W9-AXQ-051

BUILDING CONSTRUCTION

Principles, Materials, and Systems

Madan Mehta
University of Texas at Arlington

Walter Scarborough
HKS, Inc. Architecture, Engineering, Interiors

Diane Armpriest
University of Idaho

Upper Saddle River, New Jersey
Columbus, Ohio

Library of Congress Cataloging-in-Publication Data

Mehta, Madan.
Building construction : principles, materials, and systems / Madan Mehta, Walter Scarborough, Diane Armpriest.
p. cm.
Includes bibliographical references and index.
ISBN 0-13-049421-6
1. Building. I. Scarborough, Walter. II. Armpriest, Diane. III. Title.
TH146.M4288 2008
690—dc22

2006033140

Editor in Chief: Vernon R. Anthony
Associate Managing Editor: Christine M. Buckendahl
Team Assistant: Mark Fiebernitz
Editorial Assistant: Lara Dimmick
Production Editor: Stephen C. Robb
Production Coordination: Kelly Ricci/Techbooks
Art Coordinator: Karen L. Bretz
Design Coordinator: Diane Y. Ernsberger
Text Designer: Candace Rowley
Cover Designer: Candace Rowley
Cover Photo: Index Stock
Production Manager: Deidra Schwartz
Director of Marketing: David Gesell
Executive Marketing Manager: Derril Trakalo
Marketing Assistant: Les Roberts

This book was set in Optima and Garamond by TechBooks. It was printed and bound by R. R. Donnelley & Sons Company. The cover was printed by The Lehigh Press, Inc.

Figures 9.1 and 30.7(b) by Diane Armpriest. Figures 18.11, 34.3, 34.6, 35.1, 35.5, 35.6, 35.9, 35.13, 35.15, 35.18, and 35.22 by Walter Scarborough. Other photos by Madan Mehta unless indicated otherwise.

Pearson Education Ltd.
Pearson Education Singapore, Pte. Ltd
Pearson Education Canada, Ltd.
Pearson Education—Japan
Pearson Education Australia PTY, Limited
Pearson Education North Asia Ltd
Pearson Educación de Mexico, S.A. de C.V.
Pearson Education Malaysia, Pte. Ltd

10 9 8 7 6 5 4 3 2 1
ISBN-13: 978-0-13-049421-4
ISBN-10: 0-13-049421-6

FOREWORD

As the authors note in the preface of this book, building construction is a highly dynamic and complex enterprise. The range and sophistication of expertise required to design and construct contemporary buildings are increasing rapidly. As an architect for more than 30 years, I have come to appreciate the tremendous amount of knowledge today's architects, engineers, and contractors require to produce functional, economical, pleasing, and sustainable buildings. A successful design or construction professional must be prepared to cope with an exploding amount of information, mesh together the growing range of products and assemblies, coordinate diverse specialties, and manage both design and construction processes.

In *Building Construction: Principles, Materials, and Systems,* Madan Mehta, Walter Scarborough, and Diane Armpriest have combined their diverse professional and educational backgrounds to produce a resource that presents the complexity of the discipline in an accessible volume. It clearly provides the basics of building science as applied to the art of transforming materials and systems into constructible buildings. Principles that influence building performance provide the background necessary to understand why, as well as how, buildings are assembled as they are.

The book appropriately addresses each of the primary building assemblies—foundations, walls, floors, ceilings, and roofs—and how they join, seal, and integrate with other components. The performance of building enclosures and systems is reviewed in detail, which enhances the reader's understanding of the comprehensive, integrated nature of the building design and construction processes. Almost all building materials and systems have been covered in depth.

The book is unique among other available books on the subject because it is a joint effort of three authors—two of whom are engaged full time in academia, and another who has an extensive background in the profession. Together, the authors combine expertise in architecture, engineering, and construction disciplines to provide a holistic treatment of the subject. Although written primarily to educate students of architecture, engineering, and construction, the book will also serve as a reference for practitioners.

An exhaustive work, *Building Construction: Principles, Materials, and Systems* uses text and concepts and detailed drawings to convey construction assembly techniques, theory, and technology inherent in architecture. It also highlights the building industry's involvement as stewards of the environment through sustainability principles.

This volume should be read by every architect, engineer, and contractor. In my view, the book will remain the benchmark among building construction texts for a long time.

H. Ralph Hawkins

H. Ralph Hawkins, FAIA, FACHA, MPH
President and CEO
HKS Inc.

INTRODUCING A NEW, GROUNDBREAKING, COMPREHENSIVE INTRODUCTION TO CONSTRUCTION PRINCIPLES, MATERIALS, AND METHODS

A unique organization allows for an unparalleled exploration of building construction: Principles of Construction are covered in Part 1, and Materials and Systems are covered in Part 2.

CHAPTER 1 An Overview of the Building Delivery Process—How Buildings Come into Being

CHAPTER 2 Governmental Constraints on Construction

PART 2

MATERIALS AND SYSTEMS

CHAPTER 11 Materials for Wood Construction–I (Lumber)

CHAPTER 12 Materials for Wood Construction–II (Manufactured

CHAPTER 22 Masonry Materials–I (Mortar and Brick)

CHAPTER 23 Masonry Materials–II (Concrete Masonry Units, Natural Stone

FIGURE 26.38 A typical office building with precast concrete curtain wall panels. (Photo by MM.)

FIGURE 26.39 **(a)** A PC panel being unloaded from the delivery truck for hoisting into position by a crane. (Photo by MM.)

FIGURE 26.39 **(b)** PC panel of Figure 26.39(a) being hoisted to its final position. (Photo by MM.)

Hundreds of photographs and drawings illustrate the concepts and ideas presented in the text.

PRINCIPLES IN PRACTICE

How a WLF Building Resists Loads (*Continued*)

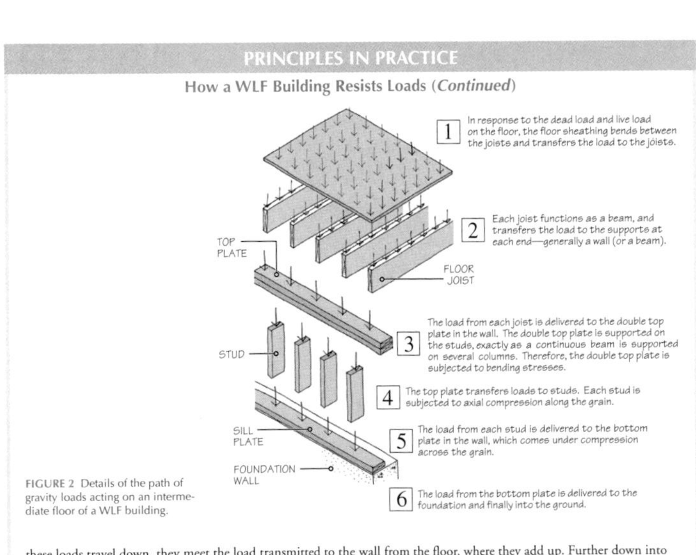

FIGURE 2 Details of the path of gravity loads acting on an intermediate floor of a WLF building.

these loads travel down, they meet the load transmitted to the wall from the floor, where they add up. Further down into the wall, the load from the next floor is added. Finally the load reaches the foundation.

Observe that the load on the wall increases as it proceeds toward the foundation. This implies that the wall should be made stronger at each lower floor. For the sake of simplicity, however, wall strength at upper floors may be kept the same as that on the lowest floor. For instance, if the first-floor walls are required to consist of 2 × 6 studs at 16 in. on center, the same stud size and spacing can also be used in the upper floors.

The load paths shown in Figure 1 indicate that a load-bearing wall at an upper floor must be aligned with that at the lower floor. Although a small offset in load-bearing walls can be managed structurally, it is advisable to plan for vertical alignment of all load-bearing walls. Figure 2 shows the details of how the gravity loads acting on an intermediate floor are transmitted to the foundation.

LATERAL LOAD PATH

In tracing the lateral loads through the structure, let us consider the loads acting on one of the walls (say, the gable-end wall) of a two-story WLF building, Figure 3. In resisting these loads, the wall behaves as a vertical slab supported by three elements—the foundation at the bottom, the intermediate floor, and the roof-ceiling assembly. The loads transmitted to each of these three elements are proportional to their respective tributary areas.

Thus, if the floor-to-floor height and the floor-to-roof height are each 10 ft, the lateral loads on the lowest 5 ft of the wall are transmitted to the foundation. The loads on the next 10 ft of the wall are transmitted to the (second) floor, whereas the loads on the remaining (uppermost) portion of the wall are transmitted to the roof.

In a WLF building, the transfer of lateral loads from the wall to the floor or the roof occurs through the studs, which function as vertical beams. Therefore, in high-wind regions, the size and spacing of studs in exterior walls may be dictated by the wind loads. In regions where the wind loads are small, gravity loads dictate the size and spacing of studs.

5.4 R-VALUE OF A MULTILAYER COMPONENT

Usually a wall or roof assembly consists of several layers of different materials, as shown in Figure 5.9. For example, an insulated brick wall consists of three layers of materials: a layer of brick, a layer of insulation, and a second layer of brick, Figure 5.10. It can be shown mathematically that the total resistance of a multilayer component, shown in Figure 5.9, is obtained by adding the R-values of individual layers:

$$R_t = R_1 + R_2 + R_3 + R_4 + \cdots \qquad \textbf{Eq. (3)}$$

where R_t is the total resistance of the multilayer component, and R_1, R_2, R_3, . . . , are the resistances of layer 1, layer 2, layer 3, and so on, respectively. Note that Equation (3) is valid only if there is no thermal **bridging** in the assembly (see Example 5 and the box entitled Thermal Bridging later in the chapter).

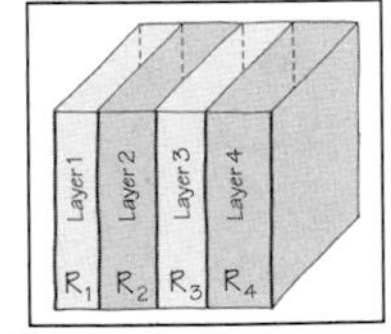

FIGURE 5.9 A multilayer assembly.

Example 2

(Estimating the R-Value of an Assembly)

Calculate the R-value of a wall assembly, which consists of a 2-in.-thick extruded polystyrene (XEPS) insulation sandwiched between an 8-in.-thick (nominal) brick wall and a 4-in.-thick (nominal) brick wall, as shown in Figure 5.10.

Note: The actual thickness of an 8-in. nominal brick wall is $7\frac{5}{8}$ in., and that of a 4-in.-thick brick wall is $3\frac{5}{8}$ in. (see Section 22.5).

Solution

From Table 5.3, the ρ-values of brick wall and extruded polystyrene insulations are 0.2 and 5.0, respectively. The total R-value of the assembly is determined as follows:

Element	R-value
4-in.-thick brick wall	3.625(0.2) = 0.725
2-in.-thick polystyrene board	2.0(5.0) = 10.0
8-in.-thick brick wall	7.625(0.2) = 1.525

$$R_t = 0.725 + 10.0 + 1.525 = 12.25$$

Because an R-value is given by the nearest whole number, the wall is an R-12 wall.

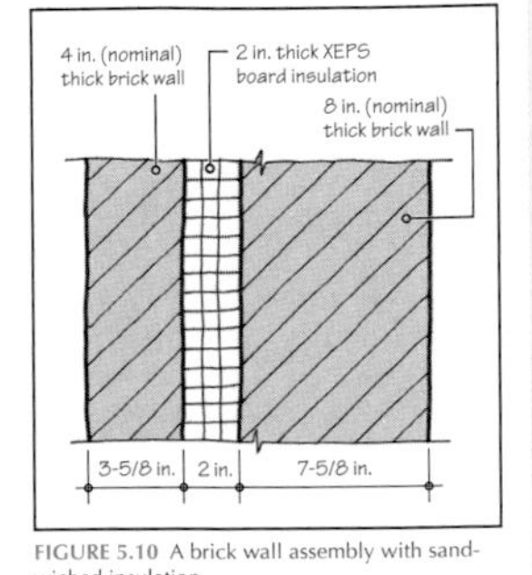

FIGURE 5.10 A brick wall assembly with sandwiched insulation.

Numerous examples provide real-world problems and their solutions.

Principles in Practice sections demonstrate practical applications of key concepts.

udinally oriented
hardwoods, have
Rays are oriented
od perpendicular
iser than the soft-
najor determinant
dwoods are gener-
oods.
m of the chemical
cture or capsule).
; trees, have broad
:iduous, shedding
lly. Oak, walnut,
y, rosewood, etc.,
luuf cells are ori-
idinal axis.
udinally oriented
ardwoods, have transverse cells, called rays. Rays are oriented
od perpendicular to the axially ogenerally denser than the soft-
najor determinant of the strength of wood, hardwoods are gen-

FOCUS ON SUSTAINABILITY

Three Tenets of Sustainability— Reduce, Reuse, and Recycle

Reduce, reuse, and recycle are considered to be the three most important tenets of sustainable construction. Their importance is in the order stated. The importance of the first tenet, in fact, far outweighs that of the other two.

While a great deal of stress is currently being placed on reusability and recyclability, the same level of concern is lacking for appropriate sizing of buildings, automobiles, and other items of human consumption. Regardless of how successful we are with reusability and recyclability, sustainability will not be achieved without seriously addressing the "reduction" tenet.

A worrying statistic about buildings is the gradual increase in the average size of new homes built in the United States during the past three decades. In 1970, a typical new house was 1,500 sq ft. In 2005, the corresponding size is 2,300 sq ft—an increase of approximately 50 percent. This, while the average family size in the same period has reduced.

CHAPTER 10 Sustainable Construction

CHAPTER OUTLINE

10.1 LIGHT-GAUGE STEEL FRAMING IN LOADBEARING APPLICATIONS

Corrosion Protection and yield Strength of Framing Members
Wall Framing in Light-Gauge Steel
Floor Framing in Light-Gauge Steel
Roof Framing in Light-Gauge Steel

10.2 ADVANTAGES AND LIMITATIONS OF LIGHT-GAUGE STEEL FRAMING

Advantages
Limitations

A chapter on "Sustainable Construction" and *Focus on Sustainability* features highlight "green" construction issues.

PRACTICE QUIZ

Each question has only one correct answer. Select the choice that best answers the question.

1. The realization of a typical building project, as described in this text, may be divided into
   a. two phases.
   b. three phases.
   c. four phases.
   d. five phases.
   e. six phases.
2. Establishing the project's economic feasibility and its overall budget is a part of the design phase of the project.
   a. True
   b. False

   c. building official of the city.
   d. architect.
   e. any one of these, depending on the type of the building.

6. The construction drawings of a building project are prepared during the
   a. SD stage of the project.
   b. DD stage of the project.
   c. CD stage of the project.
   d. preconstruction phase of the project.
   e. construction phase of the project.
7. The construction drawings of a building project are drawings that the architect uses to explain the design to the owner.
   a. True

Practice Quizzes throughout each chapter allow readers to test their comprehension.

EXPAND YOUR KNOWLEDGE

Surety Bonds

The purpose of a surety bond is to ensure that should the contractor fail to fulfill contractual obligations, there will be a financially sound party—referred to as the *surety* (also called *guarantor* or *bonding company*)—available to take over those unfulfilled obligations. The bond is, therefore, a kind of insurance that the contractor buys from a surety, generally a corporation.

There are three types of surety bonds in most building projects. A few others may be required in some special projects:

- Bid bond
- Performance bond
- Payment bond

Bid Bond
The purpose of the bid bond (also called the *bid security bond*) is to exclude frivolous bidders. It ensures that if selected by the owner, the bidder will be able to enter into a contract with the owner based on the bidding requirements and that the bidder will be able to obtain performance and payment bonds from an acceptable surety.

A bid bond is required at the time the bidder submits the bid for the project. If the bidder refuses to enter into an agreement or is unable to provide the required performance and payment bonds, the surety is obliged to pay the penalty (bid security amount), usually 5% of the bid amount, to the owner.

Performance Bond
The performance bond is required by the owner before entering into an agreement with the successful bidder. The performance bond ensures that if, after the award of the contract, the contractor is unable to perform the work as required by the bidding documents, the surety will provide sufficient funds for the completion of the project.

A performance bond protects the owner against default by the contractor or by those for whose work the contractor is responsible, such as the subcontractors. For that reason, the contractor will generally require a performance bond from major subcontractors.

Payment Bond
A payment bond (also referred to as a *labor and materials bond*) ensures that those providing labor, services, and materials for the project—such as the subcontractors and material suppliers—will be paid by the contractor. In the absence of the payment bond, the owner may be held liable to those subcontractors and material suppliers whose services and materials have been invested in the project. This liability exists even if the owner has paid the general contractor for the work of these subcontractors and material suppliers.

Pros and Cons of Bonds
The bonds are generally mandated for a publicly-funded project. In a private project, the owner may waive the bonds, particularly the bid bond. This saves the owner some money because although the cost of a bond (the premium) is paid by the contractor, it is in reality paid by the owner because the contractor adds the cost of the bond to the bid amount.

Despite its cost, most owners consider the bonds (particularly the performance and payment bonds) to be a good value because they eliminate financial risks of construction. The bid bond may be unnecessary in a negotiated bid where the owner knows the contractor's financial standing and the ability to perform. However, where uncertainty exists, a bid bond provides an excellent prequalification screening of the contractor. Responsible contractors generally maintain a close and continuous relationship with their bonding company so that the bonding company's knowledge of a contractor's capabilities far exceeds that of most owners or architects (as an owner's representative).

NOTE

Owner's Program

American Institite of Architect's (AIA) Document B141, *Standard Form of Agreement Between Owner and Architect*, defines program as "the owner's objectives, schedule, constraints and criteria including space requirements and relationships, special equipment, flexibility, expandability, systems, and site requirements."

Expand Your Knowledge features and margin *Notes* include additional information about selected topics.

KEY TERMS AND CONCEPTS

Deposition	Microstructures	Self-Tapping	Wane
Lumber	Self-Drilling	Surface Splines	Wood

REVIEW QUESTIONS

1. The two main categories under which all building loads may be classified are?
2. Which of the following can be estimated with greatest certainty?
3. The floor love load in a residential building is generally assumed as?

SELECTED WEBSITES

Steel Stud Manufacturers Association (SSMA) www.ssma.com
North American Steel Framing Alliance www.steelframingalliance.com
Structural Insulated Panel Association (SIPA) www.sips.org

FURTHER READING

1 American Wood Council (AWC) and American Forest and Paper association (AFPA): *Details for Conventional Wood Frame construction* (2001)--nearly 60-page publication that provides several standard details used in wood frame construction. It is available as free download from www.awc.org.

End-of-chapter resources include a list of *Key Terms and Concepts, Review Questions, Selected Websites,* and suggestions for *Further Reading.*

PREFACE

Building construction is a society's most dynamic enterprise, rooted in the inherent desire of humans to innovate. This dynamism is also a response to the changing socioeconomic framework caused by increasing population and declining resources. The impact of this dynamism on building construction is that new products and construction systems are added to the existing stock in a never-ending process.

In addition to new materials, new versions of traditional materials are proliferating. Consequently, today's designers and constructors face an unprecedented challenge as they strive to make well-informed decisions in the face of an expanding number of competing building products and systems available for a given application.

The most accessible source of information on building products is often those who produce them. Although product testing is generally conducted by independent agencies, the results come to us through the filter of the manufacturers' vested interests and aggressive sales strategies, slanting the information unduly in favor of a specific product.

How do we, as designers and constructors, develop a critical faculty that enables us to sift facts from exaggeration and relevance from insignificance? How do we deal with the immense regional and international diversities of construction? How do we learn to function successfully in an increasingly litigious environment?

This text is based on the premise that given a strong foundation in the fundamentals and principles of materials and systems, future designers and constructors will be well prepared to make decisions in the face of increasingly complex choices. After all, the principles will be sustained just as the conventional practices will become dated all too soon.

PARTS 1 AND 2

The study of principles alone is not sufficient to master or fully grasp a subject. The principles must be illustrated through current and traditional practices. *Building Construction: Principles, Materials, and Systems* aims to do just that. It is divided into two parts:

- Part 1, consisting of 10 chapters, deals primarily with the principles of building materials' and building assemblies' performance.
- Part 2, consisting of 25 chapters, deals primarily with specific materials and assemblies.

In our own teaching, we have found that by introducing the basic principles common to the performance of most materials early in the course, it is possible to preclude or reduce repetition when progressing from one material or system to the other. For example, thermal insulation is provided differently in different assemblies, but the thermal behavior of various assemblies is entrenched in the same basic principles. Similarly, all building assemblies must have some measure of fire endurance. Therefore, it makes sense to deal with the basics of thermal and fire-related properties in advance of their specifics.

However, there cannot be an absolute separation between the principles and practices. Some reiteration of the fundamentals as one proceeds through various materials and systems is unavoidable. Additionally, there are certain principles that are limited only to one or two materials or assemblies. Those principles must be discussed in pertinent chapters.

For instance, we had initially considered including a chapter on the durability of materials in Part 1 of the book. It did not take us long to realize that although there are a few generic aspects of durability across materials, their role is secondary. Wood's durability is limited by its combustibility, fungal decay, and termite infestation. Metals deteriorate mainly through corrosion; natural stone deteriorates through abrasion and its reaction with atmospheric acids; roofing materials deteriorate through ultraviolet radiation and hail impact, and so on. The lack of commonality means that durability attributes must be discussed in the chapter that deals with the particular material.

Water-leakage control also defies a generalized treatment because the strategies used in various building components differ from one another. For example, the principles and the details required to control water leakage through exterior walls are different from those required for basement walls, which, in turn, are different from those needed for roofs. Therefore, water-leakage control in exterior walls is covered in a separate chapter (in Part 2 of the text) before the chapters on exterior wall assemblies. Water-leakage control in basements and roofs is covered in chapter 21 and chapters 31 and 32, respectively.

For the same reasons, a few materials are covered in the "Principles" sections. It would have been incongruous to separate insulating materials from the thermal properties of materials—a chapter in Part 1. Other materials included in Part 1 are air retarders, vapor retarders, sound-absorbing materials, and joint sealants.

CHAPTER SEQUENCE

The sequence in which various materials and systems are covered (in Part 2 of the text) is based on the extent of their use rather than on a standard sequencing system, such as the MasterFormat. Wood materials and construction systems are covered first because of the extensive use of wood in North America. This is followed by coverage of steel, concrete, and masonry. The characteristics of each of these materials are discussed in one or two chapters, followed by chapters on the related construction systems.

The final 10 chapters of the text focus on finish applications. For instance, the discussion of wall cladding begins with principles of water leakage and control, followed by chapters on specific types of wall cladding systems. Some of these discussions revisit materials presented in earlier chapters, such as concrete, masonry, and stone. In other chapters, new materials (such as glass and aluminum) are covered.

This text may be used in several ways. Its size and the extent of its contents make it an ideal text for a two-semester course sequence. However, it can also be used successfully in other curricular contexts. It can be used in a one-semester introductory class to prepare beginning students for subsequent courses in structures and environmental controls and used later as a text in a more advanced construction course and design studio supplement.

SPECIAL FEATURES

In many ways, this book is similar to other books on the subject because it covers many of the same topics. However, a relatively detailed discussion of the principles and a greater emphasis on the technical aspects of construction may be considered its distinguishing features. Despite its accent on technology, the depth of coverage has been kept well within the grasp of an average undergraduate student.

Multiple-choice (Practice Quiz) questions are provided throughout each chapter. Their purpose is to help the reader gain an overall understanding of the subject and assimilate its highlights. Review questions are designed to help students develop a more detailed understanding of the subject.

Because there is a limit to the amount of information an average reader can absorb at a time, each chapter is limited in coverage. An excessively long chapter defeats that purpose. Therefore, in some cases, material is divided into two separate chapters. Chapters 11 and 12 are examples.

Several chapters conclude with a section entitled "Principles in Practice." These sections provide the opportunity for the interested reader to investigate more in-depth information on the applications of principles in architectural design and construction while maintaining the readability of the chapter.

The extent to which construction materials and systems are sustainable is an increasingly critical issue. Chapter 10 introduces the principles of sustainable construction, and this is followed up with a special section in various chapters entitled "Focus on Sustainability."

LIMITATIONS

A cursory glimpse of the chapter titles of this text will reveal that building construction is a multidimensional discipline. To develop a reasonable competence in this discipline, a fair knowledge of the fundamentals of design, history, art, and the science of building is necessary. This must be supplemented by careful and frequent observation of the construction process and interaction with contractors, material manufacturers, and assembly fabricators.

In other words, building construction cannot be learned merely by reading a book on the subject, regardless of the book's comprehensiveness. A book provides only the necessary introduction and the ability to explore and pursue the subject further. Although this book includes a large number of illustrations and construction photographs collected by the authors over the years, they should not be regarded as a substitute for the reader's own observation of materials, details, and construction processes.

Building Construction: Principles, Materials, and Systems, a joint undertaking of two educators and one practitioner, is written not only for the students of architecture and construction but also for practicing architects, engineers, and constructors. It deals with the more commonly used, time-tested materials and assemblies. It does not pretend to cover some of the so-called cutting-edge and exotic technologies because of their uncertain future. An authentic test of a building assembly's performance must come from the field, because it is virtually impossible to simulate that in a laboratory.

TEACHING RESOURCES

The authors have written a laboratory manual (0-13-243792-9) that can also be purchased from the publisher. This manual provides a set of homework exercises and projects designed to encourage critical thinking skills—exercises that typically involve the application of the information to design or construction problems.

In addition, there are a number of instruction resources available: an Instructor's Manual (0-13-049422-4), Online Instructor's Manual (0-13-243793-7), Instructor's Resource CD (0-13-230197-0) that contains PowerPoint lecture slides featuring outlines and image references to every chapter, Test Generator CD (0-13-224228-1), and Online Test Generator (0-13-230196-2). The authors hope that these resources will be a beginning of a collaborative resource developed by all who use this text in their courses.

To access supplementary materials online, instructors need to request an instructor access code. Go to **www.prenhall.com**, click the **Instructor Resource Center** link, and then click **Register Today** for an instructor access code. Within 48 hours after registering you will receive a confirming e-mail including an instructor access code. Once you have received your code, go to the site and log on for full instructions on downloading the materials you wish to use.

ACKNOWLEDGMENTS

Because building construction is a highly diverse field, the completion of this work would not have been possible without the help of a large number of experts, who have so graciously given their valuable time in reading and reviewing various chapters. The authors acknowledge with sincere gratitude the contributions of the following individuals:

James Abernethy, RA, CSI, Lawrence Technological University, Southfield, MI
Richard Atchison, Form Studios, Burleson, TX
David Biggs, P.E., Ryan-Biggs Associates, Troy, NY
J. Gregg Borchelt, P.E., Vice President, Engineering and Research, Brick Industry Association, Reston, VA
Sidney Freedman, AIA, Director of Architecture, Precast/Prestressed Concrete Institute (PCI), Chicago, IL
Thomas Gorman, Ph.D., P.E., Department of Forest Products, University of Idaho, Moscow, ID
Dennis Graber, P.E., Director of Technical Publications, National Concrete Masonry Association, Herndon, VA
Garry Heron, Redi-Mix Plant, Fort Worth, TX
Robert Hohmann, Hohmann & Barnard, Inc., Fort Worth, TX
Chris Huckabee, AIA, Huckabee and Associates, Fort Worth, TX
Jim Johnson, WJHW Acoustical Consultants, Seattle, WA
Edward Knowles, P.E., Vice President, Walters and Wolf Precast, Freemont, CA
Jerry Kunkell, P.E., University of Texas at Arlington, Arlington, TX
Joel Lewallen, Registered Roof Consultant, Fort Worth, Texas
Janice K. Means, P.E., Lawrence Technological University, Southfield, MI
Stephen Patterson, P.E., Roof Technical Services, Fort Worth, TX
Gregory Ralph, President, Dietrich Industries, Pittsburgh, PA

Mahendra Raval, P.E., TerranearPMC, LLC, Cranbury, NJ
J. Edward Sauter, AIA, Tilt-Up Concrete Association, Mount Vernon, IA
Robert Thomas, Vice President of Engineering, National Concrete Masonry Association, Herndon, VA
Jason Thomas, Structural Engineer, National Concrete Masonry Association, Herndon, VA
Van Tran, Certified Building Official, Dallas, TX
Alex Wiedenhoeft, Ph.D., Forest Products Laboratory, U.S. Department of Agriculture, Madison, WI
David Wrobel, Sharon Stairs, Medina, OH
Frank Yin, P.E., HKS Inc., Dallas, TX

The following educators reviewed the book during its development and provided numerous suggestions and ideas that the authors incorporated into the book.

Robert R. Bell, Jr., Miami University
Larry Browne, Kansas State University
Robert M. Casagrande, Southern Methodist University
David Gress, University of New Hampshire
Mark H. Hasso, Wentworth Institute of Technology
Dennis Hughes, CS Mott Community College
Ralph Johnson, Montana State University
Daphene Cyr Koch, Indiania University-Purdue University Indianapolis
Kihong Ku, Virginia Polytechnic Institute and State University
Stephen R. Lee, Carnegie Mellon School of Architecture
Thomas Mills, Virginia Polytechnic Institute and State University
Khaled Nassar, University of Maryland Eastern Shore
Ronald Nichols, Alfred State College
Craig S. Priskorn, Henry Ford Community College
Chad Richert, Henry Ford Community College
A. Paige Wyatt, Columbia Basin College
Lee Yaros, Henry Ford Community College

DISCLAIMER

The information in this book has been derived from several sources, such as the Internet, reference books, professional journals, manufacturers' literature, and the authors' professional experience. It is presented in good faith, and although the authors and the publisher have made every reasonable effort to present the information accurately, they do not assume any responsibility for its accuracy, completeness, or suitability for a particular project. It is the responsibility of the users of the book to apply their professional knowledge and consult original resources for detailed information.

Madan Mehta
Walter Scarborough
Diane Armpriest

ABOUT THE AUTHORS

Madan Mehta, B.Arch., M. Bdg. Sc., Ph.D., P.E., has been a faculty member at the School of Architecture, University of Texas at Arlington since 1985, and teaches courses in construction and structures. He was previously the Director of the Architectural Engineering Program at King Fahd University, Dhahran, Saudi Arabia. He is a Fellow of the Indian Institute of Architects and Member of the American Society of Civil Engineers.

Walter R. Scarborough, CSI, AIA, is Vice President and Director of Specifications at HKS Architects. He has almost 30 years of comprehensive experience in architecture and building construction, and has worked in contract document production, construction contract administration, and architectural specifications. A Registered Architect, he has been with HKS for more than 20 years. As director of specifications, he is responsible for product research, new master specification development, and maintaining a standard of building product quality. He also has overview responsibilities for a dozen specifiers that produce specifications for HKS projects.

Diane Armpriest, M.L.A, M.Arch., is Associate Professor in the Department of Architecture, University of Idaho. Before joining that department in 2001, she worked as an architectural project manager, and project developer and construction manager for neighborhood nonprofit housing providers. Her teaching and research interests include the pedagogy of architectural building construction technology, and the expression of structure and materials in Northwest regional architecture. Previously, she was associate professor of landscape architecture at the University of Cincinnati. Highlights of her work there include research in resource-efficient design and construction, and working with students on design-build projects.

BRIEF CONTENTS

PART 1: PRINCIPLES OF CONSTRUCTION

1 • An Overview of the Building Delivery Process—How Buildings Come into Being 3

2 • Governmental Constraints on Construction 25

3 • Loads on Buildings 53

4 • Load Resistance—The Structural Properties of Materials 79

5 • Thermal Properties of Materials 101

6 • Air Leakage and Water Vapor Control 133

7 • Fire-Related Properties 151

8 • Acoustical Properties of Materials 167

9 • Principles of Joints and Sealants (Expansion and Contraction Control) 177

10 • Principles of Sustainable Construction 197

PART 2: MATERIALS AND SYSTEMS

11 • Materials for Wood Construction–I (Lumber) 211

12 • Materials for Wood Construction–II (Manufactured Wood Products, Fasteners, and Connectors) 245

13 • Wood Light Frame Construction–I 269

14 • Wood Light Frame Construction-II 311

15 • Structural Insulated Panel System 329

16 • The Material Steel and Structural Steel Construction 341

17 • Light-Gauge Steel Construction 397

18 • Lime, Portland Cement, and Concrete 413

19 • Concrete Construction–I (Formwork, Reinforcement, and Slabs-on-Ground) 445

20 • Concrete Construction–II (Site-Cast and Precast Concrete Framing Systems) 479

21 • Soils; Foundation and Basement Construction 503

22 • Masonry Materials–I (Mortar and Brick) 541

23 • Masonry Materials–II (Concrete Masonry Units, Natural Stone, and Glass Masonry Units) 569

24 • Masonry and Concrete Bearing Wall Construction 599

25 • Rainwater Infiltration Control in Exterior Walls 631

26 • Exterior Wall Cladding–I (Masonry, Precast Concrete, GFRC, and Prefabricated Masonry) 643

27 • Exterior Wall Cladding–II (Stucco, EIFS, Natural Stone, and Insulated Metal Panels) 693

28 • Transparent Materials (Glass and Light-Transmitting Plastics) 727

29 • Windows and Doors 761

30 • Glass-Aluminum Wall Systems 791

31 • Roofing–I (Low-Slope Roofs) 817

32 • Roofing–II (Steep Roofs) 851

33 • Stairs 879

34 • Floor Coverings 903

35 • Ceilings 933

Appendix A • SI System and U.S. System of Units 949

Appendix B • Preliminary Sizing of Structural Members 955

Glossary of Terms 963

Index 977

CONTENTS

PART 1: PRINCIPLES OF CONSTRUCTION

1 • An Overview of the Building Delivery Process—How Buildings Come into Being 3

1.1 Project Delivery Phases 4 • 1.2 Predesign Phase 4 • 1.3 Design Phase 5 • 1.4 CSI MasterFormat and Specifications 9 • 1.5 Preconstruction (Bid Negotiation) Phase 11 • 1.6 Construction Phase 15 • 1.7 Construction Contract Administration 16 • 1.8 Postconstruction (Project Closeout) Phase 18 • 1.9 Alternative Project Delivery Methods 19 • 1.10 Construction Management (CM) Method 19 • 1.11 CM at Risk (CMAR) Method 21 • 1.12 Design-Build (DB) Method 22

2 • Governmental Constraints on Construction 25

2.1 Objectives of a Building Code 26 • 2.2 Enforcement of a Building Code 29 • 2.3 Prescriptive and Performance Codes 32 • 2.4 Model Codes 33 • 2.5 Contents of a Building Code 35 • 2.6 Application of a Building Code 36 • 2.7 Construction Standards 40 • 2.8 Other Major Governmental Constraints 42 • 2.9 Zoning Ordinance 43 • 2.10 Building Accessibility—Americans with Disabilities Act (ADA) 44 • Principles in Practice: Code Allowable Area and Height of Building 46

3 • Loads on Buildings 53

3.1 Dead Loads 55 • 3.2 Live Loads 55 • 3.3 Rain Loads 56 • 3.4 Wind Load Basics 57 • 3.5 Factors That Affect Wind Loads 61 • 3.6 Roof Snow Load 63 • 3.7 Earthquake Load 64 • 3.8 Factors That Affect Earthquake Loads 68 • 3.9 Wind versus Earthquake Resistance of Buildings 69 • Principles in Practice: Dead Load and Live Load Estimation 71

4 • Load Resistance—The Structural Properties of Materials 79

4.1 Compressive and Tensile Strengths of Materials 80 • 4.2 Ductility and Brittleness 83 • 4.3 Yield Strength of Materials 85 • 4.4 Elasticity and Plasticity 86 • 4.5 Modulus of Elasticity 87 • 4.6 Bending Strength of Materials 88 • 4.7 Shear Strength of Materials 92 • 4.8 Bearing Strength of Materials 93 • 4.9 Structural Failures 94 • 4.10 Structural Safety 96

5 • Thermal Properties of Materials 101

5.1 Units of Energy 103 • 5.2 Conduction, Convection, and Radiation 104 • 5.3 R-Value of a Building Component 105 • 5.4 R-Value of a Multilayer Component 109 • 5.5 Surface Emissivity 110 • 5.6 U-Value of an Assembly 113 • 5.7 Where and How Much to Insulate 116 • 5.8 Thermal Capacity 120 • 5.9 The Most Effective Face of the Envelope for Insulation 122 • Principles in Practice: Insulating Materials 123

6 • Air Leakage and Water Vapor Control 133

6.1 Air Leakage Fundamentals 134 • 6.2 Air Retarder 135 • 6.3 Water Vapor in Air 137 • 6.4 Condensation of Water Vapor 140 • 6.5 Control of Condensation 140 • 6.6 Materials Used as Vapor Retarders 141 • 6.7 Location of Vapor Retarder and Ventilation in the Space Beyond the Vapor Retarder 143 • 6.8 Vapor Retarder under a Concrete Slab-on-Ground 144 • Principles in Practice: Where Dew Point Occurs in an Assembly 146

7 • Fire-Related Properties 151

7.1 Fire Code and Building Code 153 • 7.2 Combustible and Noncombustible Materials 154 • 7.3 Products Given Off in a Building Fire 154 • 7.4 Fire-Rated Assemblies and Compartmentalization of a Building 156 • 7.5 Types of Construction 157 • 7.6 Fire-Stopping of Penetrations and Fire-Sealing of Joints 161 • 7.7 Fire-Test Response Characteristics of Interior Finishes 162 • 7.8 Role of Sprinklers 164

8 • Acoustical Properties of Materials 167

8.1 Frequency, Speed, and Wavelength of Sound 168 • 8.2 The Decibel Scale 169 • 8.3 Airborne and Structure-Borne Sounds 169 • 8.4 Airborne Sound Insulation—Sound-Transmission Class 170 • 8.5 Structure-Borne Sound Insulation—Impact Insulation Class 173 • 8.6 Sound Absorption—Noise-Reduction Coefficient 174

9 • Principles of Joints and Sealants (Expansion and Contraction Control) 177

9.1 Types of Movement Joints 178 • 9.2 Building Separation Joints and Seismic Joints 179 • 9.3 Movement Joints in Building Components 182 • 9.4 Thermal Movement 184 • 9.5 Moisture Movement 186 • 9.6 Elastic Deformation and Creep 188 • 9.7 Total Joint Dimension 189 • 9.8 Principles of Joint Detailing 190 • 9.9 Components of a Sealed Joint 191 • 9.10 Types and Properties of Joint Sealants 192

10 • Principles of Sustainable Construction 197

10.1 Fundamentals of Sustainable Architecture 200 • 10.2 Ecolabeling of Buildings 200 • 10.3 Characteristics of Green Building Products 202 • 10.4 Ecolabeling of Building Products 206

PART 2: MATERIALS AND SYSTEMS

11 • Materials for Wood Construction–I (Lumber) 211

11.1 Introduction 212 • 11.2 Growth Rings and Wood's Microstructure 215 • 11.3 Softwoods and Hardwoods 217 • 11.4 From Logs to Finished Lumber 221 • 11.5 Drying of Lumber 223 • 11.6 Lumber Surfacing 226 • 11.7 Nominal and Actual Dimensions of Lumber 227 • 11.8 Board Foot Measure 227 • 11.9 Softwood Lumber Classification 229 • 11.10 Lumber's Strength and Appearance 229 • 11.11 Lumber Grading 231 • 11.12 Durability of Wood 234 • 11.13 Fungal Decay 234 • 11.14 Termite Control 235 • 11.15 Preservative-Treated Wood 237 • 11.16 Fire-Retardant-Treated-Wood 239 • Principles in Practice: Typical Grade Stamps of Visually Graded Lumber 241

12 • Materials for Wood Construction–II (Manufactured Wood Products, Fasteners, and Connectors) 245

12.1 Glulam Members 246 • 12.2 Structural Composite Lumber—LVL and PSL 249 • 12.3 Wood I-Joists 250 • 12.4 Wood Trusses 251 • 12.5 Wood Panels 255 • 12.6 Plywood Panels 255 • 12.7 OSB Panels 257 • 12.8 Specifying Wood Panels—Panel Ratings 259 • 12.9 Fasteners for Connecting Wood Members 261 • 12.10 Sheet Metal Connectors 265

13 • Wood Light Frame Construction–I 269

13.1 Evolution of Wood Light Frame Construction 270 • 13.2 Contemporary Wood Light Frame—The Platform Frame 272 • 13.3 Frame Configuration and Spacing of Members 274 • 13.4 Essentials of Wall Framing 275 • 13.5 Framing Around Wall Openings 278 • 13.6 Essentials of Floor Framing 281 • 13.7 Roof Types and Roof Slope 285 • 13.8 Essentials of Roof Framing 287 • 13.9 Vaulted Ceilings 291 • 13.10 Sheathing Applied to a Frame 292 • 13.11 Equalizing Cross-Grain Lumber Dimensions 294 • Principles in Practice: Constructing a Two-Story Wood Light Frame Building 296 • Principles in Practice: How a WLF Building Resists Loads 303

14 • Wood Light Frame Construction-II 311

14.1 Exterior Wall Finishes in a WLF Building 312 • 14.2 Horizontal Sidings 313 • 14.3 Vertical Sidings 317 • 14.4 Finishing the Eaves, Rakes, and Ridge 318 • 14.5 Gypsum Board 320 • 14.6 Installing and Finishing Interior Drywall 323 • 14.7 Fire-Resistance Ratings of WLF Assemblies 326

15 • Structural Insulated Panel System 329

15.1 Basics of the Structural Insulated Panel (SIP) System 330 • 15.2 SIP Wall Assemblies 331 • 15.3 SIP Floor Assemblies 334 • 15.4 SIP Roof Assemblies 334 • 15.5 Advantages and Limitations of SIPS 336

16 • The Material Steel and Structural Steel Construction 341

16.1 Making of Modern Steel 344 • 16.2 Structural Steel Shapes and Their Designations 346 • 16.3 Steel Joists and Joist Girders 350 • 16.4 Steel Roof and Floor Decks 355 • 16.5 Preliminary Layout of Framing Members 360 • 16.6 Bolts and Welds 362 • 16.7 Connections Between Framing Members 364 • 16.8 Steel Detailing and Fabrication 369 • 16.9 Steel Erection 372 • 16.10 Corrosion Protection of Steel 375 • 16.11 Fire Protection of Steel 376 • Principles in Practice: Fundamentals of Skeleton Frame Construction 382

17 • Light-Gauge Steel Construction 397

17.1 Light-Gauge Steel Framing Members 399 • 17.2 Light-Gauge Steel Framing in Load-Bearing Applications 400 • 17.3 Advantages and Limitations of Light-Gauge Steel Framing 408 • 17.4 Non-Load-Bearing Light-Gauge Steel Framing 409

18 • Lime, Portland Cement, and Concrete 413

18.1 Introduction to Lime 414 • 18.2 Types of Lime Used in Construction 416 • 18.3 Portland Cement 418 • 18.4 Air-Entrained and White Portland Cements 420 • 18.5 Basic Ingredients of Concrete 421 • 18.6 Important Properties of Concrete 425 • 18.7 Making Concrete 428 • 18.8 Placing and Finishing Concrete 429 • 18.9 Portland Cement and Water Reaction 433 • 18.10 Water-Reducing Concrete Admixtures 435 • 18.11 High-Strength Concrete 435 • 18.12 Steel Reinforcement 438 • 18.13 Welded Wire Reinforcement 441

19 • Concrete Construction–I (Formwork, Reinforcement, and Slabs-on-Ground) 445

19.1 Versatility of Reinforced Concrete 448 • 19.2 Concrete Formwork and Shores 448 • 19.3 Formwork Removal and Reshoring 454 • 19.4 Architectural Concrete and Form Liners 455 • 19.5 Principles of Reinforcing Concrete 456 • 19.6 Splices, Couplers, and Hooks in Bars 459 • 19.7 Corrosion Protection of Steel Reinforcement 461 • 19.8 Reinforcement and Formwork for Columns 463 • 19.9 Reinforcement and Formwork for Walls 464 • 19.10 Types of Concrete Slabs 468 • 19.11 Ground-Supported Isolated Concrete Slab 468 • 19.12 Ground-Supported Stiffened Concrete Slab 472

20 • Concrete Construction–II (Site-Cast and Precast Concrete Framing Systems) 479

20.1 Types of Elevated Concrete Floor Systems 480 • 20.2 Beam-Supported Concrete Floors 480 • 20.3 Beamless Concrete Floors 488 • 20.4 Posttensioned Elevated Concrete Floors 491 • 20.5 Introduction to Precast Concrete 493 • 20.6 Mixed Precast Concrete Construction 494 • 20.7 Fire Resistance of Concrete Members 498

21 • Soils; Foundation and Basement Construction 503

21.1 Classification of Soils 504 • 21.2 Soil Sampling and Testing 505 • 21.3 Earthwork and Excavation 507 • 21.4 Supports for Deep Excavations 511 • 21.5 Keeping Excavations Dry 518 • 21.6 Foundation Systems 519 • 21.7 Shallow Foundations 520 • 21.8 Deep Foundations 526 • 21.9 Frost-Protected Shallow Foundations 528 • 21.10 Below-Grade Waterproofing 531 • Principles in Practice: Unified Soil Classification 535

22 • Masonry Materials–I (Mortar and Brick) 541

22.1 Masonry Mortar 543 • 22.2 Mortar Materials and Specifications 546 • 22.3 Mortar Joint Thickness and Profiles 550 • 22.4 Manufacture of Bricks 552 • 22.5 Dimensions of Masonry Units 555 • 22.6 Types of Clay Bricks 556 • 22.7 Bond Patterns in Masonry Walls 559 • 22.8 The Importance of the IRA of Bricks 562 • 22.9 The Craft and Art of Brick Masonry Construction 562 • 22.10 Efflorescence in Brick Walls 565 • 22.11 Expansion Control in Brick Walls 565

23 • Masonry Materials–II (Concrete Masonry Units, Natural Stone, and Glass Masonry Units) 569

23.1 Concrete Masonry Units—Sizes and Shapes 570 • 23.2 Concrete Masonry Units—Manufacturing and Specifications 576 • 23.3 Construction of a CMU Wall 578 • 23.4 Shrinkage Control in CMU Walls 579 • 23.5 Grout 583 • 23.6 Calcium Silicate Masonry Units 584 • 23.7 Natural Stone 585 • 23.8 From Blocks to Finished Stone 588 • 23.9 Stone Selection 590 • 23.10 Bond Patterns in Stone Masonry Walls 591 • 23.11 Glass Masonry Units 592 • 23.12 Fire Resistance of Masonry Walls 595

24 • Masonry and Concrete Bearing Wall Construction 599

24.1 Traditional Masonry Bearing Wall Construction 601 • 24.2 Importance of Vertical Reinforcement in Masonry Walls 604 • 24.3 Bond Beams in a Masonry Bearing Wall Building 605 • 24.4 Wall Layout in a Bearing Wall Building 606 • 24.5 Floor and Roof Decks—Connections to Walls 610 • 24.6 Limitations of Masonry Bearing Wall Construction 613 • 24.7 Bearing Wall and Column-Beam System 616 • 24.8 Reinforced-Concrete Bearing Wall Construction 617 • 24.9 Reinforced Concrete Tilt-Up Wall Construction 619 • 24.10 Connections in a Tilt-Up Wall Building 623 • 24.11 Aesthetics of Tilt-Up Wall Buildings 626 • Principles in Practice: The Middle-Third Rule 627

25 • Rainwater Infiltration Control in Exterior Walls 631

25.1 Rainwater Infiltration Control—Basic Strategies 632 • 25.2 Barrier Wall Versus Drainage Wall 634 • 25.3 Rain-Screen Exterior Cladding 637

26 • Exterior Wall Cladding–I (Masonry, Precast Concrete, GFRC, and Prefabricated Masonry) 643

26.1 Masonry Veneer Assembly—General Considerations 644 • 26.2 Brick Veneer with a CMU or Concrete Backup Wall 654 • 26.3 Brick Veneer with a Steel Stud Backup Wall 661 • 26.4 CMU Backup Versus Steel Stud Backup 668 • 26.5 Aesthetics of Brick Veneer 669 • 26.6 Precast Concrete (PC) Curtain Wall 670 • 26.7 Connecting the PC Curtain Wall to a Structure 673 • 26.8 Brick and Stone-Faced PC Curtain Wall 675 • 26.9 Detailing a PC Curtain Wall 678 • 26.10 Glass Fiber–Reinforced Concrete (GFRC) Curtain Wall 680 • 26.11 Fabrication of GFRC Panels 683 • 26.12 Detailing a GFRC Curtain Wall 685 • 26.13 Prefabricated Brick Curtain Wall 687

27 • Exterior Wall Cladding–II (Stucco, EIFS, Natural Stone, and Insulated Metal Panels) 693

27.1 Portland Cement Plaster (Stucco) Basics 694 • 27.2 Stucco on Steel- or Wood-Stud Walls 695 • 27.3 Stucco on Masonry and Concrete Substrates 700 • 27.4 Limitations and Advantages of Stucco 701 • 27.5 Exterior Insulation and Finish System (EIFS) Basics 702 • 27.6 Application of Polymer-Based EIFS 703 • 27.7 Impact-Resistant and Drainable EIFS 707 • 27.8 Exterior Cladding with Dimension Stone 709 • 27.9 Field Installation of Stone—Standard-Set Method 710 • 27.10 Field Installation of Stone—Vertical Support Channel Method 714 • 27.11 Prefabricated Stone Curtain Walls 717 • 27.12 Thin Stone Cladding 719 • 27.13 Insulated Metal Panels 722

28 • Transparent Materials (Glass and Light-Transmitting Plastics) 727

28.1 Manufacture of Flat Glass 730 • 28.2 Types of Heat-Modified Glass 732 • 28.3 Glass and Solar Radiation 736 • 28.4 Types of Tinted and Reflective Glass 737 • 28.5 Glass and Long-Wave Radiation 739 • 28.6 Insulating Glass Unit 740 • 28.7 R-Value (or U-Value) of Glass 742 • 28.8 Glass and Glazing 743 • 28.9 Safety Glass 745 • 28.10 Laminated Glass 746 • 28.11 Structural Performance of Glass 747 • 28.12 Fire-Resistant Glass 747 • 28.13 Plastic Glazing 750 • 28.14 Glass for Special Purposes 750 • 28.15 Criteria for the Selection of Glass 752 • 28.16 Anatomy of a Glazing Pocket 752 • Principles in Practice: Important Facts about Radiation 756 • Principles in Practice: Condensation-Resistance Factor 757

29 • Windows and Doors 761

29.1 Window Styles 763 • 29.2 Window Materials 764 • 29.3 Performance Ratings of Windows 767 • 29.4 Window Installation and Surrounding Details 769 • 29.5 Classification of Doors 773 • 29.6 Door Frames 780 • 29.7 Fire-Rated Doors and Windows 783 • Principles in Practice: A Note on Aluminum 786

30 • Glass-Aluminum Wall Systems 791

30.1 Glass-Aluminum Curtain Walls 792 • 30.2 Anchorage of a Stick-Built Glass Curtain Wall to a Structure 794 • 30.3 Stick-Built Glass Curtain Wall Details 800 • 30.4 Structural Performance of a Glass Curtain Wall 809 • 30.5 Other Performance Criteria of a Glass Curtain Wall 810 • 30.6 Nontraditional Glass Curtain Walls 812 • 30.7 Other Glass-Aluminum Wall Systems 812

31 • Roofing–I (Low-Slope Roofs) 817

31.1 Low-Slope and Steep Roofs Distinguished 818 • 31.2 Low-Slope Roof Fundamentals 819 • 31.3 Built-Up Roof Membrane 820 • 31.4 Modified Bitumen Roof Membrane 825 • 31.5 Single-Ply Roof Membrane 829 • 31.6 Rigid-Board Insulation and Membrane Attachment 832

• 31.7 Insulating Concrete and Membrane Attachment 835 • 31.8 Low-Slope Roof Flashings 837 • 31.9 Base Flashing Details 838 • 31.10 Curb and Flange Flashing Details 840 • 31.11 Protected Membrane Roof 841 • 31.12 Low-Slope-Roof Design Considerations 842 • Principles in Practice: Shingling of Built-Up Roof Felts 846

32 • Roofing–II (Steep Roofs) 851

32.1 Steep-Roof Fundamentals 852 • 32.2 Asphalt Shingles and Roof Underlayment 854 • 32.3 Installation of Asphalt Shingles 856 • 32.4 Valley Treatment in an Asphalt Shingle Roof 859 • 32.5 Ridge and Hip Treatment in an Asphalt Shingle Roof 862 • 32.6 Flashings in an Asphalt Shingle Roof 863 • 32.7 Essentials of Clay and Concrete Roof Tiles 866 • 32.8 Clay and Concrete Tile Roof Details 869 • 32.9 Types of Architectural Metal Roofs 872 • 32.10 Contemporary Architectural Metal Roofs 873

33 • Stairs 879

33.1 Stair Fundamentals 881 • 33.2 Wood Stairs 886 • 33.3 Steel Stairs 888 • 33.4 Concrete Stairs 898

34 • Floor Coverings 903

34.1 Subfloors 904 • 34.2 Selection Criteria for Floor Coverings 905 • 34.3 Ceramic and Stone Tile Flooring 906 • 34.4 Stone Panel Flooring 914 • 34.5 Terrazzo Flooring 916 • 34.6 Carpet and Carpet Tile Flooring 918 • 34.7 Wood Flooring 921 • 34.8 Resilient Flooring 924 • 34.9 Resinous Flooring 926 • 34.10 Other Floor-Covering Materials 927 • 34.11 Underlayments 927 • 34.12 Resilient Accessories—Wall Base and Moldings 928 • Principles in Practice: Showers and Tile 930

35 • Ceilings 933

35.1 Selection Criteria for Ceiling Finish Materials 934 • 35.2 No Ceiling Finish—Exposed to Above 936 • 35.3 Ceilings Attached to Building Structure 936 • 35.4 Ceilings Suspended from Building Structure 937

Appendix A • SI System and U.S. System of Units 949

Rules of Grammar in the SI System 950 • Length, Thickness, Area, and Volume 951 • Fluid Capacity 951 • Mass, Force, and Weight 951 • Pressure and Stress 951 • Unit Weight of Materials 952 • Temperature and Energy 952 • Conversion from the U.S. System to the SI System 952

Appendix B • Preliminary Sizing of Structural Members 955

Conventional Wood Light Frame (WLF) Buildings 956 • Conventional Light-Gauge Steel Frame (LGSF) Buildings 958 • Structural Steel Frame Buildings 959 • Site-Cast Concrete Frame Buildings 960 • Precast-Prestressed Concrete Members 962 • Load-Bearing Masonry and Concrete Buildings 962

Glossary 963

Index 977

PART 1

PRINCIPLES OF CONSTRUCTION

CHAPTER 1 An Overview of the Building Delivery Process—How Buildings Come into Being

CHAPTER 2 Governmental Constraints on Construction

CHAPTER 3 Loads on Buildings

CHAPTER 4 Load Resistance—The Structural Properties of Materials

CHAPTER 5 Thermal Properties of Materials

CHAPTER 6 Air Leakage and Water Vapor Control

CHAPTER 7 Fire-Related Properties

CHAPTER 8 Acoustical Properties of Materials

CHAPTER 9 Principles of Joints and Sealants (Expansion and Contraction Control)

CHAPTER 10 Principles of Sustainable Construction

• FUNCTIONALITY/OPERATIONS
• NURSING UNIT DESIGN
-EFFICIENCY
-NEW CONCEPTS

CHAPTER 1

An Overview of the Building Delivery Process—How Buildings Come into Being

CHAPTER OUTLINE

1.1 Project Delivery Phases

1.2 Predesign Phase

1.3 Design Phase

1.4 CSI MasterFormat and Specifications

1.5 Preconstruction (Bid Negotiation) Phase

1.6 Construction Phase

1.7 Construction Contract Administration

1.8 Postconstruction (Project Closeout) Phase

1.9 Alternative Project Delivery Methods

1.10 Construction Management (CM) Method

1.11 CM at Risk (CMAR) Method

1.12 Design-Build (DB) Method

A typical schematic design (SD) stage meeting of the design team. (Photo courtesy of HKS Inc.)

Building construction is a complex, significant, and rewarding process. It begins with an idea and culminates in a structure that may serve its occupants for several decades, even centuries. Like the manufacturing of products, building construction requires an ordered and planned assembly of materials. It is, however, far more complicated than product manufacturing. Buildings are assembled outdoors on all types of sites and are subject to all kinds of weather.

Additionally, even a modest-sized building must satisfy many performance criteria and legal constraints, requires an immense variety of materials, and involves a large network of design and production firms. It is further complicated by the fact that no two buildings are truly identical; each one must be custom-built to serve a unique function and respond to the uniqueness of its context and the preferences of its owner, user, and occupant.

Because of a building's uniqueness, we invoke first principles in each building project. Although it may seem that we are "reinventing the wheel," we are in fact refining and improving the building delivery process. In so doing, we bring to the task the collective wisdom of the architects, engineers, and contractors who have done so before us. Although there are movements that promote the development of standardized, mass-produced buildings, these seldom meet the distinct needs of each user.

Regardless of the uniqueness of each building project, the flow of events and processes necessary for a project's realization is virtually the same in all buildings. This chapter presents an overview of the events and processes that bring about a building—from the inception of a mere idea or concept in the owner's mind to the completed *design* by the architects and engineers and, finally, to the actual *construction* of the building by the contractor.

Design and construction are two independent but related and generally sequential functions in the realization of a building. The former function deals with the creation of the *documentation,* and the latter function involves interpreting and transforming these documents into reality—a building or a complex of buildings.

The chapter begins with a discussion of the various personnel involved in a project and the relational framework among them. Subsequently, a description of the two major elements of design documentation—construction drawings and specifications—is provided. Finally, the chapter examines some of the emerging methods of bringing a building into being and compares them with the traditional methods.

The purpose of this chapter, as its title suggests, is to provide an overall, yet distilled, view of the construction process and its relationship with design. Although several contractual and legal issues are discussed, they should be treated as introductory. A reader requiring detailed information on these topics should refer to sources such as those provided at the end of the chapter.

1.1 PROJECT DELIVERY PHASES

The process by which a building project is delivered to its owner may be divided into the following five phases, referred to as the *project delivery phases.* Although there is usually some overlap between adjacent phases, they generally follow the order listed below:

- Predesign phase
- Design phase
- Preconstruction phase
- Construction phase
- Postconstruction phase

1.2 PREDESIGN PHASE

During the *predesign phase* (also called the *planning phase*), the project is defined in terms of its function, purpose, scope, size, and economics. It is the most crucial of all the five phases, as the success or failure of the project may depend on how well this phase is defined and managed. Obviously, the clearer the project's definition, the easier it is to proceed to the subsequent phases. Some of the important predesign tasks are as follows:

- *Building program definition,* including activities, functions, and spaces required in the building, along with their approximate sizes, and relationships with each other
- E*conomic feasibility assessment,* including the project's overall budget and financing

- *Site assessment and selection,* including the verification of the site's appropriateness and determining its designated land use (Chapter 2)
- *Governmental constraints assessment,* for example, building code constraints (Chapter 2) and other legal aspects of the project
- *Design team selection*

For a house, or other small project, the program is usually simple and can be developed by the owner without external assistance. For a large project, however, where the owner may be an institution (such as a corporation, school board, hospital, church, or governmental entity), developing the program may be a complex exercise. This may be due to the size and complexity of the project or the need to involve several individuals—a corporation's board of directors for example—in decision making. These constituencies may have different views of the project, making it difficult to create a consensus.

The program development may also be complicated by situations in which the owner has a fuzzy idea of the project and is unable to define it clearly. On the other hand, experienced owners tend to have a clear understanding of the project and generally provide a detailed, unambiguous program to the architect.

It is not unusual for the owner to involve the architect and a few other consultants of the design team in preparing the program. In this instance, the design team may be hired during the predesign phase. When the economic considerations of the project are paramount, the owner may also consult a construction cost analyst.

Whatever the situation, the owner's program is the first step in the project delivery process. It should be spelled out in writing and in sufficient detail to guide design, reduce liability risk for the architect, and avoid its misinterpretation. If a revision is made during the progress of the project, the owner's written approval is necessary.

NOTE

Owner's Program

American Institite of Architect's (AIA) Document B141 *Standard Form of Agreement Between Owner and Architect,* defines program as "the owner's objectives, schedule, constraints and criteria including space requirements and relationships, special equipment, flexibility, expandability, systems, and site requirements."

1.3 DESIGN PHASE

The *design phase* begins after the selection of the architect. Because the architect (usually a firm) may have limited capabilities for handling the broad range of building-design activities, several different, more specialized consultants are usually required, depending on the size and scope of the project.

In most projects, the design team consists of the architect, civil and structural consultants, and mechanical, electrical, plumbing, and fire-protection (MEPF) consultants. In complex projects, the design team may also include an acoustical consultant, roofing and waterproofing consultant, cost consultant, building code consultant, signage consultant, interior designer, landscape architect, and so on.

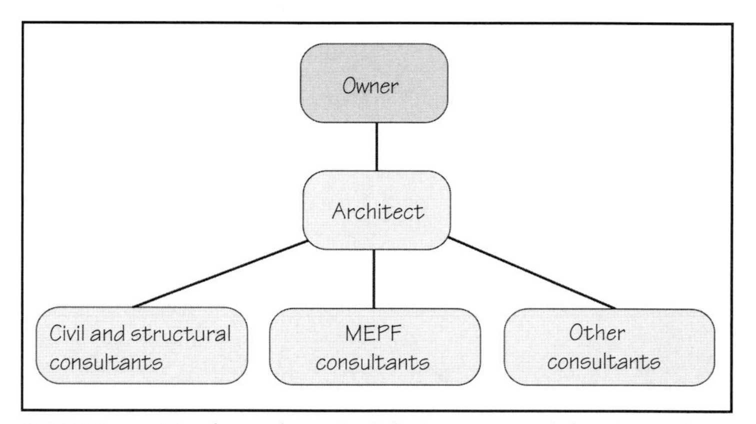

FIGURE 1.1 Members of a typical design team and their interrelationships with each other and the owner in a traditional contractual setup. A line in this illustration indicates a contractual relationship between parties.

Some design firms have an entire design team (architects and specialized consultants) on staff, in which case the owner will contract with a single firm. Generally, however, the design team comprises several different design firms. In such cases, the owner typically contracts the architect, who in turn contracts the remaining design team members, Figure 1.1.

Thus, the architect functions as the prime design professional and, to a limited degree, as the owner's representative. The architect is liable to the owner for his or her own work and also that of the consultants. For that reason, most architects ensure that their consultants carry adequate liability insurance.

In some projects, the owner may contract some consultants directly, particularly a civil consultant (for a survey of the site, site grading, slope stabilization, and site drainage), a geotechnical consultant (for investigation of the soil properties), and a landscape architect (for landscape and site design), Figure 1.2. These consultants may be engaged before or at the same time as the architect.

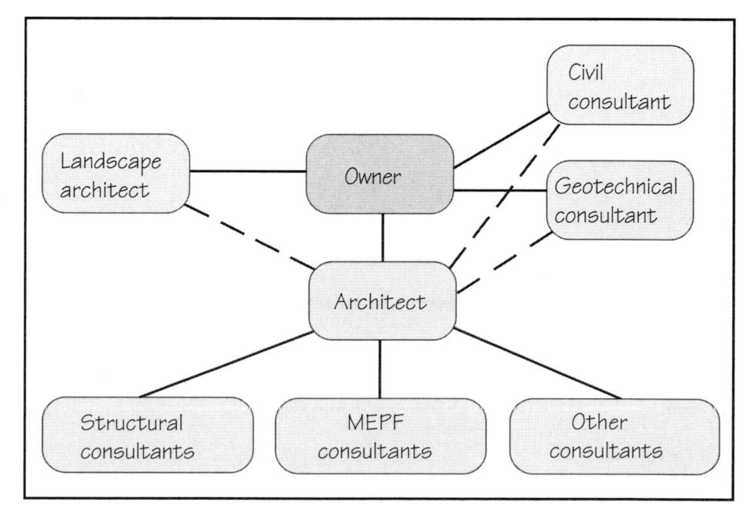

FIGURE 1.2 Members of a typical design team and their interrelationships with each other and the owner in a project in which some consultants are contracted directly by the owner. A solid line in this illustration indicates a contractual relationship between parties. A dashed line indicates a communication link, not a contract.

Even when a consultant is contracted directly by the owner, the architect retains some level of liability for the consultant's work. This liability occurs because the architect, being the prime design professional, coordinates the entire design effort, and the consultants' work is influenced a great deal by architectural decisions. Therefore, the working relationship between the architect and an owner-contracted consultant remains essentially the same as if the consultant were chosen by the architect.

In some cases an engineer or other professional may coordinate the design process. This generally occurs when a building is a minor component of a large-scale project. For example, in a highly technical project such as a power plant, an electrical engineer may be the prime design professional.

In most building projects, the design phase consists of three stages:

- Schematic design stage
- Design development stage
- Construction documents stage

Schematic Design (SD) Stage—Emphasis on Design

The *schematic design* gives graphic shape to the owner's program. It is an overall design concept that illustrates the key ideas of the design solution. The major player in this stage is the architect, who develops the design scheme (or several design options) with only limited help from the consultants. Because most projects have strict budgetary limitations, a rough estimate of the project's probable cost is generally produced at this stage.

The schematic design usually goes through several revisions, because the *first* design proposal prepared by the architect will rarely be approved by the owner. The architect communicates the design proposal(s) to the owner through various types of drawings—plans, elevations, sections, freehand sketches, and three-dimensional graphics (isometrics, axonometrics, and perspectives). For some projects, a three-dimensional scale model of the entire building or the complex of buildings, showing the context (neighboring buildings) within which the project is sited, may be needed.

With significant developments in electronic media technology, computer-generated imagery has become common in architecture and related engineering disciplines. Computer-generated walk-through and flyover simulations are becoming increasingly popular ways of communicating the architect's design intent to the owner and the related organizations at the SD stage.

It is important to note that the schematic design drawings, images, models, and simulations, regardless of how well they are produced, are not adequate to construct the building. Their objective is merely to communicate the design scheme to the owner (and to consultants, who may or may not be on board at this stage), not to the contractor.

Design Development (DD) Stage—Emphasis on Decision Making

Once the schematic design is approved by the owner, the process of designing the building in greater detail begins. During this stage, the schematic design is developed further—hence the term *design development* (DD) stage.

Although the emphasis in the SD stage is on the creative, conceptual, and innovative aspects of design, the DD stage focuses on developing practical and pragmatic solutions for the exterior envelope, structure, fenestration, interior systems, MEPF systems, and so forth. This development involves strategic consultations with all members of the design team.

Therefore, the most critical feature of the DD stage is decision making, which may range from broad design aspects to details. At this stage, the vast majority of decisions about products, materials, and equipment are made. Efficient execution of the construction documents depends directly on how well the DD is managed.

A more detailed version of the specifications and probable cost of the project are also prepared at this stage. It is not uncommon in a negotiated contract (Section 1.5) for the general contractor to become involved at this stage to provide input about the cost and constructibility of the project.

Construction Documents (CD) Stage—Emphasis on Documentation

The purpose of the construction documents (CD) stage is to prepare all documents required by the contractor to construct the building. During this stage, the engineering consultants and architect collaborate intensively to work out the "nuts and bolts" of the building and develop the required documentation, referred to as *construction documents.*

Each consultant advises the architect, but they also collaborate with each other (generally through the architect) so that the work of one consultant agrees with that of the others.

The construction documents consist of the following:

- Construction drawings
- Specifications

Construction Drawings

During the CD stage, the architect and consultants prepare their own sets of drawings, referred to as *construction drawings.* Thus, a project has architectural construction drawings, civil and structural construction drawings, MEPF construction drawings, landscape construction drawings, and so on.

Construction drawings are dimensioned drawings (usually computer generated) that fully delineate the building. They consist of floor plans, elevations, sections, and various large-scale details. The details depict a small portion of the building that cannot be adequately described on smaller-scale plans, elevations, or sections.

Construction drawings are the drawings that the contractor uses to build the building. Therefore, they must indicate the geometry, layout, dimensions, type of materials, details of assembling the components, colors and textures, and so on. Construction drawings are generally two-dimensional drawings, but three-dimensional isometrics are sometimes used for complex details. Construction drawings are also used by the contractor to prepare a detailed cost estimate of the project at the time of bidding.

Construction drawings are not a sequence of assembly instructions, such as for a bicycle. Instead, they indicate what every component is and where it will be located when the building is completed. In other words, the design team decides the "what" and "where" of the building. The "how" and "when" of the building are, however, entirely in the contractor's domain.

NOTE

Working Drawings and Construction Drawings

The term *working drawings* was used until the end of the 20th century for what are now commonly referred to as *construction drawings.*

Specifications

Buildings cannot be constructed from drawings alone because there is a great deal of information that cannot be included in the drawings. For instance, the drawings will give the locations of columns, their dimensions, and the material used (such as reinforced concrete), but the quality of materials, their properties (the strength of concrete for example), and the test methods required to confirm compliance cannot be furnished on the drawings. This information, called *specifications,* is included in the document.

Specifications are written technical descriptions of the design intent, whereas the drawings provide the graphic description. The two components of the construction documents—the specifications and the construction drawings—complement each other and generally deal with different aspects of the project. Because they are complementary, they are supposed to be used in conjunction with each other. There is no order of precedence between the construction drawings and the specifications. Thus, if an item is described in only one place—either the specification or the drawings—it is part of the project.

For instance, if the construction drawings do not show the door hardware (hinges, locks, handles, and other components) but the hardware is described in the specifications, the owner will get the doors with the stated hardware. If the drawings had precedence over the specifications, the owner would receive doors without hinges and handles.

Generally, there is little overlap between the drawings and the specifications. More importantly, there should be no conflict between them. If a conflict between the two documents is identified, the contractor must bring it to the attention of the architect promptly. In fact, construction contracts generally require that before starting any portion of the project, the contractor must carefully study and compare the drawings and the specifications and report inconsistencies to the architect.

If the conflict between the specifications and the construction drawings goes unnoticed initially but later results in a dispute, the courts have in most cases resolved in favor of the specifications—implying that the specifications, not the drawings, govern the project. However, if the owner or the design team wishes to reverse the order, it may be so stated in the owner-contractor agreement.

The Construction Document Set

Just as the construction drawings are prepared separately by the architect and each consultant for their respective portions of the work, so are the specifications. The specifications from various design team members are assembled by the architect in a single document,

EXPAND YOUR KNOWLEDGE

Relationship Between Construction Drawings and Specifications

Construction Drawings	Specifications
Design intent represented graphically	Design intent represented with words
Product/material may be shown many times	Product/material described only once
Product/material shown generically	Product/material identified specifically, sometimes proprietary to a manufacturer
Quantity indicated	Quality indicated
Location of elements established	Installation requirements of elements established
Size, shape, and relationship of building elements provided	Description, properties, characteristics, and finishes of building elements provided

called the *project manual.* Because the specifications consist of printed (typed) pages (not graphic images), a project manual is a bound document—like a book.

The major component of a project manual is the specifications. However, the project manual also contains other items, as explained later in this chapter.

The set of construction drawings (from various design team members) and the project manual together constitute what is known as the *construction document set,* Figure 1.3. The construction document set is the document that the architect uses to invite bids from prospective contractors.

OWNER'S ROLE

The owner's role in the design phase of the project may not appear as active as in the predesign phase, but it is important all the same. In fact, a conscientious owner will be fully involved in the entire breadth of the project delivery process—from the predesign phase through the project closeout phase.

PRACTICE QUIZ

Each question has only one correct answer. Select the choice that best answers the question.

1. The realization of a typical building project, as described in this text, may be divided into
   a. two phases.
   b. three phases.
   c. four phases.
   d. five phases.
   e. six phases.
2. Establishing the project's economic feasibility and its overall budget is a part of the design phase of the project.
   a. True
   b. False
3. The term MEPF stands for
   a. mechanical, electrical, piping, and foundations.
   b. mechanical, electrical, plumbing, and foundations.
   c. mechanical, electrical, plumbing, and fire.
   d. mechanical, electrical, piping, and fire.
4. The program for a building project is usually provided by the
   a. owner.
   b. general contractor.
   c. building official of the city.
   d. architect.
   e. any one of these, depending on the type of the building.
5. In a typical building project, the coordination of the building's design is done by the
   a. owner.
   b. general contractor.
   c. building official of the city.
   d. architect.
   e. any one of these, depending on the type of the building.
6. The construction drawings of a building project are prepared during the
   a. SD stage of the project.
   b. DD stage of the project.
   c. CD stage of the project.
   d. preconstruction phase of the project.
   e. construction phase of the project.
7. The construction drawings of a building project are drawings that the architect uses to explain the design to the owner.
   a. True
   b. False
8. The construction drawings of a building project are generally in the format of
   a. freehand sketches.
   b. two-dimensional plans, elevations, sections, and details.
   c. three-dimensional drawings.
   d. photographs of three-dimensional scale model(s).
   e. all the above.
9. The construction drawings for a building project generally consist of
   a. architectural drawings.
   b. structural drawings.
   c. MEPF drawings.
   d. all the above.
   e. (a) and (b) only.

Answers: 1-d, 2-b, 3-c, 4-a, 5-d, 6-c, 7-b, 8-b, 9-d.

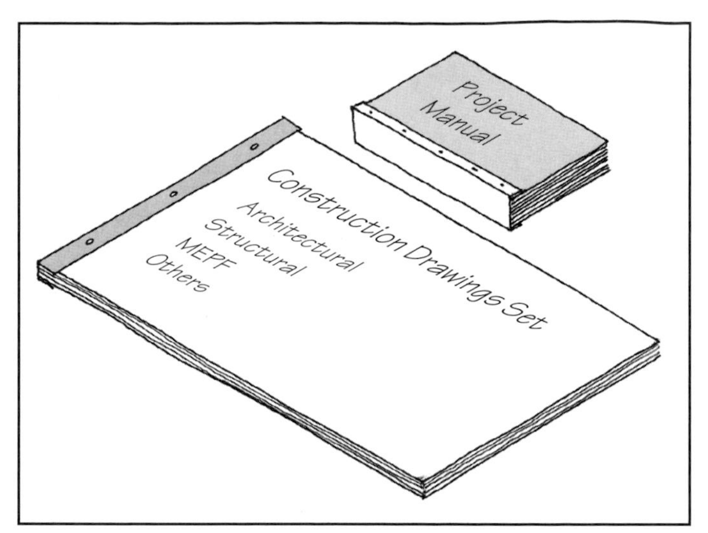

FIGURE 1.3 A construction document set consists of a set of architectural and consultants' construction drawings plus the project manual. The project manual is bound in a book format.

1.4 CSI MASTERFORMAT AND SPECIFICATIONS

The specification document for even a modest-size project can run into hundreds of pages. It is used not only by the contractor and the subcontractors, but also by the owner, material suppliers, and the site superintendent. With so many different people using it, it is necessary that the specifications he organized in a standard format, so that each user can go to the section of particular interest without having to wade through the entire document.

The standard organizational format for specifications, referred to as the *MasterFormat,* has been developed by The Construction Specifications Institute (CSI) and is the format most commonly used in the United States and Canada. The MasterFormat consists of 50 divisions, which are identified using six-digit numbers.

The first two digits of the numbering system (referred to as Level 1 digits) identify the division number. The 50 division numbers are 00, 01, 02, 03, . . . , 48, and 49. A division identifies the broadest collection of related products and assemblies.

The next two digits of the numbering system (Level 2 digits) refer to various sections within the division, and the last two digits (Level 3 digits) refer to the subsections within a section. In other words, Level 2 and Level 3 digits classify products and assemblies into progressively closer affiliations. Thus, Level 1 digits in MasterFormat may be compared to chapter numbers in a book, Level 2 digits with section numbers of a chapter, and Level 3 digits with subsection numbers of a section.

A complete list of the MasterFormat titles is voluminous. Figure 1.4 gives the division titles and also the additional details of one of the divisions—Masonry—as a broad illustration of the numbering system. Note that apart from the classification in divisions, the MasterFormat is classified in two groups: *Procurement and Contracting Group* (Division 00) and *Specifications Group* (Divisions 01 to 49).

Because the MasterFormat deals with all types of construction (new facilities, renovation, facility maintenance, services, urban infrastructural construction, equipment, and so forth), the Specification Group has been divided into four subgroups, as shown in Figure 1.4.

Recollecting MasterFormat Division Sequence

Architectural design typically involves Divisions 2 to 14 of the Facilities Construction Subgroup. Although the basis for sequencing the divisions in this subgroup is far more complicated, the first few divisions (those that are used in virtually all buildings) may be deduced by visualizing the sequence of postearthwork operations required in constructing the simple building shown in Figure 1.5. The building consists of load-bearing masonry walls, steel roof joists, and wood roof deck.

The first operation is the foundations for the walls. Because foundations are typically made of concrete, *Concrete* is Division 03. After the foundations have been completed, masonry work for the walls can begin. Thus, *Masonry* is Division 04. After the walls are completed, steel roof joists can be placed. Thus, Division 05 is *Metals.*

The installation of wood roof deck follows the joists. Hence, *Wood, Plastics, and Composites* is Division 06. Wood, plastics, and composites are grouped together because they are chemically similar hydrocarbons.

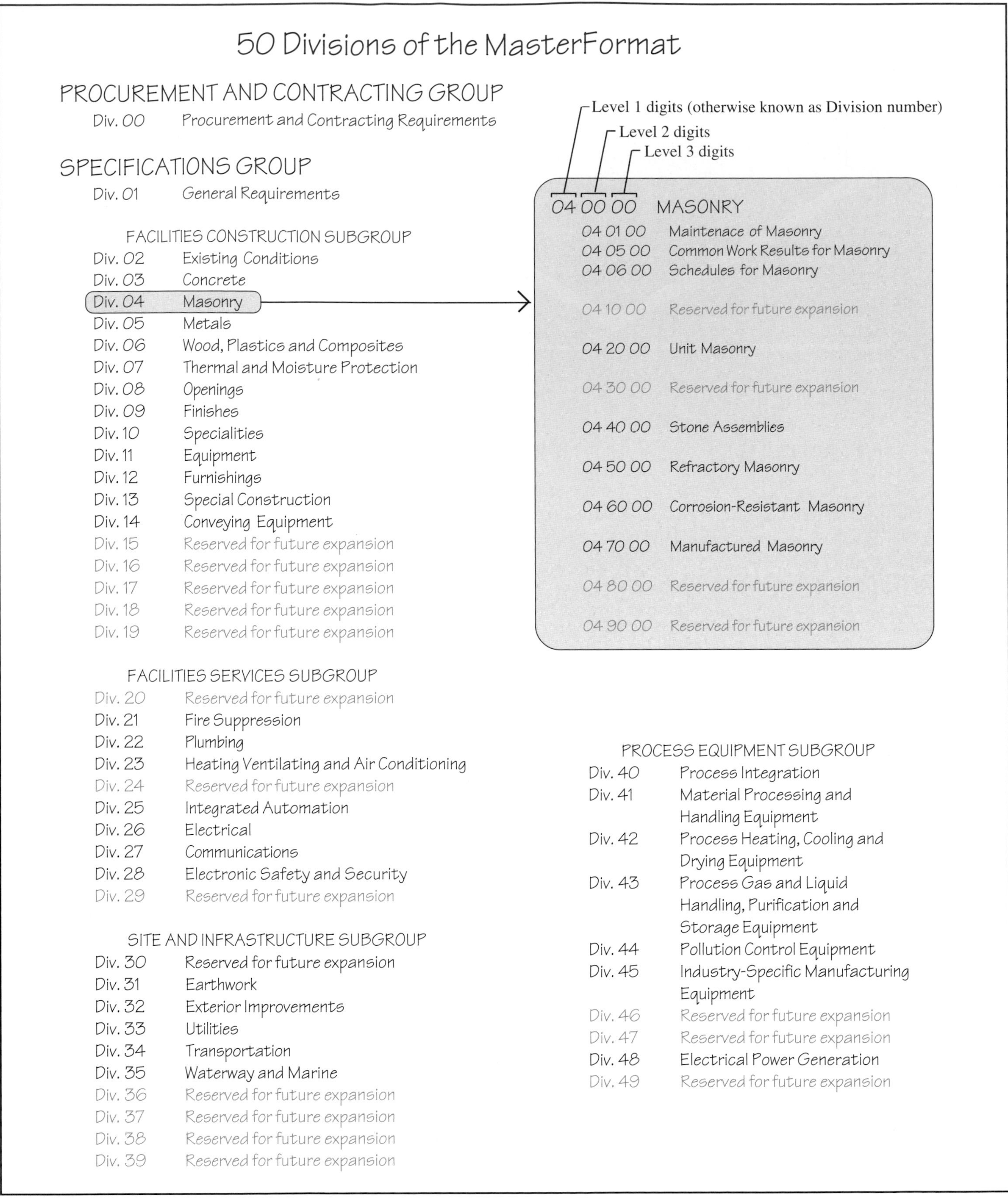

FIGURE 1.4 MasterFormat divisions. The Masonry division has been further elaborated as an illustration.

After the roof deck is erected, it must be insulated and protected against weather. Therefore, *Thermal and Moisture Protection* is Division 07. Roofing and waterproofing (of basements) are part of this division, as are wall insulation and joint sealants. The next step is to protect the rest of the envelope; hence, Division 08 is *Openings*. All doors and windows are a part of this division, regardless of whether they are made of steel, aluminum, or wood.

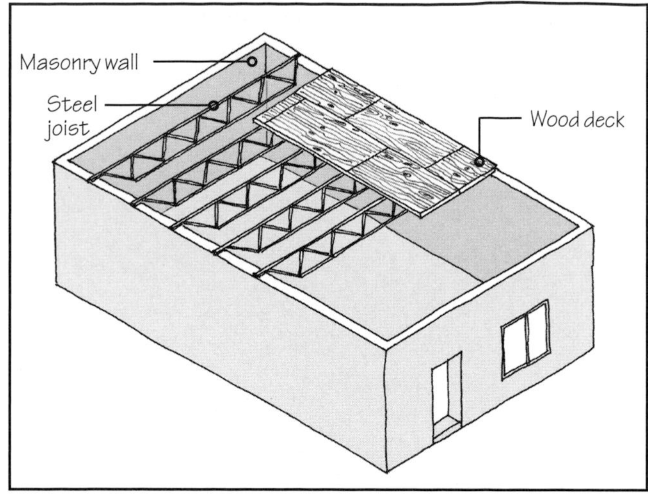

FIGURE 1.5 A simple load-bearing masonry wall building with steel roof trusses and wood roof deck, used as an aid to recalling the sequence of the first few architecturally important divisions of the MasterFormat.

With the envelope protected, finish operations, such as interior drywall, flooring, and ceiling, can begin. Thus, Division 09 is *Finishes*. Division 10 is *Specialities*, which consists of several items that cannot be included in the previous divisions, such as toilet partitions, lockers, storage shelving, and movable partitions.

Obviously, the building must now receive all the necessary office, kitchen, laboratory, or other equipment. Thus, Division 11 is *Equipment*. Division 12 is *Furnishings*, followed by *Special Construction* (Division 13) and *Conveying Equipment* (Division 14).

Before any construction operation can begin, there must be references to items that apply to all divisions, such as payment procedures, product-substitution procedures, contract-modification procedures, contractor's temporary facilities, and regulatory requirements imposed by the city or any other authority having jurisdiction. This is Division 01, called the *General Requirements*. Division 00 (*Procurement and Contracting Requirements*) refers to the requirements for the procurement of bids from prospective contractors.

CONSTRUCTION-RELATED INFORMATION

Familiarity with the MasterFormat is required to prepare the project manual and writing the specifications for the project. It is also helpful in filing and storing construction information in an office. Material manufacturers also use MasterFormat division numbers in catalogs and publications provided to design and construction professionals.

The MasterFormat is also helpful when seeking information about a construction material or system, as any serious student of construction (architect, engineer, or builder) must frequently do.

1.5 PRECONSTRUCTION (BID NEGOTIATION) PHASE

The preconstruction phase generally begins after the construction drawings and specifications have been completed and culminates in the selection of the construction team. The construction of even a small building involves so many specialized skills and trades that the work cannot normally be undertaken by a single construction firm. Instead, the work is generally done by a team consisting of the *general contractor* and a number of speciality *subcontractors*.

Thus, a project may have roofing; window and curtain wall; and heating, plumbing, ventilation, and air-conditioning (HVAC) subcontractors, among others, and so on, in addition to the general contractor. The general contractor's own work may be limited only to the structural components of the building—basements and foundations, load-bearing walls, reinforced concrete beams and columns, roof, floor slabs, and other components—with all the remaining work subcontracted.

In contemporary projects, however, the trend is toward the general contractor not performing any actual construction work but subcontracting the work entirely to various subcontractors. Because the subcontractors are contracted by the general contractor, only the general contractor is responsible and liable to the owner.

NOTE

Difference Between Specialities (Division 10) and Special Construction (Division 13)

Specialities (Division 10) includes prefabricated items such as marker boards, chalkboards, tackboards, lockers, shelves, **grilles** and screens, **louvers** and vents, flagpoles, manufactured fireplaces, and demountable partitions.

Special Construction (Division 13) includes items that are generally site fabricated but are not covered in other divisions, such as air-supported fabric structures, swimming pools, ice rinks, aquariums, planetariums, geodesic structures, and sound and vibration control.

NOTE

Important Items Included in Division 00

Procurement and Contracting Requirements

Advertisements for bids
Invitation to bid
Instruction to bidders
Prebid meetings
Land survey information
Geotechnical information
Bid forms
Owner-contractor agreement forms
Bond forms
Certificate of substantial completion form
Certificate of completion form
Conditions of the contract

In some cases, a subcontractor will, in turn, subcontract a portion of his or her work to another subcontractor, referred to as a *second-tier subcontractor,* Figure 1.6 In that case, the general contractor deals only with the subcontractor, not the second-tier subcontractor.

Whether the general contractor performs a part of the construction work or subcontracts the entire work, the key function of the general contractor is the overall management of construction. This includes coordinating the work of all subcontractors, ensuring that the work done by them is completed in accordance with the contract documents, and ensuring the safety of all workers on the site. A general contractor with a good record of site safety not only demonstrates respect for the workers but also improves profit margin by lowering the cost of construction insurance.

NOTE

Important Items Included in Division 01

General Requirements

Summary of work
Price and payment procedures
Product substitution procedures
Contract modification procedures
Project management and coordination
Construction schedule and documentation
Contractor's responsibility
Regulatory requirements (codes, laws, permits, etc.)
Temporary facilities
Product storage and handling
Owner-supplied products
Execution and closeout requirements

SELECTING A GENERAL CONTRACTOR

Several methods are used in selecting a general contractor to suit the peculiarities of the project and particular needs of the owner. Here are four commonly used methods:

- Competitive bidding method
- Invitational bidding method
- Negotiated contract method
- Multiple prime contract method

COMPETITIVE BIDDING

On most publicly funded projects, the award of a construction contract to the general contractor is based on *competitive bidding.* This refers to the process by which qualified contractors are invited to bid on the project. The invitation is generally through advertisements in newspapers, trade publications, and other public media.

The advertisement for bids includes a description of the project, its location, where to obtain the bidding documents, the price of the bidding documents, bid opening date and location, and other important information. The purpose of the advertisement is to notify and thereby attract a sufficient number of contractors to compete for the construction contract.

The general contractor's bid for the project is based on the information provided in the *bidding documents.* The bidding documents are essentially the construction document set with such additional items as the instructions to bidders, requirements with respect to the financial and technical status of bidders (see the box on surety bonds), and the contract agreement form that the successful bidder will sign when the contract is awarded. Because these additional items are text items, they are bound together as a project manual, Figure 1.7.

NOTE

Relationship Between Architect and General Contractor

A notable feature of design and construction contracts is that they are all two-party contracts, that is, owner-architect contract, architect-consultant contract, or owner general contractor contract, etc. There is no direct contractual relationship between the owner and the subcontractor. Similarly, there is no contractual relationship between the architect and the general contractor (see Figure 1.6). In the overall interest of the project, however, and as per the terms of the owner-contractor agreement, the owner generally communicates with the general contractor through the architect.

In the competitive bidding method, the bidding documents are generally given only to contractors who are capable, by virtue of their experience, resources, and financial standing, to bid for the project. Therefore, the architect (as the owner's agent) may prescreen the bidders with respect to their reputation and capability to undertake the project.

An exception to prescreening for the release of bidding documents involves projects funded by the federal, state, or local government, for which almost anyone can access the

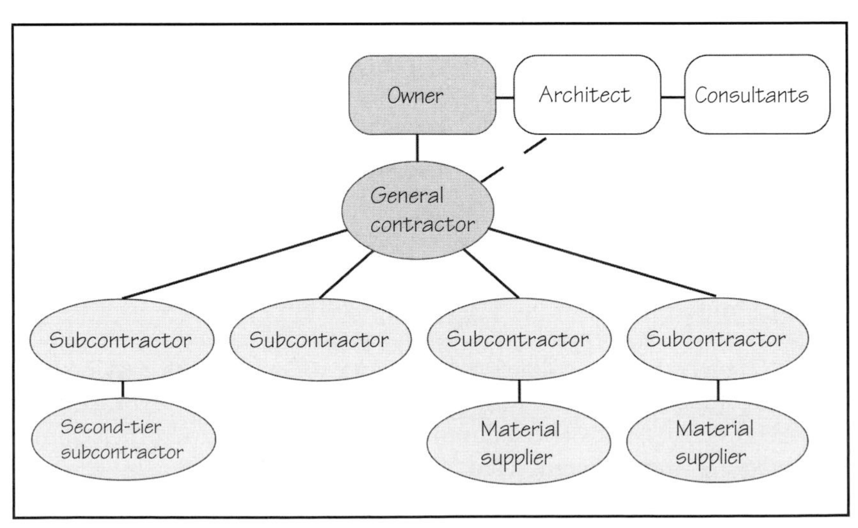

FIGURE 1.6 The construction team and their interrelationships with each other and the owner. A solid line in this illustration indicates a contractual relationship between parties. A dashed line indicates a communication link. The relationships shown here are not absolute and may change somewhat with the nature of the project.

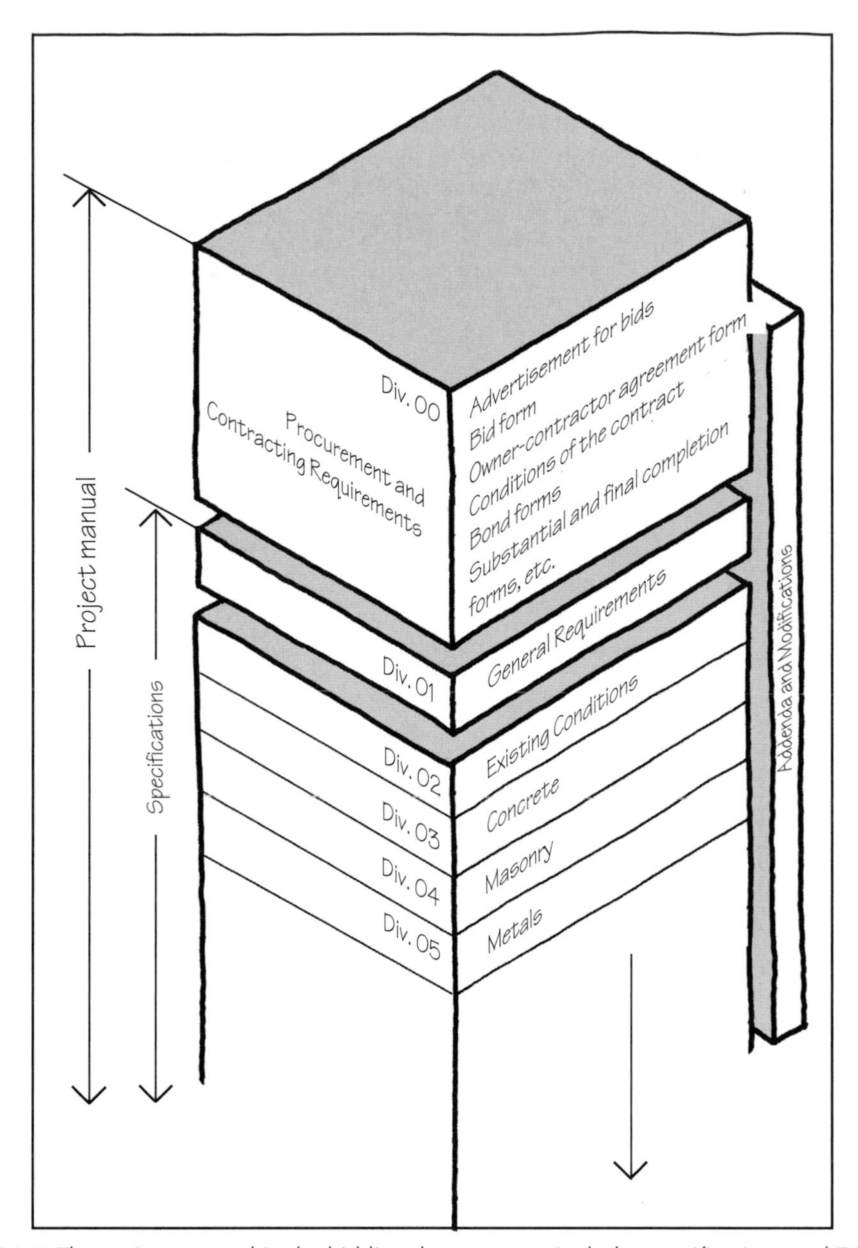

FIGURE 1.7 The project manual in the bidding document set includes specifications and Division 00. After the general contractor has been selected, the project manual (in the contract document set) generally excludes Division 00.

bidding documents. However, even in this kind of project, the number of contractors who can actually submit the bids is limited.

This limitation is generally the result of the financial security required from the bidders, known as a *bid bond*. The bidder must obtain a bid bond from a surety company in the amount specified in the bidding documents. This bond is issued based on the contractor's experience, ability to perform the work, and financial resources required to fulfill contractual obligations.

Whether or not the release of the bidding documents is restricted, the procedure stated earlier ensures that all the bidders are similarly qualified with respect to experience, technical expertise, and financial status. Because all bidders receive the same information and are of the same standing, the competition is fair; therefore, the contract is generally awarded to the lowest bidder.

INVITATIONAL BIDDING

Invitational bidding is a variation of the competitive bidding method that may be followed for some private projects. This procedure allows the owner to preselect general contractors who have demonstrated, based on their experience, resources, and financial standing, their qualifications to perform the work. The architect (as the owner's agent) may conduct the prescreening process.

NEGOTIATED CONTRACT

A method of selecting the general contractor without seeking bids is referred to as a *negotiated contract*. This method is used when the owner knows of one or more reputable, competent, and trusted general contractors. In this case, the owner simply negotiates with these

EXPAND YOUR KNOWLEDGE

Surety Bonds

The purpose of a surety bond is to ensure that should the contractor fail to fulfill contractual obligations, there will be a financially sound party—referred to as the *surety* (also called *guarantor* or *bonding company*)—available to take over those unfulfilled obligations. The bond is, therefore, a kind of insurance that the contractor buys from a surety, generally a corporation.

There are three types of surety bonds in most building projects. A few others may be required in some special projects:

- Bid bond
- Performance bond
- Payment bond

Bid Bond

The purpose of the bid bond (also called the *bid security bond*) is to exclude frivolous bidders. It ensures that if selected by the owner, the bidder will be able to enter into a contract with the owner based on the bidding requirements and that the bidder will be able to obtain performance and payment bonds from an acceptable surety.

A bid bond is required at the time the bidder submits the bid for the project. If the bidder refuses to enter into an agreement or is unable to provide the required performance and payment bonds, the surety is obliged to pay the penalty (bid security amount), usually 5% of the bid amount, to the owner.

Performance Bond

The performance bond is required by the owner before entering into an agreement with the successful bidder. The performance bond ensures that if, after the award of the contract, the contractor is unable to perform the work as required by the bidding documents, the surety will provide sufficient funds for the completion of the project.

A performance bond protects the owner against default by the contractor or by those for whose work the contractor is responsible, such as the subcontractors. For that reason, the contractor will generally require a performance bond from major subcontractors.

Payment Bond

A payment bond (also referred to as a *labor and materials bond*) ensures that those providing labor, services, and materials for the project—such as the subcontractors and material suppliers—will be paid by the contractor. In the absence of the payment bond, the owner may be held liable to those subcontractors and material suppliers whose services and materials have been invested in the project. This liability exists even if the owner has paid the general contractor for the work of these subcontractors and material suppliers.

Pros and Cons of Bonds

The bonds are generally mandated for a publicly-funded project. In a private project, the owner may waive the bonds, particularly the bid bond. This saves the owner some money because although the cost of a bond (the premium) is paid by the contractor, it is in reality paid by the owner because the contractor adds the cost of the bond to the bid amount.

Despite its cost, most owners consider the bonds (particularly the performance and payment bonds) to be a good value because they eliminate financial risks of construction. The bid bond may be unnecessary in a negotiated bid where the owner knows the contractor's financial standing and the ability to perform. However, where uncertainty exists, a bid bond provides an excellent prequalification screening of the contractor. Responsible contractors generally maintain a close and continuous relationship with their bonding company so that the bonding company's knowledge of a contractor's capabilities far exceeds that of most owners or architects (as an owner's representative).

contractors for the overall contract price, time required for completion, and other important details of the project. The negotiations are generally conducted with one contractor at a time.

A major advantage of the negotiated contract is that the general contractor can be on board during the design or predesign phase. This helps the owner ensure that the architect's design is realistically constructible. Additionally, because the contractor is the one who is most knowledgeable about construction costs, budget estimates can be obtained at various stages during the design phase. Because the vast majority of owners have to work within a limited budget, the negotiated contract is by far the most popular mode of contracting for private projects.

The negotiated contract is not devoid of all competition, because the general contractor obtains competitive bids from numerous subcontractors and material suppliers. Because the general contractor is selected during the SD or DD stage, the bids from the subcontractors can be obtained earlier, providing the potential to shorten project delivery time.

The services offered by the contractor during the design phase of a negotiated contract are referred to as contractor's *preconstruction services*. Realizing their importance, public agencies sometimes use general contracting or construction management firms to obtain preconstruction services in competitively bid projects (Section 1.10; "Construction Management Method").

Multiple Prime Contracts—Fast-Track Project Delivery

A variation of the negotiated contract, which can save project delivery time, is one in which the project is divided into multiple phases, and each phase of construction is awarded to different contractors through negotiations. The division of the project into phases is such that the phases are sequential. Thus, the first phase of the project may be site construction (site development, excavations, and foundations), the second phase may be the structural framing (columns, beams, and floor and roof slabs), the third phase may be roofing, and the subsequent phase(s) take the project to completion.

In this method of contractor selection, there is no single general contractor, the multiple contractors are each referred to as a *prime contractor*.

Sequential phasing of the project saves time because the earlier phases of the project can be constructed while the construction documents for the later phases are still in progress. Thus, the design and construction phases of the project overlap.

This project delivery method, referred to as *fast-track project delivery,* requires a great deal of coordination between phases and (several) contractors, particularly because there is no single contracting authority. It also requires a commitment from the owner that the decisions will not be delayed and, once made, will not be changed.

By contrast, the competitive (or invitational) bidding project delivery is slower because all the construction documents must be completed before the general contractor is selected. The competitive (or invitational) bidding method is also referred to as the *traditional,* or *design-bid-build,* method because each of the three phases—designing, bidding, and construction—are sequential, and one phase does not begin unless the previous phase is completed.

PRACTICE QUIZ

Each question has only one correct answer. Select the choice that best answers the question.

10. The MasterFormat has been developed by the
- **a.** Construction Specifications Institute.
- **b.** American Society for Testing and Materials.
- **c.** American National Standards Institute.
- **d.** American Institute of Architects.
- **e.** Associated General Contractors of America.

11. The MasterFormat consists of
- **a.** 20 divisions.
- **b.** 30 divisions.
- **c.** 40 divisions.
- **d.** 50 divisions.
- **e.** none of the above.

12. In the MasterFormat, Division 02 refers to
- **a.** General Requirements.
- **b.** Existing Conditions.
- **c.** Masonry.
- **d.** Metals.
- **e.** none of the above.

13. In the MasterFormat, Division 04 refers to
- **a.** General Requirements.
- **b.** Existing Conditions.
- **c.** Masonry.
- **d.** Metals.
- **e.** none of the above.

14. In the MasterFormat, windows are part of
- **a.** Division 05.
- **b.** Division 06.
- **c.** Division 07.
- **d.** Division 08.
- **e.** none of the above.

15. In the MasterFormat, roofing is part of
- **a.** Division 05.
- **b.** Division 06.
- **c.** Division 07.
- **d.** Division 08.
- **e.** none of the above.

16. In the MasterFormat, flooring is part of
- **a.** Division 05.
- **b.** Division 06.
- **c.** Division 07.
- **d.** Division 08.
- **e.** none of the above.

17. In the traditional project delivery (design-bid-build) method, the owner has separate contracts with the general contractor and the subcontractors.
- **a.** True
- **b.** False

18. Who is responsible for ensuring the safety of workers on a construction site of a typical building project?
- **a.** The architect
- **b.** The structural engineer
- **c.** The general contractor
- **d.** The owner
- **e.** A collective responsibility of all the above

19. In the traditional project delivery (design-bid-build) method for a building, there is generally
- **a.** one general contractor.
- **b.** one general contractor and one subcontractor.
- **c.** one general contractor and several subcontractors.
- **d.** several general contractors and several subcontractors.

Answers: 10-a, 11-d, 12-b, 13-c, 14-d, 15-c, 16-e, 17-b, 18-c, 19-c.

1.6 CONSTRUCTION PHASE

Once the general contractor has been selected and the contract awarded, the construction work begins, as described in the *contract documents.* The contract documents are virtually the same as the bidding documents, except that the contract documents are a signed legal contract between the owner and the contractor. They generally do not contain Division 00 of the MasterFormat.

In preparing the contract documents, the design team's challenge is to efficiently produce the graphics and text that effectively communicate the design intent to the construction professionals and the related product suppliers and manufacturing industries so that they can do the following:

- Propose accurate and competitive bids
- Prepare detailed and descriptive submittals for approval
- Construct the building, all with a minimum of questions, revisions, and changes

NOTE

Construction Documents and Contract Documents

The terms *contract documents* and *construction documents* are used interchangeably. Although essentially the same set of documents, the construction documents become contract documents when they are incorporated into the contract between the owner and the contractor.

SHOP DRAWINGS

The construction drawings and the specifications should provide a fairly detailed delineation of the building. However, they do not describe it to the extent that fabricators can produce building components directly from them. Therefore, the fabricators generate their own drawings, referred to as *shop drawings,* to provide the higher level of detail necessary to fabricate and assemble the components.

Shop drawings are not generic, consisting of manufacturers' or suppliers' catalogs, but are specially prepared for the project by the manufacturer, fabricator, erector, or subcontractors. For example, an aluminum window manufacturer must produce shop drawings to show that the required windows conform with the construction drawings and the specifications. Similarly, precast concrete panels, stone cladding, structural steel frame, marble or granite flooring, air-conditioning ducts, and other components require shop drawings before they are fabricated and installed.

Before commencing fabrication, the fabricator submits the shop drawings to the general contractor. The general contractor reviews them, marks them "approved", if appropriate, and then submits them to the architect for review and approval. Subcontractors or manufacturers cannot submit shop drawings directly to the architect.

The review of all shop drawings is coordinated through the architect, even though they may actually be reviewed in detail by the appropriate consultant. Thus the shop drawings pertaining to structural components are sent to the architect, and then to the structural consultant for review and approval. The fabricator generally begins fabrication only after receiving the architect's approval of the shop drawings.

The approval of the shop drawings by the architect is a review to check that the work indicated therein conforms with the overall design intent shown in the contract documents. Approval of shop drawings that deviate from the contract documents does not absolve the general contractor of the responsibility to comply with the contract documents for quality of materials, workmanship, or the dimensions of the fabricated components.

MOCK-UP SAMPLES

In addition to shop drawings, full-size mock-up samples of one or more critical elements of the building may be required in some projects. This is done to establish the quality of materials and workmanship by which the completed work will be judged. For example, it is not unusual for the architect to ask for a mock-up of a typical area of the curtain wall of a high-rise building before the fabrication of the actual curtain wall is undertaken. Mock-up samples go through the same approval process as the shop drawings.

XYZ
Architects
Any City

Review is only for the limited purpose of checking for conformance with information given and the design concept expressed in the Contract Documents. Review is not conducted for the purpose of determining the accuracy or completeness of other details such as dimensions or quantities or for substantiating instructions for installation or performance of equipment or systems designed by the Contractor, all of which remain the responsibility of the Contractor. Review or acceptance of a specific item shall not indicate approval of an assembly of which the item is a component. Review neither extends nor alters any contractual obligations of the Architect or Contractor and review shall not relieve the Contractor of responsibility for any deviation from the requirements of the Contract Documents.

- ☐ Accepted as Noted
- ☐ Revise, Resubmit
- ☐ Product Substitution, Submit in accordance with Section 01630
- ☐ Value Engineering, Submit in accordance with Section 01630
- ☐ Not Reviewed; Submittal not required by contract documents
- ☐ Reviewed for Architectural Issues Only
- ☐ Reviewed for Project Close Out Requirements Only

By: Date:

FIGURE 1.8 A typical stamp used by architects to indicate the result of review of shop drawings and other submittals.

OTHER SUBMITTALS

In addition to shop drawings and any mock-up samples, some other submittals required from the contractor for the architect's review are:

- Product material samples
- Product data
- Certifications
- Calculations

A typical stamp used by the architect on the submittals to indicate the outcome of the review is shown in Figure 1.8.

1.7 CONSTRUCTION CONTRACT ADMINISTRATION

The general contractor will normally have his or her own inspection process to ensure that the work of all subcontractors is progressing as indicated in the contract documents and that the work meets the standards of quality and workmanship. On smaller projects, this may be done by the *project superintendent.* On large projects, a team of quality-assurance inspectors generally assists the contractor's project superintendent. These inspectors are individuals who, by training and experience, are specialized in their own areas of construction—for example, concrete, steel, or masonry.

Additional quality control is required by the contract through the use of independent testing laboratories. For instance, structural concrete to be used on the site must be verified for strength and other properties through independent concrete testing laboratories.

Leaving quality control of materials and performance entirely in the hands of the contractor is considered inappropriate. It can leave the owner vulnerable to omissions and errors in the work and, in some instances, to unscrupulous activities. Therefore, the owner usually retains the services of the architect to provide an additional level of scrutiny to administer the construction contract. If not, the owner will retain another independent architect, engineer, or inspector to provide construction contract administration services.

OBSERVATION OF CONSTRUCTION

The architect's role during the construction phase has evolved over the years. There was a time when architects provided regular supervision of their projects during construction, but the liability exposure resulting from the supervisory role became so adverse for the architects that they have been forced to relinquish this responsibility. Instead, the operative term for the architect's role during construction is *field observation* of the work.

The observational role still allows the architects to determine that their drawings and specifications are transformed to reality just as they had conceived. It also provides sufficient safeguard against the errors caused by the contractors' misinterpretation of contract documents in the absence of the architects' clarification and interpretation.

The shift in the architect's role to observer of construction also recognizes the important and entirely independent role that the contractor must play during construction. This recognition provides full authority to the general contractor to proceed with the work in the manner that the contractor deems appropriate. Remember:

- The architect determines the *what* and *where.*
- The contractor determines the *how* and *when.*

In other words, daily supervision or superintendence of construction is the function of the contractor—the most competent person to fulfill this role. The architect provides periodic observation and evaluation of the contractor's work and notifies the owner if the work is not in compliance with the intent of the contract documents. This underscores the dividing line between the responsibilities of the architect and the contractor during construction.

Note that by providing observation, the architect does not certify the contractor's work. Nor does the observation relieve the contractor of its responsibilities under the contract. The contractor remains fully liable for any error that has not been discovered through the architect's observation. However, the architect may be held liable for all or a part of the work observed, should the architect fail to detect or provide timely notification of work not conforming with the contract documents. This is known as *failure to detect.*

Because many components can be covered up by other items in days or hours, the architect should visit the construction site at regular intervals, appropriate to the progress of construction. For example, earthwork covers foundations and underground plumbing, and gypsum board covers ceiling and wall framing. Observing the work after the components are hidden defeats the purpose of observation.

On some projects, a resident project architect or engineer may be engaged by the architect, at an additional cost to the owner, to observe the work of the contractor. Under the conditions of the contract, the contractor is generally required to provide to this person a trailer office, water, electricity, a telephone, and other necessary facilities.

INSPECTION OF WORK

There are only two times during the construction of a project (Section 1.8) that the architect makes an exception to being an observer of construction. At these times, the architect inspects the work. These inspections are meant to verify the general contractor's claim that the work is (a) substantially complete and (b) is fully performed and hence is ready for final payment. These inspections are, therefore, referred to as follows:

- Substantial completion inspection
- Final completion inspection

PAYMENT CERTIFICATIONS

In addition to construction observation and inspection, there are several other duties the architect must discharge in administering the contract between the owner and the contractor. These are outlined in the box entitled "Summary of Architect's Functions as Construction Contract Administrator." Certifying (validating) the contractor's periodic requests for payment against the work done and the materials stored at the construction site is perhaps the most critical of these functions.

A request for payment (typically made once a month unless stated differently in the contract) is followed by the architect's evaluation of the work and necessary documentation to verify the contractor's claim. Because the architect is not involved in day-to-day supervision, the issuance of the certificate of payment by the architect does not imply acceptance of the quality or quantity of the contractor's work. However, the architect has to be judicious and impartial to both the owner and the contractor and be within the bounds of the contract.

NOTE

Summary of Architect's Functions as Construction Contract Administrator

- Observe construction
- Inform the owner of the progress of work
- Guard the owner against defects and deficiencies
- Review and approve shop drawings, mock-up samples, and other submittals
- Prepare change orders, if required
- Review correspondence between the owner and the contractor and take action if required
- Prepare certificates of payment
- Make substantial completion inspection
- Make final completion inspection
- Review manufacturers' and suppliers' warranties and other project closeout documentation and forward them to the owner
- Make a judicious interpretation of the contract between the owner and the contractor when needed

CHANGE ORDERS

There is hardly a construction project that does not require changes after construction has begun. The contract between the owner and the contractor recognizes this fact and includes provisions for the owner's right to order a change and the contractor's obligation to accept the *change order* in return for an equitable price adjustment. Here again, the architect performs a quasi-judicial role to arrive at a suitable agreement and price between the owner and the contractor.

1.8 POSTCONSTRUCTION (PROJECT CLOSEOUT) PHASE

Once the project is sufficiently complete, the contractor will request the architect to conduct a *substantial completion* inspection to confirm that the work is complete in most respects. By doing so, the contractor implies that the work is complete enough for the owner to occupy the facility and start using it, notwithstanding the fact that there might be cosmetic and minor items yet to be completed.

The contractor's request for substantial completion inspection by the architect may include a list of incomplete corrective portions of the work, referred to as the *punch list.* The punch list, which is prepared by the contractor, is used by the architect as a checklist to review all work, not merely the incomplete portions of the work. If the architect's inspection discloses incomplete items not included in the contractor's punch list, they are added to the list by the architect.

The substantial completion inspection is also conducted by the architect's consultants, either with the architect or separately. Incomplete items discovered by them are also added to the list. If the additional items are excessive, the architect may ask the contractor to complete the selected items before rescheduling substantial completion inspection.

SUBSTANTIAL COMPLETION—THE MOST IMPORTANT PROJECT DATE

Before requesting a substantial completion inspection, the contractor must submit all required guaranties and warranties from the manufacturers of equipment and materials and the speciality subcontractors and installers used in the building. For instance, the manufacturers of roofing materials, windows, curtain walls, mechanical equipment, and other materials, warrant their products for specified time periods. These warranties are in addition to the standard one-year warranty between the owner and the contractor.

The warranties are to be given to the architect at the time of substantial completion for review and transmission to the owner. Because the obligatory one-year warranty between the owner and the contractor, as well as other extended-time warranties, begin from the date of substantial completion of the project, the substantial completion date marks an important project closeout event. That is why the contractor is allowed a brief time interval to complete fully the work after the successful substantial completion inspection.

Before seeking a substantial completion inspection, the contractor is generally required to secure a *certificate of occupancy* (Chapter 2) from the authority having jurisdiction over the project—usually the city where the project is permitted and built. The certificate of occupancy confirms that all appropriate inspections and approvals have taken place and that the site has been cleared of the contractor's temporary facilities so that the owner can occupy the building without obligations to any authority.

CERTIFICATE OF FINAL COMPLETION

After the contractor carries out all the corrective work identified during substantial completion inspection and so informs the architect, the architect (with the assistance of the consultants) carries out the final inspection of the project. If the final inspection passes, the *certificate of final completion* is issued by the architect, and the contractor is entitled to *final payment.*

Before the certificate of final completion is executed by the architect and, finally, the owner, the owner receives the record documents, keys and key schedule, equipment manuals, and other necessities. Additionally, the owner receives all legal documentation to indicate that the contractor will be responsible for claims made by any subcontractor, manufacturer, or other party with respect to the project.

After the certificate of final completion, the contractor is no longer liable for the maintenance, utility costs, insurance, and security of the project. These responsibilities and liabilities transfer to the owner.

RECORD DOCUMENTS

Changes of a minor nature are often made during the construction of a project. These changes must be recorded for the benefit of the owner, should the owner wish to alter or expand the building in the future. Therefore, after the building has been completed, the contractor is required to provide a set of *record drawings* (previously known as *as-built drawings*). These drawings reflect the changes that were made during the course of construction by the contractor.

In addition to record drawings, *record specifications,* as well as a set of approved shop drawings, are usually required to complete the record document package delivered to the owner.

1.9 ALTERNATIVE PROJECT DELIVERY METHODS

As stated in Section 1.5, several building projects are constructed on the basis of the owner awarding the contract to a general contractor, who, in turn, subcontracts much (or all) of the work to subcontractors. This is the traditional, design-bid-build method of project delivery, accomplished through competitive bidding, invitational bidding, or negotiated contract.

In recent times, three new methods of project delivery have become popular:

- Construction management (CM) method
- CM at risk (CMAR) method
- Design-build (DB) method

1.10 CONSTRUCTION MANAGEMENT (CM) METHOD

In the traditional design-bid-build project delivery method, the architect designs the project, prepares the bidding documents, and assists the owner in the selection of the general contractor. During the construction phase, the architect visits the site to observe the work in progress, advises the owner whether the work progress conforms with the contract documents, and takes action on the general contractor's requests for periodic payments to be made by the owner.

In other words, the architect functions in a limited sense as the owner's agent and provides professional service from the inception to the completion of the project. Various phases of this method, as described previously, are represented graphically in Figure 1.9.

In the 1970s, because of the large cost overruns and time delays on many projects, owners began to require architects to include a cost-estimating professional during the early stages of the design process. Because the best expertise in cost estimating and construction scheduling resides in the contracting community, it often meant involving the contractor during the design phase.

As the contracting community acquired the ear of the owner during the design phase, it began to influence many issues that in the traditional method were entirely within the architects' realm. The contractors' involvement in the design phase, the increased complexity of building projects, and the owners' push for timely and on-budget project deliveries made the contractors realize the need for professional management assistance in construction.

This gave birth to the full-time professional construction manager and to a new method of project delivery, referred to as the *construction management (CM) method.* In the CM method, the owner retains a construction manager as an agent to advise on such aspects as cost, scheduling, site supervision, site safety, construction finance administration, and overall building construction.

Note that the construction manager is not a contractor, but a manager who plays no entrepreneurial role in the project (unlike the general contractor, who assumes financial risks in the project). In most CM projects, the owner hires the construction manager as the first step. The CM may advise the owner in the selection of the architect and other members of the design team as well as the contracting team.

The birth of the CM delivery method does not mean that there was no construction management in the traditional, or design-bid-build method. It was there, but it was done informally and shared between the design team and the general contractor, both of whom had little formal training in the management and financial aspects of a project.

The introduction of a construction manager on the project transferred various functions of the general contractor (in a traditional method) to the construction manager. Thus, in the CM method, the *general* contractor became redundant, therefore, there is no *general* contractor in this method.

In the CM method of project delivery, the owner awards multiple contracts to various trade and speciality contractors, whose work is coordinated by the construction manager. Thus, the structural framework of the building may be erected by one contractor, masonry work done by another, interior drywall work by yet another contractor and so on.

Each contractor is referred to as the *prime contractor,* who may have one or more subcontractors, Figure 1.10. The task of scheduling and coordinating the work of all the contractors and ensuring site safety—undertaken by the general contractor in the traditional method—is, in the CM method, done by the construction manager on behalf of the owner. Additionally, the construction manager administers the contracts between the contractors and the owner.

Thus, the owner, by assuming a part of the role of the general contractor, eliminates the general contractor's markup on the work of the subcontractors. The owner may also receive some reduction in the fee charged by the architect for contract administration. Although these savings are partially offset by the fee that the owner pays to the construction manager, there can still be substantial savings in large but technically simple projects.

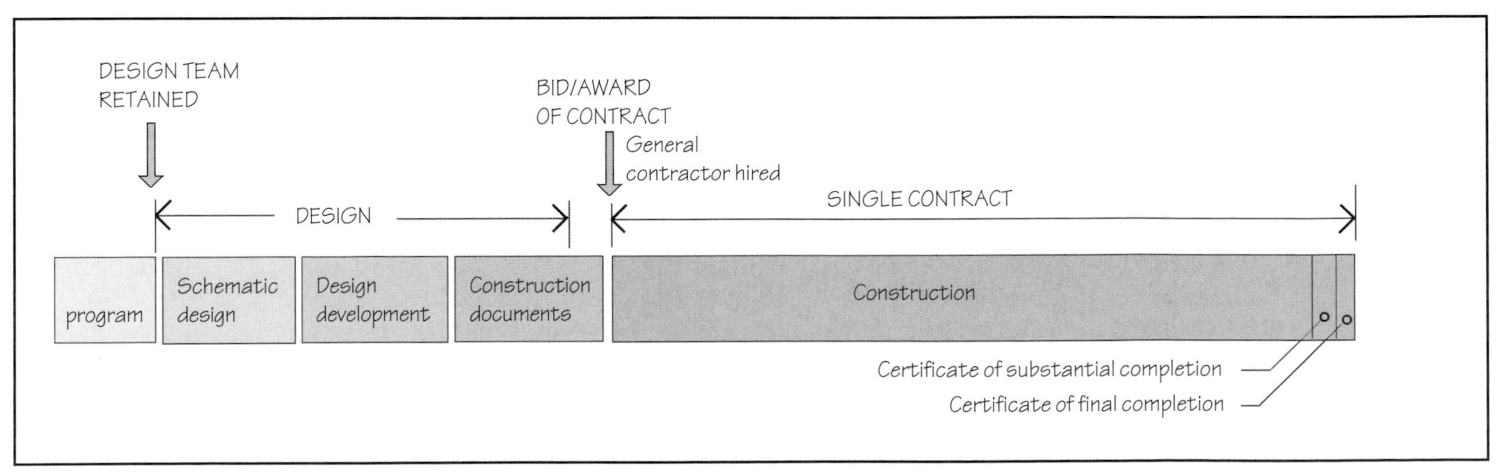

FIGURE 1.9 Sequence of operations in the traditional design-bid-build system of project delivery.

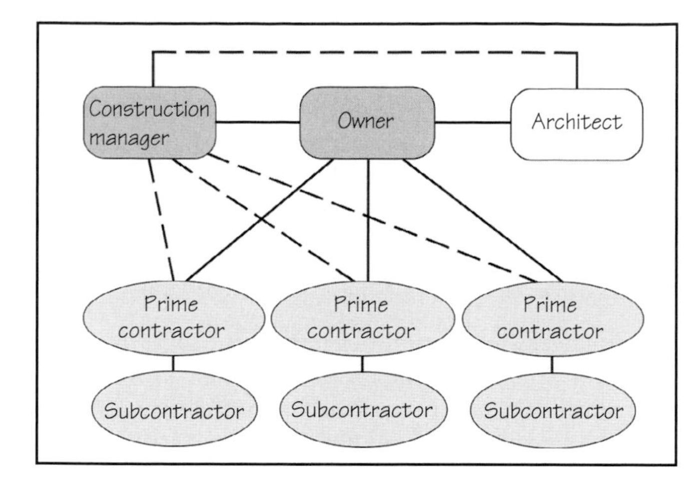

FIGURE 1.10 Contractual relationships between various parties in the CM method of project delivery. A solid line in this illustration indicates a contractual relationship between parties. A dashed line indicates only a communication link, not a contract.

The CM project delivery method is particularly attractive to owners who are knowledgeable about the construction process and can fully participate in all its aspects from bidding and bid evaluation to the closeout phase.

1.11 CM AT RISK (CMAR) METHOD

A disadvantage of the CM method lies in the liability risk that the owner assumes, which in the traditional design-bid-build method is held by the general contractor. This means that there is not the same incentive for the CM to optimize efficiency as when the CM carries financial risks.

Additionally, in the CM method, there is no single point of responsibility among the various prime contractors. Each prime contractor has a direct contract with the owner. Consequently, the CM has little leverage to ensure timely performance. The owner must, therefore, exercise care in selecting the construction manager because the cost, timeliness, and quality of the ultimate product are heavily dependent on the expertise of the construction manager.

In response to the preceding concerns, the CM method has evolved into what is known as the *CM at risk* (CMAR) method. In this method, the roles of the general contractor and the construction manager are performed by one entity, but the compensation for these roles is paid separately by the owner.

In the CMAR method, the owner contracts a CMAR company for (a) construction management services for a professional fee and (b) building the project as the general contractor. The CMAR company then works with the architect to develop construction documents that will meet the owner's budget and schedule. In doing so, the CMAR company functions as the owner's agent. Relationships between various functionaries in a CMAR project delivery method are shown in Figure 1.11.

After the drawings are completed, all the work is competitively bid by subcontractors, and the bids opened in the owner's presence. The work is normally awarded to subcontractors with the lowest bids. In working as the general contractor, the CMAR company assumes all responsibilities for subcontractors' work and site safety.

The CMAR method is being increasingly used for publicly funded projects such as schools, university residence halls, and apartments.

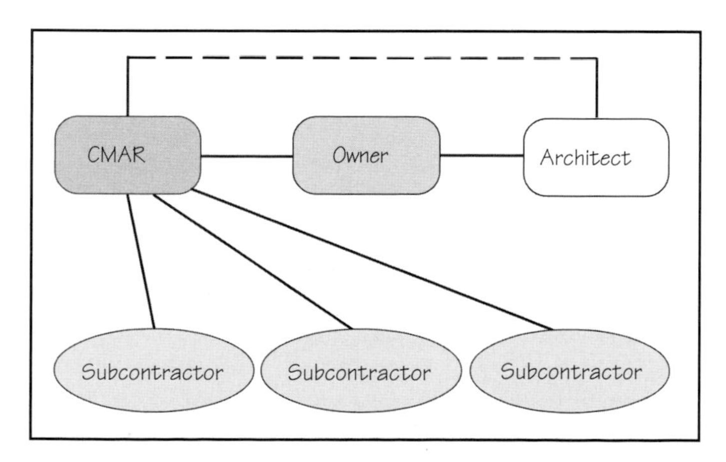

FIGURE 1.11 Contractual relationships between various parties in a CMAR method of project delivery. A solid line in this illustration indicates a contractual relationship between parties. A dashed line indicates only a communication link, not a contract.

1.12 DESIGN-BUILD (DB) METHOD

Another project delivery method that is gaining popularity is the *design-build (DB) method.* In this method, the owner awards the contract to one firm, which designs the project and also builds it, either on cost-plus-profit basis or lump-sum basis. In many ways, this method is similar to the historic *master-builder method,* in which there was no distinction between the architect and the builder. The design-build firm is usually a general contractor, who, in addition to the construction capabilities, has a design team of architects and engineers within the organization.

In this method, the architect's and engineer's responsibilities remain the same as in the traditional design-bid-build method; that is, they are liable for design errors. However, in the design-build method, their contribution is subservient to the project production schedule and budget because the owner's best interest may not always be served by the design team.

The design-build method has the advantage of an integrated design and construction team. This can provide a reduction in change orders for the owner, a faster project completion, and a single source of responsibility.

The method is not new because it has been in existence for decades in single-family residential construction. It is now being increasingly accepted in commercial construction—for both private and publicly funded projects. The establishment of the Design-Build Institute of America (DBIA) has further promoted the method.

A special version of design-build method, referred to as the *turnkey method,* consists of the design-build firm arranging for the land and finances for the project in addition to designing and constructing it.

PRACTICE QUIZ

Each question has only one correct answer. Select the choice that best answers the question.

20. A contract document set consists of
- **a.** construction drawings and specifications.
- **b.** construction drawings and project manual.
- **c.** specifications and project manual.
- **d.** specifications and bidding documents.

21. The shop drawings are prepared by the
- **a.** architect.
- **b.** structural engineer.
- **c.** mechanical engineer.
- **d.** general contractor.
- **e.** none of the above.

22. Shop drawings are generally reviewed by the
- **a.** architect.
- **b.** concerned engineering consultant.
- **c.** general contractor.
- **d.** all the above.

23. In the traditional project delivery (design-bid-build) method for a building, the day-to-day supervision of the construction is generally the responsibility of the
- **a.** architect.
- **b.** structural engineer.
- **c.** general contractor.
- **d.** all the above.

24. In the traditional project delivery (design-bid-build) method, who is typically responsible for obtaining the certificate of occupancy from the local jurisdiction?
- **a.** The architect
- **b.** The structural engineer
- **c.** The general contractor
- **d.** The owner

25. The certificate of occupancy predates substantial completion inspection of the project.
- **a.** True
- **b.** False

26. The final completion inspection of the project is generally conducted by the
- **a.** architect.
- **b.** structural engineer.
- **c.** general contractor.
- **d.** architect with the help of his/her consultants.
- **e.** local jurisdiction.

27. A record document set is generally prepared by the
- **a.** architect.
- **b.** general contractor.
- **c.** structural engineer.
- **d.** architect with the help of consultants.

28. When does the owner receive manufacturers' warranties from the general contractor?
- **a.** At the substantial completion inspection
- **b.** At the final completion inspection
- **c.** Within one year of final completion

29. In the CM method of project delivery, there is normally no general contractor.
- **a.** True
- **b.** False

30. In the CMAR method of project delivery, the construction manager
- **a.** advises the owner with respect to construction cost during the design phase.
- **b.** manages the project's construction during the construction phase.
- **c.** works as the general contractor for the project.
- **d.** all the above.
- **e.** only (a) and (b).

31. The project delivery method in which only one firm is contracted for both design and construction of the building is called
- **a.** design-bid-build method.
- **b.** design-build method.
- **c.** CM method.
- **d.** CMAR method.

Answers: 20-b, 21-e, 22-d, 23-c, 24-c, 25-a, 26-d, 27-b, 28-a, 29-a, 30-d, 31-b.

KEY TERMS AND CONCEPTS

- Certificate of final completion
- Competitive bidding
- Construction documents
- Construction drawings
- Construction manager (CM)
- Construction manager at risk (CMAR)
- Construction phase
- Construction Specifications Institute (CSI)
- Design development
- Design phase
- General contractor (GC)
- Invitational bidding
- MasterFormat
- Owner's program
- Postconstruction phase
- Preconstruction phase
- Predesign phase
- Project delivery phases
- Project manual
- Record documents
- Schematic design
- Shop drawings
- Specifications
- Subcontractors
- Substantial completion

REVIEW QUESTIONS

1. List the major phases in which the work on a traditional (design-bid-build) building project may be divided.
2. Using a diagram, show the contractual relationships between the owner, the general contractor, subcontractors, and the architect in a traditional (design-bid-build) building project.
3. List the important items contained in a project manual.
4. Explain the differences between competitive bidding and invitational bidding.
5. From memory, list the first ten divisions of the MasterFormat.
6. Explain what is included in record documents.
7. Explain the differences between construction management (CM) and construction management at risk (CMAR) project delivery methods.

SELECTED WEB SITES

American Institute of Architects (www.aiaonline.com)

Associated General Contractors of America (www.agc.org)

Construction Management Association of America (www.cmaanet.org)

Construction Specifications Institute (www.csinet.org)

Design-Build Institute of America (www.dbia.org)

FURTHER READING

1. American Institute of Architects. *General Conditions of the Contract for Construction.* AIA Document A201.
2. American Institute of Architects. *Standard Form of Agreement Between Owner and Architect.* AIA Document B141.
3. American Institute of Architects. *Standard Form of Agreement Between Owner and Contractor.* AIA Documents A101 and A114.
4. Construction Specification Institute. *The Project Resource Manual—CSI Manual of Practice*, 5th ed. Construction Specification Institute, 2005.
5. Demkin, Joseph. *The Architects's Handbook of Professional Practice.* New York: Wiley, 2001.
6. Heery, George. *Time, Cost, and Architecture.* New York: McGraw-Hill, 1975.

INTERNATIONAL PLUMBING CODE
INTERNATIONAL ZONING CODE

A Member of the International Code Family
INTERNATIONAL CODE

INTERNATIONAL PROPERTY MAINTENANCE

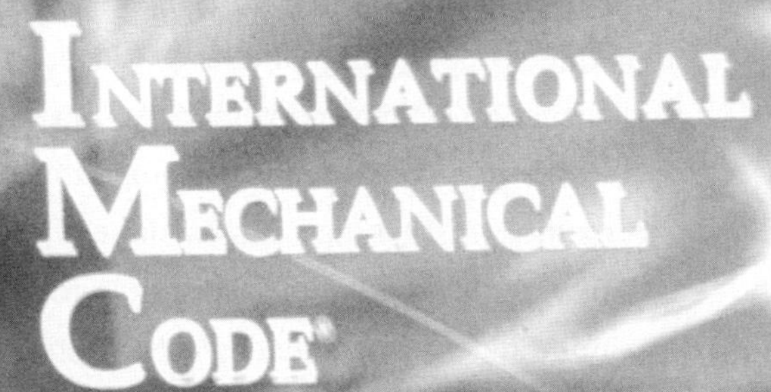
A Member of the International Code Family
INTERNATIONAL MECHANICAL CODE

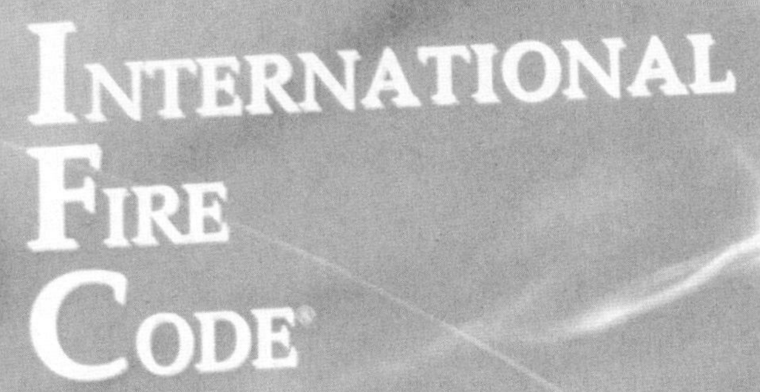
A Member of the International Code Family
INTERNATIONAL FIRE CODE

A Member of the International Code Family
INTERNATIONAL BUILDING CODE
2006
A Member of the International Code Family
INTERNATIONAL RESIDENTIAL CODE
FOR ONE- AND TWO-FAMILY DWELLINGS

CHAPTER 2

Governmental Constraints on Construction

CHAPTER OUTLINE

2.1 Objectives of a Building Code

2.2 Enforcement of a Building Code

2.3 Prescriptive and Performance Codes

2.4 Model Codes

2.5 Contents of a Building Code

2.6 Application of a Building Code

2.7 Construction Standards

2.8 Other Major Governmental Constraints

2.9 Zoning Ordinance

2.10 Building Accessibility—Americans With Disabilities Act (ADA)

PRINCIPLES IN PRACTICE: Code Allowable Area and Height of a Building

Some of the several codes referenced by a typical design team (architects, engineers and other consultants) during the design and construction of a building. (Photo by MM.)

The primary requirement of a building is that it should be safe and healthy. Visual appeal and economic viability, though important, are secondary requirements. To deliver a safe and healthy building is primarily the responsibility of the design and construction professionals—architects, engineers, and builders. However, as with all public health and safety issues, the design and construction of buildings is regulated under numerous federal, state, and local laws. The most important of these laws are contained in a document called the *building code.*

A building code is enforced by local jurisdictions such as the cities or municipalities, under police powers granted to them by the state. No building may be constructed unless it meets the requirements of the building code of the local jurisdiction.*

Although a few building code provisions are based on traditional construction practices, most provisions are firmly grounded in scientific and quantifiable data of construction performance. There is, therefore, a strong link between code requirements and the science of construction.

Ongoing technological developments in the building industry require that building codes be constantly reviewed and revised. This task is usually beyond the resources of most local jurisdictions. It is, therefore, handled by an independent agency—the model code organization—whose primary responsibility is to develop, maintain, and publish the building code and the other related codes.

This chapter begins with a discussion of the objectives of building code, followed by a description of the organizational principles and contents of the current model building code. Finally, the chapter deals with zoning ordinances and other important laws that affect building construction.

2.1 OBJECTIVES OF A BUILDING CODE

The objective of a building code is to ensure that all new construction and renovated buildings provide a minimum level of safety, health, and welfare to the occupants and public at large. Although under no legal obligation to do so, the owner or the designer may choose to exceed the requirements of the code. For the sake of economy, however, a large majority of buildings are designed to satisfy only the minimum requirements.

A building code does not regulate aspects of design that relate to a building's appearance. It deals with the issues of a building's performance. Therefore, aesthetics, color, and form-related attributes are outside the purview of building codes.

Additionally, the code protects not only the safety, health, and welfare of the owner of the project but also the general public, because the interests of the owner, the general public, and the building occupants may be at variance with one another. What may be in the best interest of the owner may not be in the best interest of the public or the occupants of the building.

It is for this reason that building construction is regulated by an impartial authority, such as the state, county, or city. It is the responsibility of the regulatory authority to ensure that the interests of all concerned are protected. Although design and construction professionals generally dislike having their work policed by an external authority, building codes have one major benefit for them all—that of liability protection. If the building has been designed and constructed in accordance with the building code and other applicable regulations, the design and construction professionals are exposed to a substantially lower liability risk.

In more specific terms, a building code regulates the following aspects of building design and construction:

- Life safety
- Fire safety
- Structural safety
- Health and welfare
- Property protection

LIFE SAFETY

Although both fire safety and structural safety are essentially life safety issues, the reverse is not always true. The term *life safety* has its own independent existence in building codes, because several safety regulations in codes are neither related to fire safety nor to structural

*There are often unregulated areas in a state or country where buildings may be constructed without conforming to any building code. Rural communities and areas outside urban boundaries are generally unregulated.

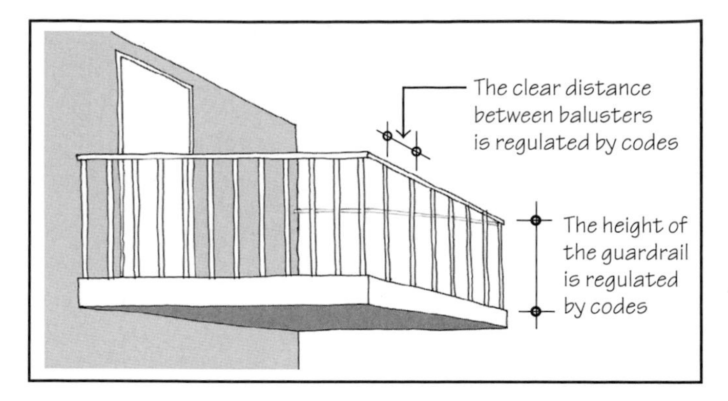

FIGURE 2.1 In addition to its structural integrity, the height of a balcony's guardrail as well as the clear space between the balusters is regulated by the codes.

safety. For example, a guardrail on a balcony, apart from being structurally adequate, must provide protection from people falling over the top of the rail or from between its vertical or horizontal members, Figure 2.1.

Thus, the regulations relating to the height of a guardrail, and the clear space between its members are life safety regulations with no relationship to fire safety or structural safety. Similarly, the handrail on a staircase, apart from being structurally strong, must provide adequate grippage. A handrail whose cross section is either too large or too small will not provide the required safety. Therefore, the building codes prescribe maximum and minimum dimensions of a handrail, in addition to its clearance from the adjacent wall, Figure 2.2. These regulations are life safety regulations, as distinct from structural safety and fire safety regulations.

The relationship between the treads and risers of a staircase, the dimensional uniformity of the treads and risers, the slope of a ramp, etc., are some other examples of life safety issues in codes, Figure 2.3. Additionally, accessibility regulations requiring that a building be easily accessible to individuals with disabilities are, in many ways, life safety regulations.

This dimension
is regulated
by codes

This dimension
is regulated
by codes

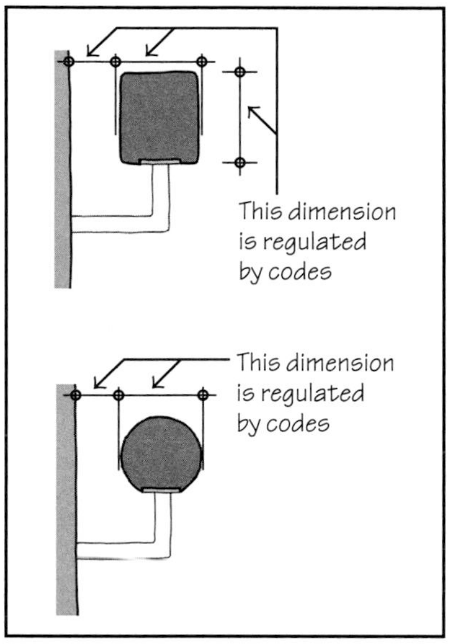

FIGURE 2.2 To ensure adequate grippage, the building code regulates various dimensions of a staircase handrail.

FIRE SAFETY

Fire safety regulations in building codes are among the most important regulations. If one were to separate structural regulations from building codes, the bulk of the remaining regulations relate, in one way or another, to fire safety issues. In fact, the history of building codes is replete with instances where the world's prominent cities either promulgated building codes for the first time or drastically revised their existing ones after the outbreak of major fires, Figure 2.4.

Obviously, fire safety regulations relate to the use of fire-resistant materials and construction. As we will see in Chapter 7, the types of building construction, as defined by the codes, are directly related to the fire resistance of the major components of the building—walls, floor slabs, roofs, beams, columns, and so

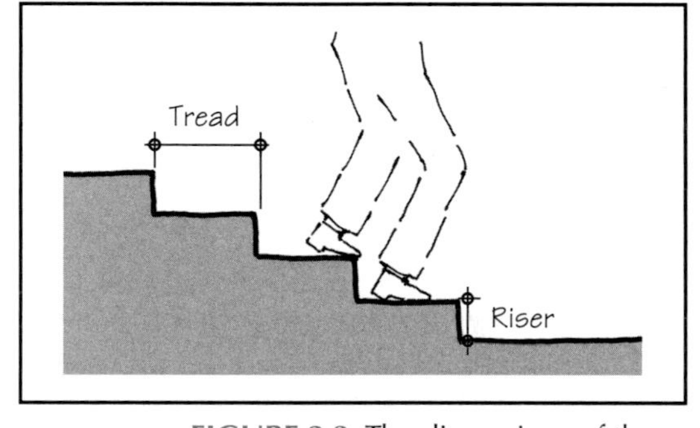

FIGURE 2.3 The dimensions of the treads and risers of a stair are regulated by the codes. Additionally, each tread must be of the same dimension and profile. The same applies to the risers (see Section 33.1).

64 A.D. The Burning of Rome
Some historians believe that it was Emperor Nero who had Rome burned. The city was rebuilt mainly with noncombustible materials and under stricter building regulations.

Great Fire of London of 1666
In 1667, the British Parliament passed a law to rebuild London with several new and more stringent building regulations.

Baltimore Fire of 1858
Baltimore established its first building code in 1859.

Chicago Fire of 1871
Chicago established its first building code in 1875.

First U.S. Model Building Code in 1905
The first model building code in the United States was published by an insurance group that insured buildings against fire damage—the National Board of Fire Underwriters. This code was later renamed as the National Building Code.

FIGURE 2.4 A few important dates that underline the importance of fire safety in building code development.

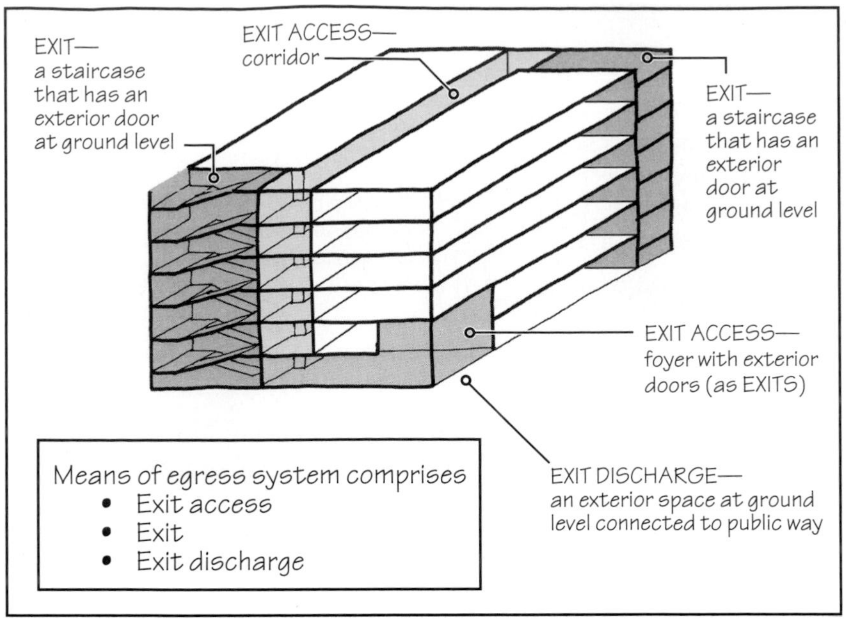

FIGURE 2.5 Means-of-egress system in a building. Building codes regulate various aspects of the means-of-egress system for fire safety reasons.

on. We will also observe in that chapter that smoke plays a predominant role in the safety of occupants during a fire. Therefore, fire safety regulations are, in fact, fire and smoke safety regulations.

One of the most important set of fire safety regulations deals with the means of egress from the building (escape routes) should a fire occur. Building codes regulate various aspects of the means of egress system in a building, including the width and height of exit enclosures, fire resistance of materials used therein, illumination levels in exit enclosures, exit signs, and so on, Figure 2.5.

Similarly, there are code regulations for exit doors, including height and width, fire resistance, panic hardware, and direction of swing, Figure 2.6. The regulations governing the provision of fire and smoke detection, fire and smoke alarms, fire suppression, fire-extinguishing systems (automatic sprinklers and stand pipes), and so on, are parts of the fire safety regulations.

STRUCTURAL SAFETY

Structural safety is obviously one of the primary objectives of a building code. A building code contains several chapters that provide detailed regulations dealing with structural design of buildings. As stated previously, structural regulations are the most numerous and account for the largest volume in the building code document.

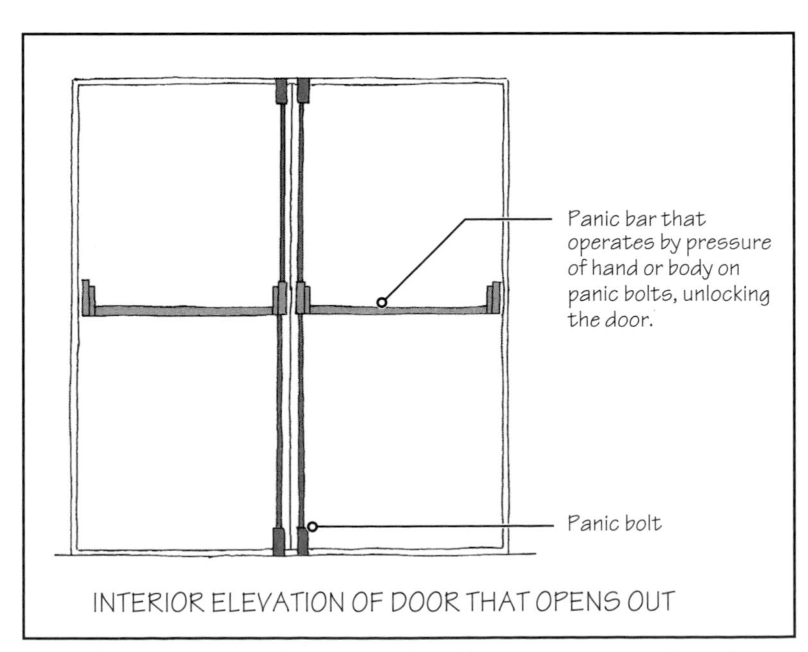

FIGURE 2.6 Panic hardware on exit doors is regulated by codes to ensure fire safety in buildings.

NOTE

Means of Egress

A means of egress in a building is a system that consists of the travel route taken by a building occupant to escape from the interior of the building to a public way. The route, which must remain unobstructed at all times, passes through the following three elements: (a) exit access, (b) exit, and (c) exit discharge.

Exit Access: An exit access is an interior space that leads to an exit. Thus, a corridor leading from a room to an exterior door at the ground level or to a staircase that has an exterior door opening at the ground level is an exit access. A room or other interior space is also part of the exit access.

Exit: An exit consists of an exterior door at the ground level or a staircase that opens at the ground level.

Exit Discharge: An exit discharge is an exterior or interior space immediately beyond the exit at the ground level that leads into a public way. It may be a courtyard, plaza, or any other open-air space adjoining an exit door. An exterior door is an exit only if it opens onto an exit discharge. An exterior door opening into an open-air courtyard that is closed from all sides is not an exit.

A portico or an open verandah at the ground floor that connects an enclosed staircase to the street is also an exit discharge.

HEALTH AND WELFARE

Although the main preoccupation of a building code is with traditional concerns of life safety, fire safety, and structural safety, it also deals with human health and welfare issues. Therefore, the building code contains provisions related to lighting, ventilation, sanitation, temperature, and noise control.

For the same reasons, the minimum dimensions of habitable rooms are regulated by the codes. Regulations pertaining to energy conservation and accessibility for individuals with disabilities are also included. In fact, energy-conservation regulations are increasingly becoming integral to building design.

PROPERTY PROTECTION

Safeguarding a building against loss or damage (property protection) is indirectly ensured through life safety and fire safety regulations, because the burning of the building or its structural collapse are two major causes of property damage. However, the codes contain several regulations that deal with property protection through requirements for materials' durability.

These durability regulations are embedded in the codes through materials and construction standards (see Section 2.7) to which the codes refer. For instance, the codes require that roof membranes and many other materials conform to relevant ASTM specifications. These specifications contain requirements for the durability of materials. Several other durability requirements, such as the decay of structural and nonstructural wood from termite and fungal attack, degeneration of materials due to freeze-thaw action, and corrosion of metals, are expressly stated in the codes.

2.2 ENFORCEMENT OF A BUILDING CODE

A building code is a legal document, enforced through the police powers of the state. It is under these powers that a state is authorized to enact legislation for the safety of its citizens. The health, traffic safety, and general welfare of the public are also promoted by the state under the same legislation.

Because it is generally agreed that the construction of buildings and the development of neighborhoods is best left under the direct control of their local citizens, the state usually delegates this power to local (city or municipal) governments. Thus, it is under the police powers delegated to them by the state legislature that local jurisdictions are able to enact, adopt, review, or change a provision of the code, subject to any overriding state or federal legislation.

BUILDING OFFICIAL

The local jurisdiction's authority to enforce and administer the code is exercised through an enforcement official, called the *building official* or the code official, who is an employee of the jurisdiction. Because the building official's authority stems from the police powers of the state, a building official has the powers of a law enforcement officer.

It is the responsibility of the building official to verify that buildings constructed in the jurisdiction are safe and comply with the provisions and the intent of the code. It is the building official's duty to prevent and/or take action to correct any violation of the code.

According to one authority [1], "A building official . . . is a highly specialized law enforcement officer whose prime mission is the prevention, correction, or abatement of violations. . . . A building official designs nothing, builds nothing, repairs nothing. His responsibility is merely to see that those persons who are engaged in these activities do so within the requirements of the law."

In order to effectively discharge duties, a building official must have a thorough knowledge of the building code, related city regulations, and the science of building construction. Law enforcement responsibilities require that a building official's dealings with the public are fair and impartial.

As the head of a city building department, the building official is usually assisted by several other functionaries (plans examiners and building inspectors) to carry out such tasks as plan reviews and field inspections. In a small city or jurisdiction, the building official may be the only person performing these tasks. An extremely small city may outsource some or all of a building official's functions to an outside consultant—a noncity employee.

NOTE

Alternative Titles for Building Official

Some jurisdictions call the building code–enforcement official the *director of building department* or *manager of building safety*. Some very small jurisdictions may have a part-time employee to enforce the codes and may designate that person as the building inspector.

BUILDING PERMIT APPLICATION

Job Address

Legal Descr.	Lot No.	Blk.	Tract

Contractor	Mail address	Zip	Phone
Permit Applicant (Engineer/Architect/Owner	Mail address	Zip	Phone

Use of Building — Number of floors

Work use (Check one)

New construction	Rehab
New constr uction-shell	Repair
Remodel	Fire repair
Interior f inish	Demolition
Addition con version	Move
	Swimming pool
	Spa/hot tub
	Misc. uses

Square feet in area	Square feet in garage	Type of sla b Post-tensioned	Rebar

NOTICE

Separate permits are required for electrical, plumbing, air conditioning and signs. This permit becomes null and void if work or construction authorized is not commenced within 120 days or if construction or work is suspended or abandoned for a period of 120 days at any time after work is commenced. Applicant is the owner and/or has the owner's consent to do the requested work. I hereby certify that I have read and examined this application and know the same to be true and correct. All provisions of law and ordinances governing this type of work will be complied with whether specified herein or not. The granting of permit does not presume to give authority to violate or cancel the provisions of any other state or local ordinances regulating construction or the performance of construction.

Plan check fee	Permit fee
Type of const.	Occupancy group &Div.
Size of bldg (Total sq ft)	No. of stories
Use zone	Max. Occ. load
Fire sprinklers reqd.	Yes No

Special approvals
Zoning
Health Dept.
Fire Dept.
Soil report
Other (specify)

WHEN PROPERLY VALIDATED (IN THIS SPACE) THIS IS YOUR PERMIT

FIGURE 2.7 A typical building permit application.

NOTE

Plans for Building Permit

The word *plans* is generally used for what really is a set of drawings that fully describe the building to be constructed. The set generally includes plans, elevations, sections, and details.

BUILDING PERMIT

The general procedure followed in administering the code is to require that the design of a proposed building complies with the provisions of the code before an official permission to commence construction is granted by the city. Before granting approval for construction in the form of a *building permit*, the city requires the submission of a building permit application, along with copies of plans for the proposed building. A typical building permit application form is shown in Figure 2.7.

If the plans are in accordance with the building code and other applicable laws, the approved building permit and plans are returned to the owner. After receiving them, the owner may commence construction. If the plans do not comply, the owner is notified by the building official, and the plans must be revised until conformance with the code is achieved.

Once construction begins, periodic and progressive inspections by the building official or the building official's representative—the building inspector—ensure that the construction meets code requirements. The building is inspected at several stages during construction. Typically, these stages are (1) prefoundation (to inspect sewage disposal and water-supply lines), (2) post-foundation, (3) after the erection of structural frame, (4) after insulation and vapor retarder installation, (5) mechanical, electrical, and plumbing (MEP) rough-in, (6) MEP final, and (7) final building completion.

Every inspection concludes with a report generated by the inspector to indicate whether or not the construction is progressing in conformance with the code. Some cities color-code the inspection result: a green tag left on the site indicates that the construction is proceeding in accordance with the code, and a red tag indicates that it has failed the inspection.

In the case of a failed inspection, the owner is obliged to make immediate corrections and request reinspection until the construction is approved. If the corrections are not made or if, in the opinion of the inspector, a gross violation of the code has been committed, a *stop work order* by the city is generally issued, Figure 2.8. Noncompliance with the stop work order results in administrative and/or legal action against the offender.

STOP WORK ORDER

City of ....

Department of Building Inspection

Address ....

You are hereby notified that a Stop Work Order is hereby issued by the Building Official of the City of .... as of .... and that no further work shall be performed on this job until such time as it is authorized under the Codes of the City of

....

Violations subject to $ 2,000.00 fine

....
Inspector

DO NOT REMOVE THIS TAG

FIGURE 2.8 A typical stop work order notice.

Certificate of Occupancy

Once the building is substantially complete (see Section 1.8), the owner may request a final inspection and apply for a *certificate of occupancy*. Permission to occupy the building is granted only after the city officials are satisfied that all work has been completed in accordance with the code.

No building may be occupied, in whole or in part, until the certificate of occupancy has been granted. Among other details, a certificate of occupancy must specify the occupancy group and division of the building, the type of construction (Section 2.6), and any other special stipulation of the code, Figure 2.9.

Certificate of Occupancy
City of ____________________
Department of Building Inspection

This certificate issued persuant to the requirements of Section ________ of the Building Code of the City of ___________ certifying that at the time of issuance this structure was in compliance with various ordinances of the city regulating building construction or use for the following:

Occupancy Group _________________ Building Permit No. ________

Division _______ Type of Construction _______ Use Zone _________

Owner of Building _______________ Address __________________

Building Address ________________ Locality _______________

By _______________________

____
Building Official

Date ____________________

POST IN A CONSPICUOUS PLACE

FIGURE 2.9 A typical certificate of occupancy.

Board of Appeals

A building code is a legal document, so it is a concise description of regulations as agreed to among the experts. Like most other laws, it does not contain any background material (except a brief commentary) that explains how decisions were made. A building code is, therefore, subject to interpretations, which may differ from jurisdiction to jurisdiction. In many cases, a clear-cut interpretation of a code provision may not even exist, because its original intent may have been either lost or not recorded.

In this situation, the owner (or the architect working as the owner's representative) may consult with the local building official for an interpretation. In the event of an unresolved difference between the interpretations of the owner and that of the building official, the building official's interpretation and decision is binding unless it is appealed.

Appeals against the building official's interpretation are referred to a *board of appeals*, which is usually a standing body of the city. The board's task is to hear appeals and adjudicate on the validity of the building official's decision. In some situations, the board may be requested by the building official to provide interpretation of an ambiguous code provision. This allows the entire board to assume responsibility for a particular interpretation, thereby reducing the building official's personal liability exposure.

The board of appeals is generally not authorized to waive any provision of the code, nor can it render judgment on the administrative provisions of the code. Its task is simply to render interpretation of the technical aspects of the code, taking into account the code's objectives and intent. The board's decision is binding on the city, and, in most cases, an appeal against the board's decision can be taken only to a court of law. The board consists of individuals who are qualified by training or experience in building construction and are not employees of the city. The building official usually serves as the ex-officio secretary of the board but does not have voting rights on the board.

PRACTICE QUIZ

Each question has only one correct answer. Select the choice that best answers the question.

1. A building code regulates
   a. the design of the building.
   b. the construction of the building.
   c. the aesthetics of the building.
   d. all the above.
   e. both (a) and (b).
2. In a building code, some life safety provisions are contained in fire safety provisions and the remaining in structural safety provisions.
   a. True
   b. False
3. The building code requirements for the dimensions of treads and risers of a staircase are
   a. fire safety issues.
   b. life safety issues.
   c. structural safety issues.
   d. health and welfare issues.
   e. property protection issues.
4. The building code requirements for the means of egress from a building are primarily
   a. fire safety issues.
   b. life safety issues.
   c. structural safety issues.
   d. health and welfare issues.
   e. property protection issues.
5. The enforcement of a building code is usually done at the level of the
   a. city in which the building is located.
   b. county in which the building is located.
   c. state in which the building is located.
   d. country in which the building is located.
6. The person in charge for enforcing the building code is generally called a
   a. code in-charge.
   b. building code official.
   c. building official.
   d. none of the above.

7. Before commencing the construction of a building, the owner must first apply to the city to obtain a
- **a.** commence work order.
- **b.** building license.
- **c.** construction permit.
- **d.** building permit.

8. To ensure that the construction of the building is being done in accordance with the building code, the city will arrange for its inspection
- **a.** only once during the construction of the building.
- **b.** two times during the construction of the building.
- **c.** three times during the construction of the building.
- **d.** several times during the construction of the building.

9. If during the construction of the building, the building inspector discovers a gross violation of the building code, he or she will generally
- **a.** invite the owner for a meeting.
- **b.** issue a warning to the owner.
- **c.** issue a stop work order to the owner.
- **d.** take the owner to the appropriate court of law.

10. In the case of a dispute between the building official and the owner in the interpretation of a building code provision, the matter is first resolved through
- **a.** the owner's architect.
- **b.** an independent arbitrator appointed by the city.
- **c.** the board of appeals of the city.
- **d.** city attorney.
- **e.** the appropriate court of law.

Answers: 1-e, 2-b, 3-b, 4-a, 5-a, 6-c, 7-d, 8-d, 9-c, 10-c.

2.3 PRESCRIPTIVE AND PERFORMANCE CODES

The older and traditional types of building codes are *prescriptive codes.* Such codes give clear prescriptions for construction systems, types of materials, and the devices to be used without permitting any alternatives. They are definitive in interpretation and are, therefore, easy to use and enforce. Their main drawback is that they cannot remain apace with the developments in building materials, technology, and safety concepts.

In a *performance code,* the performance criteria of a component are specified instead of the material or the construction system. The performance criteria are based on the function of the component. For example, a prescriptive code might require an 8-in.-thick brick wall as a party wall between two semidetached residential units, Figure 2.10. It might further specify the type of bricks and the type of mortar to be used in joints. A performance code, on the other hand, states the required properties of the wall, such as the fire resistance, sound insulation, load-carrying capacity, and durability characteristics.

In a performance code, the choice of material and the thickness of the wall is left to the discretion of the designer. As long as the wall meets the stated performance requirements, it is acceptable regardless of its material or thickness. A performance code is, therefore, more flexible and provides greater freedom for the innovation of new building products or construction systems.

Although the modern building codes have become increasingly more performance oriented, they still include a considerable degree of prescriptiveness. Several code regulations are purely performance type, several others are purely prescriptive, and many are a combination of the two, Figure 2.11. As previously stated, early building codes were primarily prescriptive. In contemporary building codes, some prescriptive provisions exist because replacing them with performance provisions would introduce unnecessary complications, whereas others serve as an alternative to the performance provisions. Where alternative provisions (of a prescriptive nature) exist, the user of the code has the option to follow either the performance provisions or the prescriptive provisions.

For instance, conventional wood frame construction, used for low-rise single or multifamily dwellings, is in the hands of small builders, who may have neither the training nor the resources to effectively use the performance-type provisions. Therefore, building codes

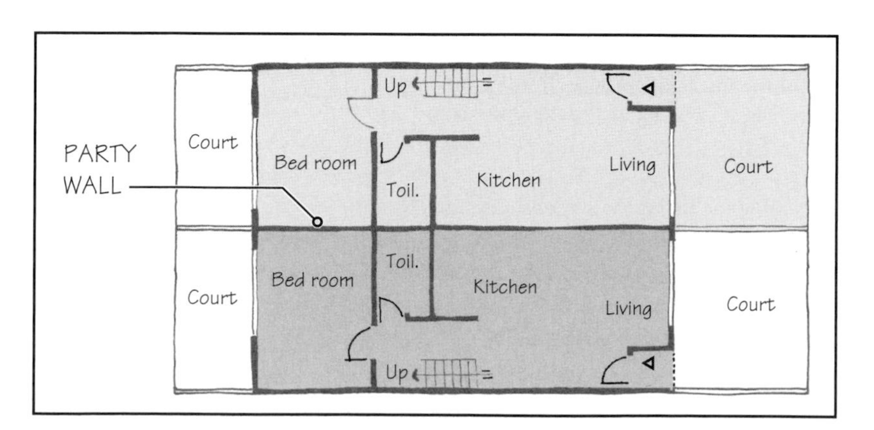

FIGURE 2.10 Two semidetached residential units with a party wall.

PERFORMANCE PROVISION	PRESCRIPTIVE PROVISION	PERFORMANCE AND PRESCRIPTIVE COMBINATION PROVISION
Section 1005.32 International Building Code "Interior exit stairways shall be enclosed. Vertical exit enclosures four stories or more shall be 2-hour fire resistance rated. Vertical exit enclosures less than four stories shall be 1-hour fire resistancerated ...."	**Section 1405.2 International Building Code** "Exterior walls shall provide weather protection for the building. The materials of the minimum thickness specified in Table 1405.2 shall be acceptable as approved weather coverings." **Table 1405.2 (part of)** Covering type — Minimum thickness (in.) Adhered masonry veneer — 0.250 Anchored masonry veneer — 2.625 Aluminum siding — 0.019 Asbestos cement board — 0.125 Asbestos shingle — 0.156	**Section 1507.3.2 International Building Code** "Clay and concrete roof tiles shall be installed on roof slopes of 2-1/2 units vertical in 12 units horizontal (21 percent slope) or greater. For roof slopes from 2-1/2 units vertical in 12 units horizontal (21 percent slope) to 4 units vertical in 12 units horizontal (33 percent slope), double underlayment application is required in accordance with Section 1507.3.3." **Section 1507.3.3, International Building Code** "Unless otherwise noted, required underlayment shall conform with: ASTM D226, Type I; ASTM D2626 Type I; or ASTM D249 mineral surfaced roll roofing.".

FIGURE 2.11 Examples of performance, prescriptive, and combination of performance and prescriptive provisions in a building code. Source: 2006 International Building Code. Copyright 2006. Falls Church, Virginia: International Code Council, Inc. Reproduced with permission. All rights reserved.

contain prescriptive (as well as performance) provisions that are applicable to conventional wood frame construction. A conventional wood frame building may comply with either the general performance-type provisions of the code or with the alternative prescriptive provisions.

2.4 MODEL CODES

There are two principal activities relating to building codes:

- Enforcement and interpretation of codes
- Formulation and updating of codes.

As stated in Section 2.2, the enforcement of a code is mainly an administrative function handled by local jurisdictions. The formulation and updating of building codes, which must be based on the latest knowledge in the realm of health safety and welfare (HSW), are, on the other hand, beyond the resources of most local jurisdictions. It is a complex activity that requires the input of a large number of technical experts, such as architects, structural engineers, fire safety engineers, chemical engineers, builders, building officials, and building material manufacturers, to name just a few.

In most countries, therefore, building codes are developed by an independent agency, which not only reduces the cost burden on local governments but also avoids unnecessary duplication of work among them. Such a code, usually referred to as the National Building Code of the country, is then adopted by local governments. For example, the National Building Code of Canada is the only model code in the country.

Although voluntary, the National Building Code of Canada has been adopted by various Canadian cities. The adoption is usually accompanied by local amendments to take into account the uniqueness of the jurisdiction. Thus, a model code is similar to a model house. A prospective home buyer may either seek to buy an exact replica of the model house or request the builder make minor changes in the model house to suit the buyer's specific requirements.

Note that a model code is not a legal document unless it is adopted by a jurisdiction through appropriate legislation. Once adopted by a jurisdiction, it becomes the (legal) building code of that jurisdiction. In the United States, as in most other countries, the power for model code adoption rests with the states. In some states, this power has been delegated to the local jurisdictions—county or city.

HISTORY OF MODEL CODES IN THE UNITED STATES

Because of the size and diversity of local conditions in the United States and because of the peculiar history in which code development took place in this country, there were three model building codes, developed by three independent model code–writing organizations, until the year 2000. Each model building code was an independent and complete document containing all necessary building code regulations. They were revised periodically, and a new edition of each code was published every three years.

The multiplicity of model codes had been a matter of great concern to the design community because of the difficulties it created for them. The designers had to be familiar with all three model codes. Visualize an architect developing a design for a national chain store. The uniformity of design might be essential to the marketing and distribution strategy of the chain. However, under the three-code setup, three separate versions of the same design were required to satisfy the requirements of three different codes, in addition to any local requirements.

Although the basic features of these three model codes were the same, each code was formulated differently. For instance, each model code prescribed different methods for determining the permissible built-up area and height of a building. None of these methods was any better than the other two; they were just different.

The multiplicity of model code organizations had some positive features, such as the competition between the three organizations as each model code organization competed with the other two to lure cities or even entire states to adopt its code. Additionally, the regional character of each model code gave an opportunity to a city or state to choose the code that best fit its needs. However, the disadvantages of multiple codes far outweighed their benefits.

THE INTERNATIONAL CODE COUNCIL (ICC)

In response to the criticism just described, the three model code organizations (ICBO, SBCCI, and BOCA) merged and jointly founded the International Code Council (ICC) in 1994. The purpose of the merger and the establishment of ICC was to discontinue production of three separate model codes and to produce a common model code. In the year 2000, the first edition of ICC's building code, called the *International Building Code* (IBC), was published. Like its predecessors, the IBC is updated every three years.

MODEL CODE ADOPTION

Although code enforcement is invariably done at the local level (city or municipality), the adoption of a model code in the United States is done either at the city level or at the state level. For instance, nearly 35 states in the United States have adopted the International Building Code for the entire state. In several other states, the International Building Code has been adopted by local jurisdictions.

There is always a time lag between the publication of a model code edition and its adoption by a state or the local jurisdictions. As stated previously, a model code is not the legal code for a city unless the city's governing council has formally adopted it through an ordinance. This process takes time, and in some cities, it may take several years. Thus, it is not

EXPAND YOUR KNOWLEDGE

U.S. Model Codes Prior to Year 2000

Uniform Building Code

The Uniform Building Code, commonly known as the UBC, was published by the International Conference of Building Officials (ICBO), headquartered in Whittier, California. The ICBO was formed in 1922; the first edition of the UBC was published in 1927 and its last edition, in 1997. It was adopted by a large number of cities, more commonly in the western United States. Because of the relative vulnerability of the western United States to earthquakes, the UBC distinguished itself from the other two model codes by its up-to-date seismic design provisions.

Standard Building Code

The Standard Building Code was published by the Southern Building Code Congress International (SBCCI), headquartered in Birmingham, Alabama. SBCCI was formed in 1940. The first edition of the Standard Building Code was published in 1945, and the last was published in 1997. Best known for its up-to-date wind-load provisions, the Standard Building Code was primarily used in the south and southeastern United States, particularly by cities lying on the Atlantic and Gulf coastlines, which are vulnerable to hurricanes.

BOCA National Building Code

The BOCA National Building Code (earlier known as the Basic Building Code) was published by the Building Officials and Code Administrators (BOCA) International, headquartered in Country Club Hills, Illinois. BOCA was founded in 1915. The first edition of this code has published in 1950, and it last appeared in 1996. It was used primarily in the central and northeastern United States.

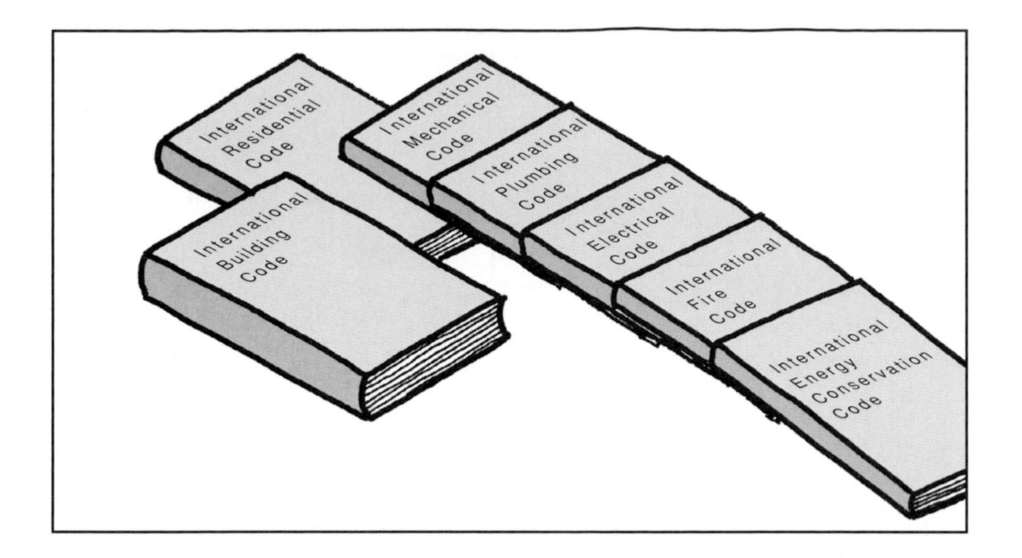

FIGURE 2.12 Some of the model codes published by the International Code Council.

uncommon to see a gap of several years between the publication of a new edition of the model code and its adoption by a state or the local jurisdictions.

Note that building activity is not regulated in every part of the country. Several rural areas in a state may be completely unregulated. Additionally, some cities in the United States, particularly the larger ones such as New York and Chicago, do not follow a model code but have their own building codes. The problems of these two cities are so different from the others and their revenues are so large that they can afford to write and constantly update their own codes. Finally, state and federal buildings are usually exempt from local building codes and follow their own codes.

In addition to the International Building Code, another model building code currently exists in the United States. This code has been developed by the National Fire Protection Association (NFPA) and is referred to as *NFPA 5000: Building Construction and Safety Code*. This code has been adopted by a very small number of local jurisdictions.

Additional Codes

A building is required to conform not only to the building code, but also to a host of other speciality codes. However, the building code is the primary code, and most design and construction professionals need to be familiar only with the building code, leaving conformance to other codes to be determined by the specialist consultants. The ICC publishes most of the speciality codes, some of which are shown in Figure 2.12 and described here:

- International Mechanical Code—relating to heating, ventilating, and air-conditioning equipment, incinerators, and other mechanical equipment in buildings
- International Plumbing Code—relating to water supply, wastewater, and storm water disposal
- International Electrical Code—relating to the electrical systems in buildings
- International Fire Code—to ensure fire-safe maintenance of buildings (also called the *fire-prevention code*)
- International Energy Conservation Code—to ensure the conservation of energy by the buildings

Note that the International Residential Code (IRC), applicable only to one- or two-family dwellings, is an alternative to the building code and the speciality codes. It is a comprehensive code that includes all provisions relating to the design and construction of a dwelling, including those related to mechanical, electrical, plumbing, etc. It is more prescriptive in nature; and hence, it is easier to use. Thus, an architect or builder involved in the design or building of one- or two-family dwellings need not refer to the building code—only to the residential code.

2.5 CONTENTS OF A BUILDING CODE

A typical building code, such as the International Building Code, may be divided under the following nine parts:

Code Administration Administrative provisions deal with the administrative aspects of the code, such as the duties and functions of a building official, plans examination,

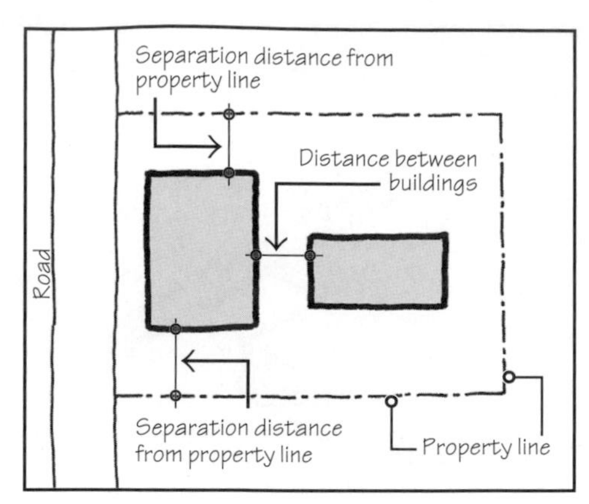

FIGURE 2.13 The separation of a building from the property lines and the adjoining buildings located within the same property provides open spaces around the building. If these spaces are accessible to a public way (or street), they can be used for fire-fighting measures. Building codes recognize accessible open spaces (referred to as frontage in determining the allowable area of the building, (see Principles in Practice at the end of this chapter).

board of appeals, issuance of a building permit, occupancy certificates, inspections, and fees. The definitions of various terms used in the code are also contained in this part.

Building Planning Building-planning provisions deal with the classification of buildings according to the occupancy and type of construction. The occupancy of the building refers to the building's use. The type of construction refers to the fire resistance of the major components of the building. These two factors—*occupancy classification* and the *type of construction*—are the two most important factors for determining the maximum allowable area of the building and its maximum allowable height.

Another factor that influences the building's allowable area is its *frontage*. The frontage of building refers to accessible open spaces around the building, such as the separation distance from the adjacent buildings and from the property lines, Figure 2.13. The term *accessible* here means the accessibility of space to a public way to allow it to be used for occupant escape and fire fighting. The greater the frontage, the lower the fire hazard the building presents to its occupants and the neighboring buildings and hence the greater the building's allowable area. For the same reason, if the building is provided with *automatic sprinklers*, its allowable area and height can be increased.

NOTE

Factors That Determine Code-Allowed Area and Height of a Building

The following factors determine the allowable area and height of a building. See Principles in Practice at the end of this chapter.

- Occupancy classification
- Type of construction
- Frontage
- Automatic sprinklers

Fire Protection Provisions deal with fire-resistive materials and construction, fire-resistive interior finishes, and fire protection systems. Fire protection systems include those that detect and suppress fires.

Occupant Needs Provisions deal with the means of egress, accessibility, and interior environment, including lighting, ventilation, sanitation, and sound control.

Building Envelope Provisions deal with the performance of the exterior envelope of the building—exterior walls, cladding, windows, roof, and so forth.

Structural Systems and Materials Provisions deal with loads on buildings, structural tests and inspections, and foundations. It is the most extensive part of the code and includes a separate chapter for each structural material—concrete, masonry, steel, and wood.

Nonstructural Materials Provisions deal with the use of nonstructural materials, such as aluminum, glass, gypsum board, and plastics.

Building Services Provisions deal with electrical, mechanical, and plumbing aspects relevant to architectural design (excluding technical aspects of these systems, which are covered in speciality codes).

Miscellaneous Provisions These deal with miscellaneous concerns, such as construction in the public right of way, site work, demolition, and existing structures.

2.6 APPLICATION OF A BUILDING CODE

Experienced design and construction professionals realize that a building code is a large and complex document. Therefore, they consider the essential features of the code early in the design process and gradually introduce its details as the design process progresses.

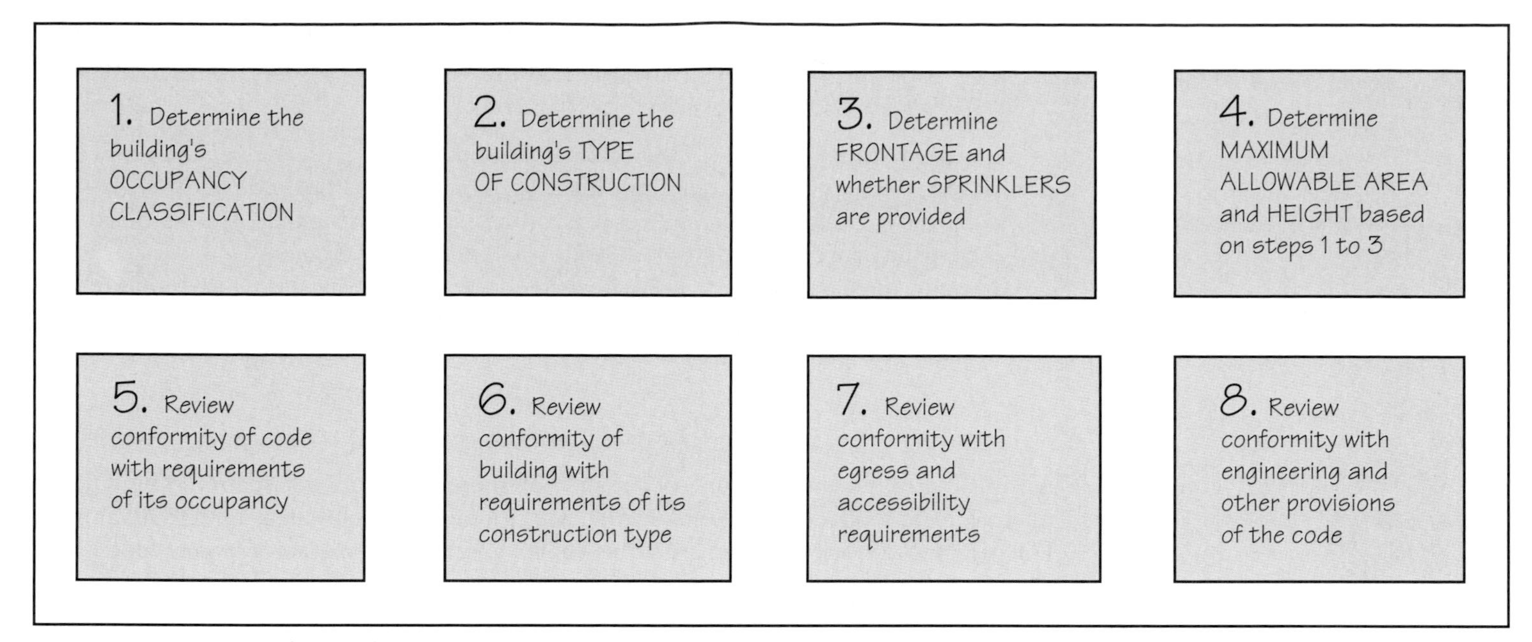

FIGURE 2.14 Steps involved in the application of a building code to a proposed building.

Because of the code's complexity, only a brief overview of how a building code is applied to a proposed building is presented here. The most important features of this process, subdivided into eight steps, are illustrated in Figure 2.14.

Step 1 in the application of the code is to determine the occupancy classification of the building. The next step (*Step 2*) is to determine the type of construction to be used. *Step 3* determines the building's frontage and whether or not automatic sprinklers will be provided.

Step 4 involves determining the maximum allowable height and the maximum allowable area of the building. This is related to the degree of hazard present in the building. The fundamental premise is that if the hazard in the building is small, the allowable area can be large, and vice versa. As stated previously, the allowable area is a function of the occupancy classification, construction type, frontage, and automatic sprinklers. The more fire-resistive the construction type and the greater the frontage, the greater the area allowed for a particular occupancy. The provision of automatic sprinklers also increases the allowable area and height of the building.

Other steps in the application of building code include reviewing the conformity of the building to the detailed provisions of its particular occupancy and type of construction and reviewing requirements for egress and accessibility. Finally, the building must be reviewed for conformity with structural engineering requirements.

Occupancy Classification

Every building presents a certain amount of hazard to its occupants. The degree of hazard depends on several factors, such as the occupancy (use) of the building, type of construction, height and number of stories, and whether any fire-detection and suppression system has been provided.

Of these factors, the occupancy of a building is one of the major hazard-determining factors. Occupancy-related hazard in a building depends primarily on the following.

Concentration of occupants Buildings that accommodate a large number of people, such as stadiums, assembly halls, auditoriums, and churches, present a greater hazard than buildings with a smaller concentration of occupants, such as individual dwellings and apartments.

Fuel content Buildings that contain flammable materials, such as automobile repair garages, paint and chemical stores, and woodworking mills, are more hazardous because they contain more combustibles than other buildings.

Mobility of occupants Buildings such as nursing homes, hospitals, day care centers, and prisons have occupants whose mobility is severely limited. Such buildings present a great deal of hazard because their occupants cannot exit the building in the event of a fire or some other emergency.

Familiarity of occupants with the building The greater the occupants' familiarity with the building, the lower the hazard. This is based on the premise that in the event of a fire or some other emergency, the occupants will be able to exit a building easily if they are familiar with it. A hotel or office building is, therefore, more hazardous than an individual dwelling.

The International Building Code classifies buildings in ten occupancy groups. Each occupancy group is given a letter designation, such as A, B, . . . , U, and is further classified into subgroups, called *divisions*. For example, the assembly occupancy is divided into five divisions, which are designated as A-1, A-2, A-3, A-4 and A-5, Table 2.1.

Generally, the hazard present in occupancies decreases with increasing division number. Thus, Occupancy A-2 is generally less hazardous than Occupancy A-1, A-3 is less hazardous than A-2, and so on.

Mixed Occupancy and Occupancy Separation

A building may consist of one occupancy or more than one occupancy. The latter is referred to as a *mixed-occupancy building*. Mixed-occupancy buildings are fairly common. For instance, a hotel may consist of bedrooms and a few conference rooms. Hotel bedrooms are Occupancy R-1, whereas the conference rooms are classified as A-3 (Table 2.1). Similarly, an office block may consist of individual offices, a day care center, a gymnasium

TABLE 2.1 OCCUPANCY CLASSIFICATIONS

	Generally decreasing hazard →				
A **Assembly**	**A-1** Movie theatre, TV and radio stations, concert hall, etc.	**A-2** Banquet hall, restaurant, nightclub, etc.	**A-3** Church, gymnasium, lecture hall, museum, etc.	**A-4** Skating rink, swimming pool, tennis court, etc.	**A-5** Stadium, grandstand, amusement park, etc.
B **Business**	Bank, professional offices (architects, attorneys, doctors, etc.), beauty shop, fire and police stations, post office, educational institution beyond grade 12, etc.		**A room or space used for assembly purposes by less than 50 persons and as an accessory to another occupancy is considered to belong to that occupancy. If the number of persons using that space is 50 or more, it will be considered as an assembly occupancy. For example, a lecture or classroom for less than 50 students in an educational building (Occupancy E if up to grade 12 or Occupancy B if beyond grade 12) is classified as Occupancy E or Occupancy B. A classroom for more than 50 students in the same building is classified as Occupancy A-3.**		
E **Educational**	Educational buildings up to grade 12 with at least 6 or more persons at any one time and day care centers for at least 5 children $2\frac{1}{2}$ years or older.				
F **Factory**	**F-1** Automobile, bicycles, bakery, clothing, furniture, etc.	**F-2** Beverages, brick and masonry, glass, gypsum, etc.			
H **Hazardous**	**H-1** Structures with explosives, unstable chemicals, etc.	**H-2** Structures with **combustible** dust, flammable gas, etc.	**H-3** Structures with aerosols, flammable solids, etc.	**H-4** Structures with toxic materials, etc.	
I **Institutional**	**I-1** Convalescent home, assisted living center, etc.	**I-2** Hospital, nursing home, 24-hour day care center, etc.	**I-3** Prison, detention center, correction center, etc.	**I-4** Adult care facility with at least 5 people, etc.	
M **Mercantile**	Department store, drugstore, market, sales room, motor vehicle service station, etc.				
R **Residential**	**R-1** Hotel, motel, boarding house, etc.	**R-2** Apartment, dormitory, monastery, fraternity, etc.	**R-3** Single-family or two-family dwelling unit, etc.	**R-4** Assisted living for more than 5 but fewer than 16 persons.	
S **Storage**	**S-1** Aircraft hangar, furniture, clothing, grain, leather, etc.	**S-2** Cement, glass, parking garage (open or enclosed), etc.			
U **Utility**	Agricultural building, carport, greenhouse, livestock shelter, etc.				

The information provided in this table is incomplete. For complete information refer to the 2006 International Building Code.

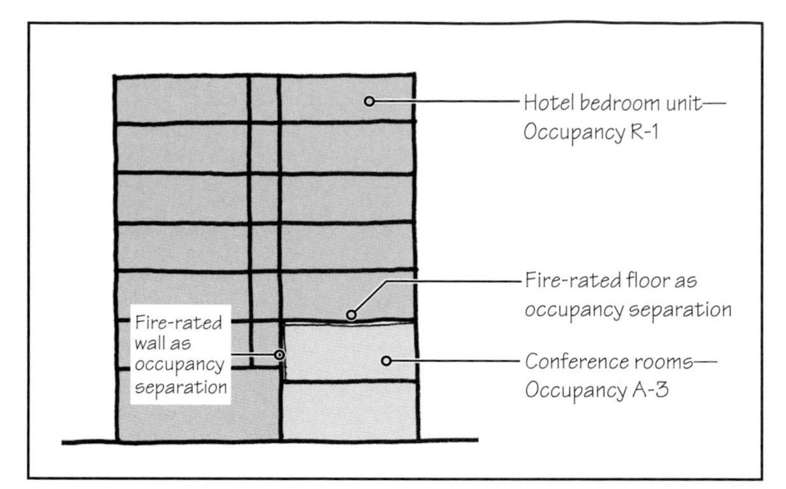

FIGURE 2.15 The separation between two different occupancies (occupancy separation) may be a wall, a floor, or both, each of which must have the minimum fire resistance rating specified by the code. If occupancy separations are not provided (i.e., if nonseparated occupancies are provided), the entire building will be treated as belonging to the more hazardous occupancy.

and a few retail shops. The individual offices are Occupancy B, a day care center is Occupancy E, a gymnasium is Occupancy A-3, and retail shops are Occupancy M.

Because of the variable degree of hazard present in a mixed occupancy, the code requires that each occupancy be separated from the other occupancy by an occupancy separation of a specified minimum fire resistance. For example, the code stipulates that in the case of a hotel building consisting of R-1 and A-3 occupancies, the occupancy separation between these two occupancies must have a minimum fire resistance of 2 h, Figure 2.15. The occupancy separation (which is a fire-rated element) may be a wall, floor, or both. The objective is to separate the two occupancies vertically as well as horizontally.

If the required occupancy separation is not provided, the entire building will be treated as belonging to the more hazardous group, adversely affecting the allowable area and height of the building. In the example of Figure 2.15, the entire building will be treated as Occupancy A-3, because this is more hazardous than Occupancy R-1. The allowable area and height of the building will be as permitted for Occupancy A-3.

If the required occupancy separation is provided, the building will be considered as one building with two distinct occupancies, and each occupancy can have its own allowable area and height. Additional details of the allowable area and height of a building (including a mixed occupancy building) are given in "Principles in Practice" at the end of this chapter.

TYPES OF CONSTRUCTION

The building's type of construction also creates its own hazard. The more fire-resistive the construction, the smaller the hazard. The International Building Code classifies construction into five basic types—I, II, III, IV, and V. This classification is based on the degree of fire resistance of various structural components of the building, such as floors, roof, columns, beams, exterior walls, and interior partitions.

Type I is the most fire-resistive construction, and Type V is the least fire-resistive. Each of these types is further subdivided in two subtypes, except Type IV (which has no subtype), Table 2.2. A detailed description of the construction types is given in Chapter 7.

TABLE 2.2 TYPES OF CONSTRUCTION

Decreasing fire resistance ↓		Decreasing fire resistance →	
	Type I	Type I(A)	Type I(B)
	Type II	Type II(A)	Type II(B)
	Type III	Type III(A)	Type III(B)
	Type IV	No subclassifications	
	Type V	Type V(A)	Type V(B)

NOTE

Accessory-Use Space

Some buildings will have one dominant occupancy and one or more minor occupancies. If the total floor area of the minor occupancies is less than 10% of the total area of the building at any floor, the minor occupancies may be treated as an *accessory-use space*. An occupancy with an accessory-use space may be considered as a single (dominant) occupancy, requiring no separation between the occupancy and the accessory-use space.

For example, consider a high-rise apartment building with a beauty shop and a restaurant at the ground floor. The dominant occupancy of the building is R-2 (Table 2.1). The beauty shop is Occupancy B, and the restaurant is Occupancy A-2. If the total area of Occupancy B and Occupancy A-2 is less than 10% of the total ground-floor area of the building, these two occupancies may be considered as accessory-use spaces.

Exceptions to this policy are classrooms or assembly spaces used for fewer than 50 persons. Such spaces need not be limited to 10% of the building's area, provided they are accessory to that occupancy.

Incidental-Use Space

Some minor occupancies may be more hazardous than the dominant occupancy. Such occupancies are treated not as accessory-use spaces but as incidental-use spaces. For example, a covered parking garage in a high-rise office building is an incidental-use space. Incidental-use spaces require a fire-rated separation from the main occupancy. The International Building Code lists the spaces that should be treated as incidental-use spaces and gives the corresponding fire-rating requirement of the separation.

PRACTICE QUIZ

Each question has only one correct answer. Select the choice that best answers the question.

11. A prescriptive building code provision is easier to enforce than a performance provision.
a. True **b.** False

12. Performance-type building code provisions allow greater innovation of building materials and construction systems than prescriptive type provisions.
a. True **b.** False

13. Writing and periodically updating the building code is generally done by:
a. the individual city.
b. the individual state.
c. the federal government.
d. a specialized independent agency.

14. The International Building Code is published by the
a. U.S. government
b. International Conference of Building Officials
c. Building Officials and Code Administrators International
d. International Code Council
e. United Nations

15. The International Building Code is generally updated every
a. Year.
b. 3 years.
c. 5 years.
d. 8 years.
e. 10 years.

16. The adoption of a model building code in the United States is done at the level of
a. each city.
b. each state.
c. either (a) or (b).

17. The only code published by the International Code Council is the International Building Code.
a. True **b.** False

18. The first step in applying the building code to a project is to
a. determine the type of construction of the building.
b. determine the width of open spaces around the building.
c. determine the occupancy classification of the building.
d. determine the number of exits from the building at the ground floor.

19. The total number of occupancies (excluding the divisions) as specified in the International Building Code are
a. 6.
b. 8.
c. 10.
d. 12.
e. none of the above.

20. Which of the following is not a recognized building code occupancy?
a. Assembly
b. Educational
c. Business
d. Office
e. Institutional

21. A hotel building is a
a. business occupancy.
b. residential occupancy.
c. hotel occupancy.
d. hospitality occupancy.
e. none of the above.

22. The International Building Code divides the types of construction into
a. Types U, V, X, and Y.
b. Types 1, 2, 3, 4, and 5.
c. Types I, II, III, IV, and V.
d. Types I, II, III, IV, V, and VI.

23. The most fire resistive type of construction is
a. Type X.
b. Type 5.
c. Type I.
d. Type V.

Answers: 11-a, 12-a, 13-d, 14-d, 15-b, 16-c, 17-b, 18-c, 19-c, 20-d, 21-b, 22-c, 23-c.

2.7 CONSTRUCTION STANDARDS

Standards are the foundations of modern building codes. They contain technical information that addresses: (1) the properties of a building product or a component, (2) test methods to determine properties of materials, and (3) the method of installation or construction.

The word *standard* is used here as a qualifier for words such as the properties, test methods, or the method of installation. Thus, a standard, in fact, refers to one of the following:

Standard specification This deals with the quality of materials, products and components.

Standard test method This determines the particular performance of a product or system through a test including methods of sampling and quality control.

Standard method of practice—construction practice, installation practice, or maintenance practice This includes specific fabrication, installation, and erection methods of a component, including its maintenance.

A standard should not be confused with a building code provision. Although a code specifies the design criteria or the required properties of a component, the standard specifies the procedures and equipment required to verify the criteria or measure the properties. For instance, a building code will give the required fire resistance (e.g., 2 h) for a wall, roof, floor, and so on. The construction standard prescribes the test procedure and equipment required to measure the fire resistance of that element.

A fully prescriptive code does not require any standards because it does not refer to any performance criterion or property. On the other hand, a performance-type code must rely heavily on the use of standards. As the trend toward performance-type codes has grown, the use of standards in building codes has increased accordingly.

Organizations That Write Standards

It is beyond the scope of most local jurisdictions or a model code organization to develop standards, because many standards are referenced by a model code. For example, the International Building Code references nearly 500 standards produced by numerous organizations. The task of formulating and updating the standards is, therefore, performed by organizations that have the necessary facilities and expertise for obtaining and evaluating performance data. These organizations may be divided into three types: trade associations, professional societies, and organizations whose primary purpose is to create, maintain, and publish standards.

Trade Associations A trade association is an association of manufacturing companies making the same product. For example, the American Wood Preservers' Association (AWPA) is an organization whose membership comprises industries that treat lumber, plywood, and other wood products with preservatives and fire-retardant chemicals. Some other important trade organizations involved in writing standards are the National Roofing Contractors' Association (NRCA), Brick Industry Association (BIA), National Concrete Masonry Association (NCMA), and Tile Council of America (TCI).

The function of a trade association is to coordinate the tasks of member companies and to protect their interests. It deals with all matters that affect the specific industry, such as collecting statistics, dealing with tariffs and trade regulations, and disseminating information about the product to design and construction professionals, particularly emphasizing its benefits over any competing or alternative product.

In order to ensure quality and competitiveness, most trade associations publish standards for the products they serve, to which all member companies must conform. Several trade associations also have certification programs to grade and stamp the product. AWPA is such an organization, and an AWPA grade stamp on wood and wood products is regarded as a mark of quality and standardization.

Professional Societies The professional societies, such as the American Society of Civil Engineers (ASCE) and the American Society of Heating, Refrigeration and Air Conditioning Engineers (ASHRAE), are another source of standards, although not extensive. These organizations develop standard specifications and testing procedures as professional obligations. For example, the American Society of Civil Engineers' publication *Minimum Design Loads for Buildings and Other Structures,* ASCE 7 Standard, is a publication that is heavily referenced by the building codes.

Standards-Producing Organizations

By far the largest and most referenced standards-producing organizations are those whose main function is to produce standards. In countries other than the United States, this task is normally handled by a single umbrella organization at the national level. In Britain, there is the British Standards Institution (BSI), in Australia, the Standards Association of Australia (SAA), and in Germany, the Deutsches Institute fur Normung (DIN).

In the United States, the picture is somewhat less clear because of the large number of organizations involved in the development of standards at the national level. Three primary standards-producing organizations in the U.S are

- American National Standards Institute (ANSI)
- American Society for Testing and Materials (ASTM)
- Underwriters Laboratories (UL)

American National Standards Institute (ANSI) The American National Standards Institute is a body similar to, but not identical with, national standards organizations in other countries, such as the BSI, DIN, and SAA. Unlike these organizations, however, ANSI generally does not develop its own standards. It functions primarily as the approving body and a clearinghouse at the national level for voluntary standards developed by other organizations in the United States. Thus, it provides an official and a nationally recognized status to standards developed by private organizations.

ANSI ensures that the standards it approves have been developed in an open and fair manner after due consultations among various interest groups, such as the producers, users, professional experts, and the general public. In case no standard exists for a product, ANSI will bring together and coordinate the expertise needed to develop the required standards.

Founded in 1918, ANSI is a nongovernmental and privately financed body, obtaining its funding primarily from membership dues and the sale of its publications. However, it represents the United States in the International Standards Organization (ISO).

American Society for Testing and Materials (ASTM) Founded in 1898, the American Society for Testing and Materials is a privately funded, nonprofit organization established for the development of standards on the characteristics and performance of materials, products, systems, and services and the promotion of related knowledge. It is by far the largest single source of standards in the world.

The Society was established as the American Section of the International Society for Testing Materials and incorporated in 1902 as the American Society for Testing Materials. In 1961, its name was changed to its present name, the *American Society for Testing and Materials,* to emphasize its interest in the materials as well as their testing.

ASTM standards are "voluntary consensus standards." The word *voluntary* implies that the standard is not mandatory, unless so required by a code or regulation that references it. In other words, the existence of an ASTM standard for a particular product (or procedure) does not mean that it cannot be produced (or used) if it does not conform to the standard. The word *consensus* implies that the standard is arrived at by a consensus of various members of the committee or subcommittee that produced it.

TABLE 2.3 LETTER DESIGNATIONS OF ASTM STANDARDS

Letter designation	Type of material or test
A	**Ferrous metals**
B	Nonferrous metals
C	Cementitious, ceramic, concrete, and masonry materials
D	Miscellaneous materials
E	Miscellaneous subjects
F	Materials of specific applications
G	Corrosion, deterioration, and degradation of materials
ES	Emergency standards
P	Proposals

ASTM standards are designated by a unique serial designation comprising an uppercase letter followed by a number (one to four digits), a hyphen, and, finally, the year of issue. The letter refers to the general classification of the material or test procedure, Table 2.3.

The year of issue refers to the year of original preparation or the year of the most recent revision of the standard. Thus, a standard prepared or revised in 2002 carries as its final number 02. If a letter follows the year, it implies a rewriting of the standard in the same year. For instance, 02a implies two revisions of the standard in 2002, 02b implies three revisions, and so on.

ASTM standards can be revised at any time by the appropriate technical committee. A standard must be reviewed every five years, which may result in its revision, reapproval, or withdrawal. In case a standard has been reapproved without any change, then the year of reapproval is placed within parentheses as part of the designation. Thus, (2002) means that the standard was reapproved in 2002. A typical ASTM standard is designated as C 136-02a: Standard Test Method for Sieve Analysis of Fine and Coarse Aggregates.

Underwriters Laboratories A different kind of standards organization is the Underwriters Laboratories (UL). It was established in 1894 by insurance companies, which were paying excessive claims caused by fires generated by the failures of new electrical and mechanical devices in buildings. Today, the UL is a self-supporting and not-for-profit organization that finances itself through the testing of various building and engineering products and the sale of its publications.

Currently, the UL maintains four testing laboratories in the United States, with its corporate headquarters in Northbrook, Illinois. It also certifies and labels products. The UL mark on a product is a certificate of its having successfully undergone some of the world's most rigorous tests for safety evaluation. In other words, the UL mark indicates that the product is safe and free from any foreseeable risk of hazard.

The Underwriters Laboratories is an independent organization, implying that it has no monetary interest in the product or its success in the marketplace. If a product passes UL's required safety testing and the manufacturer wants it to carry the UL mark, the Laboratory's field representative visits the manufacturing facilities unannounced several times during a year to ensure continued compliance of the product with the required safety standards. Certified products are listed in the UL Directories, which are published annually.

In the building industry, the UL is known mainly for testing the fire resistance rating of building components. The tests must obviously be conducted according to recognized standards. Currently, the UL has approximately 500 standards, most of which are approved as ANSI standards.

2.8 OTHER MAJOR GOVERNMENTAL CONSTRAINTS

In addition to building codes and speciality codes, building design and construction must also conform to several other legal constraints. The two most important such constraints are

- Zoning ordinances
- Accessibility standards—Americans with Disabilities Act (ADA)

In addition to the zoning ordinance and ADA regulations, building design and construction must be in conformance with several other requirements of the state or federal government. These constraints are generally specific to a building type. For example, a multifamily housing design must conform to the requirements laid down by the U.S. Department of Housing and Urban Development (HUD) in response to the *Fair Housing Act*.

Similar constraints apply to hospitals, schools, food-processing plants, and so on. Industrial and commercial buildings generally need to conform to requirements laid down by the insurance industry for coverage against property losses. Most insurance industry requirements exceed the corresponding requirements of the building code.

2.9 ZONING ORDINANCE

A *zoning ordinance* (also called the *zoning code*) of a city or county is a document containing urban planning laws. Its primary aim is to regulate the use of land under the jurisdiction of the city according to a comprehensive master plan. Thus, a zoning ordinance segregates the land of a city into different use groups and specifies the activity for which each piece of land can be used. Areas of land on which the same type of activity is allowed are called *districts*.

The number of districts and their designation varies from city to city. Usually, a city will divide its territory into four primary district groups:

- Residential district group
- Commercial district group
- Industrial district group and
- Special-purpose district group

Each of the above district groups is subdivided into several districts. For instance, the commercial district group may consist of a professional district, office district, local retail district, local business district, and general business district. Similarly, the residential district group may consist of a low-density, single-family dwelling district, medium-density, single-family dwelling district, high-density, single-family dwelling district, duplex dwelling district, and townhouse district.

To accomplish its land-use objectives, a zoning ordinance consists of two basic elements:

- Zoning map
- Zoning text

ZONING MAP

A zoning map is based on a comprehensive master plan of the city that takes into account the present and the future use of land under the city's jurisdiction. The map shows the district to which a piece of land belongs, and the zoning text specifies the types of buildings that can be built in that district. For instance, in a single-family dwelling district, the zoning ordinance will permit the construction of single-family residences and several other facilities that may be needed in that district, such as fire stations, local churches, community centers, parochial schools, playgrounds, and parks.

ZONING TEXT

The zoning text (as distinct from the graphical component—the zoning map) forms the bulk of the zoning ordinance. Apart from specifying the land use, a zoning text also gives the development standards for the city's land. These standards include the maximum ground coverage, maximum floor area ratio, minimum setbacks, maximum height, and the number of stories that may be built on a piece of land. In some cases, the types of exterior building envelope materials may be regulated.

Ground Coverage Ground coverage refers to the percentage of land area that may be covered by the building. For instance, for an office building in a downtown area, the ground coverage permitted may be as high as 100%, whereas for a single-family dwelling in a suburban location, the maximum ground coverage may be as low as 25%.

Floor Area Ratio (FAR) Floor area ratio refers to the total built-up area (sum of the built-up areas on all floors) divided by the total land area. An FAR of 2.0 means that the total allowable built-up area of the building on all floors is equal to twice the land area. Thus, an FAR of 2.0 is achieved by a two-story building covering the entire piece of land or a four-story building covering 50% of the land area, or a six-story building covering one-third the land area, and so on.

Ground coverage and FAR are powerful zoning restrictions that control the overall volume of buildings in a district without greatly jeopardizing design freedom. Volume control is an indirect means of aesthetic control of the development. It also controls population density, traffic volume, environmental pollution, and open spaces.

Restrictions placed on the height or the number of stories under a zoning ordinance also affect population density and traffic volume in the same way as ground coverage and FAR, but the main purpose of height restriction is to be sure the buildings are within the capabilities of fire-fighting equipment of the city. In some cities, height restrictions may be imposed to maintain views of existing buildings or to preserve the aesthetic character of a district.

Setbacks Setbacks refer to the distances a building must be set back from property lines, thereby specifying the minimum sizes of open areas around the building's property lines. Historically, the purpose of a front setback was to reserve land for any future widening of the street. In modern times, setbacks provide greater privacy and help to insulate buildings from traffic noise and fumes and otherwise enhance the aesthetic appeal of neighborhoods.

Zoning Ordinance Review and Administration

The formulation and enforcement of a zoning ordinance are usually delegated by the state to local authorities. Like the building code, the zoning ordinance is enforced under the police powers of the state.

Each local authority has several decision-making bodies that contribute to the formulation and review of the zoning ordinance. Theoretically, the city council is the highest zoning ordinance authority. For practical purposes, however, the zoning ordinance is under the charge of the *zoning and planning commission* of the city, which is typically a body consisting of several local citizens. The responsibilities of this commission include the formulation, development, and review of the zoning plan.

The zoning commission is (ideally) composed of individuals who have distinguished themselves by demonstrating outstanding and unselfish interest in civic affairs. They are appointed by the city council to serve for a fixed term and receive no remuneration for their work.

The initial formulation and subsequent reviews of the city's zoning plan by the zoning commission must go through a public hearing process. This allows all citizens to participate in city planning and the zoning decision-making process.

Although the zoning commission and the city council are the bodies responsible for zonal plan review, the day-to-day work related to zoning administration is carried out by city's planning department. Thus, the owner of a piece of land who desires to rezone (i.e., change the land use of) the property, would submit the request to the city's planning department, who would review it, make a recommendation, and forward it to the zoning commission for decision.

If the zoning commission agrees with rezoning, a public hearing is arranged. After the public hearing, the zoning commission makes its decision, which is usually in the form of a recommendation to city council. The city council may support the zoning commission's recommendation or hold its own public hearing on the matter before voting.

Board of Zoning Appeals

The implementation of land use control involves more than a mere formulation or review of the zoning plan. It also involves the enforcement of zoning ordinances that give rise to problems of application and interpretation. These problems are referred to the board of zoning appeals, also called the *board of adjustment*. This board consists of several members, usually appointed by the mayor or the city council.

The primary responsibility of the board of adjustment is to hear appeals on actions taken by city officials who administer zoning regulations and to make special exceptions to zoning regulations. Special exceptions generally refer to granting variances to setbacks, ground coverage, FAR, and so on, whose literal enforcement may cause unnecessary and undue hardship to the owner. Variances granted by the board must not be contrary to public interest and must not act to relieve the owner of self-created hardship. The meetings of the board are open to the public.

2.10 BUILDING ACCESSIBILITY—AMERICANS WITH DISABILITIES ACT (ADA)

The Americans with Disabilities Act (ADA) is a 1990 federal law that came into effect in January 1992. The objective of the act is to ensure equal opportunity to individuals with disabilities. A part of the act is devoted to regulating access to buildings so that they are also usable by individuals with disabilities.

The act requires that all new construction and alterations to existing construction provide barrier-free access routes on the site, into, and within all building spaces. Single-family dwellings, multifamily dwellings, and buildings funded by religious organizations, such as churches and temples, are exempt from ADA's accessibility provisions.

The concept of barrier-free access existed for nearly two decades before the promulgation of the ADA. The ANSI 117.1 standard entitled "American National Specifications for Making Buildings and Facilities Accessible to and Usable by the Physically Handicapped" had been adopted by several states and local jurisdictions.

In 1969, the U.S. federal government passed the Architectural Barriers Act (ABA), which requires that if federal funds are spent in the construction of a facility, that facility must be designed to be accessible to the physically handicapped according to the Uniform Federal Accessibility Standard (UFAS). In 1980, UFAS provisions were absorbed into ANSI 117.1. The difference between ABA and ADA is that ADA is much wider in scope than ABA. Additionally, ADA is mandated for application to all buildings except those mentioned earlier, regardless of the source of funds used in constructing them.

ADA is a comprehensive legislative act that addresses four major concerns: Title I pertains to employment discrimination of individuals with disabilities. Title II refers to the accessibility of public transportation lines, such as railroads and buses. Title III deals with accessibility in buildings, and Title IV covers accessible telecommunication facilities, such as telephones for the hearing impaired or individuals with other handicaps, and automatic teller machines.

Only Title III and Title IV of the ADA are of concern in the design of buildings. The provisions of Title III are detailed in ADA Accessibility Guidelines (ADAAG). The International Building Code also has provisions relating to accessibility (in a separate chapter). However, it must be recognized that a building code and ADA are two different legislative provisions. A building code is a local jurisdiction's law, whereas ADA is a federal law. The owner of the building is required to meet with the provisions of both laws.

ADA's accessibility requirements relate primarily to the design of entrances, doors, stairways, ramps, changes of level, sidewalks, elevators, drinking fountains, toilet facilities, door width, doorknobs, and so on.

Some states have enacted legislation and created standards that exceed the ADA's accessibility standards. In these states, the designers must comply with ADA's accessibility regulations, as well as those contained in the building codes and the state's accessibility standards.

PRACTICE QUIZ

Each question has only one correct answer. Select the choice that best answers the question.

24. A prescriptive code needs no standards.
- **a.** True
- **b.** False

25. Construction standards are typically written by
- **a.** the same organization that writes the building code.
- **b.** individual cities.
- **c.** the federal government.
- **d.** none of the above.

26. Which of the following is not a standard writing organization?
- **a.** ASTM
- **b.** ICC
- **c.** UL
- **d.** ANSI

27. Which of the following is a standards writing organization?
- **a.** ICC
- **b.** UL
- **c.** SBCCI
- **d.** ICBO
- **e.** BOCA

28. The acronym ASTM stands for
- **a.** American Standards for Testing and Manufacturing.
- **b.** Associated Society for Testing Methods.
- **c.** American Society for Testing of Materials.
- **d.** American Society for Testing and Materials.

29. The acronym UL stands for
- **a.** Universal Laboratories.
- **b.** Underwriters Laboratories.
- **c.** United Legislature.
- **d.** Union of Laborers.

30. A zoning ordinance refers to regulations that pertain mainly to
- **a.** the use of buildings within the city.
- **b.** the use of highways within the city.
- **c.** the use of land within the city.
- **d.** none of the above.

31. A zoning ordinance consists of
- **a.** the zoning map of the city.
- **b.** the zoning map and the building code of the city.
- **c.** the zoning text and the building code of the city.
- **d.** the zoning text and the zoning map of the city.
- **e.** none of the above.

32. The term FAR refers to
- **a.** factory area restrictions.
- **b.** floors as required.
- **c.** floor area ratio.
- **d.** none of the above.

33. If the total area of land under a property is 15,000 ft^2 and the total covered area on all floors of the building is 60,000 ft^2, the FAR for the property is
- **a.** 0.25.
- **b.** 1.5.
- **c.** 4.0.
- **d.** 6.75.

34. The Americans with Disabilities Act (ADA) is devoted entirely to provisions relating to accessibility within a building.
- **a.** True
- **b.** False

Answers: 24-a, 25-d, 26-b, 27-b, 28-d, 29-b, 30-c, 31-d, 32-c, 33-c, 34-b.

PRINCIPLES IN PRACTICE

Code Allowable Area and Height of a Building

During the early stages of design, an architect generally needs to obtain a rough idea of the total volume of the building, its massing, and the building's location on the site. This requires determining the maximum allowable area of the building and its maximum allowable height, as per the building code. The allowable area and height of a building are a function of the following four factors:

- Building's occupancy
- Building's type of construction
- Width of accessible open spaces around the building (frontage)
- Whether or not the building is provided throughout with automatic sprinklers

Sometimes, the total area and height of the building are already established by the building's program. In that case, the architect would determine the minimum type of construction required to satisfy the code and whether or not to provide automatic sprinklers.

The area and height calculation can be complex for some buildings. The description presented here has been simplified to provide a general understanding of the issues involved. For a comprehensive treatment, the reader should consult the jurisdiction's building code.

ALLOWABLE AREA AND ALLOWABLE HEIGHT

In determining the maximum allowable area and maximum allowable height of the building, we follow these five steps:

1. *Tabular Area and Tabular Height:* Determine the *basic allowable area per floor* and *basic allowable height* of the building. This is based only on the occupancy and the type of construction of the building, Table 1. The area so obtained is also referred to as the *tabular area*, A_t, because it is obtained directly from the table.

   The height is specified as a number of stories and also as a number of feet above the ground plane. Both restrictions must be met; that is, the number of stories must not exceed those stated and the height of the building should not exceed the allowable height in feet.
2. *Area Increase Due to Frontage:* Determine the area increase (per floor) due to accessible open spaces (referred to as *frontage*) around the building.
3. *Area Increase and Height Increase Due to Sprinklers:* Determine the area increase (per floor) and the increase in total height if the building is provided throughout with automatic sprinklers.

TABLE 1 BASIC ALLOWABLE AREAS AND HEIGHTS FOR SELECTED OCCUPANCIES

	Type of Construction	Type I A	Type I B	Type II A	Type II B	Type III A	Type III B	Type IV HT	Type V A	Type V B
Occupany group	Height (ft) ▷ Stories/Area ▽	UL	160	65	55	65	55	65	50	40
A-3	No. of stories	UL	11	3	2	3	2	3	2	1
	Area per floor (ft²)	UL	UL	15,500	9,500	14,000	9,500	15,000	11,500	6,000
B	No. of stories	UL	11	5	4	5	4	5	3	2
	Area per floor (ft²)	UL	UL	37,500	23,000	28,500	19,000	36,000	18,000	9,000
E	No. of stories	UL	5	3	2	3	2	3	2	1
	Area per floor (ft²)	UL	UL	26,500	14,500	23,500	14,500	25,500	18,500	9,500
R-2	No. of stories	UL	11	4	4	4	4	4	3	2
	Area per floor (ft²)	UL	UL	24,000	16,000	24,000	16,000	20,500	12,000	7,000
S-1	No. of stories	UL	11	4	3	3	3	4	3	1
	Area per floor (ft²)	UL	48,000	26,000	17,500	26,000	17,500	25,500	14,000	9,000

Notes: (1) This is an incomplete table. For complete information, refer to the 2006 International Building Code. Copyright 2006. Falls Church, Virginia: International Code Council, Inc. Reproduced with permission. All rights reserved.

(2) UL = unlimited. Thus, if area per floor is specified as UL, it means that the code has no limits on the allowable area. Similarly if the number of stories is specified as UL, it means that the number of stories may be unlimited. However, the city's zoning ordinance may place restrictions on the height as well as the number of stories.

(3) Area per floor = basic allowable area of one floor.

(4) No. of stories = basic number of stories allowed, which may be increased by one additional story if the entire building is sprinklered.

4. *Maximum Allowable Area per Floor:* Determine the maximum allowable area per floor, which is the sum of the areas obtained from Steps 1 to 3. We refer to this sum as $A_{per\ floor}$.
5. *Maximum Allowable Area on All Floors:* Determine the maximum total allowable area of the building on all floors, referred to as $A_{all\ floors}$.

Area Increase Due to Frontage

The basic allowable area per floor (Table 1), may be increased for frontage. The frontage increase is available only if more than 25% of building's perimeter fronts a public way or an accessible open space at least 20 ft wide. The open space must be accessible from a street or fire lane at all times. The area increase per floor due to frontage is given by the following equation:

$$\text{Area increase due to frontage} = A_t I_f \qquad \textbf{Eq. (1)}$$

where A_t = basic allowable area from Table 1
I_f = a coefficient, whose value is given by the expression

$$I_f = \left[\frac{F}{P} - 0.25\right]\frac{W}{30} \qquad \textbf{Eq. (2)}$$

where F = perimeter of a building that fronts on a public way or an accessible open space having a minimum width of 20 ft.
P = perimeter of entire building. If $(F/P) \leq 0.25$, i.e., accessible frontage is less than or equal to 25% of the total perimeter of the building, set $I_f = 0$, which implies that no area increase due to frontage is available.
W = width of public way or accessible open space, not less than 20 ft; in other words, for $W < 20$ ft, set $W = 0$. Also, if $W > 30$ ft, set $W = 30$ ft, that is, $W/30 = 1.0$. In other words, an accessible open space greater than 30 ft does not yield any additional area increase.

Area Increase Due to Sprinklers

The basic allowable area per floor (Table 1), may be increased if the entire building is provided with automatic sprinklers. The area increase per floor due to sprinklers is given by

$$\text{Area increase due to sprinklers} = A_t I_s \qquad \textbf{Eq. (3)}$$

where $I_s = 2.0$ (which implies a 200% increase) for a multistory building
$I_s = 3.0$ (which implies a 300% increase) for a single-story building

Height Increase Due to Sprinklers

If automatic sprinklers are provided throughout the building, the maximum height of the building may be increased by 20 ft, and the maximum number of stories may be increased by one. Thus, from Table 1, the maximum allowable height of Occupancy B with Type III(A) construction is 65 ft, not exceeding five stories. If this building is provided with automatic sprinklers, the maximum allowable height is 85 ft, and the maximum number of stories allowed is six.

Total Allowable Area per Floor

Thus, the total area per floor including the increases due to frontage and sprinklers, $A_{per\ floor}$, is given by

$$A_{per\ floor} = \begin{array}{l}\text{basic allowable area, } A_t \text{ (from Table 1)} + \\ \text{area increase due to frontage (from Eq. 1)} + \\ \text{area increase due to sprinklers (from Eq. 3)}\end{array} \qquad \textbf{Eq. (4)}$$

Total Allowable Area on All Floors

The total allowable area on all floors, $A_{(all\ floors)}$ is given by

$$\begin{array}{ll}(A_{all\ floors}) = 3.0(A_{per\ floor}) & \text{for a building with three or more floors} \\ (A_{all\ floors}) = 2.0(A_{per\ floor}) & \text{for a building with two floors}\end{array} \qquad \textbf{Eq. (5)}$$

(*Continued*)

PRINCIPLES IN PRACTICE

Code Allowable Area and Height of a Building (*Continued*)

Thus, if $A_{per\ floor}$ for an eight-story building = 25,000 ft², the maximum total allowable area on all floors for this building is $A_{(all\ floors)} = 75{,}000\ ft^2$. If the building is a two-story building, $A_{(all\ floors)} = 50{,}000\ ft^2$.

Examples 1 and 2 illustrate the use of this procedure.

EXAMPLE 1 • (ALLOWABLE AREA OF AN UNSPRINKLERED ONE-STORY BUILDING)

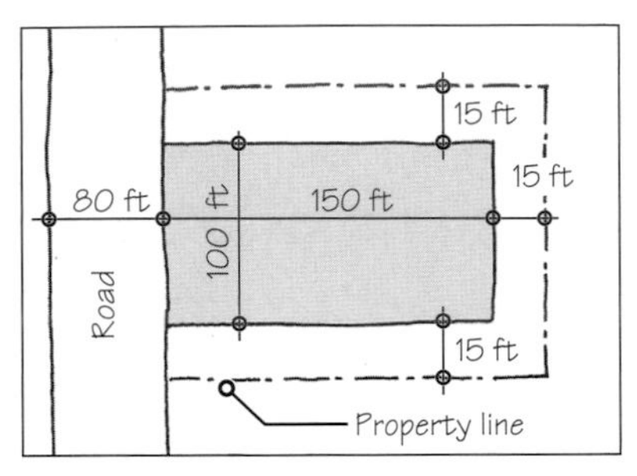

FIGURE 1 Site plan of a furniture warehouse (Example 1).

Determine if the plan of the furniture warehouse shown in Figure 1 meets the building code restrictions on allowable area and height. The building is one story high (total height = 30 ft), and the type of construction used is Type II(B). No automatic sprinklers are provided.

SOLUTION

From Table 2.1, a furniture warehouse is Occupancy S-1.

Step 1: From Table 1, the basic allowable area is $A_t = 17{,}500\ ft^2$, and basic allowable height = 55 ft.

Step 2: Area Increase Due to Frontage
Only one (100 ft) side of the building fronts a public way. Because the other three open spaces are less than 20 ft wide, they do not count as accessible open spaces. Hence

$$F = 100\ ft \quad \text{and} \quad P = 100 + 100 + 150 + 150 = 500\ ft$$

Hence, (F/P) = (100/500) = 0.2. Because (F/P) is less than 0.25, implying that less than 25% of the building's perimeter fronts an accessible open space, $I_f = 0$. From Equation (1),

Area increase due to frontage = 0

Step 3: Area Increase Due to Sprinklers
Because no sprinklers have been provided,

Area increase due to sprinklers = 0

Step 4: Allowable Area per Floor
From Equation (4),

$$(A_{per\ floor}) = 17{,}500 + 0 + 0 = 17{,}500\ ft^2$$

Step 5: Allowable Area on all Floors
Because the building is one story high, the allowable area on all floors is 17,500 ft².

Area and Height Provided
The area provided is $150 \times 100 = 15{,}000\ ft^2$, which is less than the allowable area of 17,500 ft². Therefore, the area provided is OK.

The height of the building is 30 ft, which is less than the maximum allowable height of 55 ft (Table 1), which is also OK.

EXAMPLE 2 • (ALLOWABLE AREA OF A SPRINKLERED MULTI-STORY BUILDING)

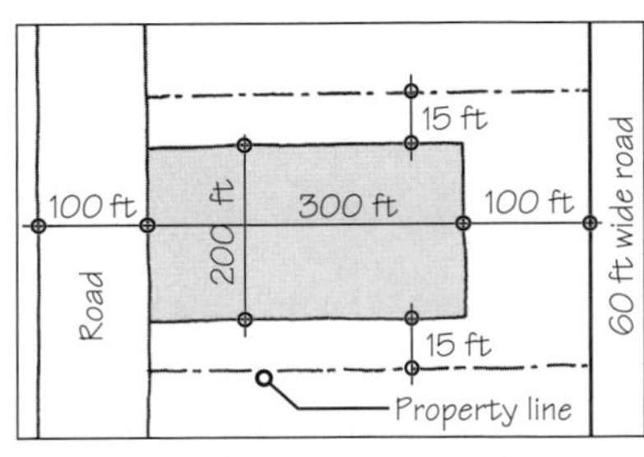

FIGURE 2 Site plan of an office building (Example 2).

An architect's site plan of a six-story office building (80 ft tall) is shown in Figure 2. Determine if the area and height of the building are as per the building code. The type of construction used is Type II(A). The entire building is provided with an approved automatic sprinkler system.

SOLUTION

From Table 2.1, an office is Occupancy B.

Step 1: From Table 1, the basic allowable area per floor is $A_t = 37{,}500\ ft^2$, basic allowable height is 65 ft, and number of stories is 5.

Step 2: Area Increase Due to Frontage
Two (200-ft) sides of the building are accessible from as public way. Because the other two open spaces are less than 20 ft wide, they are not counted as accessible open spaces. Hence,

$$F = 200 + 200 = 400 \text{ ft} \quad \text{and} \quad P = 1{,}000 \text{ ft}.$$

Hence, $(F/P) = (400/1{,}000) = 0.4$. Because $W = 100$ ft and 160 ft (>30 ft on both 200 ft sides of the building), set $W = 30$. Hence, $W/30 = 1.0$. From Equation (2):

$$I_f = (0.4 - 0.25)(1.0) = 0.15$$

From Equation (1):

$$\text{Area increase due to frontage} = (37{,}500)(0.15) = 5{,}625 \text{ ft}^2$$

Step 3: Area Increase Due to Sprinklers
Because the building is a multistory building, $I_s = 2.0$. From Equation (3),

$$\text{Area increase due to sprinklers} = (37{,}500)(2.0) = 75{,}000 \text{ ft}^2$$

Step 4: Total Allowable Area per Floor
From Equation (4):

$$(A_{\text{per floor}}) = 37{,}500 + 5{,}625 + 75{,}000 = 118{,}125 \text{ ft}^2$$

Step 5: Total Allowable Area for the Entire Building
From Equation (5), the total allowable area for the entire building is

$$A_{\text{all floors}} = 3.0(118{,}125) = 354{,}375 \text{ ft}^2$$

Area Provided
The area provided per floor is $300 \times 200 = 60{,}000$ ft^2, which is less than the allowable area ($A_{\text{per floor}}$) of 118,125 ft^2. Therefore, the area provided per floor is OK.

However, the allowable area for the entire building is 354,375 ft^2. This is less than the area provided: $6(60{,}000) = 360{,}000$ ft^2, which is no good (NG).

Therefore, the provided area must be reduced. Alternatively, the construction type may be upgraded to Type I(B); in this case there is no code-prescribed limit on the area (Table 1). The city may, however, place limitations on the area based on the city's zoning laws (Section 2.9).

Height Provided
The basic allowable height for occupancy is B = 5 stories (Table 1), with a maximum height of 65 ft. Because the building is provided with automatic sprinklers, the maximum allowable height is $65 + 20 = 85$ ft, and the number of stories allowed is $5 + 1 = 6$. Thus, the provided height of 80 ft and six stories is OK.

Allowable Area of a Mixed Occupancy Building

The allowable area and height determinations given so far refer to a single occupancy building. For a mixed occupancy building, two cases must be distinguished.

First, if the building is not provided with required occupancy separations (an *unseparated mixed occupancy building*), the allowable area and height are calculated by assuming that the entire building belongs to only one occupancy. These calculations are made for each occupancy provided. The smallest area and the smallest height obtained from the calculations are the allowable area and height of the building.

Second, if required occupancy separations have been provided, the building is treated as a duly *separated mixed occupancy building*. For such a building, the sum of respective floor areas provided for each occupancy divided by their allowable floor areas at each floor must not exceed 1.0.

For example, consider a building with three different occupancies. Let us call these occupancies Occupancy 1, Occupancy 2, and Occupancy 3. Assume that the floor area provided at any one floor for Occupancy 1 is $(A_{\text{prov}})_1$, for Occupancy 2 is $(A_{\text{prov}})_2$, and for Occupancy 3 is $(A_{\text{prov}})_3$. Assume further that the respective allowable areas for each floor for the three occupancies are $(A_{\text{allow}})_1$, $(A_{\text{allow}})_2$, and $(A_{\text{allow}})_3$; then the following equation must be satisfied.

$$\frac{(A_{\text{prov}})_1}{(A_{\text{allow}})_1} + \frac{(A_{\text{prov}})_2}{(A_{\text{allow}})_2} + \frac{(A_{\text{prov}})_3}{(A_{\text{allow}})_3} \leq 1.0 \qquad \textbf{Eq. (6)}$$

PRACTICE QUIZ

Each question has only one correct answer. Select the choice that best answers the question.

35. If a single-story building is provided throughout with automatic sprinklers, the allowable area per floor of the building may be increased by
- **a.** 100%
- **b.** 150%
- **c.** 200%
- **d.** 250%
- **e.** 300%

36. If a multistory building is provided throughout with automatic sprinklers, the allowable area per floor of the building may be increased by
- **a.** 100%
- **b.** 150%
- **c.** 200%
- **d.** 250%
- **e.** 300%

37. If automatic sprinklers are provided throughout the building, the building's height may be increased by
- **a.** 20 ft above the basic allowable height.
- **b.** 40 ft above the basic allowable height.
- **c.** 60 ft above the basic allowable height.
- **d.** 100 ft above the basic allowable height.
- **e.** none of the above.

38. If the maximum allowable area per floor for a building is 20,000 ft^2 and the maximum allowable height is 11 floors, what is the maximum allowable area for the entire building?
- **a.** 40,000 ft^2
- **b.** 60,000 ft^2
- **c.** 100,000 ft^2
- **d.** 220,000 ft^2
- **e.** none of the above

Answers: 35-e, 36-c, 37-a, 38-b.

KEY TERMS AND CONCEPTS

American National Standards Institute (ANSI)
American Society of Testing and Materials (ASTM)
Americans with Disabilities Act (ADA)
Board of Appeals
Building code
Building official
Building permit
Certificate of occupancy
Code-allowed area and height of building
Construction standards
Fire safety
Health and welfare
International Building Code
International Code Council (ICC)
International Residential Code
Life safety
Means of egress
Mixed occupancy
Model code
Occupancy classifications
Performance code provision
Prescriptive code provision
Property protection
Stop work order
Structural safety
Types of construction
Underwriters Laboratory (UL)
Zoning ordinance

REVIEW QUESTIONS

1. What is the difference between a prescriptive and a performance type code provision? Explain with the help of an example.
2. List the model code organizations that existed in the United States prior to the year 2000 and the building codes published by each.
3. Describe the relationship between a building code and construction standards. List three important standards organizations in the United States.
4. What information can you derive from a given designation of an ASTM standard such as E 119-95a?
5. List the first five steps you will follow in ascertaining that a building conforms to the provisions of the building code.
6. List at least three codes published by the International Code Council.
7. Explain the terms ground coverage and FAR.

SELECTED WEB SITES

American National Standards Institute, ANSI (www.ansi.org)

American Society for Heating, Refrigeration and Air Conditioning Engineers, ASHRAE (www.ashrae.com)

American Society for Testing and Materials, ASTM (www.astm.com)

International Code Council, ICC (www.iccsafe.org)

Underwriters Laboratory, UL (www.ul.com)

FURTHER READING

1. International Code Council: *International Building Code.*
2. International Code Council: *International Residential Code.*

REFERENCES

1. Sanderson, R: "Code and Code Administration". Country Club Hills, Il:, Building Officials and Code Administrators, 1969, p. 102.

CHAPTER 3 Loads on Buildings

CHAPTER OUTLINE

3.1 Dead Loads

3.2 Live Loads

3.3 Rain Loads

3.4 Wind Load Basics

3.5 Factors That Affect Wind Loads

3.6 Roof Snow Load

3.7 Earthquake Load

3.8 Factors That Affect Earthquake Loads

3.9 Wind Versus Earthquake Resistance of Buildings

PRINCIPLES IN PRACTICE: Dead Load and Live Load Estimation

The shallow profile of London's Millenium Footbridge provides an uninterrupted view across the Thames River, particularly of St. Paul's Cathedral (seen in the background). The design pushes the limits of possibilities for suspension bridge design and is recognized as a structural masterpiece. Engineer: Ove Arup and Partners. Sculptor: Anthony Caro. Architect: Foster and Partners. (Photo by DA.)

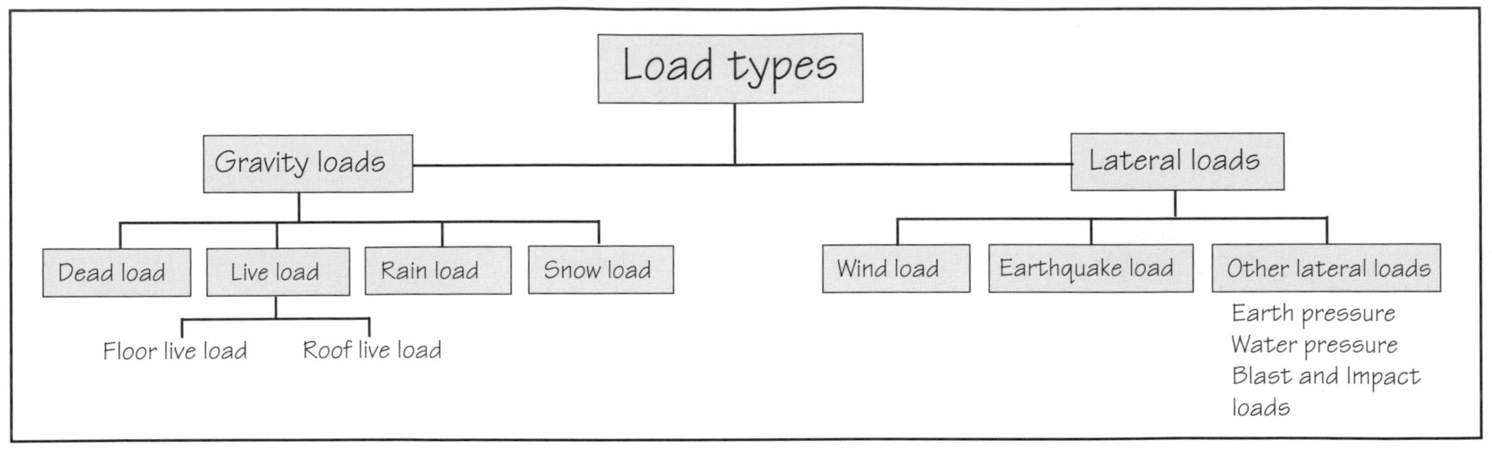

FIGURE 3.1 Some of the important loads in buildings.

Buildings are subjected to several types of loads, Figure 3.1. Two main classifications of loads are

- Gravity loads
- Lateral loads

Gravity loads are caused by the gravitational pull of the earth and act in the vertical direction. Therefore, they are also referred to as *vertical loads.* Gravity loads include the materials and components that comprise the buildings, as well as people, rainwater, snow, furniture, equipment, and all that is contained within the building. Gravity loads are further classified as *dead loads* and *live loads,* as explained in subsequent sections.

The two primary sources of *lateral loads* on buildings are wind and earthquakes. The effect of each is to create loads in the lateral (other than vertically downward) direction. For example, wind creates horizontal forces on a wall as well as vertically upward forces (suction) on a flat roof. The main effect of earthquake ground motion is to create horizontal forces in buildings, although a small amount of vertical force may also exist. Additional examples of lateral loads are earth pressure on basement walls, water pressure on tank walls, and loads caused by blasts and moving vehicles or equipment.

In this chapter, we examine the nature of different types of loads on buildings. The treatment given here is brief and qualitative and sets the stage for a detailed investigation of the topic by an interested reader.

NOTE

Loads in the SI System of Units

The unit of measurement for load in the SI system of units is the newton, which is abbreviated N. A newton is even smaller than a pound (1 lb = 4.45 N). Therefore, the unit kilonewton (kN) is commonly used.

For the load on a floor, the unit newton per square meter (N/m^2) is used instead of psf. A newton per square meter is a pascal (Pa), and a kN/m^2 is a kilopascal (kPa). Thus, 1.0 N/m^2 is synonymous with 1.0 Pa, and 1.0 kN/m^2, with 1.0 kPa. 1 psf = 47.88 Pa.

For the load on a linear element, the unit is newton per meter (N/m), or kilonewton per meter (kN/m) is used. The load on a column is expressed in newtons (N) or kilonewtons (kN).

(For conversion from the U.S. system of units to the SI System of units, and vice versa, see Appendix A.)

UNITS OF MEASUREMENT FOR LOADS

In the U.S. system of units, a load is expressed in pounds (lb). The pound is a small unit for measuring building loads. Therefore, the unit kilopounds (kips) is commonly used; 1 kip is equal to 1,000 lb. Thus, 2.5 kips = 2,500 lb.

When the load is distributed over a surface such as a floor or roof, it is generally expressed in pounds per square foot (psf), Figure 3.2. A beam is treated as a linear element, because its width is generally much smaller than its length. Therefore, the load on a beam is generally expressed in terms of pounds per foot length of the beam (lb/ft). If the load on the beam is large, it is generally stated as kilopounds per foot (kips/ft). The load on a column is expressed in terms of pounds (lb) or kilopounds (kips).

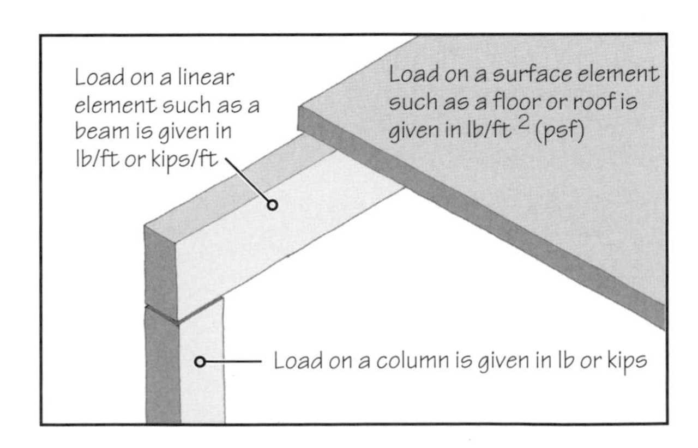

FIGURE 3.2 The loads on various elements of a building are expressed in different units.

3.1 DEAD LOADS

Dead loads are always present in a building; that is, they do not vary with time. They include the weights of materials and components that comprise the structure. The dead load of a component is computed by multiplying its volume by the density of the material. Because both the densities and the dimensions of the components are known with reasonable accuracy, dead loads in a building can be estimated with greater certainty than other load types. Densities of some of the commonly used materials are given in Table 2 in Principles in Practice at the end of this chapter.

The dead load for which a building component is designed includes the self-load of the component plus the dead loads of all other components that it supports. For example, the dead load on a column includes the weight of the column itself plus all the dead load imposed on it. In Figure 3.3, the dead load on a column is the weight of the column plus the dead load from the beams and slab resting on it. Similarly, the dead load on a beam is the weight of the beam itself plus the dead load from the slab that it supports. The dead load on the slab is only the self-weight of the slab. However, if the slab supports a floor finish, ceiling, light fixtures, or plumbing and electrical pipes, their weights must be included in the dead load acting on the slab.

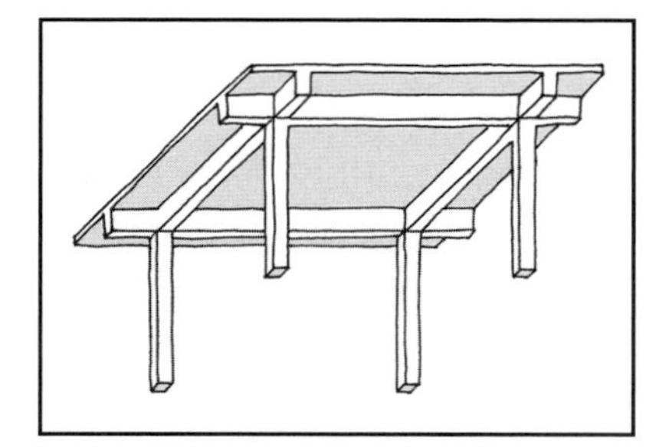

FIGURE 3.3 The dead load on the column of a building includes the weight of all components that it supports, such as the beams and floor slab plus the self-load of the column.

3.2 LIVE LOADS

Although the dead load is a permanent load on the structure, live load is defined as the load whose magnitude and placement changes with time. Such loads are due to the weights of people (animals, if the building houses animals), furniture, movable equipment, and stored materials. As shown in Figure 3.1, live loads are further divided into

- Floor live load
- Roof live load

FLOOR LIVE LOAD

Floor live load depends on the occupancy and the use of the building. Therefore, it is also called the *occupancy load,* and it is different for different occupancies. For instance, the floor live load for a library stack room is higher than the floor live load for a library reading room, which in turn is higher than the floor live load for an apartment building.

Floor live loads are determined by aggregating the loads of all people, furniture, and movable equipment that may result from the particular occupancy. Safety considerations require that the worst expected situation be considered so that the structure is designed for the maximum possible live load that may be placed on it.

Based on a large number of surveys, floor live loads for various commonly encountered occupancies, such as individual dwellings, hotels, apartment buildings, libraries, office buildings, and industrial structures, have been determined and are contained in building code tables. Table 3.1 gives floor live loads for a few representative occupancies. For instance, the floor live load for a library stack room is 150 psf, for a library reading room is 60 psf, and for an apartment building is 40 psf. Note that these are the minimum floor live loads for which the building must be designed. Even if the actual live load on a floor is smaller, it must be designed for the minimum live load specified by the building code.

Floor live load values in Table 3.1 are conservative. In most situations, the actual loads are smaller than those given. However, the architect or the engineer must recognize unusual situations that may lead to a greater actual load than the one specified in the code. In such situations, the higher anticipated load should be used. Additionally, if the live load for an occupancy not included in building code tables is to be obtained, the architect or the structural engineer must determine it from first principles, taking into account all the loads that may be expected on the structure. Most building codes require that such live load values be approved by the building official.

Floor live loads are usually assumed to be uniformly distributed over the entire floor area and are expressed in pounds per square foot (psf). However, an absolutely uniform distribution is seldom obtained in practice. Therefore, building codes recommend that the effects of live load concentration be investigated in those occupancies where it may be critical to design.

These occupancies are parking garages (where loads are concentrated on vehicle tires), libraries, offices, schools, and hospitals. Apartments, hotels, individual residences, assembly buildings (auditoriums, concert halls, grandstands), and exit areas of all buildings do not

TABLE 3.1 MINIMUM FLOOR LIVE LOADS FOR SELECTED OCCUPANCIES

Occupancy	Description	Load (psf)
Access floor system	Office use	50
	Computer use	100
Auditoriums	Fixed seating areas	60
	Movable seating areas	100
	Stage areas and enclosed platforms	125
Garages	Passenger cars	40
Hospitals	Wards and rooms	40
	Operating theaters	60
Libraries	Reading rooms	60
	Stack rooms	150
Manufacturing	Light	125
	Heavy	250
Offices	General office	50
	Corridors	80
	Lobbies	100
Residential dwellings	Dwellings and hotel guest rooms	40
	Habitable attics	30
	Corridors, public spaces, and balconies	100
Schools	Classrooms	40
	Corridors and lobbies	80
Stores	Light	125
	Heavy	250

Note: For an authoritative information, refer to the building code of the jurisdiction.

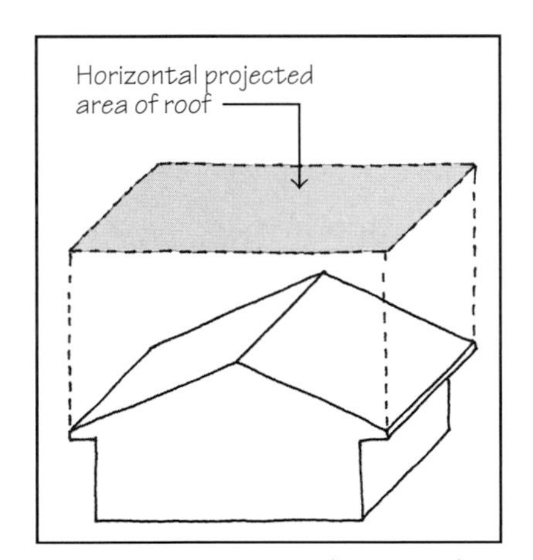

FIGURE 3.4 Horizontal projected area of a pitched roof.

require investigation for concentrated loads. (At this stage, the reader is advised to study the examples given in "Principles in Practice" at the end of this chapter.)

ROOF LIVE LOAD

The live load on a roof takes into account the weight of repair personnel and temporary storage of construction or repair materials and equipment on the roof. Roof live load is generally given as 20 psf acting on the horizontal projected area of the roof, Figure 3.4. This applies to a roof that will not be used in the future as a floor. If the building is expected to be extended vertically in the future, the roof live load is the floor live load for the anticipated future occupancy. Additionally, if the roof is to be landscaped, the load due to the growth medium and landscaping elements must be considered as dead load on the roof.

3.3 RAIN LOADS

Although roofs are designed to have adequate drainage so that no accumulation of water occurs, loads resulting from accidental accumulation of melted snow or rainwater must be considered as a possibility. Drains may be blocked by windblown debris or hail aggregating on the roof or the formation of ice dams near the drains.

Long-span, relatively flat roofs are particularly vulnerable to rainwater accumulation because, being flexible, they deflect under the weight of water. This deflection leads to yet more accumulated water, causing additional deflection, which increases accumulation. If adequate stiffness is not provided in the roof, the progressive increase of deflection can cause excessive load on the roof. Water accumulation has been the cause of complete collapse of several long-span roofs.

Generally, roofs with slope greater than $\frac{1}{4}$ in. to 1 ft (1:48 or say, 1:50 slope) are not subjected to accumulated rainwater unless roof drains are blocked. Building codes mandate $\frac{1}{4}$ in. to 1 ft as the minimum slope required for roofs, which, apart from providing positive drainage, also helps to increase the life and improve the performance of roof membranes.

Building codes also require that in addition to primary drains, roofs must be provided with secondary (overflow) drains. The secondary drains must be at least 2 in. above the primary drains so that if the primary drainage system gets blocked, the secondary system will be able to drain the water off the roof, Figure 3.5.

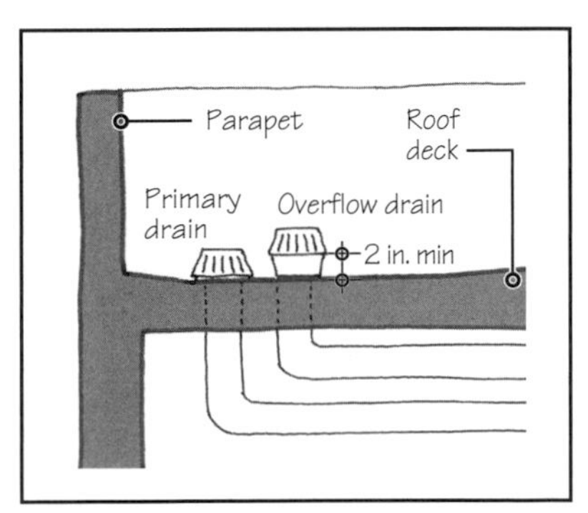

FIGURE 3.5 Roof with a parapet must be provided with two independent drainage systems.

Water accumulation generally occurs on roofs that are provided with a parapet. In the absence of a parapet, water accumulation will generally not occur. Therefore, secondary drains are not required on unparapeted roofs or on steep roofs.

PRACTICE QUIZ

Each question has only one correct answer. Select the choice that best answers the question.

1. The two main categories under which all building loads may be classified are
   - **a.** dead loads and live loads.
   - **b.** gravity loads and lateral loads.
   - **c.** terrestrial loads and celestial loads.
   - **d.** first-class and second-class loads.
2. The loads on a floor are generally expressed in
   - **a.** lb/ft^2.
   - **b.** lb/ft.
   - **c.** lb.
   - **d.** lb/ft^3.
   - **e.** none of the above.
3. The loads on a beam are generally expressed in
   - **a.** lb/ft^2.
   - **b.** lb/ft.
   - **c.** lb.
   - **d.** lb/ft^3.
   - **e.** none of the above.
4. The loads on a column are generally expressed in
   - **a.** lb/ft^2.
   - **b.** lb/ft.
   - **c.** lb.
   - **d.** lb/ft^3.
   - **e.** none of the above.
5. Which of the following loads can be estimated with greatest certainty?
   - **a.** Dead loads
   - **b.** Live loads
   - **c.** Wind loads
   - **d.** Earthquake loads
   - **e.** Snow loads
6. The floor live load in a residential occupancy is generally assumed to be
   - **a.** 40 psf.
   - **b.** 10 psf.
   - **c.** 20 psf.
   - **d.** 5 psf.
   - **e.** none of the above.
7. Which of the following occupancies has the highest floor live load?
   - **a.** Classrooms
   - **b.** Hotel guest rooms
   - **c.** Library stack rooms
   - **d.** Library reading rooms
   - **e.** Offices
8. Roof live load in a building is generally
   - **a.** less than the floor live load.
   - **b.** greater than the floor live load.
   - **c.** the same as the floor live load.
9. Roof live load is a function of the building's occupancy.
   - **a.** True
   - **b.** False
10. In which of the following roofs will you consider the effect of rain load?
    - **a.** Roof with a parapet
    - **b.** Roof without a parapet

Answers: 1-b, 2-a, 3-b, 4-c, 5-a, 6-a, 7-c, 8-a, 9-b, 10-a.

3.4 WIND LOAD BASICS

Although wind loads are primarily horizontal, they also exert an upward force on horizontal elements such as flat and low-slope roofs. Resistance against upward wind force is provided by anchoring the building to its foundations. Resistance against horizontal loads requires anchorage to foundations and the use of stiffening elements. These stiffening elements are commonly referred to as *wind-bracing elements.*

The wind-bracing requirement for a building is precisely the same as the requirement for either diagonal bracing or sheet bracing of a bookshelf. If a book rack is constructed of several shelves mounted on two side supports, it will be adequate to carry the gravity load from the books, but the assembly will be unstable. When acted upon by a horizontal force from either direction, it will move sideways (or *rack*), as shown by dotted lines in Figure 3.6(a).

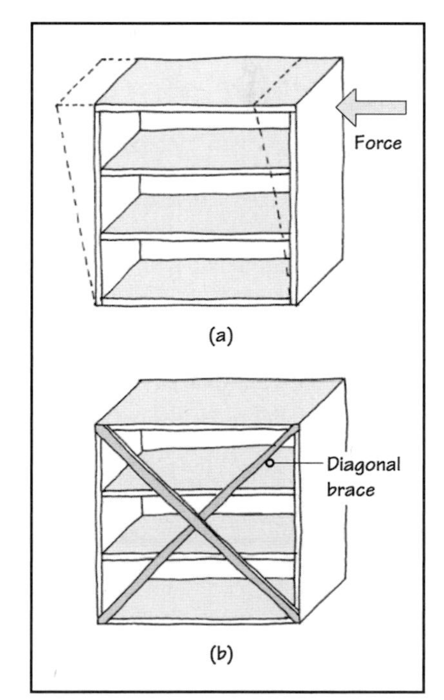

FIGURE 3.6 **(a)** When pushed by a horizontal force, a book rack constructed of several horizontal shelves and two vertical supports will move sideways (i.e., rack), as shown by dashed lines. **(b)** The racking of shelves can be prevented by diagonal braces (as shown) or by sheet bracing.

Racking of the bookshelf can be prevented by diagonally bracing one of its faces, Figure 3.6(b). An alternative method of stiffening the book rack is to apply a sheet material over the entire face instead of diagonal braces. Sheet bracing generally provides greater lateral stiffening than diagonal bracing.

For buildings, several methods of wind bracing are used, depending on the functional and aesthetic constraints of the building and the magnitude of wind loads. Because wind loads increase with the height of a building, stiffening the building against wind loads becomes increasingly important as the height of the building increases. In fact, the design of tall buildings is heavily dominated by wind-bracing requirements (see Chapter 16 for additional details).

Architects have often used this structural requirement as a forceful aesthetic expression. Two successful examples of this are the John Hancock Center and the Sears Tower, both in Chicago. The John Hancock Center uses diagonal braces on all four of its facades, Figure 3.7. The Sears Tower uses the concept of *bundled tubes* to provide lateral stiffening.

FIGURE 3.7 The wind-load resistance in John Hancock Center, Chicago, is provided by diagonal bracing on its facade.

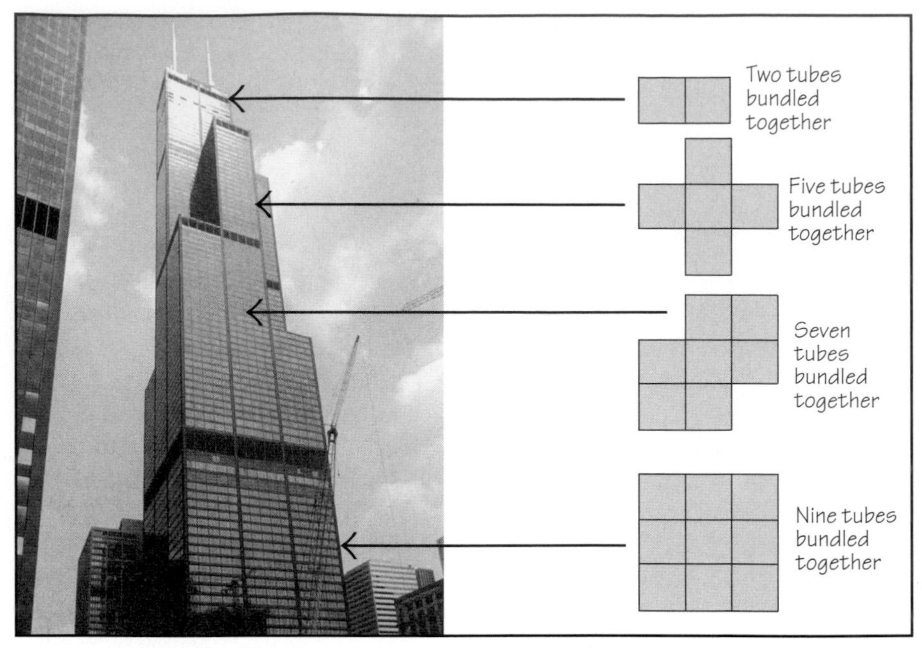

FIGURE 3.8 The wind-load resistance in Sears Tower, Chicago, is provided by using bundled tubes.

The number of tubes bundled together in the Sears Tower reduces from nine at the base of the tower to two at the top, Figure 3.8.

NOTE

A Note on Tornadoes

The suction produced in a tornado funnel lifts objects into the air like a giant suction pump. If a tornado hits a water body such as a lake or a sea, a waterspout is formed. Nearly 60% of all tornadoes are weak in intensity. However, 38% are strong enough to topple walls and blow roofs off buildings. Nearly 2% of tornadoes are so violent that they destroy virtually everything in their path.

Actual measurements of wind speeds in a tornado are not available because the area of ground struck by a tornado is relatively small. Posttornado damage observations have revealed that an average tornado has a life span of nearly 30 minutes and covers a nearly 10-mi-long by 0.25-mi-wide area of ground, i.e., 2.5 mi^2.

Thus, the probability of a tornado hitting a particular building or intercepting a permanently located wind speed–measuring instrumentation, such as that installed at an airport, is very small. Additionally, the destructive power of a tornado is so large that whenever a tornado has hit such measuring instruments, it has completely destroyed them.

TORNADOES, HURRICANES, AND STRAIGHT-LINE WINDSTORMS

The wind speed is fundamental to determining wind loads on buildings. It is the most important factor, because the wind load is directly proportional to the square of wind speed. In other words, if other factors that affect wind loads are ignored, doubling the wind speed quadruples the wind load. Tripling the wind speed makes wind load nine times larger.

Obviously, buildings must be designed for the maximum probable wind speed at that location. The highest wind speeds are usually contained in a tornado—a rotating, funnel-shaped column of air that produces strong suction, Figure 3.9.

Because of the extremely small probability that a violent tornado will hit a particular building, design wind speeds adopted by building codes for various locations do not reflect the wind speeds obtained in tornadoes. In spite of the fact that design wind speeds do not account for tornadoes, buildings have to be designed to resist tornadoes in tornado-prone locations. For this purpose, sufficient empirical information is available to design and construct tornado-resistant buildings. Figure 3.10 shows tornado-prone areas in the United States.

After tornadoes, hurricanes represent the next most severe wind phenomenon. Known as typhoons or tropical cyclones in Asia, hurricanes pose greater life safety and economic hazards to buildings and urban infrastructure than tornadoes. Like a tornado, a hurricane is also a rotating funnel-shaped wind storm, except that the funnel in a hurricane is much larger than that of a tornado.

FIGURE 3.9 Tornado in Seymour, Texas, 1979. (Photo courtesy of National Severe Storms Laboratory, National Oceanic and Atmospheric Administration, Norman, Oklahoma.)

The diameter of an average hurricane within which strong winds exist is about 500 mi and has an average lifespan of 10 days. Depending on the location, winds of up to 150 mph may occur within a hurricane. Most damage from hurricanes occurs in coastal areas and is caused primarily by the uplifting of roofs, windborne debris, and uprooted trees or branches of trees. Tidal surges caused by hurricanes can be more devastating than wind-created damage.

As a hurricane travels inland, it loses energy and degenerates into a tropical rain storm. In the United States, hurricanes develop over warm areas in the Atlantic Ocean, Caribbean Sea, Gulf of Mexico, and the northeastern Pacific Ocean. Although hurricane-induced property damage has increased over the years in the United States due to the migration of population toward coastal areas, the loss of life has decreased due to improvements in forecasting and warning systems.

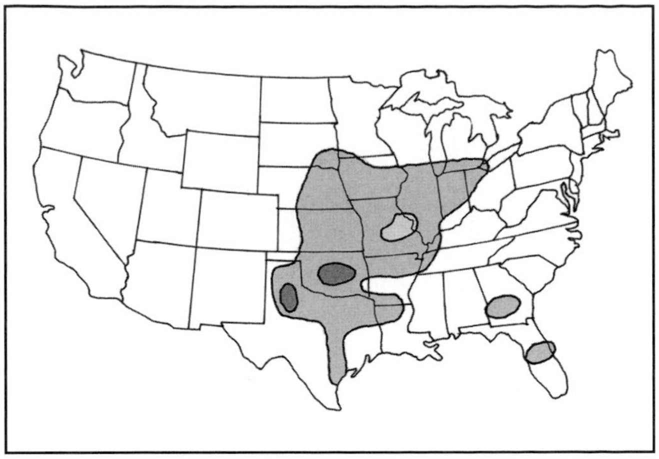

FIGURE 3.10 An approximate representation of tornado-prone regions of the United States. The darker the color of the region, the greater the tornado activity. However, no region of the United States is absolutely tornado free. For hurricane-prone regions of the United States, see Figure 3.11.

After tornadoes and hurricanes, local windstorms (e.g., thunderstorms) are the next category of violent wind activity. Because they are nonrotating or nonspinning, they are also referred to as *straight-line winds*. Being a local phenomenon, a windstorm strikes a much smaller area of ground than a hurricane. However, the frequency of occurrence of windstorms is much higher. Consequently, the damage caused by windstorms is substantial. In determining the design wind speed for a location, we consider both the hurricane activity and the straight-line wind activity.

Design (Basic) Wind Speed

Wind also tends to be gusty, implying that the wind speed varies continuously. Because wind speed–measuring instrumentation has a finite response period over which it averages the wind speed, instantaneous wind speed measurements are not possible. The shortest averaging interval used in wind speed–measuring instruments in the United States is 3 s. Because building design must be based on maximum wind speeds, we refer to the maximum wind speed as *peak 3-s gust speed.*

Peak 3-s gust speeds for several selected locations are continuously recorded in the United States. The peak 3-s gust speed that has not been exceeded at a location in the past 50 years is defined as the *basic wind speed* for that location. Basic wind speed, also called *design wind speed,* is used in determining the design wind loads on buildings. In other words, the basic wind speed has 2% annual probability of being exceeded, i.e., a 50-year recurrence value.

By international agreement, wind speeds are recorded at a height of 33 ft (10 m) above ground. Although, there are several other recording stations, wind speeds are typically recorded at local airports. A contour map giving basic wind speed values for the United States is shown in Figure 3.11. Map values are to be used as a general guide. Local building codes may specify a higher value. In that case, this higher value must be used as the basic wind speed.

Wind Direction

Although in most locations wind blows from only a few directions, in the design of buildings, it is safer to assume that it may come from any direction. Because we are interested in the worst effect produced by wind, buildings with rectilinear profiles are investigated, assuming that the wind will blow along one of the two main directions of the building.

For example, with respect to the building of Figure 3.12, the wind loads are determined based on the assumption that the wind will blow either in the X or the Y direction. If the building, as a whole, has adequate strength to resist wind loads from both the X and the Y directions, it will be adequate against wind loads from any other direction.

Additionally, no reduction in wind loads due to the shielding effect of adjacent structures is permitted. For example, in the building of Figure 3.12, the outstanding leg of the L-shaped building will tend to shield the part of the structure behind it, causing some reduction in wind loads on the shielded portion. Similar reductions may be caused by adjacent structures. Such reductions are difficult to estimate and are generally ignored.

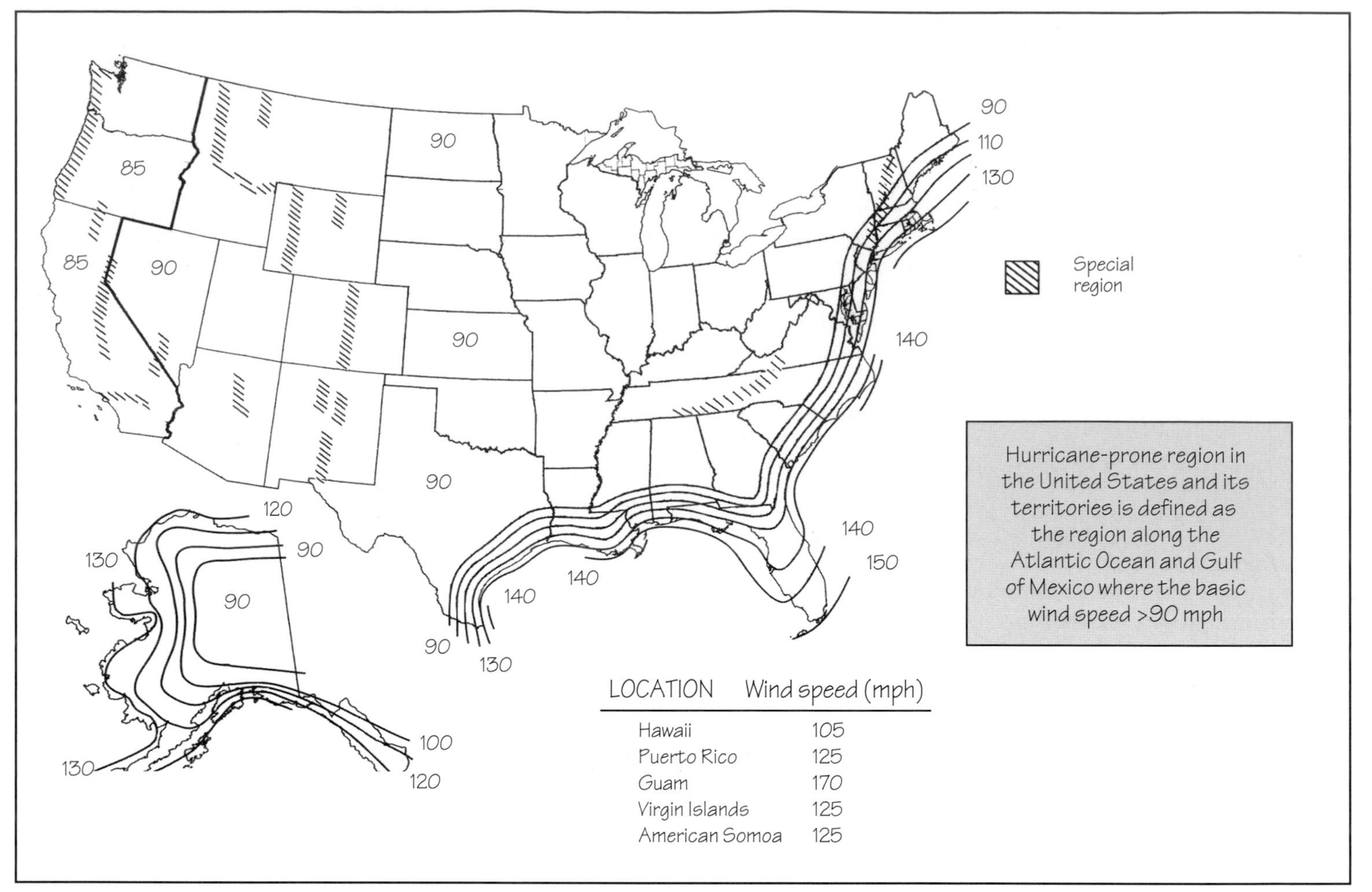

FIGURE 3.11 Three-second peak gust wind speed with 50-year recurrence interval (basic wind speed) map of the United States in miles per hour (mph).
Source: American Society of Civil Engineers: "Minimum Design Loads for Buildings and Other Structures." SEI/ASCE 7-05 (2005), with permission from ASCE.

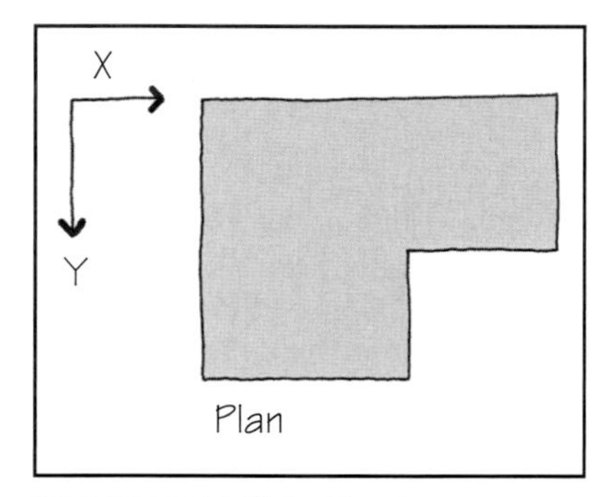

FIGURE 3.12 Wind loads on a building with a rectilinear plan are determined assuming that the wind may blow along either of the two main directions—the X or the Y direction.

Induced Pressure and Suction

Being a fluid, wind exerts pressure on building surfaces. The direction of wind pressure is always perpendicular to the building surface. Thus, on a flat roof, wind pressure is vertical. On a vertical wall, it is horizontal. Whatever the direction of pressure, it either acts toward the surface or away from the surface (suction).

On a rectangular building, wind causes inward force on the windward wall and suction on the other walls (leeward wall and side walls), Figure 3.13(a). The largest pressures occur on the windward and the leeward walls. Because a given wall may be a windward or a leeward wall, depending on wind direction, it has to be designed for both the maximum (inward) pressure and maximum suction.

Wind tunnel tests have shown that on the windward wall, wind pressure varies with height, increasing as the height increases. On the leeward wall, there is no appreciable change in pressure with respect to height. Hence, uniform pressure is assumed to act on the leeward wall, Figure 3.13(b).

As for a roof, a flat or near-flat roof is subjected to suction. A pitched roof has to be designed for suction as well as downward pressure. This is because if the wind blows parallel to the ridge, the roof is under suction, Figure 3.14(a). If the wind blows perpendicular to the ridge, the leeward slope is subjected to suction. The windward slope of the roof, however, comes under suction if the roof slope is small, which turns to downward (toward the roof) pressure as the slope increases, Figure 3.14(b).

Spatial Variation of Wind Pressure

Wind pressure on a building's facade not only varies with height above ground, but it also varies from point to point at the same height above ground. For example, significantly larger pressures are obtained at the edges and corners of a building facade than in the central part (field). In other words, a corner window is subjected to greater pressure than a window in the middle of the wall. Similarly, wind pressures on the corners of a roof are greater than along the perimeter of the roof, which in turn are greater than in the middle of the roof.

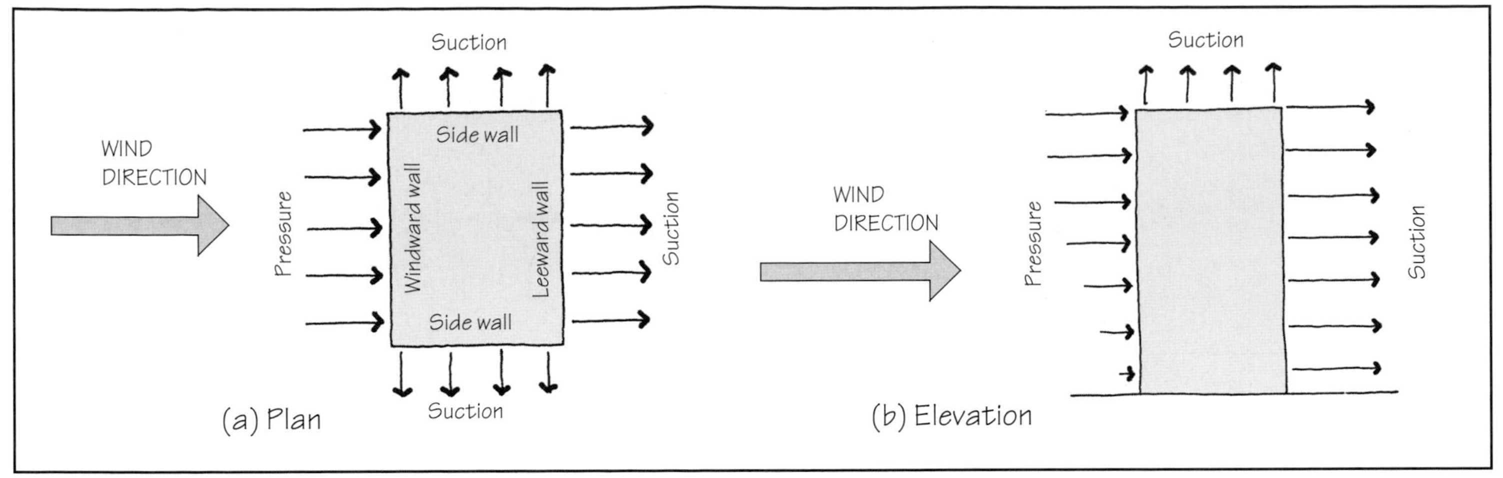

FIGURE 3.13 Wind pressures exerted on the walls of a rectilinear building: **(a)** in plan, and **(b)** in elevation. Note that on the windward wall, wind pressure increases with height. On the leeward wall, the pressure is constant over the entire height of the wall.

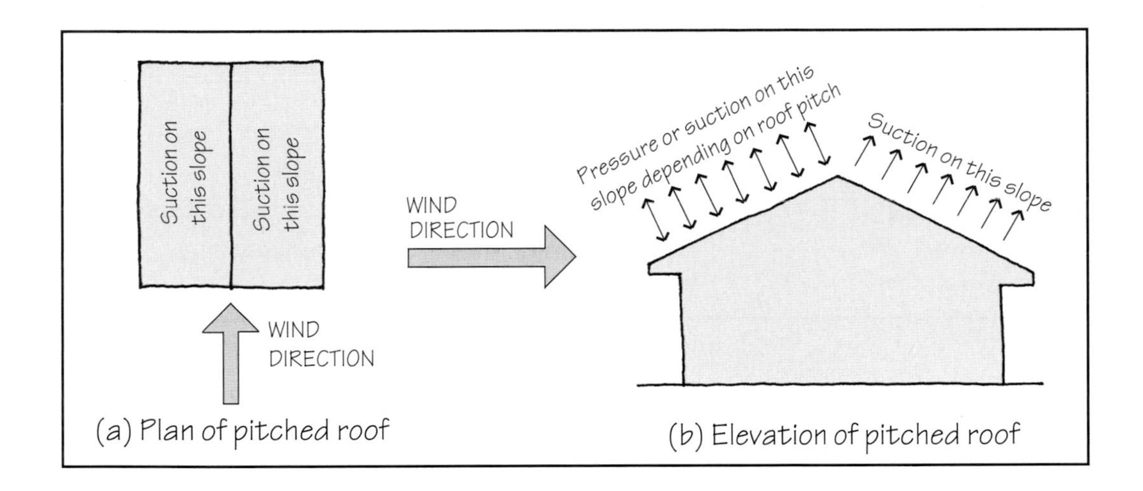

FIGURE 3.14 Wind pressures on a pitched roof.

3.5 FACTORS THAT AFFECT WIND LOADS

The predominant effect of wind speed on wind pressure has already been mentioned. Other factors that affect wind pressure are as follows:

- Height above ground
- Exposure classification of site
- Topography of site
- Enclosure classification of building
- Importance of building's occupancy

Height Above Ground

Through our daily experience, we know that wind speed increases with height above the ground. An open window on a higher floor of a multistory building feels breezier than the one on the lower floor. Consequently, wind pressures are higher on a tall building than on a short one.

Exposure Classification of Site

Wind pressures on a building also depend on the neighborhood in which the building is located. A building located in a vast open stretch of land, with no other buildings or obstructions in its neighborhood, is subjected to greater wind pressures than the same building located in a densely built-up area. This is because the wind speed in an open stretch of land is higher than in a built-up or highly wooded area.

Building sites are classified under three types, based on the roughness of the ground. The roughness of the ground is referred to as the *exposure category* of the site. The three exposure categories to which a particular site may belong are B, C, or D. Exposure Category B, or simply Exposure B, refers to an urban, suburban, or wooded area with several closely spaced obstructions, Figure 3.15. A large city center with several high-rise and closely spaced buildings also belongs to Exposure B.

Exposure A—An urban or suburban location

Exposure B— A downtown location

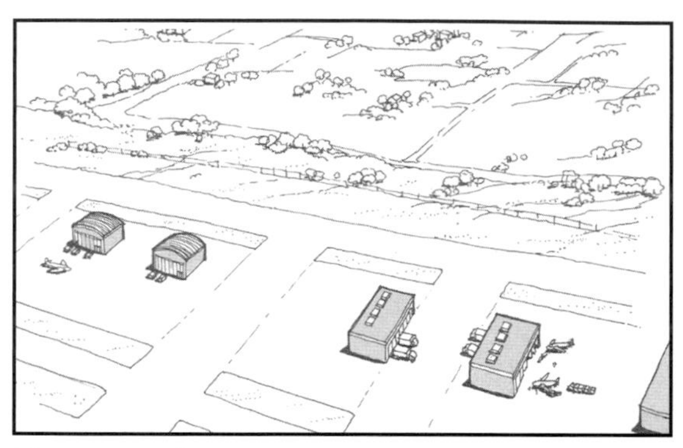

Exposure C—An open country or grassland
Wind speed measurements are made in Exposure C.

Exposure D—A flat unobstructed site facing a large water body

FIGURE 3.15 Site-exposure categories.

Exposure C refers to open terrain with scattered obstructions involving heights generally less than 30 ft. An airport is a typical example of Exposure C. Exposure D refers to flat unobstructed ground facing a large body of water (a large lake- or seafront).

TOPOGRAPHY OF SITE

If the building is located on an escarpment, as shown in Figure 3.16(a), or an isolated hill, it is subjected to greater wind pressures than if the same building is located on a flat ground.

EXPAND YOUR KNOWLEDGE

Important Facts About Wind Loads

Wind Load Is Expressed in Terms of Wind Pressure

Wind loads on buildings are generally expressed as wind pressure, that is, in pounds per square foot (psf).

Wind Pressure Is Proportional to the Square of Wind Speed

Everything else being the same, the wind pressure on a building component is proportional to the square of wind speed. More precisely,

$$p = 0.00256V^2$$

where: p = wind pressure in pounds per square foot (psf)
V = wind speed in miles per hour (mph).

Thus, if $V = 100$ mph, $p = 25.6$ psf. Because 0.00256 is approximately equal to 0.0025 or $\frac{1}{400}$, a commonly used simplification for the preceding equation is

$$p = \left[\frac{V}{20}\right]^2$$

According to this *simplified equation*, if $V = 100$ mph, $p = 25$ psf. If $V = 90$ mph, $p = 20$ psf. For $V = 10$ mph, $p = 0.25$ psf.

Magnitude of Wind Pressures on Low-Rise Buildings

Because wind speed refers to that measured at a height of 33 ft (10 m) above ground, the approximate value of wind pressure on the components of most low-rise buildings (one to three stories tall) can be obtained by the simplified equation just given. This is an important result, which we will refer to in Chapter 6 in comparing the commonly prevailing (water) vapor pressures with air pressures in buildings.

Wind Pressure on a Building Component Equals Difference of Inside and Outside Air Pressures

The wind pressure on a building component, such as a wall, window, or roof is, in fact, the difference between inside and outside air pressures. If the component is subjected to suction under wind loads, it implies that inside pressure is greater than the outside pressure. Conversely, if the outside pressure is greater than the inside pressure, the component will be subjected to push-in pressure. The inside pressure in a building is generally the atmospheric pressure. The outside pressure changes with wind speed.

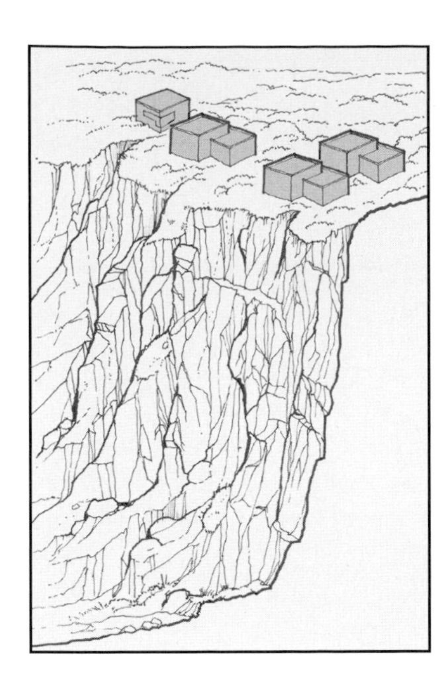

FIGURE 3.16 (a) An escarpment, as shown here, increases wind speed, which increases wind loads on buildings situated on the escarpment. See section of escarpment in (b).

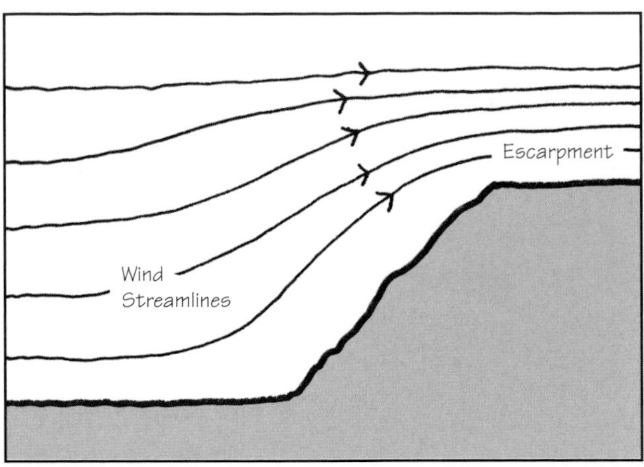

FIGURE 3.16 (b) The escarpment deflects lower wind streamlines upward, and the upper-level windstreams pass undeflected. This converges wind into a narrow region above the escarpment, increasing the wind speed. This phenomenon is known as the wind speed-up effect.

This is because as the wind approaches the escarpment or isolated hill, it has to converge or funnel into a narrow region, Figure 3.16(b). The funneling of wind increases the wind speed at the escarpment or isolated hill, increasing the wind pressure on buildings.

Enclosure Classification of Building

Building envelopes are not perfectly airtight. They are subject to air infiltration and exfiltration. Additionally, they have windows and doors, which may be partially or fully open or broken by flying debris. Because of air infiltration through the openings, the interior of a building is always subjected to some internal pressure in addition to the pressure on the exterior envelope of the building, Figure 3.17.

Internal pressures are particularly critical in buildings that have large openings on one side or two adjacent sides, because these tend to "cup" the wind, creating a "ballooning" effect, Figure 3.18. Buildings in which the ballooning effect is large are called *partially enclosed structures.* They include buildings in which the area of openings in one wall is much greater than that in the remaining walls, such as aircraft hangars, warehouses with dock doors, or garages that are open only on one side.

Importance of Building's Occupancy

Some buildings are more important than others and must be designed for the 100-year worst storm rather than the normal 50-year worst storm. Such buildings include emergency shelters, hospitals, fire stations, radio stations, and broadcasting studios. Statistical analysis shows that the 100-year worst storm causes 15% greater pressure than the 50-year worst storm. Thus, the design wind pressures on more important buildings are 15% greater than on normal buildings.

3.6 ROOF SNOW LOAD

A roof is designed for either the roof live load or the snow load, whichever is greater. The reason is that in the event of full snow load on a roof, the roof is not likely to be accessed by a repair or construction crew, who would impose additional live load on it. Like the roof

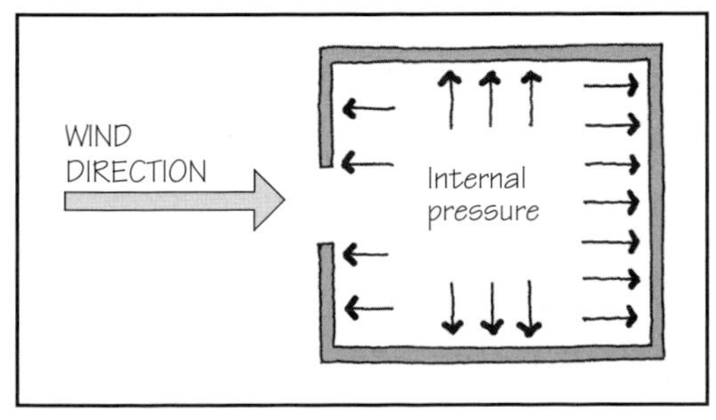

FIGURE 3.17 Internal pressures in a building due to openings or air infiltration through the envelope.

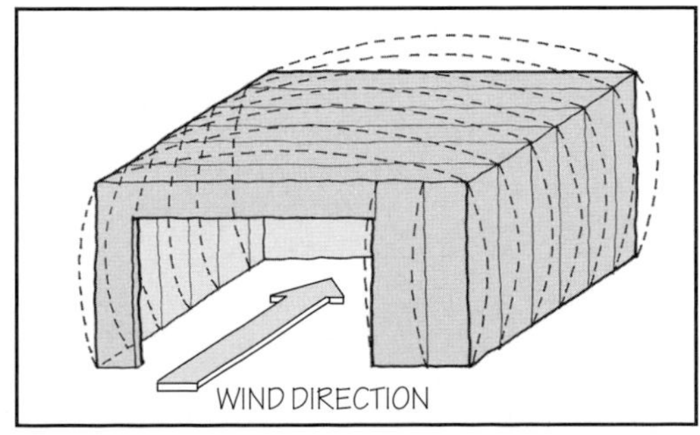

FIGURE 3.18 Ballooning of a building with a relatively large opening in one wall.

TABLE 3.2 APPROXIMATE GROUND SNOW LOADS FOR SELECTED LOCATIONS

Location	Snow load psf	Location	Snow load psf
Huntsville, Alabama	5	Jackson, Mississippi	3
Flagstaff, Arizona	48	Kansas City, Missouri	18
Little Rock, Arkansas	6	Concord, New Hampshire	63
Mt. Shasta, California	62	Newark, New Jersey	15
Denver, Colorado	18	Albuquerque, New Mexico	4
Hartford, Connecticut	33	Buffalo, New York	39
Wilmington, Delaware	16	New York City	15
Atlanta, Georgia	3	Charlotte, North Carolina	11
Boise, Idaho	9	Fargo, North Dakota	41
Chicago, Illinois	22	Columbus, Ohio	11
Indianapolis, Indiana	22	Oklahoma City, Oklahoma	8
Des Moines, Iowa	22	Portland, Oregon	8
Wichita, Kansas	14	Philadelphia, Pennsylvania	14
Jackson, Kentucky	18	Providence, Rhode Island	23
Shreveport, Louisiana	3	Memphis, Tennessee	6
Portland, Maine	60	Dallas, Texas	3
Baltimore, Maryland	22	Seattle, Washington	18
Boston, Massachusetts	34	Spokane, Washington	42
Detroit, Michigan	22	Charleston, West Virginia	18
Minneapolis, Minnesota	51	Madison, Wisconsin	35

Source: Adapted from the American Society of Civil Engineers' *Minimum Design Loads for Buildings and Other Structures*, ASCE 7-05 (2005), with permission from ASCE.

live load, snow load is also expressed in terms of the horizontal projected area of the roof (Figure 3.4).

Fundamental to determining roof snow load is the ground snow load of a location. Ground snow load values with a 50-year recurrence interval for selected U.S. locations are given in Table 3.2. These values should be treated as approximate. For authoritative data, the building code or the city's building department should be consulted.

Factors That Affect Roof Snow Load

The factors that affect snow load on a roof are

- *Ground snow load at the location*
- *Roof slope* The greater the slope of the roof, the smaller the snow load. For a flat or near-flat roof, the snow load should theoretically be equal to the ground snow load. However, records of measurements of snow deposits on roofs have shown that less snow is typically present on flat roofs than on the ground because wind blows the snow off the roofs and some snow melts due to heat escaping from the heated interiors. Snow load on flat roofs in open areas (i.e., areas that are not obstructed by wind-shielding elements such as higher structures, terrain, and trees) can be as low as 60% of that on the nearby ground.
- *Wind exposure classification of site (Exposure B, C, or D)* Everything else being the same, Exposure D has the least snow load, due to the high wind speeds, and Exposure B has the highest snow load.
- *Warm roof or cold roof* An example of a warm roof is a continuously heated greenhouse, and that of a cold roof is a typical residential structure in which the attic space is ventilated, which keeps the roof cold. An unheated building is another example of a cold roof. Snow load is smaller on a warm roof than on a cold roof.
- *Importance of building* This factor is essentially the same as for the wind loads. More important buildings, such as emergency shelters, police stations, fire stations, and radio and television studios, are designed for greater snow load than normal buildings.

3.7 EARTHQUAKE LOAD

A severe earthquake is one of the most terrifying natural events a person can experience. Regardless of where it occurs, it makes instant headlines all over the world. In the past, earthquakes killed a large number of people. Today, earthquakes pose a smaller threat to life than

Each question has only one correct answer. Select the choice that best answers the question.

11. Highest wind speeds are usually obtained in
a. hurricanes.
b. wind storms.
c. thunderstorms.
d. tornadoes.

12. An average tornado strikes a larger area on the ground than a hurricane.
a. True
b. False

13. The design wind speed is also referred to as the
a. standard wind speed.
b. official wind speed.
c. uniform wind speed.
d. straight-line wind speed.
e. basic wind speed.

14. The design wind speed for a location has a probability of occurrence of
a. once in 200 years.
b. once in 100 years.
c. once in 75 years.
d. once in 50 years.
e. once in 25 years.

15. The design wind speed for a location is the peak
a. 60-s gust speed.
b. 10-s gust speed.
c. 5-s gust speed.
d. 3-s gust speed.
e. 1-s gust speed.

16. For most of the United States, the basic wind speed is
a. 70 mph.
b. 80 mph.
c. 90 mph.
d. 100 mph.
e. 110 mph.

17. For wind blowing along one of the two major axes of a rectangular building, the windward wall is subjected to positive (toward-the-wall) pressure.
a. True
b. False

18. For wind blowing along one of the two major axes of a rectangular building, the leeward wall is subjected to positive (toward-the-wall) pressure.
a. True
b. False

19. For wind blowing along one of the two major axes of a rectangular building, the side walls are subjected to positive (toward-the-wall) pressure.
a. True
b. False

20. On a windward wall, the wind pressure
a. increases with height above the ground.
b. decreases with height above the ground.
c. is constant with height above the ground.

21. With respect to wind pressure, a building site is classified in terms of
a. Exposures A, B, C, and D.
b. Exposures A, B, and C.
c. Exposures X, Y, and Z.
d. Exposures P, Q, R, and S.
e. Exposures B, C, and D.

22. The design snow load on a roof is a function of
a. ground snow load of the location.
b. roof slope.
c. wind exposure classification of site.
d. all the above.
e. both (a) and (b).

Answers: 11-d, 12-b, 13-e, 14-d, 15-d, 16-c, 17-a, 18-b, 19-b, 20-a, 21-e, 22-d.

in the past (at least in developed nations and discounting the 2004 tsunami in the Indian Ocean) because of our improved knowledge of constructing earthquake-resistant structures.

Statistics indicate that the current annual loss of life resulting from earthquakes is much smaller than those due to hurricanes, building fires, floods, or automobile accidents. Ground shaking, the most commonly associated phenomenon with earthquakes, causes damage to buildings and other infrastructural facilities, Figure 3.19. Records show that the earthquakes of 1811 and 1812 in Missouri shifted the course of the Mississippi River considerably [1]. In addition to ground shaking, a major earthquake can produce several related problems, such as landslides, surface fractures, soil liquefaction, tsunamis, and fires.

FIGURE 3.19 A railroad deformed by an earthquake.

Soil Liquefaction

Soil liquefaction usually occurs in water-saturated, sandy soils in which the particle sizes of sand are of relatively uniform size. When such soils are shaken, the water rises to the surface, resulting in loss of foundation support to buildings constructed on them. The most notable example of soil liquefaction occurred in Nigata, Japan, during the 1964 earthquake [2]. In this earthquake several high-rise buildings simply tipped over while otherwise remaining intact.

Tsunamis

A tsunami is caused by the physical displacement of the seabed during an earthquake, which creates waves of up to 50 ft in height. So far, most tsunamis have occurred in the Pacific Ocean, with Japan witnessing the largest number of them—hence its Japanese name. However, the 2004 tsunami that occurred in the Indian Ocean was the most devastating in terms of human lives.

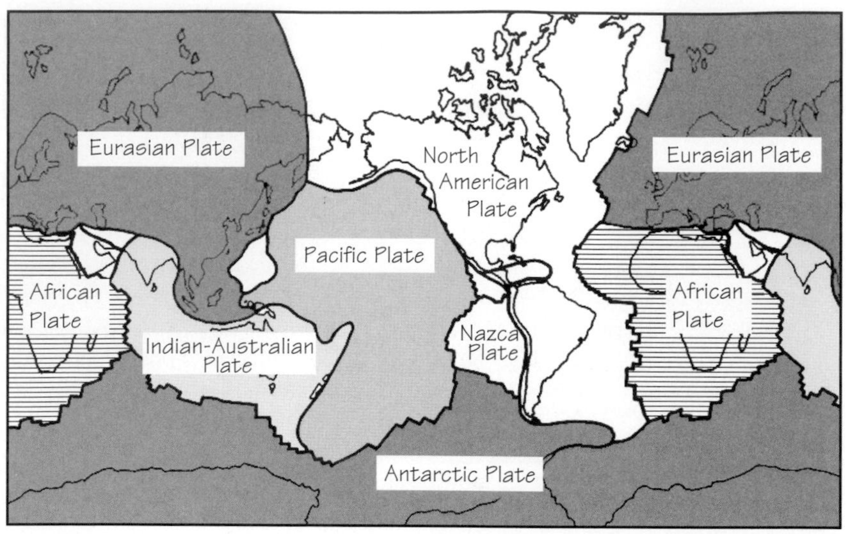

FIGURE 3.20 Map of the earth with an approximate depiction of major tectonic plates.

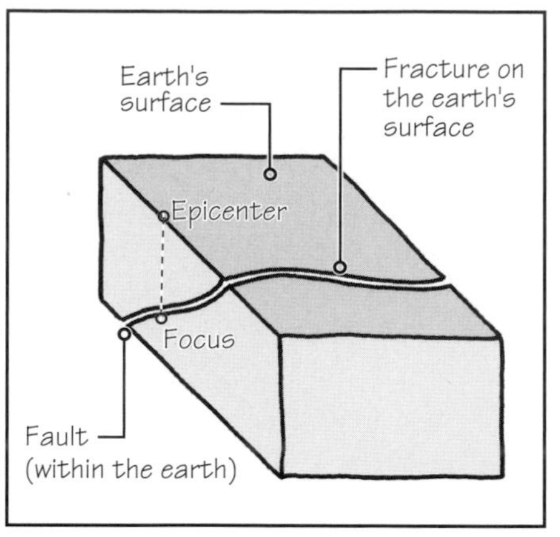

FIGURE 3.21 Fault, focus, and epicenter of an earthquake.

Geological Explanation of Earthquakes

The cause of earthquakes is fairly well established. However, we are still far from perfecting a methodology to accurately predict the location and time of earthquake occurrence.

Geologists believe that the earth's crust is divided into several individual segments called *crustal* or *tectonic plates,* Figure 3.20. These plates, floating on a molten mantle below, are constantly in motion relative to each other. The motion is so slow that it is significant only in geological time. The relative motion in the earth's crust causes compressive, tensile, or shear forces to develop, depending on whether the plates press against each other, pull apart, or slip laterally under one another. When these stresses exceed the maximum capacity of the crust to absorb them, a fracture at the crust occurs. It is this fracturing, or slippage, which is always sudden, that produces a shock wave known as seismic motion.

The plane where the fracture occurs is called a *fault.* The location where the fault originates is called the *focus,* which is generally inside the earth's crust. The point directly above the focus on the earth's surface is called the *epicenter*, Figure 3.21.

The fracture on the surface of the earth may or may not be noticeable. The 1906 San Francisco (California) earthquake left a deep ground rupture extending hundreds of miles. Known as the San Andreas fault, it stands as a distinct landmark created by an earthquake.

Frequency of Occurrence and Location of Earthquakes

An earthquake may occur at any location. In fact, mild earthquakes occur quite frequently: Several thousand a day occur in various regions of the earth, Table 3.3. They go unnoticed because they may occur in remote places or with a magnitude that is below the threshold of human perception.

The regions of the earth in proximity to plate boundaries are susceptible to more frequent and severe earthquakes. Nearly 90% of all earthquakes occur in the vicinity of plate boundaries, because the plate boundaries represent lines of greater weakness in the earth's crust. Observe in Figure 3.20 that the Pacific plate and the North American plate meet along the West Coast of the United States—a region known for its high seismic activity.

The remaining 10% of earthquakes occur because of the faults within the plates. Called *intraplate earthquakes,* they occur far away from plate boundaries. Earthquakes that have occurred in the Eastern and Midwestern United States are intraplate earthquakes. Intraplate earthquakes are infrequent but may be as severe as plate boundary earthquakes.

Because seismicity is a geographical activity, it is possible to delineate areas that are seismically more active than others. For the United States, this activity is shown in Figure 3.22, in which the degree of seismic

TABLE 3.3 WORLDWIDE FREQUENCY OF EARTHQUAKE OCCURRENCE

Descriptor	Richter magnitude	Average frequency per year
Great	≥8.0	0–1
Major	7–7.9	17
Strong	6–6.9	134
Moderate	5–5.9	1,319
Light	4–4.9	13,000
Minor	3–3.9	130,000
Very minor	2–2.9	1,300,000

Source: EQ Facts and Lists, National Earthquake Information Center, U.S. Geological Society Web Site (www.usgs.gov).

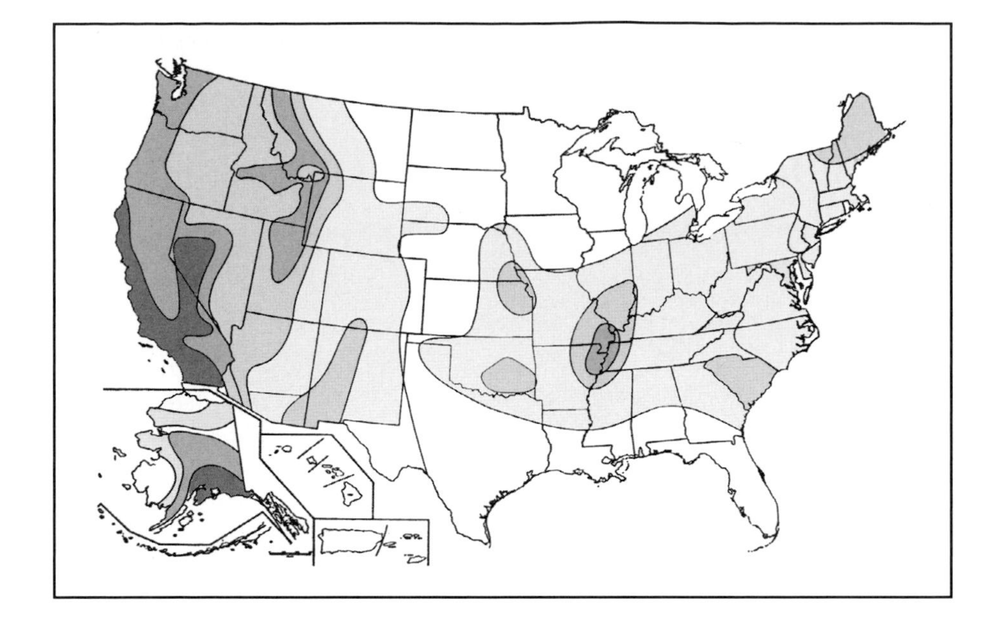

FIGURE 3.22 Map of the United States showing approximate variation of seismic activity. The darker the area, the greater the seismic activity. White areas have little or no seismic activity.

activity is denoted by the darkness of the gray color. Thus, unshaded areas of the country have little or no seismic activity, and the darkest areas are seismically most active.

EARTHQUAKE INTENSITY—RICHTER MAGNITUDE

For the general public, the scale most commonly used to describe the intensity of an earthquake is the Richter scale. The Richter scale was devised in 1935 by Professor Charles Richter of the California Institute of Technology. The scale begins at zero but has no upper limit. However, the severest earthquake recorded to date on this scale is 9.0.

According to the Richter scale, the total energy released by an earthquake (E) is proportional to Richter magnitude (R), as given by the following expression.

$$E \propto 10^{1.5\,R}$$

The symbol $\propto$ is the sign of proportionality, implying that earthquake energy is proportional to $10^{1.5R}$. Because $10^{1.5}$ is approximately equal to 32, this expression implies that an increase of 1.0 on the Richter scale gives 32 times increase in the energy output of the earthquake. Thus, an earthquake measuring 8.0 on the Richter scale is nearly 32-fold more powerful than one measuring 7.0 and nearly 1,000 times more powerful than one measuring 6.0.

An earthquake of up to 2 on the Richter scale goes unnoticed by humans; 4 to 5 may cause slight damage to poorly designed buildings. At 7, the damage is usually extensive in poorly designed buildings. An earthquake with magnitude of 6.5 and above is considered a significant earthquake. Table 3.4 gives some of the major earthquakes that have occurred in

TABLE 3.4 SOME OF THE MAJOR EARTHQUAKES SINCE 1900

Year	Location	Richter magnitude	Loss of life	Year	Location	Richter magnitude	Loss of life
1905	Kangra, India	8.6	19,000	1950	India and China	8.6	1,500
1908	Messina, Italy	7.2	100,000	1952	Hokkaido, Japan	8.6	600
1906	Columbia and Ecuador	8.9	1,000	1952	Alerce, Chile	8.5	2,000
1906	San Francisco, USA	8.3	700	1960	Alaska, USA	8.5	115
1906	South Central Chile	8.6	1,500	1964	Moro, Phillippines	8.0	6,500
1907	Karatag, Tajikistan	8.0	12,000	1976	Tangshan, China	7.5	255,000
1920	Gansu, China	7.8	200,000	1976	Sunda, Indonesia	8.0	200
1923	Kanto, Japan	7.9	143,000	1977	Mexico	8.1	5,000
1927	Tsinghai, China	7.9	200,000	1985	Western Iran	7.7	50,000
1932	Gansu, China	7.6	70,000	1990	Kobe, Japan	6.8	6,400
1933	Honshu, Japan	8.9	3,000	1995	Turkey	7.7	17,000
1935	Quetta, Pakistan	7.5	60,000	1999	India	7.7	20,000
1939	Erzincan, Turkey	8.0	33,000	2001	Sumatra (tsunami)	9.0	283,000
1945	Iran and Pakistan	8.2	4,000	2004	Pakistan	7.6	80,000
1948	USSR	7.9	110,000	2005			

Source: U.S. Geological Survey Earthquake Hazards Program (http://earthquake.usgs.gov).

various parts of the world in the past 100 years. (Major earthquakes may be measured by Richter magnitude, number of deaths, or loss of property.)

3.8 FACTORS THAT AFFECT EARTHQUAKE LOADS

The ground motion caused by an earthquake consists of random vibrations. Like any other vibration, this motion is also characterized by amplitude and frequency of vibration. Although the Richter scale gives the total energy released, it provides no information about the characteristics of ground motion (frequency and duration of vibrations) and the interaction of ground vibration with the vibrational characteristics of the building and the underlying soil. Therefore, the Richter scale cannot be used in determining earthquake loads on buildings. Instead, the following factors must be considered:

- Ground motion
- Building's mass and ductility of structural frame
- Type of soil
- Importance of the building

Ground Motion

To understand how an earthquake impacts a building, imagine a building just before the earthquake. Every part of the building is in static equilibrium. As soon as the earthquake occurs, the ground is displaced horizontally. Because the building has inertia, it takes a finite period of time for the horizontal displacement to travel to the upper parts of the building. In the meantime, the building is subjected to deformations, as indicated by its deformed shape, Figure 3.23. Thus, although there are several differences between the effects of an earthquake and a windstorm on a building (see Section 3.9), the overall building deformation produced by both is similar.

An earthquake results in ground acceleration due to an abrupt ground motion. This acceleration produces an inertial force in the building. This force, which tends to oppose the ground motion, is referred to as the *total earthquake load* on the building. The magnitude of the total earthquake load can be obtained from Newton's second law of motion. According to this law, when a body is accelerated, the inertial force (F) acting on it is equal to the product of the acceleration and the mass of the body:

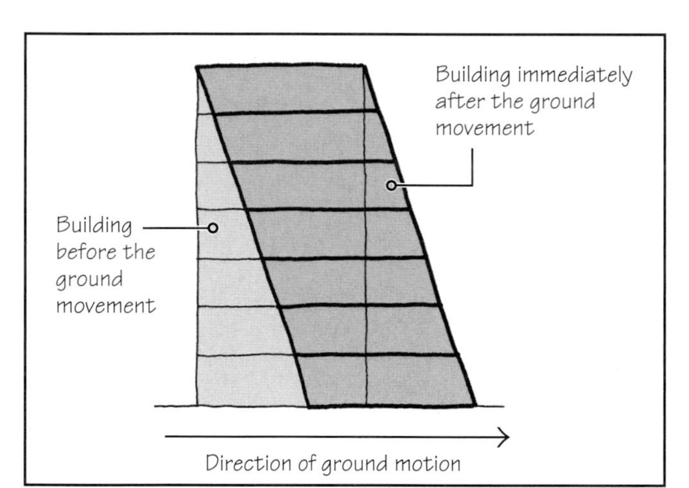

FIGURE 3.23 Deformation of a building resulting from the ground movement.

$$F = ma$$

where m is the mass of the body and a is the acceleration imparted to it.

When Newton's law is applied to buildings, *m* refers to the weight of the building and *a* refers to the acceleration of the ground. If ground acceleration is small, the earthquake load is small. This is obvious because if the ground moves very slowly (with a negligible acceleration), the building will simply go along with this motion as one unit, and the opposing inertial force (the earthquake load) will be small. In the case of a sudden and swift motion (large acceleration), the lower part of the building moves horizontally, whereas the upper part remains in its original position (see Figure 3.23). In an effort to maintain the status quo, a large opposing (inertial) force is produced in the building—that is, the total earthquake load on the building is large.

Building's Mass and Ductility of Structural Frame

From Newton's law, the other factor that affects the magnitude of the earthquake load is the weight of the building. Lighter buildings attract a smaller earthquake load than buildings constructed of heavy materials. Thus, concrete frame and masonry structures attract a greater earthquake load than wood frame or steel frame structures. Residential structures with wood siding or stucco finishes attract a smaller earthquake load than those with brick or stone veneer.

Indeed, because of its light weight, the earthquake load on a tent is negligible even in a severe earthquake, Figure 3.24. A tent structure has the additional advantage of its built-in

FIGURE 3.24 A tent resists earthquake forces admirably due to its light weight and its ability to absorb deformations.

resistance to absorb structural deformations (ductility)—an important structural property for earthquake regions.

Soil Condition—The Site Class

The type of soil on which the building rests may vary from hard rock to soft clay (Chapter 21). Studies of building damage caused by recent earthquakes has indicated the importance of the type of soil on which the building's foundations rest. A hard rock transfers earthquake ground motion to the building much more faithfully than a soft soil. Because a soft soil has its own vibrational characteristics, it modifies the ground motion, generally amplifying it.

The difference between the vibrations of hard and soft soils may be understood by imagining two bowls: one containing frozen water (representing hard soil conditions) and the other containing dry sand. When the bowls are shaken, the frozen water will vibrate in the same way as the bowl. The vibrations of the sand in the bowl, however, are more vigorous and indirectly reflect those of the bowl.

Seismic load standards classify soils in six different types, referred to as Site Classes A, B, C, D, E, and F. Site Class A represents hard rock; Class B represents rock; Class C is soft rock or dense soil, Class D is stiff soil, and Class E is soft soil. Class F represents highly sensitive soils that may experience consolidation or liquefaction as a result of ground motion. Geotechnical studies of the soil are required to determine the site class.

NOTE

Site Class

Site Class A	Hard rock
Site Class B	Rock
Site Class C	Soft rock
Site Class D	Stiff soil
Site Class E	Soft soil
Site Class F	Liquefiable soil

Importance of Building's Occupancy—The Seismic Use Group

The seismic use group is similar to the importance factor used in wind load determination. Buildings are divided into three seismic use groups—Groups I, II, and III. Most buildings belong to seismic use Group I. Group II buildings are more important than Group I buildings or contain more hazardous materials. Health care facilities, power-generating stations, elementary and secondary schools, jails, and so on, belong to Group II. Group III buildings are even more important or hazardous than Group II buildings and include police stations, fire stations, designated emergency shelters, and water-treatment facilities. Group I buildings are those that are not included in Groups II or III.

NOTE

Seismic Use Group

Group III	Most important or most hazardous buildings
Group II	Important or hazardous buildings
Group I	Normal buildings—buildings that do not belong to Group III or Group II

3.9 Wind versus Earthquake Resistance of Buildings

Buildings are generally designed to resist either earthquake loads or the wind loads, whichever causes the worst effect (greater stresses). This is based on the assumption that there is a negligible probability of maximum wind speeds occurring at the same time as an intense earthquake. In regions of low seismic activity, wind loads govern the design of lateral resistance of buildings; in regions of intense seismic activity, the reverse is the case.

The design of lightweight envelope components, such as glass curtain walls and roof membranes, are governed by wind loads even in highly seismic regions. (Remember that earthquake loads are influenced by the weight of the component).

FIGURE 3.25 Tornado damage to the glass curtain wall of Bank One Building, Fort Worth, Texas.

Because earthquake loads are mainly horizontal, the lateral bracing required to resist earthquake loads is, in general, similar to that required to resist wind loads. Nevertheless, there are several differences in details. First, an earthquake shakes not only the entire building, but also its contents. Thus, all building components, equipment, and fixtures in a building must be adequately anchored to remain intact during an earthquake.

Wind, on the other hand, acts on the building envelope. It is through the envelope that wind loads are transmitted to the building structure. If the envelope remains intact, the building will generally retain its overall structural integrity in a storm. Thus, the wind damage in a building usually begins with the damage to envelope components, such as the roof, exterior walls, exterior doors, windows, and other cladding elements, Figure 3.25. Once the building changes from fully enclosed to partially enclosed, the wind loads on the building increase due to the ballooning effect (Section 3.5), which may result in additional damage or a collapse of the structure.

Second, in determining the wind loads on a building, we calculate the maximum expected wind load. The structure is designed so that under the action of maximum wind loads, it remains elastic—that is, the structure is designed to suffer no permanent deformation under the worst expected storm.

The design approach is different for earthquake loads. The loads on a building produced by the worst expected earthquake for the location are so large that if the building were designed to remain elastic after the earthquake, it would be prohibitive in cost.

Therefore, the earthquake loads for which a building is designed may be smaller than the actual earthquake loads expected on it. The underlying design philosophy is that the building should remain elastic when resisting minor earthquakes. In the event of an intense earthquake, a part of the earthquake's energy may be dissipated in permanently deforming the building, but the building must otherwise remain intact to provide complete safety to its occupants.

Permanent deformations in a structure are possible only if the structure has the ability to sustain such deformations. As we will see in Chapter 4, materials that can sustain permanent deformation prior to failure are called *ductile* materials. Ductility is an essential property for buildings located in seismic zones but is not a requirement for resisting wind loads.

A building must not deform permanently even in the most severe windstorm. Additionally, it must be rigid enough to reduce the deflections caused by strong winds. The swaying of the upper floor of tall buildings under wind loads must be controlled to remain within acceptable limits of human tolerance.

The reverse is true for earthquake loads. Buildings must be able to deform, even permanently, to absorb the energy delivered to them by the earthquake. If brittle materials are used, they will fail under earthquake forces. Several reinforced concrete frame buildings with (brittle) unreinforced masonry infill walls between frames have collapsed or suffered serious damage under earthquake forces, Figure 3.26. Such buildings would have been unharmed by violent storms.

FIGURE 3.26 Earthquake damage to unreinforced masonry panels in buildings with reinforced concrete structural frames in Mexico earthquake, 1985.

PRINCIPLES IN PRACTICE

Dead Load and Live Load Estimation

TRIBUTARY AREA OF A BUILDING COMPONENT

In computing the dead loads and other loads on a component, the concept of tributary area is important to understand. The tributary area for a component is the area (or areas) of the building that *contributes* load on that component.

For example, in a floor system that consists of wood planks supported by parallel beams, the tributary area for a beam is the area of the floor that lies halfway to the adjacent beam on the left and halfway to the adjacent beam on the right. In Figure 1, the tributary area for beam A has been shown by the shaded area. All the load that is placed on this shaded area must be carried by beam A.

The tributary area for a column in a building is similarly obtained. In the framing plan of the building shown in Figure 2, the tributary areas of an interior column X, exterior column Y, and a corner column Z, have been shaded. Thus, column X receives all the load placed on the shaded area of the floor surrounding it. This is a rectangular area with dimensions $0.5(L_1 + L_2)$ by $0.5(W_1 + W_2)$. Thus, if L_1 and L_2 are 12 ft and 18 ft, respectively, and W_1 and W_2 are each 13 ft, the tributary area for column X is 15 ft × 13 ft.

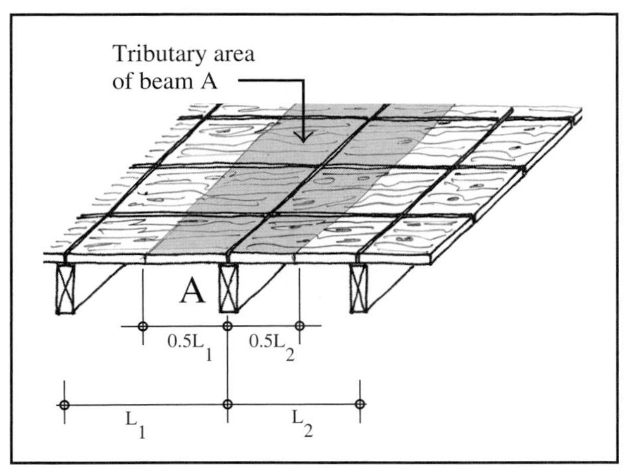

FIGURE 1 The tributary area of beam A is the shaded area. The width of this area is $0.5(L_1 + L_2)$, where L_1 and L_2 are the center-to-center distances between beams.

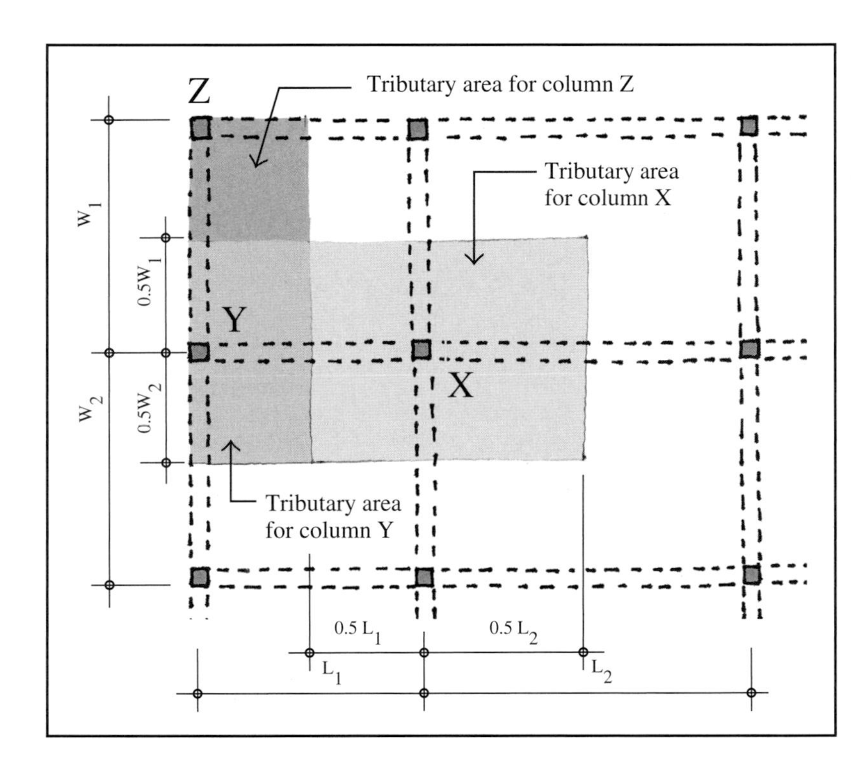

FIGURE 2 Tributary areas for columns X, Y, and Z

EXAMPLE 1 • (ESTIMATING THE DEAD LOAD ON A BEAM)

Determine the dead load on beam A of Figure 3, where each (wood) beam is 4 in. × 16 in. in cross section and is spaced 6 ft on center. The (wood) floor planks are $2\frac{1}{2}$ in. thick. Assume that the density of wood is 40 lb/ft^3 (pcf).

SOLUTION

Because the load on a beam is generally expressed in pounds/foot, we will isolate a 1-ft length of beam A and determine all the dead load on this length. What holds for this 1-ft length holds for the rest of the beam also.

$$\text{Volume of 1-ft long beam} = \left(\frac{4}{12}\right)\left(\frac{16}{12}\right)(1) = 0.44 \text{ ft}^3$$

$$\text{Weight of 1-ft long beam} = 0.44(40) = 17.6 \text{ lb}$$

(*Continued*)

Dead Load and Live Load Estimation (*Continued*)

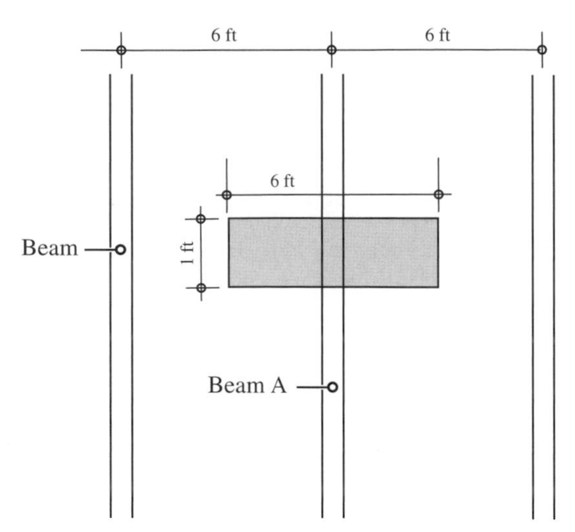

(a) Layout of 4 in. × 16 in. beams in plan

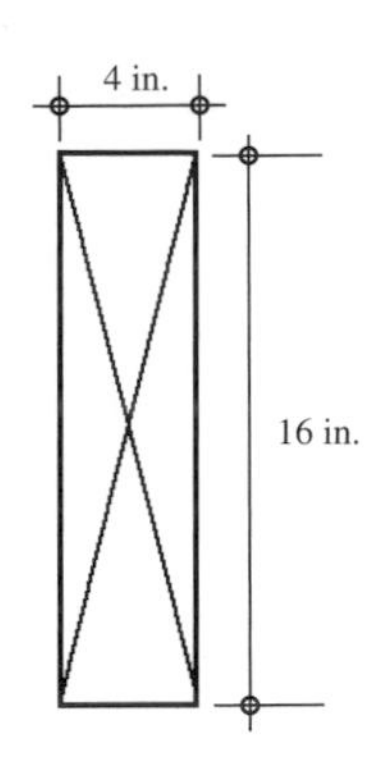

(b) Cross-section of beam

FIGURE 3 Figure for Example 1.

From Figure 3, the tributary area of a 1-ft-long beam is 1 ft × 6 ft = 6.0 ft^2. Hence, the weight of 6 ft^2 of floor planks is borne by a 1-ft-long beam.

$$\text{Volume of 6.0 ft}^2 \text{ planks} = (6.0)\left(\frac{2.5}{12}\right) = 1.25 \text{ ft}^3$$

$$\text{Weight of planks} = 1.25(40) = 50.0 \text{ lb}$$

$$\text{Total dead load on beam A} = 17.6 + 50.0 = 67.6 \text{ lb/ft}$$

EXAMPLE 2 • (ESTIMATING THE DEAD LOAD OF A TYPICAL WOOD FRAME RESIDENTIAL FLOOR)

Determine the dead load of a typical residential floor constructed of 2 × 12 (nominal) Douglas fir floor joists spaced 16 in. on center. The subfloor consists of $\frac{3}{4}$-in.-thick plywood and a $\frac{1}{4}$-in.-thick hardboard underlayment, Figure 4. Vinyl tiles are used as floor finish, and the ceiling consists of $\frac{1}{2}$-in.-thick gypsum wallboard.

Note: A (nominal) 2 × 12 floor joist has actual dimensions of $1\frac{1}{2}$ in. × $11\frac{1}{4}$ in.

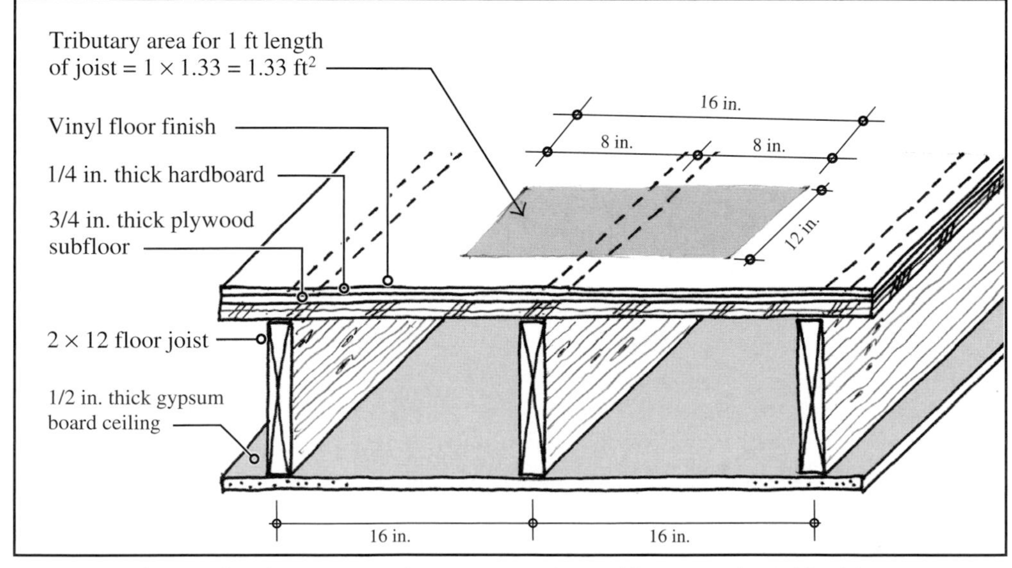

FIGURE 4 Floor and ceiling construction in a typical wood frame residential building. See Example 2.

SOLUTION

A simple method for solving this problem is to add the weights of 1 ft^2 of all sheet materials of the floor system: vinyl tiles, hardboard underlayment, plywood subfloor, and gypsum board. From Table 1, these weights are as follows:

Vinyl tiles	1.0 lb
Hardboard (2 × 0.4)	0.8 lb
Plywood (6 × 0.4)	2.4 lb
Gypsum board (4 × 0.55)	2.2 lb
Total weight	6.4 lb

Thus, the total load due to all sheet materials of the floor system is 6.4 lb/ft^2. To this weight, if we add the weight of the floor joist, we will obtain the dead load of the entire floor assembly.

The weight of a 1-ft-long floor joist can be calculated by multiplying the volume of the joist with density of Douglas fir, which is 34 lb/ft^3 (Table 2).

$$\text{Volume of 1-ft-long joist} = \left(\frac{1.5}{12}\right)\left(\frac{11.25}{12}\right)(1) = 0.12 \text{ ft}^3$$

$$\text{Weight of 1-ft-long joist} = (0.12)(34) = 4.1 \text{ lb}$$

From Figure 3.4, a 1-ft-long floor joist carries 1.33 ft^2 of floor.

Therefore, the weight of floor joist that contributes to 1 ft^2 of the floor is $\left(\frac{4.1}{1.33}\right) = 3.1$ lb.

Hence, the total dead load of 1 ft^2 of floor is $6.4 + 3.1 = 9.5$ lb.

This is an important result because the dead load of a conventional wood frame floor is assumed to be 10 psf.

TABLE 1 (SURFACE) DENSITIES OF SELECTED MATERIALS

Component	Weight lb/ft^2	Component	Weight lb/ft^2
Roof and Wall Coverings		Roof Decks	
Asphalt shingles	2.0	Metal deck, 20 gauge	2.5
Clay tiles	12–18	2-in. wood deck	5.0
Cement tile	10–16	3-in. wood deck	8.0
Wood shingles	3.0	Wood sheathing ($\frac{1}{2}$ in. thick)	1.5
5-ply gravel covered built-up roof	6.0	Fiberboard ($\frac{1}{2}$ in. thick)	1.0
Insulation (per 1 in. thickness)		Ceilings	
Cellular glass	0.7	Gypsum board (per $\frac{1}{8}$-in. thickness)	0.55
Fiberglass	0.1–0.2	Plaster on wood lath	8.0
Extruded/expanded polystyrene	0.2	Suspended steel channel system	2.0
Polyurethane/polyisocyanurate	0.2	Acoustic fiber tile	1.0
Perlite/vermiculite	2.0	Mechanical duct allowance	4.0
Floors and Floor Finishes		Clay Masonry Walls	
Hardwood flooring, $\frac{7}{8}$ in. thick	4.0	4 in. thick	39
Plywood (per $\frac{1}{8}$-in. thickness)	0.4	8 in. thick	79
Hardboard (per $\frac{1}{8}$-in. thickness)	0.4	Hollow Concrete Block Walls	
Sturdifloor (per $\frac{1}{8}$-in. thickness)	0.4	4 in thick, 70% solid, lightweight units	22
Linoleum, vinyl or asphalt tile ($\frac{1}{4}$-in. thick)	1.0	4 in. thick, 70% solid, normal weight	29
Marble or slate (per 1-in. thickness)	15	6 in. thick, 55% solid, lightweight units	27
Terrazzo ($1\frac{1}{2}$ in. thick)	19	6 in. thick, 55% solid, normal weight units	35
Frame Partitions and Walls		8 in. thick, 52% solid, lightweight units	35
Wood/steel studs + $\frac{1}{2}$ in. gypsum board both sides	8.0	8 in. thick, 52% solid, normal weight units	45
Wood/steel studs + $\frac{5}{8}$ in. gypsum board both sides + insulation + siding	11	Floor Fill	
		Sand (per inch)	8.0
Wood/steel studs + $\frac{5}{8}$ in. gypsum board both sides + insulation + brick veneer	48	Cinder concrete (per inch)	9.0

(*Continued*)

PRINCIPLES IN PRACTICE

Dead Load and Live Load Estimation (*Continued*)

TABLE 2 (VOLUME) DENSITIES OF SELECTED MATERIALS

Material	Weight, lb/ft^3	Material	Weight, lb/ft^3
Bituminous Products		Wood	
Asphalt	81	Ash	41
Tar	75	Oak	47
Metals		Douglas fir	34
Brass	526	Hem fir	28
Bronze	552	Southern pine	37
Cast iron	450	Wood Products	
Copper	556	Plywood	36
Lead	710	Particleboard	45
Steel	490	Gypsum Products	
Zinc	450	Gypsum wallboard	50
Cement and Concrete		Lath and plaster	55
Concrete (normal weight)	145	Insulating Materials	
Reinforced concrete (normal weight)	150	Fiberglass	1.0–2.5
Structural lightweight concrete	85–115	Rockwool	1.0–2.5
Perlite/vermiculite concrete	25–50	Extruded or expanded polystyrene	2.0
Portland cement	94	Polyisocyanurate	2.0
Lime	45	Glass (flat glass)	160
Masonry		Earth	
Clay brick	110–130	Clay, silt or sand (dry packed)	100
Concrete block (light weight)	105–125	Gravel	110
Concrete block (normal weight)	135	Crushed stone	100–120
Masonry mortar or grout	130	Water	63
Stone		Ice	57
Granite	165	Air	0.075
Limestone	165		
Marble	170		

EXAMPLE 3 • (ESTIMATING TOTAL GRAVITY LOAD ON A TYPICAL WOOD FRAME RESIDENTIAL FLOOR)

Determine the total gravity load (dead load + live load) on the floor of Example 2. The floor is part of an apartment building.

SOLUTION

The dead load of the floor is 10 psf, Example 2.
The live load on a residential floor is 40 psf, (Table 3.1).
Hence, the total gravity load (dead load + live load) is 10 + 40 = 50 psf.

EXAMPLE 4 • (ESTIMATING DEAD LOAD OF A WOOD FRAME RESIDENTIAL FLOOR WITH LIGHTWEIGHT CONCRETE TOPPING)

In many multifamily dwellings, a lightweight concrete topping is used over a wood floor to increase the sound insulation between floors. Determine the dead load of such a floor, assuming $1\frac{1}{2}$-in.-thick lightweight concrete topping. The density of lightweight concrete is 100 pcf.

SOLUTION

The dead load of a typical residential wood floor is 10 psf, Example 2.

Volume of 1 sq ft of concrete topping $= 1.0 \times 0.125 = 0.125$ ft^3 (Note that $1\frac{1}{2}$ in. $= 0.125$ ft).

Weight of 1 sq ft of concrete topping $= 0.125 \times 100 = 12.5$ lb

Hence, the total dead load of the floor is 10 + 12.5 = 22.5 psf.

PRACTICE QUIZ

Each question has only one correct answer. Select the choice that best answers the question.

23. According to the geogolists, the earth's surface is composed of several segments that are called
a. earth segments.
b. surface plates.
c. tectonic plates.
d. geological segments.
e. segmental plates.

24. The region of the United States that is relatively more earthquake prone is
a. northern United States.
b. southern United States.
c. eastern United States.
d. western United States.
e. none of the above.

25. The number of mild (minor to very minor) earthquakes occurring on the earth are nearly
a. 1 or 2 per year.
b. 10 to 20 per year.
c. 100 per year.
d. 1,000 per year.
e. several thousand per year.

26. An earthquake measuring 7.0 on the Richter scale is
a. twice as strong as a 6.0 earthquake.
b. 10 times stronger than a 6.0 earthquake.
c. 100 times stronger than a 6.0 earthquake.
d. 500 times stronger than a 6.0 earthquake.
e. none of the above.

27. Everything else being the same, a building constructed of heavyweight materials is subjected to a greater earthquake load than a building constructed of lightweight materials.
a. True
b. False

28. A reinforced concrete beam measures 12 in. × 24 in. in cross section. If the density of reinforced concrete is 150 lb/ft^3, the weight of beam is
a. 450 lb/ft.
b. 350 lb/ft.
c. 300 lb/ft.
d. 300 lb.
e. 300 lb/ft^3.

29. In a building whose structural system consists of columns, beams, and floor and roof slabs, the columns are spaced at 40 ft on center in one direction and 25 ft on center in the other direction. What is the tributary area for an interior column?
a. 1,000 ft^2
b. 1,000 ft
c. 500 ft^2
d. 250 ft^2
e. None of the above

30. In a building whose structural system consists of columns, beams, and floor and roof slabs, the columns are spaced at 40 ft on center in one direction and 25 ft on center in the other direction. What is the tributary area for a corner column?
a. 1,000 ft^2
b. 1,000 ft
c. 500 ft^2
d. 250 ft^2
e. None of the above

31. The total approximate dead load of a floor in a typical wood frame residential building is
a. 50 psf.
b. 30 psf.
c. 10 psf.
d. 5 psf.
e. none of the above.

32. The total approximate gravity load on a floor in a typical wood frame residential building is
a. 50 psf.
b. 30 psf.
c. 10 psf.
d. 5 psf.
e. none of the above.

Answers: 23-c, 24-d, 25-e, 26-e, 27-a, 28-c, 29-a, 30-d, 31-c, 32-a.

KEY TERMS AND CONCEPTS

Basic wind speed
Dead load
Earthquake load
Floor live load
Frequency and occurrence of earthquakes
Gravity load
Lateral load
Live load
Rain load
Richter scale
Roof live load
Seismic use group
Snow load
Soil liquefaction
Tsunami
Wind exposure classifications
Wind load

REVIEW QUESTIONS

1. What is the difference between dead load and live load? Explain with the help of examples.
2. What is the approximate wind speed in a tornado? Explain why tornado wind speeds are not considered in determining the design wind speed for a location.
3. Using sketches and notes, explain the types of wind pressures that occur on a rectangular building.

4. Using sketches and notes, explain how roof slope affects wind pressure on a roof.
5. Explain the differences between wind exposure categories B, C, and D.
6. Using sketches and notes, explain the ballooning effect in buildings.
7. What is Richter scale? What is the difference between the amounts of energy released by two earthquakes that are 8.0 and 6.0 on the Richter scale?
8. What types of construction are (a) suitable and (b) unsuitable in an earthquake-prone region?
9. Explain the major differences between designing buildings for earthquake resistance and for wind resistance.

FURTHER READING

1. American Society of Civil Engineers (ASCE) and Structural Engineering Institute (SEI). *Minimum Design Loads for Buildings and Other Structures,* ASCE 7 Document.

REFERENCES

1. Walker, B. *Earthquake.* Alexandria, VA: Time-Life Books, 1982, p. 7.
2. Botsai, E., et al. *Architects and Earthquakes.* Washington, DC: AIA Research Corporation, 1975, p. 19.

CHAPTER 4 Load Resistance–The Structural Properties of Materials

CHAPTER OUTLINE

4.1 Compressive and Tensile Strengths of Materials

4.2 Ductility and Brittleness

4.3 Yield Strength of Materials

4.4 Elasticity and Plasticity

4.5 Modulus of Elasticity

4.6 Bending Strength of Materials

4.7 Shear Strength of Materials

4.8 Bearing Strength of Materials

4.9 Structural Failures

4.10 Structural Safety

New York Times Tower, Eighth Avenue, New York City, under construction (2006)—a steel frame structure whose lateral load–resisting subsystem has been expressed on the facade. Another notable feature of the facade is the screen made of closely spaced horizontal ceramic tubes that shade the glass, conserving energy. Architect: The Renzo Piano Building Workshop in association with Fox and Fowle. (Photo by MM.)

The history of the development of architectural form coincides with the evolution of human understanding of the structural properties of materials and their application to the design and construction of buildings. Every civilization has found it necessary to understand the strength and limitations of indigenous materials in order to design safe and durable structures. The small spans of beams used in Greek temples, as shown in Figure 4.1, were due mainly to the designer's instinctive understanding that stone is a brittle material that is weak in tension and would break if the span is too long or if the beam size is too small.

FIGURE 4.1 Parthenon, Acropolis, Greece. Observe the relatively small span of beams in response to the weakness of stone in bending. (Photo courtesy of Dr. Jay Henry.)

Stone's high strength in compression was exploited in columns and walls, but the column spacings and the sizes of openings in walls had to be small. Although the main reason for providing capitals at the tops of columns in Greek temples was nonstructural, the capitals provided greater structural safety by reducing beam spans even further.

The discovery of the arch—a form that carries loads primarily in compression—enabled early builders to span greater distances in stone (or brick) than was previously possible. The arch led to the development of the vault, which is essentially a three-dimensional version of the arch, and subsequently to the dome—a form obtained by rotating the arch about a center point.

The arch, the vault, and the dome dominated architecture for centuries as the primary form-giving elements in buildings because the major construction materials available at the time, such as stone and brick, were brittle materials. The use of wood, which is less brittle than stone, was limited to either minor buildings or minor components of buildings because of its lack of durability. Sooner or later, wood structures were either consumed by fire or destroyed by termites.

FIGURE 4.2 Bourges Cathedral, Bourges, France. Buttresses and flying arches were some of the structural innovations of the time in understanding the inherent weakness of stone in tension. (Photo courtesy of Dr. Jay Henry.)

Masonry buildings, on the other hand, are extremely durable, and some have survived thousands of years. Therefore, the formal, intellectual and the technological use of brittle materials became highly perfected over time. Gothic cathedrals are excellent examples of how the limitations in the tensile strength of brittle materials were overcome to produce magnificent and technically daring structures, Figure 4.2.

The industrial revolution brought many new developments in architectural form. The availability of wrought iron and, later, cast iron had already revived the beam as a horizontal spanning element because of iron's high tensile strength. Large-size openings became possible without the use of arches, and the progression from load-bearing masonry buildings to the skeleton frame did not require much imagination. After steel became commercially available (in the midnineteenth century), the skeleton frame became the predominant structural system.

The invention of new materials—prestressed concrete, high-strength steels, reinforced concrete, and concrete masonry—led to the development of new structural forms. In fact, the modern movement and subsequent developments in architecture owe a great deal to the exploitation of the structural properties of materials.

In this chapter, we examine some of the important properties that influence the strength of materials. Emphasis is placed on properties that relate directly to basic structural materials, such as steel, concrete, brick, stone, and wood. However, the principles examined have general application.

Structural properties of materials and the behavior of structures under loads cannot be divorced from each other. They are so interrelated that often the distinction between the two is vague at best. Ductility and brittleness, elasticity, and plasticity, which are properties of materials, are also used to describe the behavior of an entire structure or components of a structure. Therefore, structural behavior is discussed in this chapter where necessary to describe the structural properties of materials.

4.1 COMPRESSIVE AND TENSILE STRENGTHS OF MATERIALS

When a force acts on a member (such as a building component), the member develops an internal resistance to the applied force. The intensity of internal resistance to the applied external force is called the *stress*. If the applied force is large, the internal resistance is large, and

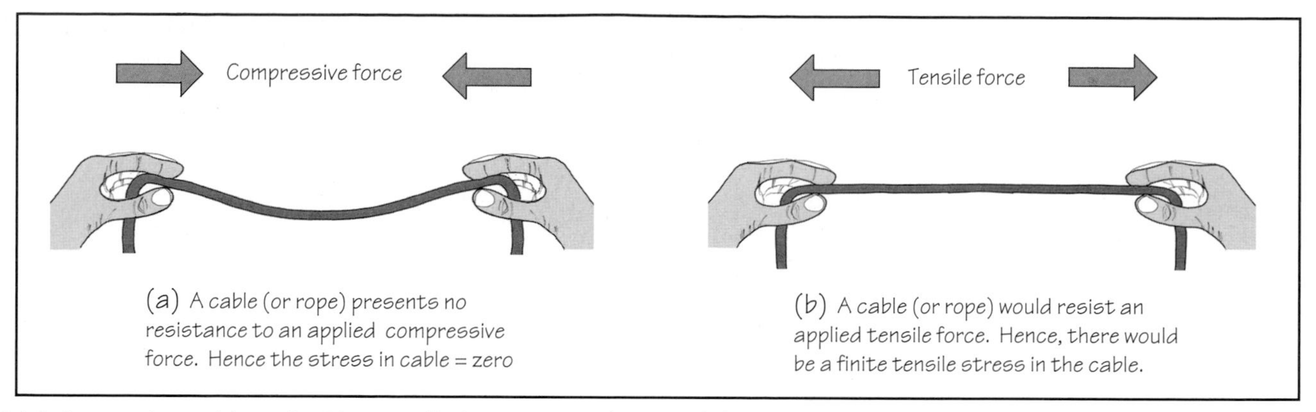

FIGURE 4.3 Stresses in a cable under **(a)** an applied compressive force, and **(b)** an applied tensile force.

so is the stress. If the applied force is small, the stress developed in the member is also small.

If a member is unable to develop any resistance to an applied force, the stress in the member is zero. For instance, if we hold a cable (or rope) between two hands and push it inward, that is, put a compressive force on the cable, we observe that it does not resist the applied force and simply gives in, Figure 4.3(a). The (compressive) stress in the cable in this case is zero.

If, on the other hand, the cable is pulled, i.e., a tensile force is applied on the cable, the cable tends to resist the force, and the cable becomes taut, implying the existence of tensile stress in the cable, Figure 4.3(b). The stress in the cable increases as the applied tensile force is increased.

From this example, we learn that the stress can either be compressive or tensile, depending on the type of external force. If the external force is compressive, the stress created in the member is *compressive stress* (or simply *compression*), and if the external force is a tensile force, the stress created in the member is *tensile stress* (or simply *tension*). A column or wall in a typical building is in compression, Figure 4.4(a). In a simple truss made of two rafters and a ceiling joist, the rafters are in compression and the ceiling joist is in tension, Figure 4.4(b). In a suspension bridge, the vertical pylons are in compression and the suspension cables are in tension, Figure 4.5.

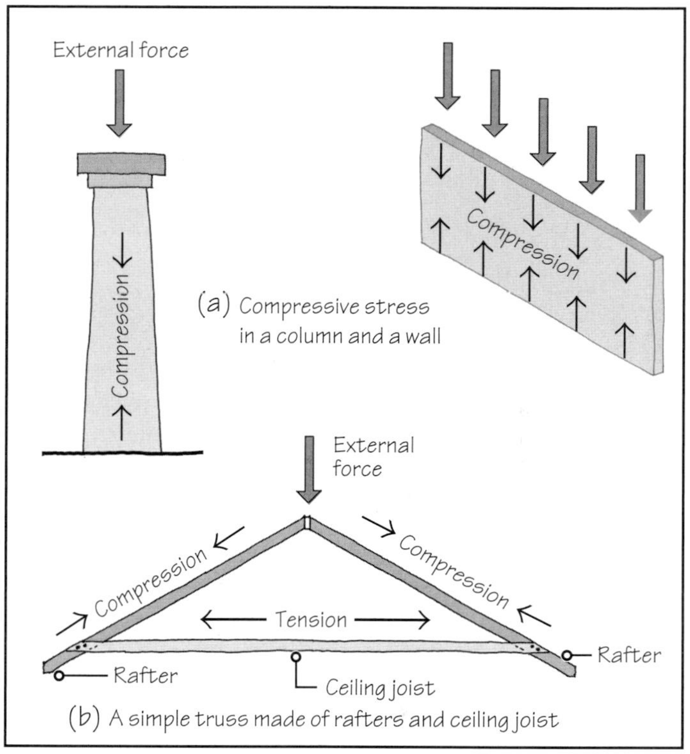

FIGURE 4.4 Compressive and tensile stresses in building components.

Quantitatively, the stress is defined as the force acting on a unit cross-sectional area (1 in.^{2} or 1 ft^2) of the member. That is why we referred to the stress as the *intensity* of internal resistance. The symbol commonly used for stress is f. Thus,

$$f = \frac{\text{force}}{\text{area}} = \frac{P}{A} \qquad \textbf{Eq. (1)}$$

where P = force and A = cross-sectional area of the member.

If the force is expressed in pounds and the cross-sectional area is in square inches, the unit of stress is pounds per square inch (psi). If the force is expressed in kips (1 kip = 1,000 lb), the unit of stress is kips per square inch (ksi).

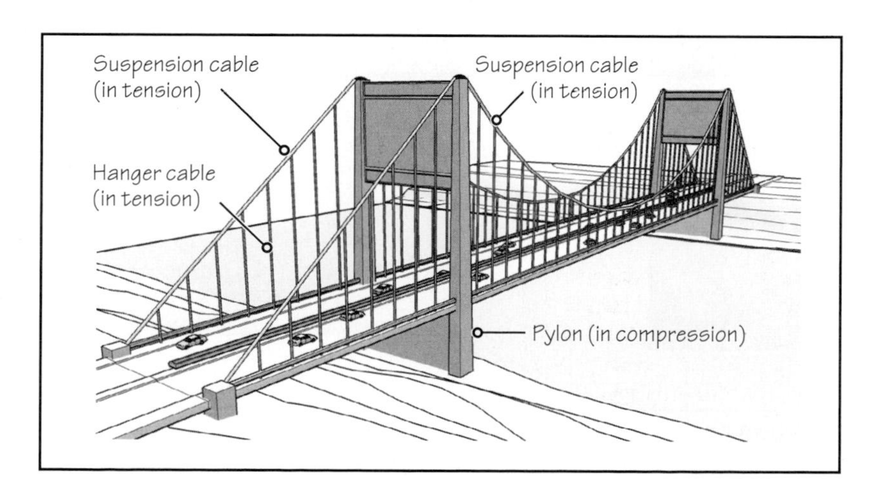

FIGURE 4.5 Tensile and compressive stresses in the members of a suspension bridge.

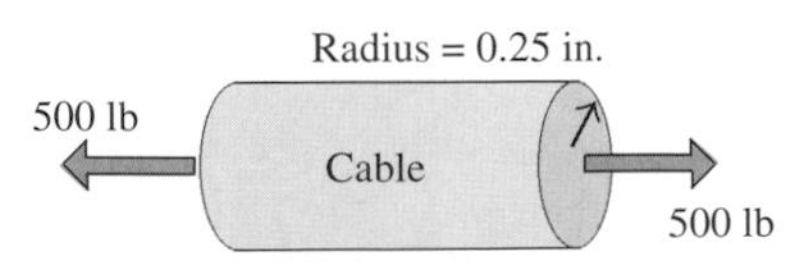

Cross-sectional area of cable = π (0.25)^2 = 0.196 in.^{2}

For instance, if the force applied on the cable of Figure 4.3(b) is 500 lb and if the diameter of the cable is 0.5 in. (cross-sectional area = 0.196 in.^{2}), the tensile stress in the cable is:

$$f = \frac{500}{0.196} = 2{,}550 \text{ psi} = 2.55 \text{ ksi}$$

ULTIMATE COMPRESSIVE STRENGTH AND ULTIMATE TENSILE STRENGTH

If the applied force on a member results in its failure, the stress in the member at failure is referred to as the *ultimate stress,* which is also known as *ultimate strength,* or simply the *strength* of the material. Thus, if the cable of Figure 4.3(b), with a cross-sectional area of 0.196 in.^{2}, fails when the applied tensile force is 1,000 lb, the *tensile strength* of the cable's material is

$$\text{Tensile strength} = \frac{1{,}000}{0.196} = 5{,}100 \text{ psi} = 5.1 \text{ ksi}$$

TESTING OF MATERIALS TO OBTAIN THEIR ULTIMATE STRENGTHS

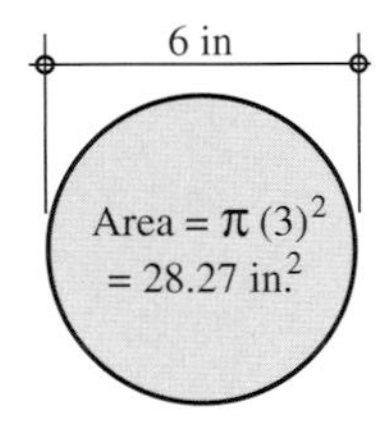

Standards such as those mentioned in Chapter 2 prescribe the approved test methods for determining the tensile and compressive strengths of various materials. For example, if we wish to determine the compressive strength of concrete, we perform a standard test. This test involves crushing a 6-in.-diameter cylinder of concrete measuring 12 in. high in a controlled setting, Figure 4.6(a) and (b). As the load is applied, it is measured and recorded continuously until the cylinder is crushed (fails), Figure 4.6(c). The ultimate compressive strength of the concrete is obtained from the load that caused the failure.

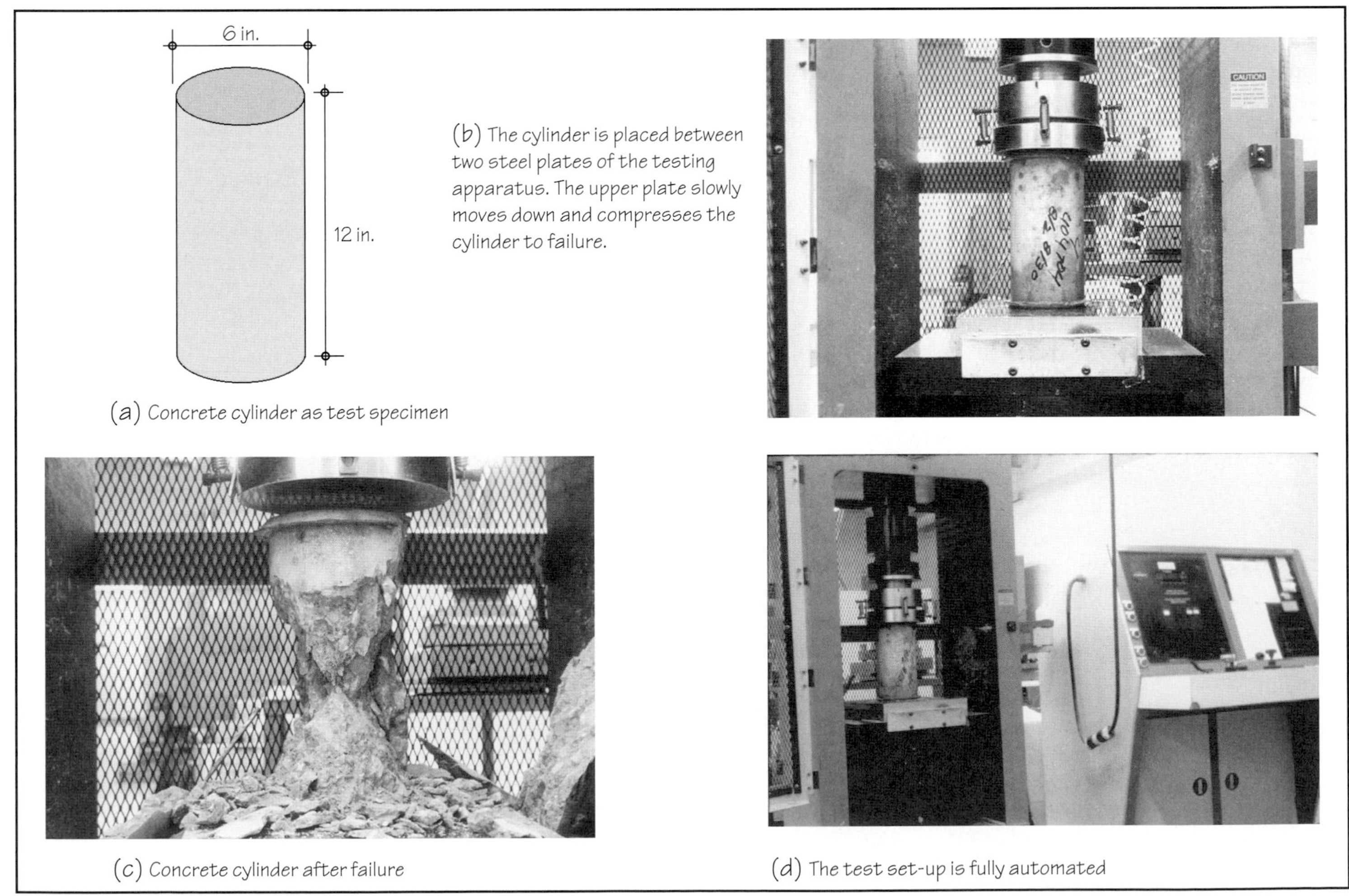

FIGURE 4.6 Apparatus and test specimen (6 in. × 12 in. concrete cylinder) used for determining the compressive strength of concrete.

For instance, in the illustration of Figure 4.6, let us assume that the cylinder failed when the load reached 115 kips. Then, the compressive strength of concrete would be

$$\text{Compressive strength} = \frac{\text{load at failure}}{\text{area of cylinder}} = \frac{115}{28.27} = 4.1 \text{ ksi}$$
$$= 4{,}100 \text{ psi}$$

The method and test setup (machine) for determining the compressive strength of brick or concrete block masonry is the same. The specimen used is a masonry prism of a specified size made from several bricks or blocks with mortar joints, Figure 4.7.

The tensile strength of concrete or masonry is very low compared with its compressive strength—approximately 10% of its compressive strength. Thus, if the compressive strength of concrete is 4,000 psi, its tensile strength is only about 400 psi. That is why the tensile strength of concrete is generally ignored in the design of concrete structures; that is, we assume that the tensile strength of concrete is zero.

We typically determine the tensile strength of steel using a round bar specially machined at the ends to provide a grip for the machine to pull the bar to failure. A commonly used tensile-testing machine is shown in Figure 4.8. The compressive strength of steel is approximately equal to its tensile strength, provided the steel member is prevented from buckling (see Section 4.9).

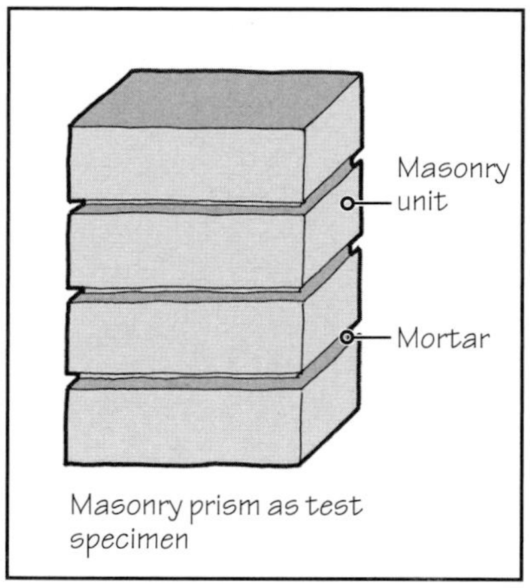

Masonry unit

Mortar

Masonry prism as test specimen

FIGURE 4.7 Test specimen used for determining the compressive strength of masonry.

4.2 DUCTILITY AND BRITTLENESS

Stress is nearly always accompanied by deformation of the member. The exception to this rule is a *rigid body*. A rigid body is defined as one that does not deform at all under the action of loads. In other words, a rigid body will not stretch, shorten, bend, or change shape regardless of the magnitude of load placed on it. In the real world, a rigid body does not exist, because all materials deform when loaded. Actual bodies, such as building components, will deform under loads, although the deformations in them are usually too small to be visible to the human eye.

The deformation caused by a compressive or a tensile stress is simply the change in the length of the member. Tensile stress causes the member to elongate, and compressive stress causes it to shorten. We are interested in both the absolute value of the change in length as well as its relative value. The relative change in length, defined as the change in length divided by the original length, is called the *strain*.

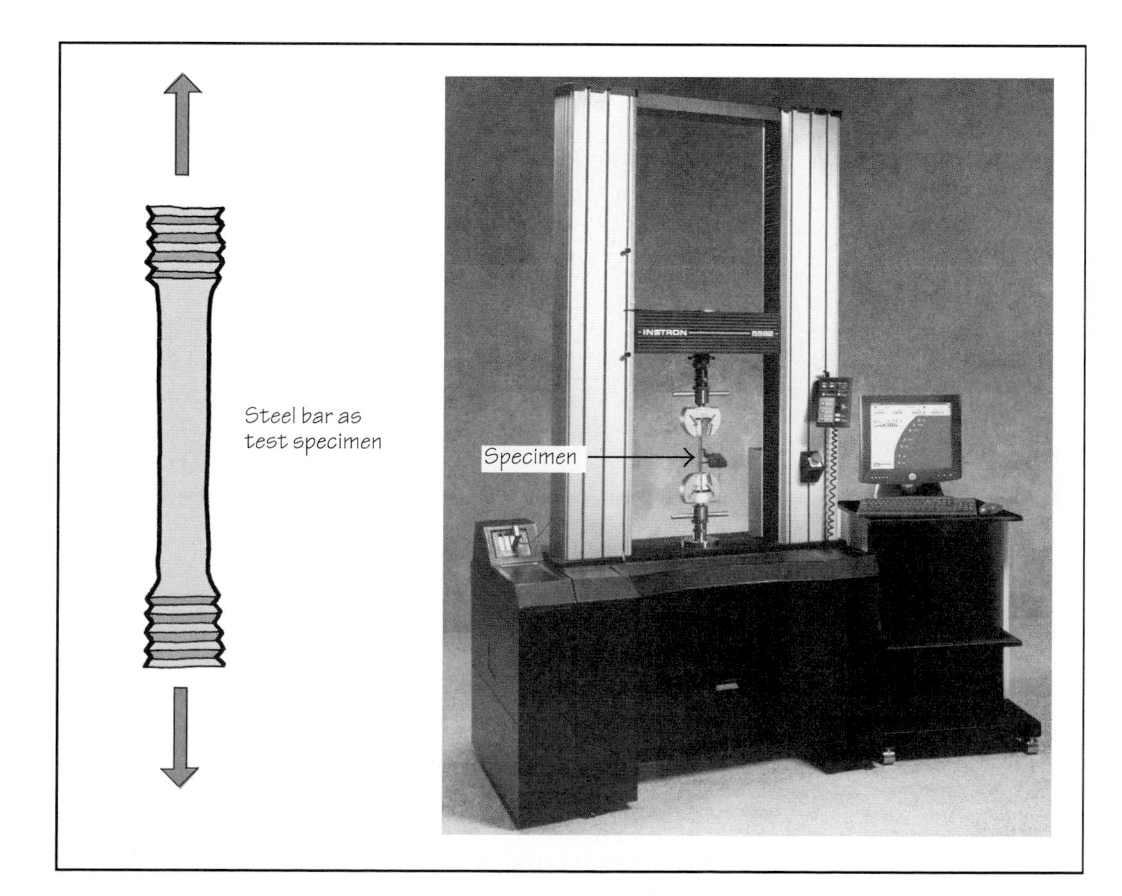

FIGURE 4.8 A typical Universal Testing Machine (UTM) measuring the tensile strength of a steel specimen. (Photo courtesy of Instron® Corporation.)

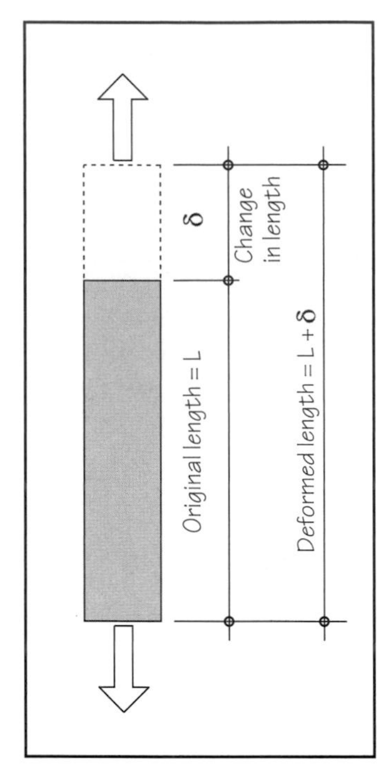

FIGURE 4.9 If a bar, whose original length is L, stretches to length L + δ under the action of a force, then the strain in bar is ε = (δ/l).

NOTE

Brittle and Ductile Materials

A brittle material is one for which ultimate strain ≤0.5%.

A ductile material is one for which ultimate strain >0.5%.

Thus, in Figure 4.9, if the original length of the bar is L, which—under the action of a tensile force—becomes L + δ, then the strain in the bar, denoted by the Greek letter ε (epsilon), is given by

$$\varepsilon = \frac{\text{change in length}}{\text{original length}} = \frac{\delta}{L} \qquad \textbf{Eq. (2)}$$

From Equation (2), if a 10-ft (120-in.) high column shortens by 0.25 in. under a load, the strain in the column is

$$\text{Strain } (\varepsilon) = \frac{0.25}{120} = 0.002$$

Because the strain is the ratio of two lengths, it is simply a number with no units of measure. The strain caused by a tensile force is called *tensile strain*; that caused by a compressive force is called *compressive strain.* The strain in the element at failure is referred to as the *ultimate strain.*

The ultimate strain for most building materials is usually small. For example, the ultimate compressive strain of concrete or masonry is approximately 0.003. It is, therefore, more convenient at times to express strain as a percentage. For example, a strain of 0.003 is equal to 0.3% strain. The ultimate strain for mild steel (the term *mild steel* is explained later) is nearly 0.35 (or 35%), which is fairly large. In other words, for mild steel, a 100-in.-long steel bar will elongate by 35 in. at failure; that is, its length will become 135 in.

Materials that produce large deformations before failure are called *ductile materials.* Conversely, materials that do not deform much before failure are called *brittle* materials. The corresponding properties are called *ductility* and *brittleness.* There is no general agreement as to the limiting value of ultimate strain that distinguishes a ductile material from a brittle material, but we generally regard a material as brittle if its ultimate strain is less than or equal to 0.5%. A ductile material is one whose ultimate strain is greater than 0.5%.

Materials such as brick, stone, concrete, and glass have an ultimate strain of less than 0.5% and are, therefore, classified as brittle materials. Approximate values of ultimate strains of the commonly used structural materials are shown in Figure 4.10.

Most metals are ductile, because they have large deformations at failure. Pure metals are more ductile than their alloys. Pure gold is seldom used in jewelry because it is very ductile and, hence, prone to damage by deformation. The same is true of pure iron. It is too ductile and soft for use as a construction material, but when alloyed with carbon to form steel, it becomes a useful construction material.

BRITTLENESS AND TENSILE STRENGTH

One of the characteristics of brittle materials is that they are stronger in compression than in tension. As stated in Section 4.1, the tensile strength of concrete is approximately 10% of its compressive strength. Brick, stone, and glass behave similarly. A ductile material such as steel has equal strength in compression and tension.

DUCTILITY AND FAILURE WARNING

Because its deformation at failure is small, a brittle material does not give any visual warning of its impending failure. Its failure is sudden. Failure of a ductile material, on the other hand, is gentle and is preceded by excessive deformation and ample warning. Structural safety concerns mandate a ductile failure in building structures. Brittle failure is not permitted. Because concrete is a brittle material and steel is a ductile material, the use of steel reinforcement helps to impart a moderate amount of ductility to reinforced concrete members.

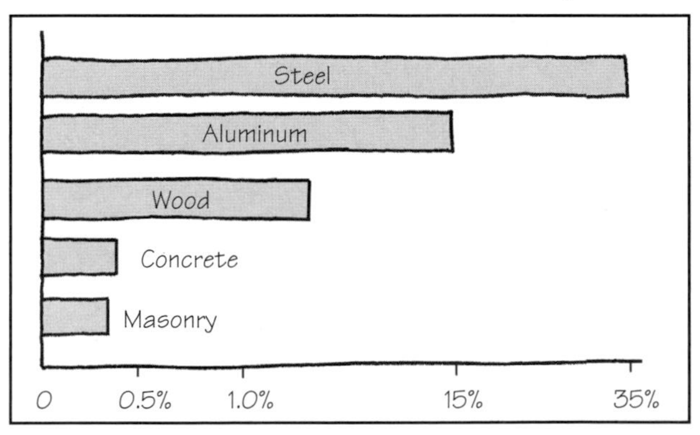

FIGURE 4.10 Approximate values of ultimate strain for selected materials (not to scale).

DUCTILITY AND MALLEABILITY

The terms *ductility* and *malleability* are used interchangeably in lay literature, although strictly speaking they represent different properties of materials. As stated previously, ductility is the property of a material to deform extensively prior to failure. Malleability, on the other hand, is the property that allows a material to be shaped by hammering, forging, pressing, and rolling. Ductile materials are generally malleable, but they need not be. For instance, cast iron is a ductile material (according to the definition of ductile material given earlier), but it cannot be converted into shapes by hammering or pressing. Cast-iron shapes are obtained by casting molten iron into molds.

PRACTICE QUIZ

Each question has only one correct answer. Select the choice that best answers the question.

1. The material in an arch is primarily in
   a. compression.
   b. tension.
   c. both tension and compression.
   d. shear.
   e. none of the above.

2. When a cable or rope is compressed by a force, what is the compressive stress in it?
   a. Equal to the force applied
   b. Force divided by the cross-sectional area of rope
   c. Zero
   d. Infinite
   e. None of the above

3. In a simple three-member truss consisting of two rafters and a tie, each member is in
   a. compression.
   b. tension.
   c. either tension or compression.
   d. shear.
   e. both tension and compression.

4. In a suspension bridge, which member(s) is (are) in compression?
   a. Suspension cable
   b. Hanger cables
   c. Support pylons
   d. Both suspension cable and hanger cables
   e. None of the above

5. A rectangular column measuring 12 in. × 12 in. in cross section carries a load of 18 kips. What is the stress in the column?
   a. 18 kips
   b. 18 ksi
   c. 2.25 ksi
   d. 125 psi
   e. None of the above

6. The test specimen used for determining the compressive strength of concrete in the United States is a
   a. cube measuring 12 in. × 12 in. × 12 in.
   b. cube measuring 6 in. × 6 in. × 6 in.
   c. prism measuring 6 in. × 6 in. × 12 in.
   d. cylinder measuring 12 in. diameter × 6 in. high.
   e. cylinder measuring 6 in. diameter × 12 in. high.

7. When we determine the strength of concrete using the test specimens, we determine the concrete's
   a. ultimate compressive stress.
   b. ultimate tensile stress.
   c. ultimate shear stress.
   d. modulus of elasticity.
   e. all the above.

8. In relation to its compressive strength, the tensile strength of concrete is
   a. much higher.
   b. slightly higher.
   c. nearly the same.
   d. much lower.
   e. slightly lower.

9. A 20-ft-high column shortens by 0.6 in. under a load. What is the resulting strain in the column?
   a. 0.0020
   b. 0.0020 in.
   c. 0.0025 in.
   d. 0.0025
   e. None of the above

10. Which gives greater warning before failure?
    a. A brittle material
    b. A ductile material

Answers: 1-a, 2-c, 3-c, 4-c, 5-d, 6-e, 7-a, 8-d, 9-d, 10-b.

4.3 YIELD STRENGTH OF MATERIALS

Because a load causes deformation, stress and strain occur simultaneously in a member. Generally, the strain increases as the stress increases. However, there are situations where stress may exist without any strain or strain without stress. For example, a beam resting on two end supports will be subjected to a change in its length as the temperature of the beam increases or decreases. In this case, the strain occurs without any stress.

But if the same beam is clamped between the two end supports so that it cannot change its length, the beam will be subjected to stresses as its temperature changes—a case in which the stress occurs without the strain. The clamped beam will be subjected to compressive stress when its temperature rises and to tensile stress when the temperature falls.

The relationship between stress and strain conveys a great deal of information about the structural properties of the material and is unique for each material. This relationship is usually given in a graphical form, called the *stress-strain diagram.* In the stress-strain diagram, the stress is generally represented on the vertical axis and the strain on the horizontal axis.

The stress-strain diagram of one of the most important structural materials—low-carbon steel—is shown in Figure 4.11. The diagram is obtained by subjecting a round steel bar of standard length and diameter to an increasing tensile force. This is done in a tensile-testing machine (Figure 4.8), which is fully automated and gives the stress-strain diagram directly, thus avoiding the need to calculate stresses and strains and to plot them manually.

The diagram of Figure 4.11 begins at the origin (point O) because the strain must obviously be zero when the stress is zero—a no-load condition. It consists of three important points; Y, U, and F.

Point Y is called the *yield point,* and the stress corresponding to this point is called the *yield stress.* Yield stress is an important property of steel. A few types of steel are distinguished from each other by their yield stress values. For instance, A-36 steel is so called because its yield stress is 36 ksi (the letter *A* in A-36 symbolizes a ferrous metal, as per

NOTE

Low-Carbon (Mild) Steel

Low-carbon steel is known as *mild steel* because of its higher ductility and the consequent ease with which it can be bent to shape and rolled into various cross-sectional shapes. High-strength steels are comparatively more brittle. They contain more carbon than mild steel. Therefore, they cannot be bent to shape or rolled into cross-sectional shapes with the same ease.

ASTM classification (Table 2.3). Steel used as reinforcing bars in concrete is called grade 40 or grade 60 steel, depending on whether its yield stress is 40 ksi or 60 ksi.

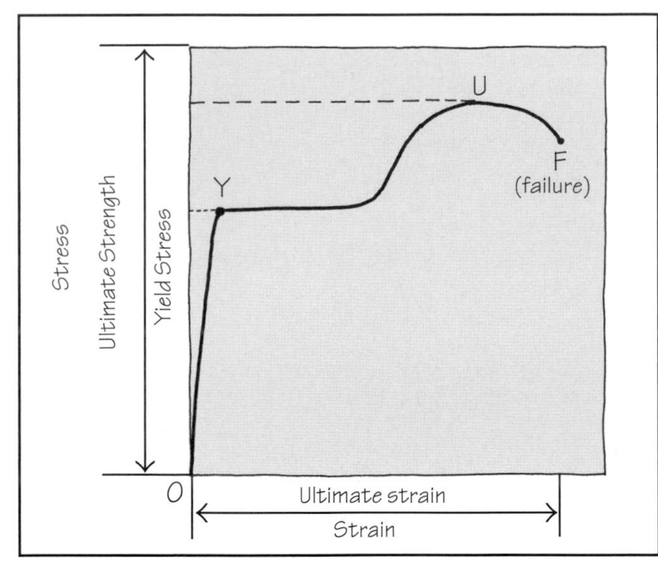

FIGURE 4.11 Stress-strain diagram for low-carbon (mild) steel in tension (not to scale).

YIELD STRESS AND ULTIMATE STRENGTH

In region OY of Figure 4.11, the relationship between stress and strain is a straight line, and stress is proportional to strain. Beyond point Y, the deformation in the bar increases substantially with little or no increase in the load, and the stress-strain diagram becomes (almost) horizontal. In other words, the steel yields after reaching point Y like a plastic material (such as modeling clay).

The yielding of steel produces a major rearrangement of the atoms in its crystalline structure, making steel stronger and harder, so the bar is able to sustain a greater load. This effect, referred to as *strain hardening,* is similar to that obtained by mechanical working on steel, such as cold rolling or hammering. We experience strain hardening when we take a piece of steel wire and bend it between our fingers. The wire becomes harder and stronger after bending, and it is more difficult to bend again at the original bend point.

Further increases in the load take the diagram of Figure 4.11 to point U, the point of maximum stress. The maximum stress is called the *ultimate stress,* or the *ultimate strength* of steel. Observe that the stress-strain diagram drops down between point U and the failure point F, indicating a loss of strength near failure. This is due to the manner in which the materials are tested and indicates the inability of testing equipment to keep the load on a rapidly disintegrating material.

Although point U represents the ultimate strength of steel, for all practical purposes it is the yield stress that is considered as the limiting stress in steel. Because steel deforms excessively after yield and because excessive deformation is to be avoided in a structure, we limit the actual stress in steel members to remain well below the yield stress. Therefore, although not absolutely correct, yield stress is also referred to as the *yield strength* of steel.

The yielding of the material is not peculiar to steel. It occurs in several metals and is an extremely useful property, because it provides ductility. Additionally, the strain hardening that follows yield makes the metal stronger.

4.4 ELASTICITY AND PLASTICITY

A material is called an *elastic material* if it is fully recoverable from deformation after the load is removed. The property responsible for the elastic behavior of the material is called *elasticity.* The converse of elasticity is called *plasticity.* Thus, a material that does not recover from its deformation after the removal of the load is called an *inelastic,* or *plastic, material.* Materials are generally elastic only up to a small strain value—approximately 0.5%. As stated earlier, the ultimate strain in brittle materials such as concrete, brick, stone, and glass is of this order (0.5%). Therefore, brittle materials are generally elastic up to failure.

Metals, on the other hand, can sustain large deformations, usually greater than 10%. They are elastic up to a certain stress level, and if loaded beyond that stress value, they are plastic—that is, their deformation beyond that point is permanent. For example, steel is elastic up to yield point (between points O and Y in Figure 4.11) but plastic thereafter. In other words, if the stress in a steel member remains below the yield stress, the member is fully recovered from any deformation after the load is removed. Materials that are elastic up to a certain stress value and plastic thereafter are called *elastic-plastic materials.* Thus, low-carbon steel is an elastic-plastic material. The same applies to aluminum—another important metal used in buildings, generally in windows and glass curtain walls.

Building structures are designed so that the stresses in materials remain within the elastic limit and so that no inelastic deformations occur under design loads. Because they are permanent, inelastic deformations affect the appearance of structures. The exception to this general philosophy is made in the design of structures located in highly seismic areas. The maximum intensity of earthquake loads in such areas is so high that if structures were designed to remain elastic, member sizes (columns, beams and floor slabs) would have to be so large as to make the structure highly uneconomical.

Thus, as stated in Section 3.9, the structural design philosophy for buildings located in seismic zones is that the building components should deform elastically under low to moderately intense earthquakes so that after the earthquake event, the building regains its original shape and size. However, under severe earthquake loads, the structure is designed to deform inelastically (plastically) without compromising its safety or integrity. The structure may become unusable and unserviceable because of plastic deformations but will remain integral to provide life safety.

This design philosophy is based on the probability that the maximum intensity of earthquake loads will occur only once in 50 years. Thus, the structure will probably suffer inelastic deformations once in 50 years, which is typically the useful life span of most buildings.

4.5 MODULUS OF ELASTICITY

The ratio of stress to strain in a material is called the *modulus of elasticity.* Thus,

$$\frac{\text{Stress}}{\text{Strain}} = E \qquad \textbf{Eq. (3)}$$

where E denotes the modulus of elasticity. A measure of the material's stiffness, the modulus of elasticity is a property of a material. The larger the value of E, the stiffer the material and the less it will deform under a given load.

On the stress-strain diagram, the value of E is represented by the slope of the diagram. The greater the slope, the larger the value of E and, hence, the stiffer the material. Thus, in Figure 4.12, the value of E is larger for Material 1 than for Material 2. For a *rigid* material, the value of E would be infinite. This is obvious from Equation (3), because for a rigid material, the strain is zero regardless of the value of stress. Therefore, the stress-strain diagram for a rigid material is a vertical line because on a vertical line, the strain is zero.

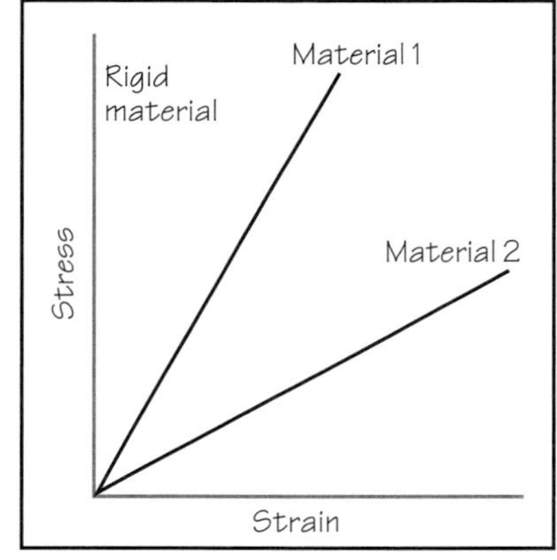

FIGURE 4.12 The slope of a material's stress-strain diagram gives the magnitude of the modulus of elasticity, E, of the material. The greater the slope, the greater the value of E.

The concept of the modulus of elasticity as a material property (constant for a material) assumes a linear stress-strain diagram. If the stress-strain diagram is a curve, the modulus of elasticity is not constant but varies with the stress level. Most construction materials fall in this category, i.e., they do not have a linear stress-strain diagram. The exception is steel, which has a linear stress-strain diagram only up to the yield point. Because yield stress is, practically, the limiting stress in steel, steel may be considered as having a linear stress-strain diagram for all practical purposes.

The modulus of elasticity of a material that does not have a linear stress-strain diagram is obtained from the straight line whose slope is approximately equal to the slope of the initial part of its stress-strain diagram. This approximation is justified on the basis that in actual buildings, the materials are usually loaded to low- or moderate-stress levels so that the slope of the initial part of the diagram is a good approximation to the modulus of elasticity of the material.

Figure 4.13 shows the stress-strain diagram for concrete. Because the diagram is nonlinear, the modulus of elasticity is given by the slope of straight line OA, where A represents 40% of the compressive strength of concrete. Thus, if the strength of concrete is 4,000 psi, point A represents a stress level of 1,600 psi.

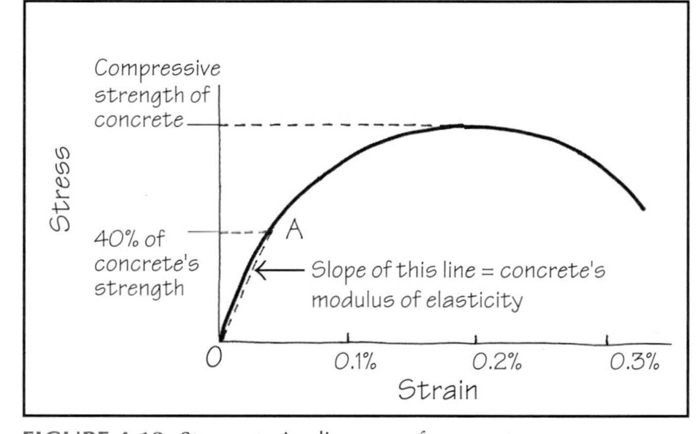

FIGURE 4.13 Stress-strain diagram of concrete.

UNITS OF MODULUS OF ELASTICITY

From Equation (3), the units of the modulus of elasticity are the same as those of stress, because strain has no units. In other words, the units of E are psi or ksi (pascals in the SI system). The E value of materials is usually a large number—in millions of pounds per square inch (gigapascals in the SI system, where 1 GPa = 10^9 Pa).

For a soft rubber, the value of E is a relatively small number. Of commonly used materials, diamond has the highest value of E (greatest stiffness). That is why diamond-tipped tools are used to cut window glass. Table 4.1 gives the modulus of elasticity values of a few selected materials.

MODULUS OF ELASTICITY AND STRENGTH OF MATERIALS

Being a measure of the stiffness of the material, modulus of elasticity determines how much a material will deform under loads. Because deformation control is a structural design criterion, modulus of elasticity is an important material property. It determines how much a beam or slab will deflect under a given load or how much a tall building will sway sideways under the action of wind.

NOTE

Artificial Diamond

The diamond commonly used on the tips of cutting tools, such as saw blades and glass-cutting wheels, is artificial. Artificial diamond, first patented by Edward Acheson in 1926, was carborundum—a trade name for silicon carbide—made from clay and carbon. Since then, several other metal carbides, such as tungsten carbide and titanium carbide, have been produced. Pure diamond is 100% crystalline carbon.

For most materials, the modulus of elasticity increases with the strength of the material. Higher-strength species of wood, for example, have higher moduli of elasticity than lower-strength species. The same is true of concrete and masonry. The greater the concrete strength, the greater its modulus of elasticity. However, steel has the same modulus of elasticity regardless of its strength. In other words, high-strength steels have the same modulus of elasticity as low-carbon (mild) steel.

TABLE 4.1 MODULUS OF ELASTICITY OF SELECTED MATERIALS

Material	Modulus of elasticity ($\times 10^6$ psi)
Wood	1.0–2.0
Brick masonry	1.0–3.0
Concrete	3.0–8.0
Steel	29.0
Aluminum alloys	10.5
Window glass	10.4
Diamond	170
Rubber	0.001

Steel is the stiffest of all construction materials; that is, it has the highest value of E (29×10^6 psi). Its modulus of elasticity is based on the straight-line portion of the stress-strain diagram, points O to Y in Figure 4.11. Because unlike other materials, the modulus of elasticity of steel does not increase with strength, high-strength steel is not as commonly used in structural steel sections as low-carbon (mild) steels.

4.6 BENDING STRENGTH OF MATERIALS

So far we have discussed situations in which either tensile or compressive stress is present in a member. For example, in the suspension bridge of Figure 4.5, the pylons are in compression, and the cables and hanger bars are in tension. In such cases, the external force acts along the axis of the member, as shown in Figure 4.14(a). The presence of the axial force may be obvious, as in Figure 4.5, or not so directly obvious, as in the rafter-joist assembly of Figure 4.4(b).

Because the external force is axial, the stress created in the member is referred to as an *axial stress*. The axial stress is uniform over the entire cross section, as shown in Figure 4.14(a), and may be either compressive or tensile. However, if the external force acts perpendicular to the axis of a member, as seen in Figure 4.14(b), the member bends, subjecting itself to both compressive and tensile stresses simultaneously.

A force acting in a downward direction on a beam that is supported at the ends bends it with a concave, *water-holding curvature*, like the inside of a saucer. An upward acting force will cause it to have a convex, *water-shedding curvature*, like the outside of a sphere. Bending (also called *flexure*) is a predominant action in building structures. Gravity loads produce bending in beams and slabs, and lateral loads cause bending in walls and columns.

Bending produces compression in one half of a member's cross section and tension in the other half. This can be demonstrated by bending a member made of a soft material such as rubber or sponge. If we mark two vertical lines, PQ and RS, on one of the longitudinal faces of a sponge beam and bend it to produce a concave curvature, as shown in Figure 4.15, we see that the lines have become closer at the top and farther apart at the bottom. Fibers of the beam represented by line PR have become shorter in length, and fibers on line QS have become longer.

Further examination of the deformed shape of rectangle PQSR shows that maximum contraction takes place along line PR, the extreme top surface of the beam. The maximum

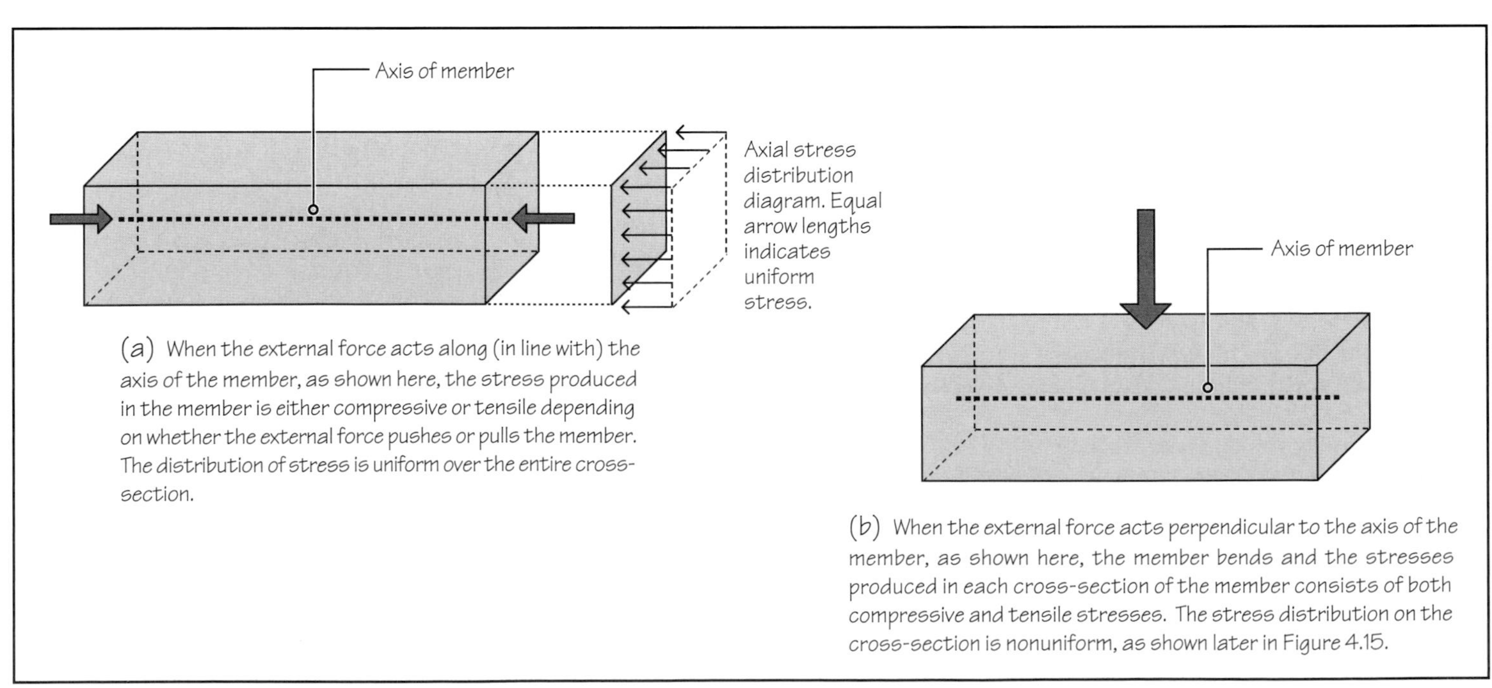

FIGURE 4.14 The orientation of the external force with respect to the axis of the member and the types of stresses produced in the member.

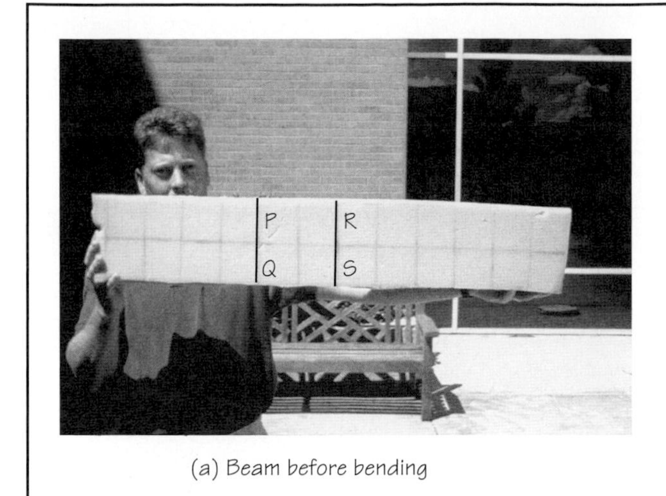

(a) Beam before bending

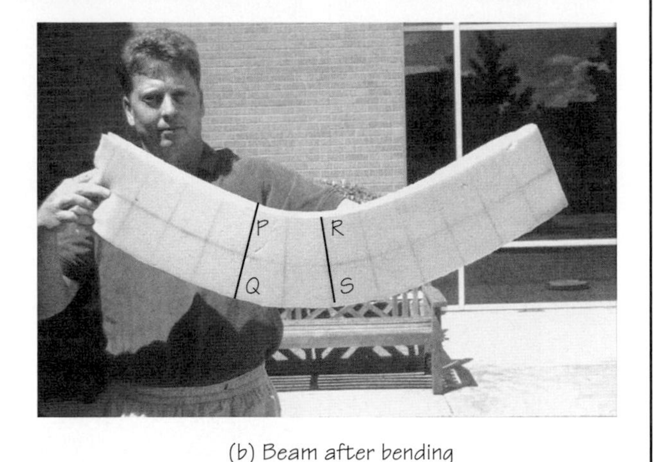

(b) Beam after bending

FIGURE 4.15 Demonstration of bending in a beam. In a beam bent with a water-holding curvature as shown in **(b)**, the upper half of the beam is in compression and the lower half of the beam in tension.

elongation is produced along line QS, the extreme bottom surface of the beam. In other words, compressive stresses are maximum at the top surface, and tensile stresses are maximum at the bottom surface of the beam.

The change from compressive to tensile stresses occurs on the surface that neither contracts nor stretches; that is, this surface maintains its original dimension and is, therefore, unstressed. This surface is the neutral plane, commonly referred to as the *neutral axis* of the beam, Figure 4.16. Thus, the neutral axis is a line on the beam's cross section where the stresses are zero.

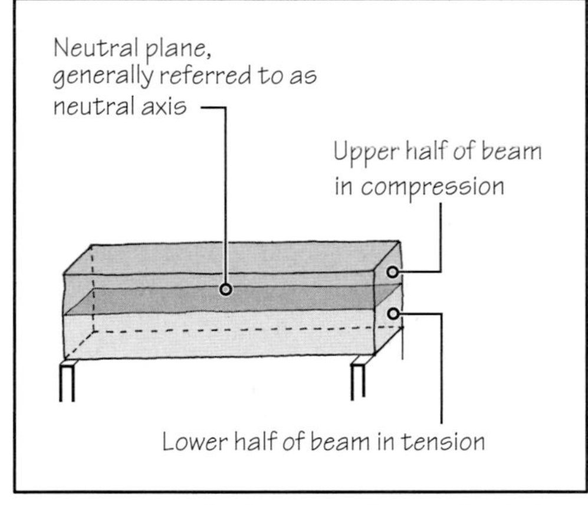

FIGURE 4.16 The location of neutral axis in a beam under bending stresses. Note that the neutral axis is, in fact, a neutral plane. It is called neutral axis because we generally draw a beam in two dimensions—in cross section—in which the neutral plane shows as a line.

Improving the Bending Strength of Members

If we isolate a small length of the beam, such as length PQSR of Figure 4.15, and examine the distribution of stresses on cross sections represented by lines PQ and RS, we see that the distribution is as shown in Figure 4.17. The term *bending stress* (or *flexural stress*) is used to describe this distribution of stresses—compressive stresses on one side of the neutral axis and tensile stresses on the other side.

The stress distribution shown in Figure 4.17 represents stresses in the beam when it bends with a water-holding curvature. If the beam bends with a water-shedding curvature, the stresses will simply be reversed; that is, tensile stresses will be created in the upper half of the beam and compressive stresses, in the lower half of the beam.

Stresses are unequally distributed throughout the beam's cross section in bending. They are concentrated at the top and bottom of the cross section, but the central part of the cross section is relatively stress free—an inefficient use of the material. By contrast, the axial stress distribution is uniform over the entire cross section, as shown in Figure 4.14(a). Thus, structural configurations that generate axial stresses in members use the material efficiently.

In suspension structures, where members are either in compression or tension (Figure 4.5), loads are carried very efficiently. That is why suspension structures are popular for long-span bridges.

NOTE

Structural Efficiency

A structurally efficient shape is one that can withstand a greater load for the same amount of material. For example, a beam with an I-section is structurally more efficient because it can withstand a greater load than a rectangular beam of the same cross-sectional area and span.

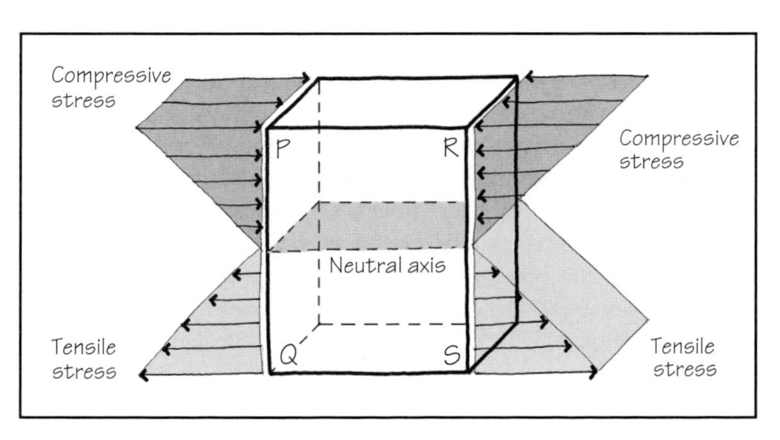

FIGURE 4.17 Stress distribution on a small length, PQSR, of the beam of Figure 4.15.

FIGURE 4.18 I-sections in steel, wood, and concrete are in common usage.

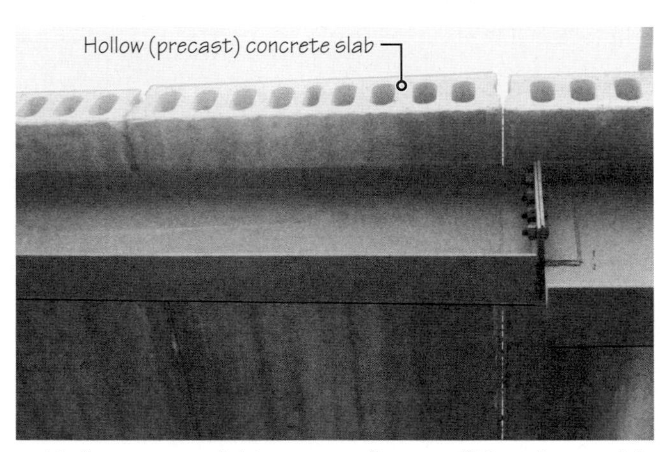

A hollow concrete slab is structurally more efficient than a solid slab because it has less material (concrete) in the center of the slab, where it is not much needed

Open-web I-section beam (called castellated beam) is more efficient in bending than an I-section beam, which in turn is more efficient than a rectangular beam with the same amount of material

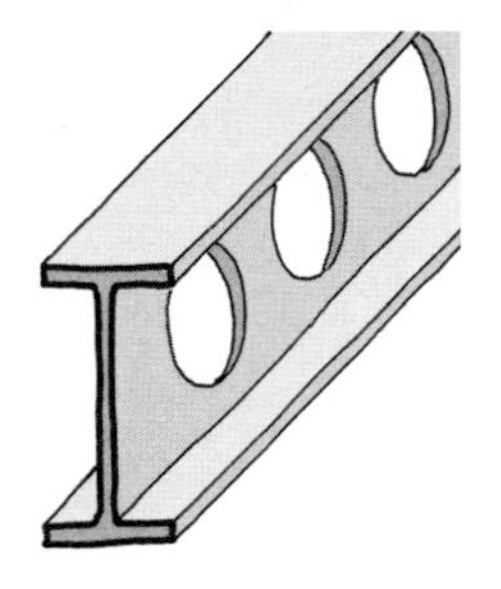

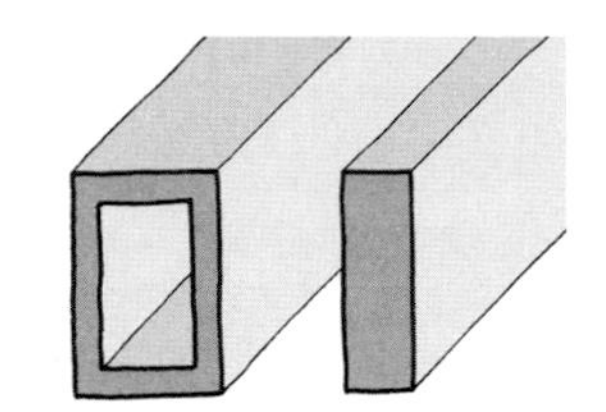

A tubular (hollow structural) section is more efficient in bending than a solid rectangular section of the cross-sectional area

FIGURE 4.19 Some strategies used to improve the structural inefficiency of members in bending.

Architects and engineers have overcome the structural inefficiency in bending by developing the I-shaped cross section. This configuration places most material at the top and bottom of the beam where stresses are greatest, thus making it more efficient than a (solid) rectangular section. I-shaped beams in steel, wood, and concrete are commonly used in buildings and civil engineering sructures, Figure 4.18.

Some of the other strategies used in response to the structural inefficiency in bending are tubular cross sections, hollow concrete slabs, and open-web-steel I-shape beams, Figure 4.19.

In fact, the structural inefficiency of an element in bending, such as a beam, is worse than that described here. The stresses are not only nonuniform on a beam's cross section but are also nonuniform along the length of the beam. From the laws of statics, it can be shown that the stress distribution on various sections along the length of a beam supported at two ends is as shown in Figure 4.20. The central cross section is most highly stressed, whereas the stresses on sections away from the center of the beam are progressively smaller.

When either the maximum tensile or maximum compressive stress on the central cross section equals the ultimate strength of the material, the beam will fail. Thus, the bending strength of the beam depends on the failure of just one (extreme) fiber—the top or the bottom fiber at the center of the beam. The rest of the beam has much lower stresses at failure.

By comparison with bending, axial compression or axial tension is a far more efficient structural action because every cross-sectional surface of the member is under equal and uniform stress distribution, as shown in Figure 4.14(a). That is why we will probably not be able to break a thin bar or rod by pulling or compressing it. But if we bend the same rod, it will

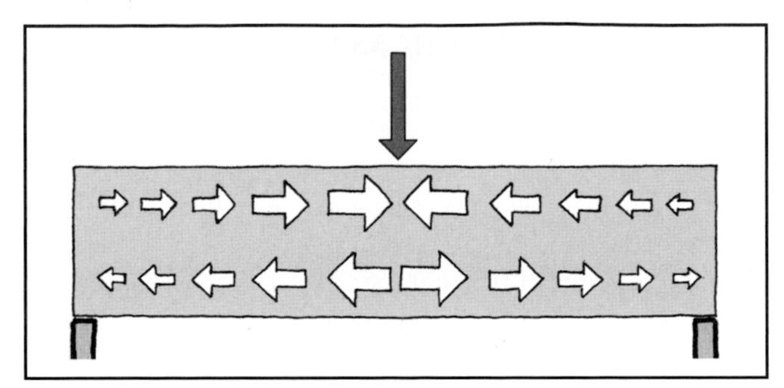

FIGURE 4.20 The distribution of stresses along the length of a member in bending. Observe that the stresses maximize in the center of the beam and progressively reduce at the supports. Thus, the material near the supports is not being fully utilized—a further illustration of the structural inefficiency of a member in bending.

break more easily. The reason is that the maximum resistance to the load in bending is being provided by only a few fibers, thereby causing the rod to break. Therefore, a relatively small load is required to break the rod in bending. By contrast, the entire member participates in resisting an axial load. Despite their structural inefficiency, beams, joists, decks, and slabs are commonly used because of the need for horizontal spanning members in buildings.

BENDING STRENGTH OF BRITTLE MATERIALS IS LOW

Because bending creates tensile as well as compressive stresses, brittle materials, being weak in tension, are also weak in bending. Ductile materials, such as steel, which have equal strengths in tension and compression, have a higher bending strength. Reinforced concrete is a composite material that exploits the moderately high compressive strength of concrete and the high tensile strength of steel.

Steel is placed in those locations of a reinforced concrete member where tensile stresses are present. For instance, in a beam supported at two ends, tensile stresses occur in the lower half of the beam cross section. Therefore, steel reinforcing bars are placed along the length of the beam at the bottom of the beam's cross section, Figure 4.21. (Also see Section 19.3 for additional details).

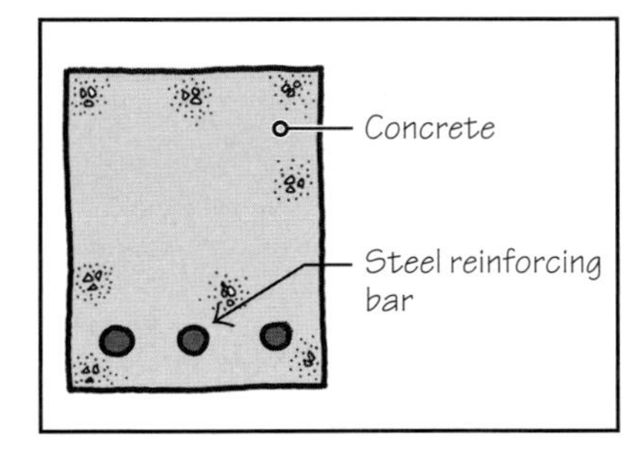

FIGURE 4.21 Section through a typical reinforced concrete beam. Note that steel reinforcing bars are used in region of concrete that is subjected to tension.

PRACTICE QUIZ

Each question has only one correct answer. Select the choice that best answers the question.

11. In a typical stress-strain diagram of a material, the stress is generally plotted along the vertical axis.
- **a.** True
- **b.** False

12. What is the yield stress of A36 steel?
- **a.** 36 lb
- **b.** 3600 ksi
- **c.** 3600 psi
- **d.** 36 ksi
- **e.** None of the above

13. The stress-strain diagram of mild steel is nearly a straight line up to the yield point.
- **a.** True
- **b.** False

14. Low carbon (mild) steel is
- **a.** an elastic material.
- **b.** a plastic material.
- **c.** an elastic-plastic material.
- **d.** none of the above.

15. Concrete is
- **a.** an elastic material.
- **b.** a plastic material.
- **c.** an elastic-plastic material.
- **d.** none of the above.

16. Modulus of elasticity refers to how
- **a.** strong a material is.
- **b.** stiff a material is.
- **c.** elastic a material is.
- **d.** serviceable a material is.
- **e.** none of the above.

17. The units of modulus of elasticity are
- **a.** psi.
- **b.** ksi.
- **c.** Pa.
- **d.** GPa.
- **e.** all the above.

18. The bending stresses along the neutral axis of a beam are
- **a.** the minimum.
- **b.** maximum.
- **c.** zero.
- **d.** none of the above.

19. Which of the following beam cross-sections is structurally more efficient in bending?
- **a.** Rectangular (solid) section
- **b.** I-section

20. Which of the following beam cross-sections is structurally more efficient in bending?
- **a.** Rectangular (solid) section
- **b.** Tubular section

Answers: 11-a, 12-d, 13-a, 14-c, 15-a, 16-b, 17-e, 18-c, 19-b, 20-b.

4.7 SHEAR STRENGTH OF MATERIALS

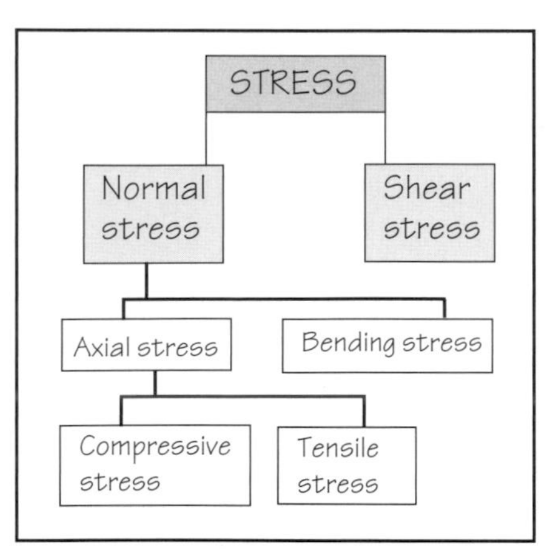

FIGURE 4.22 Types of stresses that may exist in a member.

So far we have discussed the strength of materials under axial stress and bending stress. Both these stresses—axial stress (which can be compressive or tensile) and bending stress—are *normal stresses,* implying that the stress acts perpendicular (normal) to the cross section of the member (see Figures 4.14(a) and 4.17). A second type of stress, called *shear stress,* acts tangential to the cross section of the member. In fact, normal stress and shear stress are the only two basic types of stress. Various types of stress that exist in a member belong to either normal stress or shear stress, Figure 4.22.

Shear stress is caused by a force that is tangential to the cross section of the member. Because it is tangential, a shear force tends to produce a sliding effect on the body. Shear stress is, therefore, associated with the resistance of the layers within the body to sliding over each other, Figure 4.23. Like normal stress, shear stress is also defined as the shear force divided by area and has the same units as normal stress—psi or ksi.

Consider an assembly of two wood members glued together at the interface, as shown in Figure 4.24(a). If a force pulls on the assembly, axial tension is developed in each member. In addition to the axial stress, the assembly also develops shear stress to resist sliding at the interface. The interface area of the assembly that resists shear is shown in white. For example, if this area is 2.0 in. by 2.0 in. (4.0 in.^{2}) and if the force acting on the assembly is 2.0 kips, then the shear stress developed at the interface is

$$\text{Shear stress} = \frac{\text{force}}{\text{area}} = \frac{2.0}{4.0} = 0.5 \text{ ksi} = 500 \text{ psi}$$

If the members are joined together with bolts, as shown in Figure 4.24(b), instead of being glued together at the interface, the shear force will now be resisted by the cross-sectional area of the bolts. For example, if there are two 0.5-in.-diameter bolts, the shear stress in bolts is

$$\text{Shear stress} = \frac{2.0}{2(0.196)} = 5.1 \text{ ksi}$$

LOW SHEAR STRENGTH OF BRITTLE MATERIALS

As shown in Figure 4.23, a shear force causes an angular deformation in an element, changing the element's shape, not its dimensions. Thus, a rectangular element becomes an oblique element. This introduces tension along one diagonal of the member and compression along

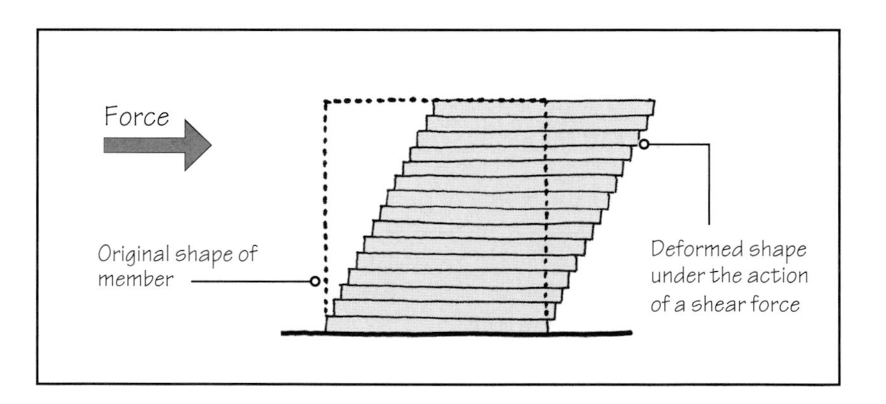

FIGURE 4.23 The effect of a shear (tangential) force on a body.

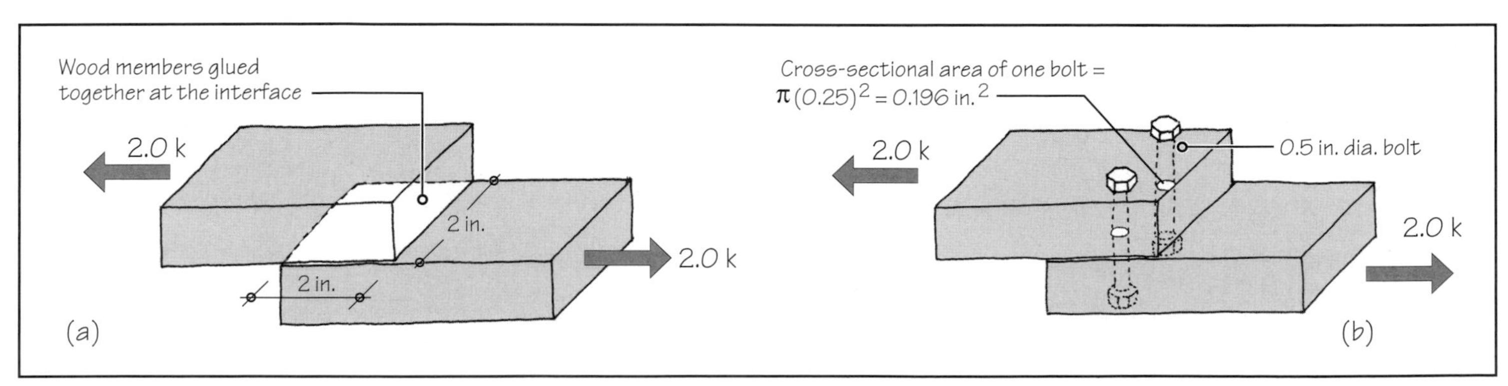

FIGURE 4.24 The magnitude of shear stress equals the shear force divided by the area that resists the shear force.

the other, Figure 4.25. Because of the tension created by shear, brittle materials—which are weak in tension—are also weak in shear. Steel, on the other hand, has a fairly high shear strength, because it has the same (high) strength in tension as well as compression.

In the concrete beam illustrated in Figure 4.21, the steel placed at the bottom of the beam is adequate to resist tensile stresses resulting from bending, but additional steel is required where shear stresses are high. Steel stirrups are therefore used in concrete beams to resist shear stresses, Figure 4.26. (See Section 19.3 for additional details.)

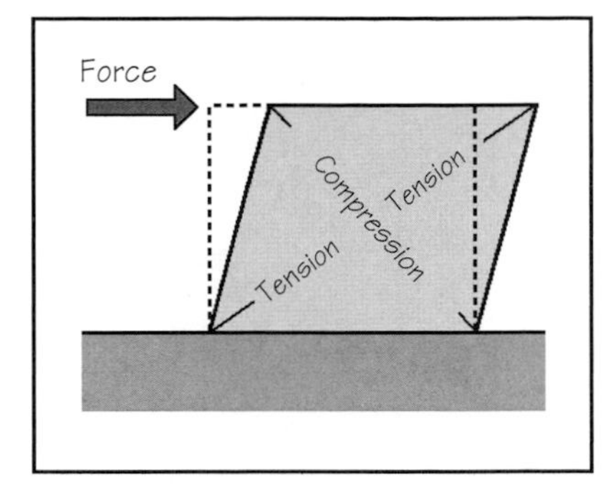

FIGURE 4.25 Shear stress is a combination of tensile stress along one diagonal and compressive stress along other diagonal.

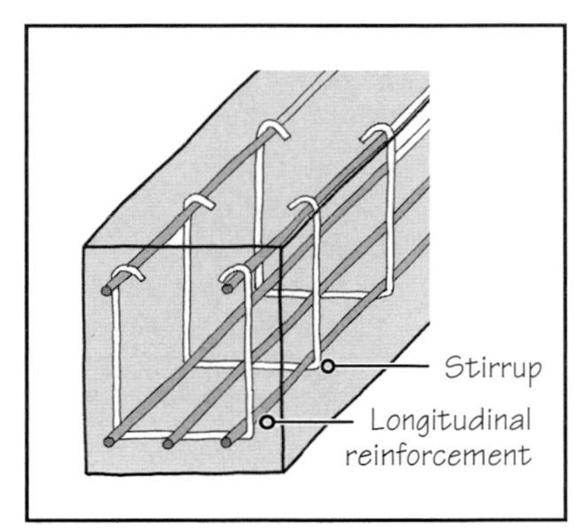

FIGURE 4.26 Steel stirrups in a reinforced concrete beam are used to increase the beam's shear strength.

4.8 BEARING STRENGTH OF MATERIALS

When we speak of the compressive strength of a material, we imply that a specimen of the material was tested to failure without any confinement in the lateral direction; that is, the test specimen was allowed to expand freely in the direction perpendicular to the load. This is how we determine the compressive strength of concrete in a standard compression test. As the cylinder is loaded to failure from the vertical direction, it is free to bulge out horizontally (see Figure 4.6).

However, if a cylinder is tested with horizontal confining stress throughout its entire height, as shown in Figure 4.27, we observe that the cylinder fails at a higher load than an unconfined cylinder made from the same concrete. For example, assume that a concrete cylinder under the standard compression test (no lateral confinement) tests to a compressive strength of 3,500 psi. If we prepare another cylinder from the same concrete and impose a lateral confining stress of 1,000 psi during the compression test, we find that the cylinder tests to a compressive strength of nearly 9,000 psi. When the lateral confinement is increased to 2,000 psi, a cylinder from the same concrete tests to a compressive strength of nearly 13,000 psi, Figure 4.28.

The state of stress in a laterally confined cylinder is referred to as *triaxial* compression, because the cylinder is being loaded three-dimensionally—from the vertical direction as well as from the horizontal directions. Triaxial compression exists in deep and thick columns and walls. In such elements, the material in the interior portions is not free to expand laterally because of the confining effect of the exterior mass. Consequently, the interior material posts a higher compressive strength than the material on the exterior of such elements.

This fact is also used in the design of a column footing. The concrete footing generally has a much larger area than the area of the column. Therefore, as the column delivers the load to the footing, the concrete immediately under the column experiences the confining effect of the mass of the concrete surrounding the column base. Thus, the concrete in the footing exhibits a greater overall compressive strength than that obtained from the standard compression test. This state of compressive stress, created at the surface of contact between the column and the footing, is referred to as the *bearing stress,* and the corresponding strength of the material is called the *bearing strength* of the material.

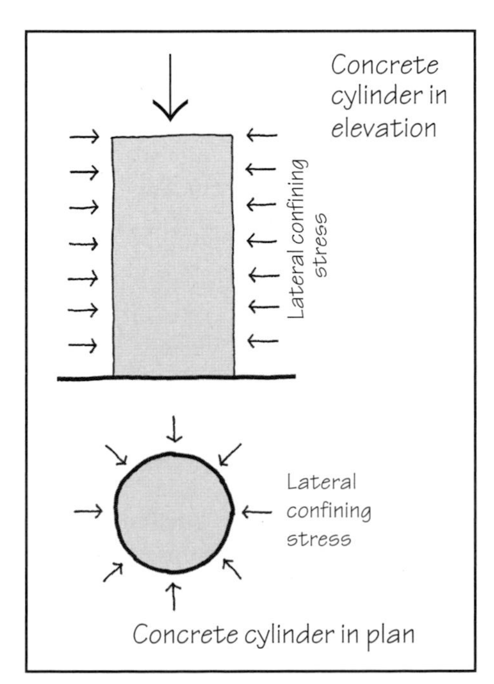

FIGURE 4.27 A concrete cylinder (shown in plan and elevation) tested to failure under lateral confinement.

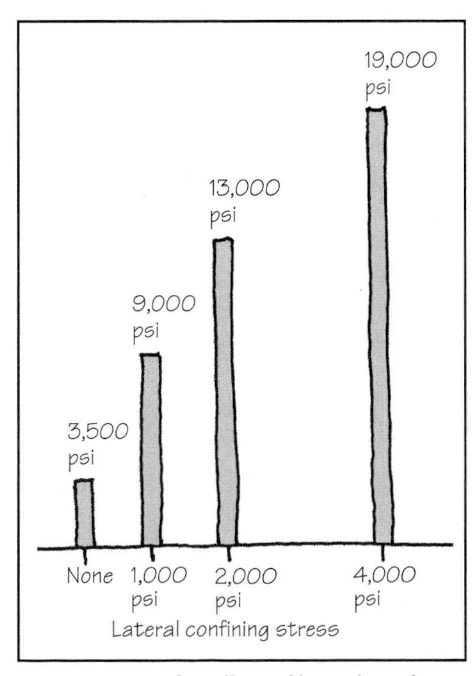

FIGURE 4.28 The effect of lateral confinement on the crushing strength of concrete.

In general, when one element compresses on another element, the stress that occurs at the interface between the two elements is the bearing stress. Therefore, bearing stress is the (compressive) load divided by the contact area between the two elements.

Bearing failure occurs when the load causes the stress in the material to exceed its bearing strength. Usually, bearing failure does not result in complete crushing of the material but a partial one, or what is referred to as *local crushing*. Local crushing in a structural member is similar to the crushing (or settlement) of ground directly under the pointed heel of a shoe.

BEARING PLATES

In building structures, a bearing plate is used to prevent local crushing that could occur when one element delivers a concentrated compressive load to another element. For example if a wide-flange steel column is placed directly on a concrete footing, a bearing failure (local crushing) in the concrete could result, because concrete's bearing strength is much less than the compressive stress in steel column.

The term *bearing* is used when one element delivers a compressive load to another element. That is why the steel plate directly under a steel column is referred to as the *bearing plate*, Figure 4.29. The area of the bearing plate is greater than the cross-sectional area of the column, which reduces the bearing stress on concrete. For the same reasons, a bearing plate is also used when a steel beam rests on a concrete column or wall, Figure 4.30.

In this situation, the column *bears* on the bearing plate, the bearing plate *bears* on concrete footing, and the footing *bears* on the soil below. The bearing stress of the materials must be checked at each interface to ensure that it does not exceed the bearing strength of the material.

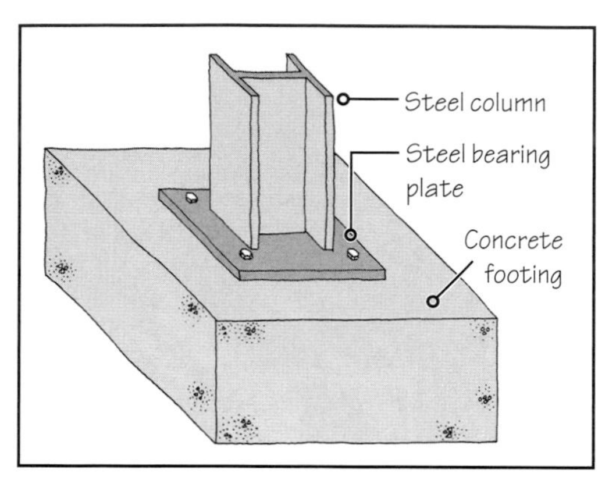

FIGURE 4.29 Bearing plate under a steel column.

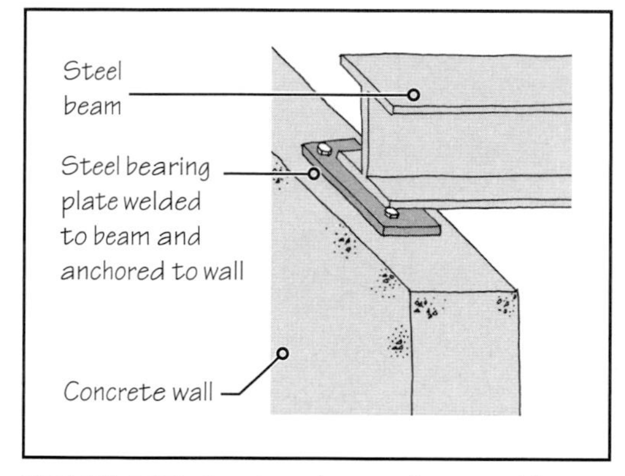

FIGURE 4.30 Bearing plate under a steel beam.

BEARING CAPACITY OF SOIL

Because the soil directly under a footing is confined by a large mass of the surrounding ground, the compressive strength of the soil is referred to as the bearing strength or, more commonly, as the *bearing capacity* of the soil (see Section 21.7). Observe that the bearing strength of a material is essentially its compressive strength plus the effect of confinement, if present. If there is no confinement, the bearing strength of the material is the same as the compressive strength of the material.

4.9 STRUCTURAL FAILURES

The structural failure of a member can occur in several ways, depending on the type of stress in the member. Because stresses in a member can either be compressive, tensile, or a combination of tensile and compressive stresses, the compressive and tensile failures are the two basic types of structural failures.

COMPRESSIVE AND TENSILE FAILURES

A compressive failure is sudden and catastrophic, Figure 4.31. A tensile failure (particularly of a ductile material) is preceded by necking and elongation of the member, giving sufficient warning of the failure, Figure 4.32.

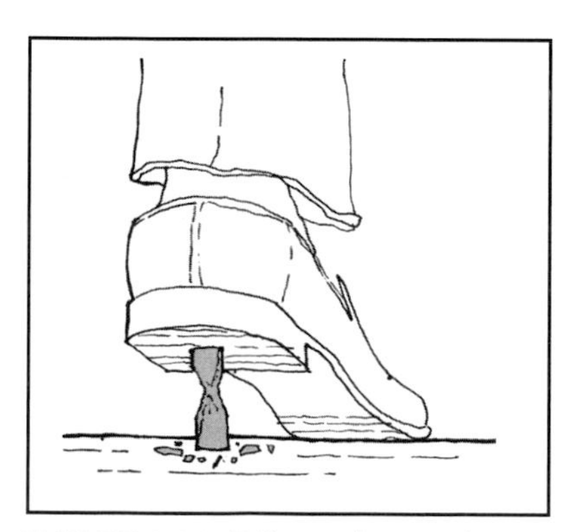

FIGURE 4.31 Failure of a member under axial compression (compressive failure). This type of failure occurs in a short, squat member made of brittle material such as concrete, brick, or stone. Compressive failure is abrupt.

BENDING FAILURE

Bending (stress) failure may either be sudden or gradual, depending on whether the member fails in compression or tension. The same applies to shear failure, because like bending stress, shear stress is also a combination of the compressive and tensile stresses.

BUCKLING FAILURE

A type of failure that is entirely different from the failures just described is buckling failure. Buckling failure occurs in a long and slender member subjected to axial compressive stress. (A slender member is one whose length is much greater than its cross-sectional dimensions.) Although a short, squat member will fail by the crushing of the material (as shown in Figure 4.31), a long, slender member will bend under an axial load. The bending of a member under an axial load is called *buckling*.

Buckling is similar to bending. The difference between bending and buckling is that bending is produced by loads that are perpendicular to the axis of the member and buckling

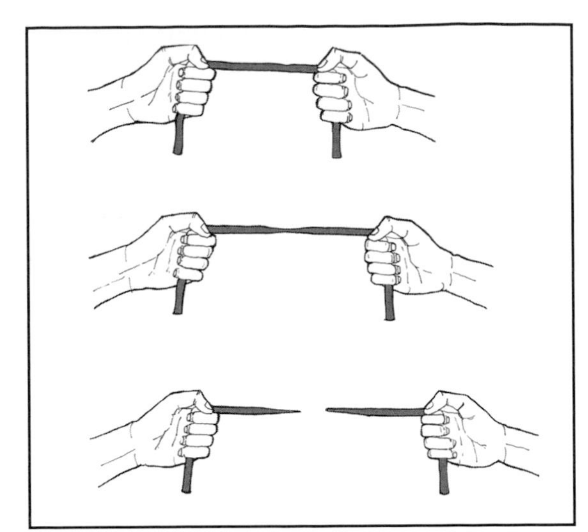

FIGURE 4.32 Failure of a member in tension—tensile failure. If the member is made of a ductile material, such as steel or aluminum, tensile failure is gradual and is generally preceded by necking of the member.

is caused by axial loads. A more significant difference between the two, however, is that the bending of a member takes place under all magnitudes of loads, however small. Buckling occurs when the load increases to a certain critical value, called the *critical load.*

Therefore, unlike bending, buckling takes place abruptly. The buckling of a member is recognized as one of the failure mechanisms, although it usually does not lead to a complete collapse of the structure. By buckling under the load, the member becomes unstable but will generally not fracture. In some situations, however, the instability caused by buckling in one member may lead to excessive stresses in other members, thereby causing a progressive collapse of the structure.

Buckling can be easily demonstrated by taking a plastic ruler and subjecting it to an increasing axial compression. The more slender the ruler, the smaller the compressive force needed to produce a buckling failure in it, Figure 4.33.

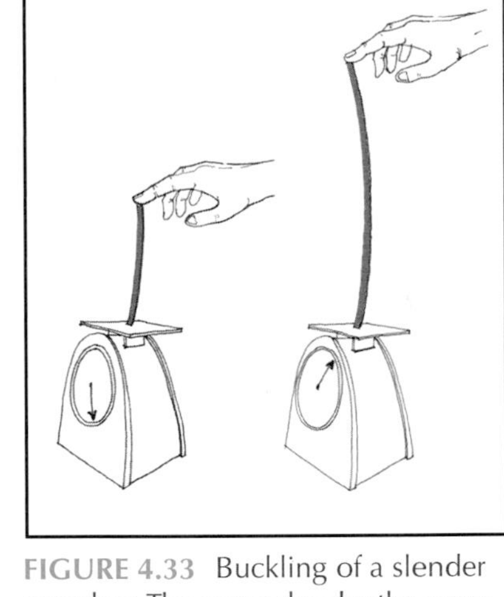

FIGURE 4.33 Buckling of a slender member. The more slender the member, the smaller its buckling load.

Buckling is an important consideration in the design of columns and load-bearing walls. Steel columns are more prone to buckling than concrete columns because they are more slender. Even slender beams and floor joists are prone to buckling because of the presence of compressive stresses resulting from their bending. Remember, bending is a combination of tensile and compressive stresses.

To strengthen a member against buckling failure, we can either reduce its slenderness—that is, increase its cross-sectional dimension or brace it at intermediate point(s) along its length or height. Usually, the latter approach (bracing the member) is used, because increasing the cross-sectional dimension is uneconomical. For instance, studs in a wood light frame wall must be solidly blocked if it is determined that they may buckle under loads, Figure 4.34. Slender beams and joists require similar bracing, Figure 4.35,

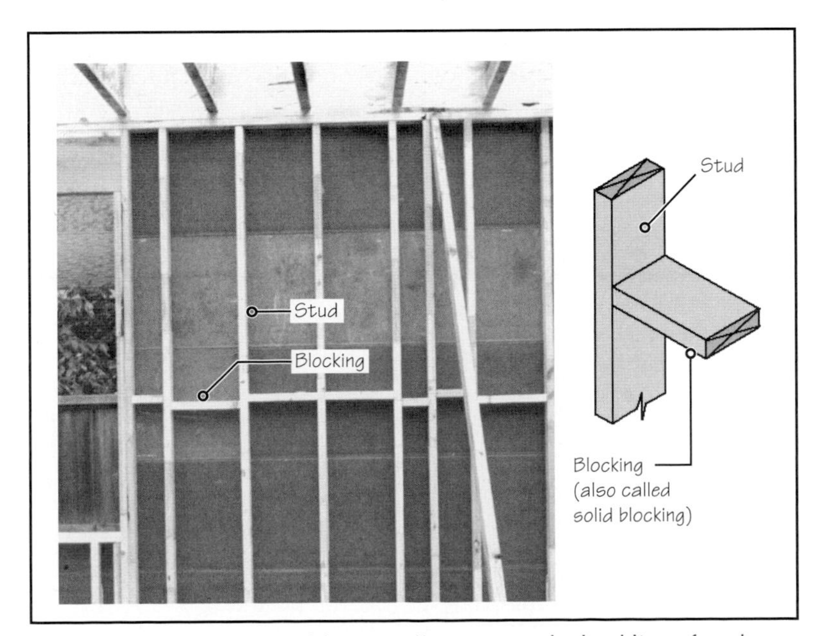

FIGURE 4.34 Solid blocking in a wood frame wall to prevent the buckling of studs.

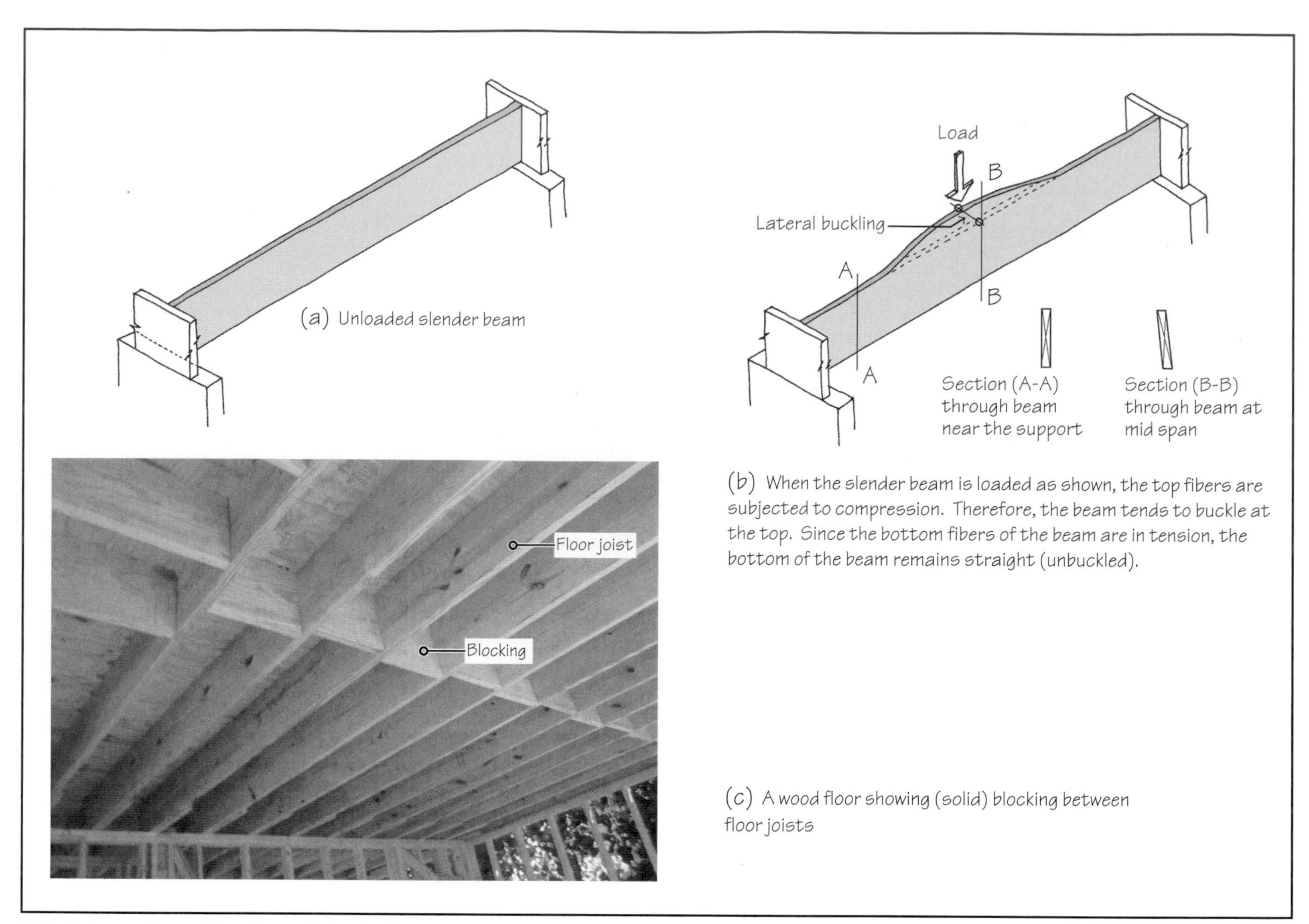

FIGURE 4.35 Lateral buckling of slender beams. In wood floor joists (which are quite slender), lateral buckling is prevented by using solid blocking between joists (see also Section 13.5).

because compression is produced in their top fibers as they bend. This compression may cause the slender beams and joists to buckle sideways, a phenomenon known as *lateral buckling*.

Lateral buckling can be verified by taking a 12-in.-long plastic ruler and subjecting it to a vertical load along its depth, as shown in Figure 4.35. We can see that the ruler will not bend downward but horizontally at the top. (Two persons, one to hold the ruler and the other to press it down, will facilitate the experiment.)

Buckling failure is not just a consideration for the entire structural member but also for its parts. Because of their slenderness, the vertical reinforcing bars in a concrete column have a tendency to buckle and break through the concrete cover under an axial load on the column. Horizontal **ties** are, therefore, provided along the height of the column to prevent this failure mode, Figure 4.36.

4.10 STRUCTURAL SAFETY

The most fundamental requirement of a structure is that it should be safe. A structure is safe if its strength is greater than the strength required to carry the loads imposed on it. The excess of the actual strength of the structure over that required to carry the loads is a measure of safety provided in the structure. Thus, the safety margin may be expressed as a ratio of actual strength to the required strength—strength required to carry the loads:

$$\text{Safety margin} = \frac{\text{actual strength}}{\text{required strength}} \qquad \textbf{Eq. (4)}$$

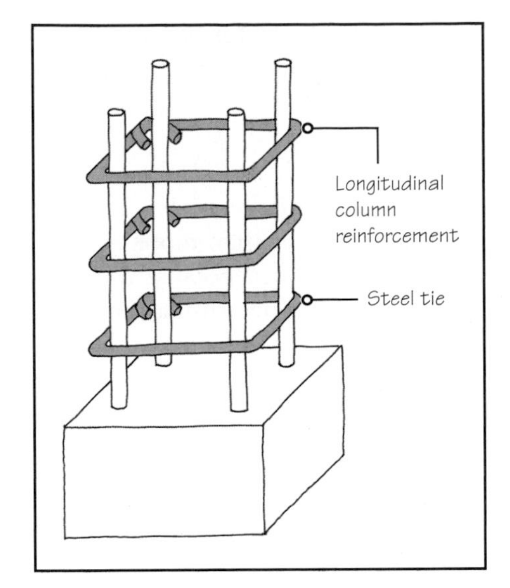

FIGURE 4.36 Horizontal ties in a reinforced concrete column to prevent the buckling of vertical reinforcement.

With this definition, the safety margin must be greater than 1.0. If we could ascertain the loads that a structure is required to support with perfect certainty, if we were absolutely certain of the strength of the material, if all materials gave sufficient warning before failure, if the strength of materials did not deteriorate with time, and if the structural design procedures used were without any ambiguity, then a safety margin slightly greater than 1.0 is all that would be needed.

The fact is, however, that none of these "ifs" can be ignored in practice. There are always uncertainties in the values of design loads used, and there is usually a great deal of variation in the strength of materials. For example, if several concrete cylinders were prepared from the same mix of concrete under identical conditions and were tested to failure using the same procedures and testing equipment, the results could vary by as much as 20%.

Additionally, materials deteriorate with time, and their deterioration is not the same under all conditions. Our design procedures are not exact and involve considerable approximation and generalization. Considering these uncertainties, it is necessary that the the safety margin be much greater than 1.0.

The greater the uncertainty, the greater the safety margin needed. In a material such as steel, which is manufactured under controlled conditions, the safety margin used is relatively low. Concrete, on the other hand, is virtually always a site-manufactured material. Its strength is dependent on placing, compacting, and curing, all of which occur on site. Therefore, the safety margin used for concrete is higher.

Similarly, because wood is a naturally grown material, there is considerable variation (uncertainty) in its strength for the same specie and grade. The safety margin required in designing with wood is even higher than that required for concrete.

The safety margin is also dependent on the type of stress present. Usually, the margin is lower for bending stress or axial tension than for axial compression, shear, bearing, or buckling. The reason is that failure of a member in axial compression, shearing, bearing, and buckling is relatively more abrupt than in bending or axial tension, requiring a higher safety margin.

A large safety margin leads to uneconomical design. A judicious compromise is, therefore, needed between adequate safety and economic considerations.

STRUCTURAL SAFETY—ALLOWABLE STRESS

There are two ways in which safety is incorporated in a structure. One way is to limit the stress in members caused by loads to a value that is less than its failure stress. This limiting stress is called the *allowable stress.* The stress in the material caused by loads must be less than or equal to the allowable stress. The ratio of the failure stress to the allowable stress is the safety margin, which is commonly called the *factor of safety:*

$$\text{Factor of safety} = \frac{\text{failure stress}}{\text{allowable stress}} \qquad \textbf{Eq. (5)}$$

For example, if the failure stress of a material is 2,000 psi and if a factor of safety of 2.0 is used, the allowable stress of that material is 2,000/2.0 = 1,000 psi.

STRUCTURAL SAFETY—LOAD FACTOR AND RESISTANCE FACTOR

An alternative method of incorporating safety in structures is to inflate the value of loads on the structure, that is, to design the structure for loads that are greater than those that will actually be present on the structure. This method is more sophisticated because the loads can be inflated based on the degree of uncertainty present in their determination. Load inflation is achieved by multiplying the loads with suitable multiplication factors, called the *load factors.*

Load factor is relatively small for dead loads, because the uncertainty in their determination is small. The load factor has the highest value for earthquake loads, because of a great deal of uncertainty in their determination.

Because there is some uncertainty in the determination of the strength of materials, the strength of the material is reduced by a factor called the *resistance factor.* Although the load factor is always greater than 1.0, the resistance factor is less than 1.0. This method of structural design is, therefore, referred to as the *load resistance factor method,* as opposed to the *allowable stress method* described previously.

PRACTICE QUIZ

Each question has only one correct answer. Select the choice that best answers the question.

21. The primary purpose of longitudinal steel bars in a reinforced concrete beam is to overcome the weakness of concrete in tension.
- **a.** True
- **b.** False

22. A material that is weak in tension is
- **a.** strong in shear.
- **b.** weak in shear.
- **c.** no relationship between the shear and tensile strengths of a material.

23. A material that is weak in tension is
- **a.** strong in bending.
- **b.** weak in bending.
- **c.** no relationship between the tensile and bending strengths of a material.

24. The stirrups in a reinforced concrete beam are used to
- **a.** increase the strength of beam in bending.
- **b.** reduce the deflection of the beam.
- **c.** increase the durability of the beam.
- **d.** increase the strength of beam in shear.
- **e.** none of the above.

25. The bearing strength of a material is closely related to its
- **a.** bending strength.
- **b.** tensile strength.
- **c.** compressive strength.
- **d.** modulus of elasticity.
- **e.** none of the above.

26. In a steel column that rests on a concrete footing, the area of the bearing plate used is generally
- **a.** greater than the cross-sectional area of the column.
- **b.** equal to the cross-sectional area of the column.
- **c.** smaller than the cross-sectional area of the column.

27. The primary purpose of ties in a reinforced concrete column is to
- **a.** prevent the buckling of vertical reinforcement in column.
- **b.** prevent the crushing of concrete.
- **c.** increase the shear resistance of the column.
- **d.** none of the above.

28. The horizontal blocking in wood studs helps prevent
- **a.** the compressive failure of studs.
- **b.** shear failure of studs.
- **c.** tensile failure of studs.
- **d.** buckling of studs.

29. The allowable stress in a material is generally
- **a.** equal to its ultimate stress.
- **b.** less than its ultimate stress.
- **c.** greater than its ultimate stress.
- **d.** none of the above.

30. In most materials, the safety margin for compressive stress and bending stress is the same.
- **a.** True
- **b.** False

Answers: 21-a, 22-b, 23-b, 24-d, 25-c, 26-a, 27-a, 28-d, 29-b, 30-b.

KEY TERMS AND CONCEPTS

Allowable stress	Buckling	Elastic-plastic material	Resistance factor
Bearing strength	Compressive strength	Load factor	Shear strength
Bending strength	Ductile material	Modulus of elasticity	Tensile strength
Brittle material	Elastic material	Plastic material	Yield strength

REVIEW QUESTIONS

1. What are the units of stress in the U.S. system of units and the SI system of units?
2. A 1-in.-diameter steel bar is subjected to a tensile force of 30 kips. Determine the stress in the bar.
3. If the bar in Problem 2 is 10 ft long, determine the elongation in the bar. The modulus of elasticity of steel is 29×10^6 psi.
4. Sketch the stress-strain diagram of low-carbon (mild) steel, showing all its important parts.
5. Explain the modulus of elasticity. Which property of the material does it represent? Give the approximate values of the modulus of elasticity of steel, concrete and wood.
6. Using sketches and notes, explain how we increase the bending efficiency of building components.
7. Explain why a brittle material is weak in shear.
8. Explain the bearing strength of a material.
9. Explain the concept of buckling. What measures do we adopt to prevent buckling failure of building components.
10. If the ultimate strength of a material is 6,000 psi and the factor of safety is 2.5, determine the allowable stress of the material.

CHAPTER 5 Thermal Properties of Materials

CHAPTER OUTLINE

5.1 Units of Energy

5.2 Conduction, Convection, and Radiation

5.3 R-Value of a Building Component

5.4 R-Value of a Multilayer Component

5.5 Surface Emissivity

5.6 U-Value of an Assembly

5.7 Where and How Much to Insulate

5.8 Thermal Capacity

5.9 The Most Effective Face of the Envelope for Insulation

PRINCIPLES IN PRACTICE: Insulating Materials

Insulated concrete forms (ICFs) are increasing in importance as a high-performance and sustainable construction system for residential and commercial buildings. The exposed forms in the upper photo are ready to receive the exterior wall cladding. The photo below shows the application of fiber cement siding as wall cladding. (Photos courtesy of American PolySteel.)

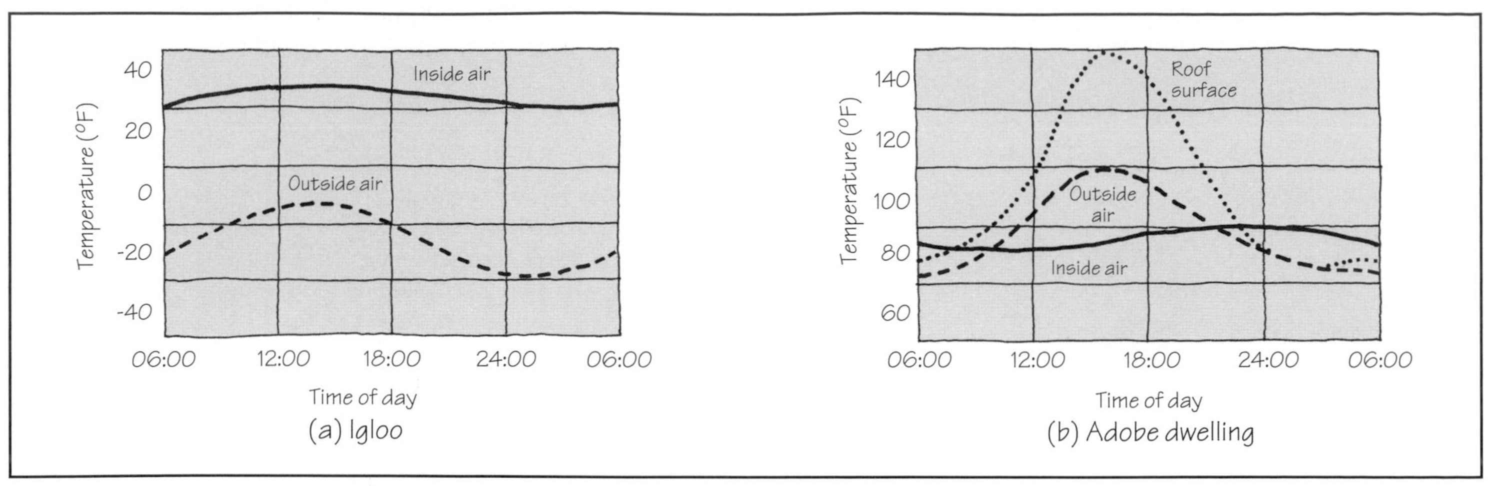

FIGURE 5.1 Diurnal variations of temperature in **(a)** an igloo and **(b)** an adobe dwelling.

In Chapters 3 and 4, we discussed loads on buildings and the structural properties of materials. The focus was on the structural system—the components that work together to support and give form to a building. The next several chapters focus on performance requirements and the characteristics of materials for the building's enclosure. (In practice, several different names are used for the building's enclosure, including the envelope or the skin. We refer to it in this text as the *building envelope.*)

The role of the building envelope is to separate the interior of the building from exterior conditions: heat and cold, wind, moisture, noise, and so on. It not only serves as the first line of defense, but it can mediate the conditions, reducing the need for mechanical services such as heating, cooling, and ventilation. In this chapter we discuss thermal properties of materials, followed by chapters on the properties related to air leakage and water vapor, fire resistance, and acoustics.

Providing thermal comfort in buildings has been an important design objective from the earliest times. The considerations of climate and manipulation of resources to produce a comfortable building have always had a profound influence on building/shelter design. The Eskimos' igloo and the adobe houses of the Middle East are some of the early examples of comfort-conscious designs in two extremes of climate.

Figure 5.1(a) and 5.1(b) shows diurnal (24-h) variations of temperatures inside a typical igloo and an adobe dwelling, along with the respective outdoor temperatures [1]. The relative stability of the indoor air temperature and the large difference between the indoor and outdoor temperatures in both the igloo and the adobe dwelling attest to the success of these designs in providing thermal comfort.

Water bodies in open courtyards of the hot, dry regions of northern India and wind scoops on roofs in the hot and humid climates of the Middle East are design solutions that provided cooling by manipulating the form and orientation of buildings to respond to the challenges of local site and climate, Figure 5.2.

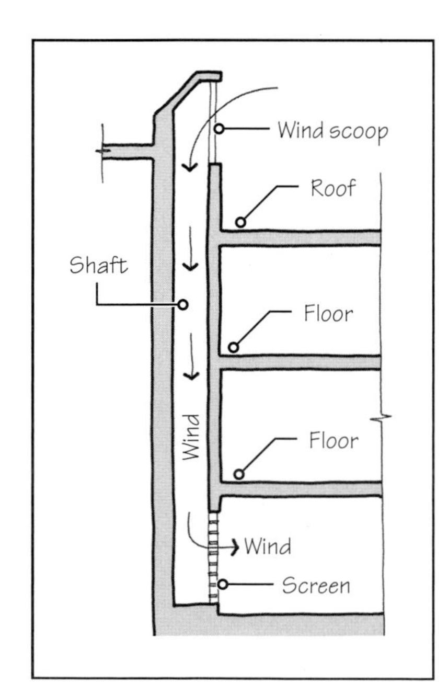

FIGURE 5.2 Section through a wind scoop showing the penetration of breeze from a vertical shaft into house interior. Wind scoops were used extensively in coastal areas of the Middle East.

Twentieth-century advances in heating and air-conditioning equipment made it possible to achieve interior comfort independent of building form and climate. Advances in materials and their mass production allowed designers to build larger spaces and taller buildings. These factors, in conjunction with modern ideas about design, resulted in buildings that conformed to specific design aesthetics rather than considerations of site and climate. Similar design solutions were used everywhere in the world, because the mechanical systems allowed any design solution to be adapted to any climate.

The resulting design freedom was fully exploited by architects to the extent that an all-glass exterior became a recognized facade formula, particularly for commercial buildings. Whereas the glass envelope was capable of resisting moisture, it tended to collect heat, increasing the requirements for cooling, even in moderate climates. The situation was acceptable because energy prices were low and the cost of operating buildings was small, but it changed dramatically in 1973 when the oil-producing and exporting (OPEC) countries placed an embargo on the sale of oil and quadrupled prices.

With the increase in energy prices, the cost of heating and cooling buildings became the single most important recurring liability. At the same time, came the realization that the earth's fossil fuel reserves were rapidly being depleted. Consequently, energy conservation became an important design objective, which led to a substantial reorientation of architectural thinking.

Consideration of thermal properties and thermal behavior of buildings assumed greater importance. The building industry reacted by producing materials and equipment that are

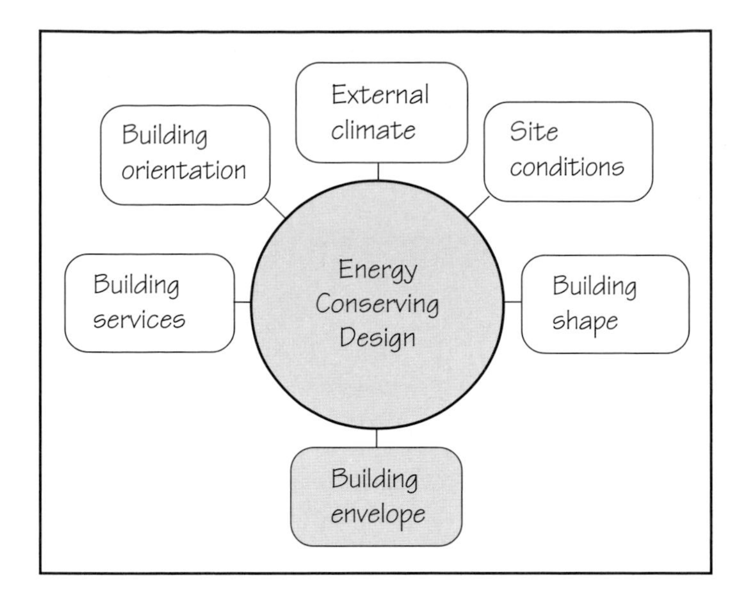

FIGURE 5.3 Factors that affect energy-conserving building design.

thermally more efficient, and the design community responded by insulating the building envelope against heat gain and heat loss. In fact, today's building envelope is a multilayered, often complex assembly of materials and components.

As shown in Figure 5.3, an energy-conserving building design is a function of several variables. It includes the consideration of the outside climate, site conditions, building orientation and shape, building services (air conditioning, water supply, lighting, and waste management), and thermal properties of the building envelope.

In this chapter, only the thermal properties of materials and assemblies that influence the design and performance of the building envelope are covered. A discussion of insulating materials is provided at the end of the chapter.

5.1 UNITS OF ENERGY

Until the middle of the nineteenth century, various forms of energy (such as the electrical energy obtained from electric generating stations, nuclear energy obtained from the splitting of atomic nuclei, and chemical energy obtained from the burning of coal, gasoline, and other combustible materials) were considered to be unrelated to each other. Therefore, different types of energy were measured in different units. Today, we know that all energy forms are interrelated by a fundamental property: a natural tendency among all forms of energy to be converted to heat. Thus, all energy forms can be measured by the amount of heat they produce. The *calorie,* (cal), a unit of heat, is, therefore, a convenient unit for the measurement of energy.

One calorie is defined as the amount of heat required to raise the temperature of 1 gram (g) of water by 1°C. In the U.S. system of units, the unit of heat is the British thermal unit (Btu), which is defined as the amount of heat required to raise the temperature of 1 lb of water by 1°F; 1 Btu is approximately equal to 252 cal.

Whereas the calorie and British thermal unit are used only with reference to heat, the unit that is used for all forms of energy (including heat energy) is the joule (J). A joule is a small unit (1 cal = 4.185 J). Therefore, kilojoules (kJ) and megajoules (MJ) are often used; 1 kJ = 1,000 J = 10^3 J, and 1 MJ = 10^6 J.

> **NOTE**
>
> **The Joule as a Unit of Energy**
>
> One joule is equal to the amount of work done by a force of 1 N (newton) in moving through a distance of 1 m.

ENERGY AND POWER

Power is defined as the rate of energy production, energy consumption or energy conversion. Therefore, the unit of power is joule per second (J/s), also called a watt (W). Because 1 J/s = 1 W, a 100-W electric lamp produces 100 J of energy (heat) per second. In 1 h, the same lamp produces or $(100 \times 60 \times 60) = 360 \times 10^3$ J/h = 360 kJ/h.

Again, the watt is a small unit. Therefore, the kilowatt (kW) is more commonly used (1 kW = 1,000 W). Because power and energy are related, we can switch back and forth between the two quantities as the situation demands. Because power is the rate of energy consumption, the total amount of energy consumed (E) in time (t) by a device whose power is P is given by

$$\text{Energy} = \text{power} \times \text{time} \quad \text{or} \quad E = P(t)$$

TABLE 5.1 RELATIONSHIP OF OTHER ENERGY UNITS TO THE JOULE	
Unit of energy	**Value in joules**
1 cal	4.185 J
1 kcal	4,185 J
1 Btu	1,050 J
1 kWh	$3.6(10)^6$ J

The kilowatt-hour (kWh), which we see in our utility bills, is the amount of energy consumed in 1 h by a device whose power is 1 kW. A 100-W lightbulb that is lit for 24 h consumes 24(100) = 2,400 Wh, or 2.4 kWh, of energy.

Electric companies charge their consumers for energy consumption in terms of the kilowatt-hour. For instance, if the price of electrical energy is 10¢/kWh, the consumer pays (10 × 2.4), or 24¢, for a 100-W lamp kept on for 24 h.

In other words, the kilowatt-hour is also an energy unit. The relationship between various commonly used energy units is given in Table 5.1.

5.2 CONDUCTION, CONVECTION, AND RADIATION

From our daily experience, we know that if two objects, which are at different temperatures, are placed in contact with each other, the temperature of the warmer object decreases and that of the cooler object increases. This phenomenon occurs because heat flows from an object at a higher temperature to that at a lower temperature, Figure 5.4. Heat transfer will continue until the temperature of two objects becomes equal. There are three modes by which heat transfer can take place: conduction, convection, and radiation.

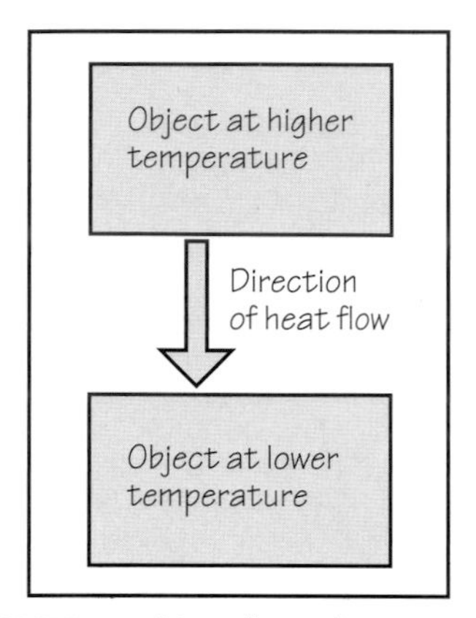

FIGURE 5.4 Heat flows from an object at higher temperature to an object at lower temperature.

CONDUCTION

Conduction is the phenomenon of heat transfer between the molecules of a substance that are in contact with each other. It is the only mode of heat transfer within solids and is characterized by physical contact between the molecules in the substance. When one end of a solid substance is heated, the molecules in the neighborhood of the heated end begin to vibrate about their mean positions with a higher average velocity. Because of the physical contact between the molecules of the substance, the increased vibrational energy in the heated molecules is transferred to adjacent cooler molecules, which in turn transfer their increased energy to still cooler molecules in their neighborhood, and so on.

CONVECTION

Convection occurs only in fluids (gases and liquids) and may be described as energy transfer by actual bulk motion of a gas (such as air) or liquid (such as water). A familiar example of convection is heating water on a stove. When water in a container is placed on a stove, the water particles nearest the burner become lighter in density as they are heated. Being lighter, these particles rise to the top and are replaced by colder (and hence heavier) particles of water from above. The colder particles are heated again and rise to the top, and the cycle continues as long as the water is being heated. This process creates motion within the fluid.

In addition to the transfer of energy by bulk motion, which creates fluid flow, a small amount of energy is also transferred by conduction in fluids because of the physical contact between fluid particles. However, the dominant mode of heat transfer in fluids is by convection, that is, the movement of fluid particles.

The process of heat transfer described by the heating of water on a stove is called *natural convection.* Natural convection results from density changes in the fluid. Convection may also occur without changes in the density of fluid, in which case it is called *forced convection.* Forced convection is produced by fans, pumps, and wind motion.

RADIATION

Radiation is the transfer of energy between two objects in the form of electromagnetic waves. Thus, the heat and light received from the sun, from a burning candle, or from an electric lamp reach us by electromagnetic radiation. Unlike conduction and convection, radiation does not require a medium to travel. This property allows radiation from the sun to reach the earth through the vast realm of space.

It is important to appreciate that heat radiation is a form of electromagnetic radiation, and all electromagnetic radiation is transformed to heat when absorbed by an object. Electromagnetic radiation can exist in a variety of wavelengths, each with different characteristics. Some of the segments of electromagnetic radiation are infrared, visible (light), ultraviolet, X-rays, and gamma rays.

Thus, although the radiation from the sun consists of ultraviolet, visible, and infrared radiations, all three components of solar radiation are converted to heat when they are absorbed by a building surface. Similarly, all radiation from an electric lamp in a room is

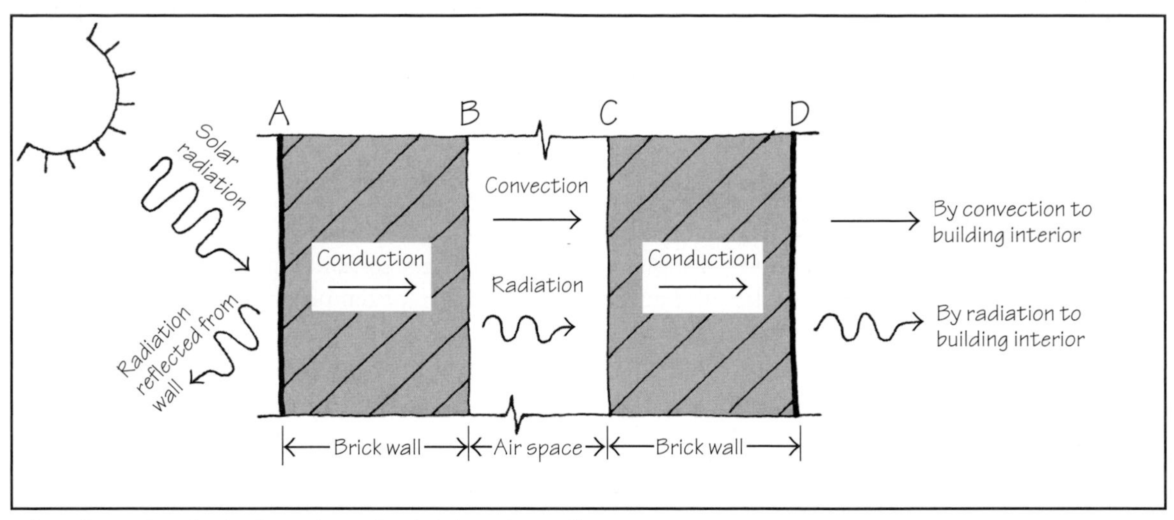

FIGURE 5.5 Modes of transfer of solar heat through a brick cavity wall.

converted to heat as it gets absorbed by the occupants, furniture, room surfaces, and room air.

Simultaneity of Conduction, Convection, and Radiation

During the process of heat transfer through a building envelope, all three modes usually come into play simultaneously. Thus, to reduce heat transfer through the envelope, all three modes must be considered, and the heat transfer potential of each mode must be reduced.

Consider the section through the brick cavity wall shown in Figure 5.5. Energy from the sun reaches the wall by radiation and is partly absorbed and partly reflected by the wall's exterior surface. The absorbed energy then travels by conduction through the wall (from surface A to surface B) until it reaches the air space.

Inside the air space (from surface B to surface C), the heat-transfer mode changes to radiation and convection. The subsequent mode is conduction through the inner layer of the wall (from surface C to surface D). Finally, the energy is transferred from the inner surface of the wall (surface D) to indoor air, room surfaces, furniture and occupants by convection and radiation.

5.3 R-VALUE OF A BUILDING COMPONENT

The term R is called thermal resistance, or simply the R-value. It is the measure of the ability of a component to resist the flow of heat through it. The R-value is related to heat flow primarily by conduction.

To understand the concept, consider a rectangular plate (such as a wall or roof) of thickness L, as shown in Figure 5.6. Let the surface temperatures of the two faces of the plate be

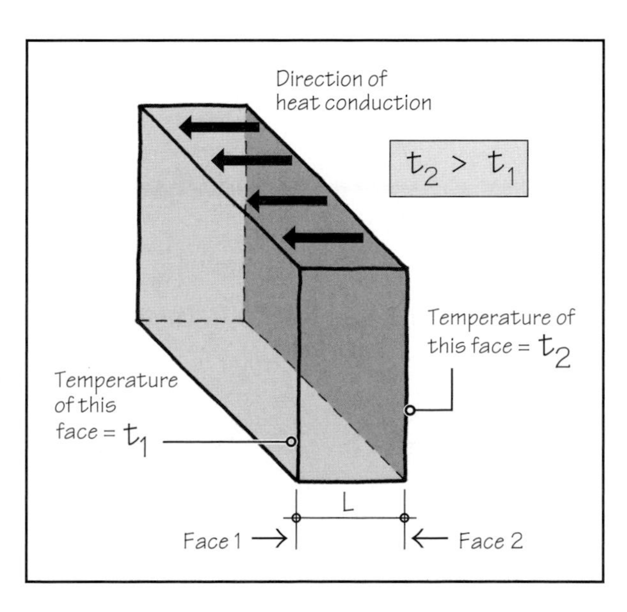

FIGURE 5.6 Heat transfer by conduction through a plate of thickness L.

(a)

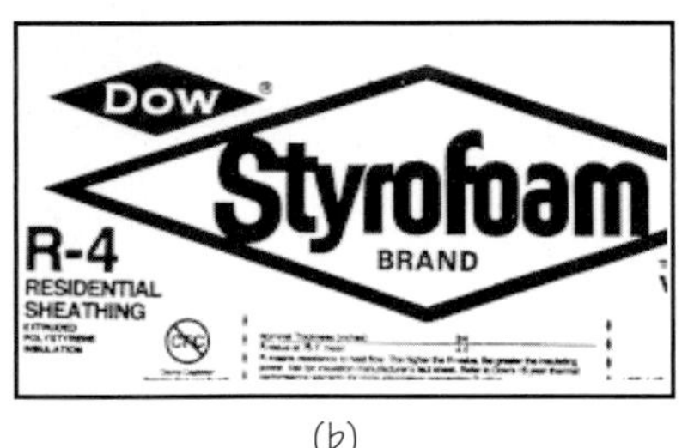

(b)

FIGURE 5.7 **(a)** R-11 fiberglass insulation in batt form. **(b)** R-4 extruded polystyrene insulation in board form.

TABLE 5.2 UNITS OF R-VALUE AND ρ-VALUE

U.S. System of units	
R-value	$\frac{ft^2\text{-}°F\text{-}h}{Btu}$
ρ-value	$\frac{ft^2\text{-}°F\text{-}h}{Btu\text{-}in.}$
SI System of units	
R-value	$\frac{m^2 \cdot °C}{W}$
ρ-value	$\frac{m \cdot °C}{W}$

To convert an R-value to an RSI-value, multiply the R-value by 0.176.

t_1 and t_2. If t_2 is greater than t_1, heat will travel from face 2 to face 1. Experiments have shown that the rate of heat transfer by conduction from face 2 to face 1 is given by:

$$q_c = \frac{(t_2 - t_1)}{\rho L} \qquad \textbf{Eq. (1)}$$

where q_c is the rate of heat conduction in Btu/h through 1 ft^2 of the plate (Btu/(h-ft^2).

Equation (1) shows that the rate of heat conduction through the plate is directly proportional to the difference between the surface temperatures of the two faces of the plate. If there is no difference between the surface temperatures—that is, if $(t_2 - t_1)$ is zero—no heat conduction will occur. Additionally, the rate of heat conduction is inversely proportional to the thickness, L, of the plate. This means that the greater the thickness of the plate, the smaller the amount of heat conducted through it.

The term ρ in Equation (1) is called *thermal resistivity.* Thermal resistivity is a property of the material and is, therefore, constant for the material. If the value of ρ is large, the rate of heat conduction through the plate is small. Such a material is called a *thermal insulator,* or an *insulating material.* Conversely, if the value of ρ is small, the rate of heat conduction through the plate is large. Such a material is called a *thermal conductor.*

If we now replace ρL in Equation (1) by R, it becomes:

$$q_c = \frac{(t_2 - t_1)}{R} \qquad \textbf{Eq. (2)}$$

Again, the term R refers to thermal resistance, and a component with a high R-value is more insulating (resistant to the flow of heat) than a material with a low R-value. It is the R-value that is generally quoted in referring to the insulating value of a component or assembly. Thus, we speak of the R-value of a component, such as R-10 or R-11. Most insulating products are labeled with their R-values, Figure 5.7.

From Equation (2), the units of R-value are (ft^2-°F-h)/Btu. The units of ρ are [(ft^2-°F-h)/(Btu-in.)], Table 5.2. Because both of these units are rather long and complex, we usually omit the units in practice—a practice that we will follow in this chapter.

When the R-value of a component is given in the SI system of units, it is usually referred to as the RSI-value. Thus, if the R-value of a component is R-10, its RSI-value is 1.76, which is referred to as RSI-1.76. Table 5.2 also gives the conversion factors for R-values and ρ-values.

DIFFERENCE BETWEEN R-VALUES AND ρ-VALUES

It is important to note again that the ρ-value is a property of the material, whereas the R-value is a property of the component of certain thickness. Because they are related to each other by the relationship $R = \rho L$, where L is the thickness of the component, the ρ-value is, in fact, the R-value of a component whose thickness is 1 in. Thus, if we know the ρ-value of a material, we can determine the R-value of a component of any thickness. For example, if the ρ-value of a material is 2.5, then the R-value of a 6-in.-thick component made of that material is $2.5 \times 6 = 15.0$; that is, the R-value of this component is 15. The R-value of a component is generally rounded to the nearest whole number (integer).

Because the ρ-value is a material property, it is the ρ-value that is used for comparing the insulating effectiveness of materials. Again, a material with a higher ρ-value is a better thermal insulator. The ρ-values of some commonly used materials are given in Table 5.3.

Example 1
(R-Value From ρ-Value)

Determine the R-value of (a) a 6-in.-thick concrete wall, (b) a 6-in.-thick fiberglass blanket, and (c) a 0.25-in.-thick glass sheet.

Solution

From Table 5.3, ρ-values of concrete, fiberglass, and glass sheet are 0.15, 3.5, and 0.14, respectively.

a. For a 6-in.-thick (normal weight) concrete wall, $R = \rho L = (0.15)(6) = 0.9$.
b. For a 6-in.-thick fiberglass blanket, $R = \rho L = (3.5)(6) = 21.0$.
c. For a 0.25-in.-thick glass sheet, $R = \rho L = (0.14)(0.25) = 0.035$.

Note that although concrete wall and fiberglass blanket are both of the same thickness, the fiberglass blanket is nearly 21.0/0.9 = 23 times more effective as a thermal insulator than concrete. In other words, it will take a nearly 23 × 6 = 138-in.-thick concrete wall to provide the same thermal insulation as a 6-in.-thick fiberglass blanket. Also note that the R-value of a sheet of glass is virtually zero.

EFFECT OF DENSITY AND MOISTURE CONTENT ON THE ρ-VALUE

As a general rule, a low-density material has a high ρ-value. Being fibrous, granular, or cellular in structure, a low-density material has air trapped within its voids. Because air has a high ρ-value, a low-density material is a better insulator than a high-density material. In fact, still air is one of the best thermal insulators available. Circulating air cannot be a good insulator because the circulation of air increases convective heat flow, thereby reducing the ρ-value.

In other words, the basic requirement of an insulating material is that the solid constituents of the material must divide its volume into such tiny air spaces that air cannot circulate within them. This characteristics ensures that convective heat transfer is virtually negligible. It also implies that all insulating materials are low-density materials.

Another important parameter that affects the ρ-value is the moisture content of the material. Water has a much lower ρ-value than air (ρ-value of air = 5.6 and ρ-value of water = 0.24, Table 5.3). Because a moist material has water trapped inside its pores in place of the air, its ρ-value is lower than that of a dry sample.

TABLE 5.3 ρ-VALUES OF COMMONLY USED BUILDING MATERIALS

Material	Resistivity (ρ-value) (ft^2-°F-h)/(Btu-in.)
Metals	
Steel	$0.0032 = 3.2 \times 10^{-3}$
Copper	3.2×10^{-4}
Aluminum	3.2×10^{-3}
Ceramic materials	
Clay bricks	0.20
Concrete (normal weight)	0.15
Concrete (structural lightweight)	0.25–0.35 (depending on density)
Insulating concrete (perlite or vermiculite)	1.70
Concrete masonry	Depends on type of concrete and cell insulation
Limestone	0.15
Sandstone	0.18
Glass	0.14
Plaster	0.35
Gypsum wallboard	0.60
Portland cement plaster	0.30
Wood and coal	
Softwoods (solid lumber or plywood)	0.9
Fiberboard	2.4
Wood charcoal	2.2
Coal	0.85
Insulating materials	
Granulated cork	3.0
Vermiculite (loose **fill**)	2.1
Perlite (loose fill, density 5.0 pcf)	3.0
Perlite (loose fill, density 10.0 pcf)	2.4
Expanded perlite board	2.8
Fiberglass (loose fill, assuming all voids are filled)	3.5
Fiberglass (batt or blanket)	3.5
Mineral wool (rock wool)	3.5
Expanded polystyrene (EPS) board (bead board)	4.0
Extruded polystyrene (XEPS) board	5.0
Polyurethane board (laminations on both sides)	6.5
Polyisocyanurate board (laminations on both sides)	6.5
Foamed-in-place polyicynene	3.6
Gases	
Air	5.6
Argon	8.9
Carbon dioxide	9.9
Chloro-fluoro-carbon (CFC) gas	16.5
Hydro-chloro-fluoro-carbon (HCFC) gas	15.0
Water	0.24

ρ-value = R-value of 1-in.-thick material

To convert a ρ-value to a ρSI-value:	**Multiply the ρ-value by 6.93.**
To convert an R-value to an RSI-value	**Multiply the R-value by 0.176.**

Values given in this table are representative values. Consult manufacturers' data for precise values.

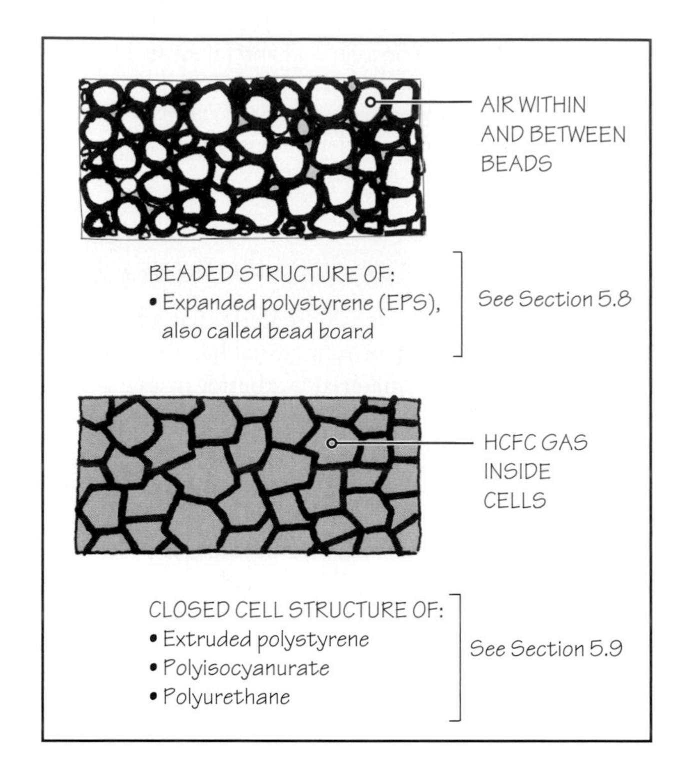

FIGURE 5.8 Cellular structure of various plastic foam insulations.

It is, therefore, important to ensure, in the detailing of buildings, that insulating materials remain dry throughout their useful life spans. This fact is important because most insulating materials, because of their location, are particularly vulnerable to water absorption from rain. In cold climates where condensation of water vapor within insulating materials is an additional hazard, vapor retarders are installed in roof and exterior wall assemblies, as discussed in Section 6.7.

Entrapment of HCFC Gas Instead of Air

Although most insulating materials contain small volumes of air to provide high insulating value, some plastic foam insulations (or, simply, *foamed plastics,* discussed at the end of the chapter) contain a gas that is more insulating than air. The gas commonly used is hydro-chloro-fluoro-carbon (HCFC) gas, which is nearly 2.5 times more effective than air as an insulation (see Table 5.3).*

Foamed plastics that entrap HCFC are manufactured by using HCFC gas as a foaming agent in a liquid polymer. From Table 5.3, the ρ-value of bead board (a foamed plastic with air inside the beads, also called *EPS board*) is 4.0, whereas the ρ-value of polyisocyanurate board, which contains HCFC, is 6.5.

Note that plastic foam insulations that trap HCFC have a closed cellular structure to retain the gas within the material, Figure 5.8. Despite their cellular structure, some of the trapped HCFC gas finds its way into the atmosphere. Therefore, air is able to permeate into the cell cavities of these foams, driving some HCFC gas out from the edges and ends of a board.

The steady outward migration of HCFC gas means that the R-value of such plastic foams decreases with time. Therefore, in referring to the R-value of a plastic foam with HCFC gas, we generally quote its *stabilized,* or *aged,* R-value.† The stabilized R-value is lower than the R-value immediately after the plastic foam's manufacture. In most cases, the R-value of such a material stabilizes after 6 months, after which the decrease in R-value is almost negligible.

Practice Quiz

Each question has only one correct answer. Select the choice that best answers the question.

1. Apart from the climate of the location, the other factors that affect energy-conscious building design are
   a. building orientation and building shape.
   b. building envelope.
   c. site conditions.
   d. all the above.
   e. (a) and (b).
2. All forms of energy eventually convert to a magnetic field.
   a. True
   b. False

*The gas used earlier as foaming agent was chloro-fluoro-carbon (CFC) gas. However, it was found to deplete the ozone layer in the upper atmosphere of the earth (see Chapter 10). The HCFC gas is less destructive of the ozone layer, and hence it has replaced CFC gas. The insulating potential of HCFC is only slightly lower than that of CFC.

†The Polyisocyanyrate Manufacturers Association (PIMA) refers to the stabilized R-value as long-term thermal resistance (LTTR).

3. The unit for the measure of energy is
   a. calorie. b. joule.
   c. Btu. d. watt-hour.
   e. all the above.

4. The unit for the measure of energy in the SI system is
   a. calorie. b. joule.
   c. Btu. d. watt.

5. The unit of power is
   a. calorie. b. joule.
   c. Btu. d. watt.

6. Within a solid, the mode of heat transfer is by
   a. conduction.
   b. convection.
   c. radiation.
   d. both (a) and (b).
   e. both (b) and (c).

7. Within an air space (cavity), the mode of heat transfer is primarily by
   a. conduction.
   b. convection.
   c. radiation.
   d. both (a) and (b).
   e. both (b) and (c).

8. Air is a good thermal insulator provided it exists in large volumes.
   a. True b. False

9. The R-value of a 2-in.-thick material is 4.0. What is the R-value of the same material if its thickness is 1 in.?
   a. 1.0 b. 2.0
   c. 6.0 d. 8.0
   e. None of the above

10. The R-value of a building assembly in the SI system is referred to as the
    a. R-value (SI). b. SI-value.
    c. RS-value. d. RSI-value.
    e. none of the above.

11. A wet or moist material provides a higher R-value than if the same material is dry.
    a. True b. False

Answers: 1-d, 2-b, 3-e, 4-b, 5-d, 6-a, 7-e, 8-b, 9-b, 10-d, 11-b.

5.4 R-VALUE OF A MULTILAYER COMPONENT

Usually a wall or roof assembly consists of several layers of different materials, as shown in Figure 5.9. For example, an insulated brick wall consists of three layers of materials: a layer of brick, a layer of insulation, and a second layer of brick, Figure 5.10. It can be shown mathematically that the total resistance of a multilayer component, shown in Figure 5.9, is obtained by adding the R-values of individual layers:

$$R_t = R_1 + R_2 + R_3 + R_4 + \cdots \qquad \textbf{Eq. (3)}$$

where R_t is the total resistance of the multilayer component, and R_1, R_2, R_3, . . . , are the resistances of layer 1, layer 2, layer 3, and so on, respectively. Note that Equation (3) is valid only if there is no thermal bridging in the assembly (see Example 5 and the box entitled Thermal Bridging later in the chapter).

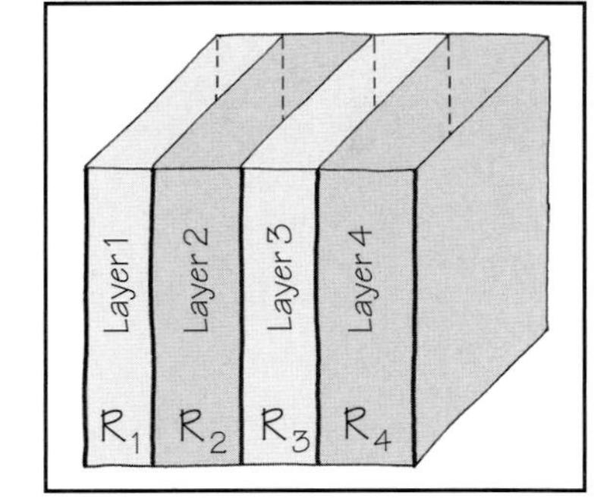

FIGURE 5.9 A multilayer assembly.

Example 2
(Estimating the R-Value of an Assembly)

Calculate the R-value of a wall assembly, which consists of a 2-in.-thick extruded polystyrene (XEPS) insulation sandwiched between an 8-in.-thick (nominal) brick wall and a 4-in.-thick (nominal) brick wall, as shown in Figure 5.10.

Note: The actual thickness of an 8-in. nominal brick wall is $7\frac{5}{8}$ in., and that of a 4-in.-thick brick wall is $3\frac{5}{8}$ in. (see Section 22.5).

Solution

From Table 5.3, the ρ-values of brick wall and extruded polystyrene insulations are 0.2 and 5.0, respectively. The total R-value of the assembly is determined as follows:

Element	R-value
4-in.-thick brick wall	3.625(0.2) = 0.725
2-in.-thick polystyrene board	2.0(5.0) = 10.0
8-in.-thick brick wall	7.625(0.2) = 1.525

$$R_t = 0.725 + 10.0 + 1.525 = 12.25$$

Because an R-value is given by the nearest whole number, the wall is an R-12 wall.

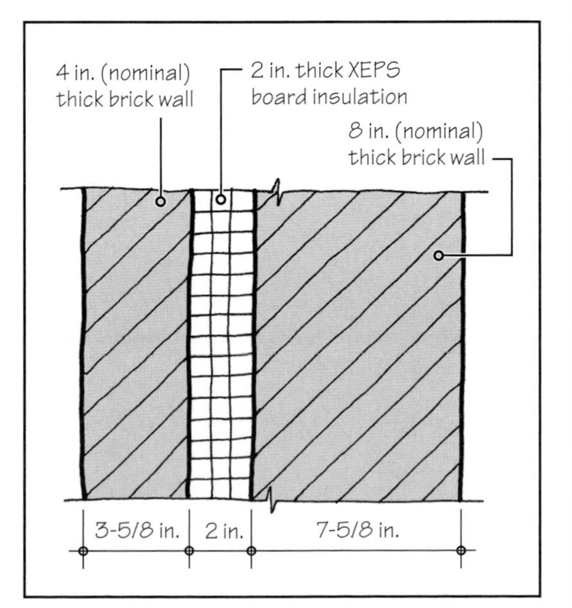

FIGURE 5.10 A brick wall assembly with sandwiched insulation.

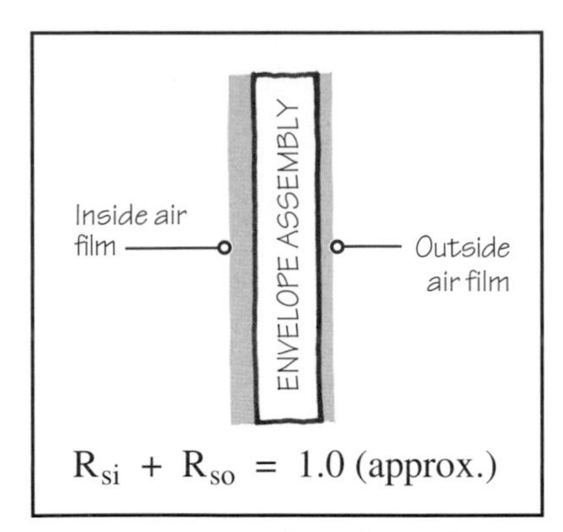

FIGURE 5.11 Inside and outside air films on a building envelope assembly.

NOTE

R-Value of a Single Sheet of Glass

The R-value of a single sheet of glass is R-1, which is due mainly to the internal and external film resistances.

NOTE

R-Value of a Wall Cavity

The R-value of a vertical (wall) cavity is approximately equal to R-1.

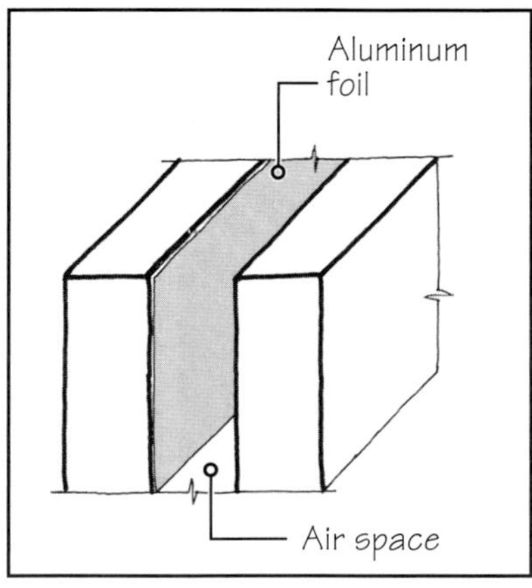

FIGURE 5.12 Radiation heat transfer across an air space is reduced if one surface of the air space is lined with a bright metal surface, such as aluminum foil. Thus, the R-value of an (aluminum) foil-lined air space is greater than that of an unlined air space.

SURFACE RESISTANCES

In addition to the layers of solid materials comprising an assembly, there are two invisible layers that contribute to its total R-value. These layers are thin films of air of nearly zero velocity that cling to the inside and outside surfaces of an assembly, Figure 5.11. Because each air film has a certain thickness, it increases the total R-value of the assembly.

The R-value of each film is called the *film resistance,* or, more commonly, the *surface resistance.* Thus, there is an inside surface resistance, R_{si}, and an outside surface resistance, R_{so}, corresponding to the inside and outside air films, respectively. Because the surface resistances are always present, Equation (3) is modified as follows:

$$R_t = R_{si} + R_1 + R_2 + R_3 + R_4 + \cdots + R_{so} \quad \textbf{Eq. (4)}$$

It is the total resistance obtained by including the two surface resistances that is responsible for *air-to-air heat transfer* through an envelope assembly. Because the inside air velocity is usually smaller than the outside air velocity, the inside film is thicker than the outside film. Hence, R_{si} is usually greater than R_{so}. In fact, R_{si} and R_{so} do not depend on the air velocity alone but also on the texture of the surface and the direction of heat flow through the assembly. For instance, an increase in surface roughness increases the surface resistance. This is because a rougher surface creates a greater drag on air movement, reducing air velocity near the surface and hence increasing the film thickness.

Although the values of R_{si} and R_{so} vary as stated, a good approximation of R_{si} is 0.7, and that of R_{so} is 0.2 (in the U.S. system of units). Thus, $R_{si} + R_{so}$ may be taken as approximately equal to 0.9, or, say, 1.0.

Because both surface resistances provide a total R-value of only 1.0, they do not significantly add to the total R-value of a typical roof or wall assembly (see Example 3 later in this chapter). However, they contribute greatly to the R-value of a glazed window or a roof skylight, because a single sheet of glass has an extremely small R-value—virtually zero. See Example 1. In fact, the total R-value of a single sheet of glass is 1.0, which is provided almost entirely by the inside and outside surface resistances.

5.5 SURFACE EMISSIVITY

So far we have discussed conductive heat transfer—the dominant mode in most building components. The other two modes, convection and radiation, occur only in air spaces that exist in wall and roof assemblies and attics. Theoretically, the concept of R-value is applicable only to solid components in which conduction is the only mode of heat transfer. However, the concept is extended to air spaces also, and an air space is assigned an R-value just like a solid component, although the primary modes of heat transfer through an air space are convection and radiation.

The R-value of an air space varies with its width, orientation (whether it is a horizontal or a vertical air space), and the direction of heat flow. The change in the R-value of an air space caused by a change in its width is a complex function of conduction, convection, and radiation heat transfers within the space.

For a vertical air space (wall cavity), the R-value increases with an increase in the width of air space, up to a value of 1.0. This value applies to an air space, which is $\frac{3}{4}$ in. wide. Beyond a width of $\frac{3}{4}$ in., the R-value of air space does not increase. Therefore, the R-value of a vertical air space is generally assumed to be 1.0. This takes into account the two film resistances within the cavity and both convection and radiation heat transfers.

Note that radiation heat transfer is unaffected by the width of the cavity. As long as there is a space between the two opposite surfaces of a cavity, radiation heat transfer will occur between them. However, radiation heat transfer can be altered (reduced) by including a layer of metal foil within the air space, Figure 5.12. This is generally achieved by laminating an aluminum foil to one of the two surfaces of the air space, referred to as *lining the air space.*

R-VALUE OF A FOIL-LINED VERTICAL AIR SPACE

Aluminum foil reduces radiation heat transfer due to its low *emissivity.* Emissivity is a property of the surface of an object that refers to its potential to emit radiation. In fact, the magnitude of radiation emitted by an object is directly related to its emissivity and its temperature.

The effect of temperature on the radiation emitted by an object is well known. If we place an iron rod inside a fire for a short while and then withdraw it, we can feel the heat emitted by the rod. If we place the same rod in the fire for a long time and then withdraw it, we see that it emits more heat. The increase in radiated heat is due to the rise in the temperature of the rod.

The effect of the emissivity of the object on the amount of radiation emitted by it is not as readily observed. However, if we take radiation measurements on a heated object, we will observe that it emits a certain amount of heat. If the same object is covered all around with aluminum foil, we will observe that the heat emitted by it is much smaller, Figure 5.13. Restaurants commonly keep a baked potato wrapped in aluminum foil so that it will emit less heat and thereby cool more slowly.* In this case, the foil reduces the emission of heat. (Note that the term *emission* is used only with radiation, not with conduction or convection.)

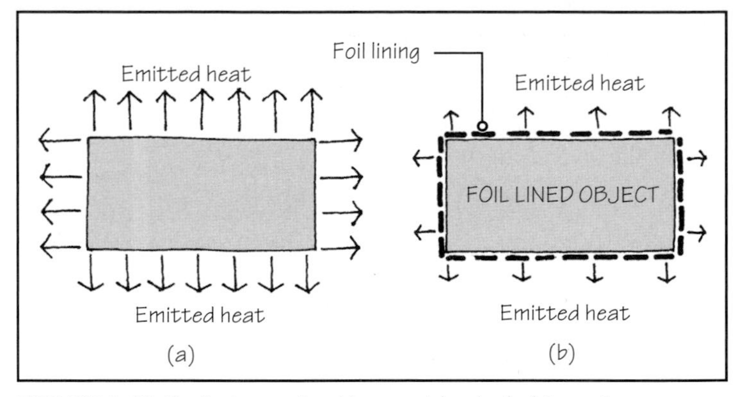

FIGURE 5.13 Radiation emitted by two identical objects that are at the same temperature. The unlined object **(a)** emits more radiation than the same object **(b)**, which is lined (wrapped) with aluminum foil.

The emissivity of a material lies between 0 and 1. Most building materials (brick, concrete, wood, plaster, gypsum board, steel, etc.) have a high emissivity—approximately 0.9. Polished metals, on the other hand, have a low emissivity. A highly polished metal, such as stainless steel or aluminum, has an emissivity of only 0.05. This means that if an object made of brick, concrete, wood, and the like, is covered with aluminum foil, the heat emitted by it will be (0.05/0.90) = 0.06 (i.e., 6%) of the heat emitted by the same object without the foil covering.

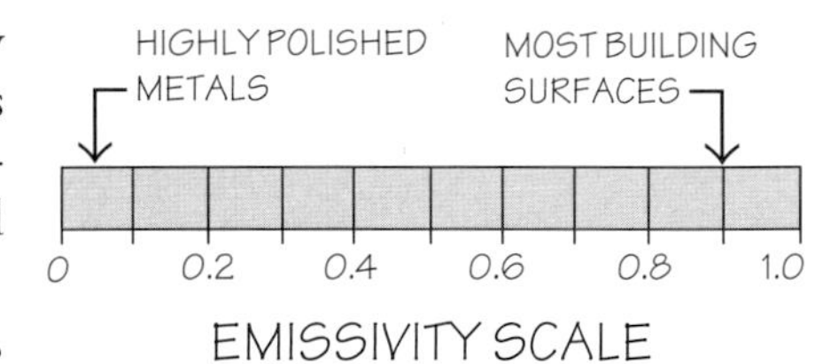

Note that emissivity is a property of the surface, not of the bulk of the material. Thus, changing the surface characteristic of an object changes its emissivity. For instance, plywood's emissivity is nearly 0.9, but if that plywood is laminated with aluminum foil, its emissivity becomes only 0.05. Aluminum foil is the metal that is commonly used in building assemblies because of its relatively low cost.

Lining a surface of the air space with foil implies that the heat emitted by that surface is small; hence, the amount of heat transferred to the opposite surface is small. This increases the R-value of the air space. Thus, the R-value of a typical foil-lined air space in a wall is approximately 2.5 (an increase of 1.5 over that of an unlined air space).

NOTE

R-Value of a Foil-Lined Wall Cavity

The R-value of a vertical (wall) cavity lined with aluminum foil is approximately equal to R-2.5.

Location of Low-Emissivity Material (Aluminum Foil) in an Air Space

A fundamental fact related to emissivity is that *the emissivity of a surface is always equal to its absorptivity*.† Because most building materials have an emissivity of 0.9, their absorptivity is also 0.9, which means that they absorb 90% of radiation falling on them. A foil-lined surface, on the other hand, absorbs only 5% of the radiation falling on it, because its absorptivity (or emissivity) is 0.05. Its reflectivity is, therefore, 0.95; that is, it reflects 95% of the radiation falling on it. That is why aluminum foil is also referred to as a *reflective insulation,* or a *radiant barrier.*

The implication of this fact is that the location of the foil in an air space—on the warmer or the cooler side of the space—has a negligible effect on the R-value of the air space. Thus, if the warmer (left-hand) surface of air space is lined, the radiation emitted by it toward the right-hand surface is small, Figure 5.14. Therefore, the heat flow across the air space is small.

If the cooler (right-hand) surface of air space is lined, it receives a much larger amount of radiation from the left-hand surface. However, because a low-emissivity material is a good reflector of radiation, the right-hand surface reflects most of the radiation back to the left-hand surface, Figure 5.15. Once again, the heat flow across the air space is small.

NOTE

R-Value of a Horizontal Cavity

The R-value of an unlined horizontal cavity, such as an attic, is approximately 2.5.

The R-value of a horizontal cavity lined with an aluminum foil is approximately 6.5.

R-Value of a Horizontal Air Space (Attic Space)

The advantage of using aluminum foil is more pronounced in an attic than in a vertical air space (wall cavity). The reason is that radiation heat transfer is much more pronounced in an attic than in a vertical air space. Although the R-value of an attic varies from summer to winter, a typical unlined attic is assumed as R-2.5, and if the same attic is lined with aluminum foil, its R-value is approximately 6.5.

*The aluminum foil also helps retain the moisture within the potato.

†Emissivity and absorptivity of a surface are equal if the radiation falling on the object is approximately of the same wavelength as the radiation emitted by it. The wavelength of radiation emitted by an object is a function of its temperature. Because the temperature range of building surfaces is small, the equality of emissivity and absorptivity of a surface is true for all practical purposes.

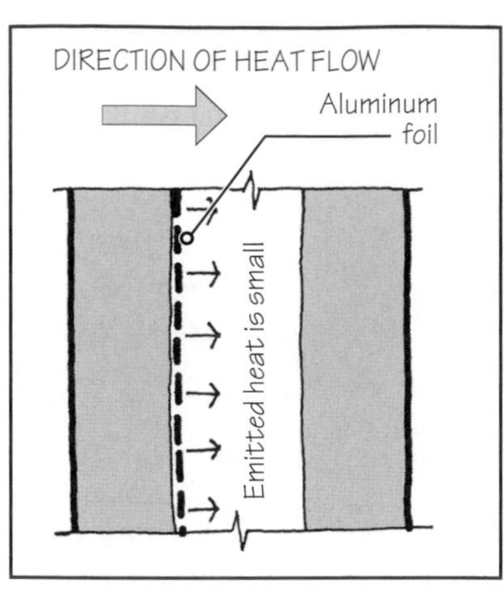

FIGURE 5.14 Radiation heat transfer through a foil-lined air space is approximately the same regardless of whether the foil is on the warmer side (as shown here) or on the colder side of the air space, as shown in Figure 5.15.

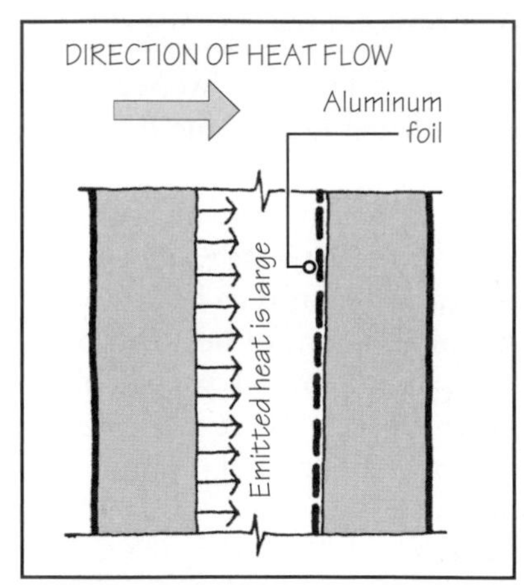

FIGURE 5.15 Radiation heat transfer through a foil-lined air space is approximately the same regardless of whether the foil is on the colder side of the air space, as shown here or on the warmer side as shown in Figure 5.14.

SURFACE EMISSIVITY AND SURFACE COLOR

In asserting that the emissivity of a building component governs the absorption or reflection of radiation that falls on it, it must be emphasized that the radiation referred to must be from a low-temperature source, such as an opposing building component or the ground. If the source of radiation is a high-temperature source, such as the sun, it is not the emissivity of the building component that governs the absorption or reflection of radiation, but the color of the building component. A light-colored component absorbs less solar radiation than a dark-colored component.

In other words, if the absorption of solar radiation by a building surface is to be reduced, the building surface should be painted with a light color, preferably white. The use of white building-envelope components in the tropics, for example, reduces the amount of solar radiation absorbed by the building.

Example 3
(R-Value of a Wall With an Air Space)

Determine the R-value of a wall consisting of 2 × 4 (actual dimensions $1\frac{1}{2}$ in. × $3\frac{1}{2}$ in.) wood studs spaced 16 in. on center. The spaces between the studs are filled with $3\frac{1}{2}$ in.-thick fiberglass insulation. Other members of the wall are shown in Figure 5.16.

Solution

The R-values of various layers of the assembly are tabulated below.

Element	R-value
Inside surface resistance	0.7
1/2 in. thick gypsum board	0.5(0.60) = 0.30
3-1/2 in. thick fiberglass insulation	3.5(3.5) = 12.25
0.5 in. thick plywood	0.5(0.9) = 0.45
2 in. wide air space	1.0
$3\frac{5}{8}$ in. thick brick veneer	3.625(0.2) = 0.73
Outside surface resistance	0.2

$$R_t = 0.7 + 0.3 + 12.25 + 0.45 + 1.0 + 0.73 + 0.2 = 15.6$$

In other words, this wall assembly is R-16 assembly.

Note that in the preceding calculations, we have ignored the presence of wood studs and assumed that the space occupied by the studs is occupied by fiberglass insulation. Because the R-value of wood studs is lower than that of fiberglass, the actual R-value of the assembly is, in fact, less than 15.6. A more precise method of determining the R-value of a component in which there are two or more materials in the same layer is explained in Example 5.

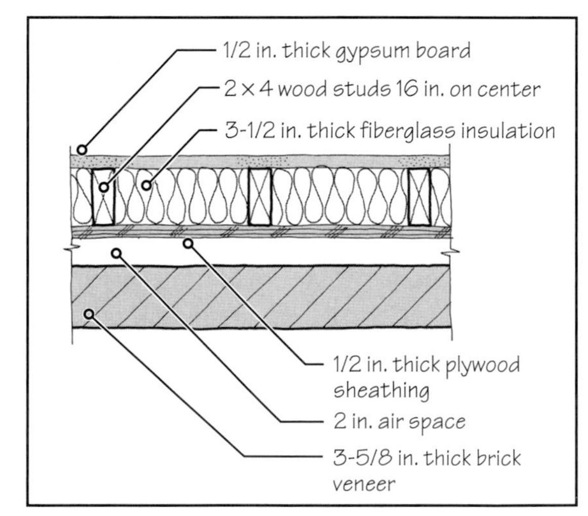

FIGURE 5.16 A brick veneer–wood stud cavity wall (in plan).

Each question has only one correct answer. Select the choice that best answers the question.

12. The R-value of a single glass sheet, without considering the film resistances, is approximately
a. 4.0. **b.** 3.0.
c. 2.0. **d.** 1.0.
e. 0.

13. The R-value of a single glass sheet, including the film resistances, is nearly
a. 4.0. **b.** 3.0.
c. 2.0. **d.** 1.0.
e. 0.

14. One of the main reasons for using HCFC gas in some insulating materials is that it
a. makes the material noncombustible.
b. makes the material more economical.
c. makes the material more resistant to decay.
d. increases the strength of the material.
e. none of the above.

15. Insulating materials that contain HCFC gas within their cells have a higher R-value than those that contain air within their cells.
a. True **b.** False

16. Which of the following insulating materials contain HCFC gas within their voids?
a. Extruded polystyrene
b. Polyisocyanurate
c. Polyurethane
d. All the above
e. Only (b) and (c)

17. For the same thickness, which of the following materials has the highest R-value?
a. Extruded polystyrene
b. Expanded polystyrene
c. Polyisocyanurate
d. Fiberglass
e. Gypsum

18. For the same thickness, which of the following materials has the smallest R-value?
a. Extruded polystyrene
b. Expanded polystyrene
c. Polyisocyanurate
d. Fiberglass
e. Gypsum

19. The R-value of a 1-in.-thick extruded polystyrene board is 5.0. If the R-value of a 1-in.-thick brick wall is 0.2, how thick should the brick wall be to equal the R-value of 1-in.-thick extruded polystyrene?
a. 10 in. **b.** 5 in.
c. 50 in. **d.** 2.5 in.
e. None of the above

20. The concept of *aged R-value* is applicable to
a. all insulating materials.
b. insulating materials that contain HCFC gas.
c. fiberglass insulation.
d. perlite board insulation.
e. all building materials.

21. The emissivity of a material lies between
a. 0 and 1. **b.** 1 and 2.
c. 1 and 10. **d.** 0 and 100.
e. none of the above.

22. Which of the following materials has the smallest value of emissivity?
a. Fiberglass
b. Highly compressed fiberglass
c. Metal
d. Highly polished metal
e. All the above have the same value of emissivity.

23. Aluminum foil is commonly used in building assemblies because it
a. lowers heat transfer due to conduction.
b. lowers heat transfer due to convection.
c. lowers heat transfer due to radiation.
d. all the above.

24. Like color, emissivity is a property of the surface of the material, not of the entire body of the material.
a. True **b.** False

25. The R-value of a building assembly is influenced by the inside and outside air films. The sum of the R-values of these two films ($R_{si} + R_{so}$) is approximately equal to
a. 0.1. **b.** 5.0.
c. 10.0. **d.** 1.0.
e. none of the above.

26. The approximate R-value of an air space in a wall is
a. R-1. **b.** R-1.5.
c. R-2.0. **d.** R-2.5.
e. none of the above.

27. The approximate R-value of an air space in a wall with an aluminum foil lining is
a. R-1. **b.** R-1.5.
c. R-2.0. **d.** R-2.5.
e. none of the above.

Answers: 12-e, 13-d, 14-e, 15-a, 16-d, 17-c, 18-e, 19-e, 20-b, 21-a, 22-d, 23-c, 24-a, 25-d, 26-a, 27-d.

5.6 U-VALUE OF AN ASSEMBLY

The thermal property, which is more directly related to heat flow through a building assembly, is called its *thermal transmittance,* or the *U-value.* The U-value is the reciprocal of the R-value. In practice, we are concerned with the air-to-air heat transmittance through a wall or a roof. Hence, the U-value includes the inside and outside surface resistances, R_{si} and R_{so}. Thus, the U-value of a component is given by

$$U = \frac{1}{R_{si} + R_1 + R_2 + R_3 + \cdots + R_{so}} \qquad \textbf{Eq. (5)}$$

If the total resistance of a component, including the inside and outside film resistances, is 10.0, its U-value is $1/10.0 = 0.10$. Architects and engineers compare the effectiveness of different construction assemblies in terms of their U-values. The real usefulness of the U-value, however, lies in heat-transfer calculations, because heat transfer is directly proportional to the U-value. The greater the U-value of a component, the greater the heat flow through it.

NOTE

U-Value and Surface Resistances

The U-value of an assembly includes the effect of internal and external surface resistances.

The U-value of an assembly can be calculated by first calculating R_t and then finding its reciprocal. Thus, the U-value of the wall assembly of Example 3 is 1/15.6 = 0.064.

NOTE

U-Value of a Single Glass Sheet

As stated in Section 5.4, the R-value of a single sheet of glass, regardless of its thickness, is approximately 1.0. Hence, its U-value is also 1.0.

U-VALUES OF A GLAZED OPENING

We saw in Example 1 that the R-value of a single glass sheet is negligible. As stated in Section 5.4, the inside and outside surface resistances provide any resistance to heat flow through a glass sheet. Because the outside and inside surface resistances add to nearly 0.9, the R-value of a single glass sheet is generally assumed to be 1.0. Because U = 1/R, the U-value of a single glass sheet is also 1.0.

The U-value of a glazed window, however, differs from that of a glass sheet because a window includes the glass as well as the frame in which the glass is held. Depending on the material used and the thickness of the frame, the U-value of the frame may be lower or higher than the U-value of glass. In general, a window with wood frame has a smaller U-value and a aluminum-framed window has a greater U-value than a single glass sheet. The U-values of glass and glazed openings are discussed further in Chapter 28.

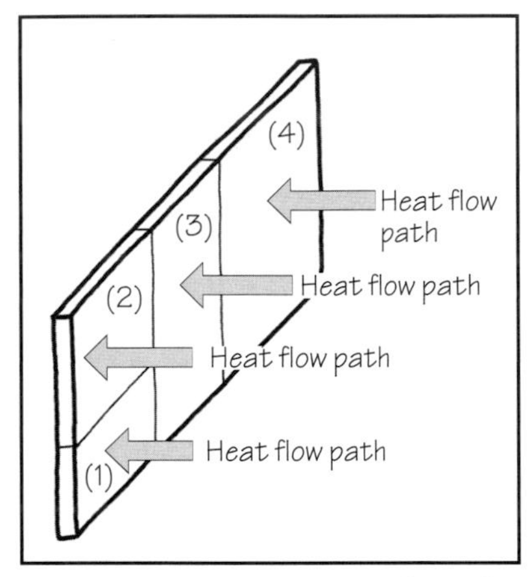

FIGURE 5.17 Heat flowing through an assembly, which has several parts with different R-values, divides itself into parallel paths. Each path represents heat flow through an individual part.

OVERALL U-VALUE OF AN ASSEMBLY—PARALLEL-PATH HEAT FLOW

The R-values of various parts of a building envelope are usually not the same. For instance, the doors and windows usually have a lower R-value than the wall itself. Similarly, a skylight has a lower R-value than the roof. So how do we obtain the overall R-value of a wall or a roof assembly that has several parts with different R-values?

The determination of the overall R-value of an assembly is simpler if we consider the U-values of the individual parts rather than their R-values. Consider a wall that consists of several parts with different U-values, as shown in Figure 5.17. As the heat flows through such an assembly, it divides itself into parallel paths. One path goes through part 1, another path through part 2, and so on. If the U-values of various parts are $U_1, U_2, U_3, \ldots$, it can be shown that the overall U-value, U_o, of the entire wall assembly is given by

$$U_o = \frac{A_1U_1 + A_2U_2 + A_3U_3 + \cdots}{A_1 + A_2 + A_3 + \cdots} \qquad \textbf{Eq. (6)}$$

where $A_1, A_2, A_3, \ldots$, are the respective areas of parts 1, 2, 3, . . . , of the assembly.

The overall R-value of the assembly, R_o, is

$$R_o = \frac{1}{U_o} \qquad \textbf{Eq. (7)}$$

Example 4
(Overall U-Value of a Wall)

Determine the overall R-value of the wall assembly of Figure 5.18. The overall dimensions of the wall are 20 ft × 12 ft, and it is composed of wood studs and brick veneer, as shown in Figure 5.16, Example 3. In other words, the R-value of the wall (excluding door and windows) is 15.6, and hence its U-value is 0.064.

The window areas are 20 ft^2 and 25 ft^2. Assume that their U-values are 1.0 and 0.7, respectively. The door area is 35 ft^2, and its U-value is 0.3.

Solution

Wall area = [(20 × 12) − (20 + 25 + 35)] = 160 ft^2. Its U-value is 0.064.
Window area = 20 ft^2. Its U-value is 1.0.
Window area = 25 ft^2. Its U-value is 0.7.
Door area = 35 ft^2. Its U-value is 0.3.

From Equation (6), the overall U-value of the wall is

$$U_o = \frac{160(0.064) + 20(1.0) + 25(0.7) + 35(0.3)}{240} = 0.24$$

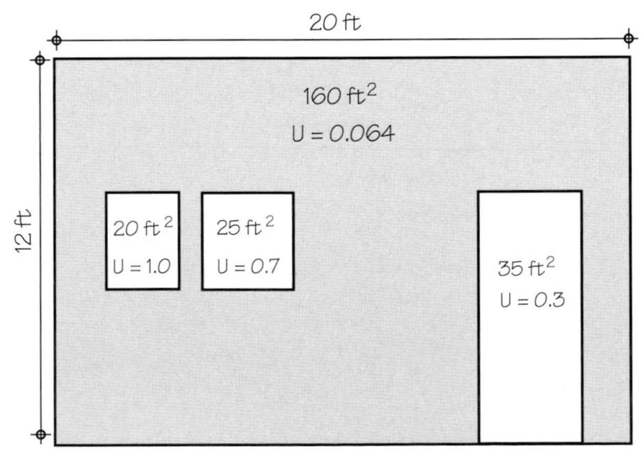

FIGURE 5.18 Wall assembly of Example 4.

From Equation (7), the overall R-value of the wall is

$$R_o = \frac{1}{0.24} = 4.2$$

Observe that although most of the wall (160 ft^2 out of a total of 240 ft^2) has an R-value of 15.6, the overall R-value of the assembly has been degraded to 4.2 because of the presence of parts with low R-values.

Example 5
(Overall U-Value of an Insulated Wood Stud Wall)

Determine the overall R-value of the wood stud and fiberglass layer in Example 3, shown in Figure 5.19.

Solution

Because the studs are spaced 16 in. on center, the width of fiberglass insulation in each stud space is 14.5 in., and the width of a stud is 1.5 in.

Using Table 5.3, the R-value of 3.5 in. thick fiberglass is 3.5(3.5) = 12.25. Hence, its U-value is 0.082.

Similarly, the R-value of the wood stud is 0.9(3.5) = 3.15. Hence, its U-value is 0.32.

The overall U-value of the layer can be obtained by considering only one stud space consisting of 14.5-in.-wide fiberglass insulation plus the 1.5-in.-wide wood stud. See Figure 5.19. From Equations (6) and (7):

$$U_o = \frac{14.5(0.082) + 1.5(0.32)}{(14.5 + 1.5)} = 0.104$$

$$R_o = \frac{1}{0.104} = 9.6$$

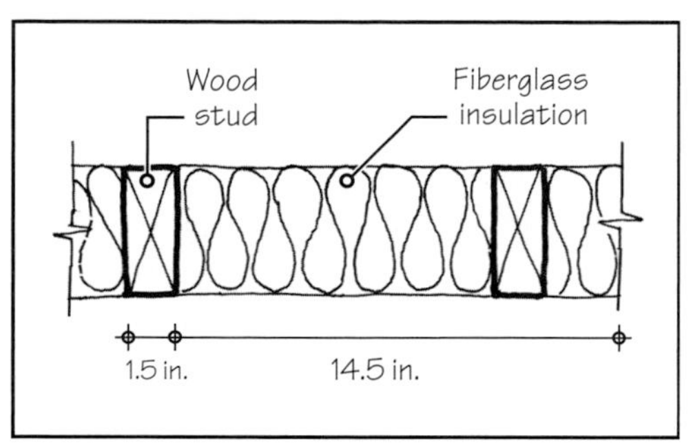

FIGURE 5.19 Stud-fiberglass layer of Example 3.

In other words, the R-value of the stud-fiberglass layer, which we had assumed to be 12.25 in Example 3, is, in fact, equal to 9.6. Note that the decrease in R-value is due to thermal bridging through the studs. If the wood studs are replaced by metal studs, thermal bridging will be more pronounced, further reducing the overall R-value of stud-fiberglass layer. Hence, a more precise total R-value of the wall assembly of Example 3 (which takes into account thermal bridging through wood studs) is not R-16 but R-13:

$$R_t = 0.7 + 0.3 + 9.6 + 0.45 + 1.0 + 0.73 + 0.2 = 13.0$$

EXPAND YOUR KNOWLEDGE

Thermal Bridging in Building Assemblies

Examples 4 and 5 indicate that the overall R-value of an assembly is governed mainly by the part(s) that have low R-value(s). This phenomenon, known as *thermal short circuiting*, or *thermal bridging*, can be further investigated by considering an assembly of two parts—A and B. Let the total area of both parts be 100 ft^2. Let part A of the assembly, with an area of 90 ft^2, have an R-value of 10.0 (U-value = 0.1). Let the area of part B be 10 ft^2 with an R-value of 1.0 (U = 1.0), Figure 1. Using Equation (6), we find the overall R-value of such an assembly to be 5.3:

$$U_o = \frac{90(0.1) + 10(1.0)}{100} = 0.19 \qquad R_o = \frac{1}{0.19} = 5.3$$

If we now increase the R-value of part A to 20 (U = 0.05), keeping part B as before, the overall R-value of the assembly becomes 6.9, an increase of 23% from the earlier value of 5.3.

A further increase in the R-value of part A to 30 (U = 0.033) changes the overall R-value of the assembly to 7.7—an increase of 12%.

This example shows that the overall R-value of the assembly is being dictated by the part that has a low R-value. It further shows that increasing the R-value of a part that already has a high R-value gives diminishing returns in increasing the overall R-value of the assembly. This fact highlights the importance of eliminating or reducing thermal bridges in the envelope.

In practical terms, it means that windows and skylights with low R-values function as efficient thermal bridges. Therefore, their areas should be limited or their R-value should be increased.

Another example of thermal bridging is an insulated stud wall assembly. Here, the studs function as thermal bridges. Thermal bridging through wood studs is generally small and is often ignored. However, it is significant through steel studs, Figure 2. This is generally countered by covering the metal studs on the outside with an insulating foam (XEPS or iso board) sheathing. The greater the R-value of exterior sheathing, the lower the thermal bridging.

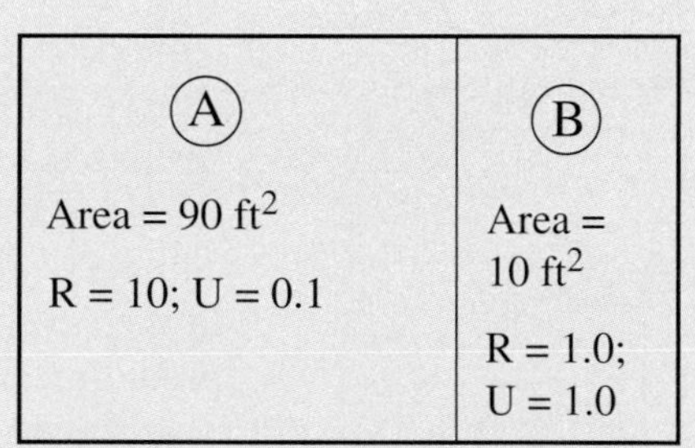

FIGURE 1 A two-part assembly with different U-values.

$$U_o = \frac{90(0.05) + 10(1.0)}{100} = 0.145$$

$$R_o = \frac{1}{0.145} = 6.9$$

$$U_o = \frac{90(0.033) + 10(1.0)}{100} = 0.13$$

$$R_o = \frac{1}{0.13} = 7.7$$

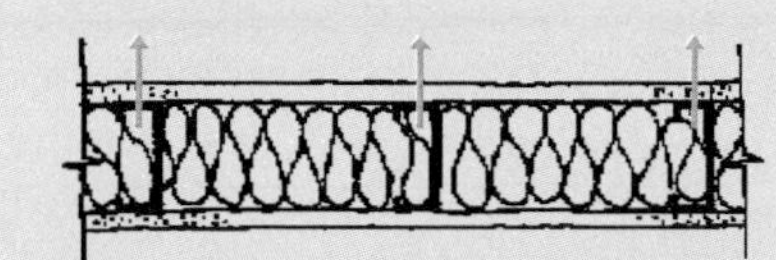

FIGURE 2 Thermal bridges in an insulated metal stud wall.

In fact, a small increase in the R-value of exterior sheathing on a steel stud wall can significantly increase the overall R-value of an insulated steel stud wall. Thus, as shown in Figure 3, the

(Continued)

EXPAND YOUR KNOWLEDGE

Thermal Bridging in Building Assemblies *(Continued)*

overall R-value of a steel stud wall with R-21 cavity insulation and R-3 exterior sheathing is equal to a wall with R-11 cavity insulation and R-5 exterior sheathing.

R-21 cavity insulation
R-3 Exterior sheathing
=
R-11 cavity insulation
R-5 Exterior sheathing

FIGURE 3 Effect of exterior insulating sheathing in metal stud assemblies.

Similar thermal bridging occurs in a concrete block wall, where the webs function as thermal bridges, Figure 4.

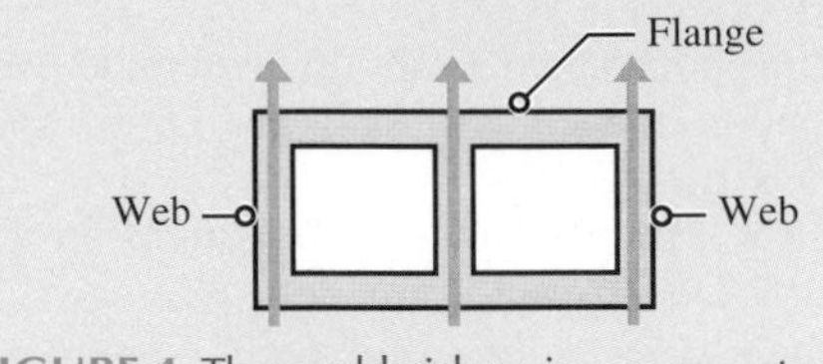

FIGURE 4 Thermal bridges in a concrete block.

NOTE

Energy Use in the U.S.

A total of 98.33 quads of energy were used in the United States in year 2003. (1 quad = 1 quadrillion Btu = 10^{15} Btu). Of this, 38.8 quads were used in buildings, representing nearly 39% of the total energy use. Industry and transportation accounted for 33% and 28%, respectively.

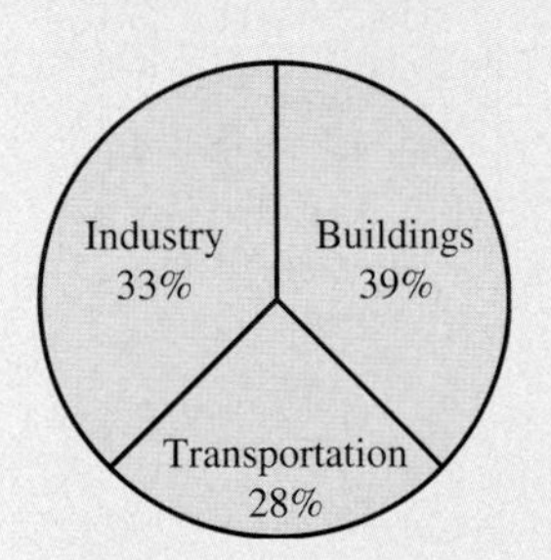

Of the total 38.8 quads of energy used in buildings in the United States in 2003, 21.3 quads were used in the residential sector. In other words, the residential sector accounted for nearly 55% of building energy use.

Source: Summary Tables, *Annual Energy Outlook 2005 With Projections to 2025.* U.S. Department of Energy, DOE/EIA Report No. 0383 (2005).

5.7 WHERE AND HOW MUCH TO INSULATE

The discussion of the thermal properties presented so far leads to two questions:

- Where to insulate a building
- How much insulation (U-value) to use.

The first question is easily answered. Obviously, the insulation is required only in the building envelope, that is, in all walls that are exposed to the outside and the roof. Interior walls and floors need no insulation unless they are exposed to an unconditioned space. For example, a floor over an unconditioned basement or crawl space should be insulated. Similarly, the wall between the garage and the house should be insulated.

The amount of insulation required is a function of two factors: (a) economics and (b) the climate of the location. From an economic standpoint, the amount of insulation required in the envelope is a function of the relationship between the price of energy and the cost of providing and installing the insulation.

If the energy prices are low, the cost of heating or cooling a building is low. In that case, a lower level of insulation is economical. On the other hand, if energy prices are high, a greater amount of insulation is justified. This is precisely why the levels of insulation in buildings increased when the world energy prices were abruptly raised in 1973. A less tangible but important value of insulation relates to the conservation of fossil fuel, extending the availability of this limited resource (see Chapter 10).

Climate affects the insulation requirement in the following way. In a more severe climate, a greater amount of insulation is needed. If the climate of a location is temperate, little or no insulation may be needed. There generally is an optimum level of envelope insulation for a particular climate. This is defined as that level of insulation at which the amount saved in energy consumption obtained over the entire life of the building exceeds the cost of providing and installing the insulation.

The severity of climate is defined by two indices: heating degree days (HDD) and cooling degree days (CDD). Simply explained, HDD represents the amount of heating that may be needed in a location over the entire year. CDD represents the corresponding figure for cooling. In the United States, insulation requirements are currently based on HDD, which varies with the location.

ENERGY CONSERVATION CODE

An important document regulating the use of energy in buildings in the United States is the *International Energy Conservation Code* (IECC). This comprehensive document deals with all aspects of energy use in buildings, including the amount of insulation in the building envelope, area of glass, level of lighting, efficiency of heating, cooling and other energy-consuming equipment, and so on.

Heating Degree Days

Crucial to the understanding of the heating degree day (HDD) concept is the *balance point temperature* of a building. The balance point temperature is defined as the outside air temperature at which the heat gain of the building from the sun and internal sources (such as lighting, cooking, and humans) equals the heat loss through the building envelope. Thus, if the outside air temperature equals the balance point temperature, no mechanical heating of the building is required.

The balance point temperature for an average building during winters in the United States is usually taken as 65°F. This implies that when the average daily outside air temperature T_o at a location is lower than 65°F, heating of the building will be required.

The difference between 65°F and T_o on a particular day gives the number of heating degree days contributed by that day. Thus, on a particular day, if the average outside air temperature at a location is 35°F, then that day will contribute (65 − 35) = 30 heating degree days to the annual HDD of the location. On the following day at that location, if the average outside air temperature is 30°F, then an additional 35 heating degree days will be added to the annual HDD of the location. The average outside air temperature for a particular day may be taken as the average of the maximum and minimum air temperatures.

We define the sum of all such heating degree days as the annual heating degree days, HDD, of a location. In other words,

$$\text{HDD} = \Sigma(65°\text{F} - T_o) \qquad \text{for each day when } T_o < 65°\text{F}$$

The annual HDD is a measure of the severity of winters of a place. The greater the value of HDD, the more severe the winter. In addition to describing the severity of winters of a location, the HDD value has been found to be directly related to the amount of energy consumed by buildings. HDD values for selected U.S. locations are given in the following table.

Approximate Annual Heating Degree Days for Selected U.S. Cities

Location	HDD	Location	HDD
Albuquerque, New Mexico	4,200	Indianapolis, Indiana	5,400
Anchorage, Alaska	9,700	Las Vegas, Nevada	2,700
Atlanta, Georgia	2,800	Little Rock, Arkansas	3,000
Baltimore, Maryland	4,500	Los Angeles, California	1,200
Birmingham, Alabama	2,800	Madison, Wisconsin	7,200
Boise, Idaho	5,500	Memphis, Tennessee	3,000
Boston, Massachusetts	5,400	Miami, Florida	200
Buffalo, New York	6,400	Minneapolis, Minnesota	7,700
Chicago (O'Hare), Illinois	6,300	New Orleans, Louisiana	1,500
Cleveland, Ohio	5,900	New York City	4,700
Dallas, Texas	2,400	Oklahoma City	3,600
Denver, Colorado	5,900	Philadelphia, Pennsylvania	4,700
Des Moines, Iowa	6,300	Phoenix, Arizona	1,100
Detroit, Michigan	6,200	Portland, Oregon	4,100
Hartford, Connecticut	5,900	San Francisco, California	2,600
Hawaii (all cities)	0	Seattle, Washington	4,400
Houston, Texas	1,500		

Note: These values are approximate. Consult the city for precise values.
Source: www.cpc.ncep.noaa.gov/products/analysis_monitoring/cdus/degree_days.

The discussion presented here is limited to the amount of envelope insulation, one of the most critical factors affecting energy use. Our particular focus is on residential occupancies, because far more energy is used in residential structures than in commercial buildings.

Insulation Requirements for Residential Structures

Energy-use regulations for residential structures are simplified in the IECC. IECC provides requirements for maximum overall U-values for walls and roofs of residential structures as a function of HDD, as shown in Tables 5.4 and 5.5. Examples 6 and 7 illustrate the use of these tables.

Insulation in Floors Over Unheated Basements or Crawl Spaces

In a cold climate, the heat loss from a ground floor can be substantial. A wood floor over a crawl space or unheated basement should be insulated between the floor joists, Figure 5.20. Note that the insulation is required under the entire floor. Table 5.6 gives the maximum U-values for such floors.

TABLE 5.4 MAXIMUM OVERALL U-VALUES FOR WALLS IN WOOD-FRAMED RESIDENTIAL BUILDINGS

	Heating degree days	U_o(max)
Detached one- or two-family residences	0	0.265
	0–2,500	$0.265 - HDD(34 \times 10^{-6})$
	2,500–7,000	$0.2188 - HDD(15.55 \times 10^{-6})$
	7,000–13,000	0.11
	>14,000	0.10
All other residential structures	0–500	0.38
	500–3,000	$0.38 - (HDD - 500)(66 \times 10^{-6})$
	3,000–6,000	0.215
	6,000–8,200	$0.215 - (HDD - 6,000)(30.5 \times 10^{-6})$
	8,200–9,500	0.148
	9,500–10,000	$0.148 - (HDD - 9,500)(55.8 \times 10^{-6})$
	>10,000	0.12

Source: 2006 International Energy Conservation Code. Copyright 2006. Falls Church, VA: International Code Council, Inc. Reproduced with permission. All rights reserved.

TABLE 5.5 MAXIMUM OVERALL U-VALUES FOR ROOF-CEILING ASSEMBLIES OF WOOD-FRAMED RESIDENTIAL BUILDINGS

Heating degree days	U_o(max)
0	0.05
0–2,500	$0.05 - HDD\ (5.6 \times 10^{-6})$
2,500–3,900	0.036
3,900–6,000	$0.036 - (HDD - 3,900)(4.78 \times 10^{-6})$
6,000–16,000	0.026
>16,500	0.025

Source: 2006 International Energy Conservation Code. Copyright 2006. Falls Church, VA: International Code Council, Inc. Reproduced with permission. All rights reserved.

Example 6
(Minimum Required R-Value of Walls)

Determine the minimum required R-value of the opaque portions of the walls of a wood-framed single-family dwelling in Philadelphia, Pennsylvania. The area of the opaque portions of the walls is 1,500 ft^2. The total glazed area is 300 ft^2. The U-value of the glazing is 0.5.

Solution

HDD for Philadelphia = 4,700. Selecting the appropriate formula from Table 5.4, the maximum overall U-value of the walls is

$$U_o(\text{max}) = 0.2188 - 4,700(15.55 \times 10^{-6}) = 0.146$$

If the U-value of the opaque portions of the walls is represented by U_{op}, then from Equation (6),

$$0.146 = \frac{U_{op}(1,500) + 0.5(300)}{(1,500 + 300)} \quad \text{and} \quad U_{op} = 0.075$$

In other words the minimum required R-value of the opaque portions of the walls is 1/0.075 = 13.3, that is, R-13.

Example 7
(Minimum Required R-Value of Roof-Ceiling Assembly)

Determine the required minimum attic (roof-ceiling assembly) insulation for a wood-framed single-family dwelling in Philadelphia, Pennsylvania.

Solution

From Table 5.5,

$$U_o(\text{max}) = 0.036 - (HDD - 3,900)(4.78 \times 10^{-6})$$
$$= 0.036 - (4,700 - 3,900)(4.78 \times 10^{-6}) = 0.032$$

Thus, the minimum overall R-value of the roof-ceiling assembly is 1/0.032 = 31.3.

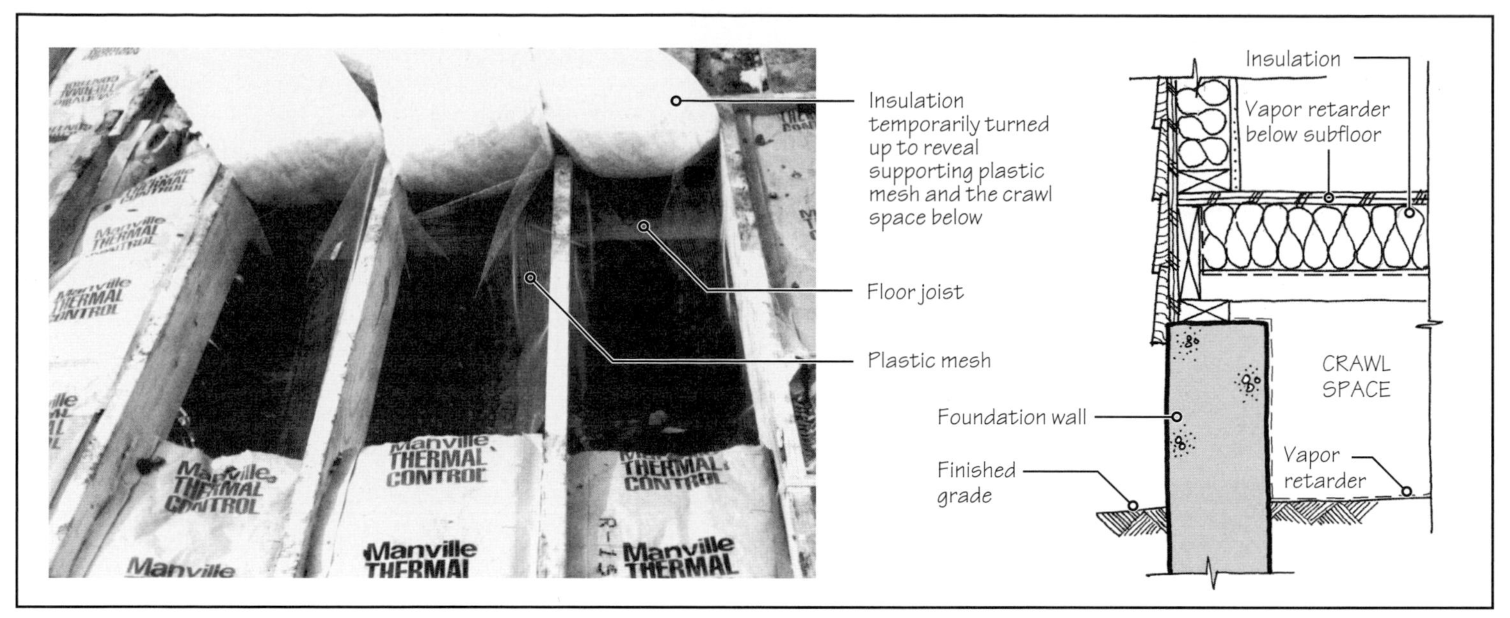

FIGURE 5.20 Insulation in a floor over a crawl space.

TABLE 5.6 MAXIMUM OVERALL U-VALUES FOR FLOORS OVER UNHEATED SPACES

Heating degree days	U_o(max)
0–1,000	0.08
1,001–2,500	0.07
2,501–15,500	0.05
15,501–16,500	$0.05 - (\text{HDD} - 15{,}500)(10 \times 10^{-6})$
>16,500	0.04

Source: 2006 International Energy Conservation Code. Copyright 2006. Falls Church, VA: International Code Council, Inc. Reproduced with permission.

Basement Wall Insulation

The maximum U-values of the walls of a heated basement are given in Table 5.7. These values can also be used for the walls of an unheated basement. If the walls are insulated, there is no need to insulate the floor over the basement. The basement walls need to be insulated only up to a depth of 10 ft below grade or to the level of the basement floor, whichever is smaller.

Insulation Under a Concrete Slab-on-Ground

If the floor is a ground-supported concrete slab (a *slab-on-ground*, also called a *slab-on-grade*), it needs to be insulated only along its exposed perimeter. There is no need to place insulation under the entire slab because the heat loss from a slab-on-ground occurs only

TABLE 5.7 MAXIMUM OVERALL U-VALUES FOR WALLS OF HEATED BASEMENTS

Heating degree days	U_o(max)
0–1,499	None required
1,500–4,500	$0.205 - \text{HDD}(233 \times 10^{-7})$
4,501–8,500	$0.11125 - \text{HDD}(25 \times 10^{-7})$
8,501–9,000	$0.6 - \text{HDD}(60 \times 10^{-6})$
>9,000	0.06

Source: 2006 International Energy Conservation Code. Copyright 2006. Falls Church, VA: International Code Council, Inc. Reproduced with permission.

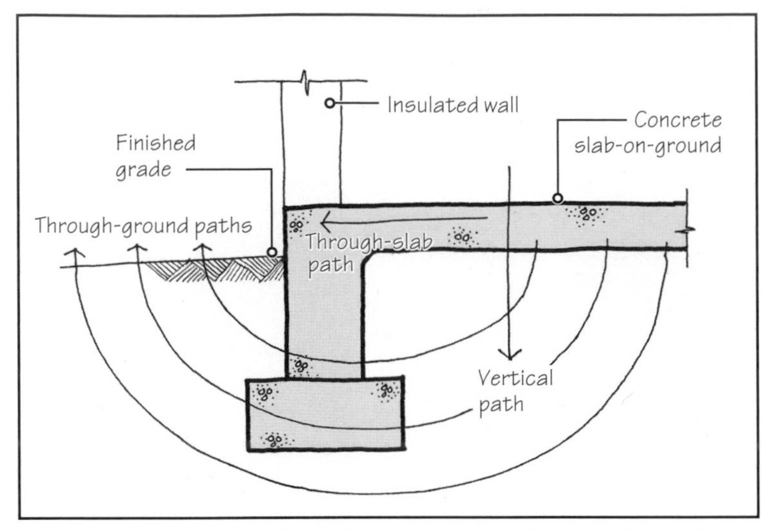

FIGURE 5.21 Various heat flow paths through a slab-on-ground (also called slab-on-grade).

from the exposed perimeter of the slab. The portion of the slab away from the exposed perimeter does not contribute much to heat loss, as explained shortly.

From Section 5.3, the thermal resistance of an element is directly proportional to its thickness L. In fact, the quantity L in Figure 5.6 refers to the length of the heat-flow path through the element. The longer the path, the greater its resistance to heat flow.

Let us now examine the heat-flow paths in a concrete slab-on-ground. A simplified visualization of heat loss from a slab-on-ground shows that there are three possible paths along which heat may flow from the slab:

- Vertical path—through the thickness of the slab and into the ground,
- Horizontal path through the slab
- Through the ground paths (shown by curved lines in Figure 5.21)

The vertical path represents an extremely large resistance to heat flow (as large as the diameter of the earth), implying that almost no heat will flow in that direction. Heat losses through the other two paths are functions of their respective lengths. Because the region of the slab away from an exposed edge has long path lengths, it does not contribute much to heat loss from the slab. Only the region that is close to the edges affects heat loss. Therefore, a slab on the ground is best insulated around its exposed perimeter. Very little is gained by insulating under the entire slab.

The standard practice is to insulate under the slab with either horizontal or vertical insulation (vertical insulation is generally preferred). Two commonly used details for insulating a slab-on-ground are shown in Figure 5.22. In both details, the location of the insulation is meant to insulate the horizontal and curved heat flow paths. The minimum R-value of the insulation is a function of the HDD of the location, as given in Figure 5.22. Note that no perimeter insulation is required in a location with HDD less than 2,500.

NOTE

Frost-Protected Shallow Foundations

The principles used in reducing the loss of interior heat through a slab-on-ground are also used in designing frost-protected shallow foundations (FSPF) in cold climates. See Chapter 21.

5.8 THERMAL CAPACITY

The concepts of heat flow and the related thermal properties of materials discussed in earlier sections apply to s*teady-state* heat transfer. A steady state is defined as the state in which the temperatures at all points within an assembly remain constant over time. The opposite of a steady state is an unsteady state, more commonly referred to as a *dynamic state.*

With particular reference to the building envelope, a steady state implies that the temperatures on both sides of the envelope are constant over time. In practice, a perfect steady state does not occur because although the inside air temperature may be kept constant by heating or cooling, the outside envelope temperature varies over time.

The outside temperature of the envelope is a function of the outside air temperature and the intensity of solar radiation. On days when the sun's rays are absent—that is, when the sky is cloudy—the diurnal variation in the outside temperature is small. Thus, an approximation of a steady state is obtained when: (1) the indoor temperature of the building is kept constant and (2) the outdoor temperature registers little daily variation.

A dynamic state is more common than a steady state. In the dynamic state, the temperature at a point within a component varies with time due to the variation of outside air and

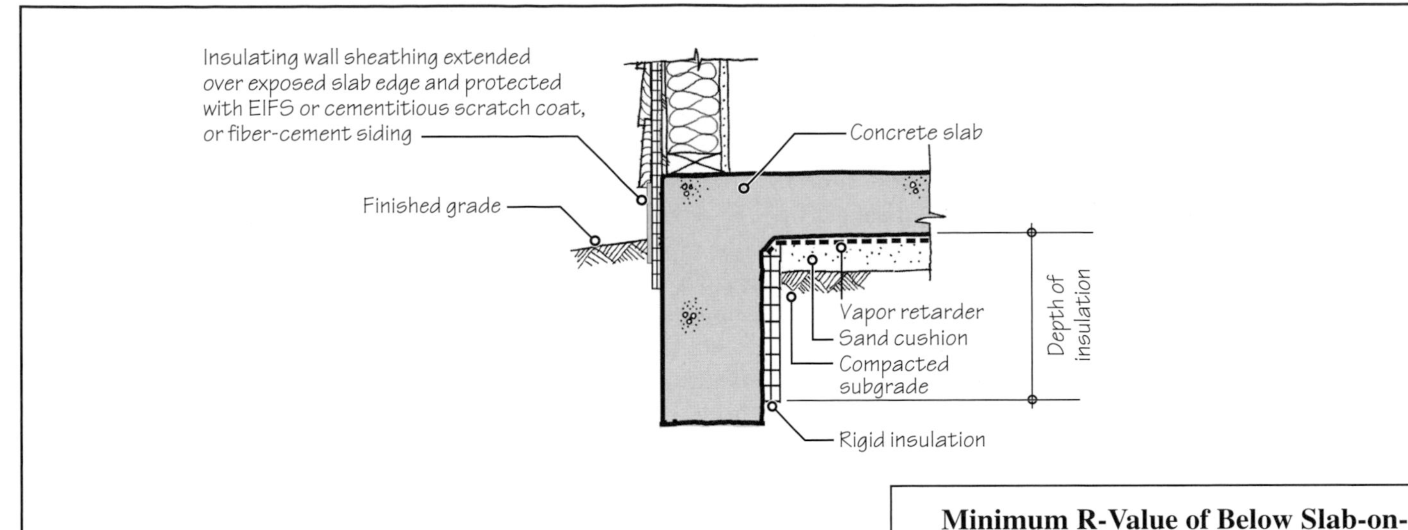

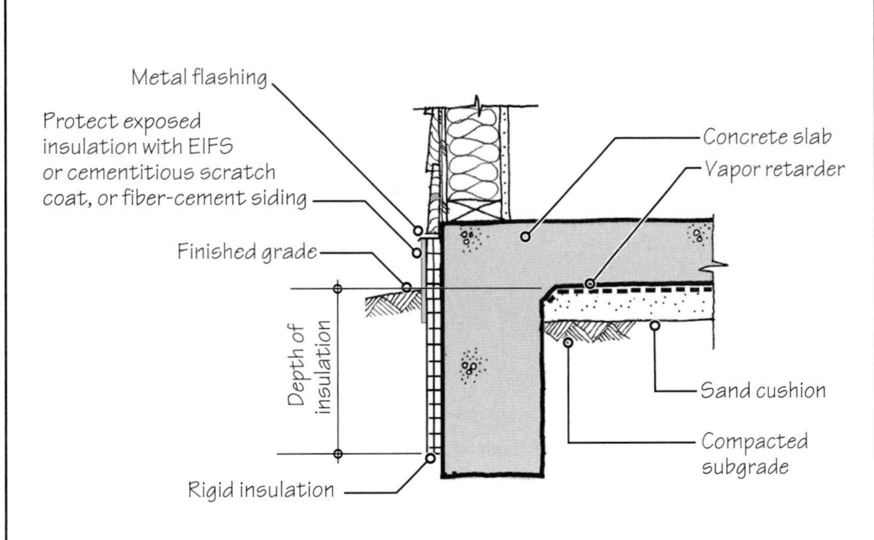

Minimum R-Value of Below Slab-on-Ground Rigid Insulation and Its Depth

HDD	Min. R-Value	Min. Depth
0 — < 2,500	None	
2,500 — < 3,500	R-4	2 ft
3,500 — < 4,500	R-5	2 ft
4,500 — < 5,500	R-6	2 ft
5,500 — < 6,500	R-9	4 ft
6,500 — < 7,000	R-11	4 ft
7,000 — < 8,500	R-13	4 ft
8,500 — < 9,000	R-14	4 ft
9,000 — <13,000	R-18	4 ft

Source: 2006 International Energy Conservation Code. Copyright 2006. Falls Church, Virginia: International Code Council, Inc. Reproduced with permission. All rights reserved.

FIGURE 5.22 Two among the several alternative details of insulating a slab-on-ground.

surface temperatures. This occurs in climates where solar radiation intensities are high, resulting in large daily and seasonal variations in the outside air temperature.

The distinction between steady and dynamic states has an important bearing on a building envelope's thermal properties. The property of the envelope that governs heat transfer under a steady state is its U-value (or R-value). In other words, under a steady state, it is the insulation in the envelope that governs the heat flow. Under a dynamic state, both the U-value (insulation) and the envelope's ability to store heat come into play, Figure 5.23.

Thermal Capacity of a Component

The ability of a component to store heat is referred to as its *thermal capacity* (TC), which is defined as the amount of heat needed to raise the temperature of 1 ft^2 of the component by 1°F. The greater the thermal capacity of the component, the greater the amount of heat it will absorb for a given rise in its temperature. *Thermal mass* and *thermal inertia* are other terms that are used synonymously for thermal capacity.

Dense materials, such as concrete, brick, stone, and adobe, have a higher thermal capacity than lightweight materials, such as wood, plastics, and insulating materials.

Thermal Capacity Versus Insulation

Because both thermal capacity and insulation affect heat transfer through the envelope, the question arises as to the relative importance of thermal capacity and the U-value of the envelope. It has been shown that the amount of energy consumed in heating or cooling buildings is independent of the thermal capacity of the envelope, provided the heat flow through the envelope remains unidirectional—that is, if heat flows from the inside to the outside or from the outside to the inside.

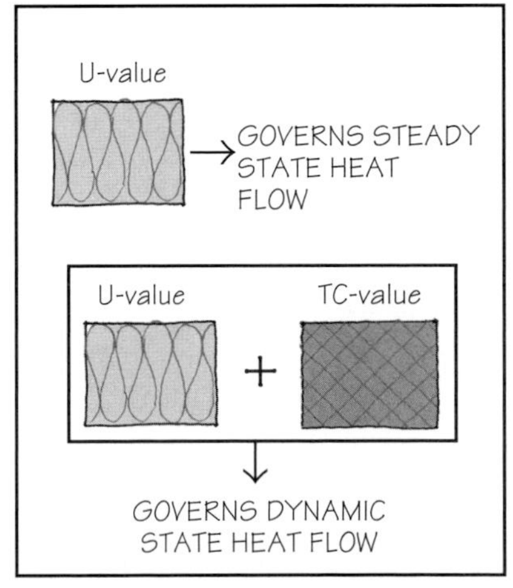

FIGURE 5.23 Envelope properties that govern heat flow under steady and dynamic states.

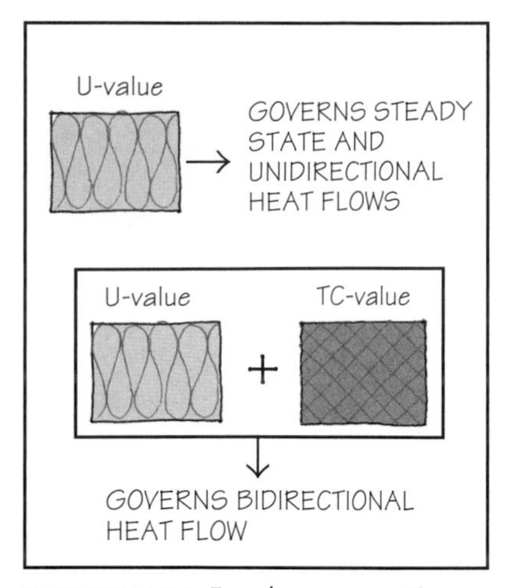

FIGURE 5.24 Envelope properties that govern unidirectional and bidirectional heat flows.

Under unidirectional heat flow, only the U-value of the envelope (i.e., the insulation) determines the rate of energy consumption. Thermal capacity of the envelope has no impact on energy consumption under unidirectional heat-flow conditions, even when unidirectional heat flow is dynamic.

On the other hand, if heat-flow direction changes over time (from the outside to the inside for some time and from the inside to the outside at other times), both the U-value and thermal capacity of the envelope determine energy consumption. The statement of Figure 5.23 can, therefore, be modified as shown in Figure 5.24.

An obvious example of unidirectional heat flow is a cold storage room, where the heat flow is generally from the outside to the inside. Thus, in cold storage buildings, the envelope's insulation determines energy consumption.

In cold climates of North America, Europe, and so on, the heat-transfer path remains unidirectional (heat flows from the inside to the outside) for most of the year. Here also, the insulation of the envelope is significant. Thermal capacity plays an insignificant role in such climates. The same is true in most warm, humid climates, where the heat flow is from the outside to the inside in air-conditioned buildings.

In some climates, where the diurnal temperature swing is extreme (say, 90°F during the day to 50°F at night), the heat-flow direction changes in a 24-h period. Thermal capacity plays a role in such climates. A high thermal capacity will stabilize indoor temperature within the comfort range, and insulation may not be required at all in a well-designed building.

5.9 THE MOST EFFECTIVE FACE OF THE ENVELOPE FOR INSULATION

The relative locations of insulation and thermal capacity in the envelope are also important. Placing the insulation on the outside of the assembly is more effective in reducing energy consumption than if the insulation is placed in the middle of the envelope section or toward the interior of the building because it eliminates or reduces thermal bridging.

The placement of high-thermal-capacity (heavyweight) materials on the interior side of the envelope helps to moderate interior temperature fluctuations. This is valid in buildings that are continuously heated or cooled. For a building that is heated or cooled intermittently, such as a church hall, the placement of high-thermal-capacity material on the interior of the building envelope implies that it will take longer to achieve comfort conditions.

PRACTICE QUIZ

Each question has only one correct answer. Select the choice that best answers the question.

28. The U-value and R-value of an assembly are related to each other by which relationship?
a. U = R
b. U + R = 1
c. (U)(R) = 1.0
d. $U^2 + R^2 = 1.0$
e. None of the above

29. A building assembly with a higher U-value is thermally more insulating than one with a lower U-value.
a. True
b. False

30. Which of the following metal stud wall assemblies gives a higher overall R-value? In both assemblies the total amount of insulation is R-15.
a. Assembly A has an exterior insulating sheathing of R-5 and within-stud insulation of R-10.
b. Assembly B has an exterior insulating sheathing of R-3 and within-stud insulation of R-12.

31. In a cold climate, a concrete slab-on-ground must be insulated under its entire area.
a. True
b. False

32. The walls of a heated basement must be insulated over the entire depth of the wall.
a. True
b. False

33. In which of the following buildings will heat flow occur unidirectionally?
a. Continuously heated or cooled buildings
b. Intermittently heated buildings
c. Intermittently cooled buildings
d. All the above
e. None of the above

34. For energy conservation, materials with a high thermal capacity are effective in climates in which the heat flow through the building envelope is
a. bidirectional within a 24-hour period during a large part of the year.
b. unidirectional within a 24-hour period during a large part of the year.
c. absent during a large part of the year.

35. Insulation is most effective if it is placed
a. on the outside face of the assembly.
b. on the inside face of the assembly.
c. in the center of the assembly.

Answers: 28-c, 29-b, 30-a, 31-b, 32-b, 33-a, 34-a, 35-a.

PRINCIPLES IN PRACTICE

Insulating Materials

A large variety of insulating materials are currently used in buildings. They can be classified in several ways, based on their configuration, physical structure or combustibility. If classified on the basis of configuration, insulating materials are classified as either *rigid insulation* or *flexible insulation.* Rigid insulation is commonly used in flat roofs and may be in the form of rigid boards or poured-in-place lightweight concrete. Flexible insulation is commonly used in stud walls and attic spaces and may be in the form of blankets, batts, or loose-fill insulation. Graphically, we distinguish between the rigid and flexible insulations, as shown in Figure 1.

A classification that is more suitable for describing various insulations is based on the insulation's physical structure. According to this classification, insulating materials may be divided into the following three categories:

- Fibrous insulation
- Granular insulation
- Foamed insulation

Commonly used materials in each category are shown in Table 1.

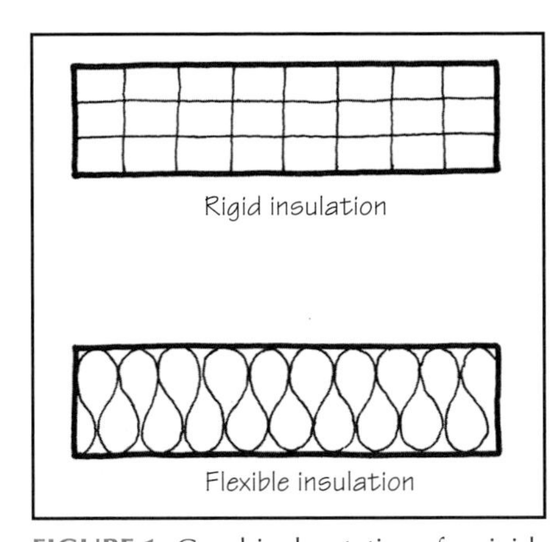

FIGURE 1 Graphical notations for rigid and flexible insulations for use in cross sections of building assemblies.

FIBROUS INSULATION

Fibrous insulating materials derive their high thermal resistivity from the air contained between the fibers. The fibers can be of mineral base or cellulosic base. Three types of mineral fibers are generally used: (a) glass fibers, called *fiberglass,* (b) fibers obtained from natural rock, called *rock wool,* and (c) fibers obtained from slag, called *slag wool.*

Fiberglass is made by fiberizing molten glass, and rock wool and slag wool are made by fiberizing molten rock and slag, respectively. The fibers are then sprayed with a binder and fabricated into the finished product. Fiberglass, rock wool, and slag wool insulations are fire resistant, moisture resistant, and vermin resistant.

Cellulosic fibers are derived from recycled newspaper, wood, and sugarcane. They are usually treated for fire resistance, minimization of smoke contribution, fungal growth, and decomposition by moisture.

NOTE

Rock Wool and Slag Wool

Rock wool and slag wool are together referred to as *mineral wool.* Although both mineral wool and fiberglass are noncombustible, mineral wool is more fire resistant than fiberglass because it has a higher melting-point temperature. Slag is a waste product obtained from the manufacture of iron, see Chapter 16.

TABLE 1 COMMONLY USED INSULATING MATERIALS

Physical structure	Configuration	Insulating material
Fibrous insulation	Batts, blankets and semirigid boards	Fiberglass Rockwool Slagwool
	Loose-fill (blown in)	Cellulosic fibers Fiberglass Rockwool
Granular insulation	Loose-fill	Expanded perlite and vermiculite granules
	Rigid boards	Perlite board Expanded polystyrene (EPS)
	Masonry inserts	EPS masonry inserts
	Insulating concrete	Perlite concrete Vermiculite concrete
Foamed insulation	Rigid boards	Plastic foams Extruded polystyrene (XEPS) Polyisocyanurate (iso board) Cellular glass
	Insulating concrete	Foamed concrete
	Foamed-in-place insulation	Foamed-in-place polyurethane

(Continued)

Insulating Materials (*Continued*)

FIGURE 2(a) Unfaced fiberglass insulation installed between studs. (Photo courtesy of CertainTeed Corporation.)

FIGURE 2(b) Kraft paper–faced fiberglass insulation installed between studs. (Photo courtesy of CertainTeed Corporation.)

Fiberglass is the most commonly used fibrous insulation and is available in the form of

- Batts
- Blankets
- Semirigid boards

Batts and blankets are normally used as wall insulation. Semirigid boards are also used in walls, but in locations where greater rigidity is needed, such as in spandrel areas of glass curtain walls. Batts and blankets are similar in appearance, composition, and density. The only difference between them is that blankets are available in rolls, whereas batts are precut from rolls into standard dimensions.

Apart from other sizes, both batts and blankets are manufactured in standard widths to fit 12-in., 16-in. and 24-in. center-to-center spacing between wood or metal studs. They are available either unfaced, Figure 2(a), or faced, Figure 2(b). Facing is usually of asphalt-impregnated paper, called *kraft paper,* which functions as a vapor retarder (see Section 6.6). Kraft paper–faced products are also available with projecting flanges for stapling to wood studs, Figure 3. Flanges may either be stapled to the faces of studs (face stapled) or to the sides of studs (inset stapled), as shown in Figure 2(b).

Unfaced insulation is particularly useful as additional insulation in attic spaces where faced insulation already exists or in ceiling spaces where rigid ceiling tiles provide support to the insulation. Unfaced batt insulation is also available to pressure fit within stud spaces, which may be used when a cover material has already been placed on the outside face of the assembly. If a vapor retarder is required in such an assembly, it must be provided separately, Figure 4.

Kraft (asphalt saturated) paper facing
Projecting flange
Fiberglass or rockwool insulation
Projecting flange

FIGURE 3 Projecting paper flanges in kraft paper–faced fibrous insulation.

Glass wool, mineral wool, or cellulosic fiber insulation may also be used as loose-fill insulation. These are usually blown into air spaces by pneumatic blowers. Blown-in insulation, Figure 5, is popular in retrofit applications where either no insulation or inadequate insulation existed previously. However, the potential of condensation within the insulation must be examined in a retrofit application, because the addition of insulation may create condensation where none existed earlier. Refer to "Principles in Practice" at the end of Chapter 6.

GRANULAR INSULATION

In materials with a granular structure, the air voids are contained inside tiny hollow beads or granules. Three types of granules are in common use

- Expanded perlite granules
- Expanded vermiculite granules
- Expanded polystyrene (EPS) granules

Perlite is a glassy volcanic rock that is expanded into granules by heat treatment. The expansion of perlite takes place due to the 2% to 6% moisture present

FIGURE 4 A clear polyethylene sheet vapor retarder being installed over unfaced fiberglass insulation between wood studs.

FIGURE 5 Blown-in insulation in attic. (Photo courtesy of CertainTeed Corporation.)

in the crude perlite rock. When heated suddenly to a temperature of nearly 1,600°F, the water in crushed perlite particles converts to steam. The pressure of the steam expands perlite particles, creating small air-filled granules, Figure 6. The process of expanding perlite is, in fact, similar to making popcorn.

Expanded vermiculite granules are made from mica. The process of making vermiculite granules is similar to that of making perlite granules. Expanded perlite and vermiculite are used as loose-fill insulation in masonry cavities, Figure 7. However, being porous, both vermiculite and perlite absorb moisture, which degrades their R-values. The freezing of trapped moisture may also lead to the deterioration of granules. Although the manufacturers of vermiculite and perlite loose-fill insulations supply these materials with a water-repellent treatment, it is important that the cavities in which they are filled remain dry.

Expanded polystyrene granules (or beads), which are commonly used as packing material, are not recommended as loose-fill insulation in masonry because they are extremely lightweight and do not completely fill air spaces. Perlite and vermiculite granules, on the other hand, flow much better and seek out even small voids. Additionally, because polystyrene is a polymer, its granules are combustible, whereas perlite and vermiculite granules are noncombustible.

Perlite Board and Expanded Polystyrene (EPS) Board

Expanded perlite is also available as a rigid board. Called *perlite board,* it is made with a combination of expanded perlite granules and mineral and cellulosic fibers. The fibers help increase the strength of the board. Asphalt is also added to the combination to improve water repellency and rot resistance. Perlite board is widely used as an insulation over flat or low-slope roof decks. It is noncombustible and can withstand the high temperature of hot asphalt.

FIGURE 6 Three stages of perlite production show the great increase in volume that takes place on expansion.

(*Continued*)

Insulating Materials (*Continued*)

FIGURE 7 Vermiculite loose-fill insulation being poured into a concrete masonry wall's cavities. (Photo courtesy of W. R. Grace & Co.)

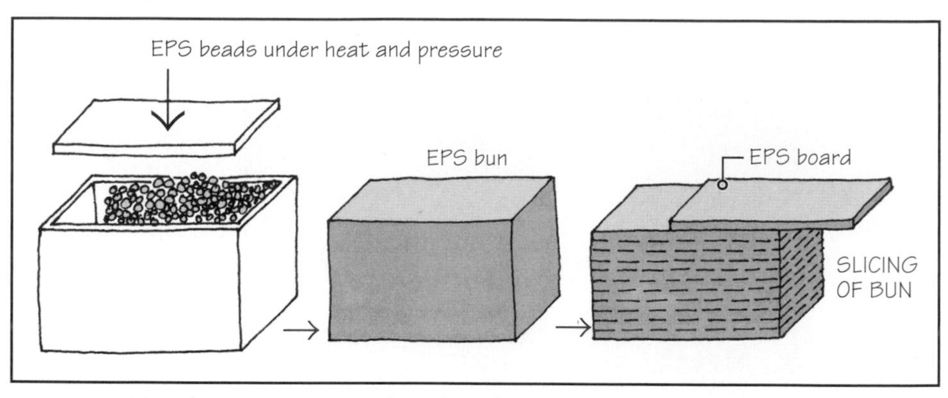

FIGURE 8 Manufacturing process of EPS boards.

Expanded polystyrene (EPS) boards—also referred to as *bead boards*—are made by packing polystyrene granules (beads), nearly $\frac{1}{8}$ in. diameter, in a mold and fusing them together under heat and pressure. The molded material is simply a large rectangular block, referred to as a *bun,* which is sliced into boards of required thickness, Figure 8. An EPS board is essentially of the same material as that used in picnic coolers.

Although the compressive strength of an EPS board is almost the same as that of a perlite board, it is not as widely used in low-slope roofs because of its combustibility and its inability to withstand high temperatures. It is, however, commonly used in EIFS (exterior insulation and finish systems) wall systems. See Chapter 27.

Because of its moldability, EPS inserts are used as an alternative to loose-fill insulation in concrete masonry walls. EPS inserts require no on-site labor, because they are placed in standard concrete blocks at the block-manufacturing plant. As Figure 9 shows, each insert, although made of one piece, has vertical slits so that it can fit tightly into the space. EPS inserts are more resistant to water absorption than loose-fill perlite or vermiculite granules.

Insulating Concrete Forms (ICF)

An interesting use of EPS is in hollow blocks, which are used as formwork for concrete. Referred to as insulating concrete forms (ICF), ICF blocks are stacked and reinforced with steel bars, and their cells are filled with concrete to form an insulated concrete wall, Figure 10.

By using EPS as permanent concrete forms, ICF construction eliminates the use of conventional wood or steel formwork. ICF manufacturers have developed several innovative products, and ICF construction is being used as an alternative to conventional wood-frame walls for residential construction.

Insulating Concrete

Another application of perlite and vermiculite granules is in (nonstructural) lightweight concrete. Vermiculite or perlite granules combined with portland cement in the ratio of 1 (portland cement): 4 to 8 (perlite or vermiculite) granules by weight, makes a lightweight concrete, which has good insulating properties. This concrete is called *insulating concrete.*

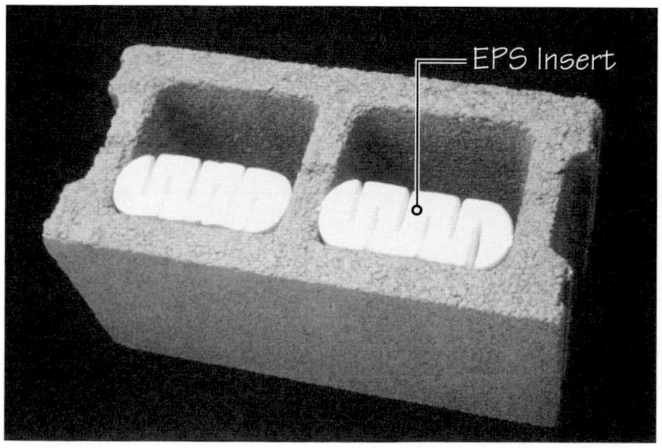

Insulating concrete is well suited for insulation on a flat roof because roof slopes can be easily created with wet concrete. The other method of providing slope on a flat roof is to use tapered board insulation, which is far more labor intensive. Because insulating concrete bonds well to most roof substrates, such as reinforced concrete slab or steel deck, it provides high-wind-uplift resistance. Additionally, because perlite and vermiculite are inorganic materials, insulating concrete is noncombustible and provides a high degree of fire resistance to the roof.

The R-value of insulating concrete is lower than other insulations (e.g., plastic foams). However, the advantages of insulating concrete, such its fire resistance, easy sloping to drains, and good bonding to substrate, can be combined with the higher

FIGURE 9 EPS inserts in concrete blocks installed at the block manufacturing plant.

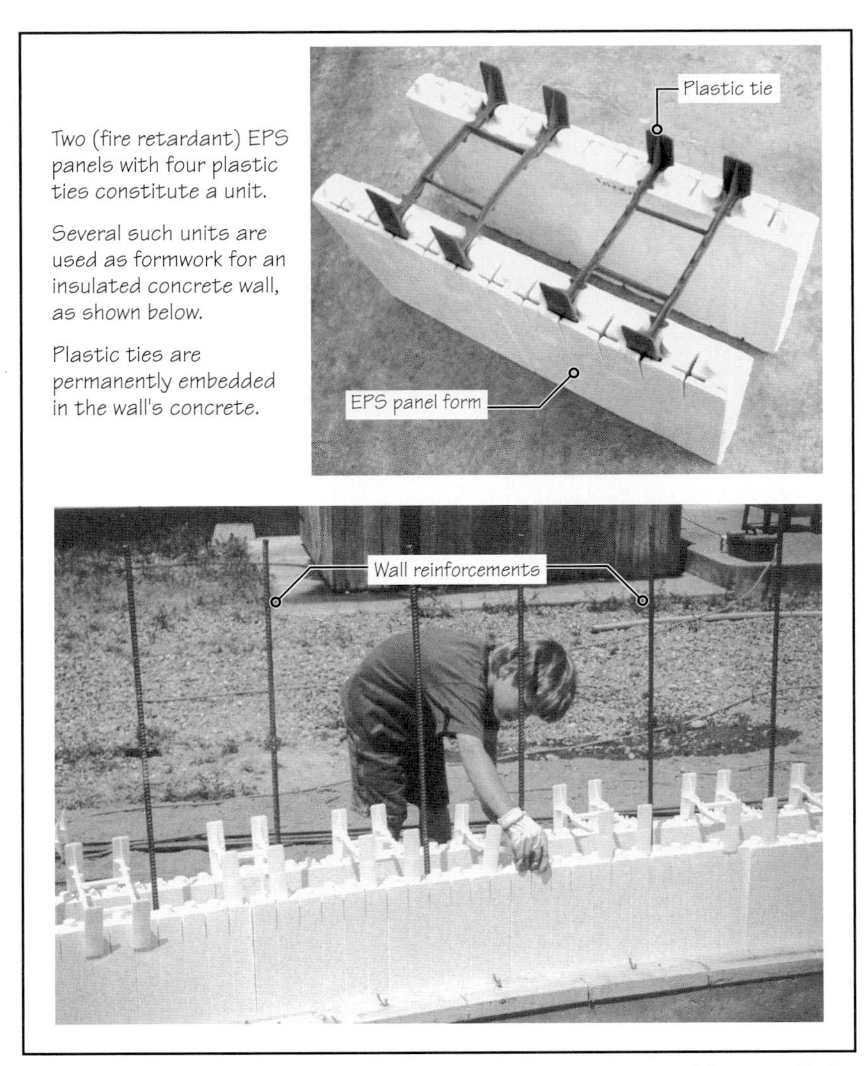

FIGURE 10 Insulating concrete form (ICF) blocks in walls are used as an alternative to wood frame walls for residential construction. Lower photo courtesy of Quad-Lock Building Systems Ltd.

R-value of EPS boards to create a hybrid assembly. In this assembly, insulating concrete is poured below and above the EPS boards so that the EPS boards are sandwiched between the two layers of concrete, Figures 11 and 12.

To structurally integrate the two layers of concrete and the EPS boards into a monolithic whole, the EPS boards are provided with holes, Figure 13. Insulating concrete is particularly well suited as insulation over steel roof decks, see Chapter 31.

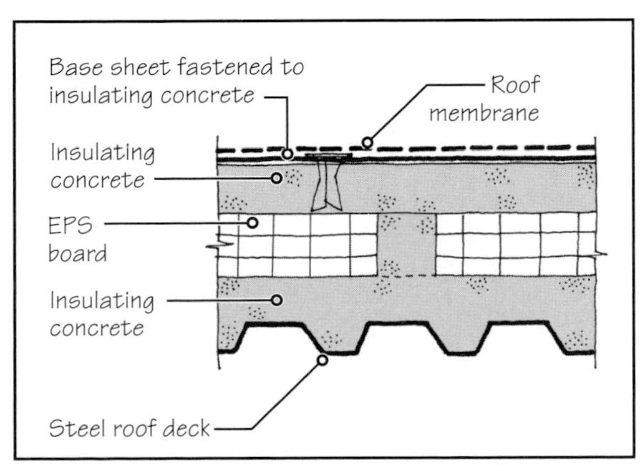

FIGURE 11 Cross section through an insulating concrete roof deck with sandwiched EPS boards.

FIGURE 12 Insulating concrete being placed over roof deck.

(*Continued*)

PRINCIPLES IN PRACTICE

Insulating Materials (*Continued*)

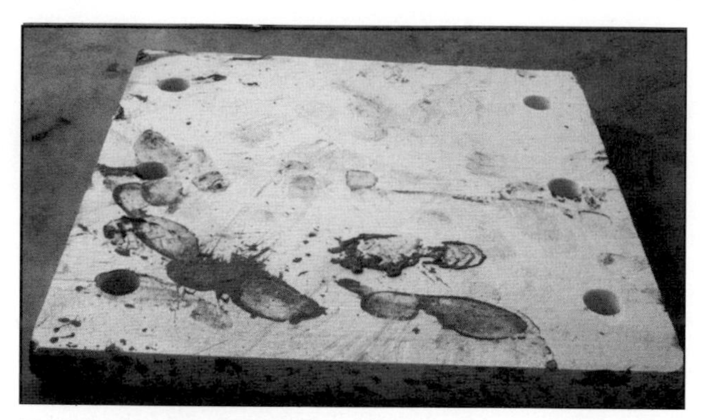

FIGURE 13 Because EPS boards are sandwiched between two layers of concrete, the holes in EPS boards structurally integrate the layers.

FOAMED INSULATION

The most commonly used insulating foams are synthetic (plastic) foams, such as *extruded polystyrene* and *polyisocyanurate*. Both are used as rigid boards, and their physical structure is cellular with closed cell cavities. The cavities contain HCFC gas.

The boards are made by the extrusion process. The extrusion process consists of forcing a semiliquid material through an aperture (die). This is precisely the same concept as forcing toothpaste out of a tube, the mouth of the tube being the die. Polystyrene board obtained from the extrusion process is called *extruded polystyrene board* (XEPS) to distinguish it from the molded EPS (beaded) board.

The extrusion process is complex and requires a large capital outlay in manufacturing. By contrast, the molding process is simpler and less expensive. That is why XEPS boards are costlier than EPS boards but provide a higher R-value and a higher compressive strength.

XEPS boards are made by mixing liquid polystyrene with liquid HCFC (as blowing agent) and forcing the mixture through a die, Figure 14. XEPS was first manufactured by the Dow Chemical Company in the United States during the 1940s under the trade name Styrofoam. However, today several manufacturers make XEPS boards and distinguish their products from each other by using different colors.

The version of HCFC gas used as a blowing agent in the production of polyisocyanurate boards (commonly referred to as *iso boards*) vaporizes at room temperature, so the liquid mixture of polymer and blowing agent is spread between the top and bottom facers to give iso boards their required shapes. Iso boards are, therefore, manufactured with facers on both sides. XEPS boards, on the other hand, do not require facers. The process of manufacturing iso boards is shown in Figure 15.

The facers used on iso boards retard the migration of HCFC gas out of the foam. Therefore, iso boards have higher thermal resistivity than XEPS boards. Facers consist of aluminum foil, glass fiber–reinforced polyester, plaster boards, wood fiber boards, and so on.

USES AND LIMITATIONS OF PLASTIC FOAM BOARDS

The closed cellular structure of XEPS and iso boards implies that they have high resistance to water and water vapor penetration. An EPS board, being a beaded product, is relatively more permeable to water and water vapor than an XEPS or iso board (Figure 5.8).

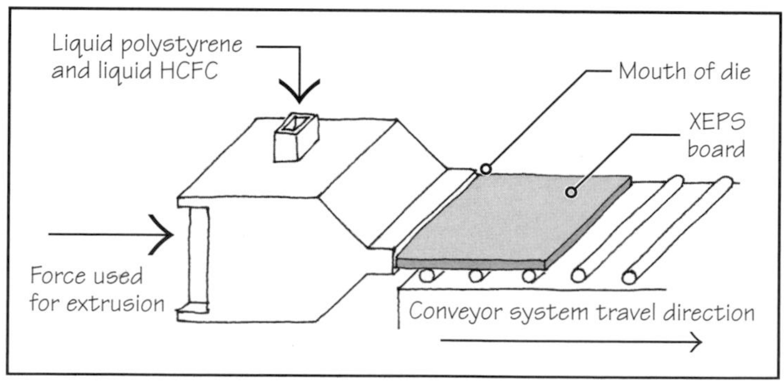

FIGURE 14 Manufacturing process of XEPS boards.

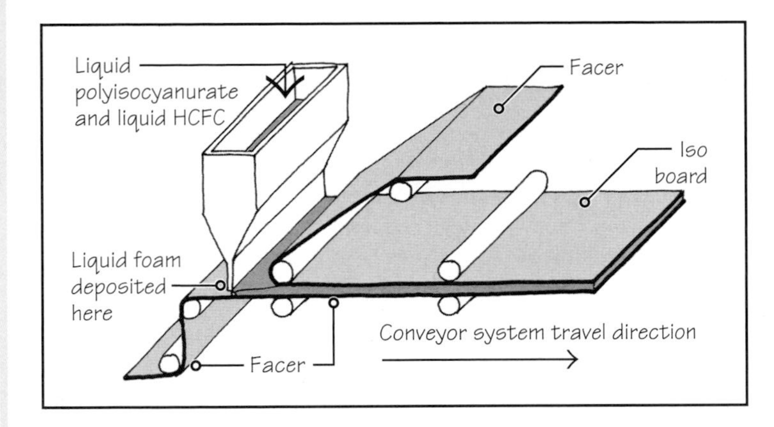

FIGURE 15 Manufacturing process of iso boards.

Plastic foams are combustible and require protection from flames and high temperatures. They should not be used toward the interior of a wall. Building codes mandate that if plastic foam is used as interior insulation, it must be covered with a minimum of $\frac{1}{2}$-in.-thick gypsum board or an equivalent thermal barrier. Plastic foams are resistant to fungal growth and chemical decomposition but can be destroyed when used below ground in heavily termite-infested soils. Polystyrene is sensitive to daylight and will deteriorate after prolonged exposure.

XEPS and iso boards are commonly used as insulation on flat or low-slope roof decks. Iso board is relatively more common as roof insulation because it can better withstand the high temperature of hot asphalt than XEPS and is also more fire resistant. Other uses of XEPS and iso boards are in wall sheathing, basement walls, and insulation under slabs-on-grade.

Foamed Glass

Another rigid-board foam product is cellular glass, which is made by expanding molten glass into a closed-cell structure and cooling it to form boards. Cellular glass boards have a much higher compressive strength than XEPS or iso boards. They are also impermeable to water vapor and water and are suitable as insulation in roofs that carry heavy loads, such as roof plazas and rooftop parking areas, and as perimeter insulation under concrete slabs-on-grade under heavy loads.

Foamed-in-Place Insulation

Foamed-in-place insulation is formed by spraying a liquid chemical (or a mixture of two liquid chemicals) in position. The liquid foams, when sprayed, expand to nearly 30 times the original volume of the liquid. After expansion, the foam solidifies into a closed-cell structure in a few minutes. The advantage of foamed-in-place insulation is that it can be provided in preexisting cavities of odd shapes where other types of insulations would be less desirable or impossible to use.

A commonly used foamed-in-place insulation, called *polyicynene* by its manufacturer, is used for insulating stud cavities, Figure 16. The open-cell structure of the foam entraps air and expands as it is sprayed into voids. The manufacturer claims that it fills the voids, providing a good air barrier.

FIGURE 16 Spraying of polyicynene insulation in stud cavities. The insulation expands and cures (hardens) quickly, then it is shaved flush with the studs. (Photo courtesy of Icynene, Inc.)

Foamed Concrete

Foamed concrete (also called *cellular concrete*) is a nonaggregate concrete that consists of portland cement, water, and a liquid foaming concentrate. The foaming concentrate creates tiny air bubbles, so that when the mix hardens, it contains a matrix of air voids separated by pure portland cement walls.

Foamed concrete is an excellent alternative to insulating concrete for roof insulation. The absence of aggregate reduces the amount of water required to pump and place the foamed concrete compared to insulating concrete. Also because of the absence of aggregate, the ρ-value of foamed concrete is slightly higher than that of insulating concrete.

PRACTICE QUIZ

Each question has only one correct answer. Select the choice that best answers the question.

36. Which of the following plastic foam insulation contains air within its voids?
- **a.** Extruded polystyrene board
- **b.** Expanded polystyrene board (bead board)
- **c.** Polyisocyanurate board (iso board)

37. Which of the following insulating materials is available in the form of rigid boards?
- **a.** Fiberglass
- **b.** Rockwool
- **c.** Insulating concrete
- **d.** Extruded polystyrene
- **e.** Vermiculite

38. Which of the following insulating materials is available in the form of batts?
- **a.** Fiberglass
- **b.** Expanded perlite
- **c.** Insulating concrete
- **d.** Extruded polystyrene
- **e.** Vermiculite

39. Which of the following insulations is combustible?
- **a.** Fiberglass
- **b.** Insulating concrete
- **c.** Perlite
- **d.** Extruded polystyrene
- **e.** Vermiculite

40. In which of the following locations is blown-in insulation most commonly used?
- **a.** Stud wall cavities
- **b.** Masonry wall cavities
- **c.** Top of a flat roof
- **d.** Attic spaces

41. Lightweight insulating concrete is generally used as insulation in
- **a.** walls.
- **b.** steep roofs.
- **c.** foundations.
- **d.** all the above.
- **e.** none of the above.

Answers: 36-b, 37-d, 38-a, 39-d, 40-d, 41-e.

KEY TERMS AND CONCEPTS

- Batt insulation
- Blanket insulation
- British thermal unit (Btu)
- Conduction
- Convection
- Fibrous insulation
- Foamed insulation
- Foamed-in-place insulation
- Granular insulation
- HCFC gas
- Heating degree days (HDD)
- Insulating concrete forms (ICF)
- Insulating materials
- International Energy Conservation Code (IECC)
- Joule (J)
- Kilowatt-hour (kWh)
- Radiation
- R-value
- Surface color
- Surface emissivity
- Thermal bridging
- Thermal capacity
- U-value
- Watt (W)
- Watt-hour

REVIEW QUESTIONS

1. Using the appropriate table, determine the R-values of the following materials.
   a. 4-in. nominal (actual $3\frac{5}{8}$-in.) thickness brick wall
   b. $\frac{5}{8}$-in.-thick gypsum board
   c. 2-in.-thick extruded polystyrene board
2. Draw a plan of the following wall assembly and then determine its R-value. Starting from the outside face of the wall, the assembly consists of the following layers.
   4-in. nominal ($3\frac{5}{8}$-in. actual) brick veneer
   2-in. air cavity
   0.75-in.-thick extruded polystyrene sheathing
   2×4 studs with 3.5-in.-thick fiberglass insulation
   0.5-in.-thick gypsum wallboard

   (Note: In calculating the total R-value, ignore the presence of wood studs and assume that the entire layer consists of fiberglass insulation).
3. What is the U-value of the assembly of Question 2?
4. What is aged R-value? To which materials does this concept apply and why?
5. Determine the overall U-value of a roof whose opaque portion, with an area of 2,500 ft^2, has an R-value of 35. There is a skylight in the roof, which has an area of 100 ft^2, and its R-value is 2.0.
6. Using the tables given in the text, determine the minimum overall R-value for the opaque walls of a single family home in
   a. Dallas, Texas.
   b. Buffalo, New York.

   Assume wood frame walls with 15% glazed area. The U-value of glazing is 0.4.

4. What is aged R-value? To which materials does this concept apply and why?
5. Determine the overall U-value of a roof whose opaque portion, with an area of 2,500 ft^2, has an R-value of 35. There is a skylight in the roof, which has an area of 100 ft^2, and its R-value is 2.0.
6. Using the tables given in the text, determine the minimum overall R-value for the opaque walls of a single family home in
   a. Dallas, Texas.
   b. Buffalo, New York.

   Assume wood frame walls with 15% glazed area. The U-value of glazing is 0.4.
7. Determine the minimum required R-values of the roof-ceiling assemblies in Question 6.
8. Using sketches and notes, explain why it is unnecessary to insulate under an entire concrete slab-on-ground.
9. What are high thermal capacity materials? In which climates do high-thermal-capacity materials help in energy conservation?

SELECTED WEB SITES

Expanded Polystyrene Molders Association (www.epsmolders.org)

Extruded Polystyrene Foam Association, XPSA (www.xpsa.com)

Insulating Concrete Forms Association (www.forms.org)

National Insulation Association, NIA (www.insulation.org)

Polyisocyanurate Insulation Manufacturers Association, PIMA (www.pima.org)

FURTHER READING

1. Nisson, Ned, and J. D. Gautam: *The Superinsulated House.* New York: John Wiley, 1985, p. 4.

REFERENCES

1. Baird, George, et al. *Energy Performance of Buildings.* CRC Press, Boca Raton, FL: 1984, p. 2.

Tyvek
Tyvek
Tyvek
Tyvek
Tyvek
Tyvek

CHAPTER 6

Air Leakage and Water Vapor Control

CHAPTER OUTLINE

6.1 AIR LEAKAGE FUNDAMENTALS

6.2 AIR RETARDER

6.3 WATER VAPOR IN AIR

6.4 CONDENSATION OF WATER VAPOR

6.5 CONTROL OF CONDENSATION

6.6 MATERIALS USED AS VAPOR RETARDERS

6.7 LOCATION OF VAPOR RETARDER AND VENTILATION IN THE SPACE BEYOND THE VAPOR RETARDER

6.8 VAPOR RETARDER UNDER A CONCRETE SLAB-ON-GROUND

PRINCIPLES IN PRACTICE: WHERE DEW POINT OCCURS IN AN ASSEMBLY

A hospital building (with site-cast reinforced-concrete structural frame) under construction showing the use of air-weather retarder in the exterior wall assembly. The scaffolding is in preparation for cladding the wall with brick veneer. (Photo by MM.)

In addition to thermal insulation and the thermal mass of the building envelope, discussed in Chapter 5, another factor that affects the heat loss or gain of a building is air leakage. The leakage of conditioned (heated or cooled) air through cracks and unsealed joints in the building envelope and its replacement by the outside air—referred to as *exfiltration* and *infiltration,* respectively—is unwanted ventilation of the building.

This increases the energy consumption of a building because the infiltrating air must be heated or cooled to the required inside air temperature. In extremely hot or cold climates where the temperature difference between the inside and the outside air is great, air leakage can add significantly to energy consumption.

Air always contains some water vapor. In fact, air is not only a mixture of nitrogen, oxygen, and carbon dioxide, but also of water vapor. The leakage of air through the building envelope is, therefore, always accompanied by water vapor.

Air leakage is not the only means by which water vapor moves in and out of a building envelope. Water vapor also moves through the envelope independent of air movement. This is due to the vapor pressure difference between the inside and the outside air. However, if the water vapor condenses within the envelope, it can cause a great deal of damage to the various components.

In this chapter, we study the mechanisms of air leakage and the movement of water vapor through envelope assemblies. We also examine materials and construction practices employed to control air leakage and the passage of water vapor and to avoid water damage resulting from condensation inside assemblies.

6.1 AIR LEAKAGE FUNDAMENTALS

FIGURE 6.1 Air infiltration and exfiltration through a building caused by the pressure differential created by wind.

It is estimated that air leakage accounts for 20% to 40% of the heat loss in buildings in North America. Two factors directly affect air leakage in a building:

- Area in the envelope that is prone to air leakage
- Air pressure difference between the inside and outside air

The greater the leakage area or the pressure difference, the greater the rate of air leakage. The leakage area is a measure of the relative tightness of the envelope. It is the area of cracks, tears, holes, and openings in the envelope and is a function of the type of construction, design, and artisanry. The leakage sites in the envelope usually occur at the joints of various components, such as exterior doors, windows, skylights, fireplaces, electrical outlets, plumbing, and duct penetrations.

The pressure difference between the inside and the outside air depends on wind speed and direction. Because wind speed increases with height above the ground, the pressure differences at the upper floors of a building are greater than those at the lower floors, causing greater air leakage at upper floors. We observed in Section 3.4 that the windward face of a building is under positive pressure, and the other faces are under suction (negative pressure). Therefore, air infiltrates through the windward face and exfiltrates through the non-windward faces of the building, Figure 6.1.

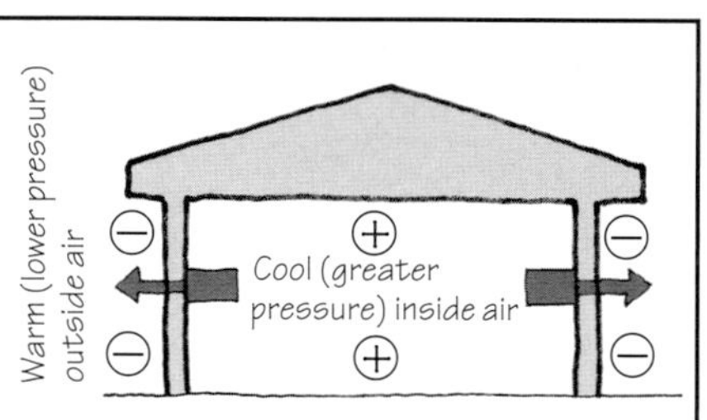

(a) During the cooling (summer) season, cool (more dense, i.e., greater pressure) inside air exfiltrates through the building envelope.

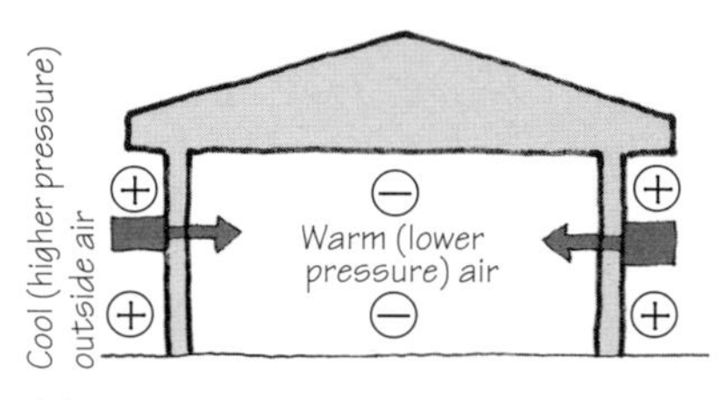

(b) During the heating (winter) season, cool (more dense, i.e., greater pressure) outside air infiltrates through the building envelope.

FIGURE 6.2 Air infiltration and exfiltration through a building due temperature differences between inside and outside air.

Another factor that affects the pressure difference is the temperature difference between inside and outside air. Because the density of air increases with decreasing temperature, cool (more dense) air exerts greater pressure than warm (less dense) air. During the cooling season, cool inside air has greater pressure than warm outside air. This leads to continuous exfiltration of cool air and infiltration of warm outside air through the envelope, Figure 6.2(a). This phenomenon is accentuated by the interior pressurization caused by the fans of the air conditioning system. During the heating season, cool outside air is at greater pressure than the warm inside air, and hence the air flow direction is reversed, Figure 6.2(b).

Because the inside-outside pressure difference is primarily a function of the outside climate (wind speed and air temperature), there is little that can be done to control it. Thus, the only means of reducing air leakage in a building is to reduce the leakage area in the building envelope. This requires sealing all joints between fixed building components (e.g., door jamb and wall) and between fixed-operable components (e.g., door jamb and door) or operable components. Sealants between fixed components are typically nonhardening synthetic compounds, (Chapter 9). *Weather stripping,* made from a resilient, compressible material, is used to seal gaps between fixed-operable or operable components.

AIR LEAKAGE RATE

Since a great deal of air leakage in buildings occurs through doors, windows, and glass curtain walls, limits are placed on their maximum permissible air leakage rates. The air leakage rate for windows (and curtain walls) is specified in terms of air leakage from a unit area of the window when the window is subjected to a standard air pressure difference (Sections 29.3 and 30.5).

6.2 AIR RETARDER

Air leakage is a particularly serious concern in wood or light-gauge steel (steel stud) frame buildings because their envelope is inherently more leaky than other types of construction, such as concrete or masonry. Additionally, as stated in Section 5.7, nearly 55% of all energy consumed in buildings in the United States is in residential occupancies, which are constructed of wood frame or light-gauge steel frame. Therefore, a small percentage decrease in energy consumption in residential buildings translates into large energy savings at the national level. The leakage sites in a typical wood frame or light-gauge steel frame building are shown in Figure 6.3.

A commonly used technique of reducing air leakage in wood or light-gauge steel construction is to wrap the exterior walls with a continuous *air retarder,* Figure 6.4. Air retarders are available in rolls of up to one story high that wrap the walls with very few joints.

An air retarder is typically made of a 5- to 10-mil-thick plastic sheet with micropores, so it allows very little air to pass through but has a high degree of permeability to water vapor. Depending on the manufacturer, it is made from plastic fibers that are either woven or spunbonded into a sheet. One manufacturer makes an air retarder out of a microperforated plastic sheet. (See Section 6.7 for the reasons why an air retarder must be perforated).

An air retarder is usually wrapped over the exterior wall sheathing and secured to the sheathing with staples. Joints in the air retarder are lapped and typically sealed with a self-adhesive tape, Figure 6.5.

The air retarder not only reduces heat gain or loss by reducing air leakage but also reduces air movement in the wall cavity in the vicinity of the insulation, thereby increasing the effectiveness of the insulation. Low-density, loose-fill, and fibrous insulations, such as fiberglass or rock wool, are more susceptible to increased convective heat transfer from infiltration than closed-cell foam insulations, such as extruded polystyrene or polyisocyanurate foam insulations.

NOTE

1 mil = $\frac{1}{1,000}$ in. = 0.001 inch

FIGURE 6.3 Typical leakage sites in a wood frame or light-gauge steel frame building.

FIGURE 6.4 An air retarder wrapped over the exterior wall sheathing of a three-story wood frame apartment building.

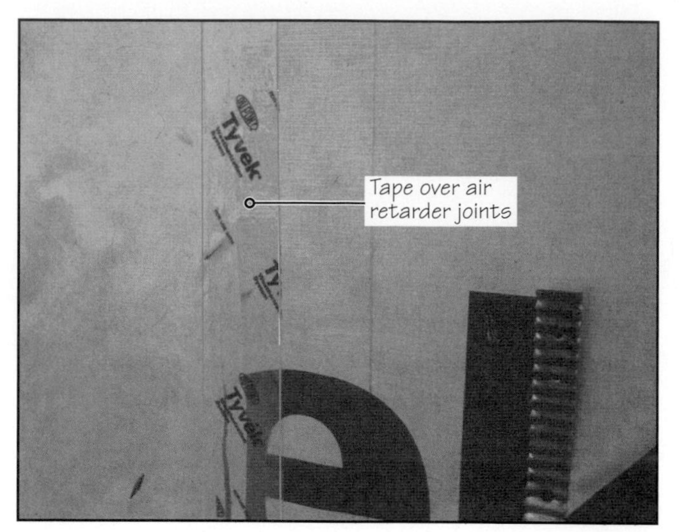

FIGURE 6.5 Joints in air retarder are taped with a self-adhering tape supplied by the air retarder manufacturer.

Note that the term *air barrier* is sometimes used in the industry for air retarder, which is incorrect because most commercially available air barriers retard the flow of air through them rather than completely stopping it.

Effectiveness of Air Retarders and Exterior Climate

The use of air retarders is standard practice in residential wood frame or light-gauge steel frame construction in climates where the winters are long and severe. This is due to the relatively large difference between the inside and the outside air temperatures in cold climates and, hence, a greater loss of energy by leakage. For example, the difference between the inside and outside air temperatures may be 60°F or more on a typical winter day in a cold climate (70°F inside air temperature and 10°F outside air temperature).

The corresponding difference in air temperatures in a warm climate is much smaller. For example, the difference in air temperatures during the summer in a hot climate seldom exceeds 30°F (70°F inside air temperature and 100°F outside air temperature). Therefore, in warm or mild climates, the use of air retarders is relatively less critical. However, if the building is located on a windy site—for example, close to a sea front or lake or on a hilltop—the use of an air retarder is advisable.

NOTE

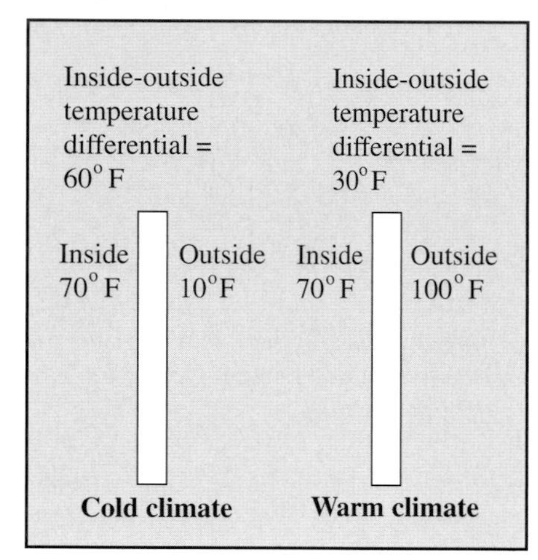

Air Retarders as Air-Weather Retarders

Air retarders also function to prevent the passage of exterior water through the wall. Therefore, they are also referred to as *air-weather retarders.* Their microperforations, although permeable to water vapor, do not allow air or (bulk) water to pass through.

Sealing Exterior Window and Door Perimeter

Despite the use of the air retarder, the joints between the wall and the window (or door) perimeter can leak air (see the dark arrows in Figure 6.3). Depending on the type of detail around the jamb, these joints can either be sealed by an elastomeric sealant (Chapter 9) or a self-adhering tape. The self-adhering tape works particularly well where the window has a nailing flange, so that the tape covers the flange and the wall, Figure 6.6.

Air Leakage in Masonry and Concrete Walls

In a well-constructed masonry building, air leakage is relatively small. However, if the mortar joints in the walls are not completely filled, leakage of air through joints may result, particularly through vertical mortar joints. A coat of cold-applied asphalt emulsion is commonly used in a wall assembly with a masonry backup wall. This coat provides weather (air and water) resistance to the backup wall (see Figure 26.21).

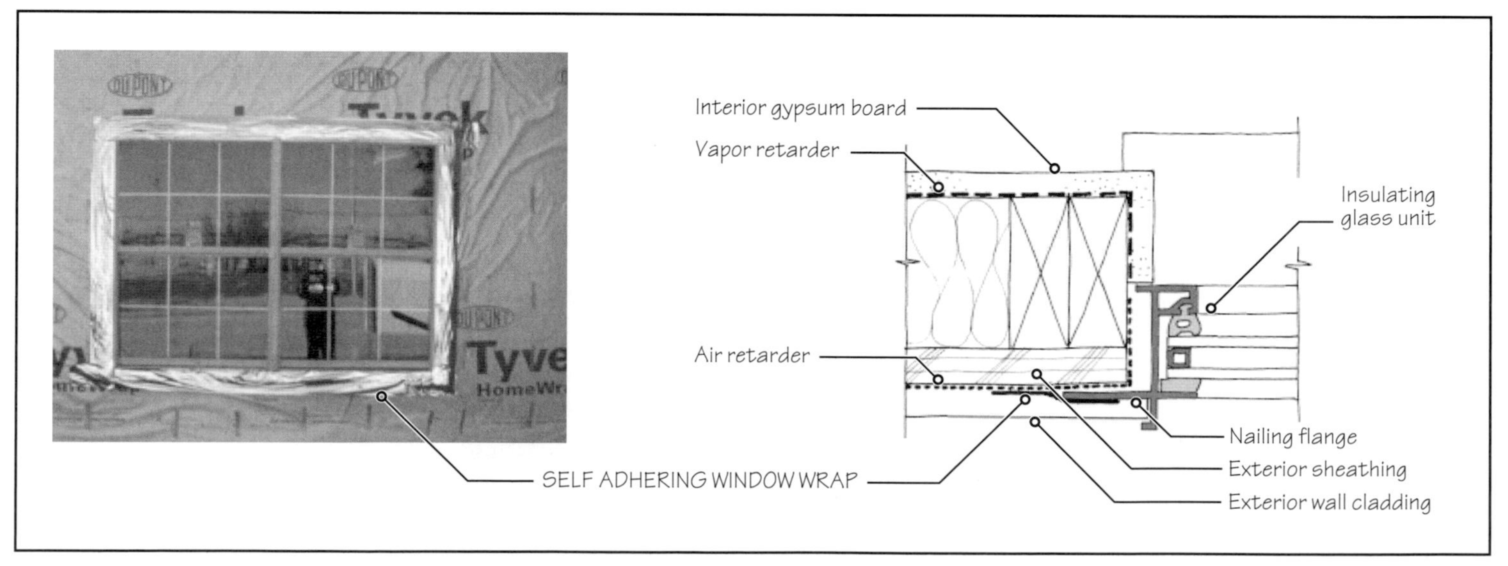

FIGURE 6.6 A proprietary self-adhering tape on the outside of a window (referred to as window wrap) to reduce air leakage through the joint between the window and the wall, photographed before the application of exterior cladding.

EXPAND YOUR KNOWLEDGE

Air Leakage and Indoor Air Quality

The quality of indoor air in buildings has come under much scrutiny in the last few years due to complaints from the occupants—of headache, eye, and nasal irritation, fatigue, and unacceptable odors. Studies conducted by the Environmental Protection Agency (EPA) in the United States have concluded that the indoor air may become excessively polluted, creating what has come to be known as *sick building syndrome.*

In addition to dust, pollen and microorganisms (fungi and mold), contaminants routinely added to indoor air are carbon dioxide, generated from combustion and human metabolism, and carbon monoxide from combustion and smoking. Several volatile organic compounds (VOC) are present due to the use of aerosols, cleaning chemicals, and disinfectants (see also Chapter 10).

Formaldehyde, a colorless gas, is a major source of indoor air pollution. Although a VOC, formaldehyde is generally listed separately because of its widespread use in building products in the form of urea formaldehyde. Urea formaldehyde is an adhesive used in wood veneer and wood fiber products such as plywood, particleboard, and fiberboard. Other sources of formaldehyde are carpet pads and underlayments, ceiling tiles, upholstery, and wall coverings.

Another dangerous source of indoor air pollution is radon, a colorless, odorless, radioactive gas present in some soils, which may leak into the interior through cracks in the floors. Radon is produced by the decay of radium, a substance that occurs in most soils. However, it is dangerous only when its concentration is high. Radon gas particles affect human lungs in the same way as smoking, that is, it increases the risk for lung cancer.

If buildings are adequately ventilated, these pollutants are generally of little concern. However, due to the relative airtightness of the envelope (doors, windows, walls, etc., and the use of air retarders) in modern energy-efficient houses, concern about indoor air quality in buildings has grown. This is particularly significant as the population of western countries ages, with a greater number of older people spending less time in fresh outdoor air.

Whereas an excessively tight envelope may save energy, it creates undesirable effects just mentioned. In order to provide the necessary amount of fresh air with a minimum impact on energy consumption, a system of controlled mechanical ventilation may be used. In this system, stale air is exhausted from one vent in the building and fresh air drawn through another vent. The heat from the stale air is used to warm the fresh air through a device called the *air-to-air heat exchanger.*

PRACTICE QUIZ

Each question has only one correct answer. Select the choice that best answers the question.

1. Air leakage through a building envelope is a function of
- **a.** inside-outside air pressure difference.
- **b.** inside-outside air temperature difference.
- **c.** outside wind speed.
- **d.** all the above.
- **e.** (a) and (c) only.

2. Which of the following is a more appropriate term?
- **a.** Air barrier
- **b.** Air retarder

3. An air retarder is available in many varieties, but in all of them, it is made from
- **a.** an unperforated plastic sheet.
- **b.** a perforated plastic sheet.
- **c.** an asphaltic felt.
- **d.** any one of the above.
- **e.** none of the above.

4. An air retarder is placed on the warm side of the envelope assembly.
- **a.** True
- **b.** False

5. An air retarder is generally used to wrap
- **a.** the walls.
- **b.** the ceiling.
- **c.** the roof.
- **d.** all the above.

6. The use of an air retarder is more critical in a
- **a.** hot, humid climate.
- **b.** hot, dry climate.
- **c.** temperate climate.
- **d.** cold climate.

7. The use of air retarders in buildings improves the quality of indoor air.
- **a.** True
- **b.** False

Answers: 1-d, 2-b, 3-b, 4-b, 5-a, 6-d, 7-b.

6.3 WATER VAPOR IN AIR

It is an observable fact that all air contains some water vapor. If water in a container is left exposed to air, the water disappears after a few days. This disappearance of water, referred to as *evaporation*, is, in fact, the conversion of water into water vapor. This property of converting water to water vapor and mixing with the surrounding air allows wet clothes and other materials to dry.

Water vapor is like steam, but at temperatures below the boiling point of water. It mixes readily with air and is invisible. In fact, air is not only a mixture of nitrogen, oxygen, and carbon dioxide, but also of water vapor. Absolutely dry air, that is, air without water vapor, is rare. It is produced only in laboratories for special purposes. Water vapor content in air is usually higher in coastal areas than inland locations.

Like air, water vapor is a gas. Consequently, it exerts pressure on the surfaces of the enclosure containing it. However, although air and water vapor are thoroughly mixed

TABLE 6.1 SATURATION WATER VAPOR CONTENT AND SATURATION VAPOR PRESSURE AS FUNCTION OF AIR TEMPERATURE

Temperature (°F)	Saturation vapor content (grains/lb of dry air)	Saturation vapor pressure (psf)
−20	2.0	1.0
−10	3.0	1.5
0	5.5	2.5
10	9.0	4.5
20	15.0	7.5
30	24.0	11.5
40	36.5	17.5
50	53.5	25.5
60	77.5	37.0
70	110.0	52.5
80	155.5	73.0
90	217.5	100.5
100	301.0	137.0
110	414.0	184.0

1 lb of water = 7,000 grains of water

together, the pressure exerted by water vapor is independent of that exerted by air.* In other words, air pressure and vapor pressure do not add up but act separately.

Vapor pressure is directly related to the amount of water vapor present in air. Air cannot contain an unlimited amount of water vapor. When the air contains the maximum amount of water vapor it can possibly hold, it is referred to as *saturated air,* and the corresponding vapor pressure is referred to as the *saturation vapor pressure.*

The amount of water (in the form of water vapor) in saturated air increases with air temperature, Table 6.1. For instance, the amount of water in saturated air at 0°F is nearly 5.5 grains per pound of dry air. At 20°F, the amount of water is approximately 15.0 grains per pound of dry air; at 40°F, the corresponding amount of water is 36.5 grains; at 60°F, it is 77.5 grains, and so on.

Because the vapor pressure is directly related to the amount of water in air, the saturation vapor pressure also increases with air temperature, as shown in Table 6.1.

RELATIVE HUMIDITY OF AIR

The occurrence of saturated air is relatively uncommon. Outside air is saturated with water vapor only during or immediately after a rain shower. Inside air is seldom saturated. In other words, the inside air at a given temperature contains less water vapor than saturated air (at that temperature).

The amount of water vapor in air is usually not given in terms of its *absolute* value, but as a *relative* amount of water vapor, referred to as the *relative humidity* (RH) of air. The relative humidity of air is expressed in percentage form by the following relationship:

$$RH = \frac{\text{weight of water (as vapor) in air}}{\text{weight of water (as vapor) in saturated air}} \qquad \textbf{Eq. (1)}$$

Because the amount of water vapor in air and the vapor pressure of air are directly related, the relative humidity is also defined by the ratios of vapor pressures:

$$RH = \frac{\text{vapor pressure of air}}{\text{vapor pressure of saturated air}} \times 100 \qquad \textbf{Eq. (2)}$$

For saturated air, RH = 100%. Similarly, a 45% RH means that the air contains 45% of the amount of water present in saturated air. More specifically, 1 lb of air at 45% RH and at 70°F contains (0.45)110 = 50 grains of water as water vapor (Table 6.1). From Equation (2), this mass of air will exert (0.45)52.5 = 23.6 psf of vapor pressure.

*Independence of pressures exerted by two gases that are mixed together is as per *Dalton's law of partial pressures.*

NOTE

Determining Vapor Pressure in Air

Example
Calculate the vapor pressure in air whose temperature is 70°F and whose relative humidity (RH) is 45%.

Solution
From Table 6.1, the vapor pressure of saturated air at 70°F is 52.5 psf. Because RH = 45%, the vapor pressure of air, from Equation (2), is

$$\left(\frac{45}{100}\right) 52.5 = 23.6 \text{ psf}$$

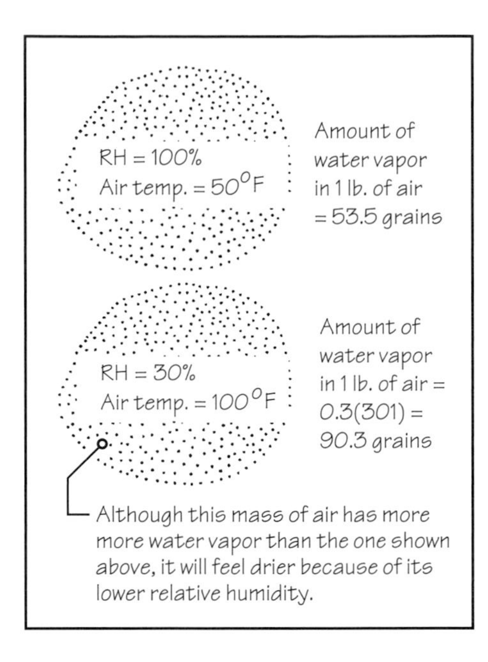

FIGURE 6.7 Human sensation of dryness or dampness of air is a function of its relative humidity, not the absolute humidity value.

The use of relative humidity of air (in place of absolute humidity) is based on the human sensitivity to water vapor content in air. Air feels drier or damper, depending on its relative humidity, not on its absolute humidity. For example, air at 30% RH and 100°F, which contains approximately 90 grains of water vapor, feels drier than air at 100% RH and 50°F, which contains 53.5 grains of water vapor, Figure 6.7.

Air with a low RH dries human skin, produces static electricity in carpeted interiors, causes respiratory health problems, and is generally uncomfortable. Air with a high RH feels moist, promotes fungal growth, and is also uncomfortable and unhealthy. The RH of mechanically conditioned air in offices is generally kept at about 45%.

MOVEMENT OF WATER VAPOR THROUGH ASSEMBLIES

Building assemblies are generally more vapor permeable than air permeable. This means that water vapor can pass through walls and roofs with greater ease than air. The reason is that the vapor pressure difference between the inside and the outside is generally much higher than the corresponding air pressure difference, as explained shortly.

In modern buildings, mechanical systems are designed to keep the inside air at constant temperature and relative humidity throughout the year—nearly 70°F air temperature and 45% RH. As determined previously, the vapor pressure of this air is 23.6 psf. The outside vapor pressure changes continuously, depending on the time of the day and the weather.

Let us now estimate the inside-outside vapor pressure differential on a typical day during the heating season when the outside air temperature is 10°F and the relative humidity is 80%.

From Table 6.1, the vapor pressure of this air is 0.8(4.5) = 3.6 psf. Therefore, the vapor pressure differential between the inside and outside is (23.6 − 3.6) = 20.0 psf, Figure 6.8(a).

This is a large pressure differential. For wind to cause the same magnitude of pressure differential (20 psf), its speed must be nearly 90 mph—hurricane wind speed—a condition that rarely occurs. On a moderately windy day, with the wind blowing at 10 mph, the

NOTE

Pressure Difference Between the Inside and the Outside Air Generated by Wind Speeds of 90 mi/h and 10 mi/h

We observed in Section 3.5 (in the box entitled "Some Important Facts About Wind Loads") that wind speed of V miles per hour creates a pressure difference of approximately $[V/20]^2$ psf. Thus, if V = 90 mph, the pressure difference between the inside and outside air is $[90/20]^2 = 20.25$ psf.

Similarly, if V = 10 mph, the inside-outside air pressure difference is $[10/20]^2 = 0.25$ psf.

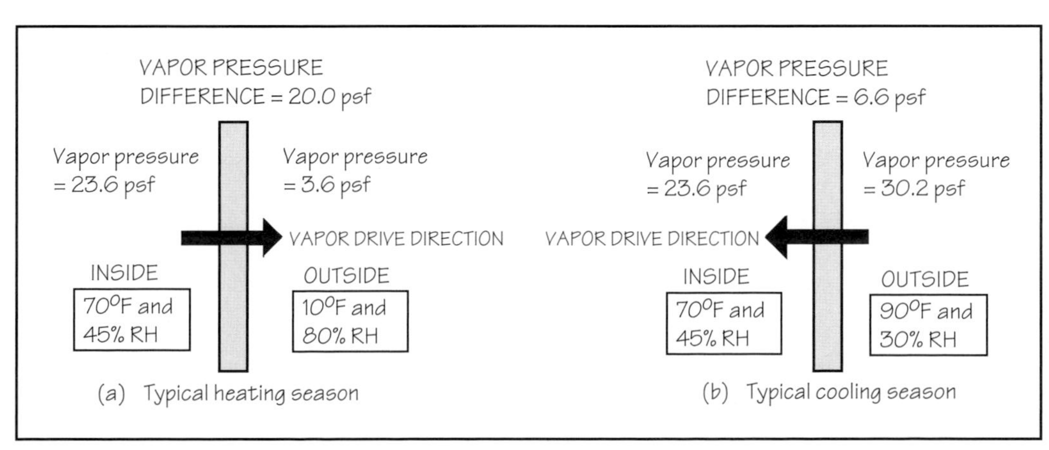

FIGURE 6.8 Differences between indoor and outdoor vapor pressures during typical summer and winter conditions.

inside-outside air pressure differential is only about 0.25 psf, which is much smaller than the vapor pressure differential of 20 psf.

Because the typical difference between inside-outside vapor pressure is much greater than inside-outside air pressure, vapor can move through building assemblies with much greater ease than air. This fact also explains why air retarders are vapor permeable but not air permeable. Because of the high inside-outside vapor pressure differential in buildings, vapor can pass through the microperforations in an air retarder with ease. On the other hand, very little air passes through them because of the low air pressure differential. (In Section 6.7, we will see why it is necessary for air retarders to be vapor permeable.)

Figure 6.8(b) shows the inside-outside vapor pressure difference that may occur on a typical day during cooling season. The corresponding vapor pressure difference (nearly 7 psf) is much smaller than that produced on a typical winter day (nearly 20 psf). This implies that vapor drive is much more forceful during the winters than the summers.

Figure 6.8 also shows the direction of vapor flow. During the heating season, the vapor moves from the inside to the outside. The direction may reverse during the cooling season. Observe that, in general:

Water vapor moves from the warm to the cold side of an assembly.

6.4 CONDENSATION OF WATER VAPOR

Consider air at a certain temperature and relative humidity. If no moisture is added or subtracted from this air and its temperature is decreased, its relative humidity will increase. If the decrease in temperature continues, a temperature will be reached at which the RH of the air is 100%. The temperature at which the air's RH becomes 100% (i.e., when the air becomes saturated) is called its *dew point temperature,* or simply the *dew point* of air. If the temperature of the air is decreased below the dew point, the water vapor in air will convert to (liquid) water—a phenomenon known as *condensation.*

Condensation occurs commonly in nature. The surface of a cup containing ice or cold water becomes wet because the temperature of the surface of the cup is below the dew point of warm and humid ambient air. As this air comes in contact with the surface of the cup, condensation of water vapor (present in air) occurs, which deposits on the surface of the cup. In heated building interiors, condensation is often observed during the winter on window glass. Such condensation is more pronounced in more humid interiors such as indoor swimming pools, aerobic centers, gymnasiums, gang showers, and hotel kitchens.

NOTE

The Requirement of Both the Existence of Dew Point and Presence of a Vapor Impermeable Surface for Condensation

Note that condensation occurs only on surfaces through which water vapor cannot pass (permeate). In other words, it will continue to move freely until it encounters a surface that interrupts its movement. At that location, condensation of vapor is possible if the temperature of the surface is less than or equal to the air's dew point. In the examples cited, the condensation occurs on the surface of the cup or window glass because they are impermeable to water vapor.

Interstitial and Surface Condensation

Apart from condensing on the surfaces of a window glass or any other cold surface, warm interior air can also condense inside a wall or roof assembly. If the water vapor can permeate into a wall or roof assembly, it will condense where the temperature of the assembly is at or below the dew point of the permeating vapor. The condensation of vapor inside an envelope assembly is referred to as *interstitial condensation,* as opposed to *surface condensation,* which occurs on the envelope's surface, such as a window glass.

Although both surface and interstitial condensation are undesirable, the latter is more so. Interstitial condensation wets the assembly, which accelerates the corrosion of metals and the decay of wood, decreases the R-value of insulation, and generally reduces the strength of materials. Interstitial condensation may also contribute to the growth of fungi and mold—a major health hazard.

Therefore, the control of condensation in buildings, particularly the interstitial condensation, is important. However, water vapor by itself is not damaging to a building assembly. It becomes damaging when it converts to water.

6.5 CONTROL OF CONDENSATION

Surface condensation can be prevented by simply increasing the R-value of the assembly. The reason condensation occurs on a single glass sheet is because the R-value of glass is quite low (R = 1). If the single glass sheet is replaced by an insulating glass unit (say, R = 3), condensation on the glass will not occur unless the interior air is very humid.

In general, if the R-value of the envelope is low, the dew point of air occurs at the surface of the envelope, leading to surface condensation. If the R-value of the envelope is raised, the dew point shifts to within the body of the envelope, Figure 6.9. This leads to interstitial

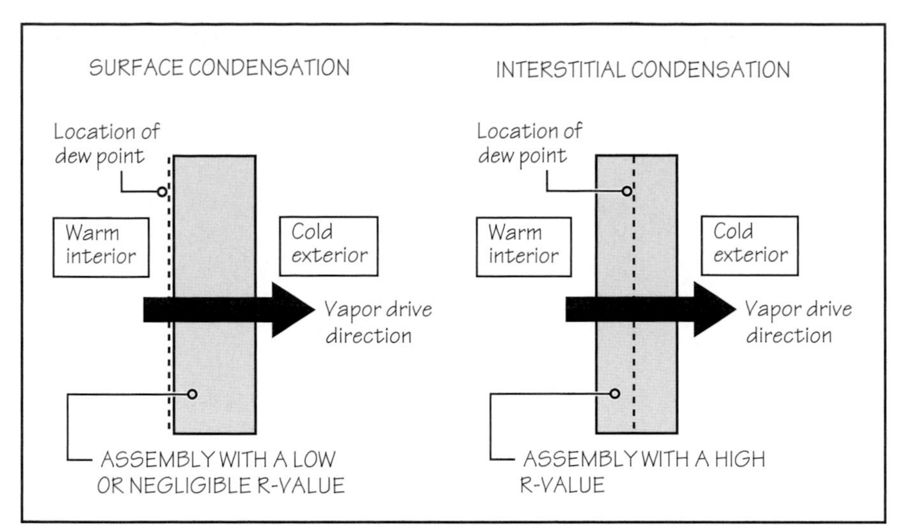

FIGURE 6.9 Effect of R-value of the assembly on the location of dew point.

condensation, which should be avoided. (See "Principles in Practice" at the end of this chapter to find how to locate the dew point in an envelope assembly).

Note that interstitial condensation became a concern only after the introduction of insulation in buildings. In earlier times, when the buildings were not insulated (low R-value), interstitial condensation did not occur, because the dew point generally occurred on the interior surface, creating surface condensation.

CONTROL OF INTERSTITIAL CONDENSATION—THE USE OF VAPOR RETARDER

As stated earlier, interstitial condensation occurs only if the water vapor is able to enter into the envelope assembly and is then unable to get out. Therefore, a two-part strategy is used to control interstitial condensation. The first part, which is discussed in this section, is to prevent the entry of water vapor into the assembly. The second part is to facilitate the exit of any vapor that has entered the assembly through ventilation, which is discussed in Section 6.7.

The entry of water vapor into an envelope assembly can take place by one or both of the following modes:

- Vapor movement through air leakage
- Vapor diffusion (due to vapor pressure differential)

Because air and water vapor are thoroughly mixed, any leakage of air through the envelope is also accompanied by water vapor. Because most air leakage takes place at cracks, penetrations, and unsealed joints, sealing the envelope prevents this mode of vapor flow.

The second mode of vapor flow takes place independently of air flow. It occurs due to the inside-outside vapor pressure differential and is called *vapor diffusion.* Thus, vapor diffuses from a region of higher vapor pressure to that of a lower vapor pressure.

How can we prevent the diffusion of water vapor into the envelope assembly? This is accomplished by the use of a *vapor barrier.* A vapor barrier is a material that is impermeable to vapor. It is also impermeable to air because if air leaks through it, so will vapor. Thus, a vapor barrier is both an air-vapor barrier. Note that the term *vapor retarder* is more appropriate, because most commercial vapor barriers are not absolutely vapor impermeable.

6.6 MATERIALS USED AS VAPOR RETARDERS

The rate at which water vapor flows (diffuses) through a material is measured by its vapor permeability. The unit of vapor permeability is called the *permeance,* or simply the *perm.* A material with a higher perm value (or rating) is more vapor permeable. If the perm rating is zero, the material is impermeable to vapor flow. Such a material is a perfect vapor retarder—a vapor barrier—provided it is free of holes, cracks, and unsealed joints.

Apart from being a property of the material, perm rating is also a function of the material's thickness. A larger thickness of the same material has a lower perm rating. Therefore, when the perm rating is given, the thickness of the component must be stated. The perm ratings of selected materials are given in Table 6.2. For instance, the perm rating of a

NOTE

Units of Permeance (Perm Rating)

U.S. System of Units

1 perm = 1 grain of water vapor passing through 1 ft^2 of a material in 1 h under a vapor pressure difference of 1 in. of mercury.

SI System of Units

1 perm = 1 mg of water vapor passing through 1 m^2 of a material in 1 s under a vapor pressure difference of 1 Pa.

1 (U.S) perm = 57.2 (SI) perm

TABLE 6.2 APPROXIMATE PERM RATINGS OF SELECTED MATERIALS (IN U.S. SYSTEM OF UNITS)

Component	Perm rating (perm)	Component	Perm rating (perm)
Aluminum foil (unpunctured)	0.0	Brick masonry, 4 in. thick	0.8
Aluminum foil on gypsum board	0.1	Concrete block masonry, 8 in. thick	2.4
Built-up roofing, 3- to 5-ply	0.0	Plaster on metal lath	15.0
15-lb asphalt felt	4.0	Building paper, grade D	5.0
PVC (plasticized), 4 mil thick	1.2	Interior primer plus 1 coat flat oil paint on plaster	1.6–3.0
Polyethylene sheet, 4 mil thick	0.08	Exterior oil paint, 3 coats on wood	0.3–1.0
Polyethylene sheet, 6 mil thick	0.06	Commercial air retarders	15–60
Polyethylene sheet, 8 mil thick	0.03		

4-mil-thick polyethylene sheet is 0.08 (in the U.S. system of units); a 6-mil-thick polyethylene sheet's perm rating is 0.06.

For a material to qualify as a vapor retarder, its perm rating must be 1.0 perm (in the U.S. system of units) or less. However, for an effective vapor retarder, a much lower value—less than or equal to 0.1 perm—is recommended.

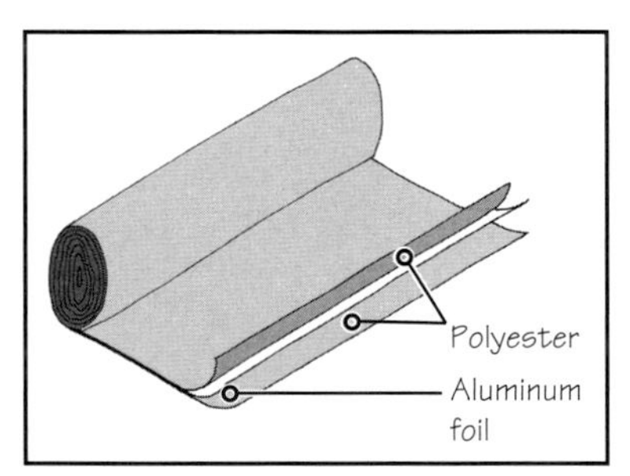

FIGURE 6.10 Aluminum foil bonded to two sheets of polyester to provide a vapor retarder with nearly zero perm rating.

Glass and metals (even a thin metal foil, such as aluminum foil) have a zero perm rating. Roof membranes—built-up roof, modified bitumen, and single-ply membranes—are excellent vapor retarders. Plastics have low perm ratings and are commonly used as vapor retarders. Asphalt-treated paper, generally referred to as *building paper* or *kraft paper*, is also a good vapor retarder (Table 6.2). Several fiberglass insulation manufacturers make fiberglass batts faced with a low-perm sheet material, such as building paper (Section 5.7).

Another commonly used vapor retarder is aluminum foil. It has a zero perm rating, provided it has no holes and (or) punctures. However, because aluminum foil does not have the required strength, it is usually laminated to another material. One manufacturer produces a vapor retarder consisting of 1-mil-thick aluminum foil sandwiched between two 0.5-mil-thick sheets of polyester, Figure 6.10. In this composite product, the aluminum foil provides zero permeance, and the polyester sheet provides tensile strength and puncture resistance.

Because the polyester sheet also has a low permeability, any small holes in the aluminum foil are sealed by the polyester sheet so that the composite product has nearly a zero perm rating. The polyester sheet also helps to reduce the corrosion of the aluminum foil.

In addition to the products just described, there are several other products with low-perm ratings, such as rubberized asphalt membranes, liquid-applied emulsions, and reinforced plastic sheets.

PRACTICE QUIZ

Each question has only one correct answer. Select the choice that best answers the question.

8. In a typical dwelling in Northern United States, the vapor pressure difference between the inside-outside air is generally
- **a.** the same during the heating and cooling seasons.
- **b.** smaller during the heating season than the cooling season.
- **c.** greater during the heating season than the cooling season.

9. The dew point of air is always
- **a.** less than or equal to the air's temperature.
- **b.** greater than or equal to the air's temperature.
- **c.** dew point is unrelated to the air's temperature.

10. In heated buildings, the vapor flow is generally
- **a.** from the inside to the outside of the envelope.
- **b.** from the outside to the inside of the envelope.

11. Which of the following materials has the smallest perm rating?
- **a.** Plastic sheet
- **b.** Metal sheet
- **c.** Brick wall
- **d.** Concrete wall
- **e.** All the above materials have the same perm rating.

12. In which of the two assemblies, one with a negligible R-value (assembly A) and the other with a high R-value (assembly B), will interstitial condensation occur?
- **a.** Assembly A
- **b.** Assembly B

13. Perm rating is a property that measures
- **a.** how effectively the air passes through the material.
- **b.** how effectively water vapor passes through the material.
- **c.** both (a) and (b).
- **d.** neither (a) nor (b).

14. In general, the greater the thickness of a material, the greater its perm rating.
- **a.** True
- **b.** False

15. For a material to be considered a vapor retarder, its perm rating in the U.S. System of units must be:
- **a.** greater than 2.0.
- **b.** less than or equal to 2.0.
- **c.** greater than 1.0.
- **d.** less than or equal to 1.0.
- **e.** none of the above.

Answers: 8-c, 9-a, 10-a, 11-b, 12-b, 13-b, 14-b, 15-d.

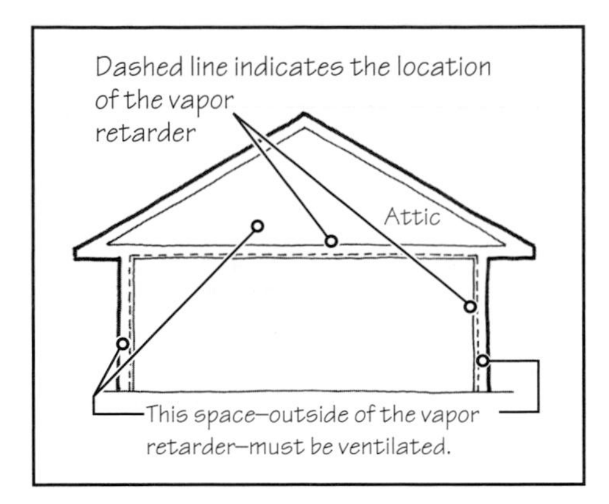

FIGURE 6.11 Location of vapor retarder in a conventional residential construction in North America.

6.7 LOCATION OF VAPOR RETARDER AND VENTILATION IN THE SPACE BEYOND THE VAPOR RETARDER

As shown in Figure 6.8, vapor flows from the inside to the outside during the winter. In summer, the direction of vapor flow may reverse (from outside to inside), particularly in warm coastal regions. The building codes require that the vapor retarder should be installed on the *warm-in-winter side of the insulation,* that is, on the interior face of the envelope.

In warm and humid climates of North America, where the higher vapor pressure may occur on the outside face of the insulation, the codes do not require the use of a vapor retarder. A vapor retarder is also unnecessary in buildings that are not insulated.

VAPOR RETARDER IN CONVENTIONAL RESIDENTIAL BUILDINGS

Because there are several vapor-producing activities in a residential building, a 4- to 6-mil-thick polyethylene sheet is typically installed in the walls and the ceiling between the interior gypsum board and the insulation, Figure 6.11 (also Figure 6.12).

In addition to using a vapor retarder, it is also important to ventilate the envelope assembly between the vapor retarder and the outside. The idea is that if any vapor enters into the assembly through the vapor retarder, it should be able to mix with the (infinite mass of) outside air to prevent its condensation within the assembly. That is why a typical attic must be ventilated. Without adequate attic ventilation, the vapor will be trapped in the attic, where it can condense.

In a wall or a flat roof assembly, vapor that passes through the vapor retarder must travel through several different layers of material before reaching outside air, and they must all be vapor permeable.

Because the dew point generally occurs within the insulation, the insulation must be vapor permeable. That is why the most commonly used insulation in residential buildings consists of fiberglass batts or blankets. For the same reason, the air retarder, which is placed on the exterior face of the walls, must also be vapor permeable and, hence, perforated. If the air retarder is not perforated, any vapor that travels through the vapor retarder will be trapped between the vapor and air retarders and will condense there.

IMPORTANCE OF ATTIC VENTILATION

Ventilating the space beyond the vapor retarder is particularly important in attics. It is also required for reasons other than condensation control. In cold climates, ventilation prevents the formation of ice dams at projecting roof eaves. If the attic is not ventilated, it remains relatively warm because of the entry of heat into the attic from the warm interior, but the eave overhangs are cold. This melts the snow in the middle of the roof, which freezes again at the eaves, forming ice dams, Figure 6.12(a). The problem is worse in an uninsulated roof-ceiling assembly.

Ice dams can substantially load the roof at eave overhang and gutter. They also prevent water from draining off the roof, which may cause roof leakage. Ceiling insulation, coupled with the ventilation of the attic, helps to eliminate ice dams as well as the buildup of water vapor, Figure 6.12(b).

Attic ventilation is also required in warm climates to reduce heat transmission from the ceiling into the interior of the building. The temperature of the air in an unventilated attic becomes much higher than the outside air because of the ability of air and the materials of the roof to store heat. This is particularly critical during the summer.

Temperature differences between the outside air and the attic air in excess of 40°F have been reported in unventilated or inadequately vented attics. The higher attic temperature

NOTE

Vapor Retarder Not Required in Warm Climates

The International Residential Code lists the counties in the United States where the use of a vapor retarder is not needed in building assemblies.

NOTE

Perm Rating of an Air Retarder

As per ASTM E1677 standard, the minimum required perm rating of an air retarder is 5.0 U.S. perms, which is five times the maximum perm rating of a vapor retarder. The perm rating of commercial air retarders is much greater than 5.0.

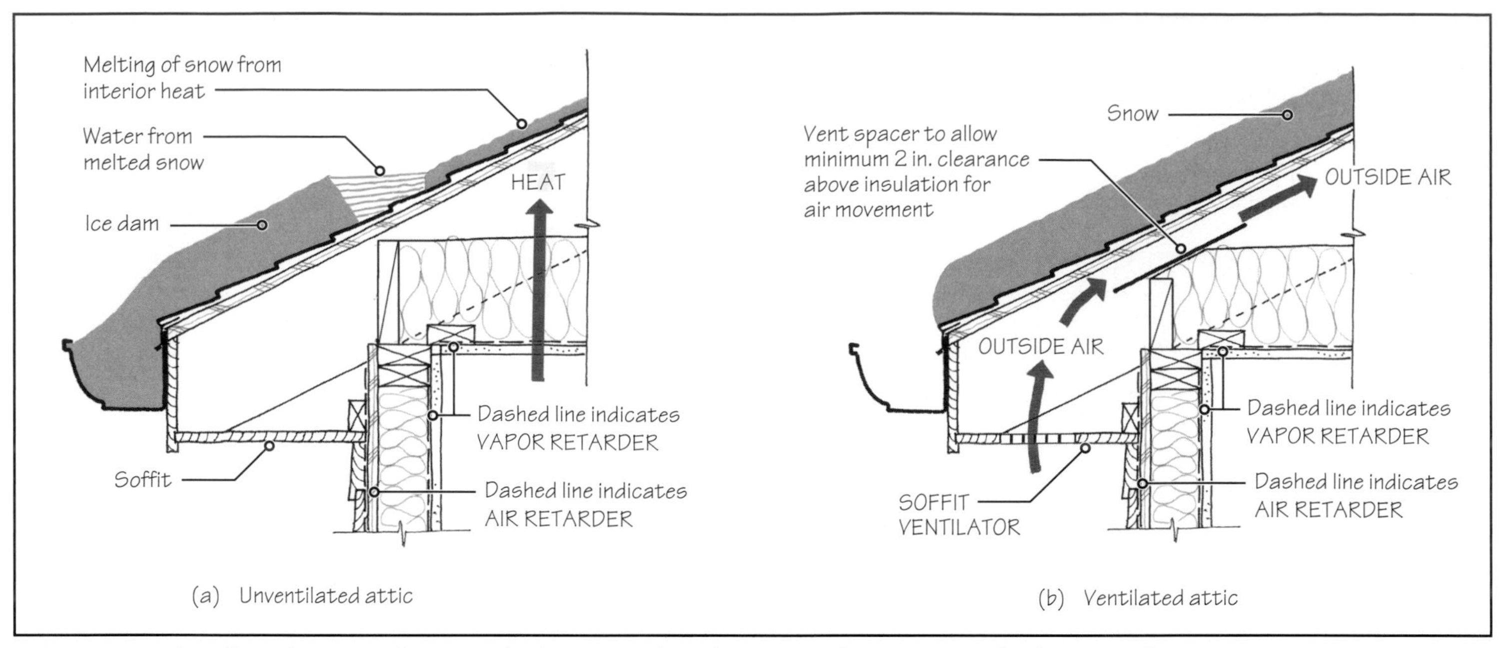

FIGURE 6.12 The effect of attic ventilation on the formation of ice dams at overhanging eave of a sloping roof.

increases heat transfer to the interior of the building. As shown in Figure 6.12(b), attic ventilation can be provided through openings at the soffit. The openings must be covered with a mesh or screen to prevent the entry of insects. Building codes mandate a certain minimum area of soffit vents.

Soffit ventilation is considered as *intake ventilation.* Adequate ventilation of the attic requires cross ventilation, using intake as well as *exhaust ventilation.* Because warm air rises, exhaust ventilation is required at a higher level in the attic. Four different alternatives are used for exhaust ventilation, Figure 6.13:

- Gable ventilators
- Ridge ventilators
- Turbine ventilators
- Gable fans

6.8 VAPOR RETARDER UNDER A CONCRETE SLAB-ON-GROUND

A vapor retarder is also required under a concrete slab-on-ground in all climates to prevent subsoil moisture from diffusing through the slab to the interior of the building. A 6-mil-thick polyethylene sheet with overlapped joints is recommended as a vapor retarder under the slab. A layer of coarse sand below the vapor retarder drains water away from the vapor retarder by eliminating capillary action. Additionally, the sand functions as a protective cushion for the vapor retarder (see Section 19.11).

A vapor retarder is not required under a slab-on-ground if the migration of moisture is not detrimental to the occupancy of the building. Such buildings include garages, carports, and other unconditioned spaces. A vapor retarder is also not required under driveways, walkways, patios, and so forth.

EXPAND YOUR KNOWLEDGE

Vapor-Tight Construction and Attic Ventilation

Although preventing condensation in an attic by providing attic ventilation is perhaps the safest approach and one that is currently mandated by building codes, it is not without disadvantages. In extremely cold climates with frequent blowing snow and rain, venting can cause serious problems by permitting snow and rain to infiltrate the vents. Experience in cold Canadian climates has demonstrated that if indoor humidity levels are controlled and the ceiling is provided with a well-installed vapor retarder with an extremely low vapor permeance (vapor-tight construction), it is possible to prevent attic condensation without providing attic ventilation.

A major advantage of eliminating attic ventilation is that it improves the effectiveness of attic insulation. However, if attic ventilation is not provided, the problem of ice dam formation should be investigated or prevented by eliminating eave overhangs.

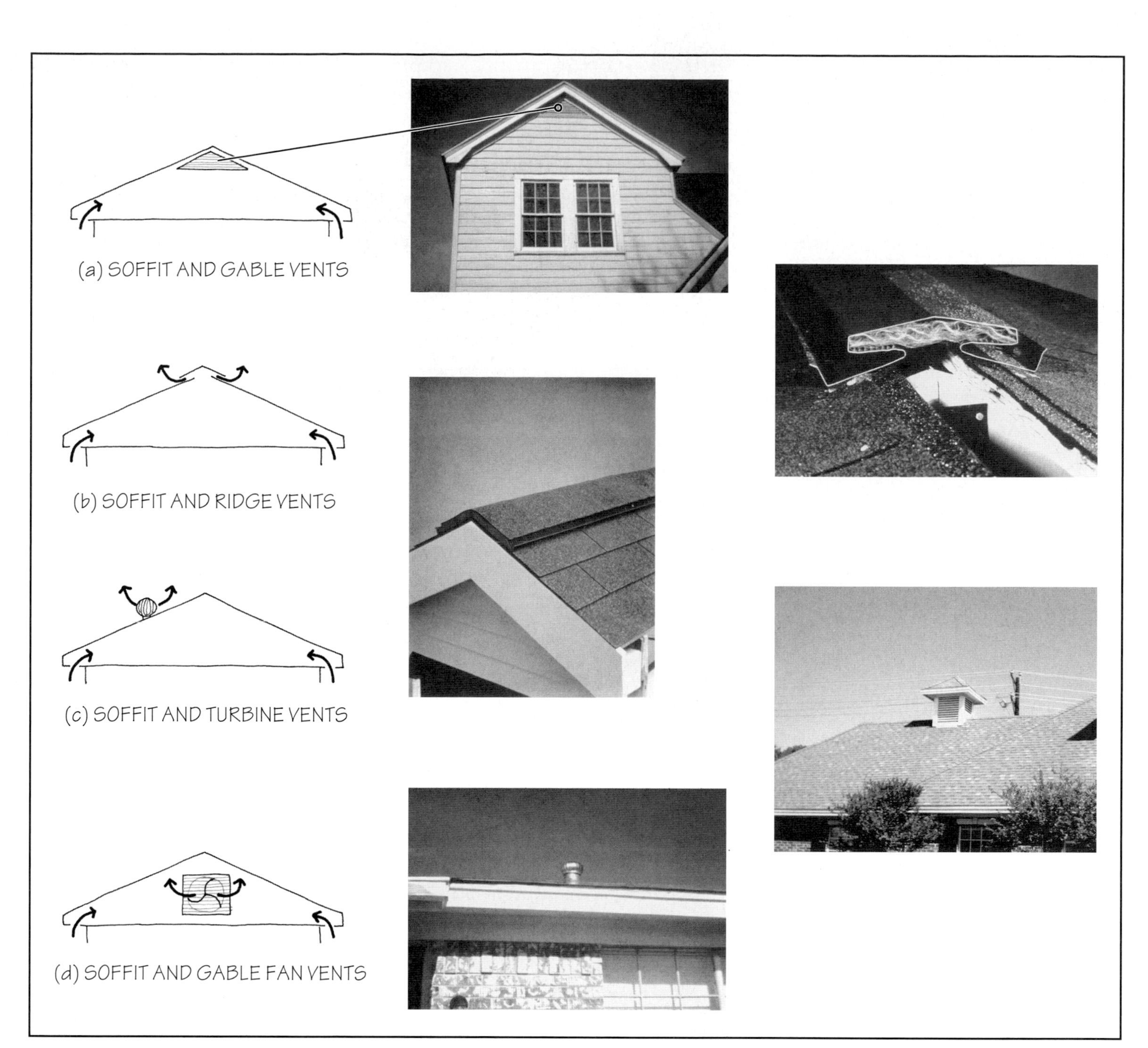

FIGURE 6.13 Various alternatives of providing attic ventilation. Soffit vents function as intake ventilators, while gable vents, ridge vents, turbine vents or gable fan vents function as exhaust ventilators. Any one of the four exhaust ventilation options may be used along with soffit vents.

PRACTICE QUIZ

Each question has only one correct answer. Select the choice that best answers the question.

16. A vapor retarder is required to be placed on the
- **a.** cold side of the assembly.
- **b.** warm side of the assembly.
- **c.** warm-in-winter side of the insulation.
- **d.** warm-in-summer side of the insulation.
- **e.** cold-in-winter side of the insulation.

17. A vapor retarder is perforated so that
- **a.** water may pass through the pores.
- **b.** water vapor may pass through the pores.
- **c.** air may pass through the pores.
- **d.** a vapor retarder is not perforated.

18. Ventilation of an attic space is generally considered important in
- **a.** cold climates only.
- **b.** moderately cold climates only.
- **c.** hot climates only.
- **d.** all climates.

19. Soffit vents alone are not adequate to provide attic ventilation.
- **a.** True
- **b.** False

20. When gable vents are provided to ventilate an attic space
- **a.** ridge vents are also necessary.
- **b.** turbine vents are also necessary.
- **c.** soffit vents are also necessary.
- **d.** no other vents are necessary.

21. An ice dam is likely to occur
- **a.** in the middle of a sloping roof.
- **b.** at the eave of a sloping roof.
- **c.** in the middle of a flat roof.
- **d.** at the edge of a flat roof.

Answers: 16-c, 17-d, 18-d, 19-a, 20-c, 21-b.

PRINCIPLES IN PRACTICE

Where Dew Point Occurs in an Assembly

TEMPERATURE GRADIENT ACROSS AN ASSEMBLY

Consider a multilayer assembly. Let the R-values of various layers of the assembly be R_1, R_2, R_3, . . ., so that $R_t = R_1 + R_2 + R_3 + \cdots$, as shown in Figure 1. Let the difference between the interior and exterior surface temperatures of this assembly be Δt. From Equation (2), Chapter 5, the heat flowing through the assembly (q) is given by

$$q = \frac{\Delta t}{R_t}$$

Because the heat flowing through each layer must be the same as that flowing through the entire assembly, it follows that:

$$q = \frac{\Delta t}{R_t} = \frac{\Delta t_1}{R_1} = \frac{\Delta t_2}{R_2} = \frac{\Delta t_3}{R_3}, \ldots$$

Thus, the temperature drop (Δt_1) across layer 1 is

$$\Delta t_1 = \frac{R_1}{R_t}\Delta t$$

Similarly, the temperature drop (Δt_2) across layer 2 is

$$\Delta t_2 = \frac{R_2}{R_t}\Delta t$$

Temperature drop across the entire assembly $= \Delta t_1 + \Delta t_2 + \Delta t_3 = \Delta t$

R_1	R_2	R_3
Temp. drop across layer $= \Delta t_1$	Temp. drop across layer $= \Delta t_2$	Temp. drop across layer $= \Delta t_3$

FIGURE 1 Temperature drops across various layers of a multilayer assembly. Temperature drop across the entire assembly is $\Delta t_1 + \Delta t_2 + \Delta t_3 = \Delta t$.

The preceding analysis indicates that the temperature drop across a layer is proportional to the R-value of that layer. Let us now apply the above analysis to an assembly consisting of three layers whose R-values are R-2, R-12, and R-1, respectively, Figure 2. Let the interior surface temperature of layer 1 be 70°F and the exterior surface temperature of layer 3 be 10°F, so that $\Delta t = 60$°F. Because $R_t = 15$, the temperature drop across layer 1 is

$$\Delta t_1 = \frac{2}{15}(60) = 8° \text{F}$$

Similarly, $\Delta t_2 = 48$°F, and $\Delta t_3 = 4$°F. The change of temperature from one side of the assembly to the other is referred to as the *temperature gradient.* For the example just discussed, the temperature gradient of the assembly is shown in Figure 2.

Because the temperature drop across each layer is proportional to the R-value of the layer, it implies that if we draw a section through the assembly, using the R-values to represent the thickness of the respective layers, the temperature gradient across the assembly will be a single straight line, Figure 3. This procedure of drawing the temperature gradient is more convenient in some situations (see Example 2).

LOCATION OF DEW POINT IN THE ASSEMBLY

In the example of Figure 2 (and Figure 3), if the relative humidity of the inside air is 50% and that of the outside air is 80%, the inside vapor pressure is much greater than the outside vapor pressure. Therefore, if the assembly is vapor permeable, the vapor will migrate from the inside to the outside.

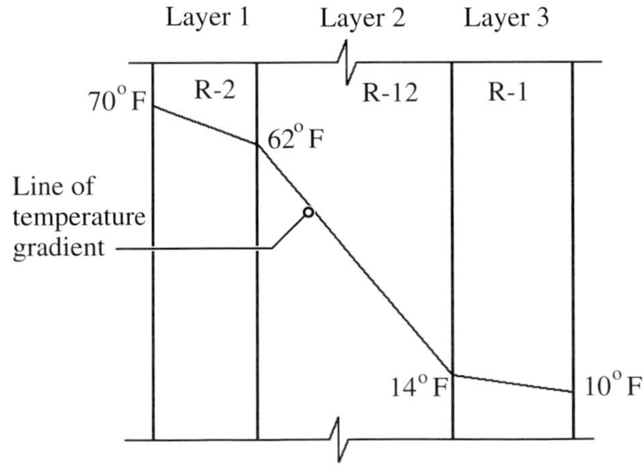

FIGURE 2 Temperature gradient superimposed over the section of a multilayer assembly.

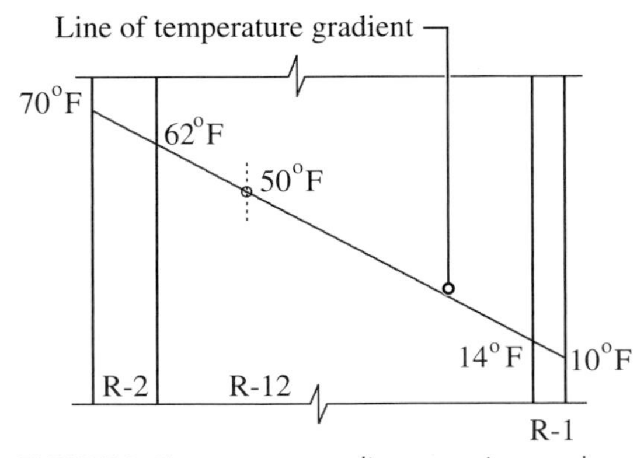

FIGURE 3 Temperature gradient superimposed over the section of the multilayer assembly (of Figure 2). In this illustration, the thickness of a layer is proportional to its R-value. In Figure 2, the thickness of a layer is its actual thickness.

TABLE 1 DEW POINT OF AIR AS A FUNCTION OF ITS TEMPERATURE AND RELATIVE HUMIDITY

	Air temperature (°F)								
RH(%)	30	40	50	60	70	80	90	100	110
100	30	40	50	60	70	80	90	100	110
90	28	37	47	57	67	77	87	97	107
80	25	34	44	54	64	73	83	93	103
70	22	31	40	50	60	69	79	88	98
60	19	28	36	46	55	65	74	83	93
50	15	24	33	41	50	60	69	78	88
40	11	18	27	35	45	53	62	71	81
30	5	14	21	29	37	46	54	62	72
20	3	8	13	20	28	35	43	52	62
10	1	5	6	9	13	20	27	34	43
0	This air will not condense.								

Table 1 gives the dew point of air as a function of its temperature and relative humidity. From this table, the dew point of inside air (70°F and 50% relative humidity) is 50°F. As the vapor migrates toward the outside, it reaches the temperature of 50°F in the middle layer, shown by the dashed line in Figure 2. It is at this location that the migrating vapor will condense.

In order to keep the analysis simple, we have so far ignored the presence of inside and outside surface resistance in the assemblies (Section 5.4). In practice, these should be taken into account, as illustrated in Examples 1 and 2.

EXAMPLE 1

Determine the location of dew point in an insulating glass unit (**IGU**). Assume that the inside surface resistances are $R_{si} = 0.7$, R-value of IGU = 1.0, and outside surface resistance = $R_{so} = 0.2$. Assume that the inside and outside temperatures are 70°F and 10°F, respectively, and the inside relative humidity is 45%.

SOLUTION

$R_t = 0.7 + 1.0 + 0.2 = 1.9$. The temperature drop across inside surface resistance is $(0.7/1.9)60 = 22$°F. Therefore, the temperature of inside surface of glass is $(70 - 22) = 48$°F. From Table 1, the dew point of inside air is 47.5°F. Because the dew point of the inside air is below the temperature of the inside surface of the glass, condensation will not occur on the glass—that is, there is no surface condensation.

Also note that because the IGU is a sealed unit, the air (or water vapor) cannot flow through it. Therefore, there will be no condensation within the IGU—that is, no interstitial condensation. Note, however, that if the IGU is replaced by a single sheet of glass, condensation on the interior surface of glass will occur under these conditions.

The temperature drop across the IGU is $(1.0/1.9)60 = 32$°F. The temperature gradient across the IGU is shown in Figure 4.

FIGURE 4 Temperature gradient through the insulating glass unit of Example 1. Note that the temperature gradient line is horizontal through both panes of glass because the R-value of a glass pane is nearly zero. Also note that the inside surface temperature of glass is 48°F and the dew point of inside air is 47.5°F. Therefore, condensation will not occur on the glass surface.

EXAMPLE 2

Determine the location of dew point in the brick veneer wall assembly of Example 3 (in Chapter 5), consisting of 2 × 4 wood studs spaced 16 in. on center. The spaces between the studs are filled with $3\frac{1}{2}$-in.-thick fiberglass insulation, as shown in Figure 5 (which is a copy of Figure 5.16 in Chapter 5).

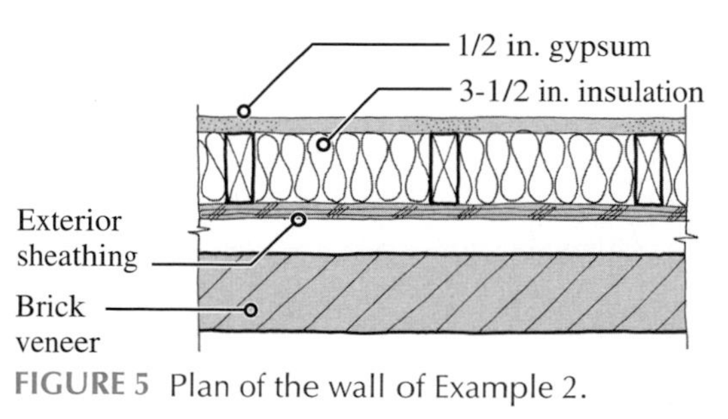

FIGURE 5 Plan of the wall of Example 2.

(*Continued*)

PRINCIPLES IN PRACTICE

Where Dew Point Occurs in an Assembly (*Continued*)

Assume that the inside and outside temperatures are 70°F and 10°F, respectively, and the inside relative humidity is 45%.

SOLUTION

From Example 3 (Chapter 5), the R-values of various layers of the assembly are

Inside surface resistance = 0.7
$\frac{1}{2}$-in.-thick gypsum board = 0.3
$3\frac{1}{2}$-in-thick fiberglass insulation = 12.2
$\frac{1}{2}$-in.-thick plywood = 0.5
2-in.-wide air space = 1.0
$3\frac{5}{8}$-in.-thick brick veneer = 0.7
Outside surface resistance = 0.2

Hence,

$R_t = 0.7 + 0.3 + 12.2 + 0.5 + 1.0 + 0.7 + 0.2 = 15.6$

The temperature gradient across the assembly, when the inside and outside air temperatures are 70°F and 10°F respectively, is shown in Figure 6. Because the relative humidity of inside air is 45%, its dew point is 47.5°F (see Table 1), which occurs within the insulation.

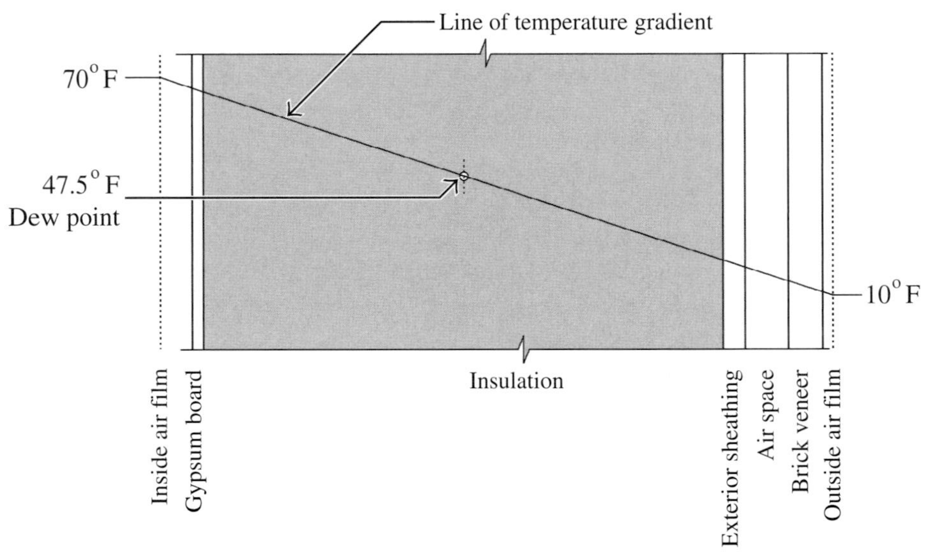

FIGURE 6 Temperature gradient across the vertical section of the wall of Example 2. The thickness of each layer is proportional to its R-value.

KEY TERMS AND CONCEPTS

Air leakage
Air pressure differential
Air retarder
Air-weather retarder
Condensation
Dew point
Interstitial condensation
Perm rating
Relative humidity of air
Surface condensation
Vapor diffusion
Vapor permeability
Vapor pressure differential

REVIEW QUESTIONS

1. Conduct a web search to identify at least two manufacturers of air retarders. How do these products differ from each other?
2. Draw a detailed plan through a typical residential wall and show the locations of the air retarder and vapor retarder.

3. Explain the concept of the relative humidity of air.
4. Explain why it is much easier for vapor to flow through building assemblies than air.

SELECTED WEB SITES

Air Barrier Association of America (www.airbarrier.org)
Building Science Corporation (www.buildingscience.com)
National Research Council of Canada—Institute for Research in Construction (www.irc.nrc-cnrc.gc.ca)
U.S. Army Corps of Engineers—Cold Region Research and Engineering Laboratory (www.crrel.usace.army.mil)
Whole Building Design Guide (www.wbdg.org)

FURTHER READING

1. Beall, Christine. *Thermal and Moisture Protection Manual for Architects, Engineers and Contractors.* New York: McGraw Hill, 1999.
2. Graham, Charles. "Use of Air Barriers and Vapor Retarders in Buildings." *Texas Architect Magazine,* September/October 2004.
3. International Code Council. *International Residential Code.*
4. Lstiburek, Joseph. "Understanding Vapor Barriers." *ASHRAE Journal,* August 2004.

CHAPTER 7

Fire-Related Properties

CHAPTER OUTLINE

7.1 Fire Code and Building Code

7.2 Combustible and Noncombustible Materials

7.3 Products Given Off in a Building Fire

7.4 Fire-Rated Assemblies and Compartmentalization of a Building

7.5 Types of Construction

7.6 Fire-Stopping of Penetrations and Fire-Sealing of Joints

7.7 Fire-Test Response Characteristics of Interior Finishes

7.8 Role of Sprinklers

Because of its noncombustibility and integral fire resistance, Type I(A) construction—the most fire resistive construction type designated by the building codes—is much easier to achieve in a site-cast reinforced-concrete building. (Photo by MM.)

Fires that could engulf an entire neighborhood or a city are rare these days.* Because of sustained improvements in fire safety and zoning regulations and advances in fire detection and suppression equipment used in buildings, fires today are generally limited to individual buildings or a small group of buildings. If fire safety considerations were less stringent than those currently in use, fire of the severity shown in Figure 7.1 would have not only caused the total destruction and collapse of this building, but also have produced a domino effect that would have burned and destroyed a large area of the city.

Despite various improvements in fire-safe design and construction, the frequency of building fires and the resulting property losses are still substantial enough to require continued research into materials and construction systems. Even though they are gradually decreasing, the statistics of deaths and injuries resulting from fires are still appalling. Fire continues to be the single largest killer of building occupants in the United States. By comparison, natural disasters, such as earthquakes, hurricanes, and floods, account for a much smaller toll of lives lost, and deaths caused by structural failures of buildings are rare these days. For instance, in the United States, an average of approximately 3,000 people (excluding firefighters) die and several times more are injured annually in building fires. Some of the injured suffer from lifelong disabilities, and many find it difficult to recover from psychological and emotional aftereffects of fire.

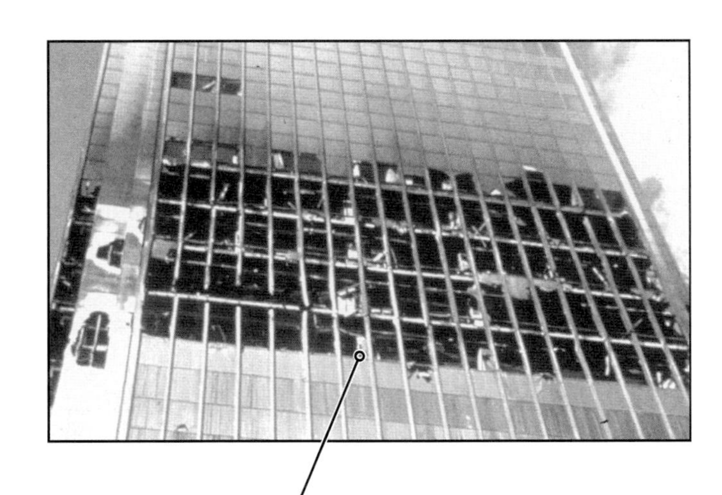

FIGURE 7.1 Four floors of the First Interstate Building in Los Angeles, California, gutted with fire. The building is still standing and is in service. Its survival is ascribed primarily to its fire-safe construction. (Photos courtesy of W. R. Grace & Co. Both photos are copyrighted by W. R. Grace & Co.)

*Although building fires are fairly common in contemporary times, fires such as the Great Fire of London (1666), which lasted several days and nights and destroyed virtually half the city of London, are, thankfully, a thing of the past.

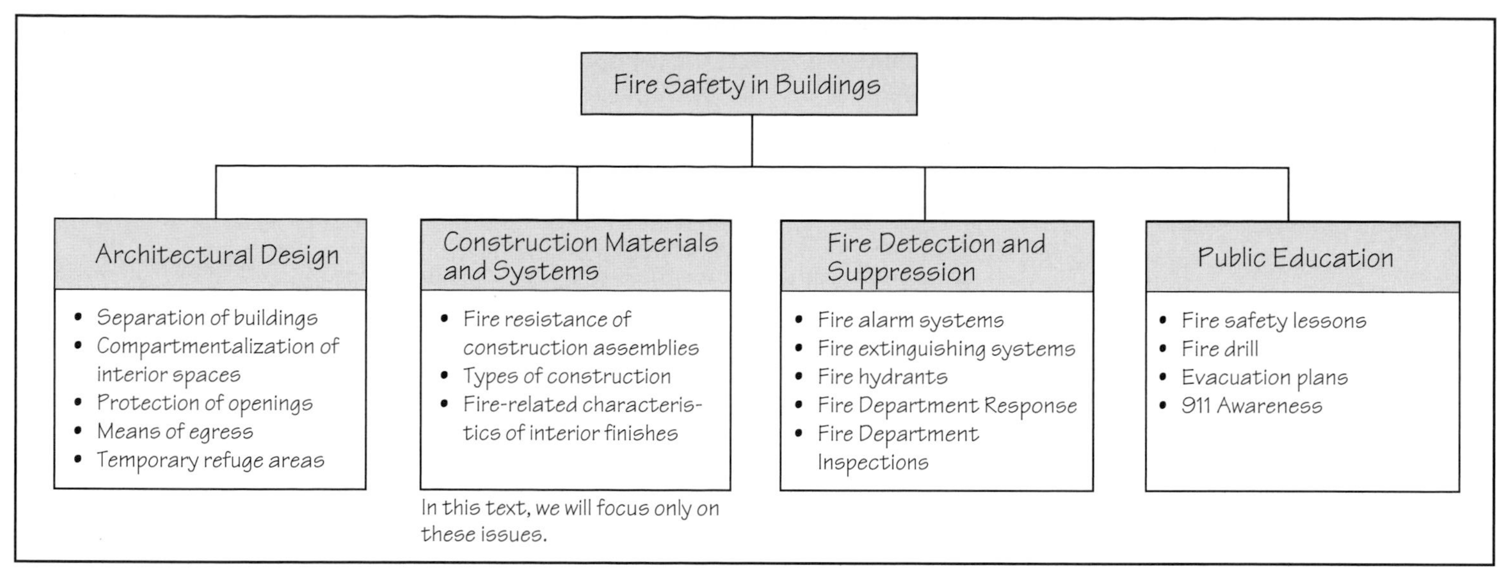

FIGURE 7.2 Factors that affect fire safety in buildings.

Because fire is the biggest hazard to life safety in modern buildings, building codes recognize this fact by making fire protection an important objective. For example, the classification of buildings in various occupancy groups, as provided in building codes, is based primarily on the degree of fire hazard present in buildings. The maximum permissible building area and height are also based on the ability of various construction assemblies to withstand fire. In fact, if the structural engineering provisions are disregarded in a building code, then of the remaining regulations, nearly two-thirds relate to fire safety considerations.

Factors That Affect Fire Safety in Buildings

Fire safety in buildings is a function of several variables, which may be grouped under the following four categories: (a) architectural design, (b) construction materials and systems, (c) fire detection and suppression, and (d) public education, Figure 7.2.

Fire-safe architectural design, an extensive area of study in itself, is outside the scope of this text. Briefly, it involves the design features of a building that reduce fire hazard and assist in speedy evacuation of people in the event of a fire, as shown in Figure 7.2.

Fire detection and suppression covers fire alarm systems, fire sprinklers, fire hydrants, standpipes, and other fire-fighting equipment. Although it is important to integrate these systems in the design of buildings, in this text, we focus on the elements of fire-safe materials and construction, including fire-test response of construction assemblies, types of construction, and fire-related characteristics of interior finishes.

7.1 FIRE CODE AND BUILDING CODE

In addition to the building code, a building is also regulated by the jurisdiction's fire code. As previously stated, fire safety is one of the prime objectives of a building code. However, once the building has been built and occupied, it must be properly maintained to remain safe against fire and other hazards. The regulations that cover aspects of fire safety in a building during its use and occupancy are included in the *fire prevention code,* or simply the *fire code.*

The fire code regulates such items as the location, maintenance, and installation of fire protection appliances, maintenance of the means of egress, and the storage of combustible and (or) hazardous materials. For example, the storage of flammable liquids in a dry cleaning establishment is regulated by the fire code, but the construction of the enclosure to store the flammable liquids and the requirement for an automatic fire suppression system are contained within the building code.

The building code and the fire code are two arms of a jurisdiction's building safety ordinances. They are enforced by the jurisdiction's building official and fire official, respectively. Because the distinction between fire-safe construction and fire prevention can be subtle and, at times, indistinguishable, there is always a certain amount of overlap between the fire code and building code provisions. The fire department and building code department of a city are expected to work in close cooperation because the aims of both are the same—to ensure public safety.

The principal fire code in the United States is the Fire Prevention Code published by the National Fire Protection Association (NFPA). However, in order to ensure compatibility and correlation between the building code and the fire code, the International Code Council publishes its own fire code, called the International Fire Code (Figure 2.12).

7.2 COMBUSTIBLE AND NONCOMBUSTIBLE MATERIALS

In the context of fire-safe construction, building materials must be distinguished either as *combustible materials* or as *noncombustible materials.* Metals used in construction (iron, steel, aluminum, copper, etc.) and concrete, brick, stone, gypsum, and so on, are noncombustible materials. Wood, paper, and plastics are combustible materials.

The distinction between a combustible and a noncombustible material is generally, although not always, obvious. For example, it is not immediately clear whether fire-retardant-treated wood is a combustible or a noncombustible material. Additionally, some materials may contain a small combustible content, which may or may not contribute appreciably to a fire, such as concrete made with polystyrene beads as aggregate. Are these materials to be considered as combustible or noncombustible materials?

A precise definition of noncombustibility is, therefore, necessary. This is provided for in ASTM E 136 Test. In this test a predried specimen of the material measuring $1\frac{1}{2}$ in. by $1\frac{1}{2}$ in. by 2 in. is placed in a 3-in.-diameter cylindrical furnace, Figure 7.3. Before placing the material in the furnace, the furnace temperature is raised to 1,382°F (750°C). The furnace has a transparent lid to allow the inspection of the specimen during the test.

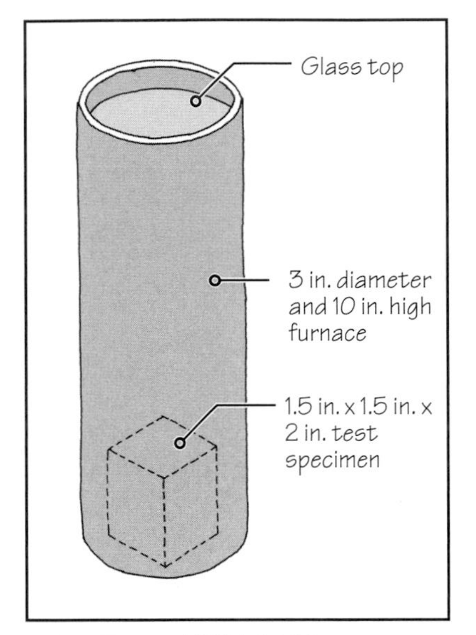

FIGURE 7.3 ASTM E 136 test setup to determine whether a material is combustible or noncombustible.

The specimen is reported as noncombustible if it satisfies the conditions laid down in the test. Fire-retardant-treated lumber and plywood do not pass the ASTM test and are, therefore, classified as combustible materials. However, recognizing that fire-retardant-treated wood makes a reduced contribution as fuel in the early stages of fire, building codes permit its use in some limited situations where only noncombustible materials are allowed.

For example, non-load-bearing partitions in Type I and Type II construction, which are otherwise required to be of noncombustible materials (metal stud and gypsum board assemblies), may be constructed of fire-retardant-treated wood framing. (See Section 7.5 for the types of construction.)

ASTM E 136 test is applicable only to elementary materials, not to composites, laminated, or coated materials. In recognition of the fact that a material with a noncombustible core (as per ASTM E 136 test) but with a thin facing of a combustible material or a combustible paint, does not contribute greatly to fire, building codes classify such a material as a noncombustible material, Figure 7.4.

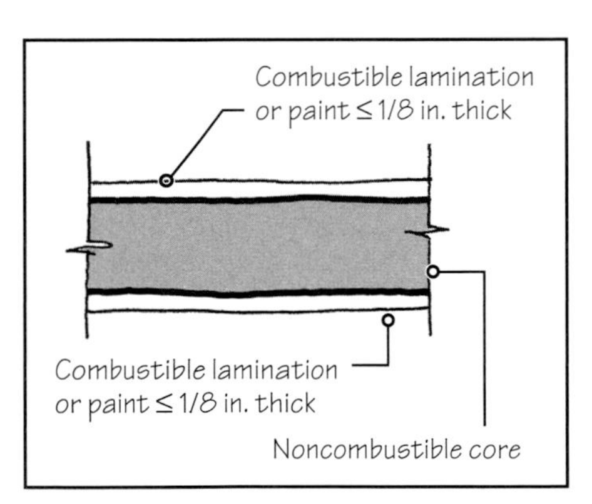

FIGURE 7.4 A material with a noncombustible core but with a combustible lamination less than or equal to $\frac{1}{8}$ in (3 mm) thick is regarded as a noncombustible material.

The constraints placed on the combustible paint or laminate are that its thickness shall not exceed $\frac{1}{8}$ in., and its flame spread rating shall not exceed 50 (see Section 7.8 for the explanation of flame spread rating). Thus, gypsum board, which has a thin paper lining, is classified as a noncombustible material because the thickness of the combustible paper lining is less than $\frac{1}{8}$ in.

NONCOMBUSTIBILITY OF A MATERIAL AND ITS ABILITY TO WITHSTAND FIRE

Note that noncombustibility refers only to the fact that a noncombustible material will not add fuel to the fire. Noncombustibility is not related to the ability of the material to withstand fire. For example, wood is a combustible material and steel is a noncombustible material, but a structure constructed of heavy sections of wood is better able to withstand fire than the one constructed of unprotected steel members, as explained later in this chapter.

7.3 PRODUCTS GIVEN OFF IN A BUILDING FIRE

The two products given off in a building fire (referred to as *products of combustion*) are heat and smoke. Smoke consists of (a) fire gases, (b) solid particulate matter (fine particles of solid matter suspended in the atmosphere) and (c) liquid particulate matter (condensed vapors dispersed in the atmosphere as tiny liquid droplets), Figure 7.5. Although fire gases in smoke are generally transparent, the solid and liquid particulate matter form an opaque cloud and diminish visibility in a fire.

Both products of combustion—smoke and heat—are responsible for human deaths and injuries. The statistics based on autopsies conducted on people who died in building fires indicate that, because of smoke's toxicity, smoke inhalation accounts for nearly 65% of all deaths. Burns account for 30%, and the remaining 5% of deaths are caused by emotional shock and heart failure, Figure 7.6.

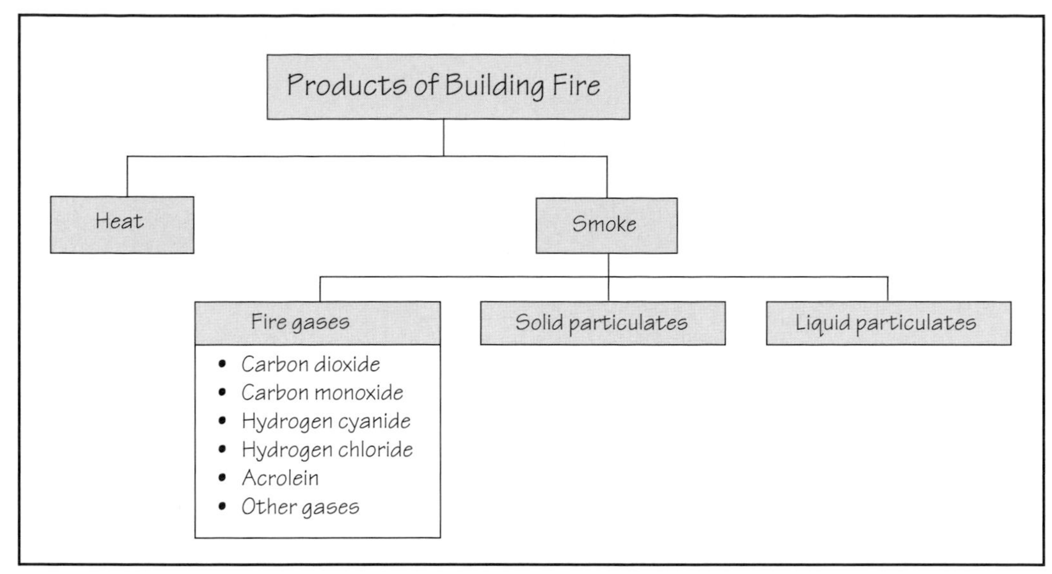

FIGURE 7.5 Products of combustion in a building fire.

The toxicity of smoke is due to gases produced in a fire. The types of gases depend on the type of material that burns. Carbon dioxide and carbon monoxide are produced in almost all fires. If a sufficient supply of oxygen (air) is available during the fire, the carbon that is present in almost every combustible material, such as wood, cotton, silk, wool, and plastics, is converted to carbon dioxide. This type of combustion is called *complete combustion.* However, most fires take place under conditions of incomplete combustion, where the amount of oxygen available is less than that required for complete combustion. In such a case, carbon monoxide (CO) is produced.

Both carbon dioxide and carbon monoxide are toxic to humans. However, carbon monoxide is considered the major threat to human life in building fires. Although less toxic than some of the other fire gases, it is generally present in such large quantities that its overall toxic effect is most devastating. It is estimated that nearly 50% of fatalities caused by smoke inhalation occur due to the effect of carbon monoxide.

Carbon dioxide is also present in large quantities in a building fire, but its toxicity is considerably lower than that of carbon monoxide. Other gases produced in a fire may include hydrogen cyanide, hydrogen chloride, or acrolein. Hydrogen cyanide is produced from the burning of wool, silk, leather, rayon, and plastics containing nitrogen. Hydrogen chloride is produced from polyvinyl chloride (PVC). Acrolein is produced from wood and paper.

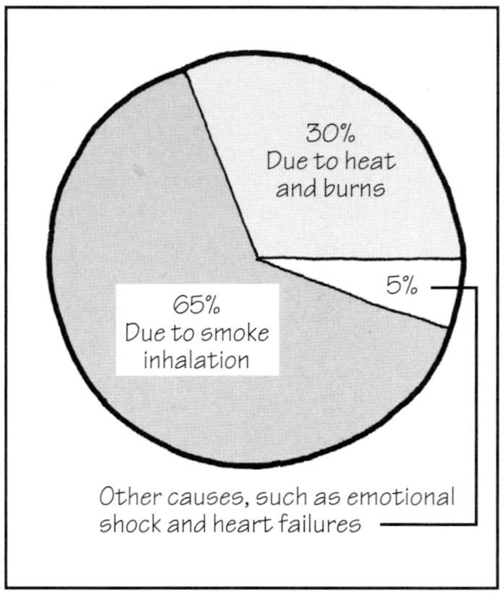

30% Due to heat and burns
5%
65% Due to smoke inhalation
Other causes, such as emotional shock and heart failures

FIGURE 7.6 Causes of deaths in a building fire.

EFFECT OF SMOKE ON VISIBILITY

The solid particulate matter in smoke consists primarily of unoxidized carbon in the form of extremely small particles suspended in the atmosphere. Incomplete combustion produces greater amounts of solid particles. The liquid particulate matter comprises mainly water vapor and tar droplets. Both the solid and liquid particles scatter light and, hence, reduce visibility.

Although reduced visibility has no direct effect on fire fatalities, it impedes the escape of a building's occupants and prolongs their exposure to toxic gases and heat. This often results in panic conditions, which are known to have caused several deaths in building fires. Smoke also has irritating effects on eyes and lungs, and this may result in serious medical problems.

PRACTICE QUIZ

Each question has only one correct answer. Select the choice that best answers the question.

1. On average, approximately how many people are killed in building fires in the United States every year?
   - **a.** 10,000
   - **b.** 1,000
   - **c.** 100
   - **d.** None of the above
2. Fire code is one of the most important chapters in a building code.
   - **a.** True
   - **b.** False
3. A material whose core is noncombustible but is covered with a combustible lamination or paint is considered a noncombustible material provided the thickness of the paint or lamination does not exceed
   - **a.** 1 in.
   - **b.** $\frac{1}{2}$ in.
   - **c.** $\frac{1}{4}$ in.
   - **d.** $\frac{1}{8}$ in.
   - **e.** $\frac{1}{16}$ in.

4. Statistics based on autopsies conducted on people who died in building fires indicate that most deaths occur due to
   a. burns.
   b. smoke.
   c. emotional shock.
   d. heart failure.
5. Which of the following properties are required of a fire-resistive building component?
   a. Structural integrity during the fire
   b. Freedom from cracks during the fire
   c. Resistance to the change of color during the fire
   d. All the above
   e. (a) and (b)
6. Fire rating is a property of
   a. the material of which a component is made.
   b. the thickness of the component.
   c. the entire assembly comprising the component.
   d. weight of the component.
   e. all the above.

Answers: 1-d, 2-b, 3-d, 4-b, 5-e, 6-c.

7.4 FIRE-RATED ASSEMBLIES AND COMPARTMENTALIZATION OF A BUILDING

One of the important design strategies employed to reduce fire hazard in a building is to subdivide it into small compartments so that the fire is limited to the compartment of its origin. This strategy is based on the fact that the fire will either extinguish itself after a while due to the lack of oxygen in the compartment or take some time to spread to other compartments. During this period, the occupants can evacuate to a place of safety.

REQUIREMENTS OF A FIRE-RATED ASSEMBLY

The concept of compartmentalization, as shown in Figure 7.7, assumes that the boundary elements of the compartment (walls, floors, and roofs) will function as barriers against the spread of fire to adjacent compartments of the building. We refer to such barriers as *fire-rated assemblies.* To meet this criterion, a fire-rated assembly must satisfy the following three requirements:

1. A fire-rated assembly should be able to perform its structural function without collapse—that is, it should be able to sustain the loads for which it has been designed throughout the duration of the fire.
2. A fire-rated assembly should remain fire-tight; that is, it should develop no cracks during the duration of the fire. The purpose of this requirement is to ensure that smoke and flames will not spread to adjacent compartments.
3. The temperature of the unexposed face of a fire-rated assembly during a fire should be so low that the heat received by radiation and (or) conduction through the assembly will not ignite combustibles in adjacent compartments.

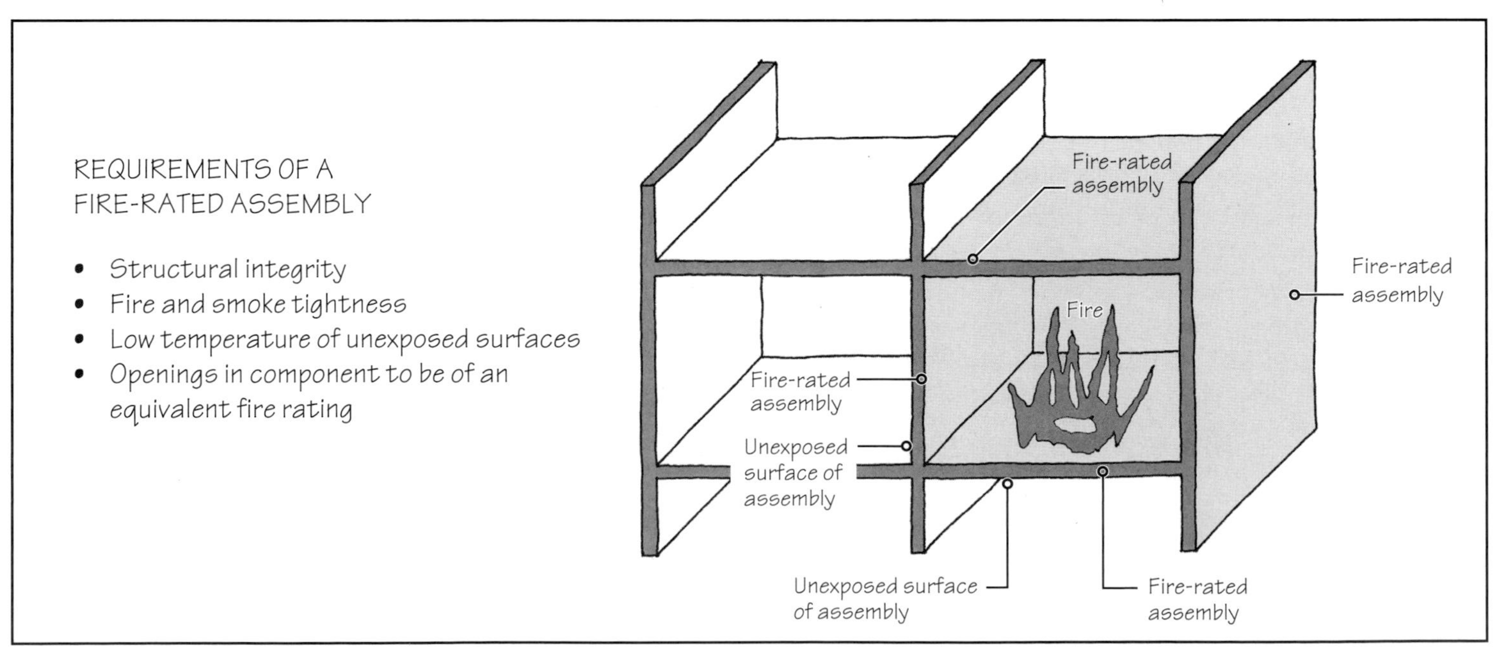

FIGURE 7.7 Compartmentalization of a building by fire-rated assemblies. A fire-rated assembly may either be a vertical or a horizontal member.

Fire-Resistance Rating

When exposed to a fire for a sufficient length of time, all building assemblies, regardless of their material, will eventually crack, disintegrate, and collapse. Thus, a fire-rated assembly can satisfy the stated requirements for a limited duration only. The ability to endure fire for the duration for which it satisfies all the stated requirements is called the *fire-resistance rating,* or simply the *fire rating* of the assembly.

In other words, the fire rating of an assembly is its ability to confine the fire to the enclosure of its origin for a particular duration. Accordingly, it is measured in units of time, that is, in hours or fractions thereof. Thus, a 1-h fire rating means that the fire-rated assembly can endure a typical building fire for at least 1 h. A 2-h fire rating means that it can endure fire for at least 2 h.

Note that by virtue of its definition, fire rating is a property of the assemblies, such as walls, floors, and roofs, and not simply of the materials of which the assemblies are constructed. Thus, the fire rating of a single material, such as concrete or wood, is meaningless.

Openings in a Fire-Rated Assembly

Obviously, a fire-rated assembly can function as a fire barrier only if the openings in the barrier (such as doors and windows) have an equivalent fire rating. Thus, in addition to the three requirements of a fire-rated assembly, the fourth requirement is that the openings therein should also be fire-rated. Equivalent fire rating of an opening does not mean that its fire rating should be equal to that of the assembly in which it is provided. Generally, the fire rating required of a door or window is less than the fire rating of the wall in which it is located.

7.5 TYPES OF CONSTRUCTION

As stated in Chapter 2, building construction is classified by building codes under the following five types:

- Type I
- Type II
- Type III
- Type IV
- Type V

The classification is based on the fire ratings of the various critical assemblies of the building. The higher the fire rating of these assemblies, the more fire resistive the type of construction. Type I construction is the most fire resistive, and Type V construction is the least fire resistive. For a given occupancy (Chapter 2, "Principles in Practice"), building codes allow greater area and greater height for a more fire-resistive type of construction.

The assemblies of a building whose fire rating is considered in assigning a type of construction classification to a building are

- Building envelope elements—exterior walls and roof
- Structural elements—floor-ceiling assemblies, structural frame including columns, beams, girders, and trusses, and bearing walls

The fire resistance of envelope elements determines the degree of hazard to which the building is subjected from an external fire, such as, from a neighboring building. The envelope elements also determine the level of fire hazard the building itself poses to the neighboring buildings. The structural elements are important because if they are not adequately protected against fire, structural failure may occur.

As Table 7.1 shows, the fire rating requirement of an assembly is generally given by the number of hours. Thus, the building codes require these assemblies to be rated as 0, 1, 1.5, 2, or 3 h. Number 0 in Table 7.1 implies that there is no requirement for fire rating; that is, the assembly may be a nonrated assembly. The maximum fire rating required of any assembly by building codes is 3 h (see Note 5 in Table 7.1).

Types I, II, III, and V are further subdivided into subtypes A and B. Thus, there are nine types of construction classifications. More precisely, therefore, Type I(A) is the most fire resistive, and Type V(B) is the least fire-resistive construction type.

The five types of construction are categorized under three groups:

- *Noncombustible group*—consisting of Types I (A and B) and Types II (A and B). In these types of construction, all envelope and structural assemblies must be of noncombustible materials (concrete, steel, or masonry).

NOTE

Types of Construction—An Overview

Noncombustible Construction–Type Group

Type I and Type II All envelope elements (exterior walls and roof), interior partitions, and the structural frame must be of noncombustible materials such as steel, concrete, or masonry.

Noncombustible/Combustible Construction–Type Group

Exterior walls must be of noncombustible materials in both Types III and IV.

Type III Structural frame and interior partitions may be of wood light-frame construction.

Type IV (HT) Structural frame must be of heavy wood sections, which is why Type IV is called heavy timber (HT) construction type. Interior partitions must be one-hour rated and may be of wood light-frame construction.

Combustible Construction–Type Group

Type V All structural elements, exterior walls, and internal partitions may be of wood light-frame construction.

TABLE 7.1 MINIMUM FIRE RATING REQUIREMENTS (IN HOURS) OF VARIOUS TYPES OF CONSTRUCTION

	Noncombustible group				Noncombustible/ combustible group			Combustible group	
	Type I		Type II		Type III		Type IV	Type V	
Building component	A	B	A	B	A	B	HT	A	B
Structural frame Columns, girders, trusses	3	2	1	0	1	0	HT	1	0
Floor construction Including slabs, beams, and joists	2	2	1	0	1	0	HT	1	0
Roof construction Including slabs, beams, and joists	1.5	1	1	0	1	0	HT	1	0
Bearing walls									
Exterior	3	2	1	0	2	2	2	1	0
Interior	3	2	1	0	1	0	1 or HT	1	0
Nonbearing Walls	The fire rating of nonbearing walls (exterior or interior) is not a determinant of the type of construction. However, building codes require a certain minimum fire rating of these walls in certain situations.								
Exterior	*Exterior nonbearing walls:* Their fire-rating requirement is a function of *fire separation distance*, type of construction and the occupancy. If the separation distance is greater than 30 ft, the required fire rating is generally zero regardless of the type of construction or the occupancy.								
Interior	*Interior nonbearing walls (partitions):* Minimum fire rating depends on the purpose they serve, i.e., as corridor walls, tenant separation walls in shopping malls, guest room separation walls in hotels, or motels.								

1. Table adapted from the 2006 International Building Code. Copyright 2006. Falls Church, VA: International Code Council, Inc. Reproduced with permission. All rights reserved.
2. Numbers in the table, such as 0, 1, 2, and 3, indicate the required minimum fire resistance rating of components in hours. Zero implies that there is no requirement for fire rating.
3. Type IV is referred to as heavy timber (HT) construction because the structural members must be of heavy wood sections, which may be left unprotected (exposed).
4. Fire separation distance is an index of the hazard that the building poses to the adjoining building. It is defined as the distance of the building to the property line or to the centerline of the street.
5. The maximum fire rating of a component in determining the building's type of construction is 3 h. The only component that may require a fire rating greater than 3 h is a fire wall. However, a fire wall is not a determinant of a building's type of construction.

- *Noncombustible/combustible group*—consisting of Type III (A and B) and Type IV. In these types of construction, the exterior walls must be of noncombustible materials, whereas other assemblies of the building may be of combustible materials. A combustible type of construction generally refers to one whose structural frame consists of wood or wood-based products.
- *Combustible group*—consisting of Types V (A and B). In these types of construction, all envelope and structural elements may be of combustible materials.

FIGURE 7.8 An example of Type II(B) construction. Notice exposed steel frame members with no fire-protective covering.

Although not expressly mentioned, Types III (A and B) and Types V (A and B) are of wood light-frame construction. The difference between them is that, in Types III (A and B), the exterior walls are required to be of noncombustible materials. In Types V (A and B), the exterior walls may be of combustible or noncombustible materials.

Note that the fire rating required of various members in Type II(B) and Type V(B) is the same—zero. The difference between the two is that in Type II(B), all envelope and structural elements must be of noncombustible materials. Generally, such construction consists of masonry walls, unprotected steel columns, beams, and roof deck, Figure 7.8. Type V(B) consists of wood light frame construction. Most single-family dwellings are constructed of Type V(B), in which the envelope and structural elements need not be fire rated (zero fire rating). A typical example of Type V(B) construction is shown in Figure 7.9.

Type IV is called *heavy timber* (HT) construction. In HT construction, floors, roofs, and the structural frame are of heavy sections of

FIGURE 7.9 Most single- and multifamily dwellings in North America are built of wood frame construction, as shown here (see also Chapters 13 and 14). The wood members are typically protected against fire with $\frac{1}{2}$-in.-thick gypsum board on both sides, which gives the wall, floor, and roof assemblies a fire rating of approximately 40 min. Because the fire rating is less than 1 h, such assemblies meet the requirements of Type V(B) construction.

wood without any added gypsum board or similar fire protection. In other words, in HT construction, the wood members are exposed without any concealed spaces such as those that occur in a conventional wood light frame building. For example, in a conventional wood light frame floor, concealed spaces occur between floor joists, Figure 7.10. Similar spaces occur in conventional wood light frame walls and attics. Such spaces are absent in HT construction.

Heavy-timber construction is slow burning and is assumed to provide an equivalent of a 1-h fire rating. Thus, the overall fire endurance of HT construction is generally considered equivalent to that of Type III(A). An example of Type IV (HT) construction is shown in Figure 7.11.

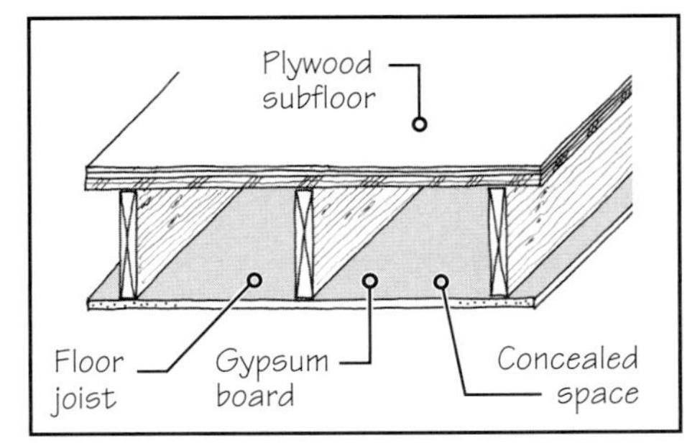

FIGURE 7.10 Concealed spaces in a conventional wood light-frame floor assembly.

Identifying the Type of Construction

For a building to be classified as a given type of construction, the fire ratings of various assemblies must either be equal to or greater than those given in Table 7.1. Note that fire ratings given in Table 7.1 are minimum fire ratings. A building cannot be designated as belonging to a particular type of construction unless it meets or exceeds all the fire-rating requirements for that type of construction.

FIGURE 7.11 An example of heavy-timber construction—Type IV. Notice heavy sections of wood with no protective covering and no concealed spaces.

Determining the Type of Construction

Example 1

Using Table 7.1, determine the type of construction of a building in which the materials and fire ratings of different components are as follows:

Structural frame—steel columns and steel girders	2-h rating
Exterior bearing walls	None provided
Interior bearing walls	None provided
Floor assembly—steel deck on steel beams with concrete topping	1-h rating
Roof assembly—steel deck	1-h rating

Solution

Because all components are of noncombustible materials, the building is either Type I or Type II construction. Comparing the fire ratings of the building's components with those of Table 7.1, the building is classified as Type II(A).

Example 2

What changes must be made in the fire ratings of components to upgrade the construction of Example 1 to Type I(B)?

Solution

Examining the fire ratings required for Type I(B) construction in Table 7.1, we see that if the fire rating of the floor assemblies of Example 1 is increased to a minimum of 2 h, it can be classified as Type I(B) construction.

Example 3

Determine the type of construction for the building in which the materials and fire ratings of different components are as follows:

Structural frame—columns and girders	None provided
Exterior bearing walls—concrete masonry	2-h rating
Interior bearing walls—wood studs and gypsum board	1-h rating
Floor assemblies—wood joists and plywood floors	1 h
Roof assembly—wood rafters and plywood sheathing	1 h

Solution

Because some components are of combustible materials, the building will be classified as Type III, Type IV, or Type V. Type IV is ruled out since the roofs and floors are not constructed of heavy timber. Comparing the given fire ratings with those of Table 7.1, we observe that the building is of Type III(A) construction. Note that the exterior bearing walls are of noncombustible materials, as required for this type of construction.

Example 4

If the exterior bearing walls of Example 3 are of wood studs and gypsum board, but with the same (2-h) fire rating, and all other components remain unchanged, what will be the revised construction type?

Solution

Because the exterior bearing walls are of a combustible material, the type of construction will be Type V(A).

PRACTICE QUIZ

Each question has only one correct answer. Select the choice that best answers the question.

7. Fire rating is measured in units of
- **a.** time.
- **b.** temperature increase per hour.
- **c.** heat released by the component in a fire test.
- **d.** pounds per square inch.
- **e.** all the above.

8. Building codes divide the types of construction into
- **a.** Types 1 to 5.
- **b.** Types I to V.
- **c.** Types 1 to 6.
- **d.** Types I to VI.
- **e.** none of the above.

9. Noncombustible types of construction are
- **a.** Types 1, 2, and 3.
- **b.** Types 1 and 2.
- **c.** Types I, II, and III.
- **d.** Types I and II.
- **e.** Type I.

10. Combustible types of construction are
- **a.** Types 4, 5, and 6.
- **b.** Types 4 and 5.
- **c.** Types IV, V, and VI.
- **d.** Type VI.
- **e.** Type V.

PRACTICE QUIZ (Continued)

11. Among the various types of construction included in a building code, which is the most fire-resistive type?
a. Type I(a)
b. Type V(a)
c. Type V(b)
d. Type HT
e. None of the above

12. The total number of the types of construction, including their subclassifications, in a building code are
a. 6.
b. 7.
c. 8.
d. 9.
e. 10.

13. Which type of construction is commonly used in the construction of a typical single-family dwelling in the United States?
a. Type I(a)
b. Type V(a)
c. Type V(b)
d. Type HT
e. None of the above

14. Fire rating of building components is generally obtained from fire tests, not through calculations.
a. True
b. False

Answers: 7-a, 8-b, 9-d, 10-e, 11-a, 12-d, 13-c, 14-a.

7.6 FIRE-STOPPING OF PENETRATIONS AND FIRE-SEALING OF JOINTS

Walls, floors, and roofs in buildings are often penetrated by HVAC ducts, pipes, electrical conduits, and communication wiring. If these assemblies (walls, floors, and roofs) are to function as effective barriers to the spread of fire, the space around the penetrations must be thoroughly sealed. The sealing of the space around a penetration is called *fire-stopping*.

An unsealed space will allow heat, smoke, and flammable gases to transmit from one side of the barrier to the other, compromising the fire rating of the barrier. The fire rating of fire-stops is generally required to match the fire rating of the barrier in which they are installed.

Fire-stopping materials must obviously be noncombustible. Several are in common use. They consist of mineral wool, sealants, and foams, Figure 7.12. Foamed fire-stop is a material that expands at room temperature and fills the space. Fire-stop materials also help to seal spaces against air and sound transmission.

Just as penetrations through a fire-rated building element are required to be fire-stopped, joints in them (such as expansion joints and control joints, Chapter 9) are also required to be fire rated.

Similarly the space between a curtain wall and a fire-rated floor or roof also needs to be fire-sealed. (See Chapter 26 for the definition of a curtain wall.) In fact, the entire spandrel assembly in a curtain wall (extending from the head of the window at a floor to the **sill** of the window above that floor) must be fire rated, Figure 7.13(a). In absence of a fire-rated spandrel and (or) a fire-rated seal in the wall-floor gap, the fire from the lower floor will not be contained within the floor of origin, Figure 7.13(b).

EXPAND YOUR KNOWLEDGE

Fire-Rating Data for Assemblies

Although procedures are available to calculate the fire rating of a few building assemblies, the most reliable approach for obtaining a fire rating is to test the assembly in the standard fire test. Several laboratories are equipped to conduct the standard test. In the United States, the most commonly used laboratory is the Underwriters Laboratories Inc., and in Canada, the Underwriters Laboratory of Canada is most commonly used.

The fire ratings of assemblies tested by the Underwriters Laboratories are given in its annual publication, entitled the *Directory of Fire Resistance*, which is the most comprehensive source available for fire ratings of assemblies. Additionally, building codes provide tables, listing the fire ratings of several selected construction assemblies. For assemblies using gypsum board, the *Fire Resistance Design Manual* published by the Gypsum Association is a valuable reference.

The test procedure that is used to measure the fire rating of an assembly is ASTM E 119, "Standard Test Methods for Fire Tests of Building Construction and Materials." In the ASTM E119 test, a full-size specimen of the assembly is exposed to fire in a furnace. A vertical furnace is used for walls and columns, and a horizontal furnace is used for floors and ceiling assemblies.

ASTM E119 Tests and World Trade Center Fires

After the collapse of the twin towers of the World Trade Center (WTC) on September 11, 2001, the reliability of ASTM E119 tests began to be widely questioned. As a result, the National Institute of Standards and Technology (NIST) of the U.S. Department of Commerce conducted four major test studies of the composite concrete-steel trussed floor system used in WTC. All four tests were conducted in the Underwriters Laboratories, two in Canada and the other two in the United States.

As a result of the NIST study (available on the Web), some changes to ASTM E119 test may be made, particularly with respect to the long-term efficacies of the fire-protective materials used on steel members.

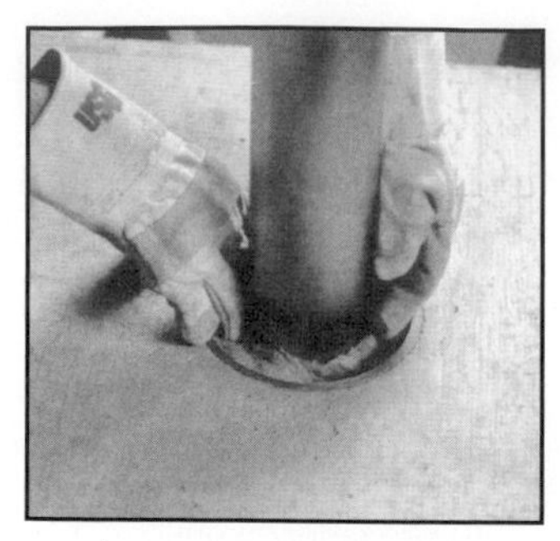
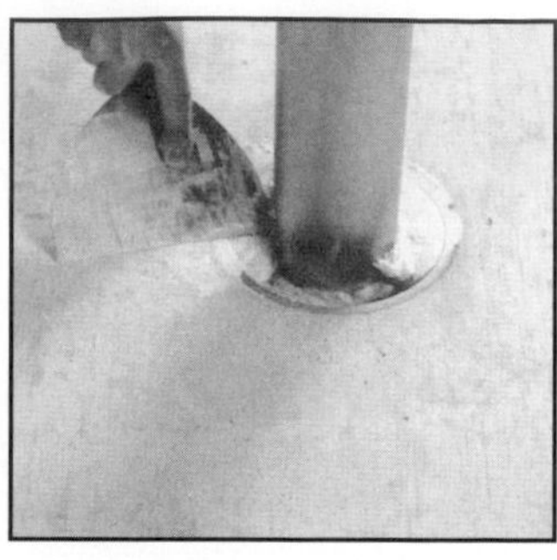
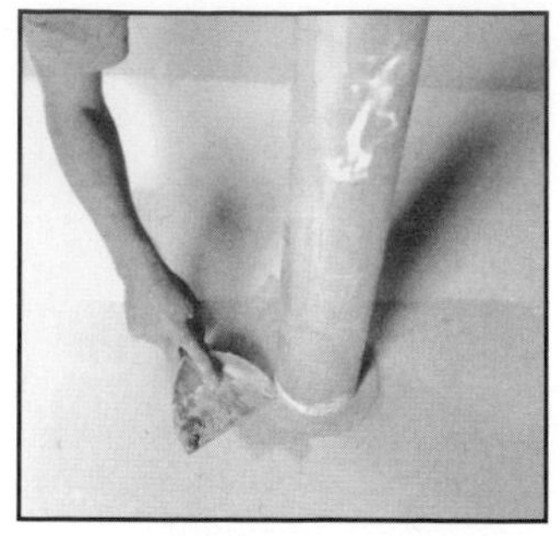

FIGURE 7.12 A typical fire-stopping around a penetration in a floor generally consists of high-density mineral wool pressure-fitted into the void, followed by a semiliquid, fire-resistant sealant troweled over mineral wool packing. (Photo courtesy of U.S. Gypsum Company.)

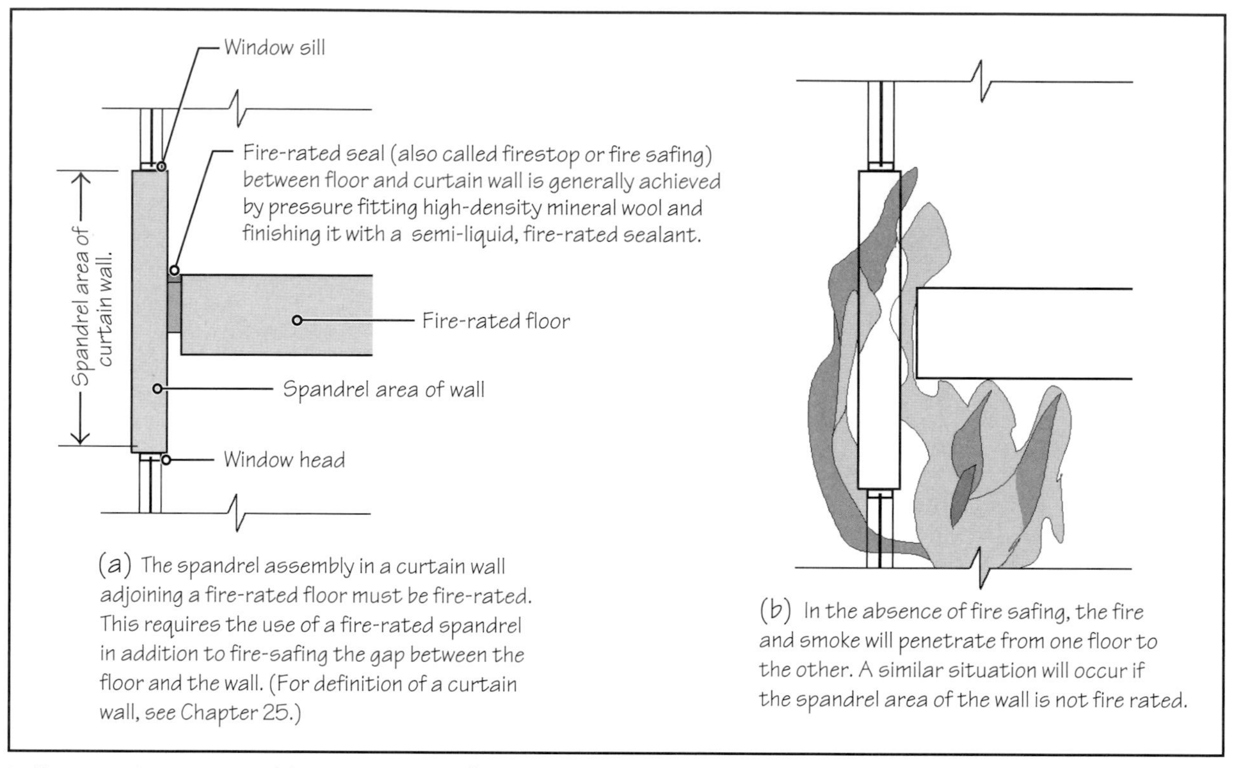

FIGURE 7.13 Fire-containment assembly in a curtain wall consists of a fire-rated spandrel and fire-rated seal between the wall and the floor.

7.7 FIRE-TEST RESPONSE CHARACTERISTICS OF INTERIOR FINISHES

Fire rating is the property of an assembly to act as a barrier to the spread of fire to adjacent compartments. Another important fire-related property of an assembly is its ability to resist the spread of fire within the compartment of origin. This property is a function of the

- Surface burning behavior of interior finish materials
- Toxicity and density of smoke generated by the burning of finish materials

Experience with actual building fires and experimental investigations with full-scale, laboratory-based fire tests have shown that apart from providing additional fuel, interior finishes contribute to fire hazard by (a) flaming of the material and (b) producing smoke. The properties that are used to assess these hazards are *flame-spread index* and *smoke-developed index,* respectively.

Flame Spread Index (FSI)

The flame-spread index (FSI) is a measure of the rate at which flames spread on the surface of an assembly or material used as an interior finish. It is an important index because a rapid spread of flames either prevents or delays the escape of occupants from the building. Its importance was realized in the 1940s, when investigations revealed that the main cause for the loss of lives in three major building fires was the rapid spread of flames on the surfaces of interior finishes.

The FSI of a finish material is obtained from the Steiner Tunnel Test and rated on a scale that begins at zero but has no upper limit. The FSI of (select-grade) red oak has been arbitrarily fixed at 100 and that of a portland cement board, at zero. In this respect, the FSI scale is

TABLE 7.2 APPROXIMATE FSI VALUES OF SELECTED MATERIALS

Interior finish material	FSI	Interior finish material	FSI
Fiberglass (kraft paper faced)	15–30	Red oak	100
Rock wool (aluminum foil faced)	0–5	Southern pine	130–190
Shredded wood fiber board (treated)	20–25	Douglas fir	70–100
Spray-on cellulosic fibers (treated)	20	Plywood paneling	75–275
Cement board	0	Fire retardant treated lumber	Less than 25
Bricks or concrete block	0	Cork	175
Concrete	0	Carpets	10–600
Gypsum board with paper lining	10–25		

similar to the Celsius scale, in which the boiling point of water and the freezing point of water are arbitrarily fixed at 100°C and 0°C, respectively.

Red oak has been chosen as the standard because the rate of flame travel on red oak is uniform, and the test results on red oak are reasonably reproducible.

A material whose FSI is 200 means that the rate of flame spread on this material is twice as rapid as that on red oak. Similarly, on a material whose FSI is 50, the rate of flame travel is half as rapid as on red oak. Table 7.2 gives typical FSI values of some commonly used building materials. Building codes limit the maximum value of FSI to 200 for an interior finish material.

Smoke-Developed Index (SDI)

The smoke-developed index (SDI) measures the visibility through the smoke resulting from burning assemblies or materials used as interior finishes. The lower the visibility through smoke, the greater the SDI value. SDI values of materials are obtained from the same test as the FSI. SDI is also measured on a scale that begins at zero but has no upper limit. Once again, the SDI value of red oak has been arbitrarily fixed at 100, and that of portland cement board is fixed at zero.

As mentioned in Section 7.3, building codes at the present time do not require the toxic effects of smoke to be rated, except prohibiting the use of some materials as interior finishes because of the excessive toxicity of their smoke. The only building code requirement of the smoke generated by materials is that the SDI value of a material used as an interior finish should not exceed 450.

Building Code Requirements for FSI and SDI

Based on FSI and SDI values, building codes classify interior finishes using three classes: Class A, Class B, and Class C, Table 7.3. The objective of this classification is to regulate the use of interior finish materials according to the fire safety requirements of spaces. Because Class A interior finishes are the least hazardous, building codes mandate that they be used in enclosed vertical exit ways—staircases and elevators. A minimum of Class B interior finish material is required in other exit ways—corridors, lounges, foyers, and so on. Class C finish may be used in areas such as individual rooms, Figure 7.14.

There are a few exceptions to these rules, for which the applicable building code should be consulted. For example, Class C materials are allowed to be used in all areas

TABLE 7.3 CLASSIFICATION OF INTERIOR FINISHES

Flame spread class	FSI	SDI
Class A	0–25	<450
Class B	26–75	<450
Class C	76–200	<450

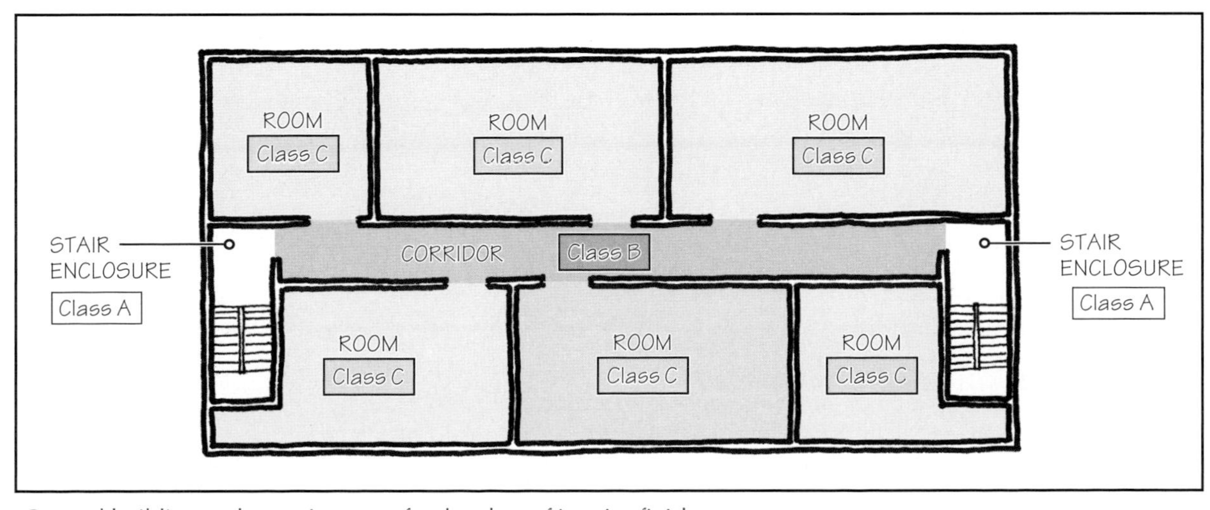

FIGURE 7.14 General building code requirements for the class of interior finishes.

of a single-family dwelling, even in stair enclosures and corridors. Additionally, the provision of automatic sprinklers in a building allows Class B finish to be used in vertical exit areas where Class A is generally needed.

No classification is given to interior finishes with an FSI greater than 200 and an SDI greater than 450, because these materials are not allowed as interior finish materials except in utility occupancies. Only finishes applied on walls and ceilings are regulated for FSI and SDI values. Floor finishes such as carpets and floor tiles are not rated for FSI or SDI. Because flames spread vertically upward, floor finishes do not get involved in a building fire until fairly late in the burning process. However, floor finishes are tested and classified for radiant flux (as per ASTM E648).

FIRE RESISTANCE AND SURFACE FLAMMABILITY

Note that the fire rating of barriers and the surface burning characteristics of interior finishes are entirely independent and unrelated properties. Whereas fire rating is a measure of the structural integrity of a barrier during a fire, FSI and SDI measure the fire hazard of an interior material. For example, heavy timber construction may give the equivalent of 1 h or more of fire rating, but it presents some degree of hazard as an interior finish. A thin steel plate, on the other hand, is a poor fire barrier with virtually zero fire rating, but it presents no hazard as an interior finish.

7.8 ROLE OF SPRINKLERS

Automatic sprinklers are an important part of fire-safe construction. Building codes encourage the use of automatic sprinklers by allowing greater area and height for a building that is sprinklered (see the Chapter 2 "Principles in Practice") in comparison with a building that is not sprinklered. For some occupancies and construction types, however, the use of automatic sprinklers is mandated by building codes.

In addition to permitting an increase in the area and height of the building, building codes allow a reduction in fire-resistance rating for some assemblies if the building is sprinklered. Similarly, lowering of the class of interior finish (from Class A to Class B or from Class B to Class C) is also permitted in some situations if the building is sprinklered.

NOTE

Sprinklers and Building Fire-Related Statistics

In 2004, there were 526,000 building fires in the United States, which caused 3,190 civilian fatalities, 82% of which (2,680) occurred in one- or two-family dwellings. Fire-related injuries during the same period were 14,175, of which 12,825 occurred in residential structures. Financial loss was reported to be $8 billion.

A building fire occurs in the United States every 60 s and in homes every 77 s. Several experts have, therefore, proposed for a long time to use building codes to mandate the provision of automatic sprinklers in U.S. homes.

Source: *Engineering News Record* (*ENR*), July 2005, pp. 11 and 88.

PRACTICE QUIZ

Each question has only one correct answer. Select the choice that best answers the question.

15. The term *fire-stopping* refers to
- **a.** a fire-rated wall with a minimum fire rating of 3 h.
- **b.** a fire-rated floor with a minimum fire rating of 3 h.
- **c.** a fire seal in a nonfire-rated wall or floor.
- **d.** a fire seal in a fire-rated wall or floor.
- **e.** none of the above.

16. The term FSI refers to
- **a.** fire safety index.
- **b.** fire and smoke index.
- **c.** fire-speed index.
- **d.** flame-spread index.

17. SDI measures
- **a.** visibility through smoke.
- **b.** toxicity of smoke.
- **c.** rate of smoke generation.
- **d.** amount of carbon dioxide in smoke.
- **e.** all the above.

18. Based on their fire properties, interior finishes are divided into
- **a.** Classes I, II, III, and IV.
- **b.** Classes A, B, C, and D.
- **c.** Classes W, X, Y, and Z.
- **d.** Classes A, B, and C.
- **e.** Classes X, Y, and Z.

19. The fire properties that determine the class of interior finishes are
- **a.** fire resistance rating and FSI.
- **b.** FSI and SDI.
- **c.** fire-resistance rating and SDI.
- **d.** fire-resistance rating, FSI and SDI.

20. The maximum value of SDI recognized by building codes is
- **a.** 450.
- **b.** 300.
- **c.** 150.
- **d.** 50.
- **e.** none of the above.

Answers: 15-d, 16-d, 17-a, 18-d, 19-b, 20-a.

KEY TERMS AND CONCEPTS

Class A interior finish
Class B interior finish
Class C interior finish
Combustible and noncombustible materials
Combustible construction types
Fire code
Fire-resistance rating
Firesealing of assemblies
Firestopping of assemblies

Flame spread index (FSI)
Heavy timber construction
Noncombustible construction types
Noncombustible/combustible construction types
Smoke developed index (SDI)
Sprinklers
Type I construction
Type II construction
Type III construction
Type IV construction
Type V construction

REVIEW QUESTIONS

1. List the factors that affect fire safety in buildings.
2. Discuss the differences between a fire code and a building code.
3. Explain why we need a test to determine whether a building material is combustible or noncombustible when we generally know which material is combustible and which is noncombustible.
4. Determine the type of construction of a building with the following data (Table 7.1):

   Structural frame; structural steel; 2-h rated
   Floors; steel joists, steel deck, and concrete topping; 1-h rated
   Roof; steel deck; 2-h rated
   Exterior bearing walls; none provided
   Interior bearing walls; none provided
5. What type of construction would the building of Problem 4 be if the floor's fire rating is increased to 2 h?

SELECTED WEB SITES

Gypsum Association (www.gypsum.org)

National Fire Prevention Association (www.nfpa.org)

Thermafiber Inc. (www.thermafiber.com)

Underwriters' Laboratories, Inc. (www.ul.com)

FURTHER READING

1. Egan, David. Concepts in Building Fire Safety. New York: John Wiley, 1978.
2. Gypsum Association. *Fire Resistance Manual.* GA Document GA-600.
3. National Fire Prevention Association. *Life Safety Code,* NFPA Document 101.

CHAPTER 8

Acoustical Properties of Materials

CHAPTER OUTLINE

8.1 Frequency, Speed, and Wavelength of Sound

8.2 The Decibel Scale

8.3 Airborne and Structure-Borne Sounds

8.4 Airborne Sound Insulation—Sound-Transmission Class

8.5 Structure-Borne Sound Insulation—Impact Insulation Class

8.6 Sound Absorption—Noise-Reduction Coefficient

Eugene McDermott Concert Hall, Morton Meyerson Symphony Center, Dallas, Texas; its acoustical quality is rated as excellent by musicians and listeners. Acoustics Consultant: ARTEC Consultants Inc., Architect: I. M. Pei. (Photo courtesy of ARTEC Consultants Inc.)

Acoustical and noise-control concerns exist in almost every modern building. An acoustical consultant is usually required for the design of an auditorium or a concert hall or for the solution of a complicated noise problem. However, acoustical issues in most commonplace buildings, such as selecting the appropriate interior finishes for a lecture room, controlling noise in a busy dining hall or isolating a hotel bedroom from street noise, are fairly elementary and are resolved by design professionals without a consultant's help.

This chapter provides an understanding of the acoustical properties of materials and assemblies that are routinely used in buildings. It begins with a discussion of basic characteristics of sound waves (frequency, speed, and wavelength), leading to the decibel scale—a unit for rating the loudness of sound. This is followed by sound-insulation and sound-absorbing properties of materials.

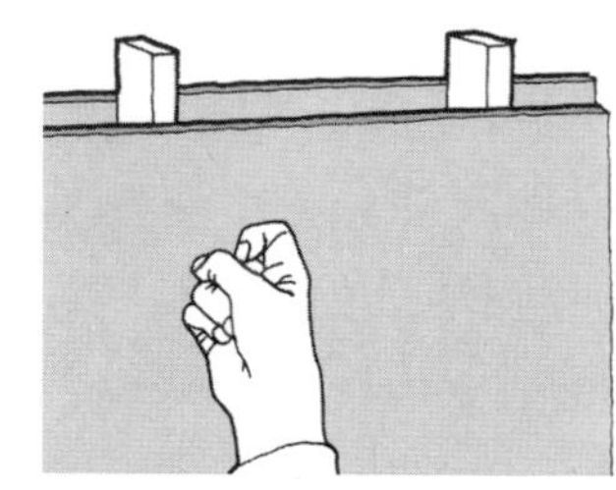

A tap on a wall makes the wall vibrate, which in turn makes the air particles vibrate, producing sound waves.

8.1 FREQUENCY, SPEED, AND WAVELENGTH OF SOUND

Sound is the human ear's response to pressure fluctuations in the air caused by vibrating objects. For example, a tap on a wall produces sound because the tap makes the wall vibrate back and forth. The back-and-forth motion of the wall is transferred to air particles in direct contact with the wall. These particles then transfer their vibratory motion to the neighboring particles, and so on.

The domino-type transfer of vibratory motion from particle to particle is what we refer to as a *sound wave*. The number of back-and-forth cycles that an object (or the air particles) moves in 1 s in a sound wave is called *sound frequency*. Its unit is cycles per second (c/s), which is also termed *hertz* (Hz) after the Austrian physicist Heinrich Hertz. Vibrations with frequencies lying between 20 Hz and 20,000 Hz (i.e., 20 kHz) are audible. This range of frequencies is called the *audible frequency range*.

Subjectively, the frequency of a sound is perceived as its pitch. A high-pitched sound means that it has a high frequency. A female voice is slightly higher pitched than the male voice.

Sounds in our environment do not generally consist of individual frequencies (a single note or pure tone), such as that produced by a tuning fork. Most sounds are complex; that is, they consist of a continuum of several frequencies. Thus, the human speech consists of all frequencies ranging from nearly 200 Hz to 5 kHz. The male voice peaks at about 400 Hz, and the female voice peaks at about 500 Hz. The range of frequencies in music is larger than the range for speech.

Frequency of sound is an important acoustical concept because the properties of building materials and construction assemblies are frequency dependent, meaning that building products vary in how they transmit or absorb sound in its many frequencies.

NOTE

Cycles Per Second and Hertz

Hertz (Hz) and cycles per second (c/s) are used interchangeably, that is,

1 Hz = 1c/s
1,000 Hz = 1 kilohertz = 1kHz

NOTE

Speech Sounds

Human speech lies in the frequency range of 200 Hz to 5 kHz. However, most of the speech energy lies in the range of 250 Hz to 2 kHz.

SPEED OF SOUND

The speed of sound in the air has been measured as 1,130 ft/s (344 m/s) and is independent of sound frequency or loudness. Because sound travels from particle to particle in a medium, a medium is required for the existence of sound. In other words, sound cannot be produced or travel in vacuum. This is in contrast with light and heat waves, which, being electromagnetic in nature, do not require a medium for travel.

NOTE

Infrasonic, Ultrasonic, and Supersonic

Frequencies below 20 Hz are called *infrasonic* frequencies. They are not heard but are perceived by humans as vibrations.

Frequencies above 20 kHz, referred to as *ultrasonic* frequencies, are also not heard by humans.

An object traveling at a speed greater than the speed of sound is said to be traveling at *supersonic* speed.

WAVELENGTH OF SOUND

As the sound wave travels, it creates excess pressure (called *compression*) and reduced pressure (called *rarefaction*) in space, just as water waves produce *crests* and *troughs*. The distance between adjacent compression peaks or adjacent rarefaction peaks in the air at an instant of time is called the *wavelength*. The frequency, wavelength, and speed of wave motion are related to each other by the following simple relationship:

$$\boxed{\text{Speed} = \text{frequency} \times \text{wavelength}} \qquad \textbf{Eq. (1)}$$

Thus, the wavelength of sound at 100 Hz is 1,130/100 = 11.3 ft. At 1 kHz, the wavelength is only 1.13 ft. At 10 kHz, the wavelength is 0.113 ft, that is, approximately 1 in. Wavelength is not as commonly used in describing material properties as frequency. However, as Equation (1) shows, it is easy to convert from frequency to wavelength, and vice versa.

8.2 THE DECIBEL SCALE

The physical quantity associated with the loudness of sound is sound pressure, which can be expressed in pounds per square foot (psf) but is generally expressed in pascals (Pa), a unit in the SI system. The minimum sound pressure to which our ear responds, the *threshold of audibility,* is 0.00002 Pa. The sound pressure that corresponds to the sensation of pain in the ear is approximately 20 Pa (0.4 psf). This is still a very small pressure as compared to the atmospheric pressure (101,300 Pa = 2,100 psf) under which we all live.

Although the sound pressures are small, their range is extremely large. Therefore, it is more convenient to use the *decibel* scale in expressing sound pressure. The decibel scale is a logarithmic scale that compares a given sound pressure with the reference sound pressure of 0.00002 Pa, the threshold of audibility.

To distinguish sound pressure (expressed in pascals) from that expressed in decibels, we call the latter the *sound pressure level* (SPL). In other words, the unit for SPL is the decibel (dB). The threshold of audibility is 0 dB and the threshold of pain is 130 dB. Figure 8.1 gives the approximate SPLs of common environmental sounds.

Obtaining convenient numbers is not the only reason for the use of the decibel scale. The decibel scale also corresponds more directly to the ear's perception of loudness. For instance, a change of 1 dB in SPL is hardly perceived by the human ear. The minimum change in SPL that is just perceptible is 3 dB. A 5-dB change is quite noticeable, and a 10-dB change is perceived as a substantial change.

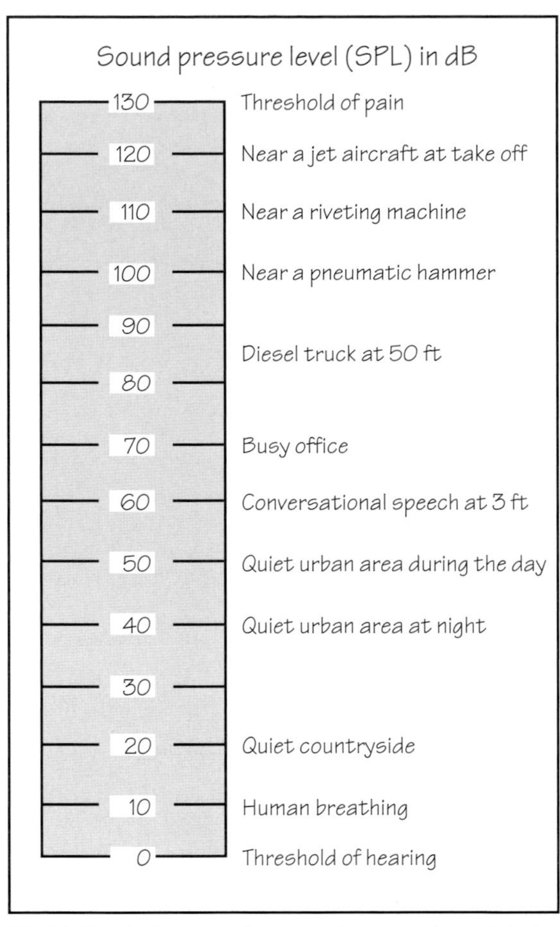

FIGURE 8.1 Commonly occurring sounds and their decibel ratings.

8.3 AIRBORNE AND STRUCTURE-BORNE SOUNDS

In building acoustics, we distinguish between two types of sounds based on the sound's origin. The reason for the distinction is that building components behave differently with respect to them. The two types of sound are

- Airborne sound
- Structure-borne sound

Most sounds in buildings are airborne sounds, such as the sounds generated by human conversation and musical instruments, Figure 8.2. Fans, motors, machinery, airplanes, and automobiles are some of the other sources that produce airborne sound.

FIGURE 8.2 Examples of airborne sound sources.

Structure-borne sound is produced by an impact of some sort on building components—walls, floors, roofs, and so on. The impact causes building elements to vibrate, and as they vibrate, they radiate sound. Because it is impact related, structure-borne sound is also referred to as *impact sound.* Thus, when a nail is driven into a wall or a person walks on a floor, structure-borne sound is produced, Figure 8.3. Other examples of structure-borne sounds are vibrating machinery rigidly connected to a floor, plumbing pipes attached to a wall, and the slamming of a door.

In other words, a structure-borne sound originates in an impact- or vibration-producing source that is in contact with a building component. The building component works as an amplifier of the sound generated by the impact or vibrating source. The impact or vibrating source itself may not create much sound.

For example, a vibrating water tap does not create much sound of its own, but if it is rigidly connected to a wall, its sound is greatly amplified because of the vibrations it produces in the wall. Similarly, the sound produced by a string instrument, such as a guitar, is amplified severalfold by the wooden body (sounding board) on which the strings are mounted.

Some sources can produce both airborne and structure-borne sounds. Once the structure-borne sound is produced by a building component, it becomes airborne sound and reaches the receiver as such.

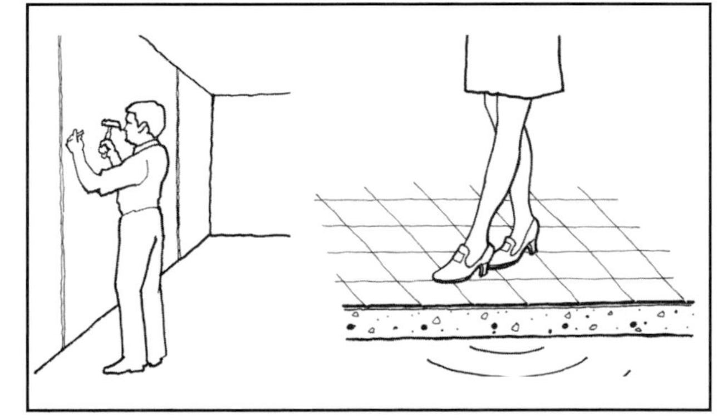

FIGURE 8.3 Examples of structure-borne sound sources.

PRACTICE QUIZ

Each question has only one correct answer. Select the choice that best answers the question.

1. Cycles per second is also called a
   - **a.** pascal.
   - **b.** newton.
   - **c.** hertz.
   - **d.** hulen.
   - **e.** none of the above.
2. Sound that is audible to human beings has a frequency lying between
   - **a.** 2 kHz and 20 kHz.
   - **b.** 200 Hz and 20 kHz.
   - **c.** 100 Hz and 800 Hz.
   - **d.** 20 Hz and 20 kHz.
   - **e.** 50 Hz and 50 kHz.
3. The pitch of a sound is related to
   - **a.** sound frequency.
   - **b.** sound energy.
   - **c.** sound speed.
   - **d.** all the above.
   - **e.** none of the above.
4. Sound speed in air is approximately
   - **a.** 10 ft/s.
   - **b.** 100 ft/s.
   - **c.** 500 ft/s.
   - **d.** 1,000 ft/s.
   - **e.** 1,100 ft/s.
5. Compared to a lightweight building element, a heavyweight element is generally a poor transmitter of airborne sound.
   - **a.** True
   - **b.** False
6. Most of the noise produced in a busy restaurant or dining hall is
   - **a.** airborne sound.
   - **b.** structure-borne sound.

Answers: 1-c, 2-d, 3-a, 4-e, 5-a, 6-a.

8.4 AIRBORNE SOUND INSULATION—SOUND-TRANSMISSION CLASS

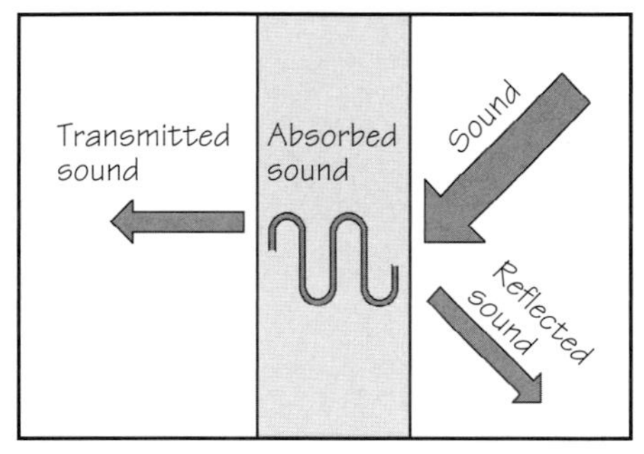

FIGURE 8.4 When a sound falls on a building assembly, part of it is reflected, part is absorbed, and the remaining is transmitted through the assembly.

When the airborne sound energy falls on a building assembly (such as a wall or ceiling), a part of the energy is reflected back into the enclosure, a part is absorbed within the material of the assembly and converted to heat, and a part is transmitted through it, Figure 8.4. An important property of the building assembly that affects sound reflection, absorption, and transmission is its surface weight (pounds per square foot).

Massive, heavyweight assemblies, such as thick masonry or concrete walls, are good sound insulators because they reflect back most of the sound. Stated differently, heavyweight assemblies are poor transmitters because a relatively small fraction of sound energy goes through them. Conversely, lightweight assemblies, such as stud walls and wood floors, are generally poor sound insulators because a relatively large amount of sound energy goes through them. However, as described later, the sound-insulating properties of lightweight assemblies can be improved by a few special provisions.

The airborne sound insulation of assemblies is measured by a quantity called *sound-transmission loss*. Sound-transmission loss (TL) is defined as the loss in sound-pressure level that occurs as the sound passes through the assembly. For example, if the sound-pressure level on the source side of a wall is 90 dB and it is 40 dB on the receiver side, the TL of the wall is 50 dB, Figure 8.5. If the sound pressure level on the source side is 70 dB, then the sound pressure level on the receiver side of the same wall is 20 dB.

The greater the TL of an assembly, the greater the sound insulation provided by it. An assembly with a TL of 50 dB is a better sound insulator than one with a TL of 40 dB. The TL of an assembly varies with frequency, generally increasing with increasing frequency. In other words, building assemblies generally provide greater sound insulation at high frequencies than at low frequencies.

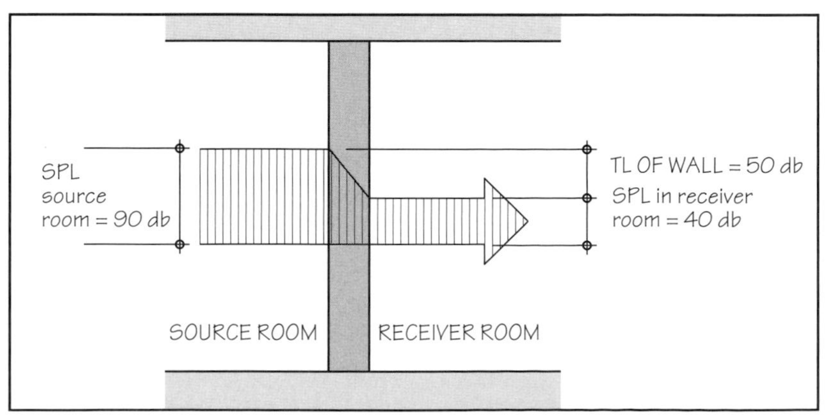

FIGURE 8.5 Definition of the transmission loss (TL) of an assembly.

Because of the variation of TL with frequency, TL cannot be used to compare the sound insulating efficacy of one assembly with another. For such a comparison, a single-number index is required. This is provided by the quantity called *sound-transmission class* (STC). The STC of an assembly is its average TL over frequencies ranging from 125 Hz to 4 kHz. The greater the STC, the more sound insulating the assembly. Note that the unit decibel is omitted in quoting the STC value in order to distinguish STC from TL. Thus, we use STC 54, not STC 54 dB, in stating the sound insulation of an assembly. Whole numbers are used in quoting the STC.

SUBJECTIVE PERCEPTION OF STC VALUES

STC is an important acoustical index. Manufacturers routinely provide the STC values of their assemblies. Building codes require that the party wall separating two dwelling units (condominiums or apartments) must be have a minimum STC of 50. A larger value is usually required as a good design practice. But how much higher? In other words, how much acoustical privacy does an STC 50 assembly or an STC 55 assembly provide? Table 8.1 attempts to answer these questions. As illustrated in this table, an assembly with an STC of 50 will allow loud speech to be heard through the assembly by a curious listener.

TABLE 8.1 SUBJECTIVE PERCEPTION OF STC VALUES

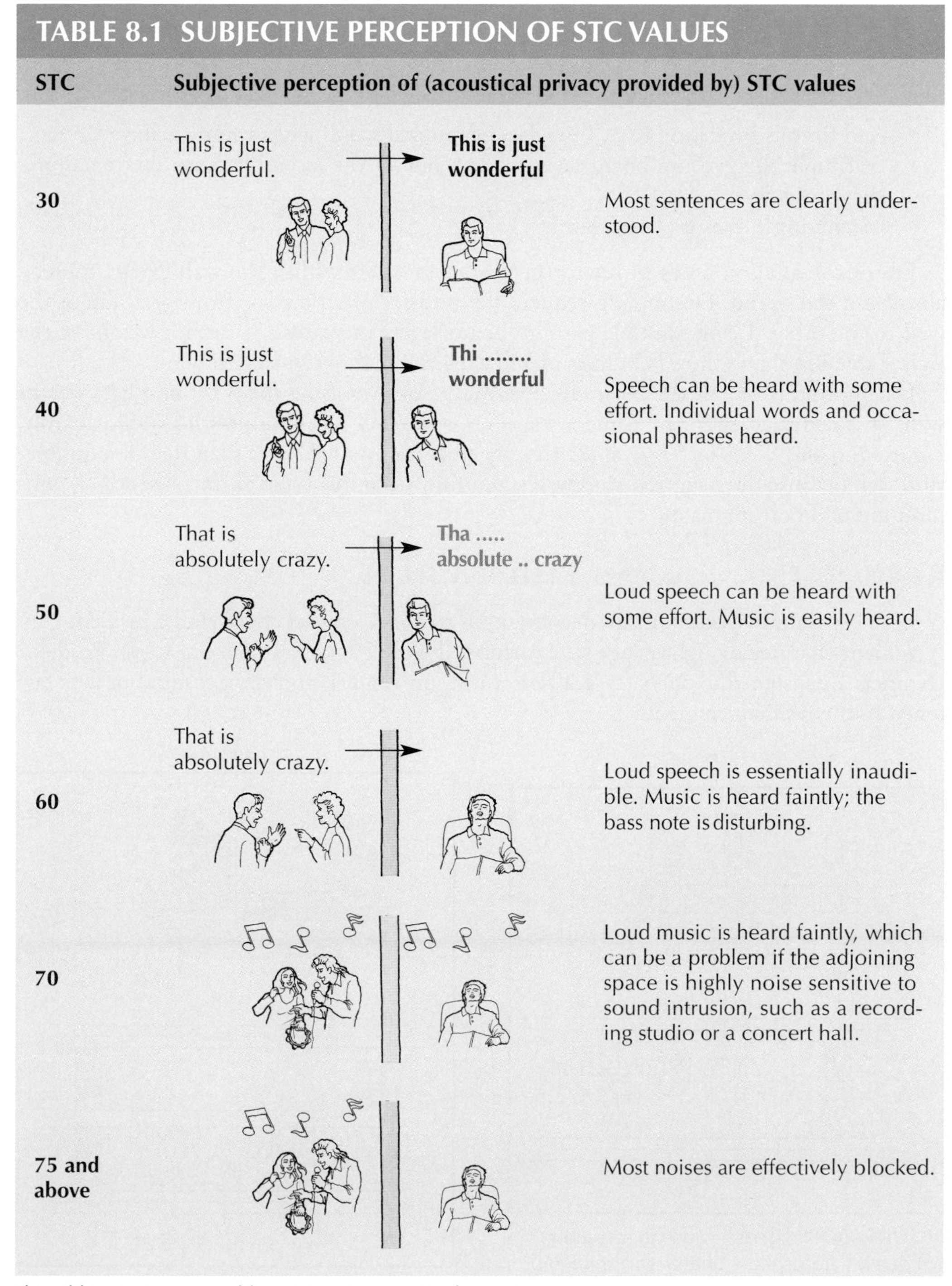

STC	Subjective perception of (acoustical privacy provided by) STC values
30	Most sentences are clearly understood.
40	Speech can be heard with some effort. Individual words and occasional phrases heard.
50	Loud speech can be heard with some effort. Music is easily heard.
60	Loud speech is essentially inaudible. Music is heard faintly; the bass note is disturbing.
70	Loud music is heard faintly, which can be a problem if the adjoining space is highly noise sensitive to sound intrusion, such as a recording studio or a concert hall.
75 and above	Most noises are effectively blocked.

This table assumes reasonably quiet environment in the receiving room.

TABLE 8.2 APPROXIMATE STC VALUES OF SELECTED WOOD STUD ASSEMBLIES WITH $\frac{1}{2}$-IN.-THICK GYPSUM BOARD ON BOTH SIDES OF STUDS

Gypsum board support system	Without cavity absorption Layers of gypsum board 1 + 1	1 + 2	2 + 2	With cavity absorption Layers of gypsum board 1 + 1	1 + 2	2 + 2
2 × 4 wood studs	37	40	43	40	43	46
2 × 4 wood studs, resilient channel (or clip) on one side	40	45	49	50	53	57
2 × 4 wood studs, resilient channel (or clip) on both sides	41	46	51	49	53	58
2 × 4 wood stud wall (staggered studs)	41	48	52	50	54	58
Double-stud wall with 1 in. space between studs	46	53	57	57	61	63

As shown in this table, using a resilient channel (or clip) on one side of a wall is adequate. Using the channel (or clip) on both sides of the wall gives a negligible increase in STC value.

Improving the STC Value of Lightweight Assemblies

As stated earlier, the TL (and hence the STC) of an assembly can be increased if it is made heavier. However, this not only adds to the cost of the assembly but also increases the dead load on the structure. That is why lightweight assemblies, such as steel or wood stud walls lined with gypsum board on both sides, are commonly used.

The STC of a wall with 2 × 4 wood studs and $\frac{1}{2}$ in. gypsum board on each side is only 37. This is is an extremely low STC value and, as shown in Table 8.1, provides little acoustical privacy. The following are commonly used methods of increasing the STC of stud-framed walls:

- Add fibrous insulation (e.g., fiberglass or mineral wool) within stud cavities
- Decouple the gypsum board layer on one side of the assembly from the remaining part of the wall
- Use multiple gypsum board layers

Fibrous insulation helps to retard the vibration of air within the wall cavity, thereby absorbing the sound. Decoupling reduces the transfer of vibrations from one side of the wall to the other. Using multiple gypsum board layers increases the surface weight of the wall. Table 8.2 shows the STC values of a few selected lightweight assemblies.

Decoupling is one of the most effective means of increasing the STC of a lightweight wall. It is achieved either by using a staggered-stud wall or a double-stud wall assembly, Figures 8.6 and 8.7. The STC value of a staggered-stud wall is lower than that of a double-stud wall because the staggered-stud wall is not fully decoupled due to the presence of common top and bottom plates.

Resilient Channels and Resilient Clips

A more commonly used means of decoupling is to use a resilient channel or a resilient clip. A resilient channel is a light-gauge steel member, Figure 8.8 (see also Figure 17.4). Resilient channels are fastened to studs (typically at 24 in. on center), and the gypsum board is fastened to the resilient channels.

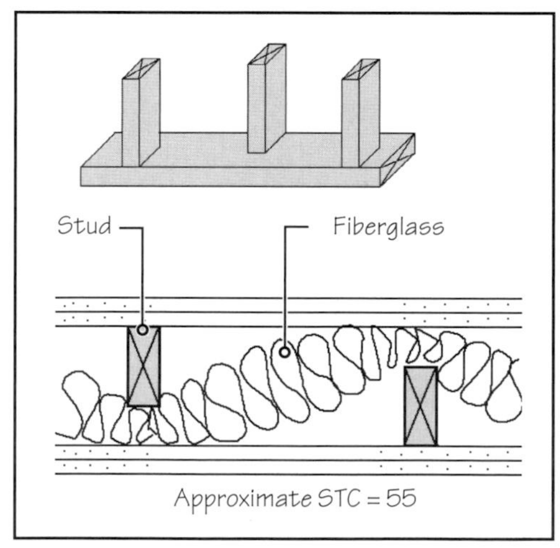

FIGURE 8.6 Staggered-stud wall assembly with cavity absorption (fiberglass or mineral wool).

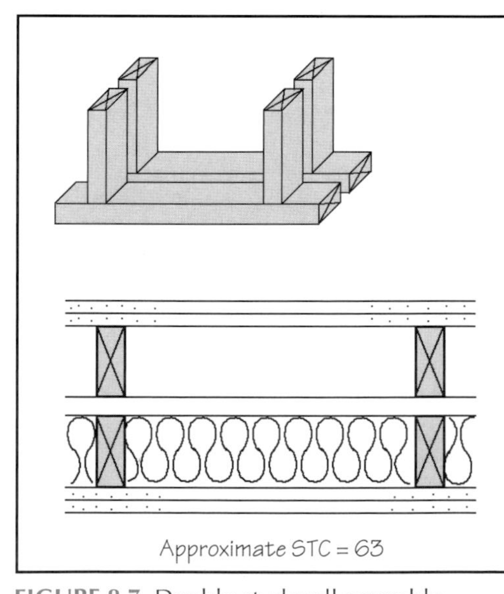

FIGURE 8.7 Double-stud wall assembly.

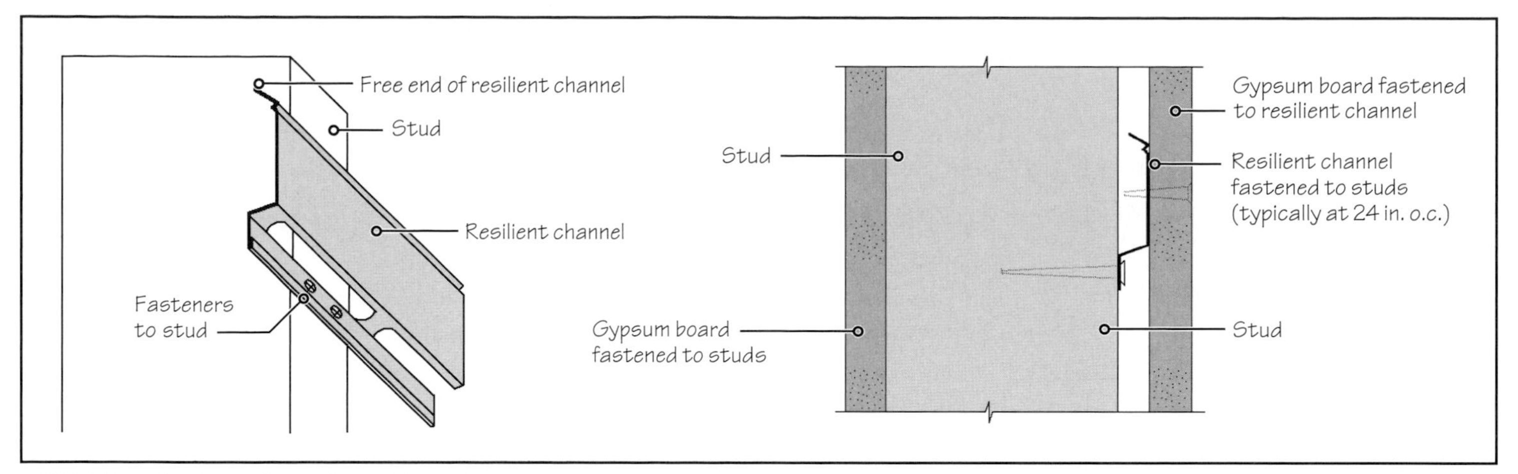

FIGURE 8.8 Resilient channel's profile and use.

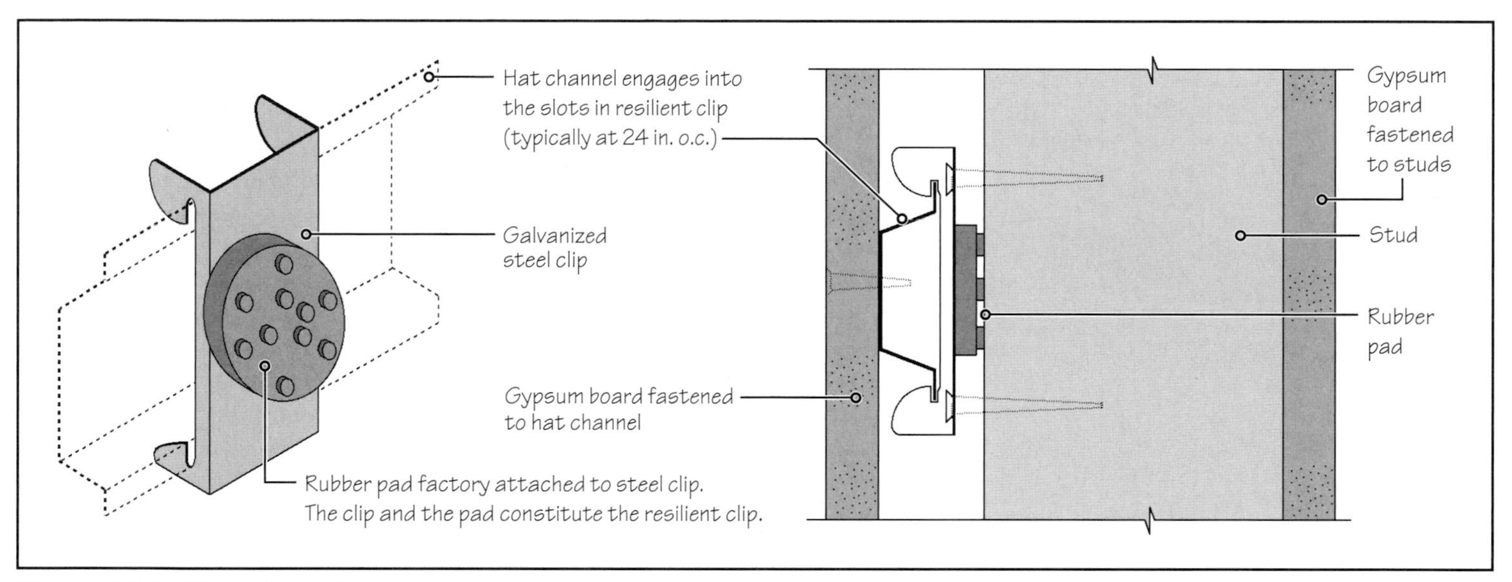

FIGURE 8.9 Resilient clip and its use.

In fastening the resilient channel, its free end must be toward the top, so that when the gypsum board is fastened, the free end pulls away from the studs. If the free end is toward the bottom, decoupling will not be achieved.

A resilient clip is a more recently introduced device as an alternative to the resilient channel. It consists of a light-gauge steel clip with a rubber pad, Figure 8.9. The rubber pad functions as a vibration absorber.

Importance of Air Seals for Airborne Sound Insulation

Because sound travels through the air, it is important that an assembly contain no voids if it is required to have a high STC. Remember that if air can go through the assembly, so will sound. It is futile to increase the surface weight of the assembly or use resilient channels, clips, or staggered studs if the is air can leak through the assembly. Therefore, the assembly should be fully sealed at all edges (perimeter) and contain no air-leakage sites.

8.5 STRUCTURE-BORNE SOUND INSULATION—IMPACT INSULATION CLASS

An effective way to reduce structure-borne sound transmission is to reduce the vibrations at the source (before they become structure-borne) by absorbing them through the use of resilient materials. For example, the most common structure-borne sound in buildings is produced by the floors as a result of footsteps, furniture movement or the vibration of machinery supported by the floor. The best way of reducing the transmission of this sound through the floor is to dampen the impact through a soft covering over the floor. The measure used to quantify structure-borne sound insulation is *impact insulation class* (IIC). As with STC values, the unit decibel is not used with IIC values.

A soft covering, such as carpet backed by a foam underlayment, increases the IIC substantially. For example, the IIC of a bare 6-in.-thick concrete floor is 25. The same floor covered with a carpet and underlayment gives an IIC of 85—an increase of 60. Note that a carpet does not improve the STC of a floor. The STC of a 6-in.-thick concrete slab is 55, regardless of whether the floor is bare or covered with a carpet.

The improvement in the IIC of wood floors by using a carpet is less pronounced because of the inherent resilience of a wood floor. For example, a typical plywood floor on wood joists and gypsum board ceiling directly attached to the studs has an IIC of 34 without the carpet and 55 with the carpet, Figure 8.10. Building codes require a minimum IIC of 50 for a floor-ceiling assembly between two independent dwelling units, such as in an apartment building.

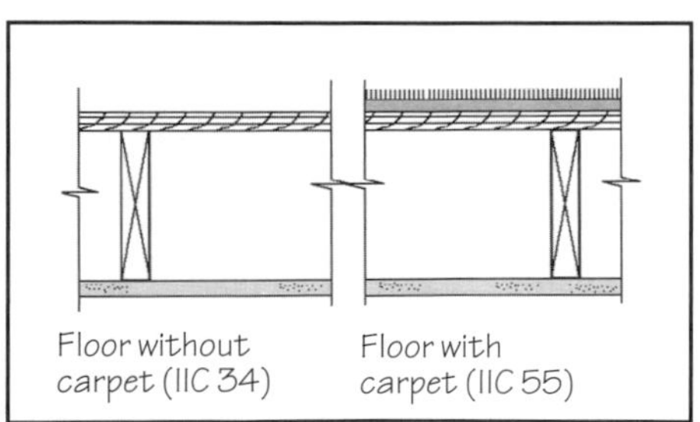

FIGURE 8.10 IIC of wood floor without and with carpet and underlayment.

8.6 SOUND ABSORPTION—NOISE-REDUCTION COEFFICIENT

While the sound originating from outside a receiving room is attenuated (reduced) by using sound-insulating construction, that originating from within the room can be reduced only by using sound-absorbing materials. All materials and objects absorb sound to some degree. However, materials whose sound-absorption coefficient is greater than 0.2 are called *sound-absorbing materials,* or *acoustical materials,* although the former term is preferable. For the same reason, the term *acoustical treatment* usually implies sound-absorptive treatment.

The sound-absorption coefficient of an assembly, α, is defined as the sound energy that is not reflected by the assembly divided by the total sound energy falling on the assembly. The term *not reflected* is used because it includes both the absorbed energy and the energy transmitted through the assembly. From the point of view of an enclosure, an open window is considered to be a perfect acoustical absorber because it does not reflect any sound.

The value of α lies between 0 and 1 and varies with the frequency of sound. For most sound-absorbing materials, we are concerned with the value of α lying between 250 Hz and 2 kHz because, as noted earlier, it is the frequency range of human speech.

Because of its variation with frequency, we cannot use the value of α to compare the sound-absorbing efficacy of one material with the other. Therefore, we use a single-number metric called *noise-reduction coefficient* (NRC), which is the value of α for a material averaged over the frequency range of 250 Hz to 2 kHz. Because NRC is the average value of α, NRC also lies between 0 and 1. The higher the NRC, the more sound absorptive the material.

The most commonly used materials for sound absorption are porous materials, called *porous sound absorbers.* They absorb sound because the air particles in the vicinity of the material go back and forth into the pores and convert their vibrational energy into heat through friction. Because the energy contained in sound is extremely small, the amount of heat so created is negligible.

For a porous material to be a good sound absorber, it is important that it has interconnected pores. Plastic foams (such as extruded polystyrene, polyisocyanurate, etc.), used as thermal insulators, are not good sound absorbers because their pores are not interconnected. Fiberglass, rockwool, and slagwool are the most commonly used sound-absorbing materials because they have interconnected pores and also because they are noncombustible.

The NRC value of a porous sound absorber is a function of its thickness. A thicker porous absorber generally has a higher value of NRC. Manufacturers routinely quote the NRC value for their sound absorptive materials. A commonly used porous absorber is the ceiling *tile,* in which the perforations can take various patterns, Figure 8.11. Other materials, particularly fiberglass wrapped in a perforated fabric, is also commonly used, Figure 8.12.

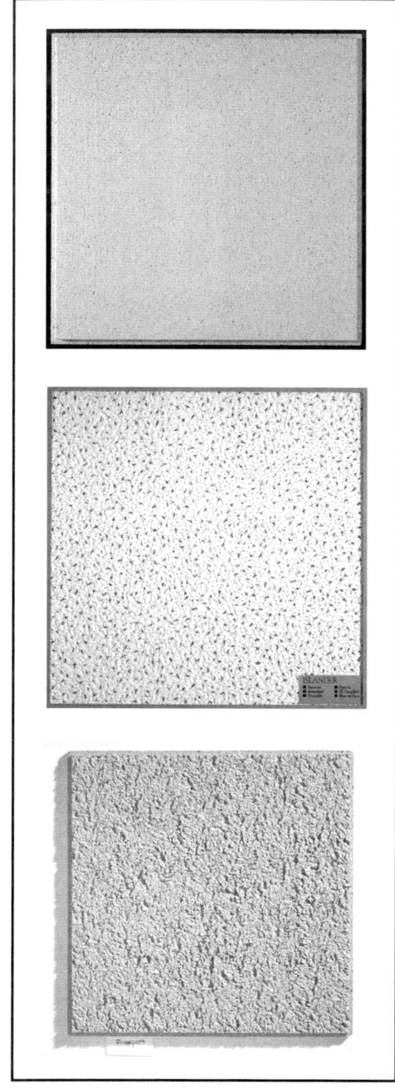

FIGURE 8.11 Various types of perforations used in ceiling tiles (see Chapter 35).

FIGURE 8.12 Low-height partitions in a large open-plan office are usually made of fabric-wrapped fiberglass to provide sound absorption. In the absence of sound absorbing material, a large office can be excessively noisy.

Each question has only one correct answer. Select the choice that best answers the question.

7. Sound transmission class (STC) is a measure of
- **a.** sound-absorbing property of a material.
- **b.** sound-insulating property of a material with respect to airborne sound.
- **c.** sound-insulating property of a material with respect to structure-borne sound.
- **d.** sound-reflecting property of a material.

8. For a party wall, between two dwelling units, such as two apartments, the building codes require its minimum STC value to be
- **a.** 100.
- **b.** 75.
- **c.** 65.
- **d.** 55.
- **e.** 50.

9. A resilient clip consists of a small light-gauge steel member with a rubber pad fastened to it.
- **a.** True
- **b.** False

10. A resilient channel is used to attenuate the airborne sound, whereas a resilient clip is used to attenuate the structure-borne sound.
- **a.** True
- **b.** False

11. The measure that is used for structure-borne sound insulation is
- **a.** transmission loss.
- **b.** impact isolation index.
- **c.** impact insulation class.
- **d.** structural insulation index.

12. A commonly used porous sound absorber is
- **a.** gypsum board.
- **b.** aluminum foil.
- **c.** microperforated plastic.
- **d.** oriented strandboard.
- **e.** fiberglass.

13. The noise-reduction coefficient of a material is evaluated between the frequency range of 250 Hz and 2 kHz because
- **a.** it is virtually impossible to evaluate this property outside the above frequency range.
- **b.** the human ear is insensitive to noise outside the above frequency range.
- **c.** most of the sound energy in the noise produced in buildings lies in the given frequency range.
- **d.** most of the sound energy in human speech lies in the given frequency range.
- **e.** none of the above.

Answers: 7-b, 8-e, 9-a, 10-b, 11-c, 12-e, 13-d.

KEY TERMS AND CONCEPTS

Airborne sound
Airborne sound insulation
Decibel scale
Frequency and wavelength of sound
Impact insulation class (IIC)
Infrasonic
Noise reduction coefficient (NRC)
Resilient channel and resilient clip
Sound absorption coefficient
Sound transmission class (STC)
Speed of sound
Structureborne sound
Structureborne sound insulation
Supersonic
Ultrasonic

SELECTED WEB SITES

Acoustical Society of America (www.asa.aip.org)
Acoustics.com (www.acoustics.com)

FURTHER READING

1. Mehta, M., Johnson, J., and Rocafort, J. *Architectural Acoustics*. Upper Saddle River, NJ: Prentice Hall, 1999.

CHAPTER 9

Principles of Joints and Sealants (Expansion and Contraction Control)

CHAPTER OUTLINE

9.1 Types of Movement Joints

9.2 Building Separation Joints and Seismic Joints

9.3 Movement Joints in Building Components

9.4 Thermal Movement

9.5 Moisture Movement

9.6 Elastic Deformation and Creep

9.7 Total Joint Dimension

9.8 Principles of Joint Detailing

9.9 Components of a Sealed Joint

9.10 Types and Properties of Joint Sealants

Detailing of the connection of copper cladding with the concrete wall in the Burton Barr Central Library, Phoenix, Arizona, was critical to the design of the building's exterior. Inset detail shows copper cladding. Architects: Will Bruder Architect and DWL Architects. (Photo by DA.)

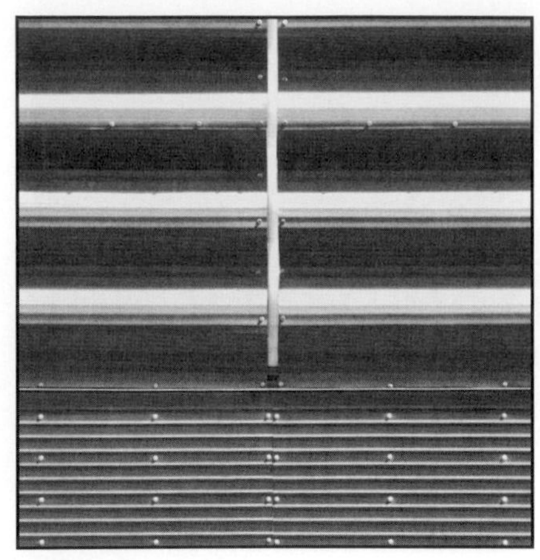

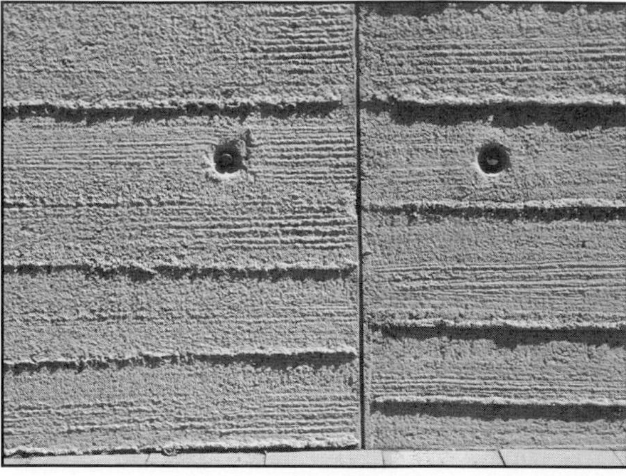

FIGURE 9.1 The fasteners used for attaching copper cladding and the control joints used in textured concrete cladding in Burton Barr Central Library, Phoenix, Arizona, contribute to the larger scale pattern in the building facade. Design Architect: Will Bruder.

Building joints result from our need to work with materials than can be easily and efficiently fabricated, transported, and assembled on site. The joints also allow us field adjustments in assemblies during construction. Additionally, visible connections resulting from the joints also provide the opportunity to develop ideas of scale, pattern, color, texture, and so on, on building planes and surfaces, Figure 9.1.

The most important issue in the design of building joints is that the dimensions of building components are constantly changing, not only with respect to their original dimensions, but also relative to each other. For example, window glass expands and contracts in response to external temperature changes. The changes in glass dimensions are usually different from the corresponding changes in the frames that hold the glass. Yet we still need to be able to make a secure connection between the window glass and frame.

Static Joints and Movement Joints

Over the years, two different strategies have been developed for joining building components. We could provide a nonmoving joint (connection) between the components, called a *static joint.* In the case of the window glass and frame, a static joint would most likely result in broken glass, broken frame, or both. The alternative is to provide a connection that allows both materials to move independently and at the same time hold the window glass securely in the frame. This movable connection is called a *dynamic joint,* or, more commonly, a *movement joint.*

Static joints typically connect materials that are the same or similar. For instance, the mortar joints between units of masonry, the joint between two sheets of gypsum wall board, and most structural joints provide nonmoving connections between components. By their nature, the movement of components in a static joint is restricted so that the stress created by such restraint must be counteracted by increased strength of the components.

A well-designed movement joint, on the other hand, has no movement stress. Therefore, it is used at the connection of materials that move at different rates. This implies that movement joints are placed where dissimilar materials meet. They may also be needed to subdivide a component into smaller sizes in order to reduce internal stresses. It is for this reason that a concrete slab-on-grade is subdivided into smaller areas with a gridwork of joints. It is also the reason a masonry wall—an assembly of static (mortar) joints—must be provided with movement joints at intervals.

The focus of this chapter is on movement joints. Because a movement joint is generally required to be sealed against air, water, and noise transmission (particularly the joints that occur in the external envelope of the building), we will also examine the commonly used sealants and sealant backup materials.

Issues related to preventing movement of rainwater through joints in wall assemblies are discussed specifically in Chapter 25.

NOTE

Construction Joint

A special type of static joint that is provided between two concrete placements is called *construction joint,* or *cold joint* (Chapter 19).

9.1 TYPES OF MOVEMENT JOINTS

Based on the purpose they serve, movement joints are classified as

- *Building joints*—joints between different parts of the building as a whole
- *Component joints*—joints between individual components of an assembly

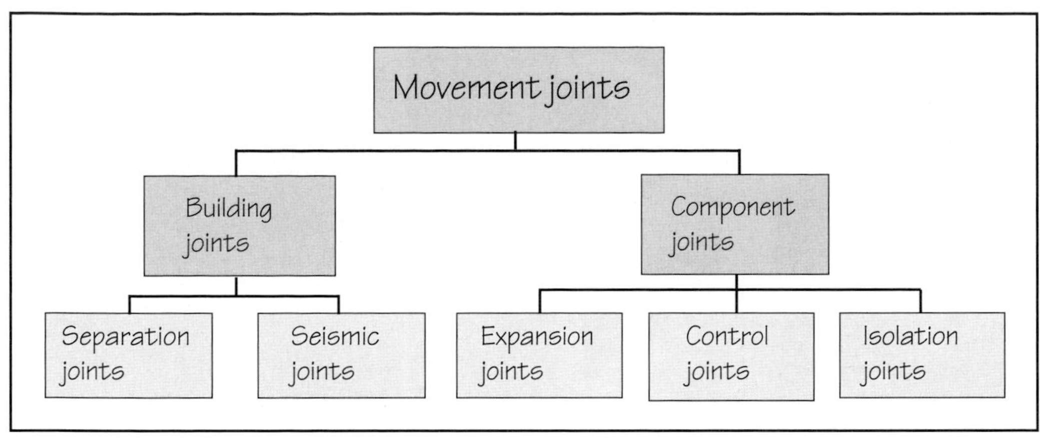

FIGURE 9.2 Types of joints in a building.

A building joint divides the entire building into two or more separate buildings. Based on the purpose for which it is provided, it is classified either as a *building separation joint* or a *seismic joint.*

A joint between two adjacent building components can be an *expansion joint,* a contraction joint (commonly called a *control joint*), or an *isolation joint.* A complete list of joint types in a building is given in Figure 9.2. Joint types are discussed in detail in the following sections.

9.2 BUILDING SEPARATION JOINTS AND SEISMIC JOINTS

A building separation joint runs continuously throughout the entire building from floor to floor and from face to face. It divides a large and geometrically complex building into smaller individual buildings, which can move independent of each other. This joint, typically 1.5 to 2 in. wide, accommodates the cumulative effect of various types of movements that occur in a building as a whole, as distinct from the movements that occur in individual components of the building.

A building separation joint prevents the stresses created in one part of the building from affecting the integrity of the other part. A building separation joint is needed in large buildings. As a rough guide, it should be provided at 250-ft (75-m) intervals, Figure 9.3.

NOTE

Building Separation Joint (or Building Expansion Joint)

The term *building expansion joint* is often used for building separation joint in some literature. This is not entirely correct, because a building separation joint accommodates more than mere expansion and contraction of a building.

NOTE

Spacing Between Building Separation Joints

The suggested spacing of 250 ft between building separation joints should be regarded as a rough guide. Most buildings of a simple rectangular shape up to 300 ft in length do not need separation joints. On the other hand, a building with major deviations from a rectangular shape may need separation joints at less than 250 ft on center.

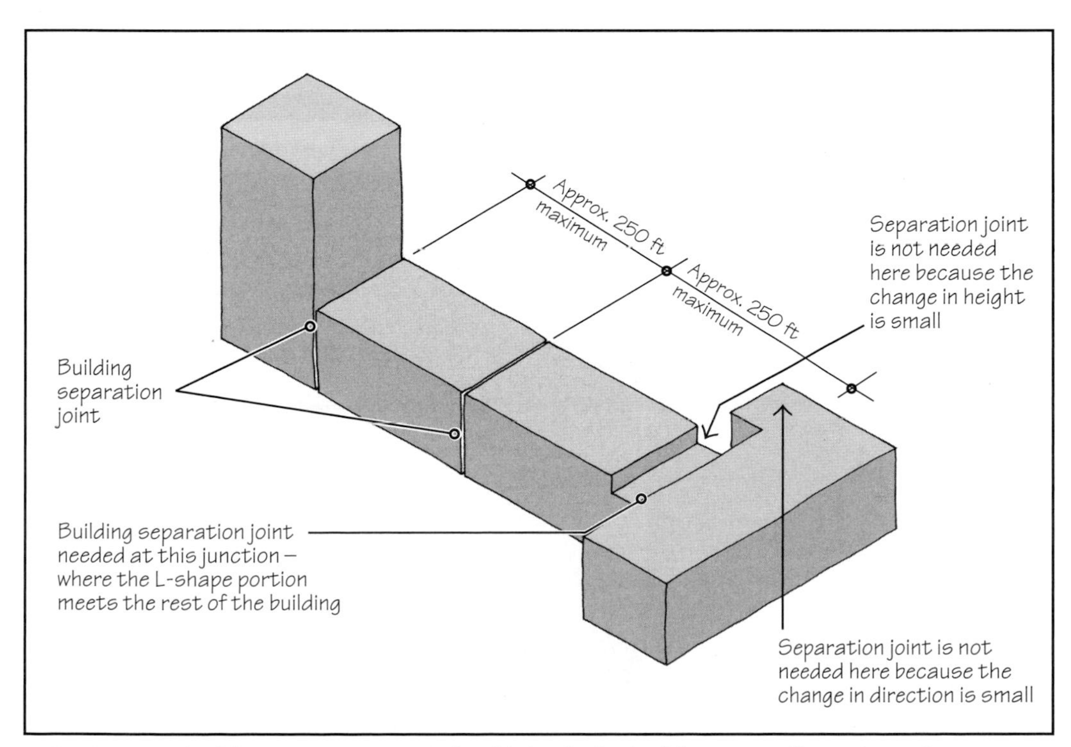

FIGURE 9.3 A building separation joint should divide the building in smaller, independent sections that are geometrically simple.

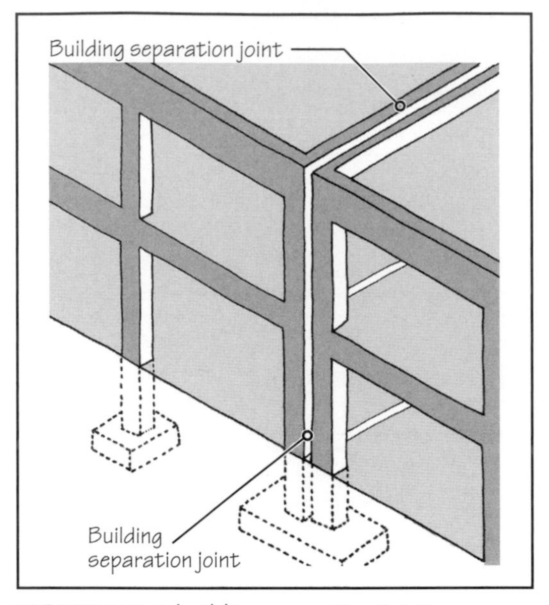

FIGURE 9.4 A building separation joint runs through the entire building from the ground floor (or the basement, if provided) up to the roof. Two columns are often used at the separation joint. However, a combined footing may be used for the two adjacent columns.

FIGURE 9.5 Building separation joint with two columns and two beams in a reinforced concrete building.

In addition, good practice requires the provision of building separation joints at the following discontinuities:

- Where a low building mass meets a tall mass. This prevents problems caused by differential expansion and contraction (including creep), in addition to any differential settlement of foundations. Minor changes in height should be disregarded.
- Where the building changes direction, such as in an L-shape or T-shape building. Minor changes in direction should be ignored.
- Where the building's structural material changes, such as, where a steel frame building meets a concrete frame. This prevents problems due to differential movement between disparate structural systems.

As far as possible, a building separation joint should divide a large building into separate, smaller buildings of simple geometries, such as rectangles. The division into separate buildings requires the use of two columns and two beams at a building separation joint, Figures 9.4 to 9.6.

The duplication of columns and beams at building separation joint is an ideal solution. An alternative solution that is commonly used is shown in Figures 9.7 and 9.8, in which a single column with a bracket is used.

FIGURE 9.6 Building separation joint with two columns and two beams in a steel frame building.

FIGURE 9.7 Building separation joint using a single column.

FIGURE 9.8 Detail of building separation joint of Figure 9.7. For additional details of the joint, see Chapter 20.

SEISMIC JOINT

A seismic joint is similar to but also different from a building separation joint. Building separation joints are introduced to accommodate moderate building movements caused by temperature and moisture changes, creep, and foundation settlements. Although they can accommodate small vertical movement, they are designed to accommodate mainly horizontal movement, that is, in the direction perpendicular to the joint.

Like building separation joints, seismic joints are also provided where there are major dissimilarities in building form. However, the purpose of seismic joints is to ensure that one section of the building does not collide with the adjacent section during an earthquake. A seismic joint must accommodate simultaneous movements in horizontal as well as vertical directions, that is, in all three principal directions.

A seismic joint is generally much wider than a building separation joint. In highly seismic locations, a seismic joint may be a few feet wide for a tall building. The width of a building separation joint is the same at each floor. The width of a seismic joint, on the other hand, should increase with height. For aesthetic and architectural reasons, however, a constant width is generally used.

Using double columns at a seismic joint is the preferred solution. However, a single column with seated connections (Figure 9.8) that are adequately restrained to prevent the seat from sliding off the support is acceptable.

COVERS OVER BUILDING SEPARATION OR SEISMIC JOINT

Building separation and seismic joints must be covered. Various proprietary products are available to cover the joints at the roof, floors, and walls. A simple floor joint cover consists of a compressible gasket between two metal sections. Each metal section is anchored to the floor slab, Figure 9.9.

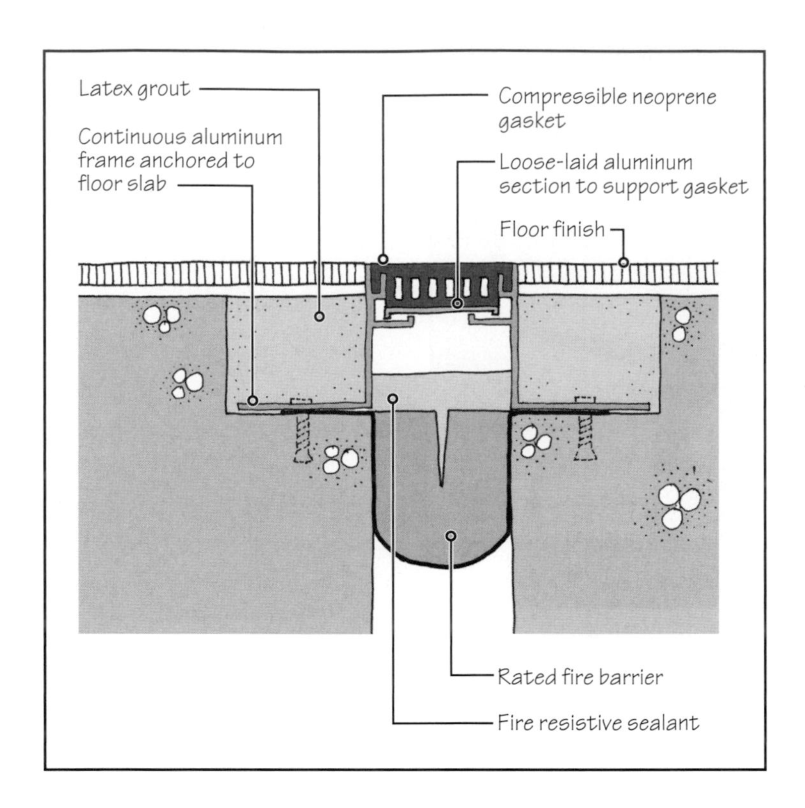

FIGURE 9.9 A typical detail of building separation joint cover at floor level. Similar covers are provided in interior and exterior walls, and roofs. Joint cover manufacturers' details should be strictly followed, particularly for fire-rated joints.

A more elaborate floor joint cover that allows greater movement is used for seismic joints. It is important to ensure that the joint has the same fire resistance as the floor, wall, or the roof in which it occurs. Once again, proprietary fire-resistive joint materials (typically consisting of ceramic insulation encapsulated in a stainless steel sheet) are available to use within the joint. Building separation joints at the roof are discussed in Chapter 31.

9.3 MOVEMENT JOINTS IN BUILDING COMPONENTS

Movement in a building component can be caused by several phenomena. Therefore, many factors must be considered in sizing and detailing movement joints in building components:

- Thermal movement
- Moisture movement
- Elastic deformation and creep
- Construction tolerances and other considerations

Although thermal movement and elastic deformation occur in all materials, the same is not true of creep and moisture movement. Steel does not creep, and its dimensions are not affected by moisture variations. Thermal movement is reversible if the component is unrestrained. Moisture movement, on the other hand, may or may not be reversible, depending on the material, Table 9.1. Creep deformation is irreversible; that is, creep is a permanent deformation.

In addition to the factors just stated, construction and material tolerances and several other factors must be considered in sizing building component joints.

Because of the multiplicity of factors and their intricate interrelationships, the theory of sizing movement joints can be fairly complex. Therefore, in practice, joint dimensions are based on well-acknowledged rules of thumb. In most situations, the rules of thumb are

TABLE 9.1 TYPES OF MOVEMENT IN SELECTED MATERIALS

Building material	Thermal	Elastic deformation	Creep	Reversible moisture	Irreversible moisture
Steel	X	X	—	—	—
Concrete	X	X	X	X	—
Concrete masonry	X	X	X	X	—
Brick masonry	X	X	X	—	X
Wood	X	X	X	X	—

X indicates the presence of particular movement in the material and — indicates the absence of such movement.

Importance of Movement Joints in Building Components

The importance of providing an adequate space for the movement of components may be appreciated by considering the expansion of a linear member such as a beam. Assume that the original length of a beam is L. Assume further that, under certain temperature and moisture changes, its length increases by an amount δ. If the beam is fully restrained at its ends so that it cannot increase in length, it will be subjected to axial compression. Consequently, the beam will tend to buckle (bend upward or downward), Figure 9.10. If the deflection of the beam is denoted by h, we find that h is several times larger than the beam's unrestrained extension, δ.

To better visualize this fact, consider the following model. An approximately 4-ft-long sheet made of a flexible material (e.g., acrylic) has been cut to fit snugly between two raised ends, Figure 9.11(a). The raised ends restrain the expansion of the sheet. If we slightly shorten the length of the sheet (say, by approximately 0.1 in.) on each end by placing a quarter between the sheet and the raised end, we find that the sheet curves up substantially, Figure 9.11(b).

Apart from the fact that such a large deformation is visually unacceptable in buildings, it also creates structural problems. A component made of a ductile material (e.g., metal) will buckle in such a situation. On the other hand, materials such as concrete, masonry, gypsum drywall, and glass will simply break under such a large deformation because, being weak in tension, these (brittle) materials cannot resist the tensile stress created by excessive bending.

This simple model highlights the importance of providing movement joints. A similar situation would occur if the beam were to contract instead of expanding. Contraction creates axial tensions, and most brittle materials will fail under relatively small contraction because their tensile strength is low.

If the component is free to move due to the provision of movement joints, no stress will be created in the component. However, although movement joints provide a convenient solution to expansion and contraction, they can be avoided (where the situation so requires) by increasing the strength and stiffness of the component. For example, the buckling of the acrylic sheet in Figure 9.11(b) can be avoided by increasing the strength (i.e., thickness) of the sheet or replacing it by a stiffer material.

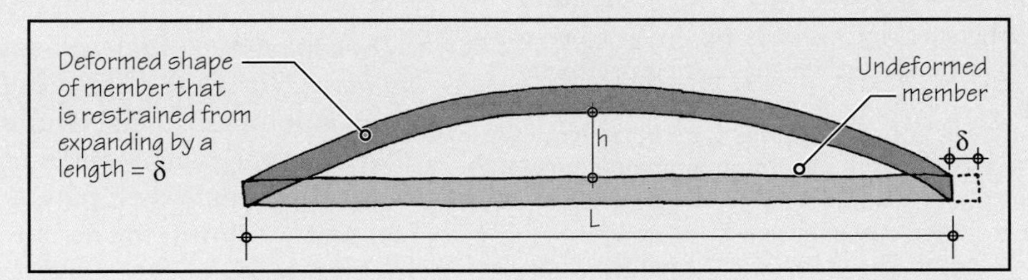

FIGURE 9.10 Deformation of a member that is restrained from expanding. Note that the dimension h is much larger than the dimension δ.

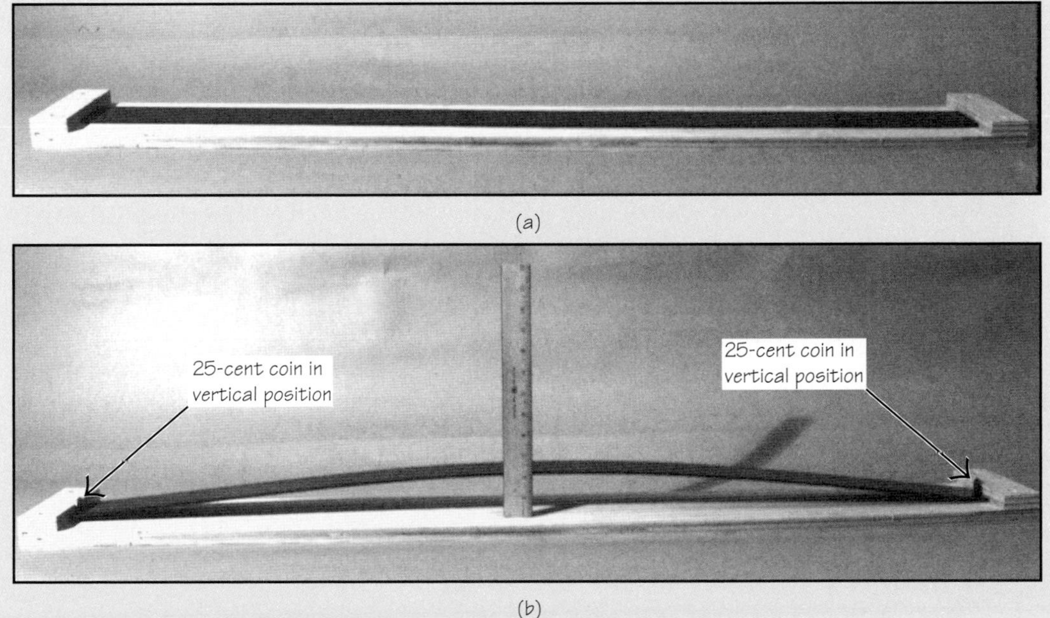

FIGURE 9.11 **(a)** An acrylic sheet in an undeformed state. **(b)** Deformation of the same sheet (nearly 4 ft long) shortened by placing a quarter at each end.

established by material manufacturing associations. For instance, the Brick Industry Association (BIA) provides rules of thumb for size and spacing of joints in brick walls. Similarly, the Portland Cement Association (PCA) and American Concrete Institute (ACI) provide rules of thumb for joints in concrete slabs and walls.

Despite the general use of rules of thumb, a basic understanding of the theory of movement of building components is important and is presented in the following sections. This information is critical when the designer must depart from the rules of thumb. It is also important in situations where rules of thumb do not exist.

NOTE

Units of α

The units of L and δ_t are inches, and the unit of (ΔT) is °F. Substituting these units in Equation (1), we get

$$\text{in.} = \alpha(\text{in.})(°\text{F})$$

So that the units on both sides of the equals sign are consistent in this equation, the units of α must be

$$\frac{1}{°\text{F}}, \quad \text{that is,} \quad °\text{F}^{-1}$$

9.4 THERMAL MOVEMENT

Thermal movement is generally the most critical movement because it occurs in all components, particularly those exposed to the exterior climate, such as exterior walls, cladding, roof membranes, slabs, and paving. It is also the largest type of movement in most components.

To determine the thermal expansion or contraction of a component, assume that its length is L and its temperature rises (or falls) by Δt (°F or °C). If the component is not restrained, its length will increase (or decrease) so that the new length will be $(L + \delta_t)$ or $(L - \delta_t)$, where δ_t is the change in length. The value of δ_t is given by the following equation:

$$\delta_t = \alpha L(\Delta t) \qquad \textbf{Eq. (1)}$$

The quantity α is a property of the material and is called the *coefficient of thermal expansion.* The greater the value of α, the greater the thermal movement. Plastics have the highest values of α, followed by metals and ceramic materials such as concrete, masonry, and glass. The units of α are $°\text{F}^{-1}$ or $°\text{C}^{-1}$. Values of α for selected materials are given in Figure 9.12.

Equation (1) can be used to determine the minimum width of joint required to prevent thermal stresses between adjacent components. The value of Δt that should be used in this calculation is a function of two factors:

- *The annual temperature range to which the component is subjected.* It is the difference between the maximum and minimum temperatures of the component. Average approximate values of annual maximum and annual minimum temperatures of vertical components for most of North America are shown in Figure 9.13.
- *The temperature of the component at the time of its construction.* Because buildings are usually not built during weather extremes, the range of ambient temperatures during which construction is undertaken is much smaller than the maximum-minimum temperature range of the component. See Figure 9.13 for typical ambient temperature range during construction in the United States.

If an exterior wall is built when ambient air temperature is 50°F, it will be subjected to a temperature gradient (Δt) of 70°F when the wall's temperature becomes 120°F.

NOTE

Component Temperature and Ambient (Air) Temperature

The temperature of building envelope components is usually not the same as the ambient air temperature. The reason is that the air temperature is recorded under shade, whereas the building envelope is exposed to sun's radiation.

Therefore, the maximum surface temperature of an envelope component is generally higher than the maximum air temperature at the location. The difference between the two temperatures depends on the degree of exposure of the component to direct sun, its material, color, amount of insulation in the component, and the wind speed.

Dark-colored components get hotter than light-colored components. Horizontal components receive solar radiation throughout the day, so they become hotter than the vertical components. Metals become hotter than nonmetals. If a component is insulated, its surface temperature is higher than an uninsulated component because an uninsulated component loses more heat to the interior.

Under a clear night sky, the minimum surface temperature of a component is lower than the minimum air temperature. This is due to radiation heat loss from the component to the sky, referred to as *nocturnal cooling.*

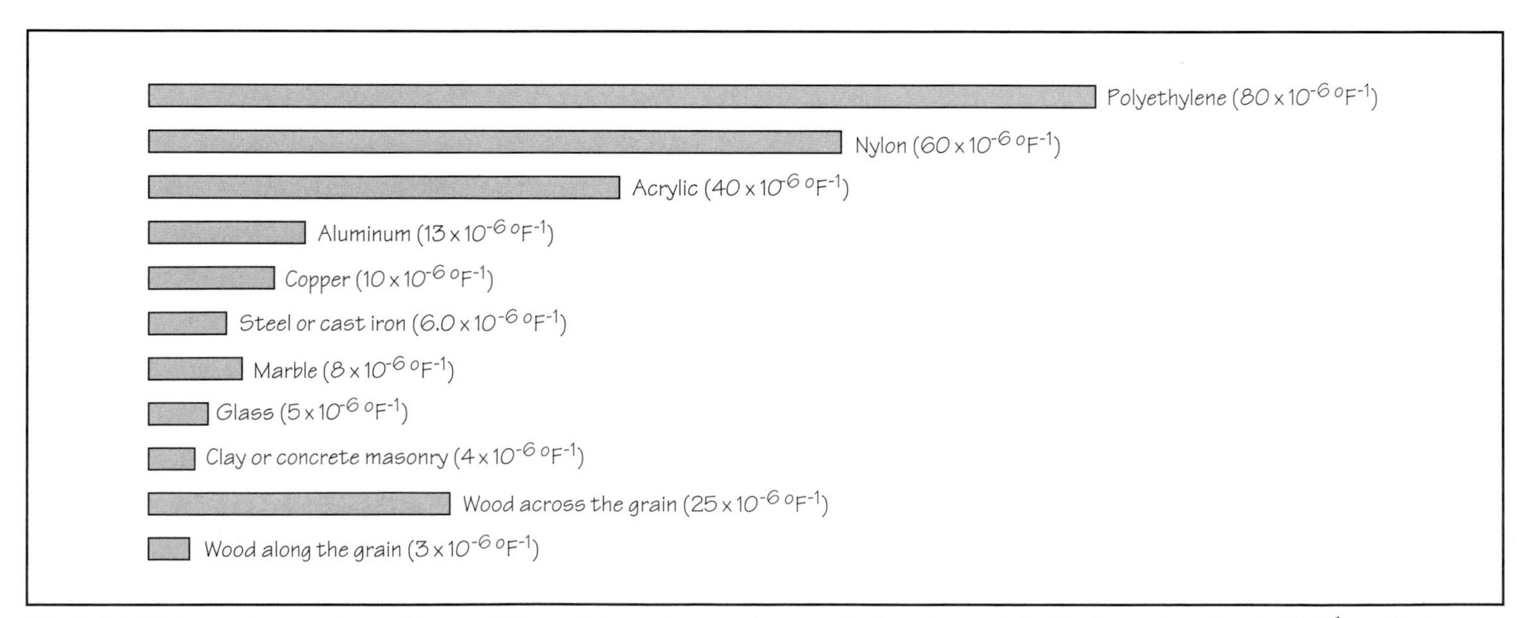

FIGURE 9.12 Approximate values of the coefficient of thermal expansion, α, of selected materials. To obtain the value in $(°\text{C})^{-1}$, multiply given values by 1.8. Thus, the value of α for clay or concrete masonry is $7.2 \times 10^{-6}\ (°\text{C})^{-1}$.

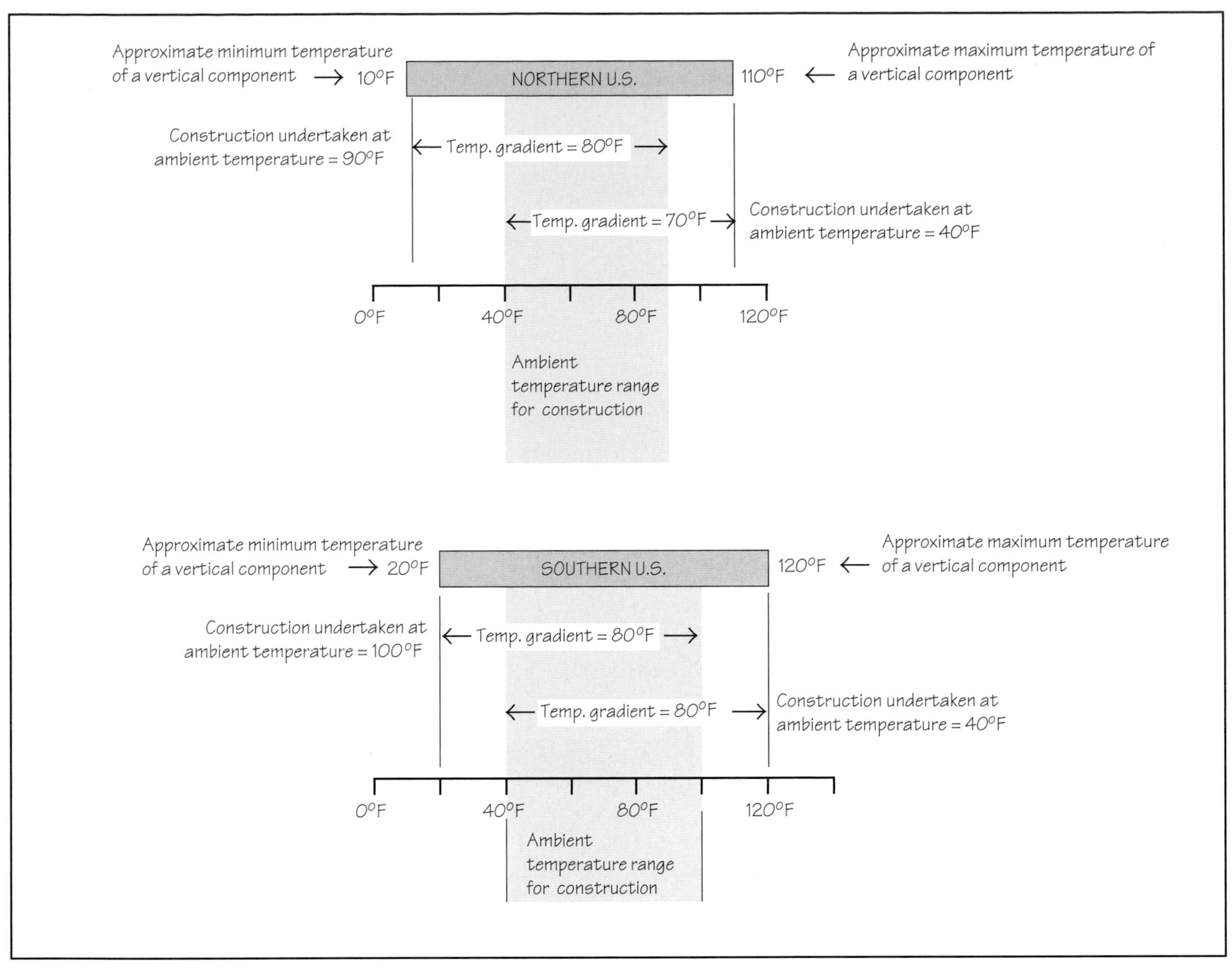

FIGURE 9.13 Typical temperature gradient to which a vertical component is subjected in the United States is nearly 80°F. This is based on the assumption that the ambient air temperature range when construction is undertaken in southern United States is between 40°F and 100°F. The corresponding range for the northern United States has been assumed to lie between 40°F and 90°F. This temperature range is particularly applicable to elements in which temperature-sensitive materials, such as portland cement and sealants, are used.

Maximum Temperature Gradient at a Location

Figure 9.13 shows that the typical maximum temperature gradient (i.e., the maximum temperature change to which a vertical component will be subjected) in the United States is nearly 80°F. For roofs and other horizontal components, the temperature gradient is generally higher.

Estimating Thermal Movement

Example 1

A residential neighborhood is to be provided with a clay (brick) masonry noise barrier wall along a busy road. It has been decided to use 100-ft-long segments of the wall with intervening joints. Determine the required width of each joint, considering only the wall's thermal movement.

0.384 in. 100 ft 0.384 in.

Minimum joint width needed to accommodate thermal movement in 100-ft wall segments of Example 1

Solution

$$L = 100 \text{ ft} = 1{,}200 \text{ in.}$$

$$\Delta t = 80°\text{F}$$

$$\alpha = 4.0 \times 10^{-6}{}°\text{F}^{-1} \quad \text{(Figure 9.12)}$$

From Equation (1), the minimum joint width (δ_t) is $(4.0 \times 10^{-6})(1{,}200)80 = 0.384$ in.

Note that this joint width does not include the provision for moisture expansion, construction, and material tolerances, which should also be included in determining the total joint width (Section 9.7).

Example 2

The masonry noise barrier wall of Example 1 has been provided with 0.384-in.-wide joints at 100 ft on center. Assuming that the wall was built when the ambient temperature was 50°F, determine the actual width of the joints when the wall temperature is (a) 90°F and (b) 10°F.

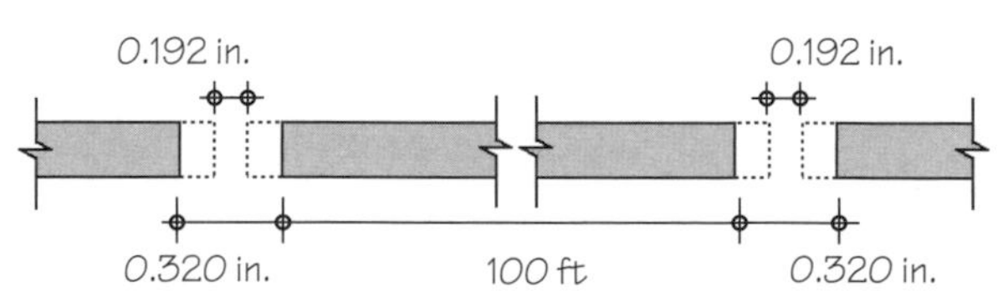

Each 100-ft wall segment of Example 2 will expand 0.096 in. on either side under a temperature rise of 40°F, giving a total expansion of 0.192 in.

Solution

$$L = 100 \text{ ft} = 1{,}200 \text{ in.}$$

$$\textit{Case (a)}: \Delta t = 90 - 50 = 40°F$$

From Equation (1), the expansion of the wall is $(4.0 \times 10^{-6})(1{,}200)40 = 0.192$ in. Thus, at 90°F, the joint width will become $0.384 - 0.192 = 0.192$ in.

$$\textit{Case (b)}: \Delta t = 50 - 10 = 40°F$$

From Equation (1), the contraction of the wall is $(4.0 \times 10^{-6})(1{,}200)40 = 0.192$ in. In other words, the gap between wall segments will become $(0.384 + 0.192) = 0.576$ in. at 10°F.

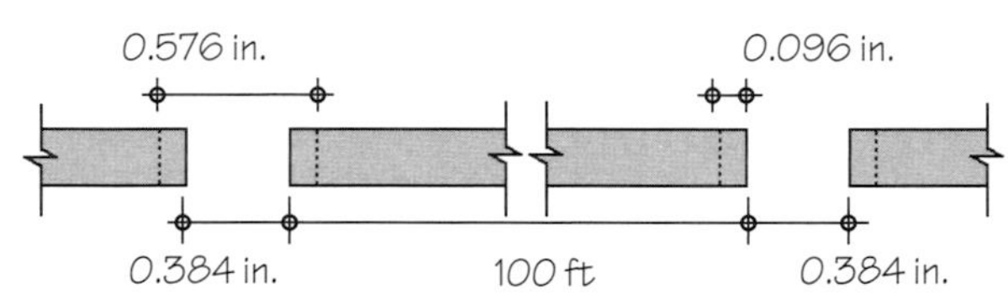

Each 100-ft wall segment of Example 2 will contract 0.096 in. on either side under a temperature decrease of 40°F, increasing the joint width to 0.576 in.

Example 3

A brick veneer wall is provided with continuous vertical expansion joints at 30 ft on center. Determine the required minimum width of each joint, considering the thermal expansion of the wall.

Solution

$$L = 30 \text{ ft} = 360 \text{ in.}$$
$$\Delta t = 80°F \quad \text{(Figure 9.13)}$$

From Equation (1), the joint width is $(4.0 \times 10^{-6})(360)80 = 0.115$ in.

9.5 MOISTURE MOVEMENT

Metals such as steel, copper, and aluminum are dimensionally stable with respect to moisture. However, portland cement–based materials shrink due to moisture loss. Because the moisture loss continues for several weeks after construction, portland cement–based components keep shrinking long after they are placed in a wall or slab. Therefore, concrete slabs, concrete and concrete masonry walls, and stucco require shrinkage joints—more commonly referred to as *control joints.*

MOISTURE MOVEMENT IN CONCRETE AND MASONRY

Clay (brick) masonry units expand on absorption of water or water vapor. This expansion, which occurs during and after construction, is irreversible; that is, clay masonry does not shrink on drying. In fact, a brick unit is smallest in size when it comes out of the kiln, after which it grows in size. Thus, brick masonry walls require *expansion joints.*

The amount of moisture movement in concrete and masonry is given by

$$\boxed{\delta_m = (\mu)L} \qquad \textbf{Eq. (2)}$$

where δ_m is the moisture expansion or contraction of a unit of length L and μ is the *moisture coefficient* of the material. The values of μ are given in Table 9.2. Note that μ has no units. It is simply a number.

TABLE 9.2 APPROXIMATE MOISTURE COEFFICIENT (μ) OF SELECTED MATERIALS

Material	Moisture coefficient (μ)
	Shrinkage
Plain concrete	0.0006
Reinforced concrete	0.0003
Concrete masonry	0.0003
	Expansion
Clay masonry	0.0003

MOISTURE MOVEMENT IN WOOD

As discussed in greater detail in Chapter 11, the microstructure of wood consists of hollow cellular tubes bundled together,

Figure 9.14. The walls of the tubes shrink and swell, depending on the ambient air's humidity and temperature, changing the dimensions of a wood member.

The wood industry has established empirical rules for providing movement joints in wood members. For instance, a 4-ft × 8-ft plywood sheet must be provided with $\frac{1}{8}$-in. (3-mm) wide joints at ends and edges in order to accommodate its movement. Moisture-related movement control in wood structures is discussed further in Section 13.10.

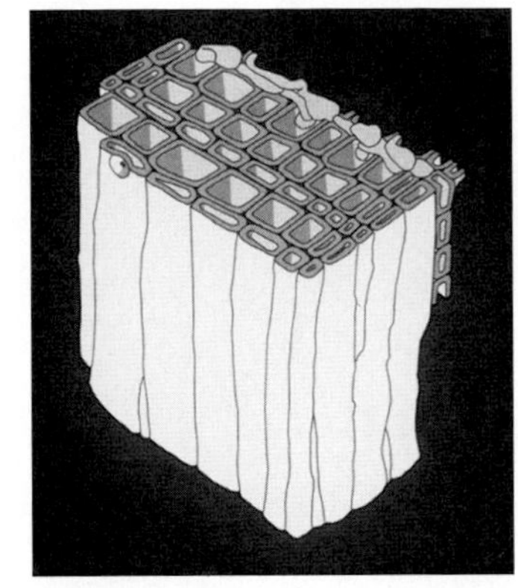

FIGURE 9.14 Microstructure of wood. See also Chapter 11.

Estimating Moisture Movement

Example 4

Expansion joints are to be provided in a brick veneer wall at 30 ft on center. Determine the moisture expansion of each 30-ft wall segment.

Solution

From Table 9.2, the moisture coefficient, μ, of clay masonry is 0.0003. Therefore, from Equation (2), the moisture expansion of each 30-ft segment is

$$0.0003(30 \times 12) = 0.108 \text{ in.}$$

Example 5

If control joints are provided in a reinforced concrete slab-on-grade at 12-ft intervals, determine the required minimum width of each joint.

Solution

From Table 9.2, $\mu = 0.0003$. From Equation (2), the minimum width of each joint is

$$0.0003(12 \times 12) = 0.043 \text{ in.}$$

PRACTICE QUIZ

Each question has only one correct answer. Select the choice that best answers the question.

1. A building separation joint in a multistory building generally
   a. runs through alternate floors, the roof, and the foundation.
   b. runs through all floors, the roof, and the foundation.
   c. runs through all floors and the roof.
   d. runs through all floors and the foundation.
   e. runs through all floors.
2. Building separation joints in a long rectilinear building are generally located at every
   a. 500 ft on centers.
   b. 400 ft on centers.
   c. 350 ft on centers.
   d. 250 ft on centers.
   e. none of the above.
3. The width of a building separation joint is generally between:
   a. 1/2 in. to 1 in.
   b. $1\frac{1}{2}$ in. to 2 in.
   c. $2\frac{1}{2}$ in. to 3 in.
   d. $3\frac{1}{2}$ in. to 4 in.
   e. $4\frac{1}{2}$ in. to 5 in.
4. A seismic joint is generally similar to the building expansion joint but smaller in width.
   a. True
   b. False
5. Which of the following materials has the highest coefficient of thermal expansion?
   a. Steel
   b. Plastics
   c. Aluminum
   d. Concrete
   e. Glass
6. The unit of measurement for the coefficient of thermal expansion (α) is
   a. °C or °F.
   b. feet or meters.
   c. watts.
   d. $(°C)^{-1}$ or $(°F)^{-1}$.
   e. none of the above.
7. The unit of measurement for the coefficient of moisture expansion (μ) is
   a. °C or °F.
   b. feet or meters.
   c. watts.
   d. $(°C)^{-1}$ or $(°F)^{-1}$.
   e. none of the above.
8. The equation used for determining the thermal movement, δ_t, of a member is
   a. $\delta_t = \alpha L(\Delta t)$.
   b. $\delta_t = L(\Delta t)$.
   c. $\delta_t = (\alpha L)/(\Delta t)$.
   d. $\delta_t = [L(\Delta t)]/\alpha$.
   e. $\delta_t = [\alpha(\Delta t)]/L$.
9. Which of the following materials is subjected to moisture movement?
   a. Wood
   b. Concrete
   c. Bricks
   d. All the above
   e. Only (a)
10. Which of the following materials has irreversible moisture expansion during service, that is, after construction.
    a. Steel
    b. Concrete
    c. Bricks
    d. Natural stone
    e. Aluminum

Answers: 1-c, 2-d, 3-b, 4-b, 5-b, 6-d, 7-e, 8-a, 9-d, 10-c.

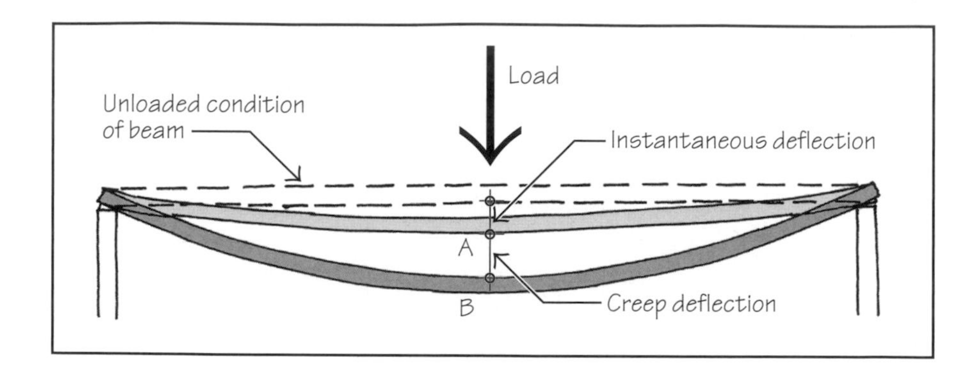

FIGURE 9.15 Instantaneous (elastic) deflection and creep (inelastic) deflection of a beam. The dashed lines show unloaded condition of the beam.

NOTE

Creep

Creep is significant in concrete and concrete masonry structures. Thus, concrete columns, particularly in high-rise buildings, keep shortening in height for up to nearly 2 years after their construction. A typical concrete column shortens by nearly $\frac{1}{8}$ in. (3 mm) per story due to creep. Fortunately, most of creep deformation occurs in the first 3 months. Therefore, the shortening of columns and the consequent lowering of the floor structure has little effect on non-load-bearing walls placed under the floors. This is because it generally takes longer than 3 months before non-load-bearing walls are erected.

NOTE

Spandrel Beam

A spandrel beam is a beam that runs on the outside edge of a floor or roof of the building, Figure 9.16.

9.6 ELASTIC DEFORMATION AND CREEP

Consider a beam of Figure 9.15, which is initially unloaded, as shown by dashed lines. If the beam is now loaded, it will deflect almost instantaneously in response to the load. Let the instantaneous deflected shape of the beam be represented by profile A. If this load is a permanent load, the beam's deflection may progressively increase and eventually assume the deflected profile B. This time-dependent additional deflection of the beam is known as *creep deflection,* and the associated phenomenon is called *creep.*

Thus, creep is the incremental deformation of a component caused by sustained loading. Among the structural materials used in buildings, creep is significant only in concrete and concrete masonry. Steel does not creep. In wood, creep deformation is relatively small.

Creep deformation in concrete increases progressively for up to nearly 2 years after the component is loaded. This implies that creep is caused only by long-term, sustained, or permanent loads. Thus, the fundamental difference between instantaneous deformation and creep deformation is that the instantaneous deformation is caused by all types of loads, whereas creep deformation is caused only by the dead load.

Because building components are designed to remain elastic under loads, the instantaneous deflection is elastic. In other words, after the load is removed, the component returns to its original undeformed state. That is why the instantaneous deformation is also referred to as *elastic deformation.* Creep deformation, on the other hand, is inelastic deformation of components and is, therefore, irreversible. In other words, even if the load is removed, creep deformation will remain.

Elastic deformations are routinely determined for all types of structures as part of structural calculations. For concrete structures, creep deformation is also determined. From the architectural viewpoint, the most critical deformation occurs in spandrel beams. Because spandrel beams have exterior walls anchored to them, it is necessary to provide flexibility in the anchorage of exterior walls so that neither the elastic nor the creep deflection of spandrel beams will cause additional stresses in exterior walls.

If a non-load-bearing wall is placed directly under a spandrel beam, an adequate space between the beam and the top of the wall should be provided to allow the beam to deflect under the loads (Figure 9.16). If adequate space is not provided, the load from the beam will be transferred to the wall, damaging it.

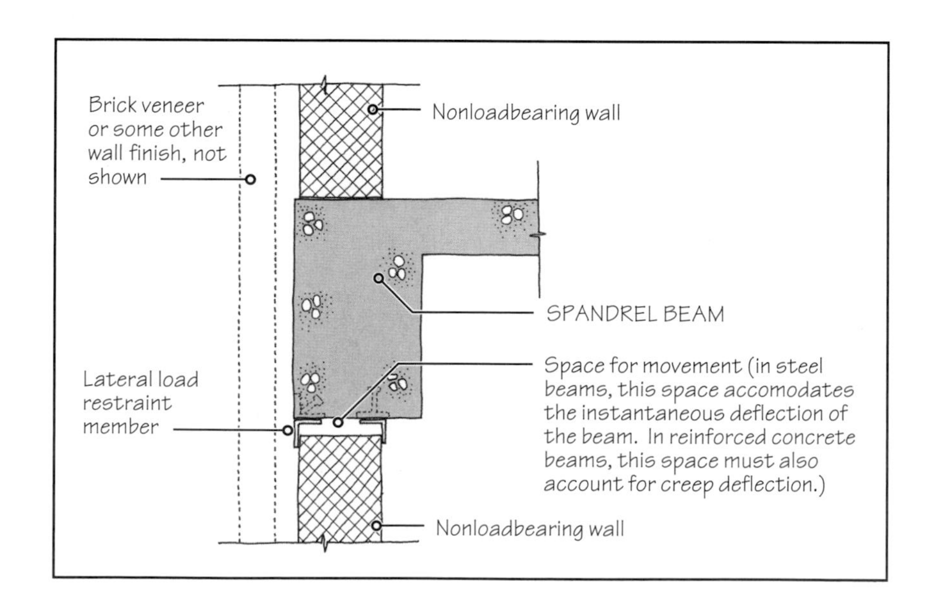

FIGURE 9.16 The connection between the spandrel beam and the nonloadbearing wall must account for the beam's deflection (see also Figure 26.22).

9.7 TOTAL JOINT DIMENSION

In addition to thermal, moisture, elastic, and creep deformations, there are several other factors that should be considered in determining the joint dimensions. One such important consideration is material tolerances to recognize that building components vary from their specified dimensions. For example, it is not uncommon to see brick or concrete masonry units that are undersized or oversized. Noticeable differences in dimensions from those stated on the drawings can occur in prefabricated, concrete curtain wall panels.

In addition to material or component tolerances, construction tolerances should also be taken into account. For instance, an expansion joint in a masonry wall may be shown as a vertical joint on the drawings, but the constructed joint is unlikely to be absolutely vertical. Construction tolerances are a function of quality control during construction. Smaller tolerances require good construction practices and supervision. A poor supervision demands larger tolerances. Therefore;

$$\text{Width of joint} = \begin{array}{c}\text{width based on}\\ \text{temperature, moisture,}\\ \text{and other movements}\end{array} + \text{tolerances} \qquad \textbf{Eq. (3)}$$

Additionally, foundation settlement and story drift (see "Principles in Practice" in Chapter 16) due to wind or earthquake loads are also important. Some chemical processes may also cause movement. Steel expands as it corrodes. Excessive corrosion of steel can cause cracking and sometime spalling of concrete. Because water expands on freezing, the freezing of water causes similar effects in materials that absorb water, such as brick, concrete, concrete masonry, and wood.

EFFECT OF SEALANT ON JOINT DIMENSION

Most joints in building envelope components are filled with elastomeric sealants. The sealants have a finite ability to cyclically stretch and compress, which is given as plus-minus movement ability of the sealant. The plus value denotes the maximum stretch the sealant can withstand, and the minus value denotes the maximum compression it can sustain.

In fact, sealants are generally classified based on their movement ability. Thus, a sealant with a movement ability of ±25% is called a *class 25 sealant.* Such a sealant is able to stretch and compress by 25% of its installed dimension. Similarly, a sealant with a movement ability of ±50% is called a *class 50 sealant.* Thus, if the width of a joint at the time the sealant is applied to it is 1 in., a class 50 sealant will safely stretch to 1.5 in. or compress to 0.5 in.

The movement ability of sealants must be included in determining the total joint width of a sealed joint. Therefore, the total joint width of a sealed joint is given by

$$\text{Joint width of sealed joint} = \frac{100}{\text{sealant class}}[\text{joint movement}] + \text{tolerances} \qquad \textbf{Eq. (4)}$$

Estimating Total Joint Width Between Components

Example 6

Continuous vertical expansion joints are to be provided in a brick veneer wall at 30 ft on center. Determine the total joint width, assuming that the joints are sealed with a class 50 sealant.

Solution

From Examples 3 and 4, the thermal and moisture movement of a 30-ft-long wall is 0.115 + 0.108 = 0.223 in. From Equation (4), total joint width is

$$\frac{100}{50}(0.223) = 0.446 \text{ in.}$$

This joint width does not include tolerances. In practice, the expansion joints in a brick veneer wall are $\frac{3}{8}$ in. (0.375 in.) wide. This width matches the thickness of mortar joints (Chapter 26). The use of a smaller than the theoretically calculated joint width induces a small amount of stress in the wall under extreme circumstances.

Example 7

Determine the required width of joints in the noise barrier wall of Example 1, considering all the necessary factors. The joints are not sealed.

Solution

From Example 1, the thermal expansion of the wall is 0.384 in. The moisture expansion of a 100-ft wall segment is $0.0003(100 \times 12) = 0.36$ in., giving a total movement of $0.384 + 0.36 = 0.744$ in. Hence, providing 1.0-in.-wide joints should take care of material and construction tolerances.

9.8 PRINCIPLES OF JOINT DETAILING

In detailing a movement joint in a component, it is important to distinguish between an *expansion joint* and a *control (shrinkage) joint.* Because the joint width in an expansion joint will become smaller over time, the filler in an expansion joint (if used) must be elastomeric to allow unrestrained movement of the components.

The filler in a control joint, on the other hand, may be elastomeric or nonelastomeric. If a nonelastomeric filler is provided in a control joint, the detailing of the joint must allow unrestrained shrinkage of the components.

The difference in the detailing of an expansion joint and a control joint can be further appreciated by examining typical joint details used in brick masonry and concrete masonry walls. As stated earlier, brick masonry walls expand over time after construction as the bricks absorb water from the mortar and the atmosphere. Therefore, brick masonry walls require expansion joints. By contrast, concrete masonry walls shrink after construction as the water used in manufacturing concrete masonry units and in the preparation of mortar evaporates. Therefore, concrete masonry walls require control joints.

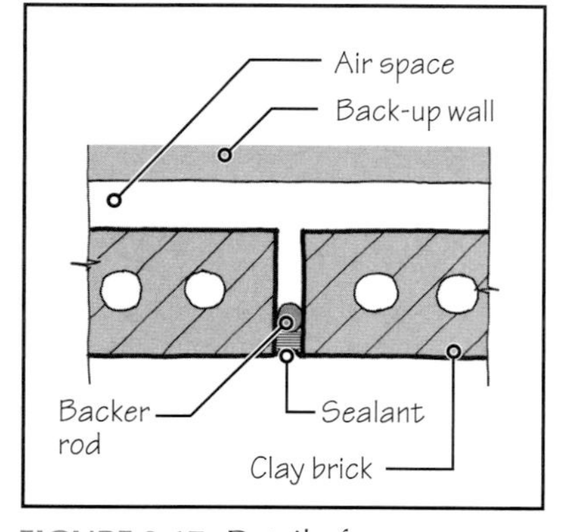

FIGURE 9.17 Detail of an expansion joint in a brick veneer wall. The joint is filled with a compressible backer rod and an elastomeric sealant (see also Section 26.1).

Figure 9.17 shows the detailing of an expansion joint in a brick veneer wall. Note that the joint is filled with a compressible backer rod and an elastomeric sealant.

Figure 9.18 shows a typical control joint detail in concrete masonry walls. The joint is filled with masonry mortar to provide shear key between the two parts of the wall. To ensure that the mortar does not adhere on both sides of the joint and restrain the shrinkage of the wall, an asphalt-saturated paper is placed on one side of the joint to break the bond.

In explaining the difference between a control joint and an expansion joint, we considered moisture-induced movement in brick and concrete masonry walls, which is largely permanent. On the other hand, to accommodate reversible movement, such as that created by temperature changes, requires either an open joint or one filled with an elastomeric sealant (Figure 9.17). In other words, an open joint or an elastomerically filled joint is a versatile joint.

Note that because temperature-induced movement is almost universal, expansion joints are provided in most components. Control joints, on the other hand, are provided in concrete and concrete masonry components. Detailed discussion of joint detailing is covered in later chapters. For example, construction joints, control joints, expansion joints, and isolation joints used in concrete construction are covered in Chapter 19.

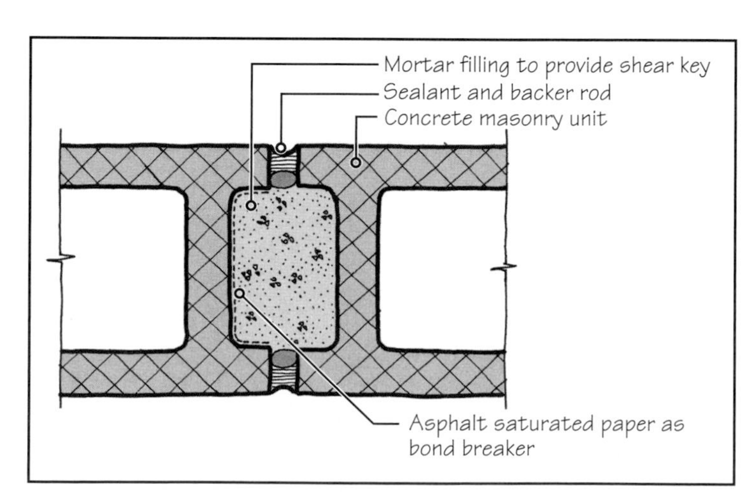

FIGURE 9.18 Detail of a control joint in a concrete masonry wall (see also Figure 23.26).

9.9 COMPONENTS OF A SEALED JOINT

The primary purpose of sealing a joint is to prevent water penetration. However, control of air leakage, dust penetration, and noise transmission are additional benefits of sealing a joint. An unsealed joint may also get filled with incompressible materials over time, resulting in joint failure.

Although joint sealant is the most important component of a sealed joint, these other components must also be carefully selected:

- Substrate
- Primer
- Sealant backup
- Bond breaker

SUBSTRATE

For a sealed joint to function effectively, it is important that the sealant be fully adhered to the surfaces of components meeting at the joint (called the *substrate*). A nonadhering sealant will obviously give a leaky joint.

The adhesion of the sealant to the substrate depends on the chemical compatibility of the sealant with the substrate material. Because all sealants are not compatible (or equally compatible) with all substrate materials, it is important to obtain compatibility information from the sealant manufacturer. This is particularly vital because substrate materials such as concrete and masonry are sometimes treated with water-repellent chemicals and metals with paints and other protective coatings. These treatments may inhibit sealant adhesion.

Some substrate materials may require special joint preparation, such as the application of primers to make the substrate compatible with the sealant. In any event, some preparation will be required on all substrates, including cleaning the surfaces of loose particles, contaminants, frost, ice, and so on.

PRIMER

The purpose of a primer is to improve the adhesion of sealant to substrate. Some sealants require primers on all types of surfaces, whereas others require primers on a few types of surfaces. Most sealant manufacturers require primers for concrete and masonry substrates to stabilize the substrate by filling the pores and strengthening weak areas.

Some sealants require different primers for different substrates, which is an important consideration in the selection of a sealant if disparate materials are to be joined. Another consideration is the length of time that must pass between the application of primer and the sealant. Obviously, a sealant that can be applied almost immediately after the application of the primer is preferable.

SEALANT BACKUP—THE BACKER ROD

A sealant backup is a compressible and resilient material, such as a plastic foam. It is usually circular in cross section; hence, it is called a *backup rod*, or, more commonly, a *backer rod*. The diameter of the backer rod should be greater than the width of the joint, so that it is held in place by compression. The backer rod performs the following functions:

- It controls the depth and shape of sealant.
- It allows the tooling of the sealant, which provides adhesion between the sealant and substrate. In an untooled joint, the sealant will not fully adhere to the substrate, resulting in a leaky joint. A properly tooled joint, on the other hand, will have full contact with the substrate, Figure 9.19. For the joint to be tooled, the backer rod must be under sufficient compression not to slide under the pressure of tooling.
- It acts as a temporary joint seal until the sealant is applied. Some backer rods can also act as secondary seals in the event of sealant deterioration due to weathering and aging (see the box entitled "Types of Backer Rods").

For the backer rod to function as a secondary seal, it must remain elastic throughout its life and not develop significant compression set. It should also be compatible with the sealant. Therefore, a backer rod approved by the sealant manufacturer should be used.

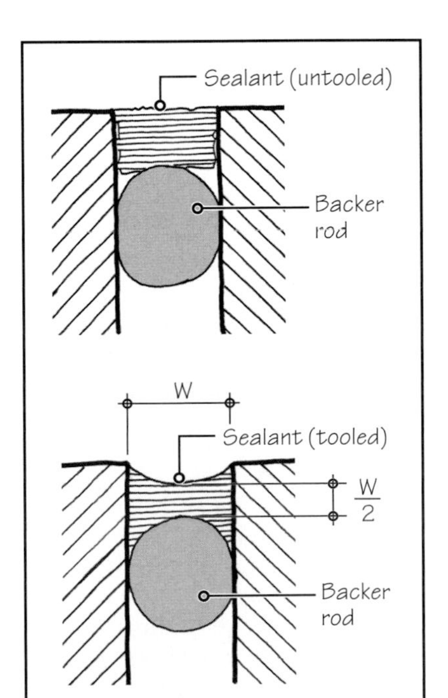

FIGURE 9.19 Untooled and tooled sealants in joints.

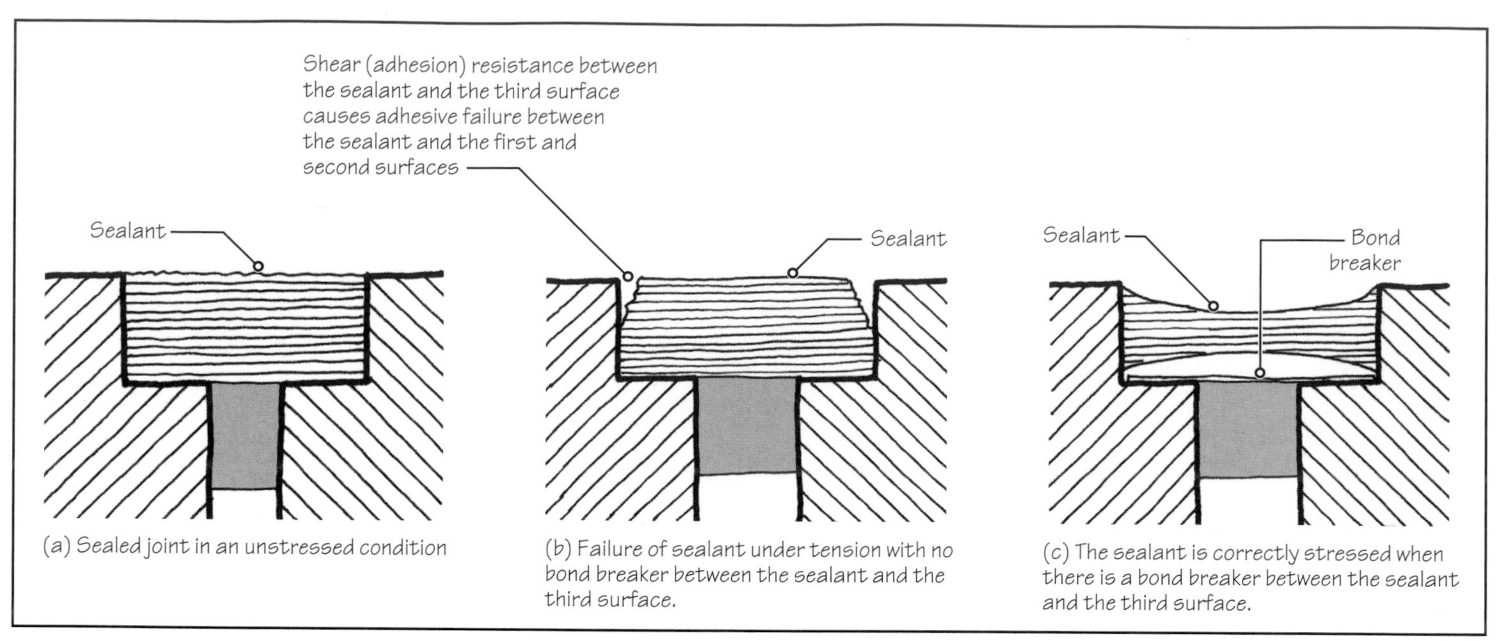

FIGURE 9.20 The effect of a bond breaker in a sealed joint.

Bond Breaker

For the sealant to function effectively, it should be bonded to only two opposite surfaces of the substrate, so that it comes under axial tension or axial compression when the joint moves. If the sealant is bonded to the third side of the joint (bottom surface), as shown in Figure 9.20, resistance to movement provided by the bond at the third surface creates stress concentration, leading to sealant detachment.

A bond breaker is required only if the third surface is hard and unyielding, such as concrete, metal, masonry, or any other inflexible backup. A bond breaker is not required with a flexible backup that will not significantly restrict the freedom of sealant movement. Thus, a bond breaker is not required in a conventional joint with a foam backer rod, such as that shown in Figure 9.19.

9.10 TYPES AND PROPERTIES OF JOINT SEALANTS

Joint sealants may be divided in three categories:

- Preformed tapes
- Caulks
- Elastomeric sealants

Preformed tapes are available in rolls. They function as seals only if they are under pressure; they are commonly used in window glazing, door jambs, and gypsum drywall.

EXPAND YOUR KNOWLEDGE

Types of Backer Rods

Backer rods are made of foamed plastics and are generally round in cross section. They are available in three types, depending on the cellular structure of the foam:

- Open-cell rods
- Closed-cell rods
- Bicellular rods

Open-cell rods are spongy and can absorb water. Closed-cell rods do not absorb water; hence, they protect the joint from water penetration until the sealant is applied. However, open-cell rods allow the sealant to cure from both the front and the back of the rod, speeding the curing process. This is an advantage because the curing of sealant, which requires the presence of air and water vapor, is a slow process that can take a few days for some sealants. Therefore, in joints that are not exposed to the weather, open-cell rods are preferred.

For exterior joints, a closed-cell rod is desirable. However, the use of a sharp instrument to squeeze the rod into the joint can puncture it, converting some of the closed cells into open cells. When the ambient temperature increases and the joint becomes narrower, the rod is set into compression. If this occurs while the sealant is still curing, the air from the punctured cells bubbles into the sealant leaving voids, a phenomenon known as *outgasing*. Note that the air can move out only into the sealant not into the rod's interior because the cells in the rod are of closed type.

More recently, bicellular rods, which do not produce outgasing, have been introduced. Bicellular rods have closed cells on the periphery but open cells in the rod's interior. This provides adequate space for the air from punctured cells to move into the rod's interior when the rod comes under compression.

Caulks are doughlike materials and are the first generation of sealing compounds. Glazing putty is an example of the earliest caulks used in buildings. It consists of nearly 12% linseed oil and 88% chalk (calcium carbonate) that hardens as it cures (oil evaporation), leading to cracking, loss of elasticity, and, hence, loss of its sealing ability. It is rarely used today.

TYPES OF ELASTOMERIC SEALANTS

Elastomeric sealants are synthetic materials (polymers) and are the ones most commonly used in contemporary construction. The five commonly used synthetic sealants are

- Polyisobutylene
- Acrylics
- Polyurethane
- Polysulfide
- Silicone

Polyisobutylene and acrylic sealants have a very low movement ability and are used as caulking materials in joints that do not move or move very little, such as the static joints between aluminum window frames. Polyisobutylene (referred to as butyl caulk) has excellent resistance to water-vapor transmission. Therefore, it is commonly used as the primary seal in an insulating glass unit. The secondary seal in an insulated glass unit is provided by the silicone sealant because of its durability and good adhesion (Section 28.6).

Polyurethane sealant has high abrasion resistance. Therefore, it is the sealant of choice in horizontal joints subjected to foot traffic. It has good compatibility with a wide variety of substrates—another factor in its favor.

Polysulfide sealant has excellent chemical and weathering resistance. It is the sealant of choice for swimming pools, wastewater treatment plants, water-treatment plants, and so on, where a durable water tight seal is required. Silicone sealant has high strength and exceptional resistance to ultraviolet. It is the only sealant that can be used in structural glazing.

Table 9.3 gives a comparative overview of three most commonly used sealants—polyurethane, polysulfide and silicone—and the following sections provide a description of the criteria that must be considered in their selection.

MOVEMENT ABILITY OF SEALANTS

The most important property of a sealant is its ability to withstand cyclic joint movements (Section 9.7). Currently available sealants are classified into three categories: low-range sealants, medium-range sealants, and high-range sealants.

A low-range sealant has limited movement ability, of the order of ±5% or less. These include (a) oil-based caulks and (b) butyl or acrylic caulks. Low-range sealants should be used only in a nonmoving joint. Their principal advantage lies in their low cost.

Medium-range sealants have a movement range of up to ±12.5%, and high-range sealants have a movement range larger than ±12.5%. Silicone sealants normally have the highest movement capability, up to ±50%.

STRENGTH AND MODULUS OF SEALANTS

A sealant must be able to withstand movement stresses. Most joint failures occur due to excessive tensile stress. Tensile failure may occur either at the substrate-sealant interface (failure due to inadequate adhesion), or within the sealant (failure due to inadequate cohesion).

Modulus refers to the modulus of elasticity of sealant, a measure of the stiffness of a material (Chapter 4). A high-modulus sealant is stiffer than a low-modulus sealant and is

NOTE

Modulus of a Sealant

From Equation (3) in Chapter 4, the stress in a material is the product of the modulus of elasticity of the material and the strain:

Stress = modulus of elasticity × strain

Thus, stress in a material is directly proportional to its modulus of elasticity. In other words, a low-modulus material will be subjected to a lower stress than a high-modulus material if the strain (deformation) of the two materials is the same.

TABLE 9.3 COMPARATIVE OVERVIEW OF COMMONLY USED SEALANTS

Sealant property	Type of sealant Polyurethane	Polysulfide	Silicone
Movement ability	Up to ±25%	Up to ±25%	Up to ±50%
Water vapor permeability	Fair	Good	Poor
Abrasion resistance	Excellent	Poor	Fair
Chemical resistance	Fair	Excellent	Fair
Durability	Good	Good	Excellent
Substrate compatibility	Excellent	Good	Incompatible with some substrates
Paintability	Good	Good	Poor

desirable where joint movement is small. Thus, a high-modulus sealant is required in structural glazing, where the stiffness of joint between the glass and the aluminum frame is important. A low-modulus sealant, on the other hand, is desirable where joint movement is large and the sealant is required purely for sealing purposes. A low-modulus sealant will be under lower stress due to movement.

NOTE

Joints in Building Components—Lapped Joints

There are several movement joints in buildings that function because of the overlap between building components. Such joints generally do not need to be sealed. Some examples of lapped joints are: joints between exterior wall siding (Chapter 14), joints between roof shingles or tiles, and joints between underlayment felts below the shingles (Chapter 32).

TOOLING TIME, CURE TIME, AND TEMPERATURE RANGE

Tooling time refers to the time (usually measured in minutes) that must be allowed to elapse before the sealant changes from liquid state to semisolid state so that it can be tooled. Cure time refers to the time it takes for sealant to be hardened to its final elastomeric state.

Application temperature range refers to the ambient air temperature range within which the sealant can be applied. Performance temperature range is the range over which the sealant will maintain its properties after it has cured.

LIFE EXPECTANCY OF SEALANTS

Life expectancy is the time (in years) after which the sealant may have to be reapplied. Most high-grade sealants, such as the structural silicones, are quoted by their manufacturers to have a life expectancy of 20-plus years.

ONE-PART OR TWO-PART SEALANTS

Sealants can either be one-part or two-part sealants. One-part sealants are easier to use but generally take longer to cure and have long tack-free time. Longer curing time means a greater delay in tooling and possible damage to the sealant if the joint moves excessively before the sealant has fully cured. Long tack-free time means that the sealants will tend to attract more dirt.

PRACTICE QUIZ

Each question has only one correct answer. Select the choice that best answers the question.

11. Creep deformation occurs in
a. concrete members.
b. steel members.
c. aluminum members.
d. all the above.

12. Creep deformation is caused by
a. live loads.
b. dead loads.
c. wind loads.
d. all the above.
e. none of the above.

13. A spandrel beam in a building is a beam that
a. has a floor slab or roof slab on both sides.
b. runs on the outside edge of a floor or roof.
c. is partially embedded in the ground.
d. is supported by a continuous wall below it.
e. supports a continuous wall above it.

14. A non-load-bearing wall that runs under a spandrel beam should be adequately separated from the beam to allow the beam to move freely in the
a. vertical direction.
b. horizontal direction.
c. both (a) and (b).

15. In determining the total width of a movement joint between building components, we account for the dimensional and construction tolerances. A poor or inadequate quality control on a construction site
a. increases construction tolerances.
a. decreases construction tolerances.
c. increases dimensional tolerances.
d. decreases dimensional tolerances.
e. none of the above.

16. A joint between two walls has been determined to be 0.5 in. if the joint is not to be filled with a sealant. If the same joint is to be filled with a Class 50 sealant, its width should be:
a. 1 in. minimum.
b. $2\frac{1}{2}$ in. minimum.
c. $\frac{1}{4}$ in. minimum.
d. $\frac{1}{4}$ in. maximum.
e. none of the above.

17. Which of the following materials does not represent a joint sealant?
a. Silicone
b. Polysulfide
c. Polyisobutylene
d. Polyvinyl chloride

18. Which of the following joint sealants has the highest abrasion resistance?
a. Polyisobutylene
b. Acrylic
c. Polyurethane
d. Silicone

19. Which of the following joint sealants has the highest movement ability?
a. Polyurethane
b. Acrylic
c. Polyisobutylene
d. Silicone

20. Backer rods are available in
a. two types.
b. three types.
c. four types.
d. five types.

21. The sealant in a joint works best when it is subjected to
a. tension, compression, and shear.
b. tension, compression, shear, and bending.
c. tension or shear only.
d. tension or compression only.

Answers: 11-a, 12-b, 13-b, 14-a, 15-a, 16-a, 17-d, 18-c, 19-d, 20-b, 21-d.

KEY TERMS AND CONCEPTS

Backer rod
Bond breaker
Building separation joint
Coefficient of moisture expansion
Coefficient of thermal expansion
Construction tolerances
Control joint
Creep
Curing of sealant
Elastic deformation
Expansion joint
Isolation joint
Joint sealant
Modulus of sealant
Movement ability of sealants
Movement joints
Sealant class
Seismic joint
Temperature gradient
Temperature range of sealant
Tooling time and cure time of sealant

REVIEW QUESTIONS

1. Using sketches and notes, explain various types of movement joints used in buildings.
2. With the help of a sketch, explain why and where building separation joints should be provided.
3. List the important properties of the following joint sealants: silicone, polyurethane and polysulfide.
4. Discuss the benefits of tooling a joint sealant.

SELECTED WEB SITES

Adhesives & Sealants Industry (www.adhesivesmag.com)
Sealant Waterproofing & Restoration Institute, SWRI (www.swrionline.org)

FURTHER READING

1. American Society for Testing and Materials. "Specifications for Elastomeric Joint Sealants," ASTM C920.
2. American Society for Testing and Materials. "Standard Guide for Use of Joint Sealants," ASTM C1193.
3. Kubal, Michael. *Construction Waterproofing Handbook.* New York: McGraw Hill, 2000.
4. Sealant Waterproofing & Restoration Institute. *Sealants: The Professionals' Guide,* 1995.

ONE WAY

CHAPTER 10 Principles of Sustainable Construction

CHAPTER OUTLINE

10.1 Fundamentals of Sustainable Architecture

10.2 Ecolabeling of Buildings

10.3 Characteristics of Green Building Products

10.4 Ecolabeling of Building Products

LEED Gold-certified Seattle City Hall. The city of Seattle now requires that all new public buildings achieve LEED certification. BCJ Architects in association with Basetti Architects. (Photo courtesy of Bruce Haguland.)

Before the beginning of the twentieth century, buildings contributed little to environmental degradation. Building construction was largely craft based, and most buildings were built through human labor or low-tech tools and occasionally using simple machines. The use of locally produced materials was the norm, and the amount of energy used in the mining of raw materials and converting them into finished products (referred to as *embodied energy*) was low.

In addition to low embodied energy, the energy used in a building's operation and maintenance (lighting, heating, cooling, water supply, and waste disposal) was also low. Air conditioning had not yet been invented. The dominant source for interior illumination was daylight, and the levels of artificial lighting in buildings were modest.

To ensure that buildings were comfortable and healthy, architects spent a great deal of design effort in creating climate-responsive buildings through appropriate site planning, building orientation, fenestration design, and the use of appropriate landscape materials (see the introduction to Chapter 5).

Around the beginning of the twentieth century, several technical discoveries changed the way in which buildings were constructed and used. The most significant change was the invention of air conditioning (by Willis Carrier in 1902). This produced a ducted system of comfort conditioning using warm or chilled air. By the middle of the twentieth century, most new buildings in North America were fully air conditioned—a technology that spread not only to buildings throughout the world, but also to forms of transportation that carried people and perishable goods (trains, buses, automobiles, etc.). The result was a hefty increase in the world's energy use.

Another major development was that of the elevator by Elisha Graves Otis, who publicly demonstrated its safety in 1854 in New York. The elevator's safety and the subsequent improvements in its efficiency and speed provided the necessary stimuli for the construction of tall buildings. Initially, Chicago and New York—and later other urban centers throughout the world—began constructing tall buildings, giving birth to an entirely new building form, the skyscraper.

The amount of energy required for the construction of a square foot of a skyscraper is far greater than that required for the construction of a low-rise building. This is due not only to the use of heavy construction equipment (excavators, bulldozers, graders, loaders, cranes, etc.) required for the construction of a skyscraper, but also to the disproportionately large amount of materials needed in its structural frame.

In addition to the large amount of embodied energy, tall buildings also consume large amounts of energy for operation and maintenance. On the other hand, because tall buildings yield denser urban habitations, they can reduce the amount of energy required for urban infrastructure—roads, water supply, sewage, and waste-disposal facilities.

The twentieth century also witnessed several major sociopolitical changes that led to further escalation in the world's energy use. The independence of several Asian and African countries and their industrialization led to the expansion of international trade and travel and increased building and infrastructural construction. The population explosion, resulting from medical breakthroughs that drastically reduced infant mortality and increased adult life spans, was another major reason for the increased use of energy.

The escalation of energy use continued unabated until the oil embargo of 1973. This event made the world suddenly realize its excessive dependence on energy and the pace at which the earth's finite energy resources were being depleted. An even more significant realization was how the world's excessive use of energy was irreparably damaging the environment and endangering people's health.

For example, the thinning of the ozone layer in the earth's atmosphere due to the use of ozone-depleting chemicals was well established during the latter part of the twentieth century. Global warming, caused by the emission of carbon dioxide and other gases, was no longer a theoretical prediction but a reality that could be testified to by irrefutable measurements and also by those that had lived through the change.

By this time (around the 1970s), the world was beginning to be aware that we were abusing the planet's resources and unless something was done soon, the damage to our ecosystem could be irreversible. We also realized that energy is only one of the several resources needed for our survival, and, for humankind to have a reasonable future, all resources, such as land (including landfills), water, air, and other materials, must be conserved and protected against depletion, pollution, and degradation.

This realization led to the concept of pursuing an *ecologically benign,* or *environmentally friendly,* existence. The term *sustainable development* eventually stuck because it included the dimension of time. Another commonly used term is *green development,* which emphasizes its association with the environment.

NOTE

Embodied Energy

Embodied Energy in a Building

The embodied energy in a building is the amount of energy consumed by all the processes involved in the production of a building—from the acquisition of raw materials to the delivery and placement of the final product into the building. It includes the energy used in mining, manufacturing, transportation, construction, and administrative functions.

Embodied Energy in a Product

The embodied energy in a material or product is the amount of energy used in its production, including the administrative functions related to its production. It does not include the energy used in the product's transportation to the construction site and its installation in the building.

EXPAND YOUR KNOWLEDGE

Ozone Depletion

The ozone layer lies in the earth's atmosphere—between 10 and 20 mi (18 and 35 km) from the earth's surface. It is in this atmospheric region that the ultraviolet component of solar radiation converts atmospheric oxygen to ozone. (An oxygen molecule consists of two atoms of oxygen, i.e., O_2, and an ozone molecule consists of three atoms of oxygen, i.e., O_3.)

The other components of solar radiation convert the ozone molecule back to an oxygen molecule, so there is a perpetual cycle of conversion from oxygen to ozone and back to oxygen. If the cycle is not disrupted, a constant thickness of ozone layer is maintained in the atmosphere at all times. The ozone layer blocks the harmful UV-B rays from reaching the earth by absorbing them. In humans, UV-B rays cause skin cancer and eye cataracts and weaken the immune system.

Substances such as chlorine aggressively react with ozone and disrupt the ozone-oxygen cycle, decreasing the thickness of the ozone layer. Several modern-day chemicals contain chlorine, such as methyl chloroform (used in cleaning solvents, adhesives, and aerosols), carbon tetrachloride (used in dry cleaning), and chloro-fluoro-carbons (CFCs), used as freon in refrigeration equipment.

Since the 1970s, scientists had warned that the ozone layer was being depleted, but it was not until 1987 that measurements revealed an unexpectedly large ozone-depleted hole over the Antarctic. In response to the revelation, new production of ozone-depleting substances was banned under the 1987 international agreement referred to as the *Montreal Protocol*. Despite this agreement (and provided all countries seriously adhere to the agreement), the chemicals that are already in use will eventually find their way into the atmosphere, causing further damage.

The first serious worldwide discourse on sustainable development is generally traced to the (1987) United Nations publication entitled *Brundtland Report* [1], named after Gro Brundtland, Prime Minister of Norway, who chaired the UN World Commission on Environment and Development. The gist of the report is best described by the following excerpt:

> . . . *the challenge faced by all is to achieve sustainable world economy where the needs of all the world's people are met without compromising the ability of the future generations to meet their needs.*

Since the Brundtland Report, sustainability has become an all-encompassing discipline because a truly sustainable development must include all facets of human activity—agriculture, manufacturing, transportation, buildings, and infrastructure.

This chapter begins with a discussion of the fundamentals of sustainability as applied to architectural design. Subsequently, it deals with the issues of sustainability, as applied to materials and construction.

It must be noted that although sustainable materials and construction practices have been growing, they are still in the early stages and are likely to go through extensive evolution as we learn from the experience. It is expected that sustainability issues will eventually assume the same importance in architecture as structural safety, fire protection, and energy conservation.

EXPAND YOUR KNOWLEDGE

Global Warming

The earth's atmosphere contains a number of gases—nitrogen (approximately 79%), oxygen (approximately 16%), carbon dioxide (approximately 4%), and several others in small amounts. Some of the atmospheric gases, particularly carbon dioxide, are responsible for producing the greenhouse effect on the earth. This allows the sun's radiation to pass through the atmosphere but traps the earth's long-wave radiation. The effect is similar to that produced by glass in a greenhouse (see the box entitled "Some Important Facts About Radiation" in Chapter 28).

In addition to carbon dioxide, there are several other gases, such as methane, nitrous oxide, and water vapor, that are responsible for the greenhouse effect. They are collectively referred to as the *greenhouse gases*. The greenhouse effect is useful because, in its absence, the earth would have been much colder, making life more difficult in several regions.

However, due to the rapidly increasing use of fossil fuels, the amount of carbon dioxide released in the atmosphere has increased substantially. Additionally, the use of nitrogen-based fertilizers for farming has increased the release of nitrous oxide. Consequently, the greenhouse effect has become more pronounced in the past century and has raised the average temperature of the earth.

The scientists believe that, if the present trend continues, the earth's temperature could rise further, melting most of the polar ice and causing the sea levels to rise so that low-lying countries, such as the Netherlands and Bangladesh, could be submerged in seawater. Cities such as New York, London, and Bangkok may also be at risk. Perhaps the most frightening part of global warming is the chain reaction that could ensue, upsetting ecosystems and causing increased flooding, severe droughts, heat waves, and so on.

In response to this scenario, 141 countries signed the *Kyoto Protocol*, which aims to reduce the emission of greenhouse gases. Several countries, including the United States, refused to sign the treaty due to their disagreement with the basic assumptions with respect to global warming. The Kyoto Treaty went into effect in February 2005.

10.1 FUNDAMENTALS OF SUSTAINABLE ARCHITECTURE

As noted earlier, sustainable building design and construction is not a new concept. It is an idea that reemerged in the 1970s and gathered momentum after the 1973 oil embargo. At that time, the focus was limited to energy conservation and the use of alternative energy sources in buildings. As awareness grew that the problem was larger than one of just energy use, a wide range of related environmental issues were also incorporated in the thinking. The work of architects and designers who propagated this thought came to be known by the more inclusive term *sustainable design.*

More specifically, sustainable design recognizes that buildings are major consumers of resources and continually generate waste and pollution. The processes of manufacturing materials and the construction of buildings also use resources and produce waste and pollution.

The land used under the buildings and related infrastructure (roads, bridges, water supply, sewage-treatment plants, etc.) disrupts the ecosystems. Several interior materials, finishes, and furnishings emit unhealthy gaseous products, causing harm to humans. The goals of sustainable architecture, therefore, are as follows.

Integrated Site Design Promote development of the built environment that minimizes its impact on the natural systems and processes of the site or restores and improves sites that have been poisoned or altered greatly over time. Strategies include minimizing site disruption, increasing development density, minimizing the buildings' footprints, using pedestrian-friendly neighborhoods, developing links to public transportation, landscaping that conserves water and reduces the heat island effect, and so on.

Water Conservation Use water-conservation strategies that reduce storm water runoff and introduce water harvesting techniques to increase local aquifer recharge; reduce or limit the use of potable water for landscaping; use low-flow plumbing fixtures, water-efficient appliances and heating, ventilation, and air-conditioning (HVAC) equipment, and so on.

Energy Conservation and Atmosphere Protection Minimize energy use through energy-efficient HVAC, lighting, and other equipment; increase the use of renewable energy sources; reduce atmospheric ozone depletion, and so on.

Resource Efficiency Reuse existing buildings; design long-lasting buildings that can be adapted for changing uses over time; reduce construction waste and implement construction waste management; increase the use of durable and reusable materials; use materials with greater recycled content; use locally or regionally produced products; and so on.

Indoor Environment Maintain good indoor air quality; increase ventilation effectiveness; reduce the emissions of volatile organic compounds (VOCs) and other contaminants by interior materials; increase the use of daylighting of interiors, and so on.

Experts [2] claim that although a sustainable building generally has a higher initial cost, the financial payback through energy savings and lower waste-disposal and water-consumption costs is rapid. Additionally, there are several financial payback features that cannot be easily quantified. These are lower employee health costs and greater productivity due to building's healthier indoor environment (cleaner air, increased amount of daylight, and greater tenant control of temperature and lighting).

NOTE

Resource Consumption by Buildings

In the United States, buildings

- Use 36% of total energy
- Use 30% of raw materials
- Use 12% of potable water
- Produce 30% of total waste
- Emit 30% of greenhouse gases

See also the box entitled "Some Interesting Energy Use Data" in Section 5.11.

Source: Kats, Gregory, et al. "The Costs and Benefits of Green Buildings." A Report to California's Sustainable Building Task Force, October 2003.

10.2 ECOLABELING OF BUILDINGS

It is difficult for a practitioner of sustainable architectural design (especially one who is new to the field) to have the knowledge required to make decisions that successfully address the numerous diverse and interrelated sustainability issues encountered in each new project. For that reason, several organizations have worked for years to develop objective, practical, and fair systems to evaluate the sustainability of buildings for a given size and building program.

The organization that has become a leader in the field is the U.S. Green Building Council (USGBC). It was formed in 1993 by a diverse group of individuals representing various interest groups, such as building owners, architects, engineers, contractors, environmentalists, building material manufacturers, utility companies, financial and insurance experts, and local, state, and federal governments.

USGBC has devised a system to rate the sustainability of a building's design. The rating system is voluntary, is based on consensus and provides a third-party, independent measure

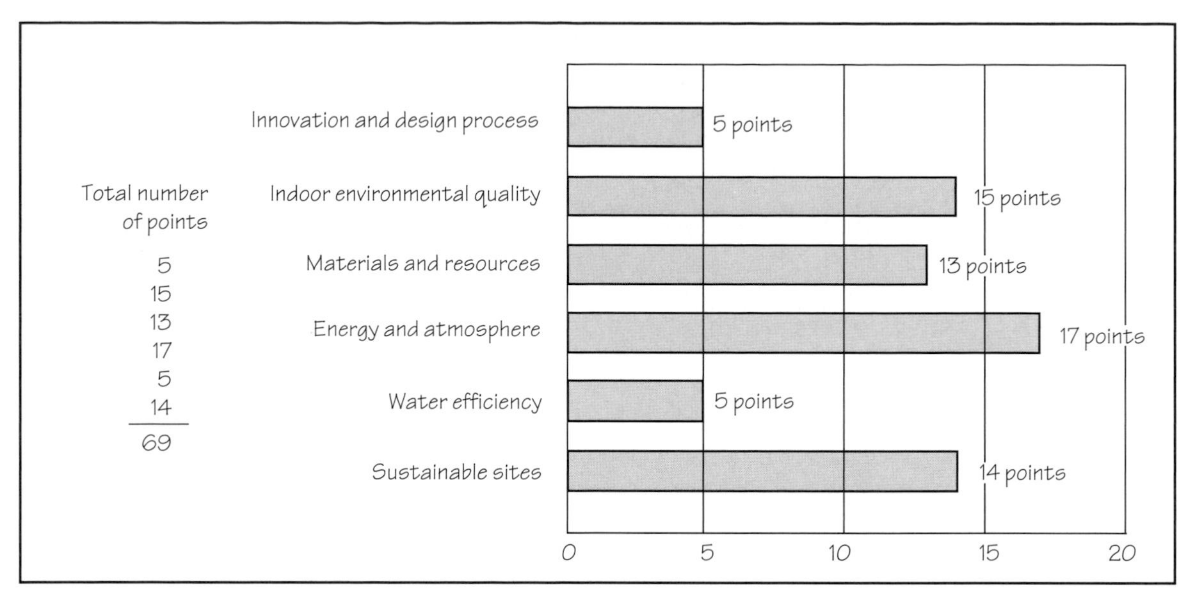

FIGURE 10.1 Relative importance of categories in LEED Green Building Rating System.

of a building's sustainability. It is referred to as the *Leadership in Energy and Environmental Design (LEED®) Green Building* rating system [3].

There are variations in the rating system to include several building types. These are called

- LEED for new construction and major renovations (LEED-NC)
- LEED for existing buildings (LEED-EB)
- LEED for commercial interiors (LEED-CI)
- LEED for homes (LEED-H)

The LEED-NC rating system is based on the performance of the building under the following six categories—five topical categories and one for innovation and design.

- Sustainable sites (maximum 14 points)
- Water efficiency (maximum 5 points)
- Energy and atmosphere (maximum 17 points)
- Materials and resources (maximum 13 points)
- Indoor environmental quality (maximum 15 points)
- Innovation and design process (maximum 5 points)

The relative importance of each category is illustrated in Figure 10.1. A building can score points under all six categories, with the total number of points in all categories adding to 69.

The measure of a building's sustainability is determined by the sum of all points it scores. Four levels of green established by the system, which are referred to as *LEED certification levels,* are given in Table 10.1. A building that scores less than 26 points is not considered a *green* building.

Each topical category is subdivided into several subcategories, which delineate the requirements of that subcategory and the number of points that can be earned by meeting the stated requirements. Additionally, each category has a few prerequisites that must be met before any points can be assigned in that category.

For example, one of the prerequisites in the Energy and Atmosphere category is that the building must meet the requirements of the latest version of ASHRAE Standard 90.1 or the local Energy Conservation Code, whichever is more stringent. (ASHRAE 90.1 is the energy performance standard for buildings, published by the American Society of Heating, Refrigerating and Air-Conditioning Engineers.)

Commissioning Authority

To ensure that the building's performance after completion is as intended and designed (particularly with respect to energy use, envelope design, water use, waste management, indoor air quality, etc.), the LEED rating system recommends that the building be commissioned. Commissioning is a systematic evaluation to verify that the basic building components and systems have been installed and calibrated to function interactively.

TABLE 10.1 LEED CERTIFICATION LEVELS AND POINT SCORES

Certification level	Points scored by building
Certified	26–32
Silver	33–38
Gold	39–51
Platinum	52–69

Engaging an independent *commissioning authority* (also called a *commissioning agent*) is preferred so that the owner obtains an objective evaluation of the architect's and contractor's work. Engaging a commissioning authority early in the project-delivery process (preferably during the design phase) ensures that the postcompletion verification strategies used by the authority are clear to the design and construction teams.

PRACTICE QUIZ

Each question has only one correct answer. Select the choice that best answers the question.

1. The embodied energy in a building is a measure of the energy consumed
   - **a.** to extract raw materials from the earth and manufacture a finished building product.
   - **b.** in the administrative functions related to raw material extraction and product manufacturing.
   - **c.** to transport the finished product to building site and install it in the building.
   - **d.** all the above.
   - **e.** (a) and (b).
2. The increase in the use of a few modern-day chemicals has increased the thickness of the ozone layer.
   - **a.** True
   - **b.** False
3. Global warming has been caused mainly by the
   - **a.** thickening of ozone layer in the atmosphere.
   - **b.** increase in the emission of greenhouse gases.
   - **c.** decrease in cloud cover over the earth.
   - **d.** deforestation of the earth's surface.
   - **e.** all the above.
4. One of the major U.S. organizations that deals with sustainable architectural design is
   - **a.** USEBC.
   - **b.** USFBC.
   - **c.** USGBC.
   - **d.** USHBC.
   - **e.** none of the above.
5. The total number of categories in LEED-NC Green Building Rating System is
   - **a.** 10.
   - **b.** 8.
   - **c.** 6.
   - **d.** 4.
   - **e.** 2.
6. As per the LEED-NC Green Building Rating System, a building may receive recognition as a green building under the following levels:
   - **a.** Platinum, Gold, Silver, and Bronze
   - **b.** Gold, Silver, Bronze, and Certified
   - **c.** Gold, Silver, and Bronze
   - **d.** Gold, Silver, and Merit
   - **e.** Platinum, Gold, Silver, and Certified
7. The three recognized tenets of sustainability are
   - **a.** health, safety, and welfare.
   - **b.** durability, life cycle assessment, and greenness.
   - **c.** reduce, renew, and reuse.
   - **d.** reduce, reuse, and recycle.

Answers: 1-d, 2-b, 3-b, 4-c, 5-c, 6-e, 7-d.

10.3 CHARACTERISTICS OF GREEN BUILDING PRODUCTS

Because building materials constitute a large part of the environmental burden created by a building, the use of green building materials and products is one of the several constituents that make a building sustainable. Extracting materials from the earth and processing them into a finished product require energy and water resources and produce waste, some of which may be hazardous.

Some products give off toxic gases after installation. Others require cleaning with chemicals that may do likewise. Postconsumer disposal of products consumes landfills, some of which may pollute groundwater.

NOTE

Current Laws on Emissions and Manufacturing

Several federal and state laws currently control toxic emissions from products during and after manufacturing. Similarly, laws also exist to control, regulate, and monitor the entire raw-material use, manufacturing, and use cycle of products. The goals of sustainability, however, aim for even higher standards.

Materials whose overall environmental burden is low are referred to as *green materials.* The relative greenness of a material is based on the same basic determinants as for the building as a whole. More specifically, the greenness of a material is a function of the following factors:

- Renewability
- Recovery and reusability
- Recyclability and recycled content
- Biodegradability
- Resource (energy and water) consumption
- Impact on occupants' health
- Durability and life-cycle assessment of greenness

RENEWABILITY

The *law of conservation of matter* states that all matter on the earth and within its atmosphere can neither be destroyed nor created. In other words, whatever existed on the earth at the dawn of time will always exist. Its physical or biological state may, however, change, either through natural or humanmade processes. The basic elements, of which a material is composed, continue to exist forever on the earth or in its atmosphere. For example, when iron corrodes, it becomes iron oxide. The amounts of iron and oxygen on the earth remain unchanged in this transformation.

The transformation of matter (contained within both physical and biological realms) from one state to the other and in various combinations is cyclical. In other words, matter cycles back and forth within physical realms and also from the physical to biological realms.

For example, there is a constant amount of water on the earth, held within the oceans, in the atmosphere, and in other terrestrial entities—both living and nonliving. The processes of evaporation, condensation, consumption, and disposal of water simply move it from one state to another and from one realm to another. There may be drought in one region and excessive rainfall in another region, but the total quantity of water on the earth and in its atmosphere remains constant.

Thus, all materials are theoretically renewable. However, some materials renew over a short-duration cycle, whereas the cycles of others extend over millions of years. Materials that have a short renewal cycle and require limited processing input to convert to a usable form are referred to as *renewable materials.* Conversely, materials that have long renewal cycles and require large processing input are called *nonrenewable materials.* A renewable material is greener than a nonrenewable material.

The renewal cycle of forests is brief—usually 25 to 50 years. To process trees to wood products requires a small amount of additional resources. Wood is, therefore, a renewable material. Perhaps one of the most renewable building materials is adobe. We can dig earth from the ground, ram it into adobe bricks, and construct buildings with them with a negligible amount of processing. When an adobe building is no longer needed, it can be demolished, and its material can be reprocessed for use in a different building with zero renewability duration. Alternatively, the material may be returned to the earth in almost the same state in which it was first obtained.

Metals, stone, glass, and plastics are examples of nonrenewable materials because they take much longer to renew and require excessive processing resources to convert into usable states. Take the example of steel again. If left unprotected, it rusts. Rust is essentially the same as iron ore but in a highly diluted form. To obtain iron from the rust, which, by mingling with the other constituents of the earth's surface, becomes further diluted in iron concentration (i.e., becomes a low-grade iron ore), requires enormous processing energy and produces enormous waste and pollution. Natural geological processes can provide iron ore with high iron concentration, but the processes would take millions of years.

FOCUS ON **SUSTAINABILITY**

Reduce, Reuse and Recycle—Three Tenets of Sustainability

Reduce, reuse, and recycle are considered to be the three most important tenets of sustainable construction. These tenets are listed in order of their importance. The first tenet (reduce), in fact, far outweighs that of the other two in importance.

Although a great deal of stress is currently being placed on reusability and recyclability, the same level of concern is lacking for appropriate sizing of buildings, automobiles, and other items of human consumption. Regardless of how successful we are with reusability and recyclability, sustainability will not be achieved without seriously addressing the *reduction* tenet.

A worrying statistic about buildings is the gradual increase in the average size of new homes built in the United States during the past three decades. In 1970, a typical new house was 1,500 ft^2. In 2005, the corresponding size is 2,300 ft^2—an increase of approximately 50% [4]. At the same time, the average family size has become smaller.

Recovery and Reusability

An effective way of greening a building is by using materials that have been recovered from the demolition of existing buildings. This reduces raw-material extraction and also reduces the burden on landfills. Obviously, salvaged materials must be obtained from nearby sources (to reduce transportation) and must be of usable quality, and their remaining life spans must meet the requirements of their reuse. The economics of recovery and reuse are more favorable with durable materials, such as bricks, natural stone, steel, and aluminum, than with materials with short life spans.

It is expected that as the use of salvaged materials becomes an accepted part of architectural practice, demolition recovery will become a more important commercial enterprise. The role of manufacturers cannot be overemphasized in this area. As the manufacturing industry moves further toward sustainability and assumes responsibility for the entire life cycle of the materials it produces, salvage and reuse could become part of the manufacturers' responsibilities.

In this scenario, the manufacturer retains the ownership of the product, which is transferred to the consumer for only a certain duration. At the end of that duration, the buyback of the product by its manufacturer is mandated, and it is to be refurbished, remanufactured,

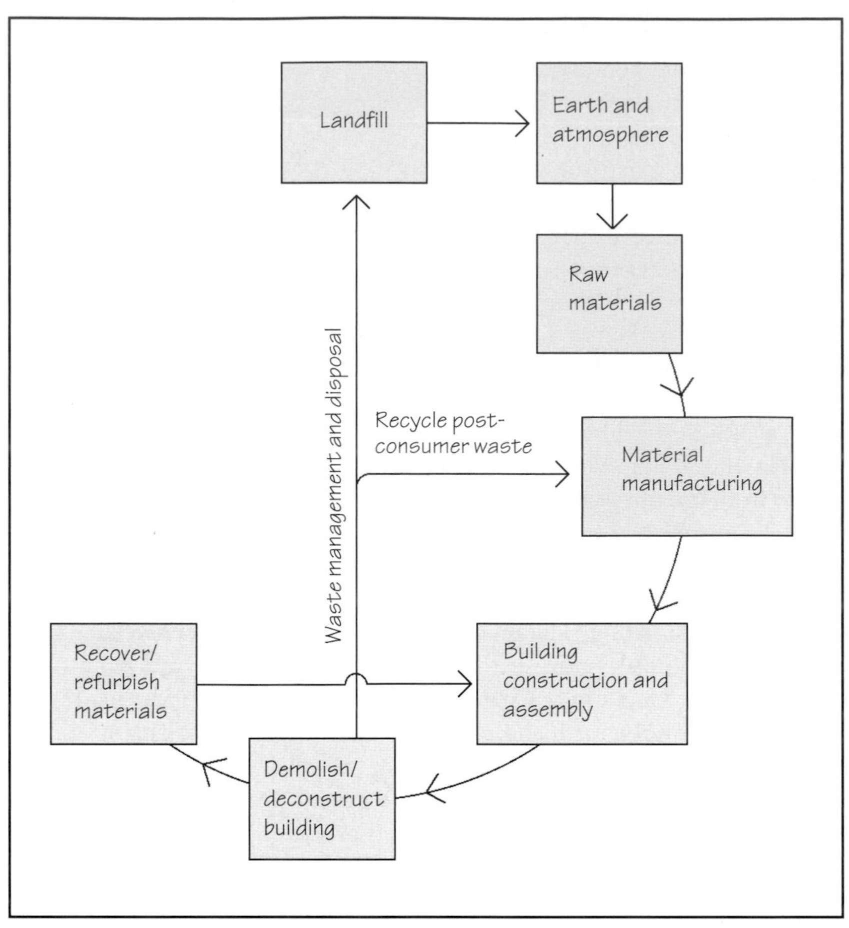

FIGURE 10.2 Closed-loop building-product life cycle. Note that after a building is demolished, a part of the products may be recycled as raw materials for manufacturing and a part may be sent to a landfill. Another part may be salvaged for use in a new building. This part may go through a few stages of recycling before it is also sent to a landfill.

NOTE

Difference Between Renewal and Recycling

The difference between the renewal and recycling of a material is subtle. Renewal refers to the process of recycling that occurs in nature. *Recycling* is a deliberate process.

or recycled as needed. The ultimate disposal of the product in a landfill also remains the manufacturer's responsibility. This should make product manufacturing and its use in buildings approach a closed-loop system, Figure 10.2.

As environmental legislation makes waste management and disposal more costly, manufacturers will be increasingly forced into making their products greener. Removal of hazardous contents, if any, before disposal and ensuring rapid degradability of materials (e.g., computers, furniture, or carpets) in landfills will be the manufacturers' responsibility.

An obvious extension of demolition recovery is the concept of building deconstruction, which implies that when a building has served its useful life, it is not demolished but deconstructed into components. The deconstructed components can be refurbished if necessary and sold for use in another building.

Conceptually, deconstruction is the reverse of construction. Only those components of a deconstructed building are disposed of as waste that cannot be reused.

RECYCLABILITY AND RECYCLED CONTENT

The greater the amount of recycled content in a material, the greener it is. Recycling is generally classified under two types: *internal recycling* and *external recycling*.

Internal recycling is the reuse of materials that are by-products or waste products from manufacturing. Brick- and glass-manufacturing industries and several others have used internal recycling for a long time. Broken or defective bricks or glass are reused as raw materials in the manufacturing process. Ready-mix concrete manufacturing plants recycle wastewater reclaimed from washing concrete mixers after they return for recharge and delivery.

External recycling, on the other hand, is reclaiming a material for use in the manufacturing process of the same or a different product after it has become obsolete or unnecessary in its present application. This is referred to as *postconsumer waste recycling*.

For example, concrete obtained from a demolished building can be crushed and used as a drainage layer under a new concrete slab-on-grade in place of crushed stone. In some cases, crushed old concrete may be used as coarse aggregate in new concrete. Similarly, reinforcing or structural steel recovered from a demolished building can be melted and processed as new steel.

The amount of recycled content in a material depends on the economics of procuring post-consumer waste. The reason that recycling of aluminum (beverage) cans became successful early in recycling history is because of the large amount of resource input (particularly energy) required in the manufacture of new aluminum from its ore, as compared with procuring recycled cans and converting them into a new or same product.

Other products with significant recycled content are steel, carpets, gypsum board, rubber, plastics, and fiberglass insulation. Increasingly larger number of products are being manufactured with recycled materials, and their recycled content is also increasing. Consequently, recycling has become a major and diverse industry that employs a large number of skilled and unskilled workers.

The number of highly skilled scientists and engineers in the recycling industry is quite large because manufacturing with recycled materials is technically as complex as (if not more complex than) manufacturing from virgin materials. For example, one of the major recycling problems in the glass industry has been to remove color from used glass—a problem that has engaged the brains of several top glass scientists.

BIODEGRADABILITY

Everything else being identical, biodegradable materials, such as adobe, wood, and paper, are greener than nonbiodegradable materials such as metals and plastics. Building product packaging, if required, consisting of paper-based products should, therefore, be preferred.

RESOURCE CONSUMPTION

Two resources that are critical in product manufacturing are energy and water. Both are, therefore, determinants of a material's greenness. As stated previously, the amount of energy used in manufacturing a material is called its *embodied energy*. Embodied energy is generally measured in megajoules per kilogram (MJ/kg) or MJ/m^3.

Table 10.2 gives approximate values of embodied energy in selected building materials [5]. Because all materials are not equal in terms of their use and properties, embodied energy cannot be used directly in comparing the energy efficiency of one material with that of the other. However, it is a good measure for comparing the energy efficiency of similar products—those that belong to the same application group. For example, embodied energy may be used in comparing plywood with oriented strandboard (OSB), one carpet with the other carpet, or one type of gypsum board with the other, and so on.

Regardless of the materials' embodied energy, the use of locally produced materials reduces the building's embodied energy by reducing transportation energy use. It also improves the local economy.

Data for total water used in manufacturing a product, similar to embodied energy data, are not generally available.

TABLE 10.2 EMBODIED ENERGY IN SELECTED MATERIALS

Material	Embodied energy MJ/kg	Embodied energy MJ/m^3
Adobe	0.42	819
Bricks	2.5	5,170
Concrete block	0.94	2,350
Concrete (4,000 psi)	1.3	3,180
Stone	0.79	2,030
Lumber	2.5	1,380
Plywood	10.4	5,720
Particle board	8.0	4,400
Aluminum (virgin)	227	515,700
Aluminum (recycled)	8.1	21,870
Steel (virgin)	32.0	251,200
Steel (recycled)	8.9	37,200
Copper (virgin)	70.6	631,164
Glass (float)	15.9	37,550
Gypsum board	6.1	5,890
Fiberglass insulation	30.3	970
Extruded polystyrene	117	3,770
Cellulose insulation	3.3	112

Impact on Occupants' Health

The relative greenness of a material is also a function of its impact on human health. A number of modern building materials, particularly adhesives used in wood products and floor finishes (carpets, linoleum, and vinyl and rubber floorings), paints, sealers, and sealants, emit volatile organic compounds (VOCs), which are harmful to humans beyond a certain level of concentration. These emissions adversely impact indoor air quality (IAQ) and, hence, the health of building occupants and are regulated by the U.S. Environmental Protection Agency (EPA).

Poorly ventilated spaces with high moisture contents and the use of fibrous materials may support the growth of mold when the materials become wet. Mold has been known to cause heath problems among building occupants.

In addition to the impact of materials on building occupants, its impact on the earth and its biosphere must also be accounted for, because some materials will either leach toxic materials during their use or degrade into hazardous substances in landfills.

Durability and Life-Cycle Assessment (LCA) of Greenness

A product's durability is closely related to its sustainability. A material with high embodied energy but with long service life may be more sustainable than one with a low embodied energy but short service life. Additionally, a material may not emit hazardous compounds when installed in the building, but the chemicals required for its manufacture or periodic maintenance may do so. Therefore, the overall ecoburden of a product should be determined based on its expected service life. In other words, a life-cycle assessment (LCA) of a products' greenness is more important than its initial greenness.

The LCA of a product evaluates the environmental impact of a product over its entire life cycle, including the product's production, transportation, use, reuse, and its final disposal. Environmental impact measures include global warming, ozone depletion, production of pollutants and toxic waste, resource consumption, land use, and so on.

10.4 ECOLABELING OF BUILDING PRODUCTS

With the increasing recognition of sustainable design and construction, most building-product manufacturers have begun to claim their products as green. The claims are generally based on only one or two green criteria that the product may satisfy.

For example, manufacturers of products made from metals (steel, aluminum, copper, etc.) claim their product as green because metals are easily recycled. Manufacturers of wood-based products claim greenness because wood is a renewable material. Masonry product manufacturers promote their products for reasons of low embodied energy and zero VOC emissions.

A comprehensive ecolabeling of building products, similar to the LEED Green Building Rating System, has not yet been developed. However, several labeling systems, which are either material specific or criteria specific, are available. A few selected, independent product-labeling systems are

- Energy Star label
- Certified Wood label
- Green Label and Green Label Plus
- Green Seal label

Energy Star Label

The Energy Star label was introduced by the U.S. Environmental Protection Agency (EPA) in 1992 to recognize energy-efficient computers. The program's success led to its extension to several other product areas and also its acceptance by several other countries. Thus, Energy Star is now an international energy-labeling system that labels home and office appliances, electric light sources, windows, and so on. More building-product manufacturers are getting their products Energy Star certified.

In collaboration with the U.S. Department of Energy (DOE), EPA is moving into other product areas. For example, the Energy Star label is now available for energy-efficient homes. Note that Energy Star label for a home is not as comprehensive as the LEED-H Green Building Rating System.

Certified Wood Label

The Certified Wood label is carried by wood products that have been produced by manufacturers according to the guidelines promulgated by the Forest Stewardship Council

(FSC). The FSC prescribes and enforces stringent environmental guidelines for forest growth and harvesting. It maintains a number of independent certifiers around the globe for product labeling. (See Chapters 11 and 12 for additional details.)

Note that FSC's certified wood label refers only to the environmental management aspects of forests. It does not include other sustainability concerns, such as embodied energy or emissions of VOCs by engineered wood products.

GREEN LABEL AND GREEN LABEL PLUS

Green Label is the mark assigned by the Carpet and Rug Institute (CRI) to carpets, rugs, and cushion materials that have low VOC emissions. Green Label Plus is a more stringent mark than Green Label for the same products.

GREEN SEAL LABEL

Green Seal label is a comprehensive, life-cycle environmental evaluation of a product based on EPA and International Standards Organization (ISO) standards. Only a few building products are currently covered by the Green Seal label.

PRACTICE QUIZ

Each question has only one correct answer. Select the choice that best answers the question.

8. Which of the following building materials is most renewable?
- **a.** Steel
- **b.** Aluminum
- **c.** Natural stone
- **d.** Wood

9. Embodied energy of a product is a good index of comparing the energy efficiency of
- **a.** one material versus the other material regardless of the type of material.
- **b.** similar materials—materials that belong to the same group.

10. Weight for weight, which of the following materials has the highest embodied energy when made from virgin raw materials?
- **a.** Steel
- **b.** Aluminum
- **c.** Natural stone
- **d.** Lumber

11. The durability and sustainability of a material are related to each other.
- **a.** True
- **b.** False

12. The certified wood label considers all aspects of sustainability of wood.
- **a.** True
- **b.** False

13. The life-cycle assessment of a product refers to its cost over its entire service life.
- **a.** True
- **b.** False

14. The Energy Star labeling program applies to
- **a.** appliances.
- **b.** products, such as windows and doors.
- **c.** homes.
- **d.** all the above.
- **e.** none of the above.

15. Green Label and Green Label Plus programs apply to
- **a.** appliances.
- **b.** products, such as windows and doors.
- **c.** homes.
- **d.** all the above.
- **e.** none of the above.

Answers: 8-d, 9-b, 10-b, 11-a, 12-b, 13-b, 14-d, 15-e.

KEY TERMS AND CONCEPTS

- Biodegradability
- Certified Wood label
- Commissioning of building
- Ecolabeling of building products
- Ecolabeling of buildings
- Embodied energy in a building
- Embodied energy in a product
- Energy Star label
- Global warming
- Green building product
- Green Label and Green Label Plus
- Green Seal label
- LEED
- LEED certification levels
- Ozone depletion
- Reduce, reuse and recycle
- Sustainable development
- USGBC

REVIEW QUESTIONS

1. In the LEED-NC Green Building Rating System, list the categories that are used to measure sustainability.
2. Explain commissioning in LEED-NC Green Building Rating System.
3. List various factors that determine the sustainability of a building product.

4. Using a sketch and notes, explain what a closed-loop product life cycle is.
5. Explain which materials are considered renewable and which are considered nonrenewable, and why.

SELECTED WEB SITES

Green Seal Products Standard and Certification (www.greenseal.org)
U.S. Environmental Protection Agency, EPA (www.epa.gov)
U.S. EPA Energy Star Program (www.energystar.gov)
United States Forest Stewardship Council (www.fscus.org)
U.S. Green Building Council, USGBC (www.usgbc.org)
University of Minnesota Center for Sustainable Building Research (www.buildingmaterials.umn.edu)

REFERENCES

1. World Commission on Environment and Development (WCED), Melbourne. *Our Common Future* (The Brundtland Report). Oxford, UK: Oxford University Press, 1987.
2. Kats, Gregory, et al. "The Costs and Benefits of Green Buildings." A Report to California's Sustainable Building Task Force, October 2003.
3. McLaughlin, Lisa. "Technology, Health, Money, Fitness, Home Trends." *Time*, March 2005, p. 74.
4. U.S. Green Building Council. *Leadership in Energy and Environmental Design Green Building Rating System.* Version 2.1, 2003.
5. Alternative Technology and Lifestyle Association (ATLA). "Embodied Energy of Materials." 7, No. 4, 1998.

PART 2

MATERIALS AND SYSTEMS

CHAPTER 11 Materials for Wood Construction–I (Lumber)

CHAPTER 12 Materials for Wood Construction–II (Manufactured Wood Products, Fasteners, and Connectors)

CHAPTER 13 Wood Light Frame Construction–I

CHAPTER 14 Wood Light Frame Construction–II

CHAPTER 15 Structural Insulated Panel System

CHAPTER 16 The Material Steel and Structural Steel Construction

CHAPTER 17 Light-Gauge Steel Construction

CHAPTER 18 Lime, Portland Cement, and Concrete

CHAPTER 19 Concrete Construction–I (Formwork, Reinforcement, and Slabs-on-Ground)

CHAPTER 20 Concrete Construction–II (Site-Cast and Precast Concrete Framing Systems)

CHAPTER 21 Soils; Foundation and Basement Construction

CHAPTER 22 Masonry Materials–I (Mortar and Brick)

CHAPTER 23 Masonry Materials–II (Concrete Masonry Units, Natural Stone, and Glass Masonry Units)

CHAPTER 24 Masonry and Concrete Bearing Wall Construction

CHAPTER 25 Rainwater Infiltration Control in Exterior Walls

CHAPTER 26 Exterior Wall Cladding–I (Masonry, Precast Concrete, GFRC, and Prefabricated Masonry)

CHAPTER 27 Exterior Wall Cladding–II (Stucco, EIFS, Natural Stone, and Insulated Metal Panels)

CHAPTER 28 Transparent Materials (Glass and Light-Transmitting Plastics)

CHAPTER 29 Windows and Doors

CHAPTER 30 Glass-Aluminum Wall Systems

CHAPTER 31 Roofing–I (Low-Slope Roofs)

CHAPTER 32 Roofing–II (Steep Roofs)

CHAPTER 33 Stairs

CHAPTER 34 Floor Coverings

CHAPTER 35 Ceilings

CHAPTER 11

Materials for Wood Construction–I (Lumber)

CHAPTER OUTLINE

11.1 Introduction

11.2 Growth Rings and Wood's Microstructure

11.3 Softwoods and Hardwoods

11.4 From Logs to Finished Lumber

11.5 Drying of Lumber

11.6 Lumber Surfacing

11.7 Nominal and Actual Dimensions of Lumber

11.8 Board Foot Measure

11.9 Softwood Lumber Classification

11.10 Lumber's Strength and Appearance

11.11 Lumber Grading

11.12 Durability of Wood

11.13 Fungal Decay

11.14 Termite Control

11.15 Preservative-Treated Wood

11.16 Fire-Retardant-Treated Wood

PRINCIPLES IN PRACTICE: Typical Grade Stamps of Visually Graded Lumber

With forest resources diminishing, new varieties of trees are cultivated and harvested as agricultural crops. The hybrid poplars shown here are harvested on approximately a 10-year cycle, and the wood is used in a variety of construction and finish products. (Photo by DA.)

There are many structural and nonstructural applications for wood products in building construction. Because wood is a general term, it is important to clarify the terminology for its various uses.

The term *lumber* applies to wood products derived directly from logs through sawing and planing operations only, with no further manufacturing except cutting to length or finger-jointing two or more pieces (see Section 11.4 for finger jointing). In other words, only solid pieces of wood are classified as lumber. Hence, *lumber, solid lumber, solid sawn lumber,* and *sawn lumber* are used synonymously.

In recent years, manufactured wood products—products manufactured with altered or transformed wood fibers—have been developed. These products are identified by their individual names and are not called lumber. Plywood was one of the earliest of these products.

Glue-laminated lumber, parallel strand lumber, and laminated veneer lumber are products that have the same or similar profiles as solid lumber. Wood trusses and I-joists are other examples of manufactured wood products. These products have been formulated primarily to conserve decreasing supplies of high-quality solid lumber members.

This and the following four chapters deal with wood construction. Properties and uses of various wood products are presented first. Lumber is discussed in this chapter and manufactured wood products, in Chapter 12. In practice, lumber and manufactured wood products are used together in the same building. Wood light frame, a common construction system utilizing lumber and manufactured wood products, is described in Chapter 13.

Chapter 14 deals with frequently used exterior and interior finishes in wood light-frame buildings. Chapter 15 deals with panelized wood frame construction in what is referred to as *structural insulated panel* (SIP) construction.

11.1 INTRODUCTION

A cursory review of architectural history reveals that stone masonry was the construction system of antiquity, yet sufficient evidence suggests that the use of wood in buildings could have preceded it. Historic stone structures survived due to their relative durability, whereas wood structures of antiquity were either consumed by fires or destroyed by wood-consuming organisms.

However, many centuries-old surviving wood buildings indicate that wood is a fairly durable material, provided it is protected from fire and biological destruction. An example illustrating wood's durability is shown in Figure 11.1. Many other examples exist in several European countries.

Wood—An Easily Renewable Material

One of the major advantages of wood as a building material is its renewability. Theoretically, all materials are renewable. However, wood is a building material that is renewed within the

FIGURE 11.1 John Ward House, Salem, Massachusetts, built in the seventeenth century and still standing. (Photo courtesy of Dr. Jay Henry.)

human time span. Trees for most softwood construction lumber are harvested within 25 to 40 years. Other naturally occurring building materials (stone, iron, aluminum, copper, etc.) are also renewed, but within geological time spans—in thousands or millions of years.

In order to optimize wood's renewablility, sufficient forest land must be available to grow trees, and the land must be carefully managed (lumber farming). Unfortunately, due to high population densities, the availability of land for lumber farming is severely limited in most parts of Africa, Asia, Europe, and Latin America. Coincidentally, forest-management skills are also relatively underdeveloped in these countries.

Wood used in buildings is, therefore, grown primarily in the less densely populated countries—the United States, Canada, Australia, New Zealand, Russia, and a few countries in Latin America. It is in these countries that wood is a major building and structural material today. The native wood in most other countries is used for nonstructural purposes, such as the doors, windows, furniture, and cabinets.*

WOOD AND OTHER STRUCTURAL MATERIALS

Many believe that wood is the weakest of the structural materials—concrete, masonry, and steel. This is true only in absolute terms. When viewed in relation to its density (unit weight), the strength of wood compares well with even the strongest of all building materials, steel. Weight for weight, wood's structural properties are almost comparable with that of steel, Table 11.1.

For instance, the average unit weight of structural wood (lumber) is nearly 35 lb/ft^3, and its average allowable compressive strength (parallel to grain) is 1.6 ksi. With a unit weight of 490 lb/ft^3, steel is nearly 14 times heavier than lumber, and its allowable compressive strength of nearly 30.0 ksi is 19 times that of the allowable compressive strength of lumber.

The strength-to-weight ratio of lumber in comparison with concrete and masonry is heavily in lumber's favor. For instance, the density of concrete (145 lb/ft^3) is approximately four times that of lumber, and the allowable compressive strength of the commonly used concrete mixes is only slightly higher than that of lumber.

The high strength-to-weight ratio of lumber not only reduces the dead loads but also precludes the need for heavy equipment for lifting and hoisting building components during construction. These and other factors, such as its lower cost and the relatively simpler construction techniques associated with it, account for wood's popularity for several building types.

In countries with abundant forest reserves, wood is used extensively for residential and low-rise commercial buildings and amusement structures. A few examples illustrating the versatility of wood for different applications are shown in Figures 11.2 to 11.4.

TABLE 11.1 COMPARISON OF LUMBER AND STEEL AS STRUCTURAL MATERIALS

Density (lb/ft^3)	
Lumber	35
Steel	490
Density ratio = 1:14	
Allowable compressive strength (ksi)	
Lumber (parallel to grain)	1.6
Steel	30.0
Compressive strength ratio = 1:19	
Allowable tensile strength (ksi)	
Lumber (parallel to grain)	1.0
Steel	30.0
Tensile strength ratio = 1:30	
Modulus of elasticity (ksi)	
Lumber	1,700
Steel	29,000
Modulus of elasticity ratio = 1:17	

Values for wood refer to the stronger lumber species and grade (such as southern pine No. 1). Note that the structural properties of lumber vary with grade, species, and cross-sectional dimensions.

Values for steel refer to steel with a yield strength of 50 ksi.

FIGURE 11.2 Millbrook Guest House in Millbrook, New York by Meditch Murphey Architects—a wood, light frame house that takes advantage of natural ventilation, sunlight, and views. It is roofed with cedar shingles and clad with stucco and limestone base, with beech and teak used throughout the interior. (Photographs © Maxwell Mackenzie.)

*In developing countries, a great deal of wood is used as household fuel.

FIGURE 11.3 Montage Apartments in Palo Alto, California, by Seidel/Holzman Architects, constructed of wood light frame and clad in combination of fiber-cement lap, fiber-cement panel, and cedar batten siding. (Photo courtesy of Tom Rider.)

FIGURE 11.4 Wood-framed skylight in Thunder Bay Regional Health Sciences Centre, Thunder Bay, Ontario, Canada, by Salter Farrow Pilon Architects (Farrow Partnership Architects; Salter Pilon Architects successors). (Photo: Peter Sellar Klik photography, provided courtesy of Farrow Partnership Architects Inc.)

Photosynthesis and Respiration

The Processes That Make Wood a Rapidly Renewable Material

The growth of trees and plants occurs due to a natural process, called photosynthesis. In this process, the water drawn from the soil and the environment reacts with the carbon dioxide in the air to produce a basic sugar (glucose), releasing oxygen. This reaction requires a tree or plant as the medium and the presence of light energy from the sun, thus the term photosynthesis, whose literal meaning is "to synthesize with the help of light." In other words

$$\text{Carbon dioxide} + \text{water} \xrightarrow[\text{energy}]{\text{Light}} \text{glucose} + \text{oxygen}$$

$$6CO_2 + 6H_2O \xrightarrow[\text{energy}]{\text{Light}} C_6H_{12}O_6 + 6O_2$$

Glucose is subsequently converted into other organic molecules that comprise the tree—cellulose, hemicellulose, and lignin. Cellulose* and hemicellulose, which are chemically similar to glucose, constitute nearly two-thirds to three-quarters the weight of wood in a tree. Cellulose and hemicellulose form the walls of wood cells, and lignin is the glue that binds the cells together.

Although cellulose, hemicellulose, and lignin constitute the bulk of the wood, small quantities of other minerals and chemicals are also transported into the tree.

Photosynthesis is the reverse of respiration—a process by which the food that the humans and animals eat is converted into sugar. This sugar is subsequently metabolized into carbon dioxide, and water, and energy is released in this process:

$$\text{Glucose} + \text{oxygen} \longrightarrow \text{carbon dioxide} + \text{water} + \text{energy}$$

$$C_6H_{12}O_6 + 6O_2 \longrightarrow 6CO_2 + 6H_2O + \text{energy}$$

Photosynthesis and respiration, being reverse of each other, form an ecological cycle on which all terrestrial life depends.

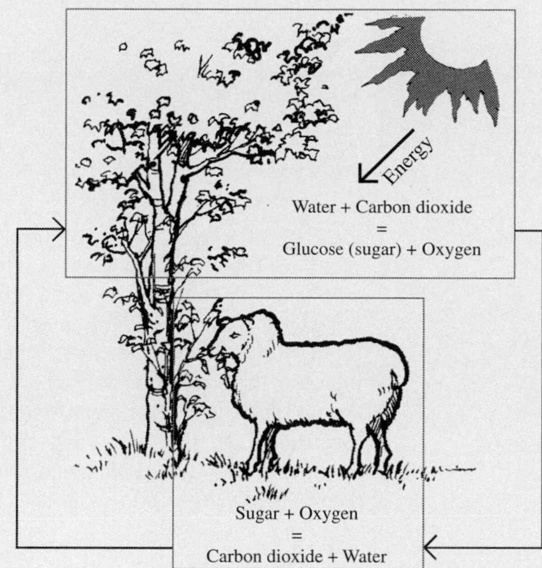

The tree obtains water from the soil and carbon dioxide from the atmosphere and converts them into sugar (contained in plant life). This reaction, in which oxygen is released, requires the presence of energy that is supplied by the sun.

Animals, humans, and other creatures consume the sugar in plants. With the help of the oxygen they breathe, they convert these substances into carbon dioxide (which they exhale), and water is given off in animal and human waste.

*Chemically, a cellulose molecule is represented by $(C_6H_{10}O_5)_n$. The subscript n indicates that a cellulose molecule is a polymer consisting of several $C_6H_{10}O_5$ monomers linked together end-to-end, as the loops in a chain. Typically, in a cellulose polymer, n may be as high as 10,000, which gives polymer lengths of nearly 0.01 mm. The orientation of the polymer chains, along the long axis of the tree, provides many of the basic properties of wood.

11.2 GROWTH RINGS AND WOOD'S MICROSTRUCTURE

A tree consists of three essential parts: (a) root structure, (b) trunk, and (c) branches and the leaf system. The root structure stabilizes the tree and the trunk provides strength required to support the branches and leaves. In addition to providing stability and strength, the other important function of the root structure and trunk is to conduct water mixed with food (nutrients) from the soil to the leaves and branches.

As the water moves through the tree and reaches the leaves, it is converted to cellulose by *photosynthesis.* The transportation of water from the soil is accompanied not only by food, but also by several other dissolved minerals and chemicals in small quantities. Water transportation requires that the microscopic structure of a tree must consist of tiny hollow tubes, referred to as *cells.* To transport food efficiently, the cells must be oriented along the long axis of the tree—vertically in the trunk and along the length of an individual branch. The cells are visible only when magnified several times under a microscope.

Figure 11.5 shows a part of the unmagnified cross section through the trunk of a tree. The outermost dark-colored layer is the bark, followed by several increasingly smaller rings toward the center of the trunk. These rings are referred to as the *annual rings,* because a new ring is typically added every year. *Growth ring* is, however, a better term because in some trees, more than one ring may be added per year.

The addition of growth rings occurs in the *cambium* layer, which lies between the bark and the outermost growth ring (Figure 11.5). The cells in the cambium layer divide and subdivide, producing the cells of the growth rings and the bark. As each growth ring is added, the girth of the tree increases.

FIGURE 11.5 A cross section through the trunk of a tree. The cambium layer, in which cell division takes place, lies between the bark and the outermost growth ring. The interior dark-colored region is the heartwood, and the light-colored region is the sapwood. The radial lines (which are rather faint in this diagram) are the rays (Section 11.3). (Photo courtesy of the Forest Products Laboratory, U.S. Department of Agriculture.)

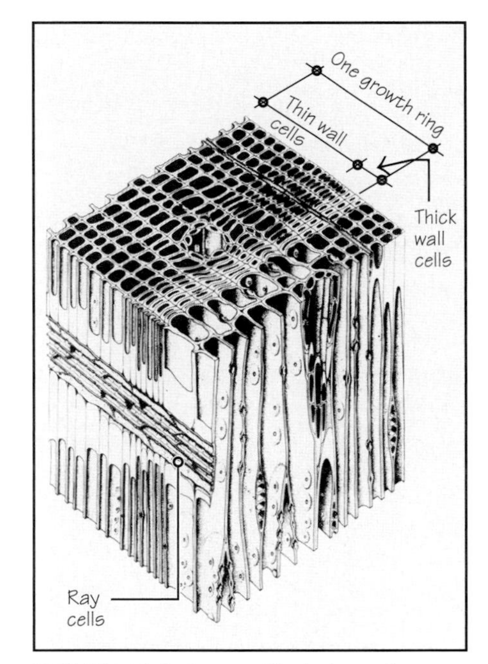

FIGURE 11.6 A magnified view of the cross section of a tree trunk. (Photo courtesy of the Forest Products Laboratory, U.S. Department of Agriculture.)

EARLYWOOD AND LATEWOOD

The cells added during the wet season have thinner walls and larger cavities. During the dry season, when the tree has to transport less food, the cell walls are thicker, and the cell cavities are smaller. The cells added during the wet and dry seasons are called *earlywood* and *latewood,* respectively.

It is the difference between the cells of earlywood and latewood that distinguishes each ring, Figure 11.6. The distinct growth rings give wood its characteristic grain structure.

In some species the difference between earlywood and latewood is less distinct than in others. For instance, in basswood all cells have a similar wall thickness and cavity size, and annual rings are indistinct. This yields wood with a relatively flat and even grain structure. By comparison, southern yellow pine, Figure 11.7, has highly differentiated cell sizes, resulting in a pronounced grain structure.

HEARTWOOD AND SAPWOOD

As the tree ages and increases in girth, not all cells are needed to conduct food. As this happens, walls and cavities of the cells formed when the tree was younger (i.e., the rings nearer the center of the trunk) are impregnated with several chemical substances. This process increases the tree's mass and strength, which is required to support its increasing height.

The obvious manifestation of the chemical deposition is a change in the color of the wood. Thus, the growth rings added in the trunk earlier during the life of the tree are darker in color than those added later (Figure 11.5).

The darker-colored region of the trunk is referred to as the *heartwood,* and the lighter-colored region is called the *sapwood.* The word *sap* is just another term for the food that the tree conducts, and hence the term *sapwood* merely expresses the function of the light-colored outer growth rings in relation to the dark-colored inner rings. The change from sapwood to heartwood is gradual, and as the tree ages, the heartwood portion in the tree increases.

NOTE

Earlywood and Latewood

The terms *springwood* and *summerwood* were earlier used for earlywood and latewood, respectively, which erroneously represented a connection with a calendar season of the year. In many tropical areas, which remain wet throughout the year, the growth of a tree occurs throughout the year.

FIGURE 11.7 A piece of southern yellow pine, illustrating its pronounced grain structure. By comparison, a piece of basswood, commonly used for architectural model making, has flat and even grain structure.

The chemical deposition in many commercial trees is toxic to fungi. Therefore, the heartwood of most commercial trees is relatively more decay resistant than the sapwood. In some species, the decay resistance of the heartwood is significant. Before the introduction of pressure-treated wood, the heartwoods of redwood and cedar were commonly specified for lumber elements resting directly on concrete or masonry foundations, such as the sill plates and sleepers.

Contemporary building codes also recognize the decay resistance of the heartwood of redwood and cedar by allowing them to be used where pressure-treated wood is required. The use of pressure-treated wood is, however, more common due to its greater decay and termite resistance.

11.3 SOFTWOODS AND HARDWOODS

Lumber is divided into two broad categories—*hardwoods* and *softwoods*. The hardwoods are generally denser than the softwoods. Because density is a major determinant of the strength of wood, hardwoods are generally stronger than the softwoods.

The distinction between softwoods and hardwoods is, however, not based on the density of the wood, because several hardwoods are lighter than the softwoods, and vice versa. For instance, balsa, among the lightest and the weakest of woods, is a hardwood. On the other hand, slash pine and longleaf pine are softwoods but are denser than many hardwoods, Figure 11.8.

The distinction between softwoods and hardwoods is based on their botanical characteristics. Softwood-producing trees, in general, do not bear flowers and have a single main stem, and most of them are evergreen, with leaves that are needlelike (i.e., conical; hence softwood trees are also called *conifers*), Figure 11.9. Pines, firs, spruces, cedars, hemlocks, and so on, are softwoods.

Hardwood-producing trees are generally flowering trees, have broad leaves, and are typically deciduous, shedding and regrowing leaves annually. Oak, walnut, birch, elm, teak, mahogany, rosewood, and so on, are hardwoods. Figure 11.10 shows a typical hardwood tree.

Softwoods in the United States come from two main regions, the western wood region and the southern pine region, and to a small extent from the northern and northeastern region, Figure 11.11. In addition to indigenous softwoods, a great deal of softwood lumber in the United States is imported from Canada. Most hardwoods in the United States are from the eastern region, from New York to Georgia.

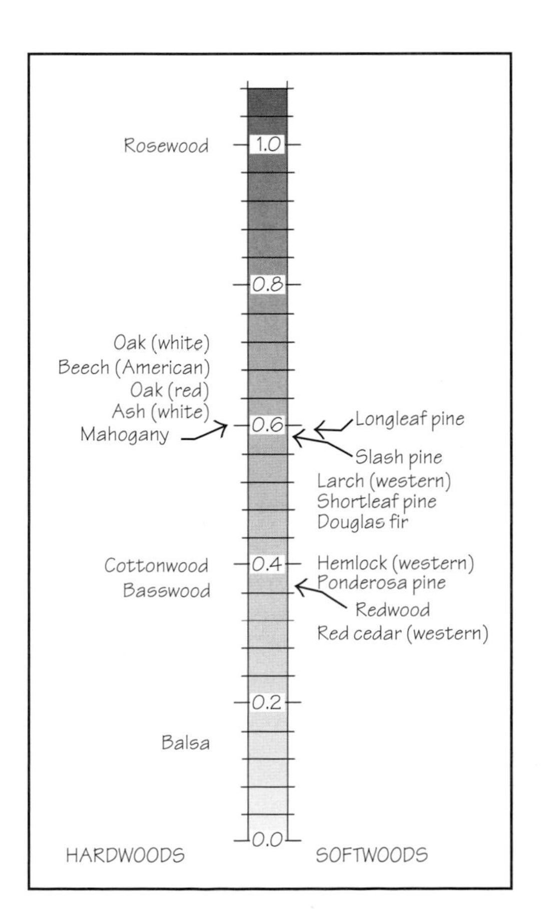

FIGURE 11.8 Specific gravities of selected softwoods and hardwoods.

NOTE

Naturally Decay-Resistant Wood

Heartwood of redwood and cedar is naturally decay resistant and is recognized as such by the building codes.

NOTE

The Term *Species*

In this text, the term *species* is used in its customary sense, not in the botanically correct sense. In the botanist's language, a particular species is designated by the combination of *genus* (generic name) and *species* (specific name). For instance, the botanical name of Eastern white pine is *Pinus strobus*, in which *Pinus* represents the genus and *strobus* represents the species. Similarly, the botanical name of Ponderosa pine is *Pinus ponderosa*. For the sake of simplicity, the term species in this text includes both botanical terms—genus and species. Note also that the term *species* is used in both singular and plural forms.

NOTE

Specific Gravity

The specific gravity of a material is an index of its density relative to the density of water, that is, the density of the material divided by the density of water. Thus, if the density of a material is less than the density of water, its specific gravity is less than 1.0. Such a material will float on water. Conversely, if the density of a material is greater than the density of water, its specific gravity is greater than 1.0, and it will sink in water.

Because the density of water is 62.5 lb/ft^3, a wood whose specific gravity is 0.5 has a density of 31.25 lb/ft^3.

FIGURE 11.9 Softwood (pine) trees in Willamette Industries forest, Ruston, Louisiana.

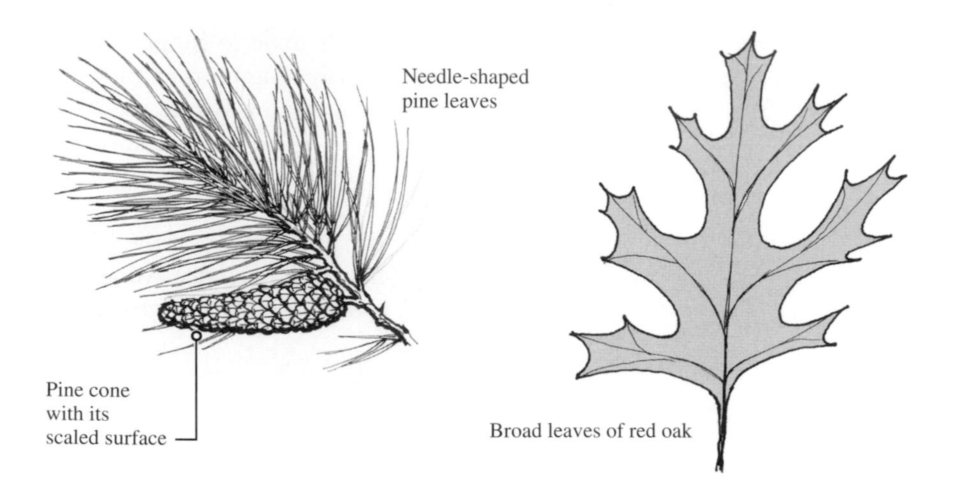

FIGURE 11.10 A hardwood tree (red oak).

MICROSTRUCTURE OF SOFTWOODS AND HARDWOODS

Although basically similar, there is some difference between the cellular structures of softwoods and hardwoods. In softwoods, all longitudinal cells are of the same type and almost the same size, Figure 11.12. These cells perform two functions: (a) provide strength to the tree and (b) conduct food from the ground up.

The cellular structure of hardwoods is more complex. Unlike softwoods, the longitudinal cells of hardwoods are of two types—one type with small cavities and the other with large cavities—with two distinct functions. The smaller cells in hardwoods provide strength to the tree, and the cells with larger cavities, referred to as *vessels,* conduct food, Figure 11.13. In many hardwoods, the vessels can be seen under an ordinary magnifying glass.

In addition to the longitudinally oriented cells, both softwoods and hardwoods have transverse cells called *rays.* Rays are perpendicular to the longitudinal cells. They provide radial transfer of food, as well as transverse strength.

Ray cells are far more prominent in hardwoods than softwoods. Due to the prominence of ray cells and the diversity in their longitudinal cells, most hardwoods show a more interesting grain structure than do softwoods.

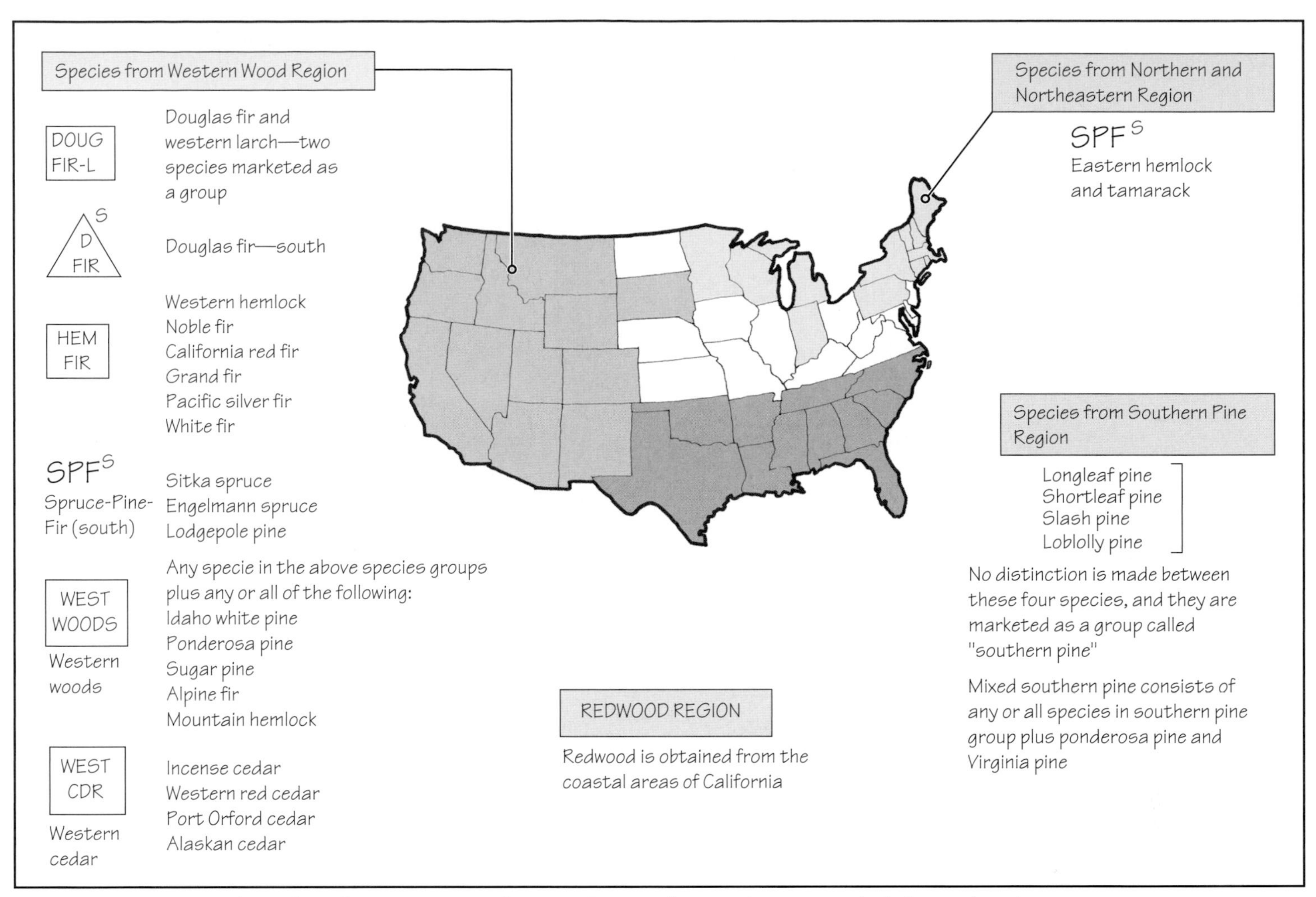

FIGURE 11.11 Major softwood producing regions in the United States, showing the states included in each region.

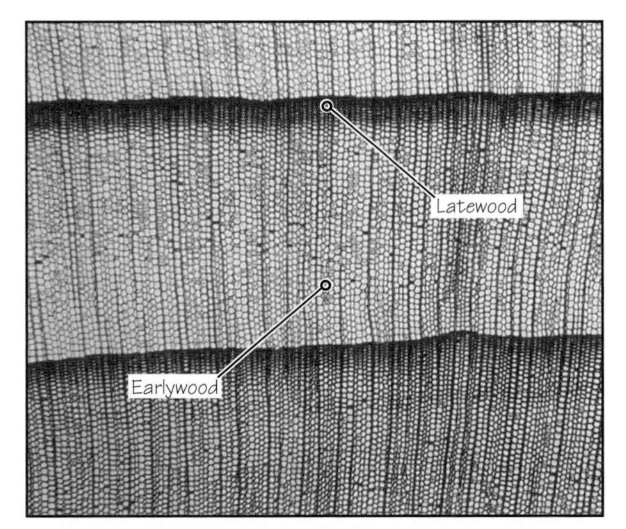

FIGURE 11.12 Magnified view of a softwood's growth rings showing the cellular structure. The lighter-colored region, with larger cell cavities and thinner walls of a growth ring, represents earlywood. The darker-colored region with smaller cell cavities and thicker walls represents latewood. (Photo courtesy of the Forest Products Laboratory, U.S. Department of Agriculture.)

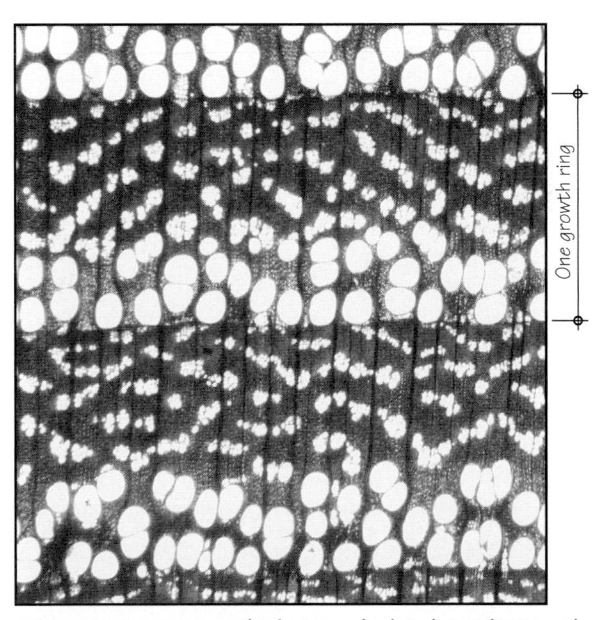

FIGURE 11.13 Magnified view of a hardwood's growth rings showing it has a more complex microstructure than that of softwood. (Photo courtesy of the Forest Products Laboratory, U.S. Department of Agriculture.)

Wood's Anisotropicity

Despite the differences previously described, the microstructures of both softwoods and hardwoods are essentially similar and consist of a cluster of longitudinally oriented hollow tubes. This cluster resembles a bundle of drinking straws glued together and reinforced with a few transversely oriented tubes (ray cells), Figure 11.14. This simplification clarifies some of the important properties of wood.

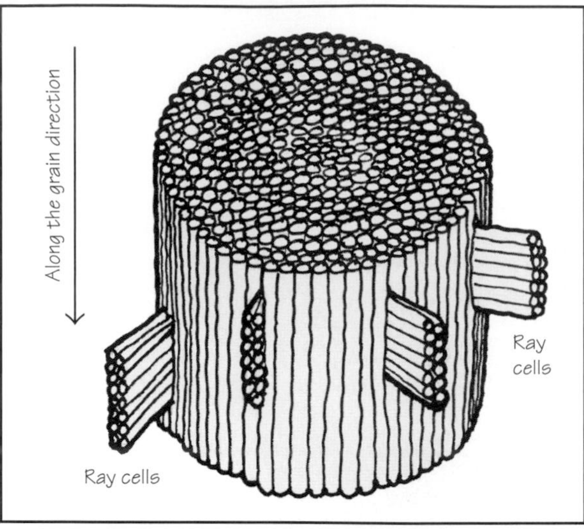

FIGURE 11.14 A highly simplified version of the cellular structure of wood—consisting mainly of hollow tubes oriented along the length of the member, resembling a bundle of straws glued together.

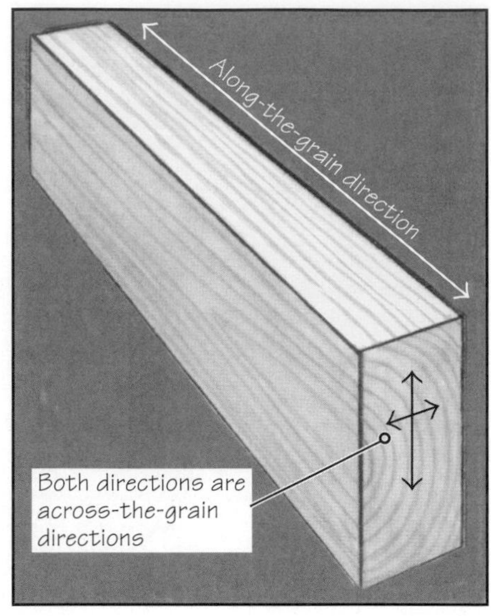

FIGURE 11.15 Along-the-grain and across-the-grain directions in lumber.

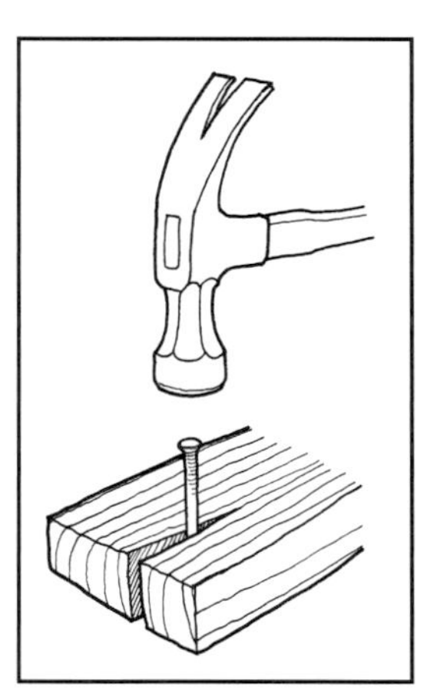

FIGURE 11.16 Lumber splits easily when nailed close to an end because of its low tensile strength across the grain, resulting from the weak glue bond between fibers.

For instance, a piece of lumber is much stronger along the grain (parallel to the axis of cells) than across the grain (perpendicular to the axis of cells), Figure 11.15. In fact, lumber's tensile strength across the grain is so small that it is neglected, as if the individual straws were not glued together. Thus, unlike steel and concrete, which are isotropic materials (having the same properties in all directions), wood is an anisotropic material.

The weak glue bond between individual cells (drinking straws) explains why a lumber piece will easily split along the grain when nailed near an end, Figure 11.16. For the same reason, wood is weaker in shear along the grain than across the grain, because individual cells can easily slip against each other.

As we will see in Section 11.5, a piece of lumber shrinks substantially across the grain due to the cell walls becoming thinner as water leaves them. However, along the grain (lengthwise), the departure of water has very little effect. Thus, lumber shrinks negligibly along the grain.

Uses of Softwoods and Hardwoods

Softwood trees mature for foresting two to three times faster than hardwood trees. Nearly 75% of North American forests contain softwoods. Because of their faster growth and, hence, their relative abundance, softwoods are commonly used for structural framing of wood buildings in North America. Thus, in most buildings, floor joists, rafters, ceiling joists, studs, sheathing, and so on, are of softwood lumber.

Hardwoods are commonly used for finish flooring, where their higher densities (giving greater abrasion resistance) are useful. Because hardwoods have a more interesting and varied grain structure, they are also used for wall paneling, trims, cabinets, furniture, and so on. Table 11.2 gives some of the major softwood and hardwood species used in building construction in North America.

TABLE 11.2 MAJOR SOFTWOOD AND HARDWOOD LUMBER SPECIES USED IN THE UNITED STATES

Softwood lumber		Hardwood lumber
Framing lumber (studs, floor and ceiling joists, rafters, headers, etc.) Southern pine, western lumber, north and northeastern lumber, Canadian lumber (See Figure 11.11)	**Roof shingles, fencing, etc.** Decay-resistant woods such as redwood, western red cedar, white cedar, and southern cypress.	**Finish flooring** Red oak and white oak are most common. Sugar maple, pecan, hickory, and teak are also used.
Siding, paneling, fascia boards, etc. Almost all species used for framing can be used for these nonstructural applications.	**Finish flooring** Denser species such as southern pine and douglas fir.	**Paneling and Molding** Ash, beech, hickory, red oak, white oak, sugar maple, pecan, black walnut, and teak.
	Doors, windows, and cabinets Ponderosa pine, sugar pine, and Idaho white pine.	

PRACTICE QUIZ

Each question has only one correct answer. Select the choice that best answers the question.

1. Which of the following two terms are used synonymously?
- **a.** Lumber and wood
- **b.** Lumber and laminated veneer lumber
- **c.** Lumber and solid sawn lumber
- **d.** All the above

2. The approximate density of lumber is
- **a.** 35 pcf.
- **b.** 35 psf.
- **c.** 55 pcf.
- **d.** 55 psf.
- **e.** 75 pcf.

3. The cellular structure of wood consists mainly of
- **a.** hollow, small, approximately spherical cells.
- **b.** solid, small, approximately spherical cells.
- **c.** hollow, small, approximately ellipsoidal cells.
- **d.** hollow, long tubular cells.
- **e.** either (c) or (d).

4. If we cut a cross section through a tree trunk, we will see a number of rings. The most appropriate term for these rings is
- **a.** annual rings.
- **b.** growth rings.
- **c.** concentric rings.
- **d.** eccentric rings.
- **e.** all the above.

5. The walls of wood cells consist primarily of
- **a.** calcium.
- **b.** glucose.
- **c.** cellulose.
- **d.** copper.
- **e.** lignin.

6. The terms *sapwood* and *heartwood* refer to
- **a.** two different wood species.
- **b.** two different subspecies of wood.
- **c.** two different types of wood cells.
- **d.** two different parts of a growth ring.
- **e.** two different parts of the same tree.

7. The difference between *softwoods* and *hardwoods* is based on
- **a.** strength of wood.
- **b.** density of wood.
- **c.** abrasion resistance of wood.
- **d.** botanical characteristics.
- **e.** decay resistance of wood.

8. Which of the following species of wood is not a softwood?
- **a.** Pine
- **b.** Fir
- **c.** Cedar
- **d.** Oak
- **e.** Redwood

9. Which of the following species of wood is not a hardwood?
- **a.** Teak
- **b.** Balsa wood
- **c.** Rosewood
- **d.** Oak
- **e.** Redwood

10. Spruce is a softwood species.
- **a.** True
- **b.** False

11. Lumber used for structural framing of buildings in North America is generally derived
- **a.** from softwood species.
- **b.** from hardwood species.
- **c.** from both softwoods and hardwoods, depending on the price of lumber.
- **d.** from both softwoods and hardwoods, depending on the time of the year.

12. Hardwoods are generally used for fine-quality flooring, furniture, and wall paneling.
- **a.** True
- **b.** False

13. In the United States, most structural lumber is grown in the
- **a.** northern region of the United States.
- **b.** western and southern regions of the United States.
- **c.** eastern and northeastern regions of the United States.
- **d.** central region of the United States.
- **e.** none of the above, because most structural lumber is imported from Mexico.

14. Redwood is grown mainly in
- **a.** Texas.
- **b.** Florida.
- **c.** Oklahoma.
- **d.** Arizona.
- **e.** California.

15. Which of the following species of wood is naturally decay resistant?
- **a.** Pine
- **b.** Fir
- **c.** Spruce
- **d.** Cedar
- **e.** Hemlock

16. In a 2 × 6 stud, the wood grain is parallel to the
- **a.** vertical direction.
- **b.** 2-in. dimension.
- **c.** 6-in. dimension.

17. Wood is
- **a.** stronger along the grain than across the grain.
- **b.** stronger across the grain than along the grain.
- **c.** equally strong in both directions of the grain.

Answers: 1-c, 2-a, 3-d, 4-b, 5-c, 6-e, 7-d, 8-d, 9-e, 10-a, 11-a, 12-a, 13-b, 14-e, 15-d, 16-a, 17-a.

11.4 FROM LOGS TO FINISHED LUMBER

The conversion of logs into finished lumber takes place in lumber mills, which are usually located close to the forests. The conversion process consists of four basic steps:

- Transportation of logs from the forest to the mill, Figure 11.17
- Debarking of logs, Figure 11.18
- Sawing the debarked logs into lumber
- Surfacing lumber members smooth

In a modern mill, the sawing operation is highly automated and is typically controlled by one person. The sawyer's chamber is equipped with computer monitors to help determine

(a)

(b)

FIGURE 11.17 **(a)** An overview of a lumber mill. **(b)** Transportation of logs from the forest to lumber mill.

(a)

(b)

FIGURE 11.18 **(a)** Logs moving on a conveyor belt for debarking. **(b)** Debarked logs moving further on the belt for sawing.

the most marketable quantity of wood from a log. The chamber's height permits the sawyer to physically observe the sawing operations.

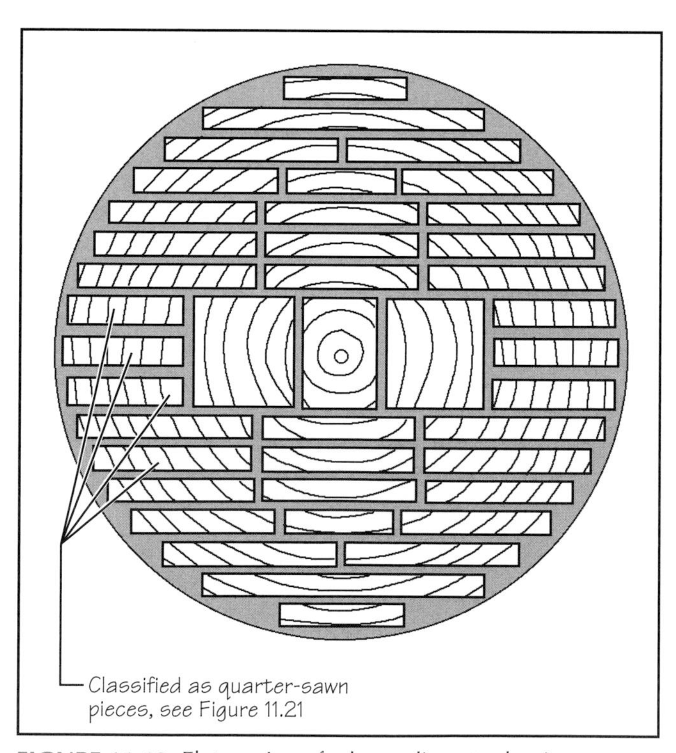

FIGURE 11.19 Flat sawing of a large-diameter log in two directions yields several pieces of dimension lumber and timbers. In a small-diameter log, the sawing may be along one direction only (through and through sawing), giving primarily dimension lumber. (See Section 11.9 for dimension lumber.)

Flat-Sawn and Radial-Sawn Lumber

Two methods are commonly used for sawing the logs: *flat sawing* and *radial sawing.* The lumber pieces so obtained are referred to as *flat-sawn lumber,* and *radial-sawn lumber,* respectively.

In flat-sawn lumber, the growth rings in some pieces run parallel to the edges of the cross section. In other pieces, the rings are diagonally oriented to the edges, and in a few other pieces, the rings are perpendicular, Figure 11.19. Another term used for flat sawing is *plain sawing.*

In radial sawing, the log is first converted into four pieces through the center of the log. That is why the term *quarter sawing* is used interchangeably with radial sawing. Each quarter is then sawed radially into required sizes, Figure 11.20. In a quarter-sawn piece, obtained from the log's center, the growth rings are perpendicular to its wider faces.

In a radial-sawn piece farther from the log's center, the angle between the rings and the wider faces of the cross section is less than 90°. Some lumber experts define a quarter-sawn piece as the one whose growth rings make an angle of 45° or more with the wider face of its cross section, Figure 11.21 (even if the piece has been obtained from flat sawing). Thus, some pieces in Figure 11.19 are classified as quarter-sawn. If the angle is less than 45°, it is classified as flat-sawn.

Quarter sawing is more complex and usually more wasteful. Hence, most framing lumber is obtained by flat sawing. As we will see later in this section, quarter-sawn lumber is more dimensionally stable. It resists wear and abrasion more evenly because denser latewood is more uniformly exposed on the surfaces of the lumber piece, Figure 11.22. Quarter-sawn lumber is commonly specified for high-grade finish floors.

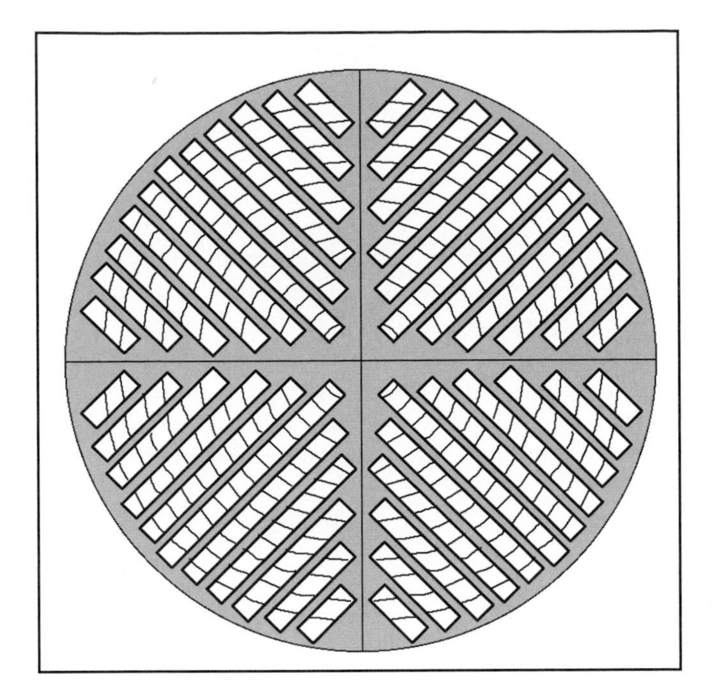

FIGURE 11.20 Quarter (radial) sawing of a log. Note that in comparison with flat sawing, radial sawing is more wasteful of wood.

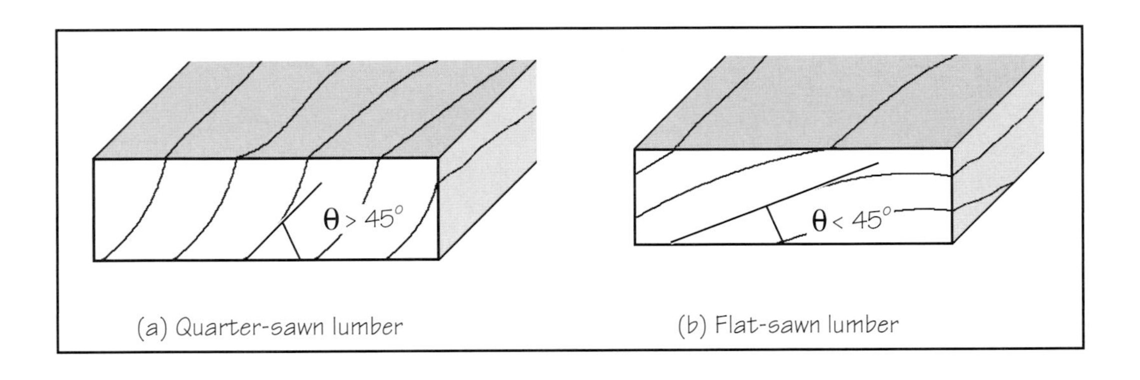

FIGURE 11.21 Theoretical definitions of quarter-sawn and flat-sawn lumber members.

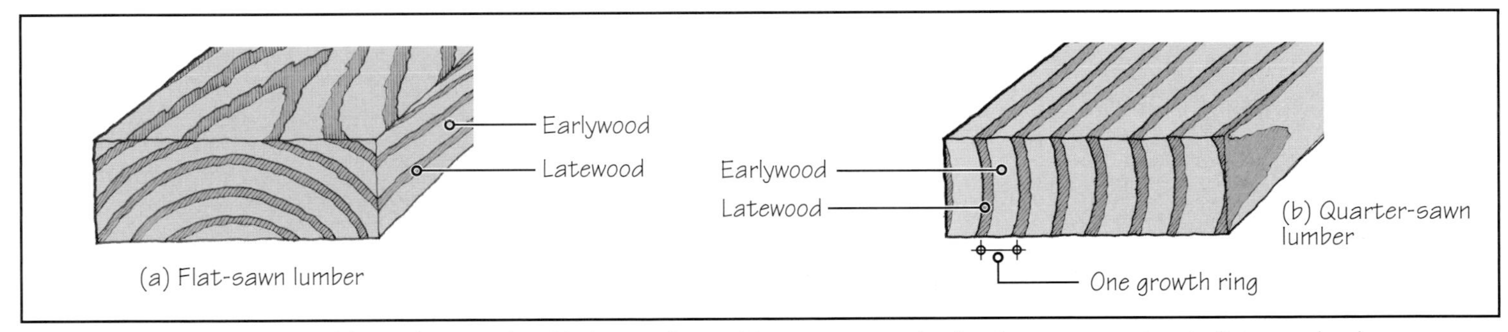

FIGURE 11.22 Distribution of denser latewood and lighter earlywood in quarter-sawn lumber is more even than in flat-sawn lumber.

11.5 DRYING OF LUMBER

After sawing, the lumber is dried and subsequently surfaced smooth before being shipped to lumber yards for sale. Although the drying of wood begins from the time the tree is cut and converted to logs, the moisture content prior to sawing far exceeds that required for use in buildings. Therefore, lumber must be dried (or seasoned). The term *seasoning* implies a controlled rate of drying. Controlled drying minimizes the separation of fibers that occurs during the drying process.

The moisture content (MC) in a piece of wood is the weight of water in wood divided by its oven-dry (completely dry) weight, expressed as a percentage:

$$MC = \frac{\text{weight of water in wood}}{\text{weight of oven-dry wood}} \times 100$$

Let us assume that a piece of wood weighs 6 lb. Now let us dry it until all its water has evaporated and then weigh it again. If the weight of this dry piece is 4.5 lb (implying that the weight of water in the wood was 1.5 lb), the moisture content in wood before drying, as obtained from the preceding equation, is

$$MC = \frac{1.5}{4.5} \times 100 = 33.3\%$$

FIGURE 11.23 Free water and bound water in wood cells. (Photo courtesy of Southern Forest Products Association.)

The moisture content in the tree before it is cut varies with the species and may be as high as 200%. A part of this water is contained in cell walls, which are in their fully swollen state, and a part of the water is contained in cell cavities. During the drying process, the cavity water, referred to as the *free water,* evaporates first.

After all the free water has evaporated, the cell wall water, called the *bound water,* begins to evaporate, Figure 11.23. The stage at which all the free water has evaporated and the bound water has just begun to evaporate is referred to as the fiber saturation point (FSP). At this point, the cell walls are in their fully swollen state, just as they were when there was free water in the cells.

GREEN VERSUS DRY LUMBER

The average moisture content in lumber at FSP is approximately 30%. It is only when the moisture content in wood falls below FSP (i.e., when the bound water begins to evaporate) that the water leaves the cell walls and the wood begins to shrink. No shrinkage of wood occurs if its moisture content is greater than fiber saturation point. Dimensional changes in lumber (shrinkage and swelling) are discussed later in this section.

Lumber is typically seasoned in the mill to a moisture content of 19% or less—a value that distinguishes *dry* lumber from *green* lumber. Thus, a piece of lumber whose moisture content is less than or equal to 19% is referred to as dry lumber, and that whose moisture content is 20% or greater is referred to as green lumber, Figure 11.24.

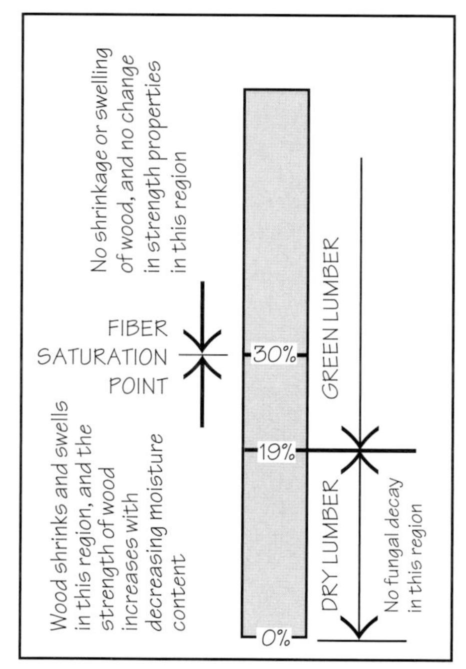

FIGURE 11.24 Definitions of dry lumber and green lumber in terms of lumber's moisture content.

The value of 19% has been selected because lumber with a moisture content of 19% or less is not susceptible to fungal decay. Additionally, the strength properties of lumber do not improve significantly below 19% moisture content. Note that the strength and stiffness of lumber increase as the moisture content decreases beyond the fiber saturation point. Seasoning also helps to reduce transportation costs because a drier piece of lumber weighs less.

AIR SEASONING VERSUS KILN SEASONING

Two processes are used to season lumber: *air seasoning* and *kiln seasoning.* In air seasoning, lumber pieces dry naturally. They are simply stacked one over the other in piles in such a way that air can circulate freely around them. The piles are kept under a roof to prevent the direct action of rain, snow, or sun.

Air seasoning is a slow process and can take several months, depending on the thicknesses and species of lumber. Kiln seasoning is much faster. The high temperature in the kiln also kills any fungus that may be present in the living tree. In kiln seasoning, the lumber is stacked in a chamber, through which warm air with a controlled amount of humidity (steam) is circulated. Figure 11.25 shows a typical seasoning kiln.

A piece of lumber (air or kiln) seasoned to a moisture content of 19% or less is identified as *surfaced dry* (S-DRY). If the moisture content at the time of surfacing is more than 19%, the lumber is identified as *surfaced green* (S-GRN). Some mills season lumber to a moisture content of 15% or less. The identification in that case is MC 15, or KD 15; KD stands for *kiln dried.* As we will see in Section 11.9, the identification of its moisture content is an important component of lumber grade.

Lumber marked S-DRY, MC 15, or KD 15 are considered to be equivalent for most structural purposes. However, for nonstructural uses such as for wall and ceiling paneling and flooring, where a greater dimensional stability is needed, a lower moisture content is desirable. Additionally, it is generally immaterial whether lumber is kiln or air dried.

NOTE

S-DRY, MC 15, and KD 15 are considered equivalent moisture-content designations for framing lumber. However, sawn lumber used in glulam beams and wood I-joists is generally seasoned to a moisture content of less than or equal to 15%.

SHRINKAGE AND SWELLING OF LUMBER

As stated previously, wood begins to shrink only when the water from cell walls begins to evaporate, that is, when the moisture content in wood is less than the fiber saturation point. The drying shrinkage of cell walls affects the dimensions of lumber's cross section only, not its length. Consequently, a vertical lumber member, such as a stud, shortens negligibly along its height on drying. Similarly, the shrinkage of a floor joist is negligible along its length. In other words, wood shrinks only across the grain, not along the grain, Figure 11.26(a).

(a)

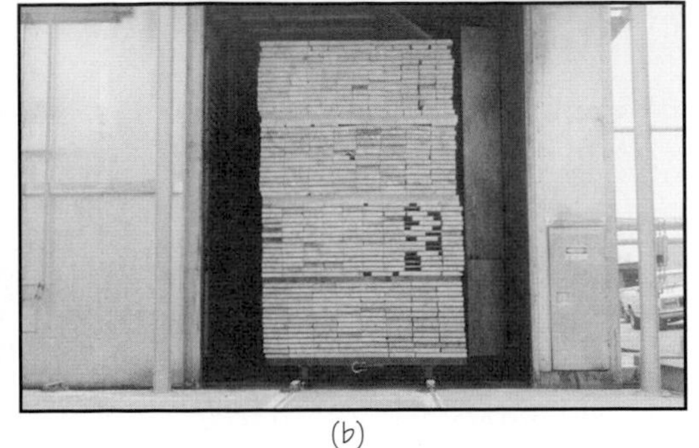
(b)

(c)

FIGURE 11.25 **(a)** A stack of unseasoned sawed lumber stacked outside the kiln, ready to be moved into the seasoning kiln seen in the background. **(b)** A view of a seasoning kiln with its doors open. **(c)** Seasoned lumber stored in the yard.

Cell walls not only shrink on drying, but also expand when moisture content increases. Being hygroscopic, cell walls respond to the water vapor in air. Therefore, a lumber cross section will generally shrink and swell due to the changes in the relative humidity of air. The shrinkage and swelling is much larger along the direction of growth rings in lumber (tangential direction) than perpendicular to the direction of growth rings (radial direction).

Consequently, quarter-sawn lumber shrinks uniformly, Figure 11.26(b) and tangentially (flat-) sawn lumber will generally cup, bow, or twist on drying. This is another reason why radially sawn lumber is preferred for wood flooring.

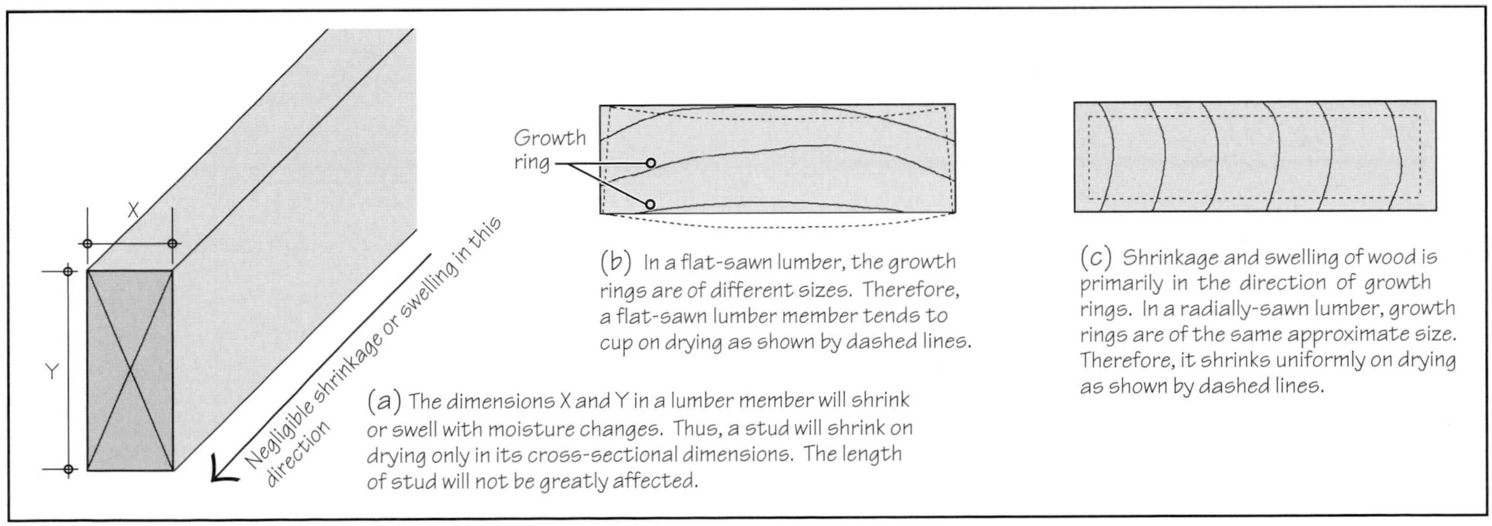

FIGURE 11.26 Drying shrinkage of a lumber member.

EXPAND YOUR KNOWLEDGE

Moisture Content of Wood and Relative Humidity of Air

A piece of lumber swells and shrinks with changes in the relative humidity (RH) of surrounding air and gradually reaches an equilibrium moisture content (EMC). Although the relationship of EMC with RH varies somewhat with the species, the following represents a good average.

Based on this relationship, the average moisture content in framing lumber lies between 5% and 14% depending on how arid or humid the location is.

RH	EMC	RH	EMC
0%	0%	75%	14%
25%	4.5%	90%	20%
50%	8.5%	100%	FSP approx. 30%

Source: Hadley, Bruce. *Understanding Wood.* Newtown, CT: Taunton Press, 1980, p. 69.

11.6 LUMBER SURFACING

Saw blades used for mass manufacturing are coarse-edged. Therefore, lumber obtained after sawing is quite rough. Because rough surfaces can cause injuries to framers, lumber used in conventional wood light framing is surfaced smooth.

Surfacing is done by high-speed planing machines—a replacement for the traditional hand-held carpenter's plane. Surfacing smooths lumber, rounds off the edges, and makes it more square, removing some of the distortions that have occurred during the seasoning process. That is why seasoning usually precedes surfacing. Lumber that has been surfaced before seasoning is identified as S-GRN (surfaced green).

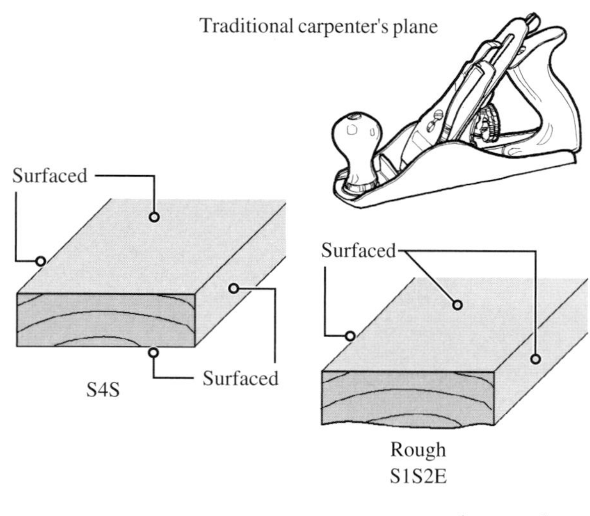

Framing lumber is surfaced on its longitudinal faces, not the ends, because the ends are usually resawn during construction. A piece surfaced on one wide face only is referred to as S1S (surfaced on one side); if surfaced on both wide faces, it is called S2S (surfaced on two sides); if surfaced on all four faces, it is called S4S.

A piece that is surfaced on one wide face and one narrow face (edge) is called S1S1E (surfaced on one side and one edge). Other designations are S2S1E and S1S2E. Lumber used for structural framing is typically surfaced S4S, fascia boards, as S1S2E, and so on.

In construction drawings, rough or surfaced lumber is graphically depicted by diagonal lines through its cross section. Two crossed diagonal lines indicate a continuous member, Figure 11.27(a), and one diagonal line indicates an interrupted (discontinuous) member, such as blocking or shims, Figure 11.27(b). Surfaced lumber is used for framing members, which are typically covered over by finish materials such as gypsum wallboard or plywood subfloor. Rough lumber is used for exposed lumber components (unprotected by gypsum board, etc.), particularly for beams and columns of large cross sections.

A *worked lumber* member is represented by growth ring symbols in the cross section, Figure 11.27(c). Worked lumber is lumber that has been dressed by additional machining to obtain the required profile. Worked lumber is used for interior finish members, such as door and window frames, trims and moldings, finish flooring, paneling, and so on. Figure 11.28 shows a typical section through the head of a door, illustrating the use of lumber's graphic symbols.

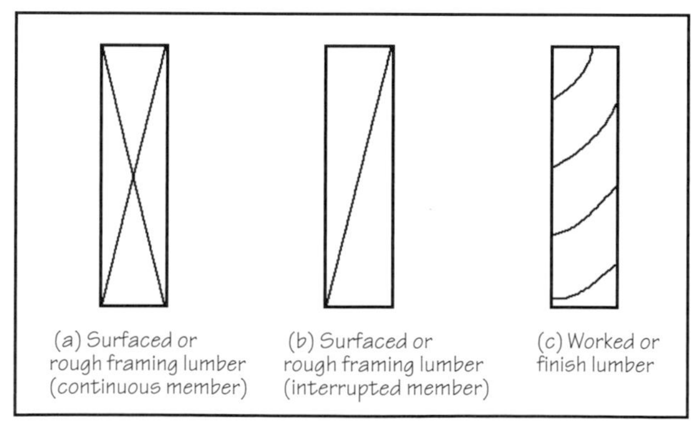

FIGURE 11.27 Graphic symbols for framing lumber and worked lumber.

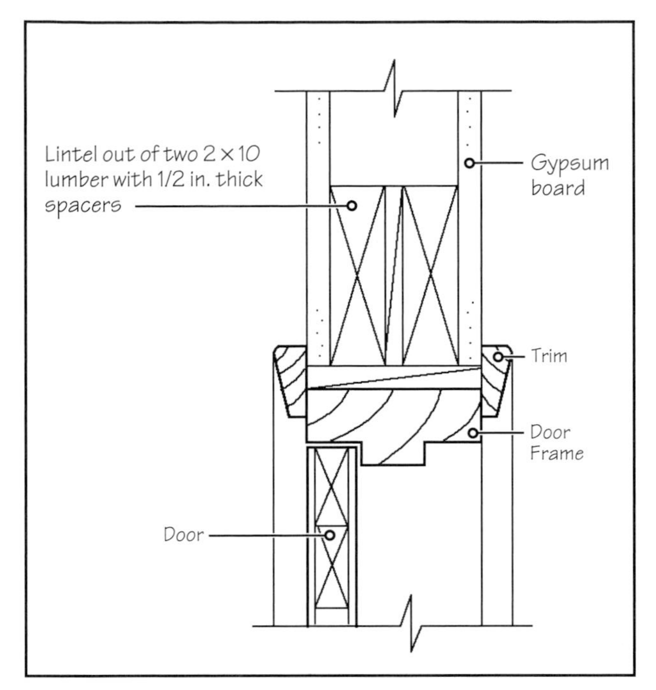

FIGURE 11.28 A section through the head of a door illustrating the use of graphic symbols for lumber members.

Finger-Jointed Lumber

Two lumber pieces can be glued together to produce a longer member using complementary finger joints on either side, Figure 11.29. By finger jointing, defects in a wood member are cut away and the good pieces are joined to obtain higher-grade lumber.

Because of high-strength and water-resistant glues, a finger-jointed member is as strong as a single-length (unjointed) member. Jointed and unjointed members can be used interchangeably.

FIGURE 11.29 Finger-jointing of lumber pieces.

11.7 NOMINAL AND ACTUAL DIMENSIONS OF LUMBER

In North America, a softwood lumber's cross section is specified by its nominal dimensions, rather than by its actual dimensions. Thus, when we say a lumber piece is 2 × 4 (two by four), we refer to its nominal dimensions and imply that its cross section measures 2 in. × 4 in. nominally. The actual dimensions of a 2 × 4 (surfaced dry) are $1\frac{1}{2}$ in. × $3\frac{1}{2}$ in.

In stating the nominal dimensions, the inch label is not used. For the actual dimensions, however, inch labels must be used. Thus, the size with and without the inch labels distinguishes the actual size from the nominal size.

Nominal dimensions correspond roughly to the wood in the log before it is sawn, seasoned, and surfaced. The difference between the two represents the amount of wood lost due to shrinkage, sawing, and surfacing. The relationship between nominal and actual dimensions of S-DRY lumber is shown in Table 11.3. Thus, a 2 × 10 (two by ten) dry lumber cross section actually measures $1\frac{1}{2}$ in. × $9\frac{1}{4}$ in.

Although the stated actual dimensions are reasonably accurate for construction purposes, wood cross sections do not have the same dimensional precision as steel cross sections, because wood swells and shrinks with changes in relative humidity in the atmosphere.

TABLE 11.3 NOMINAL AND ACTUAL SIZES OF S-DRY LUMBER

Nominal size	Actual size
1	$\frac{3}{4}$ in.
$1\frac{1}{2}$	$1\frac{1}{4}$ in.
2	$1\frac{1}{2}$ in.
3	$2\frac{1}{2}$ in.
4	$3\frac{1}{2}$ in.
5	$4\frac{1}{2}$ in.
6	$5\frac{1}{2}$ in.
8	$7\frac{1}{4}$ in.
10	$9\frac{1}{4}$ in.
12	$11\frac{1}{4}$ in.
14	$13\frac{1}{4}$ in.

11.8 BOARD FOOT MEASURE

In the United States, softwood lumber is sold by volume, similar to concrete.* The obvious volumetric measure is the cubic foot (or cubic yard; in the SI system, it is a cubic meter). However, a volumetric measure that is unique to lumber is the *board foot* (bd ft). One board foot is the amount of lumber contained in a 1-in.-thick board that measures 1 ft × 1 ft on its face. In other words, 1 ft^3 = 12 bd ft.

Thus, a 2 × 12 piece of lumber that is 20 ft long contains 40 bd ft. One hundred such pieces contain 4,000 bd ft. In calculating board feet, we use the nominal dimensions, not the actual dimensions, of lumber.

*Concrete is sold per cubic yard of green (wet) concrete.

The price of lumber is quoted in terms of 1,000 bd ft (MBF) and varies according to the lumber's cross-sectional dimensions and grade. Larger cross sections, longer lengths, and higher grades cost more. In a home-improvement building material store, lumber is typically sold per piece rather than by the board foot. In Europe and other countries, where the SI system prevails, lumber is sold by the cubic meter.

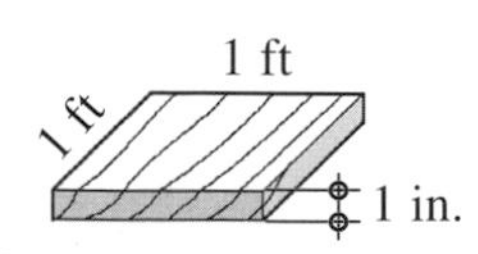

1 bd ft lumber

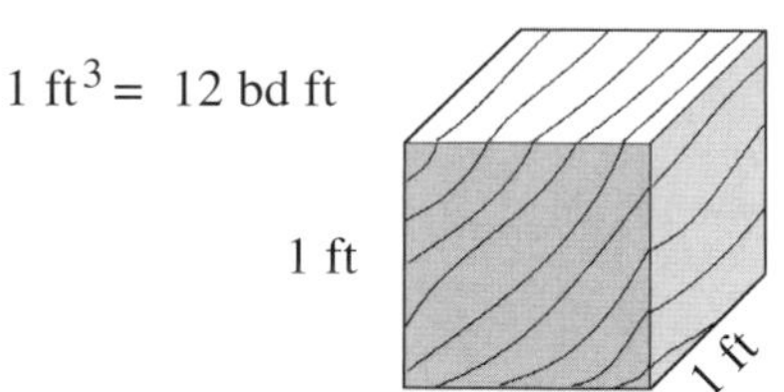

1 cubic ft lumber 1 ft

Example

Determine how much a builder will pay a lumber wholesaler for 900 pieces of 2 × 12 lumber, each 16 ft long. The price of lumber is \$800 per MBF.

Solution

$$\text{Total volume of lumber} = \frac{(16 \times 2 \times 12)(900)}{(12 \times 12)} = 2{,}400 \text{ ft}^3 = 2{,}400(12) \text{ bd ft} = 28.8 \text{ MBF}$$

$$\text{Total price of lumber} = 28.8(800) = \$23{,}040$$

PRACTICE QUIZ

Each question has only one correct answer. Select the choice that best answers the question.

18. Structural lumber is typically
a. flat-sawn.
b. quarter-sawn.
c. either (a) or (b), depending on the type of structural member.

19. Flat-sawn lumber is dimensionally more stable than quarter-sawn lumber.
a. True
b. False

20. Which of the following terms means the same as *quarter-sawn*?
a. Tangentially sawn
b. Radial-sawn

21. The weight of a piece of lumber is 2 lb. The weight of the same piece after it is fully dried is 1.8 lb. What was the original moisture content in the wood?
a. 7%
b. 9%
c. 11%
d. 15%
e. None of the above

22. Fungal decay in wood occurs when the moisture content in wood is
a. greater than or equal to 15%.
b. greater than or equal to 20%.
c. greater than or equal to 25%.
d. between 15% and 25%.
e. none of the above.

23. The moisture content in a piece of lumber has been measured to be 15%. According to wood industry, this piece of lumber will be classified as
a. dry.
b. green.
c. wet.
d. moist.
e. arid.

24. From the user's view point, the difference between MC 15 and KD 15 is generally ignored.
a. True
b. False

25. The term *S-DRY* implies that the lumber has been
a. stored in a dry climate.
b. sawed when its moisture content was ≤25%.
c. surfaced when its moisture content was ≤25%.
d. sawed when its moisture content was ≤19%.
e. surfaced when its moisture content was ≤19%.

26. The term S2S implies that the lumber has been
a. sawed twice.
b. sawed from two sides.
c. surfaced on two sides.
d. none of the above.

27. A 10-ft-long stud was installed in position when the wood's moisture content was 19%. During a long dry spell, the wood's moisture content became 8%. The new stud length will be
a. approximately 9 ft 10 in.
b. approximately 9 ft $10\frac{1}{2}$ in.
c. approximately 9 ft 11 in.
d. approximately the same as the initial stud length of 10 ft.
e. none of the above.

28. A piece of lumber has been specified as 2 in. × 4 in. in cross section. This refers to its
a. actual cross-sectional dimensions.
b. nominal cross-sectional dimensions.

29. The actual dimensions of 4 × 8 lumber are
a. $3\frac{1}{2}$ in. × $7\frac{1}{2}$ in.
b. $3\frac{1}{4}$ in. × $7\frac{1}{2}$ in.
c. $3\frac{1}{4}$ in. × $7\frac{1}{4}$ in.
d. $3\frac{1}{2}$ in. × $7\frac{1}{4}$ in.
e. none of the above.

30. The actual dimensions of a 2 × 12 lumber are
a. $1\frac{1}{2}$ in. × 11 in.
b. $1\frac{1}{2}$ in. × $11\frac{1}{2}$ in.
c. $1\frac{1}{2}$ in. × $11\frac{3}{4}$ in.
d. $1\frac{3}{4}$ in. × $11\frac{1}{4}$ in.
e. none of the above.

31. In calculating the board foot measure, we must use the actual dimensions of lumber, not its nominal dimensions.
a. True
b. False

32. A retailer of building materials purchased 1,000 pieces of 2 × 12 lumber, each 10 ft long. How many board feet did the retailer purchase?
a. 1,000 bd ft
b. 10 MBF
c. 20,000 bd ft
d. 100,000 bd ft
e. None of the above

Answers: 18-a, 19-b, 20-b, 21-c, 22-b, 23-a, 24-a, 25-e, 26-c, 27-d, 28-a, 29-d, 30-e, 31-b, 32-c.

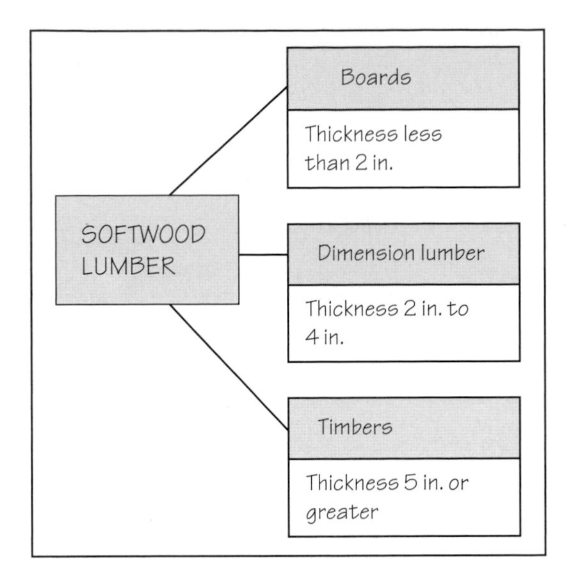

FIGURE 11.30 Size-based softwood lumber classification.

11.9 SOFTWOOD LUMBER CLASSIFICATION

A lumber piece is classified in one of the following three categories, Figure 11.30:

- Board lumber, or, simply, boards
- Dimension lumber
- Timbers

The distinction between them is based on the thickness of the member—thickness being the smaller of the two dimensions of its cross section. Thus, the thickness of a 4 × 10 member is 4 in. (nominal).

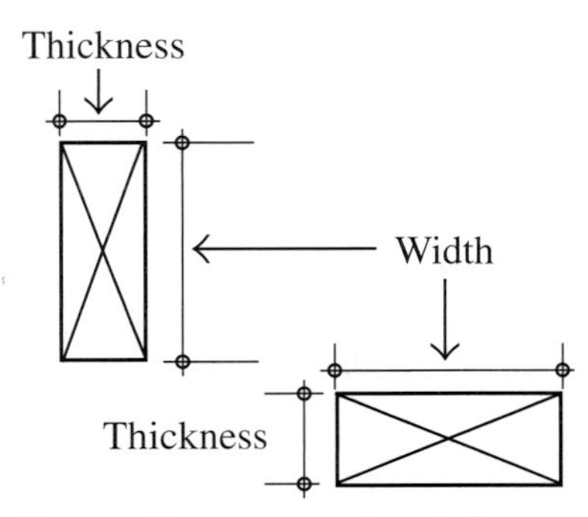

The smaller cross-sectional dimension of a lumber member is its thickness, and the larger dimension is its width.

Board lumber includes lumber whose thickness is less than 2 in. nominal. Thus, 1 × 4, 1 × 12, $1\frac{1}{2}$ × 6, and so on, are boards. Usually, 16 in. (nominal) is the maximum available width in board lumber. Board lumber is used in nonstructural applications, such as sheathing, fencing, and shelving.

Dimension lumber includes lumber whose (nominal) thickness is 2 to 4 in. The width of dimension lumber varies from 2 to 14 in. in steps of 2 in.—that is, 2 in., 4 in., 6 in., 8 in., and so on.* Thus, a 2 × 10, 3 × 6, 4 × 4, 4 × 8, 4 × 12, and so on, are all in the dimension lumber category.

Timbers includes lumber whose nominal thickness is 5 in. or greater. Thus, 5 × 8, 6 × 6, and 8 × 10 are timbers. Dimension lumber and timbers are graded based on strength.

DIMENSION LUMBER AND TIMBERS

Most of the lumber used in structural framing of wood buildings is dimension lumber. Out of this, the 2 × (referred to as "two-by,"or "2-by") lumber is most commonly used. In fact, a whole house or an apartment block can be framed, for all practical purposes, using 2-by lumber only, such as 2 × 4, 2 × 6, 2 × 8, 2 × 10, and 2 × 12. If a member thicker than 2 in. is required, two or three 2-by members can be nailed, screwed or bolted together.

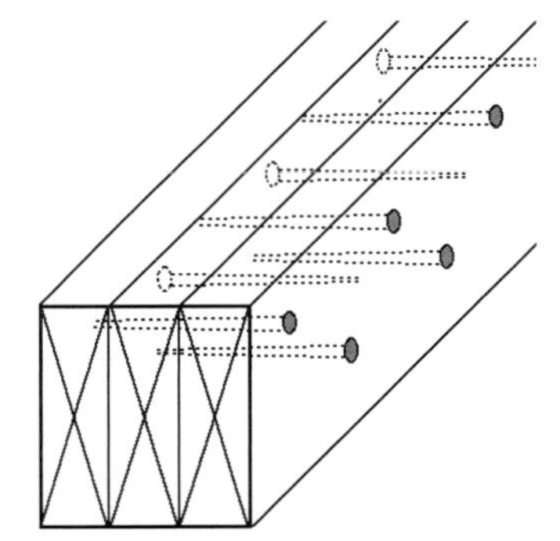

Two or three 2-by lumber members may be nailed, screwed, or bolted together to obtain a thicker member.

Dimension lumber is available in lengths of 8 ft, 10 ft, 12 ft, and so on, in steps of 2 ft up to 28 ft. Most lumber yards do not regularly stock lumber longer than 20 ft. Local availability should be checked before specifying a certain length.

Two-by lumber is typically available as S4S and in surfaced dry condition. Three- or 4-in. thick dimension lumber is available in dry or green condition. Timbers (5 in. and thicker) are generally shipped in green condition and are manufactured in S4S or rough condition.

11.10 LUMBER'S STRENGTH AND APPEARANCE

As a naturally grown material, wood does not have the same degree of uniformity as steel or concrete. Although two steel beams of the same length and cross section have identical load-carrying capacities, this is not so with lumber beams. Thus, two lumber beams of the same length and cross-sectional dimensions, even obtained from the same log, can have vastly different load-carrying capacities.

*A width greater than 12 in. may not be available in most lumber yards.

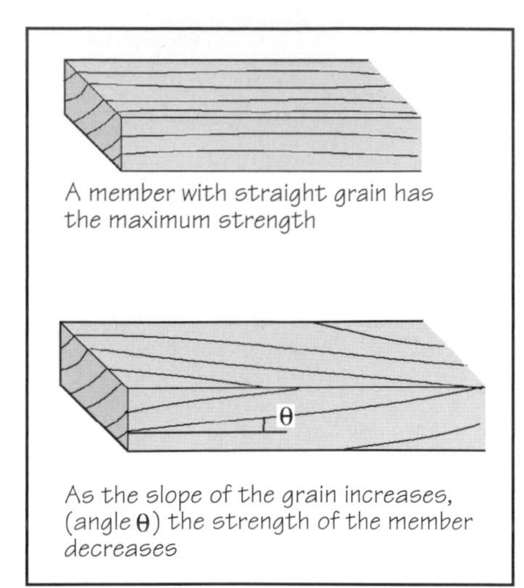

FIGURE 11.31 The slope of grain affects the lumber's strength. It is most easily observed by examing the longitudinal grain of the lumber in comparison with the edge of the member.

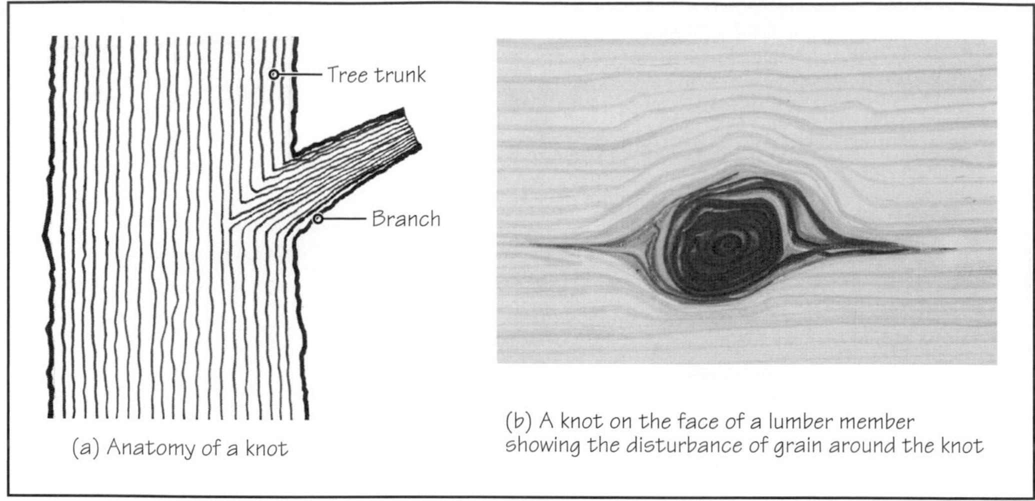

FIGURE 11.32 **(a)** Anatomy of a knot. **(b)** Grain orientation in the vicinity of a knot.

As previously discussed, the strength of a piece of lumber is affected by its species and is a function of specific gravity. Strength is also affected by its growth and manufacturing characteristics. These include slope of grain, knots, checks, shakes, splits, and wane.

Slope of Grain

An important factor that affects the strength of lumber is the slope of the grain. A piece of lumber whose grain runs parallel to the long axis of the member will have maximum strength for that species. As the slope of the grain with respect to the axis increases, the strength of the member decreases, Figure 11.31.

Knots

A knot occurs where a branch emerges from the tree trunk, Figure 11.32(a). The growth rings in the trunk of a tree are interlaced with the growth rings of its branches. Within the trunk, the growth rings of a branch form a cone whose end points toward the center of the trunk and whose diameter gradually increases toward the bark. Hence, the grain of wood in the vicinity of a branch is not straight, but highly disturbed from straightness, Figure 11.32(b). Consequently, the presence of a knot in a member reduces its strength. The more numerous the knots in a member, the lower its strength.

The adverse effect of a knot on the member's strength is more pronounced if the knot is a *loose knot.* A loose knot results when the branch dies during the tree's growth and the successive growth rings of the trunk encircle this dead branch. The death of the branch may be biologically caused, for example, by the overshadowing of the branch by the upper parts of the tree, or mechanically caused, such as by the effect of strong wind or hail storm.

A loose knot degrades the strength of wood substantially, and if rotting occurs in the knot, a *knot hole* results. A knot that is not loose, but is tightly intergrown with the adjoining tissue, is referred to as an *encased knot.*

Apart from the number, size, and type of knots (encased or loose), the location of the knot within the member also affects its strength. A knot located on or near the longitudinal axis of the member does not affect the strength as adversely as a knot located near the edges of the member, Figure 11.33. Remember that in a member subjected to bending (such as a beam, floor joist, rafter, etc.), the bending stresses are maximum at the extreme edges of the member and zero at the center (neutral axis).

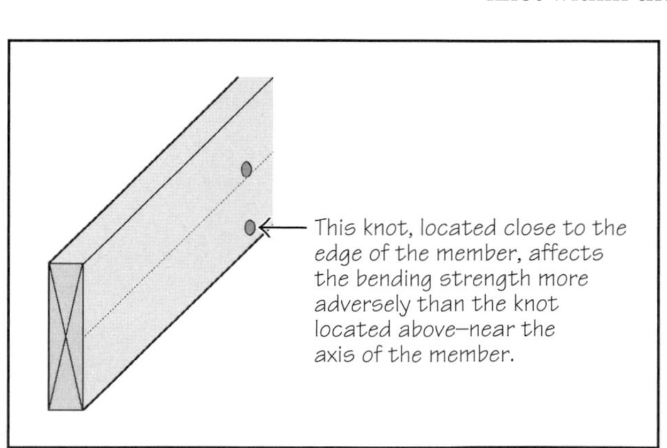

FIGURE 11.33 Effect of the location of knot on the bending strength of a member.

Checks, Shakes, and Splits

Checks, shakes, and splits are separations of wood fibers. A check is separation of wood fibers along the rays (perpendicular to growth rings). It is caused by the drying of wood and occurs at the ends of the member and also on its faces, Figure 11.34(a). It results from the fact that the surfaces of wood dry faster than its interior.

A shake is separation of wood fibers along the growth rings, Figure 11.34(b). It occurs during the growth of the tree and is not due to drying.

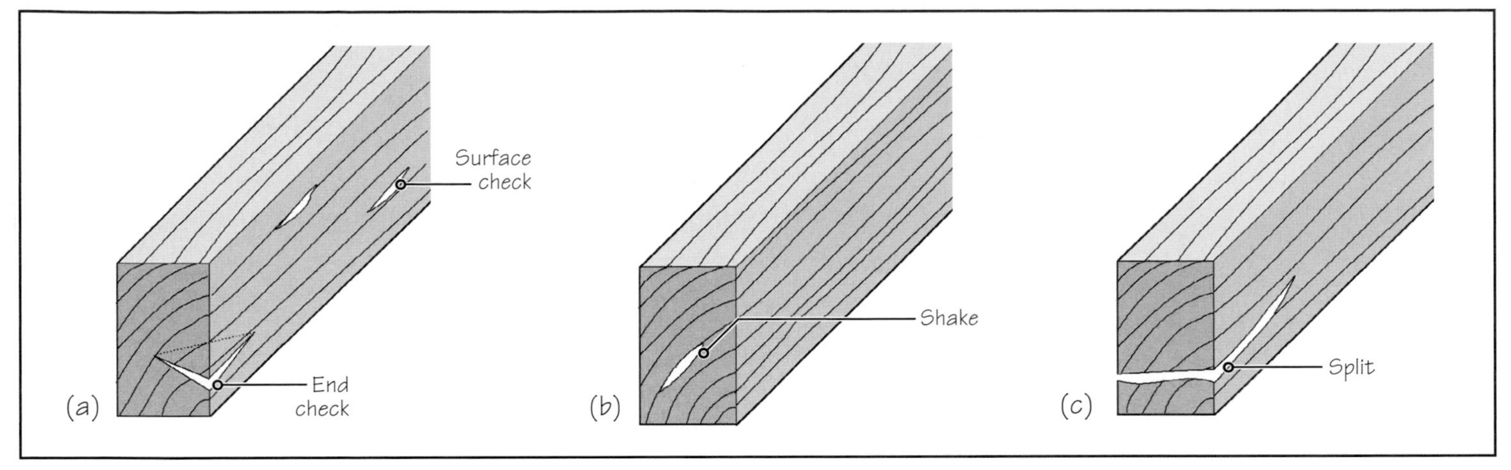

FIGURE 11.34 Check, shake, and split in lumber.

A split occurs at the ends of a member and is a complete separation of wood fibers through the entire end, Figure 11.34(c). It is believed to be caused by a weakness that occurred during the growth of the tree and was aggravated during drying.

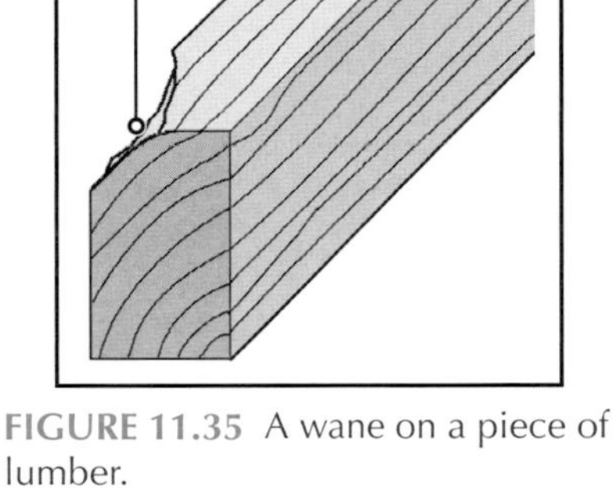

FIGURE 11.35 A wane on a piece of lumber.

WANE

A wane is the absence of wood or the presence of bark at the corner or the edge of a piece that results from the sawing process, Figure 11.35. Like the knots, checks, splits, and shakes, wanes also reduce the strength of the member. Therefore, their sizes and numbers are taken into account in the grading of lumber.

11.11 LUMBER GRADING

In North America, each piece of lumber used for structural framing is graded and so stamped, Figure 11.36. The grading is done by an independent inspection agency, which employs trained inspectors. Each piece is visually examined on all its four surfaces for growth characteristics, such as the slope of grain, knots, checks and shakes. An inspector's trained eyes can determine the grade of a piece of lumber within a few seconds. Such a grading method is referred to as *visual grading,* as opposed to *machine grading,* described later.

In visual grading, the inspection agency grades lumber according to the grading rules prescribed by a grading rules–writing agency. In North America, there are seven *grading rules–writing agencies* (six in the United States and one in Canada), which also function as inspection agencies, Table 11.4. In addition to rules–writing agencies, there are several nonrules–writing inspection agencies, which grade lumber as per the rules of the rules–writing agencies.

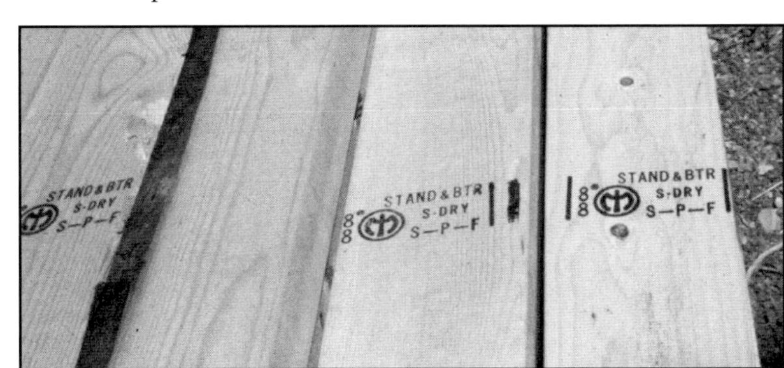

FIGURE 11.36 Grade stamps on lumber members.

VISUAL GRADING—DIMENSION LUMBER GRADES AND GRADE STAMPS

Although the seven rules–writing agencies have different grading rules for boards and timbers, their grading rules for dimension lumber are identical and are referred to as the

TABLE 11.4 GRADING RULES–WRITING AGENCIES

U.S. Agencies
Northeastern Lumber Manufacturers Association, Inc. (NELMA)
Northern Softwood Lumber Bureau (NSLB)
Redwood Inspection Service (RIS)
Southern Pine Inspection Bureau (SPIB)
West Coast Lumber Inspection Bureau (WCLIB)
Western Wood Products Association (WWPA)

Canadian Agency
National Lumber Grades Authority (NLGA)

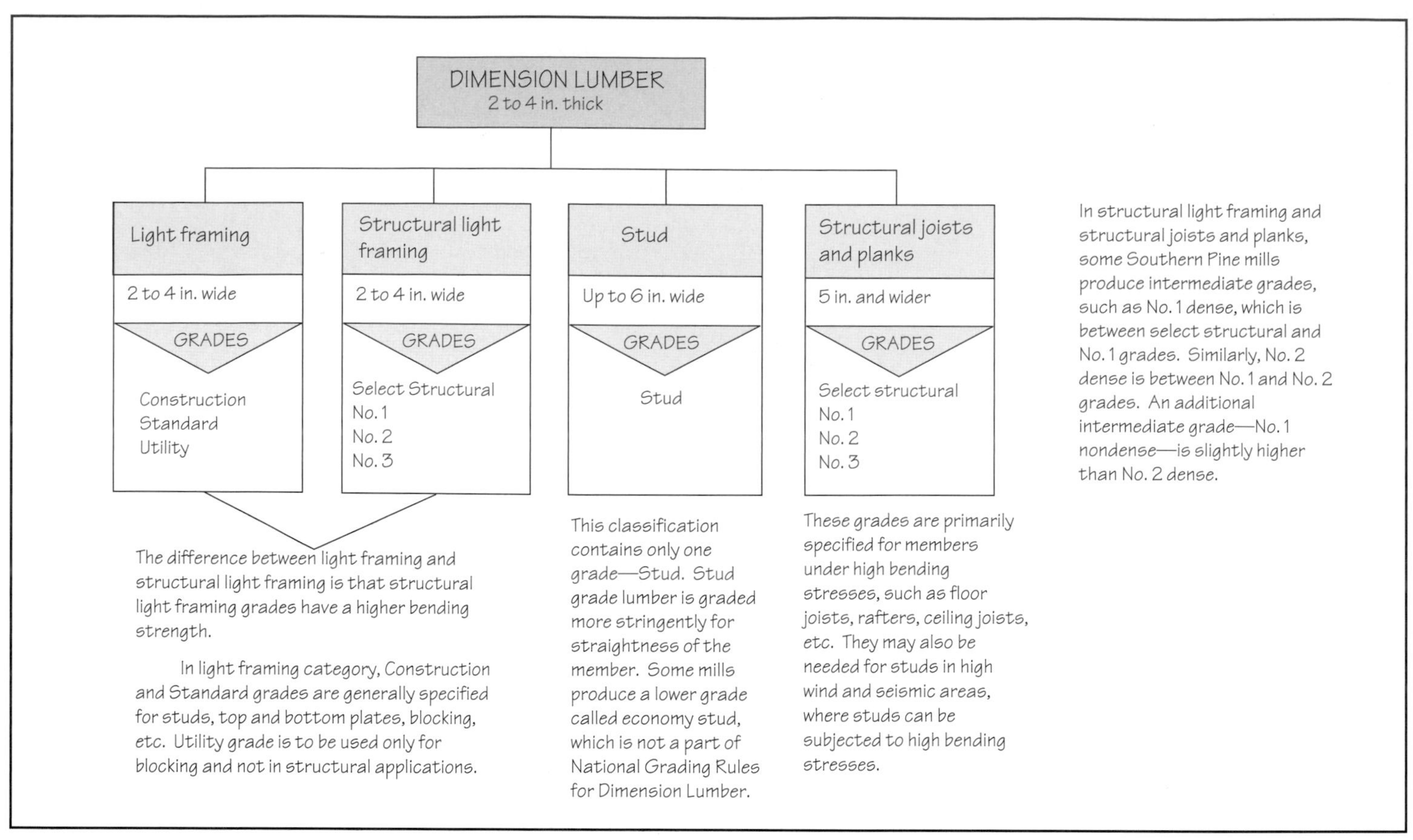

FIGURE 11.37 Various grades for dimension lumber.

National Grading Rules (NGR) for Dimension Lumber. According to these rules, dimension lumber can carry any one of the several grades shown in Figure 11.37.

Two typical grade stamps for dimension lumber are shown in Figure 11.38. Note that a grade stamp identifies the following items:

- Species of lumber
- Moisture content at the time of surfacing
- The mill that produced the piece
- The inspection (grading) agency
- The structural grade

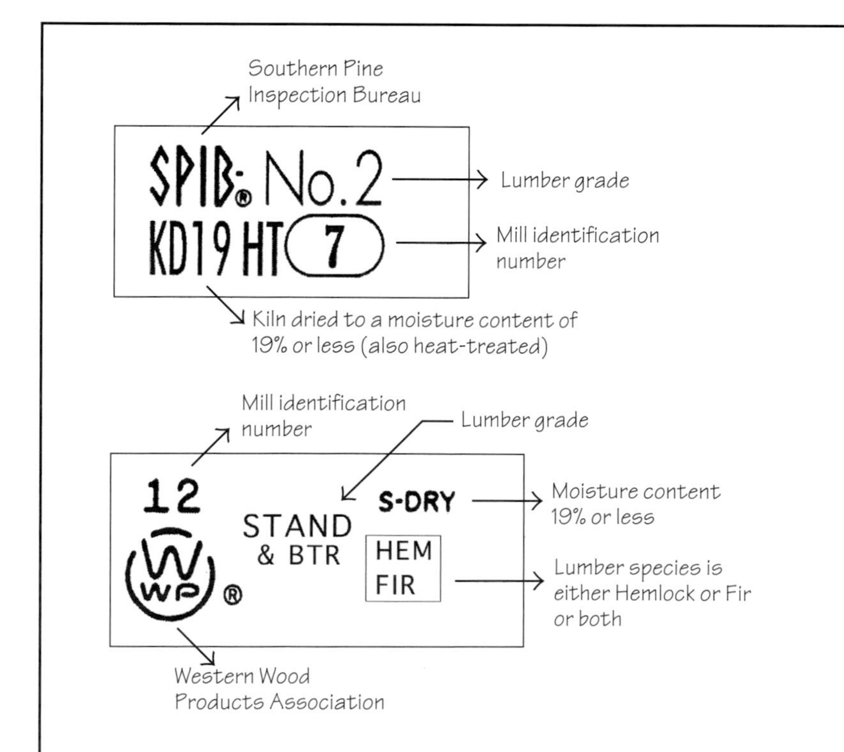

Southern pine grade stamp does not specify lumber species, since Southern pine is a mixture of four species: Longleaf pine, Shortleaf pine, Slash pine and Loblolly pine. These species are mixed together because they are almost equivalent in appearance and performance.

Grade stamps of other agencies include species information. However, two or more species may be grouped together. For example, the WWPA grade stamp given below shows that two species—Hemlock and Fir—have been combined together. Similarly, a species designation "S-P-F" means that Spruce, Pine and Fir have been grouped together.

To distinguish Canadian S-P-F designation from U.S. designation, S-P-F from the U.S. is called S-P-F (south) or S-P-F^S. Similarly, D-Fir (North) indicates that it is Canadian Douglas Fir.

Heat Treatment (HT) of Lumber is intended to eliminate pests and to make it suitable for packaging industry, which utilizes a great deal of dimension lumber. Heat treatment is not required for lumber used in construction. Heat treatment standards require that lumber be subjected to a kiln temperature of 133°F (56 °C) for a minimum of 30 minutes. Kiln drying of lumber for construction purposes is generally more stringent.

FIGURE 11.38 Typical grade stamps. A comprehensive grade stamp list of North American lumber is given at the end of this chapter.

TABLE 11.5 BASE ALLOWABLE STRESSES OF VISUALLY GRADED DIMENSION LUMBER

Species and grade	Size classification	Bending F_b	Tension parallel to grain F_t (psi)	Shear parallel to grain F_v (psi)	Compression perpendicular to grain $F_{c\perp}$(psi)	Compression parallel to grain $F_{c\parallel}$(psi)	Modulus of elasticity E (ksi)
Spruce-Pine-Fir (South)							
Select Structural	2 in.–4 in. thick	1,300	575	70	335	1,200	1,300
No. 1		850	400	70	335	1,050	1,200
No. 2		750	325	70	335	975	1,100
No. 3	2 in. and wider	425	200	70	335	600	1,000
Stud		575	250	70	335	600	1,000
Construction	2 in.–4 in. thick	850	375	70	335	1,200	1,000
Standard		475	225	70	335	1,000	900
Utility	2 in.–4 in. wide	225	100	70	335	650	900

Base Values: The values given here are base values. Design values are a function of base values and several other factors, such as lumber's moisture content, whether the member is a single-use member (such as a beam) or a repetitive-use member, type of load, and so on.

Source: American Forest and Paper Association (AF&PA) and American Wood Council (AWC): *National Design Specification (NDS) for Wood Construction.* Courtesy American Forest & Paper Association, Washington, DC, with permission.

Examples of grade stamps of various agencies for dimension lumber available in North America are given in the "Principles in Practice" at the end of this chapter.

Visual Grading—Allowable Stresses from Grade Stamp

Through extensive structural testing of solid lumber members, the allowable stress values and the modulus of elasticity of lumber of various species and grades have been compiled by the wood industry. Table 11.5 shows these values for only one species group—spruce-pine-fir. The purpose of this table is simply to illustrate the nature of the data available for the structural design of wood frame buildings.

Figure 11.39 shows the types of stresses that are critical for the structural design of various members of a conventional wood frame building.

Machine Stress-Rated Lumber

Lumber grading discussed previously is called *visual grading,* because the grading is done by visually examining the lumber piece by a trained inspector. Whereas the lumber used in the structural frame of a typical wood building is visually graded, the lumber used in more demanding structural applications, such as wood trusses, is machine graded. Machine-graded lumber is referred to as *machine stress-rated (MSR) lumber,* or simply *machine-rated lumber.*

In MSR lumber, the stiffness (modulus of elasticity) of a piece of lumber is determined by a nondestructive testing procedure, in which the piece is subjected to a given load and its

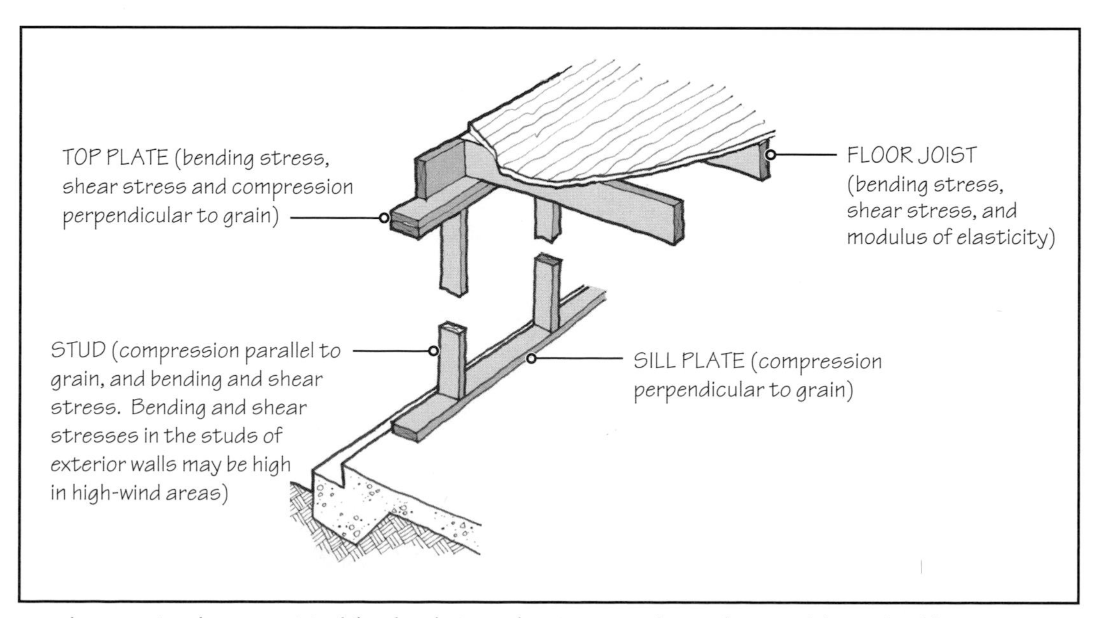

FIGURE 11.39 Structural properties that are critical for the design of various members of a wood frame building.

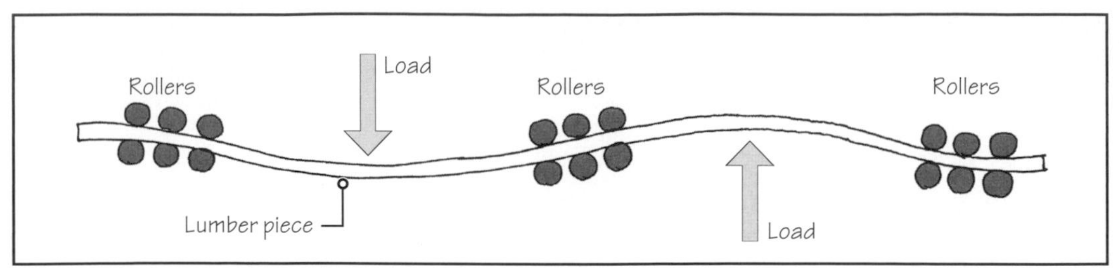

FIGURE 11.40 Outline of the equipment used for machine stress-rated lumber. Load-deflection information is fed into a computer, which automatically determines the values of the modulus of elasticity and allowable bending strength of lumber.

FIGURE 11.41 A typical grade stamp on a machine stress-rated (MSR) lumber.

deflection is noted, Figure 11.40. Because various structural properties of lumber are related to its stiffness, these can be obtained from the already-established statistical relationships.

Although a certain amount of visual inspection of lumber is necessary even in MSR lumber, machine grading is heavily automated, using high-speed equipment to obtain stiffness and other properties, including the affixing of the grade stamp.

An MSR grade stamp is similar to the grade stamp of visually graded lumber, except that in place of the grade mark in visually graded lumber, the MSR stamp gives E and F_b values. Thus, in the grade stamp of Figure 11.41, the lumber has the following properties: allowable bending stress (F_b) = 2,400 psi and modulus of elasticity (E) = 2.0×10^6 psi. Other structural properties of this piece, such as the allowable compressive stress, shear stress, and so on, are obtained from the tables prepared by the lumber industry.

11.12 DURABILITY OF WOOD

Like any other material, wood is subject to deterioration. Mechanical and chemical deterioration are the two most common ways in which a building material deteriorates. Mechanical deterioration of wood is due to physical wear, such as the abrasion caused by foot traffic or wheeling equipment on a floor. Other causes of mechanical wear, which occur primarily in wood exposed to the exterior climate, are (a) erosion of the material by wind and water,* (b) material fatigue caused by its repeated expansion and contraction, and (c) the deterioration caused by freeze-thaw cycles.

Chemical deterioration is due to the effect of chemicals (acids and alkalies) in the environment and the chemical breakdown of wood's constituents by the ultraviolet rays of the sun. Generally, mechanical and chemical deterioration of wood is of little importance, because the major cause of wood's deterioration is biological, referred to as *biodeterioration.*

BIODETERIORATION

Biodeterioration is caused by living organisms, which use wood as food. Two such groups of organisms consume wood:

- Fungi
- Insects

Deterioration of wood caused by fungi is called *fungal decay,* or, simply, the *decay of wood* (also called *rotting).* Land-based insects that consume wood are *termites,* and water-based wood-eating insects are called *marine borers.* In addition to termites and marine borers, there are carpenter ants, which, though not using wood as food, can nest inside it, thereby weakening the wood to some degree.

11.13 FUNGAL DECAY

Unlike other plants, fungi cannot manufacture their own food but live off the food produced by other plants. Because wood is mainly cellulose and lignin, it is an excellent source of food for these parasites.

Not all forms of fungi invade the cell structure of wood. Those that do are the ones that cause structural damage to wood (decay). As the decay-causing fungus consumes this food,

*The erosion due to wind is caused by dust and grit in the air. Thus, in terms of the weatherability of materials, wind behaves as a low-intensity sand blasting, and the rain serves as water blasting.

the strength of wood decreases, which is usually accompanied by a change in the external appearance of wood.

For instance, a type of fungus that consumes cellulose (white in color) makes wood turn more brown, and the fungus that consumes lignin (brown in color) makes wood turn whiter. *Brown rot* and *white rot* are the terms used for the two forms of decay.

Because fungus is a plant life, four factors are necessary for its survival and growth:

- Oxygen
- Mild temperature
- Water
- Food

When any one of these four elements is removed, fungal decay cannot take place. Thus, wood completely immersed in water at all times does not decay, due to the absence of oxygen. Similarly, fungal decay does not generally occur in extremely cold climates. The most favorable temperature range for fungal growth and survival is 70°F to 85°F (20°C to 30°C), as it is with most other plants. Thus, in a seasoning kiln, where high-temperature air is used to dry the wood, most types of fungi perish.

PREVENTING FUNGAL DECAY

Because it is not possible to control temperature and the supply of oxygen in building assemblies, the only methods available to prevent fungal decay are: (a) to control the supply of water in wood and (b) to make wood toxic to fungi.

It has been found that if the moisture content in wood is less than 20%, fungal decay does not occur. That is why 19% moisture content has been fixed as the dividing line between dry and green lumber. Specifying dry lumber and keeping it dry (below 19% moisture content) is the best insurance against decay. This fact highlights the importance of ventilating attic and crawl spaces in wood structures, because excessive moisture buildup in confined spaces can raise the moisture content of wood above 20%.

Although the lumber that is not exposed to rain or groundwater may be kept dry, this is not true with lumber exposed to a water source. Therefore, exterior lumber, lumber buried in the ground, or lumber close to or in contact with the ground will decay. Two options are available to prevent the decay of exposed lumber:

- Use of naturally decay-resistant species
- Use of preservative-treated lumber

Whereas the heartwood of a tree is generally more decay resistant than its sapwood, the heartwood of a few species, such as cedar, redwood, cypress, and black locust, have a high degree of decay resistance. These species may be used in decay-prone situations. Preservative-treated lumber is treated with chemicals that are toxic to fungi and insects.

11.14 TERMITE CONTROL

Most termite damage in buildings is caused by subterranean termites. Subterranean termites live in underground colonies. Due to their distaste for light, they normally cannot reach above-ground wood. However, they can build tubular mud tunnels over foundation walls and piers to reach the above-ground wood.

Because they can obtain water from the ground, termites do not have to depend on the water in the wood. Thus, although wood with a moisture content of less than 20% is not subject to fungal decay, it is vulnerable to termites if the termites can reach the wood.

Subterranean termites are more commonly found in warm climates; termite hazard is usually higher in warm, damp climates than in warm, dry climates. The hazard is relatively low in colder regions. Figure 11.42 shows the termite-infestation probability map of the United States.

REDUCING (OR PREVENTING) TERMITE DAMAGE

A multipart strategy is required to reduce or prevent termite damage in wood buildings. For wood members not buried in the ground, all or a combination of following strategies are generally recommended.

- Maintain distance between wood and ground
- Provide soil barrier, that is, chemical soil treatment
- Use naturally decay-resistant or preservative-treated wood

NOTE

Termites and Building Materials

Cellulose-based materials are a food for termites, but termites also attack plastics. They are known to chew through cable shields, plastic laminates, and plastic foam insulation. The use of below-ground plastic foam insulation is, therefore, not recommended in termite-infested regions.

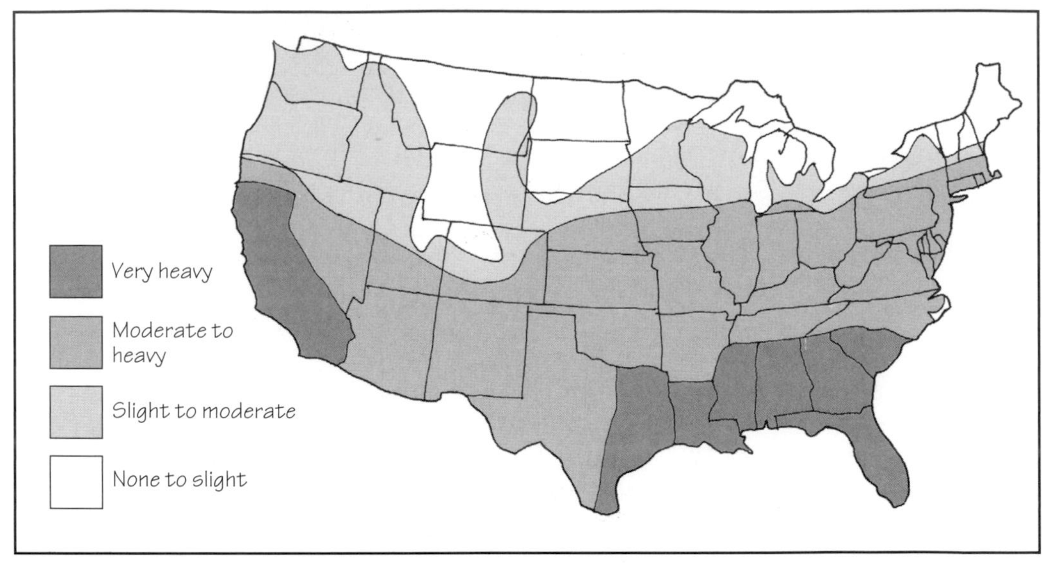

FIGURE 11.42 Termite infestation probability map of the United States. This map should be regarded only as approximate and must be verified with locally obtained information. (Map adapted from the 2006 International Residential Code. Copyright 2006. Falls Church, VA: International Code Council, Inc. Reproduced with permission. All rights reserved.)

- Use termite shield
- Inspection and remediation

The best line of defense against termites is to provide sufficient distance between the ground and the wood member, supplemented by frequent inspection to ensure that no mud tunnels exist. Figure 11.43 shows some of the commonly recommended construction details for preventing termite attack.

In most wood frame buildings, preservative-treated lumber is specified only for the sill plate, when anchored to concrete or masonry foundation. All other framing lumber is usually untreated lumber. Wood floor joists over a crawl space are required to be of preservative-treated or of decay-resistant species only if the distance between the bottom of the joists and the ground is less than 18 in. Wood beams less than 12 in. above ground must satisfy the same criterion.

Treating the soil around foundation walls and the soil below concrete slabs-on-grade or basement slabs provides additional protection against termites. Although the method of application of the chemicals (e.g., aldrin or dieldrin—0.5% in water emulsion) is fairly sim-

NOTE

Importance of a Termite Barrier

Experts estimate that termites can go through a space as small as $\frac{1}{24}$ in. Therefore, a concrete slab-on-grade should be without visible cracks. Hollow concrete masonry foundations must be fully grouted, and the termite shield must be carefully soldered at all joints.

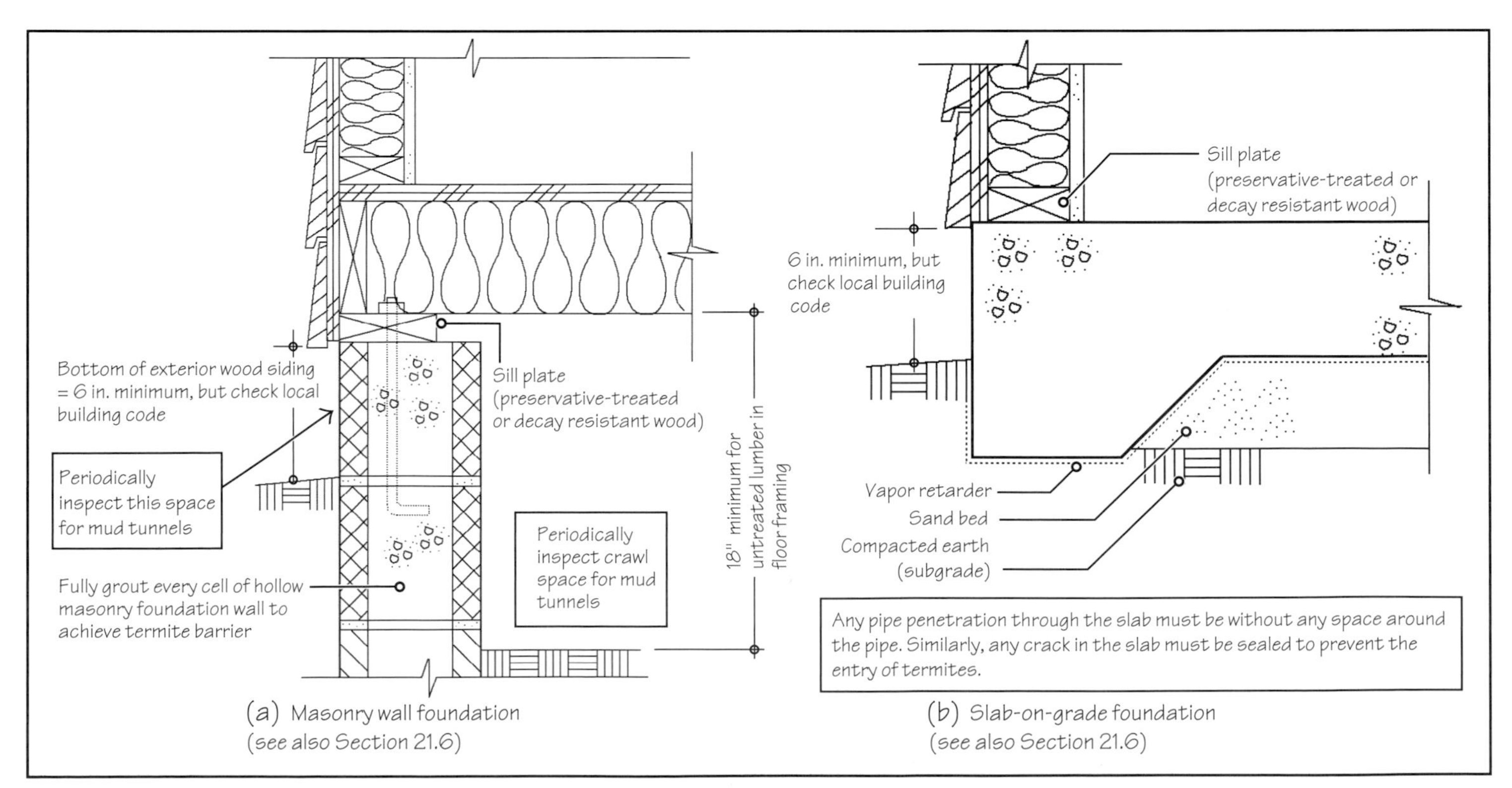

FIGURE 11.43 The use of preservative-treated or decay-resistant lumber in a conventional wood frame building.

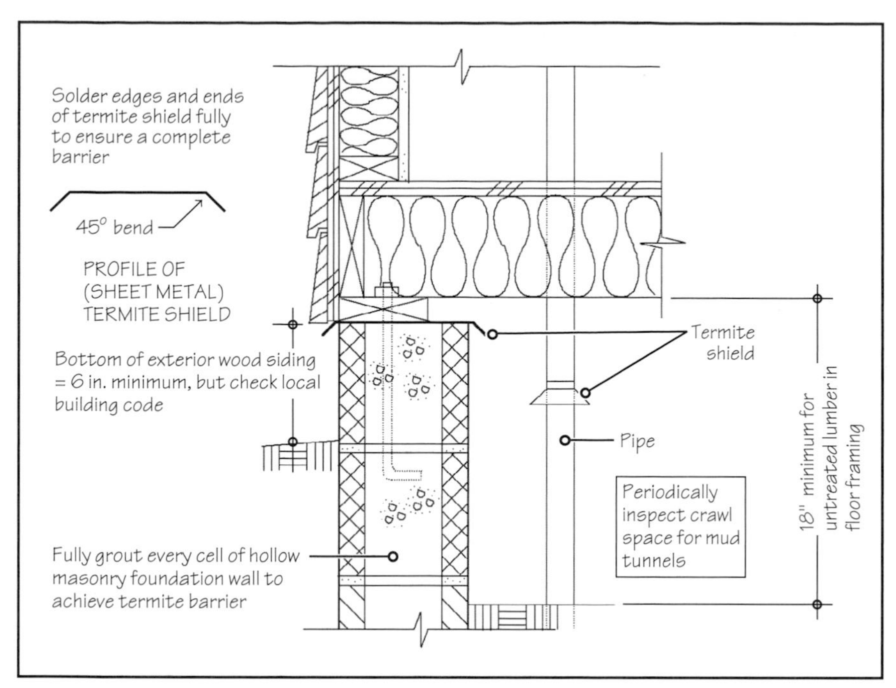

FIGURE 11.44 The use of a termite shield is recommended as an additional protection in regions with heavy infestation of termites.

ple, it is recommended that the application be done by a specialist agency. Booster applications may be needed in some locations every few years.

In locations subjected to heavy or very heavy termite hazard (based on local experience), the use of a termite shield is recommended. A termite shield is generally made of galvanized sheet steel, profiled with a drip-edge type bend. Termites are unable to maneuver this bend in building mud tunnels. All wood must be placed above termite shields, Figure 11.44.

11.15 PRESERVATIVE-TREATED WOOD

The use of naturally decay-resistant species gives limited protection against termites. Additionally, their supply is small and decreasing. Therefore, the use of decay-resistant species is infrequent.

The most effective and the most commonly used method of achieving termite protection is through the use of preservative-treated lumber. Therefore, preservative-treated lumber is commonly specified for outdoor decks, fences, verandahs, porches, and lumber buried in the ground (fence poles, wood retaining walls, wood foundations, etc.). Because the preservative is pressure injected into lumber, preservative-treated lumber is also referred to as *pressure-treated lumber*.

The preservatives used in pressure-treated lumber enable it to resist insect and fungal attack. To be effective, they must be toxic to these organisms, nondamaging to wood, and relatively safe to humans. Three types of preservatives are commonly used for the purpose:

- Creosote
- Oil-borne preservatives
- Waterborne preservatives

Creosote is the oldest and the most effective preservative against all forms of wood-eating insects, including marine borers. A distillate of coal tar, creosote is black to deep brown in color and is relatively insoluble in water. It is commonly used in railroad ties, utility poles, highway guardrail posts, marine bulkheads, and piles. Because frequent human contact with creosote is not recommended, creosote-treated lumber is not used in structural framing, decks, patios, benches, and so on. Creosote-treated lumber cannot be painted.

Pentachlorophenol (or simply penta) is the most commonly used oil-borne preservative. It is effective against fungi and land-based insects, but not against marine borers. It is commonly used for bridge timbers and utility poles.

WATERBORNE PRESERVATIVES

Wood treated with waterborne preservatives has several applications. Common applications of waterborne preservative–treated wood are outdoor decks, fences, gazebos, playground

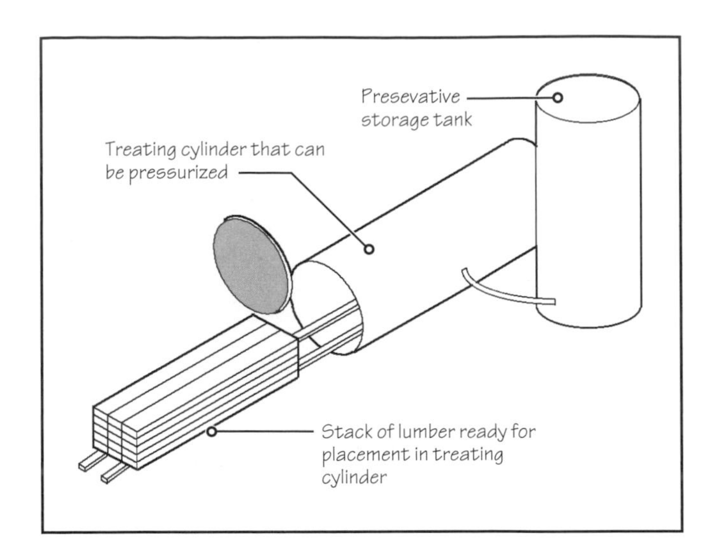

FIGURE 11.45 Outline of the equipment used for pressure treating lumber with a preservative.

equipment, structural framing, and highway noise barriers. Commonly used preservatives are

- Chromated copper arsenate (CCA)
- Alkaline copper quat (ACQ)
- Copper azole (CA)
- Sodium borates (SBX)

CCA is a waterborne mixture of the compounds of copper, chromium, and arsenic, giving the name chromated copper arsenate. Copper is toxic to fungi, and arsenic is an insecticide. Chromium is added to lock the preservative into wood.

CCA was the most widely used waterborne preservative until 2003. However, due to environmental concerns, it is no longer allowed for use in residential building components such as outdoor decks, landscape timbers, fencing, and playground structures. CCA is also not allowed in plywood, glulaminated lumber, and LVL. It will continue to be used in noise barriers, vehicular bridges, and some industrial structures.

ACQ and CA are replacing CCA for residential uses. Most wood preservers provide similar warranties of performance as they did for CCA. Both ACQ and CA have high retentivity in wood, and their leaching from wood is minimal. SBX-treated lumber is limited to indoor applications such as sill plates and interior framing lumber, because of its solubility in water.

Pressure Treatment with Waterborne Preservatives

Pressure treatment is done by placing lumber members in a cylinder, filling the cylinder with waterborne preservatives in liquid form, and applying the correct amount of pressure, Figure 11.45. The pressure helps to inject the preservative into cell cavities. The treated lumber is subsequently taken out of the cylinder and dried.

A piece of lumber kiln dried (to less than or equal to 19% moisture content) after treatment carries a stamp KDAT, meaning kiln dried after treatment. The amount of preservative injected into wood is a function of the hazard to which it would be subjected, as shown in Table 11.6.

TABLE 11.6 RETENTION OF WATERBORNE PRESERVATIVES AND LOCATION OF TREATED LUMBER OR PLYWOOD IN A BUILDING

Location of lumber or plywood	Retention (lb/ft^3)
Aboveground	0.25
Ground contact	0.40
Permanent wood foundation	0.60
Structural poles	0.60
Fresh water immersion	0.60
Salt water immersion[a]	2.50

[a]Marine borers are usually found in salt or brackish water.

FIGURE 11.46 A pressure-treated lumber pile with end tags.

TABLE 11.7 MAJOR PRESSURE-TREATABLE SOFTWOOD SPECIES

Southern pine	Douglas fir
Ponderosa pine	Hem-fir
Red pine	Western larch
Jack pine	Redwood
Sugar pine	
White Pine	

Not all wood species can be effectively pressure treated; more-permeable ones are better for treatment. Some of the commonly treated species are shown in Table 11.7. Of these, some species need incisions (slits along the grain of wood) prior to treatment to help push the preservative into the required depth of the member.

Because of the presence of copper in CCA, ACQ, and CA, pressure-treated lumber is generally greenish in color. However, a coloring agent can be added to the preservative to change the color of treated lumber to one that approaches that of untreated lumber. Treated lumber may be stained, painted, or sealed.

Treated lumber carries a treater's stamp. It is either an imprint (ink stamp) on the lumber or, more commonly, a plastic tag stapled to an end of the member, referred to as an *end tag* in the industry, Figure 11.46. Note that treated lumber carries two stamps: One stamp refers to the mill that originally produced that member and its grade, and the other stamp is the treater's stamp (or end tag). Figure 11.47 shows the information that is provided on treated lumber's end tag or stamp.

In addition to special handling precautions, treated wood has special disposal restrictions. For instance, treated wood must not be burned in stoves or fireplaces because of the toxic chemicals that it releases. The fasteners (nails and screws) and other steel components (joist hangers, etc.) used with treated wood must be of galvanized steel or stainless steel because of the corrosive effect of the preservative.

11.16 FIRE-RETARDANT-TREATED WOOD

Fire-retardant-treated wood (FRTW) is lumber or plywood that has been pressure treated with chemicals that retard the development of fire. Although not recognized as a noncombustible material (such as steel, concrete, or masonry), FRTW is permitted by building

FOCUS ON SUSTAINABILITY

Sustainability Features of Lumber

Wood is a biodegradable and rapidly renewable material. It is the only material that is theoretically inexhaustible and uses the most renewable form of energy for its production—solar energy. The only nonrenewable energy used in wood's production is in sawing, planing, and dressing of wood to its finished form. Additionally, although the production of other materials creates environmental pollution from gaseous and liquid emissions, the production of wood (the growth of trees) produces oxygen that purifies the environment. The use of wood as a building material is, therefore, a sustainable practice.

However, because forests are essential for the survival of life on earth, excessive and uncontrolled harvesting of trees to obtain wood is damaging to the environment. The wood industry has, therefore, introduced third-party certification of lumber and engineered wood products. The certification of wood is a means of ensuring that the forests are managed for sustainability through

- Maintenance of biodiversity of plant and animal species in forests
- Conservation of soil and water quality in forests
- Controlled harvesting of trees for socioeconomic benefits
- Sustained improvement in forest management practices

Certification of wood is voluntary, and a number of agencies conduct wood certification under the rules established by the Forest Stewardship Council (FSC). Certified wood is more expensive. However, it is expected that with increasing use of certified wood, a greater number of commercial forests will participate in the certification of their products, bridging the cost gap between certified and uncertified wood.

FSC is a global umbrella organization with a small permanent secretariat in Oaxaca, Mexico, that accredits wood-certifying agencies in various countries. Wood certified by an FSC-accredited certifier (e.g., SmartWood and Scientific Certification systems) carries the FSC logo.

The FSC stamp is an index of sustainable forest-management practices. It does not include other issues of sustainability, such as embodied energy and emissions of volatile organic compounds from engineered wood products.

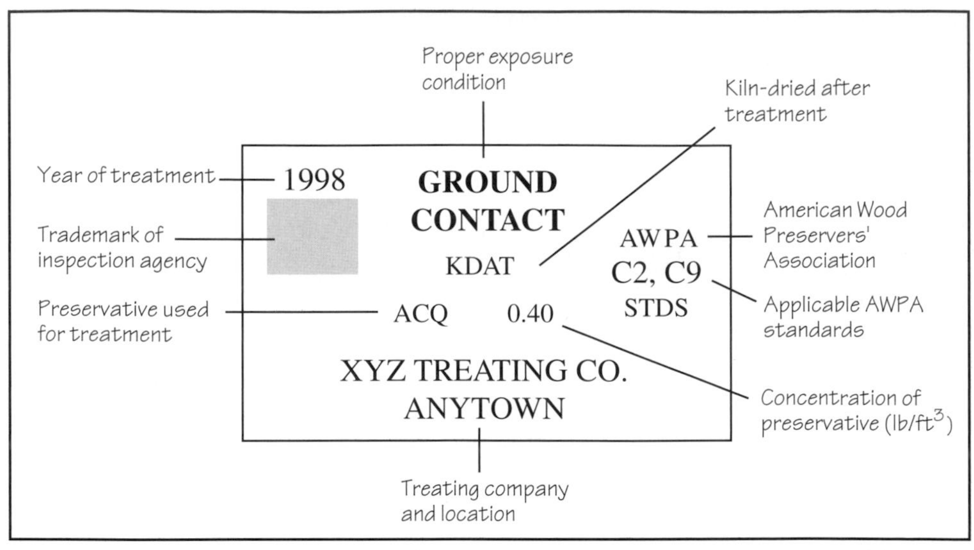

FIGURE 11.47 Information provided in ink stamp or on an end tag on pressure-treated lumber.

codes in non-load-bearing situations, where only noncombustible materials are allowed. FRTW is required to have a flame spread rating of 25 or less. Other uses of fire-retardant treatment is in wood roofing shingles and shakes. Each FRTW piece must carry the Underwriters Laboratory identification.

PRACTICE QUIZ

Each question has only one correct answer. Select the choice that best answers the question.

33. Which of the following lumber cross-sections is a dimension lumber?
- **a.** 4 × 8
- **b.** 6 × 8
- **c.** 1 × 8
- **d.** 1 × 12
- **e.** 6 × 12

34. Which of the following is not a lumber grade?
- **a.** No. 1
- **b.** No. 2
- **c.** No. 3
- **d.** No. 4
- **e.** Construction

35. The basis for distinguishing lumber as *board, dimension lumber,* and *timber* is
- **a.** the species of lumber.
- **b.** strength of lumber.
- **c.** cross-sectional dimensions of lumber.
- **d.** length of lumber.
- **e.** durability of lumber.

36. A lumber grade stamp must include information on the cross-sectional dimensions of lumber.
- **a.** True
- **b.** False

37. A lumber grade stamp must include the mill identification number.
- **a.** True
- **b.** False

38. Which of the following grades will not apply to a 2 × 8 lumber?
- **a.** No. 1
- **b.** No. 2
- **c.** No. 3
- **d.** Construction

39. Which of the following grades can only be used as blocking?
- **a.** No. 1
- **b.** No. 2
- **c.** Construction
- **d.** Utility
- **e.** Standard

40. Lumber used for structural framing members, such as studs, floor joists, and rafters, is graded using
- **a.** only visual inspection.
- **b.** only machine grading.
- **c.** both visual and machine grading must be used to arrive at lumber's grade.

41. Preservative-treated lumber is effective
- **a.** only against termite attack.
- **b.** only against fungal attack.
- **c.** against both fungi and termites.

42. A termite shield is generally made out of
- **a.** asphalt-treated felt.
- **b.** galvanized sheet steel.
- **c.** lead sheet.
- **d.** kraft paper.

43. Wood certified by Forest Stewardship Council (FSC) is available
- **a.** only in the United States.
- **b.** in the United States and Canada.
- **c.** in the United States, Canada, and Mexico.
- **d.** in (c) and several other countries.

Answers: 33-a, 34-d, 35-c, 36-b, 37-a, 38-d, 39-d, 40-a, 41-c, 42-b, 43-d.

Typical Grade Stamps of Visually Graded Lumber

U.S. Grading Agencies

Northeastern Lumber Manufacturers Association (NeLMA), Inc. Cumberland Center, Maine—a rules-writing agency and inspection agency

Northern Softwood Lumber Bureau (NSLB), Cumberland Center, Maine—a rules-writing agency and inspection agency

Pacific Lumber Inspection Bureau (PLIB), Inc., Bellevue, Washington—an inspection agency

Redwood Inspection Service (RIS), Novato, California—a rules-writing agency and inspection agency

Renewable Resource Associates (RRA), Inc., Atlanta, Georgia—an inspection agency

Southern Pine Inspection Bureau (SPIB), Pensacola, Florida—a rules-writing agency

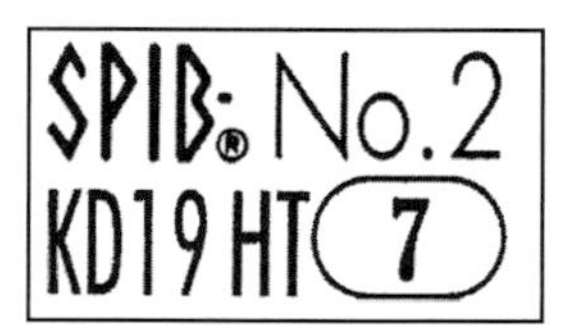

Timber Products Inspection (TPI), Conyers, Georgia—an inspection agency

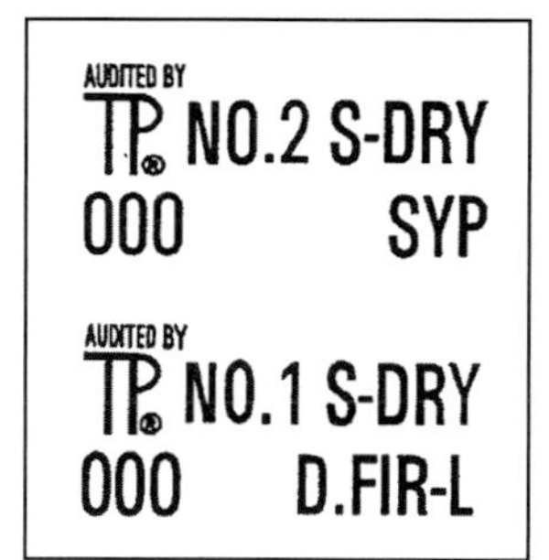

West Coast Lumber Inspection Bureau (WCLIB), Portland, Oregon—a rules-writing agency and inspection agency

(*Continued*)

Typical Grade Stamps of Visually Graded Lumber (*Continued*)

Western Wood Products Association. Portland, Oregon—a rules-writing agency and inspection agency

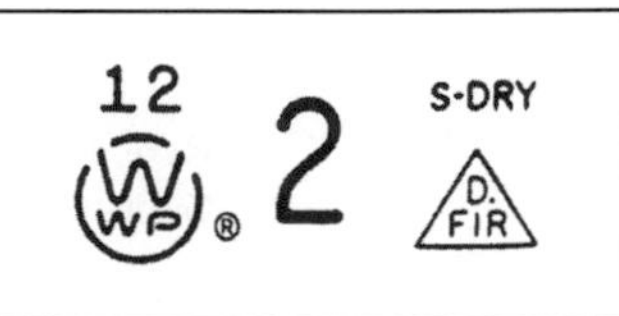

Canadian Inspection Agencies

The following Canadian inspection agencies have been certified by the American Lumber Standards Committee. There is only one rules-writing agency in Canada—the National Lumber Grades Authority (NLGA). NLGA's rules for dimension lumber are the same as those of the U.S. National Grading Rules (NGR) for Dimension Lumber.

A.F.P.A.® 00
S-P-F NLGA
KD-HT 1

Alberta Forest Products Association (AFPA), Edmonton, Alberta

CL®A
S-P-F
100
No. 2
S-GRN.

CL®A 100
SPRUCE-PINE-FIR
NO. 1 S-DRY

Canadian Lumbermen's Association (CLA), Ottawa, Ontario

CFPA® 00
S-P-F S-DRY
CONST

Central Forest Products Association (CFPA), Hudson Bay, Saskatchewan

Council of Forest Industries (COFI), Vancouver, British Columbia

MacDonald Inspection (MI), Coquitlam, British Columbia

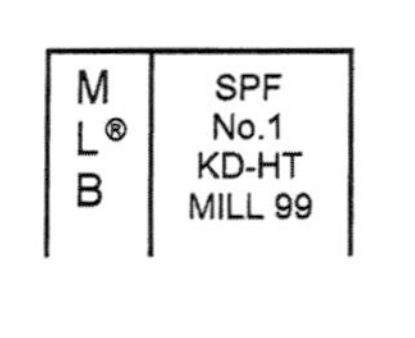

Maritime Lumber Bureau (MLB), Amherst, Nova Scotia

O.L.M.A.® 01-1
CONST. S-DRY
SPRUCE - PINE - FIR

Ontario Lumber Manufacturers Association (OLMA), Toronto, Ontario

Pacific Lumber Inspection Bureau (PLIB), Bellevue, Washington , U.S.A., British Columbia Division: Vancouver, British Columbia

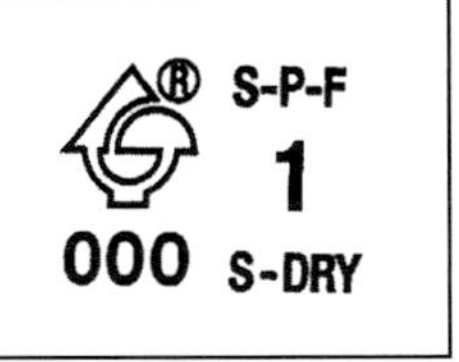

Quebec Lumber Manufacturers Association, (QFIC), Ste-Foy, Quebec

Source: These grade stamps have been adapted from the American Lumber Standards Committee's (ALSC) website (www.alsc.org), with permission. Note that this is an incomplete list. For a complete list refer to the above website.

KEY TERMS AND CONCEPTS

Actual dimensions
Air seasoning
Board foot
Board lumber
Decay resistance
Dimension lumber
Dimensional stability of lumber
Dry lumber
Finger-joint
Fire retardant treated wood
Flat sawn
Fungal decay
Glue laminated
Grade stamp
Graded lumber
Grain structure
Green lumber
Growth characteristics
Growth ring
Hardwood
Heartwood
Heavy timber
Kiln dry
Knot
Lumber
Lumber grading
Lumber surfacing
Machine grading
Machine stress rated
Nominal dimensions
Plain sawn
Planing
Preservative treated wood
Pressure treatment
Quarter sawn
Radial sawn
Sapwood
Sawing
Seasoning
Softwood
Termite control
Visual grading
Waterborne preservatives
Wood light frame
Wood microstructure
Worked lumber

REVIEW QUESTIONS

1. Describe the essential differences between softwood and hardwood trees. Give at least three commonly used species of each.
2. What is the difference between heartwood and sapwood? Explain.
3. What does the term *SPF* mean? What is the difference between SPF^{s} and SPF? Explain.
4. Using sketches and notes, explain why wood is stronger along the grain than across the grain.
5. Using sketches and notes, explain how to graphically distinguish between framing lumber and finished lumber in cross-sectional drawings.
6. What information is typically provided on the grade stamp of machine-rated lumber? Explain with the help of an example.
7. Explain how fungal decay of lumber can be prevented.

SELECTED WEB SITES

Alberta Forest Products Association (www.albertaforestproducts.ca)
American Lumber Standard Committee (www.alsc.org)
American Wood-Preservers' Association (www.awpa.com)
Southern Forest Products Association (www.sfpa.org)
Southern Pine Council (www.southernpine.com)
Western Wood Products Association (www.wwpa.org)

FURTHER READING

1. Canadian Wood Council. *Wood Reference Handbook—A Guide to the Architectural Use of Wood in Building Construction*, 1991.
2. International Code Council. *International Residential Code*. Falls Church, VA: International Code Council.

REFERENCES

1. Hoadley, Bruce. *Understanding Wood*. Newtown, CT: Taunton Press, 1980, p. 69.

CHAPTER 12

Materials for Wood Construction–II (Manufactured Wood Products, Fasteners, and Connectors)

CHAPTER OUTLINE

12.1 Glulam Members

12.2 Structural Composite Lumber—LVL and PSL

12.3 Wood I-Joists

12.4 Wood Trusses

12.5 Wood Panels

12.6 Plywood Panels

12.7 OSB Panels

12.8 Specifying Wood Panels—Panel Ratings

12.9 Fasteners for Connecting Wood Members

12.10 Sheet Metal Connectors

Glue-laminated beams support the roof and exposed OSB and plywood interiors in this sun-filled corridor in the Island Wood Outdoor Learning Center, Bainbridge Island, WA. Mithun architects + Designers + Planners. (Photo courtesy of Bruce Haguland.)

There are a growing number of wood products that are produced using techniques that extend beyond sawing and planing. One of the earliest examples—plywood—originated from a laminating process that can be traced back to ancient Egypt and China. However, it was not until the invention of waterproof adhesives in the 1930s that plywood became the first commercially produced *manufactured wood product.*

Manufactured wood products are made by bonding together lumber members, wood veneers, wood strands, wood particles, and other forms of wood fibers to produce a composite material. Their manufacturing process eliminates defects and weak points in wood (or spreads them over the product), giving a stiffer, stronger, and more homogeneous material than lumber. The process also utilizes wood materials that would have otherwise become a waste.

Two types of manufactured wood products are used in construction:

- Engineered wood (EW) products
- Industrial wood products

EW products are those that are engineered for structural applications. They include glue-laminated wood, structural composite lumber, wood I-joists, plywood, oriented strandboard, and wood trusses.

Industrial wood products include particle board, medium-density fiberboard (MDF), and high-density fiberboard (HDF), also called *hardboard.* Particle board is commonly used for cabinet work, furniture, heavy-duty shipping containers, and so on. MDF is used as a replacement for solid lumber. It has a smooth surface that allows precise machining to form complex and intricate moldings. Hardboard is commonly used as floor underlayment (Section 12.8).

This chapter begins with a discussion of various EW products. Finally, the chapter deals with fasteners and connectors used in wood frame construction.

12.1 GLULAM MEMBERS

Sawn lumber has several limitations where large cross-sectional wood members are required to span long distances between structural supports. Some of these limitations are as follows:

- Large sawn lumber cross sections can be obtained only from trees with large girths. Such trees are generally protected from harvest. Additionally, trees with excessively long life spans make lumber farming uneconomical.
- Large sawn lumber cross sections cannot be dried to an acceptable moisture content.
- Being naturally grown, there is little control over the structural properties of sawn lumber.

Glue-laminated wood (referred to as *glulam*) is the response to these limitations. It is made from individual lengths of dimension lumber that are glued together to form large cross sections. The individual lengths are joined horizontally (face laminated) as well as vertically (end jointed), Figure 12.1. The ends are generally finger jointed.

Laminating is an effective way of using short lengths of high-grade sawn lumber obtained by eliminating pieces that are of low grade due to knots, shakes, and splits. Because glulam is generally made from dimension lumber, it uses lumber that has been dried to a fairly low moisture content, generally less than 15%. Additionally, laminating allows the use of smaller trees harvested from younger forests.

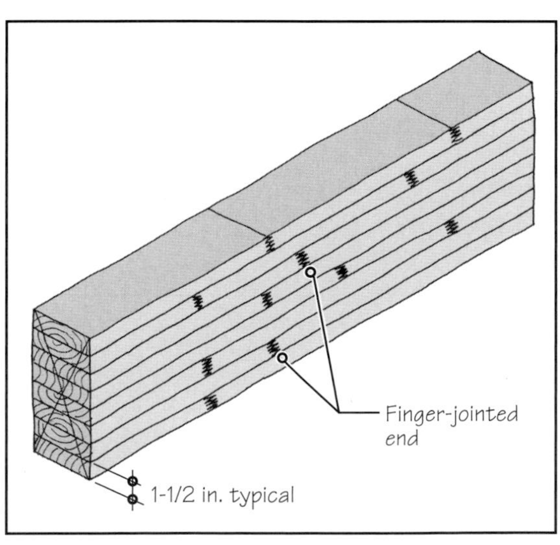

FIGURE 12.1 Anatomy of a glulam beam. A glulam beam is typically made by gluing together $1\frac{1}{2}$-in.-thick laminations of solid sawn lumber.

The adhesive used between laminations is of high strength and is fully water resistant. Consequently, a glulam member is stronger and stiffer than a sawn lumber of the same dimensions. The use of water-resistant adhesive means that a glulam member can be used in externally exposed conditions without delaminating along adhesive lines.

USES AND SIZES OF GLULAM MEMBERS

The most common use of glulam members is for long-span beams, Figure 12.2. Other uses are for heavy columns, Figure 12.3, and heavy trusses, Figure 12.4. In some situations, glulam members are specified because their large cross sections provide greater fire resistance. A fire resistance rating of up to 1 h is obtainable with unprotected glulam members.

Glulam manufacturers produce beams of standard cross sections that are cut to the required lengths. Because 2-by (i.e., $1\frac{1}{2}$-in.-thick) lumber is used in producing glulam beams, their depth is generally in multiples of $1\frac{1}{2}$ in., such as $7\frac{1}{2}$ in., 9 in., or $10\frac{1}{2}$ in.

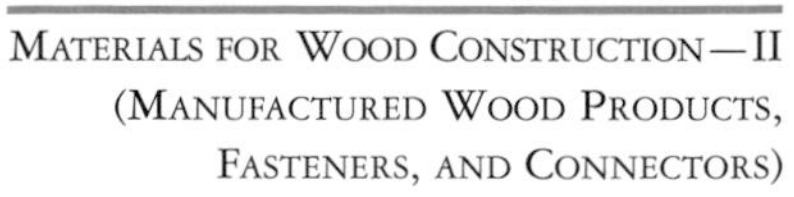

FIGURE 12.2 A glulam beam over a long span. (Photo courtesy of APA—The Engineered Wood Association.)

FIGURE 12.3 Glulam columns used in Beaverton Public Library, Beaverton, Oregon. Thomas Hacker and Associates, Architects. (Photo courtesy of APA—The Engineered Wood Association.)

FIGURE 12.4 Glulam trusses used in Goodman School, Gig Harbor, Washington. (Photo courtesy of APA—The Engineered Wood Association.)

The width of glulam beams varies depending on the width of dimension lumber used. Standard beam widths vary from nearly 3 in. to nearly 14 in. Wider beams can be custom produced. Theoretically, a glulam member can be made to any cross-sectional size and length by gluing dimension lumber side by side and face to face, Figure 12.5. However, limitations of transportation to the construction site and installing the member must be considered in choosing a large glulam member.

Because glulam is made from small wood sections under controlled conditions, it can also be curved along its length. Thus, arches and other contours can be easily obtained in glulam, Figure 12.6. In curved glulam members, laminations are generally $\frac{3}{4}$ in. thick instead of the standard thickness of $1\frac{1}{2}$ in.

FIGURE 12.5 Large cross sections of glulam members are made by gluing laminations side by side and face to face.

FIGURE 12.6 Anaheim Ice Rink, Anaheim, California. Frank Gehry Associates, Architects. (Photo courtesy of APA—The Engineered Wood Association.)

NOTE

APA, PFS, and TECO

Three U.S. organizations provide third-party certification and quality control of glulam and other engineered wood (EW) products: APA—The Engineered Wood Association, PFS Corporation, and TECO Corporation.

APA originated as the certifier of plywood products. At that time, APA was the acronym for *American Plywood Association.* With the introduction of other EW products, it has changed its name to APA—The Engineered Wood Association to better describe its new role.

PFS and TECO are sister organizations. TECO deals with the quality certification of wood panels—plywood and oriented strandboard—and PFS deals with the other EW products.

Of these three organizations, APA has by far the largest market share at the present time.

Balanced and Unbalanced Glulam Beams

Balanced glulam beams are symmetrical in lumber quality above and below the beam's mid depth. In an unbalanced beam, the quality of lumber used in upper laminations is different from those in the lower laminations to account for the fact that the lower half of the beam will be in tension, and the upper half of the beam will be in compression. Unbalanced beams are, therefore, stamped with *TOP* to ensure their correct placement, Figure 12.7. They can be used only as single-span beams, Figure 12.8.

Balanced glulam beams are more versatile because they can be used for any span condition. However, they are generally used where the beams are continuous over two or more supports, because a continuous beam is subjected to tension and compression on both faces of the beam at different locations. Similarly, a beam with an overhang (cantilever) will also require the use of a balanced glulam beam.

Specifying Glulam Members—Glulam Member Grades

Most glulam producers in the United States conform to the quality certification mechanism developed by APA—The Engineered Wood Association, which requires each glulam member to be grade stamped. Among other information, the grade stamp on a glulam

FIGURE 12.7 An unbalanced glulam beam showing the beam's top face.

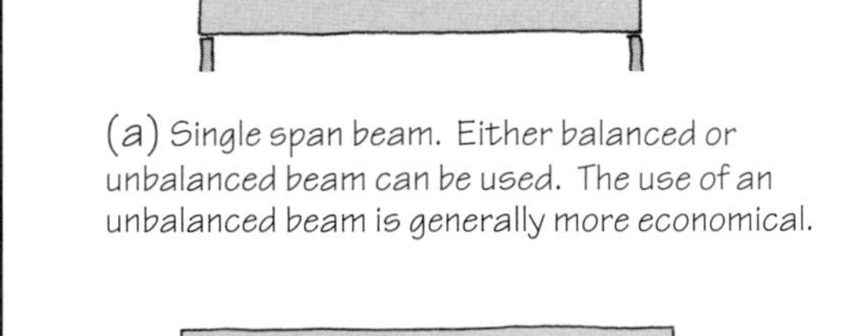

FIGURE 12.8 Single-span bean, continuous beam, and a cantilevered beam.

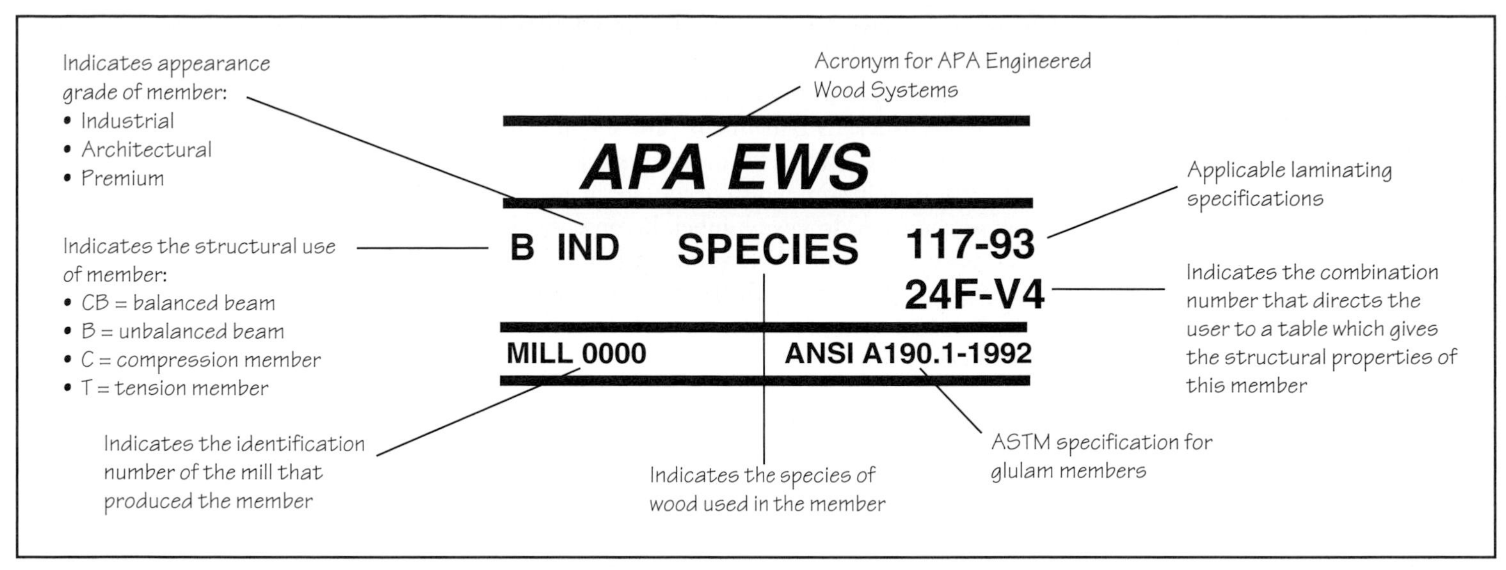

FIGURE 12.9 A typical grade stamp on a glulam member.

member provides the structural properties of the beam and its appearance. The appearance of a glulam member is classified in one of the following three grades:

- *Industrial Appearance Grade*—commonly specified in situations where the appearance of the member is relatively unimportant, such as in a warehouse, or if the member is to be covered with a finish material. Some manufacturers make industrial grade glulam beams called "headers". These are beams whose thickness is consistent with the commonly used widths of wood frame walls. Thus, glulam headers are made to widths of $3\frac{1}{2}$ in., $5\frac{1}{2}$ in., and $7\frac{1}{4}$ in. The $3\frac{1}{2}$-in.-thick header can be used over an opening in a wood frame wall with 2 × 4 studs. A $5\frac{1}{2}$-in.-thick header is used in a wall with 2 × 6 studs.
- *Architectural Appearance Grade*—an appearance grade that is intermediate between industrial and premium grades.
- *Premium Appearance Grade*—a grade commonly specified where the appearance of the beam is important.

A typical APA grade stamp on a glulam member is shown in Figure 12.9.

12.2 STRUCTURAL COMPOSITE LUMBER—LVL AND PSL

Laminated veneer lumber (LVL) is produced by gluing together dried wood veneers that are approximately $\frac{1}{8}$ in. thick. The wood grain in all veneers runs in the same direction, unlike in plywood, where the veneers are cross-grained between laminations (see Section 12.5). Thus, like sawn lumber, LVL is stronger along the grain and weaker across the grain.

LVL is generally used as floor joists and rafters. Therefore, it is usually made to a finished thickness of $1\frac{3}{4}$ in. and to depths of up to 18 in., Figure 12.10. The thickness of $1\frac{3}{4}$ in. (as compared

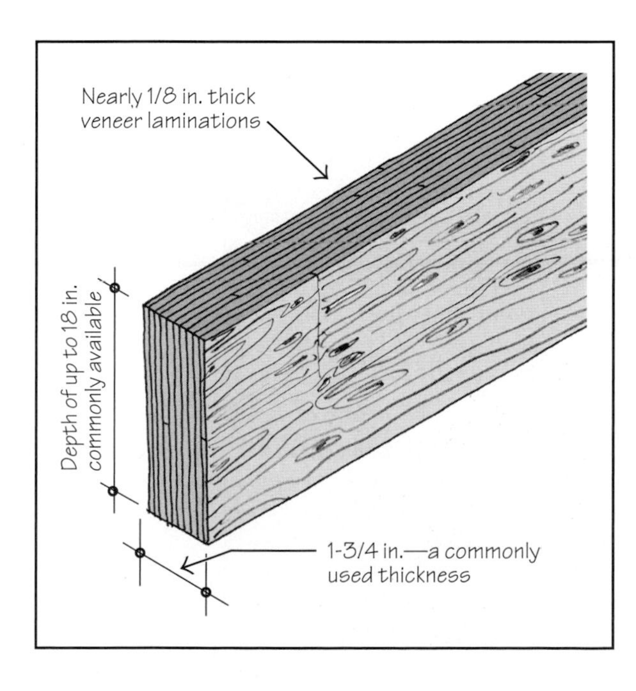

FIGURE 12.10 A typical LVL member.

to $1\frac{1}{2}$ in. for sawn lumber) provides greater fastening surface for floor or roof sheathing. Other thicknesses, such as $3\frac{1}{2}$ in. and $5\frac{1}{4}$ in., are also available. The $3\frac{1}{2}$-in.-thick LVL members are generally used as headers in 2 × 4 wood stud walls. If $3\frac{1}{2}$-in.-thick LVL is not available, two $1\frac{3}{4}$-in.-thick LVL members can be nailed together to make a $3\frac{1}{2}$-in.-thick member.

Because LVL is made from veneers that are dried to a moisture content of approximately 12% (the equilibrium moisture content in a typical building interior), it is less likely to warp or split as compared to sawn lumber. Because the defects in veneers are cut away, LVL is stronger than sawn lumber of the same dimensions. Like glulam, LVL allows the use of smaller trees.

LVL members are produced by gluing veneers to form large billets, approximately 2 ft × 4 ft in cross section and nearly 80 ft long. These billets are then sawn to yield the desired (smaller) LVL members.

PARALLEL STRAND LUMBER

A variation of LVL is called *parallel strand lumber* (PSL). PSL is made by gluing together narrow strands of veneer in place of wide veneers. The strands are made by chopping the veneers into strips, about $\frac{1}{2}$ in. wide and 8 ft long. Like the LVL, PSL is also first made into large billets, which are then sawn to required dimensions. Both LVL and PSL are proprietary to each manufacturer. Manufacturers provide structural data and other recommendations that must be adhered to in the use of LVL and PSL members. LVL and PSL are together referred to as *structural composite lumber* (SCL).

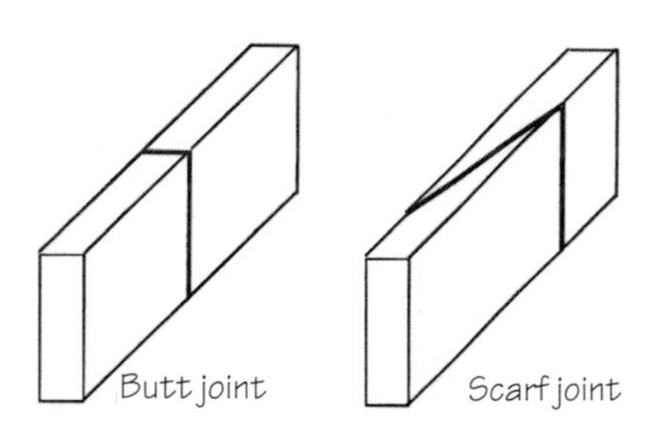

12.3 WOOD I-JOISTS

Wood I-joists (so named because of their I-shaped cross section) are made by gluing wood flanges to a wood web, Figure 12.11. The flanges are made either of sawn lumber or LVL. LVL has gained greater acceptance because I-joist flanges must be jointless, which is more easily done with LVL members because they are available in much longer lengths than sawn lumber.

The web in a wood I-joist is either of plywood or oriented-strandboard (OSB) panel. OSB is more commonly used because of the absence of core voids in OSB (Section 12.7), which gives greater shear strength than a plywood web. Joints in the web are permissible, either in the form of butt joints or scarf joints.

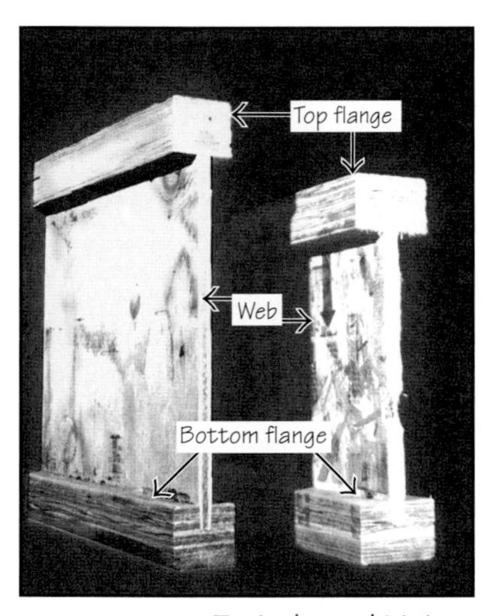

FIGURE 12.11 Typical wood I-joists.

I-joists are commonly used as floor joists, Figure 12.12, and roof rafters. They are dimensionally more stable than sawn lumber due to their plywood or OSB webs. (As we will see later in this chapter, atmospheric moisture changes do not affect plywood or OSB member dimensions as much as those of solid lumber.) That is why I-joist manufacturers claim that a floor constructed of I-joists is less prone to squeaking than the one made of sawn lumber joists.

The most common flange width of I-joists is $1\frac{3}{4}$ in., with total depth ranging from 10 in. to 18 in., Figure 12.13. Most manufacturers supply the joists to their distributors and dealers in lengths of 60 ft, which are then cut to lengths desired for the project.

Because there is less wood fiber in an I-joist, it is lighter than a corresponding sawn lumber joist. Less wood fiber means a more resource-efficient material. Another advantage of I-joists is that holes can be easily cut in webs to carry utility lines without compromising the structural capacity of the member. However, because of the proprietary and preengineered nature of the joists, manufacturers provide strict guidelines as to the size, shape, and location of holes.

FIGURE 12.12 I-joists are commonly used as floor joists because of their greater dimensional stability and stiffness and availability in longer lengths than solid sawn lumber joists.

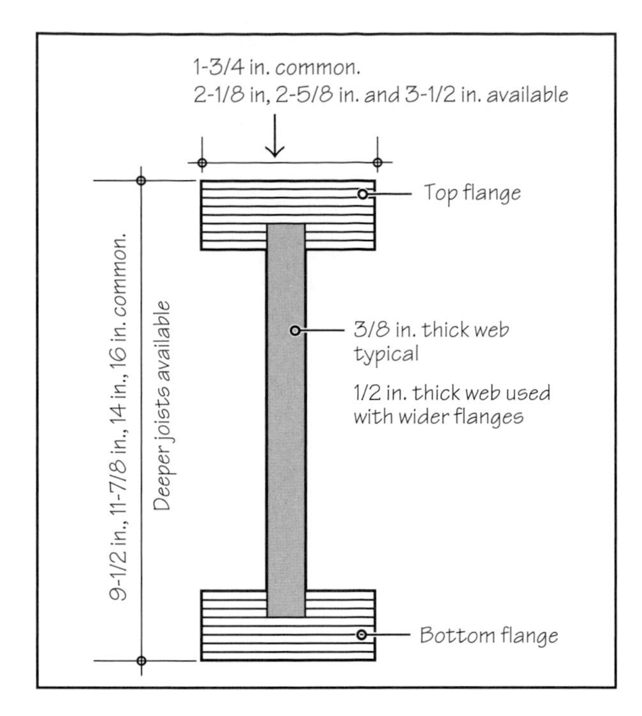

FIGURE 12.13 Commonly used dimensions of an I-joist. Manufacturer's literature must be consulted for exact dimensions.

A disadvantage of I-joists is their thin web, which makes them relatively unstable. Hence, they must be braced during construction, that is, until the installation of floor or roof sheathing.

SPECIFYING I-JOISTS

Like the structural composite lumber industry, the wood I-joist industry is also proprietary. Each I-joist manufacturer formulates its own design by testing its products. Thus, the connection between the web and the flanges is proprietary, and so is the type of adhesive. Some of the commonly used flange-web connections are shown in Figure 12.14.

I-joist manufacturers provide ready-to-use span tables and other data to help design and construction professionals select the right type of joists for a project. All important construction details are also provided.

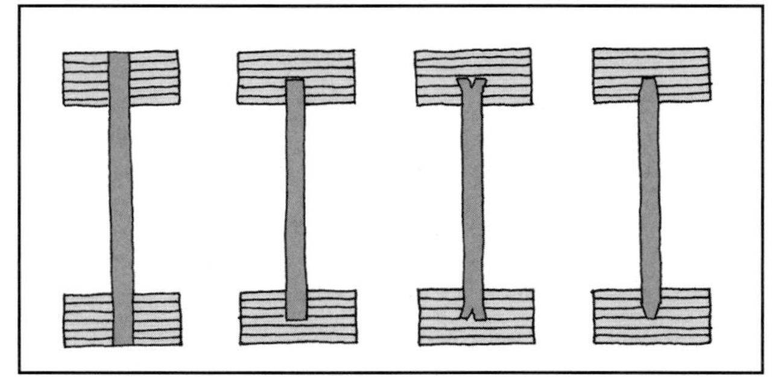

FIGURE 12.14 A few commonly used web-flange connections in I-joists.

12.4 WOOD TRUSSES

A truss consists of individual members that are joined together to form an array of interconnected triangular frames. Because a triangle is a naturally rigid geometric shape that resists being distorted when loaded from any direction, a truss is more rigid than a beam with the same amount of material. In fact, a truss is one of the most efficient means of carrying loads between supports. It is able to carry a greater load over a given span using less material than a rectangular beam or an I-beam. Figure 12.15 shows a typical wood truss, along with the important truss vocabulary.

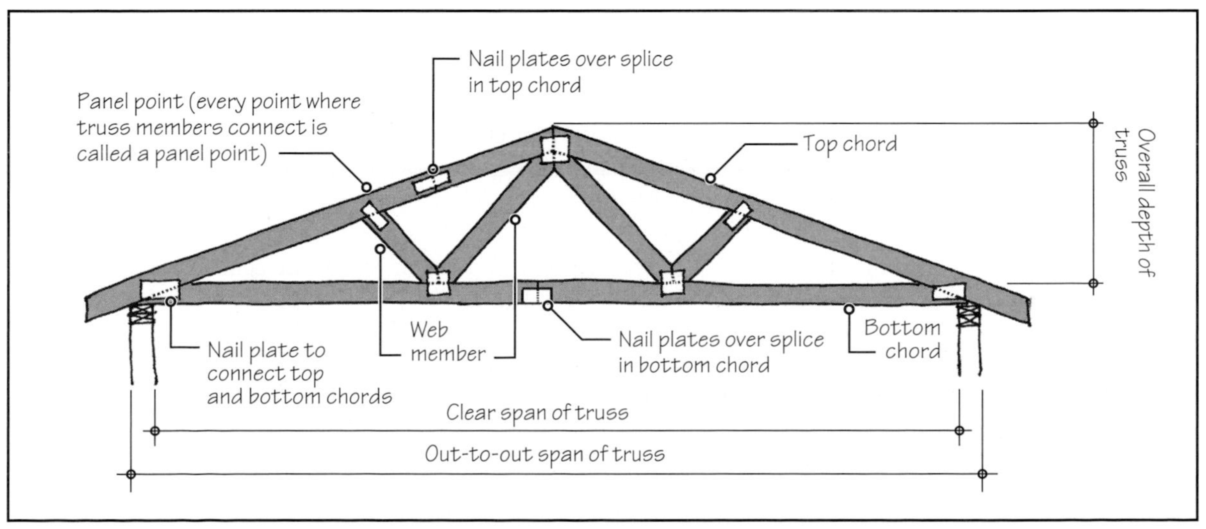

FIGURE 12.15 Important terms related to trusses.

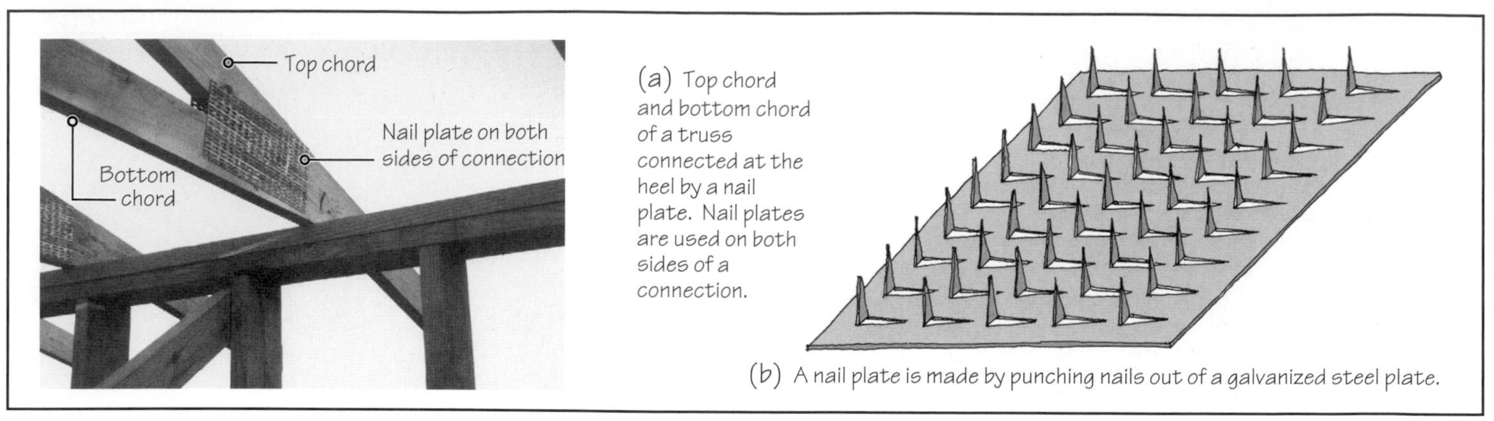

FIGURE 12.16 Metal nail plate.

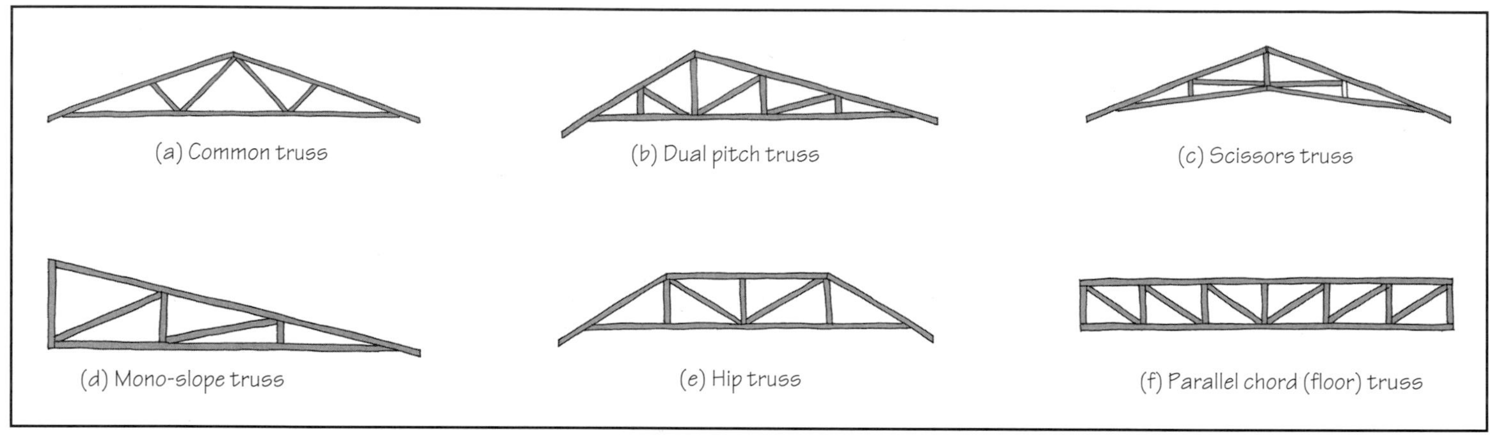

FIGURE 12.17 Commonly used shapes of wood trusses for residential and light commercial structures.

Being made of individual triangles, truss members need not be continuous. Because the members are small in size, sawn lumber (generally 2 × 4 lumber) is most commonly used for trusses meant for residential or light commercial buildings. LVL, PSL, or glulam are also used by some manufacturers as members in trusses for heavy commercial or industrial applications.

In a sawn lumber truss, the individual members are connected together by metal nail plates, Figure 12.16(a). A nail plate is generally a 16-, 18-, or 20-gauge galvanized steel plate with the nails punched out of the sheet so that the nails and the plate are of one piece, Figure 12.16(b). The nails are generally $\frac{3}{8}$ in. long, nearly 8 nails per square inch. Nail plates are machine applied in manufacturing plants so that trusses are fully fabricated before they are transported to the construction site.

Several shapes, some of which are shown in Figure 12.17, are common for wood trusses. The shape and size of a truss is limited by manufacturing and transportation capabilities. Depending on its overall shape, a wood truss is either one of the following:

- Roof truss
- Floor truss

Roof Trusses

In a roof truss, the top chord is pitched. The bottom chord is generally horizontal, but it can be sloping, as in a scissors truss, Figure 12.17(c). Roof trusses are generally designed to bear on exterior walls, with the bottom chord providing the nailing surface for the ceiling. In other words, the interior walls of a building do not carry any roof loads. This is in contrast to the stick-built rafter and ceiling joist roof, in which the ceiling joists are generally supported on interior as well as exterior walls.

The use of roof trusses, therefore, gives greater freedom in the layout of interior walls and their future reconfiguration. Wood roof trusses for residential and light commercial buildings are generally spaced 24 in. on center, Figure 12.18.

Floor Trusses

In a floor truss, the top chord is horizontal—parallel to the bottom chord. Thus, it is referred to as a *parallel chord truss.* Wood floor trusses are also called *trussed joists* because

FIGURE 12.18 Monoslope roof trusses. Two monoslope trusses laid end to end and field connected make a common roof truss. Monoslope trusses are generally used where a common truss is too large to transport to the site. Wood trusses are generally spaced 24 in. on center.

they function as floor joists, Figure 12.19. They are also used as lintel beams over openings, Figure 12.20. Trussed joists can be provided with a rectangular opening in the middle of the joist for mechanical ducts, Figure 12.21.

The versatility and the cost effectiveness of trussed joists has made them a popular choice in wood buildings with long spans (greater than 20 ft). A disadvantage of trussed joists (also of roof trusses) is that they cannot be cut to size. In other words, their use does not permit last-minute changes in design.

In addition to all-wood joists, wood-metal joists are available, in which the web members are made of hollow steel pipes and the chords are made of LVL or sawn lumber, Figure 12.22. They are generally used in industrial buildings, where the loads are greater and the spans are longer. The use of steel in webs increases the strength and stiffness of the joists and the wood chords provide a nailable surface for roof and floor sheathing and the ceiling.

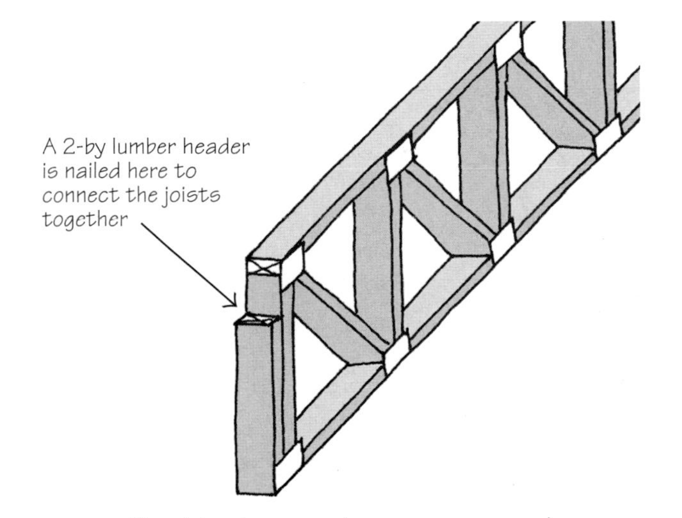

2 x 4 lumber members are commonly used to make a floor truss

Trussed joists at a building site before being hoisted into position

FIGURE 12.19 Floor trusses (trussed joists).

FIGURE 12.20 Trussed lintel over an opening.

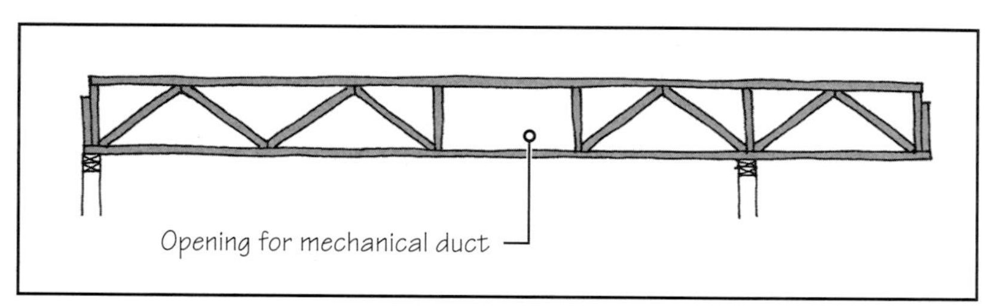

FIGURE 12.21 A floor truss with a rectangular opening to accommodate a mechanical duct.

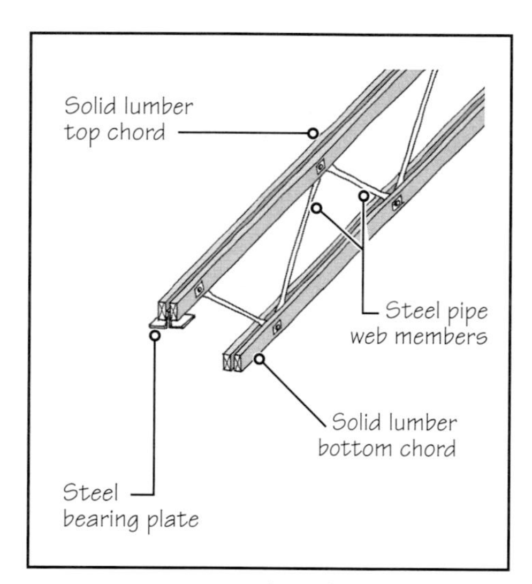

FIGURE 12.22 Wood-steel trussed joists.

SPECIFYING WOOD TRUSSES

When using trusses for a project, all that an architect, engineer, or builder has to do is to determine the load requirements, shape, span, and slope. The truss manufacturer then determines the detailed truss geometry, sizes of truss members, wood species, and so on. The truss fabricator prepares shop drawings for the trusses and sends them to the engineer, architect, or the builder for approval.

Installing trusses on the site is as important as truss design or fabrication. A crane or other lifting equipment is generally required, and installers must be skilled and experienced. During erection, wood trusses should almost always be held in an upright position—in the position in which they are to be used in the building, Figure 12.23. When held horizontally, the lateral bending of truss members during lifting may overstress the connections, causing nail plates to loosen or pop out.

FIGURE 12.23 While lifting and placing a wood truss in position, it should be held in an upright position.

PRACTICE QUIZ

Each question has only one correct answer. Select the choice that best answers the question.

1. The most common use of glulam members is in
   - **a.** structural panels.
   - **b.** long-span beams.
   - **c.** short-span beams.
   - **d.** none of the above.
2. Unprotected glulam members with a large cross section are considered to provide a fire resistance rating of up to
   - **a.** $\frac{1}{2}$ h.
   - **b.** 1 h.
   - **c.** 2 h.
   - **d.** 3 h.
   - **e.** 4 h.
3. A glulam member
   - **a.** must be made using long, continuous lengths of high-grade sawn lumber.
   - **b.** is generally stronger and stiffer than sawn lumber of the same dimensions.
   - **c.** requires complete protection from exterior elements due to its water soluble adhesive.
   - **d.** none of the above.
   - **e.** all the above.
4. Unbalanced beams can be used only as single-span beams.
   - **a.** True
   - **b.** False
5. Balanced glulam beams are mandated for
   - **a.** continuous beams.
   - **b.** single-span beams.
   - **c.** beams with an overhang.
   - **d.** (a) and (c).
   - **e.** (b) and (c).
6. Laminated veneer lumber (LVL) is
   - **a.** produced by gluing together wood veneers that are approximately $\frac{1}{8}$ in. thick.
   - **b.** generally used as floor joists.
   - **c.** glued with all veneers running in the same direction.
   - **d.** stronger along the grain and weaker across the grain.
   - **e.** all the above.
7. Wood I-joists are made of LVL or solid sawn lumber flanges and webs.
   - **a.** True
   - **b.** False
8. Compared to sawn lumber joists, wood I-joists are
   - **a.** less expensive.
   - **b.** slightly heavier.
   - **c.** dimensionally more stable.
   - **d.** all the above.
9. Characteristics of a truss include
   - **a.** greater spanning capability than a sawn lumber beam having the same amount of material.
   - **b.** individual members joined together to form an array of interconnected rectangular frames.
   - **c.** individual members joined together to form an array of interconnected triangular frames.
   - **d.** (a) and (b).
   - **e.** (a) and (c).
10. Nail plates, used to join members of a truss, are generally used only on one face of the truss.
    - **a.** True
    - **b.** False
11. Wood trusses are generally spaced at
    - **a.** 12 in. o.c.
    - **b.** 24 in. o.c.
    - **c.** 48 in. o.c.
    - **d.** 60 in. o.c.

Answers: 1-b, 2-b, 3-b, 4-a, 5-d, 6-e, 7-b, 8-c, 9-e, 10-b, 11-b.

12.5 WOOD PANELS

Wood panels are an important part of wood frame construction. They are used structurally—as floor sheathing, roof sheathing and wall sheathing—and nonstructurally—as exterior siding and interior paneling. Wood panels are divided into the following three types:

- *Veneered panels*—consisting of plywood panels.
- *Nonveneered panels*—consisting of oriented strandboard (OSB) and particle board panels. Particle board panels are generally used in shelving and furniture making and are not discussed here.
- *Composite panels*—consisting of two parallel face veneers with a nonveneer core. Their use in contemporary structural applications is limited, and at the present time, there are few U.S. manufacturers of composite panels. Therefore, only plywood and OSB panels are discussed here.

12.6 PLYWOOD PANELS

Plywood panels are made by gluing wood veneers under heat and pressure. Veneers are generally produced by a machine that holds a debarked log at two ends in a lathe and rotates the log against a stationary knife blade extending throughout the length of the log, Figure 12.24. This operation peels the log off, giving a continuous veneer about $\frac{1}{8}$ in. thick, in much the same way as one would unroll a paper towel. The veneer so obtained is subsequently cut to desired sizes.

To make a plywood panel, the defects in veneers, such as knot holes and splits, are cut away or repaired where necessary. The veneers are then dried and glued together so that the grain direction in each veneer is oriented at a right angle to the grain direction of the adjacent veneer, Figure 12.25.

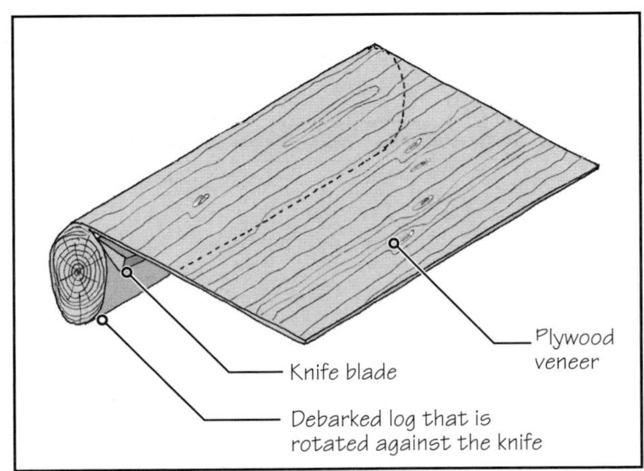

FIGURE 12.24 Rotary slicing of a log—commonly used method of making plywood veneers.

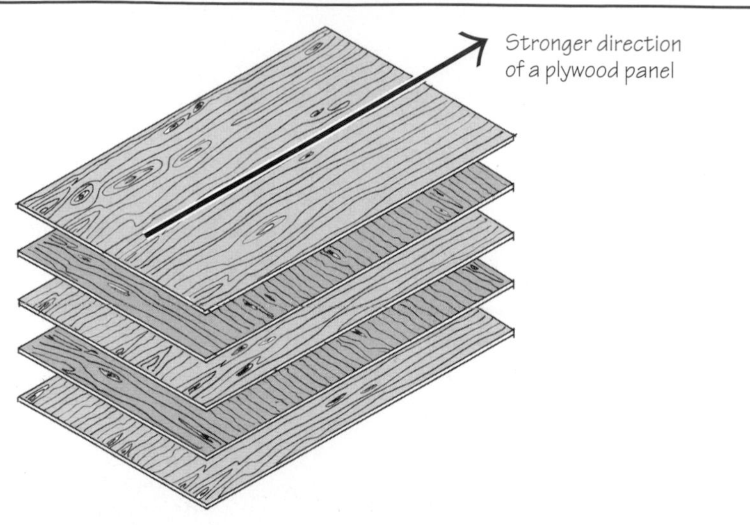

The grain direction in alternate veneer layers are at right angles to each other. Generally, this arrangement implies an odd number of veneers in a panel. However, an even number of veneers is also used in which the two innermost veneers are oriented in the same direction so that the top and bottom veneers (referred to as face veneers) are always in the same direction.

For example, if a panel is made of 6 veneers, of which the two innermost veneers are in the same direction, it is referred to as 5 layer, 6 veneer panel. The panel shown here is a 5 layer, 5 veneer panel.

Because bending stresses maximize at the top and bottom fibers in a panel and because wood is stronger along the grain than across the grain, a plywood panel is stronger in the direction of the grain of the face veneers.

FIGURE 12.25 Orientation of veneer grains in a plywood panel.

Because wood is stronger along the grain than across the grain, cross-graining tends to equalize the strength of a plywood panel in its two principal directions. It also makes a plywood panel dimensionally more stable because shrinkage and swelling of wood is high across the grain and negligible along the grain. For the same reason, a plywood panel is less likely to split than sawn lumber. Therefore, it can be nailed near its edge without splitting.

The most commonly used plywood panel size is 4 ft × 8 ft, with thickness varying from $\frac{1}{4}$ in. to 1 in. Longer plywood panels are manufactured for siding and industrial use. The panel dimension of 4 ft × 8 ft is its nominal dimension. Its actual dimension is $\frac{1}{8}$ in. smaller in length as well as its width. In other words, the actual dimensions of a 4-f × 8-ft panel are $47\frac{7}{8}$ in. × $95\frac{7}{8}$ in. This allows a panel to be installed with a $\frac{1}{8}$-in. space all around for moisture expansion, Figure 12.26.

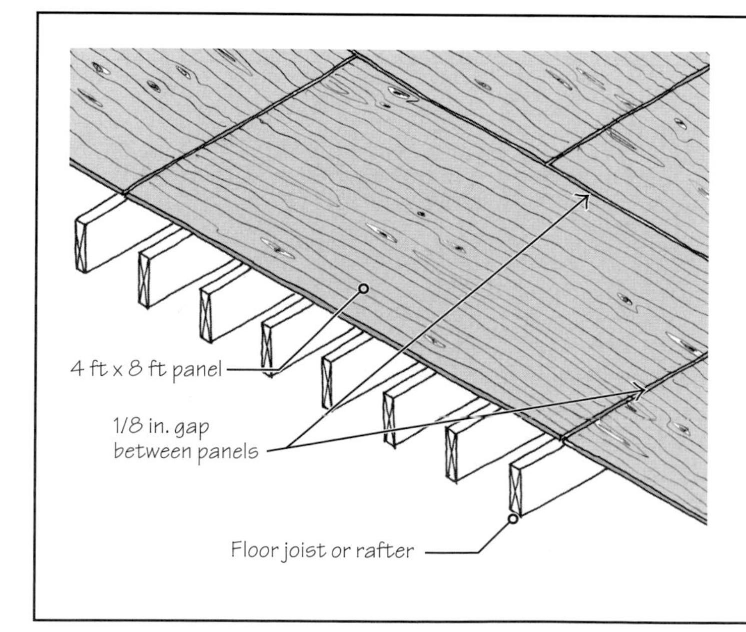

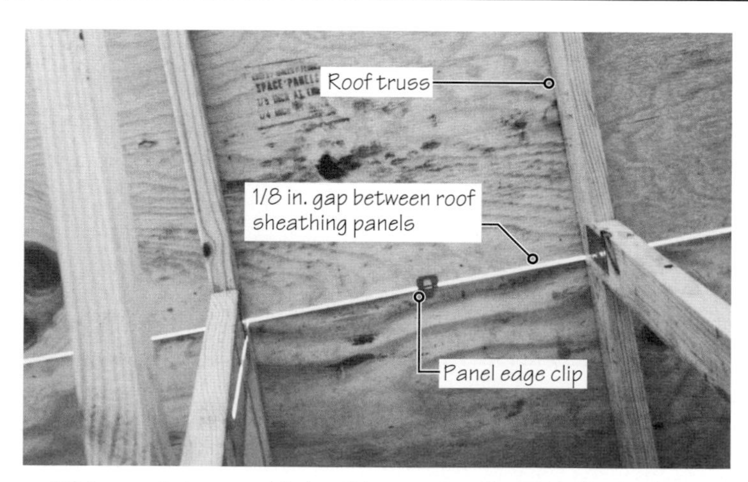

1/8 in. gap between roof sheathing panels. The white line in this image represents daylight coming through 1/8 in. gap. Note that roof sheathing requires edge clips between adjacent panels to add to their continuity.

FIGURE 12.26 Plywood panels must be oriented with their long direction perpendicular to the supporting members. Additionally, a gap of nearly $\frac{1}{8}$ in. must be left all around panels to accommodate moisture expansion.

Grain direction of face veneers of a panel are oriented along the panel's length (along the 8-ft dimension in a 4-ft × 8-ft panel), making the panel stronger in its long direction. In structural applications, the long direction of a plywood panel should, therefore, be perpendicular to the supporting joists or rafters.

Veneer Grades of Softwood Plywood

Plywood veneers used in building construction are generally obtained from softwoods. Softwood veneers are graded in five grades—A, B, C-plugged, C, and D—based on the type and size of defects such as knots and splits, Figure 12.27. Veneer grade A is the highest grade, and veneer grade D is the lowest. Thus, in veneer grade D, the number of knot holes and their sizes are higher than any other grade.

A	Smooth, paintable veneer — Not more than 18 neatly made repairs. May be used for natural finish in less demanding applications.
B	Solid surface veneer — A lower grade than grade A. Tight knots to 1 in. across grain and minor splits permitted.
C Plugged	Improved C veneer — Knot holes limited to 1/4 in. x 1/2 in. and splits limited to 1/8 in. Admits some broken grain.
C	C veneer — Tight knots to 1-1/2 in., knot holes to 1 in. across grain. Limited splits allowed.
D	D veneer — Knots and knot holes to 2-1/2 in. width across grain. Limited to Exposure 1 or Interior panels.

FIGURE 12.27 A brief description of plywood veneer grades.

Sanded, Unsanded, and Touch-Sanded Plywood Panels

The veneer grade on plywood panel faces is generally different from the veneer grades in the panel's core. Additionally, panels meant to be exposed on one side only may have a high veneer grade on one face and a lower grade on the other face. For example, a panel may be an A-C plywood panel, which indicates that one face of the panel has grade A veneer and the other face has grade C veneer.

Plywood panels with grade A or B face veneers are always sanded smooth because such panels are generally used in cabinet work, furniture work, and shelving. Panels with C-plugged, C, and D face veneers are touch-sanded to provide a more uniform panel thickness. Panels used for sheathing purposes are unsanded.

12.7 OSB PANELS

For many years, plywood panels were the only wood panels available. Recently, as a result of the desire to use resources more efficiently, technology has been developed to use shredded, wafer-thin wood strands, compressed and glued to form a panel. Panels made with

EXPAND YOUR KNOWLEDGE

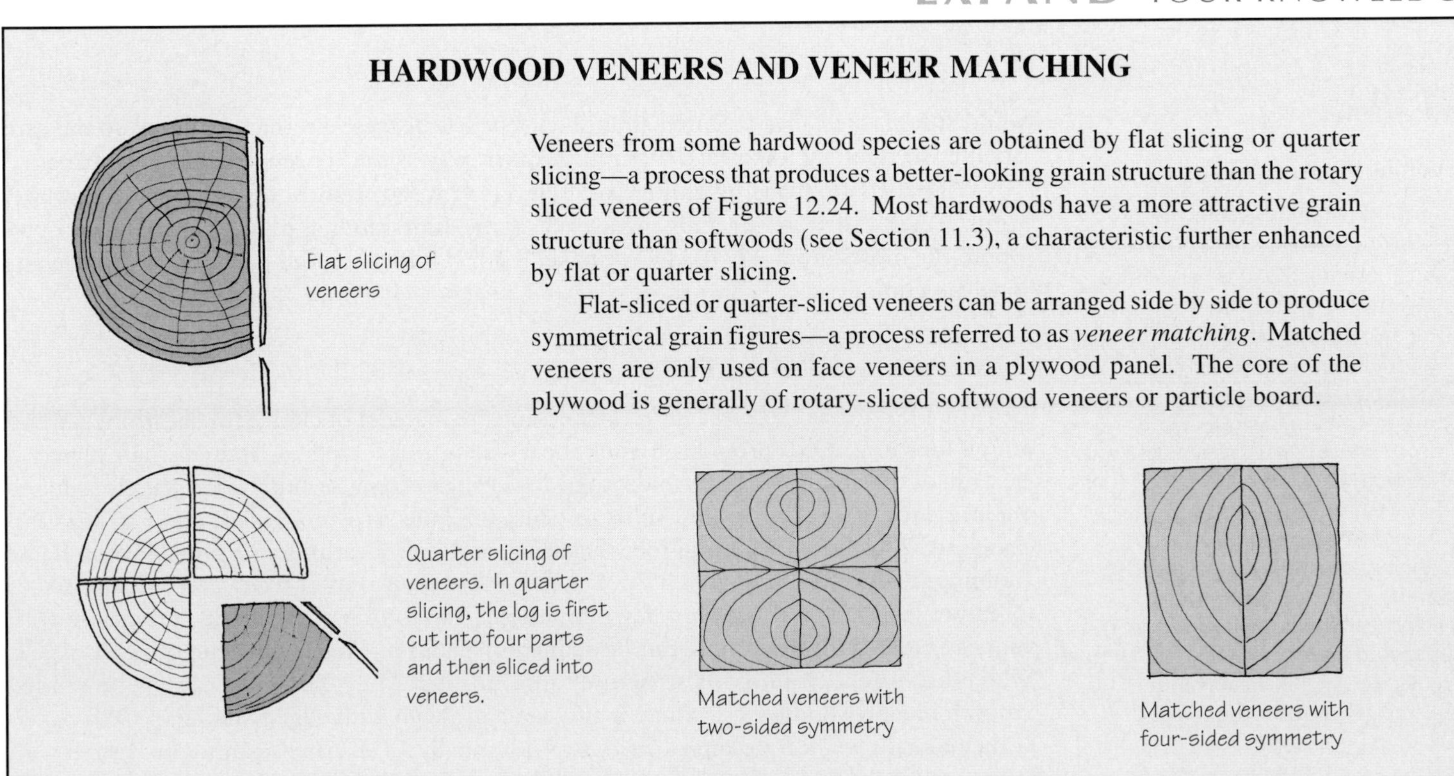

HARDWOOD VENEERS AND VENEER MATCHING

Veneers from some hardwood species are obtained by flat slicing or quarter slicing—a process that produces a better-looking grain structure than the rotary sliced veneers of Figure 12.24. Most hardwoods have a more attractive grain structure than softwoods (see Section 11.3), a characteristic further enhanced by flat or quarter slicing.

Flat-sliced or quarter-sliced veneers can be arranged side by side to produce symmetrical grain figures—a process referred to as *veneer matching*. Matched veneers are only used on face veneers in a plywood panel. The core of the plywood is generally of rotary-sliced softwood veneers or particle board.

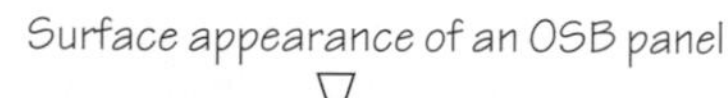

FIGURE 12.28 OSB panel manufacturing.

NOTE

Versatility of OSB Panels

For all structural sheathing (wall sheathing, floor sheathing, and roof sheathing), OSB is currently the material of choice. The use of plywood in floor and roof sheathing declined from nearly 90% in 1978 to 26% in 1999 in favor of OSB (Bruce Smith, 2001 President National Association of Home Builders, in *Builder and Developer Magazine,* 2001, an online magazine (bdmag.com)). Similarly, the use of gypsum wall sheathing has virtually disappeared and has been replaced by OSB sheathing in residential construction.

wood strands are called oriented strandboard panels because alternate layers of strands are oriented at right angles to each other in the same way as the veneers in a plywood panel.

OSB panels are made by gluing several layers of wood strands under heat and pressure, Figure 12.28. They are generally made to the same dimensions as plywood panels. Like plywood, the most commonly used OSB panel is 4 ft × 8 ft (nominal) and is stronger along its long direction.

OSB Versus Plywood

Because OSB generally costs less, it has become the material of choice for sheathing a wood frame building. OSB provides higher shear strength (racking resistance) than plywood because of the absence of core voids. (The lower shear strength in plywood panels is due to the presence of knot holes and splits in veneers.) Thus, it is not uncommon to see OSB panels used for floor sheathing, roof sheathing, and wall sheathing in a typical wood frame building, Figure 12.29.

However, OSB panels have a few limitations. Plywood panels, particularly those with higher grades on face veneers, can be stained or painted. This is not true of OSB panels, which are intended only for structural applications. They generally cannot be sanded smooth like plywood panels. There is also some problem with edge swelling in OSB panels if they remain wet for prolonged periods. Additionally, OSB panels cannot be treated with preservatives, whereas preservative-treated plywood is available.

FIGURE 12.29 Because of their lower cost and greater racking resistance, OSB panels are generally used to sheath the entire envelope of a wood frame building (see also Section 13.9).

12.8 SPECIFYING WOOD PANELS—PANEL RATINGS

Wood panels can be divided into two categories based on their end use:

- Performance-rated engineered wood panels
- Sanded and touch-sanded plywood panels

PERFORMANCE-RATED ENGINEERED WOOD PANELS

Performance-rated engineered wood panels are meant for structural applications (e.g., wall sheathing, floor sheathing, roof sheathing) as well as for nonstructural use in exterior siding. The panels are preengineered and rated accordingly. The rating provides the user with the panel's structural capacity and other performance data, such as its intended end use and durability. This simplifies the specification of the panel and relieves the user of any further investigation.

Performance rating means that as long as the panel meets the specified requirements for end use (exposure durability and span-rating), it does not matter which material—plywood or OSB—it is made of. The rating is achieved through a third-party certification process and stamped accordingly. Typical ratings (grade stamps) of panels are given in Figure 12.30. The following performance properties are of particular importance:

Intended End Use Performance-rated panels can be used in one of the following three situations:

- Sheathing (over studs, floor joists, or rafters)
- Combination floor sheathing
- Exterior siding

Combination floor sheathing is a special type of floor sheathing that replaces a two-layer covering over floor joists. If regular sheathing is used over floor joists, it requires an additional layer of panels over the sheathing panels. The additional panel layer is called *underlayment* because it functions as an underlayer for the floor finish and is typically $\frac{1}{4}$-in.-thick sanded plywood or hardboard.

Combination floor sheathing panels have been developed to act as sheathing as well as underlayment and are generally provided with tongue-and-groove (T&G) profile along the long (8-ft) edges, Figure 12.31, although square-edged panels are also available.

Whereas sheathing and combination sheathing panels may be of plywood or OSB, only plywood is used in exterior siding panels because of aesthetic and durability reasons.

Exposure Durability Engineered wood panels are produced in two exposure-durability classifications:

- Exterior
- Exposure 1

Exterior-rated panels are designed for permanent exposure to the weather—to withstand the effect of rain, humidity, and sunshine. *Exposure 1* panels are meant for use in protected situations, that is, where the panels are to be covered with an exterior facing

NOTE

Sturd-I-Floor as Combination Floor Sheathing

APA-rated Sturd-I-Floor is a commonly specified floor sheathing, ranging in thickness from $\frac{3}{4}$ in. to $1\frac{1}{8}$ in. It does not require any underlayment and is commonly used in situations where carpet is the required floor finish. According to APA, the panel surface of Stud-I-Floor has extra resistance to punch-through damage.

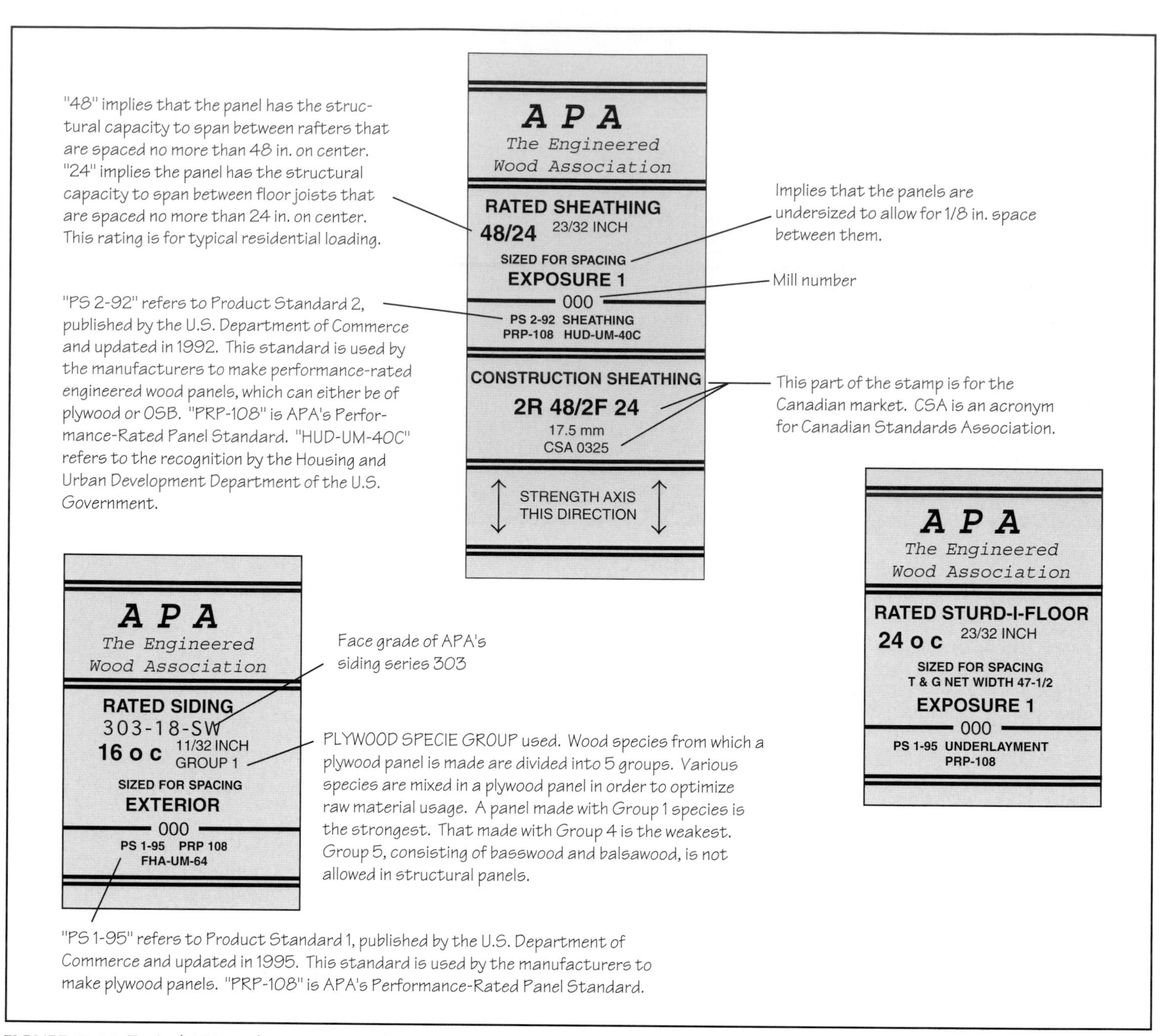

FIGURE 12.30 Typical APA grade stamps on performance-rated engineered wood panels. TECO, another grading agency, uses similar stamps.

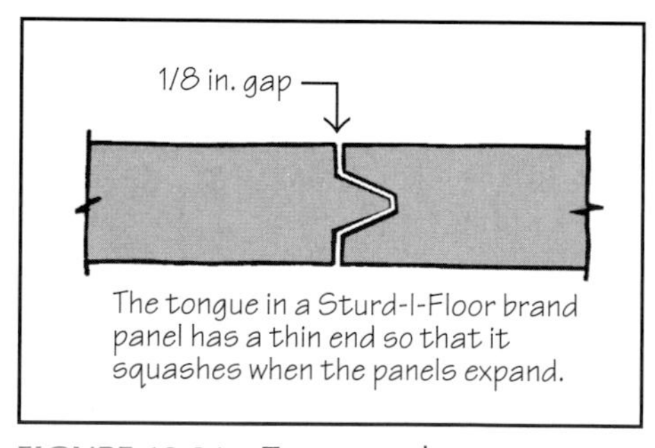

FIGURE 12.31 Tongue-and-groove joint between Sturd-I-Floor panels.

material. However, Exposure 1 panels are designed to withstand the effect of weather for several days due to construction delays. Both the Exterior and Exposure 1 panels use the same waterproof glue. The difference lies in the panels' composition.

Note that the exposure-durability classification of panels relates to the moisture resistance of the glue bond, not the fungal resistance of the panel. Thus, an Exterior-rated panel, if not used correctly, is subject to fungal attack.

Theoretically, two other durability classifications exist in the standards. These are Exposure 2 and Interior. However, at the present time, only Exterior and Exposure 1 panels are made.

Span Rating and Overall Thickness As stated previously, performance-rated panels are preengineered, which means that the architects, engineers, and builders do not need to calculate the load capacity of the panel. This information is provided by the manufacturer and is stamped on the panel, along with the panel's end use, exposure classification, and thickness. Panel thickness is given in fractions of an inch, such as $\frac{3}{8}$, $\frac{15}{32}$, and $\frac{23}{32}$. For Canadian markets, it is given in millimeters.

Sanded and Touch-Sanded Plywood Panels

Sanded and touch-sanded panels are rated for performance as well as exterior appearance. The exterior appearance refers to the grade of face veneers of the panels. A few typical grade stamps of such panels are shown in Figure 12.32. Note that the performance-rated panels are rated only for performance, not for appearance.

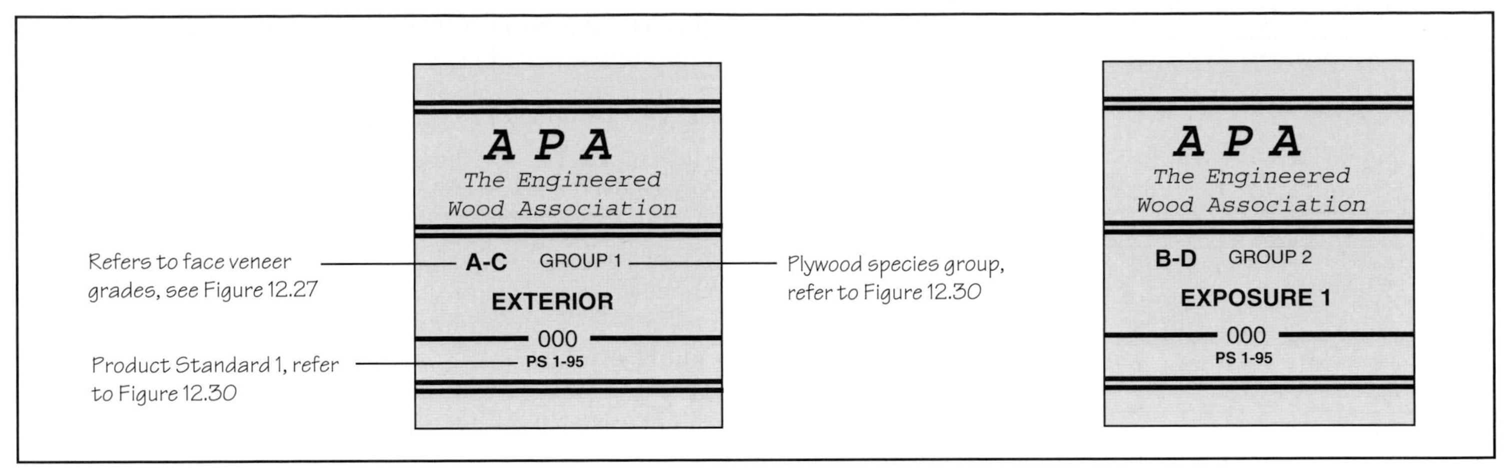

FIGURE 12.32 Typical grade stamps on sanded and touch-sanded plywood panels.

PRACTICE QUIZ

Each question has only one correct answer. Select the choice that best answers the question.

12. Plywood panels are made by gluing under heat and pressure layers of
- **a.** wood veneers so that the wood grain in all layers is in the same direction.
- **b.** wood veneers so that the wood grain in adjacent layers are perpendicular to each other.
- **c.** shredded wood strands so that the strands in all layers are in the same direction.
- **d.** shredded wood strands so that the strands in adjacent layers are perpendicular to each other.

13. Cross-graining makes a plywood panel
- **a.** dimensionally more stable.
- **b.** less likely to split than solid sawn lumber.
- **c.** stronger in the direction of face veneer grains.
- **d.** all the above.
- **e.** (a) and (c).

14. The most commonly used plywood panel (nominal) size is 4 ft × 8 ft.
- **a.** True
- **b.** False

15. Plywood panels should be placed with their longer dimension
- **a.** parallel to rafters or joists.
- **b.** perpendicular to rafters or joists.
- **c.** diagonal to rafters or joists.
- **d.** either (a) or (b).
- **e.** either (a) or (c).

16. With respect to veneer quality, softwood plywood is graded in grades
- **a.** A to C.
- **b.** B to D.
- **c.** A to D.
- **d.** A to E.
- **e.** B to F.

17. An OSB panel generally has greater shear strength than a plywood panel of the same size and thickness.
- **a.** True
- **b.** False

18. Grade stamps on engineered wood panels specify
- **a.** intended use and exposure.
- **b.** allowable spans.
- **c.** mill number.
- **d.** thickness.
- **e.** all the above.

19. Exterior siding panels may be made of OSB or plywood.
- **a.** True
- **b.** False

Answers: 12-b, 13-d, 14-a, 15-b, 16-c, 17-a, 18-e, 19-b.

12.9 FASTENERS FOR CONNECTING WOOD MEMBERS

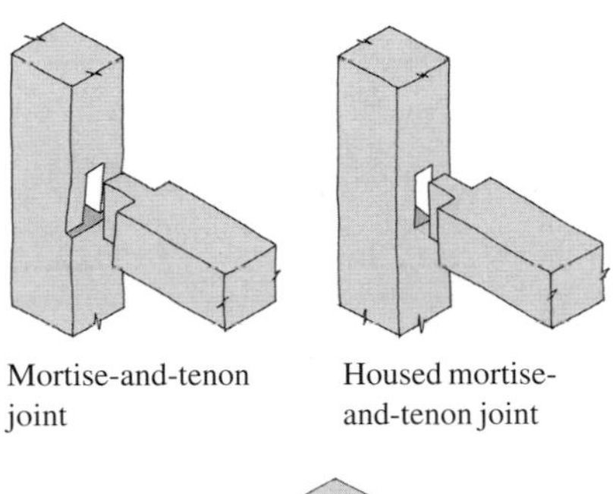

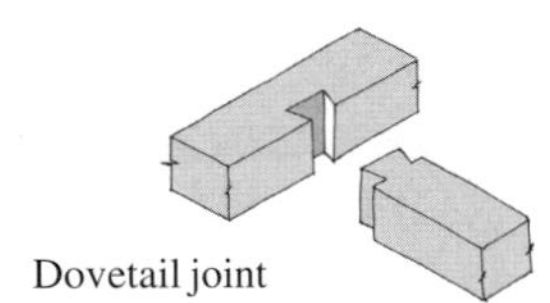

The traditional method of connecting wood members was through different types of interlocking joints, such as a mortise-and-tenon joint, housed mortise-and-tenon joint, and dovetail joint. Except for fine furniture and to some extent in heavy timber construction, these *joinery methods,* as they are referred to, have been replaced by simpler and more effective methods that often yield stronger connections. Most joints in contemporary wood construction are made by simply nailing the members together or by nailing them through sheet metal connectors. In some joints, adhesives are used in addition to nails, whereas in others, screws and bolts are necessary.

TYPES OF NAILS

A nail is generally made of low or medium carbon steel wire that is heat treated to increase its stiffness. Where increased impact resistance is needed, such as for masonry nails, steel with a higher carbon content is used. Nails made in this way without any further treatment for corrosion are called *brite nails.* In exterior siding and decks where greater corrosion resistance is needed, hot-dip galvanized nails are used. (Stainless steel nails can provide an even higher corrosion resistance but are expensive.)

For increased holding power, nails are phosphate or vinyl coated. Vinyl-coated nails produce heat due to friction when the nail is driven, melting the vinyl, which increases the bond between the wood and the nail. They have a thinner shank and are easier to drive into wood and are, therefore, called *sinker nails*. For most structural connections, however, brite (ungalvanized, uncoated) steel nails are used.

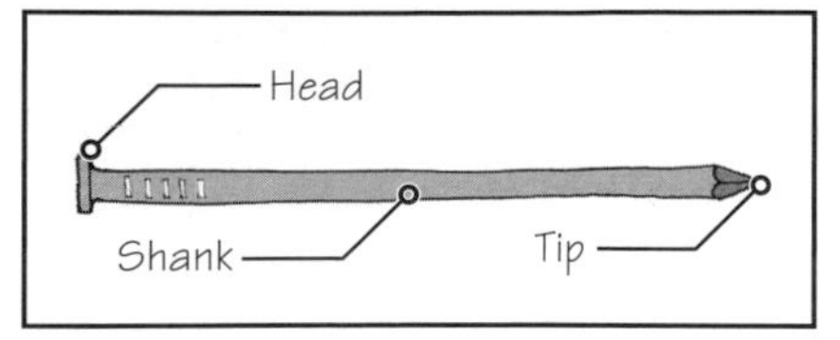

FIGURE 12.33 Parts of a nail.

A nail has three basic parts: (a) the tip, (b) the shank, and (c) the head, Figure 12.33. The tip of most nails is diamond shaped. Therefore, the nail type is distinguished by the type of its head and the type of shank. Some of the commonly used nail types in wood frame construction are shown in Figure 12.34.

For framing connections, common nails are most frequently used. Box nails are similar to common nails but have a thinner shank, which reduces wood splitting. They are generally used for attaching wood shingles. Casing nails and finish nails are used for finish carpentry. Deformed shank nails are used for attaching wood flooring or gypsum wallboard.

Nail Sizes

The length of common nails in the United States is specified by a *penny* (abbreviated as *d*) designation. For example, a 10d nail (called *a 10 penny nail*) is 3 in. long. The penny designation originated in England centuries ago when 1 poundweight of 10d nails cost 10 pence, 1 poundweight of 12d nails cost 12 pence, and so on.

Common nails are available in lengths ranging from 2d to 60d. A 2d nail is 1 in. long, and a 60d nail is 6 in. long. From 2d to 10d, the increase in length is $\frac{1}{4}$ in. per penny. Therefore, a 10d nail is 3 in. long. The next two nail sizes are 12d and 16d, with lengths of $3\frac{1}{4}$ in. and $3\frac{1}{2}$ in., respectively. A 20d nail is 4 in. long, Figure 12.35. Most commonly used nail sizes in wood frame construction are 6d, 8d, 10d, and 16d, highlighted in Figure 12.35.

Nailed Connections

Nails work best when they are subjected to shear—that is, the load is perpendicular to the length of the nails—or when they are in compression. Nails are particularly weak in withdrawal. Withdrawal resistance is needed when the load is parallel to the length of the nails, trying to pull the connected members apart, Figure 12.36. Three types of nailed connections are used in wood frame construction, Figure 12.37:

- Face-nailed connection
- End-nailed connection
- Toe-nailed connection

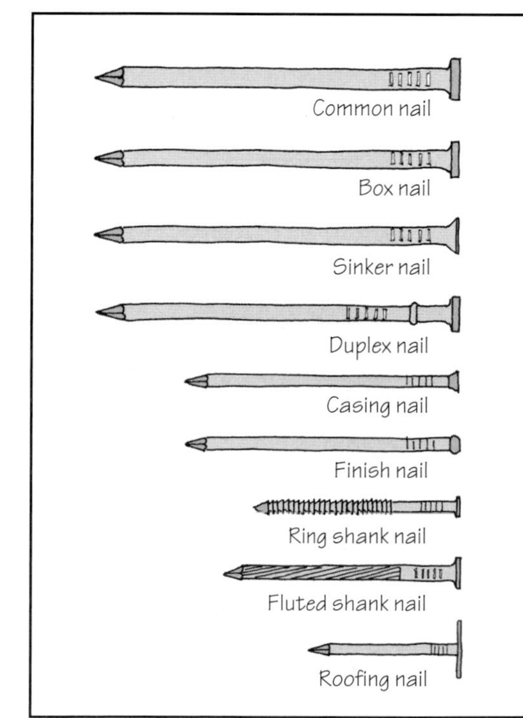

Most commonly used nail type for connecting wood frame members. Thick shank gives greater strength.

Used for attaching wood siding and shingles. Thin shank reduces wood splitting.

Tapered head that sinks into wood. Sinker nails have a thin shank like box nails and are generally vinyl coated.

Double head for temporary nailing, used in scaffolding and concrete form work.

For wood trim, window frames, casing and decks. Small head for countersinking.

For finer carpentry and finishing. Small head for countersinking.

For attaching floor sheathing and gypsum wallboard. Ring shank gives greater holding power.

For attaching wood to masonry or concrete. High carbon steel gives this nail greater impact resistance.

Large head for attaching roof shingles.

FIGURE 12.34 Commonly used types of nails in wood frame construction.

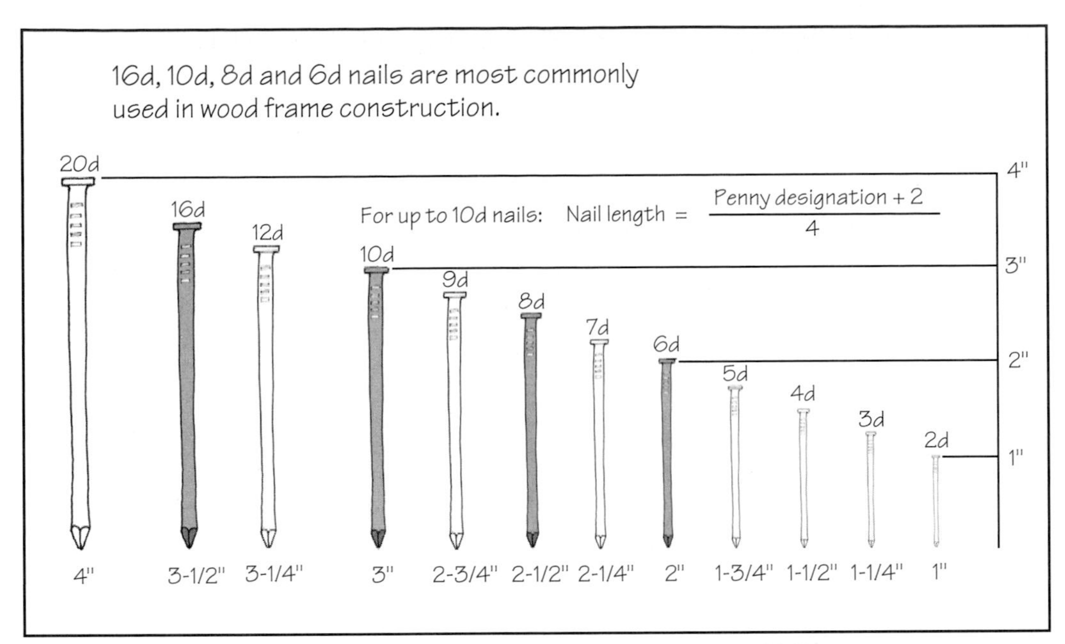

FIGURE 12.35 Standard sizes of common nails (approximately one-half full size).

A face-nailed connection is the strongest of the three because it has the highest withdrawal resistance. The withdrawal resistance of a nail is a function of the nail's orientation with respect to the grain of wood in the holding member (the member that contains the tip of the nail). If the axis of the nail is parallel to the grain of wood in the holding member, the withdrawal resistance is extremely small. In practice, it is assumed equal to zero. Withdrawal resistance is highest if the nail's axis is perpendicular to the grain of wood in the holding member, such as in a face-nailed connection.

End nailing, in which the nails are parallel to the grain in the holding member, is the weakest connection. End nailing is acceptable only in situations where the member is not subjected to withdrawal. Toe nailing is stronger than end nailing, but it is used where access for end nailing is unavailable. A fourth type of connection, referred to as *blind nailing,* is used in finished wood flooring.

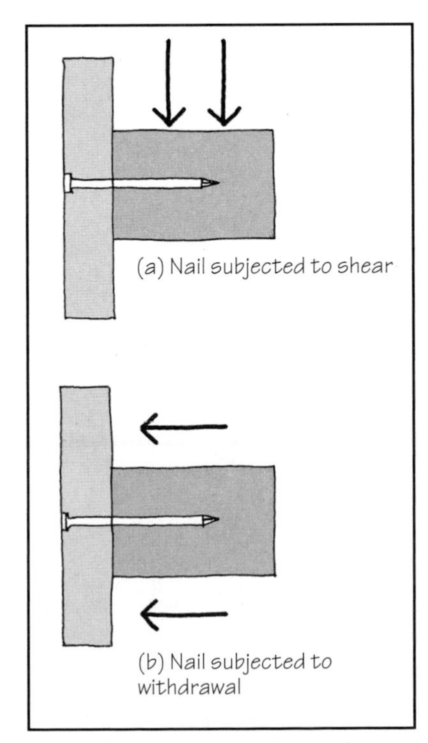

FIGURE 12.36 Nails subjected to **(a)** shear and **(b)** withdrawal.

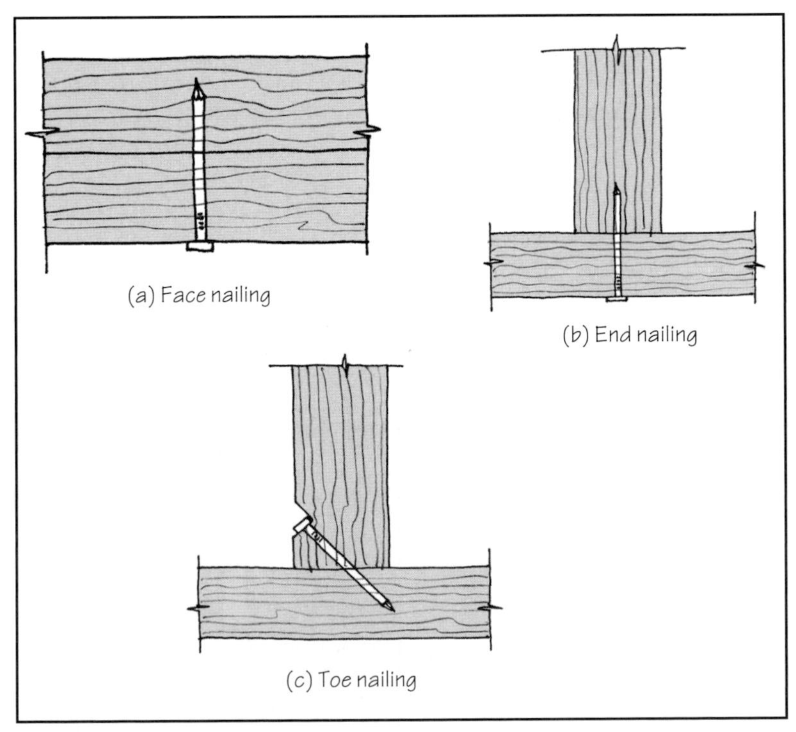

FIGURE 12.37 Types of nailed connections used in wood frame construction.

FIGURE 12.38 A typical nailing gun.

POWER-DRIVEN NAILING AND STAPLING

Although manual hammering is still used to drive nails, power-driven nailing (through pneumatic or electric nailing guns) has become the preferred method of making connections, Figure 12.38. In the power-driven system, nailing is achieved by the pull of a trigger, which not only speeds the process substantially but is also less tiring on the worker.

Nails used in nailing guns are thinner and smaller than the corresponding pennyweight sizes of hand-driven nails. Manufacturers provide the equivalence between nails used in guns and the nails used in manual hammering. Power stapling is also a recognized method of fastening wood members, generally used as an alternative to 8d or 6d nails.

NAIL POPPING

Nail popping—nails sticking out of the wood members—is a problem primarily in floor sheathing that is nailed to the joists. It occurs as the floor joists dry and shrink in size, which pulls them away from the sheathing, Figure 12.39. Because the load on sheathing is downward, the nails pop out through the sheathing.

Nail popping is particularly critical if the floor finish consists of thin, easily penetrable material, such as vinyl tiles. It can be reduced by using I-joists, trussed joists, or solid lumber joists that are relatively drier. Field gluing the sheathing to the joists before nailing them with ring shank nails further reduces nail popping, Figure 12.40. Gluing also stiffens the floor by integrating the sheathing with the floor joists to form a floor consisting of T-sections, in which the sheathing works as the flange of the T-section and the joist works as the web.

SCREWS AND BOLTS

Although screws have a much higher withdrawal resistance (holding power) than nails, they are not often used in structural wood framing because they take longer to install and are more

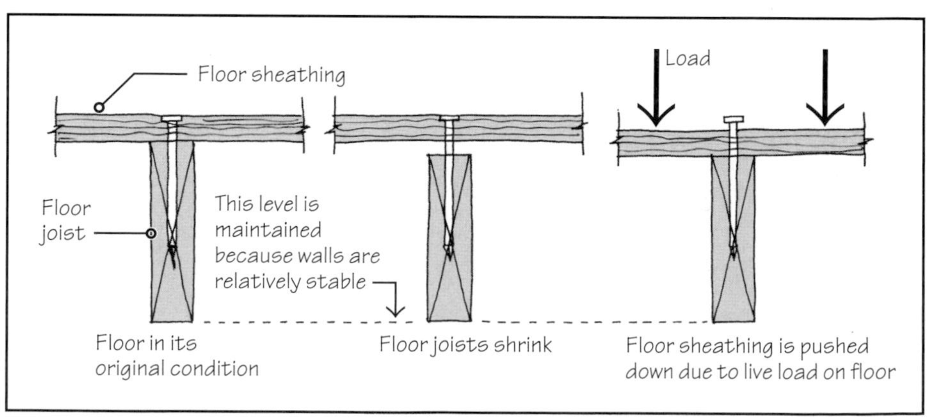

FIGURE 12.39 Mechanics of nail popping.

FIGURE 12.40 Gluing and nailing of floor sheathing to floor joists.

expensive than nails. However, they are commonly used in cabinet work, in furniture, and for fastening door and window hardware, such as hinges. Drywall screws, used for attaching wallboard, are also commonly used. In heavy structural members, lag screws or bolts are used. A lag screw has the shank of a screw but the head of a bolt, Figure 12.41.

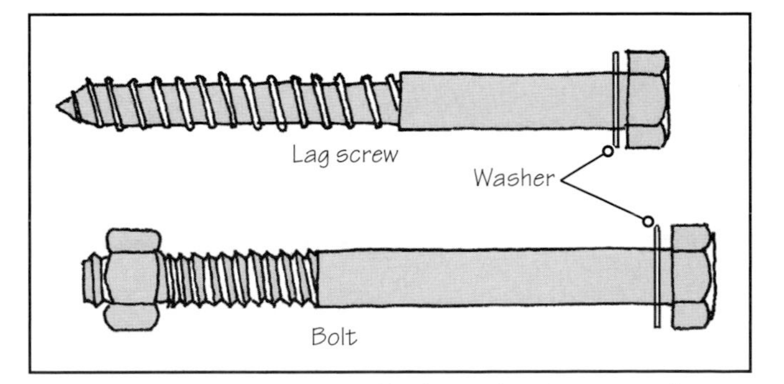

FIGURE 12.41 Lag screws and bolts used in heavy wood framing.

12.10 SHEET METAL CONNECTORS

A number of metal connectors are used in contemporary wood buildings. Nails used in conjunction with metal connectors are loaded in shear rather than in complete or partial withdrawal. Standard connectors are made of galvanized steel, but stainless steel connectors are also available. Metal connectors can be divided into two types:

- Light-gauge sheet steel connectors (generally 16 to 20 gauge) are used in conventional wood framing, such as the joints between a beam and floor joists, between rafters or trusses and top plate and between studs and bottom plate, and so on.

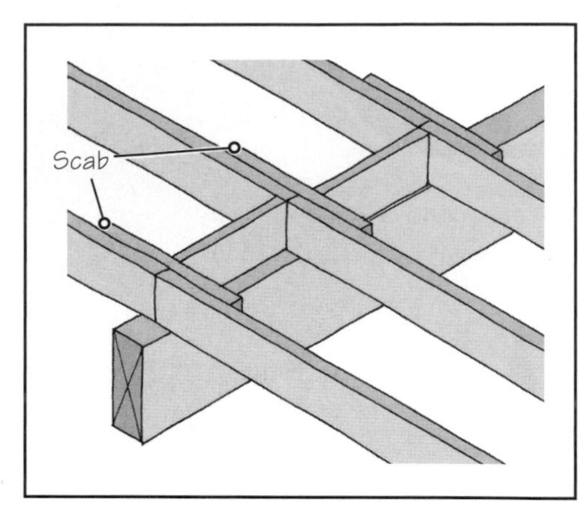

FIGURE 12.42 Floor joists resting over a supporting beam (see also Figure 14.20).

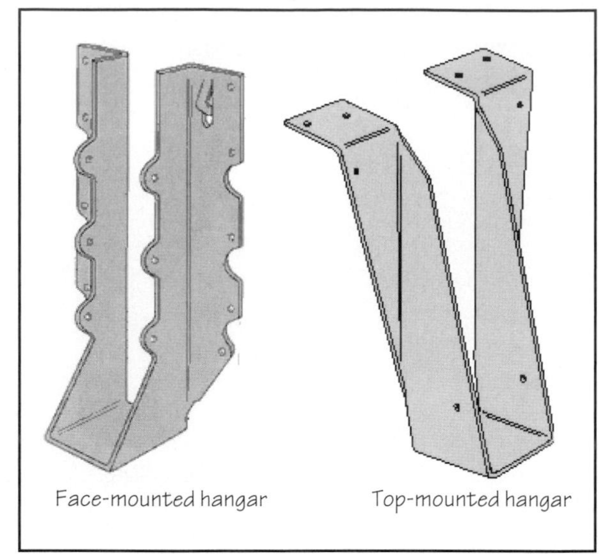

FIGURE 12.43 Typical joist hangers.

FOCUS ON SUSTAINABILITY

Sustainability in Engineered Wood Products

Like lumber, manufactured wood products are among the most sustainable of building materials, especially when produced from wood sources certified by the Forest Stewardship Council (FSC).

Many of the characteristics that make manufactured wood products highly desirable as construction materials also make them a more sustainable product. For instance, I-joists are structurally efficient and can be ordered to size and length, reducing job-site waste. They use only 50% as much wood as comparable sawn lumber members and can be manufactured using fast-growing second- or third-growth trees, mixed species, or other underutilized species. The same applies to other engineered wood products.

One of the long-standing debates related to manufactured wood products is that they are produced with formaldehyde resin adhesives. These adhesives emit a gas into the air that is known to cause adverse health effects in humans. Emission levels are highest when the product is new, but they decrease over time. In the United States, the Department of Housing and Urban Development has set standards for allowable levels of formaldehyde emissions, which manufacturers currently meet. According to the Canadian Wood Council, it is expected that standards will become increasingly more stringent in the future.

There are two types of formaldehyde commonly used. Urea formaldehyde is generally used for products designed for indoor applications. These include particle board, used as a floor underlayment and in shelving, hardwood plywood paneling, and medium-density fiberboard, used in cabinets and furniture. Some of the hazardous effects can be mitigated by encasing these products in laminates or water-based sealants. The use of adequate ventilation and the maintenance of moderate interior humidity levels also help reduce potential problems.

Phenolformaldehyde resin emits formaldehyde gas at much lower levels. It is darker in color and is used to manufacture products such as softwood plywood and oriented strandboard intended for exterior use. Designers are beginning to specify the use of these materials for interior applications as a way to reduce problems with indoor air quality.

- Heavy-gauge sheet steel connectors (generally 7 to 12 gauge) are used in heavy wood framing, such as in joints between a post and a beam, between two beams, between post and foundation, and so on. They are also used in high-wind or high-seismic regions in wood light frame buildings, which would otherwise require light-gauge connectors.

Steel connectors are integral to wood framing, but only a few commonly used connectors are discussed here. The most commonly used connector is a joist hanger, which is used for connecting floor joists with a supporting beam. Joist hangers can reduce the overall height of the floor because, without their use, the joists will have to rest over the beam, Figure 12.42.

Two types of joist hangers are commonly used: (a) face-mounted hanger and (b) top-mounted hanger, Figure 12.43. Face-mounted hangers are used to connect joists to a beam or header. Top-mounted hangers are used where face mounting is not possible, such as when connecting joists to a steel beam, Figure 12.44.

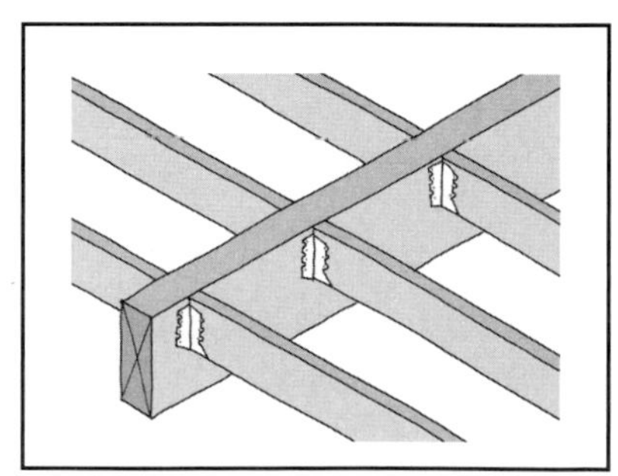

FIGURE 12.44(a) The use of face-mounted joist hangers.

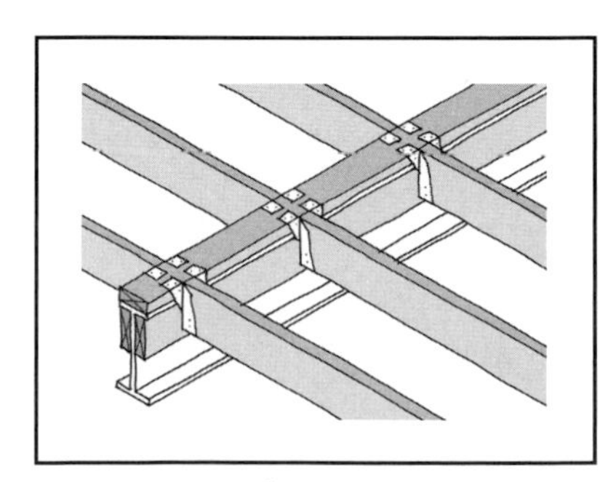

FIGURE 12.44(b) The use of top-mounted joist hangers.

PRACTICE QUIZ

Each question has only one correct answer. Select the choice that best answers each question.

20. Several different types of nails are used in wood frame construction. They are typically distinguished from each other by
- **a.** nail head and nail shank.
- **b.** nail head and nail tip.
- **c.** nail tip and nail shank.
- **d.** nail head and type of threads.
- **e.** none of the above.

21. The nails most commonly used in wood frame construction are
- **a.** 4d, 6d, 9d, and 12d.
- **b.** 6d, 12d, 18d, and 24d.
- **c.** 6d, 10d, 16d, and 20d.
- **d.** 6d, 8d, 10d, and 16d.
- **e.** 6d, 8d, 10d, and 12d.

22. Nails work best when they are subjected to
- **a.** shear.
- **b.** tension.

23. Which of the following nailed connections is the strongest?
- **a.** Face nailing
- **b.** Toe nailing
- **c.** End nailing

24. Nail popping is primarily a problem in floor sheathing and is caused by the shrinkage of floor sheathing.
- **a.** True
- **b.** False

25. Joist hangers are steel connectors
- **a.** that connect floor joists with a supporting beam.
- **b.** that connect floor joists with the supporting wall.
- **c.** that connect floor joists with floor sheathing.
- **d.** all the above.

26. The adhesives used in making engineered wood products contain
- **a.** urethane.
- **b.** styrene.
- **c.** formaldehyde.
- **d.** lignin.
- **e.** none of the above.

27. Gaseous emissions produced by engineered wood products decrease over time.
- **a.** True
- **b.** False

Answers: 20-a, 21-d, 22-a, 23-a, 24-b, 25-a, 26-c, 27-a.

KEY TERMS AND CONCEPTS

American Plywood Association (APA)
Engineered wood products
Glue laminated
Glulam
Glulam grades
Industrial wood products
Laminated veneer lumber (LVL)
Nail plate
Nail sizes and types
Nailed connections
Oriented strand board (OSB)
Panel point
Panel ratings
Parallel strand lumber (PSL)
Plywood
Sheet metal connectors
Splice plate
Structural composite lumber
Third party certification
Trussed joist
Wood fasteners
Wood floor truss
Wood I-joist
Wood panels
Wood roof truss
Wood truss
Wood veneer grades

REVIEW QUESTIONS

1. Using sketches and notes, explain the difference between a balanced and an unbalanced glulam beam and the situations in which they are specified.
2. Using sketches and notes, explain the composition of a wood I-joist. What are the advantages and disadvantages of wood I-joists as compared with solid lumber?
3. Using a sketch and notes, explain the following terms related to a wood roof truss: (a) top chord (b) bottom chord (c) web member (d) panel point (e) nail plate.
4. Using a sketch, explain the difference between a roof truss and a floor truss. In which situations, will you consider the use of wood floor trusses?
5. Which type of nail is commonly used for connections between framing members in wood light-frame structures? How do we specify the size of nails?
6. Using sketches, illustrate the following connections in wood frame members: (a) face-nailed connection, (b) end-nailed connection and (c) toe-nailed connection.

SELECTED WEB SITES

American Wood Council (www.awc.org)
APA—The Engineered Wood Association (www.apawood.org)
Canadian Sustainable Forestry Certification Coalition (www.certificationcanada.org)
Canadian Wood Council (www.cwc.ca)
Forest Stewardship Council (www.fscus.org)
Scientific Certification Systems (www.scscertified.com)
SmartWood (www.rainforest-alliance.org)
Structural Building Components Industry (www.sbcindustry.com)

FURTHER READING

1. Canadian Wood Council. *Wood Reference Handbook—A Guide to the Architectural Use of Wood in Building Construction*, 1991.
2. International Code Council. *International Residential Code*. Falls Church, VA: International Code Council.

CHAPTER 13

Wood Light Frame Construction–I

CHAPTER OUTLINE

13.1 Evolution of Wood Light Frame Construction

13.2 Contemporary Wood Light Frame—The Platform Frame

13.3 Frame Configuration and Spacing of Members

13.4 Essentials of Wall Framing

13.5 Framing Around Wall Openings

13.6 Essentials of Floor Framing

13.7 Roof Types and Roof Slope

13.8 Essentials of Roof Framing

13.9 Vaulted Ceilings

13.10 Sheathing Applied to a Frame

13.11 Equalizing Cross-Grain Lumber Dimensions

PRINCIPLES IN PRACTICE: Constructing a Two-Story Wood Light Frame Building

PRINCIPLES IN PRACTICE: How a WLF Building Resists Loads

Wood light frame (WLF) construction, invented nearly two centuries years ago, is currently the most widely used construction system for residential and low-rise commercial buildings in North America.

Wood light frame (WLF) construction is the dominant system for contemporary residential and light commercial construction in the United States and several other countries. It is examined in this chapter and the next. Chapter 13 focuses on the framing aspects of WLF construction, and Chapter 14 explores interior and exterior finishes.

This chapter begins with a brief overview of the evolution of WLF, followed by a discussion of the organizational principles of this type of construction system. Then, the essential characteristics and components of the system, including floor, wall, and roof framing and sheathing, are investigated. Concluding this chapter are two *Principles in Practice* features: The first is an example of the construction process, and the second feature describes how loads are resisted by the structure.

13.1 EVOLUTION OF WOOD LIGHT FRAME CONSTRUCTION

As the European settlers colonized America and built on the new land, they tried to re-create the ambience and spirit of their original countries. The early American buildings and construction systems were, therefore, derived mainly from European practices. Wood light frame construction, on the other hand, was invented in the United States and is uniquely American.

WLF construction was first used around 1830 in the Chicago area, which was then developing as a major urban center in the United States. There is some controversy about who should be credited with this new system. Two names—Augustine Taylor and George Washington Snow—both carpenters from the Chicago area, are said to have invented the system independent of each other.

The WLF system has evolved over nearly two centuries and has adapted marvelously to several technical innovations in buildings, such as insulation, electricity, piped water supply, and sewage disposal, which did not exist during the time of Taylor or Snow. The system's adaptability and the ease with which it could be erected led to its success and popularity. So successful has the system been that it is now the dominant system for contemporary residential and light commercial construction in the United States and several other countries.

Balloon Frame

Prior to the invention of the WLF system, most buildings in the United States were built of thick masonry walls, heavy timber posts, beams, and thick wood plank floors and roofs. The joints between heavy timber posts and beams consisted of mortise-and-tenon, or dovetail, joints (Section 12.9). Because no nails or screws were used to secure the joints, expert artisanry was needed to provide tight joints.

When the WLF system was invented to replace the heavy construction system just mentioned, it was called the *balloon frame*. It consisted of thin, closely spaced, vertical wood members (called studs) and similar floor and roof framing members (joists and rafters).

The connections between members were made with simple nails. Measuring, hammering, and sawing were, therefore, the only skills needed for the connections, making sophisticated carpentry and joinery skills unnecessary. In fact, the entire structural frame of a WLF building could be erected by a relatively unskilled crew of two or three people.

An additional advantage of the WLF system lay in its lightweight members, which needed little effort to hoist and install. Because the members were small in cross section, the total amount of wood consumed in a balloon frame building was only a fraction of that consumed in a traditionally built building of the same size. In fact, many historians believe that various U.S. cities could not have developed as rapidly as they did without the invention of the balloon frame.

As with most other inventions, the birth of the WLF system may be attributed to necessity. The necessity arose from an abundance of wood in the new country, where skilled carpenters were scarce. Immigration at that time consisted largely of poor, unskilled workers unable to find work in Europe. This was in direct contrast to a wealth of skilled carpenters in Europe, where forests were gradually depleting.

The term *balloon frame* was coined out of contempt by the critics of the system at the time, who considered it too insubstantial to withstand the forces of nature. Although time proved them wrong, the term continued to be used. In the balloon frame, the studs run the full height of the building, from the sill plate at the bottom to the top plate under the rafters. The intermediate floor joists (in a two-story building) rest on a 1×6 (or 1×4) ledger—called a *ribband*—notched into the studs, Figure 13.1.

In addition to ribbands, fire-stops are provided at floor lines in a balloon frame, Figures 13.1 and 13.2. Without the fire-stops, the air space between the studs is continuous from the top to the bottom. This creates a fire risk because the fire occurring in the stud cavity of one floor can spread to that of the other floor. The fire-stops separate the cavities and deprive the fire of the oxygen required for its growth.

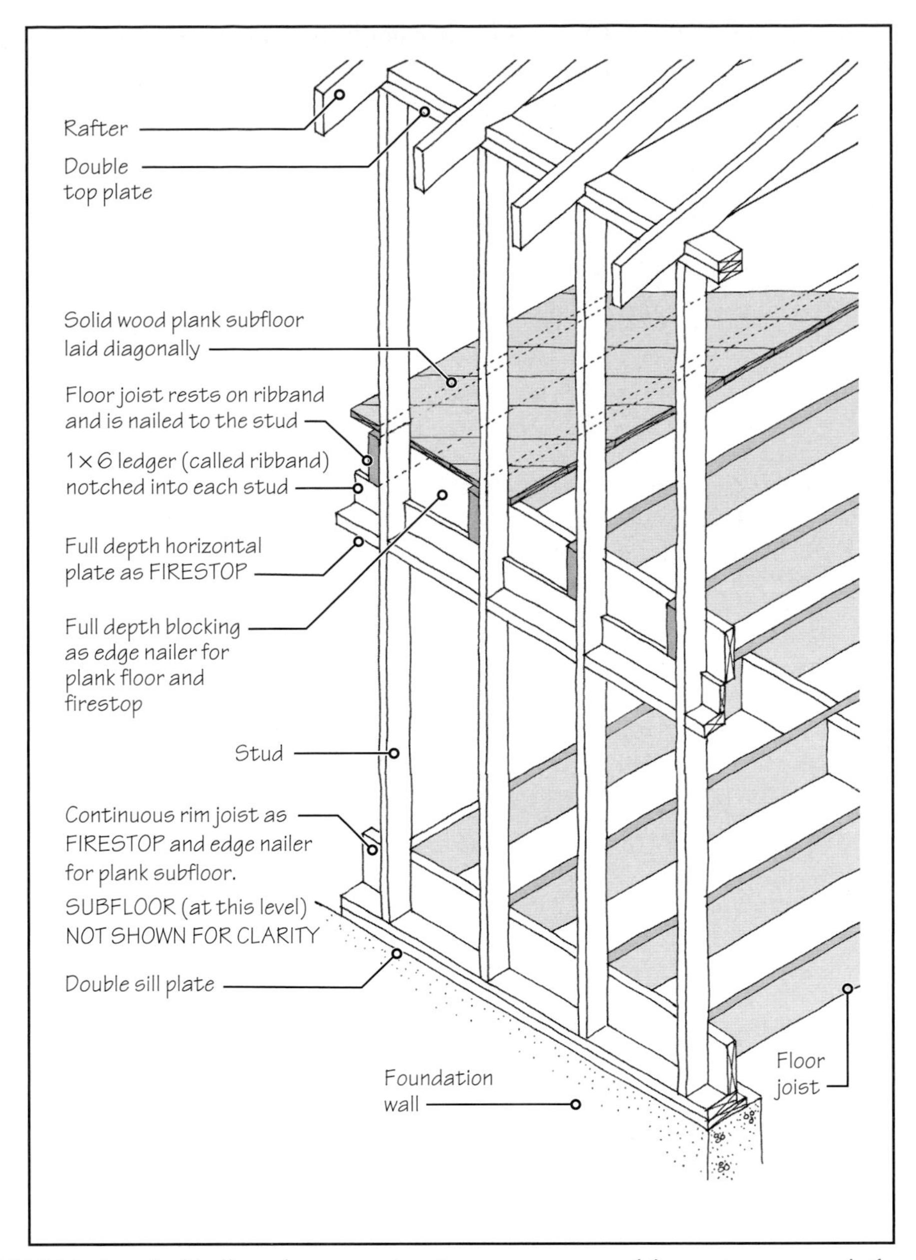

FIGURE 13.1 A typical balloon frame construction—a precursor of the contemporary platform frame.

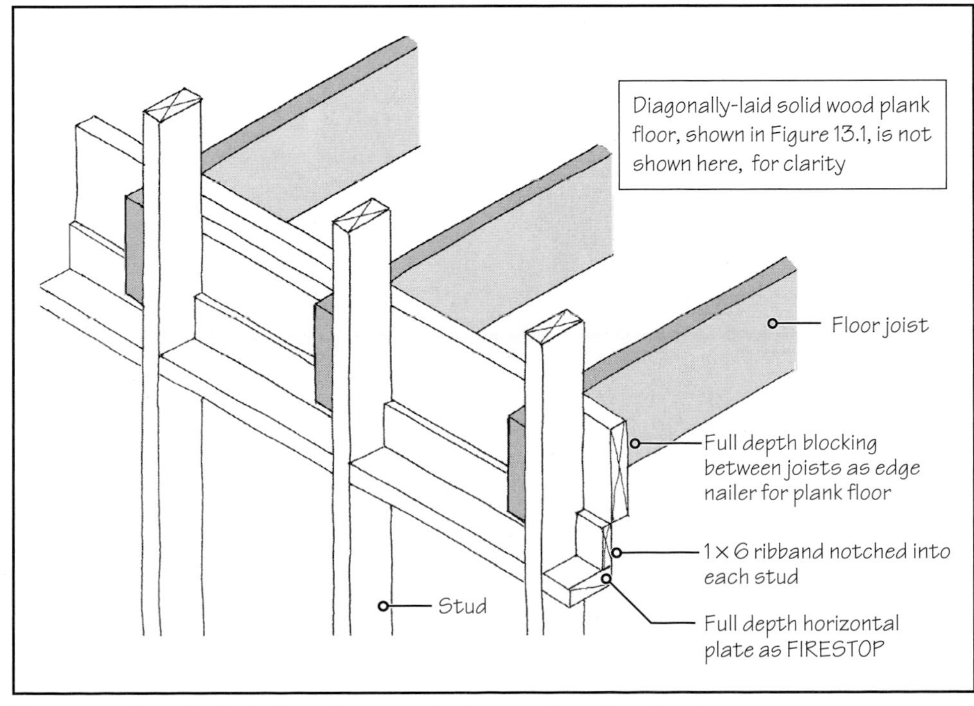

FIGURE 13.2 Detail of a balloon frame at the intermediate floor level.

The continuity of studs was recognized as a major limitation of the balloon frame. This limitation became more acute as the greater use of wood reduced the availability of long, straight members and made them more expensive. The balloon frame was, therefore, modified over time into the *platform frame,* in which the individual studs are only one story high. Contemporary WLF construction consists mainly of the platform frame, the subject of discussion in this chapter.

13.2 CONTEMPORARY WOOD LIGHT FRAME—THE PLATFORM FRAME

In the balloon frame, wall framing, floor framing, and roof framing are entirely completed before constructing the structural floor (referred to as the *subfloor,* or *floor sheathing*). In the platform frame, on the other hand, the subfloor at the first floor level is completed soon after laying the floor joists at that level. This subfloor provides a platform for the workers to stand on and to build the following story.

With the subfloor completed, the first-floor walls are then erected. These walls bear on the subfloor. Later, the floor joists for the second floor are erected, which are then sheathed with plywood or OSB panels to make the subfloor for the second story. If there is an additional story in the building, it is built exactly the same way as the lower story. Finally, the roof framing is completed. In other words, in the platform frame, the structural frame is erected story by story.

A three-dimensional view of a typical platform frame structure is shown in Figure 13.3. Because the walls are not continuous, fire-stops are automatically provided at each floor

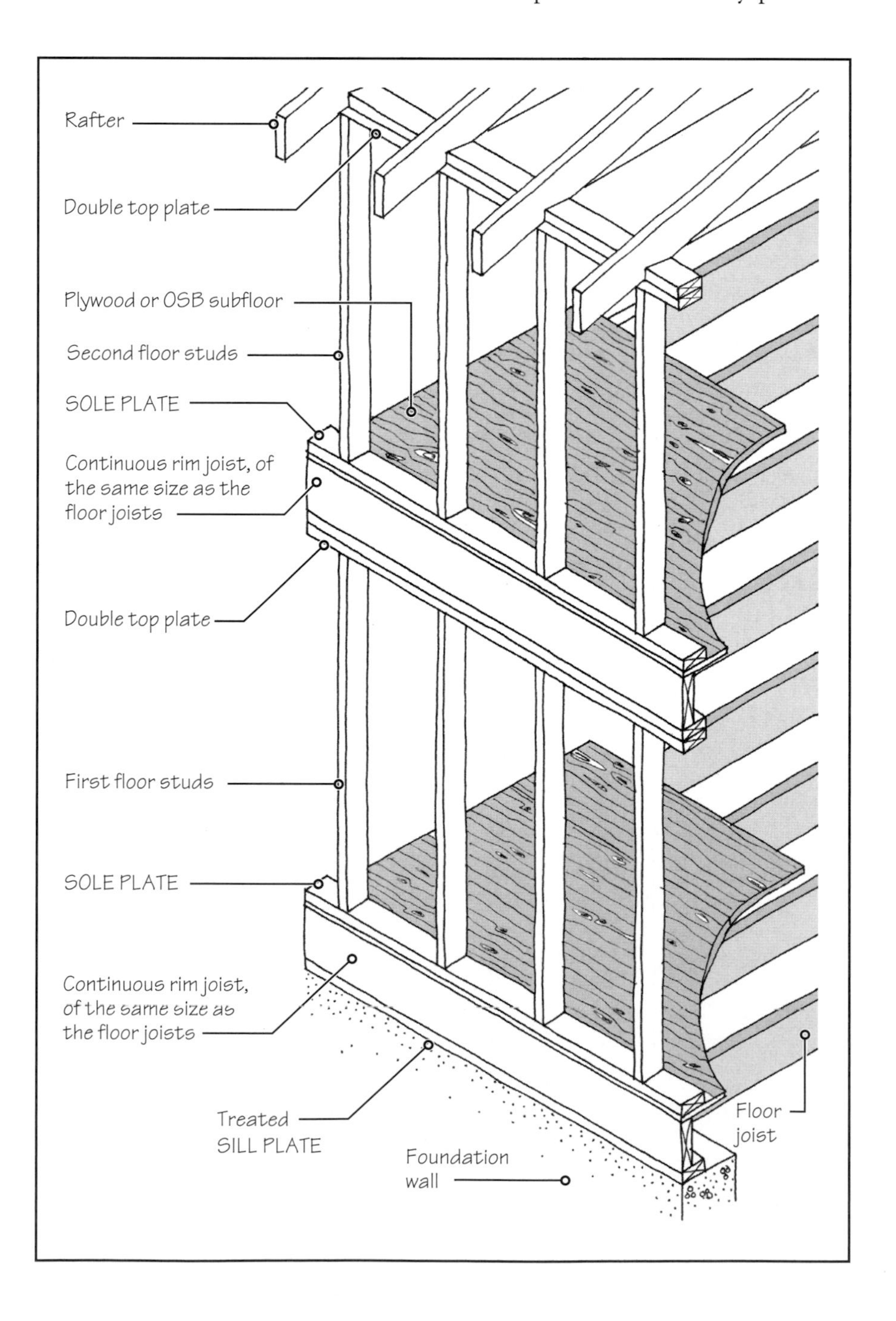

FIGURE 13.3 A typical platform frame construction.

level. Thus, there is no need to add special members for fire-stopping, as is the case in a balloon frame. Note that the roof framing is identical in both the platform frame and the balloon frame.

The platform frame system not only facilitated construction, it also increased safety for the workers because they could use the floor platform as the scaffold for constructing the subsequent floor. Exterior scaffolding and tall construction ladders are unnecessary. Additionally, it became possible to build structures of more than two stories. With the balloon frame, the practical height limitation of the building was two stories because the studs longer than two stories were—and are—not easily available. (For details on the process and sequence of the construction of a typical WLF building, see the first "Principles in Practice" at the end of the chapter.

DIMENSIONAL STABILITY OF BALLOON AND PLATFORM FRAMES

The continuity of studs in a balloon frame, however, gives it some benefits over the platform frame. For instance, the balloon frame is dimensionally more stable because the balloon frame has a much smaller overall cross-grain lumber dimension than the platform frame.

Figure 13.4 shows the cross sections through a typical platform frame and a typical balloon frame. Assuming that 2 × 12 solid lumber floor joists are used in both frames, the total cross-grain lumber in the platform frame is 33 in., as compared to 6 in. in a two-story balloon frame. Because lumber shrinks negligibly along the grain but significantly across the grain (Section 11.5), the shrinkage and swelling in a balloon frame is much smaller than in a platform frame.

Excessive shrinkage and swelling results in nail popping and cracking of brittle finishes applied to the frame, such as, gypsum board and brick veneer. The detailing of the platform frame to accommodate shrinkage and swelling effects is covered in Section 13.11.

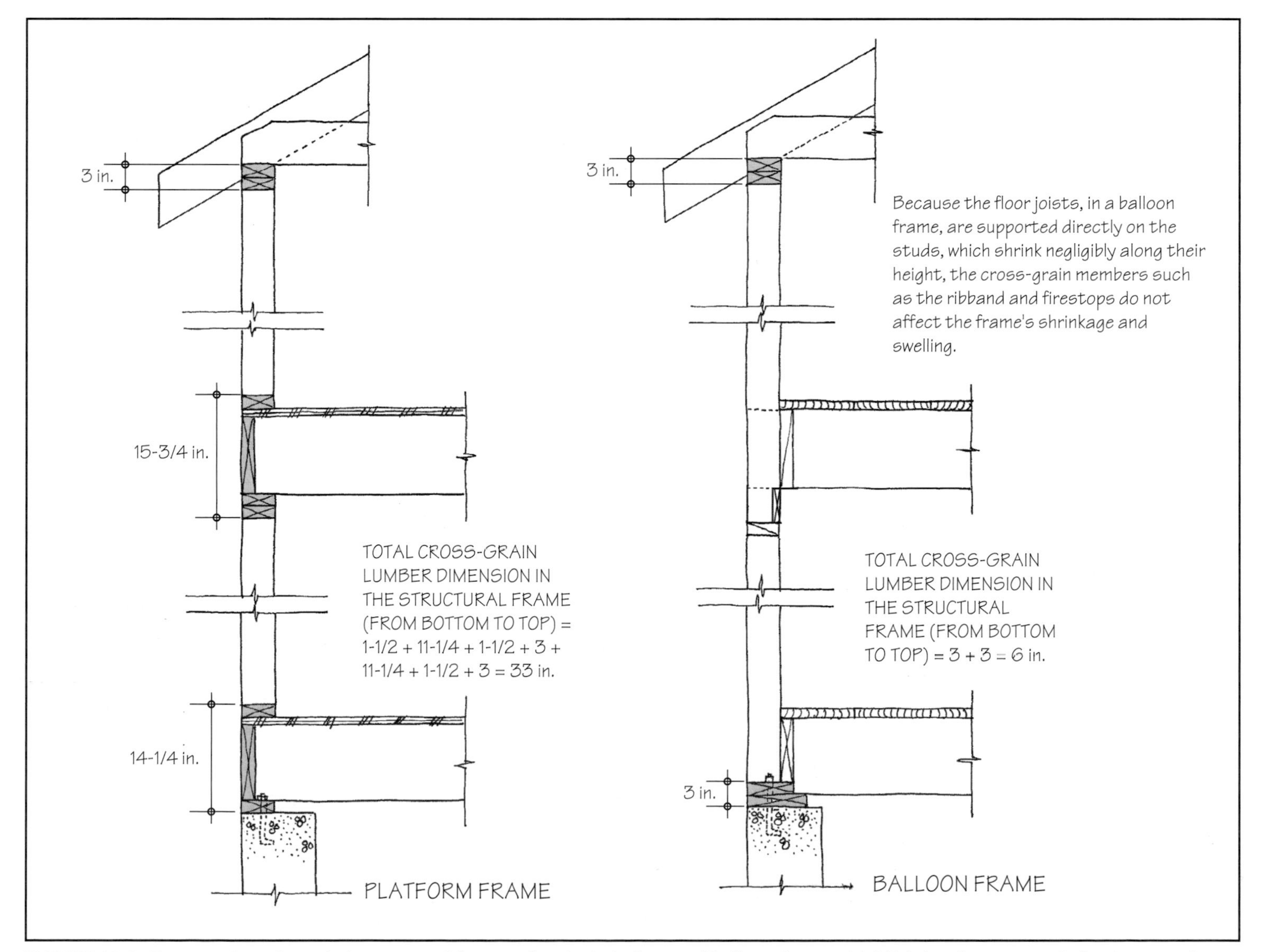

FIGURE 13.4 Cross sections through the structural frames of typical platform and balloon frames, highlighting the cross-grain lumber members in each.

STRUCTURAL STABILITY OF BALLOON AND PLATFORM FRAMES

An additional advantage of the balloon frame is its greater structural stability. Continuous studs provide a more positive connection between the first and the second floors of a balloon frame, giving greater strength and stability in resisting the lateral loads.

In the platform frame, the connection between the two floors is provided by nailing the bottom plate of walls to the underlying floor frame. Additional connection is provided through the exterior sheathing. The connection so obtained between the floors is, however, not as strong as that provided by continuous studs. Therefore, in high-wind or seismic regions, a platform frame requires metal straps and connectors to improve the connectivity between floors. See the second "Principles in Practice" at the end of this chapter.

13.3 FRAME CONFIGURATION AND SPACING OF MEMBERS

A common feature among the three principal structural elements—the walls, floors, and roof—of a platform frame building is that each of these elements is made of several parallel, closely spaced, repetitive members joined at each end to a continuous cross member that runs perpendicular to the parallel members, Figure 13.5. Cross members connect the parallel members together to form a *frame* and also play an important structural role.

In a wall assembly, the cross members are the top and the bottom plates. In a floor assembly, the cross member at each end is a continuous member of the same size as the joists, called the *rim joist,* or *band joist.* In a rafter-and-ceiling joist assembly, the top plates of the supporting walls on either side function as the cross members. The triangular

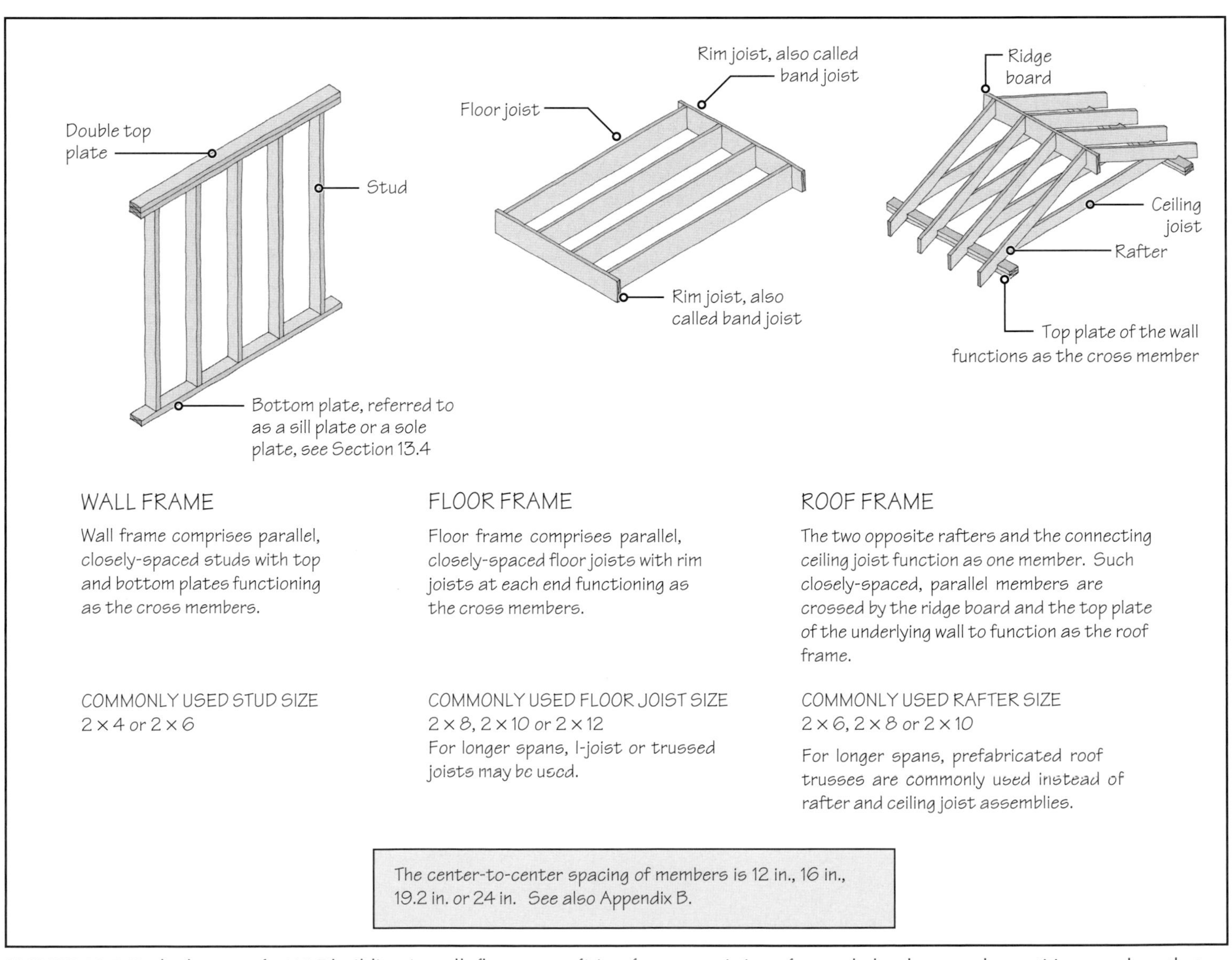

FIGURE 13.5 Each element of a WLF building (a wall, floor, or roof) is a frame consisting of several closely spaced, repetitive members that are connected at each end by a cross member.

rafter-and-ceiling joist assembly constitute repetitive members. In a trussed roof, the trusses are the repetitive members.

Most framing members in a WLF building are 2-by members (actual thickness is $1\frac{1}{2}$ in.). Generally, 2 × 4 or 2 × 6 members are used for walls; 2 × 8, 2 × 10, or 2 × 12 members are used for the floor joists; and 2 × 6, 2 × 8, or 2 × 10 are used for rafters.

The center-to-center spacing of members is either 12 in., 16 in., 19.2 in., or 24 in., as determined by loads. These dimensions are also based on the size of the panels used as *sheathing* over the framing members—plywood, OSB, or gypsum board panels (typically 4 ft × 8 ft). Although 19.2-in. spacing is not commonly used, it is recognized as a possibility because like 12-in., 16-in., and 24-in. spacings, it also divides the 8-ft (96-in.) dimension by a whole number.

PRACTICE QUIZ

Each question has only one correct answer. Select the choice that best answers the question.

1. Roughly how old is WLF construction?
a. 1,000 years
b. 800 years
c. 600 years
d. 400 years
e. 200 years

2. When WLF was first discovered, it was called a
a. balloon frame.
b. rigid frame.
c. skeleton frame.
d. platform frame.
e. portal frame.

3. WLF construction originated in
a. the United Kingdom.
b. France.
c. Australia.
d. Norway.
e. the United States.

4. In the early version of WLF construction, the studs were continuous from the foundation to the roof.
a. True
b. False

5. In time, the early version of WLF construction was modified to what is now called a
a. balloon frame.
b. rigid frame.
c. skeleton frame.
d. platform frame.
e. portal frame.

6. The type of WLF construction that has the largest amount of cross-grain lumber is a
a. balloon frame.
b. rigid frame.
c. skeleton frame.
d. platform frame.
e. portal frame.

7. The framing members in a WLF building are generally
a. 1-by solid lumber members.
b. 2-by solid lumber members.
c. 3-by solid lumber members.
d. 4-by solid lumber members.
e. none of the above.

8. The center-to-center spacings of framing members in a WLF building are generally
a. 12 in., 16 in., or 24 in.
b. 12 in., 24 in., or 36 in.
c. 2 ft, 3 ft, or 4 ft.
d. 4 ft, 6 ft, or 8 ft.
e. virtually any spacing may be used depending on architectural and structural considerations.

9. The center-to-center spacing of framing members in a WLF building is mainly based on
a. age-old practice.
b. structural considerations.
c. the size of sheathing panels.
d. insulation requirements.
e. (b) and (c).

Answers: 1-e, 2-a, 3-e, 4-a, 5-d, 6-d, 7-b, 8-a, 9-e.

13.4 ESSENTIALS OF WALL FRAMING

The top plate in a wall assembly consists of two members, each of the same size as the studs. That is why it is referred to as the *double top plate.* Doubling the top plate allows the floor joists or rafters to be placed anywhere on the top plate. If a single top plate is used, the floor joists and rafters must align with the underlying studs, Figure 13.6. In other words, the double top plate works as a beam supported on adjacent studs and transfers gravity loads from the floor (and roof) to the studs.

Another reason for doubling the top plate is to provide structural continuity in the top plate. The structural continuity in the top plate is needed because the floor (and roof) works as a diaphragm in a wood frame building (see the second "Principles in Practice" at the end of this chapter). This produces tension as well as compression in the top plate (of the exterior walls) under wind and earthquake loads.

Because a single top plate will have discontinuity at the joints, it will not be able to counteract tension. A double top plate can do so, provided the joints between the two plates are staggered. Building codes mandate a minimum 24-in. lap at the joints between the two top plates. For the same reason, the two top plates must be staggered at the corners and junctions

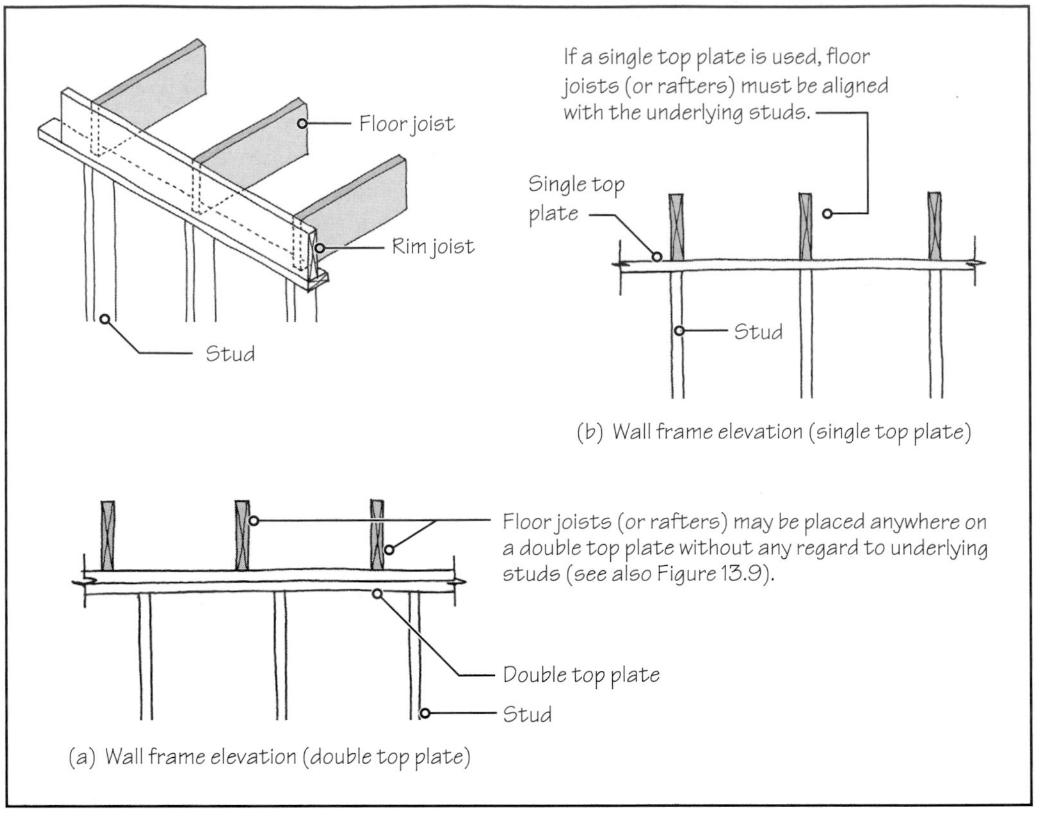

Floor joist
Rim joist
Stud
If a single top plate is used, floor joists (or rafters) must be aligned with the underlying studs.
Single top plate
Stud
(b) Wall frame elevation (single top plate)
Floor joists (or rafters) may be placed anywhere on a double top plate without any regard to underlying studs (see also Figure 13.9).
Double top plate
Stud
(a) Wall frame elevation (double top plate)

FIGURE 13.6 If a double top plate is used in a wall frame, the floor joists or roof rafters need not align with the underlying studs. See Figure 13.9 for an exception.

of the walls, Figure 13.7. If a single top plate is used, a galvanized steel plate is needed at the joints, Figure 13.8. This makes the use of the single top plate additionally inconvenient.

TRIPLE TOP PLATE

If the joists or rafters are 24 in. apart, each joist or rafter delivers greater load on the wall than if they are spaced 16 in. apart or less. Therefore, building codes require that if the supporting wall is made of 2 × 4 studs spaced 24 in. on centers, and if the spacing between joists and rafters is more than 16 in. (19.2 in. or 24 in.), each joist or rafter should be located within 5 in. of the underlying studs, Figure 13.9. If this cannot be done, a third top plate is needed.

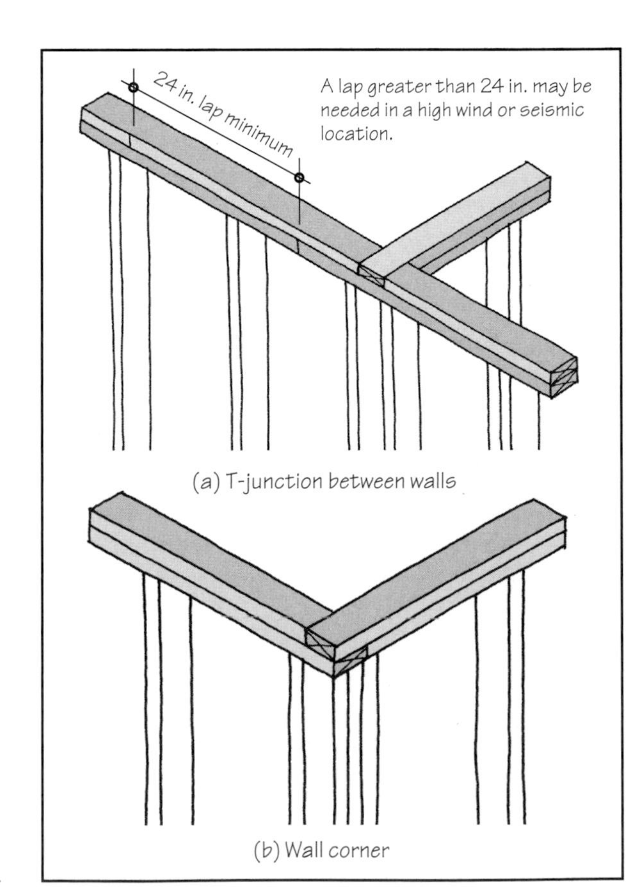

FIGURE 13.7 The joints in the two top plates in a wall must be staggered as shown. Staggered joints are also required at wall T-junctions and corners.

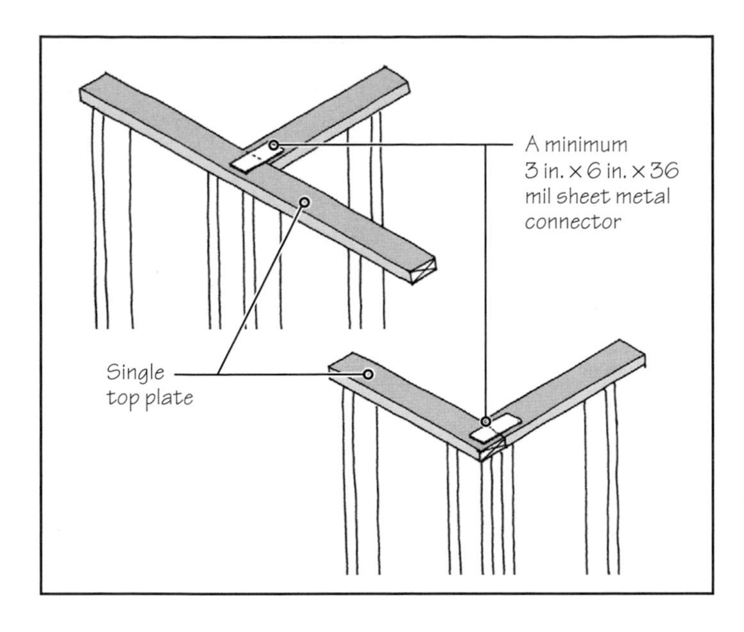

FIGURE 13.8 To provide continuity in a single top plate requires the use of sheet-metal connectors.

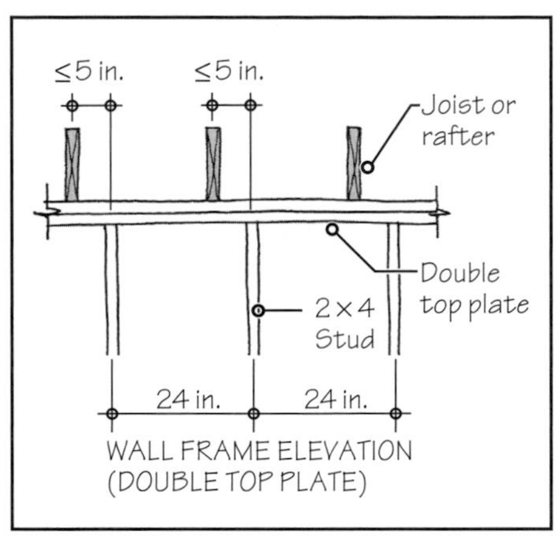

FIGURE 13.9 Restrictions on the placement of joists or rafters on a double top plate if 2 × 4 studs at 24 in. on centers are used.

If the supporting wall is framed with 2 × 6 studs at 24 in. on centers (instead of 2 × 4 studs), the 5-in. restriction or the tripling of the top plate is not required. Because of its inconvenience, a triple top plate is seldom used.

Bottom Plate in a Wall Assembly

The bottom plate in a wall is typically a single plate, called either a sill plate or a sole plate, depending on its location in the building. A *sole plate* is a plate that is not in contact with the foundation. Thus, the bottom plate of a second floor wall is called the sole plate because it is connected to the wood subfloor, not to the foundation, (Figure 13.3).

The bottom plate that is connected to the foundations is referred to as the *sill plate.* (Building foundations are discussed in detail in Section 21.6.) At this stage, it is sufficient to know that the foundations for a WLF building may consist of any one of the following three types:

- A concrete slab-on-grade.
- Reinforced concrete (or concrete masonry) foundation walls. (With such a foundation, the ground floor consists of a wood frame floor with an underlying crawl space.)
- Reinforced concrete basement walls. (The ground floor in this case is also a wood frame floor with a basement space below.)

The sill plate is also referred to in some literature as the *mud sill.* It must either be a preservative-treated wood or a naturally decay-resistant wood specie. The sill plate must be anchored to the foundations with bolts. This anchorage is critical because the entire building is secured to the foundations through these bolts.

Building codes require a minimum of $\frac{1}{2}$-in.-diameter steel bolts that must be embedded at least 7 in. into foundation concrete or masonry. The maximum spacing between bolts is 6 ft in low-wind and nonseismic locations, Figure 13.10(a). In high-wind and seismic locations, the anchor bolts are required to be larger and located closer together. Each sill plate length must have at least two bolts. Additionally, a bolt is required within 12 in. of the end of a sill plate length, Figure 13.10(b).

To reduce air infiltration, a compressible, fibrous felt may be placed between the sill and the foundation, Figure 13.11. This helps to seal the gaps between the sill plate and the uneven surface of the foundation. In termite-infested areas, the use of a continuous termite shield (Section 11.14) is a good practice, in addition to using a preservative-treated sill plate.

NOTE

Bottom Plate—Either a Sill Plate or Sole Plate

A *sill plate* is the bottom plate of a wall frame that is in direct contact with the foundation. It is generally made of preservative-treated wood, although codes allow a naturally decay-resistant wood specie to be used instead of the preservative-treated wood.

A *sole plate* is similar to the sill plate (the bottom horizontal member of a wall frame) that is not in direct contact with the foundation. A sole plate is of an untreated lumber.

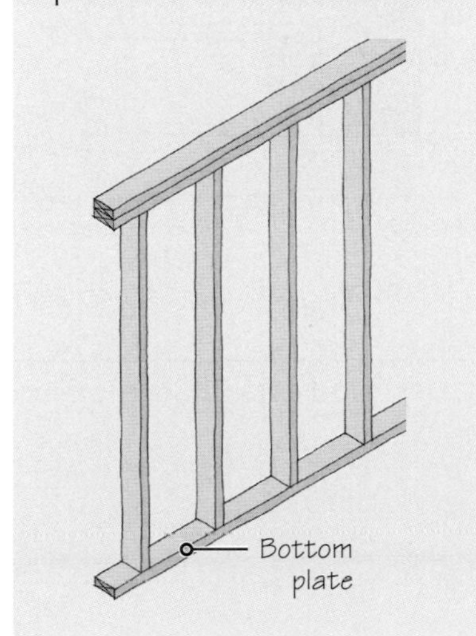

Number of Studs at a Wall Corner or a T-Junction Between Walls

An exterior wall must function as a shear wall (see the second "Principles in Practice" at the end of the chapter). A shear wall is subjected to tension at one end and compression at the other end, caused by the wall tending to overturn under the action of wind or earthquake loads. The compression caused by the lateral loads adds to the compression due to the gravity loads. Therefore, the corner of an exterior wall must be stronger than the field of the wall, requiring a minimum of three studs at the corner.

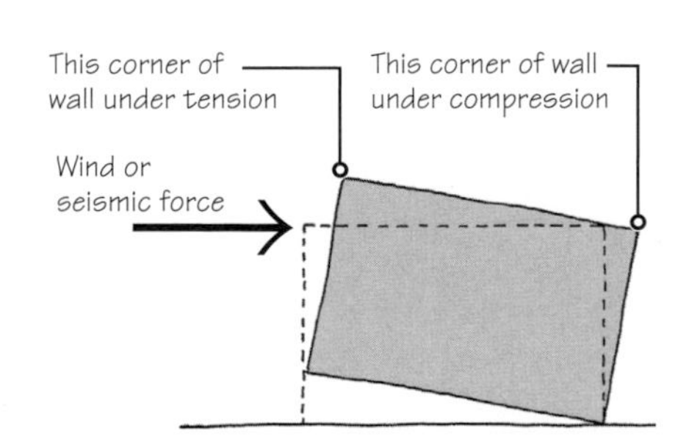

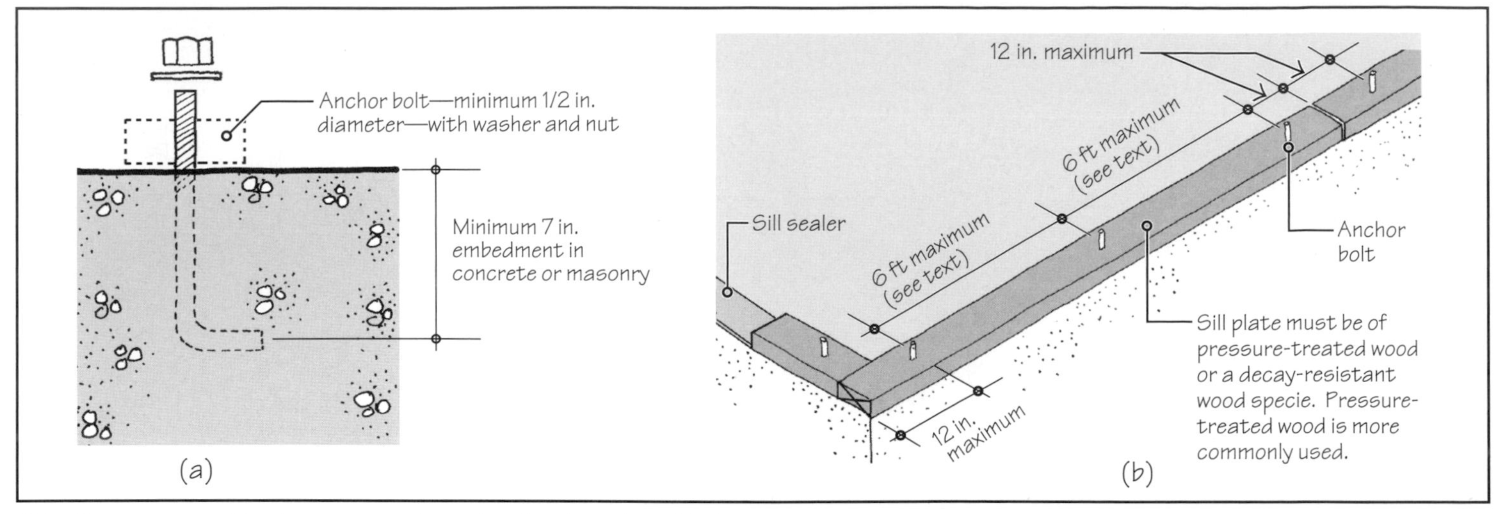

FIGURE 13.10 Specifications and spacing of anchor bolts in a sill plate.

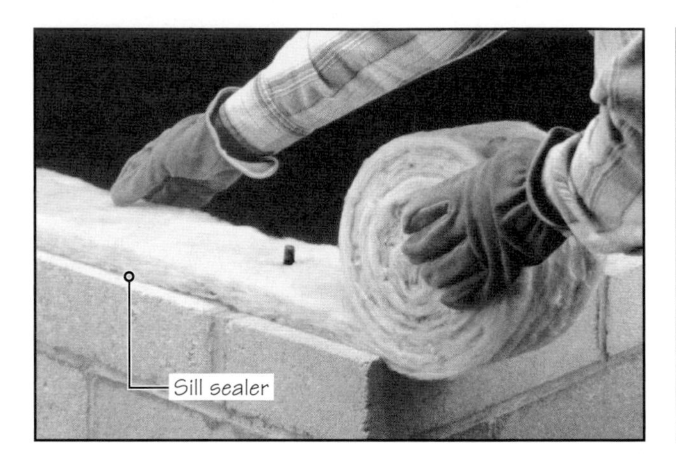

FIGURE 13.11 Sill sealer under a sill plate.

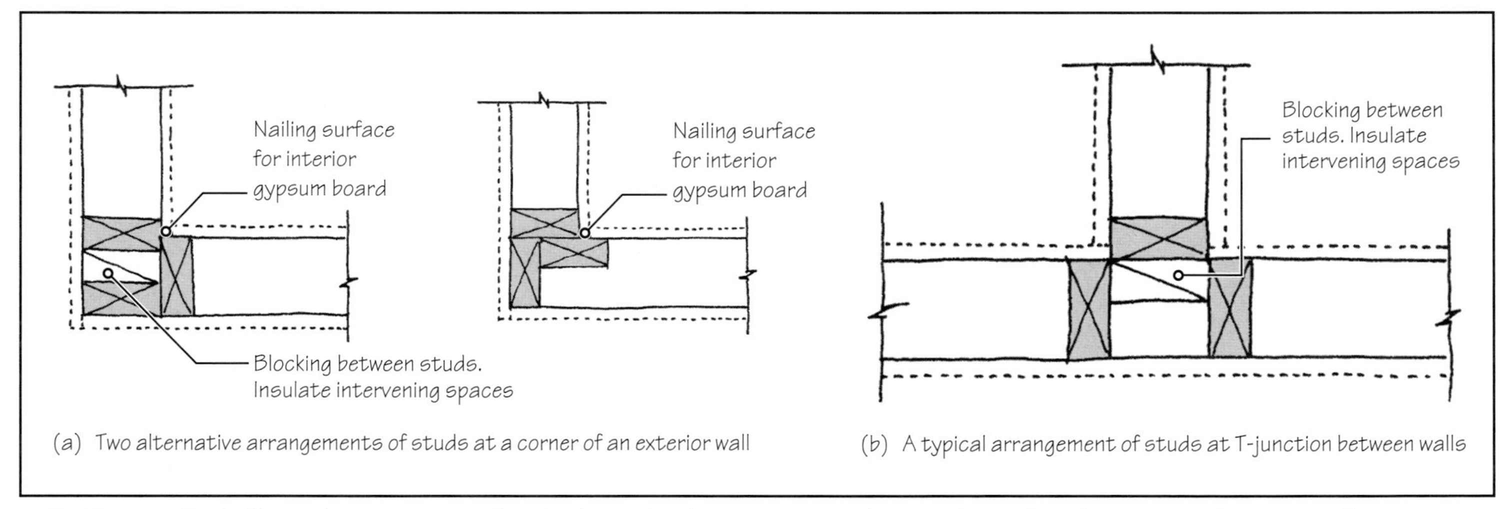

FIGURE 13.12 Typically used arrangements of studs (shown in plan) at a corner of an exterior wall and a T-junction between walls.

In addition to the structural consideration, the provision of three studs at a corner is also required to obtain adequate nailing surface for interior gypsum board and exterior sheathing. Two alternative details are typically used for a three-stud corner, Figure 13.12. This figure also shows a typical detail of studs at a wall T-junction, which also requires three studs.

13.5 FRAMING AROUND WALL OPENINGS

A typical framing around openings in a wall (doors or windows) is shown in Figure 13.13. A wall opening requires the use of *jack studs* on both sides of the opening. A jack stud is a partial-height stud that supports the lintel beam, generally referred to as *header* (or *lintel header* in some literature). One jack stud on both sides of the opening gives $1\frac{1}{2}$ in. bearing for the header. This is adequate for a small opening in a one- or two-story building. If the opening size is large or if the number of floors in the building exceeds two, the use of two or three jack studs may be needed.

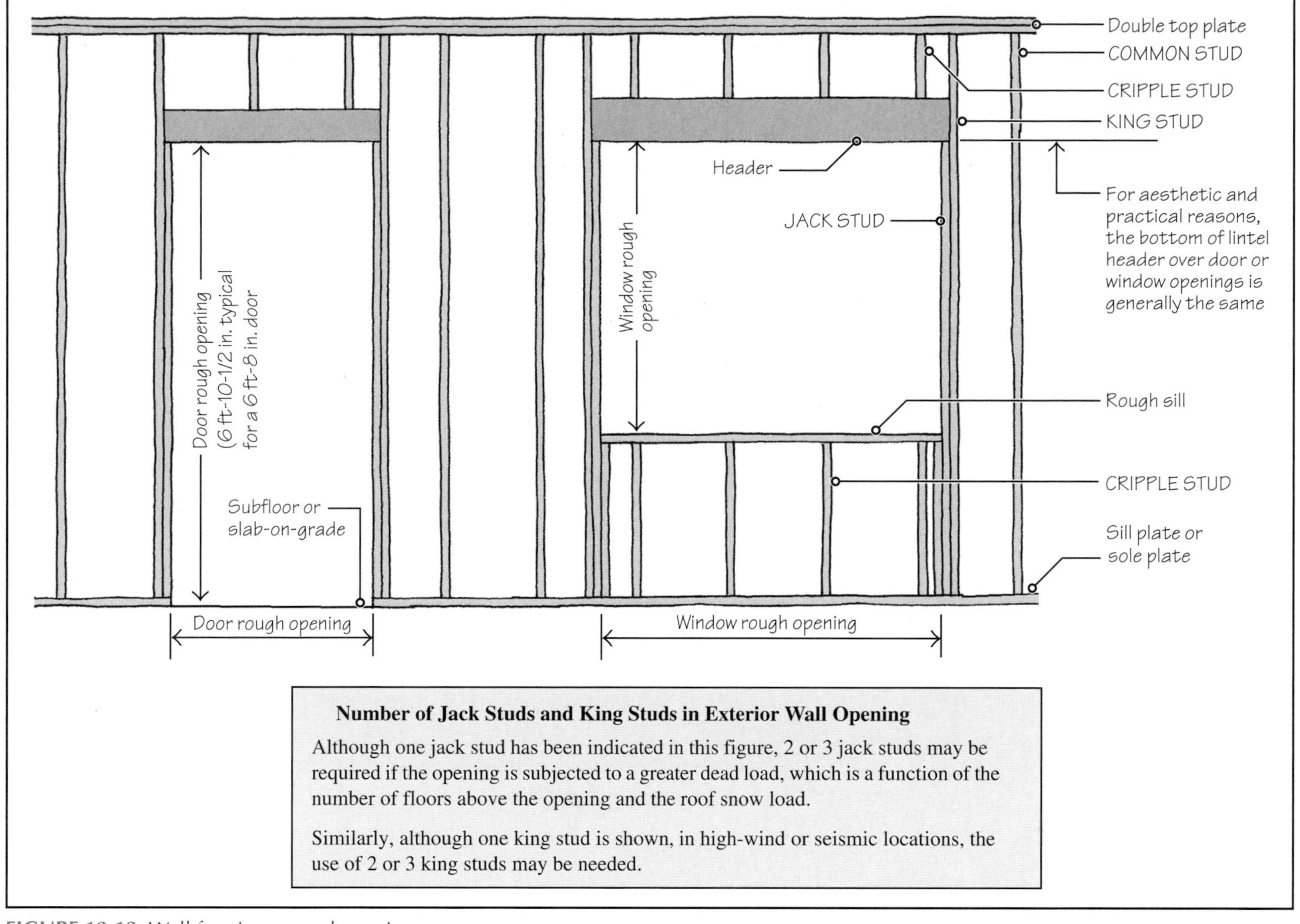

Number of Jack Studs and King Studs in Exterior Wall Opening

Although one jack stud has been indicated in this figure, 2 or 3 jack studs may be required if the opening is subjected to a greater dead load, which is a function of the number of floors above the opening and the roof snow load.

Similarly, although one king stud is shown, in high-wind or seismic locations, the use of 2 or 3 king studs may be needed.

FIGURE 13.13 Wall framing around openings.

HEADER

A header is typically made of two or three 2-by lumber members, depending on the thickness of the wall. The members are face nailed to form a beam. If the wall is framed of 2 × 4 members, two 2-by lumber members are required, with a $\frac{1}{2}$-in.-thick filler, Figure 13.14. The filler is usually a plywood or an OSB sheet. For exterior walls, a rigid

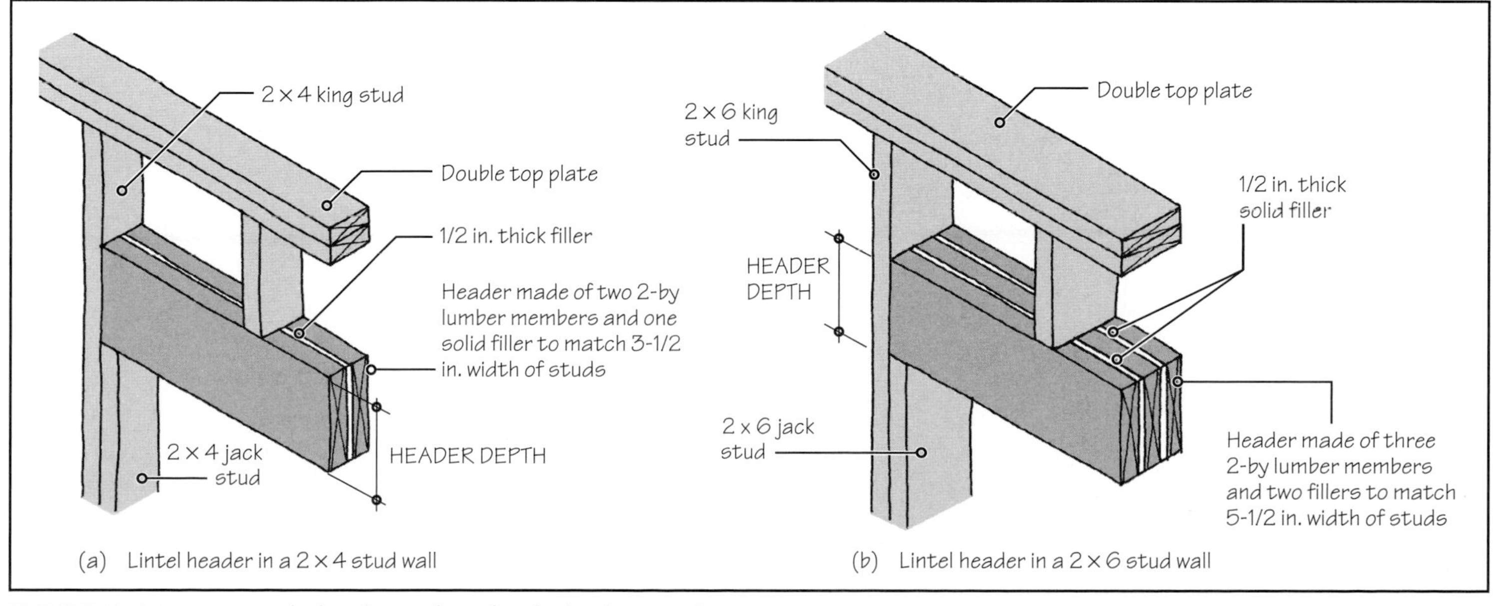

FIGURE 13.14 Anatomy of a header made with 2-by lumber members.

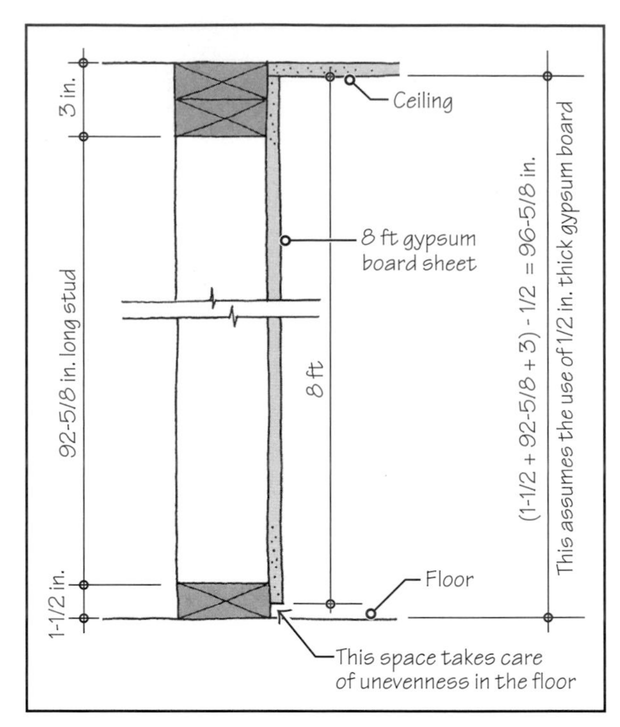

FIGURE 13.15 The use of 7 ft $8\frac{5}{8}$ in. ($92\frac{5}{8}$ in.) high studs.

plastic foam insulation is used. In a 2 × 6 wall, three 2-by lumber members are required, with two filler sheets. For large openings, trussed headers (Figure 12.20) or glulam headers are used.

PRECUT LENGTH OF STUDS AND OVERSIZED HEADERS

Lumber is usually available in lengths of 8 ft, 10 ft, 12 ft, and so on, up to a maximum length of 26 ft. (Some lumber dealers may stock lumber only up to a length of 22 ft.) A special precut length that is commonly used for studs is 7 ft $8\frac{5}{8}$ in. ($92\frac{5}{8}$ in.). The use of these studs saves on-site labor and gives a clear interior height (finished floor to ceiling) of 8 ft, Figure 13.15. The 8-ft clear height is common in multifamily dwellings, hotels, townhouses, and so on. If 2 × 12 headers are used with these studs, the opening height obtained is 6 ft $10\frac{1}{2}$ in., which is the standard lintel height for residential doors and windows, Figure 13.16.

Another commonly used special precut length of studs is $104\frac{5}{8}$ in. (8 ft $8\frac{5}{8}$ in.), which gives a floor-to-ceiling height of 9 ft. Note that 8-ft and 9-ft floor-to-ceiling heights conform to gypsum board panel sizes.

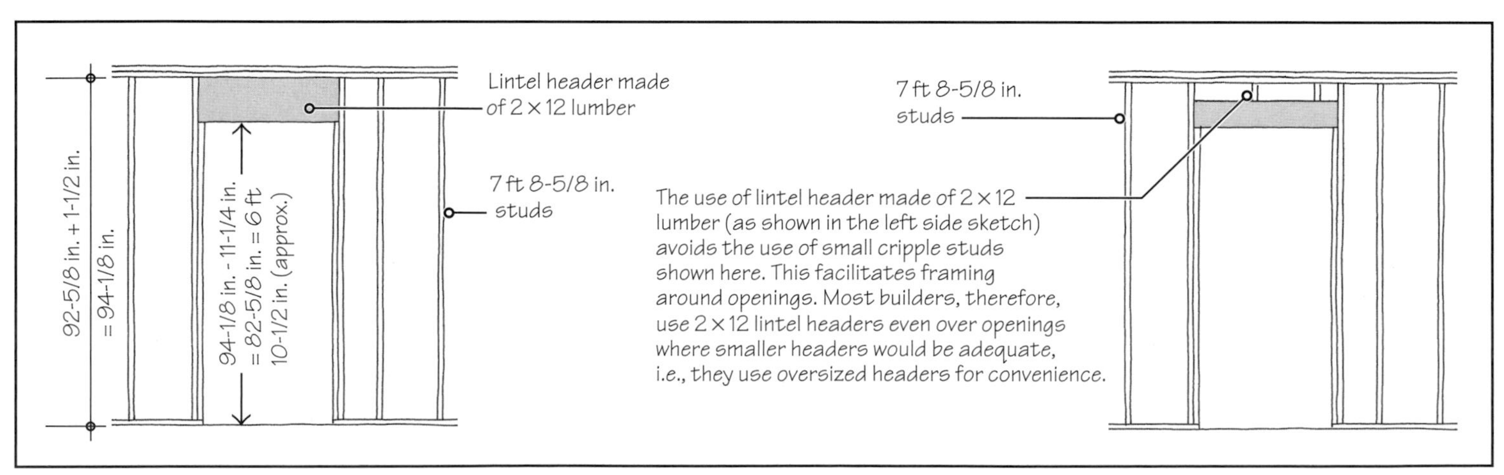

FIGURE 13.16 The use of oversized headers.

Each question has only one correct answer. Select the choice that best answers the question.

10. The top plate in a WLF wall generally consists of
- **a.** one 2-by member.
- **b.** two 2-by members.
- **c.** three 2-by members.
- **d.** any one of the above, depending on architectural considerations.
- **e.** none of the above.

11. The bottom plate in a WLF wall is called
- **a.** a sill plate.
- **b.** a sole plate.
- **c.** both (a) and (b) are synonymous terms.
- **d.** either (a) or (b), depending on the location of the wall.

12. A sill plate in a WLF wall must be anchored to the foundation with at least
- **a.** 1-in.-diameter bolts spaced a maximum of 6 ft on centers.
- **b.** $\frac{3}{4}$-in.-diameter bolts spaced a maximum of 6 ft on centers.
- **c.** $\frac{1}{2}$-in.-diameter bolts spaced a maximum of 6 ft on centers.
- **d.** $\frac{1}{4}$-in.-diameter bolts spaced a maximum of 6 ft on centers.
- **e.** none of the above.

13. A corner made by two WLF walls must have at least
- **a.** 2 studs.
- **b.** 3 studs.
- **c.** 4 studs.
- **d.** 5 studs.
- **e.** 6 studs.

14. A jack stud is always accompanied by a king stud.
- **a.** True
- **b.** False

15. In a WLF wall assembly, a cripple stud is used
- **a.** at a wall corner.
- **b.** at a wall T-junction.
- **c.** at the ends of a wall opening.
- **d.** above a header or below rough sill.
- **e.** both (a) and (b).

16. A header over an opening bears directly on
- **a.** cripple studs.
- **b.** jack studs.
- **c.** king studs.
- **d.** common studs.

17. A precut stud length equal to 7 ft $8\frac{5}{8}$ in. is commonly used because
- **a.** it provides a floor-to-ceiling height of 8 ft.
- **b.** it permits the use of $\frac{1}{2}$-in.-thick gypsum boards.
- **c.** it permits the use of OSB sheathing; otherwise more expensive plywood sheathing is needed.
- **d.** it is more energy efficient.

18. The header over an opening in a wall made of 2 × 6 studs consists of
- **a.** three 2-by lumber members nailed together.
- **b.** three 2-by lumber members nailed together with one intervening filler.
- **c.** three 2-by lumber members nailed together with two intervening fillers.
- **d.** two 2-by lumber members nailed together with three intervening fillers.
- **e.** two 2-by lumber members nailed together with two intervening fillers.

Answers: 10-b, 11-d, 12-c, 13-b, 14-a, 15-d, 16-b, 17-a, 18-c.

13.6 ESSENTIALS OF FLOOR FRAMING

The most crucial aspect of floor framing is the layout of floor joists. It is only after the floor joist layout has been finalized that we can work out the sizes of joists. The initial layout of floor joists is generally prepared by superimposing it on the architectural floor plan. In general, the joists should span between the opposite walls along the shorter of the two spans. When it is not possible to use the walls as supports, an intermediate beam is required to support the joists.

Floor Framing Plan

Figure 13.17 shows the layout of the second-floor joists (referred to as the second-floor *framing plan*) of a small (approximately 28 ft × 28 ft) building. The joists have been laid in the direction of the shorter span. However, where there is a cantilevered floor, the joists must be laid along the direction of the overhang.

Overhanging joists must bear on a support and be securely connected at the far end (end of the back span) to a wall or a beam. A beam may be a glulam beam or a built-up beam, obtained by nailing two or more 2-by members of the same size as the joists. A built-up beam in this location is referred to as a *joist header*. A glulam beam is commonly used over long spans and a built-up joist header, over short spans. In Figure 13.17, a glulam beam has been used.

Joist hangers are required where the joists are hung from a beam or joist header, Figure 13.18. The joist hangers allow the top of the beam and the joists to be at the same elevation.

A framing plan for the ground floor is needed if there is a crawl space or basement floor in the building. Figure 13.19 shows the framing plan of the ground floor of the building of Figure 13.17, assuming an underlying crawl space.

Support for Load-Bearing and Non-Load-Bearing Walls

Generally a load-bearing wall on an upper floor must lie directly over a wall on the lower floor. Additionally, all exterior walls and interior load-bearing walls should be supported on foundation walls or piers, as shown in Figure 13.18.

NOTE

What is a Load-Bearing Wall?

A load-bearing wall is defined as a wall that supports gravity loads other than its self-load. Note that lateral loads are not included in the definition because virtually all building elements (load-bearing or non-load-bearing) have to support some type of lateral load.

For instance, a window (a non-load-bearing element) has to resist lateral loads due to wind and/or earthquake. The same is true of a door or an interior partition wall. An interior partition wall must be designed to resist a minimum lateral load of 5 psf.

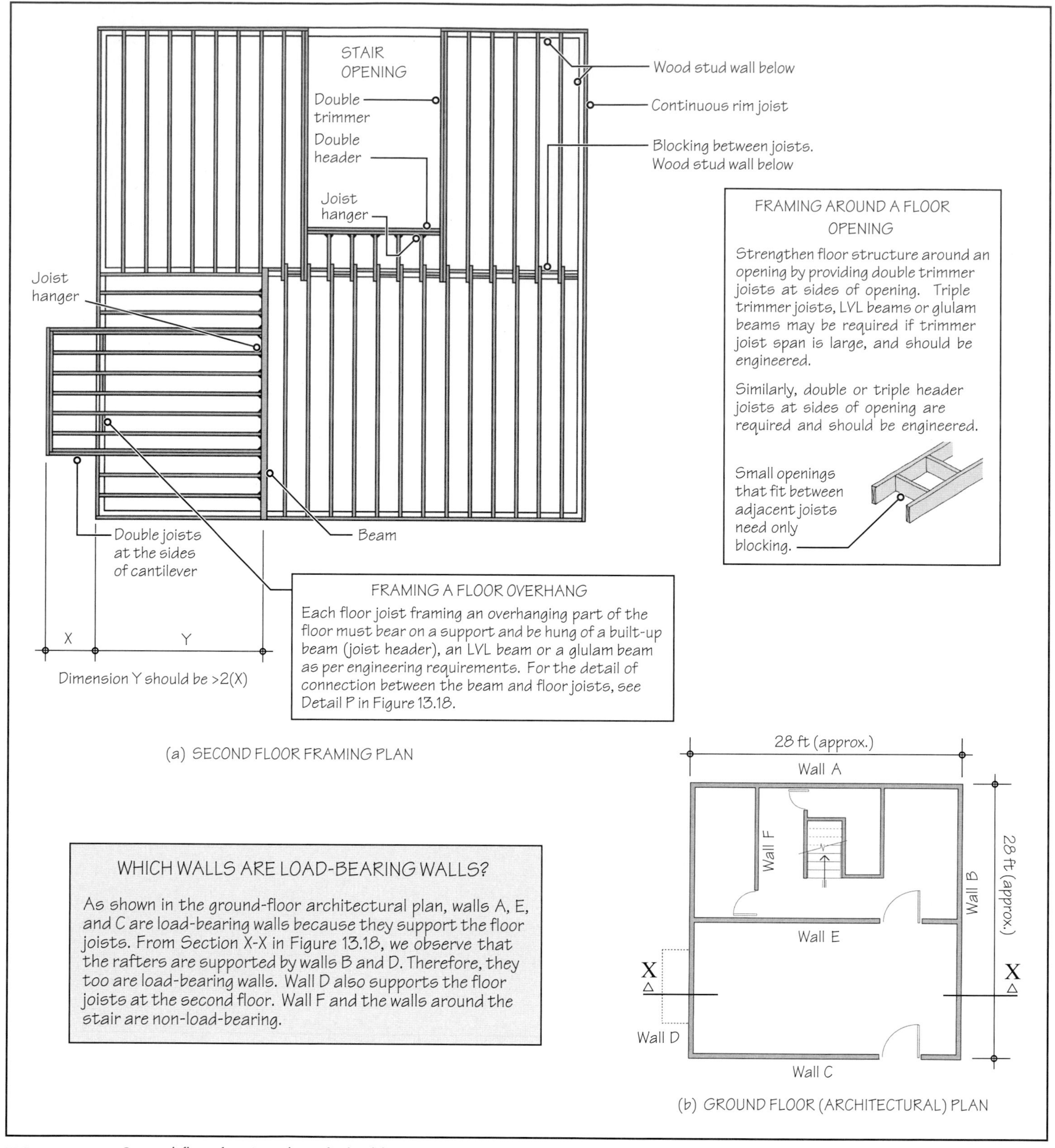

FIGURE 13.17 Second-floor framing plan of a building.

A non-load-bearing wall on an upper floor may, however, be supported by the floor. Figure 13.20 shows the typical framing enhancement under a non-load-bearing wall that runs parallel to the joists, such as wall F in Figure 13.17. If the wall runs perpendicular to the floor joists, no such reinforcement is generally needed, Figure 13.21.

Enhancement of Floor Framing

Carefully examine the layout of Figure 13.17. Observe that the joists along the two opposite ends of the cantilevered floor must be doubled. Similarly, the joists framing around an opening (in this case, a stair opening) are also required to be doubled.

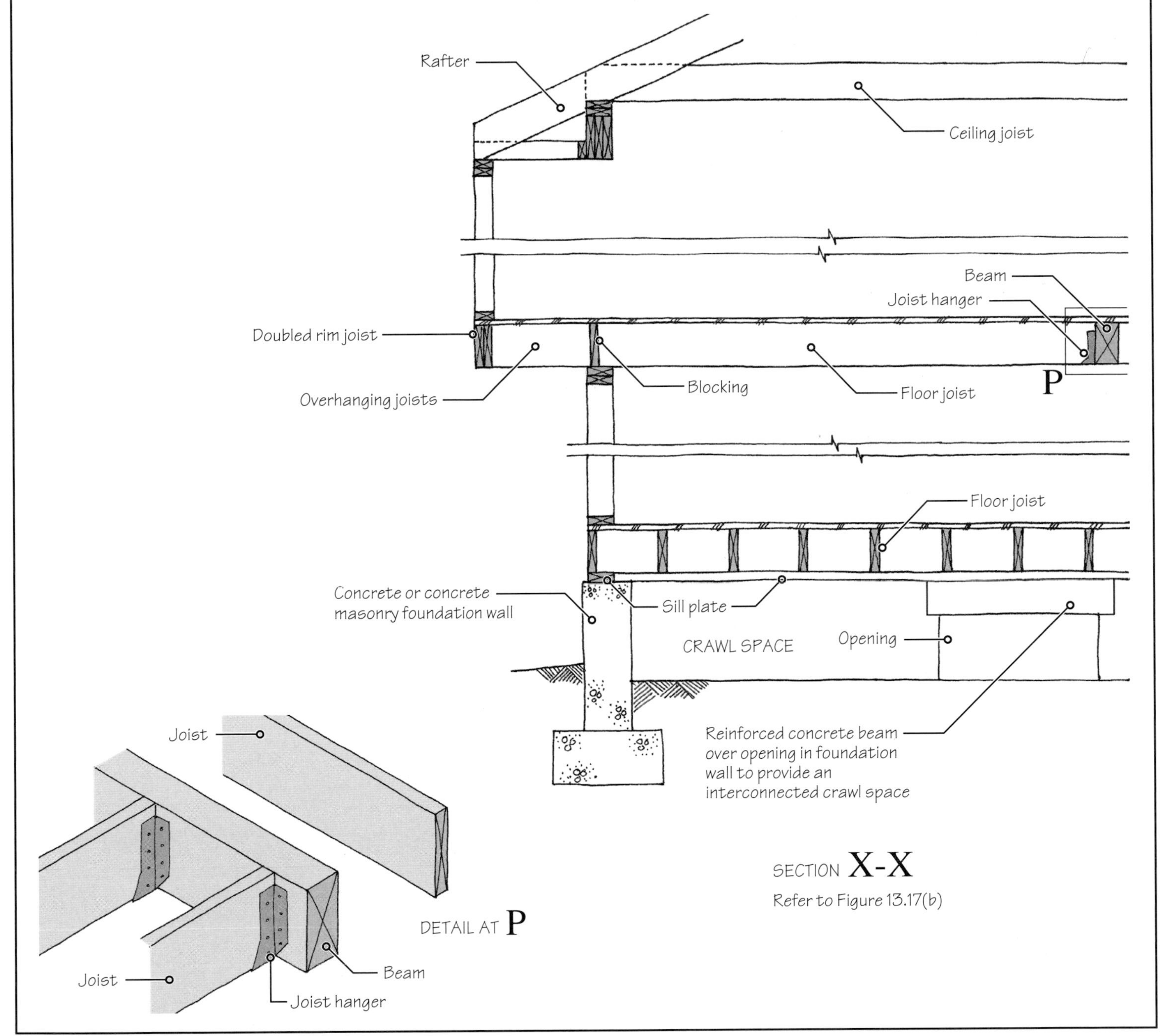

FIGURE 13.18 Section through the building of Figure 13.17. Detail P shows the connection between the overhanging floor joists at the second floor and the supporting beam.

Doubling of floor joists (or decreased spacing between joists) may also be needed in areas of the floor that carry excessive load, such as a floor topped with concrete and ceramic tiles, Figure 13.22.

Role of Rim (or Band) Joists

Because the floor joists are slender members, they are prone to lateral buckling (Section 4.9). Rim joists provide lateral restraint to the floor joists, reducing their tendency to buckle as well as connecting the ends of the joists.

For the same reasons, floor joists that have a long span may require intermediate full-depth *blocking* or *diagonal bridging*, Figure 13.23. Generally intermediate blocking or bridging is required if the floor joist depth is greater than 12 in. (nominal). However, it is prudent to provide rows of blocking or bridging for all floor joists generally at 8 ft on centers.

Diagonal bridging is generally provided through 1 × 3 lumber with bevel-cut ends to fit tight at the joists. Full-depth blocking has the advantage of not requiring bevel-cut ends. It also stiffens the floor and distributes the load between adjacent joists.

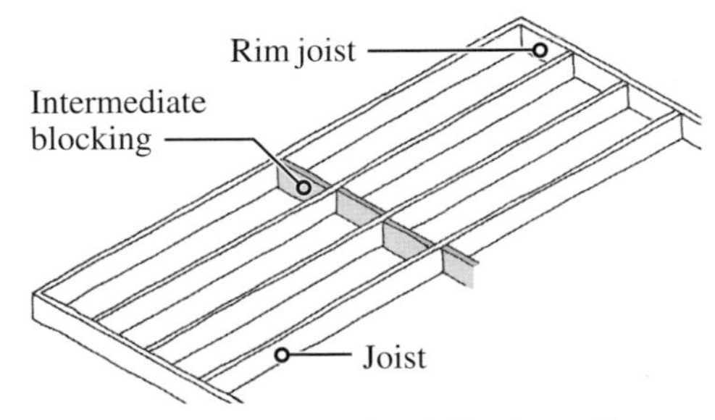

Long span floor joists should be laterally supported by intermediate full-depth blocking or diagonal bridging.

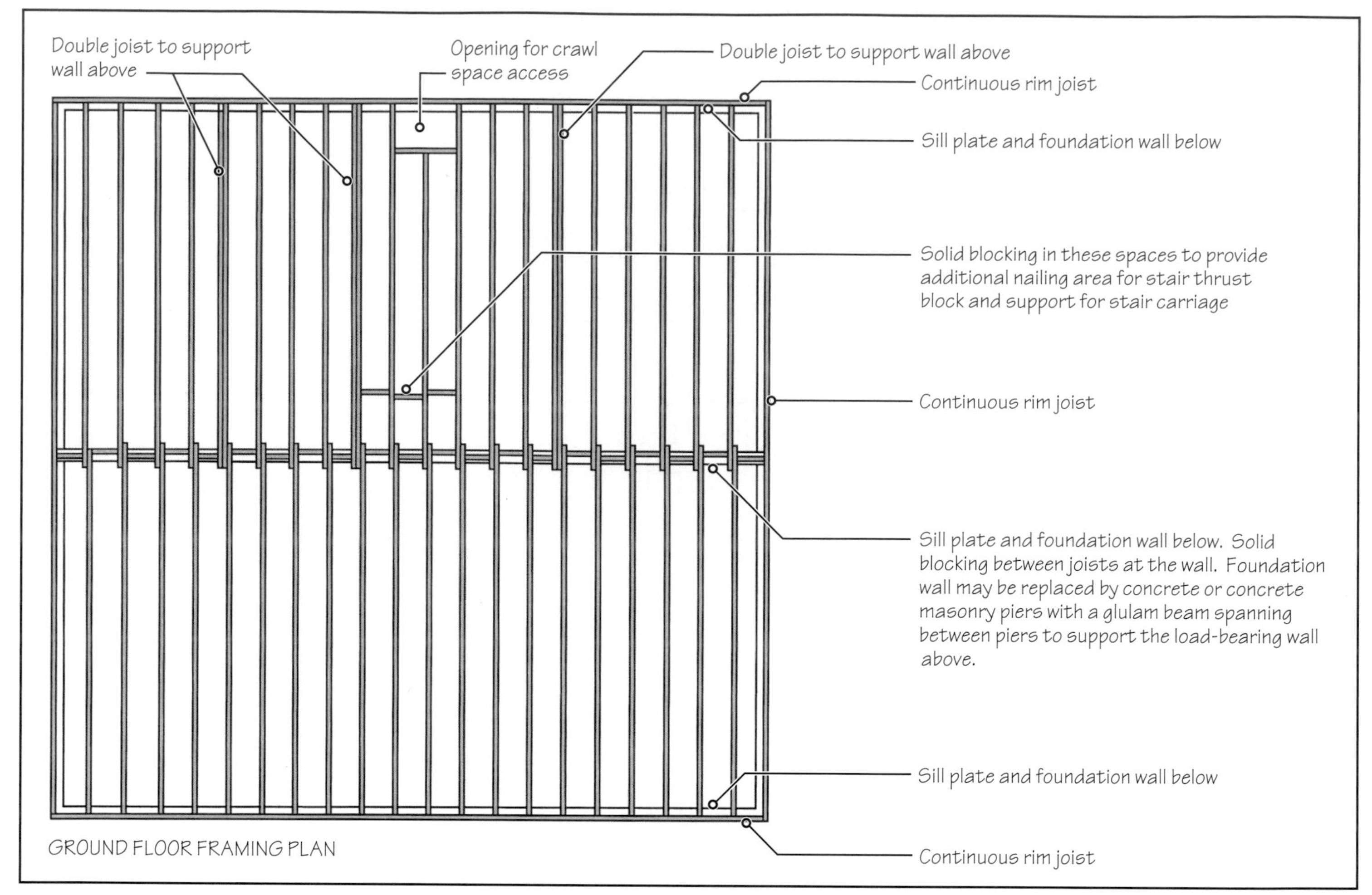

FIGURE 13.19 Ground-floor framing plan of the building of Figure 13.17 (assuming a crawl space below).

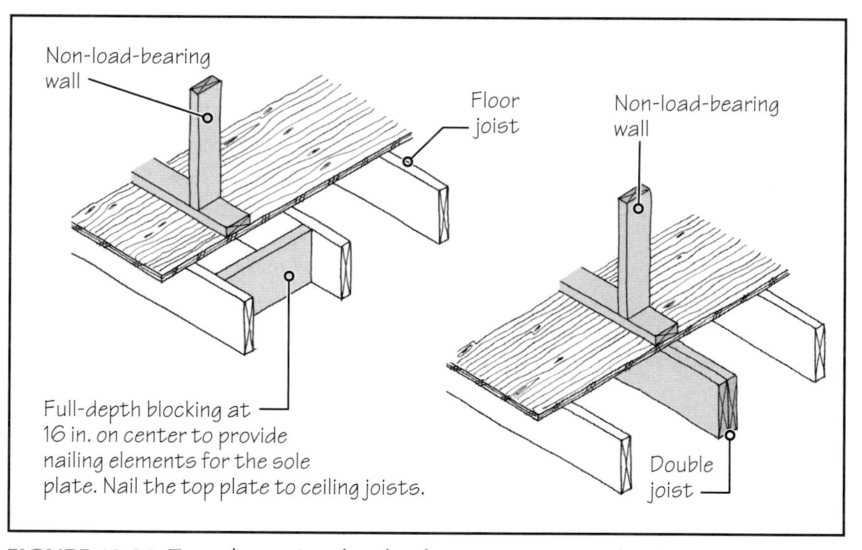

FIGURE 13.20 Two alternative details of supporting a non-load-bearing wall on a wood floor when the wall is parallel to the joists.

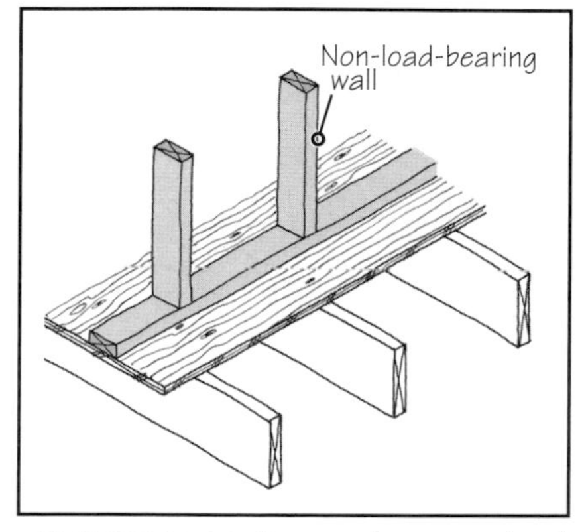

FIGURE 13.21 Support for a non-load-bearing wall on a wood floor when the wall is perpendicular to the joists.

However, full-depth blocking obstructs pipes running in the space between adjacent joists. Diagonal bridging consists of metal straps and is available from several manufacturers; it is easier to install than lumber bridging.

Structural Continuity in Floor Joists and Rim Joists

Floor joists should have continuity, provided either through overlapped ends nailed together at the supporting wall, Figure 13.24(a), or through scabs nailed to butt-jointed ends, Figure 13.24(b). Similar details are used with joists that bear on an underlying beam, Figure 13.24(c). For joists that are connected to a supporting beam with joist hangers, the continuity between joists is provided through the connections, Figure 13.24(d).

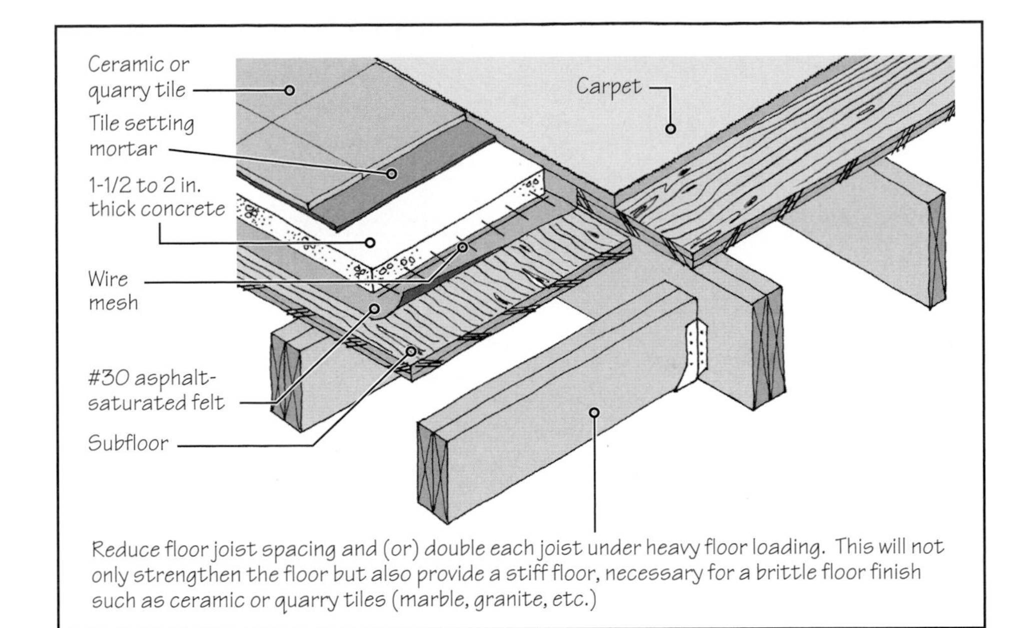

FIGURE 13.22 Strengthening of the floor frame under a heavily loaded part of the floor.

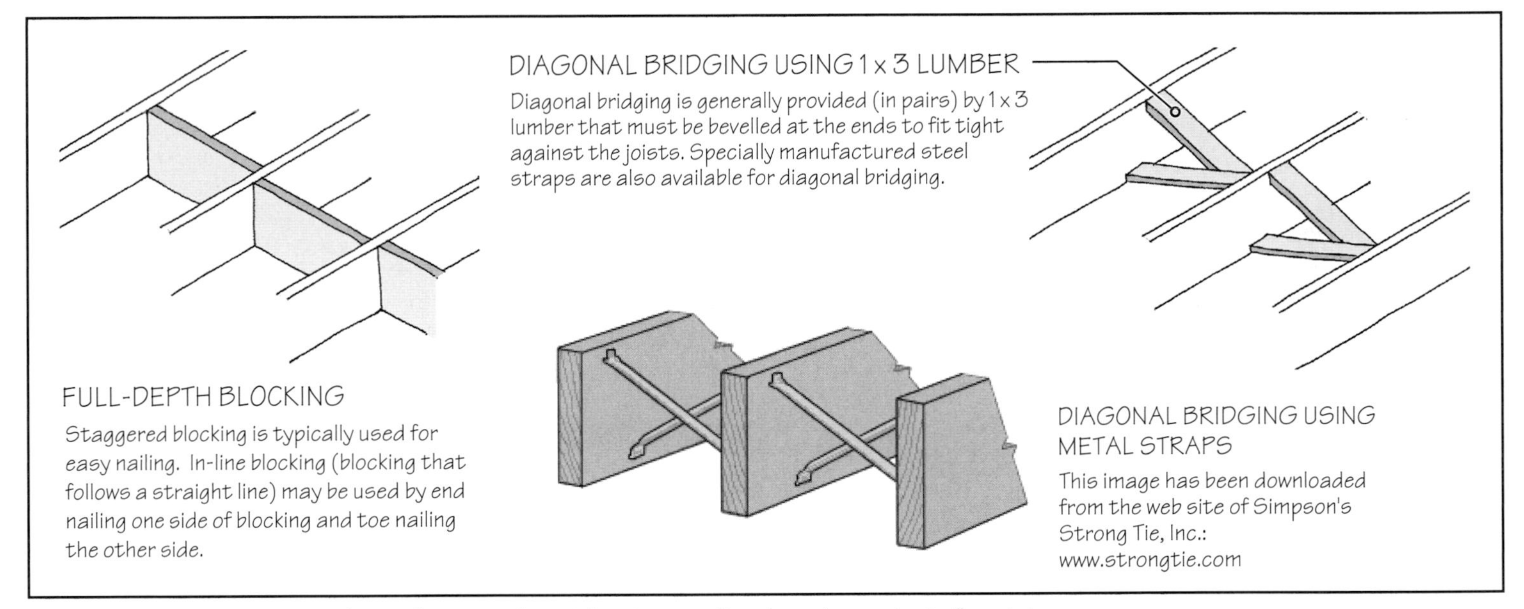

FIGURE 13.23 Three commonly used means of providing intermediate lateral restraint in floor joists.

A rim joist should also be continuous. Continuity in a rim joist adds to the continuity of the wall's top plate and is typically provided by nailing a 2-by splice member at the joint, Figure 13.25.

13.7 ROOF TYPES AND ROOF SLOPE

The roof of a WLF building is generally sloped. The more commonly used roof shapes are gable, hip, and shed roofs. Often these roof types are used in combination, Figure 13.26. Figure 13.26 also provides an introduction to some of the roof vocabulary.

The slope of a roof is not expressed in degrees but as a rise-to-run ratio. In this expression, the run is generally kept at a constant value of 12. Thus, a roof slope is expressed as 2:12, $2\frac{1}{2}$:12, 5:12, 6:12, and so on. The greater the rise, the greater the roof slope. Figure 13.27 gives a visual picture of various roof slopes. As we will observe in Chapter 31, the roofing industry divides roofs into two types, based on roof slope:

- Low-slope roof
- Steep roof

A low-slope roof is a roof whose slope is less than 3:12. A steep roof has a slope of greater than or equal to 3:12. Using a 3:12 roof slope as the dividing line is based on structural considerations, as explained later in this chapter.

An additional basis for the dividing line is roof cover considerations. Several asphalt shingle manufacturers do not warrant the use of their product on a roof with a slope of less than 3:12.

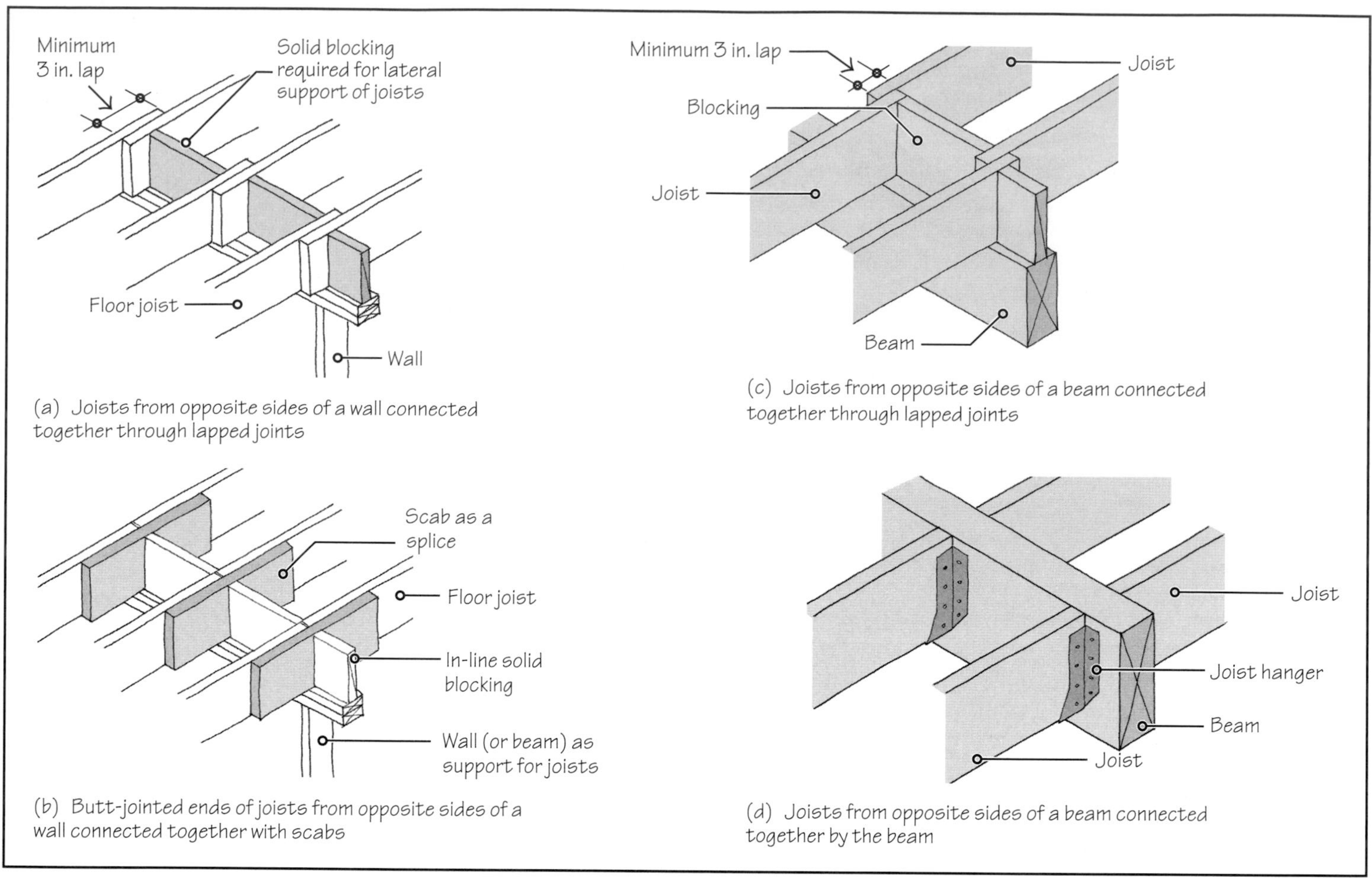

FIGURE 13.24 Details of floor joists resting on an intermediate support.

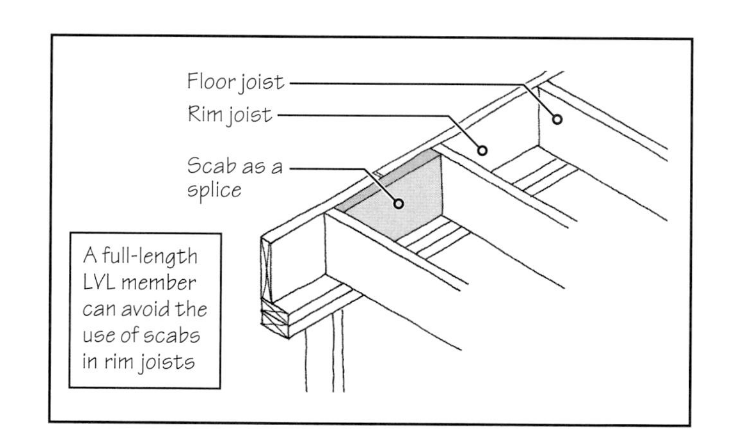

FIGURE 13.25 Detail of splice-connected rim joist.

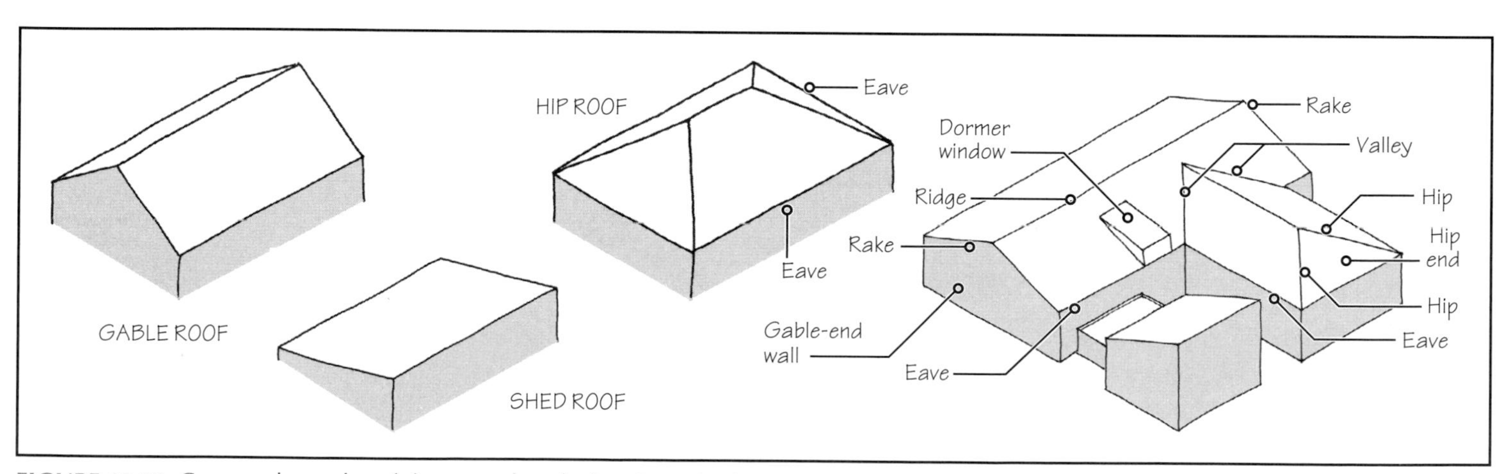

FIGURE 13.26 Commonly used roof shapes and roof-related terminology.

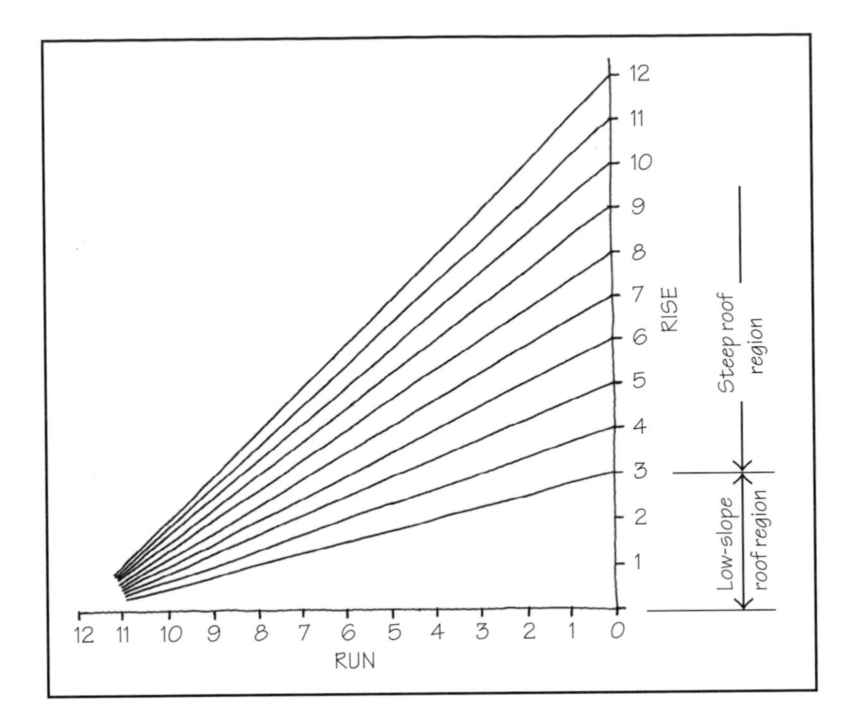

FIGURE 13.27 Roof slope.

PRACTICE QUIZ

Each question has only one correct answer. Select the choice that best answers the question.

19. In a WLF building, a floor framing plan indicates
- **a.** the layout of studs and their spacing.
- **b.** the layout of studs and exterior wall sheathing.
- **c.** the layout of floor joists and blocking.
- **d.** the layout of floor joists, floor beams and blocking.
- **e.** the layout of floor beams, studs and wall sheathing.

20. A non-load-bearing wall in a building is one that supports the loads from the roof but not from any floor.
- **a.** True
- **b.** False

21. The primary purpose of full-depth blocking in a floor frame is to
- **a.** reduce compressive stresses in floor joists.
- **b.** prevent buckling of floor joists.
- **c.** reduce tensile stresses in floor joists.
- **d.** give additional nailing surface for a gypsum board ceiling.
- **e.** give additional nailing surface for floor sheathing.

22. Blocking between floor joists is required where
- **a.** the joists bear on a wall.
- **b.** the joists bear on a beam.
- **c.** the joists are hung on a beam using joist hangers.
- **d.** all the above.
- **e.** both (a) and (b).

23. A floor in a WLF building that supports a non-load-bearing wall generally requires no additional strengthening or stiffening when
- **a.** the wall runs parallel to floor joists.
- **b.** the wall runs perpendicular to floor joists.

24. A roof with gable ends on both sides in a rectangular building has
- **a.** eaves on all four sides.
- **b.** eaves on three sides.
- **c.** eaves on two sides.
- **d.** eave on one side.
- **e.** no eaves.

25. A rectangular roof with hip ends on two sides has
- **a.** eaves on all four sides.
- **b.** eaves on three sides.
- **c.** eaves on two sides.
- **d.** eave on one side.
- **e.** no eaves.

26. In the building industry, the slope of a roof is generally expressed in
- **a.** degrees.
- **b.** radians.
- **c.** rise-to-run ratio.
- **d.** run-to-rise ratio.
- **e.** none of the above.

Answers: 19-d, 20-b, 21-b, 22-e, 23-b, 24-c, 25-a, 26-c.

13.8 ESSENTIALS OF ROOF FRAMING

The roof framing of a typical WLF building consists either of trusses or rafter-and-ceiling joist assemblies. Theoretically, there is little difference between them, except that a truss is a shop-fabricated, multitriangle frame and a rafter-and-ceiling-joist assembly is a site-fabricated, single-triangle frame.

In a rafter-joist assembly, the rafters rest on the walls at one end and are connected to the opposite rafters at the ridge. The connection between the two rafters is made through a continuous *ridge board* that runs perpendicular to the rafters.

A ridge board is generally a 2-by member that must extend from the top edge to the bottom edge of the rafter (or a little below), Figure 13.28. It has no structural function except to align the rafter ends in a straight line at the top. That is why the codes only require a 1-by member, although a 2-by member is generally used as the ridge board. Many framers use an LVL member to ensure a continuous, straight ridge.

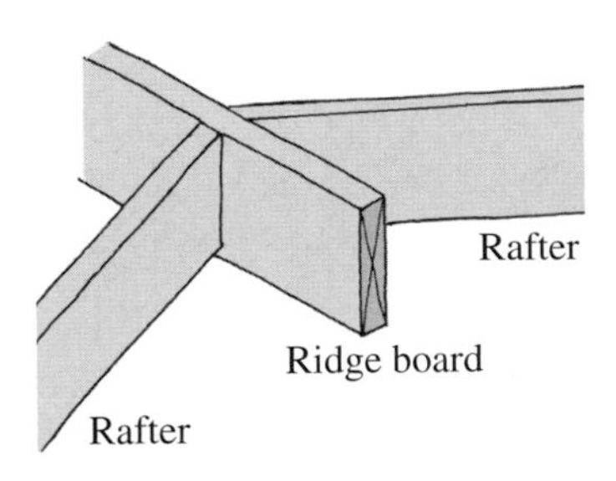

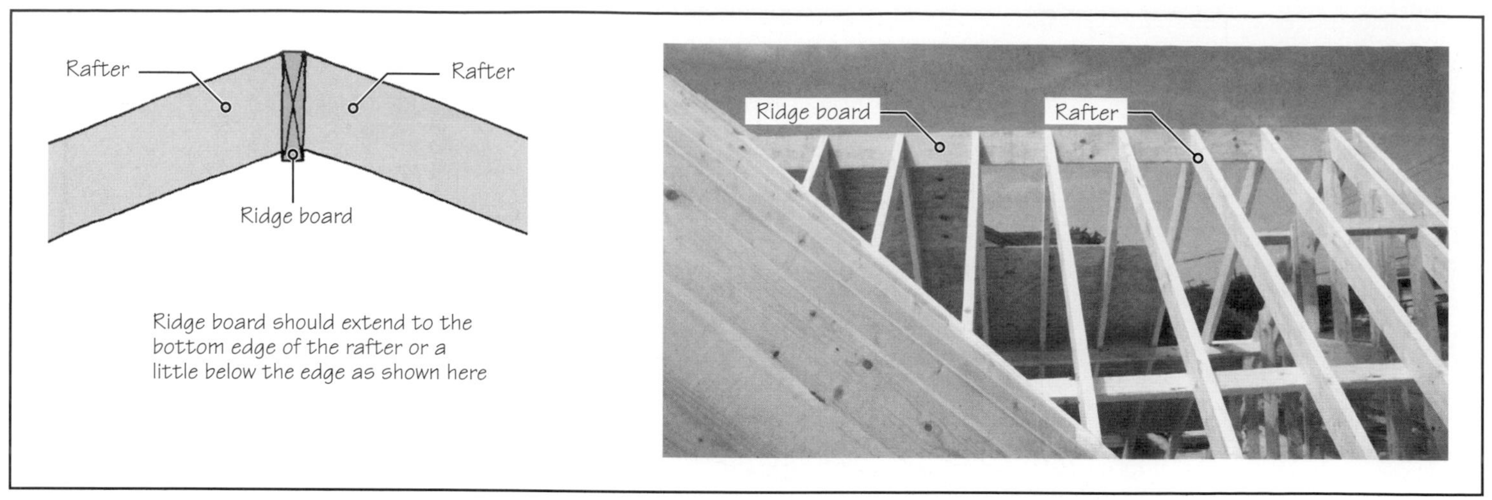

FIGURE 13.28 Ridge board and rafters.

Each rafter pair (connected at the ridge) must be tied together at the bottom to resist the outward thrust created by the gravity loads on the roof. In a typical WLF roof, this tie also functions as the ceiling joist, Figure 13.29. The rafter pair and the ceiling joist thus make a stable triangular frame, which rests on two opposite supports and delivers a vertical load at the supports with no lateral thrust.

To deliver the load vertically at the support, each rafter is cut to have a horizontal bearing on the supporting walls. Each rafter is, therefore, specially notched near the bottom. This notch is referred to as a *bird's mouth,* Figure 13.30. The horizontal part of the bird's mouth should equal the width of the supporting wall (including the thickness of sheathing) to ensure that the wall sheathing engages the top plate fully. If the sheathing is applied to the wall after the rafters are in place, it must be cut around the rafters.

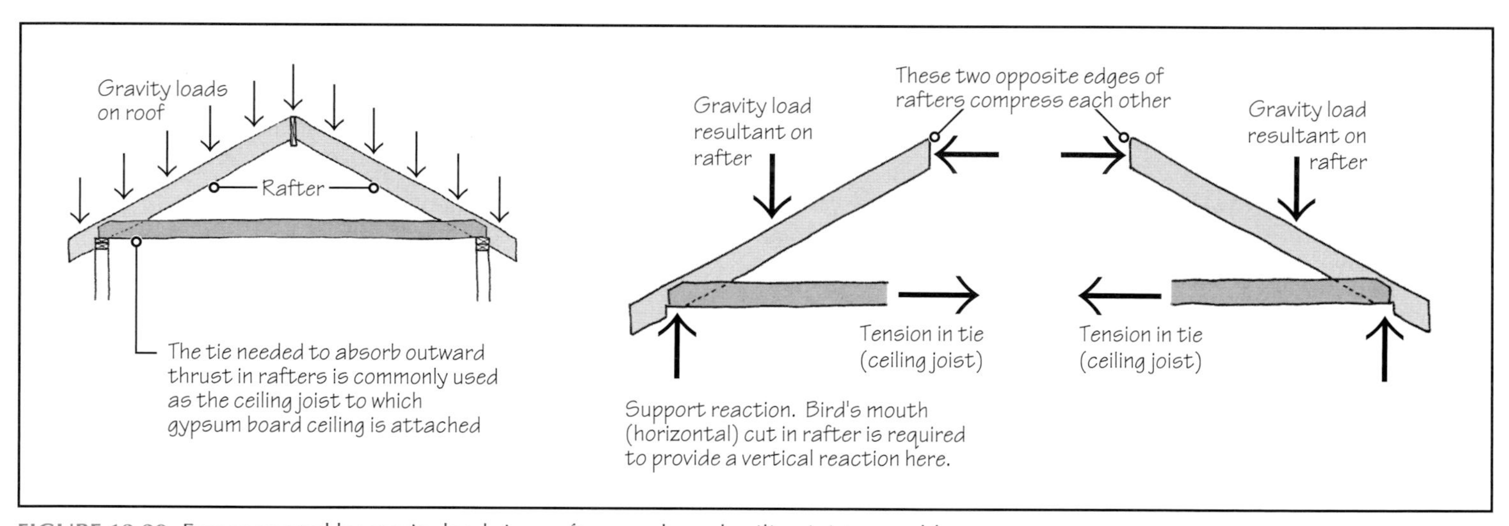

FIGURE 13.29 Forces created by gravity loads in a rafter couple and ceiling joist assembly.

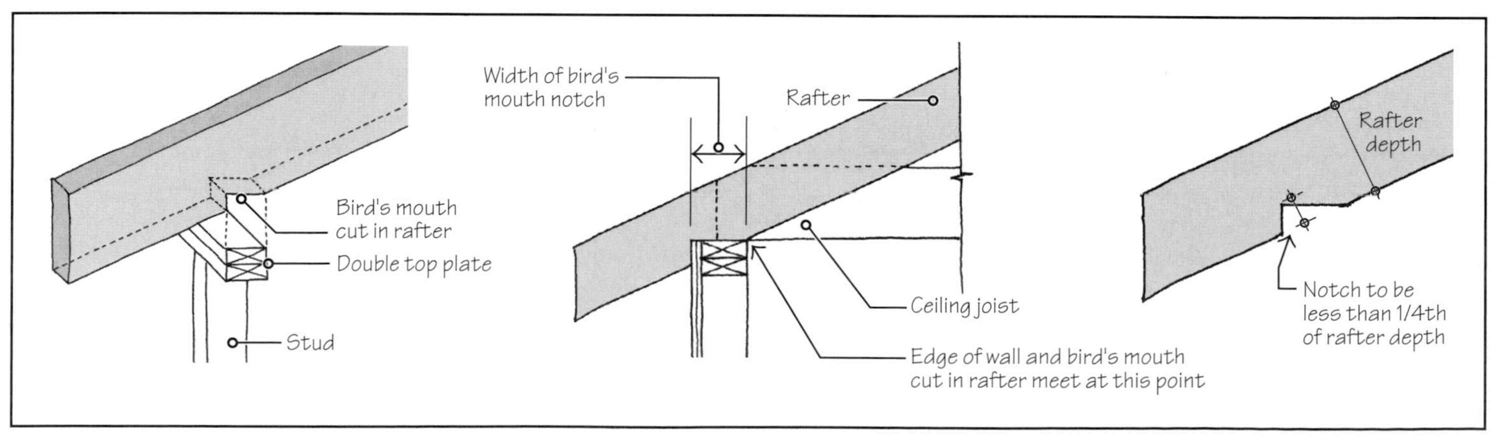

FIGURE 13.30 Bird's mouth cut in a rafter.

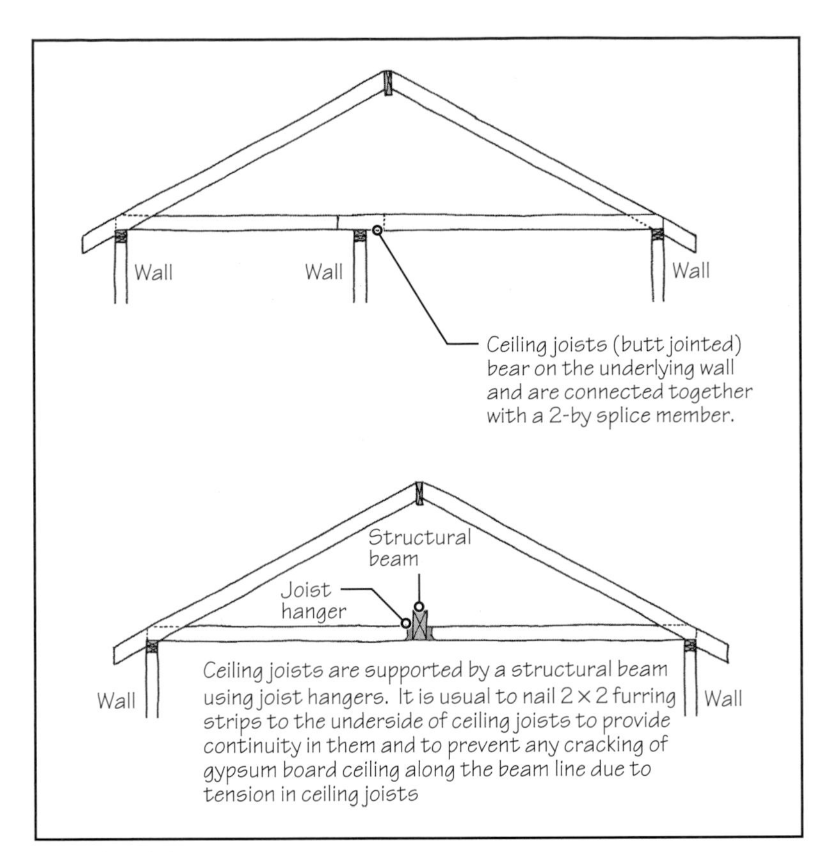

FIGURE 13.31 Ceiling joist support.

SUPPORT FOR CEILING JOISTS

Because the distance between the supporting walls is usually too large to allow the use of one-piece members as ceiling joists, the ceiling joists are generally two-piece members connected together at an intermediate support. The intermediate support is necessary because the ceiling joists are subjected not only to tension but also to bending, due to gravity loads from the attic. The intermediate support is usually an underlying wall, Figure 13.31. If a wall is not available, a beam should be introduced. Alternatively, prefabricated roof trusses may be used, which do not need intermediate supports.

PURLINS AND PURLIN BRACES AS INTERMEDIATE RAFTER SUPPORTS

Apart from providing support to ceiling joists, an intermediate wall may also be used to provide an additional support to rafters. This is done by using a purlin to support the rafters. The purlin is supported on closely spaced purlin braces, Figure 13.32. The use of purlins and braces not only reduces the size of rafters, it also increases the stiffness of the roof. This is particularly helpful in roofs supporting large gravity loads due to snow or heavy roofing materials.

COLLAR TIES

A ceiling joist tie and a bird's mouth notch are two basic features of the gravity-load resistance of a rafter-and-ceiling-joist roof assembly. Wind-load resistance in a roof assembly, on the other hand, needs additional features. Because wind loads on a roof are predominantly uplift loads (the exact opposite of the gravity loads), they produce separation of rafters at the ridge, Figure 13.33.

To restrain the rafters from separating at the ridge, collar ties are often employed. To be effective, collar ties should be located within the upper one-third of the attic. The most common location for a collar tie is about 18 in. below the ridge. A collar tie generally consists of 2 × 4 lumber.

A *ridge strap,* made of $1\frac{1}{4}$-in.-wide and 20-gauge-thick galvanized sheet steel can be used as an alternative to collar tie, Figure 13.34. Unlike the collar ties, ridge straps do not obstruct the space in the attic. However, a collar tie increases the rigidity of the

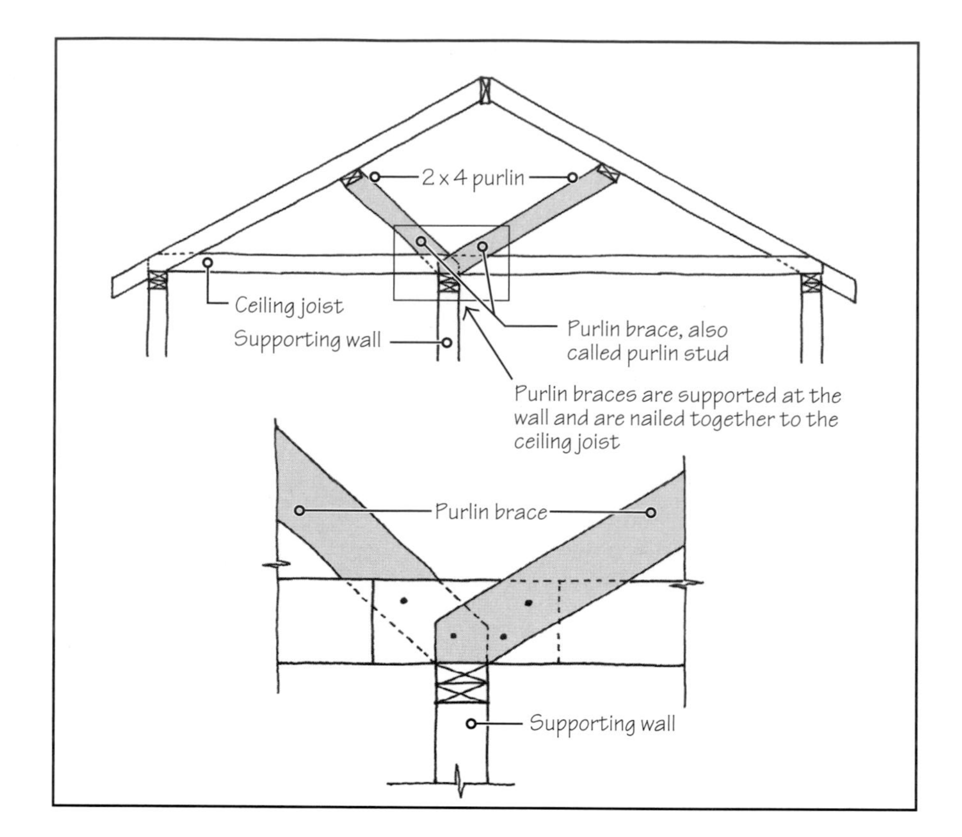

FIGURE 13.32 Purlins and purlin braces.

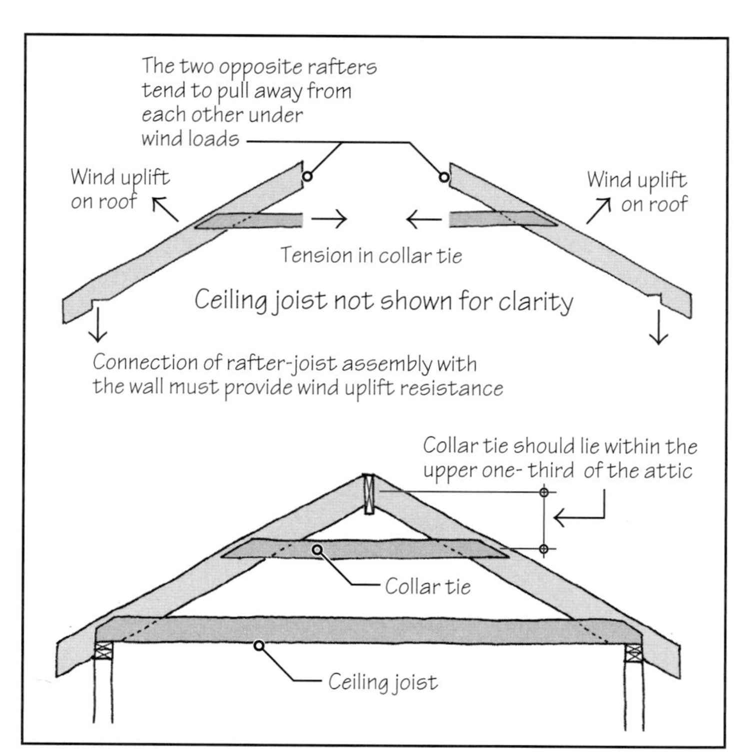

FIGURE 13.33 The use of collar ties for wind uplift resistance in a roof.

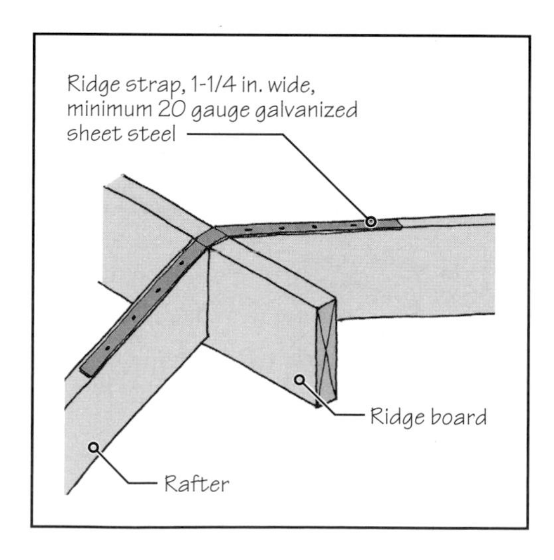

FIGURE 13.34 Ridge straps as alternatives to collar ties.

joint between the rafters at the ridge, decreasing the tension in ceiling joists under gravity loads.

Prefabricated Trusses as an Alternative to a Rafter-Joist Assembly

Wood trusses are an alternative to a stick-built rafter-joist assembly. Their use is particularly attractive for single-family dwellings in regions where prefabrication is more economical either because of the climate's severity and/or higher labor costs.

Because of the length limitations of sawn lumber, a rafter-joist assembly has limited span capabilities. When the width of the building is large (typically greater than 35 ft), the use of trusses is a more viable solution for roof framing, Figure 13.35. Wood trusses are, therefore, a norm in wood frame commercial and multifamily residential buildings. Truss manufacturers provide all the necessary engineering and details that should be used with their trusses.

FIGURE 13.35 Prefabricated wood trusses are commonly used in buildings where roof spans are large. Prefabrication reduces on-site labor and hence makes trusses a popular alternative in regions with unfavorable weather conditions or high labor costs.

13.9 VAULTED CEILINGS

The rafter-ceiling-joist assembly described previously works only if the roof slope is greater than 3:12. If the slope is less than 3:12, the tension in ceiling joists becomes too large to be handled by conventional 2-by members. In such a case, the ridge board must be replaced by a *ridge beam.* A ridge beam is a structural member, unlike the ridge board. The rafters bear on the ridge beam at the top and on the supporting walls at the eave.

Such an assembly, consisting of two opposite rafters and the ridge beam, creates no outward thrust in the rafters under gravity loads. Consequently, there is no need for ceiling joist ties. The gypsum board ceiling in such a roof may be attached directly to the rafters and ridge beam and/or the ridge beam may remain exposed. Such a ceiling is referred to as *cathedral ceiling,* or a *vaulted ceiling,* Figure 13.36. A vaulted ceiling may also be used with roof pitch greater than 3:12.

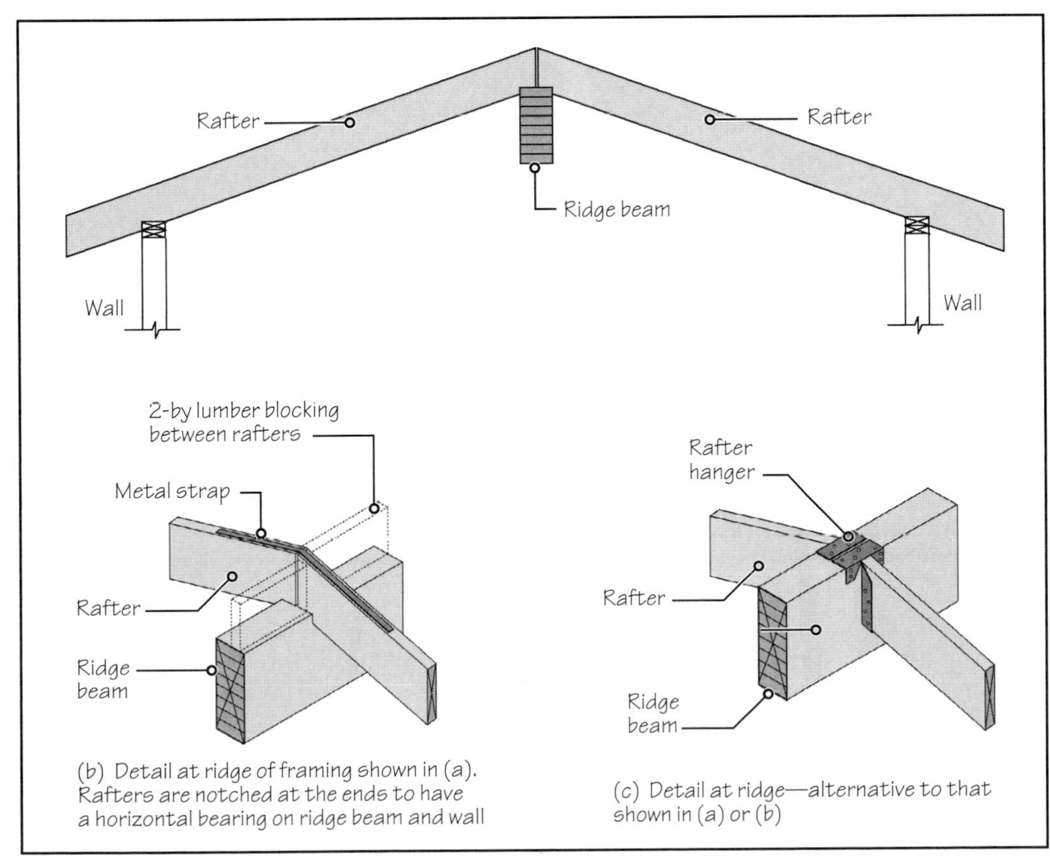

FIGURE 13.36 If the rafters are supported on a (structural) ridge beam, ceiling joists are not required to absorb the lateral thrust. Such a roof system is called a vaulted (or cathedral) ceiling.

13.10 SHEATHING APPLIED TO A FRAME

The wall, floor and roof frames must be covered with a sheathing material. Sheathing (generally a panel material) serves both structural and nonstructural functions.

WALL SHEATHING AND LATERAL BRACING OF THE FRAME

Wall sheathing furnishes a base over which exterior wall finishes are applied. It also provides a nailing base for an air-weather retarder. Structurally, wall sheathing integrates the studs into a composite wall system. It also provides bracing to the frame against lateral loads.

The most commonly used material for exterior wall sheathing (and bracing) is OSB. However, plywood, gypsum sheathing, and several other panel materials (including rigid foam insulation) are acceptable bracing alternatives. The minimum area of the exterior wall that must be covered with bracing material is a function of the number of stories of the building and the intensity of lateral load. However, as shown in Figure 12.29, the walls of most WLF buildings are often fully sheathed with OSB (excluding the openings), which generally exceeds the minimum bracing requirement.

An alternative to panel bracing is a diagonal let-in brace, Figure 13.37. This consists of a 1 × 4 lumber member fastened to studs that are notched to receive the brace. The angle of the brace with the horizontal should lie between 45° and 60°. A steel angle let-in brace may also be used in place of lumber let-in brace. This requires only a small slit cut into the studs.

The bracing strength of a let-in brace (wood or steel) is much smaller than that of plywood or OSB panel bracing. Therefore, the use of let-in brace is generally limited to situations where the lateral loads are small.

WLF buildings that have a relatively large footprint may require bracing in the interior walls in addition to the bracing provided in exterior walls. Interior-wall bracing can be provided by sheathing the interior walls on both sides with plywood or OSB and covering them with gypsum board. Alternatively, 2-by lumber (blocking) members placed diagonally through the wall may be used to obtain an interior braced wall, Figure 13.38.

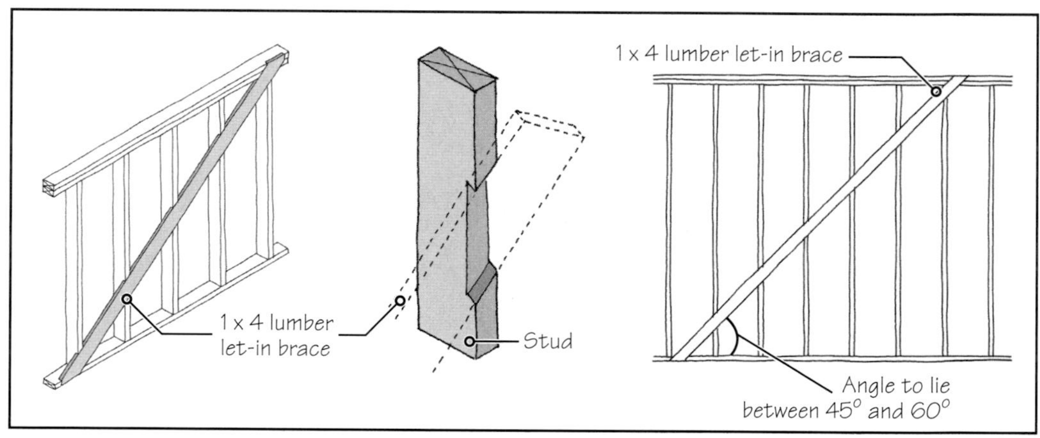

FIGURE 13.37 Lumber let-in lumber brace.

FIGURE 13.38 Diagonal bracing of an interior wall with 2-by lumber members.

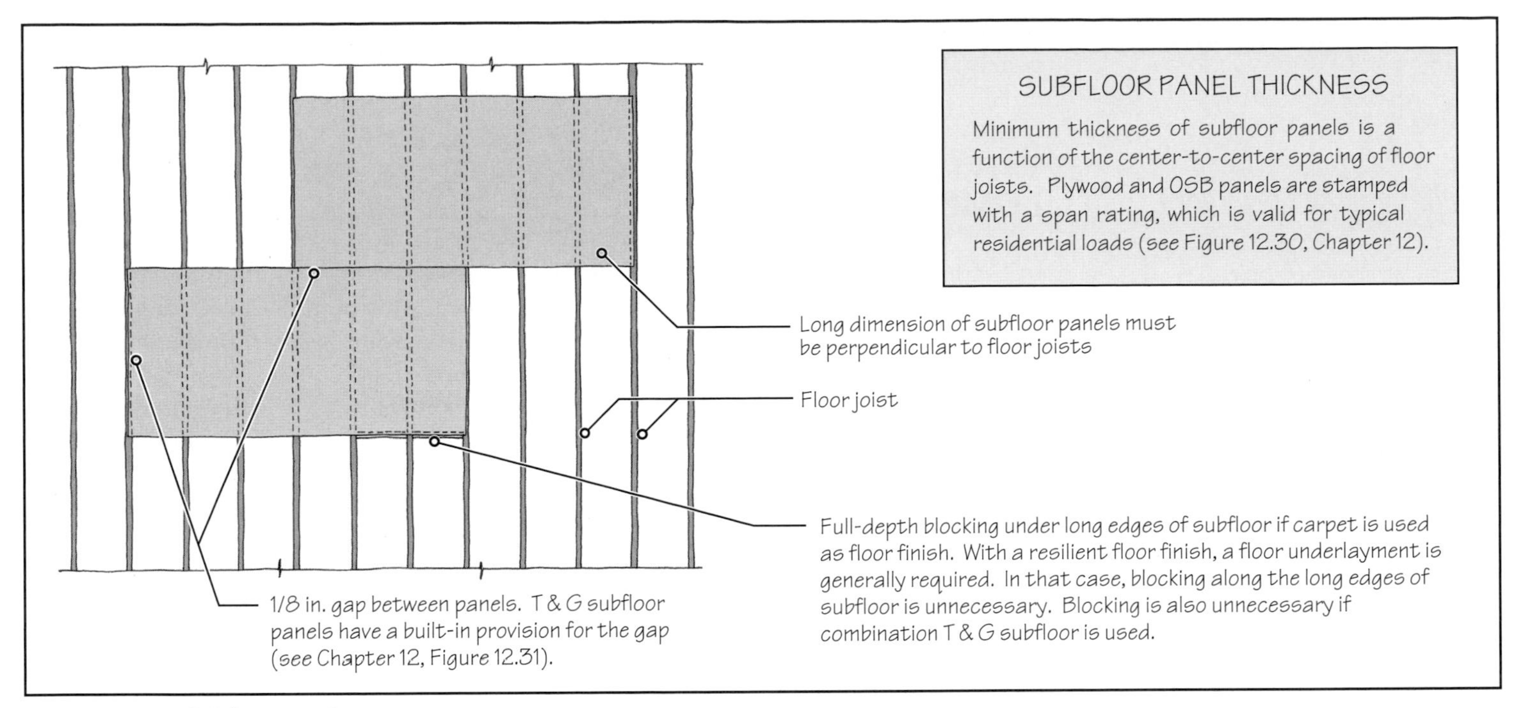

FIGURE 13.39 Subfloor requirements.

Floor Sheathing

Floor sheathing (also called *subfloor*) is a structural element because it transfers floor dead and live loads to the joists. Floor sheathing also provides diaphragm action in resisting the lateral loads. The most commonly used material for subfloor is OSB or plywood panels, generally used in 4-ft × 8-ft dimensions. Panels must be laid with the 8-ft dimensions perpendicular to the joists, Figure 13.39. Two types of subfloor panels are used, depending on the profile of the long (8-ft) edges:

- All four panel edges have a straight profile.
- Two long edges have a tongue-and-groove profile, whereas the other two edges are straight (Section 12.8).

If carpet is used as floor finish, straight-edge subfloor panels require blocking along the long edges. Resilient floor finishes (such as vinyl or linoleum tiles or sheets) generally require a smooth underlayment over an OSB or plywood subfloor (Chapter 34). When a floor underlayment is used, blocking of the long edges of the subfloor is generally not required. Blocking is also not required if tongue-and-groove panels are used for the subfloor.

Roof Sheathing

Like floor sheathing, roof sheathing is also a structural component, and OSB or plywood panels are most commonly used. The thickness of panels is based on their span rating. Roof sheathing panels do not require blocking, but metal edge clips (generally H-shaped clips) are used along the long (8-ft) edges to join adjacent panels, Figure 13.40.

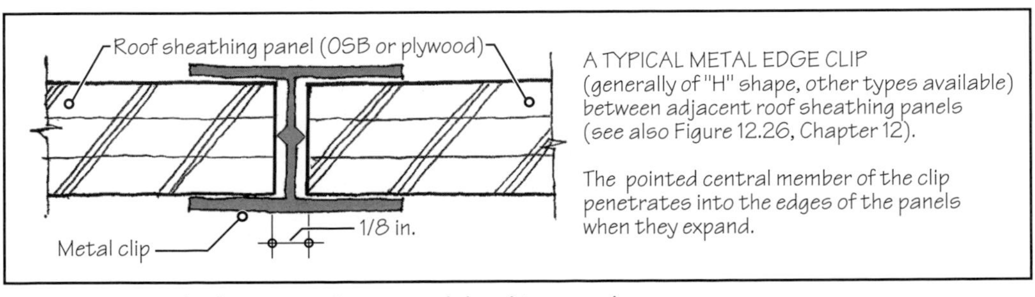

FIGURE 13.40 Clip between adjacent roof sheathing panels.

NOTE

WLF and Building Codes

As stated in Chapter 7, the building codes recognize WLF construction as Type V construction—the least fire-resistive type. Type V construction is further subclassified into Type V(A) and Type V(B). In Type V(A), all critical assemblies (such as exterior walls, floors, roofs, and interior partitions) are required to be 1-h rated. One-hour rating is typically achieved by covering the wood frame with $\frac{5}{8}$-in.-thick type X gypsum board on both sides (Chapter 14).

Type V(B) construction, commonly used for individual homes, is a nonrated WLF construction requiring no fire-rated assembly. Thus, a typical Type V(B) assembly is covered with $\frac{1}{2}$-in.-thick Type R gypsum board on the interior, whose fire rating is much less than 1 h.

(continued)

The clips automatically leave a gap of of $\frac{1}{8}$ in. between panels and allow the edges of panels to expand.

13.11 EQUALIZING CROSS-GRAIN LUMBER DIMENSIONS

In Section 13.2, reference was made to moisture-related dimensional instability of cross-grain lumber. In a WLF building, cross-grain lumber is present in floor and wall frames. In the walls, only top and bottom plates constitute cross-grain lumber. In the floors, floor joists and supporting beams constitute cross-grain lumber.

It is usually not possible to reduce cross-grain lumber dimensions in a building. All that can be done is to equalize cross-grain lumber dimensions throughout the structural frame so that the entire frame moves as a unit. Unequal cross-grain lumber dimensions lead to differential swelling and shrinkage, creating bending of framing members, which can crack brittle interior (gypsum) finishes. It may also lead to squeaking floors.

Consider the building of Figure 13.41, in which a steel beam has been used in the basement to support the overlying floor and wall. This is a commonly used detail because the steel beam is unaffected by moisture changes. The provision of a 2-by member directly above the steel beam matches the sill plate over the foundation wall, equalizing cross-grain lumber dimension. It also provides a nailing surface for the joists, Figure 13.42.

If the steel beam is replaced by a timber or glulam beam, as shown in Figure 13.43(a), cross-grain lumber dimensions will become unequal, creating dimensional instability. However, if the floor joists are hung from a timber or glulam beam using joist hangers, the dimensional inequality is substantially reduced. If the depth of the beam equals the depth of joists plus the sill plate, there is no inequality of cross-grain lumber dimensions, Figure 13.43(b).

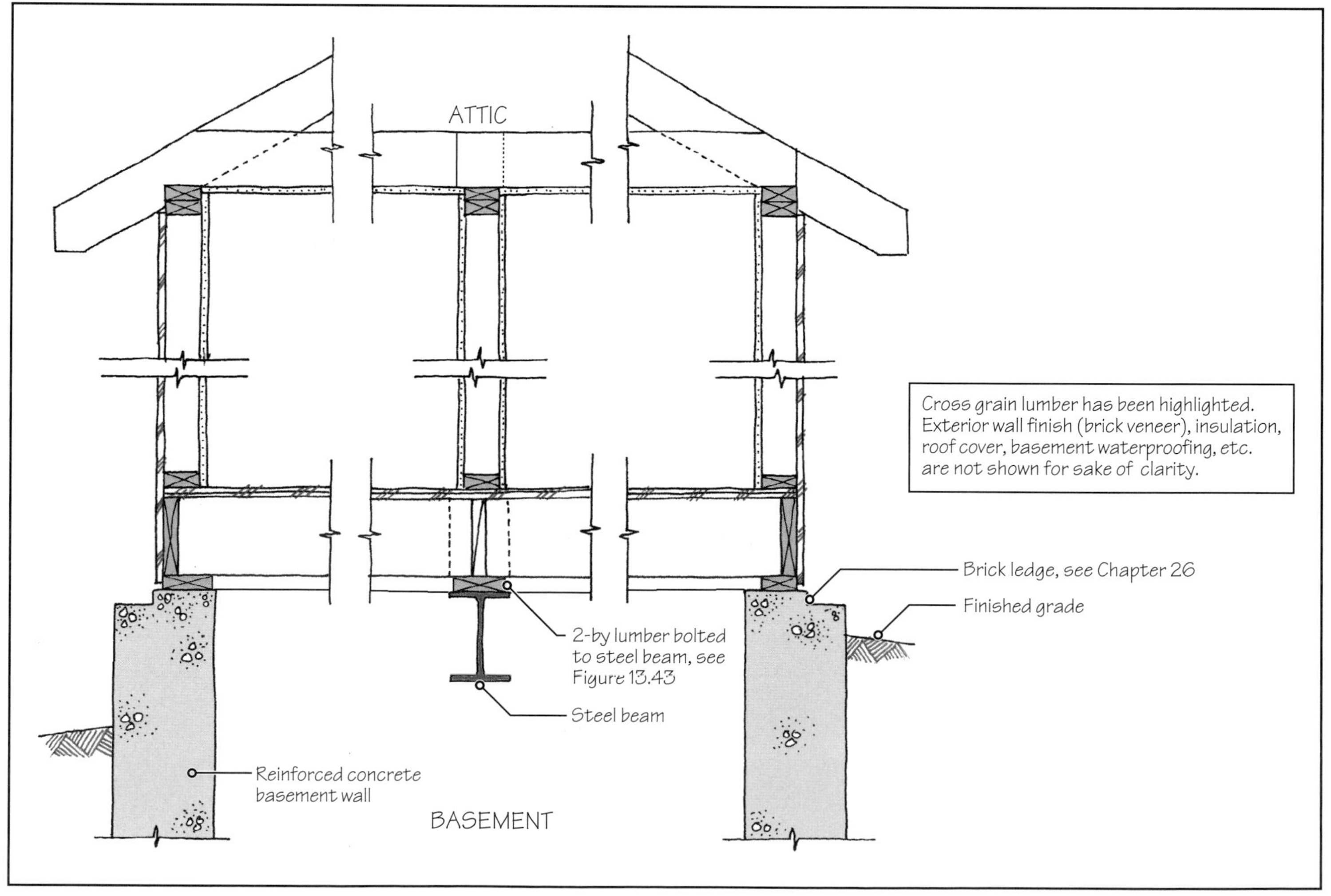

FIGURE 13.41 Equalizing cross-grain lumber dimensions in a WLF building using a steel beam.

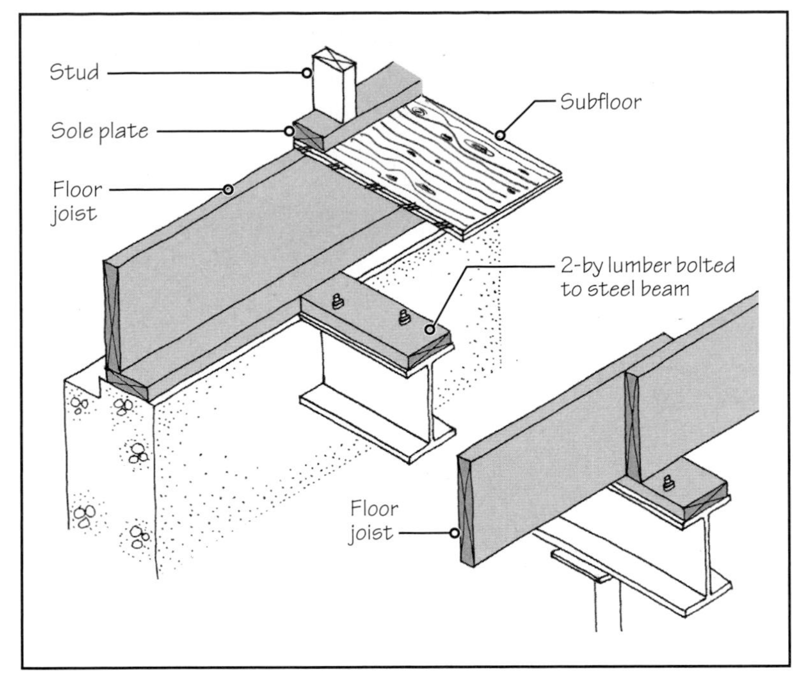

FIGURE 13.42 Floor-level detail of the structure of Figure 13.41.

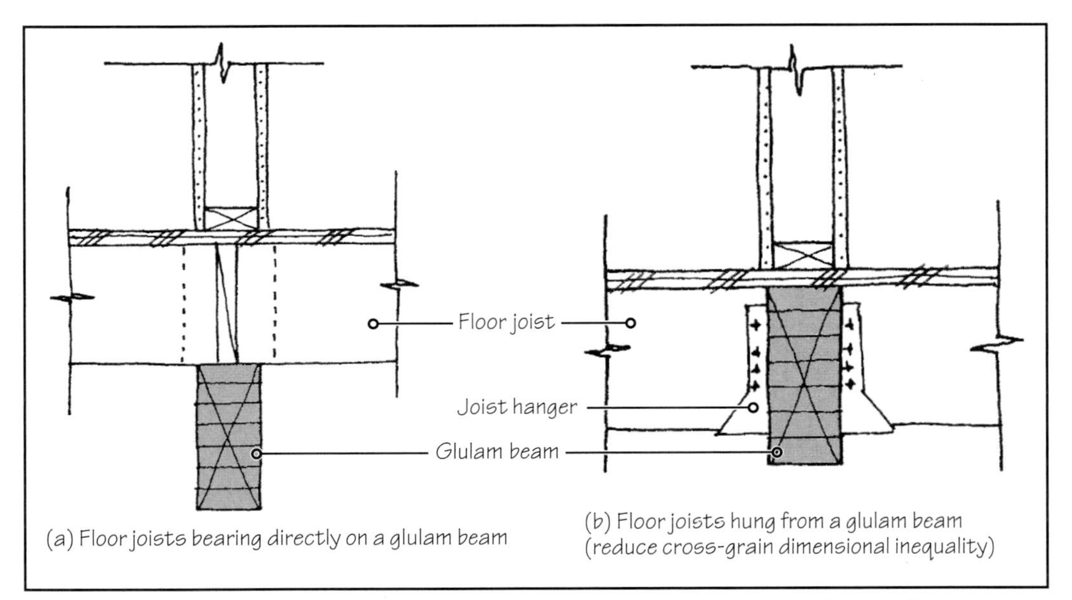

FIGURE 13.43 Use of a glulam beam and its effect on equalizing cross-grain lumber dimensions.

NOTE

WLF and Building Codes *(continued)*

The use of Type V(A) is permitted for all occupancies. With some exceptions, Type V(B) is also permitted for all occupancies. However, because of the combustibility of a wood frame, the allowable areas and heights permitted by the code for Type V buildings are fairly limited. For instance, the maximum permissible height for Type V construction used in a single-family dwelling is three stories. If automatic sprinklers are used throughout the building, the building height may be increased by one story.

PRACTICE QUIZ

Each question has only one correct answer. Select the choice that best answers the question.

27. Because a ridge board is a structural member, its size depends on roof loads.
- **a.** True
- **b.** False

28. A bird's mouth cut in rafters is made at the ridge.
- **a.** True
- **b.** False

29. In a roof with a ridge beam,
- **a.** ceiling joists are spaced at 12 in. on center maximum.
- **b.** ceiling joists are spaced at 16 in. on center maximum.
- **c.** ceiling joists are spaced at 24 in. on center maximum.
- **d.** none of the above.

30. When a collar tie is provided in a pitched WLF roof, it should be located
- **a.** in the upper one-fourth of the attic.
- **b.** in the upper one-third of the attic.
- **c.** in the lower one-fourth of the attic.
- **d.** in the lower one-third of the attic.
- **e.** in the center of the attic.

31. The primary purpose of collar ties is to
- **a.** increase the wind-uplift resistance of roof framing.
- **b.** increase the gravity-load resistance of roof framing.
- **c.** increase the wind uplift resistance of floor framing.
- **d.** increase the gravity load resistance of floor framing.

32. If the cross-grain lumber dimensions in a WLF building are the same throughout the entire building, it is not subjected to any swelling or shrinkage.
- **a.** True
- **b.** False

33. As per the building codes, the type of construction that WLF represents is
- **a.** Type I construction.
- **b.** Type II construction.
- **c.** Type III construction.
- **d.** Type IV construction.
- **e.** Type V construction.

34. The maximum number of floors allowed by building codes for a single-family dwelling built using wood light frame construction is
- **a.** one floor.
- **b.** two floors.
- **c.** three floors.
- **d.** five floors.

Answers: 27-b, 28-b, 29-d, 30-b, 31-a, 32-b, 33-e, 34-c.

PRINCIPLES IN PRACTICE

Constructing a Two-Story Wood Light Frame Building

The following text and illustrations show the process and sequence of construction of a two-story wood light frame (WLF) building. Some builders may follow a slightly different process or sequence.

Several foundation systems can be used in WLF construction, depending on soil conditions and climate (Chapter 21). In this example, a concrete slab-on-ground (also called *slab-on-grade*) foundation has been used.

STEP 1: FOUNDATION PREPARATION

If the foundation for a WLF building consists of a slab-on-grade, as shown in Figure 1, the ground must first be prepared to receive the slab. This includes clearing the site of vegetation and undesirable topsoil, and grading the site to provide appropriate drainage. Trenches are then dug for grade beams as required, as well as underground utilities (sewage, wastewater, water supply, natural gas pipes, and electrical and telecommunication lines). Trenches are backfilled after the utilities are laid and then compacted. Finally, the slab is constructed, as shown in Figure 2 (Section 19.6).

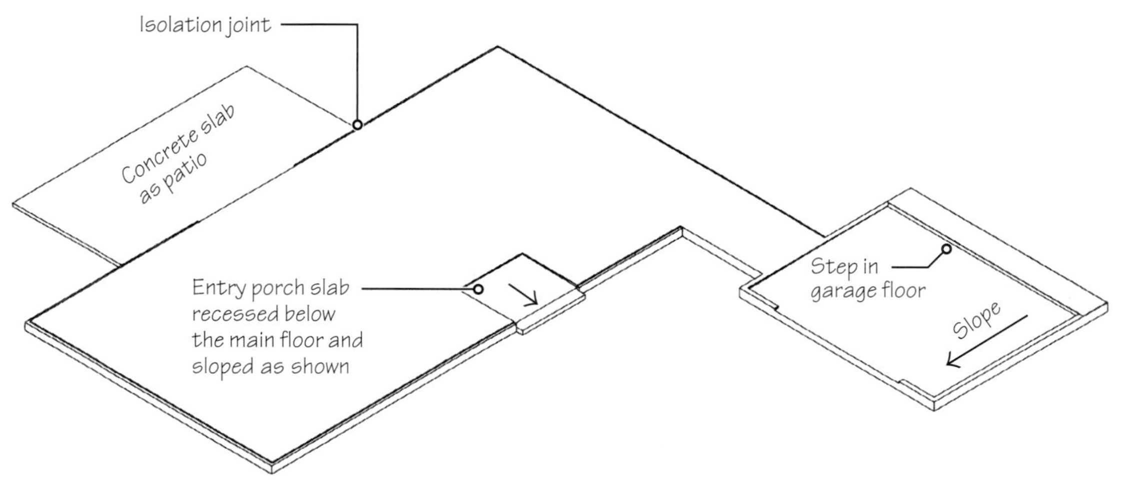

FIGURE 1 Diagram of a concrete slab-on-grade foundation.

FIGURE 2 Typical completed slab-on-grade foundation ready to receive the WLF superstructure. Note that the utilities extend (referred to as being stubbed out) above the top of the slab.

STEP 2: FRAMING THE FIRST-FLOOR WALLS

Before the wall framing for the ground floor begins, chalk lines are placed (snapped) on the slab to locate the walls per the architectural floor plan, Figures 3 and 4.

With the chalk lines in place, framing the ground floor walls can begin, Figures 5, 6, and 7. Temporary braces stabilize the walls until the frame is complete and exterior walls are sheathed. When double-height spaces are required, continuous or double studs are used to frame the high segments of the wall.

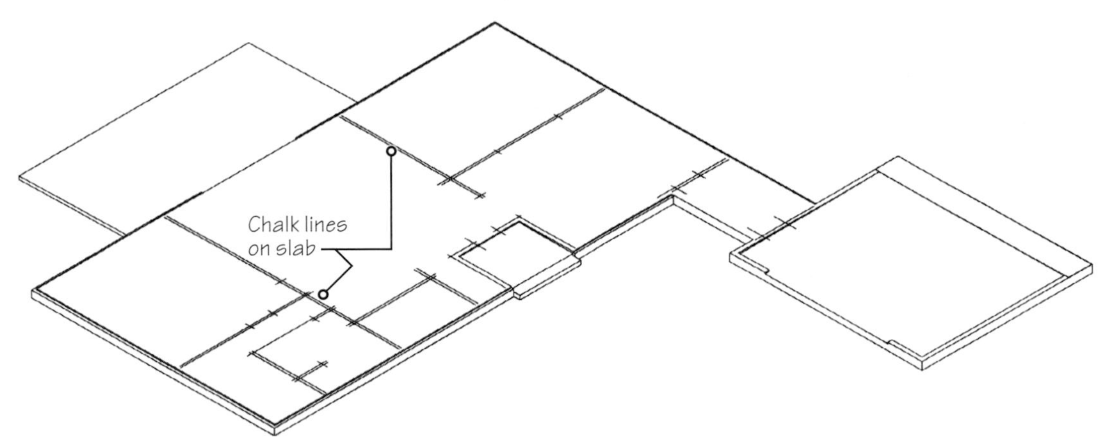

FIGURE 3 Slab-on-grade with chalk lines.

FIGURE 4 Chalk lines on the slab indicate the location of walls. Note that plumbing pipes align and are located within the depth of a wall.

FIGURE 5 Wall assembly under construction.

FIGURE 6 Walls are typically framed on the slab or floor platform and subsequently tilted up into position.

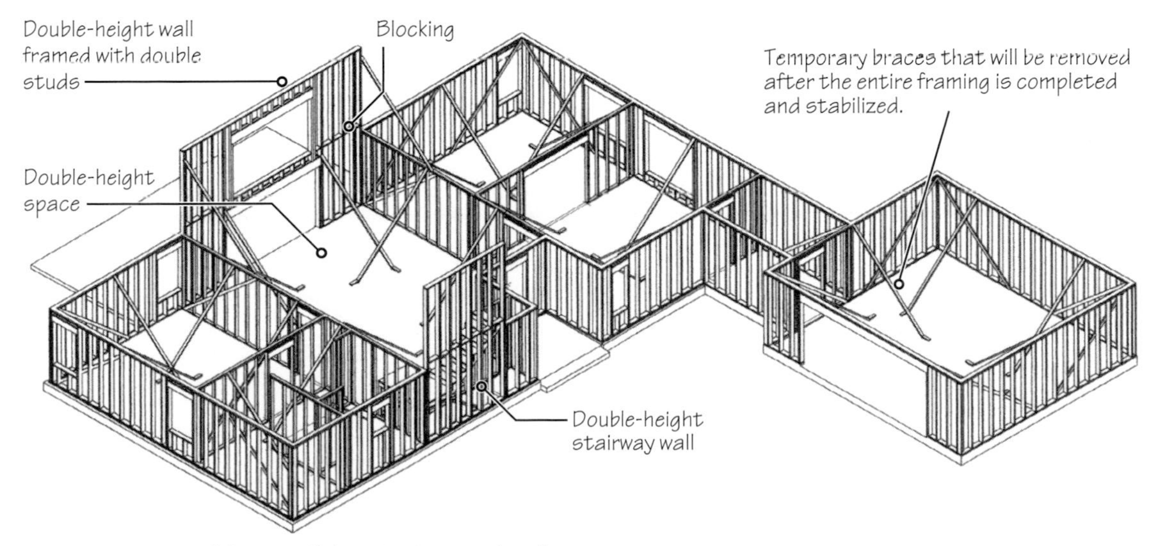

FIGURE 7 Ground floor wall framing is completed.

(Continued)

Constructing a Two-Story Wood Light Frame Building (*Continued*)

STEP 3: FRAMING THE SECOND FLOOR

Upon completion of the first-floor walls, joists can be installed for the second-story floor, Figures 8 and 9. Unless the plan dictates otherwise, the floor joists are oriented to span the short direction. At locations such as stair openings or recessed floors, floor joists may be doubled or tripled, or a beam may be used to carry the greater load.

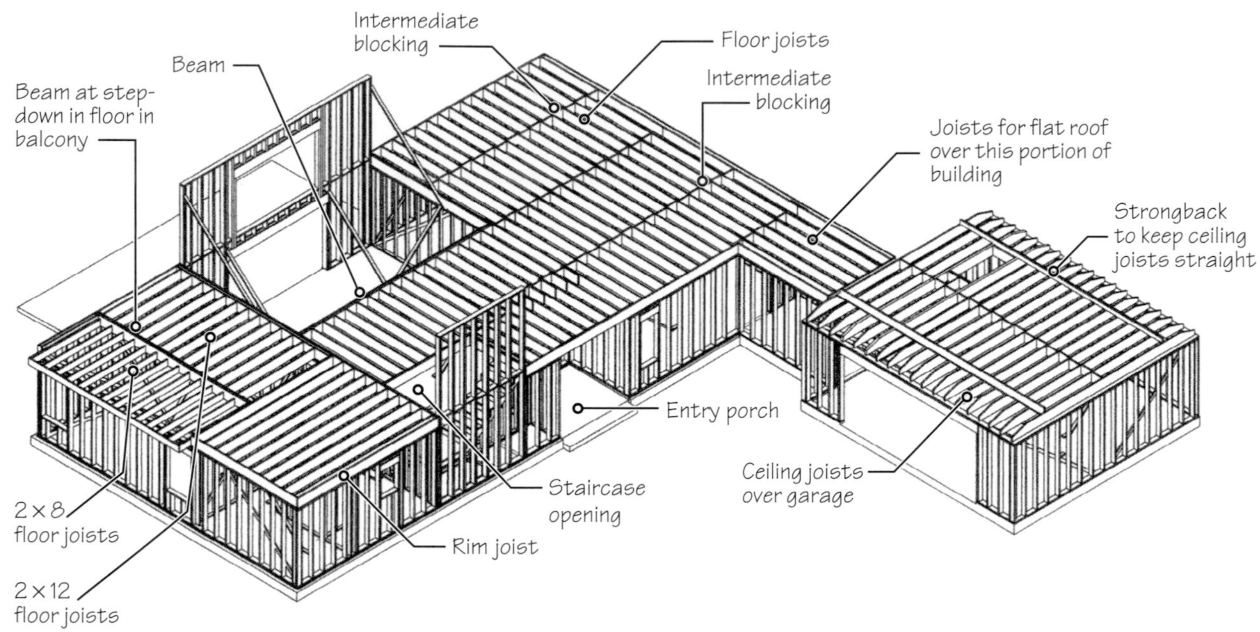

FIGURE 8 Floor framing at the second floor.

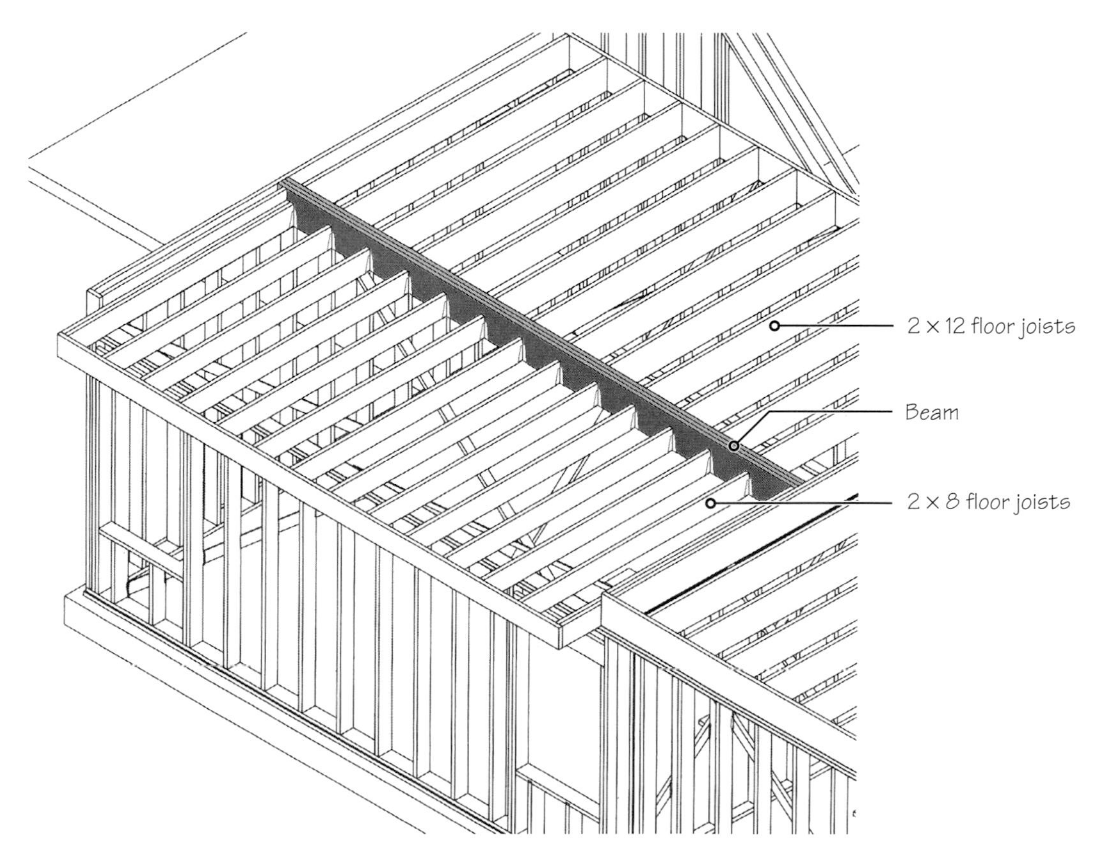

FIGURE 9 Detail of the floor framing at a stepdown in the floor (Figure 8.) Note that 2 × 12 floor joists are tripled (built up) to form a beam. The 2 × 8 floor joists supported by the beam are cantilevered beyond the supporting wall below. This is similar to the example in Figures 13.17 and 13.18

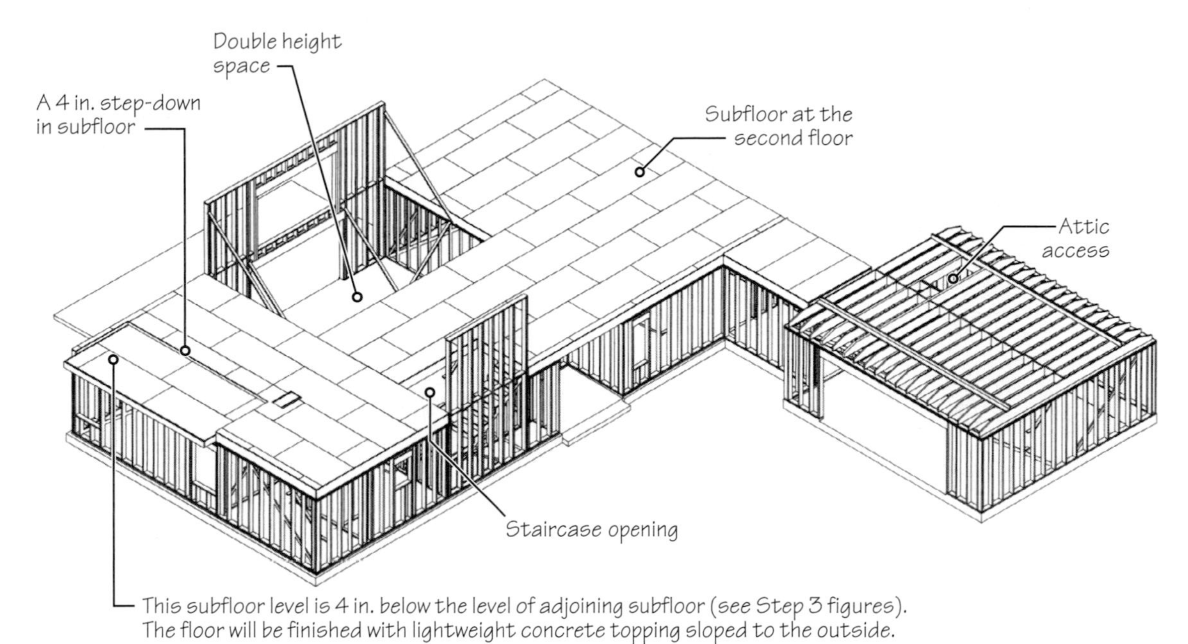

FIGURE 10 Plywood or OSB sheathing is staggered, glued, and nailed to the floor joists.

STEP 4: FLOOR SHEATHING AT THE SECOND FLOOR

Next, the sheathing is applied to the floor joists to form the second-story floor assembly, Figure 10.

STEP 5: SECOND-FLOOR WALL FRAMING

The walls for the second story are constructed on the subfloor surface, Figure 11, in a manner similar to that shown in Step 2.

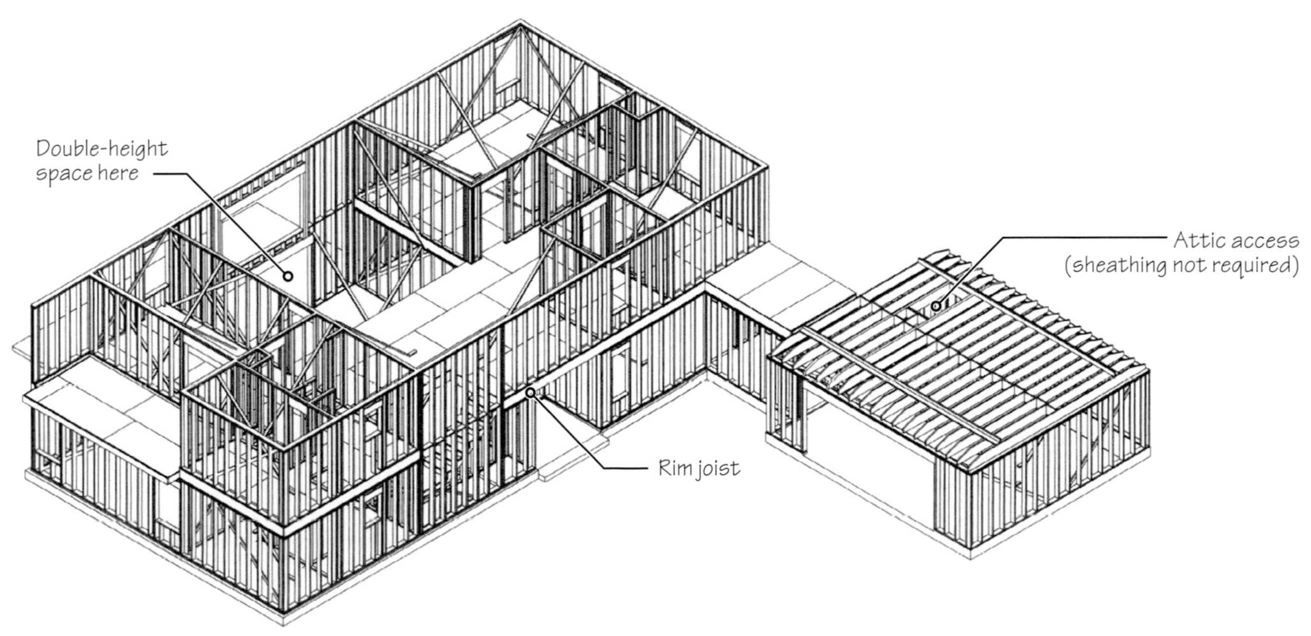

FIGURE 11 Walls are framed, erected, and braced.

STEP 6: CEILING JOIST AND ROOF FRAMING

In a stick-framed roof, ceiling joists span across the walls, Figure 12. Rafters are then set in place to complete the roof framing. When roof trusses are used, they replace the function of ceiling joists and rafters in one unit.

(Continued)

PRINCIPLES IN PRACTICE

Constructing a Two-Story Wood Light Frame Building (*Continued*)

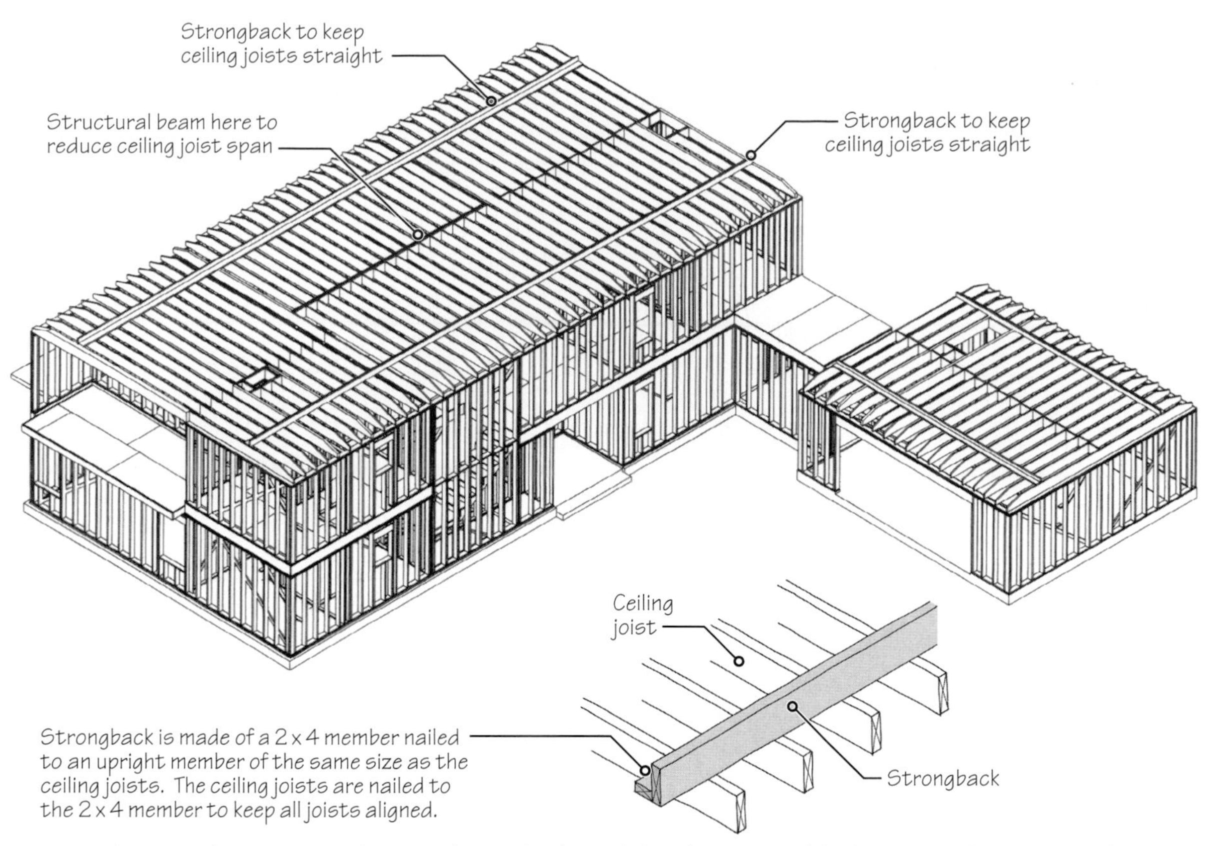

FIGURE 12 Ceiling joist framing. Note the use of strongbacks and the placement of the beam to enhance strength.

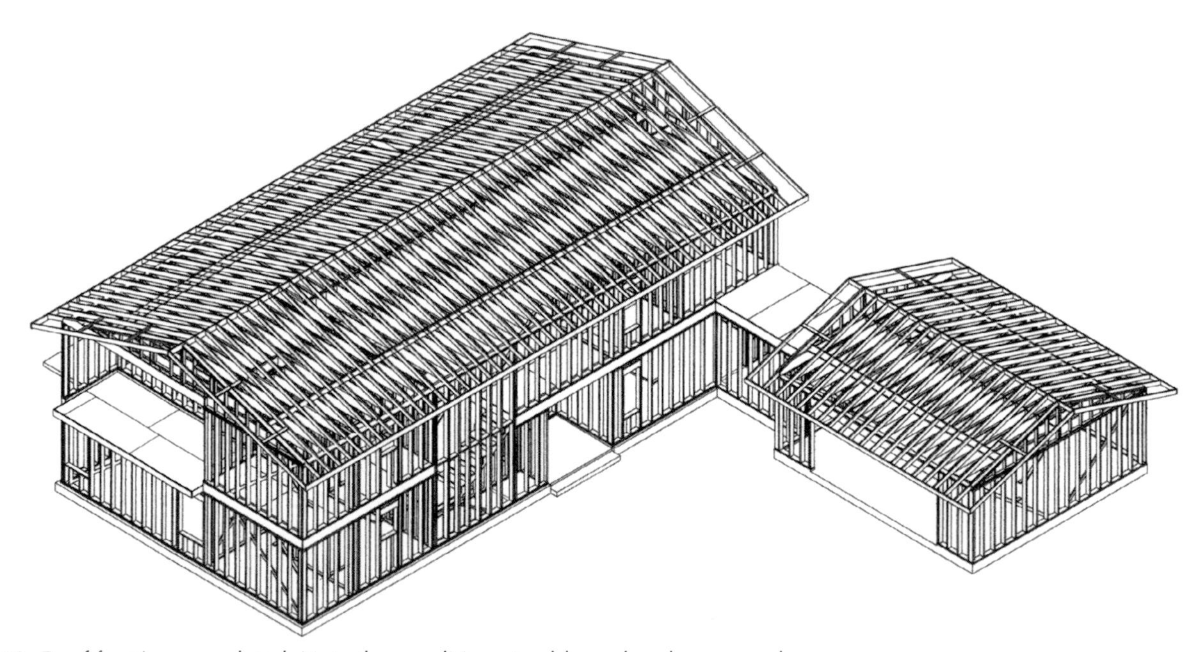

FIGURE 13 Roof framing completed. Note the condition at gable end and eave overhangs.

STEP 7: WALL AND ROOF SHEATHING

With the framing complete, Figure 13, the building is ready to receive wall sheathing, Figure 14, and then roof sheathing, Figure 15, which greatly enhances structural rigidity. Note that some framers may apply wall sheathing prior to roof framing. After the walls and roof have been sheathed, the structural frame is stable. Temporary interior wall supports can now be removed, allowing work on interior finishes.

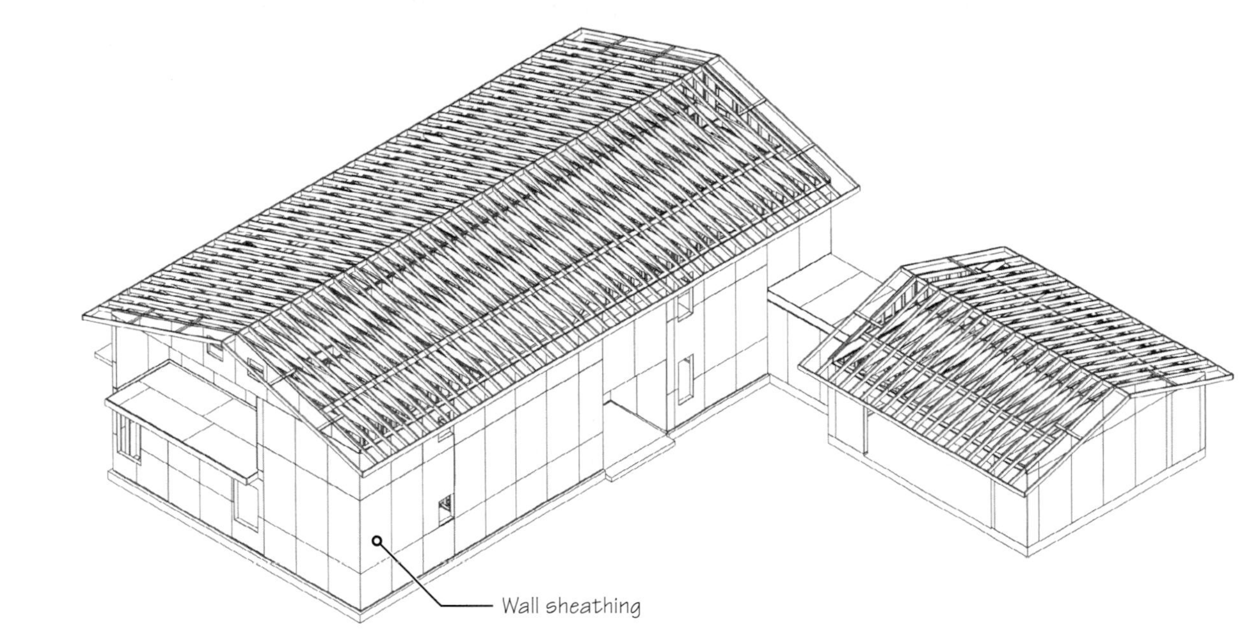

FIGURE 14 Sheathing installed on all exterior walls.

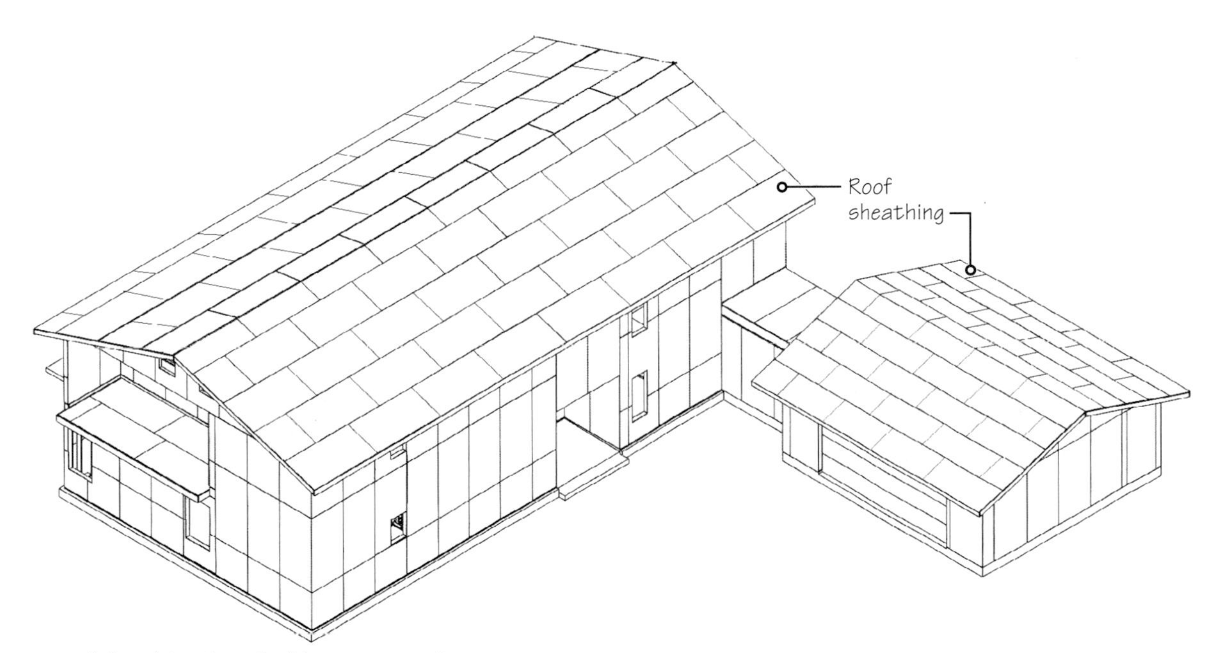

FIGURE 15 Roof sheathing installed in a staggered pattern.

STEP 8: EXTERIOR AND INTERIOR FINISHES

Roofing, including underlayment, Figure 16, and finish cover, Figure 17 (discussed in Chapter 32), are generally applied first to dry-in the interior spaces and allow inside work to progress. Interior work, including rough plumbing and electrical and wall insulation, is completed before drywall and interior finishes. Exterior work includes windows and doors, moisture/air retarder, and finish surface treatment, Figure 18.

(*Continued*)

Constructing a Two-Story Wood Light Frame Building (*Continued*)

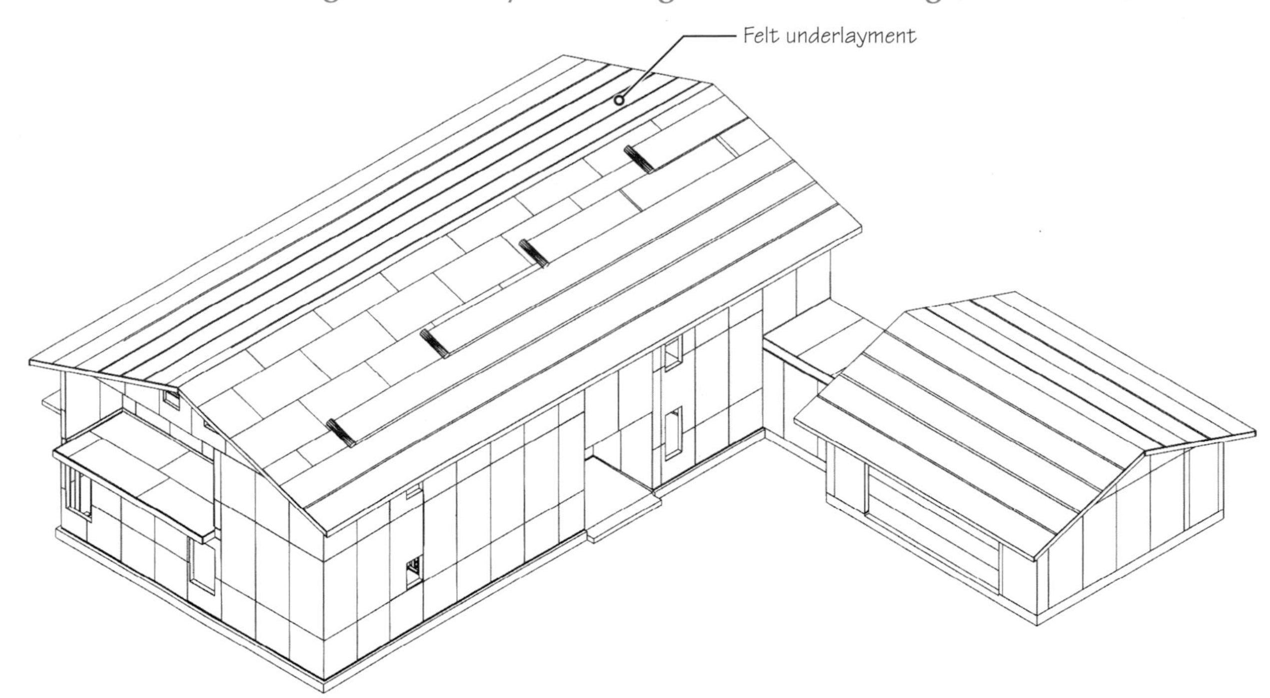

FIGURE 16 Application of felt underlayment.

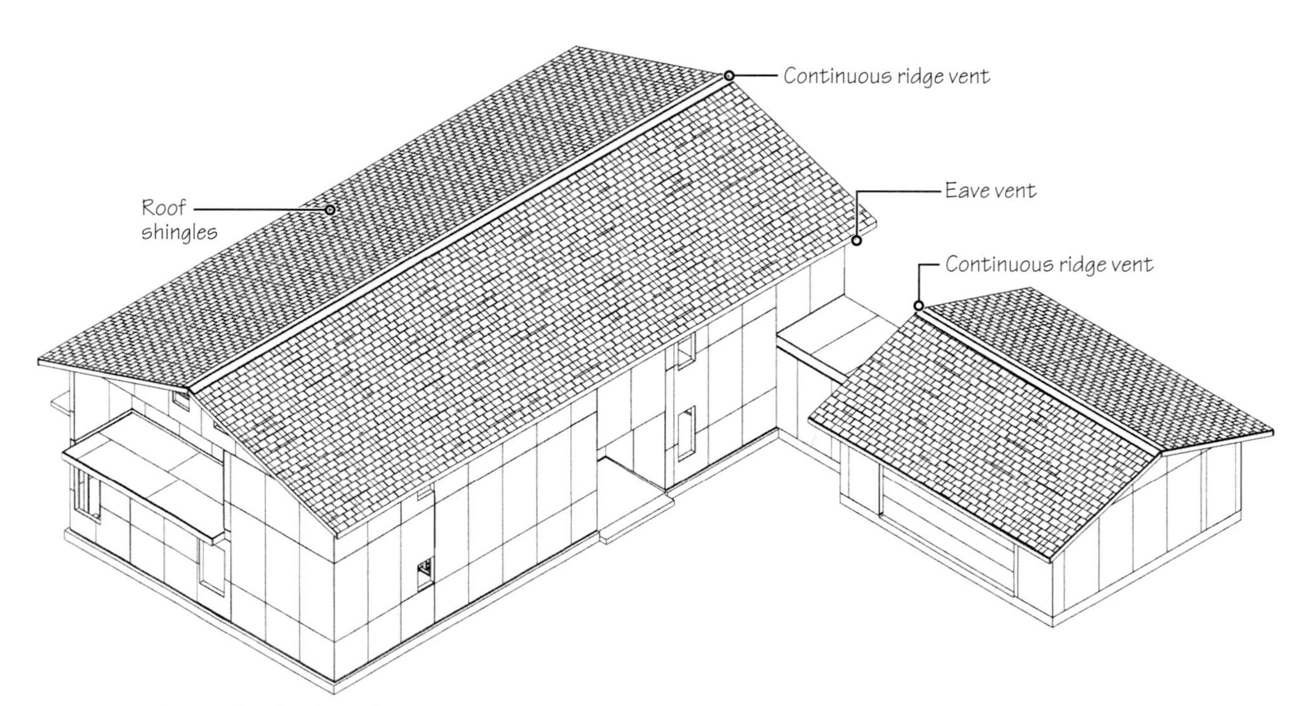

FIGURE 17 Shingles are the finish roofing cover in this example. Metal, tiles, and other options are available.

PRINCIPLES IN PRACTICE

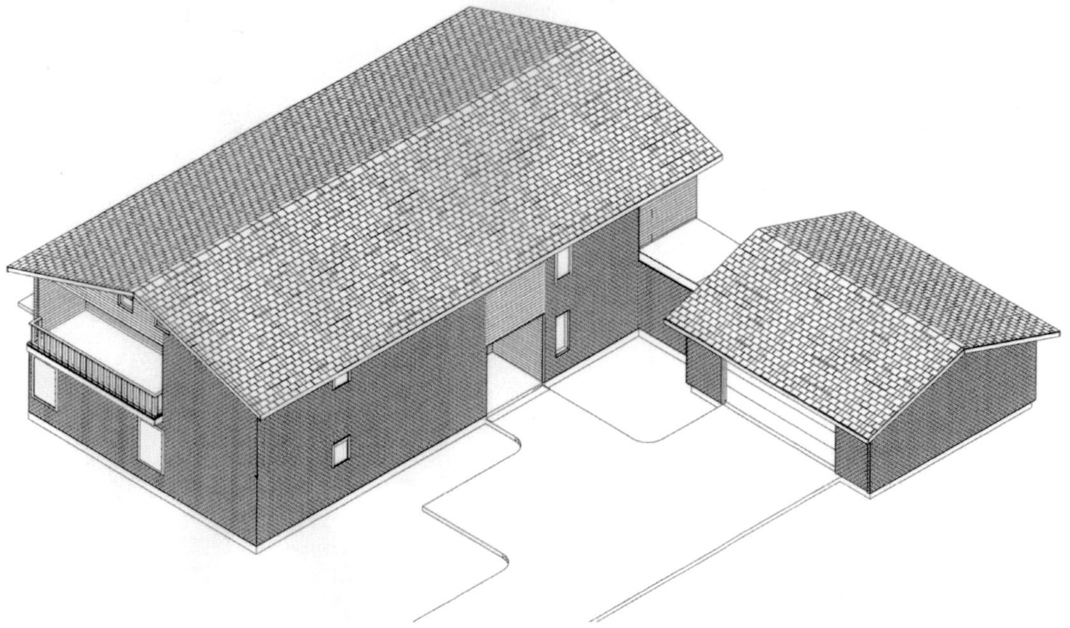

FIGURE 18 Exterior finishes include roofing and siding.

PRINCIPLES IN PRACTICE

How a WLF Building Resists Loads

All loads on a building, regardless of the type, must terminate in the ground, where they are finally absorbed. In other words, the ground serves as an infinitely large sink for all loads. In proceeding from its point of origin in the building down to the ground, a load will generally traverse through several intermediate components.

The path that a particular load takes through various components helps us understand the role of each component in resisting the loads. Additionally, it helps to determine the magnitude and types of stresses to which each component is subjected. Both gravity and lateral loads must be considered, and the stresses caused by each load are added to determine a component's size.

GRAVITY LOAD PATH

The path of gravity loads through a WLF building is relatively simple to trace and is shown in Figure 1. In tracing the gravity load path, we proceed from the top and work our way down. Rafters transmit roof loads to the supporting walls. As

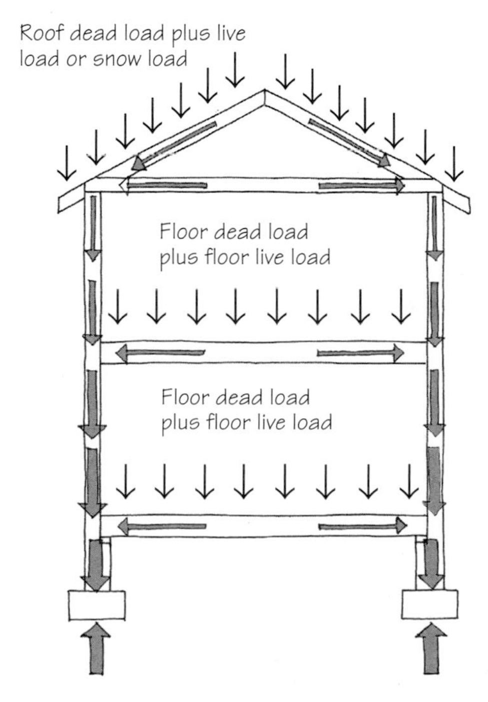

FIGURE 1 Journey of gravity loads through a WLF building.

(*Continued*)

PRINCIPLES IN PRACTICE

How a WLF Building Resists Loads (*Continued*)

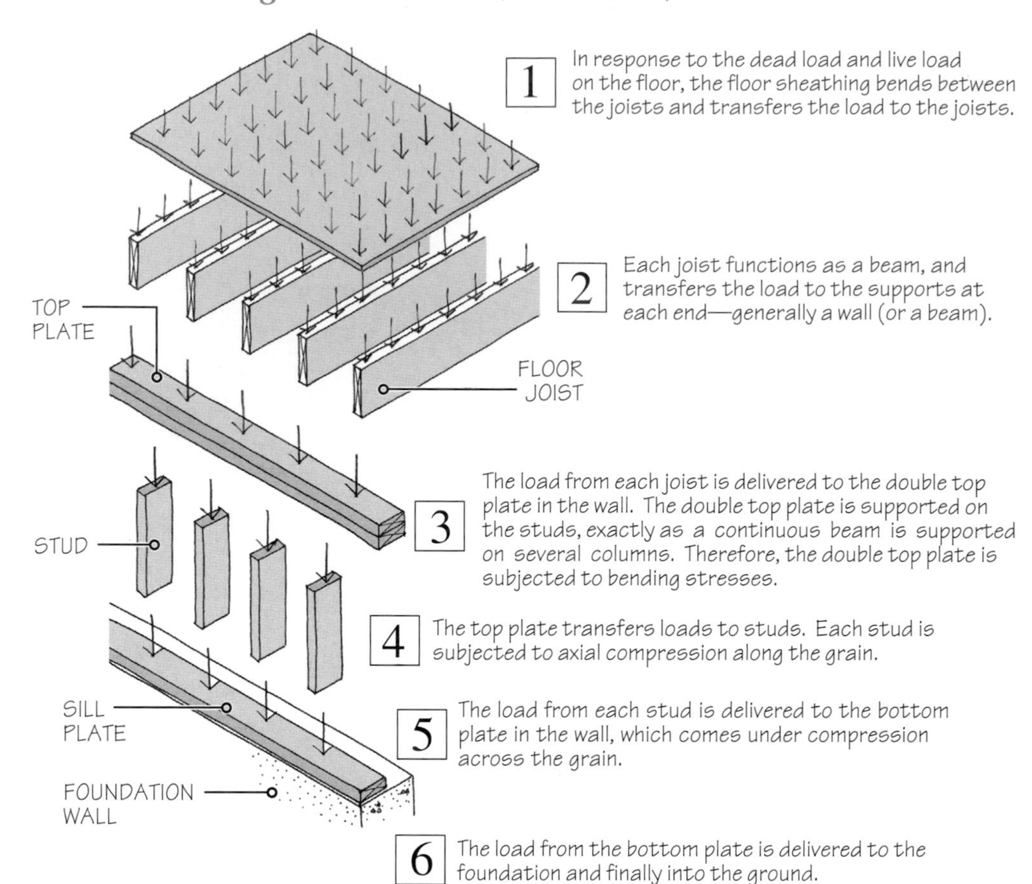

FIGURE 2 Details of the path of gravity loads acting on an intermediate floor of a WLF building.

these loads travel down, they meet the load transmitted to the wall from the floor, where they add up. Further down into the wall, the load from the next floor is added. Finally the load reaches the foundation.

Observe that the load on the wall increases as it proceeds toward the foundation. This implies that the wall should be made stronger at each lower floor. For the sake of simplicity, however, wall strength at upper floors may be kept the same as that on the lowest floor. For instance, if the first-floor walls are required to consist of 2 × 6 studs at 16 in. on center, the same stud size and spacing can also be used in the upper floors.

The load paths shown in Figure 1 indicate that a load-bearing wall at an upper floor must be aligned with that at the lower floor. Although a small offset in load-bearing walls can be managed structurally, it is advisable to plan for vertical alignment of all load-bearing walls. Figure 2 shows the details of how the gravity loads acting on an intermediate floor are transmitted to the foundation.

LATERAL LOAD PATH

In tracing the lateral loads through the structure, let us consider the loads acting on one of the walls (say, the gable-end wall) of a two-story WLF building, Figure 3. In resisting these loads, the wall behaves as a vertical slab supported by three elements—the foundation at the bottom, the intermediate floor, and the roof-ceiling assembly. The loads transmitted to each of these three elements are proportional to their respective tributary areas.

Thus, if the floor-to-floor height and the floor-to-roof height are each 10 ft, the lateral loads on the lowest 5 ft of the wall are transmitted to the foundation. The loads on the next 10 ft of the wall are transmitted to the (second) floor, whereas the loads on the remaining (uppermost) portion of the wall are transmitted to the roof.

In a WLF building, the transfer of lateral loads from the wall to the floor or the roof occurs through the studs, which function as vertical beams. Therefore, in high-wind regions, the size and spacing of studs in exterior walls may be dictated by the wind loads. In regions where the wind loads are small, gravity loads dictate the size and spacing of studs.

HORIZONTAL DIAPHRAGMS

The load transmitted from the wall to the foundation is absorbed into the ground, where all loads must eventually reach. Therefore, this load will not be traced any further.

The load transmitted to the (second) floor lies in the plane of the floor, Figure 4. In resisting this load, the floor behaves as a deep beam, with side walls functioning as its supports. One of the characteristics of a deep beam is that the tensile and compressive stresses created in it are concentrated mainly at its extreme edges.

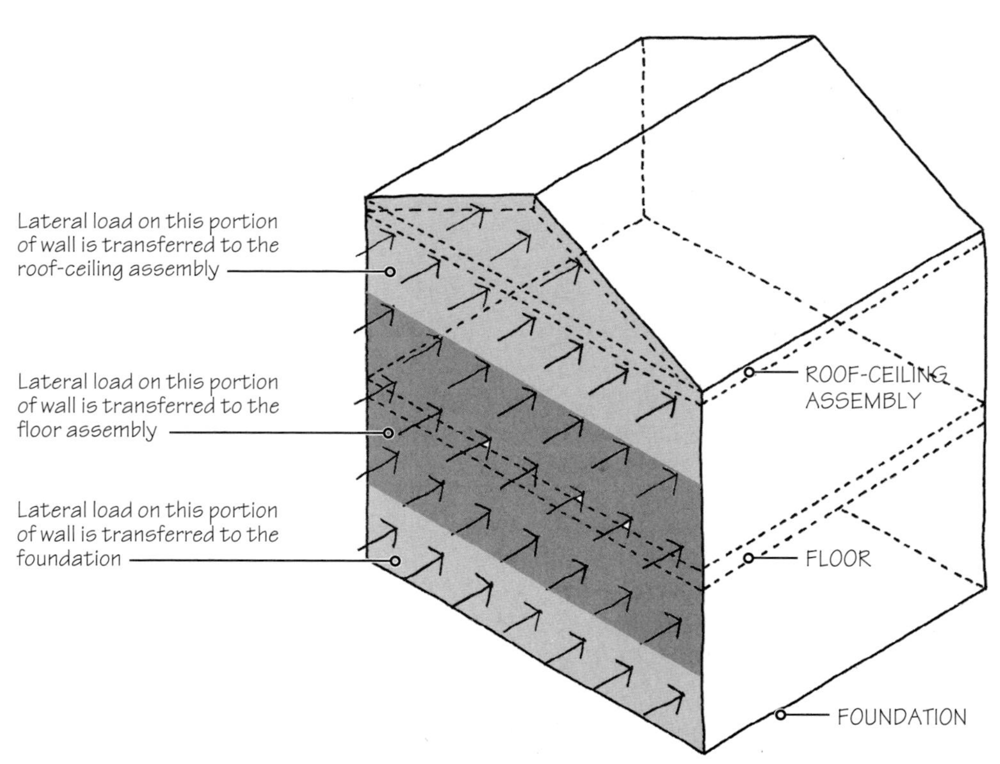

FIGURE 3 Journey of lateral loads acting on one of the walls of a WLF building. Note that the lateral loads assumed on a gable-end wall are shown as push-in loads, but they could just as well be suction loads.

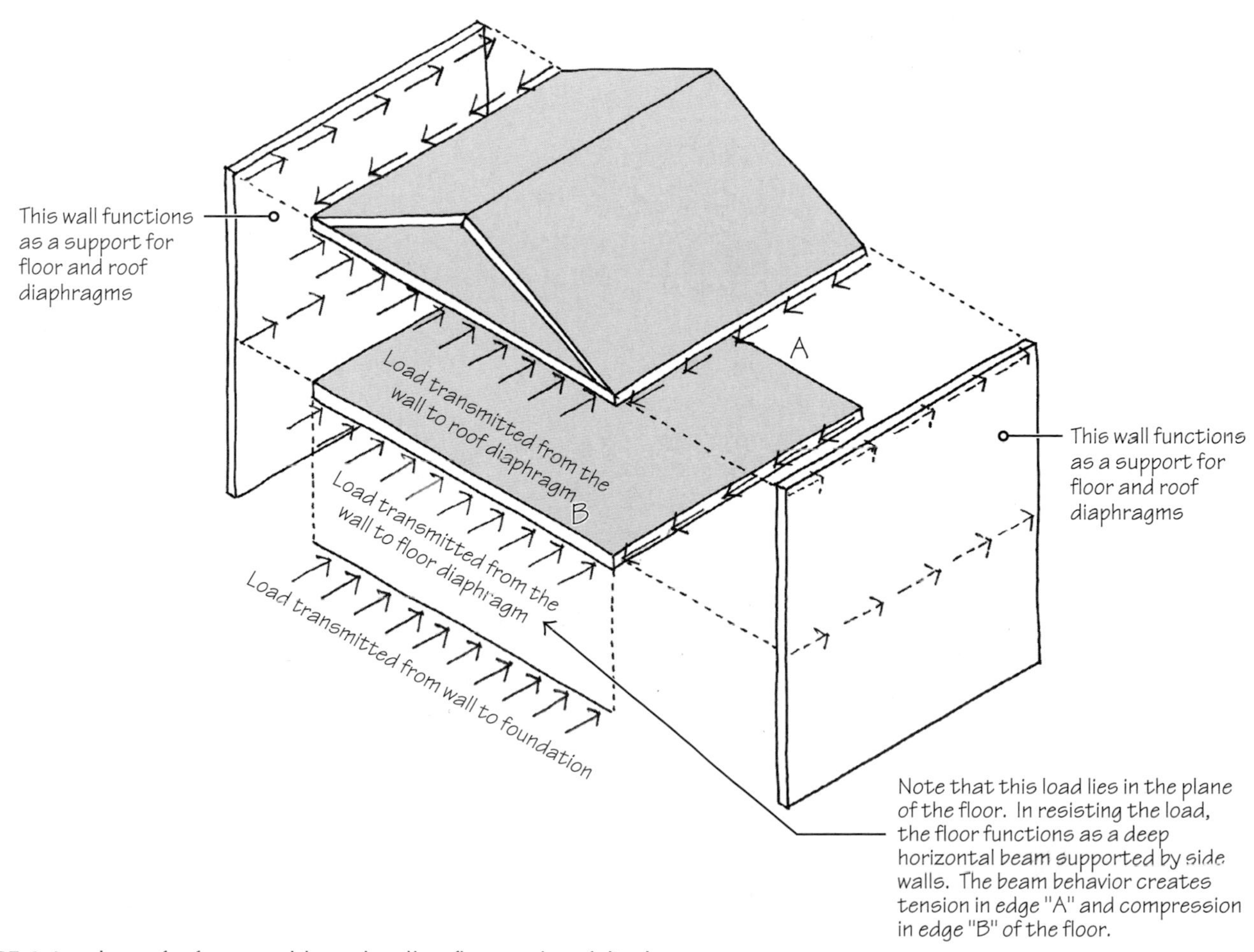

FIGURE 4 Load transfer from a gable-end wall to floor and roof diaphragms.

(*Continued*)

PRINCIPLES IN PRACTICE

How a WLF Building Resists Loads (*Continued*)

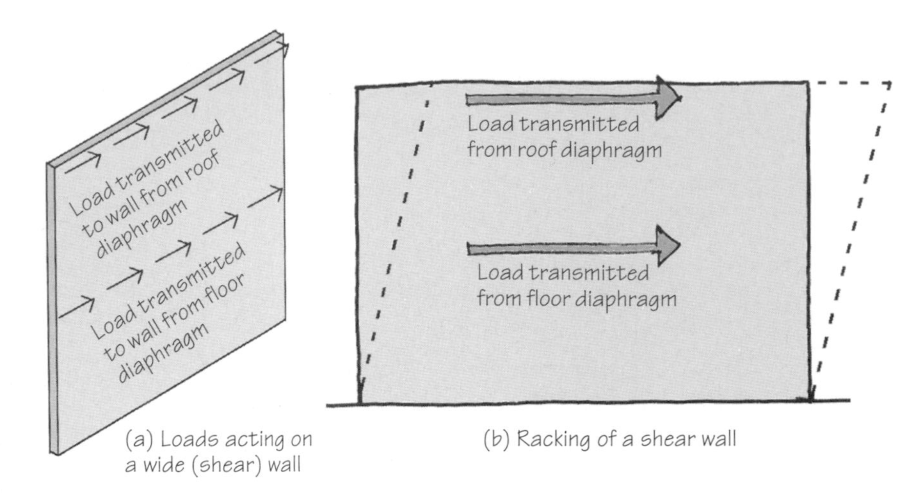

FIGURE 5 Racking in a WLF wall due to in-plane loads.

The action of the floor in transferring the lateral loads to its supports (side walls) through deep beam action is called *diaphragm action,* and the tensile and compressive stresses in it are called *diaphragm tension* and *diaphragm compression.*

In practice, we assume that diaphragm tension and compression are resisted by the double top plate supporting the floor. The presence of diaphragm tension requires that the top plate, as well as the adjacent rim joist, should be continuous.

Just as a floor works as a diaphragm in transferring lateral loads to side walls, a roof also works in the same way—although as a folded diaphragm.

SHEAR WALLS

The loads acting on a side wall are those that have been transferred to it from the floor and roof diaphragms. In resisting these loads, the wall is supported only at the foundations. Therefore, it behaves as a vertical cantilever carrying in-plane loads. This makes the wall *rack,* or deform angularly, Figure 5.

A WLF wall must, therefore, be braced against racking, which is achieved through the use of plywood, OSB sheathing, or diagonal braces (Section 13.10). Because the racking and the consequent angular deformation are caused by shear forces, a wall braced to resist racking is referred to as a *shear wall.* In a rectilinear, box-type building with four walls, all walls must be designed as shear walls to account for lateral loads from either one of the two principal directions.

In transferring the loads to the foundation, a shear wall is subjected to sliding, Figure 6(a) and overturning, Figures 6(b) and 6(c). The sliding action in a WLF building is resisted by foundation anchor bolts. The overturning action is also resisted by the same bolts if the overturning force is small. In a high-wind or seismic region, special hold-down bolts are required in each shear wall segment to counteract excessive overturning forces. A typical hold-down bolt is shown in Figure 6(d).

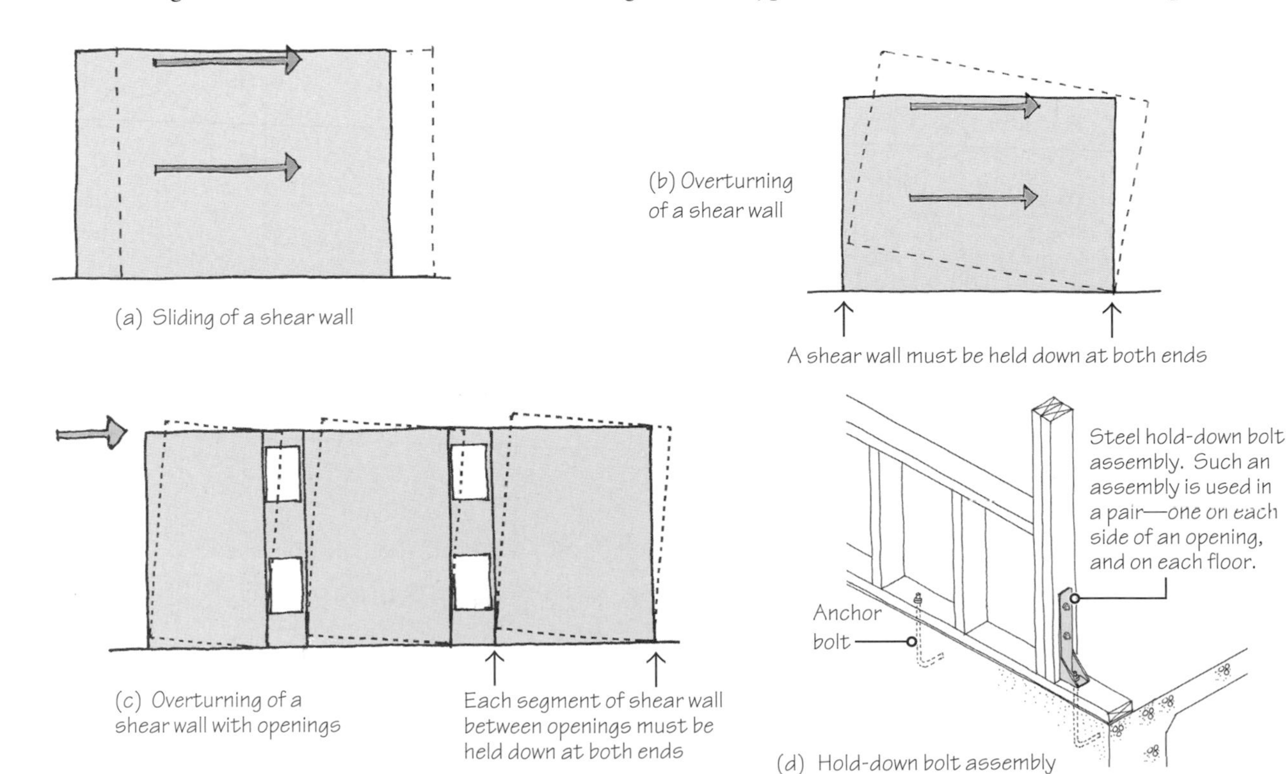

FIGURE 6 Sliding and overturning of a shear wall.

Uplift on Roofs Due to Wind or Earthquake

The interaction between horizontal diaphragms and shear walls means that if any one of these elements fails, the entire structure may collapse. Because of its light weight, a frequent cause of the failure of a WLF building is the roof blowing off due to uplift forces created by wind. Once the roof blows off, a complete collapse of the structure becomes imminent. Therefore, it is important to ensure that the roof assembly in a WLF building is adequately held down.

In other words, roof sheathing should be securely nailed to the rafters, and the rafters should be securely fastened to the wall's double top plate. In a high-wind or seismic region, the rafters are generally required to be anchored to the top plate with sheet-metal anchors, referred to as hurricane or seismic straps, Figure 7. Additional sheet steel straps may be required in a high wind or seismic region to provide a continuous load path from the roof to the foundation, Figure 8.

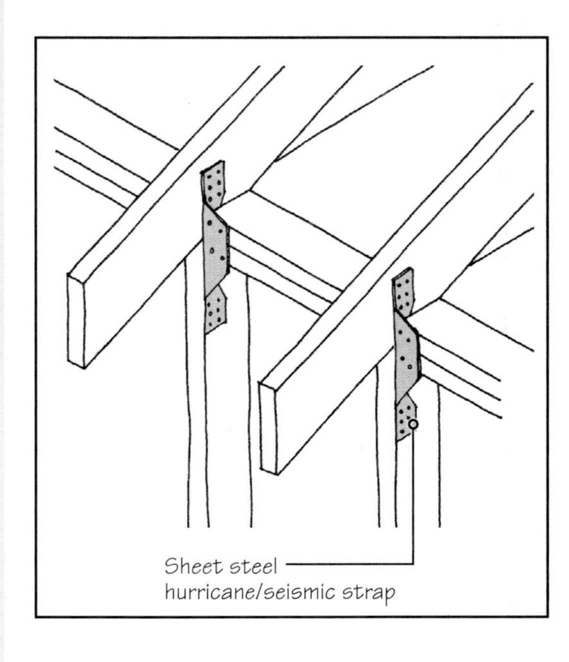

FIGURE 7 Commonly used hurricane or seismic straps for roof.

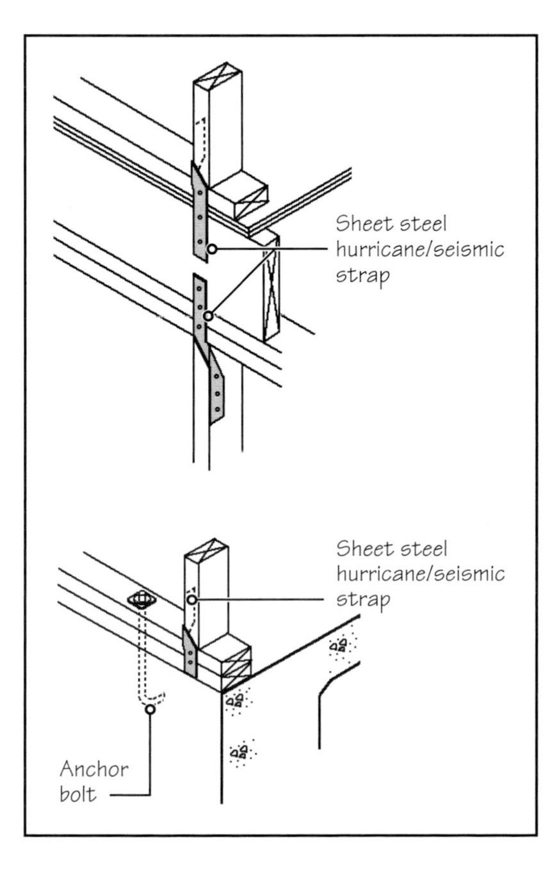

FIGURE 8 Commonly used hurricane or seismic straps to provide a continuous load path from the roof to the foundation.

(*Continued*)

PRINCIPLES IN PRACTICE

How a WLF Building Resists Loads (*Continued*)

WLF BUILDING—A LOAD-BEARING WALL STRUCTURE

The structural behavior of a WLF building just described shows that all the major elements of a building work together in countering the lateral loads. The interaction between horizontal diaphragms and shear walls in a building is not peculiar to a WLF construction. It applies to all structures—reinforced concrete frame, load-bearing masonry, or steel frame structure.

Because most walls in a WLF building double as gravity-load-resisting and shear walls, a WLF building is, in fact, a load-bearing-wall building. In other words, there is no fundamental difference between the structural behavior of a load-bearing masonry or concrete building and a contemporary WLF building. The reason for referring to it as a wood *light frame* building is that the walls consist of a frame instead of solid concrete or masonry walls.

PRACTICE QUIZ

Each question has only one correct answer. Select the choice that best answers the question.

35. In referring to a floor or roof of a building as a *horizontal diaphragm,* the refererence is with respect to
- **a.** gravity loads.
- **b.** lateral loads.
- **c.** both gravity and lateral loads.
- **d.** none of the above.

36. Hold-down bolts, when provided in a WLF wall building, are used in
- **a.** floors.
- **b.** walls.
- **c.** the roof.

37. A wood light frame building is essentially a
- **a.** load-bearing-wall structure.
- **b.** frame structure.

Answers: 35-b, 36-b, 37-a.

KEY TERMS AND CONCEPTS

Balloon frame
Blocking and bridging
Built-up beam
Cathedral ceiling
Ceiling joist
Cross-grain lumber dimensions
Eave
Floor framing
Floor sheathing
Frame configuration and spacing of members
Framing plan
Gable
Grade beam
Header
Hip
Lateral bracing of wood frame
Mud sill
Platform frame
Rafter and ceiling joist roof structure
Rake
Ridge
Ridge beam
Ridge board
Rim joist
Roof framing
Sheathing
Shed roof
Stud
Subfloor
Top plate
Valley
Vaulted ceiling
Wall framing
Wall openings
Wood light frame construction (WLF)

REVIEW QUESTIONS

1. Explain why we use 12 in., 16 in., or 24 in. as center-to-center spacing of framing members in a WLF structure.
2. Explain why the top plate in a wall assembly is generally made of two 2-by members.
3. Using sketches and notes, explain how the two top plate members are connected at a wall corner.
4. Show at least two alternative arrangements of studs at a wall corner.
5. Using sketches and notes, show the typical arrangement of studs around a wall opening. Name all the important framing members.
6. Using sketches and notes, explain the composition of a typical header over an opening.

SELECTED WEB SITES

Alberta Forest Products Association (www.albertaforestproducts.ca)

American Lumber Standards Committee (www.alsc.org)

Southern Forest Products Association (www.sfpa.org)

Southern Pine Council (www.southernpine.com)

Western Wood Products Association (www.wwpa.org)

FURTHER READING

1. American Wood Council (AWC) and American Forest and Paper Association (AFPA). *Details for Conventional Wood Frame Construction,* 2001.
2. Thallon, Rob. *Graphic Guide to Frame Construction.* Newtown, CT: Taunton Press, 2000.
3. United States Housing and Urban Development (HUD). *Residential Structural Design Guide,* 2000; updated 2004. A 400-page free downloadable publication from www.huduser.org.

CHAPTER 14

Wood Light Frame Construction–II

CHAPTER OUTLINE

14.1 Exterior Wall Finishes in a WLF Building

14.2 Horizontal Sidings

14.3 Vertical Sidings

14.4 Finishing the Eaves, Rakes, and Ridge

14.5 Gypsum Board

14.6 Installing and Finishing Interior Drywall

14.7 Fire-Resistance Ratings of WLF Assemblies

After the joints between interior gypsum board panels have been treated with tape and joint compound, the walls and ceilings in a WLF building are ready for painting. (Photo by MM.)

The previous chapter focused on the framing aspects of a WLF construction system. This chapter deals with some of the important interior and exterior finishes used in this system. It begins with a brief introduction to various exterior wall finishes typically used in a WLF building. This is followed by a detailed discussion of various types of sidings, which are unique to WLF (or light-gauge steel frame) buildings.

Other exterior wall finishes, such as stucco, EIFS, and brick veneer are described in later chapters because they are used in all types of construction systems. For the same reason, floor finishes and roof coverings for WLF buildings are discussed in later chapters—floor finishes in Chapter 34 and roof coverings in Chapters 31 and 32. Stair construction is covered in Chapter 33.

An introduction to gypsum board as an interior-wall and ceiling finish material is provided in this chapter, which concludes with a discussion of the fire ratings of WLF assemblies.

14.1 EXTERIOR WALL FINISHES IN A WLF BUILDING

Various exterior wall finishes can be applied to a WLF building. As shown in Figure 14.1, the more commonly used finishes are

- Horizontal siding
- Vertical siding (diagonal siding, used occasionally)
- Shingles

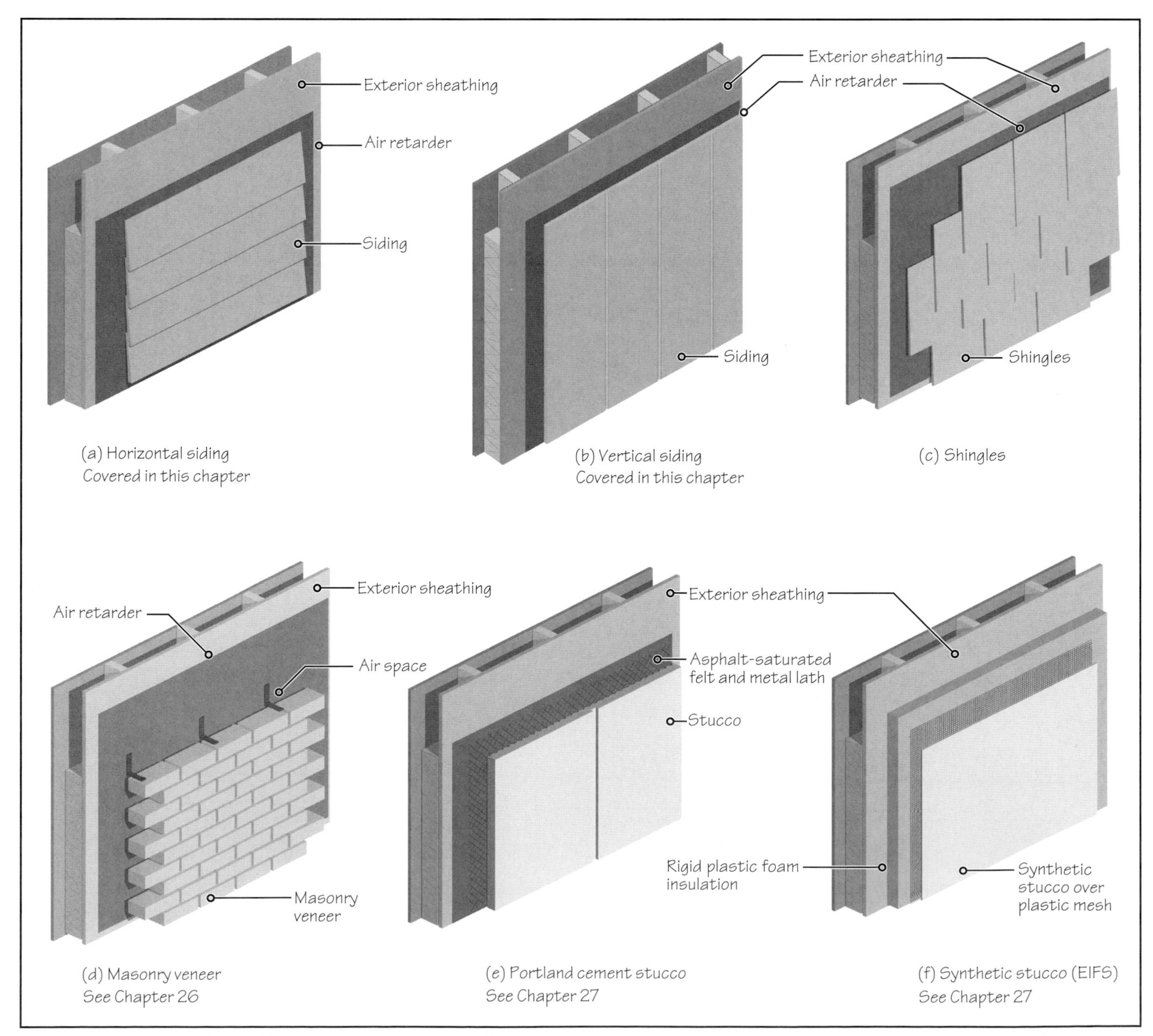

FIGURE 14.1 Commonly used finishes on WLF exterior walls.

- Masonry veneer
- Portland cement stucco
- Exterior insulation and finish system (EIFS)

Only horizontal and vertical sidings are discussed in this chapter. Masonry veneer is discussed in Chapter 26, and stucco and EIFS are discussed in Chapter 27. The use of shingles is costly and relatively uncommon these days and is not included in this text. Two or more of these finishes may be used either on the same facade or different facades of the same building.

14.2 HORIZONTAL SIDINGS

Horizontal siding, a commonly used exterior wall finish in WLF buildings, consists of long, overlapping strips of material nailed to the studs. Some siding manufacturers also allow their siding to be nailed to a nailable wall sheathing such as OSB or plywood. Because the siding strips overlap, they shed water, and the horizontal lines on the facade are accentuated under direct sunlight. Commonly used materials for horizontal sidings are

- Wood
- Plywood
- Hardboard
- Cement fiber
- Vinyl
- Metal (aluminum and steel are common)

HORIZONTAL WOOD SIDING PROFILES

Wood siding is available in various profiles and several widths (from 6 in. to 12 in. nominal) and a thickness that generally increases with the width of the siding, Figure 14.2. Lengths of up to 16 ft are common.

Wood species commonly used are cedar, redwood, and cypress because of their greater decay resistance. Other species used are sugar pine, white pine and ponderosa pine because of their resistance to warpage. Manufacturers provide the details to be used in applying their siding, which are generally similar to those used for applying beveled wood siding, described at length in this section.

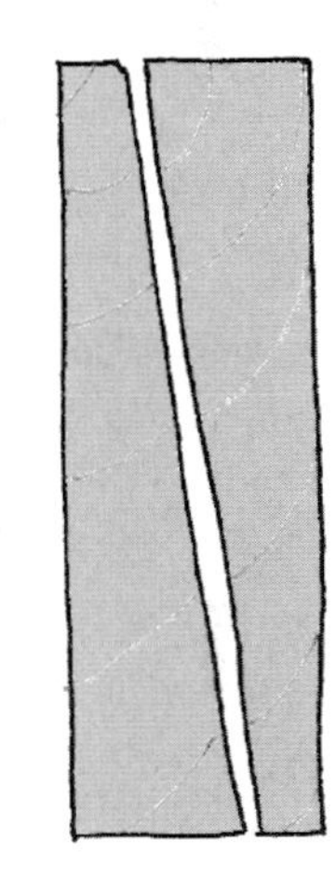

Beveled siding is made by sawing a rectangular piece of wood in two pieces.

BEVELED WOOD SIDING

The cross section of a piece of beveled wood siding (also called *clapboard*) is tapered, so that one end is thicker than the other. It is generally made by sawing a rectangular piece of lumber into two pieces. Typical siding widths range from 6 in. to 10 in. (nominal). Thicknesses range from $\frac{1}{2}$ in. to $\frac{3}{4}$ in., depending on the width of the siding.

One face of siding is generally rough, and the other face is smooth surfaced. That is why most manufacturers make beveled wood siding as S1S2E. Beveled siding is applied with the thicker side down; the upper siding overlaps the lower siding, Figure 14.3.

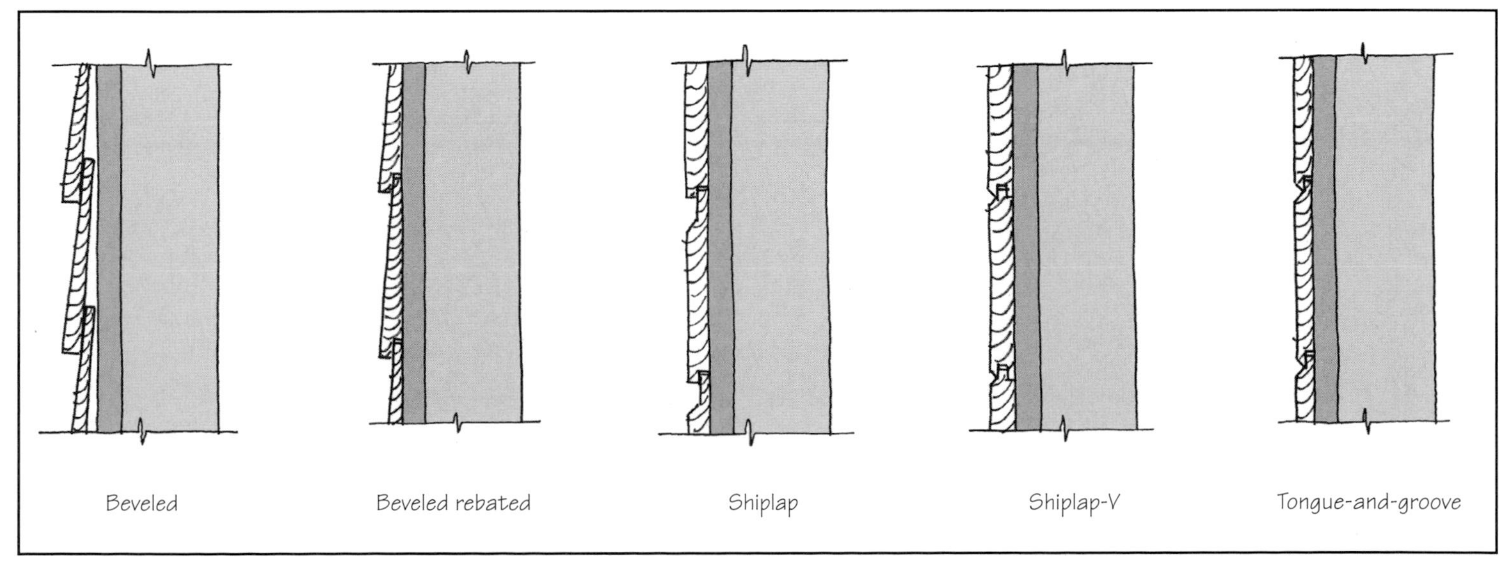

FIGURE 14.2 A few of the several available horizontal wood siding profiles.

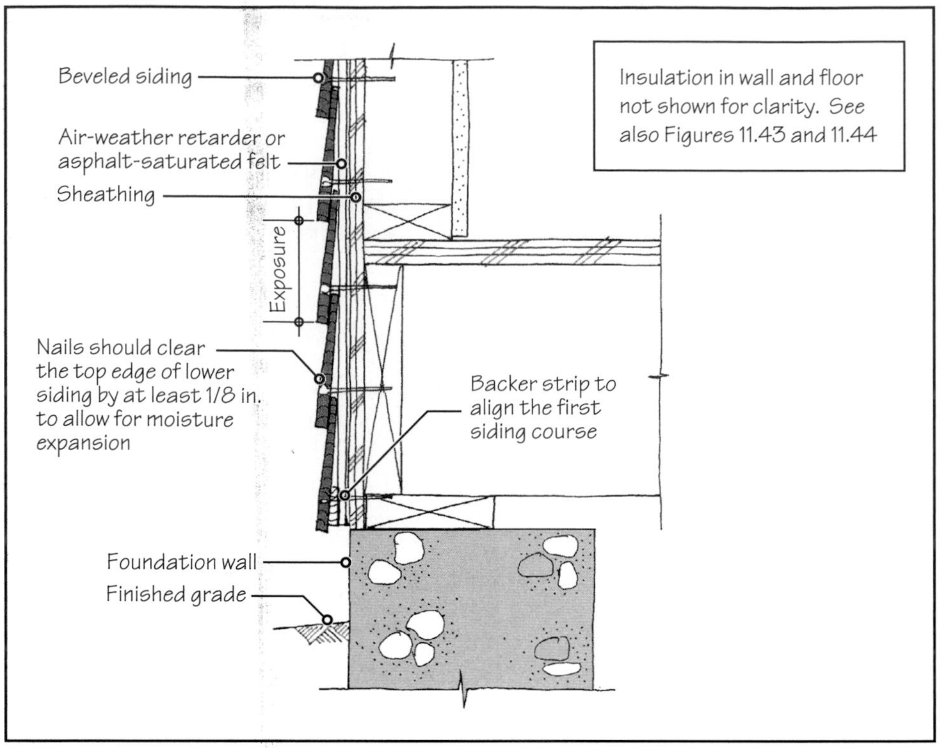

Beveled siding
Air-weather retarder or asphalt-saturated felt
Sheathing
Exposure
Nails should clear the top edge of lower siding by at least 1/8 in. to allow for moisture expansion
Backer strip to align the first siding course
Foundation wall
Finished grade
Insulation in wall and floor not shown for clarity. See also Figures 11.43 and 11.44

FIGURE 14.3 Typical section at foundation level of a wall with beveled wood siding as exterior wall finish.

The exposed width of siding is called *exposure,* which simply equals the width of siding minus the overlap. The manufacturer-recommended overlap varies from 1 to $1\frac{1}{2}$ in. The overlap may be adjusted slightly, even on the same facade, to match window sill elevation or some other elevational line. The first course of siding is backed by a backer strip to provide it with the same vertical alignment as the overlying courses.

Various grades of beveled wood siding are available, depending on the presence of knots and other defects and whether the siding is made of heartwood. Because the siding is exposed to weather, a better grade is relatively more warp resistant.

Beveled wood siding is nailed to each stud through sheathing. Nails should be of galvanized steel (or stainless steel) so that they do not stain the siding upon corrosion. The siding is nailed near its bottom, and the nails must clear the top of siding below by at least $\frac{1}{8}$ in. This allows for expansion and contraction. Thus, each siding course is held at the bottom by the nail and at the top by the overlying siding.

Providing a #15 (or #30) asphalt-saturated felt or an air-weather retarder between the sheathing and siding is recommended. The ends of siding are butt-jointed and must be staggered to avoid a continuous vertical joint, Figure 14.4. A joint should occur at a stud and is caulked. Caulking is applied to both butting ends before nailing the siding.

As an additional precaution against the entry of water, a 4-in.-wide strip of #15 (or #30) felt or air-weather retarder material is placed behind the vertical joint, as shown in Figure 14.4. The height of this strip is the same as the width of the siding. It is positioned so that it is completely covered by the siding.

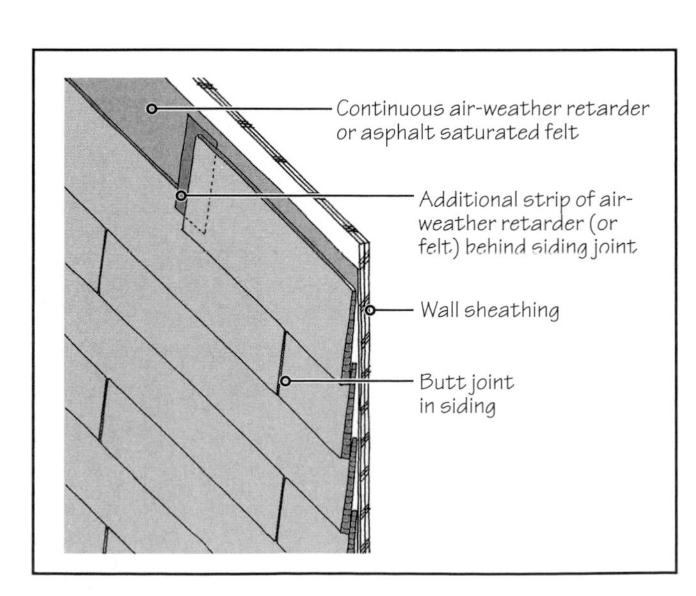

FIGURE 14.4 Use of an additional strip of air-weather retarder or asphalt-saturated felt behind a butt joint in siding.

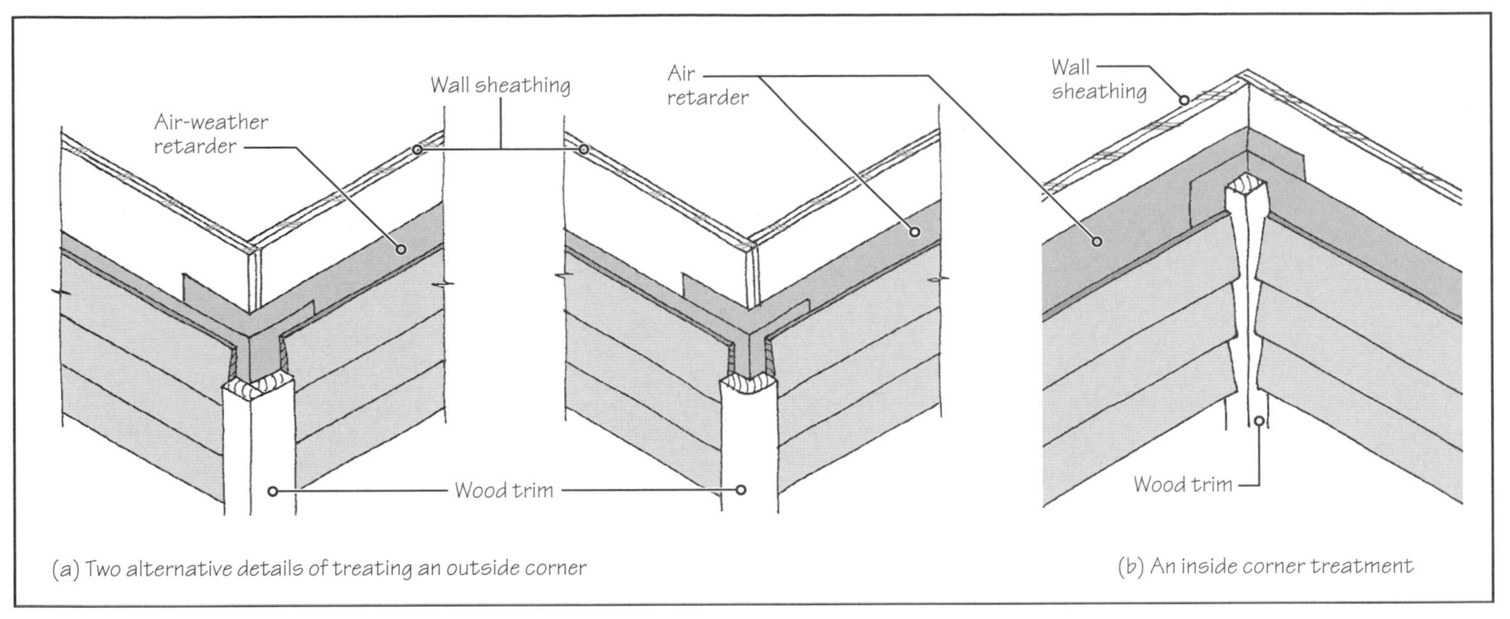

FIGURE 14.5 Typical corner-treatment details.

Beveled wood siding should be back-primed before installation. At corners and ends of walls, the siding butts against trim pieces, Figure 14.5. The trim is installed before applying the siding. The vertical joints between the siding and trims are caulked in the same way as the butt-jointed ends of siding. Doors and windows should be installed before applying siding so that the siding butts against the trim surrounding the door or window frame, Figure 14.6.

Horizontal Siding With Drainage Layer

Figure 14.7 shows a method of applying horizontal siding with an air space between the siding and the sheathing. This space is created by nailing $1\frac{1}{2}$-in.-wide and almost $\frac{1}{2}$-in.-thick vertical wood or plywood strips over the studs. The space provides a drainage layer for any water that may infiltrate behind the siding, particularly under windy conditions. Detailing the bottom of the wall with a galvanized steel flashing is recommended, Figure 14.8.

The air space between the siding and sheathing also helps to dry the siding faster—an important consideration for wood siding, which is decay prone.

Plywood and Hardboard Sidings

Plywood siding is more commonly used as vertical siding panels (described later), but strips of plywood for use as horizontal siding are available. Plywood siding strips are available as

FIGURE 14.6 Doors and windows should be installed before applying siding. In this building, an air-weather retarder has not been used, but its use is recommended.

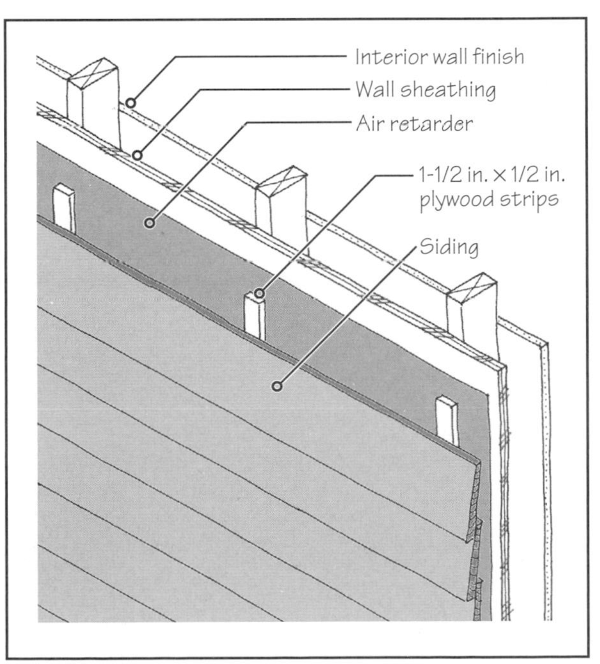

FIGURE 14.7 Horizontal siding with a drainage layer behind.

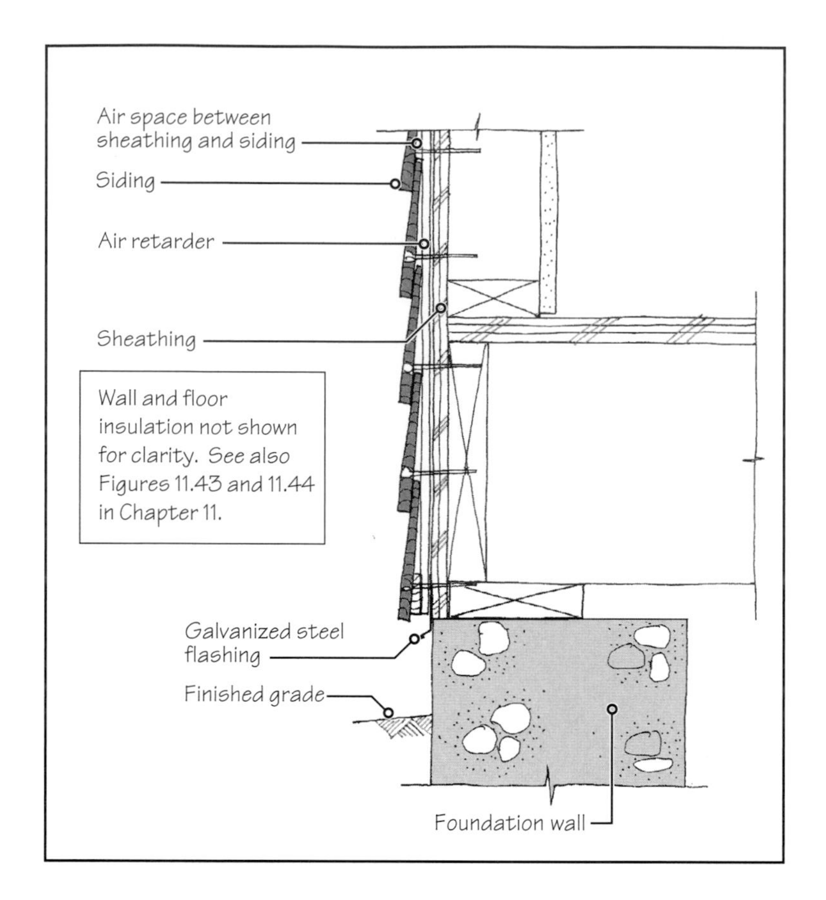

FIGURE 14.8 Section through a wall with a drainage space behind siding.

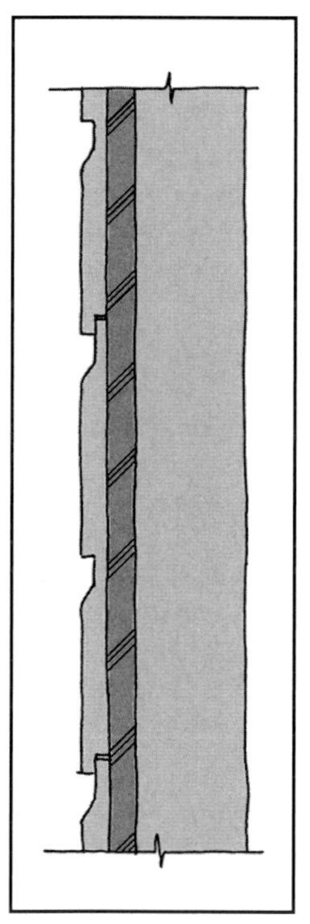

FIGURE 14.9 A commonly used hardboard horizontal siding profile.

smooth-face veneers or rough-sawn face veneers to mimic rough-sawn lumber. Generally, smooth-face veneers are not plywood veneers but medium-density or high-density wood overlays.

Hardboard is made from wood chips bonded under heat and pressure to give a high-density material. It is available in panels for vertical application or in strips for horizontal application. Hardboard siding manufacturers can provide slightly more complex profiles than wood siding, Figure 14.9. The method of application is similar to that described for beveled wood siding.

FIBER-CEMENT SIDING

Fiber-cement siding is made from portland cement, sand and wood fibers. It is noncombustible and is not prone to decay like wood, plywood, or hardboard. Horizontal siding material is almost rectangular in profile (not beveled). It is available in 6-in. to 12-in. widths and is nearly $\frac{3}{8}$ in. thick.

Manufacturers generally recommend an overlap of $1\frac{1}{4}$ in. and provide trims for use at corners and around openings to match the siding, Figure 14.10. The cutting of fiber-cement boards and siding produces cement dust. The use of a dust mask is recommended during cutting.

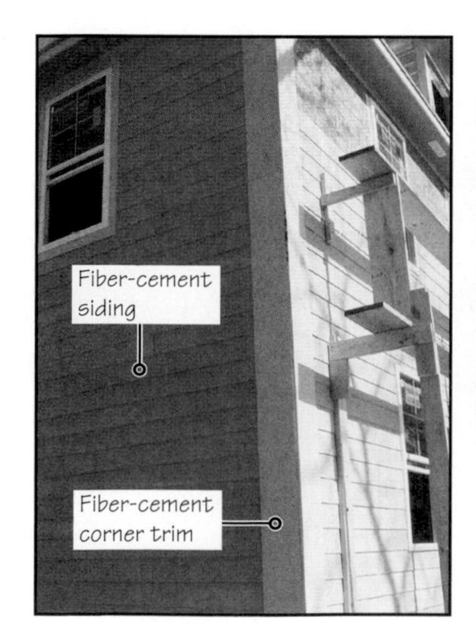

FIGURE 14.10 **(a)** Manufacturers of fiber-cement sidings provide trim materials to match their sidings. **(b)** Site cutting of cement fiber siding.

Vinyl Siding

Vinyl siding is another commonly used cladding material for WLF buildings. It is molded from polyvinyl chloride (PVC), thermoplastic polyolefin (TPO), or polypropylene, often formed to look like wood siding. It is available in lap, shingle, and board-and-batten patterns. It is assembled using siding panels, soffits, various shapes of extruded trim, and accessories and is attached with corrosion-resistant nails. The material can be easily cut with a saw.

Vinyl siding is durable and water resistant when properly detailed. It can be cleaned with water and/or a mild detergent and does not require painting. Because vinyl tends to expand and contract substantially with changes in temperature, it is critical that it be properly installed to accommodate dimensional changes. Like wood siding, vinyl siding is also combustible.

14.3 VERTICAL SIDINGS

Wood boards can be used as vertical siding, Figure 14.11(a). The most commonly used vertical siding, however, is either plywood or hardboard panels. The panels are available in 4-ft width and heights of either 8 ft or 9 ft. The panels can be used either over wall sheathing or directly over the studs. If used directly over the studs, the span rating of siding panels should not be exceeded (Figure 12.30(b)).

Several vertical plywood or hardboard siding varieties have vertical grooves to mimic wood siding boards. A few of the several patterns available are shown in Figures 14.11(b) to (d).

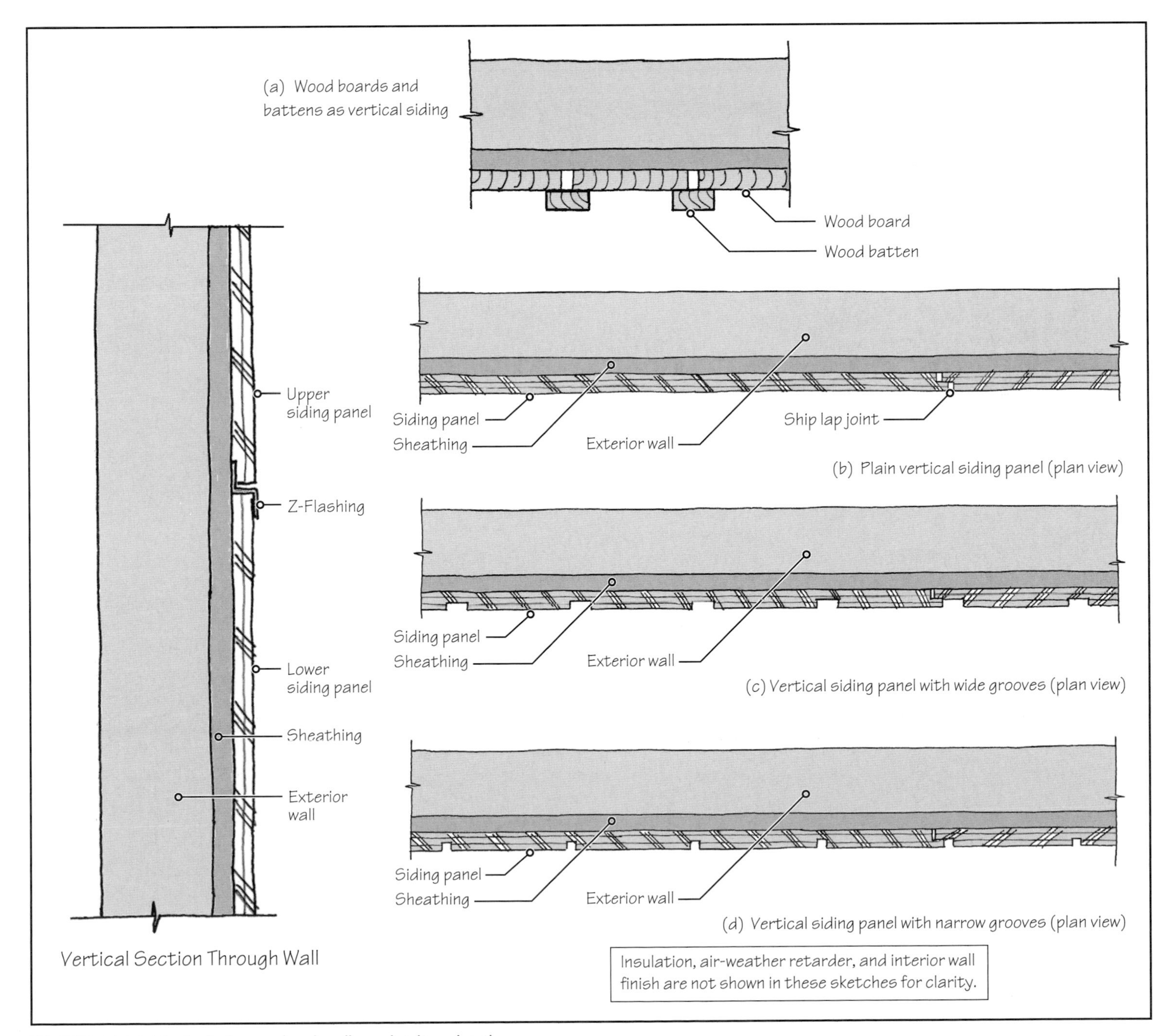

FIGURE 14.11 Vertical plywood or hardboard siding details.

Vertical joints between panels are ship-lapped. Horizontal joints require galvanized steel, aluminum, or stainless steel Z-flashing.

14.4 FINISHING THE EAVES, RAKES, AND RIDGE

The eave is the horizontal edge of the low side of a sloping roof. A hip roof has eaves on all four sides. A gable roof has eaves on two opposite sides. The sloping sides of a gable roof are referred to as *rakes* (Figure 13.26).

The eave may be almost flush with the underlying wall or project over it. A projecting eave protects the junction between the wall and the roof from rain penetration and is the one more commonly used. The projection may be a wide projection (2 to 3 ft) or a narrow projection (a few inches) and may be finished with a horizontal soffit, Figure 14.12.

If the slope of the roof is large, an eave with a horizontal soffit cannot project too far beyond the wall, or the bottom of the soffit will interfere with the door or window head, Figure 14.13. A sloping soffit, Figure 14.14, does not have this limitation. A sloping soffit has another advantage over a horizontal soffit in that the transition from rake to eave is seamless. With a horizontal soffit, the transition requires additional detailing, Figure 14.15.

RIDGE VENTILATION

In Chapter 6, the importance of attic ventilation was discussed. As stated there, attic ventilation consists of intake and exhaust ventilation in the roof. Although intake ventilation

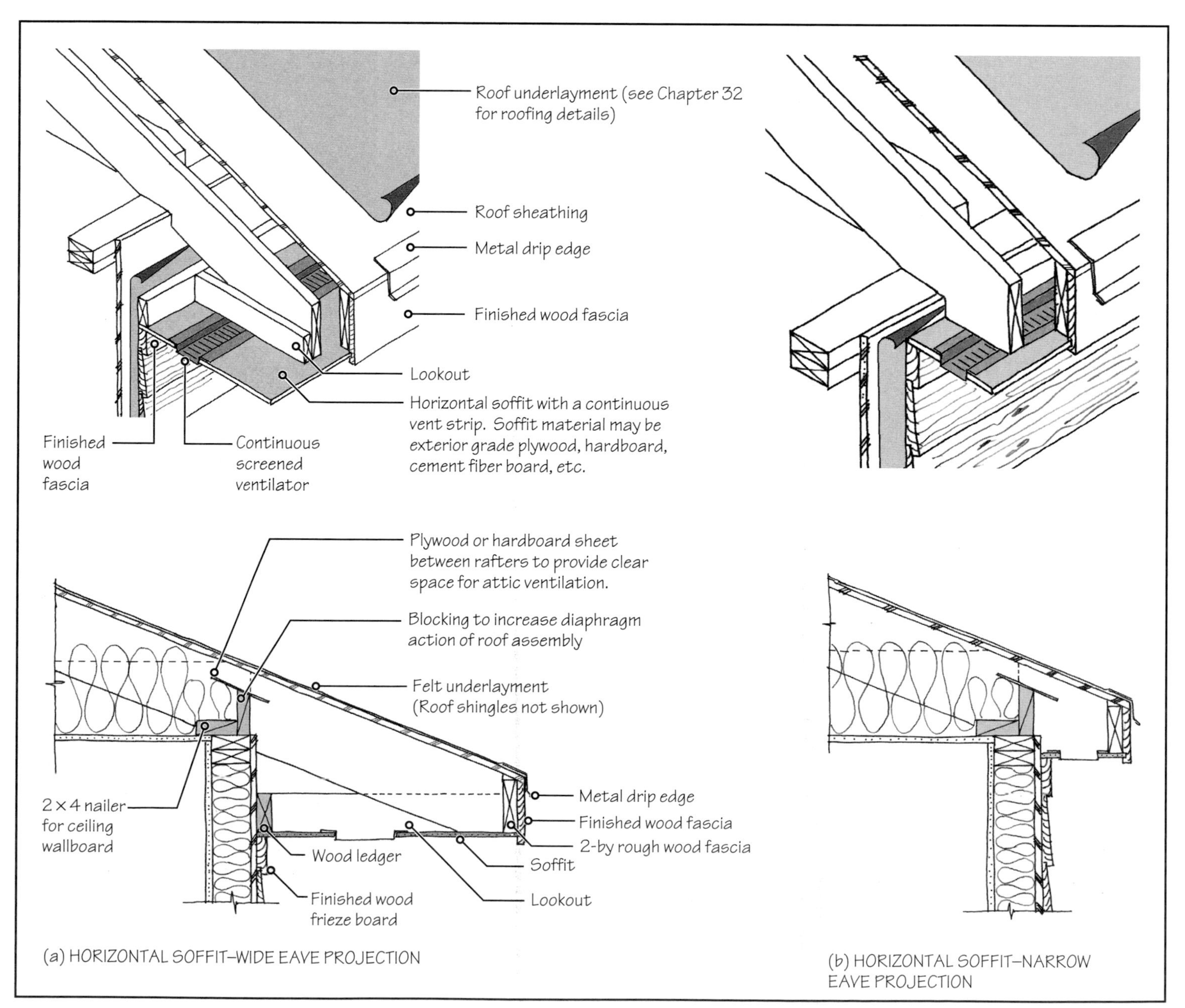

FIGURE 14.12 Details of wide and narrow eave projections.

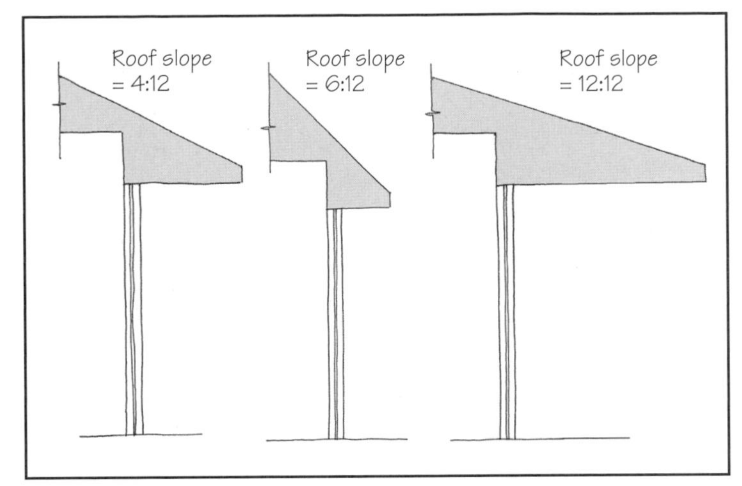

FIGURE 14.13 Effect of roof slope on eave projection with a horizontal soffit.

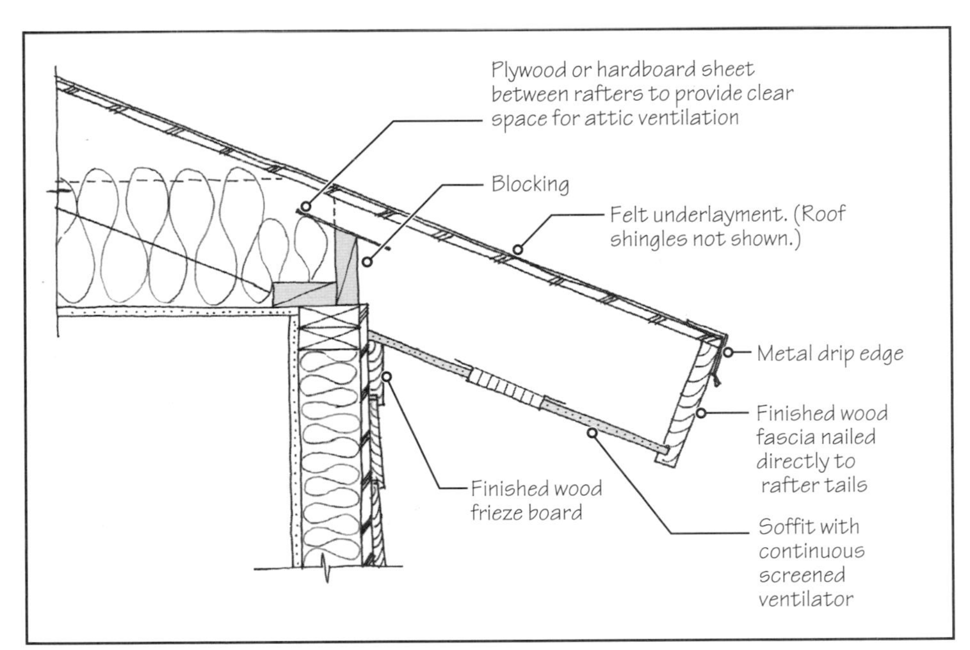

FIGURE 14.14 Details of an eave projection with a sloping soffit.

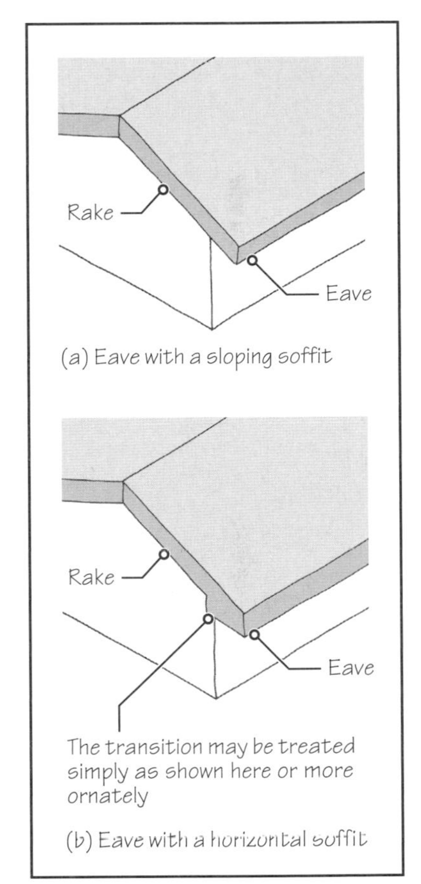

FIGURE 14.15 Eave-rake transitions.

is provided through soffit vents, exhaust ventilation can be provided in many ways (Figure 6.13).

One of the means of providing exhaust ventilation is to install a continuous ridge ventilator. Several proprietary ridge ventilators are available that can be installed over felt underlayment and covered over by roof shingles. The sheathing at the ridge is set back from the ridge board (by 1 to 2 in. on either side) to provide a clear space between rafters for ventilation, Figure 14.16.

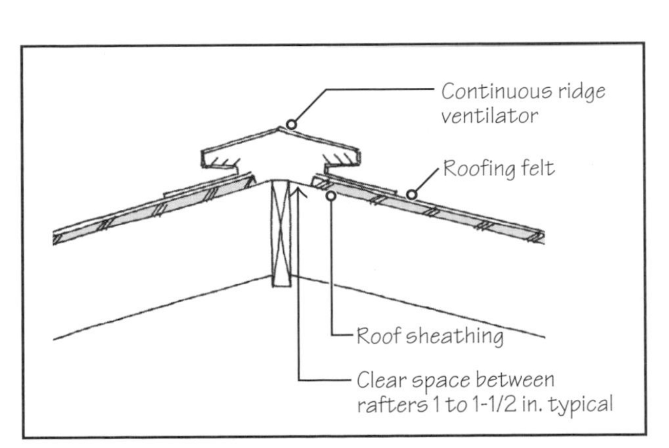

FIGURE 14.16 One of the several ways of detailing for ridge ventilation.

PRACTICE QUIZ

Each question has only one correct answer. Select the choice that best answers the question.

1. Horizontal beveled wood siding is generally made by slicing a
- **a.** solid lumber piece.
- **b.** LVL member.
- **c.** glulam member.
- **d.** OSB member.
- **e.** plywood member.

2. Horizontal siding is provided on
- **a.** the walls of a wood light frame building.
- **b.** the eave of a wood light frame building.
- **c.** the rake of a wood light frame building.
- **d.** the roof of a wood light frame building.

3. Horizontal wood siding material is generally finished smooth on the front and back faces.
- **a.** True
- **b.** False

4. Each course of beveled horizontal wood siding is nailed to the wall through
- **a.** one line of nails near the top of the siding.
- **b.** one line of nails near the bottom of the siding.
- **c.** one line of nails in the center of the siding.
- **d.** two lines of nails, one at the bottom and one at the top of the siding.

5. Vertical joints between adjacent horizontal siding boards are generally
- **a.** lapped $\frac{1}{2}$ in.
- **b.** lapped at least 1 in.
- **c.** lapped at least $1\frac{1}{2}$ in.
- **d.** butted against each other.

6. To produce a drainage layer with horizontal siding, we generally use
- **a.** 2-in.-wide horizontal strips nailed to sheathing.
- **b.** $1\frac{1}{2}$-in.-wide horizontal strips nailed to sheathing.
- **c.** 2-in.-wide horizontal strips nailed to studs before installing sheathing.
- **d.** $1\frac{1}{2}$-in.-wide horizontal strips nailed to studs before installing sheathing.
- **e.** none of the above.

7. Fiber-cement siding is
- **a.** noncombustible but prone to fungal decay.
- **b.** combustible and prone to fungal decay.
- **c.** noncombustible and not prone to fungal decay.
- **d.** none of the above.

8. Vertical siding for a wood light frame building is generally made from small strips of plywood.
- **a.** True
- **b.** False

9. An eave is generally horizontal and the rake is inclined.
- **a.** True
- **b.** False

10. The transition from a projecting eave to rake needs special detailing when the eave is provided with a
- **a.** horizontal soffit.
- **b.** sloping soffit.
- **c.** no soffit.

Answers: 1-a, 2-a, 3-b, 4-b, 5-d, 6-e, 7-c, 8-b, 9-a, 10-a.

14.5 GYPSUM BOARD

Gypsum board is the most commonly used interior wall and ceiling finish in virtually all types of buildings. It provides a strong, fire-protective cover over framing members and can be finished with paint, wallpaper, wood paneling, and so on. It is more economical and much easier to install and finish than the lath-and-plaster alternative that preceded it. Additionally, being a dry form of construction, it leaves less construction mess than lath and plaster.

Gypsum board is known by several names that are used synonymously—*drywall, wallboard, gypsum wallboard* (GWB), and *plasterboard.* The term *sheetrock* is not a generic name but is proprietary to a gypsum board manufacturer.

WHY GYPSUM IS A WONDROUS MATERIAL WITH RESPECT TO FIRE

Gypsum is a rocklike mineral found extensively on the earth's surface. Its chemical name is calcium sulfate dihydrate ($CaSO_4{\cdot}2H_2O$), which indicates that every molecule of gypsum contains two molecules of water (H_2O) chemically combined with one molecule of calcium sulfate ($CaSO_4$). The consequence of gypsum's chemical formulation is that 100 lb of gypsum rock contains 21 lb of water, that is, 21% of gypsum rock is water. (The term *di* in dihydrate means two.)

It is the water contained in gypsum that gives it the fire-protective property. Under the action of heat, the water in gypsum converts to steam. Steam not only helps to absorb the heat, but also provides a cooling effect. Because the temperature of water cannot exceed its boiling point (212°F), regardless of how long the fire acts on it, the temperature of gypsum rock cannot equal that of fire unless all the water in gypsum has converted to steam. Once all the water has turned into steam, gypsum begins to disintegrate under the effect of fire. Therefore, the presence of water in gypsum slows the disintegration process.

The temperature gradient of the wall after 2-h fire exposure, when a 6-in.-thick gypsum wall is exposed to fire in the standard test and the temperature of the wall at various points within the wall is measured, is shown in Figure 14.17. Observe that after 2 h, the temperature of the unexposed face of the wall is only 130°F, and the temperature of the wall 2 in. away from the fire side is the boiling point of water—212°F.

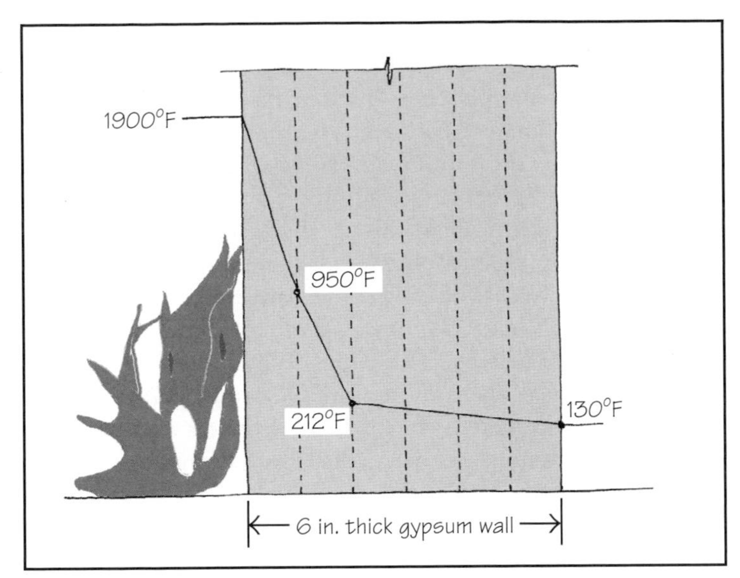

FIGURE 14.17 Temperature gradient along a 6-in.-thick gypsum wall after 2-h exposure to fire under ASTM E119 test (standard fire test).

ANATOMY AND SIZES OF GYPSUM BOARD PANELS

Gypsum board consists of a gypsum core, which is sandwiched between specially treated paper faces, Figure 14.18. The paper adds strength to the board and provides a smooth surface for finishing and painting.

The most commonly used size of gypsum board is a 4 ft × 8 ft panel. Some of the other commonly used sizes are 4 ft × 9 ft, 4 ft × 10 ft and 4 ft × 12 ft. Boards of width 4 ft 6 in. are also available and are used horizontally on walls in rooms with 9-ft-high ceilings. Longer boards are more cumbersome to install. They are heavier to carry and more difficult to maneuver in stair enclosures and other interior spaces. However, fewer joint seams in longer boards reduce cutting of boards and finishing of seams.

The most frequently used board thickness for residential buildings is $\frac{1}{2}$ in. For commercial buildings, the corresponding thickness is $\frac{5}{8}$ in. Other available thicknesses are $\frac{3}{8}$ in. and $\frac{1}{4}$ in.

The long edges of a gypsum board are tapered, as shown in Figure 14.18. The taper allows the joints to be taped over and filled with a joint compound so that a finished wall or ceiling surface appears jointless. The boards that do not require finishing (e.g., prefinished boards and gypsum sheathing panels) have square (nontapered) long edges.

The manufacturing process of gypsum boards yields an endlessly long board wrapped with paper, which is cut to the desired lengths. Therefore, the short (4-ft-wide) edges, where the cut is made, do not have paper wrapping and have smooth, square edges (not tapered).

TYPES OF GYPSUM BOARD PANELS

Several different types of gypsum board panels, with different properties, are available. The more commonly used types are as follows:

- *Type R (regular) Boards:* Type R boards are the standard boards, typically used in $\frac{1}{2}$-in. thickness.

NOTE

Plaster of Paris

When gypsum is heated to remove 75% of its water, the resulting product is calcium sulfate hemihydrate (*hemi* means *half*), referred to as *plaster of paris*. The chemical notation of plaster of paris is $CaSO_4 \cdot \frac{1}{2} H_2O$. (Ca is the symbol for calcium and SO_4 is the symbol for sulfate, so $CaSO_4$ stands for calcium sulfate.) When water is added to plaster of paris, it returns to its original hard, rocklike form—gypsum—containing calcium sulfate and two molecules of water.

Because plaster of paris reverts to gypsum with the addition of water and its subsequent setting, plaster of paris expands on setting. This is unlike portland cement, which shrinks on setting due to the evaporation of water. Thus, plaster of paris is used for casting in molds because it accurately assumes the shape of the mold.

If water is completely removed from gypsum, the resulting product, which is pure calcium sulfate, is called anhydrite. Anhydrite has a longer setting time than plaster of paris and is marketed as Keene's cement, used as **finish coat** in interior plaster.

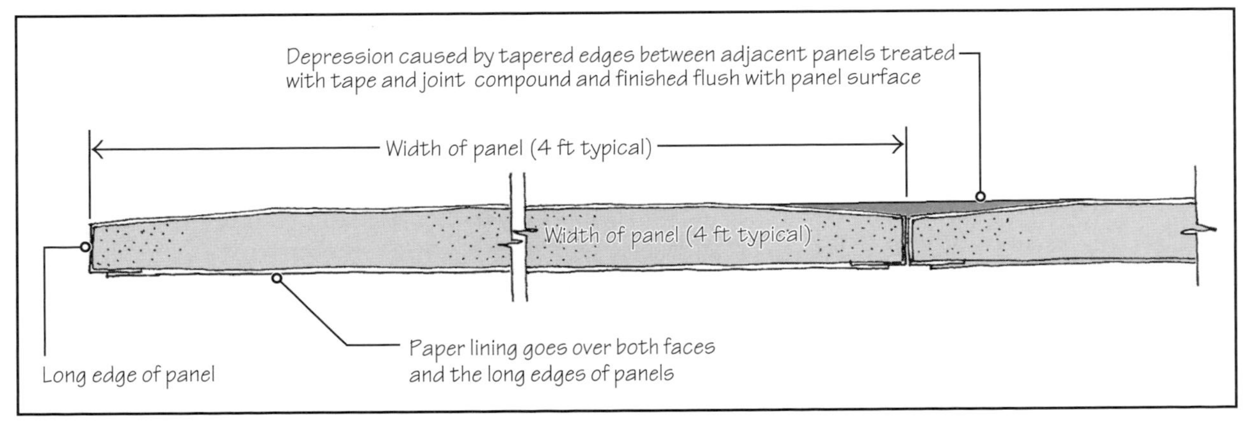

FIGURE 14.18 Section through a gypsum board panel.

- *Type X Boards:* Type X boards are more fire-resistive than Type R boards. The core in Type X boards contains noncombustible fibers (e.g., fiberglass) mixed with gypsum. The fibers provide greater strength so that the boards, if subjected to fire, maintain structural integrity for a longer time than type R boards.

  By definition, a Type X gypsum board, when used in $\frac{5}{8}$-in. thickness on both sides of a wood or metal stud, gives a minimum 1-h fire-rated wall. A $\frac{5}{8}$-in.-thick board that does not give a 1-h fire rating is not a Type X board. A $\frac{1}{2}$-in.-thick Type X panel gives a 45-min fire rating. Manufacturers call their Type X boards by different brand names, such as Firecode Core (U.S. Gypsum Company), Firebloc (American Gypsum Company), Fire-shield (National Gypsum Company), and so on, Figure 14.19.
- *Moisture-resistant (MR) Boards:* The paper facing and the core of MR boards are chemically treated to make them more water resistant than the regular boards. The paper is generally green in color. Therefore, MR boards are also referred to as *green board.* They are recommended for use in walls as backing for ceramic tiles (or other nonabsorbent finishes) in bathrooms and kitchens.

  MR boards may also be used in wall areas where there are no tiles. Therefore, MR boards have tapered edges and are available as Type R or X. MR boards, if used in ceilings, require closer spacing of ceiling joists (per the manufacturer's recommendations).
- *Mold-and-moisture-resistant (MMR) Boards:* These boards employ several different proprietary technologies to resist mold and moisture absorption. Some combine a moisture-resistant core with paper that is treated on one side to be more resistant to mold. Another uses a moisture-resistant core sandwiched between glass mats on both faces. Mold resistance is determined by an ASTM test that rates panels on a scale of 1 to 10, 1 being the least resistant and 10 being the most resistant panel. The test is conducted over a 4-week period, giving only a short-term rating. MMR boards are being increasingly used in applications where MR boards were earlier specified.
- *Flexible Boards:* These boards are more flexible than other gypsum boards and are meant for use on curved walls or ceilings. They are $\frac{1}{4}$ in. thick and have heavier paper facing to resist cracking. They are generally used in two layers or more to give the required gypsum thickness.

NOTE

Portland Cement Boards for Ceramic Tile Backing

In place of MR gypsum boards as tile backing material, portland cement boards are often specified. A portland cement board is much more resistant to water than gypsum board but is harder to install.

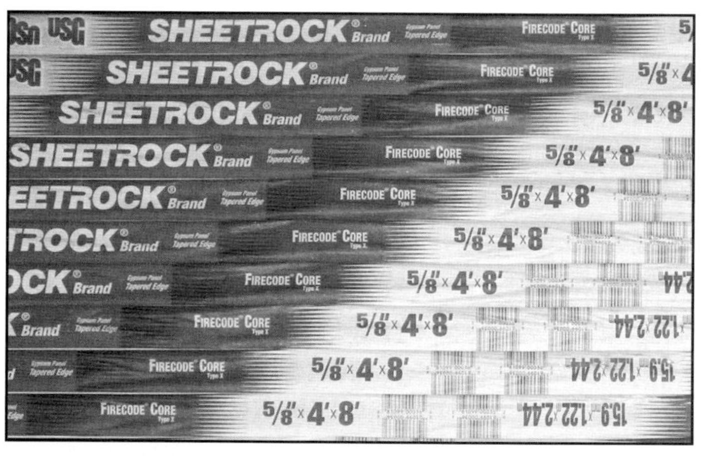

(a)

(b)

FIGURE 14.19 Type X gypsum panels: **(a)** called Firecode Core by United States Gypsum Company and **(b)** called Firebloc by American Gypsum Company.

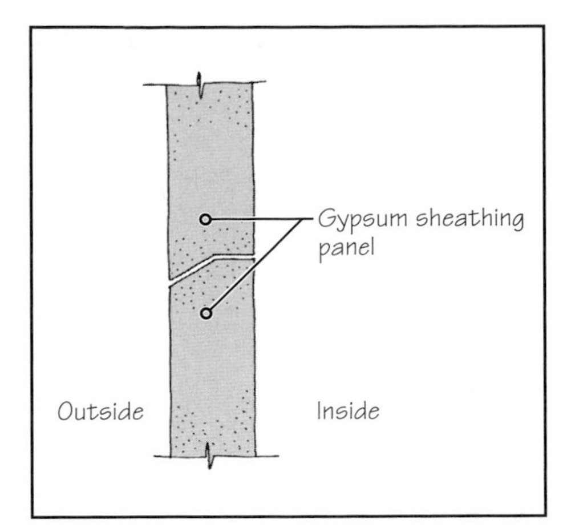

FIGURE 14.20 Vertical cross section through horizontally applied, paper-faced gypsum sheathing panels.

- *High-impact Boards:* These are boards that have a thick fiberglass mesh embedded in the core near the back to provide greater impact and penetration strength as compared with other boards. They are available as Type R or X.
- *Prefinished Boards:* These boards are covered on one face with vinyl instead of paper and do not require any finishing (tape and bed, paint, or wallpaper). Generally, both long and short panel edges are square (not tapered). Manufacturers provide special nails to match the color and texture of the finish in addition to joint-cover accessories. The vapor impermeability of vinyl covering should be taken into account before specifying the use of prefinished boards in warm, humid climates.
- *Gypsum Wall Sheathing Boards:* Although gypsum wall sheathing in wood frame buildings is used much less since the introduction of OSB boards, it is used in situations where the fire rating of the exterior walls is a concern. Gypsum wall sheathing boards have a specially treated core and water-resistant paper facings. The long edges of boards are profiled to shed water, Figure 14.20. The boards are available as Type R or X. A particularly popular gypsum sheathing for commercial construction is made from gypsum core embeded in fiberglass mats on both faces.

LIMITATIONS OF GYPSUM BOARDS

Gypsum is a water-soluble material. Therefore, it is basically an interior-use material. Gypsum boards (including MR boards) are not recommended for use in locations with direct exposure to water or continuous interior high humidity, such as in swimming pool enclosures, steam rooms, or gang shower rooms. However, gypsum boards are available for use in protected exterior locations such as for eave soffits (soffit panels) or as wall sheathing.

14.6 INSTALLING AND FINISHING INTERIOR DRYWALL

Gypsum boards are fastened to framing members using nails or screws, Figure 14.21. Screw application is preferred over nails because screws eliminate nail popping caused by wood shrinkage. On metal studs, only screws can be used. However, modern pneumatic nailers

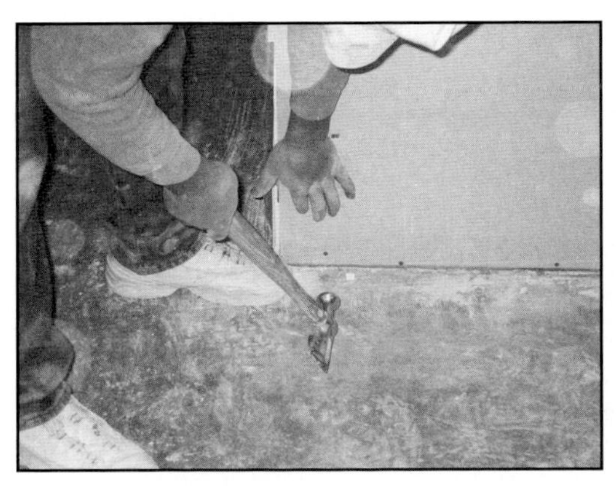

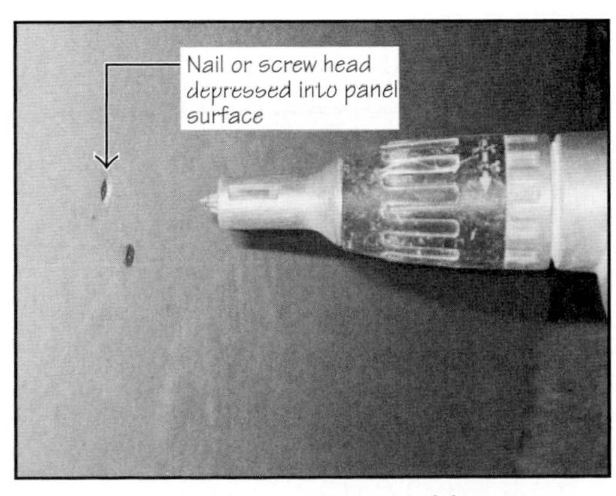

FIGURE 14.21 Gypsum panels may either be installed using nails or screws on wood framing members. On metal studs, only screws can be used. The nails and screws are slightly recessed into the panel surface so that they can be covered over by joint compound and finished smooth with panel surface. (Photos courtesy of Tommy Eaton.)

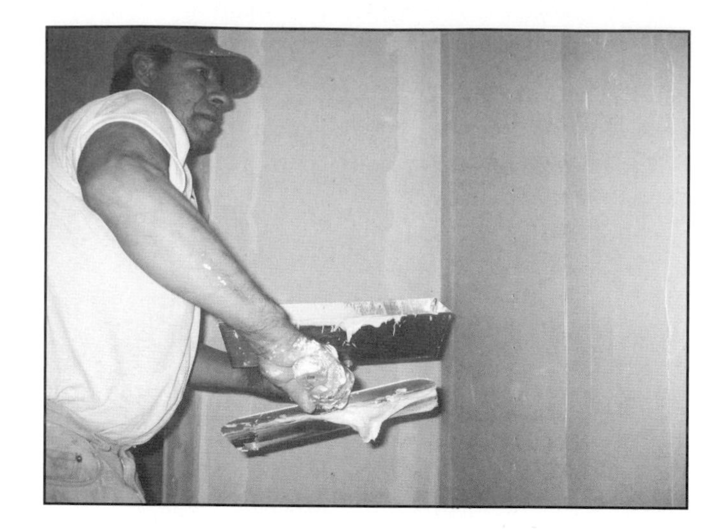

FIGURE 14.22 A drywall worker uses a pan containing the joint compound (mud) and a knife to apply it. (Photo courtesy of Tommy Eaton.)

FIGURE 14.23 Gypsum board panels for ceilings are typically attached prior to wall panels.

make nailing gypsum boards much faster than fastening with screws. The nail or screw heads are slightly depressed into the panel surface so that they can be filled with joint compound and leveled with the panel surface.

Joint compound (also referred to as *mud* in the drywall trade) has a plaster-type consistency and is available in pails or cartons in a ready-to-use formulation. Drywall finishers use a pan to hold the mud in one hand and a knife in another to apply the mud as needed, Figure 14.22.

The boards may be fastened either horizontally or vertically. They are fastened to every framing member. The fastener spacing depends on the spacing between framing members and the required fire rating of the assembly.

Ceiling boards are attached first so that the wall boards provide edge support to the ceiling boards, Figure 14.23. Joints between boards must occur on framing members. The joints are covered with a tape that reinforces the joint against movement and cracking of the joint compound. Two types of tape are used—paper tape and fiberglass scrim tape, Figure 14.24. The paper tape is adhered by spreading a thin coat of joint compound, into which the tape is lightly pressed. Fiberglass tape is self-adhering and does not need an undercoat of joint compound.

Taping of the joint is followed by an overcoat of joint compound, Figure 14.25. After the coating has dried, it is feathered smooth with a sander, and another coat of joint compound and sanding is done until the joint is ready to receive paint.

Inside corners of rooms are taped with a paper tape. The paper tape has a built-in crease, about which the tape is folded, Figure 14.26. The folded tape

FIGURE 14.24 Paper tape and self-adhering fiberglass scrim joint tape. (Photo courtesy of Tommy Eaton.)

FIGURE 14.25 Joint compound filling over joint tape. (Photo courtesy of Tommy Eaton.)

is installed in the corner, covered with joint compound, and finished smooth in the same way as on a flat joint.

Outside corners require L-shape metal or vinyl trims, which are nailed into the studs through the drywall panel. The trim is covered with three layers of joint compound coatings and finished flush with the surface of the boards, Figure 14.27. When all joints and corners are finished and the interior is ready to receive the paint, it looks something like that shown in Figure 14.28.

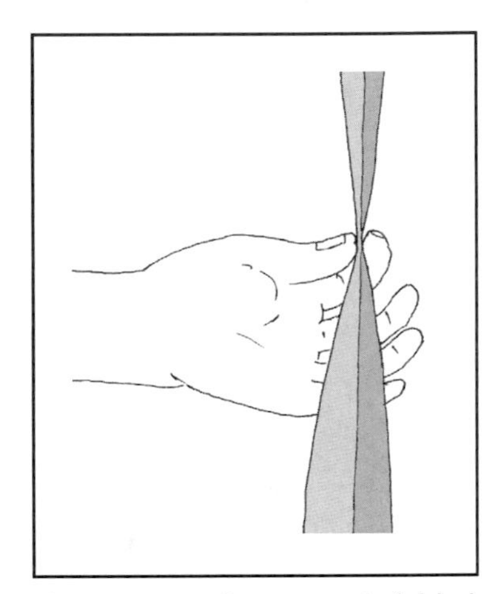

FIGURE 14.26 Paper tape is folded along the crease before installation into an inside corner.

Multilayer Drywall

Although one-layer application is the norm, sometimes two or more layers of drywall panels may be needed on one or both faces of the assembly to obtain a higher fire rating or higher sound insulation. In such an assembly, each layer may be nailed or screwed.

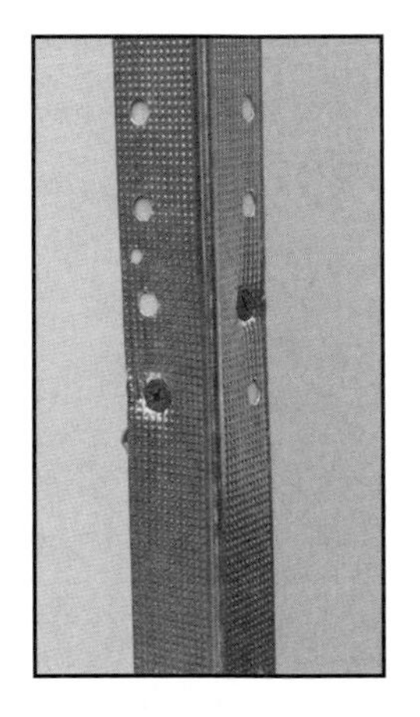

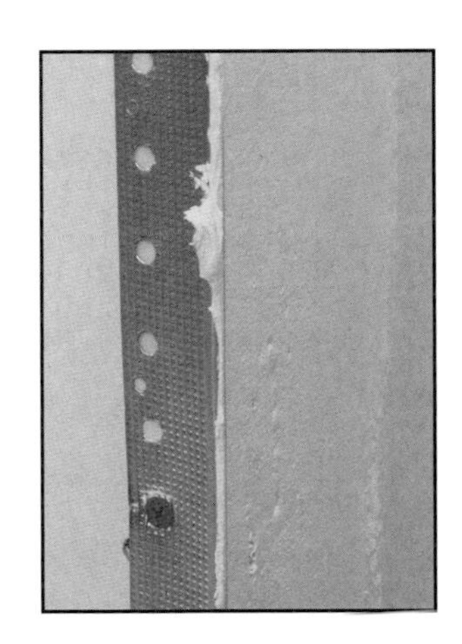

FIGURE 14.27 An outside corner is reinforced with a metal or vinyl trim before being covered by joint compound. (Photos courtesy of Tommy Eaton.)

FIGURE 14.28 A room interior ready to receive paint. (Photo courtesy of Tommy Eaton.)

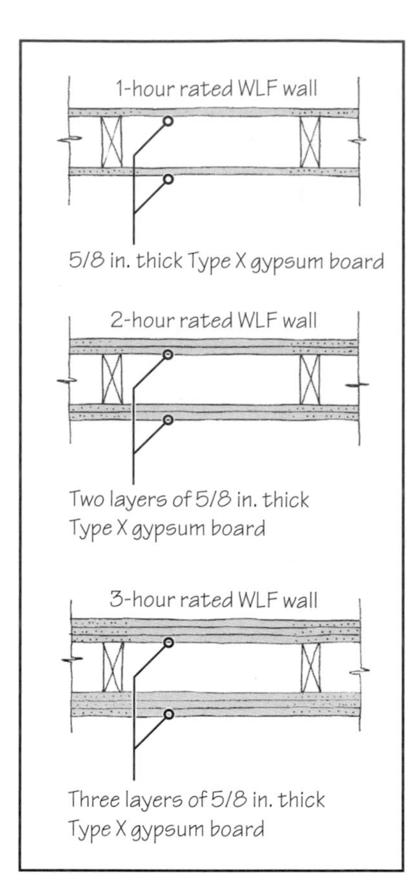

FIGURE 14.29 Fire-rated wood frame assemblies are generally obtained by using multiple layers of $\frac{5}{8}$-in.-thick Type X gypsum board.

Although the type and thickness of gypsum board is the primary factor that determines the fire rating of a wood frame assembly, a few other factors are also important: the type and spacing of nails or screws, spacing of framing members (studs or ceiling joists), the provision of insulation within framing members, and whether the panel is applied horizontally or vertically.

However, with prefinished boards, in which two-layer application is common, the base layer is fastened with nails or screws and the face layer is adhered to the base layer using manufacturer-provided adhesives. Temporary nailing is generally required on the edges of face panels to hold them in place until the adhesive has attained strength.

14.7 FIRE-RESISTANCE RATINGS OF WLF ASSEMBLIES

The fire resistance rating of a WLF wall or ceiling-floor assembly depends on the type of drywall panels. Generally, one $\frac{5}{8}$-in.-thick, Type X drywall panel on each side of a wood stud wall gives a 1-h fire-resistance rating. If two layers of the same drywall panels are used, a 2-h assembly is obtained. Three such layers give a 3-h rating, Figure 14.29.

However, the fire-resistance rating also depends on the spacing of studs, the spacing and type of nails or screws, and whether or not wall cavities are insulated. Drywall manufacturers provide the fire-rating data for various assemblies in which their product can be used. Another standard source for this information is *Fire Resistance Design Manual* published by the Gypsum Association, which is updated every few years. This publication gives the specifications and sketches of 1-h, 2-h and 3-h rated assemblies. These ratings have been established based on measurements in standard fire tests. Because the ratings are in integer hours, an assembly whose measured fire rating is 1 h 59 min is listed as a 1-h assembly. To be listed as a 2-h assembly, its fire rating must lie between 2 h and 2 h 59 min.

PRACTICE QUIZ

Each question has only one correct answer. Select the choice that best answers the question.

11. Which of the following terms for gypsum board is proprietary to a manufacturer?
- **a.** Wallboard
- **b.** Sheetrock
- **c.** Plasterboard
- **d.** Drywall

12. Gypsum board is one of the best materials for fire protection because
- **a.** it is noncombustible.
- **b.** it is heat-treated during its manufacture.
- **c.** each molecule of gypsum contains two molecules of water that retard the progress of fire.
- **d.** each molecule of gypsum contains two molecules of carbon dioxide that retard the progress of fire.

13. The most commonly used size for gypsum boards is
a. 4 ft × 8 ft.
b. 4 ft × 8 ft 6 in.
c. 4 ft 6 in. × 8 ft.
d. 3 ft × 6 ft.
e. 4 ft × 4 ft.

14. Out of Types R and X gypsum board panels, both of the same thickness, Type R is more fire-resistive.
a. True
b. False

15. One sheet of Type X gypsum board panel, $\frac{5}{8}$ in. thick, used on each side of a stud wall gives a fire rating of
a. 30 min.
b. 45 min.
c. 60 min.
d. 75 min.
e. 90 min.

16. Moisture-resistant gypsum board panels are generally
a. green in color.
b. pink in color.
c. white in color.
d. yellow in color.
e. gray in color.

17. Gypsum board panels used on curved walls are generally
a. $\frac{5}{8}$ in. thick.
b. $\frac{1}{2}$ in. thick.
c. $\frac{3}{8}$ in. thick.
d. $\frac{1}{4}$ in. thick.
e. $\frac{1}{8}$ in. thick.

18. High-impact gypsum board panels generally have a
a. steel lining at the back side.
b. aluminum foil lining at the back side.
c. fiberglass mesh at the back side.
d. plastic sheet at the back side.
e. none of the above.

19. Gypsum sheathing is generally used on
a. exterior side of an exterior stud wall.
b. interior side of an exterior stud wall.
c. exterior side of an exterior masonry wall.
d. interior side of an exterior masonry wall.

20. It is better to install ceiling gypsum board panels before installing the wall panels in a room.
a. True
b. False

21. The reason for using more than one layer of gypsum board panels on one or both faces of a wall is to obtain a stronger wall assembly.
a. True
b. False

Answers: 11-b, 12-c, 13-a, 14-b, 15-c, 16-a, 17-d, 18-c, 19-a, 20-a, 21-b.

KEY TERMS AND CONCEPTS

Board and batten
Cement fiber (fiber cement)
Clapboard
Drainage layer
Drywall
Eaves, rakes and ridge
Exterior wall finish materials
Fire resistance of WLF assemblies
Gypsum board panels
Horizontal siding
Joint compound
Plasterboard
Siding
Tongue and groove
Wallboard

REVIEW QUESTIONS

1. Draw a detail section at the foundation level through a wood light frame building whose exterior wall is finished with beveled wood siding. Assume a crawl space under the ground floor.
2. Repeat Question 1, assuming fiber-cement siding and a drainage layer.
3. With the help of a detailed section, explain the construction of a projected eave and soffit in a wood light frame building.
4. Using a sketch and notes, explain how you will detail a 2-h fire-rated WLF wall assembly.

D6

CHAPTER 15 Structural Insulated Panel System

CHAPTER OUTLINE

15.1 BASICS OF THE STRUCTURAL INSULATED PANEL (SIP) SYSTEM

15.2 SIP WALL ASSEMBLIES

15.3 SIP FLOOR ASSEMBLIES

15.4 SIP ROOF ASSEMBLIES

15.5 ADVANTAGES AND LIMITATIONS OF SIPS

A typical structural insulated panel (SIP) building under construction. (Photos courtesy of the Structural Insulated Panel Association (SIPA)).

The structural insulated panel (SIP) system of construction described in this chapter is an alternative to the conventional WLF system. It is a panelized system that reduces on-site construction time and allows the use of less-skilled labor. Because insulation is a part of the panels, insulation subtrade is eliminated, which further increases construction efficiency.

Several other advantages of the system, as well as its limitations, are described at the end of this chapter. Because it is a relatively new system that is still evolving, its current market share is small as compared with the well-established, conventional WLF system that dominates the residential and light commercial construction market in the United States.

In addition to all SIP construction, SIPs are also used as wall and roofing panels in low-rise steel frame buildings.

15.1 BASICS OF THE STRUCTURAL INSULATED PANEL (SIP) SYSTEM

The structural insulated panel (SIP) system consists of panels of sandwich composition. Each panel comprises two facing boards bonded to a core consisting of rigid plastic foam insulation, Figure 15.1. The core generally consists of expanded polystyrene (EPS). (Extruded polystyrene and polyisocyanurate cores may also be used, but they are more expensive.) The material used for the facings may be oriented strandboard (OSB) or plywood. OSB is commonly used because of its lower cost and its availability in much larger sizes. The most common thickness of OSB facing is $\frac{1}{2}$ in.

Panels are manufactured by applying structural-grade adhesive on both faces of the core and then laminating the OSB to it. The assembly is kept under pressure for the required duration. Because insulation is included in the panels, SIPs are used only in the building envelope. Therefore, in a SIP building, the interior walls are generally framed with 2-by lumber and the intermediate floors are also framed with 2-by lumber or engineered wood members. The use of SIPs for an intermediate floor is, in fact, discouraged because of its lower structureborne sound insulation. The foam in SIPs functions as an efficient acoustical bridge. SIPs may, however, be used in a floor that is required to be insulated if its lower sound-insulation value is of no concern, such as a floor over a crawl space or basement.

Although there are many similarities between the characteristics of the SIP system between manufacturers, it is a proprietary system that differs somewhat from manufacturer to manufacturer. The discussion and illustrations provided here are generic. Most SIP manufacturers have details that have been carefully developed and approved by the code authorities and should be strictly followed.

In considering the use of a SIP system, the general procedure is to send the architectural drawings of the project to the selected SIP manufacturer. Based on the analysis of the drawings, the manufacturer determines the size, shape, configuration, and thickness of panels and prepares necessary shop drawings for the fabrication of panels.

After the approval of shop drawings by the architect and the constructor, the manufacturer fabricates the panels. Each panel is uniquely identified as to its location in the building. All panels required for the project are packaged and sent to the construction site for assembly.

Because almost no cutting is required on site, the panels are shipped to the site when needed, eliminating the need for elaborate site storage facilities. The plant typically packages the panels in reverse order of their use, increasing builder's productivity. The entire fabrication work, that is, the analysis of architectural drawings, preparation of shop drawings, and fabrication of panels, is generally computerized.

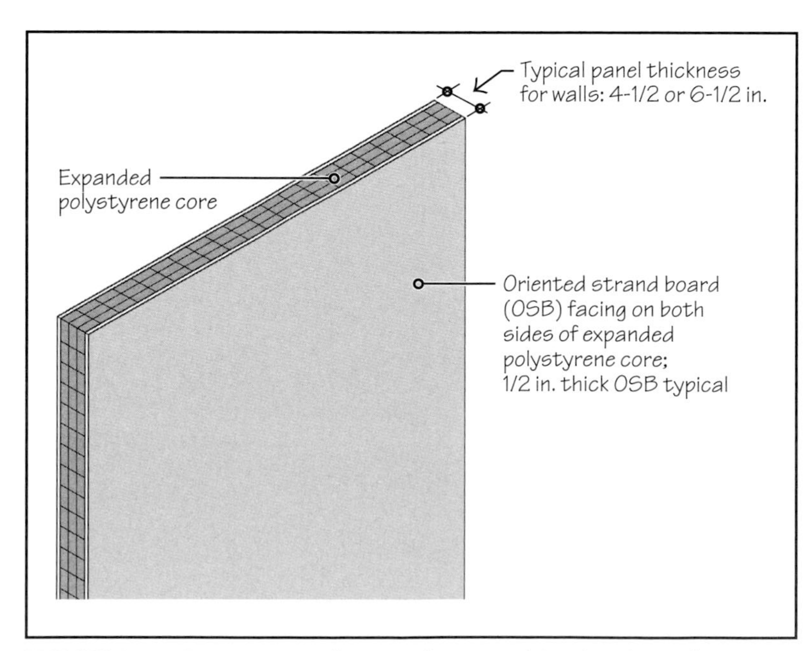

FIGURE 15.1 Composition of a typical structural insulated panel (SIP).

STRUCTURAL BEHAVIOR OF A SIP STRUCTURE

Structurally, a SIP panel functions as a composite panel in which the core integrates the two facings. Thus, under an axial compressive load, even if the load is initially delivered to one facing, both facings combine to resist the load because the core transfers the load to the other, initially unloaded facing through shear mechanism.

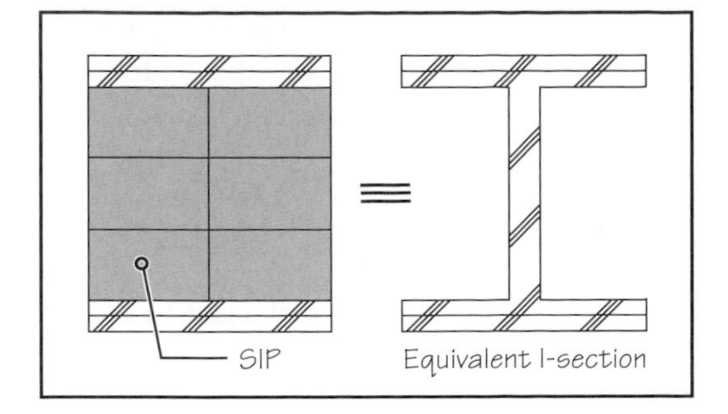

FIGURE 15.2 In carrying both the axial and the lateral loads, a SIP behaves as an I-section, where the core functions as the web of the section and the facings as its flanges.

Similar behavior comes into play when the panel is subjected to bending. Under bending, one facing is subjected to compressive stresses and the other facing is under tensile stresses. The core resists the shear stresses generated by bending. The behavior of a SIP panel is, therefore, identical to an I-section, where the facings function as the flanges of the I-section and the foam core functions as the section's web, Figure 15.2.

The core also helps to prevent the buckling of thin facings under compressive stresses, created either by the bending of the panels (in a floor or roof) or under compressive loads (in a wall). In other words, all three elements of a SIP are stressed under axial or lateral loads.

The facing-core composite behavior also comes into action in providing racking resistance to the structure, where the facings and the core act together as shear walls. In a conventional wood frame building, only the exterior sheathing and the studs provide the racking resistance.

PANEL SIZES

Panels are produced in various thicknesses, depending on whether they are used in walls, floors, or roofs. Wall panels are generally $4\frac{1}{2}$ in. or $6\frac{1}{2}$ in. thick. A $4\frac{1}{2}$-in.-thick SIP consists of a $3\frac{1}{2}$-in.-thick core, matching a wood light frame wall made of 2 × 4 studs. A $6\frac{1}{2}$-in.-thick panel (with a $5\frac{1}{2}$-in.-thick core) corresponds to a wood frame wall comprising 2 × 6 studs. Panel thicknesses of $8\frac{1}{4}$ in., $10\frac{1}{4}$ in., and $12\frac{1}{4}$ in. are used for floors or roofs.

The surface dimensions (length and width) of panels vary, depending on whether lifting and placing the panels in position at the site is done manually or using hoisting equipment. Panels of up to 8 ft × 24 ft can be produced by fabricators.

15.2 SIP WALL ASSEMBLIES

In general, wall panels are either 4 ft or 8 ft wide. They may be continuous over the entire height of the building or they may be only one floor tall, requiring floor-by-floor installation. The floor-by-floor installation of panels is similar to the construction of walls in a conventional WLF building, where the upper floor walls are installed over the subfloor at that level.

The details of the connection of wall panels to the foundation, between adjacent panels, at wall corners, and to the floors and roofs are critical. Figure 15.3 provides an overall view of a typical panel layout. The adjacent wall panels are connected together with splines, Figure 15.4.

The splines may consist of $\frac{1}{2}$-in.-thick OSB strips (referred to as *surface splines*), engineered wood studs, or double 2-by studs, depending on the requirement of the connection. Nails, screws, and adhesives are used in a connection per the manufacturer's details.

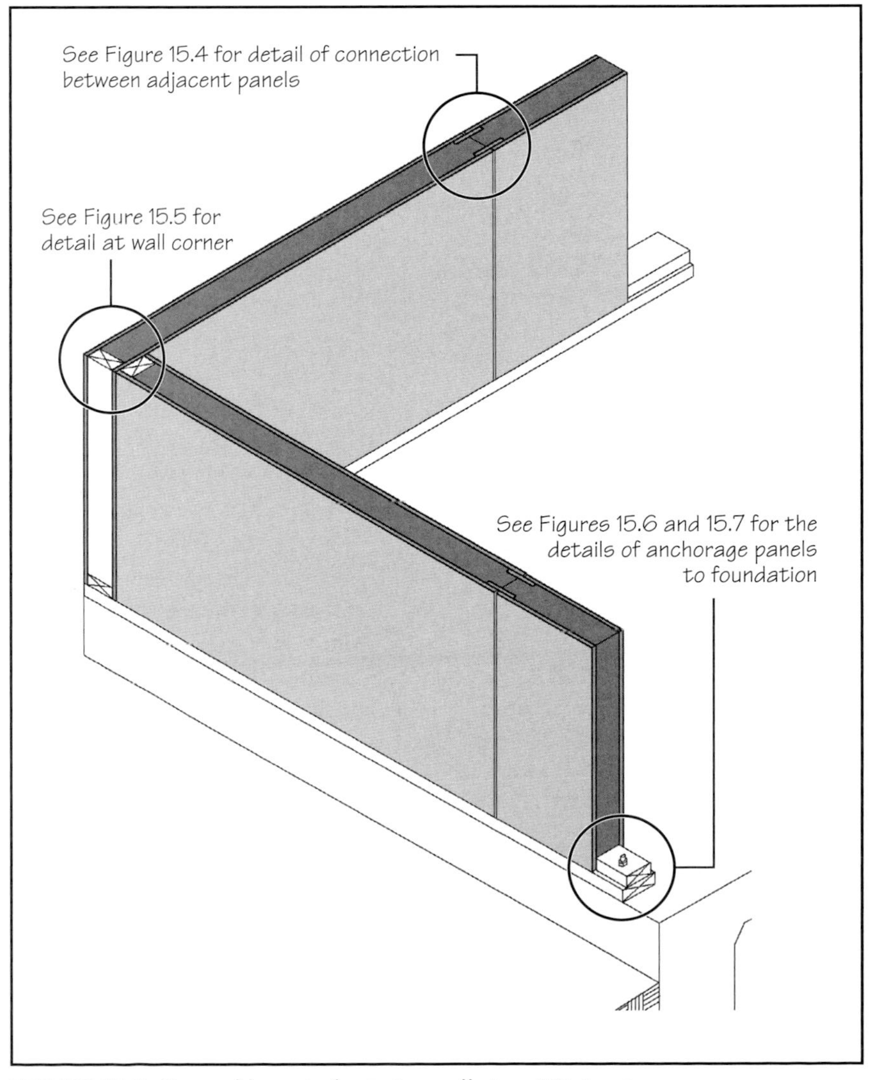

FIGURE 15.3 General layout of exterior walls in a SIP structure.

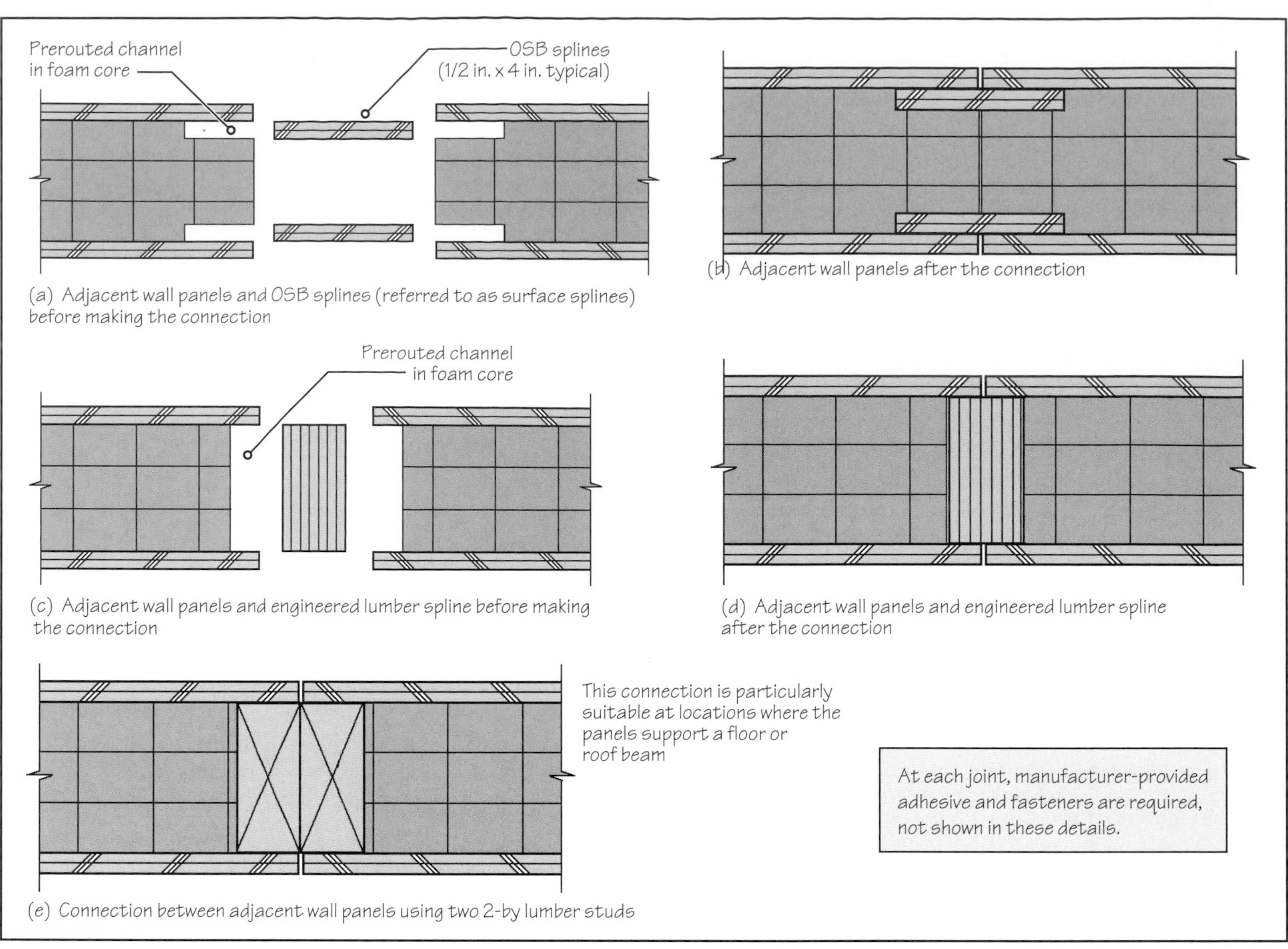

FIGURE 15.4 Three commonly used alternatives for connecting adjacent wall panels.

The use of connecting splines requires the foam core to be routed and grooved to accept the splines. This operation is performed at the panel-fabrication plant so that the panels arrive at the site already routed. Chases for electrical wiring (generally $1\frac{1}{2}$ in. in diameter) are also routed before shipping the panels to the site. Chases for other utility pipes are not included in SIPs because these pipes typically occur in the interior walls.

Figure 15.5 shows the connection between panels at a wall corner. Figures 15.6 and 15.7 show the details of the anchorage of wall panels to foundations. A two-sill-plate assembly at the

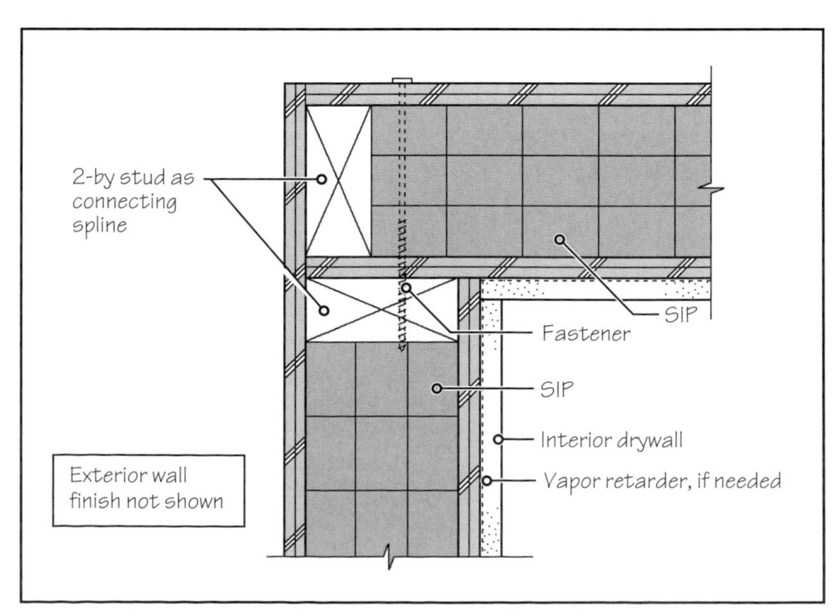

FIGURE 15.5 Typical connection at the corner of two wall panels.

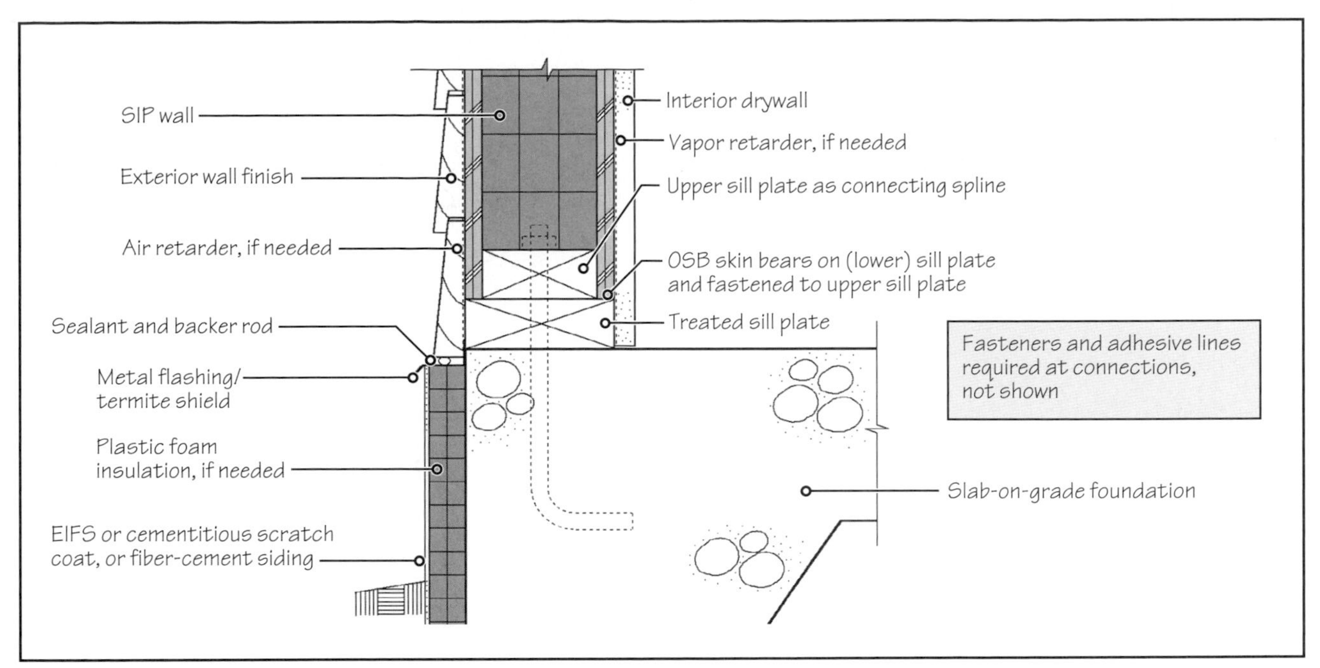

FIGURE 15.6 Typical detail of anchorage of wall panels to slab-on-grade foundation.

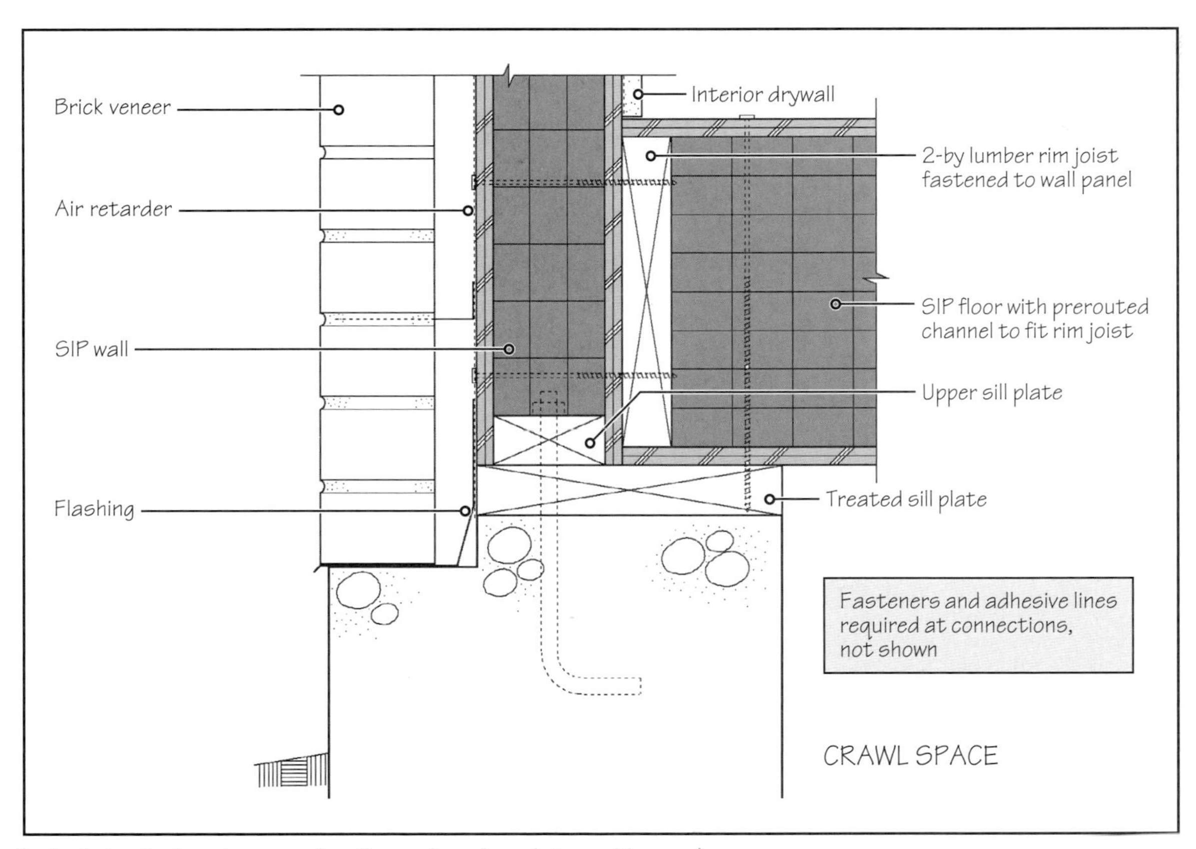

FIGURE 15.7 Typical detail of anchorage of wall panels to foundation with crawl space.

foundation is commonly recommended by the manufacturers. The lower sill plate consists of preservative-treated lumber. The upper sill plate functions as a connecting spline. The details of Figures 15.6 and 15.7 ensure that the OSB facings bear directly over the lower sill plate so that the gravity load on the wall is transferred through the facings to the foundations in bearing.

There are various ways in which windows and doors can be detailed in a SIP wall. A commonly used method is shown in Figure 15.8. As stated previously, the interior partitions in a SIP building are framed by using conventional 2-by studs. Figure 15.9 shows the detail of the junction between an exterior SIP panel and an interior wall.

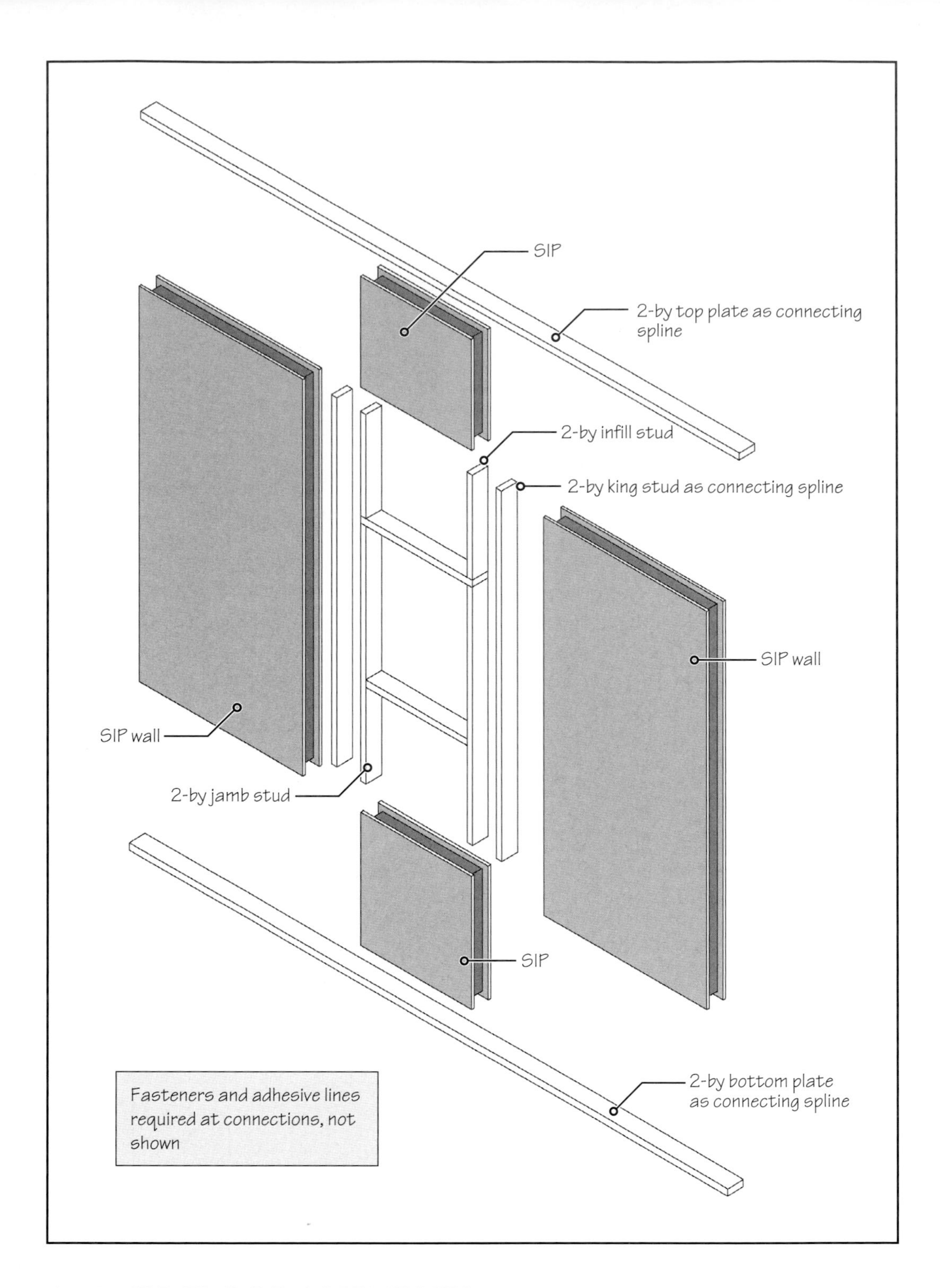

FIGURE 15.8 A typical detail of framing around a window. Framing around a door is similar.

15.3 SIP FLOOR ASSEMBLIES

As earlier stated, an intermediate floor in a SIP building is framed with 2-by lumber joists, engineered wood members (e.g., wood I-joists), or trussed joists. The floor joists may be placed on the wall panel and fastened to the top plate spline using a detail similar to that used in a conventional wood frame building, Figure 15.10.

The detail of Figure 15.10 creates a thermal bridge at the rim joist. Therefore, some SIP manufacturers recommend the detail shown in Figure 15.11, in which the floor joists are hung off the top plate spline using joist hangers. The joist hangers are top-supported, that is, fastened to the top plate in the SIPs. Where the SIPs are continuous, that is, two floors high, the floor joists are supported on joist hangers that are fastened to a ledger beam, Figure 15.12.

15.4 SIP ROOF ASSEMBLIES

Two alternative roof assemblies can be used in a SIP building. One alternative is to use any one of the various assemblies that are used in a conventional wood frame building—that is, a stick-frame rafter-ceiling assembly or wood roof truss assembly. In these assemblies, insulation is placed in the attic.

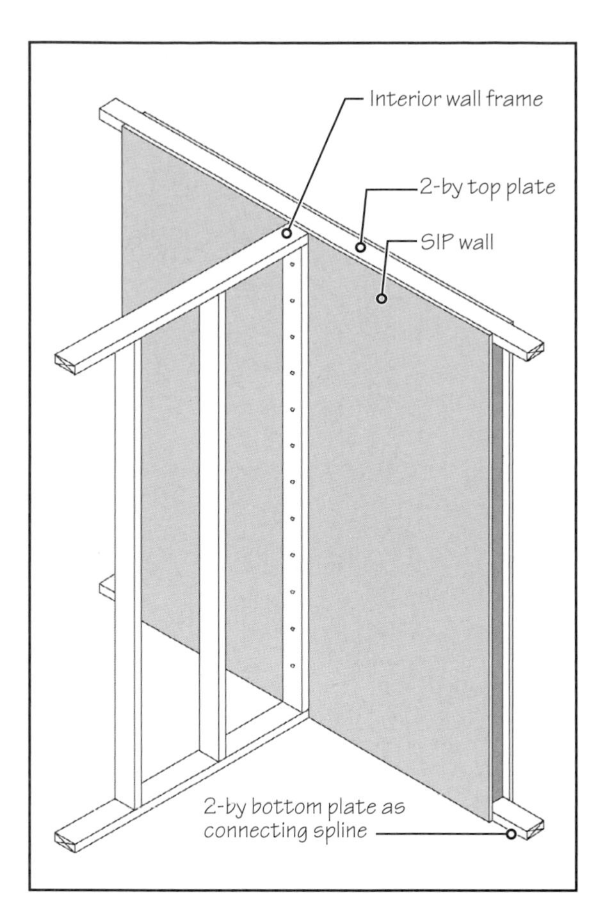

FIGURE 15.9 Connection of an interior wall to a SIP wall.

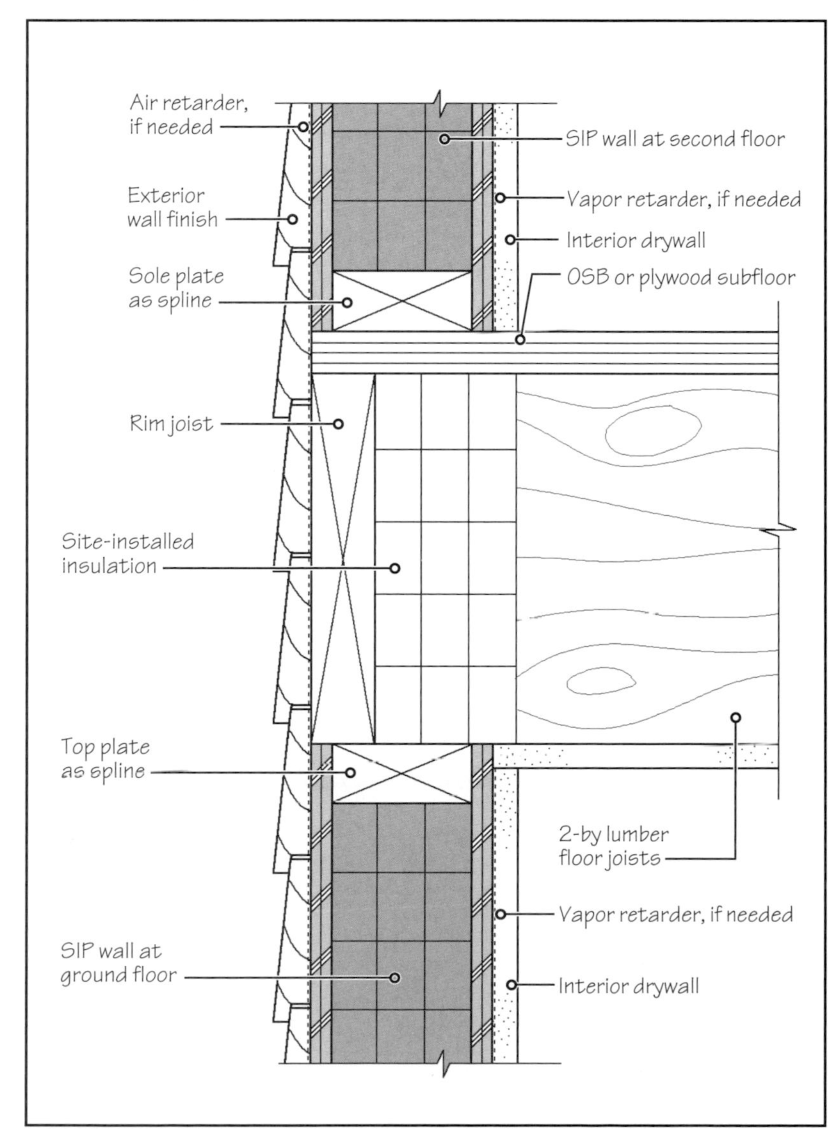

FIGURE 15.10 Detail at the junction of a SIP floor and SIP wall. The detail is similar to the one used in a conventional wood frame building. Site-installed insulation is needed behind the rim joist.

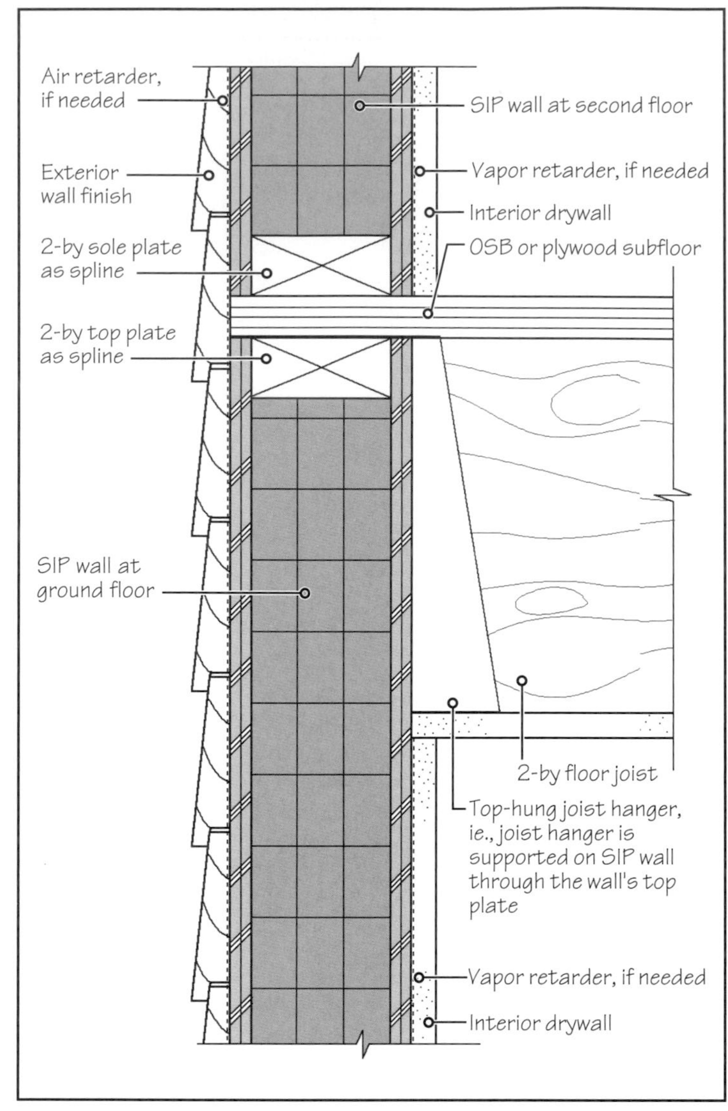

FIGURE 15.11 Detail at the intersection of floor and SIP wall—an alternative to the detail of Figure 15.10. In this detail, site-installed insulation is not needed.

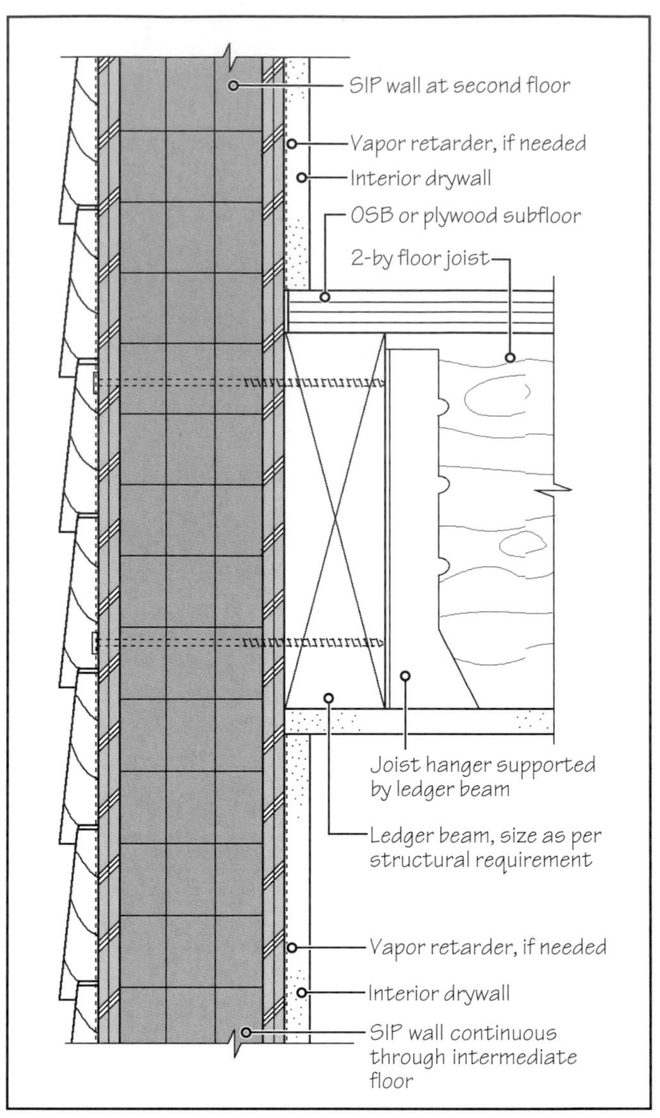

FIGURE 15.12 Detail at the intersection of floor and SIP wall, where the SIP wall is continuous through the floor.

The second alternative is to use a SIP roof, where the roof SIPs span from the ridge beam to the exterior wall, creating a cathedral ceiling. Typical details of such an alternative are shown in Figure 15.13.

15.5 ADVANTAGES AND LIMITATIONS OF SIPS

The SIP system of construction has several advantages and limitations as compared with the conventional WLF system.

ADVANTAGES

- *Panelization and Deletion of Insulation Subtrade:* As stated at the beginning of this chapter, panelization and the deletion of insulation subtrade in a SIP structure increase speed and efficiency of construction.
- *Air Retarders:* Because the panels are relatively airtight, air retarders are generally unnecessary in most climates.
- *Continuous Nailable Surface:* Because OSB facings provide a continuous nailable surface, interior drywall and exterior wall finishes can be applied without having to locate the studs.
- *Pneumatic Nailers:* Although several details require the use of screws, pneumatic nailers are used extensively.

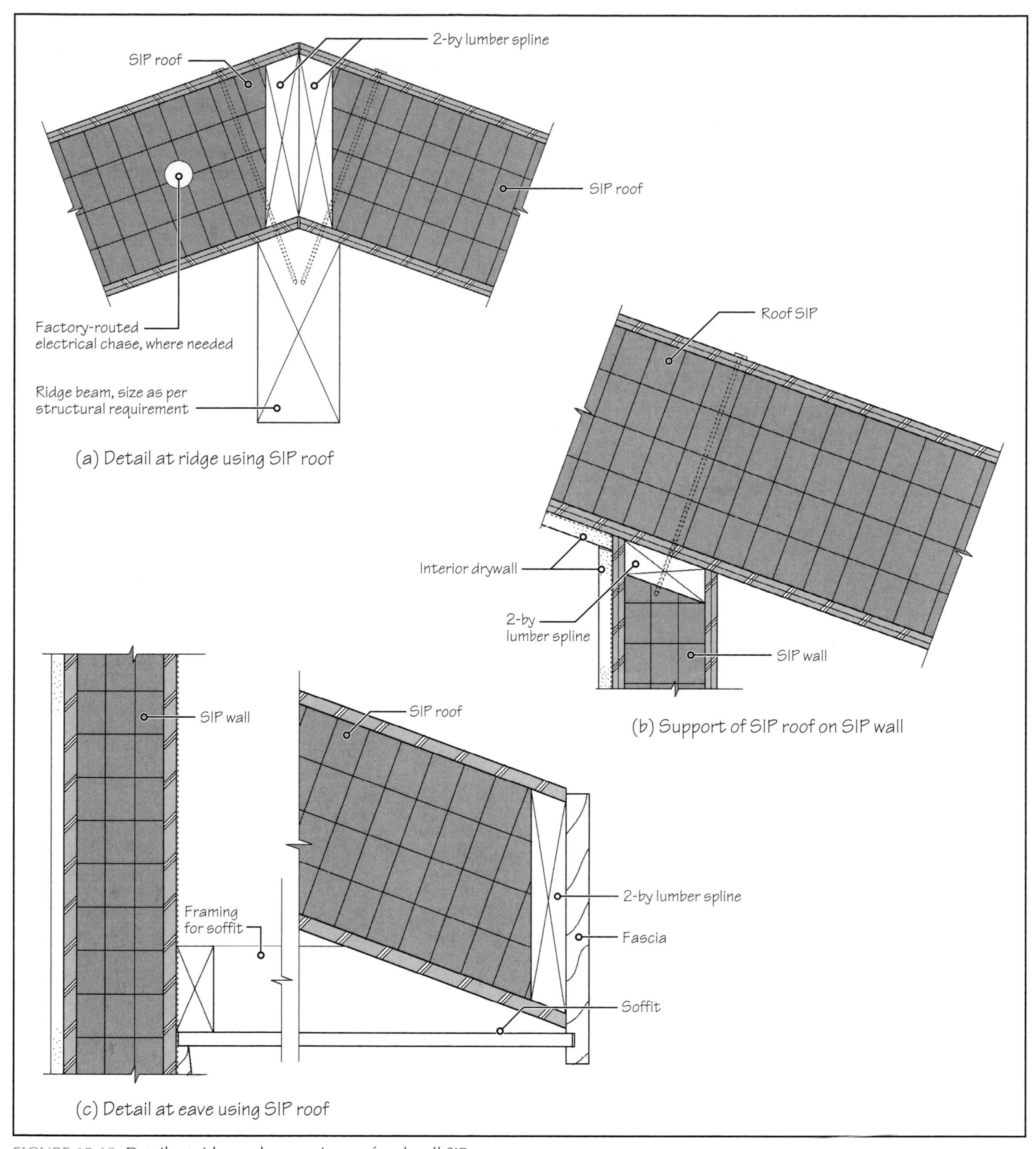

FIGURE 15.13 Details at ridge and eave using roof and wall SIPs.

- *On-site Waste:* There is very little on-site waste produced in a SIP structure because most members arrive at the site precut to size.
- *Energy Efficiency:* One of the major advantages of SIPS is its energy efficiency. A SIP structure has virtually no thermal bridges in the envelope. The use of foam core allows no air movement within the insulation, unlike that in fiberglass insulation.
- *Fire Endurance:* SIP manufacturers claim that the fire endurance of a SIP structure matches that of a conventional wood frame structure. The absence of concealed spaces in walls and roofs deprives the fire of the oxygen required for burning.

LIMITATIONS

- *Chases:* Because the electrical chases must be prerouted in the panels, a greater amount of preplanning is necessary.
- *Squareness of Panels:* Panels must be absolutely square for a successful installation.
- *Termites and Insect Attack:* Termites and insects boring through foam insulation, which can seriously impact the structural performance of the system, must be checked with the manufacturer.
- *Long-term Performance:* Long-term structural performance of the system depends on the long-term performance of adhesives.

PRACTICE QUIZ

Each question has only one correct answer. Select the choice that best answers the question.

1. In a structural insulated panel, the insulation generally consists of
- **a.** fiberglass.
- **b.** extruded polystyrene.
- **c.** polyisocyanurate.
- **d.** polyurethane.
- **e.** expanded polystyrene.

2. The facing in a structural insulated panel generally consists of
- **a.** OSB on both faces.
- **b.** OSB on the exterior face and gypsum board on the interior face.
- **c.** plywood on the exterior face and gypsum board on the interior face.
- **d.** plywood on the exterior face and OSB on the interior face.
- **e.** none of the above.

3. In a SIP structure, the interior walls are constructed of 4 ft × 8 ft SIP panels.
- **a.** True
- **b.** False

4. In a SIP building, exterior wall panels are typically
- **a.** 6 in. or 7 in. thick.
- **b.** $5\frac{1}{2}$ in. or 6 in. thick.
- **c.** 5 in. or 6 in. thick.
- **d.** 4 in. or 6 in. thick.
- **e.** $4\frac{1}{2}$ in. or $6\frac{1}{2}$ in. thick.

5. Adjacent structural insulated wall panels in a SIP building are connected through
- **a.** 2-by lumber studs.
- **b.** engineered lumber splines.
- **c.** OSB splines.
- **d.** any one of the above.
- **e.** special sheet metal connectors.

6. A corner between structural insulated wall panels in a SIP building requires a minimum of
- **a.** four 2-by lumber studs.
- **b.** three 2-by lumber studs.
- **c.** two 2-by lumber studs.
- **d.** one 2-by lumber stud.
- **e.** 2-by lumber studs are not required.

7. The connection of a wall panel to concrete foundation, as shown in this text, requires the use of
- **a.** one treated sill plate.
- **b.** two treated sill plates.
- **c.** two sill plates, one of which is treated.
- **d.** a sill plate is not needed in a SIP structure.

8. In a SIP structure, an intermediate floor may be framed with any one of the following except
- **a.** 2-by lumber members.
- **b.** wood I-joists.
- **c.** wood trussed joists.
- **d.** SIP panels.
- **e.** all the above (without exception).

9. In a SIP structure, the roof may be framed with any one of the following except
- **a.** a rafter-and-ceiling-joist assembly consisting of 2-by lumber members.
- **b.** a rafter-and-ceiling-joist assembly consisting of wood I-joists.
- **c.** wood roof trusses.
- **d.** SIP panels.
- **e.** all the above (without exception).

10. A SIP manufacturer generally supplies standard size panels, which must be cut to size at the site as required by project drawings.
- **a.** True
- **b.** False

11. Similar to the conventional WLF structure, the interior gypsum board wall finish must be nailed to the studs in a SIP structure.
- **a.** True
- **b.** False

12. The maximum size of SIP panel available is
- **a.** 4 ft × 8 ft.
- **b.** 4 ft × 10 ft.
- **c.** 8 ft × 8 ft.
- **d.** 8 ft × 16 ft.
- **e.** 8 ft × 24 ft.

Answers: 1-e, 2-a, 3-b, 4-e, 5-d, 6-c, 7-c, 8-d, 9-e, 10-b, 11-b, 12-e.

KEY TERMS AND CONCEPTS

Floor assemblies in SIP structure

Roof assemblies in SIP structure

Structural behavior of SIPS

Structural insulated panels (SIPS)

Wall assemblies in SIP structure

REVIEW QUESTIONS

1. Using a sketch and notes, explain the anatomy of a structural insulated panel. In what thickness are the panels commonly used?
2. Using sketches and notes, show how two adjacent SIP wall panels are joined.

SELECTED WEB SITES

Structural Insulated Panel Association, SIPA (www.sips.org)

FURTHER READING

1. Morley, Michael. *Building with Structural Insulated Panels.* Newtown, CT: Taunton Press, 2000.

ISUZU
NISSAN
TOYOTA
ACURA
MITSUBISHI MOTORS
TOYOTA
TOYOTA

CHAPTER 16

The Material Steel and Structural Steel Construction

CHAPTER OUTLINE

16.1 Making of Modern Steel

16.2 Structural Steel Shapes and Their Designations

16.3 Steel Joists and Joist Girders

16.4 Steel Roof and Floor Decks

16.5 Preliminary Layout of Framing Members

16.6 Bolts and Welds

16.7 Connections Between Framing Members

16.8 Steel Detailing and Fabrication

16.9 Steel Erection

16.10 Corrosion Protection of Steel

16.11 Fire Protection of Steel

PRINCIPLES IN PRACTICE: Fundamentals of Skeleton Frame Construction

The twin tubular steel arches, spanning 390 ft, create the world's largest unobstructed interior space (203,000 ft^2) in the new Exhibit Hall of the Dallas Convention Center. Architects: SOM Inc. and HKS Inc. (Photo courtesy of HKS Inc.)

FIGURE 16.1 Hong Kong and Shanghai (HSBC) Bank, Hong Kong, with its expressed structural steel frame. The frame consists of eight vertical masts, two of which are seen in this photo. Each mast comprises a cluster of four interconnected tubular steel columns that support large steel trusses every few floors. The floors between trusses are suspended from the upper truss. Architect: Sir Norman Foster; Structural Engineer: Ove Arup and Partners.

Because of its extensive use in industry, construction, and weaponry, iron is by far the most important of all metals. That is why metals are generally divided in two categories:

- Ferrous metals—metals that contain iron (the Latin term for iron is *ferrum.*)
- Nonferrous metals—metals that do not contain iron

Steel is the most important ferrous metal. Its high strength in relation to its weight makes it the material of choice for skyscrapers and long-span structures, such as sports stadiums and bridges. Its malleability and weldability allow it to be shaped, bent, and made into different types of components. These characteristics provide the versatility that architects and engineers have exploited in creating a wide range of highly expressive structures, as exemplified by Figures 16.1 to 16.3.

FIGURE 16.2 The interior of the steel dome of German Bundestag, Berlin. Architect: Sir Norman Foster.

FIGURE 16.3 Rock and Roll Hall of Fame and Museum, Cleveland, Ohio. The sloping glass skin framed with tubular steel trusses span nearly 200 ft. The steel tower structure is seen in the background. Architects: Pei, Cobb, Freed and Partners. (Photo courtesy of Dr. Jay Henry.)

The history of development of steel has been long and arduous because the fuel technology necessary to produce temperatures high enough to melt iron ore on a large scale were not available until the early eighteenth century. The melting point of iron is 2,800°F (1,540°C). By contrast, copper's melting point is approximately 2,000°F (1,100°C)—a temperature that could be obtained in a wood or charcoal fire. That is why copper and, later, bronze (an alloy of copper and tin) preceded steel by several centuries.

Wrought Iron—The Earliest Form of Iron

The earliest predecessor of steel was wrought iron. It was produced by heating iron ore in a charcoal fire. Though insufficient to melt the ore, a charcoal fire was sufficient to extract the oxygen from the ore, which consists of iron oxide mixed with some silica (sand). In this reaction, carbon dioxide and carbon monoxide are driven off, and iron oxide is reduced to iron. The heat softens the ore enough to squeeze silica out by beating the ore with a hammer—a process called *working*, which led to the use of the term *wrought iron.**

Although most of the silica is removed, a small amount (1% to 2%) remains in the form of silica fibers, formed by pounding during the working process. Pounding also converted the iron into fibers. Wrought iron is, therefore, fibrous in character, unlike steel, which is relatively homogeneous. The presence of silica also gives wrought iron the ability to resist corrosion. That is why wrought iron structures—the most notable being the Eiffel Tower in Paris (completed in 1889)—have survived without any corrosion protection.

Because of the use of charcoal, a tiny amount of carbon (nearly 0.02%) remained in iron. Thus, if we ignore the presence of the small percentage of silica, wrought iron is virtually pure iron. Its purity makes it a soft, malleable, and ductile material.

From Wrought Iron to Cast Iron

The making of wrought iron was slow and time consuming due to the repetitive working required on the ore. British industrialist Abraham Darby's discovery of the blast furnace in 1709 revolutionized iron making by providing a means of melting the ore on a large scale.

Darby used coke as the fuel to melt the ore instead of coal.† (Coal had replaced charcoal for quite some time due to the depletion of forests in England.) Coke is much stronger than coal and has a higher caloric value because it contains few or no impurities. In Darby's process, iron ore and coke were mixed together in a furnace, called the *blast furnace*. Mixing

*According to *Webster's New Collegiate Dictionary*, the term *wrought* means "worked into shape by artistry or effort, or beaten into shape by tools."

†Coke is the residue left over from heating coal in the absence of oxygen. Similarly, charcoal is the residue left over by heating wood in the absence of oxygen. Coke is sometimes called *cooked coal*.

the fuel with the ore provided the temperature required to melt the ore. The liquid iron was cooled to a solid state by casting it on a sand bed. The shape of these castings resembled a litter of newborn pigs. Therefore, the end product from the blast furnace came to be known as *pig iron*, although a more appropriate term is *cast iron*.

Darby's blast furnace replaced the manual process of making wrought iron to mass production of cast iron, thereby reducing the cost of metal dramatically. Consequently, cast iron found extensive use, initially in machines, later in railroad tracks, and finally in bridges and buildings. Cast iron's high strength and sculptability were exploited by architects in countless buildings, notably the Crystal Palace in England (1851) and several early skyscrapers in Chicago.

Because of the mixing of the ore with fuel, cast iron contains a high percentage (2.5% to 4%) of carbon. The presence of this much carbon makes iron brittle and unweldable but hard and resistant to corrosion and fire. Because of its hardness and corrosion resistance, cast iron is used in utility access hole covers and sanitary pipes. Wrought iron is no longer commercially produced.

Wrought iron (with nearly 0.02% carbon) and cast iron (with 2.5% to 4% carbon) represent two extremes of iron-carbon alloy. The "perfect" amount of carbon is from 0.1% to 1.7%, a metal that is called *steel*.

From an architect's perspective, the importance of structural steel construction lies in its versatility. Structural steel construction is suitable for almost all building types—from the extremely functional industrial warehouse to ornate and expressive museums and concert halls. It is used for 1-story to 100-story structures. In fact, for most contemporary commercial buildings, the only choice for the structural system of the building is between steel and concrete.

From a historian's perspective, the significance of steel construction is that it led to the birth of the system called *skeleton frame*, commonly known as the *frame structure*. A frame structure is defined as a structure that is made entirely of linear elements—columns and beams. Prior to the discovery of the frame structure, buildings were generally built using load-bearing masonry walls—a system that is still used (Chapter 24).

In a load-bearing wall system, the walls function as both structural and space-organizing elements. In a frame structure, on the other hand, the frame performs the structural function and the walls are used primarily for spatial organization. The separation of the two functions provided architects with a means of expression that was not available in load-bearing wall structures. (A brief introduction to the principles of skeleton frame structures is also provided in "Principles in Practice" at the end of this chapter.).

This chapter begins with a brief coverage of the production of various types of modern steels, followed by a discussion of standard structural steel shapes and various steel components, such as open-web joists and floor and roof decks. Subsequently, the chapter covers construction systems that utilize structural steel sections and components. Structural steel sections and components are so called because they are used in the structural frame of buildings.

NOTE

A Note on Metals

Metals have a characteristic "metallic" appearance. They are good conductors of heat and electricity. Unless formed into extremely thin films, they are opaque.

With the exception of gold, pure metals are not found on the earth's surface because they react readily with other elements and form compounds. For example, iron forms iron oxide when exposed to air over a period of time. Aluminum also much more readily forms aluminum oxide. Although iron oxide separates from iron in the form of rust, aluminum oxide clings to aluminum tenaciously, forming a protective skin that resists the further oxidation of aluminum.

Pure metals are soft and malleable and are, therefore, not as useful as metals that are impure (i.e., alloyed with other metals). That is why 100% pure gold is too soft to be used for jewelry. Most gold jewelry is made from 18- to 22-carat gold. Eighteen-carat gold is 75% gold and 25% other metals (18 ÷ 24 = 75%). The softness of wrought iron—also due to its excessive purity—makes it less useful than steel.

Why Some Metals Are Precious

Platinum, gold, and silver are considered precious metals primarily because of their scarcity on the earth. If aluminum was rare and gold as plentiful as aluminum, we would be using gold foil to wrap food and as a radiant barrier in buildings. In fact, when aluminum was first discovered, it was more precious than gold. It had the shine of gold, silver, and platinum and resisted corrosion, so it was, therefore, a perfect jewelry material.

As the legend has it, Napoleon Bonaparte, Emperor of France, was so impressed by this shiny, new metal that in one of his banquets, the plates for the royal guests were made from aluminum. The commoners in the banquet were served on silver plates. Bars of aluminum were exhibited alongside the French Crown jewels at the Exposition Universelle in Paris in 1855.

Once the process of extracting aluminum was perfected, the price of aluminum dropped precipitously due to aluminum's abundance on the earth's surface, and aluminum was no longer a precious metal.

16.1 MAKING OF MODERN STEEL

Extracting carbon from cast iron to obtain the so-called perfect iron-carbon alloy was a major metallurgical problem of the nineteenth century. It was finally resolved through the discovery of the basic oxygen furnace by Henry Bessemer in 1855. In the basic oxygen furnace (BOF), the excess carbon in the molten pig iron was converted to carbon dioxide by blowing oxygen through it. After the discovery of the BOF, the electric arc furnace (EAF) and open-hearth furnace (OHF) were developed, both of which obtain the same result as the BOF.

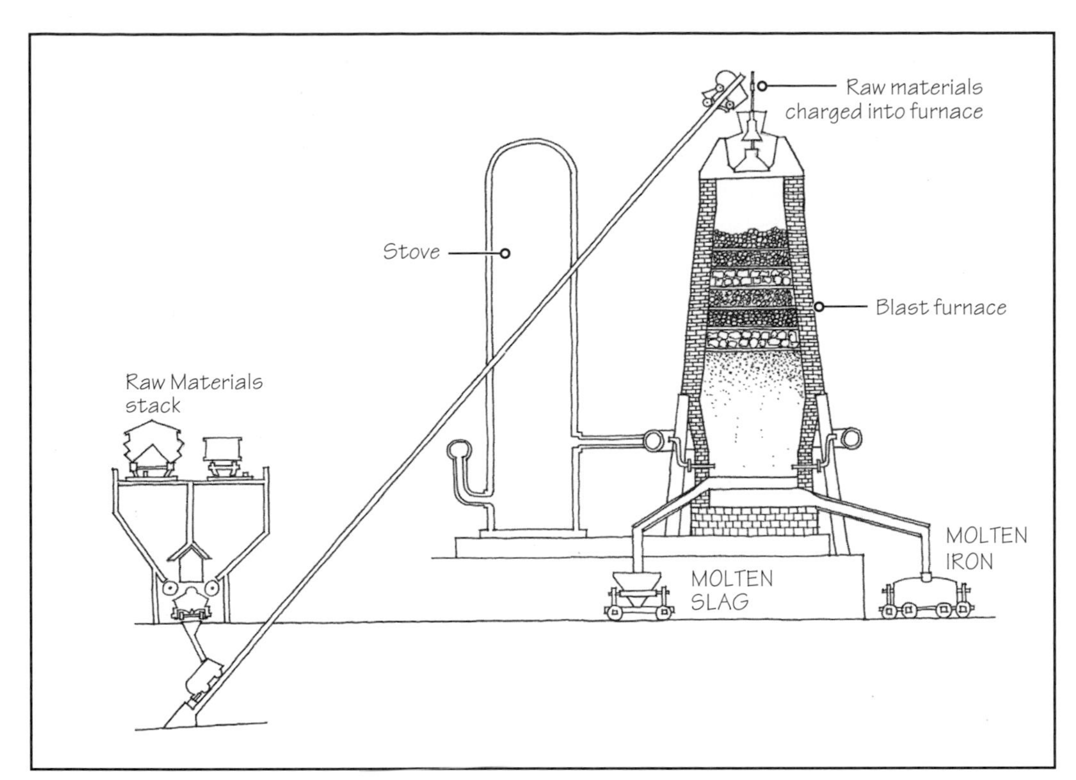

FIGURE 16.4 A typical modern blast furnace.

Modern steel can be produced in several ways. A commonly used method is the integrated method, in which a blast furnace is an important feature. A blast furnace is a huge, steel cylinder, whose interior is lined with refractory bricks, Figure 16.4. The raw materials—iron ore, coke, and limestone—are stacked in the furnace in alternate layers from the top. Limestone acts as a flux to reduce the melting point of iron. Hot air from the stove is fed into the furnace from the bottom of the stack, which ignites the coke. As the iron ore and limestone melt, they travel down the stack and settle at the bottom of the furnace. The molten material consists of two parts: molten iron and molten slag.

The slag is essentially molten limestone, but it also contains molten silica and other elements present in iron ore. Being lighter, molten slag floats over molten iron. Therefore, both slag and molten iron are drained separately from the bottom of the furnace.

Molten iron from the blast furnace is converted to molten steel by one of the three processes (BOF, EAF, or OHF), depending on the particular steel mill. At this stage, small quantities of a few other metals, such as copper, nickel, molybdenum, or chromium, are added to the molten metal to obtain the steel with the required properties. Depending on the final cross-sectional shape required, molten steel is then cast into billets, blooms, or slabs. A billet, bloom, or slab is rolled into the final structural shapes while it is still hot, Figure 16.5.

The manufacturing process just described is based on the use of virgin iron ore. A method being increasingly used is one that recycles the scrap from automobiles, transport vehicles, machinery, and so on. This method yields a more economical steel, requiring a much smaller manufacturing outlay.

COMMONLY USED STRUCTURAL STEELS

As stated previously, molten steel is alloyed with a small percentage of other metals to obtain steels that vary from each other with respect to a few important properties, such as the yield strength, tensile strength, and corrosion resistance. The most commonly used steel for framing members (W-shape columns and beams) in contemporary buildings is A992, with a yield strength of 50 ksi. The yield strength of steel is also referred to as the *steel's grade*. Thus, a steel with a yield strength of 50 ksi is called *grade 50 steel*.

Prior to the production of A992 steel, A36 steel (grade 36) was the building industry's standard. Today, A36 steel is used mainly for plates, angles, and channel sections. Corrosion-resistant steels are also available.

SLAG—A WASTE PRODUCT WITH SEVERAL USES

Slag is a waste product from the blast furnace. It has many uses, including as a lightweight aggregate, raw material in the manufacture of portland cement, and in the manufacture of insulation, called *slagwool*. Slagwool is noncombustible and is similar to fiberglass in thermal properties.

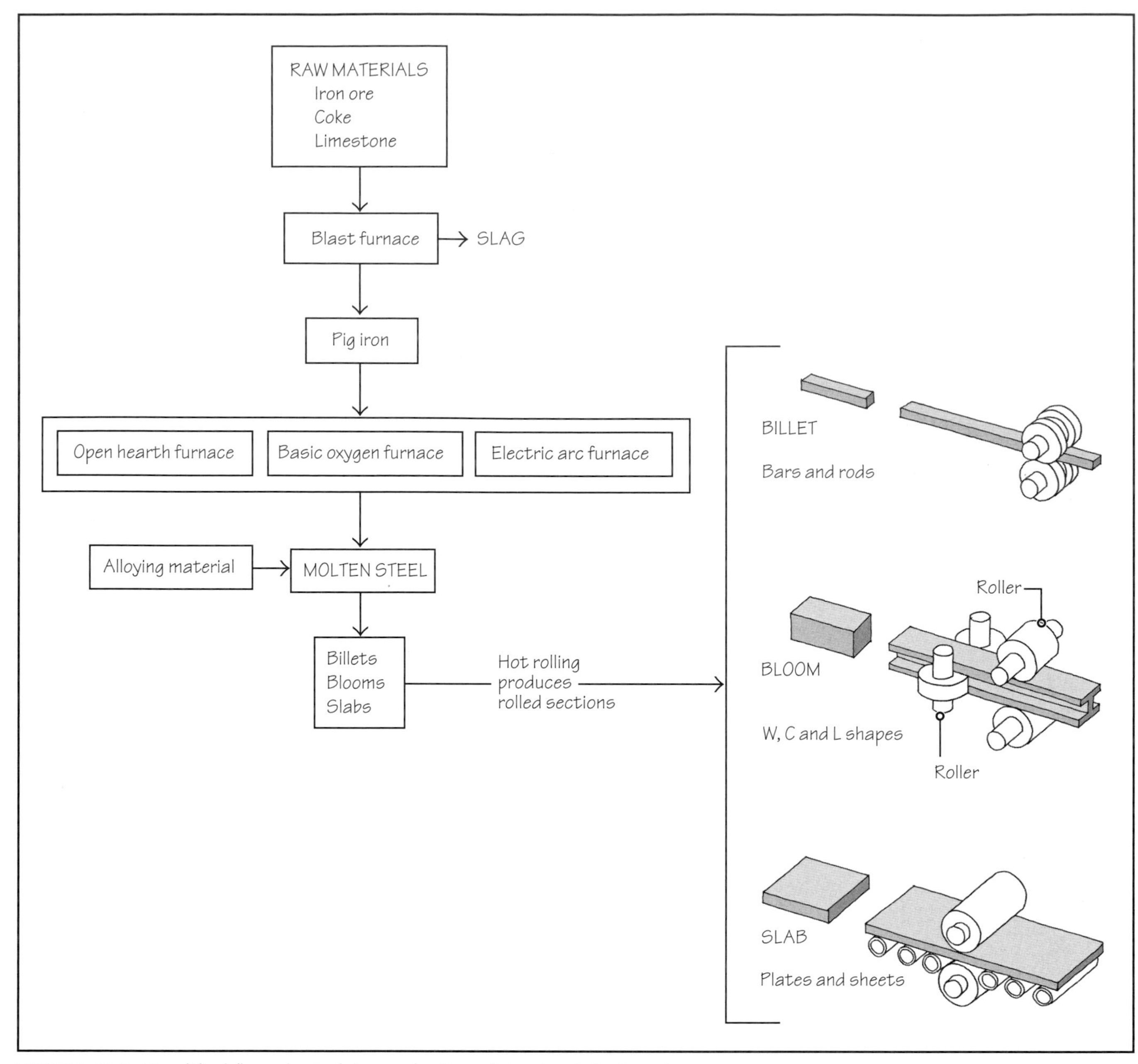

FIGURE 16.5 Simplified flow chart of the production of structural steel sections.

NOTE

Structural Steel, Reinforcing Steel, and Light-Gauge Steel

It is important to distinguish between structural steel and reinforcing steel. Steel cross sections such as I, C, L, T, HSS, plates, and pipes are called *structural steel* members.

Reinforcing steel is in the form of deformed round bars (also called *rebars*) that are used in concrete slabs, beams, and columns. Reinforcing steel is covered in Chapter 18.

Light-gauge steel members consist of thin sheets of steel that are bent to shape. Light-gauge steel construction is discussed in Chapter 17.

16.2 STRUCTURAL STEEL SHAPES AND THEIR DESIGNATIONS

Because of the large financial outlay required in producing structural steel sections, an architect or engineer must select from the standard shapes and sizes. They are available in cross-sectional shapes of I, C, L, T, pipes, tubes, round and rectangular bars, and plates, Figure 16.6. Most of these sections are produced by hot rolling (Figure 16.5).

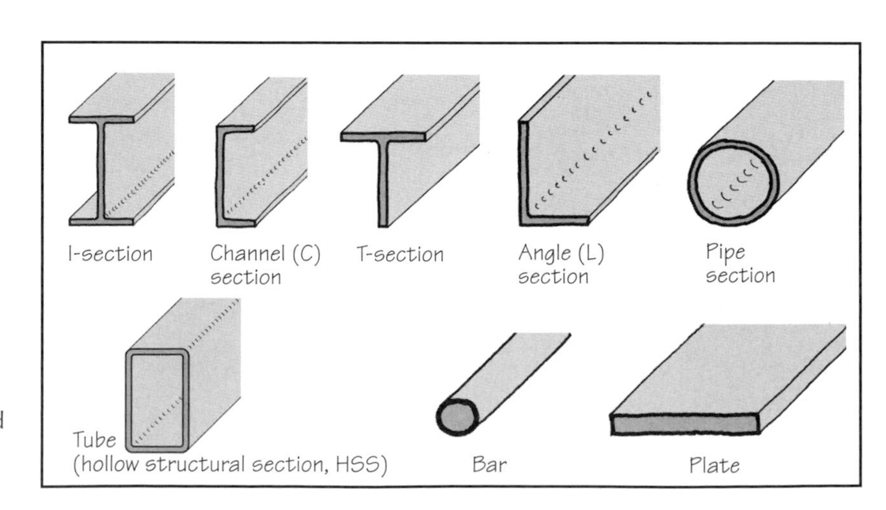

FIGURE 16.6 Commonly used structural steel sections.

I-SECTIONS

Structural steel I-sections may be classified into four shapes:

- W-shapes
- S-shapes
- HP-shapes
- M-shapes

In a W-shape, the interior surfaces of the flanges are parallel. In the S-shape, the interior surfaces of flanges are inclined at a slope of 2:12, Figure 16.7.

Another difference between W- and S-shapes is that, for the same depth, the flanges in a W-shape are wider than those in an S-shape. That is why a W-shape is called a *wide-flange section.* Because of the parallel flange surfaces and wide flanges, W shapes are the ones most used for beams and columns. S-shapes (also called *American standard* shapes) are used in cranes and rails.

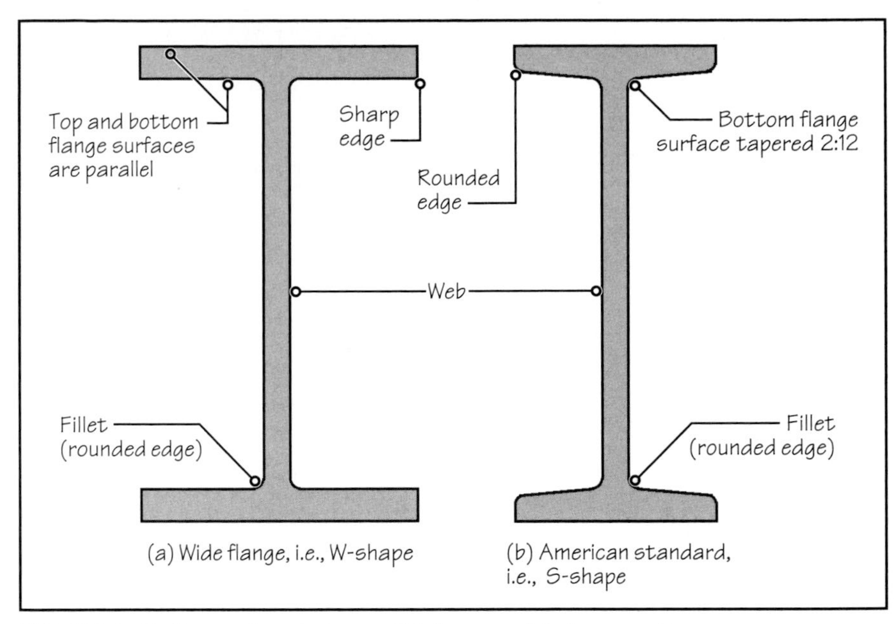

FIGURE 16.7 Comparison between W-shape and S-shape sections.

Each W-shape is designated by two numbers: the first number refers to the nominal depth of the section (in inches) and the second number gives its weight for a 1-ft length (pounds per foot). For example, W21 × 68 means that the nominal depth of the section is 21 in. and its nominal weight is 68 lb/ft. The nominal depth is the section's approximate (not exact) depth.

W-shapes are available in a large range of sizes and weights, from W40 × 503 to W4 × 13. A few manufacturers also make 44-in.-deep W-shapes. Several W-shapes share a common nominal depth but have different flange widths, flange and web thickness. Sections such as W14 × 132, W14 × 82, and W14 × 53 have different weights but the same nominal depth.

The interior flange-to-flange dimensions of most W-shapes sharing the same nominal depth designation are equal. For example, as shown in Figure 16.8, W14 × 132, W14 × 82, and W14 × 53 have two dimensional facts in common: (a) Their interior flange-to-flange dimension is the same (12.60 in.), and (b) their overall depths are approximately 14 in. All other dimensions of these sections are different from each other. Generally, W-shapes with squarish proportions (depth and flange width nearly equal), such as W14 × 132, are used as columns. Tall and narrow sections, such as W14 × 53, are used as beams.

An HP-shape is so called because it is generally used as a bearing pile and is squarish in proportion; that is, its overall depth and flange width are nearly equal, resembling the letter *H.*

NOTE

Nominal Depths of W-Shapes

W44, W40, W36
W33, W30, W27, W24, W21, W18
W16, W14, W12, W10, W8, W6, W4

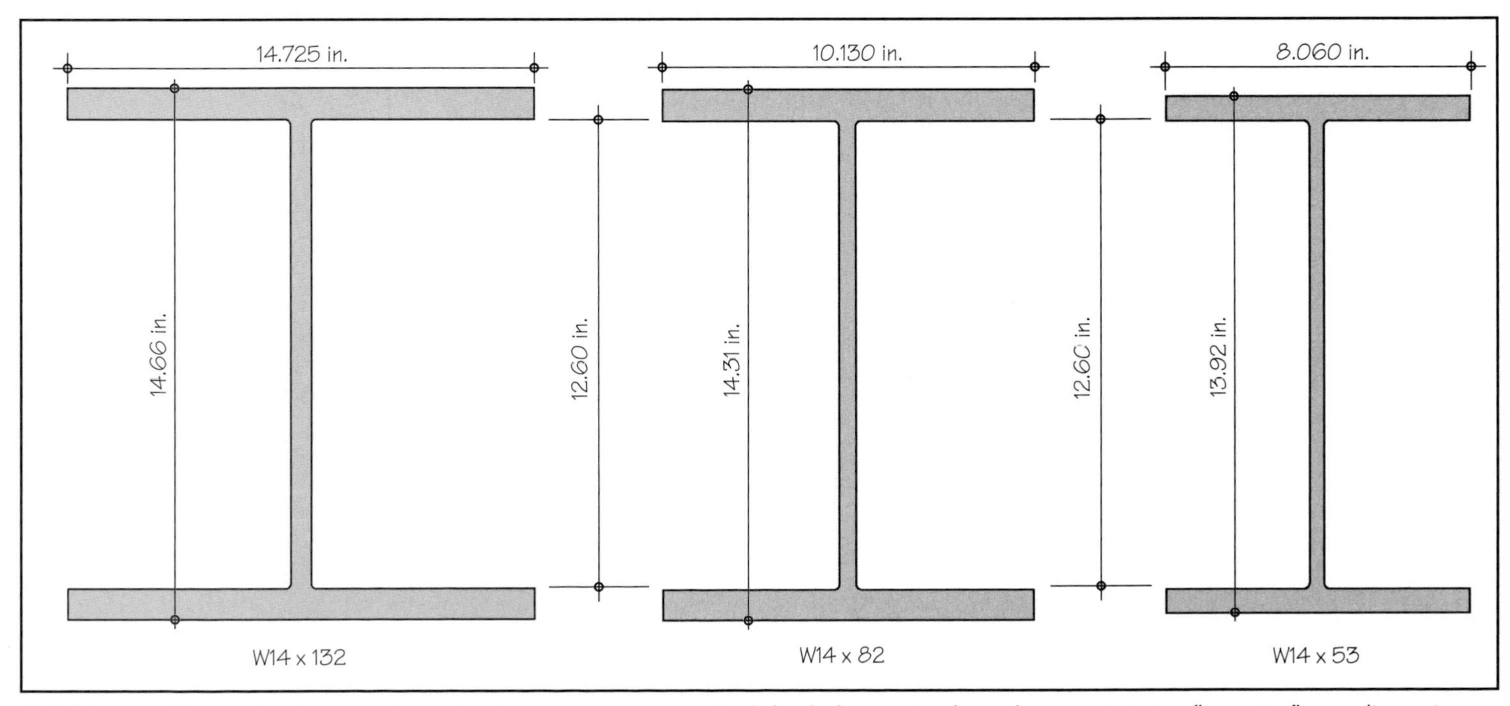

FIGURE 16.8 Most wide-flange sections that share a common nominal depth designation have the same interior flange-to-flange dimension. This illustration shows three sections, all with a nominal depth of 14 in. Their overall depths are nearly 14 in., and several other dimensions are different from each other. However, the interior flange-to-flange dimensions are 12.60 in. for all sections. Several other sections with nominal depth of 14 in. are available (Table 16.1).

TABLE 16.1 DIMENSIONAL INFORMATION ABOUT SELECTED W-SHAPES

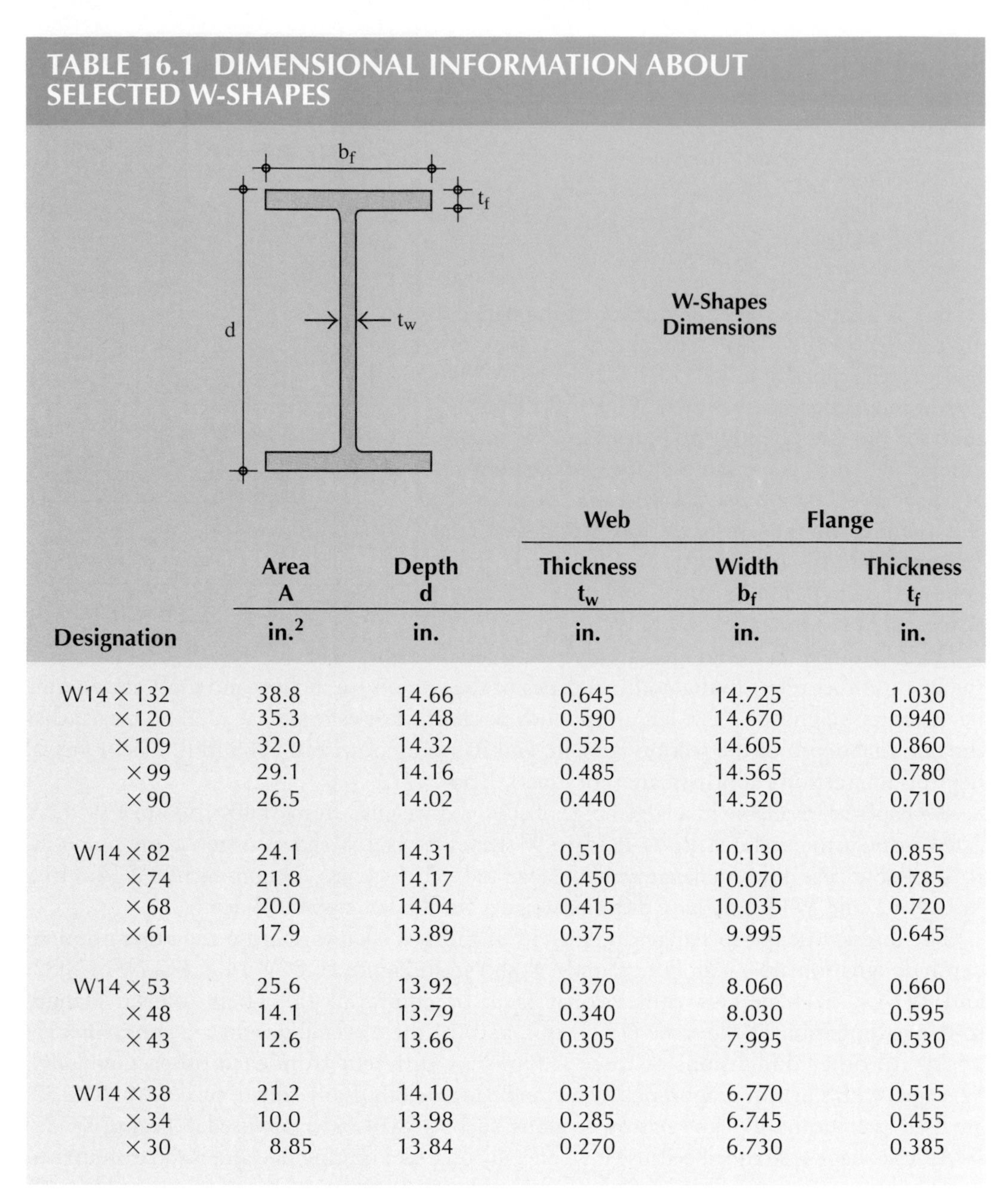

Designation	Area A	Depth d	Web Thickness t_w	Flange Width b_f	Flange Thickness t_f
	in.^{2}	in.	in.	in.	in.
W14 × 132	38.8	14.66	0.645	14.725	1.030
× 120	35.3	14.48	0.590	14.670	0.940
× 109	32.0	14.32	0.525	14.605	0.860
× 99	29.1	14.16	0.485	14.565	0.780
× 90	26.5	14.02	0.440	14.520	0.710
W14 × 82	24.1	14.31	0.510	10.130	0.855
× 74	21.8	14.17	0.450	10.070	0.785
× 68	20.0	14.04	0.415	10.035	0.720
× 61	17.9	13.89	0.375	9.995	0.645
W14 × 53	25.6	13.92	0.370	8.060	0.660
× 48	14.1	13.79	0.340	8.030	0.595
× 43	12.6	13.66	0.305	7.995	0.530
W14 × 38	21.2	14.10	0.310	6.770	0.515
× 34	10.0	13.98	0.285	6.745	0.455
× 30	8.85	13.84	0.270	6.730	0.385

This table has been adapted from *Steel Construction Manual* (13th Ed.), published by the American Institute of Steel Construction (AISC), with permission.

The table is an bridged version. A complete version includes additional dimensions and information about the engineering properties of each section, such as the moment of inertia, radius of gyration, and so on.

The thickness of the web and flanges are equal in an HP-shape. An M-shape is one that cannot be classified as W, S, or HP; the letter *M* being an acronym for *miscellaneous.*

Dimensional information and engineering properties of various structural steel shapes are available in the *Steel Construction Manual* published by the American Institute of Steel construction (AISC). Table 16.1 gives an indication of some of the information available in this publication.

C-SHAPES

Steel channels (C-shapes) are similar in profile to S-shapes, that is, their inner flange surfaces are inclined at an angle of 2:12. They are designated by two numbers after the letter C. The first number gives the overall depth of the section and the second number gives the weight of a 1-ft length of the section. Thus, C8 × 11.5 means that the channel is 8 in. deep and weighs 11.5 lb/ft.

Miscellaneous channels do not have a standard slope on the inner flange surfaces. They are designated in the same way as C-shapes, for example, MC12 × 50.

T-SHAPES

A T-shape section is made by splitting a W-shape, M-shape, or S-shape into two equal parts. It is, therefore, called WT, MT, or ST, depending on its origin. For example, two WT6 × 29 sections are obtained from one W12 × 58.

L-SHAPES, PIPES, TUBES, BARS, AND PLATES

Steel angles (L-shapes) may either be equal-leg angles or unequal-leg angles. The thickness of both legs is the same in an angle. Angles are designated by three numbers. The first two numbers give the length of each leg, and the third number gives the thickness of the legs. L4 × 4 × $\frac{1}{2}$ is an example of an equal-leg angle with legs equal to 4 in. each; the thickness of each leg is $\frac{1}{2}$ in. In an unequal-leg angle, the longer leg is mentioned first, as in L4 × 3 × $\frac{1}{4}$. Angles have various uses, such as masonry lintels and members of steel trusses.

Pipes are designated by their nominal diameter and whether the pipe is a standard weight, extra strong, or double-extra strong. These three designations refer to the pipe's wall thickness. Pipes are generally used as columns or as members of a truss.

A tube is referred to as a *hollow structural section* (HSS) and is made by bending a steel plate and welding it seamlessly. That is why the edges of a tube are rounded (Figure 16.6). An HSS may be square, rectangular, or round. Square or round HSSs are generally used as columns, and rectangular tubes are used as beams. Like pipe trusses, HSS member trusses are fairly common for long-span structures.

A rectangular or square HSS member is designated by three dimensions, all dimensions expressed in fractional numbers. Thus, an HSS 12 × 8 × $\frac{1}{2}$ is a rectangular HSS measuring 12 in. × 8 in. on the outside; wall thickness is $\frac{1}{2}$ in. A round HSS is designated by outside diameter and wall thickness, both dimensions expressed to three decimal places. Thus, an HSS 10.000 × 0.500 has an outer diameter of 10 in. and wall thickness of $\frac{1}{2}$ in.

NOTE

Bent Plates as Angles

The maximum size of the leg of a standard angle is 8 in. When an angle with a larger leg is required, it may be ordered as a plate bent to the shape of an angle. Bent-plate angles have a rounded corner, whereas a standard-size (hot-rolled) angle has a sharp corner.

Bent-plate angles are also used where an angle with odd dimensions (not available as a standard angle) is mandated by the detail.

BUILT-UP SECTIONS

When size and other requirements dictate the use of nonstandard shapes and cross-sectional dimensions, they can be obtained by welding two or more standard sections together, called *built-up sections*. A few examples of built-up sections are shown in Figure 16.9.

In some situations, it is necessary to use a W-shape beam that is deeper than the deepest (44 in.) W-shape available. Such a built-up beam can be obtained by welding steel plates together, Figure 16.10(a). Because the beam is fabricated from plates, it is also called a *plate girder*. Plate girders are fairly common in bridge structures. In buildings, they are employed to carry heavy loads over long spans, such as to support the weight of columns that must be omitted at the lower levels to achieve larger unobstructed space, Figure 16.10(b).

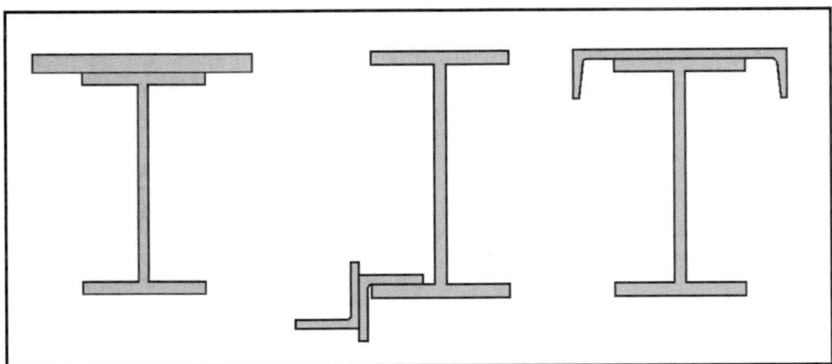

FIGURE 16.9 Some examples of built-up steel sections.

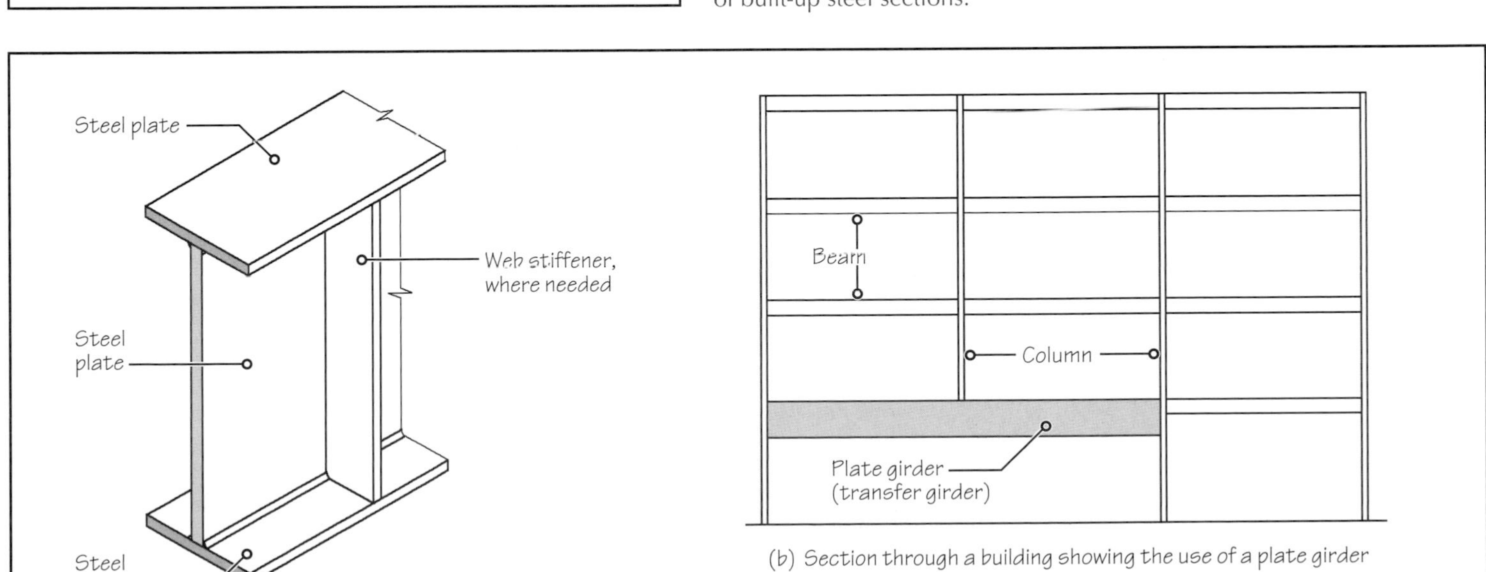

FIGURE 16.10 **(a)** Anatomy of a typical plate girder. **(b)** One of the several uses of a plate girder.

PRACTICE QUIZ

Each question has only one correct answer. Select the choice that best answers the question.

1. Which of the following ferrous metals has the largest amount of carbon?
- **a.** Steel
- **b.** Wrought iron
- **c.** Cast iron

2. Which of the following ferrous metals is least corrosion resistant?
- **a.** Steel
- **b.** Wrought iron
- **c.** Cast iron

3. The steel type most commonly used for framing members (columns and beams) of a steel building is
- **a.** steel type A36 with a yield strength of 36 ksi.
- **b.** steel type A36 with a yield strength of 360 ksi.
- **c.** steel type A992 with a yield strength of 50 ksi.
- **d.** steel type A992 with a yield strength of 40 ksi.
- **e.** none of the above.

4. Slag is
- **a.** a waste product generated from the manufacturing of iron.
- **b.** the ore used for making iron.
- **c.** the ore used for making aluminum.
- **d.** a ferrous metal that has been replaced by steel.
- **e.** none of the above.

5. A W-shape section is so called because
- **a.** it is an I-section with wide flanges.
- **b.** it is an I-section with a wide web.
- **c.** its cross-sectional shape resembles the letter W.
- **d.** it is a rectangular section whose width is larger than its depth.

6. W-shape, S-shape, HP-shape, and M-shape are all structural steel I-sections
- **a.** True
- **b.** False

7. Which of the following steel shapes is most commonly used for the structural frame of buildings?
- **a.** HP-shape
- **b.** S-shape
- **c.** M-shape
- **d.** W-shape
- **e.** C-shape

8. The designation W14 × 38 means that the
- **a.** width of the section is 14 in. and depth is 38 in.
- **b.** depth of the section is 14 in. and width is 38 in.
- **c.** depth of the section is 14 in. and its weight is 38 lb/ft.
- **d.** nominal depth of the section is 14 in. and its weight is 38 lb/ft.
- **e.** none of the above.

9. To obtain the structural engineering properties of standard steel sections, the following publication is generally used.
- **a.** Steel Fabricators' Manual
- **b.** Steel Construction Manual
- **c.** ASTM Standard on Steel Sections
- **d.** Local Manufacturers' Literature
- **e.** Steel Erectors' Manual

10. The acronym HSS stands for
- **a.** heat-strengthened steel.
- **b.** high-strength steel.
- **c.** high-strength section.
- **d.** hollow steel section.
- **e.** none of the above.

11. Structural steel shapes, such as W, C, and L, are made by
- **a.** bending and welding heated steel plates.
- **b.** cold rolling of steel blooms.
- **c.** hot rolling of steel blooms.
- **d.** casting molten steel to shape.
- **e.** any one of the above, depending on the manufacturer.

12. HSS members are made by
- **a.** bending and welding heated steel plates.
- **b.** cold rolling of steel blooms.
- **c.** hot rolling of steel blooms.
- **d.** casting molten steel to shape.
- **e.** any one of the above, depending on the manufacturer.

Answers: 1-c, 2-a, 3-c, 4-a, 5-a, 6-a, 7-d, 8-d, 9-b, 10-e, 11-c, 12-a.

16.3 STEEL JOISTS AND JOIST GIRDERS

In addition to the standard steel shapes and built-up sections, two types of prefabricated steel members are commonly used for roof and floor structures in buildings. These are trusslike, open-web members, called *joists* and *joist girders*. There is no fundamental difference between a joist and a joist girder, except that a joist girder is a heavier member and spans from column to column, whereas a joist is a lighter member that spans between the girders, Figure 16.11.

Joists and joist girders are fabricated by manufacturers according to the specifications prepared by the Steel Joist Institute (SJI). They are generally recommended for use in light, uniformly loaded roofs and floors.

STEEL JOISTS

The Steel Joist Institute classifies joists in three categories:

- K-series joists (joist depth ranges from 8 in. to 30 in.)
- LH-series joists (joist depth ranges from 18 in. to 48 in.)
- DLH-series (joist depth ranges from 52 in. to 72 in.)

K-series joists are more frequently used because they fall within the range of loads and spans that are common in buildings. Manufacturers producing K-series joists must comply with SJI's specifications, which, for the most part, are performance-oriented, giving a measure of freedom to manufacturers to select the members' sizes and shapes. However, a joist bearing the standard SJI designation must have a minimum specified load-carrying capacity.

Figure 16.12 shows a typical elevation, section, and other important details of a K-series joist. Each joist bears a designation, consisting of two numbers sandwiching the letter *K*.

FIGURE 16.11 A steel frame structure showing the use of steel joists and steel joist girders. Observe that the joists span from joist girder to joist girder. The joists bear on the girders only through their top chords (Figure 16.12**(e)** and **(f)**).

The first number gives the depth of joists in inches and the second number gives the relative stiffness of the joist. The designation system is further explained in Figure 16.12.

The load-carrying capacities of various K-series joists are tabulated as shown in Table 16.2. Thus, in practice, the joists are not structurally designed but are simply selected from SJI's load tables based on the dead load and live load that they are required to support.

LH- and DLH-series joists are designated similar to K-series joists, for example, 24LH10 and 56DLH16. The maximum allowable span of a given joist (regardless of the joist series)

NOTE

Joist Series' Nomenclature

Until 1986, SJI had three series of joists: H-series, LH-series and DLH-series. H-series joists (ranging from 8 in. to 30 in. deep) were most frequently used because their depths and load-carrying capacities were within the range commonly required. LH-series and DLH-series joists, referred to as *long-span* and *deep long-span* joists, respectively, were intended for spans and loads not covered by H-series joists.

In 1986, SJI introduced K-series joists, which replaced the earlier H-series joists primarily because of the economy resulting from the improved design concepts.

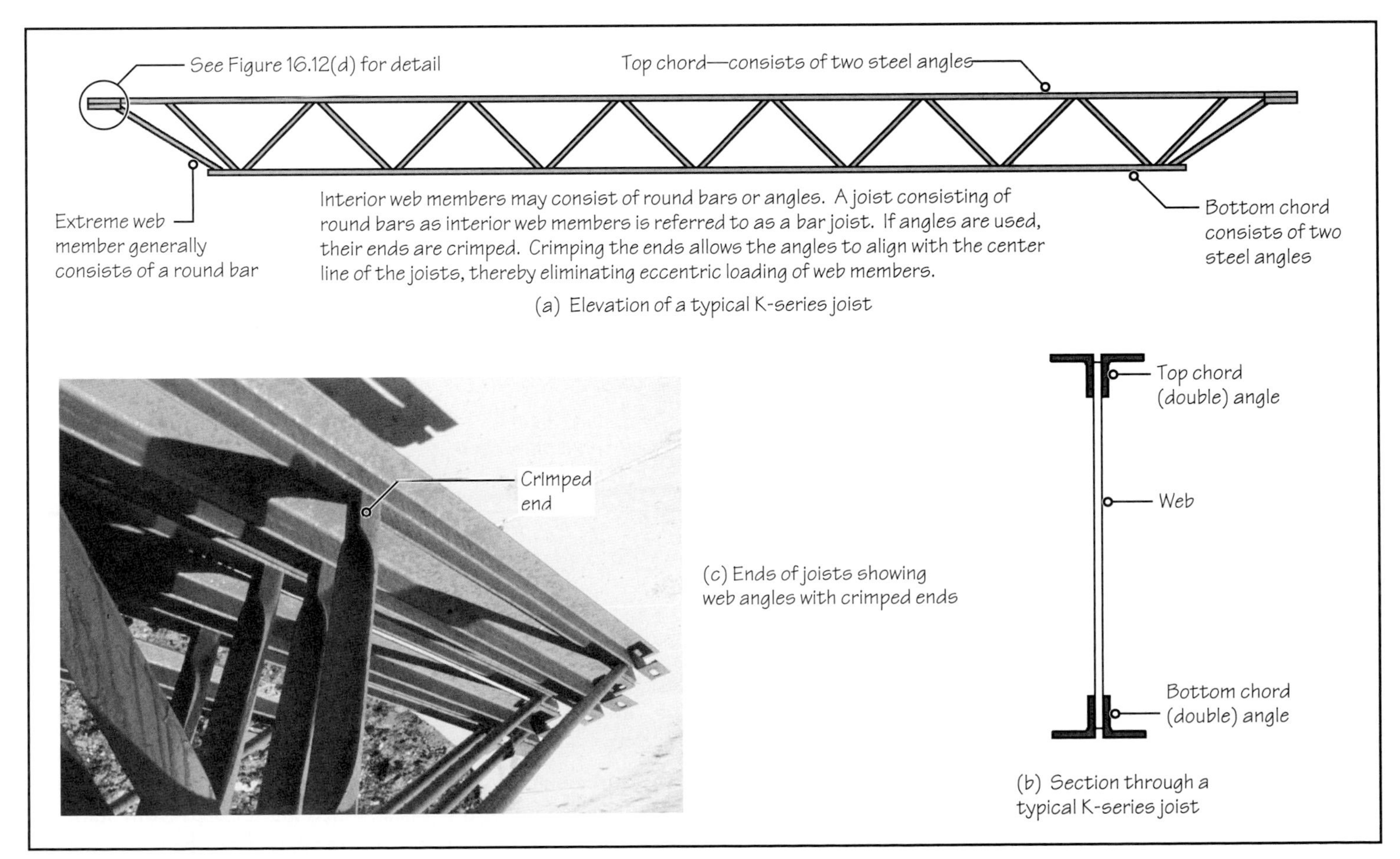

FIGURE 16.12 Essential details of K-series joists.

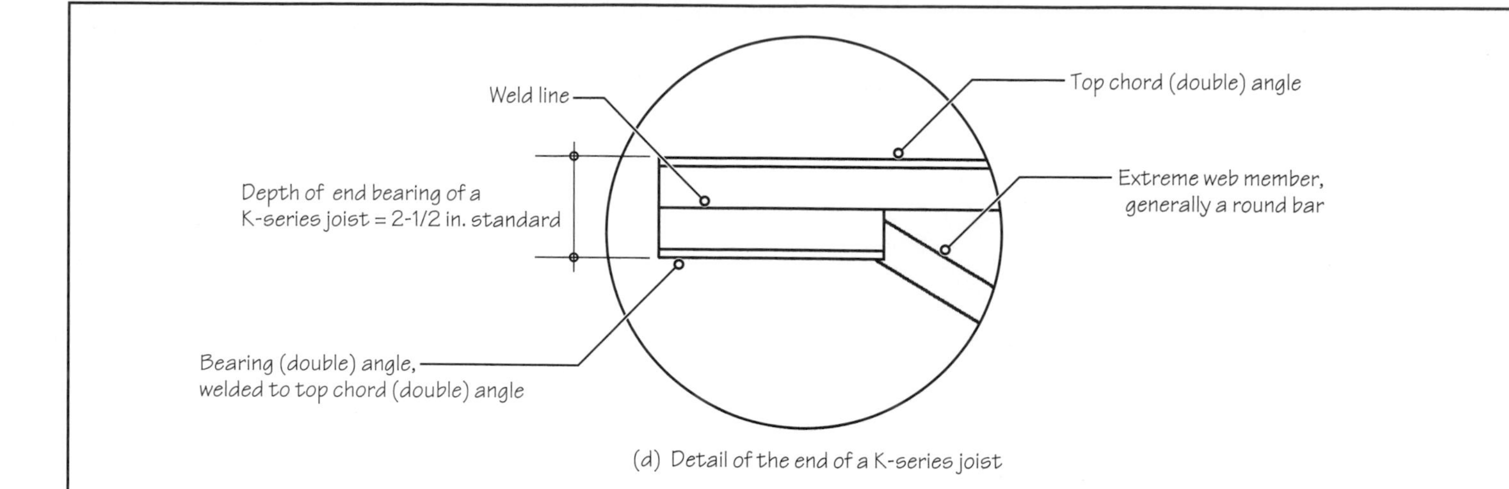

(d) Detail of the end of a K-series joist

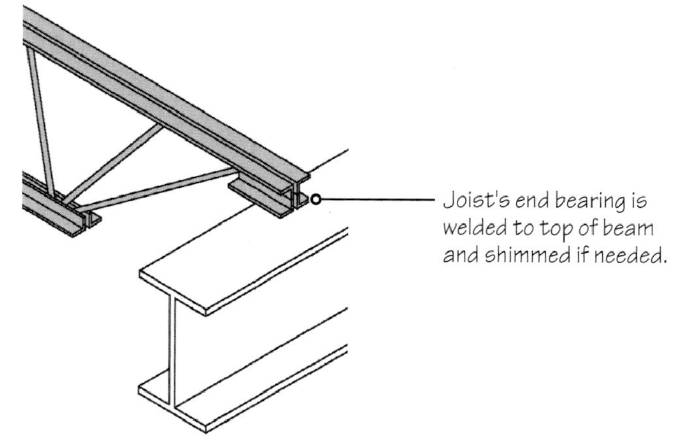

(e) Support detail of a K-series joist bearing on a wide-flange beam. Bearing of joist on a joist girder is similar.

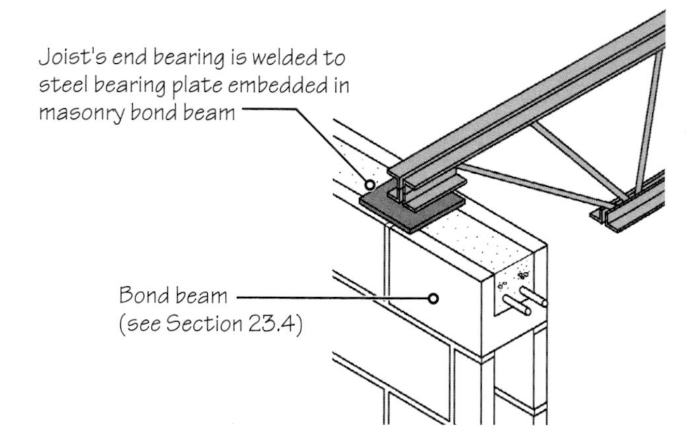

(f) Support detail of a K-series joist bearing on a masonry wall. Detail of joist bearing on a reinforced concrete wall is similar.

K-SERIES JOIST DESIGNATION

K-series joists are designated by two numbers one on each side of letter "K". The left side number gives the depth of the joist in inches and varies from 8 to 30 in 2 in. increments, i.e., the left side numbers can be 8, 10, 12, ... 30. Thus, the minimum depth of a K-series joist is 8 in. and the maximum depth is 30 in. For example, the depth of joist, designated as 16K4, is 16 in.

The right side number gives the relative weight of the joist and varies from 1 to 12 in steps of 1, i.e., the right side numbers can be 1, 2, 3, ... 12. For the same joist depth, the larger the right side number, the heavier the joist. Thus, a 16K5 joist is heavier (and, therefore, being of the same depth is stiffer) than a 16K4 joist and has a greater load carrying capacity.

HEIGHT OF END BEARING OF K-SERIES JOISTS

The height and depth of end bearing is 2-1/2 in. for all K-series joists.

TYPICAL SPACING OF K-SERIES JOISTS

Joist spacing must be determined by the engineer or achitect depending on the load-carrying capacity of floor or roof deck. It generally varies from 2 to 4 ft for floors and 4 to 6 ft for roofs .

MAXIMUM SPAN OF K-SERIES JOISTS

As Table 16.2 shows, the maximum allowable span of a joist is 24 times its depth. Thus, a 10 in. deep joist can be used for a maximum span of 240 in. = 20 ft.

FIGURE 16.12 *(Continued)* Essential details of K-series joists.

is 24 times the depth of the joist. For example, the maximum allowable span of joists designated as 24K4, 24K5, 24K6, and so on, is 48 ft.

Steel Joist Girders

As shown in Figure 16.11, joist girders are used as primary structural elements as an economical alternative to W-shape beams. Unlike the joists, which are selected from standard SJI's load tables, joist girders are designed by the manufacturers based on the specifications provided by the engineer.

TABLE 16.2 STANDARD LOAD TABLE OF SELECTED K-SERIES JOISTS									
Joist Designation	**8K1**	**10K1**	**12K1**	**12K3**	**12K5**	**14K1**	**14K3**	**14K4**	**14K6**
Depth (in.)	**8**	**10**	**12**	**12**	**12**	**14**	**14**	**14**	**14**
Approx. weight (lb/ft)	**5.1**	**5.0**	**5.0**	**5.7**	**7.1**	**5.2**	**6.0**	**6.7**	**7.7**
Span (ft)									
8	550								
	550								
9	550								
	550								
10	550	550							
	480	**550**							
11	532	550							
	377	**542**							
12	444	550	550	550	550				
	288	**455**	**550**	**550**	**550**				
13	377	479	550	550	550				
	225	**363**	**510**	**510**	**510**				
14	324	412	500	550	550	550	550	550	550
	179	**289**	**425**	**463**	**463**	**550**	**550**	**550**	**550**
15	281	358	434	543	550	511	550	550	550
	145	**234**	**344**	**428**	**434**	**475**	**507**	**507**	**507**
16	246	313	380	476	550	448	550	550	550
	119	**192**	**282**	**351**	**396**	**390**	**467**	**467**	**467**
17		277	336	420	550	395	495	550	550
		159	**234**	**291**	**366**	**324**	**404**	**443**	**443**
18		246	299	374	507	352	441	530	550
		134	**197**	**245**	**317**	**272**	**339**	**397**	**408**
19		221	268	335	454	315	395	475	550
		113	**167**	**207**	**269**	**230**	**287**	**336**	**383**
20		199	241	302	409	284	356	428	525
		97	**142**	**177**	**230**	**197**	**246**	**287**	**347**
21			218	273	370	257	322	388	475
			123	**153**	**198**	**170**	**212**	**248**	**299**
22			199	249	337	234	293	353	432
			106	**132**	**172**	**147**	**184**	**215**	**259**
23			181	227	308	214	268	322	395
			93	**116**	**150**	**128**	**160**	**188**	**226**
24			166	208	282	196	245	295	362
			81	**101**	**132**	**113**	**141**	**165**	**199**
25						180	226	272	334
						100	**124**	**145**	**175**
26						166	209	251	308
						88	**110**	**129**	**156**
27						154	193	233	285
						79	**98**	**115**	**139**
28						143	180	216	265
						70	**88**	**103**	**124**

The normal-weight numbers represent the total safe uniform load-carrying capacity of joists in pound/foot. Dead loads, including the self-load of joists, must be deducted to determine the joist's live-load capacity. Bold numbers represent the live load (in pound/foot) that will produce a deflection in joist equal to joist span divided by 360.

This table has been adapted from *Standard Specifications Load Tables & Weight Tables for Steel Joists and Joist Girders*, published by the Steel Joist Institute (SJI), with permission. For a complete version of the table, refer to this publication.

Bracing Joists and Joist Girders Against Instability

Steel joists and joist girders are slender elements and are, therefore, unstable and prone to overturning. As per SJI's specifications, the joists must be stabilized by rows of continuous horizontal members, referred to as *horizontal bridging members*. Horizontal bridging members are used in rows: One row is welded to the top chord of the joist and the other row is welded to the bottom chord, Figure 16.13.

(a) Steel frame structure with wide-flange beams as primary structural elements and steel joists as secondary elements. The steel deck (not shown here) forms the tertiary element—the three elements together typically comprise the floor or the roof structure in a steel frame building. In Figure 16.11, the primary floor elements are the joist girders. In many steel frame buildings, wide-flange beams comprise both primary and secondary elements of the floor.

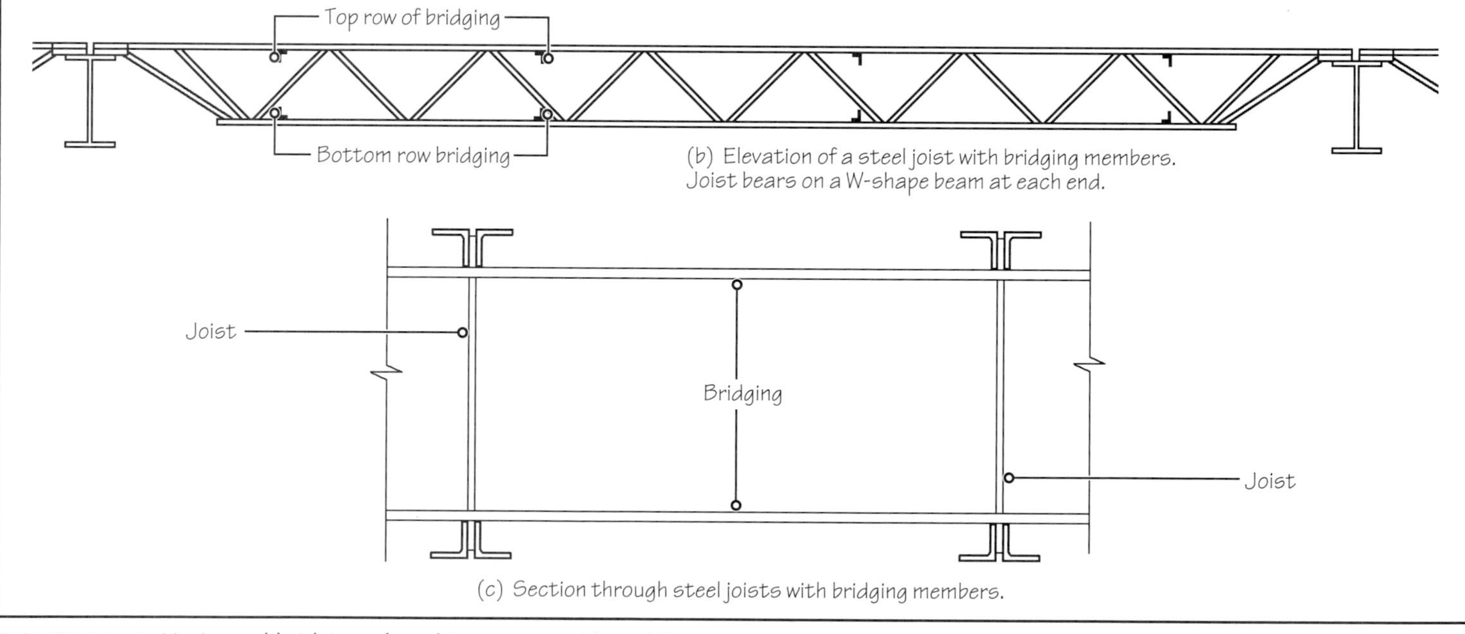

(b) Elevation of a steel joist with bridging members. Joist bears on a W-shape beam at each end.

(c) Section through steel joists with bridging members.

FIGURE 16.13 Horizontal bridging of steel joists to provide stability against overturning.

Generally, the bridging members are $1\frac{1}{4}$ in. $\times$ $1\frac{1}{4}$ in. angles but must meet the stiffness and strength requirements of SJI's specifications. An alternative to horizontal bridging is *diagonal bridging,* as shown in Figure 16.14. Regardless of which bridging alternative is used, it is important that the end spans of bridging be securely connected to the wall or spandrel beam. The spacing between the rows of bridging is also specified by SJI. Bridging is particularly important in roofs to prevent the buckling of a joist's bottom chord that comes under compression due to the uplift from the wind loads on the roof.

Unlike joists, the joist girders are stabilized by connecting their lower chords to the columns (Figure 16.11). This connection may also be designed to provide lateral-load (racking) resistance to the structural frame of the building.

EXPAND YOUR KNOWLEDGE

Steel Joists and Joist Girders Compared with W-Shapes

Advantages of Steel Joists and Joist Girders

Lighter Weight

As compared with W-shape beams, steel joists and joist girders are generally an economical alternative for roof and floor structure. The reason is that for a given span and floor/roof loads, the structural solution based on the use of joists and joist girders is lighter in weight than that obtained by using W-shape beams. A lighter-weight floor/roof structure reduces not only its own cost, but also the cost of columns and foundations.

Easier Erection

Because of their bearing-type support, steel joists are much easier and faster to erect and install than W-shape beams. W-shape beams use shear connections, which are more complex and take longer to erect. Joists also allow panelized erection, in which a number of parallel joists can be connected together with bridging members and decking and the entire assembly lifted to position.

Limitations of Steel Joists and Joist Girders

Greater Floor-to-Floor Height

The depth of a floor using W-shapes is generally smaller than that using steel joists. This is particularly helpful in a high-rise building, where even a small reduction in floor-to-floor height can translate into a substantial decrease in the overall height of the building and, hence, in the magnitude of lateral loads.

However, the open webs of steel joists allow the passage of electrical conduits, pipes, and air-conditioning ducts within the depth of the floor. Thus, although the depth of the floor using joists and joist girders may be greater than that using W-shape beams, the floor-to-floor height may not be. (Because of their open webs, joists and joist girders are also referred to as *open-web joists* and *open-web joist girders*.)

Concentrated Loads

Although the joists can be designed to support concentrated loads, they are less forgiving than W-shape beams if the location of concentrated load changes from that of its designed location. Joist girders, on the other hand, are required to support concentrated loads, particularly from the joists bearing on them.

Complex Layout of Floor and Roof Framing

Steel joists are economical for a floor or roof that is rectangular in plan, so that the layout of joists is uniform and rhythmic. They are generally not economical for a floor or roof that has nonrectangular bays with different spans for adjacent joists.

Fire Protection

Steel joists are more closely spaced and have a greater surface area than W-shape beams. Therefore, they require a larger amount of spray-on fire-protection material.

FIGURE 16.14 Diagonal bridging of steel joists is an alternative to horizontal bridging. Regardless of the bridging system used, the bridging members in the end spans must be securely connected to the exterior wall (as shown here) or the spandrel beam.

16.4 STEEL ROOF AND FLOOR DECKS

Steel decks are are made from sheet steel. All sheet-steel components are fabricated by rolling and pressing the sheets into various cross-sectional profiles at room temperature. This process is known as *cold forming* (in contrast to hot rolling, which is used in producing structural steel sections such as the W-, C-, or L-shapes). Hot rolling cannot produce sheets thin and flat enough to match the cold rolling process.

Steel decks are available in two categories: (a) *roof decks* and (b) *floor decks*. The primary difference between them is that a roof deck is generally topped with rigid insulation and roof cover, and a floor deck is topped with structural concrete fill. The terminology used to describe the deck profile is given in Figure 16.15.

The thickness of a sheet steel used in decks is expressed by its gauge number. The higher the gauge number, the thinner the sheet. Table 16.3 gives the U.S. Standard sheet-steel gauge number and the corresponding sheet thickness. Roof and floor decks are generally painted or galvanized for protection against corrosion.

Roof Decks

Commonly used steel roof decks profiles, depths, and gauges are shown in Figure 16.16. Roof decks ranging from $1\frac{1}{2}$ to $7\frac{1}{2}$ in. in depth are available. The greater the depth and thicker the gauge, the greater the spanning capability of the deck.

NOTE

Concrete Fill on Roof Decks

Some roof decks are topped with lightweight insulating concrete instead of rigid board insulation (Chapters 5 and 31).

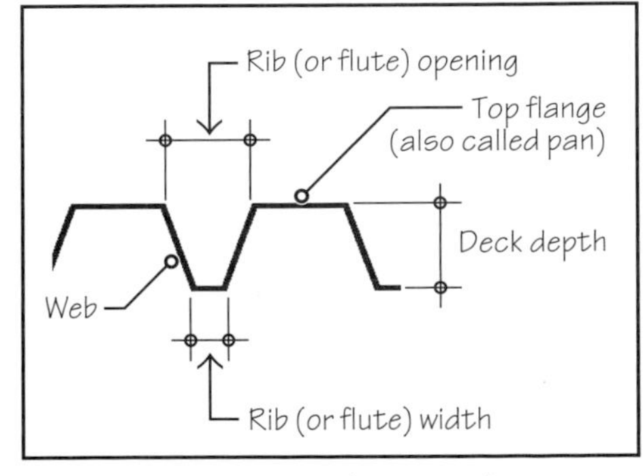

FIGURE 16.15 Deck terminology.

TABLE 16.3 SHEET-STEEL GAUGE NUMBERS AND CORRESPONDING THICKNESSES AS USED IN ROOF AND FLOOR DECKS

Gauge number	Thickness (in.)	Thickness (mm)	Gauge number	Thickness (in.)	Thickness (mm)
14	0.0747	1.88	20	0.0359	0.89
15	0.0673	1.70	21	0.0329	0.82
16	0.0598	1.50	22	0.0299	0.75
17	0.0538	1.35	23	0.0269	0.67
18	0.0478	1.19	24	0.0239	0.60
19	0.0418	1.04	25	0.0209	0.52

This is an incomplete table. Gauge numbers lower than 14 and higher than 25 are a part of the complete table.

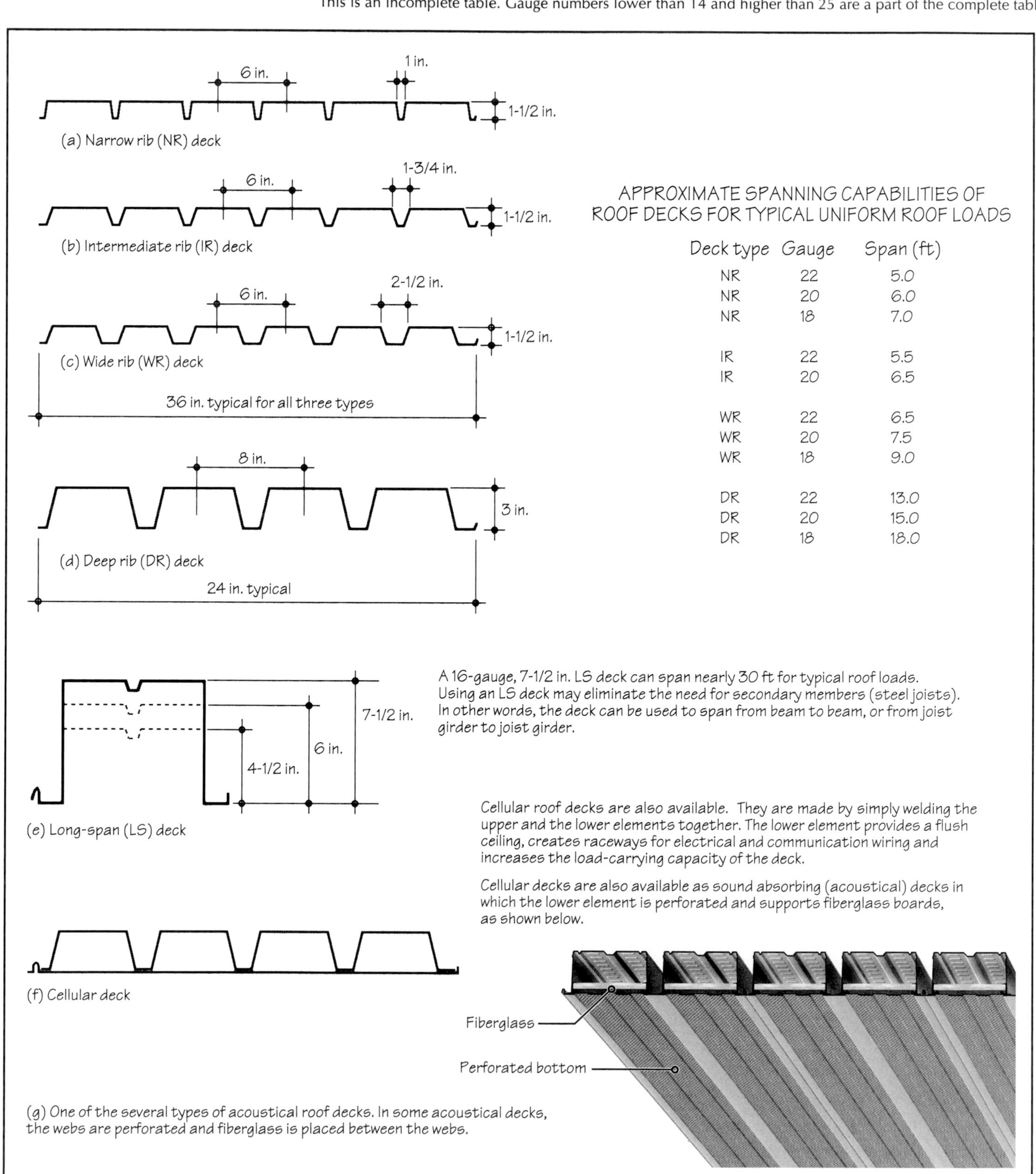

FIGURE 16.16 Commonly used roof deck types.

Because of their extensive use, $1\frac{1}{2}$-in. deep decks are available in three rib widths: narrow rib deck, intermediate rib deck, and wide rib deck. The wider the ribs, the greater the load-carrying capacity of the deck. Thus, for the same sheet thickness and deck depth, a wide rib deck has a greater load-carrying capacity than an intermediate rib deck, which in turn has a greater load-carrying capacity than a narrow rib deck.

The width of a deck panel varies from 24 to 36 in., depending on the manufacturer and the type of deck. The length of the panel is usually not fixed, and the manufacturer generally provides whatever length is specified. However, there are practical limitations on panel length, determined by transportation constraints and the ability of workers to handle long panels at the job site. Panels longer than 40 ft (12 m) are cumbersome to handle.

Steel roof decks are commonly manufactured from 22-gauge to 18-gauge sheet steel. The fastener (weld) retention of a deck increases significantly with an increase in deck thickness. Because roof decks are subjected to wind uplift, the roof deck's gauge may be governed by wind-uplift considerations rather than its capacity to carry gravity loads.

Roof decks are available with a paint finish, hot-dip galvanized, or simply prime painted. Priming of the deck gives only temporary protection. In corrosive or high-moisture environments (e.g., coastal areas), galvanized or galvanized and painted decks should be used.

Roof decks are usually anchored to supporting elements (joists or wide-flange beams) by puddle (spot) welds, Figures 16.17(a) and (b). Weld areas are field painted. As an alternative to welding, powder actuated fasteners or mechanical fasteners may be used.

Individual deck panels must be connected together at side laps, that is, along the length of the panel. This prevents differential deflection between adjacent panels and, therefore, prevents side joints from opening. Depending on the design of the deck profile, side-lap connection may be provided by screws or welds.

FLOOR DECKS

A steel floor deck functions as a working platform as well as permanent formwork for the concrete fill, so that the deck and the fill form a concrete floor slab. Two types of floor decks are used:

- In a *composite deck,* the deck also functions as steel reinforcement for the slab, reducing the need for conventional concrete reinforcing bars.

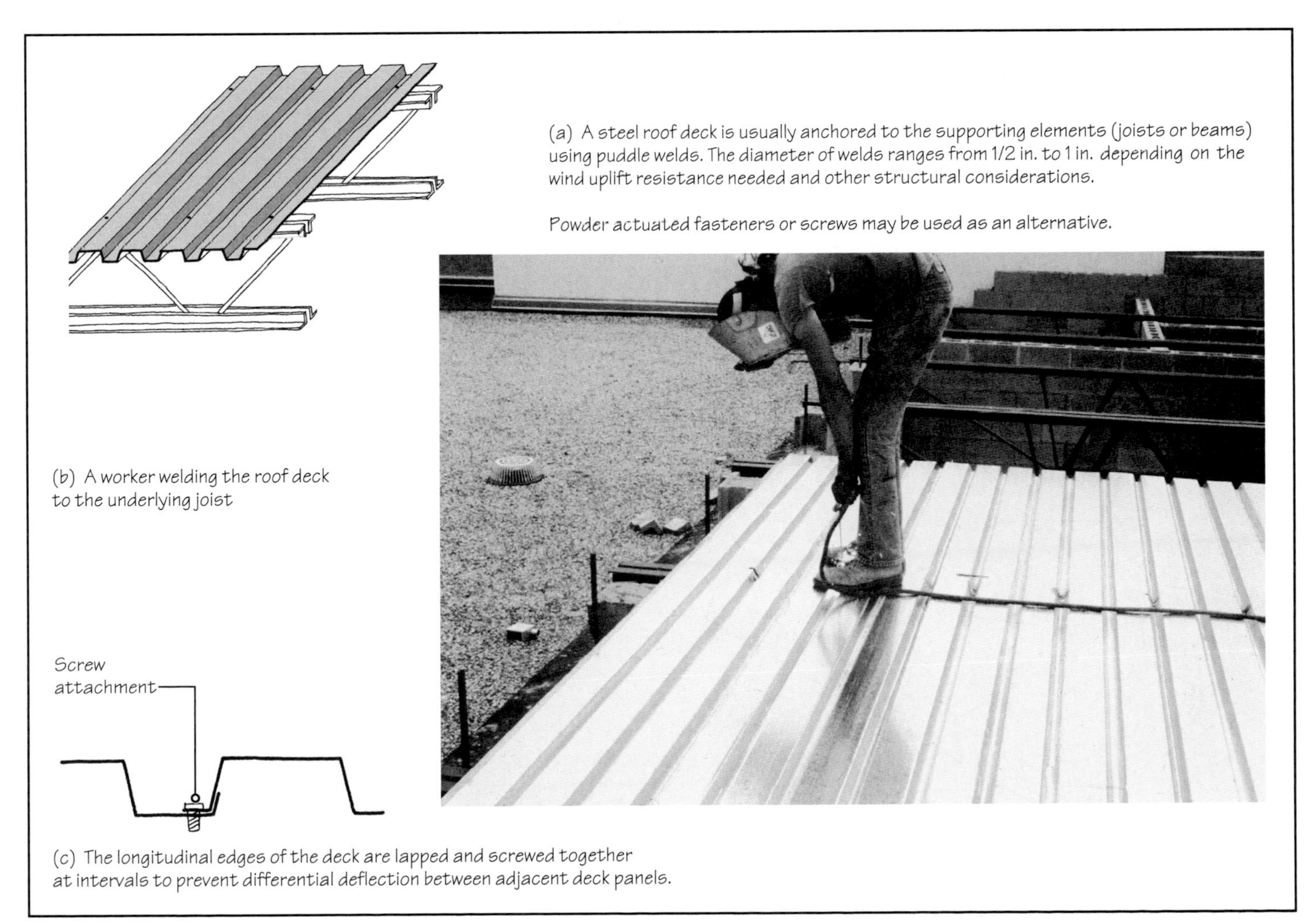

FIGURE 16.17 Anchorage of a roof deck to supporting elements.

- A *form deck* functions as a permanent formwork only. Thus, the slab must be reinforced with conventional reinforcement.

The surface of a composite floor deck is designed so that the deck bonds integrally with the overlying concrete fill. Protrusions and depressions are, therefore, stamped on the deck surface, Figure 16.18(a). Alternatively, the deck may be profiled with dovetail flutes, as shown in Figure 16.18(b), to provide the composite action.

Because a composite deck provides permanent reinforcement, the deck must be protected from corrosion. Two alternatives are generally used. The better alternative is a hot-dip galvanized finish. Another alternative is a phosphatized/painted finish, in which the exposed deck surface is painted and the unexposed surface is phosphatized. The minimum concrete cover above the deck is 2 in. A composite deck is not recommended for use in parking garages in cold regions where salt spread is used on the roads.

Because a form deck supports only the weight of wet concrete (no floor live load) and does not function as concrete reinforcement, it is usually much shallower and of a thinner gauge than a composite deck. A composite deck, on the other hand, must support the dead load of wet concrete, and when the concrete has hardened, the deck and the concrete together must support the entire load on the deck. A comparison of the typical profiles of composite and form decks is shown in Figure 16.19.

The anchorage of a floor deck to supporting joists or beams is similar to that of a roof deck, i.e., through puddle welds. The connection of the longitudinal (side) edges of adjacent deck panels is also similar.

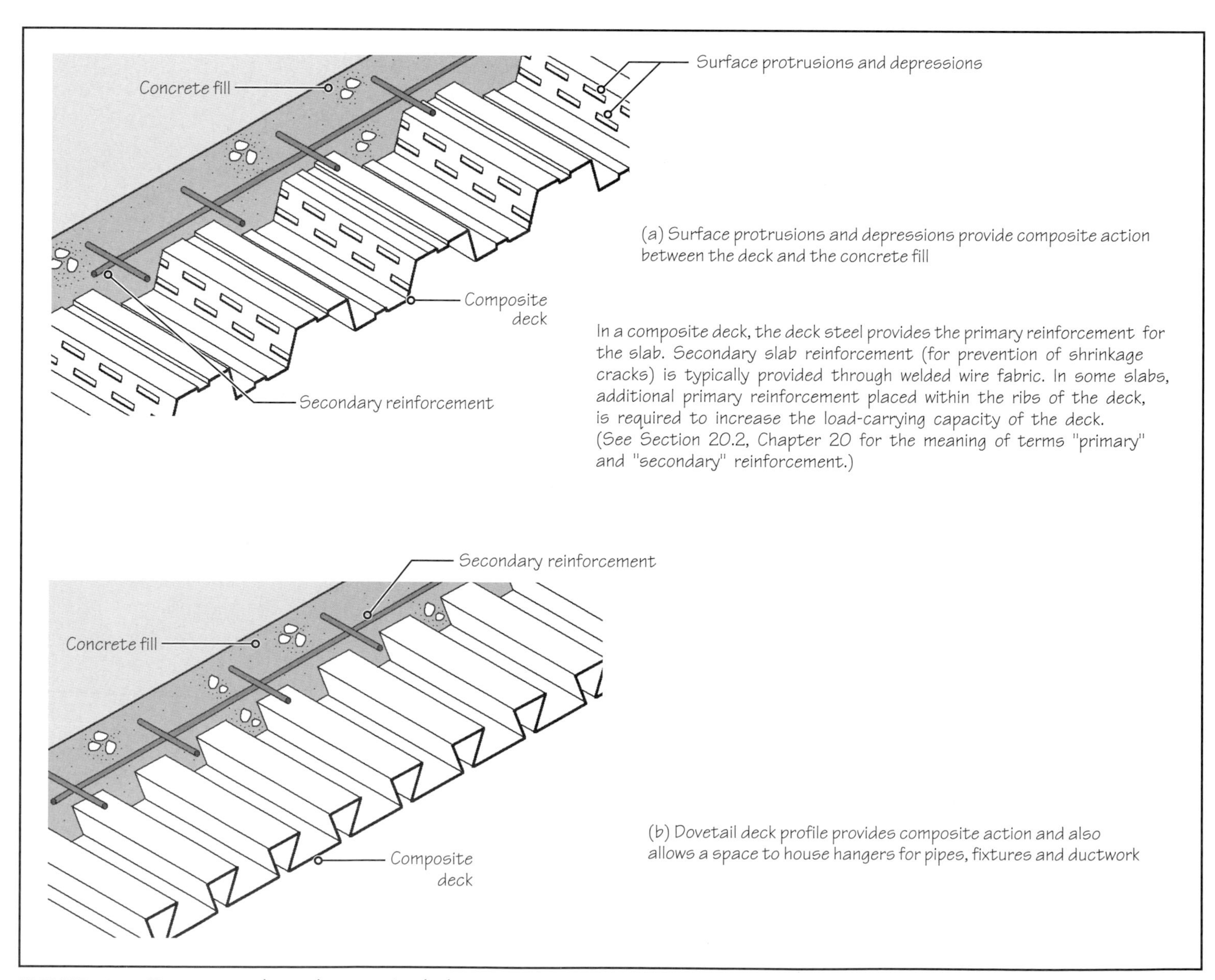

FIGURE 16.18 Two commonly used composite deck types.

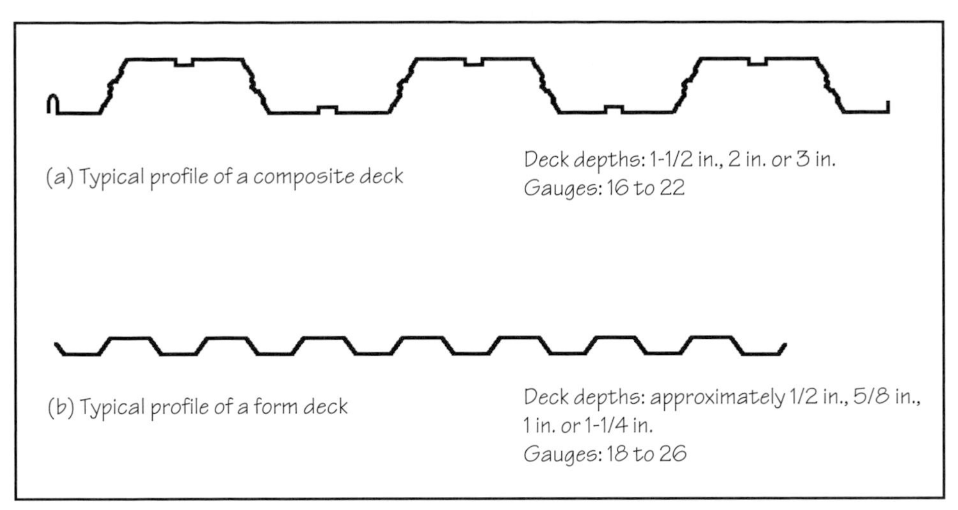

FIGURE 16.19 Comparison between the general profiles, depths, and gauges of composite decks and form decks.

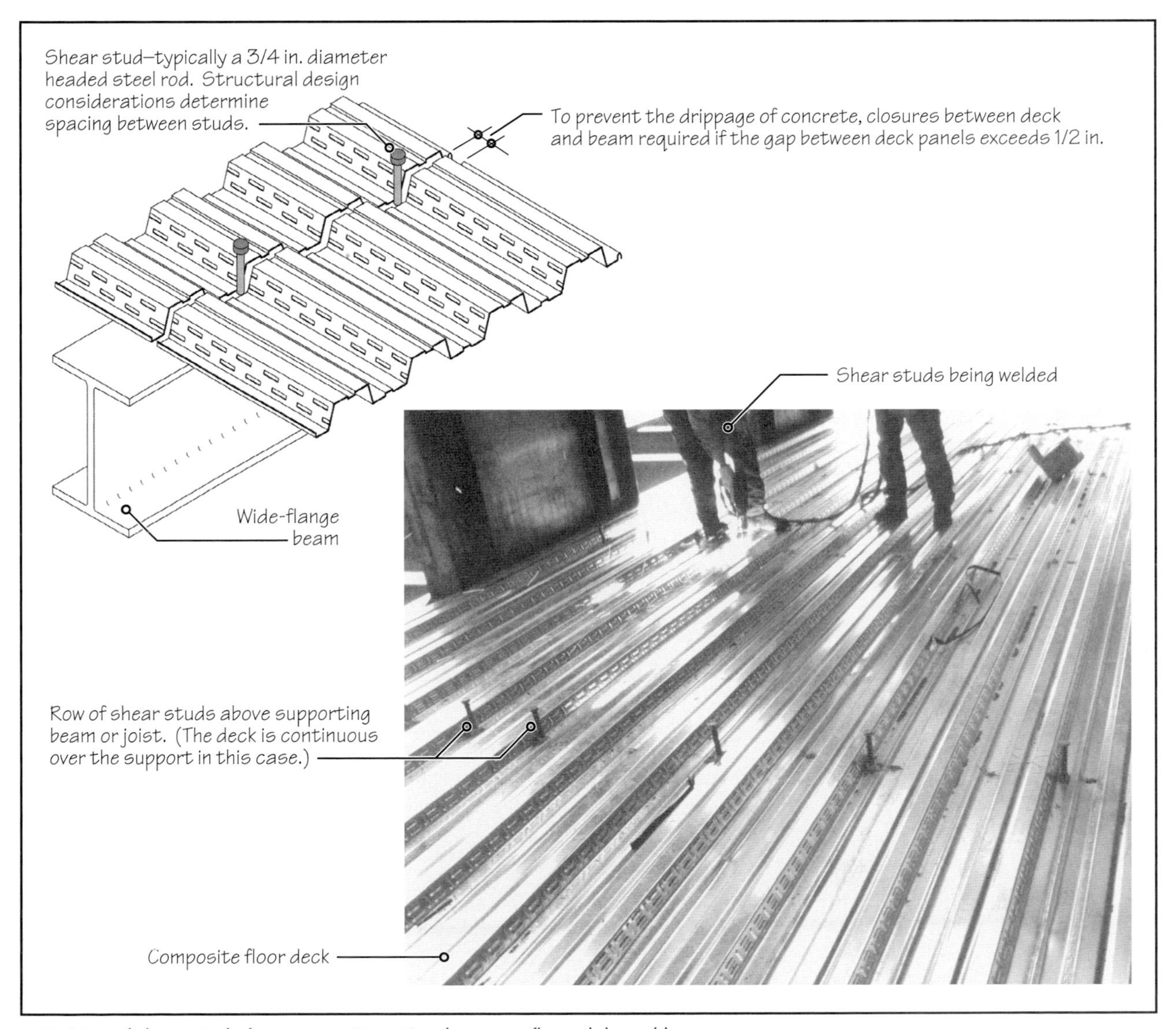

FIGURE 16.20 Use of shear studs for composite action between floor slab and beam.

Composite Action Between Floor Slab and Supporting Members

Both a composite deck and a form deck can be made to act compositely with the supporting beams by using shear studs, Figure 16.20. Shear studs prevent slippage of the deck under bending of the underlying beam. They are similar to nails that connect a plywood deck with supporting wood joists.

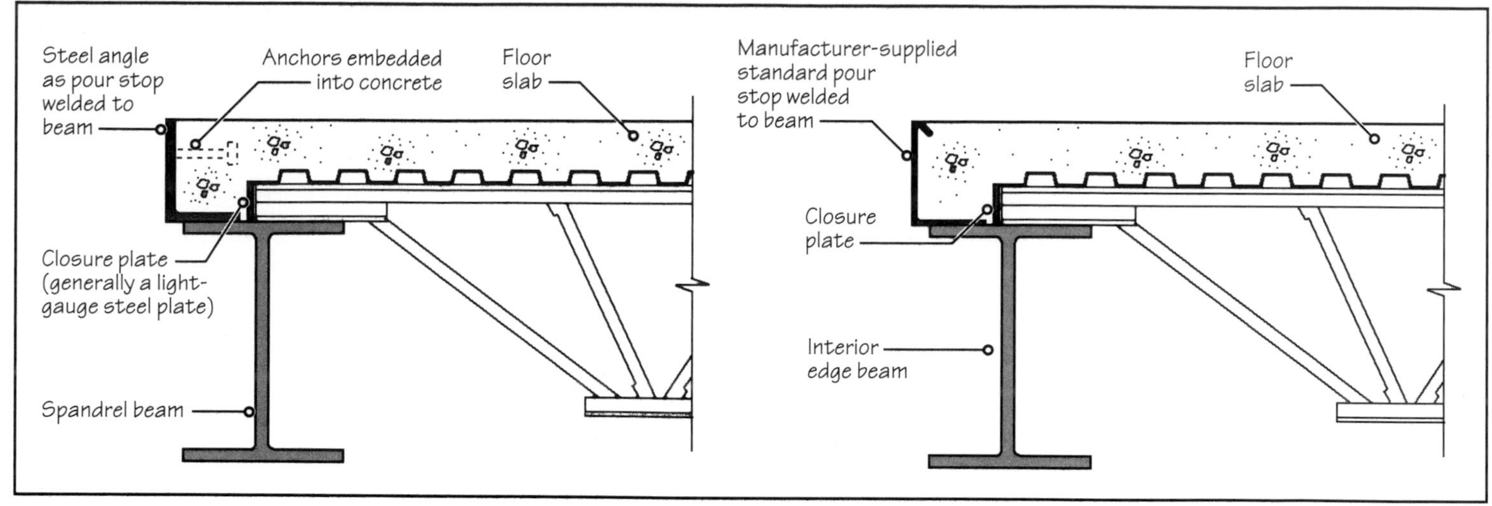

FIGURE 16.21 The use of pour stops at spandrel beams or interior mezannine floor beams.

POUR STOPS AT SPANDREL BEAMS AND EDGES OF FLOOR OPENINGS

To terminate a concrete pour at the extreme edges of a floor deck, pour stops are used. A steel angle or a manufacturer-supplied standard pour stop is used for the purpose, Figure 16.21. If the pour stop is used for anchoring another building element, such as a glass curtain wall, the structural capacity of the pour stop should be examined.

PRACTICE QUIZ

Each question has only one correct answer. Select the choice that best answers the question.

13. Which of the following does not represent a steel joist series?
- **a.** H-series
- **b.** J-series
- **c.** K-series
- **d.** LH-series
- **e.** DLH-series

14. In the steel joist designation 24LH10, the number 24 gives the
- **a.** length of the joist in feet.
- **b.** depth of the joist in inches.
- **c.** weight of the joist in pounds per foot.
- **d.** number of web members.

15. A K-series steel joist that is 20 in. deep cannot be used to span more than
- **a.** 20 ft.
- **b.** 25 ft.
- **c.** 30 ft.
- **d.** 35 ft.
- **e.** 40 ft.

16. Steel joists are braced against instability by
- **a.** horizontal bridging members at the bottom of the joists.
- **b.** horizontal bridging members at the top of the joists.
- **c.** (a) or (b).
- **d.** (a) and (b).
- **e.** none of the above.

17. Diagonal bridging may be used as an alternative to horizontal bridging to stabilize steel joists.
- **a.** True **b.** False

18. Steel joists and joist girders are best utilized in a building that has concentrated loads.
- **a.** True **b.** False

19. Steel roof decks are either form decks or composite decks.
- **a.** True **b.** False

20. Steel floor decks are either form decks or composite decks.
- **a.** True **b.** False

21. The thickness of a roof or floor deck is generally specified in terms of
- **a.** mils.
- **b.** fraction of an inch.
- **c.** gauge number.
- **d.** any one of the above.

22. Roof decks are anchored to supporting members by continuous welds along the support lines.
- **a.** True **b.** False

23. A pour stop is generally used with
- **a.** roof decks. **b.** floor decks.
- **c.** (a) and (b). **d.** none of the above.

Answers: 13-b, 14-b, 15-e, 16-d, 17-a, 18-b, 19-b, 20-a, 21-c, 22-b, 23-b.

16.5 PRELIMINARY LAYOUT OF FRAMING MEMBERS

An architect working on a steel frame building must integrate several architectural considerations along with framing considerations. The majority of buildings are rectilinear in plan. Therefore, establishing the number of bays in each direction and bay dimensions is generally the first step in preliminary design. This determines the location and center-to-center spacing of columns.

The next step is to establish the direction of primary and secondary floor and roof-framing elements and prepare preliminary framing plans for each floor and roof. The primary framing

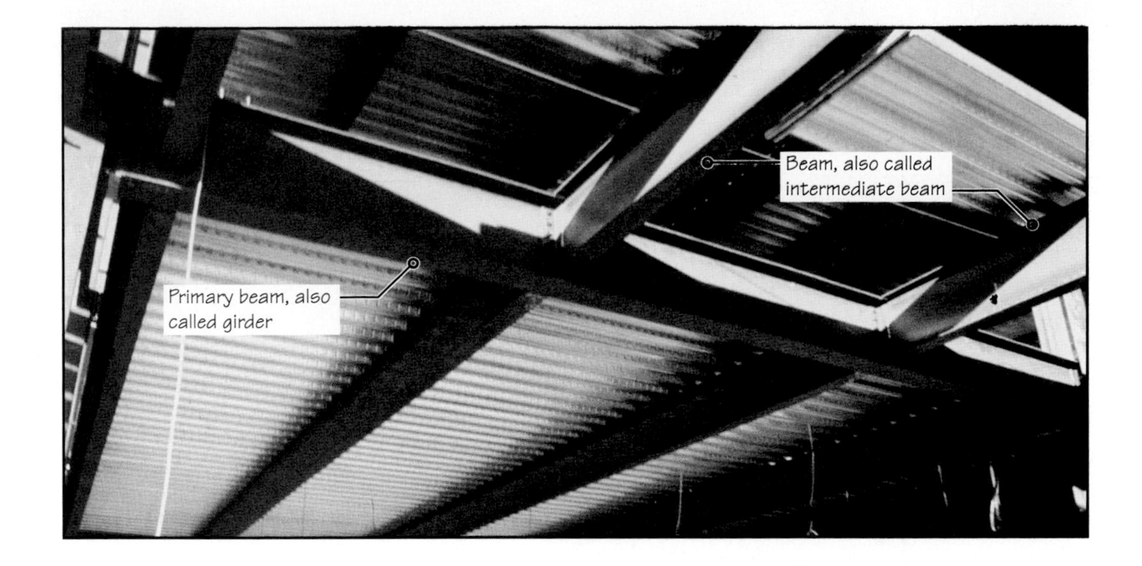

FIGURE 16.22 Underside of the floor of a building framed with W-section girders and beams.

elements support the secondary elements, which, in turn, support the floor or roof deck. In most buildings, the choice is generally between the use of W-sections or steel joists. Thus, the following variations are commonly used for roof and floor framing:

- W-sections for both primary and secondary elements, Figure 16.22. In this system, the primary elements are generally called *girders* and the secondary elements are called *beams* or *intermediate beams.*
- W-sections for primary elements and steel joists for secondary elements (Figure 16.13).
- Steel joist girders or trusses for primary elements and steel joists for secondary elements (Figure 16.11).

In some buildings, the secondary supporting elements may be omitted and the roof deck may be supported directly on the primary elements. Simultaneous with the preparation of the floor-framing plan, the architect (in consultation with the structural engineer) should determine whether to use a rigid frame, a braced frame, or a combination of the two systems.

A typical floor-framing plan of a low-rise rigid-frame building, using W-sections for both girders and intermediate beams, is shown in Figure 16.23. A version of the same framing

NOTE

A Note to the Reader

A reader not familiar with frame structures should study "Principles in Practice" at the end of this chapter.

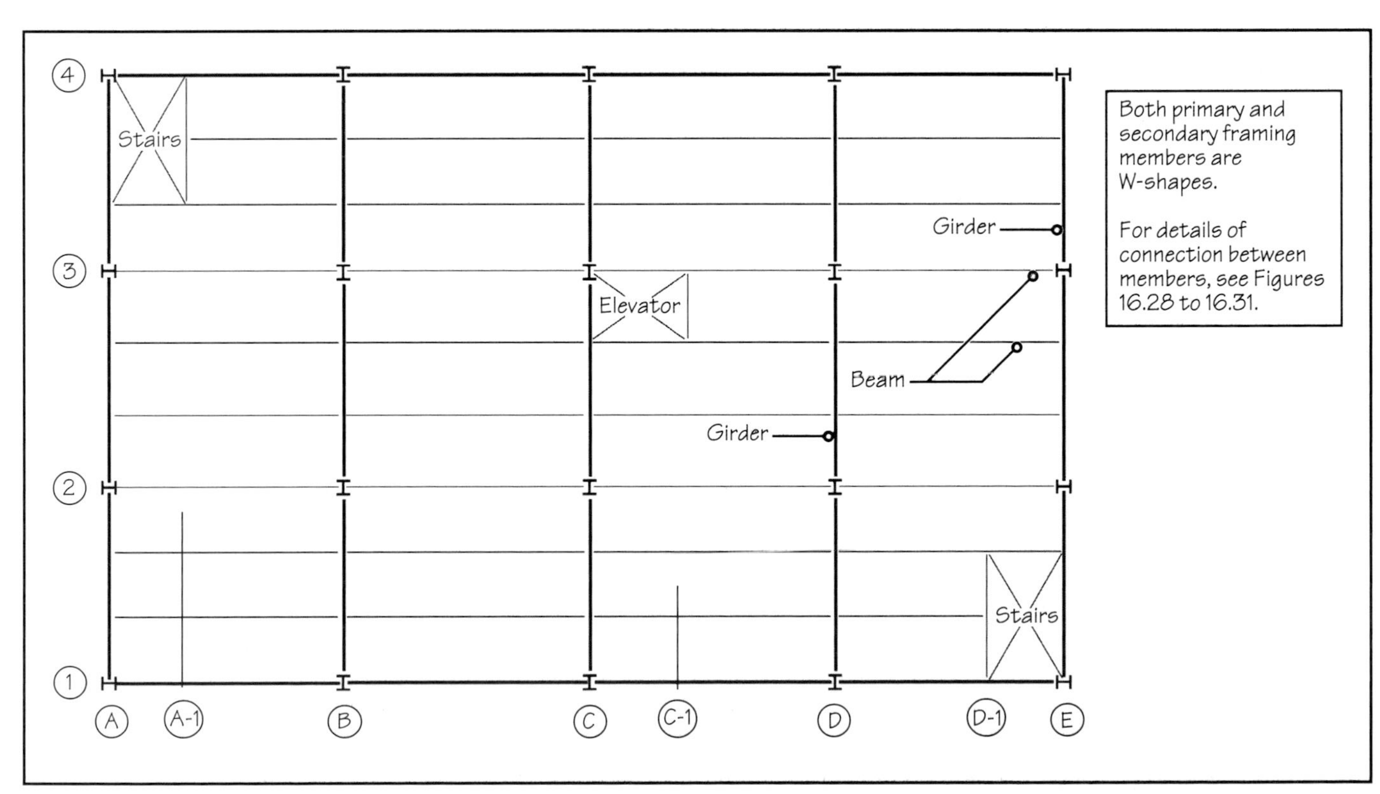

FIGURE 16.23 A typical floor-framing plan for a steel rigid frame building using W-sections for floor framing. Observe that column grid lines are identified by A, B, C, . . . , along one direction and 1, 2, 3, . . . , in the other direction. Rigid frames have been identified by heavy lines. Thus in this building, 3-bay rigid frames have been provided along grid lines A, B, C, D, and E, and 4-bay rigid frames, along grid lines 1 and 4. All other beam-column or beam-beam connections are simple (shear) connections. In a low-rise building, rigid frames along grid lines B, C, and D (or B and D) may be replaced by simple frames.

The columns along grid lines A and E have been turned through 90° to increase stiffness of frames along grid lines 1 to 4.

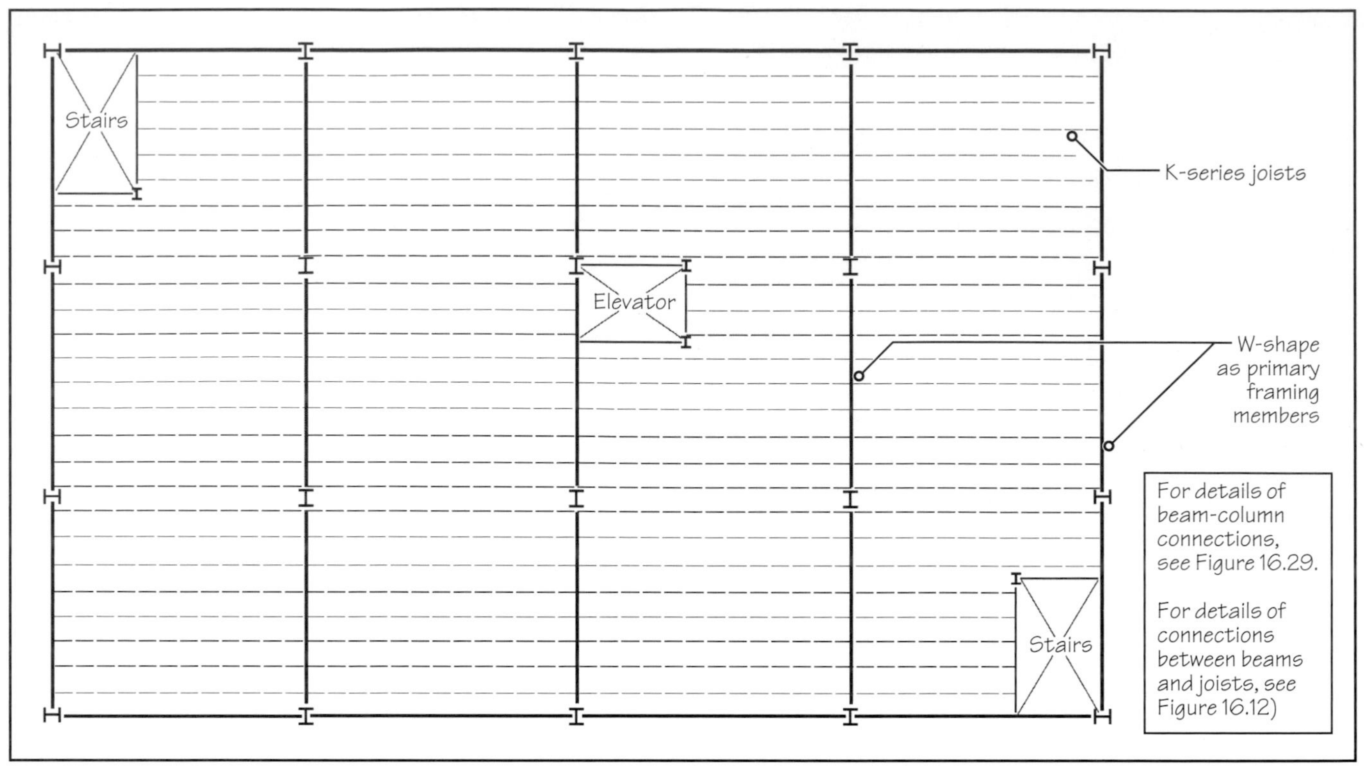

FIGURE 16.24 Floor framing plan of the building of Figure 16.23 with steel joists as intermediate members instead of W-section intermediate beams. Rigid frame lines are the same as in Figure 16.23.

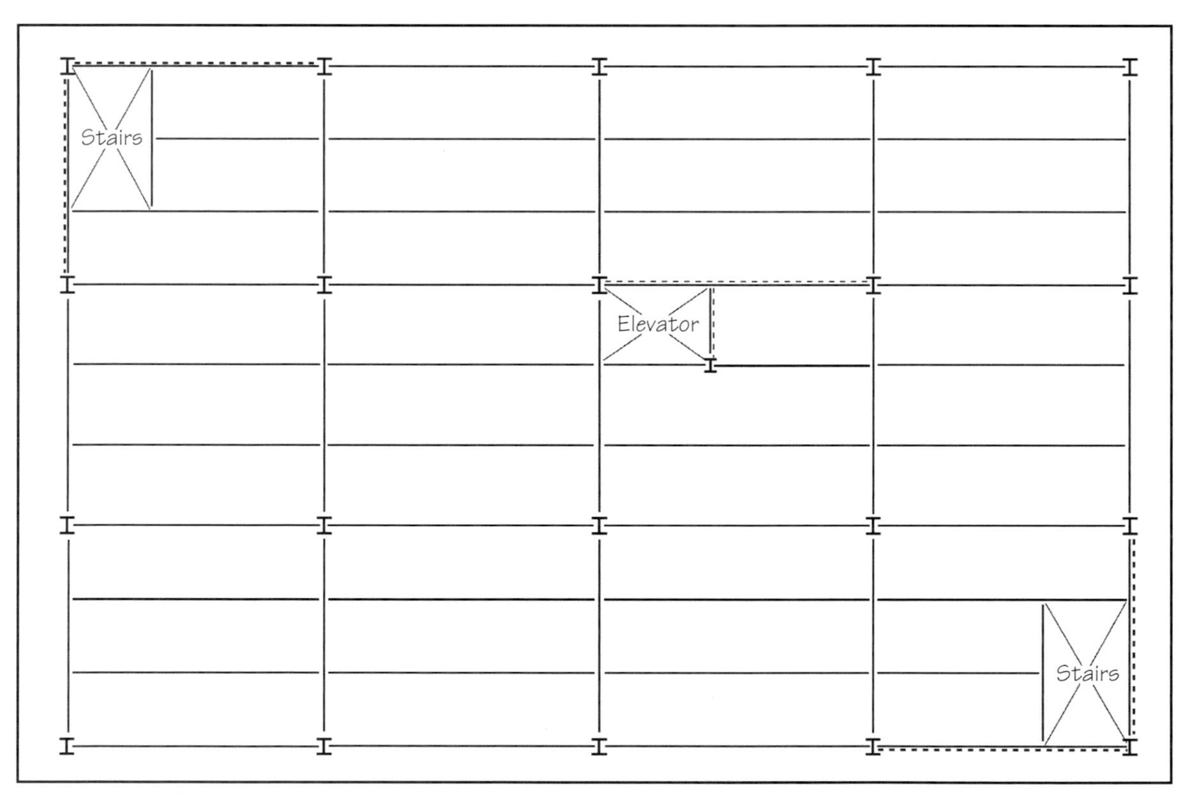

FIGURE 16.25 Framing plan of a typical braced frame building. Dashed lines indicate braced bays.

plan using steel joists as intermediate elements is shown in Figure 16.24. Figure 16.25 shows the framing plan of the same building using braced frames instead of rigid frames. (See "Principles in Practice" at the end of this chapter.)

The preliminary depth of floor- and roof-framing members should be established at this stage, in addition to the type of floor and roof decks to be used. This information helps to provide the approximate floor-to-floor height of the building. The exact dimensions of the framing elements are established after the detailed structural design of the building has been completed.

16.6 BOLTS AND WELDS

Connections between structural steel members can either be bolted or welded. Riveting, which was used extensively at one time, is no longer used.

FIGURE 16.26 A typical spud wrench for making snug-tight bolted connections.

Bolts

Steel bolts are of two types:

- *Unfinished* (or *common* or *ordinary*) *bolts* are made from carbon steel and generally have the same stress-strain characteristics as A36 steel. As per ASTM specifications, they are classified as A307 bolts. The use of A307 bolts has decreased significantly since the introduction of high-strength bolts.
- *High-strength bolts* are based on ASTM specification A325 or A490. A325 bolts are made from heat-treated carbon steel and have an approximate yield stress of 85 ksi. A490 bolts are made from a heat-treated steel alloy and have a yield strength of 120 ksi. A325 bolts are more commonly used because they cost less.

Snug-Tight Connection

For most structural steel connections, bolts are tightened to what is called *a snug-tight condition.* This condition is obtained when the connected members have been bolted together using a spud wrench with the full force of a person, Figure 16.26. The load transfer between members in a snug-tight connection (also called a *bearing connection*) occurs through shear in the bolts.

Slip-Critical Bolted Connection

Another commonly used bolted connection is a *slip-critical connection.* The bolts in a slip-critical connection are tightened to a (high) tensile stress so that shear resistance is provided through friction between the connected surfaces, not through bearing, as in a snug-tight connection. In a slip-critical connection, the bolts are first brought to a snug-tight condition. Subsequently, they are tightened further until the bolt shank is under a predetermined level of tensile stress.

There are four ways for tightening a bolt to obtain a predetermined level of tensile stress. The oldest method, which continues to be used even today, is the *turn-of-nut method.* In the turn-of-nut method, the bolt shank and nut are marked after the joint has been snugged. Then a specific amount of rotation is induced in the nut with respect to the bolt, generally from a $\frac{1}{3}$ to a full turn beyond the snug-tight condition.

The success of the turn-of-nut method depends on the correct snugging of the joint and also on ensuring that the bolt head will not turn on further tightening. Two persons are, therefore, mandatorily required to execute the tightening. One person restrains the bolt from turning and the other person turns the nut with help of a wrench. The turn-of-nut method cannot be used if the members are painted with a compressible paint.

Another commonly used method is the *twist-off bolt method.* In this method, a special electric wrench is used that induces a predetermined torque in the bolt. The torque induces the required tension, after which the splined bolt extension twists off the bolt, Figure 16.27. This connection can be made by only one person.

Two other methods of obtaining slip-critical connections are the *calibrated wrench* and *direct-tension indicator* (DTI) methods. Slip-critical connections are generally required where the members are subjected to excessive vibration and/or fatigue and where oversized holes are used for easier fit. They are more expensive than bearing-type connections. However, when made through the twist-off method, they make connection inspection and quality control considerably easier.

Spacing of Framing Members

The spacing of columns is generally a function of the architectural plan. If the dimensions of bays are unequal, the primary horizontal framing members should generally be placed along the shorter dimension, and the secondary members should be placed along the longer dimension (Figure 16.23). This strategy reduces the difference between the depths of members and also yields an economical framing system.

If the secondary members are W-shape beams, typical spacing between them is 8 to 12 ft. If joists are used as secondary members, their typical spacing is 2 to 4 ft for floors and 4 to 6 ft for roofs.

Approximate Depths of Horizontal Framing Members

Depth of joist = span/20; Depth of joist girder = span/14

Depth of W-section beam = span/22

Depth of W-section girder = span/16

See also Appendix B.

Nut and washer

1. The connection is brought to a snug tight condition using a spud wrench.

Outer socket of wrench

Inner socket of wrench

2. Tension control wrench is placed over the connection so that the inner socket of wrench engages the spline and its outer socket engages the nut.

3. When the wrench is started, the outer socket rotates the nut relative to the bolt. When the preset tension in the bolt is reached, the spline tip shears off automatically.

Ejection lever

4. After the installation is complete and the wrench is disengaged from the connection, the sheared spline is ousted by depressing the ejection lever on the wrench.

Sheared spline

FIGURE 16.27 Steps in making a slip-critical connection through the twist-off bolt method. Illustration courtesy of Nucor Fasteners with permission.

A typical pneumatic impact wrench.

WELDS

Welding is a process by which connected steel parts are brought to a plastic or fluid state through heating of the parts, allowing them to fuse together, generally with the addition of another molten metal. Welding may be done using either gas welding or arc welding. In arc welding, used more commonly today, two terminals of a high-voltage electric circuit are brought close together, creating a sustained spark across the space between the terminals. A temperature of 6,000°F to 10,000°F is produced in the arc.

One of the terminals in arc welding is the electrode, which is in the welder's hand, and the other terminal is provided by the two metals to be welded together. In the welding process, the electrode also melts and becomes a part of the connection between the two metals being fused.

WELDING VERSUS BOLTING

Welding has a much larger range of applicability than bolting. For example, hollow structural steel sections (round or rectangular) cannot be easily bolted together, but they can be easily welded.

Welded connections also eliminate the need for bolts and connection gusset plates, which can amount to a substantial saving of steel in some structures. Welding also creates continuity between the connected members, which is more difficult to obtain through bolting. However, welding requires a much greater level of skill than bolting. Welding should preferably be carried out under the controlled conditions of a shop. Good welding practice requires that the members to be welded must be dry and free from dirt and grease. Therefore, field welding is generally avoided as much as possible.

Bolting, on the other hand, is rapid, involves less-skilled labor, and can be accomplished without any surface cleaning. It is more suited to field conditions than welding because weather conditions have relatively little effect on bolting.

16.7 CONNECTIONS BETWEEN FRAMING MEMBERS

Structural steel connections are so varied and numerous that they all cannot be included here. Therefore, only the more commonly used connection types are illustrated:

- Column-to-beam connections
- Beam-to-beam connections
- Column-to-column connections (i.e., column splice)

Each one of these connections can be detailed in several ways. One of the major reasons for variations in the details for the same connection is the substitution of welds for bolts, or vice versa, particularly for the part of the connection that is shop produced. As the following illustrations show, a part of the connection between members is made in the shop, and the remaining are completed in the field.

Some fabrication shops prefer bolting to welding. Therefore, these shops generally use fully bolted connections. In other words, the shop-produced part of the connection is bolted, and because field bolting is preferred over field welding, the part of the connection completed in the field is also bolted. Fabrication shops that are better equipped for welding prefer shop-welded and field-bolted connections.

In addition to variations caused by substituting bolting for welding (and vice versa), the ease of erection also governs the choice of connection details.

NOTE

Interchangeability of Terms

With steel connections, the terms *simple connection, shear connection,* and *AISC Type II connection* are interchangeable. The terms *moment connection, rigid connection,* and *AISC Type I* are also interchangeable.

COLUMN-BEAM CONNECTIONS

The American Institute of Steel Construction (AISC) divides column-beam connections as follows:

- Rigid connection, also called moment connection or AISC Type I connection
- Simple connection, also called shear connection or AISC Type II connection
- Semirigid connection, also called AISC Type III connection

In practice, however, most connections are either simple or rigid connections. Therefore, only these two types are illustrated.

Column-Beam Simple (AISC Type II) Connections A simple connection is a flexible connection that allows the ends of the beam to rotate under the loads. (Although simple connections used in practice have some end restraint, i.e., moment resistance, the restraint is small and is, therefore, neglected.)

The most frequently used column-beam simple connection uses two angles that are shop welded to a beam and field bolted to a column flange (or web), as required, Figure 16.28(a). Where a single-angle connection is acceptable, it may be shop welded to the column and field bolted to the beam, Figure 16.28(b). A single angle may be replaced by a single plate, Figure 16.28(c).

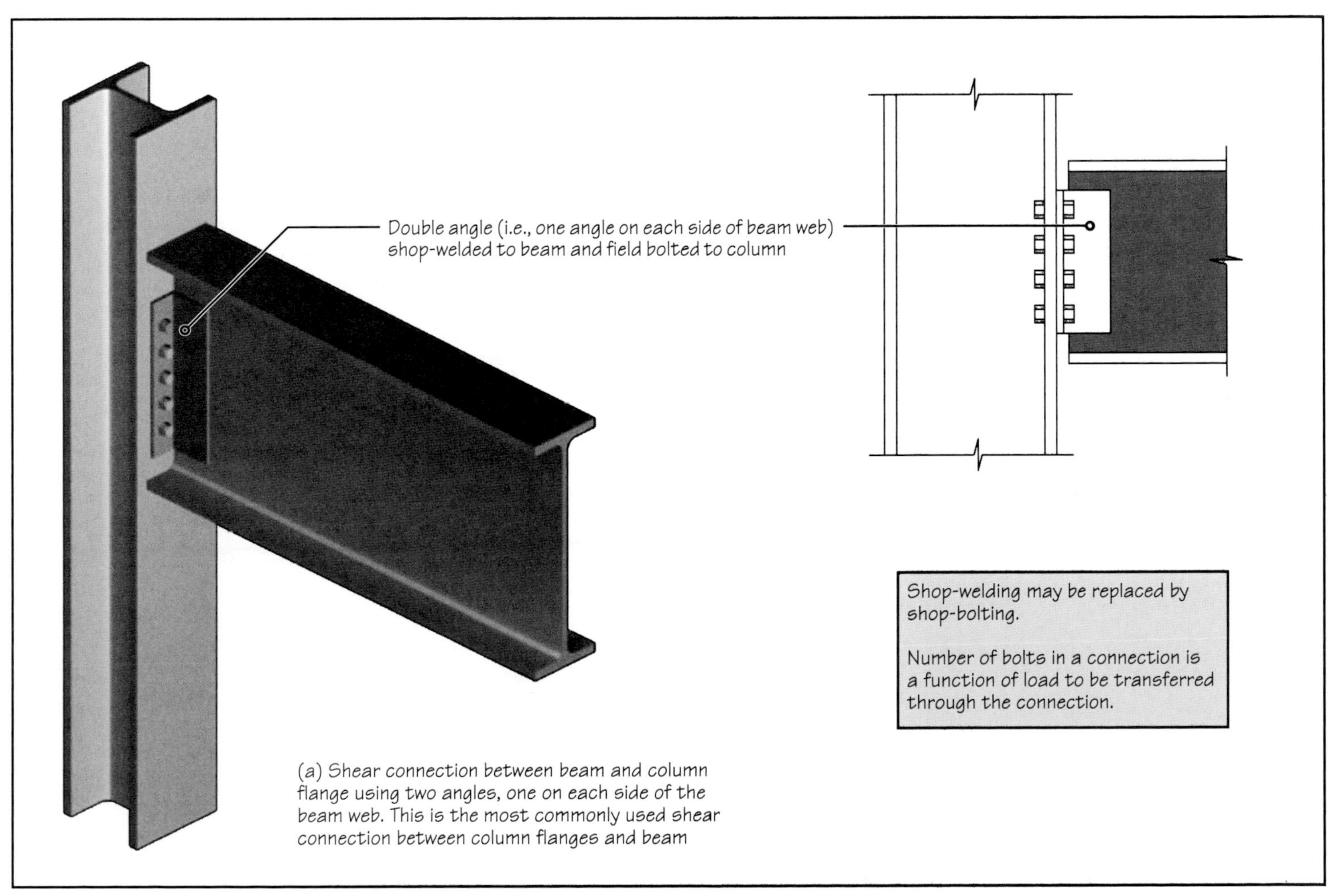

FIGURE 16.28 Typical column-beam shear (Type II) connections.

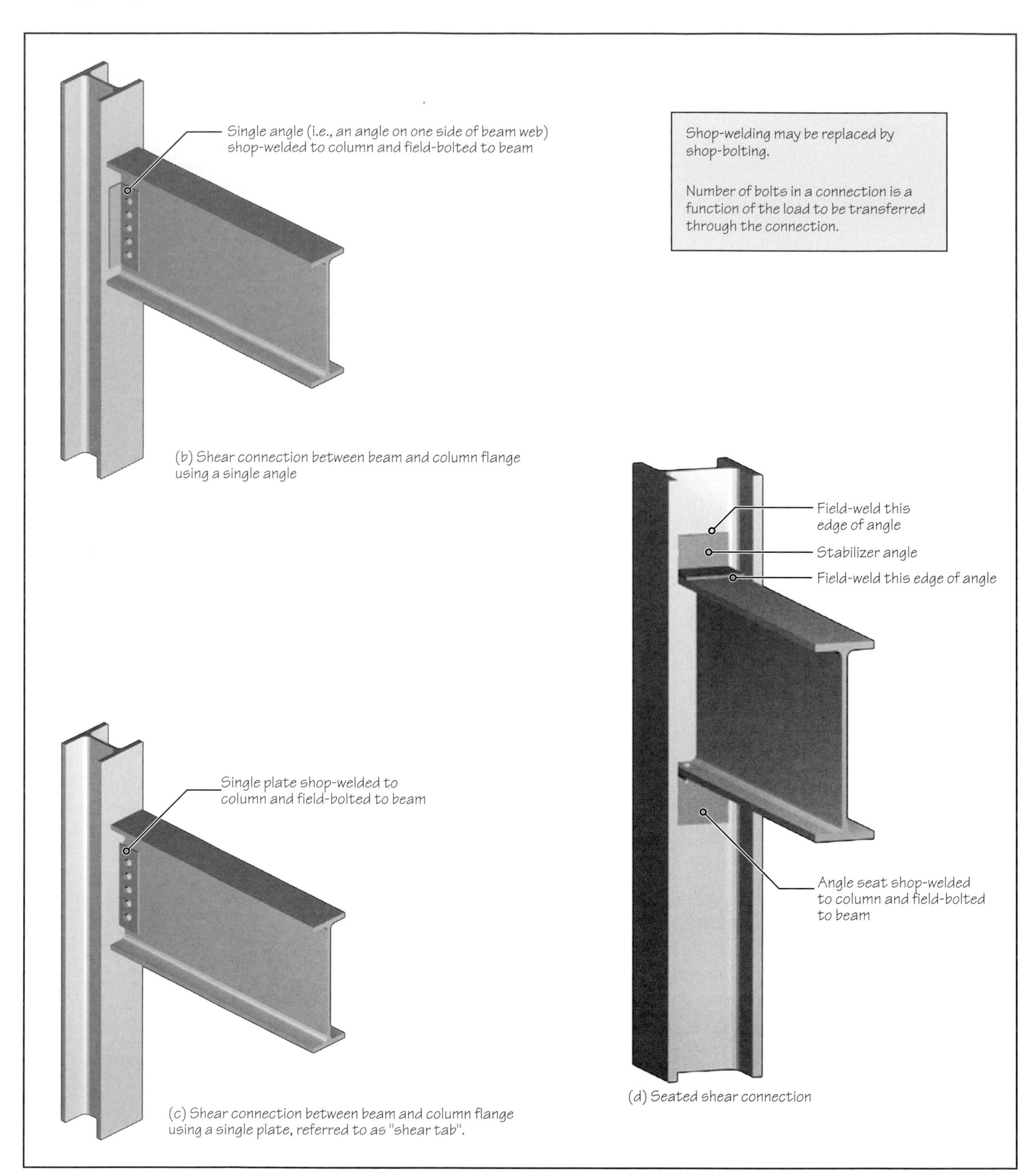

FIGURE 16.28 *(Continued)* Typical column-beam shear (Type II) connections.

A seated shear connection is well suited for connecting the beam with a column web, particularly if there is a beam on each side of the web, Figure 16.28(d). A seated shear connection consists of one angle at the bottom of the beam and the other at the top.

The top angle simply provides stability to the connection. Hence, it is called a *stabilizer angle*. To ensure that the stabilizer angle does not introduce moment resistance in the connection, it is generally field welded to both column and beam at the toes only (the edges away from the angle heel).

Column-Beam Rigid (AISC Type I) Connections Because a rigid connection must transfer tensile and compressive stresses between the beam and the column, the flanges of the beam are

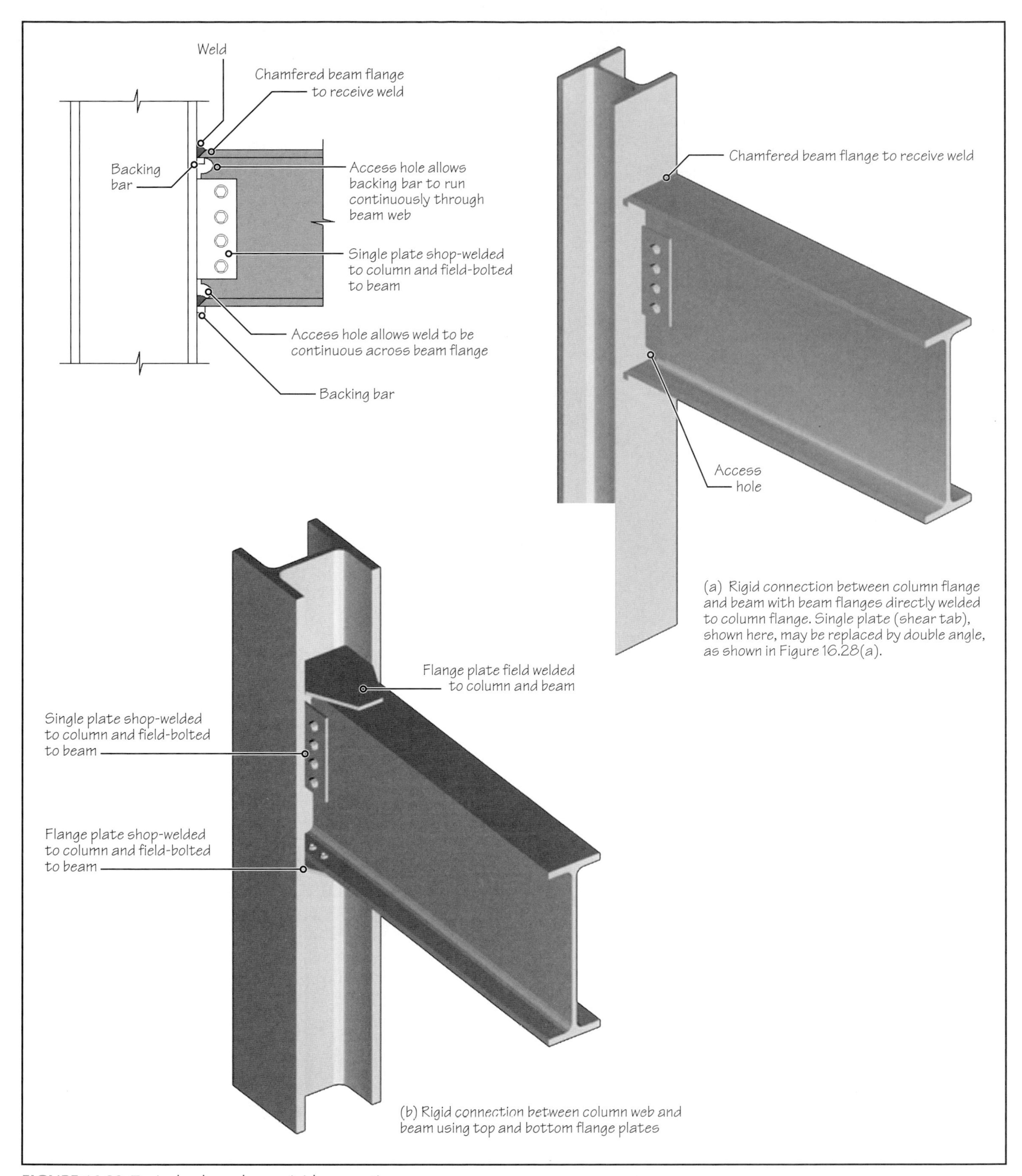

FIGURE 16.29 Typical column-beam rigid connections.

also connected to the column (in addition to the web connection). Observe from Figure 16.28 that in a shear connection, the flanges of the beam are not connected to the column.

One way of connecting beam flanges to the column is to weld them directly to the column using groove welds. This requires chamfered beam flanges, access holes, and backing bars, Figure 16.29(a). Backing bars prevent weld material from dripping down. To prevent weld material from flowing off the sides, runoff tabs are used, which are removed after making the connection. Backing bars are, however, left in place. (In seismic zones, the backing bar at the bottom flange must be removed.)

Figure 16.29(b) shows another alternative to column-beam rigid connections, in which steel plates have been used to connect beam flanges to the column instead of direct welding the beam flanges to the column. These plates are referred to as *flange plates.*

BEAM-BEAM CONNECTIONS

Beam-to-beam connections, where one beam supports another beam, are fairly common in steel floors and roofs. These connections are generally simple connections and are similar to simple column-to-beam connections, Figure 16.30.

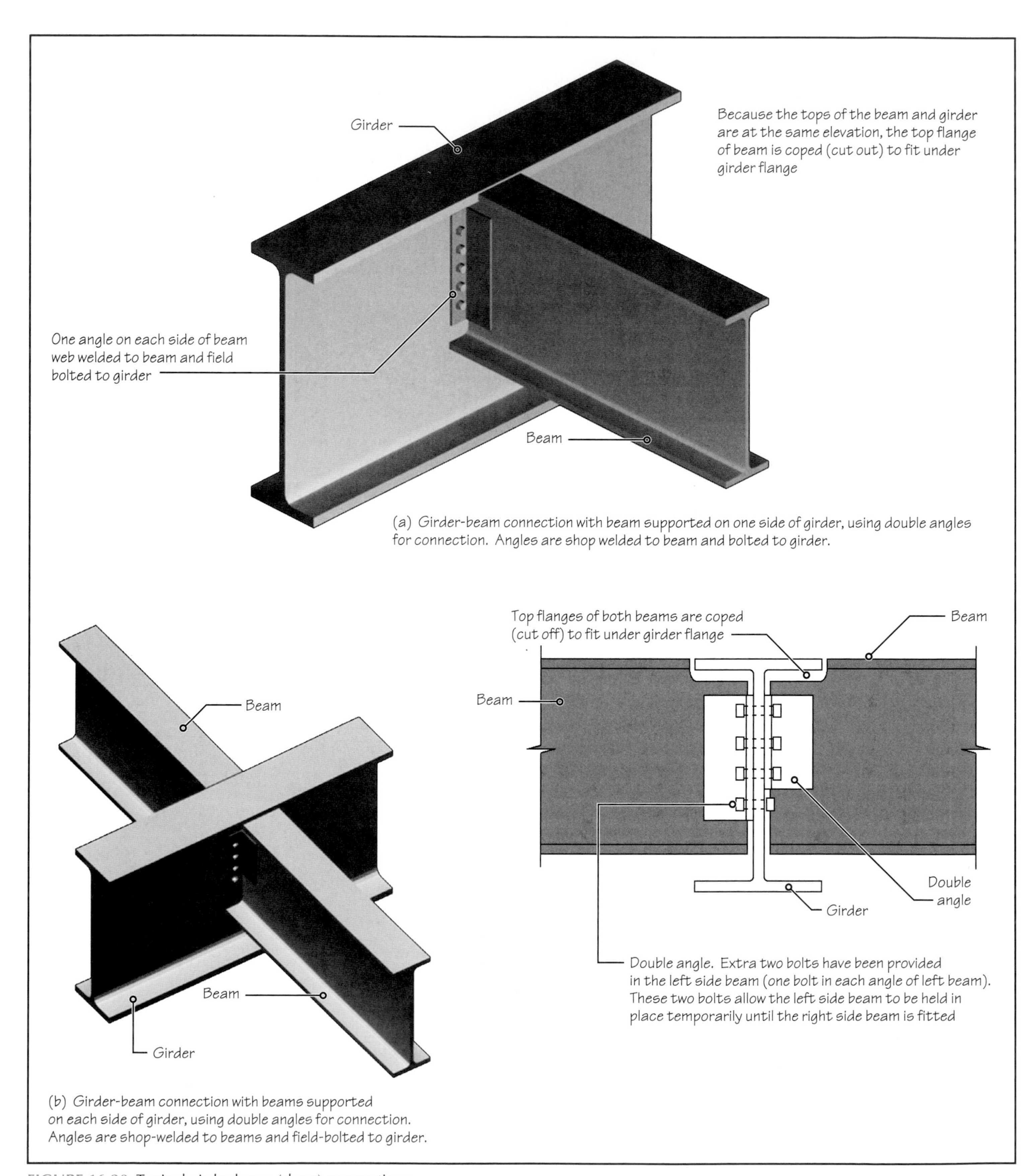

FIGURE 16.30 Typical girder-beam (shear) connections.

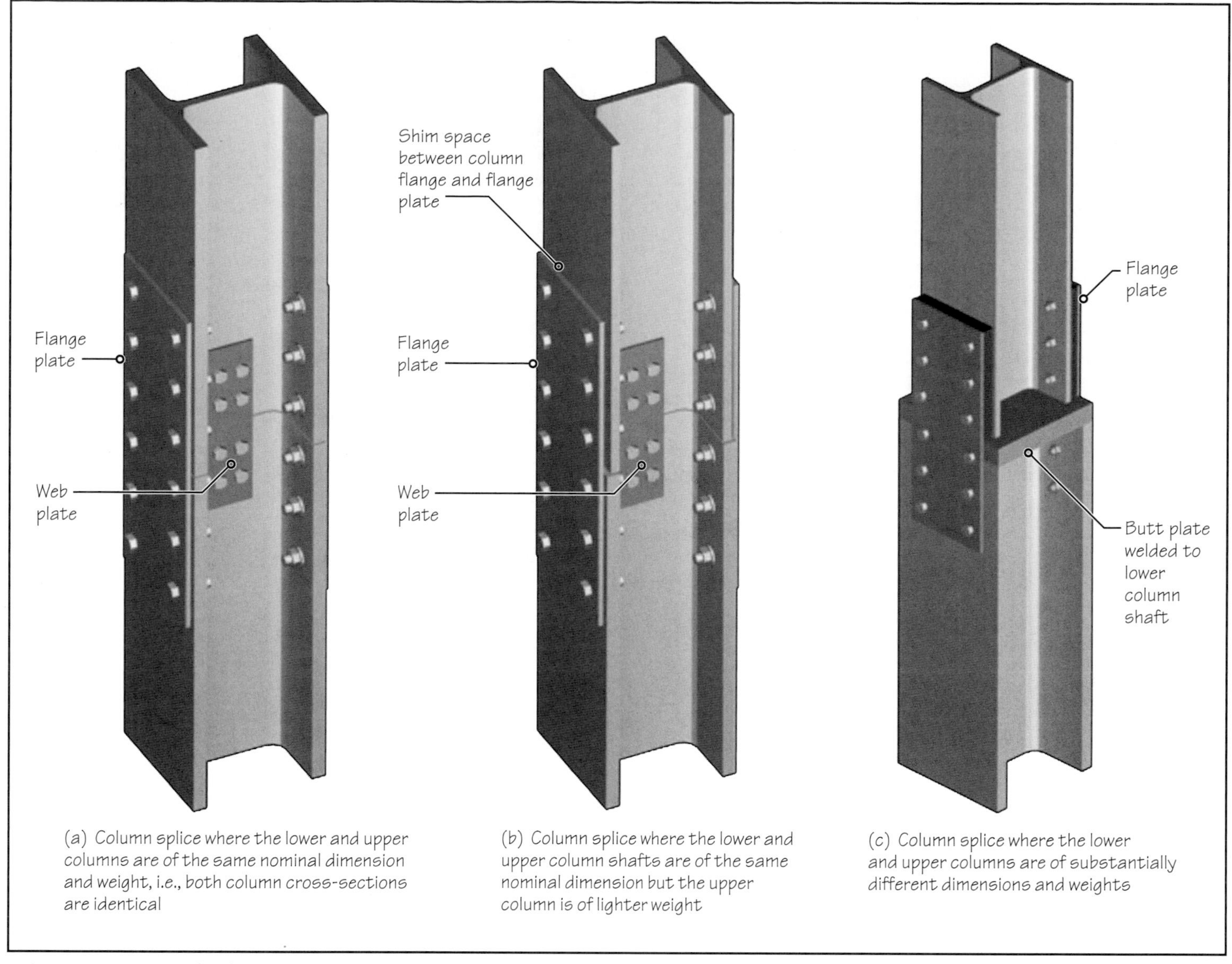

FIGURE 16.31 Typical column splices. Column ends are milled smooth giving full end bearing to achieve proper transfer of loads.

Column Splices

Column splices are a necessity in a multistory building because of the limited length of steel members. Generally, column splices are provided every two stories. In a three-story building, however, the columns can be continuous—without splices.

When provided, a column splice is located approximately 4 ft above the lower floor so that the erector can use the floor as a platform to make the connection. This is particularly important for the columns at perimeter locations, where the erector must use the lower column shaft to attach safety cables.

Where the lower and upper columns are of the same nominal dimensions, the upper column shaft bears on the lower column shaft and is connected to it with the help of flange plates, Figures 16.31(a) and (b). Flange plates may be shop welded to the lower column shaft and field bolted to the upper column shaft. However, an all-bolted connection is generally preferred. In a welded-bolted connection, the protruding flange plates can bend during transportation.

If the upper and lower columns are of slightly different nominal dimensions, a flange-plated connection may still be used because it is not necessary that the upper column must fully bear on the lower column. However, if the upper and lower columns are significantly different in dimensions, a butt plate and flange plates are used at the connection, Figure 16.31(c).

16.8 STEEL DETAILING AND FABRICATION

Unlike masonry and site-cast concrete construction, in which various components, such as walls, columns, and beams, are constructed at the site, structural steel components are brought to the site in a prefabricated and finished state, ready for erection and assembly. There is very little cutting, notching, and drilling of components at the site.

NOTE

Cambering

Camber is generally provided in floor beams to counter their deflection under dead loads. Because dead-load deflection is downward, the camber is an upward curvature in the beam. It is provided in the fabricator's shop.

The purpose of camber is to ensure that the beam becomes flat under dead loads and will be subjected to deflection only under the live loads. If camber is not provided, the beam will have a larger deflection due to the combined dead-load and live-load deflections.

Cambering adds to the cost of the beam, but the alternative is to increase the beam size, which is more expensive. However, camber should not be overspecified. Overspecification of camber not only costs more, but it may leave the top of the shear studs sticking out of the concrete topping.

Generally camber is calculated to counter the deflection created by two-thirds to three-quarters of the dead load on the beam.

The fabrication of ready-to-erect components is done in a steel fabrication shop. A fabricator, who generally works as a subcontractor to the general contractor, first prepares a set of shop drawings based on the project's contract documents. The shop drawings delineate every component of the building with its precise dimensions.

Each connection is fully detailed, giving the dimensions of all connection angles and plates, including the number of holes, bolt diameters, size and types of welds, and so on. The drawings also distinguish between shop and field welds. The type of bolted connection (snug-tight or slip-critical) is also identified. If a member is required to be cambered (based on the contract documents), the amount of camber is indicated on the shop drawings.

As described in Chapter 1, the purpose of the shop drawings is to help the fabricator fabricate various components so that they can be assembled at the site with ease and within a predetermined schedule.

STEEL DETAILING

For most buildings, the design and detailing of connections between components (such as those shown in Figures 16.28 to 16.31) are prepared by the fabricator based on the framing plans and other information provided in the project's structural drawings. Leaving connection detailing to the fabricator allows the fabricator to use the details that are most economical and best suited for erection and scheduling. Where details are aesthetically significant, the fabricator must conform to the architectural and structural requirements of the project.

Some fabricators have an in-house detailing staff. Most fabricators, however, subcontract the work to an independent steel detailer. Structural steel detailing is both a science and an art because the details must not only be able to withstand the forces (as required by the engineer's drawings) but also be economical, be easy to erect, and suit the fabrication shop's capabilities and resources.

ORDERING MATERIALS FOR FABRICATION

After the shop drawings are completed, they are first reviewed by the general contractor and then by the project architect and structural engineer. Because the modifications expected from the review process are generally minor, the fabricator orders the structural steel sections from the rolling mills while the review of shop drawings is in progress.

Most rolling mills need sufficient lead time to deliver the material, particularly the heavier structural steel sections. This is because a rolling mill rolls a certain section (and a steel grade) for a few days; then the rollers are changed to produce a different section (Figure 16.5). Thus, a given section is rolled at intervals that may range from weeks to months. Therefore, a rolling mill generally rolls a batch containing different sections on order. To hold various grades, sections, and lengths in the mill's stock waiting to receive an order is too uneconomical for most mills.

A fabricator may order heavy sections (for columns and beams) cut to the required lengths by the mill. On the other hand, the fabricator may opt to cut the sections to lengths from standard mill lengths in the shop, if that is more economical. Standard lengths are generally 30 ft, 35 ft, 40 ft, and so on, up to 60 ft. Length greater than 60 ft may be available as a special order. Smaller sections, such as plates and angles, are always cut by the fabricator.

It is important to realize that structural sections with exactly the same specified dimensions and weight may not, in reality, be of the same dimensions. Wear of the rollers, thermal distortion of hot cross sections and their subsequent cooling, and several other factors cause the dimensions of the sections with the same specifications to be different. Steel detailing and fabrication recognize this fact and provide the required connection and erection tolerances.

FABRICATION

Most large steel-fabrication shops are semiautomated, that is, computer assisted, which makes the fabrication process not only precise but also quick. Figure 16.32 shows a typical angle-shearing machine that shears the angles to size and drills holes of the required diameters and spacings without any manual labor. Whereas lighter sections, such as plates and angles, are sheared, heavier steel sections, such as beams and columns, are sawed to desired length by semi- or fully automated saws, Figure 16.33.

Welding in most fabrication shops is carried out manually. For long continuous lines of welds or repetitive welding, an automated welding system is used.

Most structural steel members do not require any surface preparation, such as blast-cleaning or a prime coat. However, if the steel has developed loose mill scale because of its extended exposure to weather, it should be blast-cleaned, particularly if it is to be treated

FIGURE 16.32 A computer-aided plate and angle-punching machine cuts plates and angles to the required size and drills holes of the required diameters and spacings without any manual labor. (Photo at Irwin Steel Fabrication Plant, Justin, Texas.)

FIGURE 16.33 A computer-aided band saw cuts heavy steel members to size. The machine is versatile enough to make skew cuts instead of a right-angle cut, as shown here. (Photo at Irwin Steel Fabrication Plant, Justin, Texas.)

with spray-on fire protection or intumescent paint or galvanized. Steel components that are to be finished with paint generally require a shop coat of primer.

As the components are being fabricated, they are identified by marks that help further fabrication and erection. Unless dispatched to the construction site immediately on fabrication, the fabricated components are stored in the fabricator's shed. The storage is planned in a way that allows easy retrieval for delivery to the site in predetermined batches.

PRACTICE QUIZ

Each question has only one correct answer. Select the choice that best answers the question.

24. In a steel frame building, the secondary framing members for a roof or floor generally consist of
- **a.** W-shape beams.
- **b.** steel joists.
- **c.** steel joist girders.
- **d.** all of the above.
- **e.** (a) and (b).

25. Which of the following is not an ASTM designation for steel bolts?
- **a.** A325 bolts
- **b.** A307 bolts
- **c.** A490 bolts
- **d.** A507 bolts

26. A snug-tight connection between framing members is obtained by
- **a.** welding the members together using arc welding.
- **b.** welding the members together using gas welding.

c. bolting the members and tightening them with a spud wrench.
d. bolting the members and tightening them with a spud wrench using full force.
e. none of the above.

27. A slip-critical connection between framing members is obtained by
a. welding the members together using arc welding.
b. welding the members together using gas welding.
c. bolting the members and tightening them with a spud wrench.
d. bolting the members and tightening them with a spud wrench using full force.
e. none of the above.

28. Field bolting of steel members is generally preferred over field welding.
a. True b. False

29. A shear connection between a steel beam and a steel column is also called
a. a simple connection. b. a rigid connection.
c. a semirigid connection. d. a AISC Type I connection.
e. none of the above.

30. A shear connection between a steel beam and a steel column can be made without connecting beam flanges to the column.
a. True b. False

31. A rigid connection between a steel beam and a steel column can be made without connecting beam flanges to the column.
a. True
b. False

32. In a multistory steel frame building, the columns are generally spliced
a. every four floors. b. every three floors.
c. every two floors. d. every floor.
e. none of the above.

33. In connecting the upper-column shaft with a lower-column shaft in a multistory steel frame building, we generally require
a. four flange plates.
b. three flange plates.
c. two flange plates.
d. one flange plate.

34. Details of connections between steel framing members are generally prepared by the building's
a. architect.
b. structural engineer.
c. general contractor.
d. steel fabricator.
e. none of the above.

35. Camber refers to
a. downward deflection of a beam.
b. upward deflection of a beam to counter downward deflection by a live load.
c. upward deflection of a beam to counter downward deflection by a dead load.
d. upward deflection of a beam to counter downward deflection by a dead load plus a live load.
e. none of the above.

Answers: 24-e, 25-d, 26-d, 27-e, 28-a, 29-a, 30-a, 31-b, 32-c, 33-c, 34-d, 35-c.

16.9 STEEL ERECTION

The erection of structural steel frame at the site may be performed by the fabricator or a separate erection company. In most cases, the general contractor will seek separate bids for fabrication and erection. If the fabricator, selected on the basis of his or her competitive bid for fabrication, also submits a competitive bid for erection, the fabrication and erection may be done by the same organization. If not, the erection contract may be given to a different entity. Fabrication and erection by the same entity is obviously preferred.

Steel frame erection begins with the erection of columns, Figure 16.34, using a crane. For a small building, a mobile crane on rubber tires is generally adequate. For a large building, a tower crane is needed in addition to one or more mobile cranes, Figure 16.35. Generally, a tower crane is built over the foundations of an elevator shaft. Climbing cranes, secured to the building's columns, can also be used. In that case, a climbing crane climbs up the columns as the building height increases. These columns are specially engineered and provided with special attachment points.

FIGURE 16.34 The erection of a steel frame begins with the erection of columns. The columns, with their shop-welded base plates, are anchored to the bolts embedded in the foundations (Figures 16.36 and 16.37) by the general contractor per the fabricator's details. In a multistory building, each column shaft is generally two stories tall.

Structural steel columns are anchored to the foundations through anchor bolts. The anchor bolts are set into column footings before pouring footing concrete. The bolts are supplied by the fabricator but embedded into the footings by the general contractor per the fabricator's details.

A steel base plate is shop welded to the column bottom with predrilled holes to penetrate the anchor bolts. Columns are leveled to ensure that they are plumb, which can be done in a number of ways. One method uses leveling bolts under the base plate, Figure 16.36(a).

A space (minimum 2 in.) between the top of the footing and bottom of the base plate is needed. After the column has been leveled and the frame squared and stabilized, this space is filled with portland cement and sand grout, Figure 16.36(b).

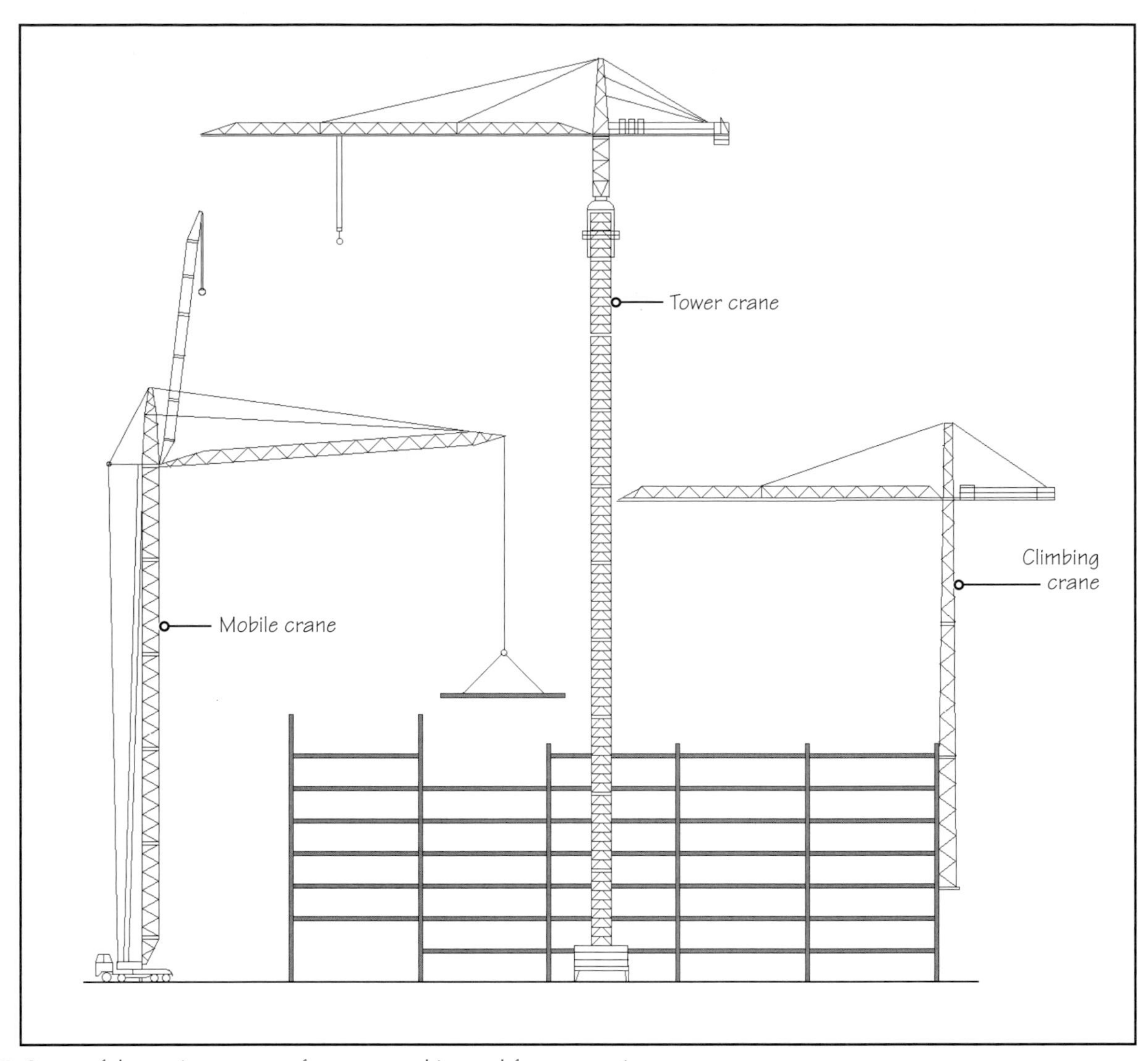

FIGURE 16.35 Some of the various types of cranes used in steel frame erection.

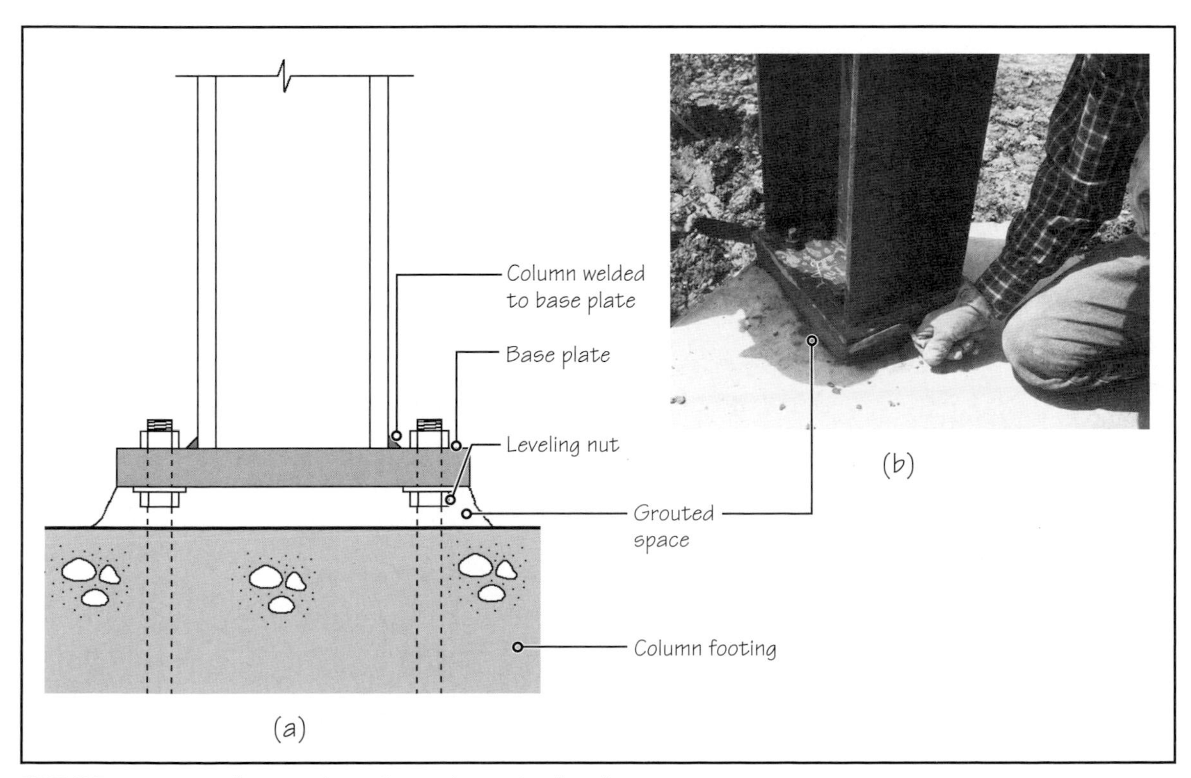

FIGURE 16.36 Leveling a column base plate using leveling nuts.

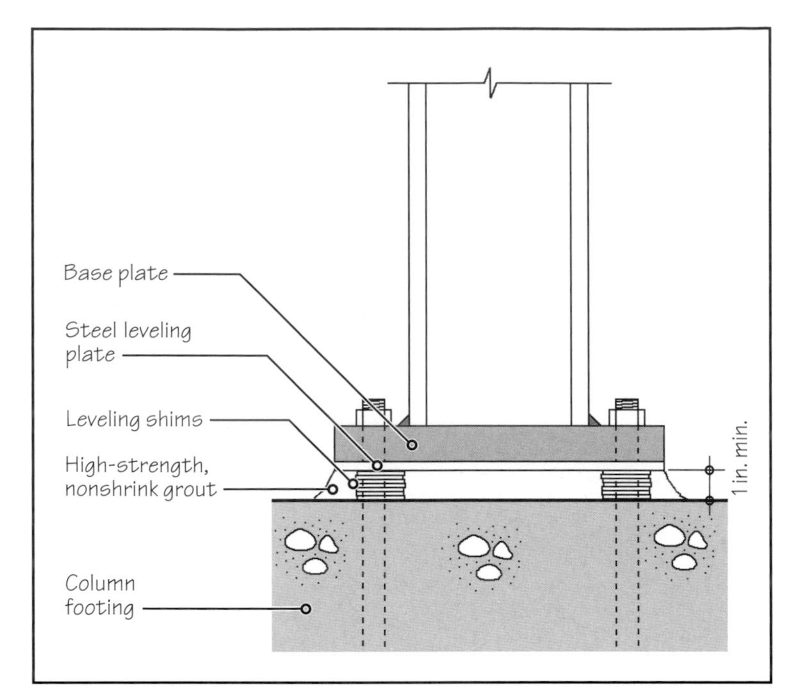

FIGURE 16.37 Leveling a column base plate using high-strength plastic shims placed under a thin steel plate.

For a wide base plate (generally more than 24 in.), grout holes in the base plate may also be needed to ensure full bearing after grouting. The grout is packed under the base plate using a trowel.

The other alternative is to level a relatively thin steel plate of the same size as the base plate using high-strength plastic shims under the plate. After the plate has been leveled, the steel column is placed over the plate and bolted into place, Figure 16.37. A space (minimum 1 in.) is needed between the top of the footing and bottom of the base plate. This space is grouted in the same way as shown in Figure 16.36(b), but the grout must be high-strength, nonshrink grout.

After a few columns for the first tier have been erected, beams are connected to them. The erection of framing members is a coordinated effort between the crane operator and the erectors. The erector must guide the member into position, initially through a tagline and finally with his or her hands, Figure 16.38.

FIGURE 16.38 A steel erector (with his body harness tied to the beam on which he sits) maneuvers the new beam into position, initially using the rope attached at the end of the beam and later by hand. The rope is referred to as a tagline in erection vernacular.

Some connections may require some forced alignment between the holes in the parts to be connected. Crowbars, hammers, drift pins, and so on, may be used to bring members into alignment, Figure 16.39.

The initial connection between members is made using some of the bolts in the connection. After most of the members have been placed in position, the remaining bolts are installed and the connection is completed by hand tightening the bolts. Finally, an impact wrench is used in case of slip-critical connections.

The placement of beams for a tier is followed by floor and roof decking. The installation of decking is done by the general contractor or the erector, depending on the contract.

FIGURE 16.39 Generally members at a connection are forced to align. In this image, a drift pin has been used to make the alignment. In some connections, crowbars or hammers may have to be used.

16.10 CORROSION PROTECTION OF STEEL

Because steel (unlike aluminum) does not automatically form a protective oxide coating, it must be protected against corrosion. However, structural steel members enclosed by the building envelope do not require any protective coating unless they are in a corrosive environment.

In other words, interior structural steel members can be left bare (mill-finished state) in most situations. This recommendation [1] is based on the surface conditions disclosed by the demolition of several long-standing structures in which steel had no protective coating. These structures suffered either no corrosion or the corrosion was localized to small areas that were affected by persistent water leakage or condensation.

Although bare steel is acceptable, almost all structural steel members receive a prime coat in the fabricator's shop before being delivered to the construction site. The prime coat (the familiar red-brown coating) provides temporary protection until the steel is wrapped by the building envelope. Structural steel that is permanently exposed to the atmosphere must, however, receive additional protection. Additionally, concealed steel components that may be subjected to wetting, such as steel anchors buried in masonry or concrete, should also be protected.

Several protective coatings are available for steel to suit different environmental conditions, aesthetics, and budget. These include acrylics, epoxies, polyurethanes, and zinc coating. For exposed structural steel members, polyurethane coatings are the hardest, toughest, and most versatile. The prime coat is generally applied in the shop, followed by one or two field-applied finish coats. A well-applied polyurethane coating provides more than 20 years protection.

For cold-formed and light structural steel members, zinc coating (referred to as *galvanizing*) is a cost-effective solution. Of various types of galvanizing, hot-dip galvanizing is by far the most reliable and the most widely used.

HOT-DIP GALVANIZING

In the hot-dip galvanizing process, the steel member is immersed in a kettle of molten zinc, which is at a temperature of about 850°F (475°C). This results in the formation of a zinc-iron alloy coating that is metallurgically bonded to steel. After the member is removed from the kettle, excess zinc is drained and shaken off. The galvanized member is then cooled in air or quenched in water. As a result of its immersion, the entire member is coated with zinc, including the protrusions, recesses, and corners.

Kettles up to 50 ft long are available. If a member is longer than 50 ft, it can be galvanized in two halves. After one half has been galvanized, it is removed, then the other half is immersed in the kettle.

Hot-dip galvanizing should be done after the member has been completely fabricated. No cutting, grinding, or process that eradicates the coating should be done after galvanizing.

The thickness of the coating influences the degree of protection. The thicker the coating, the greater the protection. The thickness of zinc coating is specified as Gxx, where G stands for galvanizing and xx indicates average coating mass in ounces per square foot (oz/ft^2) of the steel member. Thus, a G60 coating means that the member is coated on all its surfaces with an average of 0.60 oz/ft^2 of zinc. If the member is a sheet, G60 means the presence of 0.30 oz/ft^2 of zinc on each surface of the sheet. A G90 coating implies 0.90 oz/ft^2 of zinc coating on all surfaces.

In the SI system, the same zinc coating is specified as Zyy, where yy indicates average coating mass in grams per square meter (g/m^2). Because 1 oz/ft^2 = 305 g/m^2, a G60 coating is equal to Z180, and a G90 coating is equal to Z275.

NOTE

Corrosion—an Electrochemical Process

Corrosion is not simply a chemical process—it is an electrochemical process that results in the eating away of the surface of a metal when it is exposed to weather. It occurs when electrons flow through an electrolyte. (An electrolyte is a liquid that allows the flow of electrons through it.)

Water is a necessary ingredient for corrosion. Acidic water or water containing salt is a better electrolyte. That is why coastal regions are more corrosive than inland areas and deserts are less corrosive than humid regions.

EXPAND YOUR KNOWLEDGE

Corrosion of Metals

Metals, by virtue of their atomic structure, are chemically reactive materials. They react readily with the atmospheric oxygen to form metal oxides, which are chemically stable. That is why metals are generally obtained on the earth's surface in the form of their oxides, that is, iron oxide (iron ore), aluminum oxide (bauxite ore), and so on.

Atmospheric Corrosion

Because of their reactivity, metals have a tendency to return to their original (oxide) state. The oxidation reaction obviously requires atmospheric oxygen, but it also requires the presence of water. The reaction is accelerated by the presence of atmospheric impurities because they, in combination with water, can form weak acids.

For example, sulfur released from automobile and industrial exhausts dissolves into rainwater and converts to weak sulfuric acid. Sulfuric acid exists in most contemporary urban environments. It was also present in the environment before the introduction of automobiles due to the burning of coal, which contains a substantial amount of sulfur.

Water that contains salt also accelerates the oxidation of metals. Corrosion of metals is more rapid in coastal areas than inland areas. Salt spread on roads in snow-prone regions accelerates the corrosion of steel bridges and of steel reinforcement embedded in floor slabs of parking garages. The oxidation of metals, referred to as *corrosion*, begins at the surface of the metal and, if the reaction is not interrupted, it proceeds inward and can corrode the entire metal over time.

The corrosion of steel results in the formation of iron oxide—a brown-colored material called *rust*. Because rust is a loose and flaky material, it separates from the steel member and exposes the member to further corrosion. The volume of rust is greater than the original iron. That is why the corrosion of reinforcing steel not only reduces the amount of steel, but also tends to spall the concrete.

By contrast with iron oxide, aluminum oxide, formed by the oxidation of aluminum, clings securely to aluminum. Being a stable compound, aluminum oxide forms a protective layer and prevents further oxidation of aluminum. When aluminum's surface is scratched and aluminum is exposed, a new oxide layer is immediately formed over the scratch, maintaining the protection.

Galvanic Corrosion

The type of corrosion just described is called *atmospheric corrosion* because it is caused by metal's exposure to the atmosphere. An additional type of corrosion of interest to design and construction professionals is *galvanic corrosion*. Galvanic corrosion, also called *bimetallic corrosion*, is caused by the physical contact between two dissimilar metals. Galvanic corrosion is not chemically different from atmospheric corrosion; that is, both galvanic and atmospheric corrosion processes produce the same end product. The difference is that the contact between the two dissimilar metals changes the metals' respective corrosion rates. One metal corrodes more rapidly as compared with its atmospheric corrosion rate and the other metal does so more slowly. The change in their corrosion rates depends on the type of metals in contact.

The metal whose corrosion rate declines is called a *more noble* metal, and the one whose corrosion rate increases is called a *less noble* metal. The relative nobility of metals is given by the *galvanic series*, discovered by Luigi Galvani in the latter part of the eighteenth century. The galvanic series for a few selected metals is given below. Of various metals, gold and platinum are the most noble, and magnesium and zinc are the least noble.

The change in the corrosion rates of metals that are close to each other in the galvanic series is smaller. The farther apart the metals are in the series, the greater the change in their respective corrosion rates. Thus, if zinc and copper, which are far apart in the series, are placed in contact, the corrosion rate of zinc will be accelerated, and that of copper will decline substantially.

Galvanic corrosion is easily controlled by separating the metals from each other either by a plastic separator or by coating one or both metals. Coatings also inhibit atmospheric corrosion.

Galvanic Series of Selected Metals

Less noble	Magnesium
	Aluminum
	Zinc
	Steel
	Stainless steel
	Tin
	Lead
	Copper
	Bronze
	Brass
	Tiitanium
	Silver
More noble	Platinum
	Gold

Three levels of hot-dip galvanizing are commonly specified: G30 for a mildly corrosive environment, G60 for a more corrosive environment, and G90 or G120 for highly corrosive environments.

16.11 FIRE PROTECTION OF STEEL

The fire endurance of steel is poor, and its inherent incombustibility gives a false sense of security. Exposed (unprotected) steel, unless it is very thick, cannot withstand long exposure to fire. Steel's yield strength and modulus of elasticity drop to nearly 60% of their original values at about 1,100°F—a temperature that is well below the temperature used in a standard fire test.

Steel has a high coefficient of thermal expansion. When subjected to fire, an unrestrained steel member expands, pushing on the members to which it is connected. A restrained member develops high internal stresses. In either case, the consequences can be diastrous. It is important, therefore, that structural steel be protected against fire. Unprotected steel buildings may be used, but building codes severely limit the allowable areas and heights of such structures.

There are various ways to protect structural steel. The oldest method is to encase steel sections in concrete, as shown in Figure 16.40. It is generally used in columns and floor systems that consist of reinforced concrete slabs supported by steel beams so that the concrete-encased steel beams function monolithically with the slab.

The disadvantage of protecting steel with concrete encasement is the high cost of formwork and the large dead load it poses on the structure. It has, therefore, been replaced by more efficient techniques, which are lightweight and more economical, such as spray-applied fire protection and gypsum board facing.

FIGURE 16.40 A steel member encased in concrete (or masonry) is one of the oldest methods of protecting steel against fire. The use of this method is relatively uncommon in contemporary construction because of the availability of more efficient and economical alternatives.

SPRAY-APPLIED FIRE PROTECTION

Spray-applied fire protection, Figure 16.41, is the most commonly used method of protecting steel. Before spraying, dirt, loose scale, and oil are removed from the member in order to develop better adhesion. Two types of materials are used for spray-applied (or spray-on) protection:

- Mineral fiber and binder, usually fiberglass and portland cement
- Cementitious mixture containing portland cement mixed with a lightweight aggregate such as perlite or vermiculite.

In general, aggregate-based cementitious protection is more rugged, has better abrasion resistance, and is aesthetically superior to fiber-based protection. After spraying, aggregate-based protection can be finished smooth with a trowel. Because the protection is always applied before the installation of duct work, piping, and other equipment, any damage caused by such installation must be subsequently repaired.

Virtually any amount of fire resistance can be obtained by varying the thickness of spray-on protection. For a given fire rating, the thickness of spray-on protection depends on the spray-on material and the thickness of steel. The heavier the member, the thinner the required protection. However, a $\frac{5}{8}$-in. thickness per 1 h of fire rating is a good rough estimate of the required thickness of protection. Thus, a 1-h rating is obtained by applying about $\frac{5}{8}$-in.-thick material, a 2-h rating by $1\frac{1}{4}$-in.-thick material, and so on.

Fire rating is not the only criterion that spray-on protection must meet. It must also meet the minimum requirements for damageability resistance, adhesion to steel, deflection (so that the protection does not delaminate when the steel member deflects under loads), air-erosion resistance, and corrosion resistance (spray-on protection protects steel from corrosion also).

Spray-on protection must be applied on bare steel because painting or priming reduces adhesion. Additionally, it must be carried out by specialist contractors who warranty its installation, fire resistance, and other long-term performance characteristics.

GYPSUM BOARD ENCASEMENT

Spray-applied protection is relatively unattractive, which usually necessitates that the protected component be covered. An alternative to spray-on fire protection is gypsum board encasement, which provides a finished appearance to steel frame.

FIGURE 16.41 A steel-framed floor with spray-on fire protection. Observe that the steel deck may not require protection if sufficient thickness of concrete topping over the deck is available.

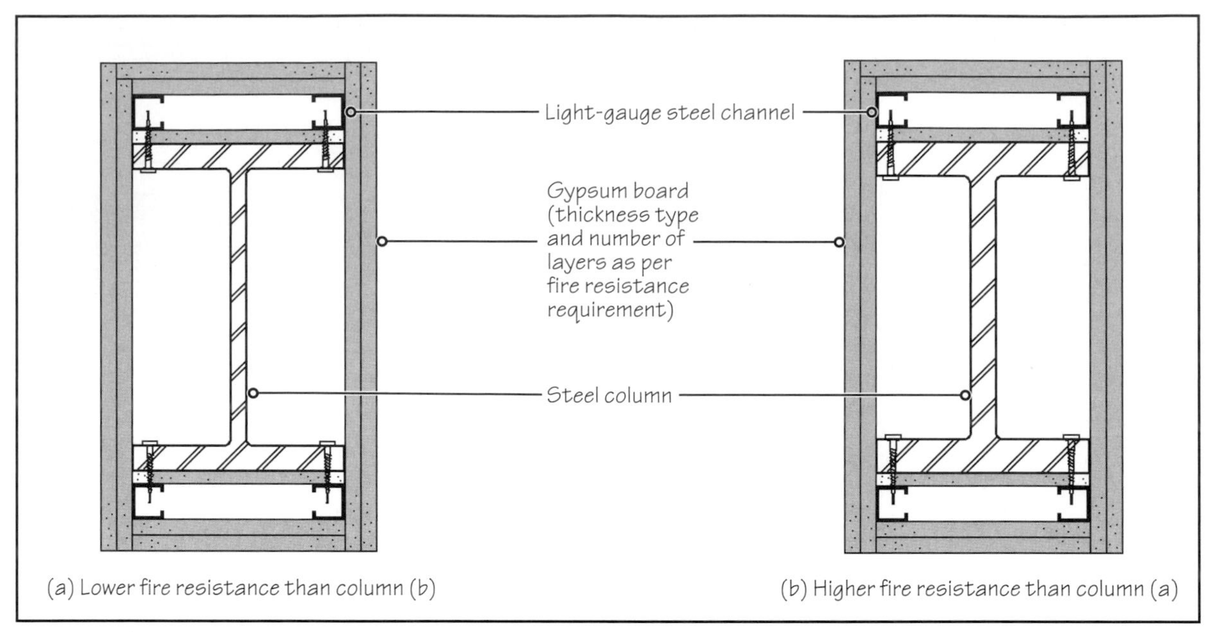

FIGURE 16.42 Details of a steel column covered with gypsum board layers. Details **(a)** and **(b)** are identical except that the column in detail **(b)** is heavier than that of the column in detail (a). Because the thickness of steel also affects the member's fire resistance, in addition to the thickness and type of gypsum board, the fire resistance rating of column **(b)** is higher than that of column **(a)**.

Gypsum board encasement consists of gypsum boards mechanically fastened to steel sections, usually screw attached to light-gauge steel studs, Figure 16.42. Almost any degree of fire rating can be obtained by varying the number of board layers. (See Chapter 14 for a discussion of gypsum boards.)

NOTE

Use of Unprotected Structural Steel Framing Members

Unprotected steel in structural framing of building is routinely used. Unprotected steel-framed buildings, defined as Type II(B) construction (Chapter 7), are allowed by building codes for all occupancy groups except I-2 occupancy group (hospitals, nursing homes, etc.).

However, if protected steel framing is used for the same building, the built-up area allowed by the building codes is greater, depending on the type of construction used. Protected steel construction types may be Type I(A), Type I(B), or Type II(A). The allowable built-up area may also be increased through the use of automatic sprinklers (see "Principles in Practice," Chapter 2).

INTUMESCENT PAINTS

An alternative to gypsum board encasement is intumescent paint on steel members. Intumescent paint is typically 20 to 50 mil (0.5 mm to 1.3 mm) thick. When exposed to the heat of fire, the paint intumesces, or swells, yielding an insulating char cover on steel that is 2 in. to 4 in. thick. It is this char layer that protects steel from fire. However, the char layer is damaged under long-term exposure to fire, and hence this technique can be used only for low levels of fire rating, generally not exceeding 2 h.

Intumescent paint is a thin film that does not alter the overall profile of steel sections. It gives the structure the look of exposed steel and is commonly used in situations where the expression of exposed steel is required from an architectural and interior design perspective, Figure 16.43.

FIGURE 16.43 A steel truss painted with intumescent paint providing a 1-h fire rating, Hospital for Sick Children, Toronto, Ontario. (Photo courtesy of AD Fire Protection Systems, Scarborough, Ontario. Architect: Zeidler Roberts Partnership. Photo: Lenscape Inc.)

Intumescent paints are available in different colors. Thus, the painted structure can be color coordinated with the overall character of the interior. Another advantage of intumescent paint protection is its abrasion resistance and relatively smooth surface (which does not easily collect dust). It is, therefore, ideal for areas where a dust-free environment is required, such as laboratories and hospitals. If a washable top coat is provided on intumescent paint, the structure can be washed with water, which is an advantage in food-processing plants.

Intumescent paint protection, however, is far costlier than conventional spray-on fireproofing. It is, therefore, not practical for buildings where economy is a major consideration. In buildings with large areas of steel deck supported on steel beams and columns, the intumescent paint may be used for beams and columns, but the deck may be protected by a more economical alternative.

UNPAINTED EXPOSED STEEL

Unpainted exposed steel has also been used, but mainly in spandrel beams. On the interior of the building, the spandrel beams are protected by conventional techniques, but on the exterior, they remain exposed. To shield the exposed faces of beams from flames, a protruding steel-sheet cover is fastened to bottom flanges of beams. For aesthetic reasons, the steel sheet is designed as an integral part of the exterior cladding of the building.

SUSPENDED CEILINGS

Suspended ceilings consisting of gypsum lath and plaster, gypsum boards, or acoustical tiles are also used to provide fire protection to otherwise unprotected steel beams or trusses in roof-ceiling or floor-ceiling assemblies. The ceiling grid may be directly attached to, or hung from, the bottom flanges of beams or bottom chords of trusses.

The effectiveness of such a system depends on how well the ceiling continues to perform as a barrier. Often the maintenance is not adequate, resulting in a defective ceiling tile or gypsum board not being replaced properly, so that hot gases enter the plenum space to attack the unprotected steel.

FOCUS ON SUSTAINABILITY

Sustainability Features of Steel

Steel can be easily recycled by melting old steel and reforming it. Large quantities of new steel are being made from steel recovered from old cars, washing machines, refrigerators, structural members from old buildings, and so on. The steel industry claims that along with aluminum, steel is the most recycled building material. Because the industry is moving toward sustainable practices, new steel contains increasingly larger percentages of recycled content.

In the United States, all structural steel sections, such as the W-shape, C-shape, and L-shape, are produced from the electric arc furnace (EAF) process, which utilizes mostly recycled steel.* The basic oxygen furnace (BOF) system of making steel, which uses nearly 75% virgin ore, is primarily used in nonstructural applications.

As mentioned earlier, steel does not corrode if used in interior locations. It is, therefore, a durable material even without any corrosion protection. Another sustainable feature of steel is that it lends itself readily to deconstruction, particularly if mechanical fasteners are used for component assembly. In a well-designed assembly, steel components can be disassembled, refabricated, and reassembled at a new location (and new use) without the need to recycle.

Steel structures, being lighter in weight (as compared with concrete and masonry structures) require smaller foundations, conserving on the use of concrete and excavations.

Although the manufacturing of steel produces pollution, there are no pollution issues with steel during its service. Unlike engineered wood components, there are no outgassing issues with steel. However, steel members protected by fibrous fire protection may dispel stray fibers in the air.

*Hewitt, Christopher. "Sustainable Steel," *Modern Steel Construction*, September 2003.

A Building Separation Joint in a Steel Frame Building

As discussed in Chapter 9, continuous building separation joints are required in many buildings. A preferred way is to use two columns and two beams at the joint. Often for the sake of economy, however, a single column with two beams is used. The following detail explains the construction of such a joint in a building with W-shape columns and beams. (See Chapter 20 for a similar detail in a reinforced concete building.)

Details of a building separation joint for a single-story building, consisting of joists and joist girders, are shown in the accompanying art.

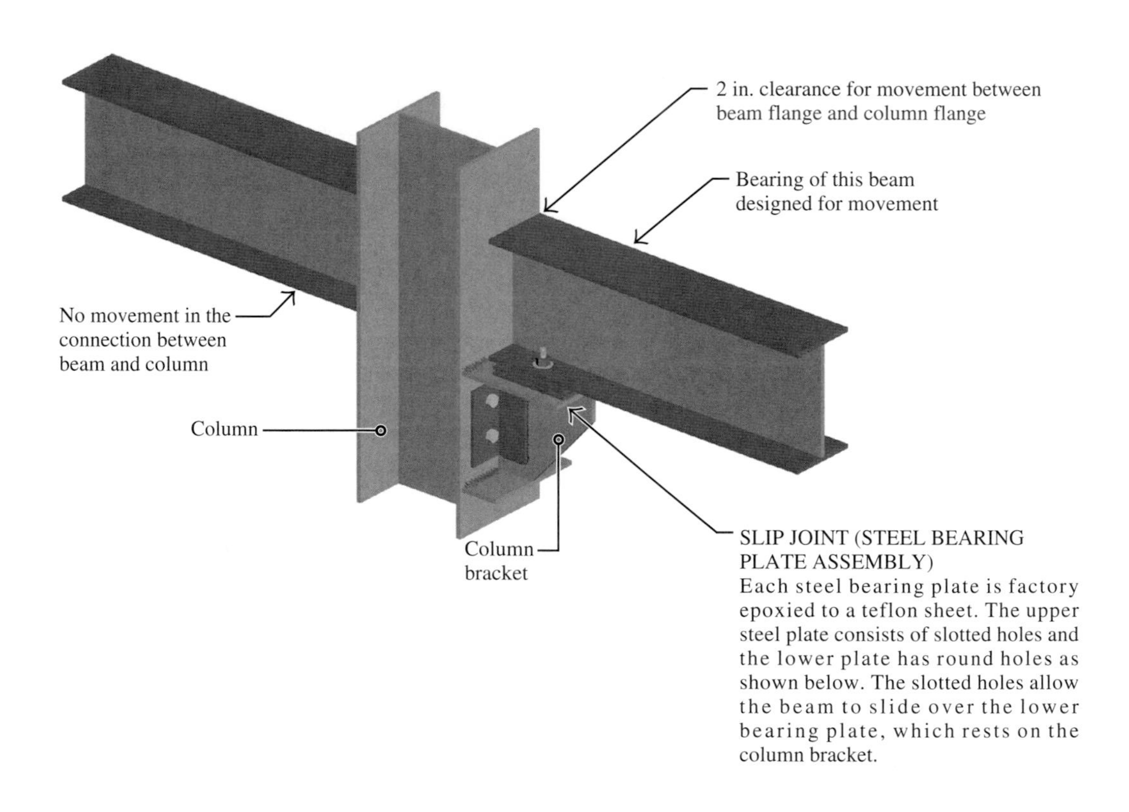

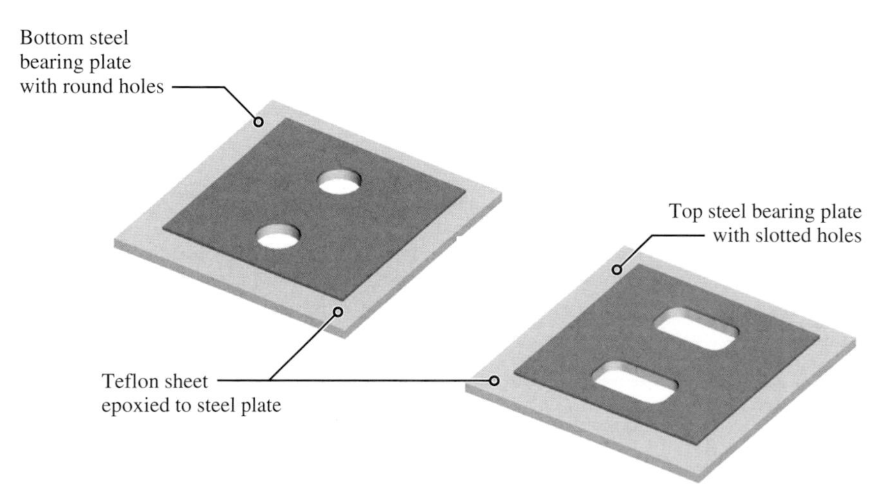

STEEL BEARING PLATES
Although the plates have been shown separated from each other, they are placed over each other with teflon surfaces in contact. (For properties of teflon sheet, see "Building Separation Joint in a Sitecast Concrete Building" in Chapter 20.)

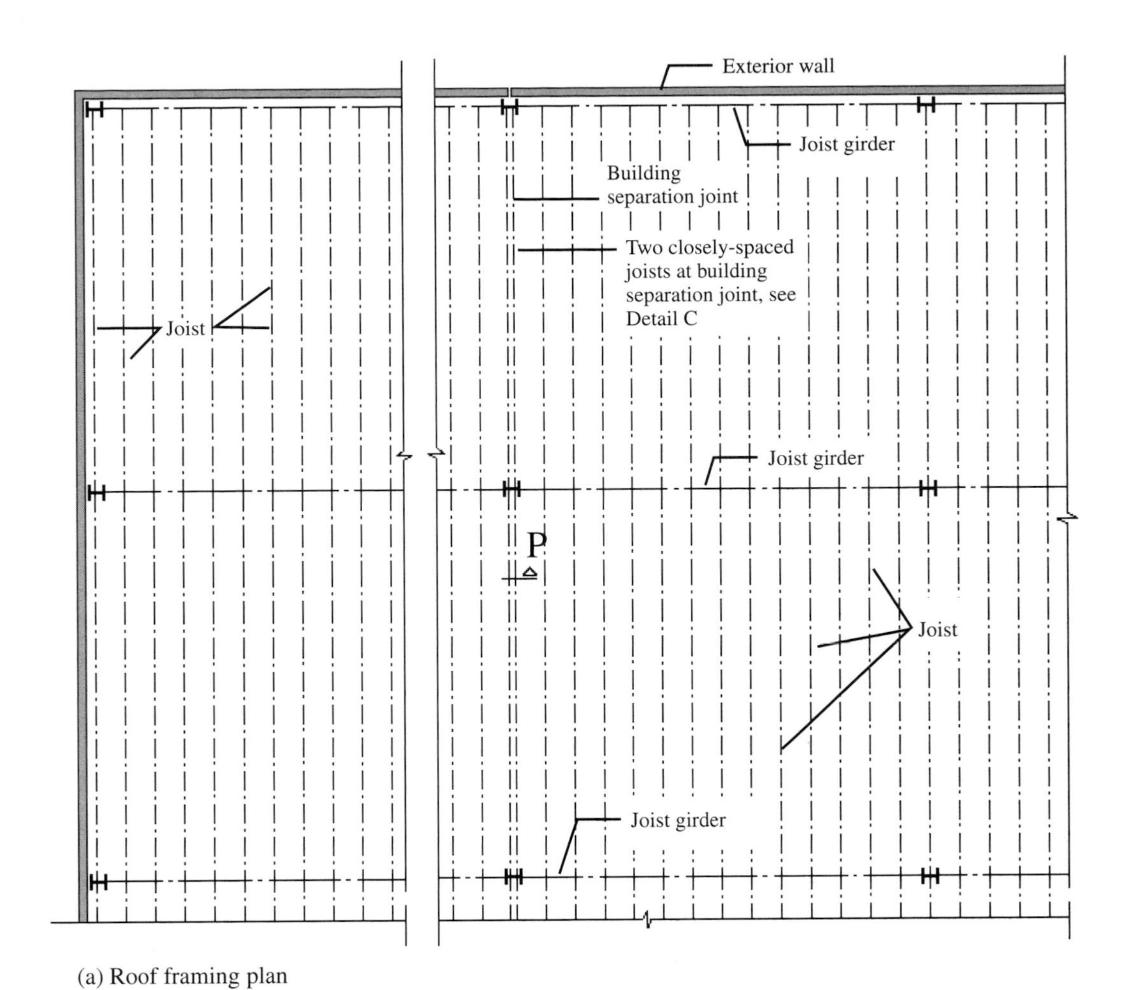

(a) Roof framing plan

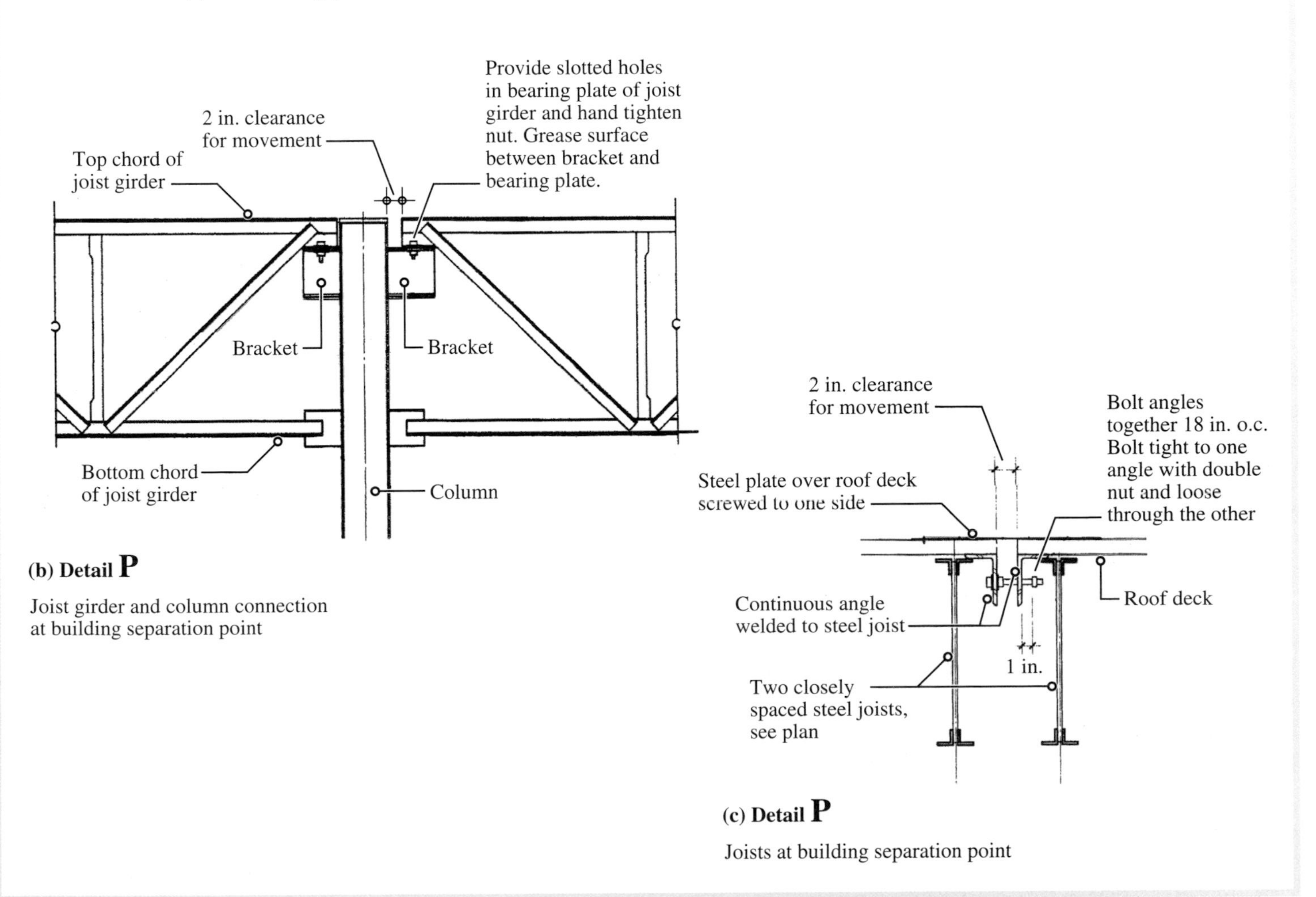

(b) Detail P

Joist girder and column connection
at building separation point

(c) Detail P

Joists at building separation point

PRINCIPLES IN PRACTICE

Fundamentals of Skeleton Frame Construction

As mentioned in the introduction to this chapter, a frame structure is defined as a structure made entirely of linear elements—beams and columns. The earliest predecessor of the frame structure is the *post-and-beam* (also called *post-and-lintel*) structure. A post-and-beam structure consists of columns (posts) and beams (lintels) with little or no connection between the posts and the beams. It was among the earliest structural systems devised by humankind, Figure 1. Ancient Egyptians, Greeks, and Romans used the post-and-beam system extensively to produce architectural masterpieces that stand until this day (Figure 4.1).

A post-and-beam structure works well for gravity loads. Under gravity loads, the beam is subjected to bending. Because there is no connection between the posts and the beam, the bending in the beam is not transferred to the posts, Figure 2. Consequently, the posts are under pure compression—no bending.

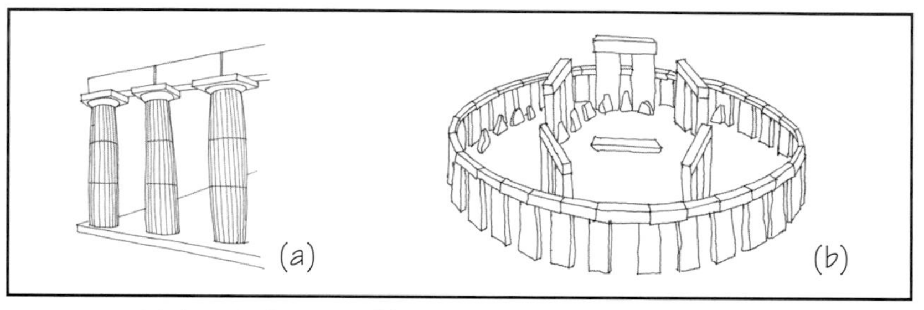

FIGURE 1 **(a)** A typical post-and-beam structure used by ancient Egyptians, Greeks and Romans (Figure 4.1). **(b)** Stonehenge (U.K., approximately 2500 to 1700 BC) is generally cited as the earliest surviving example of the post and beam structure.

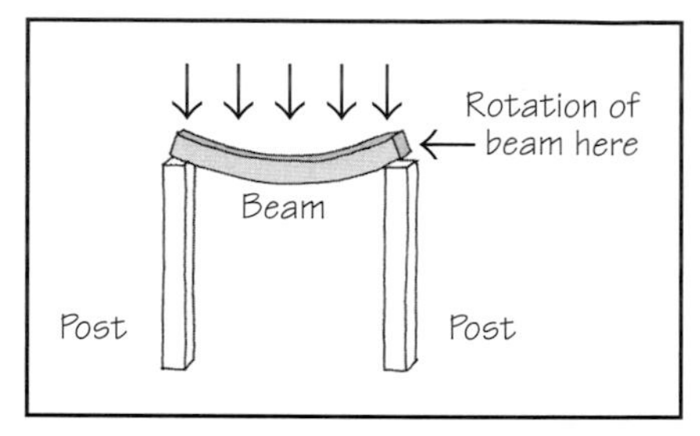

FIGURE 2 Under gravity loads, the beam in a post-and-beam structure is subjected to bending, which causes it to rotate at the joint. Because of the absence of a connection between the post and the beam, the rotation of the beam at the joint is not transferred to the post. The post is, therefore, subjected to compression only.

Under lateral loads (wind and earthquake), the post-and-beam structure is theoretically unstable. When the lateral loads are within the plane of the structure, the beam tends to slide over the column, and the column tends to overturn or slide along its base, Figure 3(a). If the lateral loads are perpendicular to the plane of the post-and-beam structure, the entire structure tends to overturn, Figure 3(b).

Early post-and-beam structures used large and hefty stones. Stability of the structure against overturning and sliding was achieved through large dead load of members and the friction between their surfaces.

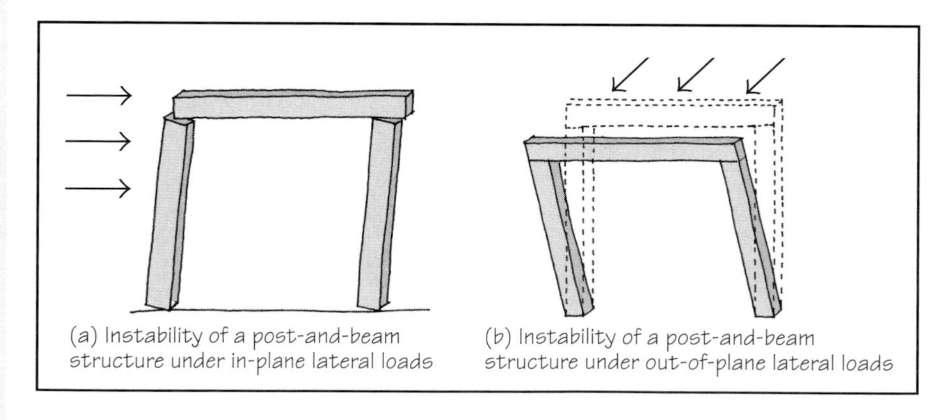

FIGURE 3 Under the action of both in-plane or out-of-plane lateral loads, a post-and-beam structure is unstable.

A frame structure is similar to post-and-beam structure, except that the members of the frame (the columns and beams) are connected together. The connection between a column and a beam can be a *pin connection,* giving a pin-connected frame, or it can be a *rigid connection,* giving a rigid frame.

A pin-connected column-beam frame can be one of the following two types:

- Pin-connected cantilevered frame
- Pin-connected braced frame

PIN-CONNECTED CANTILEVERED FRAME

A pin connection is defined as a connection that allows the members to rotate at the connection independently of each other. In other words, the rotation of one member at the connection is not transferred to the other member. The simplest pin-connected frame consists of a single-bay frame (i.e., one beam and two columns) in which the beam is connected to each column through a single pin (a nail, screw, or a bolt), Figure 4.

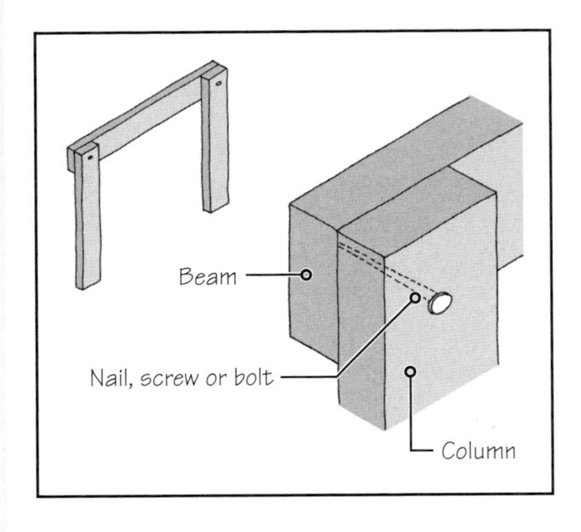

FIGURE 4 A single-bay pin-connected frame in which the beam and column are connected together with a single nail, screw, or bolt. A pin connection is also referred to as shear connection because the gravity load from the beam is transferred to the column creating shear in the nail, screw, or bolt.

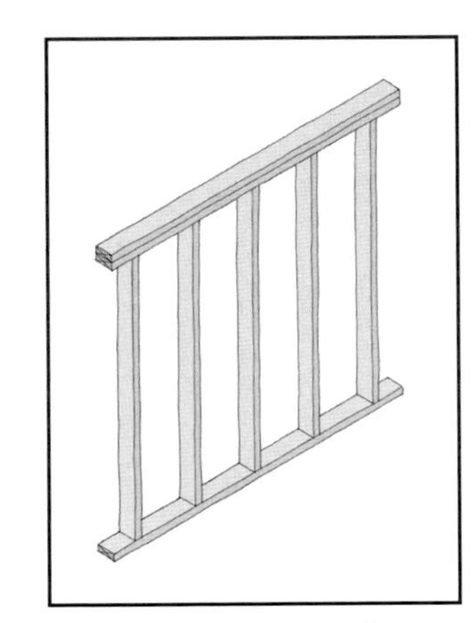

FIGURE 5 A connection between a stud and the top (or bottom) plate of a wood light frame, typically using two or three nails, is considered a pin connection.

This connection allows the the beam to bend and rotate under gravity loads without transferring the bending to the column. The connection, however, transfers the gravity load on the beam through shear created in the pin. A pin connection is, therefore, also referred to as a *shear connection.* Another term used for this connection is a *simple connection* (Section 16.7).

In practice, a shear (pin or simple) connection need not have lapped members, as shown in Figure 4. Additionally, welds or multiple nails, screws, or bolts may be used to achieve a pin connection. Thus, the connections between the studs and the top and bottom plates of a wood light frame, which are typically joined with two or three nails at each connection, behave as pin connections, Figure 5.

Similarly, the joints between wood truss members, connected together with nail plates (Chapter 12) are also considered as pin connections. The same is true of the joints between the members of a steel joist or joist girder in which the members are welded together (Figures 16.11 to 16.13). Although such joints are not true (100%) pin connections, they provide adequate rotation of members at a joint to be considered as pin connections.

In buildings, a joint detailed as a *true* pin connection is used for both structural and ornamental reasons, Figure 6. True pin connections are more common in civil engineering structures, such as waterway, highway, or railroad bridges.

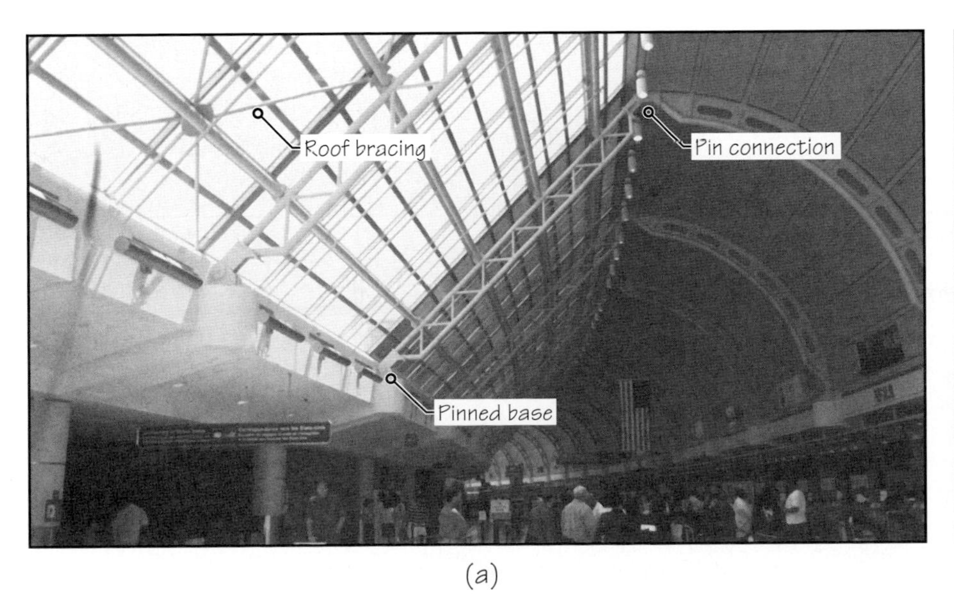

(a)

(b)

FIGURE 6 **(a)** A glass roof at Toronto Airport supported by three-pin frames. Each frame consists of two members, a linear trussed member and a curved hollow-web member. The members are joined at the top with a pin connection. **(b)** A pin connection has also been used to support each member at the base. All three pins in each frame have been detailed as true pins.

(Continued)

Fundamentals of Skeleton Frame Construction (*Continued*)

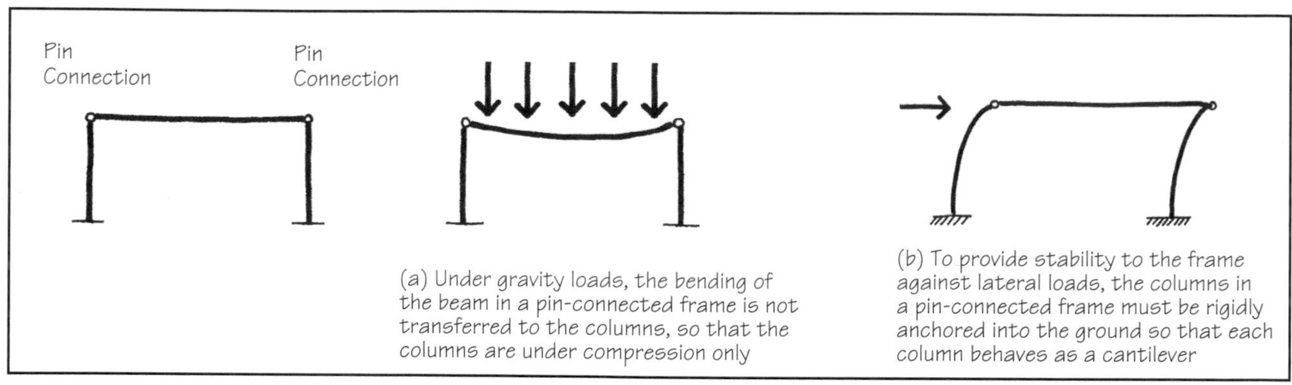

FIGURE 7 In a conceptual line diagram of a pin-connected frame, a pin is represented by a circular dot, and the frame members are represented by single heavy lines.

Details of commonly used shear connections between structural steel beams and columns were illustrated in Section 16.7.

In a conceptual line diagram of a pin-connected frame, a pin connection is represented by a hollow dot, and the columns and the beams are represented by single heavy lines, Figure 7. If the beam of a pin-connected frame is subjected to gravity loads, it will bend as shown in Figure 7(a). The bending of the beam (which causes the beam to rotate at the joint) is not passed on to the column. Therefore, the columns in a pin-connected frame come under pure compression, exactly as in a post-and-beam structure.

One way to ensure the stability of the frame of Figure 7 against lateral loads is to anchor the columns rigidly into the ground so that they are cantilevered from the ground like utility poles. Under the action of a lateral load, both columns will deflect horizontally, as shown in Figure 7(b)—a tendency called *racking*. Excessive racking of the cantilevered frame of Figure 7 is one of its major limitations. Racking can be reduced only by increasing the size and stiffness of columns—a solution that becomes increasingly inefficient as the column height increases.

Note that because of the pin connection between the beam and the columns, the bending in the columns of the cantilevered frame is not transferred to the beam. Thus, in transferring the lateral load from one column to the other, the beam of Figure 7 comes under compression.

Another limitation of the cantilevered frame of Figure 7 is the need for rigid column bases. The rigidity at the base of each column can be achieved by anchoring the column to foundation, that is, a concrete footing. The footing must be heavy and sufficiently large so that the bending in the columns will not rotate the footing.

A shallow, isolated column footing is generally not rigid enough to resist rotation of the footing, particularly if the footing is small and the soil is compressible. Such a footing simply rotates with the column, Figure 8(b), representing a *pinned column base*. A deep, isolated column footing, Figure 8(c), or a mat footing in which several columns share a common footing (Chapter 21), is treated as a *rigid* (or *fixed*) *column base.*

NOTE

Racking of Frame

The term *racking* represents lateral deflection (angular deformation) of a frame under lateral loads and is an important design consideration. Building codes limit the amount of racking in buildings. Excessive racking causes occupant discomfort, damage to nonstructural components, such as doors and windows, and fatigue in structural components, which may lead to structural failure over a period of time.

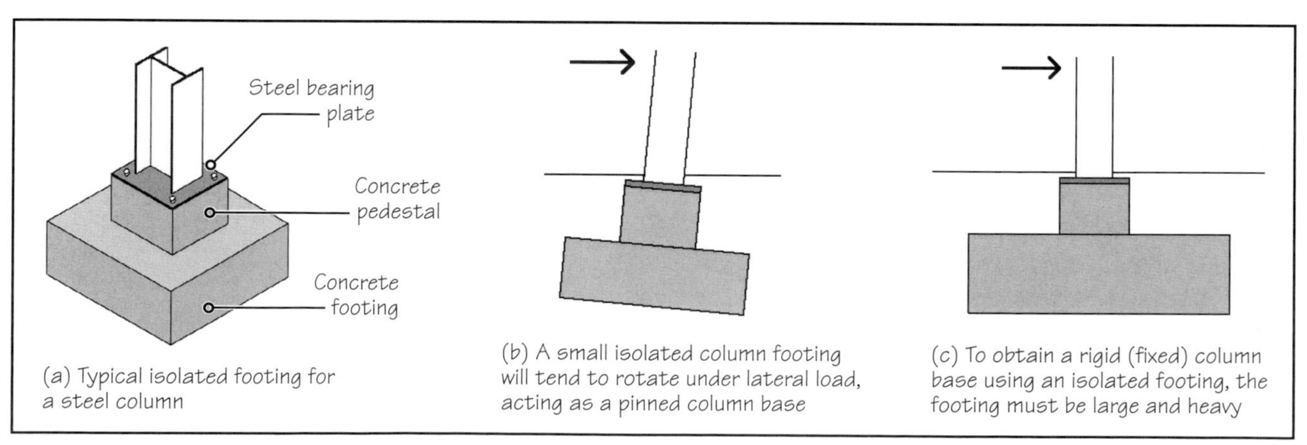

FIGURE 8 **(a)** A typical isolated footing for a steel column. **(b)** A small column footing is regarded to provide a pinned column base because it will tend to rotate under the action of lateral load. **(c)** To obtain cantilever behavior from a column with an isolated footing, the footing must be large and heavy.

PIN-CONNECTED BRACED FRAME

In the conceptual line diagram, a pinned column base is represented in the same way as a pin connection, that is, by a circular dot. Thus, a single-bay, pin-connected frame in which the columns have pinned bases is a 4-pin frame, Figure 9. Such a frame is obviously unstable as the frame will simply fold under lateral loads. A 4-pin frame is also unstable under an unsymmetrical gravity load on the beam. In fact, a 4-pin frame is not a structure, but a *mechanism.*

A 4-pin frame can be stabilized, however, by bracing the frame in its own plane. The most efficient and commonly used bracing system is a set of diagonal braces, also referred to as *X-bracing,* or *cross bracing,* Figure 10. Under lateral loads, one diagonal brace is subjected to compression and the other to tension. The brace subjected to compression buckles under the load and becomes ineffective. Stability of the frame is, therefore, provided by the brace that is in tension. Because steel is strong in tension, X-bracing members need not be heavy. Therefore, X-bracing generally consist of steel rods, hollow pipes, plates, or angles. In high-rise structures, wide flange or hollow structural steel sections are used.

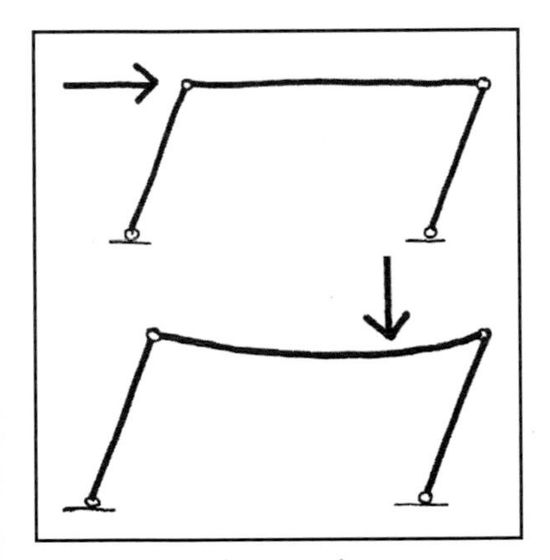

FIGURE 9 A four-pin frame is not a structure but a mechanism because it is unstable under both lateral loads and unsymmetrical gravity loads.

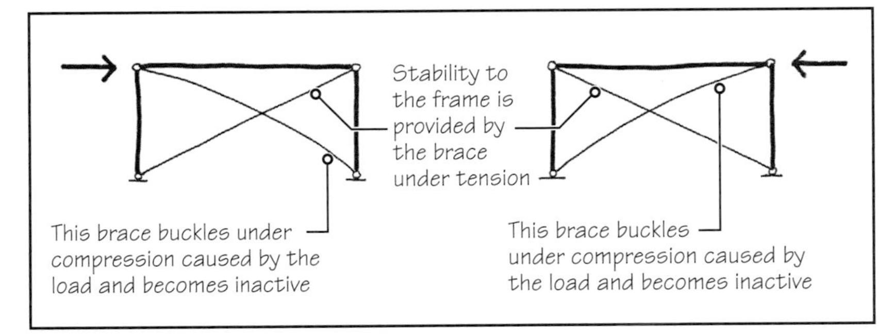

FIGURE 10 A four-pin frame can be stabilized by using X-bracing. Note that only one of the two X-braces is active under a lateral load.

Other bracing alternatives are the *K-brace* and *eccentric K-brace,* Figure 11. K-bracing and eccentric K-bracing are generally used in multistory buildings, where substantial openings (doors or windows) are required in braced bays. The difference between K-brace and eccentric K-brace lies in their details. The braces in an eccentric K-brace do not meet at a point, leaving a small length of beam between the connections. An eccentric K-brace is generally recommended for use in seismic areas due to its ability to provide ductility of the frame.

Because a K-brace (or an eccentric K-brace) is connected to the center of the overlying beam, it must carry gravity loads on the beam, which halves the beam's span, reducing beam size. Gravity load support implies that K-braces are heavier than X-braces. Thus, unlike X-braces, in which only one brace is active at a time, both K-braces are designed to resist tension as well as compression.

In low-rise buildings, one diagonal brace may be used, which must be heavy enough to resist compression as well as tension, Figure 12(a). A particularly interesting bracing alternative for a pin-connected frame is a *panel brace* in which the entire space within the frame is filled with a panel, Figure 12(b). In concrete or steel structures, the panel brace consists of a masonry or concrete infill wall, referred to as *infill brace,* or *shear wall.*

RIGID FRAME

In a rigid frame the connections between the beam and the columns are rigid. Unlike a pin connection, a *rigid connection* retains a 90° angle between the connected members on deformation under the loads. Rigidity implies that the bending of one member is transferred through the joint to the other member. Therefore, a rigid connection is also called a *moment-resisting connection,* or, simply, a *moment connection.*

Because both the beam and the columns bend together, all three members of a rigid frame function collectively in resisting a load. Consequently, when the beam of a rigid frame is subjected to gravity loads, its deformation is smaller than that of the beam in a pin-connected frame. In the conceptual line diagram, a moment connection is represented by a small diagonal thickening of the joint, as shown in Figure 13.

(*Continued*)

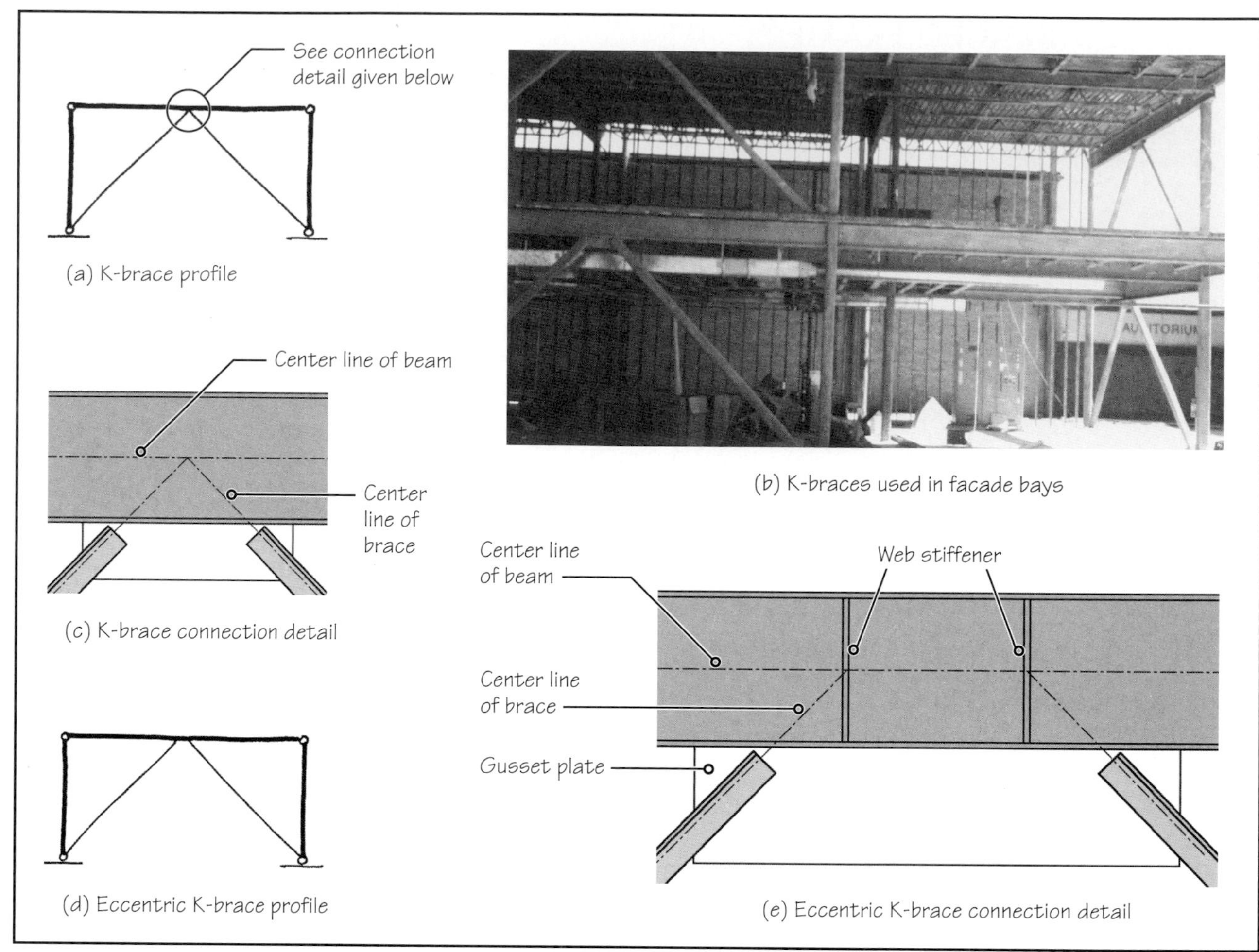

FIGURE 11 K-braces and eccentric K-braces.

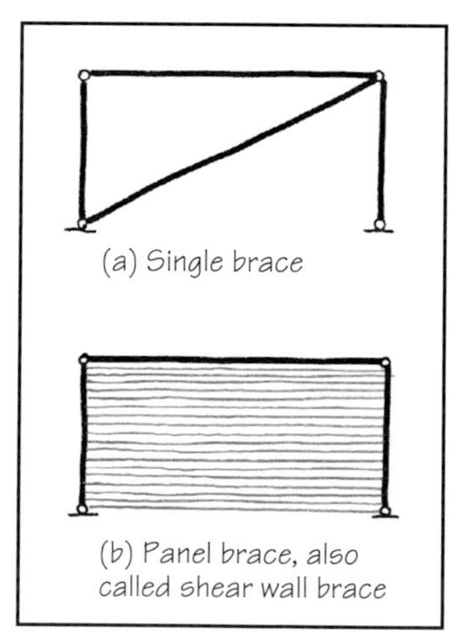

FIGURE 12 **(a)** If a single brace is used, it must be sufficiently heavy against buckling under compression. **(b)** Shear wall bracing.

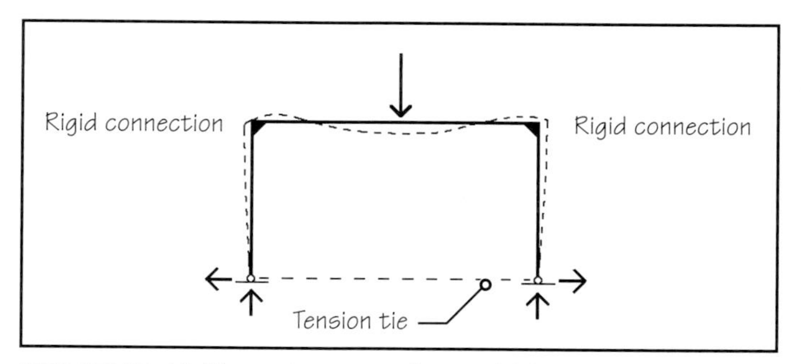

FIGURE 13 Unlike a pin connection, a rigid conection retains its 90° angle between the connected members on deformation. Consequently, a rigid frame resists gravity loads by creating bending in all three frame members. Because bending is structurally inefficient, a rigid frame is generally less economical than a comparable braced frame.

Note that a rigid frame under gravity loads experiences an outward thrust in column bases, similar to that of an arch.

In a single-line diagram of the frame, a rigid connection is denoted by a diagonal thickening of the joint.

RIGID FRAME VERSUS BRACED FRAME

A rigid frame does not require bracing for lateral stability, providing an unobstructed space. However, a rigid connection is more difficult to make than a shear connection. In a wood rigid frame, joint rigidity may be obtained by securely fastening a plywood or OSB gusset plate on both sides of the joint, Figure 14. Compare this with simply nailing the members together to obtain a pin connection.

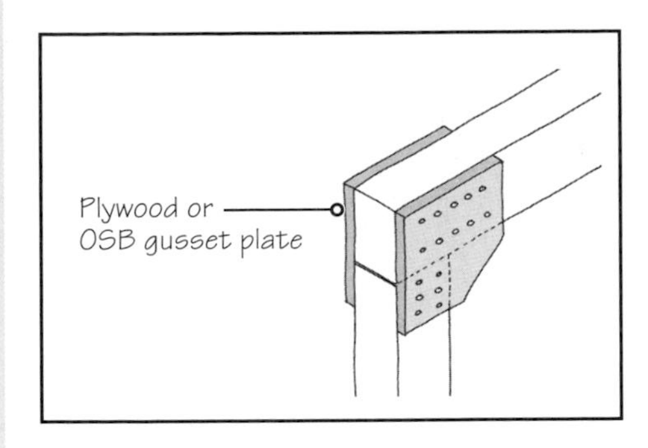

FIGURE 14 A rigid connection between two wood frame members can be obtained by nailing a plywood or OSB gusset plate on both sides of the joint. A rigid connection between two steel members is more complex (Figure 16.29).

As illustrated later, the difference between a rigid connection and a shear connection is even more pronounced in a steel frame structure. Simplicity of connections is one reason why braced steel frames are preferred over rigid steel frames.

Braced frames are generally less expensive and provide for faster erection. Rigid frame structures are used only where braced frames are architecturally unacceptable because of the obstruction caused by the braces. Some of the other advantages of a braced frame over a rigid frame are as follows:

- The columns in a rigid frame are subjected to bending under lateral loads (Figure 13). In a braced frame, the columns are subjected to axial stresses only. Because bending is an inefficient structural action, the columns in a rigid frame are heavier than the columns in a braced frame. This difference becomes progressively more pronounced as the height of the structure increases.
- The racking of a rigid frame is larger than that of a comparable braced frame, Figures 15(a) and (b). In fact the term *rigid* is a misnomer because a braced frame is much stiffer than a corresponding rigid frame. The racking of a rigid frame can be decreased somewhat by using fixed (rigid) column bases, but the increase in stiffness so obtained does not match the inherent stiffness of a braced frame, Figure 15(c). Racking of a rigid frame also occurs under unsymmetrical gravity loads, Figure 16.
- Another limitation of a rigid frame is its arching action, which makes the columns spread out under gravity loads, requiring a heavier footing or a tie that connects the column footings (Figure 13).

Because of its several limitations, a rigid-frame solution for a building that is more than six to seven floors in height becomes too uneconomical to justify its use.

MULTIBAY SINGLE-STORY FRAMES

The fundamentals discussed with respect to single-bay frames also apply to multibay frames. However, it is not necessary to brace every bay of a multibay frame. Sufficient frame stiffness can be achieved by bracing a few selected bays in a single-story, multibay frame building. For example, in a 4-bay, single-story frame, either two end bays or two central bays may be braced,

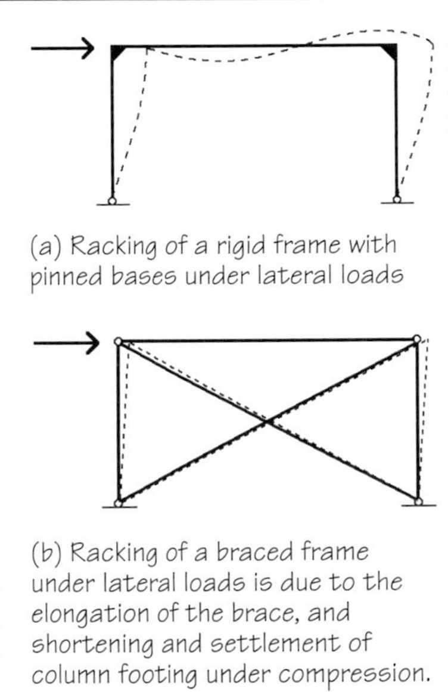

(a) Racking of a rigid frame with pinned bases under lateral loads

(b) Racking of a braced frame under lateral loads is due to the elongation of the brace, and shortening and settlement of column footing under compression.

The racking of a braced frame is much less than that of a rigid frame even when the rigid frame is supported on fixed column bases. A braced frame is, therefore, much stiffer than a corresponding rigid frame.

A K-braced frame is stiffer than a corresponding X-braced frame because of its shorter diagonals.

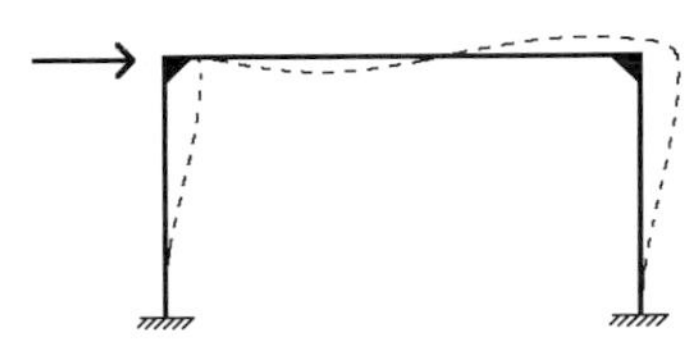

(c) Racking of a rigid frame with fixed bases under lateral loads is smaller than that of a rigid frame with pinned bases.

FIGURE 15 Comparison of the racking of a rigid frame and braced frame.

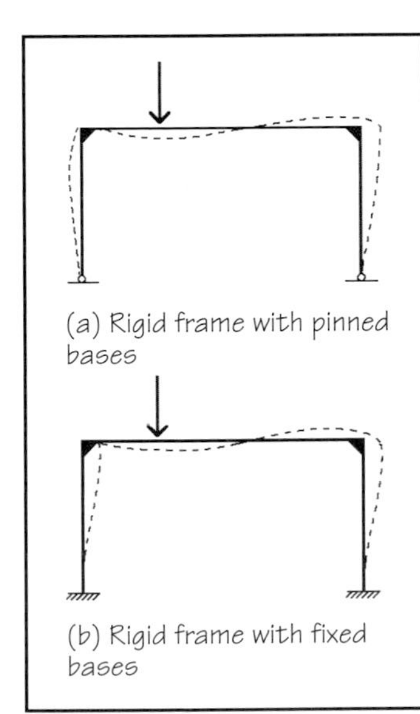

(a) Rigid frame with pinned bases

(b) Rigid frame with fixed bases

FIGURE 16 Racking of the frame is produced even under unsymmetrical gravity loads, which is more pronounced in a rigid frame than in a braced frame.

(*Continued*)

Fundamentals of Skeleton Frame Construction (*Continued*)

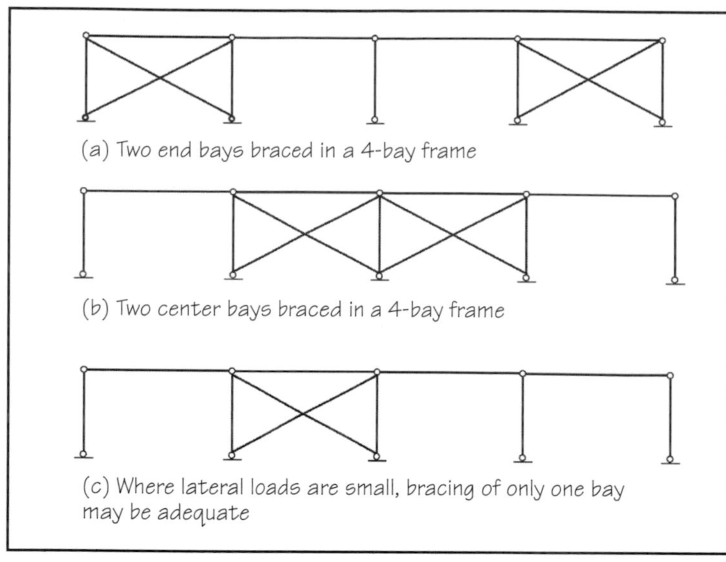

FIGURE 17 Some of the ways of bracing a 4-bay, single-story frame.

Figures 17(a) and (b). Where lateral loads are small, bracing of only one bay may be sufficient, Figure 17(c).

The three-dimensional nature of the building requires that the frames in both principal directions be braced, Figure 18(a). As far as possible, braced bays should be symmetrically located in the building. Asymmetrical locations of braced bays are structurally inefficient but may be used where architectural considerations dictate their use. A combination of a rigid frame and braced frame may also be used. In a common braced and rigid-frame combination, the rigid frame is used along one principal direction of the building and braced frame is used along the other direction, Figure 18(b).

Just as every bay of a multibay braced frame structure need not be braced, every bay of a multibay rigid frame need not be a rigid frame. Sufficient rigidity may be obtained by using rigid column-beam connections for a few selected bays and leaving the others as pin connections, Figures 19(a) and (b).

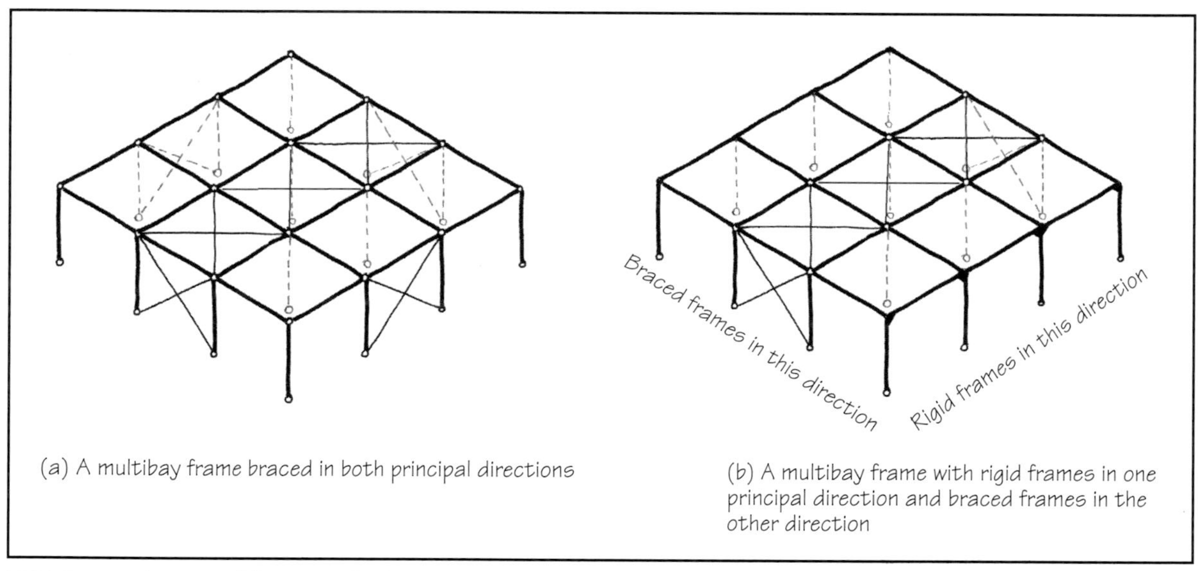

FIGURE 18 **(a)** Braced frames in both principal directions. **(b)** Braced frames in one direction and rigid frames in the other direction.

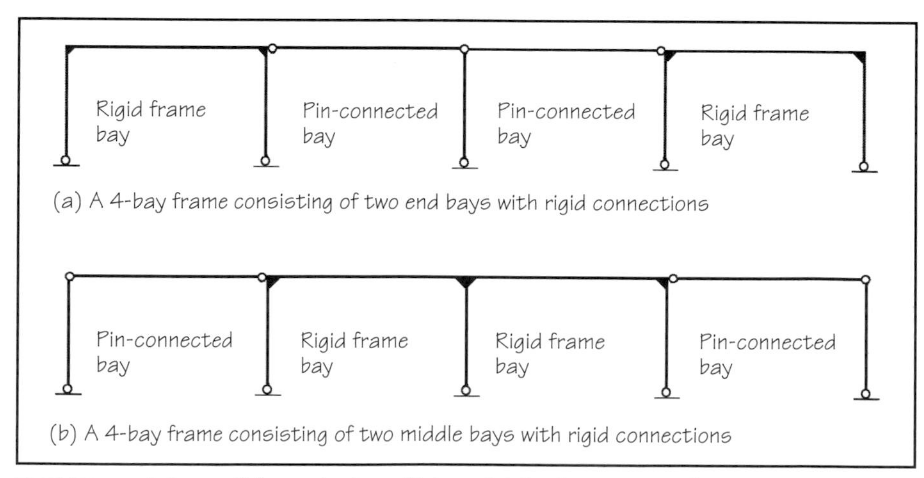

FIGURE 19 It is possible to obtain sufficient rigidity by using rigid connections in a few bays of a multibay frame.

STIFFNESS OF HORIZONTAL DIAPHRAGMS (FLOORS AND ROOF)

Resistance to lateral loads in a building requires that various components of the building function interactively. This implies that the horizontal diaphragms (floors and roof) of the building should be sufficiently stiff so that they can transfer lateral loads from vertical elements on one face of the building to the vertical elements of other faces.

Generally, reinforced concrete floors, concrete-filled steel deck floors and insulation-topped steel roof decks have sufficient stiffness for the purpose (referred to as providing *diaphragm action*). Where diaphragm action is not available, bracing of roofs may be needed, as shown in Figure 18. (See also Figure 6, where the large glazed roof is braced by diagonal braces.)

MULTISTORY STEEL FRAMES

Although the principles of framing a multistory building are similar to those of a multibay, single-story building, a much larger number of solutions exist as the building height increases. Complete coverage of the framing solutions for high-rise buildings is outside the scope of this text. The coverage provided here is limited to buildings up to 30 to 40 stories in height.

For a low-rise multistory building, a steel rigid frame may be used. Note once again that a rigid-frame solution becomes progressively uneconomical with increasing building height due to the inordinate increase in column and beam sizes. However, because of the absence of braced bays, it provides greater adaptability in organizing the interior spaces and exterior elevations. A typical steel rigid frame structure is shown in Figure 20.

A braced frame structure is more commonly used for both low-rise and high-rise buildings. The bracing may be incorporated in the exterior bays, where it is generally most efficient. In a large building, a few interior bays may also need to be braced, Figure 21. In a braced steel frame building, all column-beam connections are simple connections.

FIGURE 20 A typical steel rigid frame building. Observe the absence of braced bays that gives greater freedom in organizing the interior spaces and building's facade.

FIGURE 21 A typical braced frame building. Observe the provision of braced bays in the interior as well as the exterior of the building.

(*Continued*)

Fundamentals of Skeleton Frame Construction (*Continued*)

TALL BUILDINGS IN STEEL

The ideal locations for interior bracing are around staircase shafts, elevator shafts, restrooms, and so on, because such spaces do not require large openings. A frequently used bracing system for tall buildings (up to 30 or 40 stories) is to brace only a few interior bays in the center of the building, giving a rigid central *shear core*, Figure 22. By providing the shear core, all the lateral loads are resisted by the core. Therefore, the columns, which are not part of the core (e.g., perimeter columns), do not resist any lateral loads. They are designed to resist gravity loads only. Additionally, all column-beam connections are simple connections.

The core is used to enclose functions such as stairs, elevators, restrooms, and HVAC shafts. For an office building, the core may occupy as much as one-third the total floor area of the building at each floor. The core also permits the perimeter of the building to be fully glazed.

An alternative to a braced shear core is to use a reinforced concrete shear core, Figures 23 and 24. The reinforced concrete core provides rigidity to the building by behaving as a tube. The tube may have perforations to accommodate doors

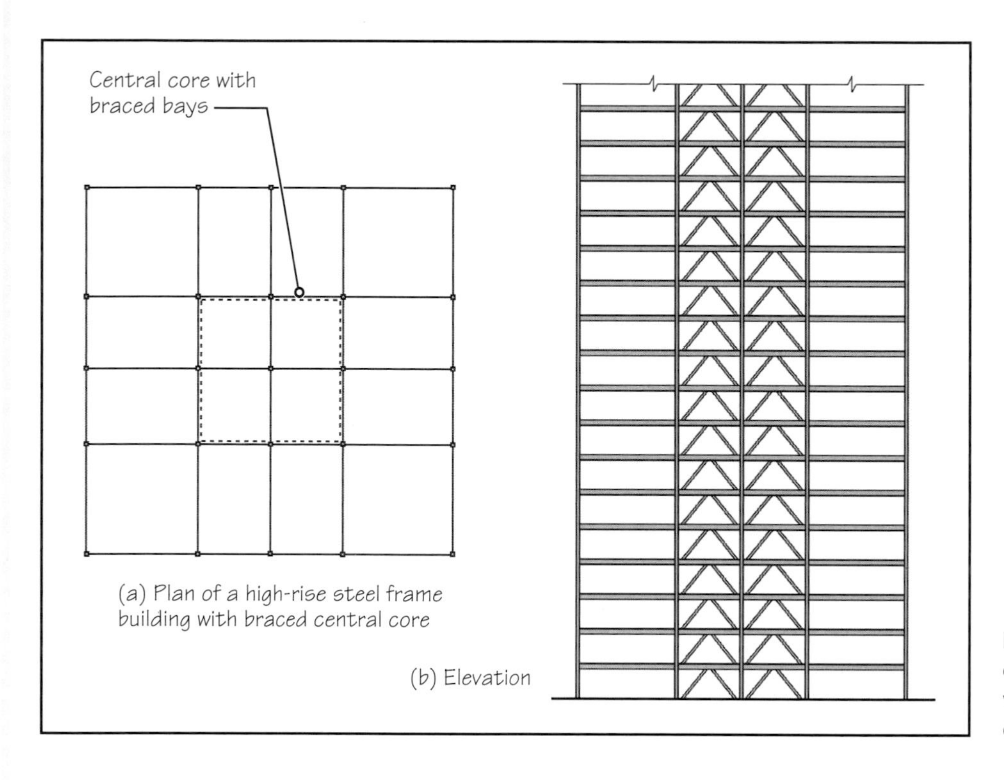

FIGURE 22 Plan and elevation of a high-rise steel frame building with braced steel frame shear core.

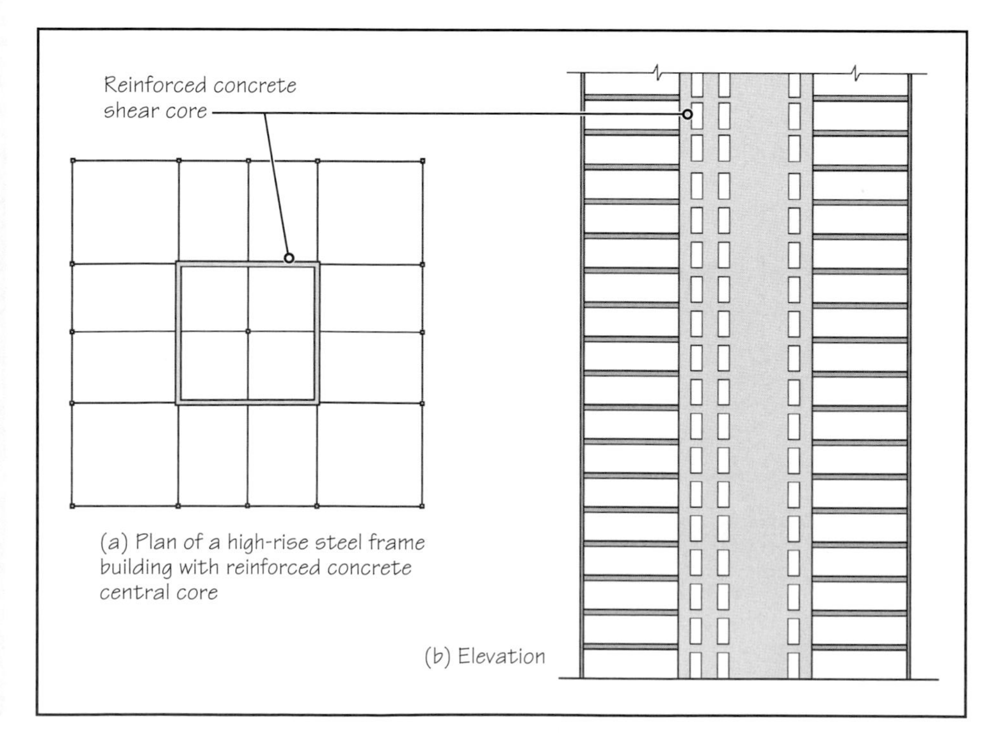

FIGURE 23 Plan and elevation of a high-rise steel frame building with reinforced concrete shear core.

FIGURE 24 A steel frame building with reinforced concrete central core under construction in Hong Kong.

and other openings, as long as the perforations are not too large. A reinforced concrete core has an advantage over a braced steel frame core because of its inherent high fire resistance, which is generally mandated of stair and elevator shafts. A particularly elegant building consisting of a reinforced concrete core is the 12-story office building shown in Figure 25. In this building, there are no perimeter columns, and the floors, consisting of steel wide-flange beams, have been suspended from the top of the core with steel cables.

FIGURE 25 Twelve-story West Coast Transmission Company Building, Vancouver, Canada, consisting of a reinforced concrete shear core. There are no columns in the building. The floors are framed out of steel wide-flange beams and are suspended at the perimeter from the core with steel cables. The absence of columns gives a large, column-free space at the ground. Similar buildings (with floors suspended from a concrete or steel-framed central core) have been constructed in several places [2].

(*Continued*)

Fundamentals of Skeleton Frame Construction (*Continued*)

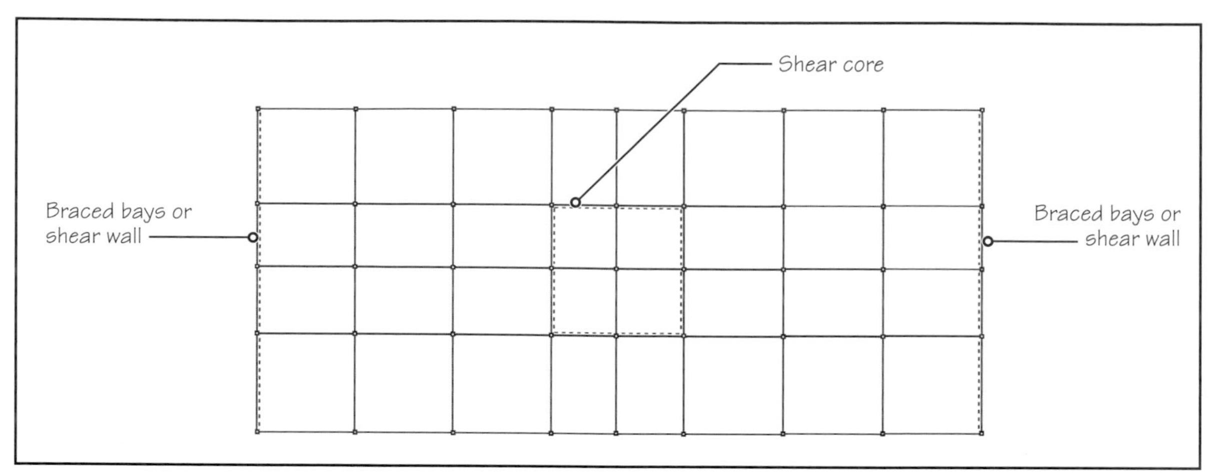

FIGURE 26 A commonly used method of framing a tall building that is much longer in one direction than the other is to use a central core and two end shear walls (or braced bays).

In a high-rise building, which is much longer in one direction than the other direction, a combination of a shear core and braced end shear walls is often used, Figure 26.

LONG-SPAN STEEL BUILDINGS

Just as steel is the only structural material available for very tall buildings (skyscrapers), it is also the only material suitable for long-span structures (spans greater than 100 ft). Long-span structures are typically required for sports facilities, convention centers, atriums, and other assembly spaces.

For spans of up to about 200 ft, the use of steel joists provide an economical solution, Figure 27. For longer spans, two-way trusses, two-way W-shape grid structures, and steel space frames are often used, Figures 28 to 31.

FIGURE 27 The use of steel trusses in the University of Houston Athletic Center (truss span is approximately 200 ft; truss depth approximately 10 ft at midspan). (Photo courtesy of Nucor-Vulcraft Group, with permission.)

FIGURE 28 The use of two-way steel trusses in main hall, Convention Center, Arlington, Texas (truss spans are 300 ft × 400 ft).

FIGURE 29 Atrium space consisting of a two-way grid of W-shape beams, King Saud University, Riyadh, Saudi Arabia.

FIGURE 30 Arched steel space frame used in William Beaumont Hospital, Royal Oak, Michigan. (Photo courtesy of Novum Structures, Inc., with permission.)

(*Continued*)

PRINCIPLES IN PRACTICE

Fundamentals of Skeleton Frame Construction (*Continued*)

FIGURE 31 Steel space frame used in the canopy of Steven F. Udvar-Hazy Center, National Air & Space Museum, Dulles International Airport, Chantilly, Virginia. (Photo courtesy of Novum Structures, Inc., with permission.)

PRACTICE QUIZ

Each question has only one correct answer. Select the choice that best answers the question.

36. Two dissimilar metals are generally separated from each other by a separator, which helps to prevent
- **a.** atmospheric corrosion.
- **b.** galvanic corrosion.

37. In a galvanized steel plate that carries the designation G60, number 60 means that
- **a.** the total mass of zinc coating on both surfaces of the plate is 60 oz/ft^2.
- **b.** the total mass of zinc coating on each surface of the plate is 60 oz/ft^2.
- **c.** the total mass of zinc coating on both surfaces of the plate is 60 g/ft^2.
- **d.** the total mass of zinc coating on each surface of the plate is 60 g/ft^2.
- **e.** none of the above.

38. Spray-on fire protection materials for steel contain
- **a.** portland cement with mineral fibers or lightweight aggregate.
- **b.** polymer-based cement with mineral fibers or lightweight aggregate.
- **c.** portland cement with normal-weight aggregate.
- **d.** polymer-based cement with normal-weight aggregate.
- **e.** any one of the above, depending on the manufacturer.

39. Fire-protection of structural steel framing members is mandated by building codes for all buildings.
- **a.** True
- **b.** False

40. A braced frame structure is
- **a.** generally more economical than a comparable rigid frame.
- **b.** has a smaller amount of racking than a comparable rigid frame.
- **c.** has a greater amount of racking than a comparable rigid frame.
- **d.** (a) and (b).
- **e.** (a) and (c).

41. Eccentric K-bracing is generally recommended for use
- **a.** because it is more economical than other forms of bracing.
- **b.** because it reduces the racking of the frame more than other forms of bracing.
- **c.** in hurricane-prone regions.
- **d.** in seismic regions.
- **e.** all of the above.

42. A steel frame building may consist of either a rigid frame or braced frame. A combination of the two systems (braced- and rigid-frame systems) cannot be used in the same building.
- **a.** True
- **b.** False

43. In a braced-frame steel structure, the braced bays
- **a.** should be provided on the exterior facade of the building.
- **b.** should be provided in the interior of the building.
- **c.** should be provided on both exterior facade and the interior of the building.
- **d.** may be provided on the exterior facade or the interior of the building.
- **e.** may be provided on the exterior facade, the interior of the building, or both.

44. A multistory steel frame structure must be provided with a central shear core.
- **a.** True
- **b.** False

45. The central shear core in a steel frame building can be constructed of reinforced concrete walls.
- **a.** True
- **b.** False

Answers: 36-b, 37-e, 38-a, 39-b, 40-d, 41-d, 42-b, 43-e, 44-b, 45-a.

KEY TERMS AND CONCEPTS

Atmospheric corrosion
Billet
Bloom
Bolts and bolted connections
Bridging
Built-up sections
Camber
Cast iron
Coke
Column splice
Composite deck
Corrosion protection
Decking
Erection process
Floor deck
Form deck
Frame structure
Galvanic corrosion
Galvanizing
Girder
Gusset plate
Hollow structural section
Intumescent paint
Open-web steel joist
Pour stop
Rigid (moment) connection
Rigid frame
Roof deck
Shear stud
Shim
Simple (shear) connection
Slag
Space truss
Spandrel beam
Spray-applied fire protection
Steel
Steel detailing
Steel fabrication process
Steel grades
Steel joists and joist girders
Structural steel shapes
Welding
Wide-flange
Wrought iron
Yield strength

REFERENCES

1. American Institute of Steel Construction. *A Guide for Architects*, 1988, p. 85.
2. Zahner, L. W. *Architectural Metals*. New York: John Wiley, 1995, p. 30.

FURTHER READING

1. American Institute of Steel Construction (AISC). *Designing With Steel—A Guide for Architects*, 1998.
2. American Institute of Steel Construction (AISC). *Steel Construction Manual*, 13th ed.

SELECTED WEB SITES

American Institute of Steel Construction, AISC (www.aisc.org)
American Iron and Steel Institute, AISI (www.steel.org)
Steel Deck Institute, SDI (www.sdi.org)
Steel Joist Institute, SJI (www.steeljoist.org)

REVIEW QUESTIONS

1. What is the important difference between the W-shapes used for beams from those used for columns.
2. Using sketches and notes, explain what is a plate girder and where it is used.
3. Sketch the elevation of a typical K-series joist.
4. Sketch the end of a K-series joist in three dimensions. What is the standard end-bearing depth of a K-series joist?
5. Using sketches and notes, explain how the joists and joist girders are stabilized against overturning.

CHAPTER 17

Light-Gauge Steel Construction

CHAPTER OUTLINE

17.1 Light-Gauge Steel Framing Members

17.2 Light-Gauge Steel Framing in Load-Bearing Applications

17.3 Advantages and Limitations of Light-Gauge Steel Framing

17.4 Non-Load-Bearing Light-Gauge Steel Framing

Light-gauge steel has a long history of use for the framing of interior partitions and backup exterior walls of commercial buildings (shown here). The use of light-gauge steel for the framing of load-bearing walls, floors, and roofs of residential buildings is relatively new, but it is gradually gaining in popularity. (Photo by MM.)

In the previous chapter, the shapes, sizes, properties, and applications of structural steel were covered. This chapter is devoted to the use of light-gauge steel in buildings.

Light-gauge steel framing has long been used in commercial buildings for non-load-bearing walls. More specifically, light-gauge steel is used extensively for interior partitions and exterior non-load-bearing walls in Types I and II construction, where the building codes mandate the use of noncombustible materials (see Chapter 7 for the definition of Types I and II construction). Because light-gauge steel is both lightweight and noncombustible, it is the material of choice in these applications.

More recently, the light-gauge steel frame has begun to compete with wood light frame (discussed in Chapter 13) in some areas of the United States. In this application, light-gauge

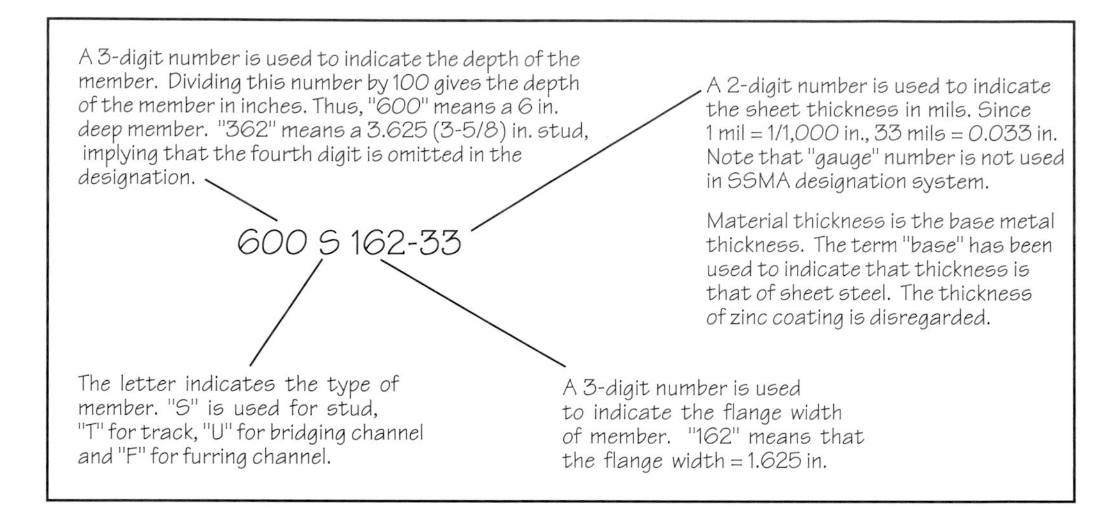

FIGURE 17.1 Designation system for identifying light-gauge steel members, established by the Steel Stud Manufacturers Association (SSMA).

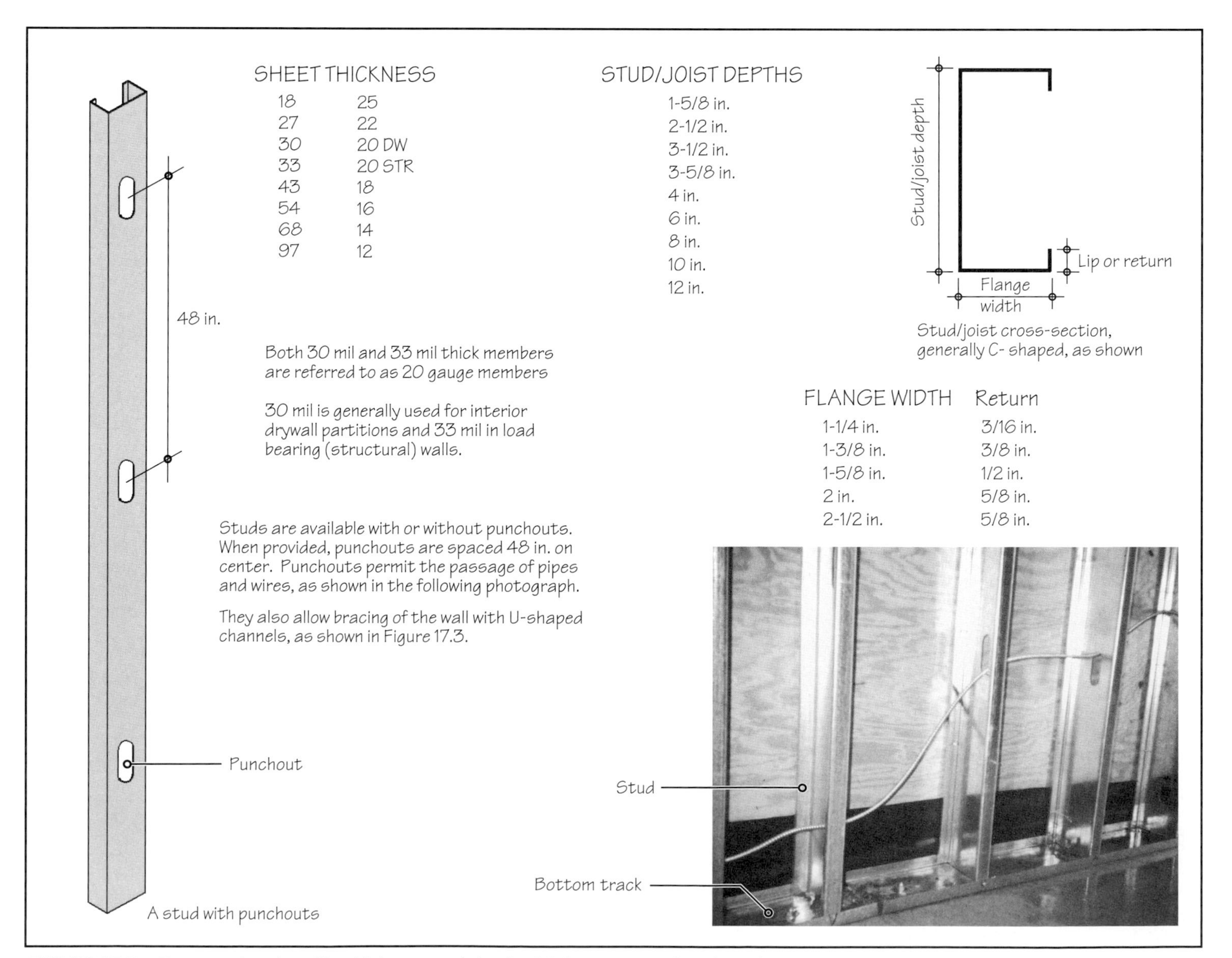

FIGURE 17.2 Cross-sectional profile, thickness, and depth of light-gauge steel studs and joists.

steel is used in load-bearing elements—walls, floors, and roofs—in the same way as wood light frame is used.

One region that has seen an increasing use of light-gauge steel in load-bearing applications is the western United States, where the higher strength, greater ductility, and lower dead load of light-gauge steel gives it an advantage over wood frame in resisting seismic forces. Other areas of growth are warm, humid areas that are infested with termites and/or susceptible to mold growth. Durability against termites is the primary reason for the popularity of light-gauge steel in the structural framing of homes in Hawaii, Florida, Georgia, and Texas.

17.1 LIGHT-GAUGE STEEL FRAMING MEMBERS

Light-gauge steel (also called *cold-formed steel*) framing members are made from sheet steel and are virtually always hot-dip galvanized. Four types of light-gauge steel framing members are commonly used:

- Studs and joists (symbolized by *S* in the SSMA designation)—generally C-shaped
- Tracks (symbolized by *T* in the SSMA designation)—U-shaped
- Bridging channels (symbolized by *U* in the SSMA designation)
- Furring channels (symbolized by *F* in the SSMA designation)

The Steel Stud Manufacturing Association (SSMA) uses a standard designation system to identify its four products, as shown in Figure 17.1.

Light-gauge steel studs are available in several depths, flange widths, and gauges, Figure 17.2. Interior partitions commonly use $3\frac{5}{8}$-in.-deep studs (20 or 22 gauge). Deeper and heavier studs may be required for load-bearing walls and exterior non-load-bearing walls.

Light-gauge steel tracks match the studs so that the studs fit within the tracks, Figure 17.3. The tracks run at the top and bottom of the studs, simulating the top and bottom plates in

NOTE

Gauge Versus Mil as Sheet Thickness of Light-Gauge Steel Members

In specifying sheet thickness of light-gauge steel members, SSMA encourages the use of *mil* instead of the old *gauge* number. However, both units are currently used. Figure 17.2 shows the conversion from mil to gauge.

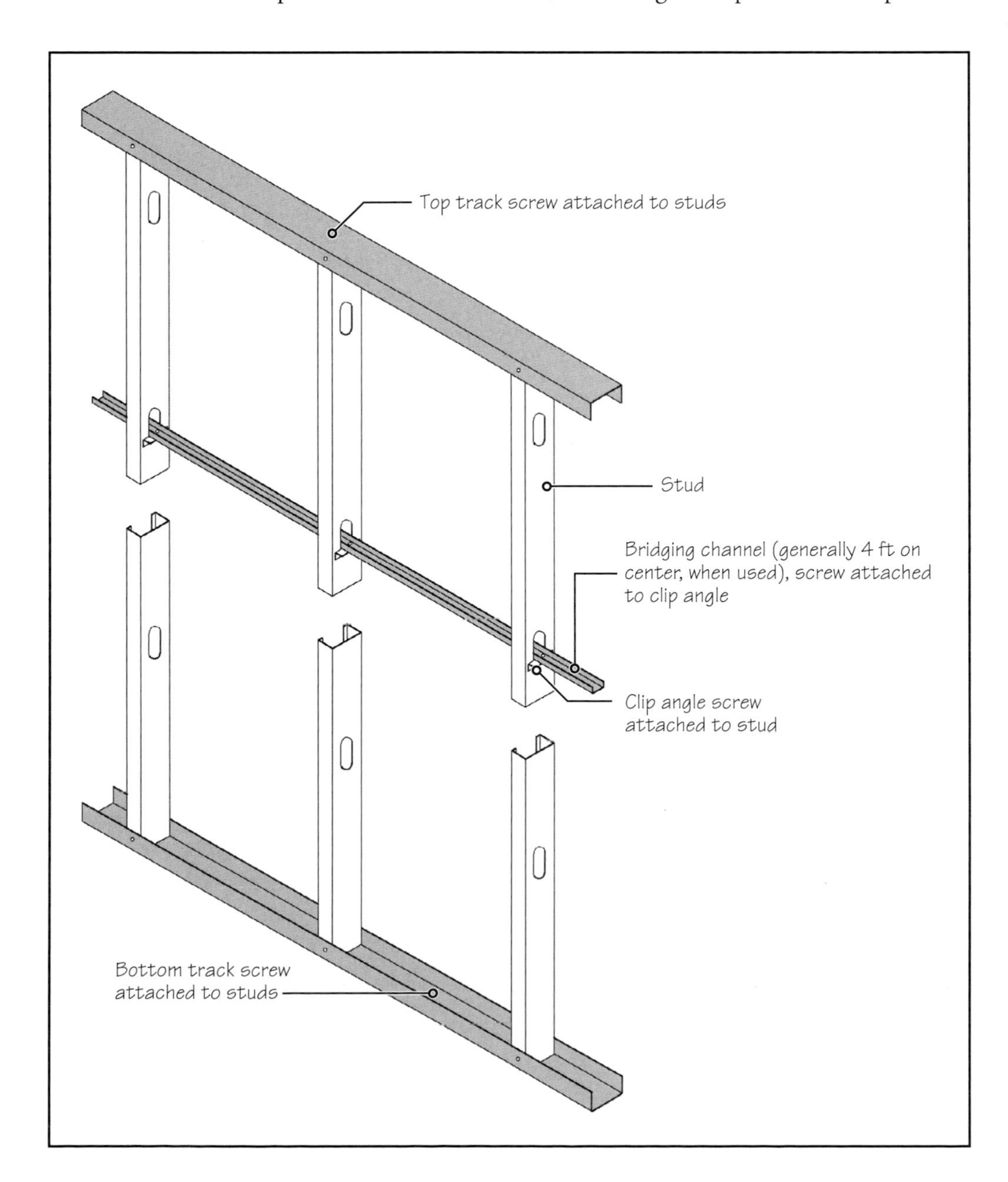

FIGURE 17.3 Typical framing of a wall using light-gauge steel members. Note that bridging channels can be used only with punched studs. They are used to brace a non-load-bearing wall but do not provide adequate bracing for a load-bearing wall.

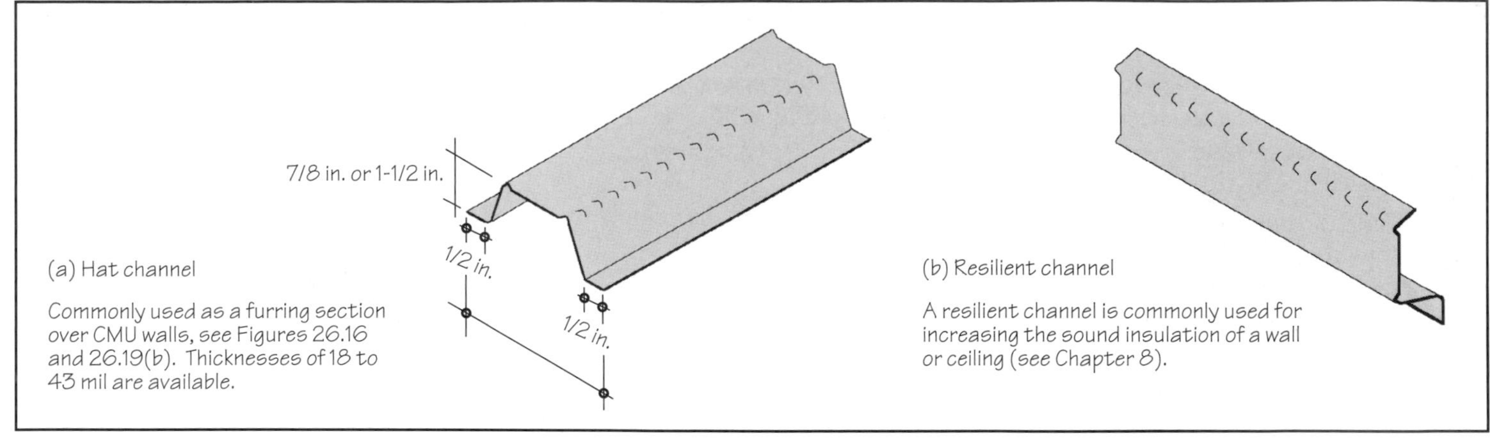

FIGURE 17.4 Two commonly used light-gauge steel furring channels.

wood light frame construction. Light-gauge steel bridging channels can be used with punched studs to brace the wall, similar to blocking in wood frame construction. When required, they are generally provided at 48 in. on center.

Two types of furring channels are in common use: (a) hat channels and (b) resilient channels, Figure 17.4. A hat channel is generally used on the interior face of a masonry wall with drywall finish (Figure 26.19). A resilient channel is used to increase the airborne sound insulation of a wall or floor (Chapter 8).

PRACTICE QUIZ

Each question has only one correct answer. Select the choice that best answers the question.

1. A load-bearing light-gauge steel frame structure is generally lighter than a comparable WLF structure.
   a. True **b.** False
2. Because steel is a strong material, light-gauge steel frame walls in load-bearing applications do not require bracing and blocking.
   a. True **b.** False
3. The amount of corrosion protection required in light-gauge steel frame members is generally
   a. G30.
   b. G45.
   c. G60.
   d. G90.
   e. none of the above.
4. The yield strength of light-gauge steel used in load-bearing applications is generally
   a. 33 ksi.
   b. 40 ksi.
   c. 50 ksi.
   d. either (a) or (b).
   e. either (a) or (c).
5. Punchouts in light-gauge steel studs are generally spaced at
   a. 48 in. o.c.
   b. 40 in. o.c.
   c. 24 in. o.c.
   d. 16 in. o.c.
   e. none of the above.

Answers: 1-a, 2-b, 3-c, 4-e, 5-a.

17.2 LIGHT-GAUGE STEEL FRAMING IN LOAD-BEARING APPLICATIONS

Load-bearing light-gauge steel frame (LGSF) construction is far more recent in origin than WLF construction (in use for nearly two centuries). In its present stage of development, it is almost identical to and is modeled after WLF construction. A one-for-one equivalence generally exists between the framing members of WLF and LGSF structures.

Beyond the structural frame, however, both LGSF and WLF systems are identical; that is, there is no difference in sheathing materials (for walls, floors, and roof), exterior and interior finishes, or in the installation of service and utility items between buildings using one or the other system. Thus, plywood and OSB are used as wall sheathing, subfloor, and roof sheathing in structures framed with light-gauge steel in the same way as they are used in wood frame structures. (When LGSF is used as structural frame in noncombustible construction—Type II Construction—gypsum sheathing is used instead of plywood or OSB.)

Figures 17.5 and 17.6 show the exterior and interior views of a typical load-bearing LGSF, highlighting its likeness with WLF. To a large extent, the likeness is intentional because it makes it easier for the builders to switch between the systems, increasing versatility and reducing the investment in the training of workers.

FIGURE 17.5 The exterior of a residential structure in which all load-bearing and non-load-bearing walls, floors, and roof are framed with light-gauge steel members. Observe the similarity of framing of this building with a wood light frame building. (Photo courtesy of Dietrich Metal Framing.)

FIGURE 17.6 The interior of a dwelling with light-gauge steel frame. In this structure, interior shear walls have been provided by using 20-gauge steel straps fastened to sheet-steel gusset plates, which in turn, are fastened to studs. Light-gauge steel framed building generally requires interior bracing of walls, whereas a similar wood framed building may not. (Photo courtesy of Dietrich Metal Framing.)

If the construction process and framing details were vastly different, it would have discouraged builders, who have been used to and trained in WLF, from attempting the use of LGSF. This would have increased the cost of LGSF and also impeded its development. As the builders get familiar with LGSF, its use will increase. This should encourage the development of new applications that are uniquely suited to LGSF.

Corrosion Protection and Yield Strength of Framing Members

As per the International Residential Code, the minimum corrosion protection required on all LGSF members is G60 (see Chapter 16 for the meaning of this designation). The minimum yield strength of steel for LGSF members is 33 ksi. Generally, 33-ksi or 50-ksi steel is used.

Wall Framing in Light-Gauge Steel

Figures 17.7(a) to (c) show various stages in the framing of the walls of a light-gauge steel structure. Like WLF, LGSF walls are assembled on the floor and tilted into place. Even though the organization of LGSF is similar to WLF, connections and fasteners are quite different. Generally, the studs are fastened to the top and bottom tracks with screws.

The screws used in assembling light-gauge members are *self-drilling, self-tapping* screws. Self-drilling implies that the screws drill their own holes and self-tapping implies that they form (i.e., *tap*) their own thread.

The International Residential Code provides prescriptive design tables, which can be used to select the size of studs for given loads and the building's length and width. Manufacturers of LGS members also provide similar tables in addition to providing typical

NOTE

Stud Size and Spacing for LGSF Residential Framing

As explained earlier, 350S162 means a $3\frac{1}{2}$-in.-deep stud with a flange width of $1\frac{5}{8}$ in. Dimensionally, this is similar to a 2×4 wood stud. Similarly, 550S162 is similar to a 2×6 stud.

Sheet thickness of steel used in light-gauge studs is 33, 43, 54, or 68 mils (i.e., 20, 18, 16, or 14 gauge) depending on the gravity and lateral loads on the wall. The spacing of light-gauge steel studs is generally 24 in.

See Figure 17.11 for blocking of studs

See Figure 17.8 for anchorage of bottom track to foundation

FIGURE 17.7(a) A preassembled wall is tilted up and plced over foundation anchor bolts. Other wall assemblies, seen in the foreground, will be similarly placed and anchored to the foundation. (Photo courtesy of Dietrich Metal Framing.)

FIGURE 17.7(b) This image shows the progress in framing of walls of the structure shown in Figure 17.7(a). (Photo courtesy of Dietrich Metal Framing.)

FIGURE 17.7(c) Light-gauge steel wall framing showing temporary wall bracing, which is similar to that used in a wood frame structure. (Photo courtesy of Dietrich Metal Framing.)

framing details. Some of the manufacturers' details use proprietary members that are unique to the manufacturer.

The size of studs commonly used in exterior walls of a typical single-family residence is generally 350S162 or 550S162 with sheet thickness of 33, 43, 54, or 68 mil (i.e., gauges 20, 18, 16, or 14), depending on the gravity, lateral loads, and yield strength of steel.

The center-to-center spacing of studs may be 12 in., 16 in., or 24 in. However, because of the high strength of steel, 24-in. spacing is more common.

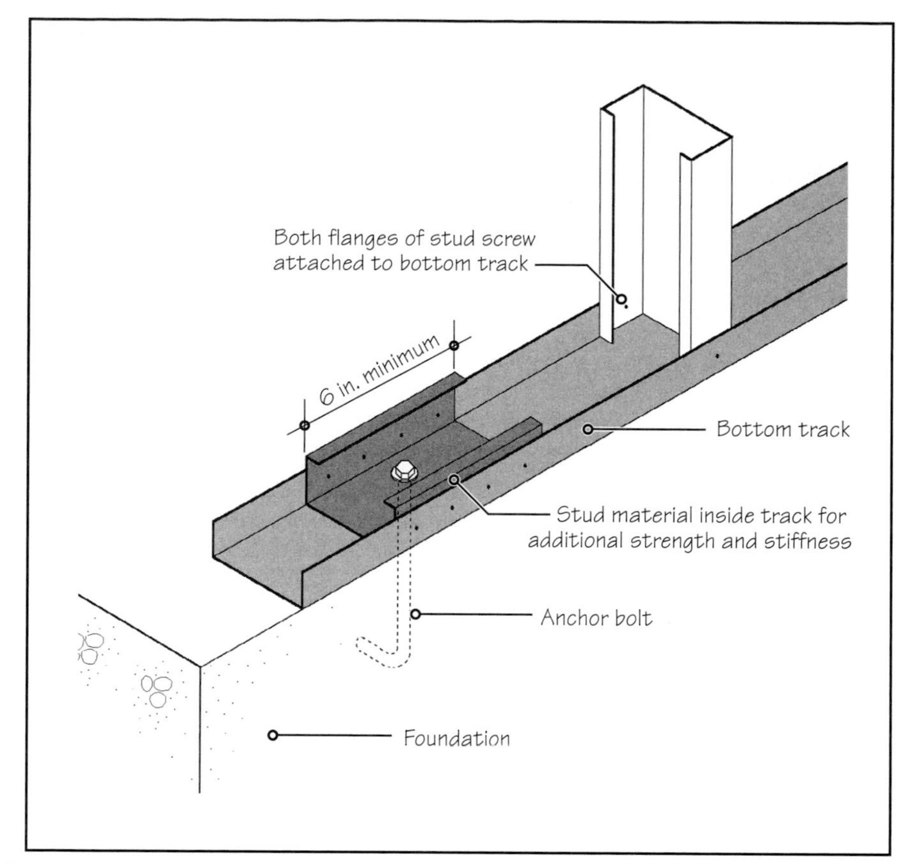

FIGURE 17.8 Anchorage of light-gauge steel frame wall to foundation.

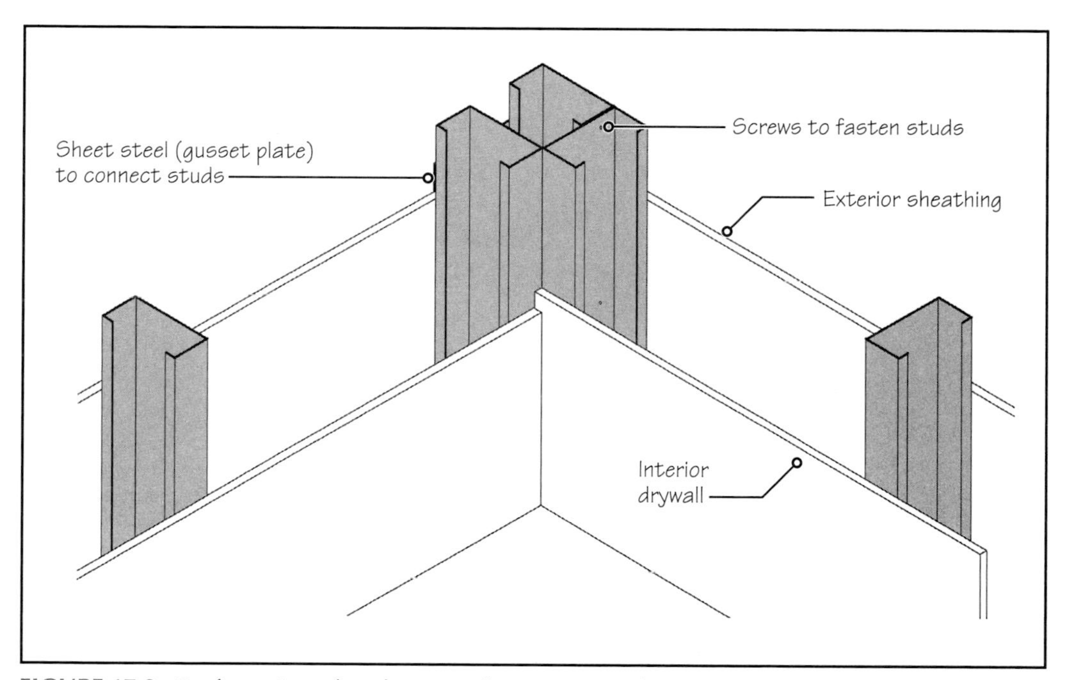

FIGURE 17.9 Configuration of studs at a wall corner. Note that a minimum of three studs are required at a corner, exactly as with a WLF wall. A sheet steel gusset plate is used to connect studs. In a WLF wall, 2-by lumber blocking is used between the studs for the connection (Figure 13.12).

Figures 17.8 to 17.12 provide a few important details of wall framing and connections between members showing the similarities and differences between LGSF and WLF construction. Details given here are generic and are merely representative of the system. Manufacturer's details should be referenced for more specific information.

Floor Framing in Light-Gauge Steel

Figure 17.13 shows the framing of a typical light-gauge steel floor. A major difference between a light-gauge steel floor and a wood floor framing is that the light-gauge steel floor joists must align with the underlying studs. This is because the commonly used top track in

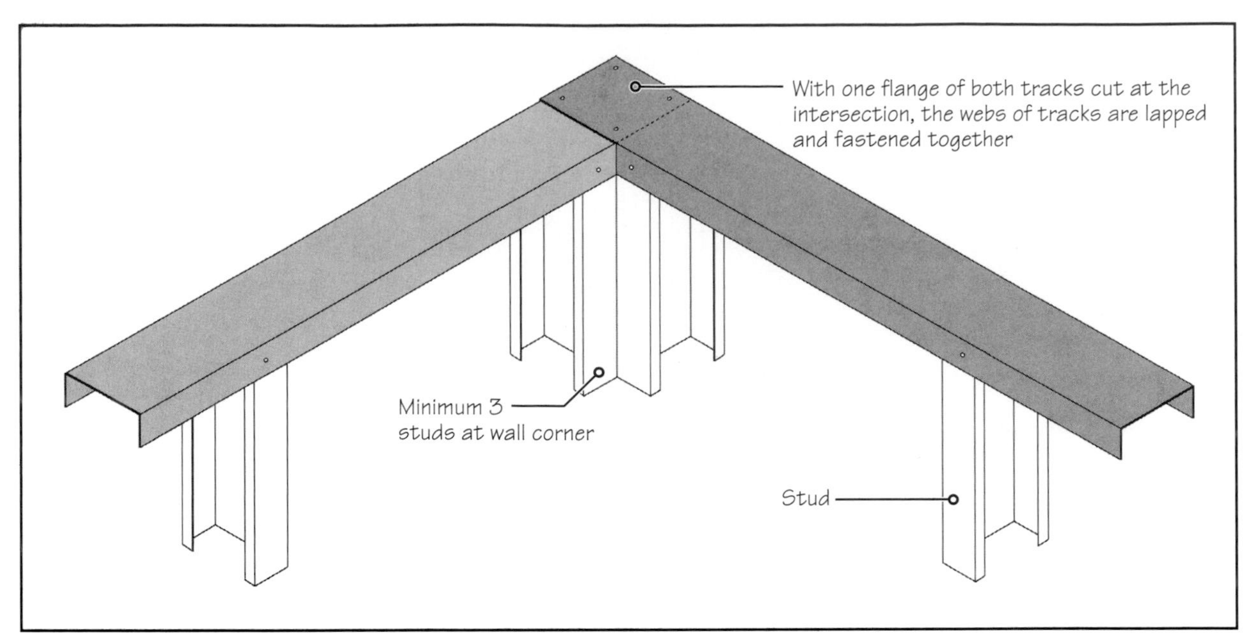

FIGURE 17.10 Detail of connection between intersecting tracks at the top of a wall corner.

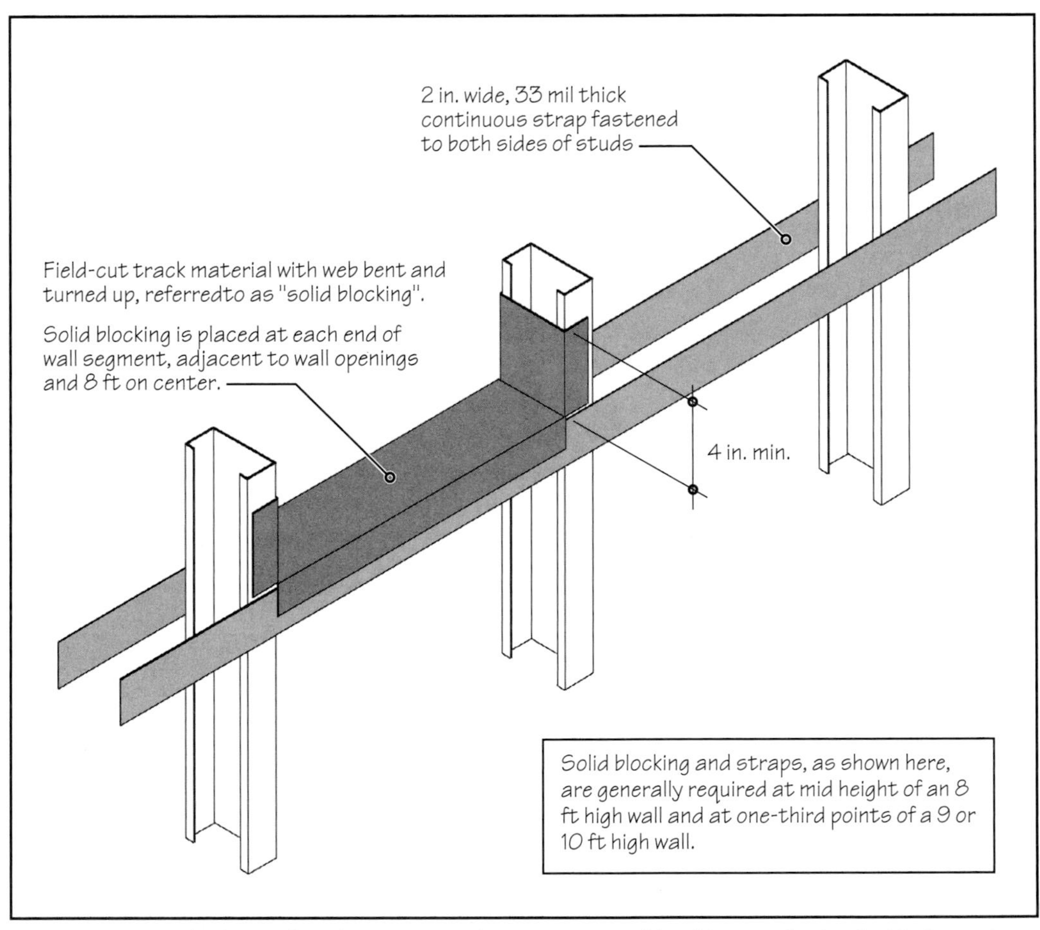

FIGURE 17.11 Blocking of studs to prevent their rotation and buckling under loads. U-channel bridging through punchouts (shown in Figure 17.3) does not generally provide sufficient bracing for studs in load-bearing applications. The detail has been adapted from Dietrich Metal Framing.

NOTE

In-Line Framing

In-line framing is mandated if the structure is designed using the prescriptive code provisions. Where in-line framing is inconvenient, the top track and studs should be structurally engineered.

a LGSF wall is unable to function as a bending member, unlike the double top plate in a wood frame wall.

In-line framing (as it is generally referred to in the industry) also applies to rafters and roof trusses, so that the load from the floor or roof is transferred directly to the studs. A maximum tolerance of $\frac{3}{4}$ in. between the center lines of studs and joists or rafters is permitted, Figure 17.14.

Joist depths of $7\frac{1}{4}$ in., $9\frac{1}{4}$ in., and $11\frac{1}{4}$ in., which match the depths of standard wood joists, are available in addition to the depths of 8 in., 10 in., 12 in., and 14 in. Light-gauge

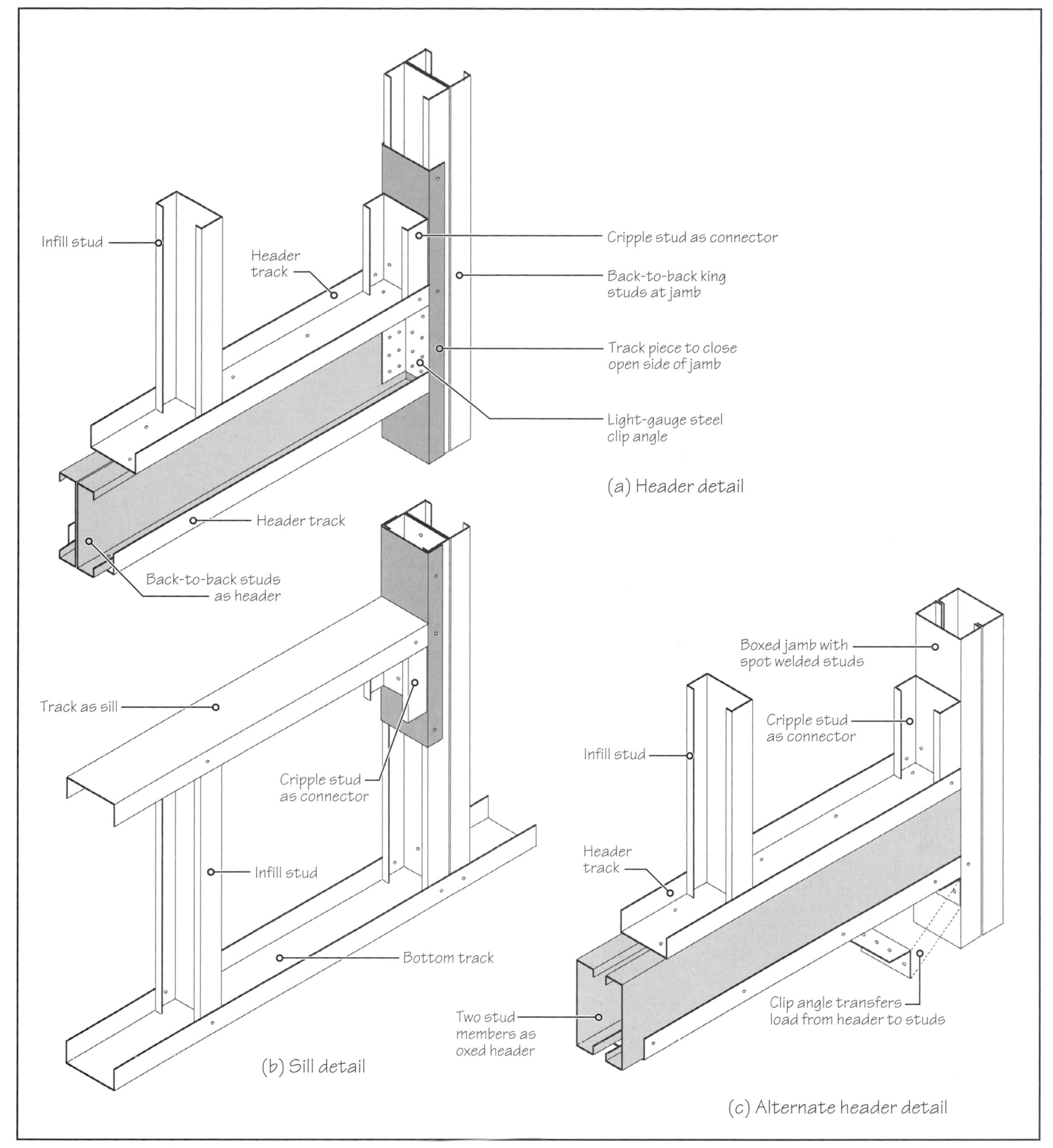

FIGURE 17.12 Details of a window header and sill in a load-bearing wall. Door header detail is similar.

steel joists typically include holes in their webs to accommodate HVAC, plumbing, and other utility lines, Figure 17.15.

Once the floor joists are in place, the plywood or OSB subfloor for the next floor is laid in exactly the same way as in a WLF building. After the completion of the subfloor, the walls for the next floor are erected, Figure 17.16.

Roof Framing in Light-Gauge Steel

Figure 17.17 shows the framing of a typical light-gauge steel roof and Figure 17.18 shows the framing detail at an eave. Although the roof can be framed using individual

FIGURE 17.13 Layout of floor joists in a typical light-gauge steel structure. (Photo courtesy of Dietrich Metal Framing.)

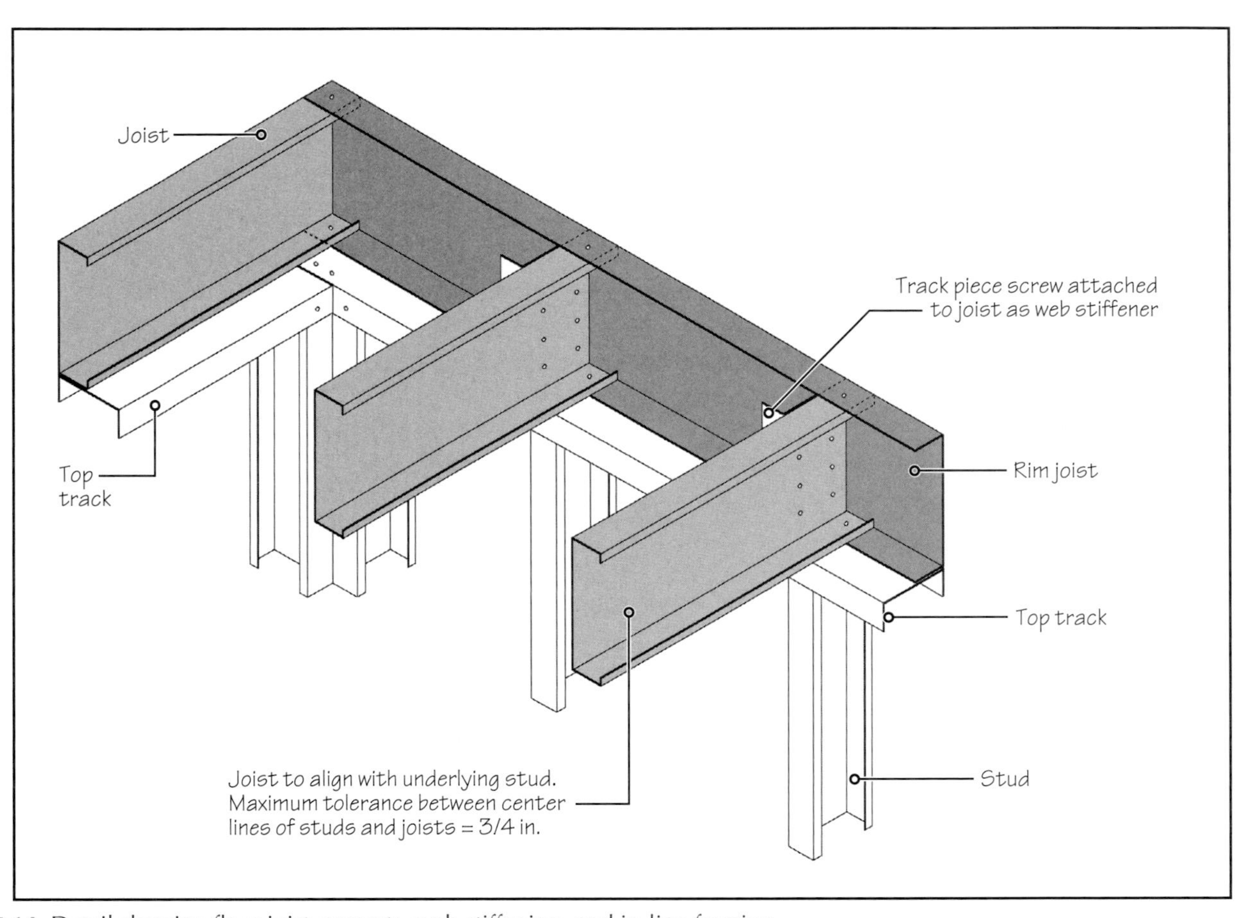

FIGURE 17.14 Detail showing floor joist supports, web stiffening, and in-line framing.

FIGURE 17.15 Floor joists with extruded holes allow water supply, plumbing, and other utility pipes. (Photo courtesy of Dietrich Metal Framing.)

FIGURE 17.16 A worker erects the second-floor walls in a light-gauge steel structure. The wall is erected over a plywood or OSB subfloor in exactly the same way as in a wood frame building. (Photo courtesy of Dietrich Metal Framing.)

FIGURE 17.17 Use of light-gauge steel roof trusses. (Photo courtesy of American Residential Steel Technologies.)

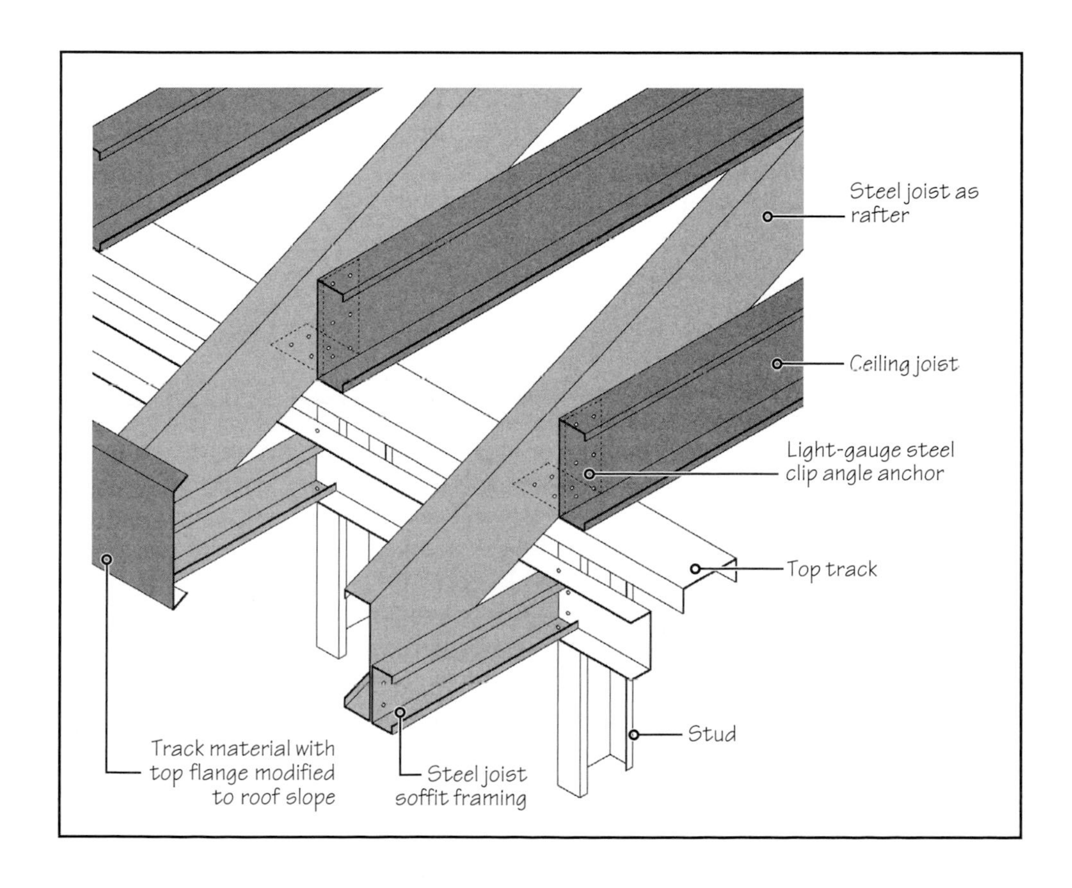

FIGURE 17.18 A typical eave detail in a light-gauge steel stick-built roof.

rafters and ceiling joists (stick-framed roof-ceiling assembly), builders find it more convenient to use light-gauge steel trusses because of the ease with which they can be fabricated.

Being lighter than corresponding wood trusses, light-gauge steel trusses can generally be placed in position without special lifting and hoisting equipment. Additionally, the joints between members of a light-gauge steel truss are more rigid, so that the trusses are easier to handle and more forgiving during their placement.

17.3 ADVANTAGES AND LIMITATIONS OF LIGHT-GAUGE STEEL FRAMING

Because LGSF and WLF systems are similar, they share several attributes and characteristics with each other:

- Both are dry systems of construction.
- Both systems provide spaces between framing members that allow infill insulation and passage of utility pipes.
- Both systems are lightweight, which eliminates the need for heavy construction and hoisting equipment.
- Both systems require simple, inexpensive tools for connecting the members. Welding is generally unnecessary in LGSF structures.
- Both systems accept the same interior and exterior finishes.

In addition to the shared characteristics, a LGSF structure has several advantages and disadvantages vis-à-vis the WLF structure.

ADVANTAGES

- *Lighter weight:* LGSF members are much lighter than corresponding WLF members, reducing workers' fatigue.
- *Higher strength:* For the same overall cross-sectional dimensions, steel frame members are stronger and more ductile than WLF members. Higher strength and ductility is the reason for the increased use of LGSF for residential framing in seismic and high-wind areas.
- *Dimensional stability:* Steel is not subjected to dimensional instability—shrinkage, swelling, twisting, and warpage caused by moisture changes. Steel frame members are straight and true.
- *Uniform quality:* Unlike wood, steel is a manufactured material with uniform and consistent properties.
- *Durability:* Steel is not subject to fungal decay or termite attack.
- *Noncombustibility:* A major advantage of steel as compared to wood is steel's noncombustibility.
- *Waste recycling:* LGSF manufacturers generally provide framing members cut to size to reduce on-site labor and waste.

LIMITATIONS

- *Corrosion:* Although LGSF members are galvanized, they can be subjected to corrosion in a highly corrosive environment. A thicker than G60 zinc protection (e.g., G90) may be needed in some areas, which increases cost substantially.
- *Thermal bridging:* A major disadvantage of LGSF is the low R-value of framing members, which reduces the effectiveness of infill insulation. Although this is offset somewhat by a 24-in. center-to-center spacing of studs and the use of an insulating exterior sheathing, it increases the cost of the building.
- *Bracing and web stiffening:* Because LGSF members are thin and lightweight, a greater number of bridging and bracing lines and web-stiffening measures are required to prevent twisting and buckling of framing members.
- *Cost:* The cost of materials for an LGSF is comparable with that of a WLF. However, an LGSF has a higher labor cost due to the use of screw guns instead of the pneumatic nailing used in a WLF. The development of improved fastening systems for LGSF may, however, change the situation in the future.

NOTE

Dust Marking

If a light-gauge steel frame structure is not properly detailed, a phenomenon known as *ghosting*, or *dust marking*, may occur due to the temperature differential between the framing members and the rest of the wall. The phenomenon, which generally occurs in cold climates, creates dust deposits on the interior drywall, highlighting the framing members.

17.4 NON-LOAD-BEARING LIGHT-GAUGE STEEL FRAMING

As stated earlier, LGSF has virtually no competition for the framing of interior partitions, suspended ceilings, and exterior backup walls in buildings of Types I and II construction. Whereas, the interior partitions are used as space-dividing elements, an exterior backup wall provides lateral load support for exterior wall finishes, such as masonry veneer, stucco, EIFS or natural stone.

Light-Gauge Steel Interior Partitions

Figure 17.19 shows the use of LGSF for an interior partition. Generally, 20-gauge, $3\frac{1}{2}$-in.- or 4-in.-wide (350S125-30 or 400S125-30) studs are used for partitions. Because steel is a strong material, the size and spacing of studs for walls is governed by their deflection due to lateral loads rather than the strength of the studs.

Building codes require that an interior partition should be able to withstand a minimum lateral load of 5 psf. (This value may be exceeded in seismic zones.) Manufacturers provide tables for selecting stud size and spacing for a given partition height without structural calculations.

Light-Gauge Steel Back-Up Exterior Walls

Figure 17.20 shows the framing for a backup exterior wall in a low-rise building, and Figure 17.21 shows similar framing for a high-rise building. The size of studs and their spacing in a backup exterior wall is generally based on the wind load. Manufacturers provide tables to size the studs for a given height of the backup wall as a function of wind load, similar to the tables for interior partitions. Additional details of LGSF backup exterior walls are covered in Chapter 26.

(a) Studs terminate at ceiling height

To track fastened to floor (see Chapter 26, Figure 26.28, if a slotted track is required)

Full-height stud

Intermediate track to engage ceiling and drywall; drywall terminates here

Bottom track fastened to floor

(b) Studs terminate at underside of upper floor

FIGURE 17.19 Two commonly used framing alternatives for a light-gauge steel interior partition wall. The interior partitions in most commercial buildings terminate at the ceiling level, leaving a gap between the underside of the floor (or roof) and the ceiling. The gap allows the passage of HVAC ducts and other utility lines. Such a wall is acoustically more transparent than a wall that extends up to the underside of floor or roof. **(a)** One way to support such a wall at the top is to use light-gauge steel braces. **(b)** Alternatively, the studs extend to the underside of the upper floor. Intermediate track blocking is required for anchoring drywall and ceiling.

FIGURE 17.20 A light-gauge steel exterior backup wall framing in a low-rise (one- or two-story) building. The wall is a non-load-bearing wall, which functions as a backup for the exterior wall finish and also transfers lateral loads to the structural frame of the building. In this wall, the studs extend continuously from the ground floor to the top of the parapet. The connection between the wall and the floor (or roof) must be a slip connection to allow the floor (or roof) to move (deflect) without stressing the wall (Figure 26.31).

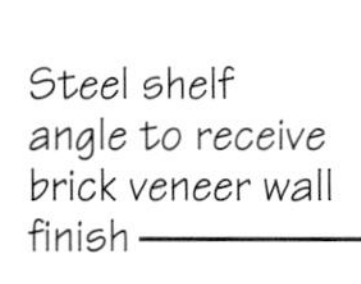

FIGURE 17.21 A light-gauge steel exterior backup wall framing in a high-rise building. The connection of the top of the wall (i.e., at the underside of each floor) must allow the floor to deflect without stressing the wall (Figure 26.27).

PRACTICE QUIZ

Each question has only one correct answer. Select the choice that best answers the question.

6. Unlike WLF walls, LGSF walls cannot be assembled on the floor and subsequently tilted and moved into position.
   a. True
   b. False
7. The anchorage of the bottom track of a LGSF wall to concrete footing
   a. requires 6-in.-long stud material to strengthen the connection.
   b. requires 6-in.-long additional track material to strengthen the connection.
   c. requires 6-in.-long light-gauge furring section to strengthen the connection.
   d. requires no additional strengthening.
8. The corner of a load-bearing LGSF wall requires a minimum of three studs (similar to a WLF wall).
   a. True
   b. False
9. The top track in a load-bearing LGSF wall must be doubled in the same way as the top plate in a WLF wall.
   a. True
   b. False
10. *In-line framing* in an LGSF structure means that
    a. the studs must be continuous across an intermediate floor.
    b. the floor joists must align with the studs in the underlying supporting wall.
    c. the floor joists from two opposite directions of an underlying supporting wall must be in line, not staggered.
    d. floor-to-floor height in a two- or three-story structure must be the same.
11. The subfloor material commonly used with an LGSF structure is a light-gauge steel deck.
    a. True
    b. False
12. Spot welding is the most common method of joining LGSF members.
    a. True
    b. False
13. The benefit of using load-bearing LGSF instead of load-bearing WLF in seismic zones is the
    a. lighter weight of LGSF.
    b. ductility of steel.
    c. thermal bridging.
    d. (a) and (b).
    e. (b) and (c).

14. Per the building codes, interior partitions must be designed for a minimum lateral load of
a. 20 psf.
b. 15 psf.
c. 10 psf.
d. 5 psf.
e. none of the above.

15. Sheet thickness of commonly used light-gauge steel studs for interior partitions is
a. 30-mil-thick sheet.
b. 35-mil-thick sheet.
c. 40-mil-thick sheet.
d. 45-mil-thick sheet.
e. none of the above.

Answers: 6-b, 7-a, 8-a, 9-b, 10-b, 11-b, 12-b, 13-d, 14-d, 15-a.

KEY TERMS AND CONCEPTS

- Anchorage and connections
- Blocking and straps
- Bridging channels
- Furring channels
- Header
- Light gauge framing members
- Punch out
- Roof framing
- Soffit
- Stud
- Track

REVIEW QUESTIONS

1. Discuss why the details of light-gauge steel framing used in load-bearing applications are similar to those used in wood light frame structures.
2. In which regions of the United States does LGSF have advantages over WLF, and why?

SELECTED WEB SITES

North American Steel Framing Alliance (www.steelframingalliance.com)
Steel Stud Manufacturers Association (SSMA) (www.ssma.com)

FURTHER READING

1. American Iron and Steel Institute (AISI). *Standard for Cold Formed Steel Framing—A Prescriptive Method for One or Two Family Dwellings,* 2001.
2. International Code Council (ICC). *International Residential Code,* 2006.

CHAPTER 18

Lime, Portland Cement, and Concrete

CHAPTER OUTLINE

18.1 Introduction to Lime

18.2 Types of Lime Used in Construction

18.3 Portland Cement

18.4 Air-Entrained and White Portland Cements

18.5 Basic Ingredients of Concrete

18.6 Important Properties of Concrete

18.7 Making Concrete

18.8 Placing and Finishing Concrete

18.9 Portland Cement and Water Reaction

18.10 Water-Reducing Concrete Admixtures

18.11 High-Strength Concrete

18.12 Steel Reinforcement

18.13 Welded Wire Reinforcement

Two Union Square Tower in Seattle, WA (completed in 1989), in which 19,000-psi concrete was used for the first time (in tubular steel and concrete composite columns). Architects: NBBJ Group. (Photo by MM.)

This chapter begins with a discussion of lime, which, until the middle of the nineteenth century, was the only cement available for use in masonry mortar, plaster, and concrete. Today, lime's use is limited to a small component in mortar, stucco, whitewash, and soil stabilization.

Portland cement, whose discovery led to the replacement of lime in most applications, is covered next. Without portland cement, modern architecture, as we know it, would not be possible. In fact, it is virtually impossible to construct a building today without using portland cement in some form or another.

Portland cement is the main ingredient in concrete. Therefore, a discussion of concrete follows portland cement. Concrete is covered here as a material. Its applications, that is, concrete construction systems, are discussed in the following three chapters.

Because concrete is generally used in conjunction with steel reinforcement (as reinforced concrete), we also cover steel reinforcement here. Steel reinforcement consists of round bars (referred to as *rebars*) or steel wires welded together to form a mesh, referred to as *welded wire reinforcement* (WWR).

NOTE

A Note on Quicklime

Historical records indicate that British soldiers used quicklime to incapacitate their French enemies in the war of 1217 by throwing it in their faces—one of the earliest records of chemical warfare.

The heat produced in the hydration of 1 lb of quicklime is nearly 450 Btu. Because 1 Btu is the amount of heat required to raise the temperature of 1 lb of water by 1°F (Chapter 5), 450 Btu will raise the temperature of 3.2 lb of water from 70°F to 212°F—the boiling point of water. In other words, if 1 lb of quick lime is mixed with 3.2 lb of water initially at room temperature, that water will begin to boil.

Wood box carts carrying quicklime have ignited through accidental penetration of rainwater through leaks in the carts. Criminals misused quicklime to dispose of the corpses of their victims by dumping them in a pit, covering them with quicklime, and backfilling with wet soil.

18.1 INTRODUCTION TO LIME

Lime is a cement (binder) and has been used for thousands of years in masonry mortars to bind the stone and brick units in walls. Until the discovery of portland cement, lime-sand mortar was the only masonry mortar available. Egyptians used lime-sand mortar in the construction of pyramids, and virtually all historic stone buildings in Europe used lime-sand mortar. In lime-sand mortar, lime is the binder and sand is the filler.

Lime is made from limestone, one of the most abundant rocks on the earth's crust. Chemically, limestone is calcium carbonate. Lime is produced by simply heating limestone—a process known as *calcining.*

Calcining is typically done by first crushing limestone rock into approximately 2-in.-size particles and then heating them in a kiln to a temperature of approximately 1,800°F (1,000°C). Calcining expels the carbon dioxide out of calcium carbonate, leaving calcium oxide. (It is an endothermic reaction, implying that heat is required for the reaction.)

Calcium oxide is called *quicklime.* Quicklime is a caustic substance that corrodes metals and causes severe damage to human skin. However, it reacts readily with water to form calcium hydroxide. Calcium hydroxide is called *hydrated lime,* or *slaked lime,* because it contains water that is chemically combined with calcium oxide. It is a relatively benign material and is the one that is commonly used in building construction.

The reaction of water with calcium oxide is an exothermic reaction, implying that heat is produced during the reaction. In fact, the amount of heat produced is so large that if 1 lb of quicklime is mixed with approximately 3 lb of water, the water boils.

Both quicklime and hydrated lime are white in color. Hydrated lime is generally used in powder form. The amount of water used to hydrate quicklime is much greater than that required for hydration reaction. Therefore, after the completion of hydration reaction, the quicklime and water mixture is dried to evaporate the excess water. The dried hydrated lime is then pulverized to powder form. In the United States, hydrated lime is sold in 50-lb bags, Figure 18.1.

FIGURE 18.1 A typical lime bag.

CARBONATION OF LIME

The hardening property of lime results from its ability to react with atmospheric carbon dioxide to form calcium carbonate—the raw material used in making lime. Thus, the chemical reaction of lime with carbon dioxide returns lime to its parent material—limestone—which is a hard and strong material. Stated differently, the calcination of limestone resulting in quicklime, the hydration of quicklime to produce hydrated lime, and the carbonation of hydrated lime back to limestone is a cycle, Figure 18.2.

As the atmospheric carbon dioxide begins to react with lime-sand mortar in a wall, it converts the exposed areas of the mortar into a hard carbonate crust. This crust hinders further penetration of carbon dioxide into the interior volume of mortar. Therefore, the carbonation reaction—the setting and hardening of lime-sand mortar—is a slow reaction. The slow setting is also due to the fact that only a small fraction (less than 4%) of air is carbon dioxide. Thus, it is difficult to lay more than a few masonry courses per day using lime-sand mortar because the weight of additional courses squeezes the mortar out of masonry.

The cementing property of lime is not simply the result of its hardening but is due more to the absorption of water in mortar by brick or stone units. As the water evaporates and the

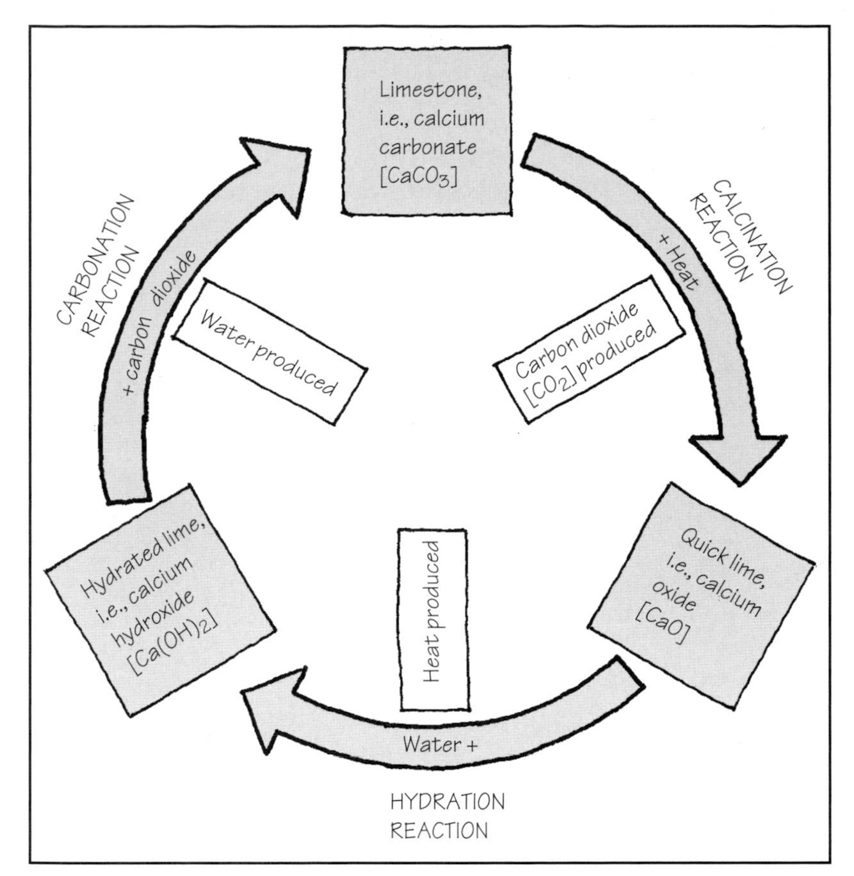

FIGURE 18.2 Limestone cycle—calcination, hydration, and carbonation.

mortar shrinks, it draws the units toward itself. The tiny pores, pits, and crevices on the surfaces of the units, which get filled with fresh mortar, add to lime's cementing property.

POZZOLANIC REACTION—ROMAN CONCRETE AND MORTAR

In addition to its hardening caused by carbonation, lime experiences another type of hardening. This is caused by lime's reaction with amorphous (noncrystalline, i.e., glassy) silica, referred to as a *pozzolanic reaction.* Although not understood scientifically at the time, the pozzolanic reaction was first discovered by ancient Romans around the first or second century BC.

Romans found that if lime and volcanic ash were mixed, the mixture, when used with sand and water, gave a mortar that set more quickly, was stronger, and was more durable than lime-sand mortar. The greater durability was due to the pozzolanic reaction that occurs when lime and volcanic ash are mixed with water. The mixture produces a water-resistant, or *hydraulic cement*—a cement that does not dissolve in water. Lime by itself (i.e., free lime, defined as lime in the absence of pozzolanic reaction) dissolves slowly in water over a period of time.

The main ingredient of volcanic ash is mainly microscopic silica (with a small percentage of microscopic alumina). As the volcanic lava jets out of the ground, the carbon in the rock and other organic matter is converted to carbon dioxide gas, which carries with it particles of rock that have melted and fused into glassy (amorphous) particles. (Soil and most rocks consist mainly of oxides of silicon and aluminum, called *silica and alumina,* respectively.)

Due to the pressure of volcanic eruption, molten glass particles are converted to tiny beads of glass, which we call *volcanic ash.* It is both the microscopic size and amorphous nature of silica and alumina in volcanic ash that give it the pozzolanic property. Therefore, volcanic ash is called a *pozzolanic material,* or simply a *pozzolana.*

Ancient Romans perfected the art and science of making hydraulic cement from lime and volcanic ash and used it to produce concrete, known as *Roman concrete,* which is not too different from present-day concrete. Romans used two closely spaced stone or brick walls as the permanent formwork, within which the concrete was placed. In many such buildings, the facing walls have disintegrated

NOTE

Chemistry of Lime

Calcination of Limestone

Limestone $\xrightarrow{\text{Heat}}$ lime + carbon dioxide

$CaCO_3 \xrightarrow{\text{Heat}} CaO + CO_2$

Hydration of Quicklime

Calcium oxide + water $\longrightarrow$ Calcium hydroxide + heat

$CaO + H_2O \longrightarrow Ca(OH)_2$ + heat

Carbonation of Hydrated Lime

Calcium hydroxide + carbon dioxide $\longrightarrow$ Calcium carbonate + water

$Ca(OH)_2 + CO_2 \longrightarrow CaCO_3 + H_2O$

NOTE

Roman Concrete and Mortar

Historians conjecture that the Romans discovered lime from cooking and heating hearths, which were generally made of limestone slabs. The fire from the hearths calcined the limestone, and subsequent rain showers hydrated it. Volcanic ash was abundantly available in the region because it had a number of active volcanoes at the time.

In fact, the term *pozzolana* is derived from Pozzuoli, a town in southern Italy that had several active volcanoes during the Roman period. Evidence of extinct volcanoes is abundantly available on the outskirts of modern Pozzuoli.

FIGURE 18.3 Temple at Palestrina, Italy, built around the first century BC, showing concrete walls. In this temple, the facing stone veneer, used as permanent formwork for the concrete walls, has disintegrated and disappeared. (Photo courtesy of Professor Michael Yardley.)

FIGURE 18.4 A typical aqueduct in ancient Rome that supplied water to the city of Rome. (Photo courtesy of Professor Michael Yardley.)

NOTE

Pozzolana

By definition, a pozzolanic material (also called *pozzolana*) is a material that, when mixed with lime and water, converts the mixture into a hydraulic cement. For it to be a pozzolana, the material must be in extremely small (microscopic) in particle size and contain amorphous silica. (See the introduction to Chapter 28 for the meaning of the term *amorphous.*)

A commonly used present-day pozzolana is fly ash—a waste product from coal-fired electricity generating stations. Like volcanic ash, fly ash contains microscopic and amorphous silica and alumina. The average size of particles in fly ash is less than 20 μm [1] (1 μm, called a micrometer is one millionth of a meter, or 0.001 mm). By comparison, the average portland cement particle size is 45 μm.

Another contemporary pozzolana is silica fume—a waste product from the silicon industry. The particle size of silica fume is even smaller than that of fly ash (Section 18.11).

The reaction between lime and a pozzolana yields a complex form of calcium silicate hydrate, which is the main constituent produced when portland cement is mixed with water. Because portland cement is a much better hydraulic cement, lime and pozzolanas are not used directly in contemporary construction.

or a part of the wall has fallen, revealing the interior concrete, Figure 18.3.

Because their cement was hydraulic, the mortar used in Roman masonry was also hydraulic. They exploited this property to build aqueducts to supply water to the city of Rome, Figure 18.4. Architectural historians claim that the 150-ft diameter dome of the Pantheon, Rome, built around AD 120 and still standing, is made of concrete using lime, volcanic ash, and broken stone, Figure 18.5.

18.2 TYPES OF LIME USED IN CONSTRUCTION

The composition of lime outlined previously assumed that limestone consists only of calcium carbonate. This is only partially correct because limestone occurs in two types:

- *High calcium limestone* consisting of approximately 95% calcium carbonate
- *Dolomitic limestone* consisting of approximately 60 to 80% calcium (Section 23.7)

FIGURE 18.5 Interior of the dome of Pantheon, Rome built around AD 120. (Photo courtesy of Dr. Jay Henry.)

Dolomitic limestone shares most of its properties with high-calcium limestone; that is, it can be calcined to obtain quick lime. Quick lime obtained from dolomitic limestone is calcium oxide chemically combined with magnesium oxide ($CaO \cdot MgO$). Although dolomitic quick lime also has great affinity to water, it does not hydrate as readily as calcium oxide.

Higher temperature and pressure are generally needed to fully hydrate all magnesium oxide. Unhydrated oxides in lime are problematic (particularly in plaster work) because, given right temperature, pressure, and humidity, they can hydrate abruptly well after the construction is complete. Because the hydration of lime is exothermic, blistering and popping generally occurs.

The lime industry produces the following two types of hydrated lime:

- Type N (normal) hydrated lime
- Type S (special) hydrated lime

Type S hydrated lime has maximum (8%) unhydrated oxides. In Type N, there is no such limitation. Type S lime is more expensive and is preferred over Type N lime, not simply because of the limited amount of unhydrated oxides but also because it gives greater workability and plasticity.

Lime used in masonry mortar and plastering is generally Type S hydrated lime. Manufacturers identify Type S lime by so labeling the bags, Figure 18.6. A bag not so labeled is usually Type N lime, as shown in Figure 18.1.

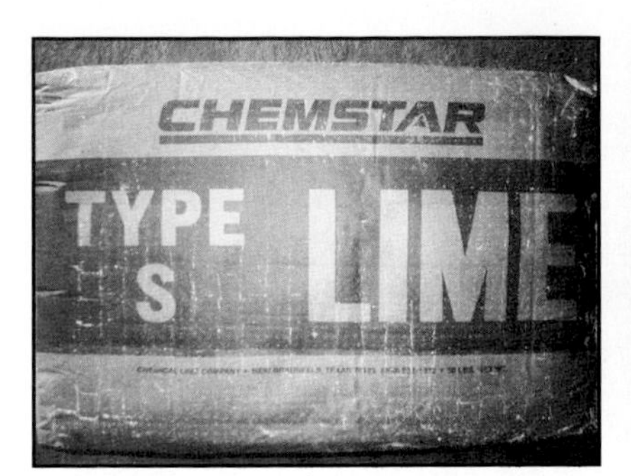

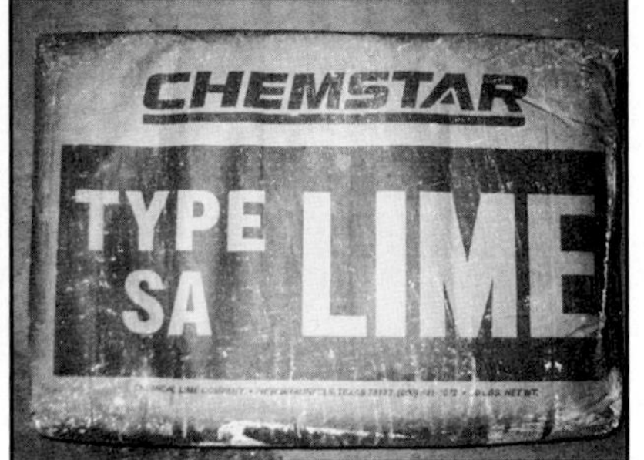

FIGURE 18.6 Type S hydrated lime and Type SA hydrated lime bags.

Type S lime can be produced from either dolomitic or high-calcium limestone. In the United States, dolomitic limestone is preferred because it gives greater workability to lime—a quality that masons and plasterers require. Lime is also produced with air entraining agents mixed with it. Air entrainment increases the resistance of lime mortar against the freeze-thaw effect, but decreases its strength (Section 18.4). Air-entrained lime is labeled by manufacturers as NA lime or SA lime.

USES OF LIME

Lime is used in construction as well as nonconstruction activities. Some of the nonconstruction uses of lime are in water purification, water softening, sewage sludge treatment, making of plastics and paints, and so on. Lime is used in municipal water-treatment plants because of its disinfecting property. It removes hazardous compounds, such as lead, from water and also softens hard water. In construction activities, lime is principally used in

- Masonry mortar (Chapter 22)
- Plaster and stucco (Chapter 27)
- Soil stabilization

NOTE

Workability and Plasticity

The term *workability* is used with reference to concrete, mortar, and plaster mixes. With concrete, it is somewhat quantified by slump measure (Section 18.6). With mortar and plaster mixes, there is no such quantification of workability. However, a good mason can judge workability while spreading the mortar or plaster mix with a trowel.

Basically, a workable mortar or plaster will spread more easily—a difference similar to spreading a crunchy peanut butter or creamy peanut butter on a piece of toast.

The term *plasticity* is related to workability. It is a measurable term with reference to mortar and plaster mixes. Workability includes *plasticity* and *water retentivity*. A mortar mix with a higher degree of plasticity and water retentivity is a more workable mix. Water retentivity is described in Section 22.1.

PRACTICE QUIZ

Each question has only one correct answer. Select the choice that best answers the question.

1. Lime is made by
   - **a.** heating crushed limestone and then pulverizing it.
   - **b.** mixing crushed limestone with water and then pulverizing the mixture.
   - **c.** mixing crushed limestone with sulfuric acid and then pulverizing the mixture.
   - **d.** mixing crushed limestone with nitric acid and then pulverizing the mixture.
2. Quicklime reacts readily with water.
   - **a.** True
   - **b.** False
3. Hydrated lime is obtained by treating
   - **a.** slaked lime with water.
   - **b.** quicklime with water.
   - **c.** limestone with water.
   - **d.** slaked lime with portland cement.
   - **e.** quicklime with portland cement.
4. Quicklime is most commonly used in building construction.
   - **a.** True
   - **b.** False
5. Hydrated lime–water paste sets and hardens due to the
   - **a.** chemical reaction of water with lime.
   - **b.** chemical reaction of the oxygen in air with lime.
   - **c.** chemical reaction of the carbon dioxide in air with lime.
   - **d.** chemical reaction of the nitrogen in air with lime.
   - **e.** chemical reaction of the water in lime with the carbon dioxide in air.
6. Which of the following materials is a pozzolana?
   - **a.** Portland cement
   - **b.** Lime
   - **c.** Fly ash
   - **d.** Sand
   - **e.** Concrete
7. One of the important cementitious ingredients in making Roman concrete was
   - **a.** clay.
   - **b.** volcanic ash.
   - **c.** sand.
   - **d.** portland cement.
   - **e.** none of the above.
8. Hydrated lime is available in two types, which are referred to as
   - **a.** Types N and S lime.
   - **b.** Types I and II lime.
   - **c.** Types X and Y lime.
   - **d.** Types A and B lime.
   - **e.** Types 1 and 2 lime.
9. A pozzolanic material must be mainly composed of
   - **a.** microscopic silica.
   - **b.** microscopic and amorphous silica.
   - **c.** microscopic and crystalline silica.
   - **d.** silica and alumina.
   - **e.** amorphous silica and alumina.

Answers: 1-a, 2-a, 3-b, 4-b, 5-c, 6-c, 7-b, 8-a, 9-b.

18.3 PORTLAND CEMENT

Portland cement, a hydraulic cement, was patented in 1824 by Joseph Aspidin, a British stonemason. Aspidin burnt finely ground limestone and clay in a kiln and found that the resulting product was a hydraulic cement, which set and gained strength much more quickly than lime.

Aspidin called his new product *portland cement* because the concrete that he made from it resembled, in color and strength, a highly sought-after natural limestone quarried from the Isle of Portland, off the British coast. Note that portland cement is no longer a brand name but a generic one; therefore, a large number of manufacturers produce portland cement.

Modern portland cement is made from almost the same basic raw materials and virtually by the same basic process that Aspidin used, that is, by heating a mixture of limestone and clay. Limestone comprises approximately two-thirds of the raw material required for making portland cement.

Another major raw material required is clay or shale. (Shale is a highly compressed form of clay.) Clay consists of the oxides of silicon, aluminum and iron—called *silica, alumina,* and *iron oxide,* respectively. Some clays can provide all the silica. alumina and iron oxide required in the manufacture of portland cement. Other clays may be supplemented by sand (to provide silica), bauxite (to provide alumina), and iron ore (to provide iron oxide). Generally, therefore, one or two raw materials, in addition to limestone and clay, are needed in portland cement's manufacture.

Manufacturing Process of Portland Cement

The manufacture of portland cement begins with quarrying of limestone, which is later crushed to approximately $\frac{5}{8}$-in.-size particles. Crushed limestone and other raw materials (clay, sand, and iron ore) are stored in silos for processing as needed. Figure 18.7 is a conceptual diagram of the entire manufacturing process.

The raw materials are mixed together in required proportions and ground to a fine powder in a grinding mill. Powerful fans draw the powder from the grinding mill into a dust

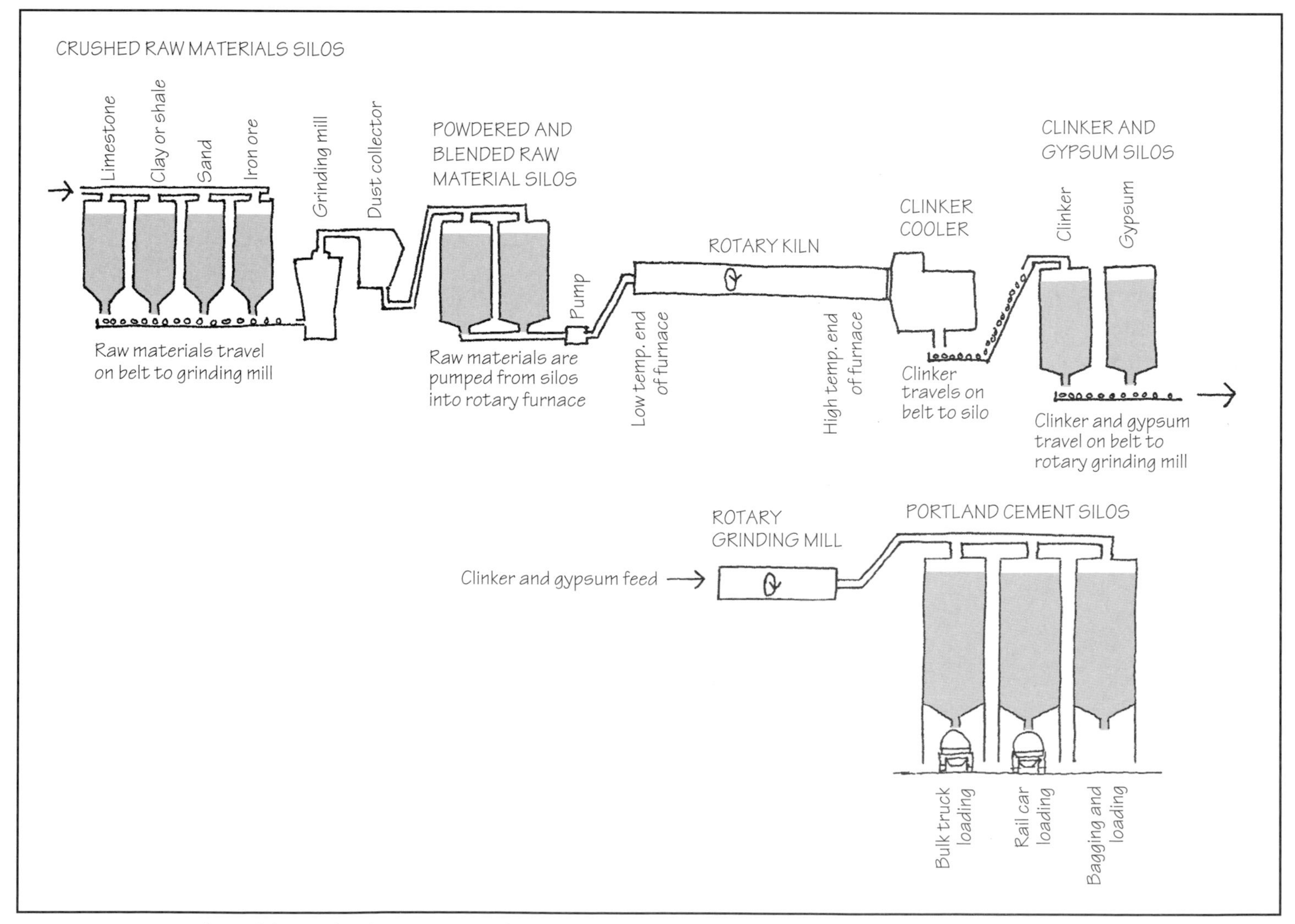

FIGURE 18.7 An outline of the portland cement–manufacturing process.

collector, and from there it is taken to storage silos. From the silos, the powder goes into a rotating, or rotary, kiln.

The rotary kiln—a huge steel cylinder lined inside with refractory bricks—is at the heart of the manufacturing process. It is considered to be the heaviest rotating industrial equipment. It is slightly sloped to allow the material to tumble forward by gravity. The kiln temperature increases as the material moves forward. In the initial section of the kiln, where the temperature is around 1,800°F (1,000°C), calcium carbonate is converted to calcium oxide, similar to the manufacturing of lime. The discharge of carbon dioxide is, therefore, a major environmental pollutant in the neighborhood of a portland cement plant.

Toward the far end, where the temperature reaches up to approximately 3,400°F (1,900°C), calcium oxide reacts with other raw materials to form complex calcium compounds. Materials become partially molten and coalesce into approximately $\frac{1}{2}$-in.-size nodules, called *clinkers,* Figure 18.8. The clinker is cooled and stored in silos.

In the final manufacturing stage, a small quantity of gypsum is added to clinkers to control the setting properties of portland cement. Clinkers and gypsum are ground and blended into a very fine powder, which is considered portland cement. It is then stored in silos for transportation in trucks or railroad cars or to a bagging facility. Portland cement is gray in color, with an average particle size of approximately 45 μm.

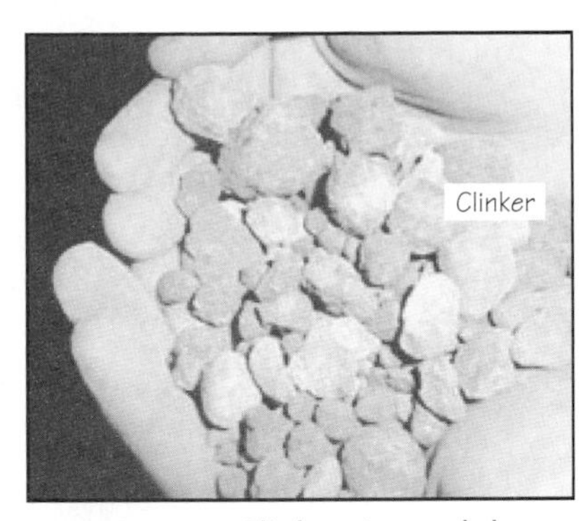

FIGURE 18.8 Clinker size and shape.

TYPES OF PORTLAND CEMENT

Most portland cement is used in making concrete, required for buildings, pavements, roads, bridges, and dams. Other uses of portland cement are in masonry mortar, plaster, stucco, and flooring. Because requirements for various concretes and other mixes vary, portland cement is manufactured in five different types—Types I, II, III, IV, and V. These types are distinguished from each other by small changes in their chemical composition, giving them different properties, Table 18.1.

NOTE

1 μm (called a micrometer) = 10^{-6} m = 10^{-3} mm. Therefore, 45 μm = 0.045 mm.

Type I is a *general-purpose* portland cement. Type III portland cement is *high-early-strength* portland cement. It develops strength at a faster rate than other types in the initial stage, but its final strength is the same as that of the other types. As described in Section 18.9, portland cement develops its ultimate strength over several months.

Type III portland cement is generally used in making precast concrete elements, such as hollow-core slabs, double-tees, and concrete pipes. Economic considerations require that the formwork for precast elements be used as frequently as possible. Generally, these elements are cast one morning and stripped off the form the following morning, giving a 24-h turnover cycle (Chapter 20).

Type IV is *low-heat-of-hydration* portland cement. It is meant for use in massive concrete structures (dam walls or bridge piers), where the temperature rise due to heat generated from hydration must be minimized. Due to its limited use, Type IV cement is produced by manufacturers only on special request.

Type V is *sulfate-resistant* portland cement. Portland cement can be affected adversely by the presence of sulfur. In fact, concrete made out of Type I portland cement will decompose into small fragments (spall) after just a few years in the presence of high-sulfur-containing environments. Some soils and groundwater have high sulfur content. The use of Type V portland cement is recommended in such environments.

Type II is *moderate-sulfate-resistant and moderate-heat-of-hydration* portland cement. It combines the properties of Types IV and V to a moderate degree. Therefore, several manufacturers

TABLE 18.1 PORTLAND CEMENT TYPES AND THEIR USES

Portland cement type	Description and use
I	**General-purpose** portland cement, used where there is no special requirement
II	**Moderate-sulfate-resistant and low-heat-of-hydration** portland cement
III	**High-early-strength** portland cement, used in precast elements and where high early strength is required, such as in cold weather
IV	**Low-heat-of-hydration** portland cement, used in massive civil engineering structures
V	**Sulfate-resistant** portland cement, used where high sulfate resistance is required

EXPAND YOUR KNOWLEDGE

Composition of Portland Cement

Complex calcium compounds are formed during the heating of raw materials in a rotary kiln. These compounds impart different properties to portland cement, and their relative proportions determine the type of portland cement. These compounds are

Tricalcium silicate (C_3S)—called *alite*—approximately 50%
Dicalcium silicate (C_2S)—called *belite*—approximately 25%
Tricalcium aluminate (C_3A)—called *celite*—5 to 12%
Tetracalcium-alumino-ferite (C_4AF)—called *iron*—up to 8%

Alite hydrates rapidly. It is responsible for initial set and early strength (from 1 to 7 days). Belite hydrates slowly and contributes to later strength gain (28+ days).

Celite liberates a great deal of heat in the initial few days and contributes somewhat to early strength. Too much celite can reduce sulfate resistance of portland cement.

Iron has little impact on portland cement properties but is needed as a flux to reduce energy consumption during manufacturing. A flux reduces the temperature at which the raw materials begin to melt and coalesce. The characteristic gray color of portland cement is due to the presence of iron and a small amount of manganese. White portland cement is made by reducing the iron and manganese contents.

Gypsum (calcium sulfate) is added to increase the setting time of portland cement. Setting is stiffening (early hardening) of the mix, that is, when the mix becomes semisolid. The mix should neither set too rapidly nor too slowly. A rapid set does not give enough time for placing and finishing the mix. A slow set delays finishing operations, such as troweling and floating. Set control is, therefore, important. The amount of gypsum is carefully controlled.

Portland cement has a small quantity of alkalis—sodium and potassium oxides—which can react adversely with some forms of aggregates used in concrete.

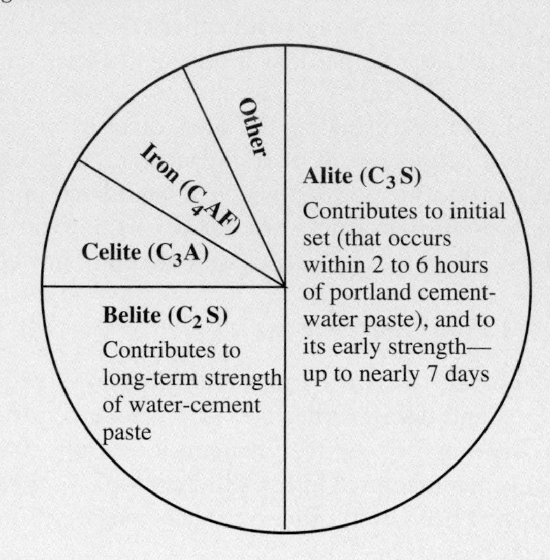

FIGURE 18.9 Type I/II portland cement. A portland cement bag generally weighs 94 lb.

make portland cement that meets the requirements of both Types I and II and label the product as Type I/II portland cement, Figure 18.9. Approximately 90% of all portland cement used is Type I (or Type I/II) followed by Type III (approximately 5%).

The particle size of portland cement affects its performance. The finer the particles, the greater the surface area available for water to coat cement particles. This increases the rate of hydration leading to higher early strength of portland cement paste. Type III portland cement, therefore, has finer particles than the other types.

18.4 AIR-ENTRAINED AND WHITE PORTLAND CEMENTS

Portland cement types previously described are the basic types. These types are also available with an integrally combined air-entraining agent. Air entrainment in concrete, mortar, or plaster increases the durability of these materials against freezing and thawing.

The process of placing concrete in forms requires a much greater amount of water in a mix (called the *water of convenience*) than that required for the chemical reaction between portland cement and water (Section 18.9). An inadequate amount of water in concrete does not allow concrete to be fluid enough to flow into the form and fill it completely. This results in honeycombed concrete. The excess water in a concrete mix evaporates as it stiffens, leaving voids. These voids are in addition to the voids between concrete ingredients—portland cement and aggregate—and those caused by lack of (or poor) consolidation. These voids contain *entrapped air*.

Because of the entrapped air, concrete is porous and tends to absorb rainwater. During freezing weather, the absorbed water turns into ice and expands. If a concrete is critically saturated and undergoes cycles of freezing and thawing, it will spall unless air voids are present to dissipate the pressure caused by the increased volume. To reduce freeze-thaw damage, tiny particles of air are introduced in a concrete mix, referred to as *air entrainment*. As the absorbed water in concrete expands on freezing, the entrained air relieves the pressure. Air-entrained concrete is commonly specified in exposed concrete elements subjected to freeze-thaw cycles.

Air entrainment can be achieved in two ways. The preferred way is to add an air-entraining chemical in the mix, referred to as an *air-entraining admixture*. The alternative is to use an air-entrained

NOTE

Entrapped Air Versus Entrained Air in Concrete

Entrapped air in concrete does not provide freeze-thaw resistance because it is not uniformly distributed in concrete and the air particle size is too large to be effective (generally 1 mm or larger in size). Additionally, entrapped air voids are irregular in shape and are generally interconnected.

Entrained air voids, on the other hand, are much smaller (10 to 100 μm in size), almost spherical in shape, and uniformly distrubuted in concrete, not interconnected.

A typical (non-air-entrained) concrete has an air content of approximately 1.5%. An air-entrained concrete requires about 6% air in order to provide high freeze-thaw resistance.

portland cement. Air-entrained portland cement has an air-entraining chemical as its integral part. All five basic types of portland cement are available as air-entrained portland cements. They are identified as Types IA, IIA, IIIA, IVA, and VA. With air-entraining admixture, normal portland cement (Type I, II, III, IV, or V) is used.

WHITE PORTLAND CEMENT

The more commonly used portland cement is gray in color. For aesthetic reasons, white portland cement is preferred, particularly for terrazzo flooring, stucco, and architectural concrete. Colored aggregates in a terrazzo floor show much better against the background of white portland cement. Additionally, when pigments are added to white portland cement, the resulting color is better than that obtained with the use of gray portland cement.

White portland cement is made by reducing the iron and manganese contents. This complicates the manufacturing process and increases the energy cost. Consequently, white portland cement is approximately twice the cost of normal (gray) portland. However, because portland cement is only one part of a concrete or stucco mix, the final product may only be 50% more expensive.

Like the gray cement, white portland cement is available in bags, Figure 18.10, or in bulk. It is generally produced as Type I or Type III. These types have the same properties as the gray Type I and Type III portland cement, respectively.

FIGURE 18.10 A white portland cement bag.

18.5 BASIC INGREDIENTS OF CONCRETE

The art and science of making concrete virtually disappeared after the fall of the Roman Empire until the discovery of portland cement approximately 1,400 years later. As stated previously, modern concrete is similar to Roman concrete, except that in modern concrete, portland cement is used as the glue to bind the stone aggregate together instead of a mixture of lime and volcanic ash.

The basic ingredients of modern concrete are portland cement paste (i.e., portland cement and water) and aggregate. Together, these materials provide a hard, rocklike substance, which is durable, fire resistant, and relatively inexpensive. It is used universally because virtually every region of the earth has the raw materials necessary to produce it. Additionally, the technology associated with its production and use is fairly simple.

Unlike other structural materials, concrete can be formed to any desired shape. In fact, an entire structure, regardless of its shape or size, can be formed monolithically—as if it were one object (*mono* means *one,* and *lithos* means *piece of stone*). Some pioneers of the modern movement in architecture (e.g., Le Corbusier and Frank Lloyd Wright) and several structural engineers (e.g., Robert Maillart and Pier Nervi, and more recently, Santiago Calatrava) have exploited the sculptural property of concrete, Figure 18.11.

NOTE

What is Concrete?

Concrete consists of

- Aggregate (coarse and fine aggregate) as matrix or filler
- Portland cement–water paste as an adhesive

Any other material used in concrete is called a ***concrete admixture*** (Sections 18.10 and 18.11).

FIGURE 18.11a Chapel at Ronchamp, France, by architect Le Corbusier. Le Corbusier was among the pioneers in exploiting the structural properties of concrete.

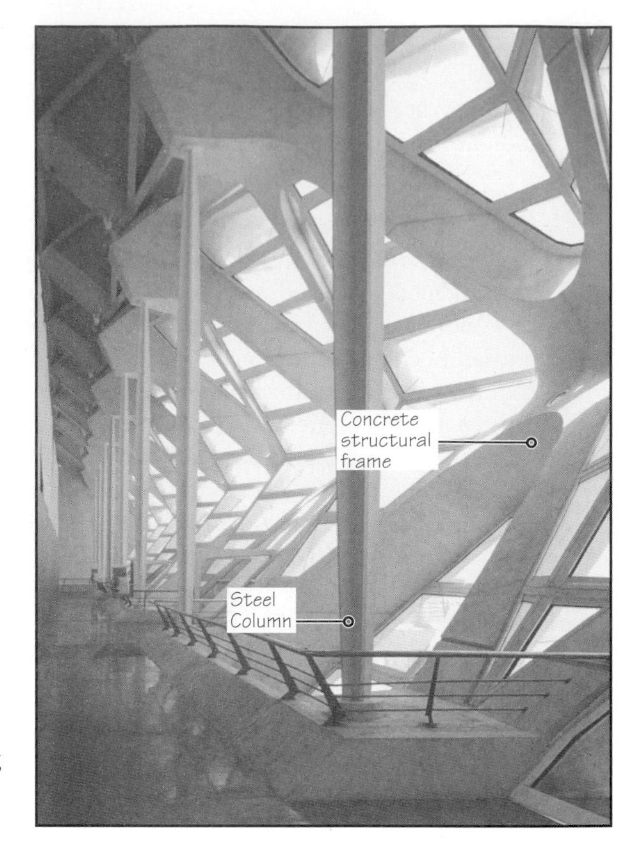

FIGURE 18.11b Concrete structural framework in the City of Arts and Sciences, Valencia, Spain by architect-engineer, Santiago Calatrava, illustrating the sculptural properties of structural concrete.

SIZE OF THE AGGREGATE

Aggregate in concrete occupies 65% to 80% of the volume of concrete, the remainder being portland cement. The aggregate must consist of different-size particles, referred to as *size gradation.* Proper gradation ensures that smaller particles fit within the voids created by larger particles, so that the entire mass of concrete is relatively dense, Figure 18.12(a).

A well-graded aggregate (implying that it consists of particles of various sizes) not only gives a stronger concrete, but also reduces the amount of portland cement necessary to wrap the particles and fill spaces between them. An aggregate that consists of only one or two sizes of particles has a higher percentage of voids and, therefore, requires a much larger amount of portland cement, Figure 18.12(b). Because portland cement is far more expensive than aggregates, this gives an uneconomical concrete.

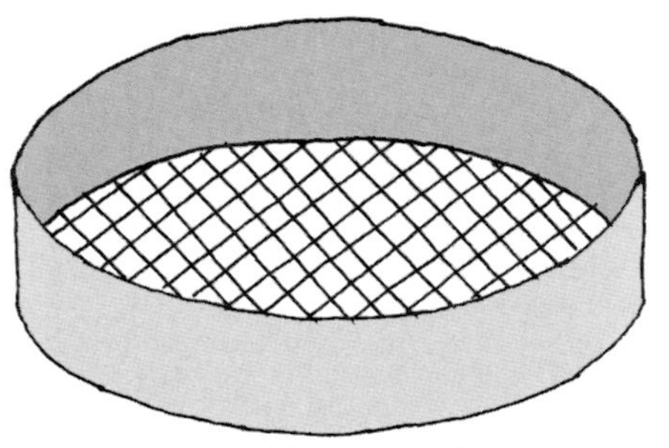

A typical sieve used for grading of aggregates in concrete laboratories. A No. 4 sieve has wires spaced at 1/4 in. (0.25 in.) on center. A No. 10 sieve has wires at 1/10 in. (0.1 in.) on center, etc. Clear opening between wires in a No. 4 sieve = 4.75 mm and that of a No. 10 sieve = 2.00 mm, see Chapter 21, Section 21.1.

In general, therefore, the aggregate in a concrete mix consists of several sizes. However, the concrete industry divides the aggregate into two size groups:

- Fine aggregate
- Coarse aggregate

Fine aggregate is generally sand, but more precisely it is that material of which 95% passes through a No. 4 sieve. A No. 4 sieve consists of a wire mesh with wires spaced at $\frac{1}{4}$ in. on center. Because the wires have a certain standard thickness, the largest particle size of fine aggregate that can pass through a No. 4 sieve is slightly smaller than $\frac{1}{4}$ in.

Fine aggregate needs to be graded from a No. 4 sieve down to a No. 100 sieve. Recommended grading is shown in Table 18.2.

Coarse aggregate is that aggregate of which 95% is retained on a No. 4 sieve. It consists of either crushed stone or gravel. Gravel has several advantages over crushed stone, but crushed stone is commonly used because it is more economical.

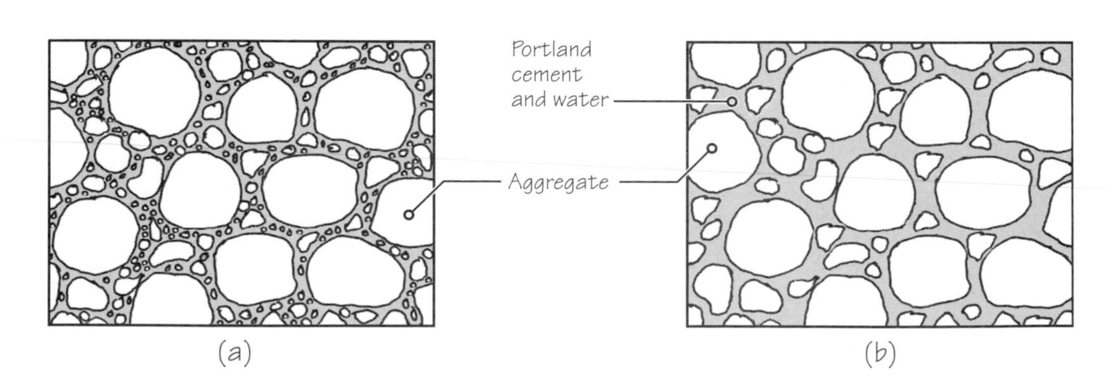

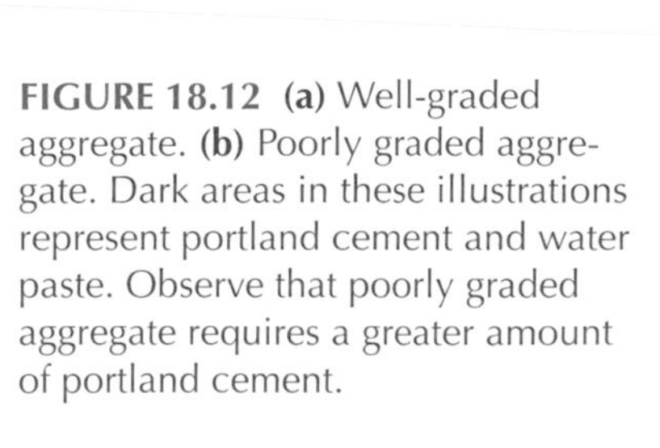

FIGURE 18.12 **(a)** Well-graded aggregate. **(b)** Poorly graded aggregate. Dark areas in these illustrations represent portland cement and water paste. Observe that poorly graded aggregate requires a greater amount of portland cement.

TABLE 18.2 RECOMMENDED GRADING OF FINE AGGREGATE FOR CONCRETE SLABS

Sieve size	Percent of aggregate passing sieve
$\frac{3}{8}$ in.	100
No. 4	95–100
No. 8	80–90
No. 16	50–75
No. 30	30–50
No. 50	10–20
No. 100	2–5

Although the minimum size of coarse aggregate is limited by the No. 4 sieve, its maximum size is a function of the smallest dimension of the building element and the spaces between steel reinforcement. For example, the American Concrete Institute (ACI) Code requires that for a concrete wall, the maximum coarse aggregate size should be smaller than one-fifth the thickness of the wall. Obviously, if the size of coarse aggregate is too large for the space in which it is placed, it will be difficult to consolidate, giving nonhomogeneous and inconsistent concrete.

Similarly, if the steel reinforcement is closely spaced, the maximum coarse aggregate size must be small. The objective is that the concrete should flow easily around the reinforcement. In most concrete used in buildings, the maximum coarse aggregate is $\frac{3}{4}$ in. Concrete used in thick walls and heavy foundations may have aggregate up to $1\frac{1}{2}$ in. in size; it may be up to 6 in. in dam walls.

The size of coarse aggregate has a bearing on the concrete's cost. Larger particles have a smaller surface area for a given volume. Therefore, a concrete made with large particles requires less portland cement, giving a more economical concrete, Figure 18.13.

NOTE

Gravel

Gravel consists of naturally occurring particles of rock, 0.2 in. to 3 in. size (Figure 21.1). Because gravel is formed by erosion and weathering of rocks, it is round in shape, in contrast to crushed stone. The term *pea gravel* refers to gravel that is approximately $\frac{1}{4}$ in. to $\frac{3}{8}$ in. size.

NORMAL-WEIGHT AND LIGHTWEIGHT COARSE AGGREGATES

Concrete may either be structural concrete or nonstructural concrete. Nonstructural concrete is generally insulating concrete, which is used as roof insulation (Chapters 5 and 31). Structural concrete is used in a structural frame—beams, slabs, columns, and walls.

The most commonly used structural concrete has a compressive strength of 4,000 psi. However, any concrete with a compressive strength of 2,000 psi or above is considered to be structural concrete. Structural concrete may either be normal-weight or lightweight concrete.

The American Concrete Institute Code defines lightweight structural concrete as one that, in its hardened state, weighs from 85 to 115 lb/ft^3. Normal-weight concrete weighs

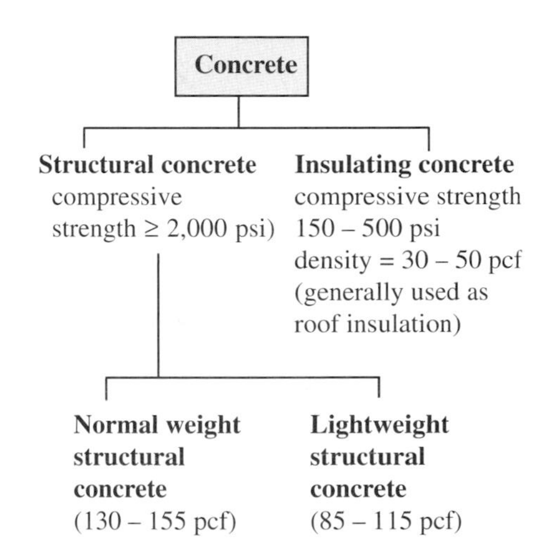

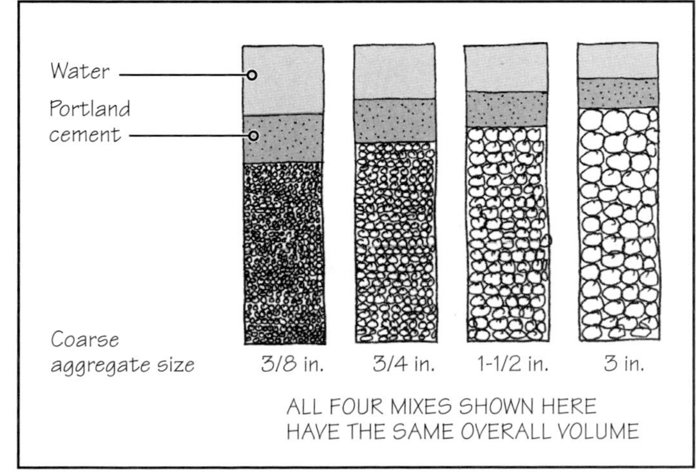

FIGURE 18.13 Effect of aggregate size on the amount of cement-water paste. Four mixes shown here have the same overall volume. Note that as the size of aggregate increases, the amount of cement-water paste needed for a given volume of concrete decreases. Because portland cement is the major cost-contributing ingredient in a mix, a concrete with larger aggregate is generally more economical.

NOTE

In a concrete mix, portland cement paste coats the surface of aggregates. If the surface area of aggregates is small, the amount of cement paste needed is small. Large aggregates have a small surface area for a given volume. Therefore, everything else being the same, a mix consisting of larger aggregates is more economical.

That the surface area of smaller particles contained in a given volume is greater than those of larger particles can be explained by considering 2-in. and 1-in. cubes. The surface area of a 2-in. cube is 24 in.^{2}, and its volume is 8 in.^{3}.

Now consider 1-in. cubes. To obtain a volume of 8 in.^{3} using 1-in. cubes, 8 cubes are needed. The surface area of these 8 cubes is $8 \times 6 = 48$ in.^{2}—double that of a 2-in. cube. Thus, the use of half-size particles has doubled the surface area, although the space occupied by the particles has not changed.

FIGURE 18.14 Two commonly used aggregates in concrete—crushed limestone as normal-weight aggregate and expanded shale as lightweight aggregate. The quantities shown here are such that both aggregate piles have the same approximate weight. Expanded shale, being fired clay, is generally brown in color, similar to that of a fired clay brick.

NOTE

Carbonate Versus Silicious Aggregates

Concrete aggregates are generally obtained from limestone, granite, quartz, sandstone, and so on. Limestone (calcium carbonate) aggregate gives a concrete that is slightly more fire resistant than the concrete obtained from aggregates produced from granite, quartz, sandstone, and so on.

In general, carbonate-containing aggregates give a concrete that is more fire resistant than that obtained from silicious aggregates (see Table 20.1).

NOTE

Wash Water

Concrete mixers are washed after delivery and before being recharged with the next batch. Ready-mix plants are increasingly using the wash water in the mix—a more sustainable alternative.

from 130 to 155 lb/ft^3. The difference between normal-weight and lightweight structural concretes is only in the type of coarse aggregates used. The fine aggregate in both is the same.

Normal-weight concrete is obtained using normal-weight aggregate. Commonly used normal-weight aggregates are crushed limestone, granite, quartz, and so on. Lightweight structural concrete is obtained using lightweight aggregate.

One commonly used lightweight aggregate is expanded shale, obtained by heating crushed shale to a high temperature. This converts any water present in a shale particle into steam, which—in trying to exit—expands the particle, reducing its density. Because it is a baked material and because shale is a compressed form of clay, the color of expanded shale is similar to that of brick. Figure 18.14 portrays the difference between normal-weight limestone aggregate and expanded shale.

Lightweight structural concrete costs more than normal-weight concrete because of the higher cost of producing the aggregate. However, it can yield savings in the overall cost of the structure because it reduces dead loads. This is particularly attractive in a high-rise structure, where any reduction in the dead load of floor slabs reduces the size of supporting beams. This, in turn, reduces the size of columns, which leads to lighter foundations.

Lightweight structural concrete has, however, limitations as to strength. High-strength concrete is difficult to obtain using lightweight aggregates. However, the commonly used concrete strength of 4,000 psi is easily obtained using lightweight aggregates. Therefore, in a high-rise structure, lightweight concrete is used in floor beams and slabs, and normal-weight concrete is used in columns. The reason is that, whereas the columns have to support the load of all the upper floors, the floor beams and slab are required to support only the load of that floor.

QUALITY OF WATER

Water is an important component of concrete. Portland cement derives its cementing property from its reaction with water. Water used in concrete must be clean. A rule of thumb in the concrete industry is that if the water is fit for drinking, it is fit for use in concrete. Because seawater contains salts, its use leads to corrosion of reinforcing steel. It also modifies the portland cement–water reaction and is, therefore, not appropriate for use.

PRACTICE QUIZ

Each question has only one correct answer. Select the choice that best answers the question.

10. The main ingredient used in the manufacture of portland cement is
- **a.** marble.
- **b.** granite.
- **c.** limestone.
- **d.** sand.
- **e.** iron ore.

11. ASTM specifications classify portland cement in five basic types, referred to as
- **a.** Types A, B, C, D, and E.
- **b.** Types I, II, III, IV, and V.
- **c.** Types PC1, PC2, PC3, PC4, and PC5.
- **d.** Types X, Y, Z, P, and Q.
- **e.** none of the above.

12. The most commonly specified type of portland cement is
- **a.** Type C.
- **b.** Type III.
- **c.** Type PC1.
- **d.** Type I/II.
- **e.** Type II/III.

13. Which type of portland cement is commonly used in making precast concrete members?
- **a.** Type C
- **b.** Type III
- **c.** Type PC1
- **d.** Type I/II
- **e.** Type II/III

14. White portland cement is generally weaker than normal (gray-colored) portland cement.
- **a.** True
- **b.** False

15. The aggregate for concrete is generally subdivided into coarse aggregate and fine aggregate. The distinction between the two types of aggregates is based on the
- **a.** aggregate's strength.
- **b.** aggregate's surface texture.
- **c.** aggregate's density.
- **d.** aggregate's particle size.
- **e.** aggregate's parent rock.

16. In a No. 4 sieve, the dimensions of voids are approximately
- **a.** 4 in. × 4 in.
- **b.** $\frac{1}{4}$ in. × $\frac{1}{4}$ in.
- **c.** 2 in. × 2 in., giving an area of approximately 4.0 in.^{2}.
- **d.** 1 in × 4 in.
- **e.** none of the above.

17. A well-graded coarse aggregate is one in which the size of all aggregate particles is approximately the same.
- **a.** True
- **b.** False

18. Everything else being the same, a concrete with a larger coarse aggregate requires less portland cement and water paste to give the same concrete strength.
- **a.** True
- **b.** False

19. Lightweight structural concrete is heavier than insulating concrete.
- **a.** True
- **b.** False

Answers: 10-c, 11-b, 12-d, 13-b, 14-b, 15-d, 16-b, 17-b, 18-a 19-a.

18.6 IMPORTANT PROPERTIES OF CONCRETE

Properties of concrete that are of interest to architects, engineers, and builders may be divided under the following two heads:

- Fresh (plastic-state) concrete properties
- Hardened concrete properties

FRESH CONCRETE PROPERTIES—WORKABILITY (SLUMP) OF CONCRETE

A good concrete mix should be able to retain its homogeneity (nonsegregation of particles) during all operations involved in transporting the concrete from the mixer to the form and placing, compacting, and finishing the concrete. Additionally, it should be possible to perform these operations with relative ease and in a reasonable amount of time. In other words, a good concrete should be easy to pump (when required), place, and compact, and it should set within reasonable time so that finishing can be done without delay.

Whether a concrete meets these requirements or not can often be judged visually by an experienced concrete technician. However, a measure commonly used to describe them collectively is called *workability*. Workability of concrete is the ease with which a concrete can be placed and compacted in the form with minimum loss of consistency and homogeneity. A concrete that is not workable is referred to as a *harsh* concrete. To obtain a workable concrete,

- The aggregates should be well graded.
- Everything else being the same, large aggregates reduce workability. Adequate fine material is important in providing a workable concrete. That is why the addition of fly ash or silica fume, for example, in a mix improves its workability. Because the particle size of portland cement is small, increasing portland cement also increases workability. However, excessive portland cement in a mix makes it too gluey and tacky, reducing workability.
- The aggregates should not be too angular in shape. Gravel, because of its naturally round particles, gives greater workability than crushed stone. However, crushed stone, if properly graded, yields good workability. Similarly, sand obtained from river beds or mined from sand pits gives greater workability than that obtained by crushing and pulverizing stone.
- An adequate amount of water is necessary for workability.

A commonly used measure for workability is *slump*. Although slump does not evaluate all aspects of workability, it is its best available measure. The slump of concrete is measured using a truncated cone. The cone is open at the top and bottom and has handles and foot tabs. It measures 4 in. at the top and 8 in. at the bottom and is 12 in. high, Figure 18.15.

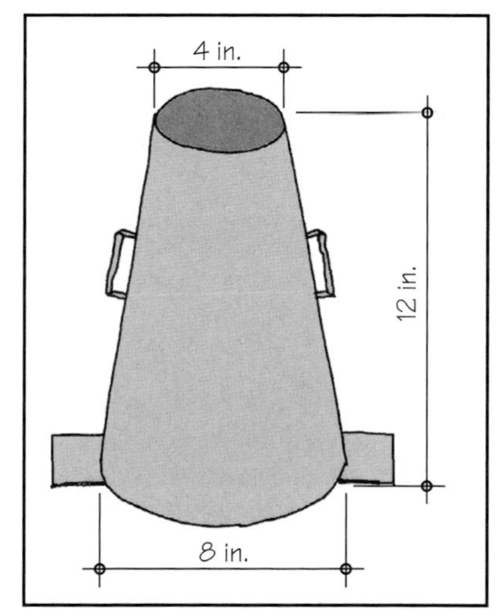

FIGURE 18.15 A slump cone is open at top and bottom.

(a) A technician filling slump cone and rodding the concrete inside the cone with a steel rod.

(b) After the cone is full, the technician lifts the slump cone and inverts it to measure concrete's slump using a float to line with cone's top.

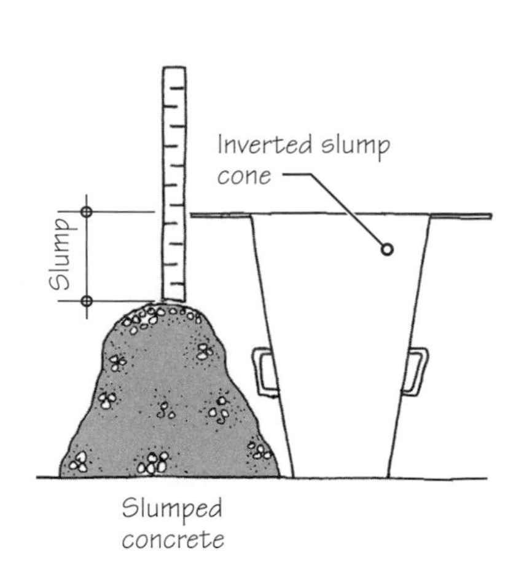

FIGURE 18.16 Measurement of the slump of concrete.

NOTE

Approximate Required Slump Values

Walls and columns	4–5 in.
Beams	4–5 in.
Slabs	3–4 in.

Slump measurements are performed by a qualified concrete technician at the construction site immediately after the concrete is received. To measure slump, the inside surface of slump cone is dampened with water, and any old concrete sticking to its surface is removed.

The cone is placed on a rigid, flat, smooth, and nonabsorbent surface. It is held firmly in place by standing on foot tabs, Figure 18.16(a). Concrete is filled in the cone in three layers, each 4 in. high (as per ASTM C143). Each layer is tamped 25 times through its depth by a $\frac{5}{8}$-in.-diameter smooth steel rod—a process also referred to as *rodding*.

After the last layer has been fully rodded, the excess concrete is struck off so that the concrete is flush with the top of the cone. The cone is then removed with a slow upward motion. As soon as the cone is removed, the concrete settles. The slump of concrete is the height by which the top of concrete has settled. It is measured in inches using a ruler over the inverted slump cone, Figure 18.16(b).

A workable concrete will slump uniformly, retaining its overall conical profile. A poorly graded concrete will slump nonuniformly.

Different slumps are required for different applications of concrete. A low value of slump gives a concrete that will not completely flow in the form and will require careful compaction.

In some situations, a high-slump concrete is desired, such as in the grout used in masonry walls. In most situations (unless a water-reducing admixture has been used, Section 18.10), a high value of slump means the use of excess water. This, as we will see later, reduces concrete strength. Generally for concrete walls, columns, and beams, a slump of 4 to 5 in. is required. For slabs, a slump of 3 to 4 in. is generally adequate.

Fresh Concrete Properties—Particle Segregation and Bleeding of Concrete

If concrete is not carefully placed in the forms or if it is compacted excessively, larger aggregates settle down and the smaller ones rise to the top. The segregation of particles gives a nonhomogeneous concrete, reducing its strength. It is important, therefore, that concrete be placed in position carefully and not be thrown from a large distance.

Segregation of aggregates is generally accompanied by bleeding. Bleeding is the rising of water to the top. This is similar to the rise of water that occurs in a heap of wet sand on a beach when tapped a few times. Bleeding occurs naturally when the mix is compacted or floated. Therefore, a certain amount of bleeding is inherent during placing, compacting, and finishing. Excessive bleeding should be avoided.

FIGURE 18.17 Casting concrete test cylinders.

Hardened Concrete Properties—Compressive Strength of Concrete

The two most important properties of hardened concrete are durability and compressive strength. The first step in determining concrete's strength is to cast test cylinders from the concrete received at the job site. The casting of cylinders and the slump test are generally performed at the same time.

A concrete cylinder is 6 in. in diameter and 12 in. high. Building code and standards specify the number of test cylinders that must be cast for a certain amount of concrete used. A concrete test cylinder is made by placing concrete in a mold in three layers, similar to the slump test. Each layer is rodded 25 times, and finally the excess concrete is struck off, leaving it flush with the top of the mold, Figure 18.17.

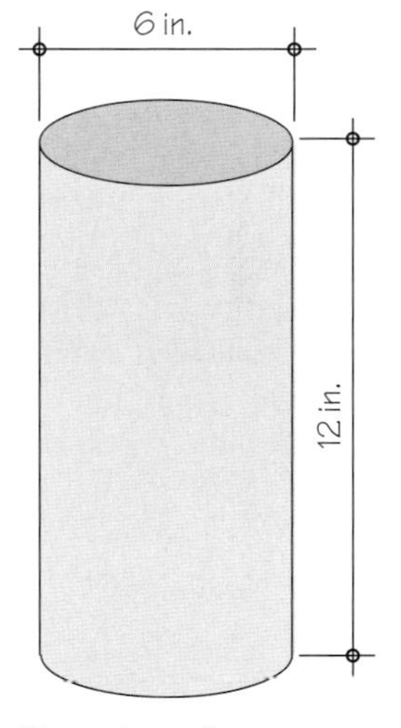

Dimensions of a concrete test cylinder

The molds for cylinders are generally of plastic, steel, or coated cardboard. Plastic and steel molds are slightly oiled before filling them with concrete. Cardboard molds have an integral waterproofing on the inside surface. The outside of a mold is identified with an indelible marker as to the the job site and the date of casting. After the molds have been cast, their tops are covered with a plastic sheet to retain the water in the concrete. They are kept at the construction site for 24 h after casting, care being taken to ensure that they remain undisturbed for this period. After 24 h, they are carefully transported to a concrete testing laboratory, where the molds are removed. At the time of removing the molds, the identifying information is transferred to the cylinders, and they are then placed in a moist chamber. The air in a moist chamber is fully saturated at approximately 73°F. The cylinders remain in the moist chamber until they are tested to failure. Some research laboratories that do not have a moist chamber use a water tank in which the cylinders are kept immersed.

> **NOTE**
>
> **Field Curing of Concrete Cylinders**
>
> To obtain the compressive strength of concrete that is representative of the field conditions, cylinders may be field cured.

As described later, concrete develops its strength over a long period of time. Typical concrete strength used for design purposes is its 28-day strength. Cylinders are, therefore, generally tested 28 days after casting. To determine the strength of concrete, the cylinders are removed from the moist chamber (or water tank) and crushed to failure, Figure 18.18. The compressive strength of concrete is simply the failure load divided by the cross-sectional area of the cylinder (Chapter 4).

Because the top and bottom surfaces of a concrete cylinder are rough, they are capped with a smooth-surfaced material before placing the cylinder in the test machine. Smooth cylinder ends give a more reliable test result. Rough ends lead to local crushing of protruding

FIGURE 18.18 A concrete test cylinder in position in the testing machine, ready to be compressed to failure (see also Figure 4.6).

concrete particles, falsifying the result. Rough ends also misalign the cylinder; the cylinder must be in a vertical position in the machine. The capping material commonly used is hot sulfur because it sets quickly on cooling and gives a strong and smooth surface.

Concrete is the only building material whose quality is verified after it has been used in the building. Cylinder and slump tests verify that the concrete supplied by the concrete manufacturer has the strength and slump specified for the project.

18.7 MAKING CONCRETE

Concrete is made by mixing coarse and fine aggregates, portland cement, and water. A small amount of concrete for a do-it-yourself job may be made by mixing various ingredients with a shovel, adding water, and mixing the ingredients further until all materials have blended. Alternatively, bags of dry concrete mix (consisting of premixed aggregates and portland cement) may be obtained from a building material store. A dry concrete mix needs only the addition of water.

FIGURE 18.19 A transportable mixer for on-site mixing of small quantities of concrete.

If a slightly larger quantity of concrete is required, an on-site mobile concrete mixer can be used, Figure 18.19. However, where even a small degree of control on the quality of concrete is required, the concrete should be obtained from a ready-mix plant. A ready-mix plant is a concrete-manufacturing (or making) facility. Approximately 95% of all concrete used in contemporary building construction is obtained from such a facility. Most cities in the United States have at least one ready-mix plant within the city or close by.

An important part of concrete manufacturing is *mix design*. Mix design involves determining the correct amounts of various ingredients to give a concrete that has the required durability, strength, workability, and any other property specified by the architect or engineer. Mix design is fairly complex and is covered in detail in a text devoted entirely to the subject. One reason for its complexity is the enormous variability in the quality of aggregates, even when they are obtained from the same source. A typical ready-mix plant has technical personnel with expertise on mix design.

Figure 18.20 shows an overview of a typical ready-mix plant. The most prominent and visible part of the plant is a set of cylindrical silos. The silos contain coarse aggregate, fine

FIGURE 18.20 An overview of a ready-mix plant.

FIGURE 18.21 Stockpiles of sand and coarse aggregates in a ready-mix plant. Observe the separation between different aggregates that ensures correct proportioning of the mix.

aggregate, and portland cement. The aggregates are initially stockpiled in an open yard within the plant compound, Figure 18.21. From the stockpiles, the aggregates are conveyed into the silos as needed. The space directly under the silos is designed to hold a concrete mixer truck. Correct amounts of various materials are fed by gravity from the silos into the mixer, Figure 18.22. Weighing the materials and charging the truck mixer is fully automated and is controlled from a central location that overlooks the charging operation, Figure 18.23. The first material fed into the mixer is usually water, followed by aggregates and portland cement. A small amount of water may be added at the construction site if needed. During travel, the mixer drum is set to rotate, so that when the concrete arrives at the site, it is almost ready for discharge (with some site remixing, as needed). A typical concrete mixer truck holds 10 yd^3 of fresh concrete.

FIGURE 18.22 A truck mixer being charged with a mix.

18.8 PLACING AND FINISHING CONCRETE

Because concrete begins to set within a few hours after the addition of water to the dry mix, it is a perishable material. Therefore, it must be placed in the desired position soon after being received at the construction site. Placing concrete requires transporting the concrete from the mixer truck to its final destination, that is, into the form of the building component. If the mixer truck can be parked fairly close to where the concrete is to be placed, the concrete can be delivered using a chute. A chute is simply an open steel channel that extends from the discharge end of the mixer to the concrete's destination, Figure 18.24. Almost every mixer has a few lengths of chute attached to it. A chute transports concrete by gravity and can only be used if the concrete's final destination is below the discharge end of the mixer.

FIGURE 18.23 A typical control room in a ready-mix plant.

Another method for placing concrete is to use buckets, Figure 18.25. Buckets are typically hoisted by cranes, but helicopters have been used in remote or demanding situations. The bucket method of transporting concrete is generally used for small quantities of concrete, where pumping is uneconomical.

Transporting concrete using a pump is common. It is particularly attractive where the construction site is spread over a large area and transporting concrete by chutes is not possible. Pumping is also used for high-rise structures in place of buckets. Pumping concrete is usually achieved by first transferring the concrete from the mixer into a pump truck. From the pump truck, the concrete is piped to the destination, Figure 18.26. The pump truck consists of a hopper into which the concrete is deposited. The hopper consists of a screen to ensure that undesirable materials that may clog the pipe or the pump are arrested by the screen, Figure 18.27.

FIGURE 18.24 Concrete being brought into the formwork of a grade beam through an open chute from the mixer.

CONCRETE CONSOLIDATION

Once the concrete has been placed in the form, it must be consolidated. Consolidation is the process of compacting concrete to ensure that the concrete is without voids and air pockets. On a small job, consolidation can be done manually with the help of a steel rod, wherein the worker simply rods into the concrete—up and down and with some sideways motion. It is, however, more common to employ a high-frequency power-driven vibrator. This is typically an internal (or immersion) vibrator inserted into the concrete. An external vibrator is one that vibrates the form and is more commonly used in precast concrete plants.

FIGURE 18.25 Concrete being filled in the bucket, which is hoisted to position using a crane. Bucket method of transporting concrete is generally used for small quantities of concrete, where pumping of concrete will be uneconomical. Note that a large quantity of concrete is required to fill a pump line, which is wasteful if the quantity of concrete to be placed is small.

FIGURE 18.26 Pumping of concrete from mixer truck into a pump truck and then to its final destination, which, in this case, is a slab-on-grade.

(a)

(b)

FIGURE 18.27 **(a)** Concrete being transferred from the mixer truck to the hopper of a pump truck. **(b)** Details of a pump hopper, which has a wire screen to arrest any undesirable element in concrete that may clog the pump or the pipeline.

An internal vibrator consists of a rod (1-in. to 3-in. diameter) connected to a flexible shaft, Figure 18.28. The vibrator is inserted into freshly placed concrete. As the concrete in a particular location is compacted, the vibrator is moved to the next location. Because the vibrator has a finite area of influence, the new insertion point must be fairly close to the previous insertion point. Excessive vibration must be avoided because it leads to particle separation and bleeding of concrete. For the same reason, concrete must be carefully placed in the forms, not dropped from an excessive height.

FIGURE 18.28 Compaction of concrete using a power-driven vibrator. Note proper vertical placement of vibrator.

FIGURE 18.29a Striking concrete in a beam with a wood straightedge.

FIGURE 18.29b Striking concrete in a slab with a wood straightedge.

FINISHING THE CONCRETE SURFACE

After the concrete has been compacted, its exposed surfaces are finished while the concrete is still plastic. The exposed surfaces are those that are not covered by the formwork. The finishes include

- Strikeoff (screeding)
- Floating (darbying)
- Troweling

The purpose of strikeoff (also called *striking,* or *screeding*) is to level the concrete surface. It is done with a wood straightedge, generally a 2×4 member, Figure 18.29. The straightedge removes excess concrete as it is moved from one point to another. It also achieves some surface compaction. Low spots behind the straight edge are filled using a shovel or hand trowel, and high spots are struck off.

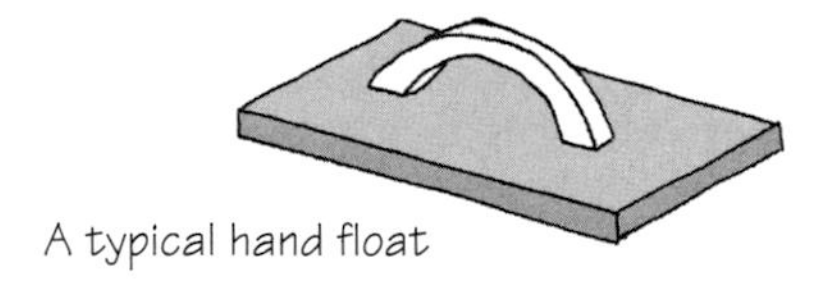

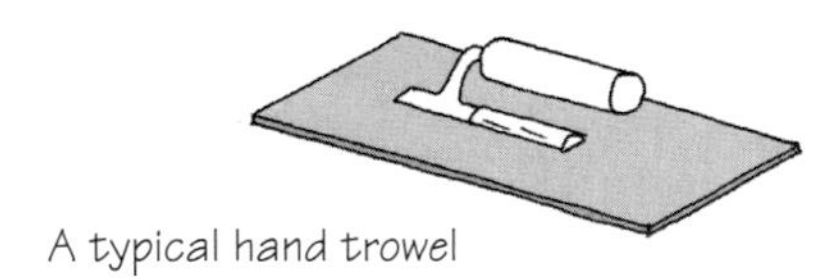

Immediately following the strikeoff operation, a concrete surface is floated. Floating (also called *darbying*) is usually done by a hand float or a bull float, which smoothes the surface further, Figure 18.30. For non-air-entrained concrete, floats are usually made of wood. For air-entrained concrete, they are generally made of aluminum or magnesium alloys.

Striking and floating are the only operations needed for most concrete that does not need any additional finishing. However, an interior concrete slab that is to be covered with a carpet or thin floor tiles must be troweled to achieve a smooth surface. Troweling is done after the concrete has stiffened somewhat (reached its initial set point), which may be an hour or longer after floating, depending on the ambient temperature. (A general rule of thumb is that concrete should sustain foot pressure with a maximum $\frac{1}{4}$ in. indentation before beginning troweling.) Troweling may be done using a hand trowel, which consists of

FIGURE 18.30 Floating of a small concrete surface with a hand float and a large surface with a long-handle bull float.

FIGURE 18.31 Troweling of concrete with a power-driven troweling machine, which is available in two different types, as shown. Each machine has metal blades that rotate over the concrete surface.

a steel plate attached to a wood handle, or by a power-driven rotary trowel, Figure 18.31. Where a rough surface is required, troweling is replaced by brooming. This is achieved through the use of steel wire brooms or fiber brooms.

PRACTICE QUIZ

Each question has only one correct answer. Select the choice that best answers the question.

20. The slump of concrete is measured
a. when the concrete is fresh.
b. 28 days after casting the concrete in a test specimen.
c. 14 days after casting the concrete in a test specimen.
d. 7 days after casting the concrete in a test specimen.
e. 1 day after casting the concrete in a test specimen.

21. The slump of concrete is a measure of the
a. compressive strength of concrete.
b. tensile strength of concrete.
c. modulus of elasticity of concrete.
d. workability of concrete.

22. The test specimen used for determining the compressive strength of concrete in the United States is a
a. 6-in. diameter, 6-in.-high cylinder.
b. 6-in. diameter, 12-in.-high cylinder.
c. 6-in. diameter, 18-in.-high cylinder.
d. 6-in. diameter, 12-in.-high cone.

23. The compressive strength of concrete is generally measured
a. when the concrete is fresh.
b. 28 days after casting the concrete in a test specimen.
c. 14 days after casting the concrete in a test specimen.
d. 7 days after casting the concrete in a test specimen.
e. 1 day after casting the concrete in a test specimen.

24. In a high-rise building, the concrete may be brought from the ground to the location of placement by
a. chutes.
b. buckets.
c. pumping.
d. (a) or (b).
e. (b) or (c).

25. After concrete has been placed in the form, it should be compacted as much as is economically practical.
a. True **b.** False

26. The consolidation and compaction of concrete on a construction site is generally done
a. manually using a 2-in.-diameter, 3-ft-long steel rod.
b. manually using a 1-in.-diameter, 3-ft-long steel rod.
c. using a power-driven rod vibrator.
d. using a power-driven plate vibrator.

27. The finishing operations of a concrete surface usually involve striking, floating, and troweling. The sequence in which these three operations are generally performed is
a. striking, floating, and troweling.
b. floating, striking, and troweling.
c. striking, troweling, and floating.
d. troweling, floating, and striking.

Answers: 20-a, 21-d, 22-b, 23-b, 24-e, 25-b, 26-c, 27-a.

18.9 PORTLAND CEMENT AND WATER REACTION

Concrete and other mixes made from portland cement gain their strength due to the reaction of portland cement with water, referred to as the *hydration of portland cement.* The amount of water required for complete hydration is about 40% of the weight of portland cement. In other words, for complete hydration, the water-cement ratio (referred to as the *w-c ratio*) should be 0.40. Often however, a larger quantity of water is needed to provide the requisite workability of concrete.

To obtain a normal-weight concrete with a slump of 4 in. or so (used for beams and columns), a w-c ratio of 0.45 to 0.55 is usually needed, which exceeds that needed for complete hydration. In some concretes, the w-c ratio required may be as high as 0.7. The excess water eventually evaporates, leaving air voids in hardened concrete, which reduce concrete strength.

Experiments have indicated that the strength of concrete is inversely proportional to the w-c ratio, Figure 18.32. Therefore, a concrete should contain the minimum amount of water that gives it the required workability.

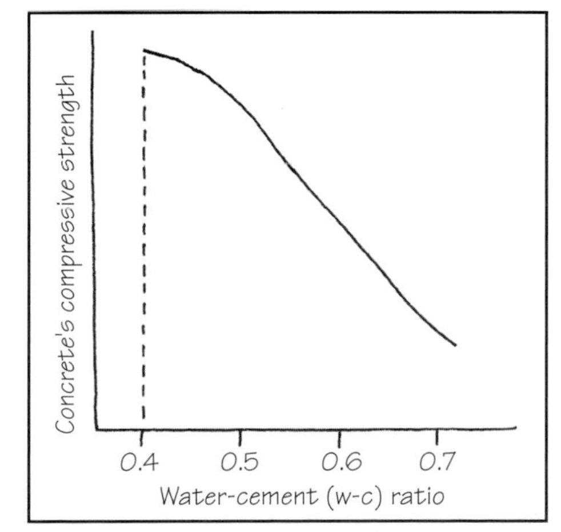

FIGURE 18.32 Strength of concrete as a function of the w-c ratio. For a given concrete (type of aggregates and the amount of portland cement), its strength increases as the amount of water used is reduced.

DURATION OF HYDRATION REACTION—THE CURING OF CONCRETE

The hydration reaction begins as soon as water and portland cement come into contact, but the rate at which this reaction proceeds is extremely slow. It takes up to 6 months or longer for concrete to gain its full strength, Figure 18.33. However, approximately 80% of concrete strength develops in 28 days.

Approximately two-thirds of the 28-day strength is obtained in the first 7 days, and approximately half, in the first 3 days. This is true only if sufficient water and favorable temperature are available for the hydration reaction to continue. That is why concrete test cylinders are kept in a moist chamber until tested. Providing moisture to concrete continuously for hydration is called *curing of concrete.*

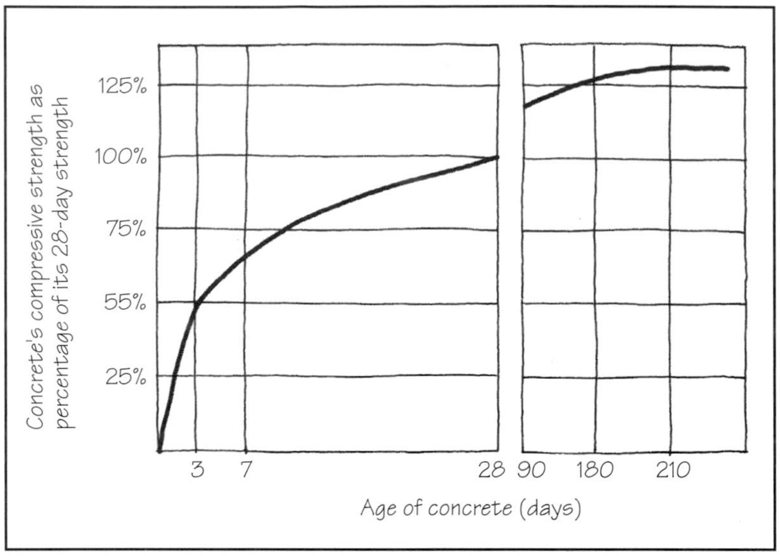

Concrete's compressive strength as percentage of its 28-day strength
125%
100%
75%
55%
25%
3 7 28 90 180 210
Age of concrete (days)

FIGURE 18.33 Compressive strength of concrete as a function of its age. Observe that concrete keeps gaining strength well beyond 28 days. Because we generally use concrete's 28-day strength as the design strength, the additional strength adds to the safety of a concrete structure.

Curing not only increases the strength of concrete, but also reduces its permeability to water. A well-cured concrete is denser and, hence, stronger and more durable. On construction sites, curing is begun as soon as the concrete has fully set (solidified), which is generally 12 to 24 h after placing the concrete. Curing in the initial stages of hardening is extremely important and should continue for as long as possible, not less than 7 days. The following methods of field curing are common:

- *Keeping concrete wet with water:* This is the most effective curing method, which includes covering the exposed surfaces of concrete with water absorptive materials such as burlap cloth, rags, or blankets, and sprinkling water on them, Figure 18.34. If it is possible to continuously sprinkle water on concrete, cover materials are not necessary.
- *Covering concrete with a plastic sheet:* Because concrete usually has more water than needed, covering concrete with a plastic sheet retains its moisture. However, in warm and (or) windy climates, where the evaporation of water is rapid, this method alone is not adequate.
- *Curing compounds:* This consists of spraying concrete with a liquid membrane. A liquid membrane is like a plastic sheet. However, it becomes an integral part of the concrete surface. If the concrete is to receive a subsequent surface treatment, such as a floor or other finish that requires bonding with concrete, curing compounds should not be used. Curing compounds are often used on street pavements and bridge decks, which do not have a source of continuous water supply available.

INFLUENCE OF TEMPERATURE ON HYDRATION

Hydration of portland cement is temperature sensitive. The rate of hydration increases as the ambient air temperature increases, and vice versa. Under 55°F, the rate of hydration decreases significantly. The use of Type III portland cement helps under these circumstances. However, low-temperature concreting should be avoided because significant reduction in concrete strength occurs if the water in concrete freezes within a few hours of the concrete's placement (generally before it attains a compressive strength of 500 psi). That is

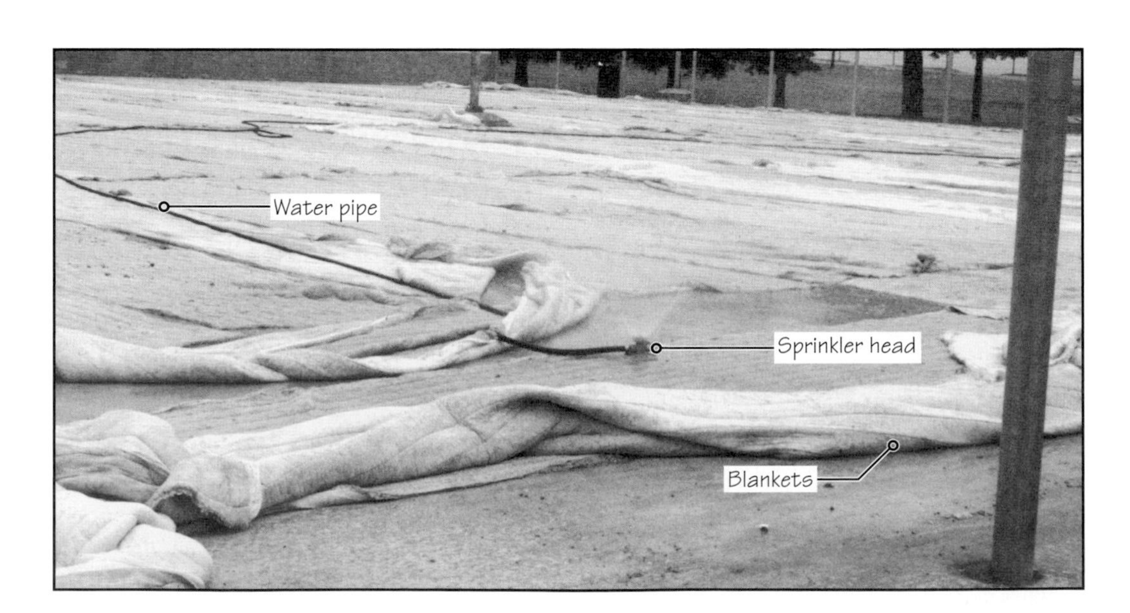

FIGURE 18.34 Curing of a recently poured concrete slab-on-grade. Notice blankets, water pipe, and sprinkler.

why concrete is generally not placed if the temperature of the surrounding air is below 40°F or is expected to fall below 40°F within a day or so.

If concrete must be placed when the temperature is under 40°F, one or more of the following precautions must be taken: warming of aggregates and water, insulating the forms, surrounding the concrete in a heated enclosure, or use of Type III portland cement.

Special precautions are also required if concreting is to be done in hot weather—generally, if the air temperature exceeds 100°F. However, low humidity and windy conditions may necessitate these precautions below 100°F. Generally, wetting the aggregates with water to cool them evaporatively, building a protective cover to shield against direct solar radiation and wind, concreting after sunset, or using chilled water or crushed ice are some of the precautions that should be taken.

NOTE

Key Temperature: 40°F and Rising

It is common not to place concrete if the ambient air temperature is 40°F and decreasing. In other words, concrete should be placed only when the ambient air temperature is at least 40°F and rising, unless necessary precautions are taken to prevent damage to fresh concrete due to its freezing.

18.10 WATER-REDUCING CONCRETE ADMIXTURES

Ever since it became known that reducing water content in concrete increases its strength, the concrete industry began to find ways of reducing the amount of water without decreasing the workability of concrete. This finally became possible with the discovery of chemicals known as *plasticizers,* or, more commonly, as *water-reducing admixtures* (WRA).

A WRA acts on the surfaces of portland cement particles. The commonly used concrete mixes do not contain a WRA. In the absence of a WRA, the particles of portland cement tend to cluster together. To overcome the clustering (flocculation) tendency, a larger amount of water is needed to make concrete workable. A WRA reduces flocculation and disperses cement particles in the mix so that each particle is covered with a sheath of water around it, rather than the particles of cement. This provides greater lubrication, increasing workability. WRA increases concrete slump by 10% to 15%. Additionally, each cement particle, being surrounded by water, is able to hydrate more fully, which increases concrete strength.

The discovery of WRAs led to the production and use of stronger concrete. However, the real breakthrough in producing high-strength concrete came with the discovery of superplasticizers, also known as *high-range water reducers* (HRWR). An HRWR, which is chemically different from a WRA, works on the same principles as a WRA but reduces water requirement substantially, up to about 30%.

The introduction of HRWRs has produced concrete strengths that were considered impossible a couple of decades ago. Concrete up to a compressive strength of approximately 50,000 psi has been produced in laboratories and up to 20,000 psi has been used in buildings [2, 3]. This is a major leap forward considering that until 1960, the highest concrete strength used in buildings was 5,000 psi.

18.11 HIGH-STRENGTH CONCRETE

The realization of high concrete strengths means that concrete competes with steel for the structural frame of tall buildings. Until recently, steel was the only material that could be used for buildings taller than 40 stories or so. Today, concrete has been used in the structural frame of some of the tallest buildings in the world. The commonly used yield strengths of steel, 36,000 psi and 50,000 psi, are well above the (current) maximum possible concrete compressive strength of 20,000 psi. However, concrete competes favorably with steel because of its two major advantages:

- The first advantage is concrete's inherent fire resistance. Steel must be protected through an additional fire-protective treatment. A fire-protective treatment adds to the cost of a steel structure in two ways. The first cost is that of the treatment. The second cost is in the drop ceilings under the floors to conceal the protective treatment. Fire protective treatment is unsightly, fibrous, and easily scrapable. Without a drop ceiling, scraped fibers can get into the environment, with obvious health concerns.
- The second advantage of concrete over steel is that the modulus of elasticity of concrete increases with increasing strength. Steel's modulus of elasticity is constant at 29,000 ksi, regardless of steel strength. Modulus of elasticity is an important structural property, particularly for tall buildings. Remember (Chapter 4) that the modulus of elasticity signifies the stiffness of a material. A tall building, which functions structurally as a vertical cantilever such as a flagpole, is subjected to deflection and side sway caused by wind and earthquake. Therefore, stiffness is an important requirement of the material used in a tall building.

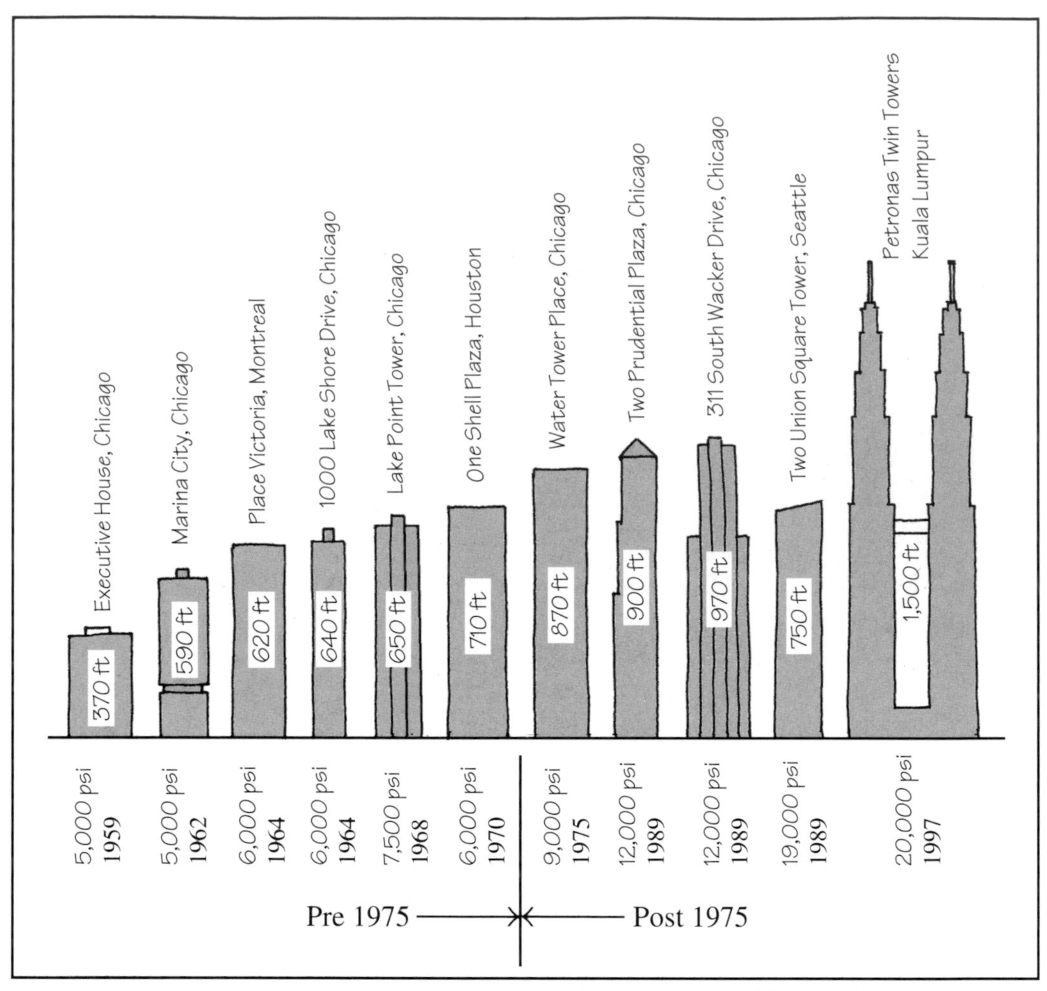

FIGURE 18.35 Progressive increase in concrete strength in the structural frame of buildings.

Although the modulus of elasticity of even high-strength concrete is well below that of steel, increasing concrete strength gives dividends in terms of increased modulus of elasticity. This is not so for steel, because of its constant modulus of elasticity. Thus, although steel is produced in strengths much higher than 50 ksi, its use in the structural frames of buildings is rare.

The consequence of these facts is that the heights of buildings with a concrete structural frame have progressively increased over the years, Figure 18.35. Until 1959, the tallest building with a concrete structural frame was 370 ft high, in which 5,000-psi concrete was used. In 1997, the 1,500-ft-high Petronas Towers were built in Kuala Lumpur utilizing steel-encased concrete columns with concrete strength of 20,000 psi. The petronas Towers are taller than the steel-framed Sears Tower in Chicago. (Taller buildings, in China and Dubai, are currently under construction.)

NOTE

Concrete Admixtures

A concrete admixture is defined as a material other than portland cement, coarse and fine aggregates, and water.

A commonly used concrete admixture is fly ash. Air-entraining admixture is used to increase concrete's durability against the freeze-thaw effect. It is more commonly used in concrete pavements and bridges in colder climates. The following other admixtures are used as needed:

- WRA
- HRWR
- Silica fume
- Set accelerator
- Set retarder

OTHER INGREDIENTS OF HIGH-STRENGTH CONCRETE

Concrete is a material that can be produced with compressive strength ranging from 2,000 psi to 20,000 psi on construction sites. This is unlike steel, which is available in two or three standard strengths. The concrete industry has, therefore, divided concrete into two classifications based on its strength:

- Conventional concrete—compressive strength $<$ 6,000 psi
- High-strength concrete—compressive strength $\geq$ 6,000 psi

This division is totally arbitrary because no one factor affects concrete strength abruptly. Generally, conventional concrete may be produced without the use of admixtures. High-strength concrete requires the use of admixtures. A *concrete admixture* is defined as a material other than portland cement, coarse and fine aggregates, and water. WRAs and HRWRs are not the only materials required for the production of high-strength concrete. High-strength concrete also requires the use of other admixtures, such as fly ash and silica fume.

Fly Ash—A Sustainable Concrete Ingredient

As mentioned previously, fly ash is a waste product produced from coal-fired power stations. It is used in both high-strength and conventional concretes. Because fly ash particles are microscopic in size, they densify concrete, filling in voids between portland cement particles. The densification of concrete increases its strength and durability. Small particles of fly ash also increase concrete's workability and reduce permeability and creep of concrete (Section 9.6).

However, the main benefit of fly ash is in its pozzolanic property. During the manufacture of portland cement, a certain amount of lime (calcium hydroxide) does not convert to complex portland compounds—alite, belite, celite, and so on. This lime is referred to as *free lime.* Fly ash reacts with free lime, converting it into a hydraulic cement, and adds to concrete strength.

Silica Fume—A Sustainable Concrete Ingredient

Silica fume is a waste product from the silicon industry. It is an extremely fine and amorphous silica (sand) with average particle size of 0.1 μm [4]. In other words, more than 100 silica fume particles occupy the same space as one particle of portland cement. Silica fume fills the voids that other concrete materials cannot. It is an extremely lightweight material and is available in bags, Figure 18.36. A large silica fume bag weighs less than 10 lb. Silica fume is an expensive concrete admixture. Therefore, it is not commonly used in conventional concrete. It has excellent pozzolanic properties. Like fly ash, it reacts with free lime to convert it into a hydraulic cement.

High-Strength Concrete and Portland Cement Quantity

Because portland cement is the binding agent, its contribution to concrete strength is obvious. In general, the greater the amount of portland cement, the greater the concrete's strength. However, increasing the amount of portland cement in a mix yields exponentially smaller increases in its strength. In general, more than 900 lb of portland cement per cubic yard does not increase concrete strength significantly.

Disadvantages of High-Strength Concrete

The cost of concrete increases with increasing strength. At 20,000 psi or so, the mix is not only very expensive but requires a great deal of quality control at the site, which increases the cost further. Remember, concrete strength depends on the quality of aggregates and admixtures, grading of aggregates, and the quantity of portland cement. It also depends on placing, compacting, and curing. An ultra-high-strength concrete requires that all these factors must be very tightly controlled.

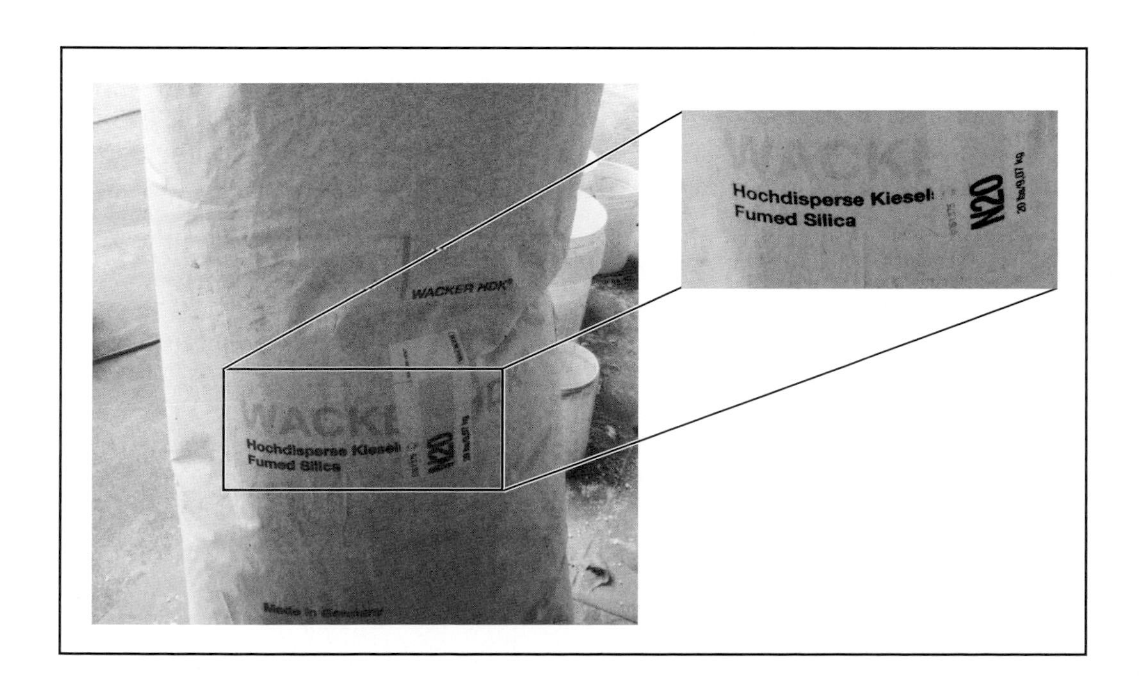

FIGURE 18.36 A silica fume bag.

18.12 STEEL REINFORCEMENT

Concrete is much weaker in tension than compression. Its tensile strength is approximately 10% of its compressive strength. Therefore, concrete is generally used in conjunction with steel reinforcement, which provides the tensile strength in a concrete member. The use of plain concrete—concrete without steel reinforcement—is limited to pavements and some slabs-on-ground.

Steel is the ideal material to complement concrete because the thermal expansion of both materials is the same. In other words, when heated or cooled, both steel and concrete expand or contract equally. Consequently, no stress is caused by differential expansion or contraction. Composite materials that expand differentially are subjected to such stresses.

FIGURE 18.37 Deformations in steel bars.

Steel also bonds well with concrete. In a composite material, the bond between two materials is necessary for it to function as a single material. The bond between steel and concrete is due to the chemistry of the two materials, which produces a *chemical bond* between them. Additionally, as water from concrete evaporates, it shrinks and grips the steel bars, giving a *mechanical bond.*

The mechanical bond is enhanced by using reinforcing bars, or *rebars,* that have surface deformations, Figure 18.37. Because mechanical bond is a function of the area of contact between the two materials, surface deformations increase that area, thereby increasing the bond. For the same reason, a rebar that has a light, firm layer of rust bonds better with concrete. Rust that is produced by leaving rebars outside on a construction site for a few days or weeks is not objectionable as long as the rust is not loose or flaky. Loose and flaky rust should be scraped using burlap or a piece of cloth. Excessively rusted rebars should not be used.

Virtually all reinforcement in a concrete member consists of deformed bars. Plain (smooth, undeformed) bars are used only as dowels at expansion joints. At expansion joints, rebars are oiled (or covered with a sleeve) before being embedded in concrete to allow movement at the joint. This requires the use of plain bars. Plain rebars are also used as spiral reinforcement in round columns.

Reinforcing Steel Grade

As we learned in Chapter 4, yield strength is an important steel property. Rebars are hot rolled from steels of the following different yield strengths:

- 40,000 psi—referred to as *grade 40 steel*
- 60,000 psi—referred to as *grade 60 steel*

The cost of both steel grades is approximately the same. Grade 60 steel is $1\frac{1}{2}$ times stronger than grade 40. Therefore, it is by far the most frequently used steel grade. Grade 40, which is more ductile, that is, more easily bendable, is used where rebars need to be field bent. Grade 50 is also available but is rarely used.

Diameter and Length of Bars

Eleven different diameters of rebars are available, from $\frac{3}{8}$ in. to $2\frac{1}{4}$ in., Table 18.3. The diameter of a bar is generally stated in terms of a number. Thus, a No. 3 bar (commonly

TABLE 18.3 STANDARD BAR DIAMETERS AND AREAS

Bar size and designation	Diameter (in.)	Cross-sectional area (in.^{2})
#3	0.375	0.11
#4	0.500	0.20
#5	0.625	0.31
#6	0.750	0.44
#7	0.875	0.60
#8	1.000	0.79
#9	1.128	1.00
#10	1.270	1.27
#11	1.410	1.56
#14	1.693	2.25
#18	2.257	4.00

stated as a #3 bar) has a diameter of $\frac{3}{8}$ in. A #4 bar has a diameter of $\frac{1}{2}$ in., a #5 bar has a diameter of $\frac{5}{8}$ in., and so on, up to a #8 bar, which has a diameter of 1 in.

According to this numbering system, a #9 bar should have a diameter of $\frac{9}{8}$ in. (1.125 in.). However, its actual diameter is 1.128 in., providing a cross-sectional area of 1.0 in.^{2}. In structural calculations, it is the cross-sectional area of a bar that is more important than the bar diameter. Therefore, rationalizing area is more important than the diameter. That is why a #18 bar has a diameter of 2.257 in., which gives an area of 4.0 in.^{2}. Note that $\frac{18}{8} = 2.25$.

Deformed rebars are available from #3 upward. A $\frac{1}{4}$-in.-diameter bar is available only as a plain bar, but it is not used as concrete reinforcement except in welded wire reinforcement, as described later. The diameter of a deformed bar is its nominal diameter. The nominal diameter is the diameter that gives the same cross-sectional area as a plain bar.

Bar lengths of 20 ft and 40 ft are standard stock sizes. Lengths up to 60 ft are generally available as special order. The 60-ft length is based more on transportation limitations than on manufacturing.

BAR IDENTIFICATION MARKINGS

Each rebar in the United States carries an identification marking, Figure 18.38. The marking consists of four pieces of information in the following order:

- Producing mill's identification number or symbol.
- Type of steel. The most commonly used steel is new billet steel. Its identification mark is *N*. Mark *S* is used for special billet steel. Mark *R* is used for rail steel. Some mills close to old, unused rail lines produce this steel. Mark *A* stands for axle steel.
- Bar size.

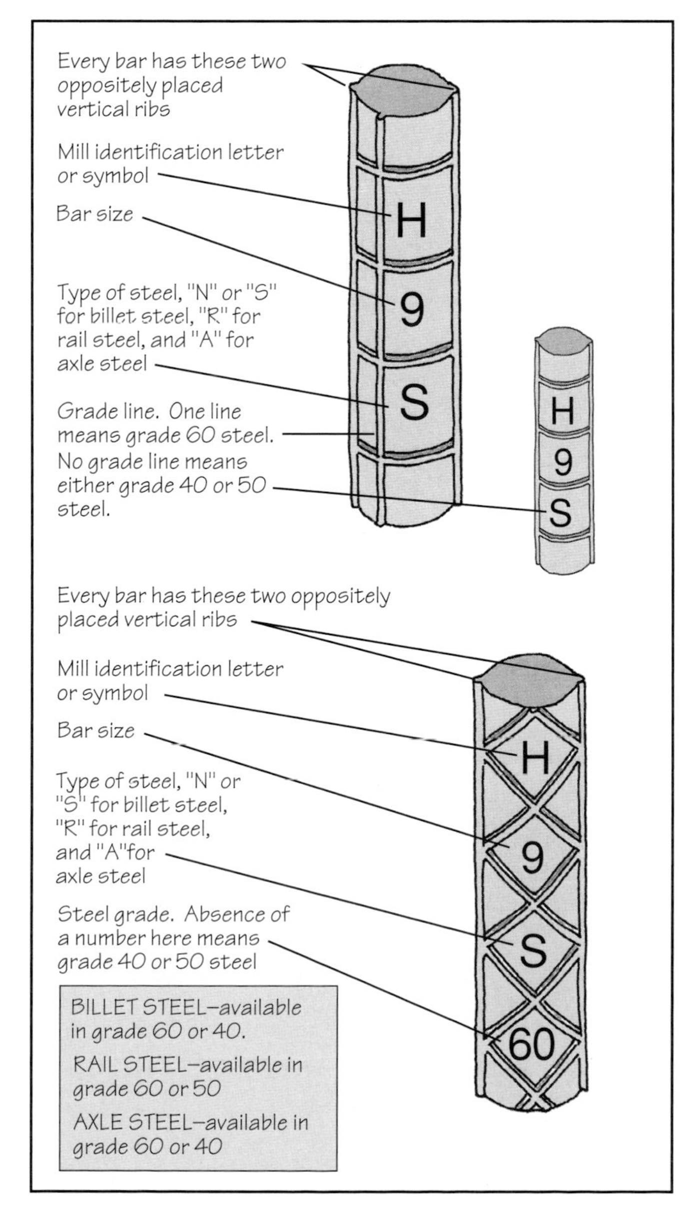

FIGURE 18.38 Two commonly used identification markings on reinforcing bars.

FIGURE 18.39 Bars generally arrive at a construction site bent to shape and cut to required lengths. Plastic tags attached to the bundles identify the building component (beam, column, or slab) to which the bars belong.

- Steel grade. Each bar carries at least two vertical lines. These are simply deformation lines. An additional vertical line on a bar is the grade line, implying that it is grade 60 bar. No grade line means that it is grade 40 or 50 bar. Bars that have diagonal deformations have the grade stamped on them during manufacturing if they are of grade 60. No grade stamp means a grade 40 or 50 bar.

BENDING AND TYING OF BARS

In a concrete member (beam, slab, or column), some rebars need to be straight, whereas others need to be bent to required profiles. The cutting and bending of rebars are done in a rebar fabrication plant. Rebars arrive at a construction site prebent and in the required lengths, Figure 18.39. Rebar bundles arrive at the construction site with identification tags that identify the structural members to which the bars belong.

After the bars are received at the site, they are assembled in beam-and-column cages. The cage assembly requires tying the bars together, Figure 18.40. Tie wires are approximately $\frac{1}{20}$-in.-diameter steel wires, which can be easily bent with the hands.

EPOXY COATED BARS

Most reinforcing bars in concrete members are black bars, that is, they do not have any surface coating. However, concrete members that are exposed to corrosive atmospheres require coated bars. Fusion-bonded epoxy coated bars are commonly used in concrete bridges and parking garages exposed to deicing salts, where the salts gradually penetrate through concrete to corrode the bars. Epoxy-coated bars may also be used in other corrosive atmospheres.

(a)

(b)

FIGURE 18.40 **(a)** Bars being assembled in cages at the construction site, which requires tying them together with a steel wire. **(b)** Close-up of tied bars, showing the wire.

FIGURE 18.41 A heap of welded wire fabric matts on a construction site.

18.13 WELDED WIRE REINFORCEMENT

Welded wire reinforcement (WWR) is a prefabricated reinforcing steel available in rolls or mats, Figure 18.41. Rolls come in widths of 5 to 7 ft. Mats come in various dimensions. They are commonly used in ground-supported slabs (or pavement) and steel deck–supported slabs, where reinforcement requirements are marginal, that is, much smaller than those needed for reinforced concrete suspended slabs.

A WWR roll or mat consists of a rectangular mesh of steel wires. Longitudinal wires are spot welded to transverse wires at each intersection. If conventional rebars are used in place of a WWR, each bar would need to be tied manually at intersections. The use of a WWR greatly speeds reinforcement assembly at the site.

Designation of a WWR Roll or Mat

Both plain and deformed bars are used in a WWR, in steel grades of 40 and 60. A WWR roll or mat is designated by either *W* or *D*, implying the use of plain wires or deformed wires, respectively. Additional designation includes the cross-sectional areas of wires and their spacing.

For example, a WWR roll designated as 6 × 8-W2.9 × W2.9 means that the longitudinal wires are spaced 6 in. on centers, and the transverse wires are at 8 in. on center. *W* indicates plain wires, and the cross-sectional area of the longitudinal and transverse wires is 2.9/100, that is, 0.029 in.^{2}. A 6 × 6-D4.0 × D4.0 means that it is a fabric consisting of deformed wires spaced 6 in. on center in both directions. The wires have a cross-sectional area of 0.04 in.^{2}.

FOCUS ON SUSTAINABILITY

Sustainability Considerations in Concrete

Like other materials, concrete has a few inherent sustainability features, which include durability, easy recycling, use of waste materials from other industries, no VOC emissions, and a high surface reflectivity of the finished material.

Durability

Concrete does not burn, corrode, or decay. Consequently, it is a durable material and, hence, environmentally responsible. Concrete foundations, floor slabs, structural frame members, roads, and pavements last a long time.

Recycling

Concrete obtained from old, demolished buildings and other structures can be broken or crushed and can then be used as aggregate in subgrade bases under building foundations, slabs-on-grade, roads, and pavements.

Use of Waste Materials

Most modern concrete utilizes fly ash, which is a waste product obtained from the manufacture of steel.

VOC Emissions

Concrete does not emit any volatile organic compounds (VOC), unlike plywood, OSB, carpets, and so on.

High Surface Reflectivity

Compared to several other materials, a concrete surface has reflectivity. Therefore, roads and pavements built out of concrete contribute less to urban-heat island effect as compared with asphalt roads and pavements.

Portland Cement and Sustainability

A major criticism with respect to concrete is the amount of energy used and the amount of carbon dioxide released in the production of portland cement. In recognition of this fact, the industry is embracing initiatives that will make the manufacturing process more environmentally friendly.

PRACTICE QUIZ

Each question has only one correct answer. Select the choice that best answers the question.

28. Portland cement sets and gains strength by virtue of its chemical reaction with
- **a.** water.
- **b.** oxygen in air.
- **c.** carbon dioxide in air.
- **d.** aggregates.
- **e.** none of the above.

29. The rate at which concrete gains strength increases as the ambient air temperature increases.
- **a.** True
- **b.** False

30. High-strength concrete is defined as a concrete whose compressive strength is greater than or equal to
- **a.** 4,000 psi.
- **b.** 5,000 psi.
- **c.** 6,000 psi.
- **d.** 8,000 psi.
- **e.** 10,000 psi.

31. A high-strength concrete generally requires a
- **a.** water-increasing agent.
- **b.** water-reducing agent.
- **c.** sand-reducing agent.
- **d.** sand-increasing agent.

32. Steel reinforcing bars, generally used in concrete, have a yield strength of
- **a.** 60 ksi.
- **b.** 80 ksi.
- **c.** 100 ksi.
- **d.** 120 ksi.

33. A #8 reinforcing bar has a diameter of
- **a.** $\frac{1}{8}$ in.
- **b.** 8 in.
- **c.** 1 in.
- **d.** none of the above.

34. In a welded wire fabric mat that has a designation of 8×8-D4.0 $\times$ 4.0, the number 4.0 implies that the wires are
- **a.** deformed wires, each with a diameter of 4.0 mil.
- **b.** deformed wires, each with a diameter of 0.04 in.
- **c.** deformed wires, each with a cross-sectional area of 0.04 in.^{2}.
- **d.** deformed wires, each with a cross-sectional area of 0.4 in.^{2}.
- **e.** deformed wires, each with a cross-sectional area of 4.0 in.^{2}.

35. Epoxy-coated bars are generally used in
- **a.** high-rise buildings built in warm climates.
- **b.** parking garages built in warm climates.
- **c.** high-rise buildings built in cold climates.
- **d.** parking garages built in cold climates.
- **e.** none of the above.

Answers: 28-a, 29-a, 30-c, 31-b, 32-a, 33-c, 34-c, 35-d.

KEY TERMS AND CONCEPTS

Aggregate
Air entrained portland cement
Concrete admixtures
Curing concrete
Finishing concrete
Floating
Fly ash
Formwork
Fresh concrete
High strength concrete
Hydrated lime
Hydration reaction
Ingredients of concrete
Lime
Placing concrete
Portland cement
Portland cement and water reaction
Pozzolanic reaction
Properties of concrete
Rebar
Reinforced concrete
Silica fume
Slump test
Steel reinforcing bars (rebar)
Striking
Troweling concrete
Welded wire fabric
Welded wire reinforcement
White portland cement
Workability of fresh concrete

REVIEW QUESTIONS

1. Describe the difference between quicklime and slaked lime.
2. Describe how lime-sand mortar hardens and functions as an adhesive.
3. List various types of portland cement and where they are typically specified.
4. What is air-entrained portland cement? Where is it typically specified?
5. What is lightweight structural concrete? How does it differ from normal weight concrete? What are the advantages and disadvantages of using lightweight structural concrete?
6. Using sketches, explain how the compressive strength of concrete is determined.
7. Discuss the importance of the water-cement ratio in concrete.
8. Discuss the commonly used concrete admixtures.

SELECTED WEB SITES

American Concrete Institute, ACI (www.aci-int.org)
Concrete Construction Magazine (www.concreteconstruction.com)

Concrete International Magazine (www.concreteinternational.com)

National Lime Association, NLA (www.lime.org)

World of Concrete (www.worldofconcrete.com)

FURTHER READING

1. American Concrete Institute (ACI). *Building Code Requirements for Structural Concrete and Commentary.* ACI 318-05, 2005.
2. American Concrete Institute (ACI). *Manual of Concrete Practice* (6-volume set), 2005.

REFERENCES

1. Kosmatka, S. H., and Panarese, W. C. *Design and Control of Concrete Mixtures,* 13th ed. Skokie, IL: Portland Cement Association, 1994, pp. 21 and 68.
2. Harriman, M. S. "Structural Concrete: High Strength." *Architectural Record,* October 1990, pp. 85–92.
3. Godfrey, K. A. "Concrete Strength Record Jumps 36%," *Civil Engineering,* October 1987, pp. 82–97.
4. Kosmatka, S. H., and Panarese, W. C. *Design and Control of Concrete Mixtures,* 13th ed. Skokie, IL: Portland Cement Association, 1994, p. 69.

CHAPTER 19

Concrete Construction–I (Formwork, Reinforcement, and Slabs-on-Ground)

CHAPTER OUTLINE

19.1 Versatility of Reinforced Concrete

19.2 Concrete Formwork and Shores

19.3 Formwork Removal and Reshoring

19.4 Architectural Concrete and Form Liners

19.5 Principles of Reinforcing Concrete

19.6 Splices, Couplers, and Hooks in Bars

19.7 Corrosion Protection of Steel Reinforcement

19.8 Reinforcement and Formwork for Columns

19.9 Reinforcement and Formwork for Walls

19.10 Types of Concrete Slabs

19.11 Ground-Supported Isolated Concrete Slab

19.12 Ground-Supported Stiffened Concrete Slab

Walkways of the King Saud University campus, Riyadh, Saudi Arabia. Reinforced concrete has been used for most buildings on this campus because of concrete's durability and the easy availability of raw materials in this desert landscape. (Photo by MM.)

As stated in Chapter 18, concrete is weak in tension. On average, the tensile strength of concrete is 10% of its compressive strength. The low tensile strength of concrete is offset by combining it with steel, which is strong in tension. In this concrete-steel combination, steel is fully embedded in the concrete so that both concrete and steel function integrally.

Plain concrete (i.e., concrete without steel reinforcement) may be used only in situations where tensile stresses are minimal, such as, in pavements, ground-supported slabs and lightly loaded wall footings on stable and uniformly compacted soils. Even in these situations, a nominal amount of steel is desirable to resist the tensile stresses caused by the shrinkage of concrete or unforeseen bending. There are two ways in which steel is used to strengthen concrete.

REINFORCED CONCRETE

In one option, steel reinforcing bars (rebars), discussed in Chapter 18, are used. In most concrete members subjected to bending (e.g., beams and slabs), reinforcing bars are positioned in locations where tensile stresses are likely to be produced. Because steel is also much stronger than concrete in compression, reinforcing bars are also used to provide compressive strength (e.g., in concrete columns, walls, and some beams).

A concrete member containing reinforcing bars is called a *reinforced concrete member.* If an entire structure consists of reinforced concrete members, which is fairly common, the structure is referred to as a *reinforced concrete structure.*

PRESTRESSED CONCRETE

The alternative to the use of reinforcing bars is to introduce a permanent initial compression in a concrete member, giving a *prestressed concrete member.* The compression is introduced in regions of the member where tension is expected to be produced by the loads. The magnitude of compression produced by prestressing is controlled to ensure that it reduces or cancels the tension created by the loads.

Prestressing of a member is obtained using prestressing cables, called *strands,* which are made of high-strength steel wires. The most commonly used strands have a yield strength of 270 ksi, which is 4.5 times stronger than the most commonly used reinforcing bars (Grade 60 steel; see Chapter 18). Prestressing can be accomplished in two ways:

- Pretensioning
- Posttensioning

In pretensioning, the strands are tensioned between two fixed abutments using hydraulic jacks, Figure 19.1. Subsequently, concrete is cast in the form enclosing the tensioned strands and allowed to cure, creating a bond between the strands and concrete. When the concrete has developed sufficient strength, the strands are cut free at the ends of the member. Because the cut strands tend to relax and attempt to return to their original untensioned state (but are unable to do so because of their bond with concrete), they compress the member.

If the strands are placed in the center of the member's cross section, a uniform compressive stress is produced over the entire cross section. Generally, however, the strands are placed eccentrically so that the prestress developed in the member is the reverse of the stresses that are expected to be produced by the loads.

For example, for a single-span beam supported at two ends, the strands are placed near the bottom of the beam, Figure 19.1(b), so that after the strands are released, compressive stresses are created at the bottom of the beam and an upward curvature (camber) is developed, Figure 19.1(c). When loads are applied, the beam deflects downward and becomes straight.

The strands shown in Figure 19.1 are straight. A more effective use of the strands is to profile them so that their eccentricity changes along the length of the member in response to the stresses created by the loads. In posttensioned concrete members, the profiling of strands (called *draped strands*) is in the shape of a smooth curve (Figure 20.19). Due to the fabrication process, draped strands in pretensioned concrete members, however, must have abrupt changes in their profiles.

Posttensioning differs from pretensioning in that in posttensioning, the strands are tensioned after the concrete has been placed. The strands for posttensioning are encased in sleeves that are placed in the form before placing the concrete. After the concrete has gained sufficient strength, the strands are tensioned. Because the strands are free to move within the sleeves, they can be tensioned without simultaneously tensioning the concrete. After tensioning the strands to the extent required, they are anchored in that position by mechanical wedges at the member's ends, which restrains them from returning to their original

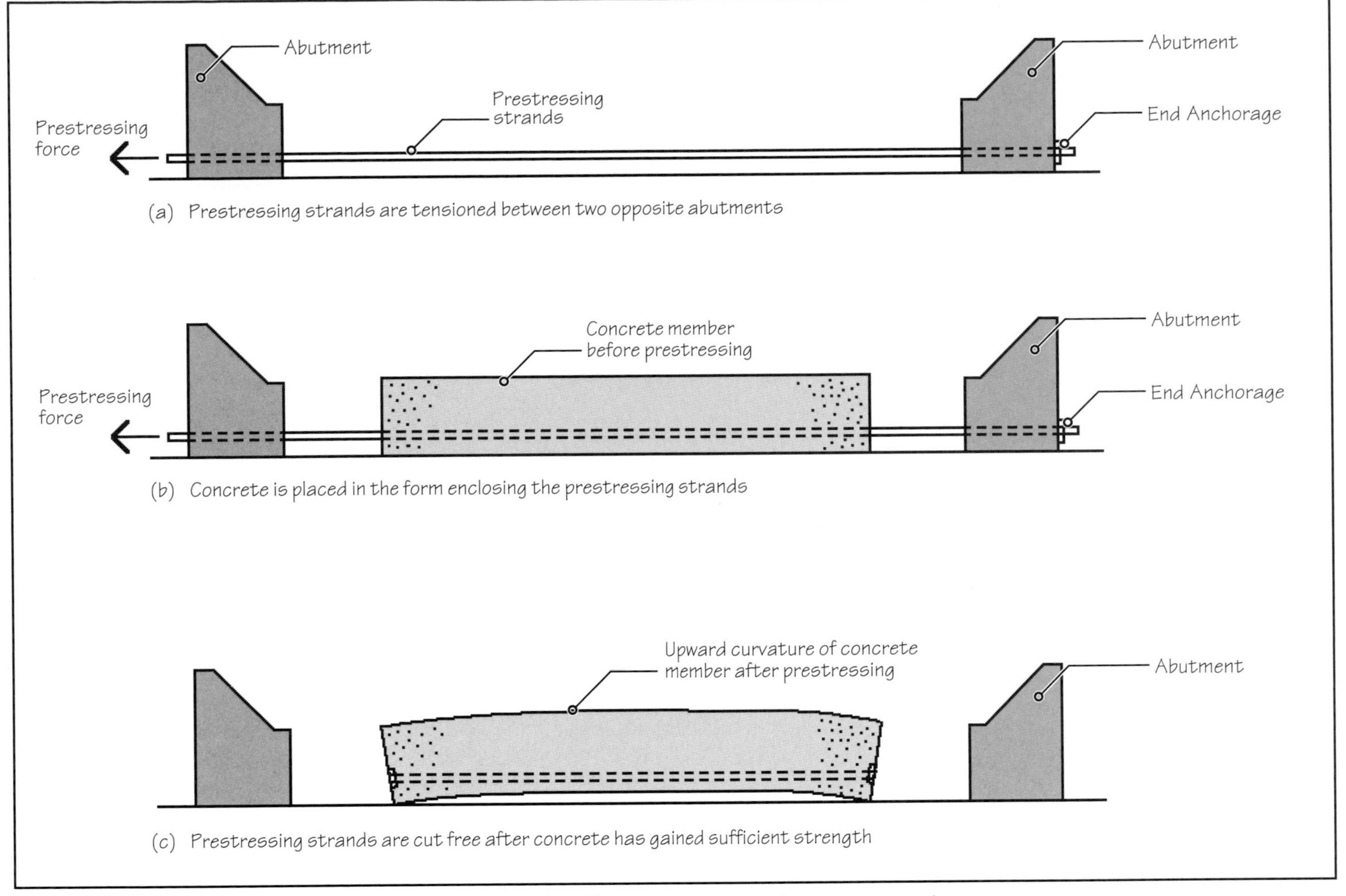

FIGURE 19.1 Three important stages in making a pretensioned (precast, prestressed) concrete member.

(untensioned) state. The prestress created in the member by posttensioning is identical to that created by pretensioning.

Strands and Tendons

The combination of strands, sleeves, end anchorages, and so on, is referred to as *tendons.* The tendons are laid within the concrete forms in the same way as the reinforcing bars. A tendon may consist of one strand (Figure 19.43) or multiple strands. (Because a strand is the most important component of as a tendon, the strand is sometimes referred to as a tendon.)

Advantages and Disadvantages of Prestressing

There are several benefits of prestressing. The two most important ones are reduced cracking in concrete and the reduction in the dimensions of structural members, resulting in smaller dead loads and overall economy. The disadvantages of prestressing are the higher cost of tendons as compared with reinforcing bars and the need for greater quality control and a more skilled labor force.

Site-cast Concrete and Precast Concrete

Pretensioning is generally used for precast concrete members and is accomplished at the precaster's plant. Therefore, a pretensioned member is also called a *precast, prestressed member.* Precast, prestressed concrete members are transported to the construction site and assembled in a manner similar to that of structural steel members.

Posttensioning, on the other hand, is generally used for site-cast members and is done at the construction site. Thus, site-cast (also called *cast-in-place* or *CIP*) *concrete* may either be reinforced concrete or posttensioned concrete. (As illustrated in Chapter 20, reinforced concrete structures can also be posttensioned, in which prestressing tendons and reinforcing bars are used in the same component).

This and the following chapter are devoted to concrete construction. This chapter deals with various aspects of site-cast concrete construction, such as formwork and shoring, principles of

EXPAND YOUR KNOWLEDGE

Some Important Terms

Reinforced-Concrete and Nonprestressed Concrete

In concrete literature, the term *reinforced concrete* is also referred to as *nonprestressed concrete.* Similarly, steel reinforcement (steel rebars) is referred to as *nonprestressing reinforcement.* This recognizes that rebars and prestressing strands are two different types of concrete reinforcement. It also recognizes the growing use of prestressed concrete in buildings, so that referring to steel rebars as *conventional reinforcement* is becoming less common.

Site-Cast Concrete, Cast-in-Place Concrete and Precast Concrete

Although the terms *site-cast concrete* and *cast-in-place concrete* are generally used interchangeably, there is a subtle difference between them. The terms cast-in-place, cast-in-situation, or cast-in-situ are synonymous terms, where the member is cast in the location where it is supposed to remain. This distinguishes it from a site-precast concrete member, which although cast on site, is not cast in-situ. For example, concrete tilt-up walls are site-precast. Concrete tilt-up wall construction is covered in Chapter 24.

reinforcing, and concrete surface finishes. Construction of concrete slabs-on-ground are also covered in this chapter. Other issues of concrete construction, such as site-cast and precast concrete framing systems, are covered in Chapter 20.

19.1 VERSATILITY OF REINFORCED CONCRETE

Reinforced concrete can be used for any structure, small, big, low-rise or high-rise—in buildings, roads, bridges, tunnels, retaining walls, dams, and so on. In one form or another, reinforced concrete is used in almost every contemporary building, regardless of the geographical location.

The materials for reinforced concrete (crushed stone, sand, and water) are relatively inexpensive and locally available. In most locations, steel required for reinforcing bars and portland cement are also locally produced. Additionally, the skill and training required for quality control in reinforced-concrete construction can be imparted to a local labor force far more easily than that required for steel construction (steel erection, welding, bolting, fire-retardant treatment, etc.).

The integral fire resistance of reinforced concrete and various types of finishes that can be achieved add to its versatility, making it the most popular construction system. In many developing countries, reinforced concrete is used in almost every element of an architecturally significant building—structural frame, exterior and interior walls, basements, foundations, water-containing structures, pavements, street furniture, and so on, Figure 19.2.

The versatility of reinforced concrete is also due to the fact that the two structural materials (concrete and steel) are complementary, which allows variation in the size of a reinforced-concrete member for a given load and span. Thus, the size of a beam or column, providing the same strength, can be varied (to some extent) by varying the relative amounts of the two materials—a flexibility not available in steel or wood construction.

For instance, by increasing the amount of steel in a beam, it is possible to reduce the amount of concrete needed, or vice versa. The reduction in the amount of concrete results in a smaller beam size. (Note that the amount of steel even in a highly reinforced concrete beam seldom exceeds 2% of the beam's cross section. Therefore, a change in the amount of steel alters the beam strength but does not impact the physical dimensions of the beam.)

NOTE

Concrete Construction and Building Codes

Because concrete is inherently fire enduring, concrete members can easily achieve the fire ratings required of Type I(A) construction—the most fire-resistive construction. For most occupancies, the building codes allow an unlimited area and an unlimited height if the building is built with Type I(A) construction. Unlike steel, concrete elements do not require add-on fire protection to meet the fire-rating requirements of a given type of construction.

19.2 CONCRETE FORMWORK AND SHORES

In reinforced-concrete construction, concrete and reinforcement must be contained in molds, referred to as *formwork.* The formwork for elevated slabs and beams must be supported on vertical supports, referred to as *shores.* Both formwork and shores are temporary structural elements that are removed after the reinforced concrete members are able to support themselves.

For a reinforced-concrete slab-on-ground, the formwork is simple and generally consists of 2-by solid lumber (or plywood) edge forms braced by stakes, Figure 19.3. Shores are not required in this case because the concrete is ground supported.

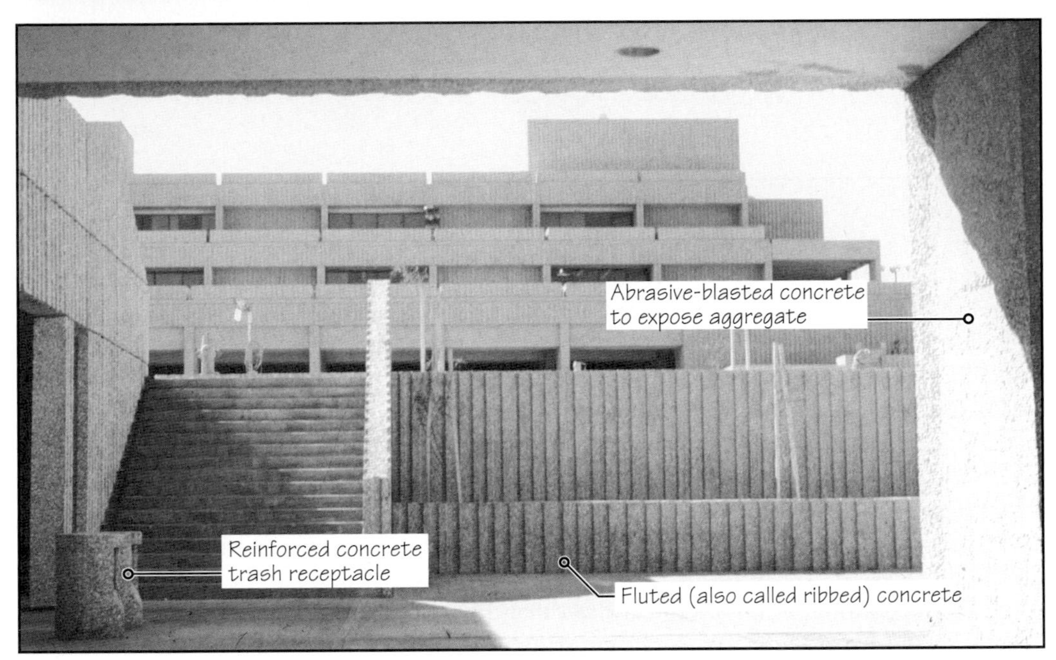

FIGURE 19.2 Almost all buildings on the campus of King Fahd University of Petroleum and Minerals in Dhahran, Saudi Arabia, are constructed of reinforced concrete. It has been used in the walls, structural frame, roof and floor slabs, stairs, water tower, roads and pavements, fountains, street furniture, and so on, highlighting the versatility of this material.

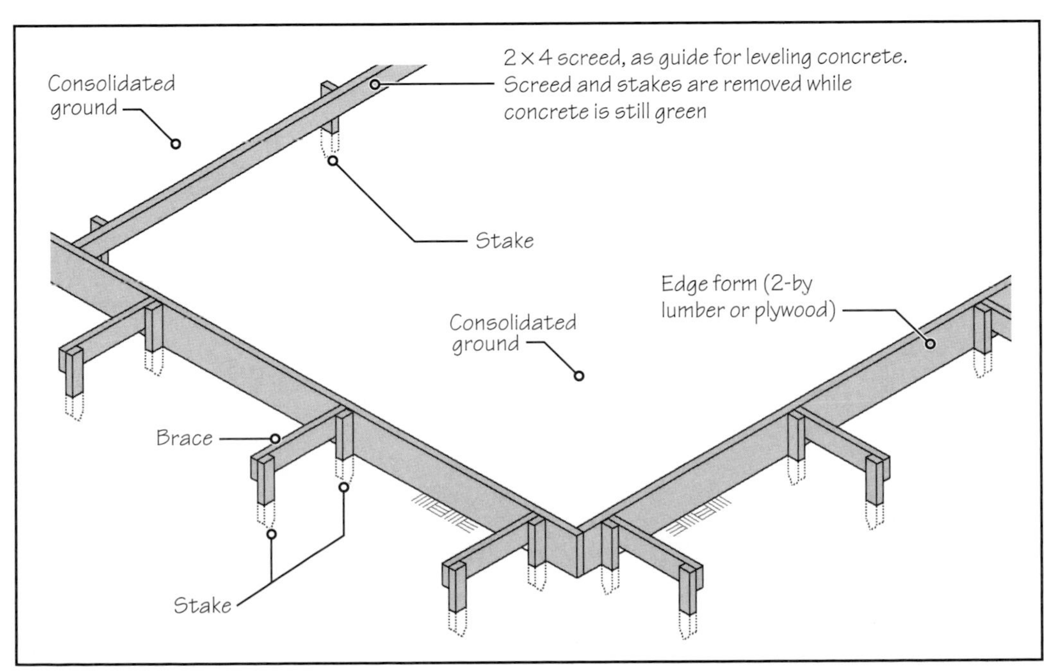

FIGURE 19.3 Formwork for a ground-supported slab.

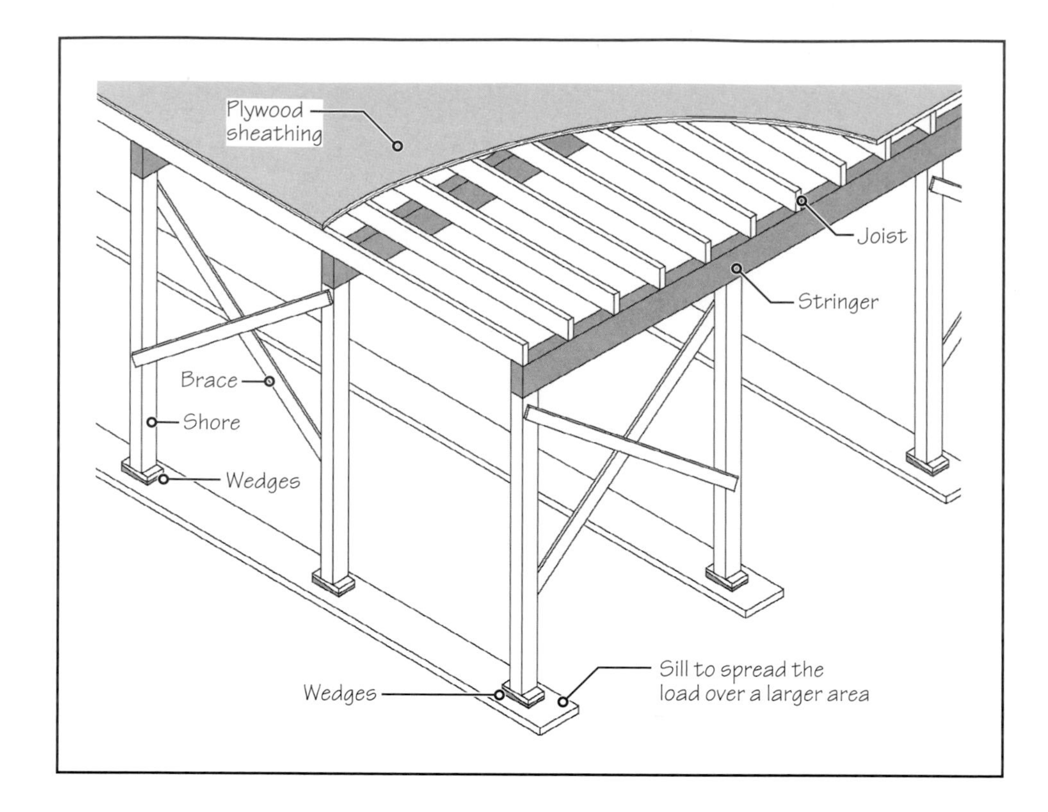

FIGURE 19.4 Formwork and shores for an elevated concrete slab. In this illustration, the shores and formwork are constructed of plywood sheathing and dimension lumber members. As shown in Figure 19.11, steel and aluminum are also used as formwork and shoring materials.

For an elevated concrete slab, the formwork and shores are more complex. Figure 19.4 shows the formwork and shores for a floor slab without beams. The formwork for a complex floor system is more complicated. That is why the formwork in a typical reinforced-concrete building is generally the single most expensive item.

On average, the concrete costs nearly 25% to 30% of the cost of the structure (including placing and finishing of concrete), and steel reinforcement costs approximately 20% to 25% (including laying reinforcement in the forms and making reinforcement cages, etc.). The remaining 50% to 55% of the cost of the structure is in formwork.

For structures with complex geometries, the cost of formwork may exceed 75% of the total cost. The excessive cost of formwork is the primary reason why concrete shell roofs, Figure 19.5, are uncommon. This is despite the fact that the amount of concrete and steel required in them is relatively small due to the inherent structural efficiency obtained from their geometry.

Formwork economy is, therefore, considered early in the design of a reinforced-concrete building. As far as possible, structural patterns and shapes that allow simple, standard, and

FIGURE 19.5 Reinforced concrete shell roof in the Oceonographic Museum in the City of Arts and Sciences, Valencia, Spain, by Architect Santiago Calatrava.

Observe the large roof span, constructed of extremely thin concrete. The amount of reinforcement required is also small. Concrete shell structures are highly efficient in the use of materials because of their geometry, which induces mainly compression and very little bending stress in the structure. However, their geometry makes formwork extremely expensive. The overall cost of concrete shell roofs is quite high.

reusable formwork are the best. Sizes of beams, columns, and other components should be repeated as much as possible.

For example, the size of columns should be kept the same from floor to floor by varying the strength of concrete and (or) the amount of reinforcement. Similarly, the width and depth of beams should also be kept the same from bay to bay.

MATERIALS USED FOR FORMWORK AND SHORES

Criteria in the selection of materials for formwork are cost, strength, reusability, durability, ease of assembly, and weight. The following materials generally satisfy the criteria.

Wood and Plywood Dimension lumber and plywood are the most commonly used form materials because of their cost, strength, nailability, and reusability. Lumber is commonly used in the supporting frame, and plywood is used as horizontal sheathing for floors and vertical sheathing for beams, walls, and columns. Engineered lumber, such as laminated veneer lumber, is also used when its higher cost over dimension lumber is justified.

Form plywood is a special type of plywood that can withstand repeated wetting from concrete. It can also be obtained with an overlay lamination of smooth-surfaced materials, such as high-density overlay (HDO) and medium-density overlay (MDO). Laminations increase the durability and allow for easier stripping and cleaning of plywood for reuse.

In addition to several patented fasteners and clamps, common nails and duplex nails are also used for fastening wood shores, braces, and forms. Wood shores may be supported on (hardwood) wedges that allow height adjustment and dismantling. Sill supports are required to spread the load on a larger area.

Several proprietary systems are available for adjusting the height of wood shores as alternatives to wedges. One system uses two overlapped shores clamped together with a patented device, Figure 19.6. Another system uses a screw-based steel assembly that is nailed to the bottom of a wood shore, Figure 19.7.

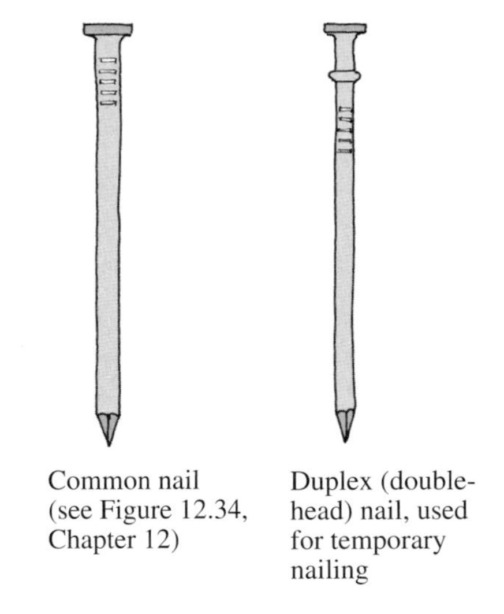

Common nail (see Figure 12.34, Chapter 12)

Duplex (double-head) nail, used for temporary nailing

Steel Steel is also a commonly used form and shore material because of its high strength. Steel angles, channels, and prefabricated steel joists are used in the supporting frame for slab forms and shores. The use of steel joists can substantially reduce the congestion of shores that typically occurs with lumber-based formwork.

Steel pipe shores can be used as individual shores, but they are more commonly used as scaffold-type framed shores. They are generally provided with screw-type adjustment at the bottom and an adjustable head at the top, Figure 19.8.

Aluminum as Form Material Aluminum alloys (which do not react with fresh concrete) are being increasingly used as form material because of their lighter weight and high strength-to-weight ratio. Generally, aluminum angles are used in trusses or in extruded, one-piece joists with U-shaped flanges to provide nailing and bolting capabilities, Figure 19.9.

FIGURE 19.6 Shores for a reinforced concrete floor slab in which two dimension lumber members are lapped and clamped together with a patented device. The lapped and clamped assembly provides adjustment in the height of a shore.

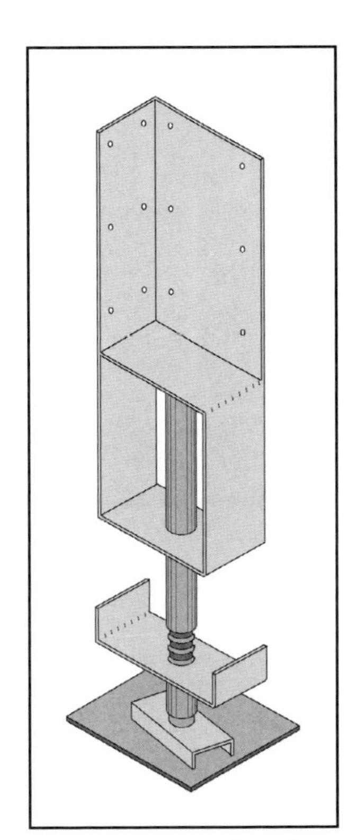

FIGURE 19.7 A screw-based metal assembly that can be nailed to one end of a lumber shore to give it height adjustability.

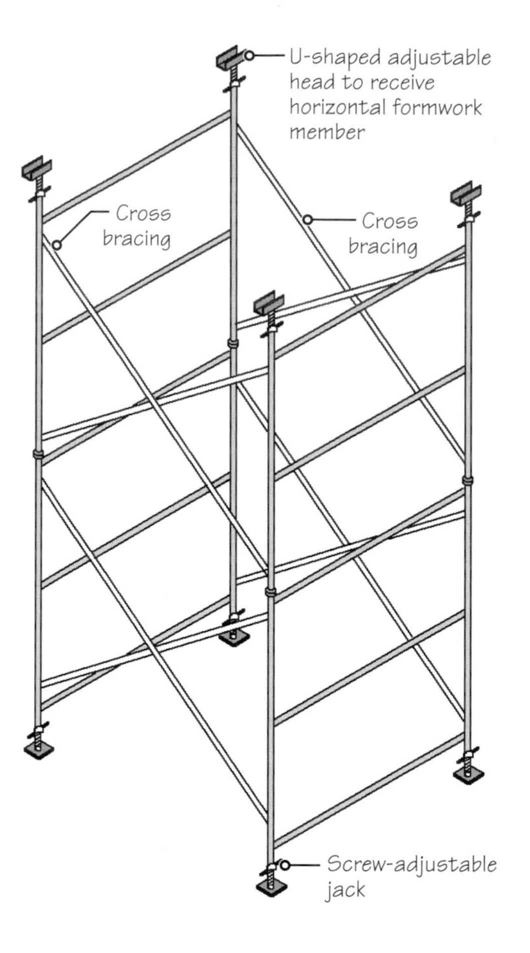

FIGURE 19.8 Steel pipe scaffold shores are available in panel frames in widths of 2 to 5 ft and heights of 4 to 6 ft. Panel frames can be assembled vertically and braced diagonally to make a stable shore frame.

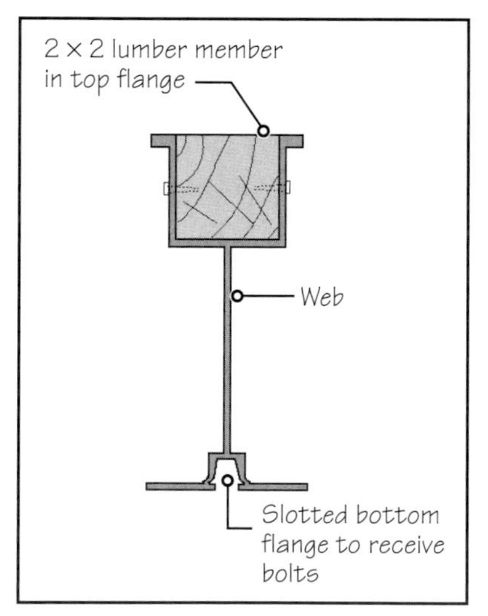

FIGURE 19.9 A typical extruded aluminum section used as joists for formwork.

FIGURE 19.10 Formwork for an entire bay between columns can be assembled as a large panel and flown into position by a crane. Referred to as flying forms, they can be demounted with the help of jacked shores and lifted to the next floor for reuse. Flying forms are particulaly convenient in a high-rise building with identical floor slab component dimensions. In this image, the right-hand side panel is being lowered into position by the crane.

FIGURE 19.11 Details of the panelized flying form in Figure 19.10. The panel consists of steel joists that are supported on two aluminum trusses, one at each end. The vertical members of trusses extend down to form shores, which are provided with adjustable jacks.

The steel joists support lumber planks, which, in turn, support plywood sheathing.

Glass Fiber–Reinforced Plastic (GFRP) as Form Material Because it can be molded into shape and has a smooth surface, GFRP is commonly used for pans and domes required for one-way or two-way joist floor systems (Chapter 20).

PREFABRICATED PANEL FORMS

Because the major cost of formwork in a multistory building is in the floors, it is fairly common to prefabricate the formwork and shores for each bay of the floor as a complete unit and "fly" it into position with a crane, Figure 19.10. The form units, referred to as *flying forms,* are generally complete, freestanding units with their own adjustable shores designed for easy removal and hoisting to the next level, Figure 19.11.

The term *flying form* is typically used with floor slab forms, but other large prefabricated panel forms (called *gang forms*) are also used for walls and other reinforced concrete elements.

PRACTICE QUIZ

Each question has only one correct answer. Select the choice that best answers the question.

1. According to building codes, plain concrete (concrete without steel reinforcement) is not permitted in new construction.
   - **a.** True
   - **b.** False
2. The yield strength of prestressing steel is
   - **a.** 40 ksi.
   - **b.** 50 ksi.
   - **c.** 60 ksi.
   - **d.** 100 ksi.
   - **e.** none of the above.
3. The term *prestressed concrete* refers to both pretensioned and posttensioned concrete.
   - **a.** True
   - **b.** False
4. Which of the following statements is true?
   - **a.** The term *strands* includes tendons.
   - **b.** The term *tendons* includes strands.
   - **c.** The term *tendons* refers to reinforcing bars that have been bent to required shapes.
   - **d.** The term *strands* refers to reinforcing bars that have been bent to required shapes.
   - **e.** none of the above.
5. Which of the following statements is true?
   - **a.** Pretensioning is generally done at the construction site and yields a precast, prestressed concrete member.
   - **b.** Pretensioning is generally done at precaster's plant and yields a precast, prestressed concrete member.
   - **c.** Posttensioning is generally done at the construction site and yields a precast, posttensioned concrete member.
   - **d.** Posttensioning is generally done at precaster's plant and yields a precast, posttensioned concrete member.
6. In a typical sitecast reinforced concrete structure, the cost of formwork and shores is
   - **a.** approximately equal to the combined cost of concrete and steel.
   - **b.** approximately 75% of the combined cost of concrete and steel.
   - **c.** approximately 50% of the combined cost of concrete and steel.
   - **d.** approximately 25% of the combined cost of concrete and steel.
   - **e.** none of the above.

7. Concrete shell roofs are not commonly used because
a. they require a relatively large amount of concrete per square foot of floor area.
b. they require a relatively large amount of steel reinforcement per square foot.
c. they require prestressing, which substantially raises their cost.
d. all the above.
e. none of the above.

8. In site-cast concrete construction, shores are used
a. in ground-supported concrete slabs.
b. in elevated concrete slabs.
c. as formwork for concrete walls.
d. as formwork for concrete columns.
e. none of the above.

9. Height adjustability in shores made of dimension lumber is provided through
a. lapped members clamped together with specially made clamps.
b. screw-based metal assembly.
c. (a) or (b).
d. height adjustability is not provided in shores.

10. Scaffold-type shores are generally made from
a. dimension lumber.
b. plywood.
c. steel pipes.
d. steel plates.
e. all the above.

11. High-density overlay (HDO) on form plywood
a. improves the strippability of forms.
b. allows easier cleaning of forms.
c. increases reusability of forms.
d. all the above.
e. none of the above.

12. Aluminum is commonly used as form and shore material because
a. it is chemically compatible with wet concrete.
b. it is an easily nailable material.
c. its strength-to-weight ratio is low, which reduces the weight of forms and shores.
d. its strength-to-weight ratio is high, which reduces the weight of forms and shores.
e. all the above.

13. The term *flying form* refers to
a. a large formwork assembly for floor slabs.
b. a large formwork assembly for roof slabs.
c. a large formwork assembly for floor and roof slabs.
d. a large formwork assembly for concrete walls.
e. a large formwork assembly for columns.

Answers: 1-b, 2-e, 3-a, 4-b, 5-b, 6-a, 7-e, 8-b, 9-c, 10-c, 11-d, 12-d, 13-c.

19.3 FORMWORK REMOVAL AND RESHORING

To speed construction, forms and shores are removed as soon as possible. Another advantage of early removal of formwork is that if repair or patching of a concrete surface is needed, it can be accomplished while the concrete is in the early stages of curing, favorable to a good bond.

However, a premature removal of forms may not only be dangerous but may also affect the quality of finish. The time between the end of concreting and the removal of forms (referred to as *stripping time*) is a function of the ambient temperature, strength of concrete, and the type of component. Higher ambient temperature and higher strength of concrete require less stripping time. Forms from vertical surfaces of members, such as column forms, the sides of beams, and slabs, can be removed earlier. After their removal, concrete surfaces are generally wrapped with blankets or burlaps and moist cured.

Forms from the horizontal surfaces can be removed only when the concrete has gained sufficient strength to support itself, which is generally assumed to be 70% of the specified strength of concrete. Stripping time is typically specified by the project's structural engineer.

RESHORING

After the formwork from the horizontal surfaces of the structure (beams and slabs) is stripped and moved to the next higher level, the surfaces are supported by shores, a process called *reshoring*. Reshoring facilitates maximum reuse of formwork by taking advantage of the partial strength of the lower floors to support the weight of wet concrete on the floors above. Other objectives of reshoring are

- To prevent deflection and creep of concrete (a concrete that has not gained sufficient strength creeps and deflects more)
- To reduce hairline cracks in concrete

In a multistory building, where one floor is generally cast 7 to 10 days apart, one shored story and two reshored stories are generally required below the story on which wet concrete is placed. This requirement is relaxed if there is a longer gap between the construction of the floors, Figure 19.12. Reshores are generally placed in the same vertical position from floor to floor.

FIGURE 19.12 In this building under construction, the concreting for the top floor has been completed and the forms for the next floor are being installed. The two underlying floors have been reshored, and the reshores from the ground floor have been removed to allow interior construction. As far as possible, reshores must align in the same vertical position from floor to floor.

FORM RELEASE AGENTS

Before placing concrete in the forms, their interior surfaces are coated with release agents to allow the forms to be stripped without damage to both the forms and the concrete. Release agents lubricate the surfaces and, in the case of wood and plywood forms, they also seal the surfaces against water absorption, which increases form life. Various proprietary release agents are available, which are either oil- or wax-based coatings. Better-quality release agents allow several reuses of forms between coatings.

RESPONSIBILITY FOR FORMWORK DESIGN

The structural safety of formwork and shores is obviously vital. Because construction site safety is in the realm of the general contractor (Chapter 1), the structural design of formwork is also the general contractor's responsibility. Thus, the general contractor is free to choose the forming and shoring systems that will yield an economical structure within the requirement of the finished product specified by the architect/engineer.

19.4 ARCHITECTURAL CONCRETE AND FORM LINERS

In the early days of its development, concrete was considered only an engineering material, suitable for gutters, sewers, pavements, roads, and structural frames of buildings. It was considered unsuitable for building facades because of its dull gray, unappealing appearance, which deteriorates with time only because of shrinkage cracks, attracting dust and moss.

Le Corbusier was the among the pioneers who, apart from exploiting its sculptural properties, made concrete count as an architectural material—suitable for building facades and interiors. In most of Le Corbusier's buildings, *smooth, off-the-form concrete* was used; that is, concrete was left exposed (in its natural state) without any additional surface treatment.

An acceptable surface quality straight from the forms can be obtained only if the formwork surface is good, the concrete has an optimum water-cement ratio, and the concrete is vibrated sufficiently to avoid honeycombing. Excessive vibration can cause bleeding or separate the ingredients.

SMOOTH, OFF-THE-FORM CONCRETE

Although smooth, off-the-form concrete provides the most economical surface finish, it is generally not recommended for contemporary facades or interiors because

- Color variations are difficult to avoid between batches of concrete, particularly with gray portland cement. They are relatively easier to control with (more expensive) white portland cement and are less noticeable with aesthetic (relief) joints on the surface.

- Smooth forms should be free from surface imperfections. Additionally, with smooth forms, the vibration of forms is recommended unless care is taken to ensure that internal vibrators do not damage form surfaces.
- Repair of smooth-surface concrete after removing the formwork is generally noticeable.

TEXTURED CONCRETE SURFACES

Because of the concerns with smooth, off-the-form concrete, the use of textured concrete is more common for building facades. (Interior application of textured concrete is relatively uncommon because of its propensity to collect dust.) Textured concrete surfaces can be obtained by (a) surface treatment after formwork removal, (b) using form liners, or (c) both.

The following are commonly used surface treatments on concrete after removing the forms:

- *Exposed aggregate finish by water washing:* This is achieved by applying chemical retarders on formwork, which retards the setting of concrete near the surface. After removing the form, the surface layer of concrete (cement-and-fine-aggregate paste) is carefully removed by pressure washing with water to expose the coarse aggregates. Retarders that expose coarse aggregates to various depths (generally up to one-third of its size) are available.
- *Abrasive blasting:* Depending on the type of abrasive and the amount of pressure applied, abrasive blasting can be: light blasting, medium blasting, or heavy blasting. The difference between the three types is easily noticeable (Figure 26.40). Blasting removes the natural gloss of aggregates and provides a mat (nonglossy) appearance, as compared with that obtained from chemical retarders.
- *Acid etching:* Acid etching involves removing surface cement paste with dilute hydrochloric acid, which reveals the sand and a small percentage of coarse aggregate. In this application, acid-resistant sand and coarse aggregate is recommended (such as quartz or granite). Carbonate aggregates, such as marble, limestone and dolomite, are not acid resistant (Chapter 23).

ARCHITECTURAL CONCRETE THROUGH FORM LINERS

A large variety of textures and patterns can be created on concrete by lining the formwork with the required pattern. The use of form liners is generally more economical than modifying the concrete surface after removing the forms. A few patterns obtained through the use of form liners are shown in Figure 19.13. Form liners, which are generally made from polymeric materials, are available for single or repeat applications (Figure 26.48). They are attached to the formwork or casting beds (for tilt-up concrete walls or precast concrete panels).

19.5 PRINCIPLES OF REINFORCING CONCRETE

A reinforced-concrete member must obviously have an adequate amount of steel. In addition, the reinforcement must also be placed in the correct location. Sometimes a seemingly trivial error in the location of reinforcement can have serious consequences. For example, because an overhanging portion of a concrete slab is subjected to tension at its top surface (and compression at the bottom), the reinforcement must be located near the top surface. If the reinforcement is inadvertently located at the bottom of the slab (as is generally the case with slabs without overhangs), collapse of the slab may occur when the formwork is removed. Such mistakes, though not common, have occurred in buildings with poor inspection protocol.

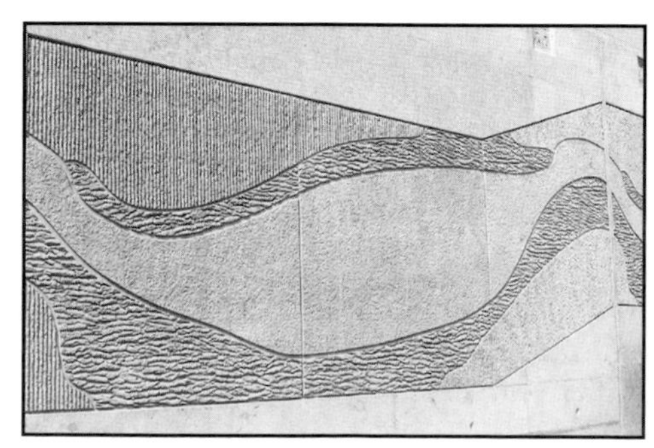

FIGURE 19.13 Some of the several patterns and textures that can be obtained in architectural concrete through the use of form liners. (Photos courtesy of The Greenstreak Group.)

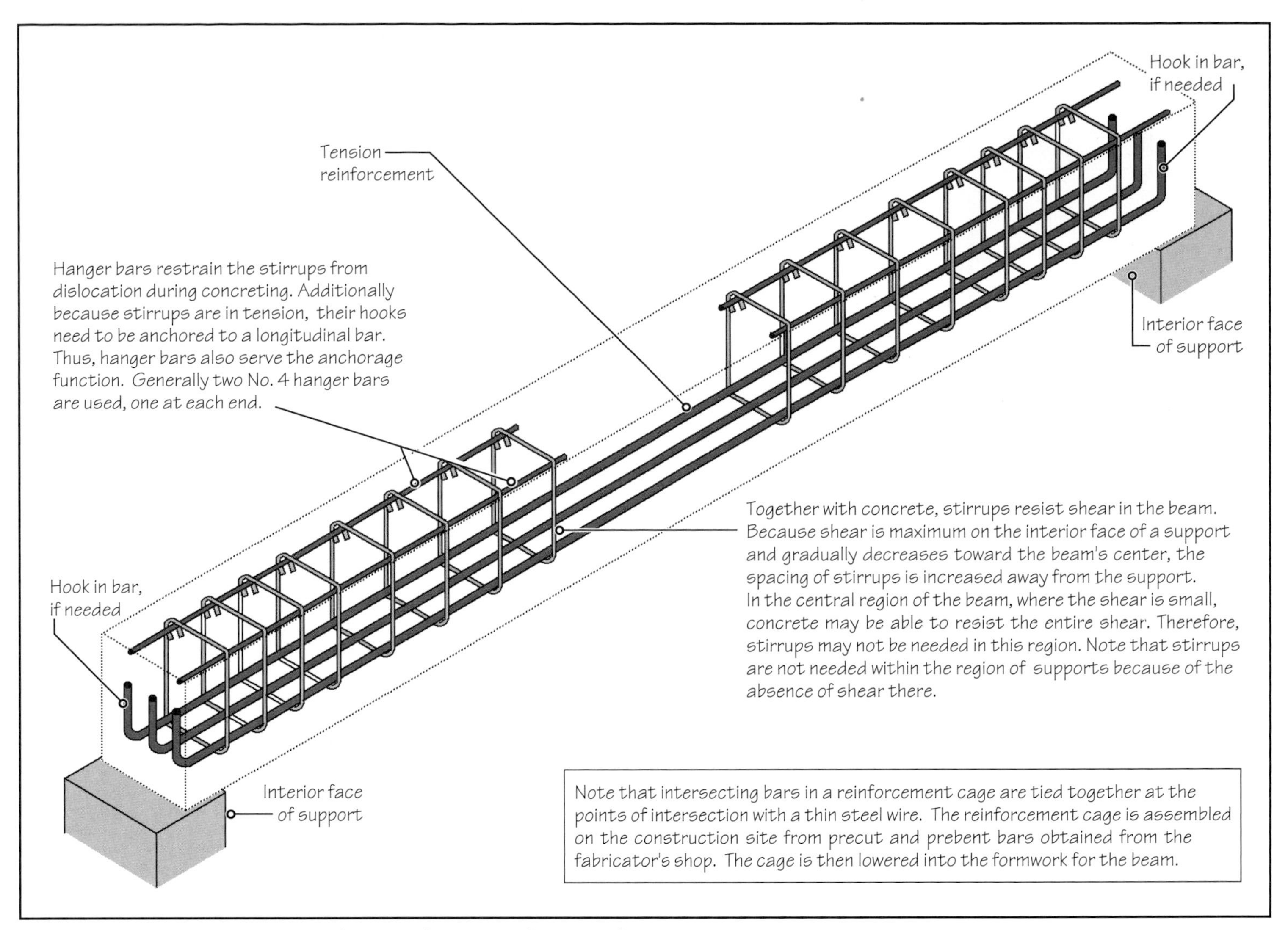

FIGURE 19.14 Reinforcement cage for a simply supported concrete beam.

In addition to the amount and location, there are several other issues related to reinforcement. Consider a typical reinforcement cage for a *simply supported beam,* Figure 19.14.

TENSION REINFORCEMENT, STIRRUPS, AND HANGER BARS

Under gravity loads, a *simply supported beam* is subjected to tensile stresses at the bottom and compressive stresses at the top. Reinforcement is, therefore, required at the beam's bottom, referred to as *tension reinforcement.* No bars are required at the top of this beam unless it also has compression reinforcement.

In addition to tension reinforcement, stirrups are also generally required in a beam to resist shear. Also referred to as *shear reinforcement,* stirrups are typically made from No. 3 or No. 4 bars. In response to the decreasing shear force in a beam away from the supports, stirrup spacing is increased from the face of the support toward the beam's center. No change is made in the diameter of stirrup bars.

Stirrups generally consist of a loop with two vertical legs. Because only the vertical legs of a stirrup resist shear, stirrups can be open-loop stirrups, Figure 19.15(a). Open-loop stirrups reduce reinforcement congestion and allow easier placement of concrete. Closed-loop stirrups are more commonly used, Figures 19.15(b) and (c), and are mandated in seismic zones, where stress reversal may occur, and in beams with compression reinforcement.

BEAMS WITH COMPRESSION REINFORCEMENT (DOUBLY REINFORCED BEAMS)

A reduction in the size of a reinforced-concrete beam for a given span and load can be obtained by increasing the amount of tension reinforcement. If the tension reinforcement cannot be accommodated in a single layer (Figure 19.14), it may be provided in two layers, Figure 19.16.

However, the American Concrete Institute (ACI) Code does not allow an increase in tension reinforcement beyond a certain percentage. This is to ensure a ductile failure of the beam under an overload. For further reduction in beam size, additional steel may be used in

NOTE

Simply Supported Beams and Continuous Beams

A simply supported beam is a beam that bears on two end supports. When the beam bends under loads, the ends of the beam will rotate freely; that is, the supports do not present any restraint to end rotation.

A beam whose ends are framed into a column is unable to rotate freely at the ends, such as the continuous beam of Figure 19.18.

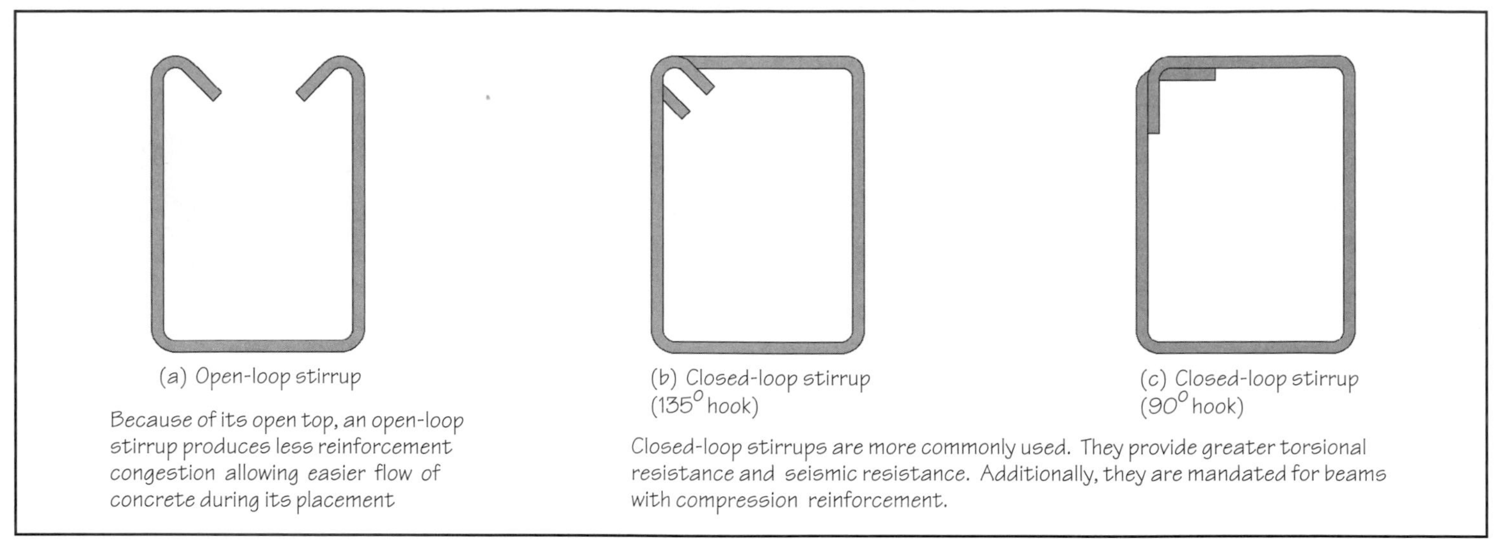

FIGURE 19.15 Commonly used stirrup shapes.

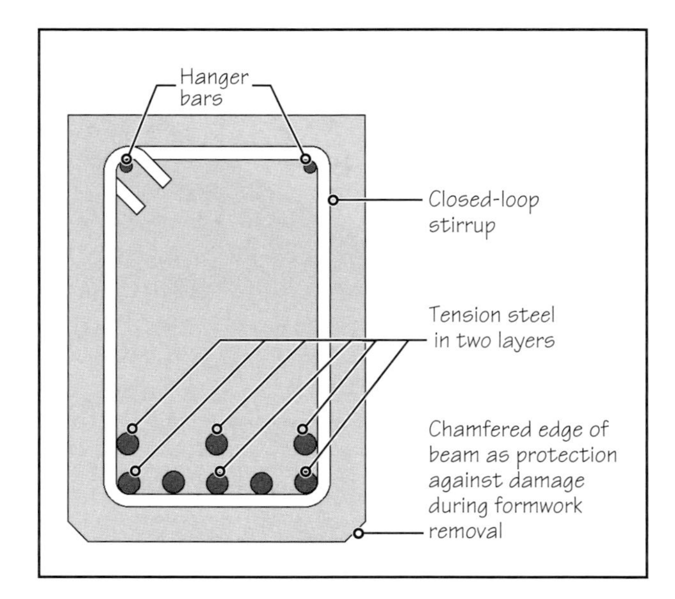

FIGURE 19.16 A concrete beam with tension steel in two layers.

the compression region, that is, at the top of a simply supported beam, referred to as *compression reinforcement.* When compression reinforcement is used in a simply supported beam, it runs continuously from support to support, like the tension reinforcement. Hanger bars are not required in this beam type, and stirrups are required throughout the beam's length.

A beam provided with compression reinforcement is called a *doubly reinforced beam,* Figure 19.17. A beam with with tension reinforcement only, such as that shown in Figure 19.16, is called a *singly reinforced beam.*

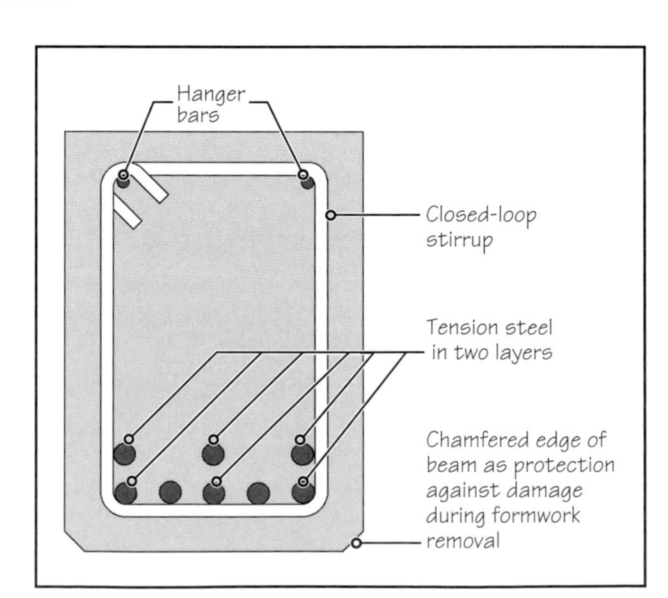

FIGURE 19.17 A section through a doubly reinforced beam. The beam shown in Figure 19.16 (and also Figure 19.14) is a singly reinforced beam.

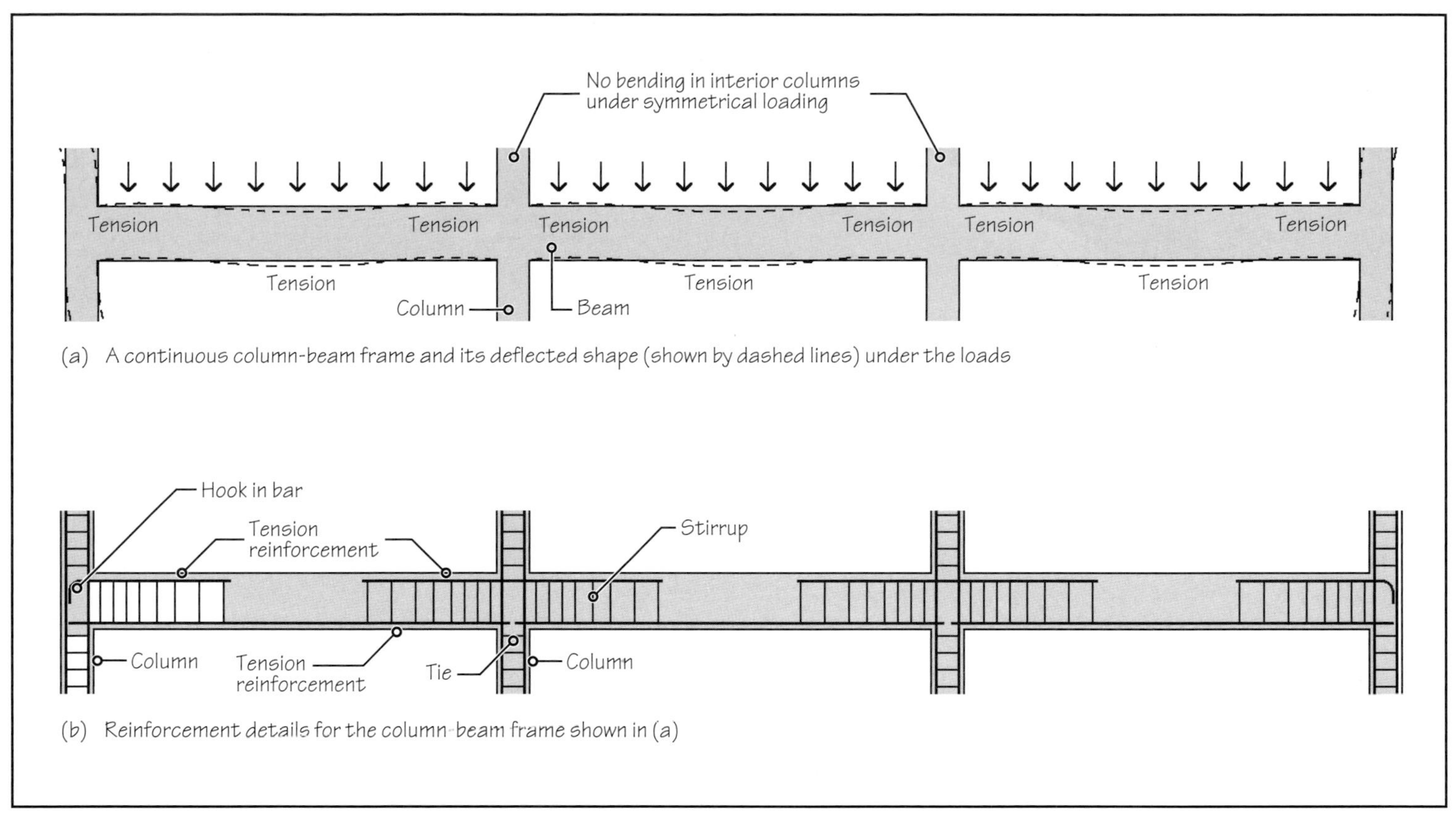

FIGURE 19.18 Reinforcement details in a column-beam frame in response to the stresses created by the loads. Note that the longitudinal bars at both top and bottom of the beam are tension-reinforcing bars.

Continuity in Reinforced-Concrete Members

A single-span, simply supported beam (Figure 19.14) is seldom used in site-cast reinforced-concrete construction. Instead, reinforced-concrete members are generally continuous, which creates a stress pattern that is different from that of a simply supported beam. For instance, in the three-span beam-column frame of Figure 19.18(a), tension is created at the bottom of the beam in the middle region and at the top of the beam near the supports. The reinforcement in such a beam is, therefore, more complex, Figure 19.18(b). Note that both top and bottom longitudinal bars in the beam are tension-reinforcing bars.

19.6 SPLICES, COUPLERS, AND HOOKS IN BARS

Reinforced concrete can function as a structural material only if there is a perfect bond (adhesion) between the concrete and reinforcing bars. This bond allows two lengths of reinforcing bars to function as one continuous bar through lap splices. Because reinforcing bars are available in finite lengths (generally up to a maximum of 60 ft), lap splicing allows reinforcement to extend continuously in long beams, large slabs, and multistory columns.

Because concrete works as glue, the stress in a bar is transferred to concrete and from there to the lapped adjacent bar, Figure 19.19(a). This is similar to joining two strips of cardboard by lapping them and placing glue in the lap to make them function as one continuous strip. The greater the force to be transferred from one strip to the other, the greater the required lap length and/or the need for stronger glue, Figure 19.19(b).

Because reinforcement is designed to be under maximum stress (60 ksi for Grade 60 steel), the lap splice lengths have been standardized for bars of different diameters, concrete strength, and steel grade. Splices are generally located in places of minimum stress in the member. In columns, however, they generally are located at the floor level for convenience, Figure 19.20. One way to splice column reinforcement is to bend the top of the lower column bars inward to accommodate the lapped bars from the upper column, Figure 19.21.

Mechanical Couplers for Reinforcing Bars

Lap splices are the most common means of splicing bars. However, they produce reinforcement congestion, which is problematic in heavily reinforced columns. Therefore,

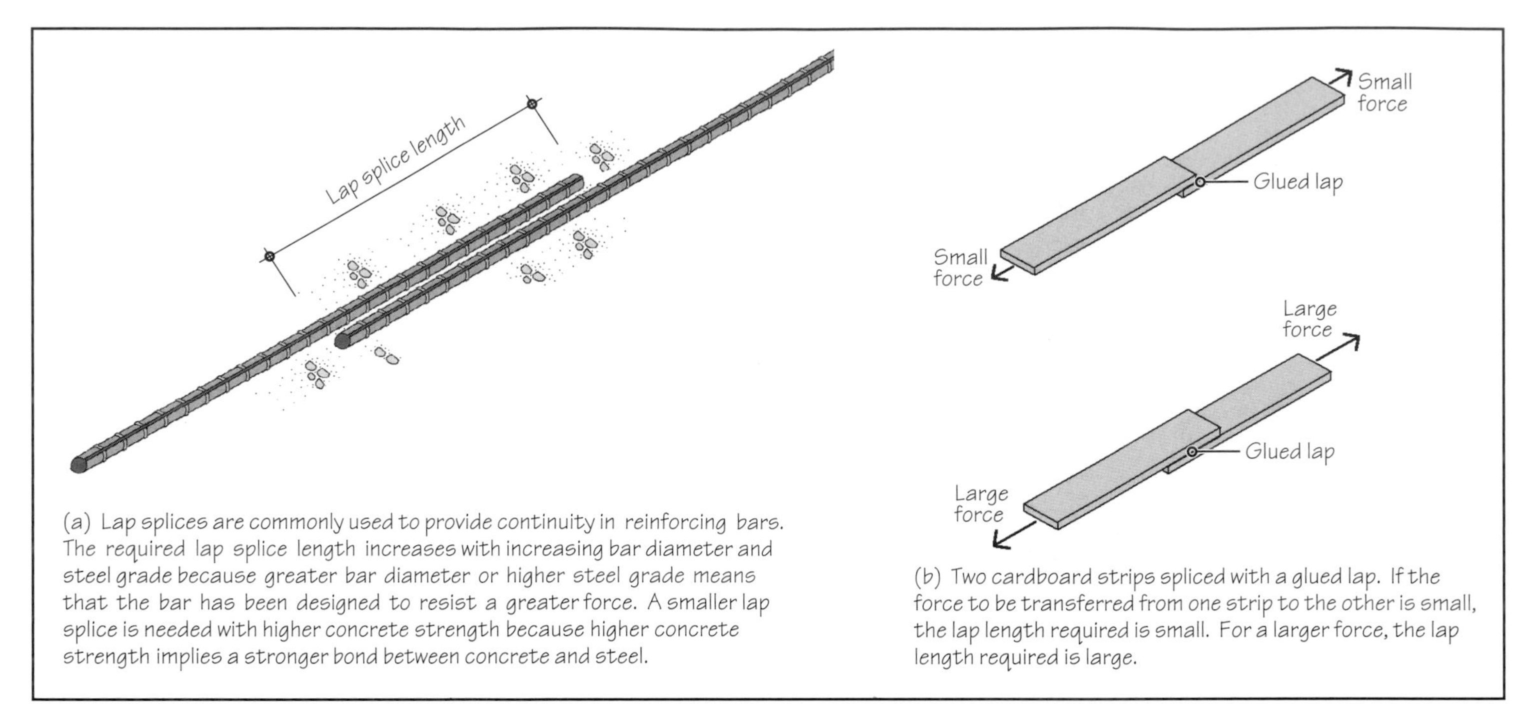

FIGURE 19.19 Principles of lap splicing of reinforcing bars.

FIGURE 19.20 Lap splicing of column bars at the floor level. The lapped bars are generally placed adjacent to each other and tied together. A maximum spacing of 6 in. between lapped bars is, however, permitted by the code.

FIGURE 19.21 Column bars bent inward to receive bars for the upper floor.

mechanical couplers are used to splice bars, Figure 19.22. Various types of proprietary couplers that have been tested for structural adequacy in gripping the bars from both ends are available.

Use of Hooked Longitudinal Bars

For a reinforcing bar to develop the design tensile stress (60 ksi for Grade 60 steel), the bar must be embedded in concrete for sufficient length. If the embedment length of the bar is inadequate, the bar will slip due to loss of anchorage, which may cause the member to fail.

Consider the simply supported beam in Figure 19.23. If the length of a bar from the point of maximum tensile stress (the beam's center in this case) to the end is inadequate, the bar will lose its anchorage, causing it to slip. To increase anchorage in such a case, hooks are provided at the ends of bars. If sufficient embedment length can be provided through straight bars, hooks are not needed. Thus, the provision of a hook is tantamount to increas-

FIGURE 19.22 Reinforcement cage of a heavily reinforced column showing the use of couplers for splicing of bars. Note that the couplers have been staggered in adjacent bars.

ing the embedment length of a bar. For bars used in a beam or slab, 90° or 180° hooks are used, Figure 19.24.

Hooks are generally required at the junction of a beam meeting the last column in a multicolumn bay because sufficient space is not available there for straight bars (Figure 19.18(b)). Note that hooks are effective only for bars in tension, not for bars in compression.

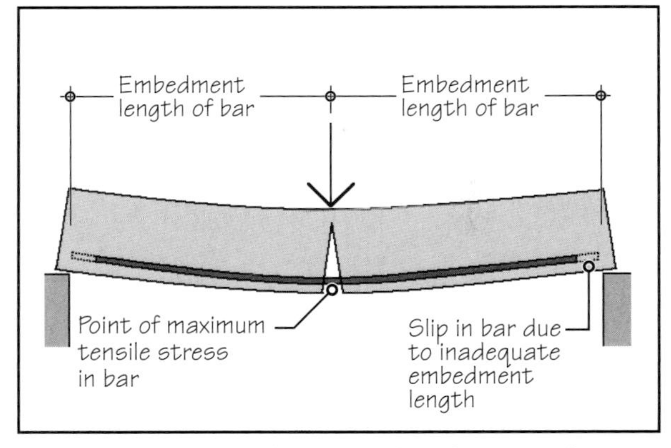

FIGURE 19.23 Bond failure of a bar due to inadequate embedment length.

19.7 CORROSION PROTECTION OF STEEL REINFORCEMENT

To protect reinforcement from corrosion, a minimum amount of concrete cover is required. For reinforcement in footings, the minimum required cover is 3 in. For interior beams, the minimum cover required is 1.5 in., and exterior beams require a 2-in. cover, Table 19.1.

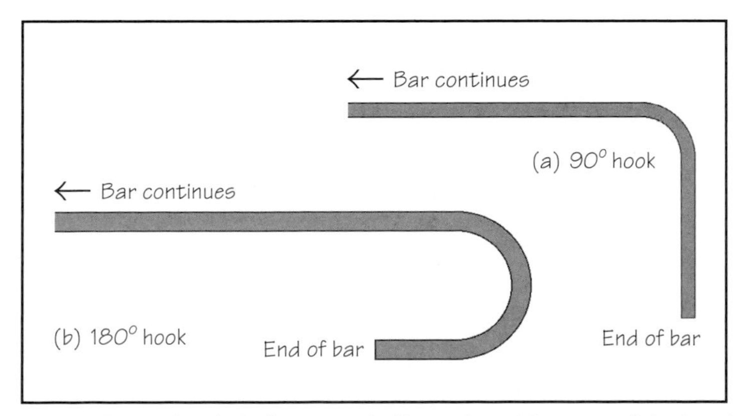

FIGURE 19.24 90° and 180° hooks in bars. Hook dimensions (diameter of the bend and the extension of the bar from the bend to the end of the bar) have been standardized by the American Concrete Institute (ACI) Code so that both hooks have the same anchorage capacity and can be used interchangeably.

TABLE 19.1 MINIMUM CONCRETE COVER FOR REINFORCEMENT

Exposure of concrete	Minimum cover (in.)
Cast against and permanently exposed to earth	3.0
Exposed to earth or weather	2.0
Not exposed to earth or weather	
Beams and columns	1.5
Slabs and walls	0.75

Source: *Building Code Requirements for Structural Concrete,* ACI 318-05.

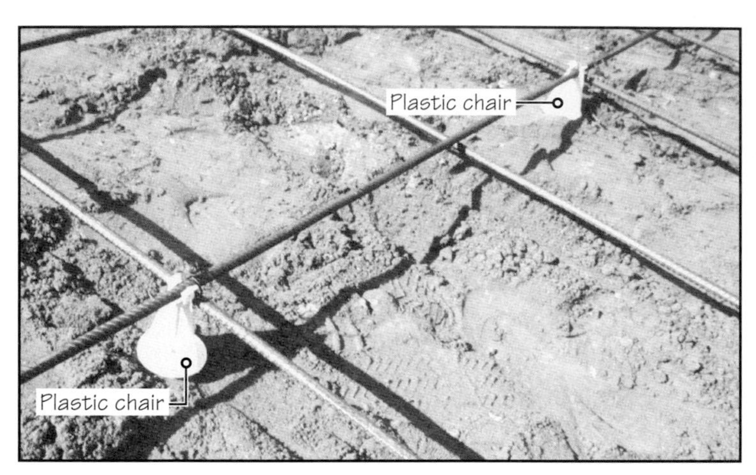

(a) Plastic chairs as bar support. Note that the bars from one direction are supported on chairs and the bars from the other direction are tied to the supported bars.

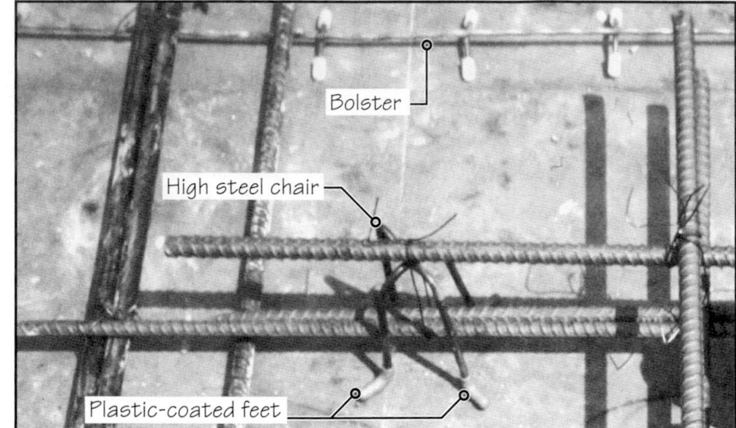

(b) Steel chairs are bolsters. Note plastic covered feet of chairs and bolsters.

FIGURE 19.25 Chairs and bolsters as bar supports. A bolster is an assembly of several steel chairs joined together with a steel bar.

The cover is measured from the exposed concrete surface to the nearest edge of reinforcement. Therefore, in a beam, the cover is measured from the outer edge of stirrups (Figure 19.17), and in a column, from the outer edge of ties (Figure 19.27(b)).

To ensure the required cover in a beam or slab, bar supports are used. These are available either in the form of individual *chairs* or bolsters. Chairs may be of molded plastic, Figure 19.25(a) or of steel wire, Figure 19.25(b). Steel chairs have plastic-coated feet to prevent corrosion. A bolster is made from steel wire and several chairs, connected together in a long length. Both chairs and bolsters are available in various heights.

PRACTICE QUIZ

Each question has only one correct answer. Select the choice that best answers the question.

14. Reshoring concrete structures is necessary where excessive deflection of floor slabs or foundation settlement is anticipated on removal of shores.
- **a.** True
- **b.** False

15. The structural design of formwork design is the responsibility of the
- **a.** general contractor.
- **b.** structural engineer.
- **c.** structural engineer, but the design must be approved by the building official.
- **d.** general contractor, but the design must be approved by the structural engineer.
- **e.** none of the above.

16. Form liners are used to
- **a.** ensure easy strippability of forms.
- **b.** allow acid etching of concrete surface after stripping the forms.
- **c.** allow acid etching of concrete form before placing concrete.
- **d.** produce textured concrete surfaces.
- **e.** allow water washing of concrete after stripping the forms.

17. The primary purpose of steel reinforcement is to increase the tensile strength of concrete elements. However, steel reinforcement is also used to increase the shear strength and compressive strength of concrete elements.
- **a.** True
- **b.** False

18. Stirrups are used in a concrete beam
- **a.** to increase the beam's compressive strength.
- **b.** to provide tensile strength in beam since concrete's tensile strength is negligible.
- **c.** to increase the tensile strength of beam above that provided by concrete.
- **d.** to provide shear strength in beam because concrete's shear strength is negligible.
- **e.** to increase the shear strength of beam above that provided by concrete.

19. Stirrups are generally made of
- **a.** No. 3 or No. 4 bars.
- **b.** No. 4, No. 5, or No. 6 bars.
- **c.** No. 6 or No. 7 bars.
- **d.** bars of any diameter, as needed.

20. A doubly reinforced beam is one in which
- **a.** reinforcing bars are provided in bundles of two bars.
- **b.** reinforcing bars are provided at the top and bottom of the beam.
- **c.** reinforcing bars are provided in two layers at the bottom of the beam.
- **d.** reinforcing bars are provided in two layers at the top of the beam.
- **e.** none of the above.

21. Mechanical couplers are used
- **a.** to lap splice reinforcing bars.
- **b.** as an economical alternative to lap splicing of reinforcing bars.
- **c.** where the lap-splice length of reinforcing bars is excessive.
- **d.** where lap splicing of reinforcing bars will produce excessive congestion.
- **e.** none of the above.

22. Hooks are used
- **a.** to lap splice reinforcing bars.
- **b.** where lap-splice length is excessive.
- **c.** where lap splicing of bars will produce excessive congestion.
- **d.** all the above.
- **e.** none of the above.

23. Hooks in bars are produced by turning the bars through
- **a.** 30° or 60°.
- **b.** 45° or 90°.
- **c.** 90° or 180°.
- **d.** 180° or 270°.

24. Steel reinforcement in concrete elements that are permanently exposed to the earth, such as in footings, must have a minimum concrete cover of
- **a.** 5 in.
- **b.** 4 in.
- **c.** 3 in.
- **d.** 2 in.
- **e.** 1 in.

25. The chairs used for supporting reinforcing steel bars are generally
- **a.** plastic chairs.
- **b.** steel-encased plastic chairs.
- **c.** steel chairs with plastic-coated feet.
- **d.** (a) and (b).
- **e.** (a) and (c).

Answers: 14-b, 15-a, 16-d, 17-a, 18-e, 19-a, 20-b, 21-d, 22-e, 23-c, 24-c, 25-e.

19.8 REINFORCEMENT AND FORMWORK FOR COLUMNS

The reinforcement in a column is primarily to resist compressive forces. However, because columns are also subjected to bending, the reinforcement must be located on the column periphery. Another reason for peripheral location is that the reinforcement cage of a column confines the concrete within the cage, even after the concrete has been crushed to failure, resulting in a more gradual ductile column failure.

A column reinforcement cage consists of *longitudinal bars* and *ties*, Figure 19.26. The total area of the longitudinal reinforcement for a column is based on structural requirements but must lie between 1% and 8% of the gross area of the column. Generally, however, it is difficult to accommodate more than 4% reinforcement area unless the bars are welded together at splices or coupled with mechanical couplers.

Ties are similar to stirrups but have a twofold structural function. One is to prevent the buckling of the longitudinal reinforcement (which consists of slender bars under compression). If the bars are not prevented from buckling, they will burst out of the concrete. Their other function is to provide shear resistance to columns subjected to bending. Ties also help in the formation of reinforcement cage.

Ties are generally No. 3 or No. 4 bars. Each longitudinal column bar must be enclosed by a tie corner (see Figure 19.27(a) for exception). Several tie shapes may be needed in the same column, Figure 19.27. In round columns, round ties may be used, but it is more common to use a continuous helical spiral, Figure 19.28. A column with a spiral reinforcement is referred to as a *spiral column*. A column with ties is called a *tied column*.

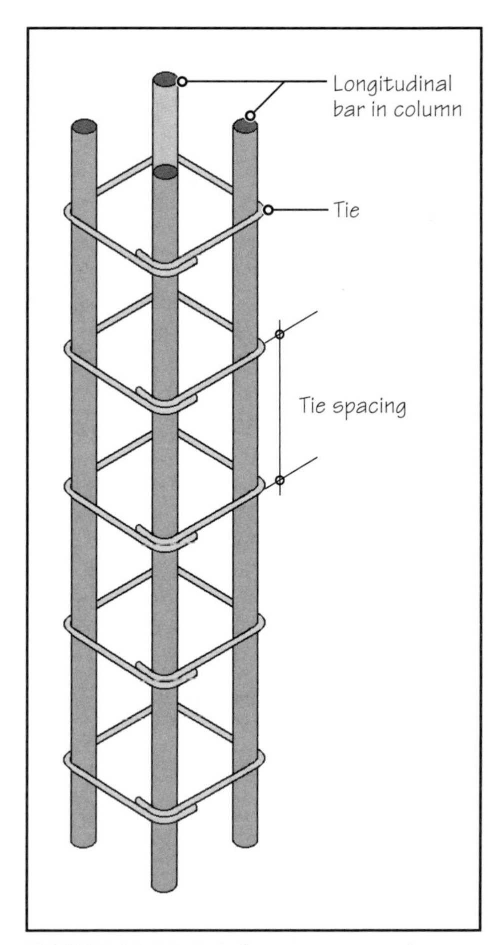

FIGURE 19.26 Reinforcement cage for a column consists of longitudinal bars and ties. The minimum number of longitudinal bars in a square or rectangular column is four (as shown here). In a column with a larger number of bars, the tie pattern is more complex, as shown in Figure 19.27.

Column Forms

Column forms for rectangular or square columns are generally made of wood or steel. A typical column form consists of four panels, braced by a steel angle frame on all sides. The angle frame is hinged at one end and clamped at the opposite end so that the form can be stripped by simply unclamping and rotating the unclamped side, Figure 19.29.

Round column forms may be of steel plate or waterproof fiberboard, Figure 19.30. Steel forms are reusable and are generally semicircular and of short lengths that are bolted together to obtain the correct height. Fiber forms are one piece and meant for one-time use.

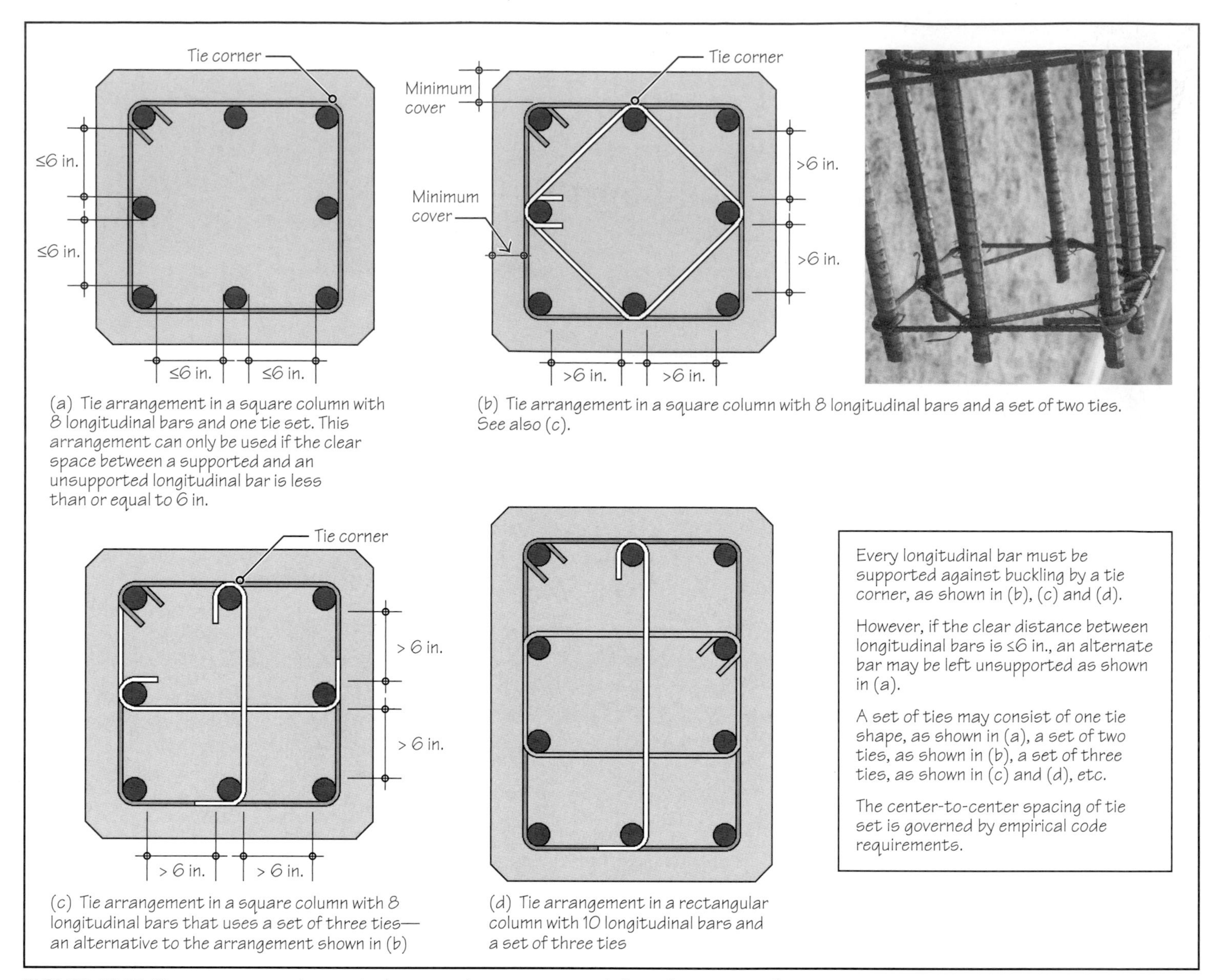

FIGURE 19.27 Tie arrangements in a typical rectangular or a square reinforced concrete column.

FIGURE 19.28 Reinforcement cage for a spiral column consisting of longitudinal bars and a continuous helical spiral.

19.9 REINFORCEMENT AND FORMWORK FOR WALLS

Concrete walls are commonly used in buildings as retaining walls, basement walls, shear walls, or load-bearing walls. The essential difference between the reinforcement in a wall and a column is that ties are not required in a wall, unless the wall supports a large gravity load.

The reinforcement in a wall consists of a mesh formed by vertical and horizontal bars, provided in one central layer. In a wall 10 in. or thicker, two layers of reinforcement are required close to each face of the wall, but separated from the face by minimum required cover, Figure 19.31.

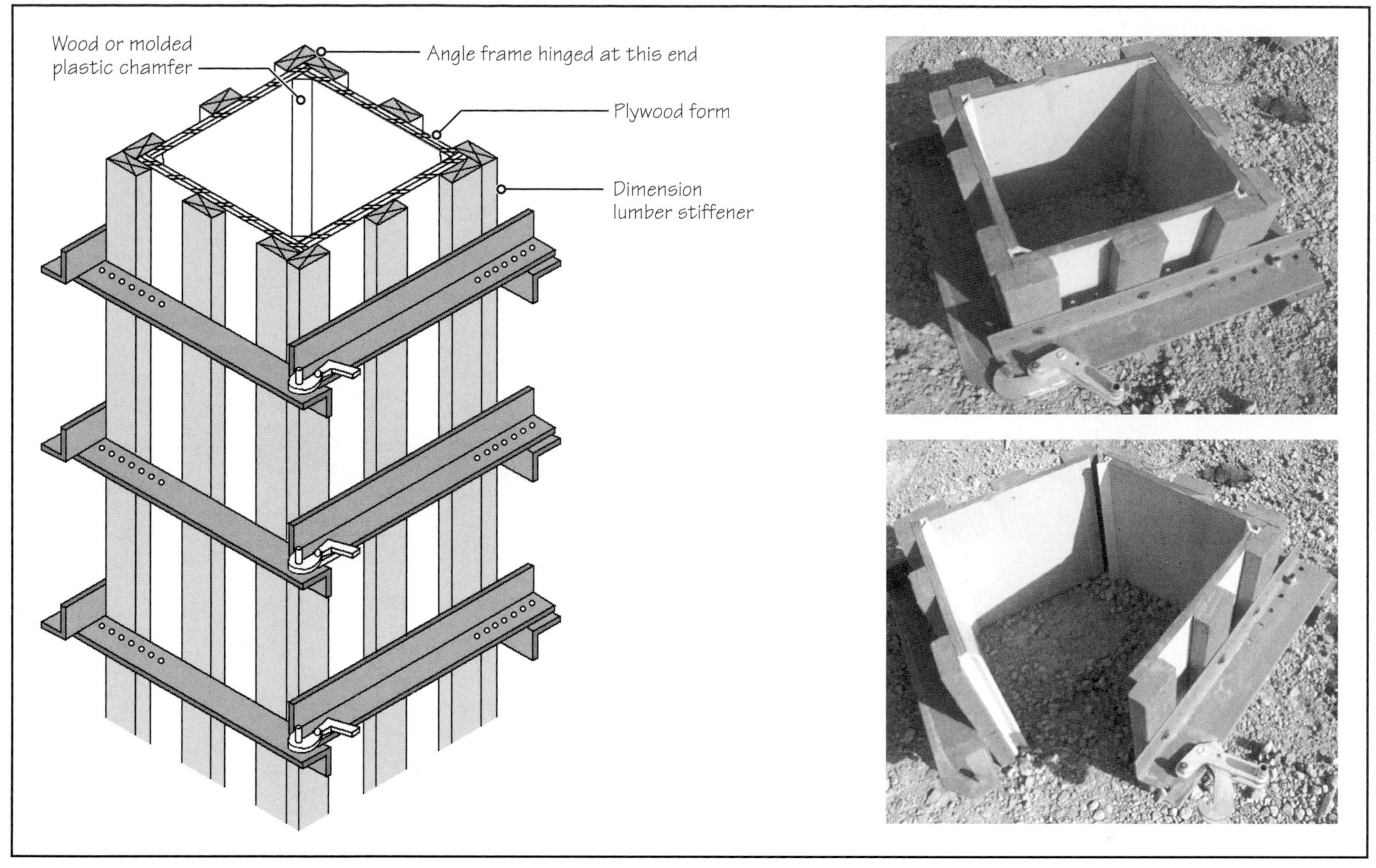

FIGURE 19.29 A typical rectangular or square column form. The photographs show a demonstration of a hinged column form in a formwork fabricator's yard.

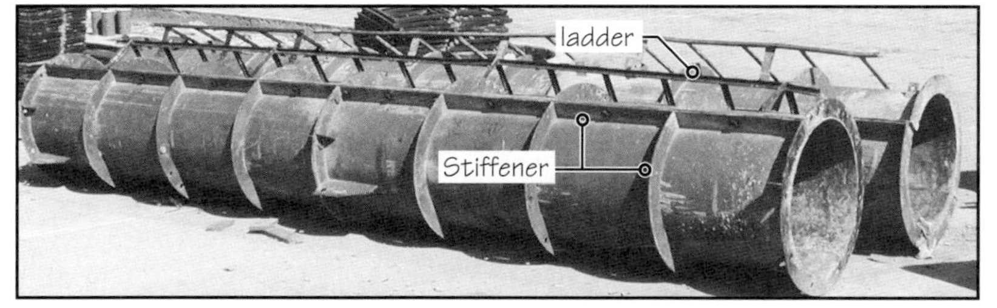

(a) Formwork for a concrete column made of 2 ft long semicircular plates with end stiffeners that are bolted together at the stiffeners.

(b) A typical fiber form.

FIGURE 19.30 Two types of forms for round columns.

Formwork for Walls

The formwork for each face of a wall is generally made of plywood panels (generally $\frac{3}{4}$ in. thick) supported by horizontal members (generally 2 × 4 lumber) called *walers,* Figure 19.32. The walers are supported by vertical lumber, called *stiffbacks*. The stiffbacks are supported by diagonal braces, Figure 19.33. The formwork for each face is separated by *form ties*—specially shaped steel wires that separate and hold the formwork from two opposite faces, and resist the pressure of wet concrete.

Form ties are available in various types. One commonly used type is a *snap tie,* which consists of a plastic cone and a loop at each end. The end-to-end distance between cones represents the wall thickness, Figure 19.34.

The wire of a snap tie under each cone has a narrow, brittle section that allows it to be snapped there by a plier-assisted twist of the end loop. The snapping is done after the formwork has been stripped. After removing the end loops and the plastic cones, the pockets left by the cones are filled with a sealant.

The entire formwork assembly (shown in Figure 19.33) is held together with patended clamps that do not require nails or fasteners. One manufacturer's clamp is shown in Figure 19.35.

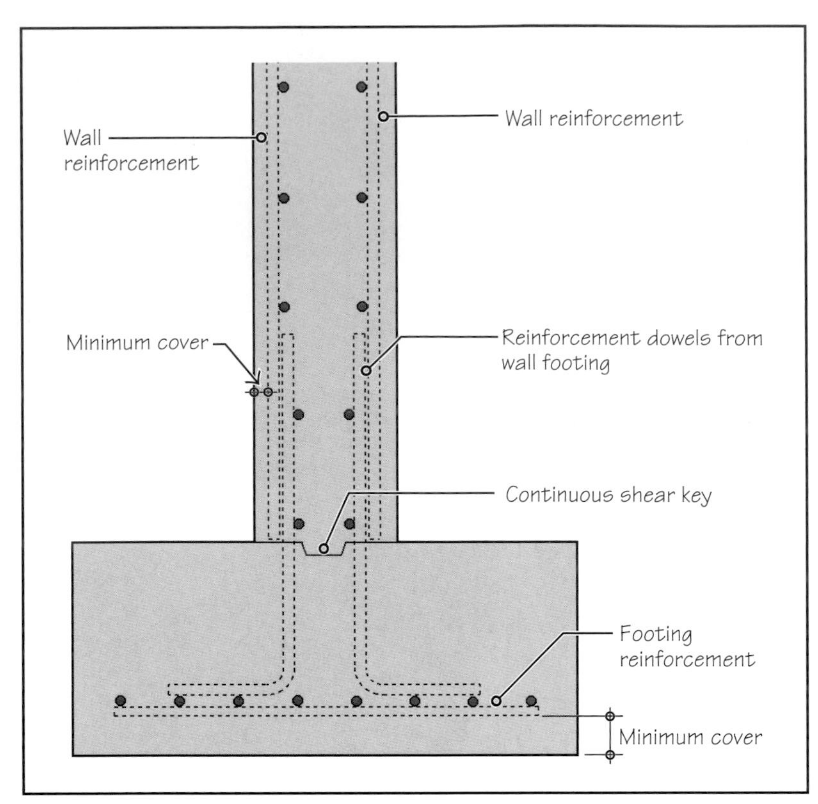

FIGURE 19.31 A section at foundation level through a reinforced concrete wall with two-layer reinforcement. Note that the footing of the wall is constructed first with reinforcing dowels projecting out of the footing. [This is similar to constructing the footing for a column. See "Other Examples of Construction Joints" (Section 19.11).]

FIGURE 19.32 Workers equipped with a safety harness erect the formwork for a reinforced concrete wall.

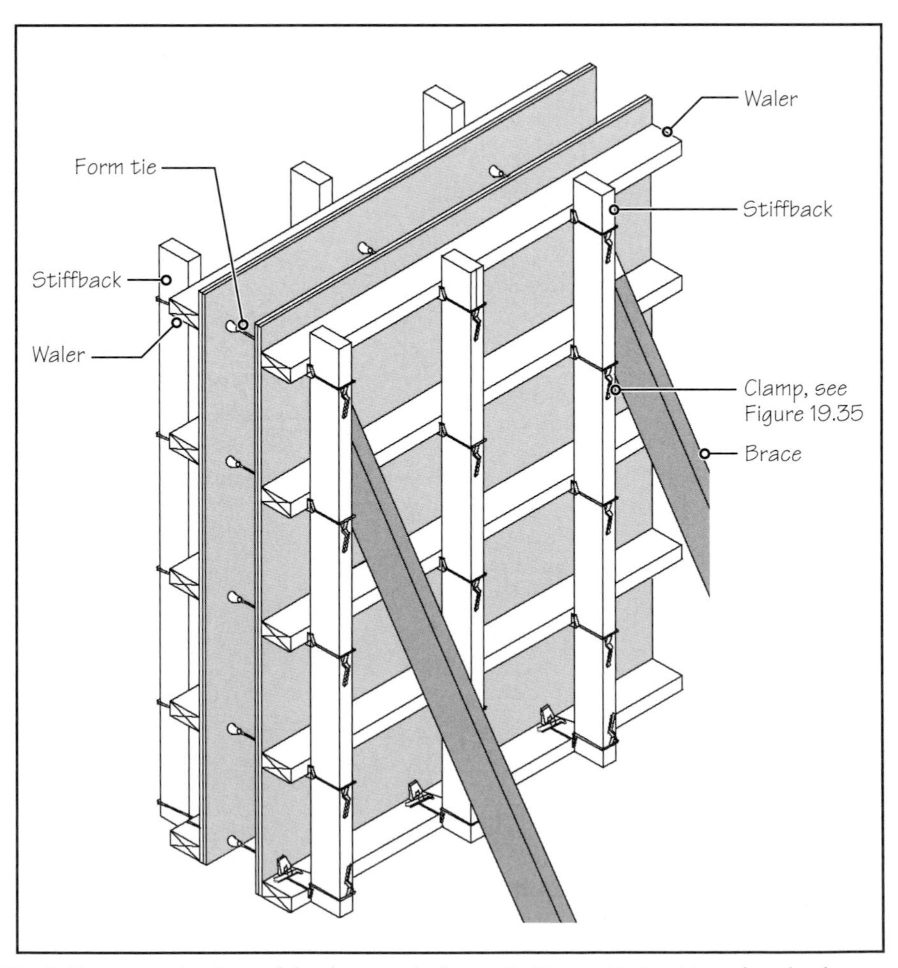

FIGURE 19.33 Isometric view of the formwork shown in Figure 19.32. Note that the formwork is made from plywood and dimension lumber. The entire assembly is fastened together with removable clamps (Figure 19.35). Form ties between opposite plywood panels separate them and hold them against the pressure of wet concrete.

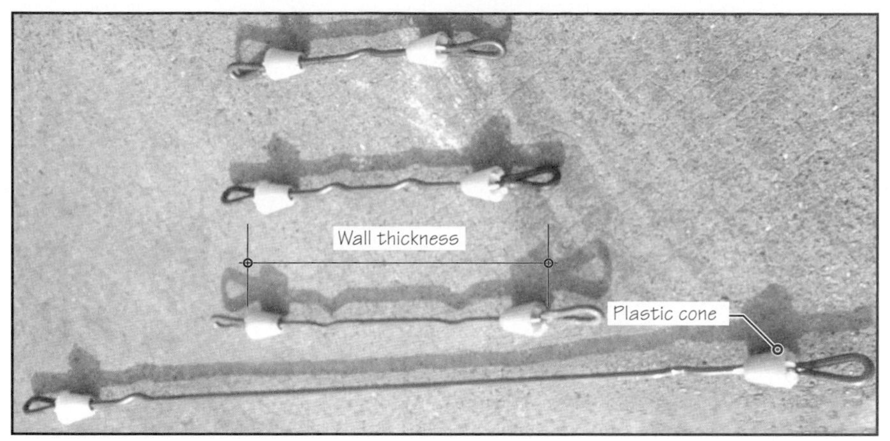

FIGURE 19.34 **Snap ties** are available in various sizes to suit various wall thicknesses. The out-to-out dimension between plastic cones represents wall thickness. Several other form tie designs are available.

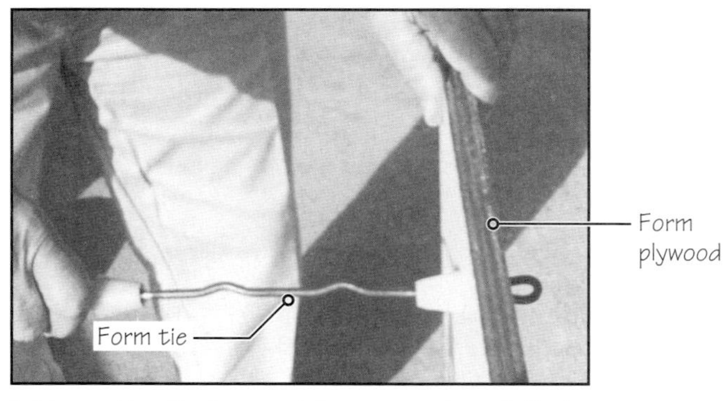

(a) In erecting the formwork for a concrete wall, first the ties are inserted from the inside of plywood form through pre-drilled holes.

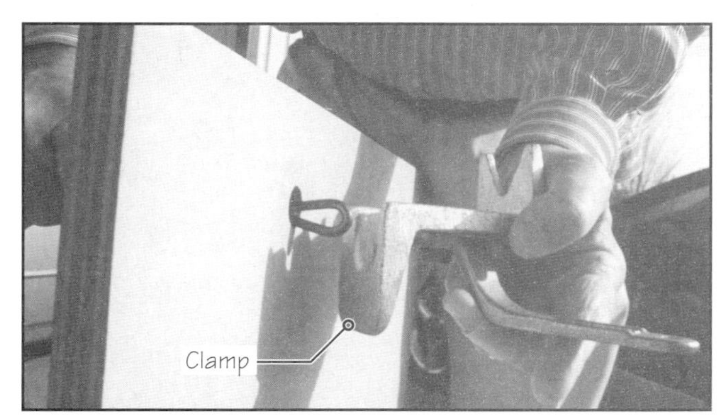

(b) Next, a patented clamp is placed over the endloop of each tie on the outside of plywood.

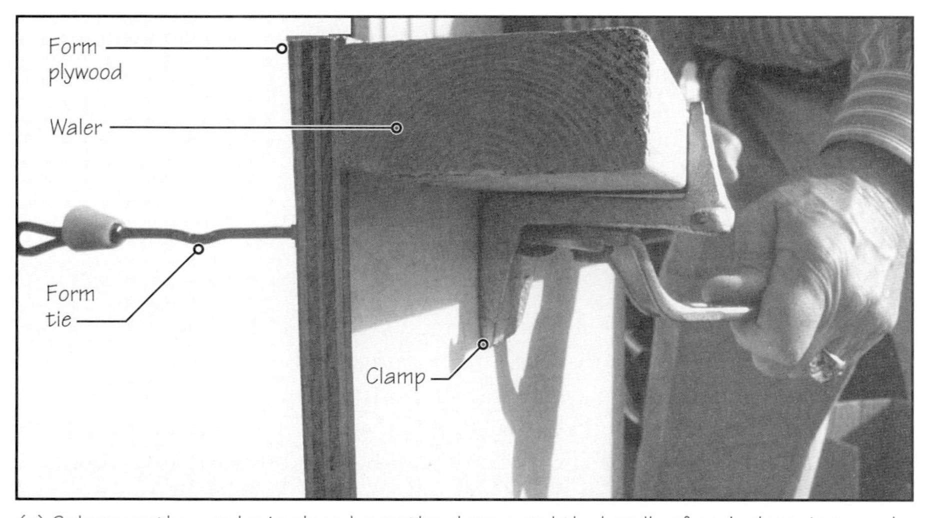

(c) Subsequently, a waler is placed over the clamps and the handle of each clamp is turned to complete the fastening of form tie with the waler.

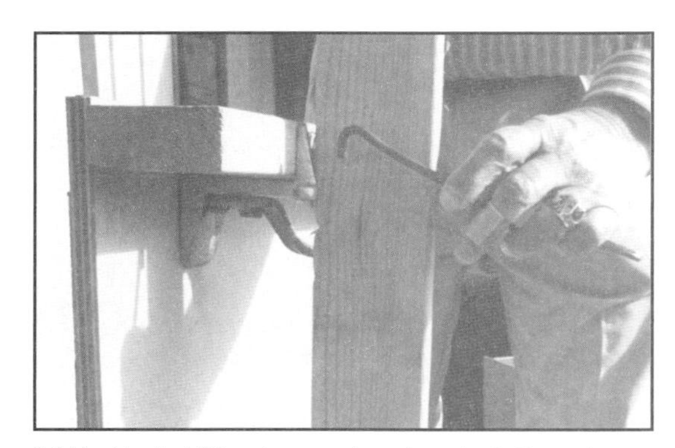

(d) Vertical stiffbacks are placed against the walers and another mating clamp is engaged into each previously installed clamp.

(e) Clamp handle is turned down to complete the connection of stiffback with the waler.

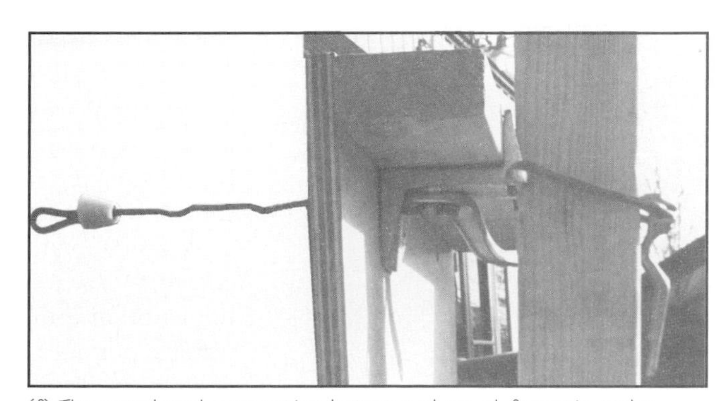

(f) The completed connection between plywood, form tie, waler, and stiffback.

FIGURE 19.35 Details of fastening the form ties, form plywood, walers, and stiffbacks with a patented clamping device.

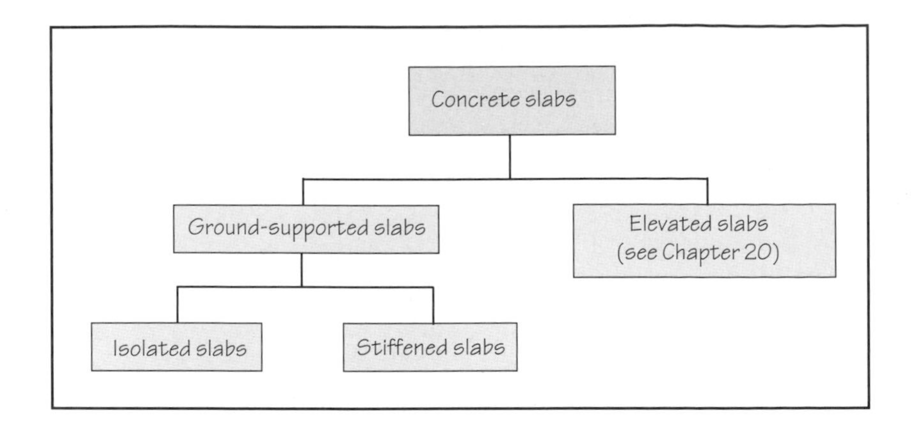

FIGURE 19.36 Types of concrete slabs.

19.10 TYPES OF CONCRETE SLABS

As shown in Figure 19.36, site-cast concrete slabs can be of two types:

- *Ground-supported slabs* (also called *slabs-on-ground* or *slabs-on-grade*) bear directly on a compacted ground (grade) with organic top soil removed. Construction of concrete slabs-on-ground is covered in this chapter (Sections 19.11 and 19.12).
- *Elevated slabs* rest on and are part of the structural frame of the building. Therefore, they are also referred to as *framed slabs*. Elevated concrete slabs are used at second and higher floors of the building. However, where the soil conditions are unfavorable (or if the building has a basement), they are also used at the first (ground) floor. Construction of elevated slabs is covered in Chapter 20.

GROUND-SUPPORTED CONCRETE SLABS

Ground-supported concrete slabs are further subdivided into

- Isolated concrete slabs
- Stiffened concrete slabs

Isolated concrete slabs are those that are separated from the building's foundation with an isolation joint. A stiffened concrete slab is generally designed to function as slab-and-foundation combination for wood light (or light-gauge steel) frame buildings and is particularly well suited for such buildings on expansive soils.

NOTE

Classification of Ground-Supported Slabs as Types I, II, and III

A classification of ground-supported concrete slabs similar to that given in Figure 19.36 is given by the Building Research Advisory Board (BRAB). In this classification, isolated slabs are called Type I or Type II, respectively, depending on whether they are unreinforced or reinforced with nominal reinforcement. A ground-supported stiffened slab is called slab Type III [1].

19.11 GROUND-SUPPORTED ISOLATED CONCRETE SLAB

An isolated ground-supported concrete slab is commonly used on stable, undisturbed soils (with all organic material removed) or on select fills compacted to provide a uniformly strong base. As long as the bearing capacity of the supporting ground is not too low, it is generally not a critical issue because concrete slabs are fairly rigid. Therefore, concentrated loads on the slab (from foot traffic, moving furniture, forklifts, and other items) are spread over a large area. Consequently, the stress on the ground due to the loads on the slab is relatively small.

VAPOR RETARDER

A vapor retarder (generally a 6-mil-thick polyethylene sheet with lapped joints) is placed immediately below an isolated slab. It prevents the passage of subsoil water and water vapor through the slab into the building's interior. In the absence of a vapor retarder, the vapor from below the slab may condense under an object covering the slab, such as a rug, furniture legs, or boxes.

A vapor retarder is particularly important if the interior of the building is heated or cooled and/or if the slab is covered with a vapor-impermeable floor finish such as vinyl tiles, linoleum, or thin-set terrazzo. A vapor retarder is not required under a slab if the passage of water vapor does not affect the building's interior, such as a garage or an open shed. It is also not required under driveways, pavements, or patios, for example.

A secondary advantage of a vapor retarder is that it works as a slip membrane between the slab and the subbase so that the slab moves more freely as it shrinks. In the absence of a

NOTE

Grade and Ground

The terms *grade* and *ground* are generally used interchangeably. Thus, *slab-on-ground* and *slab-on-grade* imply the same thing. However, *slab-on-ground* is more commonly used in the literature published by the American Concrete Institute (ACI), Portland Cement Association (PCA), and Post-Tensioning Institute (PTI).

On the other hand, terms such as: natural grade, existing grade, finished grade, compacted grade, above grade, below grade, grade beams, grading lines (contour lines), and so on, are in common use.

slip membrane, the shrinkage of the slab is restrained, which induces tensile stresses in the slab. These stresses are in addition to the tensile stresses caused by shrinkage. Because the reinforcement in an isolated slab-on-ground is provided to resist shrinkage stresses, a vapor retarder reduces reinforcement requirements.

A disadvantage of using a vapor retarder is that it aggravates differential shrinkage between the top and the bottom of the slab. Water is lost by evaporation from the top of the slab but remains relatively intact at the bottom, leading to a dishlike curling of the slab at the edges, Figure 19.37. Curling can be reduced by using concrete with a low water-cement ratio and proper curing to ensure the availability of sufficient moisture at the top of the slab.

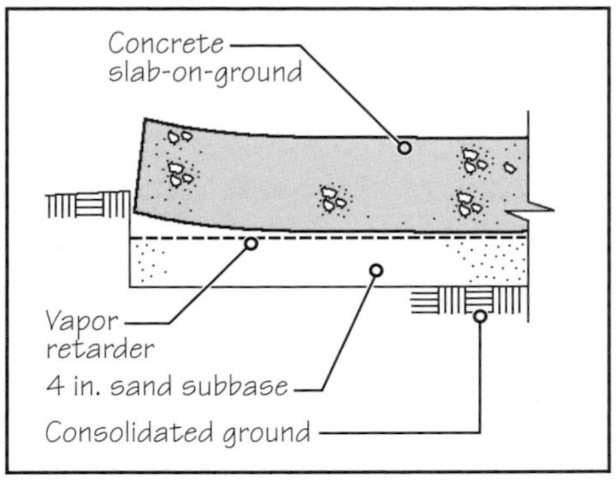

Concrete slab-on-ground

Vapor retarder

4 in. sand subbase

Consolidated ground

FIGURE 19.37 Curling of a concrete slab-on-ground due to differential drying shrinkage.

SUBBASE AS THE DRAINAGE LAYER

A subbase—a layer of granular material placed immediately above the top of compacted ground (below the vapor retarder, if used)—is generally required. Although a properly placed and consolidated concrete slab is resistant to infiltration of subsoil water, a subbase provides additional protection by providing a capillary break. The subbase also equalizes slab support by filling any uneven spots in the ground.

The required subbase thickness is generally 4 in. A thicker subbase does not yield greater benefit. Subbase material can be crushed stone, gravel, or sand. Sand is preferred in slabs provided with a vapor retarder because, unlike crushed stone or gravel, it does not puncture the vapor retarder.

CONCRETE STRENGTH AND THICKNESS

A concrete strength of 3,000 psi for residential slabs and 4,000 psi for commercial and light-industrial slabs is generally recommended. A higher strength is needed only for more heavily loaded slabs or where greater wear resistance is needed. The thickness of an isolated reinforced concrete slab is also a function of the load it must carry with a typical minimum of 4 in.

AMOUNT OF STEEL REINFORCEMENT

A concrete slab supported on stable, compacted soil may not require steel reinforcement. However, a small (i.e., nominal) amount of reinforcement is generally provided in an isolated slab, Figure 19.38. Its purpose is to reduce the size of shrinkage cracks in the slab. (See [2] for determining the amount of reinforcement.)

Note that reinforcement does not prevent the slab from cracking but reduces the width of each individual crack, keeping the overall crack width unaffected. Note also that reinforcement does not increase the load-carrying capacity of the slab, which is primarily a function of slab thickness. Because the size of individual cracks are reduced, control joints can be spaced farther apart in a reinforced slab than in an unreinforced slab. It follows, therefore, that if control joints in a slab are unacceptable, the slab must be provided with a greater amount of reinforcement.

NOTE

Fibrous Concrete Reinforcement in Slabs-on-Ground

Because the purpose of reinforcement in an isolated concrete slab-on-ground is to control shrinkage cracks, synthetic (polypropylene, nylon, and polyester) fibers are being increasingly used as replacement for steel reinforcement. They are mixed with other concrete ingredients.

Synthetic fibers are generally 0.5 to 1.5 in. long and approximately 0.04 in. (1 mm) thick. Shorter fibers are effective during the early stages of concrete hardening, and the longer fibers increase the tensile strength of hardened concrete. Fiber-reinforced concrete is generally more economical than hand-tied steel bars.

FIGURE 19.38 An isolated slab being inspected before concreting. Observe the vapor retarder and reinforcing bars, supported on chairs. Also observe the blockouts for columns, which isolate the slab from the building's structure.

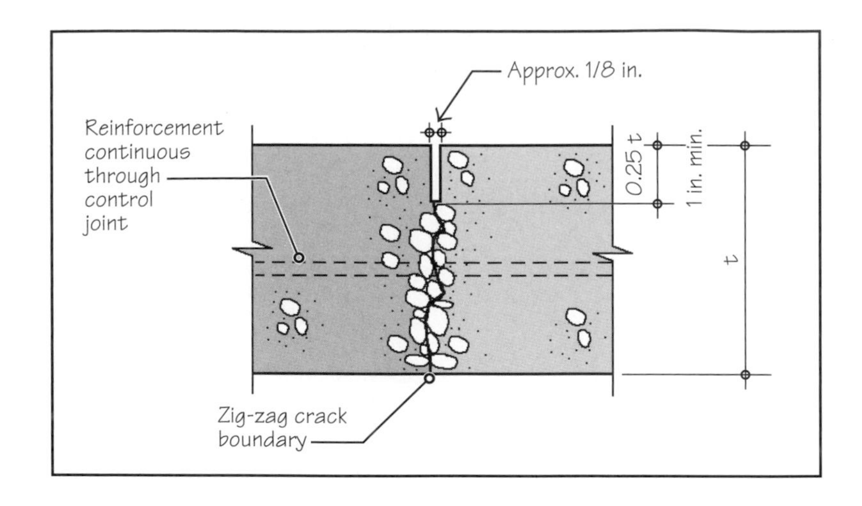

FIGURE 19.39 Section through a control joint in slab. The slab generally cracks through its entire thickness under a control joint. Shear transfer between adjacent sections of the slab at a control joint is provided by the zigzag crack along aggregate boundaries.

LOCATION OF STEEL REINFORCEMENT IN AN ISOLATED SLAB

Because the purpose of reinforcement in an isolated, ground-supported concrete slab is to control shrinkage cracks, it should be located as close to the top of the slab as possible, keeping in view the requirements of cover and surface finish.

JOINTS IN AN ISOLATED CONCRETE SLAB-ON-GROUND

An isolated concrete slab-on-ground requires the following types of joints:

- Control joints
- Isolation joints
- Construction joints

CONTROL JOINTS

Control joints accommodate the shrinkage of concrete. Their purpose is to provide weakness in the slab at predetermined locations to force the slab to crack there. In the absence of control joints, the slab will crack in a random, haphazard pattern. Control joints are generally provided by sawing the slab at intervals to a depth of 0.25 times the thickness of the slab, Figure 19.39. The width of saw-cut joint is approximately $\frac{1}{8}$ in.

Because concrete begins to shrink and crack as it hardens, the slab must be cut as soon as it is hard enough to provide a clean, unraveled joint (usually within 48 h of concrete placement). Control joints are spaced at 24 times the thickness of slab in both directions, but not more than 15 ft apart. (In exterior concrete walkways and pavements, control joints may be provided through saw-cut joints or tooled joints. A tooled joint is a shallow groove, made while the concrete is green.)

The crack that develops under a control joint follows a zigzag pattern along the boundaries of coarse aggregate particles. The zigzag crack, shown in Figure 19.39, provides shear interlock between adjacent sections of the slab, referred to as *aggregate interlock.*

ISOLATION JOINTS

Unlike control joints, *isolation joints* in a concrete slab extend the entire thickness of the slab. They are typically $\frac{1}{2}$ in. thick and are provided to ensure that the slab is isolated from the building's structural components so that their movement (creep, foundation settlement, etc.) is not transferred to the slab. The joint space is generally filled with asphalt-saturated fiberboard and covered with a sealant. Figure 19.40 shows the location of control joints and isolation joints in a typical isolated interior slab.

CONSTRUCTION JOINTS

A *construction joint* (also called a *cold joint*) is a nonmovement joint in a large concrete slab. This is used when the concrete cannot be placed in one continuous operation. The time interval between two concrete placements may be a few hours or several days.

A construction joint is detailed as if the component is monolithic across the joint. In other words, the reinforcement goes through the joint undisturbed. However, if the interval

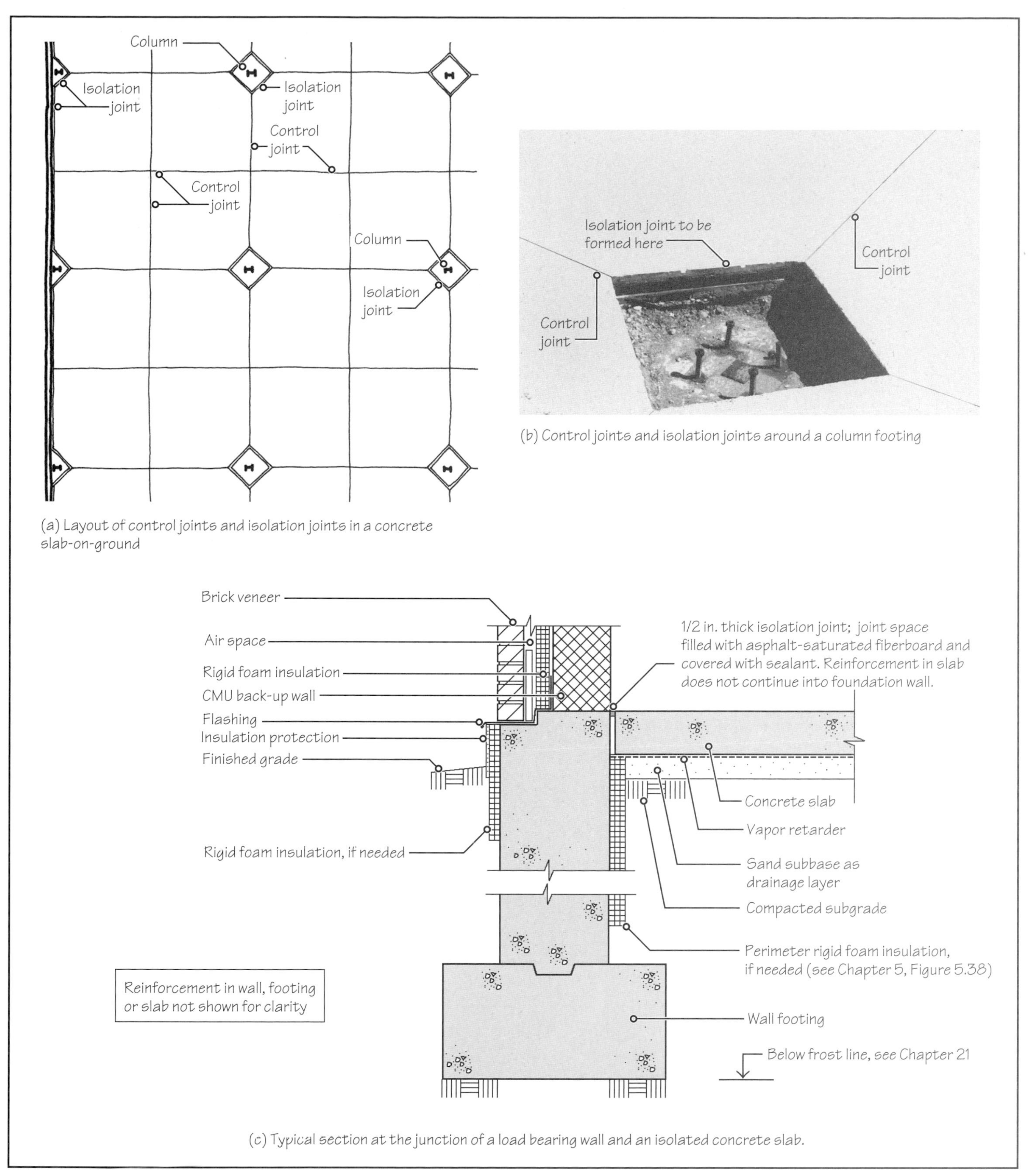

FIGURE 19.40 Isolation joints and control joints in an interior isolated concrete slab-on-ground.

between the concrete placements is long, the surface of contact is cleaned with a stiff wire brush before placing the next batch. This improves the bond between the two placements, enhancing the monolithicity between them.

A shear key is generally provided at a construction joint in slab, which distributes the load across the joint, Figure 19.41. The shear key prevents vertical differential movement of adjacent slab sections, similar to that provided by aggregate interlock across a control joint.

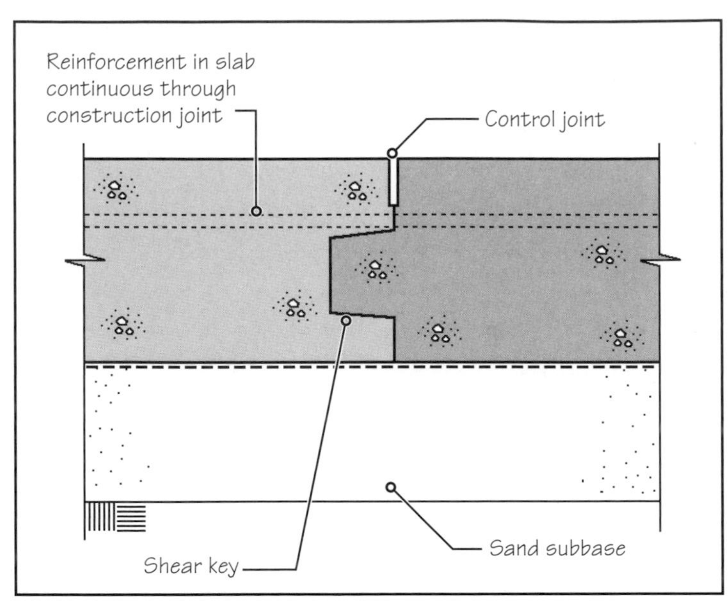

FIGURE 19.41 Detail of a keyed construction joint. A control joint is generally provided at a construction joint in a slab-on-ground.

EXPAND YOUR KNOWLEDGE

Other Examples of Construction Joints

Another example of a commonly used construction joint is the joint between a concrete column and its footing, Figure 1. Practical considerations require that the footing must be constructed first, and after the concrete in the footing has hardened, the formwork for the column is erected. Subsequently, the concrete for the column is placed.

Additional construction joints in a column are placed between the column top and the bottom of the beam or slab that rests on the column. These joints are not visible and, therefore, they do not need any aesthetic treatment.

Horizontal construction joints in a concrete wall must, however, be provided with a relief joint because (a) it is difficult to match colors between two concrete placements, and (b) the top surface of the lower concrete may not be perfectly level, resulting in an irregular construction line, or it may be slightly out of plane with the adjacent concrete. A relief joint creates a shadow line on the surface and helps to obscure the visual imperfections of the joint, Figure 2.

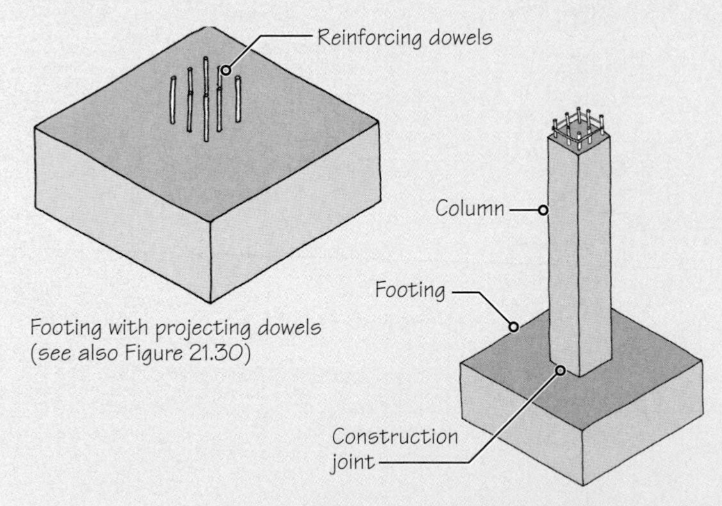

FIGURE 1 Construction joint between a column and its footing.

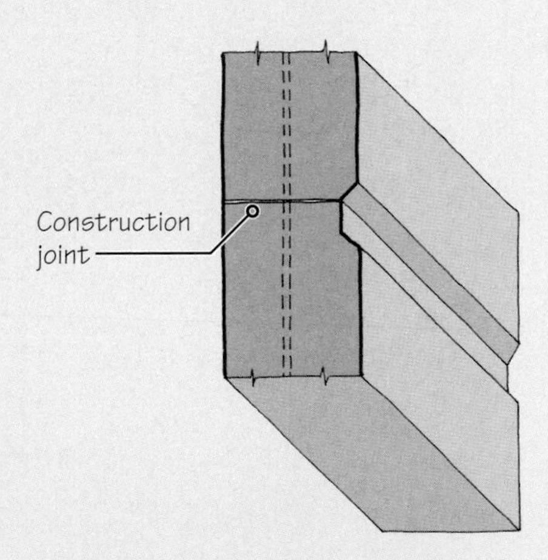

FIGURE 2 Horizontal construction joint in a concrete wall.

19.12 GROUND-SUPPORTED STIFFENED CONCRETE SLAB

A ground-supported, stiffened concrete slab is similar to an isolated concrete slab, except that the slab is stiffened with perimeter and interior beams (ribs) in both directions, giving a *ribbed concrete slab*. It is often used as a slab-and-foundation combination for wood light frame (or light-gauge steel frame) buildings. The ribs not only contribute to the stiffness of the slab, but also function as footings for load-bearing walls.

A stiffened slab is designed to withstand loads without excessive deflection, even when it loses ground support at some locations. It is, therefore, well suited as a slab-and-foundation combination for light frame buildings on soils subjected to swelling and shrinkage so that the slab moves up and down uniformly with movement of the ground. Inadequate stiffness of the slab can produce differential movement, which can produce cracking in brittle elements supported by the slab, such as masonry walls, or gypsum board.

A stiffened slab-on-ground can be designed as either as one of the following:

- Ribbed reinforced-concrete slab (i.e., with reinforcing bars)
- Ribbed posttensioned (PT) concrete slab (i.e., with prestressing steel)

A PT slab is more commonly used because of its relative economy and the reduction in the size and number of cracks. Additionally, a PT slab generally does not require control joints. The construction of a PT slab is discussed here in detail. However, the principles of constructing a ribbed, reinforced slab-on-ground are identical to those of a ribbed, PT slab-on-ground, except for the difference in the type of steel used. However, a PT slab requires greater skill and technology for construction. Additionally, if a hole is to be cut into a PT slab after its construction, it must be ensured that a prestressing tendon is neither cut nor damaged by it.

PRESTRESSING TENDONS AND THEIR LAYOUT

Figure 19.42 shows a typical PT slab during concrete placement. The prestressing tendons typically consist of an individual $\frac{1}{2}$-in.-diameter, 7-wire strand, called a *monostrand* (multistrand tendons are used in some elevated slabs). The strand is coated with corrosion-inhibiting grease and encased in an extruded plastic sleeve (sheathing), Figure 19.43.

FIGURE 19.42 A PT concrete slab during concrete placement. Observe prestressing tendons supported by plastic chairs and a vapor retarder that runs continuously under the slab. Observe the provision of ribs in both directions. Note that the tendons placed at the bottom of ribs are transitioned up at the ends.

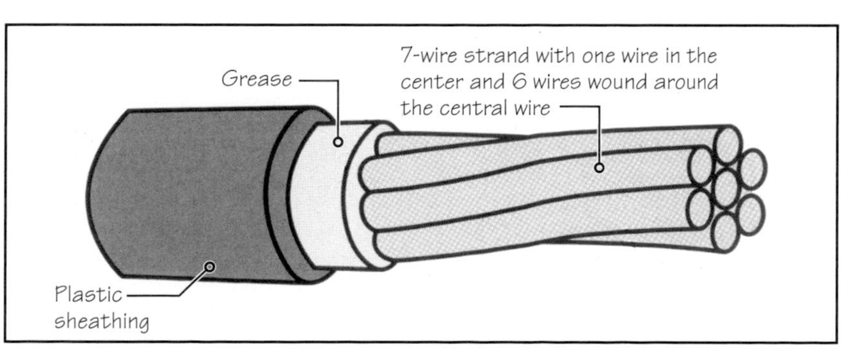

FIGURE 19.43 Anatomy of a monostrand. (Photo courtesy of Jack W. Graves, Jr.)

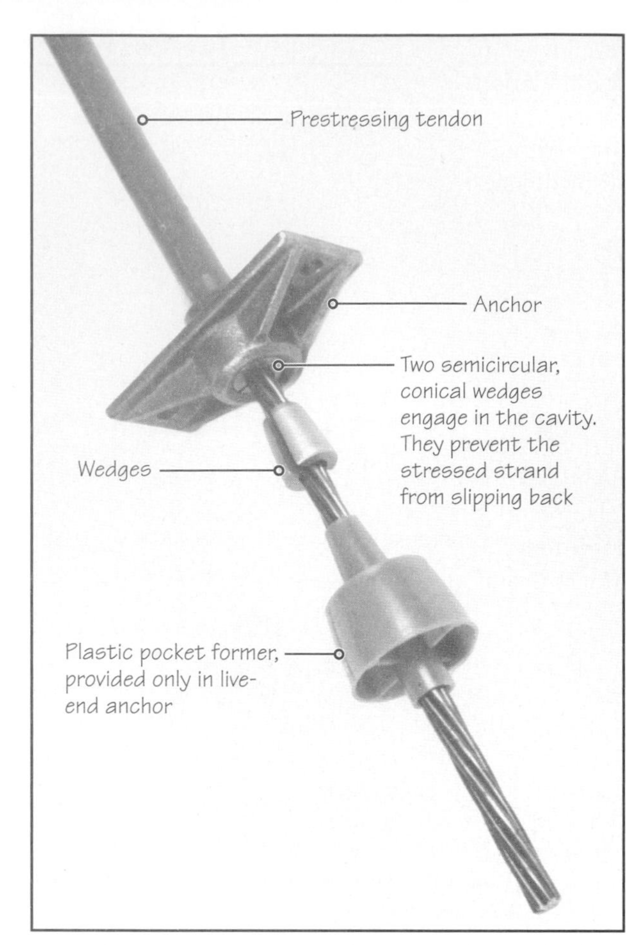

FIGURE 19.44 Anatomy of a live-end anchor. A dead-end anchor is similar (Figure 19.49). (Photo courtesy of Jack W. Graves, Jr.)

Each individual tendon has two anchor assemblies, one at each end. One end has a fixed anchor assembly, called a *dead end.* The opposite end has the stressing-end assembly, called a *live end.* The tendon is stressed from the live end. Dead-end and live-end anchors are almost identical. The anchor contains a cavity to accommodate two semicircular conical steel wedges that prevent the stressed tendon from slipping back, Figure 19.44.

The only difference between a live end and a dead end is that the live-end anchor is provided with a plastic pocket former, which gets embedded into concrete. The pocket former is removed before stressing the tendons and grouting the pocket, as described later.

A PT ribbed slab-on-ground is typically 4 to 5 in. thick. The ribs are typically 10 to 12 in wide and 24 to 36 in. deep, spaced a maximum of 10 to 15 ft on centers. Figure 19.45 shows the plan and section of one such slab and Figure 19.46 shows the details of dead end and live end anchors.

Stressing of Tendons

A concrete strength of 3,000 psi is generally used for a PT slab. The concrete is placed, vibrated, finished, and cured in the same way as for any other slab. After the concrete has hardened (generally 24 to 48 h after its placement), the edge forms from the slab are removed. This allows an easy removal of pocket formers while the concrete is still green.

The tendons are stressed after the concrete has gained sufficient strength (approximately 2,000 psi). However, the stressing should not be delayed excessively to forestall the development of shrinkage cracks in the slab. Generally, an interval of 3 to 10 days is used between concrete placement and stressing, depending on factors such as ambient temperature, water-cement ratio of concrete, and admixtures used in concrete.

The prestressing equipment consists of a jack connected to a hydraulic pump, Figure 19.47. The effective compressive prestress introduced in the slab is 50 to 100 psi, which is determined by the gauge readings on the pump and

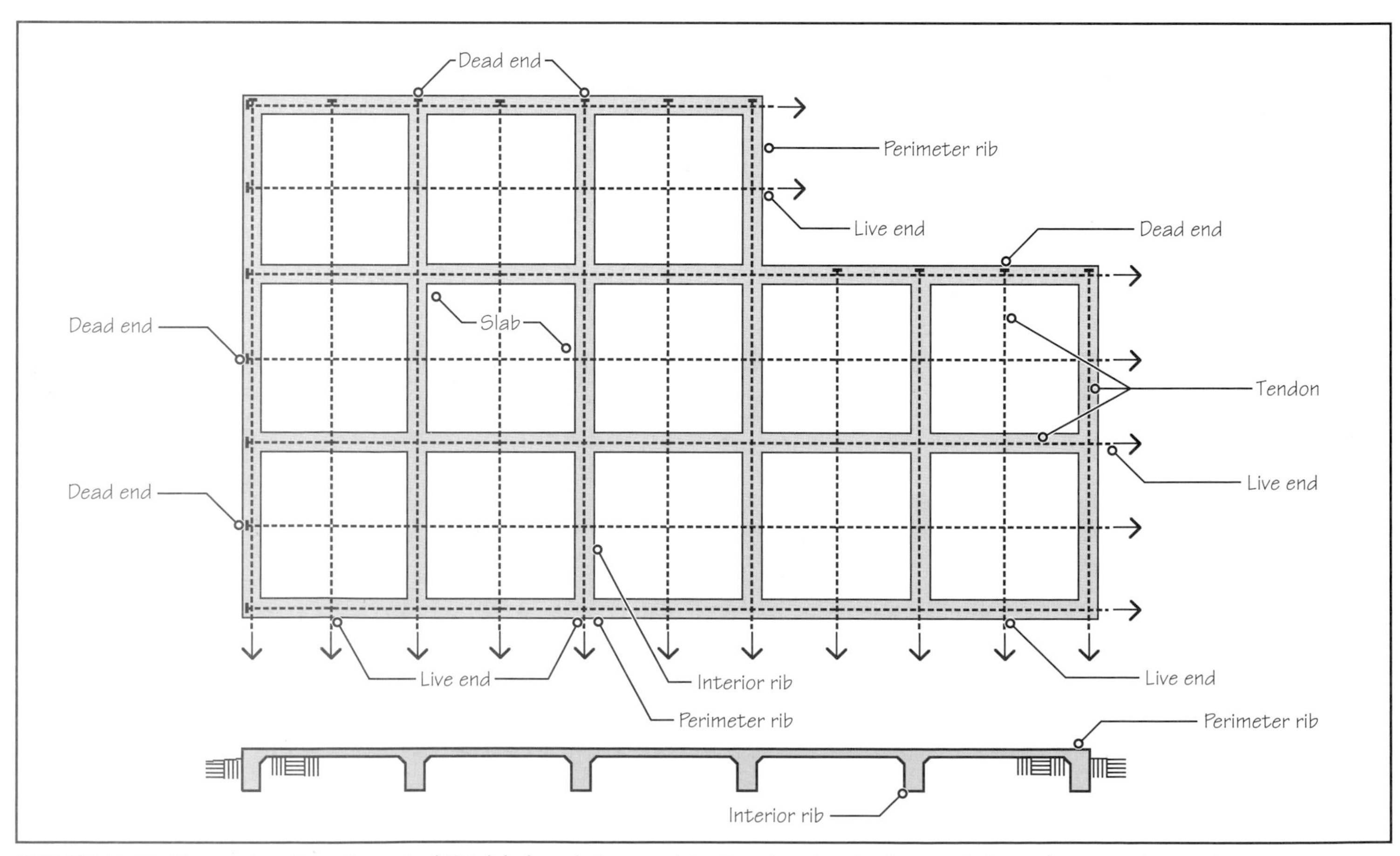

FIGURE 19.45 Plan and section of a typical PT slab-foundation combination showing the layout of ribs and prestressing tendons.

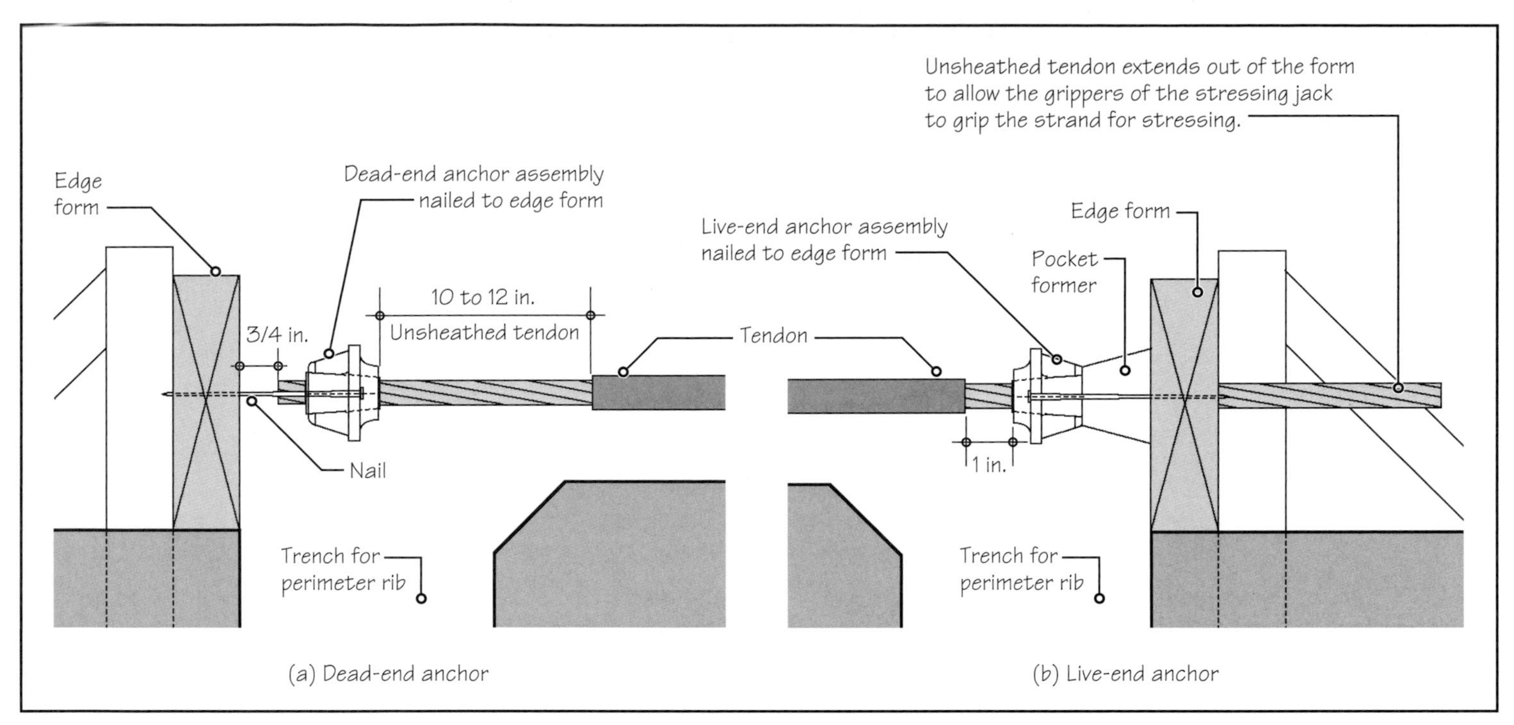

FIGURE 19.46 Details of a dead end and a live end.

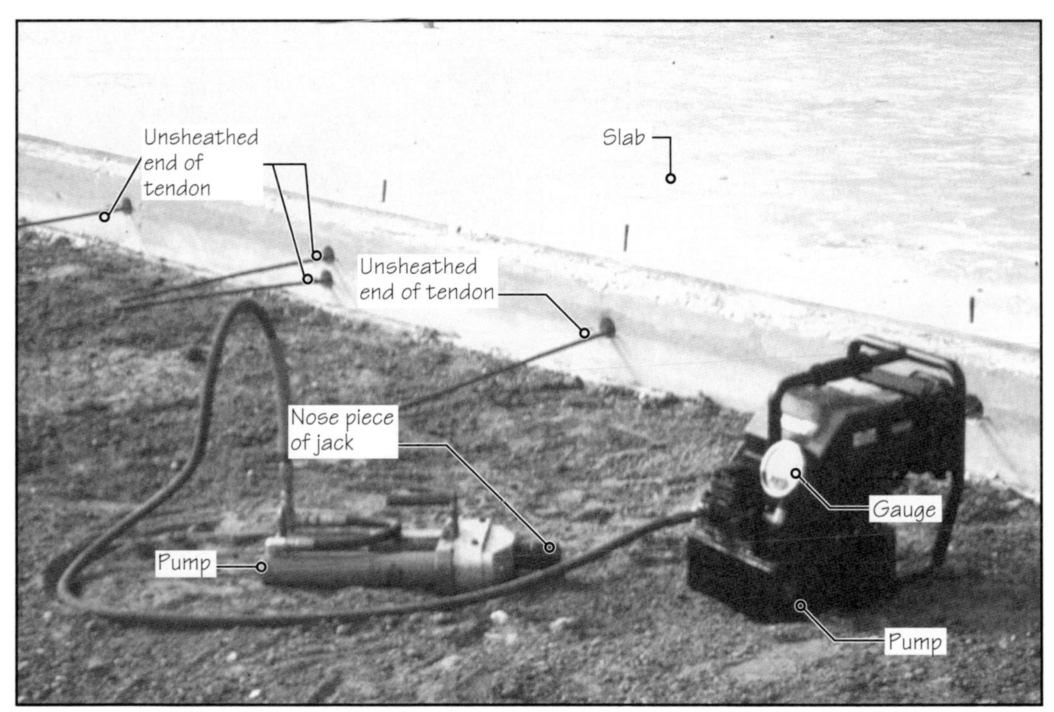

FIGURE 19.47 A PT concrete slab with tendons projecting at the live ends ready for stressing. Seen in the foreground is prestressing equipment, consisting of a jack that is connected to a hydraulic pump. (Photo courtesy of Jack W. Graves, Jr.)

verified by measuring the elongation in tendons after stressing, Figure 19.48(a). Therefore, before stressing begins, each tendon is reference marked with paint at the live end at a constant distance from the slab edge, Figure 19.48(b).

After the tendon has been tensioned by the required force, the excess length of tendon is either saw-cut or flame-cut, and the pocket left by the pocket former is patched with non-shrink grout, Figure 19.49.

Posttensioned Slab-on-Ground to Eliminate Control Joints

Although posttensioned, ground-supported slabs are generally used as a slab-foundation combination, they are also used in situations where they do not function as foundations, such as, where control joints are unacceptable and the cracking of the slab is to be minimized. In such a slab, posttensioning of the slab is done in two stages. The first stage of posttensioning is done within 36 to 48 h of concrete placement to prevent any initial cracking of the slab, followed by the second stage a few days later.

NOTE

Ribbed Slab Versus Constant-Thickness Slab

A constant-thickness PT slab-on-ground is often used in place of a ribbed PT slab-on-ground. Referred to as *California slab,* it is 7.5 in. or thicker (without any beams). Another version of a constant thickness slab is a 5-in.-thick slab with perimeter beams and no intermediate beams.

Source: Engineering News-Record (www.enr.com), September 20, 2004, p. 8.

(a) A tendon being stressed by the jack. The amount of prestressing force in the tendon is indicated by the gauge reading and verified by a predetermined elongation of the tendon.

The distance between the edge of slab and the paint mark on tendon is maintained as a constant in all strands.

(b) Stressed and unstressed tendons

FIGURE 19.48 Procedure used for verifying the elongation in tendons. (Photo courtesy of Jack W. Graves, Jr.)

(a) After stressing, the tendon is cut with a saw (or a torch).

(b) After cutting the tendon, the pocket is filled with a nonshrink grout to prevent corrosion of tendon.

FIGURE 19.49 Cutting of tendon and patching the pocket with grout. (Photo courtesy of Jack W. Graves, Jr.)

PRACTICE QUIZ

Each question has only one correct answer. Select the choice that best answers the question.

26. The most important reason for providing ties in a concrete column is to
- **a.** prevent the buckling of longitudinal column bars.
- **b.** increase the compressive strength of a column.
- **c.** increase the shear strength of a column.
- **d.** allow the formwork to be tied to the longitudinal bars in the column.
- **e.** Ties are generally not provided in a concrete column.

27. The most important reason for providing ties in a concrete wall is to
- **a.** prevent the buckling of longitudinal bars in wall.
- **b.** increase the compressive strength of wall.
- **c.** increase the shear strength of wall.
- **d.** allow the formwork to be tied to the longitudinal bars in wall.
- **e.** Ties are generally not provided in a concrete wall.

28. Formwork for a concrete wall generally consists of
- **a.** plywood, form ties, and walers fastened together with duplex nails.
- **b.** plywood, form ties, walers, and stiffbacks fastened together with duplex nails.
- **c.** plywood, form ties, walers, and stiffbacks fastened together with clamps.
- **d.** plywood and stiffbacks fastened together with clamps.

29. A vapor retarder is required under
- **a.** all concrete slabs-on-ground.
- **b.** all exterior concrete slabs-on-ground.
- **c.** all interior concrete slabs-on-ground.
- **d.** under some interior concrete slabs-on-ground.

30. The strength of concrete in a residential concrete slab-on-ground is generally
- **a.** 1,000 psi.
- **b.** 2,000 psi.
- **c.** 3,000 psi.
- **d.** 4,000 psi.
- **e.** 5,000 psi.

31. An isolated concrete slab-on-ground is generally
- **a.** used on expansive soils.
- **b.** provided with perimeter beams.
- **c.** provided with perimeter and intermediate beams.
- **d.** unreinforced.
- **e.** none of the above.

32. A construction joint in a concrete slab-on-ground is required if the concrete in the slab cannot be placed in one continuous operation.
- **a.** True
- **b.** False

33. The tendons used for prestressing a PT concrete slab-on-ground generally consist of
- **a.** a single, $\frac{1}{2}$-in.-diameter, 9-wire strand.
- **b.** a single, $\frac{1}{2}$-in.-diameter, 7-wire strand.
- **c.** two-strand tendons, each strand made of 7 wires.

d. three-strand tendons, each strand made of 7 wires.
e. none of the above.

34. In a PT concrete slab-on-ground, the tendons are stressed from the
a. dead ends.
b. live ends.
c. upper ends.
d. lower ends.
e. any one of the above.

35. Pocket formers in a PT concrete slab-on-ground are used at
a. dead ends.
b. live ends.
c. upper ends.
d. lower ends.
e. any one of the above.

36. A PT concrete slab-on-ground must have perimeter and intermediate ribs.
a. True
b. False

Answers: 26-a, 27-e, 28-c, 29-d, 30-c, 31-e, 32-a, 33-b, 34-b, 35-b, 36-b.

KEY TERMS AND CONCEPTS

Abrasive blasting
Acid etching
Architectural concrete
Bolster
Cast in place concrete
Chair
Cold joint
Column forms
Compression reinforcement
Construction joint
Control joints
Elevated slabs
Exposed aggregate
Flying forms
Form liners
Form ties
Formwork and shores
Gang forms
Glass fiber-reinforcement
Ground supported stiffened slabs
Isolation joint
Monostrand
Plain concrete
Post-tension
Pre-cast, prestressed concrete
Prestressed concrete
Pre-tension
Principles of reinforcing concrete
Reinforced concrete
Reinforcement and formwork for columns
Reinforcement and formwork for walls
Reinforcement cage
Reinforcement in slabs
Reshoring
Shear reinforcement
Shores
Slab-on-ground (slab-on-grade)
Smooth, off-the-form concrete
Snap ties
Splices, couplers and hooks in bars
Stiffbacks
Stirrup
Strands
Stripping time
Tendon
Tension reinforcement
Waler

REVIEW QUESTIONS

1. Sketch commonly used open-loop and closed-loop stirrups.
2. Using sketches and notes, explain the typical layout of reinforcement in a simply supported, reinforced concrete beam.
3. Sketch a typical
   a. control joint in a slab-on-ground.
   b. construction joint in a slab-on-ground.
   c. isolation joint in a slab-on-ground.
4. Sketch in section a typical ribbed concrete slab-on-ground.

SELECTED WEB SITES

American Concrete Institute, ACI (www.aci-int.org)
Ceco Concrete (www.cecoconcrete.com)
Concrete Construction Online (www.concreteconstruction.net)
Portland Cement Association, PCA (www.cement.org)
Precast/Prestressed Concrete Institute, PCI (www.pci.org)
World of Concrete (www.worldofconcrete.com)

FURTHER READING

1. Dobrowolski, J. A. *Concrete Construction Handbook.* New York: McGraw-Hill, 1998, Portland Cement Association (PCA), 2000.
2. Hurd, M. K. *Formwork for Concrete.* American Concrete Institute (ACI), 1995.
3. Post-Tensioning Institute (PTI): *Construction and Maintenance Procedures Manual For Post-Tensioned Slabs-On-Ground,* 1998.

REFERENCES

1. Building Research Advisory Board (BRAB), National Academy of Sciences/National Research Council. *Criteria for Selection and Design of Residential Slabs-on-Ground.* Washington DC, 1968.
2. American Concrete Institute (ACI). *Design and Construction of Concrete Slabs On Grade,* 1986, Seminar Course Manual/SCM-11(86).

CHAPTER 20

Concrete Construction–II (Site-Cast and Precast Concrete Framing Systems)

CHAPTER OUTLINE

20.1 Types of Elevated Concrete Floor Systems

20.2 Beam-Supported Concrete Floors

20.3 Beamless Concrete Floors

20.4 Posttensioned Elevated Concrete Floors

20.5 Introduction to Precast Concrete

20.6 Mixed Precast Concrete Construction

20.7 Fire Resistance of Concrete Members

The floors and roof in most contemporary site-cast reinforced-concrete buildings are posttensioned for structural efficiency. Note the posttensioning tendons projecting from floor slabs, whose ends will be cut and grouted to prevent corrosion. (Photo by MM.)

This chapter is a continuation of Chapter 19. It begins with a discussion of commonly used types of elevated site-cast concrete slabs. (Slabs-on-ground were covered in Chapter 19.)

This is followed by precast, prestressed concrete elements such as hollow-core slabs, solid planks, double tees, and inverted tees. Construction systems related to these elements are also discussed.

20.1 TYPES OF ELEVATED CONCRETE FLOOR SYSTEMS

Site-cast reinforced-concrete framing systems consist of horizontal elements (elevated floor/roof slabs and beams) and vertical elements (columns and walls). Approximately 80% to 95% of the cost of materials and formwork of a concrete structural frame is in the horizontal framing elements of the frame. Consequently, the choice of the elevated floor system constitutes the most important item in a concrete structure.

Elevated concrete floor systems can be classified as: (a) beam-supported floors and (b) beamless floors. They are further divided into several types, Figure 20.1.

20.2 BEAM-SUPPORTED CONCRETE FLOORS

A reinforced-concrete floor slab with beams on all four sides can either be a *one-way slab* or a *two-way slab*. They are also called *one-way solid slabs* or *two-way solid slabs* to distinguish them from one-way joist slabs or two-way joist slabs, which are not completely solid.

The transfer of loads from a solid slab to the four supporting beams is represented approximately by 45° lines originating from the slab corners, Figure 20.2(a). If the slab panel is a square, each supporting beam receives the same amount of load, Figure 20.2(b). If the slab panel is rectangular, one pair of beams carries a greater load than the other pair.

ONE-WAY SOLID SLAB

If the ratio of the long dimension to the short dimension of a slab panel is greater than or equal to 2.0, most of the load on the slab is transferred to the long pair of beams, that is, the load path is along the short dimension of the slab panel, Figure 20.2(c). The load path along the long dimension of the slab is negligible.

Because the load is effectively transferred along one direction in Figure 20.2(c), the slab behaves as a one-way slab. The reinforcement in a one-way slab is placed along the short direction, referred to as the *primary reinforcement* to distinguish it from the nominal reinforcement placed in the perpendicular direction, called the *secondary reinforcement*. The purpose of secondary reinforcement is to resist stresses caused by concrete shrinkage and thermal expansion and contraction of the slab.

TWO-WAY SOLID SLAB

If the ratio of the long to short dimension of a slab panel is less than 2.0, the slab is considered to behave as a *two-way slab*. However, real two-way slab behavior occurs when the ratio of the two dimensions is as close to 1.0 as possible (between 1.0 and 1.25). In a two-way slab, both directions participate in carrying the load. Reinforcement is, therefore, provided in both directions as primary reinforcement. Although not common, both one-way and two-way slabs may occur in the same floor, Figure 20.3.

NOTE

Preliminary Thickness of One-Way and Two-Way Slabs

Estimate the thickness of a one-way slab by dividing slab span by 24. Thus, if the slab span is 12 ft, use a slab approximately 6 in. thick.

Two-way slab thickness is span/36. Use longer of the two spans. Thus, if a slab panel measures 16 ft × 21 ft, the slab thickness is approximately 21/36 = 7/12 ft, or a 7-in.-thick slab.

Also see Appendix B, and consult a structural engineer for exact sizes.

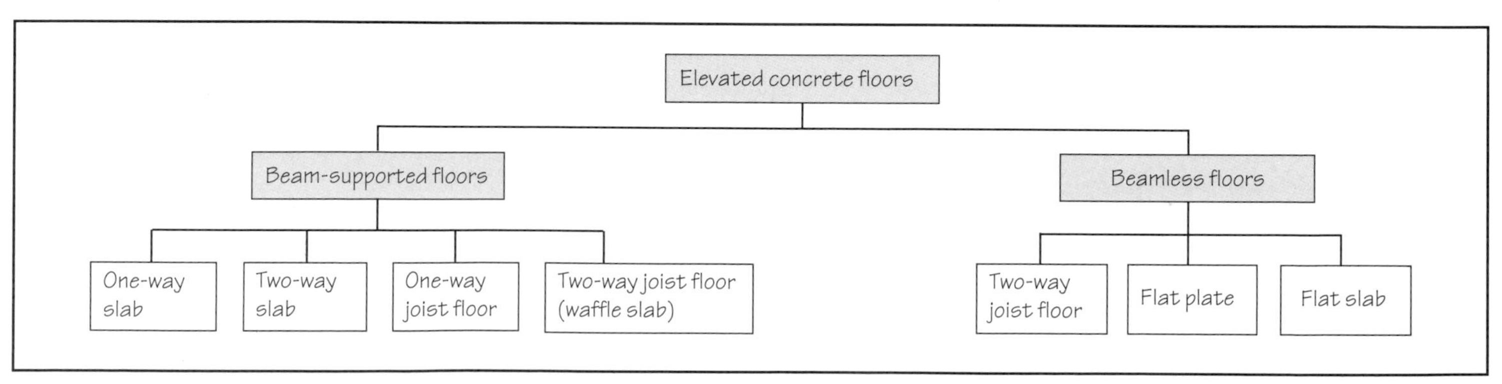

FIGURE 20.1 Types of elevated concrete floors.

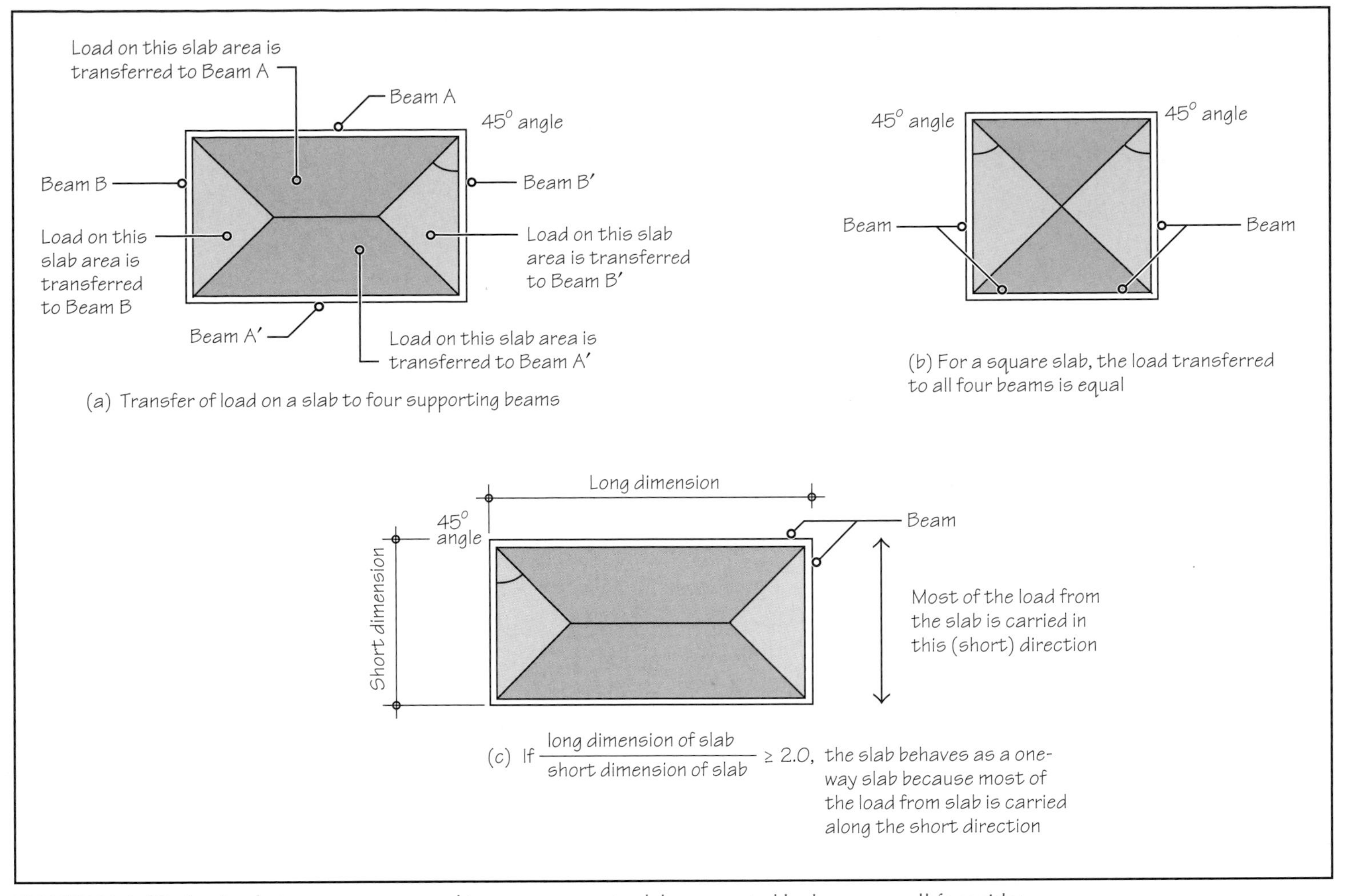

FIGURE 20.2 Distinction between one-way and two-way concrete slabs supported by beams on all four sides.

Beam and Girder Floors

One-way and two-way solid slabs become increasingly thick and hence uneconomical as their span increases. Generally, a slab thicker than 8 in. is discouraged because it creates a large dead load on the floor. For a one-way slab, an 8-in. slab thickness is reached with a span of approximately 16 ft. For a square two-way slab, a span of approximately 24 ft requires an 8-in.-thick slab.

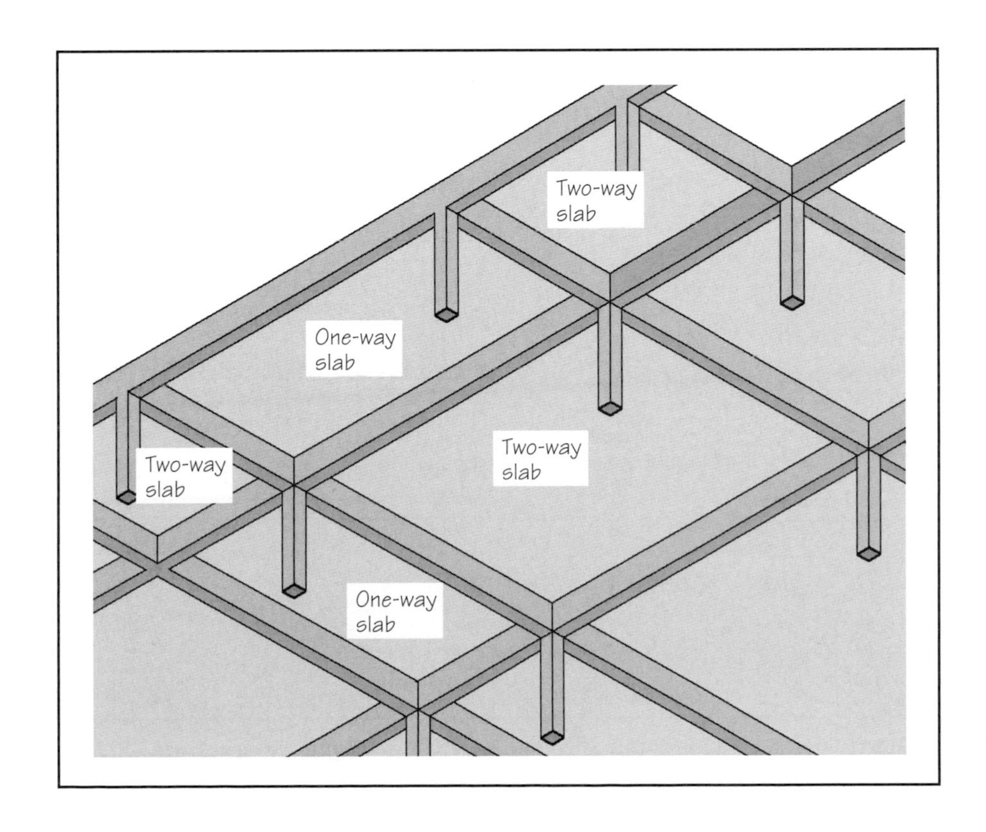

FIGURE 20.3 Although not common, one-way and two-way slabs can occur in the same floor.

Because 16-ft and 24-ft dimensions are relatively small for column spacing, one-way and two-way slabs are generally used in a *beam-and-girder floor,* Figure 20.4(a), or in a *two-way beam-and-girder floor,* Figure 20.4 (b).

BAND BEAM FLOOR

A concrete floor that cannot be constructed with a flat form deck becomes uneconomical. Therefore, the floor systems shown in Figure 20.4 are relatively uncommon because of the complexity of formwork resulting from deep beams around slab panels.

A one-way slab floor with wide and shallow, continuous beams, referred to as *band beams* (in contrast with the conventional *narrow beams*), gives a more economical formwork, Figure 20.5. Because the beams are wide, the slab span is reduced, reducing slab thickness. Additionally, because the beams are shallow, the floor-to-floor height is smaller, reducing the height of columns, interior partitions, and exterior cladding. A smaller number of stories also reduces the overall height of the building, which reduces the magnitude of lateral loads on the building.

ONE-WAY JOIST FLOOR

A concrete floor that results from an extremely economical formwork consists of closely spaced, narrow ribs in one direction supported on beams in the other direction, Figure 20.6. Because the ribs are narrow and closely spaced, the floor resembles a wood joist floor. It is, therefore, called a *joist floor,* or a *ribbed floor,* but it is more commonly known as *one-way joist floor* to distinguish it from the *two-way joist floor* described later.

A one-way joist floor is constructed with U-shaped pans as formwork placed over a flat-form deck. The gap between pans represents the width of joists, which can be adjusted by placing the pans closer or farther apart, Figure 20.7. The pans are generally made of glass fiber–reinforced plastic (GFRP) and can be used repeatedly.

The vertical section through a pan tapers downward for easy stripping and has supporting lips at both ends. The pan widths and heights have been standardized to give two categories of one-way joist floors:

- Standard-module one-way joist floor
- Wide-module one-way joist floor

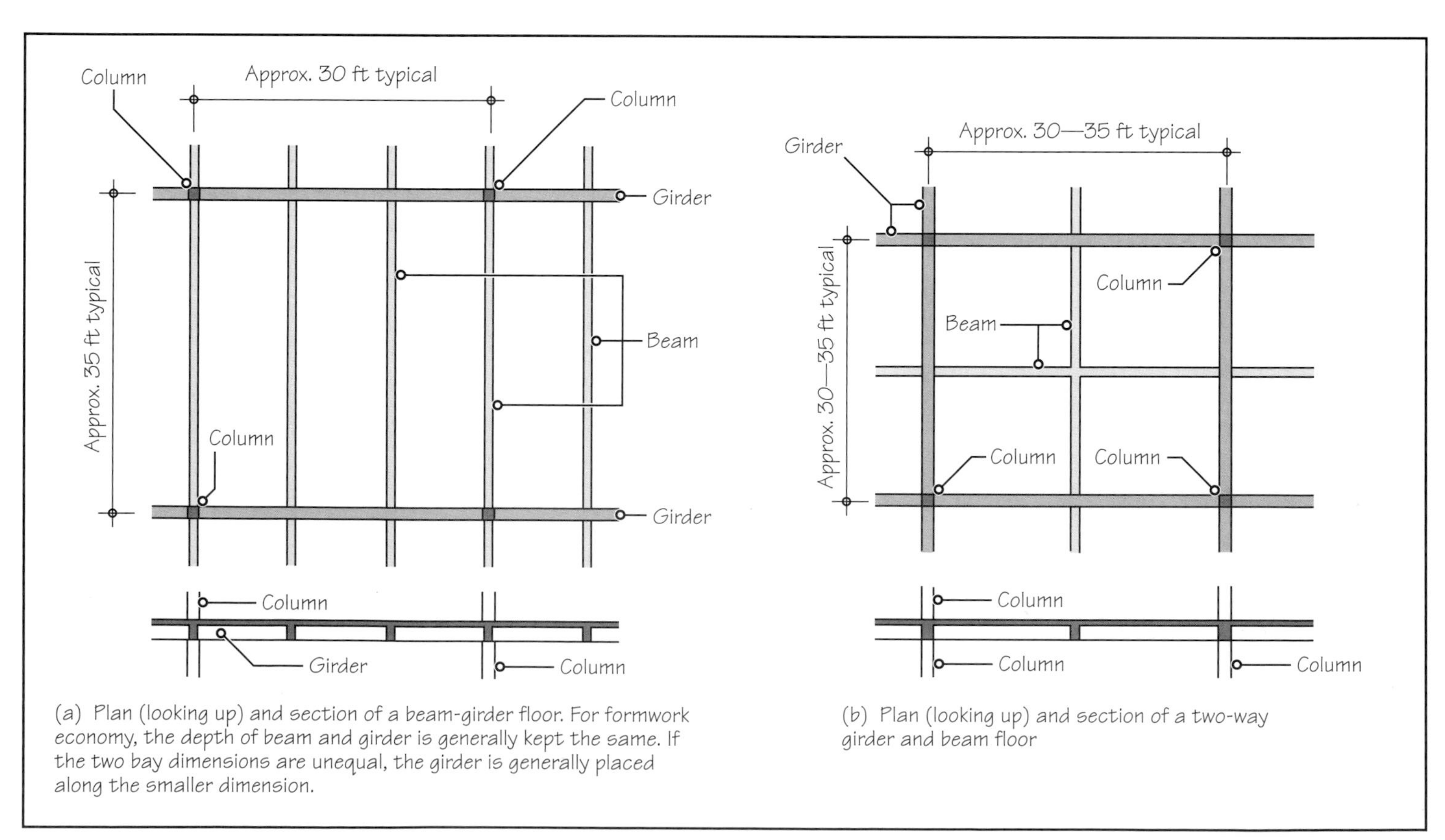

FIGURE 20.4 Plan and section through beam-girder floors.

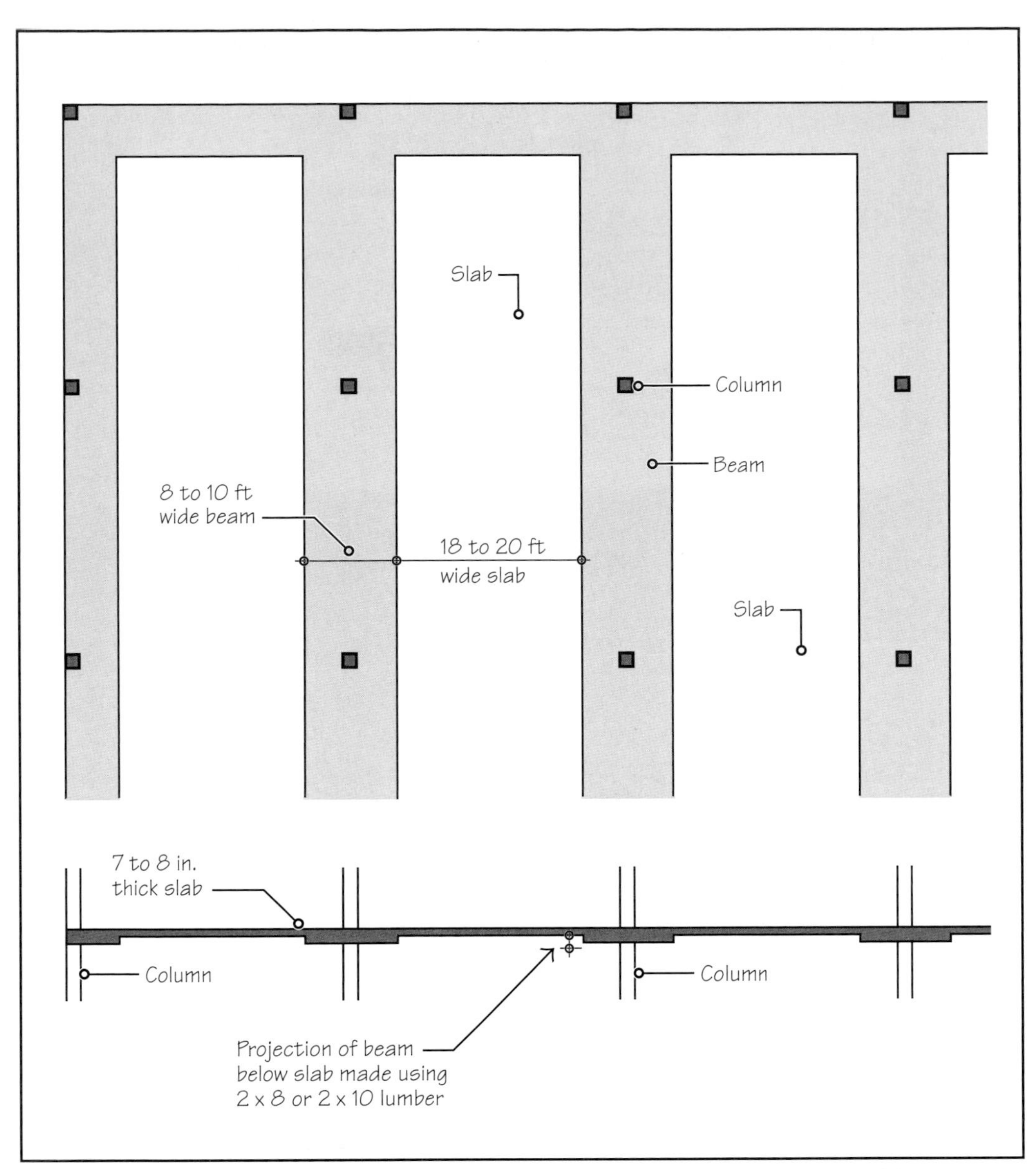

FIGURE 20.5 Plan (looking up) and section through a typical banded slab, which yields a more economical formwork than the beam-and-girder system of Figure 20.4. The projection of the beam below the bottom of the slab is generally obtained by using dimension lumber, such as 2 × 8 or 2 × 10. This illustration also shows typical spans and member thicknesses (also see Appendix B).

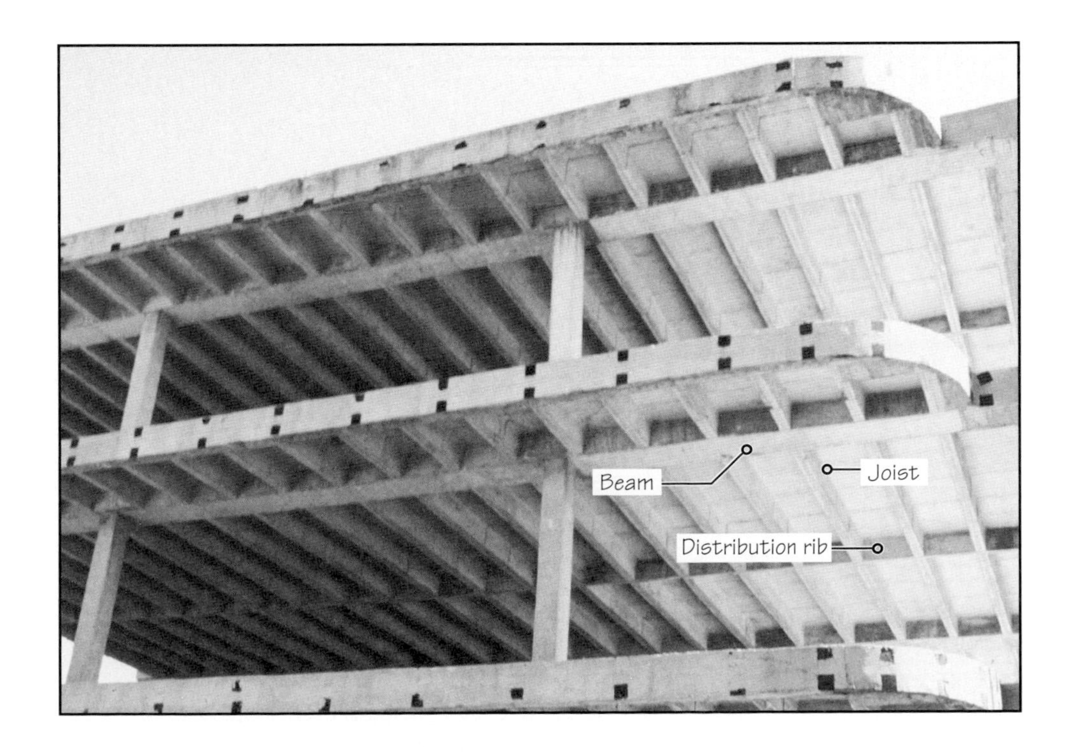

FIGURE 20.6 A one-way joist floor. Observe that the bottom of beams and joists is at the same level to allow the entire slab to be formed on a flat-form deck.

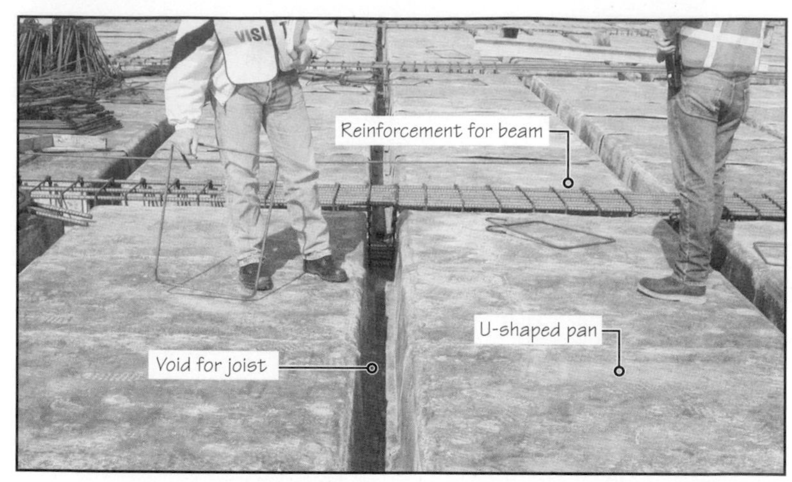

FIGURE 20.7 Formwork for a one-way joist floor showing U-shaped pans. The space between pans represents the width of joists. Notice that the reinforcement for beams has already been laid. Reinforcement for joists and the slab is yet to be completed.

STANDARD-MODULE ONE-WAY JOIST FLOOR

Standard-module pans are 20 in. and 30 in. wide, Figure 20.8. These dimensions have been standardized so that, with 4-in.- and 6-in.-wide joists, the center-to-center spacings between joists are 2 ft and 3 ft, respectively. A slab thickness of $3\frac{1}{2}$ in. is often used (to provide a 1-h fire-rated floor), although structural considerations require a minimum thickness of only 2 in. with 20-in. pans and $2\frac{1}{2}$ in. with 30-in. pans. The slab is designed as a one-way slab resting on the joists, which, in turn, are designed as beams.

Pans of various depths are available to create joists of various depths to suit different joist spans. Because the load on each joist is small, reinforcement requirements are also small. No stirrups are generally used in joists, thus requiring the concrete to resist the entire shear. The increase in shear near-joist supports is resisted, if needed, by increasing joist width, Figure 20.9. Special pans with closed ends on one side are available to produce the widening of joists.

A 4- to 5-in.-wide load-distribution rib is often provided in the center of joists.

WIDE-MODULE ONE-WAY JOIST FLOOR

Wide-module pans are available in 53-in. and 66-in. widths, Figure 20.10. They are generally used with 5-ft and 6-ft center-to-center joist spacings, giving joist widths of 7 in. and 6 in., respectively. However, almost any joist width can be created to suit load and span conditions. Joist widths of 8 and 9 in. are also common.

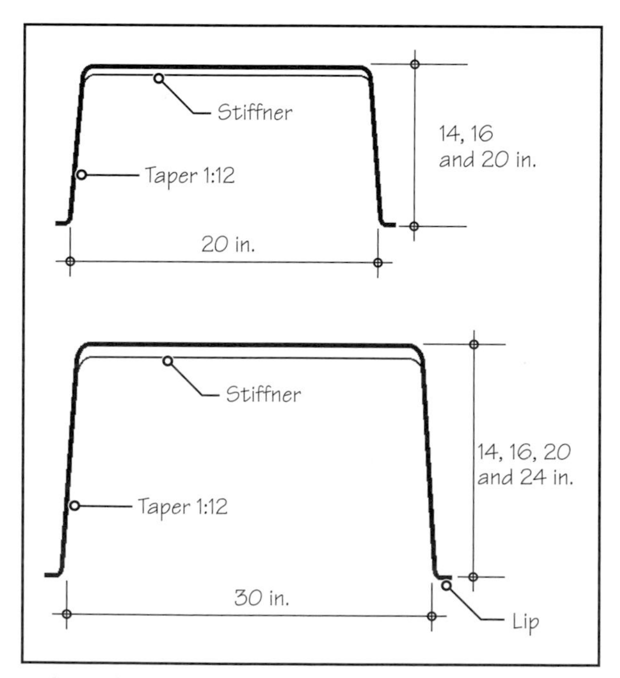

FIGURE 20.8 U-shaped standard-module pans (consult the manufacturer for available sizes). Note that the pans have open ends on both sides, except the pans used adjacent to the beam or the distribution rib, which have closed ends (Figure 20.9).

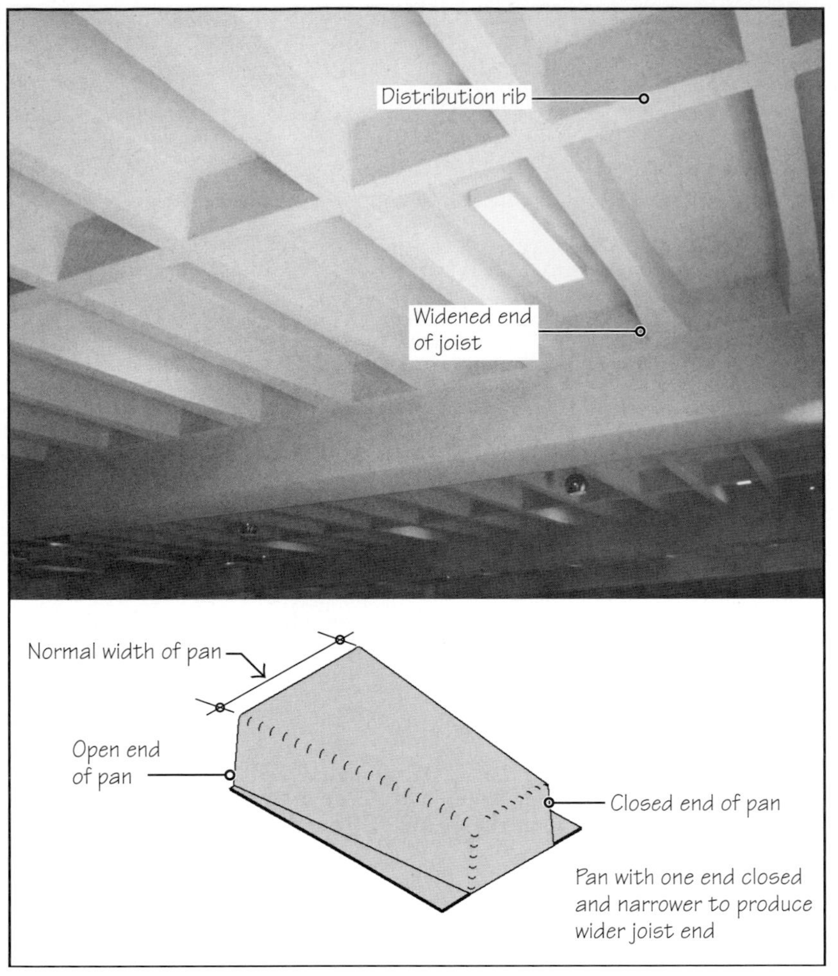

FIGURE 20.9 A one-way joist floor with standard pans. Observe that the joist width has been increased near the beams to accommodate greater shear in the joists at that location. The beams in this slab are deeper than the joists. This is unusual because it increases the cost of formwork as compared with that of a slab with the same depth of joists and beams (Figure 20.6). A distribution rib has been provided in the center of the slab.

Because a wide-module floor has a larger spacing between joists, the slab thickness required is large. Therefore, a wide-module floor is generally used where a minimum fire rating of 1 h (required slab thickness = $3\frac{1}{2}$ in.) or 2 h (required slab thickness = $4\frac{1}{2}$ in.) is needed. A thick slab also allows electrical conduits to be embedded within the slab, Figure 20.11.

NOTE

Preliminary Depth of a One-Way Joist Floor

Estimate the depth of a one-way joist floor by dividing joist span by 18. Thus, if the span of joists is 36 ft, use 24-in.-deep joists, which includes the thickness of the slab. Using a $4\frac{1}{2}$-in.-thick slab and 20-in.-deep pans for a wide-module floor gives a floor depth of $24\frac{1}{2}$ in.

Where the spacing of columns in either direction is unequal, the supporting beams should preferably be oriented along the shorter direction and the joists in the longer direction.

Also see Appendix B, and consult a structural engineer for exact sizes.

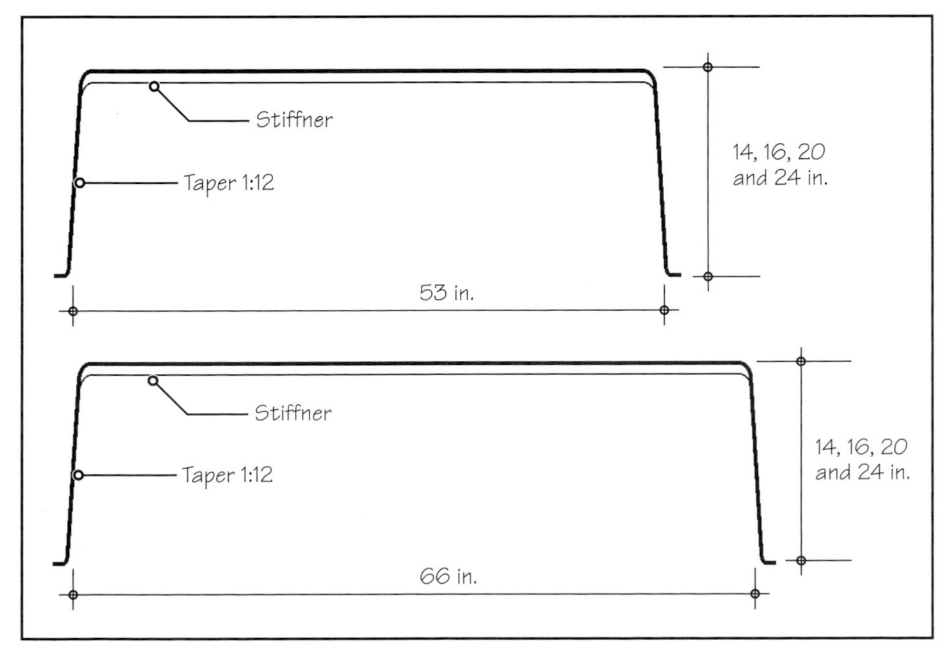

FIGURE 20.10 Wide-module pan dimensions. Consult manufacturers for available sizes.

FIGURE 20.11 A thick slab in a one-way joist floor gives a floor with a higher fire rating and also allows electrical conduits and junction boxes to be embedded in the concrete.

The joists in a wide-module floor are generally designed with shear reinforcement. Therefore, the widening of joist ends at beams is unnecessary. Special end caps are available to close the open ends of pans near the beams, Figure 20.12.

TWO-WAY JOIST FLOOR (WAFFLE SLAB)

A two-way joist floor, also called a *waffle slab,* consists of joists in both directions, Figure 20.13. For the same depth of joists, a waffle slab yields a stiffer floor than a one-way joist floor. It is, therefore, used where column-to-column spacing lies between 35 and 50 ft. A waffle slab is best suited for square or almost square column-to-column bays. When left exposed to the floor below, the waffle slab provides a highly articulated ceiling.

FIGURE 20.12 Wide-module pans and end caps. Because both ends of the pans are open, end caps are needed to close the gap to form a void for the beam.

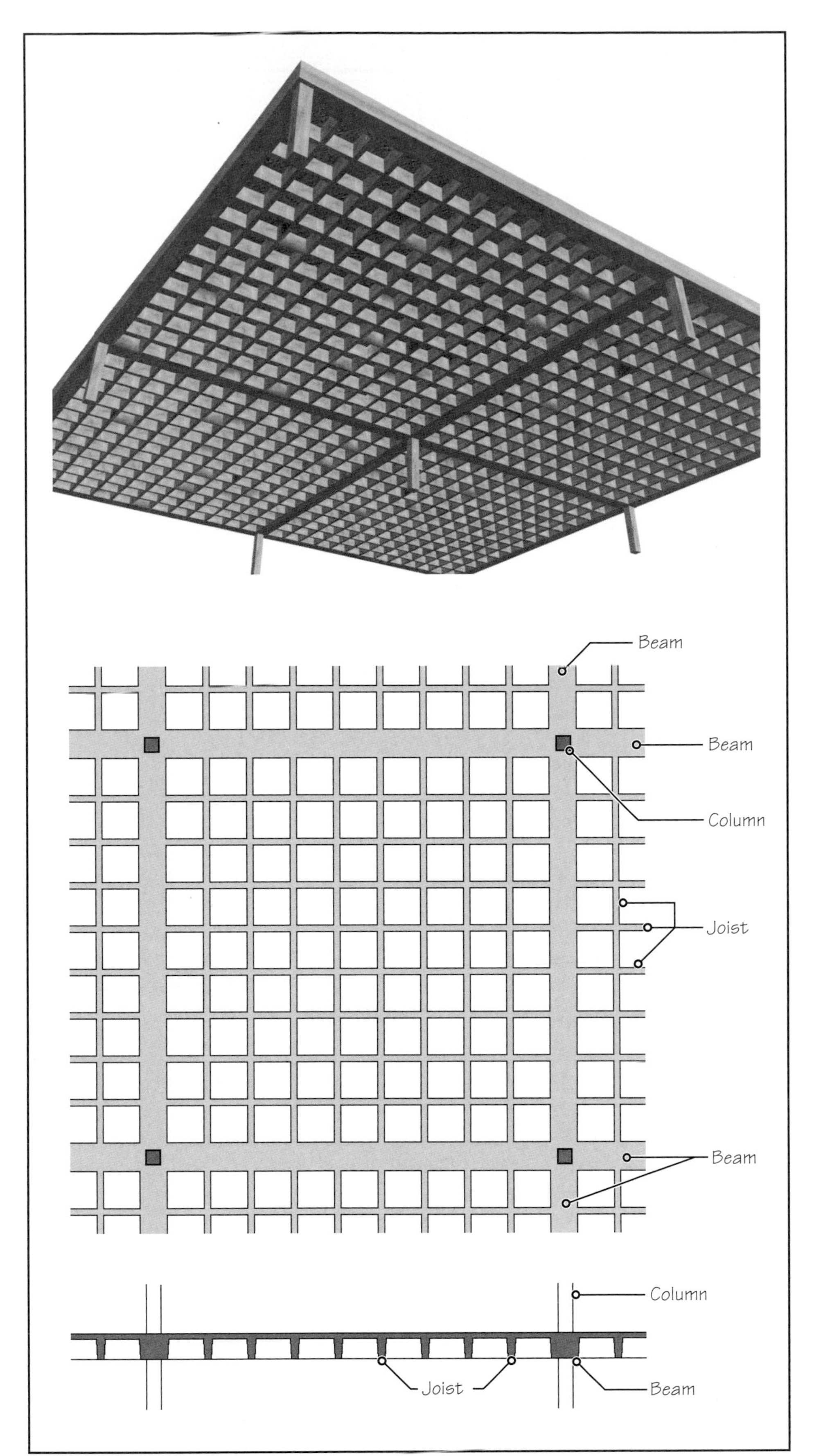

FIGURE 20.13 Two-way joist floor (waffle slab) supported on beams on all sides.
(a) Isometric from below. **(b)** Plan (looking up) and section through slab.
Note that a waffle slab can also be constructed without beams (Figure 20.15).

NOTE

Preliminary Depth of Waffle Slab

Estimate the depth of a waffle slab by dividing the longer span by 24. Thus, if the span is 40 ft, use 20-in-deep waffles that includes the thickness of slab. Using a $4\frac{1}{2}$-in-thick slab for a 2-h rated floor and 16-in.-deep domes gives a floor depth of $20\frac{1}{2}$ in.

Also see Appendix B, and consult a structural engineer for exact sizes.

The formwork for the slab consists of glass fiber–reinforced plastic domes placed on a flat form deck with lips butting each other, Figure 20.14. Like the pans, the domes can withstand repeated use. Dome dimensions have been standardized to produce 3-ft, 4-ft, and 5-ft center-to-center distances between domes in a variety of depths. The domes have a wide supporting lip on all sides and are laid on a flat-form deck.

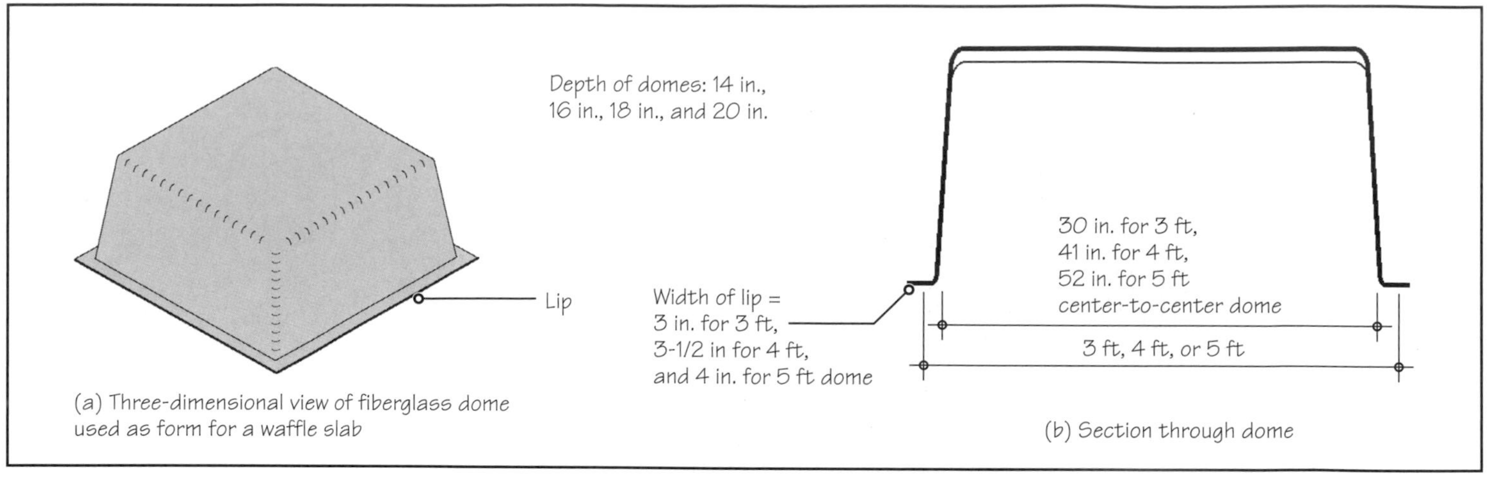

FIGURE 20.14 Standard GFRP dome sizes for a waffle slab. Consult the manufacturer for available sizes.

PRACTICE QUIZ

Each question has only one correct answer. Select the choice that best answers the question.

1. A reinforced concrete slab panel with beams on all four sides is
   - **a.** a one-way slab.
   - **b.** a two-way slab.
   - **c.** a four-way slab.
   - **d.** (a) or (b).
   - **e.** (b) or (c).
2. A reinforced-concrete slab panel with beams on all four sides and measuring 15 ft × 22 ft is
   - **a.** a one-way slab.
   - **b.** a two-way slab.
   - **c.** a four-way slab.
   - **d.** (a) or (b).
   - **e.** none of the above.
3. A reinforced-concrete slab panel with beams on all four sides and measuring 15 ft × 32 ft is
   - **a.** a one-way slab.
   - **b.** a two-way slab.
   - **c.** a four-way slab.
   - **d.** (a) or (b).
   - **e.** none of the above.
4. In a band beam reinforced-concrete floor, the beams are
   - **a.** narrow and deep.
   - **b.** wide and shallow.
   - **c.** located at a spacing of 8 to 10 ft on centers.
   - **d.** unnecessary.
5. A band beam floor is generally more economical than a beam-and-girder floor because
   - **a.** it requires a smaller quantity of concrete.
   - **b.** it requires a smaller quantity of reinforcement.
   - **c.** its formwork is more economical.
   - **d.** none of the above.
6. A standard-module one-way joist floor is constructed using pans that are
   - **a.** 20 in. wide.
   - **b.** 30 in. wide.
   - **c.** 40 in. wide.
   - **d.** (a) or (b).
   - **e.** (b) or (c).
7. Based on structural considerations only, the thickness of slab in a standard-module one-way joist floor need not exceed
   - **a.** 2 in. to $2\frac{1}{2}$ in.
   - **b.** 3 in. to $3\frac{1}{2}$ in.
   - **c.** 4 in. to 5 in.
   - **d.** 5 in. to 6 in.
8. Pans used as formwork for one-way joist floors have a
   - **a.** U-shaped profile.
   - **b.** Z-shaped profile.
   - **c.** dome profile.
   - **d.** any one of the above.
   - **e.** none of the above.
9. A wide-module one-way joist floor is constructed using pans that are
   - **a.** 66 in. wide.
   - **b.** 60 in. wide.
   - **c.** 53 in. wide.
   - **d.** (a) or (b).
   - **e.** (a) or (c).
10. In a one-way joist floor, the supporting beams run in one direction and the joists run in the other direction. If the spans in two directions are unequal, the beams should preferably span along the
    - **a.** shorter direction.
    - **b.** longer direction.
11. The formwork used for a waffle slab has a
    - **a.** U-shaped profile.
    - **b.** Z-shaped profile.
    - **c.** dome profile.
    - **d.** any one of the above.
    - **e.** none of the above.
12. Using standard formwork components for a waffle slab, the center-to-center distance between waffles (voids) is
    - **a.** 3 ft, 4 ft, or 5 ft.
    - **b.** 2 ft, 4 ft, or 6 ft.
    - **c.** 4 ft, 5 ft, or 6 ft.
    - **d.** 5 ft, 6 ft, and 7 ft.
    - **e.** none of the above.

Answers: 1-d, 2-b, 3-a, 4-b, 5-c, 6-d, 7-a, 8-a, 9-e, 10-a, 11-c, 12-a.

20.3 BEAMLESS CONCRETE FLOORS

A waffle slab is more commonly constructed as a beamless slab, Figure 20.15. In a beamless waffle slab, a few domes on all sides of a column are omitted so that the thickness of slab at the columns is the same as that of the joists. The thickening of the slab at the columns provides shear resistance (against the slab punching through the columns).

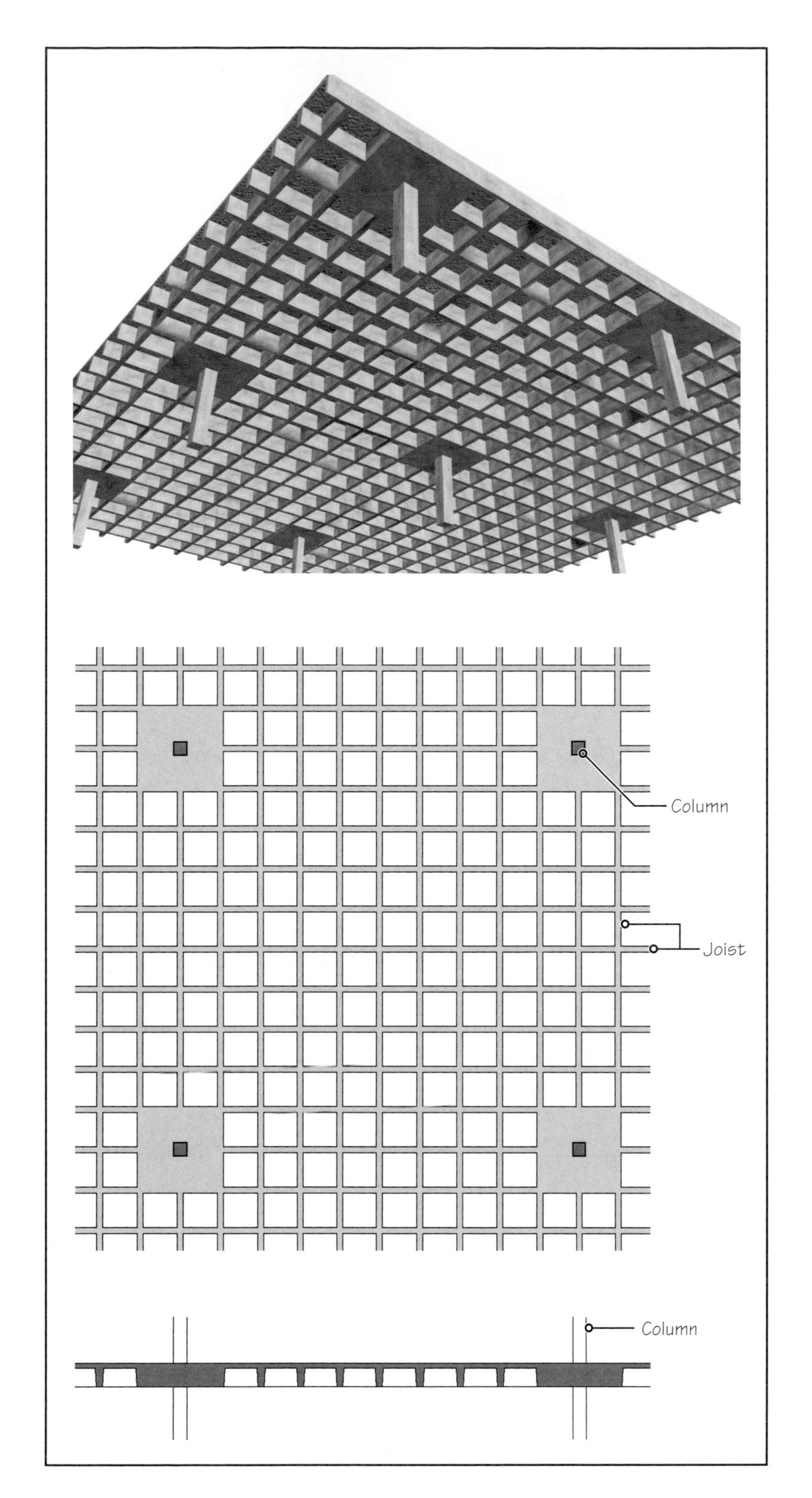

FIGURE 20.15 A beamless waffle slab.

Isometric, plan (looking up), and section through slab.

Note that a waffle slab can also be constructed with beams (Figure 20.13). A beamless waffle slab may be provided with spandrel beams, if needed.

FLAT PLATE

A flat plate consists of a solid slab supported directly on columns, Figure 20.16. A flat plate is similar to a two-way banded slab, except that the beam bands in both directions are concealed within the thickness of the slab. Therefore, the spans that can be achieved economically with a flat-plate floor are smaller than those obtained from one-way or two-way joist floors.

Flat-plate slabs are suitable for occupancies with relatively light live loads, such as hotels, apartments, and hospitals, where small column-to-column spacing does not pose a major design constraint. Additionally, a drop ceiling is not required in these occupancies, and HVAC ducts can be run within the corridors, where a lower ceiling height is acceptable.

FIGURE 20.16 A flat-plate slab in a high-rise hotel-condominium building under construction. Observe the use of perimeter beams, which may be required with heavy exterior wall cladding. Flat-plate slabs without perimeter beams are common (Figure 20.21). Additionally, columns need not be round. Rectangular or square columns may be used with flat-plate slabs.

NOTE

Preliminary Depths of Flat-Plate and Flat-Slab Floors

Estimate the depth of a flat-plate floor by dividing the longer clear distance between columns by 30. Thus, if the distance between columns is 20 ft, use an 8-in.-thick slab.

For a flat-slab floor, estimate the thickness of the slab by dividing the longer clear distance between columns by 35. Thus, if the distance between columns is 24 ft, use an 8-in.-thick slab.

Also see Appendix B, and consult a structural engineer for exact sizes.

A flat-plate slab results in a low floor-to-floor height, and its formwork is economical. Because the beams are concealed within the slab thickness, columns need not be arranged on a regular grid—a major architectural advantage. However, a flat plate is a two-way system; hence, the column spacing in both directions should be approximately the same. A slab thickness of approximately 6 in. is generally needed for 15-ft × 15-ft column bays and approximately 8 in. for 20-ft × 20-ft bays with residential loads.

FLAT SLAB

A *flat slab* is similar to a flat plate, but it has column heads, referred to as *drop panels,* Figure 20.17(a). The primary purpose of drop panels is to provide greater shear resistance at the columns, where the shear maximizes.

Structurally, the drop panel must extend a minimum of $\frac{1}{6}$ of the slab span in each direction, and its drop below the slab must at least be 25% of the slab thickness, Figure 20.17(b). For formwork economy, the drop depth is also based on lumber dimensions, Figure 20.17(c). With round columns, however, manufacturers supply column forms that have built-in drop panels and column capitals, Figure 20.18.

A flat slab is generally used where the live loads are relatively high, such as in parking garages or storage or industrial facilities.

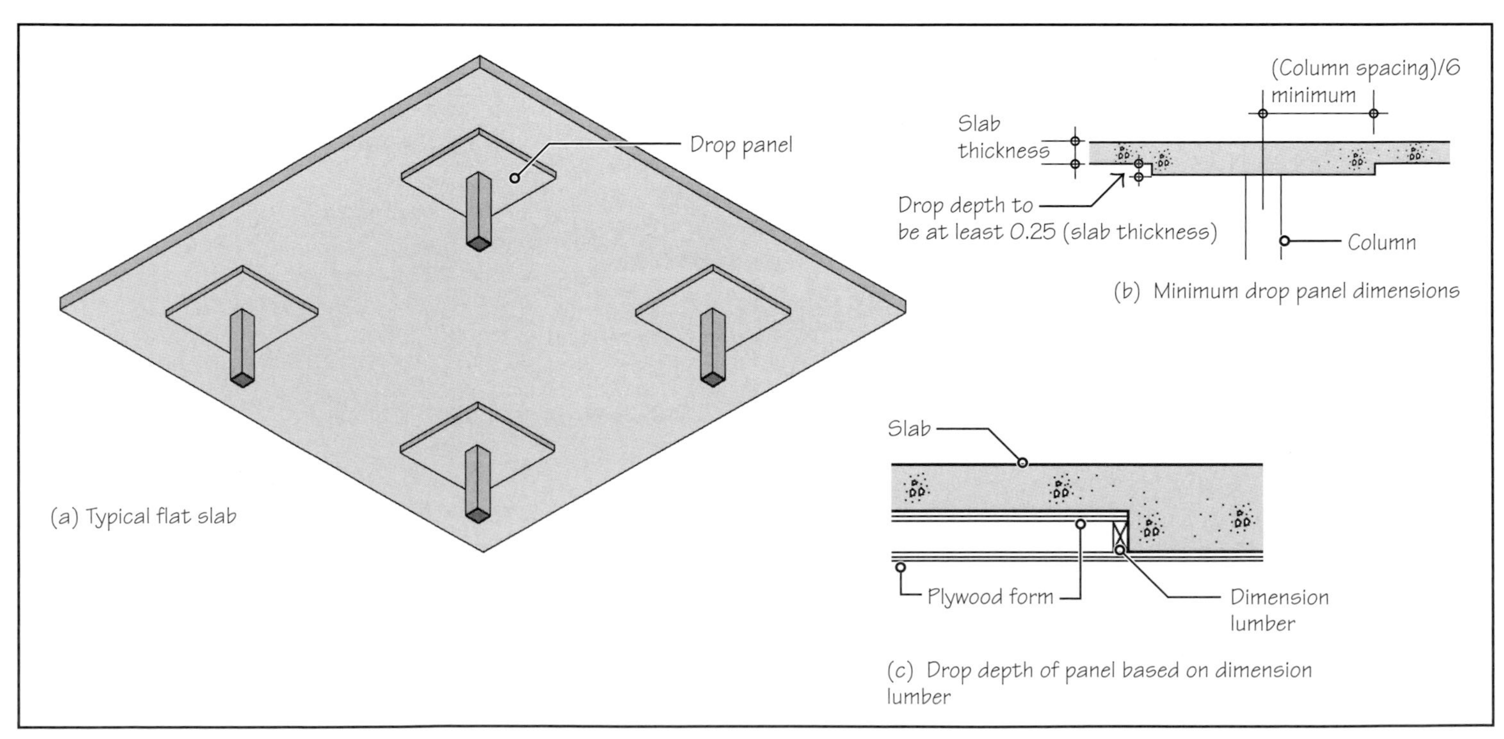

FIGURE 20.17 Typical details of a flat slab and minimum code requirements for drop panel's dimensions.

FIGURE 20.18 A flat slab with (round) drop panels and column capitals.

20.4 POSTTENSIONED ELEVATED CONCRETE FLOORS

The reinforced-concrete elevated-slab systems described previously can also be posttensioned. The posttensioning of slabs reduces slab and beam dimensions, reducing the dead load of the floor. This is particularly helpful in seismic areas, where a lower dead load imparts a lower seismic load on the structure. Smaller slab and beam dimensions also result in a lower floor-to-floor height, reducing the cost of exterior cladding. Column and foundation sizes are also reduced.

Because the code places a limit on the maximum amount of prestress that can be introduced, an elevated, posttensioned concrete slab has a significant amount of conventional reinforcement as well, Figure 20.19. However, the amount of conventional reinforcement in a posttensioned slab is lower than that required if the slab were not posttensioned, reducing reinforcement congestion in long-span and/or heavily loaded members.

FIGURE 20.19 A cage for a concrete beam consisting of conventional reinforcement and prestressing tendons ready to be lowered into beam form. Observe the drape (generally parabolic) provided in prestressing tendons to balance the stresses ceated by gravity loads on the beam. The steel frames shown are used to make the cage. They will be removed after the cage has been placed in the form. Reinforcement for the slab is yet to be completed. Figure 20.20 shows the same beam after its reinforcement has been lowered and the reinforcement and prestressing tendons for the slab are in place.

FIGURE 20.20 Layout of reinforcement and prestressing tendons in the slab of a building. The reinforcement and prestressing tendons for the beam are shown in Figure 20.19. Note the banded arrangement of tendons in the slab.

NOTE

Two Most Commonly Used Concrete Floor Systems

Although various site-cast concrete floor systems have been described in this text, the two most economical floor systems are the *flat-plate floor* and the *one-way joist floor*. For typical live loads of 50 psf, the one-way flat-plate floor is economical for spans of up to approximately 30 ft. For spans between 30 ft and 50 ft, a one-way joist floor is economical [1].

Generally, the prestressing tendons in a slab are distributed in a *banded arrangement*, rather than being individually distributed throughout the span in a *basket-weave arrangement*. Generally, the tendons are grouped together along the support lines in one direction and bands of tendons are placed at equal spacings in the transverse direction, Figure 20.20.

Posttensioned concrete slabs are being increasingly used for all types of structures. They have long been used in parking garages because of their long column-to-column spacings and are being increasingly favored for high-rise office buildings, hospitals, and condominiums, Figure 20.21. The process of posttensioning in an elevated structure is similar to that of a posttensioned slab-on-ground, covered in Chapter 19.

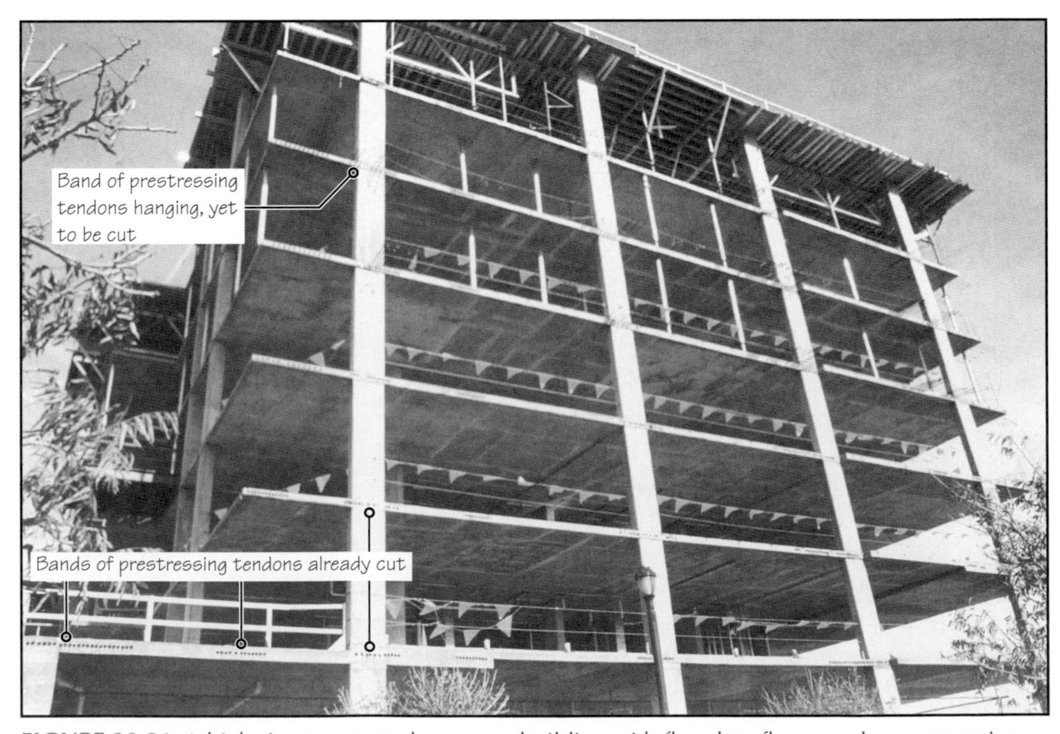

FIGURE 20.21 A high-rise prestressed concrete building with flat-plate floors under construction. Notice the prestressing tendons that have been flame-cut at the lower floors. At the upper floors, the prestressing tendons can be seen hanging and are yet to be cut.

20.5 INTRODUCTION TO PRECAST CONCRETE

As stated in the introduction to Chapter 19, precast concrete members (except concrete tilt-up walls) are fabricated in a precast plant and transported to the construction site for assembly. Precasting, which is generally done in covered or sheltered spaces, is particularly helpful in climates that limit the use of site-cast concrete.

Additionally, because precasting is done at the ground level, the cost of formwork and shoring is considerably reduced. Formwork cost reduction is also achieved through the use of standard-size elements cast in permanent forms, which are reused several times more than the formwork used for sitecast members. Precasting also allows greater quality control in the strength of concrete and surface finishes. Most surface finishes are more easily obtained in a precast plant than at the site—often several floors above ground.

Precast elements used as horizontal framing members are generally prestressed (i.e., pretensioned). To ensure rapid reuse of forms and prestressing equipment, Type III (high early strength) concrete is generally used with a relatively higher strength concrete (5,000 to 6,000 psi). To accelerate the setting and hardening of concrete, members are steam cured. Consequently, most precast concrete members are made on a 24-h cycle.

Precast concrete also has many disadvantages. The main disadvantage of precast concrete lies in the cost of transportation. Although precast members are generally lighter than corresponding site-cast members (because of prestressing), they are still fairly heavy. Transportation also limits the length and width of precast members.

Another disadvantage of precasting is the need for heavier hoisting equipment at the construction site and additional safety measures that must be observed during erection. Erection and assembly at the site also introduces the need for a more skilled workforce than that needed for site-cast concrete construction. Architecturally, the most limiting factor in the use of precast concrete is the difficulty in sculpting concrete at a large scale, which is more easily realized with site-cast concrete. Precast elements are generally straight with standard profiles.

ARCHITECTURAL PRECAST AND STRUCTURAL PRECAST CONCRETE

Precast concrete members are classified as (a) *architectural precast concrete* and (b) *structural precast concrete.* Architectural precast refers to concrete elements that are used as nonstructural, cladding elements. Their most common use is in precast concrete curtain walls (Chapter 26).

Structural precast concrete, which is covered in this chapter, includes all elements of a building's structural frame (floor/roof slabs, columns, and walls). Although an entire room or assembly of rooms can be precast, most structural precast is used in standard elements that are assembled on site to form spaces.

STRUCTURAL PRECAST—ALL PRECAST AND MIXED PRECAST CONSTRUCTION

A building can be constructed of all precast concrete members in which columns, load-bearing walls, and floor and roof elements are of precast concrete. This system is referred to as *all-precast concrete construction.* However, the use of *mixed precast construction,* in which only some elements of the building are of precast concrete, is by far more common. Only mixed precast construction is discussed here.

PRACTICE QUIZ

Each question has only one correct answer. Select the choice that best answers the question.

13. A flat slab floor consists of
- **a.** a slab of constant thickness.
- **b.** a slab with beams on all four sides of slab panel.
- **c.** a slab with drop panels at each column.
- **d.** none of the above.

14. A flat plate floor consists of
- **a.** a slab of constant thickness.
- **b.** a slab with beams on all four sides of slab panel.
- **c.** a slab with drop panels at each column.
- **d.** none of the above.

15. Commonly used beamless concrete-floors are
- **a.** one-way and two-way joist floors.
- **b.** flat-plate and flat-slab floors.
- **c.** beam-and-girder floors.
- **d.** all the above.

16. In a posttensioned elevated concrete floor, prestressing tendons are combined with conventional steel reinforcement.
- **a.** True
- **b.** False

17. The primary reason for using posttensioned concrete floors is to
- **a.** reduce or prevent the cracking of concrete.
- **b.** improve safety against failure.

c. reduce floor depth.
d. reduce or prevent corrosion of reinforcement.
e. reduce formwork cost.

18. The strength of concrete used in precast concrete members is generally
a. 1 to 2 ksi.
b. 2 to 3 ksi.
c. 3 to 4 ksi.
d. 4 to 5 ksi.
e. 5 to 6 ksi.

19. The concrete used in precast concrete members is generally
a. Type V.
b. Type III.
c. Type I.
d. none of the above.

20. Architectural precast concrete refers to concrete members used in building interiors.
a. True
b. False

Answers: 13-c, 14-a, 15-b, 16-a, 17-c, 18-e, 19-b, 20-b.

NOTE

Approximate Spanning Capability of Hollow-Core Slabs

To determine the approximate thickness of a hollow-core slab floor, divide the span by 40. Thus, if the distance between supporting members is 25 ft, the thickness of hollow-core slab is (25 × 12)/40 = 7.5 in. Hence, use an 8-in. slab. A 12-in. slab has an approximate spanning capability of 40 ft.

Due to lighter loads on roofs, slab thickness for roofs may be determined by dividing the span by 50.

Also see Appendix B, and consult a structural engineer for exact sizes.

20.6 MIXED PRECAST CONCRETE CONSTRUCTION

In mixed precast concrete construction, precast concrete floor and roof members are used with a steel frame, site-cast concrete frame, or a load-bearing masonry structure. Prestressed floor and roof elements consist of the following:

- Hollow-core slabs
- Solid planks
- Double-tee units
- Inverted-tee beams

HOLLOW-CORE SLABS AND SOLID PLANKS

Hollow-core slabs are prestressed concrete slabs that contain voids in their central region. The voids reduce the dead load of the slab by 40% to 50%, as compared with a site-cast concrete slab for the same span.

Slabs are produced in thicknesses of 4 in., 6 in., 8 in., 10 in., 12 in., and 14 in., with the 8-in.-thick slab the most commonly used. The width of the slabs is generally 4 ft. Narrow-width slabs are produced in the plant by cutting standard 4-ft slabs along their length to fit a space that is not a multiple of 4 ft. In addition to 4-ft-wide slabs, some plants make slabs that are 3 ft 4 in. and 8-ft wide.

Figure 20.22 shows a typical section through an 8-in.-thick, 4-ft-wide slab. The number of voids and their cross-sectional profiles are manufacturer specific. Manufacturers provide the load capacity of their slabs, along with suggested details for connecting the slabs with supporting members.

A site-cast concrete topping may be used over the slabs. The topping provides structural integration of slab units and increases the floor's fire resistance and sound insulation. It also functions as a leveling bed, particularly with units with uneven camber. In many buildings, however, topping is omitted for the sake of economy. The topping, when used, is generally 2 in. to $2\frac{1}{2}$ in. thick and is generally reinforced with welded wire reinforcement.

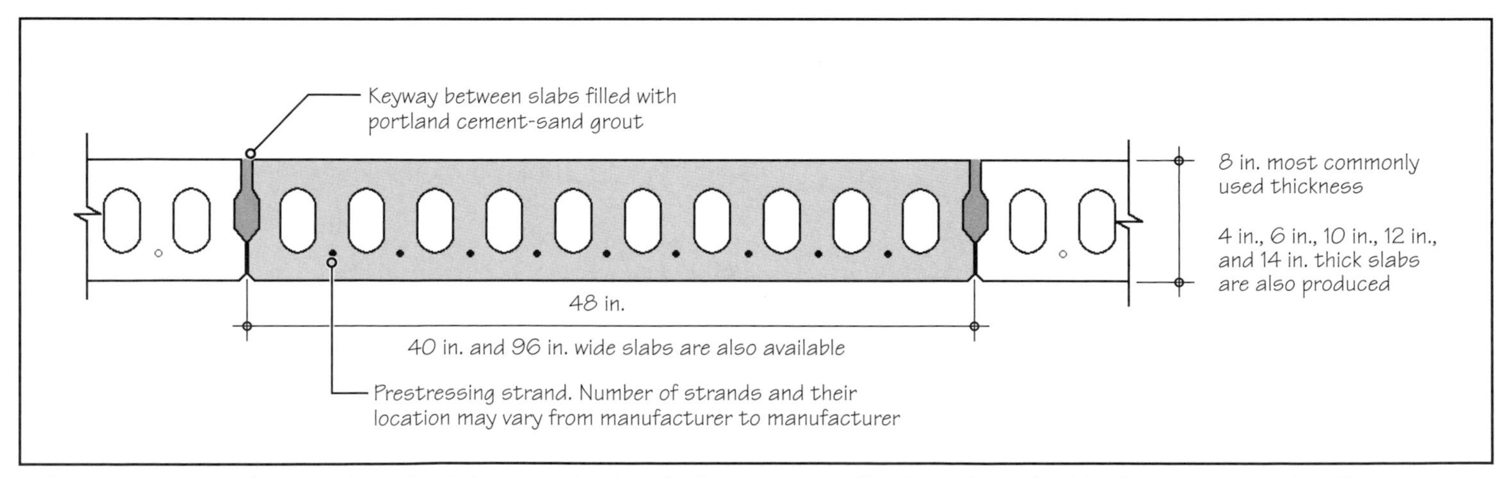

FIGURE 20.22 Typical section through a 4-ft-wide, 8-in.-thick hollow-core slab. The dimensions of voids, their number, and profile are manufacturer specific.

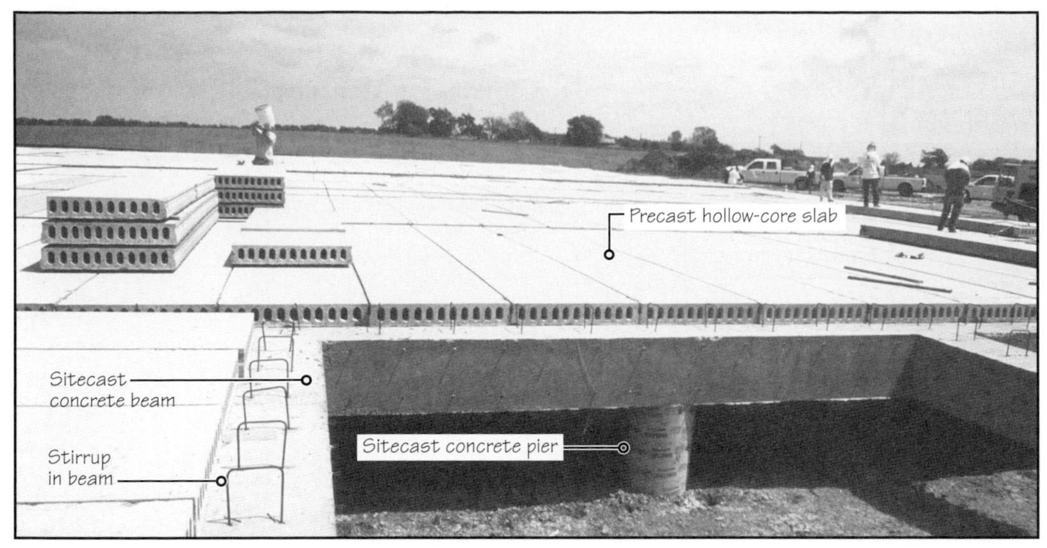

FIGURE 20.23 Mixed precast concrete construction consisting of site-cast concrete beams and piers with precast concrete hollow-core slabs. This image also exemplifies an alternative for the construction of the ground-floor concrete slab of a building on poor soil conditions. Drilled concrete piers have been used to transfer the load to deeper, stable, high-bearing-capacity soil strata.

After all hollow-core slabs have been laid, welded wire reinforcement will be placed on them followed by approximately 2-in.-thick site-cast concrete. This will structurally integrate the slabs and the supporting beams (observe the steel stirrups projecting from the beams).

(See Chapter 21 for other alternatives of providing ground floor concrete slabs on unsatisfactory soils.)

Figures 20.23 and 20.24 show hollow-core slabs with a site-cast reinforced concrete frame and a steel frame structure, respectively. Typical details of connections of slab units with the supporting frames are shown in Figure 20.25.

Hollow-core slabs are produced through a long line extrusion process. The machine, housed in a precasting enclosure, lays concrete on a casting bed over prestressing tendons, which have already been pretensioned to the required level of stress. It forms continuous voids as it travels along the length of the bed. The concrete used is high-strength, zero-slump concrete.

After casting, the slab is steam cured and cut to customer-specified lengths. The cut lengths of slabs are removed from the bed the morning after casting and the process just described is repeated on a 24-h basis.

Solid precast concrete planks are similar to hollow-core planks and are generally used where superimposed loads are relatively light.

DOUBLE-TEE UNITS

A typical section through a double-tee unit is shown in Figure 20.26. The width of units is generally 8 ft with center-to-center distance of 4 ft between beams. Double-tee units with widths of 9 ft, 10 ft, and 12 ft are also available from some manufacturers. Overall depths of double-tee units vary from 16 in to 36 in. to give different spanning capabilities.

NOTE

Approximate Spanning Capability of Double-Tee Units

To determine the approximate depth of a double-tee unit floor, divide the span by 28. Thus, if the span of units is 60 ft, the approximate depth of the floor is $(60 \times 12)/28 = 26$ in.

Standard available depths of double-tee units are generally 14 in., 18 in., 24 in., and 30 in. The manufacturer should be consulted for more precise information.

Also see Appendix B, and consult a structural engineer for exact sizes.

FIGURE 20.24 Mixed precast concrete construction consisting of steel frame structure with precast concrete hollow-core slabs.

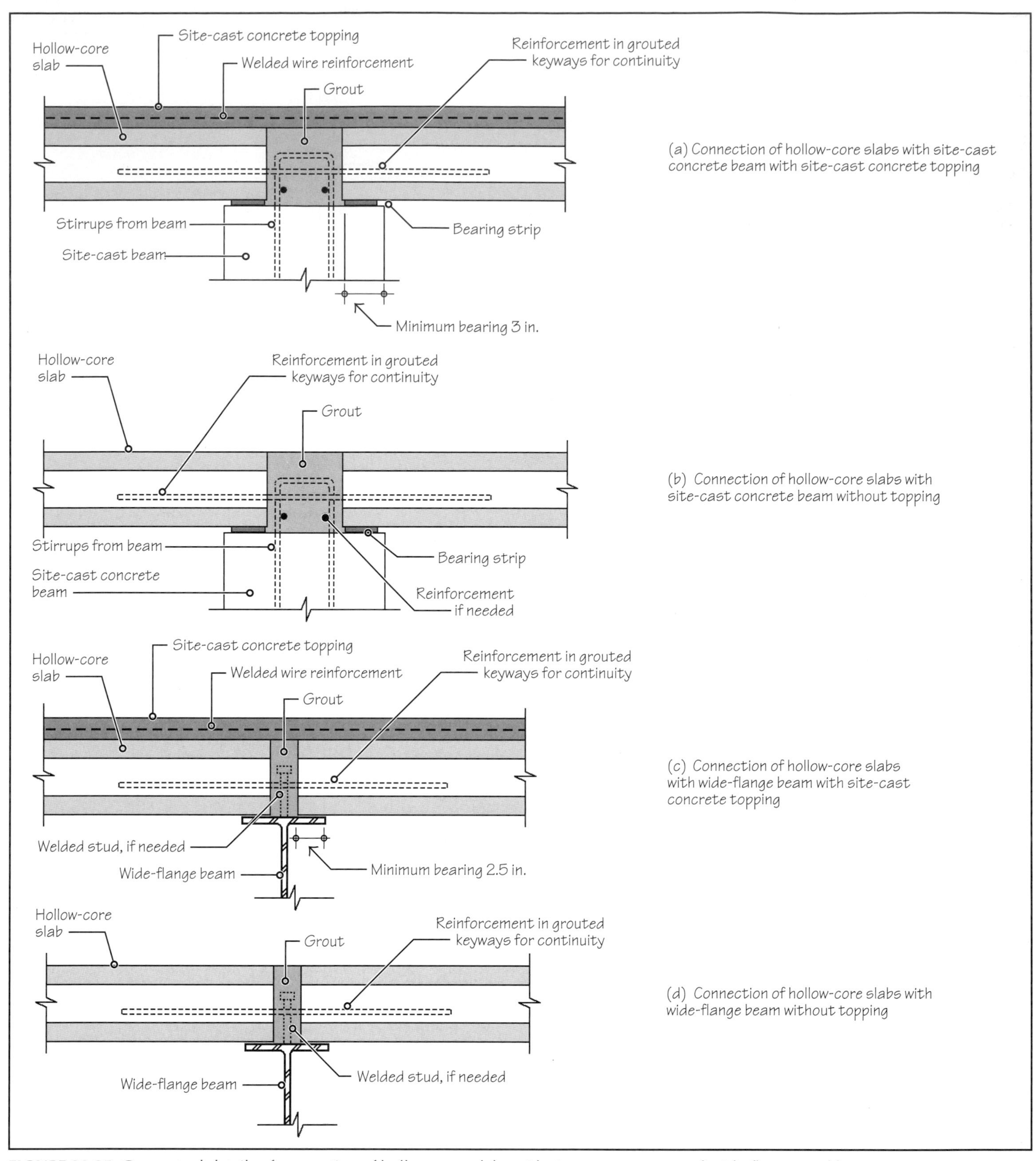

FIGURE 20.25 Conceptual details of connection of hollow-core slabs with site-cast concrete and wide-flange steel beam. (See Chapter 24 for the connection of hollow-core slabs with load-bearing masonry walls.)

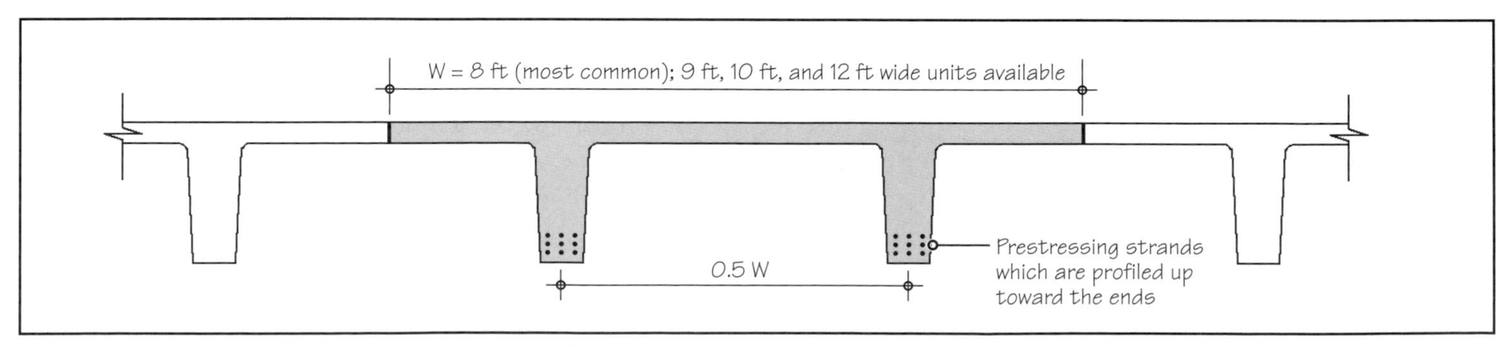

FIGURE 20.26 Typical section through a double-tee floor.

FIGURE 20.27 Use of double-tee units for a parking garage. In this case, the units are supported on precast inverted T-beams.

Double-tee units are used where the spans are large and cannot be provided economically with site-cast concrete construction or hollow-core slabs. They are commonly used for hotel and bank lobbies. As with hollow-core slabs, a topping of concrete and welded wire reinforcement (WWR) may be used on double-tees for structural integration and leveling.

Another common use of double-tees is in multistory parking garages, where a minimum distance of 60 ft between columns is generally necessary. In garages, double-tees are generally supported on site-cast or precast inverted T-beams, Figure 20.27, and the continuous topping is omitted. Structural integration between units and the supporting beam is obtained through a band of site-cast concrete topping and WWR at the ends of the units, Figure 20.28.

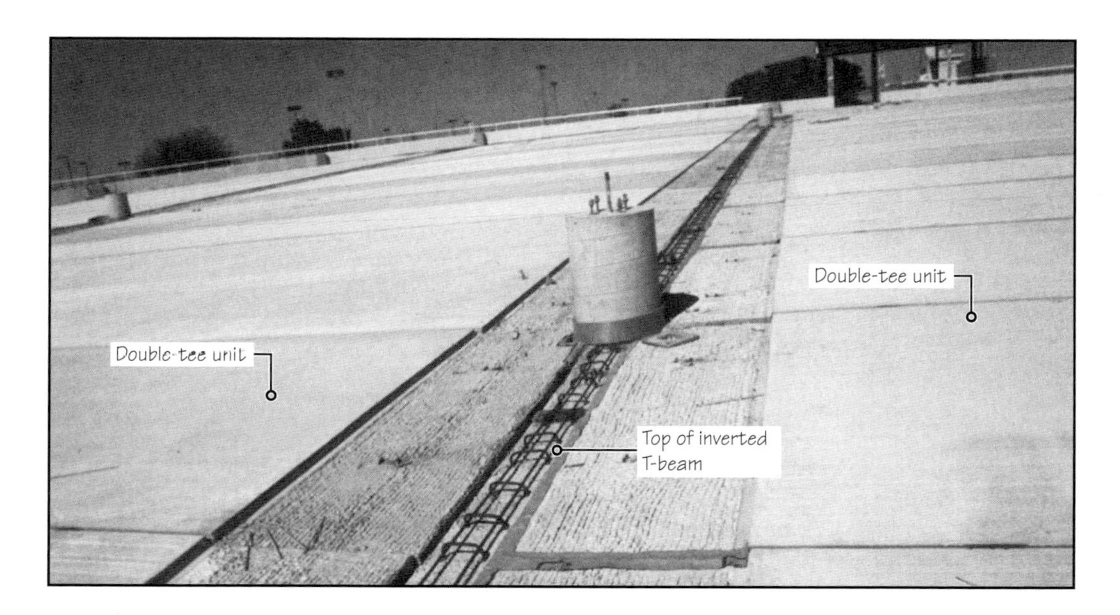

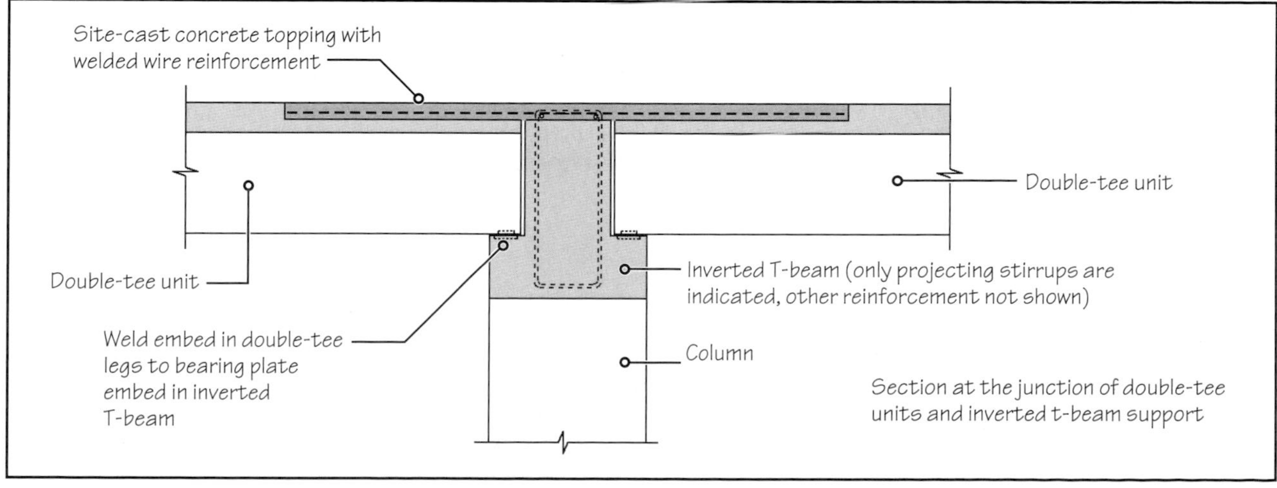

FIGURE 20.28 Band of site-cast concrete topping that integrates double-tees at the supports. Observe the depressed top of double-tees near the top of the inverted T-beam, and the roughness of the surface. Also observe stirrups projecting out of the inverted T-beam. Welded wire reinforcement (WWR) will be placed over the depressed band before placing concrete, as shown in the section.

FIGURE 20.29 Formwork for fabricating double-tee units. Observe the bulkhead separators between units.

FIGURE 20.30 Inverted-tee beams. Observe the marks left by torch-cutting the prestressing strands.

NOTE

Single-Tee Units

Single-tee units were used (generally for long spans), but they are not commonly used today because of erection instability problems.

Double-tee units are cast on a long line of steel formwork. Several units are made in one casting operation by dividing the bed with separators. Figure 20.29 shows a typical formwork set up for their fabrication.

INVERTED-TEE BEAMS

As shown in Figure 20.28, precast inverted-tee beams are generally used as supporting members for hollow-core slabs or double-tee units. They are prestressed. Steel embeds required for site welding are provided in them at the time of fabrication. Figure 20.30 shows typical inverted tee beams in a precast plant.

20.7 FIRE RESISTANCE OF CONCRETE MEMBERS

The fire resistance of a concrete member is a function of three variables:

- Thickness of the member
- Type of coarse aggregate
- Cover over reinforcement and prestressing tendons

A thicker member has a greater fire resistance. For the same member thickness, lightweight aggregate gives a greater fire resistance than a normal-weight aggregate. Among the normal-weight aggregates, carbonate aggregate (i.e., limestone) gives a greater fire resistance than noncarbonate (i.e., siliceous) aggregates.

Local building codes prescribe the minimum thickness of members to achieve a given fire resistance, Table 20.1. In addition to minimum thickness requirements, the member must also satisfy minimum cover requirements. The minimum cover required for corrosion protection (Table 19.1) is adequate to satisfy fire-resistance requirements in most situations. However, the local building code must be referenced for precise information.

TABLE 20.1 FIRE RESISTANCE OF CONCRETE MEMBERS

Member	Aggregate type	Fire resistance 1-h	2-h	3-h
Minimum thickness (in.) of floor slabs or walls	Lightweight	2.5	3.6	4.4
	Carbonate	3.2	4.6	5.7
	Siliceous	3.5	5.0	6.2
Minimum thickness (in.) of columns	Lightweight	8	9	10.5
	Carbonate	8	10	11
	Siliceous	8	10	12

PRACTICE QUIZ

Each question has only one correct answer. Select the choice that best answers the question.

21. In mixed precast concrete construction, precast concrete members are generally used for
- **a.** shear walls.
- **b.** load-bearing walls.
- **c.** columns.
- **d.** floors.
- **e.** none of the above.

22. A 12-in.-thick hollow-core floor slab will generally span up to approximately
- **a.** 40 ft.
- **b.** 30 ft.
- **c.** 25 ft.
- **d.** 20 ft.
- **e.** none of the above.

23. When a concrete topping is used over hollow-core slabs, its thickness is approximately
- **a.** 1 in.
- **b.** $1\frac{1}{2}$ in.
- **c.** 2 in.
- **d.** 3 in.
- **e.** 4 in.

24. The purpose of concrete topping on hollow-core slabs is to
- **a.** structurally integrate the slabs.
- **b.** level the top of slabs.
- **c.** increase the sound insulation of the floor.
- **d.** increase the fire rating of the floor.
- **e.** all the above.

25. The most commonly used width of a double-tee floor unit is
- **a.** 8 ft.
- **b.** 6 ft.
- **c.** 4 ft.
- **d.** none of the above.

26. Double-tee floors are commonly used
- **a.** in residential occupancies.
- **b.** where spans lie between 30 ft to 40 ft.
- **c.** where spans lie between 40 ft to 50 ft.
- **d.** spans are in excess of 50 ft.

27. The fire resistance rating of concrete members is a function of
- **a.** member thickness.
- **b.** type of coarse aggregate in concrete.
- **c.** concrete cover for corrosion protection.
- **d.** all the above.

Answers: 21-d, 22-a, 23-c, 24-e, 25-a, 26-d, 27-d.

EXPAND YOUR KNOWLEDGE

Building Separation Joint in a Site-Cast Concrete Building

As discussed in Chapter 9, continuous building separation joints are required in many buildings. A preferred method is to use two columns and two beams at the joint. Often for the sake of economy, however, a single column with two beams is used. The following detail explains the construction of such a joint in a site-cast reinforced-concrete building. (See Chapter 16 for a similar detail in a steel frame building.)

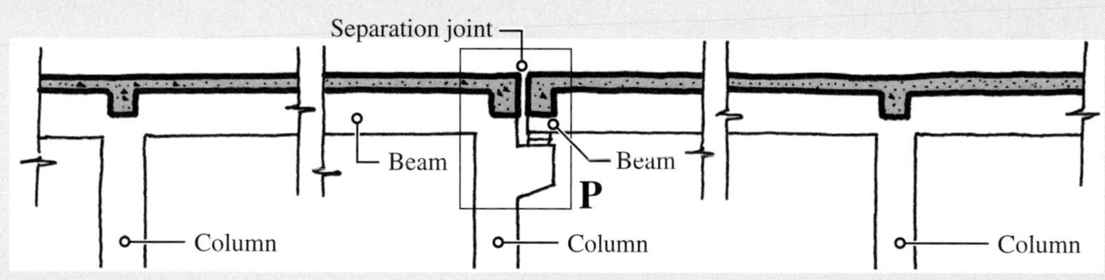

FIGURE 1 Section through a concrete floor at a building separation joint using one column at the joint (also see Chapter 9).

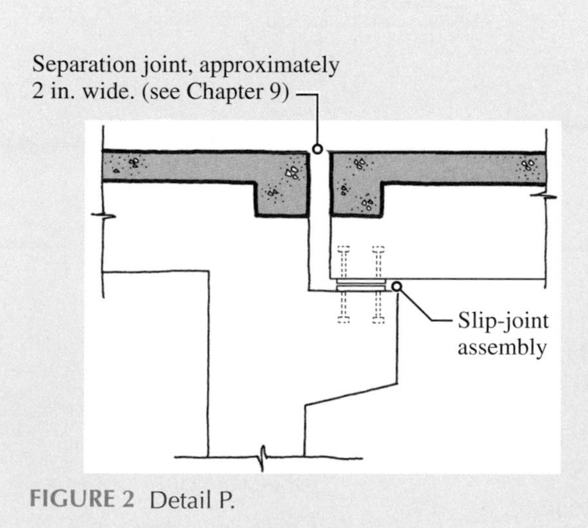

FIGURE 2 Detail P.

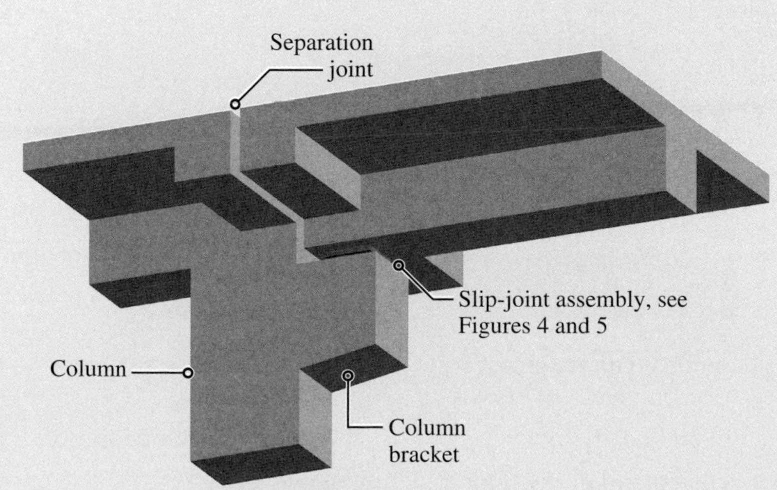

FIGURE 3 Three-dimensional view of Detail P.

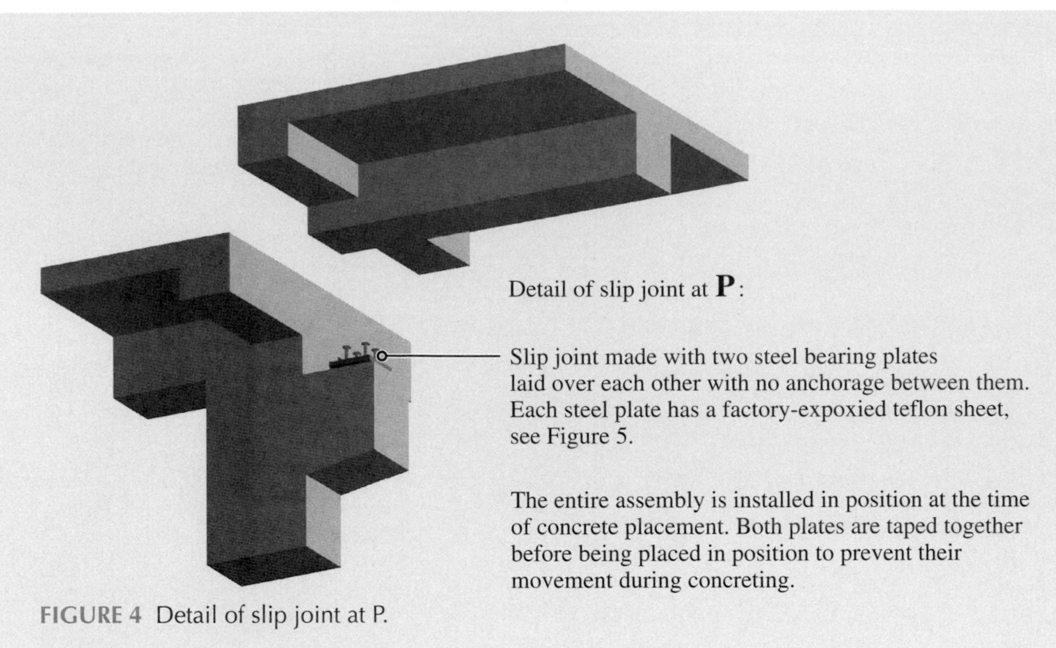

FIGURE 4 Detail of slip joint at P.

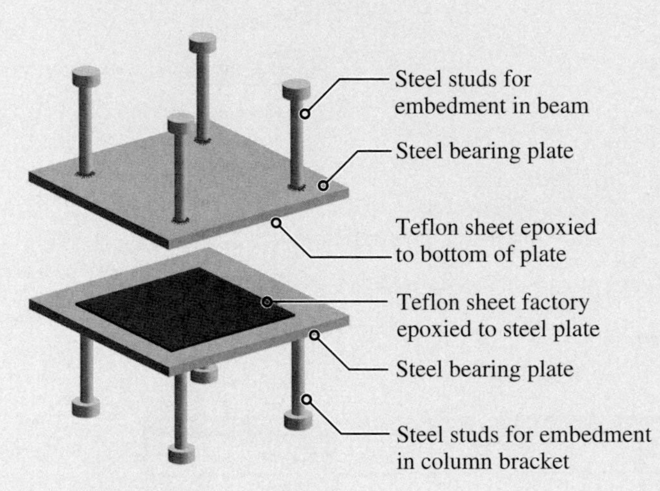

FIGURE 5 Slip-joint assembly.

Coefficient of Friction

Assume two rectangular blocks are laid, one over the other. A weight, P, is placed on the upper block, which is then made to slide over the lower block with the help of a horizontal force, F. If F is the horizontal force that is just able to slide the upper block, then the **coefficient of friction**, ϕ, between the surfaces of the blocks is given by

$$\phi = \frac{F}{P}$$

If the force required to slide is small, ϕ is small. The value of ϕ between two teflon sheets is nearly 0.05 (as reported by a separation joint assembly manufacturer, Conserve Inc., Georgetown, South Carolina).

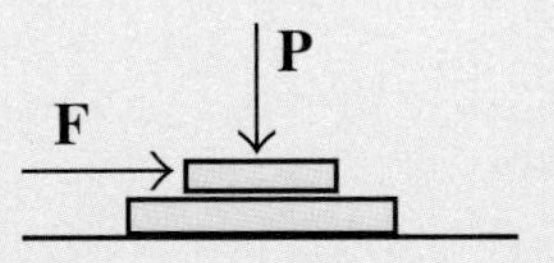

KEY TERMS AND CONCEPTS

- Architectural precast concrete
- Band beam floor
- Beam and girder floor
- Beamless floor
- Beam-supported concrete floors
- Building separation joint
- Double tee
- Drop panels
- Elevated concrete floor systems
- Elevated, post tensioned concrete floor
- Flat plate floor
- Flat slab floor
- Hollow core slab
- Inverted T-beam
- One-way joist floor
- One-way slab floor

Pan formwork
Pre-cast concrete
Primary reinforcement
Ribbed floor
Ribbed slab
Secondary reinforcement
Simply supported beam
Site-cast concrete
Solid concrete plank
Structural precast concrete
Two-way joist floor
Two-way slab
Waffle slab

REVIEW QUESTIONS

1. Explain the difference between one-way slab and two-way slab actions.
2. Using sketches, explain the commonly used beamless concrete-floor systems.
3. Sketch the following precast concrete structural members.
   a. Hollow-core units
   b. Solid planks
   c. Double-tee units
   d. Inverted-tee beams

SELECTED WEB SITES

American Concrete Institute, ACI (www.aci-int.org)
Ceco Concrete (www.cecoconcrete.com)
Concrete Construction Online (www.concreteconstruction.net)
Portland Cement Association, PCA (www.cement.org)
Precast/Prestressed Concrete Institute, PCI (www.pci.org)
World of Concrete (www.worldofconcrete.com)

FURTHER READING

1. Dobrowolski, J. A. *Concrete Construction Handbook*. New York: McGraw-Hill, 1998.
2. Fanella, D. A. *Concrete Floor Systems—Guide to Estimating and Economizing*. Skokie, IL: Portland Cement Association (PCA), 2000.
3. Hurd, M. K. *Formwork for Concrete*. Farmington Hills, MI: American Concrete Institute (ACI), 1995.
4. Post-Tensioning Institute (PTI). *Construction and Maintenance Procedures Manual For Post-Tensioned Slabs-on-Ground,* 1998.

REFERENCES

1. Alsamsan, I., and Kamara, M. *Simplified Design of Reinforced Concrete Buildings of Moderate Size and Height.* Skokie, IL: Portland Cement Association, 2004.

CHAPTER 21

Soils; Foundation and Basement Construction

CHAPTER OUTLINE

21.1 Classification of Soils

21.2 Soil Sampling and Testing

21.3 Earthwork and Excavation

21.4 Supports for Deep Excavations

21.5 Keeping Excavations Dry

21.6 Foundation Systems

21.7 Shallow Foundations

21.8 Deep Foundations

21.9 Frost-Protected Shallow Foundations

21.10 Below-Grade Waterproofing

PRINCIPLES IN PRACTICE: Unified Soil Classification

Excavation of a basement in progress. The vertical cut of the excavation is supported by soldier piles and lagging. Additional lengths of lagging will be installed as the excavation proceeds to a greater depth. (Photo by MM.)

A well-designed and constructed foundation is essential for a successful building project. Defects in foundations are extremely difficult to correct. Therefore, foundation design and construction set the standard for the building.

Before the construction of foundations can begin, the site must be prepared to receive them. These activities include excavation, site stabilization, and the control of subsurface water. Other subgrade activities may include infrastructure access, drainage, and thermal and moisture protection. Building foundations are typically constructed of site-cast reinforced concrete. Alternative materials are reinforced masonry (which has the same essential characteristics as concrete) and preservative-treated timbers or other wood products.

This chapter begins with a discussion of the importance of soil investigation on foundation design, including an introduction to the classification and testing of soil. This is followed by methods of excavation and earthwork, excavation-support systems, and excavation-dewatering methods. A general discussion of shallow and deep foundation systems is followed by a more detailed investigation of footings for shallow foundations and deep foundations. The chapter concludes with foundation details and a discussion of methods of frost protection, drainage, and waterproofing.

21.1 CLASSIFICATION OF SOILS

If organic matter, which is present only in small quantities in most locations, is ignored, the earth's crust consists mainly of mineral (i.e., inorganic, noncombustible) matter. The earth's mineral matter is generally classified as *rocks* or *soils*.

In rocks, the mineral particles are firmly bonded together. Soil particles exist either as individual particles or as a conglomerate of several easily separable particles. In most locations, the top layer of the ground consists of soil; rock generally occurs deep under the surface. The foundations of most buildings are, therefore, soil supported, except where the soil's strength is inadequate and the foundations must bear on bedrock. (Rocks are classified as *igneous rocks, sedimentary rocks,* and *metamorphic rocks* (Chapter 23).)

GRAVELS, SANDS, SILTS, AND CLAYS

An important parameter used for classifying soils is the size of soil particles. The size of soil particles is measured by passing a dried soil sample through a series of sieves, each with a standardized opening size (Figure 21.4). Soil particles that are retained in a No. 200 sieve are called *coarse-grained soils* (>0.075 mm), and soils that pass through this sieve are considered to be *fine-grained soils*.

As shown in Figure 21.1, coarse-grained soils are further divided into *gravels* or *sands*. Gravels have particles that are retained on a No. 4 sieve but are smaller than 3 in. Particles larger than 3 in. are called *cobbles,* or *boulders*. Sand particles pass through a No. 4 sieve but are retained on a No. 200 sieve.

Fine-grained soils are divided into *silts* and *clays*. Particles smaller than 0.075 mm but larger than 0.002 mm are classified as silt. Particles smaller than 0.002 mm are classified as clay.

UNIQUENESS OF CLAY

Apart from the particle size, a major factor that distinguishes clay from other soil constituents is particle shape. Gravel, sand, and silt particles are approximately equidimensional; that is, they are approximately spherical or ellipsoidal in shape. This is because gravels, sands, and silts are the result of mechanical weathering.

Clay particles, on the other hand, are the result of chemical weathering. They are, therefore, not equidimensional but have flat, platelike shapes. Because of their flat particle shape, the surface-area-to-volume ratio of clays is several hundred or thousand times greater than the corresponding ratio for gravels, sands, or silts.

The behavior of clayey soils is greatly influenced by the electrostatic forces that develop between platelike surfaces. In presence of water, these forces are repulsive, which increases the space between plates. Therefore, in the presence of water, clayey soils swell, and as water decreases, they shrink.

COHESIVE AND NONCOHESIVE SOILS

One factor that distinguishes fine-grained from coarse-grained soils is the property of cohesion. Fine-grained soils are cohesive—that is the particles adhere to each other in the presence of water. For this reason, they are called *cohesive soils*. Coarse-grained soils are typically single grained, lacking cohesiveness, and are, therefore, referred to as *noncohesive soils*.

NOTE

Rock-Soil Cycle

In the early stages of the cooling of the earth from its molten state, the earth's crust consisted only of rocks. Abrasion caused by the movement of water and air on the earth's surface disintegrated rocks into smaller soil particles. Fatigue resulting from thermal expansion, contraction, freezing, and thawing of rocks enhanced their disintegration.

In addition to *mechanical weathering,* as just described, rocks also went through *chemical weathering,* caused by their reaction with atmospheric oxygen, plant life, and various chemicals transported by water.

Both weathering processes continue to this day. As the disintegrated particles are carried by water and deposited into valleys, lakes, and deltas, they are eventually converted back to rock due to the pressure of overlying material. Thus, rock and soil formations are parts of a perpetual cycle, referred to as the *geological cycle*.

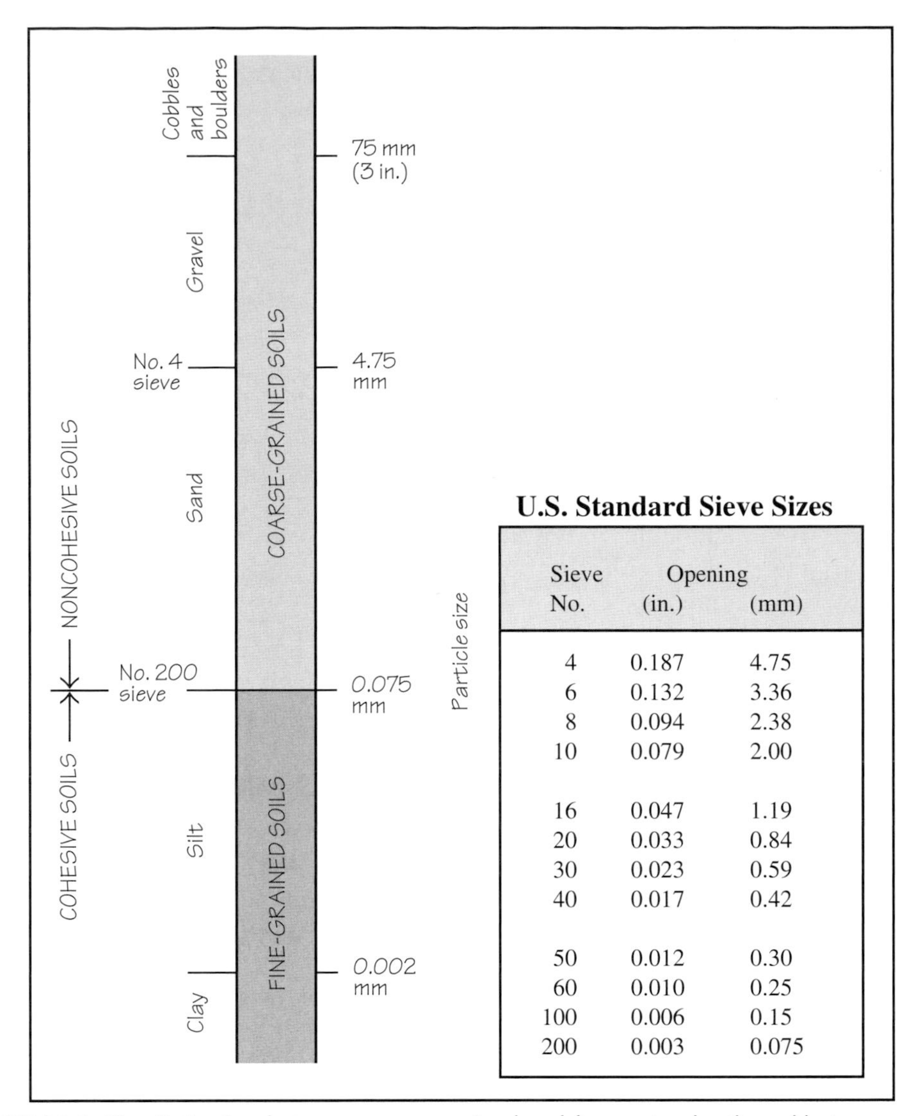

U.S. Standard Sieve Sizes

Sieve No.	Opening (in.)	Opening (mm)
4	0.187	4.75
6	0.132	3.36
8	0.094	2.38
10	0.079	2.00
16	0.047	1.19
20	0.033	0.84
30	0.023	0.59
40	0.017	0.42
50	0.012	0.30
60	0.010	0.25
100	0.006	0.15
200	0.003	0.075

FIGURE 21.1 The distinction between coarse-grained and fine-grained soils and between gravels, sands, silts, and clays. The term cobbles is used for pieces of naturally rounded rocks between 3 and 12 in. Boulders describe similar material greater than 12 in. For U.S. standard sieve sizes, see [1].

Between silts and clays, clayey soils are more cohesive. It is due to their cohesiveness that a potter can mold objects from wet clay, and excavated (cut) faces in a clayey soil can remain vertical without collapsing for a greater depth than other soil types. Cut faces in noncohesive soils are sloped for stability or shored by a support system.

ORGANIC SOILS

Soil consisting mainly of organic matter is called *organic soil.* The constituents of organic soil are fully or partially decayed plant matter, called *peat,* with little or no mineral constituents. Organic soils are highly compressible and unsuitable for building foundations. They are typically dark in color and have a characteristic odor of decay. Additionally, they may have a fibrous texture due to bark, leaves, grass, branches, or other fibrous vegetable matter.

Organic soils form the top layer of soil and are typically removed before construction begins. If the soil characteristics are desirable, they are often stockpiled and redistributed after construction is complete.

21.2 SOIL SAMPLING AND TESTING

Field exploration of below-ground soil conditions is an important first step before designing a building's foundations. The objectives of this exploration and sampling are to determine the

- *Engineering properties of soil* at various depths, such as shear strength, unconfined compression, triaxial compression, and so on, that help determine the soil's bearing capacity.
- *Particle-size distribution of soil* to assess the stability of cut faces, required excavation support, and the soil's drainage characteristics.

- *Nature of excavation* that will suit the soil. (For instance, excavation in rock is more complicated and expensive than excavating in an ordinary soil.)
- *Depth of water table* to plan for dewatering of excavations (if required) and its effect on existing construction.
- *Compressibility of soil* to assess foundation settlement.

Two methods are generally used for field exploration; (a) the test pit method and (b) the test boring method.

TEST PIT METHOD

The test pit method consists of digging pits or trenches that are large enough for visual inspection of the soil and for procuring samples for laboratory testing by getting inside the pit or trench. It is considered to be the more reliable method because it allows direct inspection and assessment of several soil properties of the undisturbed soil through field tests with simple instruments.

There are several limitations to the test pit method. It is more expensive, and the cost increases substantially as the depth of pit increases. Also, if the water table is high, the depth of inspection is limited.

TEST BORING METHOD

The test boring method, which is more commonly used, involves obtaining soil samples by boring into the ground using a truck-mounted, power-driven hollow stem, fitted at the end either with a cutting bit or a fluted drilling auger. A typical drilling operation is shown in Figures 21.2.

When the boring reaches the depth at which a sample is required, the cutting bit is withdrawn and a standard steel sampling tube (called a *Shelby tube*) is pushed into the soil through its full height and the core sample is removed. (A tube that opens up into two semicircular halves, called a s*plit-spoon,* is also used instead of the Shelby tube. From a rocky soil, rock cores are cut out.)

(a)

(b)

FIGURE 21.2 **(a)** A truck-mounted drilling machine that bores a hole in the ground with a cutting tip at the end of a hollow pipe stem. During the drilling operation, compressed air is pushed through the hollow stem to suck the drilled soil out to keep the boring clean. **(b)** A hollow pipe stem with a typical cutting bit.

Soil specimens obtained from sampling tubes are wrapped in plastic, labeled with the required information, and transported to the testing laboratory, Figure 21.3. Several test borings are generally required across a large site, and the data obtained from various borings is coordinated. The boring method also allows determination of soil's density and strength by impacting the bottom of boring with a standard amount of impact energy. The test, called the *standard penetration test,* is performed at several elevations in a boring.

Laboratory Testing of Soil Samples

Soil samples obtained from pits or borings are subjected to a number of laboratory tests. The sieve analysis determines the particle-size distribution of the soil. This is used to classify the soil (see "Principles in Practice" at the end of this chapter). It also provides important information about several engineering properties of the soil. Sieve analysis is performed on an oven-dried part of the specimen by placing the sample on the uppermost sieve of a sieve stack, consisting of sieves of different sizes, Figure 21.4.

Other tests include determining the soil's moisture content, dry density, liquid limit, plastic limit, compressive strength (unconfined, i.e., axial compression, and confined, i.e., triaxial compression), shear strength, and so on.

The data obtained during boring—from visual inspection of the soil, a standard penetration test, water table elevation, and so on—are immediately recorded on the testing laboratory's standard data sheet, referred to as the *boring field log.* After subsequent laboratory testing of samples, a final *boring log* is prepared. Each boring, therefore, results in its own boring log. One such log is shown in Figure 21.5.

FIGURE 21.3 Soil samples obtained from borings through the use of a sampling tube are wrapped in plastic, labeled with date of sampling, location of boring at the site, depth, and so on, and sent to the soil testing laboratory.

21.3 EARTHWORK AND EXCAVATION

Site preparation in advance of construction generally includes

- Fencing the site from adjacent public or private property
- Locating and marking existing utility lines so that they will not be damaged during construction
- Demolishing unneeded existing structures and utility lines
- Removing unneeded trees, shrubs, topsoil, and extraneous landfill, if present

Excavation is the first step of construction. It refers to the process of removing soil or rock from its original location, typically in preparation for constructing foundations, base-

NOTE

Standard Penetration Test

The standard penetration test consists of impacting the sampling tube with a hammer that delivers a standard amount of energy with each blow. The number of blows required to drive the sampling tube 12 in. into the soil is called *standard penetration resistance,* N.

The value of N is used as a rough index of the soil's bearing capacity, particularly for sandy soils. For a precise determination of bearing capacity, other soil properties must be used along with the value of N.

(a)

(b)

FIGURE 21.4 **(a)** A typical stack of sieves used to determine grain-size distribution of oven-dried soil specimens. The bottom of the stack is not a sieve but a container. The sieve immediately above the container is generally a No. 200 sieve. The sieve stack is shaken per the standard procedure, either manually or by placing it on a motorized shaker. The amount of soil retained on each sieve is weighed. The amount obtained from the bottom container is the amount that passes No. 200 sieve. A graph showing the percentages of soil passing through various sieves is plotted. **(b)** After performing sieve analysis for a specimen, the sieves are cleaned and restacked for analyzing the next specimen. (Photos courtesy of the Geotechnical Laboratory of the University of Texas at Arlington.)

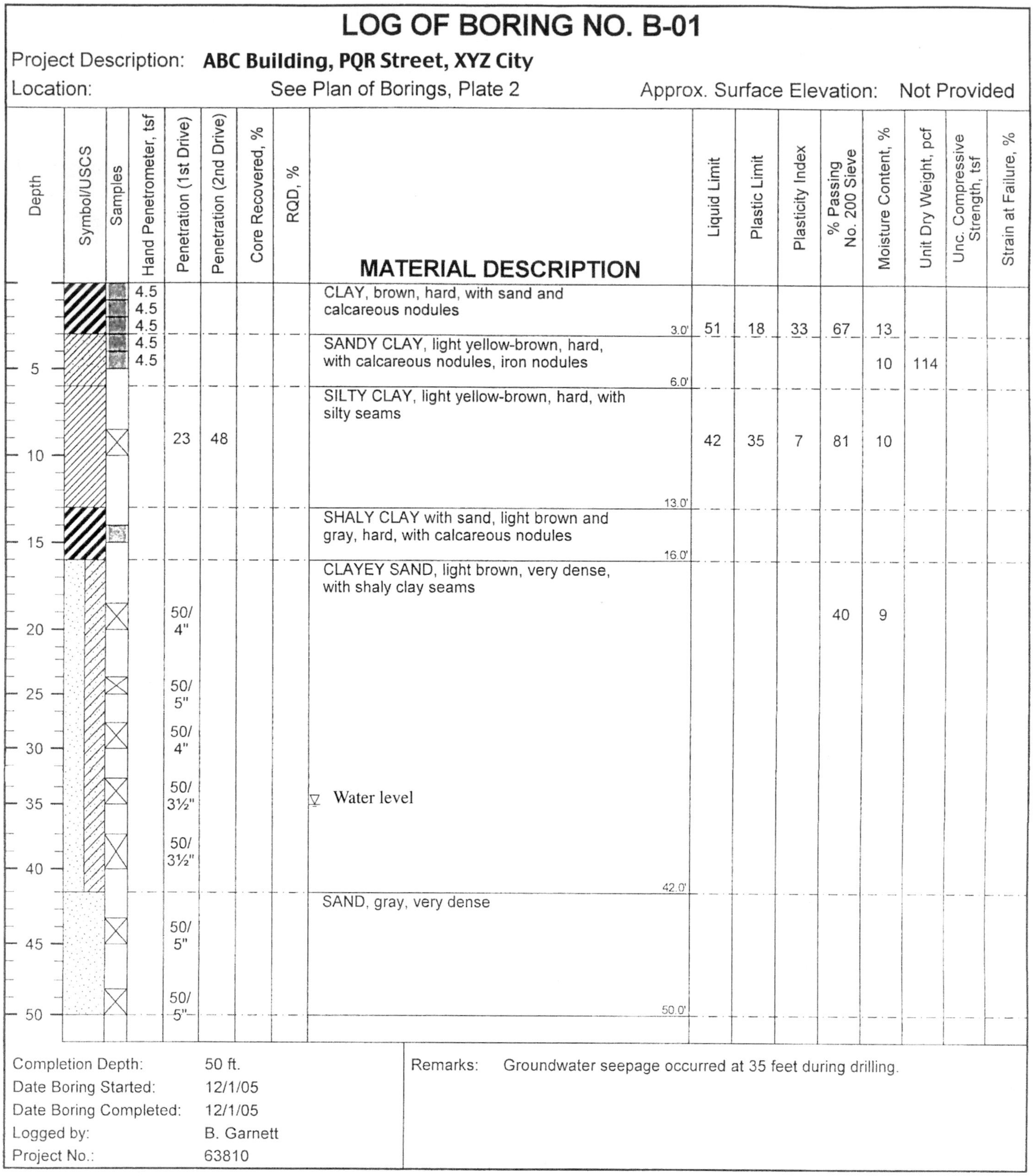

LOG OF BORING NO. B-01

Project Description: **ABC Building, PQR Street, XYZ City**

Location: See Plan of Borings, Plate 2 — Approx. Surface Elevation: Not Provided

Depth	Symbol/USCS	Samples	Hand Penetrometer, tsf	Penetration (1st Drive)	Penetration (2nd Drive)	Core Recovered, %	RQD, %	MATERIAL DESCRIPTION	Liquid Limit	Plastic Limit	Plasticity Index	% Passing No. 200 Sieve	Moisture Content, %	Unit Dry Weight, pcf	Unc. Compressive Strength, tsf	Strain at Failure, %
0–3.0'			4.5 4.5 4.5					CLAY, brown, hard, with sand and calcareous nodules	51	18	33	67	13			
3.0'–6.0'			4.5 4.5					SANDY CLAY, light yellow-brown, hard, with calcareous nodules, iron nodules					10	114		
6.0'–13.0'				23	48			SILTY CLAY, light yellow-brown, hard, with silty seams	42	35	7	81	10			
13.0'–16.0'								SHALY CLAY with sand, light brown and gray, hard, with calcareous nodules								
16.0'–42.0'				50/ 4"				CLAYEY SAND, light brown, very dense, with shaly clay seams				40	9			
25				50/ 5"												
30				50/ 4"												
35				50/ 3½"				▽ Water level								
40				50/ 3½"												
42.0'–50.0'				50/ 5"				SAND, gray, very dense								
50				50/ 5"				50.0'								

Completion Depth: 50 ft.
Date Boring Started: 12/1/05
Date Boring Completed: 12/1/05
Logged by: B. Garnett
Project No.: 63810

Remarks: Groundwater seepage occurred at 35 feet during drilling.

FIGURE 21.5 A typical boring log. Courtesy of Kleinfelder, Fort Worth, Texas. (See "Principles in Practice" at the end of this chapter for an explanation of some of the terms used in this log.)

ments, and the grading of the ground. Excavated material required for backfill or fill is stockpiled on the site for subsequent use. Unneeded material is removed from the site for appropriate disposal. Excavations are generally classified as

- Open excavations
- Trenches
- Pits

Open excavations refer to large (and often deep) excavations, such as, for a basement. *Trenches* generally refer to long, narrow excavations, such as, for footings under a wall. *Pits* are excavations for the footing of an individual column, elevator shaft, and so on.

(a) Excavator and hauler

(b) A small excavator as a trencher

(c) Front-end loader spreads and distributes the soil.

(d) Loader and backhoe

(e) Compactor

(f) Mini compactor (rammer) generally used for compacting small areas

FIGURE 21.6 Some of the various equipment used for earthwork (excavation, hauling, and compaction).

NOTE

Backfill, Fill, Select Fill, and Engineered Fill

Backfill refers to the material excavated from the site that is used to fill the excavated trench or pit after constructing the foundations or basement. Unless the excavated material is unsuitable, backfill involves returning the excavated material to its original location.

Fill is soil that is used for grading work to provide surface drainage, landscaping, erosion control, and so on, as per the site design. Generally, the material not used as backfill is used as fill, unless it is unsuitable. *Select fill* is a soil with required properties that is imported to a given site, placed, and compacted to give the desired result. For example, a select fill may consist of a coarse-grained, stable soil to replace the existing expansive (clayey) soil at a site.

Engineered fill is soil that is specially prepared, placed, compacted, and supervised per the geotechnical engineer's design and specifications. For example, it may be soil mixed with crushed angular aggregate of a required grading to give a controlled void space to increase stormwater retention and reduce runoff.

The depth of excavation depends on the type of foundation. For a slab-on-ground, the depth of excavation is small, whereas the depth of excavation for a basement is directly related to the number of basement floors. For trenches and pits, the excavation must reach the soil with sufficient bearing capacity to support the load. If the soil with the required bearing capacity is not present at a reasonable depth, deep foundation elements, which do not require conventional excavation (Section 21.8), are used.

Excavation requires various types of power equipment, such as excavators, compactors, and earth-moving equipment (front-end loaders and backhoes), some of which are shown in Figure 21.6.

OPEN EXCAVATIONS

Excavated (cut) faces generally require some temporary support to prevent cave-ins while the foundation system or basement walls are constructed. The simplest support system is to

FIGURE 21.7 An example of a self-suporting open excavation for a basement.

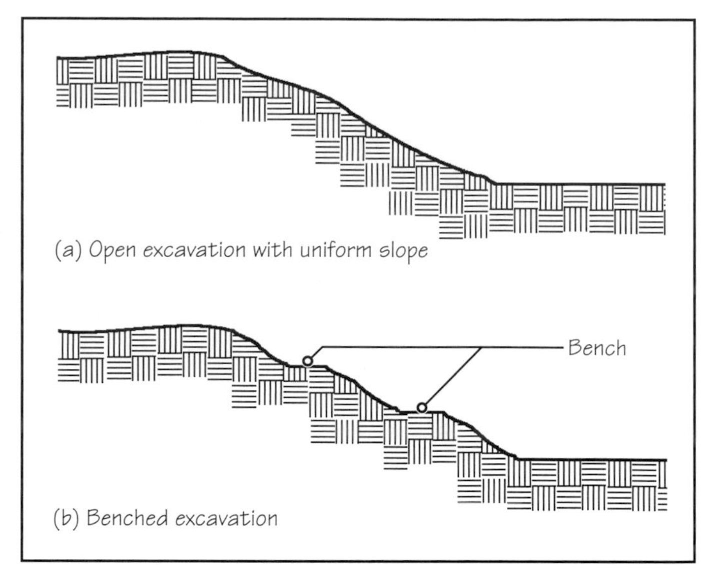

FIGURE 21.8 Two alternatives used for open excavations.

provide adequate slope in the cut face so that it is able to support itself, Figure 21.7. This is feasible only if the site is large enough to accommodate sloped excavations. Because the amount of excavated material is large, sloped excavations also require a larger site area for stockpiling the material until used as fill or backfill.

The maximum natural slope at which a soil will support itself must be determined from soil investigations. Excavation in coarse-grained soils require shallower slope than in fine-grained soils. A sloped excavation may either be uniformly sloped or stair-stepped, called *benched excavations,* Figure 21.8. Benches increase slope stability. They are also easier to compact. Compacted benches further enhance slope stability.

PRACTICE QUIZ

Each question has only one correct answer. Select the choice that best answers the question.

1. Soil and rock consist essentially of the same matter.
  a. True
  b. False

2. In the commonly used soil classification system, *sand and silt* are classified as
  a. coarse-grained soils.
  b. fine-grained soils.
  c. organic soils.
  d. inorganic soils.
  e. none of the above.

3. The four commonly used classifications of soils arranged in the order of decreasing particle size are
  a. gravel, sand, silt, and clay.
  b. gravel, silt, sand, and clay.
  c. gravel, silt, clay, and sand.
  d. gravel, sand, clay, and silt.
  e. none of the above.

4. Of the four commonly used classifications of soil particles
  a. clay particles have a unique shape.
  b. gravel particles have a unique shape.
  c. sand particles have a unique shape.
  d. silt particles have a unique shape.

5. Cohesive soils are those in which the dry soil particles tend to separate from each other when mixed with a small amount of water.
  a. True
  b. False

6. To obtain soil-sampling tubes from below ground, which soil testing method is commonly used?
  a. Test pit method
  b. Test boring method
  c. Test pressure method
  d. All the above
  e. None of the above

7. During soil testing, soil samples are obtained from
  a. one central location on the site but from different depths below ground.
  b. several locations on the site but from same depth below ground.
  c. several locations on the site, and from several depths below ground.
  d. none of the above.

8. All testing of soil samples is generally done at the site
  a. because modern soil sampling and testing equipment is fully equipped and automated.

b. because the delay in bringing soil samples to the laboratory falsifies test results.
c. because the vibration of soil samples caused by transportation falsifies test results.
d. all the above.
e. none of the above.

9. Engineered fill refers to the soil that is
a. specially formulated to provide the required properties.
b. placed per the geotechnical engineer's specifications.
c. compacted per the geotechnical engineer's specifications and inspection.
d. all the above.
e. none of the above.

10. A benched excavation is generally used
a. on open, suburban sites.
b. on tight, downtown sites.
c. where the excavation depth is less than 5 ft.
d. where dewatering of the site is not required.

Answers: 1-a, 2-e, 3-a, 4-a, 5-b, 6-b, 7-c, 8-e, 9-d, 10-a.

21.4 SUPPORTS FOR DEEP EXCAVATIONS

Self-supporting sloped excavations cannot be provided where the site area is restricted or adjoining structures are present. In these cases, the excavation must consist of vertical cuts. In cohesive soils, shallow vertical cuts (generally 5 ft or less in depth) may be possible without any support system. Deeper vertical cuts must be provided with a support system.

SHEET PILES

For depths of up to about 15 ft, vertical sheets of steel, referred to as *sheet piles,* can be driven into the ground before commencing excavations. Sheet piles consist of individual steel sections that interlock with each other on both sides. The interlocks form a continuous barrier to retain the earth. Structurally, sheet piles function as cantilevers and, therefore, must be buried into the soil for sufficient depth below the bottom of the excavation. (Sheet piles are also used as permanent barriers against water, e.g., harbor walls, flood-protection walls, etc.)

Sheet piles are available in many cross-sectional profiles. The most commonly used profile is a Z-section, Figure 21.9. The sections are driven into the ground one by one using either hydraulic hammers or vibrators, Figure 21.10.

For deeper excavations (generally greater than 15 ft), sheet piles are braced with horizontal or inclined braces or anchored with tiebacks, Figure 21.11. (Tiebacks used with sheet piles are similar to those shown in Figure 21.14.) Sheet piles are removed after they are no longer required or can be left in place if needed.

SOLDIER PILES, TIEBACKS, AND LAGGING

One of the disadvantages of sheet piles is the noise and vibration created in driving them, particularly in stiff soils where the vibratory method is ineffective and hydraulic hammers must be employed. The soldier piles and lagging system is an alternative to sheet piles and is frequently used in urban areas, Figure 21.12.

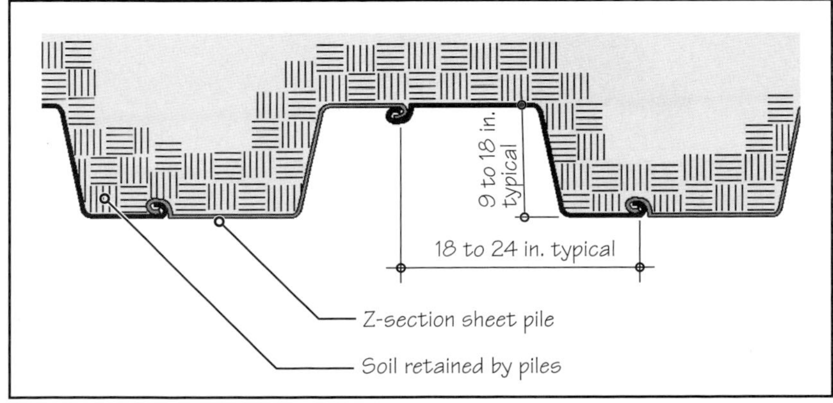

FIGURE 21.9 Steel sheet piles with a Z-shaped profile. Adjacent sections interlock with each other. Several other sectional profiles are also available.

FIGURE 21.10 Steel sheet piles being driven into the soil using a diesel pile driver. (Photo courtesy of R. I. Carr.)

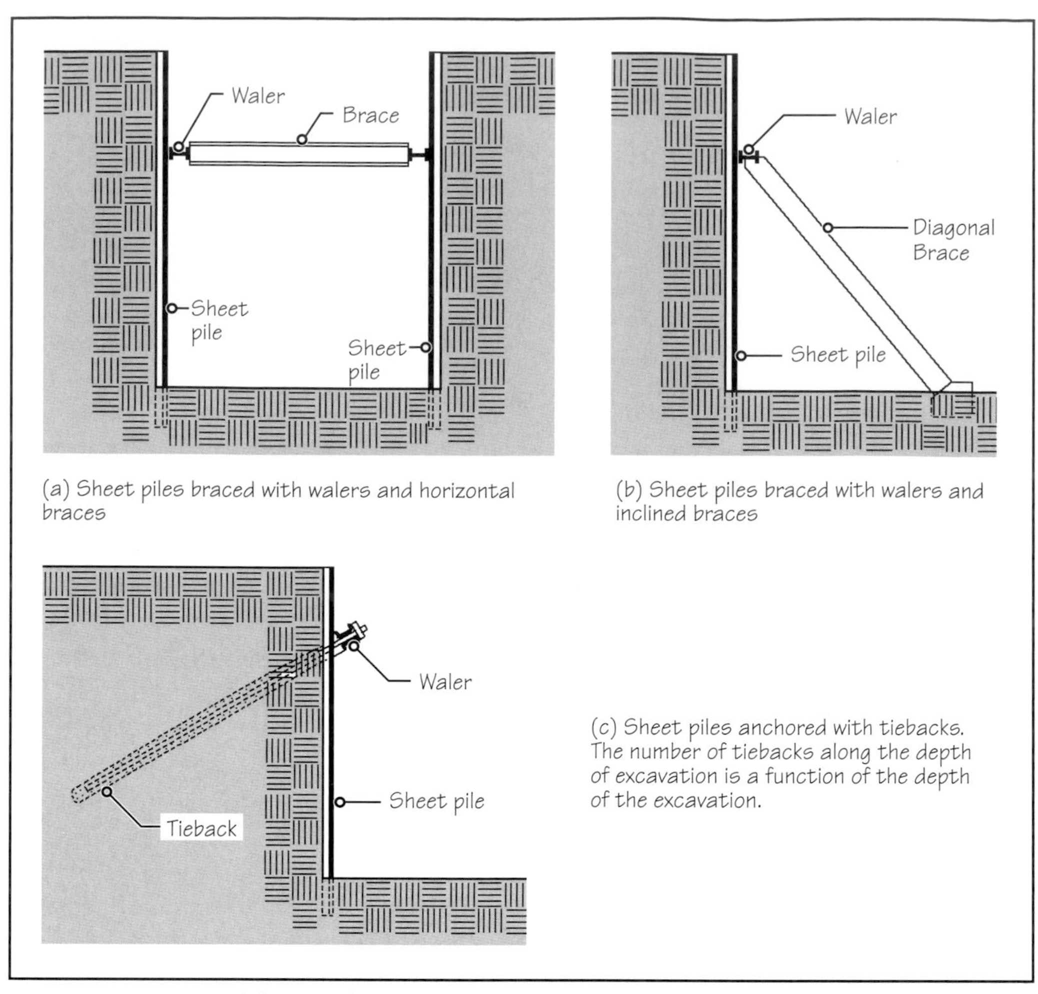

FIGURE 21.11 **(a)** and **(b)** Two alternatives of bracing sheet piles in deep excavations. **(c)** Anchored sheet piles See Figure 21.14 for tieback details.

FIGURE 21.12 A bird's eye view of cut faces for a deep basement supported by soldier piles and lagging.

The system consists of rough-sawn, pressure-treated horizontal lumber members, called *lagging*, and vertical steel members (W, HP, or twin C-sections), called *soldier piles*. In this system, soldier piles are placed in predrilled holes, approximately 6 to 10 ft on centers. After placing the piles, the holes are filled with lean concrete. Excavation is commenced after the concrete around the piles has gained sufficient strength.

As the excavation proceeds, piles are braced by inclined tiebacks, Figure 21.13. Drilling for tiebacks is done through the space between the twin C-sections of soldier piles, Figure 21.14.

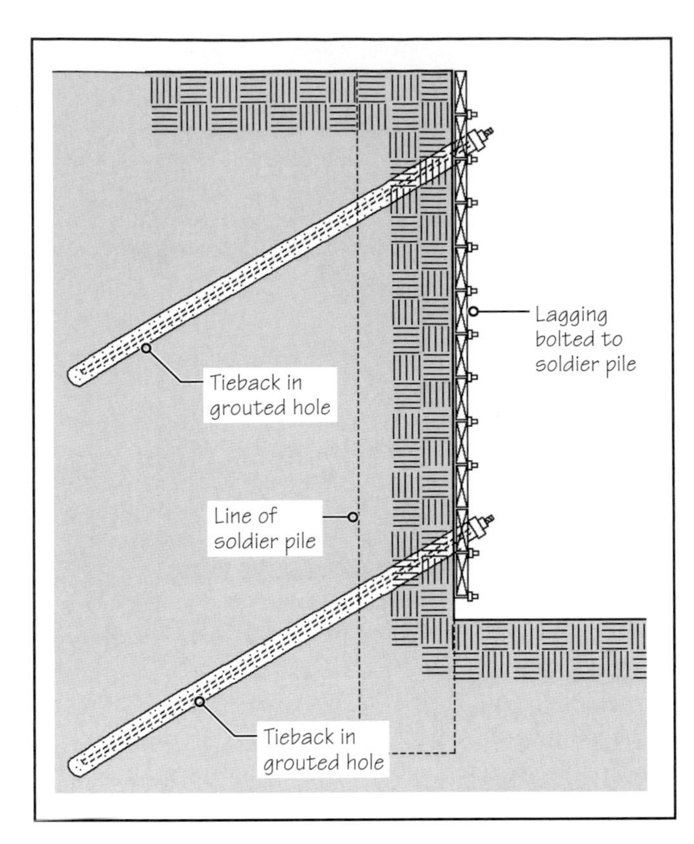

FIGURE 21.13 A section through an excavation supported by soldier piles and lagging. Tiebacks are installed as the excavation proceeds (see Figure 21.14). This is followed by bolting the lagging to soldier piles (see Figure 21.16). The number of tiebacks required is a function of the depth of excavation and the type of soil.

(a)

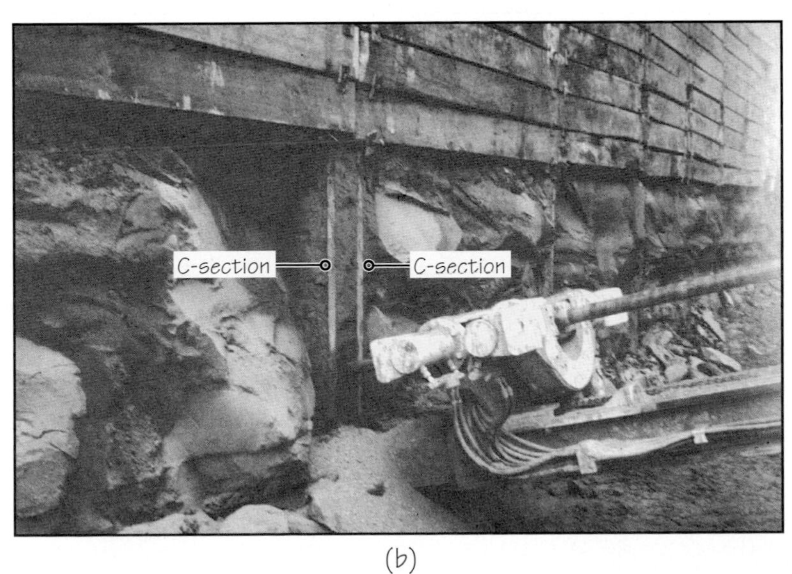

(b)

FIGURE 21.14 **(a)** Drilling of a tieback hole in progress. **(b)** Close-up of the same operation. Note that the hole is drilled in the space between the twin C-sections of the piles (see also Figures 21.15 and 21.16). The concrete around soldier piles is lean (weak) concrete to allow easier drilling of tiebacks.

FIGURE 21.15 After the tieback hole has been drilled, a high-strength steel tendon is pushed into the hole and the hole is pressure grouted with concrete. After the concrete has gained sufficient strength, the tendon is stressed (posttensioned) and then anchored to the pile.

After a tieback hole has been drilled, high-strength steel tendons are placed in the hole, and the hole is pressure grouted with high-slump concrete, Figure 21.15. After the grouted concrete has gained sufficient strength, the tendon is stressed (posttensioned) and anchored to the pile.

The attachment of lagging to piles involves (a) scraping the lean concrete around piles, (b) welding threaded bolts to piles, and (c) placing steel washers and nuts over adjacent laggings, Figure 21.16.

Whereas the sheet pile method of excavation support is relatively water resistant, the soldier pile and lagging system allows a much greater amount of water to percolate through the lagging. Therefore, the pile and lagging system is used in situations where water percolation is relatively small. In both cases, however, dewatering of the excavation is generally required to keep the excavation dry (Section 21.5).

Contiguous Bored Concrete Piles

In situations where the excavation is close to an adjacent building or the property line, tiebacks cannot be used. In this situation, closely spaced, reinforced concrete piles, called *contiguous bored piles* (CBPs), are often used, Figure 21.17. Each pile is made by screwing an auger into the ground. The auger, called a *continuous flight auger* (CFA), has a hollow stem in the middle of a continuous spiral drill.

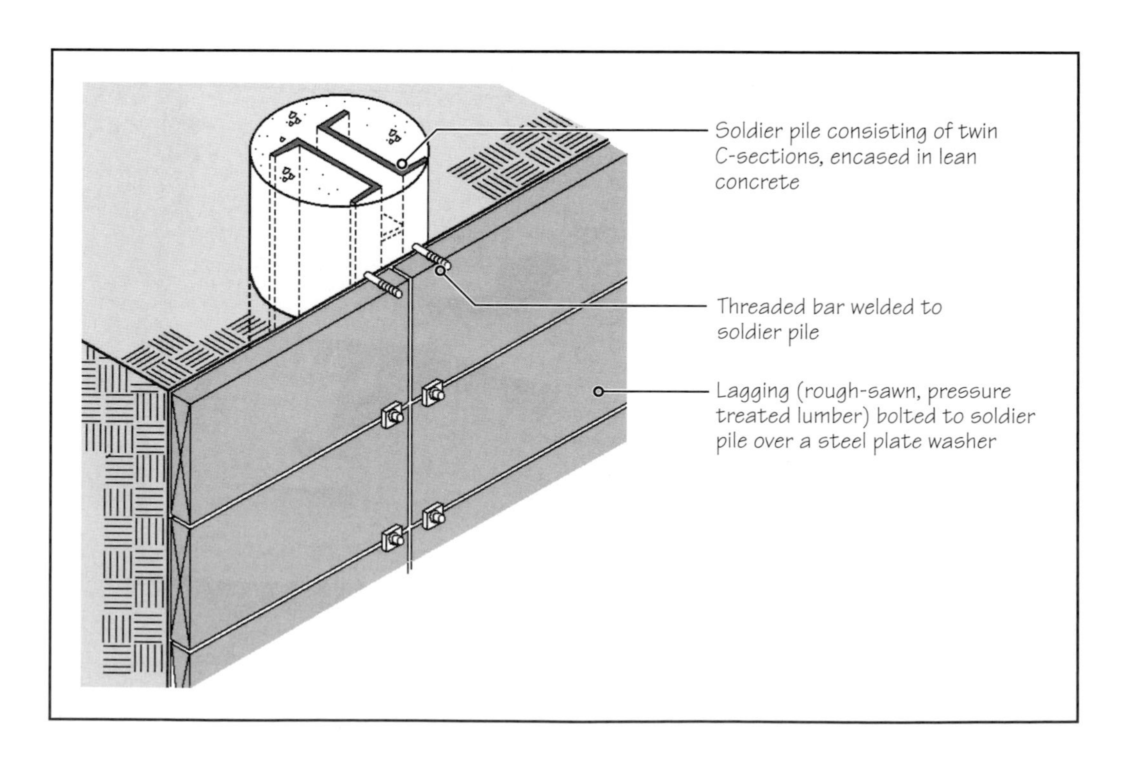

FIGURE 21.16 Connection of lagging with soldier piles.

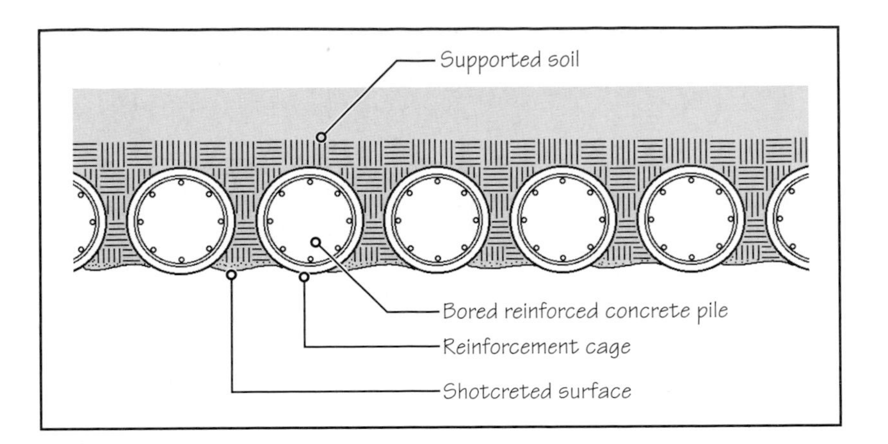

FIGURE 21.17 Closely speced drilled (bored) concrete piles, shown in plan.

Once the bottom drill has reached the required depth below the ground, high-slump concrete is pumped down the hollow stem of the auger to the bottom of the bore. Once the pumping starts, the auger is progressively withdrawn. The withdrawing auger brings the soil from the bore to the surface, where it is removed. Thus, the sides of the bore are supported at all times by the soil-filled augur or concrete. Immediately after the entire bore has been concreted, a reinforcement cage is lowered in the concrete-filled bore.

The depth of CBPs is limited by the length of the reinforcement cage that can be lowered through the concrete. (The reinforcement cage is specially designed to have adequate stiffness to penetrate into fresh concrete.) CBPs up to 100 ft deep have been constructed. They are generally 18 to 36 in. diameter, depending on the depth of the excavation.

After the piles have been constructed and the excavation proceeds, the piles and the small soil spaces between them are shotcreted to reduce water percolation through the cut face, Figure 21.18.

Secant Piles

A major shortcoming of CBPs is the gaps between piles and the consequent lack of water resistance of the excavation support. This is overcome by the use of the modified version of CBPs called *secant piles.*

Secant piles essentially consist of two sets of interlocking contiguous piles. The first set, called the *primary piles,* is bored and concreted in the same way as the CBPs. The center-to-center distance between the primary piles is slightly smaller than twice their diameter.

After constructing the primary piles, the *secondary piles* are bored at middistance between the primary piles, which also bores through a part of the primary piles, Figure 21.19(a). The boring for secondary piles is done before the concrete in the primary piles has gained full strength to reducc wear on the blades of the auger. The secondary piles are concreted

FIGURE 21.18 Contiguous bored piles used as supports for a deep basement excavation in a tight downtown location. In areas farther away from the existing building or property line, the soldier pile and lagging system is used. Note that as the excavation proceeds, the lower part of the drilled concrete piles will also be shotcreted, as shown here for the upper part.

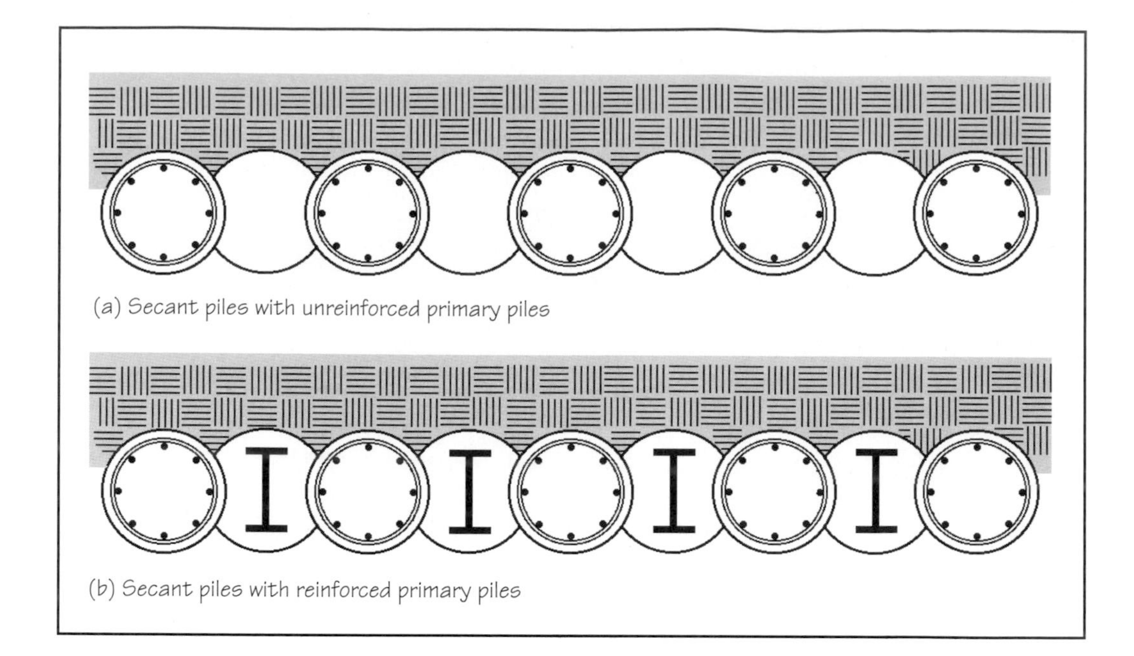

FIGURE 21.19 Excavation support through secant piles.

and reinforced in the same way as the CBPs. In most situations, only the secondary piles are reinforced, Figure 21.19(a). This precludes accidental cutting of any misaligned reinforcement in the primary piles. Thus, with nonreinforced primary piles, the primary piles provide a water-resistant layer between the secondary piles.

Where greater bending strength of the support system is required, the primary piles are also reinforced, generally with a wide-flange steel section, and the secondary piles are reinforced with a circular steel reinforcement cage, Figure 21.19(b).

REINFORCED CONCRETE WALL USING BENTONITE SLURRY AS TRENCH SUPPORT

Another excavation support system, commonly used in situations where the underground water table is relatively high, is a reinforced concrete wall. Construction of such walls is done by excavating 10-ft- to 15-ft-long discontinuous trench sections down to bedrock, called *primary panels.* The width of trench sections is the required thickness of concrete wall. So that the soil does not collapse, the trench is continuously kept filled with bentonite slurry as the excavation proceeds. (Bentonite slurry is a mixture of water and bentonite clay, which pressurizes the walls of the trench sufficiently to prevent their collapse during excavation.)

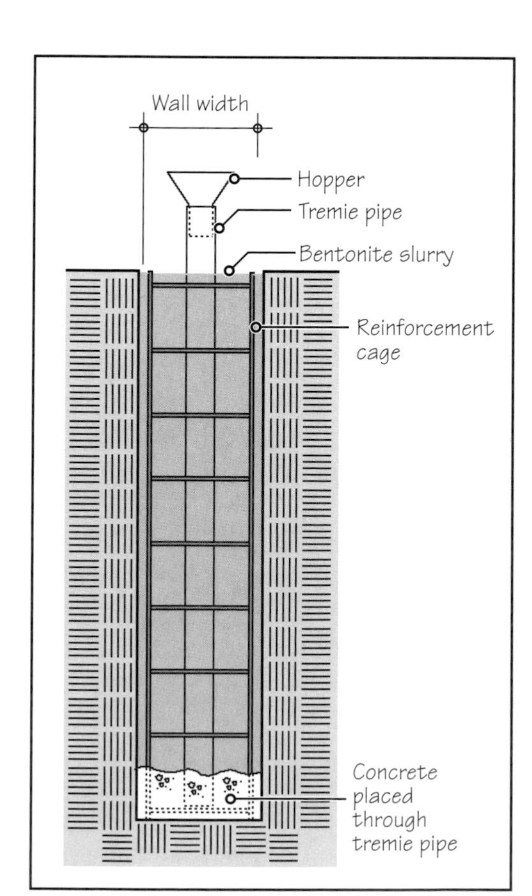

FIGURE 21.20 Trench support through bentonite slurry.

Special excavation equipment is used to extract soil through the slurry-filled trench. After the excavation for the entire primary panel is complete, a reinforcement cage is lowered in the trench. Concrete is then placed in the trench panel using two or more *tremie pipes,* typically one at each end of the panel. Concrete is placed from the bottom up, and the discharge end of tremie is always buried in concrete. A tremie pipe is generally an 8-in.- to 10-in.-diameter steel pipe with a hopper at the top, Figure 21.20.

As concreting proceeds, the slurry is pumped out from the top of the trench and stored for later use. After the primary panels have been constructed, excavation for secondary panels (between the primary panels) is undertaken in the same way as for the primary panels. To provide shear key and water resistance between primary and secondary panels, a steel pipe is embedded at the end of each primary panel prior to its concreting. These pipes are removed after the concrete in the primary panels has gained sufficient strength.

The tremie pipe method of concrete placement requires great care and expertise, particularly the initial placement of concrete, which is generally a richer mix. It also must be placed slowly so that it does not get too diluted by the slurry.

SOIL NAILING

Soil nailing is a means of strengthening the soil with closely spaced, almost horizontal steel bars that increase the cohesiveness of the soil and prevent the soil from shearing along an inclined plane. In other words, the steel bars connect imaginary, inclined

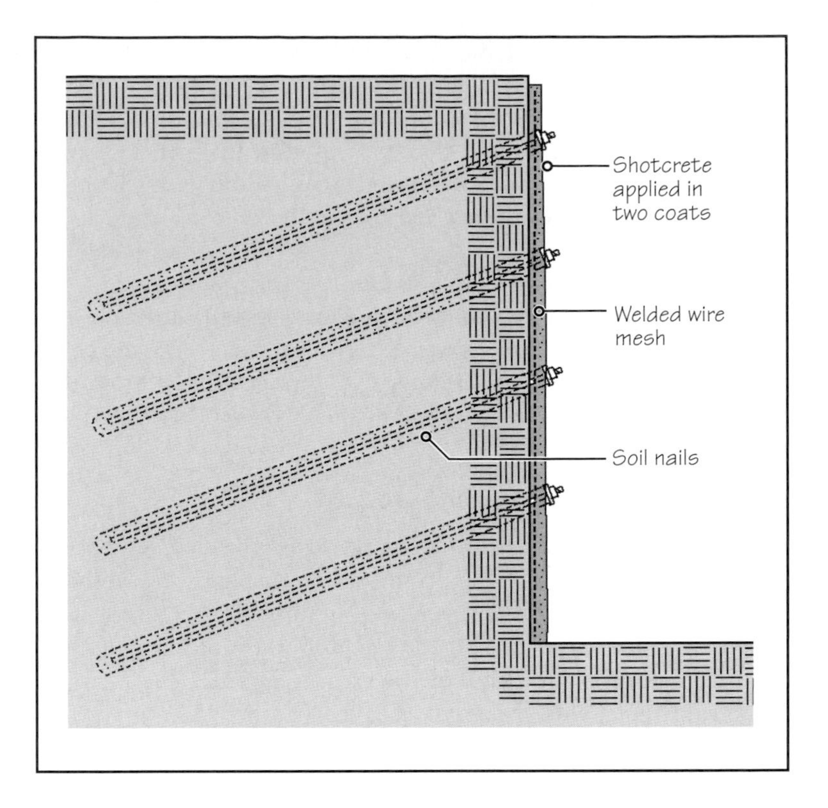

FIGURE 21.21 Section through a soil-nailed excavation support.

layers of the earth into a thick block that behaves as a gravity-retaining wall when excavated. The process of soil nailing consists of the following steps:

- The soil is first excavated 5 to 7 ft deep, depending on the ability of the cut face to remain vertical without supports.
- Holes are drilled along the cut face at 3 to 4 ft on centers so that one hole covers approximately 10 to 15 ft^2 of the cut face.
- Threaded steel bars (approximately 1 in. in diameter) are inserted in the holes. The length of bars is a function of the soil type but is approximately half the final depth of excavation. The bars protrude a few inches out of the holes.
- The holes are grouted with concrete.
- Welded wire reinforcement is placed over the wall and tied to the protruding bars.
- A layer of shotcrete is applied to the mesh.
- Plates and washers are inserted in the protruding bars and locked in position with a nut.
- A second layer of shotcrete is applied, Figure 21.21.
- These steps are repeated with the next depth of cut.

Figure 21.22 shows a soil-nailed excavation. Note that unlike tiebacks, soil nails are not posttensioned. Additionally, soil nailing can be used for both temporary and permanent excavations.

FIGURE 21.22 View of a soil-nailed excavation. (Photo courtesy of Schnabel Foundation Company.)

21.5 KEEPING EXCAVATIONS DRY

It is important to keep excavations free from groundwater during excavation. Even a small amount of water logging in excavated areas impedes further excavation, is unsafe for workers, and is harmful to equipment and machines. Additionally, if the amount of groundwater is reduced, so is the pressure exerted by the soil on excavation supports—sheet piles, soldier piles, and so on.

Groundwater control in an excavation consists of two parts: (a) preventing surface water from entering the excavation through runoff and (b) draining (dewatering) the soil around the excavations so that the groundwater level falls below the elevation of proposed excavation. Two commonly used methods of dewatering the ground are *sump pumps* and *well points.*

DEWATERING THROUGH SUMPS

Sump dewatering consists of constructing pits (called *sumps*) within the enclosure of the excavation. The bottom of sumps must be sufficiently below the final elevation of the excavation. As the groundwater from surrounding soil percolates into the sump, it is lifted by automatic pumps and discharged away from the building site, Figure 21.23. The number of required sumps is a function of the excavation area. The method of discharge of pumped water must meet with the approval of local authorities.

DEWATERING THROUGH WELL POINTS

Sump dewatering works well in cohesive soils (where the percolation rate is slow) and where the water level is not much higher than the final elevation of the base of excavation. A more effective dewatering method uses forced suction to extract groundwater. This works by sinking a number of vertical pipes with a screened end at the bottom (called *well points*) around the perimeter of the excavation. The well points reach below the floor of the excavation and are connected to large-diameter horizontal header pipes at the surface.

The header pipe is connected to a vacuum-assisted centrifugal pump that sucks water from the ground for discharge to an appropriate point. For a very deep excavation, two rings of well points may be required. The well points in the ring away from the excavation terminate at a higher level than those close to the excavation, Figure 21.24.

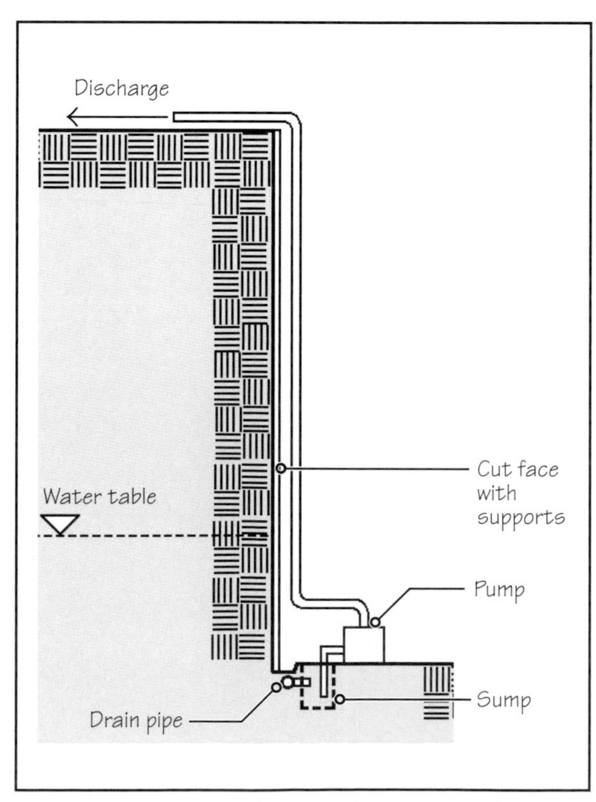

FIGURE 21.23 Sumps for ground water control in an excavation.

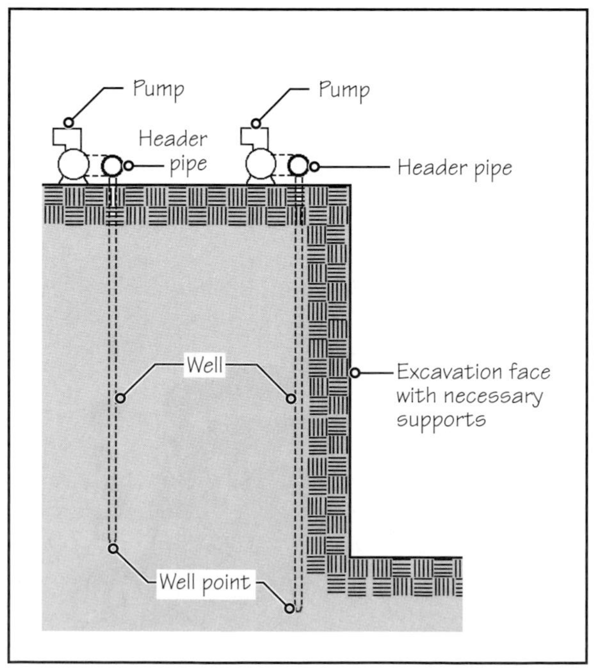

FIGURE 21.24 Section through an excavation showing two rings of well points.

Whereas the sump method of dewatering does not greatly affect the existing water table, dewatering by well points can lower the water table considerably. The effect of this on the adjoining buildings must be considered because it can cause consolidation and settling of foundations of existing buildings on some types of soils.

Dewatering of excavations can be fairly complicated and generally requires an expert dewatering subcontractor for large and complicated operations.

PRACTICE QUIZ

Each question has only one correct answer. Select the choice that best answers the question.

11. Sheet piles are used
- **a.** as shallow foundations in buildings.
- **b.** as deep foundations in buildings.
- **c.** as formwork for concrete walls.
- **d.** as excavation supports.
- **e.** none of the above.

12. Sheet piles are generally made of
- **a.** plywood.
- **b.** steel.
- **c.** aluminum.
- **d.** cast iron.

13. In the soldier pile and lagging system, soldier piles consist of
- **a.** sheet steel.
- **b.** sheet aluminum.
- **c.** structural steel sections.
- **d.** reinforced concrete.
- **e.** precast concrete.

14. In the soldier pile and lagging system, soldier piles are oriented
- **a.** horizontally.
- **b.** vertically.
- **c.** both horizontally and vertically.
- **d.** inclined at about 45° to the horizontal.
- **e.** none of the above.

15. In the soldier pile and lagging system, lagging generally consists of
- **a.** pressure-treated lumber.
- **b.** pressure-treated plywood.
- **c.** structural steel sections.
- **d.** sheet steel.
- **e.** none of the above.

16. Tiebacks used with the soldier pile and lagging system are constructed from
- **a.** prestressing tendons in grouted holes.
- **b.** structural steel sections.
- **c.** reinforcing steel bars in grouted holes.
- **d.** prestressing tendons and reinforcing steel bars.
- **e.** none of the above.

17. Contiguous bored piles are constructed using
- **a.** structural steel sections.
- **b.** site-cast reinforced concrete.
- **c.** precast concrete.
- **d.** precast, prestressed concrete.
- **e.** none of the above.

18. In terms of design and construction, secant piles are related to
- **a.** sheet piles.
- **b.** soldier piles.
- **c.** contiguous bored piles.
- **d.** none of the above.

19. The center-to-center distance between secant piles is generally
- **a.** 5 to 8 ft.
- **b.** 8 to 12 ft.
- **c.** 10 to 15 ft.
- **d.** 15 to 20 ft.
- **e.** none of the above.

20. Soil nailing refers to
- **a.** a cut face of excavation supported by closely spaced steel bars or angles.
- **b.** a cut face of excavation supported by closely spaced steel bars or angles and shotcreted.
- **c.** a cut face of excavation supported by closely spaced steel bars or angles and shotcreted over welded wire reinforcement.
- **d.** a cut face of excavation supported by closely spaced steel bars or angles in grouted holes and shotcreted over welded wire reinforcement.
- **e.** none of the above.

21. Dewatering of an excavation, as described in this text, can be done by the
- **a.** pit method.
- **b.** sump method.
- **c.** well-point method.
- **d.** (a) and (b).
- **e.** (b) and (c).

Answers: 11-d, 12-b, 13-c, 14-b, 15-a, 16-a, 17-b, 18-c, 19-e, 20-d, 21-e.

21.6 FOUNDATION SYSTEMS

A building may be considered to comprise a

- Superstructure
- Substructure
- Foundation system

In this classification, the *superstructure* generally refers to the part of the building that is above the ground level. The *substructure* refers to basements, if provided, and the *foundation system* refers to the structure that lies below the substructure. Where a basement does not exist, a building has only a superstructure and foundations.

SHALLOW AND DEEP FOUNDATIONS

Foundations are further classified as being either shallow or deep. *Shallow foundations* are used for all types of buildings. They extend a relatively short distance below ground and bear directly on the upper soil levels, Figure 21.25.

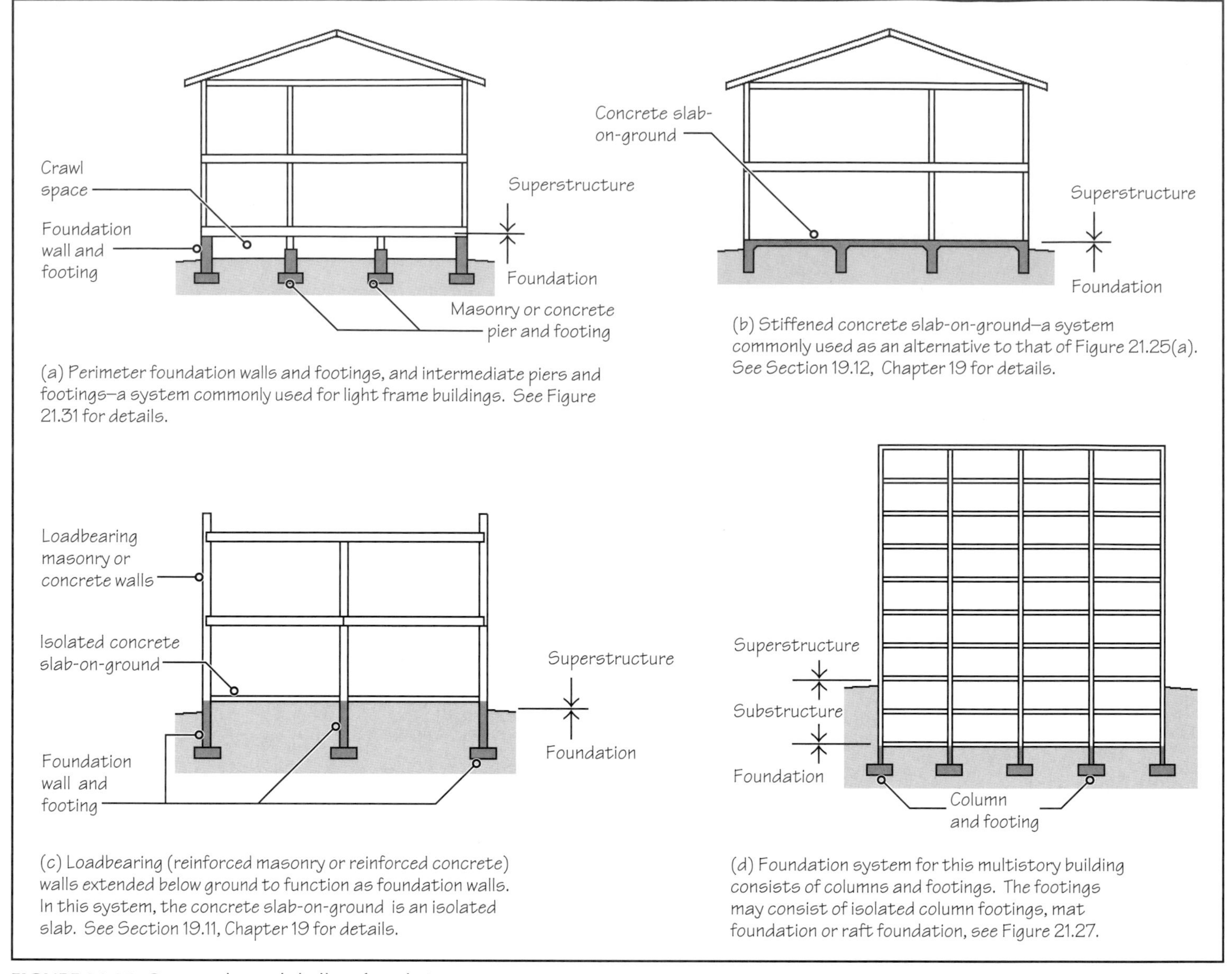

FIGURE 21.25 Commonly used shallow foundation systems.

Deep foundations are used where soil of adequate strength is not available close enough to the surface for the use of shallow foundations. They are also used where the soil near the surface is unstable (subject to swelling and shrinkage). Deep foundations consist of either piles or piers, Figure 21.26.

21.7 SHALLOW FOUNDATIONS

Except for a concrete slab-on-ground, all shallow foundations consist of footings. A footing is wider than the wall or column it supports. This is because the soil is weaker than the material used in walls or columns. Thus, the primary purpose of a footing is to distribute the superimposed load on a large area of the soil. Most footings are constructed of reinforced concrete, but plain concrete footings may be used for lightly loaded walls.

Types of Footings

As shown in Figure 21.27, footings are generally classified as follows:

- *Wall footings,* also called *strip footings,* are commonly used for load-bearing wood, masonry, or concrete walls, Figure 21.27(a).
- *Isolated* (independent) *column footings* are used where columns are lightly loaded or bear on soils with a high bearing capacity, Figure 21.27(b). Sometimes, particularly with steel columns, a stepped footing (with a pedestal) is used to save concrete, Figure 21.27(c).

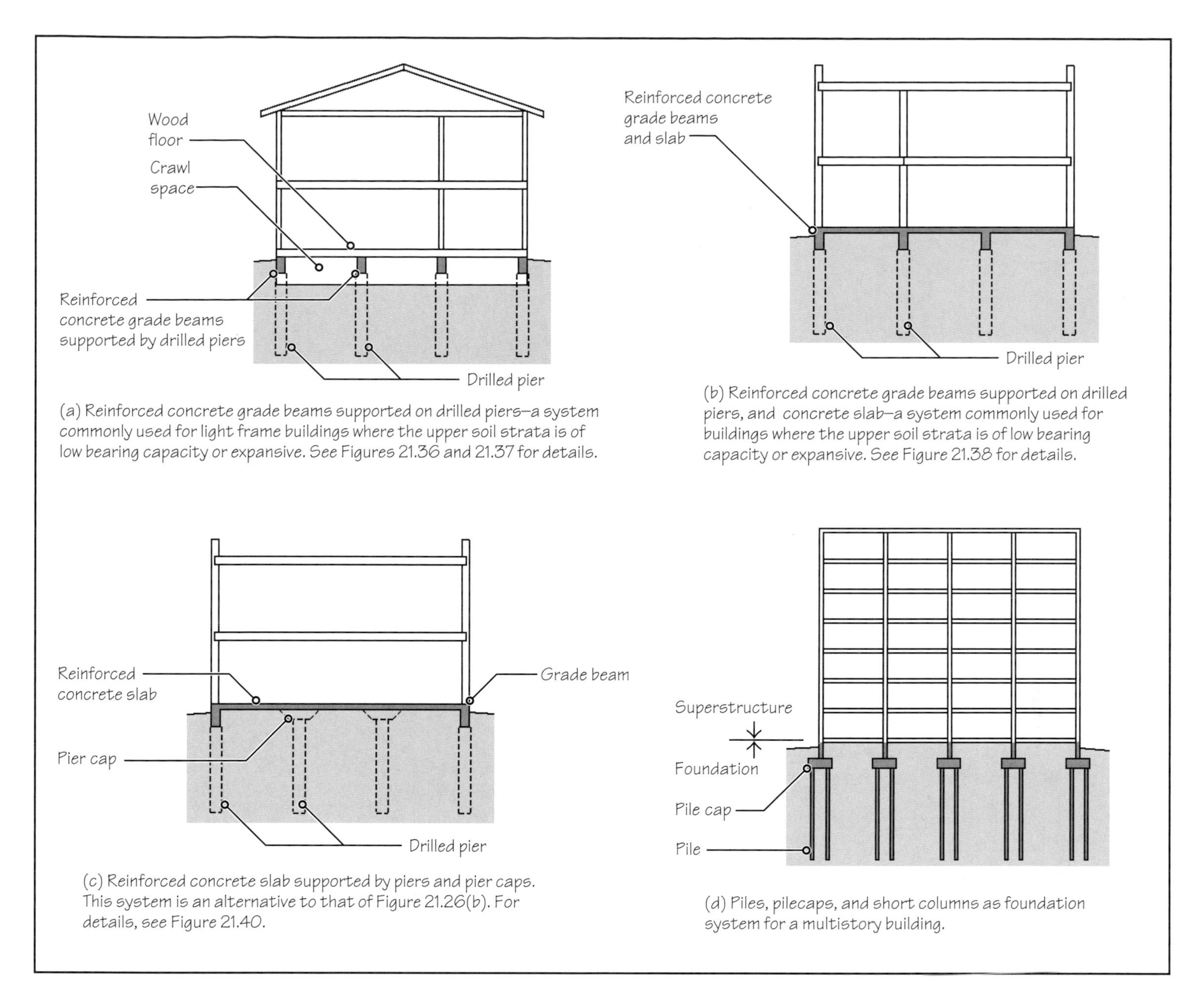

FIGURE 21.26 Commonly used deep foundation systems.

- A *combined footing is* a combination of two isolated column footings. They are generally used when a column must be placed on a property line or adjacent to an existing building, Figure 21.27(d).

  An alternative to a combined footing is a *strap footing,* Figure 21.27(e), or a *cantilevered footing,* Figure 21.27(f), in which a beam is used between the two isolated footings.

  A combined footing is also used when two or more adjacent columns are closely spaced and heavily loaded. Combining them into one footing saves excavation cost and distributes the load on a larger area.
- In a *mat footing* (also called *mat foundation*), all columns and walls of a building bear on one large and thick (usually several feet thick) reinforced concrete slab, Figure 21.27(g). It is used where the bearing capacity of the soil is low (even at a large depth below ground), so that independent column footings become so large that only small unexcavated areas are left between footings. If the excavation required for isolated footings in a building is more than 50% of the footprint of the building, it is generally more economical to use a mat foundation.
- A *raft,* or *floating, foundation* is a type of mat foundation. It consists of a hollow mat formed by a grid of thick reinforced concrete walls between two thick reinforced concrete slabs, Figure 21.27(h). In this case, the weight of the soil excavated from the ground is equal to the weight of the entire building, so that the pressure on the soil is unchanged from the original condition, making the building float on the soil.

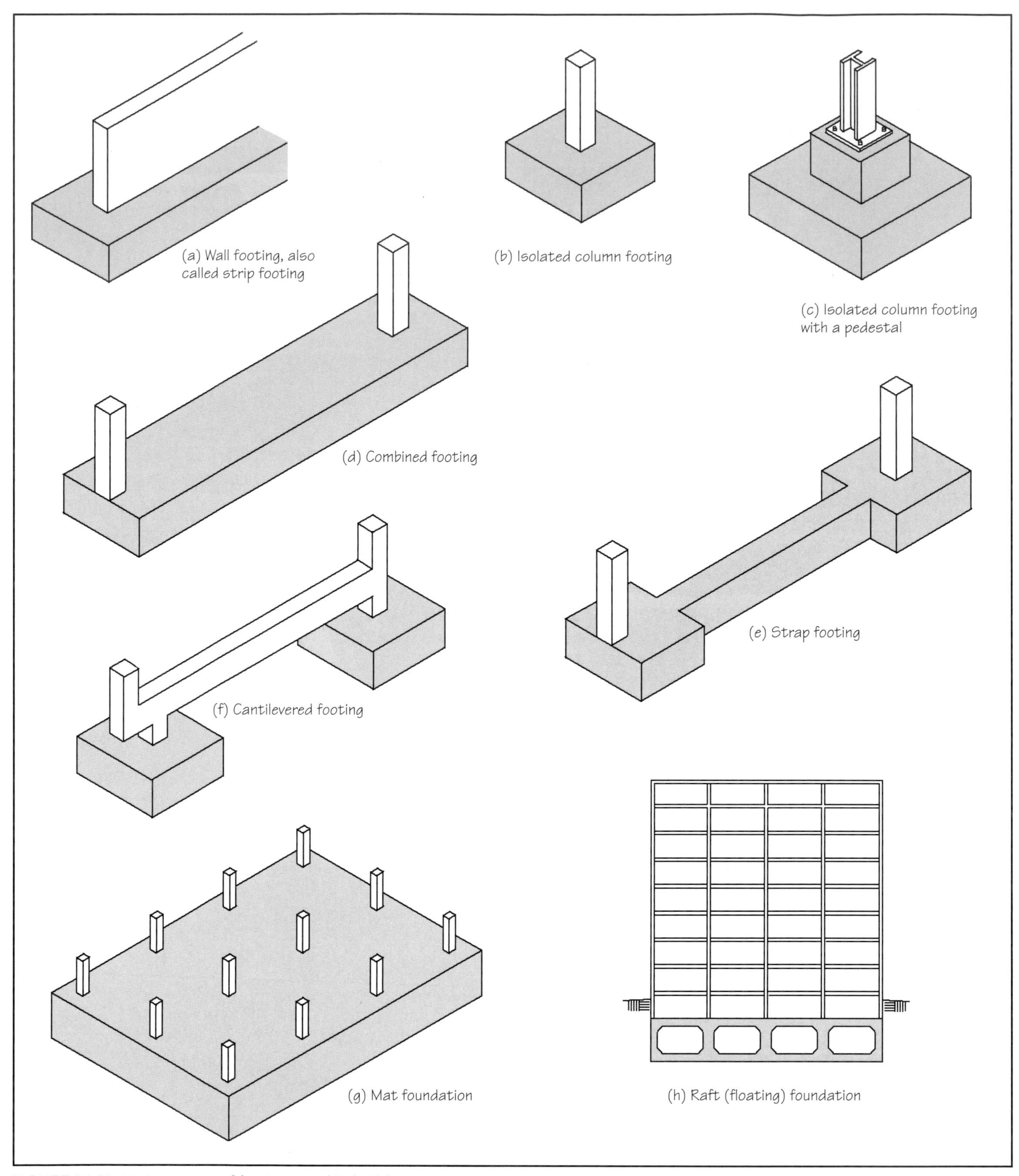

FIGURE 21.27 Various types of footings used in buildings.

Design Parameters of a Footing

Four important design parameters of a footing, as shown in Figure 21.28, are the

- Area of the footing (length × width)
- Thickness of the footing
- Depth of the bottom of the footing below grade
- Steel reinforcement

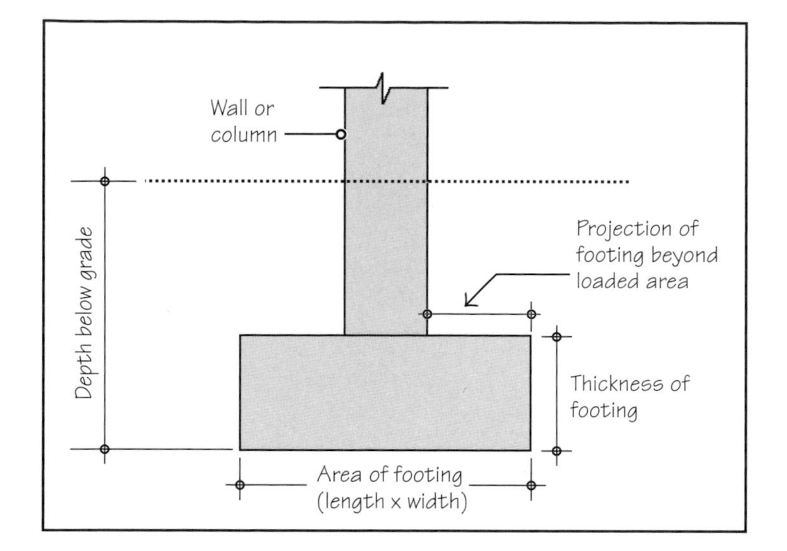

FIGURE 21.28 Four major design parameters of a footing are

- Spread (area) of footing
- Thickness of footing
- Depth below grade
- Amount and location of steel

The *area of a footing* is a function of soil strength (soil's allowable bearing capacity). The greater the allowable bearing capacity of the soil, the smaller the footing area for a given load on the footing.

In order to distribute the load on the soil through its entire area, the footing must be sufficiently rigid—a function of footing thickness. Additionally, the column or the wall supported by the footing must not punch through the footing. Therefore, the *thickness of footing* is determined by the load on the footing and the strength of concrete used in footing.

The *depth below grade* of the bottom of the footing is determined by the soil bearing capacity and the depth of the frost line. The footing must rest on undisturbed soil (or engineered fill) and extend deep enough to reach soil with the required bearing capacity.

In addition, the bottom of the footing must lie below the frost line—the depth below which freeze-thaw conditions do not occur. This prevents the footing from heaving (lifting) as the soil freezes and expands and settling as the soil thaws and shrinks. This requirement does not apply if the footing bears on rock or in locations of permanently frozen soil (permafrost).

The depth of the frost line in the continental United States ranges from zero in southern Florida to as deep as 8 feet in the Northeast. The exact depth of frost line is available from the local building department.

In most cold northern climates, the depth of the frost line determines footing depth, whereas soil bearing capacity is the most likely governing factor where freeze-thaw conditions are minimal.

The *amount of steel and its location* in the footing is a function of gravity load, lateral load, and the projection of the edge of the footing from the loaded area of the footing. Figure 21.29(a) shows reinforcement for a lightly loaded isolated column footing, and Figure 21.29(b) shows reinforcement for a mat foundation to support a tower crane (to assist in the construction of a high-rise building).

(a)

(b)

FIGURE 21.29 **(a)** Steel reinforcement in an isolated reinforced-concrete column footing. Note that the reinforcement is located near the bottom of the footing and is oriented in both directions (with the required minimum cover, generally 3 in.). Shown also are vertical reinforcing dowels, which will be lap spliced with column reinforcement after the concrete in the footing has been placed. Note that the dowels are bent at the bottom to help them stay in position during concrete placement. **(b)** Steel in the footing for the central crane used for the construction of a high-rise building.

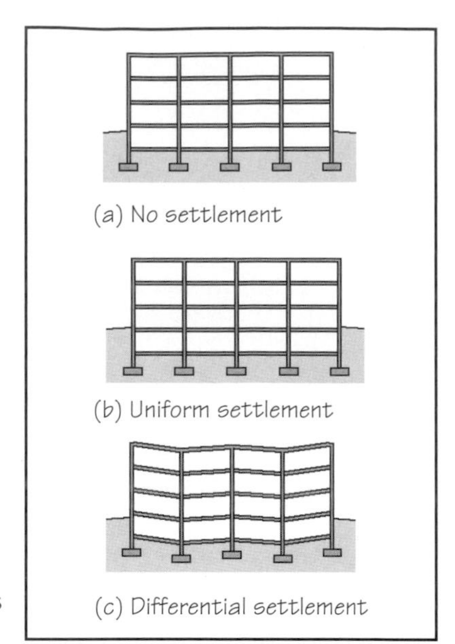

FIGURE 21.30 Effects of uniform and differential settlements of footings.

SETTLEMENT OF FOOTINGS

Another design parameter for footings is the amount of settlement. Settlement occurs due to the deformation of the soil under loads. In coarse-grained soils, the settlement is immediate; that is, it occurs as the load is applied. Therefore, most of the settlement in coarse-grained soils occurs during the construction of the building.

In fine-grained (particularly clayey) soils, part of the settlement is immediate, and the remainder (called *consolidation settlement*) occurs over a period of several months or years. Consolidation settlement occurs when the water held between clay particles is squeezed out and the soil consolidates and shifts.

Settlement may be either uniform or differential. In uniform settlement, the entire structure settles by the same amount, and in cases of excessive settlement this may result in the ground floor of a building sinking to below the finished grade, Figure 21.30(b).

With differential settlement, which is a more critical consideration, the footings may settle by different amounts, Figure 21.30(c). Whereas some differential settlement is unavoidable, excessive differential settlement must be avoided.

The settlement of a building's foundation must be predicted and accounted for in foundation design. Differential settlement in excess of L/300, where L is the center-to-center distance between columns, may cause cracking in brittle materials and may cause structural damage or failure when in excess of L/150 [2].

ALLOWABLE BEARING CAPACITY OF SOIL

As the preceding discussion shows, a soil's allowable bearing capacity is fundamental to the design of footings. The allowable bearing capacity of a soil is obtained by dividing its ultimate (failure) bearing capacity by a factor of safety. The ultimate bearing capacity of soil is determined from several properties, which are obtained through soil testing. Allowable bearing capacity of a soil may also be obtained from the local building code, Table 21.1.

TABLE 21.1 PRESUMPTIVE ALLOWABLE BEARING CAPACITY OF SOILS

Class of material	Allowable bearing capacity
Crystalline bedrock	12.0 ksf
Sedimentary and foliated rock	4.0 ksf
Soil types GW and GP	3.0 ksf
Soil types SW, SP, SM, SC, GM, and GC	2.0 ksf
Soil types CL, ML, MH, and CH	1.5 ksf

1 ksf (kips per square feet) = 1,000 psf (pounds per square feet).
For symbols of soil types, refer to the Unified Soil Classification System in "Principles and Practice" at the end of this chapter.

Source: International Code Council. *International Building Code,* 2006, Chapter 18.

The values given in the codes are highly conservative and are used for the design of small buildings where soil investigation may not be economically justifiable.

CONSTRUCTION DETAILS OF SHALLOW FOUNDATIONS

Several construction details for shallow foundations that were conceptually diagrammed in Figure 21.25 can be found in Chapter 19. A few additional details (used for light-frame buildings) are shown in Figure 21.31.

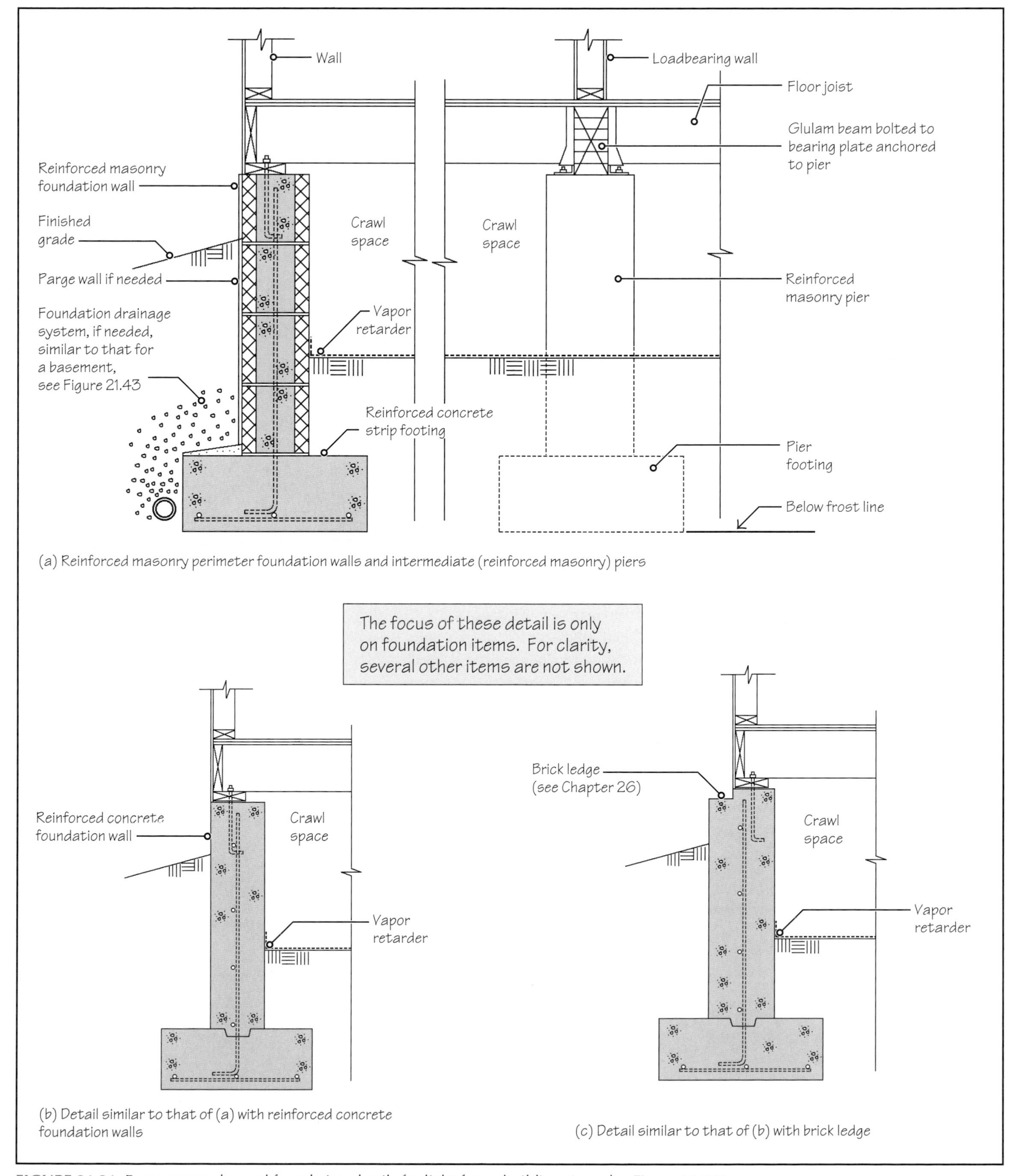

FIGURE 21.31 Few commonly used foundation details for light-frame buildings (see also Figure 21.25).

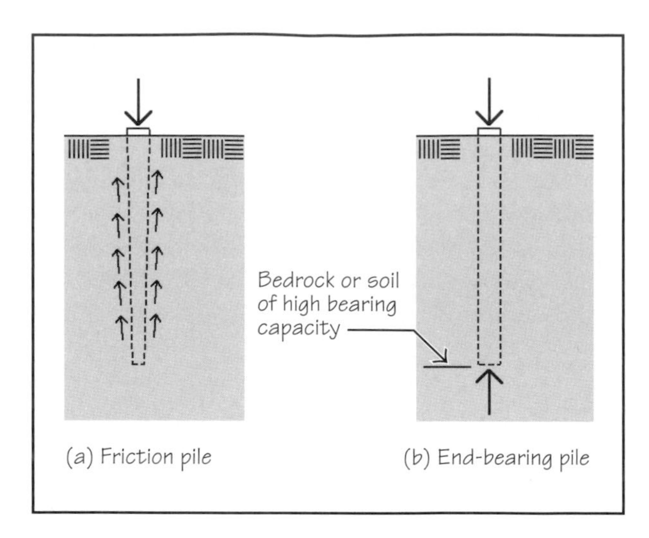

FIGURE 21.32 Friction piles and end-bearing piles.

21.8 DEEP FOUNDATIONS

Deep-foundation elements comprise piles and piers and are like slender columns buried in the ground. They are prevented from buckling because they are confined by the soil. Piles and piers transfer the load either to bedrock or to soil of high bearing capacity while passing through unsuitable soil.

As previously stated, deep foundations are used where shallow foundations cannot be used because the topsoil is expansive or has a low bearing capacity or where the frost line is deep. They are also used in buildings that house items or instrumentation sensitive to settlement of shallow foundations.

Piles are generally driven into the ground, except concrete piles, which can also be site-cast in predrilled holes. Site-cast piles are called *piers,* or *caissons.* (Thus, contiguous bored concrete piles, described in Section 21.4, can be called piers. In this section, the term *pile* means a *driven pile.*)

PILES

Steel, reinforced concrete, prestressed concrete, and wood are commonly used as pile materials. Steel and concrete piles are generally used under heavier loads. Wood piles are limited in their load capacity because of the nature of the material and the limitations on the cross-sectional area of tree trunks.

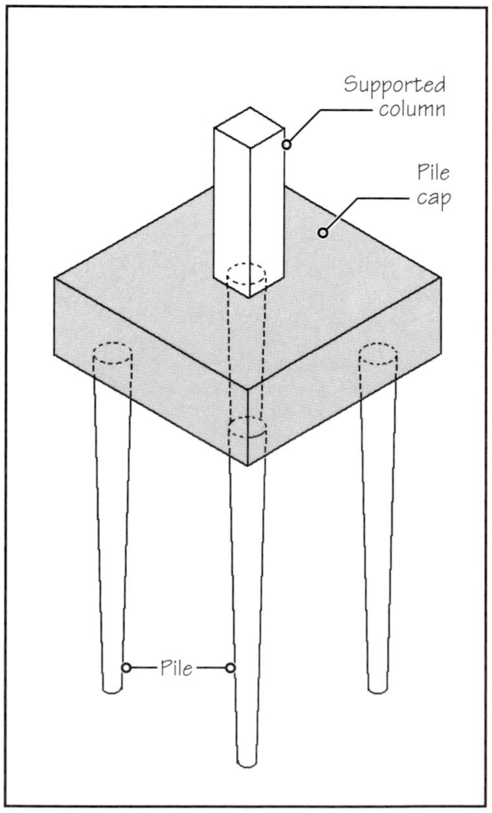

Steel piles consist of H-shape, or hollow, pipes. Hollow steel piles are filled with concrete after being driven. Steel piles are susceptible to corrosion in some environments.

Precast concrete piles are generally solid, but hollow precast concrete sections have been used. Concrete piles are subject to attack by sulfur present in some soils, requiring the use of sulfate-resistant portland cement (Chapter 18). Wood piles are treated with creosote or CCA as a preservative.

The selection of piles for a building is a function of several factors, such as availability, cost, below-grade environment, load capacity of piles, and the equipment required to drive them.

Piles that transfer the load through friction created between the surface of the pile and the soil are called *friction piles.* They are generally tapered to a narrow cross section at the bottom to facilitate driving and to increase friction. Piles that transfer most of the load to the bottom strata, with very little through friction, are called *end-bearing piles,* Figure 21.32.

Because they are driven, piles are generally smaller in cross section than piers, and the load capacity of an individual pile is generally small. Piles are, therefore, often used in clusters of three or more at 2 to 3 ft on centers and support a reinforced concrete cap, called a *pile cap.* A pile cap functions like an isolated column footing, Figure 21.33. Two or more columns can be placed on a large pile cap.

FIGURE 21.33 Pile cap supported by a cluster of four piles. A minimum of three piles are generally required for a pile cluster.

PIERS

Piers are often used individually (as opposed to cluster of piles) because they can be made almost as large as required. Pier diameter generally varies between 18 and 36 in., but 6-ft-diameter piers have been used [3]. Another reason for using individual piers

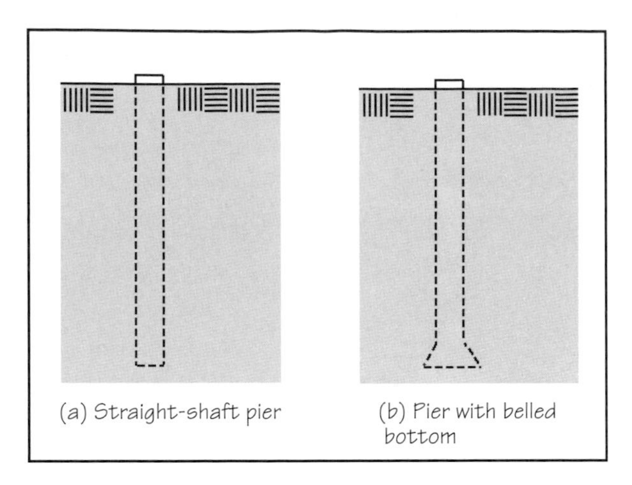

FIGURE 21.34 Straight-shaft pier and pier with belled bottom.

is that their vertical alignment is almost assured. (Because piles are driven, alignment is uncertain and, therefore, an individual pile is used only under light loading.)

A pier may also be designed as a *friction pier* or an *end-bearing pier* or as both. Piers may either have a straight shaft or be belled at the bottom, Figure 21.34.

A pier is generally reinforced with a steel cage before concreting. When piers are used as supports for another concrete element (a grade beam or slab), steel reinforcing dowels that extend beyond the top of the pier are placed in them before concreting is completed, Figure 21.35.

FIGURE 21.35 **(a)** A truck-mounted auger drills the hole for a pier. **(b)** A reinforcement cage is lowered in the hole after the drilling is complete. **(c)** Concrete is placed in pier hole through a chute. **(d)** Steel dowels are placed before finishing the top of the pier. If the pier were to support a wood or steel column, steel anchor bolts would have replaced the dowels (Figure 16.36).

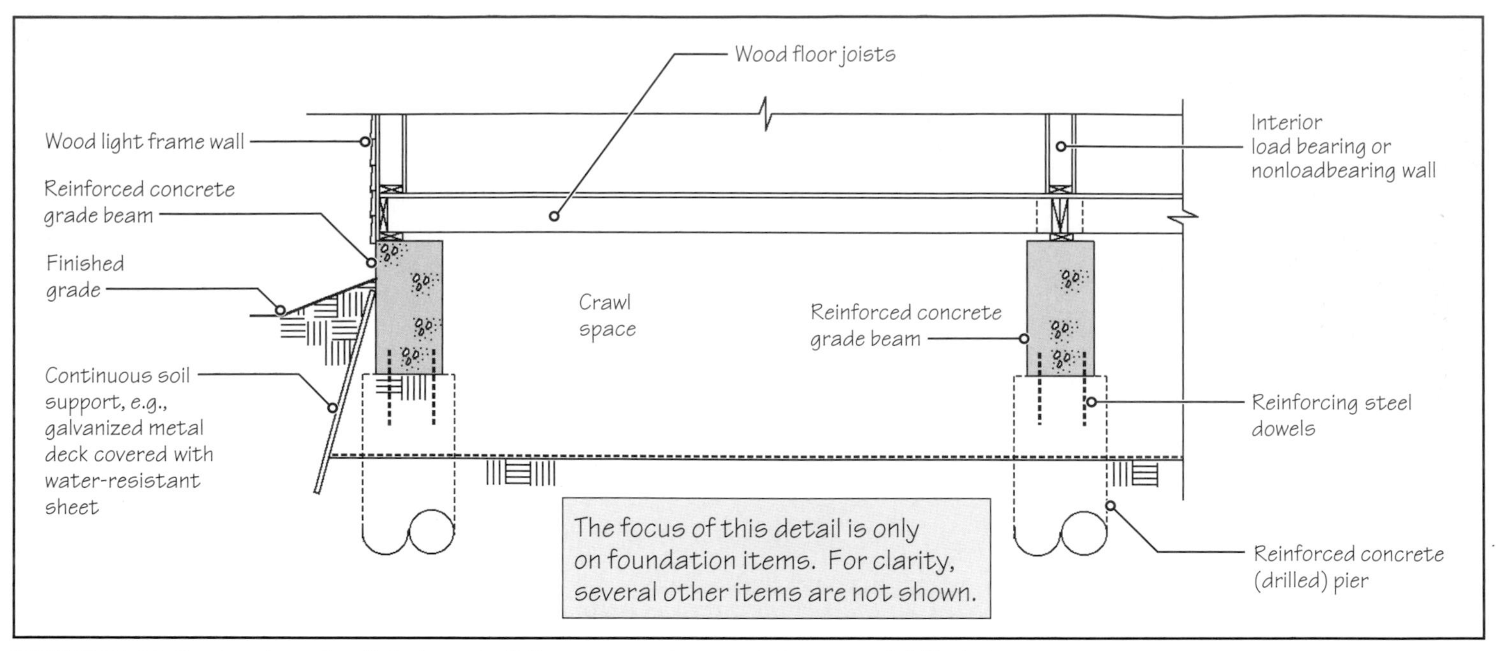

FIGURE 21.36 Section through a light-frame building supported on grade beams and drilled piers.

Construction Details for Deep Foundations

Figures 21.36 to 21.41 provide details of some of the deep-foundation systems that were conceptually diagramed in Figure 21.26.

21.9 FROST-PROTECTED SHALLOW FOUNDATIONS

Until recently, a concrete slab-on-ground could not be used in cold climates because of the need to place the bottom of the foundation below the frost line. Thus, the foundations, even for light-frame buildings, had to be deeper than that required by the soil's bearing capacity considerations.

The introduction of the *frost-protected shallow foundation* (FPSF) system has made it possible to use a concrete slab-on-ground foundation in any climate. An FPSF system incorporates insulation in the vicinity of the slab foundation and uses the heat escaping from the ground-floor slab of a heated building to keep the foundation perimeter sufficiently warm.

(a)

FIGURE 21.37 **(a)** Construction of reinforced grade beams supported on drilled piers (Figure 21.36).

(b)

(c)

FIGURE 21.37 (*Continued*) **(b)** Grade beams after the formwork has been stripped off. **(c)** Construction of wood floor is in progress.

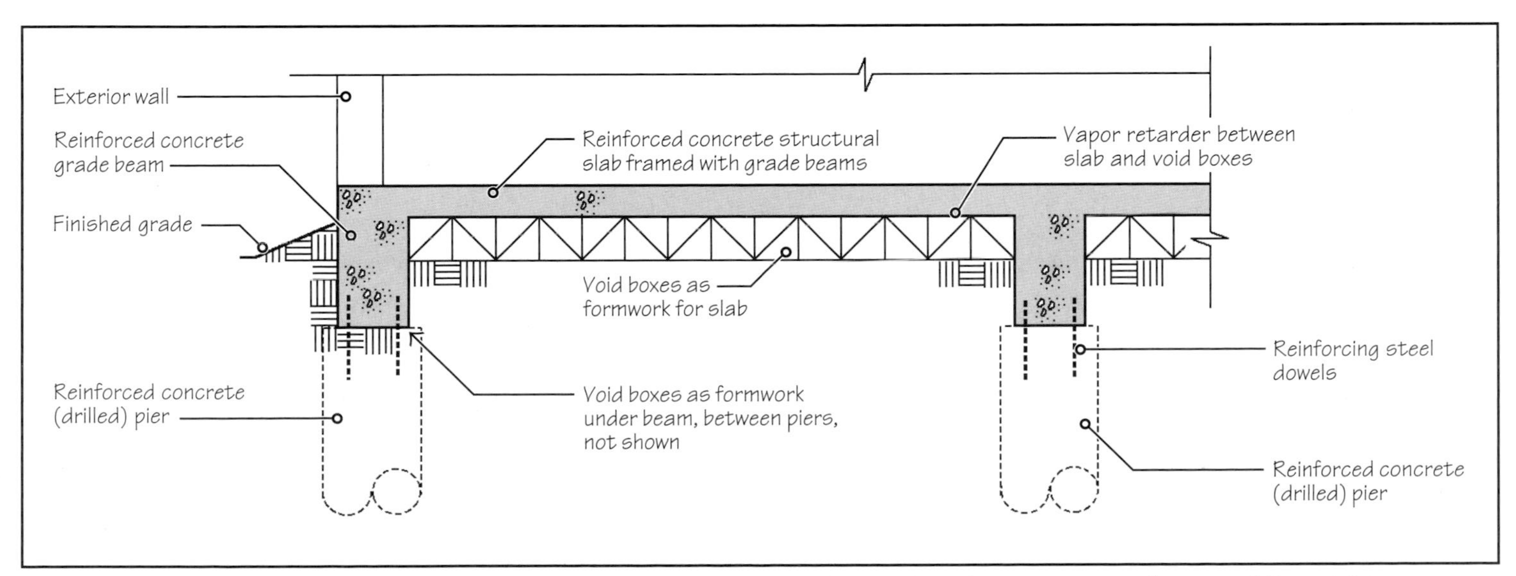

FIGURE 21.38 Section through a structure supported on grade beams and drilled reinforced-concrete piers. The ground-floor slab is not ground-supported but is framed with grade beams as a structural slab. Void boxes, made of cardboard, used as formwork under slab and beams, decompose shortly after construction.

FIGURE 21.39(a) This image shows the construction of the system in Figure 21.38. The grade beams, supported by drilled piers, have been partially concreted. Void boxes will now be placed in the space between grade beams (Figure 21.39(b)). The remaining part of beams will be concreted along with the slab.

FIGURE 21.39(b) Void boxes are being placed between grade beams. These will then be covered with vapor retarder prior to setting reinforcement and placing concrete for the structural slab and remaining volume of the grade beams.

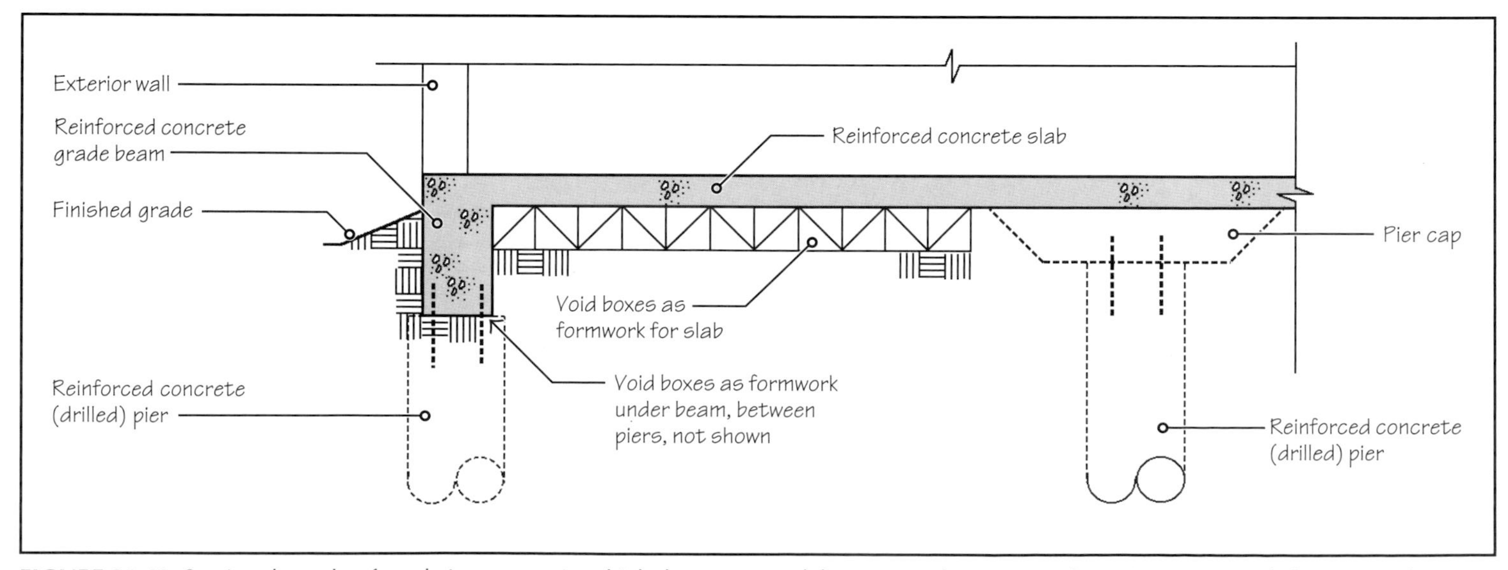

FIGURE 21.40 Section through a foundation system in which the concrete slab-on-ground is supported on pier caps. Grade beams, as shown here, may be provided to support heavy exterior walls. Structurally, this ground floor slab is similar to elevated flat slab described in Chapter 20.

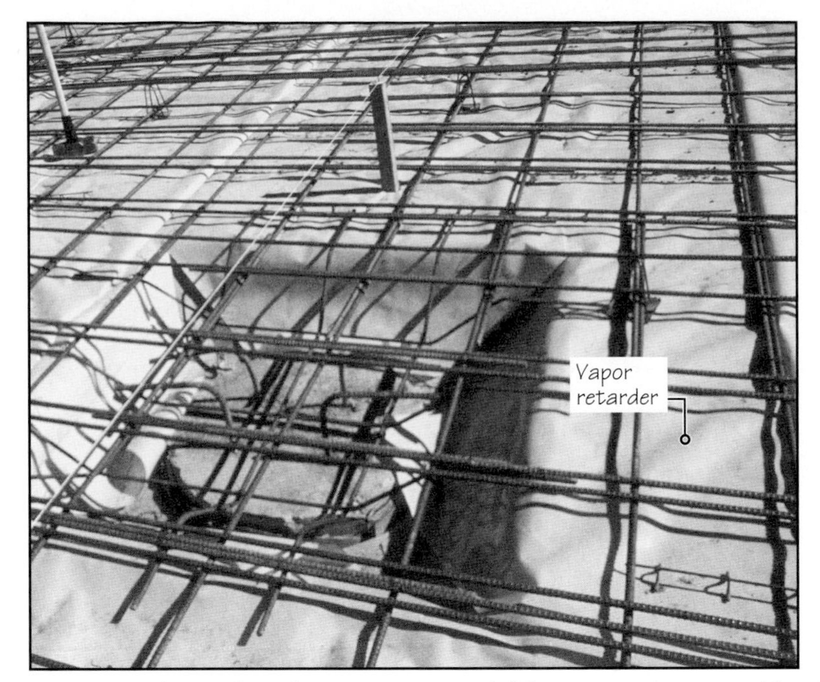

FIGURE 21.41 A view of a reinforced concrete structural slab-on-ground supported by pier caps before the placement of concrete. Note the depression around the pier to form a pier cap and the vapor retarder provided under reinforcing bars.

In other words, an FPSF system effectively raises the depth of frost line near a building's foundation.

The use of insulation in an FPSF is similar to insulating the perimeter of a concrete slab-on-ground for energy conservation reasons (Section 5.11). The difference, however, is that in FPSF, additional horizontal insulation is required outside the building. The details of a typical FPSF are shown in Figure 21.42.

The height and width of insulation and the associated R-values are a function of the local climate [4]. In a properly designed FPSF, the depth of foundation need not exceed 16 in., even in the coldest parts of North America. Note that the FPSF system does not apply to permanently frozen grounds (permafrost).

21.10 BELOW-GRADE WATERPROOFING

The first line of defense against below-grade water leakage in buildings is to prevent water from reaching the foundation by directing all surface storm-water runoff away from the building. This is done by sloping the finish grade away from the building at a gradient of $\frac{1}{2}$ in. per foot (1:25) and by discharging rainwater from roofs directly into a storm-water system or sump. Additional methods to prevent water from percolating into the soil near the foundation include using low-permeability topsoil for plantings, concrete, or pavers adjacent to the building perimeter.

The second line of defense against below-grade leakage is to waterproof the below-grade structure—basement walls and basement floor. Waterproofing consists of (a) application of a waterproofing layer to basement walls and floor and (b) incorporating a drainage system.

WATERPROOFING LAYER

The waterproofing layer is applied to the outside surface of the basement wall and under the basement floor, so that it forms a continuous, unbroken barrier to the entry of water. Layers of hot coal-tar pitch or hot asphalt alternating with fiberglass felt to yield a 3- to 5-ply membrane used to be the most common basement-waterproofing material. This technique was precisely the same as that used for built-up roofing (Chapter 31).

However, although a hot-tar or asphalt system is easily applied on a horizontal surface, its application on a vertical surface is difficult. Safety concerns are critical in underground waterproofing because of the limited working space available. Therefore, a hot-applied waterproofing system has largely been replaced by cold-applied systems.

Commonly used cold-applied, below-grade waterproofing systems are similar to single-ply roof membranes and consist of rubberized asphalt or thermoplastic sheets. The sheets, usually 60 mil (1.5 mm) thick, are available in self-adhering rolls with a release paper.

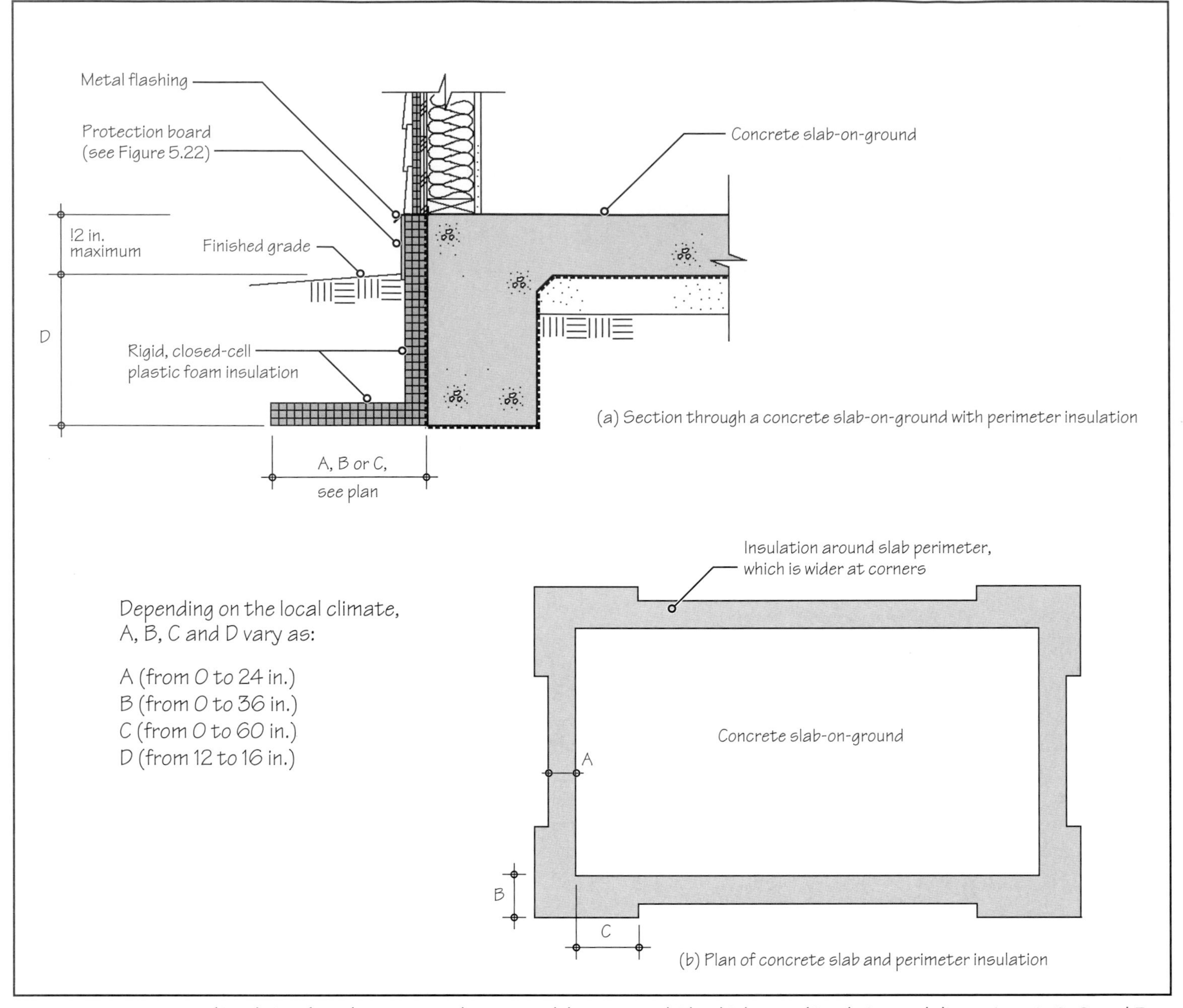

FIGURE 21.42 Layout of insulation for a frost-protected concrete slab-on-ground. The thickness of insulation and dimensions A, B, C, and D are a function of the local climate (see [4] and the local building code).

Before applying the sheets, the basement wall and floor surfaces are cleaned and primed with the manufacturer's primer. Because the sheets are self-adhering, they form a continuous waterproof membrane.

In place of membrane waterproofing, liquid-applied, elastomeric compounds are also used, which can be sprayed on, rolled on, or brushed on. These are particularly attractive for complex surface formations but are prone to application errors.

Drainage System

The drainage system is an important component of below-grade waterproofing. Its purpose is to collect, drain, and discharge subsoil water away from the building. The drainage system consists of: drainage mats and a drain pipe. The drainage mats are installed against the waterproofing layer or against the insulation layer, where needed. Some manufacturers provide the drainage mats laminated to extruded or expanded polystyrene board. Drainage mats have a thick open-weave structure that allows the subsoil water to drain downward by gravity. They eliminate or reduce water pressure acting on a below-grade wall and protect the waterproofing layer from damage that might be caused by the backfill.

Drain pipe is a 4- to 6-in.-diameter perforated pipe set within a bed of crushed stone that allows the water to seep into the pipe. Drain pipes are most commonly made of PVC or corrugated polyethylene. A synthetic filter fabric is used as a cover over crushed stone, which captures fine soil particles and prevents them from clogging the pipe perforations.

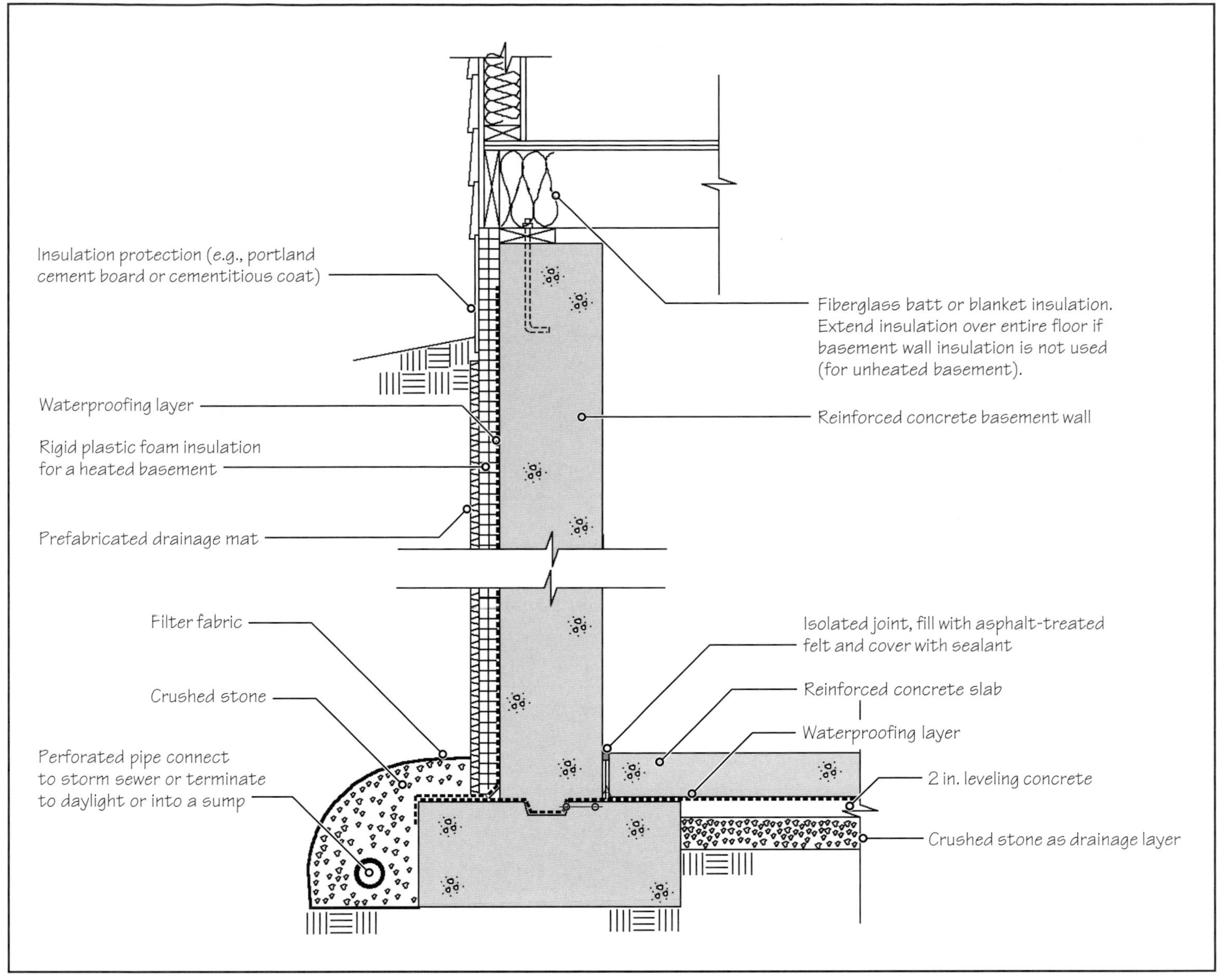

FIGURE 21.43 A section through a typical residential basement. Although a reinforced-concrete wall has been shown here, it can be replaced by a reinforced-masonry wall.

The drain pipe runs all around the basement and is installed a little above the bottom of the wall footing. (If installed below the foundation, it may tend to wash away the soil below the foundation, causing foundation settlement or foundation failure.) The water collected by the drain must be discharged to a lower ground or drained into a sump pit. From here, a self-actuated pump lifts the water into a storm drain. A section through a typical residential basement is shown in Figure 21.43.

Positive-Side, Negative-Side, and Blind-Side Waterproofing

The waterproofing layer shown in Figure 21.43 is called *positive-side waterproofing* because it has been applied on the side of the wall with direct exposure to water. *Negative-side waterproofing* refers to waterproofing that is applied from the opposite side, that is, from the interior. Negative-side waterproofing is generally used in remedial applications.

Positive-side waterproofing is more effective and is more commonly used. It also protects the wall against water seepage, growth of mold, corrosion of metals, and so on. It requires adequate space outside a below-grade wall for applicators to work, which is later backfilled. Positive-side waterproofing, therefore, remains accessible after its completion.

Blind-side waterproofing refers to waterproofing applied to the outside of the wall that becomes inaccessible after the construction. It is used in situations where the basement wall is supported by an excavation-support system such as soldier piles and lagging, which are left in place. In this situation, the waterproofing layer is applied to the excavation-support system and the basement wall is constructed against it. Figures 21.44 and 21.45 illustrate blind-side waterproofing application.

FIGURE 21.44 Waterproofing of a reinforced-concrete basement wall cast against a soldier pile and lagging system of excavation support (blind-side waterproofing).

(a)

(b)

FIGURE 21.45 **(a)** In this deep basement, a part of the cut face has been shotcreted. The shotcreted face is treated with drainage mat and waterproofing. **(b)** After the installation of drainage mat and waterproofing, reinforcement is laid for the basement wall. This will be followed by wall formwork and concrete placement.

Each question has only one correct answer. Select the choice that best answers the question.

22. Which of the following foundations is a deep foundation?
- **a.** Concrete slab-on-ground
- **b.** Strip footing
- **c.** Combined footing
- **d.** Mat foundation
- **e.** None of the above

23. A foundation system consisting of site-cast, reinforced concrete grade beams supported by drilled piers is considered a
- **a.** shallow-foundation system.
- **b.** deep-foundation system.

24. A ground-floor, site-cast reinforced concrete slab formed on void boxes is generally used where
- **a.** the soil is stable and has a high bearing capacity.
- **b.** the upper soil strata is expansive or has a low bearing capacity.
- **c.** the upper soil strata consists of rock.
- **d.** the water table is high.

25. A foundation system for wood light frame buildings, consisting of reinforced-concrete grade beams supported on drilled piers, is generally used where
- **a.** the soil is stable and has a high bearing capacity.
- **b.** the upper soil strata is expansive or has a low bearing capacity.
- **c.** the upper soil strata consists of rock.
- **d.** the water table is high.

26. A ground-floor, site-cast reinforced-concrete slab formed on void boxes and supported by pier caps is structurally similar to a(n)
- **a.** girder-and-beam elevated floor.
- **b.** elevated banded slab.
- **c.** elevated flat slab.
- **d.** elevated flat plate.

27. A mat foundation is generally
- **a.** a long and deep beam-type foundation strip that typically supports a row of columns.
- **b.** a large and deep slab-type foundation that typically supports a number of columns.
- **c.** a large and shallow slab-type foundation that typically supports a number of columns.
- **d.** a group of isolated column footings.

28. The values of allowable bearing capacity of soils given in building codes are generally lower than those obtained from soil tests.
- **a.** True
- **b.** False

29. When drilled piers are used as foundation support, they are generally provided in a cluster of three or more.
- **a.** True
- **b.** False

30. Frost-protected shallow foundations (FSPF) are generally provided with
- **a.** horizontal perimeter insulation under the slab.
- **b.** horizontal perimeter insulation outside the slab.
- **c.** vertical insulation and horizontal perimeter insulation under the slab.
- **d.** vertical insulation and horizontal perimeter insulation outside the slab.
- **e.** none of the above.

31. Negative-side waterproofing for a basement wall is provided from the interior face of the wall.
- **a.** True
- **b.** False

32. Which of the following waterproofing terms is incorrect?
- **a.** Positive-side waterproofing
- **b.** Negative-side waterproofing
- **c.** Neutral-side waterproofing
- **d.** Blind-side waterproofing

Answers: 22-e, 23-b, 24-b, 25 b, 26-c, 27-b, 28-a, 29-b, 30-d, 31-a, 32-c.

PRINCIPLES IN PRACTICE

Unified Soil Classification

Soil classification described in Section 21.1 is too broad to be useful in building construction, where a more detailed classification, called the *Unified Soil Classification System* (USCS), is employed. In USCS, the distinction between coarse and fine-grained soils is as follows.

- *Coarse-grained soils* are soils that contain more than 50% of material retained on a No. 200 sieve.

  Coarse-grained soils are further subdivided into *gravelly soils* and *sandy soils*. Gravelly soils are those in which 50% or more of coarse fraction is retained on a No. 4 sieve. In sandy soils, more than 50% of coarse fraction passes through a No. 4 sieve.

  Gravelly soils are further subdivided as *clean gravels* or *dirty gravels*. Similarly, sandy soils are further subdivided as *clean sands* or *dirty sands*.
- *Fine-grained soils* are soils that contain 50% or more of material passing through a No. 200 sieve. Fine-grained soils are further subdivided into Group L soils and Group H soils. This distinction is based on the plasticity of the soil, which is measured by the soil's *liquid limit*. A soil with a *low* plasticity (liquid limit $\leq$50%) belongs to Group L soils. A soil with a *high* plasticity (liquid limit $>$50%) is classified as a Group H soil.

Additional details of USCS are given in Table 1.

(*Continued*)

Unified Soil Classification (*Continued*)

TABLE 1 UNIFIED SOIL CLASSIFICATION SYSTEM (USCS)

Major division			Group symbols	Group description
Coarse-grained soils >50% material retained on No. 200 sieve	**Gravelly soils** >50% of coarse fraction retained on No. 4 sieve	**Clean gravels** Fines <5%	GW	Well-graded gravels and gravel-sand mixtures, little or no fines
		In field test, no stain on wet palm	GP	Poorly graded gravels and gravel-sand mixtures, little or no fines
		Dirty gravels Fines >12%	GM	Silty gravels, gravel-sand-silt mixtures
		In field test, stain on wet palm	GC	Clayey gravels, gravel-sand-clay mixtures
	Sandy soils ≥50% of coarse fraction passes No. 4 sieve	**Clean sands** Fines <5%	SW	Well-graded sands and gravelly sands, little or no fines
		In field test, no stain on wet palm	SP	Poorly graded sands and gravelly sands, little or no fines
		Dirty sands Fines >12%	SM	Silty sands, sand-silt mixtures
		In field test, stain on wet palm	SC	Clayey sands, sand-clay mixtures
Fine-grained soils ≥50% material passes No. 200 sieve	**Silts and clays** Liquid limit ≤50%		ML	Inorganic silts, very fine sands, rock flour, silty or clayey fine sands
			CL	Inorganic clays of low to medium plasticity, gravelly clays, sandy clays, silty clays, lean clays
			OL	Organic silts and organic silty clays of low plasticity
	Silts and clays Liquid limit >50%		MH	Inorganic silts, micaceous or diatomaceous fine sands or silts, elastic silts
			CH	Inorganic clays of high plasticity, fat clays
			OH	Organic clays of medium to high plasticity
Highly Organic Soils			Pt	Peat, muck, and other highly organic soils

Note: Gravelly or sandy soils with fines lying between 5% and 12% carry a hyphenated designation such as GW-GM, GW-GC, and so on.

Source: Adapted from American Society for Testing and Materials (ASTM): "Standard Practice for Classification of Soils for Engineering Purposes (Unified Soil Classification System)", ASTM D2487. "Classification of Soils for Engineering Purposes (Unified Soil Classification System," with permission.

SYMBOLS USED IN USCS

USCS uses the following symbols to designate various soils:

G for gravel
S for sand
M for silt
C for clay

W for well-graded particles, that is, when particles of various sizes are present, giving a relatively denser soil

P for poorly graded soils, that is, soils with very little (i.e., almost uniform) gradation of particle sizes. (Such soils contain more void space and are, therefore, less dense.)

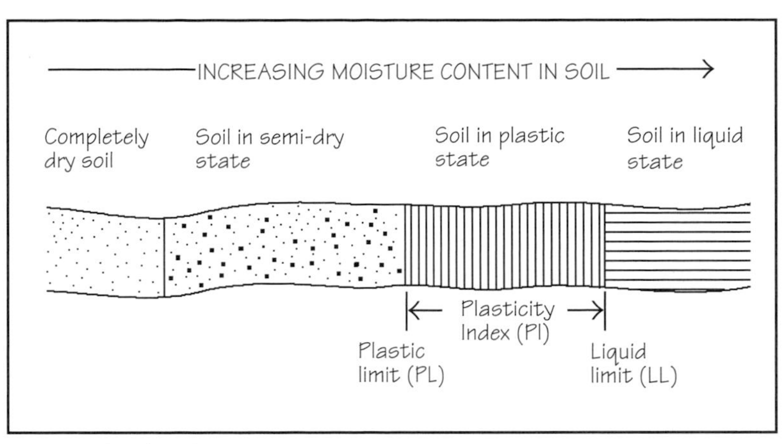

FIGURE 1 Four states of a fine-grained soil with increasing moisture content.

L for low plasticity (liquid limit $\leq$50%)
H for high plasticity (liquid limit $>$50%)
O for organic soils

Two symbols are used in conjunction to designate a soil. Thus, GW designates a well-graded gravel, GP designates a poorly graded gravel, ML designates a silty soil with liquid limit less than or equal to 50%, and so on.

PLASTIC LIMIT AND LIQUID LIMIT OF FINE-GRAINED SOILS

The liquid limit of soil (a measure of the soil's plasticity) indicates the behavior of soil in the presence of water. If we take a sample of completely dry fine-grained soil and put a little water in it, it will become slightly moist, or semidry.

If the amount of water in soil sample is increased, it will become plastic (puttylike), implying that it can be rolled into a rope between the palms of both hands. The ratio of the weight of water to the dry weight of the soil (expressed as a percentage moisture content) when a soil just changes from a semidry state to the plastic state is called the *plastic limit* of the soil.

If we keep adding water to the same soil sample, a stage is reached when the soil loses its plasticity and turns into a slurry, that is, a liquid, Figure 1. The moisture content of soil when it just turns into a liquid is called its *liquid limit.*

Gravels and sands do not become plastic. They change from a solid state to a liquid state. That is why liquid limit (LL) and plastic limit (PL) concepts apply only to fine-grained soils—silts and clays.

Another important soil property is plasticity index (PI), defined as

$$PI = LL - PL$$

For example, if a soil has a PL of 35% and LL of 60%, its PI is $60 - 35 = 25\%$. Between two soils with the same LL, the one with a larger value of PI retains more water and is, therefore, more susceptible to swelling and shrinkage due to changes in moisture.

EXPANSIVE SOILS

Some fine-grained soils expand with an increase in moisture content and shrink when the moisture content decreases. Foundations in such soils, referred to as *expansive soils*, are constructed to ensure that the structure is not adversely affected by soil movement.

A rough guide to swell-shrink susceptibility of a fine-grained soil is provided by the following indices:

Highly expansive soil (high swell-shrink potential): LL $>$ 50% and PI $>$ 30%
Medium expansive soil (medium swell-shrink potential): LL between 25% and 50% and PI between 15% and 30%.
Stable soil (low swell-shrink potential): LL $<$ 25% and PI $<$ 15%.

Expansive soils are found in many regions of the United States, such as California, Nevada, Texas, Arizona, and Maryland.

(*Continued*)

Unified Soil Classification (*Continued*)

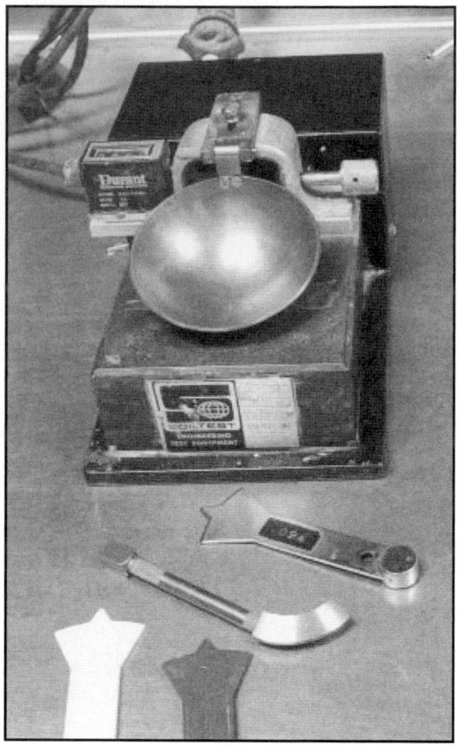

TESTING FOR LIQUID LIMIT OF FINE-GRAINED SOILS

A liquid-limit testing apparatus consists of a metal cup mounted on a hard rubber base, Figure 2. The motor attached to the cup raises it to a prescribed height and releases it for a free fall on the rubber base.

To determine the liquid limit, the soil specimen is mixed with different amounts of water to give several batches of the specimen. A batch is placed in the cup and flattened to the required spread and thickness in the prescribed manner. A groove is then cut in the flattened batch with a standard grooving tool (shown in the foreground).

The number of falls of the cup that close the groove in a batch are recorded. A graph relating the moisture content of the soil with the number of falls for each batch is plotted. From this graph, the liquid limit is obtained by reading the amount of moisture content that would be required to close the groove with 25 falls.

FIGURE 2 A liquid-limit testing apparatus. (Photo courtesy of the Geotechnical Laboratory of the University of Texas at Arlington.)

TESTING FOR PLASTIC LIMIT OF FINE-GRAINED SOILS

A plastic-limit test involves taking a prescribed weight of dry sample and mixing it with water until it is almost saturated. The sample is rolled back and forth between two plastic plates until a thread of nearly $\frac{1}{4}$ in. in diameter is formed, Figure 3.

The thread is then balled up and rolled again and then again. Each rolling operation is performed under a fan so that the specimen gets progressively drier. A stage is reached when the thread breaks into short lengths, Figure 4. The percentage moisture content in the specimen at this stage is the plastic limit of the soil.

Note that both liquid-limit and plastic-limit tests are performed on the fraction of fine-grained soils that passes through a No. 40 sieve.

FIGURE 3 Rolling of moist soil into a thread. (Photo courtesy of the Geotechnical Laboratory of the University of Texas at Arlington.)

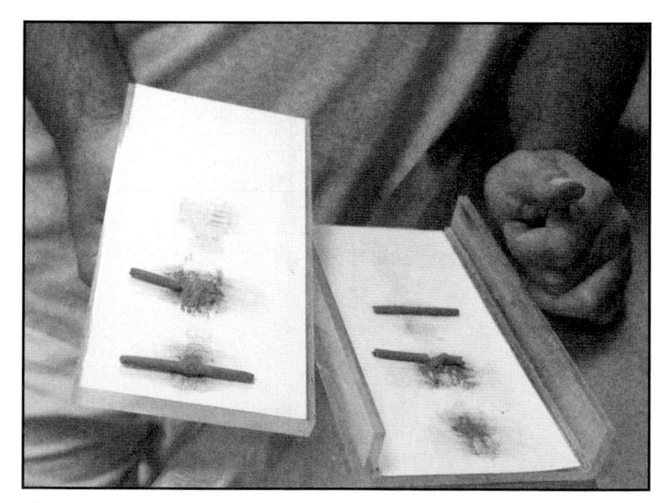

FIGURE 4 As the moist soil thread is progressively dried while rolling, it breaks. (Photo courtesy of the Geotechnical Laboratory of the University of Texas at Arlington.)

Each question has only one correct answer. Select the choice that best answers the question.

33. In the Unified Soil Classification System, the symbol M stands for
- **a.** medium-grained soil.
- **b.** moist soil.
- **c.** soil of medium plasticity.
- **d.** none of the above.

34. In the Unified Soil Classification System, the symbol GP stands for
- **a.** poorly graded gravel.
- **b.** pea gravel.
- **c.** peat and gravel.
- **d.** soil with good performance.
- **e.** none of the above.

35. Which of the following soils has the lowest liquid limit?
- **a.** Sand
- **b.** Silt
- **c.** Clay

36. The expansiveness of a soil is determined from the values of
- **a.** plastic limit and liquid limit.
- **b.** liquid limit and plasticity index.
- **c.** plastic limit and plasticity index.
- **d.** none of the above.

Answers: 33-d, 34-a, 35-a, 36-b.

KEY TERMS AND CONCEPTS

Bearing capacity of soil
Below-grade waterproofing
Bentonite slurry
Blind-side waterproofing
Boring log
Classifications of soils
Clay
Cohesive soil
Combined footing
Contiguous bored concrete piles
Deep excavations
Deep foundations
Dewatering
Drilled piers
Driven piles
Earthwork
Expansive soils
Footings
Frost-protected shallow foundations (FPSF)
Gravel
Mat footing
Negative-side waterproofing
Noncohesive soil
Open excavations
Organic soil
Positive-side waterproofing
Raft (floating) foundation
Sand
Secant piles
Shallow foundations
Sheet piles
Silt
Soil nailing
Soldier piles
Test boring
Test pit
Tremie pipe
Unified soil classification system (USCS)

REVIEW QUESTIONS

1. Using sketches and notes, explain the soldier pile and lagging system of excavation support.
2. Using sketches and notes, explain the contiguous bored pile system of excavation support.
3. Using sketches and notes, explain the secant pile system of excavation support.
4. Using sketches and notes, suggest two alternatives for constructing a ground-floor concrete slab on a soil that is expansive.
5. Using sketches and notes, explain
   a. strap footing.
   b. mat foundation.
   c. raft foundation.

FURTHER READING

1. Kubal, Michael. *Construction Waterproofing Handbook*. New York: McGraw Hill, 2000.
2. Puller, Malcolm. *Deep Excavations—A Practical Manual*. London: Thomas Telford, 2003.

REFERENCES

1. American Society for Testing and Materials (ASTM). "Standard Specifications for Wire Cloth and Sieves for Testing Purposes," ASTM E11-04.
2. Cernica, John. *Soil Mechanics*. New York: John Wiley, 1995, p. 251.
3. Olin, Harold, Schmidt, John, and Lewis, Walter. *Construction—Principles, Materials, and Methods*, 6th ed. New York: Van Nostrand Reinhold, 1995, p. 153.
4. National Association of Home Builders (NAHB). *Design Guide for Frost Protected Shallow Foundations*, 1994.

CHAPTER 22 Masonry Materials–I (Mortar and Brick)

CHAPTER OUTLINE

22.1 MASONRY MORTAR

22.2 MORTAR MATERIALS AND SPECIFICATIONS

22.3 MORTAR JOINT THICKNESS AND PROFILES

22.4 MANUFACTURE OF BRICKS

22.5 DIMENSIONS OF MASONRY UNITS

22.6 TYPES OF CLAY BRICKS

22.7 BOND PATTERNS IN MASONRY WALLS

22.8 THE IMPORTANCE OF THE IRA OF BRICKS

22.9 THE CRAFT AND ART OF BRICK MASONRY CONSTRUCTION

22.10 EFFLORESCENCE IN BRICK WALLS

22.11 EXPANSION CONTROL IN BRICK WALLS

Brick, one of the oldest construction materials, is still valued for its high compressive strength, sound insulation, fire resistance, durability, and appearance. These qualities were exploited in the interior of the high school building shown. Architect: Huckabee Architecture, Engineering, Construction Management. (Photo courtesy of Michael Lyons, Architectural Photographer.)

Masonry is one of the oldest building materials. Sun-dried clay (adobe) bricks were used as early as 8,000 BC. The origin of stone masonry is generally traced by historians to the early Egyptian and Mesopotamian civilizations, which existed around 3,000 BC. Indeed, until steel and portland cement were discovered in the mid-nineteenth century, stone was the only building material available for the construction of large building structures and bridges.

The history of architecture is replete with examples of magnificent buildings in which dressed (and partially dressed) stones were used in almost every building element—walls, columns, beams, arches, roofs (vaults and domes), and floors. In several buildings, the sizes of stones were so large that historians have differing theories as to the hoisting apparatus used by the builders at the time, Figure 22.1.

The history of architecture is also a testament to the durability and aesthetics of stone. It is these properties that render stone a matchless material even for present-day buildings, particularly those that require durable and maintenace-free facades, Figure 22.2. In fact, one of the most-used material for the exterior facades of contemporary skyscrapers is stone, Figure 22.3.

FIGURE 22.1 Parthenon, Acropolis, Greece (fifth century BC). The stones at the column bases are nearly 6 ft in diameter and 3 ft 6 in. high. The sizes of stones in the Egyptian pyramids are much larger. (Photos courtesy of Dr. Jay Henry.)

FIGURE 22.2(a) Split-face (cleft) Italian travertine used as cladding material in the J. Paul Getty Center, Los Angeles, California (a complex of several buildings). The building in the background is clad with metal panels. Architect: Richard Meier. (Photo courtesy of Dr. Jay Henry.)

FIGURE 22.2(b) Smooth-surfaced limestone cladding, Morton Meyerson Symphony Center, Dallas, Texas. Architect: I. M. Pei and Partners.

Unlike in the past, when stone was used both as structural and a finish material, its contemporary use is mainly in the form of thin slabs—as wall cladding and floor-finish material. However, some present-day architects, in an effort to create the ambience of the old times, have used stone in much the same way as used centuries ago, Figure 22.4. Such use is, however, quite limited.

Whereas stone is uniquely suited for cladding high-rise and significant buildings, other masonry materials, such as brick and block, are more economical facade alternatives and are widely used in buildings. If we lump all masonry materials together, we observe that the use of masonry on contemporary building facades exceeds that of all other materials combined.

FIGURE 22.3 One of the several towers of the World Financial Center, New York, clad with granite and glass. The proportion of granite (in relation to glass) decreases with increasing height. Architect: Cesar Pelli. (Photo courtesy of Dr. Jay Henry.)

UNIT MASONRY

Because bricks are generally made from clay, brick masonry is also referred to as *clay masonry*. Block masonry is called *concrete masonry* because the blocks are made from concrete. Because masonry is laid unit by unit (e.g., brick by brick or block by block), it is also referred to as *unit masonry*. Bricks and blocks are, therefore, called clay masonry units and concrete masonry units (CMU), respectively.

Masonry units are bonded together with mortar to yield a composite building component—generally a wall. Thus, mortar is the common ingredient in all masonry construction. Mortarless masonry, although possible, is uncommon and has relatively few applications; it is not discussed in this text. Because mortar is common to virtually all masonry, we discuss it first, followed by bricks. Other masonry materials, including concrete masonry units, stone, and glass masonry, are discussed in Chapter 23. Masonry construction systems are discussed in Chapters 24 and 26.

22.1 MASONRY MORTAR

Mortar consists of a binder (cementitious material), a filler, and water. Portland cement and hydrated lime comprise the binder, and the filler is sand. When these three elements are mixed together with the required quantity of water, mortar results.

The primary function of masonry mortar is to bond the masonry units into an integral masonry wall. Because mortar, in its plastic state, is pliable, it molds itself to the surface profile of the units being mortared. This not only helps seal the wall against water and air infiltration, it also provides a cushion between the units.

The mortar cushion also compensates for size variations between individual units. Another important role of mortar is to provide surface character to masonry through shadow lines at mortar joints and color intervention between the units.

WORKABILITY AND WATER RETENTIVITY OF MORTAR: THE ROLE OF LIME

Although portland cement is the primary cementitious material in mortar, lime imparts several useful and important properties to both the plastic as well as the hardened mortar. In the plastic (wet) state of mortar, lime improves its workability and water retentivity. A mortar

FIGURE 22.4(a) Cistercion Church Addition, Irving, Texas. Architect: Gary Cunningham, Dallas, Texas. Some of the split-face limestone blocks, used as load-bearing elements in this building, are 10 ft × 3 ft × 2 ft, weighing more than 4 tons each.

FIGURE 22.4(b) Interior view of Cistercion Church addition, Irving, Texas.

NOTE

Mortarless Masonry Walls

A common contemporary use of mortarless masonry is in segmental retaining walls. Mortarless masonry is extremely uncommon in a conventional wall.

If a mortarless masonry wall (also called dry-stacked masonry) is used in a conventional wall, the units may need grinding to provide a smooth bed surface. To provide bending resistance in a mortarless masonry wall, the voids in the units may be grouted and reinforced, and/or both surfaces of the wall may be coated with a surface-bonding material, such as plaster.

comprising portland cement only (without lime) is coarse and hence less workable. In the hardened state, lime improves the water resistance of the wall. A wall built with portland cement and lime mortar is more watertight than a wall built with only portland cement mortar.

In Chapter 18, we observed that the workability of concrete is a quantifiable property. This is not the case with mortar. The workability of mortar is difficult to quantify because it is a function of several interdependent factors. However, a mason with even a limited amount of experience and training can easily distinguish between a workable and nonworkable mortar.

A workable mortar is cohesive and spreads easily on the units using a trowel. Because of its cohesiveness, it clings to the vertical surfaces of the units and the trowel without sliding down. It extrudes easily so that excess mortar in the joints can be troweled off without the mortar dropping off or smearing the units. A lay (and rather crude) explanation of the difference between a workable and nonworkable mortar is the difference between spreading a creamy (more workable) and a crunchy (less workable) peanut butter on a piece of toasted bread.

Another important property of plastic mortar is its water retentivity. This is the ability of mortar to retain water without letting it bleed out. A mortar with good water retentivity remains soft and plastic for a long period of time and allows only a limited amount of water to be absorbed by the units.

Water retentivity and workability are directly related to each other. Extremely fine sand particles, air-entraining agents, and lime increase workability and water retentivity of mortar. Because a certain amount of water absorption by the units is necessary for the bond between the mortar and the units, excessive water retentivity is to be avoided because it reduces the bond.

NOTE

Lime for Masonry Mortar

The lime recommended for use in masonry mortar is Type S hydrated lime (Chapter 18).

NOTE

Flexural Tensile Bond Strength

A wall subjected to bending (flexure) experiences tension as well as compression. These stresses are referred to as *flexural tension* and *flexural compression*, respectively. Because unreinforced masonry is relatively stronger in compression than tension, the *flexural tensile strength* of masonry is more critical than flexural compressive strength.

Because flexural tensile strength (of an unreinforced masonry wall) is due to the bond between the units and the mortar, it is called *flexural tensile bond strength*, or simply the *bond strength* (Figure 22.6). As stated in the text, the bond strength is not relevant in a masonry wall provided with steel reinforcing bars to resist bending (Figure 22.7).

WATERTIGHTNESS OF A MASONRY WALL: THE ROLE OF LIME

Lime also improves the elasticity of hardened mortar, that is, it makes the mortar more rubbery. In other words, a lime-based mortar is able to flex somewhat in its hardened state. This reduces the cracks caused by the bending of the wall under lateral loads. Lime also provides an *autogenous healing* property to mortar. Autogenous healing refers to the self-sealing of small cracks produced either within the mortar or at the interface between the mortar and the units. The cracks may result either from the bending stresses in masonry or the drying shrinkage of portland cement in mortar.

MORTAR STRENGTH: THE ROLES OF PORTLAND CEMENT AND LIME

Two strength properties of mortar are generally of interest:

- Compressive strength
- Flexural tensile bond strength

Although several factors affect the compressive strength of mortar, the most important factor is the mortar's cementitious content. As we will observe later, the total amount of cementitious content (portland cement plus lime) in various types of mortar is roughly constant with respect to the amount of sand. The relative proportions of portland cement and lime are, however, different.

Increasing the amount of portland cement with respect to lime increases the mortar's compressive strength. Conversely, increasing the amount of lime with respect to portland cement decreases the mortar's compressive strength. Because mortar is an integral part of a masonry wall, its strength affects the compressive strength of the wall, Figure 22.5.

EXPAND YOUR KNOWLEDGE

A Historical Note on Cementitious Materials in Mortar

Before the discovery of portland cement (PC), masonry mortar consisted of lime (as the cementitious material) and sand. After the discovery of portland cement, the masons discarded the use of lime and made the mortar using only portland cement and sand. The change from lime mortar to PC mortar was instinctive and spontaneous because portland cement gives a much stronger mortar than lime. Additionally, portland cement sets and hardens far more rapidly than lime, allowing the masons to lay more masonry units per day than using the lime mortar, increasing the masons' productivity.

Remember from Chapter 18 that the setting and hardening of lime is due to its carbonation reaction with carbon dioxide in air. Because the air contains only about 4% carbon dioxide, the setting and hardening of lime is slow. The setting and hardening of portland cement, on the other hand, is due to portland cement's reaction with water, whose quantity is generally far greater than that needed for the hydration reaction.

A few decades later, however, it was discovered that the walls made with PC mortar leaked far more than those made with lime mortar. The investigation of the walls' permeability to water has led us to our understanding of the importance and benefits of using lime in contemporary masonry mortar.

Unlike the compressive strength of mortar, which is a property only of the mortar, bond strength is a property of the masonry wall. It is a measure of the bond between the masonry units and the mortar. It comes into play when an unreinforced masonry wall is subjected to bending (flexure), Figure 22.6.

The bond between a masonry unit and the mortar is both a chemical and a mechanical bond. Therefore, the bond strength of masonry is a function of several factors, such as the types of units, surface roughness of units, artisanry (such as the pressure applied between the units at the time of mortaring), curing conditions (air temperature, wind and humidity), and so on.

Another important factor that affects the bond between the mortar and the units is the amount of water in mortar. The amount of water in mortar must be sufficiently high so that the mortar can flow and be sucked into the minute crevices in masonry units. This develops and improves the bond between the units and the mortar. Everything else being the same, increasing the amount of portland cement in mortar with respect to lime increases the bond strength of masonry.

The bond strength of masonry is pertinent primarily in an unreinforced masonry wall, such as masonry veneer, or in an unreinforced masonry backup wall. In a vertically reinforced masonry wall, Figure 22.7, steel reinforcement resists flexural tension. Therefore,

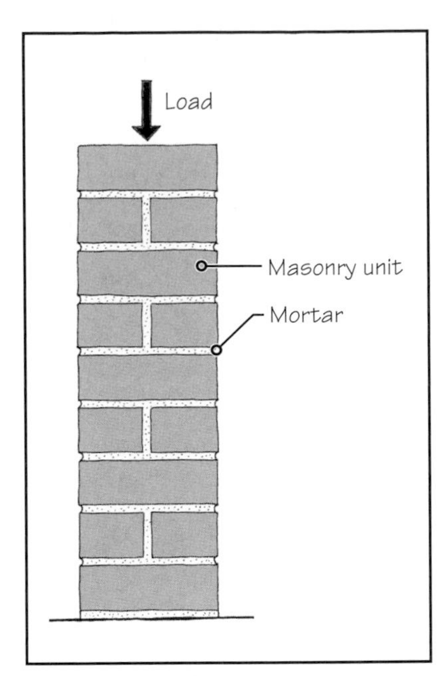

FIGURE 22.5 Because a masonry wall is composed of masonry units and mortar, the compressive strength of the wall is a function of the strength of the units and the strength of the mortar.

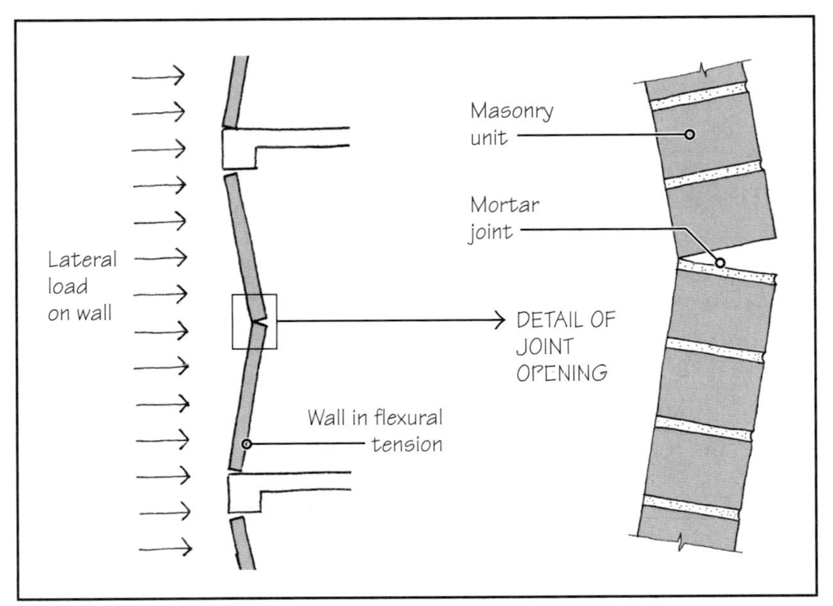

FIGURE 22.6 A wall is subjected to flexural tension when it is subjected to lateral loads (wind, earthquake, etc.). The flexural tension can lead to the wall's cracking. Because the bond between the units and the mortar is much weaker than the tensile strength of the units, the wall generally cracks at the mortar joints, that is, the mortar joints open up.

This illustration shows the opening up of horizontal mortar joints caused by the vertical bending of the wall. If the wall bends horizontally, vertical joints will tend to open.

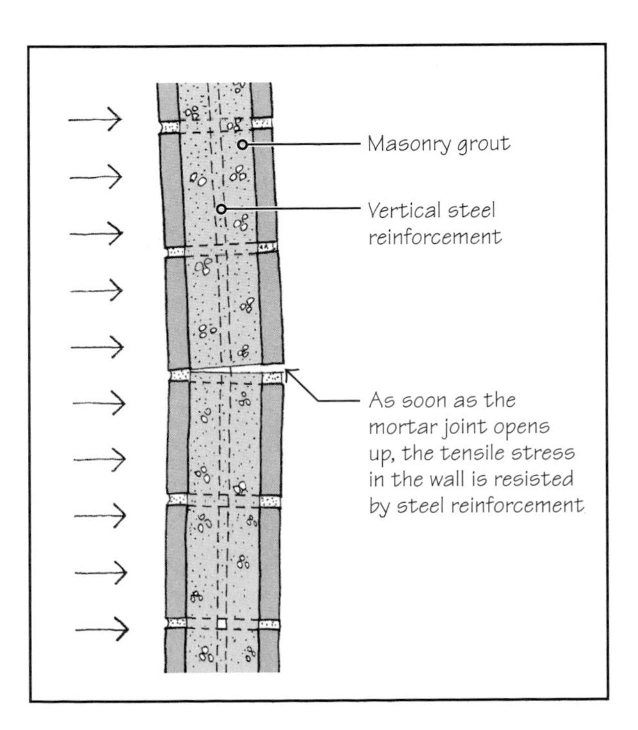

FIGURE 22.7 Flexural tensile stress in a reinforced (and grouted) masonry wall is resisted by the bond between the masonry and mortar only until the mortar joints have opened. This is the case with relatively small lateral loads. As the load increases and a mortar joint opens, the tensile stress is transferred to steel reinforcement. The bond between the masonry and the mortar is now irrelevant. For a three-dimensional illustration of a (vertically) reinforced masonry wall, see Figure 23.7.

NOTE

Autogenous Healing

Because a lime-based mortar is somewhat elastic, it is less liable to crack. However, if cracks are formed, the carbonation of lime fills them. In the carbonation process, lime absorbs atmospheric carbon dioxide (Chapter 18), which increases its mass and volume. The increase helps to fill the voids caused by drying shrinkage of portland cement and by flexural cracking of masonry—a process known as *autogenous healing*.

The term autogenous healing has been borrowed from medicine and refers to the natural healing process of human bones. When a bone in our body fractures, a surgeon simply aligns the fractured parts to their original positions and sustains them in that position using a cast. The fractured pieces fuse together automatically in due course. Although the healing of a bone fracture is complex, it is primarily due to the calcium in our bones—similar to the calcium in lime.

from a purely structural view point, the bond strength is not relevant in a reinforced masonry wall (Figure 22.7).

REQUIRED STRENGTH OF MASONRY MORTAR

The foregoing discussion indicates that the strength of a masonry wall is directly related to the strength of the mortar. It also indicates that an increase in lime (with respect to portland cement) decreases the strength of mortar and, hence the strength of a masonry wall, but it increases the wall's watertightness.

A wall's watertightness is just as important as its strength. In fact both watertightness and strength are somewhat interrelated. Water penetrating a wall corrodes steel reinforcement, ties, and other embedded accessories, which may ultimately reduce the wall's strength. The masonry industry, therefore, favors a relatively low-strength mortar for most masonry walls.

Another reason for choosing a low-strength mortar is to ensure that if the cracking of a masonry wall is to occur, it occurs in mortar joints, not the units, because it is much easier to repair a broken mortar bed than a broken masonry unit. Because a low-strength mortar is more workable, it provides better artisanry and full coverage of joints and, hence, a more watertight wall. In fact, a general recommendation for masonry walls is to

Specify the weakest mortar that will give the required performance.

However, there are situations where the strength of masonry is more important than its watertightness, such as the walls located in high-wind or seismic zones and heavily loaded interior walls. In these situations, a high-strength mortar is recommended.

NOTE

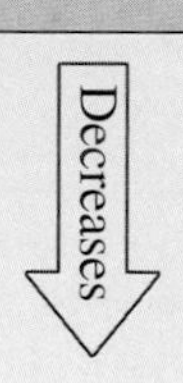

- Workability of mortar
- Water rentivity of mortar
- Watertightness of masonry wall

Lime

Decreases

- Strength of mortar and, hence, the strength of masonry wall, particularly the wall's flexural tensile bond strength

22.2 MORTAR MATERIALS AND SPECIFICATIONS

Mortar is prepared at the construction site in a small mixer (Figure 18.19). Hand-mixing with a shovel is appropriate only for a very small job. The amount of water needed in a mortar mix is not controlled by specifications but left to the discretion of the mason to obtain the required workability. Water for mortar must be clean, potable, and free of impurities.

MATERIALS FOR MASONRY MORTAR

Portland cement (PC) used in mortar is the same as that used in making concrete (Chapter 18). Generally Type I/II portland cement is used. Lime for mortar is Type S lime (Chapter 18).

The sand used in mortar is referred to as *mason's sand.* It may be manufactured sand (pulverized stone) or mined from natural sand deposits. It must conform to ASTM specifications, which require specific grading (particle size variation). This grading gives the mason's sand nearly 30% voids within a given volume of sand. The voids are the spaces between individual sand particles, which are filled by the binder (PC and lime).

PROPORTION SPECIFICATION—PORTLAND CEMENT AND LIME (PCL) MORTAR

Because the cementitious materials fill the voids between the sand particles, the total amount of cementitious compounds in a masonry mortar is almost fixed. Mortar specifications require the amount of sand in a mortar to lie between $2\frac{1}{4}$ to 3 times the amount of cementitious compounds—portland cement + lime (PCL). The sand : PCL ratio is a volumetric ratio, not a weight ratio. In other words

$$\frac{\text{Volume of sand (ft}^3\text{)}}{\text{Volume of PCL (ft}^3\text{)}} = 2.25 \text{ to } 3.0$$

The range allows the masons to adjust the mix for workability and water retentivity of the mortar.

Varying the relative proportions of portland cement and lime in a mortar mix provides different types of mortar—referred to as Types M, S, N, and O, Table 22.1. Different mortar types have different recommended applications.

Mortar Type M has the maximum amount of portland cement and hence the least amount of lime. It gives the strongest mortar. Because of its low lime content, Type M mortar is the least workable, is less elastic in its hardened state, exhibits more shrinkage, and,

TABLE 22.1 TYPES OF PORTLAND CEMENT PLUS LIME (PCL) MORTARS BASED ON MATERIAL PROPORTIONS

	Mortar type	Materials to Be Proportioned by Volume			Recommended applications
		Portland cement (PC)	Lime (L)	Sand	
Strongest Mortar →	M	1	$\frac{1}{4}$	As per ASTM specifications, the amount of sand for all types of mortar is $2\frac{1}{4}$ to 3 times the amount of cementitious materials (PC + L). Generally, however, the amount of sand used is 3 times (PC + L). Thus, in Type S mortar, if PC = 1 ft^3 and L = 0.5 ft^3, the amount of sand is 3(1 + 0.5) = 4.5 ft^3.	Interior masonry floors, foundation walls, parapet walls, and so on
	S	1	Over $\frac{1}{4}$ to $\frac{1}{2}$		Load-bearing and non-load-bearing walls in seismic or high-wind regions, chimneys, and so on
	N	1	Over $\frac{1}{2}$ to $1\frac{1}{4}$		General-purpose mortar for exterior load-bearing and non-load-bearing walls
Weakest Mortar →	O	1	Over $1\frac{1}{4}$ to $2\frac{1}{2}$		Interior non-load-bearing walls in nonseismic and low-wind regions

The amount of water should be as much as that required for workability by the mason. This recommendation is unlike that in concrete, where, for strength reasons, the minimum amount of water is recommended to produce the required workability and consistency.

therefore, results in a wall with greater potential for leakage. It is, therefore, not recommended for use in walls, except below-grade (foundation) walls. It may be used in interior masonry floors for its greater strength and abrasion resistance. (Exterior masonry paving is generally mortarless.)

Type O mortar has the least amount of portland cement and the maximum amount of lime. It is, therefore, the weakest mortar. Because of its low strength (and hence low durability), Type O mortar is not generally recommended for use in contemporary exterior walls. The most common use for type O mortar is for tuck-pointing repairs of old, historic masonry.

Type S mortar is commonly recommended for exterior (load-bearing or non-load-bearing) walls in seismic or high-wind regions. It is preferred over Type N mortar because of its higher bond strength, although it is less water resistant. Type N mortar is the general-purpose mortar, used in most non-load-bearing exterior walls, such as masonry veneer walls. It may also be used in load-bearing walls. However, where a greater strength and durability is needed, Type S mortar should be considered.

Proportion Specification Using Masonry Cement or Mortar Cement

Although PCL mortar is commonly used, a preblended cementitious mix for mortar (in one bag) is available. It is sold as *masonry cement* and is available as Types M, S, and N, Figure 22.8. Thus, to produce a Type S mortar, the mason simply mixes 1 ft^3 of Type S masonry mortar with $2\frac{1}{4}$ to 3 ft^3 of sand. There is no need to use separate portland cement and lime bags.

The use of one-bag masonry cement precludes the need to measure and mix the cementitious materials on site, thus simplifying the preparation of mortar. It also gives a mortar that is more consistent in quality and appearance than a PCL mortar. In masonry cement, cementitious materials are factory blended under a stricter quality-control environment than a site-mixed PCL mortar.

EXPAND YOUR KNOWLEDGE

Letter Designations (M, S, N, and O) for Mortar Types

The letters M, S, N, and O are the alternate letters in the word *masonwork:*

MaSoNwOrK

Although letter *K* does not represent any current ASTM mortar type, it is used in preservation work (as a more appropriate substitute for Type O mortar) to simulate historic masonry, where only lime mortars were used. Type K mortar has 1 part portland cement and between $2\frac{1}{2}$ to 4 parts lime.

The choice of M, S, N, and O has been preferred over Types I, II, III, and IV (or Types 1, 2, 3, and 4) because no type is better than the other. Each type has its own unique set of applications.

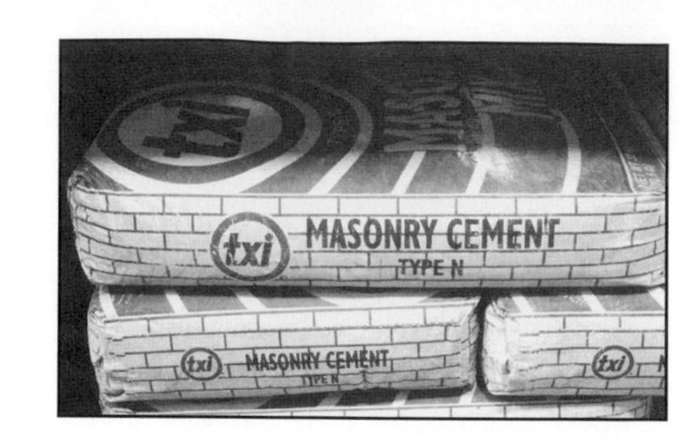

FIGURE 22.8 A Type N masonry cement bag.

Masonry cement may contain cementitious materials other than portland cement may contain and pulverized limestone in place of lime. To improve the workability of mortar, some manufacturers use air-entraining agents in their masonry cement. This reduces the bond between the units and the mortar. Codes recognize these facts by reducing the allowable bond strength of masonry built with masonry cement mortar.

Another preblended cementitious mix for mortar is called *mortar cement.* Like masonry cement, it is also a one-bag mix, but it differs from masonry cement in that it gives the same bond strength as PCL mortar. Masonry codes treat mortars made with mortar cement or PCL as being equivalent. Like masonry cement, mortar cement bags are available in Types M, S, and N.

PROPERTY SPECIFICATION OF MORTAR

An alternative way to specify mortar is by its properties, similar to how concrete is specified. The most distinguishing mortar property is its compressive strength, determined by crushing 2-in. × 2-in. cubes after 28 days.

By testing the cubes, the amounts of cementitious materials can be established in the laboratory to give the required mortar strength. The proportions so established are used in producing the mortar at the construction site. The sand content in a mix obtained from property specification is generally greater than that used in the proportion specification ($2\frac{1}{4}$ to 3 times the cementitious materials).

Mortar produced in the laboratory using the property specification is referred to as Type M mortar if its 28-day compressive strength is at least 2,500 psi or Type S mortar if its compressive strength is at least 1,800 psi, and so on, as shown in Table 22.2.

NOTE

A Note on the Strength of Mortar

The compressive strengths of mortar given in Table 22.2 are 28-day strengths, obtained from tests of 2-in. mortar cubes prepared in the laboratory. The actual compressive strengths of the corresponding mortars in the wall is much higher than the tabular values because of the small ($\frac{3}{8}$-in.) thickness of mortar and the confinement provided by the units.

PROPERTY SPECIFICATION VERSUS PROPORTION SPECIFICATION OF MORTAR

The use of the property specification in place of the proportion specification leads to greater economy in the use of cementitious materials. In other words, a mortar type based on proportion specification generally has greater strength than that given in Table 22.2, but it also uses a greater amount of cementitious materials.

Because the property specification requires laboratory preconstruction testing, it is more cumbersome to follow and enforce. Therefore, the use of property specification is relatively uncommon. It is generally used in situations where a unique mortar type is needed or where the available materials do not meet required specifications (e.g., sand does not meet gradation requirements).

TABLE 22.2 STRENGTHS OF MORTAR TYPES BASED ON PROPERTY SPECIFICATIONS

Mortar type	Compressive strength (psi)
M	2,500
S	1,800
N	750
O	350

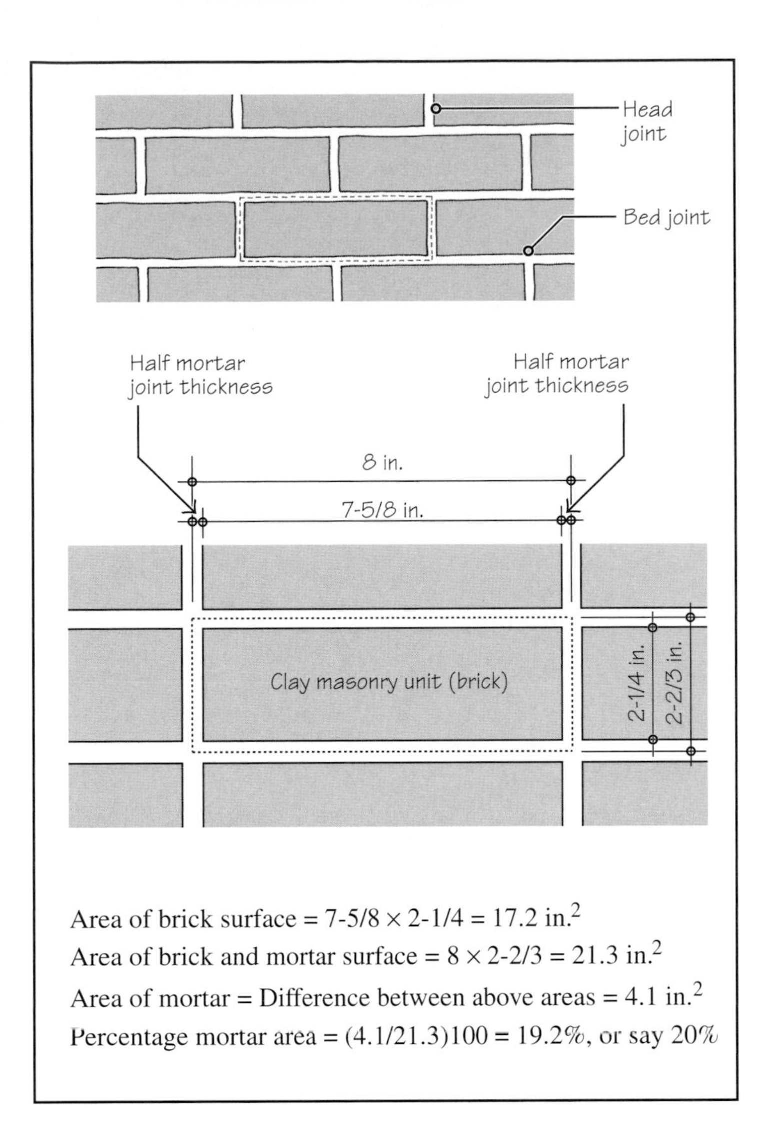

FIGURE 22.9 In a typical brick wall in running bond, mortar occupies nearly 20% of the wall's elevation.

In any case, the specifier must use only one type of specification for a project, either property specification or proportion specification, not both. If neither is specified, the proportion specification governs by default.

WHITE AND COLORED MORTARS

In a typical brick masonry wall (in running bond, see Section 22.7), mortar occupies nearly 20% of the wall's facade area; the remaining 80% is occupied by the bricks, Figure 22.9. Therefore, the color of mortar and the profile of mortar joints contributes greatly to the aesthetics of a masonry wall.

White mortar joints against dark-colored masonry units provide handsome contrast. White mortar against white or lightly colored units gives pleasant harmony. White mortar is obtained by using white portland cement and white sand. White masonry cement is also available, Figure 22.10.

Color pigments can also be added to mortar mix to produce a colored mortar. In colored mortar, white portland cement (or white masonry cement) and white sand are generally used to provide a greater control on mortar color.

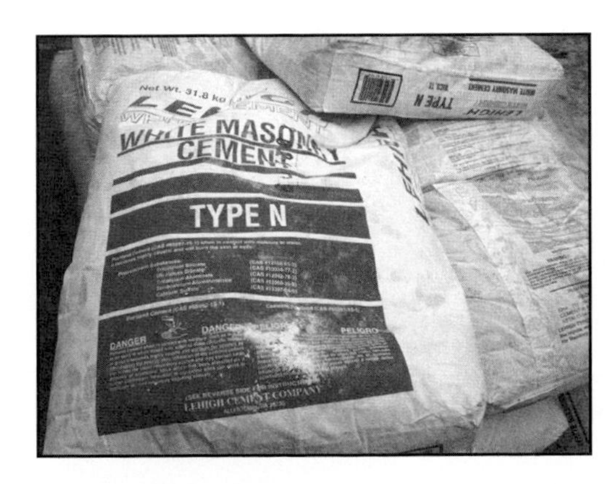

FIGURE 22.10 A Type N white masonry cement bag.

PRACTICE QUIZ

Each question has only one correct answer. Select the choice that best answers the question.

1. The term CMU is generally used as an acronym for
a. clay masonry unit.
b. concrete masonry unit.
c. clay masonry underbed.
d. concrete masonry underbed.
e. cement mortar underbed.

2. The cementitious material(s) in contemporary masonry mortar consist(s) of
a. portland cement.
b. lime.
c. portland cement and sand.
d. portland cement and lime.
e. portland cement, lime, and sand.

3. Which of the following materials increases the mortar's cohesiveness?
a. Mortar joint profile
b. Portland cement
c. Sand
d. Lime
e. All the above

4. Which of the following materials increases the mortar's compressive strength?
a. Mortar joint profile
b. Portland cement
c. Sand
d. Lime
e. All the above

5. Which of the following materials increases the mortar's workability?
a. Mortar joint profile
b. Portland cement
c. Sand
d. Lime
e. All the above

6. A wall made of portland cement mortar (without lime) is more watertight than a wall made of portland cement and lime mortar.
a. True
b. False

7. The flexural tensile bond strength of a masonry wall is relevant in
a. an unreinforced masonry wall subjected to gravity loads.
b. an unreinforced masonry wall subjected to lateral loads.
c. a reinforced masonry wall subjected to gravity loads.
d. a reinforced masonry wall subjected to lateral loads.
e. all the above.

8. The flexural tensile bond strength of a masonry wall is a function of
a. the compressive strength of mortar.
b. the type of masonry units.
c. quality of artisanry.
d. curing of mortar.
e. all the above.

9. For a masonry wall, the compressive strength of mortar should be as high as economically feasible.
a. True
b. False

10. Masonry mortar types are classified as
a. Types M, S, N, and O.
b. Types I, II, III, and IV.
c. Types 1, 2, 3, and 4.
d. Types P, Q, R, and S.
e. either (b) or (c).

11. The mortar with the highest compressive strength is
a. Type M.
b. Type I.
c. Type 1.
d. Type P.

12. The mortar most commonly used in an exterior masonry veneer wall is
a. Type S.
b. Type N.
c. Type I.
d. Type II.
e. Type P.

13. The mortar commonly recommended for use in masonry walls in seismic regions is
a. Type I.
b. Type II.
c. Type P.
d. Type S.
e. Type N.

14. Mortar may be specified either by proportion specification or by property specification. Which of these two is more commonly used?
a. Proportion specification
b. Property specification

15. When mortar is specified using proportion specification, various solid materials that constitute masonry mortar are proportioned based on their
a. wet weights.
b. dry weights.
c. wet volumes.
d. dry volumes.
e. none of the above.

16. Preblended cementitious mix for masonry mortar is available as
a. PCL cement.
b. masonry cement.
c. mortar cement.
d. (a) and (b).
e. (b) and (c).

Answers: 1-b, 2-d, 3-d, 4-b, 5-d, 6-b, 7-b, 8-e, 9-b, 10-a, 11-a, 12-b, 13-d, 14-a, 15-d, 16-e.

NOTE

The diameter of wire used as joint reinforcement is limited to a maximum of half the mortar joint thickness, that is, to a diameter of $\frac{3}{16}$ in. in a $\frac{3}{8}$-in.-thick mortar joint. See Section 23.4 for a description of joint reinforcement.

22.3 MORTAR JOINT THICKNESS AND PROFILES

A masonry wall consists of horizontal and vertical mortar joints, referred to as *bed joints* and *head joints,* respectively (Figure 22.9). For aesthetic and practical reasons, both joints are generally of the same thickness.

The most commonly used mortar joint thickness is $\frac{3}{8}$ in. This thickness has been established after considering several factors, including the embedment of joint reinforcement in mortar. The mortar's ability to provide a cushion between the units and to absorb their dimensional variations implies that the mortar joints cannot be too thin.

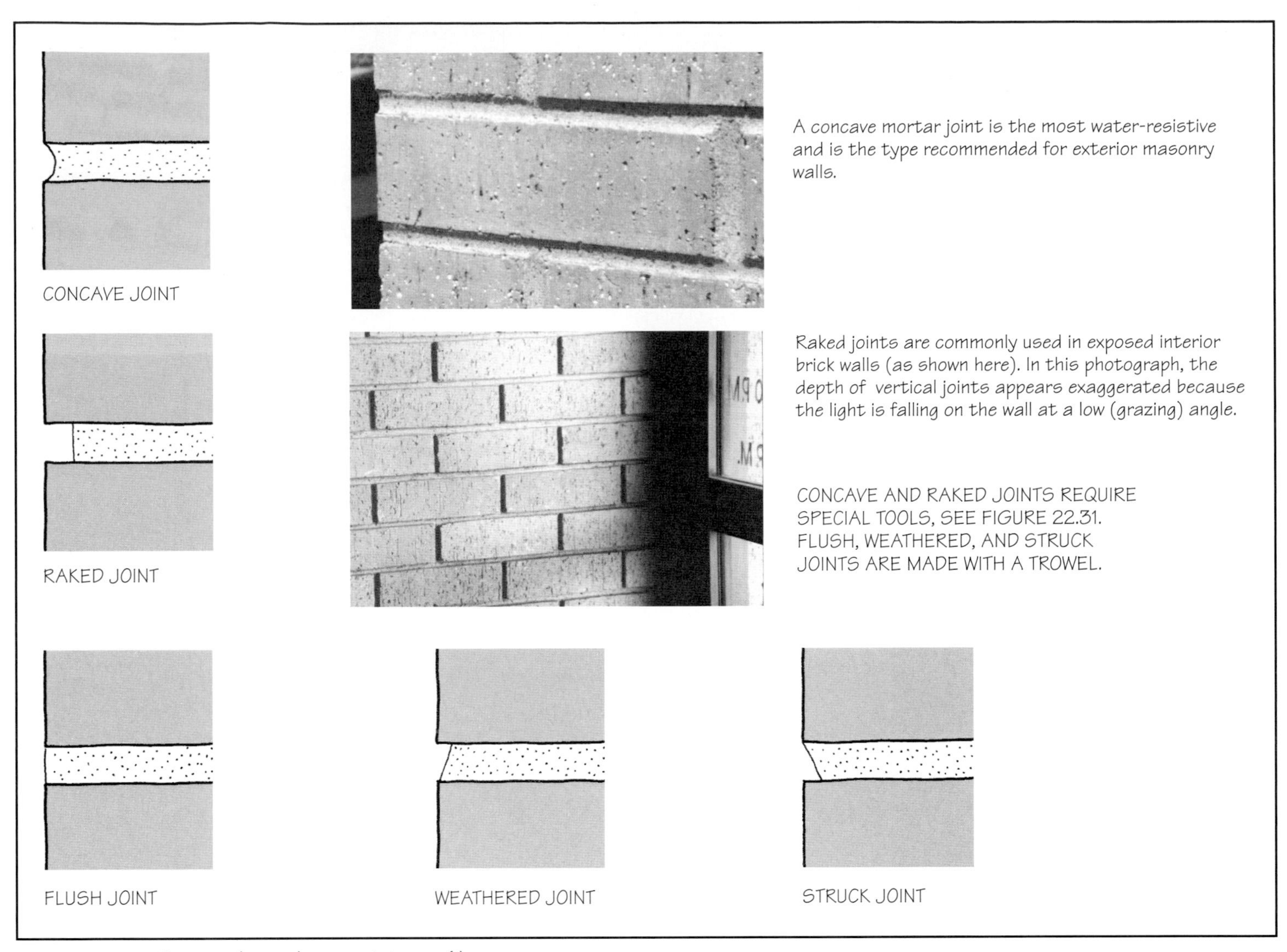

FIGURE 22.11 Commonly used mortar joint profiles.

Mortar Joint Profiles

The mortar joint profile not only affects the aesthetics of the wall, but also the water resistance of the wall. Some commonly used mortar joint profiles, as shown in Figure 22.11, are

- Concave joint
- Raked joint
- Flush joint
- Weathered joint
- Struck joint

The most watertight joint profile is the concave joint. It is obtained by tooling the joint with a cylindrical tool that compresses the mortar against itself and the units (Figure 22.31). Because most water penetrates a masonry wall through the unit-to-mortar interface (relatively little through the units or the mortar), compressing the mortar improves the seal between the units and the mortar. This makes the wall with concave mortar joints more resistant to water penetration than a wall with any other joint profile. A concave joint is also more resistant to freeze-thaw damage, Figure 22.12.

The compression and the resulting seal provided by a concave joint are not provided by other joint profiles. That is why a tooled concave joint profile is generally recommended for all exterior masonry.

For interior masonry, raked joints and flush joints are common. Raked joints highlight the units by providing deeper shadows within the mortar. This effect is dramatized if the light falls at a low angle to the wall. The disadvantage of raked joints is that they collect dust. Flush joints are used for interior masonry where dust is a concern. Between the weathered and the struck joints, the weathered joint gives greater water resistance by directing the water away from the joint.

FIGURE 22.12 Spalling of bricks due to freeze-thaw damage in a brick wall. In freeze-thaw damage, the water absorbed by the units and the mortar expands on freezing, producing compressive stresses in the wall. Over repeated freeze-thaw cycles, the mortar spalls (i.e., crumbles and falls off).

22.4 MANUFACTURE OF BRICKS

In North America, brick (clay) masonry is generally used either as an exterior or interior wall cladding material, for example, brick veneer (Chapter 26). The high strength of bricks makes them durable against freezing and thawing and slow erosion by rainwater and wind—important requirements for an exterior finish. Several brick manufacturers make their bricks with a compressive strength of 6,000 to 10,000 psi. Clay masonry is also suitable as an exterior cladding due to its mass, which provides high fire resistance and sound-insulating properties (Chapter 8). Finally, constructing a brick wall requires a high level of craft. This gives brick facades an aesthetic character not available in other facade materials, such as site-cast concrete, or precast concrete, or insulated metal panels.

The appearance of a brick facade conveys an image of permanence and stability; brick facades are used frequently in significant civic buildings, schools, and college campuses. In fact, as a facade material, brick is by far the most widely used material in contemporary buildings.

The high strength of bricks should make them a logical choice for load-bearing wall applications. However, this is not the case. The reason is that it is more difficult to insulate a brick wall or reinforce it with steel bars than a concrete masonry wall. Therefore, in load-bearing masonry construction, the load-bearing component is generally concrete masonry, with brick as a facade material (Chapter 26).

BRICK MANUFACTURING

The primary raw materials for making bricks are clay and shale. The two main constituents of clay and shale are the oxides of silicon and aluminum. Some minor components are iron and other metal oxides, which are particularly responsible for giving brick its red-brown color. White or light-colored bricks are made by using clay that is naturally deficient in metal oxides and removing whatever metal oxides are present in it. White bricks are generally more expensive than the normal (red-brown) bricks.

Although modern technology has substantially changed the details of brick manufacturing, it is conceptually simple and consists of the following six operations:

- Mining clay from the ground
- Grinding and sieving clay to a fine powder
- Mixing water with sieved clay
- Forming wet clay into the desired brick shape (green bricks)
- Drying green bricks
- Firing dried bricks in a kiln

Brick shapes can be formed by one of the following two methods:

- Extrusion of wet clay through a die—*extruded bricks*
- Molding wet clay—*molded bricks*

NOTE

Difference Between Clay and Shale

Clay is available at the surface of the earth. Therefore, clay and *surface clay* are synonymous terms. *Shale* is also clay but is available deep in the ground. Because of the pressure of the overlying material, shale has hardened to a high compressive strength, almost equalling that of stone.

EXTRUDED BRICKS

The extrusion of wet clay is done by forcing it through a die, which yields a column of clay that slides over a moving belt, Figure 22.13. The process is similar to the extrusion of toothpaste through a tube. The die consists of conical rods, which create core holes in the clay column. The cross-sectional dimensions of the clay column determine the length and width of brick.

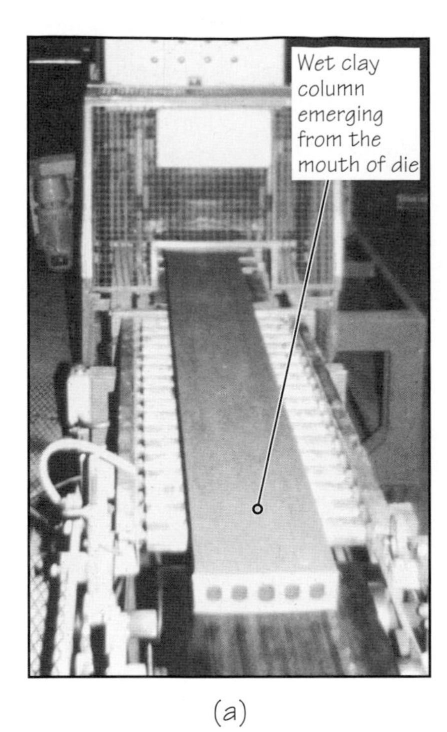

FIGURE 22.13 Manufacture of extruded clay bricks. **(a)** A wet clay column emerging from the die. **(b)** Die mouth with conical rods. (Photos courtesy of Acme Brick Company, Fort Worth, Texas.)

As the clay column moves forward, a wire cutter, consisting of a number of parallel wires, cuts it into individual bricks, Figure 22.14. The spacing between wires is the brick height. If any surface texture is to be applied to bricks, it is applied before the cutting operation, Figure 22.15. If no surface texture is applied, the brick has a smooth surface finish, resulting from the pressure applied by the steel die as the clay passes through it. That is why the smooth texture is referred to as *die skin texture.* A few other commonly used surface finishes are shown in Figure 22.16.

After the bricks have been cut, they are transported to a drying chamber, in which the temperature ranges from 100°F to 400°F. Bricks must be dried before firing them in the kiln to prevent cracking of green bricks. The heat used in the drying chamber is the exhaust heat from the kiln, where the bricks are fired.

The kiln used in modern brick manufacturing plants is a long, tunnel kiln, Figure 22.17. Cars of dried bricks move slowly through the kiln on a rail. The kiln temperature varies from nearly 400°F at the entry point of the dried bricks to a maximum of about 2,000°F. At nearly 2,000°F, clay particles soften (virtually melt) and fuse together, creating a hard and strong brick. Fired bricks are removed from the kiln, sorted, strapped in cubes, and stacked in the yard until delivery to the construction site, Figure 22.18.

CORE HOLES IN EXTRUDED BRICKS

Extruded bricks have through-and-through *core holes.* The primary reason for core holes is that they lead to more uniform drying and firing of bricks. Although the core holes reduce

FIGURE 22.14 A rotary wire cutter cuts the wet clay column into several bricks in one pass. (Photo courtesy of Acme Brick Company, Fort Worth, Texas.)

FIGURE 22.15 Some manufacturers make bricks with special surface textures that are applied to the wet clay column with rotating, textured drums. (Photo courtesy of Acme Brick Company, Fort Worth, Texas.)

(a) Smooth textured brick

(b) Wire-cut texture

(c) Tumbled-brick texture

(d) Rock-face texture

FIGURE 22.16 A few of the several textures available on extruded bricks.

FIGURE 22.17 A typical tunnel kiln. (Photo courtesy of Acme Brick Company, Fort Worth, Texas.)

FIGURE 22.18 A typical storage yard at a brick manufacturing plant.

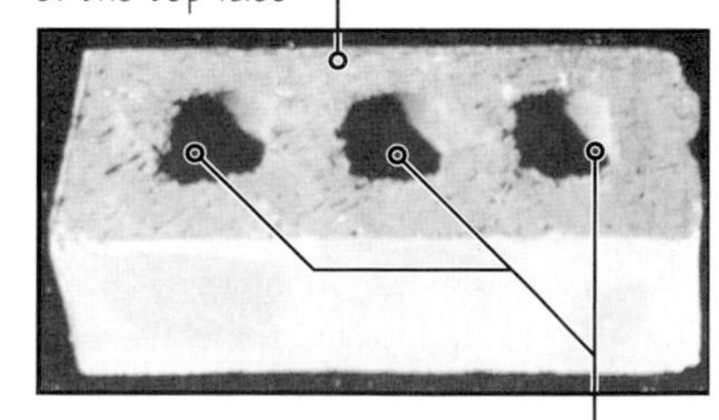

the bearing area of a brick and, hence, the compressive strength of masonry, the improved unit strength resulting from uniform firing compensates for the bearing area reduction. Core holes also improve the bond between the bricks and the mortar. The number and sizes of core holes is manufacturer specific, depending on the size of the brick.

MOLDED BRICKS AND FROGS

Molded bricks are made by force dropping individual lumps of wet clay into brick molds, Figure 22.19. The excess clay is scraped off the molds and green bricks are removed from

FIGURE 22.19 Individual lumps of wet clay are used in forming molded bricks. (Photo courtesy of Acme Brick Company, Elgin, Texas.)

FIGURE 22.20 Several hundred bricks are molded in one single pass in the molding machine. (Photo courtesy of Acme Brick Company, Elgin, Texas.)

the mold immediately thereafter. Modern brick manufacturing is highly mechanized, and several bricks can be molded in one single pass rather than the historic molding of individual bricks, Figure 22.20. In both extruded and molded-brick manufacturing, there is little, if any, manual handling.

A typical molded brick

Because there is little pressure applied in forming the molded bricks, the clay must contain a greater amount of water than that used in the extruded bricks. As the water evaporates, it leaves voids in bricks. Consequently, a molded brick is generally weaker and softer than an extruded brick. However, molded bricks are preferred by some designers and owners because they do not have the machinelike (more precise) appearance of extruded bricks.

Whereas an extruded brick has through-and-through core holes, a molded brick contains a depression, referred to as a *frog*. The frog can take many shapes, depending on the manufacturer. Two such frogs are shown in Figure 22.21. In laying the molded bricks in a wall, the frog faces downward, so that any water entering the wall will not be held by the frog recess.

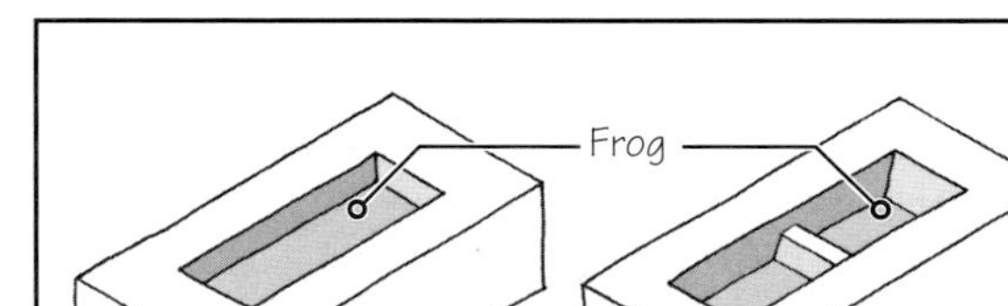

FIGURE 22.21 Two of the several shapes of frogs used in molded bricks.

22.5 DIMENSIONS OF MASONRY UNITS

A masonry unit (both clay brick and concrete block) has three types of dimensions:

- Specified dimensions
- Nominal dimensions
- Actual dimensions

The *specified dimension* of a masonry unit is the finished dimension that the specifier has requested and the manufacturer desires to achieve. However, because the manufacturing process is not perfect, the *actual dimensions* of a unit are different from the specified dimensions. The difference between the specified and the actual dimensions of a masonry unit must lie within the *dimensional tolerance* established by the industry for that product.

The *nominal dimension* of a unit is the specified dimension plus one mortar joint thickness. The nominal dimensions of a unit refers to the space occupied by one unit (and the associated mortar) in the wall. Thus,

$$\text{Nominal dimension} = \text{specified dimension} + \text{one mortar joint dimension}$$

Because the standard mortar joint thickness is $\frac{3}{8}$ in., the nominal dimension of a masonry unit is $\frac{3}{8}$ in. greater than the specified dimension. Thus, if the specified length of a masonry unit is $7\frac{5}{8}$ in., its nominal length is 8 in.

In practice, the nominal dimensions of units are generally given, and in doing so, the inch labels are not used. Inch marks are used with the specified dimensions. Thus, if

$$\text{A unit's nominal dimensions} = 4 \times 2\tfrac{2}{3} \times 8,$$
$$\text{Its specified dimensions} = 3\tfrac{5}{8}\text{ in.} \times 2\tfrac{1}{4}\text{ in.} \times 7\tfrac{5}{8}\text{ in.}$$

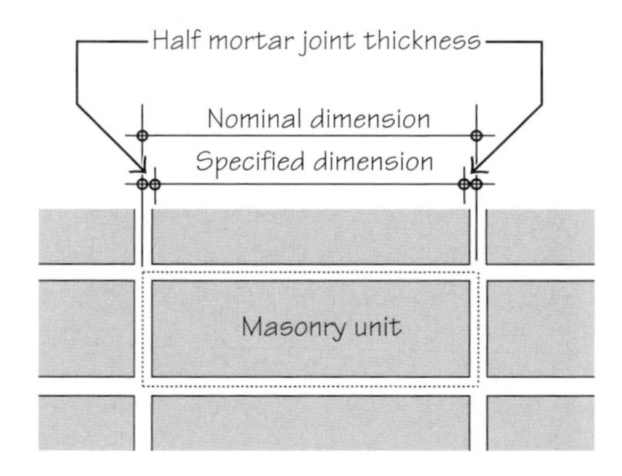

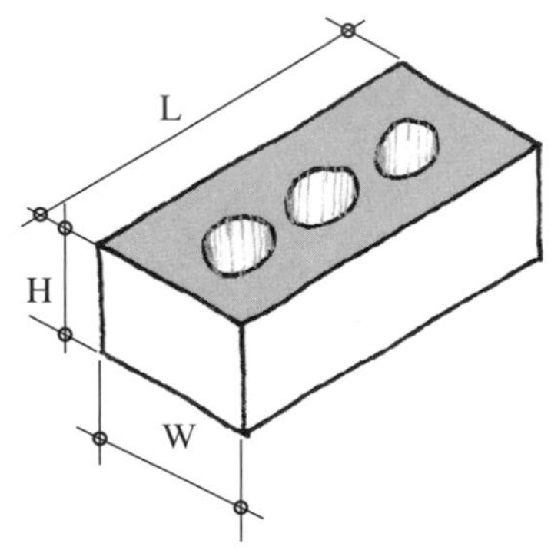

Masonry unit dimensions must be stated in the following sequence: W × H × L.

Sequencing Masonry Unit Dimensions

Observe the sequence in which the unit's dimensions have been stated. The width (W, i.e., *through-the-wall* dimension) is stated first, followed by the face dimensions—height (H) × length (L). Specifying unit's dimensions using a sequence other than W × H × L will confuse the subcontractor, manufacturer, or supplier.

Hollow Versus Solid Masonry Units

Both bricks and concrete blocks generally contain voids. In the masonry industry, a *solid* masonry unit is less than 25% hollow (≥75% solid cross-sectional bearing area). A *hollow* masonry unit is 25% or more hollow.

Most bricks are solid units because their total core area is less than 25%. A brick without cores must be specified as *100% solid.* A 100% solid brick is generally used for paving or as coping at the top of walls. Because brick-manufacturing machinery is set up to make cored or frogged bricks, 100% solid bricks are special bricks.

PRACTICE QUIZ

Each question has only one correct answer. Select the choice that best answers the question.

17. The most commonly used mortar joint thickness is
- **a.** $\frac{3}{4}$ in.
- **b.** $\frac{5}{8}$ in.
- **c.** $\frac{1}{2}$ in.
- **d.** $\frac{3}{8}$ in.
- **e.** $\frac{1}{4}$ in.

18. Which of the following mortar joint profiles should be specified for an exterior wall?
- **a.** Concave joint
- **b.** Raked joint
- **c.** Flush joint
- **d.** Weathered joint
- **e.** Struck joint

19. Bricks that have core holes are
- **a.** extruded bricks.
- **b.** molded bricks.
- **c.** (a) and (b).

20. Bricks that have frogs are
- **a.** extruded bricks.
- **b.** molded bricks.
- **c.** (a) and (b).

21. The primary reason for providing core holes in bricks is to
- **a.** reduce the weight of bricks.
- **b.** provide uniform drying and firing of bricks.
- **c.** increase the shear resistance of masonry walls.
- **d.** none of the above.

22. The nominal dimension of a masonry unit whose actual dimension equals $7\frac{5}{8}$ in. is
- **a.** 7 in.
- **b.** $7\frac{1}{4}$ in.
- **c.** $7\frac{1}{2}$ in.
- **d.** $7\frac{3}{4}$ in.
- **e.** none of the above.

23. A masonry unit that has no voids (e.g., no core holes) is referred to as a
- **a.** solid unit.
- **b.** 100% solid unit.
- **c.** (a) or (b).

24. A masonry unit that has 20% voids (e.g., 20% core hole area) is referred to as a
- **a.** solid unit.
- **b.** hollow unit.
- **c.** partially hollow unit.
- **d.** none of the above.

25. In specifying the dimensions of a masonry unit, which dimension is stated first?
- **a.** Length, L, of unit
- **b.** Height, H, of unit
- **c.** Width, W, of unit

26. In specifying the dimensions of a masonry unit, which dimension is stated last?
- **a.** Length, L, of unit
- **b.** Height, H, of unit
- **c.** Width, W, of unit

Answers: 17-d, 18-a, 19-a, 20-b, 21-b, 22-e, 23-b, 24-a, 25-c, 26-a.

22.6 TYPES OF CLAY BRICKS

Bricks are available in various sizes. The most commonly used brick is the (extruded) *modular brick,* which measures $3\frac{5}{8}$ in. × $2\frac{1}{4}$ in. × $7\frac{5}{8}$ in. Its nominal dimensions are $4 \times 2\frac{2}{3} \times 8$. Some of the other commonly used brick sizes are shown in Figure 22.22.

Types of Clay Bricks Based on Use

Bricks used in different applications must obviously be different. The bricks used as face veneer must have lower dimensional tolerances and be more weather resistant than bricks used in backup brick walls. Similarly, the bricks used in paving and flooring must be more abrasion resistant than facing bricks or backup bricks. Some of the commonly specified brick types are

- Facing bricks—solid or hollow
- Building bricks—solid or hollow
- Paving bricks—100% solid

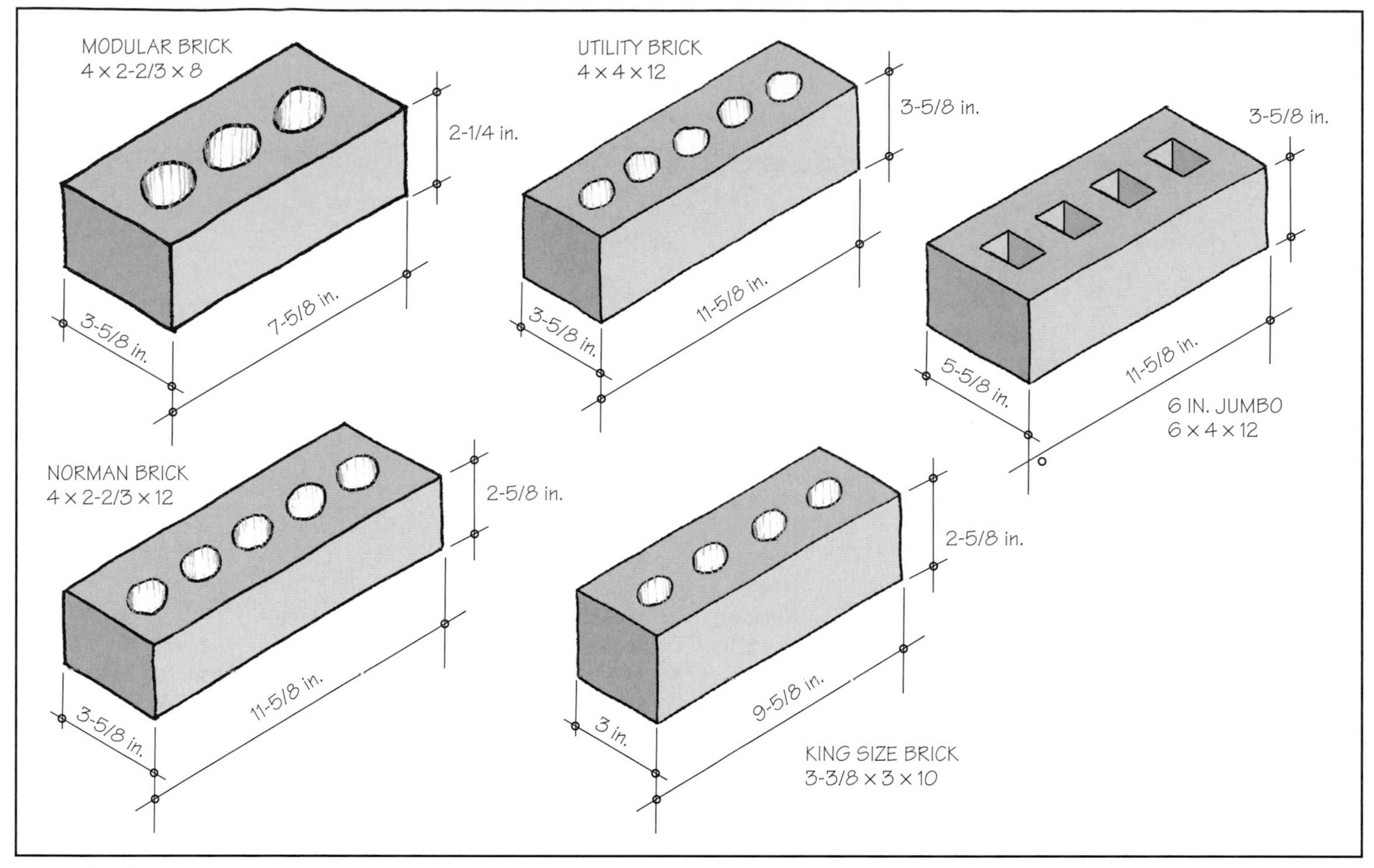

FIGURE 22.22 Commonly used brick types.

Because clay masonry units are generally used in face veneers, the use of facing bricks is among the most common, followed by paving bricks and building bricks. We discuss these three types here. Information about other types may be obtained from applicable ASTM standards.

FACING BRICKS

Facing bricks are used in exposed exterior or interior walls and are classified as follows:

- *Type FBS,* in which the dimensional tolerances, chippage, and warpage are *standard* for the industry (hence the letter *S* in the type designation). Type FBS is the default specification for facing bricks. In other words, where no type is specified, Type FBS is provided.
- *Type FBX,* in which the lowest dimensional tolerances, chippage, and warpage are permitted. In other words, the dimensional tolerances in Type FBX are smaller than those permitted in Type FBS. The letter *X* in the type designation stands for *extra* special.
- *Type FBA,* which is specially manufactured with large variations in dimensions, textural effects, and warpage to produce architectural facing bricks (hence the letter *A* in the type designation).

All facing bricks must meet durability requirements of the location. The durability of facing bricks is represented by their grade, which may either be

- Grade SW or Grade MW

Thus, an FBS (or FBX or FBA) brick can either be of Grade SW or MW. Grade SW bricks can withstand severe weathering conditions, and Grade MW can withstand medium weathering conditions. Table 22.3 gives the durability requirements for brick grades. The durability of a brick refers to its freeze-thaw resistance, which is a function of the compressive strength of brick and the amount of (cold and boiling) water absorbed by the brick.

The selection of a particular grade is based on the ability of the brick to resist damage caused by freeze-thaw cycling when it is wet. One method to assess the freeze-thaw exposure that a brick will experience is to use weathering index (WI) of the location, and the position of the brick in the building, Figure 22.23. WI is an estimate of the freeze-thaw potential of the geographic location.

NOTE

ASTM and Facing Brick

The most referenced standard for facing bricks is ASTM C216, "Standard Specification for Facing Brick."

TABLE 22.3 PHYSICAL REQUIREMENTS FOR VARIOUS GRADES OF BRICKS

Grade	Minimum compressive strength (psi) based on gross area (brick flatwise)		Maximum water absorption (%) 5-h boiling		Maximum saturation coefficient	
	Average of 5 bricks	Individual brick	Average of 5 bricks	Individual brick	Average of 5 bricks	Individual brick
SW	3,000	2,500	17.0	20.0	0.78	0.80
MW	2,500	2,200	22.0	25.0	0.88	0.90
NW	1,500	1,250	No limit	No limit	No limit	No limit

All three grades are used for building bricks. Facing bricks are made in SW and MW grades only.

Source: Brick Industry Association (BIA). *Technical Notes on Brick Construction (9A)*, "Manufacturing, Classification and Selection of Brick."

If WI ⩾ 500, the location has a *severe* freeze-thaw potential. If WI lies between 50 and 500, the location has *medium* freeze-thaw potential. If WI < 50, the location has *negligible* freeze-thaw potential. Most brick manufacturers make only SW bricks to give their product the versatility of use in any climate and any position in a building.

NOTE

ASTM and Building Bricks

The most referenced standard for building bricks is ASTM C 62; "Standard Specification for Building Brick."

BUILDING BRICKS

Building bricks are similar to facing bricks except that they do not have any limitations on dimensional variations, warpage, or chippage. Thus, building bricks are rated for durability and strength only; that is, they carry a grade—Grade SW, MW, or NW. The specifications

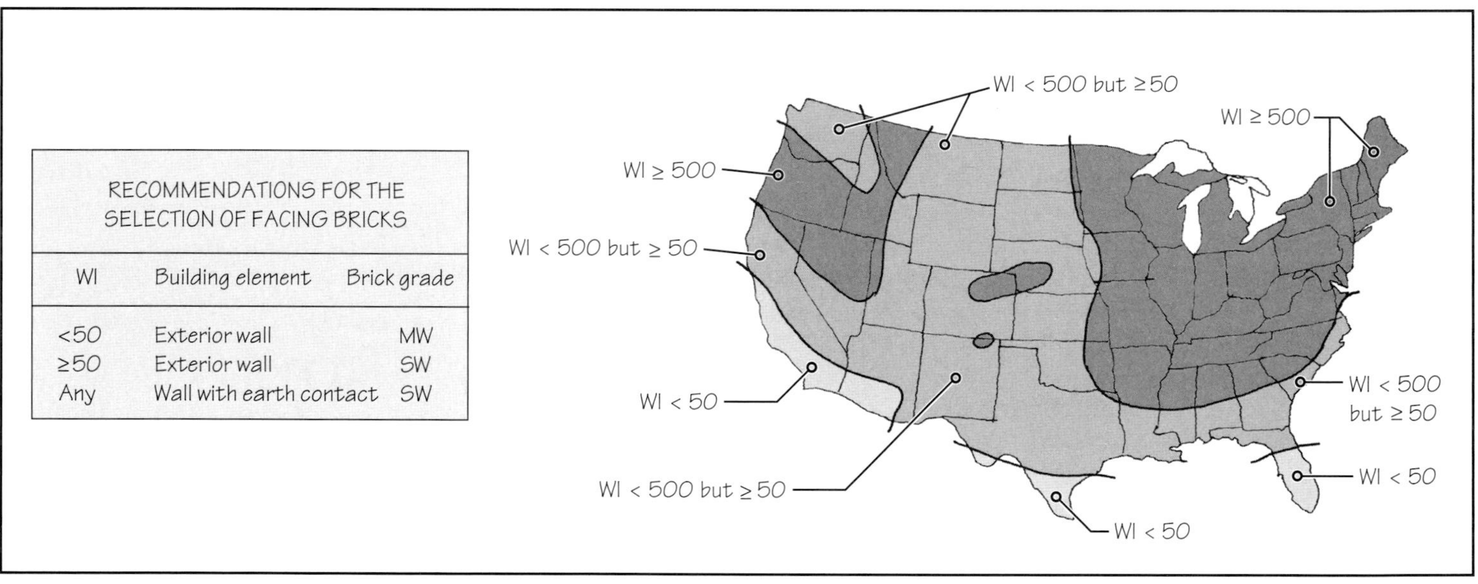

FIGURE 22.23 Weathering index (WI) map of the United States.

EXPAND YOUR KNOWLEDGE

Boiling Water Absorption, Saturation Coefficient, and Freeze-Thaw Resistance

To determine the saturation coefficient of a brick, we first immerse it in cold, that is, room-temperature, water for 24 h. The amount of water absorbed by the brick in 24 h is measured. We then boil the same brick for 5 h and determine the amount of water absorbed by the brick. The ratio of cold-water absorption (C) to boiling-water absorption (B) is defined as the saturation coefficient (SC) of the brick; that is, SC = C/B. Because the brick generally absorbs more water on boiling, B ⩾ C, that is, SC ⩽ 1.0.

For example, assume that a dry brick weighs 4.5 lb, which after 24-h immersion in cold water weighs 5.0 lb. Therefore, C = 5.0 − 4.5 = 0.5 lb. Assume further that when the same brick is boiled for 5 h and weighed thereafter, it weighs 5.2 lb, giving B = 5.2 − 4.5 = 0.7 lb. The saturation coefficient of brick is

$$SC = \frac{C}{B} = \frac{0.5}{0.7} = 0.71$$

The greater the difference between the two absorptions (i.e., the smaller the SC), the greater the ability of the brick to allow for the absorbed water to expand within the brick on freezing.

Although not perfect, the boiling-water absorption, the saturation coefficient, and the compressive strength of a brick are the best-known measures at the present time of a brick's freeze-thaw resistance.

Weathering Index of a Location

The weathering index (WI) of a location is a measure of the freeze-thaw severity of a location, and is defined as:

WI = (average annual freezing-cycle days) (average annual winter rainfall)

Thus, if the number of annual freezing-cycle days for a location is 24 and the total annual winter rainfall for that location is 20 in., WI = 24(20) = 480.

A *freezing-cycle day* is defined as the day on which freeze and thaw occur. In other words, a freezing cycle day is a day when the air temperature passes through 32°F. It is not necessary to examine maximum and minimum temperatures of each day of the year to determine the number of feezing-cycle days of a location. Instead, the number of freezing-cycle days can be obtained by (number of days when the minimum temperature ≤32°F) − (number of days when the maximum temperature ≤32°F). See the adjacent sketch.

Annual winter rainfall is the total rainfall (in inches) between the first killing frost in the fall and the last killing frost in the spring.

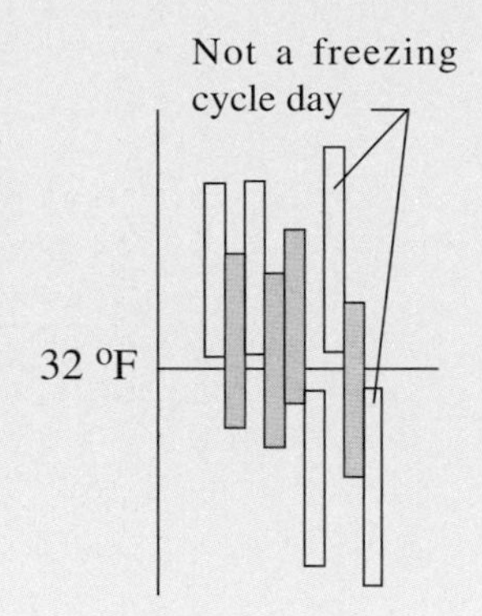

for Grade SW and MW in both facing bricks and building bricks are identical. Building bricks have an additional grade NW (Table 22.3).

Building bricks are generally used in brick walls that are covered with another facing material. However, they may be used in face veneers if their warpage, chippage, and dimensional variations are acceptable.

Paving Bricks

Paving bricks are graded for freeze-thaw as Class SX, MX, or NX—from most freeze-thaw resistant to least freeze-thaw resistant. They are also graded for abrasion resistance as Types I, II, and III—from the highest abrasion resistance to least abrasion resistance.

NOTE

ASTM and Paving Bricks

The most referenced standard for paving bricks is ASTM C 902, "Specification for Pedestrian and Light Traffic Paving Bricks" or ASTM C 1272, "Specification for Heavy Vehicular Paving Bricks."

22.7 BOND PATTERNS IN MASONRY WALLS

Bricks can be assembled in a wall in several patterns, referred to as *bond patterns,* or simply as *bonds.* The purpose of a bond is functional as well as aesthetic. Functionally, the bond is meant to stagger the units so that the load on one unit is shared by an increasing number of underlying units. A one-*wythe* masonry wall (a wall whose thickness equals the width, W, of one unit), built from whole (uncut) units, can have two types of bonds, Figure 22.24:

- Running bond
- Stack bond

A stack bond is used primarily for aesthetic reasons, Figure 22.25. As explained in Section 22.4, the (horizontal) bending strength of a stack-bond wall is lower than that

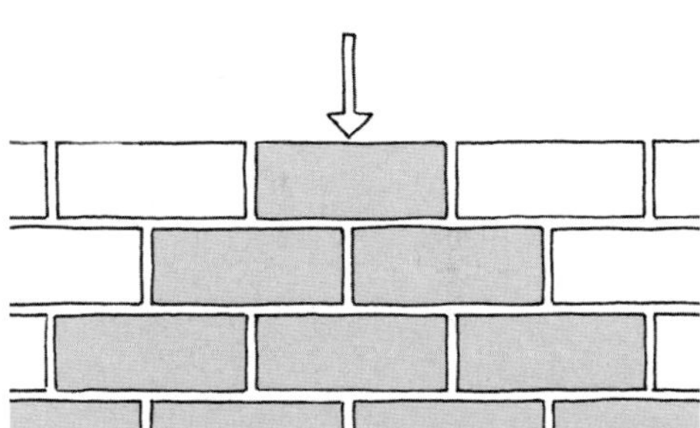

The staggering of masonry units and gravity load distribution. Also see the box "Arching Action in Masonry Walls" in Chapter 23 (Section 23.1).

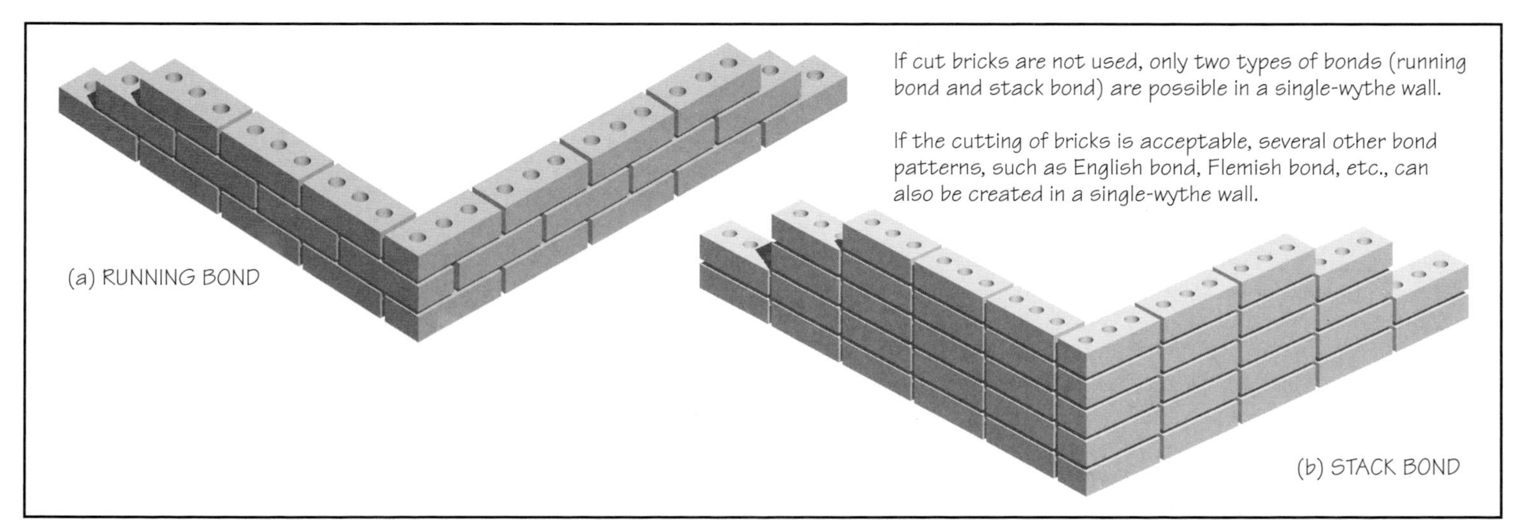

FIGURE 22.24 Single-wythe walls in **(a)** running bond and **(b)** stack bond.

FIGURE 22.25 Stack-bonded walls.

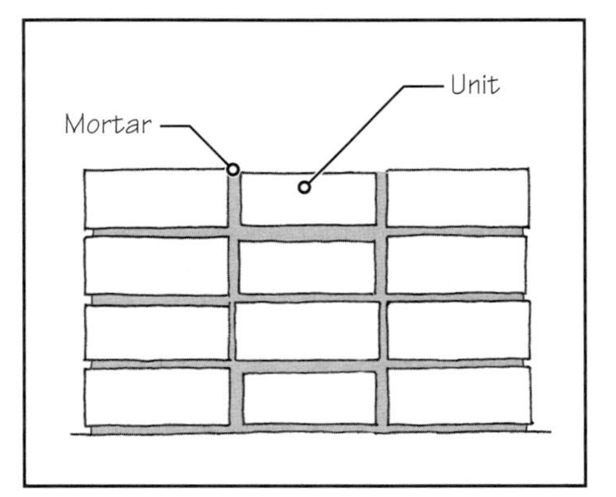

FIGURE 22.26 A stack-bonded wall with large variations in masonry unit sizes is visually unacceptable.

of a running-bond wall. Therefore, a stack-bond wall must be provided with horizontal reinforcement.

The bricks used in a stack-bond wall should have as much dimensional uniformity as possible so that the mortar joints have the least variation in width, Figure 22.26.

BRICK ORIENTATIONS ON WALL ELEVATION

In Figure 22.24, the bricks have only one orientation. The orientation refers to the exposed face of the brick. The exposed face of the brick in Figure 22.24 is called the *stretcher face,* or simply the *stretcher*. Being a six-faced figure, a brick can have six different face orientations on a wall elevation, Figure 22.27:

- Stretcher
- Header
- Rowlock
- Soldier
- Shiner
- Sailor

When a course of masonry consists of stretchers only (as in running-bond and stack-bond walls of Figure 22.24), it is called a stretcher course. If it consists of headers only, it is called a header course, and so on. Single-wythe walls are generally made of all stretcher courses (Figure 22.24) and are often terminated at the top with one or more soldier courses to cover the core holes, Figure 22.28.

ENGLISH, FLEMISH, AND COMMON (AMERICAN) BONDS

As stated previously, contemporary brick masonry is used mainly as cladding. Therefore, most brick walls are one-wythe walls in running bond. However, fence walls are generally two-wythe walls or sometimes three-wythe walls. In two- or multiple-wythe walls, several bond patterns are possible.

NOTE

The Terms: Masonry Wythe and Masonry Course

A masonry *wythe* is a vertical section of masonry that is one unit thick. The walls shown in Figure 22.24 are both one-wythe walls. Thus, if a wall is built out of modular bricks, a wall that is 4 in. (nominal) thick is a one-wythe wall. If the wall is 8 in. (nominal) thick, it is a two-wythe wall. A 12-in.-thick wall is a three-wythe wall, and so on. The walls shown in Figure 22.29 are two-wythe walls.

A masonry *course* is a horizontal layer of masonry units. The English and Flemish bond walls of Figures 22.29(a) and (b) consist of 4 courses. The wall of Figure 22.29(c) consists of 14 courses.

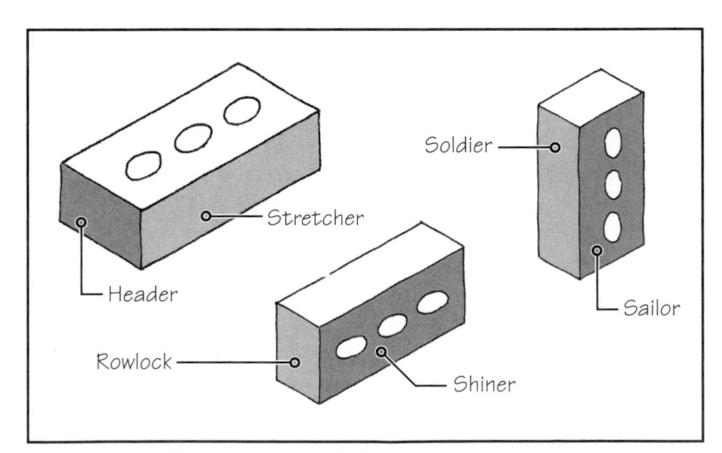

FIGURE 22.27 Orientations of bricks on a wall elevation.

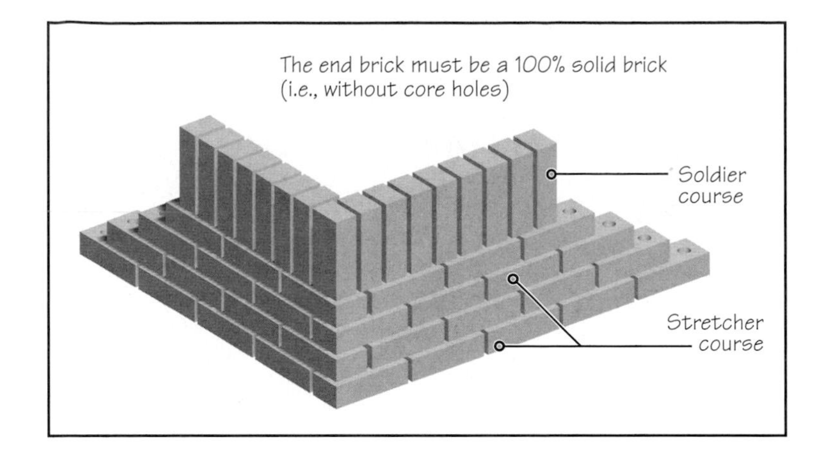

FIGURE 22.28 A single-wythe wall is often terminated with a soldier course at the top to cover the core holes in stretcher courses.

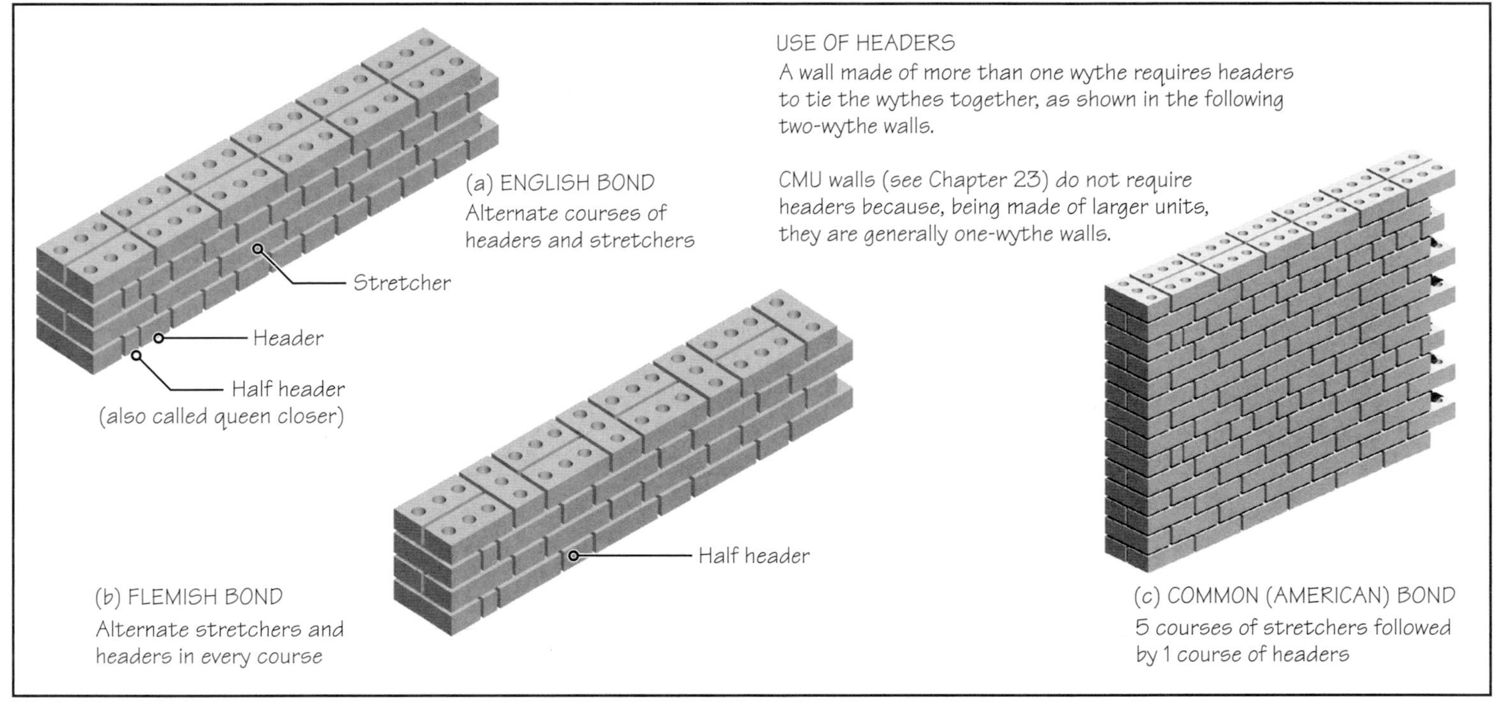

FIGURE 22.29 Two-wythe brick walls—in English, Flemish, and Common bonds.

Three commonly used bonds are the *English bond, Flemish bond,* and *American bond,* Figure 22.29. An English bond consists of alternate courses of headers and stretchers. In Flemish bond, there are alternate headers and stretchers in each course. In common (American) bond, there is a header course after every five courses of stretchers.

Just as a single-wythe wall is terminated with a soldier course to cover the core holes, a two-wythe wall is terminated with a rowlock course, Figure 22.30. Alternatively, a precast concrete or stone coping may be used as termination.

The bonds shown in Figure 22.29 use headers to tie the two wythes of masonry. If a header-tied masonry wall is used in a heated or cooled building where the interior and exterior faces of the wall have a large temperature difference, the headers can break. Therefore,

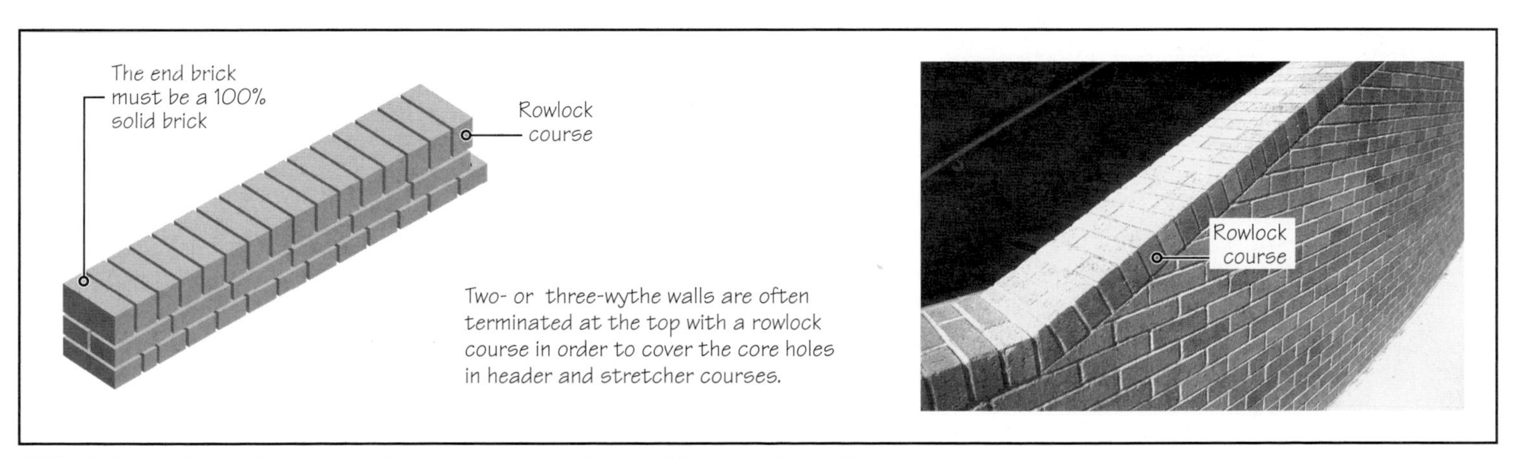

FIGURE 22.30 Rowlock course as the terminal course in a multiwythe brick wall.

in contemporary masonry, English, Flemish, or American bonds, are normally used in fence walls in which the opposite faces are not exposed to temperature differentials.

22.8 THE IMPORTANCE OF THE IRA OF BRICKS

If bricks are too porous, they can suck too much water out of the mortar, leaving insufficient water for the hydration reaction (between portland cement and water). This can weaken the mortar and, hence, the wall. Bricks that are highly absorptive must, therefore, be wetted with water prior to being laid in the wall. The wetting of bricks should be such that the interior of the brick is moist but the exterior surfaces of the brick are dry.

The preferred way of wetting absorptive bricks is to let water run on the pile of bricks so that the bricks become wet. Depending on the weather, this may either be done the evening before (or a few hours before) the bricks are laid. This ensures that the interiors of the bricks are wet but the exteriors are dry. Bricks that are wet on the surface will tend to float on the mortar bed and may also bleed water out of the mortar.

Bricks that are not highly absorptive do not need wetting and can be laid dry. Brick specifications use a measure referred to as the *initial rate of absorption* (IRA) to determine whether the bricks require wetting or not. If IRA $>$ 30 g/min/30 in.^{2}, the bricks should be wetted prior to laying; otherwise, not. (Note that in colder regions, wetting of bricks may not be appropriate.)

22.9 THE CRAFT AND ART OF BRICK MASONRY CONSTUCTION

The construction of a brick wall can appear deceptively simple to a lay observer, particularly in the hands of an expert mason. As compared to wood framing or concrete placing, however, the construction of brick walls requires much greater skill and craft. Additionally, it is

The mason spreads the mortar over a previously-laid brick course. In the case of a brick veneer wall (shown here), the mason ensures that little or no mortar falls in the cavity space behind the veneer.

The mason picks up a brick to be laid, and mortars ("butters"—a term used by the masons for mortaring) the head joint of the brick before laying it in the wall. Some masons will fully butter the brick to be laid. Other masons will fully mortar the head joint of the already-laid brick, and only butter the heads of the brick to be laid. In any case, all joints must be fully mortared.

Using the trowel handle, the mason presses the freshly-laid brick against the adjacent bricks to give a better bond and a more watertight mortar joint.

FIGURE 22.31 Steps involved in constructing a single-wythe brick wall to the string line.

EXPAND YOUR KNOWLEDGE

IRA and Its Measurement

The initial rate of absorption (IRA) is obtained by immersing a dry brick in water to a depth of $\frac{1}{8}$ in. The immersion is done for only 1 min, after which the brick is removed. The brick is weighed before and after immersion. If the weight of dry brick is W_d and of the wet brick is W_w, then IRA is given by

$$\text{IRA} = \frac{W_w - W_d}{A_n}(30 \text{ in.}^2) \quad \text{where } A_n = \text{the immersed area of the brick}$$

If IRA > 30 g/min/30 in.^{2} of the brick's immersed surface area, the bricks should be wetted; otherwise, not. A surface area of 30 in.^{2} has been chosen because it gives a convenient figure of 1 g of water absorption per square inch of a brick's surface.

slow and labor intensive. To speed construction, several masons work simultaneously on building a wall. This requires coordination and quality control to ensure uniformity of artisanry.

The first step in constructing a brick wall is to lay the corners. The string line is then stretched between the corners and the bricks are laid to the string line, maintaining a level and plumb wall. In addition to the string line, the primary tools used by a mason are a trowel and level. Some masons use a story pole to give uniformity in course heights.

The mortar joints are tooled after the mortar has become sufficiently hard but is still pliable (referred to as *thumbprint hard*). Figure 22.31 shows a typical sequence of steps involved in constructing a brick wall.

Using the trowel, the mason removes the extra mortar that has been squeezed out of the joint. Because the mortar has good water retentivity, very little mortar sticks to the brick face. However, the brick face is cleaned with a brush after the mortar has somewhat dried.

After the mortar has stiffened (referred to as "thumb-print hard" in the industry), the mason tools the joints. In this photograph, the tooling is being done to obtain concave joints.

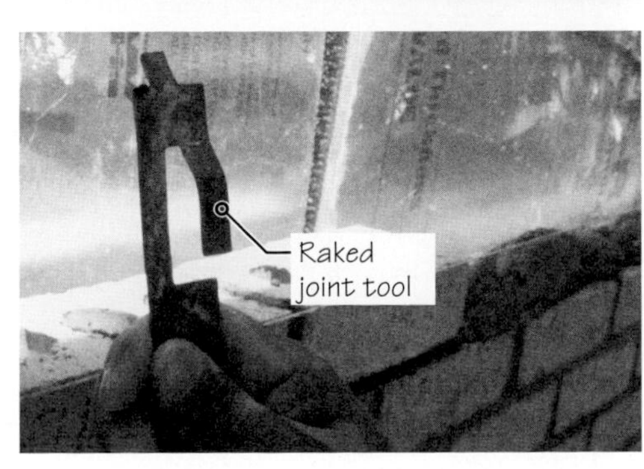

The left-side image shows a concave joint tool and the right-side image shows a raked joint tool.

FIGURE 22.31 (*Continued*) Steps involved in constructing a brick wall.

FIGURE 22.32 Brick walls even in routine buildings have some measure of interest, created through accent bands consisting of bricks with different bond patterns, colors, and textures, bricks laid in different planes, or different masonry units.

NOTE

Retempering Mortar

Retempering mortar (the addition of water to unused mortar that has somewhat dried), is acceptable provided the mortar has not stiffened excessively. In general, mortar should be prepared immediately prior to its use. If needed, the plastic life of unused mortar can be extended somewhat by covering it with a plastic sheet.

Retempering reduces the strength of mortar, but if it is done within $2\frac{1}{2}$ h of initial preparation, the results are acceptable. Therefore, mortar prepared more than $2\frac{1}{2}$ h (depending on the weather) prior to its use should be discarded. Retempering colored mortar may result in undesirable color variation.

Because brick masonry is handcrafted, it lends itself easily to artistic expression. This ease is due to the large number of bond patterns in which the standard bricks can be arranged in a wall, the availability of bricks in different colors and textures, and the brick's potential to harmonize aesthetically with other masonry materials. In fact, brick walls, even in commonplace buildings, are seldom prosaic and monotonous, Figure 22.32.

Bricks have also been used to create sculptures and wall murals, Figure 22.33. The realization of such works of art requires full-size drawings, sculpting of each brick in wet clay by the sculptor, and firing the sculpted bricks in a kiln. Finally, the fired bricks are assembled in the wall by the sculptor as per the drawings.

Repointing of Mortar Joints

Mortar may deteriorate due to freeze-thaw action and/or erode from water. This leads to receding joint depth, resulting in a leaky wall. The recessed joints need to be refilled with mortar—a process referred to as *repointing*.

Repointing of mortar joints used to be far more common when lime mortar was used. It is relatively rare in contemporary masonry because of the use of some portland cement in mortar. However, there are situations, particularly with Type O mortar, that the joints in some masonry walls may need repointing. Repointing involves raking the existing joints and filling them with mortar.

Raking involves grinding into the joint to obtain adequate depth (typically twice the mortar joint thickness) and filling the joint with the new mortar, Figure 22.34. Before filling the joint, the joint is wetted to improve the bond between the existing and the new mortar.

FIGURE 22.33 A small part of a large brick mural.

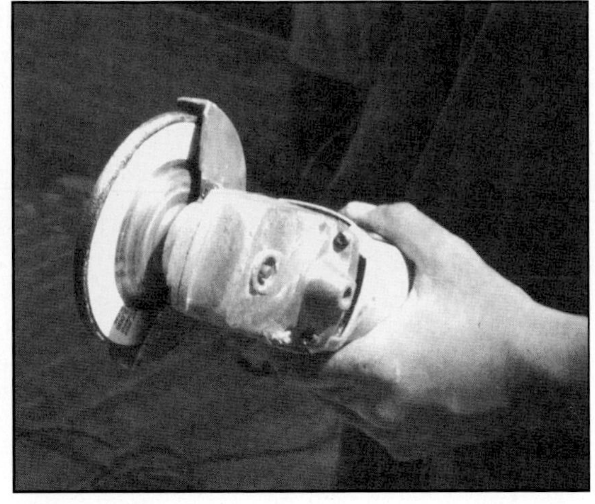

FIGURE 22.34 Grinding of mortar joints in preparation for repointing. The photo at right shows the grinding tool.

22.10 EFFLORESCENCE IN BRICK WALLS

Effloresecence is the deposit of a white substance on a masonry wall. This substance consists of water-soluble salts, present in masonry units, in the mortar, or both, which migrate to the outside of the units as the water evaporates, Figure 22.35. Efflorescence does not generally create any structural or sanitation problem, but it is unsightly.

Although efflorescence may occur in both concrete and clay masonry, it is more common and more obvious in brick masonry walls because of the dark color of bricks. The likelihood of efflorescence can be reduced by selecting masonry units that have been tested for the absence of soluble salts (per ASTM C 67).

Efflorescence may also originate from mortar due to soluble salts in water or sand. Minor efflorescence can be removed by washing the wall, allowing it to dry, and then brushing off the dry salt. Repeating this cycle a few times should leach out all the salts. Efflorescence that appears several years after the wall was built could be due to unclean water penetrating the wall, such as from a lawn sprinkler system, or leakage from a roof.

22.11 EXPANSION CONTROL IN BRICK WALLS

As described in Chapter 8, brick walls expand irreversibly due to the absorption of moisture. They also expand or contract due to temperature changes and foundation settlements. Therefore, brick masonry is divided into segments, with continuous vertical joints between the segments. Because expansion of brick walls is more dominant than contraction, the vertical joints are detailed as expansion joints.

FIGURE 22.35 Efflorescence (white spots) on a brick wall.

Most brick masonry is used as veneer in buildings whose structural system consists of a wood, steel, or concrete frame or load-bearing reinforced-concrete or concrete masonry walls. Therefore, we focus only on the discussion of expansion joints in brick veneer walls, which is covered in Section 26.1.

PRACTICE QUIZ

Each question has only one correct answer. Select the choice that best answers the question.

27. Per ASTM standards, facing bricks are classified as Types FBS, FBX, and FBA. Which of these types has the most stringent requirements for dimensional tolerance, chippage, and warpage?
- **a.** FBS
- **b.** FBX
- **c.** FBA

28. The freeze-thaw resistance of bricks has been found to be a function of the
- **a.** compressive strength of bricks.
- **b.** boiling-water absorption of bricks.
- **c.** saturation coefficient of bricks.
- **d.** all the above.
- **e.** (a) and (b).

29. Weathering index of a location is a function of the
- **a.** average annual freezing cycle days.
- **b.** average annual winter rainfall of the location.
- **c.** average annual air temperature of the location.
- **d.** all the above.
- **e.** (a) and (b).

30. In a running-bond wall
- **a.** each course consists of headers.
- **b.** each course consists of stretchers.
- **c.** each course consists of alternate headers and stretchers.
- **d.** the courses alternate between headers and stretchers.
- **e.** none of the above.

31. In an English-bond wall,
- **a.** each course consists of headers.
- **b.** each course consists of stretchers.
- **c.** each course consists of alternate headers and stretchers.
- **d.** the courses alternate between headers and stretchers.
- **e.** none of the above.

32. In a Flemish-bond wall,
- **a.** each course consists of headers.
- **b.** each course consists of stretchers.
- **c.** each course consists of alternate headers and stretchers.
- **d.** the courses alternate between headers and stretchers.
- **e.** none of the above.

33. IRA is a measure of the tensile strength of a brick.
- **a.** True
- **b.** False

34. Bricks that are highly absorptive may need to be wetted before being laid in the wall.
- **a.** True
- **b.** False

35. *Retempering* of mortar means
- **a.** adding portland cement to mortar.
- **b.** adding lime to mortar.
- **c.** adding sand to mortar.
- **d.** adding water to mortar.

36. Retempering of mortar, if needed, should be generally done within
- **a.** $5\frac{1}{2}$ h of the initial preparation of mortar.
- **b.** 4 h of the initial preparation of mortar.
- **c.** $2\frac{1}{2}$ h of the initial preparation of mortar.
- **d.** 1 h of the initial preparation of mortar.
- **e.** none of the above.

37. Efflorescence in masonry refers to
- **a.** excessive chippage of masonry units.
- **b.** excessive warpage of masonry units.
- **c.** white spots in masonry walls.
- **d.** yellow spots in masonry walls.
- **e.** none of the above.

Answers: 27-b, 28-d, 29-e, 30-b, 31-d, 32-c, 33-b, 34-a, 35-d, 36-c, 37-c.

KEY TERMS AND CONCEPTS

Actual dimensions of masonry unit
American bond
Autogenous healing
Bond strength of masonry wall
Building bricks
Clay masonry unit
Concrete masonry unit
Core holes in bricks
Efflorescence
English bond
Extruded bricks
Facing bricks
Flemish bond
Freeze-thaw resistance of bricks
Lime and portland cement for mortar
Masonry course
Masonry mortar
Masonry wythe
Modular brick
Molded bricks
Mortar joint profiles
Nominal dimensions of masonry unit
Paving bricks
PCL mortar
Property specification of mortar
Proportion specification of mortar
Running bond
Specified dimensions of masonry unit
Stack bond
Stretcher, header, rowlock, shiner, soldier and sailor faces
Unit masonry
Water retentively of mortar
White and colored mortar
Workability of mortar

REVIEW QUESTIONS

1. Describe the importance of lime in mortar.
2. Describe the importance of portland cement in mortar.

3. Give the ASTM classification of mortar types and their respective applications.
4. Explain why the masonry industry recommends the use of the weakest mortar that will give the required performance.
5. Using sketches and notes, describe various commonly used mortar joint profiles. Which profile gives the most watertight wall and why?
6. Explain the difference between the actual and the nominal dimensions of masonry units.
7. Bricks may carry a grade of SW, MW, or NW. Which factors determine the grade?
8. Explain what efflorescence is and how it can be mitigated.

SELECTED WEB SITES

Brick Industry Association, BIA (www.bia.org)

International Masonry Institute, IMI (www.imiweb.org)

Masonry Construction (www.masonryconstruction.com)

FURTHER READING

1. Beall, Christine. *Masonry Design and Detailing for Architects, Engineers and Builders.* New York: McGraw-Hill, 1997.
2. Brick Industry Association (BIA). *Technical Notes on Brick Construction.* (These notes are continuously updated at different times.)

CHAPTER 23

Masonry Materials–II (Concrete Masonry Units, Natural Stone, and Glass Masonry Units)

CHAPTER OUTLINE

23.1 CONCRETE MASONRY UNITS–SIZES AND SHAPES

23.2 CONCRETE MASONRY UNITS–MANUFACTURING AND SPECIFICATIONS

23.3 CONSTRUCTION OF A CMU WALL

23.4 SHRINKAGE CONTROL IN CMU WALLS

23.5 GROUT

23.6 CALCIUM SILICATE MASONRY UNITS

23.7 NATURAL STONE

23.8 FROM BLOCKS TO FINISHED STONE

23.9 STONE SELECTION

23.10 BOND PATTERNS IN STONE MASONRY WALLS

23.11 GLASS MASONRY UNITS

23.12 FIRE RESISTANCE OF MASONRY WALLS

Load-bearing CMU walls support steel trusses in this auditorium building. (Photo by DA.)

This chapter is a continuation of the previous chapter and deals with masonry materials not covered there—concrete masonry units, natural stone, and glass masonry units. Because joint reinforcement and grout are an integral part of most concrete masonry walls, these are also covered in this chapter.

Steel reinforcement, used in reinforced masonry, is the same as that used in reinforced concrete construction and was covered in Chapter 18. Masonry construction systems are discussed in Chapters 24 and 26.

23.1 CONCRETE MASONRY UNITS—SIZES AND SHAPES

Concrete masonry units (CMU), also called *concrete blocks,* are more versatile and hence more complex than clay bricks. A much larger number of surface finishes and colors can be obtained in CMU than in brick.

Additionally, a CMU is a much larger unit than a brick. A typical CMU occupies 12 times the space of a modular brick. Thus, laying one CMU is equivalent to laying 12 bricks, which reduces the labor cost in erecting a wall. Because of its much larger size, a CMU contains large voids.

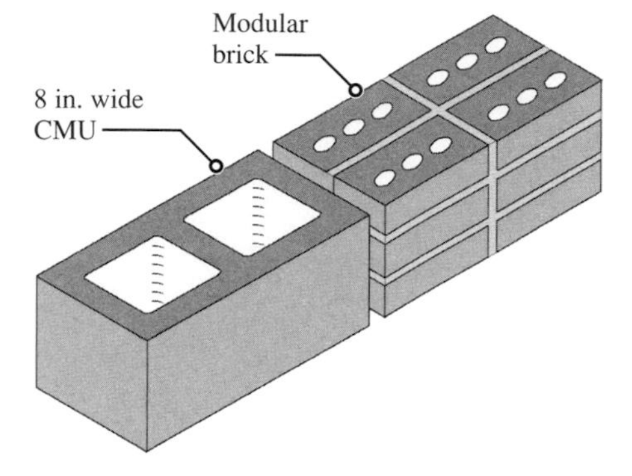

Relative sizes of CMU and bricks.
Volume of 1 CMU = Volume of 12 modular bricks

As shown later in this chapter, the voids in a CMU wall are continuous through its height. This arrangement allows the voids to be filled with grout (which is like concrete) and reinforced vertically—a requirement for load-bearing walls and walls required to resist large lateral loads. Thus, unlike clay bricks, which are generally used in face veneers, CMU are used in both structural and face veneer applications.

SIZES OF CONCRETE MASONRY UNITS

Concrete masonry units are available in several different sizes. The shape of a typical unit is shown in Figure 23.1. The length and height of all units is generally the same: length = 16 in. nominal ($15\frac{5}{8}$ in. specified) and height = 8 in. nominal ($7\frac{5}{8}$ in. specified). The width varies from 4 in. to 12 in. nominal in steps of 2 in., that is, width = 4, 6, 8, 10, 12, or 14 in. Because the length and height of a CMU is generally fixed, a CMU's size is generally specified by its width. An 8-in.-wide unit is most commonly used.

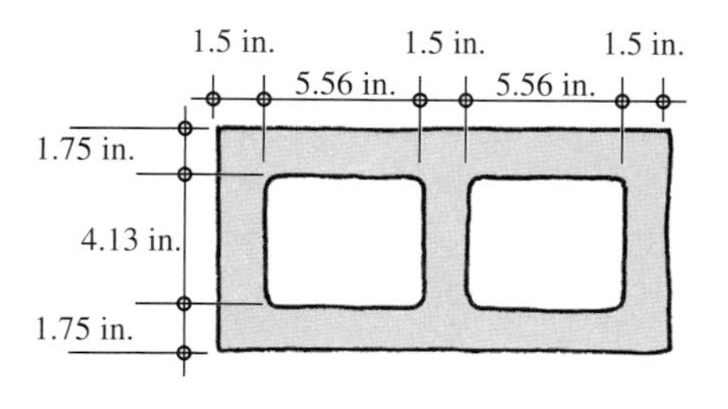

Assuming that the average thickness of face shells and webs is 1.75 in. and 1.5 in. respectively, the cells in a typical 8 in. unit are 4.13 in. × 5.56 in. = 23 in.^{2}.

The face shells and the webs of a unit are generally flared at the top of the unit as laid in the wall. Thus, a unit has a top and a bottom. The taper helps the mason in lifting and placing the unit in a wall. ASTM specifications provide minimum thickness of face shells and webs for CMU of different sizes. For example, an 8-in. CMU must have a minimum face shell thickness of $1\frac{1}{4}$ in. at the thinnest point and 1 in. minimum thickness for the webs.

The voids in a CMU are called *cells.* Most manufacturers make a 2-cell CMU. The longitudinal walls of a CMU are called *face shells,* and the transverse walls are called *webs.* Thus, a 2-cell unit has three webs. In an 8-in. unit, the cells are approximately 23 in.^{2}—

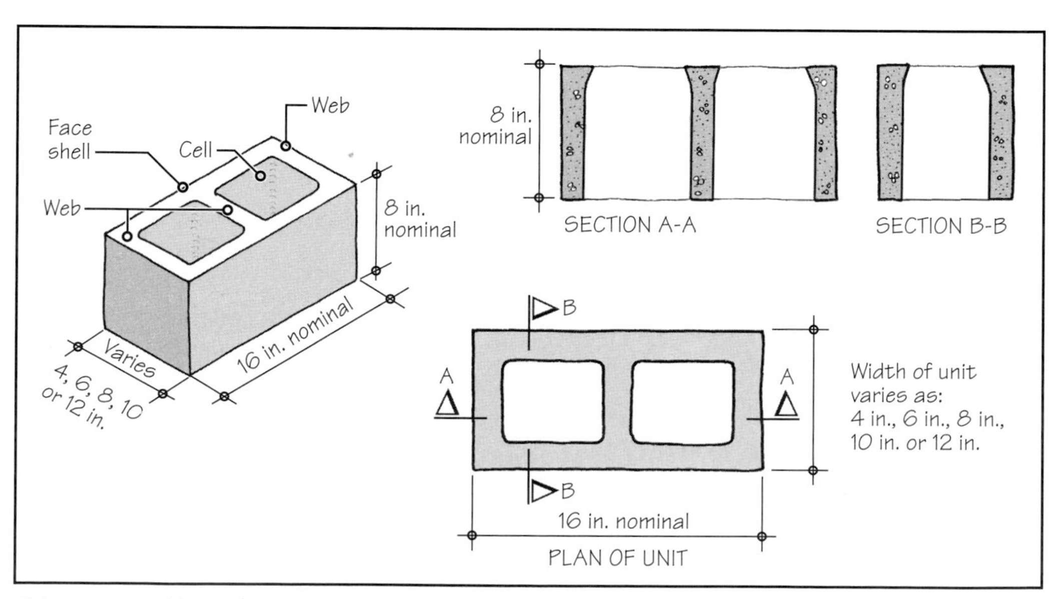

FIGURE 23.1 Sizes of concrete masonry units.

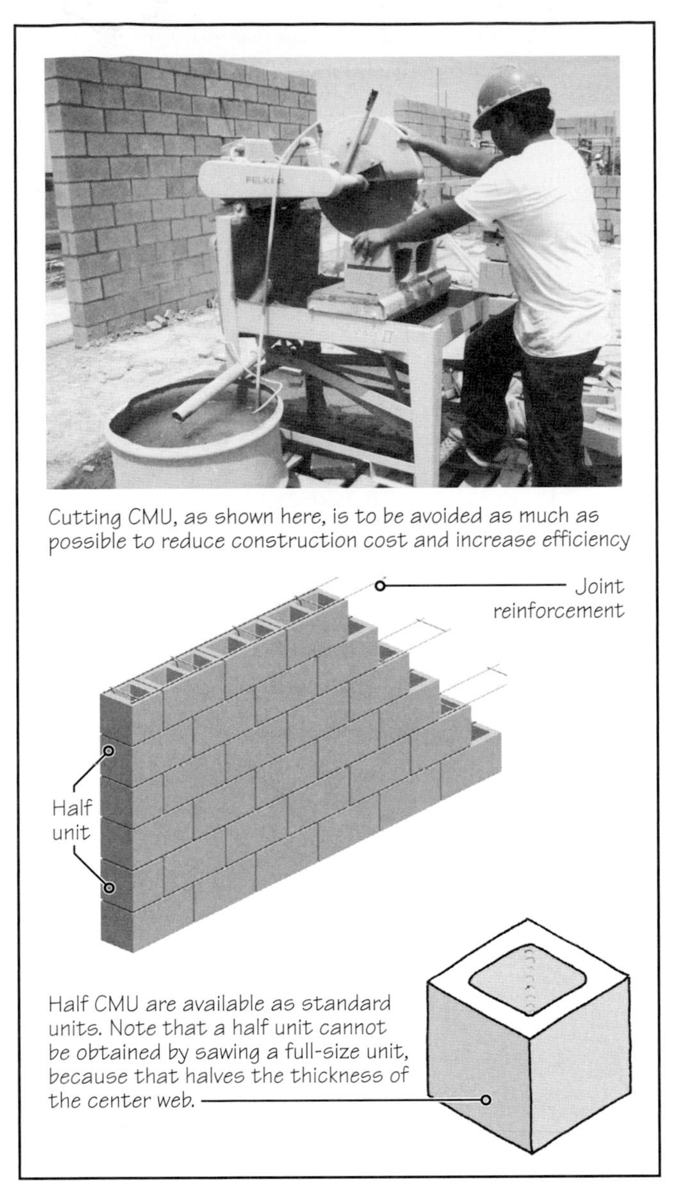

FIGURE 23.2 Half-units and their use in a wall.

a space that is large enough for reinforcing and grout (see Section 23.5 for a discussion of grout).

SCORED, BULLNOSE, AND SASH UNITS, AND UNITS WITH PROJECTING FACE SHELLS

Various shapes of CMUs are available to serve different functions in a wall. A typical manufacturer may make 20 or more different shapes and several different finishes. When all sizes, shapes, surface finishes, and colors are considered, a CMU manufacturer may have hundreds of different products.

Although the unit shape of Figure 23.1 constitutes the bulk of a CMU wall, a half-unit is required at the end of a wall, Figure 23.2. Therefore, a half-unit is generally provided as a standard unit by the manufacturers. Where an odd-dimension unit is needed, a full-size unit may be sawed at the site. Sawing of units is time and labor intensive, and, hence, it raises the cost of construction.

Figure 23.3 shows a bullnose unit and a scored unit. Bullnose units are used where a sharp wall corner is to be avoided. When scored units are used, the wall has the appearance of a stack-bond wall but the strength of a running-bond wall, Figure 23.4 (see also Figure 23.24).

The units shown in Figure 23.5(a) have projecting face shells at one or both ends. They are generally used interchangeably with the flush-end units of Figure 23.1. However, as described later, a concrete masonry unit is generally mortared on face shells only. The projecting face shells make it more convenient to mortar the head joints. Units with projecting face shells are also used in control joints (Figure 23.26 and Figure 9.18). Control joints can also be made with sash units, Figure 23.5(b).

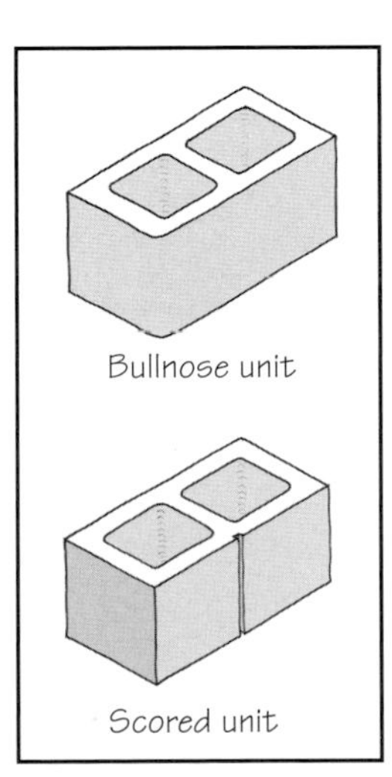

FIGURE 23.3 Bullnose and scored units.

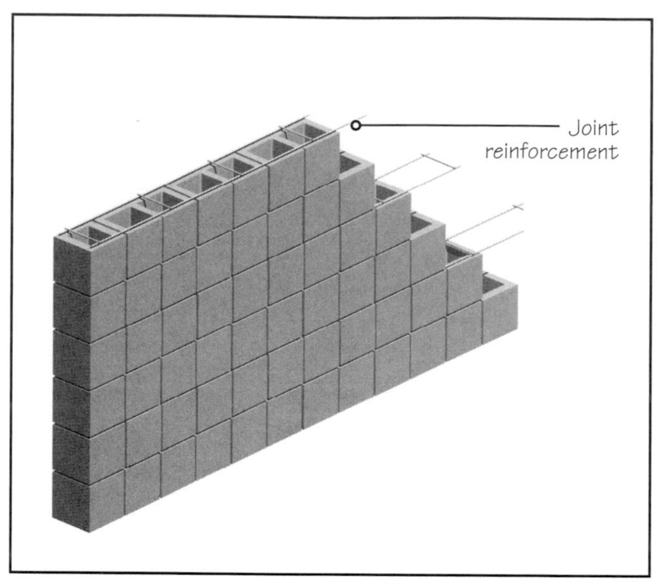

FIGURE 23.4 The use of scored units in a wall.

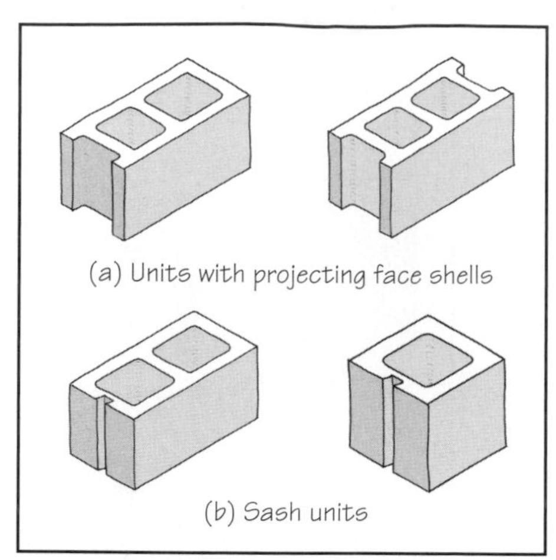

FIGURE 23.5 CMU with projecting face shells and sash units. CMUs with projecting face shells on one or both ends can be substituted for the plain-ended units of Figure 23.1. These units can also be used in control joints (Figure 23.26).

LINTEL UNIT

A lintel unit is a U-shaped unit, Figure 23.6. When grouted and reinforced horizontally, several lintel units in the same course function like a concrete beam that can be used to span an opening. For a large opening requiring a deeper lintel, some manufacturers make a 16-in.-high (2-course-high) unit.

A 16-in.-deep lintel can also be constructed by using an 8-in.-high lintel unit and fully grouting the lintel course and the course immediately above it. Similarly, a 24-in.-deep lintel can be obtained by grouting two courses of masonry above the standard 8-in.-high lintel.

Lintels in a CMU wall may also be made of steel, precast concrete, or site-cast concrete. Concrete masonry lintels have the advantage of maintaining the bond pattern, color, and surface texture of the wall.

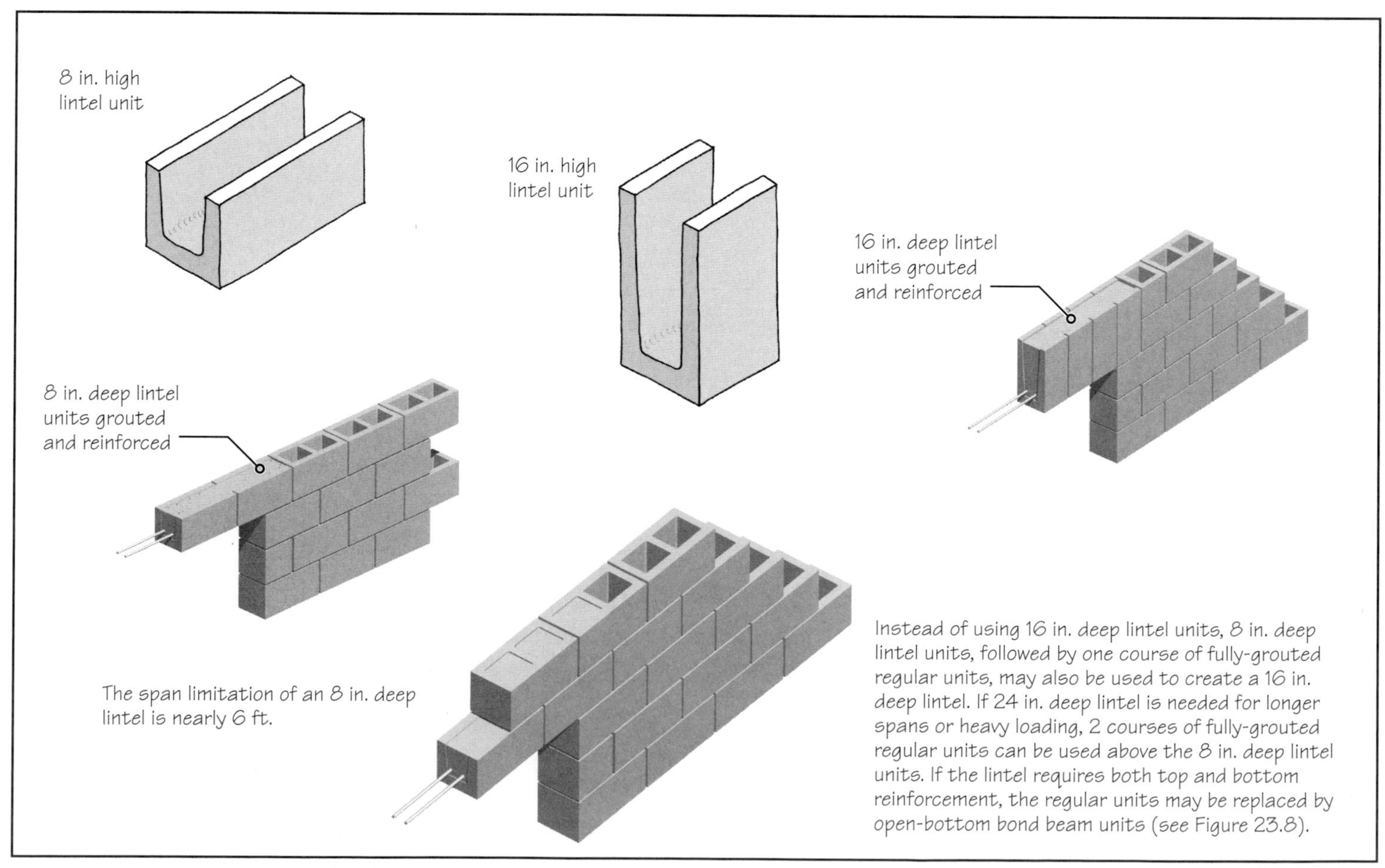

FIGURE 23.6 Concrete masonry lintels.

Arching Action in Masonry Walls

Masonry walls are generally built in running bond, English bond, Flemish bond, and so on. In all these bond patterns, the head joints are staggered so that a concentrated gravity load, as it travels down toward the foundation, is distributed on an increasingly larger number of masonry units, Figure 1. In other words, a vertical load placed on a wall is directed downward at a 45° angle on either side of the load, Figure 2.

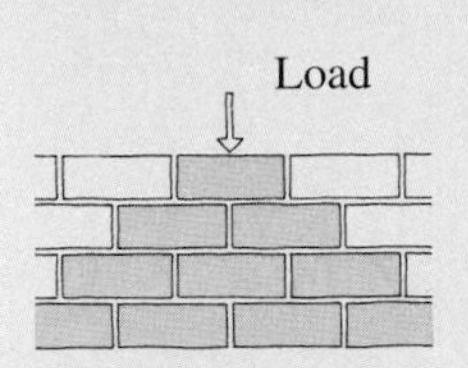

FIGURE 1

The 45° flaring of the load in masonry walls is referred to as the *arching action* because the phenomenon is similar to the behavior of an arch, which transfers the gravity loads placed on it to side abutments. Arching action occurs in all masonry walls in which the head joints in one course are (sufficiently) staggered from those in the upper and lower courses. Arching action is altogether absent in a dry-stacked, stack-bond wall. In a mortared stack-bond wall, some arching action occurs due to vertical shear transfer from one unit to the other through mortar joints, but it is ignored.

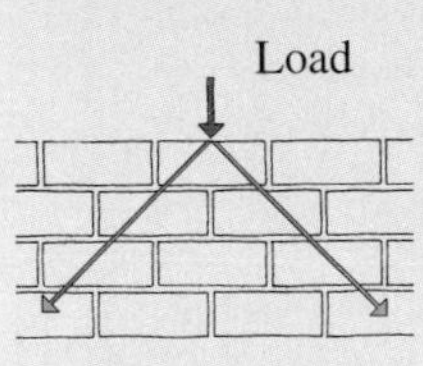

FIGURE 2

It is because of the arching action that sizable holes can be created in an existing masonry wall without causing the wall to collapse. The practical consequence of arching action is that a lintel beam (or simply a lintel) spanning an opening may not receive all the load placed directly above it, Figure 3. If the height of masonry above the lintel is more than half the width of opening (to enable a 45° triangle to be formed within the wall), the load carried by the lintel is only the load of the masonry within the 45° triangle, Figure 4.

In practice, a 12-in. depth of masonry above the apex of a 45° triangle is needed to assume a triangular load on the lintel. In other words, if the height of masonry above the lintel is half the lintel span plus 12 in., arching action is assumed and the lintel is designed for triangular loading plus the self-load of the lintel. An additional requirement for arching action is that abutments on either side of the opening are sufficiently thick to absorb the lateral thrust caused by the arching action.

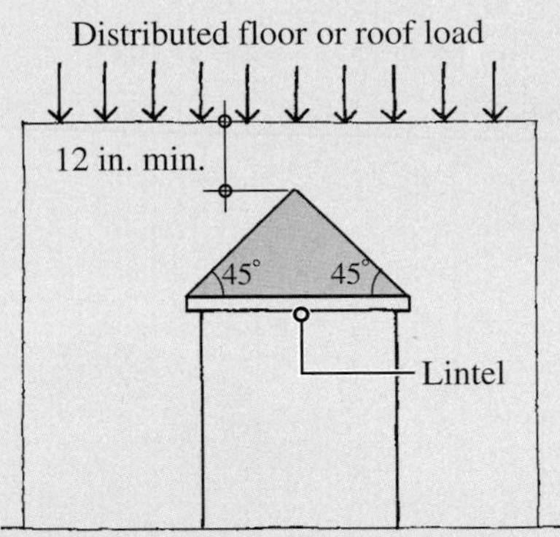

FIGURE 4 The lintel is designed to carry this triangular load of masonry plus the self-load of the lintel. The distributed superimposed (floor or roof) load on the top of the wall is not carried by the lintel. However, a concentrated load on the top of the wall, lying directly above the opening, must be considered as a distributed load on the lintel.

If the wall of Figure 4 carries a distributed superimposed load (due to an overlying floor or roof), it is not carried by the lintel but by the wall on either side of the opening. Because this condition occurs commonly, the lintels over openings in masonry walls need only a small amount of steel reinforcement.

However, a concentrated load directly above the opening should be considered as a concentrated load on the lintel.

In the absence of arching action, the lintel must be designed for the rectangular load of the wall plus any distributed and concentrated superimposed load above the lintel opening, Figure 5.

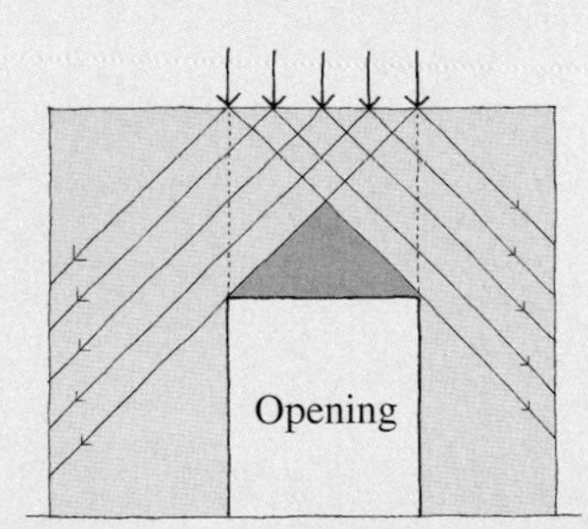

FIGURE 3 A load placed on the top of a masonry wall is directed downward flaring at 45° from the vertical. Therefore, the lintel over the opening receives the load from the triangular (highlighted) area only.

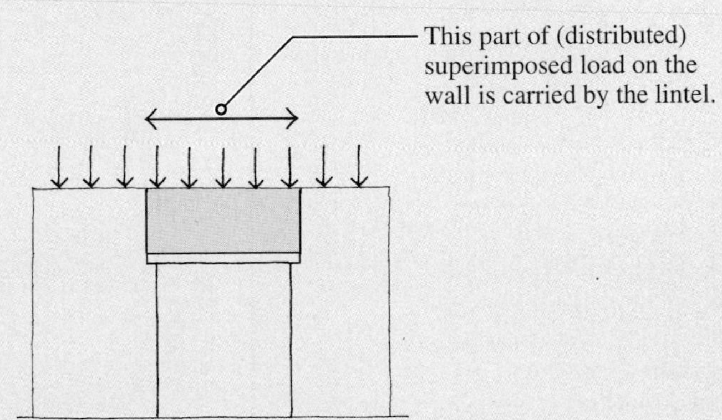

FIGURE 5 In absence of arching action, the lintel must be designed to carry this rectangular load of masonry plus the self-load of the lintel and any (distributed and concentrated) superimposed wall load directly above the opening.

A-UNIT, H-UNIT, AND BOND BEAM UNIT

An A-unit is used in a reinforced concrete masonry wall, Figure 23.7. The use of an A-unit eliminates the need to place (string) the unit over the reinforcement or an electrical conduit. An H-unit is used where every cell is reinforced. Because the cells in a concrete masonry wall are 8 in. on center, reinforcing each cell implies that the reinforcement in the wall is 8 in. on centers. A concrete masonry wall with reinforcement at 8 in. on centers is a fully grouted wall (no ungrouted cells). Note that a reinforced concrete masonry wall can have reinforcement only at 8 in., 16 in., 24 in., 32 in., and so on, on centers.

Some other commonly used concrete masonry units are bond beam units and pilaster units, Figure 23.8. The use of bond beam units is illustrated later in this chapter (Section 23.4).

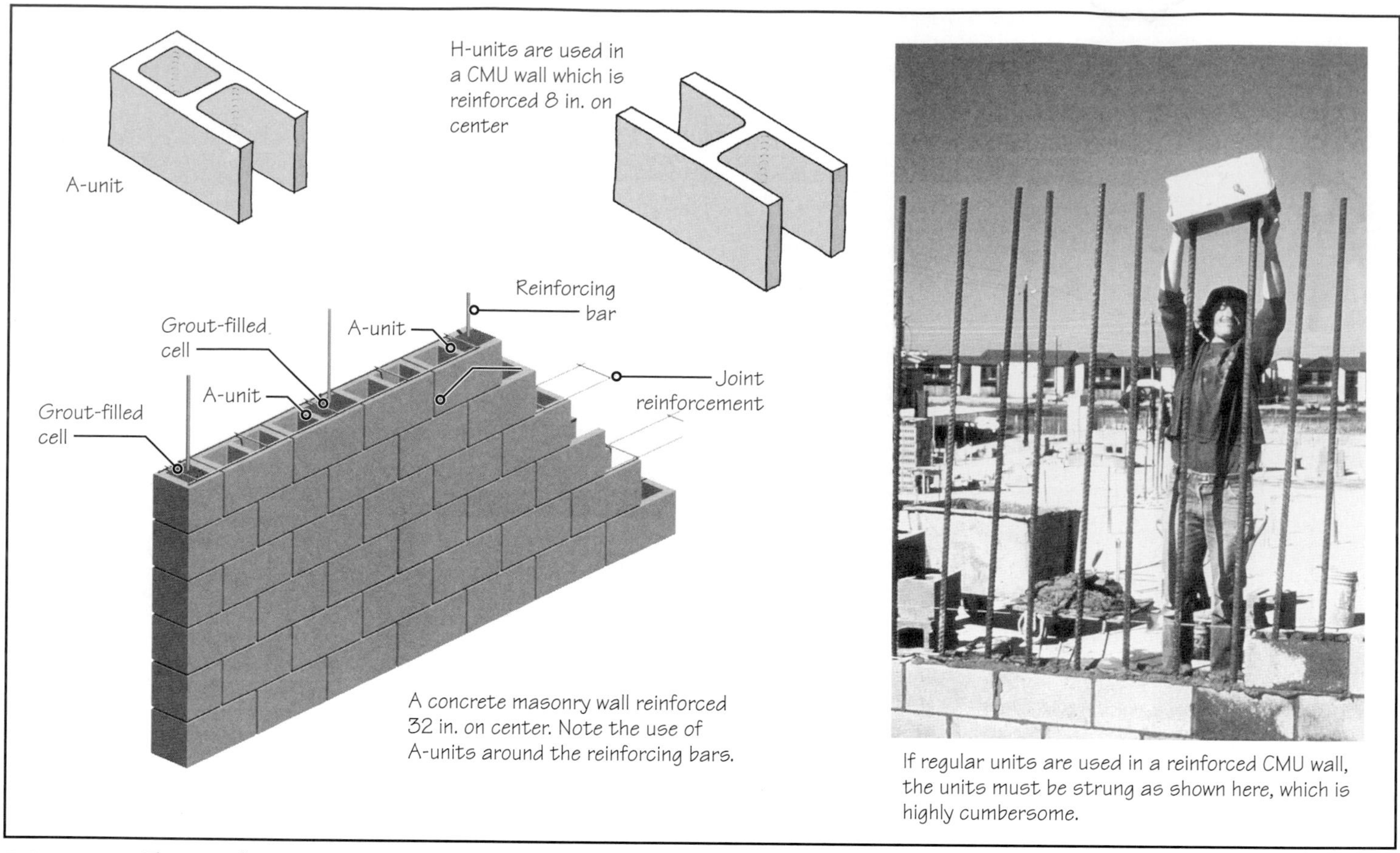

FIGURE 23.7 The use of A-units and H-units in a reinforced CMU wall. (Photo courtesy of the Brick Industry Association, Reston, Virginia.)

NOTE

Pilasters, Columns, and Piers

Pilasters, columns, and piers are similar elements. However, a column refers to an independent, isolated, vertical load-bearing masonry or nonmasonry member. The term *pier* is generally used for a masonry column that is short in height, generally used as foundation for wood frame buildings with an underlying crawl space.

A pilaster is a column formed by thickening a small area of a masonry wall, which may project on one or both sides of the wall. Like a reinforced concrete column, a reinforced masonry column requires ties. A pilaster does not require ties unless it is provided with compression reinforcement.

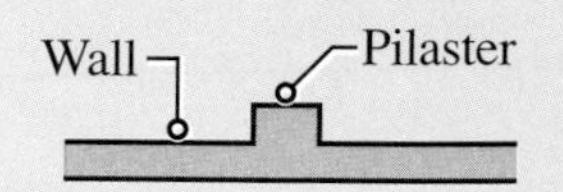

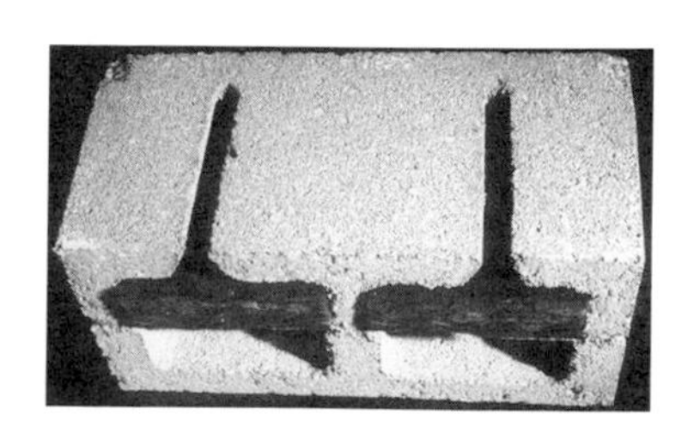

Accoustical unit

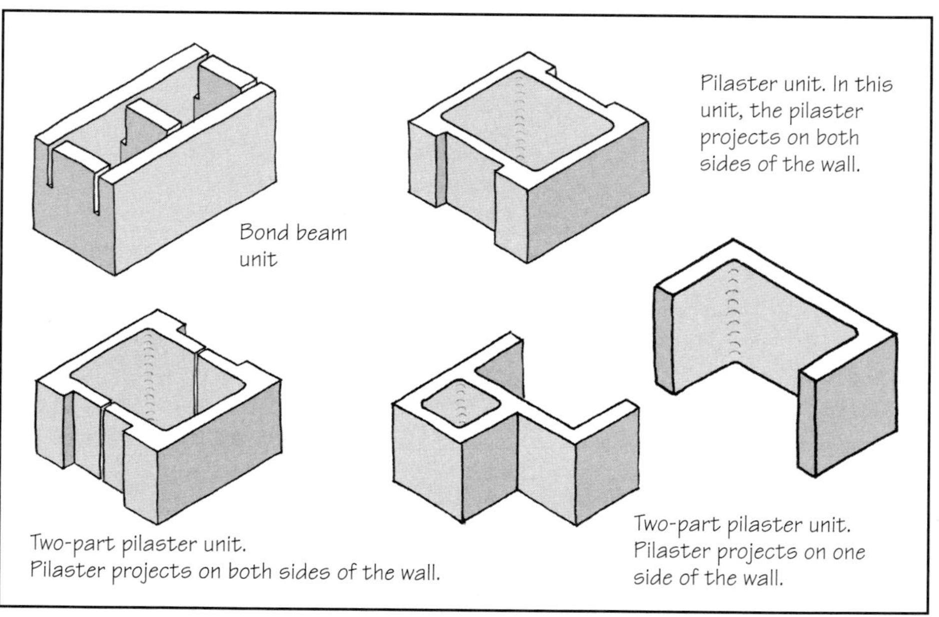

FIGURE 23.8 Bond beam unit and pilaster units.

Acoustical Units

An acoustical unit is commonly used where a concrete masonry wall with sound absorptive properties is needed. It has fiberglass embedded behind open slits.

Surface Textures—Split-Face Unit and Ribbed Unit

Concrete masonry units can be provided with surface textures other than the (standard) smooth texture. A fairly rough texture is obtained in a split-face unit, Figure 23.9. A split-face unit mimics a rough, stonelike texture and is produced by fracturing a fully hardened double CMU with a guillotine, which produces two split-face units. Split-face units are available with rough texture on one face or two adjacent faces (corner unit). Half-units are also available.

FIGURE 23.9 Split face units.

FIGURE 23.10 Ribbed units in Crawford Housing, New Haven, Connecticut. Architect: Paul Rudolph. (Photo courtesy of Dr. Jay Henry.)

Another highly textured unit is the ribbed unit, which has several vertical ribs. A ribbed unit is also available as a split-face unit so that the ribs are not smooth. When ribbed units are laid in a running-bond pattern, the ribs align to produce continuous vertical lines, Figure 23.10.

SURFACE FINISHES—BURNISHED UNIT AND GLAZED UNIT

The exposed surface of a concrete masonry unit can be ground smooth to reveal the aggregate. Variations in aggregate color, size, and type and the use of integral pigments can produce a surface that resembles a smooth-surfaced granite. Ground-face concrete masonry units are called *burnished units* and are often used in interiors where no additional finish is required. They can also be used in an exterior wall, often as accent bands in a wall made with split-face units, Figure 23.11.

A glazed unit has a facing of a glazing material bonded to one or more faces of the unit. The facing, approximately $\frac{1}{10}$ in. thick, is applied after the block has been made. The facing extends a little beyond the CMU face on all sides, Figure 23.12.

A glazed CMU surface is impervious to moisture and dust collection and is easy to clean, so it is sanitary. Glazed CMU, which are available in various colors, are often used in hospital interiors, public kitchens, and so on—wherever surface cleanliness is important. The suitability of glazed CMU on exteriors must be verified with the manufacturer to ensure color fastness and resistance to delamination by freeze-thaw action.

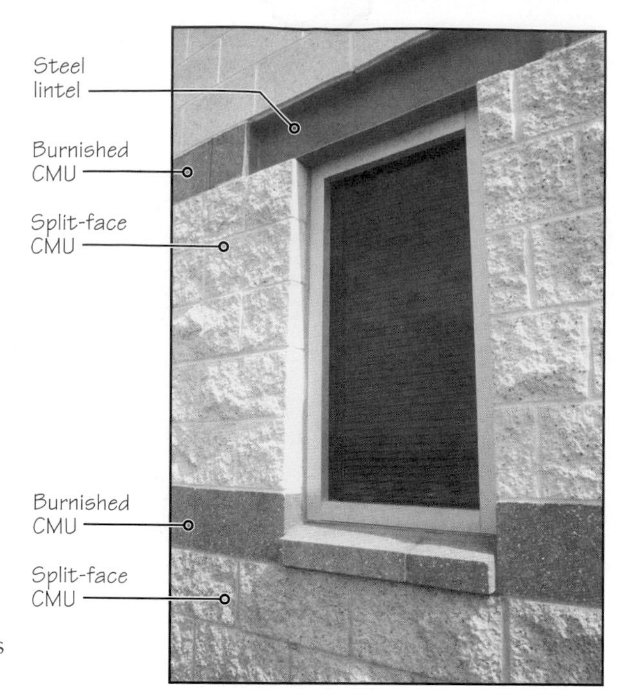

FIGURE 23.11 Use of burnished units as accent bands.

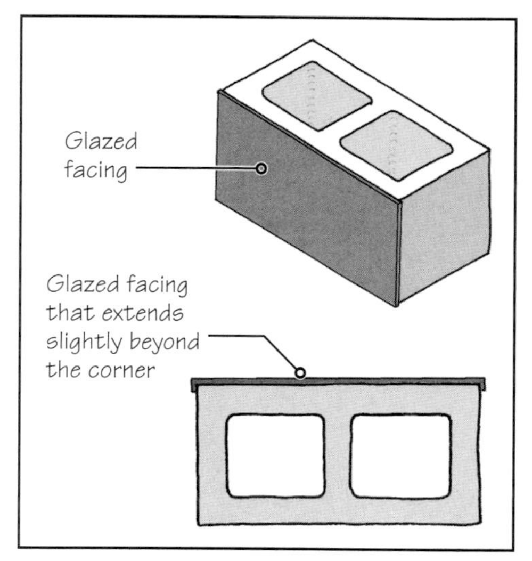

FIGURE 23.12 A glazed concrete masonry unit.

23.2 CONCRETE MASONRY UNITS—MANUFACTURING AND SPECIFICATIONS

Concrete masonry units are made from concrete. Therefore, a CMU-manufacturing plant has concrete-making equipment in addition to the equipment required to manufacture the units. Briefly, CMU manufacturing involves casting fairly dry (zero-slump) concrete mix in molds of the desired shape and compacting the concrete.

Because of the zero-slump concrete, the unit has adequate strength to be removed from the mold immediately after it has been cast. It then travels on a belt until it is removed and stacked on a multistack cart, Figure 23.13.

The multistack cart, which is generally mounted on rails, is sent to a curing chamber, where the units undergo accelerated curing by warm, supersaturated air, Figure 23.14. The curing is generally a 24-h cycle, so that the units are removed from the chamber the following day and stored in the yard for shipping to the construction site, Figure 23.15.

SPECIFICATIONS FOR CONCRETE MASONRY UNITS

Virtually all shapes of CMU can be made from lightweight or normal-weight concrete. A lightweight CMU is made from lightweight coarse aggregate, and normal-weight units are

FIGURE 23.13 Freshly made (green) CMUs travel on a conveyor belt to be stacked on a cart that delivers them to the curing chamber.

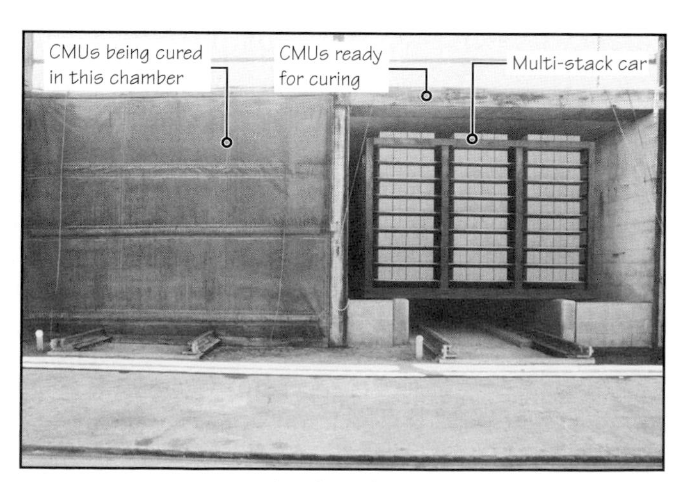

FIGURE 23.14 Curing chambers for accelerated curing of CMUs.

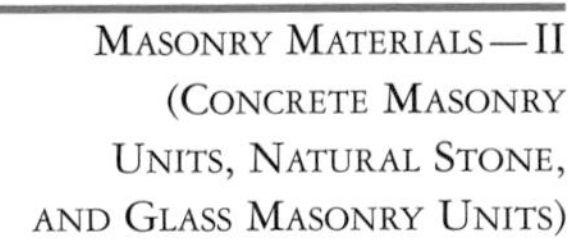

FIGURE 23.15 A storage yard of a typical CMU manufacturing plant.

made from normal-weight coarse aggregate. The weight classification is based on the following concrete densities.

- *Lightweight CMU*—dry concrete density < 105 lb/ft^3
- *Normal-weight CMU*—dry concrete density ≥ 125 lb/ft^3

A *medium-weight* unit is also made by some manufacturers, with a concrete density of 105 to 125 lb/ft^3. A typical-8 in. (2-cell) lightweight unit weighs approximately 28 lb, and a normal-weight unit weighs approximately 38 lb. Lightweight units are preferred because they reduce the dead load on a structure, and they are easier to work with. The masons may charge less for laying them. However, because lightweight aggregate costs more (Chapter 18), lightweight units are generally more expensive than the normal-weight units.

Lightweight units have the necessary strength for most situations. Where a high unit strength is needed, normal-weight units should be considered. In any case, the minimum compressive strength of a CMU used in load-bearing applications is 1,900 psi (based on the net bearing area of the unit).

PRACTICE QUIZ

Each question has only one correct answer. Select the choice that best answers the question.

1. The voids in concrete masonry units are called
a. voids.
b. cores.
c. cells.
d. frogs.
e. none of the above.

2. The nominal length of a typical CMU is
a. 18 in.
b. 16 in.
c. 12 in.
d. 8 in.
e. variable.

3. The nominal height of a typical CMU is
a. 18 in.
b. 16 in.
c. 12 in.
d. 8 in.
e. variable.

4. The nominal width (through-wall thickness) of a typical CMU is
a. 18 in.
b. 16 in.
c. 12 in.
d. 8 in.
e. variable.

5. A typical CMU has
a. one web.
b. two webs.
c. three webs.
d. four webs.
e. five webs.

6. A typical CMU has
a. one face shell.
b. two face shells.
c. three face shells.
d. four face shells.
e. five face shells.

7. Which of the following CMU shapes is used to simulate a stack-bond wall, although the wall is, in fact, made with a running-bond pattern?
a. Lintel
b. Bullnose unit
c. Sash unit
d. Unit with projecting face shells
e. Scored unit

8. Which of the following CMU shapes has a rounded corner?
a. Lintel
b. Bullnose unit
c. Sash unit
d. Unit with projecting face shells
e. Scored unit

9. The CMUs with projecting face shells can be used on both sides of a control joint in a wall. The other unit that is commonly used in the same situation is a
a. scored unit.
b. bullnose unit.
c. sash unit.
d. lintel unit.

10. Which of the following CMU shapes has a U-shape profile in its vertical cross section?
- **a.** Lintel unit
- **b.** Bullnose unit
- **c.** Sash unit
- **d.** Unit with projecting face shells
- **e.** Scored unit

11. The arching action occurs in a masonry wall with
- **a.** a running-bond pattern.
- **b.** a stack-bond pattern.
- **c.** both (a) and (b).
- **d.** neither (a) nor (b).

12. A lintel over an opening in all masonry walls is designed to carry the load from a triangular shape of superimposed masonry.
- **a.** True
- **b.** False

13. If a CMU wall is vertically reinforced at 16 in. on centers, which of the following units will you recommend for use in the wall?
- **a.** A-unit
- **b.** H-unit
- **c.** Standard unit with three webs
- **d.** Sash unit
- **e.** Bond beam unit

14. If a CMU wall is vertically reinforced at 8 in. on centers, which of the following units will you recommend for use in the wall?
- **a.** A-unit
- **b.** H-unit
- **c.** Standard unit with three webs
- **d.** Sash unit
- **e.** Bond beam unit

15. If a CMU wall is vertically reinforced at 24 in. on centers, which of the following units will you recommend for use in the wall?
- **a.** A-unit
- **b.** H-unit
- **c.** Standard unit with three webs
- **d.** Both (a) and (b)
- **e.** Both (a) and (c)

16. The center-to-center spacing of vertical reinforcement in a CMU wall cannot be less than 8 in.
- **a.** True
- **b.** False

17. Which of the following CMUs has slotted webs that are broken before being laid in the wall?
- **a.** A-unit
- **b.** H-unit
- **c.** Standard unit with three webs
- **d.** Sash unit
- **e.** Bond beam unit

18. The terms *pilaster* and *pier* are synonymous.
- **a.** True
- **b.** False

19. A split-face CMU is made by treating a standard CMU with oxyacetylene torch to produce a highly smooth surface on one or more faces of the unit.
- **a.** True
- **b.** False

20. A burnished CMU is made by treating a standard CMU with oxyacetylene torch to produce a highly smooth surface on one or more faces of the unit.
- **a.** True
- **b.** False

Answers: 1-c, 2-b, 3-d, 4-e, 5-c, 6-b, 7-e, 8-b, 9-c, 10-a, 11-a,12-b, 13-a, 14-b, 15-e, 16-a, 17-e, 18-b, 19-b, 20-b.

23.3 CONSTRUCTION OF A CMU WALL

The construction of a CMU wall differs from that of a brick wall in the following ways:

- Because a CMU is much heavier than a brick, it requires two hands to lay instead of the one-hand operation used in laying bricks, Figure 23.16. Therefore, it takes greater effort and time to lay a CMU than a brick, but because of a CMU's much larger size, the construction of a CMU wall is much faster.
- Unlike bricks, whose bed and head joints are fully mortared, a CMU is generally mortared on its exterior periphery. The masonry industry refers to it as *face-shell mortaring* of the units, Figure 23.17. In face-shell mortaring, the webs of a CMU are not mortared. The exceptions are stack-bond walls and partially grouted CMU walls. Mortaring the webs on both sides of a grouted cell in a partially grouted wall prevents

FIGURE 23.16 The laying of CMUs is a two-hand operation versus the one hand used in laying bricks.

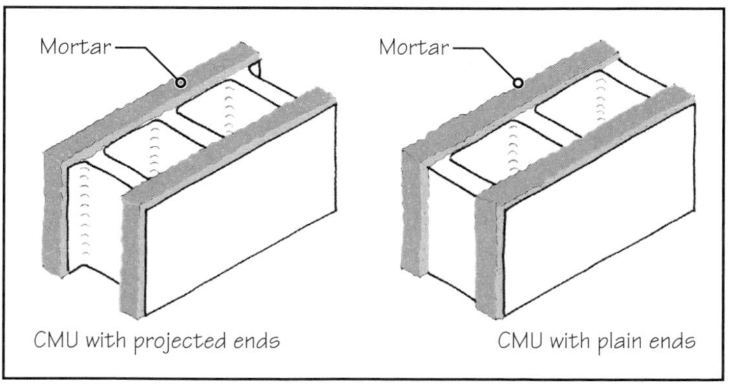

FIGURE 23.17 In a CMU wall, only face shells are generally mortared; that is, the webs are not mortared and the head joints are mortared to the depth of the face shells.

FIGURE 23.18 **(a)** A reinforced CMU wall (shown on the left) is grouted only in the cells that are reinforced. The other cells are generally left ungrouted, unless the grouting of unreinforced cells is required for greater fire resistance and (or) sound insulation of the wall. Note that in this photograph the CMU wall is reinforced 32 in. on center.

(b) A reinforced brick masonry wall must be fully grouted in the space between the two wythes of the wall.

FIGURE 23.19 **(a)** The unit is aligned and pressed downward and sideways using the trowel.
(b) The mortar is brushed off the face of the walls when it has dried sufficiently.
(c) The concave joint tool used with CMUs is longer than the one used with brick masonry (Figure 22.31).

the grout from flowing into ungrouted cells. The first course of a CMU wall is laid over a full bed of mortar regardless of the type of wall.
- A typical CMU wall contains joint reinforcement for shrinkage control, not required in brick walls other than a stack-bonded brick wall.
- A reinforced CMU wall is generally *grouted partially* because only cells that contain reinforcement need the grout. Other cells are left ungrouted unless fire-resistance and/or sound-insulation requirements of the wall require a *fully grouted* wall. A reinforced brick wall, on the other hand, must be fully grouted in all cases—the grout being provided in the space between two masonry wythes, Figure 23.18.

The images of Figure 23.19 highlight some of the similarities and differences between the construction of brick and CMU walls.

23.4 SHRINKAGE CONTROL IN CMU WALLS

As stated in Chapter 9, clay bricks expand during service, whereas CMU shrink. As with concrete, the shrinkage of CMU continues for several months after their manufacture. Although ASTM specifications require that the manufacturers limit the maximum shrinkage potential of their units to the specified value, it is not economical to let CMU stabilize their dimensions fully before selling them.

In other words, a CMU that arrives on the construction site is still shrinking. Therefore, some shrinkage of units is to be expected after they are placed in the wall. Two means are employed to control the shrinkage of CMU walls:

- Providing horizontal reinforcement in the wall.
- Providing control (shrinkage) joints so that a long CMU wall is divided into smaller segments.

NOTE

Why Only Face-Shell Mortaring of Concrete Masonry Units?

1. When a wall bends, the tensile and compressive stresses are concentrated primarily at the front and the back faces of a wall. Mortaring the webs, therefore, does not significantly increase a wall's bending strength.
2. Because CMUs are laid in running bond, the webs of a CMU in one course do not fully align with the webs of CMUs in the lower course. Therefore, if the mortar is placed on the webs, it will not serve any purpose. Additionally, not mortaring the webs speeds construction.

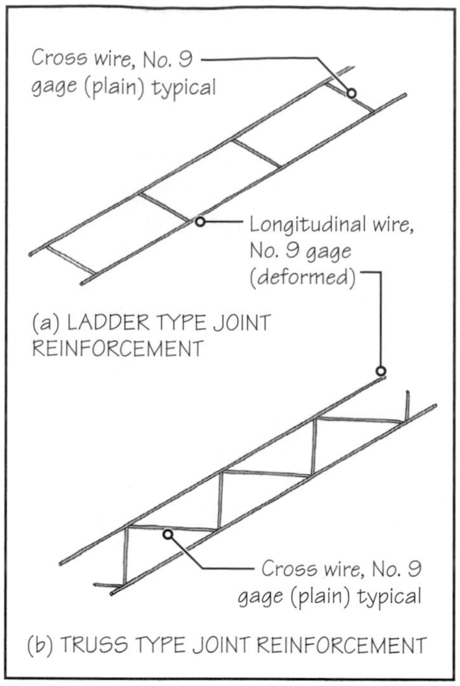

Cross wire, No. 9 gage (plain) typical
Longitudinal wire, No. 9 gage (deformed)
(a) LADDER TYPE JOINT REINFORCEMENT
Cross wire, No. 9 gage (plain) typical
(b) TRUSS TYPE JOINT REINFORCEMENT

FIGURE 23.20 Types of joint reinforcement commonly used. Although 9-gage wires are most common, other sizes are also available. The maximum diameter of wire used in joint reinforcement is half the mortar thickness.

The purpose of horizontal reinforcement is to resist tensile stresses in the wall caused by the shrinkage of units. Note that the horizontal reinforcement cannot prevent the formation of cracks. Nor can it reduce the total width of cracks in a given length of a wall. However, reinforcement distributes the cracks in the wall by increasing their number and reducing the width of individual cracks—the total width of cracks remaining unchanged. Thus, instead of one or two large cracks, numerous small cracks are formed in the wall if joint reinforcement is used. Small cracks are more water resistant, heal more easily, and can be sealed with a coating.

HORIZONTAL REINFORCEMENT—JOINT REINFORCEMENT

Horizontal reinforcement is generally provided by steel wire reinforcement placed in the mortar joints, referred to as *joint reinforcement.* Joint reinforcement consists of two parallel longitudinal wires welded to cross wires. Two types of cross-wire arrangement are used, Figure 23.20.

- In *ladder-type joint reinforcement,* cross wires are perpendicular to the longitudinal wires.
- In *truss-type joint reinforcement,* the cross wires are diagonal to the longitudinal wires.

As we will see in Chapter 26, both ladder- and truss-type reinforcement can have additional wires welded to them to tie a masonry veneer wall to a CMU backup wall. The longitudinal wires in joint reinforcement are embedded in the mortar. This is achieved by placing the joint reinforcement on the masonry bed and then placing the mortar over it, Figure 23.21. Because of face-shell mortaring, the cross wires are not covered by mortar.

The tensile stresses caused by the shrinkage of a CMU wall are resisted by the longitudinal wires. The cross wires simply hold the longitudinal wires together. The maximum permissible diameter of the longitudinal wires is half the mortar joint thickness (i.e., $\frac{3}{16}$ in.), but the smaller-diameter W1.7 wire (referred to as No. 9 gauge wire) is most commonly used.

The longitudinal wires are slightly deformed (textured) to improve their bond with mortar. The cross wires (also No. 9 gauge) are generally plain, spaced no more than 16 in. on

NOTE

Steel Wire Designation

A cold-drawn steel wire is designated by the cross-sectional area of the wire (see also Chapter 18). Thus, a W1.7 wire has a cross-sectional area of 0.017 in.^{2} (diameter = 0.147 in., or nearly $\frac{1}{7}$ in.). It is also referred to as No. 9 gauge wire.

A W2.1 wire has an area of 0.021 in.^{2}. It is also referred to as No. 8 gauge wire. Some other commonly used wires are

W1.1—gauge No. 11, area = 0.011 in.^{2}
W2.8—$\frac{3}{16}$-in. diameter, no gauge designation, area = 0.028 in.^{2}
W4.9—$\frac{1}{4}$-in. diameter, no gauge designation, area = 0.049 in.^{2}

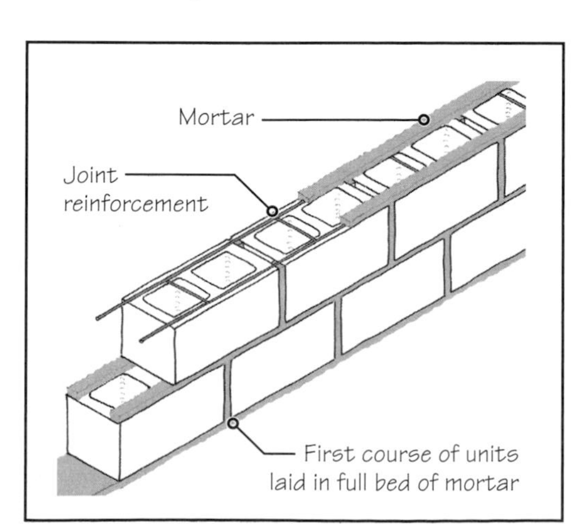

FIGURE 23.21 To embed the joint reinforcement in mortar, the standard practice is to place the joint reinforcement on the bed surface of the units first and then lay the mortar over it.

centers. Hot-dip galvanizing, in which the reinforcement is dipped in a hot bath of zinc, is required for corrosion protection. The thickness of galvanizing (zinc) film required depends on the exposure of the wall to corrosion. Where corrosion protection is critical, stainless steel joint reinforcement is recommended.

The fabrication of joint reinforcement (welding of cross wires with longitudinal wires and cutting to required lengths) should precede galvanizing. The alternative, in which the reinforcement is fabricated with galvanized wires, leaves the welded spots without any protection. If the latter is the mode of fabrication, the welded spots and sheared ends should be coated with zinc.

Joint reinforcement is generally available in 10-ft lengths. The distance between longitudinal wires has been standardized for CMU walls of different thickness. The standard requires that the longitudinal wires be spaced as far apart as possible. However, a minimum $\frac{5}{8}$-in. cover must be provided by the mortar, Figure 23.22. Therefore, for use in an 8-in. CMU wall, the out-to-out distance between wires is $(7\frac{5}{8} - \frac{5}{8} - \frac{5}{8}) = 6\frac{3}{8}$ in.

The vertical spacing of joint reinforcement in a wall is a function of the diameter of longitudinal wires and the thickness of the wall. With No. 9 gauge wires, the recommended spacing is 16 in. on center. This amounts to placing the joint reinforcement every other course of masonry. If $\frac{3}{16}$-in. longitudinal wires are used, the vertical spacing of joint reinforcement can be increased.

More stringent horizontal reinforcement requirements (wire diameter and vertical spacing of joint reinforcement) than those given here for crack control may be needed to resist high-wind and seismic loads.

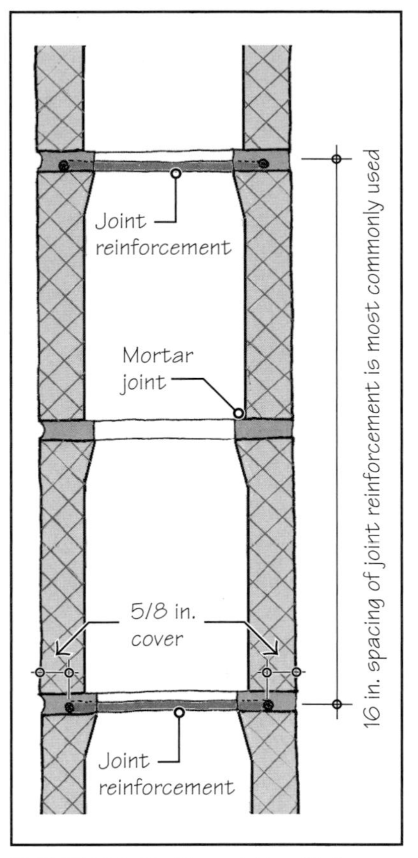

FIGURE 23.22 The width of joint reinforcement for CMU walls is standardized to provide $\frac{5}{8}$ in. mortar cover to longitudinal wires.

HORIZONTAL REINFORCEMENT—BOND BEAMS

Horizontal reinforcement in CMU walls may also be provided through bond beams. A bond beam is a CMU beam formed by grouting the bond beam units. The slotted webs of bond beam units are broken by the mason before laying them, providing a continuous space for horizontal reinforcement and grouting.

Because the bond beam units are open at the top and the bottom, a plastic mesh fabric is placed below the units to prevent the grout from leaking into the lower cells. For crack control, the bond beams are generally spaced 4 ft on center vertically with only one No. 4 reinforcing bar ($\frac{1}{2}$-in. diameter) in each bond beam, Figure 23.23.

Bond beams are more cumbersome to provide than joint reinforcement. However, the reinforcement in a bond beam is better protected, and a bond beam can be reinforced more heavily to resist high-wind and seismic forces. As we will observe in Chapter 24, bond beams are required for reasons other than crack control in load-bearing masonry structures.

HORIZONTAL REINFORCEMENT AND STACK-BOND WALLS

A stack-bond wall is much weaker in horizontal bending than a running-bond wall because the continuous head joints open easily, presenting little resistance to horizontal bending, Figure 23.24(a). In a running-bond wall, the masonry units have to overcome the shear (sliding) resistance at the bed joints before the head joints can open, Figure 23.24(b).

Horizontal reinforcement substantially increases the horizontal bending strength of a stack-bond wall. Tests have shown that if a CMU stack-bond wall is provided with joint reinforcement at 16 in. on centers, its bending strength is 20% greater than that of an

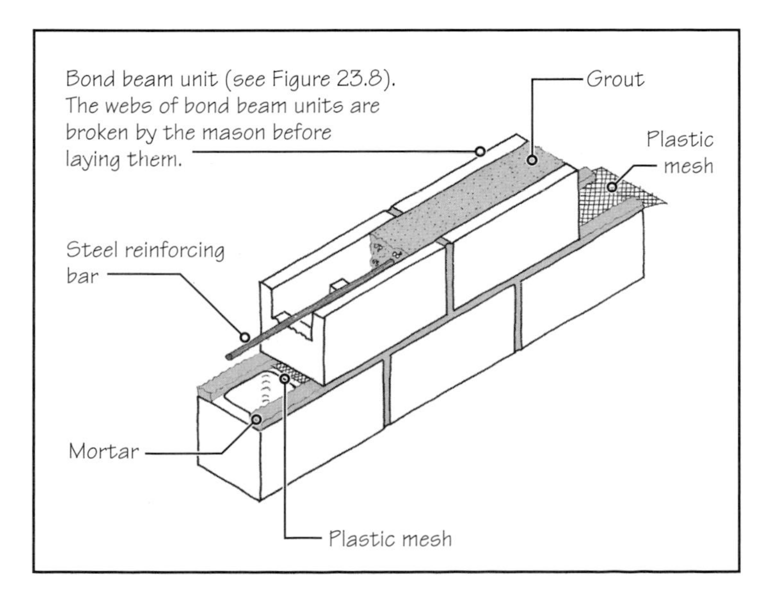

FIGURE 23.23 The use of bond beams for crack control.

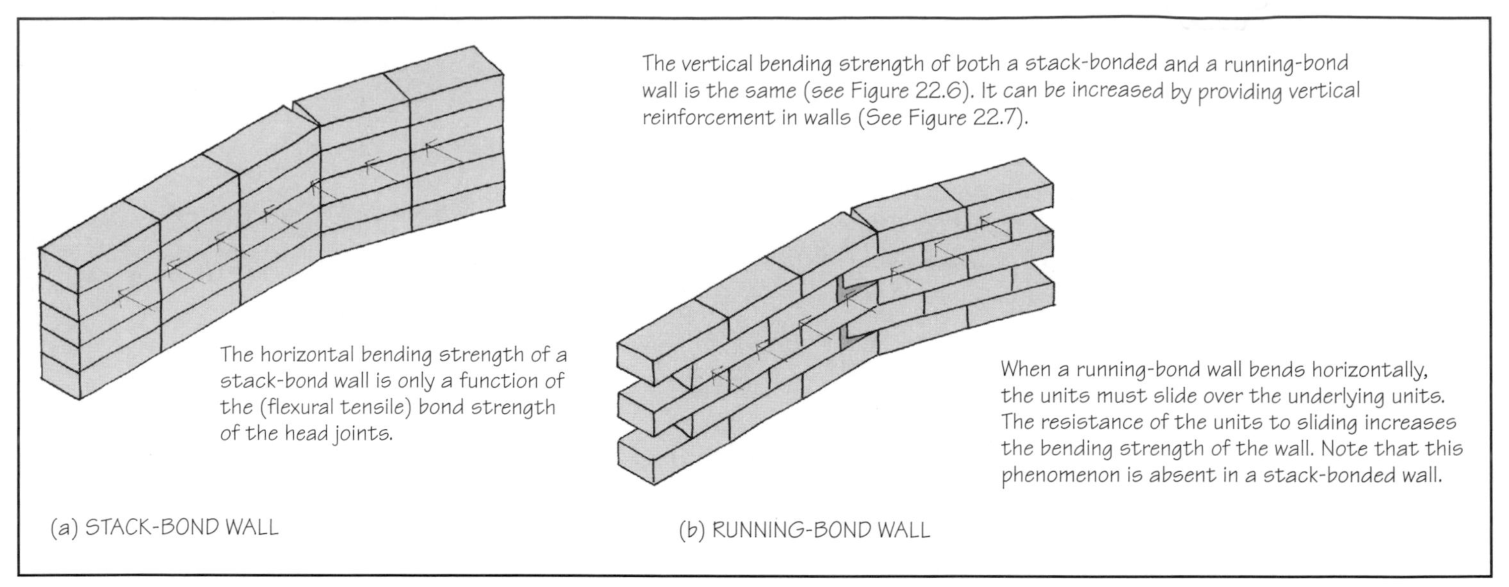

FIGURE 23.24 **(a)** The horizontal bending strength of a stack-bond wall is low because it is only a function of the flexural tensile bond strength of vertical joints. **(b)** The horizontal bending strength of a running-bond wall is greater than that of a stack-bonded wall because of the shear (sliding) resistance provided by the units.

unreinforced running-bond wall. If the running-bond wall is also reinforced horizontally in the same way as a stack-bond wall, both the stack-bond and the running-bond wall are equal in horizontal bending strength. Horizontal reinforcement in bond beams (4 ft on center) achieves the same result as the joint reinforcement laid 16 in. on center.

Control Joints

In addition to the horizontal reinforcement, crack control in CMU walls requires the provision of continuous vertical control joints. The width of a control joint is usually the same as the thickness of a mortar joint—$\frac{3}{8}$ in. The recommendations for control joint spacing are as follows:

- Length-to-height ratio of a wall between control joints should be less than or equal to 1.5. In any case, the distance between control joints should not exceed 25 ft. Thus, if the height of a wall (from floor to floor or from floor to roof) is 10 ft, the maximum distance between control joints is 15 ft. However, if the wall height is 20 ft, the maximum distance between control joints is 25 ft.
- In addition to meeting the first provision, several other provisions should also be met, as shown in Figure 23.25.

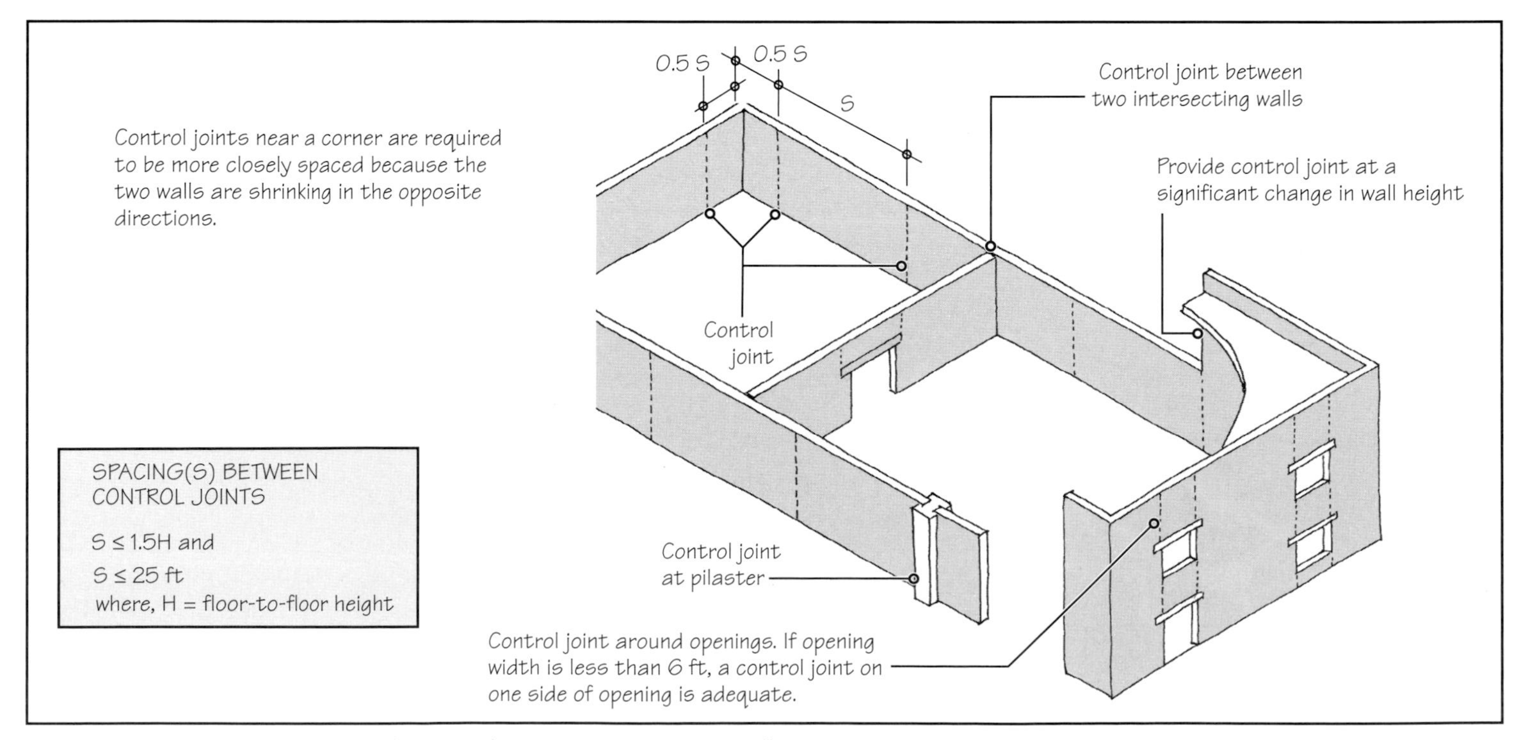

FIGURE 23.25 Recommendations for control joint spacing in CMU walls.

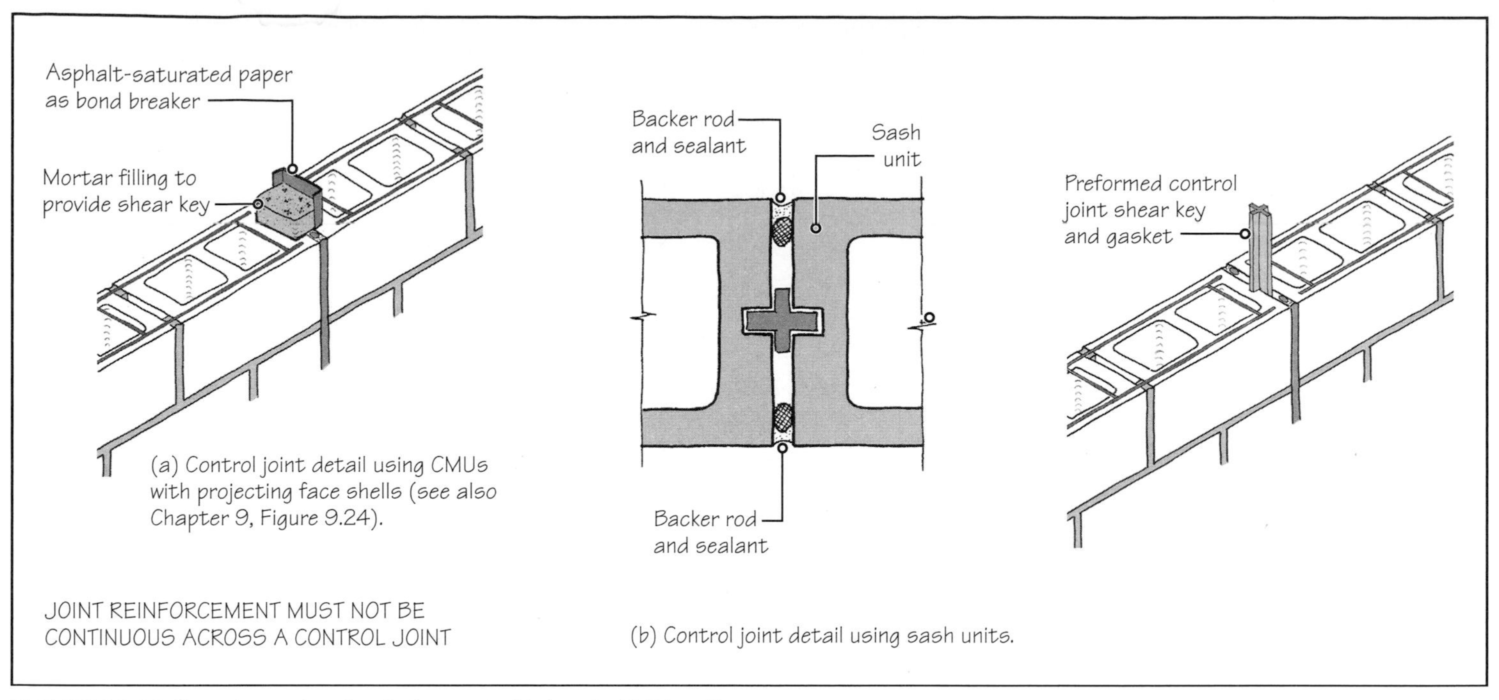

FIGURE 23.26 Two alternative details for control joints in CMU walls.

The detailing of a control joint must ensure free horizontal movement of wall segments in addition to the transfer of the lateral load between adjacent segments. A typical control joint detail is shown in Figure 23.26(a). An alternative detail, which uses a sash unit on both sides of a control joint, is shown in Figure 23.26(b). A cross-shaped PVC gasket is used in sash unit recesses. The parts of the gasket within the sash units are required to have adequate hardness to function as shear key—to transfer the lateral loads across the joint.

23.5 GROUT

The purpose of masonry grout is to fill the voids in masonry walls so that the grout, the masonry units, and the reinforcement are integrated into a composite whole. Grout is, therefore, a cementitious mix, in many ways similar to concrete. Important differences between grout and concrete, however, do exist. For example, the durability of concrete is important, but this not a concern with grout because it is protected by the units. A small amount of lime is permitted in grout, but not in concrete.

Because grout is placed in masonry voids, it contains more water than concrete. In other words, grout is a "soupy" mix to allow it flow down the voids, which are relatively small in size. The required slump of grout is between 8 to 11 in., depending on the absorption characteristics of masonry units, void dimensions, and the ambient temperature and humidity conditions at the time of placing the grout.

Types of Grout

Void dimensions also affect the type of aggregate used in grout. Grout is classified as *fine grout* and *coarse grout.* In fine grout, aggregate consists of sand only. In coarse grout, the aggregate consists of sand and coarse aggregate. The maximum size of coarse aggregate is generally limited to $\frac{3}{8}$ in.

Fine grout is used where void dimensions are small. If void dimensions are large (greater than 1.5 in. $\times$ 3 in.) coarse grout is preferred because it shrinks less than fine grout. In any case, a shrinkage compensating admixture may be specified for grout. Excess shrinkage (a result of the large slump of grout) can leave voids, reducing the bond between the units, grout, and the reinforcement. Proper consolidation will, however, reduce shrinkage to some extent.

The compressive strength of grout must be at least equal to the compressive strength of masonry, but not less than 2,000 psi. The strength of grout is established by testing specimens after 28 days of casting them. As an alternative to testing, grout may be proportioned by the requirements of Table 23.1, which results in a grout with a compressive strength of approximately 2,500 psi.

Two methods are used to pour grout in masonry walls—pumping and bucketing, Figure 23.27. Pumping is more commonly used in a *high-lift grouting* operation. High-lift grouting is used where the entire height of wall (up to a maximum of 24 ft) is grouted in

TABLE 23.1 PROPORTIONS OF MATERIALS IN A STANDARD GROUT MIX

Grout type	Portland cement (PC)	Proportions By Volume		
		Lime (L)	Aggregate: Fine	Aggregate: Coarse
Fine grout	1	$0–\frac{1}{10}$	2.25–3.0 times the sum of PC and L	—
Coarse grout	1	$0–\frac{1}{10}$	2.25–3.0 times the sum of PC and L	1 to 2 times the sum of PC and L

NOTE

ASTM and CSMUs

The most referenced standard for calcium silicate masonry units is ASTM C73, "Standard Specification for Calcium Silicate Brick (Sand-Lime Brick)."

FIGURE 23.27 Two alternative methods of pouring grout are pumping or bucketing. In these images, the masons are pouring the grout in the cells of underlying CMUs. The CMUs showing at the top of the wall are bond beam units.

one operation. Bucket placement is economical in small jobs or where *low-lift grouting* is specified. In low-lift grouting, the wall is grouted as it is constructed, generally up to 4 ft in height. Some architects and engineers prefer low-lift grouting because it gives better quality control, particularly if the wall is heavily reinforced.

FIGURE 23.28 Smooth-surfaced and rough-surfaced CSMUs used in the cladding of the York Administration Building, New Market, Ontario, Canada. (Photo courtesy of Arriscraft International.)

23.6 CALCIUM SILICATE MASONRY UNITS

Calcium silicate masonry units (bricks and blocks), abbreviated CSMU, are made from fine sand and lime with no portland cement. They are also called *sand-lime bricks and blocks.* They are pressure molded like concrete blocks but are cured in an autoclave, under pressure (several atmospheres) and steam (300°F to 400°F). Autoclaving accelerates the pozzolanic reaction between sand (silica) and lime, producing a hydraulic cement (Chapter 18).

Because lime is white in color, the natural color of a CSMU is wheatish or pearl gray, depending on the color of the sand. Other colors, obtained by adding mineral pigments, are available.

The compressive strength of CSMU can be as high as 8,000 psi. Their high compressive strength implies that the units have relatively low water absorption and can, therefore, be used either as a facing material, for which they are commonly used, or in load-bearing walls. Several important buildings have been designed with CSMU as the cladding material, Figure 23.28.

CSMU are graded for durability as SW and MW grades—the same as the grades for facing (clay) bricks. The absence of coarse aggregate in the units implies that the units can be sawed to size while maintaining a fine surface texture, similar to the smooth texture obtained by sawing a stone slab. Some CSMU manufacturers, therefore, promote their product as a synthetic stone.

The absence of portland cement in CSMU implies that the shrinkage of units is small. Therefore, the joints in CSMU walls are located using the same rules as for CMU walls but are detailed to allow for both expansion and contraction.

PRACTICE QUIZ

Each question has only one correct answer. Select the choice that best answers the question.

21. In a typical CMU wall, face-shell mortaring is used. This implies that the mortar is placed
- **a.** only on the horizontal surfaces of the face shells.
- **b.** on the horizontal and vertical surfaces of the face shells.
- **c.** on the horizontal surfaces of the face shells as well as the webs.
- **d.** on the vertical surfaces of the face shells only.

22. The grouting of the cells of a CMU wall is required where
- **a.** the cells are provided with steel reinforcement.
- **b.** a greater fire resistance of the wall is required.
- **c.** a greater sound insulation of the wall is required.
- **d.** all the above.
- **e.** both (a) and (b).

23. The joint reinforcement refers to
- **a.** the vertical reinforcement in a masonry wall.
- **b.** the horizontal reinforcement in a masonry wall.
- **c.** both vertical and horizontal reinforcements in a masonry wall.

24. Joint reinforcement in CMU walls generally consists of
- **a.** a single wire.
- **b.** two parallel wires connected together by cross wires.
- **c.** three parallel wires connected together by cross wires.
- **d.** two parallel wires without cross wires.
- **e.** none of the above.

25. The primary purpose of joint reinforcement in a CMU wall is to
- **a.** increase the compressive strength of the wall.
- **b.** increase the shear strength of the wall.
- **c.** absorb the compressive stress in the wall caused by the expansion of units.
- **d.** absorb the tensile stress in the wall caused by the expansion of units.
- **e.** absorb the tensile stress in the wall caused by the shrinkage of units.

26. Bond beams in a CMU wall can be used to serve the same purpose as joint reinforcement.
- **a.** True
- **b.** False

27. The spacing of control joints in a CMU wall
- **a.** should not exceed 25 ft.
- **b.** is a function of the length to height ratio of wall.
- **c.** is a function of the average annual air temperature of location.
- **d.** (b) and (c).
- **e.** (a) and (b).

28. Masonry grout is a mixture of
- **a.** portland cement and water.
- **b.** portland cement, aggregate, and water.
- **c.** portland cement, lime, and water.
- **d.** lime, aggregate, and water.
- **e.** none of the above.

29. Masonry grout has a slump of approximately
- **a.** 2 to 3 in.
- **b.** 3 to 4 in.
- **c.** 4 to 5 in.
- **d.** 5 to 6 in.
- **e.** none of the above.

30. Calcium silicate masonry units are made from
- **a.** portland cement, sand, and water.
- **b.** lime, sand, and water.
- **c.** portland cement, lime, sand, and water.
- **d.** limestone, sand, and water.

Answers: 21-b, 22-d, 23-b, 24-b, 25-e, 26-a, 27-e, 28-e, 29-e, 30-b.

23.7 NATURAL STONE

Natural stone is obtained from rocks that constitute the earth's crust. Rock and stone are essentially the same materials, except that after the rock has been quarried, it is called stone. Stone that is used in buildings is called *building stone.* Clay and sand present on the earth's surface are the result of the physical and chemical disintegration of rock by rainwater and air.

The term *dimension stone* refers to stone that has been fabricated to required dimensions, texture, surface finish, and so on, and meets performance requirements for durability, strength, water absorption, and the like. The term includes stone cladding panels, veneer stone, counter- and tabletops, wall copings, stair treads and risers, and balusters. It specifically excludes broken or crushed stone.

Stone is a natural material, so its characteristics (properties and appearance) are inconsistent. There is a great deal of variability even in slabs or blocks obtained from the same quarry pit. It is, therefore, too complex a material to describe fully in a few pages.

However, to be able to select and specify it intelligently for a project, architects, engineers, and builders require a working knowledge of the material. The description given here is an attempt to meet that modest requirement and dwells mainly on the properties and surface finishes of commonly used building stones:

- Granite
- Limestone
- Marble
- Travertine
- Sandstone
- Slate

GRANITE

Granite is an *igneous rock.* It is the strongest and densest of building stones. It weathers far more slowly than other stones and takes an extremely good polish. Therefore, it is commonly used in the exterior cladding of significant buildings.

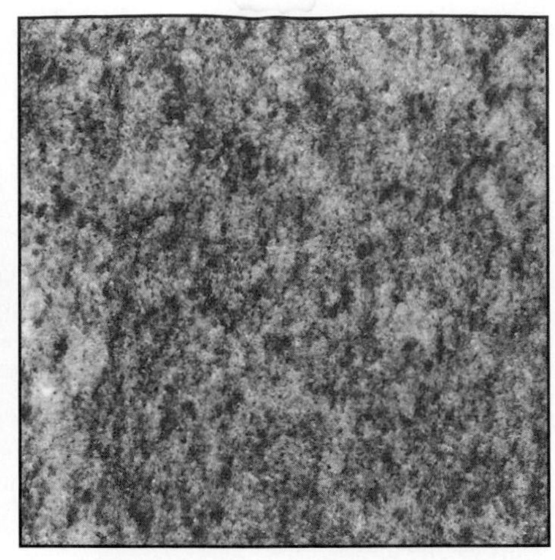

FIGURE 23.29 Speckled (granular) surface of granite.

Chemically, granite is a fusion of three minerals, feldspar (silicate of aluminum, calcium, sodium, and potassium), quartz (silicon dioxide), and mica (potassium aluminum silicate). These minerals are present as *granules,* which give granite its characteristic speckled surface appearance and name (*granite* from *granular*), Figure 23.29.

In general, granite contains 25% to 40% quartz and 3% to 10% mica; the remainder (50% or more) is feldspar. The relative proportions of these minerals and the size of their granules determine the strength, color, and surface appearance of granite. Quartz is the strongest and more durable of the three minerals. Therefore, a greater amount of quartz gives a stronger and harder granite that is more difficult to process—sawing, profiling, and grinding.

Being silicon dioxide, quartz is chemically the same as sand and, hence, is white to light pink in color. A granite low in quartz is generally darker in color. Commercial black granite is extremely low in quartz, approaching 0%, and is really not granite, but a stone called *basalt* by the geologists. Black granite (basalt), which has a handsome finish, is commonly used as table- or countertops. Being relatively weaker, it is not favored for use where high strength or abrasion resistance is necessary, such as floors and stair treads.

EXPAND YOUR KNOWLEDGE

Geological Classification of Rocks

The earth's surface consists of the following three types of rocks:

- Igneous rock
- Sedimentary rock
- Metamorphic rock

As the earth's crust began to cool from its original molten state to its present solid state, the first rock to form was igneous rock. In its molten form, the earth's surface was essentially the same as the one that is present deep inside the earth today. This molten rock, which is forced out during a volcanic eruption, is referred to as lava or magma.

The texture of *igneous rock* depends mainly on how slowly or rapidly the molten mass cools. Slow cooling allows the molecules to arrange themselves in crystals. If the molten mass cools rapidly, the molecules do not have the time to arrange themselves in crystals. Thus, a rapidly cooled volcanic rock is noncrystalline (nongranular) and glassy—brittle and relatively weak.

The igneous rock that exists deeper inside the earth's crust cooled rather slowly because of the insulating effect of the overburden. The coarse-grained crystalline structure of granite—an igneous rock—is the result of the slow cooling process that took place over many years. The crystalline structure of granite gives it the hardness and strength for which it is well known.

Sedimentary rock began to form as the igneous rock disintegrated due to erosion by water. As the eroded particles were carried by water, a great deal of marine and other life forms were also carried with it. These substances were deposited in seas, lakes, valleys, and deltas and were compressed by the pressure of overlying material and bonded together, forming a different kind of rock, called *sedimentary rock.*

Limestone, travertine, and sandstone are sedimentary rocks. The difference between them is the chemical composition of the sediment beds and the material that cemented the beds. Limestone and travertine consist primarily of calcium carbonate. Sandstone consists of silicon dioxide.

Metamorphic rocks originate from either igneous or sedimentary rock that has been altered (morphed) in its chemical structure by the action of heat and pressure. Marble, quartzite, and slate are metamorphic rocks. Marble is morphosed from limestone; quartzite, from sandstone; and slate, from shale. Because of the action of pressure and heat, a metamorphic rock is generally stronger than the rock from which it originated.

Chemical Classification of Rocks

In terms of their chemistry, rocks may be classified as:

- Siliceous rocks
- Calcareous rocks
- Argillaceous rocks

Siliceous rocks, such as sandstone, quartzite, and granite, are rich in silica (silicon dioxide). Calcareous rocks, such as limestone, marble, and travertine, are rich in calcite (calcium oxide). Argillaceous rocks, such as slate, are rich in alumina (aluminum oxide). Calcareous rocks are generally more prone to disintegration by acids but are slightly more fire resistive than silicious rocks.

Both feldspar and mica present weakness in granite, particularly mica because it decomposes more easily. Quartzite, a stone that is almost 100% quartz, is extremely strong. It is commonly used as an aggregate for ultra-high-strength concretes.

LIMESTONE

Limestone is a *sedimentary rock,* consisting primarily of the carbonates of calcium and magnesium, with small amounts of clay, sand, and organic material, such as seashells and other fossils. Limestone consisting of approximately 95% calcium carbonate and 5% impurities is called *calcite limestone.* That consisting of 60% to 80% calcium carbonate and 20% to 40% magnesium carbonate is called *dolomitic limestone.* Dolomitic limestone is generally stronger than calcite limestone.

Unlike granite, limestone is generally nongranular, with a relatively uniform surface appearance. It is softer than both marble and granite and hence easier to quarry, saw, and shape. It ranges in color from white to gray and does not take polish.

Calcium carbonate reacts with acids. The presence of acid in several foods (citric acid in lemons, limes, oranges, etc., and acetic acid in vinegar and pickles) reacts with limestone. That is why limestone is not recommended for use as kitchen or dining tabletops. Limestone facades in areas with serious acid rain problems have deteriorated. However, because acid rain is a relatively recent problem (related primarily to the emissions from the use of petroleum and coal), the deterioration of limestone is also a recent problem.

Several historic buildings with limestone facades have performed quite well in the absence of reactive atmospheres. Some of the important buildings with limestone facades are the Empire State Building and Rockefeller Center in New York City, the Pentagon in Washington, D.C., and the Chicago Tribune Tower in Chicago, Illinois. Limestone is also commonly used as concrete aggregate.

MARBLE

Geologically, marble is different from limestone because it is a *metamorphic rock.* Chemically, however, marble is similar to limestone. In fact, marble is limestone, which under centuries of high pressure and heat in the earth's crust morphed (changed) from a sedimentary rock to a metamorphic rock.

Because of the pressure and heat, marble is stronger and denser than the original limestone but weaker than granite. Like granite, it takes a good polish. Commercially, denser varieties of limestone that accept polish are sold as marble.

Marbles vary in surface appearance, which can be without patterns or with patterns—veiny, or mottled, or both, Figure 23.30. The presence of veins often indicates the presence of faults, weakness, or cracks in marble. The use of veiny and variegated marble is generally discouraged for exterior applications. Marbles vary greatly in color—from white to black, pink, and so on. Being chemically identical to limestone, marble is also vulnerable to acid attack.

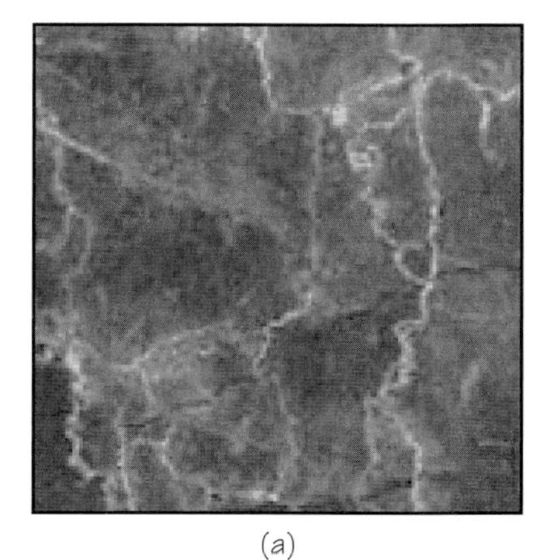

(a)

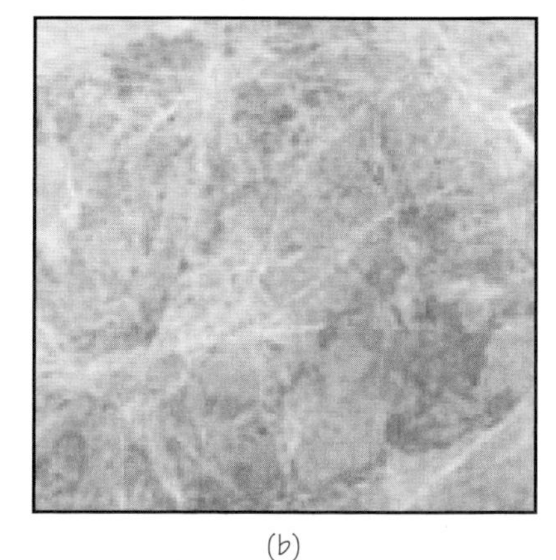

(b)

FIGURE 23.30 **(a)** Highly veiny marble. **(b)** Mottled as well as veiny marble.

TRAVERTINE

Travertine is a *sedimentary rock* obtained from the sediments of limestone dissolved in springwater. Springwater (particularly from hot springs) running over limestone deposits dissolved the limestone, which subsequently sedimented (i.e., deposited) in a nearby location. Sometimes the sediment trapped some water, which eventually evaporated, leaving voids in the rock. Travertine is, therefore, a porous stone, and travertine slabs are pitted with voids.

Travertine is closely related to limestone, but because it did not morph like marble, it is a softer stone. Most travertine varieties, therefore, do not take polish. Denser varieties, which take good polish, are referred to as *travertine marble.*

Because of its porous structure, travertine does not have the durability required to be used in thin slabs for exterior cladding. However, architects, who like the surface texture created by the pits, have used travertine in building exteriors as (thick) masonry walls. When used as a flooring material, its pitted surface can collect dirt, requiring greater maintenance unless the pits are filled with a portland cement (or an epoxy) and ground smooth.

SANDSTONE

Sandstone is a *sedimentary rock* formed by layers of sand (quartz) particles with oxides of calcium, silicon, and iron as cementing agents. If the cementing agent consists primarily of the oxide of silicon, sandstone is light in color and strong. If cemented by iron oxide, the sandstone is brown or red in color and softer. Sandstone that has a large amount of calcium oxide as cement is relatively more vulnerable to disintegration.

SLATE

Slate is a *metamorphic rock* formed by the morphosis of clay and mica sediments. Slate has a nongranular, smooth texture and is characterized by distinct cleavage planes that permit easy splitting in slabs as thin as $\frac{1}{4}$ in. That is why slate is used as roofing and flooring material. It is available in various colors, ranging from black to pink and light green.

23.8 FROM BLOCKS TO FINISHED STONE

Natural stone is procured by stone fabricators from quarries in the form of large blocks. Because of the way they are extracted from the rocks, blocks are of irregular sizes, Figure 23.31. The blocks are converted into slabs and other cross-sectional profiles in stone-fabrication plants. The conversion is done by sawing the blocks—a process similar to sawing lumber, except that water is used continuously during the sawing process to keep the saw blades cool, Figure 23.32.

The saw shown in Figure 23.32 uses one reciprocating blade that moves both horizontally and vertically so that only one slab is sliced at a time. In more sophisticated machines, several saw blades are ganged together to give multiple slabs in one pass.

Cross-sectional profiles for round columns, column caps, balusters, wall copings, window sills, and so on, are obtained by machining the sawn material to the desired shape, Figure 23.33. Different types of power-driven tools or tool attachments are available for different profiles. Although most stone fabrication is automated, complex ornamental work requires working with hand tools such as chisels, picks, and hammers, Figure 23.34.

FIGURE 23.31 Blocks of stone outside a stone fabricating plant. (Photo at Texas Quarries, Austin, Texas.)

FIGURE 23.32 Conversion from stone blocks to slabs through sawing. (Photo at Texas Quarries, Austin, Texas.)

FIGURE 23.33 Sawing and machining to convert stone blocks into special cross-sectional profiles. (Photo at Texas Quarries, Austin, Texas.)

FIGURE 23.34 Complex stonework requires working with hand tools. (Photo courtesy of Texas Quarries, Austin, Texas.)

FINISHES ON STONE SLABS AND PANELS

The surfaces of stone slabs and panels can be finished in several ways. The finish not only affects the surface appearance of the stone but also its durability. A honed or polished finish is generally more durable because it facilitates the drainage of water from the surface. The following are some of the commonly used finishes on stone slabs and panels:

- *Sawn finish:* If the stone is not finished beyond sawing, the surface is called a sawn finish. A sawn finish has visible saw marks.
- *Honed finish:* When a sawn finish is ground smooth with an abrasive material, a honed finish is obtained. Honing requires repeated grinding with increasingly finer abrasives. Water is used continuously during the honing process to control dust, Figure 23.35.
- *Polished finish:* Conceptually, there is no difference between a honed and a polished finish. A honed finish is smooth but with a matt appearance. A polished finish is obtained by grinding the stone surface beyond the honed finish with finer abrasives and finally buffing it with felt until the surface develops a sheen. A polished finish brings out the color of stone to its fullest extent by reflecting light like a mirror. The

FIGURE 23.35 Honing (grinding) of stone slabs. (Photo at Texas Quarries, Austin, Texas.)

difference in colors of a rough and polished finish on the same stone is easily noticeable, often significant.

A clear penetrating sealer that adds to the sheen is generally applied to the surface of a polished stone. The sealer increases the durability of stone by sealing the pores, adding resistance to chemical attack and the formation of stains. Only dense stones, such as granite, marble, and dense varieties of limestone, travertine, and slate, can develop polish.

- *Flame-cut finish:* Flame-cut finish, also referred to as *thermal finish,* is a rough finish obtained by torching the stone surface with a natural gas or oxyacetylene torch. Before torching, the stone is thoroughly wetted. The heat from the torch expands the absorbed water into steam, which breaks loose the surface particles in the stone, leaving behind a rough surface. Generally, a flame-cut finish is used only on granite because other stones are too porous to break only at the surface.

  The roughness of a flame-cut finish makes it ideal for floors, particularly those that are subject to frequent wetting. Often, the flame-cut finish is used in the treads of granite-topped stairs and polished finish for the risers, which gives the desired contrast between the treads and the risers.
- *Bush-hammered finish:* Bush-hammered finish is also a rough finish and is obtained by hammering off the surface of stone with picks.
- *Split-face (cleft) finish:* Like CMU, stone can also be split through one of its faces, yielding two split-face slabs. Splitting is easier in slate which has natural cleavage planes, in which case the stone is referred to as *cleft-finished.* In the stone industry, however, the terms *cleft* and *split-face* are used interchangeably.
- *Sandblasted finish:* Although not as commonly used as other finishes, sandblasting of stone yields a rough surface.

23.9 STONE SELECTION

The selection of stone for a particular use is a function of several factors. Budget and aesthetics (color, pattern, and surface appearance) are the two most important factors to be considered for stone used in building interiors. For exterior use, the history of performance of a stone in the local environment (durability) is obviously another important factor.

EXPAND YOUR KNOWLEDGE

Granite Versus Marble

Of the various building stones, granite and marble are the most important. Both are dense and strong stones and take good polish. In many buildings, the final choice of stone for a particular application is between granite and marble.

Both granite and marble have their own beauty and individual personalities. In most situations the selection is based on personal preference, cost, and availability. Marble comes in a much wider range of colors than granite and offers interesting veins. Veins are streaks of a different color than the main body of marble—much like the grain in wood (Figure 23.30). Veins may be streaks of weakness and cracks. Marble is commonly used for bathroom countertops and walls.

Granite is also available in many colors, and its surface is spotty and speckled. It is the material of choice for kitchen countertops. Because of its hardness, it can better withstand damage from kitchen knives than any other stone. Its abrasion resistance and resistance to acids are also factors in its favor. The reasons that make granite most suited for use in kitchens also make it most suited for exterior cladding of buildings. Its high density and low water absorption further add to its durability, particularly against freeze-thaw. However, marble also performs well in exterior cladding and has been used extensively for centuries.

Both marble and granite are suitable for the cladding of interior walls and interior floors.

TABLE 23.2 COMMON APPLICATIONS OF SELECTED STONES

Application	Commonly used stones
• Exterior wall cladding	Granite, marble, and limestone
• Interior wall cladding	Granite, marble, and limestone
• Interior flooring	Granite, marble, and slate
• Stair treads and risers	Granite
• Kitchen counter top	Granite
• Bathroom counter top	Granite and marble
• Wall copings and balusters	Granite, marble, and limestone
• Roofing	Slate

Where the material has no track record, the physical properties of the stones being evaluated should be compared. Generally, the following properties are of importance:

- Density
- Water absorption
- Compressive strength
- Flexural strength (modulus of rupture)
- Abrasion resistance

Of these properties, density and water absorption are the most important factors because they affect the durability of stone against freeze-thaw and chemical attack. Abrasion resistance is important in floors, but it is generally related directly to density. Flexural resistance of stone is important in wall-cladding applications. Stone quarries and/or stone suppliers provide the required data for their stones.

Table 23.2, based on common uses of various stones, provides a rough guide in stone selection.

23.10 BOND PATTERNS IN STONE MASONRY WALLS

In contemporary buildings, natural stone is generally used as thin slabs. For exterior or interior wall cladding, slabs vary in thickness from $\frac{3}{4}$ in. to 2 in. For flooring, slab thickness can be as low as $\frac{3}{8}$ in. The thinner the stone, the smaller the size of slab in which it is available. A slab thickness of less than $\frac{3}{4}$ in. is referred to as a *tile* in stone industry.

Stones used in exterior-wall veneers are generally 3 to 4 in. thick. Those used in load-bearing stone walls are thicker. In some cases, the stones are so large and thick that they cannot be laid by hands but require mechanical hoists.

Stone veneer and load-bearing stone walls are referred to as *stone masonry* to distinguish them from thin stone cladding. Stone masonry walls are laid with mortar, stone by stone, in the same way as bricks and CMUs. In this section, stone masonry construction is discussed. Thin stone cladding, which is constructed differently, is covered in Chapter 27.

Because natural stone is not available in uniform sizes, as are bricks and CMUs, the bond patterns in stone masonry walls are different from those used in brick or CMU walls. Two basic patterns used in stone masonry walls are

- Rubble masonry
- Ashlar masonry

Rubble masonry walls are made from stones whose sides are irregular (i.e., not at right angles to each other). Rubble masonry is further subdivided as *random rubble* and *coursed rubble.* In random rubble masonry, the mortar joints are irregular. A random rubble wall may consist either of stones obtained from the quarries or rounded riverbed boulders. In coursed rubble, the bed joints line up after every few pieces of stone, Figure 23.36(d). Therefore, the mason has to select the stones in the field (or shape them using a pointed hammer) so that they fit in the available spaces.

In *ashlar masonry* the sides of the stones are dressed square (at right angles to each other). The front and back faces of the stone may, however, be dressed or undressed. Like rubble masonry, ashlar masonry is also divided into *random ashlar* and *coursed ashlar,* Figure 23.37.

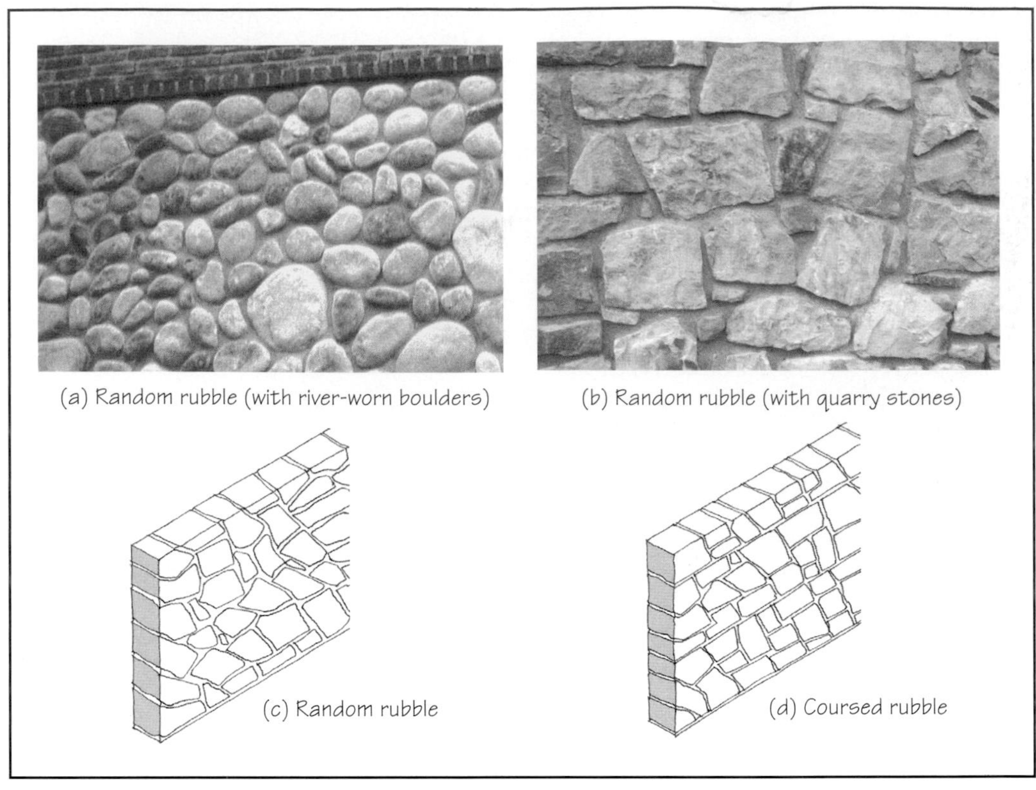

FIGURE 23.36 Random rubble and coursed rubble masonry walls.

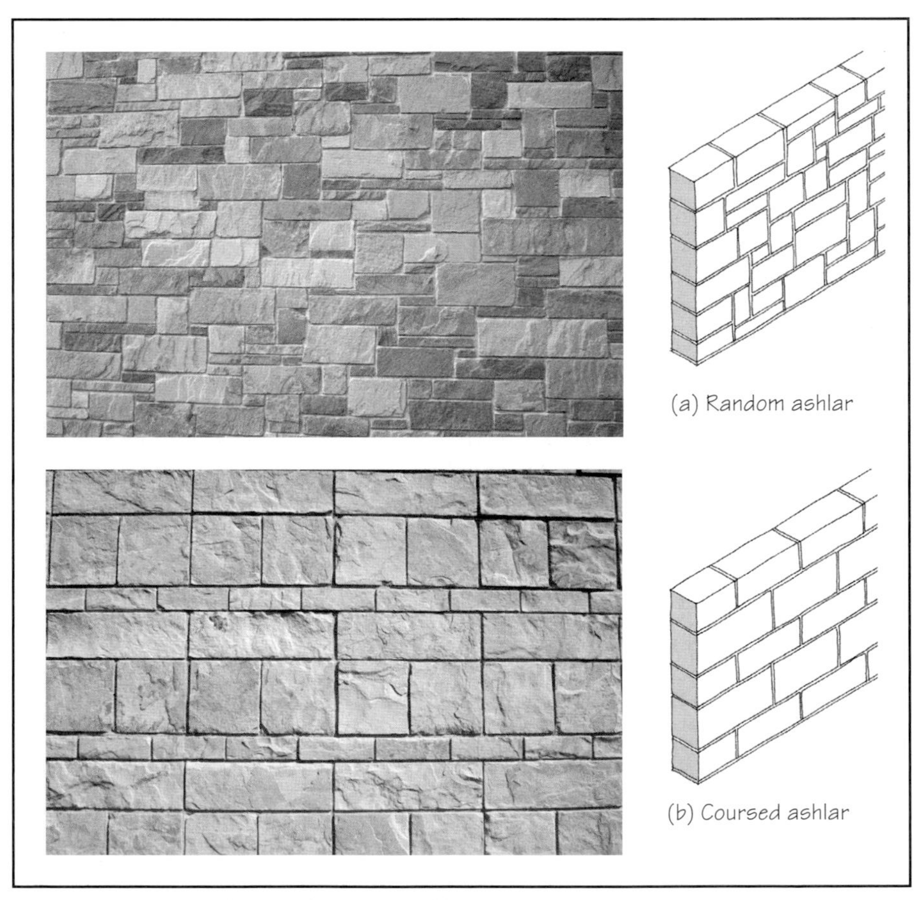

FIGURE 23.37 Random ashlar and coursed ashlar masonry walls.

23.11 GLASS MASONRY UNITS

Glass masonry units (also called glass blocks, or GMUs) are used as non-load-bearing walls in virtually all types of projects—commercial and residential—on the exterior as well as in the interior of buildings. By combining the modularity of masonry units and the transparency of glass, glass masonry units give designers a means of expression that is not available in other materials, Figure 23.38.

FIGURE 23.38(a) Glass masonry units used in the exterior walls, University of New Hampshire campus building. (Photo courtesy of Pittsburgh Corning, Pittsburgh, Pennsylvania.)

FIGURE 23.38(b) Glass masonry units used in the exterior walls, Ventura Plaza, Troy, Michigan. (Photo courtesy of Pittsburgh Corning, Pittsburgh, Pennsylvania.)

FIGURE 23.38(c) Glass masonry shower enclosure. (Photo courtesy of Pittsburgh Corning, Pittsburgh, Pennsylvania.)

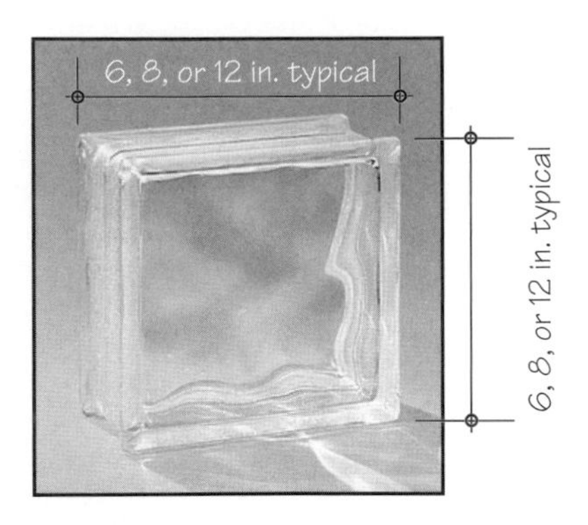

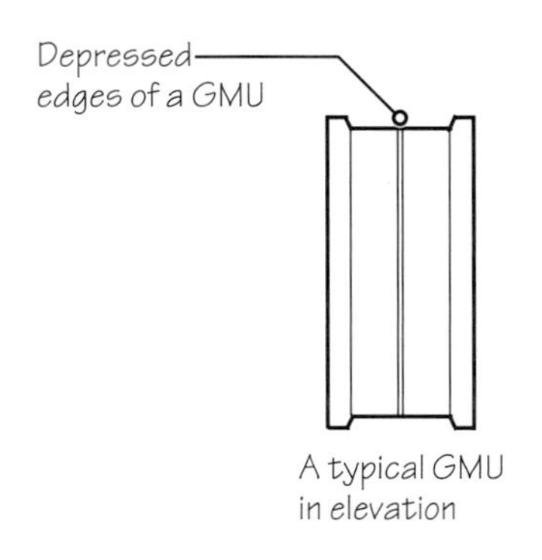

Glass masonry units are manufactured in several sizes. The most common face dimensions of units are 6 in. × 6 in., 8 in. × 8 in. and 12 in. × 12 in. with thicknesses of 3 to 4 in. The units are generally hollow with air trapped within, although solid units are also available. Because of the trapped air, the R-value of a glass masonry unit is nearly 2.0 (approximately the same as that of an air-filled insulating glass unit (Section 28.7).

Glass blocks are generally translucent. Thus, a glass block wall provides light similar to a frosted window glass. However, a glass block wall gives greater privacy, more security, and greater sound insulation as compared with a glass window or glass curtain wall. Where greater transparency is needed, transparent units are used.

Glass masonry walls also have a higher fire-resistance rating than conventional glass walls. A 45-min rating is easily achievable, and a higher rating is available.

CONSTRUCTION OF A GLASS MASONRY WALL

The construction of a glass masonry wall is similar to other masonry walls. Glass masonry units are generally laid in stack bond with portland cement–lime mortar. The joints are fully mortared, as with clay bricks.

The cross-sectional profile of a glass masonry unit is slightly depressed in the interior so that the mortar is thicker in the interior than on the face of the units. The face thickness of mortar is generally $\frac{1}{4}$ in., so that the difference between nominal and actual dimensions of units is $\frac{1}{4}$ in. In other words, an 8 in. × 8 in. unit is $7\frac{3}{4}$ in. × $7\frac{3}{4}$ in actual dimensions.

Glass masonry units are nonstructural. Therefore, a glass masonry wall must be treated as a non-load-bearing wall. It should not be designed to support any gravity load except its self-load. However, it must resist lateral (wind and earthquake) loads and be able to transfer them to the structural frame.

A single glass masonry panel is treated as a window unit and can be provided in a wood, masonry, or concrete wall. A lintel is required over such a panel to absorb the gravity loads in the same way as that required above a window or door opening.

A large glass masonry wall is treated as a combination of panels held between structural steel or reinforced-concrete framing members, Figure 23.39. Manufacturers provide the data to facilitate the structural design of a panel, in addition to some engineering support.

Figure 23.40 shows a detail commonly used for the construction of a glass masonry panel. Steel anchors are used at the top and the sides (jambs) to transfer the lateral load from the panel to the supporting members. The long leg of an anchor is embedded in the mortar joint, and the short leg is fastened to the jamb or the head of the supporting frame.

Both vertical and horizontal anchors are generally used at 16 in. on center. However, a horizontal anchor is required above the first course at the bottom of the panel, and another anchor is needed below the first course at the top of the panel.

FIGURE 23.39 Staircase enclosure, Lake Dallas High School, Dallas, Texas. (Photo courtesy of Pittsburgh Corning, Pittsburgh, Pennsylvania.)

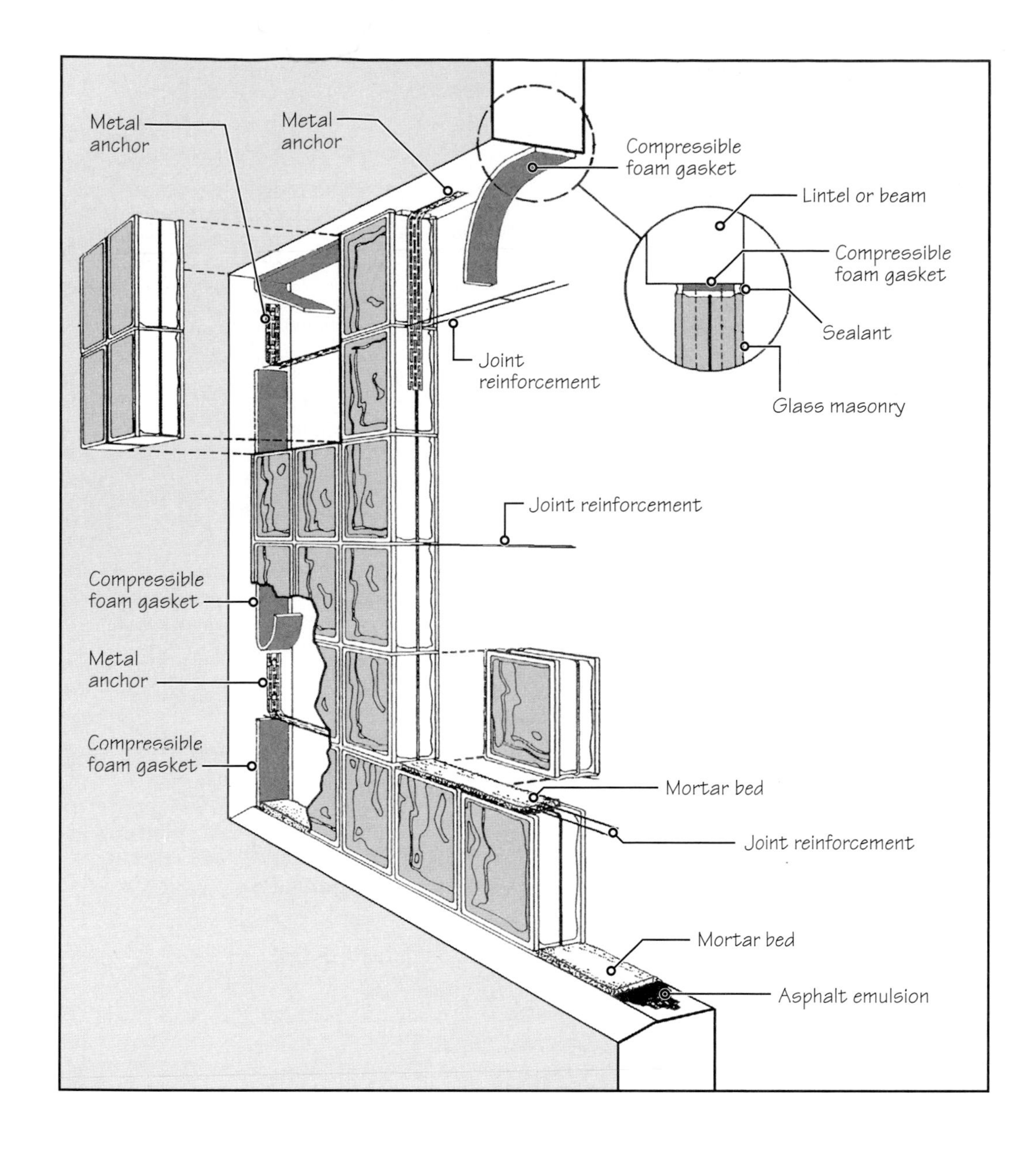

FIGURE 23.40 A typical detail used for the construction of a glass masonry panel. Detail adapted from that provided by Pittsburgh Corning, Pittsburgh, Pennsylvania.

Joint reinforcement is needed to stiffen the panel so that the panel as a whole is able to transfer the lateral load to the structure. Generally, the joint reinforcement is placed in the same course as the horizontal anchors.

23.12 FIRE RESISTANCE OF MASONRY WALLS

Fired clay, concrete, and stone are noncombustible and inherently fire enduring. Therefore, the fire-resistance ratings of masonry assemblies is generally high. One of the important factors that affects the fire-resistance rating of a masonry wall is the amount of solid content in them. The greater the solid content, the greater the fire-resistance rating of a wall. Therefore, a grouted wall has a higher fire-resistance rating than an ungrouted wall.

Because CMUs are generally hollow, the fire resistance rating of a CMU wall is given in terms of the wall's equivalent (solid) thickness. Equivalent thickness is the thickness of the wall excluding the cells (voids). Thus, if a wall is made of $7\frac{5}{8}$-in.-thick (8-in. nominal) units that are 60% solid (40% hollow), the wall's equivalent thickness is 7.625(0.6) = 4.575 in. Note that $7\frac{5}{8} = 7.625$, and 40% hollow means 60% solid. The equivalent thickness of a fully grouted wall is the thickness of the wall itself.

Another factor that influences the fire resistance rating of a CMU wall is the type of aggregate used in the CMUs. A wall made with CMUs containing lightweight aggregate gives a higher fire-resistance rating for the same equivalent thickness than a wall with CMUs containing normal-weight aggregate.

Table 23.3 gives the approximate fire-resistance ratings of selected masonry assemblies. For design purposes, more authoritative publications should be referenced. A comprehensive reference is that published by the Underwriters' Laboratories (UL). Another good source is the International Building Code.

FOCUS ON **SUSTAINABILITY**

Sustainability Features of Masonry

The most important sustainability feature of masonry is its durability. Masonry structures can stand and remain serviceable for a very long time, requiring little maintenance. Their durability is a result of several factors. First, they degrade slowly because they are resistant to physical, biological, and chemical deterioration. Second, they are noncombustible. The following are additional sustainability features of masonry:

Local production: Clay masonry and concrete masonry units are generally produced in most communities.

Reusability: Masonry materials can be easily reused. Salvaged bricks and stones are available in most communities. Some owners and architects prefer to use salvaged masonry. Interlocking concrete masonry pavers or units used in retaining walls do not require any mortar and can be disassembled and reused

Recyclability: Masonry materials can also be easily recycled. Crushed and pulverized bricks (referred to as *grog*) can be mixed with clay and used to manufacture new bricks. Stone can be used as aggregate for concrete and crushed concrete masonry units can be used as under bed for concrete footings and slabs-on-grade.

Absence of VOC Emissions: Masonry materials do not emit any volatile organic compounds. Although radon gas has been associated with certain soils, brick manufacturers ensure the absence of radon in the clay used in the manufacture of bricks.

Thermal Mass: All masonry materials provide thermal mass, which may help conserve energy in some climates (Chapter 5).

Embodied Energy: The embodied energy in stone and CMU is lower than clay brick, which has much lower embodied energy than metal or glass.

Negative sustainability aspects of masonry include soil erosion and habitat loss caused by the mining of clay and stone. However, like the other industries, the masonry industry is becoming increasingly aware of its responsibilities toward the environment. Brick manufacturers are increasingly converting the abandoned pits into lakes for use by the community.

TABLE 23.3 APPROXIMATE FIRE-RESISTANCE RATINGS OF SELECTED MASONRY WALLS

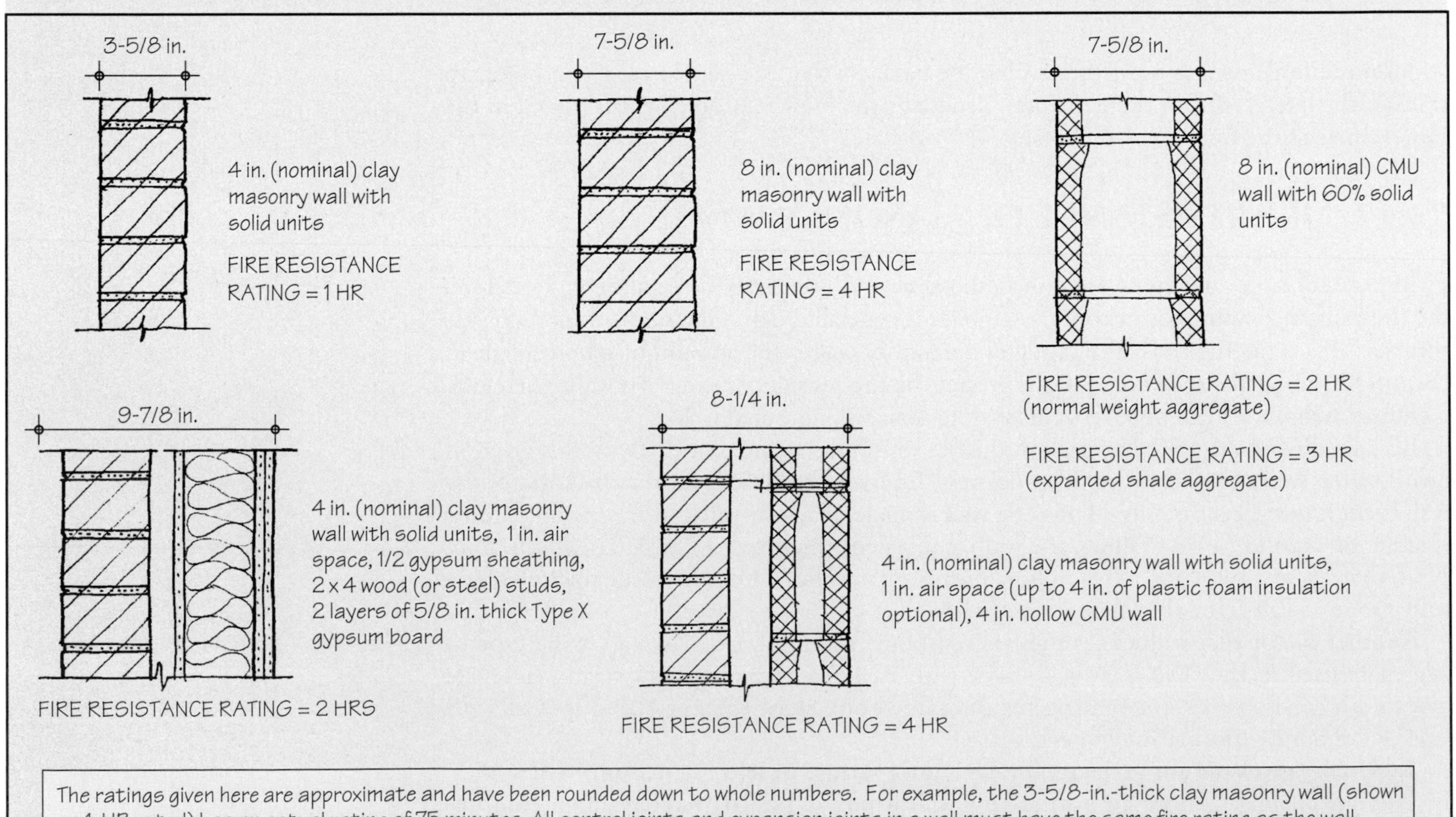

The ratings given here are approximate and have been rounded down to whole numbers. For example, the 3-5/8-in.-thick clay masonry wall (shown as 1-HR rated) has an actual rating of 75 minutes. All control joints and expansion joints in a wall must have the same fire rating as the wall.

PRACTICE QUIZ

Each question has only one correct answer. Select the choice that best answers the question.

31. Geologically, the earliest rock on the earth's surface was
- **a.** igneous rock.
- **b.** sedimentary rock.
- **c.** metamorphic rock.
- **d.** none of the above.

32. Marble is
- **a.** an igneous rock.
- **b.** a sedimentary rock.
- **c.** a metamorphic rock.
- **d.** none of the above.

33. Granite is
- **a.** an igneous rock.
- **b.** a sedimentary rock.
- **c.** a metamorphic rock.
- **d.** none of the above.

34. Slate is
- **a.** an igneous rock.
- **b.** a sedimentary rock.
- **c.** a metamorphic rock.
- **d.** none of the above.

35. Which of the following stones is generally more vulnerable to disintegration by an acidic atmosphere?
- **a.** Granite
- **b.** Marble

36. In general terms, which of the following stones is considered most durable?
- **a.** Marble
- **b.** Granite
- **c.** Limestone
- **d.** Sandstone
- **e.** Slate

37. The bond patterns that are used in brick masonry, such as running bond, English bond, and Flemish bond, are also used in stone masonry.
- **a.** True
- **b.** False

38. Dressed stones are required to be used in
- **a.** rubble masonry.
- **b.** ashlar masonry.
- **c.** CMU masonry.
- **d.** CSMU masonry.

Answers: 31-a, 32-c, 33-a, 34-c, 35-b, 36-b, 37-b, 38-b.

KEY TERMS AND CONCEPTS

Ashlar masonry
A-unit, H-unit, split-face unit, ribbed unit, acoustical unit
Burnished unit, glazed unit
Calcium silicate masonry unit
Concrete masonry unit (CMU)
Coursed rubble
Face-shell mortaring
Flame-cut finish
Glass masonry unit (GMU)
Granite, marble, limestone, slate, sandstone and travertine
Honed finish
Joint reinforcement
Lightweight unit, normal-weight unit
Masonry grout
Natural stone
Polished finish
Random rubble
Rubble masonry
Sawn finish
Scored unit, bullnose unit, sash unit, lintel unit, bond beam unit
Split-face finish

REVIEW QUESTIONS

1. Explain the differences between lightweight and normal-weight CMUs, including their uses.
2. Explain why CMUs are generally face-shell mortared. In which situations, full-bed mortaring of CMUs is used?
3. Explain the purpose of joint reinforcement in CMU walls.
4. With the help of sketches, explain why a running-bond wall has a higher horizontal bending strength than a stack-bond wall, whereas the vertical bending strength of both walls is equal.
5. What measures are used to respond to the shrinkage of concrete masonry walls?
6. Explain the differences between fine grout and coarse grout and where each is used.
7. Explain why granite is the most durable of building stones. Where would you recommend the use of granite?
8. Explain the differences between marble and granite.

SELECTED WEB SITES

Indiana Limestone Institute of America, ILIA (www.ilia.com)
Magazine of Masonry Construction (www.masonryconstruction.com)
Marble Institute of America, MIA (www.marble-institute.com)
National Concrete Masonry Association, NCMA (www.ncma.org)
Northwest Granite Manufacturers Association, NWGMA (www.nwgma.com)
Pittsburgh Corning Glass Block, Pittsburgh Corning Corporation (www.pittsburghcorning.com)

FURTHER READING

1. Beall, Christine. *Masonry Design and Detailing for Architects, Engineers and Builders.* New York: McGraw-Hill, 1997.
2. National Concrete Masonry Association (NCMA). *NCMA TEK Notes.* (These notes are continuously updated at different times.)
3. Marble Institute of America (MIA). *Dimension Stone Design Manual,* 2003.
4. Marble Institute of America. *Dimension Stones of the World,* 1992.

CHAPTER 24 Masonry and Concrete Bearing Wall Construction

CHAPTER OUTLINE

24.1 Traditional Masonry Bearing Wall Construction

24.2 Importance of Vertical Reinforcement in Masonry Walls

24.3 Bond Beams in a Masonry Bearing Wall Building

24.4 Wall Layout in a Bearing Wall Building

24.5 Floor and Roof Decks—Connections to Walls

24.6 Limitations of Masonry Bearing Wall Construction

24.7 Bearing Wall and Column-Beam System

24.8 Reinforced-Concrete Bearing Wall Construction

24.9 Reinforced-Concrete Tilt-Up Wall Construction

24.10 Connections in a Tilt-Up Wall Building

24.11 Aesthetics of Tilt-Up Wall Buildings

PRINCIPLES IN PRACTICE: The Middle-Third Rule

Jubilee Church, Rome, under construction. Architect: Richard Meier and Partners. The three doubly curved shell walls of this building are of precast concrete using white portland cement and white aggregate. To ensure that the concrete remains white and dirt-free throughout its service life the concrete mix contains a photocatalytic admixture that helps decompose the organic and inorganic substance on concrete surface using light energy. (Photo courtesy of Dr. Jay Henry.)

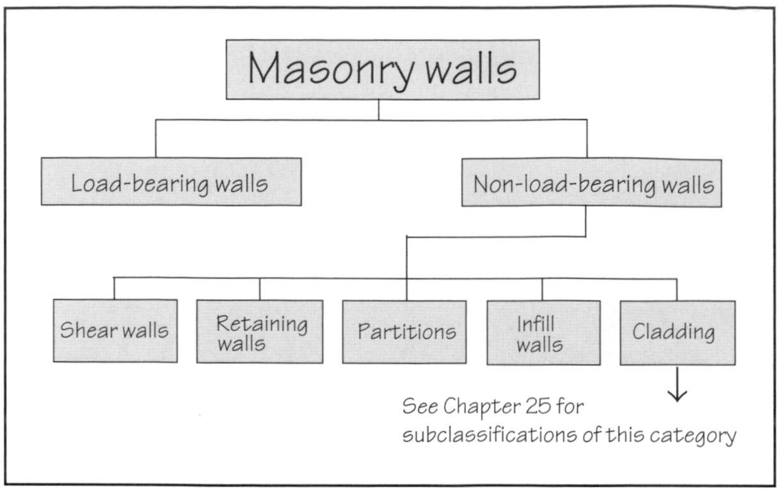

Masonry walls

- Load-bearing walls
- Non-load-bearing walls
  - Shear walls
  - Retaining walls
  - Partitions
  - Infill walls
  - Cladding

See Chapter 25 for subclassifications of this category

FIGURE 24.1 Various applications of masonry walls.

The use of masonry in contemporary buildings is primarily in the walls. Minor uses occur in interior or exterior paving. Although masonry roofs (vaults, domes, and shells) were frequently constructed in the past, they are relatively uncommon in modern buildings.

Masonry walls may either be load-bearing or non-load-bearing walls. Non-load-bearing masonry walls include retaining walls, exterior infill and cladding, and interior partitions, Figure 24.1. Masonry cladding functions as an exterior weather-resistant cover and is similar to the exterior wall sheathing in a wood light frame building. Masonry cladding is discussed in Chapter 26.

Masonry infill occupies the space between the concrete or steel frame members. It may be used to resist the racking of the frame—that is, serve as a shear wall. In that case, the infill must completely fill the space between the columns of the frame so that the lateral load is transferred from the columns to the infill, and vice versa, Figure 24.2(a).

If the infill is not used as a shear wall, adequate separation between the columns and the infill is required so that the lateral load is not transferred from the columns to the infill, Figure 24.2(b).

ADVANTAGES OF A MASONRY BEARING WALL BUILDING

A load-bearing wall (by definition) is a wall that supports gravity loads in addition to its self-load. Load-bearing walls may also act as shear walls. (However, as shown in Figure 24.1, some shear walls may be non-load-bearing walls, i.e., they may not support any gravity load other than their self-loads.)

A major advantage of a load-bearing masonry structure, also called a *masonry bearing wall structure,* is that the walls that are required for load-bearing purposes may also function as shear walls in addition to enclosing and dividing spaces. This contrasts with a steel or concrete frame structure, where the frame serves the structural function only. Space-enclosing and

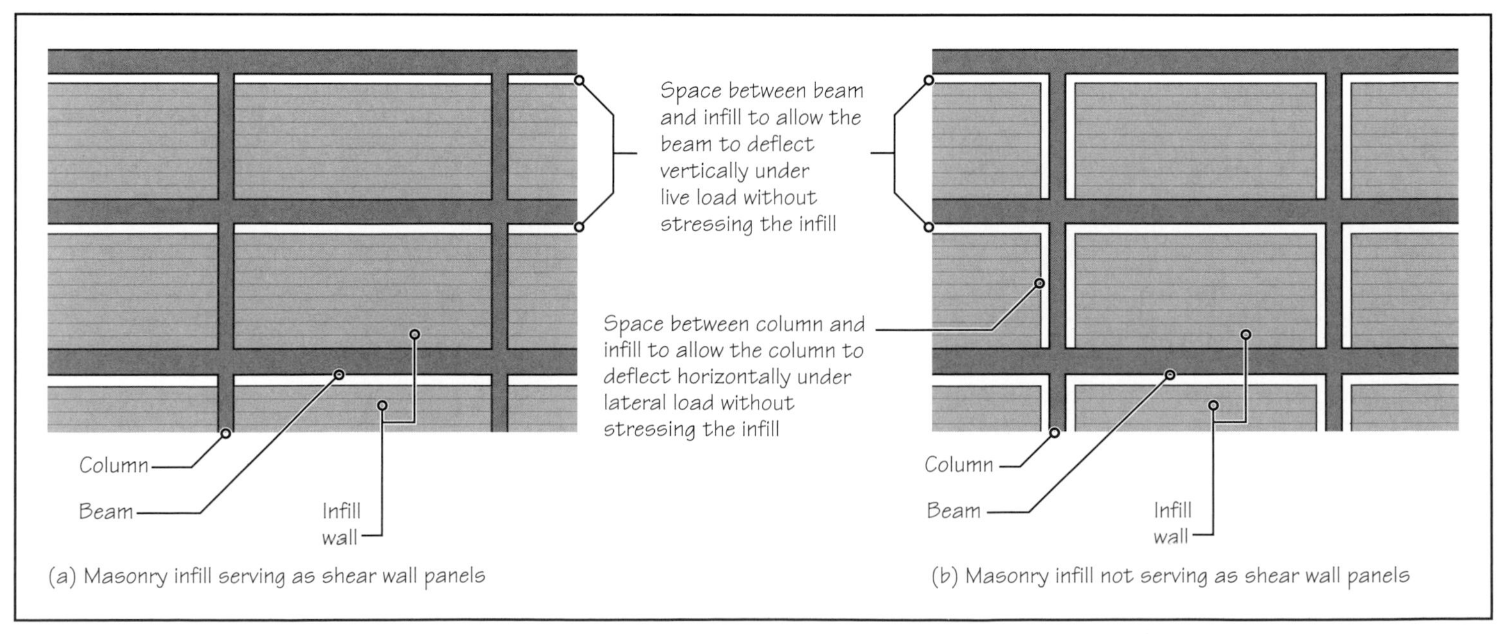

FIGURE 24.2 Masonry infill between the members of a structural frame may or may not be used to shear (racking of the frame).

dividing functions must, therefore, be performed by nonstructural walls. The separation of the structural and nonstructural elements in a frame structure requires careful detailing of the connections between them. Such details are unnecessary in a load-bearing masonry structure.

Constructed of an inherently fire-resistive, mold-free, sound-insulating, relatively inexpensive, and durable material, load-bearing masonry has a favorable life-cycle cost index. That is why masonry experts claim that load-bearing masonry is a more sustainable alternative to a concrete or steel frame structure in many applications.

BEARING WALL STRUCTURES IN REINFORCED MASONRY AND REINFORCED CONCRETE

Although masonry has been used historically without any steel reinforcement, in contemporary times, the use of plain (unreinforced) masonry is less common. Reinforced masonry has several advantages over plain masonry at little additional cost. As discussed in Section 24.2, reinforcement increases a masonry wall's flexural and shear strengths. Additionally, because steel is a ductile material, a reinforced-masonry wall is better able to absorb structural deformations by the yielding of steel—an important requirement for structures located in seismic regions. Plain masonry's ability to absorb deformations is limited due to the brittle nature of masonry.

Poor performance of unreinforced-masonry structures has been demonstrated again and again through studies of postearthquake and posthurricane building damage all over the world. However, it must be added that low-rise, plain masonry bearing wall structures have excellent performance records in locations that are not subjected to earthquakes or extreme winds.

This chapter begins with an introduction to the traditional load-bearing masonry, followed by a discussion of its contemporary alternative. Because there is a great deal of similarity between bearing wall construction in reinforced masonry and reinforced concrete, reinforced-concrete bearing wall construction is also covered in this chapter.

Reinforced-concrete bearing wall structures can be constructed of site-cast concrete or precast-concrete. The precast-concrete bearing wall system is typically a concrete tilt-up wall system, which is also discussed in this chapter.

24.1 TRADITIONAL MASONRY BEARING WALL CONSTRUCTION

The masonry bearing wall system is one of the oldest construction systems. Until the invention of the frame structure, the masonry bearing wall system was the only system available for constructing major buildings. Although several large and tall masonry structures were built, their design was based on arbitrary rules deduced from an intuitive rather than a scientific basis. Gothic cathedrals symbolize the daring to which such intuitive understanding was extended.

Over time, the intuitive understanding was embedded into the building codes as empirical rules for the design of masonry bearing walls. These rules required that the thickness of exterior masonry walls progressively increase toward the lower floors. The increase in wall thickness was necessary to accommodate greater gravity loads at lower floors.

Another important reason for progressively increasing wall thickness was to ensure stability against overturning by wind loads. (Very little was known at the time about seismic loads.) The code-required thickness was based on the assumption that in resisting the wind loads, each exterior wall functions as a freestanding wall, behaving as a vertical cantilever fixed into the ground—like a flagpole.

The cantilever action produces extreme compression on one face of the wall and extreme tension on the opposite face. Masonry is strong in compression. Therefore, compression was not of much concern. However, plain (unreinforced) masonry is weak in tension, and its tensile strength is governed by the flexural bond between masonry units and the mortar, which is quite low. When the tension in masonry exceeds its flexural bond strength, the mortar joints open up (Figure 22.6).

Because only plain masonry was used at the time, the wall thickness was based on the premise that masonry should not be subjected to any tension. Stated differently, the premise assumed that the tensile strength of masonry was negligible. This ensured some redundancy because masonry has a some (although low) flexural tensile strength.

THE MIDDLE-THIRD RULE

A simple law of statics, called the *middle-third rule* (fairly well understood at the time) indicated that to ensure the absence of tensile stress in masonry, the resultant of the gravity load

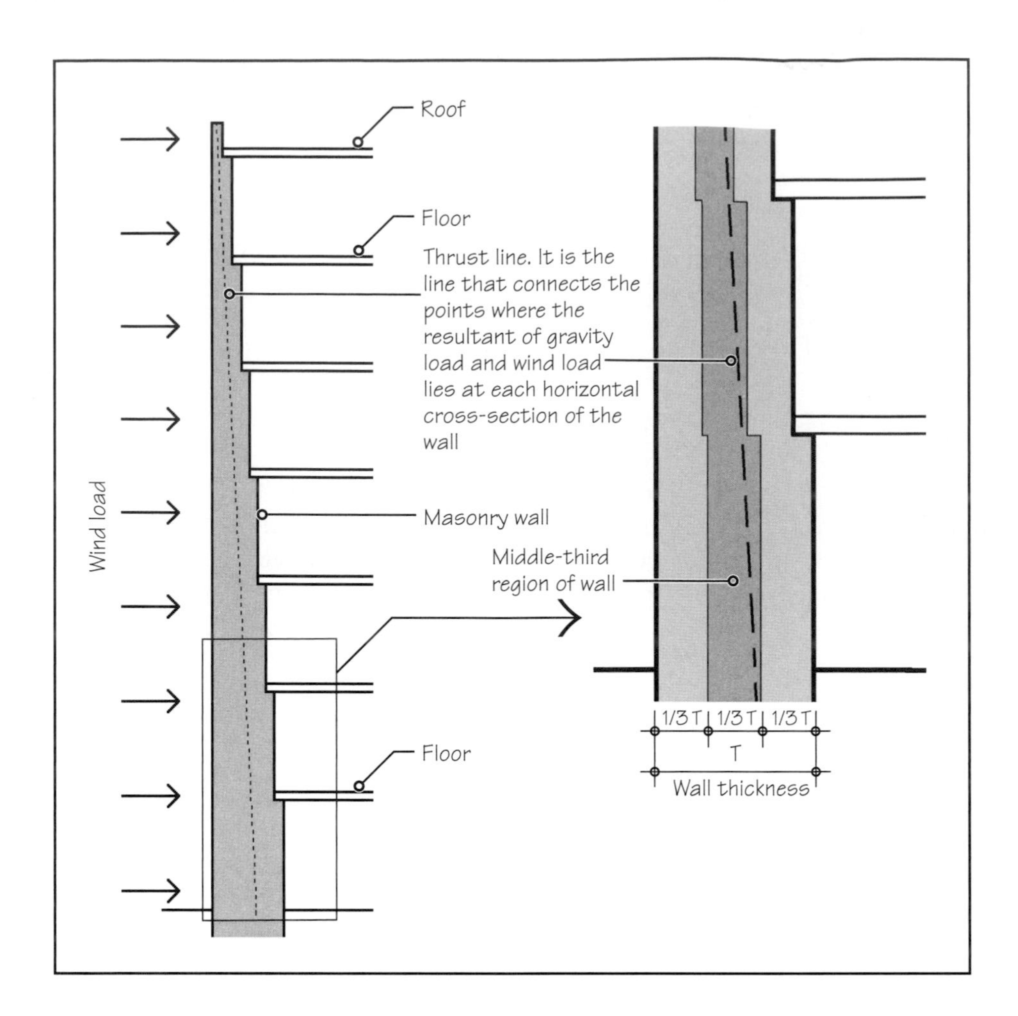

FIGURE 24.3 Middle-third rule applied to an exterior freestanding load-bearing masonry wall.

FIGURE 24.4 Monadnock Building, Chicago, completed in 1891, preserved as the last high-rise traditional bearing wall structure. Architects: Burnham and Root. (Photo courtesy of Dr. Jay Henry.)

and the wind load at any horizontal section of the wall must remain within the middle third of the section, Figure 24.3. For instance, if the wall thickness at a section is 18 in, the resultant of the gravity and the lateral load at that section must lie within the central 6 in. thickness of the wall.

In other words, the middle-third rule states that if the resultant of the gravity and wind loads lies within the middle third of a section, then that section is subjected to compressive stresses only. If the resultant lies outside the middle third, a part of the wall will be subjected to tension (see "Principles in Practice" at the end of this chapter).

HEIGHT LIMITATION OF TRADITIONAL MASONRY BEARING WALL SYSTEM

The assumption that each exterior masonry wall functions as a freestanding vertical cantilever, unable to resist tension, made the building code requirement highly conservative, necessitating extremely thick walls at the base of a tall building. That is why the exterior brick walls of the 215-ft-high, 16-story Monadnock Building (completed in 1891 in Chicago) are approximately 6 ft thick at the ground level, Figure 24.4.

Because the lower floors of a building are the most usable and profitable, it was apparent that high-rise masonry structures were economically unviable because of an unduly large percentage of floor area occupied by the walls. In fact, the Monadnock Building was the last high-rise masonry bearing wall structure in North America until the mid-twentieth century, when the system was revived using a more efficient structural design methodology.

In general, the interior structure of most masonry bearing wall buildings consisted of columns and beams. The columns were of cast iron and the beams were of wrought iron. In later structures, steel was substituted for both cast iron and wrought iron. In several masonry buildings, (heavy) timber columns and beams were used in the interior structure. For example, in the seven-story brick bearing wall building housing the Sixth Floor Museum in Dallas, Texas, completed in 1901 and now preserved for its historic importance, approximately 15-in.-thick timber columns were used to support each

The sixth floor of the Dallas County Administration Building (formerly the Texas School Book Depository) which was used to assasinate President John F. Kennedy in November, 1963. The floor was converted to a museum in 1989, called the Sixth Floor Museum.

Office space at the second floor of the building currently used by the Dallas County Administration, showing (heavy) timber columns and beams

FIGURE 24.5 The seven-story load-bearing brick building in Dallas, Texas, housing the Sixth Floor Museum.

floor, Figure 24.5. The brick walls are approximately 30 in. thick at the ground level, and the floor structure consists of timber beams and lumber planks.

Because the wind load was resisted only by the exterior walls, designed as freestanding elements, the interior structure of a traditional masonry bearing wall building was designed for gravity loads only; that is, wind loads were ignored in the design of interior walls, columns, and beams.

BEARING WALL SYSTEM—ITS ABANDONMENT AND REVIVAL

The use of the masonry bearing wall system for high-rise buildings went out of favor as the skeleton (iron and later steel) frame became popular. Steel is much stronger than masonry. Therefore, the columns in a frame structure were much smaller than masonry bearing walls. Because the walls in a frame structure are nonstructural, their thickness was considerably reduced. This conserved valuable floor space. A secondary advantage of the skeleton frame was the speed with which it could be erected as compared to thick masonry walls.

The beginning of the twentieth century saw major advances in the science of structural engineering, particularly in the area of design and analysis of frames and continuous beams. Advances were also made in the science of materials related to steel and reinforced concrete. Because masonry does not lend itself to frame construction, it was largely displaced from the realm of structural materials.

CONTEMPORARY BEARING WALL CONSTRUCTION

The situation began to change with reconstruction work in Europe after World War II, which required the construction of large-scale and relatively inexpensive housing. It was also the time when the structural design theories of reinforced concrete had become fairly sophisticated. The extension of these theories to masonry, which, as a material, is not vastly different from concrete, was only logical. Additionally, architects and engineers began to understand the interactive structural behavior of the walls and horizontal components (floors and roof) of a building.

During the 1950s several high-rise masonry bearing wall buildings were built, initially in Europe and subsequently in North America. The design was based on a rational engineering approach rather than on the earlier empirical rules.

The structural behavior of a contemporary masonry bearing wall building is similar to that of conventional wood light frame construction (see "Principles in Practice" in Chapter 13). However, the construction details required to respond to the behavior are not identical. The reason is that in a masonry bearing wall building, the walls, floors, and roof are solid components, whereas in a wood light frame building; they consist of linear frame members with intervening cavity spaces.

24.2 IMPORTANCE OF VERTICAL REINFORCEMENT IN MASONRY WALLS

Using the contemporary, rational design approach, it is possible to design masonry bearing wall residential structures (apartments, hotels, and motels) of approximately 20 stories using 8- to 10-in.-thick walls. In fact, an 8-in.-thick CMU wall is the industry's standard for masonry bearing wall structures. For long-span buildings with relatively high walls, such as those used in gymnasiums, warehouses, and large assembly spaces, 10- or 12-in.-thick CMU walls may be needed. CMU walls thicker than 12 in. are rare.

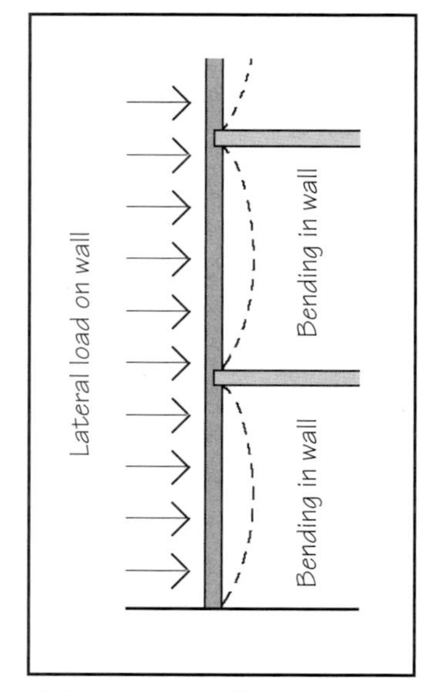

FIGURE 24.6 Bending in a wall caused by eccentric gravity load and lateral load on wall.

For a one- or two-story building in a low-wind or a nonseismic region, masonry walls may not require any vertical reinforcement. In a taller building or a building in a high-wind or seismic region, vertical reinforcement is generally needed. (Masonry walls without vertical reinforcement are referred to as *plain-masonry* walls, and those with vertical reinforcement are called *reinforced-masonry* walls.)

The benefits of reinforced masonry over plain masonry are as follows:

- Reinforcement increases the strength of a wall against bending caused by (a) eccentric gravity loads on the wall and (b) lateral loads perpendicular to the wall, Figure 24.6.
- Reinforcement increases the sliding resistance of the wall to in-plane lateral loads (as in a shear wall), reducing horizontal displacement between floors, Figure 24.7. In other words, reinforcement stitches the floors together. Additionally, it provides positive anchorage between the wall and the foundation.
- Reinforcement helps to resist the tension caused by overturning of the wall, Figure 24.8 (and functions like hold-down bolts in a wood light frame building; see "Principles in Practice" in Chapter 13). Therefore, both ends of a masonry wall are typically reinforced. Reinforcement is also provided near both ends of a door or window opening. Reinforcement at ends of an opening also helps to absorb the impact from closing and opening of a door or window.

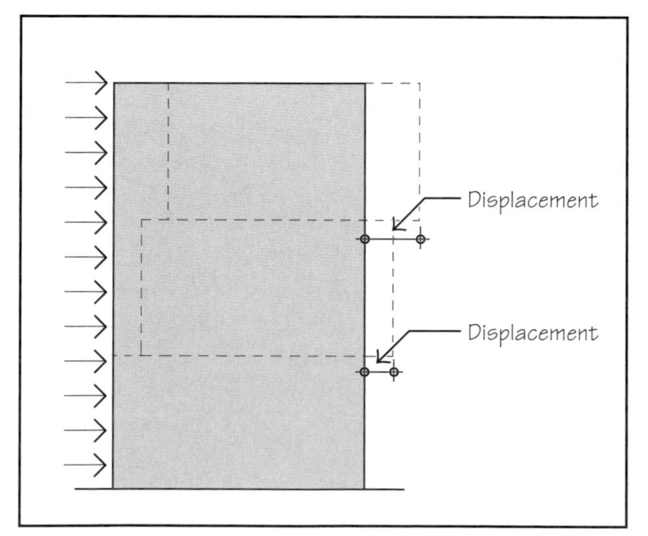

FIGURE 24.7 Vertical reinforcement helps stitch floors together, reducing horizontal displacement between them caused by in-plane lateral loads on a wall.

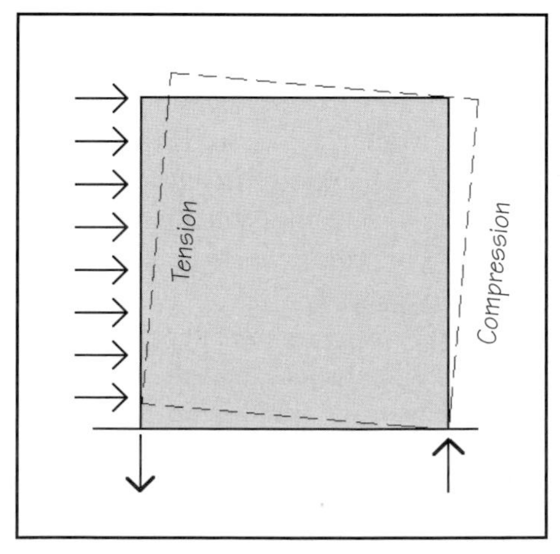

FIGURE 24.8 Vertical reinforcement helps resist tension caused by overturning. Because lateral loads can act on either side of a wall, both ends of a wall require reinforcing.

24.3 BOND BEAMS IN A MASONRY BEARING WALL BUILDING

In most masonry bearing wall structures, the floors consist of precast concrete hollow-core slabs, metal deck, or wood deck. Being individual elements, they do not have the continuity to resist diaphragm tension (see "Principles in Practice" in Chapter 13). Therefore, they are anchored to masonry beams, called *bond beams.* Because a bond beam is reinforced and continuous, it is able to resist the diaphragm tension.

A bond beam is provided at each floor level and roof level in all exterior walls and at all interior bearing walls and shear walls, Figure 24.9. A bond beam is also provided at the top of the parapet. A parapet-level bond beam does not have a structural function but instead facilitates the anchorage of coping on the wall.

Floor- and roof-level bond beams are similar to the bond beams used for shrinkage control in CMU walls (Figure 23.23). However, the floor- and roof-level bond beams are generally more heavily reinforced than those used for shrinkage control.

Generally, a bond beam is only one unit (8 in.) deep, but, if needed, its depth can be increased to 16 in. by grouting the regular concrete masonry units below (or above) the bond beam units, Figure 24.10. The reinforcement in a bond beam must be continuous to be able to effectively absorb diaphragm tension. It must, therefore, be adequately lap spliced between ends of bars, particularly at wall corners and intersections.

Because a bond beam is a grouted element, it provides a suitable base on which the roof or floor trusses and joists can be supported and anchored. Some architects and engineers prefer to use site-cast reinforced-concrete bond beams, Figure 24.11. However, because a site-cast concrete bond beam requires formwork, its construction is a little more cumbersome than that of a masonry bond beam.

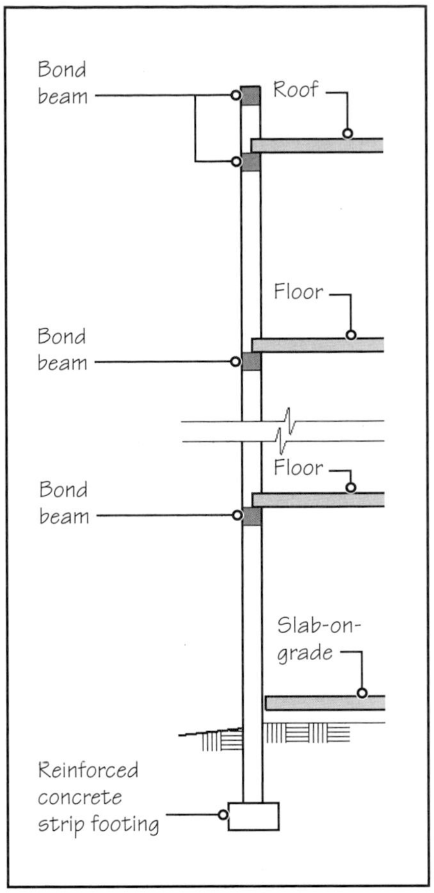

Bond beam
Roof
Floor
Bond beam
Floor
Bond beam
Slab-on-grade
Reinforced concrete strip footing

FIGURE 24.9 Locations of bond beams in a masonry bearing wall structure.

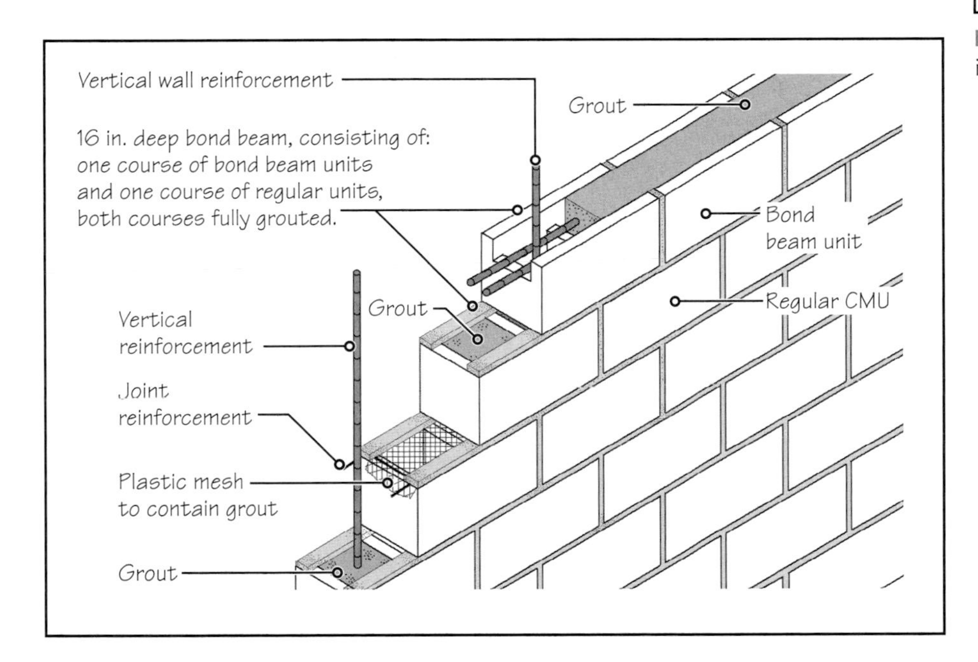

FIGURE 24.10 Detail of a 16-in.-deep bond beam in a CMU wall (Figure 23.23).

FIGURE 24.11 A site-cast reinforced-concrete bond beam at the second-floor level of a load-bearing masonry building.

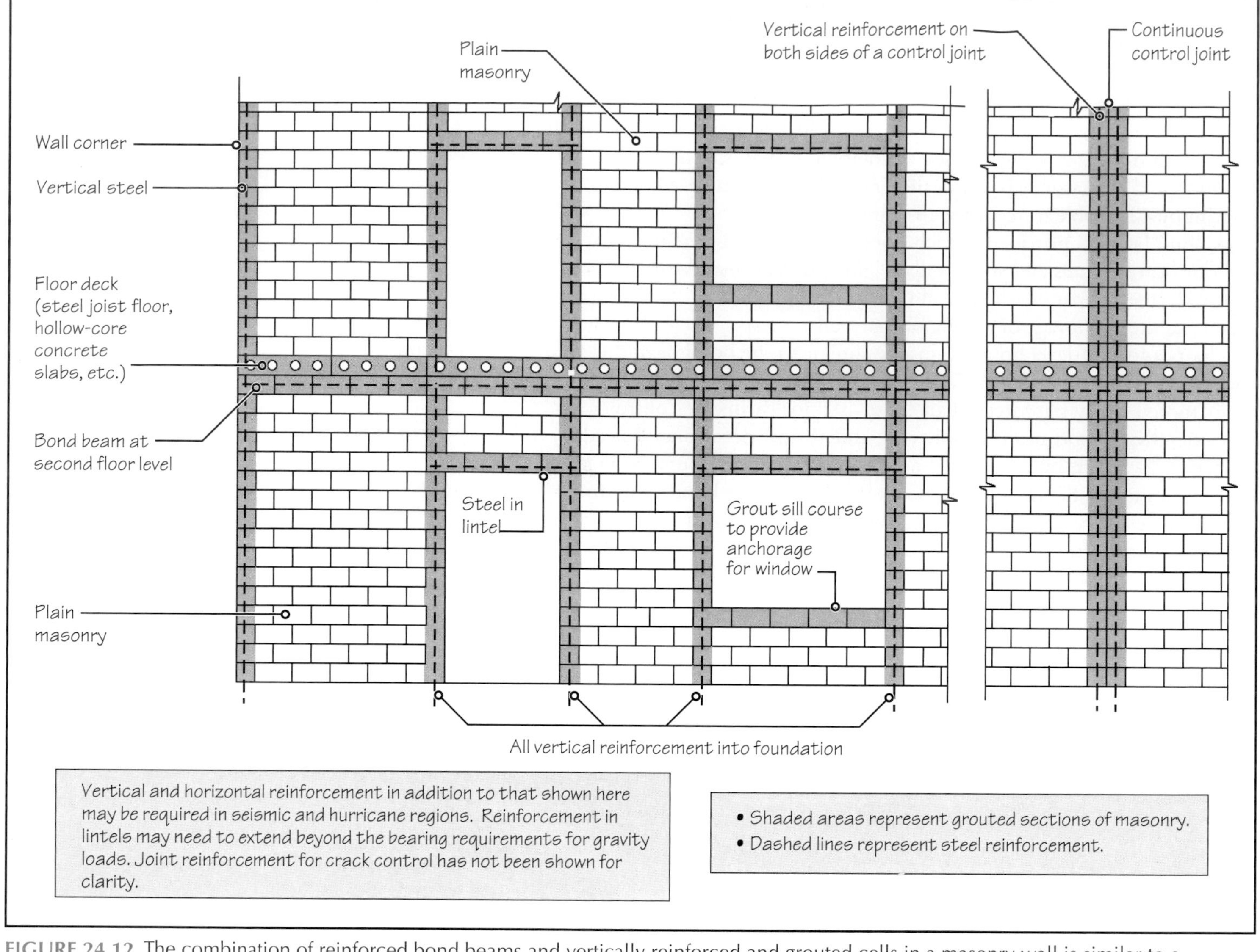

FIGURE 24.12 The combination of reinforced bond beams and vertically reinforced and grouted cells in a masonry wall is similar to a reinforced-concrete frame with intervening plain masonry infill.

Vertical Reinforcement and Bond Beam Combination—A Reinforced Masonry Frame

The combination of bond beams and vertically reinforced (and grouted) CMU cells forms a reinforced-masonry frame, Figure 24.12. The frame ties the intervening plain-masonry into an integral mass, increasing the wall's resistance to lateral loads. The system is identical to a reinforced-concrete frame with plain-masonry infill that is a common form of construction in many countries.

24.4 WALL LAYOUT IN A BEARING WALL BUILDING

For a bearing wall building to be structurally efficient, a sufficient number and adequate lengths of walls must be provided in both principal directions of a building, yielding a *cellular-type* plan. The simplest cellular-type plan is a single-story, rectangular (box-type) structure, commonly used for long-span bearing wall buildings (gymnasiums, lecture halls, and other large assembly or storage spaces), Figure 24.13.

For a multifloor bearing wall structure, a cellular-type structure is best achieved when the walls at an upper floor are at the same location as the ones on the lower floor, giving repetitive floor plans, Figure 24.14. Repetitive floor plans are not only structurally more efficient but also more easily constructible. They allow greater prefabrication and familiarity with the construction process, giving economy.

A multifloor bearing wall structure is, therefore, ideal for occupancies in which economy is a major consideration and cellular-type, repetitive floor layout is intrinsic to the occupancy. Such occupancies are generally residential—apartments, hotels, motels, student dormitories, correction facilities, hospital wards, and so on.

FIGURE 24.13(a) A gymnasium with 12-in.-thick CMU bearing walls (in a box-type floor plan) and steel joist roof. The joists span nearly 100 ft between walls and are approximately 60 in. deep.

FIGURE 24.13(b) Support for joists of the building of Figure 24.13(a). Voids, which contain steel bearing plates, are left in the wall at support locations during the construction of the wall. A continuous masonry bond beam is provided under the voids.

FIGURE 24.13(c) Acoustical roof deck in the building of Figure 24.13(a). Observe the daylight penetrating through the deck due to the perforations in the webs of the deck. The perforations expose the fiberglass sound-absorbing infill to be placed between the flutes (see Figure 16.5 for deck terminology).

Repetitiveness of floor layout requires that all load-bearing or shear walls are continuous to the foundation. This is particularly important in high-rise bearing wall structures, where there is a temptation to provide larger uninterrupted spaces at the first-floor or the basement level.

The discontinuity of walls at foundations, referred to as a *soft story,* is structurally feasible but requires a heavy transfer structure, Figure 24.15. The transfer structure carries the superimposed gravity loads and provides the required lateral load resistance. A soft story is particularly problematic in seismic zones.

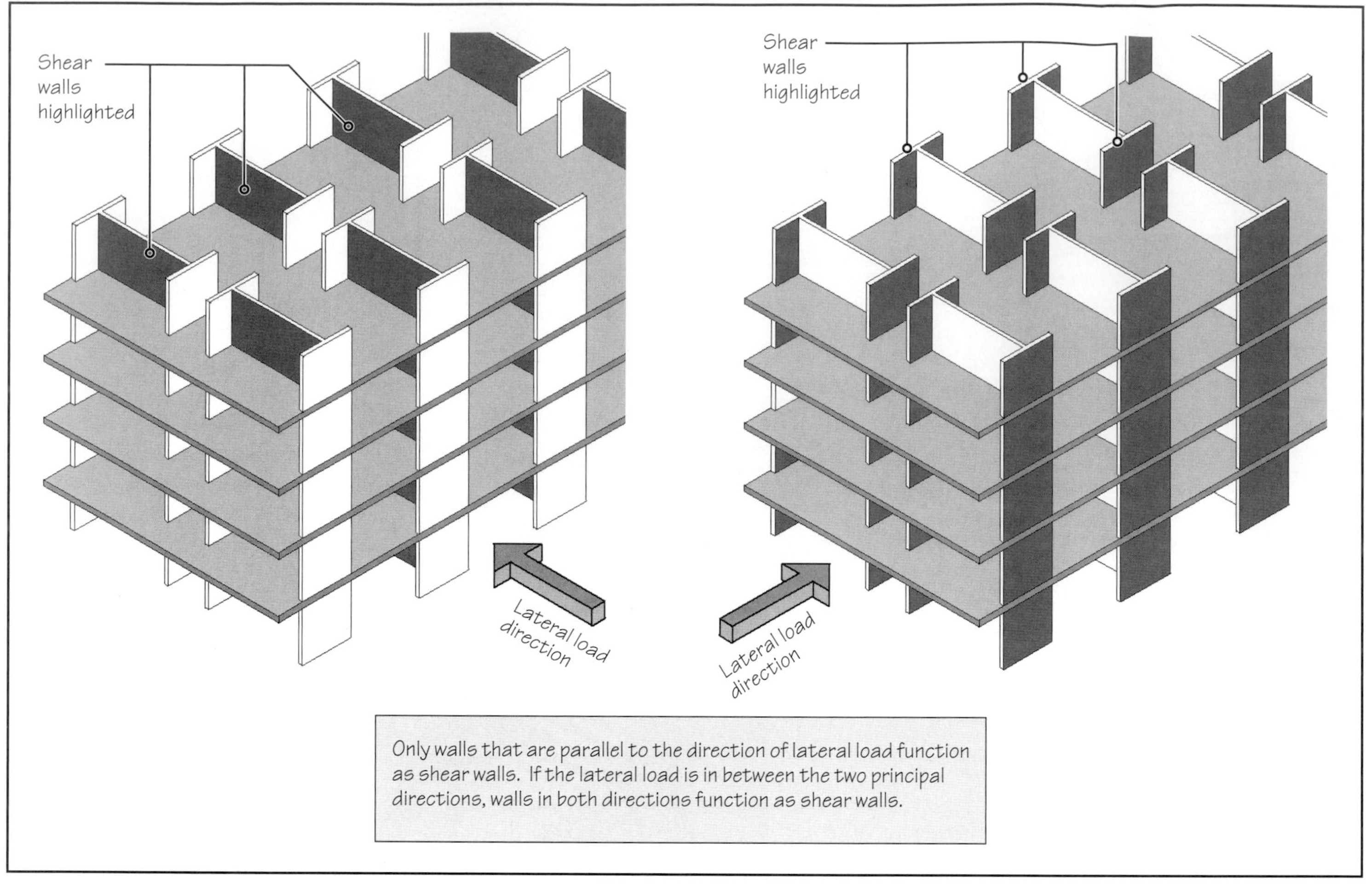

FIGURE 24.14 A bearing wall structure should have lateral load resisting (shear) walls in both principal directions. Each figure shows the walls that function as shear walls when subjected to lateral load in one principal direction.

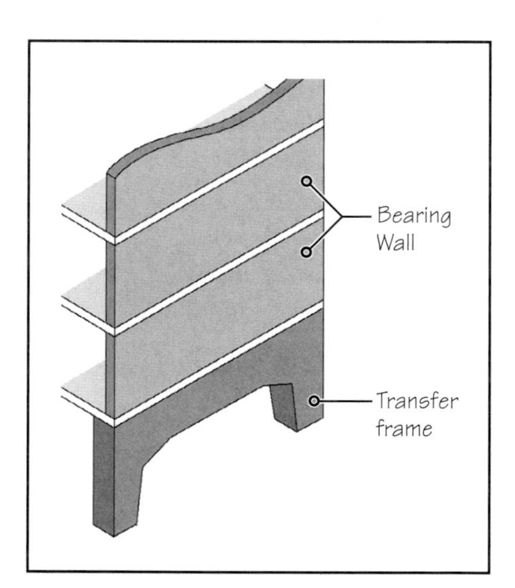

FIGURE 24.15 Transfer frame at the first floor or basement level of a multistory bearing wall structure, resulting in a soft story.

Another desirable attribute of a bearing wall structure is symmetry in wall layout. An asymmetrical wall layout leads to rotation of building (torsion) under lateral loads. The greater the asymmetry, the greater the torsion created, which increases the cost of the structure.

Role of Floor and Roof Decks on Wall Layout

Floor and roof decks can either be one-way decks or two-way decks (Chapter 20). In a one-way deck, the deck spans between two opposite walls; that is, the gravity load on the deck is carried in one direction across the two opposite walls. In a two-way deck, the load is transferred to all four walls. Two-way decks are generally of site-cast concrete.

One-way decks, which are more commonly used, may be of steel, precast concrete, or wood. The use of one-way decks yields the following two types of bearing wall plans:

- Cross bearing wall plan
- Longitudinal bearing wall plan

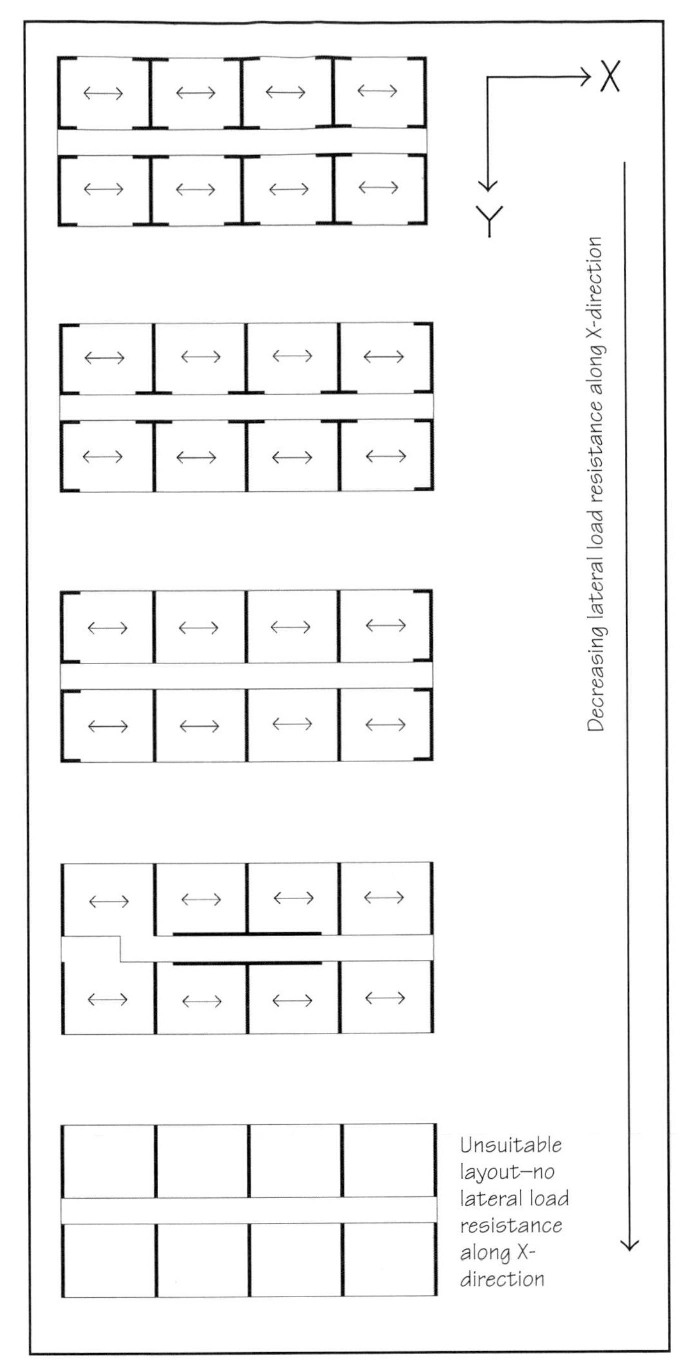

FIGURE 24.16 Relative suitability of various cross bearing wall plans (adapted from [1]).

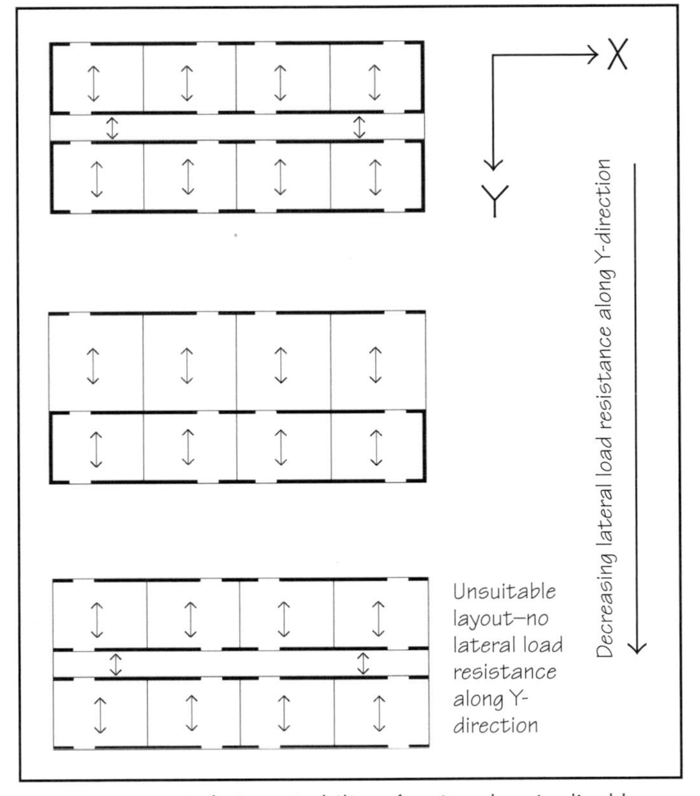

FIGURE 24.17 Relative suitability of various longitudinal bearing wall layouts (adapted from [1]). Note that a longitudinal bearing wall plan provides larger, unobstructed interior spaces as compared with a cross bearing wall plan.

In a cross bearing wall plan, the bearing walls are transverse to the main axis of the building. Figure 24.16 compares the suitability of a few cross bearing wall plans. A similar analysis for longitudinal bearing wall plans is presented in Figure 24.17. Larger openings are generally possible with a cross bearing wall plan. It is, therefore, more commonly used. A longitudinal bearing wall plan provides larger, unobstructed interior spaces. It is used where interior walls are architecturally undesirable.

PRACTICE QUIZ

Each question has only one correct answer. Select the choice that best answers the question.

1. Historical load-bearing masonry structures were designed assuming that
   - **a.** all exterior walls resisted the lateral loads through interaction with each other.
   - **b.** all exterior walls resisted the lateral loads individually.
   - **c.** both interior and exterior walls resisted the lateral loads.
   - **d.** interior and exterior walls and floor and roof diaphragms resisted the lateral loads collectively.
   - **e.** none of the above.

2. Use of the load-bearing masonry wall system became extinct for a while and was revived
   - **a.** around the early nineteenth century.
   - **b.** around the mid-nineteenth century.
   - **c.** around the late nineteenth century.
   - **d.** after World War I.
   - **e.** afer World War II.

3. The structural behavior of a contemporary load-bearing masonry wall building is similar to that of
- **a.** a conventional wood light frame building.
- **b.** site-cast reinforced-concrete frame building.
- **c.** precast-concrete frame building.
- **d.** steel frame building.
- **e.** none of the above.

4. In referring to a masonry wall as a reinforced-masonry wall, it is implied that the wall contains
- **a.** horizontal reinforcing bars.
- **b.** joint reinforcement.
- **c.** vertical reinforcing bars.
- **d.** bond beams.
- **e.** none of the above.

5. Masonry walls without joint reinforcement are called plain masonry walls.
- **a.** True
- **b.** False

6. A bond beam in a masonry bearing wall building is required for structural reasons
- **a.** above all openings in exterior walls.
- **b.** at each floor level.
- **c.** at each floor level and roof level.
- **d.** at each floor level, roof level, and top of parapet.
- **e.** none of the above.

7. A bond beam in a masonry bearing wall building must be a
- **a.** steel, reinforced-concrete, reinforced-masonry, or wood beam.
- **b.** steel, reinforced-concrete, or reinforced-masonry, beam.
- **c.** steel or reinforced-concrete beam.
- **d.** reinforced-concrete or reinforced-masonry beam.
- **e.** none of the above.

8. A bond beam in a masonry bearing wall building
- **a.** is preferably used along the shorter span.
- **b.** is preferably used along the longer span.
- **c.** may be used along the shorter or longer span.
- **d.** is embedded in the walls.
- **e.** none of the above.

9. A typical bond beam is provided with
- **a.** horizontal reinforcement.
- **b.** stirrups.
- **c.** horizontal reinforcement and stirrups.
- **d.** none of the above.

10. A load-bearing wall structure works best when the floor plan of the building has
- **a.** walls that are distributed almost uniformly in both principal directions.
- **b.** walls at an upper floor that align with the walls at the lower floor.
- **c.** walls that are continuous up to the foundations.
- **d.** all the above.

11. A cellular-type floor plan in a multistory building is generally inherent in the following occupancies.
- **a.** Residential occupancies
- **b.** Business occupancies
- **c.** Educational occupancies
- **d.** Mercantile occupancies
- **e.** none of the above.

12. In a cross-bearing wall structure, the load-bearing walls are
- **a.** perpendicular to the main axis of the building.
- **b.** parallel to the main axis of the building.
- **c.** (a) or (b).
- **d.** (a) and (b).

13. Compared to a longitudinal bearing wall floor plan, a cross–bearing wall floor plan generally gives larger exterior wall openings.
- **a.** True
- **b.** False

Answers: 1-b, 2-e, 3-a, 4-c, 5-b, 6-c, 7-d, 8-d, 9-a, 10-d, 11-a, 12-a, 13-a.

24.5 FLOOR AND ROOF DECKS—CONNECTIONS TO WALLS

The structural integrity of a bearing wall building depends on the connections between the walls and the floor or roof decks. As shown in Figure 24.18, a connection between a floor or roof and a wall must be adequate to sustain the following three forces:

- Gravity load from the floor or roof deck to the wall
- Out-of-plane shear between the wall and the floor or roof deck
- In-plane shear between the wall and the floor or roof deck

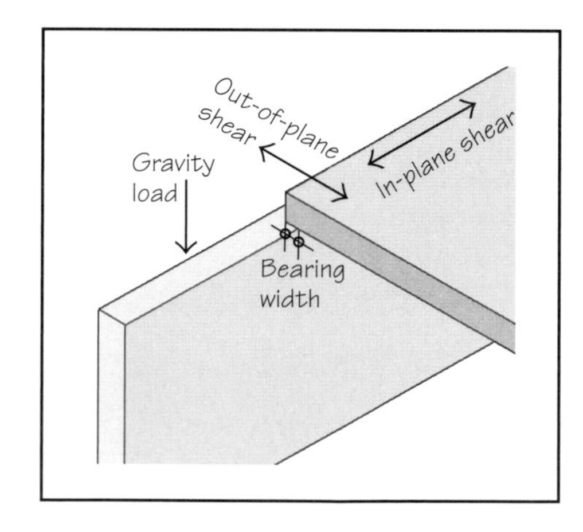

FIGURE 24.18 Connection forces between a floor (or roof) deck and a bearing wall structure.

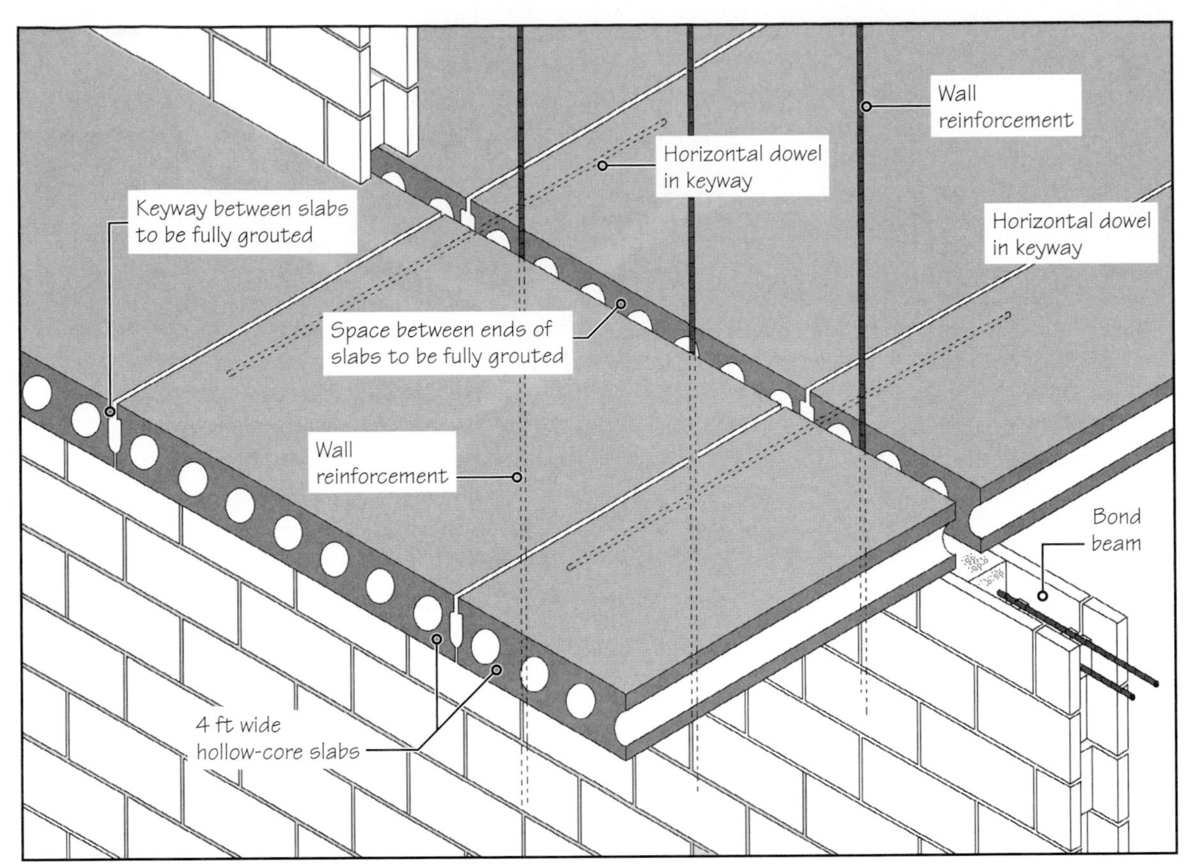

FIGURE 24.19 Schematic connection detail between an interior concrete masonry bearing wall and a 4-ft-wide hollow-core precast concrete slabs. Detail is similar if the hollow-core slabs are 8 ft wide (Figure 24.27(d)).

Figures 24.19 to 24.23 show typical connection details between masonry walls and various commonly used floor and roof decks—precast concrete, steel deck, and wood. Precast-concrete hollow-core slabs are most commonly used in mid-to-high-rise masonry bearing wall residential buildings. As stated in Chapter 20, precast hollow-core slabs are produced in various widths (widths of 4 ft and 8 ft are more common). Steel and wood decks are common in low-rise buildings.

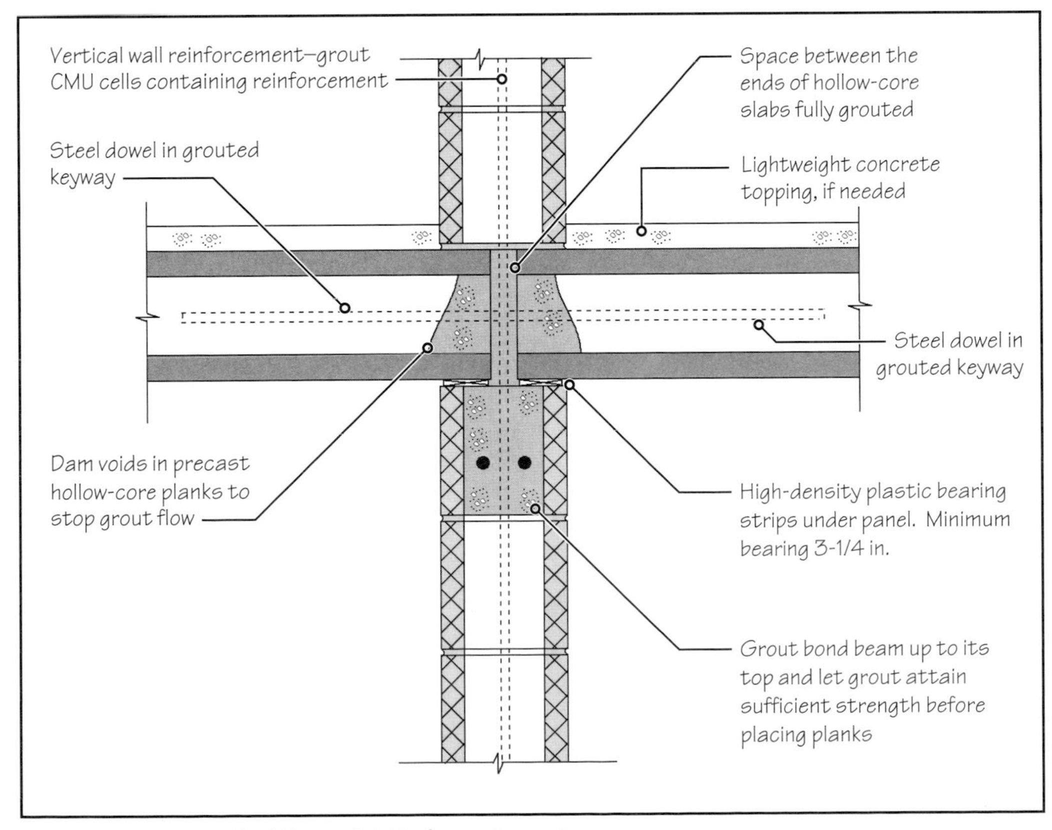

FIGURE 24.20 Detail of Figure 24.19 shown in section.

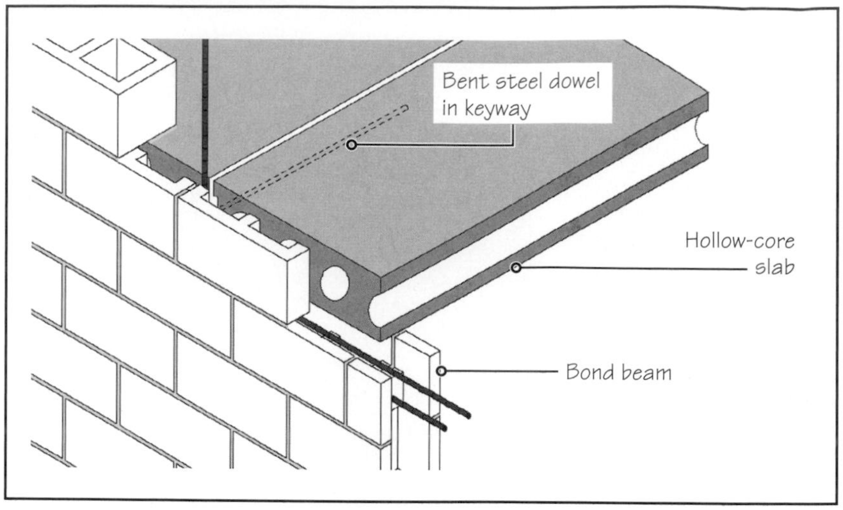

FIGURE 24.21 Schematic connection detail between an exterior concrete masonry bearing wall and hollow-core precast concrete slabs.

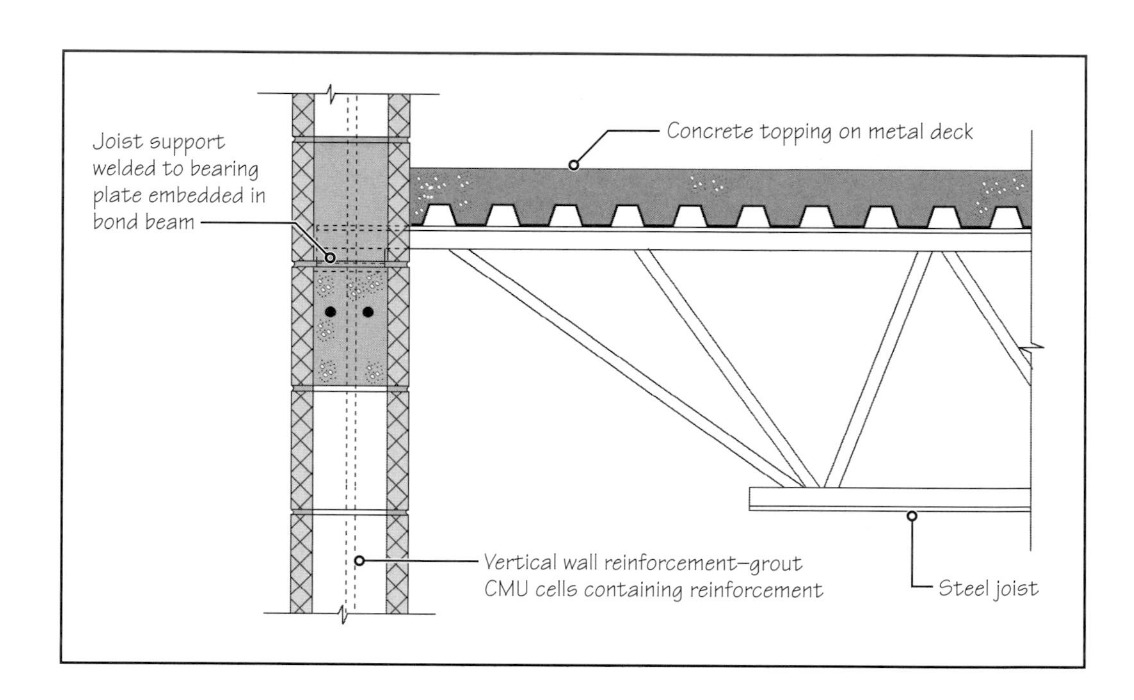

FIGURE 24.22 A schematic connection detail between a load-bearing masonry wall and metal deck floor.

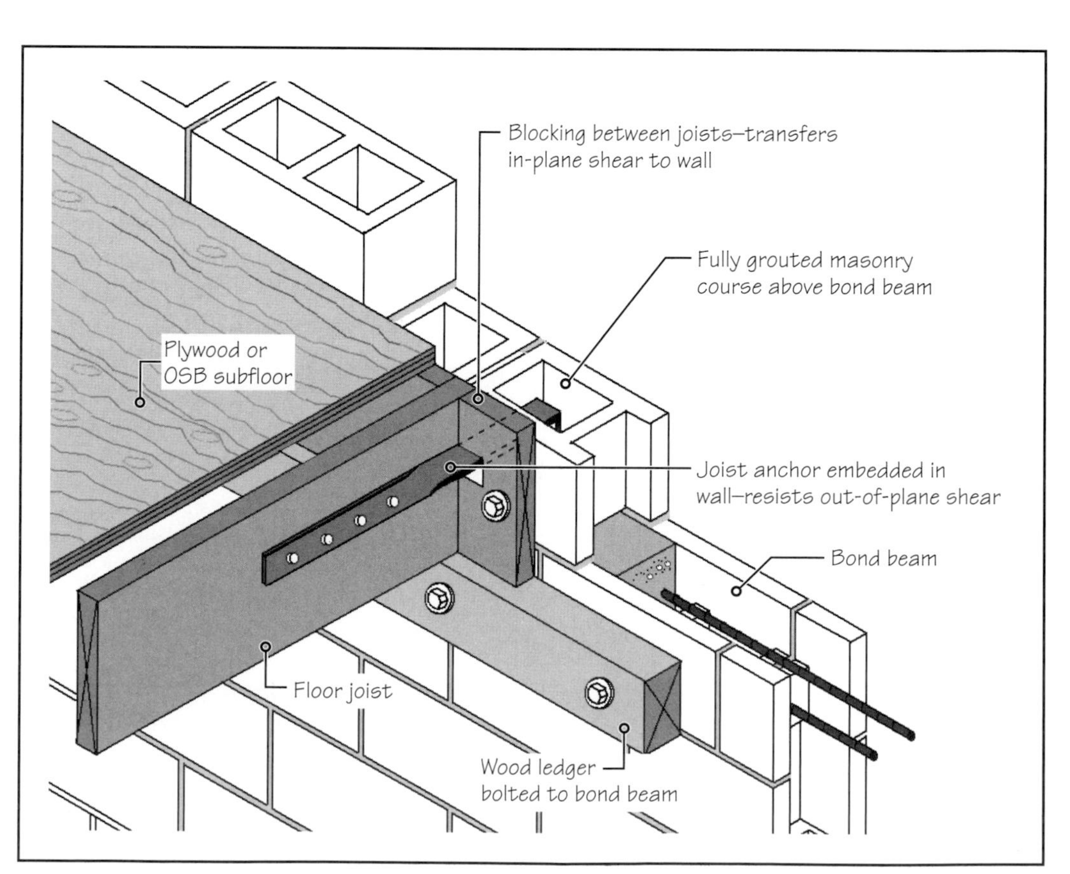

FIGURE 24.23 A schematic connection detail between a load-bearing masonry wall and wood light frame floor.

24.6 LIMITATIONS OF MASONRY BEARING WALL CONSTRUCTION

As stated in Section 22.1, the cutting of masonry units increases the cost of construction and also slows it down. That is why the cutting of units is generally avoided in masonry structures. Consequently, the length of masonry walls is based on modular dimensions. In CMU walls, the module is 8 in. Therefore, the length of CMU walls and opening widths are multiples of 8 in.

With a mortar joint thickness of $\frac{3}{8}$ in., the clear size of an opening is $\frac{3}{8}$ in. greater than that given by the multiple of 8 in. In other words, the rough opening width is 3 ft $4\frac{3}{8}$ in. ($40\frac{3}{8}$ in.), not 3 ft 4 in. (40 in.), Figure 24.24(a).

Similarly, the width of the masonry wall between openings is $\frac{3}{8}$ in. less than that given by the multiples of 8 in. Thus, the length of masonry between openings is 1 ft $3\frac{5}{8}$ in. ($15\frac{5}{8}$ in.), not 1 ft 4 in. (16 in.).

The 8-in. module also applies to height dimensions. Thus, the floor-to-ceiling heights in a CMU masonry wall building are generally 8 ft, 8 ft 8 in., 9 ft 4 in., and so on. Similarly, a rough door-opening height of 7 ft $4\frac{3}{8}$ in., shown in Figure 24.24(b), is commonly used. This dimension is (slightly) excessive for a standard 7-ft-0-in.-high door.

Where standard-height doors are desired, a starter course of saw-cut masonry units may be used. (However, note that in most commercial construction, doors and windows are custom fabricated to opening dimensions. Therefore, standard door and window dimensions generally do not apply to commercial construction.)

NOTE

The dimension of a masonry opening is $\frac{3}{8}$ in. greater than the multiples of masonry unit module. The dimension of the masonry between openings is $\frac{3}{8}$ in. less than the multiples of the masonry unit module.

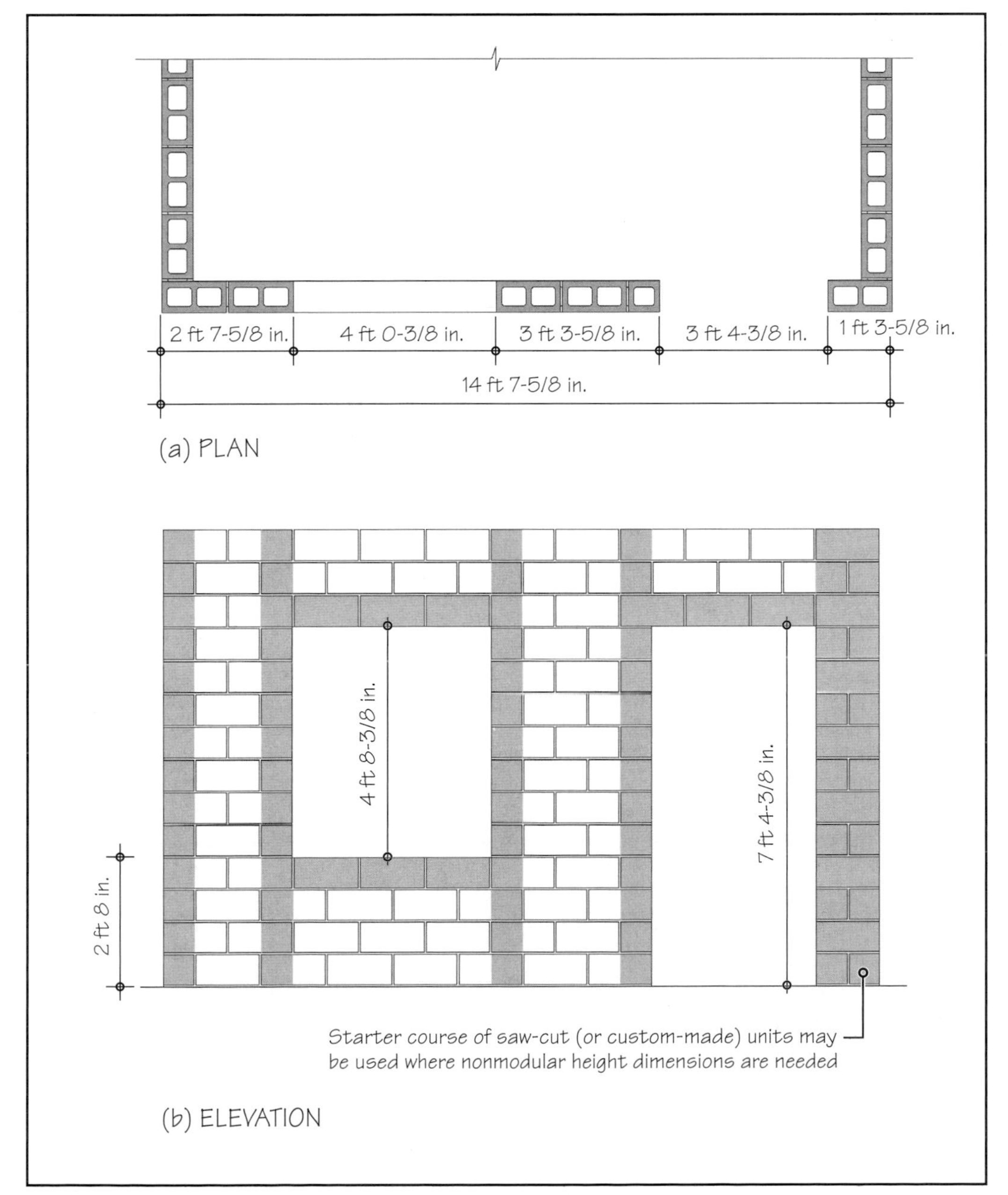

FIGURE 24.24 Effect of modularity of units on the dimensions of walls and openings.

Water-Resistive Exterior Finish Required

Because contemporary masonry bearing walls are relatively thin, they require some form of weather-resistive exterior coating, (clear coating or paint), plaster (stucco or EIFS), or some form of cladding to prevent water from entering the interior (Chapters 26 and 27). Exterior walls in traditional masonry structures did not require such cladding because their thickness made them inherently water resistive.

Coordination Between Trades During Construction

Electrical conduit and junction boxes and other utility lines need to be built into walls as the construction progresses. Door frames are generally erected in position and the walls are built around them. Therefore, greater coordination between various trades is needed during the construction of a masonry building than is required for a steel or concrete frame building. Where precast-concrete floor slabs are used, detailed shop drawings are required to indicate the location of openings for vertical continuity of service lines.

Figures 24.25(a) to (f) illustrate the highlights of the construction process of a typical high-rise masonry bearing wall building, and Figure 24.25(g) shows the same building in finished stage.

FIGURE 24.25(a) This and the following few images show the construction of an eight-story load-bearing masonry hotel building. The building, as shown, is in its early stages of construction.

FIGURE 24.25(b) Masonry wall construction at a typical floor (photograph taken at the fifth floor). Observe CMUs stacked on scaffolding in preparation for laying the walls.

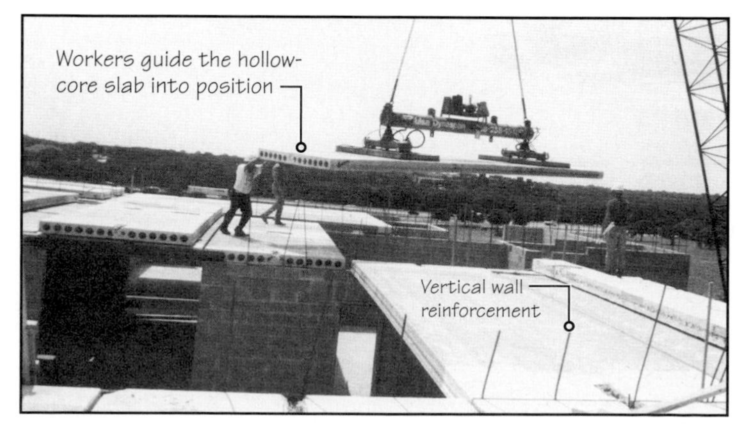

FIGURE 24.25(c) An 8-ft-wide precast-concrete hollow-core slab being flown into position. Observe vertical reinforcement projecting from lower level walls.

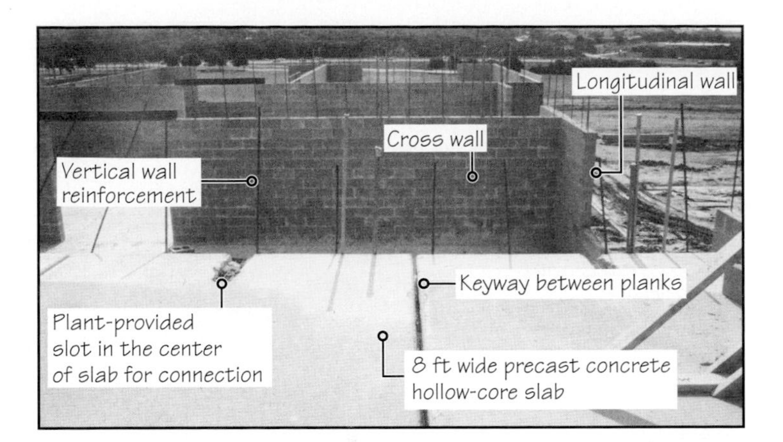

FIGURE 24.25(d) Keyway between 8-ft-wide planks. Additional connection slots have been provided in the center of each slab. Observe that the floor of plan of the building is a cross–bearing wall plan (Figure 24.16).

FIGURE 24.25(e) Connnection between longitudinal and cross walls is provided through 24-in. × $1\frac{1}{2}$-in. × $\frac{1}{4}$-in. steel plates with ends of plates bent up 2 in. Bent ends of plates are embedded in grouted CMU cells.

FIGURE 24.25(f) Electrical conduits and other utility lines must be built into walls during construction.

FIGURE 24.25(g) The eight-story hotel residential block with masonry bearing wall structure (shown under construction in (a) to (f)). The office and conference areas in this building have steel frames because of their large, unobstructed spaces.

24.7 BEARING WALL AND COLUMN-BEAM SYSTEM

Although masonry bearing wall construction is most suited for mid- to high-rise residential occupancies, as previously described, it is also well suited for several low-rise building types. Chief among them are school and college buildings, offices, shopping centers, and religious buildings.

However, such buildings generally consist of a hybrid construction system, in which all exterior walls are masonry bearing walls and the interior structure is a combination of masonry bearing walls, drywall partitions, and a column-beam frame. The lateral load resistance is generally designed into the masonry walls.

The column-beam frame may consist of cast-in-place concrete or steel. Concrete is better suited because both concrete and concrete masonry have similar properties, giving little differential-movement problems. However, a steel frame is generally, more economical and is, therefore, widely used.

Differential-movement problems between steel columns and masonry walls must be considered; they are generally more easily managed in low-rise buildings. Figure 24.26 shows a typical building that combines load-bearing masonry with steel frame.

FIGURE 24.26(a) A school building with a combination of load-bearing masonry walls and steel frame, under construction.

FIGURE 24.26(b) School building consisting of load-bearing masonry walls and structural steel frame. Architects: Huckabee, Inc. (Photo courtesy of Huckabee, Inc., and Chaplo Architectural Photographers, Inc.)

24.8 REINFORCED-CONCRETE BEARING WALL CONSTRUCTION

Site-cast reinforced-concrete bearing walls can be used in place of masonry bearing walls. When used in conjunction with site-cast reinforced-concrete floor slabs, they provide a robust structure because of the inherent continuity between the vertical and horizontal elements of the building.

In other words, the joints between the walls and the floor slabs of a site-cast reinforced-concrete bearing wall structure are tougher than those in masonry wall and precast-concrete hollow-core slab structure. Site-cast-concrete bearing wall structures are, therefore, better able to resist lateral loads. Additionally, site-cast reinforced-concrete walls can seamlessly integrate with a site-cast reinforced-concrete column-beam frame in case a hybrid (bearing wall and frame) system is used.

In general, a site-cast-concrete bearing wall structure can be designed with thinner walls than a similar masonry bearing wall structure. Figure 24.27 shows the construction of a 22-story site-cast reinforced-concrete bearing wall residential building where the walls are only 6-in.-thick.

FIGURE 24.27 A 22-story site-cast reinforced-concrete bearing wall apartment building under construction. The walls in this building are only 6-in.-thick.

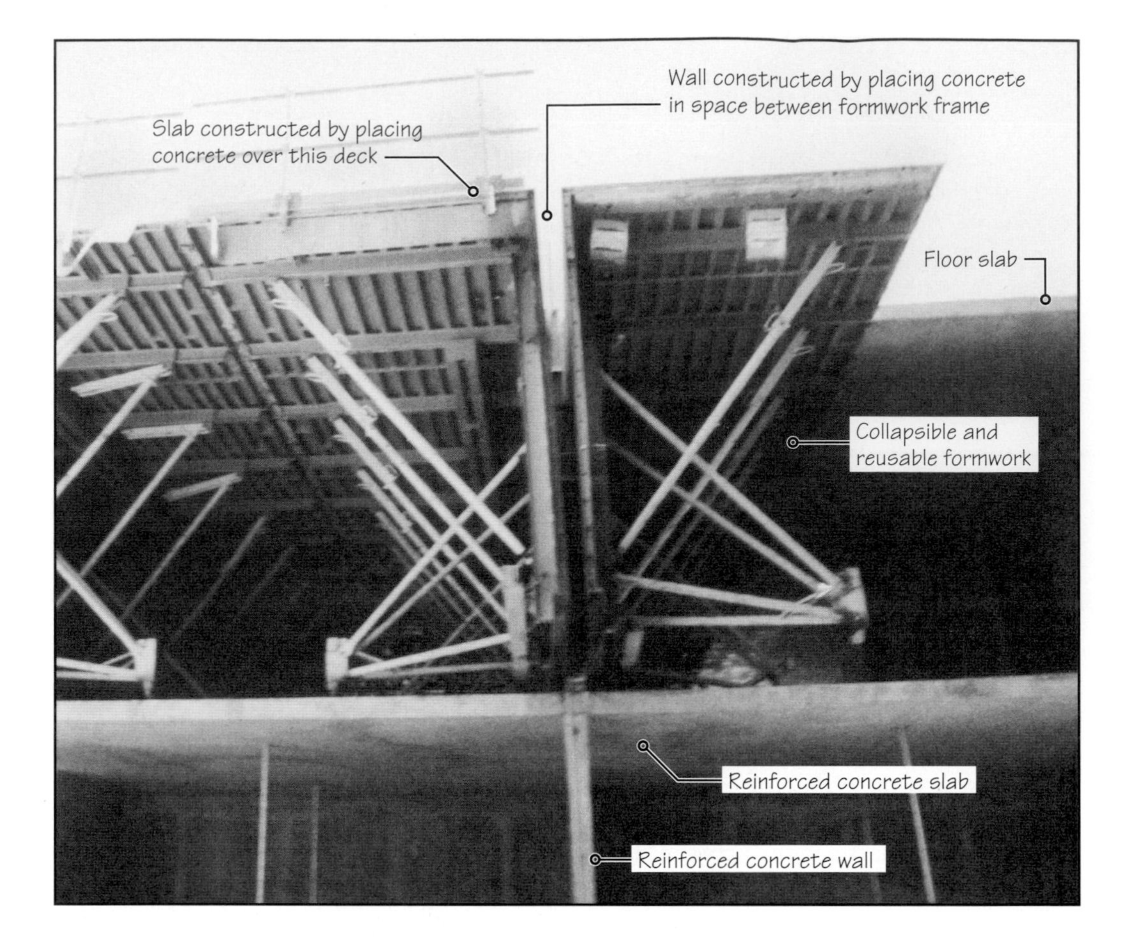

FIGURE 24.28 Prefabricated, collapsible tunnel forms, commonly used for constructing high-rise residential, load-bearing reinforced-concrete wall buildings.

As stated in Chapter 19, a site-cast-concrete structure is economical only if the cost of formwork is controlled. A type of formwork that has gained favor in constructing high-rise reinforced-concrete bearing wall structures is the *tunnel form,* Figure 24.28. A tunnel form consists of prefabricated and collapsible inverted-L forms that allow the walls and slabs for a floor to be cast simultaneously, usually on a 7-day cycle.

Virtually any thickness of walls and slabs can be constructed by increasing or decreasing the distance between tunnel forms. However, the system is feasible only with a modular and repetitive floor plan, such as that obtained in residential structures.

A reinforced-concrete bearing wall building has the same advantages as a masonry bearing wall building, such as good sound insulation, high fire resistance, and mold-free structure. The disadvantages are also the same, consisting mainly of construction that is inflexible in spatial organization. Additionally, like their masonry counterpart, exterior site-cast-concrete walls generally require a water-resistive cladding.

PRACTICE QUIZ

Each question has only one correct answer. Select the choice that best answers the question.

14. The commonly used floor system in high-rise and midrise masonry bearing wall buildings in North America is
- **a.** cast-in-place reinforced concrete slabs.
- **b.** precast-concrete hollow-core slabs.
- **c.** precast-concrete double-tee units.
- **d.** all the above.
- **e.** none of the above.

15. The keyway between precast-concrete hollow-core slabs runs perpendicular to the bearing walls.
- **a.** True
- **b.** False

16. The keyway between precast-concrete hollow-core slabs is generally
- **a.** grouted.
- **b.** reinforced with horizontal steel dowels.
- **c.** reinforced with T-shaped steel dowels.
- **d.** (a) and (c).
- **e.** (a) and (b).

17. Where the cutting of masonry units is discouraged, the wall dimensions are based on the masonry module. The module used in CMU walls is
- **a.** 4 in.
- **b.** 8 in.
- **c.** 12 in.
- **d.** 16 in.
- **e.** none of the above.

18. The clear height of openings in CMU walls is
- **a.** $\frac{3}{8}$ in. greater than the multiples of the masonry unit module.
- **b.** $\frac{3}{8}$ in. less than the multiples of the masonry unit module.
- **c.** $\frac{1}{8}$ in. greater than the multiples of the masonry unit module.
- **d.** $\frac{1}{8}$ in. less than the multiples of the masonry unit module.
- **e.** none of the above.

19. In high-rise residential buildings with site-cast reinforced-concrete bearing walls, the floors and roofs are generally constructed of
- **a.** steel joists and metal deck.
- **b.** wide-flange steel beams and metal deck.
- **c.** precast-concrete hollow-core slabs.
- **d.** site-cast concrete slabs.
- **e.** none of the above.

20. The formwork used in constructing high-rise residential buildings in site-cast reinforced-concrete bearing walls generally consists of
- **a.** large, prefabricated, horizontal plywood forms that allow walls and slabs to be cast simultaneously.
- **b.** large, prefabricated, vertical plywood forms that allow walls and slabs to be cast simultaneously.
- **c.** large, prefabricated, tunnel forms that allow walls and slabs to be cast simultaneously.
- **d.** none of the above.

Answers: 14-b, 15-a, 16-e, 17-b, 18-a, 19-d, 20-c.

24.9 REINFORCED-CONCRETE TILT-UP WALL CONSTRUCTION

Another widely used bearing wall system is the reinforced-concrete tilt-up wall system. In this system, the walls are precast on site, generally over a concrete slab-on-ground, which also serves as the ground-floor slab of the building, Figure 24.29. A form-release agent is used between the slab-on-ground and the walls.

If the ground-floor slab of the building is too small to accommodate the precasting of all walls (i.e., a building with a small footprint and relatively tall walls), *stack casting* of walls is used. In stack casting, one or more walls are cast on the lower wall using a form-release agent between walls.

After the walls have attained sufficient strength, they are lifted with a crane from their horizontal position and placed into required locations, Figure 24.30. Because of the limited lifting capacity of the crane and the stresses produced in the wall during the lift, tilt-up walls cannot be too high or too long. A concrete tilt-up wall system is, therefore, limited to low-rise buildings, generally up to 50 ft high (one to four stories). The walls in a typical tilt-up wall building consist of several panels, each panel extending to its full height.

FIGURE 24.29 Tilt-up wall panels are cast as reinforced-concrete slabs on a slab-on-ground, which forms the ground-floor slab of the building.

After being placed in position, a tilt-up wall panel is temporarily supported by steel pipe braces. A brace is anchored near the top (generally at two-thirds height) of the panel. The bottom of a brace is anchored either to the interior slab-on-ground or to a concrete pad outside the building, Figure 24.31. The braces hold the panel plumb and resist the lateral loads during the building's construction. They are removed only when the building has attained lateral stability, provided by the roof structure and the connection of panels to the slab-on-ground.

The cost of a tilt-up wall system compares favorably with that of a masonry bearing wall system. It is generally used for the same occupancies as the masonry bearing wall system—warehouses, distribution centers, low-rise offices, religious buildings, and shopping centers.

Most concrete tilt-up wall buildings are hybrid structures in which the exterior walls are tilt-up walls and the interior structure consists of steel columns and beams and drywall partitions, Figure 24.32. The roof and floor structure may consist of open-web steel joists, wide flange steel beams, and so on.

FIGURE 24.30 Lifting of a tilt-up wall panel and setting it in position. Note inclined braces placed on the interior of this single-story building.

FIGURE 24.31 A typical tilt-up wall building consists of several panels, each extending from the bottom to the top. The braces may be placed on the outside of the building (as shown here) or the inside of the building (in a single-story building).

FIGURE 24.32 Most tilt-up wall buildings use tilt-up exterior bearing walls with interior steel beams and columns.

THICKNESS OF TILT-UP WALLS AND CONCRETE STRENGTH

The thickness of tilt-up walls is determined from structural considerations. However, the following rule of thumb gives a rough estimate of the wall thickness for a single-story tilt-up wall:

$$\text{Panel thickness} = \frac{\text{unsupported height of panel}}{48} \qquad \text{(generally not less than 6 in.)}$$

where the unsupported height is the distance between the slab-on-ground and the roof/floor, Figure 24.33. Thus, if the unsupported height of panels is 24 ft, the panel thickness is approximately 6 in. The strength of concrete customarily used for panels is 4,000 psi.

SHAPE AND MAXIMUM SIZE OF PANELS

A tilt-up wall panel may be with or without openings. Openings in a panel represent areas of weakness that can damage the panel during lifting. Generally, a minimum wall width of 18 in. is required around an opening, Figure 24.34(b). Large openings may be created by using two inverted-L panels, Figure 24.34(c), or using a lintel panel that bears on the two adjacent panels, Figure 24.34(d).

Temporary structural steel bracing members can also be used to strengthen an opening. Referred to as *strongbacks,* they are attached to a panel before lifting and removed after the panel's erection, Figure 24.34(e). The maximum size of panels depends on the lifting capacity of the crane. Panels greater than 800 ft^2 are possible but are less practical.

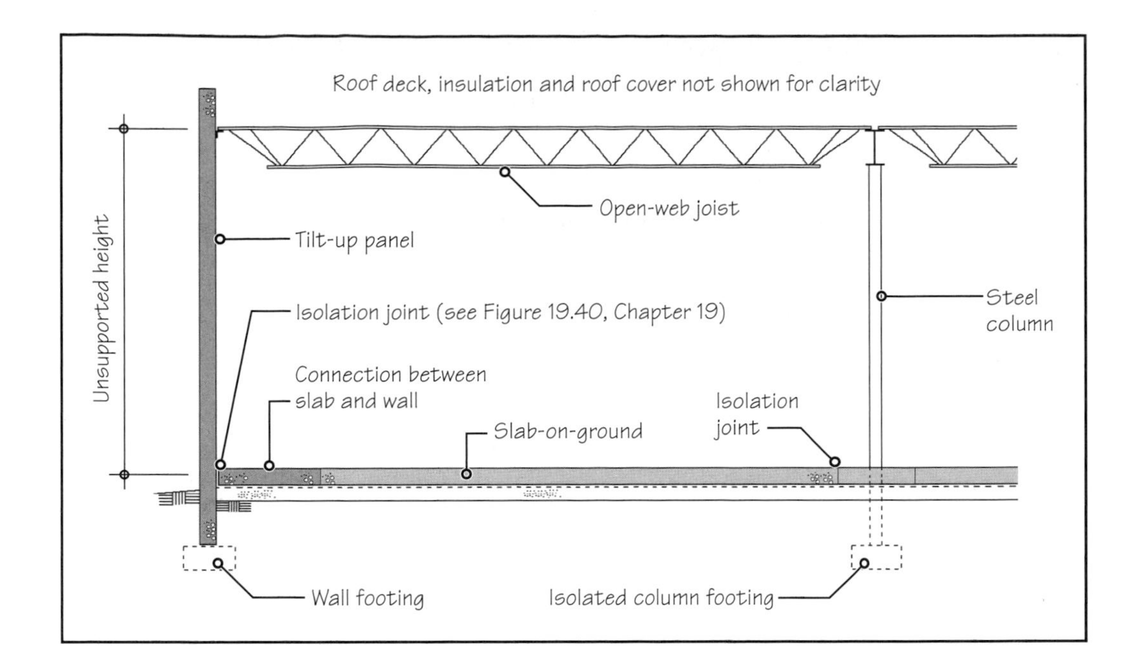

FIGURE 24.33 Partial section through a single-story tilt-up wall building.

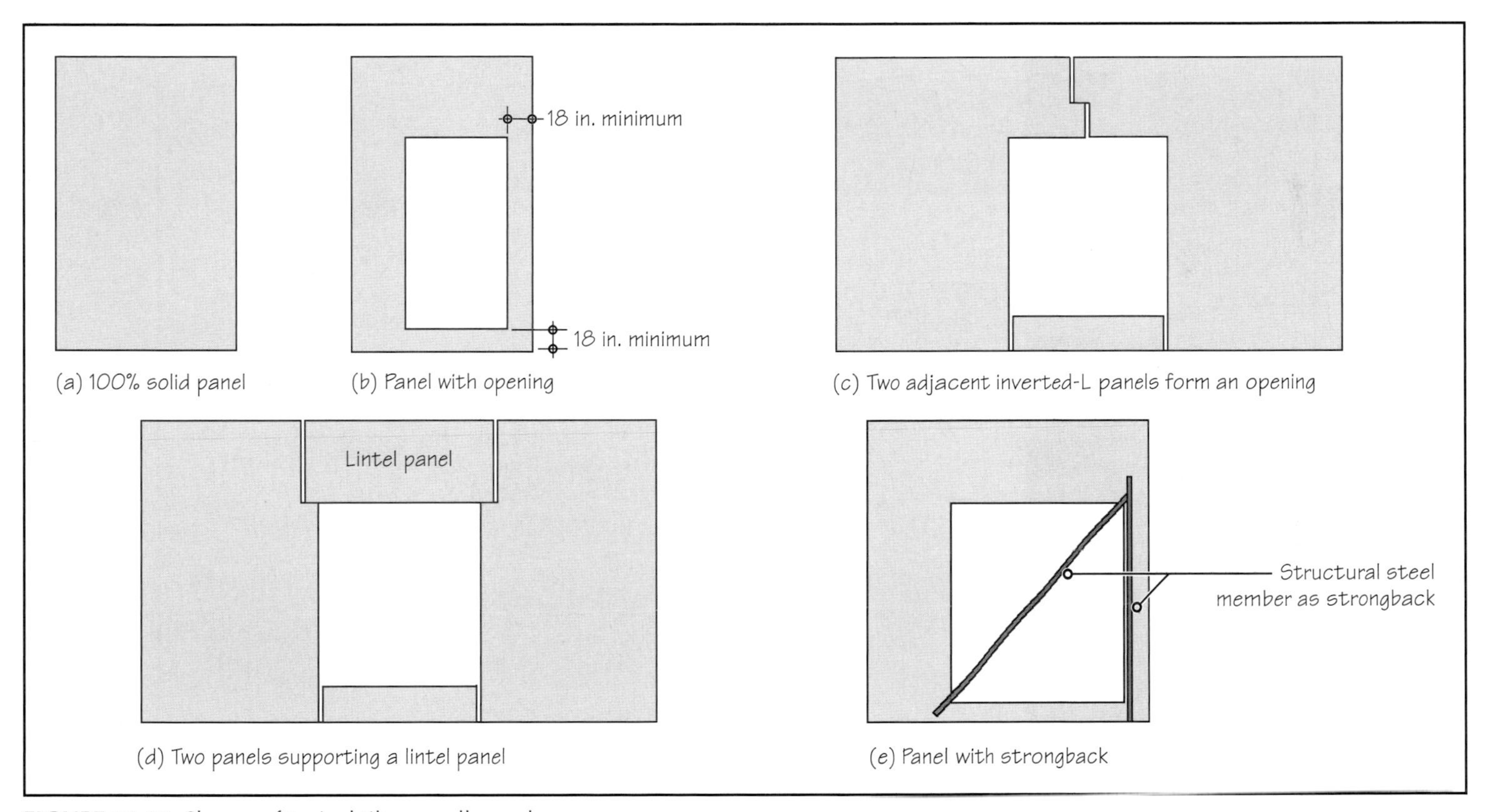

FIGURE 24.34 Shapes of typical tilt-up wall panels.

SLAB-ON-GROUND, CONCRETE STRENGTH, AND REINFORCEMENT

The Tilt-up Concrete Association (TCA) recommends a minimum concrete strength of 3,000 psi for the slab. However, the use of 4,000-psi concrete is customary. The thickness of the slab-on-ground in a tilt-up wall building depends on its occupancy and the load imposed on it during construction. Generally, the load on the slab from construction traffic is greater than that of the occupancy load, Figure 24.35. TCA recommends a minimum slab thickness of 5 in. regardless of the loads.

FOUNDATIONS FOR TILT-UP WALLS

The foundations for tilt-up wall panels generally consist of the following types, Figure 24.36:

- Continuous strip footing
- Isolated pad footing
- Drilled pier footing

FIGURE 24.35 Typical construction traffic on the slab-on-ground of a tilt-up wall building.

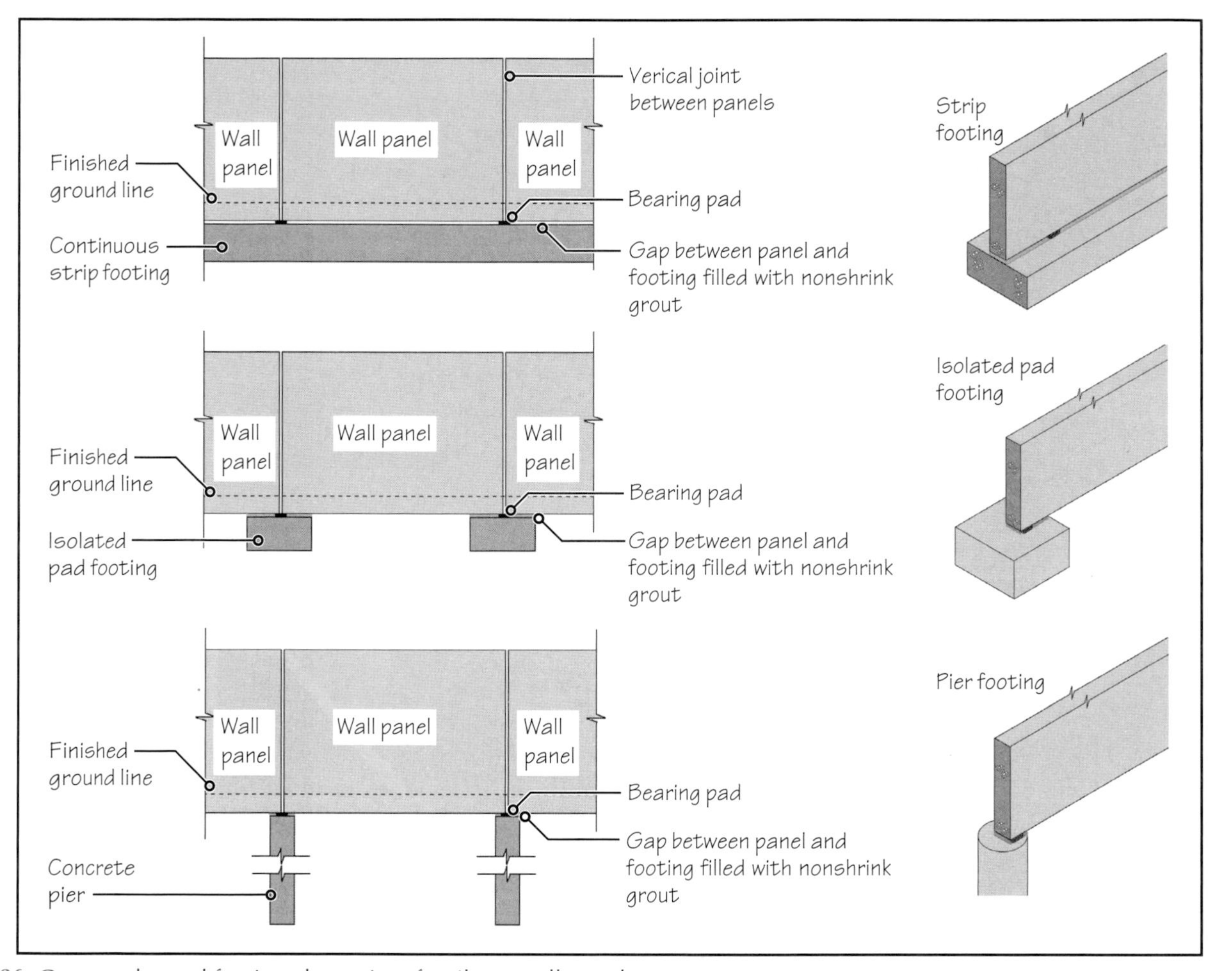

FIGURE 24.36 Commonly used footing alternatives for tilt-up wall panels.

On a strip footing, a panel has a continuous bearing. On an isolated pad or pier footing, the panel bears only at its ends, carrying the load as a deep beam between the footings. Pier footing is generally used in weak or expansive soils (Chapter 21).

Regardless of the type of footing, the bottom of the panel is typically about 2 in. above the top of the footing. This gap is produced by setting the panel on high-strength bearing pads, generally one set of bearing pads at each end of the panel. The bearing pads help to level the panel and ensure that it is set at the required elevation. A panel is released from the crane only after it has been leveled.

Two adjacent panels generally bear on the same bearing pad, Figure 24.37. The gap between panels and the footings is filled with high-strength, nonshrink grout so that in the completed building, the load from the panel is transferred over a greater area than the area of the bearing pads.

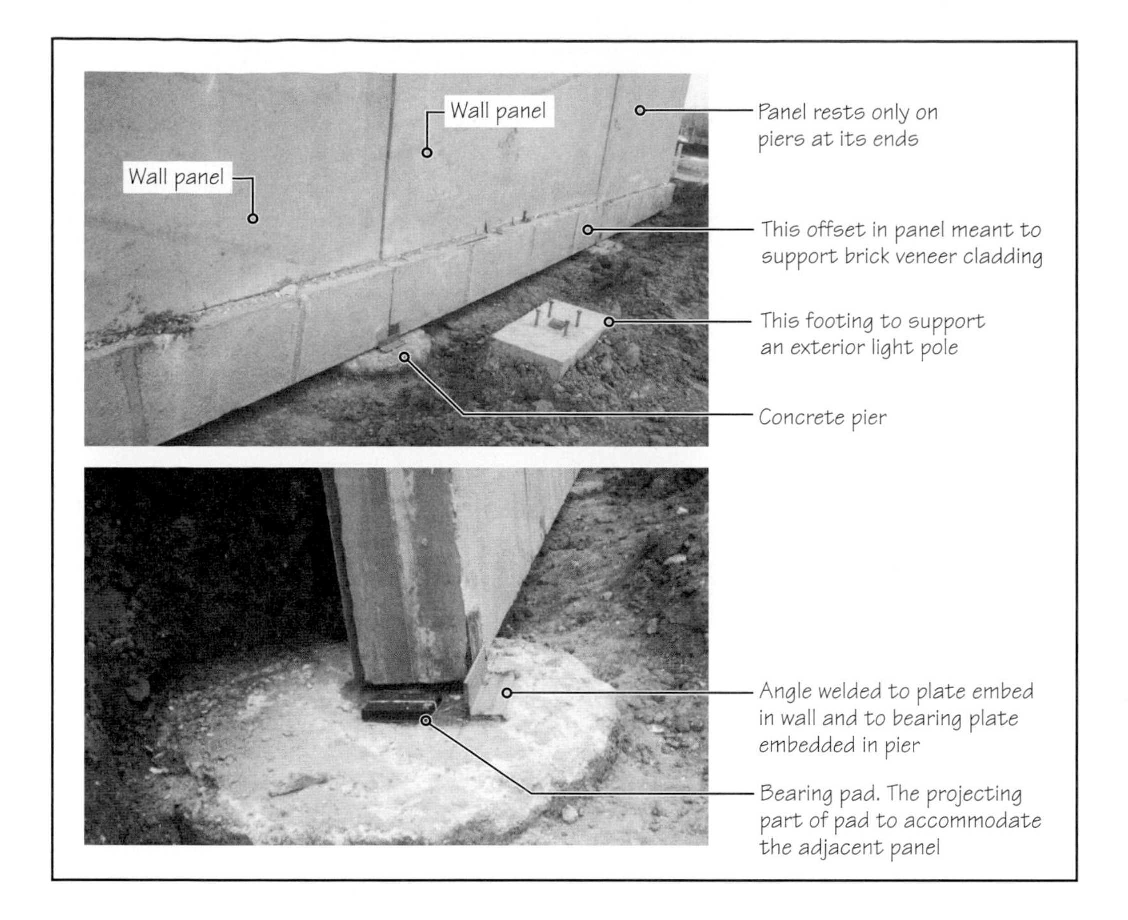

FIGURE 24.37 Detail of pier footing for a tilt-up wall panel.

24.10 CONNECTIONS IN A TILT-UP WALL BUILDING

Connections between various elements of a tilt-up wall building are critical for the overall structural integrity of the building. The connection between the slab-on-ground and the panels, and between the panels and the roof are particularly important.

CONNECTION BETWEEN SLAB-ON-GROUND AND PANELS

A tilt-up wall does not have any continuity with the foundations. This is in contrast with a masonry or cast-in-place concrete wall, in which the vertical reinforcement is continuous between the walls and the footing. Instead, the tilt-up walls are connected to the slab-on-ground by one of the following two methods:

- A continuous *closure strip,* Figure 24.38
- Steel embeds in wall and slab welded together with steel plates, Figure 24.39

A closure strip is a part of the slab-on-ground, and is typically 3 ft to 5 ft wide. The slab-on-ground for a tilt-up wall building is poured in two stages. The first stage, comprising the bulk of the slab, forms the casting bed for the wall panels. The closure strip, a strip of slab that lies between the wall panels and the casting-bed slab, is poured in the second stage—after the panels have been placed in position.

The connection between the wall panels and the slab is provided through two sets of overlapping reinforcing bars, one set projecting out of the panels and the other set projecting from the slab. After the panels have been erected, the bars projecting out of the panels, are straightened to lap the bars in the (yet unpoured) closure strip. Subsequently, concrete is poured in the closure strip, as shown in Figure 24.38.

NOTE

Connection Between Slab-On-Ground and Wall Panels

In some regions, the wall panels are connected to the continuous footing in lieu of the slab-on-ground. This type of connection is preferred in cold climates to ensure that the connection is below the frost level.

CONNECTION BETWEEN PANELS AND ROOF OR FLOOR

The most important connection in a tilt-up wall building is that between the panels and the roof and floor structure. The details of roof and floor connections vary with the roof and floor structure, but the basic design philosophy is essentially the same in all of them. Important aspects of a few such details are shown in Figures 24.40 and 24.41.

PANEL-TO-PANEL CONNECTIONS AND JOINT SEALS

The issue of connections between adjacent wall panels is controversial. Some experts suggest that two or three welded connectors between all adjacent panels should be provided.

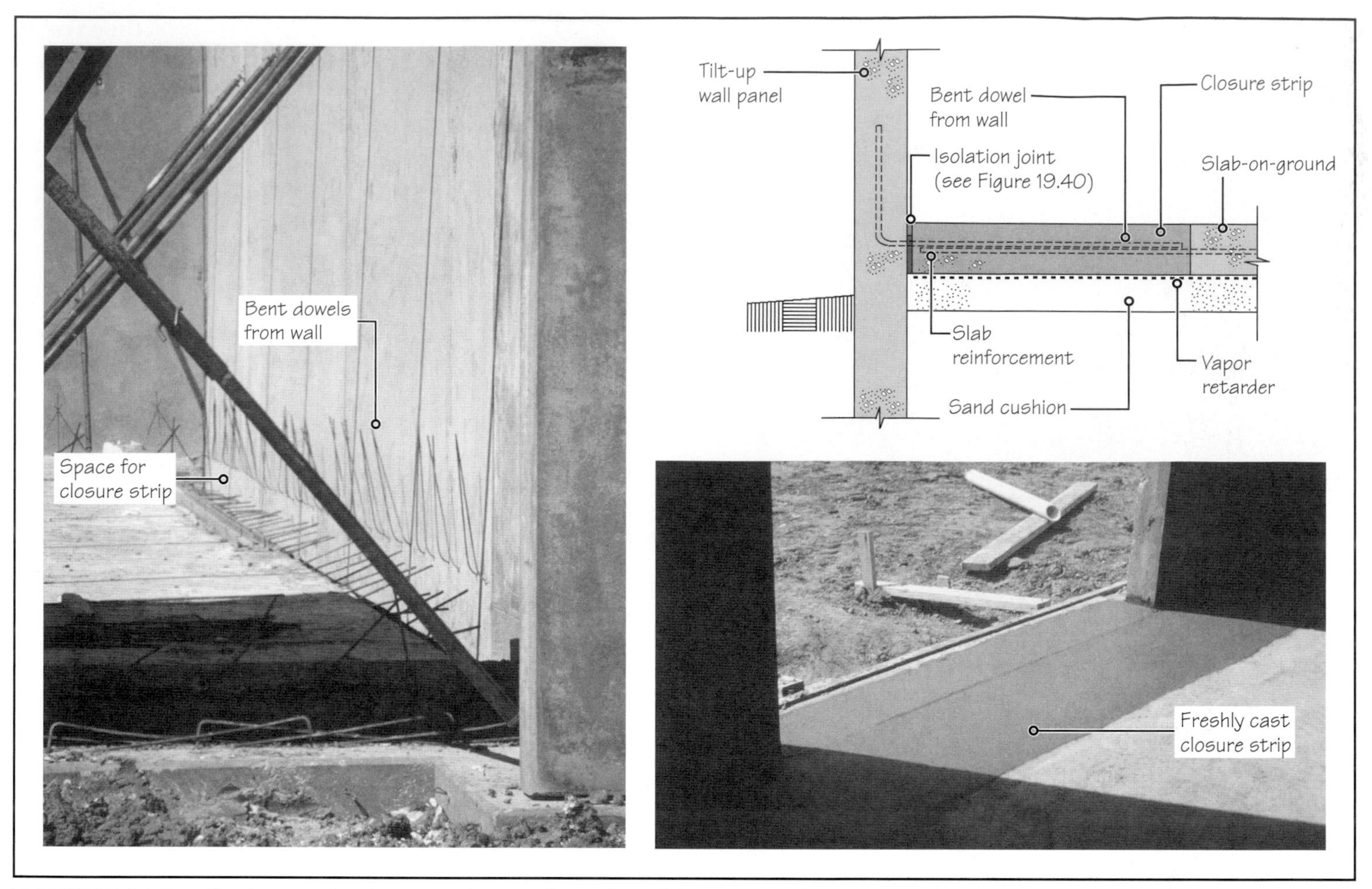

FIGURE 24.38 Details of connection between tilt-up wall panels and the concrete slab-on-ground through a continuous closure strip.

FIGURE 24.39 Details of the connection between tilt-up wall panels and the concrete slab-on-ground using individual steel embeds in panels and the slab.

Others believe that connectors restrain the movement of panels and should not be provided, except between panels meeting at a corner. Connections between adjacent panels in seismic zones are generally required.

Panels should be double sealed, preferably from the interior and exterior, for water and air tightness, Figure 24.42. The outer seal should include weep holes.

Exterior Wall Finish

Tilt-up wall panels can be finished on the exterior in the same way as masonry walls. Cladding materials such as masonry veneer, portland cement stucco, and EIFS may be

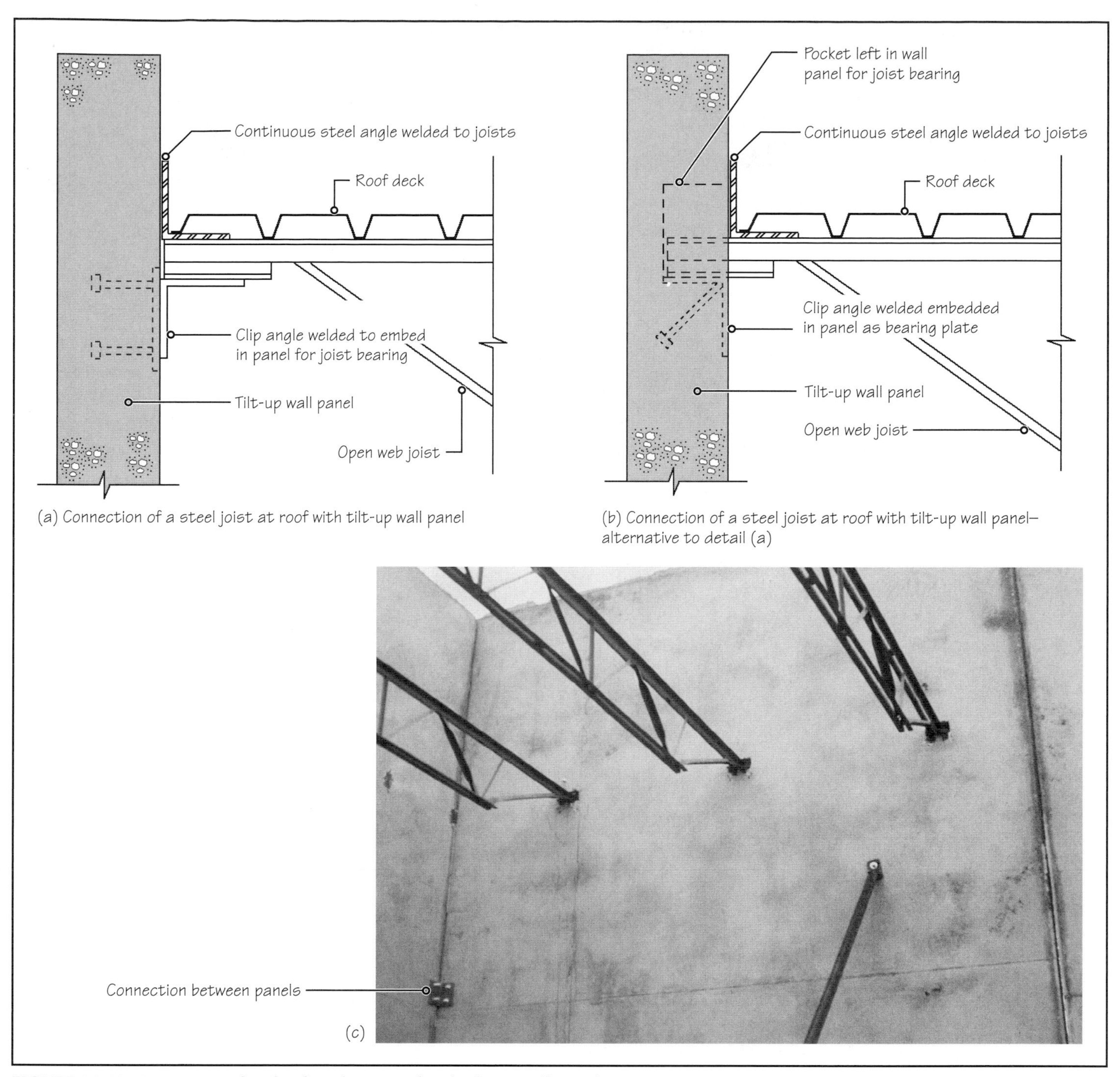

FIGURE 24.40 Connection details of steel joist roof with tilt-up wall panels.

FIGURE 24.41 Connection details of W-shape floor beams with tilt-up wall panels.

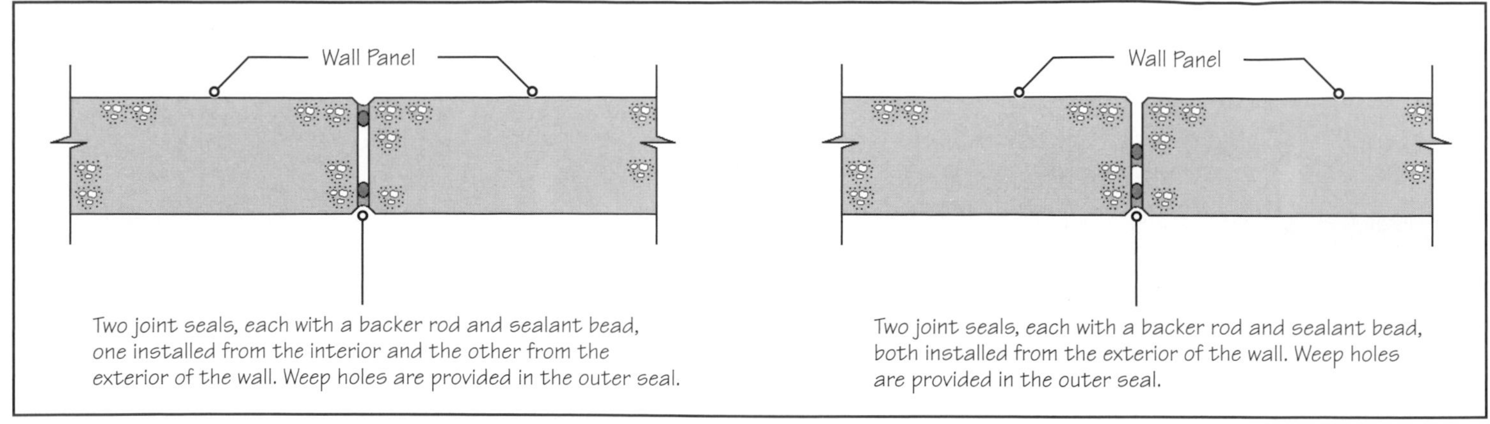

FIGURE 24.42 Joint seals between panels.

used to cover the panels. However, because concrete is relatively more water resistant than many other exterior wall materials, some designers leave tilt-up wall buildings unclad where economy is important. Unclad tilt-up wall panels are either painted or left in exposed concrete finish.

24.11 AESTHETICS OF TILT-UP WALL BUILDINGS

The discussion presented in this chapter may give an impression that tilt-up wall buildings are inherently prosaic, box-type buildings, lacking aesthetic appeal. Although it is true that a tilt-up wall building does not provide the same degree of design freedom as a frame structure, several architects have been able to successfully balance creativity with economics, as shown in the images of a few selected buildings that follow, Figure 24.43.

FIGURE 24.43(a) Airport Corporate Center, Santa Rosa, California. (Photo courtesy of Tilt-Up Concrete Association.)

FIGURE 24.43(b) West Springs Church, Ballwin, Montana. (Photo courtesy of Tilt-Up Concrete Association.)

FIGURE 24.43(c) Beacon Pointe Offices, Weston, Florida. (Photo courtesy of Tilt-Up Concrete Association.)

Each question has only one correct answer. Select the choice that best answers the question.

21. The concrete tilt-up wall construction system is generally used in
- **a.** high-rise buildings—20 stories or more.
- **b.** midrise buildings—5 to 15 stories.
- **c.** low-rise buildings—1 to 4 stories.
- **d.** all the above, depending on the local expertise and economy.

22. In a tilt-up wall construction system, the walls are
- **a.** cast in place at the construction site.
- **b.** precast in a plant and transported to the construction site for erection.
- **c.** precast at the construction site.
- **d.** (b) or (c), whichever is more economical.

23. The temporary braces that support the concrete tilt-up walls are typically removed when
- **a.** all the walls of the building have been erected.
- **b.** all the walls have been anchored to the slab-on-ground.
- **c.** the roof structure is complete.
- **d.** all the above.
- **e.** none of the above.

24. The most commonly used strength of concrete for tilt-up walls is
- **a.** 6,000 psi. **b.** 5,000 psi.
- **c.** 4,000 psi. **d.** 3,000 psi.
- **e.** 2,000 psi.

25. The most commonly used strength of concrete for the slab-on-ground in concrete tilt-up wall buildings is
- **a.** 6,000 psi. **b.** 5,000 psi.
- **c.** 4,000 psi. **d.** 3,000 psi.
- **e.** 2,000 psi.

26. The foundations for concrete tilt-up walls consist of
- **a.** continuous strip footings.
- **b.** isolated pad footings.
- **c.** drilled pier footings.
- **d.** any one of the above.
- **e.** none of the above.

27. During the erection of concrete tilt-up walls, a gap is left between the bottom of a wall panel and the top of the footing. This gap is approximately
- **a.** $\frac{1}{2}$ in. **b.** 2 in.
- **c.** 4 in. **d.** none of the above.

28. Tilt-up concrete walls are connected to the slab-on-ground with
- **a.** a continuous closure element that typically consists of a steel angle.
- **b.** a continuous concrete closure strip in the slab-on-ground.
- **c.** steel weld plates or angles welded to walls and the slab-on-ground.
- **d.** (a) or (c).
- **e.** (b) or (c).

Answers: 21-c, 22-c, 23-d, 24-c, 25-c, 26-d, 27-b, 28-e.

PRINCIPLES IN PRACTICE

The Middle-Third Rule

The middle-third rule can be derived fairly easily using the principles of statics. However, it can also be demonstrated experimentally, as shown in Figures 1 to 4. If a rigid rectangular block is placed on a soft, compressible base and loaded vertically, we see that the block will tend to settle into the base.

When the load is in the center (middle) of the block, the settlement of the block is uniform, indicating that the compressive stress created by the load is uniformly distributed over the block, Figure 1. If the load is moved slightly off center (say to the right) but still remaining within the middle third of the block, as shown in Figure 2, the block will settle more to the right than the left, indicating that the compressive stress under the right side is greater than under the left side of the block.

If the load is now moved to the extreme right end of the middle third, the right edge of the block will settle further and the left edge of the block will just touch the base, as shown in Figure 3. This indicates that the stress under the left edge of the block is zero.

If the load is moved to lie within the outer third, as shown in Figure 4, the left edge of the block will lift above the base, indicating the presence of tension under the left edge.

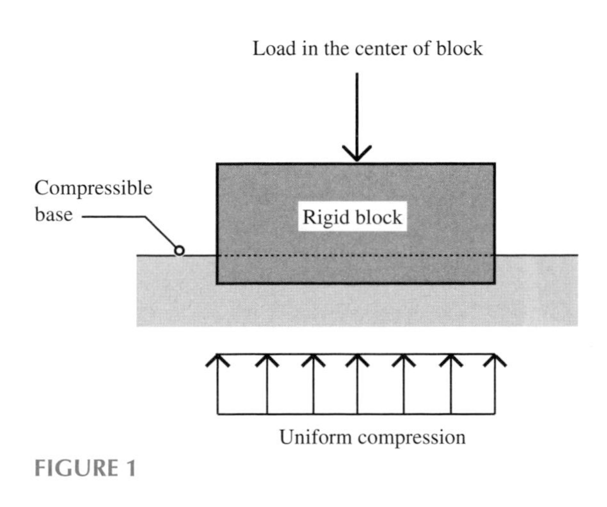

FIGURE 1

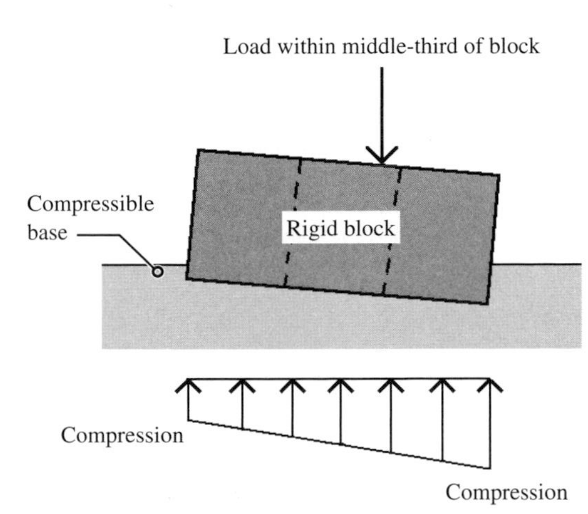

FIGURE 2

(Continued)

PRINCIPLES IN PRACTICE

The Middle-Third Rule (*Continued*)

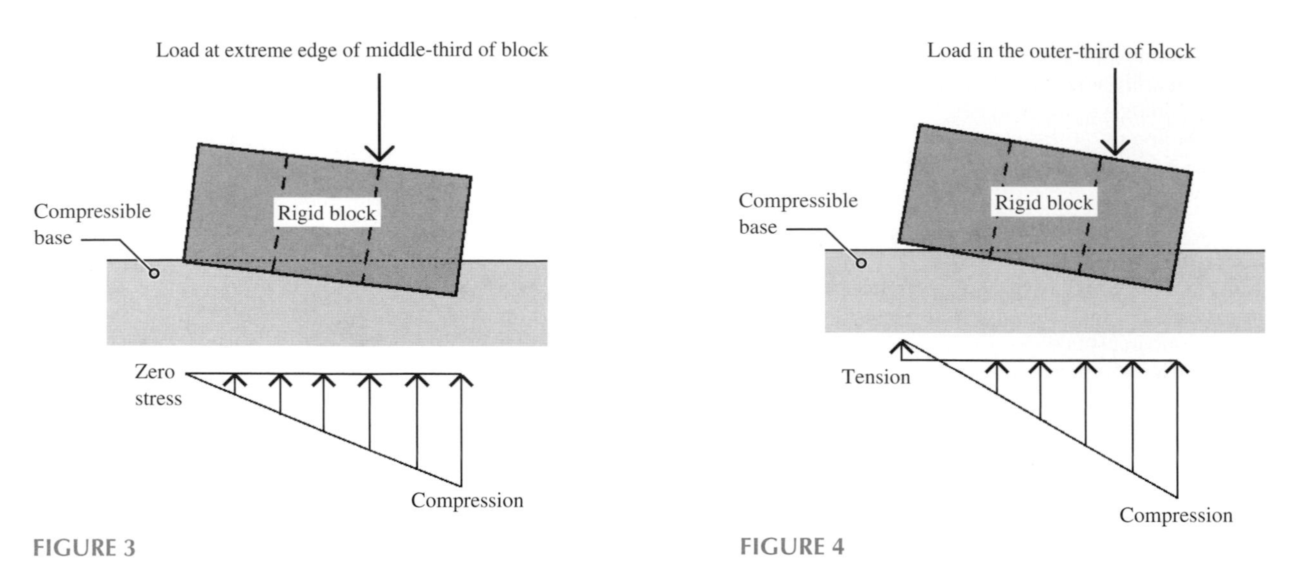

FIGURE 3

FIGURE 4

Figures 1 to 4 illustrate that as long as the load on the block lies within the middle third, no tension will be created at the interface between the block and the base.

MIDDLE-THIRD RULE AND GOTHIC CATHEDRALS

The middle-third rule has been used in the design of structures since the Gothic times. Although understood only intuitively at the time, the designers of the Gothic cathedrals used it with great success to ensure that the resultant of the lateral thrust from the flying arches (originating from the stone vaults) and the gravity load of the vertical buttresses remains in the middle third of the horizontal section of the buttresses, Figure 5. Such a design ensured that masonry walls remained free of tensile stresses.

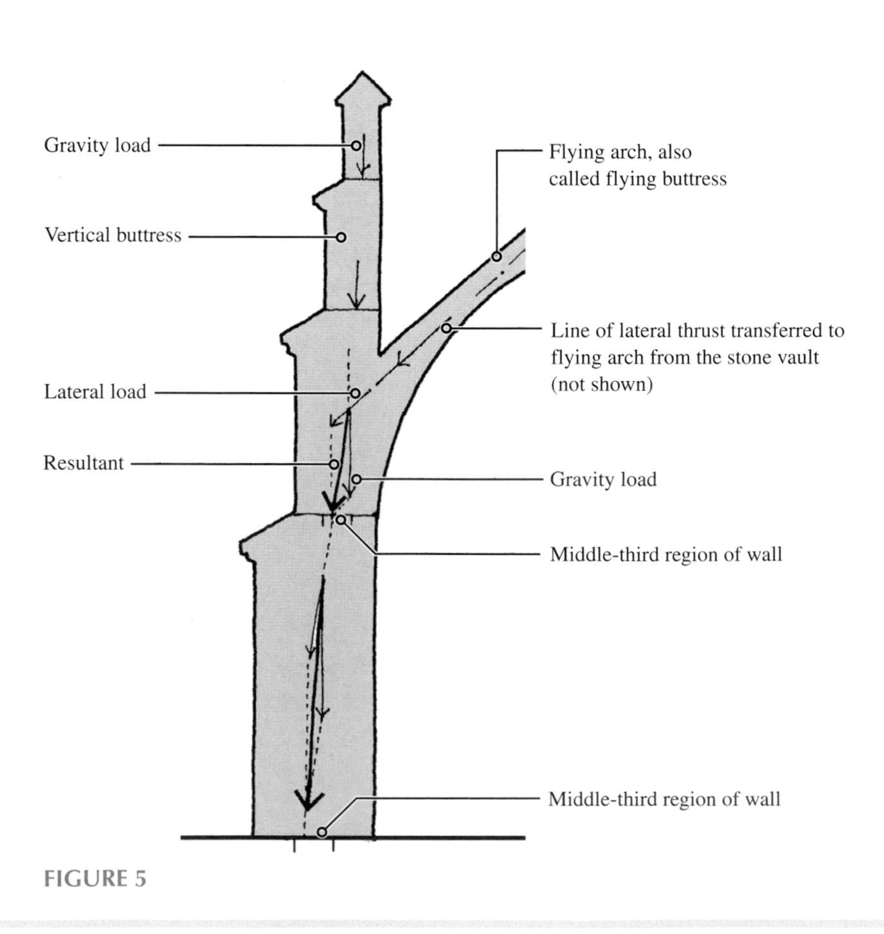

FIGURE 5

KEY TERMS AND CONCEPTS

Bond beams

Closure strip

Contemporary masonry bearing wall construction

Cross-bearing wall plan

Foundations for tilt-up walls

Importance of vertical reinforcement in masonry walls

Longitudinal-bearing wall plan

Modularity of units and masonry dimensions

Reinforced concrete bearing wall construction

Reinforced concrete tilt-up wall construction

Strength of concrete in tilt-up walls

Traditional masonry bearing wall construction

REVIEW QUESTIONS

1. With the help of sketches and notes, explain why the walls in a traditional masonry bearing wall structure had to be excessively thick.
2. With the help of sketches and notes, explain the middle-third rule.
3. Explain the importance of vertical reinforcement in load-bearing masonry walls.
4. With the help of sketches and notes, explain why it is desirable to reinforce the ends, corners, and the jambs around openings in the walls of load-bearing masonry buildings.
5. What is the purpose of bond beams, and where are they required in a typical load-bearing masonry structure?
6. Explain why the loads on the slab-on-ground in concrete tilt-up wall building are generally higher during the construction of the building than after the building is complete.

SELECTED WEB SITES

International Masonry Institute, IMI (www.imiweb.org)

Masonry Construction (www.masonryconstruction.com)

National Concrete Masonry Association, NCMA (www.ncma.org)

Tilt-Up Concrete Association, TCA (www.tilt-up.org)

FURTHER READING

1. American Concrete Institute (ACI). "Tilt-Up Concrete Structures." *Manual of Concrete Practice*, 2006.
2. Beall, Christine. *Masonry Design and Detailing for Architects, Engineers and Builders.* New York: McGraw-Hill, 1997.
3. National Concrete Masonry Association (NCMA). *NCMA TEK Notes.* Herndon, VA: NCMA. (These notes are continuously updated at different times).
4. Tilt-Up Concrete Association (TCA) and American Concrete Institute (ACI). *The Tilt-Up Design and Construction Manual*, 5th ed.

REFERENCES

1. International Masonry Institute (IMI). *The Planning and Design of Loadbearing Masonry Buildings.*

CHAPTER 25

Rainwater Infiltration Control in Exterior Walls

CHAPTER OUTLINE

25.1 Rainwater Infiltration Control—Basic Strategies

25.2 Barrier Wall Versus Drainage Wall

25.3 Rain-Screen Exterior Cladding

This image shows the interior of the air space between a CMU backup wall and limestone cladding, where the air space functions as a drainage cavity, protecting the backup wall from water intrusion. (The photo was taken during the installation of cladding at the wall's corner.) (Photo by MM.)

Exterior walls are one of the major determinants of the appearance of a building. They convey images such as strength or solidity (brick- or stone-clad walls), lightness or openness (glass-metal curtain walls), or a sense of movement or activity (bright, glistening metal curtain walls). In terms of the building's performance, the importance of exterior walls is even greater. Together with the roof, they constitute the building's envelope, providing the separation between inside and outside and serving to maintain an acceptable interior environment.

The exterior walls can serve the envelope function only if they are able to perform well under a range of environmental conditions. They must

a. Prevent water infiltration from rain and snow
b. Control heat loss and heat gain
c. Control air leakage and water vapor transmission
d. Resist fire
e. Control sound transmission
f. Accommodate movement due to thermal, moisture, and other causes

Factors (b) to (f) are covered in Chapters 5–9. Water infiltration is, however, the most critical of all concerns. Its importance is universal because no building can be considered a "shelter" unless its envelope is water resistant. Additionally, a water-resistant envelope has been the fundamental requirement of buildings through the ages—well before other environmental factors emerged as design concerns.

This chapter focuses on the principles of water-infiltration control in exterior-wall assemblies. The construction and detailing of a variety of these assemblies is covered in Chapters 26 and 27. The principles of water-infiltration control in basement walls are covered in Chapter 21, and roofs are discussed in Chapters 31 and 32.

25.1 RAINWATER INFILTRATION CONTROL—BASIC STRATEGIES

Three major forces that affect water infiltration through exterior walls are: gravity, wind and surface tension. Because these forces frequently act on an assembly simultaneously, multiple strategies must be used in the same assembly to counter water infiltration.

GRAVITATIONAL FORCE–INDUCED INFILTRATION

Gravity generally affects water penetration through an exposed horizontal surface of a wall and can be countered by providing a nominal slope in the surface. The purpose of slope (usually 1:12) is to drain water away from vulnerable parts of a building. Thus, a roof overhang protects the upper portion of a wall, and the coping on the top of a masonry fence wall is sloped on both sides of the wall, Figure 25.1.

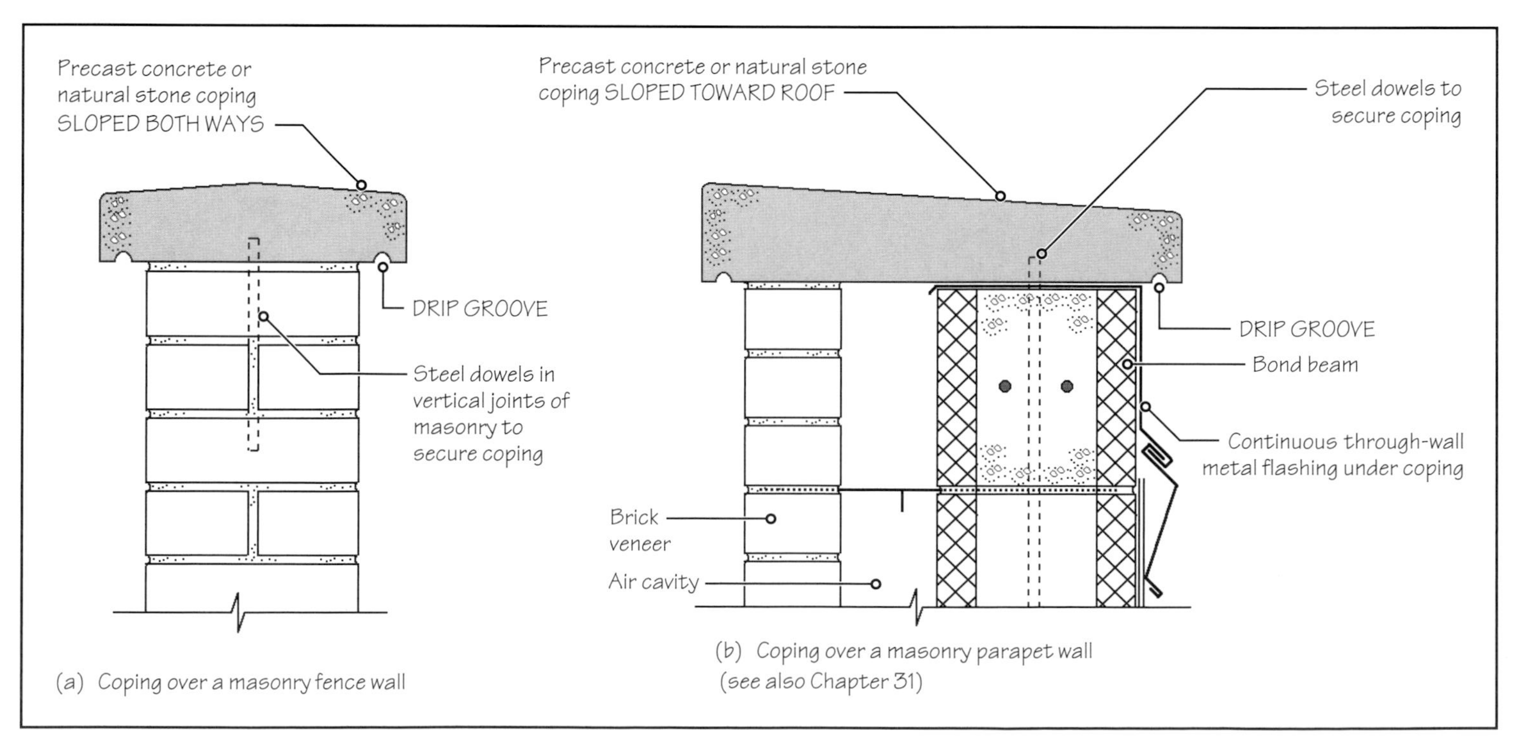

FIGURE 25.1 The importance of sloping the exposed horizontal surfaces of a wall.

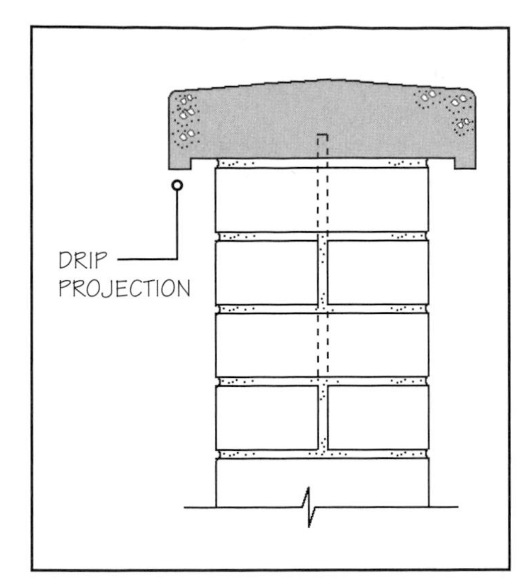

FIGURE 25.2 Detail of Figure 25.1(a) redrawn with a drip projection in coping instead of a drip groove.

The coping over a roof parapet is generally sloped toward the roof so that the water drains on the roof. In exterior window sills and thresholds under entrance doors, the slope is directed away from the building.

Surface Tension–Induced Infiltration

Surface tension creates forces of adhesion between the building surface and water, allowing a thin film of water to travel horizontally along the soffit instead of dropping vertically. A drip mechanism, which consists either of a continuous groove or a vertical projection, counteracts this phenomenon, Figures 25.1 and 25.2.

Surface tension is also responsible for creating capillary forces. Capillary forces are suction forces that occur in tiny spaces between two surfaces in very close proximity to each other. In this case, the water is sucked between the surfaces.

Water can also be sucked into the tiny pores of porous materials such as brick, CMU, or stone. Coating the surface of a porous material with a sealer helps close the pores. Because the sealer degrades rapidly due to ultraviolet radiation, abrasion, and other physical processes, the use of sealers is not a durable means of waterproofing a porous surface. To be effective, sealers on exterior walls need frequent reapplication.

Capillary suction can also occur in a joint between two nonporous materials if the joint is narrow. In narrow joints, capillary suction can be prevented by introducing a larger space, nearly $\frac{1}{4}$ in. or wider, immediately beyond the capillary space. This space is referred to as the *capillary break,* Figure 25.3.

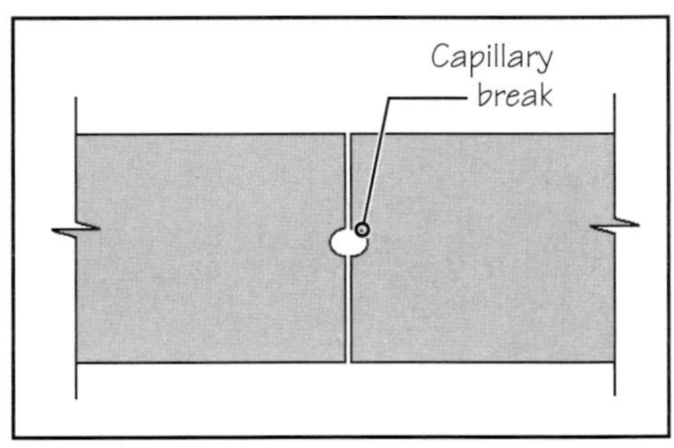

FIGURE 25.3 A narrow vertical butt joint can be made more water resistant by providing a capillary break.

Wind-Induced Infiltration

Providing an *overlap* between members meeting at a joint is another commonsense measure. The force of wind can impart kinetic energy to water, causing the water to travel horizontally as well as vertically in the joint. Therefore, the overlap must be sufficiently large to present a reliable baffle against wind-driven rain, Figure 25.4. A lapped horizontal joint

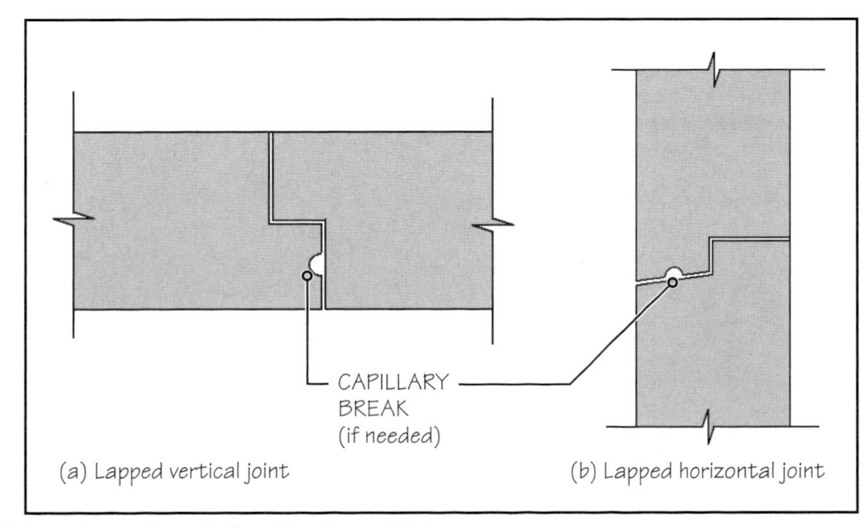

FIGURE 25.4 Lapped vertical and horizontal joints.

NOTE

What Is Surface Tension?

Surface tension is the tension that exists at the exterior surface of a liquid because of the missing bond between molecules. A molecule in the interior of a liquid is subjected to forces of attraction from adjacent molecules. Because the molecule is surrounded from all sides, it is subjected to a uniform attractive force from all directions. Consequently, the resultant force on an interior molecule is zero.

However, the molecules of liquid lying at an exterior surface are subjected to forces of attraction from the interior only. Consequently, the molecules at the exterior surface are pulled toward the interior, causing tension at the exterior surface of the liquid.

Surface tension is responsible for the force of adhesion between water and the underside of a horizontal surface and for capillary suction through tiny spaces.

performs better than a lapped vertical joint, because water has to work against the force of gravity to get across the overlap in a horizontal joint. Capillary breaks can also be included in a lapped joint.

A horizontal butt joint can be made infiltration resistant by incorporating an L- or Z-flashing, Figure 25.5. L- or Z-flashing consists of an impervious membrane that provides a barrier to water and directs it to the exterior surface. Various types of sheet materials—galvanized steel, stainless steel, copper, lead, and polyvinyl chloride (PVC)—are used as flashing materials. Other uses of flashing are shown in Figures 25.6 and 25.8.

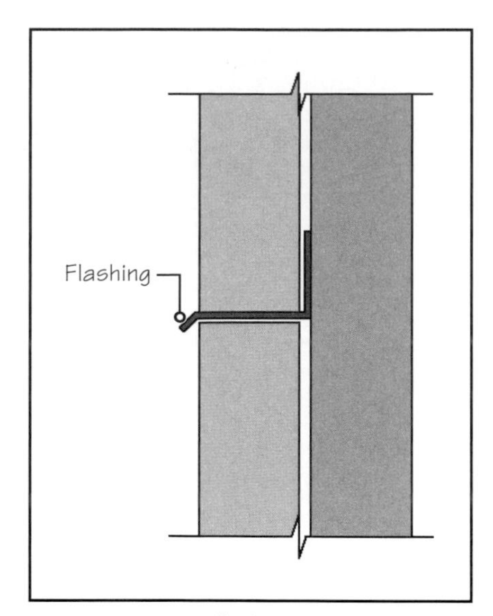

FIGURE 25.5 L-flashing at a horizontal butt joint. If the projecting part of flashing is turned down at a 90° angle instead of 45° angle, it is referred to as Z-flashing.

THE IMPORTANCE OF JOINT SEALANTS

Filling joints with sealants (Chapter 9) is another commonly used strategy to keep water from penetrating through joints between building components. Because sealants degrade over time from exposure to environmental factors, particularly solar radiation and water, they should not be relied upon as the only water-resisting element. A certain amount of redundancy must be built into the detail in case the sealant fails. Usually all or most of the preceding strategies are used in a single detail, as shown in a section at the sill level of a typical wood window, Figure 25.6.

25.2 BARRIER WALL VERSUS DRAINAGE WALL

Water-resistant exterior walls may be divided into three types:

- Walls with overhangs
- Barrier walls
- Drainage walls

WALLS WITH OVERHANGS

This type of wall system depends on protective floor and roof overhangs to prevent water infiltration, Figure 25.7. It has limited application because the overhangs are required to shield the wall completely from wind-driven rain and, therefore, must be quite deep. In addition, the overhangs pose economic as well as design constraints on the exterior envelope, particularly that of a multistory building. However, a roof overhang is commonly used in a low-rise building (eave projection) and can provide substantial protection from water penetration at the wall-roof junction.

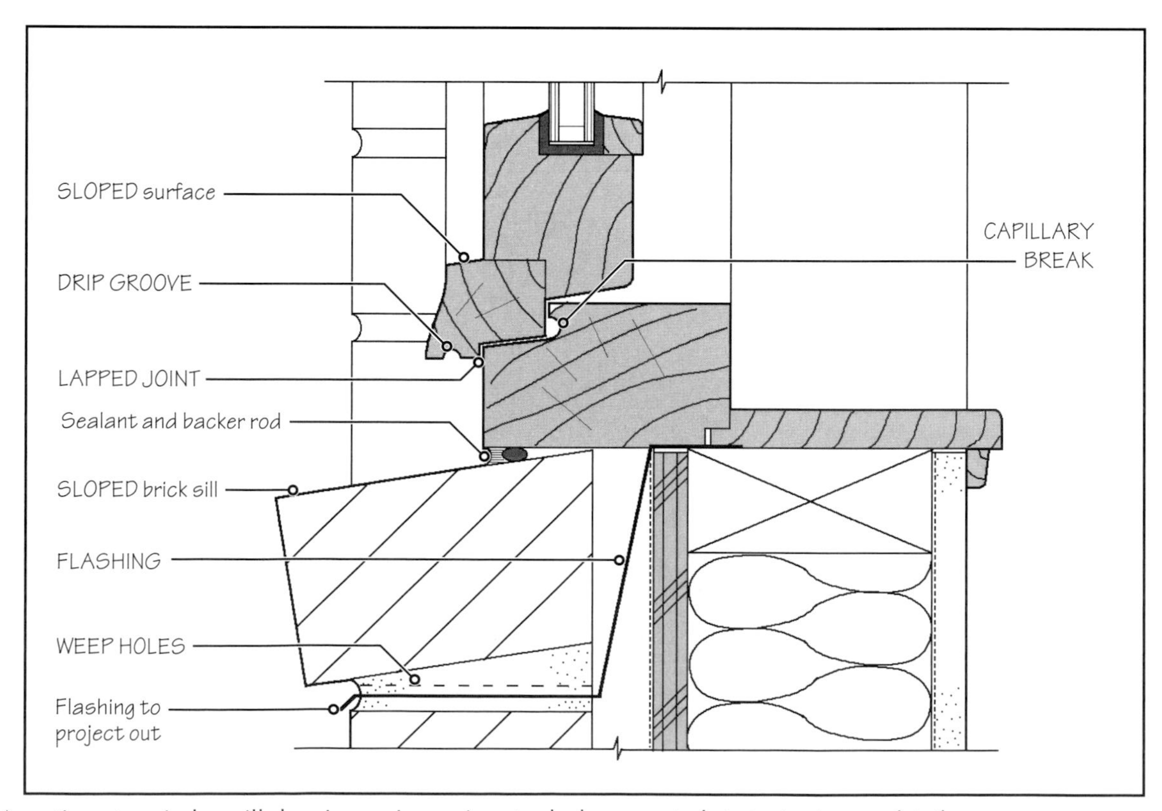

FIGURE 25.6 A section at a window sill showing various rainwater-leakage-control strategies in one detail.

BARRIER WALL

A barrier wall is a wall that functions as the primary structure and, at the same time, performs as the building envelope. Historic buildings such as the Monadnack Building and contemporary CMU load-bearing structures discussed in Chapter 24 are examples of this wall type. This wall type resists water infiltration either (a) by providing an impervious barrier to water or (b) by functioning as a water reservoir.

An example of a barrier wall that uses impervious materials as a barrier is a CMU or concrete wall system with an applied water-resistant veneer, such as EIFS or stucco (Chapter 27). On the other hand, an example of a barrier wall that acts as a reservoir is a porous masonry or concrete wall that is so thick that any absorbed water is not able to migrate to the wall's interior surface. The thicker the wall, the larger the reservoir. In this type of wall, only the outer layer of the wall gets wet, even during a long rainy spell; given a long dry spell, the entire wall will dry again.

It is very important that a barrier wall does not develop through-wall cracks that may be caused by weathering, expansion or contraction, foundation settlement, and so on. If cracks develop, water infiltration will result. Careful detailing and attention to water resistance in materials are therefore essential to the long-term performance of a barrier wall.

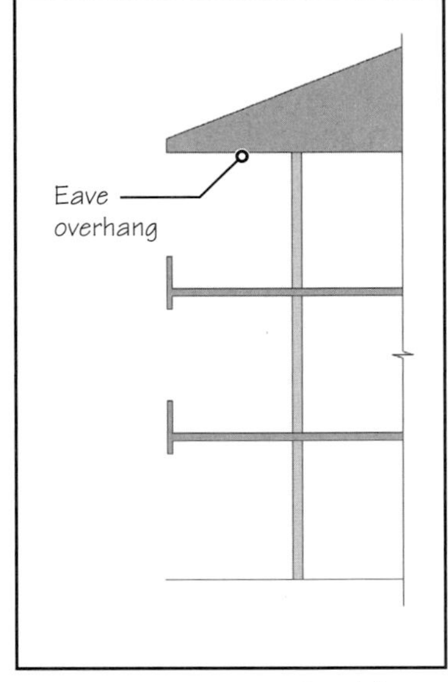

FIGURE 25.7 Deep roof and floor ovehangs can substantially reduce water leakage from wind-driven rain through walls.

DRAINAGE WALL

Many contemporary exterior-wall assemblies are designed as drainage walls. A drainage wall is a wall that consists of an exterior cladding, an inner backup wall, and an intervening air space between the cladding and the backup.

In a drainage wall, the exterior cladding is the first defense against water infiltration. Therefore, it is made as water resistant as possible. However, some water will inevitably penetrate through the cladding. Any water that leaks through the cladding collects in the air space and is drained out through small openings at the bottom of the cladding, called *weep holes,* Figure 25.8.

IMPORTANT FEATURES OF A DRAINAGE WALL

A drainage wall must have sufficient built-in redundancy to remain infiltration-free for the entire life of the building and includes the following features:

- Exterior cladding
- Air space
- Flashing
- Water-resistant backup
- Weep holes

Flashing in a drainage wall is a continuous waterproof membrane, which provides a continuous barrier that originates at the backup wall and penetrates through the air space

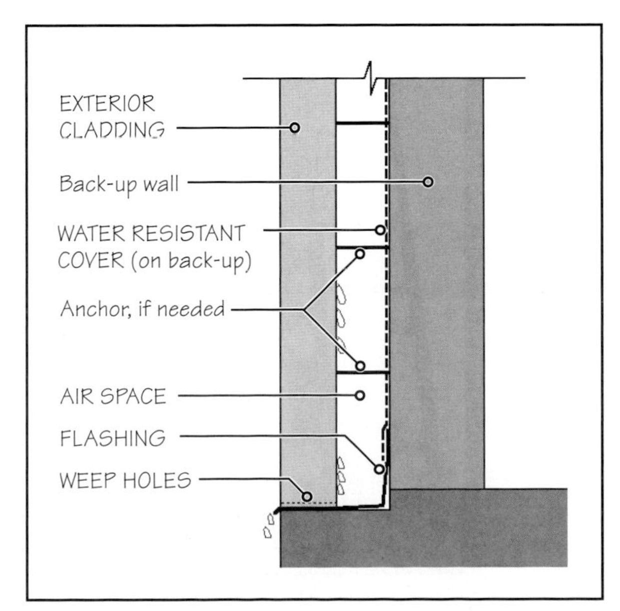

FIGURE 25.8 A section through a drainage wall showing its important water-infiltration control features. (Note that a drainage wall may also have insulation, which is not a part of water-infiltration control.)

and cladding. The purpose of flashing is to collect and channel any water that gets into the air space back to the exterior. It is attached to the backup wall at the bottom of a section of a drainage wall and works in conjunction with weep holes, which allow water to escape.

The *water-resistant layer* on the backup wall protects it from water that may accidentally reach the backup by traveling over metal anchors (where used) or unintentional obstructions in the air space. This layer also functions as an air barrier. Because the water-resistant layer is shielded by the cladding, it does not degrade as rapidly as an exposed surface.

Although the drainage wall concept was initially introduced with reference to a porous exterior cladding consisting of brick, CMU, concrete, or stone, it applies equally to an impervious exterior cladding, such as one of glass or metal. Note that the underlying philosophy of the drainage wall is the redundancy of its two water-resistant layers. If the exterior layer is made of an impervious material, the redundancy is that much greater.

EXPAND YOUR KNOWLEDGE

Exterior-Wall Cladding Systems: Curtain Wall, Veneer Wall, and Infill Wall

The primary exterior-wall types in use today are *curtain walls, veneer walls,* and *infill walls.*

Curtain Walls

As the name suggests, a *curtain wall* forms a curtain on the exterior face of a building. It is used with a frame structure and covers the building's structural frame and is directly suspended from it. In a curtain wall, the wind loads are transferred directly from the curtain wall to the building's structural frame.

A 2-in. minimum separation is generally required between the curtain wall and the building's structural frame to accommodate unintended dimensional irregularities in the frame.

A typical example of a curtain wall is a *glass-aluminum curtain wall* (Chapter 30). *Opaque curtain walls* consist of precast concrete panels, glass fiber–reinforced panels, metal panels, and so on. A backup wall may be used in an opaque curtain wall to provide an interior wall finish and to incorporate insulation, electrical, and other utility conduits within the wall. Because the wind load is resisted by the curtain wall, the backup wall in this assembly (if provided) does not experience any wind loads.

Veneer Walls

Another exterior wall cladding, which is similar to an opaque curtain wall is a *veneer wall.* A veneer wall can be used with both a frame structure or a bearing wall structure. A backup wall is always required in a veneer wall. The wind loads that act on the veneer are transferred to the backup wall. The backup wall then transfers them to the building's structural frame. A veneer wall may be of two types:

- Anchored veneer
- Adhered veneer

In an anchored veneer, the anchors connect the veneer to the backup; therefore, they participate in transferring the wind loads from the veneer to the backup wall. An anchored veneer wall is typically designed as a drainage wall with a minimum 2-in. air space between the veneer and the backup wall.

A veneer may also be applied without anchors, in which case it is adhered to the backup wall. Such a veneer is called an *adhered veneer.* An adhered veneer wall is a barrier wall, unlike the anchored veneer. Stucco and EIFS wall assemblies are examples of adhered veneers.

Infill Walls

Like a curtain wall and a veneer wall, an infill exterior wall is also a non-load-bearing wall assembly. It occupies the space between the building's columns and beams, leaving the structural frame exposed to the outside. An infill wall can be used only with a frame structure. It can be designed either as an anchored infill wall or an adhered infill wall.

Before energy-use concerns became critical, most exterior-wall assemblies were designed as infill walls, Figure 1. In these buildings, the steel or concrete structural frame, being of an extremely low R-value, contributed to substantial thermal short-circuiting. In contemporary buildings, the use of curtain walls or veneer walls is the norm because they cover the building's structural frame and, therefore, can be designed with greater thermal efficiency.

Exterior Cladding

The term *exterior-wall cladding* (or simply *cladding*) is a general term that is used for all exterior wall finishes.

FIGURE 1 A typical example of exposed structural frame and exterior infill walls.

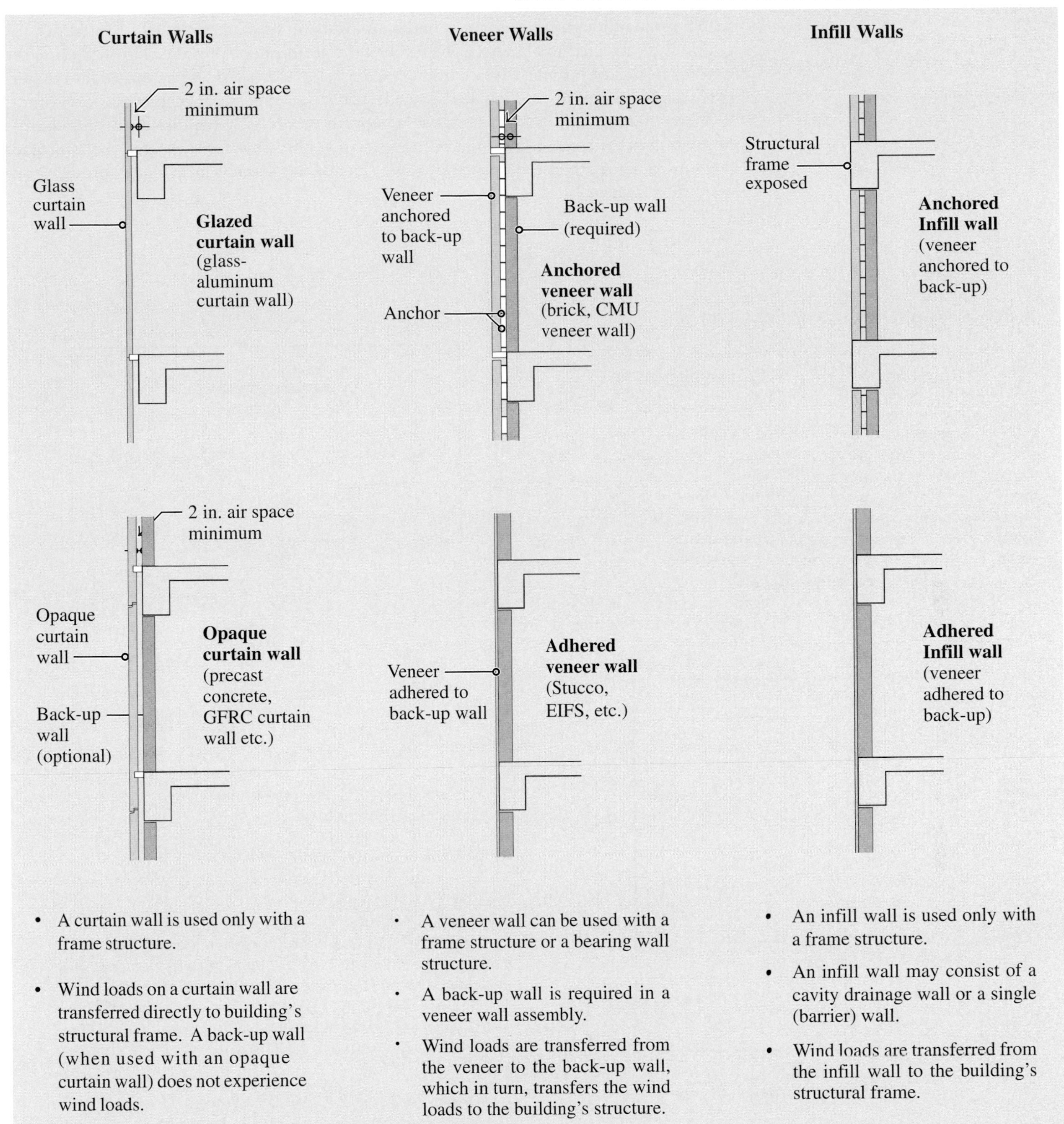

- A curtain wall is used only with a frame structure.
- Wind loads on a curtain wall are transferred directly to building's structural frame. A back-up wall (when used with an opaque curtain wall) does not experience wind loads.

- A veneer wall can be used with a frame structure or a bearing wall structure.
- A back-up wall is required in a veneer wall assembly.
- Wind loads are transferred from the veneer to the back-up wall, which in turn, transfers the wind loads to the building's structure.

- An infill wall is used only with a frame structure.
- An infill wall may consist of a cavity drainage wall or a single (barrier) wall.
- Wind loads are transferred from the infill wall to the building's structural frame.

25.3 RAIN-SCREEN EXTERIOR CLADDING

A rain-screen cladding system includes the basic principles of the drainage wall system, but it also addresses the issue of water penetration due to unequal distribution of air pressure on the exterior of the wall and in the airspace between the cladding and backup wall. This is required when weep holes and joints that provide for drainage become sources of water penetration by suction under conditions of wind-driven rain.

In order to reduce the suction of water in a drainage wall, the pressure between the airspace and the outside should be equalized as much as possible. Pressure equalization is accomplished by not completely sealing the cladding. In fact, providing weep holes and purposely incorporating other openings in the cladding help to create pressure equalization.

If the openings in the cladding of a drainage wall, such as the weep holes, are few and small in area, the pressure between the airspace and the exterior cladding will not be equalized. In such a wall, although the airspace is under atmospheric pressure, the outside surface of the cladding is subjected to a pressure greater than the atmospheric pressure. This will cause the water to be sucked into the airspace through weep holes. Water will also be sucked into the airspace through joints in cladding that are inadequately sealed or have become leaky over time.

Water suction will take place on the windward facade only because at this facade, the pressure in the airspace is lower than the outside pressure. On the other facades, the airspace pressure is higher than the outside pressure. Hence, no suction takes place through non-windward facades.

EXPAND YOUR KNOWLEDGE

Rain-Screen and Pressure Equalization

To fully appreciate the phenomenon of suction of water into the airspace in a drainage wall, the fundamentals of wind loads on buildings must be reviewed. When there is no wind (zero wind speed), the air pressure on the inside and outside surfaces of the walls of an enclosure is the same, equal to the atmospheric pressure—2,100 psf. Because the atmospheric pressure works on both sides of a wall equally, the wall is under perfect equilibrium, and there is no wind load on the wall, Figure 1.

Equilibrium is disturbed by wind, which creates additional pressure on the outside surface of a windward wall, whereas the pressure in the enclosure is equal to the atmospheric pressure. For example if the wind speed is 100 mph, the wind load on the wall is nearly 25 psf (Section 3.5). This means that the outside surface of the windward wall is under a pressure of 2,125 psf, but the enclosure is under a pressure of 2,100 psf, Figure 2.

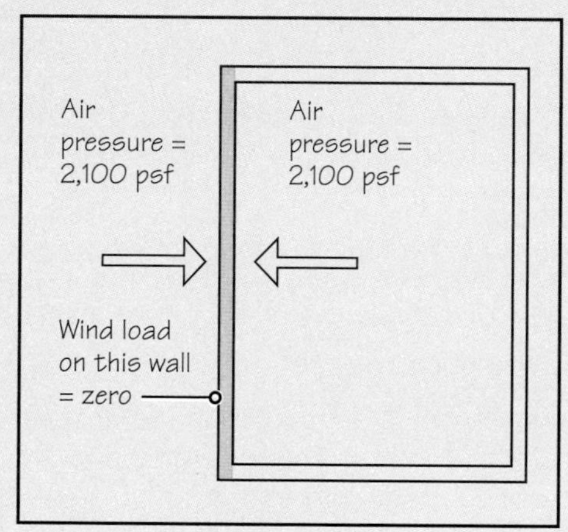

FIGURE 1 Outside and inside air pressures on a wall under zero wind speed.

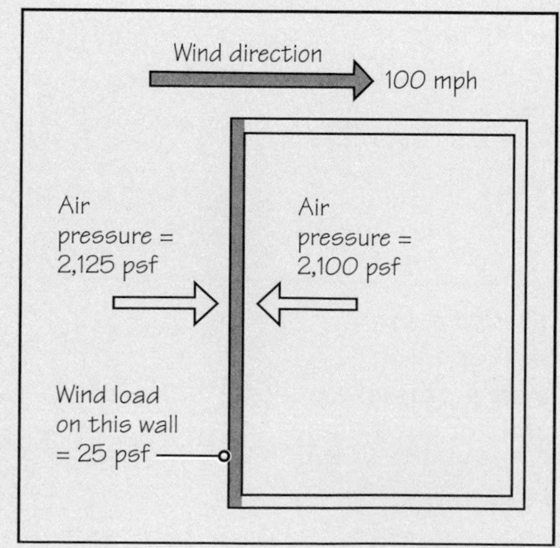

FIGURE 2 Outside and inside air pressures on a windward wall of a fully enclosed building.

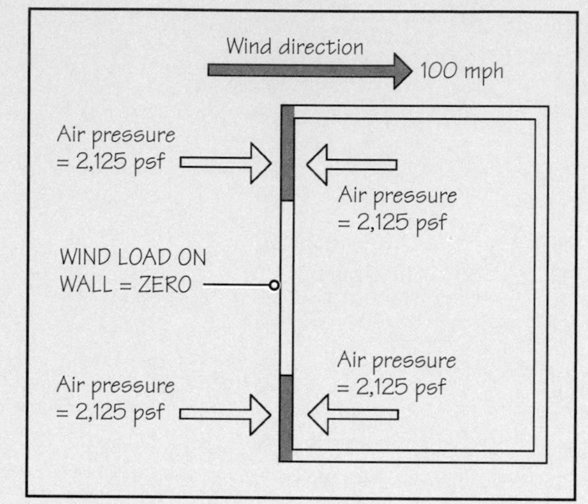

FIGURE 3 Outside and inside air pressures in an enclosure that has a large opening on the windward wall.

It is this difference between the inside and outside air pressures that we refer to as the *wind load* (Section 3.5). It is also this pressure difference that causes suction of water through the windward cladding because the pressure in the (drainage) airspace is atmospheric pressure, whereas the pressure on the exterior face of cladding is greater than the atmospheric pressure.

If the windward wall has large openings and the rest of the enclosure has no openings at all, the wind will move into the enclosure and the air pressure inside the enclosure will be equal to the outside pressure, as shown in Figure 3. In the case of perfect equalization of pressures, the windward wall will not be subjected to any wind load, and there will be no suction created at the windward wall.

Atmospheric Pressure Equalization and Human Bodies

The atmospheric air pressure is due to the weight of the column of air on the earth's surface. This equals the pressure that a 29.92-in. (say, 30 in. = 2.5 ft) column of mercury exerts at the column's bottom. It equals 14.7 psi, or approximately 2,100 psf.

A pressure of 2,100 psf is the same pressure that a 14-ft-thick concrete slab will exert at its base. In other words, human bodies are subjected to the same pressure that is exerted by a 14-ft-thick concrete slab at its base. We are not squashed by this pressure (i.e., there is no wind load on our body) because the interior of our body also contains air, and there is air pressure equalization between the inside and outside of our bodies. (Because concrete weighs about 150 pounds per cubic foot (pcf), a 14-ft-high concrete column exerts a pressure of about 2,100 psf at its base. Mercury weighs approximately 850 pcf. Therefore, only a 2.5-ft-high mercury column exerts the same pressure.)

WIND LOAD ON THE CLADDING OF A PRESSURE-EQUALIZED WALL

Because there is no pressure differential between the inside and the outside faces of cladding in a pressure-equalized drainage wall, the cladding is not subjected to any wind load. In such a wall, all the wind load acts on the backup, and the cladding functions as a nonstructural element whose primary function is to control water penetration by allowing free movement of air for pressure equalization. Therefore, the exterior cladding in a pressure-equalized wall is referred to as a screen—or, more commonly, as the *rain screen.*

PRACTICAL LIMITATIONS OF RAIN-SCREEN WALL

Theoretically, the exterior cladding in a pressure-equalized drainage wall should not be subjected to any wind load. In practice, however, it is not possible to achieve a completely load-free cladding. The reason is that pressure equalization does not occur instantaneously. It takes some time for pressure equalization to take place, depending on how rapidly the outside air can move into the intervening air space. If the area of openings in the wall is large, pressure equalization will be rapid. In the case of a small opening area, pressure equalization will be slow.

In other words, there is a time lag between a change in the outside pressure and the corresponding change in airspace pressure. During this period (when pressure equalization is taking place), the cladding is subjected to wind loads. If there is no further change in outside pressure, the airspace and the outside will continue to be under equal pressures.

However, wind in a storm or hurricane occurs in gusts, so exterior air pressure changes almost continuously. This disturbs pressure equalization in a drainage wall. Therefore, the cladding is constantly subjected to wind loads, although the magnitude of wind load on the cladding of a pressure-equalized wall is smaller than on a nonpressure-equalized wall.

AIRTIGHT BACKUP WALL

Once the pressure in the airspace is equalized with the pressure on the wall's facade, no air movement will occur in the space (because any movement of air implies the existence of pressure differential). The consequence of this fact is that the rain-screen principle works only if the backup is completely airtight. If the backup has openings, air will move from the airspace into the enclosure, implying the existence of a pressure differential (hence, the absence of pressure equalization).

If the air is able to move into the enclosure through the backup wall, water may be sucked into the enclosure. Therefore, in a pressure-equalized wall, the backup wall must be made airtight. This is usually achieved by sealing all the joints in the backup wall and/or providing an air barrier. The air barrier is usually placed on the outside face of the backup wall so that it can also function as the damp-proofing layer.

COMPARTMENTALIZATION OF AIRSPACE

Another cause of air movement in an intervening air space is shown in Figure 25.9. Under the action of wind, the windward facade is under positive pressure, and the side facades are

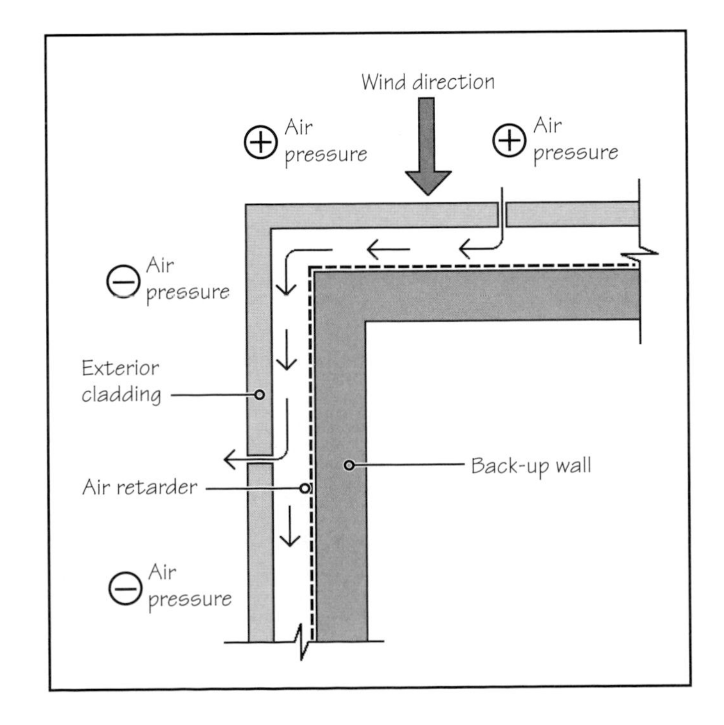

FIGURE 25.9 Because wind produces positive pressure on the windward wall and negative pressure on other walls of a building, the air in a continuous airspace will circulate and move in and out through weep holes and other openings (joints, cracks, etc.) in the cladding. This makes it impossible to achieve pressure equalization between the outside and the air space.

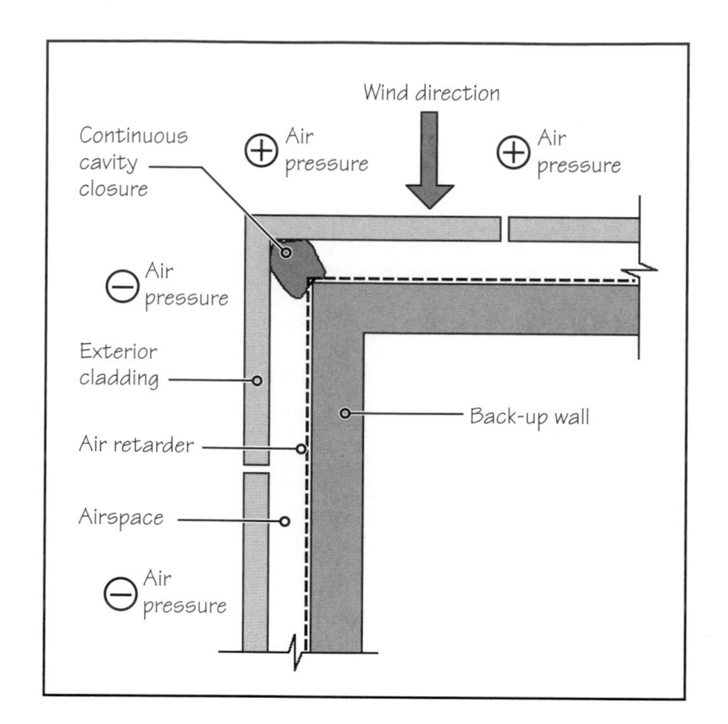

FIGURE 25.10 To prevent air movement within the airspace, the airspace should be divided into small sections with continuous cavity closures, particularly at corners and floor levels.

under negative pressure. If the air space is continuous from one facade to the other, air will move through the space as shown, negating attempts at pressure equalization.

Therefore, the airspace of a pressure-equalized drainage wall must be closed at wall corners. This is usually achieved by using a continuous vertical closure at the corners of the airspace, Figure 25.10. A closed-cell compressible filler, such as neoprene sponge, may be used as the closure material.

In addition to providing vertical closures at wall corners, the airspace should also be closed by horizontal closures. In fact, the airspace should be subdivided into independent compartments. The compartmentalization is in response to the variation of wind pressure over the building facade. Wind pressure is greater at upper floors of a building than at lower floors and is greater at edges and corners of the building than in the middle (Section 3.5). If the airspace is not compartmentalized, the pressure inequality over the facade will generate air movement inside the space.

Each compartment should have openings in the cladding at the top and bottom so that pressure equalization can take place. The openings should preferably be far enough away from each other to prevent airflow short-circuiting in the compartment. Each compartment should also be relatively small and should be independently drained by weep holes. Although preferable, it is not necessary that airspace closures must form absolutely airtight compartments. In summary, an ideal rain-screen wall should consist of the following three water-infiltration-control features:

- Voids in cladding for pressure equalization in the airspace
- Compartmentalization of the airspace
- Airtight backup wall

PRACTICE QUIZ

Each question has only one correct answer. Select the choice that best answers the question.

1. The coping on a roof parapet is generally
   - **a.** sloped toward the interior (roof).
   - **b.** sloped toward the outside.
   - **c.** sloped on both sides.
   - **d.** either (a), (b), or (c).
   - **e.** dead flat.

2. The coping on a masonry fence wall is generally
   - **a.** sloped toward the interior.
   - **b.** sloped toward the outside.
   - **c.** sloped on both sides.
   - **d.** either (a), (b), or (c).
   - **e.** dead flat.

3. A drip mechanism at the underside of a projecting horizontal surface counters the effect of
   - **a.** gravity.
   - **b.** surface tension.
   - **c.** kinetic energy.
   - **d.** capillary force.
   - **e.** all the above.

4. Capillary break is a
   - **a.** vertical barrier between two abutting surfaces of different materials.
   - **b.** horizontal barrier between two abutting surfaces of different materials.

c. wider space in an otherwise narrow joint.
d. narrow space in an otherwise wide joint.
e. none of the above.

5. Z-flashing is effective in
a. a horizontal butt joint in cladding.
b. a vertical butt joint in cladding.
c. a horizontal lapped joint in cladding.
d. a vertical lapped joint in cladding.
e. all the above.

6. In terms of water infiltration, the exterior walls in historical buildings functioned as
a. barrier walls. b. drainage walls.

7. The important feature of an (exterior) drainage wall is
a. exterior cladding.
b. waterproofing of the backup wall.
c. flashing.
d. weep holes.
e. all the above.

8. Metal anchors are needed to transfer lateral loads to the backup wall in
a. some veneer walls. b. all veneer walls.

9. A curtain wall is separated from the structural frame of the building by a minimum of
a. 1 in. b. 2 in.
c. 3 in. d. 4 in.
e. none of the above.

10. A curtain wall is typically used in buildings whose structural system consists of a
a. load-bearing masonry wall structure.
b. load-bearing reinforced-concrete wall structure.
c. reinforced-concrete or steel frame structure.
d. steel frame structure.
e. all the above.

11. A backup wall is required in a
a. curtain wall.
b. veneer wall.
c. infill wall.
d. all the above.

12. Suction that pulls wind-driven rain into the airspace of a drainage wall takes place on the windward face of a wall.
a. True
b. False

13. In an ideal pressure-equalized drainage wall,
a. the cladding resists all the wind load.
b. the backup resists all the wind load.
c. the backup is provided with openings to equalize air pressure in the airspace.
d. the joints in the cladding are sealed to control water infiltration.
e. all the above.

14. An ideal pressure-equalized drainage wall is also referred to as
a. pressure-screen wall.
b. wind-screen wall.
c. no-pressure wall.
d. rain-screen wall.
e. all the above.

15. For a pressure-equalized drainage wall to function effectively, the airspace must be continuous from floor to floor.
a. True b. False

Answers: 1-a, 2-c, 3-b, 4-c, 5-a, 6-a, 7-e, 8-a, 9-b, 10-c, 11-b, 12-a, 13-b, 14-d, 15-b.

KEY TERMS AND CONCEPTS

Barrier wall
Drainage wall
Flashing
Rainscreen exterior cladding
Surface tension-induced infiltration
Weep holes
Wind-induced infiltration

REVIEW QUESTIONS

1. Using a sketch and notes, explain what a capillary break is and where it is commonly used.
2. Explain the differences between a barrier wall and a drainage wall.
3. Using sketches and notes, explain the salient features of a pressure-equalized drainage wall.
4. Explain why the airspace in a pressure-equalized drainage wall must be compartmentalized.

SELECTED WEB SITES

Canadian Mortgage and Housing Corporation (www.cmhc-schl.gc.ca)
National Research Council, Canada (Institute for Research in Construction) (www.irc.nrc-cnrc.gc.ca)

FURTHER READING

1. Brock, Linda. *Designing the Exterior Wall.* New York: John Wiley, 2005.
2. Canadian Mortgage and Housing Corporation (CMHC): *Rain Penetration Control—Applying Current Knowledge.* Ottawa, Ontario, 2000.
3. Nashed, Fred. *Time-Saver Standards for Exterior Walls.* New York: McGraw-Hill, 1996.

CHAPTER 26 Exterior Wall Cladding–I (Masonry, Precast Concrete, GFRC, and Prefabricated Masonry)

CHAPTER OUTLINE

26.1 Masonry Veneer Assembly—General Considerations

26.2 Brick Veneer with a CMU or Concrete Backup Wall

26.3 Brick Veneer with a Steel Stud Backup Wall

26.4 CMU Backup versus Steel Stud Backup

26.5 Aesthetics of Brick Veneer

26.6 Precast Concrete (PC) Curtain Wall

26.7 Connecting the PC Curtain Wall to a Structure

26.8 Brick and Stone-Faced PC Curtain Wall

26.9 Detailing a PC Curtain Wall

26.10 Glass Fiber–Reinforced Concrete (GFRC) Curtain Wall

26.11 Fabrication of GFRC Panels

26.12 Detailing a GFRC Curtain Wall

26.13 Prefabricated Brick Curtain Wall

Brick veneer buildings, illustrating one of the most widely used cladding systems in contemporary US construction, line this New York City street. Note glass-clad Bloomberg Tower (by Cesar Pelli and Associates with Schuman, Lichtenstein, Claman and Effron) in the background. (Photo courtesy of Emporis.)

This is the first of the two chapters on exterior wall finishes, it includes masonry veneer, precast concrete, glass fiber–reinforced concrete (GFRC), and prefabricated masonry panels. Other exterior wall finishes—stucco, exterior insulation and finish systems (EIFS), stone cladding, and insulated metal panel walls—are discussed in the next chapter.

26.1 MASONRY VENEER ASSEMBLY—GENERAL CONSIDERATIONS

Among the most commonly used veneer walls is a single wythe of brick (generally 4 in. nominal thickness), referred to as *brick veneer*. The backup wall used with brick veneer may be load bearing or non–load bearing and may consist of one of the following:

- Wood or light-gauge steel stud
- Concrete masonry
- Reinforced concrete

A wood stud (or steel stud) load-bearing backup wall, Figure 26.1, is generally used in low-rise residential construction. Concrete masonry, non–load-bearing steel stud, and reinforced concrete backup walls are generally used in commercial construction. In fact, brick veneer with concrete masonry backup is the wall assembly of choice for many building types, such as schools, university campus buildings, and offices.

The popularity of brick veneer lies in its aesthetic appeal and durability. A well-designed and well-constructed brick veneer wall assembly generally requires little or no maintenance. The discussion in this chapter refers to brick veneer. It can, however, be extended to include other (CMU and stone) masonry veneers with little or no change.

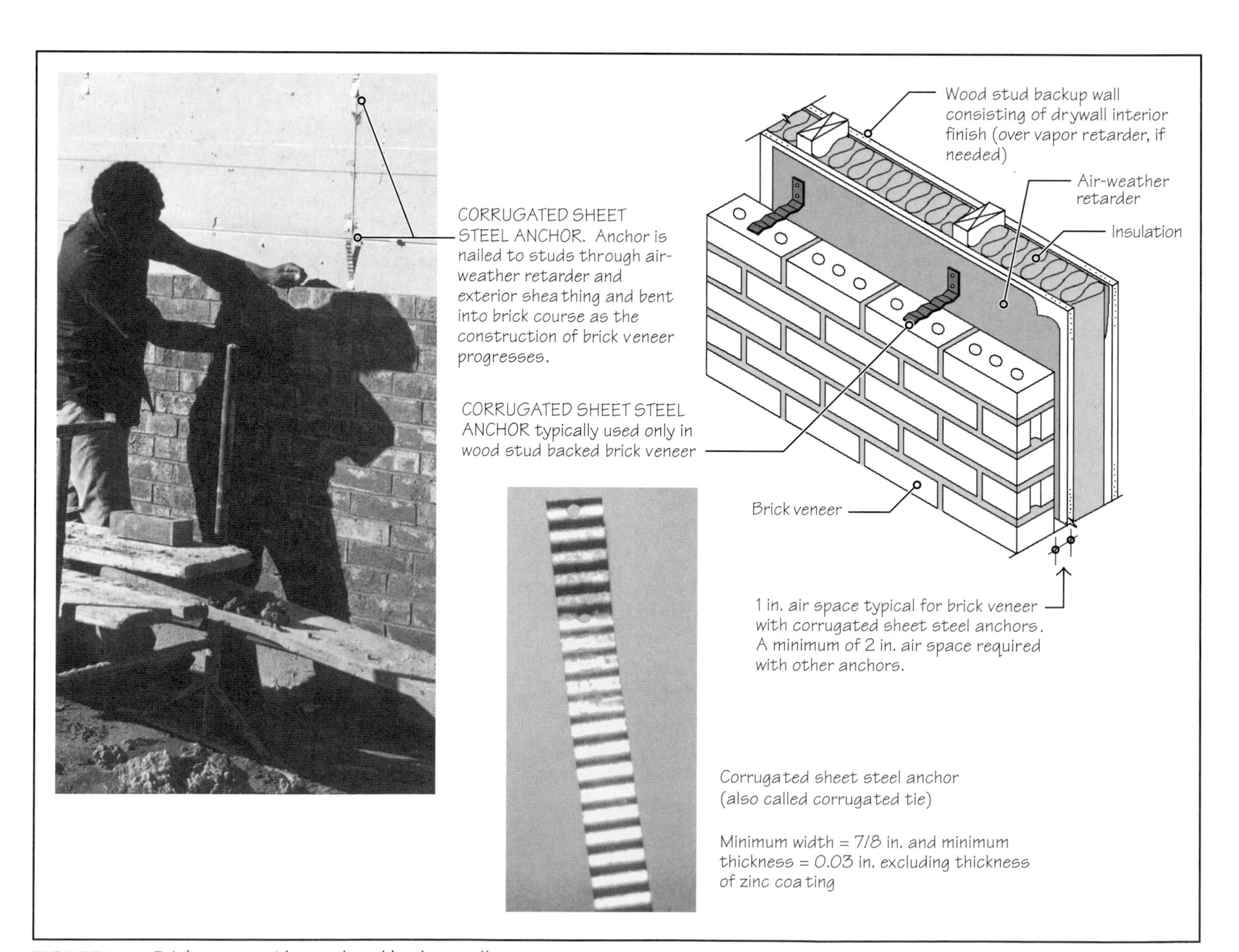

FIGURE 26.1 Brick veneer with wood stud backup wall.

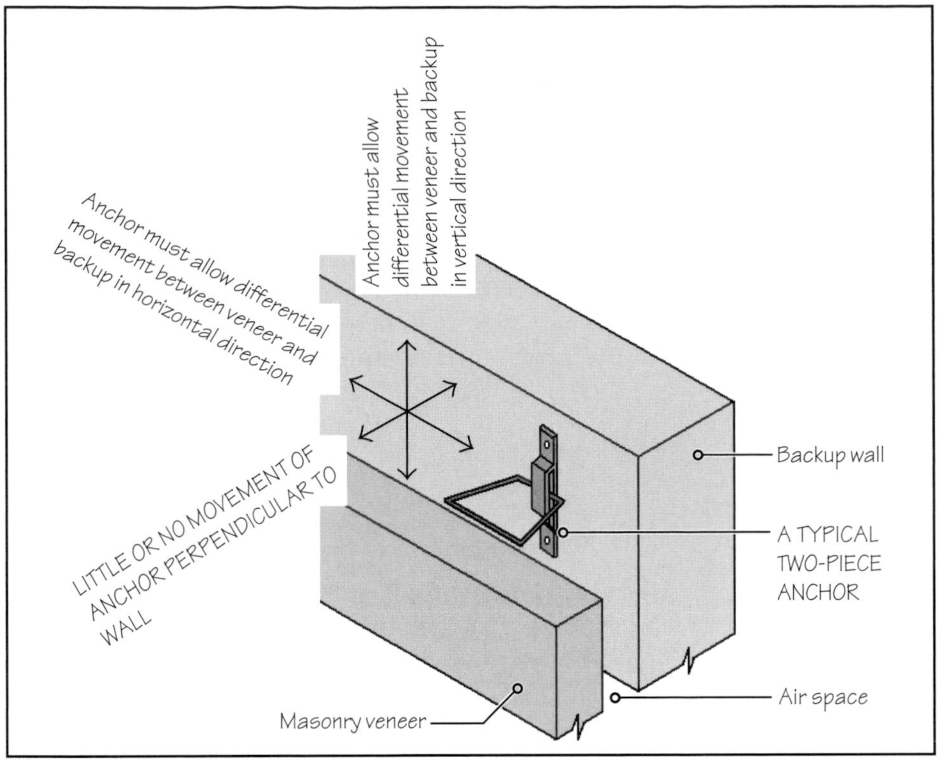

FIGURE 26.2 Adjustability requirements in various directions of a two-piece anchor.

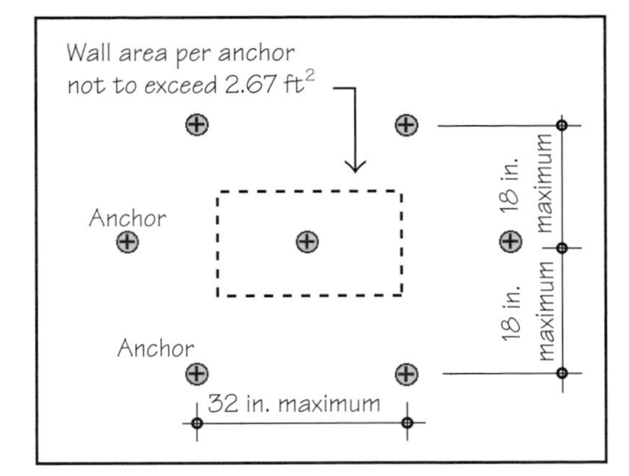

FIGURE 26.3 Anchor spacing should be determined based on the lateral load and the strength of anchors. However, for a one-piece, corrugated sheet anchor or a two-piece, adjustable wire anchor (wire size W1.7), the maximum spacing allowed is as shown. W1.7 means that the cross-sectional area of wire is 0.017 in.^{2} (Section 18.13). Anchors are generally staggered, as shown.

ANCHORS

In a brick veneer assembly, the veneer is connected to the backup wall with steel anchors, which transfer the lateral load from the veneer to the backup wall. In this load transfer, the anchors are subjected to either axial compression or tension, depending on whether the wall is subjected to inward or outward pressure.

The anchors must, therefore, have sufficient rigidity and allow little or no movement in the plane perpendicular to the wall. However, because the veneer and the backup will usually expand or contract at different rates in their own planes, the design of anchors must accommodate upward-downward and side-to-side movement, Figure 26.2.

Anchors for a brick veneer wall assembly are, therefore, made of two pieces that engage each other. One piece is secured to the backup, and the other is embedded in the horizontal mortar joints of the veneer. An adjustable, two-piece anchor should allow the veneer to move with respect to the backup in the plane of the wall, but not perpendicular to it.

An exception to this requirement is a one-piece sheet steel corrugated anchor (Figure 26.1). The corrugations in the anchor enhance the bond between the anchor and the mortar, increasing the anchor's pullout strength. But the corrugations weaken the anchor in compression by making it more prone to buckling. A one-piece, corrugated anchor is recommended for use only in low-rise, wood, light-frame buildings in low-wind and low-seismic-risk locations.

Galvanized steel is commonly used for anchors, but stainless steel is recommended where durability is an important consideration and/or where the environment is unusually corrosive.

The spacing of anchors should be calculated based on the lateral load and the strength of the anchor. However, the maximum spacing for a one-piece, corrugated anchor or an adjustable, two-piece wire anchor (wire size W1.7) is limited by the code to one anchor for every 2.67 ft^2 [1]. Additionally, they should not be spaced more than 32 in. on center horizontally and not more than 18 in. vertically, Figure 26.3.

AIR SPACE

An air space of 2 in. (clear) is recommended for brick veneer. Thus, if there is $1\frac{1}{2}$-in.-thick (rigid) insulation between the backup wall and the veneer, the backup and the veneer must be spaced $3\frac{1}{2}$ in. apart so that the air space is 2 in. clear, Figure 26.4.

In a narrower air space, there is a possibility that if the mortar squeezes out into the air space during brick laying, it may bridge over and make a permanent contact with the backup wall. A 2-in. air space reduces this possibility. In a wood-stud-backed wall assembly with one-piece corrugated anchors, however, a 1-in. air space is commonly used (Figure 26.1).

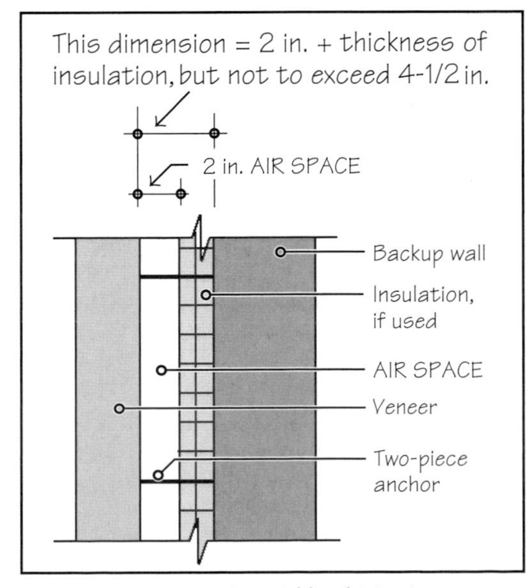

FIGURE 26.4 A cavity width of 2 in. is recommended, except in low-rise residential buildings, where a 1-in. cavity width is common. The gap between the veneer and the backup wall should not exceed $4\frac{1}{2}$ in.

The maximum distance between the veneer and backup wall is limited by the masonry code to $4\frac{1}{2}$ in. unless the anchors are specifically engineered to withstand the compressive stress caused by the lateral load. With a large gap between the veneer and the backup wall, the anchors are more prone to buckling failure.

SUPPORT FOR BRICK VENEER—SHELF ANGLES

The dead load of brick veneer may be borne by the wall foundation without any support at intermediate floors up to a maximum height of 30 ft above ground. Uninterrupted, foundation-supported veneer is commonly used in a one- to three-story wood or light-gauge steel frame buildings, Figure 26.5(a). In these buildings, the air space is continuous

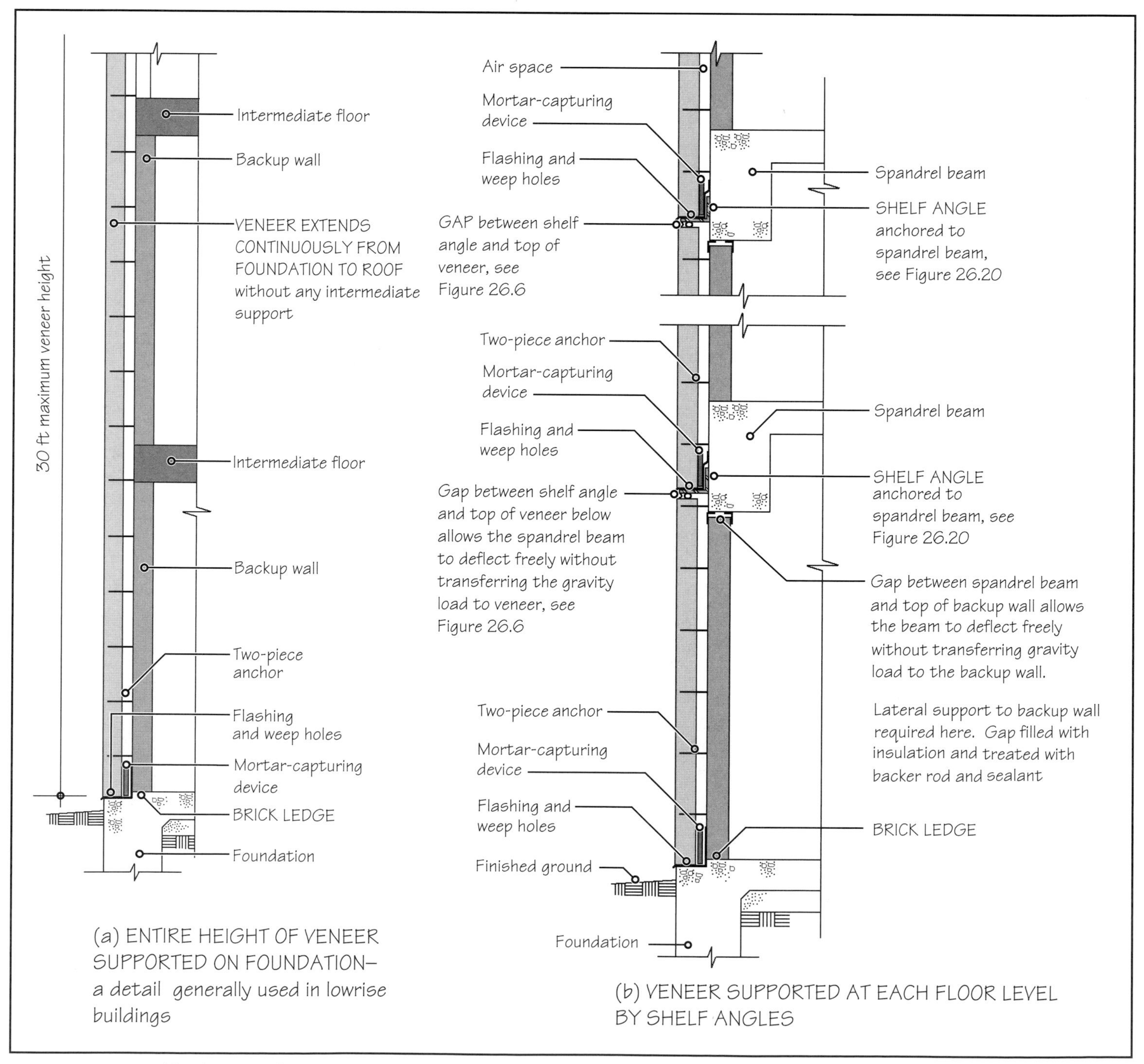

FIGURE 26.5 Dead load support for veneer.

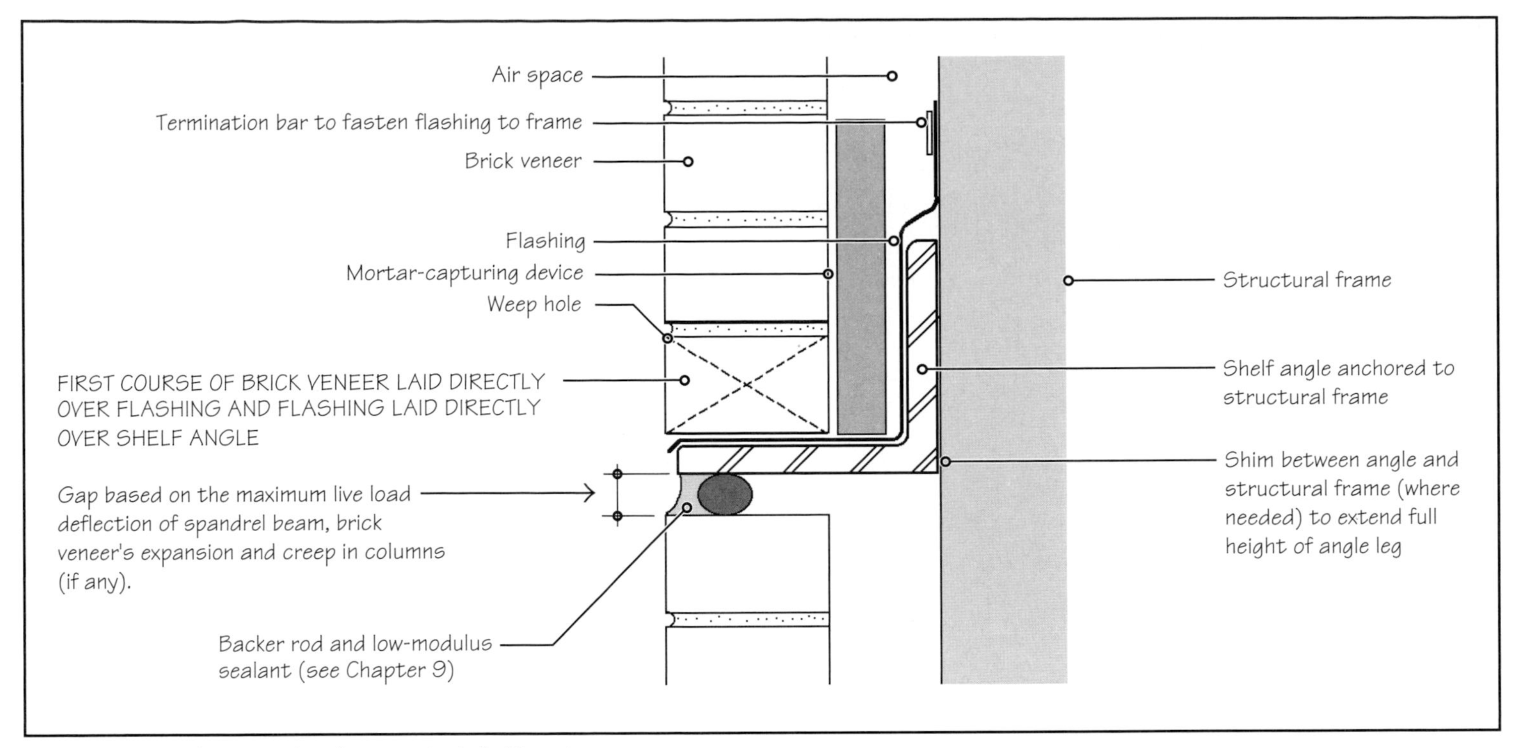

FIGURE 26.6 Schematic detail at a typical shelf angle.

from the foundation to the roof level, and the entire load of the veneer bears on the foundation. A $1\frac{1}{2}$-in. depression, referred to as the *brick ledge,* is commonly created in the foundation to receive the first course of the veneer.

In mid- and high-rise buildings, the veneer is generally supported at each floor using (preferably hot-dip galvanized) steel *shelf angles* (also referred to as *relieving angles*). Shelf angles are supported by, and anchored to, the building's structure. In a frame structure, the shelf angles are anchored (welded or bolted) to the spandrel beams, Figure 26.5(b). In a load-bearing wall structure, the shelf angles are anchored to the exterior walls. The details of the anchorage of a shelf angle to the structure are given later in the chapter.

A gap should be provided between the top of the veneer and the bottom of the shelf angle. This gap accounts for the vertical expansion of brick veneer (after construction) and the deflection of the spandrel beam under live load changes. The gap should be treated with a backer rod and sealant, Figure 26.6. The veneer may project beyond the shelf angle, but the projection should not exceed one-third of the thickness of the veneer.

Shelf angles must not be continuous. A maximum length of about 20 ft is used for shelf angles, with nearly a $\frac{3}{8}$-in. gap between adjacent lengths to provide for their expansion. The gap should ideally be at the same location as the vertical expansion joints in the veneer.

NOTE

Brick Ledge

A brick ledge refers to the depression in concrete foundation at the base of brick veneer and is generally $1\frac{1}{2}$ in. deep, because it is formed by a 2-by lumber used as a blockout when placing concrete (Figure 26.5). The depression further prevents water intrusion into the backup wall.

LINTEL ANGLES—LOOSE-LAID

Whether the veneer is supported entirely on the foundation or at each floor, additional dead load support for the veneer is needed over wall openings. The lintels generally used over an opening in brick veneer are of steel (preferably hot-dip galvanized) angles, Figure 26.7. Unlike the shelf angles, lintel angles are not anchored to the building's structural frame but are simply placed (loose) on the veneer, Figure 26.8.

To allow the lintel to move horizontally relative to the brick veneer, no mortar should be placed between the lintel bearing and the brick veneer. Flashing and weep holes must be provided over lintels in exactly the same way as on the shelf angles.

LOCATIONS OF FLASHINGS AND END DAMS

As shown in Figure 26.7, flashings must be provided at all interruptions in the brick veneer:

- At foundation level
- Over a shelf angle
- Over a lintel angle
- Under a window sill

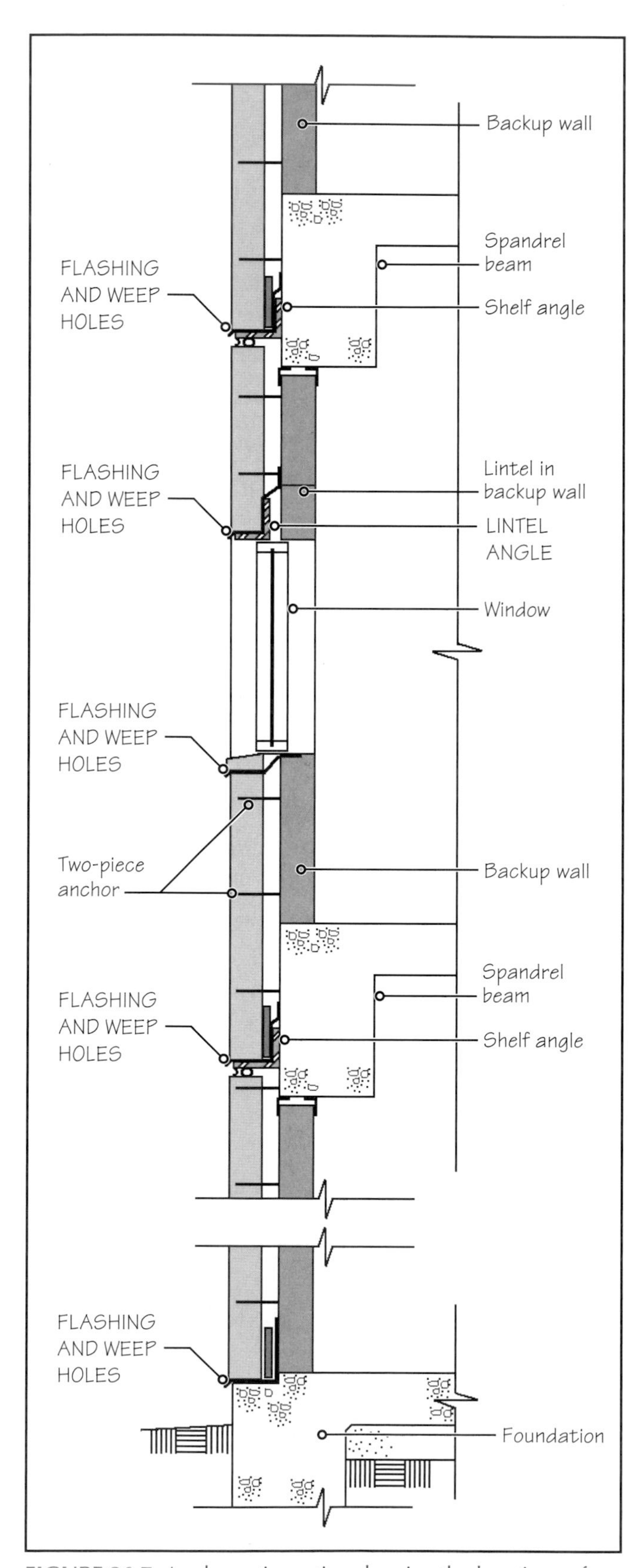

FIGURE 26.7 A schematic section showing the locations of shelf angles and lintel angles in a typical brick veneer wall assembly.

The backup wall in this building consists of metal studs with gypsum sheathing. The tape covers the joints between sheathing panels. If an air-weather retarder was used, taping of joints would be unnecessary.

Lintel angle is simply placed on the veneer with no anchorage to building's structure. No mortar bed should be used under lintel bearing.

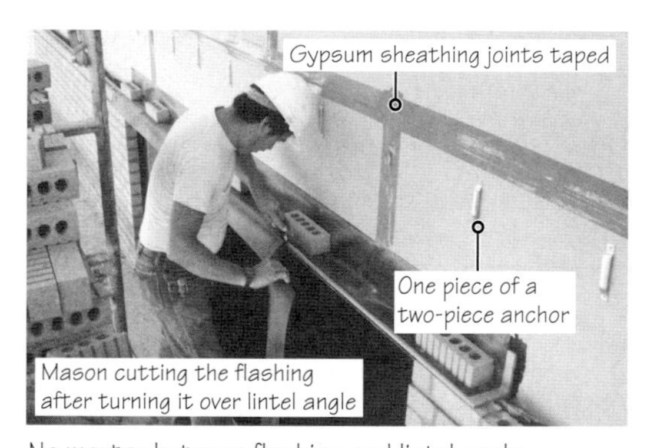

No mortar between flashing and lintel angle, see Figure 26.6.

The first course of veneer over flashing is laid without any mortar bed between bricks and flashing, see Figure 26.6.

FIGURE 26.8 Lintel angles in a brick veneer wall.

FIGURE 26.9 End dams in flashing. (Photo courtesy of Brick Industry Association, Reston, Virginia.)

Joints between flashings must be sealed, and all flashings must be accompanied by weep holes. The flashing should preferably project out of the veneer face to ensure that the water will drain to the outside of the veneer (Figure 26.6). Where the flashing terminates, it must be turned up (equal to the height of one brick) to form a dam to prevent water from entering the air space, Figure 26.9.

Flashing Materials

Flashing material must be impervious to water and resistant to puncture, tear, and abrasion. Additionally, flashing must be flexible so that it can be bent to the required profile. Durability is also important because replacing failed flashing is cumbersome and expensive. Therefore, metal flashing must be corrosion resistant. Resistance to ultraviolet radiation is also necessary because the projecting part of the flashing is exposed to the sun. Commonly used flashing materials are as follows:

- Stainless sheet steel
- Copper sheet
- Plastics such as
  - Polyvinyl chloride (PVC)
  - Neoprene
  - Ethylene propylene diene monomer (EPDM)
- Composite flashing consisting of
  - Rubberized asphalt with cross-laminated polyethylene, typically available as self-adhering and self-healing flashing
  - Copper sheet laminated on both sides to asphalt-saturated paper or fiberglass felt

Copper and stainless steel are among the most durable flashing materials. Copper's advantage over stainless steel is its greater flexibility, which allows it to be bent to shape more easily. However, copper will stain light colored masonry because of its corrosion, which yields a greenish protective cover (patina). Copper combination flashing, consisting of a copper sheet laminated to asphalt-saturated paper, reduces its staining potential.

A fairly successful flashing is a two-part flashing, comprising a self-adhering, self-healing polymeric membrane and stainless steel drip edge. The durability and rigidity of stainless steel makes a good drip edge, and the flexible, self-adhering membrane simplifies flashing installation.

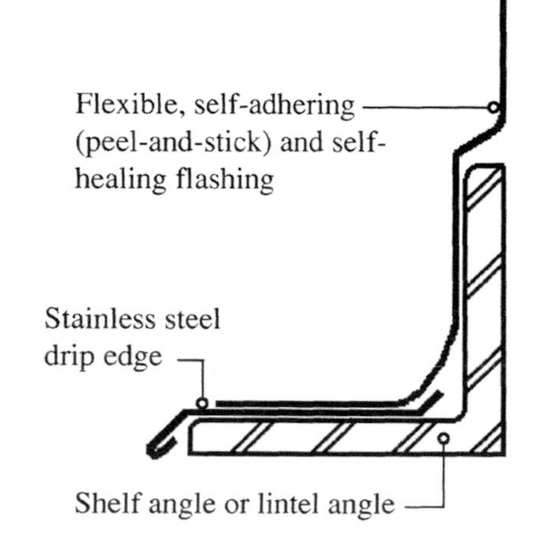

Construction and Spacing of Weep Holes

Weep holes must be provided immediately above the flashing. There are several different ways to provide weep holes. The simplest and the most effective weep hole is an open, vertical mortar joint (open-head joint) in the veneer, Figure 26.10.

FIGURE 26.10 Open head joints as weep holes.

NOTE

Weep Hole Spacing

A center-to-center spacing of 24 in. is generally used for weep holes when open-head joints are used. A spacing of 16 in. is used with weep holes consisting of wicks or tubes.

To prevent insects and debris from lodging in the open-head joint, joint screens may be used. A joint screen is an L-shaped, sheet metal or plastic element, Figure 26.11(a). Its vertical leg has louvered openings to let the water out, and the horizontal leg is embedded into the horizontal mortar joint of the veneer. The joint screen has the same width as the head joints. An alternative honeycombed plastic joint screen is also available, Figure 26.11(b).

Instead of the open-head joint, wicks or plastic tubes ($\frac{3}{8}$-in. diameter) may be used in a mortared-head joint. Wicks, which consist of cotton ropes, are embedded in head joints, Figure 26.12. They absorb water from the air space by capillary action and drain it to the outside. Their drainage efficiency is low.

Plastic tubes are better than wicks, but they do not function as well as open-head joints. They are placed in head joints with a rope inside each tube. The ropes are pulled out after the veneer has been constructed. This ensures that the air spaces of the tubes are not clogged by mortar droppings.

A sufficient number of weep holes must be provided for the drainage of the air space. Generally, a weep hole spacing of 24 in. is used with open-head joints; 16-in. spacing is used with wicks or tubes.

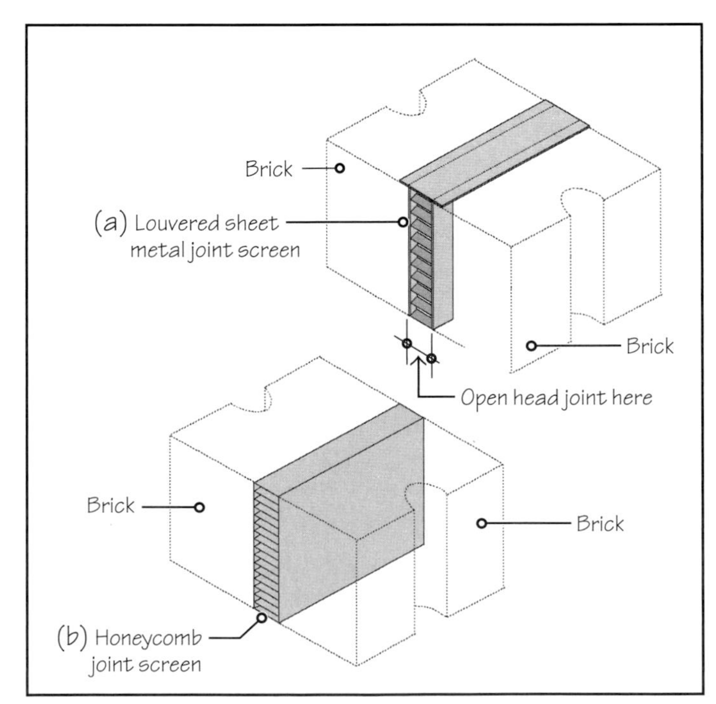

FIGURE 26.11 Two alternative screens used with weep holes consisting of open-head joints.

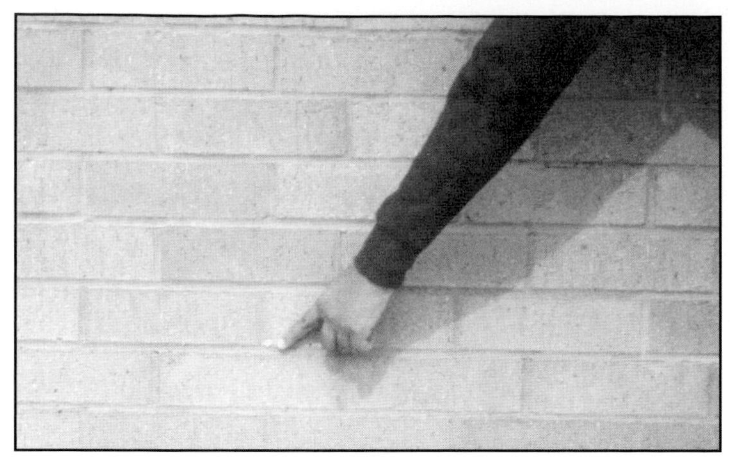
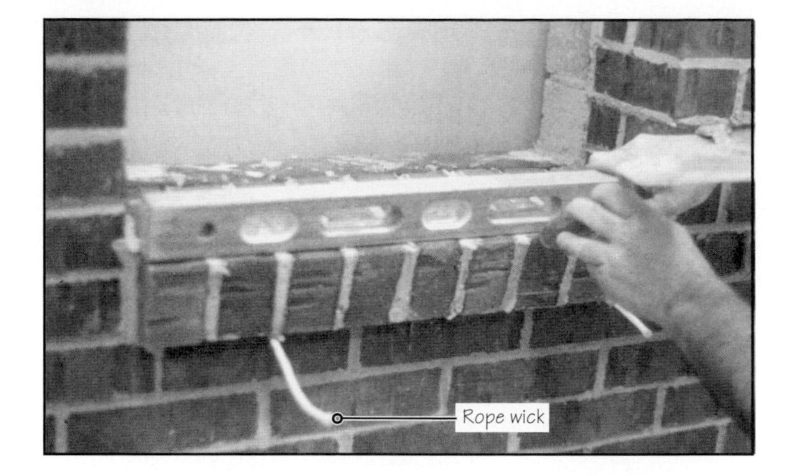

FIGURE 26.12 Wicks as weep holes.

Mortar Droppings in the Air Space—Mortar-Capturing Device

For the air space to function as an effective drainage layer, it is important to minimize mortar droppings in the air space. Excessive buildup of mortar in the air space bridges the space. Additionally, the weep holes function well only if they are not clogged by mortar droppings. Poor bricklaying practice can result in substantial accumulations of mortar on the flashing. Care in bricklaying to reduce mortar droppings is therefore essential.

Additional measures must also be incorporated to keep the air space unclogged. An earlier practice was to use a 2-in.-thick bed of pea gravel over the flashing. This provides a drainage bed that allows the water to percolate to the weep holes.

A better alternative is to use a *mortar-capturing device* in the air space immediately above the flashing. This device consists of a mesh made of polymeric strands, which trap the droppings and suspend them permanently above the weep holes, Figure 26.13. The use of a mortar-capturing device allows water in the air space to percolate freely through mortar droppings to reach the weep holes.

Continuous Vertical Expansion Joints in Brick Veneer

As stated in Section 9.8, brick walls expand after construction. Therefore, a brick veneer must be provided with continuous vertical expansion joints at intervals, Figure 26.14.

The maximum recommended spacing for vertical expansion joints is 30 ft in the field of the wall and not more than 10 ft from the wall's corner [2]. The joints are detailed so that sealant and backer rods replace mortar joints for the entire length of the continuous vertical expansion joint, allowing the bricks on both sides of the joint to move while maintaining a waterproof seal, Figure 26.15. The width of the expansion joint is $\frac{3}{8}$ in. (minimum) to match the width of the mortar joints.

With vertical expansion joints and the gaps under shelf angles (which function as horizontal expansion joints), a brick veneer essentially consists of individual brick panels that can expand and contract horizontally and vertically without stressing the backup wall or the building's structure.

Mortar Type and Mortar Joint Profile

Type N mortar is generally specified in all-brick veneer except in seismic zones, where Type S mortar may be used (see Section 22.2). A concave joint profile yields veneer with more water resistance (see Section 22.3).

(a) Mortar capturing device, as installed in air space.

(b) Mortar capturing device with mortar droppings, which allows the water to find its way to the weep holes even with the droppings.

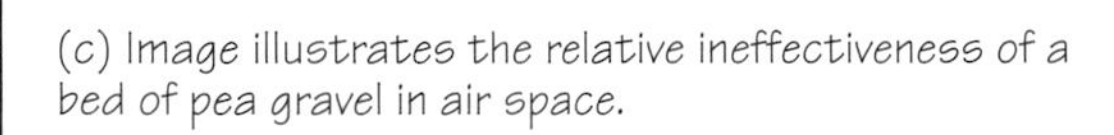

(c) Image illustrates the relative ineffectiveness of a bed of pea gravel in air space.

FIGURE 26.13 Mortar-capturing device. (Photos courtesy of Mortar Net USA Ltd., producers of The Mortar Net™.)

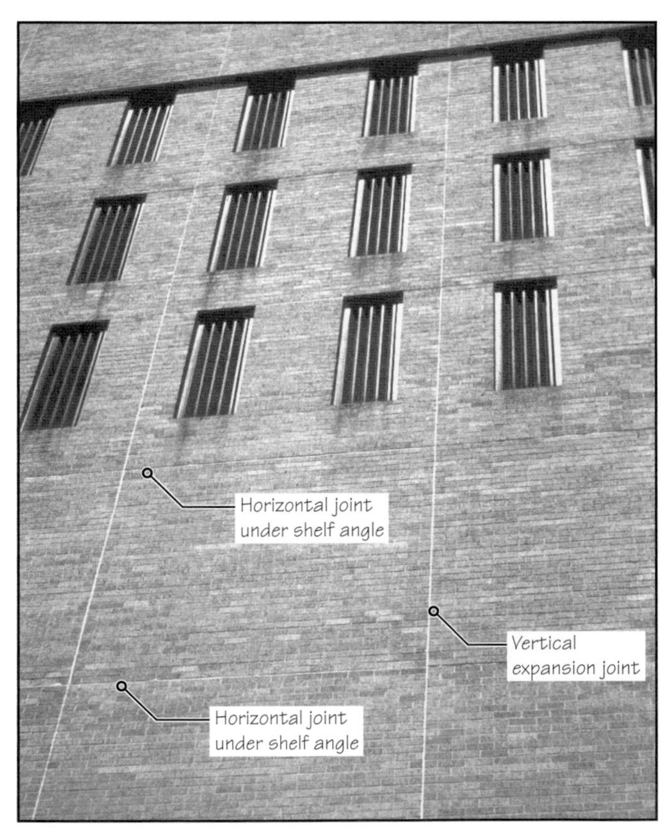

FIGURE 26.14 Continuous vertical expansion joints in brick veneer. Note continuous horizontal joints under shelf angles.

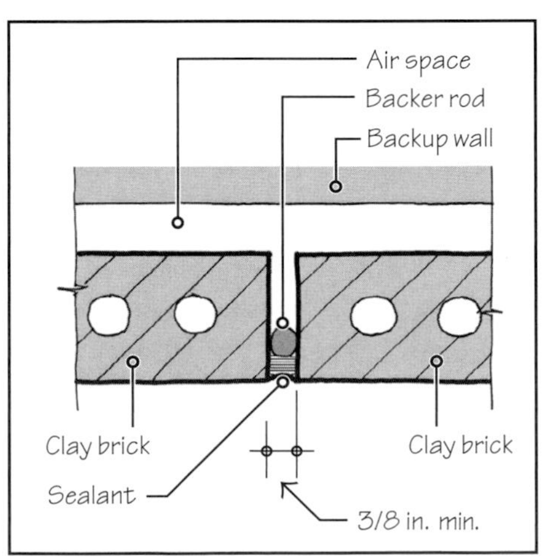

FIGURE 26.15 Detail plan of a vertical expansion joint in brick veneer. This illustration is the same as Figure 9.23.

Each question has only one correct answer. Select the choice that best answers the question.

1. The backup wall in a brick veneer wall assembly consists of a
- **a.** reinforced concrete wall.
- **b.** CMU wall.
- **c.** wood or steel stud wall.
- **d.** all the above.
- **e.** (b) and (c) only.

2. In a brick veneer wall assembly, the wind loads are transferred directly from the veneer to the building's structure.
- **a.** True
- **b.** False

3. The anchors used to anchor the brick veneer to the backup wall are generally two-piece anchors to allow differential movement between the veneer and the backup
- **a.** in all three principal directions.
- **b.** perpendicular to the plane of the veneer.
- **c.** within the plane of the veneer.
- **d.** none of the above.

4. The anchors in a brick veneer wall assembly provide
- **a.** gravity load support to both veneer and backup.
- **b.** lateral load support to both veneer and backup.
- **c.** gravity load support to the veneer.
- **d.** lateral load support to the veneer.

5. The minimum required width of air space between brick veneer and CMU backup wall is
- **a.** 1 in.
- **b.** $1\frac{1}{2}$ in.
- **c.** 2 in.
- **d.** 3 in.
- **e.** none of the above.

6. The minimum width of air space generally used between brick veneer and wood stud backup wall is
- **a.** 1 in.
- **b.** $1\frac{1}{2}$ in.
- **c.** 2 in.
- **d.** $2\frac{1}{2}$ in.
- **e.** 3 in.

7. A steel angle used to support the weight of brick veneer over an opening is called a
- **a.** Lintel angle.
- **b.** Shelf angle.
- **c.** Relieving angle.
- **d.** all the above.
- **e.** (a) or (b).

8. A shelf angle must be anchored to the building's structural frame.
- **a.** True
- **b.** False

9. A lintel angle must be anchored to the building's structural frame.
- **a.** True
- **b.** False

10. In a multistory building, shelf angles are typically used at
- **a.** each floor level.
- **b.** each floor level and at midheight between floors.
- **c.** at foundation level.
- **d.** all the above.
- **e.** (a) and (c).

11. For a brick veneer that bears on the foundation and continues to the top of the building without any intermediate support, the maximum veneer height is limited to
- **a.** 40 ft.
- **b.** 35 ft.
- **c.** 30 ft.
- **d.** 25 ft.
- **e.** 20 ft.

12. A shelf angle in a brick veneer assembly must provide
- **a.** gravity load support to the veneer.
- **b.** lateral load support to the veneer.
- **c.** gravity load support to both veneer and backup.
- **d.** lateral load support to both veneer and backup.

13. In a brick veneer assembly, flashing is required
- **a.** at foundation level.
- **b.** over a lintel angle.
- **c.** over a shelf angle.
- **d.** under a window sill.
- **e.** all the above.

14. In a brick veneer assembly, weep holes are required at
- **a.** each floor level.
- **b.** each alternate floor level.
- **c.** immediately above the flashing.
- **d.** immediately below the flashing.
- **e.** 2 in. above a flashing.

15. The most efficient weep hole in a brick veneer consists of a
- **a.** Wick.
- **b.** Plastic tube.
- **c.** Open head joint.
- **d.** None of the above.

16. A mortar capturing device in a brick veneer assembly is used
- **a.** at each floor level.
- **b.** at each alternate floor level.
- **c.** immediately above a flashing.
- **d.** immediately below a flashing.
- **e.** 2 in. above a flashing.

17. In a brick veneer assembly, vertical expansion joints should be provided at a maximum distance of
- **a.** 40 ft in the field of the wall and 40 ft from a wall's corner.
- **b.** 30 ft in the field of the wall and 30 ft from a wall's corner.
- **c.** 30 ft in the field of the wall and 20 ft from a wall's corner.
- **d.** 30 ft in the field of the wall and 10 ft from a wall's corner.

Answers: 1-d, 2-b, 3-c, 4-d, 5-c, 6-a, 7-a, 8-a, 9-b, 10-a, 11-c, 12-a, 13-e, 14-c, 15 c, 16-c, 17-d.

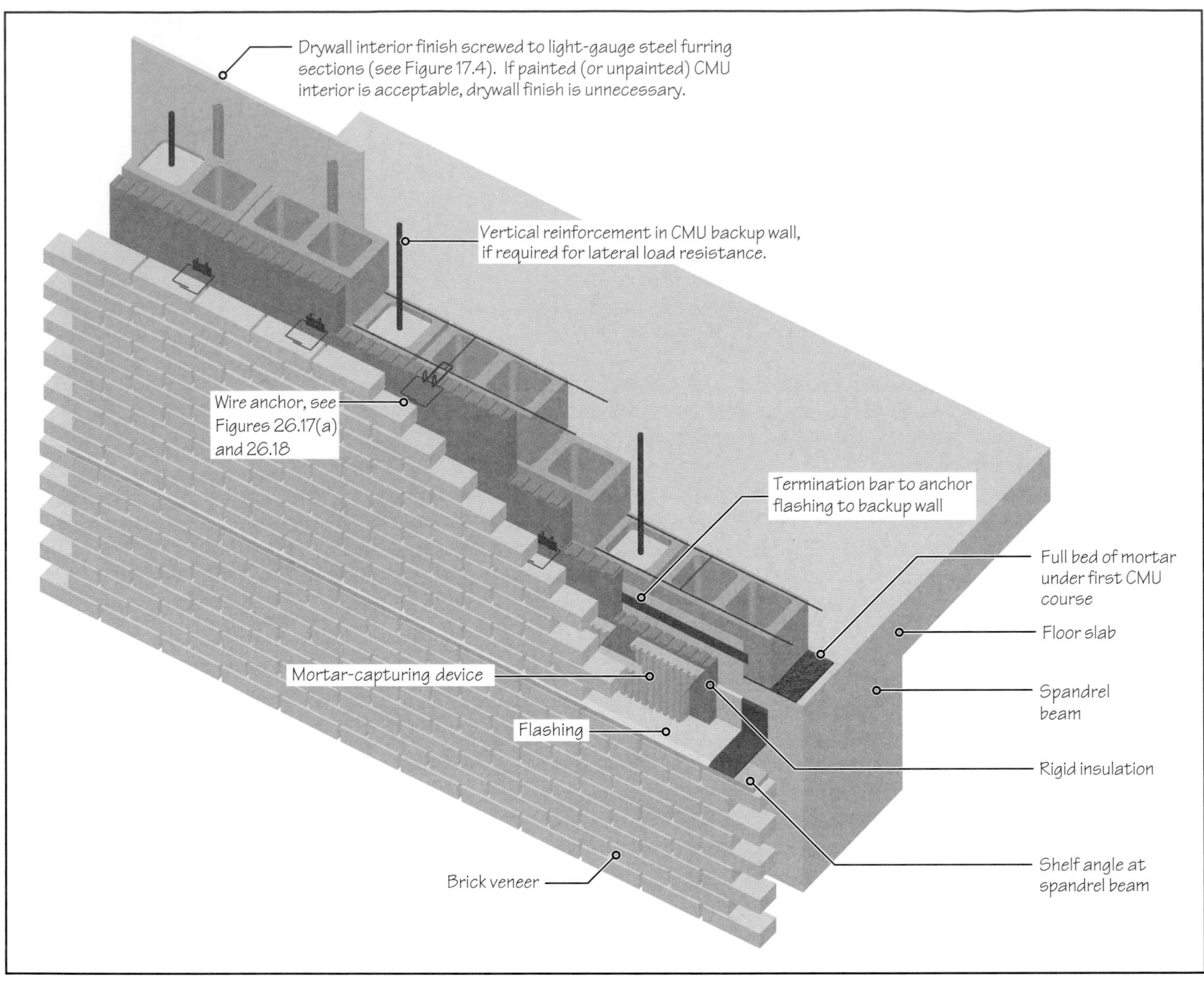

FIGURE 26.16 Brick veneer with CMU backup.

26.2 BRICK VENEER WITH A CMU OR CONCRETE BACKUP WALL

Figure 26.16 shows an overall view of a brick veneer wall assembly with a concrete masonry backup wall. The steel anchors that connect the veneer to the CMU backup wall are two-piece wire anchors. One piece is part of the joint reinforcement embedded in the CMU walls, see Figure 26.17(a). The other piece fits into this piece and is embedded into the veneer's bed joint.

Several other types of anchors used with a concrete masonry backup wall are available, such as that shown in Figure 26.17(b). Figure 26.17(c) and (d) shows typical anchors used with reinforced-concrete members.

In seismic regions, the use of seismic clips is recommended. A seismic clip engages a continuous wire reinforcement in brick veneer. Both the seismic clip and wire are embedded in the veneer's bed joint. A typical seismic clip is shown in Figure 26.18. [See Figure 26.17(d)

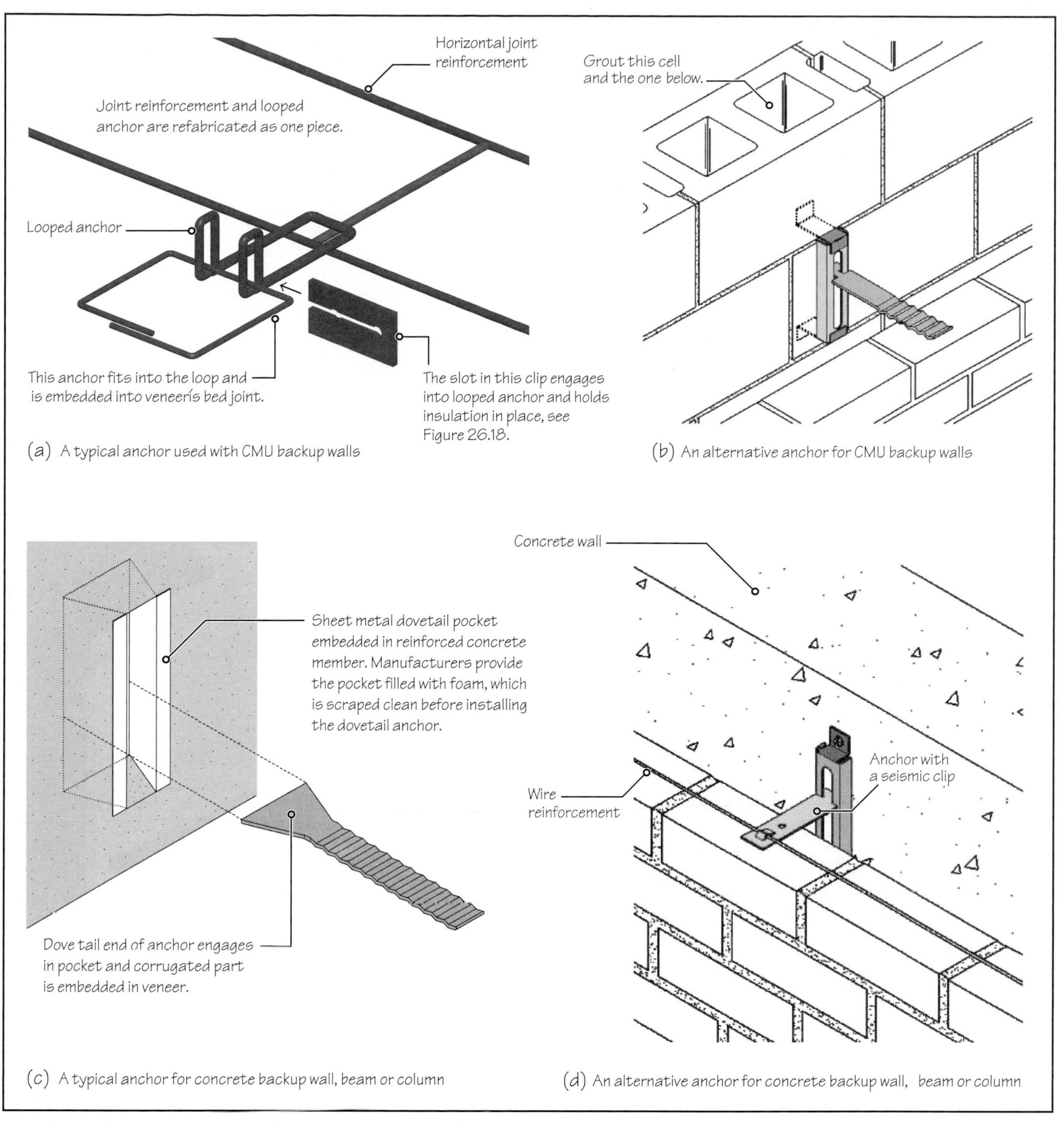

FIGURE 26.17 Typical anchors used with CMU and concrete backup walls.

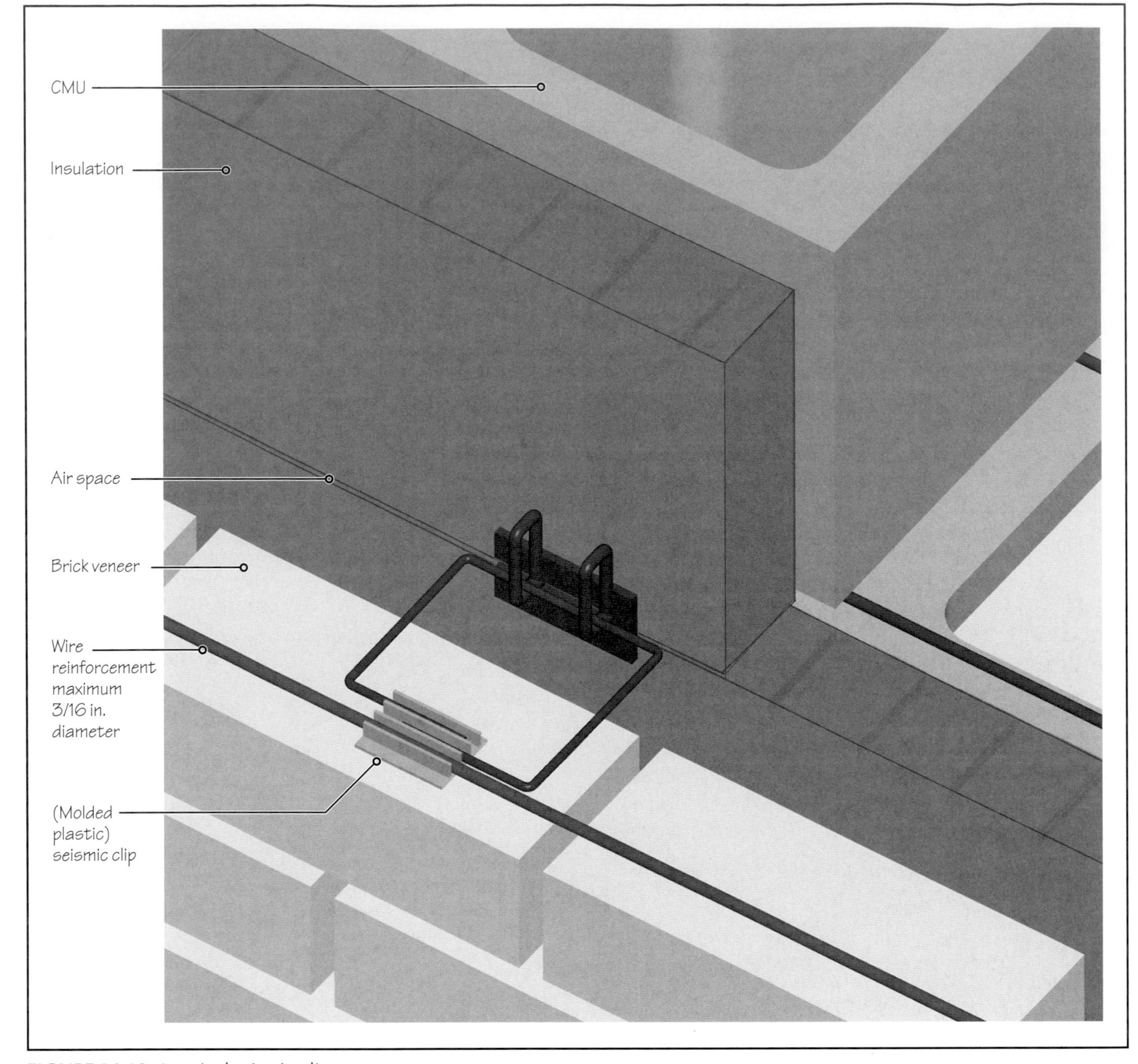

FIGURE 26.18 A typical seismic clip.

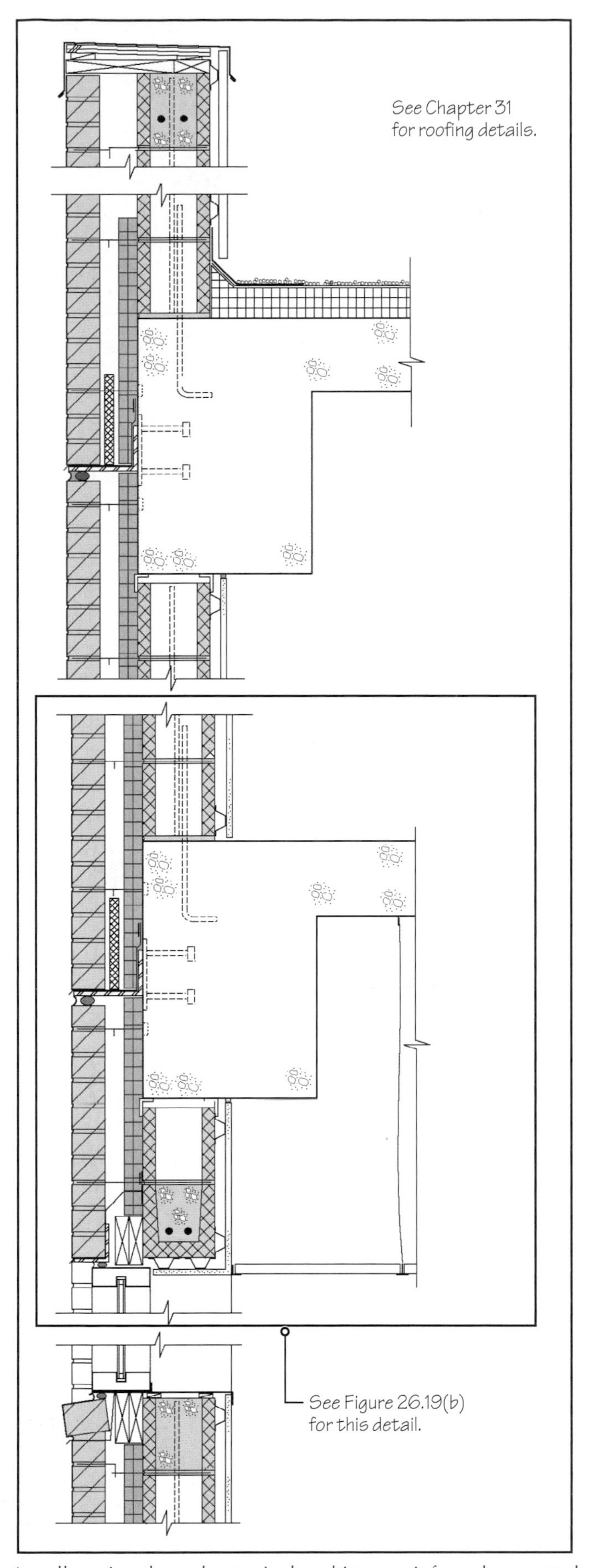

FIGURE 26.19(a) A wall section through a typical multistory reinforced concrete building with brick veneer and CMU backup wall.

for an alternative seismic clip.] Figures 26.19 to 26.22 show important details of brick veneer construction and a CMU backup wall.

The shelf angle and lintel angles can be combined into one by increasing the depth of the spandrel beam down to the level of the window head, Figure 26.23—a strategy that is commonly used with ribbon windows and continuous brick veneer spandrels, Figure 26.24.

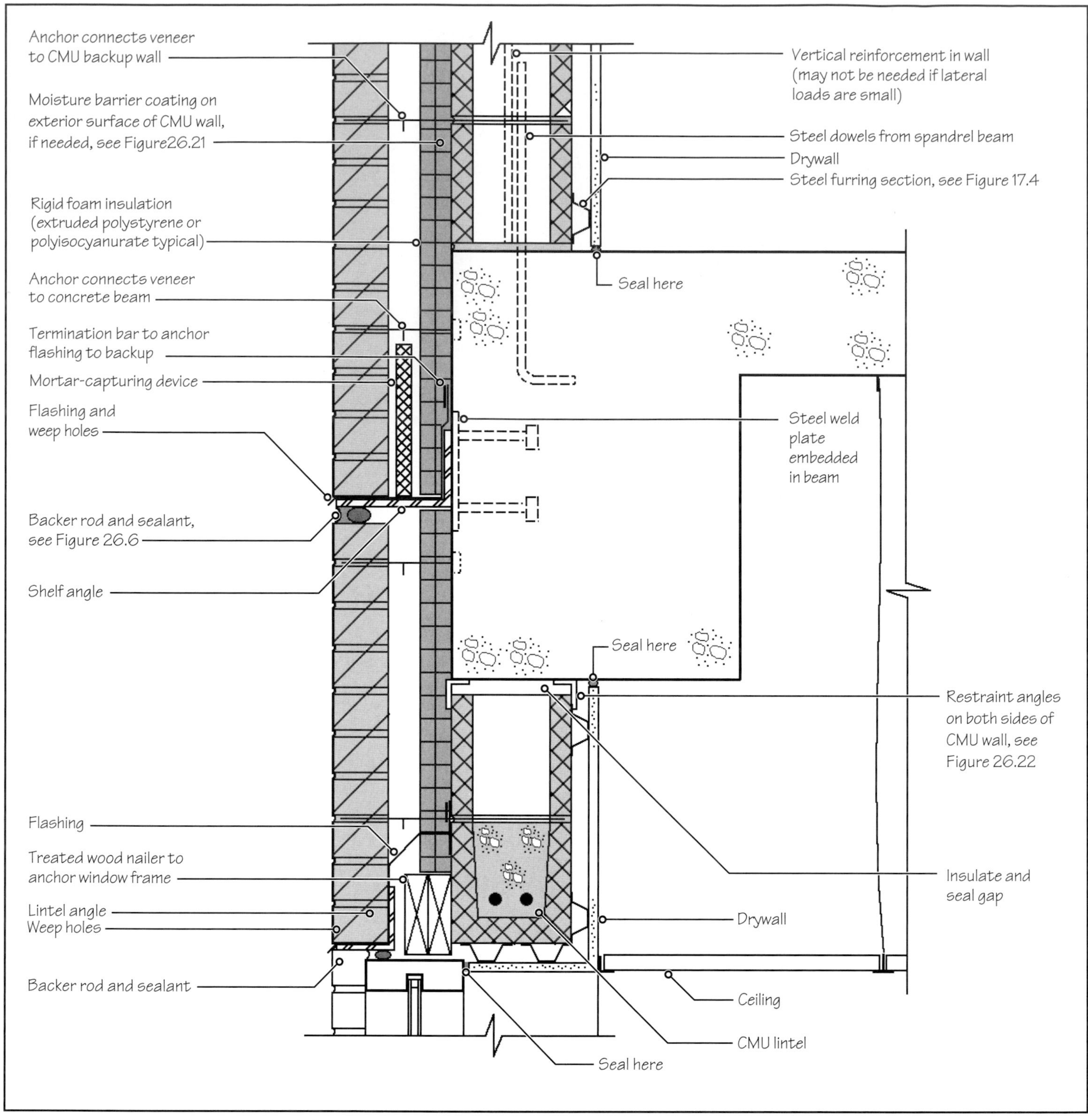

FIGURE 26.19(b) Detail of Figure 26.19(a).

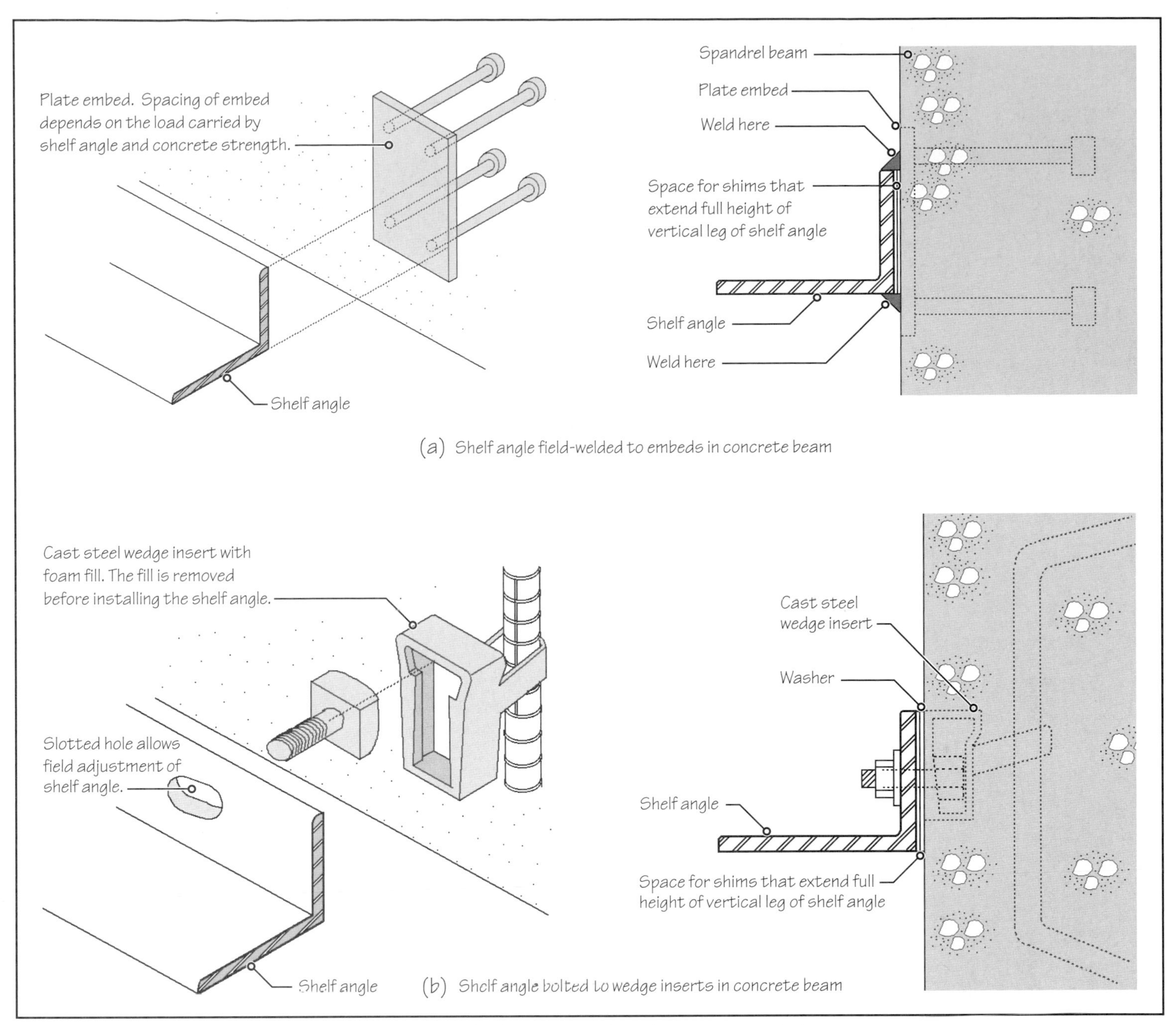

FIGURE 26.20 Two alternative methods of anchoring a shelf angle to reinforced concrete spandrel beam.

FIGURE 26.21 The exterior surface of a CMU backup wall may be treated with a water-resistant coating before constructing brick veneer.

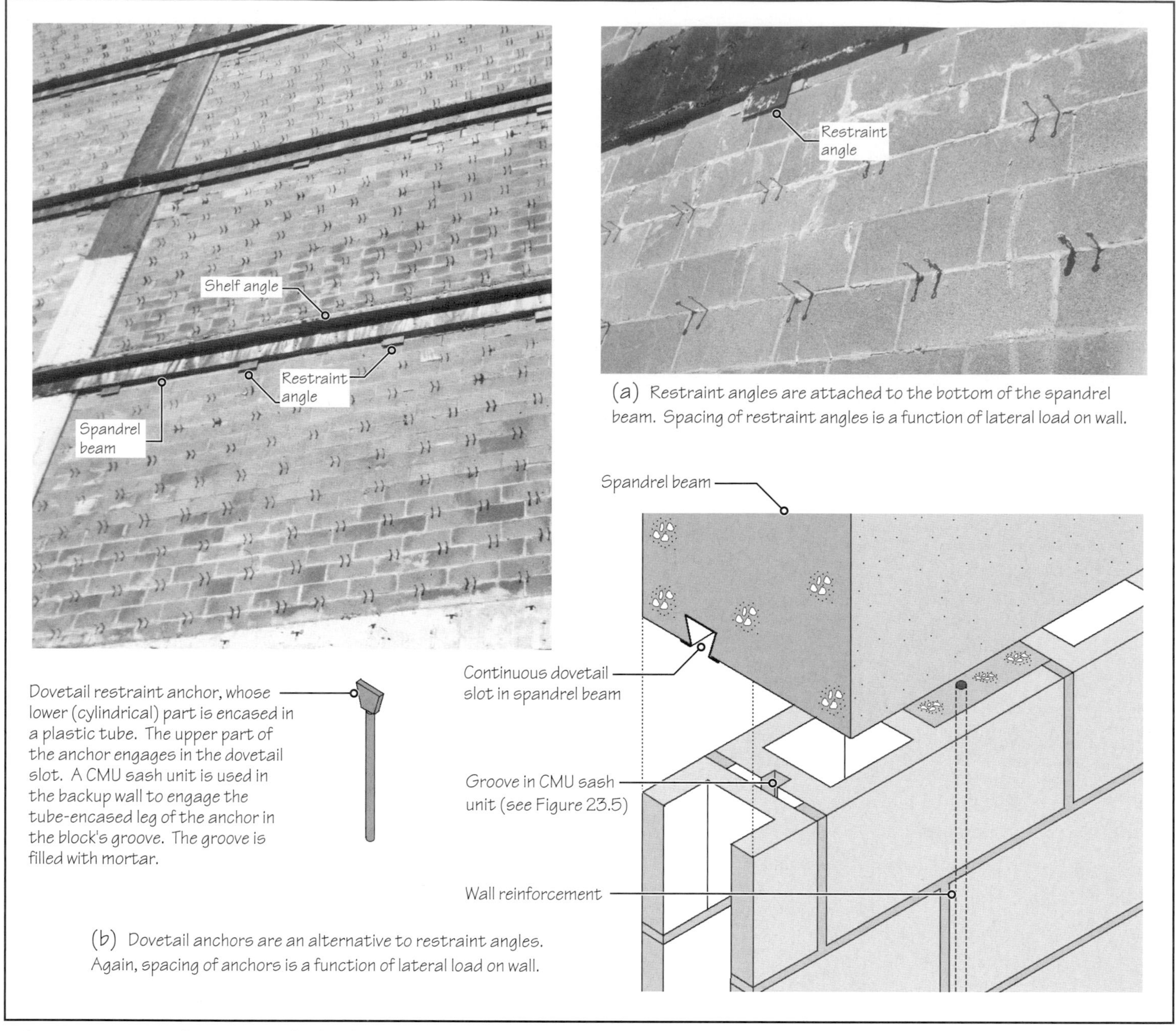

FIGURE 26.22 Two alternative methods of providing lateral load restraint at the top of a CMU backup wall.

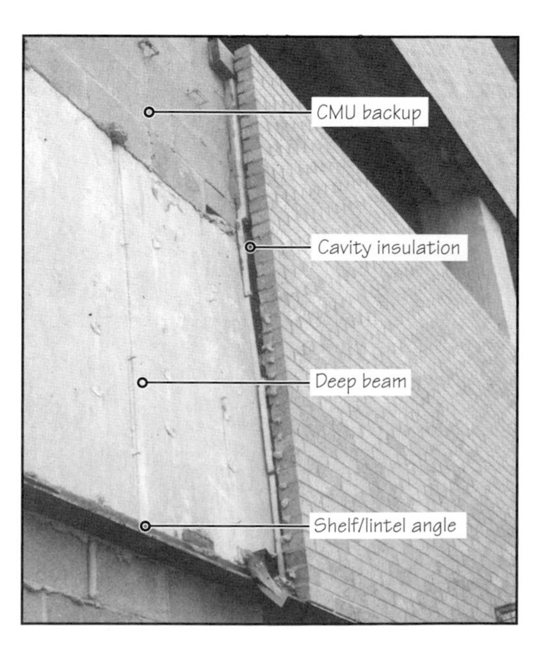

FIGURE 26.23 (Left) With a deep spandrel beam that extends from the top of window head in the lower floor to the sill level of the window at the upper floor, the shelf angle also serves as lintel angle.

FIGURE 26.24 (Right) A building with brick veneer spandrels and ribbon windows.

26.3 BRICK VENEER WITH A STEEL STUD BACKUP WALL

The construction of brick veneer with a steel stud backup wall differs from that of a CMU-backed wall mainly in the anchors used for connecting the veneer to the backup. Various types of anchors are available to suit different conditions. The anchor shown in Figure 26.25 is used if the air space does not contain rigid foam insulation so that it is fastened to steel studs through exterior sheathing.

The anchor shown in Figure 26.26 is used if rigid foam insulation is present in the air space. The sharp ends of the prongtype anchor pierce into the insulation (not the sheathing) and transfer lateral load to the studs without compressing the insulation.

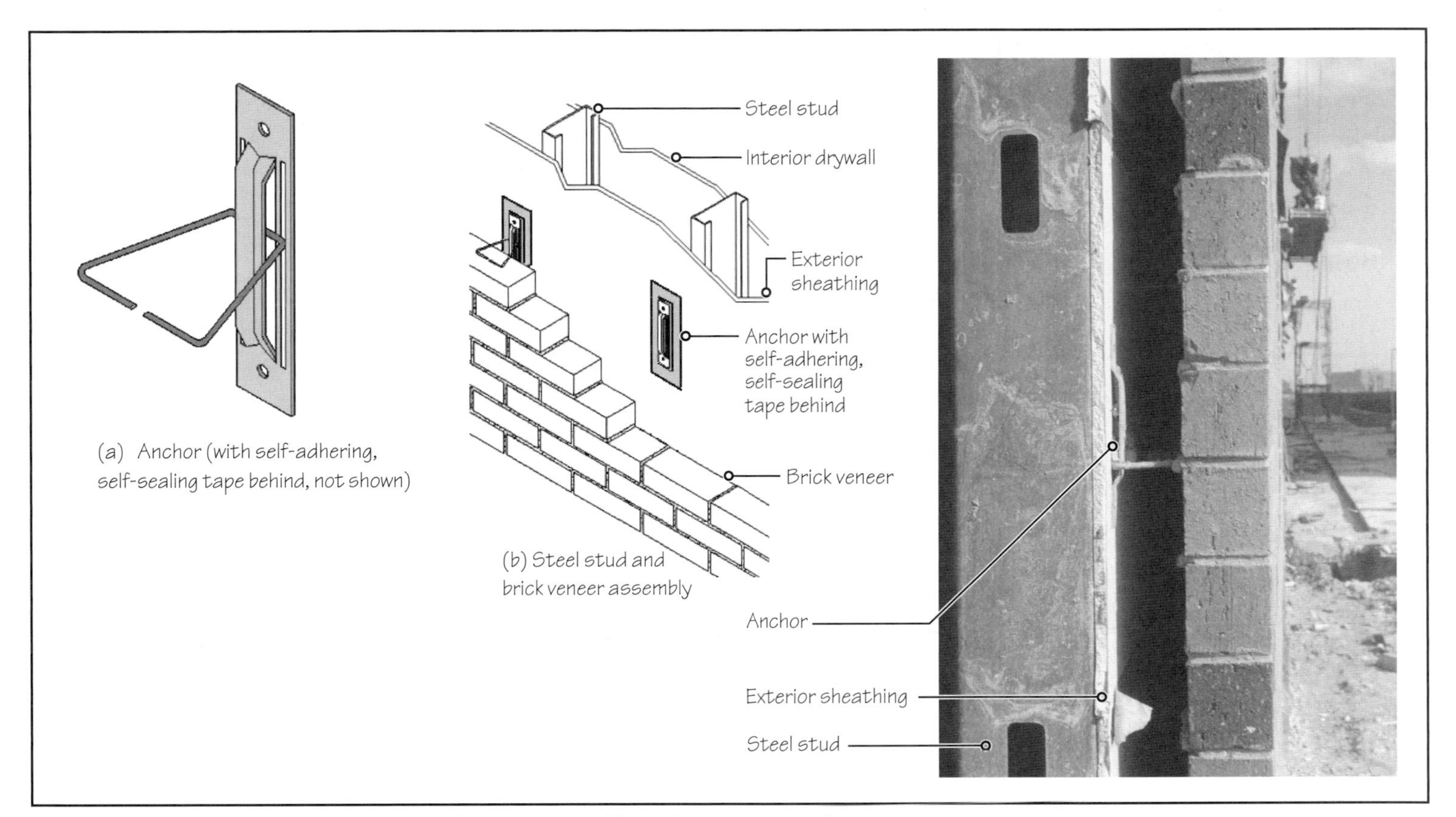

FIGURE 26.25 A typical steel stud and brick veneer assembly without cavity insulation.

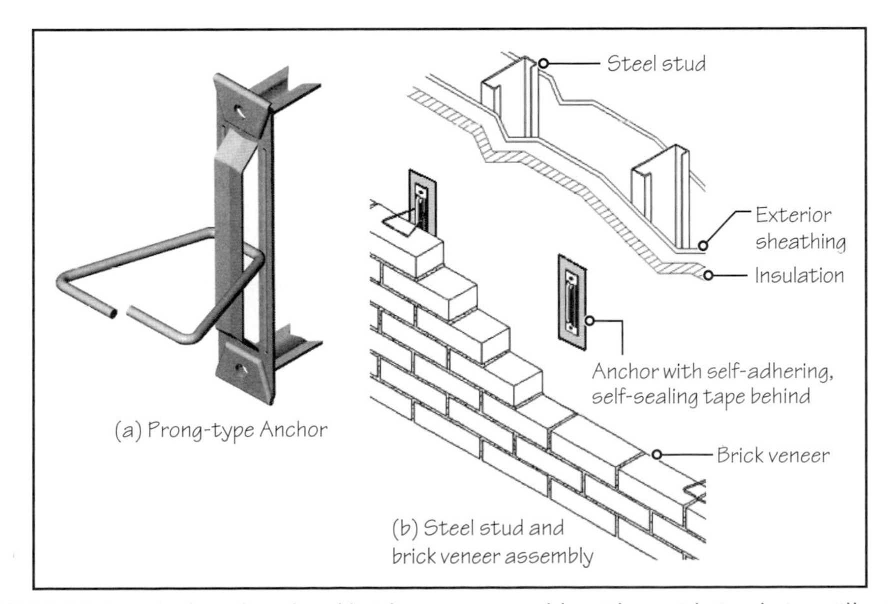

FIGURE 26.26 A typical steel stud and brick veneer assembly with outside insulation. (Illustration courtesy of Hohmann and Barnard, Inc.)

STEEL STUD BACKUP WALL AS INFILL WITHIN THE STRUCTURAL FRAME

Figure 26.27 shows a detailed section of brick veneer applied to a steel stud backup wall with the reinforced concrete structural frame. The deflection of the spandrel beam is accommodated by providing a two-track assembly consisting of a deep-leg track and a normal track, Figure 26.28(a). The upper track of this slip assembly is fastened to the beam. The studs and the interior drywall are fastened only to the lower track, which allows the upper track to slide over the lower track.

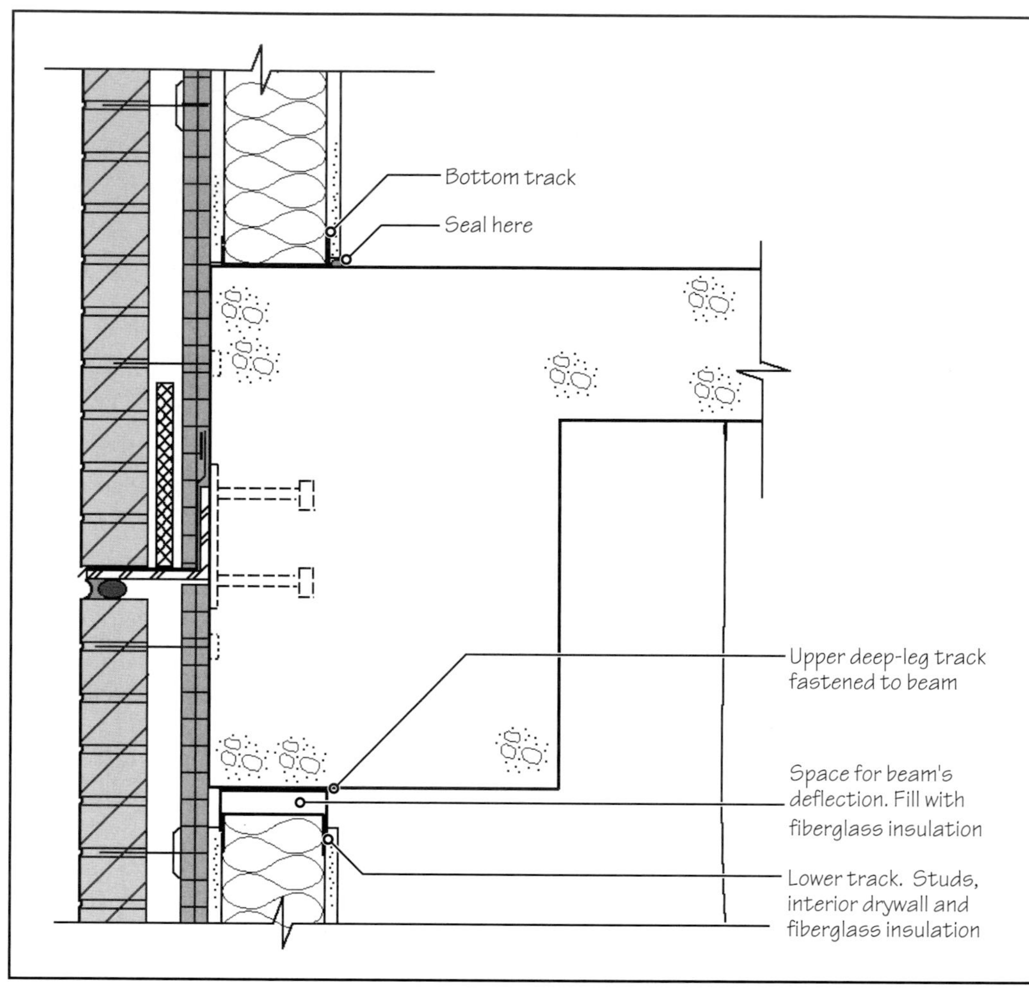

FIGURE 26.27 Detail of a typical steel stud and brick veneer assembly in a reinforced concrete structure.

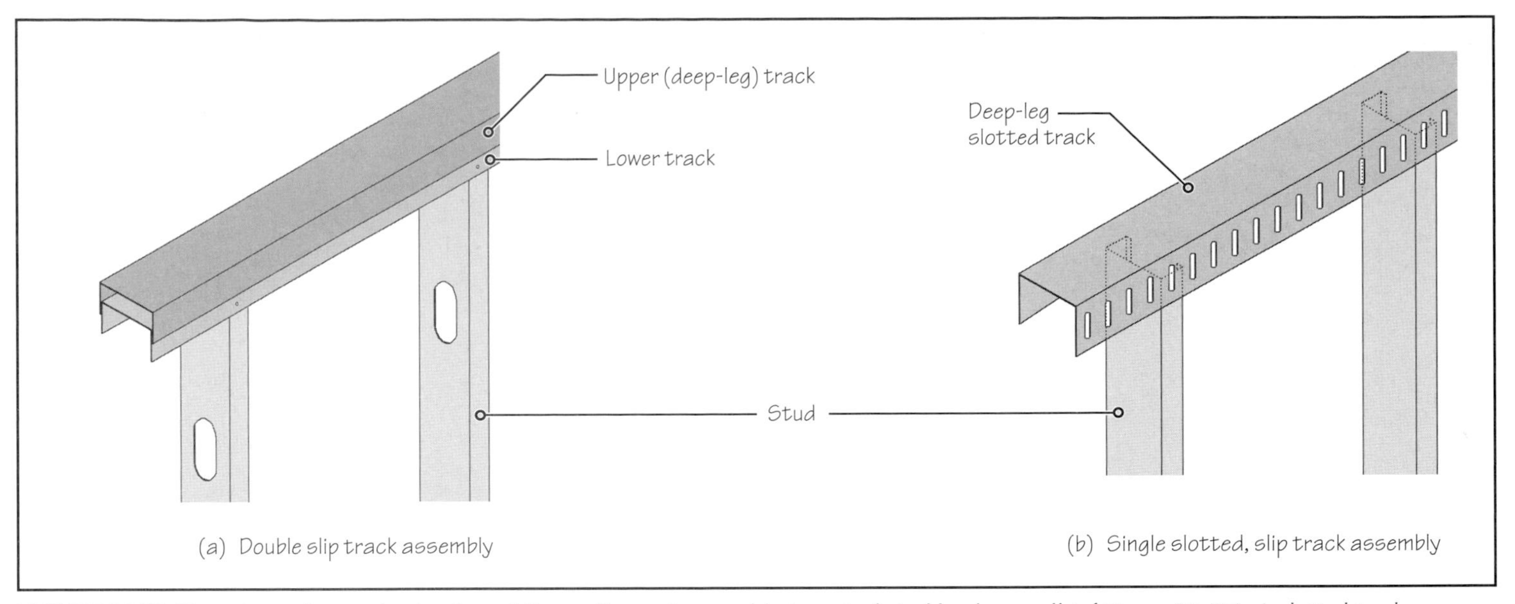

FIGURE 26.28 Two alternative methods of providing a slip-track assembly in a steel stud backup wall (of Figure 26.27). A slotted track assembly provides a positive connection between the studs and the track.

Alternatively, a single, slotted, deep-leg slip-track assembly may be used. See Figure 26.28(b). The studs are loose fastened to the top track through the slots. The drywall is not fastened to the slotted track. The slotted track assembly is more economical and also provides a positive connection between the track and the studs.

Steel Stud Backup Wall Forward of the Structural Frame

In low-rise buildings (one or two stories), putting the steel stud backup on the outside of the structure allows it to cover the structural frame. Thus, the studs are continuous from the bottom to the top, requiring no shelf angles, Figures 26.29 and 26.30.

Slip connections must be used to connect the studs with the floor or roof so that the structural frame and the wall can move independently of each other. Two alternative means of providing slip connections are used, depending on the profile of the studs, Figure 26.31.

FIGURE 26.29 In a one- or two-story building with steel stud backup wall and steel frame structure, the studs are generally continuous from the bottom to the top. The vertical continuity of studs across a floor and (or) roof also helps to reduce their size in resisting the lateral loads.

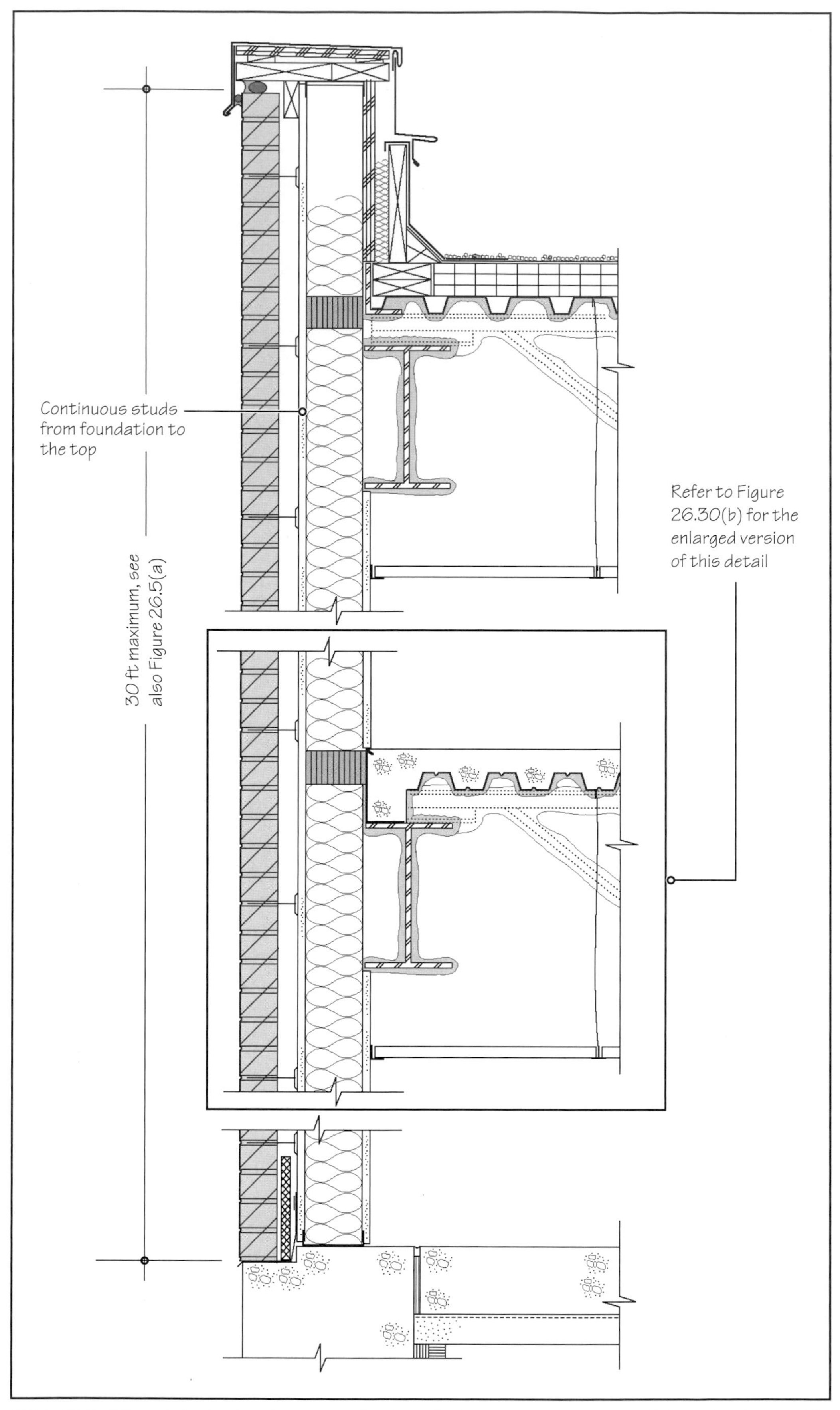

FIGURE 26.30(a) A typical section through a low-rise (one- or two-story) steel-frame building with a brick veneer and steel stud backup wall assembly. Brick veneer is continuous from the foundation to the underside of parapet coping.

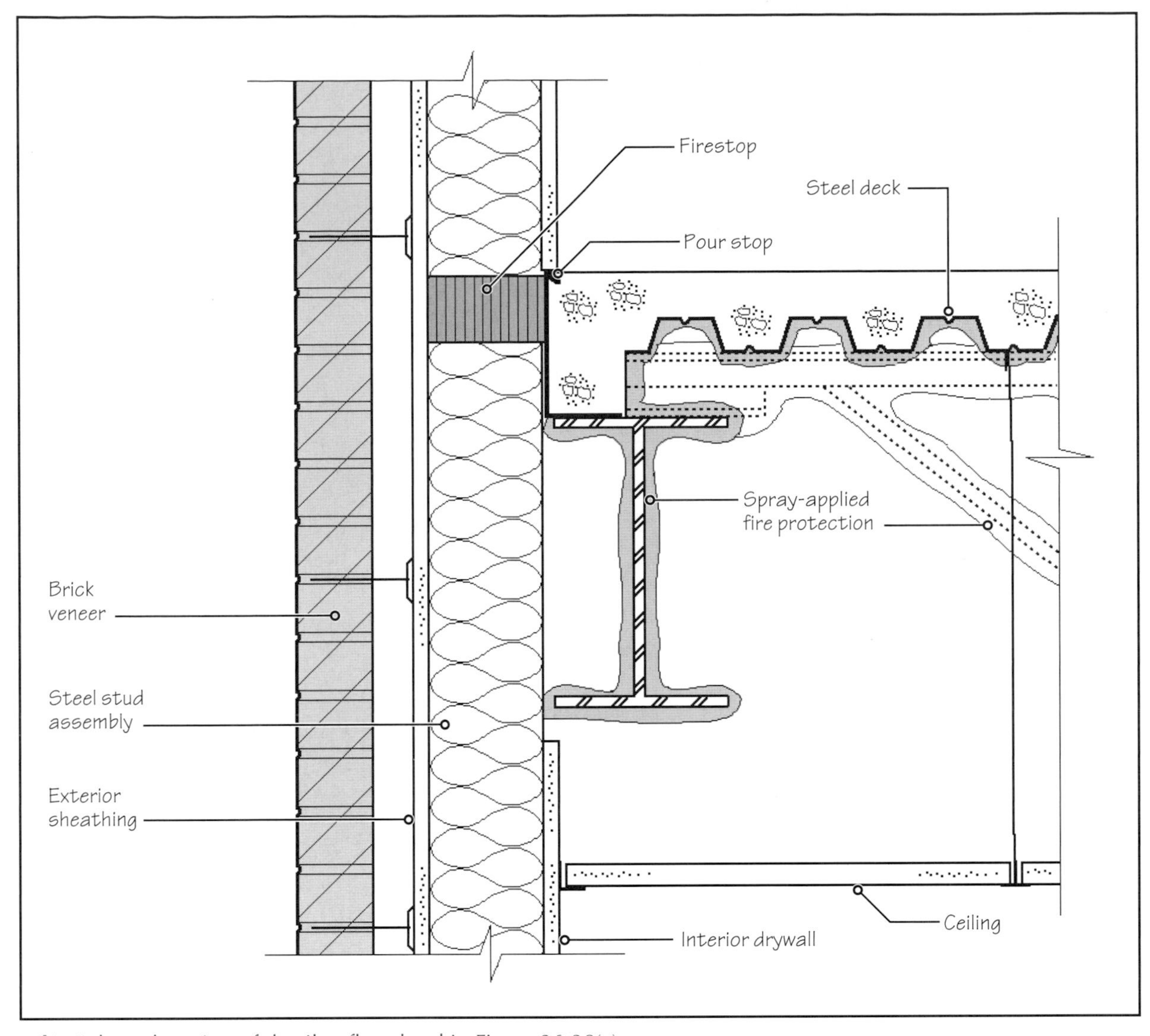

FIGURE 26.30(b) Enlarged version of detail at floor level in Figure 26.30(a).

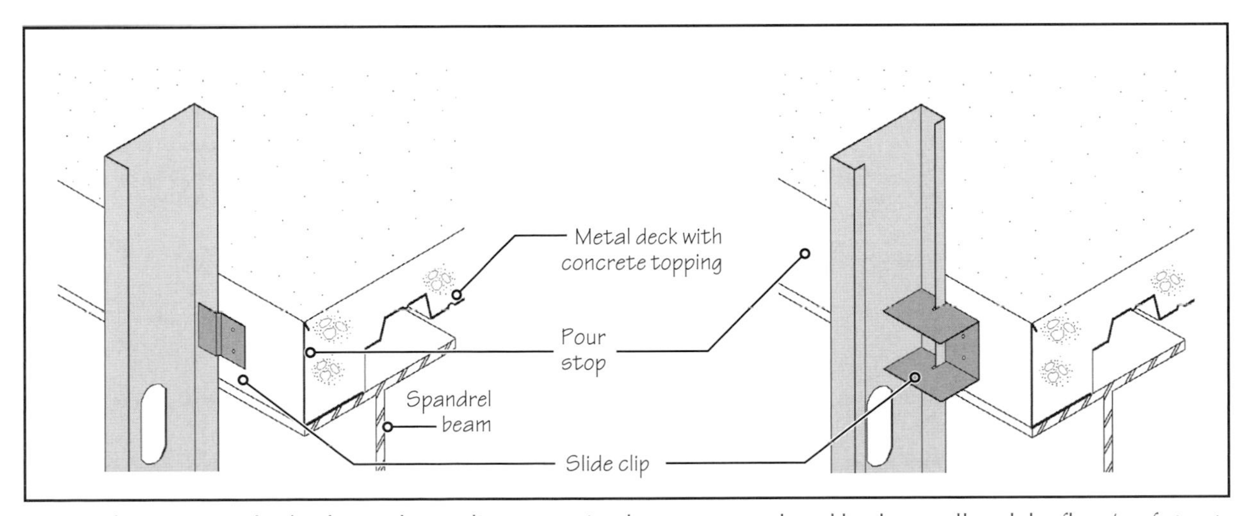

FIGURE 26.31 Two alternative methods of providing a slip connection between a steel stud backup wall and the floor/roof structure, depending on the profile of studs.

A brick veneer attached to a steel stud backup wall forward of the structure can also be used in mid- and high-rise buildings. Two alternative details commonly used for buildings with ribbon windows and brick spandrels are shown in Figures 26.32 and 26.33.

WEATHER RESISTANCE OF A STEEL STUD BACKUP WALL

Weather resistance of exterior sheathing on steel studs may be accomplished by one of the following two means:

- Using a membrane that covers the entire sheathing, such as an air-weather retarder
- Using a water-resistant tape on the joints between sheathing panels

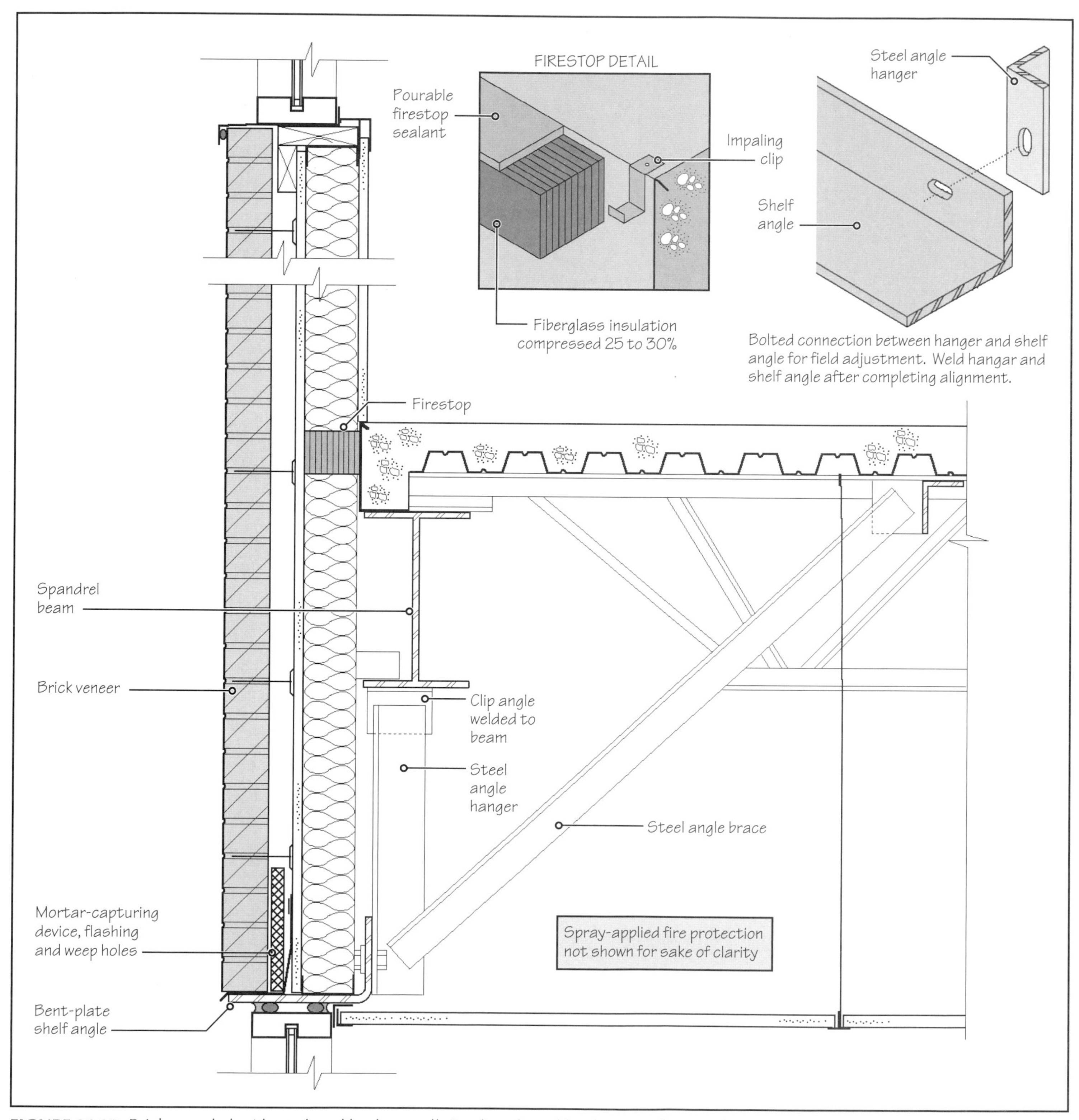

FIGURE 26.32 Brick spandrel with steel stud backup wall. Steel studs and brick veneer bear on bent-plate shelf angle.

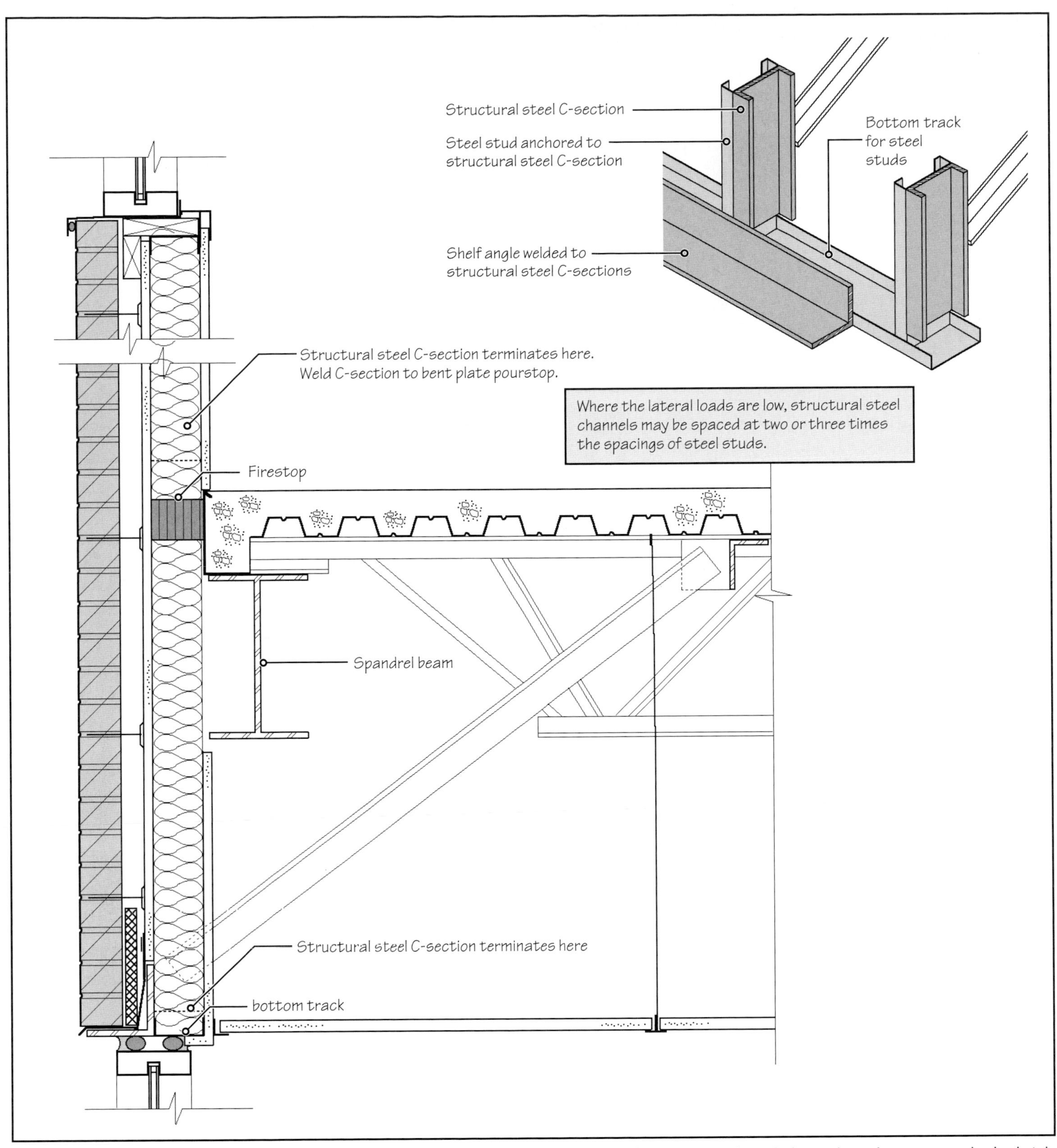

FIGURE 26.33(a) Brick spandrel with steel stud backup wall. The shelf angle is hung from structural steel channels and supports only the brick veneer. See Figure 26.33(b).

(b)

(c)

FIGURE 26.33 **(b)** Structural frame for brick spandrel with steel stud backup wall. The shelf angle is hung from vertical structural steel channels and supports only the brick veneer. **(c)** Wall framing showing steel studs anchored to structural steel channels.

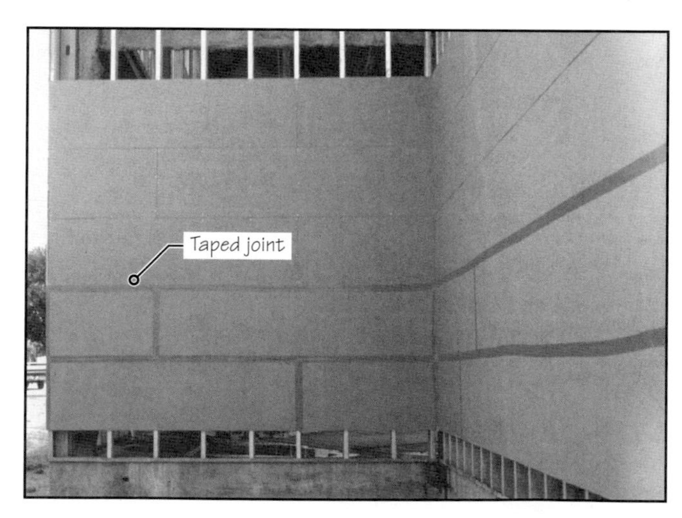

FIGURE 26.34 A partially taped exterior sheathing over steel studs, photographed while the taping of sheathing was in progress.

The use of a continuous membrane conceals the joints between siding panels, making it more difficult for the masons to locate the studs to which the fasteners must be anchored. With taped joints, it is easier to locate the studs, Figure 26.34.

26.4 CMU BACKUP VERSUS STEEL STUD BACKUP

A major benefit of a steel stud backup wall in a brick veneer assembly (as compared with a CMU backup wall) is its lighter weight. For a high-rise building, the lighter wall not only reduces the size of spandrel beams but also that of the columns and footings, yielding economy in the building's structure. However, this benefit is accompanied by several concerns.

Steel studs can deflect considerably before the bending stress in them exceeds their ultimate capacity. Brick veneer, on the other hand, deflects by a very small amount before the mortar joints open. Open mortar joints weaken the wall and also increase the probability of leakage, corroding the anchors.

Thus, the steel stud backup and brick veneer assembly performs well only if the stud wall is sufficiently stiff. To obtain the necessary stiffness, the deflection of studs must be controlled to a fairly small value. In fact, the design of a steel stud backup wall to resist the lateral loads is governed not by the strength of studs but by their deflection.

The Brick Industry Association (BIA) recommends that the lateral load deflection of steel studs, when used as backup for brick veneer assembly, should not exceed

$$\frac{\text{stud span}}{600}$$

where the stud span is the unsupported height of studs. For example, if the height of studs (e.g., from the top of the floor to the bottom of the spandrel beam in Figure 26.7) is 10 ft, the deflection of studs under the lateral load must be less than

$$\frac{(10 \times 12)}{600} = 0.5 \text{ in.}$$

Increasing the stiffness of studs increases the cost of the assembly. Another concern with steel stud backup is that the veneer is anchored to the backup only through screws that engage the threads within a light-gauge stud sheet. Over a period of time, condensation can corrode the screws and the corresponding holes in studs, making the screws come loose. Condensation is, therefore, an important concern in a steel stud–backed veneer. A more serious concern is that the anchor installer will miss the studs.

By comparison with steel studs, anchoring of brick veneer to a CMU backup does not depend on screws; hence it is more forgiving. Additionally, the anchors in a CMU backup wall are embedded in the mortar joints, and if they are made of stainless steel, their corrosion probability is extremely low.

Another advantage of a CMU backup wall is its inherent stiffness. To obtain a steel stud backup wall of the same stiffness as a CMU backup wall substantially increases the cost of the stud backup.

NOTE

Deflection of a Steel Stud Wall Assembly

The design criterion of deflection not exceeding span divided by 600, suggested by BIA, is the minimum requirement. For critical buildings, a more stringent deflection criterion, such as span divided by 720 or span divided by 900, is recommended by some experts.

Steel stud manufacturers generally provide tables for the selection of their studs to conform to the deflection criteria.

26.5 AESTHETICS OF BRICK VENEER

It is neither possible nor within the scope of this text to illustrate various techniques used to add visual interest to brick veneer facades. However, a few examples are provided:

- Use of recessed or projected bricks in the wall, Figure 26.35.
- Use of bricks of different colors or combining clay bricks with other masonry materials or cast concrete, Figure 26.36.
- Warping the wall, Figure 26.37.

FIGURE 26.35 Use of recessed or projected bricks with different hues. The projections must be small so that the core holes in bricks are not exposed.

FIGURE 26.36 Use of precast (white) concrete bands in veneer.

FIGURE 26.37 Warped brick veneer used in Frank Gehry's Peter B. Lewis Building, Cleveland, Ohio. (Photo courtesy of Mario Segovia.)

PRACTICE **QUIZ**

Each question has only one correct answer. Select the choice that best answers the question.

18. A typical anchor used with brick veneer and CMU backup wall assembly consists of a
- **a.** two-piece anchor, of which one is embedded in CMU backup and the other is embedded in the veneer's mortar joint.
- **b.** three-piece anchor, of which two are fastened to the CMU backup and the other is embedded in the veneer's mortar joint.
- **c.** two-piece anchor, of which one is an integral part of the joint reinforcement in CMU backup and the other is embedded in the veneer's mortar joint.
- **d.** (a) and (b).
- **e.** (a) and (c).

19. When lintel angles and shelf angles are combined in the same angle in a brick veneer clad building, the angle should be treated as a
- **a.** lintel angle anchored to the structural frame of the building.
- **b.** lintel angle loose-laid over underlying veneer.
- **c.** shelf angle loose-laid over underlying veneer.
- **d.** shelf angle anchored to the structural frame of the building.

20. In general, a CMU-backed brick veneer is more forgiving of construction and workmanship deficiencies than a steel stud–backed brick veneer.
- **a.** True
- **b.** False

21. In a brick veneer assembly with a steel stud backup wall, the design of studs is generally governed by
- **a.** the compressive strength of studs to withstand gravity loads.
- **b.** the shortening of studs to withstand gravity loads.
- **c.** the bending strength of studs to withstand lateral loads.
- **d.** the deflection of studs to withstand lateral loads.
- **e.** any one of the above, depending on the wall.

Answers: 18-e, 19-d, 20-a, 21-d.

26.6 PRECAST CONCRETE (PC) CURTAIN WALL

Unlike brick veneer, which is constructed brick by brick at the construction site, a precast concrete (PC) curtain wall is panelized construction. The panels are constructed off-site, under controlled conditions, and transported to the site in ready-to-erect condition, greatly reducing on-site construction time.

Although the PC curtain wall system is used in all climates, it is particularly favored in harsh climates, where on-site masonry and concrete construction are problematic due to freeze hazard and slow curing rate of portland cement. Panelized construction eliminates scaffolding, increasing on-site workers' safety. Because panel fabrication can be done in sheltered areas, it can be accomplished uninterrupted, with a higher degree of quality control.

PC curtain walls are used for almost all building types but more often are used for mid- to high-rise hospitals, apartments, hotels, parking garages, and office buildings, Figure 26.38.

PC curtain wall panels are supported on and anchored to the building's structural frame and are hoisted in position by cranes, Figure 26.39(a) and (b). The panels are fabricated in a precast concrete plant and transported to a construction site.

The structural design of PC curtain wall panels is generally done by the panel fabricator to suit the fabrication plant's setup and resources and to provide an economical product. A typical precast concrete plant generally has in-house structural engineering expertise.

In a PC curtain wall project, an important role for the architect is to work out the aesthetic expression of the panels (shapes, size, exterior finishes, etc.). This should be done in consultation with the precast plant. A great deal of coordination between the architect, engineer of record, general contractor, precast plant, and erection subcontractor is necessary for a successful PC curtain wall project.

Because of the sculptability of concrete and the assortment of possible finishes of the concrete surface (smooth, abrasive-blasted, acid-etched, etc.), PC curtain walls lend themselves to a variety of facade treatments. The use of reveals (aesthetic joints), moldings, and colored concrete further add to the design variations, Figure 26.40.

FIGURE 26.38 A typical office building with precast concrete curtain wall panels.

Panel Shapes and Size

Another key design decision is the size, shape, and function of each panel. These include window wall panels, spandrel panels, and column covers, and so on, Figure 26.41. The panels are generally one floor high, but those spanning two or (occasionally) three floors may be used.

PC wall panels are generally made as large as possible, limited only by the erection capacity of the crane, transportation limitations, and gravity load delivered by the panel to the structural frame. For structural reasons, the panels generally extend from column to column. Smaller panel sizes mean a larger number of panels, requiring a greater number of support connections, longer erection time, and, hence, a higher cost. If the scale of large panels is visually unacceptable, false joints can be incorporated in the panels, as shown in preceding images.

FIGURE 26.39(a) A PC panel being unloaded from the delivery truck for hoisting into position by a crane.

Concrete Strength

PC curtain wall panels are removed from the form as soon as possible to allow rapid turnover and reuse of the formwork. This implies that the 28-day concrete strength should be reasonably high so that when the panel is removed from the form, it can resist the stresses to which it may be subjected during the removal and handling processes.

The required concrete strength is also a function of the curtain wall's exposure, durability requirements, shape, and size of the panels. Flat panels may require higher strength (or greater thickness) as compared to ribbed or profiled panels. Therefore, the strength of concrete must be established in consultation with the precast plant supplying the panels.

The most commonly used 28-day strength of concrete for PC curtain wall panels is 5,000 psi. This relatively high strength gives greater durability, greater resistance to rainwater penetration, and an improved in-service performance. In other words, the panels are better able to resist stresses caused by the loads, building movement, and volume changes induced by thermal, creep, and shrinkage effects.

For aesthetic and economic reasons, a panel may use two mixes—a face (architectural) mix and a backup (structural) mix. In this case, the two mixes should have nearly equal expansion and contraction coefficients to prevent undue bowing and warping of the panel. In other words, the strength, slump, and water-cement ratios of the two mixes should be nearly the same.

Because panels are generally fabricated face down on a flat formwork, the face mix is placed first followed by the backup mix. The thickness of the face

FIGURE 26.39(b) PC panel of Figure 26.39(a) being hoisted to its final position.

FIGURE 26.40(a) Lightly abrasive blasted panels with a great deal of surface detailing.

FIGURE 26.40(b) The part of this panel on the left side of the reveal is lightly abrasive blasted, and the right side part is medium abrasive blasted.

FIGURE 26.40(c) Panel with exposed aggregate finish.

mix is a function of the aggregate size but should not be less than 1 in. The precaster's experience should be relied upon in determining the thicknesses and properties of the two mixes.

Panel Thickness

The thickness of panels is generally governed by the handling (erection) stresses rather than the stresses caused by in-service loads. A concrete cover on both sides and the two-way reinforcement in a panel generally gives a total of about 4 in. of thickness. Add to this the

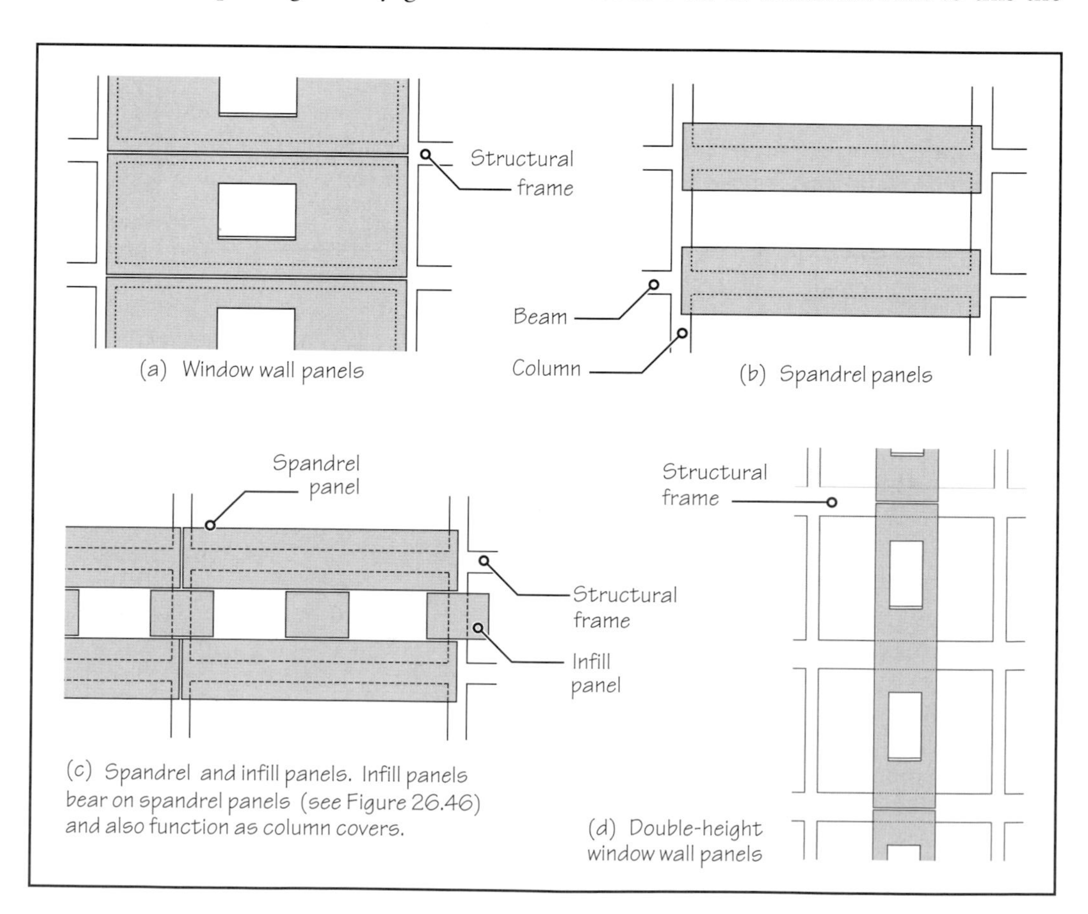

FIGURE 26.41 A few commonly used panel shapes. Panels should preferably extend from column to column so that their dead load (which is delivered through two supports only) is transferred to the beam as close to the column as possible (see Section 26.7).

thickness that will be lost due to surface treatment, such as abrasive blasting and acid etching, and the total thickness of a PC wall panel cannot be less than 5 in.

However, because panel size is generally maximized, a panel thickness of less than 6 in. is rare. A thicker panel is not only stronger but is also more durable, is more resistant to water leakage, and has higher fire resistance. Greater thickness also gives greater heat-storage capacity (Chapter 5), making the panel less susceptible to heat-induced stresses.

MOCK-UP SAMPLE(S)

For PC curtain wall projects, the architect requires the precast plant to prepare and submit for approval a sample or samples of color, texture, and finish. The *mock-up* panels, when approved, are generally kept at the construction site and become the basis for judgment of all panels produced by the precast plant.

26.7 CONNECTING THE PC CURTAIN WALL TO A STRUCTURE

The connections of PC curtain wall panels to the building's structure are among the most critical items in a PC curtain wall project and are typically designed by the panel fabricator. Two types of connections are required in each panel:

- Gravity load connections
- Lateral load connection

There should be only two gravity load connections, also referred to as *bearing supports*, per panel located as near the columns as possible. The lateral load connections, also referred to as *tiebacks*, may be as many as needed by structural considerations, generally two or more per panel, Figure 26.42.

NOTE

The connection system of a PC curtain wall resembles that of brick veneer connection system and is common to all types of curtain walls. The shelf angles in brick veneer provide the gravity load connection, and the anchors between the veneer and the backup provide the lateral load connection.

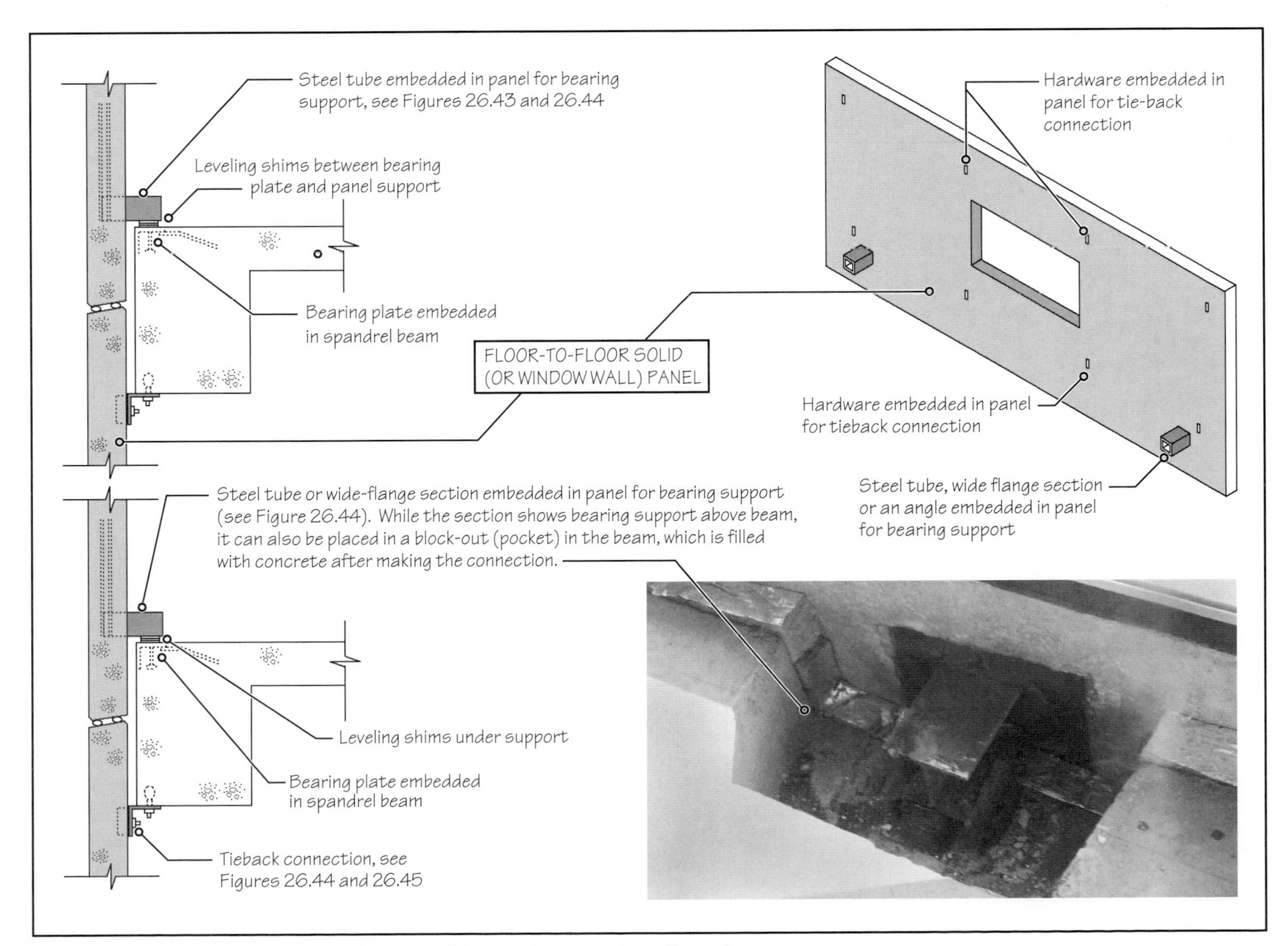

FIGURE 26.42 Support connections for a typical floor-to-floor curtain wall panel.

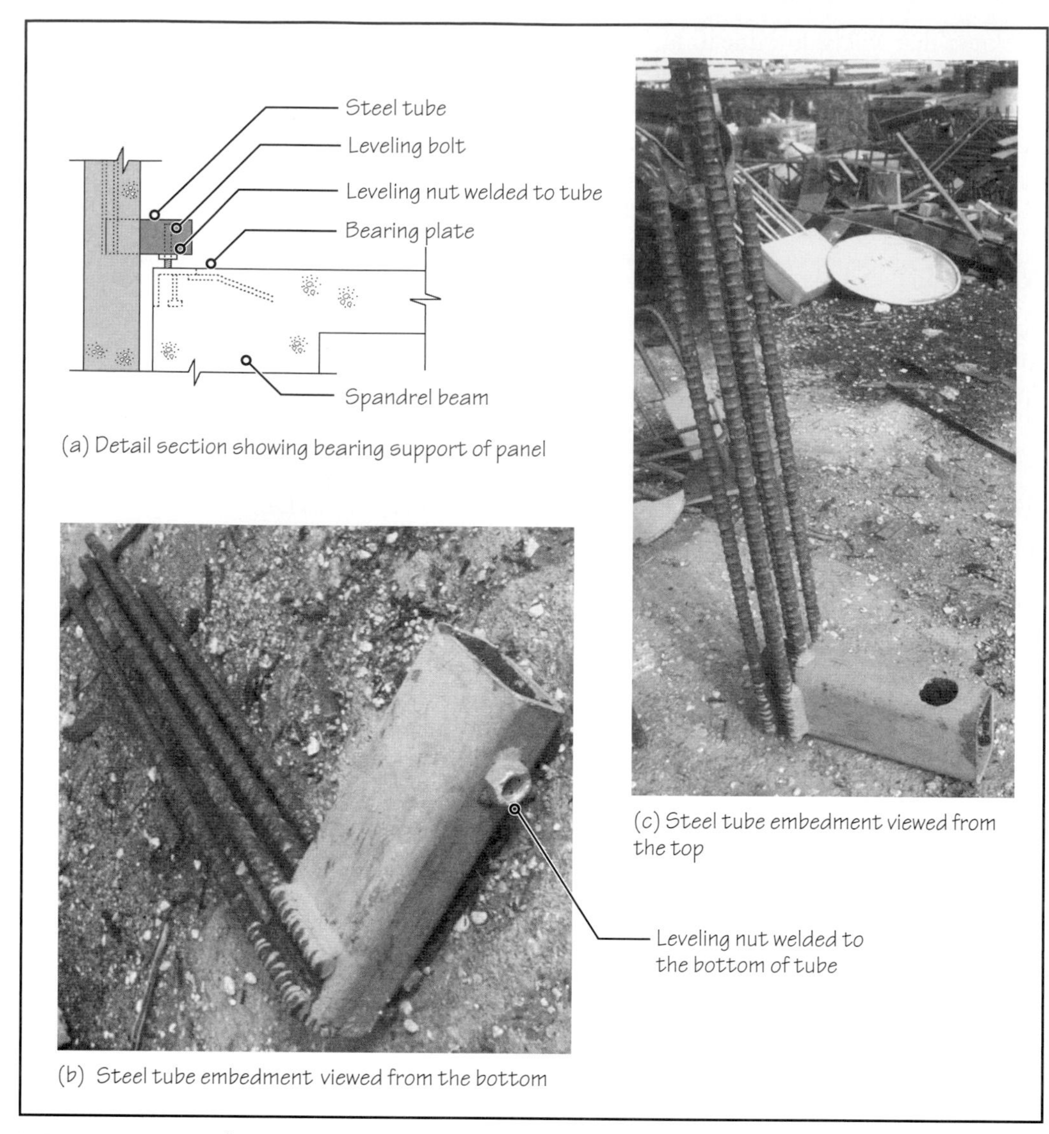

FIGURE 26.43 Leveling nut in a steel tube used for bearing support.

BEARING SUPPORTS

A commonly used bearing support for the floor-to-floor panels is provided through a section of steel tube, a part of which is embedded in the panel and a part of which projects out of the panel. The projecting part rests on the (steel angle) bearing plate embedded at the edge of the spandrel beam. Dimensional irregularities, both in the panel and in the structure, require the use of leveling shims (or bolts) under bearing supports during erection.

After the panels have been leveled, the bearing supports are welded to the bearing plate. The bearing support system is designed to allow the panel to move within its own plane so that the panel is not subjected to stresses induced by temperature, shrinkage, and creep effects.

In place of leveling shims in a bearing support, a leveling bolt is often used, Figure 26.43. The choice between the shims and the bolt is generally left to the preference of the precast manufacturer and the erector. Alternatives to the use of steel tube for bearing supports are steel angles or a wide-flange (I-) section, Figure 26.44.

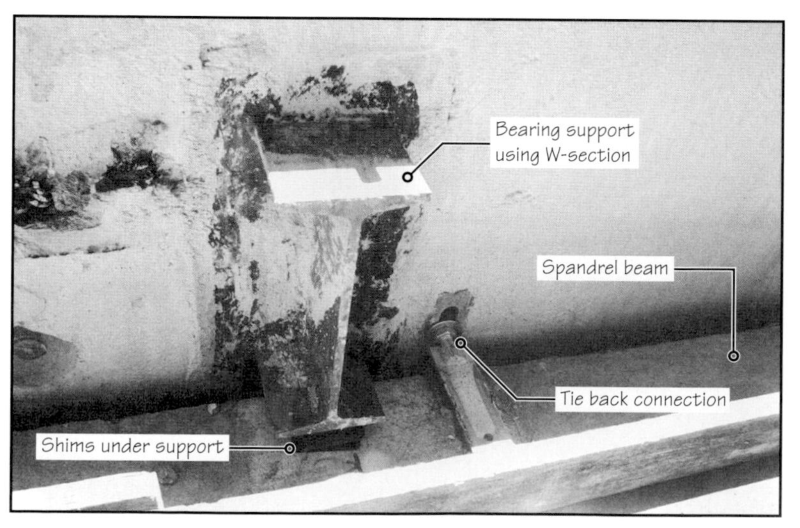

FIGURE 26.44 Wide-flange section used as bearing support in a PC panel.

TIEBACKS

A lateral load connection (tieback) is designed to resist horizontal forces on the panel from wind and/or earthquake and due to the eccentricity of panel bearing. Therefore, it must be able to resist tension and compression perpendicular to the plane of the panel.

A tieback is designed to allow movement within the plane of the panel. The connection must, however, permit

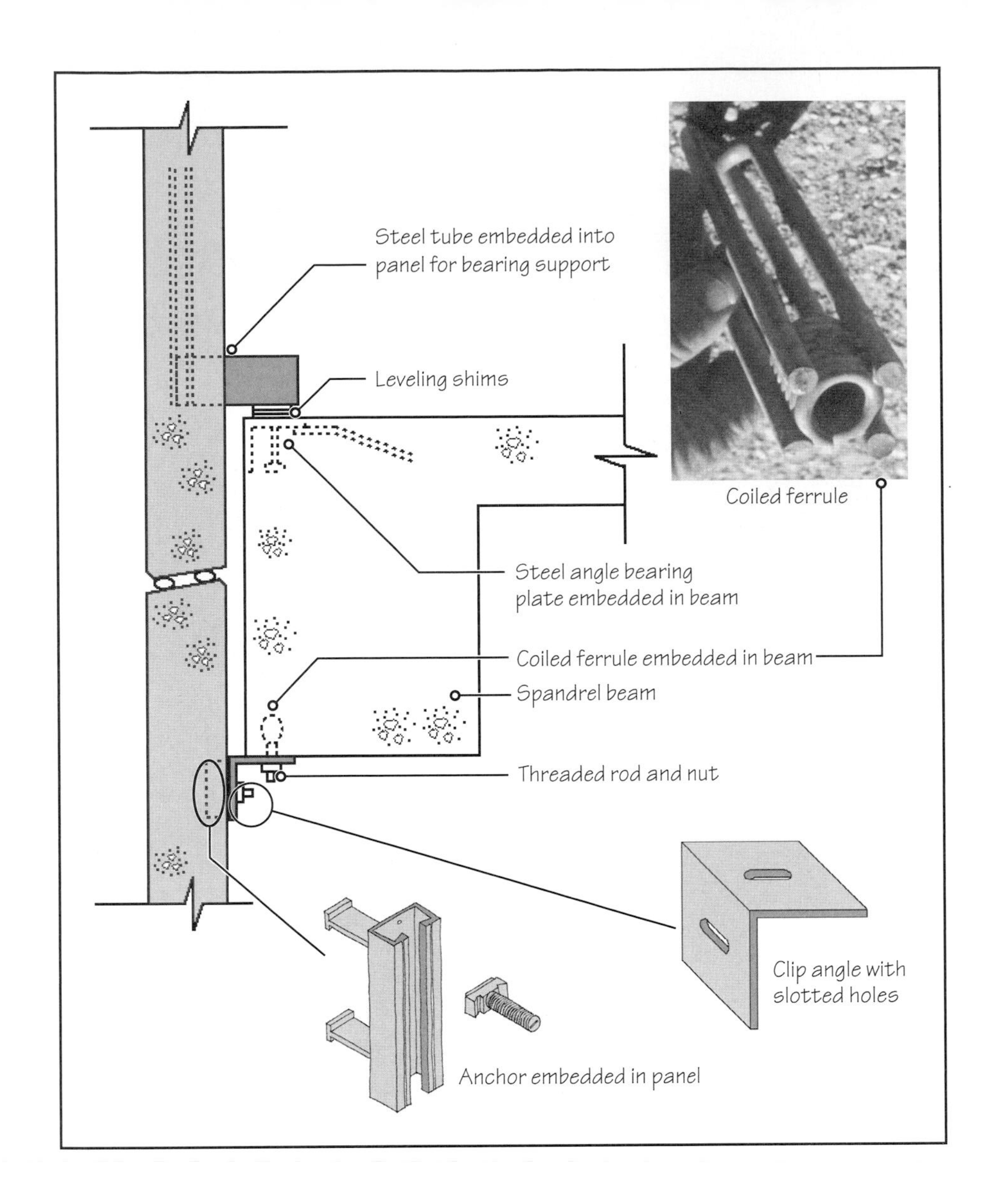

FIGURE 26.45 A typical tieback connection that allows three-way field adjustment during panel erection in addition to in-service vertical deflection of spandrel beam and thermal expansion/contraction of panel.

adjustment in all three principal directions during erection. A typical tieback is shown in Figure 26.45 (see also Figure 26.44).

SUPPORT SYSTEMS FOR SPANDREL PANELS AND INFILL PANELS

Support systems for a curtain wall consisting of spandrel and infill panels are shown in Figure 26.46; see also Figure 26.41(c).

PANELS AND STEEL FRAME STRUCTURE

PC wall panels, which create eccentric loading on the spandrel beams, create torsion in the beams. Due to the lower torsional resistance of wide-flange steel beams, PC panels used with a steel frame structure are generally designed to span from column to column and made to bear directly on them. Tiebacks, however, are connected to the spandrel beams. (Note that the rotation of the spandrel beam caused by the eccentricity of panel bearing is partially balanced by reverse rotation caused by the the floor's gravity load on the spandrel beam.)

CLEARANCE OF PANELS FROM THE STRUCTURAL FRAME

The Precast/Prestressed Concrete Institute (PCI) recommends a minimum horizontal clearance of 2 in. of precast panels from the building's structural frame.

26.8 BRICK AND STONE-FACED PC CURTAIN WALL

PC curtain wall panels may be faced with thin (clay) bricks at the time of casting the panels. Generally, $\frac{3}{4}$- to 1-in-thick bricks are used. They are available in various shapes, Figure 26.47. The bricks are placed in the desired pattern in the form, and the concrete is

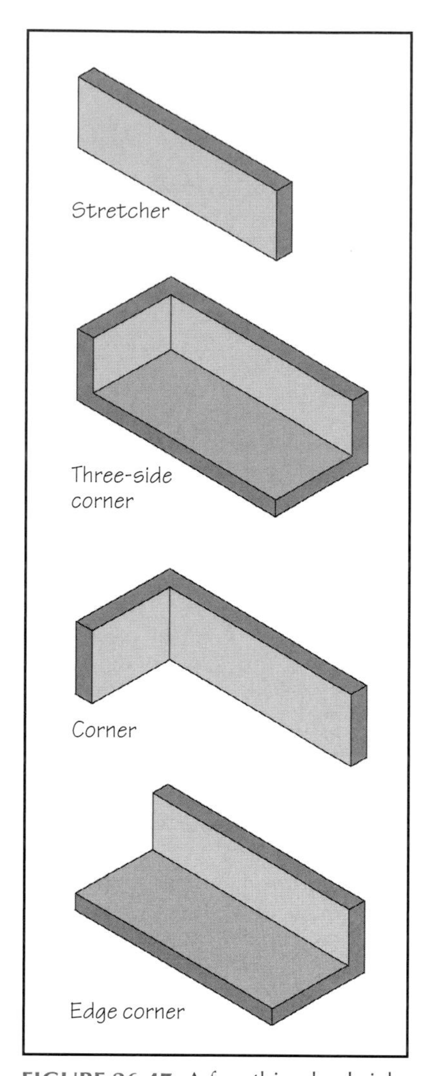

FIGURE 26.47 A few thin clay brick shapes available.

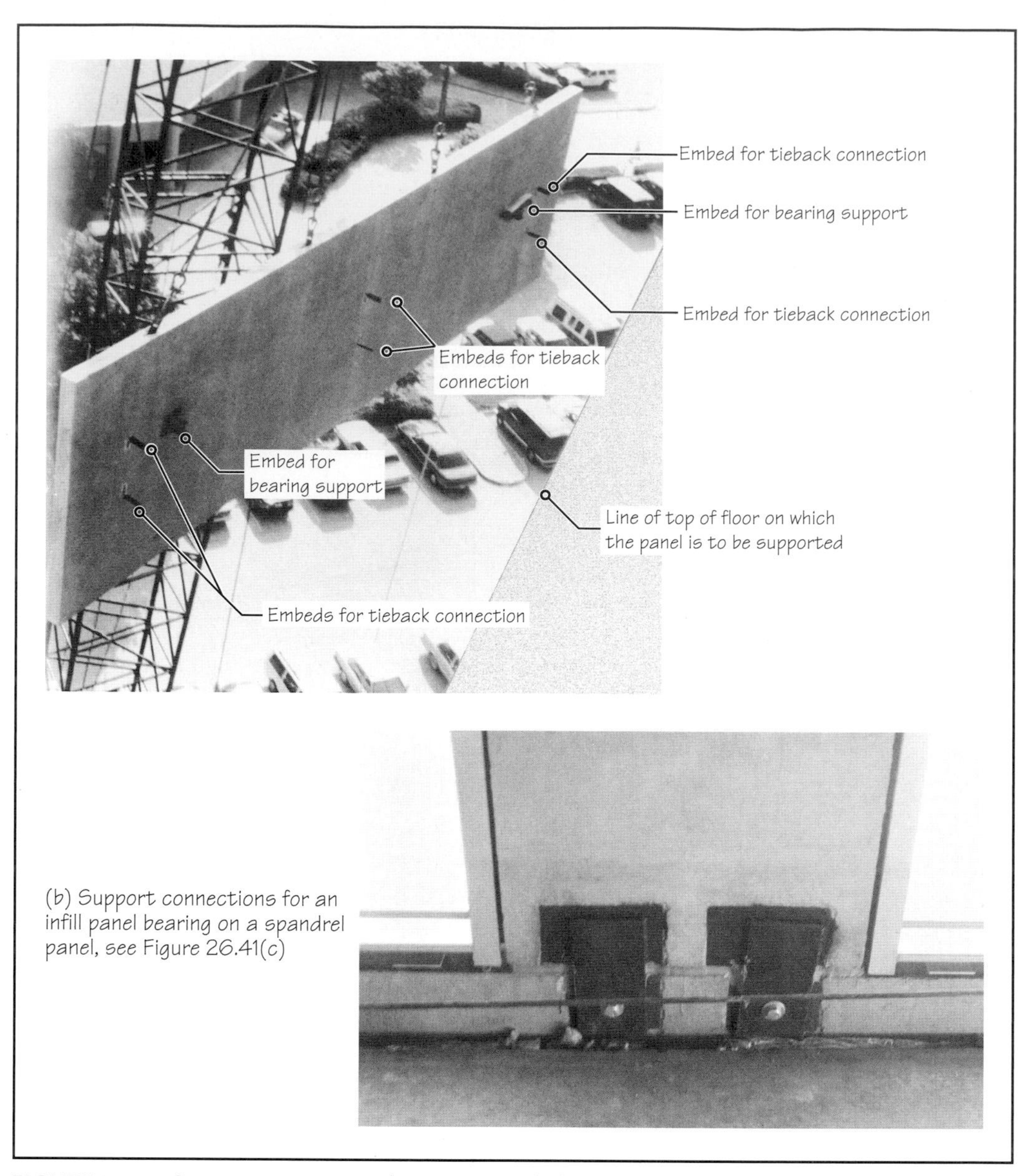

FIGURE 26.46 Support connections for: **(a)** a spandrel panel, **(b)** a column cover bearing on spandrel panels.

placed over them. To prevent the bricks from shifting during the placing operation, a rubber template is used, within which the bricks are placed, Figure 26.48. The template aligns the bricks and allows the concrete to simulate the mortar joints.

In well-designed and well-fabricated panels, it is generally difficult to distinguish between a site-constructed brick veneer wall and a brick-faced PC curtain wall. Brick-faced panels have the same advantages as other PC curtain wall panels—that is, no on-site construction and no scaffolding.

Thin bricks need to be far more dimensionally uniform than full-size bricks. With full-size bricks, dimensional nonuniformity is compensated by varying the mortar thickness.

Because clay bricks expand by absorbing moisture from the atmosphere (moisture expansion), thin bricks should be allowed to season for a few weeks at the precaster's plant before being used. This reduces their inherent incompatibility with concrete, which shrinks on drying.

Although moisture expansion and drying shrinkage are the major causes of incompatibilities between bricks and concrete, other differences between the two materials, such as the coefficient of thermal expansion and the modulus of elasticity, must be considered. Because of these differences, a brick-faced concrete panel is subject to bowing due to differential expansion and contraction of the face and the backup.

Bowing can be reduced by increasing the stiffness of the panel. Using two layers of reinforcement is encouraged when the thickness of the panel permits. Additionally, some precasters use reverse curvature in the panels.

FIGURE 26.48 Rubber template form liner used with thin brick-faced panels.

The bond between concrete and bricks is obviously very important for a brick-faced concrete wall panel. The back surface of bricks used in brick-faced panels should contain grooves, ribs, or dovetail slots to develop an adequate bond. The bond between concrete and brick is measured through a shear strength test. The architect should obtain these results from the brick manufacturer before specifying them for use.

The bond between concrete and bricks is also a function of how absorptive the bricks are. Bricks with excessively high or excessively low water absorption give a poor bond. Bricks with high water absorption are subjected to freeze-thaw damage.

Stone Veneer–Faced PC Panels

PC curtain wall panels can also be faced with (natural) stone veneer. The thickness of the veneer varies with the type of stone and the face dimensions of the veneer units. Granite and marble veneers are recommended to be at least $1\frac{1}{4}$ in. (3 cm) thick. Limestone veneer should be at least 2 in. thick.

The face dimensions of individual veneer units are generally limited to 25 ft^2 for granite and 15 ft^2 for marble or limestone. Thus, a typical PC curtain wall panel has several veneer units anchored to it.

The veneer is anchored to the concrete panel using stainless steel flexible dowels, whose diameter varies from $\frac{3}{16}$ to $\frac{5}{8}$ in. Two dowel shapes are commonly used—a U-shaped (hairpin) dowel and a pair of cross-stitch dowels, Figure 26.49. The dowels are inserted in holes drilled in the veneer. The dowel holes, which are $\frac{1}{16}$ to $\frac{1}{8}$ in. larger in diameter than the diameter of the dowels, are filled with epoxy or an elastic, fast-curing silicone sealant. Unfilled holes will allow water to seep in, leading to staining of the veneer and freeze-thaw damage.

Because the dowels are thin and flexible, they allow relative movement between the veneer and the backup. To further improve their flexibility, rubber washers are used with the dowels at the interface of the veneer and the backup.

The depth of anchor in the veneer is nearly half the veneer thickness. Their anchorage into concrete varies, depending on the type of stone and the loads imposed on the panel.

There should be no bond between stone veneer and the backup concrete to prevent bowing of the panel and cracking and staining of the veneer. To prevent the bond, a bond breaker is used between the veneer and the backup. The bond breaker is either a 6- to 10-mil-thick polyethylene sheet or a $\frac{1}{8}$- to $\frac{1}{4}$-in.-thick compressible, closed-cell polyethylene foam board. Foam board is preferred because it gives better movement capability with an uneven stone surface.

A PC curtain wall panel may be either fully veneered with stone or the veneer may be used only as an accent or feature strip on a part of the panel.

Other Form Liners

Because concrete is a moldable material, several geometric and textured patterns can be embossed on panel surface using form liners. An architect can select from a variety of standard form liners available or have them specially manufactured for a large project (Section 19.4).

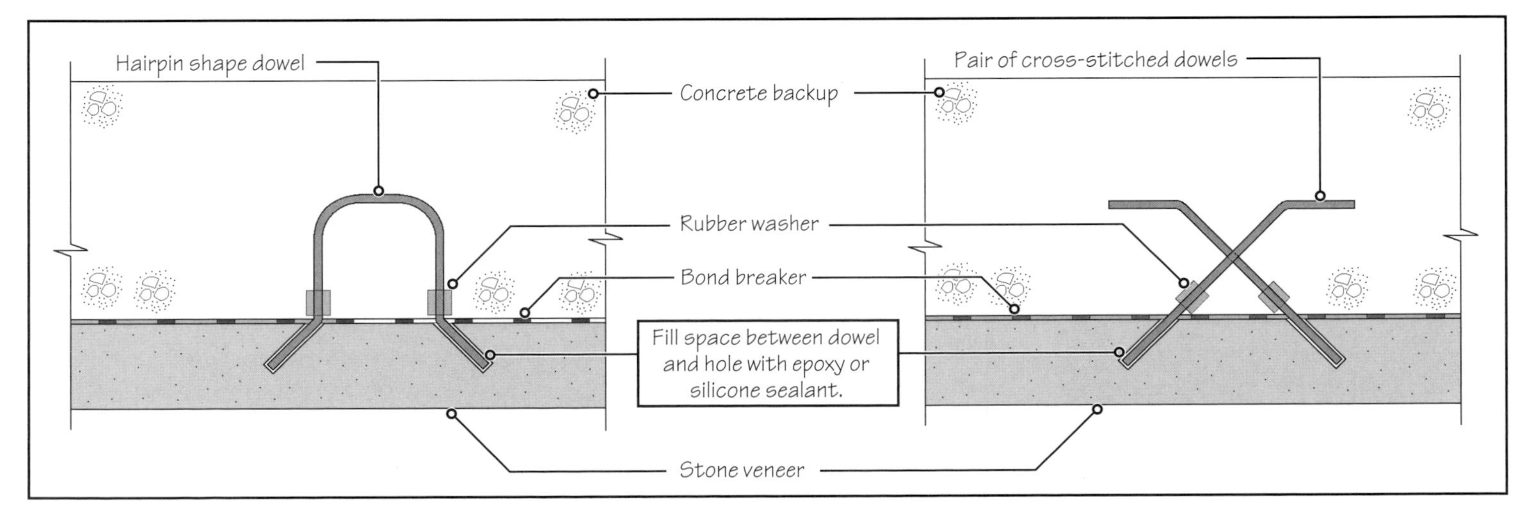

FIGURE 26.49 Two commonly used dowel anchors.

26.9 DETAILING A PC CURTAIN WALL

A typical PC curtain wall is backed by an infill steel stud wall. The stud wall provides the interior finish and includes the insulation and electrical and other utility lines. Because the backup wall is not subjected to wind loads, it needs to be designed only for incidental lateral loads from the building's interior (generally 5 psf). Therefore, a fairly lightweight backup wall is adequate. This contrasts with a backup stud wall in a brick veneer wall, which must be designed to resist wind loads and be sufficiently stiff to have a relatively small deflection.

Figure 26.50 shows a representative detail of a PC curtain wall with a steel stud backup wall. The space between the panels and the backup wall may be filled with rigid insulation, if needed.

Joints Between Panels

Detailing the joints between panels is also a critical aspect of a PC curtain wall project. A minimum joint width of 1 in. between panels is generally recommended. Although a single-stage joint is commonly used, the preferred method is to use a two-stage joint system, consisting

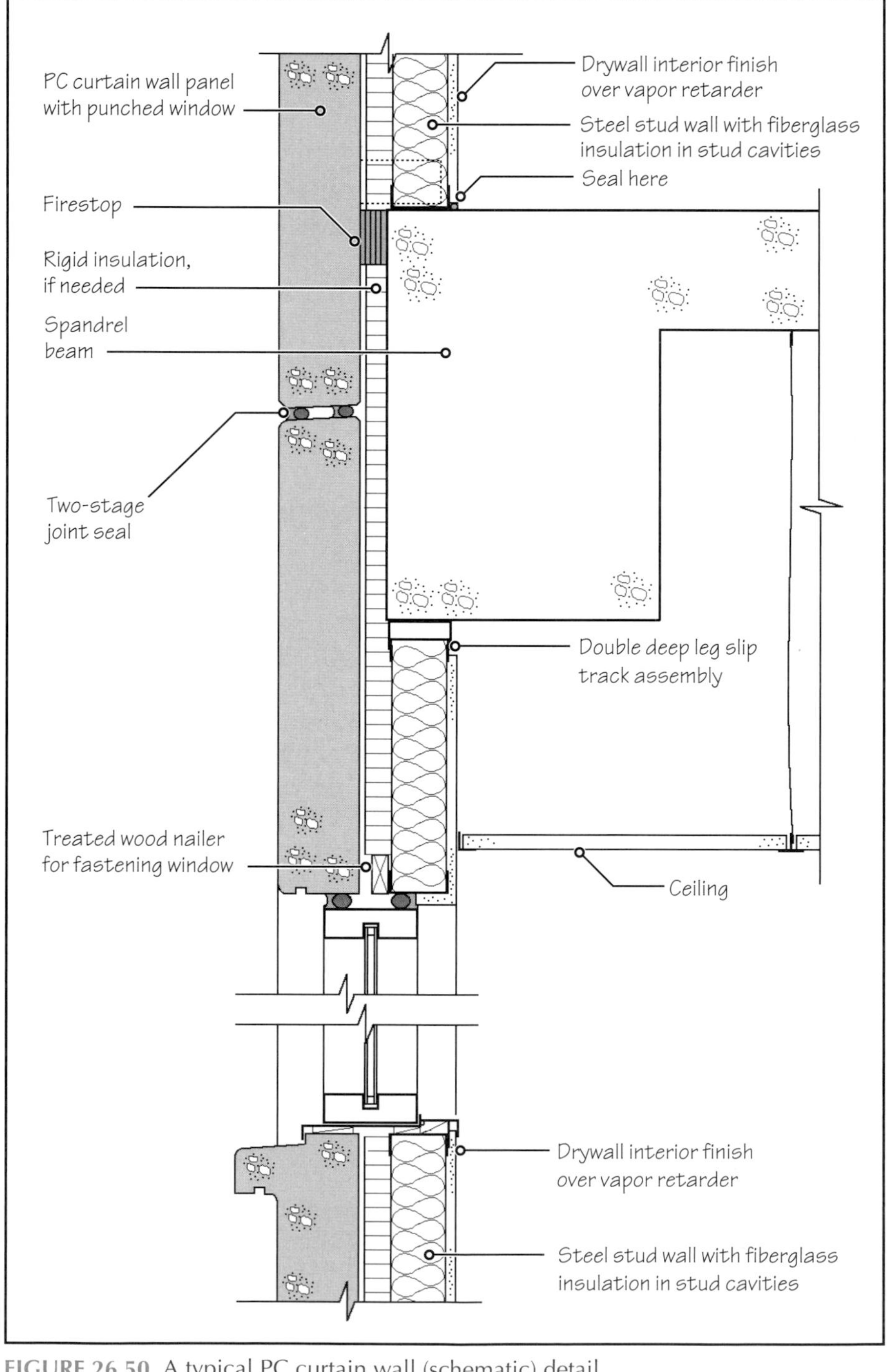

FIGURE 26.50 A typical PC curtain wall (schematic) detail.

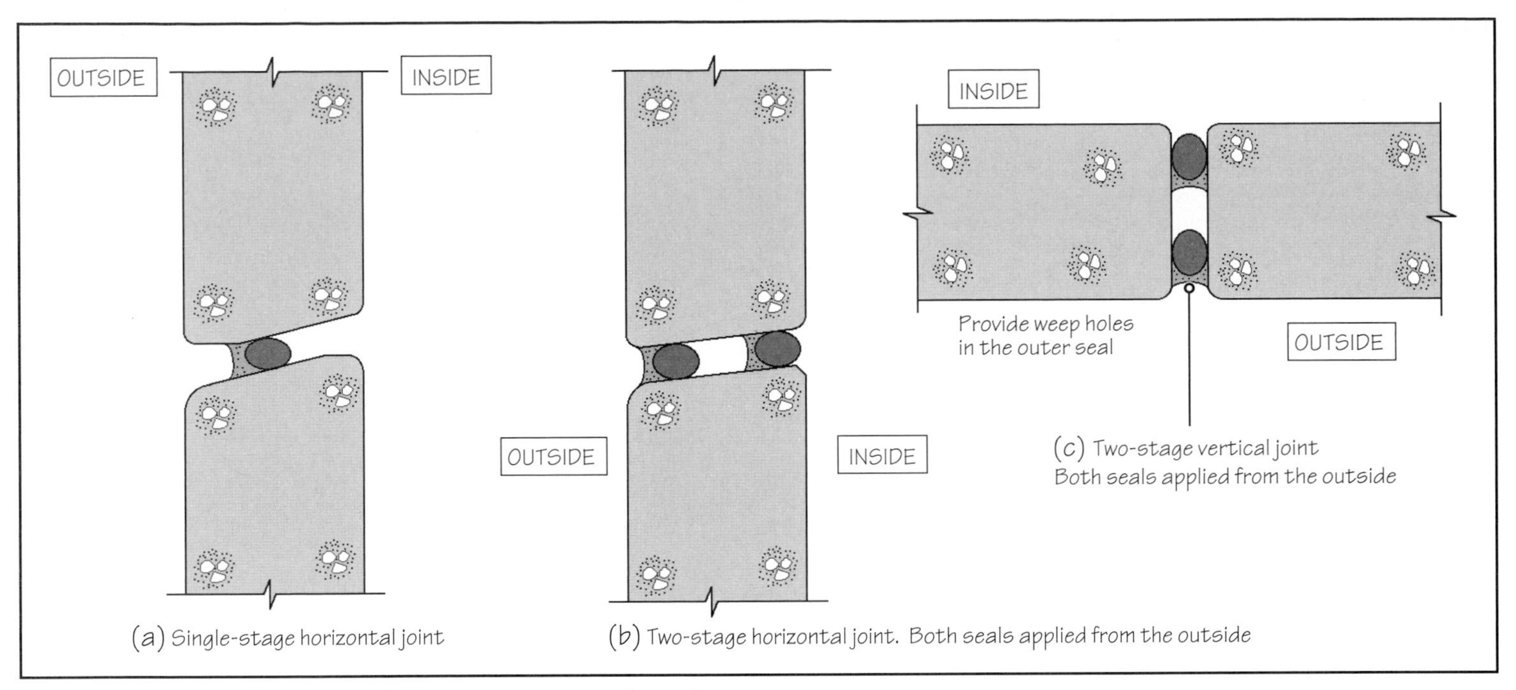

FIGURE 26.51 Treatment of joints between PC curtain wall panels.

of a pair of backer rod and sealant bead combination. One backer rod and sealant bead combination is placed toward the outer surface of the joint and the other is placed toward the inner surface, Figure 26.51.

The outer seal provides a weather barrier and contains weep holes. The inner seal is continuous without any openings and is meant to provide an air barrier. The air barrier must extend continuously over a panel and across the joint between panels. The provision of the air barrier and the openings in the exterior seal makes the two-stage joint system function as a rain screen (Chapter 25).

Both inner and outer seals should be applied from the outside to avoid discontinuities at the spandrel beams and floor slabs. This requires a deep-stem roller to push the backer rod deep into the joint and a long-nozzle sealant gun.

Insulated, Sandwiched Panels

PC curtain wall panels with rigid plastic foam insulation sandwiched between two layers of concrete may be used in cold regions. A sandwich panel consists of an outer layer of concrete and an inner layer of concrete, in which both layers are connected together with ties through an intermediate layer of rigid plastic foam insulation.

The outer layer is a nonstructural layer, whereas the inner layer is designed to carry the entire load and transfer it to the structural frame. The panel is fabricated by first casting the concrete for the nonstructural layer. This is followed by embedding the ties and placing the insulation boards over the ties. The insulation is provided with holes so that the ties project above the insulation and are embedded in the concrete that is cast above the insulation.

PRACTICE QUIZ

Each question has only one correct answer. Select the choice that best answers the question.

22. The strength of concrete used in precast concrete curtain walls is generally
- **a.** greater than or equal to 3,000 psi.
- **b.** greater than or equal to 4,000 psi.
- **c.** greater than or equal to 5,000 psi.
- **d.** none of the above.

23. Precast concrete curtain wall panels are generally
- **a.** cast on the construction site.
- **b.** fabricated by a readymix concrete plant and transported to the construction site.
- **c.** (a) or (b).
- **d.** none of the above.

24. The number of bearing supports in a precast concrete curtain wall panel must be
- **a.** one.
- **b.** two.
- **c.** three.
- **d.** four.
- **e.** between three and six.

25. The structural design of precast concrete curtain wall panels is generally the responsibility of the
- **a.** structural engineer who designs the structural frame of the building in which the panels are to be used.
- **b.** structural engineer retained by the panel fabricator.
- **c.** structural engineer retained by the general contractor of the building.
- **d.** structural engineer specially retained by the owner.
- **e.** none of the above.

26. During the erection of precast concrete curtain wall panels, the panels are leveled. The leveling of the panels is provided in the
a. panels' bearing supports.
b. panels' tieback connections.
c. (a) or (b).
d. (a) and (b).

27. The bricks used in brick-faced concrete curtain wall panels are generally
a. of the same thickness as those used in brick veneer construction.
b. thicker than those used in brick veneer construction.
c. thinner than those used in brick veneer construction.
d. any one of the above, depending on the building.

28. A bond break such as a polyethylene sheet membrane is generally used between the concrete and bricks in a brick-faced precast concrete curtain wall panel.
a. True
b. False

29. The connection between natural stone facing and the backup concrete in a stone-faced precast concrete curtain wall panel is obtained by
a. The roughness of the stone surface that is in contact with concrete.
b. nonmoving steel dowels.
c. flexible steel dowels.
d. any one of the above.

Answers: 22-c, 23-d, 24-b, 25-b, 26-a, 27-c, 28-b, 29-c.

26.10 GLASS FIBER–REINFORCED CONCRETE (GFRC) CURTAIN WALL

As the name implies, glass fiber–reinforced concrete (GFRC) is a type of concrete whose ingredients are portland cement, sand, glass fibers, and water. Glass fibers provide tensile strength to concrete. Unlike a precast concrete panel, which is reinforced with steel bars, a GFRC panel does not require steel reinforcing.

The fibers are 1 to 2 in. long and are thoroughly mixed and randomly dispersed in the mix. The random and uniform distribution of fibers not only provides tensile strength, but also gives toughness and impact resistance to a GFRC panel. Because normal glass fibers are adversely affected by wet portland cement (due to the presence of alkalies in portland cement), the fibers used in a GFRC mix are alkali-resistant (AR) glass fibers.

GFRC Skin and Steel Frame

A GFRC curtain wall panel consists of three main components:

- GFRC skin
- Light-gauge steel backup frame
- Anchors that connect the skin to the steel backup frame

Of these three components, only the skin is made of GFRC, which is generally $\frac{1}{2}$- to $\frac{3}{4}$-in.-thick. The skin is anchored to a frame consisting of light-gauge (galvanized) steel members, Figure 26.52. In this combination, the GFRC skin transfers the loads to the frame, which, in turn, transfers the loads to the building's structure. The size and spacing of backup frame members depends on the overall size of the panel and the loads to which it is subjected.

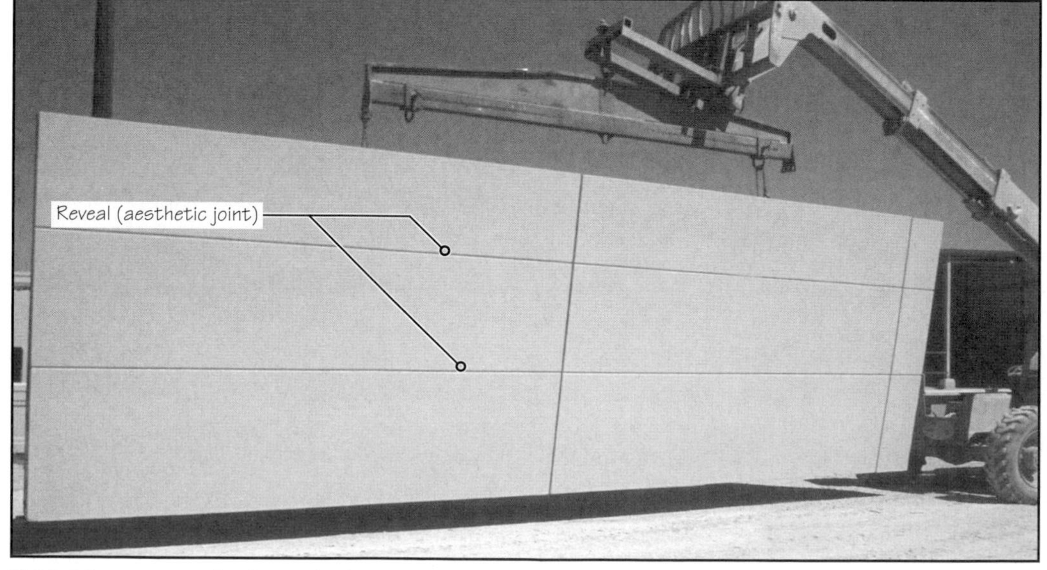

FIGURE 26.52(a) The front of a large GFRC panel after light abrasive blasting. It is being lifted for storage in the fabricator's yard.

FLEX ANCHORS

The skin is hung 2 in. (or more) away from the face of the frame using bar anchors. The gap between the skin and the frame is essential because it allows differential movement between the skin and the frame, particularly during the period when the fresh concrete in the skin shrinks as water evaporates.

The anchors are generally $\frac{3}{8}$-in.-diameter steel bars, bent into an L-shape. To provide corrosion resistance, cadmium-plated steel is generally used for the anchors. One end of an anchor is welded to the frame, and the other end is embedded in the skin. The skin is thickened around the anchor embedment. The thickened portion of the skin is referred to as a *bonding pad*, Figure 26.53.

The purpose of the anchors is to transfer both gravity and lateral loads from the skin to the frame. In doing so, the anchors must be fairly rigid in the plane perpendicular to the panel; that is, they should be able to transfer the loads without any deformation in the anchors.

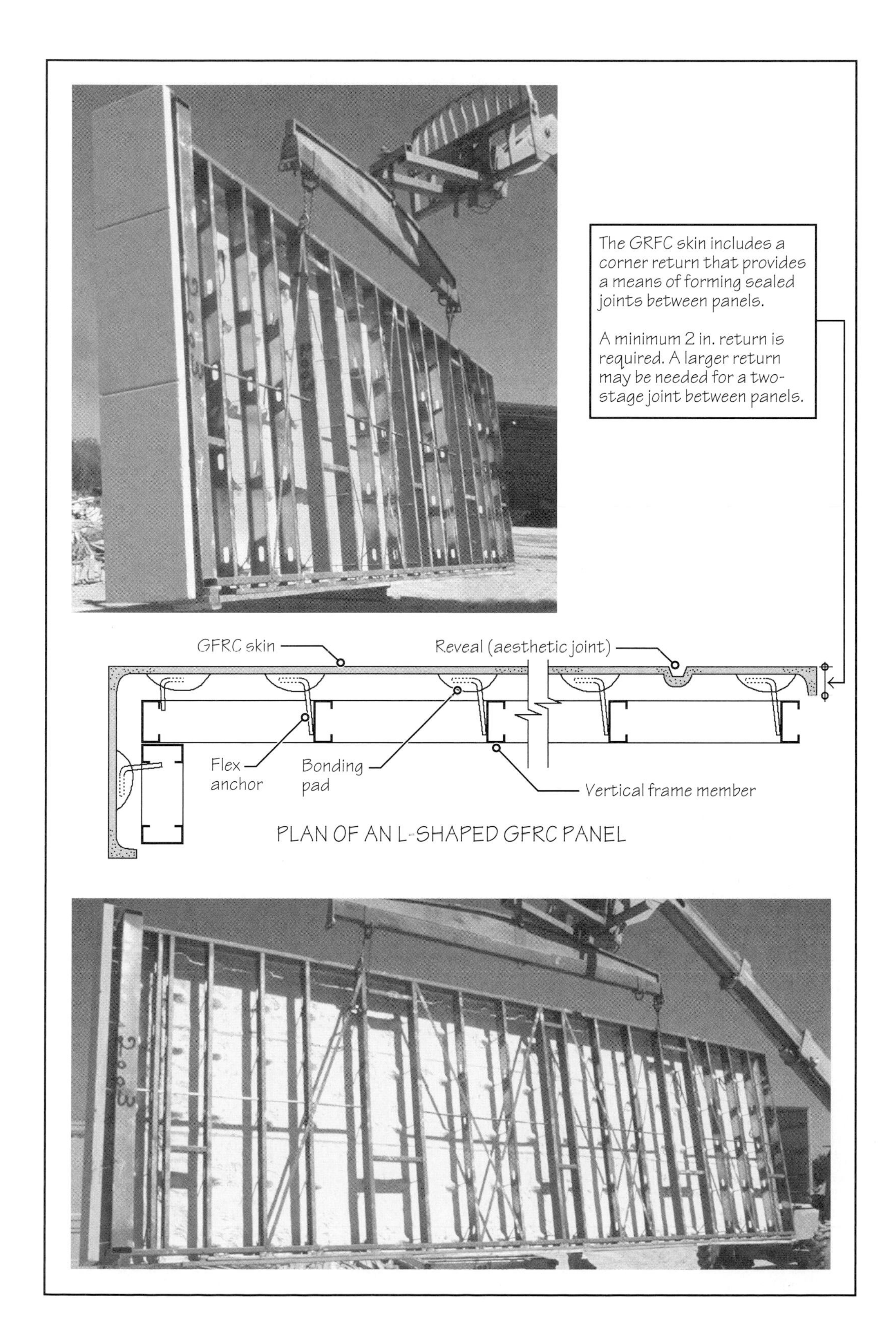

FIGURE 26.52(b) The side and back of an L-shaped GFRC panel showing the supporting steel frame.

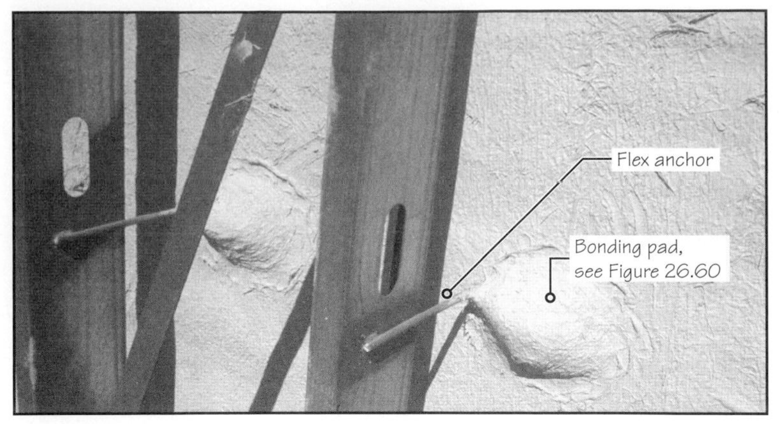

FIGURE 26.53 Bonding pads and flex anchors.

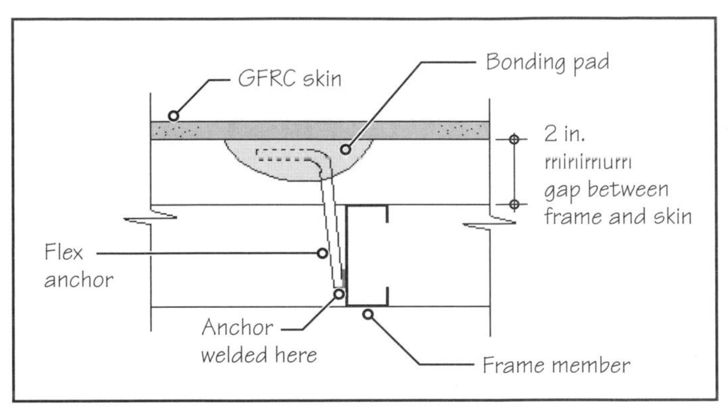

FIGURE 26.54 Flexibility of flex anchors.

However, the skin will experience in-plane dimensional changes due to moisture and temperature effects. The anchor must therefore have sufficient flexibility to allow these changes to occur without excessively stressing the skin. One way of achieving this goal is to flare the anchor away from the frame and to weld it to the frame member at the far end, Figure 26.54. The term *flex anchor* underscores the importance of anchor flexibility.

Panel Shapes

Like the precast concrete curtain wall panels, GFRC curtain wall panels can be made into different shapes, depending on the design of the building's facade. The most commonly used panels are floor-to-floor solid panels (Figure 26.52), window wall panels, Figure 26.55, and spandrel panels. A spandrel panel is essentially a solid panel of smaller height. Because the panels can be configured in several ways for a given facade, the architect must work with the GFRC fabricator before finalizing panel shapes.

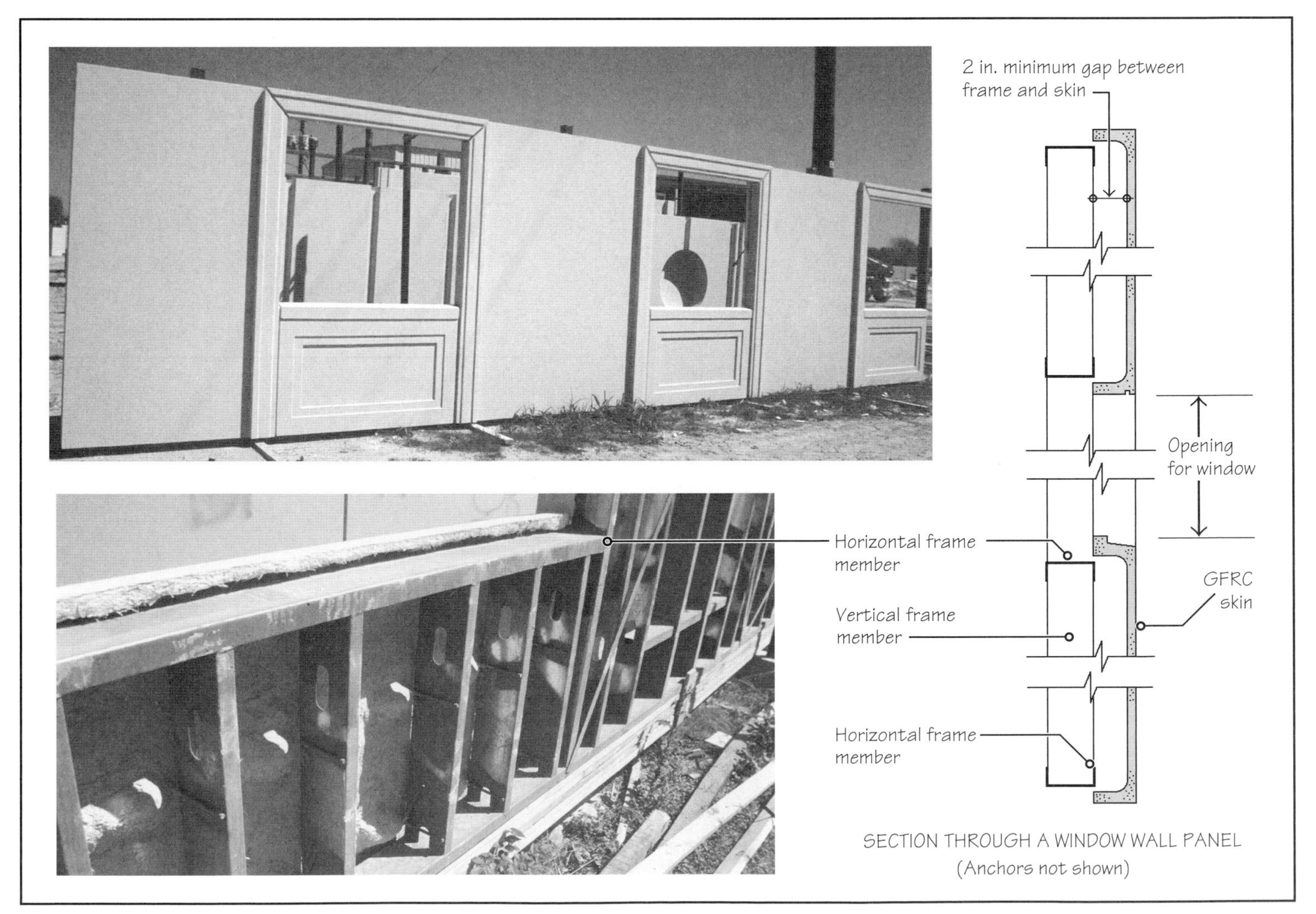

FIGURE 26.55 The front and back of a typical window wall panel.

26.11 FABRICATION OF GFRC PANELS

Figures 26.56 to 26.60 show the six-step process of fabricating a GFRC panel.

- *Preparing the Mold:* The mold must be fabricated to the required shape, Figure 26.56. The mold generally consists of plywood, but other materials, such as steel or plastics, may be used. A form-release compound is applied to the mold before applying the GFRC mix to facilitate the panel's removal from the mold.
- *Applying the Mist Coat:* Before GFRC mix is sprayed on the panel, a thin cement-sand slurry coat, referred to as a *mist coat*, is sprayed on the mold, Figure 26.57. Because the mist coat does not contain any glass fibers, it gives a smooth, even surface to the panel. The thickness of the mist coat is a function of the finish the panel face is to receive. If the panel face is to be lightly abrasive-blasted, a $\frac{1}{8}$-in.-thick mist coat is adequate.
- *Applying the GFRC Mix:* Soon after the mist coat is applied, GFRC mix is sprayed on the mold. See Figure 26.58(a). GFRC mix consists of (white) portland cement and sand slurry mixed with about 5% (by weight) of glass fibers. Because air may be trapped in the mix during spraying, the mix is consolidated after completing the spray application by rolling, tamping, or troweling, Figure 26.58(b).
- *Frame Placement:* After the GFRC spray is completed, the steel backup frame is placed against the skin, leaving the required distance between the skin and the frame, Figure 26.59.
- *Bonding Pads:* Finally, bonding pads are formed at each anchor using the same mix as that used for the skin, Figure 26.60.
- *Removing the Panel from the Mold and Curing:* The panel (skin and frame) is generally removed from the mold 24 h after casting and subsequently cured for a number of days. Because the panel has not yet gained sufficient strength, special care is needed during the panel's removal.

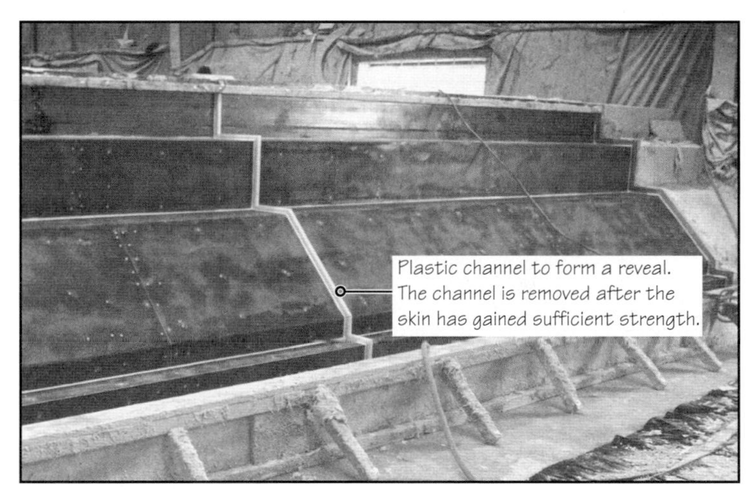

FIGURE 26.56 The mold of a roof cornice panel.

FIGURE 26.57 Application of mist coat.

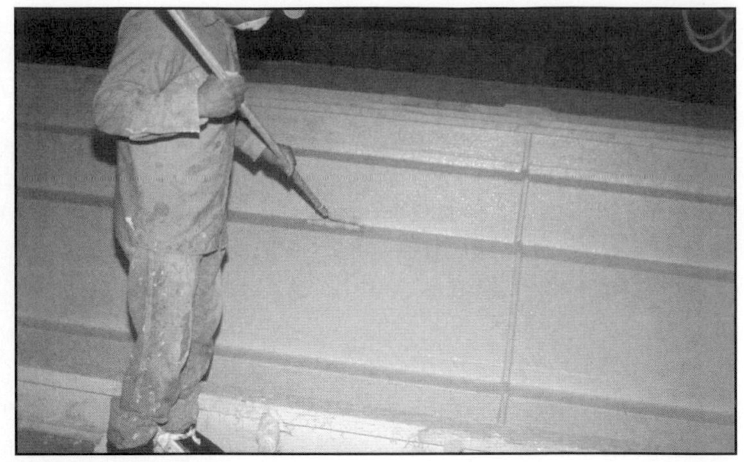

FIGURE 26.58 **(a)** Spraying GFRC mix over the mist coat. **(b)** Consolidating GFRC mix using a roller around bends, edges and corners.

FIGURE 26.59 Placing a frame against the skin.

FIGURE 26.60 Making of a bonding pad (see also Figure 26.53).

Surface Finishes on GFRC Panels

The standard finish on a GFRC panel is a light abrasive-blasted finish to remove the smoothness of the surface obtained from the mist coat. However, a GFRC panel can also be given an exposed aggregate finish, which results in a surface similar to that of a precast concrete panel.

To obtain an exposed aggregate finish, the mist coat is replaced by a concrete layer, referred to as a *face mix*. The thickness of the face mix is about $\frac{3}{8}$ in., and the aggregate used in the face mix is between $\frac{3}{16}$ in. and $\frac{1}{4}$ in. The exposure of aggregate is obtained by abrasive blasting or acid etching.

26.12 DETAILING A GFRC CURTAIN WALL

A GFRC panel is connected to the building's structure in a system similar to that of a precast concrete panel. In other words, a GFRC panel requires two bearing connections and two or more tieback connections. However, because GFRC panels are much lighter than the precast concrete panels, their connection hardware is lighter.

As with precast concrete curtain walls, the structural design of panels and their support connections is generally the responsibility of panel fabricator. Figures 26.61 and 26.62

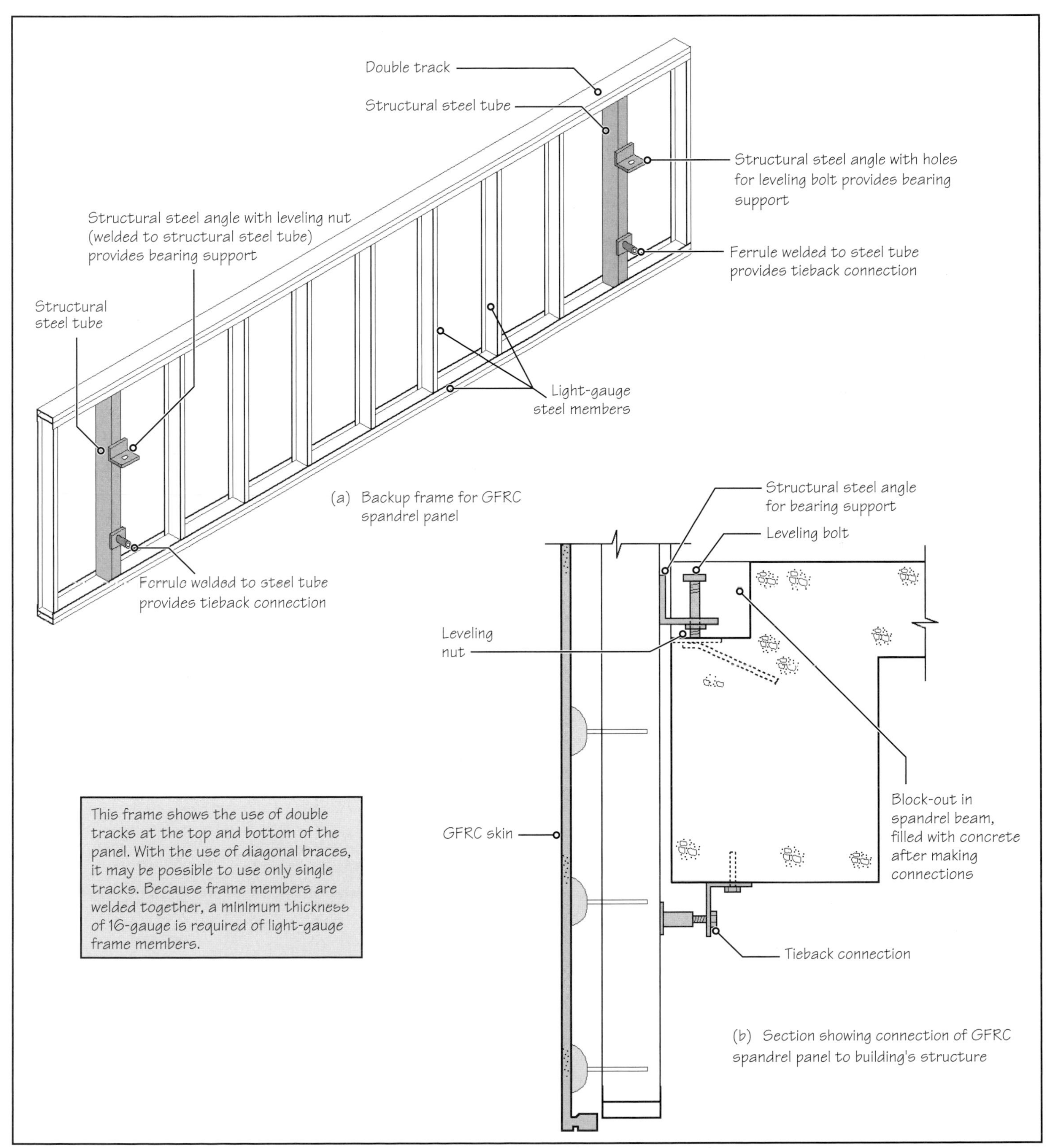

FIGURE 26.61 Schematic details of bearing and tieback connections in a spandrel GFRC panel (see also Figures 26.62 and 26.63).

show two alternative schematics for the connection of a GFRC spandrel panel to the building's structure.

Figure 26.63 shows a typical GFRC curtain wall detail. Because GFRC skin is thin, some form of water drainage system from behind the skin should be incorporated in the panels. Additionally, the space between the GFRC skin and the panel frame should be freely ventilated to prevent the condensation of water vapor. This is generally not a requirement in precast concrete curtain walls. A two-way joint sealant system may be used at panel junctions similar to that for PC curtain walls (Figures 26.50 and 26.51).

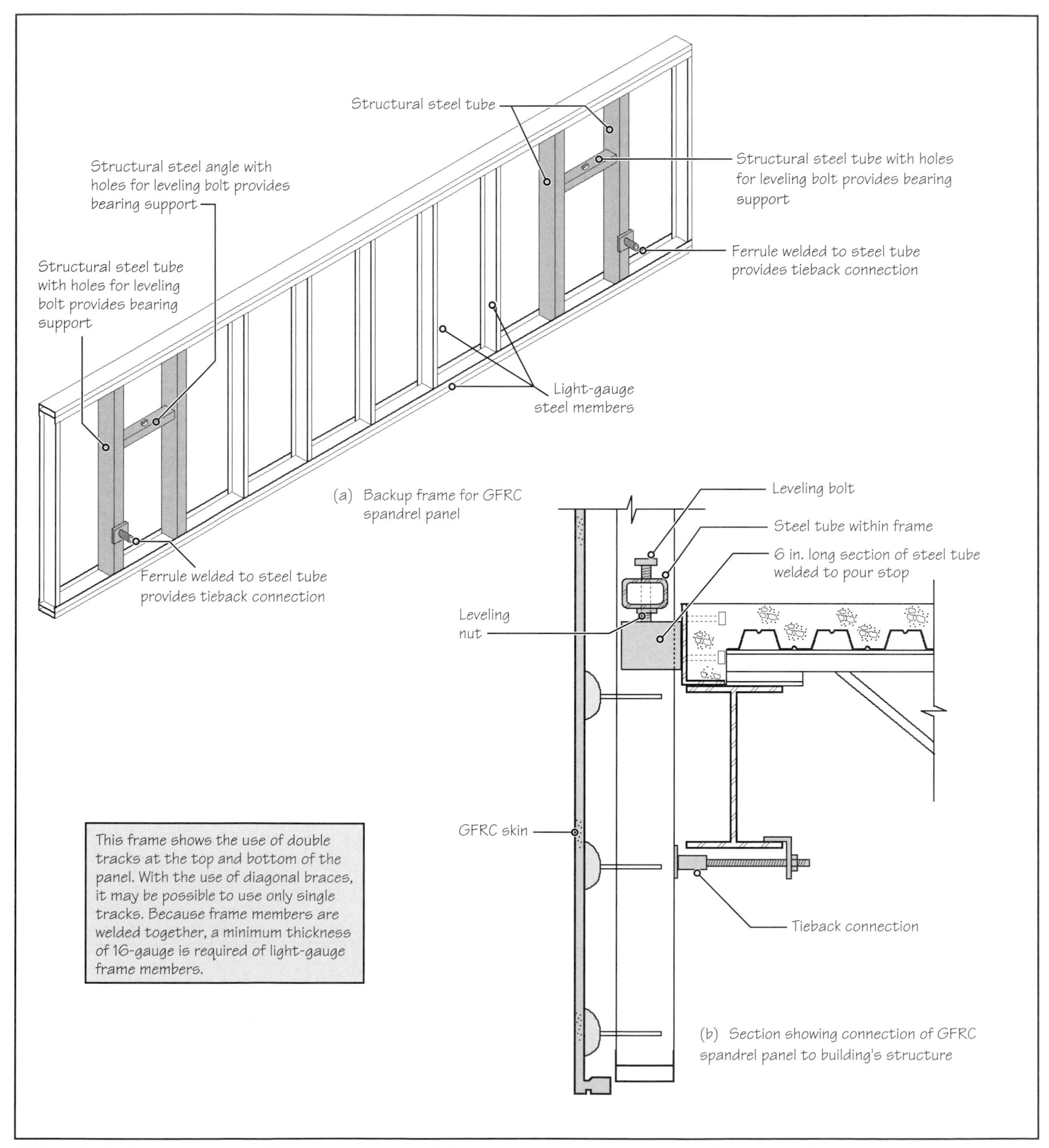

FIGURE 26.62 Schematic details of bearing supports and tieback connections in a spandrel GFRC panel (see also Figures 26.61 and 26.63).

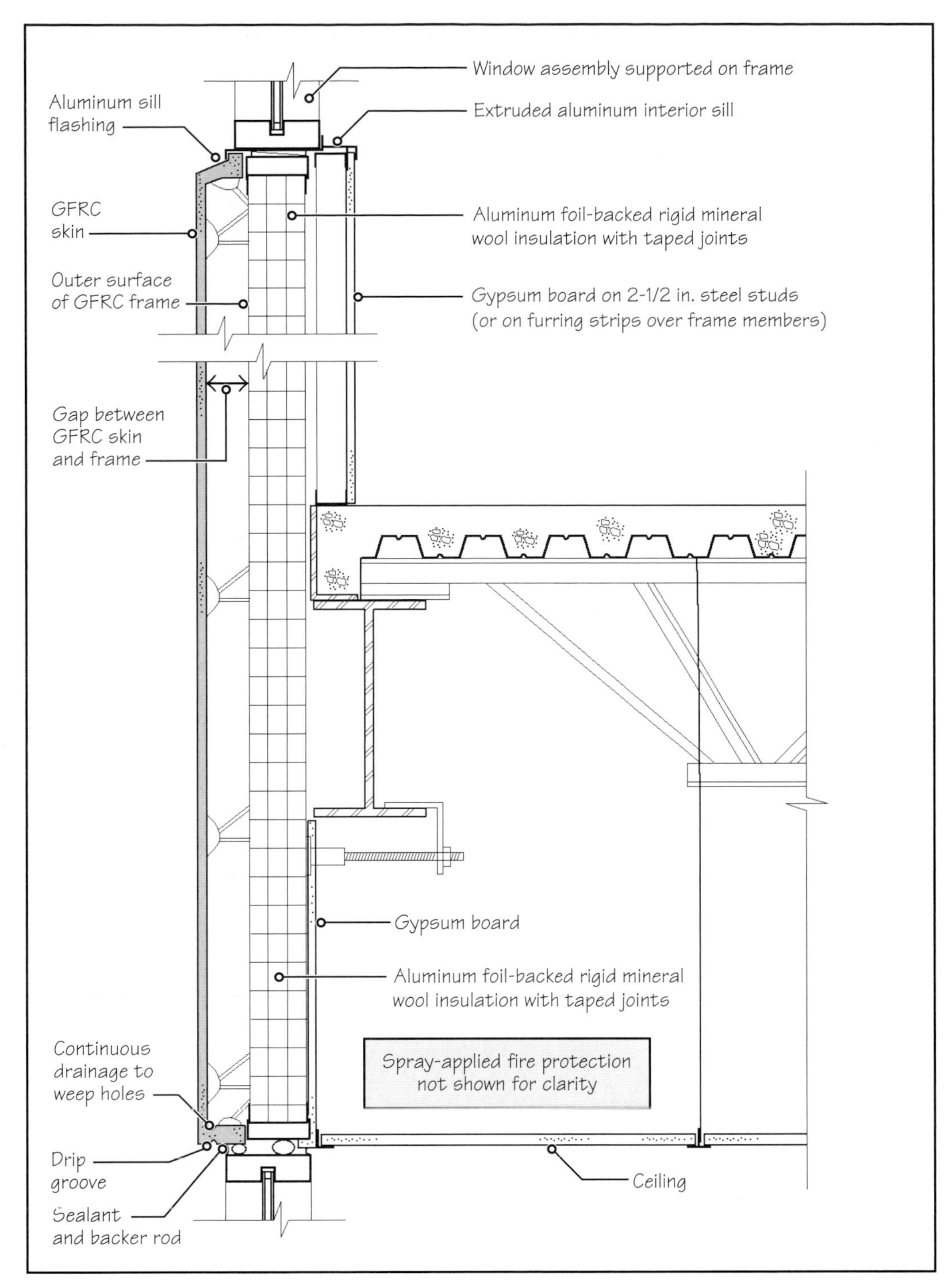

FIGURE 26.63 Schematic detail of an exterior wall with GFRC spandrel panels and ribbon windows.

26.13 PREFABRICATED BRICK CURTAIN WALL

Prefabricated brick panels using full-thickness bricks can also be used as curtain wall panels. The facade of a building with prefabricated brick panels looks similar to that of a building with site-constructed brick veneer. In a facade with prefabricated brick panels, however, the exterior soffit can also be of bricks, giving an all-brick appearance.

Figure 26.64 shows a completed building with prefabricated exterior brick panels. The panels are L-shaped so that the wall and the spandrel and soffit elements are integrated into one panel. Figure 26.65 shows the details and various stages of erection of the panels.

Brick panels are constructed in the masonry contractor's fabrication yard (or indoor plant) and transported to the site. They are reinforced in both directions. Reinforcement in one direction is provided by epoxy-coated steel bars placed into the core holes of bricks. The core holes are filled with portland cement–sand grout. In the other direction, galvanized steel joint reinforcement placed within the bed joints of panels is used.

In addition to reinforcing bars and joint reinforcement, a brick curtain wall panel generally requires two structural steel frames integrated into the brick panel to provide bearing

FIGURE 26.64 A building with prefabricated brick panels. (Photo courtesy of Dee Brown, Inc.)

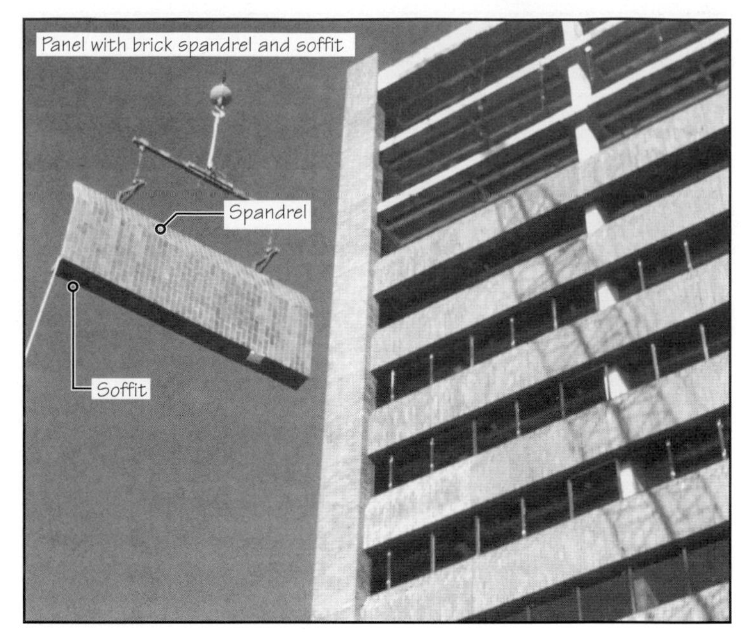

FIGURE 26.65 Erection of prefabricated brick panels in the building of Figure 26.64. (Photo courtesy of Dee Brown, Inc.)

FIGURE 26.66 A close-up of the prefabricated brick panel used in the building shown in Figure 26.64. (Photo courtesy of Dee Brown, Inc.)

supports and tieback connections. The details of the structural frame, used in the panels of the building of Figure 26.64, are shown in Figures 26.66 and 26.67.

A steel stud backup wall incorporating insulation, vapor barrier, and drywall is required with prefabricated brick curtain wall. Additionally, some means must be provided for the drainage of water from behind the panel.

PRACTICE QUIZ

Each question has only one correct answer. Select the choice that best answers the question.

30. The term GFRC is an acronym for
- **a.** glass fiber–reinforced concrete.
- **b.** glass fiber–reinforced cement.
- **c.** glass fiber–restrained concrete.
- **d.** glass fiber–restrained cement.
- **e.** none of the above.

31. A GFRC curtain wall panel consists of
- **a.** a GFRC skin.
- **b.** a light-gauge steel frame.
- **c.** anchors.
- **d.** all the above.
- **e.** (a) and (b).

32. The GFRC skin is typically
- **a.** 2 to 3 in. thick.
- **b.** $1\frac{1}{2}$ to $2\frac{1}{2}$ in. thick.
- **c.** 1 to 2 in. thick.
- **d.** $\frac{1}{2}$ to $\frac{3}{4}$ in. thick.
- **e.** None of the above.

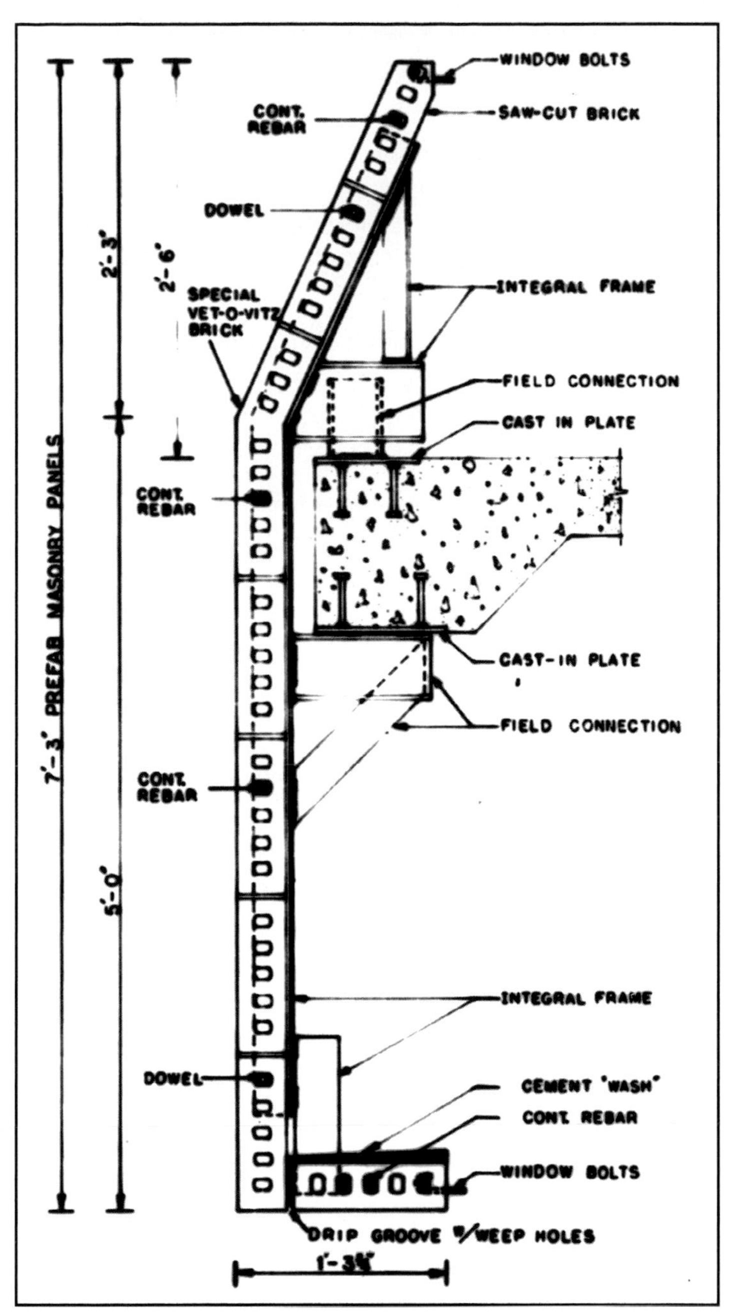

FIGURE 26.67 Shop drawings of the prefabricated brick panel used in the building of Figure 26.64. (Illustration courtesy of Dee Brown, Inc.)

PRACTICE QUIZ *(Continued)*

33. The glass fibers used in GFRC skin are referred to as AR fibers. The term AR is an acronym for
- **a.** alkali resistant.
- **b.** acid resistant.
- **c.** alumina resistant.
- **d.** aluminum reinforced.
- **e.** none of the above.

34. The anchors used in GFRC panels are typically
- **a.** two-piece anchors.
- **b.** one-piece anchors.
- **c.** either (a) or (b), depending on the thickness of GFRC skin.
- **d.** either (a) or (b), depending on the lateral loads to which the panel is subjected.
- **e.** none of the above.

35. The GFRC skin is obtained by
- **a.** casting the GFRC mix in a mold similar to casting a precast concrete member.
- **b.** applying the mix with a trowel.
- **c.** spraying the mix over a mold.
- **d.** any of the above, depending on the panel fabricator.

36. GFRC curtain wall panels are generally lighter than the corresponding precast concrete curtain wall panels.
- **a.** True
- **b.** False

37. In prefabricated brick curtain wall panels, the bricks used are
- **a.** generally of the same thickness as those used in brick veneer construction.
- **b.** generally thicker than those used in brick veneer construction.
- **c.** generally thinner than those used in brick veneer construction.

38. Prefabricated brick curtain wall panels are generally fabricated
- **a.** in a brick-manufacturing plant.
- **b.** in a masonry contractor's fabrication yard.
- **c.** in a precast concrete fabricator's plant.
- **d.** at the construction site.

Answers: 30-a, 31-d, 32-d, 33-a, 34-b, 35-c, 36-a, 37-a, 38-b.

KEY TERMS AND CONCEPTS

Adjustability requirements of two-piece anchor
Air space
Anchors for masonry veneer
Brick ledge
Brick veneer
Brick veneer with CMU backup wall
Brick veneer with steel stud backup wall
Brick veneer with wood stud backup wall
Brick-faced PC curtain wall
Corrugated sheet steel anchor
Expansion joints in brick veneer
Flashing
GFRC skin
Glass fiber-reinforced concrete (GFRC) curtain wall
Lintel angle
Mortar-capturing device
PC curtain wall bearing supports
PC curtain wall tiebacks
Precast concrete (PC) curtain wall
Prefabricated brick curtain wall
Shelf angle
Stone-faced PC curtain wall
Weep holes

REVIEW QUESTIONS

1. Using sketches and notes, explain the adjustability requirements of anchors used in a brick veneer wall assembly.
2. Using sketches and notes, explain the functions of a shelf angle and a lintel angle in a brick veneer assembly.
3. With the help of sketches and notes, explain various ways by which weep holes can be provided in a brick veneer assembly.
4. Discuss the pros and cons of using a CMU backup wall versus a steel stud backup wall in a brick veneer wall assembly.
5. With the help of sketches and notes, explain the support system of a precast concrete curtain wall panel.
6. With respect to a GFRC panel, explain (a) bonding pad and (b) flex anchor.

SELECTED WEB SITES

Brick Industry Association, BIA (www.bia.org)
International Masonry Institute, IMI (www.imiweb.org).
Magazine of Masonry Construction (www.masonryconstruction.com)
Precast/Prestressed Concrete Institute, PCI (www.pci.org)

FURTHER READING

1. American Concrete Institute (ACI). "Tilt-Up Concrete Structures," *Manual of Concrete Practice.*
2. Brick Industry Association (BIA). *Technical Notes on Brick Construction.*
3. Brock, Linda: *Designing the Exterior Wall,* New York: John Wiley & Sons, 2005.
4. Canadian Mortgage and Housing Corporation (CMHC). *Architectural Precast Concrete Walls—Best Practice Guide,* 2002.
5. Precast/Prestressed Concrete Institute (PCI). *Architectural Precast Concrete,* 1989.
6. Precast/Prestressed Concrete Institute (PCI). *Recommended Practice for Glass Fiber Reinforced Panels.*
7. Tilt-Up Concrete Association (TCA) and American Concrete Institute (ACI). *The Tilt-Up Design and Construction Manual.*

REFERENCES

1. American Concrete Institute: *Building Code Requirements and Commentary for Masonry Structures and Specifications for Masonry Structures and Related Commentaries,* ACI530/530.1.
2. Brick Industry Association (BIA): "Brick Veneer Steel Stud Walls," *Technical Notes on Brick Construction* (28B).

CHAPTER 27 Exterior Wall Cladding–II (Stucco, EIFS, Natural Stone, and Insulated Metal Panels)

CHAPTER OUTLINE

27.1 Portland Cement Plaster (Stucco) Basics

27.2 Stucco on Steel- or Wood-Stud Walls

27.3 Stucco on Masonry and Concrete Substrates

27.4 Limitations and Advantages of Stucco

27.5 Exterior Insulation and Finish System (EIFS) Basics

27.6 Application of Polymer-Based EIFS

27.7 Impact-Resistant and Drainable EIFS

27.8 Exterior Cladding with Dimension Stone

27.9 Field Installation of Stone—Standard Set Method

27.10 Field Installation of Stone—Vertical Support Channel Method

27.11 Prefabricated Stone Curtain Walls

27.12 Thin Stone Cladding

27.13 Insulated Metal Panels

Curved copper panels clad the Coucil Chamber at the Redmund, WA, City Hall. Architect: Mulvanny/G2 Architecture. (Photo by DA.)

This chapter continues the discussion of exterior-wall cladding systems. Topics discussed in this chapter are portland cement plaster (stucco), exterior insulation and finish systems (EIFS), stone cladding, and insulated metal wall panels.

27.1 PORTLAND CEMENT PLASTER (STUCCO) BASICS

Plaster has been used for centuries as an exterior and interior wall and ceiling finish. Apart from rendering the wall and ceiling surfaces smooth and paintable, plastering makes them more resistant to water and air infiltration and increases sound insulation and resistance to fire.

A plaster mix is similar to masonry mortar mix and consists of cementitious material(s), sand, and water. In some plasters, a fibrous admixture is also used. Prior to the discovery of portland cement and gypsum, lime was the only cementitious material available for plaster. In contemporary construction, gypsum and portland cement are the primary cementitious materials. Because gypsum is not a hydraulic cement (i.e., it will dissolve in water), gypsum plaster is suitable only for interior applications, not subjected to high humidity levels or wetting.

As noted in Chapter 14, gypsum plaster has largely been replaced by (prefabricated) gypsum boards. Therefore, we limit the discussion to portland cement plaster. Unlike gypsum plaster, portland cement plaster can be used on both interior and exterior surfaces.

The predominant use of portland cement plaster in contemporary buildings is as an exterior wall finish—the topic of discussion in this chapter. Its use as an interior finish is limited to situations where high humidity levels or wetting of the plastered surfaces occur, such as in saunas, public shower rooms, and commercial kitchens. In most parts of the United States, exterior portland cement plaster is referred to as *stucco.*

Although stucco can be applied on adobe walls, it is more commonly used on

- Light-gauge steel stud walls
- Wood stud walls
- Masonry walls
- Concrete walls

Because stucco is a portland cement–based material, the application of stucco requires appropriate temperature conditions. Generally, stucco should be applied if the ambient air temperature is at least 40°F (and rising).

MIX COMPOSITION FOR STUCCO COATS

Stucco is typically applied sequentially in three coats over a wall. The first two coats constitute the *base coat* for the application of the third coat, called the *finish coat.* The mix for both base coats is essentially the same, containing portland cement, lime, sand, and water. These ingredients are site blended in the same way as masonry mortar.

Portland cement is the glue that bonds all constituents of the mix, which eventually cure into a strong and rigid surface. Lime imparts plasticity and cohesiveness to the mix. Plasticity implies that the mix can be spread easily, and cohesiveness implies that the mix will hold and not sag on a vertical surface during its application. Because masonry cement gives essentially the same properties as lime (Section 22.2), lime may be replaced by masonry cement.

Unlike the base coat, the finish coat is generally factory blended with integral color and is available in bags, Figure 27.1. Water is the only additional material needed to prepare the finish coat for application. A factory-blended mix (in place of a site-blended mix) has the advantage of providing consistent quality from batch to batch. Even with factory-blended mix, the finish coat is applied continuously, with interruptions only at the control and expansion joints.

Two types of factory-blended mixes for the finish coat are available:

- Portland cement–based mix
- Acrylic polymer–based mix

The use of an acrylic polymer–based finish coat is more common because it is more flexible, reduces cracking of the stucco surface, and provides consistent and relatively nonfading colors. However, portland cement finish coat is more breathable than polymer-based finish.

FIGURE 27.1 Bags containing factory-blended finish coat mix.

27.2 STUCCO ON STEEL- OR WOOD-STUD WALLS

The anatomies of wood and light-gauge steel frame walls with stucco finish are essentially identical, as shown in Figure 27.2(a). Both wall assemblies require an exterior (gypsum, plywood, OSB, or cement board) sheathing, a water-resistant membrane, and a self-furring metal base.

The water-resistant membrane is the second line of defense against water intrusion. The first line of defense is the stucco finish itself. In most projects, one layer of asphalt-saturated paper (kraft paper) is used as the water-resistant membrane. It is applied horizontally with laps between sheets. Where greater redundancy is required or where wood-based sheathing

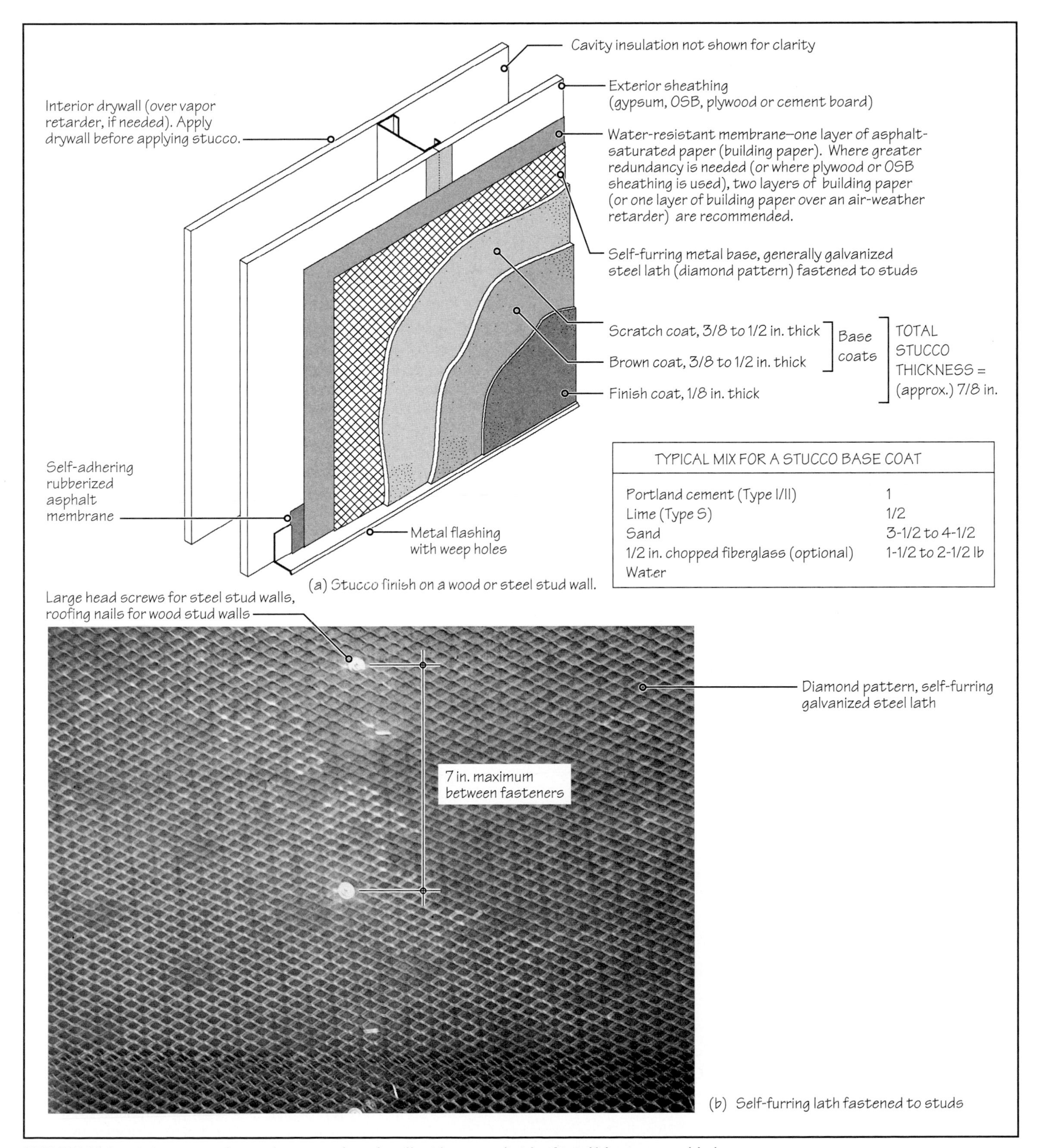

TYPICAL MIX FOR A STUCCO BASE COAT	
Portland cement (Type I/II)	1
Lime (Type S)	1/2
Sand	3-1/2 to 4-1/2
1/2 in. chopped fiberglass (optional)	1-1/2 to 2-1/2 lb
Water	

FIGURE 27.2 **(a)** Anatomy of a steel- or wood-stud wall with stucco finish. **(b)** Self-furring metal lath.

NOTE

Building Paper Backing for Stucco—The Drainage Plane

The building paper absorbs a certain amount of water during the application of stucco. As the paper and stucco dry, the paper wrinkles and the stucco shrinks, leaving small vertical drainage channels behind the stucco. Although the drainage channels are small (and unlike an air space in a drainage wall), they help in keeping a stucco wall dry. The multitude of drainage channels are referred to as a *drainage plane.*

A proprietary air-retarder-type backing for stucco is also available that has a specially made crinkled surface, providing a drainage plane similar to the one that develops in a paper-backed stucco.

(OSB or plywood) is used, two layers of kraft paper or one layer of kraft paper backed by an air retarder is recommended.

The metal base generally consists of self-furring galvanized steel lath (diamond pattern), Figure 27.2(b). The lath is fabricated from steel sheets, which are slit at regular intervals and then stretched. For this reason, the lath is also known as *expanded metal lath.* The lath sheets are finally hot-dip galvanized, as needed.

The lath provides a mechanical key to which the stucco bonds. The self-furring character of the lath is obtained by incorporating dimples or other means in the lath during the stretching process that hold it about $\frac{1}{4}$ in. away from the substrate. Thus, when the stucco coat is applied, the lath is embedded within it, becoming an integral part of the stucco, much like the reinforcing bars in a reinforced concrete slab.

Self-furring lath is also available with a continuous backing of kraft paper integral with the lath. This combination increases work efficiency, particularly in large stucco projects.

In a wood-stud wall, the lath is fastened to the studs with large-head nails. In a light-gauge steel-stud wall, the lath is anchored to the studs with self-drilling, self-tapping screws with large (nearly $\frac{1}{2}$-in.-diameter) heads. Because the lath is anchored to the studs, the stucco is structurally engaged to the studs.

APPLICATION OF STUCCO

Of the two base coats, the first is called the *scratch coat* and the second is called the *brown coat.* The scratch and brown coats are each nearly $\frac{3}{8}$ in. thick and the finish coat is nearly $\frac{1}{8}$ in. thick, giving a total stucco thickness of approximately $\frac{7}{8}$ in.

A base coat may be applied by hand using a traditional hawk and a trowel. For medium- and large-size jobs, however, the mix is typically sprayed on a wall, Figure 27.3.

For the scratch coat, the sprayed-on material is troweled with sufficient pressure to squeeze it through the lath so that the lath is embedded completely within the scratch coat. Generally, one plasterer sprays the mix, followed by another plasterer hand troweling the sprayed material. After the troweling operation, the material is scratched (hence the name *scratch coat*).

A notched trowel is used for both operations. Its flat surface is used to squeeze the coat into the lath, and its notched edges are used to scratch the surface. On a wall (vertical surface), horizontal grooves are scratched, which serve as the mechanical key for the following base (brown) coat, Figure 27.4.

The mix for the brown coat is is also sprayed on the surface in the same way as the scratch coat. However, the sprayed-on material is brought to an even plane using a wood or metal float, which also densifies the applied material, Figure 27.5. After floating, the surface is troweled with a steel trowel, which smoothes the surface further and prepares it for the finish coat.

The time interval between the scratch and brown coats depends on the ambient temperature and humidity conditions. It is important that the scratch coat develop sufficient strength and rigidity to carry the weight of the brown coat without damaging the grooves.

FIGURE 27.3(a) Spraying of stucco base coat over metal lath.

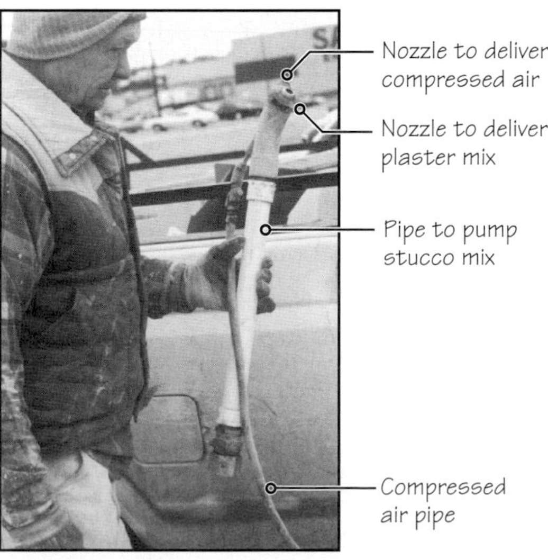

FIGURE 27.3(b) The pipe used to spray stucco mix ends into a nozzle. The spray action for the mix is provided by a separate compressed-air pipe that also ends into a nozzle close to the plaster mix nozzle.

FIGURE 27.4 Application of scratch coat. Observe that one plasterer sprays the mix, followed by another plasterer, who trowels and scores the surface.

FIGURE 27.5 Application of the brown coat.

At the same time, the scratch coat should not set and dry so much that its bond with the subsequent coat is compromised.

A 24- to 48-h interval between the scratch and brown coats is commonly used in most climates. If a larger interval is used, the scratch coat should be moist cured until the application of the brown coat.

The finish coat is obviously the most important component of the stucco finish because it provides the required color and texture. It can be applied by hand or sprayed. The sprayed finish may be left as is or textured using a sponge float or other suitable means. A longer time interval (generally, 7 to 10 days) between the brown coat and the finish coat is required to allow the base (scratch and brown) coats to stabilize from shrinkage.

Some finish coat manufacturers require two applications of their material. The first application is that of a primer, which prepares the base coat surface to receive the finish coat and improves adhesion between the base coat and the finish coat. The finish coat develops the final stucco color and texture.

CONTROL JOINTS AND EXPANSION JOINTS

Because of the presence of portland cement, shrinkage is an inherent feature of a stucco surface, which leads to its cracking. Although the cracking of stucco cannot be fully eliminated, it can be controlled by providing closely spaced control joints.

ASTM standards require that the area between control joints in stucco applied on a stud-wall assembly not exceed 144 ft^2, with neither dimension of this area exceeding 18 ft. As far as possible, control joints should also be provided around openings in walls, Figure 27.6.

FIGURE 27.6 Control joints on a stucco facade.

In addition to control joints, expansion joints are also needed in a stucco finish. Although control joints are a means of controlling the shrinkage of stucco finish, the expansion joints respond to large movements in the building structure. Thus, expansion joints should be provided at each floor level in the exterior wall in a multistory building to absorb the movement in the spandrel beam, Figure 27.7. Expansion joints are also needed where there is a major change in building elevation or where a stucco-finished wall abuts a wall made of a different material.

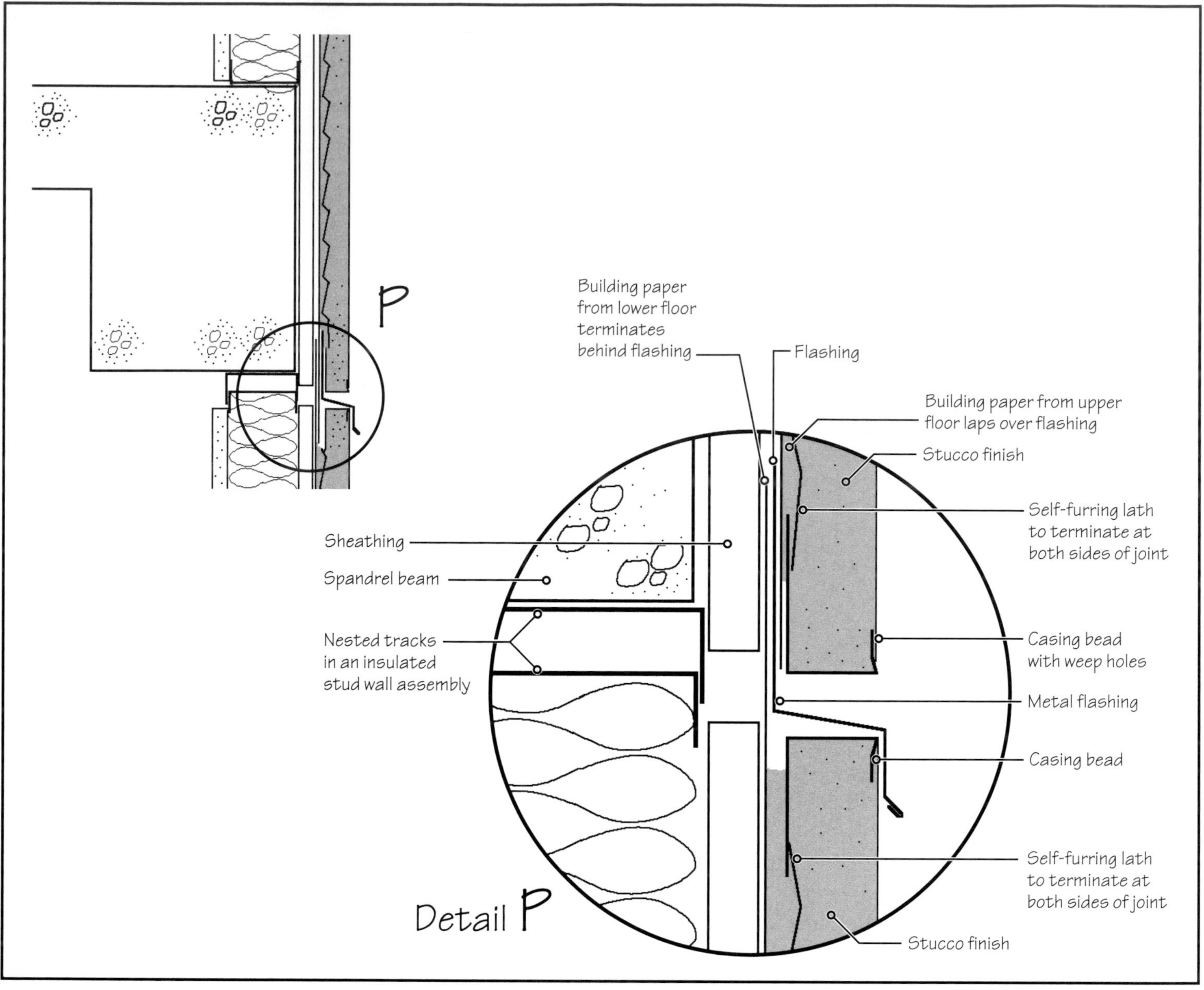

FIGURE 27.7 A typical horizontal expansion joint detail in a stucco-clad exterior stud wall. For a vertical expansion joint, a two-piece expansion-joint accessory may be used (Figure 27.9(c)).

NOTE

Control Joint Locations on Stucco Applied on Wood- or Steel-Stud Walls

- The maximum area between control joints is 144 ft^2.
- The maximum length or width of an area between control joints is 18 ft.
- Length-to-width ratio of an area between control joints should lie between 1 and 2.5.

If the movement of the spandrel beam is not transferred to the wall, the wall needs only control joints. This is shown in Figure 27.8, which illustrates the detailing of a typical control joint and the termination of stucco at the foundation level.

Control joints and expansion joints are constructed using accessories, Figure 27.9. The flanges of joint accessories consist of expanded metal or a slotted material so that they can integrate with the lath and be fastened to the studs in the same way as the lath.

A control joint accessory consists of a single piece, whereas an expansion joint is a two-piece accessory. In the case of an expansion joint, the lath must terminate on both sides of the joint. The lath should, however, be continuous under a control joint.

Apart from the control and expansion joint accessories, several other accessories and trims, such as casing beads, corner beads, and flashing, are required in a typical stucco project. A casing bead is used at the terminal edge of stucco surface. Therefore, it is also referred to as *stucco stop*.

An exterior corner bead provides a straight vertical or horizontal intersection between two surfaces. It also guards against chipping at corners from impact and establishes the thickness of the stucco finish. An interior corner bead is generally not needed with stucco.

The materials used for joint accessories and trims are zinc, galvanized steel, and PVC. In a highly corrosive environment, the use of zinc or PVC should be considered, depending on the local experience. Where accessories abut, the joints at the ends of two lengths or their intersections should be sealed with exterior-grade sealant.

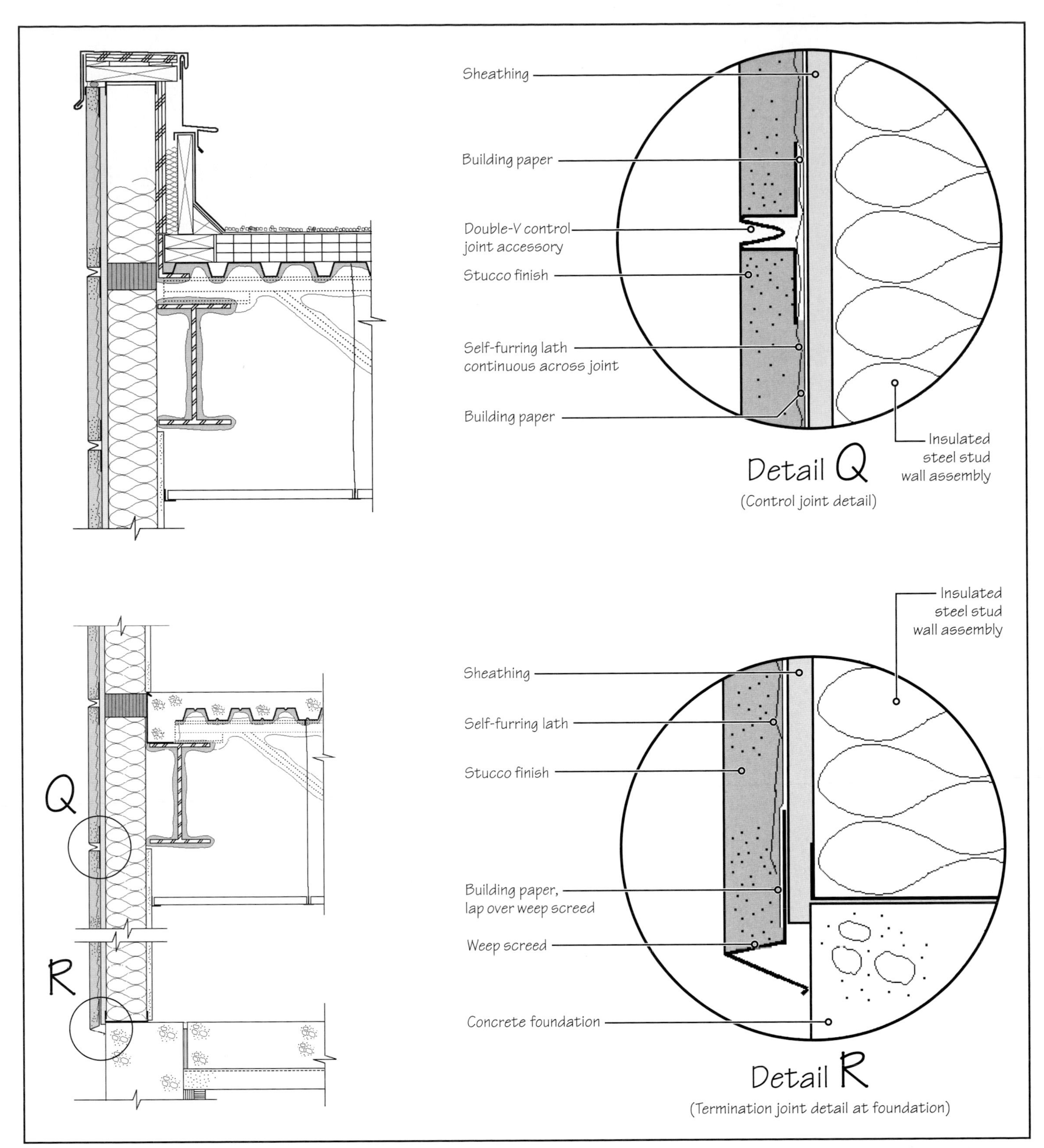

FIGURE 27.8 A typical wall section and corresponding details of a low-rise steel-frame building with stucco finish on the exterior wall.

Rigidity of Stud Wall Assembly

Because a fully cured stucco surface is relatively thin ($\frac{7}{8}$ in.) and brittle, it is important that the back-up wall assembly the reasonably rigid. A flexible assembly will aggravate cracking, leading to rapid deterioration of the wall from water penetration, freeze-thaw damage, and so on.

Building codes and standards require that the deflection of wood- and steel-stud wall assemblies clad with stucco be controlled to a maximum of span divided by 360 (L/360). More-stringent deflection-design criterion (e.g., maximum deflection not to exceed L/420) will obviously give a better-performing stucco wall.

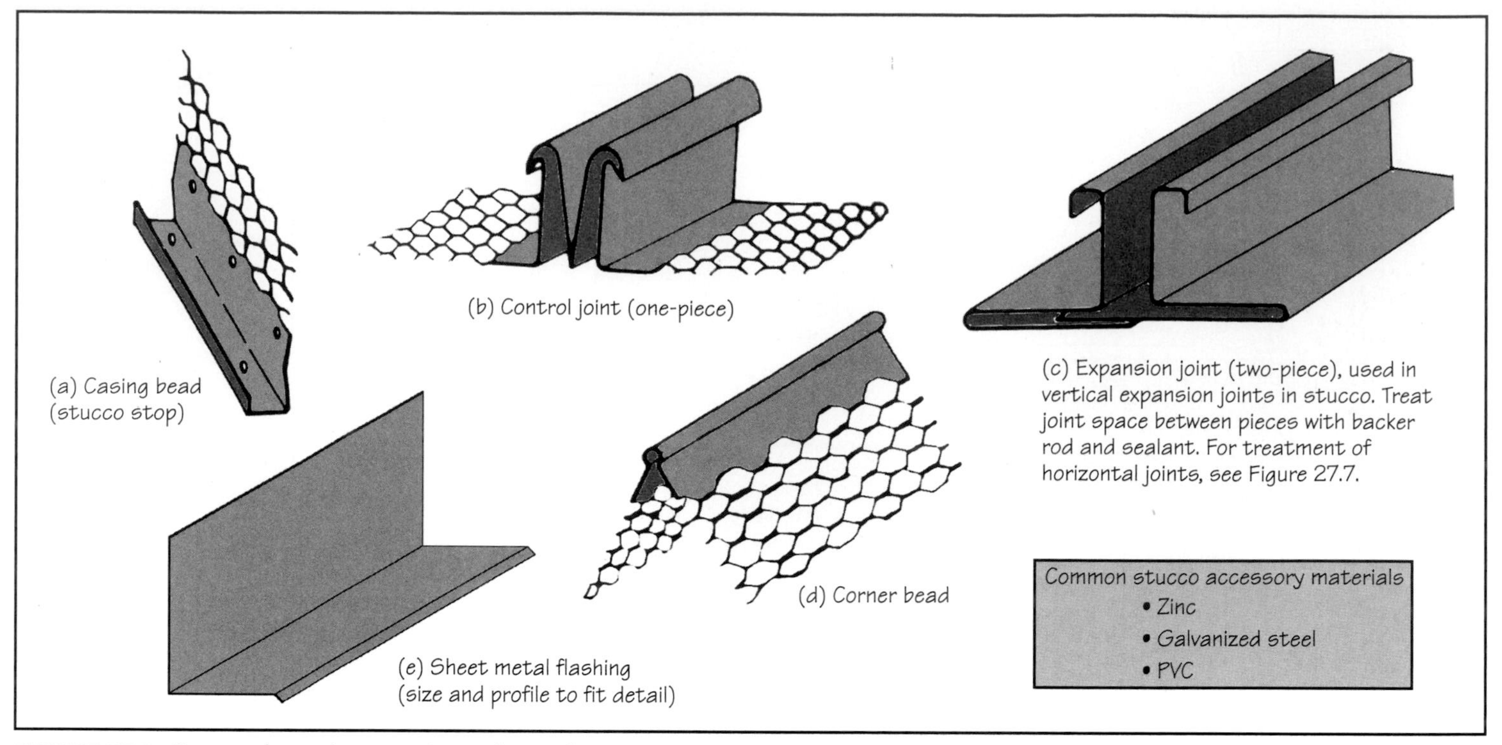

FIGURE 27.9 Commonly used accessories and trims for stucco.

NOTE

Deflection of Stud-Wall Assemblies

The deflection of stud walls that receive stucco finish should not exceed the span divided by 360, that is, L/360 caused by lateral loads. Here, L is the vertical span of studs. For steel studs, L is generally the distance between the bottom track and the top track in the assembly.

Steel-stud manufacturers provide tables for the selection of their studs to conform to the deflection criteria. See Section 26.4 for the corresponding deflection criterion for brick veneer backed by steel studs.

For wood studs, L refers to the distance between the bottom plate and the top plate in the wall assembly.

One-Coat Stucco

Several manufacturers have developed a one-coat stucco, in which the traditional two base coats (scratch coat and brown coat) are replaced by one base coat. The mix for the base coat, which is proprietary to the manufacturer, includes glass fibers in addition to portland cement and lime. Sand and water are the only ingredients required to prepare the base coat material.

The base coat for one-coat stucco is generally $\frac{1}{2}$ in. thick and is applied on the lath in the same way as the scratch coat but is finished smooth like the brown coat. A thicker base coat may be needed to meet fire-resistance rating, as required by the applicable building code. The finish coat is applied on one-coat stucco in the same way as on the traditional two-coat stucco.

One-coat stucco reduces labor and time. The glass fibers in the base coat help reduce cracking by increasing its flexural strength and impact resistance.

27.3 STUCCO ON MASONRY AND CONCRETE SUBSTRATES

Masonry is an excellent substrate for stucco because it is far more rigid than a stud wall. Additionally, the surface roughness and porosity of masonry yields a good bond for the stucco finish. Therefore, stucco applied on a masonry wall does not need lath. (Remember that a self-furring lath was required on a stud wall to develop a surface to which stucco would bond. The bond between stucco and concrete masonry is particularly strong because both are portland cement–based materials.)

Stucco applied on a masonry wall usually consists of two coats (a base coat to smoothe any irregularities on the wall's surface and a finish coat) with a total thickness of $\frac{5}{8}$ to $\frac{3}{4}$ in., Figure 27.10. To retain the natural roughness of masonry, the mortar joints in masonry are left flush, i.e., they are not tooled. Obviously, the masonry surface must be clean and free from defects that may compromise the bond between stucco and masonry.

Because masonry is porous, it may suck water from the mix, leaving insufficient water in stucco. Therefore, a masonry surface may need prewetting before applying the base coat.

A concrete wall is neither as absorptive nor as rough as a masonry wall. Therefore, light sand blasting followed by a liquid bonding coat is generally required on a concrete wall, as recommended by the stucco manufacturer. The total thickness of stucco on a concrete wall is nearly the same as on masonry wall.

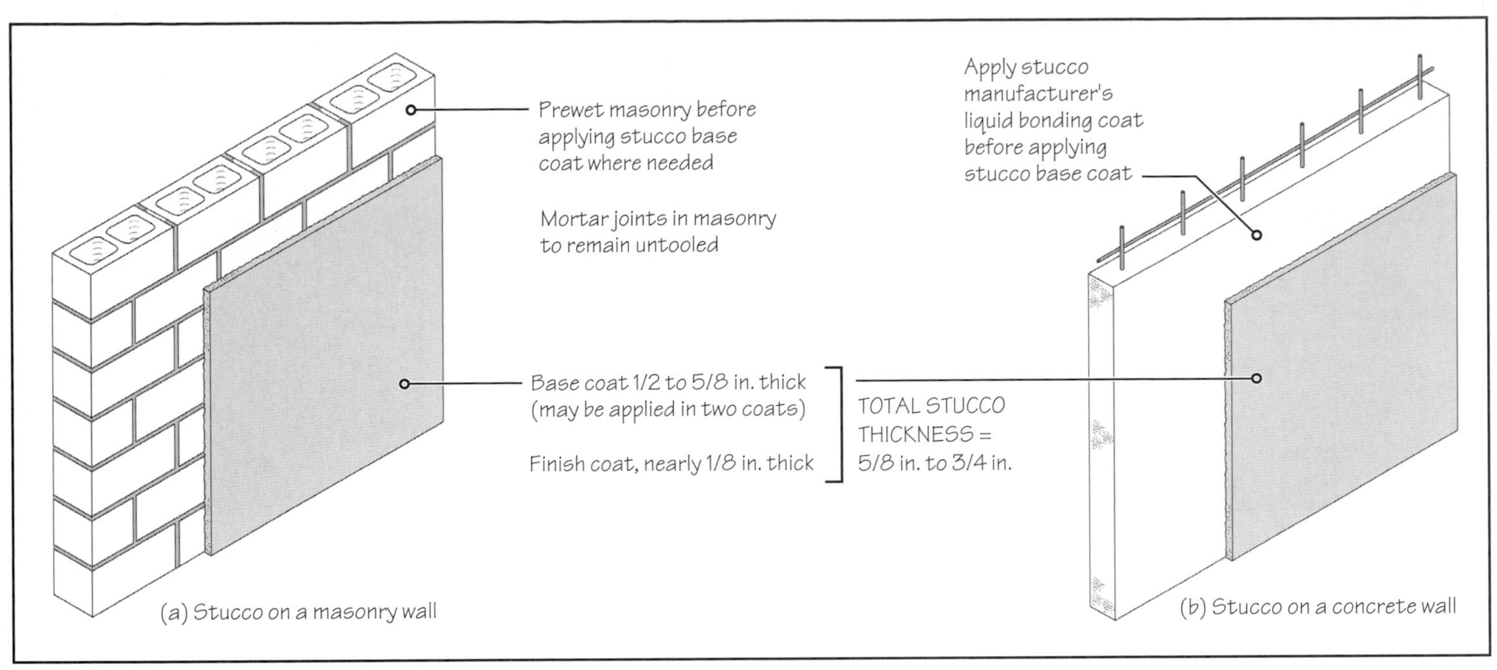

FIGURE 27.10 Anatomies of stucco-finished **(a)** masonry wall and **(b)** concrete wall.

JOINTS IN STUCCO-FINISHED MASONRY AND CONCRETE WALLS

Because masonry and concrete walls are more rigid than wood- or steel-stud walls, the control joints in stucco applied on them can be spaced farther apart. The recommended maximum area of a stucco panel between control joints on masonry or concrete substrates is 250 ft^2.

As far as possible, control joints and expansion joints in stucco should be in the same locations as the corresponding joints in substrate masonry or concrete. The control joints and other accessories are fastened to masonry or concrete using masonry or concrete nails.

NOTE

Control Joint Locations on Stucco Applied on Masonry or Concrete Walls

- Maximum area between control joints = 250 sq. ft.
- Other requirements are the same as for the stud walls.

27.4 LIMITATIONS AND ADVANTAGES OF STUCCO

A stucco-clad masonry or concrete wall is a barrier wall (Chapter 25). It does not have any means of draining rainwater out of the assembly should rainwater permeate through the stucco. A stucco-clad steel- or wood-stud framed wall, on the other hand, is not a barrier wall.

Although the anatomy of a stucco-clad framed wall does not qualify it to be called a drainage wall, water penetration tests have indicated that if the water intrudes through stucco cladding under a long spell of wind-driven rain, it is able to drain out from over the water-resistant barrier behind stucco (i.e., through the *drainage plane*). However, good artisanry, rigidity of substrate, use of control joints, and expansion joints and sealants are essential for a well-performing stucco assembly, particularly in wet climates.

Additionally, because stucco is a portland cement–based material, it is a breathable surface. Should water permeate stucco, it will begin to evaporate as soon as the rain stops and the wall begins to dry. (With an acrylic polymer–based finish coat, the evaporation of water may be slower.)

Although its relative thinness and light weight may present some challenges in keeping a stucco-clad wall dry, it is an advantage in seismic zones. Remember from Section 3.7, the earthquake load on a building component is directly related to its dead load. Consequently, a stucco-finished wall is subjected to a smaller earthquake load than a wall clad with heavier finishes, such as masonry veneer or concrete curtain wall.

Additionally, because stucco is an adhered finish (in contrast with anchored masonry veneer), it is better able to resist the vibrations caused by an earthquake. A stucco finish is often recommended for use in relatively dry and seismically active regions.

Another advantage of stucco is the clarity of form that results from its use. The variety of colors that are available add to the aesthetics of a stucco-finished building. Architects have exploited this attribute of stucco, as shown in the two images of Figure 27.11.

(a)

FIGURE 27.11(a) Beach Apartments, Albuquerque, New Mexico, Architect: Antoine Predock.

(b)

FIGURE 27.11(b) Visual Arts Center, College of Santa Fe, New Mexico. Architect: Ricardo Legoretta.

PRACTICE QUIZ

Each question has only one correct answer. Select the choice that best answers the question.

1. In addition to water, a stucco mix consists of
- **a.** gypsum, lime, and sand.
- **b.** gypsum, portland cement, and sand.
- **c.** portland cement, lime, and sand.
- **d.** lime, glass fibers, and sand.

2. On wood-stud and light-gauge steel-stud wall assemblies, stucco is typically applied in three coats. Beginning from the first to the last, the coats are generally called
- **a.** brown coat, scratch coat, and finish coat.
- **b.** scratch coat, brown coat, and finish coat.
- **c.** prime coat, scratch coat, and finish coat.
- **d.** prime coat, brown coat, and finish coat.
- **e.** prime coat, base coat, and finish coat.

3. On wood-stud and light-gauge steel-stud wall assemblies, stucco is typically applied in three coats. The total typical thickness of three coats is
- **a.** $1\frac{1}{2}$ in.
- **b.** $1\frac{1}{4}$ in.
- **c.** $1\frac{1}{8}$ in.
- **d.** 1 in.
- **e.** $\frac{7}{8}$ in.

4. The required color of a stucco surface is
- **a.** obtained by spray painting the surface with the required color after the finish coat has fully cured.
- **b.** obtained by spray painting the surface with the required color immediately following the finish coat.
- **c.** obtained by using colored portland cement and lime.
- **d.** integral to the mix of all coats used in stucco.
- **e.** integral to the mix of the finish coat.

5. Metal lath is required when stucco is applied on
- **a.** stud-wall assemblies.
- **b.** concrete walls.
- **c.** masonry walls.
- **d.** all the above.
- **e.** none of the above.

6. Metal lath, when used in stucco, is installed on the wall
- **a.** before the application of the first stucco coat.
- **b.** sandwiched between the first and second stucco coats.
- **c.** directly under the finish coat.
- **d.** between the second and the finished stucco coats.

7. Control joints on a stucco surface are provided primarily to
- **a.** control thermal expansion and contraction of metal lath.
- **b.** control thermal expansion and contraction of stucco surface.
- **c.** control differential movement between stucco surface and the substrate.
- **d.** control drying shrinkage of stucco.

8. Control joints on a stucco surface are formed by using accessories, which consist of
- **a.** galvanized steel.
- **b.** zinc.
- **c.** PVC.
- **d.** any one of the above.

9. Control joints are required to be more closely spaced on a stucco surface backed by a stud wall than on a stucco surface backed by a masonry wall.
- **a.** True
- **b.** False

10. The total thickness of stucco on a concrete or concrete masonry wall is generally
- **a.** the same as that on a stud-backed wall.
- **b.** greater than that on a stud-backed wall.
- **c.** smaller than that on a stud-backed wall.

11. When control joints are provided in a stucco clad wall, expansion joints are unnecessary.
- **a.** True
- **b.** False

Answers: 1-c, 2-b, 3-e, 4-e, 5-a, 6-a, 7-d, 8-d, 9-a, 10-c, 11-b.

27.5 EXTERIOR INSULATION AND FINISH SYSTEM (EIFS) BASICS

Exterior insulation and finish system (abbreviated EIFS and generally pronounced as "eefs") consists of a layer of a rigid polystyrene foam insulation, a fiberglass-reinforcing mesh, a polymer-based base coat, and a polymer-based finish coat. The reinforcing mesh is embedded in the base coat.

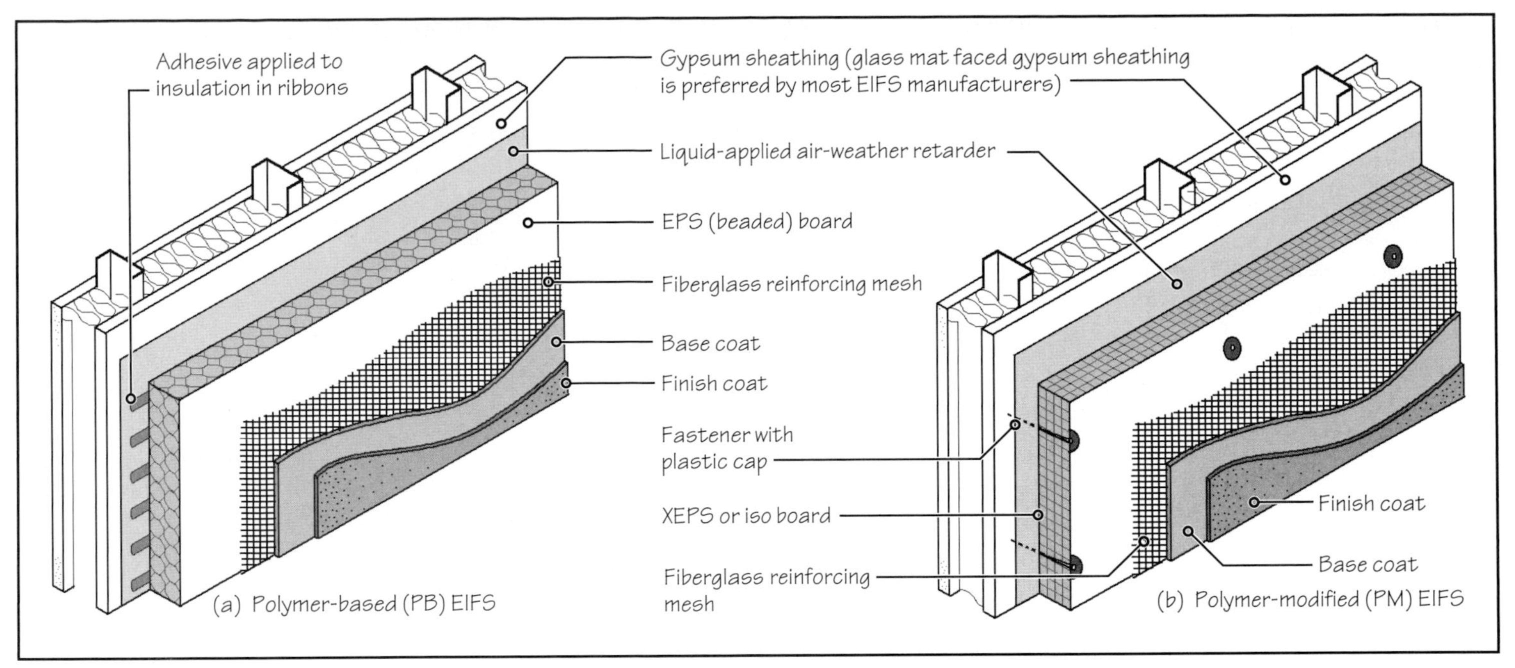

FIGURE 27.12 Difference between PB and PM EIFS.

Because, in its finished appearance, an EIFS-clad facade looks like a stucco facade, it is also referred to as *synthetic stucco* to distinguish it from portland cement stucco—the *conventional stucco.*

The description of EIFS just given is that of a basic system. More sophisticated (and, hence, more complex) systems are available (Section 27.7). The important fact to recognize is that an EIFS assembly is a system because it comprises several chemically complex materials.

Compatibility between system parts and also among the substrate, trims and joint accessories, flashings, sealants, and so on, must be ensured. Therefore, all materials for the assembly must be obtained from one manufacturer and applied per the manufacturer's instructions. The manufacturer's warranty for the system is generally contingent on the use of its approved material distributor, certified applicator, and recommended construction details.

The EIFS Industry Members Association (EIMA), a trade organization representing EIFS manufacturers, classifies an EIFS under two categories:

- Polymer-based (PB) EIFS, also called *soft-coat* EIFS
- Polymer modified (PM) EIFS, also called *hard coat* EIFS

In the PB system, the insulation consists of expanded polystyrene (EPS, i.e., molded and beaded) boards, which are adhered to the substrate, Figure 27.12. The total thickness of PB EIFS lamina (mesh, base coat, and finish coat) is approximately $\frac{1}{8}$ in.

The PM system uses extruded polystyrene (XEPS) or polyisocyanurate (iso) boards, which are anchored to the substrate using steel screws and plastic caps. The base coat on a PM system consists of a polymer-modified portland cement and is at least $\frac{1}{4}$ in. thick.

The PB system is far more commonly used and is the one described here in further detail. Although most manufacturers' systems are similar, there are differences between them. Therefore, the manufacturer's literature should be consulted for more authoritative information.

27.6 APPLICATION OF POLYMER-BASED EIFS

EIFS is a versatile cladding and is used for all types of projects—low-rise, midrise and high-rise buildings—in commercial as well as residential projects. It can be used over wood frame, light-gauge steel-stud frame, concrete, or masonry walls.

The use of rigid foam insulation in EIFS has two major advantages. First, it is energy efficient because placing insulation on the exterior of a wall assembly results in a higher R-value than when it is placed within or toward the interior of the assembly (Section 5.9).

Second, the foam insulation allows a great deal of detail work to be easily incorporated on the facade. The foam can be molded to shape, it can be cut out to provide grooves for surface relief, and its thickness can be varied to give accent bands. The surface detail remains virtually unchanged after the application of the base and finish coats because of their relatively small total thickness.

FIGURE 27.13(a) To make a selection, three alternative facade alternatives were requested by the architect from the EIFS subcontractor for use in a highrise condominium project (Figure 27.13(b)).

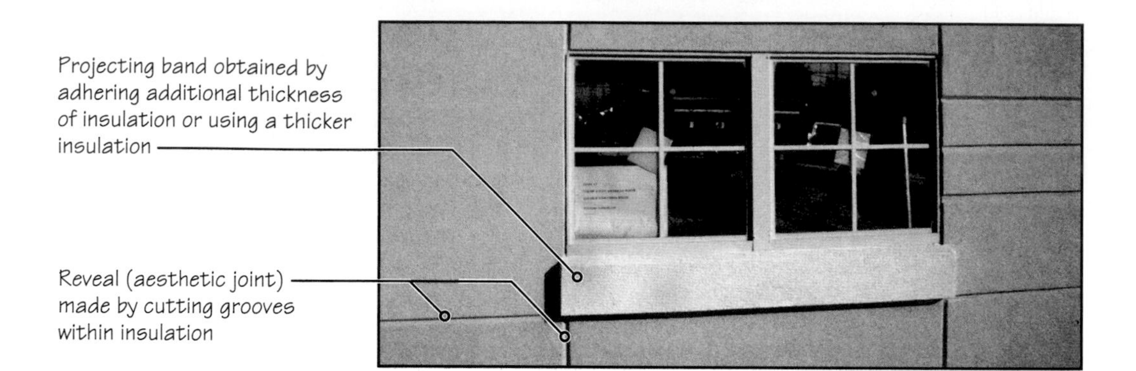

FIGURE 27.13(b) A close-up view of one of the mock-up panels shown in Figure 27.13(a).

Third, a large variety of colors are available in an EIFS finish coat. Most EIFS-clad facades are richly detailed and variously colored. Few other cladding materials lend themselves to such ornateness and bright, relatively fade-resistant colors as EIFS. Therefore, a mock-up panel of EIFS cladding is all the more important in most projects, Figure 27.13.

Fourth, EIFS does not require control joints, such as those used with stucco, adding to its versatility. Fifth, EIFS is one of the least expensive exterior wall cladding available today—less than any other system discussed in this and the previous chapter.

EIFS is, however, not without problems. Its low impact resistance and the lack of ability to breathe (in contrast with portland cement stucco) create problems (Section 27.7).

Installation of Insulation

The first step in the application of PB EIFS is to adhere the insulation over the liquid-applied air-weather retarder placed over the substrate. The adhesive is applied to the back of the insulation in ribbons using a notched trowel, Figure 27.14. (Alternatively, adhesive daubs and perimeter strips may be used, Figure 27.15(a).) After applying the adhesive, the board is pressed against the substrate, to which it clings instantaneously because of its light

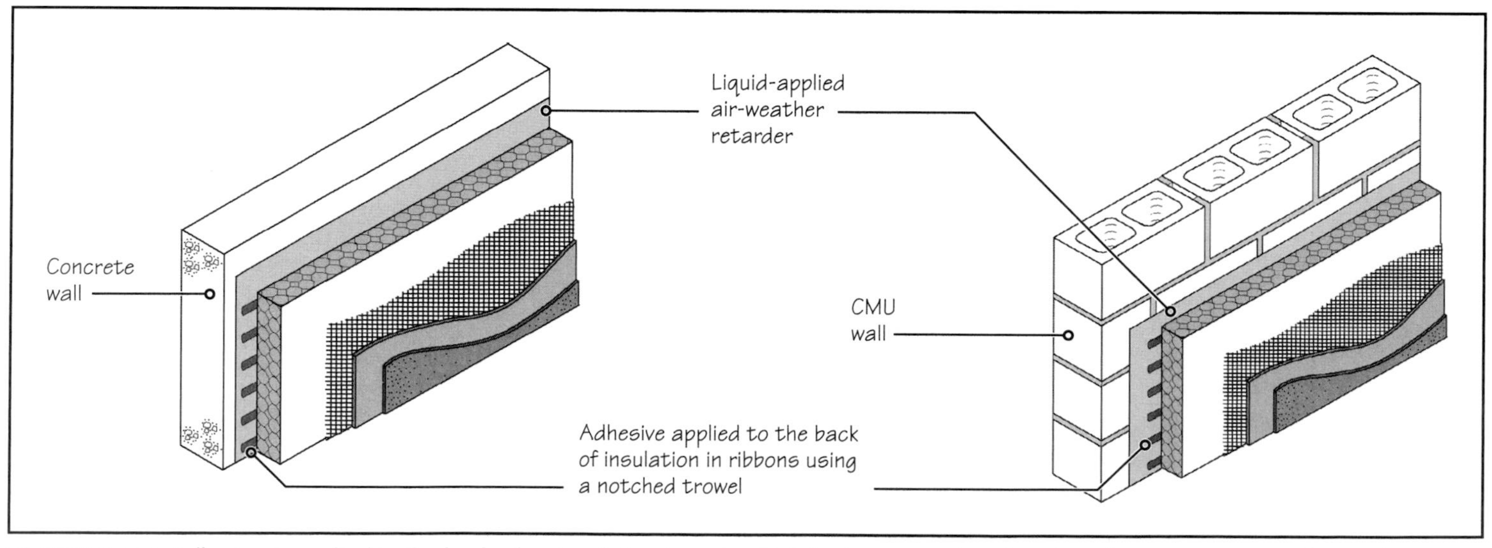

FIGURE 27.14 Adhesive is applied to the back of an insulation board, which is then pressed against the substrate—gypsum, plywood, or OSB sheathing, concrete, or CMU wall.

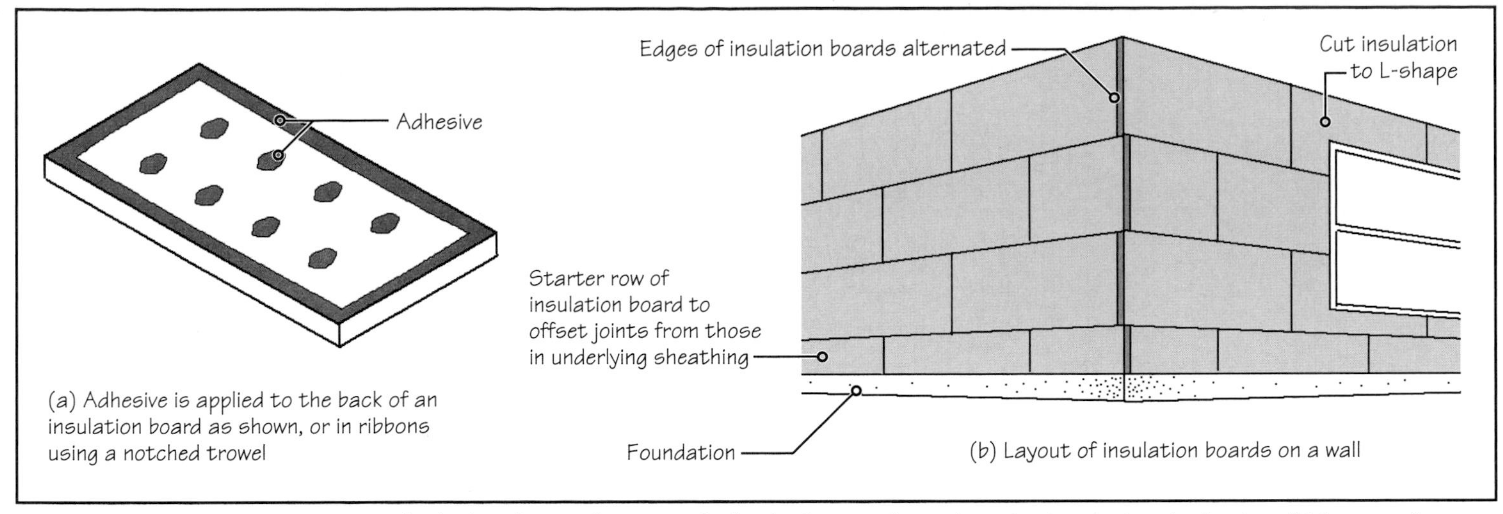

FIGURE 27.15 **(a)** An alternative method of applying adhesive to the back of an insulation board using daubs of adhesive. **(b)** Layout of insulation boards on a wall.

weight. The boards are arranged in a running bond pattern to avoid continuous joints, Figure 27.15(b).

Once the insulation boards have been installed, they are sanded smooth, a process referred to as *rasping*. Rasping removes surface irregularities resulting from an uneven substrate or uneven insulation boards, Figure 27.16. Generally more than one sander grade is needed to obtain the required surface. If aesthetic grooves are needed, they are cut into the insulation boards at this stage, Figure 27.17. The minimum thickness of insulation behind the groove must at least be $\frac{3}{4}$ in.

FIGURE 27.16 These images show the process and tools used to rasp insulation boards after they have been adhered to the substrate in this highrise condominium project.

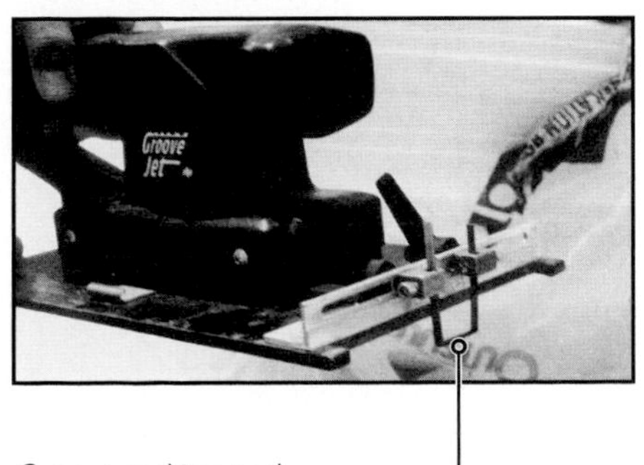

FIGURE 27.17 Cutting of reveals aesthetic grooves in insulation after it has been adhered to the substrate with the help of groove-making tool and a level.

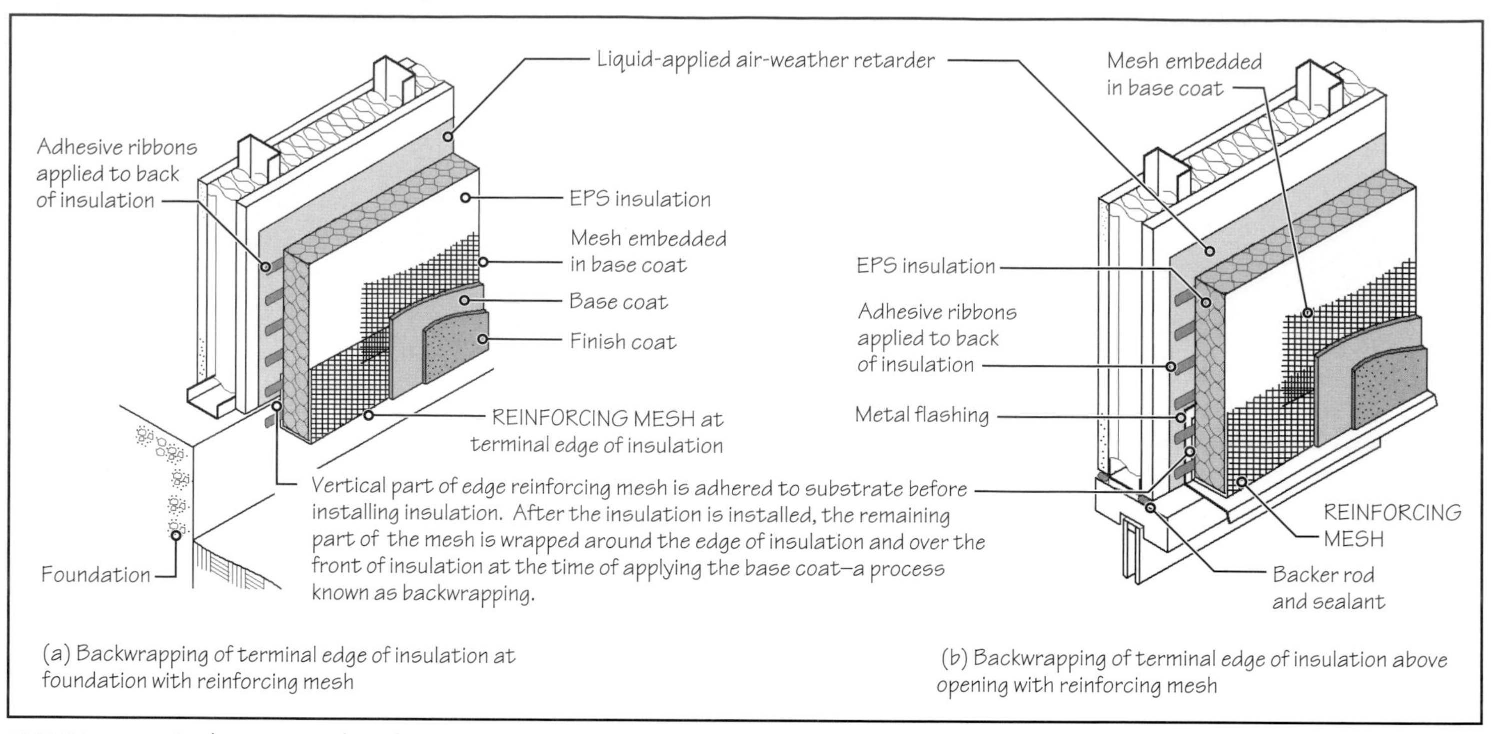

FIGURE 27.18 Backwrapping of insulation—wrapping of the terminal edges of insulation with reinforcing mesh.

Application of EIFS Lamina (Base Coat, Mesh, and Finish Coat)

Before applying the base coat, the terminal edges of insulation boards—at foundation and around openings—are wrapped with *reinforcing mesh,* a process known as *backwrapping,* Figure 27.18. The reinforcing mesh is embedded in the base coat. Backwrapping strengthens the edges and functions like casing bead in conventional stucco.

After wrapping the edges, the base coat is applied. Then immediately after, the mesh is unrolled over the base-coat and, using the trowel and some additional base-coat material, the mesh is fully embedded into the base coat, Figure 27.19. The finish coat is either sprayed or troweled on the base coat.

Movement Control in an EIFS Wall

Because both the base coat and the finish coat are polymer based, the EIFS lamina is a relatively flexible membrane. Therefore, an EIFS-clad wall does not require any control joints. Expansion joints are, however, required where large movements in the structure or substrate are expected. These locations are (a) at the spandrel beam level, (b) where an EIFS-clad wall abuts a wall of a different material, (c) at a major change in wall elevation, and (d) at a building expansion joint. Figure 27.20 shows the detailing of a typical expansion joint in an EIFS wall.

FIGURE 27.19 The mesh is unrolled over the base coat layer and, using a trowel and additional base coat material from the hawk, the applicator embeds the mesh into the base coat.

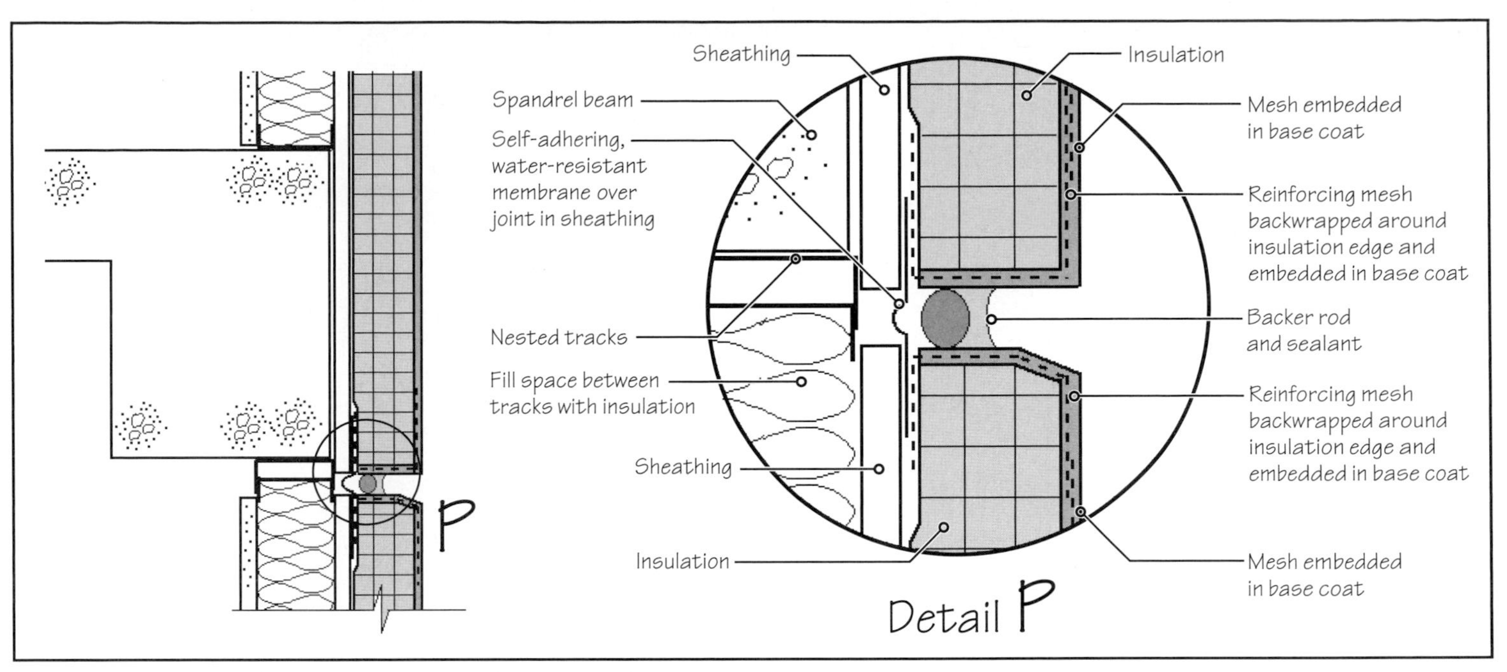

FIGURE 27.20 A typical expansion joint detail in an EIFS-clad wall.

27.7 IMPACT-RESISTANT AND DRAINABLE EIFS

As stated earlier, there are two major concerns in the standard EIFS cladding system described in the previous section. The first concern is its low resistance to impact. This is due to the low compressive strength of EPS insulation and the thinness of EIFS lamina. Damage by hailstorms has been reported on EIFS cladding, particularly on horizontal surfaces, such as aesthetic bands and window sills.

The second concern is that an EIFS-clad wall is a barrier wall. Being polymer based, it does not breathe as well as a stucco-clad wall. Any water that permeates through an EIFS cladding will generally take a long time to evaporate. This leads to an overall deterioration of the wall assembly and may result in the growth of mold in some walls. Therefore, good artisanry and quality control during application and good detailing with respect to flashing and sealants, particularly around openings and terminations, is extremely important in an EIFS-clad wall. To address these concerns, EIFS manufacturers have introduced an impact-resistant EIFS and a drainable EIFS.

Impact-Resistant EIFS

Impact-resistant EIFS consists of two layers of base coat. The first layer is embedded within a thicker reinforcing mesh, and the second layer has the same mesh as the standard EIFS, Figure 27.21. Although the impact-resistant EIFS has higher compressive strength than the

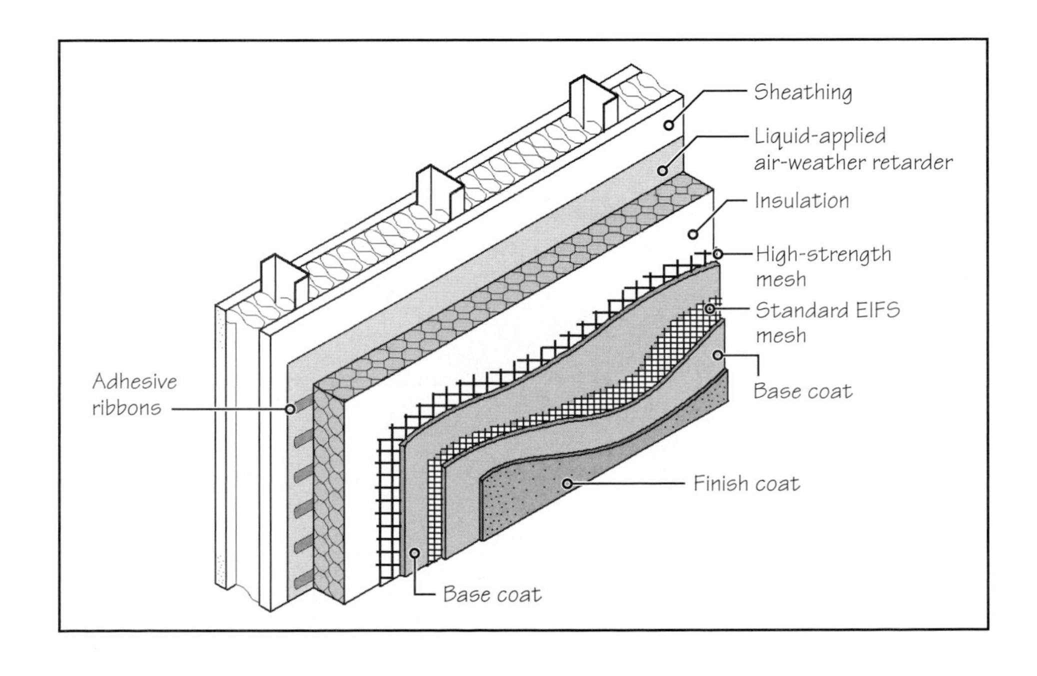

FIGURE 27.21 Anatomy of an impact-resistant EIFS assembly.

FIGURE 27.22 In this high-rise residential project, masonry veneer has been used in the lower floors and EIFS has been used in the upper floors because of the low impact resistance of EIFS.

standard EIFS, it is not equivalent to the compressive strength of most other cladding systems. Therefore, in many projects, the first few floors of the building are clad with a high-compressive-strength cladding, such as masonry veneer, concrete, and stucco, and the higher floors are clad with EIFS, Figure 27.22.

Drainable EIFS

Two types of drainable EIFS wall systems are available. The system with insulation adhered to the substrate has vertical drainage grooves at the back of the insulation, Figure 27.23.

The other system, in which the insulation is mechanically anchored to the substrate, has a thin drainage mat and building paper behind the insulation, Figure 27.24. The drainage mat is typically made of interwoven plastic strands. As an alternative to the mat and building paper, one manufacturer uses a layer of synthetic material with drainage lines.

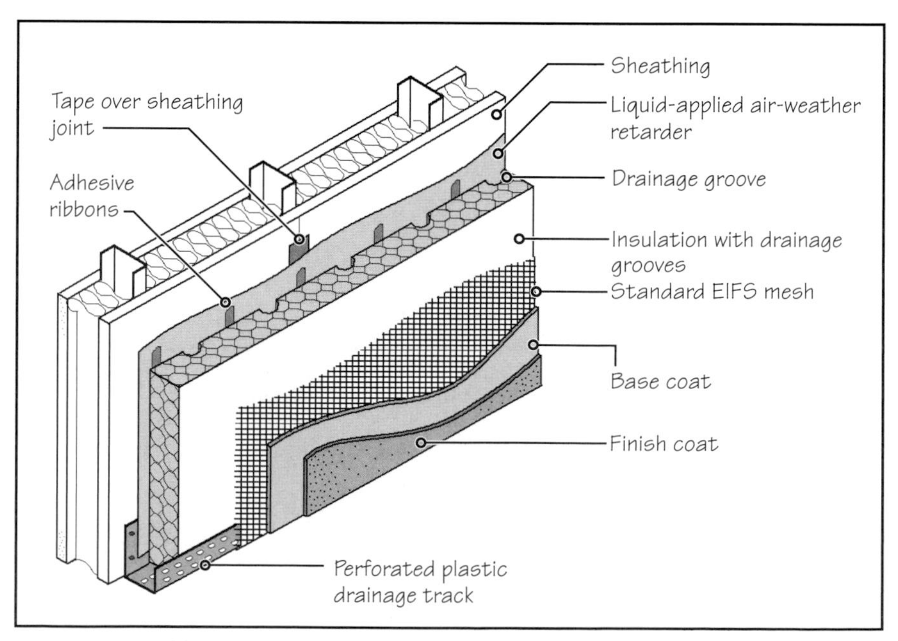

FIGURE 27.23 Drainable EIFS system in which the insulation is adhered to the substrate and the drainage is provided by vertical grooves in insulation.

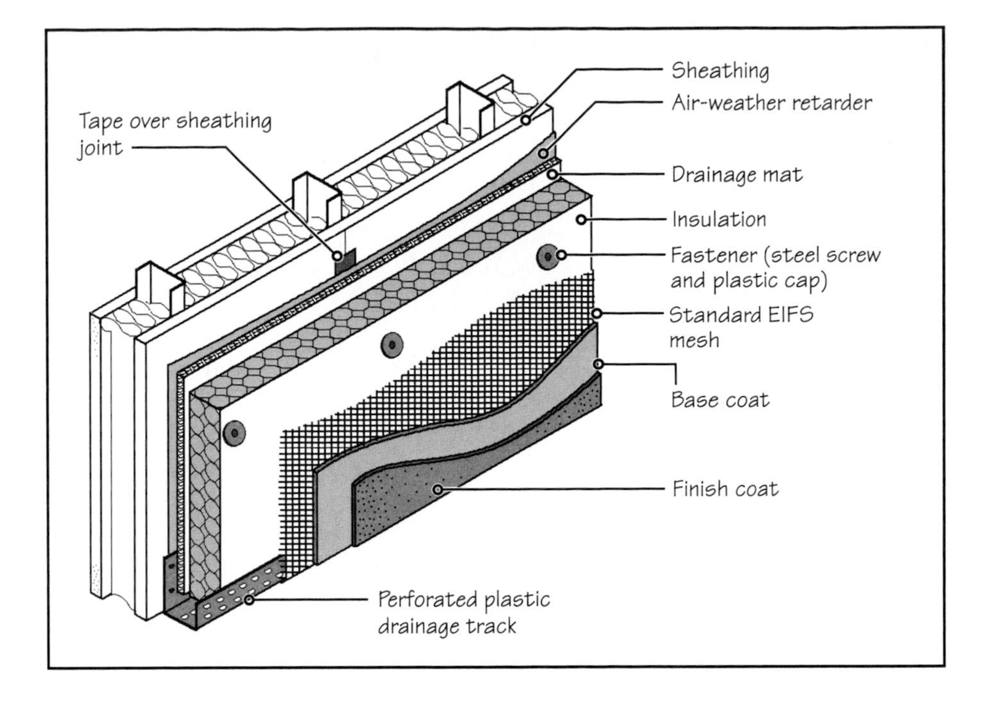

FIGURE 27.24 Drainable EIFS system in which the insulation is mechanically fastened to wall-framing members and the drainage is provided by a drainage mat, made of interwoven plastic strands.

Both systems require a perforated plastic track at the bottom to allow water to weep at the base. (The bottom track is similar to a casing bead in a portland cement stucco wall.) In a multistory building, the tracks are provided at each floor level. Both systems (shown in Figures 27.23 and 27.24) can be modified to include pressure-equalizing accessories to make the EIFS cladding perform as a rain screen.

PRACTICE QUIZ

Each question has only one correct answer. Select the choice that best answers the question.

12. The term *EIFS* is an acronym for
- **a.** exterior insulation and finished stucco.
- **b.** exterior insulation and finish system.
- **c.** externally insulated finish system.
- **d.** envelope insulation with finished stucco.

13. EIFS is classified in two categories: polymer-based (PB) EIFS and polymer-modified (PM) EIFS. Which of these two is more commonly used?
- **a.** PB EIFS
- **b.** PM EIFS

14. In PB EIFS, the insulation commonly used is
- **a.** fiberglass.
- **b.** polyisocyanurate.
- **c.** extruded polystyrene.
- **d.** expanded polystyrene.
- **e.** none of the above.

15. In PB EIFS, the EIFS lamina consists of
- **a.** base coat, insulation, and finish coat.
- **b.** base coat, mesh, and finish coat.
- **c.** insulation, mesh, and finish coat.
- **d.** insulation, lath, base coat, and finish coat.

16. The thickness of the lamina in PB EIFS is approximately
- **a.** 1 in.
- **b.** $\frac{1}{2}$ in.
- **c.** $\frac{1}{4}$ in.
- **d.** $\frac{1}{8}$ in.
- **e.** $\frac{1}{16}$ in.

17. Backwrapping in an EIFS-clad wall refers to
- **a.** wrapping the edges of the openings in the wall with a water-resistant tape.
- **b.** wrapping the terminal edges of EIFS insulation with EIFS mesh.
- **c.** wrapping the joints between EIFS insulation with EIFS mesh.
- **d.** wrapping the joints between EIFS insulation with a water-resistant tape.

18. The requirements for control joints in an EIFS wall are
- **a.** generally the same as in a stucco-clad wall.
- **b.** more stringent than in a stucco-clad wall.
- **c.** less stringent than in a stucco-clad wall.
- **d.** none; control joints are not required in an EIFS-clad wall.

19. Impact-resistant EIFS generally has
- **a.** one base coat and one finish coat.
- **b.** one base coat and two finish coats.
- **c.** two base coats and two finish coats.
- **d.** two base coats and one finish coat.
- **e.** three base coats and one finish coat.

20. Expansion joints are required in an EIFS-clad wall.
- **a.** True
- **b.** False

Answers: 12-b, 13-a, 14-d, 15-b, 16-d, 17-b, 18-d, 19-d, 20-a.

27.8 EXTERIOR CLADDING WITH DIMENSION STONE

Granite, marble, and limestone are the three stones commonly used for exterior cladding. The minimum recommended thickness for exterior granite cladding slabs is $1\frac{1}{4}$ in. (3 cm) with panel size of 20 ft^2 or less. The corresponding thickness for marble and limestone is 2 in. (5 cm). Greater thickness is used for larger slab sizes or for greater durability.

Stone cladding can either be field installed slab by slab at the construction site or prefabricated into curtain wall panels. Field installation can be done by one of the following two methods:

- Standard-set installation
- Vertical channel support installation

NOTE

Quarter-Point Supports

Quarter-point dead-load supports for slabs give uniform center-to-center spacing between supports and also almost produce least bending stresses in slabs.

27.9 FIELD INSTALLATION OF STONE—STANDARD-SET METHOD

In the standard-set method, each stone slab is directly anchored to the backup wall with its own dead-load and lateral-load supports. Two dead-load supports are required for each slab, which are provided by stone liner blocks. The liner blocks are bolted to the slab at a stone-fabrication plant with stainless steel bent bolts set in epoxy resin (typically at quarter points), Figure 27.25.

In installing stone slabs, each liner block is made to bear on a J-shaped shelf angle clip that is anchored to a CMU or reinforced-concrete backup wall, Figures 27.26 and 27.27. A setting pad that functions as a cushion and a shim is typically used under each liner block.

Tieback anchors provide lateral load supports and consist of split-tail anchors, whose tails engage in the slots (kerfs) provided in the cladding slabs, Figure 27.28. Split-tail anchors

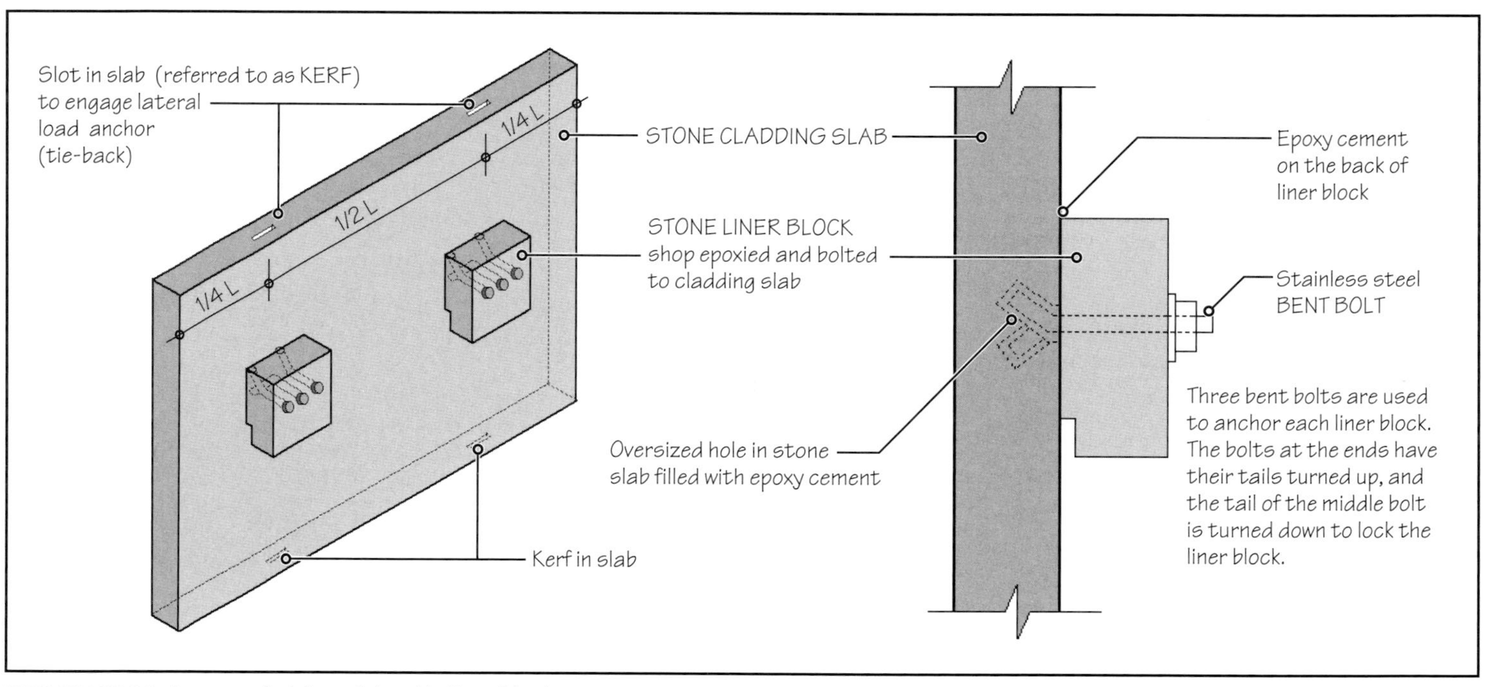

FIGURE 27.25 A stone cladding slab with liner blocks at quarter points.

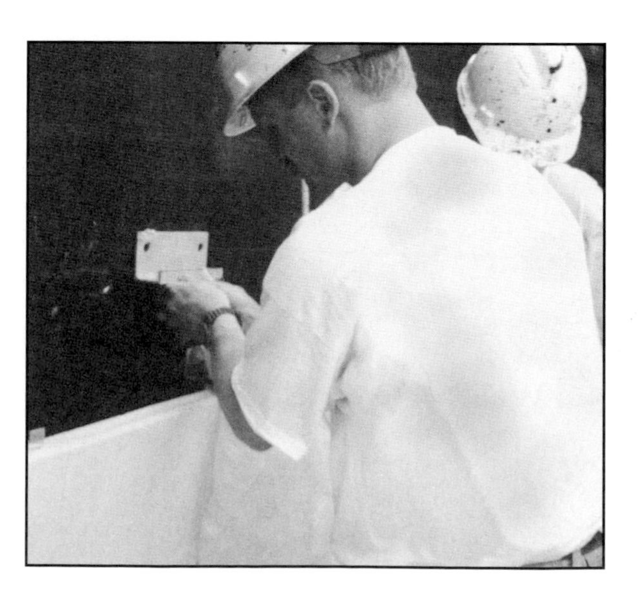

FIGURE 27.26 Anchorage of a J-shaped shelf angle clip to a CMU backup wall whose surface has been treated with liquid-applied air-weather retarder.

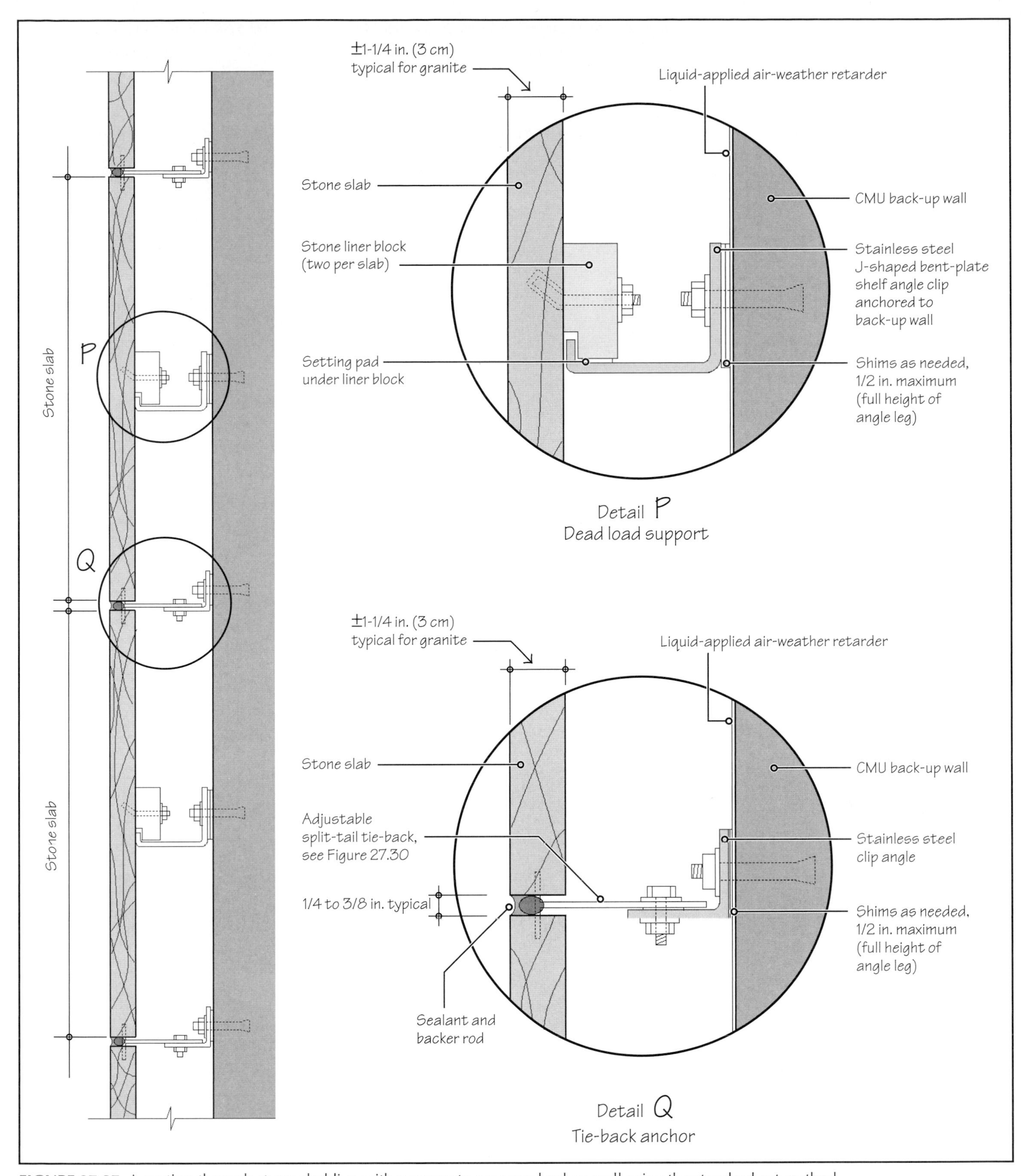

FIGURE 27.27 A section through stone cladding with a concrete masonry backup wall using the standard-set method.

may be secured to the backup wall directly, as shown in Figure 27.29. Alternatively, they may be secured through a clip angle anchored to the backup wall, Figure 27.30. Securing the split-tail anchors through an angle provides greater adjustability in their location.

The kerfs in stone slabs must be filled completely with fast-curing sealant before inserting the anchors, Figure 27.31. An incomplete seal may lead to intrusion of rainwater in them, causing freeze-thaw damage to stone slabs.

The number of tieback anchors is determined by the load resistance of each anchor and the magnitude of lateral loads. However, a minimum of four anchors are required for a slab area of up to 12 ft^2, with additional anchors for a larger area, as needed.

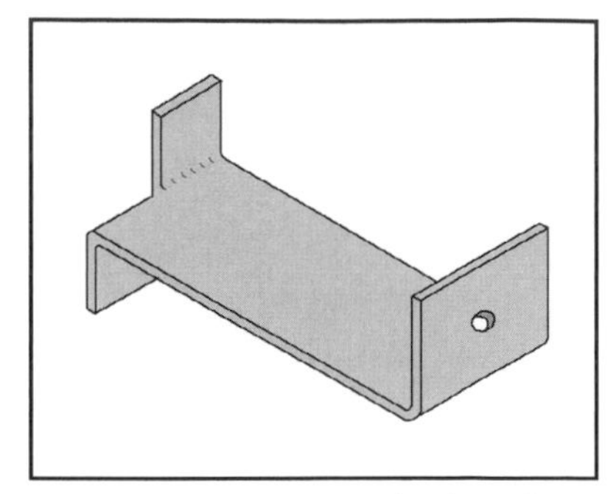

FIGURE 27.28 A typical split-tail anchor.

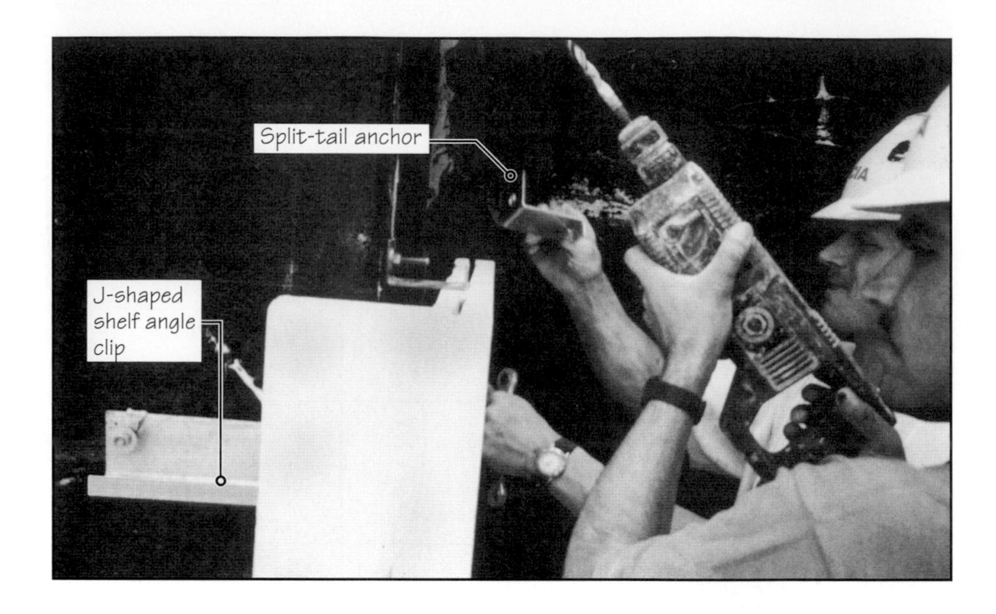

FIGURE 27.29 A split-tail anchor secured directly to a backup wall.

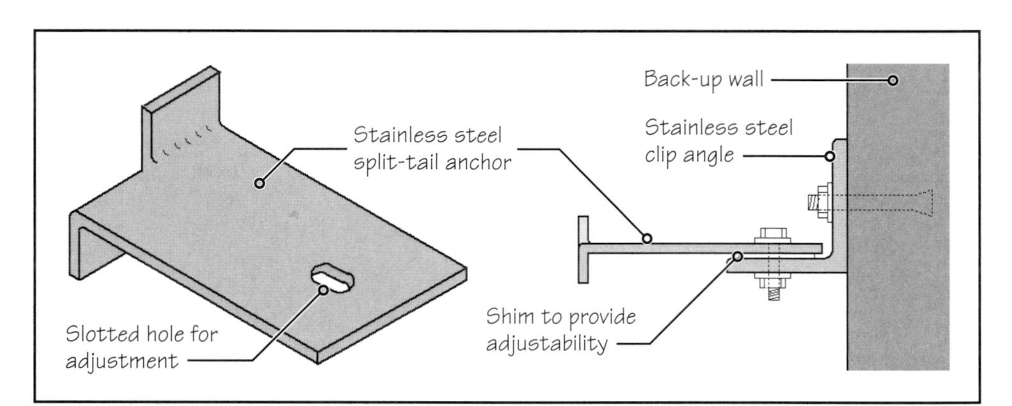

FIGURE 27.30 An adjustable split-tail tieback secured to the backup wall through a clip angle.

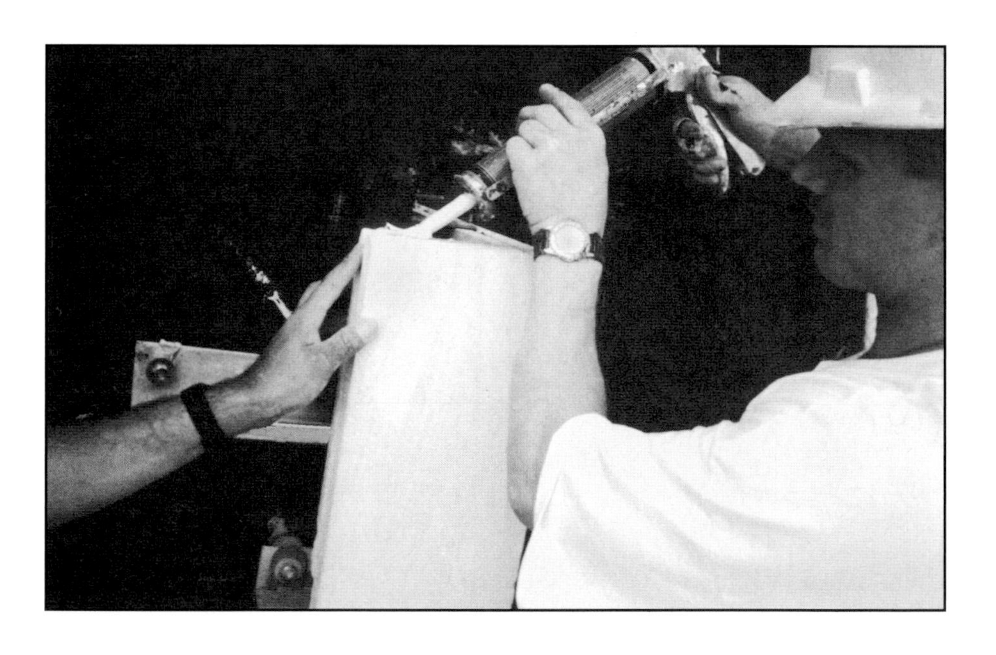

FIGURE 27.31 Applying sealant in kerfs of a stone slab before inserting tiebacks.

Bent-Plate Clips as an Alternative to Stone Liner Blocks

A commonly used substitute for a stone liner block is a stainless steel bent-plate clip, referred to as a *bayonet clip,* Figure 27.32. Slabs with bayonet clips have less stone and, therefore, a smaller dead load than slabs with liner blocks. Bayonet clips (two per slab) are bolted and epoxied to the cladding slab in the same way as the liner blocks.

Another advantage of using bayonet clips is that they can be anchored to the cladding slab using a large round-head bolt, referred to as a *bolt anchor*. A bolt anchor is inserted through a specially cut trench in the slab, Figure 27.33. Unlike a bent bolt, a bolt anchor is not required to be embedded in epoxy cement. Epoxy-free bolting is preferred by some stone experts due to the uncertainty in the long-term performance of epoxy cements.

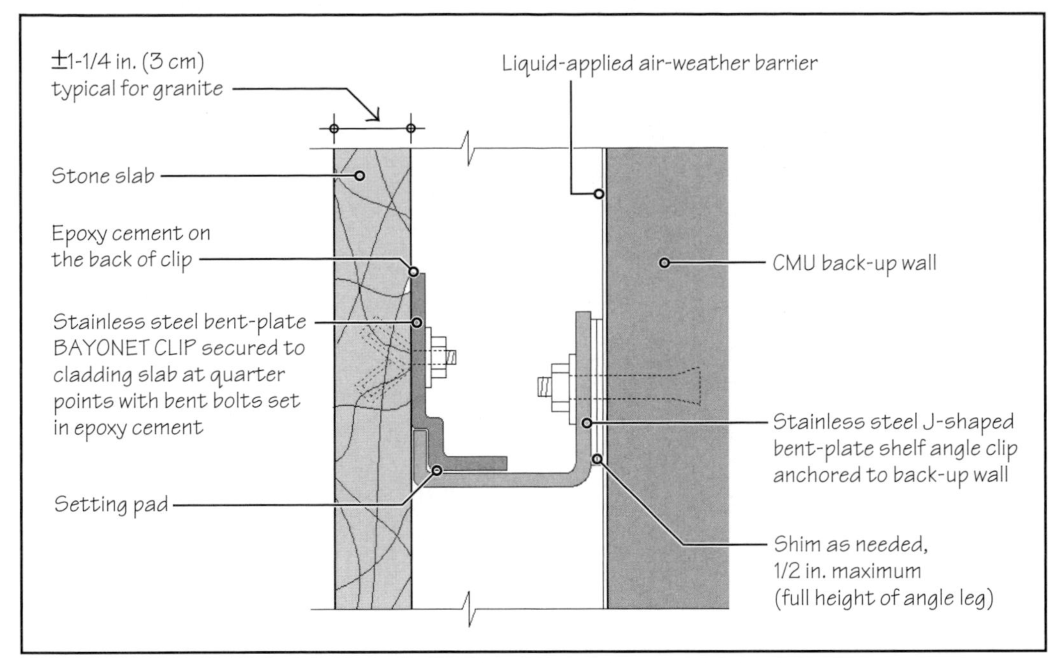

FIGURE 27.32 Bayonet anchor clips as an alternative to liner blocks for dead-load supports.

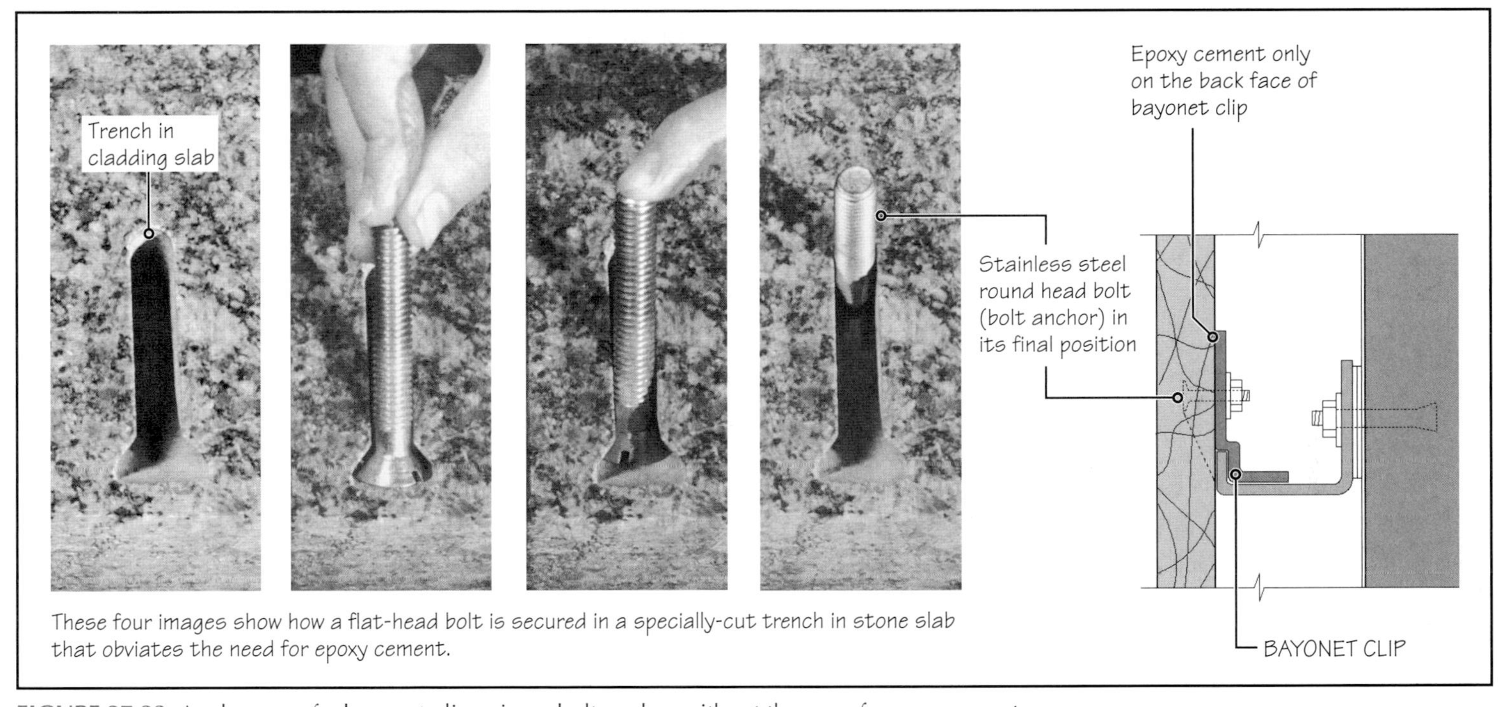

FIGURE 27.33 Anchorage of a bayonet clip using a bolt anchor without the use of epoxy cement.

DEAD-LOAD SUPPORT AND TIEBACK ANCHOR IN ONE CLIP ANCHOR

The dead-load supports and tieback anchors for stone slabs can be combined in one clip anchor, which is fabricated from two stainless steel members—a bent plate and a flat plate, Figure 27.34. The use of one anchor substantially facilitates the installation of cladding. At locations where flashing is required, dead-load supports and tieback clip anchors need to be separated, as shown in Detail Q of Figure 27.35.

With rigid foam insulation provided between the cladding and CMU backup wall, the insulation must be cut around dead-load supports and tieback anchors, Figure 27.36. In such a case, the anchorage clips may be required to be heavier or braced.

The details shown previously can be modified if the backup wall consists of a light-gauge steel-stud wall instead of a CMU backup wall, Figure 27.37.

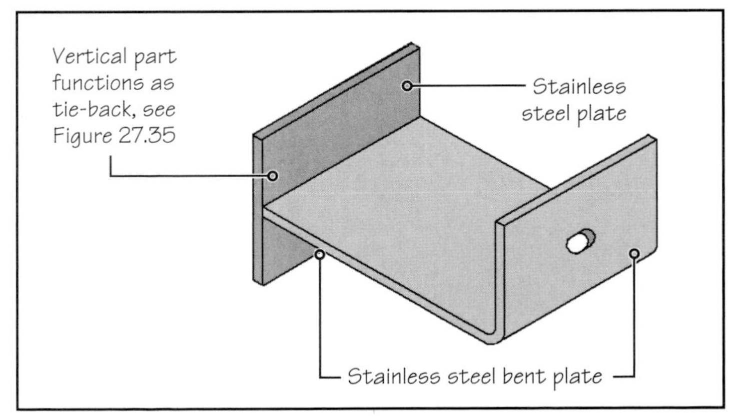

FIGURE 27.34 A combined dead-load support and tieback anchor formed by welding a stainless steel flat plate to a stainless steel bent plate.

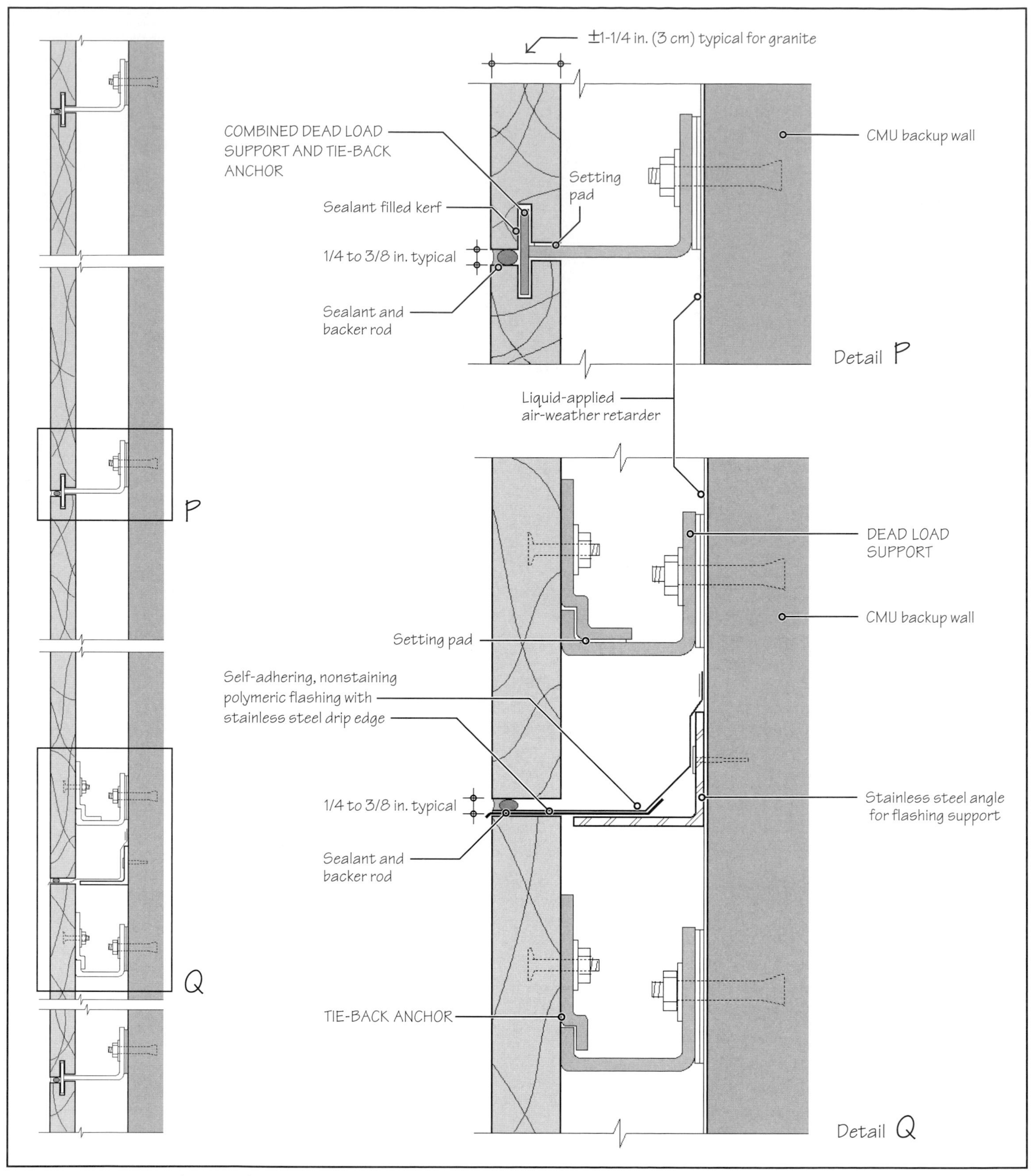

FIGURE 27.35 A typical section through a stone-clad exterior wall using combined dead load and tie back anchors. Separate dead-load supports and tiebacks have been used at locations of flashing and weep holes.

27.10 FIELD INSTALLATION OF STONE—VERTICAL SUPPORT CHANNEL METHOD

The anchorage of stone slabs is substantially simplified by using continuous vertical support channels, Figure 27.38. The manufacturers of channels provide various accessories to anchor the channels to the backup wall.

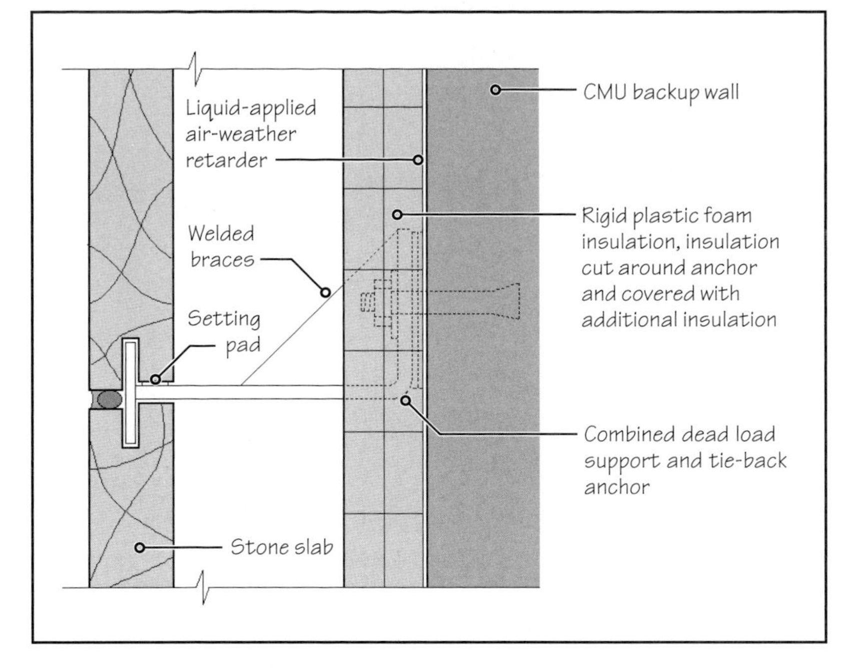

FIGURE 27.36 An alternative to detail P (in Figure 27.35) if the air space in front of the CMU backup wall is provided with insulation.

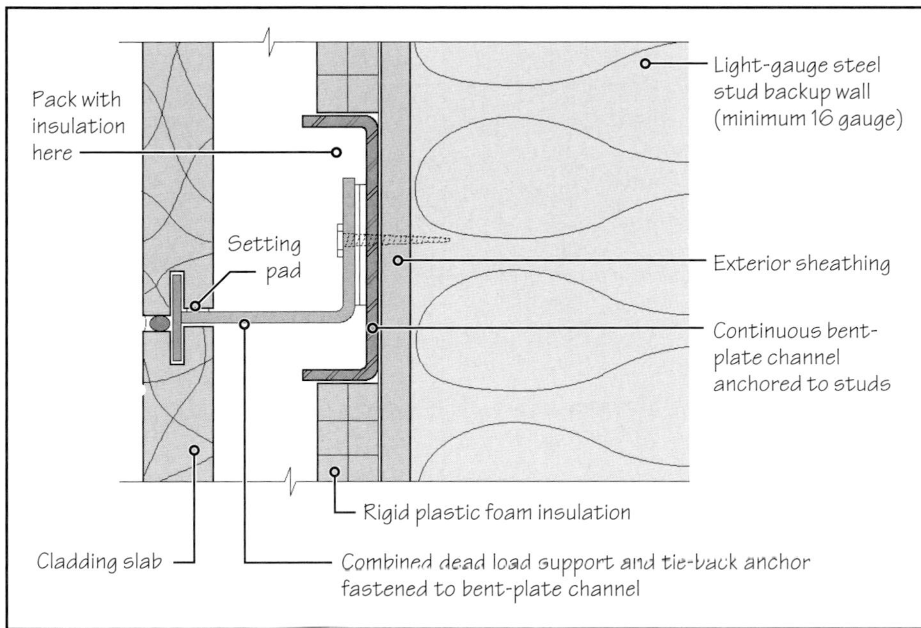

FIGURE 27.37 A typical stone cladding detail with a steel-stud backup wall.

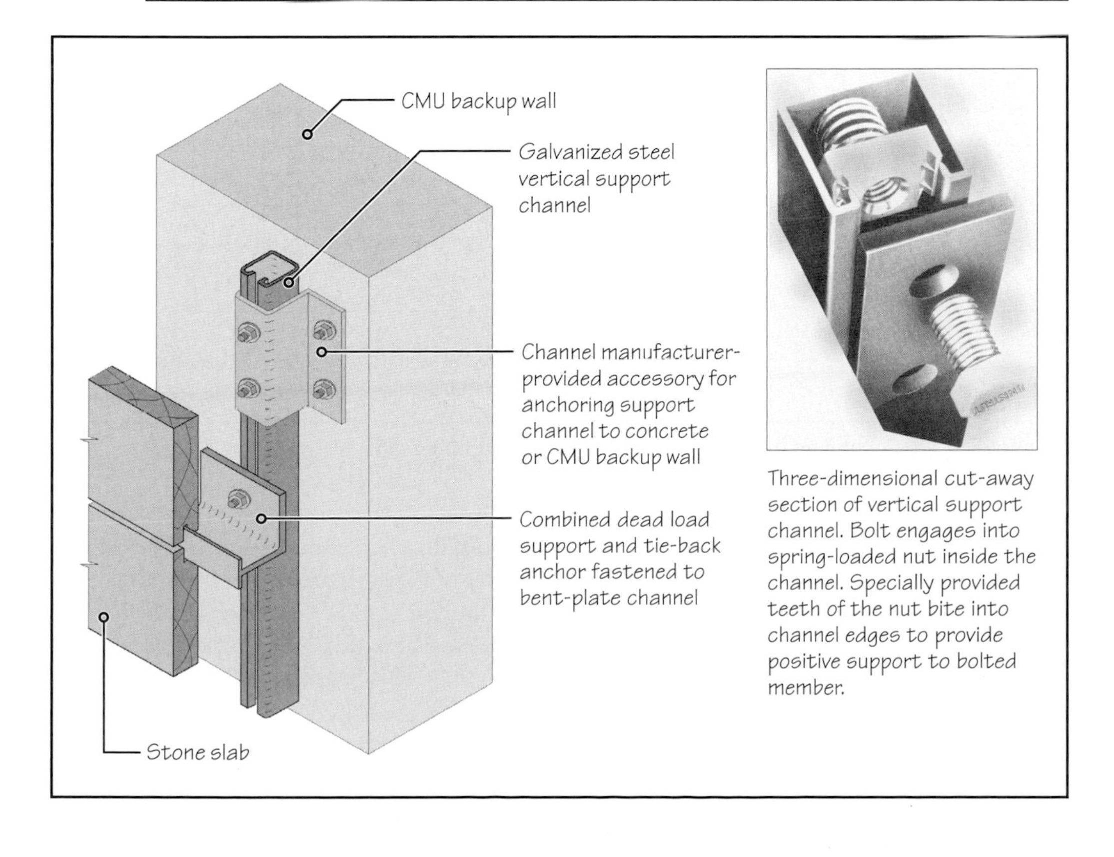

FIGURE 27.38 Adjustable anchorage of a vertical channel to a concrete or CMU backup wall.

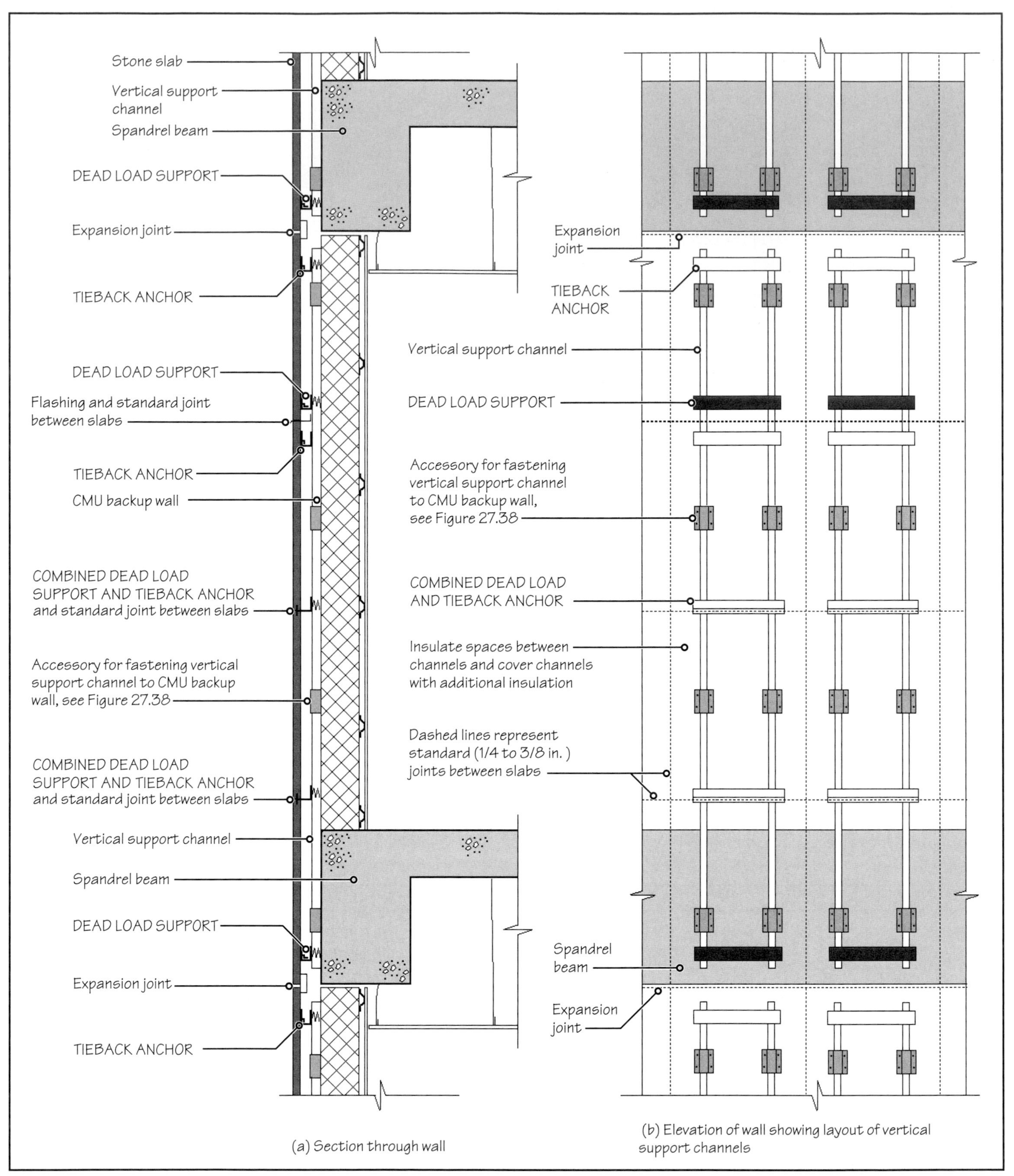

FIGURE 27.39 A typical section and elevation showing the anchorage of stone cladding using support channels.

The channels are spaced at quarter points of the slabs and extend from floor to floor, with breaks at each floor level, Figure 27.39. With a CMU backup wall, the support channels are anchored directly to it, as shown in Figure 27.38. With a steel-stud backup wall, a continuous galvanized steel plate is fastened to the studs at suitable intervals, to which the channels are anchored, Figure 27.40.

Where a short length of stone return (e.g., at a soffit) is required, it is obtained through the use of stainless steel clip angles bolted to slabs, Figure 27.41.

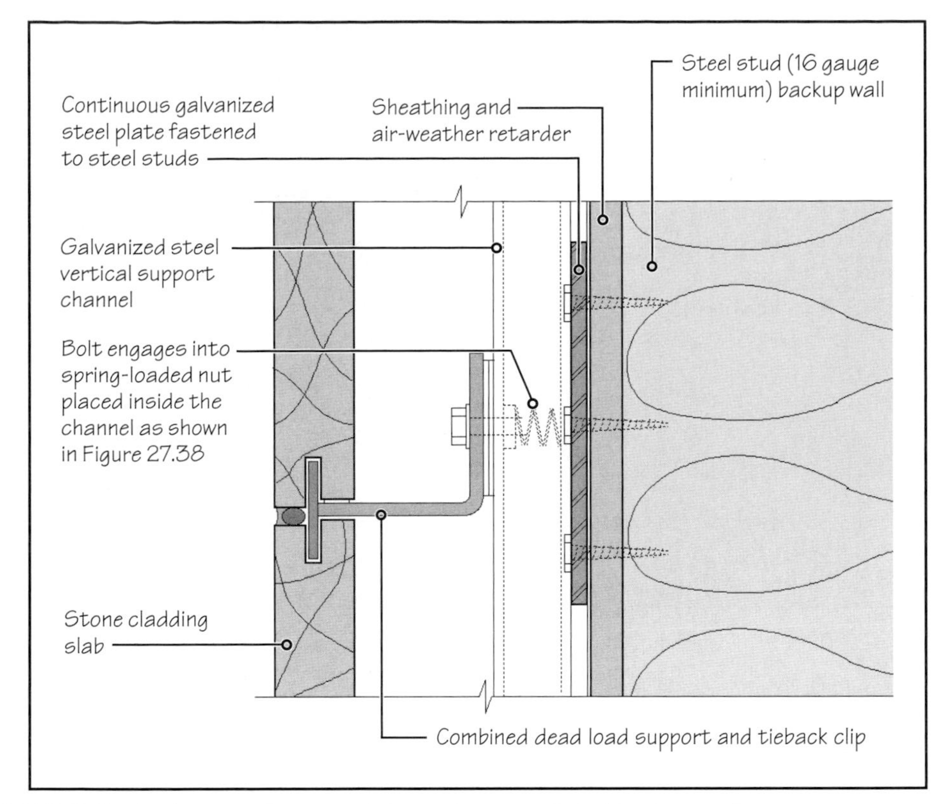

FIGURE 27.40 Anchorage of a vertical channel to a steel-stud backup wall.

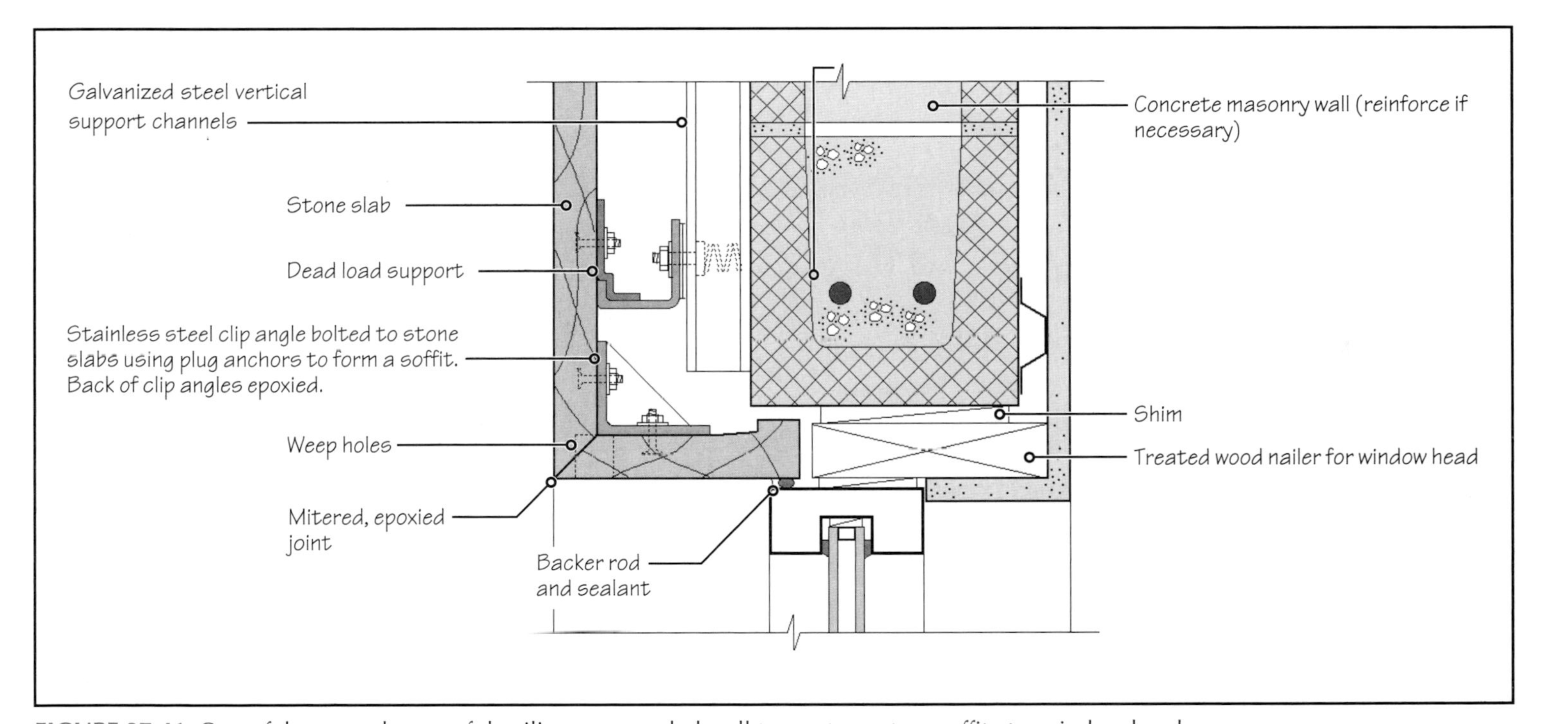

FIGURE 27.41 One of the several ways of detailing a stone-clad wall to create a return soffit at a window head.

27.11 PREFABRICATED STONE CURTAIN WALLS

Instead of installing stones slab by slab to a backup wall, they can be anchored to a steel truss frame, generally at the construction site. The stone-frame assembly forms a panel, which is lifted to position by a crane and hung from the building's structure like any other curtain wall panel (precast concrete or GFRC panel).

Generally, each panel extends from column to column and is supported on them. The panelized system is preferred for use in locations where labor costs are high, unfavorable weather conditions exist, or the site is unsuitable for the construction of scaffolding. The construction images of Figure 27.42 illustrate the panelized stone cladding.

(a) Panel being fabricated at the plant. Note that a panel consists of a number of stone slabs

(b) Worker measuring the distance between dead load support anchors for a slab

(c) Back of a prefabricated stone-clad spandrel being lifted for anchorage to the building's structure

(d) Prefabricated stone clad spandrel ready for anchorage to building's structure

(e) Installation of a tie-back anchor for the slab

(f) Dead load support for panel

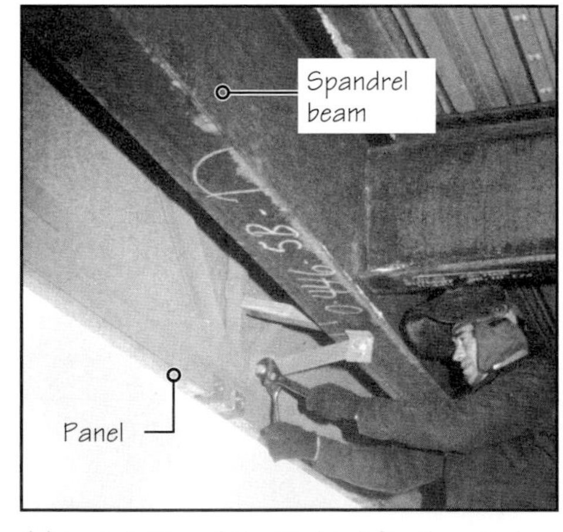

(g) Installation of the tie-back for the panel

FIGURE 27.42(a)–(g) Fabrication and installation of prefabricated stone curtain wall panels. (Photos courtesy of Dee Brown, Inc.)

PRACTICE QUIZ

Each question has only one correct answer. Select the choice that best answers the question.

21. The dead-load supports for each stone slab (in a stone-clad wall) are generally
- **a.** two, placed in the center of the slab, one above the other.
- **b.** two, placed at the same level at one-third points.
- **c.** three, placed at the same level at one-third points.
- **d.** two, placed at the same level at quarter points.
- **e.** four, placed at the same level at quarter points.

22. A bayonet clip in the stone cladding industry refers to
- **a.** dead-load support.
- **b.** wind-load support.
- **c.** flashing retainer.
- **d.** none of the above.

23. A combined dead-load and lateral-load support tieback anchor for stone cladding consists of
- **a.** a stainless steel L-shaped bent plate.
- **b.** a stainless steel J-shaped bent plate.
- **c.** a stainless steel H-shaped bent plate.
- **d.** a stainless steel L-shaped bent plate welded to a stainless steel flat plate.
- **e.** none of the above.

24. The term *kerf* in stone cladding industry refers to
- **a.** a slot in a stone slab.
- **b.** dead-load support for a stone slab.
- **c.** tieback anchor for a stone slab.
- **d.** none of the above.

25. When vertical support channels are used in stone cladding, dead-load supports and tiebacks are not required for stone slabs.
- **a.** True **b.** False

26. The thickness of granite used in wall cladding is generally
- **a.** $1\frac{1}{4}$ in. **b.** $1\frac{1}{2}$ in.
- **c.** 2 in. **d.** $2\frac{1}{4}$ in.
- **e.** none of the above.

27. In the prefabricated, panelized stone cladding system, stone slabs are backed by
- **a.** a frame consisting of light-gauge steel members.
- **b.** a frame consisting of laminated veneer lumber members.
- **c.** a ribbed steel deck.
- **d.** a truss consisting of structural steel members.
- **e.** any one of the above depending on the building.

28. In the prefabricated, panelized stone cladding system, stone slabs are anchored to the backup with
- **a.** portland cement–sand mortar.
- **b.** epoxy cement.
- **c.** epoxy-sand adhesive.
- **d.** none of the above.

Answers: 21-d, 22-a, 23-d, 24-a, 25-b, 26-a, 27-d, 28-d.

27.12 THIN STONE CLADDING

Another form of panelized stone cladding uses an extremely thin (nearly $\frac{1}{4}$ in. thick) stone veneer bonded to an aluminum honeycomb backing. The panels are manufactured by epoxy-cementing aluminum honeycomb on both sides of an approximately $\frac{3}{4}$-in.-thick stone slab (generally granite or marble). The honeycomb-stone-honeycomb combination is sawn through the middle, producing two identical panels, Figure 27.43.

After sawing, the stone facing on each panel is finished as needed (e.g., polished, honed, abrasive blasted, bush-hammered, etc.). The honeycomb backing is $\frac{3}{4}$ in. thick, giving an overall finished panel thickness of approximately 1 in. (For a panel intended for interior use, e.g., for lining the walls of elevator lobbies, foyers, and ceilings, the honeycomb backing is only $\frac{3}{8}$ in. thick.)

The standard treatment on an exposed edge of the panel is a small return, as shown in Figure 27.44. Where a larger return is required (e.g., to obtain a deeper soffit at a window head), it is obtained by cementing a continuous aluminum angle to the honeycomb, Figure 27.45.

The standard size to which the panels are made is 4 ft × 8 ft, but other sizes are available, with a maximum of 5 ft × 10 ft. The light weight of panels makes their installation convenient, particularly in high-labor areas. A 1-in.-thick stone-honeycomb panel weighs only 3.3 psf, which is approximately the weight of $\frac{1}{4}$-in.-thick glass.

The bending strength of a stone-honeycomb panel is fairly high because of the honeycomb backing and the fiber-reinforced epoxy skin bonded to it. The composition also gives a great deal of ductility to the panel so that it is able to flex under lateral loads.

The panels' light weight, high ductility, and bending strength make it ideal for use in seismic areas, where the aesthetics of natural stone cladding without its heavy weight are required. These are some of the reasons cited for their use in the Courthouse in Anchorage, Alaska, Figure 27.46, and the International Business Center, Moscow, Russia, Figure 27.47.

ANCHORAGE OF STONE-HONEYCOMB PANELS

The most commonly used method of anchoring the panels to the steel stud, concrete, or concrete masonry backup wall uses two interlocking channels. One of these channels is shop installed to the back of the panel and the other channel is field anchored to the backup wall, Figure 27.48. Figure 27.49 shows a typical detail of the use of the panels in a building.

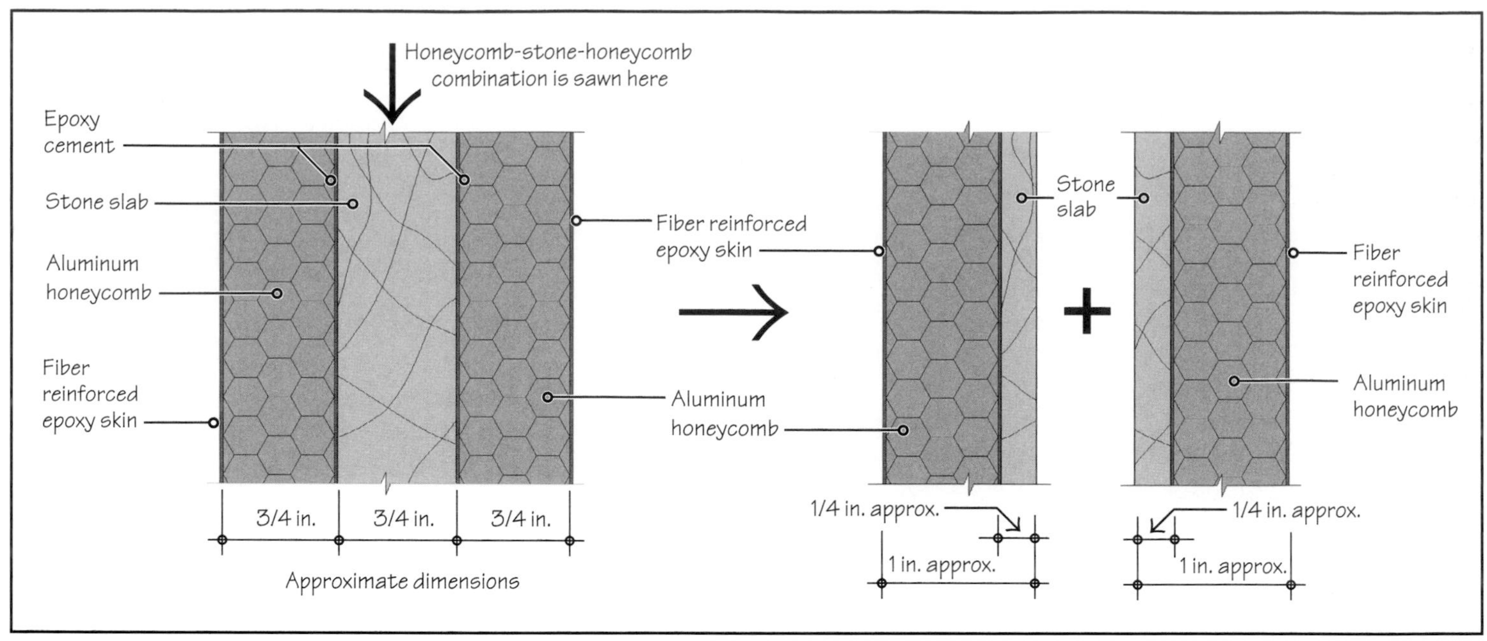

FIGURE 27.43 Making of a thin stone panel.

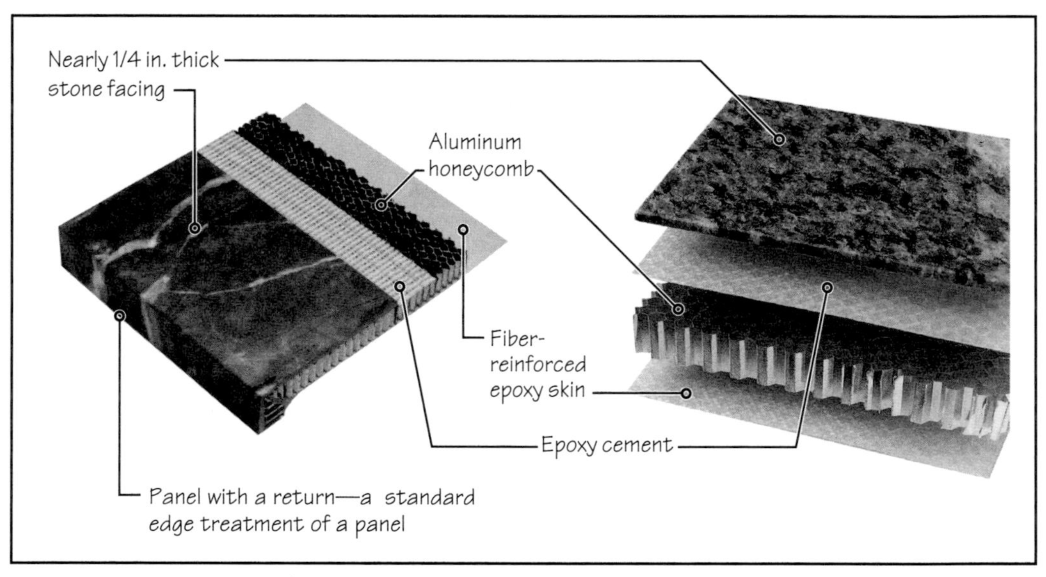

FIGURE 27.44 Anatomy of a stone-honeycomb panel showing the standard treatment of its exposed edge. (Photos courtesy of Stone Panels, Inc.)

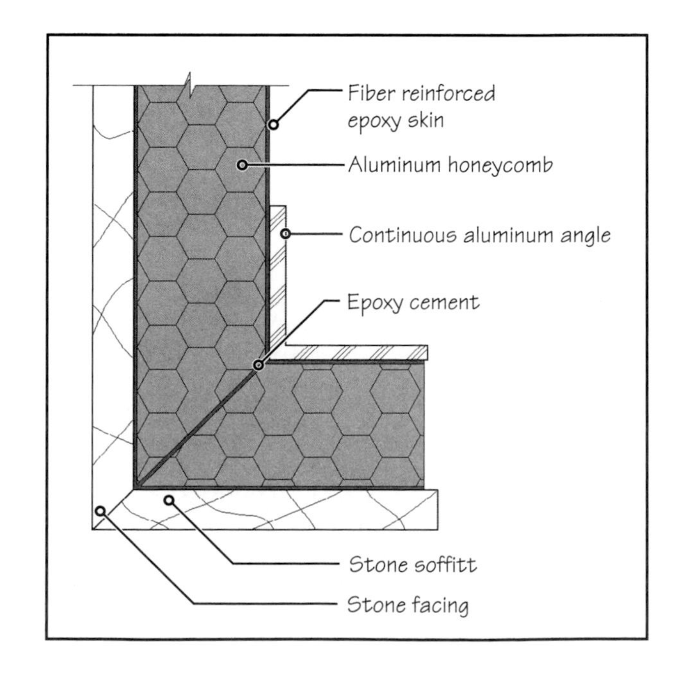

FIGURE 27.45 Anatomy of a stone-honeycomb panel with deep soffit.

FIGURE 27.46 Courthouse, Anchorage, Alaska, with stone-honeycomb exterior-wall cladding. (Photo courtesy of Stone Panels, Inc.)

FIGURE 27.47 International Business Center, Moscow, Russia, with stone-honeycomb exterior-wall cladding. (Photo courtesy of Stone Panels, Inc.)

Prefabricated Stone-Honeycomb Curtain Wall Panels

Stone-honeycomb panels can also be prefabricated into curtain wall panels, which generally extend from column to column and are hung from the building's structure like other curtain wall panels, Figure 27.50.

27.13 INSULATED METAL PANELS

Another lightweight exterior cladding system consists of metal (typically steel) panels with 2- to 3-in. factory-installed polyurethane foam insulation between metal skins. The panels are available factory painted in various colors, in galvanized steel and in stainless steel. Several surface finishes, such as smooth, embossed, and precast concrete like texture, are available.

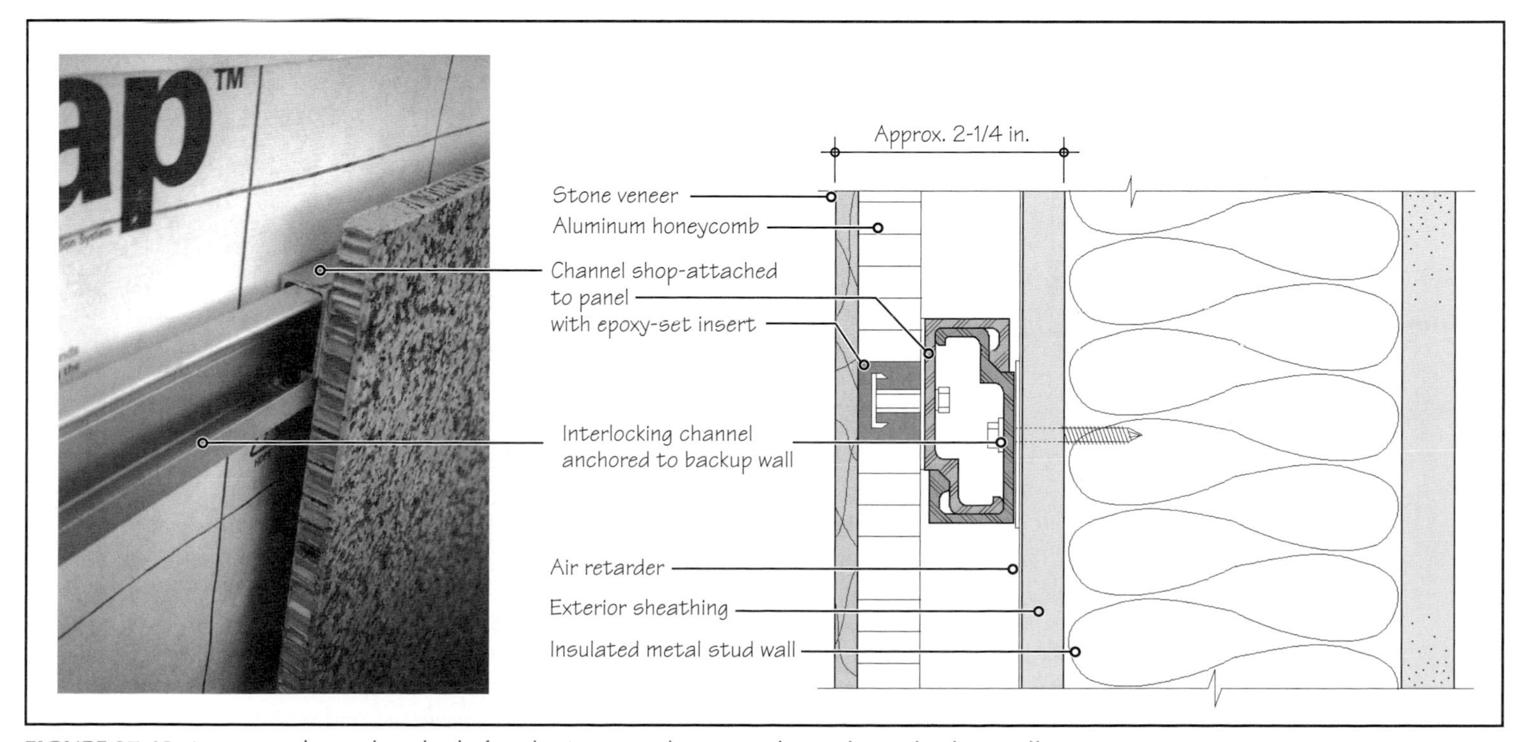

FIGURE 27.48 A commonly used method of anchoring stone-honeycomb panels to a backup wall.

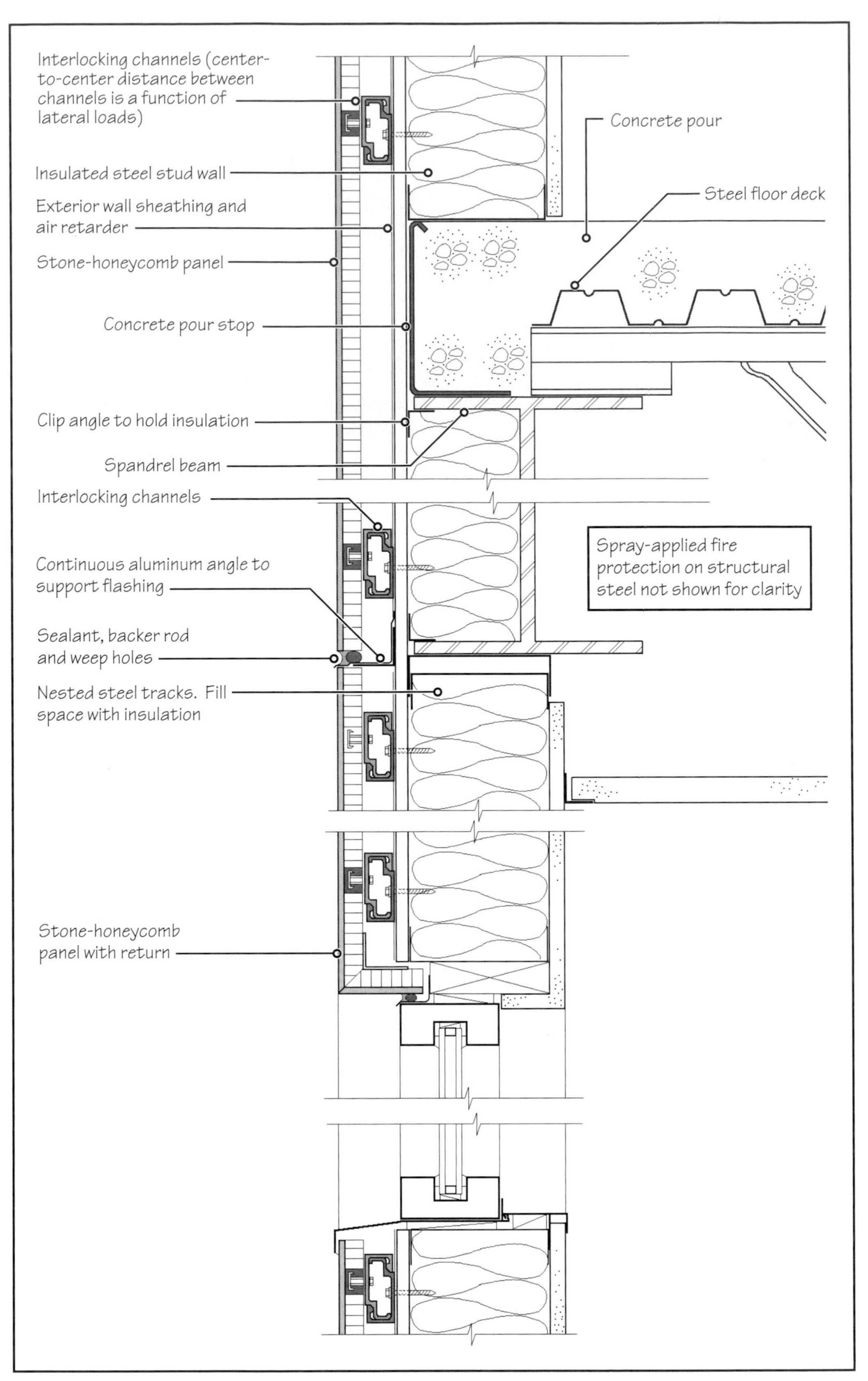

FIGURE 27.49 A typical wall section through a steel-frame building clad with stone-honeycomb panels.

Because of the insulating core, the panels have a high R-value. Therefore, additional insulation in the wall may not be needed. The joinery between panels has been perfected by manufacturers to provide concealed fasteners and variable joint widths, Figure 27.51.

The panels weigh less than 3 psf and are available in widths of 2 to 4 ft and in lengths of up to approximately 30 ft. The panels can be installed horizontally on a metal-stud wall or vertically from spandrel beam to spandrel beam with intermediate horizontal supports. When installed over a metal-stud wall, the exterior sheathing may be omitted because the panels provide a weather barrier. The panels can be integrated with windows and other openings with manufacturer-provided accessories and detailing assistance, Figure 27.52.

FIGURE 27.50 Curtain wall panels fabricated from stone-honeycomb panels. (Photos courtesy of Stone Panels, Inc.)

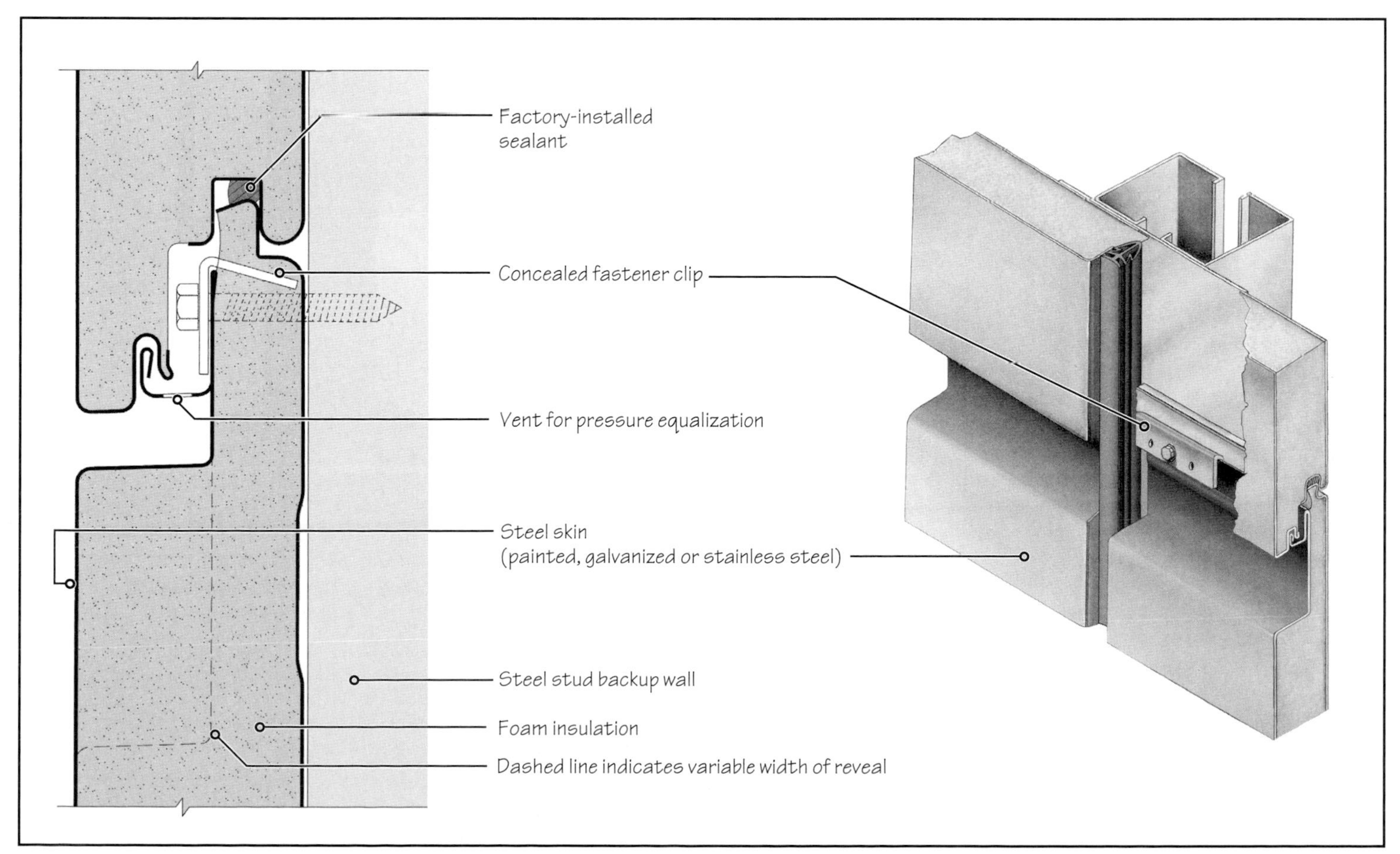

FIGURE 27.51 A section and a cut-away section showing the anatomy of a typical insulated metal panel. (Images courtesy of Centria Architectural Systems, Inc.)

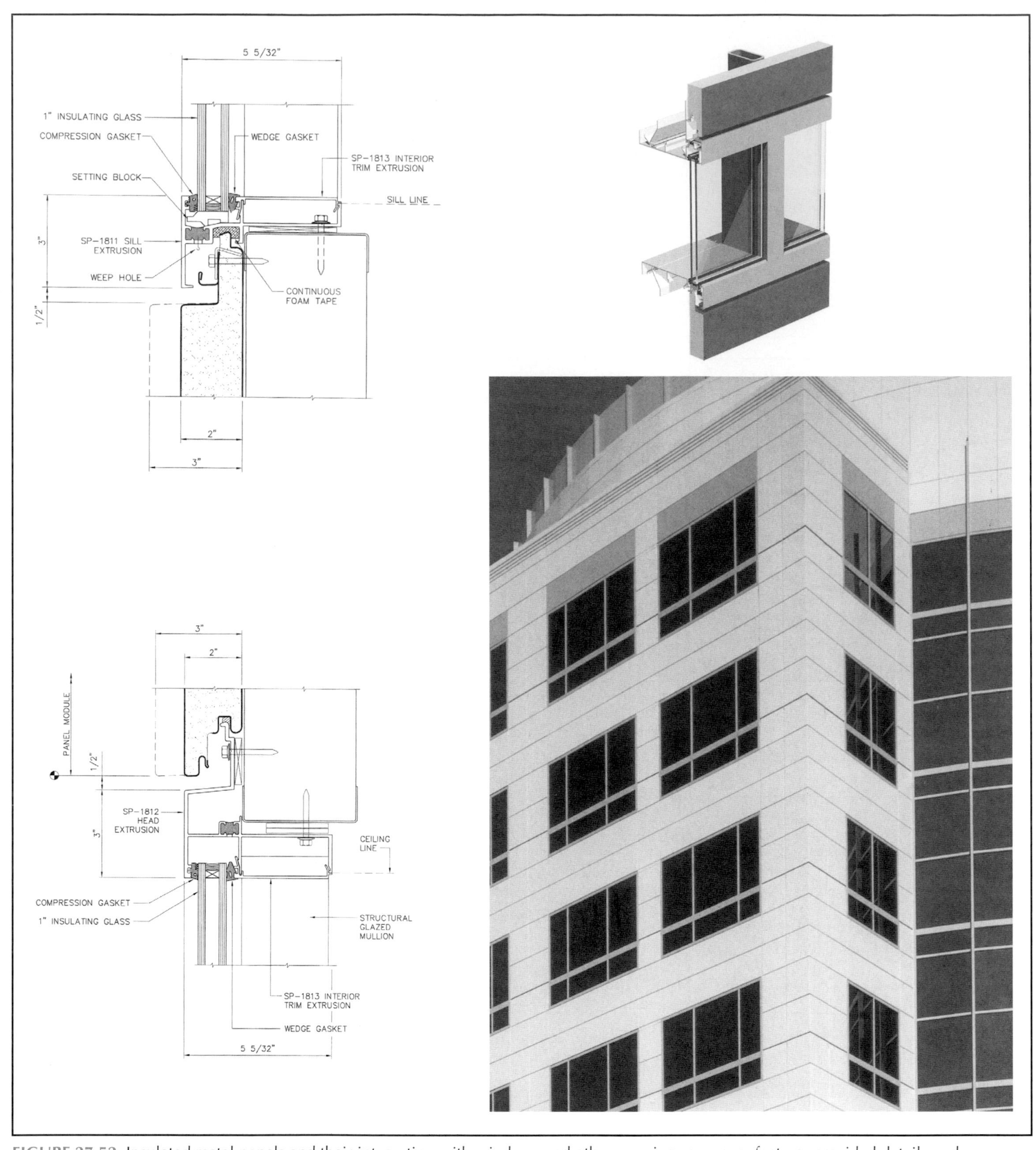

FIGURE 27.52 Insulated metal panels and their integration with windows and other openings per manufacturer-provided details and accessories. (Images and details courtesy of Centria Architectural Systems, Inc.)

PRACTICE QUIZ

Each question has only one correct answer. Select the choice that best answers the question.

29. In a stone-honeycomb wall cladding panel, the thickness of stone is approximately
- **a.** 1 in.
- **b.** $\frac{3}{4}$ in.
- **c.** $\frac{1}{2}$ in.
- **d.** $\frac{1}{4}$ in.
- **e.** $\frac{1}{8}$ in.

30. In a stone-honeycomb wall cladding panel, the honeycomb is generally made of
- **a.** stainless steel.
- **b.** aluminum.
- **c.** titanium.
- **d.** copper.
- **e.** none of the above.

31. A stone-honeycomb panel is anchored to the backup wall using
- **a.** epoxy cement.
- **b.** separate dead-load supports and tiebacks.
- **c.** two interlocking metal channels.
- **d.** stainless steel bolts.

32. In an insulated wall metal panel, the insulation is sandwiched between
- **a.** a metal sheet in the front and a fiberglass scrim at the back of the panel.
- **b.** a metal sheet on both sides.
- **c.** a metal sheet in the front and an acrylic sheet at the back of the panel.
- **d.** none of the above.

Answers: 29-d, 30-b, 31-c, 32-b.

KEY TERMS AND CONCEPTS

Backwrapping
Brown coat
Control joints in stucco
Drainable EIFS
Drainage plane for stucco
EIFS lamina
Expansion joints in stucco
Exterior insulation and finish system (EIFS)
Finish coat
Impact-resistant EIFS
Insulated metal panels
One-coat stucco
Polymer-based EIFS
Polymer-modified EIFS
Portland cement plaster (stucco)
Prefabricated stone curtain walls
Reinforcing mesh
Scratch coat
Self-furring lath
Stone cladding—standard-set method
Stone cladding—vertical-support channel method
Stone-honeycomb panels
Stucco accessories
Stucco thickness
Thin stone cladding

REVIEW QUESTIONS

1. Using a sketch, explain the anatomy of stucco applied over a metal-stud wall assembly showing all components.
2. Using a sketch, explain the difference between PB EIFS and PM EIFS.
3. Using a sketch, explain the anatomy of a stone-clad CMU wall.
4. Using a sketch, explain the anatomy of a stone-clad light-gauge steel wall.

FURTHER READING

1. ASTM Standard C926. "Standard Specification for Application of Portland Cement-Based Plaster."
2. ASTM Standard C1063. "Standard Specification for Installation of Lathing and Furring to Receive Interior and Exterior Portland Cement Plaster."
3. ASTM Standard C1397. "Standard Practice for Application of PB Exterior Insulation and Finish System."
4. EIFS Manufacturers Association (EIMA). *Guide to EIFS Construction.* Morrow, Georgia, 2000.
5. Marble Institute of America. *Dimension Stone Design Manual.* Los Angeles, CA, 2003.
6. Portland Cement Association. *Plaster/Stucco Manual.* Skokie, Illinois, 2003.

CHAPTER 28 Transparent Materials (Glass and Light-Transmitting Plastics)

CHAPTER OUTLINE

28.1 Manufacture of Flat Glass

28.2 Types of Heat-Modified Glass

28.3 Glass and Solar Radiation

28.4 Types of Tinted and Reflective Glass

28.5 Glass and Long-Wave Radiation

28.6 Insulating Glass Unit

28.7 R-Value (or U-Value) of Glass

28.8 Glass and Glazing

28.9 Safety Glass

28.10 Laminated Glass

28.11 Structural Performance of Glass

28.12 Fire-Resistant Glass

28.13 Plastic Glazing

28.14 Glass for Special Purposes

28.15 Criteria for the Selection of Glass

28.16 Anatomy of a Glazing Pocket

PRINCIPLES IN PRACTICE: Important Facts About Radiation

PRINCIPLES IN PRACTICE: Condensation-Resistance Factor

Lites of clear glass supported by a structural steel frame in the Science Museum of the City of Arts and Sciences, Valencia, Spain. Architect: Santiago Calatrava. (Photo by MM.)

The introduction of daylight into buildings has always been an important design requirement. In early buildings, this requirement was satisfied by providing voids in walls. Subsequently, in an effort to control daylight, ventilation, and security, windows with wood shutters were developed. The use of oiled-paper panels or muslin cloth in wooden frames, which provided some light without opening the windows, was the next step in providing transparency in buildings.

With the discovery of glass, the window's performance (as well as its appearance) changed dramatically. It was now possible to obtain daylight effectively without admitting other environmental elements, such as wind, rain, dust, and insects. The earliest use of glass in openings is usually traced to the Roman Empire, where it was employed in small sizes.

The first significant use of glass, however, occurred during the Middle Ages as stained glass windows gained popularity in the church. Staining (or painting) of the glass was used as decoration to hide defects such as air bubbles and colored impurities, which were integral to the glass made at that time. During the Middle Ages, church windows were particularly large. They not only provided daylight, but also used dramatic pictures made from brightly stained glass to explain Biblical stories to people who could not read.

The transparency of glass improved greatly with the perfection of the *crown glass* process around the fourteenth century [1]. This method provided glass with fewer air bubbles and other defects. In this process, molten glass was first converted to a globe by blowing through a long pipe, then transferring the globe from the blow pipe to a rod, and subsequently cutting the globe open. The cut globe was finally flattened into a disc by repeated heating and vigorous spinning, Figure 28.1.

In a similar process, called the *cylindrical process,* molten glass was first blown into a globe and subsequently enlarged into a cylinder by swinging the pipe back and forth. The cylinder was later cut along its length, flattened by heating, and converted to sheets, Figure 28.2.

Clear glass, as we know it today, came into commercial use at the end of the seventeenth century with the invention of the glass-making process by Bernard Perrot [2]. In Perrot's process, molten ingredients of glass were first cast in a mold and later spread into sheets by rollers. When the glass solidified fully, it was ground and polished on both sides with abrasives. The mechanization of rolling and grinding operations brought down the price of glass sufficiently so that it could be used commercially in mirrors and store fronts. Although the process of manufacturing modern glass is substantially different from that of the cast or cylindrical glass of several centuries ago, the ingredients are virtually the same.

The transparency of glass, coupled with modern manufacturing techniques that provide large sheets of glass inexpensively, are responsible for the extensive use of glass in modern buildings. These developments have also led to the use of glass as the primary material for the cladding of many building types, Figures 28.3 and 28.4.

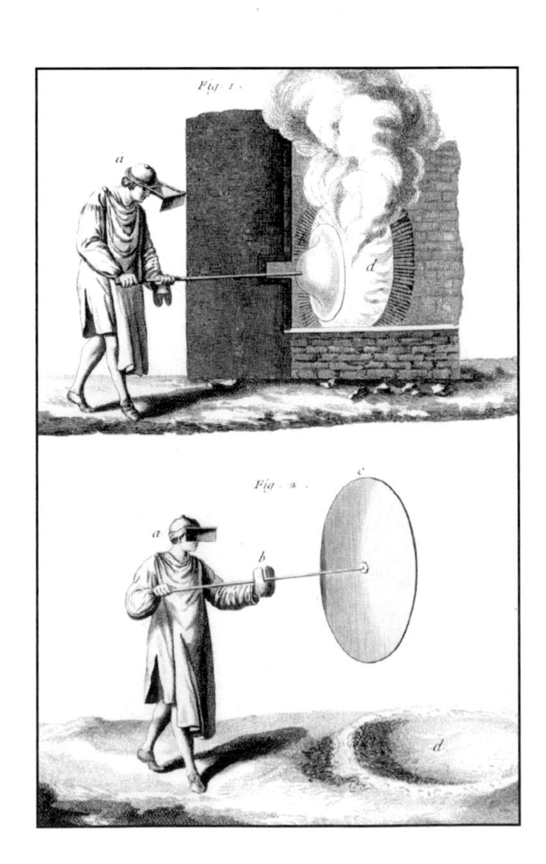

FIGURE 28.1 Crown glass–making process. (Photo courtesy of Corning Museum of Glass, Corning, New York.)

FIGURE 28.2 Cylindrical glass–making process. (Photo courtesy of Corning Museum of Glass, Corning, New York.)

FIGURE 28.3 The glass pyramid admits light into an underground addition to the Louvre Museum, Paris, France. Architect: I. M. Pei and Partners. (Photo courtesy of Dr. Jay Henry.)

FIGURE 28.4 A transparent cube of glass supported by a light-steel structural backup encloses the Rose Center for Earth and Space, New York, New York (see also Figure 30.15). Architects: Polshek Partnership.

EXPAND YOUR KNOWLEDGE

Why Glass Is Transparent

Although glass is a solid substance, its molecular structure resembles that of a liquid. A typical solid is composed of numerous tiny crystals. Each crystal has a definite geometric form at the atomic level, which reflects the arrangement of its constituent atoms. The crystals are packed together in a regular manner to form a repeating network or lattice.

Crystals are formed when the material changes from its molten state to a solid state. Crystallization occurs abruptly at a specific temperature for a given material, called the *freezing point* (or the *crystallization temperature*). In the molten (liquid) state, a material has an amorphous (noncrystalline) structure. An amorphous structure is one in which the constituent atoms are joined to one another but not in a regular three-dimensional pattern—rather in a random pattern.

Glass is a material that has never crystallized. It becomes hard while still retaining its liquid structure. Thus, glass is sometimes called an *amorphous solid* or a *supercooled liquid.* Because of its amorphous structure, glass does not have a definite melting point like a crystalline solid.

The amorphous structure of glass is responsible for its transparency. All solids except glass and clear plastics are opaque to light. The opaqueness of a crystalline solid is explained by the fact that light is reflected at each crystal boundary as it passes through the solid. At each reflection, some light is lost.

Because there are numerous such crystal boundaries, even within a small thickness of a solid material, light goes through a large number of reflections, losing some light at each reflection. This makes the material behave as an opaque material. Glass is virtually one large crystal containing no internal boundaries, and that is why it is transparent. Several crystalline impurities, however, can make glass translucent or opaque.

The amorphous structure of glass is not merely a property of the materials of which the glass is made because many other materials can also be made to produce an amorphous structure on solidification. The amorphous structure of a solid material is also a property of the rate at which molten ingredients are cooled. If cooled slowly, the atoms in the molten mass have sufficient time to organize into a regular pattern to become a crystalline solid.

Therefore, in the manufacturing of glass, the ingredients must be cooled rapidly to below the crystallization temperature to prevent crystallization.

In this chapter, the modern process of manufacturing glass as well as its various types and their properties are described. Also discussed are light-transmitting plastics, which are sometimes used in lieu of glass.

28.1 MANUFACTURE OF FLAT GLASS

NOTE

Silica Glass

Glass made only from silica (silicon dioxide), with no other additives, is used in situations where its greater resistance to heat and higher transmissivity to radiation is required, such as in mercury vapor lamps and telescope mirrors. Glass made with pure silica is called *silica glass.*

The primary raw material for making *flat glass* (i.e., window glass, to distinguish it from glass block and other glass products) is sand. Chemically, sand is silicon dioxide, also called *silica.* Sand used for glass manufacturing is obtained from sandstone deposits. (Seashore sand is unsuitable for glass making because it has too many impurities.)

Although silica is all that is needed to make glass, other ingredients are added to modify several properties. The two major ingredients added to silica for making flat glass are sodium oxide and calcium oxide (lime). That is why flat glass is sometimes referred to as *soda-lime glass.* Soda-lime glass consists of approximately 72% silica, 15% sodium oxide, 9% calcium oxide, and 4% other minor ingredients.

One of the minor impurities in the ingredients for making glass is iron oxide, which occurs naturally in sand. Iron oxide gives a clear glass sheet its bluish-green tinge when viewed from its edge.

Sodium oxide works as a *flux.* A flux is an additive that lowers the melting point of the main ingredient—in this case, the silica. The melting point of pure silica is very high—nearly 3,100°F. When sodium oxide is added to silica, the mixture melts at a much lower temperature, which reduces the cost of glass making.

Pure silica in its molten form is highly unworkable. It is so viscous and tacky that any bubbles of air or gas produced during melting do not readily escape from the molten mass. Sodium oxide also makes the molten mixture less viscous and, hence, more workable.

The mixture of silica and sodium oxide yields glass that is not too durable. It slowly dissolves in water and has a low resistance to chemical attack. The addition of calcium oxide stabilizes the mix, resulting in a glass that is durable and more easily worked.

The Float Glass Process

Although the raw materials for making flat glass have remained virtually unchanged, the manufacturing process has evolved considerably over the years. In the process that is commonly used today, the raw materials are granulated, mixed together, and loaded into a furnace, where they melt.

Broken pieces of glass from an earlier batch and scrap glass are also loaded in the furnace at this stage. From the furnace, the molten material goes to a molten glass tank and then to a bed of molten tin. Because the density of molten tin is higher than that of molten glass, the latter floats on the molten tin bed.

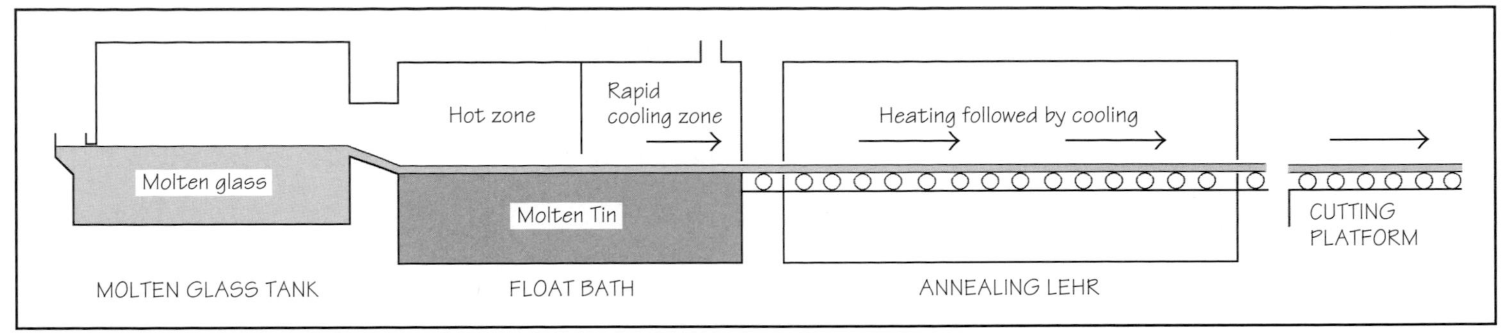

FIGURE 28.5 Outline of float glass–manufacturing process.

The process is controlled so that a predetermined thickness of molten glass travels continuously over the tank of molten tin, called the *float bath,* Figure 28.5. As the molten glass moves to the end of the float tank, it is cooled rapidly to prevent crystallization. There, it solidifies into a sheet and then travels over rollers to the annealing chamber, called the *annealing lehr.*

The glass that enters the annealing lehr has internal stresses locked within its body. This is due to the thermal gradient that is set up between the external surfaces and the interior of the glass due to rapid cooling in the float bath.

In the annealing lehr, the glass is first heated sufficiently to relieve any stresses created during the solidifying process. It is then cooled very slowly so that every glass particle cools at the same rate to ensure that no stresses are frozen in the glass. The annealing lehr is, therefore, several hundred feet long.

At the end of the annealing lehr, the glass sheet emerges as a continuous ribbon at room temperature, free of any internal stresses, Figure 28.6. It is then cut into desired lengths by automatic cutters, packed and transported to its destination, Figure 28.7.

NOTE

Invention of Float Glass Process

The float glass process was first used in 1959 by Pilkington Brothers Limited near Liverpool, England, and has since become a worldwide standard for the manufacturing of glass.

FIGURE 28.6 In the float glass–manufacturing process, glass emerges as an endless ribbon at room temperature. (Photo courtesy of PPG Industries Inc.)

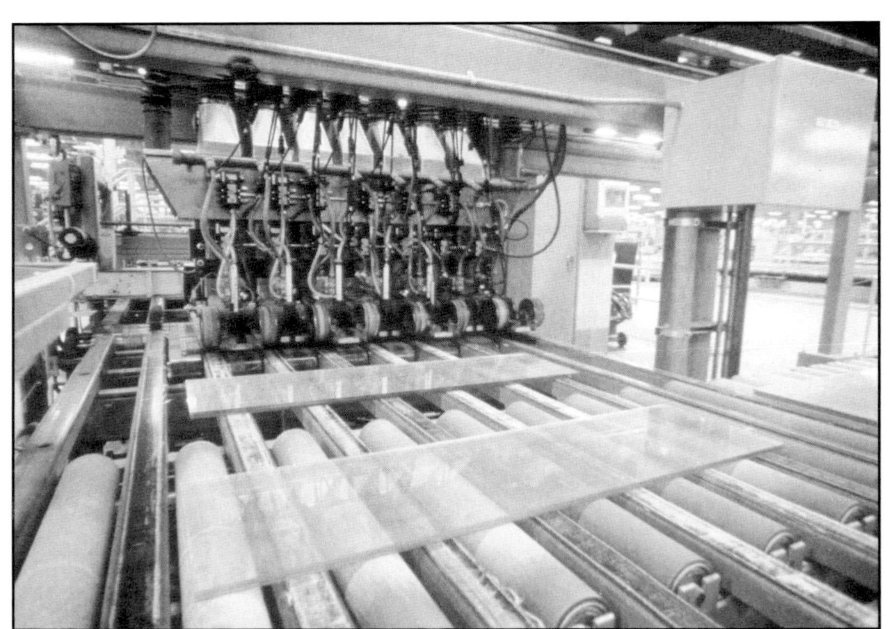

FIGURE 28.7 Glass is cut by automatic cutters to required sizes before packing and shipping. (Photo courtesy of PPG Industries Inc.)

NOTE

Nominal Versus Actual Thickness of Flat Glass

Glass thickness is given in terms of its nominal thickness. The actual glass thickness is its nominal thickness plus or minus a tolerance. For example, the actual thickness of a $\frac{1}{4}$-in. 6-mm-thick glass sheet must lie between 0.219 in. (5.56 mm) and 0.244 in. (6.20 mm).

NOTE

Annealed, Tempered, Heat-Soaked, and Heat-Strengthened Glasses

Annealed glass is the basic glass, obtained directly from the float glass process. *Tempered glass, heat-soaked glass,* and *heat-strengthened glass* are obtained by heat-treating annealed glass.

Because the top surface of a liquid must be uniformly horizontal due to gravity, the floating of molten glass over molten tin ensures that the top and bottom surfaces of glass are parallel. This provides a uniform thickness of glass sheet—an important requirement to ensure distortion-free vision through glass.

The glass obtained from this process is called *float glass* to distinguish it from sheet glass and plate glass, which were the two types of flat glass used before the discovery of the float glass manufacturing process.

Virtually all flat glass manufactured today is produced by the float glass process. With a few exceptions, sheet glass and plate glass are no longer commercially produced.

TYPICAL FLAT GLASS THICKNESS

The commonly used nominal thicknesses of flat glass range from $\frac{3}{32}$ in. to 1 in. Glass of $\frac{3}{32}$-in. thickness is referred to as *single-strength (SS) glass,* and that of $\frac{1}{8}$-in. thickness is called *double-strength (DS) glass.* Manufacturers must be consulted for the availability of certain thicknesses before specifying it for a project.

28.2 TYPES OF HEAT-MODIFIED GLASS

Flat glass obtained from the float glass process, without any further treatment, is the basic glass, referred to as *annealed glass.* Annealed glass, which may be clear or tinted, is the most commonly used glass type in buildings. However, where a stronger glass is required, annealed glass is heat treated before use. Heat treatment increases the bending strength and the temperature resistance of glass and makes it more suitable for applications where annealed glass is inadequate. The following list identifies three types of heat-modified glass:

- Tempered glass, also referred to as *fully tempered (FT) glass*
- Heat-soaked tempered glass
- Heat-strengthened (HS) glass

TEMPERED GLASS

Tempering is the exact opposite of annealing. Whereas annealing (the process of slow cooling after heating) reduces or eliminates locked-in stresses, tempering produces them. Tempering involves heating the glass below its softening point, to approximately 1,300°F, and suddenly cooling (quenching) it by blowing a jet of cold air on all surfaces of the glass simultaneously. This causes the outer layers of glass to harden quickly while the interior of the glass is still soft. As the interior begins to cool, it tends to shrink but is prevented from doing so by the already-hardened outer glass surfaces, Figure 28.8. Consequently, the exterior of the glass comes under a state of compression and the interior, under a state of compensating tension.

Like all brittle materials, glass is weak in tension. Therefore, a glass sheet is weak in bending also, because bending creates tensile (and compressive) stresses in the sheet. The locked-in compressive stresses in the outer layers of a tempered glass cancel (or reduce) the tensile stresses produced by bending. Consequently, tempered glass is approximately four times stronger than annealed glass in bending. It can also withstand greater deflection than annealed glass of the same thickness, Figure 28.9, and is far more resistant to impact and thermal stresses.

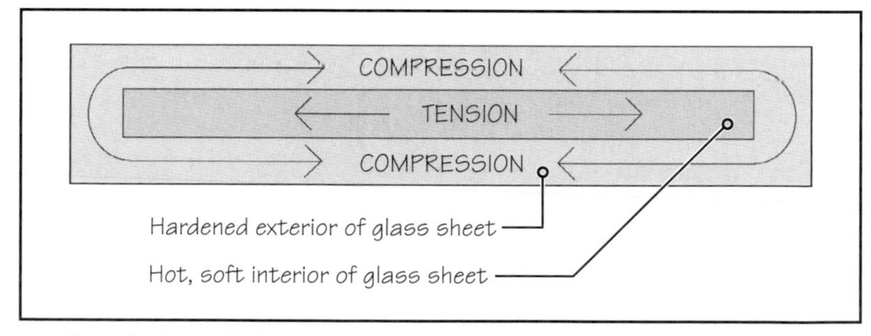

FIGURE 28.8 When the heated glass is quenched with cold air during the tempering process, the outside layers of glass become hard, whereas the interior remains hot and soft. As the interior of glass cools, it shrinks and tends to pull the hardened outside layers inward, producing compressive stresses in them. Because the outside layers have already hardened, they resist the inward pull. Consequently, the interior of glass comes under tension. These stresses are permanently locked in a tempered glass sheet.

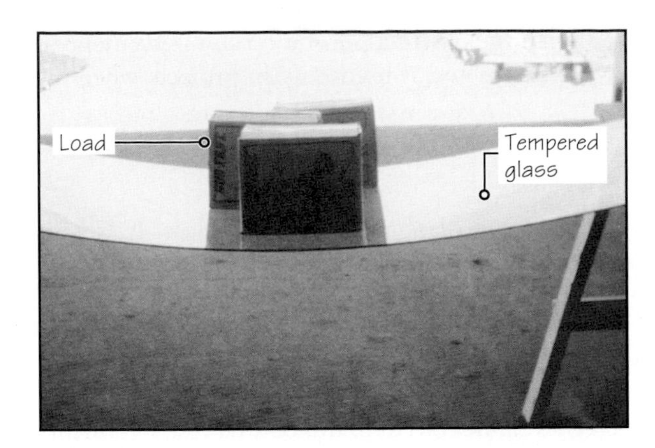

FIGURE 28.9 Deflection of a tempered glass sheet under load. Observe the ductility of tempered glass, that is, its ability to deflect substantially prior to failure.

Tempering does not affect other properties of glass, such as solar heat gain, U-value, or the color of glass. However, because of the bowing and warping caused by the shrinkage of glass during heat treatment, tempered glass may contain a slightly noticeable distortion of view under some light conditions, which is generally not objectionable.

Tempering must be done after the glass has been cut to size. A tempered glass sheet must not be cut, drilled, or edged because these processes release the locked-in stresses, causing the glass to disintegrate abruptly. Abrasive blasting and etching may be done with some care. However, both abrasive blasting and etching reduce the thickness of outer compressed layers, reducing the effectiveness of tempering.

USE OF TEMPERED GLASS

When tempered glass breaks, it breaks into tiny, square-edged, cubicle-shaped granules—a breakage pattern usually referred to as *dicing,* Figure 28.10. Annealed glass, on the other hand, breaks into long, sharp-edged pieces. Tempered glass is, therefore, used in hazardous locations provided it meets the requirements of safety glazing, as described in Section 28.9.

SPONTANEOUS BREAKAGE OF TEMPERED GLASS

Nickel sulfide and certain other impurities present in glass ingredients do not fully melt during glass manufacture. They are known to expand after days or even years of glass manufacture when subjected to extreme thermal stresses or thermal shock (e.g., sudden cooling of sun-heated glass by rainwater). When this expansion occurs in tempered glass, it creates excessive tensile stress in the glass, resulting in its spontaneous breakage.

This breakage, although rare, occurs abruptly and can be a safety hazard. Building codes require that if there is a walking or other usable surface under a tempered glass opening, a protective screen must be provided below it to prevent injuries to humans.

HEAT-SOAKED TEMPERED GLASS

When it is necessary to reduce or eliminate spontaneous breakage of tempered glass, heat-soaked tempered glass is used. In the heat-soaking process, tempered glass panes are subjected to cyclical heating and cooling to simulate long-term field conditions. Glass that does not break in the process is safe against spontaneous breakage and is called *heat-soaked tempered glass.*

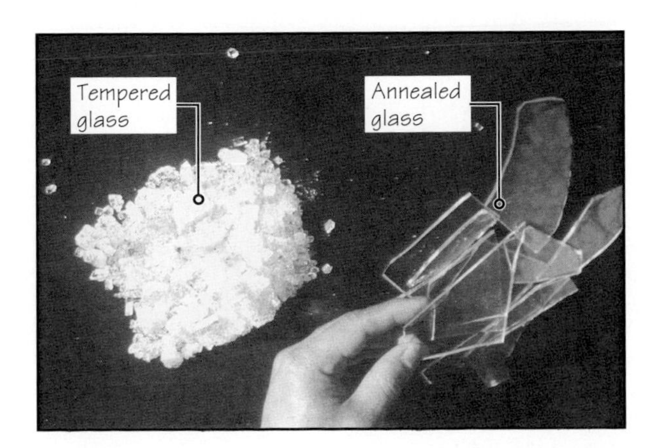

FIGURE 28.10 Breakage pattern of tempered glass and annealed glass. Tempered glass breaks into tiny, square-edged granules. Annealed glass breaks into, large, sharp-ended pieces.

FIGURE 28.11 An example of a typical all-glass curtain wall building.

NOTE

Ceramic Frit on Glass

A ceramic frit coating is deposited on glass in individual dots or lines and then fused (baked) into glass under high temperature. The coating can be of different colors and can be opaque or transluscent. Fritted glass can use any pattern to produce images or graphic designs on glass.

Because the testing is expensive and results in the destruction of a certain percentage of tempered glass panes, heat-soaked glass is expensive and is used only in projects where its higher cost can be justified.

HEAT-STRENGTHENED GLASS

Heat-strengthened glass falls in between annealed glass and tempered glass. It is heat treated in exactly the same way as tempered glass but to a lower temperature of approximately 1100°F. It is nearly twice as strong as annealed glass in bending.

Heat-strengthend glass breaks into pieces that are sharper than those of tempered glass, but more blunt than those obtained from the breakage of annealed glass. In other words, heat-strengthened glass does not "dice" on breaking. Therefore, heat-strengthened glass is not a safety glass. Like tempered glass, heat-strengthened glass cannot be cut or drilled after heat treatment.

USE OF HEAT-STRENGTHENED GLASS

The primary use of heat-strengthened glass is in situations where the glass is subjected to large thermal stresses. One such situation is the spandrel area of an all-glass curtain wall, Figure 28.11. A spandrel area is an area of a wall that is between the head of a window on one floor and the sill of the window on the floor above. This is the part of the building facade that includes edge beams and the floor slab. The purpose of spandrel glass panes is to hide the structural components behind them. Thus, an all-glass curtain wall consists of two distinct areas of glass: vision glass and spandrel glass, Figure 28.12.

In colder regions of North America, it is common practice to use monolithic (single) sheets of glass in spandrel areas and insulating glass units in vision areas. (Refer to Section 28.6 for the description of an insulating glass unit.) To prevent seeing through, the spandrel glass is opacified either with a ceramic frit coating or a polyester film.

To improve the energy performance of the glass curtain wall, spandrel areas are generally insulated. At least a 1-in. space is generally recommended between the insulation and glass to simulate the depth of view of vision glass. Using the same type of glass in both the vision and spandrel areas of the wall (particularly a tinted-reflective glass) adds to the blend between the vision and the spandrel panels.

Because of the presence of insulation, the spandrel glass in an all-glass curtain wall does not dissipate heat as readily as the vision glass. Consequently, the spandrel glass is subjected to much greater thermal stress than the vision glass. Because of its greater strength as compared to annealed glass, heat-strengthened glass is usually specified for the spandrel panels.

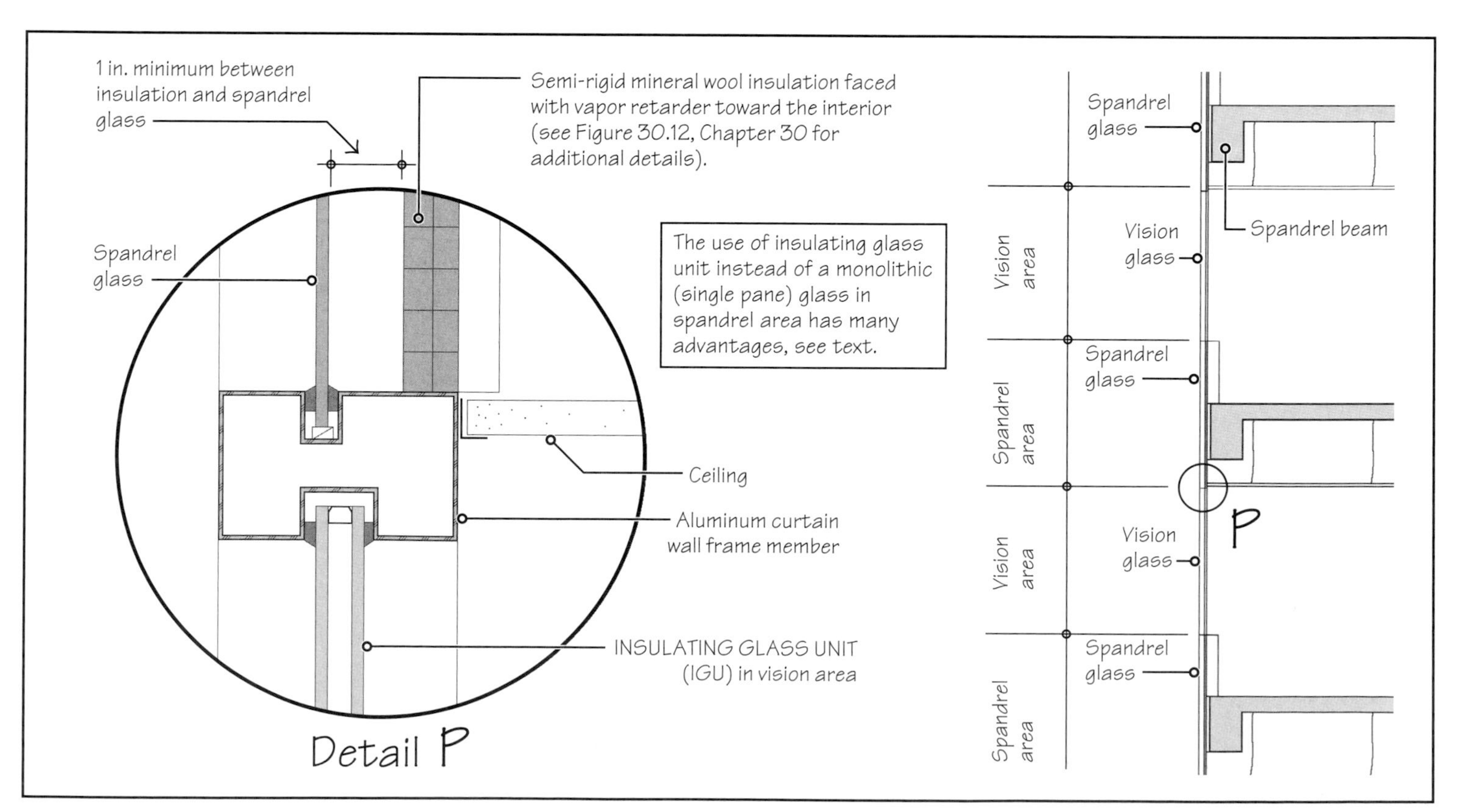

FIGURE 28.12 A section through an all-glass curtain wall showing spandrel glass and vision glass areas (see also Figure 7.13). Although a monolithic glass has been shown in the spandrel area (Detail P), the same insulating glass unit used in the vision area may be used there.

The space between the insulation and spandrel glass is particularly susceptible to condensation, which may lead to wetting of insulation and corrosion of metals. Therefore, the entry of water vapor in this space should be prevented through the use of a good vapor retarder that is fully sealed toward the interior (see Chapter 30 for additional details). Condensation effects can be reduced by the use of insulating glass units in spandrel areas.

The use of insulating glass units in spandrel areas in place of monolithic glass is fairly common in milder climates of North America, which gives a better blend between vision and spandrel areas and may preclude the use of spandrel insulation. The inward glass pane of such an insulating glass unit in the spandrel is opacified and heat strengthened. In situations where a glass pane is subjected to uneven solar shading, the vision areas may also require the use of a heat-strengthened glass.

HEAT-STRENGTHENED GLASS VERSUS TEMPERED GLASS

Although tempered glass is nearly twice as strong as the heat-strengthened glass, heat-strengthened glass is preferred over tempered glass in most situations, except those that mandate the use of tempered glass (e.g., as safety glazing). The reasons are as follows:

- Spontaneous breakage is less of a problem in heat-strengthened glass.
- When heat-strengthened glass breaks, most of it stays within the opening, similar to annealed glass. Tempered glass, on the other hand, fractures in small pieces and tends to evacuate the opening.
- There is generally less optical distortion in heat-strengthened glass.

BENT GLASS

Bent glass is made from float glass that has been heated sufficiently to become plastic so that it can be bent to shape. Various forms of bent-glass shapes have been used in windows and skylights. Bent tempered glass is also available and is commonly used in revolving doors, curved glass handrails, elevator cars, and office partitions.

PRACTICE QUIZ

Each question has only one correct answer. Select the choice that best answers the question.

1. The microscopic (atomic) structure of glass resembles that of
- **a.** solids.
- **b.** liquids.
- **c.** gases.
- **d.** none of the above.

2. To make glass transparent, the molten glass ingredients must be
- **a.** cooled very slowly during manufacture.
- **b.** cooled slowly during manufacture.
- **c.** cooled rapidly during manufacture.
- **d.** any one of the above.

3. The glass obtained from the float glass process without any further treatment is called
- **a.** sheet glass.
- **b.** annealed glass.
- **c.** tempered glass.
- **d.** heat-strengthened glass.
- **e.** plate glass.

4. If glass is $\frac{1}{8}$ in. (3 mm) thick, it is also called
- **a.** single-layer glass.
- **b.** single-strength glass.
- **c.** double-strength glass.
- **d.** basic glass.
- **e.** monolithic glass.

5. Which of the following types of glass is generally used as safety glass?
- **a.** Sheet glass
- **b.** Annealed glass
- **c.** Tempered glass
- **d.** Heat-strengthened glass
- **e.** Plate glass

6. Which glass type is commonly specified for the spandrel areas of an all-glass curtain wall?
- **a.** Sheet glass
- **b.** Annealed glass
- **c.** Tempered glass
- **d.** Heat-strengthened glass
- **e.** Plate glass

7. Which of the following glass types has locked-in stresses?
- **a.** Tempered glass
- **b.** Heat-soaked tempered glass
- **c.** Heat-strengthened glass
- **d.** All the above
- **e.** None of the above

8. To obtain heat-soaked tempered glass, the glass must be tempered first and then
- **a.** annealed.
- **b.** etched.
- **c.** heat strengthened in a furnace.
- **d.** subjected to cyclic heating and cooling.
- **e.** all the above.

9. Which of the following glass types, when broken, breaks into tiny, square-edged granules?
- **a.** Tempered glass
- **b.** Heat-strengthened glass
- **c.** Annealed glass
- **d.** None of the above

10. Spontaneous breakage generally occurs in
- **a.** annealed glass.
- **b.** heat-strengthened glass.
- **c.** heat-soaked tempered glass.
- **d.** tempered glass.
- **e.** laminated glass.

Answers: 1-b, 2-c, 3-b, 4-c, 5-c, 6-d, 7-d, 8-d, 9-a, 10-d.

28.3 GLASS AND SOLAR RADIATION

When solar radiation (i.e., direct solar beam) falls on a glass surface, a part of it is transmitted through glass, a part is reflected, and a part is absorbed by the glass. A $\frac{1}{8}$-in.-thick clear glass sheet transmits approximately 85%, reflects 10%, and absorbs nearly 5% of solar radiation. In other words, the transmissivity, reflectivity, and absorptivity of clear glass for solar radiation are 0.85, 0.10, and 0.05, respectively. Thus, if 100 units of solar energy fall on a clear glass sheet, 85 units are transmitted to the interior, 10 units are reflected to the exterior, and 5 units are absorbed by the sheet, Figure 28.13.

The 5 units of solar radiation absorbed by a clear glass sheet increase its temperature, and the glass becomes a low-temperature (long-wave) radiator. It releases (by radiation and convection) R_i units of absorbed heat to the inside and R_o units to the outside, so that $R_i + R_o = 5$ units.

The relative values of R_i and R_o depend on the internal and external air temperatures. A greater amount of heat is released toward the cooler side. Thus, if the inside air is cooler than the outside air, $R_i > R_o$ (e.g., R_i may equal 3.0 units, so $R_o = 2.0$ units). Conversely, if the outside air is cooler than the inside air, $R_o > R_i$.

When inside and outside air temperatures are the same, $R_i = R_o = 2.5$ units. In such a case, the total amount of solar radiation penetrating the glass will equal $85 + 2.5 = 87.5$ units. In other words, 87.5% of solar radiation incident on the glass penetrates the glass.

SOLAR HEAT-GAIN COEFFICIENT (SHGC)

A commonly used measure of how well a glass sheet performs with respect to direct solar radiation is the *solar heat-gain coefficient.* It is defined as the amount of solar radiation that penetrates a given glass (including the absorbed radiation that is subsequently released to the interior) divided by the solar radiation incident on the glass:

$$\text{SHGC} = \frac{\text{solar energy gain through a glass}}{\text{solar energy incident on glass}}$$

Thus, as shown in Figure 28.13, SHGC = 0.875 for a $\frac{1}{8}$-in.-thick clear glass. The smaller the value of SHGC, the smaller the heat gain through the building. For warm (southern) regions of the United States, SHGC should be as low as possible. For cold (northern) regions of the United States, a higher SHGC value is desirable because it may provide daytime heating of the building through glazed openings.

SHADING COEFFICIENT (SC)

Another measure closely related to SHGC is the *shading coefficient (SC).* SC is defined as the solar heat gain through a given glass divided by the solar heat gain through an unshaded, clear $\frac{1}{8}$-in.-thick glass under the same internal and external conditions. In other words, SC of glass is a ratio of SHGC of glass and SHGC of clear $\frac{1}{8}$-in.-thick clear glass:

$$\text{SC} = \frac{\text{SHGC of glass in question}}{\text{SHGC of } \frac{1}{8} \text{ in. clear glass}}$$

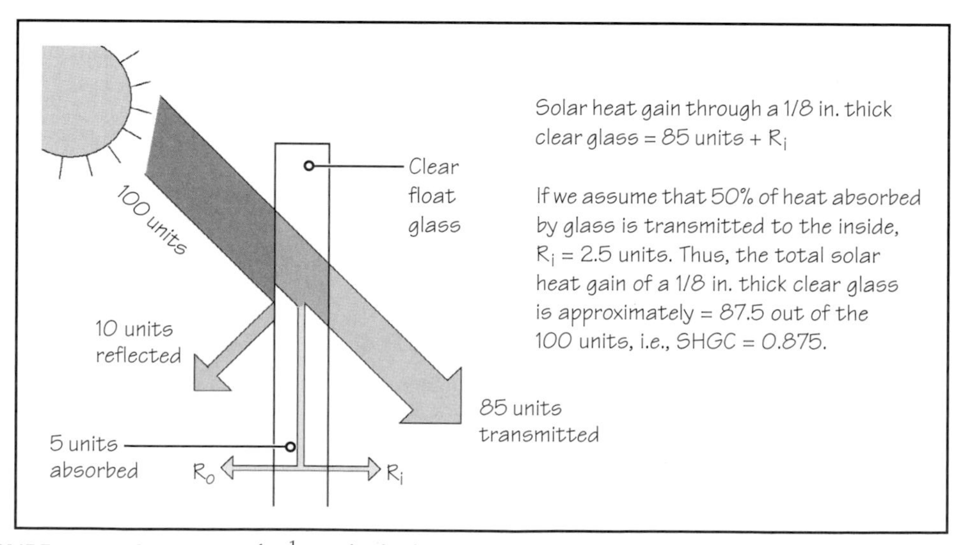

FIGURE 28.13 Properties of a $\frac{1}{8}$-in.-thick clear glass with respect to solar radiation.

From the preceding definition, the SC of a $\frac{1}{8}$-in.-thick clear glass is 1.0. This is practically the maximum possible value of the shading coefficient. The shading coefficient of a clear $\frac{3}{32}$-in. glass is nearly 1.01.

Visible Transmittance (VT) and LSG Index

Because transparency is the primary reason for using glass in buildings, another important property of glass is its ability to transmit light (the visible part of solar radiation; see "Principles in Practice" at the end of this chapter). Visible transmittance (VT) is defined as the percentage of the visible part of solar radiation transmitted through glass. For $\frac{1}{8}$-in.-thick clear glass, VT is approximately 0.90.

The LSG (light-to-solar gain) index of a glass is the ratio of VT and SHGC:

$$\text{LSG index} = \frac{\text{VT of glass}}{\text{SHGC of glass}}$$

The greater the value of LSG index, the more efficient the glass is with respect to reducing the solar heat gain and increasing light transmission. Because VT and SHGC for a $\frac{1}{8}$-in.-thick clear glass are 0.875 and 0.90, respectively, the LSG index for this glass is 0.90/0.875 = 1.03. Glass manufacturers can provide glass with a high LSG index by using spectrally selective, low-E coating on glass (Section 28.5).

Ultraviolet Transmittance (UVT)

Radiation, in general, has an adverse effect on human skin and eyes. It is also responsible for fading the colors of fabrics, carpets, paintings, and art works. Ultraviolet radiation has a higher energy content than radiation at longer wavelengths. Thus, although it is only 3% of the total solar radiation, the material degradation potential of ultraviolet radiation is far greater than that of visible, or long-wave, radiation. Manufacturers of glass quote the ultraviolet transmittance of their products. A glass with lower ultraviolet transmittance is generally preferable.

28.4 TYPES OF TINTED AND REFLECTIVE GLASS

Two commonly used glass products that are specially produced to lower solar heat gain are tinted glass and reflective glass. A glass may be either tinted, reflective, or tinted and reflective. Tinted and reflective glass is commonly used in the curtain walls of an all-glass building to reduce unwanted heat gain through the building.

Tinted (or Heat-Absorbing) Glass

Glass can be tinted to a desired color by adding metallic pigments to molten constituents of glass during its manufacture. Tinted glass is also called *heat-absorbing glass* because it absorbs more solar radiation than clear glass under identical conditions.

A thicker glass, made with exactly the same concentration of pigment and the same batch of molten constituents as a thinner glass, appears to be deeper in color because a greater amount of light is absorbed by a thicker glass. For the same reason, a thicker tinted glass absorbs more heat than a thinner tinted glass.

As a result of greater heat absorption, the temperature of tinted glass is higher than that of clear glass when direct solar radiation is incident on the glass. A thick tinted glass sheet, exposed to intense solar heat, should be heat strengthened to withstand greater thermal stresses.

The most commonly used colors in tinted glass are bronze, gray, and blue green. If everything else is the same, a green tinted glass has a higher LSG index (i.e., is more efficient in reducing solar heat gain and increasing light transmission) than other tinted glass.

Reflective Glass

Reflective glass works as a mirror from the outside during the day, hiding interior activity. At night, however, the interior activity is visible from the outside when the interior space is lit.

Reflective glass is made by bonding metal or metal oxide coatings to one surface of a clear or tinted glass. Some of the metals used are chrome, stainless steel, titanium, and copper oxide. The coating is extremely thin so that sufficient light can pass through the glass and is deposited on the glass by one of the two methods:

- Magnetic sputtering
- Pyrolytic deposition

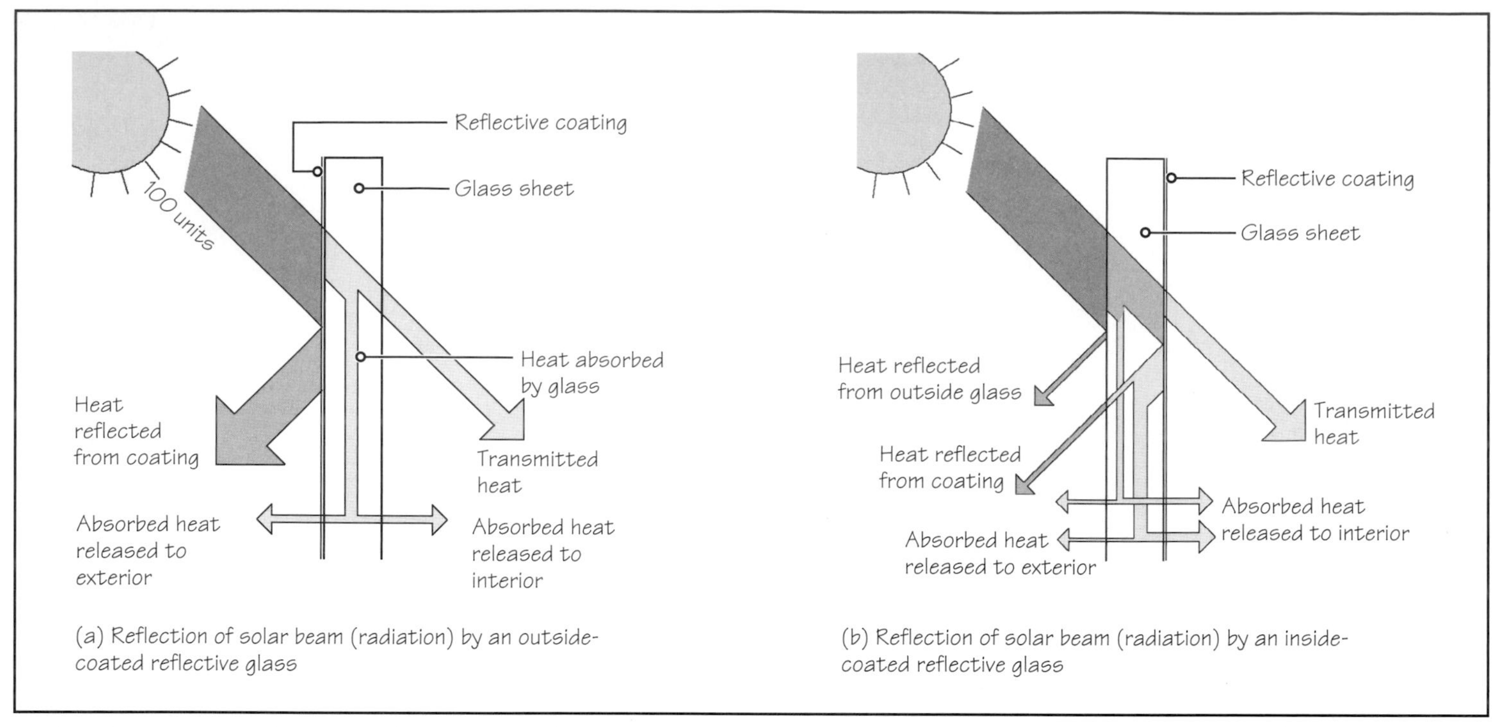

FIGURE 28.14 An outside-coated reflective glass reflects a greater amount of solar beam than an inside-coated reflective glass but is less durable.

Magnetic sputtering requires a large vacuum chamber in which atoms of metal or metal oxide are dislodged from the original material by an induced electrical charge. These atoms impinge on the surface of the glass placed in the chamber, creating a thin layer of coating. If the glass is required to be tempered or heat strengthened, it is heat treated prior to the application of the sputtered coating.

In the pyrolytic deposition process, a metal or metal oxide coating is applied to hot glass. This is accomplished either in the heat-strengthening furnace (for a heat-treated glass), at the hot end of the annealing lehr, or at the cool end of the float bath for annealed glass. Thus, the reflective coating is virtually fused into the glass sheet because the glass constituents are in a semisolid state when the coating is applied. Consequently, a pyrolytically deposited coating is more durable than a magnetically sputtered coating.

Reflective Glass and Solar Radiation

A reflective coating on the outside surface of a glass is more effective in reflecting solar radiation, as compared with the coating on the inside surface, Figure 28.14. This is due to the fact that in an inside-coated glass, the solar radiation passes through the glass twice, and hence it goes through two absorption cycles instead of one (in an outside-coated glass). Thus, an inside-coated glass will become warmer (and be subjected to greater thermal stress) than an outside-coated glass under the same conditions and will generally require heat strengthening. This is particularly true if the glass is also tinted.

The durability of the coating in an outside-coated glass must be examined. A sputtered coating is too soft to withstand the abrasive effects of wind and rain. Therefore, a sputtered coating should be exposed only to the interior. A more protected location, such as surface 2 or surface 3 of an insulating glass unit, is preferred for sputtered coating (see Section 28.6 for the description of insulating glass unit). A pyrolytic coating may be exposed to the exterior, but the cleaning of such a glass must be done per the manufacturer's recommendations.

Reflective glass reduces solar heat gain but also reduces visible transmittance (VT). Because lower VT is generally undesirable, the use of reflective glass has decreased since the introduction of new spectrally selective, low-E coatings for glass, which provide low solar heat gain without greatly reducing VT (Section 28.5).

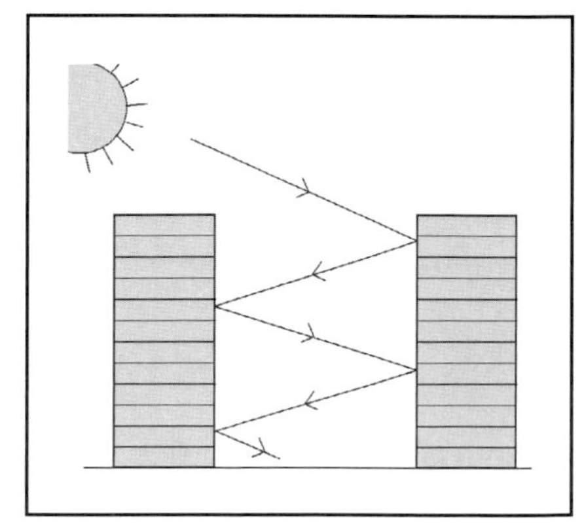

FIGURE 28.15 The use of reflective glass on the opposite facades of all-glass high-rise buildings can make a narrow street become unduly warm, particularly during summer time.

Another disadvantage of reflective glass is that it reflects the solar radiation toward the street and the surrounding buildings. The glare created by such reflection is known to have temporarily blinded passing motorists. Several high-rise, all-glass buildings located on the opposite sides of a narrow street with reflective glass facades may make a street (particularly a narrow street) unduly warm in the summer due to multiple reflections of solar radiation, Figure 28.15.

28.5 GLASS AND LONG-WAVE RADIATION

The properties of glass (SHGC, VT, and UVT) discussed before are relevant only with respect to solar radiation (see Principles in Practice at the end of the chapter). The properties of glass with respect to long-wave radiation are shown diagrammatically in Figure 28.16. Of the 100 units of long-wave radiation incident on glass, 90 units are absorbed, 10 units are reflected, and practically nothing is transmitted through glass. In other words, the long-wave absorptivity of glass is 0.90, reflectivity is 0.10, and transmissivity is zero. Because the absorptivity is equal to emissivity (see Section 5.5), the emissivity of glass is also 0.90.

Of the 90 units of long-wave radiation absorbed by glass, a part is radiated and convected toward the interior and a part, toward the exterior. If the outside air temperature is low—a typical winter condition—most of the absorbed heat is released toward the outside. In other words, the glass is soaking up 90% of the radiant heat incident on it from the enclosure, most of which is being dissipated to the outside, Figure 28.17.

At this stage, it is important to appreciate that this mode of heat transfer is not the only mode by which heat from the inside is being lost to the outside. The other important mode of heat transfer is due to the temperature difference between the inside and outside air.

LOW-EMISSIVITY (LOW-E) GLASS

A simple way to reduce the loss of interior heat absorbed by glass to the exterior is to coat the surface of the glass with a low-emissivity film. Because the emissivity of a surface is equal to its absorptivity, the lowering of emissivity implies lowering its (long-wave) absorptivity and, hence, increasing the surface's (long-wave) reflectivity.

Consider once again 100 units of interior heat incident on a glass. If the interior surface of this glass is coated with a low-emissivity coating with an emissivity of 0.1, 10% of the

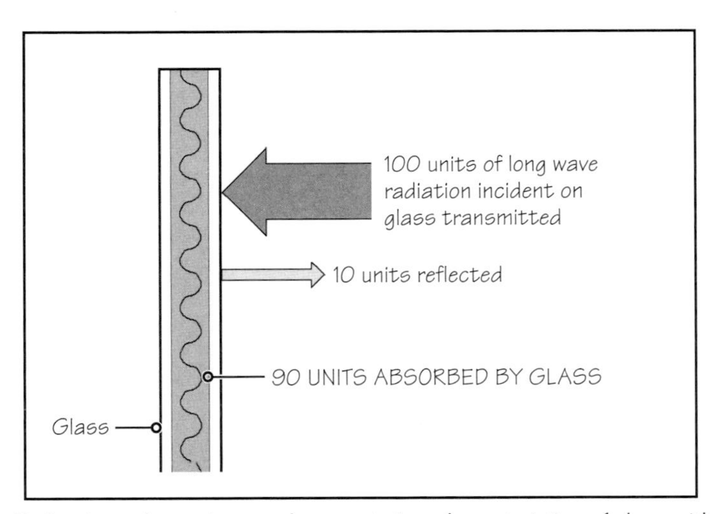

FIGURE 28.16 Reflection, absorption, and transmission characteristics of glass with respect to long-wave radiation.

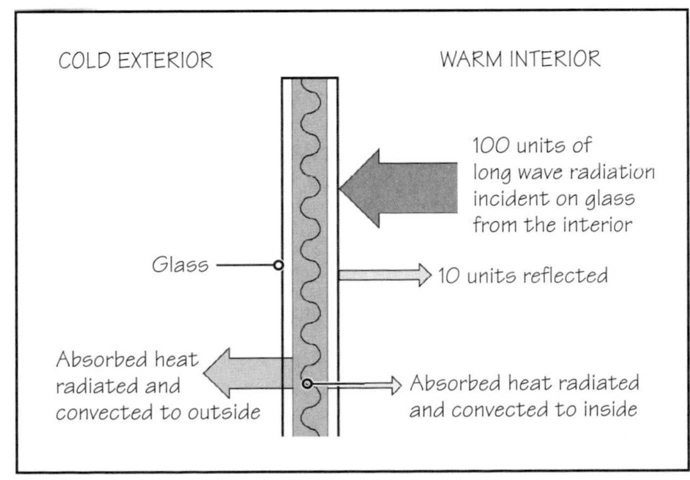

FIGURE 28.17 Long-wave radiation absorbed by float glass from a warm interior and its dissipation by radiation and convection to a cold exterior. Observe that the bulk of interior heat is dissipated to the outside; very little returns to the interior.

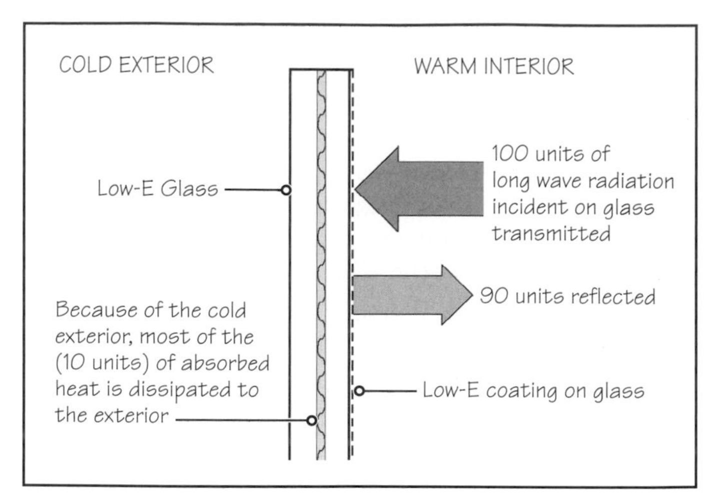

FIGURE 28.18 Long-wave radiation absorbed by low-E glass from a warm interior and its dissipation. Observe that the bulk of interior heat returns to the interior.

heat from the room interior will be absorbed and 90% will be reflected back to the interior, Figure 28.18. A glass with a low-emissivity coating is called a *low-E glass*.

Thus, low-E glass considerably reduces radiant heat loss through the glass and is extremely useful in cold climates. In other words, the behavior of low-E glass to long-wave radiation is similar to the behavior of reflective glass to solar radiation. Just as reflective glass reflects solar radiation back to the exterior, low-E glass reflects long-wave interior heat back to the interior. Low-E glass is also effective in warm climates, although less so (see the box entitled "Location of Low-E Coating in an IGU" later in this chapter).

A low-E coating is almost invisible to the eye because it transmits most of the visible solar radiation. A *spectrally selective low-E coating* is similar to low-E coating in that it transmits most of the visible radiation, but it is more efficient in blocking the transmission of the infrared and ultraviolet parts of solar radiation by absorbing and/or reflecting them. Both a reflective coating or tinting of the glass, on the other hand, are clearly visible because they reflect or absorb the visible solar radiation, respectively.

The process of producing low-E glass is exactly the same as producing a reflective glass, that is, by depositing a metal or a metal oxide coating on one of its surfaces, either by magnetic sputtering or pyrolytic deposition.

The sputtered coating has lower abrasion resistance than the pyrolytically deposited coating but gives a lower emissivity value—nearly 0.10. Low-E glass produced by the pyrolytic process has an emissivity value of nearly 0.3. Because of its lower abrasion resistance, sputtered coating is referred to as *soft-coat low-E* coating and pyrolytically coated glass as *hard-coat low-E* coating. Sputtered-coated low-E glass can be used only in an insulating glass unit where the coating is applied to either surface 2 or surface 3 of the unit.

28.6 INSULATING GLASS UNIT

In Section 5.4, it was noted that the R-value of a glass sheet is approximately 1.0, which is provided almost entirely by the inside and outside surface resistances. This is an extremely low value as compared to the opaque portions of a wall. That is why most heat loss or gain in a building occurs through glazed areas.

Because air is a good insulator, the R-value of glass can be increased by providing a cavity space between two sheets of glass. To avoid problems with the cleaning of the surfaces of glass facing into the cavity, it is necessary that the cavity space be hermetically sealed. An unsealed cavity traps dust and moisture, both of which fog the view through the glass.

A sealed assembly consisting of two glass sheets with an intervening cavity space is called an *insulating glass unit* (IGU). The most commonly used IGU in commercial construction is a 1-in. unit, in which the total width of the unit (thickness of two glass sheets plus the width of the cavity) is 1 in. In this unit, the width of the cavity varies, depending on the thickness of the glass. If the two sheets of glass are each $\frac{1}{4}$ in. thick, the width of cavity is $\frac{1}{2}$ in. If the thickness of each glass sheet is $\frac{3}{16}$ in., the width of the cavity is $\frac{5}{8}$ in. In residential windows, a $\frac{5}{8}$-in.-thick IGU (instead of a 1-in.-thick IGU) is generally used.

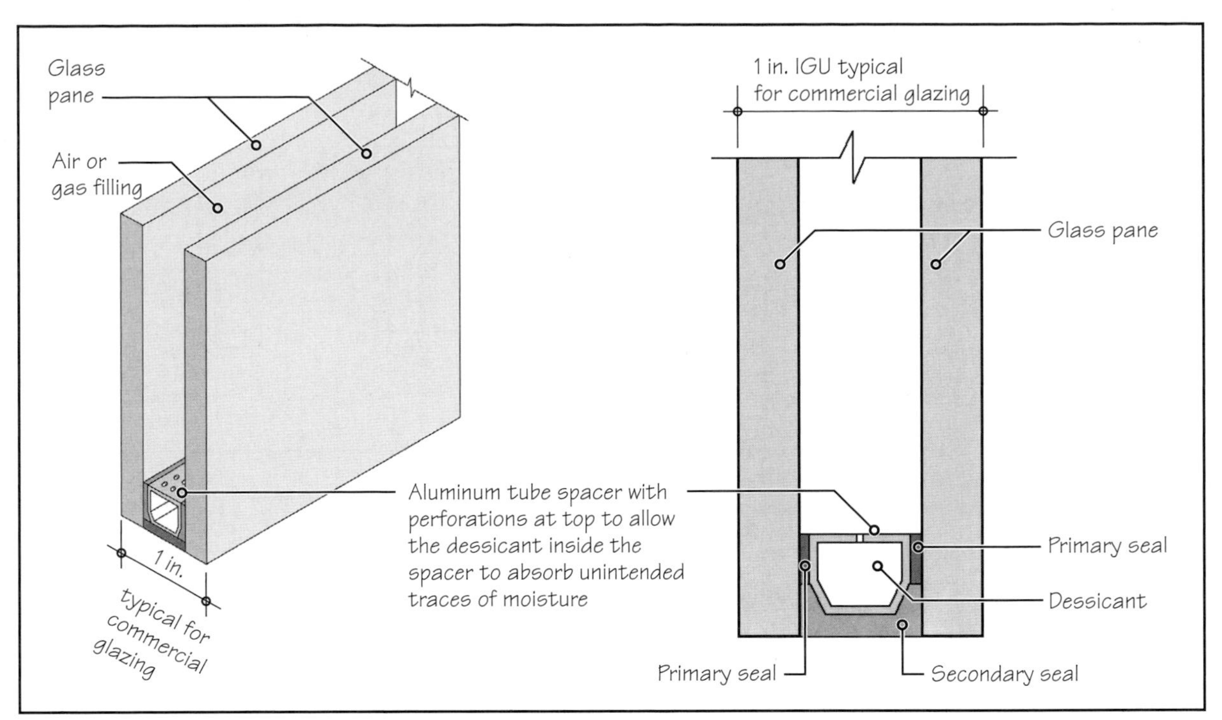

FIGURE 28.19 Anatomy of an insulating glass unit (IGU). Although an aluminum spacer is shown here, less-conducting spacers are available that increase the R-value of the IGU.

The construction of an IGU is shown in Figure 28.19. It consists of two sheets of glass with a spacer, which is sealed to the glass at the sides—the primary seal. The entire assembly is further sealed around all four edges—the secondary seal. The cavity space is filled with dry air or a gas.

However, as an extra precaution, the metal spacer is filled with a desiccant to absorb any incidental moisture that may not have fully evacuated from the cavity. The spacer may be of a clear silvery finish or any other required color.

Because argon has a higher insulating value than air (Table 5.3), many manufacturers provide argon-filled units. Krypton is another gas that is used as a filling in IGUs. Krypton has better insulating properties than argon but is more expensive. Some IGU manufacturers use a krypton-argon mixture.

Air, argon, or krypton are filled in the cavity with the pressure of 1 atm to prevent any loading of glass due to pressure differences. Note that if a vacuum were to replace the air or gas filling, the glass would be subjected to the atmospheric pressure (nearly 2,100 psf), which is several times greater than the wind load expected on a window glass, even under the worst possible hurricane.

EXPAND YOUR KNOWLEDGE

Standard Surface Designation of Glass

Because contemporary window and curtain wall glass is generally not a single sheet of glass but a fabricated product with coatings, laminations, and gas or air filling, it is necessary to have a standard numbering system for various surfaces. The glass industry has adopted the numbering system for the surfaces of various glass products, as shown in the following images. Note that the number sequence starts from the exterior surface of glass. Only the glass surfaces carry a number. The surfaces of laminations or coatings are not numbered.

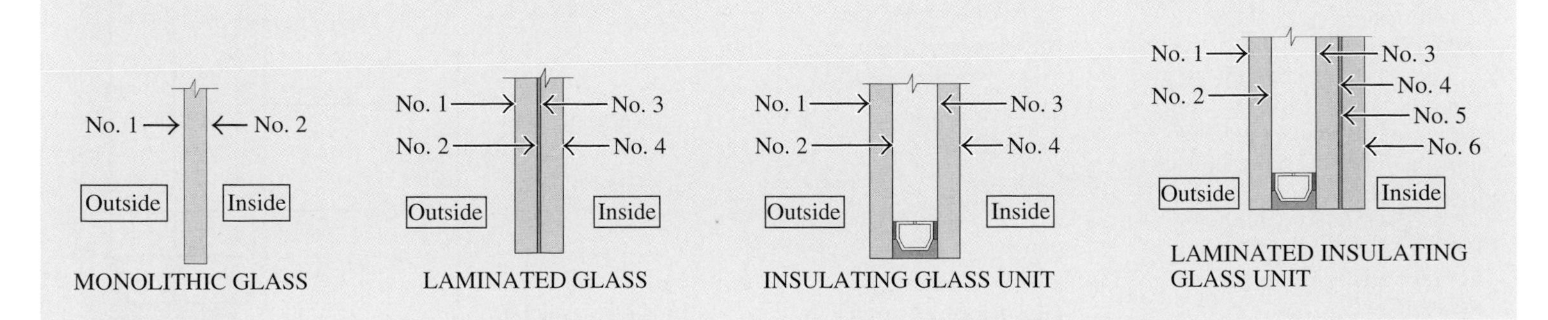

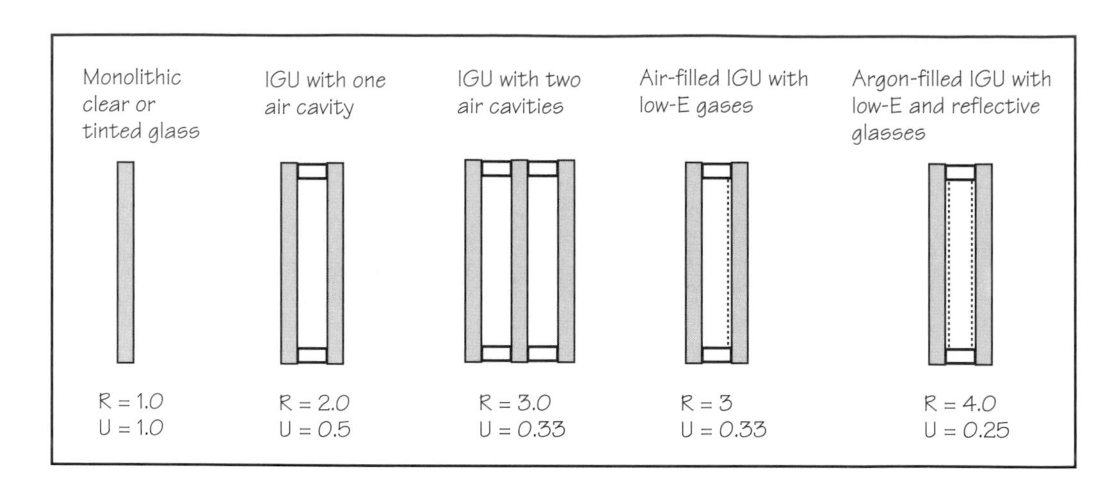

FIGURE 28.20 Approximate R-values (and U-values) of various types of insulating glass units. Note that the U-value is the reciprocal of the R-value.

28.7 R-VALUE (OR U-VALUE) OF GLASS

Knowing that the R-value of a single sheet of glass is approximately 1.0, the approximate R-values (or U-values) of other glass products can be obtained with reference to Table 28.1. Manufacturer's data must be consulted for precise values.

From Table 28.1, one air space adds 1.0 to the R-value of glass. Thus, an IGU with one air space has an R-value of nearly 2.0 (1.0 for the R-value of inside and outside surface resistances and 1.0 for the R-value of the air space. The two sheets of glass add negligibly to the total R-value of the unit). The U-value of this unit is approximately 0.5, Figure 28.20. The R-value of an IGU with two air spaces is approximately 3.0 (U = 0.33).

A low-E coating adds an R-value of approximately 1.0 to the unit. Thus, a low-E coated insulating glass unit with one air space has an R-value of nearly 3.0 (1.0 for the R-value of inside and outside surface resistances, 1.0 for the R-value of the air space, and 1.0 for low-E coating). The U-value of this unit is 0.33.

A reflective coating also helps to reduce the U-value of glass to a small extent because a reflective coating is of a similar material as the low-E coating. From Table 28.1, the contribution of a reflective coating to the R-value is approximately 0.2. Thus, a reflective-coated monolithic glass will have an R-value of nearly 1.2 (U-value = 0.83). Note that the tinting of glass does not change its U-value.

TABLE 28.1 ADDITION TO THE R-VALUE OF GLASS BY VARIOUS ITEMS

Item	Contribution to R-value of glass
Air-filled cavity	1.0
Argon-filled cavity	1.6
Low-E coating	1.0
Reflective coating	0.2
Glass tint	0.0

Note that a tinted glass has the same R-value as a clear glass and a reflective coating improves the R-value of a glass only slightly.

Argon-filled units have lower U-values as compared with units that contain air. Argon filling adds an R-value of approximately 1.6 to the unit. Thus, an argon-filled, low-E coated reflective insulating unit has a total R-value of nearly 3.8 (1.0 for internal and external surface resistances, 1.6 for argon-filled space, 1.0 for low-E coating, and 0.2 for reflective coating).

An increase in the R-value of a glass product beyond 4.0 can be achieved only by adding a second cavity space in the insulating unit, but if this is accomplished by adding a third sheet of glass, it makes the unit heavier and costlier. One manufacturer makes a low-E insulating unit by using a patented system in which a clear plastic film is suspended halfway into the cavity space, Figure 28.21.

The film divides the cavity in two spaces, decreasing the U-value of the unit without making it heavier. The low-E coating is placed on the plastic film. Being transparent, the

NOTE

Location of Low-E Coating in an IGU

A low-E coating is generally placed on the interior glass pane of an IGU (on surface 3). This location works well for cold climates because the low-E coating reflects the interior heat back.

For warm climates, a low-E coating is generally placed on the exterior glass pane of an IGU (on surface 2), which reduces the inward emission of solar heat absorbed by the exterior pane. However, if the IGU must also have reflective coating, it is placed on surface 2 and the low-E coating, on surface 3. In this location, the low-E coating reflects the heat from the cavity space back to the exterior.

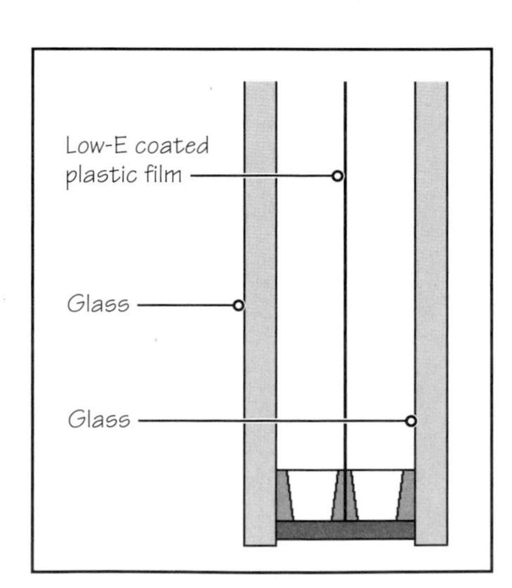

FIGURE 28.21 An IGU with a low-E coated plastic film in cavity space.

film is not visible, so that the unit is virtually indistinguishable from the standard one cavity unit. The R-value of this unit with argon filling is reported to be approximately 5.0, and it is 6.0 with krypton filling.

Because of the intervening plastic, this unit has a limitation on its ability to withstand high temperature. Insulating glass units with two low-E coated plastic films and krypton filling (giving three cavity spaces) are available with an R-value of approximately 8.0.

U-VALUES FOR SUMMER AND WINTER CONDITIONS

In the previous discussion, we have assumed that the U-value of glass is constant for a given type of glass or an IGU. In fact, the U-value of glass depends, to some extent, on the external and internal environmental conditions. Thus, glass manufacturers quote separate U-values for summer and winter conditions. The difference between the two values is significant only if a detailed analysis of the energy performance of a building is needed.

PRACTICE QUIZ

Each question has only one correct answer. Select the choice that best answers the question.

11. Tinted glass absorbs more solar radiation than clear glass.
- **a.** True
- **b.** False

12. If the reflective coating on tinted glass is on the interior surface of the glass, the solar radiation absorbed by the glass is larger than if the same coating is on the outside surface.
- **a.** True
- **b.** False

13. When solar radiation falls on a glass, a certain amount of solar energy enters the enclosure through the glass. A commonly used measure of solar heat gain through the glass is called
- **a.** VT.
- **b.** SHGC.
- **c.** UVT.
- **d.** CRF.
- **e.** none of the above.

14. Before applying reflective coating on a glass, the glass must first be cut to size.
- **a.** True
- **b.** False

15. Which of the following indices measures the effectiveness of the transparency of glass, in relation to the solar heat gain through glass?
- **a.** UVT
- **b.** SC
- **c.** SHGC
- **d.** LSG
- **e.** None of the above

16. A reflective coating increases the R-value of glass only marginally.
- **a.** True
- **b.** False

17. An IGU is
- **a.** an assembly of two glass sheets.
- **b.** an assembly of two glass sheets with an intervening space.
- **c.** a sealed assembly of two glass sheets with an intervening space.
- **d.** a sealed assembly of two glass sheets with an intervening space and a spacer.
- **e.** none of the above.

18. Which type of gas is used in an IGU?
- **a.** Air
- **b.** Argon
- **c.** Krypton
- **d.** All the above
- **e.** None of the above

19. To obtain an IGU with a high R-value, the space between glass sheets should be
- **a.** air filled.
- **b.** argon filled.
- **c.** hydrogen filled.
- **d.** none of the above.

20. Surface 3 in an IGU is
- **a.** the surface of exterior glass sheet facing the exterior.
- **b.** the surface of interior glass sheet facing the interior.
- **c.** the surface of exterior glass sheet facing the cavity space.
- **d.** the surface of interior glass sheet facing the cavity space.

21. The approximate R-value of a clear glass sheet is
- **a.** 1.0.
- **b.** 2.0.
- **c.** 3.0.
- **d.** 4.0.
- **e.** 5.0.

22. The approximate R-value of a tinted glass sheet is
- **a.** 1.0.
- **b.** 2.0.
- **c.** 3.0.
- **d.** 4.0.
- **e.** 5.0.

23. The approximate R-value of an air-filled IGU is
- **a.** 1.0.
- **b.** 2.0.
- **c.** 3.0.
- **d.** 4.0.
- **e.** none of the above.

24. The approximate R-value of an air-filled IGU with a low-E coated glass is
- **a.** 1.0.
- **b.** 2.0.
- **c.** 3.0.
- **d.** 4.0.
- **e.** none of the above.

Answers: 11-a, 12-a, 13-b, 14-b, 15-d, 16-a, 17-d, 18-d, 19-b, 20-d, 21-a, 22-a, 23-b, 24-c.

28.8 GLASS AND GLAZING

The discussion of R-values given in the previous sections refers to glass only. In practice, however, we are interested in the R-value of glazing, which includes the glass and the frame in which the glass is held. The R-value of framing material (aluminum, vinyl, or wood) is not the same as glass. Therefore, the R-value of a glazing (window or a curtain wall) is different from that of glass alone.

The overall R-value of glazing is obtained by averaging the R-values of its three parts: the center of glass, the edge of the glass, and the frame. The center of the glass constitutes most of the glazing, Figure 28.22. The R-value of the center of the glass is the R-value of glass itself. The approximate R-values given in Figure 28.20 refer to the R-values of the center of the glass.

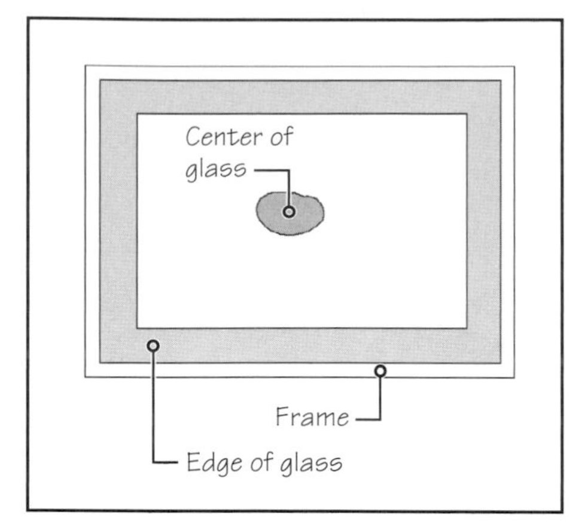

FIGURE 28.22 Three parts of a typical glazing (center-of-glass, edge-of-glass, and frame) that contribute differently to heat flow.

The edge of glass is an area surrounding the frame, which has a width of approximately 2.5 in. A greater amount of heat is conducted through this area than the center of the glass because of its proximity to the frame (and the presence of the aluminum spacer in an IGU).

Therefore, the size of glazing affects its R-value. A small aluminum window generally has a lower R-value than a large window, although the materials used in both windows may be the same. This is due to the relatively larger area of frame and edges in a small window. In commercial curtain walls, where the framing members are spaced farther apart, the difference between the R-values of the glass and the glazing is smaller than that of a window.

THERMAL BREAK IN ALUMINUM FRAMES

The R-value of the frame depends on the dimensions of the frame, its material, and its configuration (i.e., whether or not there is a thermal break in the aluminum frame). A thermal break consists of an insulating connector (usually polyurethane) that connects the two parts of the aluminum frame, Figure 28.23. Polyurethane has excellent rigidity and strength to function as framing material.

The frames of most aluminum windows used in contemporary commercial construction are thermally broken. The use of thermal break in aluminum window and glass curtain wall frames has improved the R-values of glazing in addition to reducing condensation potential, Figure 28.24. Thermally broken window (or glass curtain wall) frames are more important for cold, northern U.S. climates than for warm, southern climates.

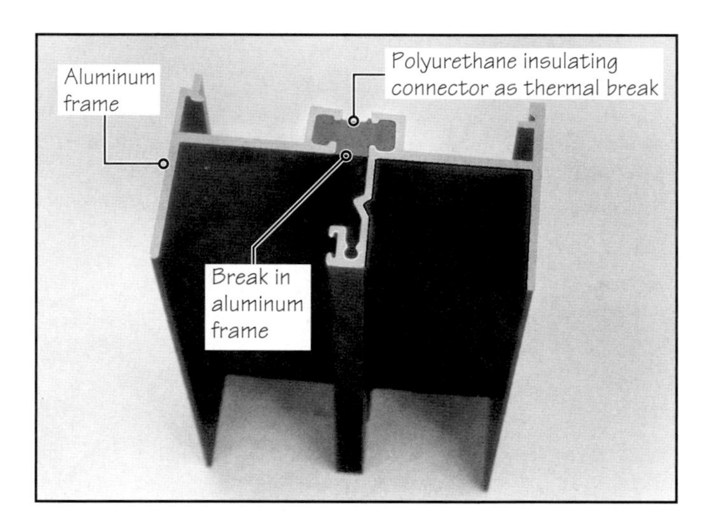

FIGURE 28.23 A polyurethane insulating connector (as a thermal break) between inside and outside parts of an aluminum window frame.

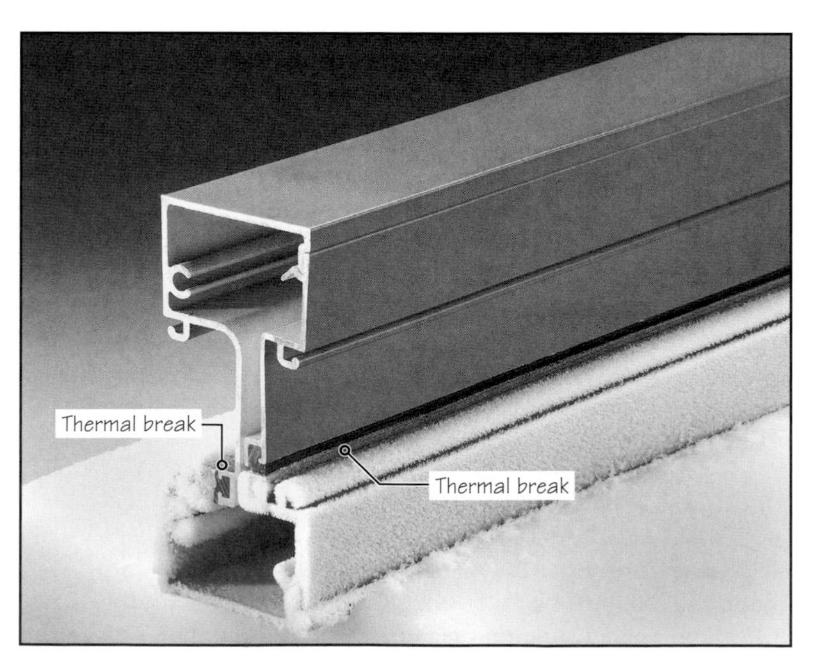

FIGURE 28.24 Illustration of the effectiveness of a thermal break in an aluminum frame. (Photo courtesy of Vitawall Architectural Products.)

LOW-CONDUCTIVITY SPACER IN AN INSULATING GLASS UNIT

Another factor that affects the R-value of glazing is the type of spacer used in an IGU. Traditionally, the spacer in an IGU consisted of a perforated rectangular aluminum tube. The perforations and the dessicant material within the tube help absorb any unintended moisture in the unit (Figure 28.19). Recently, spacers, which are thermally less conducting than the aluminum tube, have been introduced. These consist of a U-shaped aluminum spacer (instead of a tube), plastic, or a plastic foam spacer. The technology related to producing more efficient spacers is called *warm edge technology* (WET).

CONDENSATION RESISTANCE FACTOR (CRF) OF GLAZING

Because the R-value of glazing is fairly low as compared with the opaque portions of a wall, condensation on the interior surface of the glass or its frame can readily occur when the outside temperature is low and the interior relative humidity is moderate to high. A measure that rates a glazing for its condensation potential is called the *condensation resistance factor (CRF);* see "Principles in Practice" at the end of this chapter.

28.9 SAFETY GLASS

Because of the transparency of glass, a large glazed opening without an intermediate horizontal framing member may be mistaken for a clear (unglazed) opening. In this situation, there is a likelihood that a person may unwittingly attempt to walk through the glass thinking it is a clear opening. In this and several other situations, the use of annealed glass is hazardous because of the sharp pieces it produces on breaking.

Building codes require the use of safety glass in locations where accidental human impact is expected. Such locations are called *hazardous locations.* Some examples of hazardous locations are shown in the adjacent box. Two types of glass are considered to be safety glass:

- Fully tempered glass
- Laminated glass

They provide safety in different ways. On impact, tempered glass breaks into small, blunt, and relatively harmless pieces. Laminated glass, on the other hand, is safe because it stays in place after breaking due to the plastic interlayer.

Not all tempered or laminated glass is safety glass. To be recognized as such, the glass must comply with the appropriate test standards. In this test, the glass specimen is mounted in a test frame, Figure 28.25. The test frame has an impactor suspended from the top. The impactor, which weighs 100 lb, consists of a leather punching bag filled with lead shots. The impact on the test specimen is produced by moving the impactor away from its rest position and releasing it.

The test simulates the impact of a person running into a glazing. Although 100 lb is less than the average weight of a human being, it is considered adequate because, in an impact with a glazing, the entire weight of the person is not likely to be involved. The glass is assumed to have passed the test provided the specimen satisfies the breakage criteria outlined

NOTE

Hazardous Locations

Some examples of hazardous locations in buildings requiring safety glass are given here. For a comprehensive list and description, refer to the building code.

- Glass in a swinging door
- Glass in panels of a sliding door and bifold closet door
- Glass in a door and enclosure for a hot tub, whirlpool, sauna, steam room, bathtub, and shower.
- Glass in some fixed conditions adjacent to a glass door.
- Glass in guards and railings.

(a) Specimen glass in test frame, which is about to receive the impact

(b) Glass after the test. The size of broken glass pieces (which must be small enough and within the limits of specified in the test standard) indicate that the glass has passed the test.

FIGURE 28.25 Testing a tempered glass specimen to verify that it conforms to the requirements for a safety glass.

in the test standard. Building codes require that every pane of tempered glass used in hazardous locations bear a label identifying its compliance with the test.

28.10 LAMINATED GLASS

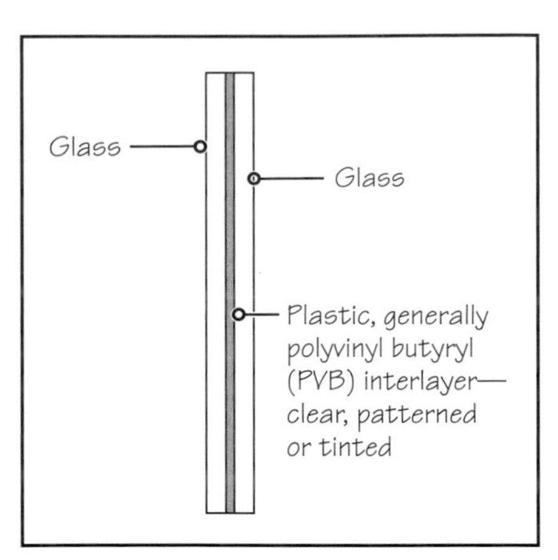

FIGURE 28.26 Section through a laminated glass sheet.

In its simplest form, laminated glass is made from two layers of glass laminated under heat and pressure to a plastic interlayer so that all three layers are fused together, Figure 28.26. In the event of an impact, the glass layer in laminated glass may break, but it will not fall out of the frame. The glass tends to remain bonded to the interlayer, minimizing the hazard of shattered glass.

Laminated glass has been used for automobile windshields for a long time. In architectural applications, it is the product of choice for skylights, sloped and overhead glazing, zoos, and aquariums. Because the glass remains in the opening even after breakage, laminated glass is also recommended for windows in earthquake-prone areas to reduce injury from abrupt shattering of glass.

The plastic interlayer, generally polyvinyl butyryl (PVB), may be clear, patterned, or tinted. It comes in three standard thicknesses: 15 mil, 30 mil, and 60 mil. Where greater safety or security is required, a greater thickness of the interlayer (90 mil or thicker) or multiple layers of thinner sheets may be used.

A major advantage of the plastic interlayer is that it also blocks out most of the ultraviolet radiation. The plastic interlayer also improves sound-transmission properties of glass. Any type of glass—annealed, heat-strengthened, or tempered—can be used in making laminated glass. Laminated glass can also be used in IGUs.

LAMINATED GLASS AS SAFETY GLASS IN HAZARDOUS LOCATIONS

If used in hazardous locations, laminated glass must pass the same impact-resistance test as tempered glass. Additionally, it must also pass the boil test. The boil test determines the ability of glass to withstand resistance to high temperature. In this test, 12-in. by 12-in. specimens of laminated glass are immersed in boiling water for 2 h. The specimens are deemed to have passed the test if no bubbles or other defects more than $\frac{1}{2}$ in. from the outer edges of the glass develop.

HURRICANE-RESISTANT AND BLAST-RESISTANT GLASS

With a suitable thickness of PVB lamination, laminated glass can be made to resist the impact of wind-borne missiles and is mandated for openings in hurricane-prone coastal regions. Blast- (explosion-) resistant laminates are also available, and blast-resistant glass is commonly specified to meet higher security requirements, Figure 28.27. Tempered glass can be laminated, and laminated-tempered glass can be used in IGUs.

FIGURE 28.27 U.S. Federal Courthouse in Jacksonville, Florida, combines stringent security requirements (including blast protection) and the requirements for light and openness of the facade through the use of laminated glass. Architects: HLM Design, Jacksonville, Missisipi. (Photo courtesy of Solutia, Inc.)

28.11 STRUCTURAL PERFORMANCE OF GLASS

Glass is a brittle material; that is, a glass pane will fail at a relatively low deflection. As previously stated, glass is weak in tension and bending but relatively strong in compression.

Although glass is sometimes used in structural applications, such as small floor slabs and stair treads, its primary use is in nonstructural applications, such as windows, skylights, and guardrails. Therefore, glass is not considered a structural material like steel, wood, concrete, and masonry.

Glass in windows is required to resist wind, earthquake, and thermal loads. Glass in skylights must be able to withstand snow loads, loads due to repair personnel, and wind and earthquake loads. Glass in the rear walls of squash and racquetball courts is subjected to impact loads, and the glass used in aquariums must be able to withstand hydrostatic pressure. All these loads cause bending in glass. Thus, the most important structural property of glass that is of interest to us is its bending strength.

Unlike other materials, the strength of glass is statistically complex because every glass sheet has numerous randomly distributed microscopic internal cracks and other flaws. Furthermore, the edge conditions of glass also affect its strength. Clean-cut edges provide the maximum strength. If the edges are damaged, the ability of the glass to resist the load is reduced. Thus, the actual breakage of glass results from a complex interaction between the size, orientation, distribution, and severity of cracks and edge defects.

Consequently, the strength of an individual glass pane cannot be predicted with any reasonable accuracy. In other words, if several identical panes of glass are tested, there will be an unusually large variation in their breaking loads. Therefore, the strength of a glass pane is expressed in terms of the *probability of breakage* under a given load, that is, the number of panes that are likely to break per 1,000 panes. Thus, a probability of breakage of $\frac{3}{1,000}$ (three per thousand) means that 3 out of a total of 1,000 panes will probably break when a specified load is placed on the panes for the first time.

The maximum probability of breakage allowed by building codes for determining glass thickness in windows and curtain walls to withstand wind loads is $\frac{8}{1,000}$. A lower probability of breakage may be required for glass in high-rise and other critical buildings.

ASTM Standard E 1300, "Standard Practice for Determining Load Resistance of Glass in Buildings," provides an authoritative procedure for determining glass thickness to resist a given wind load. A representative ASTM chart (that applies to a $\frac{1}{4}$-in.-thick annealed glass supported on all four sides) is shown in Figure 28.28. An example of the use of the chart is also given in this figure.

Higher Strengths of Heat-Strengthened and Tempered Glass

As noted previously, tempered glass is nearly four times stronger than annealed glass. This implies that a tempered glass pane can withstand four times as much load as an annealed glass pane of the same thickness and size. However, the modulus of elasticity of tempered glass (and heat-strengthened glass) is the same as that of annealed glass. This implies that the deflection of tempered glass and annealed glass panes of the same size and thickness will be equal under a given load. Thus, if tempered glass is made to withstand a greater wind load because of its higher strength, a greater deflection will be produced in the panes. This will place greater pressure on gaskets and seals, leading to their premature failure. It may also expel the glass out of the opening.

Therefore, in practice, tempered glass and heat-strengthened glass are designed to withstand the same wind load as annealed glass. In other words, the higher strengths of tempered glass and heat-strengthened glass are usually not exploited. Tempered glass and heat-strengthened glass are used primarily for their greater impact resistance, greater thermal stress resistance, and safer breakage pattern.

28.12 FIRE-RESISTANT GLASS

Ordinary glass (whether annealed, tempered, or laminated) is not resistant to fire. When subjected to a typical building fire, it shatters in 2 or 3 min. The types of glass commonly used as a fire-resistant glass are

- Wired glass
- Intumescent multilaminate glass
- Intumescent gel-filled glass units

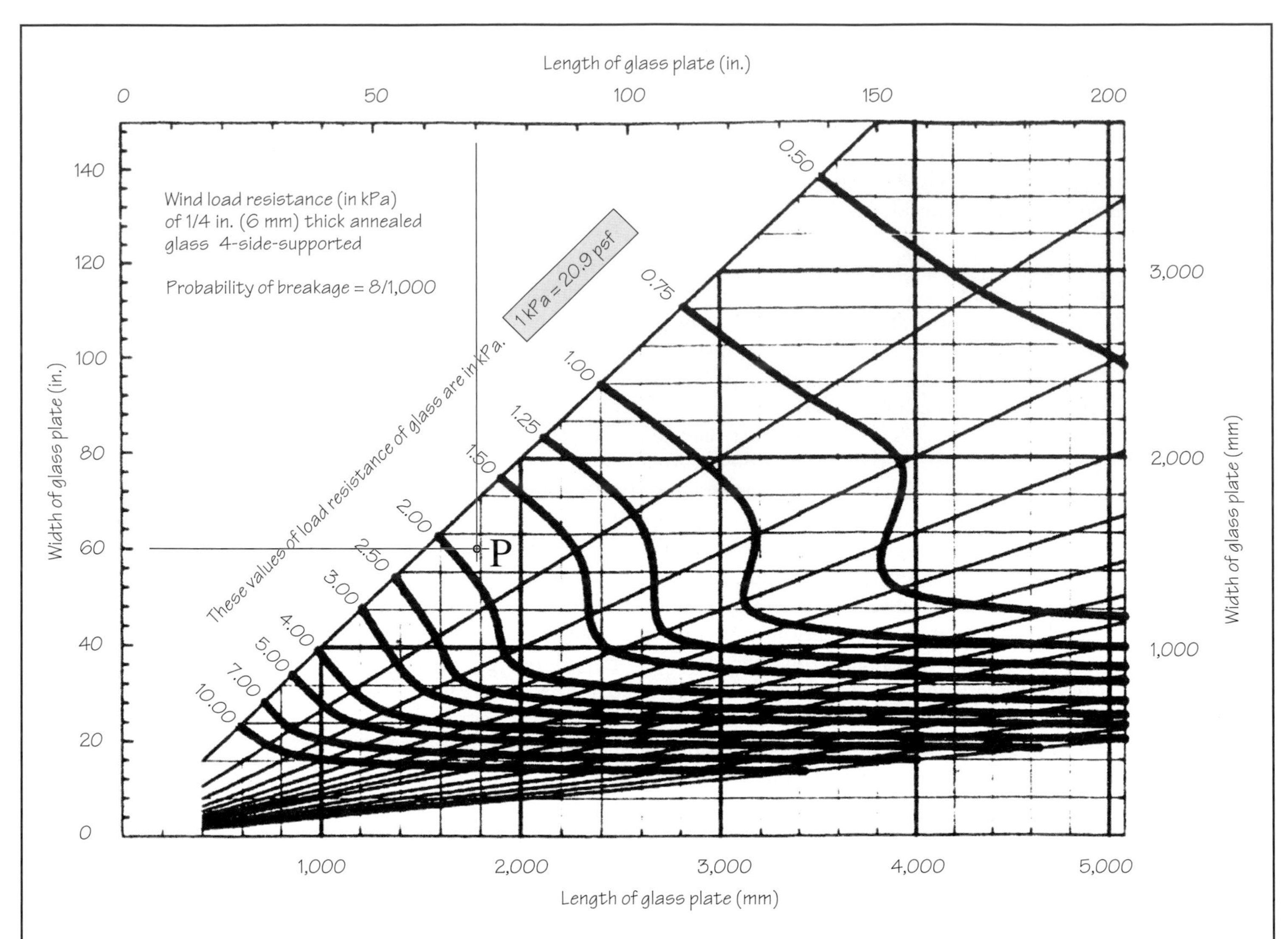

Notes:

1. Wind load resistance refers to 3-second gust loads.
2. Curved lines represent constant wind load resistance contours.
3. Inclined lines refer to constant aspect ratio (ratio of length to width of glass plate)
4. Length of glass plate is its larger dimension and the width is its smaller dimension.

Example: Determine the magnitude of wind load that a 1/4 in. (6 mm) thick annealed glass pane (plate) can resist. The glass pane is 70 in. long and 60 in. wide and is supported on all four sides.

Solution: Draw a vertical line to correspond to 70 in. length and a horizontal line to correspond to 60 in. width of pane. Call the point of intersection of these two lines as P.

Point P identifies the load resistance contour. To interpolate between contours, draw the (inclined) line from the origin to meet P. Interpolate along the inclined line. From the above figure, the load resistance of glass = 1.85 kPa = (1.85 × 20.9) = 38.7 psf.

The probability of 8/1,000 implies that when a load of 38.7 psf occurs for the first time on 1,000 panes (each 70 in. × 60 in.), 8 panes will likely break.

FIGURE 28.28 Wind-load resistance of $\frac{1}{4}$-in.-thick annealed glass.
(Source: American Society for Testing and Materials, Philadelphia, Pennsylvania."Standard Practice for Determining Load Resistance of Glass in Buildings," ASTM E 1300.)

Wired Glass

Wired glass is made by the rolling process, not by the float glass process. During the rolling process, welded wire mesh (which looks like chicken wire mesh) is embedded in the middle of the glass thickness so that the resulting product is a steel wire–reinforced glass, Figure 28.29. The wire diameter is nearly 0.020 in.

The minimum thickness of glass recognized as fire-rated glass is $\frac{1}{4}$ in. When subjected to fire, wired glass also breaks in 2 to 3 min, but unlike annealed glass, which falls out of the frame, wired glass is held within the opening because of the wire mesh.

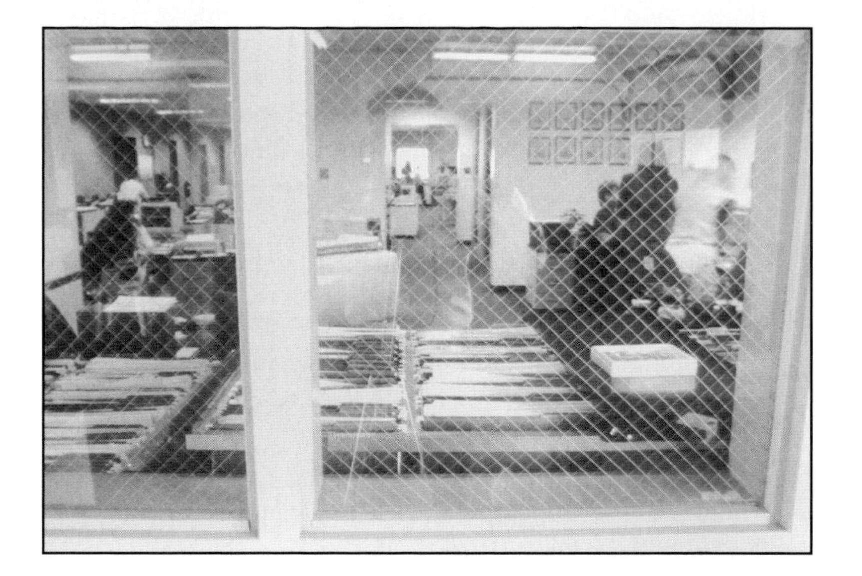

FIGURE 28.29 Wired glass in a fire-rated window between a parking garage and an office room.

Because of the embedded wires, wired glass is often confused with safety glass, which it is not. Unlike tempered glass, wired glass is not impact resistant. Additionally, its bending strength is only half that of annealed glass. On impact, wired glass breaks into sharp pieces, and the broken wires will generally project out of the glass, causing injury to a person. The wires can even act as a spider web on human impact and catch a victim rather than permit the person to pass through safely. Thus, although wired glass is fire resistant, it is not a safety glass.

INTUMESCENT MULTILAMINATE GLASS

Intumescent multilaminate glass (IMG) consists of multiple sheets of annealed glass alternated with intumescent interlayers. The number of glass and intumescent layers determine the fire rating. A fire rating of up to 2 h is attainable. When fire occurs, the intumescent layers become opaque and expand to provide the fire resistance.

INTUMESCENT GEL-FILLED UNITS

Intumescent gel-filled glass is an insulating glass unit in which the intervening cavity is filled with a clear gel. In the case of fire, the gel absorbs the heat, transforming the glass into a heat-insulating opaque crust.

A unit consisting of two sheets of $\frac{1}{4}$-in.-thick tempered glass with $\frac{3}{8}$-in. gel provides a $\frac{1}{2}$-h fire rating. The same construction with $1\frac{1}{8}$ in. gel provides a 1-h rating. A 2-h rating is obtained by a unit with two gel spaces of $1\frac{1}{4}$ in. and three sheets of glass. Under normal conditions (when there is no fire), this product looks like any other insulating glass unit with clear glass panes, Figure 28.30. These units can also be obtained to serve as safety glazing.

NOTE

Fire-Rating Limitations of Wired Glass

Building codes place severe limitations on the use of wired glass as a fire-rated glass: $\frac{1}{4}$-in.-thick wired glass can be used in an opening whose fire rating is between 60 and 90 min, provided the area of wired glass is less than 100 in.^{2}. If the fire rating of the opening is 45 min, the maximum permissible area of wired glass is 9 ft^2 (1,296 in.^{2}). If the fire rating of a window is 20 min or less, the area of wired glass is not limited by the codes.

For a fire-rated opening, both the wired glass and the opening must be so labeled by the manufacturers. All $\frac{1}{4}$-in.-thick wired glass is not a fire-rated glass.

FIGURE 28.30 Lobby at the Red Lion Inn, Costa Mesa, California. Glazed walls provide $1\frac{1}{2}$-h rating. (Photo courtesy of SAFTI FIRST, a division of O'Keeffe's Inc.)

28.13 PLASTIC GLAZING

Clear plastic sheets (referred to as *light-transmitting plastics*) have gained popularity in some applications as an alternative glazing material. Three important advantages of plastic over glass are as follows:

- Plastic can be bent to curves far more easily than glass. Plastic is the material of choice in curved skylights, particularly doubly curved shapes such as domes.
- Plastic is several times stronger than glass of the same thickness and is also more impact resistant. It does not shatter or crack like glass. It is, therefore, specified in fenestration where glass breakage due to vandalism is a concern or where a high degree of security is required.
- Plastic glazing is lighter than glass.

However, the disadvantages of plastic glazing far outweigh its advantages. That is why its use is limited to such applications where glass cannot be used. Some of the disadvantages of plastic glazing are as follows:

- Plastic has a much greater coefficient of thermal expansion than glass. If the movement of plastic glazing is restricted, it will visibly bow out in the direction of higher temperature. Framing details must, therefore, allow linear movement as well as the rotation of glazing. Consequently, framing details are more complex with plastic glazing, which further adds to its cost.
- Humidity changes also affect the dimensional stability of plastics. Plastic expands with increasing humidity, creating the same problems as thermal expansion.
- Plastic is not abrasion resistant, although abrasion-resistant coatings on plastic glazing considerably improve the abrasion resistance of plastics.
- Plastic is vapor permeable, so that sealed insulating units are not possible.
- Plastic yellows with age (ultraviolet degradation), reducing its clarity and light transmission.
- The most serious disadvantage of plastic is its combustibility. It will contribute fuel in a building fire. Therefore, building codes place severe limitations on the area of plastic glazing and the height of the building above which it cannot be used.

NOTE

Building Codes and Plastic Glazing

Building codes place stringent restrictions on the use of light-transmitting plastics in wall and roof openings. Less stringent restrictions apply to buildings of Type V(B) construction (see Section 7.5 for an explanation of the types of construction).

ACRYLIC VERSUS POLYCARBONATE GLAZING

At the present time, two types of plastics are used for glazing: acrylic and polycarbonate. Acrylic has better resistance to ultraviolet radiation than polycarbonate. It yellows more slowly and has greater resistance to abrasion and scratching than polycarbonate. It also has greater resistance to chemicals such as household and glass cleaners.

Polycarbonate, on the other hand, has greater impact resistance than acrylic. It can withstand 30 times as great an impact as acrylic and 250 times as great an impact as glass, making it the most vandal-resistant glazing. Polycarbonate is, therefore, commonly specified for bus shelters and school windows where vandalism is a problem.

Polycarbonate can also withstand higher temperatures than acrylic. Therefore, polycarbonate is commonly used in signs that are exposed to high temperatures from light sources. Acrylic is commonly used in curved skylights and greenhouses, where greater durability and weatherability is required.

NOTE

Acrylic and Polycarbonate

Acrylic—more weather and chemical resistant; commonly specified for skylights and greenhouses

Polycarbonate—more vandal and heat resistant; commonly specified in bus shelters and glazing subjected to vandalism

28.14 GLASS FOR SPECIAL PURPOSES

There have been major advancements in glass and glazing technology in recent years. A few of the glass types used for special purposes are described next.

SECURITY GLAZING

Security glass refers to a glass that is resistant to burglary, forced entry, and bullet penetration. Because the safety against these threats is a function of both glass and the frame in which the glass is held, the term *security glazing* is commonly used rather than security glass. Typically, security glazing is composed of single or multiple panes of glass fused with single or multiple sheets of plastic, usually held in a steel frame.

Security glazing can be designed to withstand different levels of threats, including repeat assaults. For example if the threat level is burglary, the design must consider repeated impacts from hand-held weapons such as hammers, crowbars, or bricks.

Ballistic threats of various levels can be accommodated, depending on the power of the firearms to be resisted. Ballistic gazing is designed to be shatterproof on the victim's side, because the shattering of glass on the victim's side can cause serious injuries. Ballistic glazing is generally specified for drive-in bank windows, armored vehicles, and any other application where a high degree of security is required.

Radiation-Shielding Glass and Plastic

The ability of a building material to shield against radiation is generally a function of its density. The greater the density and the thicker the material, the greater its radiation-shielding capability. Metals, concrete, and masonry materials (provided there are no cracks in concrete and masonry) provide excellent radiation protection.

However, lead—because of its highest density among building materials and its flexibility—is the most commonly used radiation shielding material. Where transparency is not required, lead curtains are commonly used, and lead aprons are worn for additional protection.

Radiation-shielding glass is made by adding lead (usually in the form of lead oxide) to the same ingredients as those of soda-lime glass. Lead acrylic is made from lead-acrylic polymer. Lead glass and lead acrylic have much greater radiation-shielding potential than ordinary soda-lime glass.

Dimmable Glass

Dimmable glass refers to glass in which light-transmission characteristics change from an opaque to a transparent glass with the flick of a switch. It consists of two layers of glass with a plastic interlayer that encapsulates polymer-dispersed liquid crystals (PDLC). When an electric field is applied to PDLC, the crystals orient themselves to let the light through. In its natural state (i.e., without the electric current), the glass is opaque, Figure 28.31.

Also referred to as *electrochromic glass,* it is used in situations where privacy requirements are changeable, such as conference rooms or executive offices. Although far more expensive than normal glass, it may be used in residential and office glazing for privacy and energy-conservation reasons. The switch mechanism can also be made automatic by a photovoltaic mechanism that responds to the ambient light environment. If the ambient light is excessive (e.g., due to direct solar radiation on the glass), the glass turns opaque. When the ambient light is low, the glass turns transparent.

Self-Cleaning Glass

A special oxide coating applied to one surface of the glass can make it self-cleaning. The coating breaks down organic dirt (due to trees, leaves, bird droppings, etc.) on the glass so that the glass is cleaned when rainwater falls on it. The coating does not affect inorganic dirt (e.g., plaster and paint droppings), which must be removed by conventional means. (The coating becomes active due to ultraviolet rays and may take nearly 1 week of ultraviolet exposure for a complete breakdown of organic dirt.)

FIGURE 28.31 Example of the use of dimmable glass, showing a glass pane that is transparent in the left-hand image and is opaque in the right-hand image. (Photo courtesy of Polytronix, Inc.)

28.15 CRITERIA FOR THE SELECTION OF GLASS

With so many factors to consider, the selection of glass, particularly for windows and curtain walls, is not an easy task. The selected glass must obviously be structurally adequate on its own. Additionally, the frame in which the glass is held must be adequate, so that the entire glazing system is structurally adequate.

The safety of glass against human impact, the resistance of glass against wind-borne missiles, and the security of glass against blasts or bullets are other important considerations. These considerations apply to the glass as well as to the frame.

The energy performance of glass also plays an important role in its selection. As with structural and safety considerations, the energy performance of the entire glazing must be evaluated. To help designers, manufacturers of windows and curtain walls provide energy performance ratings of their products (Chapters 29 and 30).

A checklist of factors that must be considered in the selection of glass for windows, curtain walls, and skylights follows:

- Structural adequacy—wind-load resistance, earthquake resistance (where needed), snow-load resistance (for skylights)
- Energy performance (U-value, SHGC, VT, LSG index, air infiltration, CRF, as they apply to the entire glazing, are some of the important factors)
- Resistance to temperature-related stress
- Fire resistance, where needed
- Safety glass, where needed
- Blast resistance, where needed
- Missile-impact resistance and cyclic wind-load resistance in hurricane-prone areas
- Impact resistance in sports facilities (e.g., in racquetball and squash courts)
- Radiation shielding, where needed

28.16 ANATOMY OF A GLAZING POCKET

In most glazing systems, the glass is captured by the framing members within a U-shaped pocket that is continuous on all four edges of the glass, referred to as the *glazing pocket,* Figure 28.32. Glazing pockets provide bearing and lateral load supports to glass. Additionally, they collect any incidental rainwater that might go past the glazing seals and weep it out.

Because the coefficient of thermal expansion of glass is different from that of the framing members, sufficient space must be left between the edges of the glass and the edges of the glazing pocket to allow the glass to move. Additionally, as the glass moves within the opening, it must not directly touch the frame but must be able to "float" within the opening. Important components of a glazing pocket are

- Setting blocks
- Edge blocks and (in seismic zones) corner blocks
- Glazing seals
- Weep holes

SETTING BLOCKS

Each pane of glass is supported on two identical *setting blocks,* generally at quarter points, Figure 28.33. Setting blocks serve two important functions: (a) they provide a hard, but resilient, bearing support to the glass, and (b) they lift glass up from the bottom of the glazing

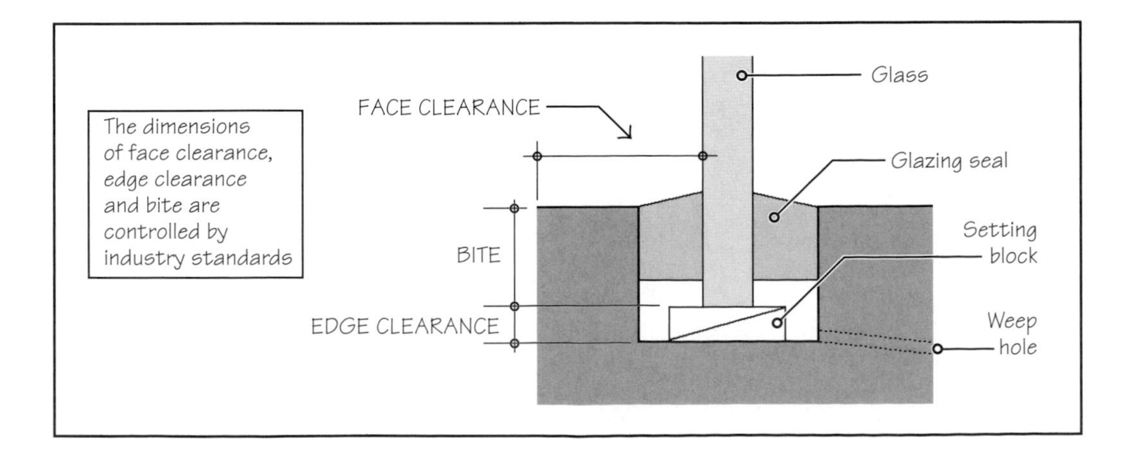

FIGURE 28.32 Glazing pocket and its components.

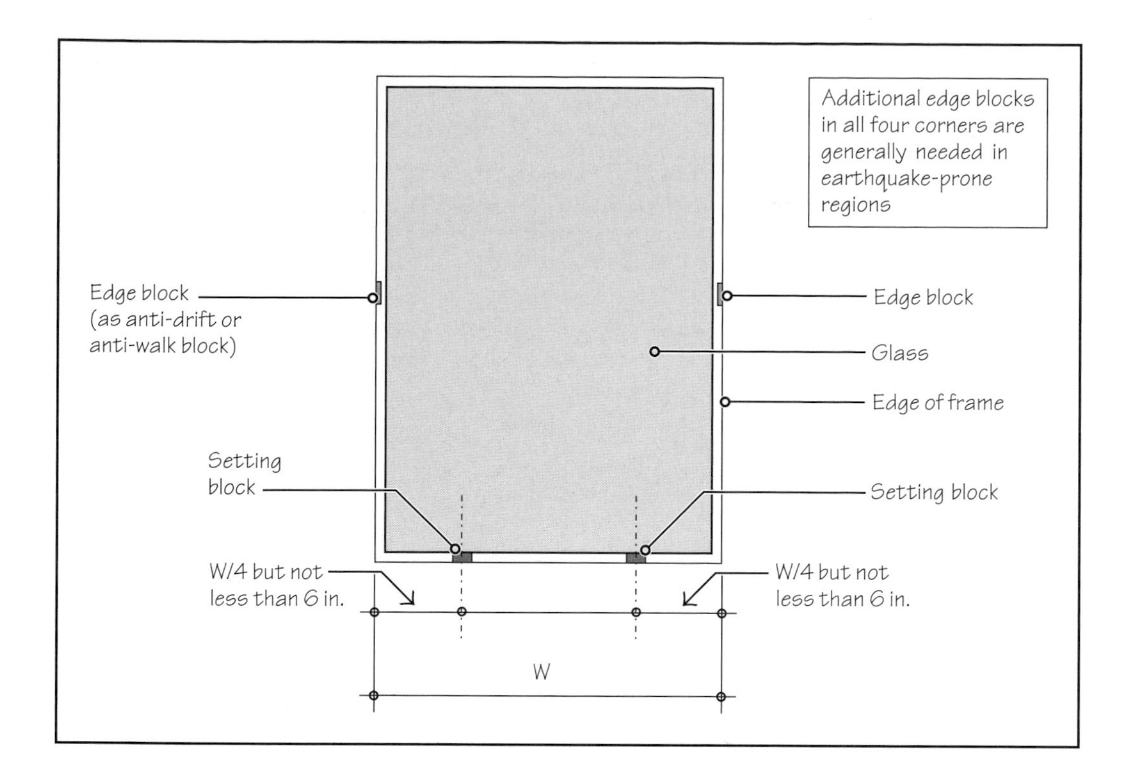

FIGURE 28.33 Setting blocks and edge blocks. Both are made of the same material.

pocket to allow water to drain out from underneath and to protect the glass from coming in direct contact with water. (The seals in an IGU and the PVB layer in a laminated glass are susceptible to damage by sustained contact with water.)

Neoprene, EPDM, or silicone are the materials commonly used for the setting blocks. Generally, the setting blocks are rectangular in shape, but other profiles may be used as long as they allow sufficient clearance around the blocks for the passage of water.

EDGE BLOCKS

Edge blocks are used to separate the vertical edges of glass from the vertical frame members. Also referred to as *antiwalk blocks,* they cushion the glass from the frame and allow the glass to move within the opening without touching the frame. The movement of glass may be due to temperature changes or the effect of wind or earthquake forces. Corner blocks are required in addition to edge blocks in seismic zones.

GLAZING SEALS—WET-GLAZED AND DRY-GLAZED SEALS

Glazing seals are used within the glazing pockets to cushion the glass and also help seal the glazing against air and water infiltration. One of the earliest glazing seals was *glazing putty,* also called *glazing compound.* Glazing putty is made from linseed oil and powdered limestone. It was generally used in a rabbeted frame to function both as a seal as well as a stop, Figure 28.34.

Glazing putty has largely been replaced by better glazing seals. Based on the glazing seal used, contemporary glazing is referred to as one of the following systems:

- Wet-glazed system
- Dry-glazed system
- Wet/dry-glazed system

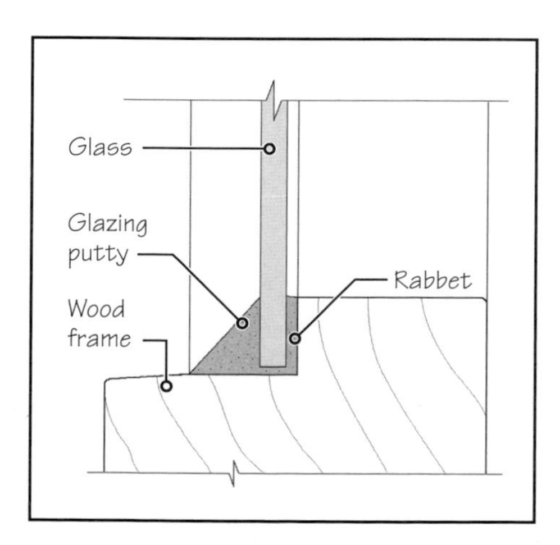

FIGURE 28.34 In earlier wood and steel windows, glazing putty served both as a *stop* and also as a *seal.*

The term stop in glazing literature is used for a member that holds the glass against the rabbet in the frame, providing lateral support to glass.

The recess in the frame against which the glass is held is called a *rabbet.* The stop and the rabbet constitute the glazing pocket.

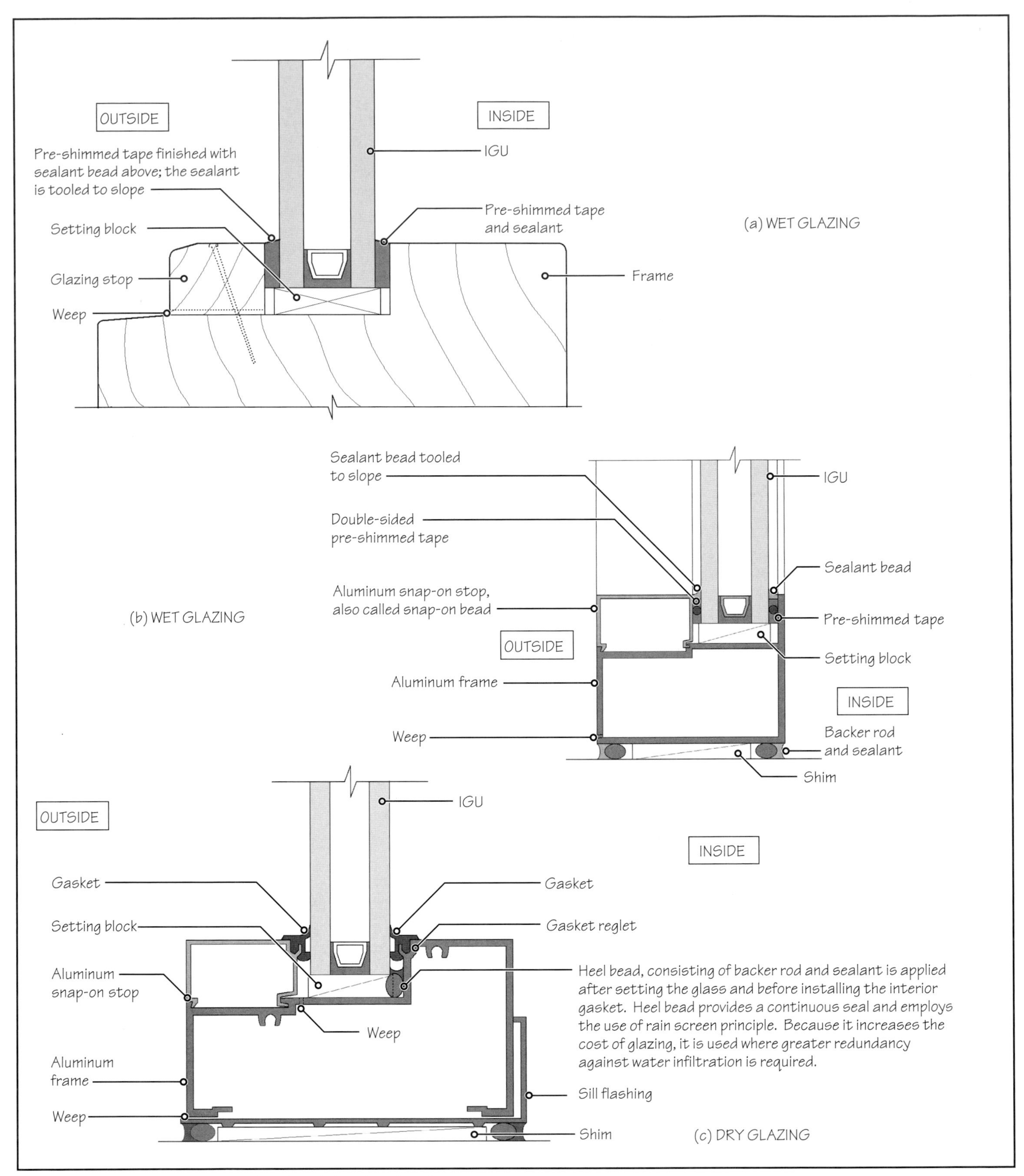

FIGURE 28.35 Examples of wet-glazed and dry-glazed seals.

A wet glazing seal consists of preshimmed double-sided tape, finished with elastomeric sealant bead applied above the tape. The sealant is tooled to slope away from the glass, Figures 28.35(a) and (b). Glazing putty (of earlier times) was an example of a wet glazing seal.

Dry glazing seals consist of *gaskets*. A gasket is generally made of neoprene, EPDM, or silicone rubber composite. Its material, size, and profile make it elastomeric so that after installation, a gasket is under compression, clinging to the glass and the reglet to provide a seal. Therefore, it is also called a *compression gasket*.

Gaskets need some form of engagement into the frame. Therefore, gaskets are generally used with aluminum frames because, aluminum being a soft metal, can be easily profiled

TABLE 28.2 RELATIVE ADVANTAGES AND DISADVANTAGES OF DRY AND WET GLAZING SEALS

Type of seal	Advantages	Disadvantages
Wet glazing seal	Better resistance to water infiltration	Requires exterior and interior installation Performance dependent on artisanry Installation weather dependent More expensive
Dry glazing seal	Can be installed from the interior Less weather dependent Less dependent on artisanry Less expensive	Less watertight Gasket can shrink and create openings

into a complex section to capture the gaskets. The part of the aluminum section that captures the gasket is referred to as *gasket reglet,* Figure 28.35(c).

Dry glazing is more commonly used in contemporary glazing than wet glazing because the performance of dry glazing is less affected by artisanry, weather, and compatibility issues, Table 28.2. A wet/dry-glazed system is a combination of the two systems.

Rain Screen Principle Applied to Glazing Pocket

Regardless of whether dry glazing or wet glazing is used, some water may enter the glazing pocket due to poor artisanry or the degradation of the seals. Some window and curtain wall manufacturers recommend the use of the rain-screen principle to glazing pockets.

Because an air seal in the backup is an important component of a rain screen, the use of an additional continuous sealant bead and backer rod in the glazing pocket, referred to as *heel bead,* is recommended, Figure 28.35(c). Heel bead is applied in the glazing pocket after setting the glass and before applying the interior gasket.

PRACTICE QUIZ

Each question has only one correct answer. Select the choice that best answers the question.

25. The R-value of an aluminum-framed glazing is generally
- **a.** higher than that of glass.
- **b.** lower than that of glass.
- **c.** equal to that of glass.

26. Laminated glass is
- **a.** an assembly of two glass sheets welded (fused) together into one sheet.
- **b.** an assembly of two glass sheets with a cavity.
- **c.** a sealed assembly of two glass sheets with a cavity.
- **d.** an assembly of two glass sheets fused together with a plastic interlayer.
- **e.** none of the above.

27. Which of the following statements is true about plastic glazing?
- **a.** It is stronger than glass.
- **b.** It is more vapor permeable than glass.
- **c.** It is less scratch resistant than glass.
- **d.** All the above.
- **e.** None of the above.

28. The primary reason for the limited use of plastic glazing in buildings is that it is more expensive than glass.
- **a.** True
- **b.** False

29. Which type of glass is commonly used in drive-in bank windows?
- **a.** Annealed glass
- **b.** Tempered glass
- **c.** Laminated glass
- **d.** Heat-strengthened glass

30. Glass can be tempered as well as laminated; that is, laminated, tempered glass is available.
- **a.** True
- **b.** False

31. A glazing stop is used
- **a.** as a bearing support for glass.
- **b.** to seal the glass against water and air infiltration.
- **c.** to provide lateral support to glass.
- **d.** to provide a cushion between the glass and the glazing pocket.
- **e.** none of the above.

32. If the glass is sealed with preshimmed tape and sealant, the glazing is referred to as
- **a.** dry glazed.
- **b.** wet glazed.
- **c.** dry/wet glazed.
- **d.** wet/wet glazed.
- **e.** dry/dry glazed.

33. A heel bead is used in glazing pocket.
- **a.** True
- **b.** False

34. A heel bead is used to
- **a.** increase the R-value of glazing.
- **b.** decrease solar heat gain through glass.
- **c.** decrease water infiltration through glazing.
- **d.** increase wind-load resistance of glazing.
- **e.** none of the above.

Answers: 25-b, 26-d, 27-d, 28-b, 29-c, 30-a, 31-c, 32-b, 33-a, 34-c.

PRINCIPLES IN PRACTICE

Important Facts about Radiation

Radiation is emitted by all objects. Consequently, all objects receive radiation from other objects in their surroundings. In other words, there is a constant radiation exchange occurring between objects that have a direct line of sight between them (i,e., that can see each other).

As the temperature of an object increases, the wavelength of radiation emitted by it becomes smaller. An object at a high temperature emits most of its radiation in short wavelengths. Conversely, an object at a low temperature radiates in long wavelengths.

A type of high temperature radiation of interest is solar radiation. Although the actual temperature of the interior of the sun is several million degrees Fahrenheit, solar radiation received on the earth's surface has the same characteristics as the radiation from an object whose temperature is approximately 11,000°F (6,000°C).

Low-temperature radiation is present everywhere because it is emitted by objects on the surface of the earth, such as the ground, sky, landscape elements, building surfaces, and furniture. The temperature of these objects seldom exceeds 150°F (60°C).

Because solar radiation is a high-temperature radiation, it is referred to as *shortwave radiation.* The radiation from earthly objects, on the other hand, is *long-wave radiation.* This difference is shown in Figure 1, which gives the wavelength composition of solar radiation and radiation from an object at 100°F.

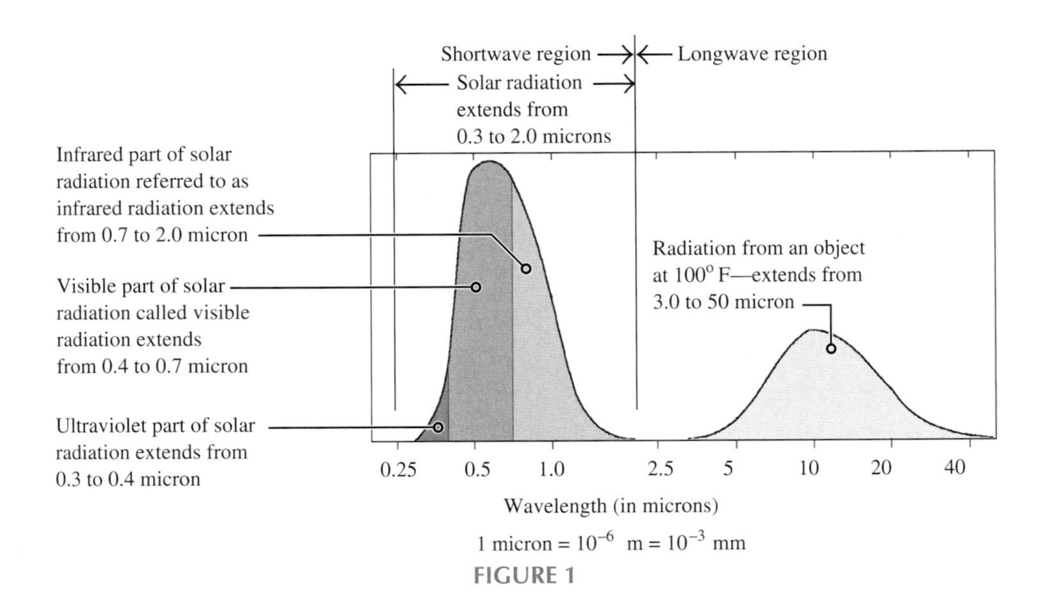

FIGURE 1

COMPONENTS OF SOLAR RADIATION (ULTRAVIOLET, VISIBLE, AND INFRARED RADIATION)

Observe that solar radiation extends from a wavelength of 0.3 μm to 2.0 μm. Approximately 3% of solar radiation lies in the ultraviolet region (wavelength less than 0.4 μm), 50% lies in the visible region (0.4 to 0.7 μm), and the remaining 47% lies in the infrared region. Radiation from an object at 100°F extends from a wavelength of nearly 3 to 50 μm.

The wavelength region up to 2.0 μm is generally referred to as the *shortwave region,* and beyond 2.0 μm, is called *long-wave region.* Building surfaces receive shortwave radiation only from the sun. However, they receive long-wave radiation from all surrounding objects, including the sky, because they are at low temperature.

The radiation from the sky is a low-temperature radiation because it is due to the heat contained in the atmosphere (air, dust, and water vapor particles). It is also referred to as *diffuse solar radiation.* The term *solar radiation* refers to the radiation contained in direct solar beams.

SURFACE EMISSIVITY, SURFACE COLOR, AND LONG-WAVE RADIATION

Building surfaces and objects behave differently to shortwave and long-wave radiation. As shown in Figure 28.13, a clear glass sheet transmits virtually all (85%) shortwave solar radiation. Having penetrated the glass, solar radiation heats room surfaces and the contents of the room. These objects, being at low temperature, emit long-wave radiation. Glass is virtually opaque to long-wave radiation because it absorbs 90% of it (Figure 28.16). Having absorbed the interior heat, glass dissipates a part of it to the interior and a part to the exterior by convection and radiation. In other

PRINCIPLES IN PRACTICE

words, glass does not allow the interior heat to escape as readily as it admitted the solar heat. This is the well-known *greenhouse effect.*

From Section 5.5, we know that emissivity of a surface—a concept that applies only to long-wave radiation—is equal to its absorptivity. Thus, the reason glass absorbs nearly 90% of long-wave radiation is because its emissivity is 0.9.

A low-emissivity (low-E) surface absorbs little and reflects most of long-wave radiation. A low-E coating on glass changes the behavior of glass in response to long-wave radiation. A reflective coating on glass, on the other hand, changes the behavior of glass with respect to solar radiation. It does not greatly affect the behavior of glass with regards to long-wave radiation.

Similarly, long-wave radiation does not discriminate between colors of surfaces. Solar radiation does. A white or light-colored surface is more reflective of solar radiation than a dark-colored surface. Long-wave radiation does not differentiate between the various colors.

PRINCIPLES IN PRACTICE

Condensation-Resistance Factor

The condensation-resistance factor (CRF) has been defined such that the higher its value, the better the performance of glazing with respect to condensation. The CRF of a glazing is obviously related to its R-value. The higher the R-value of glazing, the higher its CRF. Thus, a window with an IGU has a higher CRF than one with a single sheet of glass.

The CRF of a window or curtain wall is obtained by measuring it in a test chamber. It is given in whole numbers. Manufacturers provide CRF values for their products. In choosing a glazing based on CRF value, Table 1, which gives the minimum recommended CRF value of glazing based on the interior humidity and outside winter air temperature, is generally invoked. The inside temperature is assumed as 68°F (20°C).

For example, according to this table, if the minimum winter air temperature at a location is 20°F and if the interior relative humidity of a building is 40% (typical of an office building), a minimum CRF value of 48 should be specified for the glazing in that building. In obtaining CRF values from the Table 1, the outside air temperature is the design winter temperature of the location, defined as the temperature that is exceeded for more than 97.5% of the time during December, January, and February.

Specifying a CRF value based on Table 1 does not ensure that no condensation will occur on the interior surface of glazing. It simply means that condensation will be minimal and within acceptable limits. CRF values below 35 are generally not specified.

TABLE 1 RECOMMENDED MINIMUM CRF VALUES OF GLAZING

This table is based on American Architectural Manufacturers Association (AAMA) Standard 1503-98 (revised 2004), "Voluntary Test Method for Thermal Transm-ittance and Condensation Resistance Factor of Windows, Doors, and Glazed Wall Sections."

The entry *X* in the table indicates that no test data are available in this standard. Underlined numbers indicate extrapolated values from the test data.

The table is based on the interior temperature of 68°F (20°C) and outside wind speed of 15 mph.

		Inside relative humidity					
		15%	20%	25%	30%	35%	40%
	Outside air temperature	**Minimum CRF values**					
	−20°F	45	52	57	62	X	X
Minneapolis →	−10°F	40	47	53	58	65	X
Chicago →	0°F	32	40	47	52	58	63
Boston → New York →	10°F	17	30	37	45	52	57
Atlanta →	20°F	X	X	23	33	40	48

PRACTICE QUIZ

Each question has only one correct answer. Select the choice that best answers the question.

35. Building surfaces emit
- **a.** shortwave radiation.
- **b.** ultraviolet radiation.
- **c.** long-wave radaition.
- **d.** all the above.
- **e.** none of the above.

36. Glass is virtually opaque to
- **a.** shortwave radiation.
- **b.** ultraviolet radiation.
- **c.** long-wave radiation.
- **d.** all the above.
- **e.** none of the above.

37. Solar radiation is
- **a.** shortwave radiation.
- **b.** long-wave radiation.

38. A low-E glass is a poor reflector of long-wave radiation.
- **a.** True
- **b.** False

39. The condensation resistance factor (CRF) of glazing is directly related to
- **a.** the material used in framing members of glazing.
- **b.** whether or not reflective glass has been used in the glazing.
- **c.** whether or not tempered glass has been used in the glazing.
- **d.** the R-value of the glazing.
- **e.** all the above.

40. The minimum value of the condensation-resistance factor (CRF) generally specified is
- **a.** 55.
- **b.** 45.
- **c.** 35.
- **d.** 25.
- **e.** 15.

Answers: 35-c, 36-c, 37-a, 38-b, 39-d, 40-c.

KEY TERMS AND CONCEPTS

- Acrylic and polycarbonate glazing
- Annealed glass
- Condensation resistance factor (CRF)
- Dimmable glass
- Dry glazing
- Edge blocks
- Fire-resistant glass
- Flat glass
- Float glass process
- Gasket
- Glazing pocket
- Glazing seals
- Heat-modified glass
- Heat-soaked glass
- Heat-strengthened glass
- Insulating glass unit
- Laminated glass
- Low-E glass
- LSG index
- Plastic glazing
- Pre-shimmed tape
- Radiation shielding glass
- Reflective glass
- Self-cleaning glass
- Setting blocks
- Shading coefficient (SC)
- Solar heat gain coefficient (SHGC)
- Tempered glass
- Tinted glass
- Visible transmittance (VT)
- Wet glazing
- Wired glass

REVIEW QUESTIONS

1. Using sketches and notes, explain the process of manufacturing flat glass.
2. Using a sketch, explain why a tempered glass sheet is stronger than an annealed glass sheet.
3. What is heat-soaked glass and where is it used?
4. Using a sketch, explain the anatomy of an insulating glass unit.
5. Using a sketch, explain the standard surface designation system for (a) a monolithic glass and (b) an IGU.
6. Explain how a low-E coating increases the R-value of glass.
7. Which type of glass would you recommend for the following locations?
   a. Sliding-glass entrance door
   b. Spandrel area of an all-glass curtain wall
   c. Teller window of a motor bank
   d. Rear wall of a racquet ball court

SELECTED WEB SITES

American Architectural Manufacturers Association, AAMA (www.aamanet.org)

Glass Association of North America, GANA (www.glass.website.com)

Insulating Glass Manufacturers Association (www.igmaonline.org)

National Fenestration Rating Council, NFRC (www.nfrc.org)

FURTHER READING

1. American Society for Testing and Materials. "Standard Practice for Determining Load Resistance of Glass in Buildings," ASTM Standard E 1300.
2. Amstock, Joseph. *Handbook of Glass Construction.* New York: McGraw-Hill, 1997.
3. Flat Glass Manufacturers Association (FGMA). *Sealant Manual,* 1990.
4. Glass Association of North America (GANA). *Glazing Manual,* Topeka, KS: 2004.
5. Glass Association of North America (GANA). *Laminated Glass Design Guide,* 1994.

REFERENCES

1. Newman, Harold. *An Illustrated Dictionary of Glass.* London: Thames and Hudson Publishers, 1977, p. 82.
2. Peter, John. *Design With Glass,* New York: Reinhold, 1964, p. 9.

CHAPTER 29 Windows and Doors

CHAPTER OUTLINE

29.1 Window Styles

29.2 Window Materials

29.3 Performance Ratings of Windows

29.4 Window Installation and Surrounding Details

29.5 Classification of Doors

29.6 Door Frames

29.7 Fire-Rated Doors and Windows

PRINCIPLES IN PRACTICE: A Note on Aluminum

Tilt-up concrete load-bearing walls support a lightweight steel roof In St. Ignatius Chapel, Seattle, WA. Architect: Steven Holl. (Photo by DA.)

Not long ago, windows and doors could be made only from wood. Ornate carpentry was often used, and the architects spent an inordinate amount of time and effort detailing the cross-sectional profiles of window and door frames, window sashes, door panels, and the related trims.

Because of the lack of automation and manufacturing technology, the carpentry and joinery related to window and door fabrication was carried out at the construction site. The performance requirements of windows and doors were modest at best and were directly related to the skill of the craftsperson building the window (or door); good artisanry yielded higher-performing windows (or doors). For windows, letting the daylight in and keeping the rainwater out were the only requirements. Because large sheets of glass were not available at that time, window glazing consisted of small panes of glass held by narrow horizontal and vertical wood members. For doors, the primary performance requirement was security.

Like most other building components, windows and doors have evolved over the years. Windows are now made not only from wood but also from aluminum, steel, vinyl, and fiberglass. Additionally, two materials are sometimes combined in the same window to take advantage of their different properties.

Complex cross-sectional profiles for frames and sashes, sophisticated glazing, gaskets, weatherstripping, and hardware are required to satisfy more-demanding performance requirements for windows. Consequently, contemporary windows are almost always shop fabricated, and, in some cases, they are shipped preglazed to the construction site. This means that the only operation required at the site is to secure them into the opening and seal the joints between the window and the wall opening.

Due to their sophisticated fabrication process and the rigorous performance testing to which they are subjected, the architect's design of contemporary windows is generally limited to establishing their size, shape, and relationship with the building facade and selecting the manufacturers and window material that will satisfy the performance and aesthetic requirements. The detailing of windows is limited to designing their surrounds, that is, detailing at the jambs, head, and sill.

This chapter begins with an introduction to the commonly used window styles, materials used in window and sash frames, performance ratings of windows, and window installation.

A similar discussion of door types then follows. Because of their immense variety, this chapter focuses on the more commonly used door types (interior wood and steel doors). A discussion of the fire rating of windows and doors is presented next, and the chapter concludes with the frequently used window and door terminology. The reader may choose to browse through the terminology before reading the main text.

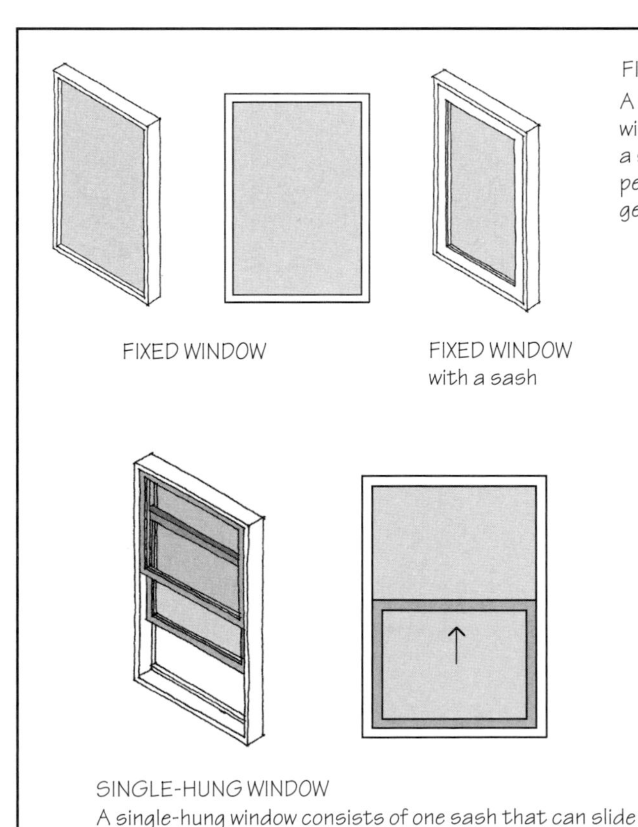

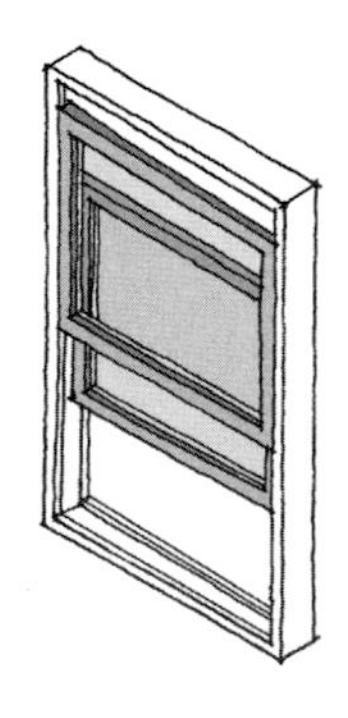

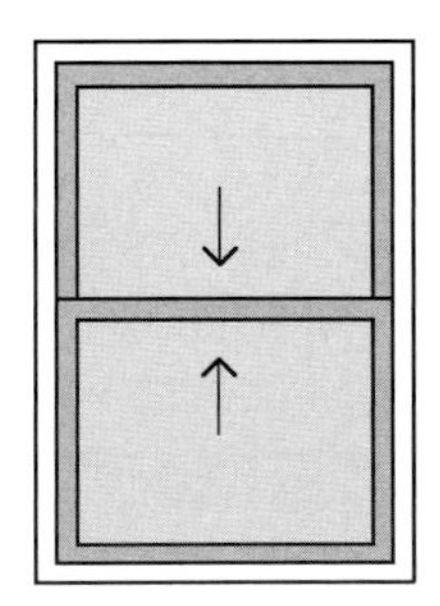

FIGURE 29.1 Commonly used window styles.

29.1 WINDOW STYLES

Figure 29.1 shows various commonly used window styles that are classified under two broad categories:

- Fixed windows
- Operable windows

NOTE

Window Terminology

A reader may wish to brush up on window terminology given at the end of the chapter before proceeding with this section.

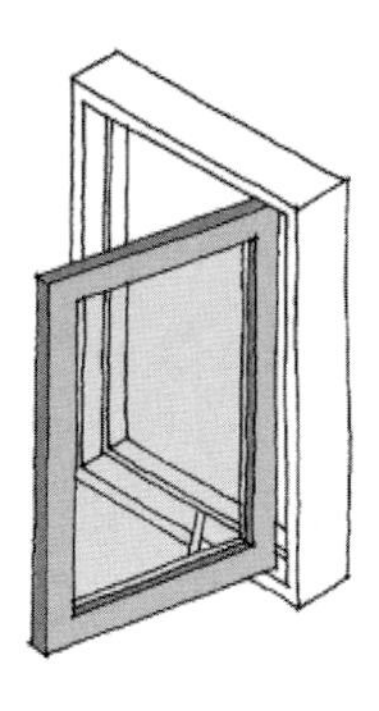
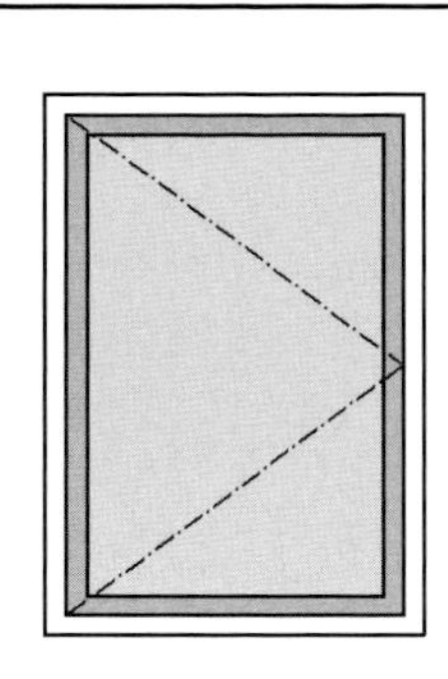

CASEMENT WINDOW

A casement window (also called a side-hung window) may consist of one operable sash, as shown here or two sashes that close on each other with or without a center mullion. Because the sash closes on the frame with pressure providing a compression seal, a casement window is generally less prone to air leakage and, hence, can be more energy efficient than single-, double-hung or sliding windows. A casement window can provide up to 100% ventilation.

Screen units with a casement window are generally provided on the inside with sashes opening out.

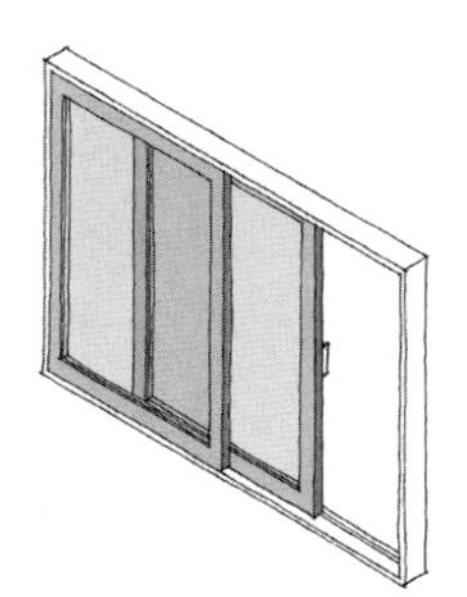
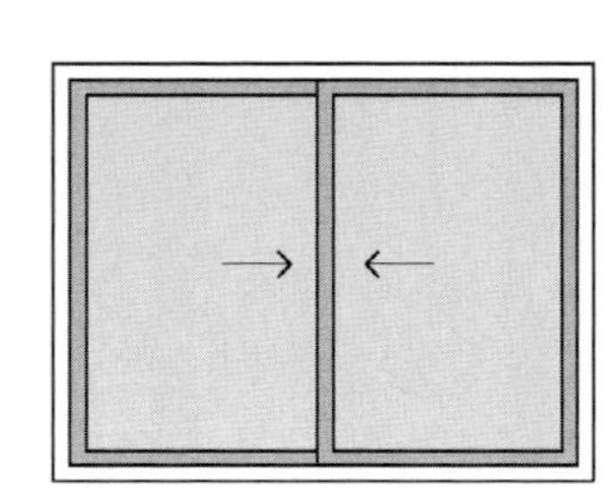

SLIDING WINDOW

A sliding window consists of one sash that slides horizontally over a fixed sash. Alternatively, both sashes may slide with respect to each other, as shown here. A sliding window can provide up to 50% ventilation. A sliding window is more prone to air leakage than other window styles.

An insect screen unit can be easily included toward the exterior with a sliding window.

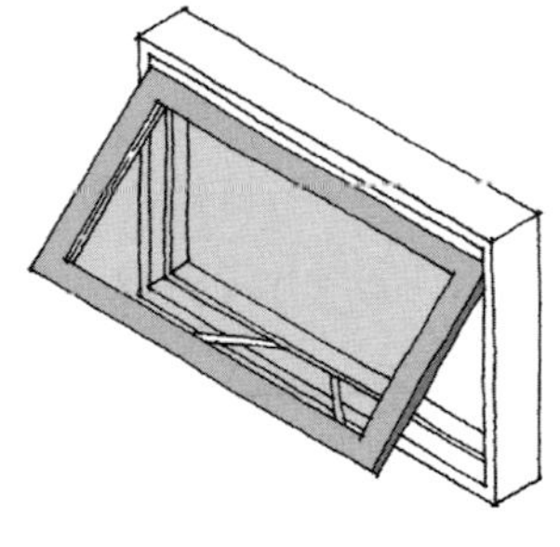
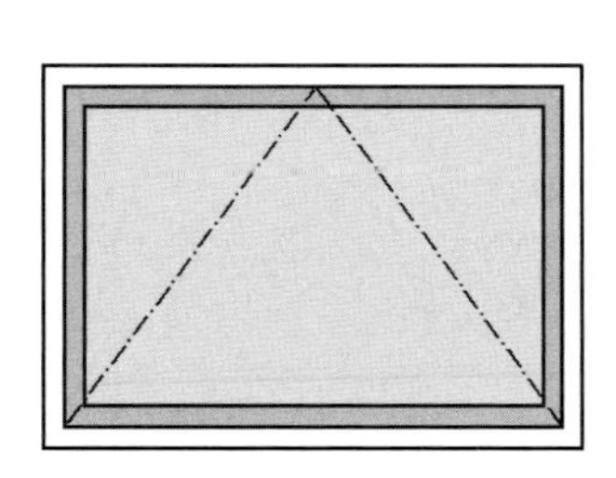

AWNING WINDOW

An awning window (also called a top-hung window) is similar to a casement window but provides a degree of rain protection when the window is partially open. It can provide up to 100% ventilation. An insect screen unit can only be provided toward the interior of an awning window.

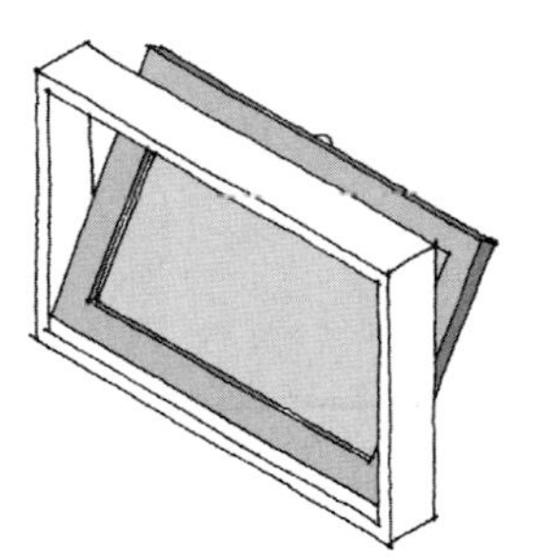
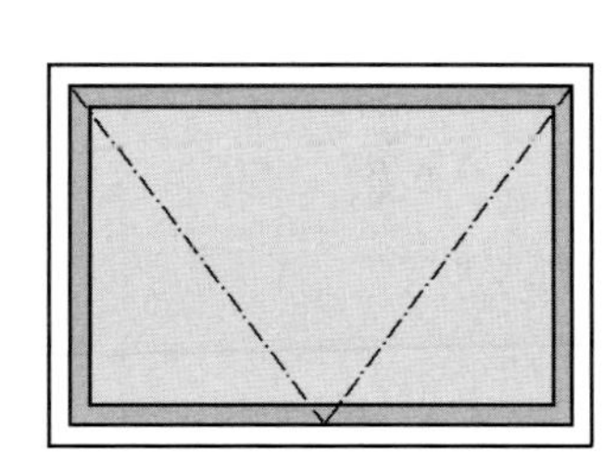

HOPPER WINDOW

A hopper window (also called a bottom-hung window) is similar to an awning window but opens inward at the top. An insect screen unit can be provided toward the outside. Like an awning window, it can provide up to 100% ventilation.

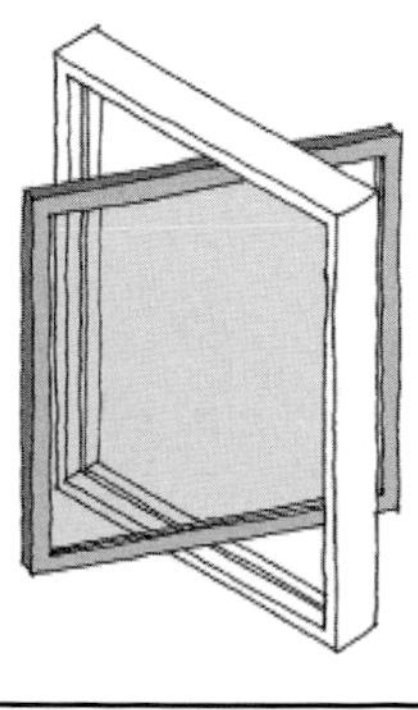
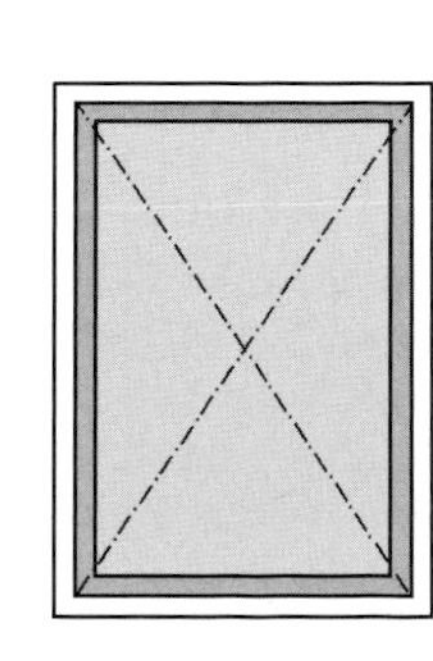

PIVOTING WINDOW

A pivoting window may be pivoted at the center or off center. It allows easy cleaning of the window from the inside and can provide up to 100% ventilation. It has the ability to direct the flow of ventilation. A screen unit cannot be provided with a pivoting window.

FIGURE 29.1 (***Continued***) Commonly used window styles.

FIGURE 29.2 Wood windows provide warmth and beauty that other window materials generally cannot provide. (Photo courtesy of Marvin Windows and Doors.)

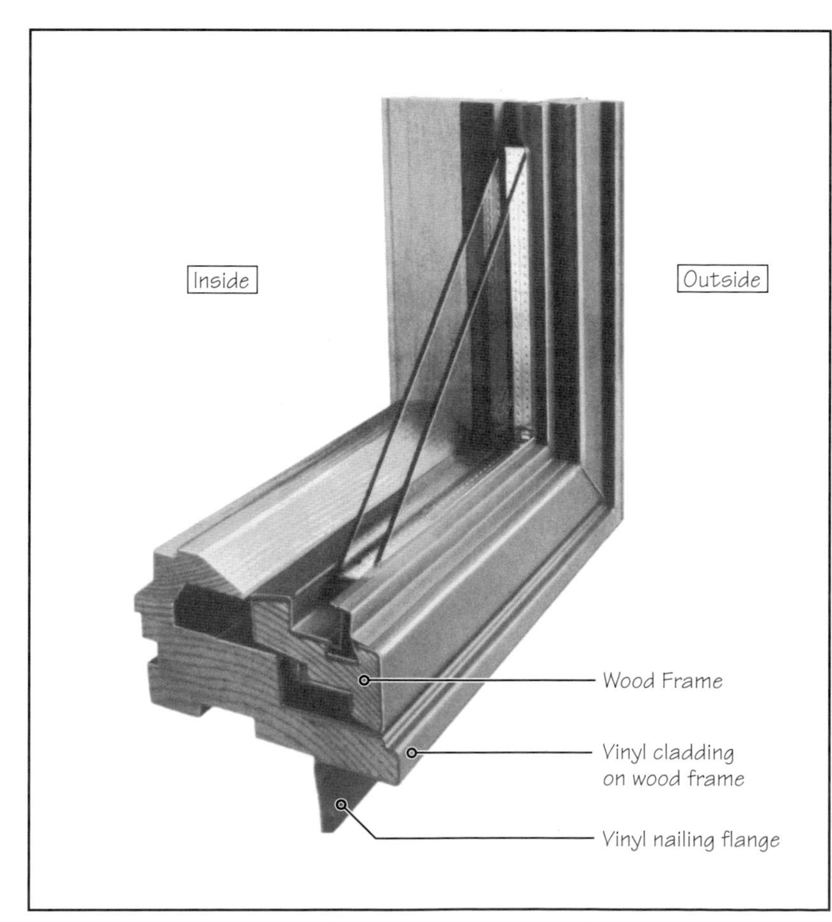

FIGURE 29.3 Cross section through a vinyl-clad wood window. (Photo courtesy of Andersen Corporation.)

Fixed windows are the simplest of all windows. Because they do not have (operable) sashes, they are more economical and are less likely to leak air or water. They are commonly used in nonresidential buildings with highly controlled heating, cooling, and ventilating systems. A fixed window is also referred to as *direct-glaze window*. The term *picture window* is used for a fixed window whose width is larger than its height to provide a panoramic view.

Operable windows are commonly used in residential occupancies (homes, apartments, and hotels) and sometimes in patient rooms of a hospital, and so on. In many situations, fixed and operable units are combined in the same window.

29.2 WINDOW MATERIALS

As stated previously, five materials are used in contemporary windows: wood, aluminum, steel, vinyl, and fiberglass. The strengths and weaknesses of these materials are discussed in this section.

Wood

Wood is the oldest material used for window frames and sashes. However, because of its swelling and shrinkage problems, fungal decay, and termite vulnerability and the need for periodic staining or painting, all-wood windows are not often used today. The advantage of wood is its high R-value. Additionally, its warmth and beauty are unmatched by any other window material, Figure 29.2. Therefore, wood windows are generally used in high-end homes and offices requiring cozy and homey interiors.

To improve their durability and eliminate periodic painting, manufacturers provide wood windows whose frames and sashes are clad on the exterior with vinyl or aluminum, Figure 29.3. Unclad wood windows are generally used in historical buildings where restoration to their original design is required.

Aluminum

Aluminum is by far the most commonly used window material today. It is the only window material that is also used in glass curtain walls and storefronts. It is not subject to moisture shrinkage or swelling. Because the anodized or painted finish on aluminum is virtually permanent, aluminum windows require almost no maintenance.

The malleability and flexibility of aluminum is perhaps the most important reason for its extensive use as a window and curtain wall material. Because of its malleability, aluminum can be extruded into complex cross sections, which are specially designed to facilitate the joinery of window and curtain wall components (see "Principles in Practice" at the end of this chapter). For example, continuous round, hollow intrusions, referred to as *screw splines*, are commonly provided in aluminum cross sections to join horizontal and vertical frame members, Figure 29.4.

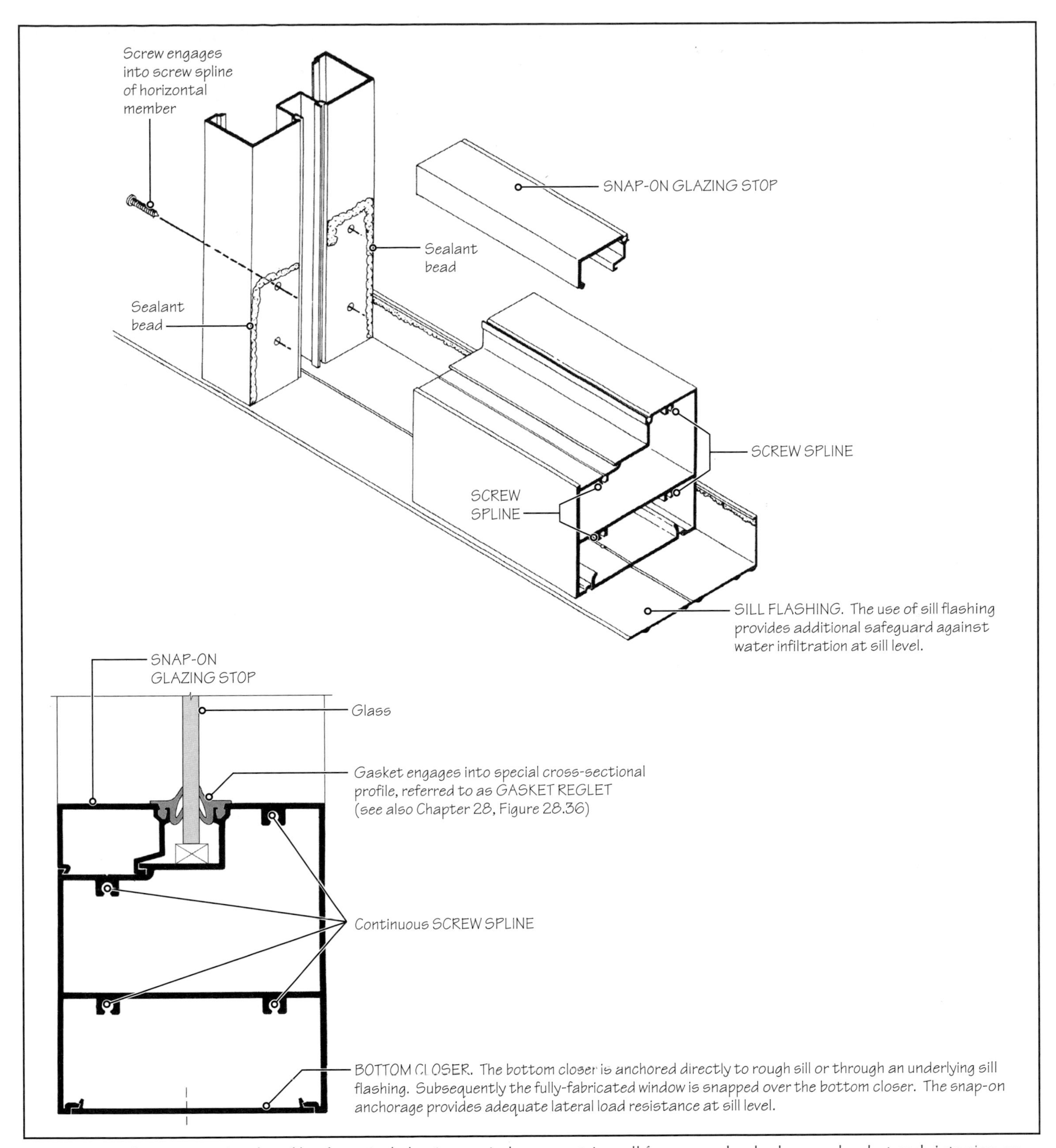

FIGURE 29.4 The cross-sectional profile of a typical aluminum window or curtain wall frame member looks complex, but each intrusion or protrusion has a special purpose. Observe the use of screw splines to connect horizontal and vertical frame member and snap-on cross-sectional profiles that allow members to be connected together without any fasteners.

The flexibility of aluminum allows the use of *snap-on* joinery, whereby two separate aluminum sections can be joined together without any fasteners. For example, in aluminum windows and curtain walls, glazing stops are simply snapped on (no need for screws or other fastening material) (Figure 29.4). Additionally, two aluminum windows are routinely snap-connected by using mating mullion profiles to provide a larger window with a common mullion—without the use of fasteners, Figure 29.5.

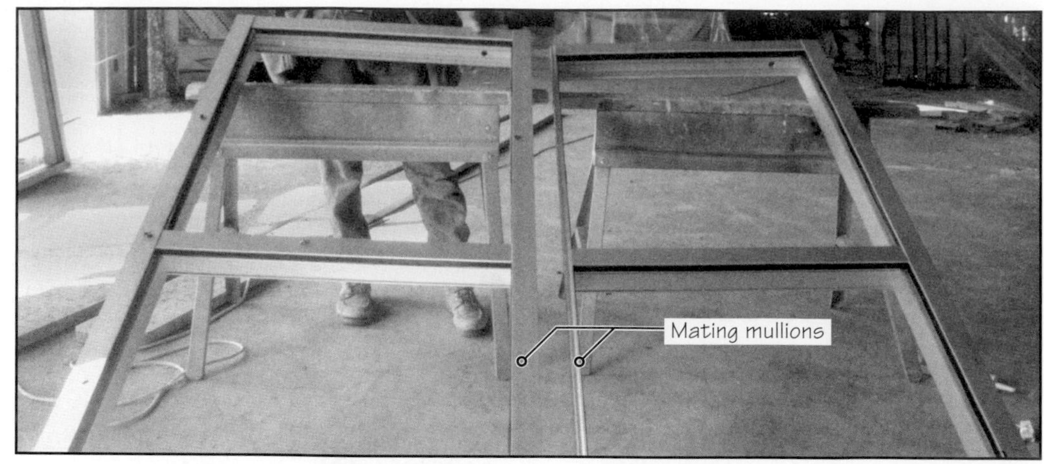

(a) Two separate aluminum windows ready to be connected together at mating mullions. The right-hand window has a narrow mullion and the left-hand window has a broad mullion, so that the overall width of the mated mullions is the same as the other mullion

(b) Windows being snapped together to form one window.

(c) The two windows have become one window with a common mullion without the use of fasteners.

FIGURE 29.5 Two aluminum windows are snap connected at mating mullions. Several windows can be connected together in this manner.

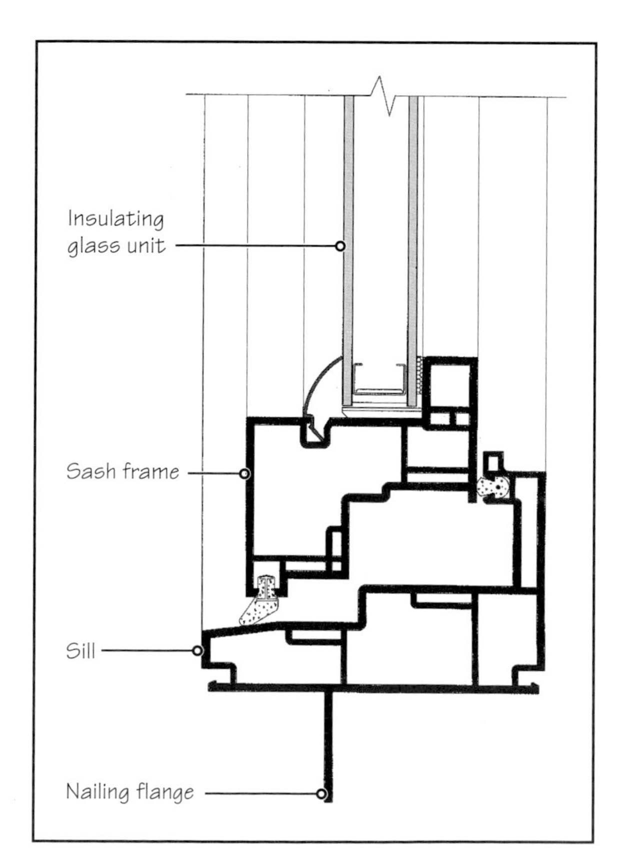

FIGURE 29.6 Typical cross-sectional profile of a vinyl window.

The disadvantage of aluminum is its low R-value. However, with the introduction of *thermally improved* also called *thermally broken* (Figure 28.23) aluminum frames, the situation has substantially improved.

Steel

Before the introduction of aluminum for windows, steel was commonly used as an alternative to wood. Problems with steel corrosion and the need for frequent painting have substantially reduced the use of steel windows. However, because steel is a strong material, the frame members can be narrower than those of other materials, giving a sleeker window appearance. Because of steel's strength, steel windows are often specified for prisons and other locations where greater durability and greater resistance to lateral loads (e.g., blast-resistant glazing) are required.

Vinyl and Fiberglass

Vinyl (an abbreviation for polyvinyl chloride, PVC) frames and sashes are almost maintenance free because vinyl does not corrode or degrade due to fungal or termite attack. Vinyl windows require no painting and are available in a variety of nonfading colors. Vinyl frame and sash cross sections are made from the extrusion process and are generally hollow, Figure 29.6. They provide excellent R-value.

A major disadvantage of vinyl and of plastics in general is their high coefficient of thermal expansion. The introduction of fiberglass in windows has substantially reduced this problem. The coefficient of thermal expansion of fiberglass is compatible with that of glass. Thus, fiberglass frames and sashes are dimensionally stable, strong, and available in different colors.

Each question has only one correct answer. Select the choice that best answers the question.

1. The term *direct-glaze window* implies
- **a.** an operable window.
- **b.** a fixed window.
- **c.** a fixed window whose length is greater than its height.
- **d.** a fixed window whose height is greater than its length.
- **e.** none of the above.

2. The term *picture window* implies
- **a.** an operable window.
- **b.** a fixed window.
- **c.** a fixed window whose length is greater than its height.
- **d.** a fixed window whose height is greater than its length.
- **e.** none of the above.

3. Which of the following window types can provide up to 100% ventilation when fully open?
- **a.** Single-hung window
- **b.** Double-hung window
- **c.** Sliding window
- **d.** Casement window
- **e.** All of the above

4. Which of the following windows is top-hung?
- **a.** Awning window
- **b.** Hopper window
- **c.** Casement window
- **d.** Pivoting window
- **e.** None of the above

5. In vinyl (or aluminum) clad windows, the cladding is used on exterior
- **a.** window frame members.
- **b.** sash frame members.
- **c.** both window frame and sash frame members.

6. Screw splines are provided in
- **a.** window and curtain wall frames made from wood.
- **b.** window and curtain wall frames made from steel.
- **c.** window and curtain wall frames made from vinyl.
- **d.** window and curtain wall frames made from aluminum.
- **e.** all the above.

7. Screw splines are round hollow intrusions in frame members that facilitate the connection of horizontal and vertical members of a
- **a.** wood window.
- **b.** aluminum window.
- **c.** steel window.
- **d.** vinyl window.
- **e.** all the above.

8. Snap-on joinery is generally used in
- **a.** wood windows.
- **b.** aluminum windows.
- **c.** steel windows.
- **d.** all the above.
- **e.** none of the above.

Answers: 1-b, 2-c, 3-d, 4-a, 5-c, 6-d, 7-b, 8-b.

29.3 PERFORMANCE RATINGS OF WINDOWS

Apart from the aesthetic considerations, such as shape, style, frame color, and finish, several performance characteristics must be considered in selecting a window for a project. Two rating systems, one relating to a window's energy performance and the other relating to a window's structural and water-leakage performance, help architects in this task. Both regimes are material neutral—that is, they apply uniformly to windows of different materials. Additionally, they apply to the entire window assembly, consisting of the frame, sash, glass, gaskets, and so on.

ENERGY-PERFORMANCE RATING OF A WINDOW

The energy-performance rating of a window assembly is provided by the National Fenestration Rating Council (NFRC), which is backed by the U.S. Government's Energy Star initiative. NFRC-rated windows are labeled for their U-values and solar heat–gain coefficient (SHGC), Figure 29.7. Labeling for other energy-related criteria, such as visible transmittance (VT), air leakage, and the condensation-resistance factor (CRF), is at the discretion of the manufacturer. The current energy-conservation code requires conformance only with the U-value and SHGC of windows.

World's Best Window Co.
Millennium 2000+
Vinyl-Clad Wood Frame
Double Glazing • Argon Fill • Low E
Product Type: **Vertical Slider**
(per NFRC 100-97)

ENERGY PERFORMANCE RATINGS

U-Factor (U.S./I-P)	Solar Heat Gain Coefficient
0.35	**0.32**

ADDITIONAL PERFORMANCE RATINGS

Visible Transmittance	Air Leakage (U.S./I-P)
0.51	**0.2**

Manufacturer stipulates that these ratings conform to applicable NFRC procedures for determining whole product performance. NFRC ratings are determined for a fixed set of environmental conditions and a specific product size. Consult manufacturer's literature for other product performance information.
www.nfrc.org

FIGURE 29.7 A typical label on an NFRC-rated window. (Illustration courtesy of National Fenestration Rating Council.)

STRUCTURAL, WATER-LEAKAGE, AND OTHER PERFORMANCE RATINGS OF A WINDOW

In addition to being energy efficient, a window assembly must be strong enough to resist wind loads, it should not leak water or air, and the window's locking mechanism should prevent forced entry. The window manufacturing industry (comprising the manufacturers of wood, aluminum, vinyl, and fiberglass windows) has jointly established standards by which their windows are tested. All windows must be tested for

- Resistance to wind loads
- Resistance to water leakage

- Resistance to air leakage
- Resistance to forced entry

Based on the tests, a window is assigned a rating designation that includes the following information:

- *Window style:* whether the window is a fixed, casement, sliding, or awning window. The acronyms for the commonly used window styles are given in Table 29.1.
- *Window performance class:* the five performance classes, coded as R, LC, C, HC or AW, Table 29.2.
- *Window performance grade:* the allowable (or design) wind load on the window in psf (pounds per square foot). A factor of safety of 1.5 is used so that if the window assembly resists a wind load of 45 psf, its design wind load is 30 psf (i.e., the window's performance grade is 30).

  The window is also tested for water leakage under a static water pressure given in Table 29.2. Under this pressure, the assembly should not leak any water. Thus, water leakage is a pass/fail test. The window's air leakage should not exceed 0.3 cfm/ft^3 (cubic feet per minute per square foot)—another pass/fail test. The window should also pass the forced-entry resistance test as specified in the standard, once again a pass/fail test.
- *Window size:* an important part of the designation system. It is stated as "window width × window height" in inches. Generally, if everything else is the same, a smaller window generally performs better than a larger window under the tests. Therefore, the standard specifies a certain minimum window size for each performance class and performance grade. The manufacturer can exceed the specified size requirement, and if that is the case, it can be so included in the designation, as explained in Figure 29.7.

TABLE 29.1 ABBREVIATIONS FOR COMMONLY USED WINDOW STYLES IN THE WINDOW-RATING DESIGNATION

Window style	Code
Fixed	F
Single/Double/Triple hung	H
Casement	C
Horizontal sliding	HS
Awning/Hopper	AP
Vertically pivoted	VP
Horizontally pivoted	HP

OTHER PERFORMANCE CHARACTERISTICS OF WINDOWS

The four performance characteristics just described are the minimum requirements for windows. Generally, a manufacturer will have its windows tested for several other characteristics, such as the operating force test (force required to open and close windows), life-cycle test (which determines the damage over time to hardware parts used in the window), and the vertical deflection test (which determines the deflection of a sash in a casement window). Acoustical windows are tested for their sound insulation property.

WINDOWS AND BUILDING CODES

Windows in habitable rooms of dwelling units must satisfy the building code's requirement for a certain minimum amounts of natural light and ventilation. They are also required to

TABLE 29.2 PERFORMANCE-CLASS AND PERFORMANCE-GRADE REQUIREMENTS OF WINDOWS

Performance class and abbreviation	Performance grade (design wind load) (psf)	Minimum wind load resistance (psf) [1]	Water-leakage test pressure (psf) [2]	Air-leakage test pressure (psf)	Air-leakage rate ft^3/(min. ft^2)
Residential (R)	15	22.5	2.86	1.57	0.3
Light commercial (LC)	25	37.5	3.75	1.57	0.3
Commercial (C)	30	45.0	4.50	1.57	0.3
Heavy commercial (HC)	40	60.0	6.00	1.57	0.3
Architectural (AW)	40	60.0	8.00	6.24	0.3

[1] The design wind load equals two-thirds of the load a window resists under the test. For example, if the window resists 60 psf under the test, the design wind load for the window is 40 psf. In other words, the window may be subjected to a maximum wind load of 40 psf in service. Hence, the window's performance grade is 40.

[2] The water leakage of a window is determined by subjecting it to a static pressure that is 15% of its design wind load. Thus, if the design wind load for a window is 40 psf, it must pass a water-leakage test under a pressure of 6.0 psf. AW windows are tested under a pressure that is 20% of the design wind load (in place of 15%) and R windows are tested under 19% design wind-load pressure.

Observe that HC and AW windows have the same performance grade of 40, but AW windows must resist water leakage and air leakage under higher pressures.

Window style:
(see Table 29.1)

"HS" = horizontal sliding window

Window performance class and grade:
(see Table 29.2)

LC25 indicates that:

(a) the window is classified as "light commercial";

(b) the design wind load on window should not exceed 25 psf;

(c) it has passed the water leakage test under its grade;

(d) its air infiltration rate is ≤0.3 cfm;

(e) it has passed the forced entry test.

Manufacturer's maximum window size

This dimension indicates that the manufacturer has tested the maximum window size for this class and grade at 120 in. × 59. As stated previously, each performance class and grade must be tested to a "minimum" window size. For example, HS-LC25 must be tested to a minimum window size of 69 in. × 54 in.

The manufacturer can exceed the size requirement of the standard, as indicated by this label, since 120 in. × 59 in. is greater than the required minimum size of 69 in. × 54 in.

As per the standard, the designation system recognizes a manufacturer exceeding the minimum requirements for a performance class and grade. For example, for a horizontal sliding window to receive a grade of HS C30 must be at least 71 in. × 59 in. and must resist a design wind load of at least 30 psf. If a manufacturer's window of this size tests to a design wind load of 35 psf, it can be designated as HS C35. This higher test wind load of (35 × 1.5 = 52.5 psf) also requires that the window must pass under a correspondingly higher water leakage test pressure.

FIGURE 29.7 Explanation of a typical label on a window tested under ANSI/AAMA/NWWDA Standard 101/I.S.2.

provide a means of emergency egress from the dwelling in the event of fire. For example, building codes generally require that the clear area of at least one window in each sleeping room of a dwelling, when open, must be at least 5.7 sq ft.

29.4 WINDOW INSTALLATION AND SURROUNDING DETAILS

Contemporary windows are generally brought to the site fully assembled and glazed, which simplifies the installation process. On-site glazing is generally limited to fixed windows with large glass panes.

A window may be installed before or after the exterior wall cladding is in place. Most commercial window manufacturers recommend the installation of their windows after the exterior-wall cladding is in place to prevent physical damage to windows, such as the breakage of glass and scratching of coatings. Chemical damage is also a concern. For example, fresh mortar can adversely react with aluminum. If window installation must precede the exterior wall finish, adequate precautions must be taken to protect the windows from damage, Figure 29.8.

Installation of Windows With Nailing Flanges

Windows with nailing flanges are generally used in buildings with exterior wood siding, stucco, and EIFS cladding. They can also be used in buildings with masonry veneer cladding,

FIGURE 29.8 Aluminum windows were installed in this high-rise condominium building before applying the exterior (EIFS) wall finish. Windows were, therefore, protected from damage, as shown here.

but the detailing is more complicated. The installation of a window with nailing flanges is relatively simple because it is anchored to the wall through the flanges.

Because the flanges project over the entire window perimeter, they seal the rough opening, providing a built-in flashing. However, additional waterproofing of the rough opening is generally recommended to provide a second line of defense against water and air leakage, Figure 29.9.

(a) A factory-glazed aluminum window with nailing flanges and simulated divided lites (grilles within insulating glass unit).

(b) Window being brought into the opening. Observe the self-adhering waterproof tape around rough opening.

(c) A window must be leveled before it is anchored to the wall. In this image, one installer holds the level while the other installer is getting ready to screw the window to the wall through the nailing flange.

FIGURE 29.9 A few steps in the anchorage of a window with nailing flanges into a stud wall.

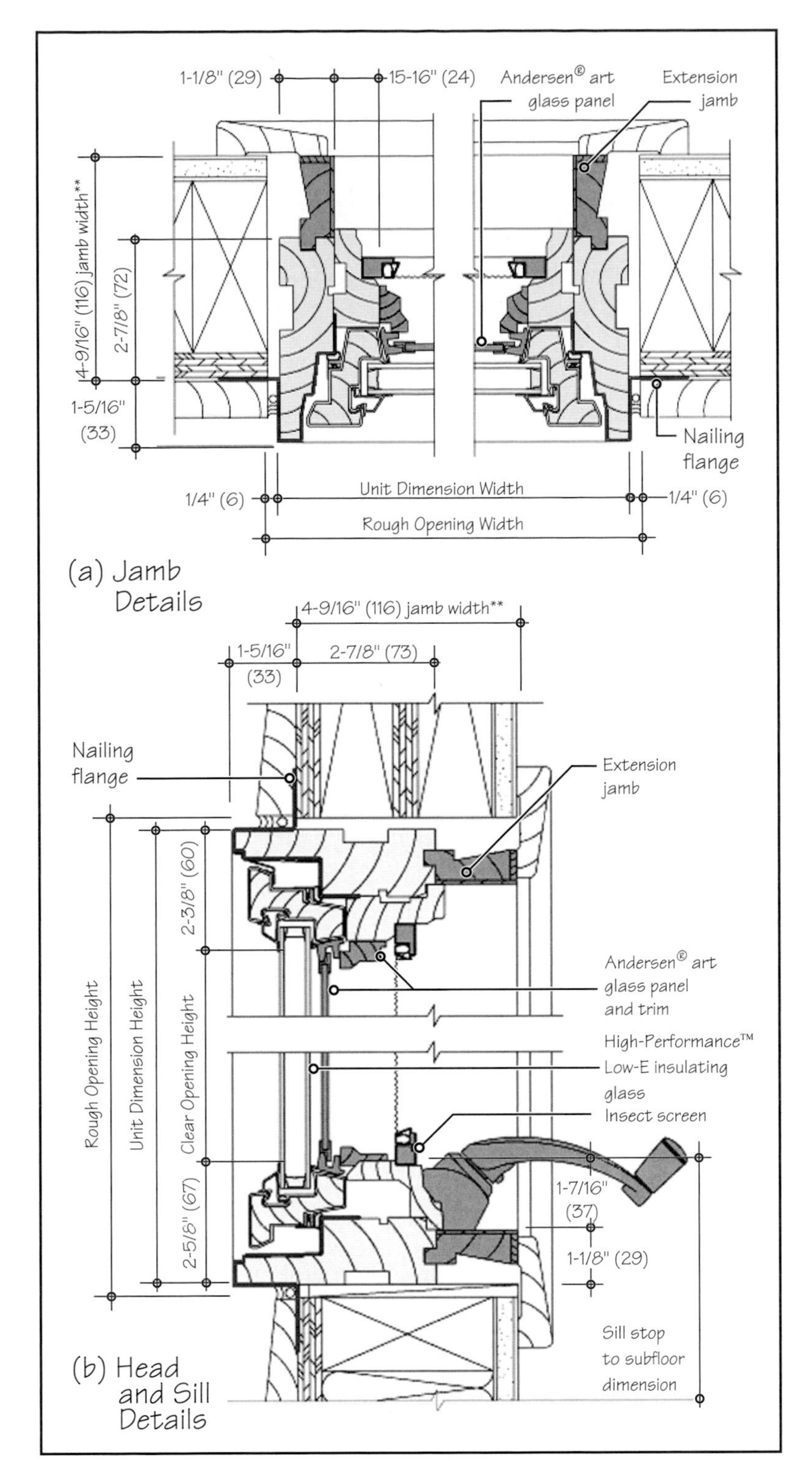

FIGURE 29.10 Suggested jamb, sill, and head details of a vinyl-clad wood window by Andersen Corporation. Observe that the window is anchored to the wall through vinyl nailing flanges. (Illustration courtesy of Andersen Corporation.)

Manufacturers typically provide recommendations for anchorage and installation of their windows and detailing at the jambs, head, and sill, which an architect can use as a guide, Figure 29.10.

Installation of Windows Without Nailing Flanges

Windows without nailing flanges can be installed in several ways. Two commonly used methods for aluminum windows are shown in Figures 29.11 and 29.12. Figure 29.11 shows the use of a perimeter receptor frame comprising a sill receptor (also called subsill), two jamb receptors, and a head receptor. The receptors are first anchored to the rough opening with anchorage points fully sealed. Subsequently, the window (complete with frame, sashes, glass, hardware, etc.) is placed within the receptor frame.

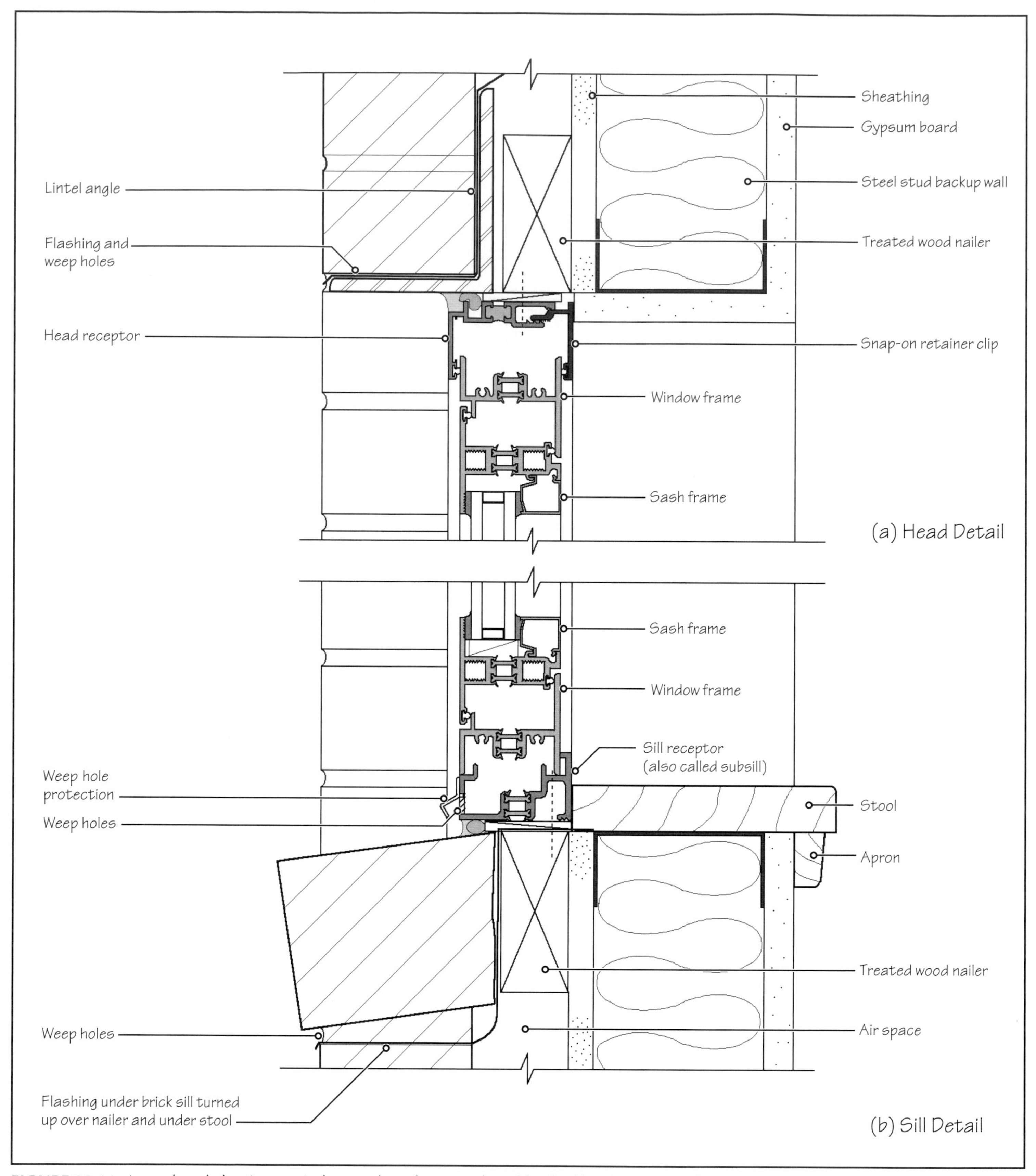

FIGURE 29.11 A preglazed aluminum window anchored to a steel-stud-backed brick veneer wall using receptors and retainer clips.

Lateral load resistance to the window is provided by aluminum retainer clips that are snapped onto the receptors from the inside. By changing their thickness, cross-sectional profile, and locking striations, retainer clips can provide the required lateral load resistance. Interior finishes around the jambs, sill, and head are now installed. In this method of installation, the window is not directly anchored to the rough opening.

Any incidental water that leaks into the window is collected in the subsill. Therefore, the subsill is provided with end dams to prevent its runoff from the ends and also with weep holes to let the water drain out.

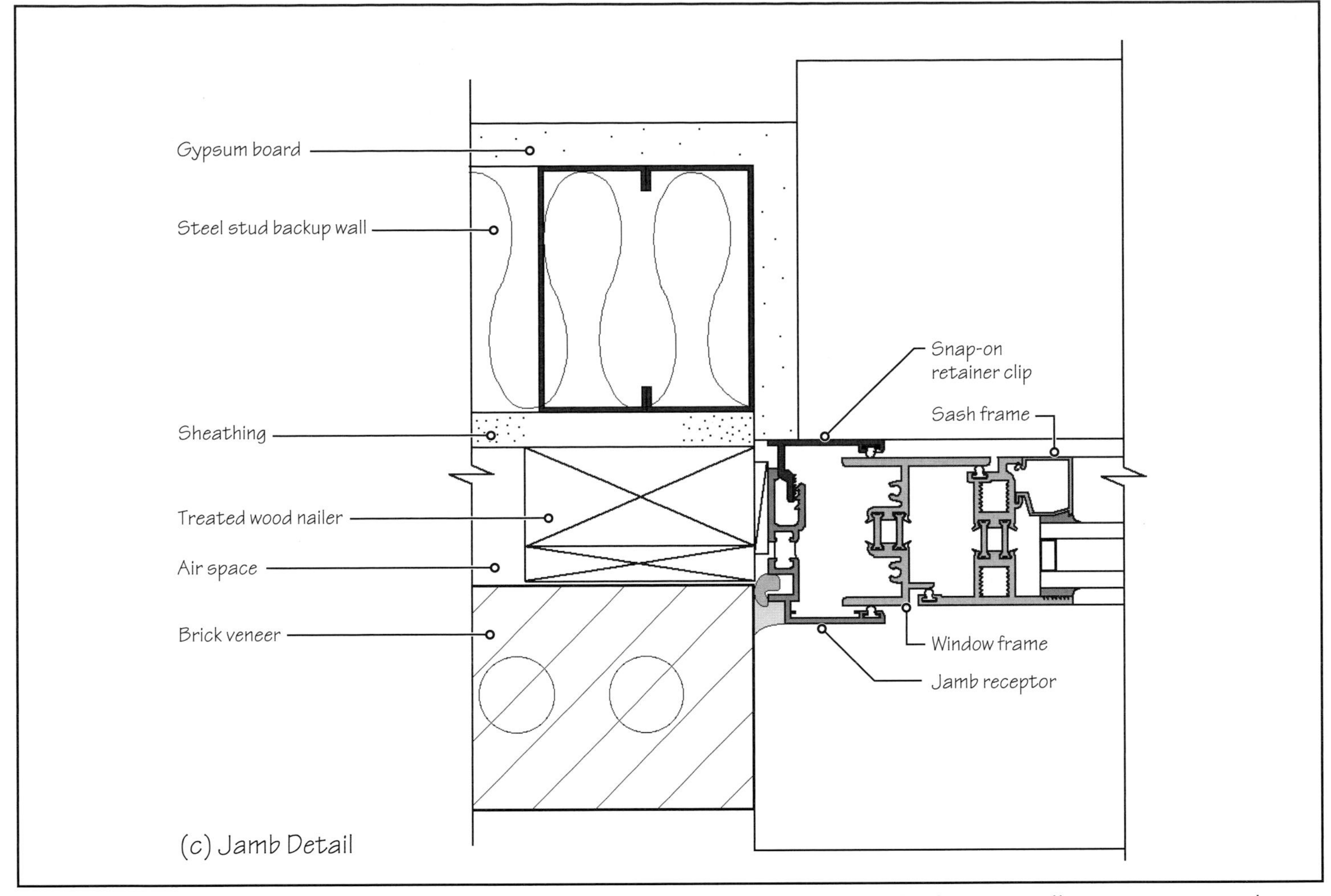

FIGURE 29.11 (*Continued*) A preglazed aluminum window anchored to a steel-stud-backed brick veneer wall using receptors and retainer clips.

Figure 29.12 shows an alternative anchorage method, which utilizes two-piece interior perimeter trims. In this method, the window is first placed into the opening, squared, and leveled. The first (L-shaped) trim piece is then anchored to the rough opening on all four sides and also to the window. Subsequently, the second trim piece is snapped on the first trim piece. Interior finishes can then be applied to the opening.

The details shown in Figures 29.11 and 29.12 can be modified in several ways, depending on the wall construction and the exterior finish. For example, an aluminum sill can be used instead of a brick sill, Figure 29.13. Manufacturers have several standard components for use with different interior and exterior requirements. They can also provide custom-made trims and accessories with little or no additional cost.

29.5 CLASSIFICATION OF DOORS

Because of the large variety of doors used in contemporary buildings, they can be classified in several ways. The simplest classification is to group them as: *interior* and *exterior doors.* Exterior doors include entry doors and garage and overhead doors. Entry doors in residential buildings are generally insulated and rated for energy performance.

Classification Based on Door Material

Doors can also be classified based on the materials of which they are made. Wood and metal (steel, stainless steel, and aluminum) are the commonly used door materials. Fiberglass-reinforced plastic (FRP) is a more recently introduced material.

Wood doors are the most popular interior doors. Among the metal doors, steel doors are used where greater security, fire resistance, rot resistance, blast resistance, and wind-load resistance are required. Stainless steel doors are preferred for food-processing plants, freezer rooms, commercial kitchens, and so on.

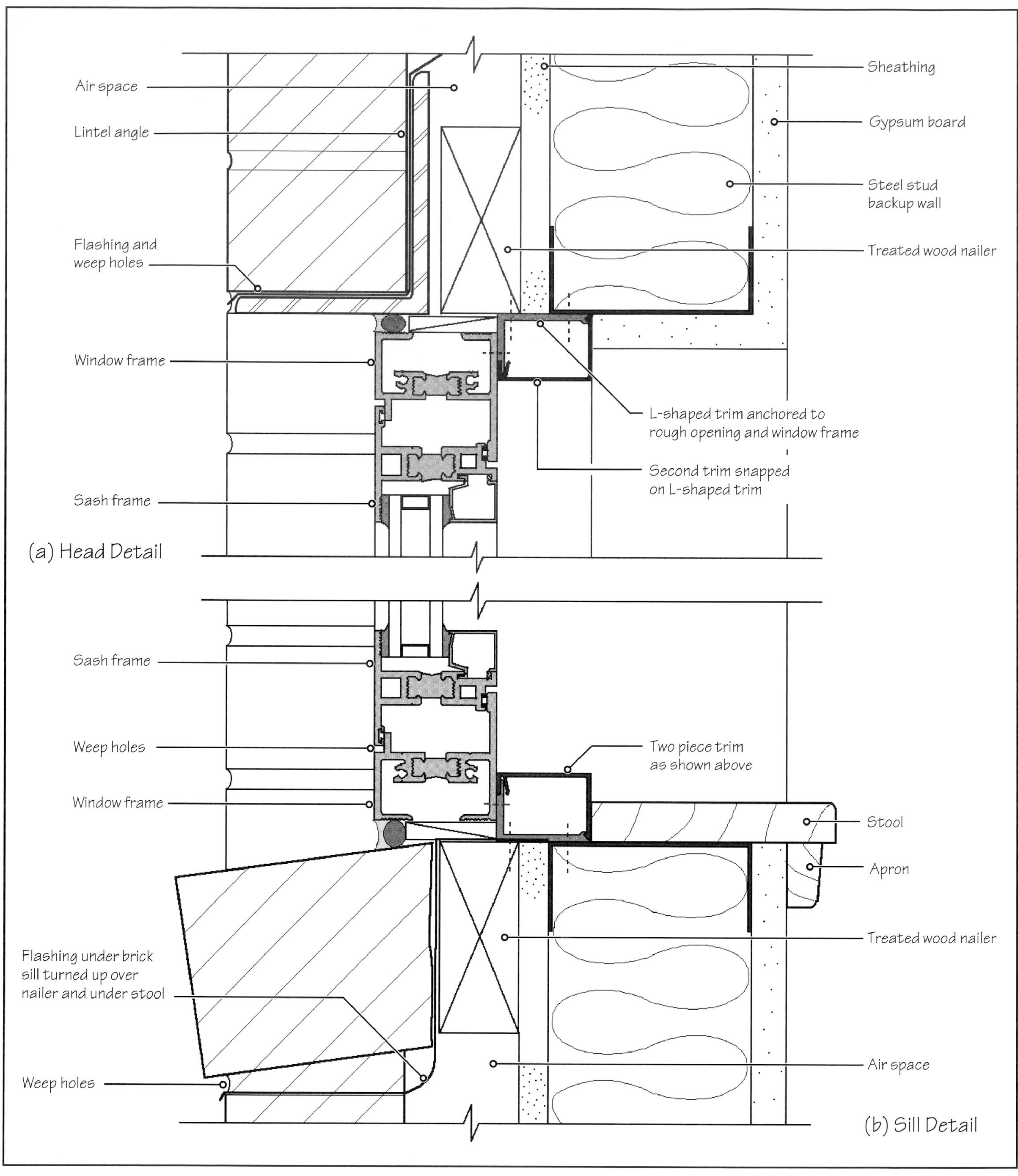

FIGURE 29.12 A preglazed aluminum window anchored using two-piece trims.

Aluminum doors are generally glazed and are commonly used in public buildings. Aluminum flush doors and panel doors are specified where corrosion resistance is important, such as water-treatment plants, swimming pools, and pumping stations.

Classifications Based on Door Operation and Door Styles

A helpful classification of doors is based on their mode of operation, as shown in Figure 29.14. Of these, the single-leaf hinged door is most commonly used. Another classification is based on door style (i.e., door elevation), Figure 29.15.

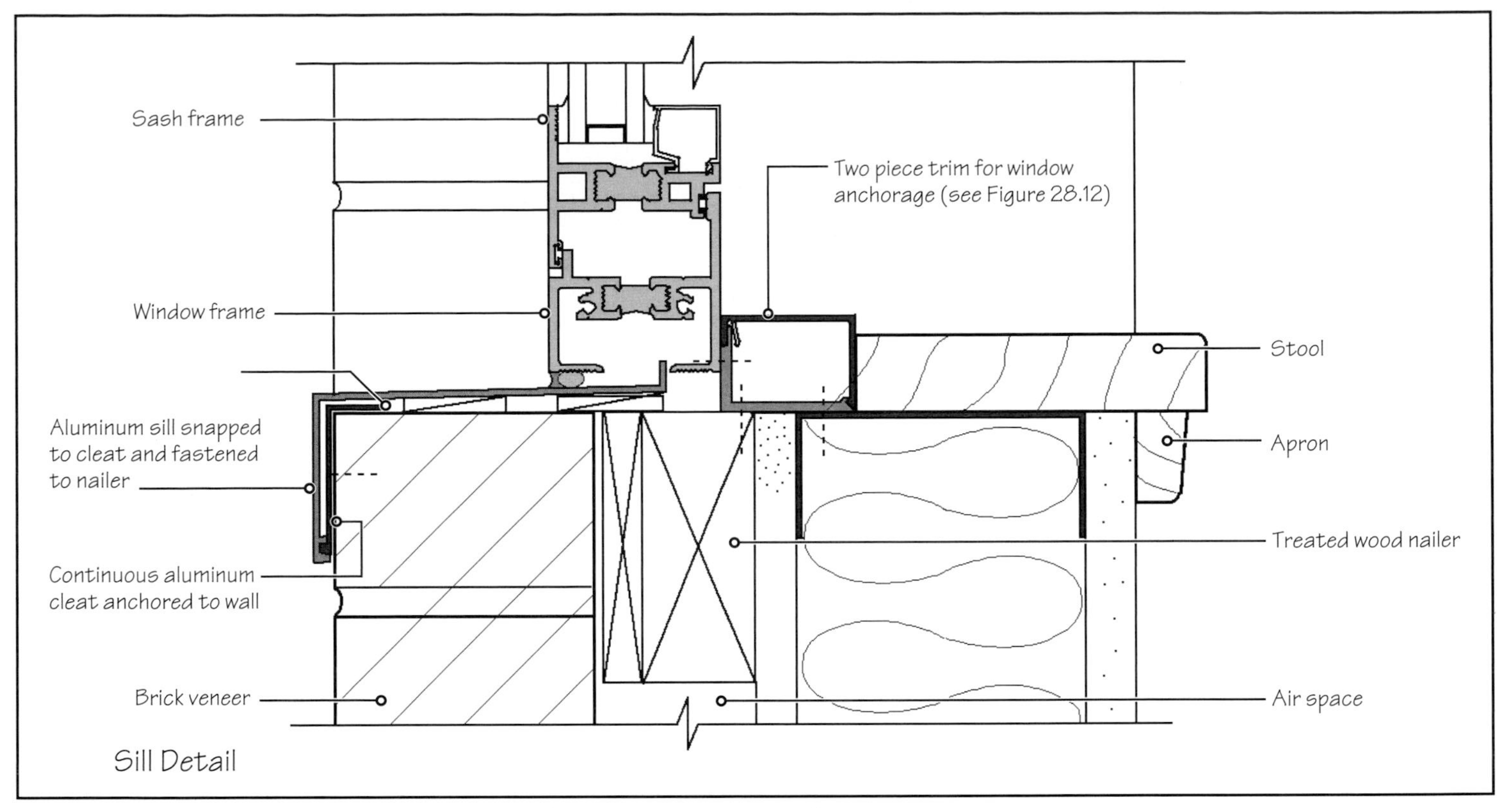

FIGURE 29.13 A preglazed aluminum window with an exterior aluminum sill.

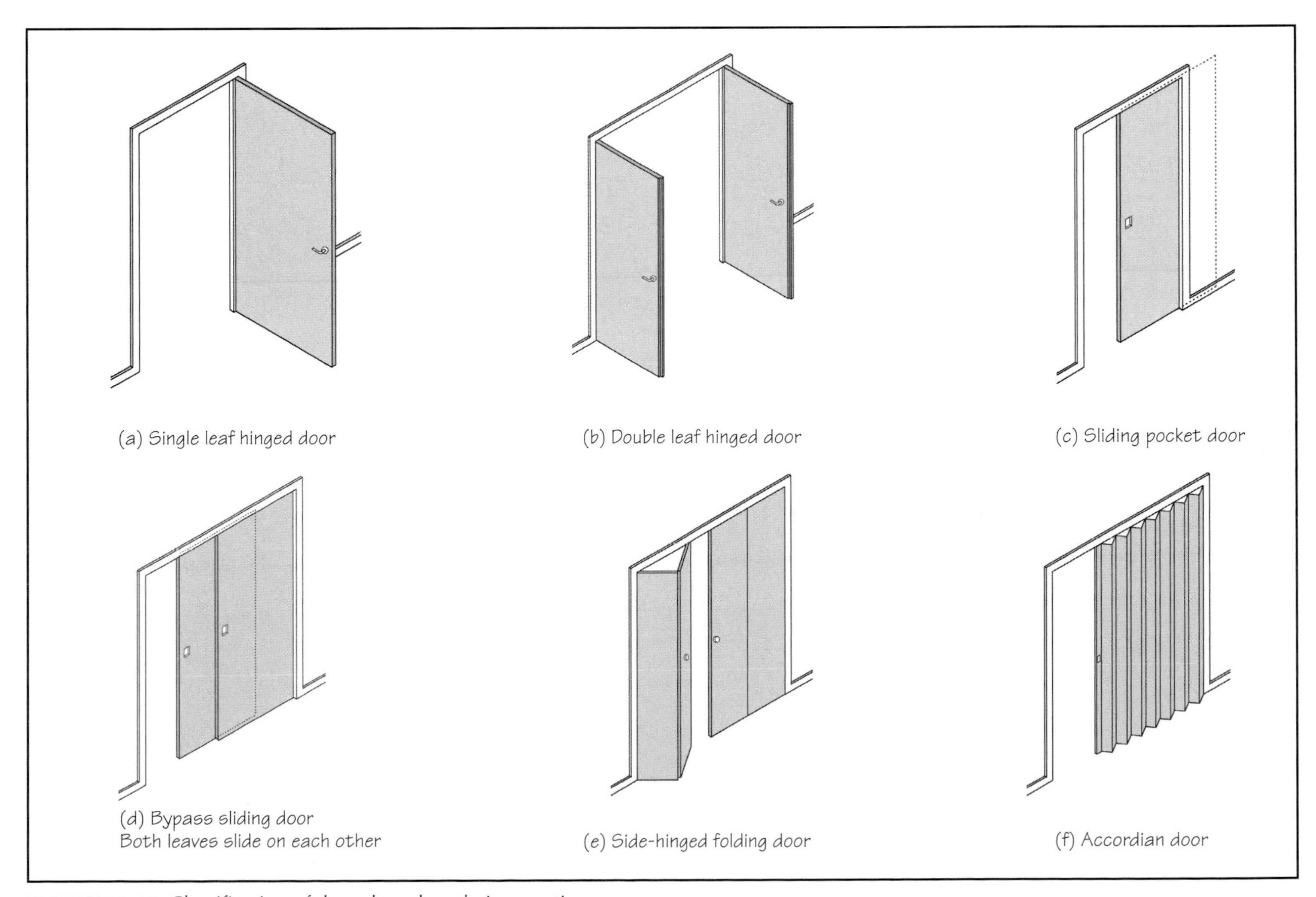

FIGURE 29.14 Classification of doors based on their operation.

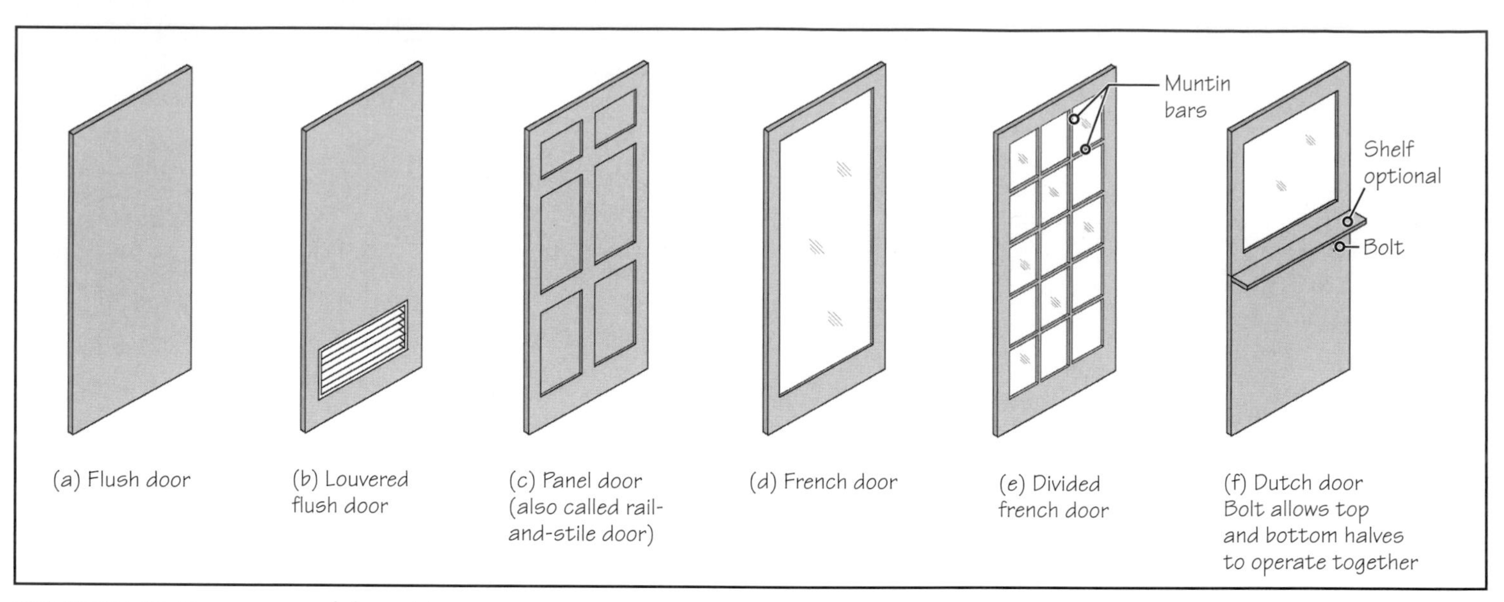

FIGURE 29.15 Common wood door styles.

Of the various styles shown in Figure 29.15, flush doors and panel doors are more popular. Construction details of wood flush doors and panel doors are illustrated in Figures 29.16 and 29.17. Figure 29.18 gives standard door sizes for wood doors, their hardware, and handling.

Metal (steel) doors are generally hollow-core doors and are available in both flush and panel door styles. Construction of metal hollow-core flush doors is illustrated in Figures 29.19 and 29.20.

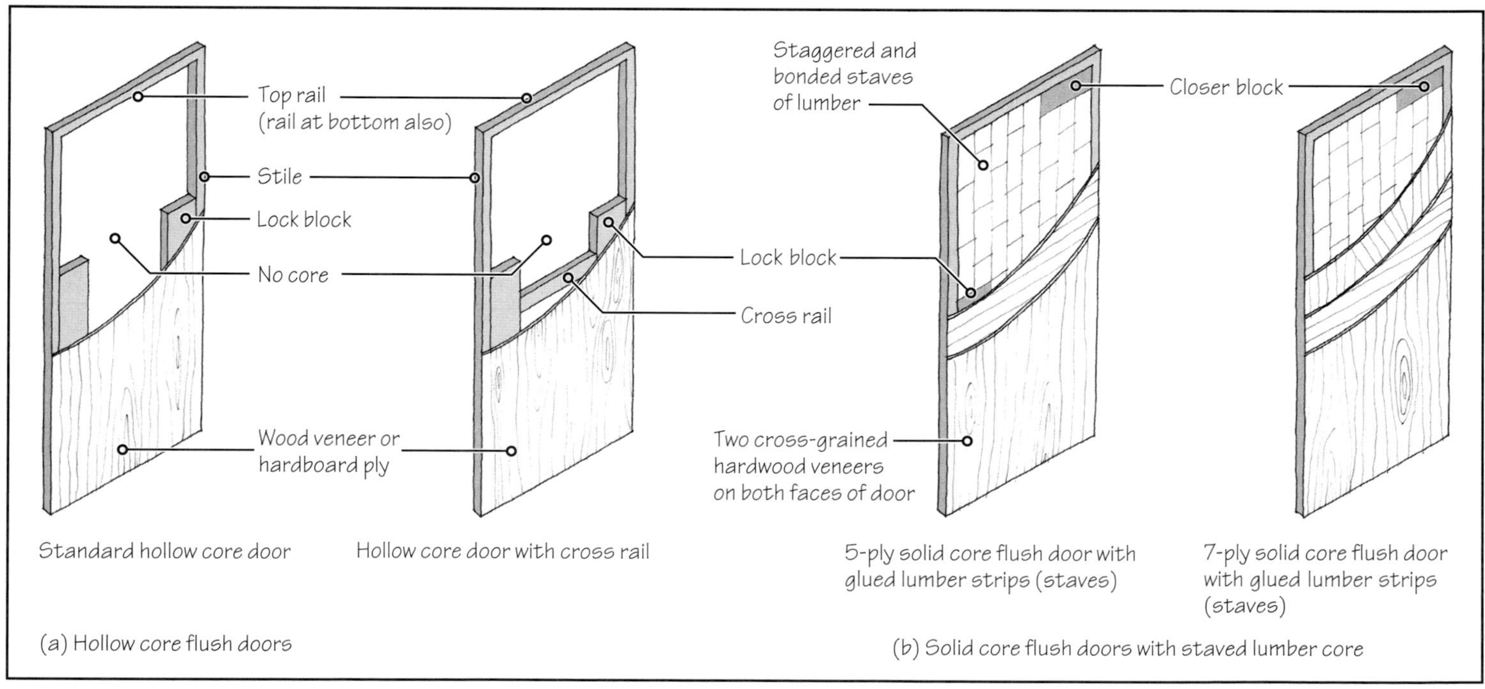

FIGURE 29.16 Construction details of wood (hollow core and solid core) flush doors.

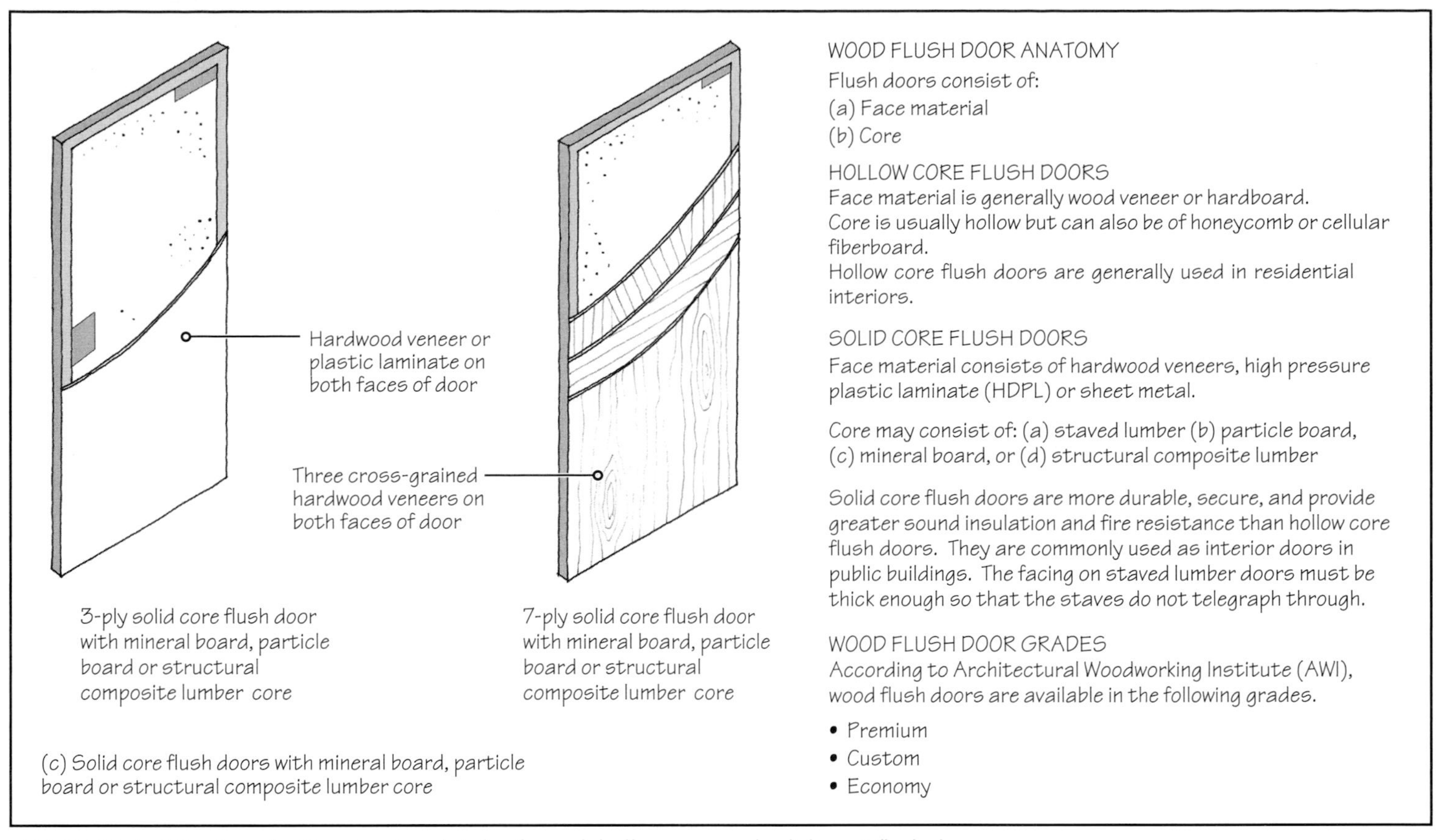

FIGURE 29.16 ***(Continued)*** Construction details of wood (hollow core and solid core) flush doors.

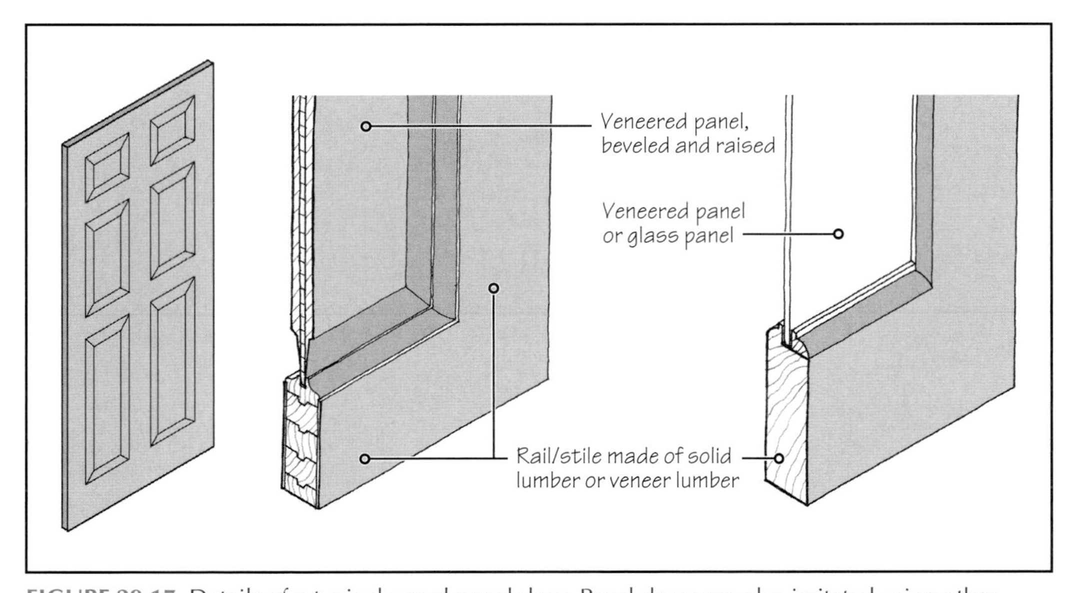

FIGURE 29.17 Details of a typical wood panel door. Panel doors are also imitated using other materials.

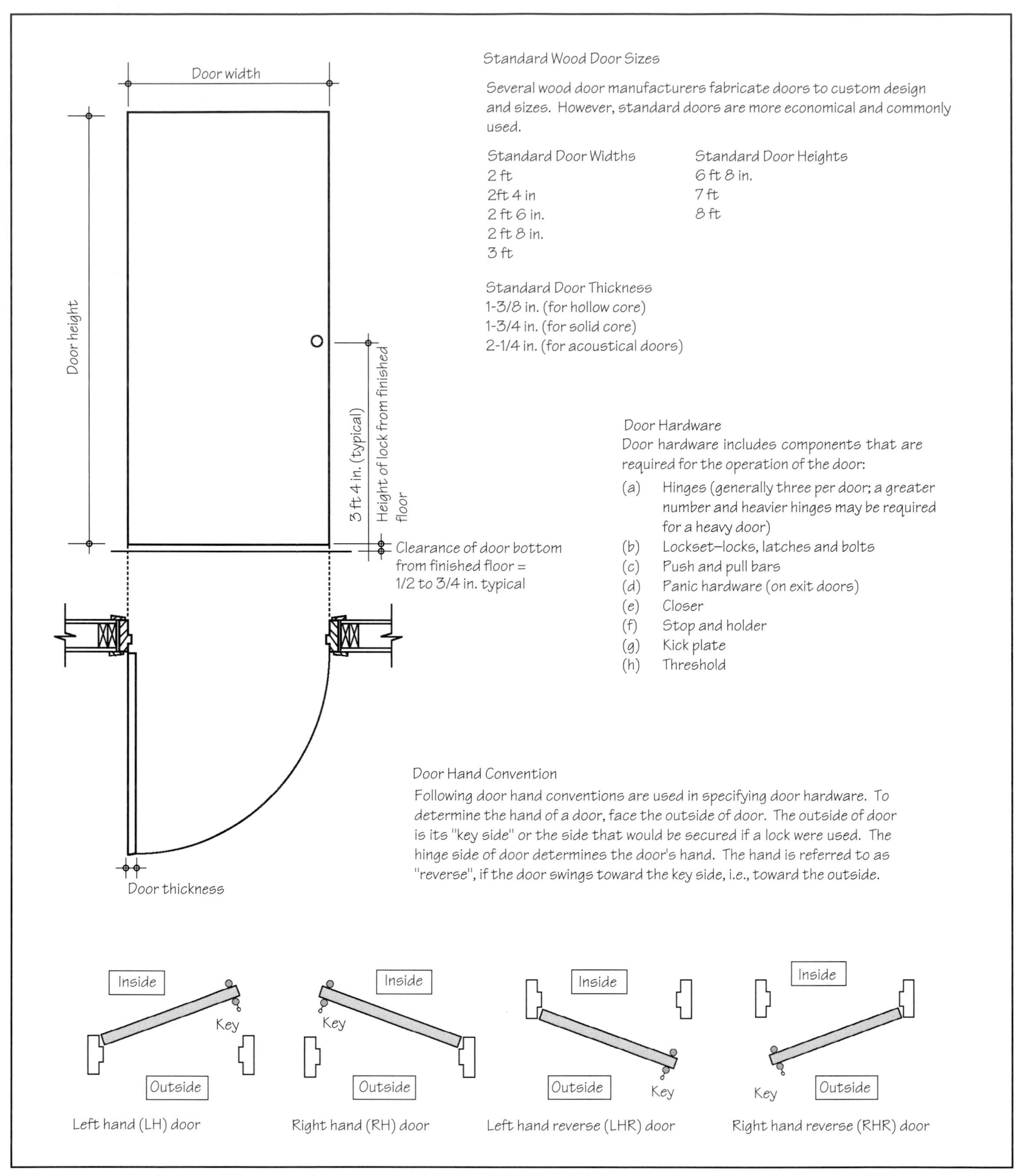

FIGURE 29.18 Standard wood-door sizes, hardware, and door-hand conventions.

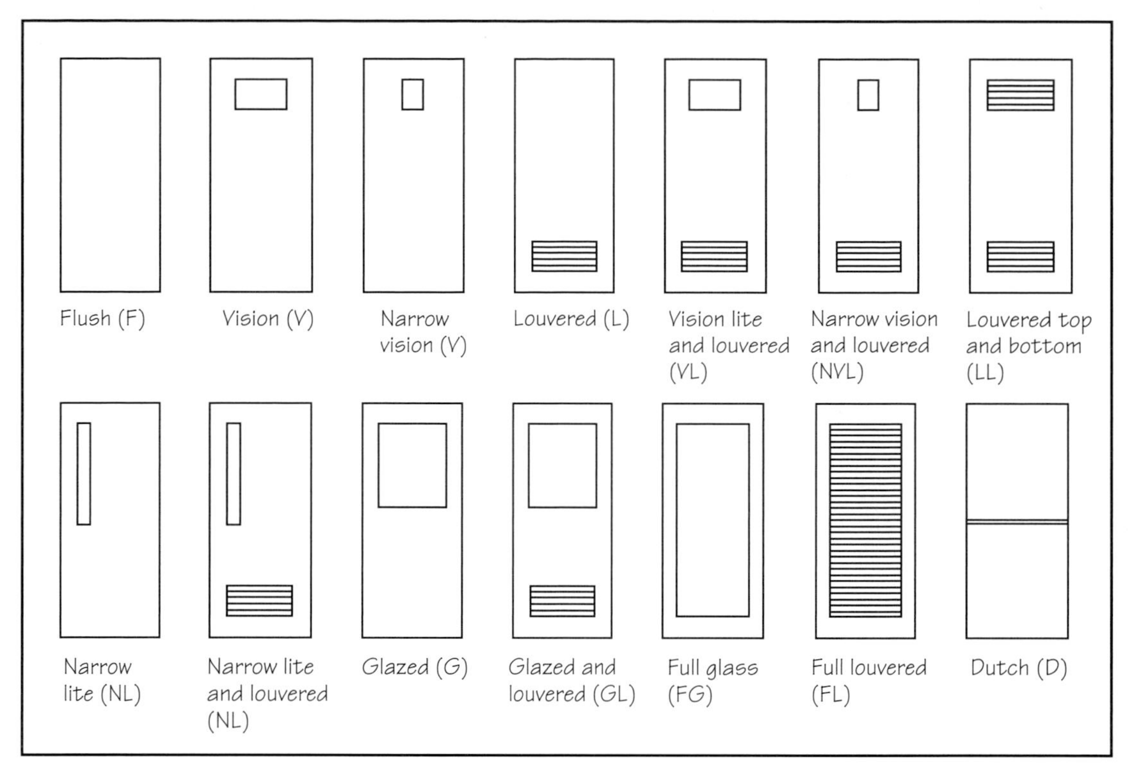

FIGURE 29.19 Commonly used hollow metal flush-door styles.

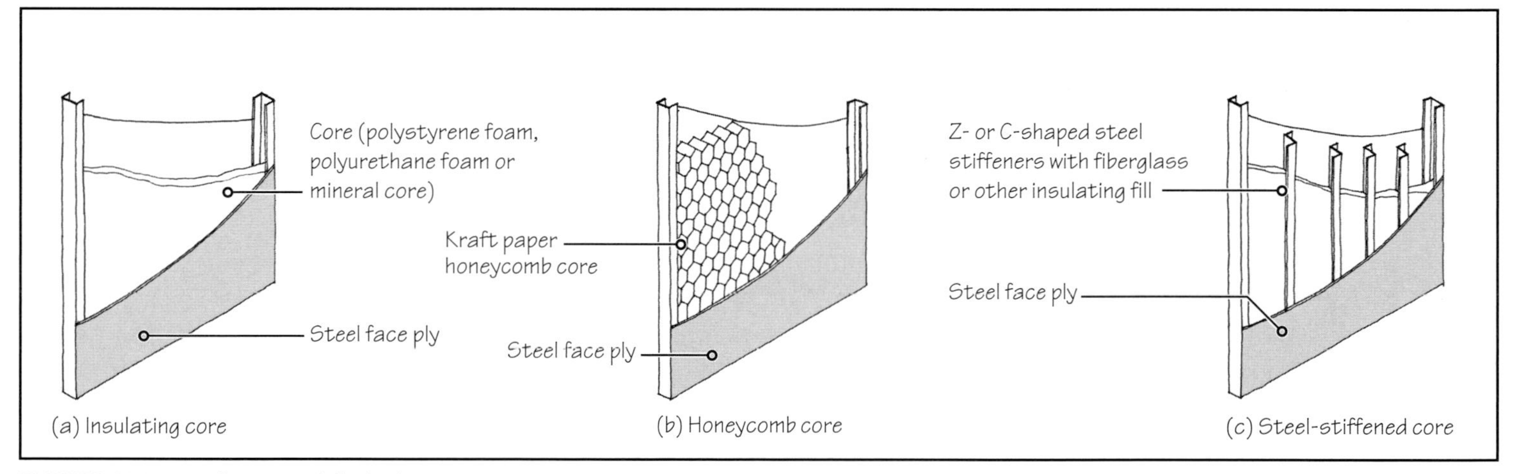

FIGURE 29.20 Hollow metal flush-door construction.

29.6 DOOR FRAMES

Wood door frames are used with wood, fiberglass, and metal doors in non-fire rated applications. Metal door frames, however, are used with metal doors or fire-rated wood doors. Figure 29.21 shows commonly used wood and metal door-frame shapes, and Figures 29.22 and 29.23 show typical construction details of wood and metal door frames, respectively.

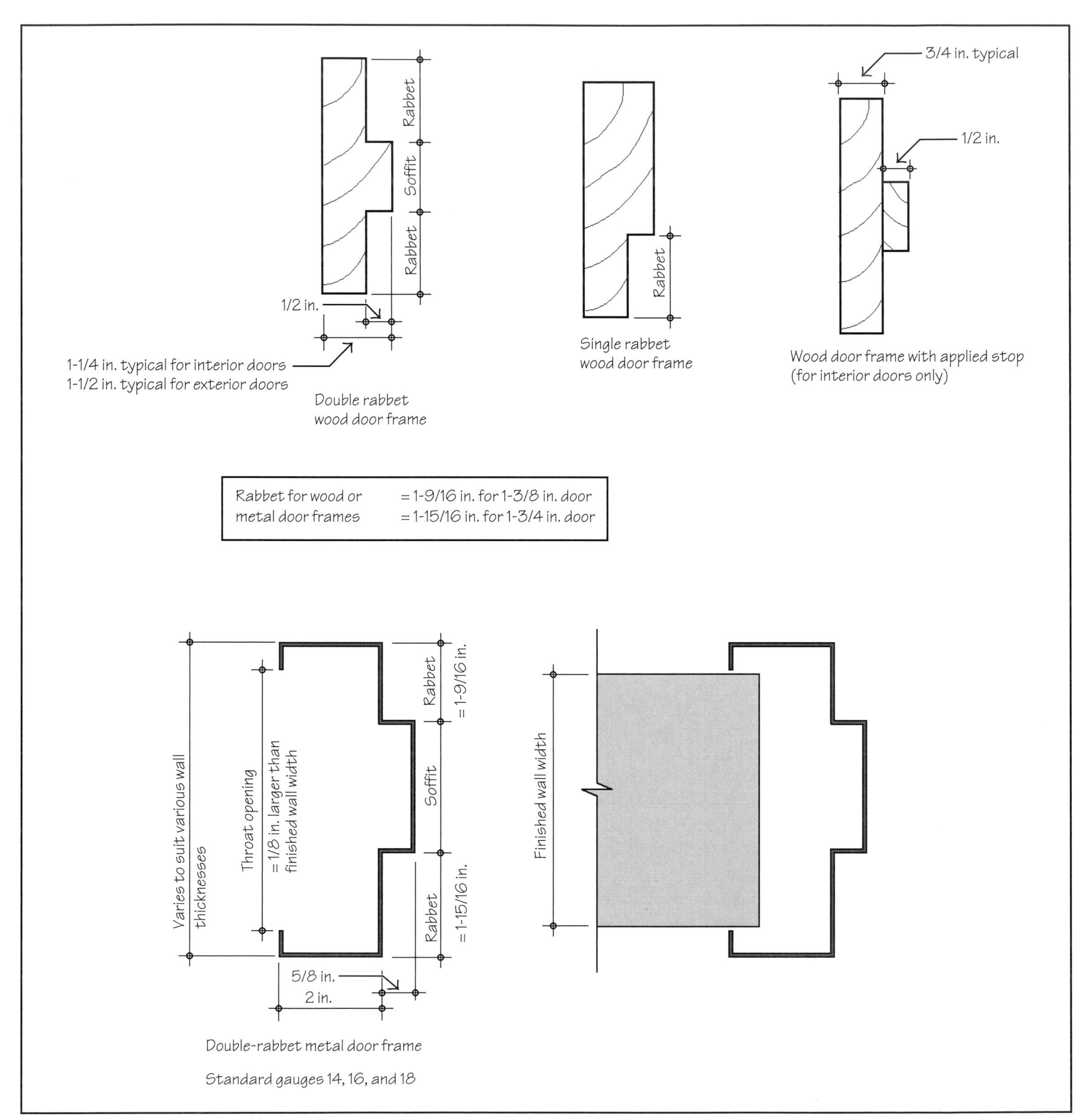

FIGURE 29.21 Typical wood and metal door-frame shapes.

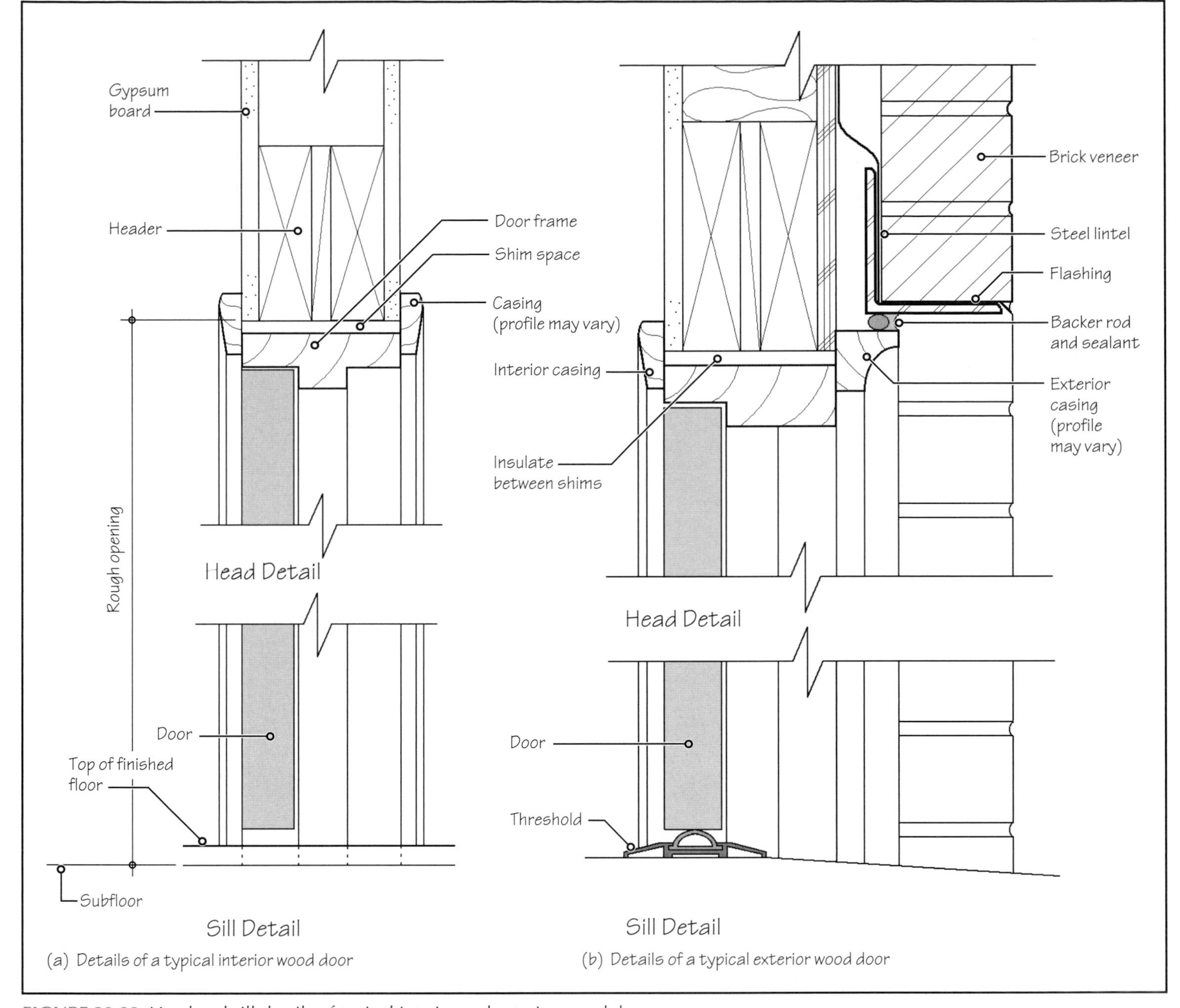

FIGURE 29.22 Head and sill details of typical interior and exterior wood doors.

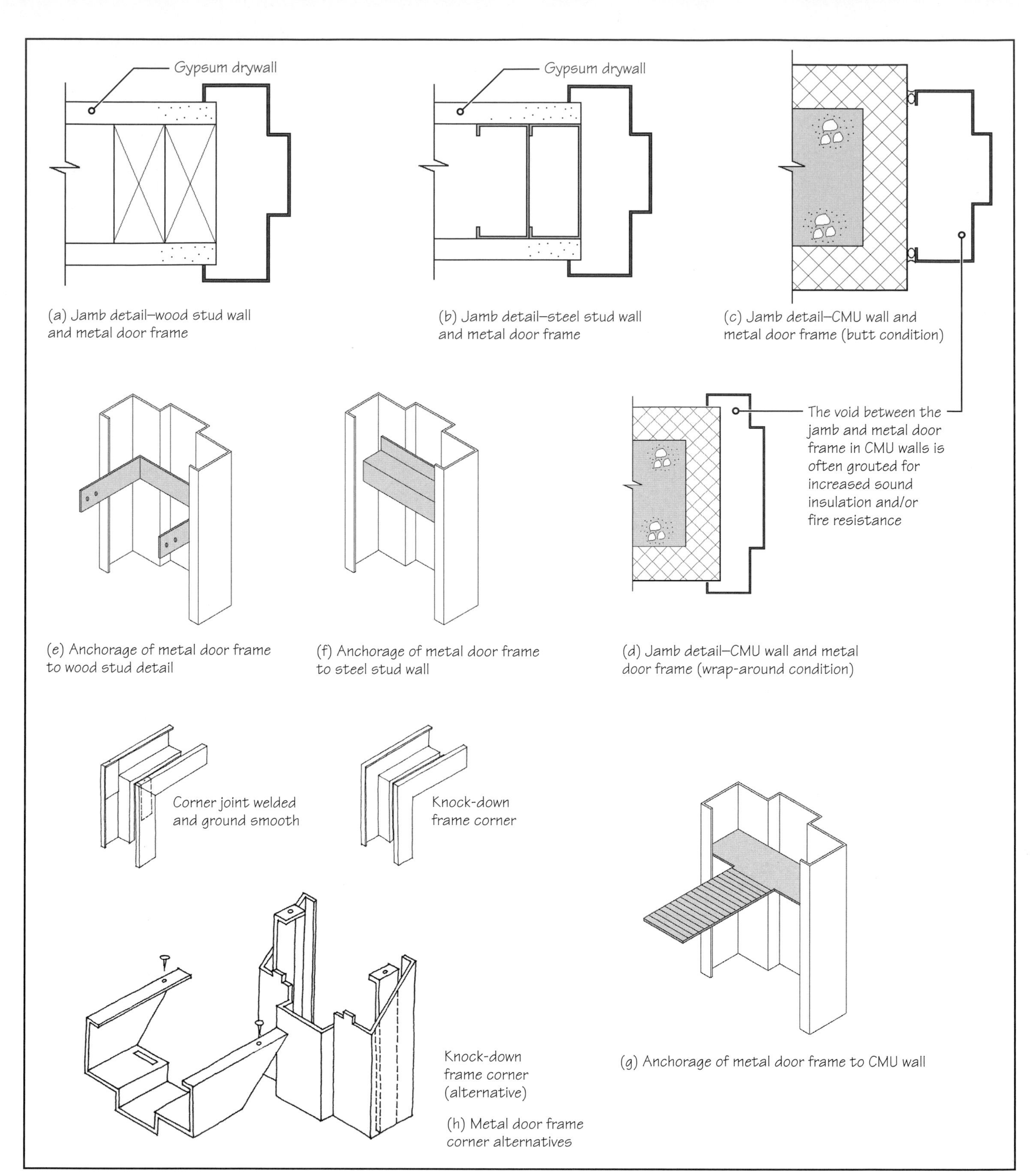

FIGURE 29.23 Metal door-frame details.

PRACTICE QUIZ

Each question has only one correct answer. Select the choice that best answers the question.

9. The label on an NFRC-rated window, must include values for
- **a.** U-value and shading coefficient.
- **b.** U-value and air infiltration.
- **c.** U-value and solar heat–gain coefficient.
- **d.** U-value and visible transmittance.
- **e.** Solar heat–gain coefficient and air infiltration.

10. In a window that carries a performance label of HS-C30 71 × 59, the number 30 indicates that the window
- **a.** will leak a maximum of 30% of room's air per hour.
- **b.** conforms to the applicable standard up to a size of 30 in. × 30 in..
- **c.** should not be subjected to a maximum wind load of 30 psf.
- **d.** should not be subjected to a maximum combined wind load and vapor pressure of 30 psf.
- **e.** none of the above.

11. In a window that carries a performance label of HS-C30 71 × 59, the letter C indicates that the window
- **a.** belongs to a performance class called *commercial*.
- **b.** is a casement window.
- **c.** is a C-grade window.
- **d.** none of the above.

12. Nailing flanges are only provided in steel windows.
- **a.** True
- **b.** False

13. The use of nailing flanges in windows also requires the use of
- **a.** retainer clips.
- **b.** two L-shaped trims.
- **c.** none of the above.
- **d.** both (a) and (b).
- **e.** either (a) or (b).

14. Most windows are shop-glazed so that they arrive at the construction site in a fully assembled condition.
- **a.** True
- **b.** False

15. A hollow metal door frame is obtained from
- **a.** tubular steel.
- **b.** sheet steel that is pressed into shape.
- **c.** tubular steel and sheet steel welded together.
- **d.** none of the above.

16. A hollow metal door frame is installed so that it
- **a.** buts against the jambs of a CMU wall.
- **b.** wraps around the jambs of a CMU wall.
- **c.** (a) or (b).
- **d.** (a) and (b).

Answers: 9-c, 10-e, 11-a, 12-b, 13-c, 14-a, 15-b, 16-c.

29.7 FIRE-RATED DOORS AND WINDOWS

A fire-rated wall can serve as a barrier to the spread of fire only if the openings (doors, shutters, and windows) installed in them also provide the same (or nearly the same) degree of fire protection. Building codes require a certain minimum level of fire protection of openings provided in fire-rated walls.

The *fire-protection ratings* of openings is usually allowed to be lower than the fire-resistance ratings of walls in which they are installed. For example, the fire rating of a door in a wall is 75% of the fire rating of the wall. Thus, in a 2-h-rated wall, a $1\frac{1}{2}$-h-rated door is considered adequate. An opening with a lower fire-protection rating than the wall in which it is installed does not greatly compromise the overall integrity of the wall. This is true because, under normal conditions of use, there is usually a lower fire hazard in the vicinity of an opening as compared with the rest of the wall.

For example in the case of a door, there is a clear space on both sides of the door to provide for unobstructed traffic. Hence, there is less fuel available in close proximity to the door, and the resulting fire hazard is less than in the rest of the wall. Another reason is that the total opening area in fire-rated walls is usually so limited by building codes that a lower protection of openings does not greatly reduce the overall effectiveness of the wall.

Reference to a *rated opening* implies that the entire assembly is rated. The assembly consists of the door (or window), door frame (or window frame), anchorages, sill, handles, latches, hinges, and so on. Building codes require that a fire-rated door or window and its frame bear an approved label.

Note that in referring to the fire rating of openings, the term *fire-protection rating*, not fire-resistance rating, has been used. This is because the fire rating of openings is not determined by the same test as for the fire rating of walls (ASTM E 119 test).

FIRE DOORS

A fire door may be a 3-h-, 90-min-, 60-min-, 45-min-, or 20-min-rated door. A 20-min-rated door assembly is, in fact, a smoke-and-draft-control assembly. Its main purpose is to minimize the transmission of smoke from one side of the door to the other. A 20-min door is usually required in a 1-h-rated corridor wall. A $1\frac{3}{4}$-in.-thick solid-bonded wood-core door usually provides a 20-min protection. However, to be accepted as a 20-min door, it must pass the required test to substantiate performance.

Fire Windows

Where the walls (interior or exterior) are required to be fire rated, a glazed window placed in such walls must provide at least a 45-min protection. A glazed window with a 45-min protection is referred to as a *fire window*. Because 45-min protection is relatively low for a protected opening, building codes place restrictions on the allowable areas of fire windows.

EXPAND YOUR KNOWLEDGE

Window Terminology

Two important components of a window are (a) window frame and (b) window sash. The terminology related to these and other components of a window is given next.

Frame

- *Head*—the top horizontal member of the frame
- *Jamb*—a side vertical member of the frame
- *Sill*—the bottom horizontal member of the frame
- *Mullion*—an intermediate vertical member of the frame
- *Rail*—an intermediate horizontal member of the frame
- *Nailing flange*—Some manufacturers provide windows with a nailing flange. A nailing flange runs continuously over the outside of a window frame and functions as a flashing. It also provides a means of anchoring the window to wall opening.

Sash

- *Stile*—a vertical sash member
- *Rail*—a horizontal sash member
- *Muntins*—thin horizontal and vertical dividers, commonly used in early windows when large sheets of good quality glass were not available (In some contemporary windows, large sheets of glass are used, but to simulate a divided window, grilles are used within the cavity of an insulating glass unit. Alternatively, a grill may be placed over the glass pane, either on the inside face or the outside face or on both faces. The grilles are removable for easy cleaning of glass. A window with imitation muntins is referred to as comprising *simulated divided lites*. A window with true dividers is referred to as a window with *authentic*, or *true, divided lites*.)
- *Weatherstripping*—a strip of resilient material that provides a seal between the sash and the frame to reduce air and water leakage
- *Glazing stop*—a feature that holds the glass against the rabbet in the sash or frame
- *Rabbet*—a step in a sash or frame cross section against which the glass is held
- *Gasket*—a strip of resilient material between the glass and the glazing pocket (Double-sided tape is used by some window manufacturers instead of gasket.)
- *Daylight opening* (DLO)—visible glass area in a window (It is the total area of glass minus the invisible area, i.e., the area of glass that is buried within glazing pockets.)
- *Hardware*—hinges, latches, locks, levers, and so on.

Window Surrounds (Interior)

- *Stool*—a horizontal trim member that abuts against the window sill and covers the rough sill
- *Jamb extension*—a horizontal or vertical trim that covers the rough head and jamb and extends the depth of window frame (Note that the term *jamb extension* also refers to the extension of the head because the head is sometimes referred to as *head jamb*.)
- *Apron*—the trim installed on the wall under the stool.
- *Casing* (or trim)—decorative members that cover the joints between adjacent materials.

Window-Opening Configurations

- *Rough opening*—the opening within which the window is placed (Rough opening dimensions are at least $\frac{1}{2}$ in. greater

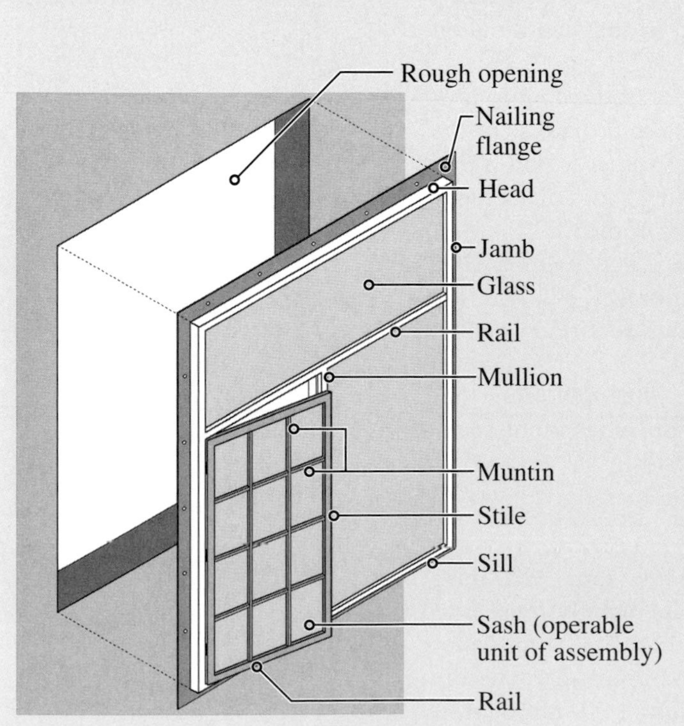

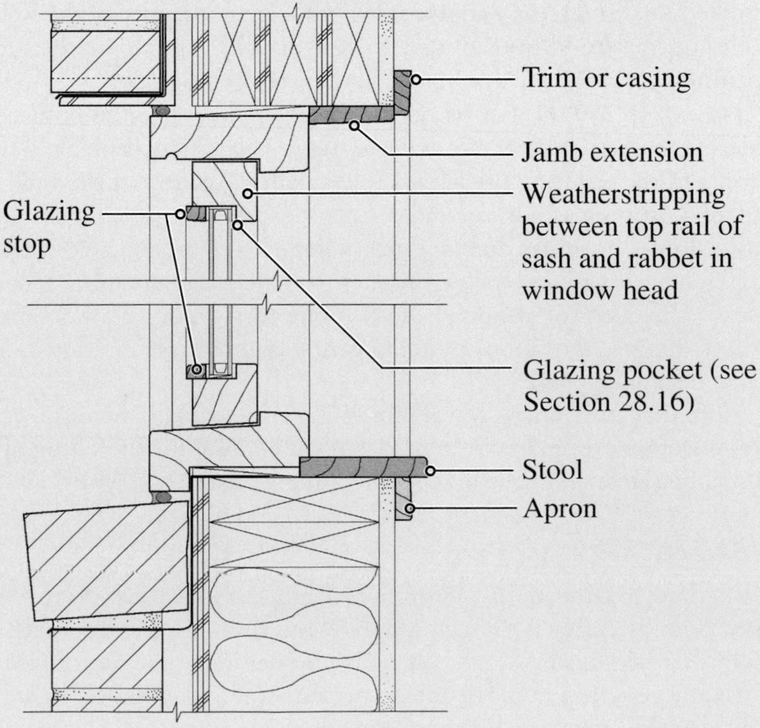

Window Terminology (*Continued*)

than the actual window unit dimensions to allow an easy fit and alignment of the window in the opening. The terms *rough sill, rough head,* and *rough jamb* refer to the respective areas of a rough opening

- *Punched window and punched glazing*—a window with an opaque wall around it (The term *punched glazing* is generally used for large, fixed glazing in which the glass is site installed. In terms of appearance, both punched windows and punched glazing are the same.)
- *Strip windows* and *glazing*—an array of windows placed side by side to form a horizontal strip (or ribbon) window system (The term *strip glazing* is used with large, fixed glazing units in which the glass is site installed.)
- *Window wall and glass curtain wall*—a wall in which the windows extend from floor to roof, or from floor to floor, called a window wall (A window wall must be distinguished from a glass curtain wall, which hangs as a curtain from the building's structure and bypasses the intermediate floors or roof (Chapters 25 and 30).)
- *Projected window*—includes windows whose sashes project out of the window plane when open, such as casement, awning, hopper, and pivoted windows (Projected windows are generally more airtight in closed position than hung or sliding windows because of the pressure seal between the frame and the sash.)

Punched windows

Strip (or ribbon) windows

EXPAND YOUR KNOWLEDGE

Door Terminology

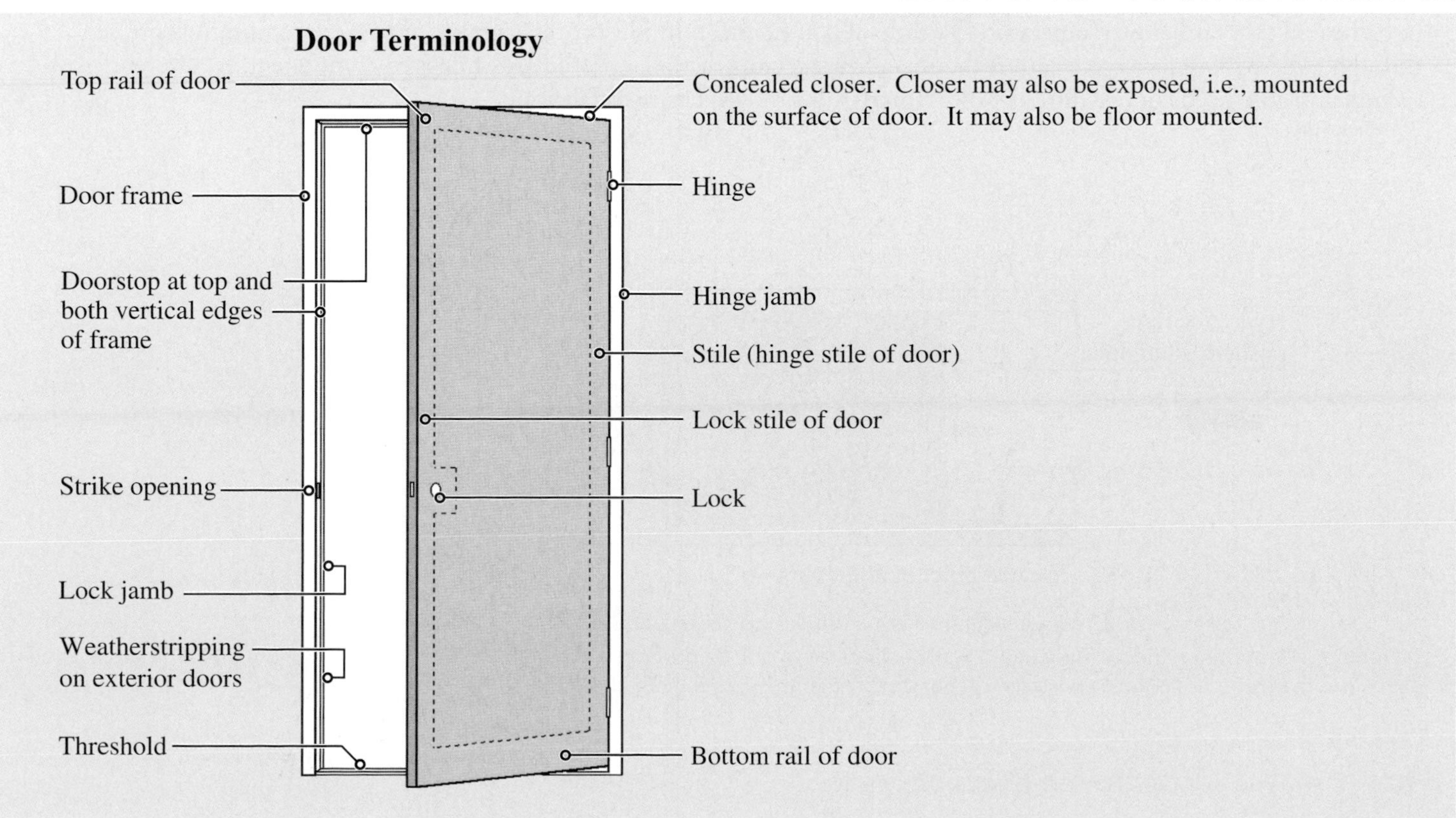

PRINCIPLES IN PRACTICE

A Note on Aluminum

Aluminum is a metal whose appearance resembles and corrosion resistance matches that of stainless steel. It is, however, much lighter, softer, and more flexible than steel or stainless steel. Its softness allows it to be easily extruded into complex cross-sectional profiles. Its flexibility allows aluminum components to be joined together by the snap-on technique, requiring no fasteners. Its corrosion resistance implies that it can be left exposed to the atmosphere without any protective coating.

Corrosion resistance, flexibility, and softness are the reasons aluminum is the preferred metal for glass curtain walls, storefronts, and windows. Nearly one-third of the entire production of aluminum is used in these components.* Aircrafts and automobiles also consume a large portion of the aluminum produced.

Aluminum, in the form of aluminum oxide, is far more abundant on the earth's crust than any other metal oxide, and most rocks and soils contain some aluminum oxide. However, the energy required to extract aluminum from common rocks and soils is very high. The only economical method of producing aluminum is from the ore called *bauxite,* named after Baux, a town in Southern France. The bauxite ore contains nearly 60% aluminum oxide and 40% of other compounds, such as iron oxide and silicon oxide. Producing aluminum, even from the bauxite ore, is an energy-intensive process. This explains why, weight for weight, aluminum is more expensive than steel, despite the aluminum's relative abundance on the earth. It also explains why the aluminum-manufacturing industry resorted to recycling of cans and other scrap well before other industries contemplated sustainable manufacturing.

Like other pure metals, pure aluminum is too weak and soft to be used as a building material. Small amounts (altogether up to 5%) of other elements, such as copper, zinc, silicon, iron, magnesium, and manganese, alloyed with aluminum make it a more practical metal. Numerous aluminum alloys are available to suit different applications. However, pure aluminum is more corrosion resistant than its alloys. Therefore, where corrosion resistance is critical, anodized or painted aluminum is used.

The higher corrosion resistance of pure aluminum is exploited in the metal called *alclad.* Alclad contains an aluminum alloy core that is metallurgically bonded to pure aluminum on the surface. It is a more expensive process than anodizing and is commonly used in aircrafts.

ALUMINUM EXTRUSIONS

Manufacturing aluminum from bauxite ore has several similarities with manufacturing iron from its ore. The end product is cast pig aluminum, which is converted to aluminum billets, bloom, or slabs. Aluminum sections can be made by hot-rolling billets in the same way as steel sections.

Sheets obtained from aluminum slabs can be cold-formed into simpler profiles, similar to light-gauge steel members. However, extrusion is a cheaper and more versatile process for aluminum than hot rolling or cold forming. Additionally, fairly complex cross-sectional shapes which cannot be obtained from hot rolling, can be produced by extruding. Therefore, except sheets, most architectural aluminum components are produced from the extrusion process, Figure 1.

The melting point of most aluminum alloys is nearly 1,100°F (600°C). Therefore, extrusion requires aluminum billets to be heated to a fairly low temperature—nearly 800°F (450°C). In this process, a billet of heated aluminum is pressed through a die, whose face has a cutout that matches the cross-sectional profile of the desired component. A different cross-sectional profile needs only a different die, which is not too expensive to fabricate.

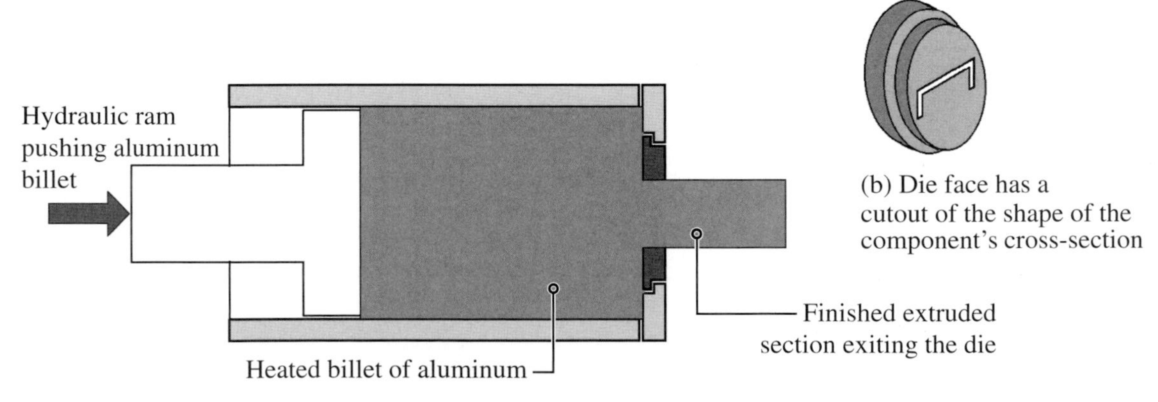

FIGURE 1 Aluminum window and curtain wall sections are made by pushing a heated aluminum billet through a die, a process called extrusion. Extrusion is similar to pushing toothpaste through the opening of a toothpaste tube.

*Zahner, L. W. *Architectural Metals.* New York: John Wiley, 1995, p. 30.

PRINCIPLES IN PRACTICE

ANODIZED ALUMINUM

Aluminum obtains its corrosion resistance because of the aluminum oxide film it develops on its surface on exposure to the atmosphere. This occurs fairly quickly. In fact, the extrusion emerging out of the die mouth is already covered with an oxide film. Like the chromium oxide film on stainless steel, aluminum oxide film is also self-repairing.

The aluminum surface obtained from the die is referred to as *mill finish*. Mill-finished aluminum has a bright, silvery (the familiar aluminum foil) finish, which turns dull gray after a few years. The number of years depends on the amount of pollution and the amount of rainfall. If water from rain or condensation is allowed to lodge on the surface, black streaks may develop on the surface.

Most architectural aluminum is, therefore, either anodized or painted. Anodization essentially preoxidizes the aluminum. The only difference between the natural mill-finished aluminum and anodized aluminum is in the thickness of the oxide film. The natural oxide film is nearly one-millionth of an inch (0.0001 mil) thick. The anodic coating is several thousand times thicker—nearly 0.2 to 0.7 mil thick. A thicker coating is selected for exterior exposures. Coatings greater than 3 mil fracture and chip easily.

The anodic coating may either be clear or colored. A clear coating leaves aluminum looking its natural silvery color. It is also the more common and economical finish and has been used in several outstanding building facades.

Colors available in anodized aluminum are fairly limited and generally restricted to black, dark bronze, medium bronze, light bronze, and white and can have slight variations between pieces.

PAINTED ALUMINUM

Two types of paint finishes are commonly used on aluminum:

- Baked-enamel coating
- Fluoropolymer coating

Baked-enamel coating requires pretreatment of mill-finished aluminum with required chemical(s), followed by spray application of thermosetting, modified acrylic enamel coating, followed by oven baking. Baked-enamel finish is generally recommended for interior applications. The colors obtained are uniform and opaque.

A fluoropolymer coating involves pretreatment of mill-finished aluminum with pretreatment chemicals, followed by one to four coats of a thermocured product system containing not less than 70% of polyvinylidene fluoride (PVDF) resin, known in the industry as KYNAR 500®.

PVDF finish is strong and durable and is primarily suited for exterior applications. A wide range of colors, including exotic colors, is available. Two types of durability ratings with respect to PVDF finish are one that gives 5 years of performance at the test requirements and another that gives 10 years.

PRACTICE QUIZ

Each question has only one correct answer. Select the choice that best answers the question.

17. Which of the following terms is used with fire-rated windows or doors?
a. Fire-resistance rating
b. Fire-protection rating

18. Fire-rated doors are rated as
a. 3-h, 90-min, 60-min, 45-min, or 20-min.
b. 3-h, 120-min, 60-min, or 30-min.
c. 4-h, 3-h, 2-h, 1-h, or 30-min.
d. 4-h, 3-h, 2-h, or 1-h.
e. none of the above.

19. Which of the following terms describes an intermediate vertical member of a window frame?
a. Mullion
b. Stile
c. Rail
d. Jamb
e. Muntin

20. Which of the following terms describes an intermediate horizontal member of a window frame?
a. Mullion
b. Stile
c. Rail
d. Jamb
e. Muntin

21. Which of the following terms refers to thin and narrow horizontal and vertical members of a window sash?
a. Mullion
b. Stile
c. Rail
d. Jamb
e. Muntin

22. A stool is used in a window in the building's interior.
a. True
b. False

Answers: 17-b, 18-a, 19-a, 20-c, 21-e, 22-a.

KEY TERMS AND CONCEPTS

Apron
Awning window
Casement window
Double-hung window
Double-rabbet door frame
Energy-performance rating of windows
Fixed window
Glazing stop
Hollow-core flush door
Hollow-metal flush door
Hopper window
Jamb extension
Lock stile
Metal door frame
Mullion
Muntin
Nailing flanges
National Fenestration Rating Council (NFRC)
Pivoting window
Rail
Rough opening
Sash
Screw spline
Single-hung window
Single-rabbet door frame
Sliding window
Snap-on joinery
Snap-on retainer clip
Solid core flush door
Stile
Stool
Threshold
Window head
Window performance class
Window performance grade
Window sill
Wood panel door

REVIEW QUESTIONS

1. Sketch the following window styles: (a) casement window, (b) double-hung window, (c) single-hung window, (d) awning window, and (e) hopper window.
2. A window has been rated as HS-C35 71 × 59. Explain each term of the rating.
3 Sketch in plan a hollow metal door frame and its junction with a metal stud wall.

SELECTED WEB SITES

American Architectural Manufacturers Association, AAMA (www.aamanet.org)
Architectural Woodwork Institute, AWI (www.awinet.org)
Door and Hardware Institute, DHI (www.dhi.org)
Hollow Metal Manufacturers Association, HMMA (www.hollowmetal.org)
National Fenestration Rating Council®, NFRC (www.nfrc.org)
Steel Door Institute, SDI (www.steeldoor.org)
Window & Door Manufacturers Association, WDMA (www.wdma.com)
Woodwork Institute, WI (www.wicnet.org)

CURB
LANE
BUSES
ONLY

CHAPTER 30 Glass-Aluminum Wall Systems

CHAPTER OUTLINE

30.1 Glass-Aluminum Curtain Walls

30.2 Anchorage of a Stick-Built Glass Curtain Wall to a Structure

30.3 Stick-Built Glass Curtain Wall Details

30.4 Structural Performance of a Glass Curtain Wall

30.5 Other Performance Criteria of a Glass Curtain Wall

30.6 Nontraditional Glass Curtain Walls

30.7 Other Glass-Aluminum Wall Systems

Double-skin glass curtain wall at the Seattle Justice Center, Seattle, Washington. NBBJ Architects. (Photo courtesy of Tisha Egashira.)

A glass-aluminum wall system is a special type of exterior wall in which the wall consists of glass panes (also called *lites*) held within vertical and horizontal aluminum framing members. It shares some of the characteristics with its smaller counterpart—the aluminum windows discussed in Chapter 29. However, there are many differences between the two: scale, aesthetic character, performance properties, design, detailing, and installation.

Three commonly used glass-aluminum wall system systems are

- Glass-aluminum curtain walls
- Punched and strip glazing systems
- Storefront systems

A vast majority of contemporary buildings include one or more of these systems in the same building. The reasons include the unparalleled opportunity provided by them to obtain the maximum amount of daylight and view, the economy of cost as compared to other exterior wall cladding systems, and the recent technological advances in the thermal and structural performance of glass wall systems.

Of these three systems, the most used and the most complex is the glass-aluminum curtain wall system, which is presented in detail. The other two systems (strip system and storefronts) are discussed to the extent that they differ from the curtain wall system.

30.1 GLASS-ALUMINUM CURTAIN WALLS

Because of their increasing use, glass-aluminum curtain walls (or simply *glass curtain walls*) are constantly evolving in their design and performance. Therefore, a succinct classification that includes all contemporary glass curtain walls is impossible. The American Architectural Manufacturers Association (AAMA), an association of the manufacturers of windows and curtain walls, however, classifies glass curtain wall systems into five types based on their anatomy:

- Stick-built (or, simply, stick) systems
- Unitized systems
- Unit and mullion systems
- Panel systems
- Column cover and spandrel systems

These systems are illustrated in Figure 30.1. The stick system is the oldest and the most widely used system. The remaining four systems are different from the stick system because they consist of prefabricated wall units, similar to the (opaque) curtain wall panels.

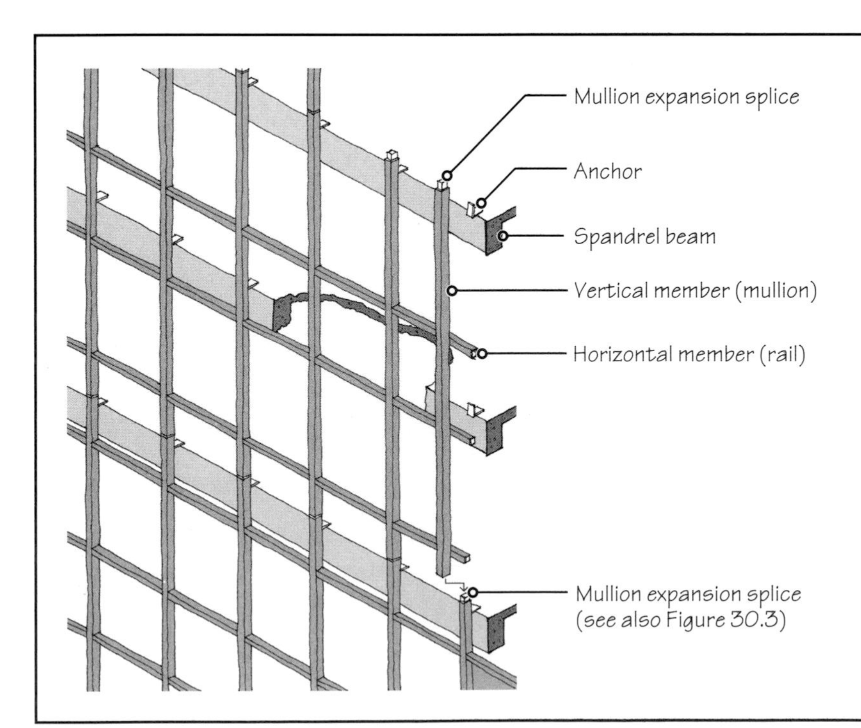

(a) STICK SYSTEM

In the stick system, the curtain wall is installed piece by piece. Generally, the mullions are installed first, followed by the rails. Subsequently the glass panes are installed within the mullion-rail framework.

The anchorage of the wall to the structural frame is through the mullions. The mullions may span from floor to floor or over two floors. Thermal expansion and contraction of mullions is accommodated through expansion joints in mullions.

The system components are shop-fabricated and shipped to the construction site in knocked down (KD) version. Therefore, it has relatively low shipping costs and also permits a greater degree of on-site adjustment, as compared to the other systems.

Its disadvantage include longer on-site assembly time and more on-site labor than the other systems.

FIGURE 30.1 Types of glass curtain walls—the stick-built system. (Illustration adapted from AAMA, *Curtain Wall Design Guide*, 1996, with permission.)

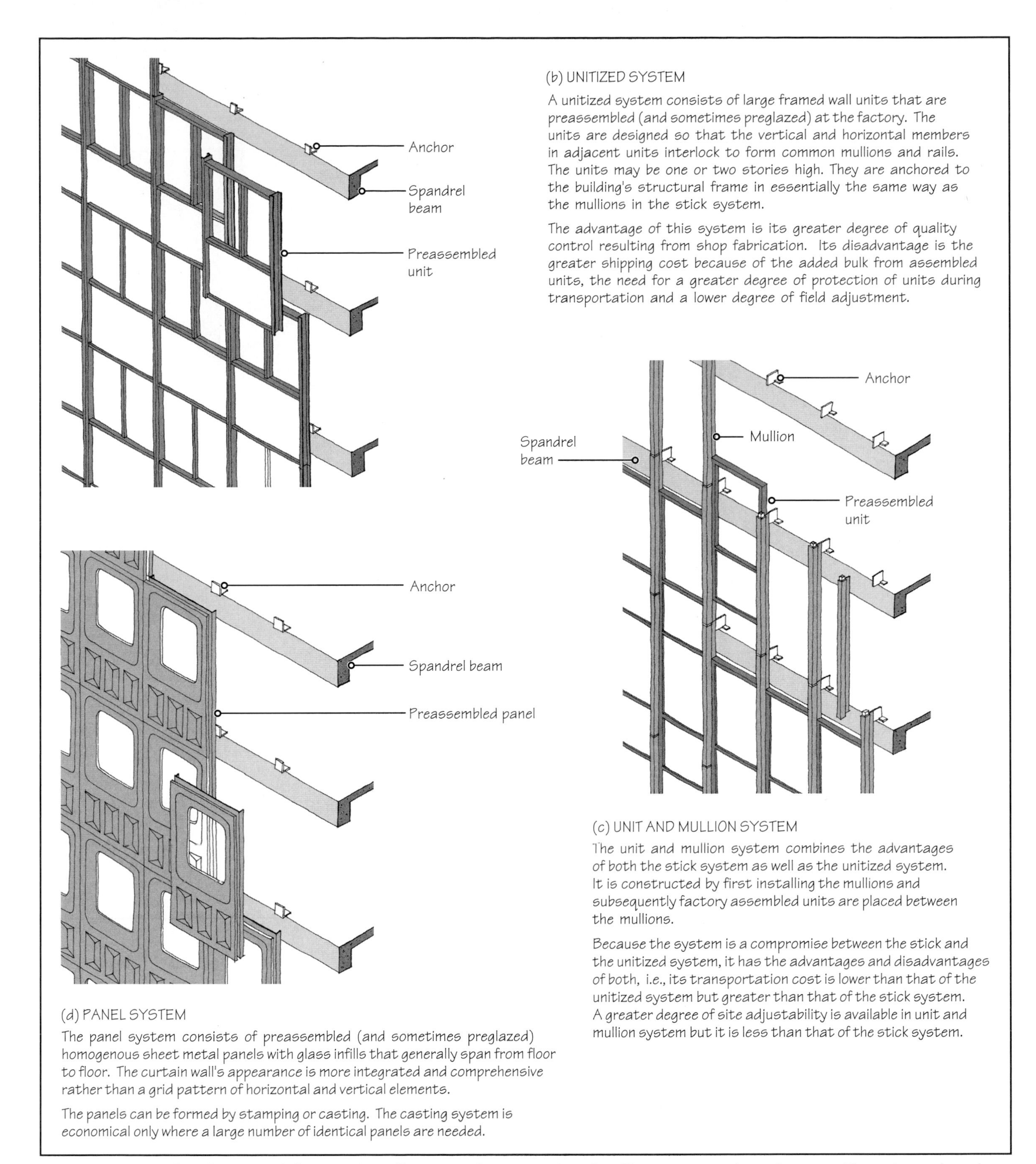

FIGURE 30.1 (***Continued***) Types of glass curtain walls—unitized system, unit and mullion system, and panel system. (Illustrations adapted from AAMA, *Curtain Wall Design Guide*, 1996, with permission.)

Standard and Custom Wall Systems

Most major glass curtain wall manufacturers have their own facility for extruding the aluminum sections of commonly used curtain wall facades. Walls constructed from a manufacturer's standard off-the-shelf aluminum sections are referred to as *standard walls.*

Custom curtain walls utilize cross-sectional shapes extruded specifically for a project in response to an architect's design. Because the cost of dies and other equipment required to extrude custom cross sections can be recovered from just one fair-size project, custom curtain

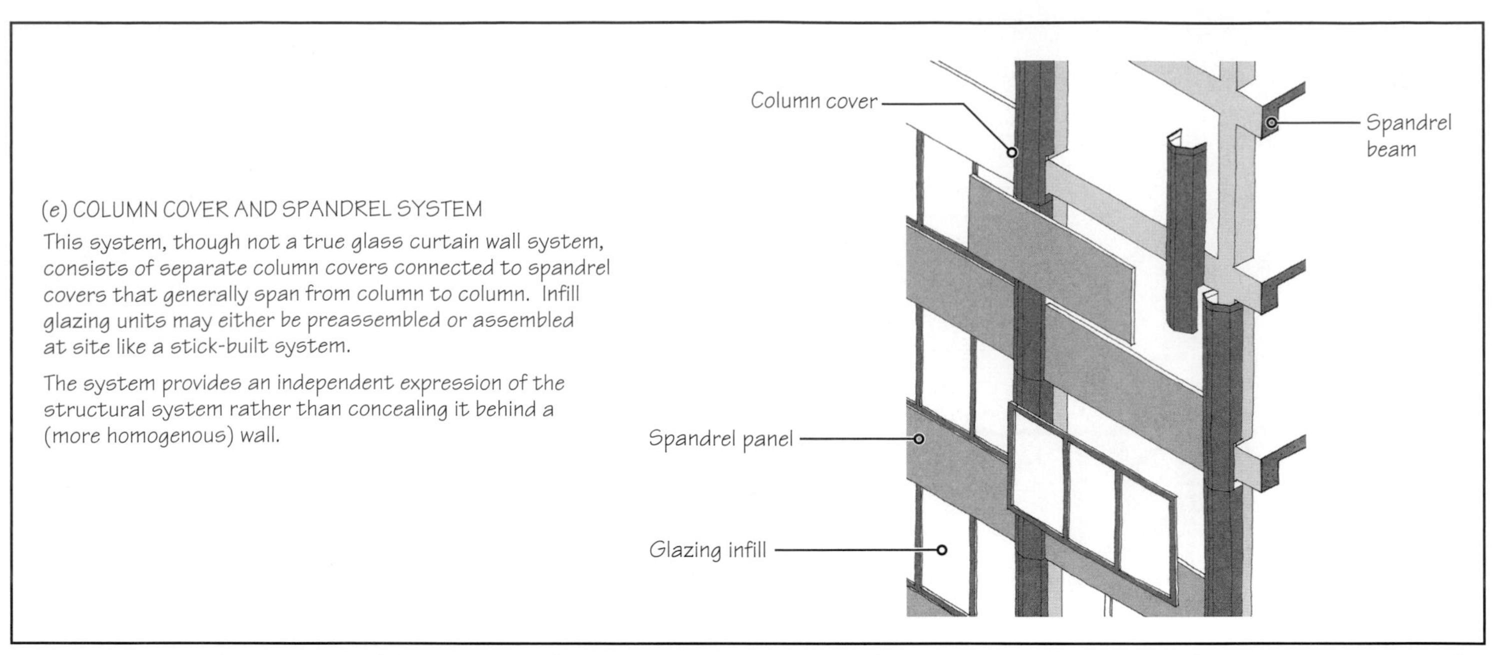

FIGURE 30.1 (***Continued***) Types of glass curtain walls—column cover and spandrel system. (Illustrations adapted from AAMA, *Curtain Wall Design Guide*, 1996, with permission.)

walls are fairly common. Custom walls should, however, be tested for performance before they are used in a project. Performance data for standard walls are available from the manufacturers.

Walls made from standard components are obviously more economical. However, this does not imply that the standard components yield only one type of wall design. In fact, the components are generally quite adaptable, and the manufacturers can provide a few custom components for a standard system, so that the facade expressions obtained from the use of standard components can be numerous. If the number of custom components in a wall becomes excessive, the cost of a standard wall may approach that of a custom wall.

30.2 ANCHORAGE OF A STICK-BUILT GLASS CURTAIN WALL TO A STRUCTURE

As with other curtain walls, a glass curtain wall must be spaced away from the building's structural frame to account for the slight dimensional variations (within the allowed tolerances) in the structural frame. A 2-in. space is generally the minimum requirement. A wider space may be required for tall buildings.

Dead-Load Anchors and Expansion Anchors

As shown in Figure 30.1(a), a stick-built glass curtain wall consists of vertical members (*mullions*) and horizontal members (*rails*). The profiles of both mullions and rails are almost identical and are tubular in cross section.

The wall is anchored to the building's structural frame through the mullions. All mullions in a wall are installed first; then the rails are inserted between them. Three rails are commonly used per floor to create two separate areas of glass at each floor—vision glass and spandrel glass.

In a building, (or part of a building) where there is no vision glass, such as in a multistory parking garage, intermediate rails are needed only to reduce the size of glass panes. Two rails per floor are commonly used in that situation, Figure 30.2. The center-to-center spacing between mullions is generally 4 to 6 ft, depending on the lateral load intensity and the desired appearance of the facade.

To allow for the expansion and contraction of mullions caused by temperature changes, each mullion must be provided with expansion joints. Thus, the mullions consist of short lengths (one or two floors tall) that terminate in expansion joints at both ends, Figure 30.3.

An expansion joint also absorbs the creep in concrete columns and the live-load deflection of the spandrel beam to which the mullions are anchored. Therefore, the expansion

FIGURE 30.2(a) The progress in the installation of a stick-built glass curtain wall on an office building in which the lower floors are parking floors and the upper floors are office floors. See also Figure 30.2(b).

FIGURE 30.2(b) The progress in the installation of a stick-built glass curtain wall on an office building in which the lower floors are parking floors and the upper floors are office floors. See also Figure 30.2(a).

joint width must be determined on a project-by-project basis. Note that an expansion joint allows movement in the vertical direction only.

Because all loads on a wall are transferred to the structural frame through the mullions, each mullion is provided with a dead-load support anchor (or, simply, a *DL anchor*) designed to carry the weight of the respective portion of the curtain wall.

A DL anchor fully restrains the movement of a mullion; that is, the mullion is immobile in all three principal directions at a dead-load support. Therefore, a DL anchor transfers both dead loads and the lateral loads on a mullion to the building's structural frame.

Two types of mullion spans are generally used in a stick-built glass curtain wall, Figure 30.4:

- Single-span mullion systems
- Twin-span mullion systems

In a single-span mullion system, each mullion extends only over one floor. DL anchors are, therefore, required at every floor, except at the ground floor, where the building's foundation provides dead-load support to the first mullion length, Figure 30.4(a).

In a twin-span mullion system, the mullions extend over two floors. Because a mullion can have only one dead-load support, DL anchors are provided at alternate floors, Figure 30.4(b). Another difference between a single-span and a twin-span system is that in a

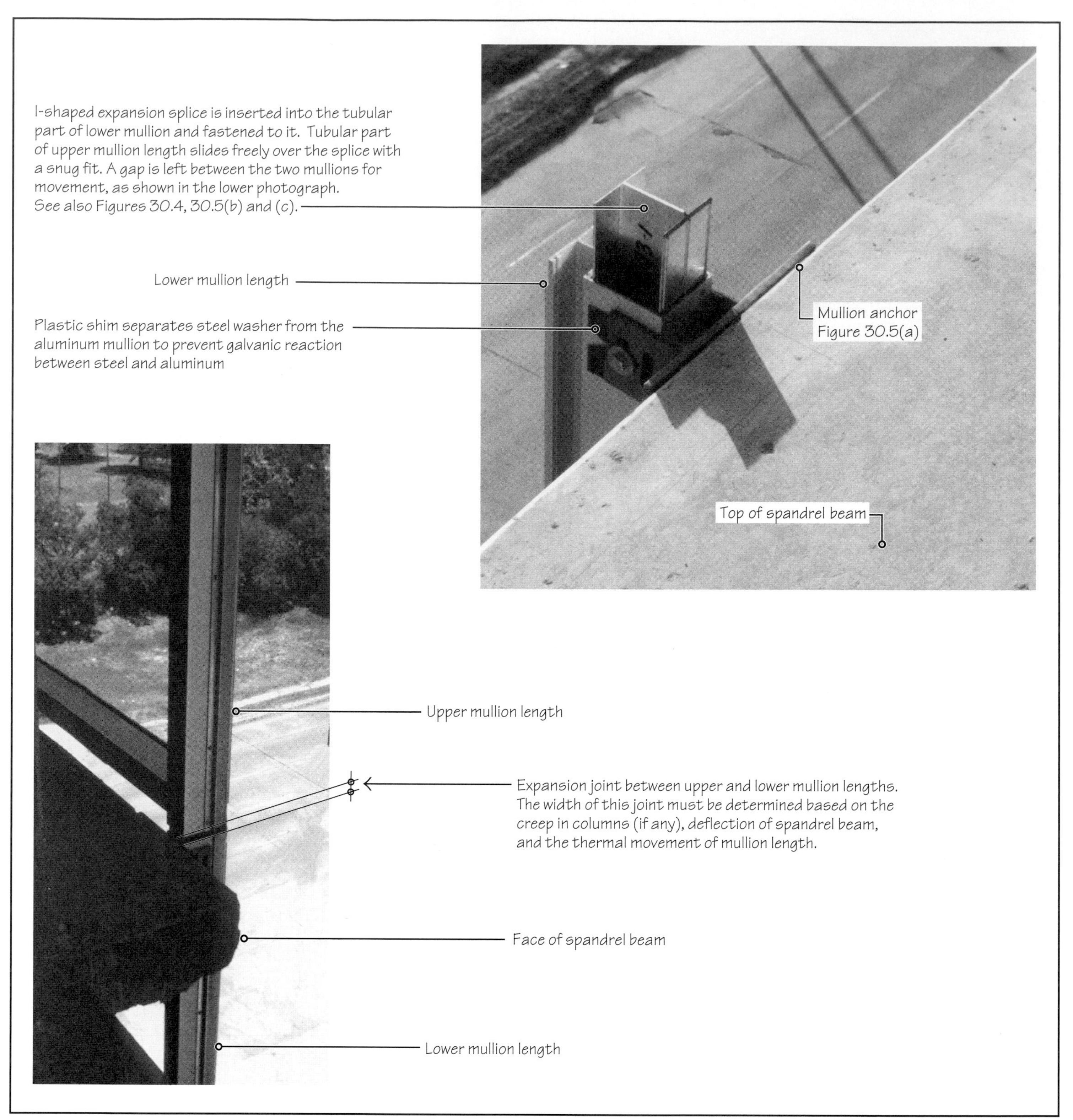

FIGURE 30.3 A typical expansion joint between two mullion lengths.

twin-span system, expansion anchors (or, simply, *EX anchors*) are also required at alternate floors. In a single-span system, an EX anchor is required only at the second floor of the building.

DL anchors and EX anchors are steel members to which the mullions are bolted. As shown in Figure 30.4(b), they are almost identical. The only difference between them is that in a DL anchor, the upper pair of holes is round, and in an EX anchor, the upper pair of holes has vertically slotted holes that allow vertical movement.

Anchoring a Mullion to a DL or EX Anchor

Figure 30.5 shows the anchorage details of a mullion to a DL anchor and an EX anchor. Anchoring a mullion to a DL anchor (or an EX anchor) is a two-step process. The first step includes providing a temporary connection between the mullion and the anchor,

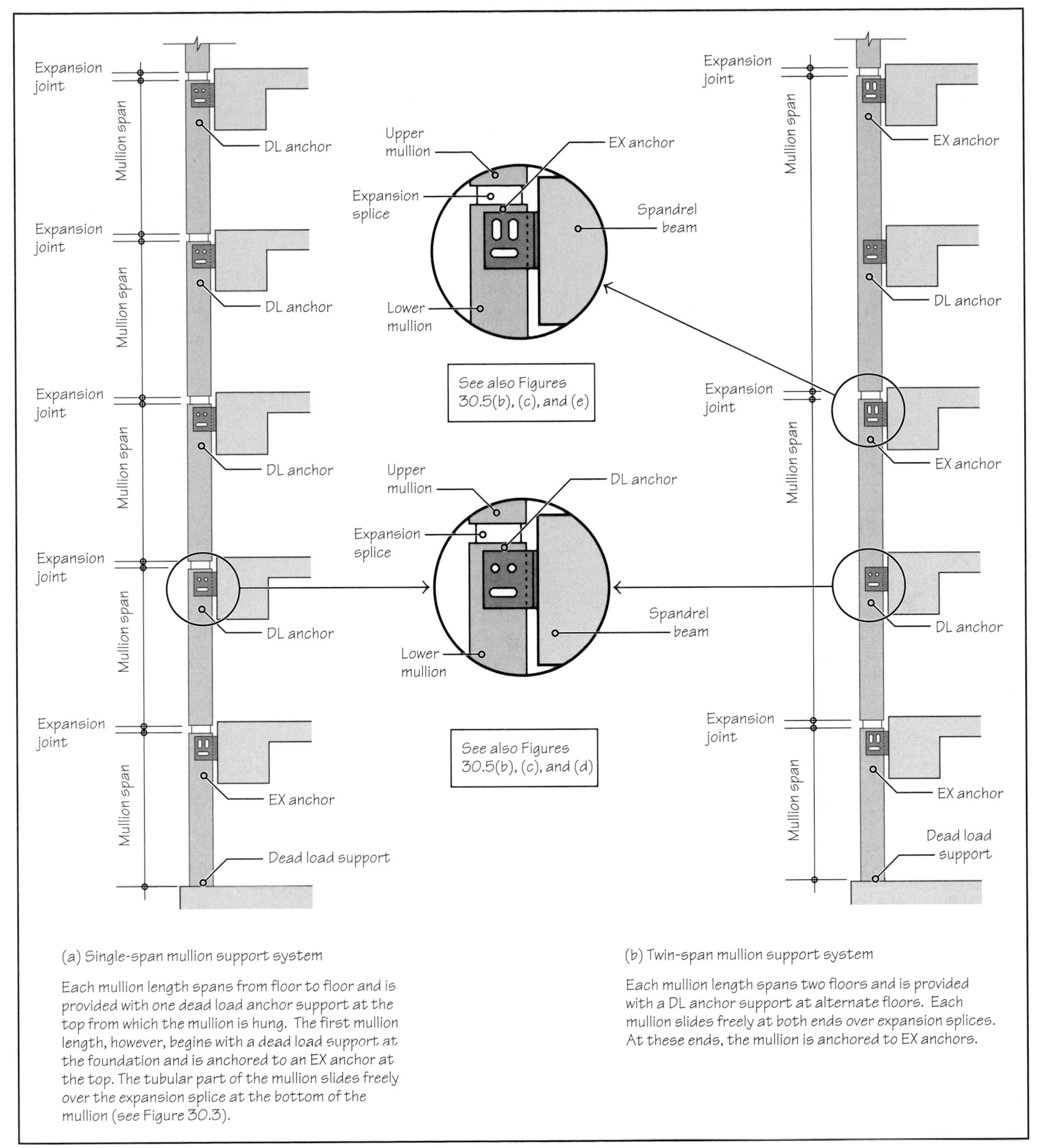

FIGURE 30.4 Support systems for single-span and twin-span curtain wall mullions. Observe that each mullion has only one dead-load support.

Figure 30.5(b). After all mullion lengths are correctly aligned, permanent connections between the mullion and the anchors is made.

A permanent connection requires field drilling into the mullion through predrilled holes in the anchors, Figures 30.5(c), (d), and (e). Predrilled holes in anchors provide for field adjustment to cater to the (allowed) dimensional variations in the structural frame of the building.

As with other curtain walls (precast concrete, GFRC, natural stone, etc.), the anchorage system of a glass curtain wall to the building's structure is typically provided by the curtain wall's supplier.

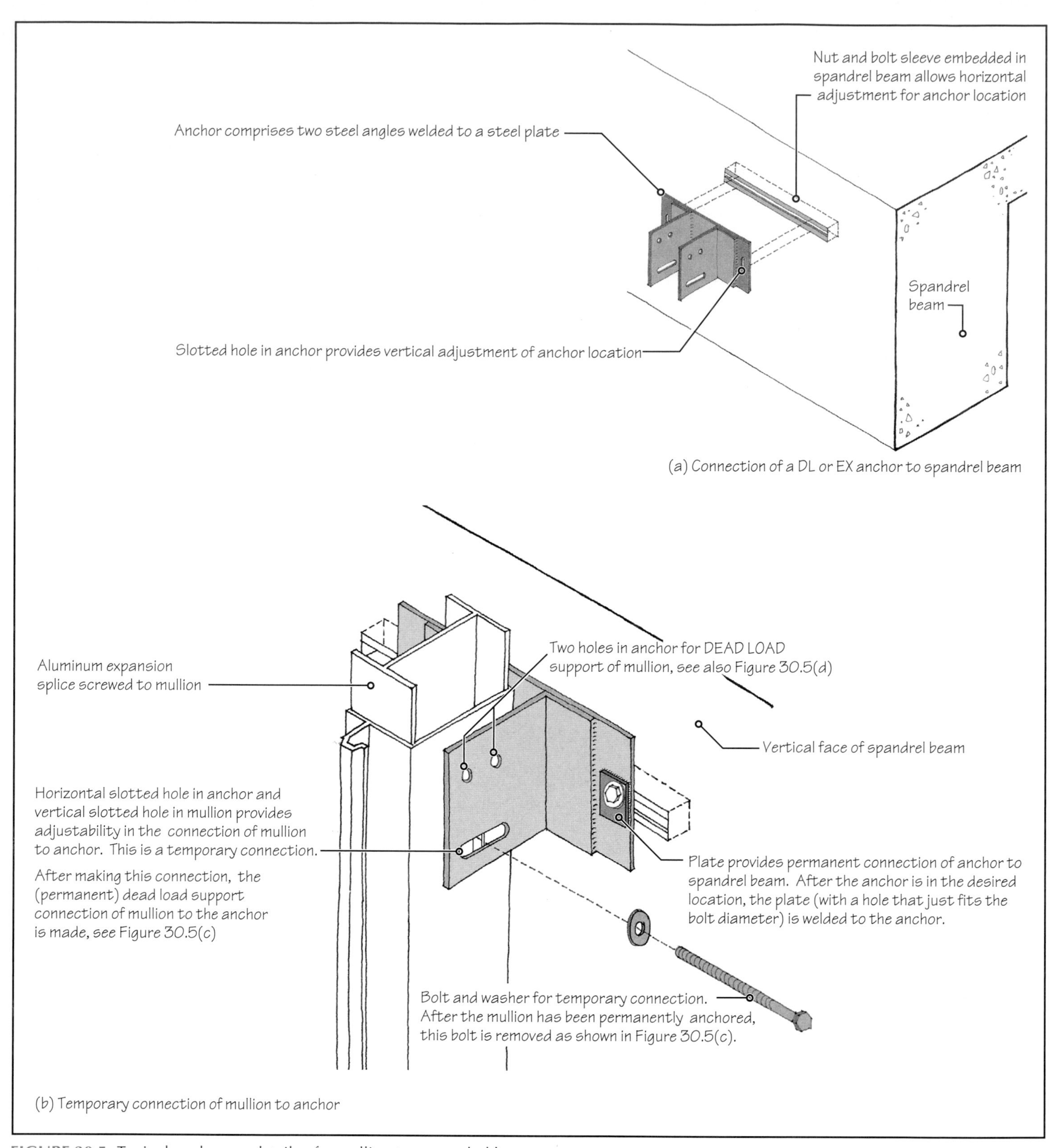

FIGURE 30.5 Typical anchorage details of a mullion to a spandrel beam.

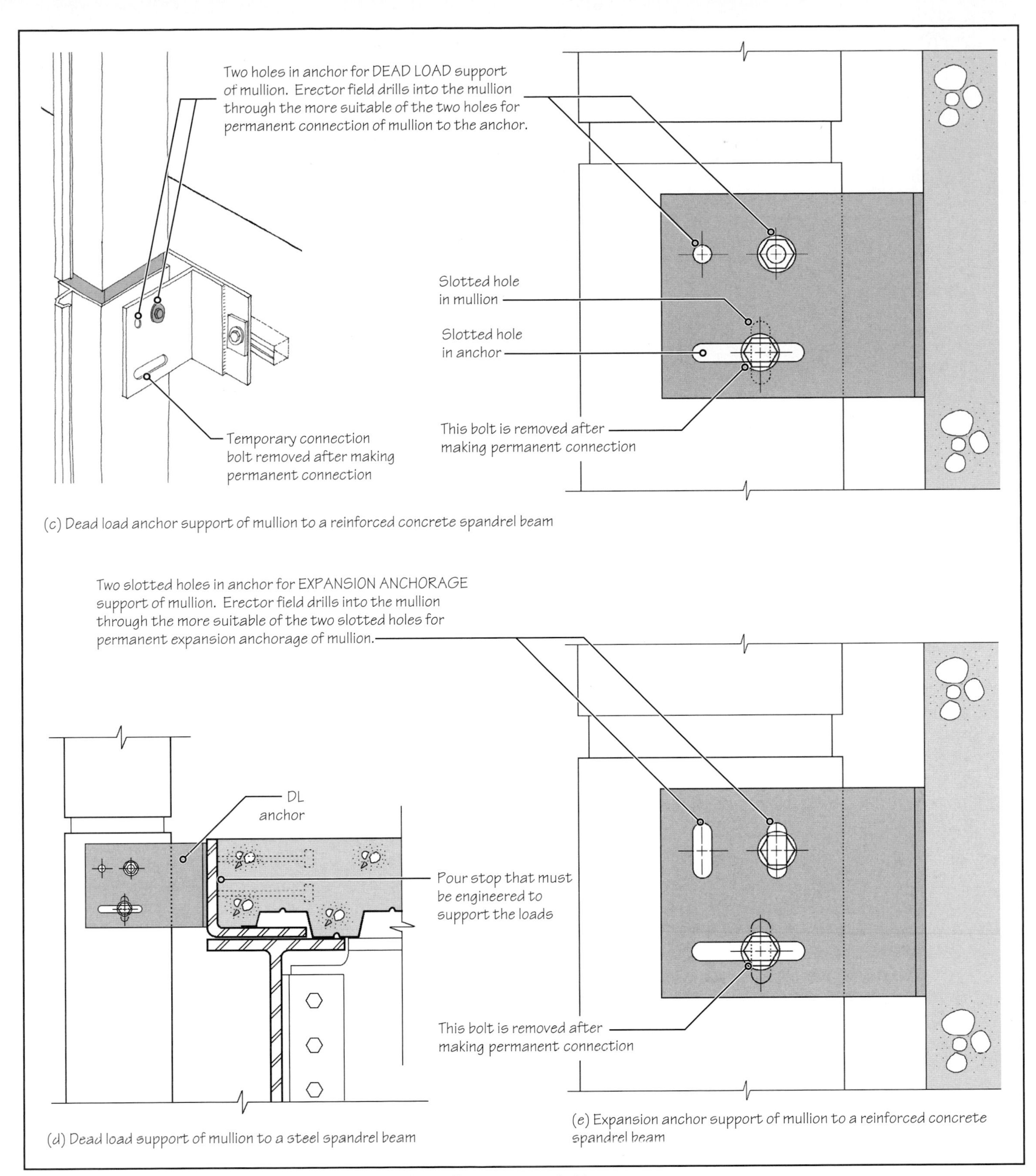

FIGURE 30.5 (*Continued*) Typical anchorage details of a mullion to a spandrel beam.

PRACTICE QUIZ

Each question has only one correct answer. Select the choice that best answers the question.

1. A stick-built glass curtain wall consists of
- **a.** mullions and glass.
- **b.** rails and glass.
- **c.** mullions, rails, and glass.
- **d.** mullions, rails, and preassembled units.
- **e.** preassembled units and glass.

2. A unitized glass curtain wall consists of
- **a.** mullions and glass.
- **b.** rails and glass.
- **c.** mullions, rails, and glass.
- **d.** mullions, rails, and preassembled units.
- **e.** preassembled units and glass.

3. Glass curtain walls in which aluminum framing sections are specially profiled for a particular project are
- **a.** rare because of the prohibitive cost of manufacturing custom profiles.
- **b.** uncommon because of the extremely high cost of manufacturing custom profiles.
- **c.** not uncommon because the cost of custom profiles can be recovered from a few repeat mid- to large-size projects.
- **d.** fairly common because the cost of custom profiles can be recovered from one large-size project.

4. In a stick-built glass curtain wall, the mullions are typically spaced at
- **a.** 2 ft to 4 ft o.c.
- **b.** 4 ft to 6 ft o.c.
- **c.** 6 ft to 10 ft o.c.
- **d.** 10 ft to 15 ft o.c.
- **e.** as needed for the project.

5. In a stick-built glass curtain wall, the rails are typically spaced at
- **a.** 2 ft to 4 ft o.c.
- **b.** 4 ft to 6 ft o.c.
- **c.** 6 ft to 10 ft o.c.
- **d.** 10 ft to 15 ft o.c.
- **e.** as needed for the project.

6. A stick-built glass curtain wall is anchored to the building's structure through
- **a.** mullions.
- **b.** rails.
- **c.** both mullions and rails.
- **d.** none of the above.

7. In a single-span mullion system for a glass curtain wall,
- **a.** one dead-load anchor is provided at every floor level.
- **b.** two dead-load anchors are provided at every floor level.
- **c.** one dead-load anchor is provided at every alternate floor level.
- **d.** two dead-load anchors are provided at every alternate floor level.
- **e.** only one dead-load anchor is provided for the entire height of the wall.

8. In a twin-span mullion system for a glass curtain wall,
- **a.** one dead-load anchor is provided at every floor level.
- **b.** two dead-load anchors are provided at every floor level.
- **c.** one dead-load anchor is provided at every alternate floor level.
- **d.** two dead-load anchors are provided at every alternate floor level.
- **e.** only one dead-load anchor is provided for the entire height of the wall.

9. In a single-span mullion system for a glass curtain wall,
- **a.** one expansion anchor is provided at every floor level.
- **b.** two expansion anchors are provided at every floor level.
- **c.** one expansion anchor is provided at every alternate floor level.
- **d.** two expansion anchors are provided at every alternate floor level.
- **e.** only one expansion anchor is provided for the entire height of the wall.

10. In a twin-span mullion system for a glass curtain wall,
- **a.** one expansion anchor is provided at every floor level.
- **b.** two expansion anchors are provided at every floor level.
- **c.** one expansion anchor is provided at every alternate floor level.
- **d.** two expansion anchors are provided at every alternate floor level.
- **e.** only one expansion anchor is provided for the entire height of the wall.

11. The width of an expansion joint between adjacent mullion lengths in a typical stick-built glass curtain wall is generally
- **a.** 1 in. standard.
- **b.** $\frac{1}{2}$ in. standard.
- **c.** $\frac{1}{4}$ in. standard.
- **d.** $\frac{1}{16}$ in. standard.
- **e.** must be determined on project-by-project basis.

Answers: 1-c, 2-e, 3-d, 4-b, 5-e, 6-a, 7-a, 8-c, 9-e, 10-c, 11-e.

30.3 STICK-BUILT GLASS CURTAIN WALL DETAILS

After the mullions have been anchored to the structural frame, the remaining items in the wall's erection require

- Connecting the rails to the mullions
- Installing the glass.

RAIL-TO-MULLION CONNECTION

Manufacturers use various methods to connect the rails to the mullions. A commonly used method involves short aluminum extrusions, called *shear blocks,* which are fastened to the mullions with screws. Subsequently, the rails are snapped over the shear blocks, one shear block at each end of a rail, Figure 30.6. Thus, no fasteners are used between the rail and the shear blocks.

Because the length of each rail is small (4 ft to 6 ft is typical), a fairly small space is required for the expansion or contraction of a rail. In general, the length of a rail is $\frac{1}{16}$ in. shorter than the clear distance between mullions.

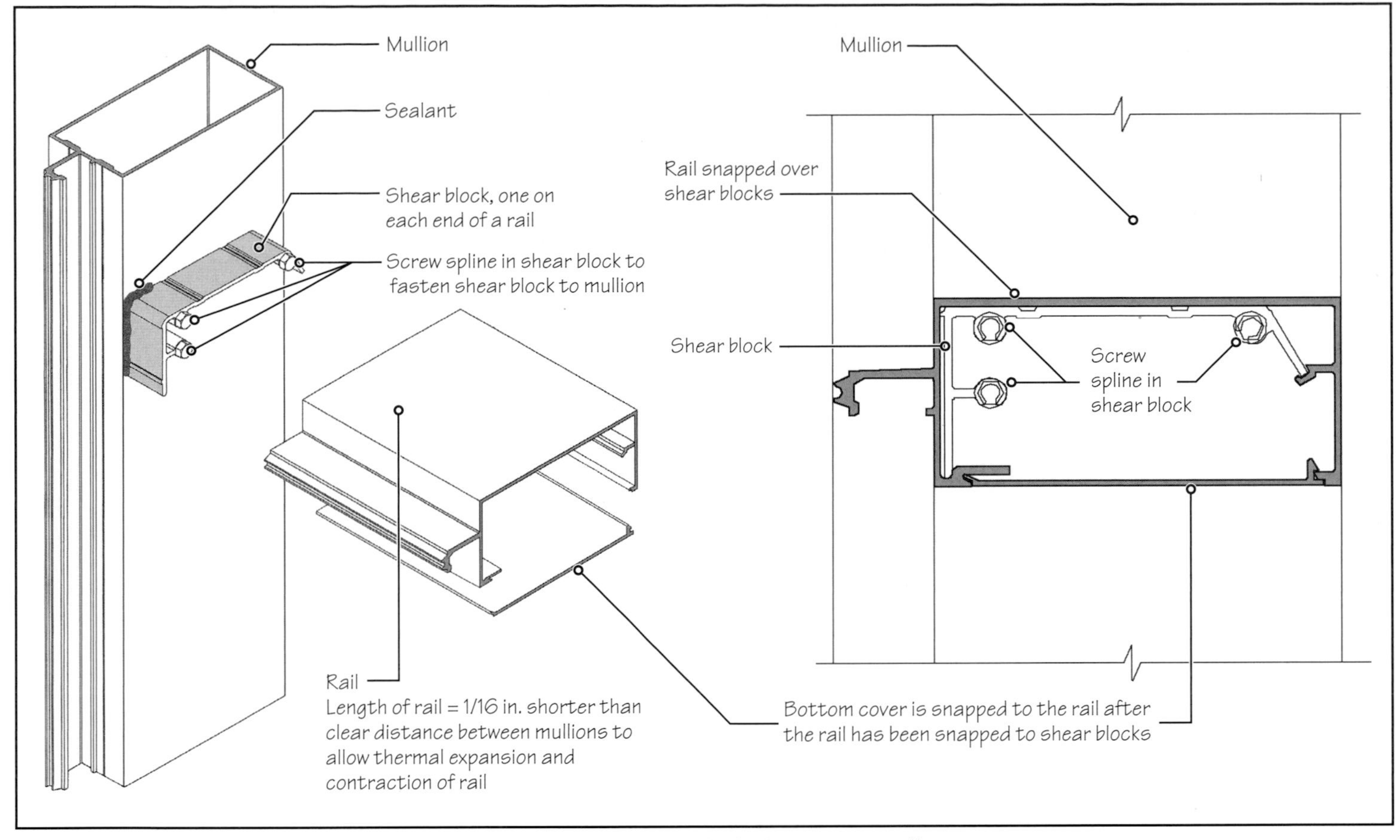

FIGURE 30.6 Typical connection between rails and mullions of a stick-built glass curtain wall.

Outside-Glazed and Inside-Glazed Curtain Walls

One of the factors that determines the cross-sectional shapes of mullions and rails is whether the glass in the wall is to be installed from the outside or the inside of the building, referred to, respectively, as

- Outside-glazed curtain walls
- Inside-glazed curtain walls

In an *outside-glazed* wall, the glass panes are installed from the outside of the building by workers standing on a scaffold or staging. This method of installing glass is less efficient and more expensive due to the cost of scaffolding or staging. It is generally used for low- to midrise buildings. The glass in an outside-glazed wall can be secured in two ways:

- Pressure plate–captured glass (Figures 30.7 to 30.9)
- Structural silicone sealant–adhered glass (Figure 30.10)

In an *inside-glazed* wall, the glass is installed by workers standing on the appropriate floor of the building. The system is more efficient because it does not require scaffolding or staging. It is the system of choice for high-rise buildings. However, the cross-sectional shapes of mullions and rails for the inside-glazed system are relatively more complex than the corresponding shapes for the outside-glazed system.

Outside-Glazed Walls (Pressure Plate–Captured Glass)

In an outside-glazed curtain wall, the glass is held by horizontal and vertical pressure plates, which are fastened to the mullions and rails with screws. A plastic insert is used between the pressure plate and the mullion (or the rail), which functions as a thermal separator. The pressure plates are finally covered with snap-on covers, Figure 30.7.

Because the covers are the only externally visible part of the curtain wall frame, they have a major influence on the curtain wall's appearance. The covers can be profiled into various shapes, Figure 30.8.

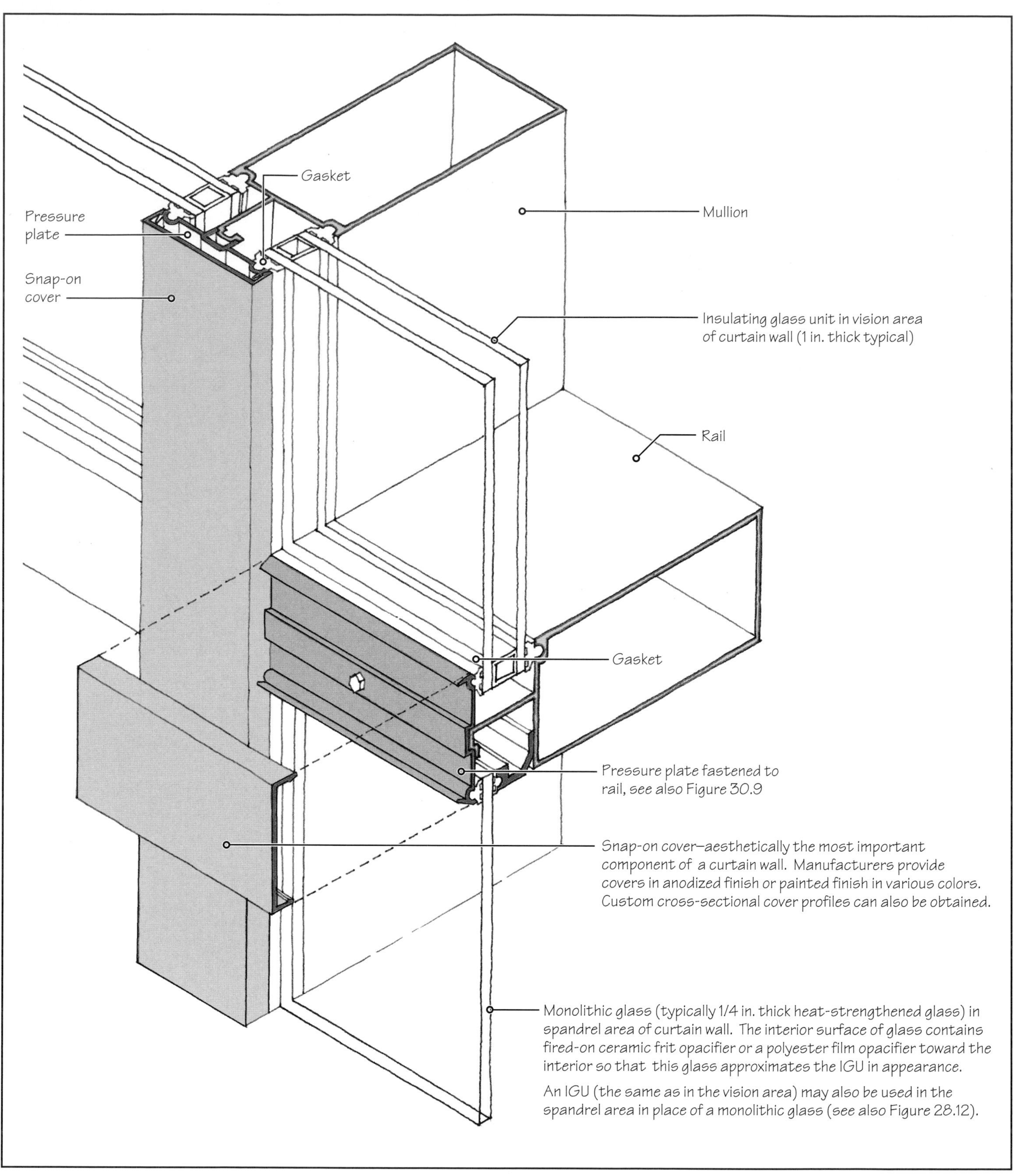

FIGURE 30.7(a) Anatomy of an outside-glazed glass curtain wall (pressure plate–captured glass).

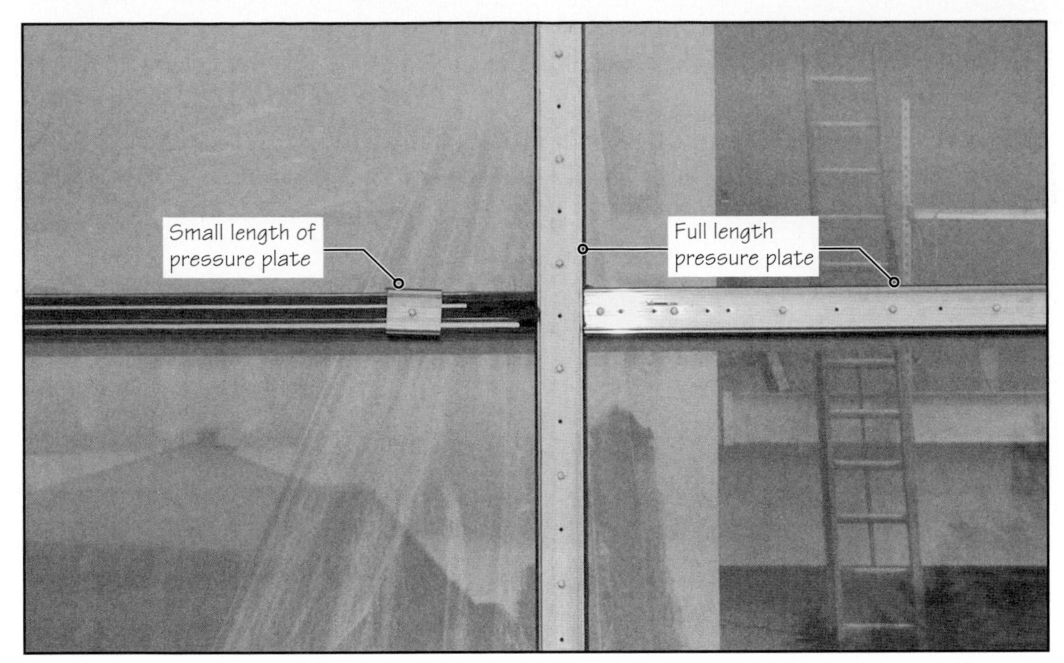

FIGURE 30.7(b) In fastening the pressure plates to curtain wall framing members, the glass is temporarily held by small pressure plate members. After several glass panes are in position, the temporary pressure plates are removed and replaced by full-length pressure plates.

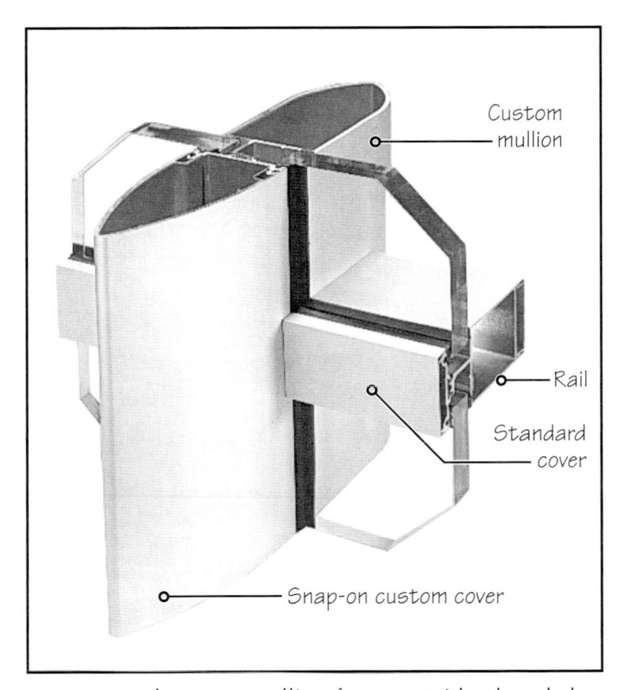

FIGURE 30.8 A custom cover and custom mullion for an outside-glazed glass curtain wall. (Photo courtesy of Vistawall Architectural Products.)

The exterior and interior gaskets should prevent water from leaking through the wall. However, a curtain wall system typically includes accommodations for the drainage of water, should it penetrate beyond the gaskets. This is accomplished through drainage weep holes in the pressure plates and the covers. Thus, in a typical curtain wall, each glass-pane frame is drained independently. Figure 30.9 shows typical sections through a pressure plate–captured outside-glazed curtain wall.

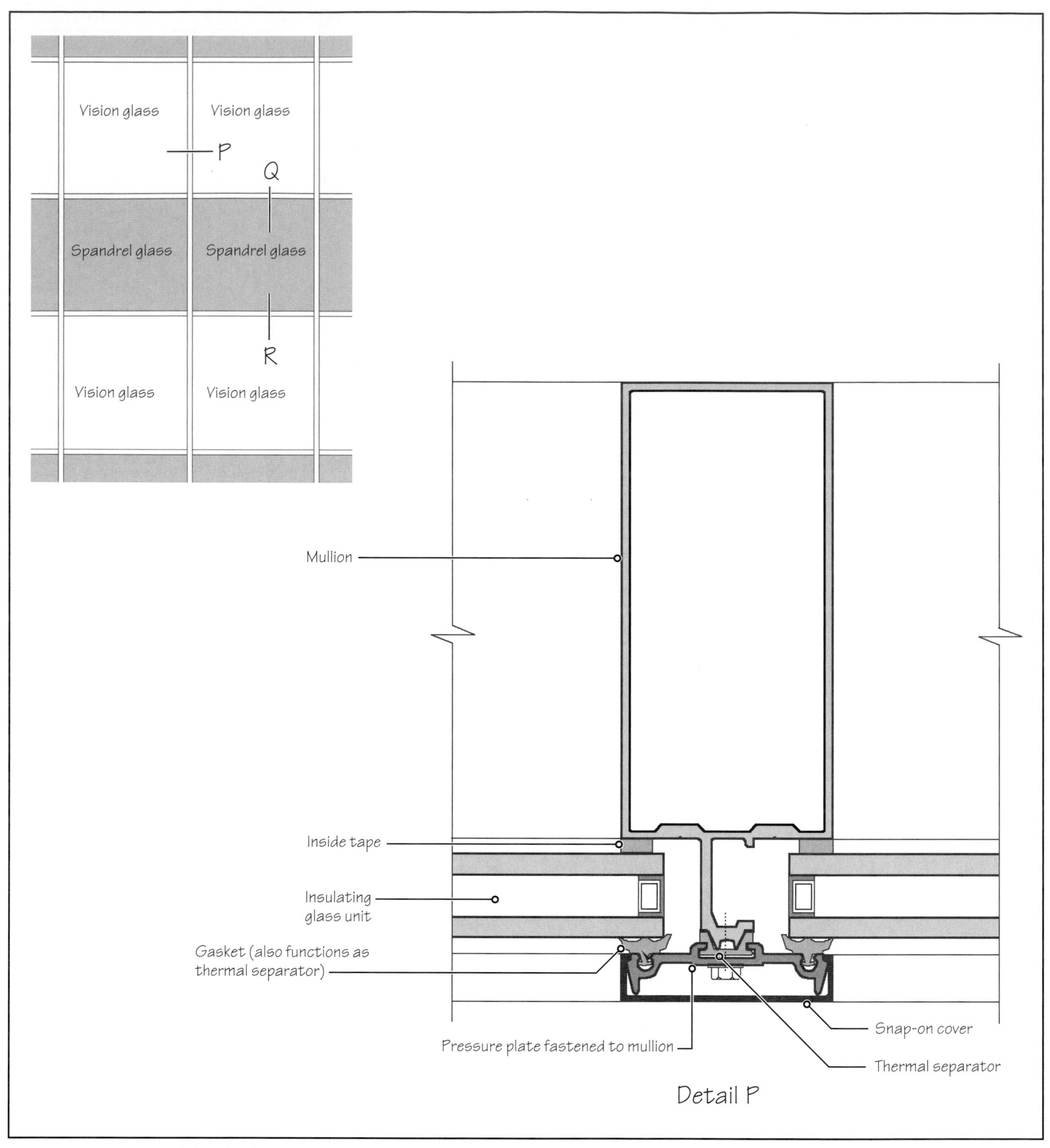

FIGURE 30.9(a) Typical details of an outside-glazed glass curtain wall (pressure plate–captured glass). Aluminum sections used in the details are by Vistawall Architectural Products. Other manufacturers provide similar sections.

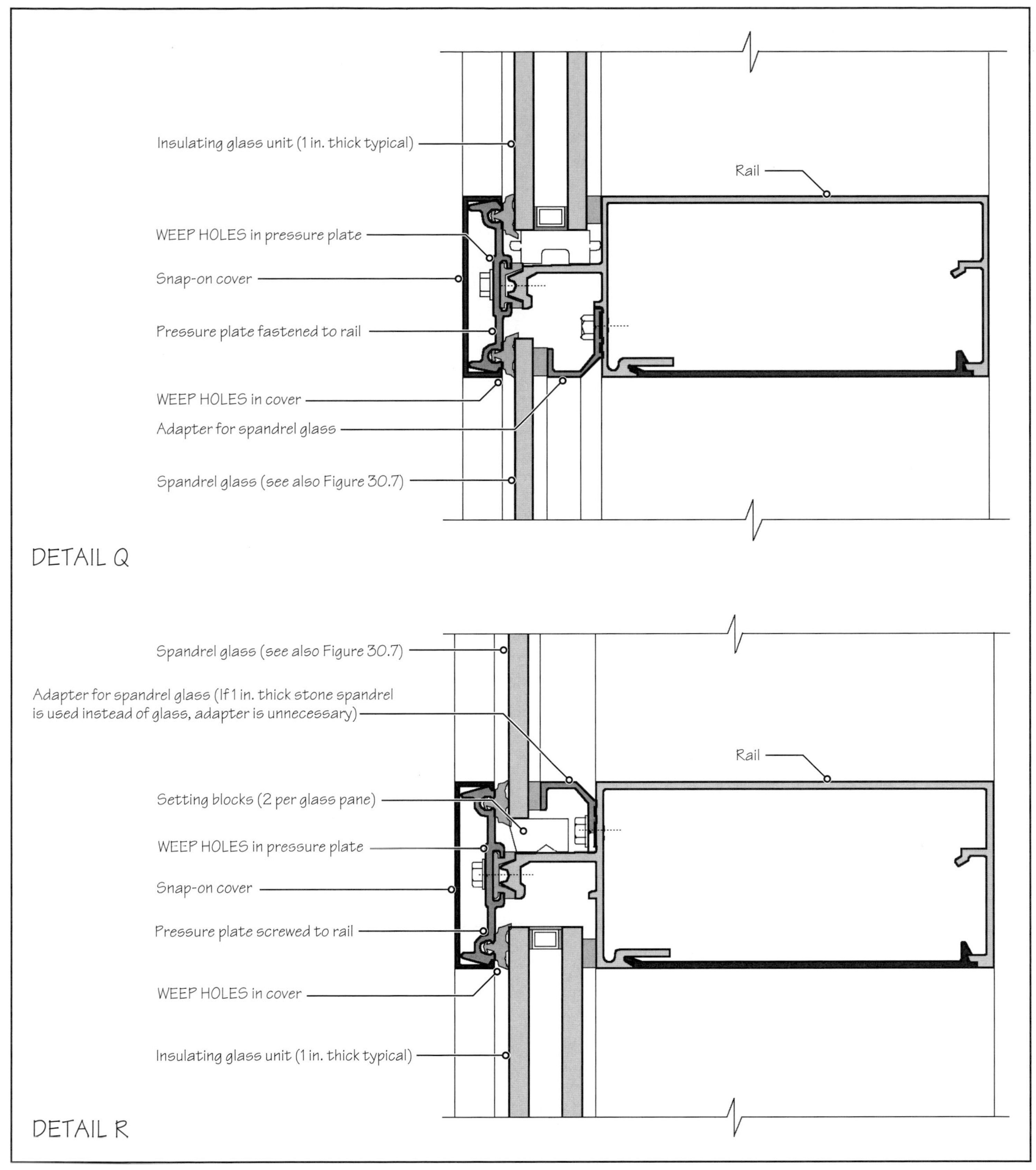

FIGURE 30.9(b) Typical details of an outside-glazed glass curtain wall (pressure plate–captured glass). Aluminum sections used in the details are by Vistawall Architectural Products. Other manufacturers provide similar sections.

Outside-Glazed Walls (Structural Silicone Sealant–Adhered Glass)

Another version of an outside-glazed curtain wall is one in which the glass is held by structural silicone sealant. In this type of system, the vertical edges of a glass pane are adhered to

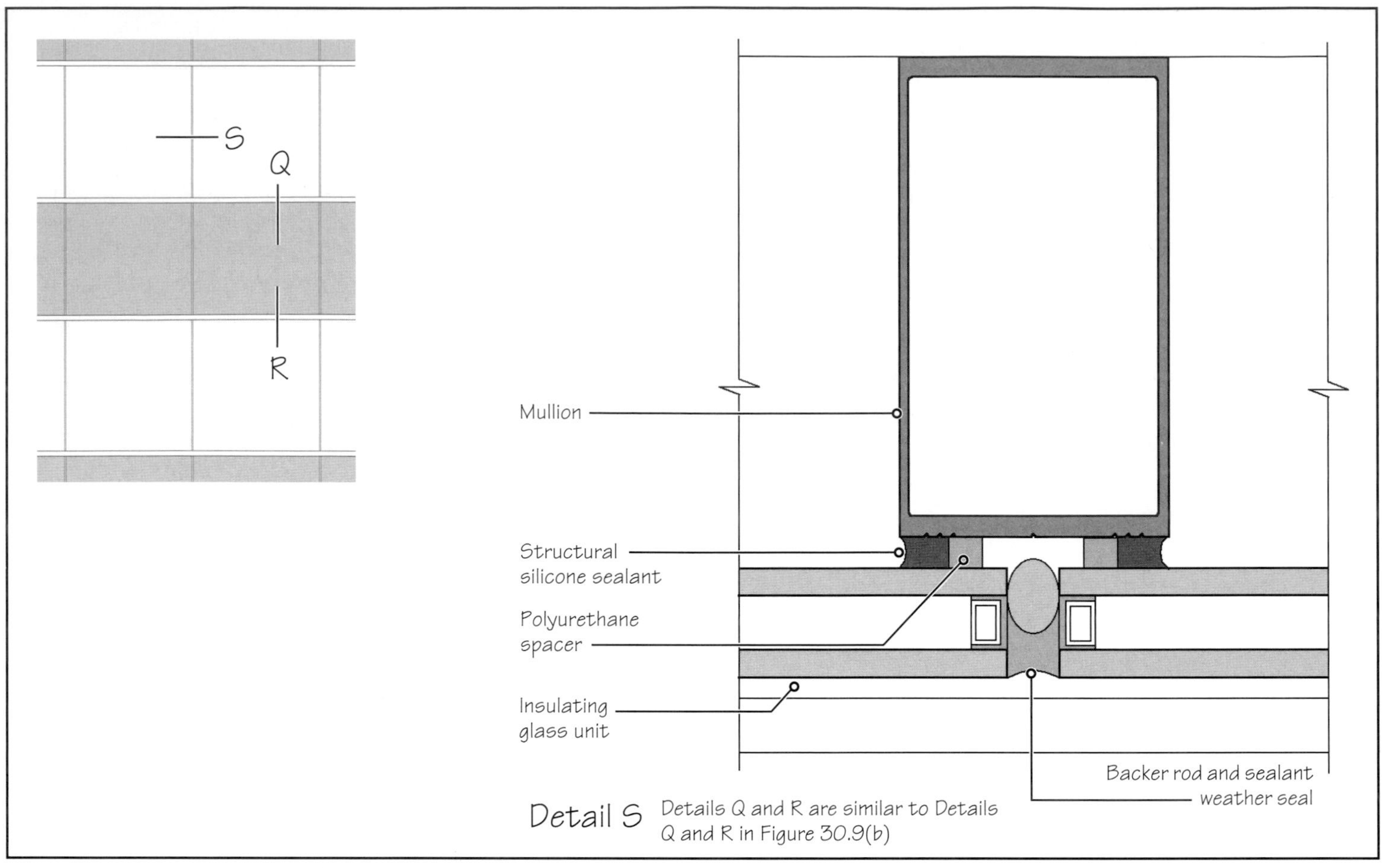

FIGURE 30.10 Typical details of an outside-glazed curtain wall (structural silicone sealant–adhered glass). Aluminum section used in the detail is by Vistawall Architectural Products. Other manufacturers provide a similar section.

the mullions with beads of structural silicone sealant. The mullions in this wall are similar to those of an outside-glazed wall without the mullion nose. The horizontal edges of the glass are supported on rails and anchored to them through standard pressure plates, Figure 30.10. The absence of vertical pressure plates in the system accentuates the horizontality of the covers.

INSIDE-GLAZED WALLS

In an inside-glazed wall, pressure plates are not used. Instead, the mullions and rails include glazing pockets on three sides (on both vertical sides and at the bottom of a glass pane), in which the glass is inserted from inside the building, Figure 30.11. After the glass has been placed in the correct position, a snap-on glazing stop at the top secures it to the frame.

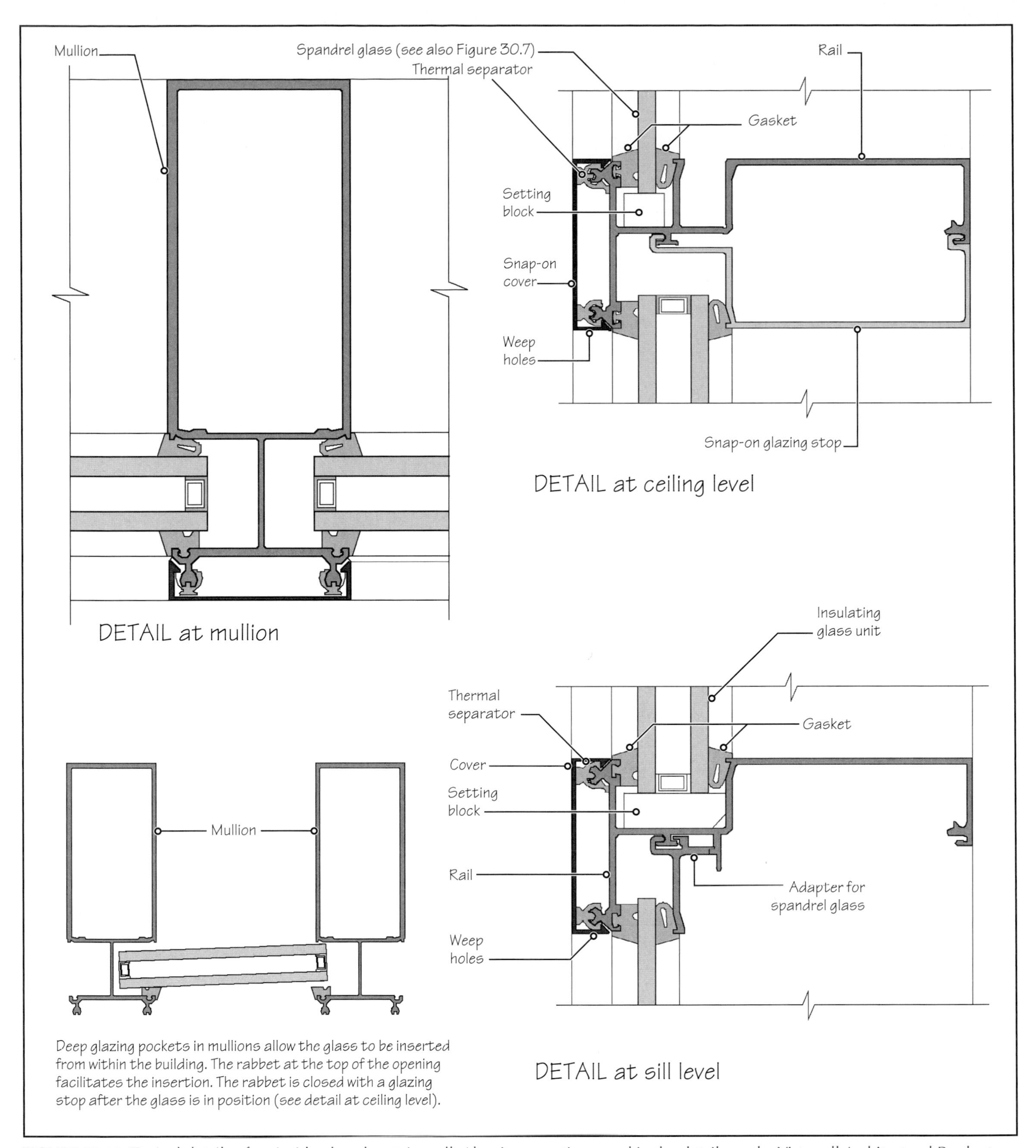

FIGURE 30.11 Typical details of an inside-glazed curtain wall. Aluminum sections used in the details are by Vistawall Architectural Products. Other manufacturers provide similar sections.

Detailing a Multistory Glass Curtain Wall

Figure 30.12 shows the details of a typical multistory (outside-glazed) glass curtain wall at the sill, floor, and ceiling levels. At the sill level, an aluminum stool provides the interior finish. It is snapped to a continuous clip on one side, and its vertical leg is fastened to a treated wood nailer on the other side.

Usually a heat-strengthened glass pane (with a fired-on ceramic frit opacification or a polyester film opacification on the interior surface of glass to prevent seeing through it) is used in the spandrel area (Figure 30.7). A fire-containment assembly is generally required to prevent the passage of fire and smoke between adjacent floors of the building. The assembly

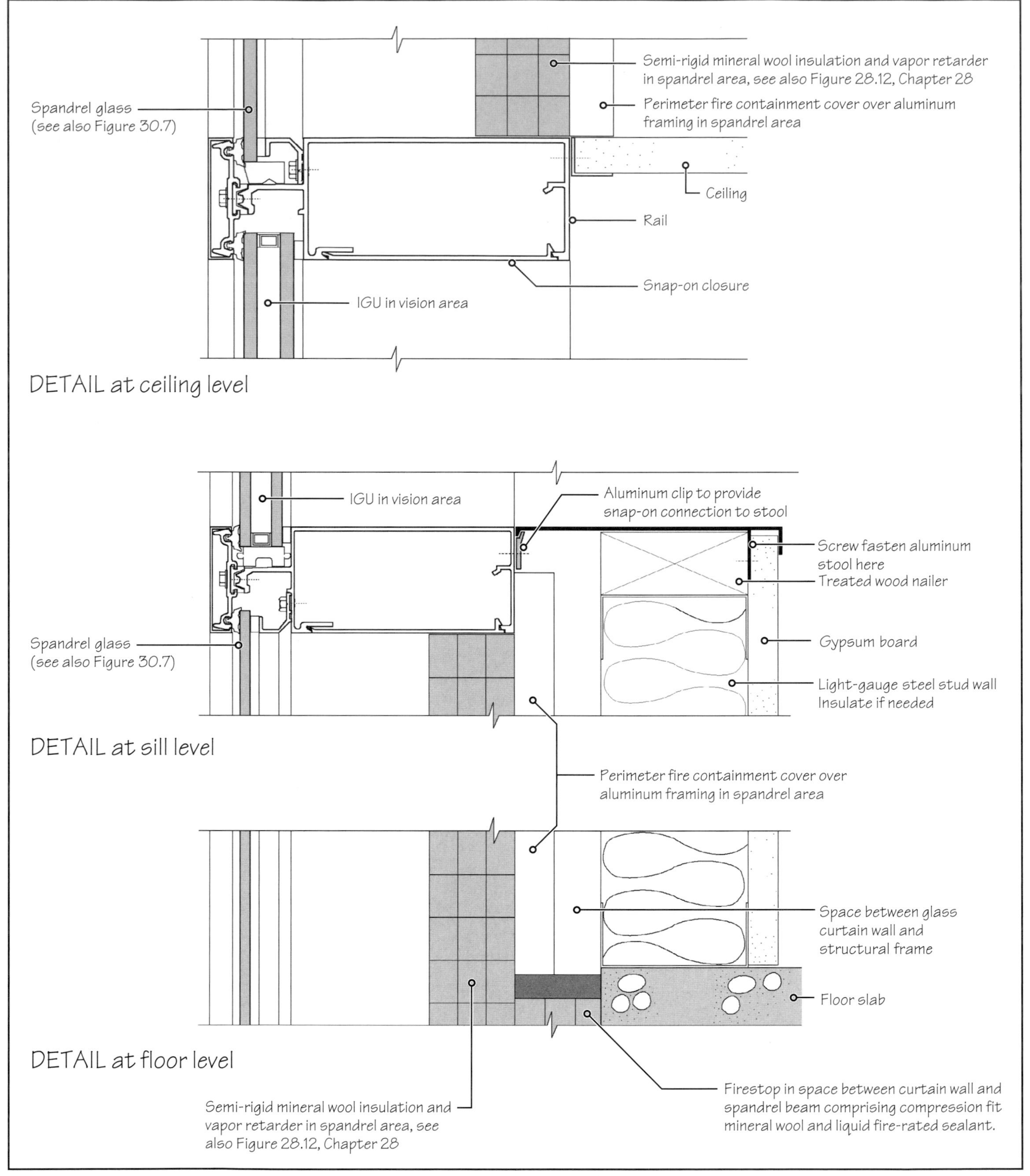

FIGURE 30.12 Typical floor-level, sill-level, and ceiling-level details of an outside-glazed glass curtain wall. Note insulation and fire-stopping.

consists of semirigid mineral wool insulation set between mullions and rails. The insulation should cover the curtain wall framing in the spandrel area. The gap between the curtain wall and the floor is closed with fire-stopping (mineral wool with a liquid-applied seal above).

It is important to control condensation in the space between the spandrel glass and the backup insulation because the condensed vapor will not only corrode fasteners, but it may also leak into the building interior as bulk water. Therefore, a vapor-diffusion analysis should be made and if condensation is an issue, mineral wool insulation should be faced with an aluminum foil scrim toward the building interior. Where a high degree of condensation potential exists, a metal panel (referred to as *metal back pan*) may be specified in place of an aluminum foil scrim.

Each question has only one correct answer. Select the choice that best answers the question.

12. The width of an expansion joint in the rails of a typical stick-built glass curtain wall
a. is 1 in. standard.
b. is $\frac{1}{2}$ in. standard.
c. is $\frac{1}{4}$ in. standard.
d. is $\frac{1}{32}$ in. standard.
e. must be determined on project-by-project basis.

13. In a stick-built glass curtain wall, the mullions are erected first and the rails are inserted between them.
a. True **b.** False

14. A shear block is used
a. at a mullion expansion joint to allow the mullion to move.
b. at the dead-load anchor of a mullion.
c. at the expansion anchor of a mullion.
d. to connect a rail to adjacent mullions.
e. none of the above.

15. A glass curtain wall in which the glass is pressure plate captured is glazed from the
a. building's interior.
b. building's exterior.
c. either (a) or (b).

16. As shown in this text, a glass curtain wall in which the glass is structural silicone adhered is glazed from the
a. building's interior.
b. building's exterior.
c. (a) or (b).

17. In a glass curtain wall with pressure plate–captured glass there are
a. no covers.
b. covers in both horizontal and vertical directions.
c. covers only in the horizontal direction.
d. covers only in the vertical direction.

18. As shown in the text, a glass curtain wall with structural silicone–adhered glass
a. has no exterior covers.
b. has exterior covers in both horizontal and vertical directions.
c. has exterior covers only in the horizontal direction.
d. has exterior covers only in the vertical direction.

19. A typical glass curtain wall is provided with weep holes to drain infiltrating water even though it is sealed from both the inside and the outside with gaskets or tapes.
a. True
b. False

Answers: 12-d, 13-a, 14-d, 15-b, 16-b, 17-b, 18-c, 19-a.

30.4 STRUCTURAL PERFORMANCE OF A GLASS CURTAIN WALL

The most important structural requirement of a glass curtain wall is its ability to resist lateral loads (particularly wind loads), including missile-impact resistance in hurricane-prone regions. Just as the design of the curtain wall's anchorage to the structure is accomplished by the curtain wall manufacturer, the curtain wall's lateral-load-resistance design is also provided by the manufacturer, based on the lateral-load intensities provided by the project architect or structural engineer.

Manufacturers generally have several standard sections designed to suit various lateral-load intensities. For high lateral-load intensities, a strategy often used is to enclose structural steel (or aluminum) sections within the mullions, Figure 30.13. The enclosed steel sections

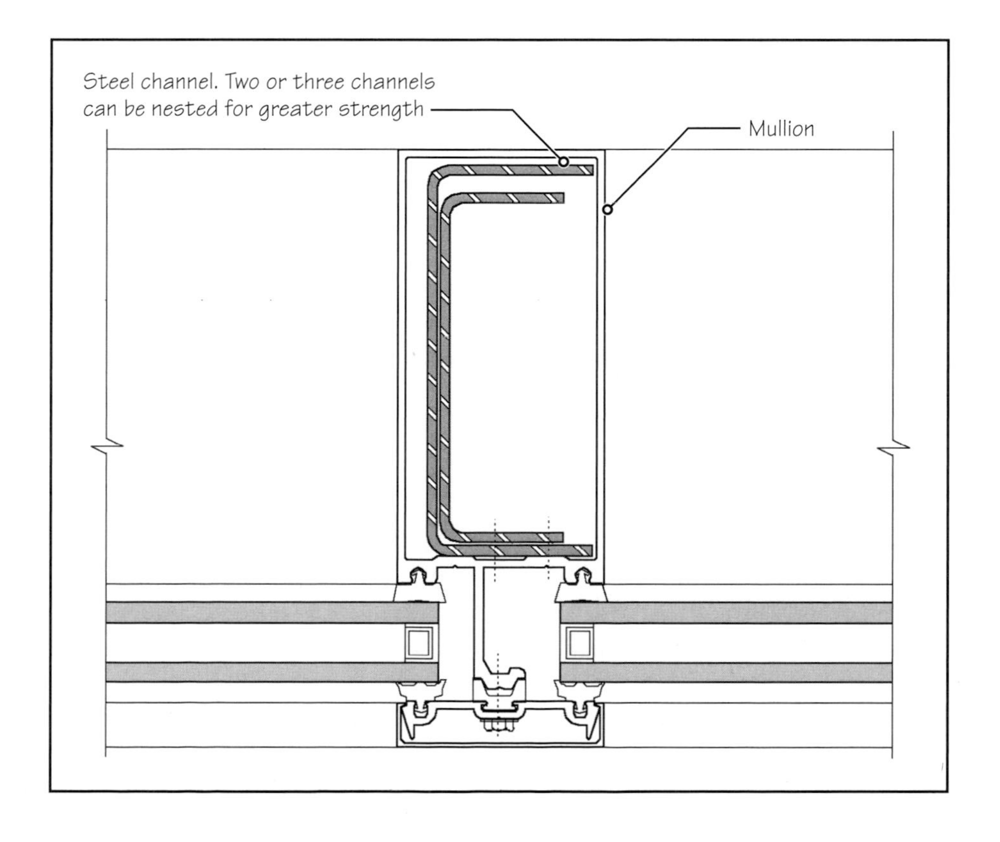

FIGURE 30.13 One of the ways to increase the lateral-load resistance of aluminum mullions is to enclose structural steel sections within them.

In this curtain wall, the supporting structure for aluminum sections consists of tubular steel columns. In a very tall curtain wall, vertical steel trusses and horizontal steel members or a steel space frame may be used.

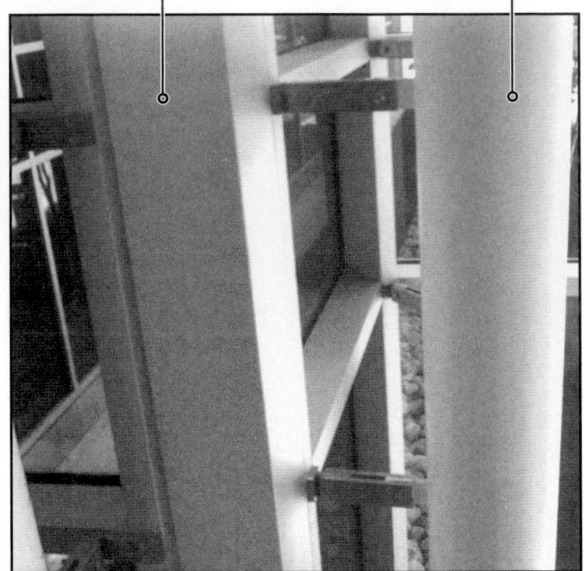

FIGURE 30.14 A tall glass curtain wall with standard curtain wall sections anchored to an interior structural steel frame.

and the mullions are fastened together to produce a composite action between them. Structural C- or I-sections are commonly used as enclosed sections. Channels provide the advantage of nesting, so that two or three channels may be used within the same mullion. The enclosed steel sections are suitably coated to prevent galvanic action between aluminum and steel.

An alternative to enclosed steel sections is to anchor the mullions to an exposed steel structural frame, Figure 30.14. This strategy is generally used in a tall curtain wall where the mullions do not have intermediate supports to reduce their span, such as those provided by the floor structure in a multistory curtain wall.

30.5 OTHER PERFORMANCE CRITERIA OF A GLASS CURTAIN WALL

Nonstructural performance of a glass curtain wall is just as important as its structural performance. Among the important nonstructural design criteria for a glass curtain wall are

- Air-infiltration control
- Rainwater- and meltwater-penetration control
- U-value
- Solar heat gain
- Condensation resistance
- Vapor diffusion
- Acoustical transmission
- Hurricane resistance
- Seismic resistance
- Thermal and structural movement
- Glass-cleaning-equipment load

For standard curtain walls, manufacturers provide the values for these criteria based on the tests conducted by recognized third-party laboratories. For custom walls, technical design support is generally available from the manufacturers. For a complicated wall design, the architect may need additional help from a curtain wall design consultant and a specialized testing laboratory to determine the wall's performance.

AIR-INFILTRATION CONTROL

In the United States, the maximum air infiltration allowed through a glass curtain wall is typically 0.06 cfm/ft^2 under an inside-outside air pressure difference of 1.57 psf. In Canada, the requirement is three times more stringent—that is, 0.02 cfm/ft^2 under an air pressure difference of 1.57 psf.

Where lower air infiltration is required, curtain wall systems, which provide a rate of up to 0.01 cfm/ft^2 under an inside-outside air pressure difference of 6.24 psf, are available. (Note that a 1.57-psf air pressure difference translates to that exerted by wind speed of 25 mph. Similarly a 6.24-psf air pressure difference is equivalent to that created by a 50-mph wind speed (Chapter 3).)

Air-infiltration control not only conserves energy but also reduces ice buildup on the exterior of curtain wall components. Ice buildup is caused by the condensation of water vapor that escapes from the building interior along with air. When the ice melts, the meltwater may leak into the building interior. Therefore, a more stringent air-infiltration-control criterion is generally needed in colder climates.

NOTE

cfm is an acronym for cubic feet per minute.

RAINWATER- AND MELTWATER-PENETRATION CONTROL

Water-penetration control (of both rainwater and meltwater) is perhaps the most important nonstructural performance requirement of a glass curtain wall. Glass curtain wall systems are generally designed to ensure no water penetration when tested under a static air pressure difference (between the inside and the outside) that is at least 20% of the inward structural design wind load on the curtain wall. Thus, if the inward structural design wind load on the wall is 50 psf, the system is tested for water penetration under a static air pressure difference of at least 10 psf. A more stringent water-penetration criterion is required for a building located in areas subjected to frequent and intense wind-driven rain. In addition to conforming to the static pressure test, a glass curtain wall is also required to conform to dynamic pressure test criterion.

Water-penetration control is accomplished in different ways by system manufacturers; it typically includes adequate drainage in the aluminum joinery and glazing pockets. For example, in the stick-built glass curtain wall described earlier, weep holes are provided in the pressure plates and snap-on covers that drain the water to the outside. The architect's details must also ensure the management of water into the curtain wall from a nonglass wall assembly that is above, where such a wall exists.

NOTE

AAMA and Water Penetration

The American Architectural Manufacturers Association (AAMA) defines water penetration as the appearance of uncontrolled water other than condensation on the interior face of any part of the curtain wall.

U-VALUE, SOLAR HEAT GAIN, AND CONDENSATION RESISTANCE

These three interrelated criteria are a function of the type of glass, type of aluminum framing (thermally improved or not), and the center-to-center spacing of framing members, as explained in Chapter 28. The architect must specify their values in consultation with the heating, ventilating, and air-conditioning (HVAC) consultant (who also needs these values to design the building's HVAC system and to meet the energy code requirements).

WATER VAPOR ANALYSIS

As previously stated, interior water vapor may result in ice buildup on curtain wall framing and condensation of water vapor in the building interior. These issues are particularly critical in cold climates, where a vapor analysis of the curtain wall system is generally required.

ACOUSTIC TRANSMISSION

Glass curtain walls in buildings located in areas where high levels of exterior noise are present (e.g., near airports or busy highways) may need a higher sound-transmission-loss specification than other areas.

HURRICANE IMPACT RESISTANCE AND SEISMIC RESISTANCE

Because of Florida's experience with extensive wind damage to glass curtain walls from hurricanes, several coastal cities in the United States are requiring that glass curtain walls be missile-impact resistant, particularly in the lower floors of the building. Similarly, because of California's experience with earthquake damage to glass curtain walls, resistance to shaking, glass drifting, and horizontal movement of components is required for buildings in seismically active areas.

THERMAL AND STRUCTURAL MOVEMENT

Aluminum-framing members and glass expand and contract due to temperature changes and the sudden cooling effects of precipitation. Glass curtain walls require sufficient expansion and contraction control built into them to allow thermal movement and the movement of spandrel beams due to live load deflection. The expansion splice required in a stick-built curtain wall (Figure 30.3) accounts for this requirement.

GLASS-CLEANING-EQUIPMENT LOAD

High-rise curtain walls include provisions for periodic cleaning. This means that they must include anchorage points for the staging of cleaning equipment that is lowered from the roof to the front of the curtain wall. These anchors add point loads on the wall, and this information needs to be communicated by the architect to the system manufacturer.

30.6 NONTRADITIONAL GLASS CURTAIN WALLS

The glass curtain wall types discussed so far are traditional types that have evolved over a long period. Two recently introduced systems that warrant discussion are

- Point-supported (mullionless) glass curtain walls
- Double-skin curtain walls

POINT-SUPPORTED GLASS CURTAIN WALLS

A sophisticated and unique version of a glass curtain wall that is not as common (due to its cost) is the mullionless glass curtain wall developed by Pilkington Company. In this system, each glass pane is supported at four corners by stainless steel *spider-shaped connectors*. The connectors are held by horizontal stainless steel tension cables and compression struts that, in turn, are held by vertical steel frames, Figure 30.15. The glass used in the system is generally heat-soaked tempered glass.

DOUBLE-SKIN GLASS CURTAIN WALLS

Another innovative glass curtain wall system is the double-skin curtain wall system, also referred to as a *bioclimatic glass curtain wall*. In this system, two glass curtain walls, separated by 1 to 5 ft of air space, are used. The air space serves as a buffer between the two skins, tempering the outdoor air, and serves as a plenum for the building's HVAC system. The outer skin may include computer-controlled operable glazing and solar shading devices.

Although the primary benefit of a double-skin system is energy conservation, it also provides effective water-penetration control, better air-infiltration control, higher sound insulation, and so on. The proponents of the system, which has been used extensively in Europe, claim it to be more sustainable under the present energy prices in Europe, which are considerably higher than in the United States.

As smart glazing systems with environment adaptive technologies (e.g., electrochromic glass) and energy-generating capabilities (e.g., photovoltaic glass) evolve further, double-skin glass wall systems may become more popular. In that scenario, the outer glass skin may not only help conserve energy but also generate some energy to power the building.

30.7 OTHER GLASS-ALUMINUM WALL SYSTEMS

In addition to curtain walls, two additional glass-aluminum wall systems commonly used are

- Punched and strip glazing
- Storefront system

PUNCHED AND STRIP GLAZING

Punched glazing is similar to a punched window (see the window terminology in Chapter 29), except that the glass in punched glazing is generally fixed and site installed due to its large size. By contrast, a punched window is generally shop glazed and may contain operable

A typical spider connector. Each connector holds one corner of all four glass panes that meet at the connector. An elastomeric sealant seals the edges between glass panes.

FIGURE 30.15 Pilkington's Planar Glass Curtain Wall System used in the Rose Center for Earth and Space, New York (Figure 28.4). Architects: Polshek Partnership.

sashes. The frame for punched glazing may either be shop assembled or stick-built on site. The frame is anchored to the opening (jambs, head, and sill) instead of being anchored to the building's structural frame (as in a curtain wall).

Strip (also called ribbon) glazing is similar to punched glazing, with several glazing units placed in linear alignment. Strip glazing is also anchored to the head and sill of an opening.

STOREFRONT SYSTEM

A storefront is a large glazing unit that is generally one story or two stories tall and extends from the ground up. The primary difference between a curtain wall and a storefront is in the glazing system's performance for nonstructural criteria (air infiltration, water penetration, CRF, etc.). Most storefront systems available from manufacturers conform to the low end of nonstructural performance criteria or none at all. A standard storefront is generally used when it can be protected by an overhang to control water leakage through the system.

Another difference between a curtain wall and a storefront is that rainwater that enters a curtain wall is drained by the horizontal member of each individual lite. In a storefront system, the water entering through a lite must travel vertically down the mullion and be drained at the weep holes near the ground level.

PRACTICE QUIZ

Each question has only one correct answer. Select the choice that best answers the question.

20. Air infiltration rate through a glass curtain wall is specified in terms of
- **a.** pounds per square feet.
- **b.** cubic feet per minute.
- **c.** cubic feet under a given inside-outside air pressure difference.
- **d.** cubic feet per minute under a given inside-outside air pressure difference.
- **e.** none of the above.

21. A point-supported glass curtain wall has
- **a.** no supporting mullions and rails.
- **b.** only vertical mullions.
- **c.** only rails.
- **d.** both mullions and rails.

22. In a point-supported glass curtain wall, the connectors that hold the glass are
- **a.** concave shaped.
- **b.** convex shaped.
- **c.** spider shaped.
- **d.** beetle shaped.
- **e.** none of the above.

23. The performance requirements for a storefront are generally much higher than those for a glass curtain wall.
- **a.** True
- **b.** False

Answers: 20-d, 21-a, 22-c, 23-b.

KEY TERMS AND CONCEPTS

Dead load anchors and expansion anchors
Double-skin glass curtain wall
Double-span mullion system
Glass-aluminum curtain wall
Inside-glazed wall
Outside-glazed wall
Performance criteria of glass-aluminum curtain walls
Point-supported glass curtain wall
Pressure plate
Punched glazing
Rail-mullion connection
Single-span mullion system
Snap-on cover
Standard and custom wall systems
Stick-built glass-aluminum curtain wall
Storefront
Strip glazing

REVIEW QUESTIONS

1. Using sketches and notes, explain the differences between a single-span mullion support system and a twin-span mullion support system for a glass curtain wall.
2. Use three-dimensional sketches to illustrate each of the following.
   a. A typical dead-load anchor used in a glass curtain wall
   b. A typical expansion anchor used in a glass curtain wall
3. Sketch in three dimensions a typical spider connector.

SELECTED WEB SITES

American Architectural Manufacturers Association AAMA (www.aamanet.org)

FURTHER READING

1. American Architectural Manufacturers Association. "Standard Test Method for Exterior Windows, Curtain Walls and Doors for Water Penetration Using Dynamic Pressure," AAMA Standard 501.1.
2. American Society for Testing and Materials. "Standard Test Method for Determining Rate of Air Leakage Through Exterior Windows, Curtain Walls, and Doors Under Specified Pressure Differences Across the Specimens," ASTM Standard E 283.
3. American Society for Testing and Materials. "Standard Test Method for Water Penetration of Exterior Windows, Skylights, Doors, and Curtain Walls by Uniform Static Air Pressure Difference," ASTM Standard E 331.

CHAPTER 31 Roofing–I (Low-Slope Roofs)

CHAPTER OUTLINE

31.1 Low-Slope and Steep Roofs Distinguished

31.2 Low-Slope Roof Fundamentals

31.3 Built-Up Roof Membrane

31.4 Modified Bitumen Roof Membrane

31.5 Single-Ply Roof Membrane

31.6 Rigid-Board Insulation and Membrane Attachment

31.7 Insulating Concrete and Membrane Attachment

31.8 Low-Slope Roof Flashings

31.9 Base Flashing Details

31.10 Curb and Flange Flashing Details

31.11 Protected Membrane Roof

31.12 Low-Slope-Roof Design Considerations

PRINCIPLES IN PRACTICE: Shingling of Built-Up Roof Felts

The underside of a parapeted low-slope roof showing two entirely independent (primary and secondary) roof drainage systems. (Photo by MM.)

The roof is one of the most critical components of a building. It is the primary source of construction litigation and building owners' complaints. One estimate indicates that nearly 65% of all lawsuits brought against architects originate in roofing problems [1]. This statistic becomes significant considering that the roof covering of a typical building is replaced every few years—12 to 14 years on the average. Would we accept replacing other components of a building envelope so frequently—doors, windows, curtain walls, and so on? A major reason for this situation is the severity of physical and chemical degradation, by solar radiation, rainwater, snow, hail, and windstorm, to which a roof is subjected. Another reason is the ever-increasing complexity of a contemporary building and, more specifically, of a contemporary roof.

Until the middle of the twentieth century, roofing alternatives were few, and roof design was relatively simple. For a flat roof, the roofing membrane consisted of either a coal tar or an asphalt built-up roof on a concrete or wood deck. For a steep roof, clay tiles or wood shingles were the major alternatives. This is not true today. A relatively sudden spurt in the gross national products of Western economies after World War II introduced large industrial buildings with relatively light steel roof decks. A roof that performed well on a stiff concrete or wood deck failed prematurely on a light and flexible steel deck.

The advent of air conditioning and more efficient heating systems led to urban development in extremely warm and cold climates. Roof insulation that was seldom used earlier became the norm. The greater degradation of roofs in extremely hot and cold climates and the effect of roof insulation introduced factors whose impact was not realized until several years later.

The oil embargo of 1973 led to significant increases in the cost of built-up roofs, which made single-ply roofs economically competitive for the first time. This, combined with several other market forces, resulted in a significant increase in the market share of single-ply roofs. Some of the early single-ply roofs had a fairly brief history of use, which further increased roofing problems. Although roofing products are generally subjected to several laboratory tests before being introduced in the market, none of these tests can fully simulate the actual field conditions.

This chapter is the first of the two chapters on roofing. It begins with a general introduction to roofing, including the basis for distinguishing between low-slope and steep roofs. Subsequently, various aspects of low-slope roofs are covered. Steep roofs are discussed in the following chapter.

31.1 LOW-SLOPE AND STEEP ROOFS DISTINGUISHED

A discussion of roofing must begin with its classification under two types:

- Low-slope roof
- Steep roof

A low-slope roof is a *water-resisting* roof and consists of a continuous roof membrane over a relatively flat roof surface. Although, water ponding is generally to be avoided, a low-slope roof has to be designed for a certain depth of rainwater accumulation that might occur in the event of blockage of the roof's drainage system. The depth of rainwater accumulation is predetermined based on roof geometry and the design of its drainage system. The roof membrane in a low-slope roof must, therefore, act as a waterproofing membrane and be able to resist rainwater load and water leakage until the drainage system is able to function again. This period may be as long as a few hours or a few days.

A steep roof, also called *water-shedding* roof, typically consists of small individual roofing units (shingles) that overlap each other. The roof surface must be sloped so that the water is shed off the roof by gravity. The slope must generate adequate gravitational force to overcome the forces produced by wind, head pressure, and capillary action that might push the water up the slope between adjacent shingles and cause the roof to leak. Double coverage in shingles and adequate overlap are necessary to prevent roof leakage.

The rate of discharge of water from the roof is not only critical in keeping the roof and the building's interior dry, it is also one of the most important factors in establishing the durability of the roof. There is sufficient long-term data to confirm that—everything else being the same—the durability of a roof increases directly with the slope of the roof. Steep roofs, in general, outsurvive the low-slope roof varieties.

A low-slope roof is is defined as a roof with a slope of less than 3:12 (25% slope) and greater than $\frac{1}{4}$:12 (2% slope). A dead-flat roof is not permitted by the codes. A steep roof is defined as a roof with slope greater than or equal to 3:12.

NOTE

Low-Slope Roof and Steep Roof

Low-slope roof	Slope $< 3{:}12$
Steep roof	Slope $\geq 3{:}12$

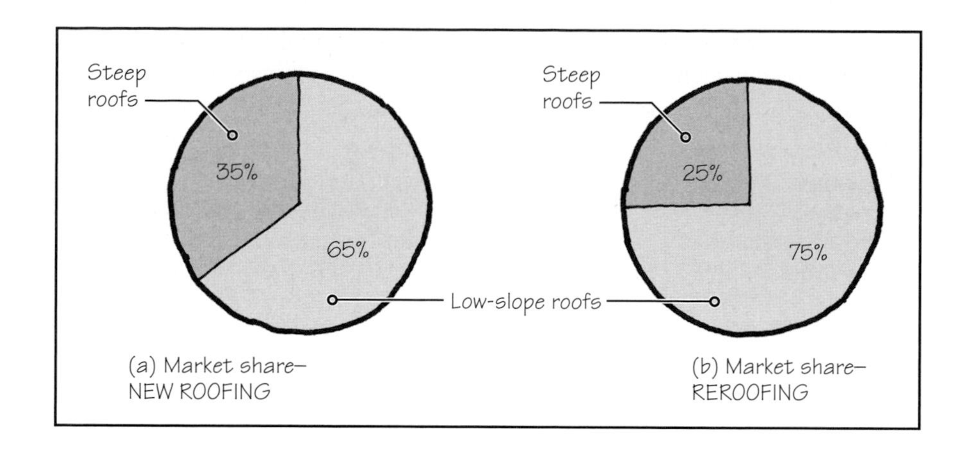

FIGURE 31.1 Approximate market shares of low-slope and steep roofs.

Despite the greater durability of a steep roof, the market share of low-slope roofs is nearly twice as large as that of steep roofs, Figure 31.1 [2]. Low-slope roofs are generally used for commercial and industrial buildings and are mostly constructed to a minimum mandated slope of $\frac{1}{4}$:12. This slope, which yields an almost flat roof, makes the structural framing of the roof and the supporting structure most economical because an increase in roof slope increases roof area, the dimensions of framing members, and related structure. Additionally, a flat roof provides an easily accessible space for installing HVAC and other equipment. Buildings with flat roofs generally have their HVAC equipment mounted on the roofs. Steep roofs, which are generally used for low-rise residential buildings (single-family homes, apartments, motels, etc.) have their HVAC equipment placed on the ground.

A low-slope roof also makes maintenance and reroofing, particularly on high-rise buildings, relatively easy. Imagine reroofing or repairing a steep roof a few hundred feet above the ground!

31.2 LOW-SLOPE ROOF FUNDAMENTALS

A low-slope roof has the following components, Figure 31.2:

- Roof membrane, including a protective cover or coating above the membrane, where needed
- Insulation
- Roof deck
- Flashings

In some roofs, particularly in cold regions or over high-humidity enclosures, a vapor retarder may also be needed.

Roof Membranes

The roof membrane is the most important component of a roof because it is the waterproofing layer. Being the top layer, it is constantly subjected to the weathering effects of sun, rain, snow, hail, and wind. Roof membranes are divided into three general categories:

- Built-up roof membrane
- Modified bitumen roof membrane
- Single-ply roof membrane

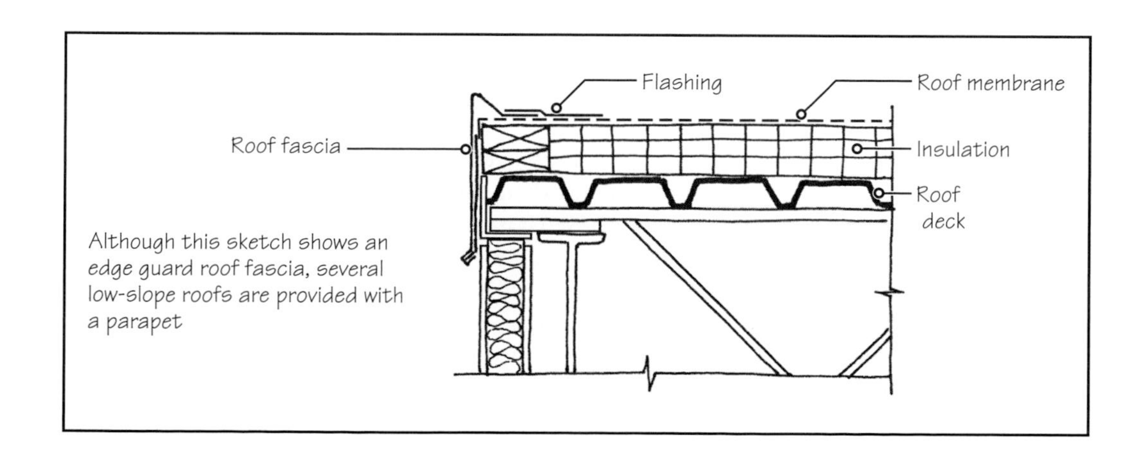

FIGURE 31.2 Components of a low-slope roof.

A *built-up roof (BUR) membrane* consists of 3 to 5 plies of felts with intervening moppings of bitumen (asphalt or coal tar). Because both asphalt and coal tar are adversely affected by ultraviolet radiation, a built-up roof membrane is generally protected by aggregate cover. The aggregate cover also increases the roof's fire resistance. Of the two, asphalt built-up membranes are far more common than coal tar built-up membranes.

A *modified bitumen roof* (MBR) membrane is similar to a built-up roof membrane and comprises two to three plies of modified bitumen sheets with intervening moppings of bitumen. A protective aggregate cover or a mineral granule–surfaced cap sheet is required.

A *single-ply roof (SPR) membrane* consists of only one sheet of a synthetic polymer (plastic) and does not require any protective cover. The market share of BUR, MBR, and SPR is approximately the same—nearly one-third each.

ROOF INSULATION

In a typical low-slope roof, the insulation is sandwiched between the membrane and the deck. Apart from reducing energy costs, insulation provides a suitable substrate for the roof membrane. Over a steel roof deck, insulation also provides the necessary surface flatness.

ROOF DECK

A low-slope roof deck may be of steel, wood, cast-in-place concrete, precast concrete, gypsum concrete, or wood fiber–cement deck. Steel and concrete roof decks are more commonly used in low-slope roofs.

ROOF FLASHINGS

Flashings are an integral part of a low-slope roof. They are required at roof terminations, such as the free edges of a roof, at parapets, and roof expansion joints. Flashings are also required at penetrations, such as around interior roof drains, and around pipe or tubular supports for rooftop equipment.

LOW-SLOPE ROOF—A SYSTEM OF COMPATIBLE COMPONENTS

Each roof component must obviously serve its own function, but it must also be compatible with other roof components. For instance, the insulation must not only give the required R-value, it must also provide a rigid substrate against hail and foot traffic. It should also have a long-term chemical compatibility with the deck and the membrane.

Similarly, the roof deck must not only provide the required structural support for the entire assembly and other loads, it should also be dimensionally stable to prevent the overstressing of the insulation and the membrane. It must provide a suitable base for the anchorage of insulation and the membrane and have adequate fire resistance.

With several factors that affect its performance and several components and subcomponents that constitute a low-slope roof, roof design must be considered as a system. It should consider the interaction between roof components and the effects of external factors such as the external climate, solar radiation, wind, rain, hailstorm, fire, and rooftop traffic (necessitated by the equipment's and roof's maintenance operations). The selection of all components of the system, including fasteners and adhesives, from the same manufacturer is, therefore, important in the case of a low-slope roof. Where a manufacturer does not make all components of the system, components approved by the primary manufacturer must be used to ensure their compatibility.

NOTE

Metal Panel Roofs and Protected Membrane Roofs

In a typical membrane-covered low-slope roof assembly, the membrane deteriorates rapidly due to its exposure to weathering elements. In response to this weakness, two other low-slope roofs have gradually increased their market shares:

- Metal panel roofs
- Protected membrane roofs

A metal panel roof consists of thin sheets of metal bent into panels that are laid over roof insulation and anchored to the deck. They may be used in low-slope or steep roofs, although they function better on steep slopes (slope greater than 3:12). The initial cost of metal panel roofs (a nonmembrane roof assembly) is higher than membrane roofs. Metal panel roofs are discussed in Chapter 32.

In a protected membrane roof, the insulation is loose laid over the deck to which the membrane is attached directly. Protected membrane roof is discussed later in this chapter.

Fluid-Applied Membrane Roofs

Fluid-applied membrane roofs consist of polyurethane foam insulation sprayed over a low-slope roof deck. The insulation is covered with a (fluid) coating that cures into a membrane. It is commonly used in south and southwestern United States with a relatively small overall market share.

31.3 BUILT-UP ROOF MEMBRANE

A typical built-up roof membrane consists of a felt laid over a mopping of bitumen, followed by a second mopping of bitumen, and then the second felt, and so on. In other words, a number of felt layers (called plies), separated by interply moppings of bitumen, constitute a built-up roof membrane.

The greater the number of plies, the thicker and, hence, stronger and more durable the membrane. Three to 5 plies is generally the norm. The last ply is typically covered with a surfacing material to protect the entire membrane from the effects of weather and external fire. The most common surfacing material is stone aggregate laid over a flood coat of bitumen. Because the quantity of bitumen for the flood coat is much greater than that required

FIGURE 31.3(a) With a bucket full of hot bitumen and a mop, a roofer begins to lay a built-up roof.

FIGURE 31.3(b) Although the bitumen can be applied with mops, it is more convenient to use a bitumen dispenser on large roofs. The unrolling of felt over bitumen is followed by pressing the felt down with a squeegee.

FIGURE 31.3(c) This image shows the roof after all built-up felt plies have been applied. The roof is now ready to receive the flood coat of bitumen and stone aggregate surfacing, as shown in the following two images.

FIGURE 31.3(d) The bitumen under aggregate surfacing is poured over the deck, not mopped. That is why it is called a *flood coat*.

FIGURE 31.3(e) The bitumen flood coat is covered with stone aggregate. For a small roof, shovels are used to spread the aggregate.

FIGURE 31.3(f) Although the bitumen for a flood coat can be poured with buckets, as shown previously, it is usually poured on large roofs using a bitumen dispenser followed by aggregate spreader.

for interply moppings, it is poured over the roof, whereas the interply bitumen is simply mopped on. Figures 31.3(a) to (f) illustrate a few stages in the laying of a typical built-up roof membrane.

In the alternate layers of felts and bitumen in a built-up roof, the bitumen is the waterproofing material. However, bitumen alone cannot be used, because it is a thermoplastic material. It becomes soft at high temperatures and begins to flow. At low temperatures, it becomes hard and brittle and cracks. Thus, bitumen does not have the requisite tensile strength to withstand stresses imposed by the changes in temperature, deck movement, foot traffic, hailstorm, and so on.

The felts work as reinforcing material, giving the required tensile strength to the membrane, Figure 31.4. The felts also stabilize the bitumen against flow, because the interwoven felt fibers form mini-receptacles within which the bitumen is held,

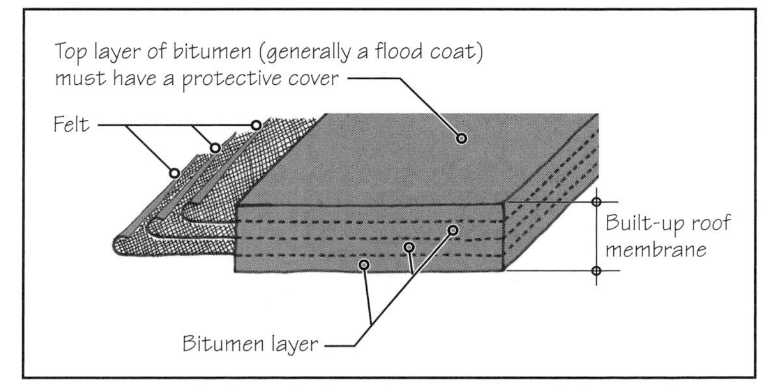

FIGURE 31.4 A built-up roof membrane consists of alternate layers of bitumen moppings and felts. The felts provide tensile strength to the membrane, in which the bitumen is the waterproofing agent. The greater the number of felts, the stronger the roof membrane.

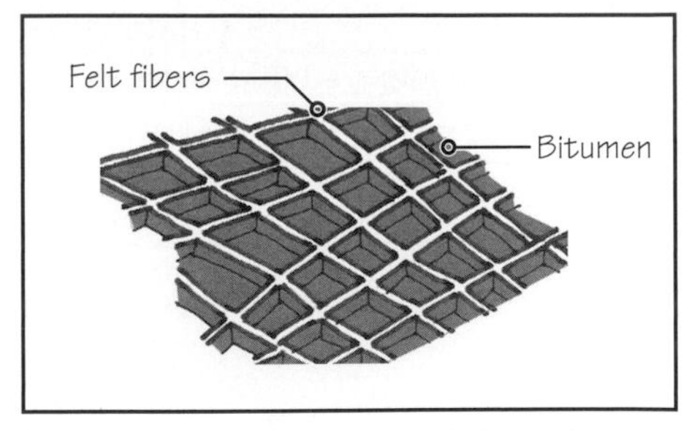

FIGURE 31.5 Interwoven felt fibers form miniature containers to hold the bitumen, preventing its flow.

preventing its flow, Figure 31.5. A heavy mopping of bitumen without felt simply cracks in cold weather, due to the lack of tensile strength, and flows like a thick paste during hot weather, due to the lack of containment. Thus, the felts allow for a more significant buildup of bitumen, which increases the weatherproofing and waterproofing of the membrane.

Organic and Fiberglass Roof Felt

A built-up roof felt consists of fibers (or strands) pressed into a sheet. A felt may be either an *organic felt* or an *inorganic felt,* depending on whether the fibers are organic or inorganic. Organic felt is made from paper or wood fibers or a combination of both. Inorganic felt is made from glass fibers.

The manufacturing process for both organic and fiberglass felts is similar to paper making, in that a mass of fibers is pressed under rollers to give a thin, flat sheet. In fact, untreated organic felt (without the bitumen treatment) is virtually indistinguishable from thick handmade paper.

Untreated fiberglass felt, on the other hand, looks like a woven mat. The weave in fiberglass felt is so sparse that light and air can easily pass through it. By comparison, organic felt is thick, nonporous, and opaque to light, Figure 31.6.

After the rolling process, felt is treated with bitumen. Because both coal tar and asphalt are black in color, treated felt is black in color. Treated felt is used in a built-up roof. Therefore, the term *felt* in our discussion generally implies a bitumen-treated felt. The treatment consists of simply covering the felt with bitumen.

Although treatment with bitumen reduces the porosity of fiberglass felt, treated fiberglass felt is still highly porous. Treated organic felt is nonporous and, hence, water resistant, Figure 31.7. Generally, organic felt is specified as an underlayment for roof shingles and tiles. Because of its porosity, fiberglass felt cannot be used in such applications (Chapter 32).

Porosity of fiberglass felt is, however, an advantage in built-up roof applications. Because the asphalt at the time of application is at a high temperature (nearly 400°F), any air trapped during mopping expands and forms blisters under the felt, if not allowed to escape. If there is any moisture in the substrate, it turns into steam and also forms blisters if unable to escape. Blisters obviously weaken the roof membrane, because they represent areas where the roof is unattached to the substrate. The pressure exerted on the membrane by water vapor and air inside the blister can exceed the tensile strength of the membrane, causing its rupture. The inherent porosity of fiberglass felt allows trapped air and vapor to escape. Therefore, fiberglass felt is commonly used in built-up roofs.

Felt is typically manufactured in 36-in.-wide rolls, although metric (1-m-wide) felt is also available from a few manufacturers. Because bitumen is a highly tacky substance, the felt is surfaced with fine mineral sand or other release agent to prevent its adhesion inside the roll.

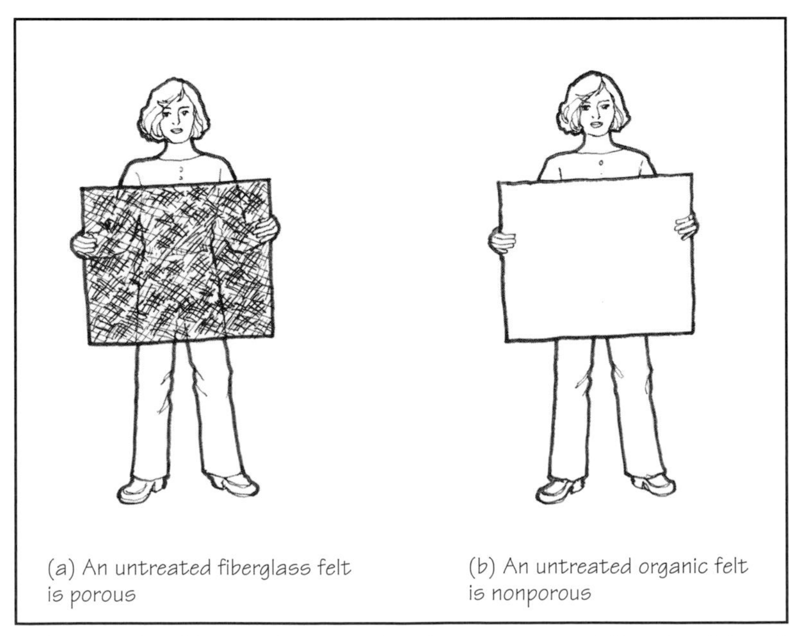

FIGURE 31.6 The partial visibility of the person holding an untreated fiberglass felt indicates its sparse weave. By comparison, an organic felt is opaque and nonporous.

FIGURE 31.7 Although less so than an untreated fiberglass felt, a treated fiberglass felt is fairly porous. A treated organic felt is nonporous and, hence, useful as a water-resistant underlayment in steep roofs (Chapter 32).

TYPES OF FIBERGLASS FELT

Asphalt-treated fiberglass felt: Asphalt-treated fiberglass felt is classified either Type IV or Type VI, having a tensile (or tear) strength of 44 lb/in. and 60 lb/in., respectively. Type VI, the stronger felt, is recommended where the membrane is subjected to a high tensile stress. These stresses may be caused by a high daily or annual temperature differential, a relatively flexible deck, or excessive impact on a roof due to hailstorms and/or foot traffic. In addition to using Type VI felt, a thicker (4- to 5-ply) built-up roof membrane is recommended for the preceding situations, since a thicker membrane is stronger.

A Type VI felt roll.

Coal tar-treated fiberglass felt: The use of tar-treated fiberglass felt has been decreasing over time. Coal tar fumes can produce extreme discomfort for the installers. Additionally, due to the porosity of fiberglass felt and the cold-flow property of tar, the tar filters through the felt, so that the felt tends to sink to the bottom of the membrane. This separates the felt from the bitumen.

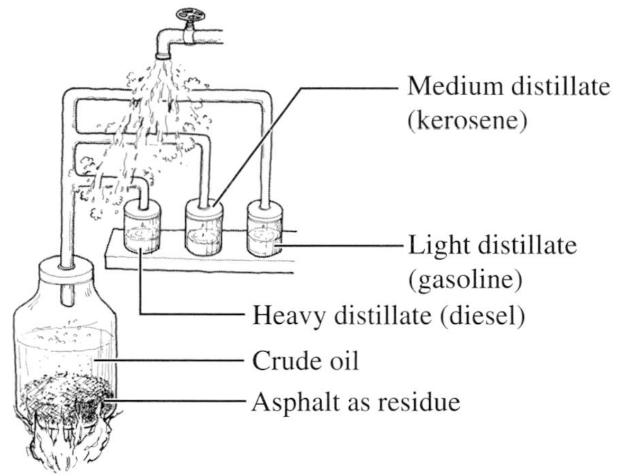

Distillation of crude oil

BITUMEN—ASPHALT AND COAL TAR

Two types of bitumen—*asphalt* and *coal tar*—are used in built-up roofing. Asphalt is the waste product (residue) obtained from the distillation of crude oil at petroleum refineries. Coal tar (also called *tar,* or *pitch*) is obtained from the distillation of coal. Although the distillate in this process is coal tar, the residue is coke. Coke is used in steel manufacturing industry (Section 16.1).

Although both asphalt and tar are highly resistant to water, coal tar is slightly more so. Coal tar built-up roof manufacturers claim their roofs withstand ponded water, a claim not advanced by any other roofing membrane manufacturer. However, asphalt built-up roofs are far more commonly specified than coal tar built-up roofs, although asphalt appeared on the scene much later. The primary reason is the significantly lower health risk to roofers from the use of asphalt compared to tar. The other benefits of asphalt are its greater availability and lower cost, compared to tar.

TYPES OF ROOFING ASPHALT

ASTM specifications divide roofing asphalts into four types: Types I, II, III, and IV. The most important property that distinguishes one type from the other is the *softening-point temperature* (SPT) of asphalt. SPT is the temperature at which asphalt begins to flow. It is directly related to the weathering characteristics of asphalt. The lower the SPT, the more durable the asphalt and the more easily the asphalt will heal any cracks caused by expansion or contraction in roof membrane. Figure 31.8 gives the SPT ranges of the four asphalt types.

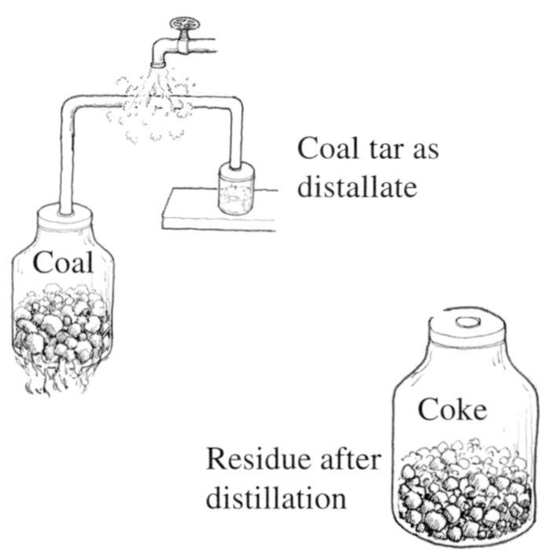

Distillation of coal

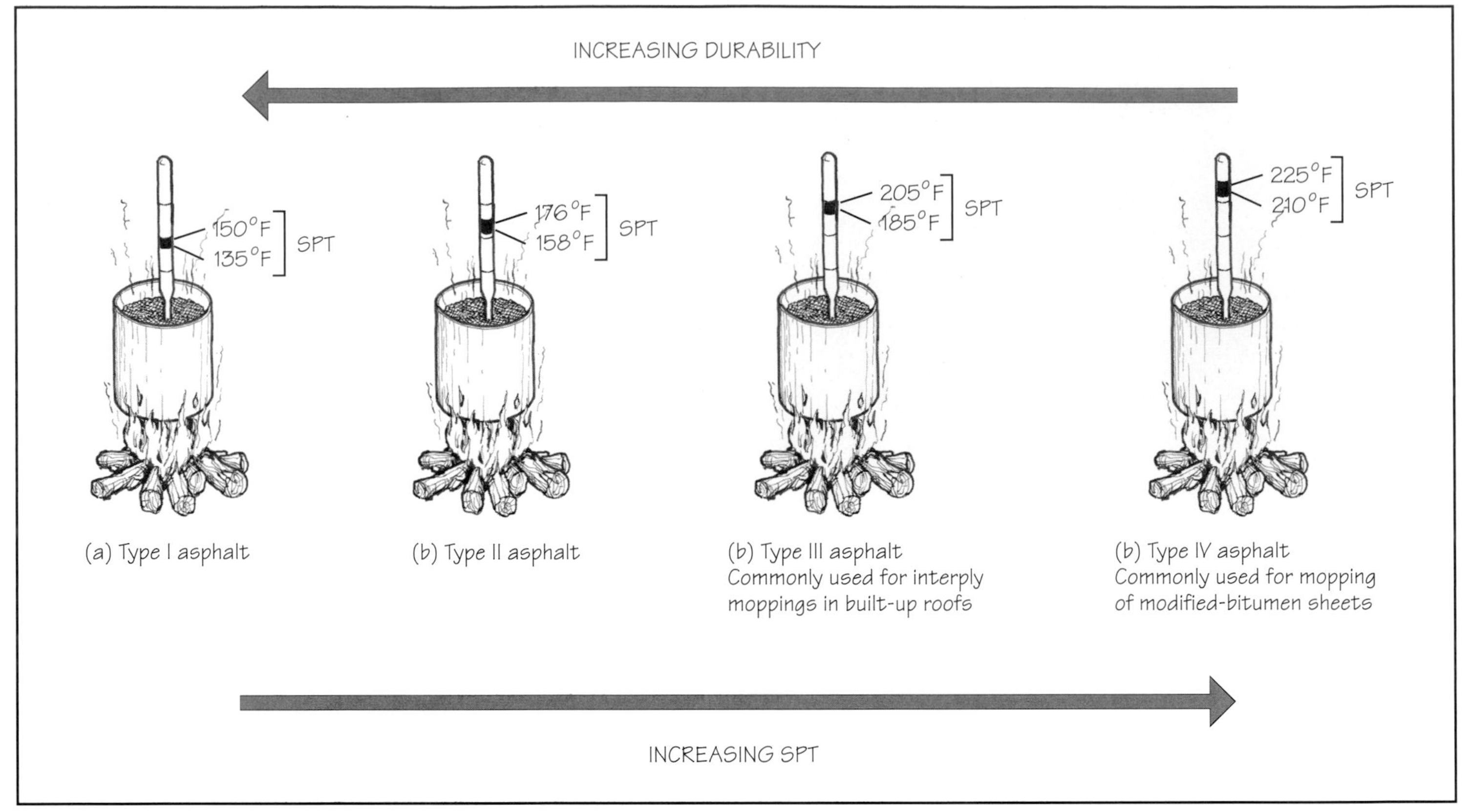

FIGURE 31.8 Types of roofing asphalts and their softening-point temperatures (SPT) and relative durability. Type III asphalt is most commonly used in built-up roofs, and Type IV is used in mop-applied modified bitumen roofs.

FIGURE 31.9 Types III and IV asphalts are available in paper-wrapped kegs, as shown.

The most commonly used asphalt is Type III. It is used as an interply adhesive and as an adhesive to bond insulation to the deck or to bond two layers of insulation together. Because of its lower viscosity, Type III (and also Type IV) can be used to adhere bituminous flashings on vertical surfaces.

Type IV asphalt performs better than Type III, but its lower durability is a deterrent. Type III asphalt is also specified for the flood coat for roofs in warm climates. However, in colder climates (and roof slope permitting), Type I or Type II asphalt should be considered for the flood coat due to its better weathering characteristics. Types III and IV asphalts are available in paper-wrapped kegs, Figure 31.9. Types I and II are available in metal containers because of their lower SPT.

Amount of Asphalt Required

Approximately 25 lb of asphalt per roof square is recommended for interply moppings and 60 lb per roof square is suggested for the flood coat. The flood coat is poured on the roof, not mopped on, to provide the large quantity of asphalt required.

Built-Up Roof Surfacing

Because both coal tar and asphalt degrade over time due to their exposure to ultraviolet radiation, the topmost layer in a built-up roof must be protected. Stone aggregate is by far the most commonly used protective covering. Apart from protecting the bitumen from ultraviolet radiation, an aggregate surface increases the fire resistance of the roof, protects the membrane from hail impact, and adds weight over the membrane, increasing its resistance to wind uplift.

The aggregate must be a graded aggregate so that most particles sink into the bitumen. The average recommended aggregate size is $\frac{3}{8}$ in., with a maximum aggregate size of $\frac{3}{4}$ in. A minimum aggregate weight of 400 lb per square is generally recommended.

Water-worn river gravel is ideal surfacing material because it does not damage the membrane under occasional foot traffic, but crushed stone may be used if the traffic on the roof is minimal. Blast-furnace slag (Section 16.1) also makes a good surfacing material. Lightweight aggregate, which may be dislodged under high wind or rainstorm, is generally discouraged. It may clog the drains and disintegrate under freeze-thaw action.

NOTE

Roof Square

Built-up roof felt rolls are 36 in. wide and 36 ft long. Thus, the area of a roll is 108 ft^2. With 2-in. head lap and 6-in. lap at the end of a roll, one roll covers a roof area of 34 in. × 35 ft 6 in., which is approximately equal to 100 ft^2.

In the roofing industry, 108 ft^2 is referred to as a *factory square* and 100 ft^2 is called *roof square*.

NOTE

Weight of Aggregate Surfacing

The weight of aggregate surfacing on a buit-up roof is generally 400 lb per roof square, that is, 4 lb/ft^2.

PRACTICE QUIZ

Each question has only one correct answer. Select the choice that best answers the question.

1. A low-slope roof is defined as a roof whose slope is less than
   - **a.** 1:12.
   - **b.** 1.5:12.
   - **c.** 2:12.
   - **d.** 2.5:12.
   - **e.** 3:12.
2. Low-slope roofs are more commonly used than steep roofs.
   - **a.** True
   - **b.** False
3. Because of the general trend toward *sustainable* and *organic* construction, organic felts are increasingly replacing fiberglass felts in built-up roofs.
   - **a.** True
   - **b.** False
4. Bitumen-treated fiberglass felt
   - **a.** is less porous than bitumen-treated organic felt.
   - **b.** is more porous than bitumen-treated organic felt.
   - **c.** has the same porosity as bitumen-treated organic felt.
   - **d.** none of the above.
5. Which of the following represents an asphalt-treated fiberglass felt?
   - **a.** Type II
   - **b.** Type III
   - **c.** Type IV
   - **d.** No. 15
   - **e.** No. 30
6. A built-up roof membrane is generally finished with
   - **a.** single-ply roof membrane.
   - **b.** aggregate surfacing.
   - **c.** a flood coat of bitumen.
   - **d.** all the above.
   - **e.** (b) and (c).
7. A roof square implies
   - **a.** 1 ft^2 of roof area.
   - **b.** 10 ft^2 of roof area.
   - **c.** 100 ft^2 of roof area.
   - **d.** 1 m^2 of roof area.
   - **e.** 10 m^2 of roof area.
8. Asphalt is a waste product obtained from
   - **a.** the manufacturing of coal tar.
   - **b.** the manufacturing of rubber.
   - **c.** petroleum refining.
   - **d.** manufacturing of steel.
   - **e.** none of the above.
9. The weight of aggregate surfacing on a built-up roof is generally
   - **a.** 5 lb per roof square.
   - **b.** 15 lb per roof square.
   - **c.** 25 lb per roof square.
   - **d.** 40 lb per roof square.
   - **e.** 400 lb per roof square.
10. Which of the following built-up roof membranes is more commonly used?
    - **a.** Asphalt built-up roof membrane
    - **b.** Tar built-up roof membrane
11. Which of the following roofing asphalts is commonly used for interply mopping?
    - **a.** Type I
    - **b.** Type II
    - **c.** Type III
    - **d.** Type IV
    - **e.** None of the above

Answers: 1-e, 2-a, 3-b, 4-b, 5-c, 6-e, 7-c, 8-c, 9-e, 10-a, 11-c.

31.4 MODIFIED BITUMEN ROOF MEMBRANE

A modified bitumen sheet is similar to built-up roof felt because the waterproofing agent is the bitumen, to which polymers have been added to modify the bitumen's properties. The polymer's addition to the bitumen improves its characteristics so that the modified bitumen is more pliable and elastomeric as compared to the (unmodified) bitumen. The polymer also increases the ultraviolet radiation resistance of the bitumen. Modified bitumen is, therefore, more resistant to cold temperature and to ultraviolet radiation than the bitumen used in a built-up roof membrane.*

Out of the two modified bitumens—modified asphalt and modified coal tar—modified asphalt is far more commonly used. The two most common asphalt modifiers are

- Styrene butadiene styrene (SBS)
- Attactic polypropylene (APP)

The SBS modifier is a synthetic rubber. Therefore, an SBS-modified sheet is more flexible and more resistant to cold temperature than an APP-modified sheet. APP-modified asphalt has greater resistance to ultraviolet radiation, which gives an APP sheet greater weatherability than an SBS sheet.

A modified bitumen sheet consists of a reinforcing mat, called the *carrier,* which is impregnated and coated with modified bitumen on both sides. The carriers commonly used are polyester or fiberglass scrims or both. A sheet that has both fiberglass and polyester reinforcements is referred to as a *dual-carrier* sheet, Figure 31.10. Most manufacturers

NOTE

SBS- and APP-Modified Bitumen Sheets

SBS-Modified Bitumen Sheet

- Greater flexibility
- Greater resistance to cold environment

APP-Modified Bitumen Sheet

- Greater resistance to ultraviolet radiation

*Regular (unmodified) ashphalt and coal tar become relatively brittle in a cold environment. However, this fact does not greatly influence the performance of a built-up roof in cold climates as sometimes believed. For instance, the long-term performance of built-up roofs in Canada and the northern United States compares favorably with those of modified bitumen or single-ply roofs.

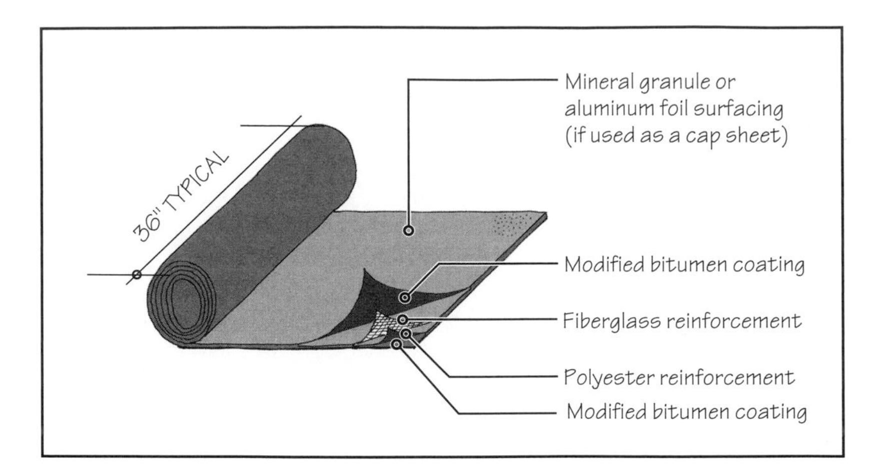

FIGURE 31.10 Anatomy of a modified bitumen membrane with fiberglass and polyester scrims.

make modified bitumen sheets in 36-in. widths, like the built-up roof felt, although the use of 1-m-wide felt is gradually increasing.

Polyester, which is itself a polymer, has high elongation and thus gives pliability to the sheet. It also adds puncture and tear resistance to the sheet, which is important in resisting roof-top traffic and damage caused by roofers' and repairpersons' tools. Fiberglass has very little elongation but provides tensile strength and increases the sheet's fire resistance.

The thickness of a modified bitumen sheet varies from about 100 to 200 mil. Therefore, a modified bitumen sheet is much thicker than an asphalt felt, which is only about 30 mil thick. It is also nonporous, unlike the asphalt-treated felt. A thicker modified bitumen sheet is selected if greater strength is required to withstand greater impact and foot traffic. Because of its greater thickness, a modified bitumen roof needs to consist of only two to three sheets to give almost the same performance as 4 to 5 plies of a built-up roof.

A modified bitumen sheet may either be smooth or surfaced with mineral granules or metal foil. A smooth-surfaced sheet is generally used as an intermediate ply or as a base sheet in a 2- or 3-ply modified bitumen membrane. A mineral granule or foil-faced sheet is used as a cap sheet. The commonly used metal foils are aluminum, copper, and stainless steel.

A cap sheet performs the same function as the flood coat and surfacing in a built-up roof. Mineral granules (typically $\frac{1}{10}$-in. stone chips) and metal foil function as protective covers in a cap sheet. If a smooth-surfaced modified bitumen sheet is used as a cap sheet, it must be protected by a bitumen flood coat and aggregate, like a typical built-up roof.

A typical SBS-modified granule-surfaced cap sheet roll.

SBS-Modified Bitumen Membrane

An SBS sheet is typically mop applied with hot asphalt. SBS sheets requiring torch application and, more recently, self-adhering membranes are also available. The low-temperature flexibility (elastomeric nature) of an SBS sheet provides some advantage in colder climates over a conventional built-up roof or an APP-modified bitumen roof.

With hot-asphalt-mopped SBS sheets, Type IV roofing asphalt is commonly used. Type IV asphalt has an application temperature of nearly 500°F—a temperature high enough to soften the SBS-modified asphalt in SBS sheets. (SBS-modified asphalt softens at approximately

EXPAND YOUR KNOWLEDGE

The Terms: Membrane, Felt, and Sheet

As mentioned in Section 31.3, the individual sheets between bitumen moppings in a built-up roof are referred to as *felts,* and the entire assembly of felts and bitumen moppings is referred to as a *membrane.* When modified bitumen sheets were first introduced (mid-1960s), it was thought that only one modified bitumen sheet would perform as well as an assembly of felts and bitumen moppings in a built-up roof—the built-up roof membrane. Therefore, the modified bitumen sheets were referred to as *membranes* and were even included in the single-ply membrane family.

It was subsequently realized that a single sheet of modified bitumen is inadequate and that two or more sheets (plies) are generally needed to provide a reasonable roof. However, each such sheet is referred to as a membrane in some roofing literature, which is an incorrect use of the term *roof membrane.* A membrane is the entire waterproofing layer on the roof. It may consist of 1 ply, as in a single-ply roof membrane; an assembly of 4 to 5 plies, as in a built-up roof; or 2 to 3 plies, as in a modified bitumen roof.

Similarly, the term *felt,* in roofing vernacular, is reserved for either an organic or fiberglass sheet, typically used in a built-up roof, which together with a bitumen mopping constitutes one built-up roof ply. Other roofing sheets are not referred to as felts. For instance, in a commonly used modified bitumen membrane assembly, consisting of two plies, the first ply is called a *base sheet,* and the second ply is called a *cap sheet.* The base sheet is usually a smooth-surfaced modified bitumen sheet, and the cap sheet is a granule-surfaced modified bitumen sheet.

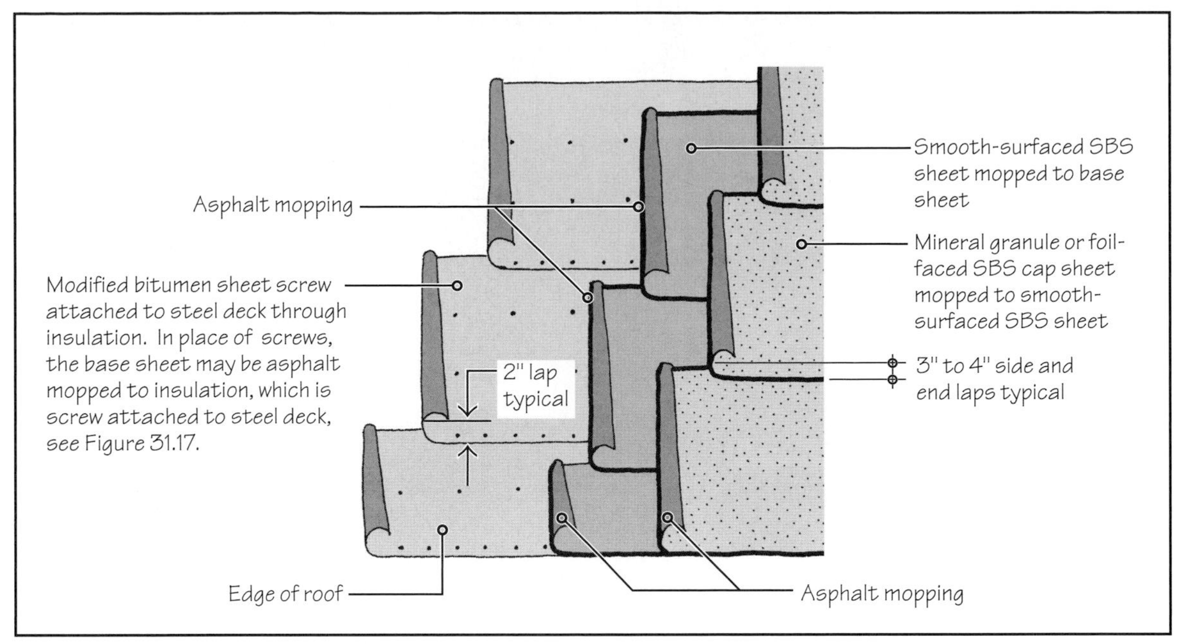

FIGURE 31.11 A 3-ply SBS-modified roof membrane, consisting of granule- or foil-faced SBS-modified bitumen cap sheet and two smooth-surfaced SBS sheets.

350°F.) The amount of asphalt required is the same as for interply moppings in a built-up roof, that is, nearly 25 lb per roof square.

A 2- or 3-ply SBS membrane gives good performance under most conditions. A 2-ply membrane may consist of a smooth-surfaced SBS base sheet covered by a granule-surfaced SBS cap sheet. A foil-faced SBS cap sheet may be substituted for a granule-surfaced cap sheet. A 3-ply SBS membrane consists of two smooth-surfaced sheets, followed by a mineral or foil-faced SBS cap sheet, Figure 31.11.

Built-Up and SBS Hybrid Membrane

An SBS-modified bitumen sheet can also be combined with a built-up roof membrane to give an SBS and built-up hybrid. Such a membrane may consist of a 2- to 4-ply conventional built-up roof followed by a granule or foil-faced SBS cap sheet. In place of an SBS cap sheet, a smooth-surfaced SBS sheet with asphalt flood coat and gravel surfacing may be used, as in a built-up roof, Figure 31.12.

The asphalt for the flood coat is generally the regular (unmodified) roofing asphalt, although SBS-modified asphalt can also be used. SBS-modified asphalt is compatible with (regular) roofing asphalt and can be used with built-up roof felts or with SBS sheets.

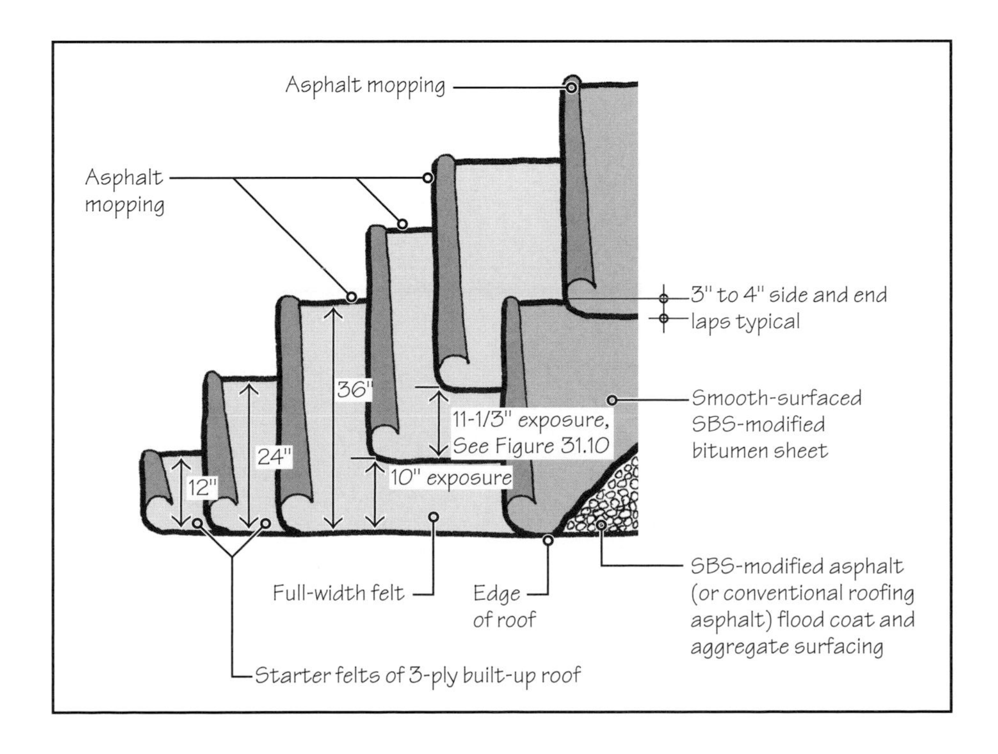

FIGURE 31.12 A built-up and SBS hybrid consisting of smooth-surfaced SBS-modified bitumen sheet and aggregate-covered flood coat over a 3-ply built-up roof. Note that built-up roof felts are installed in shingled format (see "Principles in Practice" at the end of this chapter).

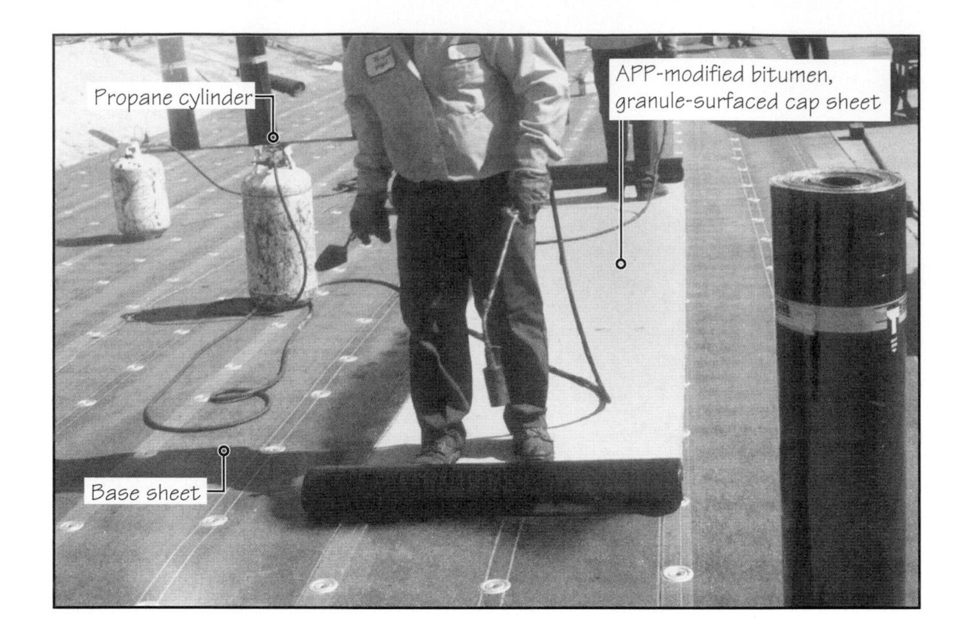

FIGURE 31.13 Granule-surfaced, APP-modified bitumen cap sheet being adhered to a base sheet with the help of a propane torch. The base sheet has been attached to the underlying steel deck through rigid-board foam insulation using screws.

APP-Modified Bitumen Membrane

Just as the addition of an SBS modifier raises the softening-point temperature of asphalt, so does the addition of an APP modifier. In fact, the addition of an APP modifier raises the softening-point temperature of the APP-modified asphalt so much that an APP-modified membrane cannot be hot mopped like the SBS sheet or a built-up roof felt. The application temperature of even type IV asphalt (nearly 500°F) is not high enough to melt the asphalt in an APP sheet for good adhesion.

FIGURE 31.14 A dragon wagon with its multiple torches. (Photo courtesy of Siplast Inc.)

The most common method of installing an APP sheet is through the open flame of a handheld propane torch, Figure 31.13, which has a temperature of more than 1,000°F. In this application, the flame is applied to the sheet roll from one end to the other. As the flame melts the modified asphalt on the underside of the sheet, the roll is unwound and pressed down. In this way, the sheet is fully adhered to the substrate. The sheets are usually factory laminated with a thin polyethylene release sheet to prevent adhesion in the roll. The torch burns the polyethylene sheet. Some manufacturers use sand dusting in place of a polyethylene sheet.

An alternative to the hand-held torch is a *dragon wagon,* which consists of a series of propane torch heads mounted on a wheelable cart that also carries the APP sheet roll, Figure 31.14. A dragon wagon speeds up the installation of an APP sheet but should be used only on a large open roof with few penetrations.

Torched Membrane's Suitability

A torched membrane is suitable in situations where the hot bitumen used in a conventional built-up roof (or with an SBS roof) cannot be pumped from the ground to the roof, such as on the roof of a high-rise building. The torched system has similar advantage in cold climates, where keeping the asphalt hot enough between the kettle and the point of application is a problem.

In fact, the torch-adhered, SBS-modified bitumen roof may be preferred for a colder region for the preceding reasons and also because of the greater flexibility of an SBS roof as compared with an APP or a built-up roof. However, the open torch used in laying a torchable membrane presents a fire hazard, which must be considered before specifying a torched membrane. Fire extinguishers must be available on the roof for any emergency, and personnel need to be left on the roof for "fire watch" for a few hours after the day's work.

Although an APP sheet has a degree of ultraviolet resistance, a surfacing over it gives additional protection. Mineral granule or foil surfacing on an APP cap sheet adds to its durability. Where practical, an APP roof should be flood coated and covered with gravel for greater durability and fire resistance.

APP and Built-Up Roof Hybrid

A two-layer APP membrane—the first layer, a smooth-surfaced APP sheet and the top layer, a mineral-surfaced APP cap sheet—can provide an adequate membrane in most situations. Alternatively, an APP cap sheet may be used over 2 to 3 plies of built-up roof felt, giving an APP and built-up roof hybrid membrane.

31.5 SINGLE-PLY ROOF MEMBRANE

Although a built-up roof membrane (and, to a smaller extent, a modified bitumen membrane) is constructed on the roof felt by felt, a single-ply roof has only a 1-ply membrane and does not require the cumbersome use of hot bitumen. It is easier to lay because it has only 1 ply.

A single-ply membrane is made of a polymeric material. Because a polymer is a synthetic material, it is capable of a great deal of chemical manipulation by minor changes in one or more of its constituents. Therefore, a large number of polymers are used as roof membranes. However, polymers (plastics), in general, can be divided into two broad categories:

- *Thermosetting polymer (unweldable plastic):* a polymer that does not soften on heating once it is cured into hardness, like a boiled egg. Therefore, the seams in a thermosetting membrane cannot be heat welded but must be adhered together by an adhesive or a double-sided seam tape. The most commonly used thermosetting membrane is EPDM, an acronym that stands for ethylene propylene diene monomer.
- *Thermoplastic polymer (heat-weldable plastic):* a polymer that softens on heating and hardens on cooling over and over again, like butter. Asphalt and coal tar behave like a thermoplastic. A thermoplastic material is heat weldable, which makes seaming operation relatively easier. However, thermoplastic polymer is not as stretchable as thermosetting polymer. Currently, the most commonly used thermoplastic roof membranes are PVC (polyvinyl chloride) and TPO (thermoplastic polyolefin).

ADVANTAGES AND DISADVANTAGES OF A SINGLE-PLY MEMBRANE

By virtue of the material of which it is made, a single-ply membrane is highly flexible and can withstand large elongation before failure. For example, a typical EPDM membrane will stretch 300% to 500% of its original length before tearing apart, as compared to 50% for an SBS-modified bitumen sheet and 4% for a fiberglass-reinforced built-up roof felt.

For a variety of reasons, a typical single-ply membrane comes in much wider and longer rolls than a built-up roof felt or a modified bitumen sheet, Figure 31.15. Although advantages (because it reduces the number of lap splices, which reduces lap-failure possibilities), a large roll size also means that roof penetrations are more difficult to work around. Therefore, a single-ply membrane has an advantage on a large roof with few penetrations.

The relative flexibility and elongation characteristic of a single-ply membrane is useful in climates with a large annual or daily temperature variation. A single-ply membrane (particularly EPDM) retains its elastomeric character even in subzero Fahrenheit temperatures. Another advantage of a single-ply membrane is that it does not require any aggregate surfacing.

A major disadvantage of a single-ply roof membrane is the absence of second, third, or additional plies. This implies a lack of redundancy in a single-ply roof—a redundancy that is helpful in any waterproofing application. A puncture or split results in a roof leak more

FIGURE 31.15 Although this is a 30-ft-wide EPDM membrane (with a center fold so that the roll is only 15 ft wide), single-ply membrane rolls are available in widths larger than 30 ft, shown here.

readily in a single-ply roof than in a built-up roof because of the greater thickness, multiple plies, and self-healing property of bitumen in a built-up roof.

As stated previously, the overall market share of single-ply membranes is approximately the same as that of built-up or modified bitumen membranes—that is, roughly one-third each. Because EPDM, PVC, and TPO are the more commonly used single-ply membranes, the following discussion relates to them only, although it is general enough to apply to other single-ply membranes also.

EPDM MEMBRANE

Because EPDM is a synthetic rubber, it is essentially similar to the material used in automobile tires. Firestone is, therefore, one of the major EPDM membrane manufacturers. The typical thickness of an EPDM membrane is 45 to 60 mil, although some manufacturers make a thicker membrane for use where a higher puncture and tear resistance is needed.

EPDM polymer lacks inherent fire resistance, but fire-resistive EPDM membranes are available. However, with a loose-laid, ballasted EPDM roof, a non-fire-resistive membrane may be specified because the ballast will provide the fire resistance. Being a synthetic rubber (and the consequent flexibility), the use of EPDM is more common in colder climates of North America. In warmer climates and in response to sustainability considerations, several architects and roof consultants specify white membranes, which are produced more economically in PVC or TPO.

PVC MEMBRANE

PVC (polyvinyl chloride), the most versatile plastic, is available in two types. The rigid PVC is used for pipes, siding, and window sashes. The soft and pliable type is used in shower curtains, raincoats, electrical wire insulation, and roof membranes. Softness and flexibility—properties that are necessary for a roofing membrane—are obtained through the addition of a plasticizer. However, despite the addition of a plasticizer, a PVC membrane, though flexible, is far less stretchable than an EPDM membrane.

The primary advantage of a PVC membrane is that the seams in the membrane can be heat fused. A simple tool, such as that shown in Figure 31.16, which supplies hot com-

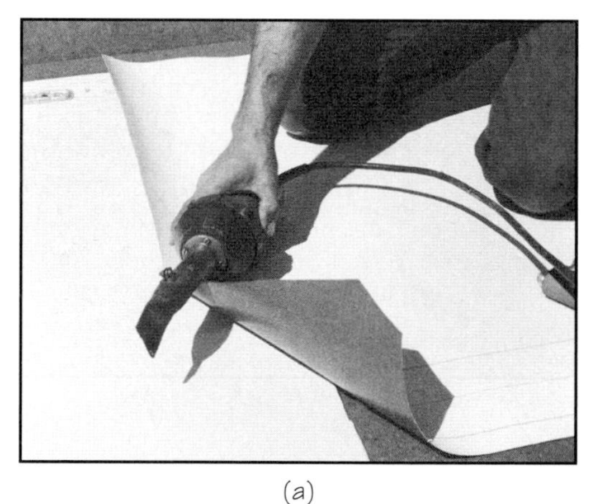

(a)

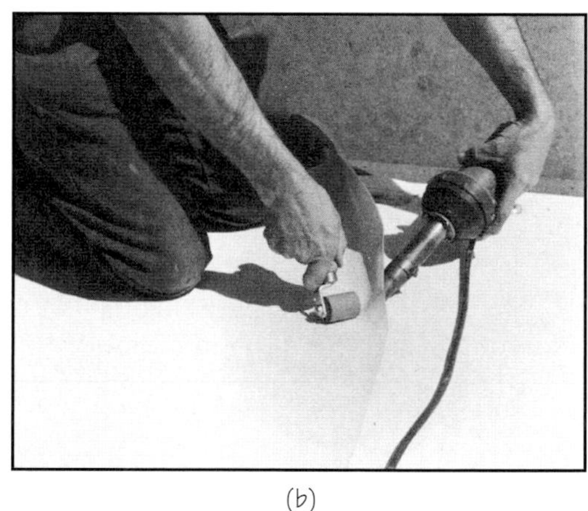

(b)

(c)

FIGURE 31.16 **(a)** Hand-held welding tool that supplies hot compressed air through a flat nozzle. **(b)** A roofer welds the seams of a white PVC membrane and presses down the fused seams with a roller. **(c)** Self-propelled hot-air welding machine that welds and presses the seams—an alternative to the hand-held welding tool. The machine speeds the welding operation.

EXPAND YOUR KNOWLEDGE

EPDM, PVC and TPO Membranes Compared

Criterion	EPDM	PVC	TPO
Lap-seam attachment	Two-sided tape or contact adhesive	Seams heat welded	Seams heat welded
Fire resistance	Fire-resistive EPDM available	Inherently good	Fire-resistive TPO available
White-sheet availability	Available	More easily made	More easily made
Resistance to UV	Good for black EPDM	Good	Good
Compatibility with insulation	Compatible with most insulations	Incompatible with polystyrene foam due to plasticizers	Compatible with most insulation
Compatibility with asphalt	Incompatible	Incompatible	Incompatible

pressed air, is all that is needed to fuse the seams. This makes roof installation much easier and more economical as compared to EPDM roofs, in which the seams must be joined together with a double-sided tape.

Another advantage of PVC is its inherent fire resistance, which is provided by the chlorine atom in polyvinyl chloride. PVC membranes are generally used in 45- to 80-mil thickness. They are available in several colors, although white and gray are most common.

TPO Membrane

Like PVC, TPO is also thermoplastic. Therefore, it has the same advantages over EPDM as PVC, that is, the heat weldability of its seams. Although less flexible than EPDM, TPO is inherently more flexible than PVC, not requiring the addition of plasticizers in its chemical formulation.

This makes TPO more economical as compared with PVC; also, there is no concern with the gradual evaporation of plasticizers (as with PVC), which causes PVC membrane to loose its flexibility over time. Thus, TPO combines the advantages of EPDM as well as of PVC, which explains the increasing popularity of TPO among the single-ply membranes.

PRACTICE QUIZ

Each question has only one correct answer. Select the choice that best answers the question.

12. Which of the following modified bitumen membranes is more elastomeric?
- **a.** SBS mod bit
- **b.** APP mod bit

13. Which of the following modified bitumen membranes is generally torch applied?
- **a.** SBS mod bit
- **b.** APP mod bit

14. Which of the following roofing asphalts is commonly used for mopping an SBS-modified bitumen membrane?
- **a.** Type I
- **b.** Type II
- **c.** Type III
- **d.** Type IV
- **e.** SBS-modified asphalt

15. A dragon wagon is used in laying
- **a.** a single-ply roof membrane.
- **b.** a modified bitumen roof membrane.
- **c.** a built-up roof membrane.
- **d.** all the above.
- **e.** none of the above.

16. EPDM is a thermosetting plastic.
- **a.** True
- **b.** False

17. The lap seams in an EPDM roof membrane are typically sealed by
- **a.** heat welding.
- **b.** liquid adhesives.
- **c.** single-sided tape.
- **d.** double-sided tape.
- **e.** any one of the above.

18. The lap seams in a PVC roof membrane are typically sealed by
- **a.** heat welding.
- **b.** liquid adhesives.
- **c.** single-sided tape.
- **d.** double-sided tape.
- **e.** any one of the above.

19. The lap seams in a TPO roof membrane are typically sealed by
- **a.** heat welding.
- **b.** liquid adhesives.
- **c.** single-sided tape.
- **d.** double-sided tape.
- **e.** any one of the above.

20. Of the two single-ply roof membranes, EPDM and PVC, PVC is generally
- **a.** more flexible and more fire resistant.
- **b.** less flexible but more fire resistant.
- **c.** more flexible but less fire resistant.
- **d.** less flexible and less fire resistant.

Answers: 12-a, 13-b, 14-d, 15-b, 16-a, 17-d, 18-a, 19-a, 20-b.

31.6 RIGID-BOARD INSULATION AND MEMBRANE ATTACHMENT

A type of insulation commonly used in a low-slope roof consists of semirigid boards. Batt and blanket insulations are not suitable for low-slope roofs because of their low compressive strength. Two commonly used semirigid board insulations (both are plastic foams, see "Principles in Practice," Chapter 5) are

- Polyisocyanurate boards (also called *iso boards*)
- Extruded polystyrene boards (also called *XEPS boards*)

Iso boards have a much wider applicability than XEPS boards. An XEPS board is chemically incompatible with some membranes, particularly a PVC membrane.

Between the roof membrane and the insulation, an intervening medium (called a *cover board*) is commonly used to provide compatibility between the membrane and the insulation. With hot-applied (hot asphalt or torched) membranes, such as built-up and modified bitumen membranes, the most commonly used cover board is perlite board (typically 1 in. thick). Perlite board also adds to the R-value of the system, although it is not as effective an insulator as iso board or an XEPS board.

With single-ply membranes, fiberglass-reinforced gypsum board (typically $\frac{1}{4}$ in. thick) is commonly used. Apart from providing compatibility, gypsum board adds to the rigidity to the roof system, providing a better bearing surface for foot traffic and greater resistance to hail damage.

NOTE

Cover Board

With hot-applied membranes, the cover board is perlite board. If the asphalt is applied directly over iso board or XEPS board, both of which are sensitive to the high temperature of asphalt, the membrane tends to blister. Perlite board, which has a good high-temperature resistance, has been found to be a good cover board with hot-applied membranes.

With single-ply membranes, the use of cover board is not required but is preferred. The commonly used cover board with single-ply membranes is fiberglass-reinforced gypsum board. It is generally not recommended under a hot-applied membrane because of its chemically combined water (Chapter 14). Under the high temperature of asphalt (approximately 400°F), the water in gypsum converts to steam. This causes blisters to form under the membrane. Additionally, the high temperature of asphalt may result in the disintegration of gypsum.

INSULATION ATTACHMENT—BUILT-UP AND MODIFIED BITUMEN MEMBRANES

The insulation used with built-up and modified bitumen membranes is iso board covered with perlite board. The insulation and cover-board combination also helps prevent thermal bridging through insulation joints. Therefore, the joints between the two layers are generally staggered.

Typical thickness of perlite board is 1 in. The thickness of iso board varies, depending on the the required R-value of the roof assembly. Figure 31.17 illustrates the arrangements for the attachment of insulation to steel and concrete decks and that of membrane to the insulation.

As Figure 31.17 shows, the insulation is attached to a steel deck with screws. To a concrete deck, the insulation may be adhered with bitumen mopping, provided the deck is dry. If the deck is not fully dry, the moisture in the deck will blister through bitumen as steam, reducing attachment strength. Nonbituminous adhesives are available that do not require a fully dry concrete deck. Spot or strip mopping is generally used with such adhesives, Figure 31.18.

INSULATION ATTACHMENT—SINGLE-PLY MEMBRANES

With a single-ply membrane, the combination of iso board and cover board in staggered layers is generally used. Whereas the attachment of built-up and modified bitumen membranes

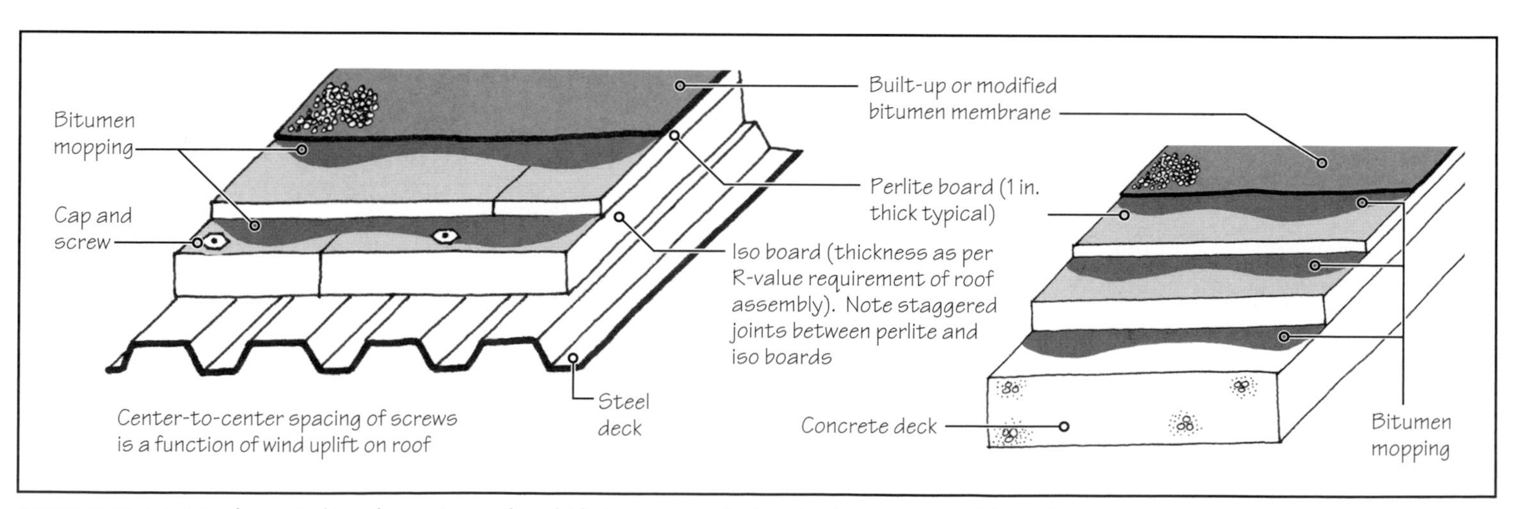

FIGURE 31.17 Attachment of insulation to steel and (dry) concrete decks—built-up and modified bitumen membranes.

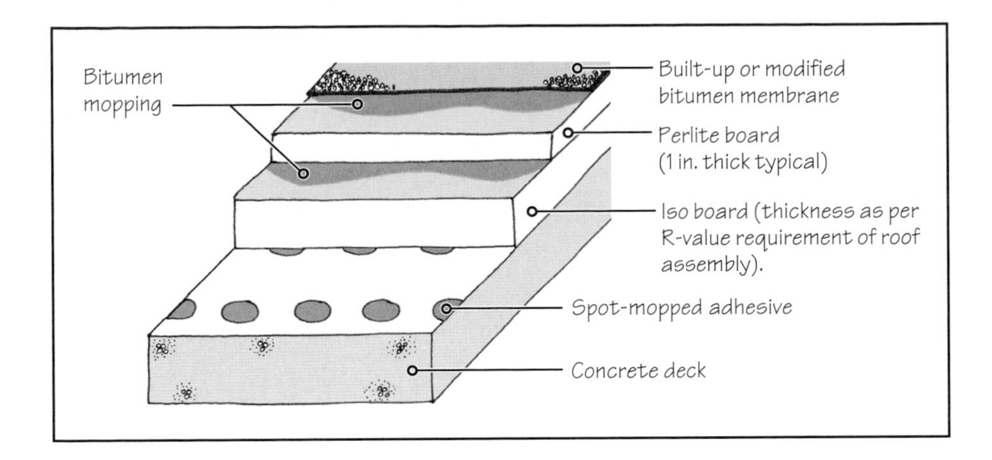

FIGURE 31.18 Attachment of insulation to a concrete deck with nonbituminous spot- or strip-mopped adhesive, generally used with a concrete deck that may not be fully dry at the time of insulation attachment.

is through bitumen mopping, a single-ply membrane (EPDM, PVC, and TPO) is attached using one of the following three systems:

- Fully adhered
- Mechanically fastened
- Ballasted

FULLY ADHERED SYSTEM OF SINGLE-PLY MEMBRANE ATTACHMENT

With a fully adhered single-ply membrane, manufacturer-provided adhesive is applied to the underside of the entire membrane, Figure 31.19(a). After adhering the membrane, the laps between adjacent membranes are joined using a double-sided tape, embedded in lap seams, Figure 31.19(b).

A fully adhered, single-ply membrane system is labor intensive compared to a mechanically fastened or a ballasted single-ply system. However, the system is particularly well suited to high-wind regions.

MECHANICALLY FASTENED ATTACHMENT SYSTEM FOR SINGLE-PLY MEMBRANE

In a mechanically fastened system, the membrane is laid over the insulation, which has already been fastened to the deck. (The fastening of insulation to the deck is nominal—to hold the insulation in place until the membrane has been anchored.) The spacing of fasteners for the anchorage of membrane to the deck is a function of the wind uplift on the roof.

The membrane-fastening system consists of screws and a continuous bar (referred to as a *batten bar* in the industry), through which the screws are applied, Figure 31.20(a). The batten bar, typically made of metal or plastic, is covered over by double-sided tape in a lap seam. Intermediate fasteners and the batten bar, which are not within the seams, are covered with self-adhesive splice tape, Figure 31.20(b). Some manufacturers provide individual plates instead of batten bars for their systems.

Under the effect of wind, a mechanically fastened system is subjected to concentrated loads at the fasteners. This is unlike a fully adhered system, in which the loads are distributed over the entire membrane. Wind may make the mechanically fastened membrane flutter, causing its premature failure at the fasteners. Therefore, care should be taken in specifying a mechanically fastened system in a high-wind location.

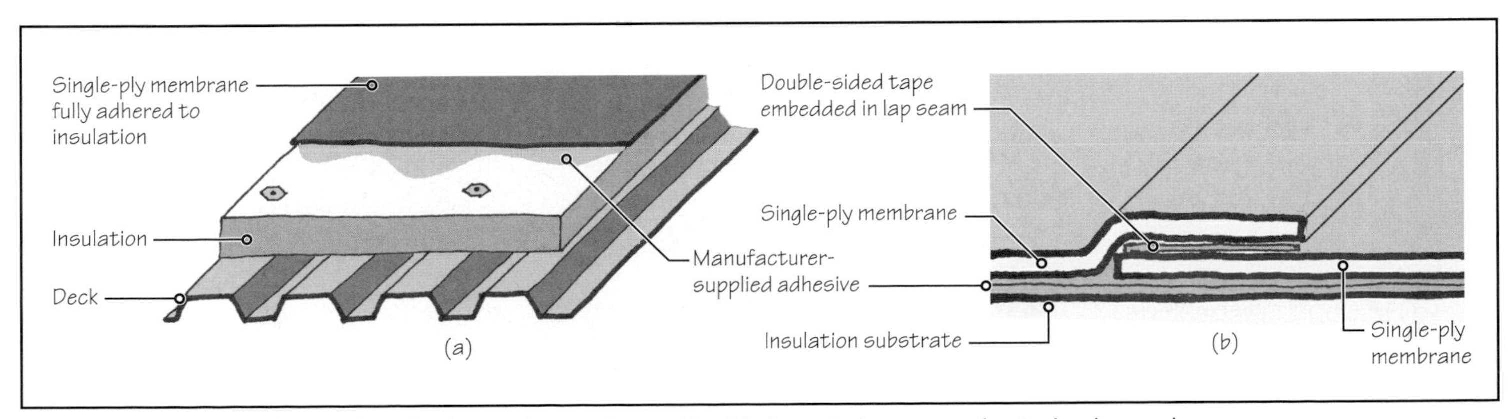

FIGURE 31.19 **(a)** Fully adhered single-ply membrane. **(b)** Double-sided tape in lap seams of a single-ply membrane.

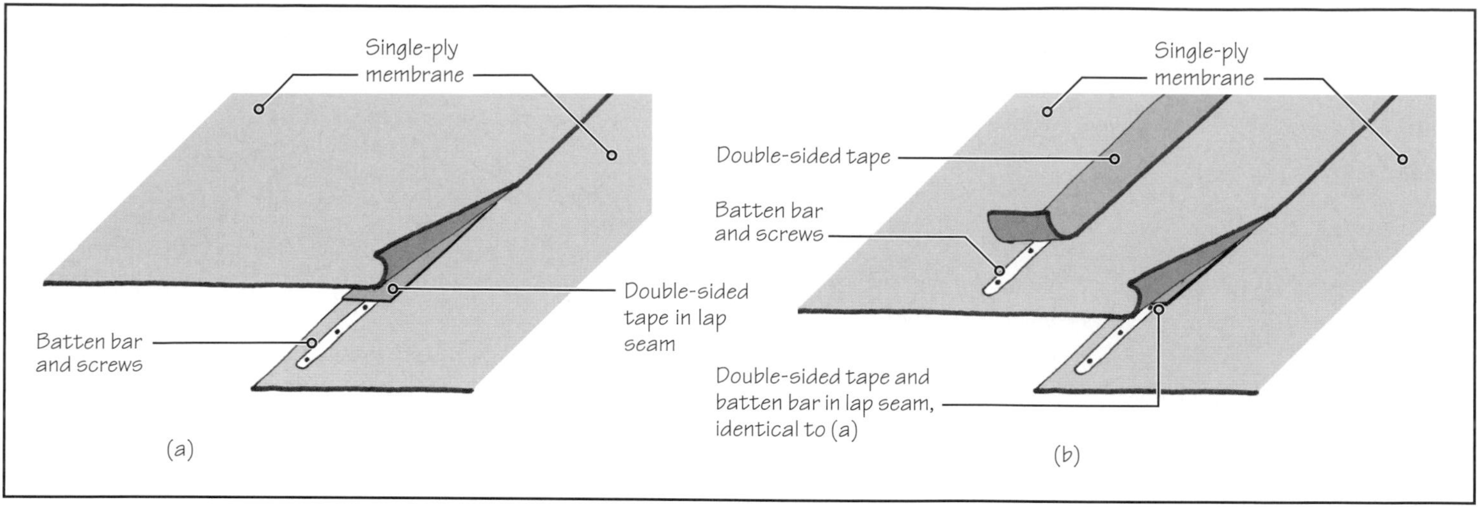

FIGURE 31.20 **(a)** The anchorage of a mechanically fastened single-ply membrane utilizes a batten bar and screws in lap seams of membranes. **(b)** Where the membrane requires fastening in areas other than the seams, a batten bar, screws, and double-sided tape are used.

Loose-Laid, Ballasted System of Single-Ply Membrane Attachment

In a loose-laid, ballasted system, the entire membrane is laid loose over the insulation, followed by a loose-laid single-ply membrane, Figure 31.21. The lap seams are adhered together using a double-sided tape (thermosetting membrane) or heat-welded thermoplastic membrane. The membrane is anchored to the deck only at the roof perimeter, at curbs and penetrations. The entire membrane is then covered with ballast, Figure 31.22.

Ballast generally consists of river-worn gravel ($1\frac{1}{2}$-in. to 2-in. particle size). The use of crushed stone is discouraged because it can puncture the membrane under foot traffic. The weight of gravel required varies with the wind uplift on the membrane. Concrete pavers may also be used instead of gravel.

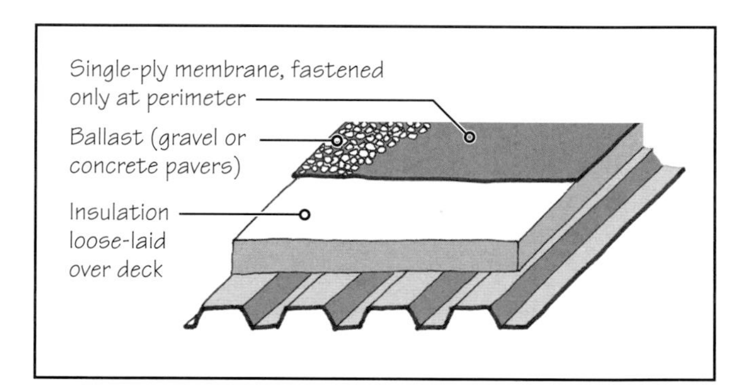

FIGURE 31.21 Loose-laid, ballasted single-ply membrane system. Note that insulation is not fastened to the roof deck.

FIGURE 31.22 In a loose-laid, ballasted, single-ply membrane system, the ballast is generally brought on the roof using a crane and buckets. Ballast spreaders are then used to spread the ballast on the roof.

The ballasted system is the most economical single-ply system. The ballast adds to fire resistance and weatherability. The system is recommended only for roofs of low-rise buildings in low-wind regions. Because the dead load of ballast is high, the deck must be structurally adequate to withstand the ballast load.

31.7 INSULATING CONCRETE AND MEMBRANE ATTACHMENT

A low-slope roof insulation (other than the rigid boards) is lightweight insulating concrete or foamed concrete. As described in in Chapter 5 ("Principles in Practice"), insulating concrete or foamed concrete is poured over a roof deck in wet format. When the concrete hardens, it becomes monolithic with the deck.

Insulating Concrete

Where insulating concrete is used over a steel deck, the use of a slotted (i.e., perforated) deck is recommended. The perforations in the deck allow the moisture in the concrete to gradually vent out from the underside of the deck by the building's heating and cooling system. Therefore, roof membrane can be installed as soon as the concrete has become sufficiently hard to hold the fasteners, even though it may not have fully dried.

After the insulating concrete has hardened (typically after 3 to 5 days), an asphalt-treated base sheet (preferably a mineral granule faced base sheet) is fastened to the concrete as a substrate for a built-up or modified bitumen roof. Special fasteners have been developed by the industry to fasten the base sheet, which can simply be driven into the insulating concrete, Figure 31.23. The fasteners flare as they penetrate into the concrete, Figure 31.24.

After the base sheet has been fastened, the built-up roof or modified bitumen membrane is mopped to the base sheet. A single-ply membrane may also be installed over insulating concrete, but the manufacturers of insulating concrete recommend consulting with manufacturers of single-ply membrane for compatibility, and so on.

(a)

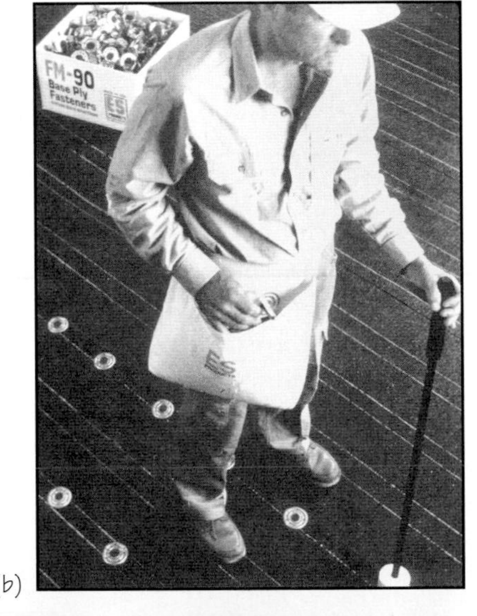

(b)

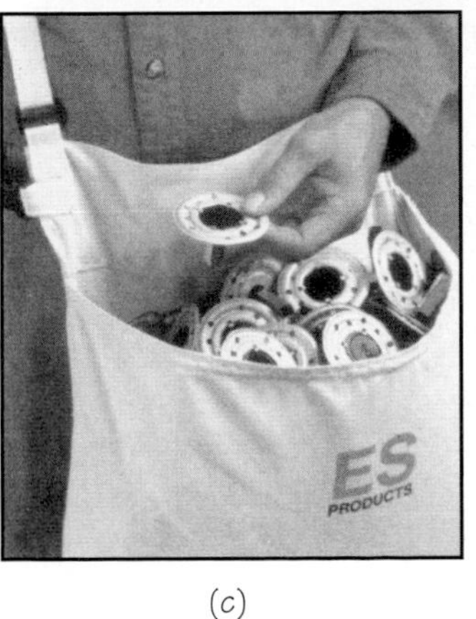

(c)

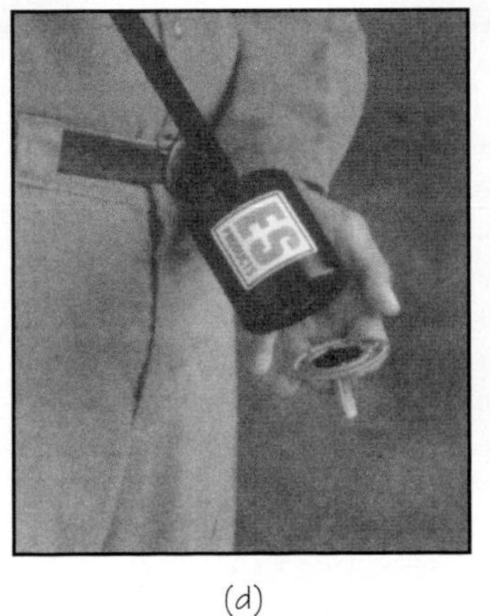

(d)

(e)

FIGURE 31.23 **(a)** A typical self-locking sheet-metal fastener used for anchoring the base sheet to insulating or foamed concrete. **(b)** A roofer fastens the base sheet with a magnetic driver. **(c)** A roofer with a pouch full of fasteners. **(d)** Embedded magnets in the head of the driver orient and hold the fastener perpendicular to the roofing surface. **(e)** A vertical "tamping" motion sets the fastener securely into the concrete. (Photos courtesy of ES Products, Inc.)

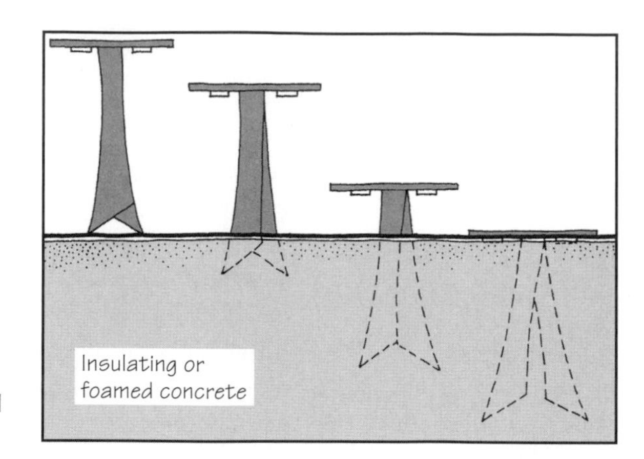

FIGURE 31.24 The fastener flares as it is driven into the concrete. (Sketch adapted from that of ES Products, Inc.)

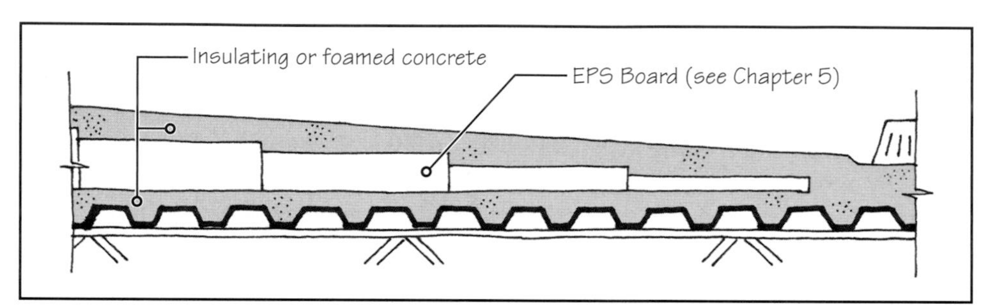

FIGURE 31.25 Section through a roof showing stair-stepped EPS board pattern to achieve a slope on a dead-level roof deck. See also Figures 11 to 13 ("Principles in Practice," Chapter 5).

Because below-the-deck venting cannot be provided in a reinforced concrete or a precast concrete deck, the use of insulating concrete on such decks requires that the insulating concrete be so dry that it does not create water vapor problems in the roofing membrane.

A major advantage of insulating concrete is that roof slope can easily be provided in the insulation, even if the roof deck is dead level. This is done by using expanded polystyrene (EPS) boards of different thicknesses to create a stair-stepped pattern, Figure 31.25. Providing slope over a dead-level deck with tapered rigid board insulation is expensive and complicated. Insulating concrete is also well suited with decks that have a complex slope geometry.

FOAMED CONCRETE

As explained in Chapter 5 ("Principles in Practice"), foamed concrete, also called *cellular concrete,* is a nonaggregate (insulating) concrete that consists of portland cement, water, and a liquid foaming concentrate. Foamed concrete requires only one-third the amount of water used in insulating concrete. Therefore, foamed concrete does not need a slotted metal deck, although its use is preferred. The other details, such as fastening a base sheet, stair-stepping EPS boards for roof slope, and so on, are similar to those used with insulating concrete.

PRACTICE QUIZ

Each question has only one correct answer. Select the choice that best answers the question.

21. Perlite board (as cover board), sandwiched between the insulation and the roof membrane, is generally used with
- **a.** asphalt-mopped membrane.
- **b.** fully adhered single-ply membrane.
- **c.** mechanically fastened single-ply membrane.
- **d.** loose-laid, ballasted single-ply membrane.

22. The cover board generally used between a single-ply roof membrane and rigid board insulation is
- **a.** iso board.
- **b.** specially treated perlite board.
- **c.** glass-fiber-reinforced gypsum board.
- **d.** extruded polystyrene board.
- **e.** none of the above.

23. A mechanically fastened single-ply roof membrane is generally used with
- **a.** steel roof decks.
- **b.** reinforced concrete decks.
- **c.** precast-concrete decks.
- **d.** none of the above.

24. A loose-laid, ballasted roof is generally used with
- **a.** built-up roof membrane.
- **b.** mod-bit membrane.
- **c.** single-ply membrane.
- **d.** all the above.

25. For a building located in a high-wind region, which of the following roof systems is most suitable?
a. Loose-laid ballasted single-ply system
b. Fully adhered single-ply system
c. Mechanically fastened single-ply system
d. Gravel-covered built-up roof
e. Protected membrane system

26. With a steel deck topped with lightweight insulating concrete, the commonly used board insulation is
a. iso board.
b. polyurethane board.
c. expanded polystyrene (EPS) board.
d. not used with insulating concrete.

27. A slotted steel roof deck is mandated for use with
a. all rigid-board insulations.
b. lightweight insulating concrete.
c. foamed concrete.
d. none of the above.

Answers: 21-a, 22-c, 23-a, 24-c, 25-b, 26-c, 27-b.

31.8 LOW-SLOPE ROOF FLASHINGS

The termination of a roof, which generally occurs over a wall, requires careful detailing. The wall either stops below the roof, in which case the roof terminates into a free edge (requiring a guard edge), or the wall projects above the roof as a parapet wall. In either case, the differential movement between the roof and the wall and the higher wind loads at roof terminations increase the stresses in a roof membrane at a wall-roof junction. Additional strength is, therefore, required at a roof-wall junction, which is provided by a special sheet material called *flashing*. Flashing also helps to integrate the roof membrane with the wall and further waterproofs the wall-roof junction, which is more vulnerable to leakage than the field-of-roof, which is the area of the roof that is not a roof perimeter, corner, or termination.

Flashing is also required around penetrations in the roof, such as roof drains, gutters, vent and plumbing stacks, curbs below skylights and roof-top equipment, expansion joints, and area dividers.

Flashings may be divided into three basic types, Figure 31.26:

- Base flashings
- Curb flashings
- Flange flashings

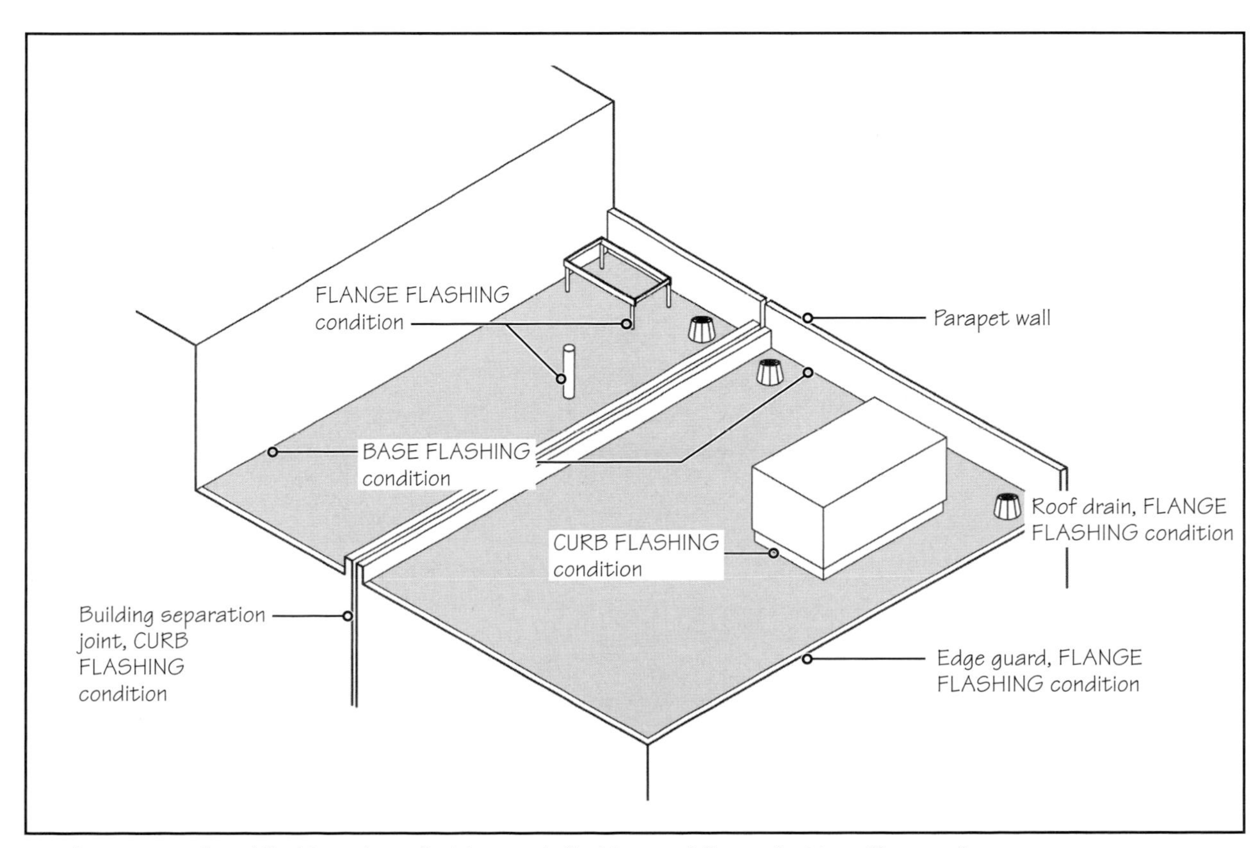

FIGURE 31.26 Three types of roof flashings; base flashing, curb flashing, and flange flashing, illustrated.

31.9 BASE FLASHING DETAILS

Base flashing occurs at the junction of a roof and a wall that rises above the roof, such as a parapet wall. Five important components of base flashing details are

- Membrane flashing
- Wood nailers
- Cants (required only with built-up and modified bitumen membranes)
- Counterflashing
- Parapet coping

Figure 31.27 illustrates the principles of detailing a base flashing against a low-height parapet (up to about 2 ft high) where the roof deck is supported by a bearing wall.

Figure 31.28 illustrates the same principles, but the roof deck is supported on a structural frame. In this case, the detail must allow differential movement between the exterior wall and the roof, which is provided for by separating the roof from the wall at base flashing level. Base flashing details for a high parapet are essentially similar but are a little more complex [1].

Membrane Base Flashing

Membrane base flashing provides additional reinforcement to the roof membrane and extends the waterproofing ability of the roof to the highest anticipated water level (8-in. minimum is recommended). Membrane flashing also provides additional waterproofing at the roof-wall junction, which, being under greater stress, is more prone to leakage than the field-of-roof.

Membrane base flashing must have adequate strength and pliability to withstand minor movement between the wall and the roof. Therefore, the most commonly used membrane base flashing material with a built-up roof is an SBS-modified bitumen sheet. An SBS membrane base flashing is also used with an SBS-modified bitumen roof. An SBS-modified

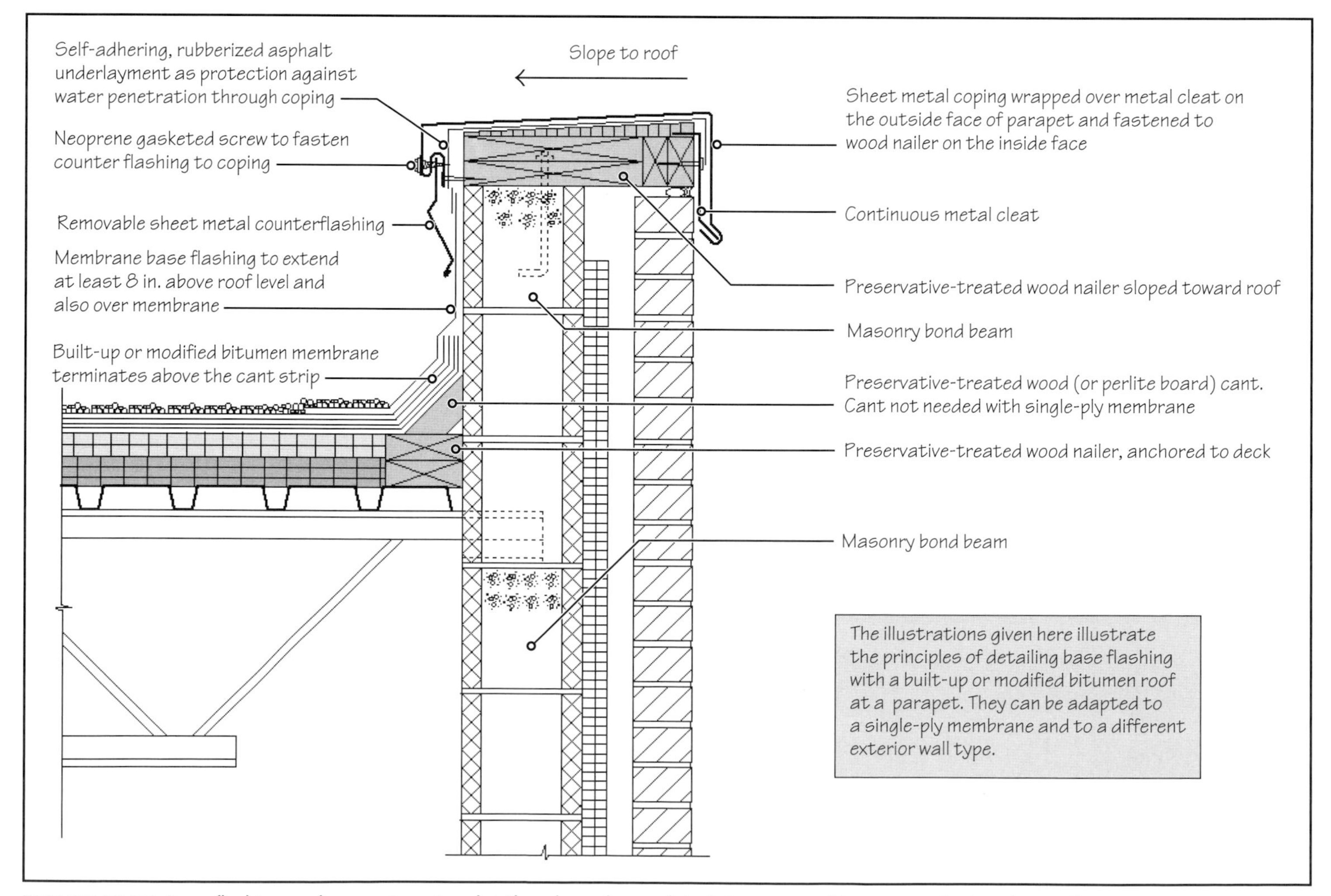

FIGURE 31.27 Base flashing and parapet coping details, where the roof deck is supported by an exterior bearing wall.

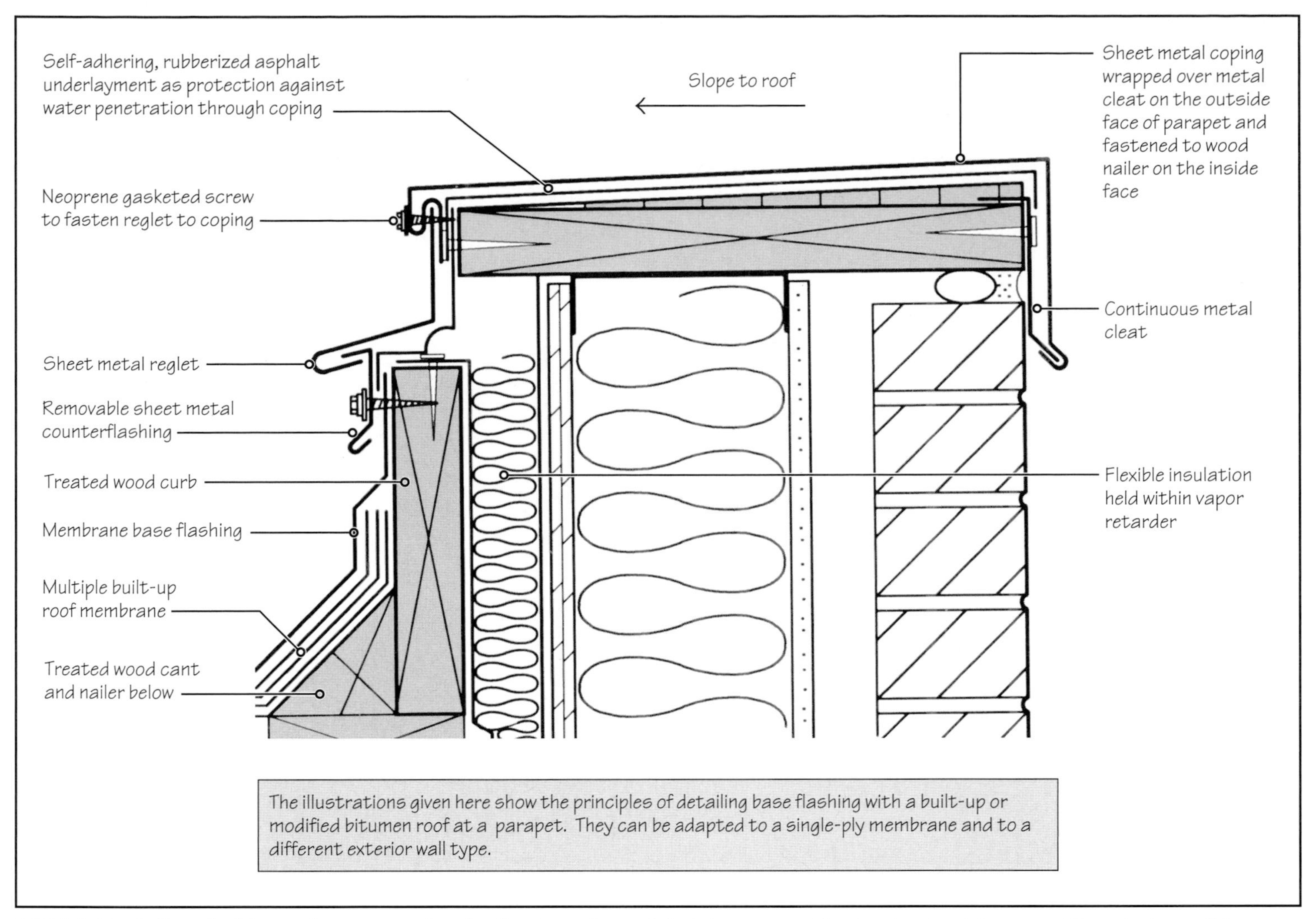

FIGURE 31.28 Base flashing and parapet coping details where the roof deck is supported by the structural frame and the exterior wall is nonbearing.

membrane base flashing is generally a mineral-granule-surfaced sheet. Mineral granules protect the flashing against weathering. This is helpful because gravel protection is not available on the vertical portion of membrane base flashing.

Torch-applied flashing sheets are also used, typically with a torch-applied roof membrane. A single-ply flashing material is used with a single-ply roof membrane.

Perimeter Wood Nailers and Cants

Preservative-treated wood nailers are generally used at the base. Nailers add strength and provide a nailable surface to which the membrane and flashing are fastened.

Because bituminous materials are relatively brittle, they cannot be bent to form a 90° angle because they will crack during service. A cant strip is, therefore, provided at the base. Cant strips may be of pressure-treated wood or of a rigid insulating material, such as perlite board. Perlite-board cant is generally adhered to the nailer.

Counterflashing

Counterflashing, generally made of sheet metal, is a removable and reusable component that allows for reroofing without demolishing other components of the detail. Galvanized steel is most commonly used for counterflashing, but aluminum, copper, or stainless steel are specified for greater durability. Counterflashing laps over the top of membrane flashing to prevent the entry of water between the membrane flashing and the parapet wall.

Metal Coping

The coping on the top of a parapet wall is generally made of sheet metal. The coping wraps over a continuous metal cleat on the outside face of the parapet and is fastened to

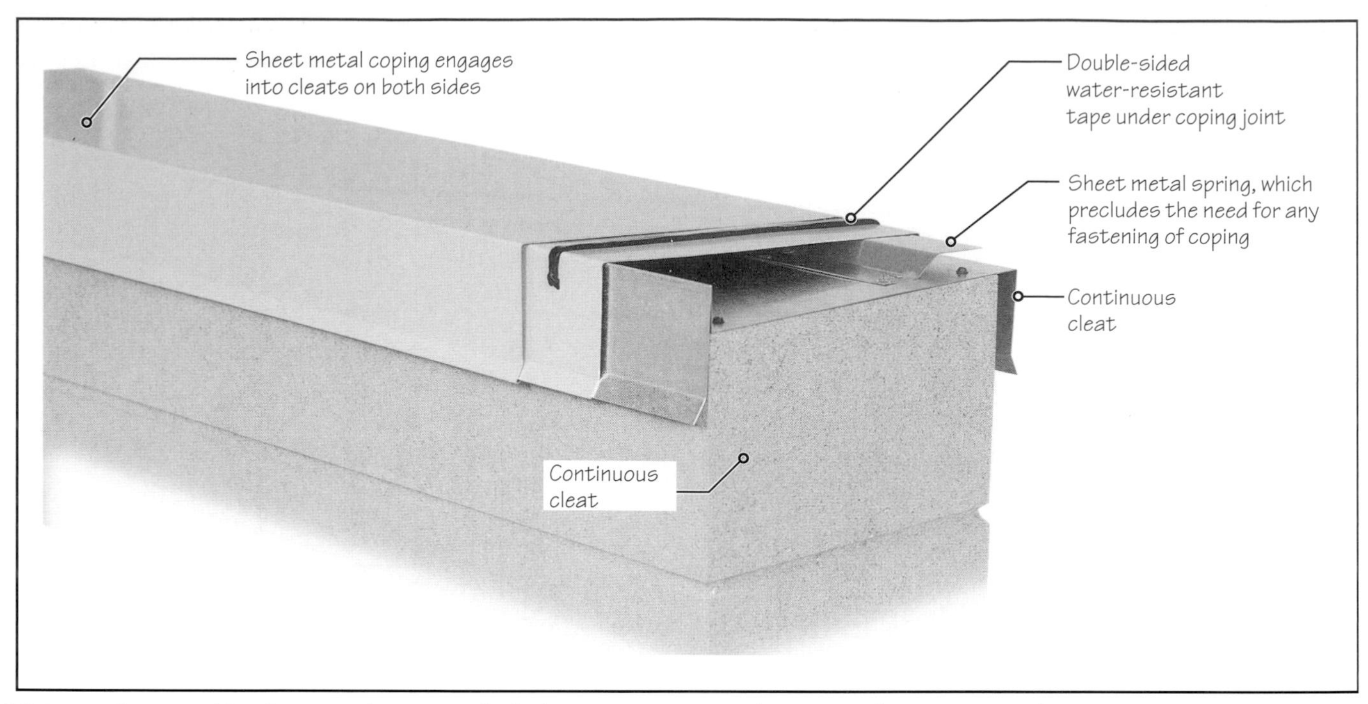

FIGURE 31.29 Snappped-in-place metal coping, which does not require any fasteners. (Photo courtesy of MM Systems Corporation.)

the parapet nailer with high-domed, neoprene-gasketed fasteners on the inside face of parapet.

In place of the metal copings fabricated in the contractor's shop (shown in Figures 31.27 and 31.28), manufacturer-fabricated sheet-metal copings may be specified. Most manufacturers' copings are spring loaded, so that they can be snapped to continuous cleats on both sides of coping, precluding the need for external fasteners, Figure 31.29.

31.10 CURB AND FLANGE FLASHING DETAILS

Curb flashing occurs at a curb placed at a building expansion joint, around a roof opening such as skylight, or under rooftop equipment, Figure 31.30. Like base flashing, curb flashing must also extend vertically up the curb, terminating at least 8 in. above the roof level. Figures 31.31 and 31.32 illustrate the principles of curb flashing at an expansion joint.

Flange flashing does not extend vertically but lies in the plane of the roof. It occurs at a free edge of the roof, such as, at an edge guard, around a roof drain or vent stack, or around a pipe

FIGURE 31.30 Examples of curb flashing.

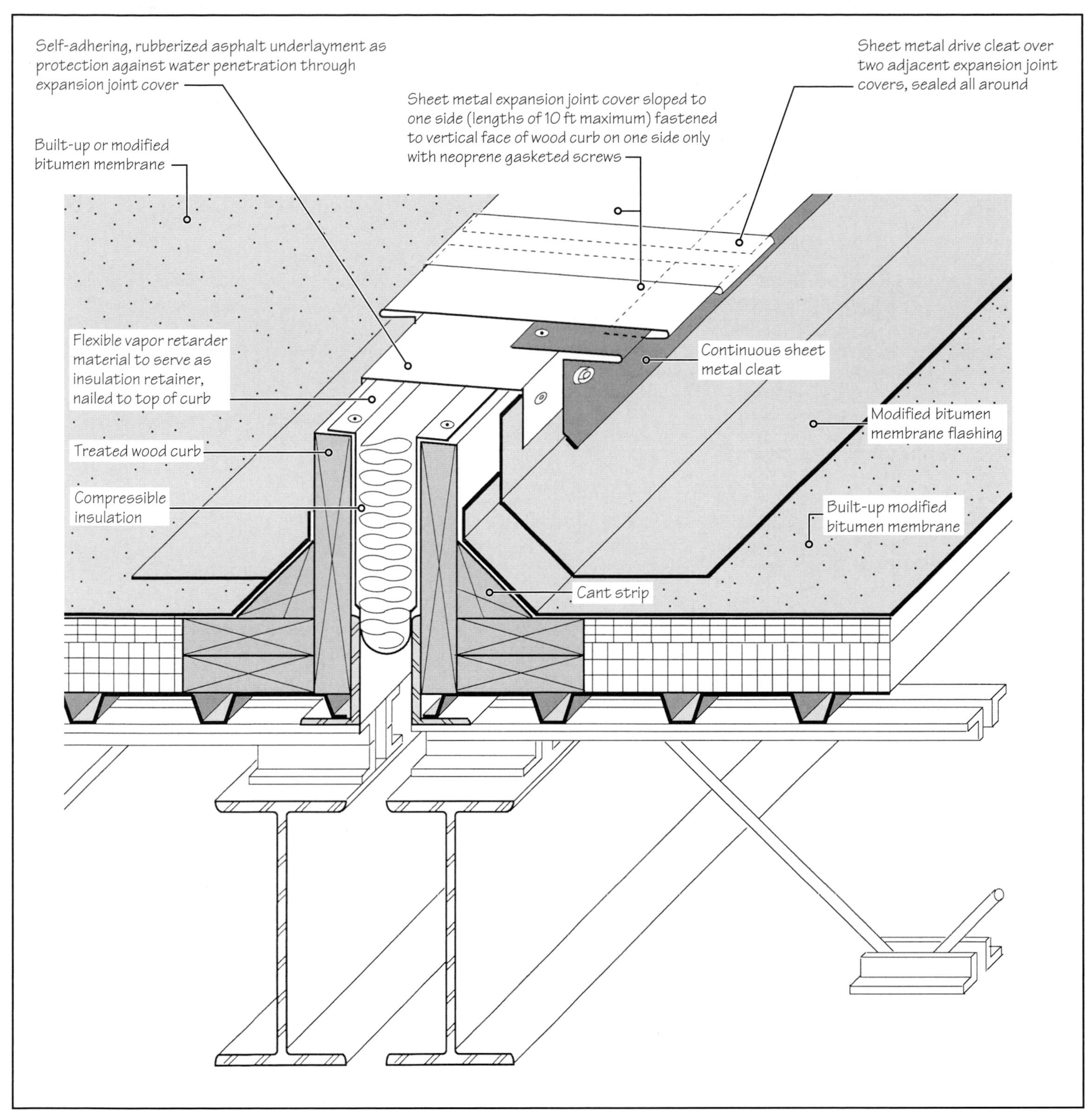

FIGURE 31.31 A typical roof expansion joint detail with a site-fabricated sheet-metal expansion joint cover.

used for supporting a structural frame under rooftop equipment. Figure 31.33 illustrates a typical edge-guard fascia detail and Figure 31.34 shows flashing around a pipe penetration.

31.11 PROTECTED MEMBRANE ROOF

In a protected membrane roof (PMR), the insulation is laid loose over the roof membrane, which is generally installed straight on the deck, Figure 31.35. In the case of a steel deck, a leveling board (typically a low-insulation board) is used under the membrane to provide a flat and rigid surface. The type of membrane and its installation is similar to that in a conventional roof assembly. Any one of the various membranes (built-up roof, modified bitumen, or single-ply membrane) can be used in a PMR.

After the installation of the membrane and the associated flashings at roof perimeter and penetrations, the membrane is covered with insulation, which is then covered with ballast. The ballast may consist of river gravel or precast-concrete pavers. Pavers make the entire

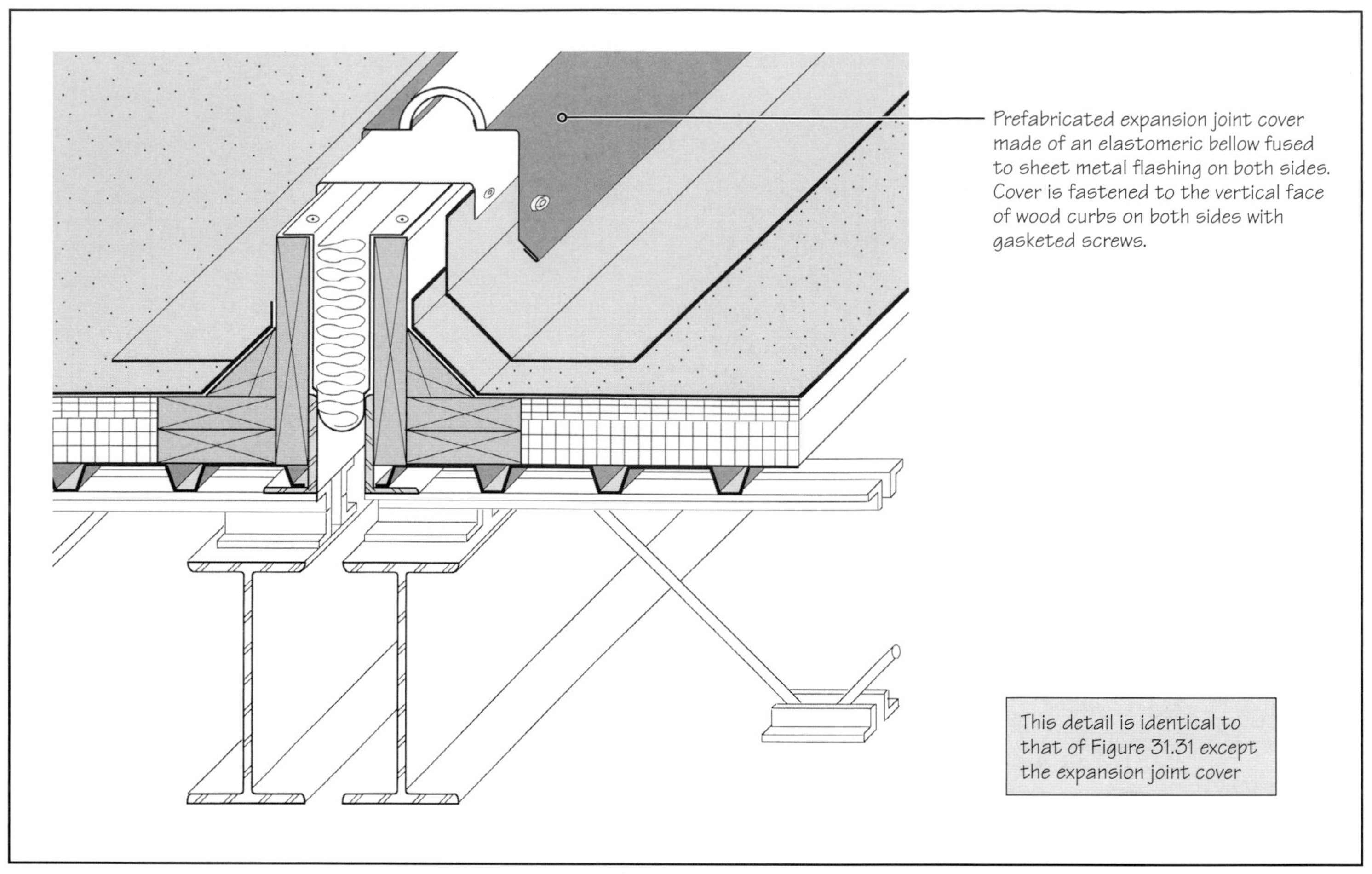

FIGURE 31.32 A typical roof expansion joint detail with prefabricated expansion joint cover.

roof surface walkable, in addition to providing an easily maintainable roof. Pavers also make the roof look cleaner compared to gravel or crushed stone. Because the order of insulation and roof membrane is reversed in a PMR, it is also referred to as an *inverted roof.*

With the protection provided to it by the insulation and the ballast cover, the roof membrane is shielded from weathering elements, such as ultraviolet radiation and hail and traffic impact. The membrane is also protected from thermal cycling—a major cause of a roof membrane's deterioration in a conventional roof assembly where the membrane is the uppermost layer.

Insulation in a PMR

Because the temperature cycling and weathering elements are borne by the insulation in a PMR, insulation becomes a more critical component of the assembly. PMR insulation must be water impermeable, have high resistance to freeze-thaw cycles, have high compressive strength, and have a high dimensional stability. Among the commonly used low-slope roof insulations, extruded polystyrene is the only insulation that meets these and the other requirements of a PMR.

Filter Fabric

Another important component in a PMR is a filter fabric, which consists of a porous plastic membrane laid over the insulation. Its purpose is to keep silt and small debris from flowing into insulation joints and restricting the insulation's expansion and contraction. The filter fabric also protects roof drains against blockage by such materials.

31.12 LOW-SLOPE-ROOF DESIGN CONSIDERATIONS

With so many different types of low-slope roofing systems available (membranes, insulation types, and decks), which system is most appropriate for a project? Some of the major factors that affect the selection of an appropriate roof system are:

- *Locally Available Roofing Technology* In areas where labor is relatively less expensive, built-up and modified bitumen roofs are more common.

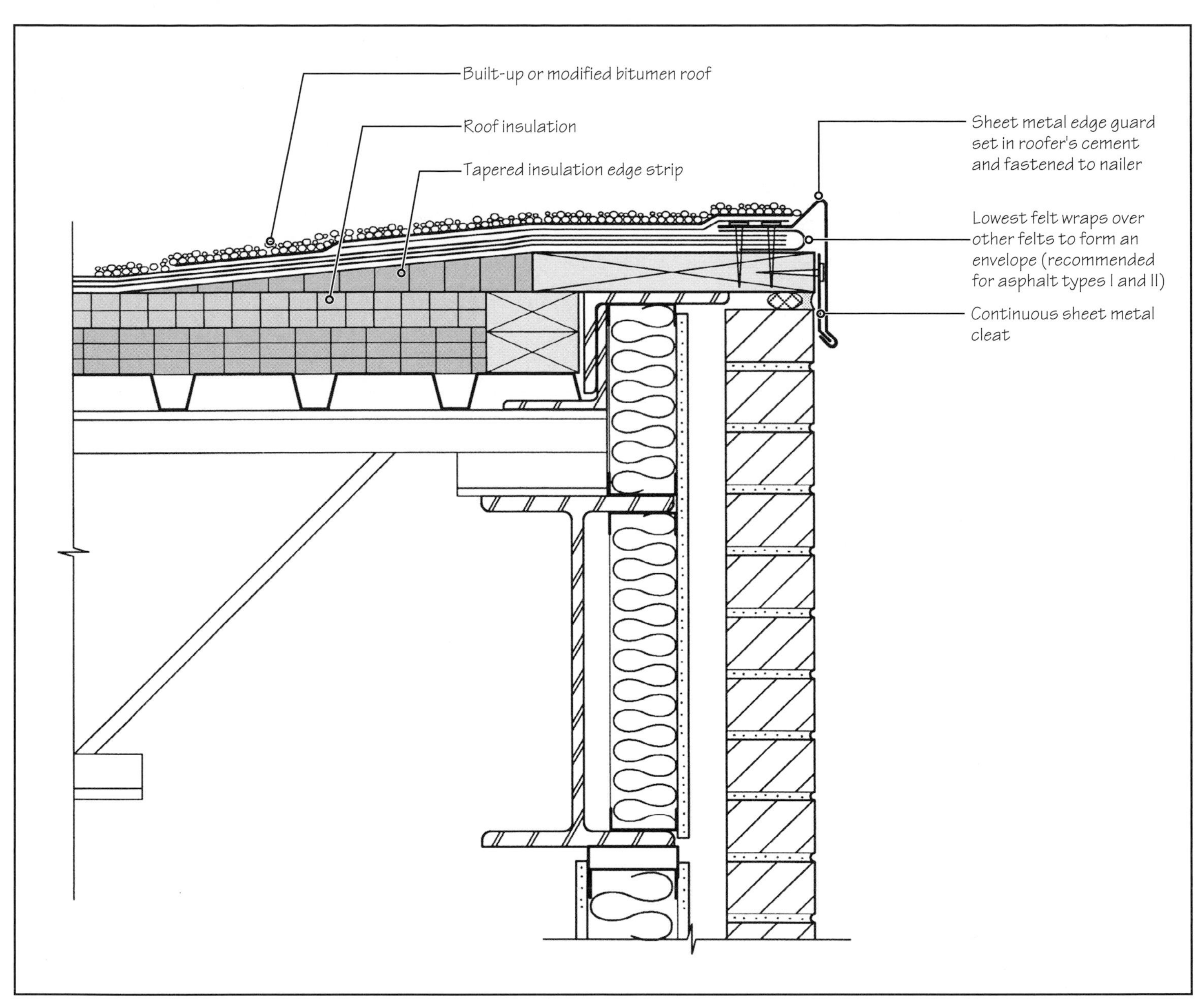

FIGURE 31.33 Flange flashing at an edge guard. This detail shows the use of a built-up roof membrane. Detail using a single-ply membrane is similar.

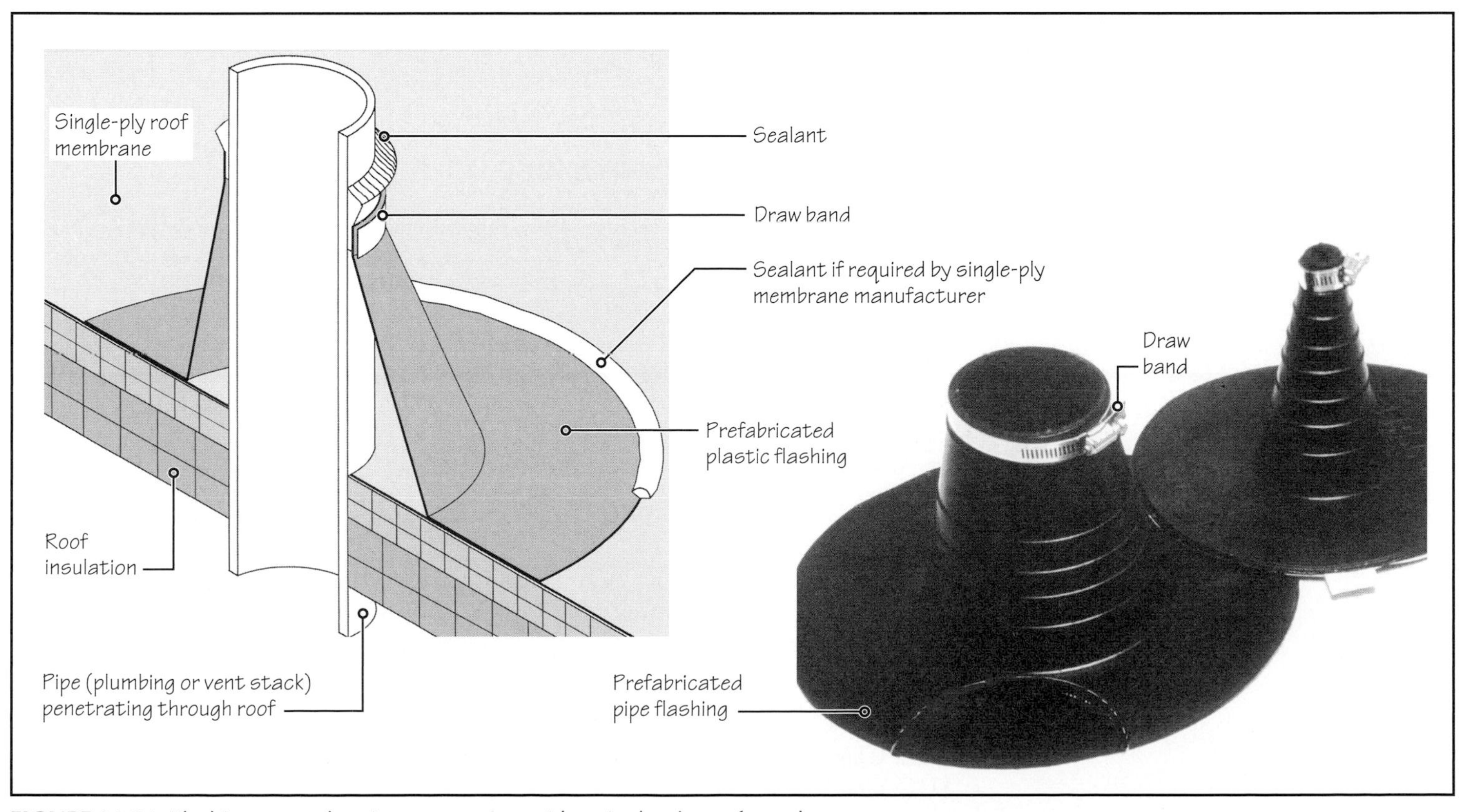

FIGURE 31.34 Flashing around a pipe penetration with a single-ply roof membrane.

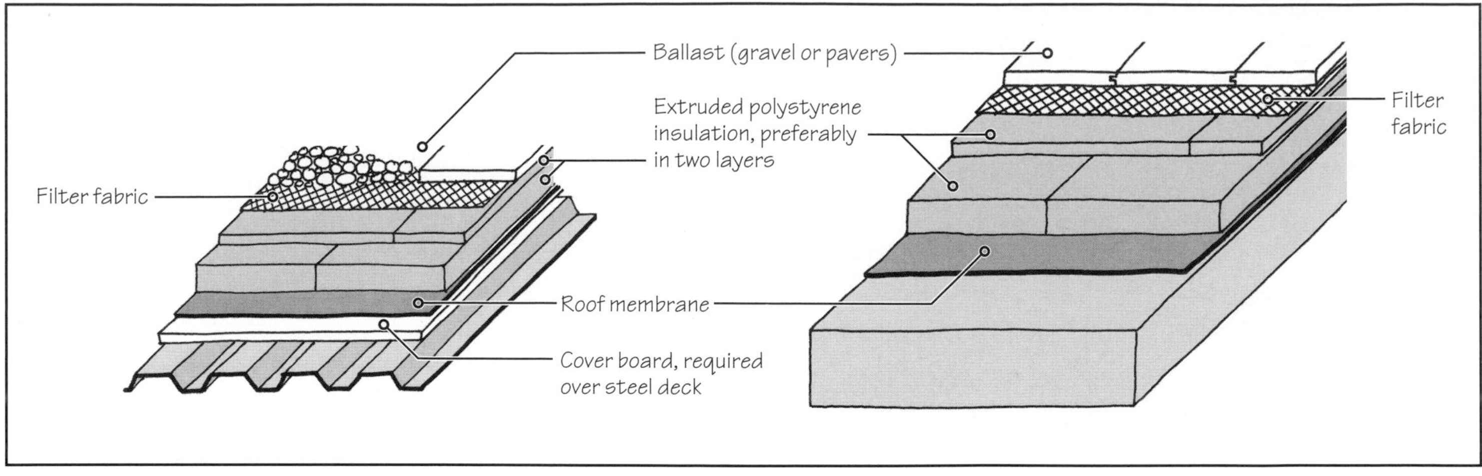

FIGURE 31.35 Components of a protected roof membrane.

- *System's Suitability to Local Climate* In extremely cold climates, a mechanically fastened or ballasted single-ply system may be more appropriate than a fully adhered single-ply or hot-asphalt-applied membrane system because both asphalt and single-ply membrane adhesives require additional precautions when applying in a cold environment.

  In a high-wind region, fully adhered systems are more appropriate because the uplift stresses are distributed over the entire roof membrane instead of having the stress concentration that occurs at the fasteners in mechanically fastened systems. A ballasted system should be avoided, and a mechanically fastened system should be detailed with care.
- *Curved Roofs and Roofs with Large Slopes* Such roofs do not lend themselves to the application of hot bitumen. A fully adhered, single-ply membrane is generally more appropriate for such projects.
- *Chemical Environment on the Roof* Roof membranes on industrial facilities and commercial kitchens must be examined to ensure the membrane's compatibility with the environment on the roof. Roofing-system manufacturers generally provide the required information.
- *Roof Warranty* Everything else being the same, the manufacturer's warranty may impact the system's selection.
- *Long-Term Cost and Sustainability Aspects of the System* Although the initial cost of the system is important, the long-term life-cycle cost of the system should not be overlooked. The effect of the system on the environment during and after the roof's installation should also be considered.

Life-Safety Issues in Low-Slope Roof Design

In addition to the factors just mentioned, the following three life-safety issues impact a roof system's selection.

- Fire-safety considerations
- Wind-uplift and hail-impact resistance
- Drainage design

Fire Safety

Roof systems are separated into three classes, in order of decreasing fire safety. Data obtained from the tests conducted by the Underwriters Laboratory (UL) on full-size roof assembly specimens are considered to be most authoritative. UL classification is based on the exposure of a roof to an exterior fire. (The fire-resistance rating of construction assemblies (Chapter 7) are based on fire assumed to originate from the interior of the building.)

- (UL) Class A roof
- (UL) Class B roof
- (UL) Class C roof

Class A is the most fire resistive and Class C is the least fire resistive roof. Generally, the fire rating of a given roof system decreases with increasing slope because the exterior fire spreads more easily on a roof with a larger slope. Additionally, gravel-covered, ballast-covered, or foil-covered roofs are generally more fire resistive than those without such covers.

WIND-UPLIFT AND HAIL-IMPACT RESISTANCE

A low-slope roof system must be designed to withstand twice the maximum-expected design wind uplift on a roof. In other words, if the maximum expected wind uplift on a roof is 75 psf, the roof system must be able to withstand at least 150 psf, which implies a minimum safety factor of 2.0.

The Factory Mutual Research Corporation (FMRC) is the most recognized establishment for rating of roof assemblies for wind-uplift and hail-impact resistance. FMRC data, obtained from tests conducted on full-size roof-assembly specimens, are provided in the following format:

Class 1-60 and Class 1-SH

In this classification, 1 implies that the assembly has passed the calorimeter test—a pass-fail test that measures the amount of combustible content in the assembly. An assembly that exceeds the combustible content limit fails the test and is not tested for any other performance (wind or hail). An assembly that passes the calorimeter test is called an *approved assembly*.

Thus, Class 1-60 implies an approved assembly with wind-uplift resistance of 60 psf. Such an assembly can be used on a roof designed to withstand a maximum wind uplift of 30 psf. Similarly, 1-240 means an assembly can be used where the design wind uplift on roof is 120 psf.

Class 1-SH implies an approved assembly that can withstand severe hail impact. In terms of hail resistance, a roof assembly is rated either as 1-SH or 1-MH. A 1-MH rating implies an approved assembly that has medium hail-impact resistance.

Manufacturers of roof systems provide UL and FMRC ratings of their assemblies—important considerations in roof system's selection.

DRAINAGE DESIGN

After the appropriate roof system has been selected, a roof's drainage design is completed. Like fire and wind, drainage design is an important life-safety issue because a poor drainage design can lead to the collapse of a low-slope roof, which is generally an abrupt collapse.

FOCUS ON SUSTAINABILITY

Sustainability Considerations in Low-Slope Roofs

In response to sustainability concerns, manufacturers of roofing systems have introduced a number of ecofriendly initiatives.

Recycling

Recycling has been initiated throughout the roofing industry. Built-up roofing and modified bitumen roofing products are being recycled for use in asphalt pavements for roads and highways. Prior to this initiative, nearly 9 million tons of asphalt roofing ended up in landfills in the United States.

Manufacturing of Roofing Materials From Industrial Wastes

New roofing products are being developed from waste obtained from old single-ply roofs, automobile tires, shoes, and several other products. The industry is conscious of making recycling and reuse economically viable.

Roof Surface Reflectivity

An effective means of increasing a roof's sustainability is to improve its energy efficiency through a reflective cover. A black surface absorbs 70% to 80% of solar energy incident on the roof. If the roof surface is white, its solar energy absorption reduces to 20% to 30%, which can reduce roof surface temperature by as much as 15°F to 20°F, reducing the cooling loads substantially during the summers.

Additionally, the use of a reflective surface does not significantly increase heating load in winters. The reason is that the heat loss through a roof is typically much smaller than the heat gain. For example, when it is 30°F outside and 70°F inside, the temperature differential between inside and the outside is 40°F. It may even be less on a sunny day because the roof's surface temperature will be greater than 30°F. In a warm climate, a dark roof's surface temperature may be 140°F or more, giving a temperature differential of 70°F.

Lower roof surface temperature also decreases the membrane's degradation, increasing the service life of the roof. Roofing membrane manufacturers are developing and promoting white roofs through white single-ply membranes or membranes coated with reflective surfaces. Some cities require the use of white roofs for city-owned buildings to obtain LEED certification and lower utility bills. White aggregate- or ballast-covered roofs generally do not qualify as white roofs because they loose their reflectivity fairly rapidly.

Green Roofs

Landscaped roofs, consisting of vegetation planted on a lightweight soil or other growth medium, have received considerable attention. It not only reduces energy use by the building but also reduces urban heat island effect.

A landscaped roof can also help control storm water runoff by temporarily storing water on the roof and releasing it slowly into underground or a surface storm water drainage system.

PRINCIPLES IN PRACTICE

Shingling of Built-Up Roof Felts

The recommended application procedure of a built-up roof membrane is by shingling of felts, Figure 1. Layer-by-layer application (i.e., phased application) of felts is discouraged.

In a shingled application, the upper felt overlaps the lower felt by a constant dimension. In a 3-ply roof membrane, there are at least 3 felts under any roof point. Similarly, a 4-ply roof must have 4 felts at every point.

FIGURE 1 Successive built-up roof felts are generally laid in shingle fashion, with the upper felt overlapping the lower felt by a constant dimension. The alternative, applying the felts layer by layer (i.e., in a phased manner) has several disadvantages and is not used.

In a shingled built-up roof (BUR), the lap at the exposed end of a felt with respect to the lowest felt under that end, referred to as the *head lap,* is 2 in. as the industry standard. Under the head lap, there is one additional felt. Thus, there are 4 felts under a head lap in a 3-ply membrane, Figure 2. The dimension EXP is called the *exposure* of the felt. EXP is related to the number of plies by the following relationship:

$$\text{EXP} = \frac{\text{felt width} - \text{head lap}}{\text{number of BUR plies}}$$

Thus, with a felt width of 36 in., EXP (in a 2-ply membrane) = 17 in. In a 3-ply membrane, EXP = $11\frac{1}{3}$ in., and so on. Built-up roof felts are manufactured with 2-, 3- and 4-ply lines marked on them, as shown in Figure 3, to help roofers lay the felts correctly.

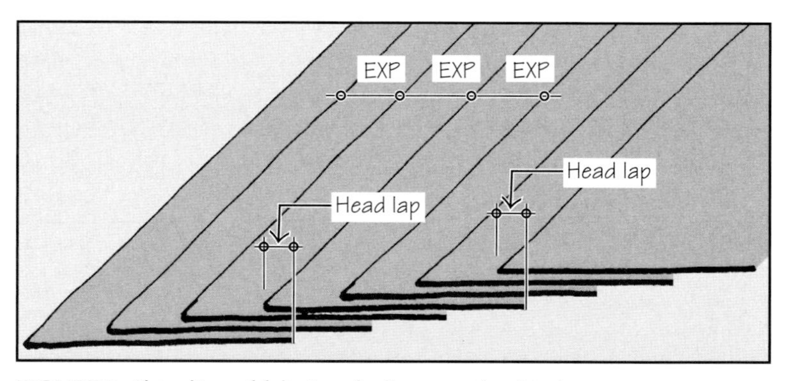

FIGURE 2 Shingling of felts in a built-up roof, with the terms *exposure* and *head lap* defined. The industry standard for head lap is 2 in. This sketch shows a 3-ply built-up roof membrane.

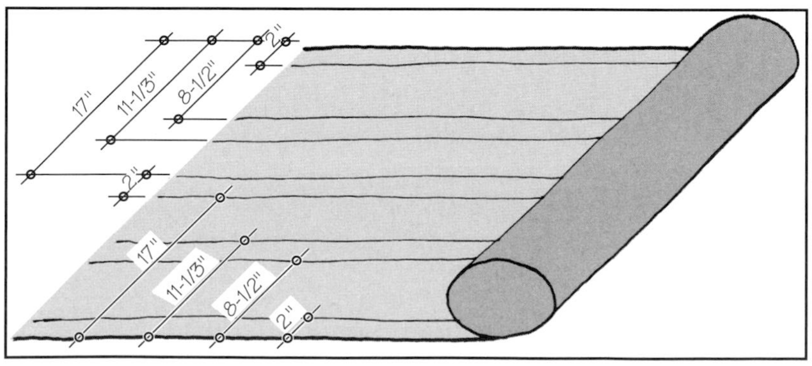

FIGURE 3 Factory-applied lines on a built-up roof felt.

PRINCIPLES IN PRACTICE

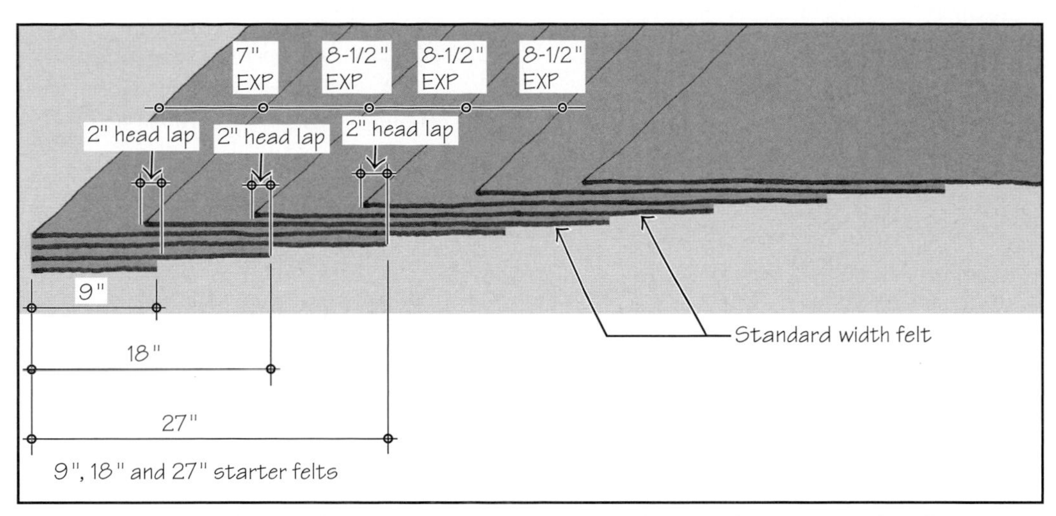

FIGURE 4 The use of starter felts at the starting edge of a built-up roof membrane. This illustration shows a 4-ply membrane.

To provide the necessary number of felts at the starting edge of the roof (usually the roof's lowest edge), the first few felts are not the full width of 36 in. The width of the starter felts is obtained simply by dividing the full width by the number of built-up roof plies.

Thus, in a 4-ply membrane, there are three partial-width felts at the starting edge—9 in., 18 in., and 27 in. wide, Figure 4. In a 3-ply membrane, there are two partial width felts—12 in. and 24 in. wide. In a 2-ply built-up roof membrane (not usually specified), there is only one partial width felt—18 in. wide.

Advantages of Shingling

Shingling of felts results in completed sections of the roof at the end of each day. Thus, the membrane remains protected from interply moisture penetration during interruptions in roofing operations. If the felts were laid in a phased manner, there would be a possibility of moisture being trapped between felt layers, either from rain or overnight condensation. The entrapment of debris in a phased application is also of concern.

Another advantage of shingling is that it provides better adhesion of plies compared to the phased application. This is due to a slower cooling of bitumen in a shingled membrane. In a shingled membrane, there are three to four layers of bitumen applied within a short time span. In a phased application, the previous layer of bitumen has virtually cooled off when the subsequent layer is applied.

PRACTICE QUIZ

Each question has only one correct answer. Select the choice that best answers the question.

28. The flashing around a pipe penetrating through the roof is
- **a.** base flashing.
- **b.** curb flashing.
- **c.** flange flashing.
- **d.** none of the above.

29. The flashing around a roof drain is
- **a.** base flashing.
- **b.** curb flashing.
- **c.** flange flashing.
- **d.** none of the above.

30. A cant strip is not required in a base flashing detail with a single-ply roof membrane.
- **a.** True
- **b.** False

31. Counter flashing is generally
- **a.** made of PVC and is permanently anchored to the deck or parapet wall.
- **b.** made of PVC and is a removable item.
- **c.** made of sheet metal and is permanently anchored to the deck or parapet wall.

PRACTICE QUIZ (*Continued*)

d. made of sheet metal and is a removable item.
e. none of the above.

32. In a protected membrane roof, the insulation is
a. below the deck.
b. between the deck and the membrane.
c. above the membrane and fully adhered to it.
d. above the membrane but loose laid over it.
e. not used.

33. FM 1-120 represents
a. the fire rating of a roof assembly.
b. wind-uplift resistance of roof assembly.
c. hail-impact resistance of roof assembly.
e. none of the above.

34. UL Class B represents:
a. the fire rating of a roof assembly.
b. wind-uplift resistance of roof assembly.
c. hail-impact resistance of roof assembly.
e. none of the above.

35. FM MH represents
a. the fire rating of a roof assembly.
b. wind-uplift resistance of roof assembly.
c. hail-impact resistance of roof assembly.
e. none of the above.

36. The exposure of felts in a 4-ply built-up roof is
a. 2 in.
b. 4 in.
c. 6 in.
d. 8 in.
e. none of the above.

37. The head lap in a 4-ply built-up roof is
a. 2 in.
b. 4 in.
c. 5 in.
d. 6 in.
e. none of the above.

Answers: 28-c, 29-c, 30-a, 31-d, 32-d, 33-b, 34-a, 35-c, 36-e, 37-a.

KEY TERMS AND CONCEPTS

APP sheet
Bitumen
Built-up roof membrane
Cant strip
Coal tar
Coping
Counterflashing
EPDM membrane
Flood coat
Loose-laid, ballasted single-ply roof
Low-slope roof
Low-slope roof flashings
Low-slope roof insulation
Metal cleat
Modified-bitumen roof membrane
Protected membrane roof (PMR)
PVC membrane
Roofing asphalt
SBS sheet
Single-ply roof membrane
Steep roof
TPO membrane
Water-holding roof
Water-shedding roof

REVIEW QUESTIONS

1. Using a sketch, explain the essential components of a low-slope roof.
2. Discuss the differences between SBS-modified and APP-modified bitumen roof membranes.
3. Using sketches, explain the anatomies of the following roofs. Show how the insulation and roof membrane are attached.
   a. Steel deck, rigid-board insulation, and built-up roof membrane.
   b. Reinforced-concrete deck, rigid-board insulation, and SBS-modified bitumen membrane
   c. Steel deck, foamed-concrete insulation, and built-up roof membrane
   d. Concrete deck, rigid-board insulation, and loose-laid, ballasted single-ply membrane
4. With the help of a sketch, explain a protected membrane roof.

SELECTED WEB SITES

Asphalt Roofing Manufacturers Association, ARMA (www.asphaltroofing.org)
The Institute of Roofing, Waterproofing, and Building Envelope Professionals, RCI (www.rci-online.org)
National Roofing Contractors Association, NRCA (www.nrca.net)
Single Ply Roofing Institute (www.spri.org)

FURTHER READING

1. National Roofing Contractors Association (NRCA). *The NRCA Roofing and Waterproofing Manual.*
   This $8\frac{1}{2} \times 11$, approximately 1500-page book is a comprehensive reference resource for all aspects of roofing.
2. Patterson, S. and Mehta, M. *Roofing Design and Practice.* Upper Saddle River, NJ: Prentice Hall, 2001.
   This $8\frac{1}{2} \times 11$, approximately 300-page book is a primer on roofing and is aimed primarily at the architects and architecture students. It covers both low-slope- and steep-roof design.

REFERENCES

1. Patterson, S., and Mehta, M. *Roofing Design and Practice.* Upper Saddle River, NJ: Prentice Hall, 2001.
2. Kane, K. "Another Strong Year for the Industry," *Professional Roofing,* 1999, p. 14.

CHAPTER 32 Roofing–II (Steep Roofs)

CHAPTER OUTLINE

32.1 Steep-Roof Fundamentals

32.2 Asphalt Shingles and Roof Underlayment

32.3 Installation of Asphalt Shingles

32.4 Valley Treatment in an Asphalt Shingle Roof

32.5 Ridge and Hip Treatment in an Asphalt Shingle Roof

32.6 Flashings in an Asphalt Shingle Roof

32.7 Essentials of Clay and Concrete Roof Tiles

32.8 Clay and Concrete Tile Roof Details

32.9 Types of Architectural Metal Roofs

32.10 Contemporary Architectural Metal Roofs

Architectural metal panel roofs (upper photo) and slate and clay tiles (lower photo) are among the more commonly used steep roofs in high-end residential and low-rise commercial buildings, but asphalt shingles are the most widely used steep-roofing material overall. (Photos by MM.)

As a continuation of the previous chapter, in which low-slope roofs were discussed, this chapter deals with steep roofs—roofs that have a slope of greater than or equal to 3:12. It begins with an introduction to the components of a steep roof and then discusses the types of steep roofs commonly used. Of the various types of steep roofs, asphalt shingle roofs are most commonly used. Therefore, asphalt shingle roofs are covered in greater detail than the other roof types.

As mentioned in Chapter 31, a roof type that is used for both low-slope and steep roofs is the architectural sheet-metal roof. Architectural sheet-metal roofs are becoming increasingly popular because of their greater durability and the availability of a vast array of non-fading colors. Architectural sheet-metal roofs are covered at the end of the chapter.

32.1 STEEP-ROOF FUNDAMENTALS

The basic components of a steep roof are illustrated in Figure 32.1:

- Roof shingles
- Roof underlayment
- Ice dam protection membrane, where needed
- Roof deck
- Flashings

ROOF SHINGLES

Roof shingles can be of various types. The more commonly used shingle varieties are

- Asphalt shingles
- Slate shingles
- Concrete and clay tiles
- Wood shingles
- Metal shingles

As stated earlier, asphalt shingles (also called *composition shingles*) have the largest steep-roof market share because of their low cost and acceptable life span. They are available in different colors and textures and have a satisfactory aesthetic. With shadow lines added to their surface, some varieties can mimic a wood shingle roof. Another advantage of asphalt shingles is their light weight, which reduces labor and transportation costs and also the cost of the structural components of the roof.

Slate shingles have one of the longest recorded history of use. They have been used in many buildings of historical importance in Europe and the United States. Because they are a hard and dense natural stone, slate roofs last a long time. They perform quite well when subjected to repetitive freeze-thaw cycles. Well constructed and detailed, a slate roof can last 75 years or more. Obviously, they are expensive. Imitation slates are also available, but their performance records must be checked before specifying them.

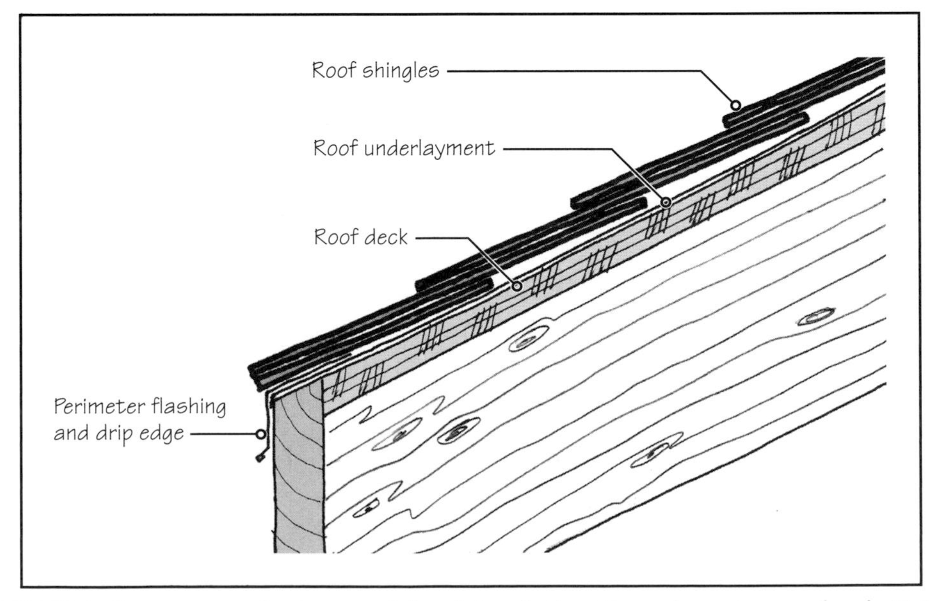

FIGURE 32.1 Components of a steep roof. (Insulation in the attic under a steep roof and attic's ventilation may also be considered as components of a steep roof. See Chapter 5 and Section 6.7.)

Clay tiles create a handsome roof due to the availability of a large variety of tile shapes. Like slate roofs, clay tile roofs are quite durable. However, it is important to understand that there can be significant differences in the quality of tiles, which can affect the durability of a clay tile roof.

Concrete tile roofs, though fairly durable, are generally less so than clay tile roofs. There can also be problems with long-term fading of colors of concrete tiles.

Wood shingles and shakes, once extensively used in the United States, have a decreasing market share today because of their fire hazard. Although fire-retardant-treated wood shingles are available, their long-term performance history is uncertain. A large variety of metal shingles are used in contemporary steep roofs. Like wood shingles, their market share is also small.

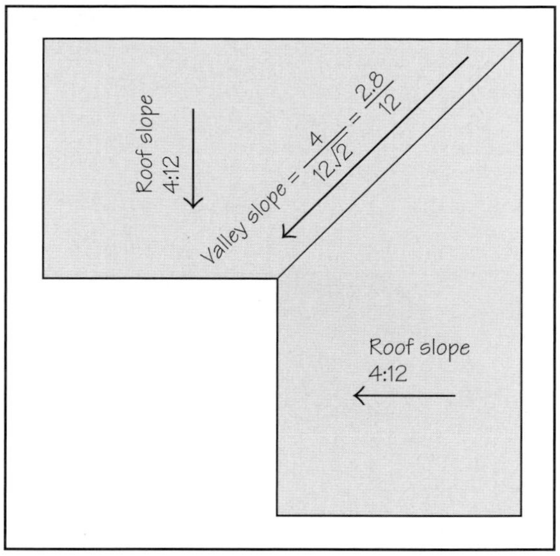

FIGURE 32.2 Roof slope in a valley is much lower than the rest of the roof. Therefore, a valley requires additional valley underlayment (Figure 32.14).

Roof Underlayment

In a properly designed and constructed steep roof, rainwater should normally not get under the shingles. However, a water-resistant layer, called a *roof underlayment,* is required under the shingles throughout the roof as a second line of protection against water leakage. The requirements for the type of roof underlayment varies with the type of shingles and roof slope. (Wood shingles do not require underlayment in some situations.)

A valley requires an additional roofing layer, referred to as a *valley underlayment.* A valley is more vulnerable to water leakage than the field of the roof because its slope is smaller than the roof itself. For example, the slope of a valley created by two intersecting planes, each with a slope of 4:12, is only 2.8:12—a 30% reduction in slope, Figure 32.2.

A significant head of water can be present in a long and flat valley. In addition to a smaller slope, the water running through the valley experiences more turbulence because it arrives there from two opposite directions.

Ice Dam Protection Membrane

Another special type of underlayment is called an *ice dam protection membrane.* It is required for use at the eaves in climates where there is a potential for the formation of ice at an overhanging eave.

The formation of an ice dam can make water back up under the shingles (Figure 6.12). Ice dam protection is recommended in regions where average daily temperature in January is less than 25°F, Figure 32.3. However, many roofers and roofing consultants recommend the provision of an ice dam protection membrane at the eaves in all locations.

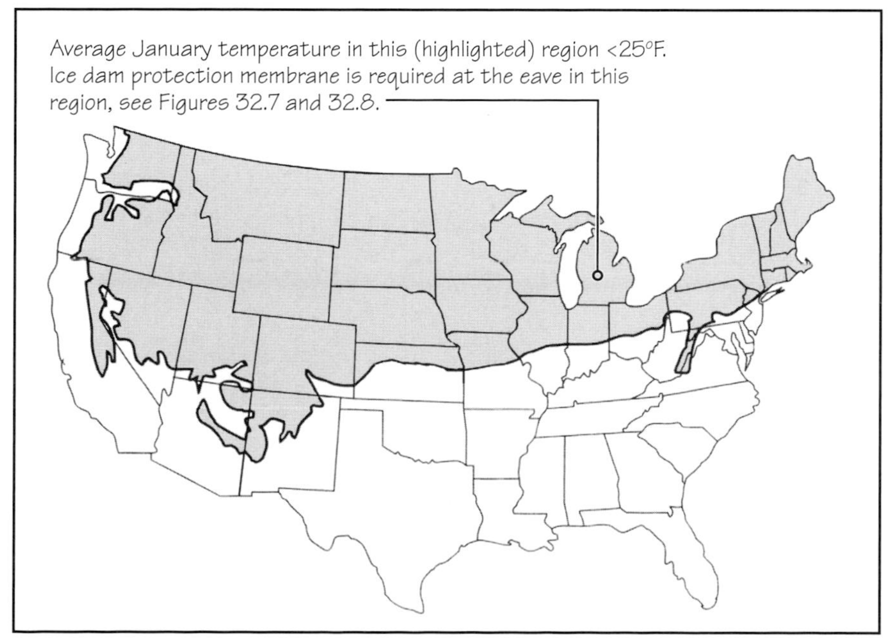

FIGURE 32.3 Approximate areas of the United States where the use of ice dam protection membrane at the eave is recommended (see [1] in Chapter 31). Local building code should be consulted for precise information.

ROOF DECK

The deck, being a structural component of the roof, is its most important element. In a contemporary steep roof, the deck generally consists of plywood or oriented strandboard panels. A spaced deck, consisting of wood lath (typically 1 × 4 or 1 × 6 wood members spaced according to the length of shingles) may be used with some steep-roof types, but its use is limited today.

FLASHINGS

As in low-slope roofs, flashings are additional waterproofing components used to create a watertight seal at terminations and transitions in the roof. Critical areas for flashings are valleys, eaves, rakes, chimneys, and other penetrations.

Flashings around roof penetrations, such as a chimney or a pipe, are particularly critical. The water is not only interrupted by a penetration but is redirected and compressed on a smaller area of the roof as it flows around a penetration.

FIGURE 32.4 Asphalt shingles with color variations in its granule surface, which give shadow lines to mimic wood or slate roofing.

32.2 ASPHALT SHINGLES AND ROOF UNDERLAYMENT

Asphalt shingles are manufactured in strips measuring approximately 36 in. × 12 in., with up to five cutouts. The cutouts separate a shingle strip into tabs, making the strip appear to be composed of several smaller shingles, which improves the shingle's appearance.

Asphalt shingles vary in weight from about 200 lb to nearly 400 lb per roof square. The heavier varieties, which consist of two or more layers of shingles laminated together, are generally more durable and aesthetically more pleasant.

Each shingle strip is made from a fiberglass mat, which is saturated and coated with asphalt and surfaced with mineral granules. The granules give the desired surface color to a shingle, which may vary from white, gray, and black to various shades of red, brown, and green, Figure 32.4.

Three most commonly used asphalt shingle profiles are shown in Figure 32.5. Each shingle is coated with a line of self-sealing adhesive, which helps to adhere that shingle with the overlying shingle. The adhesion is activated by the sun's heat after the shingles have been installed on the roof. The adhesive line increases the wind resistance of shingles. Out of the three self-sealing shingle varieties shown in Figure 32.5, the 3-tab shingle is the most commonly specified.

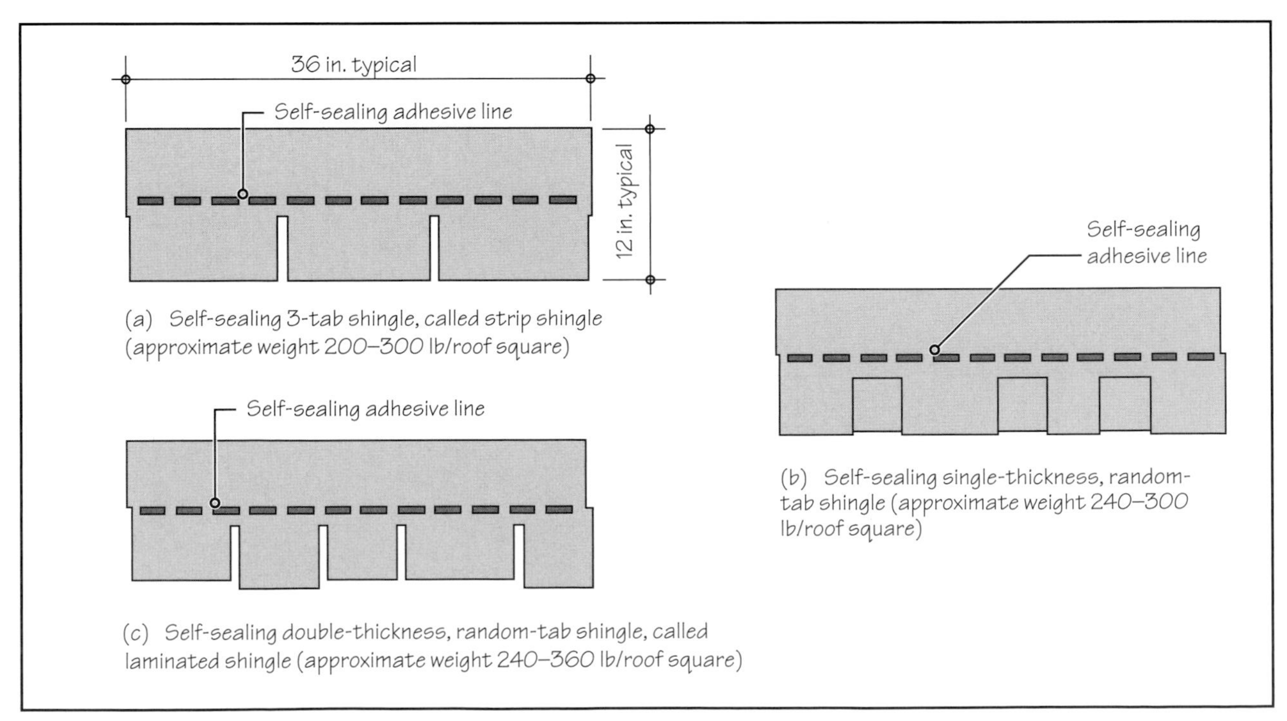

FIGURE 32.5 Commonly used asphalt shingle profiles.

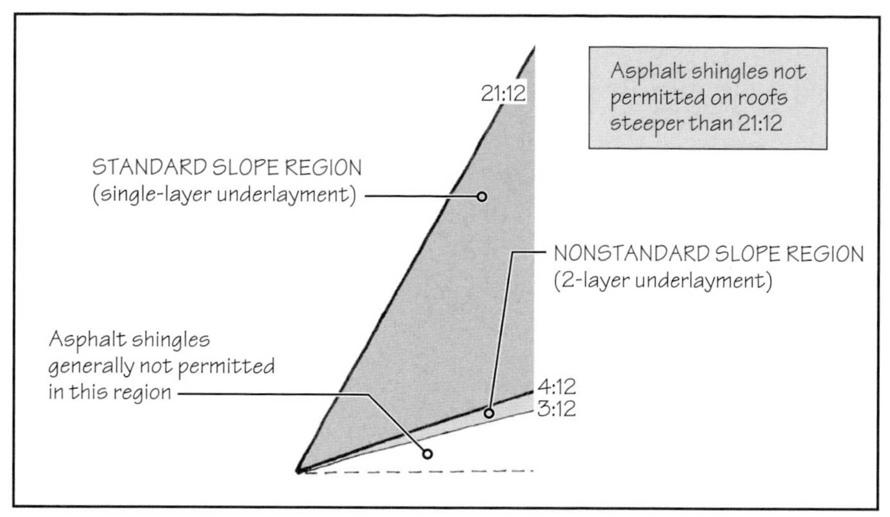

FIGURE 32.6 Slope and underlayment requirements for asphalt shingle roofs.

Roof Underlayment and Ice Dam Protection for a Standard-Slope Asphalt Shingle Roof

The recommended roof slope for asphalt shingles is between 4:12 and 21:12. This is referred to as the *standard slope* for asphalt shingles. A slope lower than the standard slope may also be used but requires special treatment, Figure 32.6.

The recommended roof underlayment for a standard-slope roof is No. 15 or heavier asphalt-treated organic felt. The underlayment is applied with 2-in. laps at the edges of the roll and 6-in. laps at the ends of the roll, Figure 32.7. It is nominally nailed or stapled to the deck—enough to hold it in place until the shingles are installed. It is essential that the underlayment is turned up the abutting walls or curbs (Figures 32.22 and 32.24).

The ice dam protection membrane on a standard-slope roof is typically a proprietary product, which is generally a self-adhering (peel-and-stick) modified bitumen sheet. It should extend a distance of at least 24 in. from the exterior wall toward the interior of the building.

Roof Underlayment and Ice Dam Protection for a Nonstandard-Slope Asphalt Shingle Roof

Asphalt shingles may be used on a roof slope lying between 3:12 and 4:12, but generally the roof's performance is not as good as that on a standard slope. Note, however, that some asphalt shingle manufacturers permit the use of their product on a slope as low as 2:12. The underlayment on a nonstandard-slope (or low-slope) roof must be a two-layer felt underlayment,

NOTE

Asphalt-Treated Organic Felts as Underlayment

As stated in Chapter 31, asphalt-treated organic felts are used as steep-roof underlayment. They are available in two types:

- No. 15 felt
- No. 30 felt

When first developed, a No. 15 felt weighed 15 lb per roof square and No. 30 felt weighed 30 lb per roof square. Currently, as per ASTM D226 Standard, No. 15 felt and No. 30 felt weigh 11.5 lb and 26 lb per roof square, respectively.

Because of its greater thickness and weight, a No. 30 felt is more durable and performs better than a No. 15 felt.

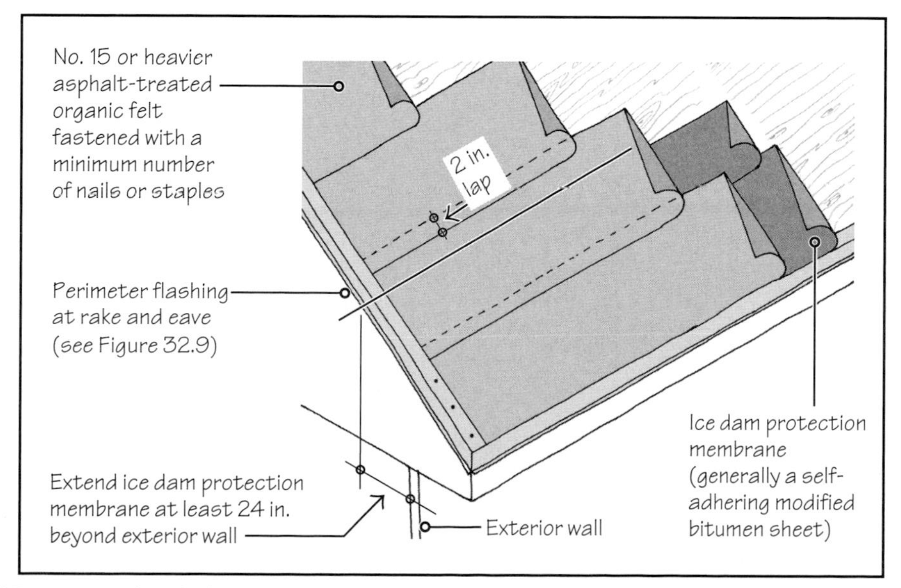

FIGURE 32.7 Roof underlayment and ice dam protection membrane on a standard-slope (≥4:12) asphalt shingle roof.

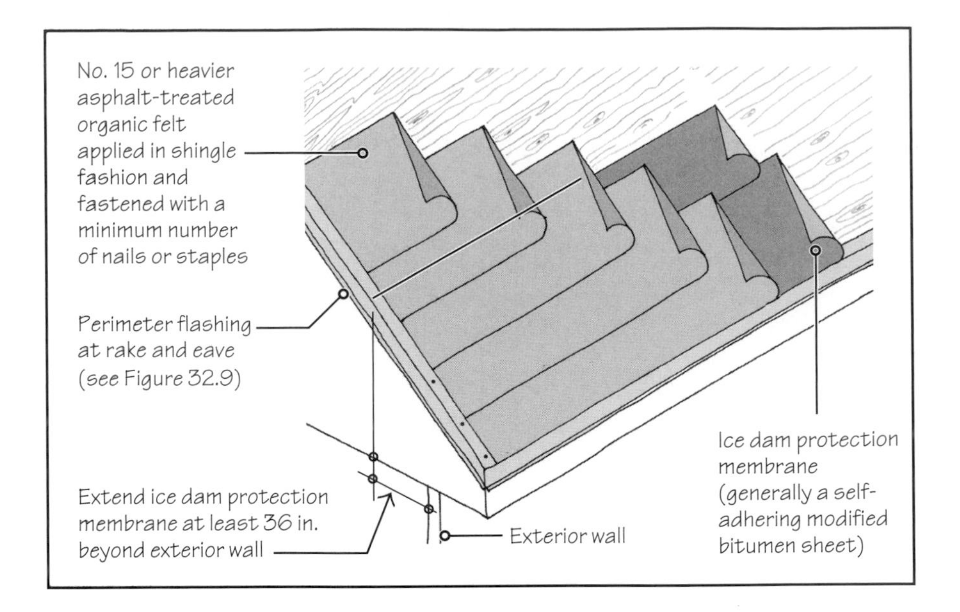

FIGURE 32.8 Roof underlayment and ice dam protection membrane on a nonstandard-slope asphalt shingle roof (slope between 3:12 and 4:12).

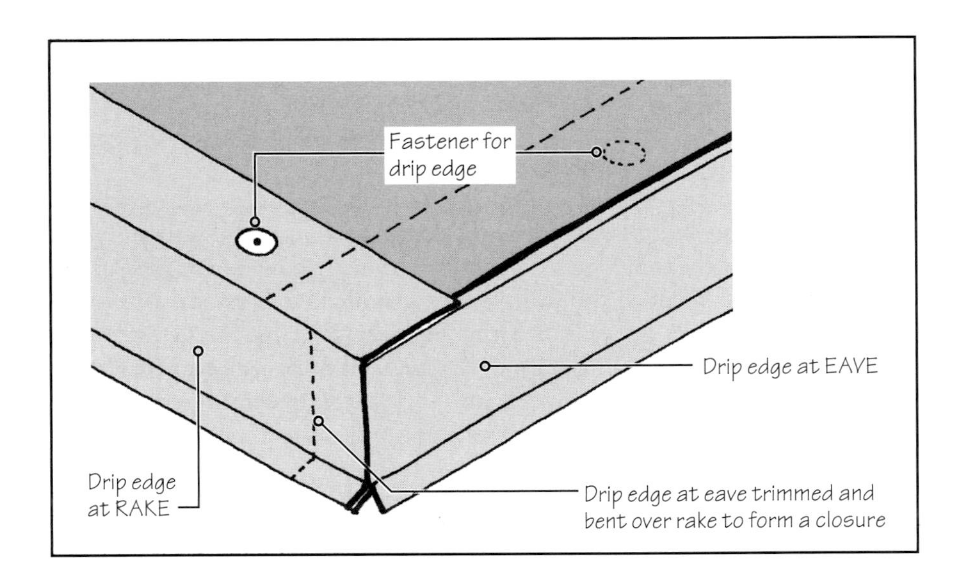

FIGURE 32.9 Drip edge at eave and rake of an asphalt shingle roof.

applied in shingle fashion, Figure 32.8. The ice dam protection membrane should extend a distance of at least 36 in. from the exterior wall toward the interior of the building.

Perimeter Flashing at Eave and Rake—The Drip Edge

A sheet-metal drip edge should be provided at all eaves and rakes and mechanically fastened to the deck. Galvanized steel or prefinished metal (minimum 28 gauge) is the most commonly used metal for a drip edge, but stainless steel, copper, or aluminum may also be used for higher durability. On an eave, the drip edge is applied before laying the underlayment. On a rake, on the other hand, it is applied over the underlayment, Figure 32.9.

32.3 INSTALLATION OF ASPHALT SHINGLES

After the roof underlayment and ice dam protection membrane have been installed, the roof is ready to receive the shingles. The procedure to install various types of asphalt shingles (Figure 32.5) is almost identical. Because 3-tab shingles are most commonly used, the installation procedure described here refers to their installation.

The installation of 3-tab shingles begins with the starter course at the eave, which is applied directly over the underlayment. A starter course consists of the standard 3-tab shingles, whose tabs have been cut off, Figure 32.10(a). The remaining part of each shingle, which is nearly 7 in. wide, is nailed to the deck, with the self-sealing adhesive line placed along the outside edge of the eave, Figure 32.10(b).

The self-sealing adhesive line of the starter course bonds with the overlying shingles (of the first course), improving wind performance of the roof. The starter course is laid from

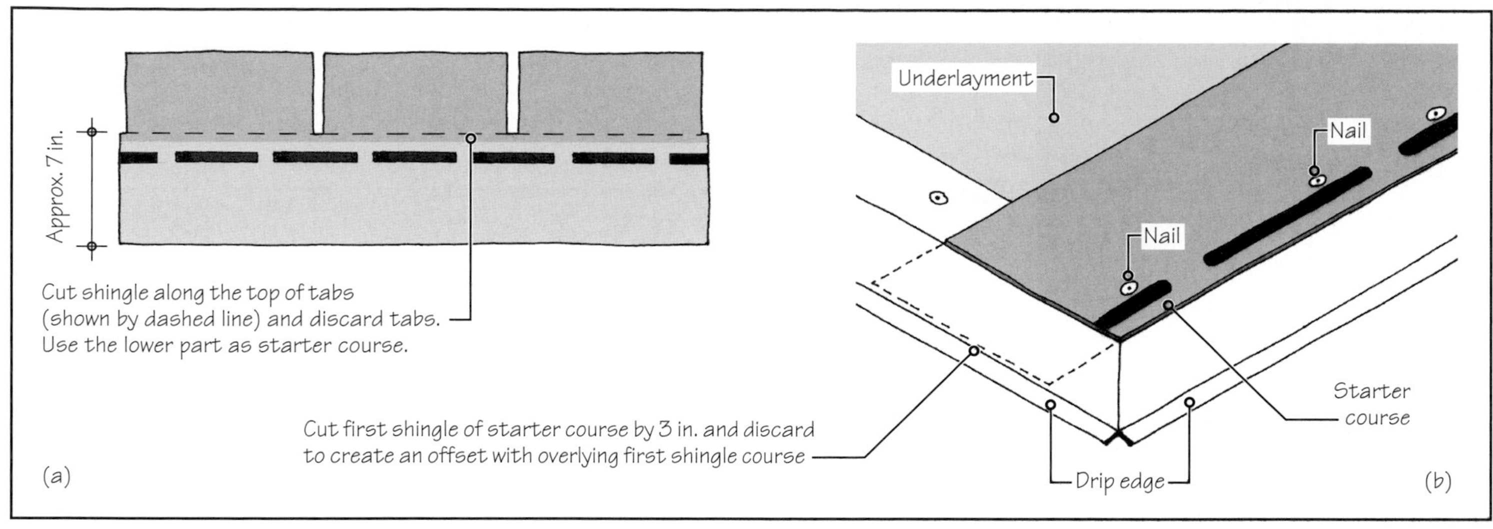

FIGURE 32.10 Installation of starter course in an asphalt shingle roof.

one end of the roof to the other. The first shingle in the starter course is cut by 3 in. along the shingle's length, so that the joints in the starter course are offset from the joints in the first course of shingles.

Next the first course of shingles is laid. The first course of shingles fully covers the starter course, and each shingle is a full-length shingle. The second course of shingles is laid so that only the tabs of the first course are exposed, and so on. In other words, asphalt shingles have a double coverage—two shingle thicknesses under every point, except under the headlap, where there are three shingle thicknesses.

The first shingle of the second course is cut in length by $\frac{1}{2}$ tab (nearly 6 in.). This cut breaks the joints in the upper and lower course of shingles, Figure 32.11. The first shingle of the third course is cut by 1 full tab, nearly 12 in. The first shingle of the fourth course is cut by $1\frac{1}{2}$ tabs, and so on. The pattern of shingles generated by this half-tab offset procedure is referred to as the *6-in. offset method,* because the shingles are offset by $\frac{1}{2}$ tab (6 in.). In addition to the 6-in. offset method, the *3-in. offset method, 4-in. offset method,* and *5-in. offset method* are also used.

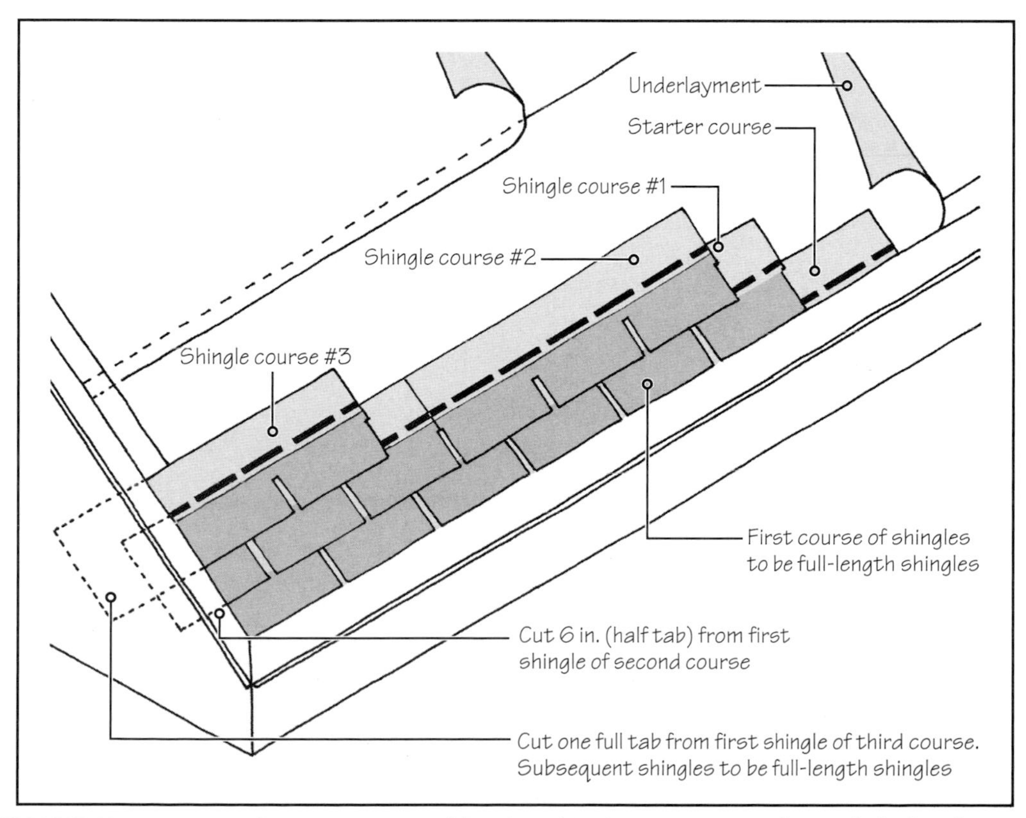

FIGURE 32.11 Layout of starter course and first few shingle courses. Note that asphalt shingles are laid with a double coverage—2-shingle thickness under every point except under the headlap, where there are 3 shingle thicknesses.

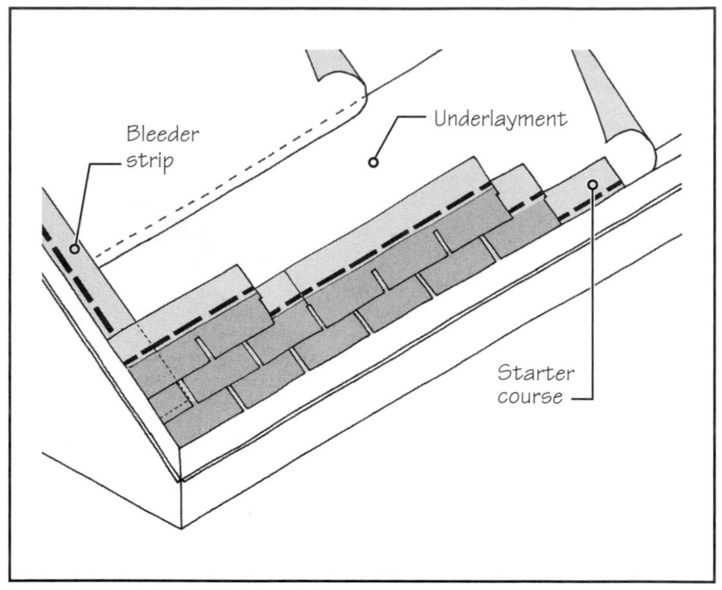

FIGURE 32.12 The use of a bleeder strip at the rake. A bleeder-strip shingle is identical to the starter-strip shingle at the eave.

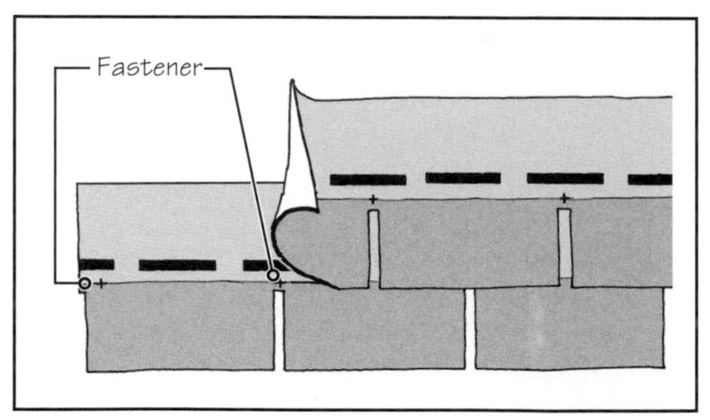

FIGURE 32.13 Standard fastening pattern requires 4 nails per shingle—one fastener at each end (nearly 1 in. from end) and one nail between each tab. In high-wind regions, a larger number of nails are needed.

To improve the performance of the roof, it is desirable to include a course of bleeder strips at the rake, Figure 32.12. A bleeder strip is similar to the starter course.

FASTENERS FOR SHINGLES

Asphalt shingles are fastened to the deck with 11- or 12-gauge galvanized steel roofing nails with $\frac{3}{8}$-in.- to $\frac{7}{16}$-in.-diameter nail heads (Figure 12.34). Each shingle strip requires a minimum of four nails, placed as shown in Figure 32.13. In high-wind regions and on steep slopes, a greater number of nails may be needed.

PRACTICE QUIZ

Each question has only one correct answer. Select the choice that best answers the question.

1. Which of the following steep roof materials has the longest history of use?
   - **a.** Asphalt shingle
   - **b.** Clay tile
   - **c.** Concrete tile
   - **d.** Slate roofs
   - **e.** Wood shingle
2. Which of the following steep roof materials is most commonly used in contemporary buildings?
   - **a.** Asphalt shingle
   - **b.** Clay tile
   - **c.** Concrete tile
   - **d.** Slate
   - **e.** Wood shingle
3. In a steep roof, the ice dam protection membrane is generally provided at the
   - **a.** ridge.
   - **b.** eave.
   - **c.** rake.
   - **d.** valley.
   - **e.** hip.
4. A double-layer roof underlayment is required in an asphalt shingle roof if the roof slope is
   - **a.** between 3:12 and 4:12.
   - **b.** between 3:12 and 5:12.
   - **c.** between 3:12 and 6:12.
   - **d.** between 3:12 and 9:12.
   - **e.** none of the above.
5. An asphalt shingle roof is generally not permitted if roof slope exceeds
   - **a.** 12:12.
   - **b.** 15:12.
   - **c.** 18:12.
   - **d.** 21:12.
   - **e.** 24:12.
6. Roof underlayment commonly used in steep roofs is
   - **a.** No. 1 asphalt-treated fiberglass felt.
   - **b.** No. 15 asphalt-treated organic felt.
   - **c.** No. 30 asphalt-treated organic felt.
   - **d.** (a) and (b).
   - **e.** (b) and (c).
7. At the eave of an asphalt shingle roof,
   - **a.** a drip edge is installed before installing roof underlayment.
   - **b.** a drip edge is installed after installing roof underlayment.
   - **c.** a drip edge is not required.
8. On the rake of an asphalt shingle roof,
   - **a.** a drip edge is installed before installing roof underlayment.
   - **b.** a drip edge is installed after installing roof underlayment.
   - **c.** a drip edge is not required.
9. Asphalt shingles are generally laid in
   - **a.** single coverage with a 5-in overlap between top and bottom shingles.
   - **b.** double coverage with a 2-in. overlap between top and bottom shingles.
   - **c.** double coverage with a 2-in. head lap between top and bottom shingles.
   - **d.** triple coverage with no head lap.
   - **e.** none of the above.

Answers: 1-d, 2-a, 3-b, 4-a, 5-d, 6-e, 7-a, 8-b, 9-c.

32.4 VALLEY TREATMENT IN AN ASPHALT SHINGLE ROOF

A valley occurs where two pitched roofs meet and water converges. In asphalt shingle roofs, any one of the following three types of valley treatments may be used:

- Open valley
- Woven valley (also called closed valley)
- Closed-cut valley

Regardless of the type of valley, each valley requires

- Valley underlayment
- Valley flashing

Although the flashing material varies with the type of valley, the underlayment requirement for each valley is the same. A valley underlayment consists of a full-width (36-in.-wide) asphalt-treated organic felt (No. 15 or heavier) laid with the center of the roll laid on the centerline of the valley, Figure 32.14. A self-adhering ice dam protection membrane below the valley underlayment is helpful because valleys are common sources of roof leaks.

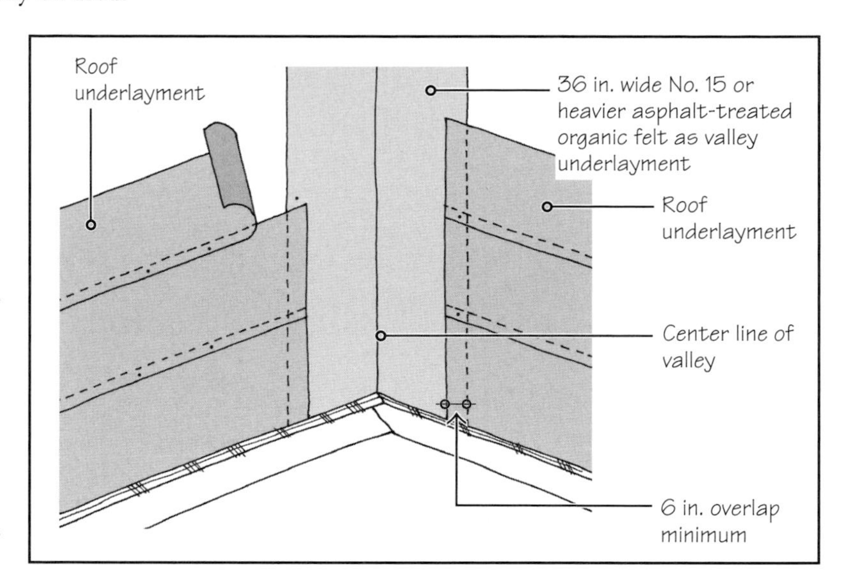

FIGURE 32.14 Valley underlayment and roof underlayment.

Like the roof underlayment, the valley underlayment is also secured with a minimum number of nails. Roof underlayment from the field of the roof is cut so that it overlaps the valley underlayment by a minimum of 6 in. Some roofers prefer the use of a self-adhering ice dam protection membrane as the valley underlayment.

OPEN VALLEY—ROLL ROOFING MATERIAL AS VALLEY FLASHING

In an open valley, the shingles are held away from the centerline of the valley so that the valley flashing is exposed to view. An open valley provides a relatively smooth and more rapid discharge of water.

The flashing in an open valley may consist of either roll roofing material or sheet metal. The construction of an open valley using roll roofing material as valley flashing is shown in Figure 32.15. Roll roofing material is an asphalt-treated, organic felt with one smooth surface and the other surface covered with mineral granules.

Roll roofing valley flashing consists of two layers of felt. The first layer, which is laid immediately over the valley underlayment, is an 18-in.-wide roll roofing felt with granule surface facing down. The second layer consists of the same material, but it is 36 in. wide. It is laid over the first layer with the mineral granule surface facing up. Both layers of roll roofing are secured with a minimum number of nails or staples.

The width of the exposed part of the flashing should increase at the rate of $\frac{1}{8}$ in. per ft toward the bottom of the valley. In practice, this is achieved by placing chalk lines over the valley flashing that converge upward. The minimum width of the exposed part of valley at the top is 6 in. Thus, if the valley is 24 ft long, the minimum width of the exposed part of valley flashing at the bottom is 9 in.

The shingles are trimmed along chalk lines. The end and corner of each shingle adjacent to the valley is cut and adhered with roofer's cement to prevent water entering under the shingles.

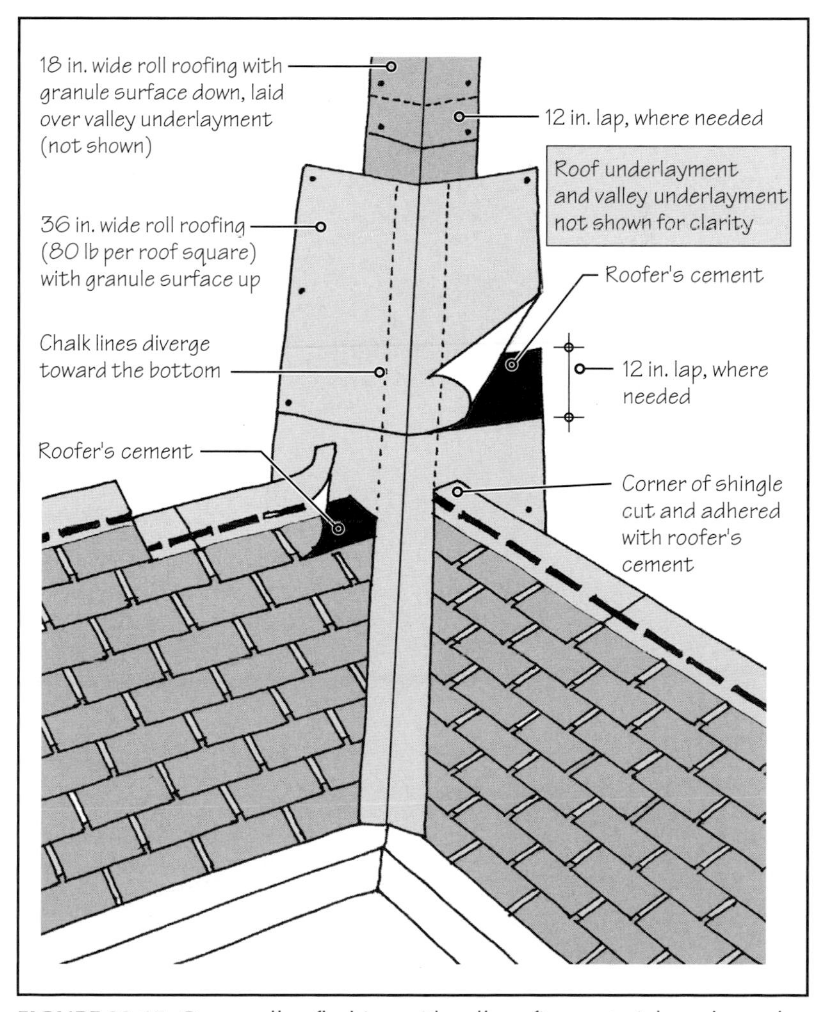

FIGURE 32.15 Open-valley flashing with roll roofing material used as valley flashing.

OPEN VALLEY—SHEET-METAL VALLEY FLASHING

The construction of an open valley using sheet-metal flashing is similar to roll roofing flashing. The sheet metal generally consists of a minimum of 24-gauge galvanized steel and must be at least 24 in. wide. It is placed over the valley underlayment, Figure 32.16(a). It is typically not nailed, but secured with special metal clips that allow the sheet metal to expand and contract, Figure 32.16(b). Sheet-metal flashing should be used in lengths of no more than 12 ft to reduce their expansion and contraction.

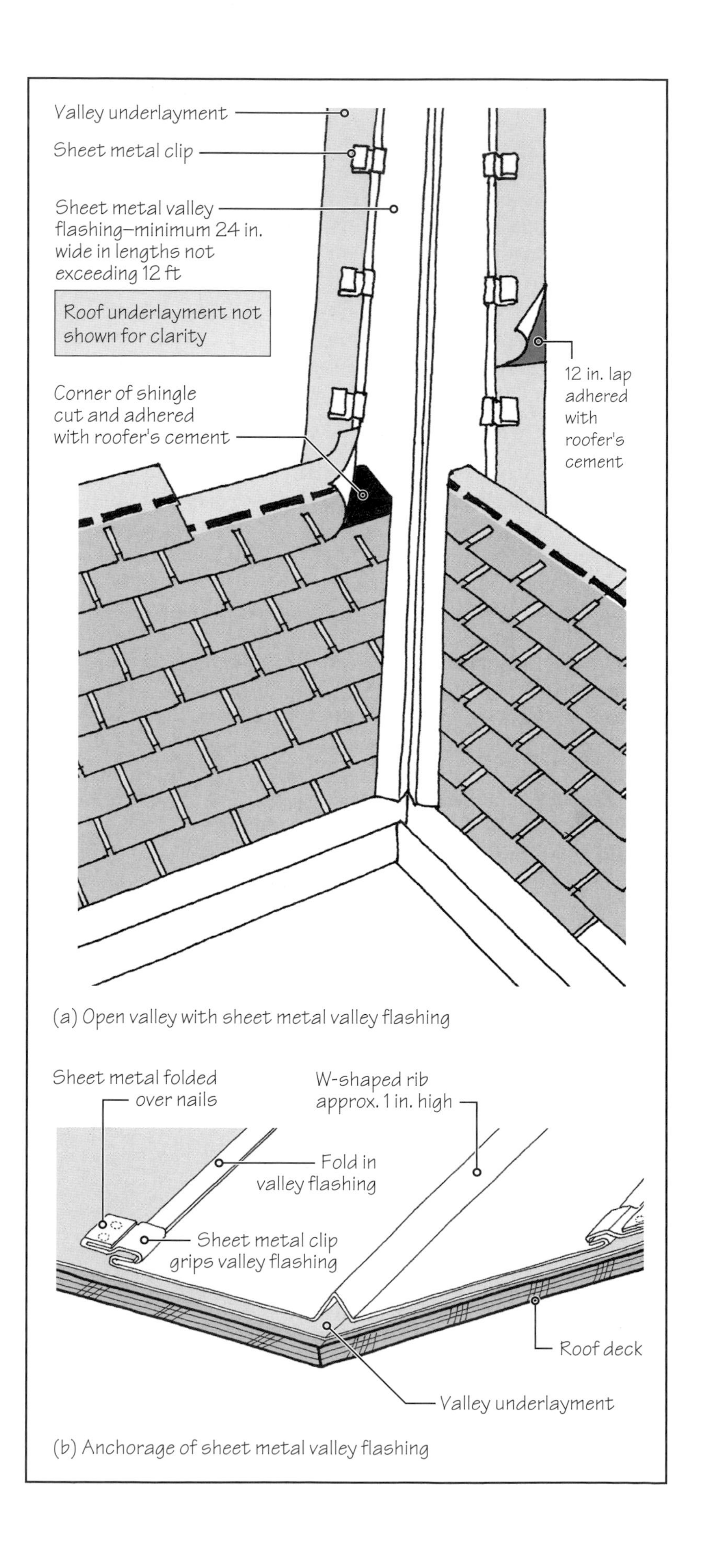

FIGURE 32.16 Open valley in an asphalt shingle roof with sheet-metal valley flashing.

The metal clips, which are located at 8 in. to 24 in. on center, grip the sheet-metal flashing at one end. The other end of a clip is nailed to the deck and then folded over to cover the nails. This protects the shingles against damage by any backout of the nails.

Although the fastening of metal flashing with clips is generally recommended, the clips can telegraph through a lightweight shingle roof, adversely affecting the roof's appearance and performance. In such a case, sheet-metal flashing may be nailed at the roof's outer edges. The outer edges may then be stripped in with a self-adhering ice dam protection membrane.

However, nailing sheet-metal flashing can be problematic in regions with large daily or annual temperature variations. That is why an open-valley sheet-metal flashing is generally not recommended in such areas.

Sheet-metal flashing is generally profiled to form a W-shaped rib in the center. This reduces the crossover of water from one side of the roof to the other.

Woven Valley

In a woven valley, shingles from the two adjacent roofs weave into each other, covering the valley flashing, Figure 32.17. Because of the weaving of shingles, the thickness of roofing material in a woven valley is greater than in an open valley. Therefore, a woven valley is more durable. It can better withstand the traffic of roofers—they tend to walk in the valley because the valley's slope is lower than the adjoining areas of the roof. A woven valley, however, does not work well with heavier-weight laminated shingles because these shingles do not bend as much.

The valley flashing in a woven valley consists of roll roofing material and is identical to that provided in an open valley. However, the 18-in.-wide roll roofing sheet shown in Figure 32.15 may be omitted, using only the 36-in.-wide sheet. Sheet-metal valley flashing is not to be used in a woven valley.

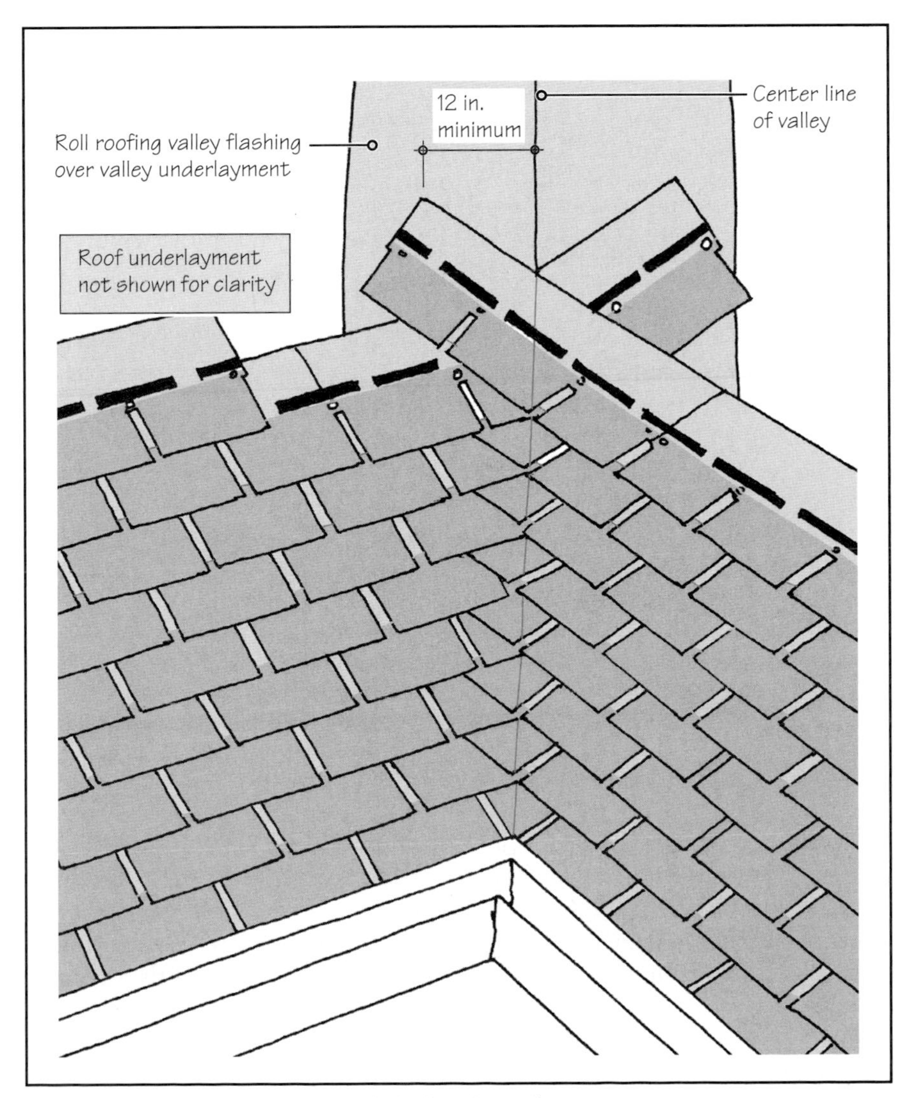

FIGURE 32.17 A woven valley in an asphalt shingle roof.

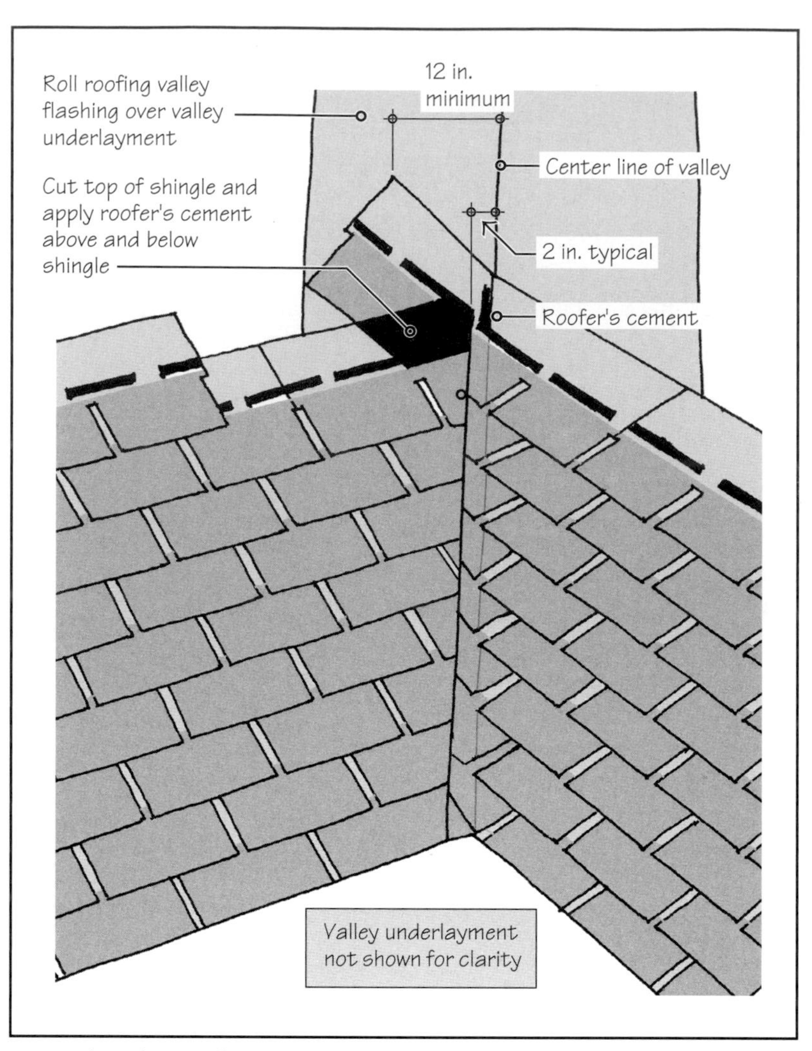

FIGURE 32.18 A closed-cut valley.

Once the valley flashing has been installed, shingles are laid, starting from the eave upward. The shingles extend to the opposite side of the valley by at least 12 in.

CLOSED-CUT VALLEY

In a closed-cut valley, the shingles from one side of the valley are cut parallel to the line of the valley so that the valley is partially closed, Figure 32.18. A closed-cut valley combines the advantages and disadvantages of the other two valley types and is more commonly used. It is more resistant to roofers' traffic than an open valley but less than that of a woven valley.

Valley flashing in a closed-cut valley is identical to that in a woven valley. The shingles from one side of the valley extend to the opposite side by a minimum of 12 in. However, the shingles from the other side are trimmed parallel to the centerline of the valley, leaving an overlap of 2 in.

32.5 RIDGE AND HIP TREATMENT IN AN ASPHALT SHINGLE ROOF

Asphalt shingles from both sides of a ridge are butted against each other at the ridge, which is then capped with ridge shingles. Some manufacturers provide special ridge shingles, but these can also be prepared on site from standard shingles. To do so, a shingle is field trimmed along dashed lines, Figure 32.19. The dashed lines taper slightly away from the cutouts. The trimming converts a 3-tab shingle in three nearly 12-in. × 12-in. shingles.

These smaller shingles are now bent and nailed over the ridge as caps, using a 5-in. exposure, Figure 32.20. In cold weather, the shingles may be warmed somewhat to increase their pliability, because fiberglass shingles are generally brittle.

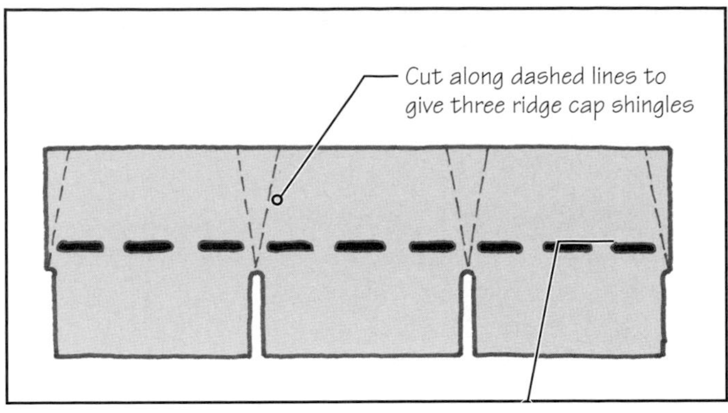

FIGURE 32.19 Creating ridge (or hip) caps from a standard 3-tab shingle.

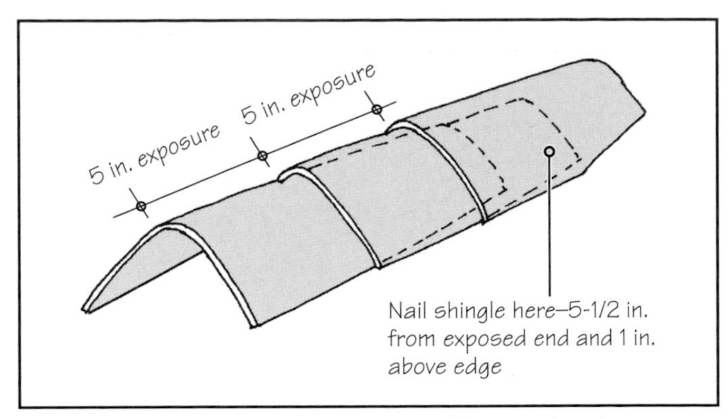

FIGURE 32.20 Layout and nailing of ridge cap shingles.

FIGURE 32.21 An asphalt shingle roof showing hip cap shingles and ridge cap shingles. Note that ridge cap shingles and hip cap shingles are identical.

The treatment of a hip is similar to that of a ridge. The hip is capped with the same shingles as a ridge, and the shingles are installed from the bottom of the hip to the top, Figure 32.21.

32.6 FLASHINGS IN AN ASPHALT SHINGLE ROOF

A sloping roof plane that abuts a vertical side wall is flashed with metal shingles placed over the end of each asphalt shingle course—a flashing system referred to as *step flashing*. Metal shingles are made of galvanized or prepainted sheet steel with a minimum thickness of 24 gauge.

The profile of a metal shingle is an L-shape, each leg at least 5 in., so that the shingle extends 5 in. in both horizontal and vertical directions. The length of the shingle is 2 in. greater than the exposure of roof shingles. Thus, with the most commonly used 3-tab roof shingles that have a 5-in. exposure, 7-in.-long metal shingles are commonly used. With its 7-in. length and 5-in. exposure, each metal shingle overlaps the underlying metal shingle by 2 in., Figure 32.22.

Each metal shingle is placed slightly up slope of the exposed edge of the roofing shingle that overlaps it. In other words, the horizontal surface of metal shingles is not visible, and each metal shingle is completely covered by the overlying roof shingle. Each metal shingle is secured to the deck by two nails placed at its top edge; the vertical leg of a metal shingle is not fastened. Metal shingles are counterflashed by siding that terminates 2 in. above the roof surface.

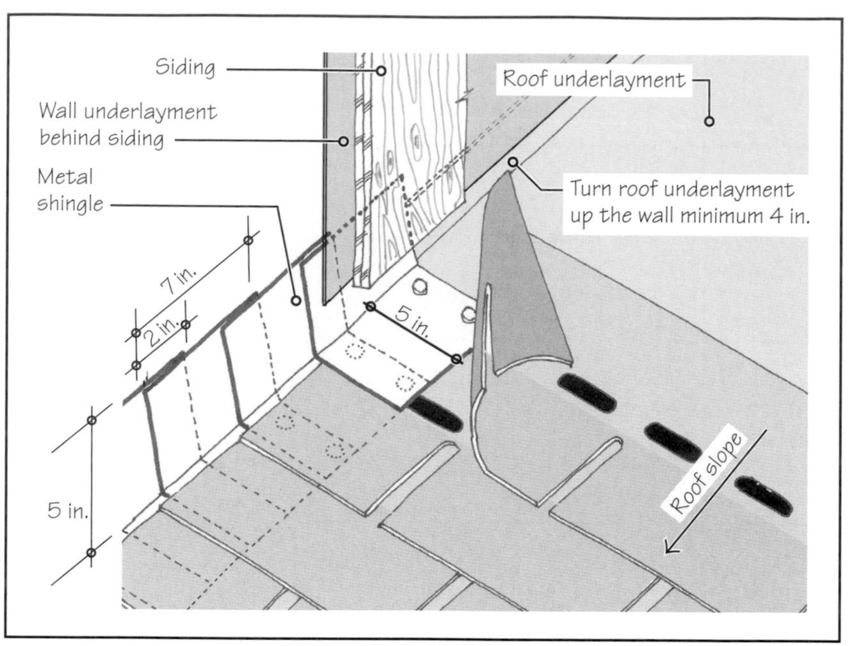

FIGURE 32.22 Step flashing against a wall that abuts the sloping side of an asphalt shingle roof.

Flashing Around a Masonry Chimney

Flashing around a masonry chimney consists of the following four-component sheet-metal flashings, where the sheet metal is generally prepainted steel, 24 gauge or thicker, or any other compatible metal:

- Apron flashing
- Step flashing
- Cricket flashing
- Counterflashing

Sheet-metal apron flashing is placed over the downslope side of chimney and tightly wrapped around it, Figure 32.23(a). This requires prefabricating the flashing to a profile, shown in Figure 32.23(b), and adhering it to the chimney with roofer's cement.

After the apron flashing has been applied, step flashing is installed on the side walls of the chimney in the same way as shown in Figure 32.22. Note that the first step-flashing shingle wraps over the apron flashing at the corner by at least 2 in., Figure 32.24. Step flashing is also set with roofer's cement applied to chimney walls.

The upslope side (backside) of a chimney requires a cricket. Thus, before installing any metal flashing, roof shingles, or a roof underlayment around a chimney, a cricket is formed with plywood or oriented strandboard, Figure 32.25. Roof underlayment is subsequently installed on the deck and over the ice dam protection membrane. (Because a chimney is one of the primary sources of leaks on an asphalt shingle roof, a self-adhering ice dam protection

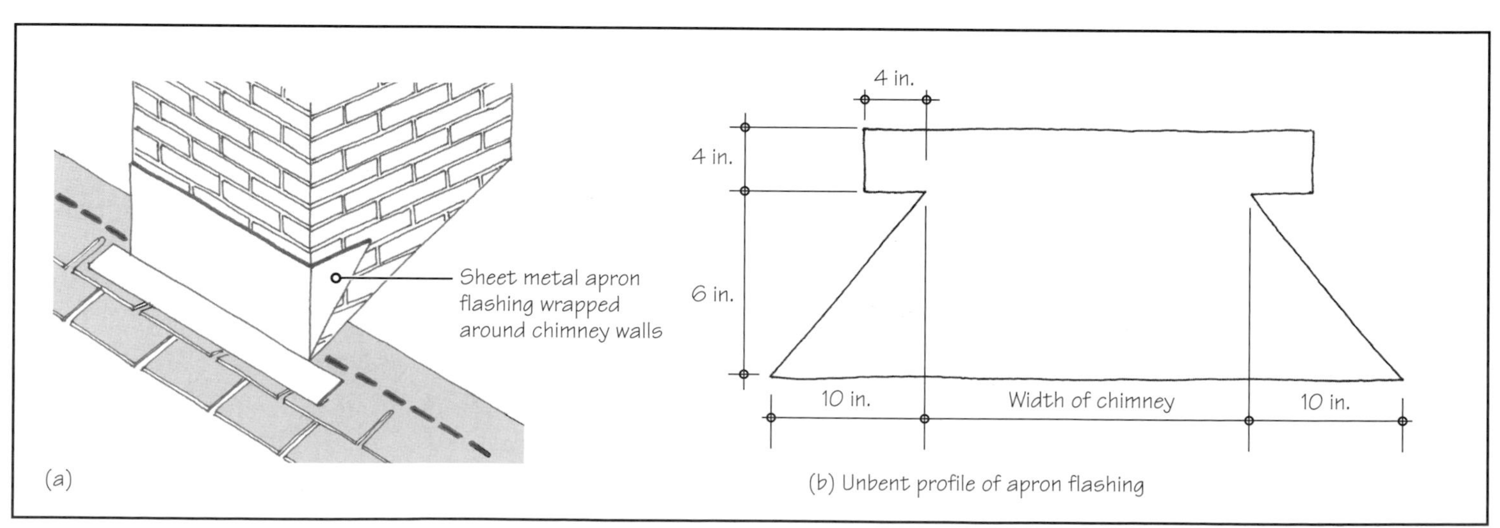

FIGURE 32.23 Apron flashing on the front face of chimney.

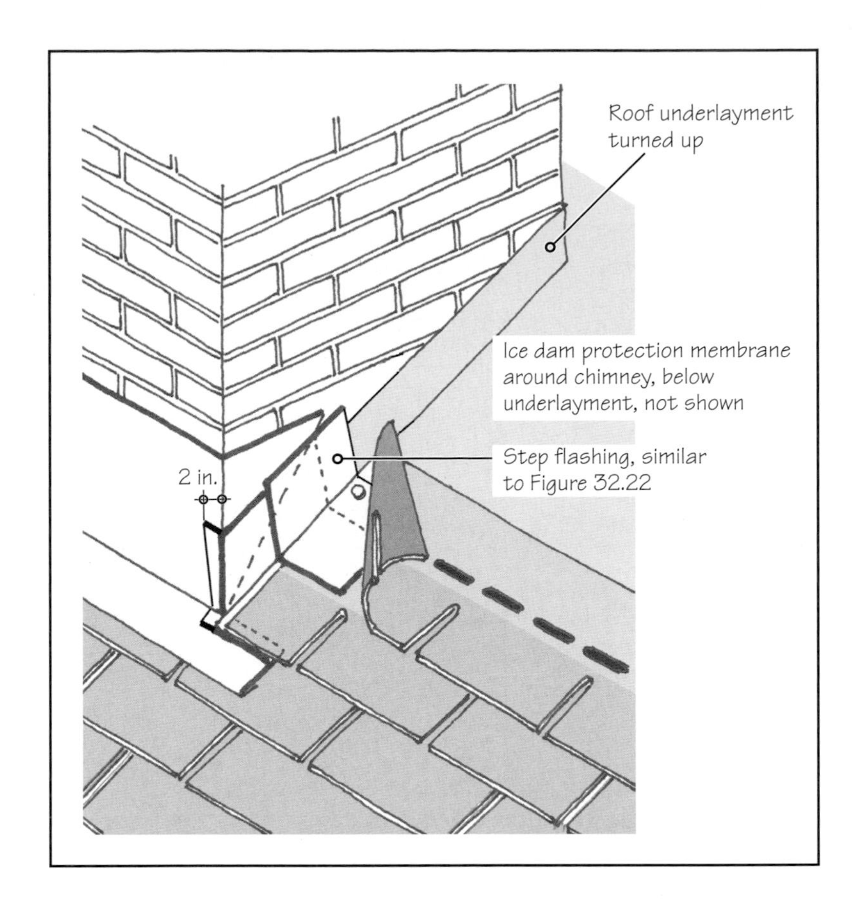

FIGURE 32.24 Step flashing on the side walls of a masonry chimney.

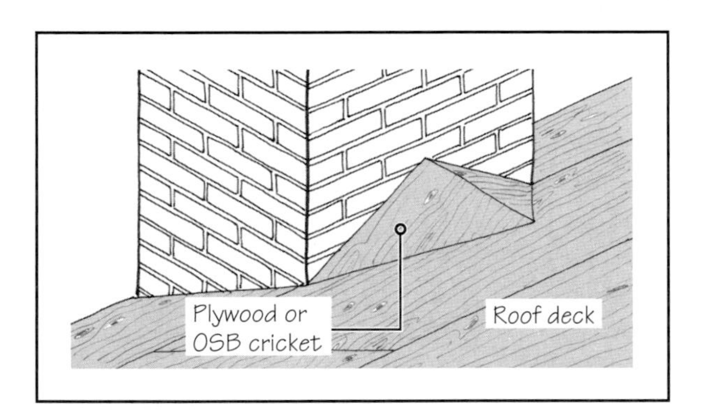

FIGURE 32.25 Formation of plywood or OSB cricket on the upslope side (back) of chimney wall.

membrane should preferably be installed all around the base of the chimney, with its edges turned up the chimney walls.)

The flashing over the cricket consists of a prefabricated sheet-metal flashing, Figure 32.26. The joints in cricket flashing are thoroughly soldered. The chimney is then ready to receive sheet-metal counterflashing, Figure 32.27.

Counterflashing is generally of the same metal as the flashing. One edge of the counterflashing laps over the flashing, and the other edge is inset into masonry mortar joints,

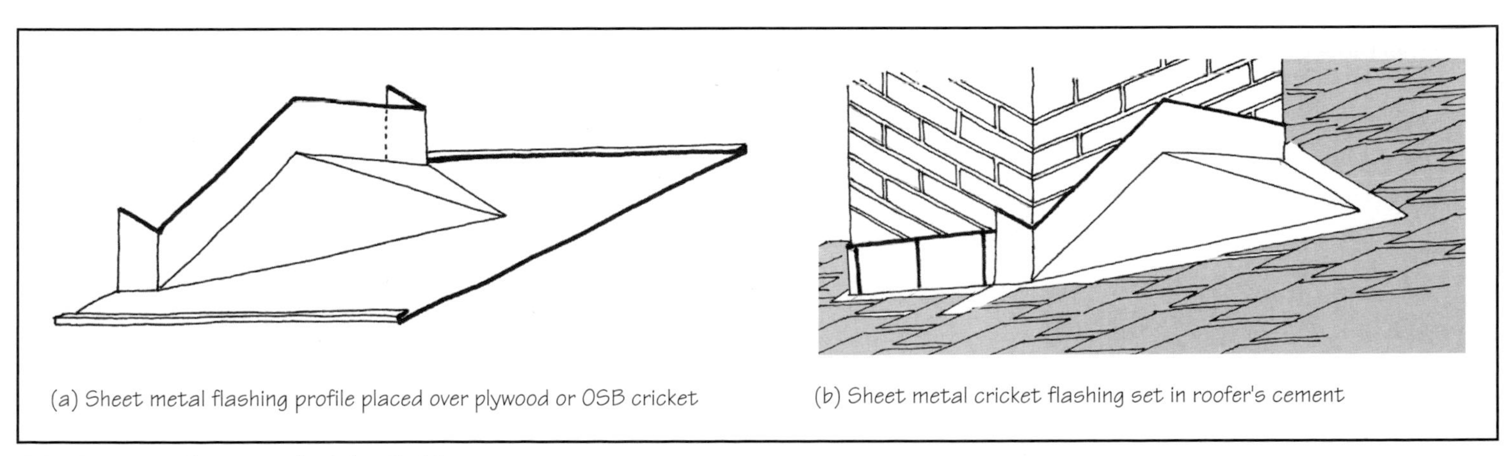

FIGURE 32.26 Sheet-metal cricket flashing.

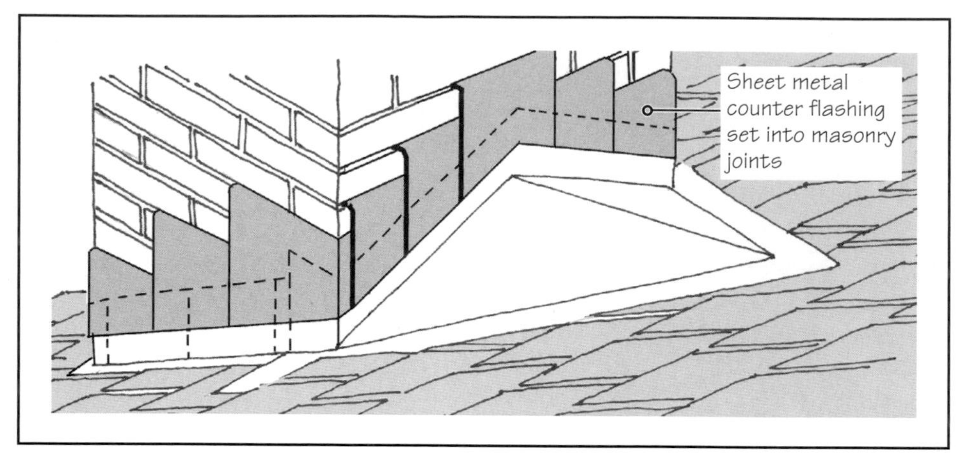

FIGURE 32.27 Counterflashing over chimney wall, set into masonry joints, as shown in Figure 32.28.

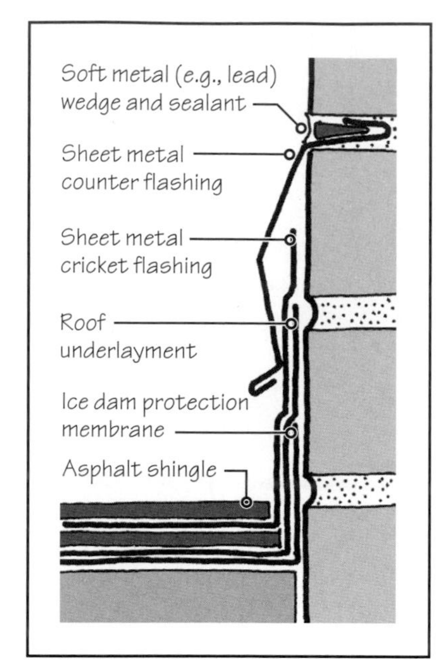

FIGURE 32.28 Section through counterflashing.

Figure 32.28. This requires cutting the appropriate mortar joints to a depth of nearly 1.5 in.—a cut that is commonly referred to as a *raggle*. The inset edge of counterflashing is profiled to fit into the raggle, giving a friction fit. Counterflashing is further secured by driving a soft metal wedge into the raggle and finishing the joint with an elastomeric sealant.

PRACTICE QUIZ

Each question has only one correct answer. Select the choice that best answers the question.

10. In an asphalt shingle roof, roof underlayment is installed first followed by valley underlayment.
- **a.** True
- **b.** False

11. In an asphalt shingle roof, either valley underlayment or valley flashing is used.
- **a.** True
- **b.** False

12. In an asphalt shingle roof, sheet-metal valley flashing is used in an
- **a.** open valley.
- **b.** closed-cut valley.
- **c.** woven valley.
- **d.** all the above.
- **e.** (a) and (b) only.

13. In an asphalt shingle roof, the ridge and hip are capped with shingles. The ridge cap and hip cap shingles can be used interchangeably.
- **a.** True
- **b.** False

14. The junctions between the walls of a chimney and a steep roof need to be flashed. Which of the following flashings is required as chimney flashing?
- **a.** Step flashing
- **b.** Apron flashing
- **c.** Cricket flashing
- **d.** All the above
- **e.** Only (a) and (c)

15. In an asphalt shingle roof, which of the following flashings is interwoven with shingles?
- **a.** Step flashing
- **b.** Apron flashing
- **c.** Cricket flashing
- **d.** All the above
- **e.** Only (a) and (c)

16. In a steep roof, which of the following flashings is made from sheet metal?
- **a.** Step flashing
- **b.** Apron flashing
- **c.** Cricket flashing
- **d.** All the above
- **e.** Only (a) and (c)

17. Which of the following flashings in a steep roof has the shape of a small steep roof?
- **a.** Step flashing
- **b.** Apron flashing
- **c.** Cricket flashing
- **d.** All the above
- **e.** Only (a) and (c)

Answers: 10-b, 11-b, 12-a, 13-a, 14-d, 15-a, 16-d, 17-c.

32.7 ESSENTIALS OF CLAY AND CONCRETE ROOF TILES

The service life spans of concrete and clay tile roofs are significantly longer than asphalt shingle roofs. Because of their color, texture, and tile profile, they produce handsome-looking roofs. That is why in modern construction, concrete and clay tiles are extensively used in low-rise commercial structures and higher-priced homes and apartment buildings.

Types of Clay and Concrete Roof Tiles

Tiles come in a large variety of profiles, sizes, and colors, depending on the manufacturer. The clay tile's color is primarily a function of the chemical composition of the clay from which it is made. If any additional color-modifying agent is used, it is added to the clay before firing. Therefore, the color becomes integral to the tile. With its integral color, the color of clay tile weathers only slightly with time, caused mainly by the pollutants in the air. Glaze-coated clay tiles are also available. The glaze is applied to the already-fired tiles, which are then refired to obtain a durable glaze.

The color of concrete tiles is a function of the color of the portland cement, aggregates, and any pigments, if used. The gradual erosion of the surface of the tile due to running water on the roof exposes the aggregates in a concrete tile, which can create a noticeable change in the tile's color over time. Initial and periodic sealing of the tiles' surface helps retain the color longer.

Concrete tiles can also be factory painted on their exposed surface. Several manufacturers make concrete tiles with the characteristic (brown) terracotta color painted on them to mimic clay tiles. Being an applied finish, the paint tends to fade over time.

Clay and Concrete Roof Tile Profiles

Because of the enormous variety in the profiles of concrete and clay tiles, it is impossible to discuss all available profiles. Although some profiles are used only with clay tiles and others, with concrete tiles, most tile profiles are essentially similar. Except for flat tile, all tile profiles provide for water channels that direct water down the roof. That is why roof tiles are generally laid in single coverage with a simple overlap of 3 in. between succeeding courses. This is unlike asphalt shingle roofs, which are laid in double coverage with a 2-in. headlap.

One of the oldest and a commonly used profile in clay tiles is the *mission tile,* which consists of an almost semicircular barrel profile. The tiles are laid alternately, one tile in a water-holding setting (pan) and the other tile in a water-shedding setting (cover), Figure 32.29. Both pan and cover tiles are provided with one hole to receive a fastener at the head of each tile.

Another commonly used clay tile profile is a derivative of the mission tile. Called *S-shape tile,* a *one-piece barrel,* or an *S-shape mission tile,* it integrates the pan and cover of the mission

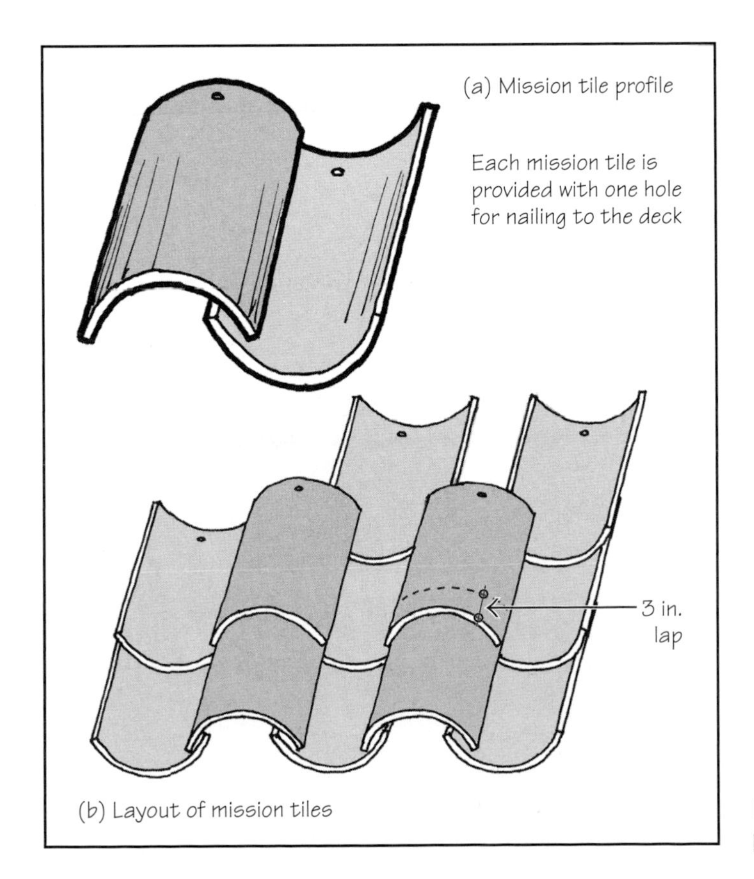

FIGURE 32.29 Mission tile profile and layout.

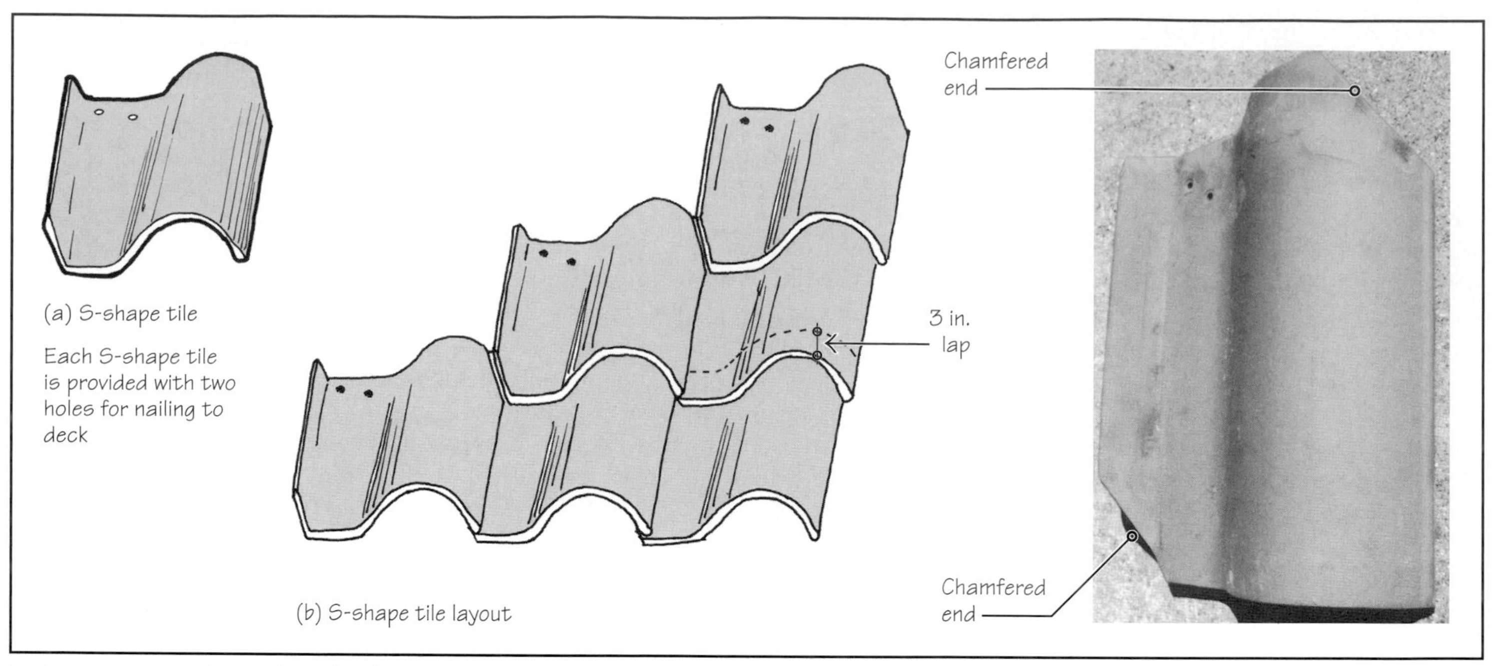

FIGURE 32.30 S-shape tile and its layout.

tile into one unit, Figure 32.30. The S-shape tile generally comes with its two opposite ends chamfered to allow for easier drainage of water. The chamfer is generally of the same size as the overlap needed between the adjacent tiles to give a better, more exact fit between them. Each tile is provided with two fastener holes at its head.

Other profiles include a flat tile. Flat tiles are laid like asphalt shingles or slates—in double coverage and with a headlap of 2 in., Figure 32.31. That is why they are also referred to as *shingle tiles*. Because of its double coverage, the weight of a shingle tile roof is higher than the roofs constructed of curved roof tiles.

Water Absorption and Strength of Tiles

Unlike asphalt shingles, concrete and clay tiles absorb water. Water absorption increases the weight of the tile roof and leads to freeze-thaw damage of tiles. It also aggravates the color-degradation tendency as the water-soluble pollutants enter the body of the tile. Less-porous tiles are obviously better but are more expensive.

The porosity of clay tiles varies from 2% to 10%; that of concrete tiles, from 3% to 20%. Sealers are generally used to reduce the effect of porosity. The service life of the sealers and the cost of repeat applications must be considered in tile selection. It may be more economical in the long run to specify a denser, albeit more expensive, tile.

Some tiles are graded (Grades 1, 2, or 3) for their freeze-thaw resistance. Grade 1 is typically specified in severe freeze-thaw climates, Grade 2, in moderate freeze-thaw climates, and Grade 3, in climates with negligible freeze-thaw activity.

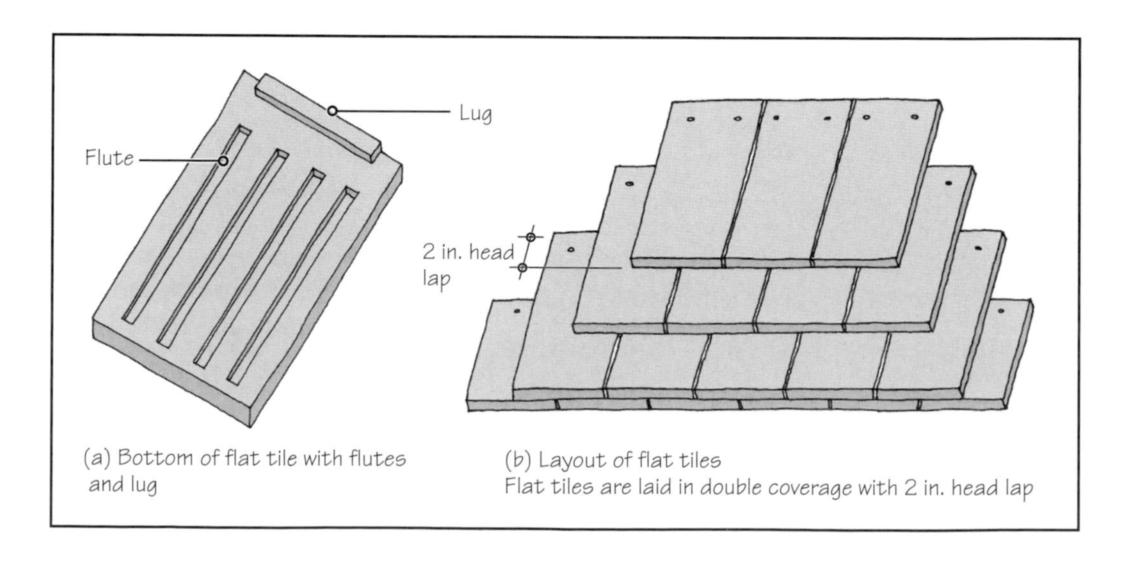

FIGURE 32.31 Flat tile and its layout.

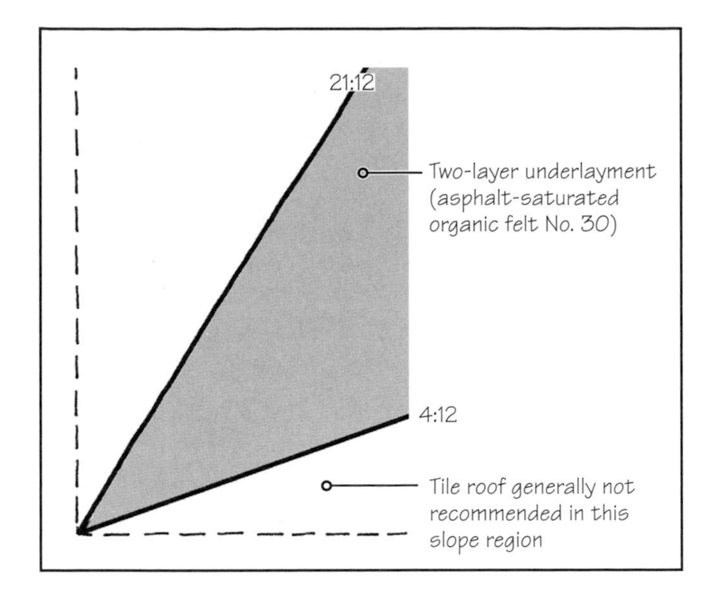

FIGURE 32.32 Roof slope and recommended underlayment for tile roofs.

Porosity is not merely an index of the tile's water absorptivity, it also indicates the tile's strength. A more porous tile is weaker. Strength is an important selection criterion for concrete and clay tiles, because a stronger tile is generally more durable.

ROOF UNDERLAYMENT AND TILE ROOFS

Because a tile roof is a relatively long-lasting roof, the underlayment should match the durability of the tiles. The following underlayment specifications is generally recommended:

- *For roofs with a slope greater than 4:12:* A minimum two layers of No. 30 asphalt-saturated felt laid in shingle fashion, Figure 32.32.
- *For roofs with a slope less than 4:12:* Tile roofs are generally not recommended in this slope region, although some manufacturers allow their product for use in this region.

ICE DAM PROTECTION MEMBRANE

An ice dam protection membrane at the eave should be specified in all climates. However, it is particularly important in cold climates. The minimum requirement for an ice dam protection membrane is the same as for the asphalt shingle roofs.

32.8 CLAY AND CONCRETE TILE ROOF DETAILS

The easiest and most commonly used method of fastening the tiles is to nail them to the substrate. The nails should be of stainless steel or copper to match the long life of a tile roof. Except for mission tiles, other tiles can be nailed straight to the deck. However, the best installation of tiles is achieved if a network of battens and crossbattens of treated wood is nailed to the deck over the underlayment, Figure 32.33. The tiles are then nailed to the (top) battens, which run horizontally. The horizontal battens should be a minimum of 1 in. × 2 in. (nominal), and the vertical battens should be a minimum of $\frac{1}{2}$ in. × $1\frac{1}{2}$ in. nominal.

The vertical battens are generally laid 24 in. on center. The spacing of the horizontal battens is a function of the length of the tile. Thus, with a 3-in. lap between successive courses, the center-to-center spacing of horizontal battens equals the tile length minus 3 in.

The batten network creates a space under the tiles, which allows for a clear drainage of water should the water seep under the tiles. Because the tiles are generally laid in single coverage with only a 3-in. lap, the seepage of water under the tiles in severe weather is not uncommon.

The space under the tiles also permits a more rapid drying of the underlayment. Battens also help to hang the tiles that consist of lugs. A major advantage of a batten and crossbatten network is that such a system can be used with slopes less than 4:12, provided the underlayment is treated like a built-up roof. In that case, the tiles serve simply as a decorative roof covering.

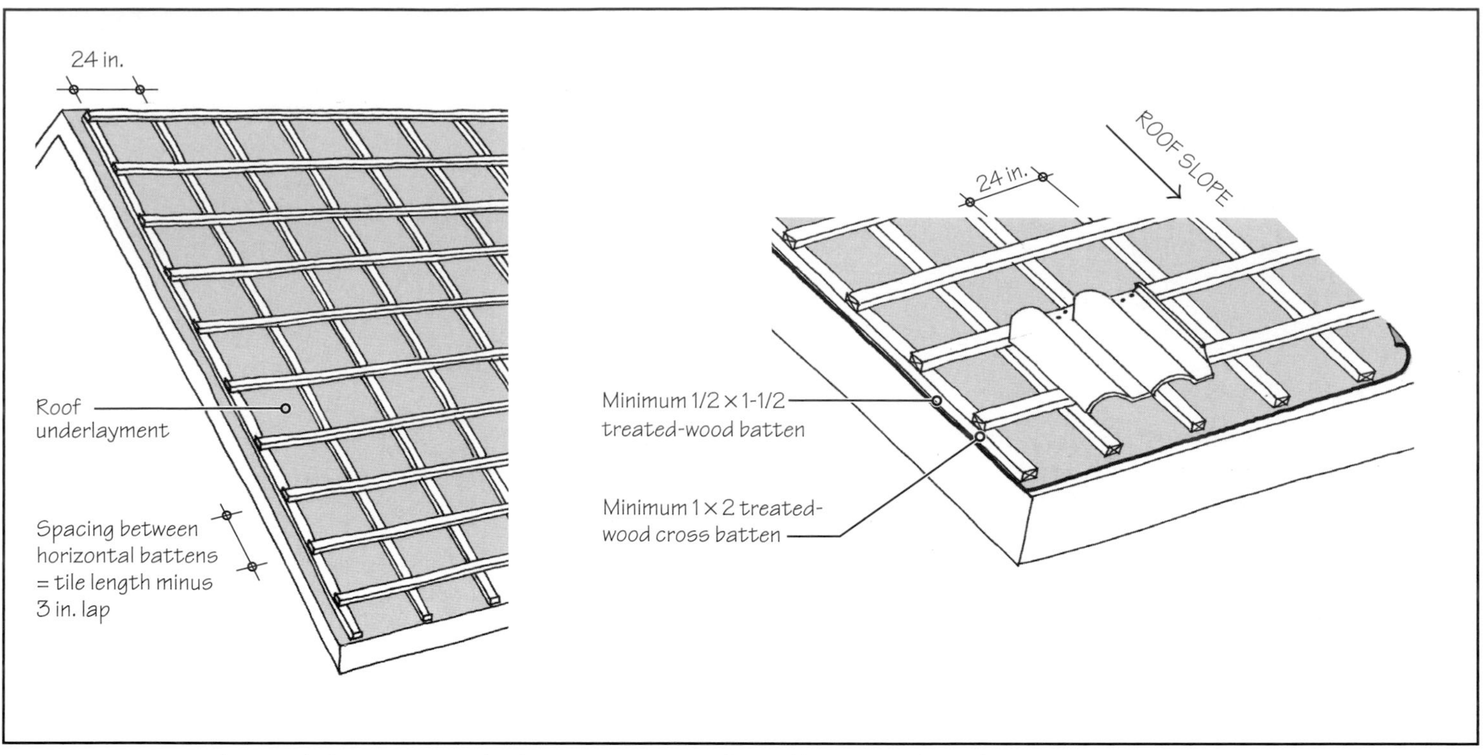

FIGURE 32.33 Installation of tiles on a network of preservative-treated wood battens and crossbattens.

In place of the batten and crossbatten network, a system of one-way horizontal battens may be used, Figure 32.34. The battens are generally laid in 4-ft lengths, separated by a minimum of a $\frac{1}{2}$-in. space. The $\frac{1}{2}$-in. separation allows the drainage of any water trapped by the battens. However, this system does not perform as well as the crossbatten system because the dust and debris block the drainage, causing leaks.

In addition to nailing each tile, additional anchorage may be required in a high-wind or high-seismic region, Figure 32.35.

Eave Treatment

Unlike an asphalt shingle roof, the drip edge is not necessary with a tile roof because the eave tiles can be made to project sufficiently beyond the fascia. However, if a drip edge is used, it should be of a durable metal (e.g., 24-oz copper) to match the long life of a tile roof. A treated-wood edge strip is also required at the eave to give the correct alignment to the first tile, Figure 32.36. Tile manufacturers also make starter tiles that preclude the need of an edge strip.

Eave-closure pieces are required to fill the space between the tiles and the underlying construction. (With flat tiles, eave-closure pieces are obviously not needed.) Closure pieces can

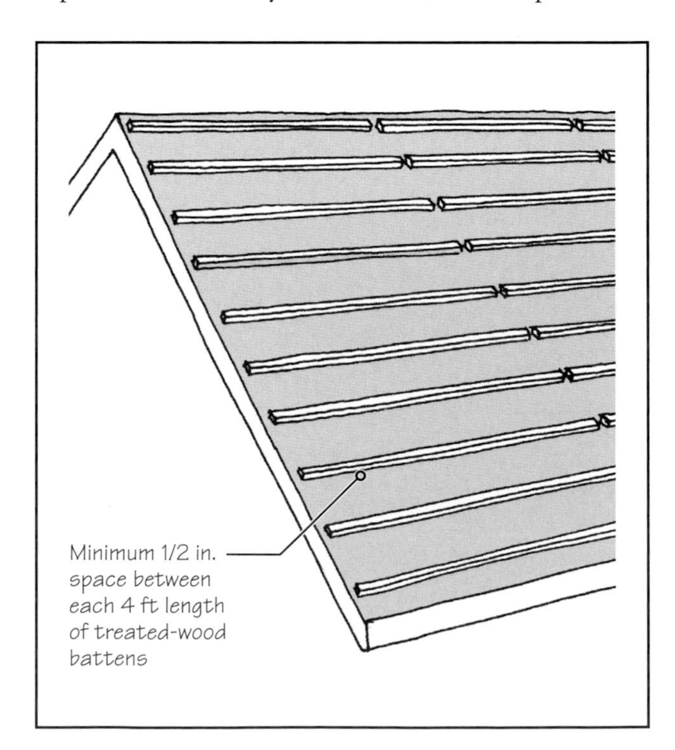

FIGURE 32.34 In place of batten and crossbatten arrangement of Figure 32.33, one-way battens may be used.

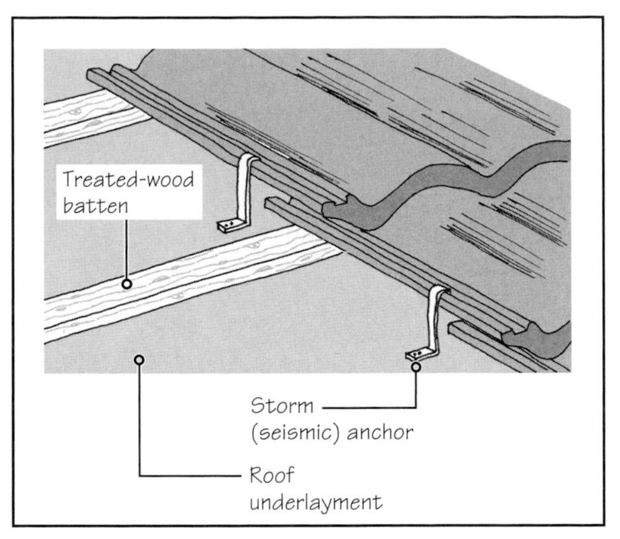

FIGURE 32.35 Storm or seismic tile anchors.

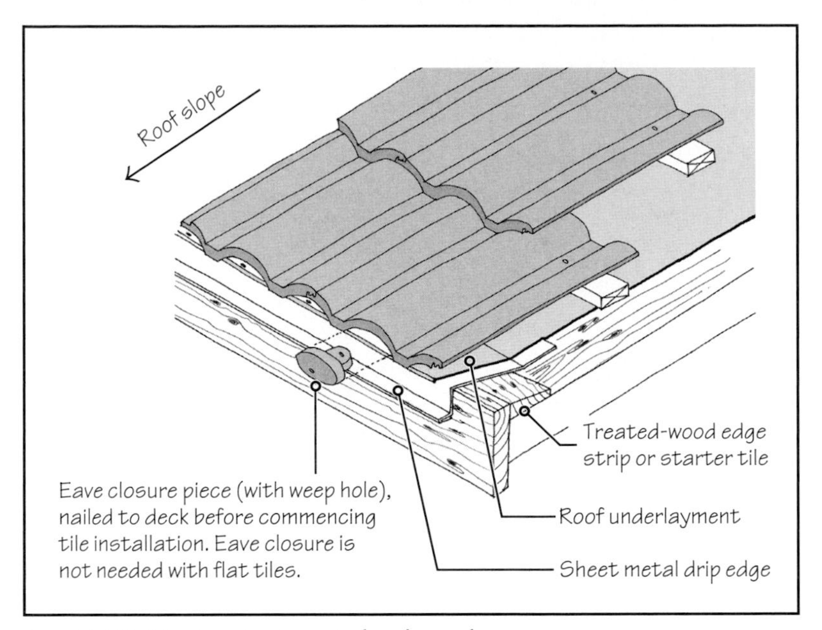

FIGURE 32.36 Eave treatment of a tile roof.

be of clay (with clay tiles), concrete, or plastic. They are provided with weep holes to allow the escape of water that may seep under the tiles. The closure pieces are nailed to the deck before installing the tiles. The closing of the space under the tiles prevents the nesting of birds and insects. Portland cement mortar may also be used to fill the gap between the eave tiles and the underlying part of the roof, if weep holes are provided in each mortar filling.

RAKE, RIDGE, AND HIP TREATMENTS

With curved tiles, specially prepared rake tiles are generally used to weatherproof the rake, Figure 32.37(a). The construction of a ridge and a hip are similar. With curved tiles, a

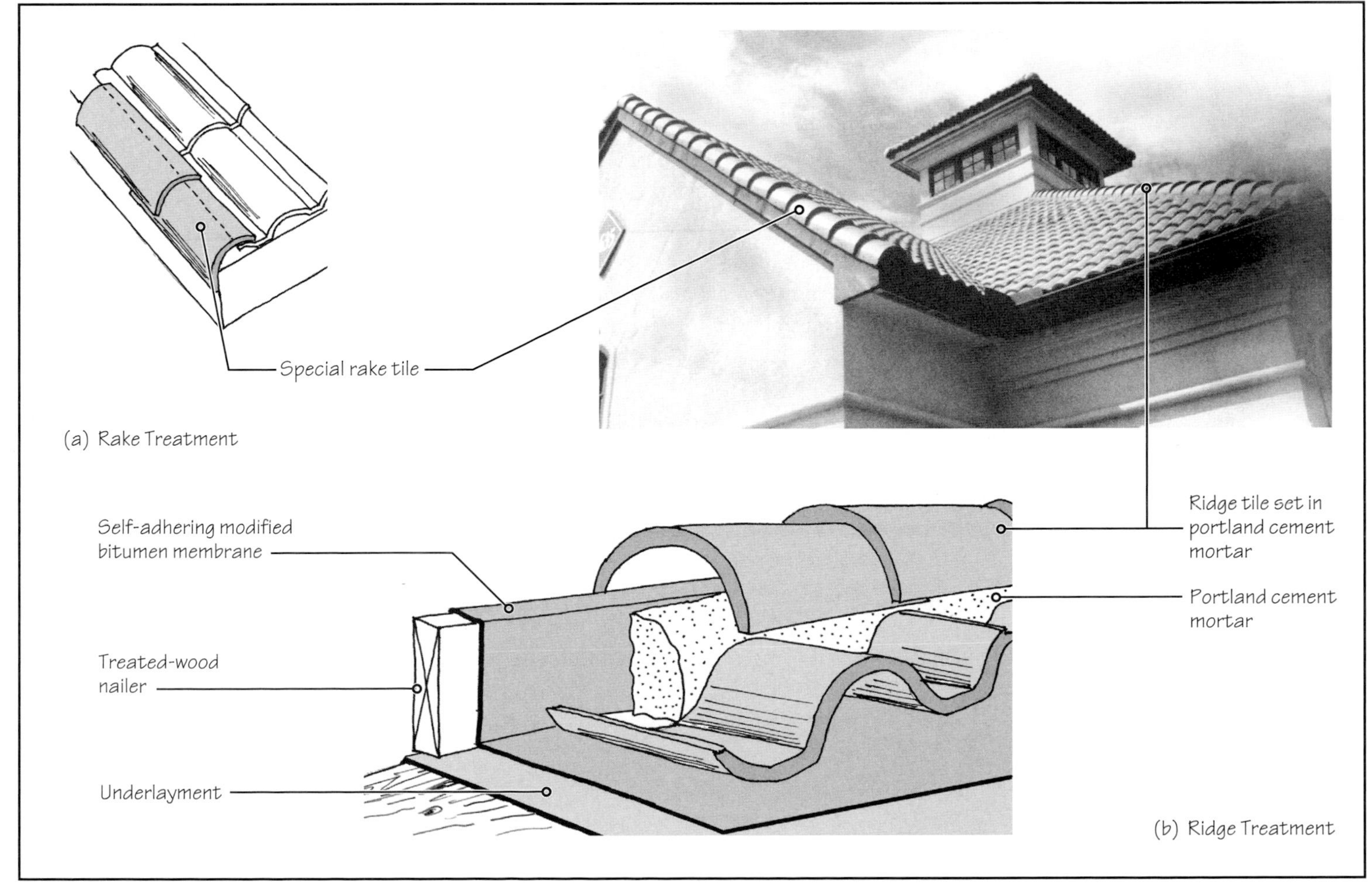

FIGURE 32.37 Rake and hip treatment in a tile roof. Ridge treatment is similar.

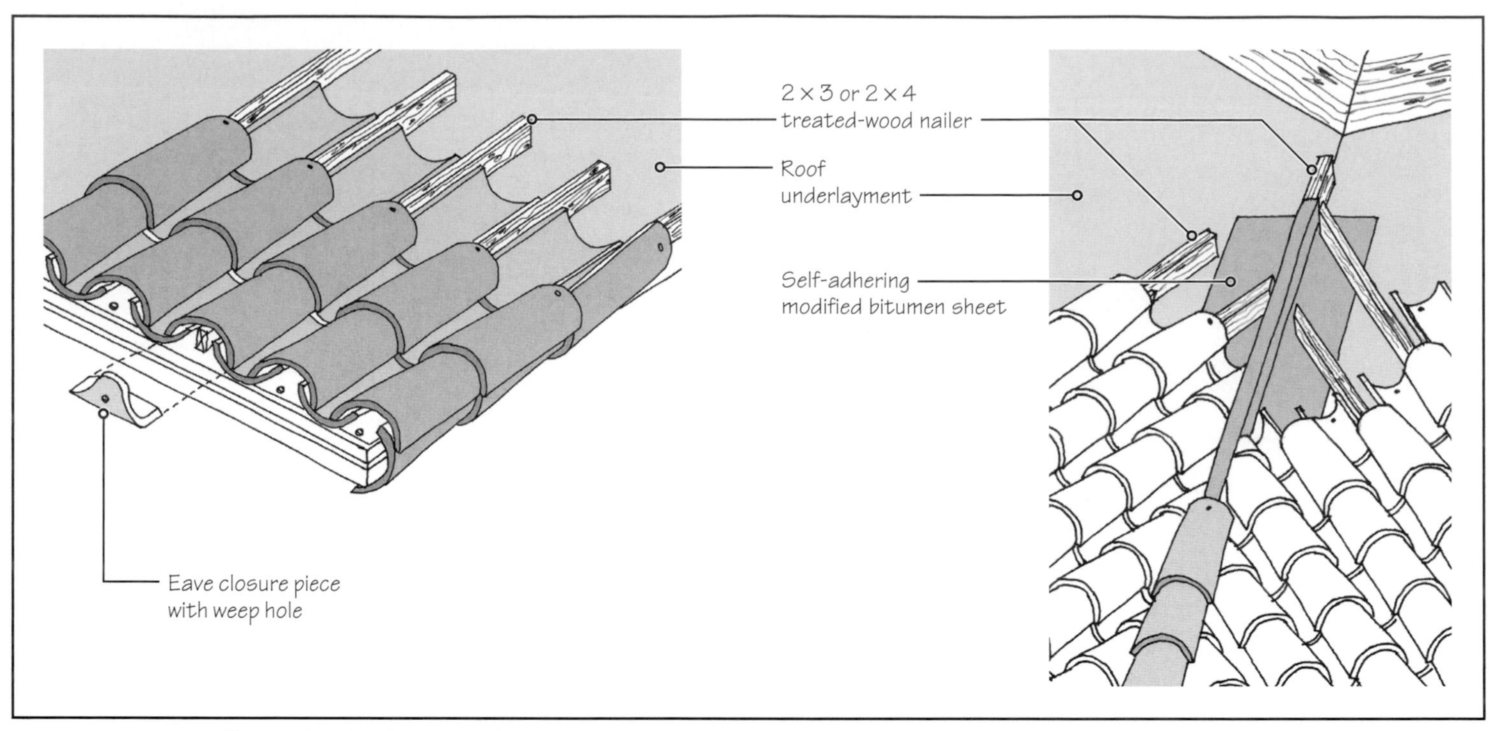

FIGURE 32.38 Installation details of mission tiles.

treated-wood nailer is installed at the ridge (or the hip), and a self-adhering modified bitumen sheet is wrapped over the nailer. Field tiles are brought close to the nailer, and the nailer is covered with special ridge (or hip) tiles set in portland cement mortar. The mortar holds the ridge (or hip) tiles and also closes the gap between the ridge tiles (or hip tiles) and the field tiles, Figure 32.37(b). Each ridge (or hip) tile is nailed to the nailer. Although these nails are protected by the overlapping ridge (or hip) tile, it is necessary to seal the nail with roofer's cement or sealant.

Installation of Mission Tiles

Mission tiles require an entirely different installation system. A treated-wood nailer, aligned along the slope of the roof, is placed under every cover piece, and the cover piece is nailed to the nailer, Figure 32.38(a). The nailer is usually either 2 × 3 or a 2 × 4 (nominal size), depending on the height of the cover piece. The nailer not only helps to secure the tile, it also reduces the possibility of tile breakage when roofers walk on the tiles. The tiles are laid with 3-in. standard overlap, as with other tiles.

Eave-closure pieces are required, and the rake is covered with the field tiles and rake cover tiles. Additional nailers are needed at the rake, Figure 32.38(b). The construction of a ridge or a hip is similar to that of other tile roofs.

32.9 Types of Architectural Metal Roofs

Sheet-metal roofs are increasingly becoming a part of the urban landscape. The development of durable coatings on sheet steel, coupled with a vast array of nonfading coating colors, has increased the popularity of metal roofs. Metal roofs are spared the rapid atmospheric and solar degradation to which other commonly used low-slope roofs (e.g., built-up, modified bitumen, or single-ply roofs) are subjected.

Some owners of commercial buildings, unhappy with frequent replacement of their existing low-slope roofs, are switching to metal roofs when reroofing is needed. This is despite the fact that a change from a low-slope roof to a metal roof on an existing building requires a great deal of additional under-roof components. Sheet-metal roofs can be divided under two types:

- Structural sheet-metal roofs
- Architectural sheet-metal roofs

Structural sheet-metal roofs double as a roof covering and also as a structural deck. A separate deck is not provided with such roofs, and the roof panels are supported directly on purlins, which are typically spaced 4 to 5 ft on center. Structural metal roofs are commonly used in low-cost industrial buildings in what are generally referred to as *preengineered metal buildings*. Preengineered metal buildings use highly standardized manufacturer-specific components and assemblies and are beyond the scope of this text.

Architectural sheet-metal roofs, which are discussed further here, require a supporting deck and an underlayment. Historically, they were limited for use in steep slopes (slope ≥ 3:12), but recently, some manufacturers allow their use in slopes less than 3:12 if sealants are used between overlapping panels. Architectural sheet-metal roofs may be further subdivided into the following types:

- *Traditional metal roofs* are typically made from sheet metals that can be soldered and are inherently durable in their natural state, not requiring any protective finish, such as copper, lead-coated copper, or terne-coated stainless steel.
- *Contemporary metal roofs* are typically made from sheet steel and aluminum, which have been treated with protective finishes. Thus, the metals commonly used in such roofs are painted steel, metallic-coated steel, and aluminum.

Traditional metal roofs have a long history, because they originated in an era when building construction relied entirely on manual fabrication and assembly. Although adapted to modern industrialized modes of construction, traditional metal roofs still rely to some extent on hand-detailed artisanry. That is why traditional metal roofs are also referred to as *custom metal roofs* and are the only metal roofs that can be used on complex roof geometries, including domes and cupolas.

Contemporary metal roofs are manufacturer specific, with hardly any custom-made components. The introduction of industrialized methods of metal forming and the discovery of durable paints and metallic coatings have made contemporary metal roofs an economical alternative to the traditional copper or lead-coated copper roofs. In fact, the durability and colorfastness of paints and coatings are responsible for the acceptance of sheet steel and aluminum as roofing material.

The discussion provided here is brief and applies to contemporary architectural metal roofs. For additional details, the reader should refer to references given at the end of the chapter.

32.10 CONTEMPORARY ARCHITECTURAL METAL ROOFS

As stated earlier, architectural metal roofs require a deck and an underlayment. The deck can be any one of the deck types used in low-slope roof applications, such as, concrete, steel, or plywood (or oriented strandboard). In commercial buildings, in which steel or concrete decks are common, a typical metal roof assembly consists of the structural deck, one or two layers of insulation, a layer of plywood (or OSB), an underlayment, and the metal roof, Figure 32.39.

In this assembly, plywood (or OSB) works as a nail base to which the metal panels are fastened. The underlayment provides the secondary waterproofing and is generally one layer of No. 30 asphalt-saturated organic felt. In some applications, however, the nail base may be omitted, and the metal panels may be fastened to the deck through the insulation.

An ice dam protection membrane is also needed, as in other steep roofs. The recommendation for an ice dam protection membrane is the same as for other steep roofs. In areas with high snow loads, an ice dam protection membrane should be installed over the entire roof. Additionally, continuous ventilation under the plywood nail base should be provided, and the roof should be designed as a cold roof.

Fabrication of Contemporary Architectural Metal Roof Panels

One of the several contemporary architectural metal roofs is of the standing seam type, in which each panel has two upturned legs, one at each longitudinal edge of the panel. The upturn is generally $\frac{3}{4}$ in. to $1\frac{1}{2}$ in. high. A larger upturn is needed where the rainfall intensity is high. The width of the panel varies from 10 in. to 24 in., Figure 32.40. The upturned edges of adjacent panels meet to form a seam, which can be waterproofed in several ways.

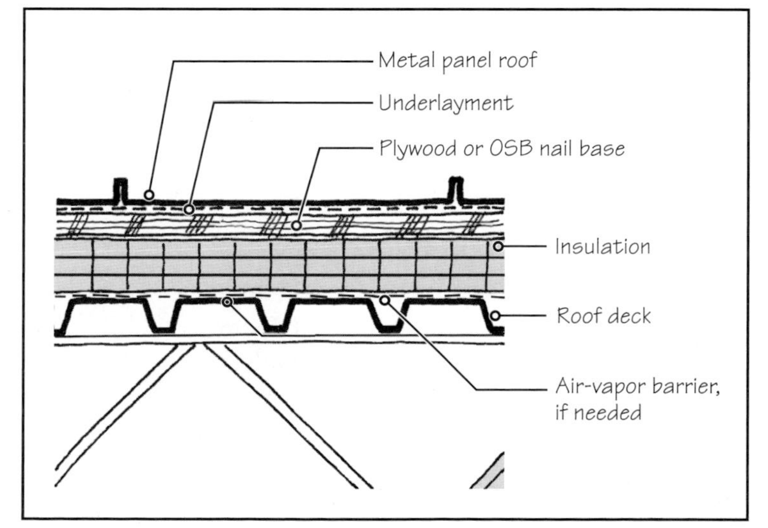

FIGURE 32.39 Typical anatomy of a metal-panel roof.

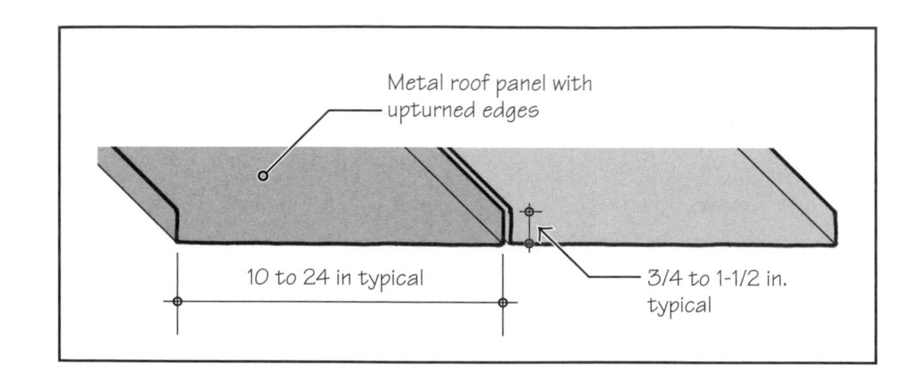

FIGURE 32.40 Typical metal-panel roof profile.

The panels are generally fabricated to the required profile at the construction site from a coil of (prefinished) sheet metal, Figure 32.41. This means that panels of almost any length can be used. The longer the panel, the smaller the number of transverse seams.

In many cases, only one length of panel on each side of a sloping roof is needed, completely precluding the use of transverse seams. In such a case, there is only one joint between the two panels from the two opposite slopes of the roof—at the ridge. The length of a panel is limited by its handleability. Panels of 100 ft in length are not uncommon.

Site fabrication of panels reduces the cost of transporting long panels from the factory to the site. In tall buildings, the panel-fabricating equipment is generally placed on the flat portion of the roof, which avoids lifting panels from the ground to the roof.

ANCHORING THE PANELS

The panels are fastened to the substrate by metal cleats, whose design varies from manufacturer to manufacturer. One such fastening method is shown in Figure 32.42. The center-to-center

FIGURE 32.41 Most architectural metal roof panels are fabricated at the construction site from a sheet-metal roll using a simple panel-forming machine. Panel length is restricted only by the handling capability. Large lengths are particularly difficult to handle on a windy day.

distance between cleats is a function of the wind uplift on the roof and may vary from 12 in. (or less) to 5 ft. The cleat shown is of a *fixed type* that does not allow any movement in panels. Generally, if panels are longer than 40 ft, *expansion cleats* are required. Expansion cleats allow the panels to move—an important requirement for long panels.

After the panels have been fastened, each seam is covered with a snap-on cover piece, Figure 32.43. The snap-on cover piece has a factory-applied sealant throughout the length of the cover piece, which squeezes over and around the panel seams, providing additional waterproofing, Figure 32.44.

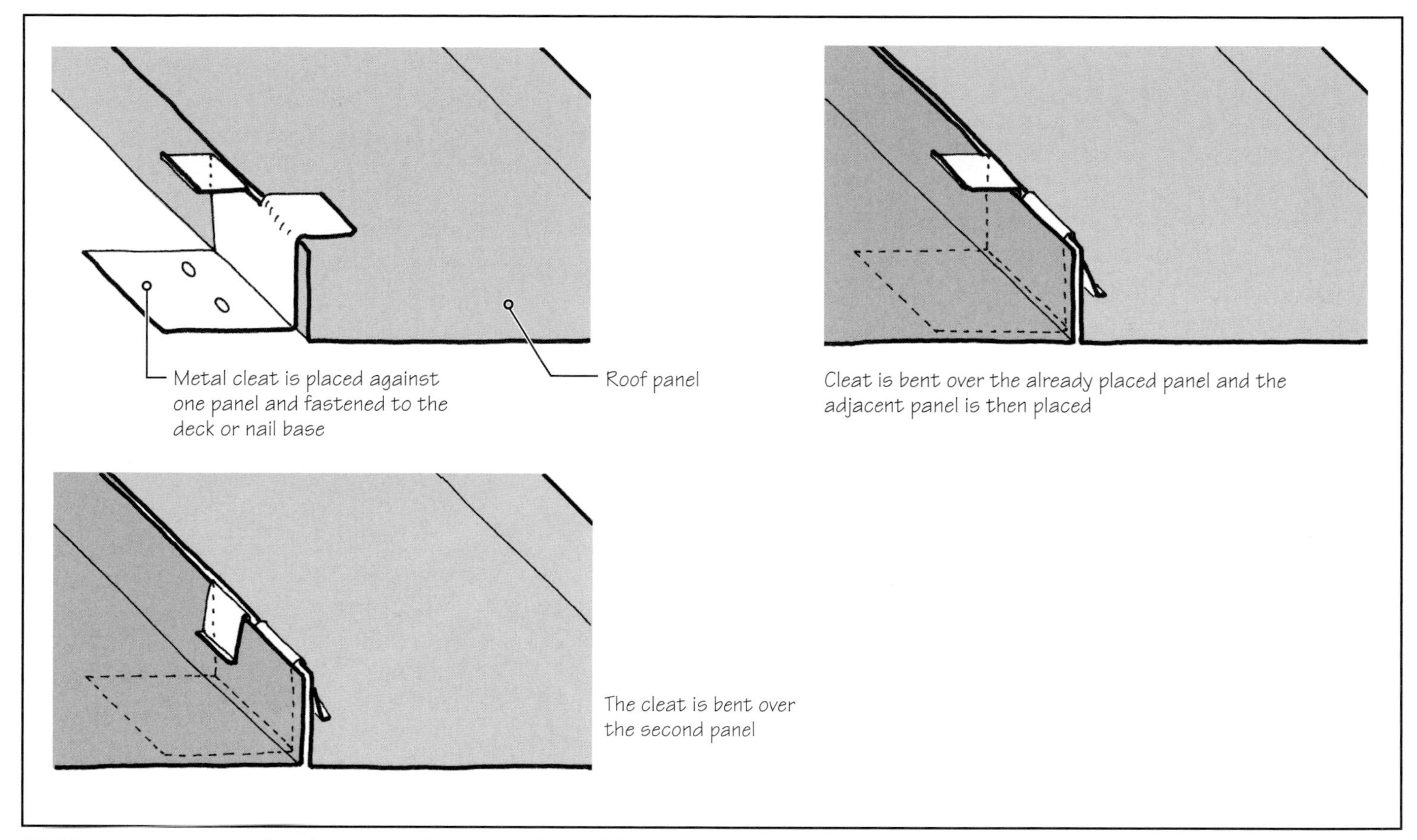

FIGURE 32.42 Details of anchoring metal panels to the roof deck (or nail base) using sheet-metal cleats.

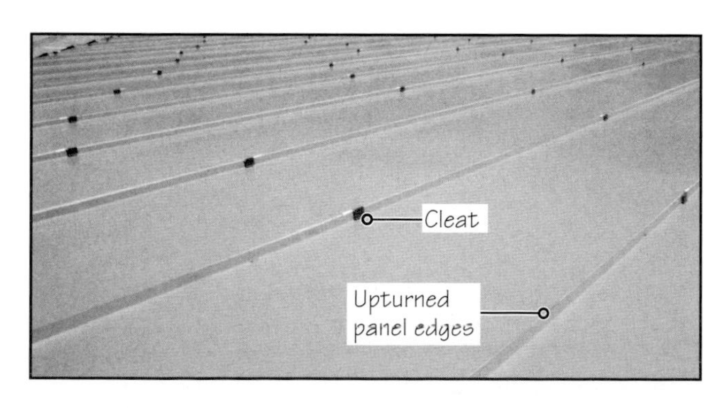

A metal panel roof showing anchorage cleats at the upturned edges (seams) of two adjacent panels. The cleats are covered by snap-on covers as shown below.

Installation of snap-on covers. The cover engages into the cleats, as shown in Figure 32.44.

FIGURE 32.43 Installation of snap-on cover pieces over panel seams.

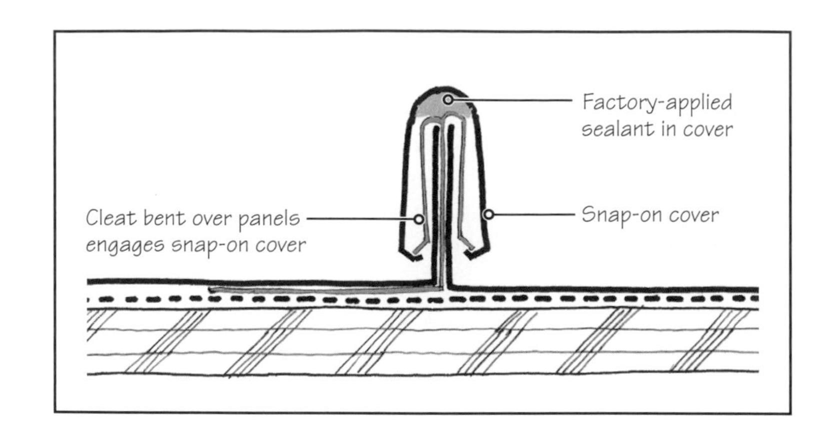

FIGURE 32.44 Section through a seam showing the cleat and snap-on cover.

RIDGE AND EAVE DETAILS

A typical ridge detail, which consists of a continuous Z-shape sheet metal closure set in sealant and soldered to the underlying metal panels, is shown in Figure 32.45. A preformed ridge cap engages into the closure. An ice dam protection membrane is adhered over the ridge directly below the roof panels. Figure 32.46 shows a typical eave detail. Note that the flat portion of the roof panels is turned over the eave flashing.

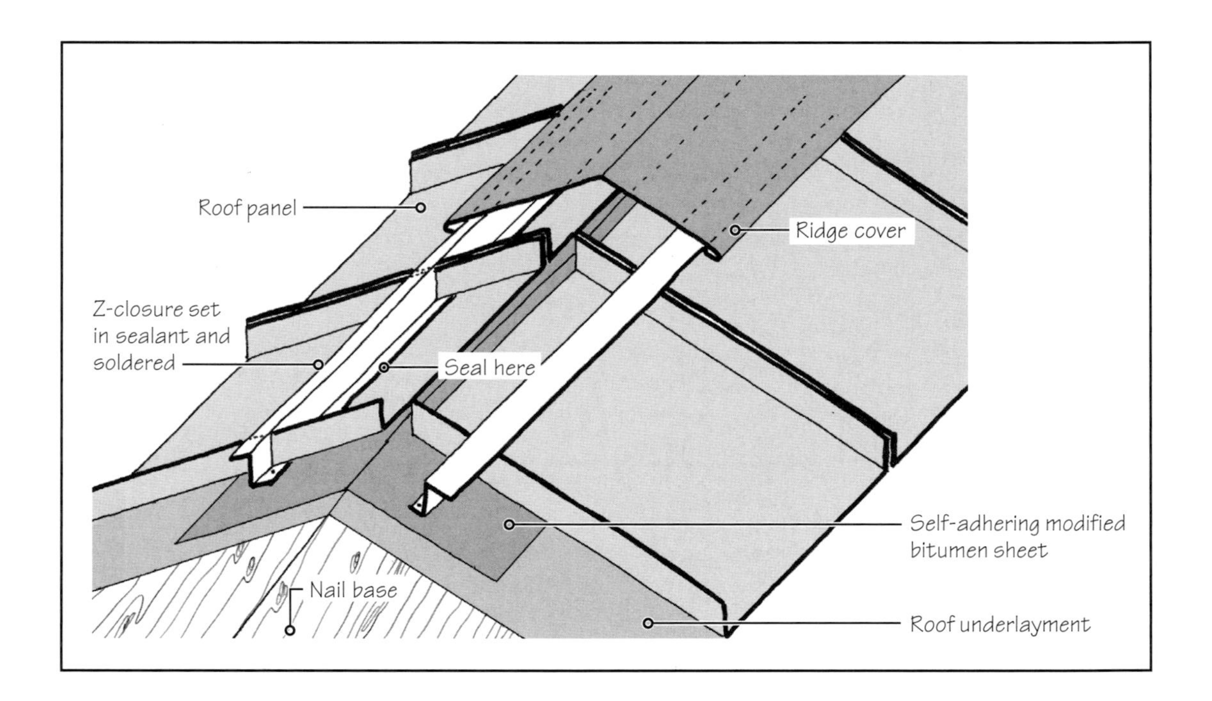

FIGURE 32.45 A typical ridge detail.

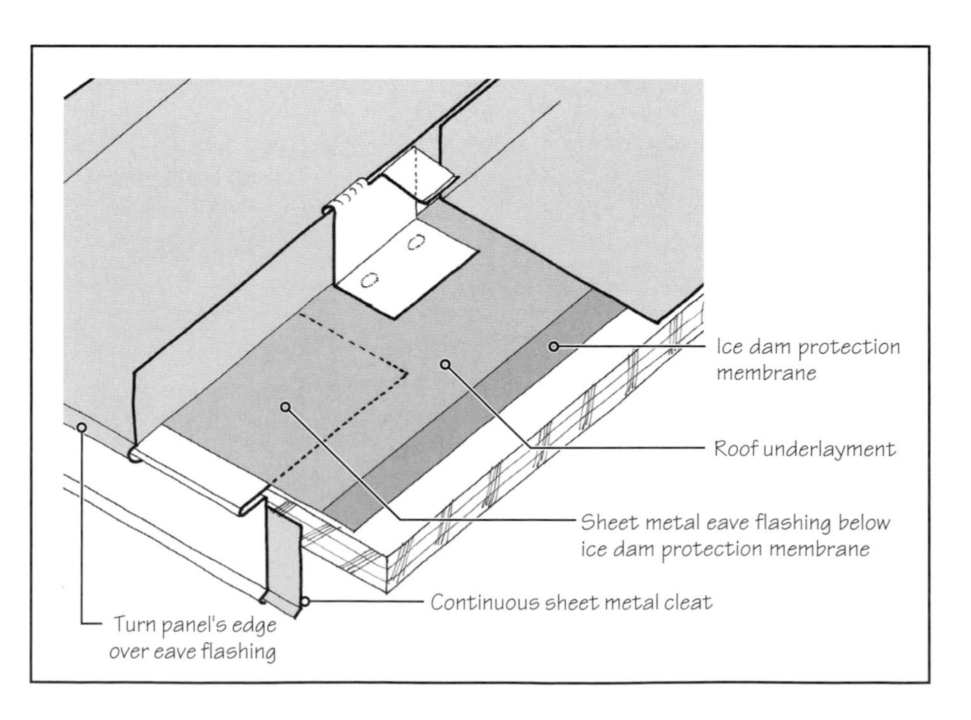

FIGURE 32.46 A typical eave detail.

Each question has only one correct answer. Select the choice that best answers the question.

18. Clay and concrete tiles are generally laid in
- **a.** single coverage with a 5-in overlap between top and bottom shingles.
- **b.** double coverage with a 2-in. overlap between top and bottom shingles.
- **c.** double coverage with a 2-in. head lap between top and bottom shingles.
- **d.** triple coverage with no head lap.
- **e.** none of the above.

19. A roof tile profile that resembles a barrel is generally referred to as a
- **a.** barrel tile.
- **b.** mission tile.
- **c.** semicircular tile.
- **d.** all the above.

20. Because a clay or concrete tile roof is more durable, the requirement for roof underlayment in a concrete or clay tile roof is less stringent than in an asphalt shingle roof.
- **a.** True
- **b.** False

21. Everything else being the same, which of the following clay or concrete tile roofs gives best performance?
- **a.** Tiles laid over two-way batten system
- **b.** Tiles laid over one-way batten system
- **c.** Tiles without any batten system

22. The batten system in a clay or concrete tile roof is installed
- **a.** between the roof deck and roof underlayment.
- **b.** between the deck and clay (or concrete) tiles.
- **c.** between the roof underlayment and clay (or concrete) tiles.
- **d.** none of the above.

23. The primary purpose of a batten system in a clay (or concrete) tile roof is to
- **a.** provide a nail base for tiles.
- **b.** provide a cushion to the tiles.
- **c.** provide a drainage layer under tiles.
- **d.** none of the above.

24. Mission tile roofs require a treated-wood nailer under each row of tiles. The nailer is generally
- **a.** 2 × 2 lumber.
- **b.** 2 × 3 lumber.
- **c.** 2 × 4 lumber.
- **d.** (a) or (b).
- **e.** (b) or (c).

25. An architectural metal roof serves both as roof deck and roof cover.
- **a.** True
- **b.** False

26. An architectural metal roof can be used as a low-slope or steep roof.
- **a.** True
- **b.** False

27. In a standing-seam metal roof, the roof panels are
- **a.** shop formed and site finished.
- **b.** shop formed and shop finished.
- **c.** site formed from a prefinished sheet metal roll.
- **d.** any one of the above.

Answers: 18-e, 19-b, 20-b, 21-a, 22-c, 23-c, 24-e, 25-b, 26-a, 27-c.

KEY TERMS AND CONCEPTS

- Architectural metal roof
- Asphalt shingles
- Asphalt-treated organic felt
- Cricket
- Drip edge
- Ice dam protection membrane
- Metal shingles
- Mission tiles
- Open valley
- Roof underlayment
- Slate shingles
- Step flashing
- Valley flashing
- Wood shingles
- Woven valley

REVIEW QUESTIONS

1. Using a sketch, explain the essential components of a steep roof.
2. Explain the pros and cons of using an open valley, closed-cut valley, and woven valley in an asphalt shingle roof.
3. Using a sketch, explain the layout of clay or concrete tiles using a treated-wood cross-batten system.
4. Using a sketch, explain the layout of mission tiles.

SELECTED WEB SITES

Asphalt Roofing Manufacturers Association, ARMA (www.asphaltroofing.org)

Copper Development Association, CDA (www.copper.org)

The Institute of Roofing, Waterproofing, and Building Envelope Professionals, RCI (www.rci-online.org)

National Roofing Contractors Association, NRCA (www.nrca.net)

FURTHER READING

1. National Roofing Contractors Association (NRCA). *The NRCA Roofing and Waterproofing Manual.*
   This $8\frac{1}{2} \times 11$, nearly 1,500-page book is a comprehensive reference resource for all aspects of roofing.
2. Patterson, S., and Mehta, M. *Roofing Design and Practice.* Upper Saddle River, NJ: Prentice Hall, 2001.
   This $8\frac{1}{2} \times 11$, nearly 300-page book is a primer on roofing and is aimed primarily at the architects and architecture students. It covers both low-slope and steep-roof design.

CHAPTER 33

Stairs

CHAPTER OUTLINE

33.1 Stair Fundamentals

33.2 Wood Stairs

33.3 Steel Stairs

33.4 Concrete Stairs

A stair with an expressed structure used as a feature in a large atrium. (Photo by MM.)

A stair, defined as a series of ascending (or descending) steps, is an important accessory in a building. However, because a stair provides repetition and rhythm, it also lends itself to a strong artistic expression, as illustrated by the images in Figure 33.1.

Architectural historians claim that the stair remained a purely functional element (without artistic overtones) until the end of the fifteenth century. The beginning of the sixteenth century, inspired by Leonardo da Vinci's sketches, however, "signaled a new era of expression for the staircase" [1]. From then on, the staircase played an increasingly important visual role, often becoming a sculptural feature in a space, an imperial entrance to a public building or a significant facade element.

The birth of the elevator—and, subsequently, the escalator—reduced the functional importance of the stair. More recently, the requirement to make buildings equally accessible to persons with disabilities further eroded the significance of stairs.

FIGURE 33.1 A few images that signify the creative potential of contemporary stairs. **(top)** A reinforced concrete stair in an entrance lobby. (Photo courtesy of HKS Inc.) **(bottom)** An all-glass entry stair that winds around a cylindrical glass elevator, Apple Store, New York City.

FIGURE 33.1 *(Continued)* A few images that signify the creative potential of contemporary stairs. An all-glass staircase (glass treads and glass supporting walls), Apple Store, Soho, New York City.

Because a staircase cannot be used by people in a wheelchair, it is no longer a mandatory feature of an entrance lobby. (Increasingly, entrance lobbies in contemporary public buildings are designed without a staircase.) Consequently, the staircase is reverting to its purely functional role—fulfilling the requirement as exit stair or standby vertical circulation in the event of electrical outage or mechanical interruption.

This chapter begins with a general introduction to stairs, followed by the details of construction of simple wood, steel, and concrete stairs.

33.1 STAIR FUNDAMENTALS

Because a stair provides vertical transportation, it is a part of the means-of-egress (exit) system of a building. It is also a relatively hazardous element because injuries due to falls from stairs are common. For this reason, stair design is stringently controlled by the building codes.

TREAD, RISER, AND NOSING

There are two main components of a stair: *treads* and *risers*. A tread is the horizontal surface on which one walks. The riser is the vertical component that separates one tread from another. Generally, a stair has several treads and risers. For the sake of safety, the dimensions of treads and risers must be uniform in a stair. Building codes allow only a small dimensional variation because a perfect uniformity is unachievable.

In walking on a horizontal or an inclined surface, an average person can comfortably traverse a distance of 24 to 25 in. in one step. Therefore, a rule of thumb generally used in proportioning the treads and risers of a stair is

$$\boxed{2(\text{riser height}) + \text{tread width} = 24 \text{ to } 25 \text{ in.}}$$

Thus, if the risers in a stair are each 5 in. high, the tread width should lie between 14 and 15 in. The most commonly used dimensions for an interior public stair are 12- to 13-in. treads and 6-in. risers. Outdoor stairs generally have a smaller riser and hence a wider tread. Building codes generally require a riser height between 4 in. and 7 in. and a minimum tread width of 11 in.

In most stairs, the tread is a simple, flat surface, and the riser is a solid vertical surface, Figure 33.2(a). Where the space is limited, the effective tread width can be increased somewhat by inclining the risers, Figure 33.2(b), or by projecting the front edge of the tread beyond the riser, Figure 33.2(c). The front edge of a tread is referred to as the *nosing*.

NOTE

Treads and Risers of a Stair

The treads and risers of a stair must meet the following dimensional requirements:

Minimum tread width = 11 in.

Riser height = 4 to 7 in.

Residential Stair

Building codes are less restrictive for a stair within a dwelling unit:

Minimum tread width = 10 in.

Riser height = 4 to $7\frac{3}{4}$ in.

A 10-in. tread is allowed in a dwelling unit stair, provided a nosing projection of $\frac{3}{4}$ to $1\frac{1}{4}$ in. is used. If the nosing projection is not used, an 11-in. minimum tread width is required.

Some local building codes may allow a riser height of greater than $7\frac{3}{4}$ in. for a residential basement stair.

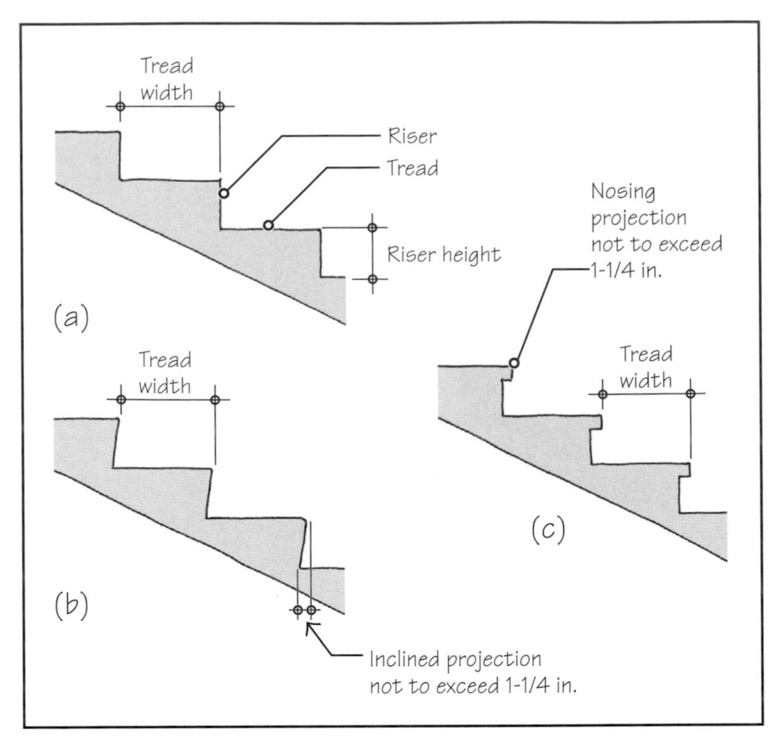

FIGURE 33.2 Tread, riser, and nosing configurations in a stair.

When an inclined riser or a projected nosing is used, the code-required minimum width of a tread does not change. In other words, the width of a tread is considered as the horizontal distance between the vertical planes of the foremost projections of adjacent treads, as shown in Figures 33.2(b) and (c).

The nosing of a tread is subjected to the maximum abrasion. In public stairs with heavy traffic, the treads should consist of a strong and dense material such as granite, high-strength concrete, or steel. Alternatively, a separate nosing (approximately $2\frac{1}{2}$ in. wide) consisting of an abrasion-resistant and skid-resistant material is epoxied or embedded into the tread.

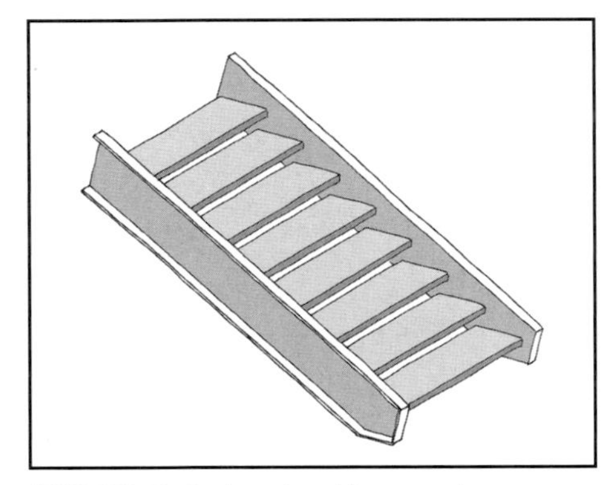

FIGURE 33.3 A stair with open risers.

Stairs can also be constructed without risers, referred to as *open-riser stairs,* Figure 33.3. Because of safety concerns, open-riser stairs are subject to more stringent code restrictions than stairs with solid risers. For example, open-riser stairs are generally not allowed as exit stairs. Additionally, the clear vertical distance between the treads of open-riser stairs cannot exceed 4 in.

STAIR SHAPES

The most commonly used stair shape is a U-shape stair (in plan). It consists of two flights of stairs between floors with a midfloor landing (or simply *midlanding,* or *landing*), Figure 33.4. In addition to the U-shape stair, some of the other commonly used stair shapes are

- *Straight-run stair* with one or two flights, Figure 33.5: A straight-run stair with more than two flights can be used but is uncommon.
- *L-shape stair,* Figure 33.6(a): Where the space is limited, the landing of an L-shape stair can be used for steps, yielding trapezoidal (pie-shaped) treads, referred to as *winders,* Figure 33.6(b). Stairs with winders are not as safe as those with rectangular treads and their use in an exit stair is strictly controlled by the codes.
- *Circular stair:* A circular stair may consist of all winders and can take many shapes. A spiral stair is a special type of circular stair in which the treads twist around a central column and are cantilevered from it, Figure 33.7. It is generally an open-riser stair. Again, the building code has several restrictions on the use of a spiral stair. A helical stair is a circular stair without a central supporting column (described at the end of the chapter).

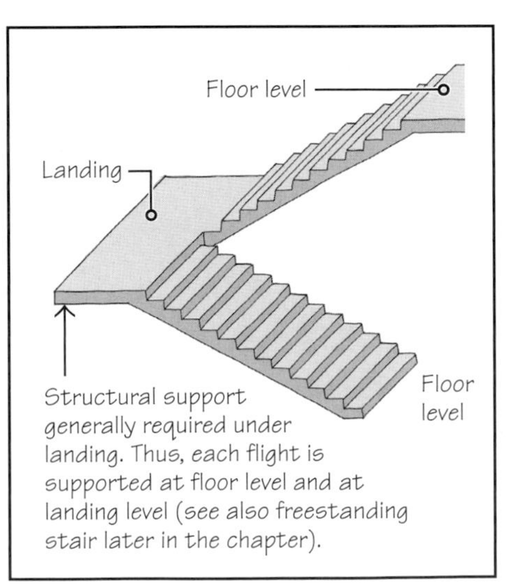

FIGURE 33.4 A U-shaped stair.

OPEN AND CLOSED STAIRS

Stairs are also described as either open or closed. An open stair is exposed to below on one or more sides, whereas a closed stair is fully enclosed with a stair enclosure (stair shaft) and is usually accessed through a doorway.

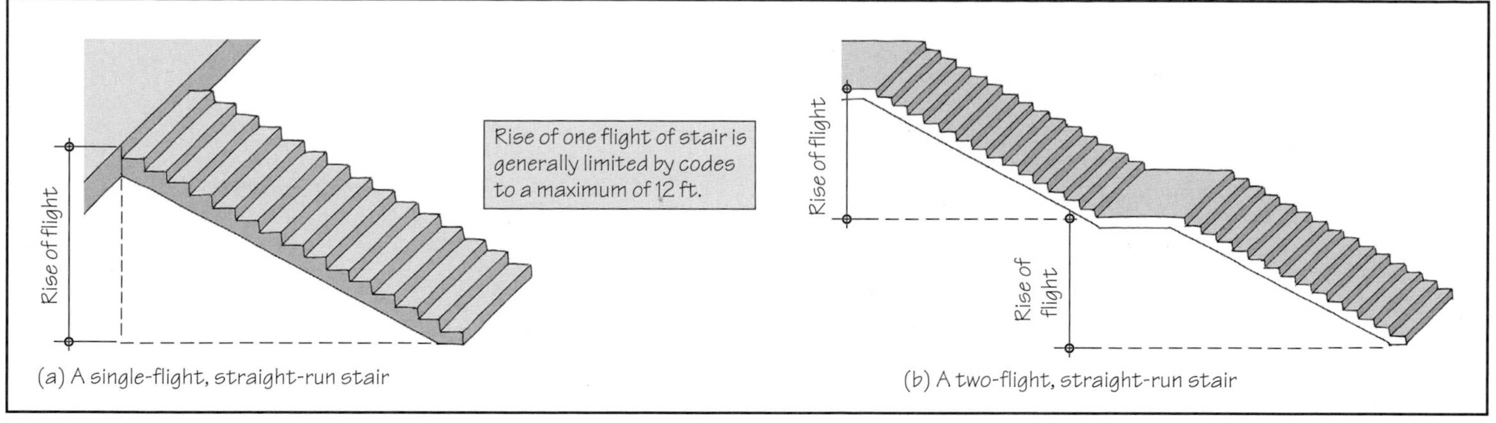

FIGURE 33.5 Straight-run stairs.

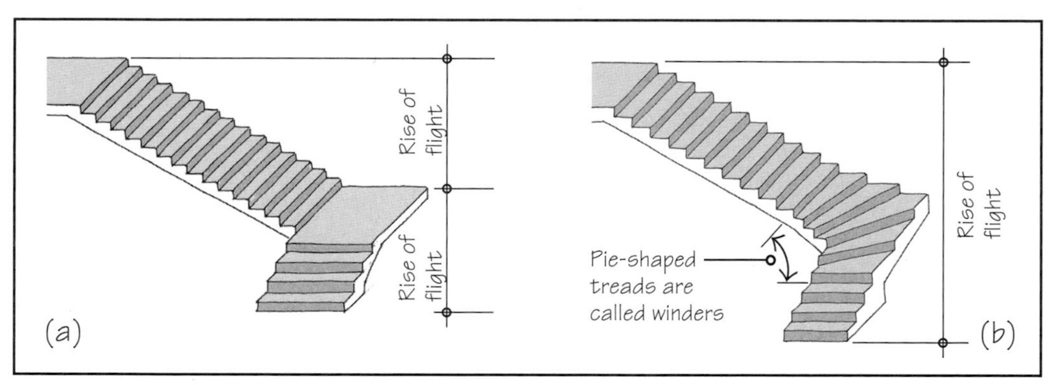

FIGURE 33.6 L-shape stairs **(a)** without winders and **(b)** with winders.

FIGURE 33.7 A spiral stair.

Width of Stair

The minimum width of a stair is determined by its purpose. When used as an exit stair, its width depends on the number of occupants it serves (occupant load) but is not less than 44 in. clear (between handrail to handrail) for an open exit stair or 48 in. for an enclosed exit stair. An exit stair for an occupant load of less than 50 or a stair within a dwelling unit has a minimum width of 36 in.

Head Room

The head room in a stair is the minimum clearance between a tread and a projection above, Figure 33.8. Building codes generally require the head room to be a minimum of 80 in. at any point on the stair.

Guard Unit, Handrail, Balusters, and Newel Post

The edge of a stair exposed to a change in height (i.e., not protected by the wall of the enclosure) must have a *guard unit* to protect against falling. The minimum height of a guard unit is 42 in., Figure 33.9(a). The clear distance of openings in a guard unit must not exceed 4 in.

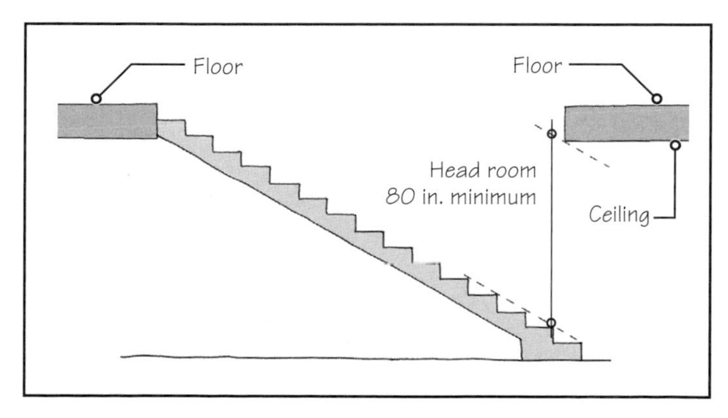

FIGURE 33.8 Headroom in a stair.

EXPAND YOUR KNOWLEDGE

Summary of Stair-Design Criteria

1. Tread depth and riser height must be dimensionally uniform throughout.
2. Minimum tread width = 11 in. Riser height = 4 to 7 in. (See Section 33.1 for exceptions for residential stairs.)
3. Nosing projection $\leq 1\frac{1}{4}$ in.
4. Stair width = function of occupant load, but not less than 48 in. for enclosed exit stair, 44 in. for open exit stair, or 36 in. for a stair serving an occupant load of less than 50 or a residential stair.
5. Width of landing $\geq$ stair width.
6. Rise of one flight $\leq$ 12 ft.
7. Headroom $\geq$ 80 in.
8. Use of winders restricted.
9. Height of guard rail = 42 in. minimum. Height of handrail = 34 to 38 in.
10. Handgrip portion of handrail must have a circular cross section between $1\frac{1}{4}$ in. to $2\frac{5}{8}$ in. Noncircular profiles must provide equivalent graspability.

The height of a handrail in a stair is generally required to lie between 34 in. and 38 in. The cross-sectional profile of a handrail is controlled by codes to give it the required graspability.

In some wood stairs, the first and/or the last vertical member of the guard unit (referred to as a *baluster*) is highlighted by using a more ornate design. Such a baluster is referred to as a *newel post,* Figure 33.9(b).

Stair Layout and Stair Plan

In preparing a stair layout, we first determine the floor-to-floor height and then calculate the number of risers and treads. Assume that the floor-to-floor height in a building is 10 ft 8 in., that is, 128 in. Assume further that we would like the riser height to be approximately 6 in. Dividing 128 by 6 gives us the number of risers:

$$\text{Number of risers} = \frac{128}{6} = 21.3$$

Because the number of risers must be a whole number, assume 21 risers. Dividing 128 in. by 21 gives the exact riser height, 6.1 in. From the tread-riser relationship given earlier, the tread width is:

$$24 \text{ (or 25)} - 2(6.1) = 11.8 \text{ to } 12.8 \text{ in.}$$

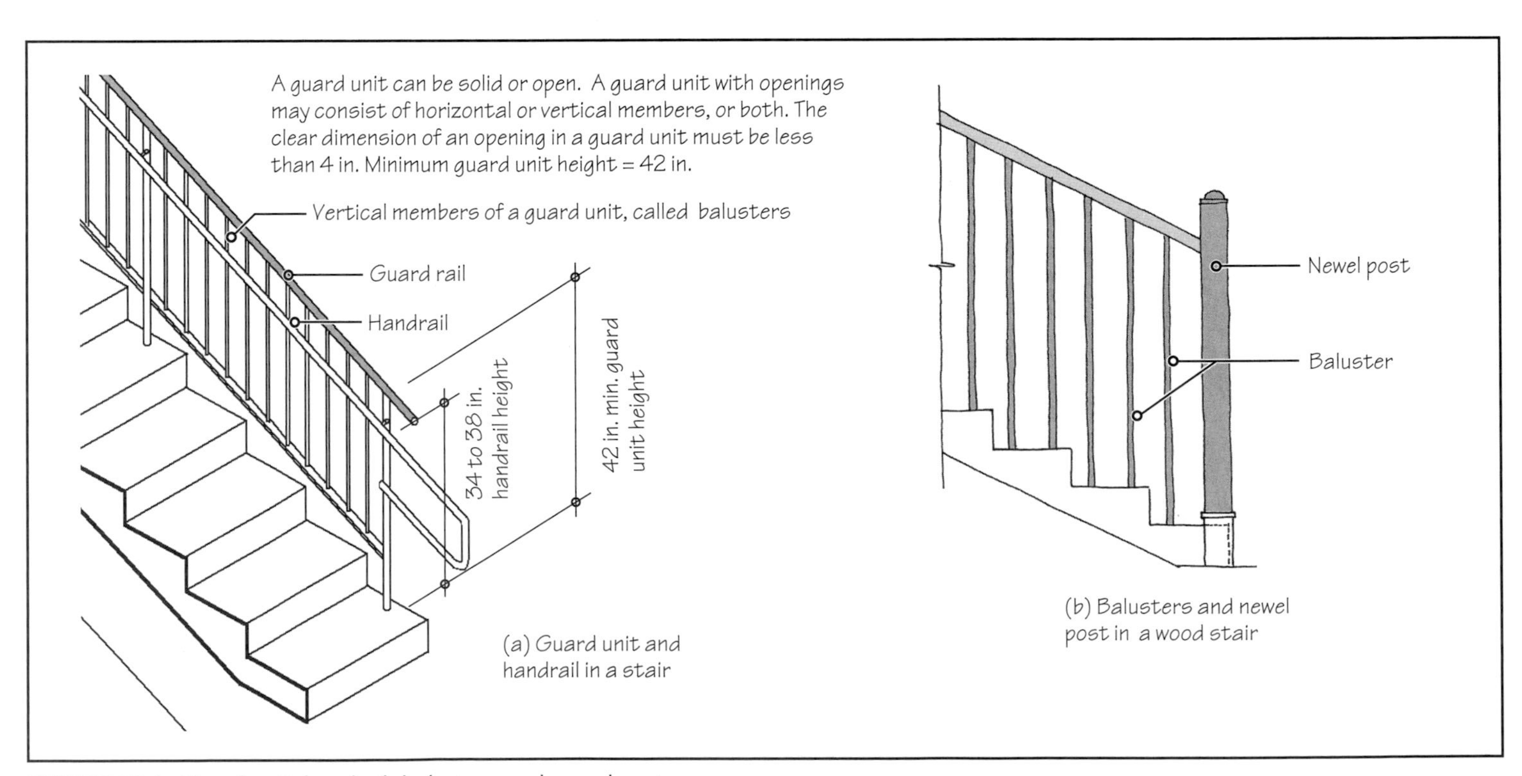

FIGURE 33.9 Guard unit, handrail, balusters, and newel post.

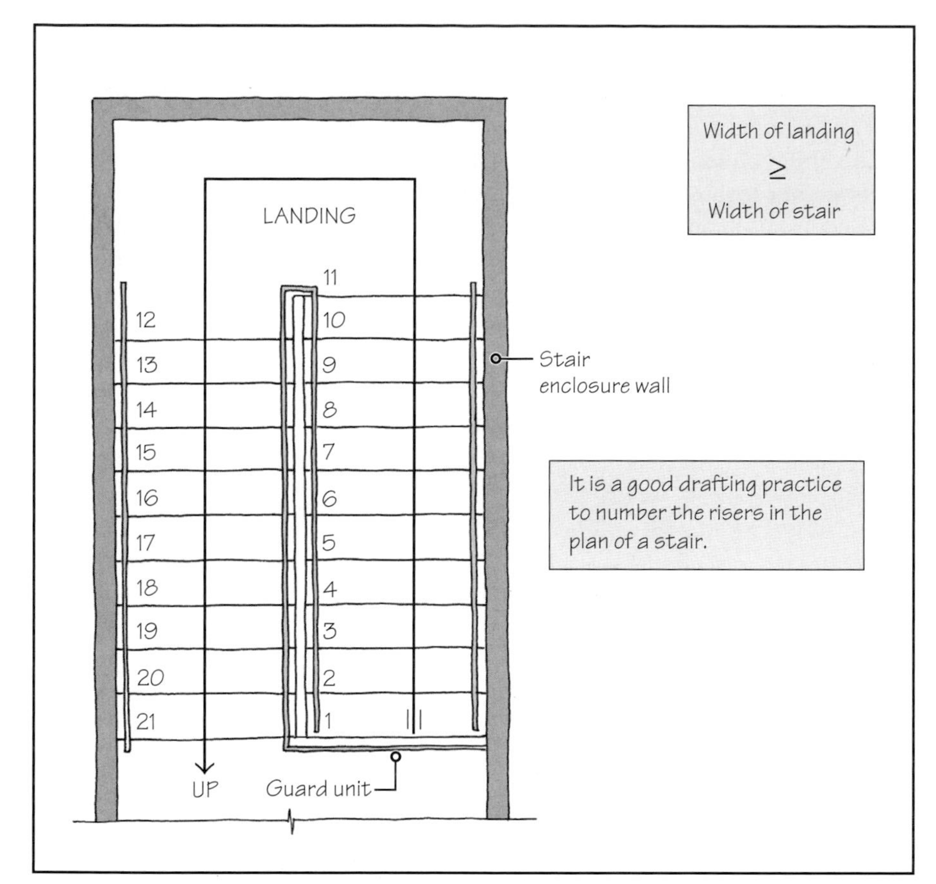

FIGURE 33.10 Plan of a U-shaped stair (at second-floor level) that extends only from the first floor to the second floor.

We will use a tread width of 12.0 in. Assume further that a U-shape stair is desired and the width of the stair is 4 ft. By code, the minimum width of landing must be the same as the the width of stair, that is, 4 ft. With these data, a plan of the stair can be drawn, as shown in Figure 33.10.

Stair-Drawing Conventions

There are standard conventions for how stairs are shown in building plans. The sketch of Figure 33.10 is the plan at the second-floor level of a U-shape stair that extends from the first to the second floor only. The plan of the same stair at the first-floor level is shown in Figure 33.11(a). If the same stair were to extend over several floors, then the plan of the stair at a typical floor is generally drawn as shown in Figure 33.11(b).

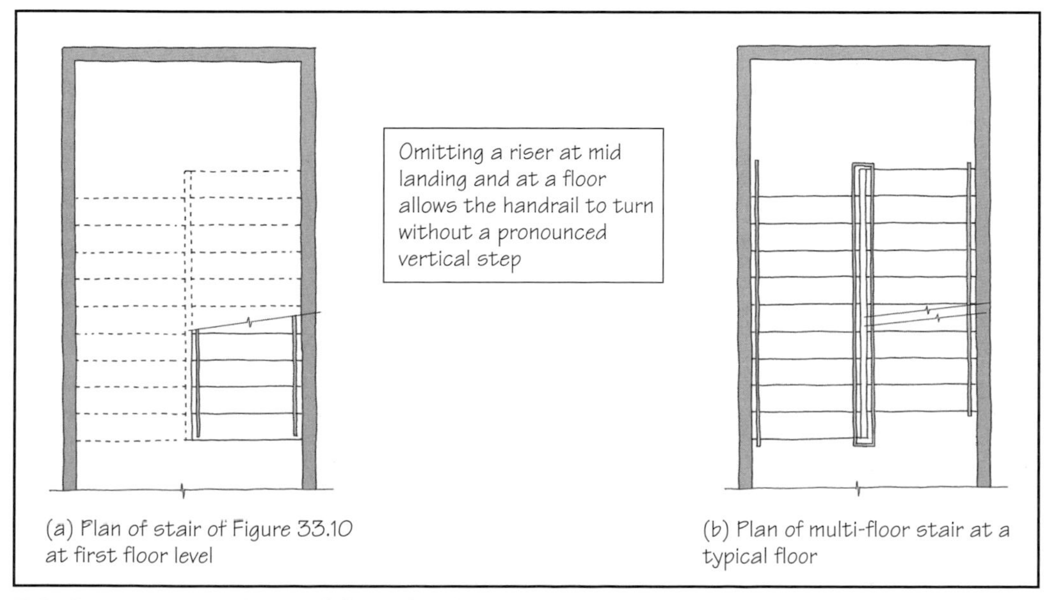

FIGURE 33.11 Stair plans at different levels.

FIRE-RATED STAIR (SHAFT) ENCLOSURE

Because most stairs in a building are used as exit stairs, they need to be enclosed by vertical enclosures (also referred to as *shafts,* or *shaft enclosures*). Generally, a shaft enclosure is required to be 1-h rated for a building up to 4 stories tall and 2-h rated for a building with 5 stories or more. Shafts are not required for individual single-family dwellings (up to 4 stories tall).

Building codes also contain several other exceptions to the requirement of shaft enclosures. For example, shafts are not required if the stair connects only two floors and is not used as an exit stair.

PRACTICE QUIZ

Each question has only one correct answer. Select the choice that best answers the question.

1. An approximate formula generally used in determining the tread dimension (T) and riser dimension (R) in a stair is
- **a.** 2T + R = 24 to 25 in.
- **b.** 2T + 2R = 24 to 25 in.
- **c.** 2R + T = 24 to 25 in.
- **d.** R + T = 24 to 25 in.
- **e.** none of the above.

2. The minimum tread width required by building codes for a nonresidential stair is
- **a.** 11.0 in.
- **b.** 11.5 in.
- **c.** 12.0 in.
- **d.** 12.5 in.
- **e.** 13.0 in.

3. The minimum riser height required by building codes for a nonresidential stair is
- **a.** 6.0 in.
- **b.** 5.5 in.
- **c.** 5.0 in.
- **d.** 4.5 in.
- **e.** 4.0 in.

4. The maximum nosing projection allowed for a stair is
- **a.** 3.0 in.
- **b.** $2\frac{1}{2}$ in.
- **c.** 2 in.
- **d.** $1\frac{1}{2}$ in.
- **e.** $1\frac{1}{4}$ in.

5. A riser must be vertical. It cannot be inclined.
- **a.** True
- **b.** False

6. A U-shape stair has been provided between the first floor and the second floor of a building with a midlanding. This stair has
- **a.** one flight.
- **b.** two flights.
- **c.** three flights.
- **d.** four flights.

7. The rise of one flight of stair is generally limited by building codes to
- **a.** 7 ft.
- **b.** 8 ft.
- **c.** 10 ft.
- **d.** 12 ft.

8. Given a multistory building with a floor-to-floor height of 10 ft and an optimal riser height of 7 in., how many treads would you use for a U-shaped stair with a midlanding between floors? (The landing is not counted as a tread.)
- **a.** 15
- **b.** 16
- **c.** 17
- **d.** 18
- **e.** None of the above

9. A handrail and guardrail in a stair are synonymous.
- **a.** True
- **b.** False

10. The minimum height of a guardrail in a stair is
- **a.** 34 in.
- **b.** 36 in.
- **c.** 38 in.
- **d.** 40 in.
- **e.** none of the above.

11. A stair constructed without risers is generally called a
- **a.** no-riser stair.
- **b.** closed-riser stair.
- **c.** open-riser stair.
- **d.** hollow stair.

12. A stair with treads cantilevered from a central column is a
- **a.** circular stair.
- **b.** U-shape stair.
- **c.** L-shape stair.
- **d.** spiral stair.
- **e.** helical stair.

13. The minimum width of a stair in a dwelling unit is
- **a.** 2 ft 6 in.
- **b.** 3 ft.
- **c.** 3 ft 6 in.
- **d.** 4 ft.
- **e.** none of the above.

Answers: 1-c, 2-a, 3-e, 4-e, 5-b, 6-b, 7-d, 8-a, 9-b, 10-e, 11-c, 12-d, 13-b.

33.2 WOOD STAIRS

The most important parts of a wood stair are the *carriages* (also called *rough stringers*). Carriages are the structural elements of a stair and are specially cut to support the treads. Figure 33.12 shows a commonly used method of framing a wood stair.

PREFABRICATED WOOD STAIRS

There are several manufacturers who supply prefabricated wood stairs per the architect's design. Prefabricated wood stairs are usually transported in a knocked-down (KD) version,

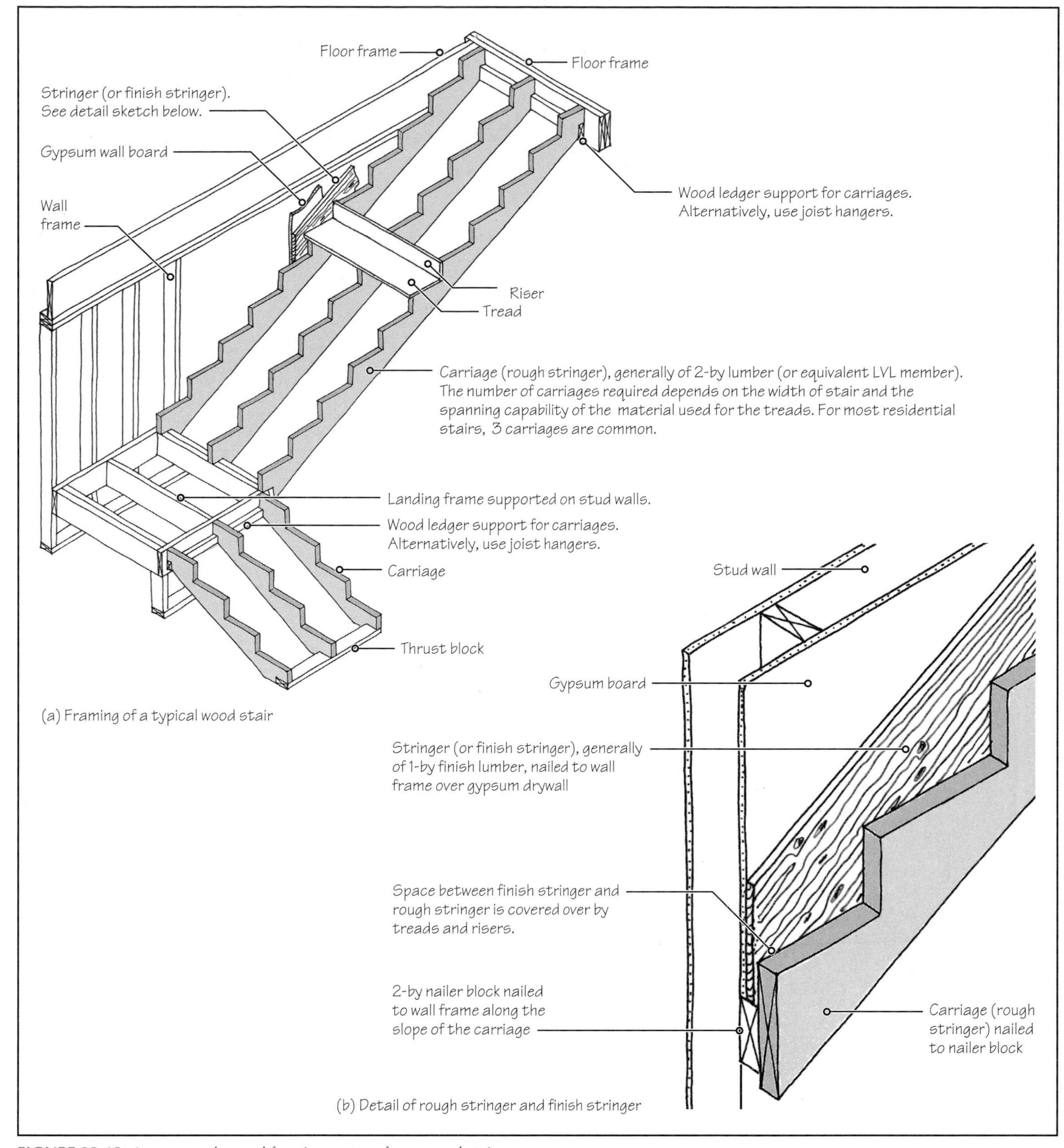

FIGURE 33.12 A commonly used framing system for a wood stair.

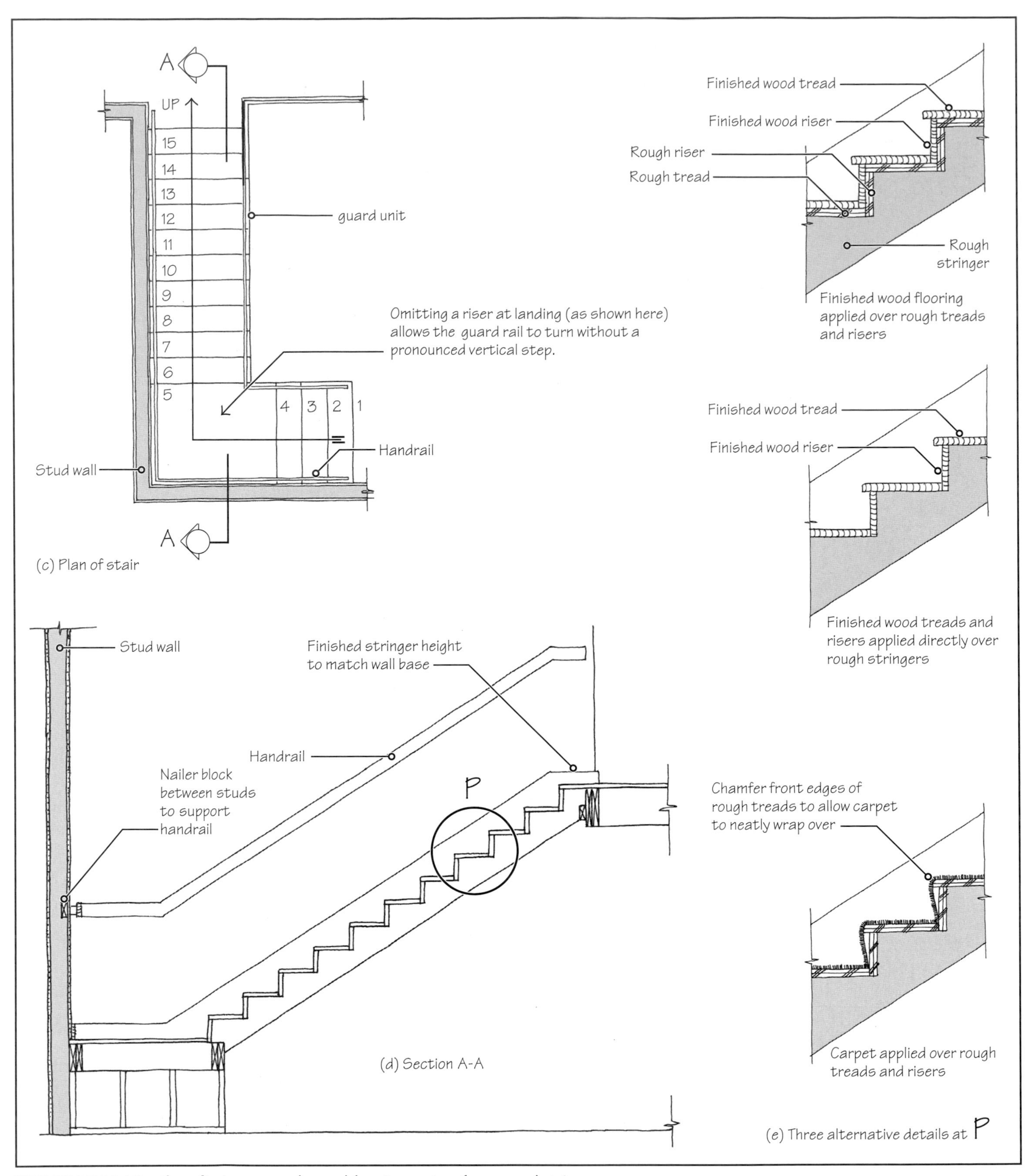

FIGURE 33.12 (*Continued*) A commonly used framing system for a wood stair.

where each part is uniquely numbered for assembly on site. They are commonly used for more-ornate stairs requiring detailed millwork and craft, not usually possible at the site.

33.3 STEEL STAIRS

Stairs in public buildings are generally constructed of steel or concrete. Because steel stairs can be shop fabricated and brought to the site ready for installation, they are more commonly used than concrete stairs.

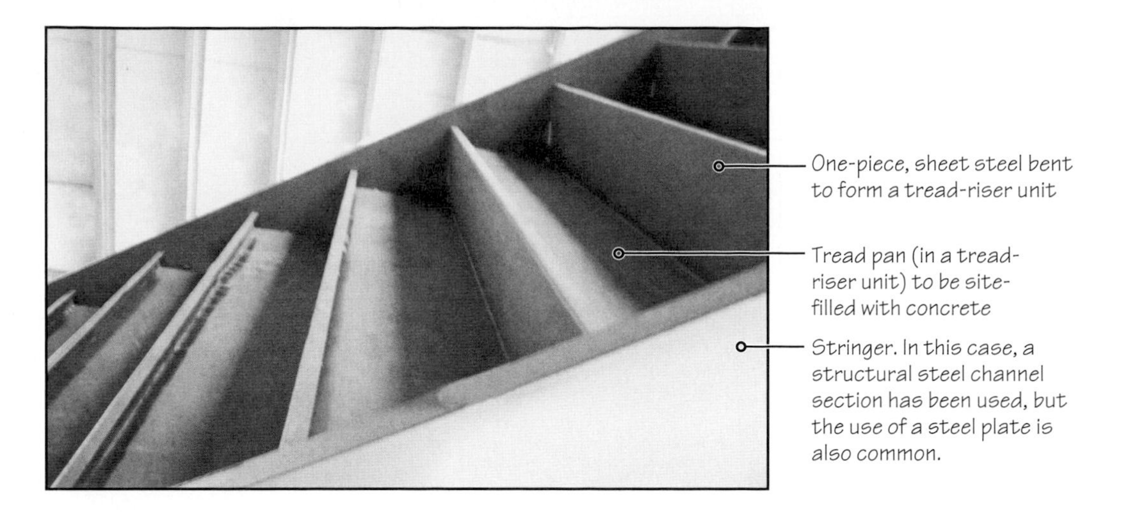

FIGURE 33.13 A typical prefabricated steel stair consists of two stringer beams (stringers) to which tread-riser units made from sheet steel are welded; see Figure 33.14 for detail. (In this stair, the guard unit and handrail have not yet been installed.)

Prefabricated steel stairs are used in all types of public buildings, that is, steel- and concrete-frame buildings and load-bearing masonry buildings. They are particularly popular for exit stairs.

A typical prefabricated flight of a steel stair consists of two stringer beams (stringers) to which tread-riser units made of sheet steel are welded, Figure 33.13.

The tread pan is generally site filled with concrete, Figure 33.14. For good wear resistance, a concrete strength of 5,000 psi is generally specified. Other tread finishes include a precast-concrete *drop-in tread* with a slip-resistant broom finish, Figure 33.15(a), and sheet steel with a raised, diamond-shaped checkered pattern, Figure 33.15(b). Factory-installed epoxy-aggregate fill or wear- and slip-resistant coatings can also be used.

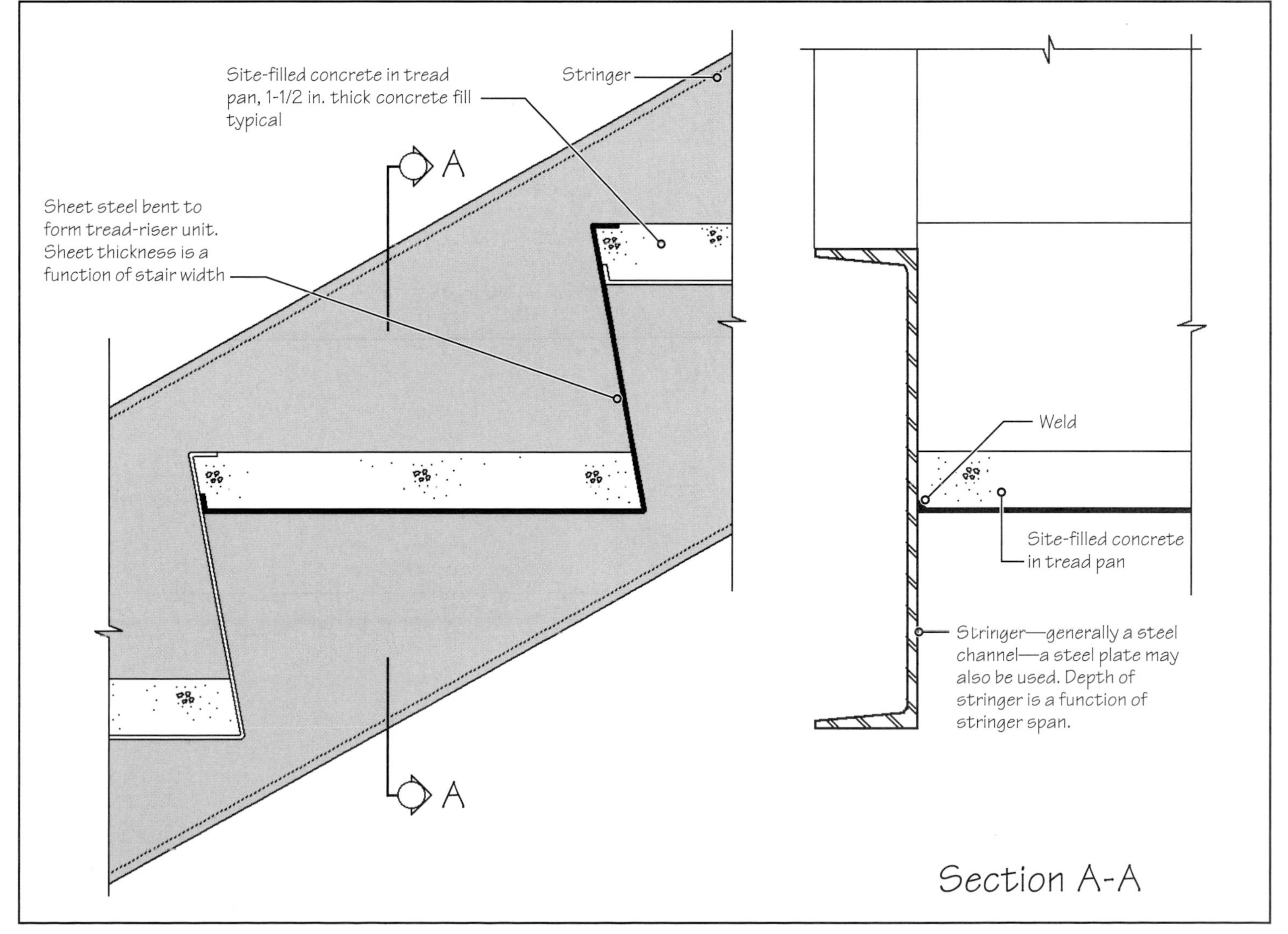

FIGURE 33.14 Typical detail of tread-riser units welded to stringers. In this detail, tread pans are filled with concrete at the site (see Figure 33.15 for alternatives).

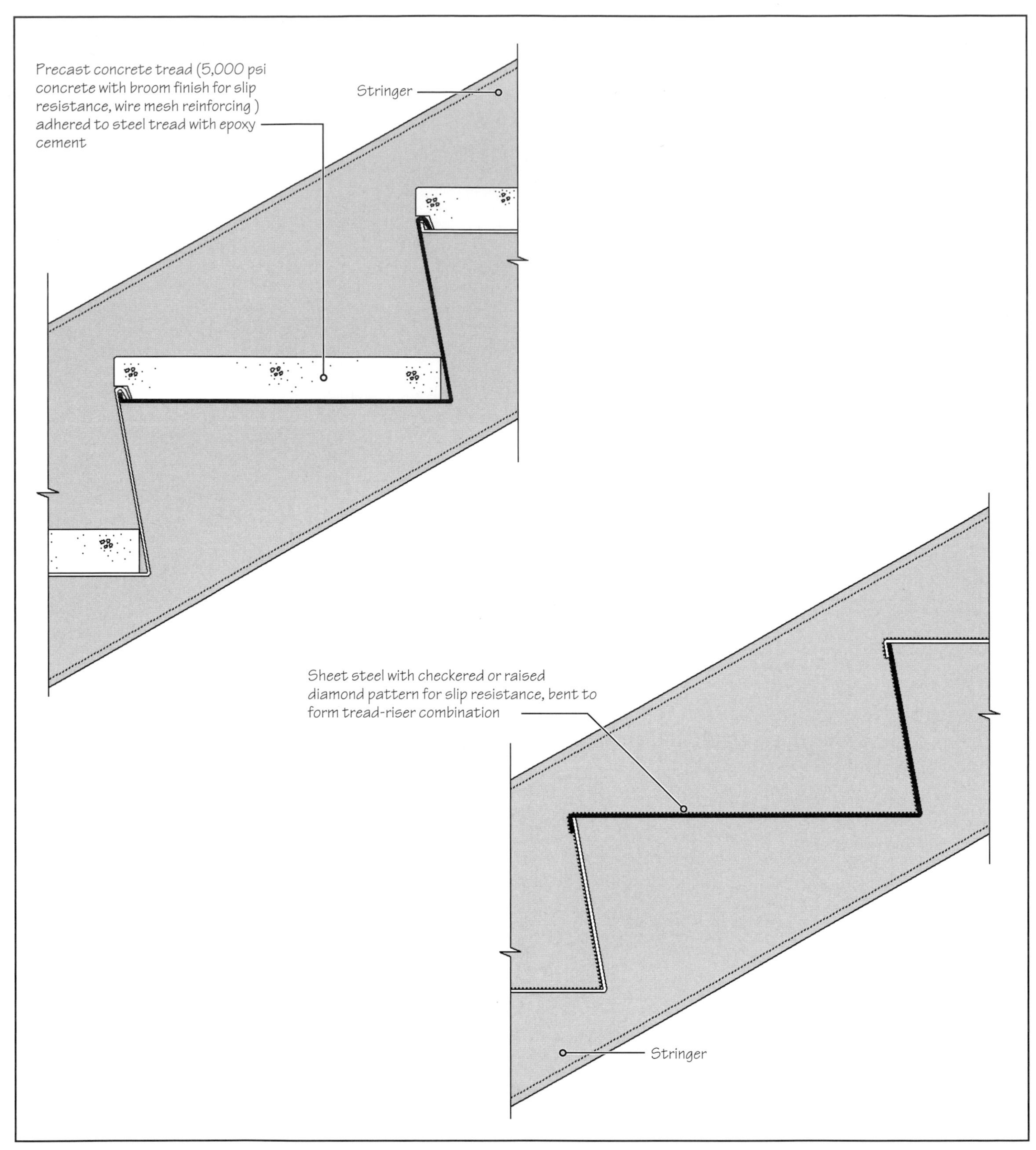

FIGURE 33.15 Typical detail of tread-riser units in a steel stair with precast-concrete and steel treads.

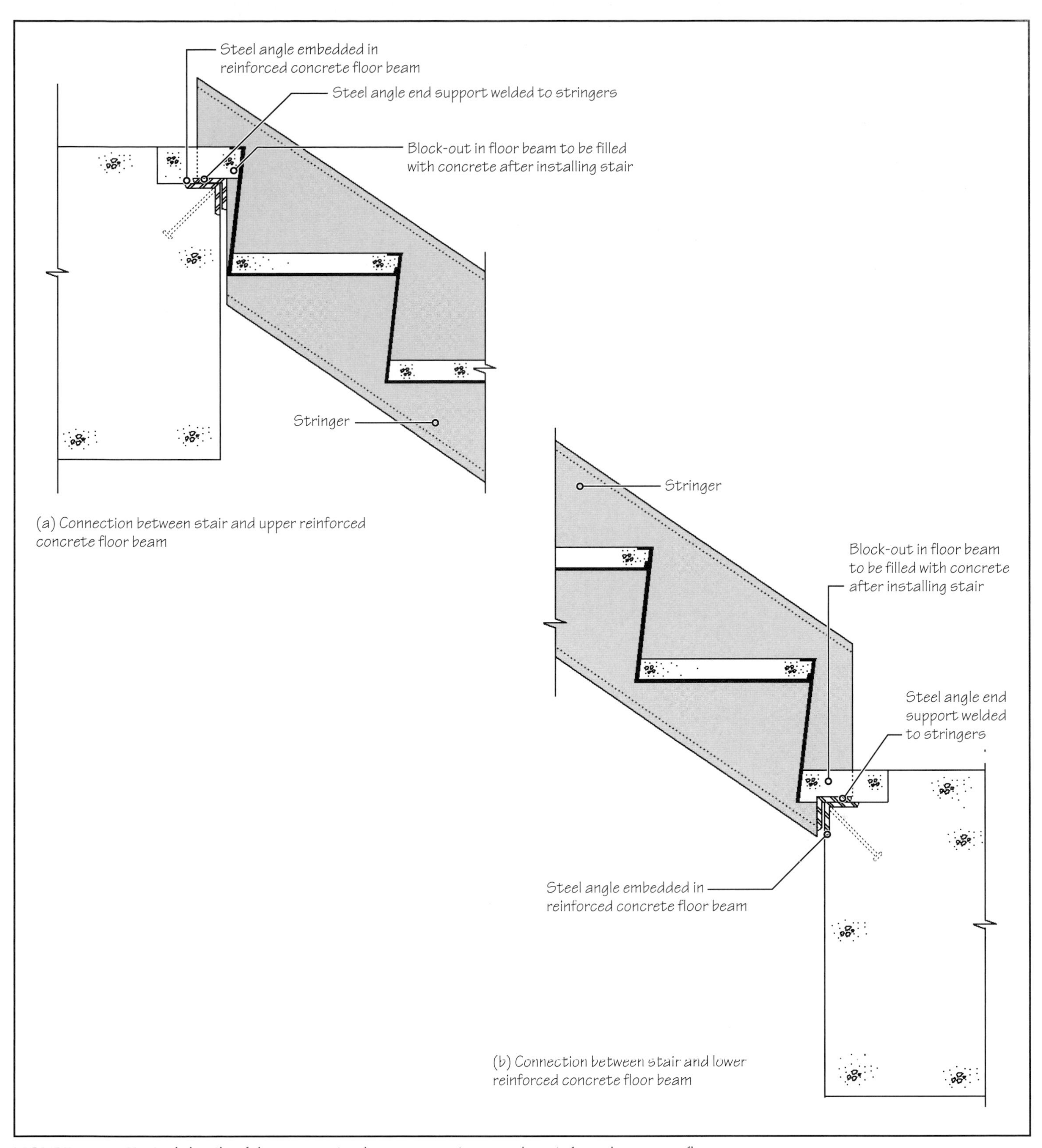

FIGURE 33.16 Typical details of the connection between a stringer and a reinforced-concrete floor.

Stringers

Stringers in a steel stair function as beams, spanning from the floor to the landing and from the landing to the next floor. They generally consist of a structural-steel channel or steel plate ($\frac{3}{16}$ in. or $\frac{1}{4}$ in. thick is typical). The depth of stringers is a function of stringer span. The tread-riser units span between the stringers. Figures 33.16 and 33.17 show typical details of support connections between the stringers and the floor of the building.

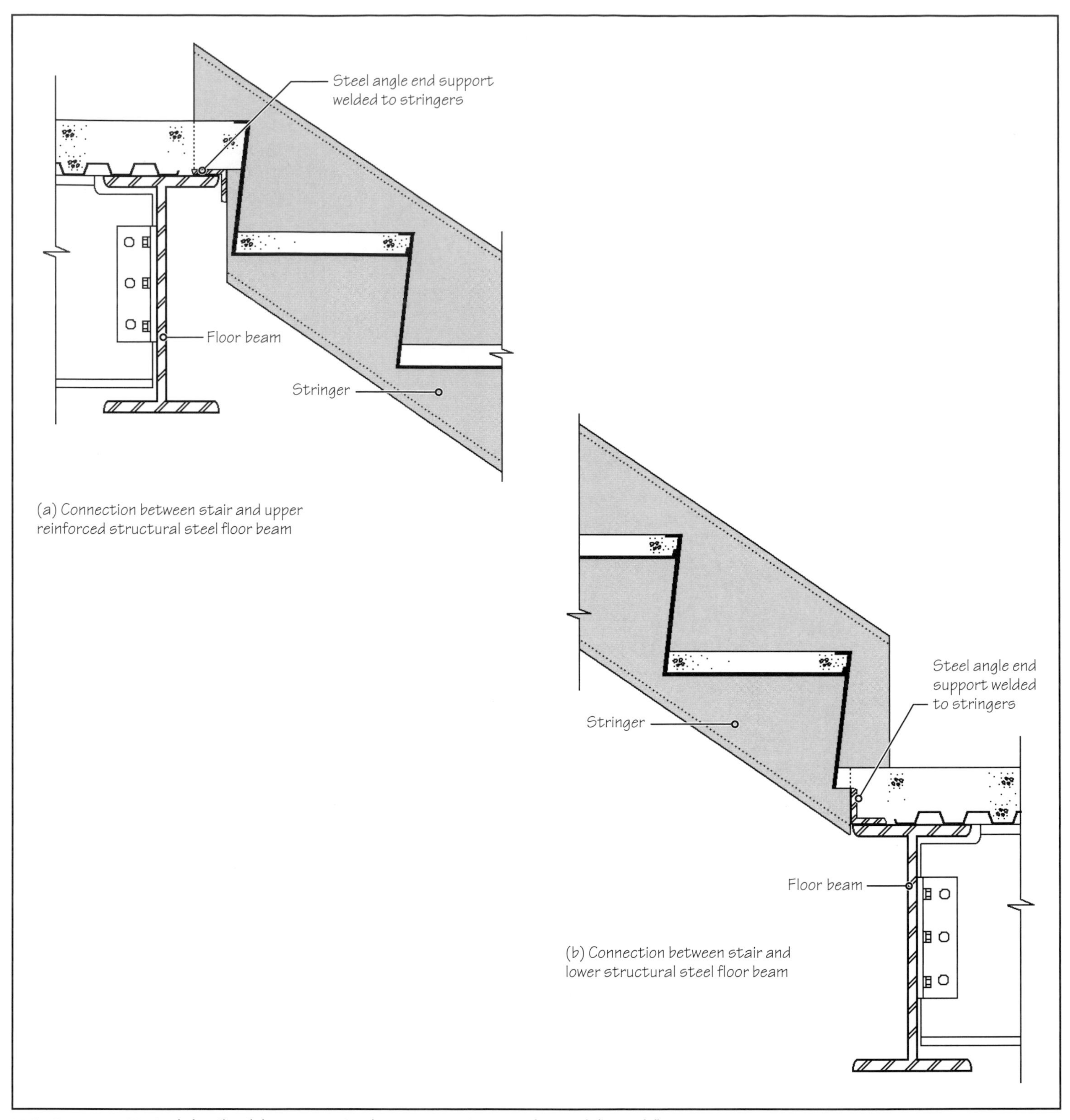

FIGURE 33.17 Typical details of the connection between a stringer and a steel-framed floor.

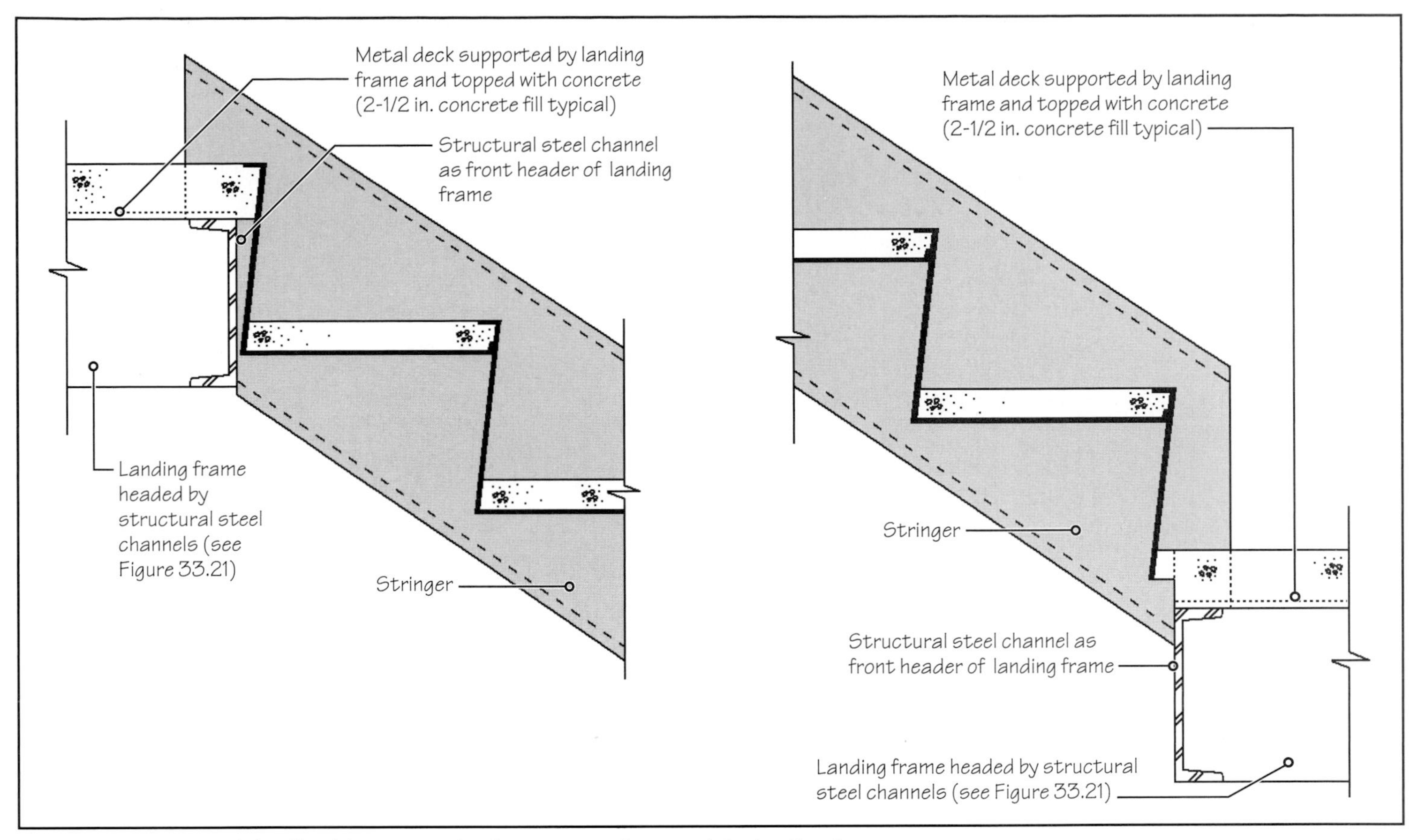

FIGURE 33.18 Typical details of the connection between a stringer and the landing frame.

LANDING FRAME

The landing of a steel stair is generally framed with structural steel members as a unit, called a *landing frame* (Figure 33.22). Typical details of connections between stringers and landings are shown in Figure 33.18. The finish on the landing is generally the same as that on the treads. Thus, where site-cast concrete is used on treads, the landing is also topped with concrete.

The landing frame is supported by suspending it from the upper-level floor beams with hanger bars, Figure 33.19, on tubular steel posts bearing on the lower-floor

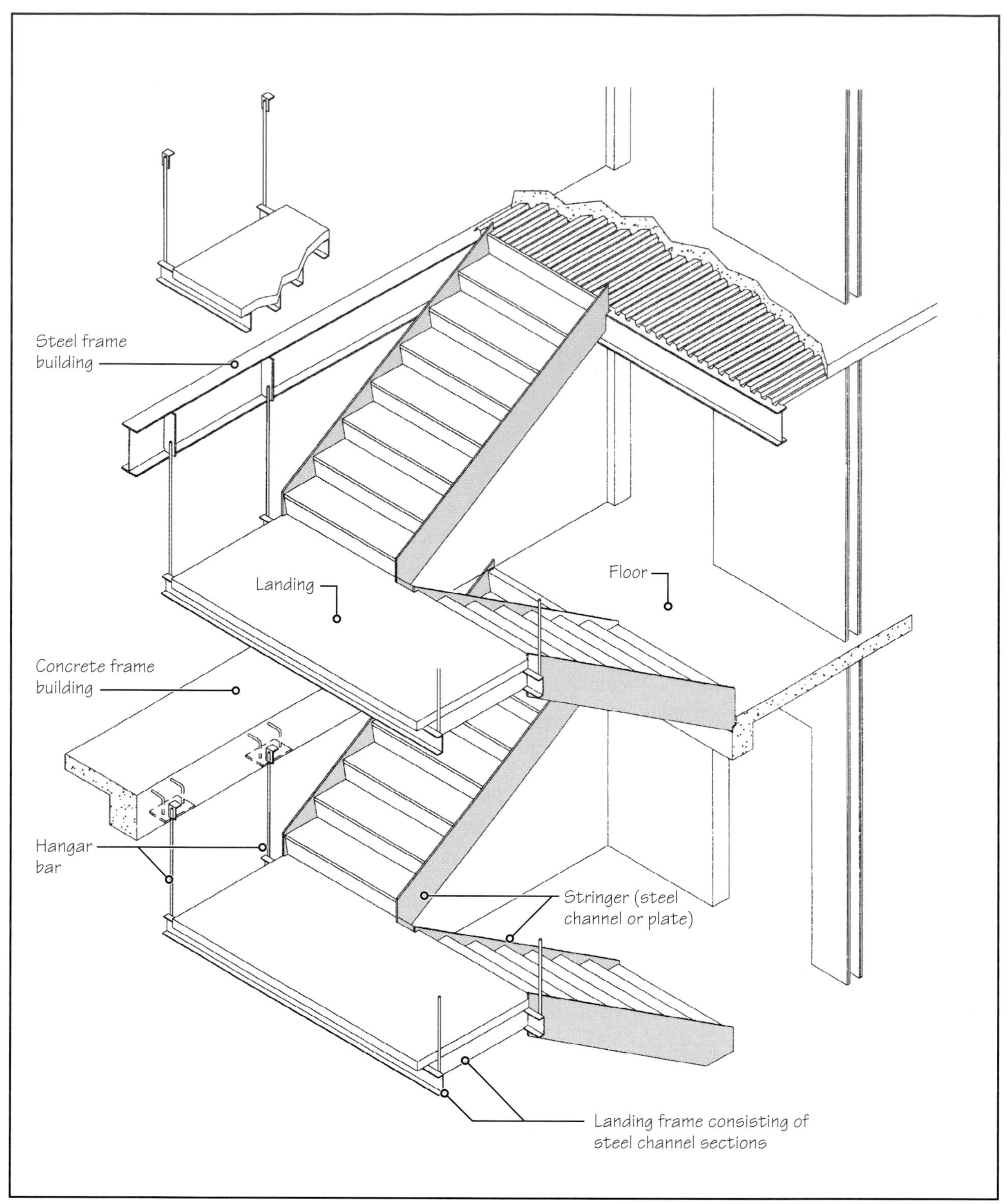

FIGURE 33.19 Landing frame of a steel stair suspended from the upper-floor beams in a steel- or concrete-frame building with drywall stair enclosure. (Illustration courtesy of The Sharon Stair Systems.) (Guard units and handrail have not been shown for clarity.)

beams, Figure 33.20, or on load-bearing (masonry or concrete) stair-enclosure walls, Figure 33.21.

The use of suspended landings, Figure 33.22, is more common in both steel- and concrete-frame buildings because the height of the landing can be field adjusted with a few turns of the nuts. Additionally, the entire stair can be erected before constructing the walls of the stair enclosure.

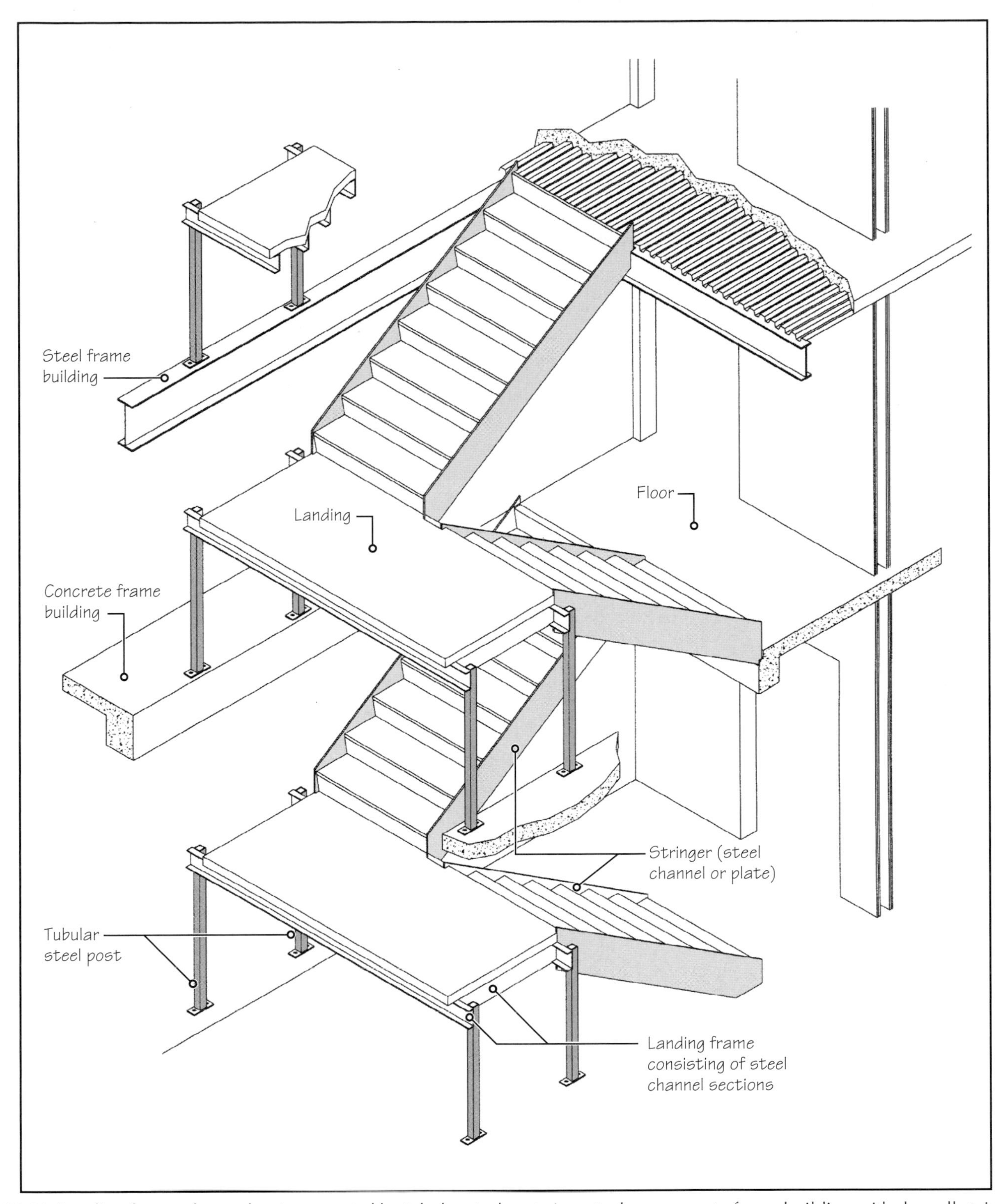

FIGURE 33.20 Landing frame of a steel stair supported by tubular steel posts in a steel- or concrete-frame building with drywall stair enclosure—an alternative to suspended landings. (Illustration courtesy of The Sharon Stair Systems.) (Guard units and handrail have not been shown for clarity.)

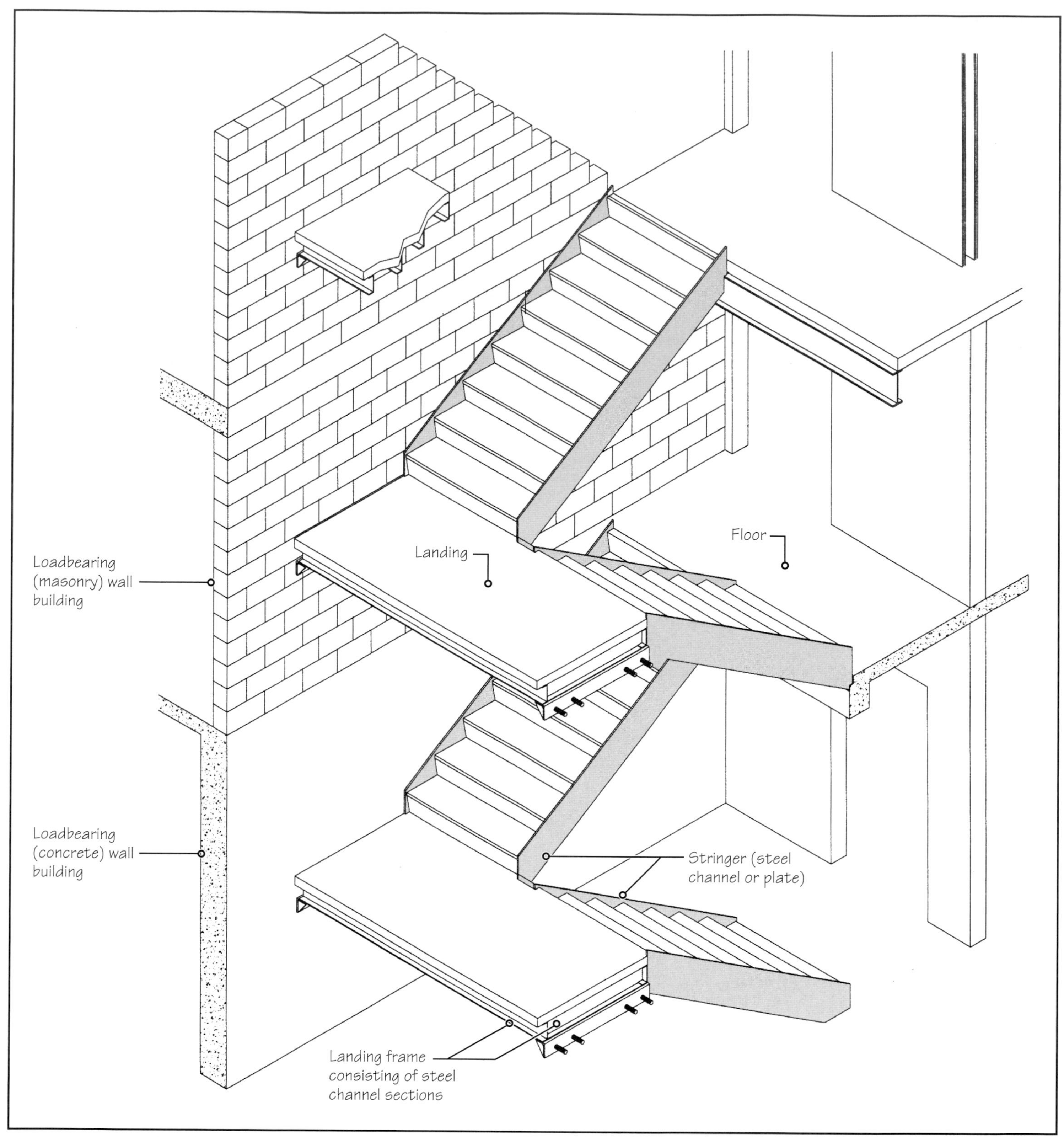

FIGURE 33.21 Landing frame supported by masonry or concrete walls of stair enclosure. (Illustration courtesy of The Sharon Stair Systems.) (Guard units and handrail have not been shown for clarity.)

Freestanding, Self-Supporting Stairs

In steel stairs discussed so far, each flight is supported between a floor and the landing. Steel and reinforced-concrete stairs can also be constructed without any supports at the landings (designed as cantilevers and supported only at the floors). Cantilevered stairs, also referred to as *self-supporting stairs,* can either be U-shaped or circular in plan.

A self-supporting, U-shaped steel stair with three flights and two intermediate landings between each floor is shown in Figure 33.23. In this stair, the stringers (of structural-steel

FIGURE 33.22 A typical steel stair with a suspended landing—a commonly used landing-support system in concrete- and steel-frame buildings. (Guard units and handrail have not yet been installed.)

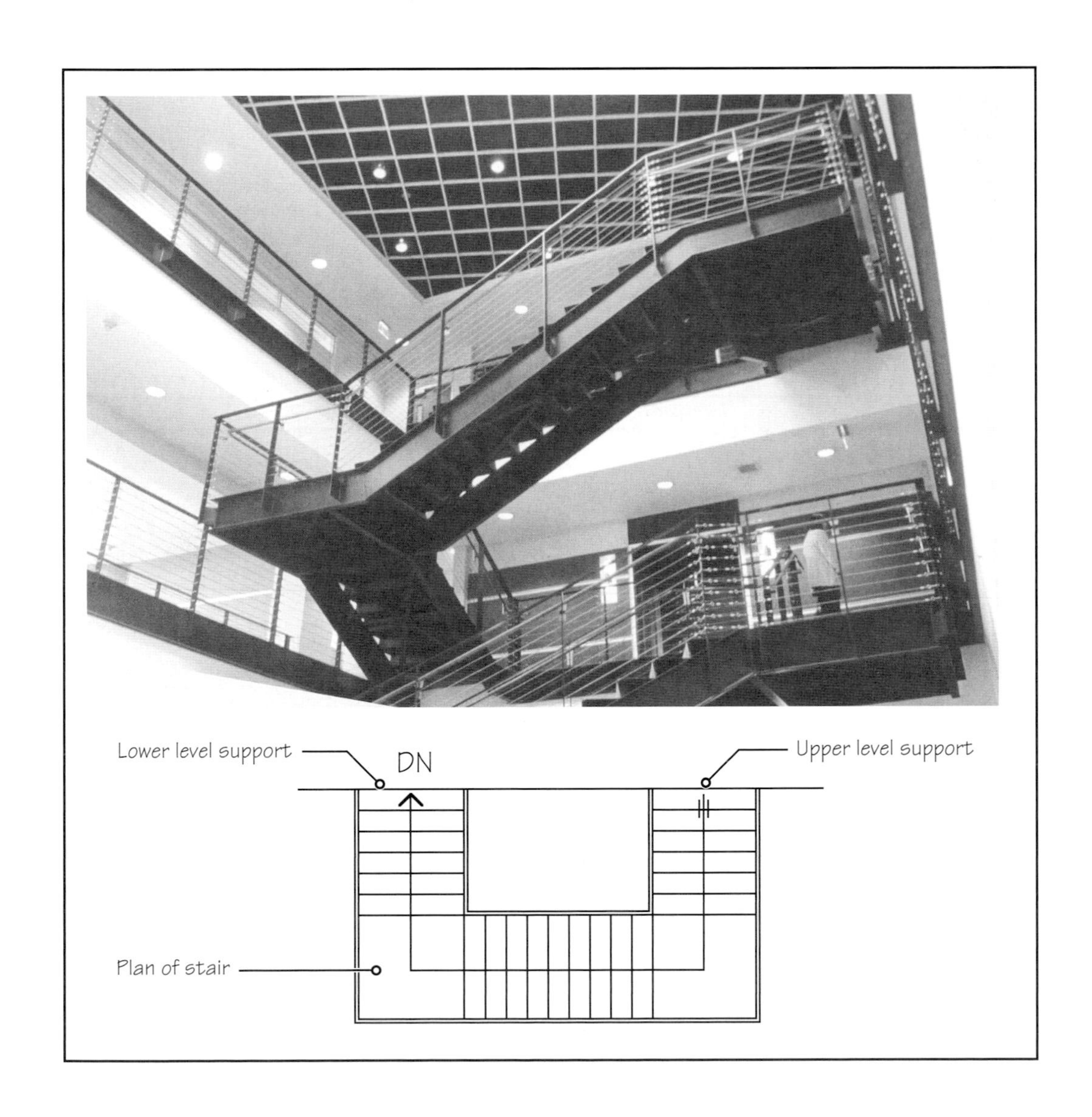

FIGURE 33.23 A self-supporting, U-shaped steel stair. See also Figure 33.24.

channels) function as continuous, spatially bent beams that are rigidly connected to the floor beams at both floors. Tread-riser units that span between stringer beams are made of structural-steel plate, Figure 33.24.

Self-supporting, circular steel stairs can be constructed with or without landings, Figure 33.25. Called *helical* (or *helicoidal*) stairs, they are fairly common in steel, concrete, and wood. A helical stair is similar to a spiral stair but without a central column support.

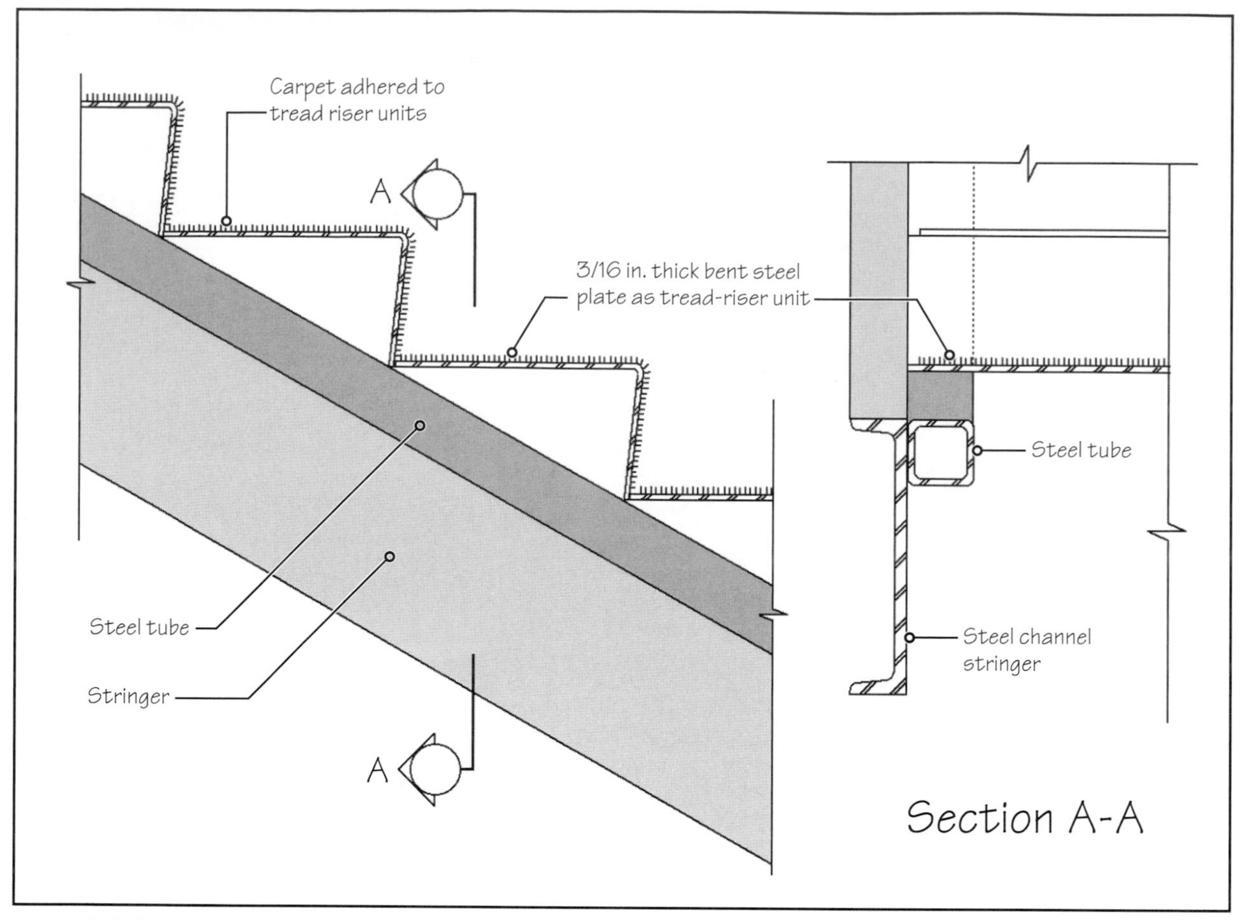

FIGURE 33.24 Each flight of the stair shown in Figure 33.23 consists of a frame made from 3-in. square steel tubes. The frame is rectangular and is cross-braced within its plane. Tread-riser units are welded to the tubular frame, completing a flight. Each such prefabricated flight unit has been welded to stringers.

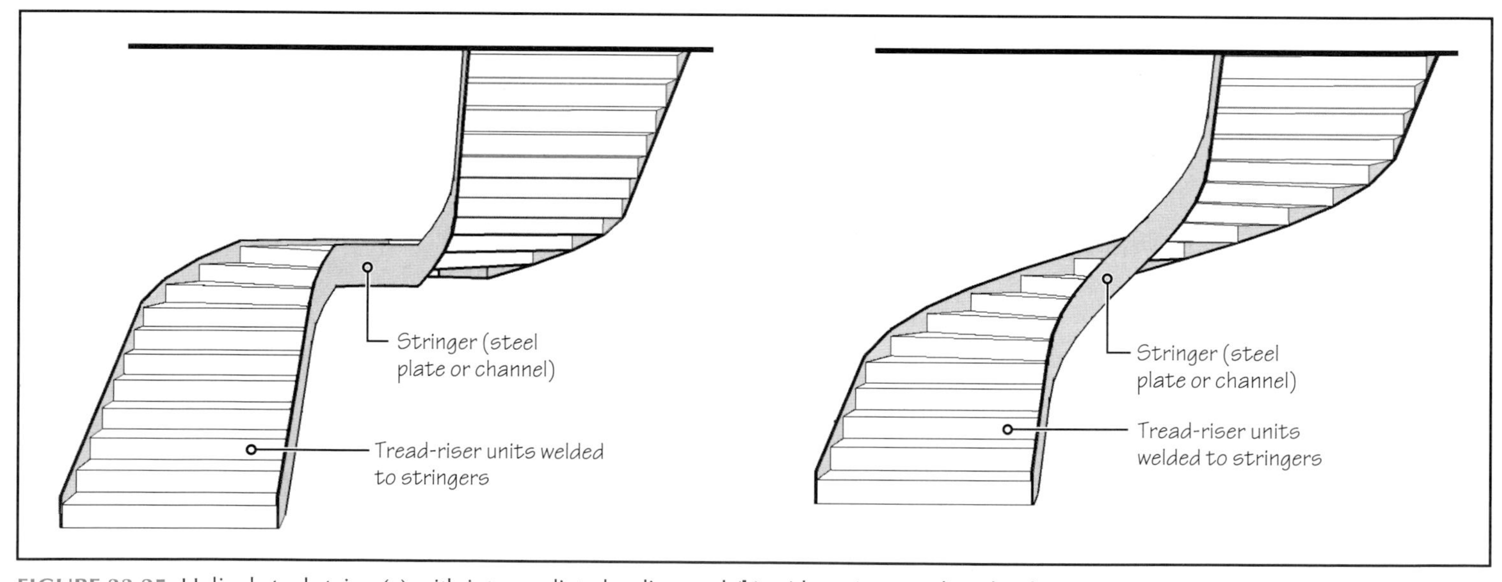

FIGURE 33.25 Helical steel stairs: **(a)** with intermediate landing and **(b)** without intermediate landing.

33.4 CONCRETE STAIRS

Although concrete stairs can be precast and prefabricated, their use is limited because they are heavy, which increases the cost of transportation and installation. Most concrete stairs are site cast. A typical site-cast concrete stair behaves as a slab supported between a floor and the landing. A section through such a stair is shown in Figure 33.26.

As previously noted, concrete stairs can also be self-supporting, Figure 33.27. Helical concrete stairs are also common.

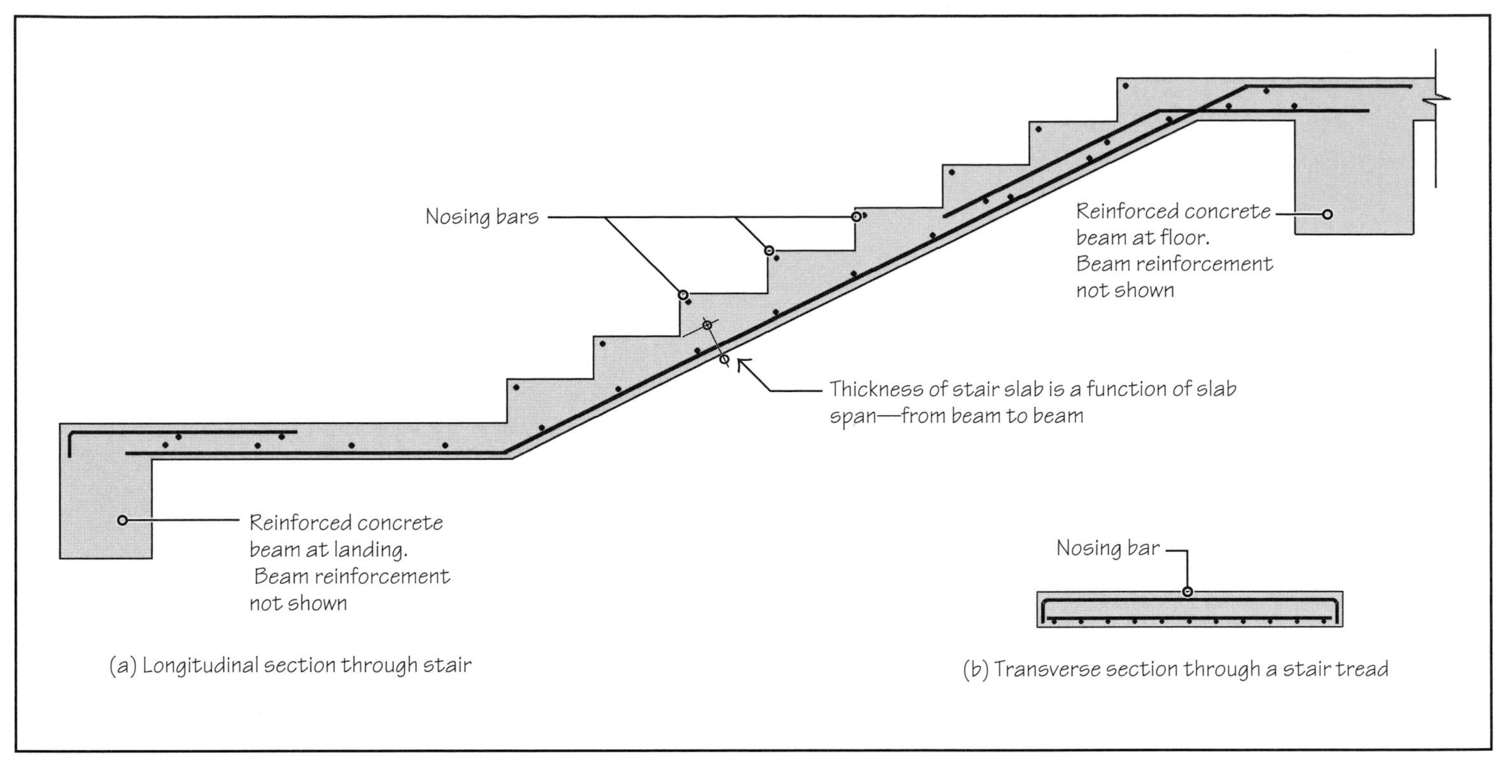

FIGURE 33.26 Sections through a typical site-cast concrete stair showing steel reinforcement.

FIGURE 33.27 A U-shaped, self-supporting reinforced-concrete stair.

PRACTICE QUIZ

Each question has only one correct answer. Select the choice that best answers the question.

14. In a typical wood stair, inclined beams that are cut to allow for the support of treads are called
a. rough stringers.
b. finish stringers.
c. balusters.
d. all the above.
e. none of the above.

15. In a typical wood stair, the number of carriages required is determined by
a. width of stair.
b. spanning capacity of carriage material.
c. floor-to-floor height.
d. all the above.
e. none of the above.

16. In a prefabricated steel stair, treads and risers are generally two separate components.
a. True
b. False

17. In a typical prefabricated steel stair, the number of stringers required is determined by the
a. width of stair.
b. spanning capacity of carriage material.

- **c.** floor-to-floor height.
- **d.** all the above.
- **e.** none of the above.

18. In a typical prefabricated steel stair, the stringers are cut to accommodate treads and risers
- **a.** True
- **b.** False

19. The stringers in a typical prefabricated steel stair are generally made of
- **a.** wide-flange sections.
- **b.** channel sections.
- **c.** plates.
- **d.** (a) and (b).
- **e.** (b) and (c).

20. The landing frame in a typical prefabricated steel stair can be hung from the building's structural frame.
- **a.** True
- **b.** False

21. In a self-supporting, freestanding stair, landings need no supports.
- **a.** True
- **b.** false

22. A self-supporting, freestanding stair can be made only of steel.
- **a.** True
- **b.** False

Answers: 14-a, 15-d, 16-b, 17-e, 18-b, 19-e, 20-a, 21-a, 22-b.

KEY TERMS AND CONCEPTS

Baluster
Circular stair
Flight
Freestanding (self-supporting) stair
Guard unit
Handrail
Landing
Nosing
Riser
Spiral stair
Straight-run stair
Stringer
Tread
Tread-riser relationship
U-shape stair

REFERENCE

1. Spens, Michael. *Staircases*. London: Academy Editions, 1995, p. 7.

CHAPTER 34 Floor Coverings

CHAPTER OUTLINE

34.1 Subfloors

34.2 Selection Criteria for Floor Coverings

34.3 Ceramic and Stone Tile Flooring

34.4 Stone Panel Flooring

34.5 Terrazzo Flooring

34.6 Carpet and Carpet Tile Flooring

34.7 Wood Flooring

34.8 Resilient Flooring

34.9 Resinous Flooring

34.10 Other Floor-Covering Materials

34.11 Underlayments

34.12 Resilient Accessories—Wall Base and Moldings

PRINCIPLES IN PRACTICE: Showers and Tile

Main entry foyer of Luis Obici Memorial Hospital, Suffolk, VA, showing the use of terrazzo flooring. Architects: HKS Inc. (Photo courtesy of HKS Inc.)

Floors have had a broad range of functions and finishes down through the history of architecture, ranging from basic and utilitarian to complex and ornate. They have been ignored, and they have been treated as the canvas for some of the most enduring artistic expressions. It should be noted that 'looking down' after entering a building is an important consideration that designers should not ignore.

The commitment to floors has always been based on the project budget. If funds are not available, floors are left basic, and if funds are available, finishes tend to be of a higher quality. Some people still ignore floors and treat them as utilitarian surfaces. However, because there are so many material, color, pattern, and covering options available, floors are still valued as surfaces worthy of financial investment and creative expression.

This chapter explores many of the most frequently used floor covering materials, including ceramic and stone tiles, stone panels, terrazzo, carpet and carpet tiles, wood, and resilient and resinous coverings.

34.1 SUBFLOORS

The top surface of the structural floor is called the subfloor. Subfloors are constructed early in the building process. Therefore, it should be expected that the subfloor surface will become damaged due to the construction that takes place above it, necessitating some subfloor surface repairs before the floor covering is laid.

TYPES OF FLOOR STRUCTURES

There are generally two types of subfloor structures over which floor coverings can be installed:

- *Concrete Subfloors:* Elevated concrete floors (cast-in-place concrete and concrete on metal deck) and concrete slabs-on-grade, Figure 34.1.
- *Wood Subfloors:* Wood panels supported by wood light frame or light-gauge steel frame members, Figure 34.2.

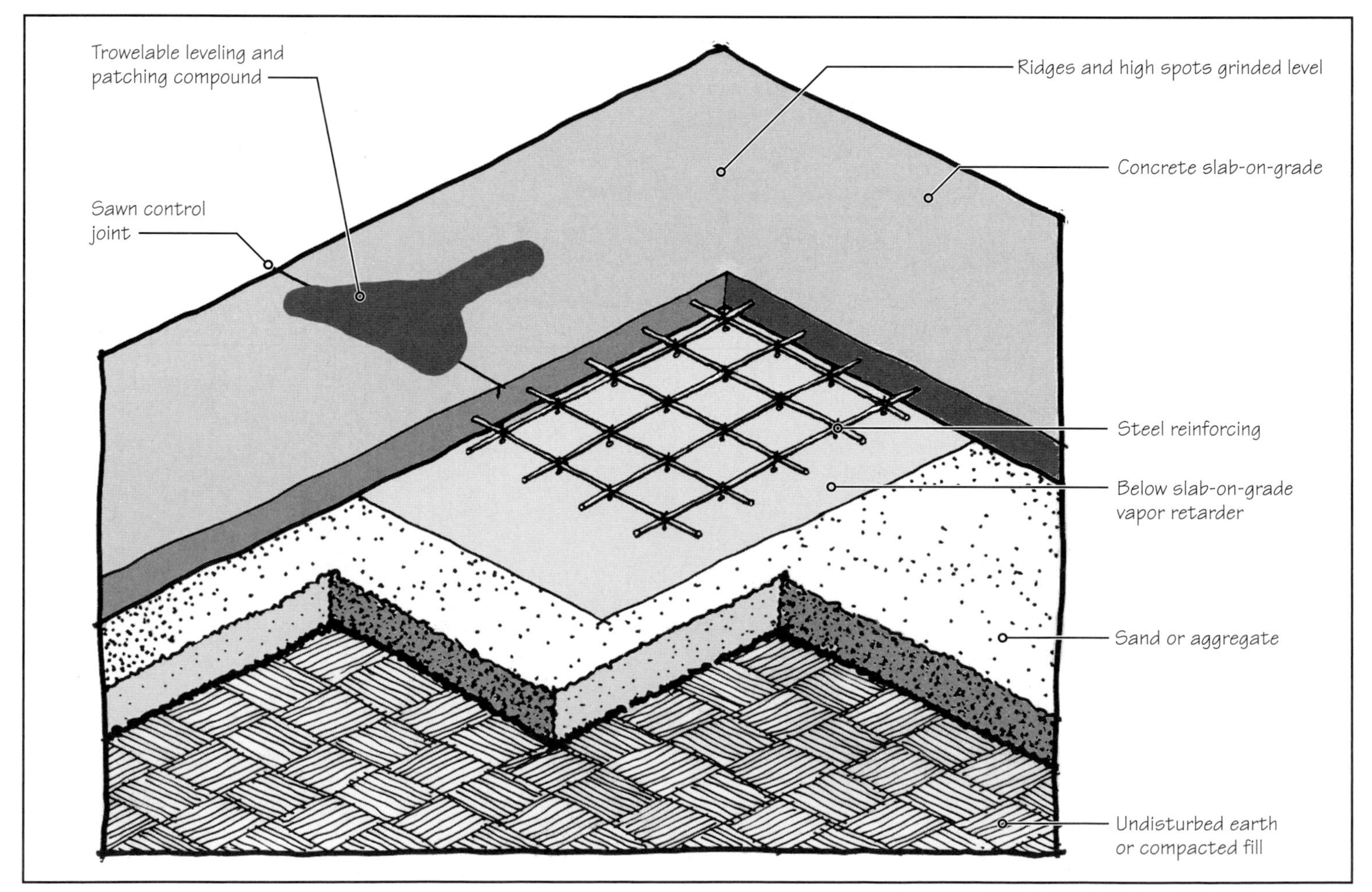

FIGURE 34.1 Typical concrete slab-on-grade components including the leveling and patching compound used before laying the floor covering.

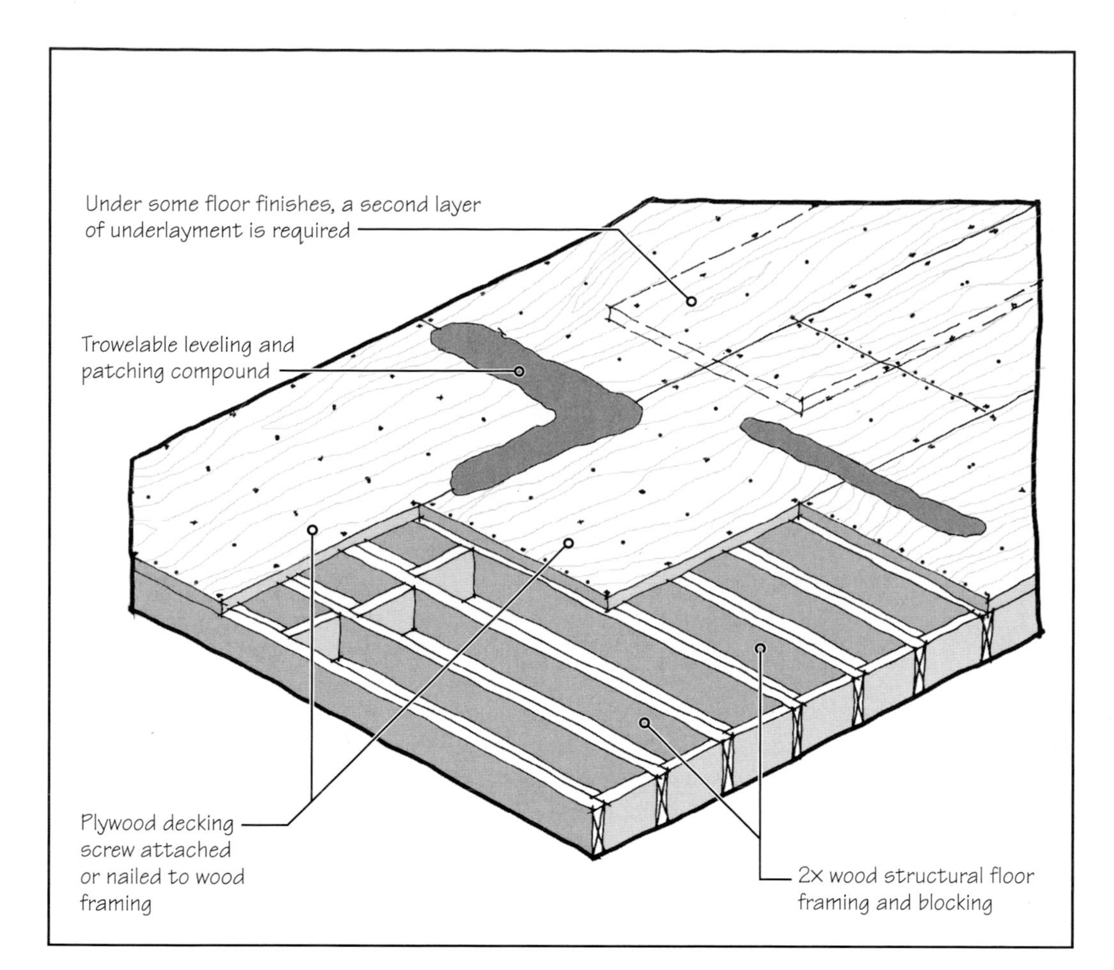

FIGURE 34.2 Typical wood framed floor structure illustrating preparation required prior to laying the floor covering.

SUBFLOOR CHARACTERISTICS

The long-term durability of floor coverings depends on the quality of the subfloor conditions. A good subfloor will have some of the following characteristics:

- *Sound:* Concrete should be finished smooth without ridges, ripples, and imperfections, and wood structures should be properly attached, solid and tight, with nails, screws, or adhesive.
- *Dry:* Subfloor must be sufficiently dry because adhesives will not adhere properly to a wet or partially wet surface. Additionally excessive movement of subfloor surface will damage floor coverings as the moisture leaves the subfloor.
- *Flat, Smooth, and Level:* Subfloors should be flat, smooth, level, and free from irregularities such as rough spots, cracks, depressions, ridges, and holes. Many floor coverings are thin and flexible, and imperfections in the subfloor will telegraph through the product and become permanent imperfections in the floor covering, Figure 34.3. Generally, subfloors should be level with no more than $\frac{1}{8}$ to $\frac{3}{16}$ in. of vertical change in any 10-ft horizontal dimension.
- *Clean and Free of Foreign Materials:* There should be no foreign materials, including dust, solvents, paint, wax, oil, grease, concrete curing compounds, sealers, hardeners, alkaline salts, excessive concrete carbonation or laitance, gypsum dust, mold, mildew, and residual adhesive and adhesive removers in the case of floor covering replacements.
- *Deflection:* As demonstrated by Figure 4.17, deflection in a subfloor structure places floor coverings in compression, resulting in cracks in tile and terrazzo and puckers and wrinkles in resilient tiles and sheets. The choice of floor coverings should be coordinated with the deflection of the building structure to avoid long-term defects.

34.2 SELECTION CRITERIA FOR FLOOR COVERINGS

There are an almost limitless number of floor covering products and materials available today. The design and articulation of floor surfaces are limited only by the designer's imagination. In fact, it is not unusual for the floor coverings to be the dominant element for the interior design of a building, so designers must consider the design intent as well as the manner in which people will interact with the floors.

NOTE

The dryness of a concrete slab-on-grade is particularly important if the floor covering is polymer-based (e.g., vinyl tiles, linoleum, thin-set terrazzo, and resinous flooring). Because these floor coverings are not vapor permeable, they do not allow water vapor to pass through them. Therefore, as the water in the slab converts to water vapor, it delaminates the floor covering from the slab. (Remember from Chapter 6 that water vapor pressure is generally very high.) The use of a vapor retarder under a concrete slab-on-grade is important for the same reason.

The dryness of a concrete slab is not of much concern if the floor covering and adhesive are portland cement–based, such as portland cement terrazzo.

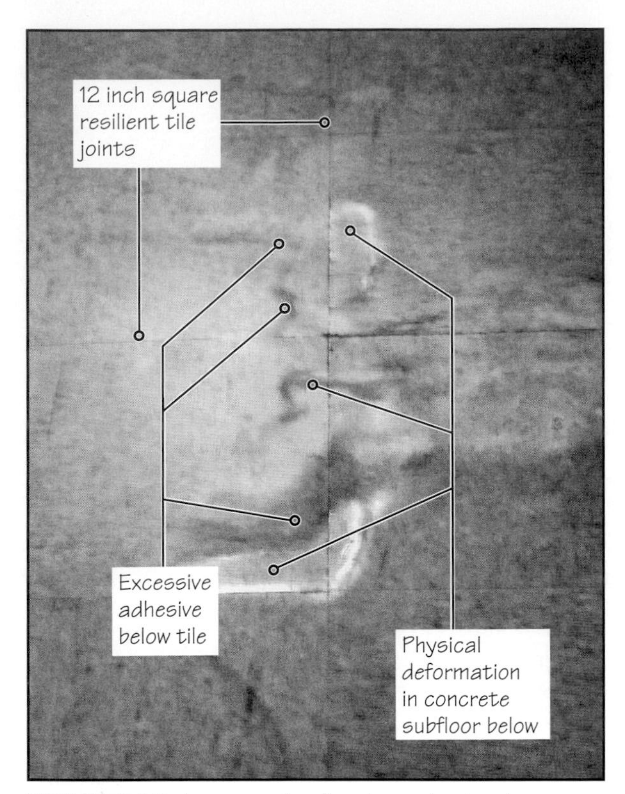

FIGURE 34.3 An example of an imperfection in a concrete subfloor that, with time and traffic, has telegraphed through the vinyl composition tile flooring to create a permanent deformation.

There are many performance factors that should be considered when selecting floor coverings. Some of the following factors are critically important and necessary, whereas others are optional and discretionary (not all apply to every situation):

- *Slip Resistance:* Floor coverings should be resistant to people slipping and falling. The slip resistance of a floor covering is measured by laboratory testing that establishes a numerical value known as the *coefficient of friction.* The Americans with Disabilities Act (ADA) recommends values under dry conditions of 0.6 for level surfaces and 0.8 for ramps.
- *Types, Amount, and Frequency of Traffic:* Floor coverings should be suitable for the abrasive impact pedestrian foot traffic will have on it. Foot traffic impact can be measured by the number of crossings per day. One manufacturer classifies frequency in the following manner: light (less than 100), moderate (100 to 1,000), heavy (1,000 to 10,000) and extra heavy (more than 10,000).
- *Durability and Longevity:* Floor coverings should be durable enough to endure any expected use and provide an adequate service life. They should be selected to resist staining from liquids, chemicals, and reagents. Also, exposure to large amounts of ultraviolet radiation, as through skylights or large windows, may fade some floor coverings.
- *Flammability:* Building codes require resistance to fire propagation (flame spread and smoke development) for some applications.
- *Sound Absorption:* Floor coverings that absorb sound can lower the airborne noise level within rooms.
- *Sound Isolation:* Floor coverings can improve the impact insulation class (IIC) when building codes mandate minimum sound isolation requirements for floors (Chapter 8).
- *Hygienic Qualities:* In some applications a floor covering must not give sanctuary to dirt and bacteria; resistance to mold and mildew growth is especially important.
- *Walking Comfort:* When people will be constantly standing on or walking over a floor covering, underfoot comfort is very important. Hard, unyielding surfaces can cause walker fatigue and, in some cases, lead to foot, leg, or back stress problems.
- *Sustainability:* Floor coverings and adhesives that do not out-gas volatile organic compounds are an important part of sustainability criteria included in the LEED rating system.
- *Maintenance Requirements:* For every floor covering, there is a prescribed maintenance procedure. Some are more time consuming and expensive to maintain than others. The long-term maintenance goals and budgets for the building must be considered, because maintenance represents a continuous expense.

34.3 CERAMIC AND STONE TILE FLOORING

Ceramic and stone tile are among the most versatile floor coverings available. The composition and installation of ceramic tile has changed little over the decades. Stone can be fabricated into either tiles, considered in this section, or panels, considered in Section 34.4. Tile is adhered with mortar, and the voids between tiles are then filled with grout. They are especially suited for installations that do not require thick beds of mortar.

Characteristics of Ceramic Tile

Ceramic tile is made from natural clay, porcelain, or other ceramic materials. The exposed face is either glazed or left unglazed and then fired to a temperature sufficient to produce the necessary physical properties.

Glazed tiles have glassy, or glossy, exposed surfaces. The glaze protects the tile body against water absorption and provides for a wide range of colors. Glazes can be opaque, transparent clear, or colored clear, and sheens may be mat (low or no gloss), semimat (semigloss, which is a moderate gloss), or bright (high gloss). One of the weaknesses of glaze is that it can be scratched, which changes its appearance and performance characteristics.

Stone Tile Characteristics

Stone tile, also known as cut stone, has many of the same characteristics as stone panel flooring, which is discussed later in this chapter. In contrast to panels, the back of tiles are gauged, or cut flat, to establish uniform thicknesses.

TABLE 34.1 PERMEABILITY AND APPLICATION OF VARIOUS TILES

Tile type	Permeability	Application
Impervious	0.5% or less	Frequent or regular water contact
Vitreous	0.5 to 3.0%	Occasional water contact and/or when water is quickly removed
Semivitreous	3.0 to 7.0%	
Nonvitreous	Above 7.0%	No water contact except maintenance

Physical Properties of Ceramic and Stone Tile

Ceramic and stone tile have a variety of physical properties, which influence the selection of the appropriate tile for a particular use.

- *Quality and Uniformity:* Consistent tile size, color, pattern, and texture are important for the finished construction. Because ceramic tiles are fired products, facial and structural defects may develop that disqualify them for installation. Crazing, the development of tiny cracks that develop in the surface of a tile or in the glazing, also allows premature deterioration.
- *Shapes and Dimensions:* Dimensional consistency is important for proper installation. The width of the grout joint is determined by the amount of variation that is allowed in the width, length, and thickness of a group of tiles. Tight control over dimensional variation allows smaller grout joints, whereas looser control necessitates wider joints.
- *Warpage:* If tiles are not uniformly flat, they will not lay flat on the subfloor and will break under heavier loads because portions of the tile are not adequately supported. In addition, after installation warped tiles that are slightly higher than an adjoining tile, called lippage, could cause people to trip and fall.
- *Water Absorption:* Tiles do not absorb atmospheric moisture as wood does; however they will absorb water when in direct contact with it. The tile industry classifies tiles according to permeability, Table 34.1.
- *Breaking Strength:* Tiles must maintain their integrity when subjected to loads. The breaking strength of a tile is a good indication of how well a tile will or will not perform. Tiles should resist breaking when objects are dropped on the floor.
- *Abrasion Hardness:* Tiles should be hard enough to resist the abrasion that will occur over the service life of the tile.

Types, Shapes, and Sizes of Ceramic Tiles

There are a wide variety of ceramic tile types, shapes, and sizes available today, Table 34.2 and Figures 34.4, 34.5, and 34.6.

Setting Methods

Before discussing the various tile-setting products, the three tile-setting methods are explained. The primary difference between them is the thickness of the mortar bed and the type of mortar used.

- *Thick-set (or thick-bed):* Usually requiring a 2- to 3-in.-thick mortar bed, this method is necessary where (a) the floor tiles are large in size, such as large panels of natural stone, (b) the floor must be sloped to floor drains, (c) there is excessive variation in the thickness of tiles, as is generally the case with natural stone panels, or (d) the subfloor has surface irregularities. The assembly consists of the following parts (see "Principles in Practice" at the end of this chapter):

  First, a mortar bed is set. If waterproofing is required, the mortar bed is placed in two applications with a waterproof membrane in the middle. If the floor will be subjected to a considerable load or if the floor structure deflection will be more than allowed by the tile installation method, the mortar bed should be reinforced with wire mesh or metal lath. Additionally, a cleavage (bond-breaking) membrane is sometimes necessary to prevent bonding with the subfloor. Reinforcing and cleavage membranes are not needed if the mortar bed is to be bonded with the subfloor.

  Second, tile is set on the mortar bed while the mortar is still green, that is, not cured. If the tile is set after the bed has cured, a portland cement bond coat is required between the tile and the mortar bed.

- *Medium-set (or medium-bed):* Although not officially recognized by the several industry standards, the medium-set method actually involves thin-set mortars

TABLE 34.2 TILE TYPES, GENERAL DESCRIPTIONS, AND DIMENSIONAL CHARACTERISTICS OF THE VARIOUS CERAMIC TILE CLASSIFICATIONS

Tile type	General description of tile	Impervious tile	Vitreous tile	Semivitreous tile	Nonvitreous tile	Typical sizes in inches				
						$\frac{1}{4}$ in. thick	$\frac{5}{16}$ in. thick	$\frac{3}{8}$ in. thick	$\frac{1}{2}$ in. thick	$\frac{3}{4}$ in. thick
UNGLAZED CERAMIC MOSAIC TILE	• Formed by dust-pressed or plastic method • Either porcelain or natural clay composition • Plain or abrasive mixture throughout • Homogeneous color throughout body	X				1 × 1 1 × 2 2 × 2 3 × 3				
QUARRY TILE	• Glazed or unglazed • Formed by extrusion method • Either natural clay or shale composition	X	X	X				6 × 6 8 × 8 8 × 4	3 × 3 4 × 4 6 × 6 8 × 8 6 × 3 8 × 4	6 × 6 8 × 4
PAVER TILE	• Glazed or unglazed • Formed by dust-pressed method • Either porcelain or natural clay composition	X	X	X		3 × 3		4 × 4 8 × 8 12 × 12 8 × 4	4 × 4 6 × 6 8 × 8 8 × 4	
PORCELAIN TILE	• A ceramic mosaic or paver tile • Formed by dust-pressed method • Composition resulting in a dense, smooth, fine-grained tile with a face that is sharply formed	X				1 × 1 3 × 3 6 × 6 12 × 12 18 × 18 20 × 20 24 × 24 12 × 18 12 × 24				
GLAZED WALL TILE	• Not expected to withstand excessive impact • Not subject to freezing and thawing conditions • Not suitable for floors				X		$4\frac{1}{4} \times 4\frac{1}{4}$ $6 \times 4\frac{1}{4}$ 6 × 6 3 × 6			
GLASS MOSAIC TILE		X				$\frac{3}{8} \times \frac{3}{8}$ 2 × 2	1 × 1 2 × 2 $4\frac{1}{4} \times 4\frac{1}{4}$ 3 × 6			

that can be applied thicker than traditional thin-set mortars. Allowing a bed of $\frac{1}{4}$ to $\frac{3}{4}$ in. in thickness, this method provides extra setting space when the subfloor is not properly prepared or when large-format tiles (18 to 24 in. in one dimension) are used.

- *Thin-set (or thin-bed):* More popular than the other methods (less material and labor), this method is used where the tiles are (a) small in size, generally less than 12 in. × 12 in., (b) no slope to floor drains is required, (c) the tile thickness is relatively uniform, or (d) the subfloor does not have excessive surface irregularities. The mortar bed is generally $\frac{1}{8}$ in. thick, generally consisting of polymer-based adhesives, Figures 34.9. 34.10, and 34.11.

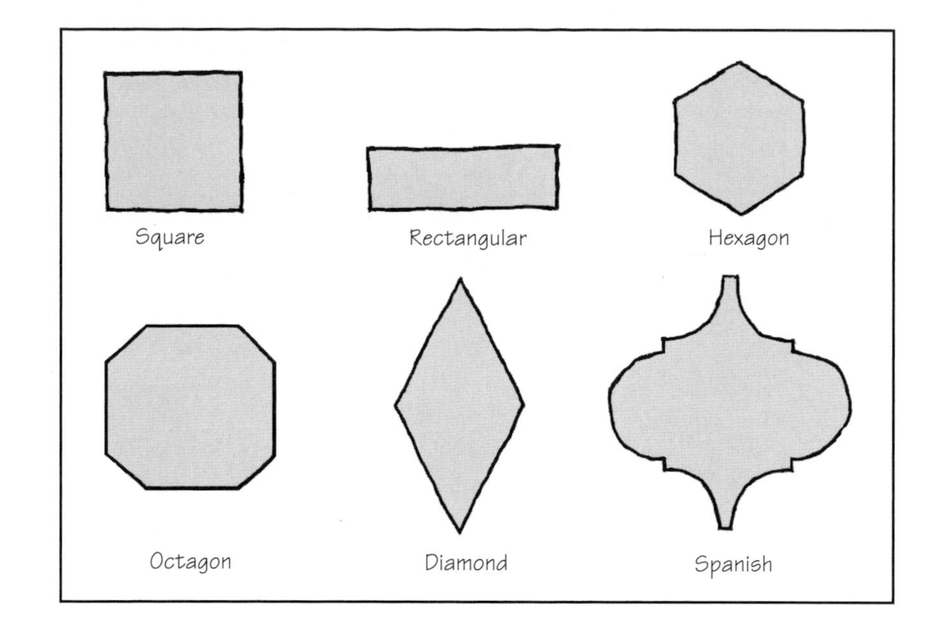

FIGURE 34.4 Several typical ceramic tile shapes, all available in many sizes.

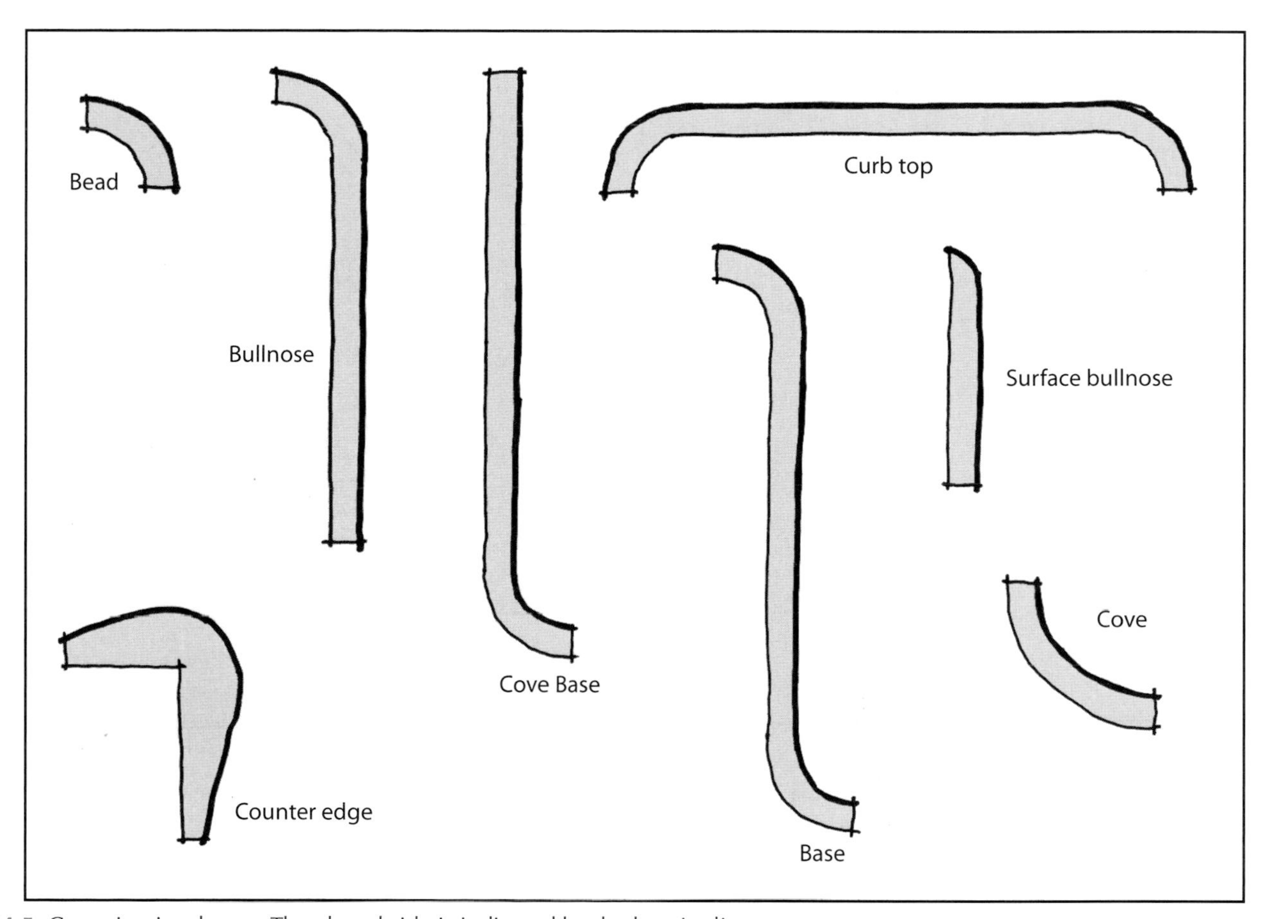

FIGURE 34.5 Ceramic trim shapes. The glazed side is indicated by the heavier line.

Setting Materials—Mortars, Adhesives and Epoxies

A wide variety of setting materials are available to adhere ceramic and stone tile to the subfloor using any of the three setting methods. The Tile Council of North America (TCNA) publishes the TCA *Handbook for Ceramic Tile Installation,* which contains installation methods for every type of mortar and the various applications in which each can be used (Figures 34.7 and 34.8).

- *Organic Adhesives:* Usually ready-to-use liquid or powdered water-emulsion latex products that cure by evaporation, adhesives are typically for light-duty installations and for interior use only, not suitable for high temperatures.
- *Cement Mortars:* Typically for general-duty installations, cement mortars consist of mixtures of portland cement, sand, water, and water-retentive additives for dry-set cement mortars or a latex polymer additive for modified cement mortars.
- *Water-Cleanable Epoxies:* Suitable for heavy-duty installations, for high temperature conditions, or for specific functions (not for general use), epoxies are composed of an epoxy resin and a hardener.

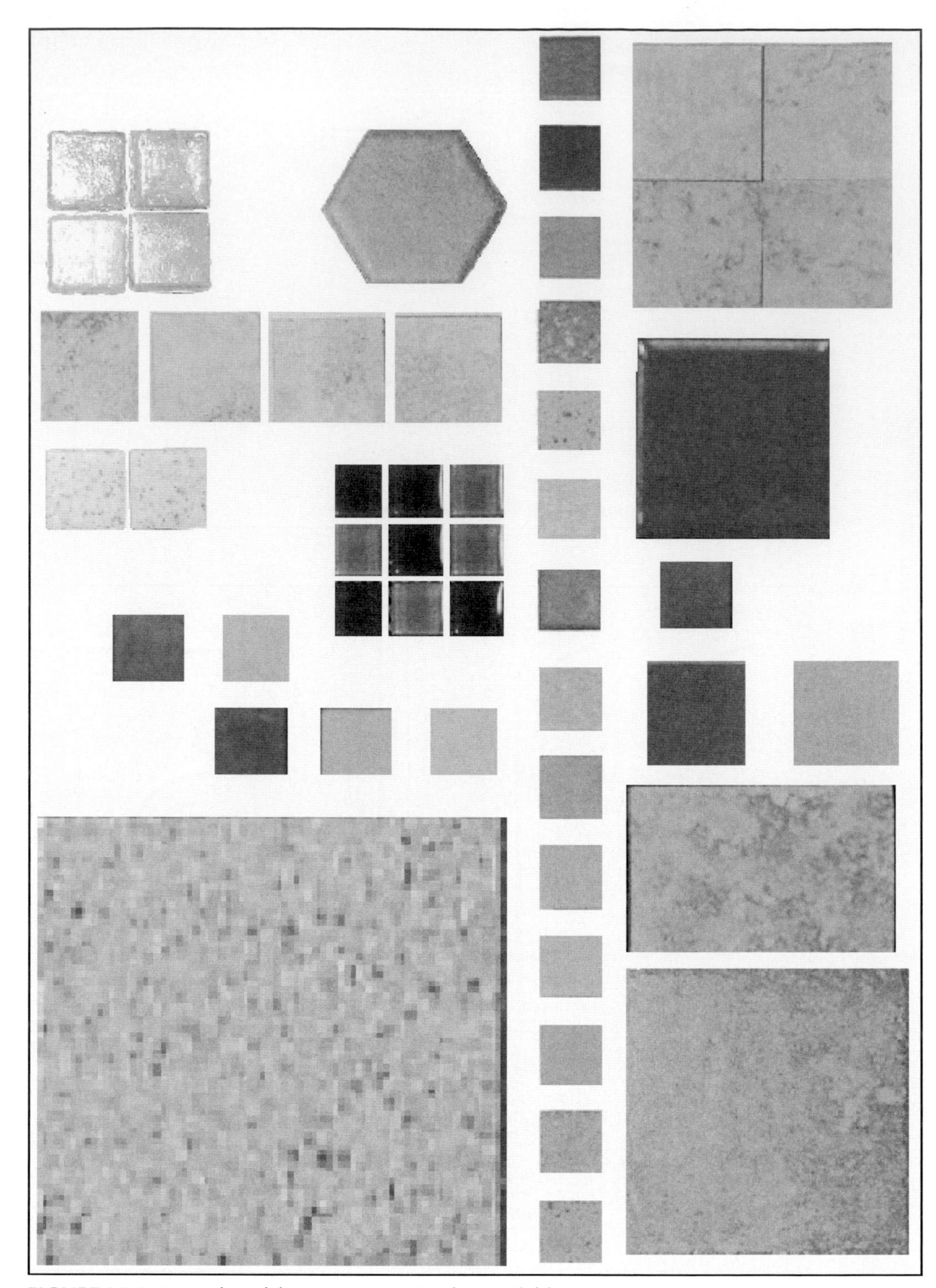

FIGURE 34.6 Examples of the many ceramic tiles available.

- *Furan Resin Mortars:* Formulated for resistance to chemicals, this mortar consists of furan resin, powder containing carbon or silica fillers, and an acid catalyst.

SETTING MATERIALS—GROUTS

After the tiles have been set and the mortar has cured, grout is used to fill the joints between the tiles. The TCA *Handbook* contains installation methods for every type of grout and the various applications in which each can be used. These include:

- *Jobsite-Mixed Sand–Portland Cement Grouts:* On-the-job mixtures of portland cement, sand, lime, and water.
- *Premixed Standard Cement Grouts:* Unsanded cement grouts contain water-retentive additives and are for joints $\frac{1}{8}$ in. wide or less, whereas sanded cement grouts contain sand and other ingredients for joints $\frac{1}{8}$ in. wide or greater.
- *Polymer-Modified Cement Grouts:* An improvement over standard cement grouts, these grouts possess increased color stability, good flexural and bond strengths, stain resistance, and lower moisture absorption, so they resist frost damage.

- *Water-Cleanable Epoxy Grouts:* These grouts are essentially the same as epoxy mortars.
- *Furan Resin Grouts:* These grouts are essentially the same as furan resin mortars.

INSTALLATION

Architects rely on industry standards to determine the appropriate installation method for an application: TCA *Handbook* and ANSI A108/A118/A136 Series, *American National Standard Specification for the Installation of Ceramic Tile.* Commonly used TCA *Handbook* guidelines for tile floor installations are shown in Figure 34.7 for concrete subfloors and Figure 34.8 for wood panel subfloors. Other installation methods in the TCA *Handbook* include ceilings, swimming pools, countertops, stairs, steam rooms, shower stalls, bathtub

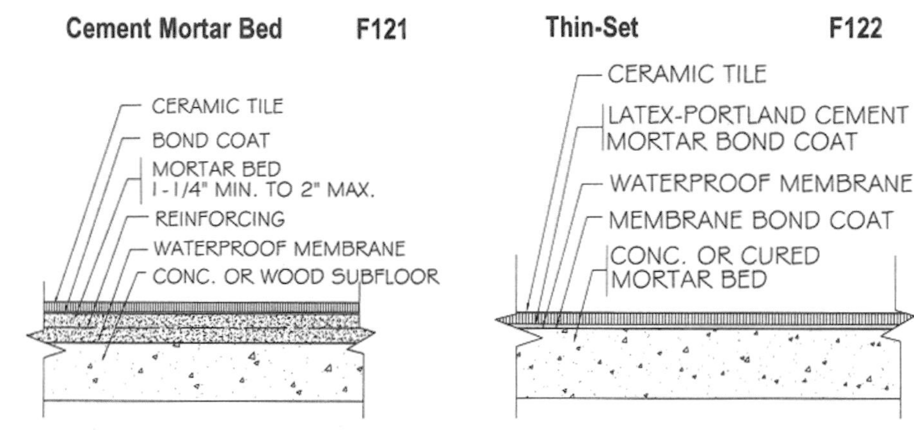

Cement Mortar Bed F121

Recommended Uses:
- wherever a waterproof interior floor is required in conjunction with ceramic tile installed on a portland cement mortar bed.
- see page 17 NOTE for exterior uses.

Limitations:
- deflection not to exceed 1/360 of span.
- not recommended for severe chemical exposure.

Requirements:
- design floor areas over which tile is to be applied to have deflection not greater than 1/360 of the span. Make allowance for live load and impact as well as all dead load, including weight of the tile and setting bed.
- mortar bed thickness—1-1/4" min. to 2" max.
- mortar beds in excess of 2"-thick shall be detailed by the architect.

Materials:
- waterproof membrane—ANSI A118.10.
- mortar—ANSI A108.1A.

Preparation by Other Trades:
- maximum variation in subfloor shall not exceed 1/4" in 10'-0" from the required plane.
- slope subfloor 1/4" per foot to drain.

Preparation by Tile Trade:
- waterproof membrane—install to comply with pertinent codes and manufacturer's directions.

Tile Installation and Movement Joints:
- follow Method F111 or F141.

Thin-Set F122

Recommended Uses:
- wherever a waterproof interior floor is required in conjunction with ceramic tile installed in a thin-set method.
- see page 17 NOTE for exterior uses.

Limitations:
- deflection not to exceed 1/360 of span.
- not recommended for severe chemical exposure.

Requirements:
- design floor areas over which tile is to be applied to have deflection not greater than 1/360 of the span. Make allowance for live load and impact as well as all dead load, including weight of the tile and setting bed.

Materials:
- waterproof membrane—ANSI A118.10.
- latex-portland cement mortar—ANSI A118.4.
- polymer modified tile grout—ANSI A118.7.

Preparation by Other Trades:
- maximum variation in subfloor—1/4" in 10'-0" from the required plane.
- slope subfloor 1/4" per foot to drain.

Preparation by Tile Trade:
- waterproof membrane—install to comply with pertinent codes and manufacturer's directions.

Membrane Installation:
- membrane manufacturer's directions.
- membrane—ANSI A108.13.

Installation Specifications:
- tile—ANSI A108.5.

Movement Joint (architect must specify type of joint and show location and details on drawings):
- movement joints—mandatory according to Method EJ171

FIGURE 34.7 Two commonly used installation methods for setting tile on concrete subfloors. F-121 is used for thick-set applications, and F-122 is used for thin-set applications. (Used with permission of Tile Council of America, Inc.)

Cement Mortar F141

CERAMIC TILE
BOND COAT
MORTAR BED
1-1/4" MIN. TO 2" MAX.
REINFORCING
CLEAVAGE MEMBRANE
SUBFLOOR

Recommended Uses:
- over all wood floors that are structurally sound.
- where radiant heat pipes are laid over the wood subfloor, screed fill flush to top of pipes before placing a cleavage membrane and reinforced mortar bed.
- see page 17 NOTE for exterior uses.
- where waterproof floor is required, use waterproof membrane meeting ANSI A118.10.
- contact waterproof membrane manufacturer for specific detail.

Requirements:
- cleavage membrane—ANSI A-2.1.8.
- reinforcing mandatory.
- design floor areas over which tile is to be applied to have a deflection not greater than 1/360 of the span when measured under 300 lb. concentrated load (see ASTM C627).
- mortar beds in excess of 2"-thick shall be detailed by the architect.

Materials:
- mortar bed, reinforcing, and cleavage membrane—ANSI A108.1A.
- bond coat—portland cement paste on a mortar bed that is still workable, or dry-set mortar or latex-portland cement mortar on a cured bed.
- grout—ANSI A118.3, A118.6, A118.7, or A118.8.

Preparation by Other Trades:
- subfloor—19/32" tongue and groove plywood with 1/8" gap between sheets or 1" nominal boards when on joists 16" o.c.
- depressing floor between joists on ledger strips permissible in residential use.

Organic Adhesive F142

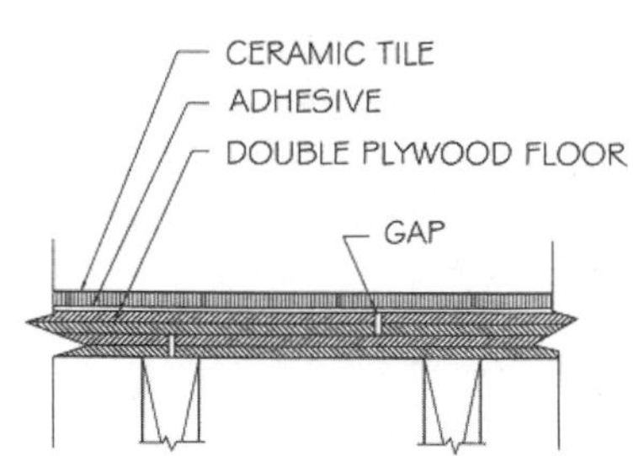

Recommended Uses:
- over wood floors exposed to residential traffic only. For heavier service, select Methods F141, F143, or F144.

Limitations:
- will not withstand high-impact or wheel loads.
- interior dry applications.

Requirements:
- design floor areas over which tile is to be applied to have a deflection not greater than 1/360 of the span when measured under 300 lb. concentrated load (see ASTM C627).
- double plywood floor—clean and free of dirt, dust, and oily film.

Materials:
- organic adhesive—ANSI A136.1 Type 1 floor type.
- grout—ANSI A118.3, A118.6, A118.7. or A118.8.

Preparation by Other Trades:
- subfloor—19/32" tongue and groove plywood with 1/8" gap between sheets or 1" nominal boards when on joists 16" o.c.
- underlayment—19/32" min. exterior glue plywood with 1/8" gap between sheets.
- max. variation in the plywood surface—1/4" in 10'-0" from the required plane. Adjacent edges of plywood sheets—max. 1/32" above or below each other.
- offset end and edge joints of the underlayment panels by at least two inches from the joints of subfloor panels; they should not coincide with framing below.
- underlayment fasteners should not penetrate joists below

Movement Joint (architect must specify type of joint and show location and details on drawings):
- movement joints—mandatory according to Method EJ171, page 68.

Installation Specifications:
- tile—ANSI A108.4.
- grout—polymer modified, ANSI A108.10; epoxy, ANSI A108.6.

FIGURE 34.8 Two commonly used installation methods for setting tile on wood subfloors. F-141 is used for thick-set applications, and F-142 is used for thin-set applications. (Used with permission of Tile Council of America, Inc.)

walls, refrigerated rooms, new tile over existing tile, window stools, and thresholds. It also contains guidelines for installing tile over sound-rated floors. Examples of the installation process are shown in Figures 34.9, 34.10, and 34.11. One of the most challenging installations for tile is in showers. For a closer look at using the thick-set method in a shower, see Principles in Practice.

One of the most overlooked aspects of successful tile installation for large floor areas is the need for movement joints. Movement joints are typically filled with a pedestrian traffic grade urethane or silicone sealant over either a foam backer rod or bond-breaking tape. The following general guidelines are helpful for locating movement joints in a tile installation:

- For interior installations, 20 to 25 ft in each direction when not exposed to direct sunlight and 8 to 12 ft in each direction when exposed. For exterior installations, 8 to 12 ft in each direction.

(a)

(b)

(c)

FIGURE 34.9 Installation of tile using the thin-set method at an exterior building entrance. **(a)** Latex-modified portland cement mortar being spread on a concrete slab using a notched trowel for a thin-set application. **(b)** A tile being set into the mortar bed. The purpose of the notched trowel is so that the proper amount of mortar will be applied—when the tile is pressed down the mortar becomes evenly distributed. **(c)** Grount being applied to the tile joints, usually 24 h after tile has been set.

- When tile adjoins other restraining structural members, such as columns, curbs, walls, and ceilings; also at changes in floor elevation.
- At subfloor construction, contraction, and expansion joints.
- Widths of joints should never be less than the joint in the subfloor below. As a general guideline, joints should be $\frac{3}{8}$ in. when joints are 8 ft on center, $\frac{1}{2}$ in. in southern climates, and $\frac{3}{4}$ in. in northern climates when 12 ft on center. Joint widths should be increased $\frac{1}{16}$ in. for each 15°F increase when the difference between the summer and winter surface temperature is more than 100°. Quarry and paver tile joints should never be less than $\frac{1}{4}$ in., and ceramic mosaic tile and glazed wall tile should never be less than $\frac{1}{8}$ in.

FIGURE 34.10 A closer view of mortar being spread using a notched trowel (in this case an epoxy mortar). (Photo courtesy of Mapei Corp.)

34.4 STONE PANEL FLOORING

Stone panels for flooring, also known as dimension stone, are natural stones that have been selected and fabricated (cut and trimmed) to specific shapes and sizes, with or without mechanical dressing of one or more surfaces.

Many of the physical properties of natural stone were discussed in Chapter 23. It is important to note that the properties of stone critical for selecting stone for exterior cladding may not be important (or as important) for the stone used in floors. For example, weatherability and flexural strength of stone (important for exterior cladding) are less important for flooring, where stone's compressive strength and abrasion resistance are extremely important.

Natural stone used for panel flooring include granite, marble, limestone, slate, and other quartz-based stones such as sandstones, bluestones, and quartzities.

Stone panels are uniformly fabricated on five faces, with the backs being left ungauged (cut in a way that does not allow for uniform thicknesses). The panels are actually slabs of stone and usually have large dimensions in one or both directions. Because the panels are not uniformly thick, they must be installed over a thick-set mortar bed.

FIGURE 34.11 A closer view of setting larger-format tile in mortar. The purpose of the notched trowel is so that when the tile is set and pushed into place, the mortar will redistribute itself into a uniform bed. (Photo courtesy of Mapei Corp.)

Physical Properties of Stone Panels

Most stones are more than adequate for use as a floor covering because they have the ability to resist abrasion, wear, and absorption. When used in exterior applications, damage is possible due to water permeability, inelasticity, or low compressive strength. Some fabrication options, such as microfractures caused by thermal finishing, may render stone unsuitable for exterior applications.

Patterns and Finishes for Stone Panel Flooring

Virtually any stone panel floor pattern can be used, limited only by the shape and size of the individual stone panels that can be fabricated.

Whereas there are more suitable finishes for some stones, such as polished for granite and honed for marble, generally any of the following mechanical finishes can be used with most stones:

- *Polished:* Finished to a reflective sheen and resistant to wear (granites more so than marbles), a polished finish can be scratched and dulled by abrasive materials, such as sand on the shoes of people that walk over it.
- *Honed:* Finished to a uniformly matte sheen, a honed finish can be used to mask wear.
- *Thermal:* Exposure to an open flame, which essentially burns off the immediate surface, leaving a slightly roughened surface, so thermal finishing provides improved slip resistance.

Installation

The usual installation method for stone panel flooring is the thick-bed method, Figure 34.12. Because the backs of stone panels are usually ungauged, the panels are larger than tiles, and there is slight variation in thickness among the panels, the thin-set installation method cannot be used.

The subfloor is not covered with a cleavage membrane when bonding between the subfloor and the stone panel flooring is desired. A setting bed composed of a damp mix of portland cement and sand that is reinforced for unbonded assemblies or unreinforced for bonded assemblies is placed and then leveled. Prior to setting the stone panels, the setting bed is sprinkled with water to start the cement-setting process.

(c) (d)

FIGURE 34.12 **(a)** Natural stone panel flooring being set by the thick-bed method. Once the exact panel locations are determined, several rows are set as the guide for the remainder of the installation. Note the water-sprinkling can for sprinkling the setting bed to start the setting process. Also note this is a bonded installation without reinforcing or cleavage membrane. **(b)** The thick-bed mortar is being readied for the installation of the next stone panel. The rubber mallet is used to adjust the panels. **(c)** Plastic shims are used to keep the joints between the stone panels uniformly the same width. The circular stickers in the corner of the panels are panel numbers that are coordinated to the shop drawings. **(d)** Finished stone flooring after grouting and polishing.

While the setting bed is still plastic, the backs of the stone panels are fully buttered with a bond coat of cementitious materials and then set in place. Care must be taken to maintain uniform joint widths and consistent elevations of the various panels.

After the mortar has cured sufficiently, the joints are grouted. It should be noted that because of the uniformity of panel sizes, the joints of stone panel flooring may be as small as $\frac{1}{8}$ in. in width.

PRACTICE QUIZ

Each question has only one correct answer. Select the choice that best answers the question.

1. The architect should expect the subfloor to sustain some damage during the construction process.
   - **a.** True
   - **b.** False
2. The two most common materials used as a subfloor are
   - **a.** steel decking and wood plank.
   - **b.** concrete and steel.
   - **c.** concrete and wood panel.
   - **d.** wood light frame and plywood panels.
   - **e.** carpet and vinyl flooring.
3. Subfloors should be free of moisture because
   - **a.** moisture can cause wood to swell, damaging floor coverings.
   - **b.** moisture in the subfloor causes water-based adhesives to debond.
   - **c.** moisture can make floor surfaces slippery.
   - **d.** all the above.
   - **e.** (a) and (b).

4. Slip resistance is measured through laboratory tests that determine
   a. the coefficient of resistance.
   b. slip resistance.
   c. emulsification.
   d. the coefficient of friction.
   e. the impact frequency.
5. In applications where hygiene is the most important consideration, a floor covering that is resistant to mold and mildew growth is required. The floor covering should also
   a. provide sanctuary to dust and dander.
   b. resist penetration by dust and bacteria.
   c. have a low impact resistance.
   d. be in a color that camouflages any dirt that may accumulate.
   e. be inexpensive.
6. Stone panel is another name for stone tile.
   a. True
   b. False
7. Ceramic tile is available in many different shapes and
   a. one size.
   b. two sizes.
   c. three sizes.
   d. many sizes.
   e. all sizes.
8. The thick-set method of setting tile is used when
   a. a slope is required to drain water.
   b. thin ceramic tile is used.
   c. irregular tile is used.
   d. (a) and (b).
   e. (a) and (c).
9. Epoxy mortar and epoxy grout are similar in chemical composition.
   a. True
   b. False
10. Stone panel flooring is laid using the
    a. deep-set method.
    b. medium-set method.
    c. thick-set method.
    d. grout mix.
    e. thin-set method.

Answers: 1-a, 2-c, 3-e, 4-d, 5-b, 6-b, 7-d, 8-e, 9-a, 10-c.

34.5 TERRAZZO FLOORING

Terrazzo flooring has much in common with concrete. Like concrete, a terrazzo binding matrix is mixed with several aggregates and placed, wet and plastic, in its final location. Then, unlike concrete, the exposed surface is ground and polished after it cures to expose the binding matrix and aggregates, thus revealing a smooth and colorful finish. No longer limited to utilitarian purposes as in the past, the use of terrazzo has become much more common; it is available in an explosive variety of creative designs, Figure 34.13.

Characteristics of Terrazzo Flooring

The binding matrix is the material that holds the aggregate chips in position. Until about the end of the twentieth century, the most common binding matrix was cementitious, combining portland cement (usually white), aggregate chips, and pigments if needed (either alkali resistant or synthetic powdered, inorganic substances) to form a uniform matrix color. This binding matrix is now being replaced with a resinous epoxy, most commonly polyester, polyacrylate-modified cement, or polyurethane.

The creative design of terrazzo flooring is the primary criterion for selecting aggregate chips. The defining quality for these chips is their ability to be ground and polished. Marble chips are the most common aggregate; however, granite, onyx, travertine, or glass chips are also used, and mother-of-pearl is especially favored. Like other aggregates, chips are graded in increments of $\frac{1}{8}$ in. according to size from Number 0 (passing a $\frac{1}{8}$-in. screen but retained on a $\frac{1}{16}$-in. screen) to Number 8 (passing a $1\frac{1}{8}$-in. screen but retained on a 1-in. screen). Chips are selected based on the relationship between the chip size and the overall thickness of the terrazzo installation. For example, a Number 6 chip (between $\frac{7}{8}$ and $\frac{3}{4}$ in.) would not be used for a $\frac{3}{8}$-in.-thick epoxy terrazzo installation. Venetian toppings use large aggregate chips, usually $\frac{3}{4}$ in.

FIGURE 34.13 The creative use of divider strips and colored terrazzo at the entry of a hospital. (Photo courtesy of HKS Inc.)

Metal divider and control strips are used to control cracking and to create designs. Cementitious terrazzo requires closely spaced strips to control cracking. Resinous terrazzo does not crack, so the strips are used to create decorative designs. Usually made of white zinc alloy, brass, or plastic, strips are available in standard, K and L shapes with top widths of $\frac{1}{8}$, $\frac{1}{4}$, and $\frac{1}{2}$ in.

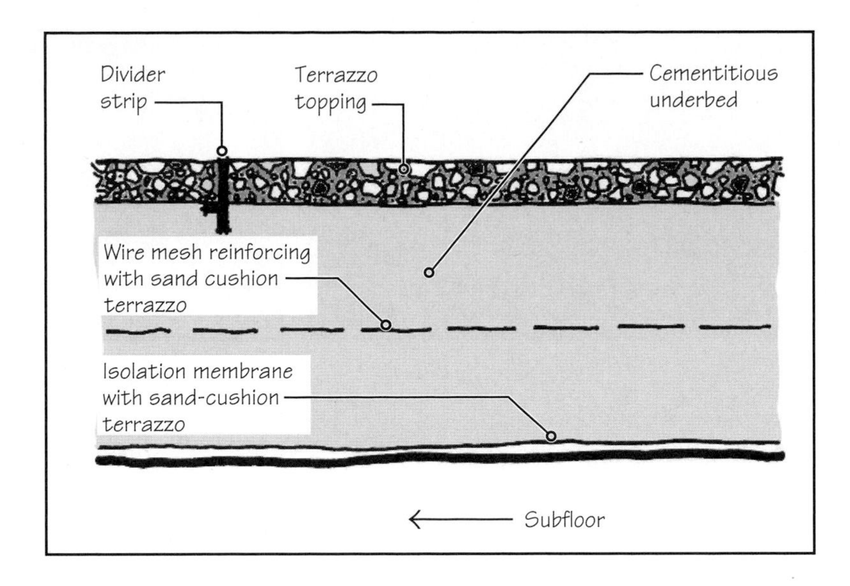

FIGURE 34.14 A section through cementitious terrazzo.

TYPES AND INSTALLATION OF TERRAZZO FLOORING

As with tile, there are industry standards for the various types of terrazzo. These are defined in the National Terrazzo and Mosaic Association's (NTMA) *Terrazzo Ideas and Design Guide.* Terrazzo types are distinguished according to the manner in which they are installed.

- *Cementitious Terrazzo:* This type requires a recessed subfloor of up to 3 in. deep, Figure 34.14.
  - *Sand-Cushion Cementitious Terrazzo:* This type consists of a $\frac{1}{2}$-in.-thick standard terrazzo topping over a $2\frac{1}{2}$-in.-thick reinforced cementitious underbed that is separated from—and not allowed to bond with—the subfloor by a $\frac{1}{4}$-in-thick sand bed. Total thickness ranges from $2\frac{3}{4}$ to 3 in. Shrinkage is relieved by the divider strips, which are required at closely spaced intervals. Substrate cracks do not telegraph through to the topping.
  - *Bonded Cementitious Terrazzo:* This type consists of a $\frac{1}{2}$-in.-thick standard terrazzo over an unreinforced cementitious underbed that is bonded to the subfloor. Total thickness ranges from $1\frac{3}{4}$ to $2\frac{1}{4}$ in. Crack control is very important, and the installation should include divider strips at closely spaced intervals.
  - *Structural Terrazzo:* Combining a standard topping with a concrete slab-on-grade structural slab, this type is used when single responsibility is desired. The total thickness ranges from $4\frac{1}{2}$ to 6 in.
  - *Rustic Terrazzo:* When grinded surfaces will be too smooth, such as in exterior locations, rustic terrazzo is used. Rather than grinding and polishing, rustic terrazzo is washed with water or otherwise treated to expose the aggregate. Variations include unbonded, bonded, monolithic, and structural.
- *Epoxy Terrazzo:* Combining the assets of cementitious systems with the dramatic improvements made in epoxy resins, epoxy terrazzo is lighter and more flexible than cementitious terrazzo, Figure 34.15. It can develop higher bond strengths and is resistant to mild acids, impact indentations, concentrated load impressions, and staining. Epoxy terrazzo requires a thin veneer ($\frac{1}{4}$ to $\frac{3}{8}$ in.), eliminating the need for floor recesses. Unless required for a design, divider strips may be located at larger intervals than with cementitious systems. The time from placement to finish grinding is shorter than with cementitious systems. When combined with crack isolation underlayment, epoxy terrazzo can have the highest resistance to crack telegraphing of any system.

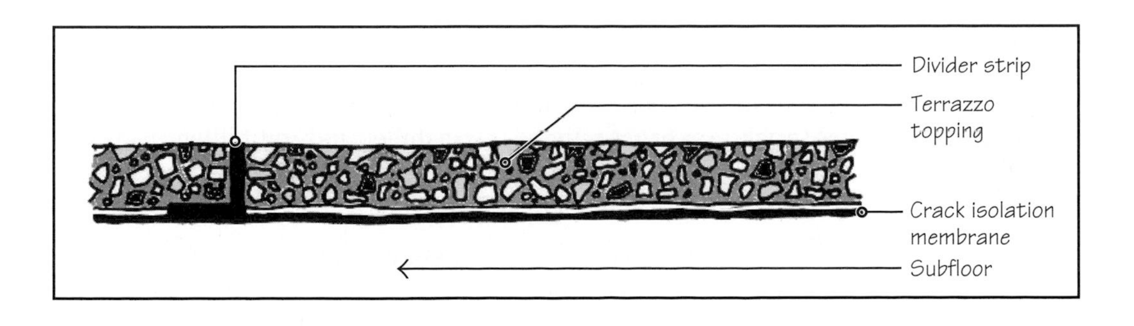

FIGURE 34.15 A section through epoxy terrazzo.

(a)

(b)

(c)

FIGURE 34.16 **(a)** Cracks in concrete being filled with epoxy prior to setting divider strips. **(b)** Divider strips being installed using an adhesive. **(c)** Epoxy terrazzo is ground after it has cured. A dust collection is connected to the grinding machine by a hose significantly reducing airborne dust. Safety precautions are required for breathing.

- *Other Thin-Set Terrazzo Systems:* Used when specific performance is required, these include conductive, polyester-resin, and polyacrylate-modified cement types.
- *Precast Terrazzo:* Terrazzo can be formed at a manufacturing plant into a variety of precast shapes, including floor tiles, sloped shower-stall floors (complete with upturned base and cutout for the plumbing drain), stair treads and risers, and wall base strips.

INSTALLATION

When the subfloor or cementitious underbed has been prepared (depending on the system used), including filling cracks, Figure 34.16(a), divider strips are attached, Figure 34.16(b); then the terrazzo topping is placed in a wet and plastic condition. It is leveled, rolled, compacted, and troweled to a dense, uniform, and flat surface that reveals the divider strips. The topping is allowed to cure until it develops sufficient strength to prevent chips from being lifted or pulled out by the grinding machine. It is then ground with a series of various grit stones, progressing from rough to fine; imperfections are grouted; and, finally, it is polished to the desired sheen, Figure 34.16(c).

34.6 CARPET AND CARPET TILE FLOORING

The most tactily comforting floor covering is carpet. With a centuries-long history of exquisite rugs and carpets, carpet continues to be the covering of choice due to its softness, contribution to quietness, and emotional feeling of comfort. Carpet is also used extensively for rooms where a quiet environment is necessary or desired.

CHARACTERISTICS OF CARPET AND CARPET TILE

All carpet is manufactured in continuous lengths. The essential difference between carpets is in their construction and how they are used in a building. There are two basic types of carpet available: rolled goods and tiles. Broadloom, or rolled goods, dominates the carpet industry (rolls are typically 6 or 12 ft in width and seams are treated so they appear invisible). Carpet tiles, usually used in commercial buildings, have rubber backings and can be easily changed when a tile becomes stained or otherwise damaged, Figure 34.17.

FIGURE 34.17 Carpet tiles of several colors to create a pattern in the waiting room of a hospital. (Photo courtesy of HKS Inc.)

FIBERS AND YARNS

Fibers, the basic component of carpet, are either natural or synthetic. Fibers must be of particular elasticity, fineness, uniformity, durability, weight (called denier, which is a measure of weight to unit length), luster (brightness or reflectivity), cleanability, and soil hiding (ability to hide the presence of soil). There are three fibers that are used to make the yarn that is used to make carpet:

- *Nylon:* Proven to be among the most durable fibers developed, nylon is the most popular fiber commercially and, consequently, dominates the market. Developed as a petrochemical based fiber by DuPont in 1938, nylon can be dyed to achieve bright colors. It is resistant to crushing, has good fade resistance, and can be easily cleaned. One of its weaknesses is that it is prone to static electricity buildup; however, that can be alleviated with antistatic treatments.

- *Polypropylene:* Known by the trade name of Olefin, polypropylene is a by-product of the gasoline-refining process, has superior soil-resistant properties, and has low water-absorption rates. Weaknesses are that colors tend to be dull because it does not accept water-based dyes or stains, has poor resiliency, has a lower melting point, and has poor texture retention.
- *Wool:* Wool is the original carpet fiber and is generally the most expensive. It has an outstanding ability to be dyed in deep, rich colors. It has exceptional memory and will return to its original state, without rupture or tearing, after being stretched, and it can also shed water. Although not commonly used in commercial buildings, wool is especially suited for hospitality and entertainment areas, and it is common in residential applications.

Fibers are grouped together in various ways to make yarn, which is then used to construct the exposed face of carpet, Table 34.3. There are several ways to achieve colors and patterns in carpet and carpet tile. In some cases the fiber is dyed before being made into yarn (synthetic fibers are solution-dyed as the filament is extruded), in some cases the yarn is stock-dyed before spinning or yarn-dyed after spinning, and in other cases the carpet is dyed after it is constructed (similar to printing).

Construction and Physical Properties of Carpet and Carpet Tiles

There are different ways of constructing carpet, each with particular advantages and disadvantages. As the yarns are constructed into a carpet, there are several face constructions that are available, Table 34.4. Carpet construction is detailed, Figure 34.18, in terms of density (weight of the pile in a unit volume), pile height (dimension between the top of the primary backing and the finished top of the yarn), rows or wires (number of pile yarn tufts for each 1 in.), stitches (number of yarn tufts for each 1 in. of length), gauge (number of yarn ends

TABLE 34.3 CARPET CONSTRUCTION TYPES

TUFTED	Accounting for the vast majority of broadloom carpets, this inexpensive and fast construction method involves hundreds of needles on tufting machines, which essentially sew, or punch, multiple rows of yarns into and through a primary backing. The back is coated with latex to stabilize the tufts, and is then backed with a secondary backing to give the completed construction dimensional stability.
FUSION BONDED	Dominating the carpet tile industry, fusion bonding consists of a spun yarn bundle that is heat-fused into a liquid vinyl compound in a sandwich like configuration. A knife then splits the sandwich into two carpets; thus cut pile is the only face construction that can be produced.
NEEDLE-PUNCHED	Hundreds of barbed needles punch individual loose synthetic fibers through a woven blanket of synthetic fibers that forms a dense, homogeneous sheet without pile, similar to a heavy felt.
KNITTED	Similar to textile knitting, various yarns are knitted into connected loops in differing directions to accomplish a sheet that does not require backings.
WOVEN	Historically, carpet and rugs were woven by hand on looms. Today, machines weave pile yarns and backing yarns simultaneously. Although requiring more time than the tufting process, woven carpet is distinguished by its tailored textures and intricate patterns. Two of the most important assets of woven carpets are that they are more dimensionally stable and more resistant to wear than other carpet constructions.

TABLE 34.4 CARPET FACE CONSTRUCTIONS

	CUT PILE Tops of the yarn loops are cut to reveal the ends of fibers, creating a plush appearance.
	TEXTURED SAXONY Tops of yarns are evenly cut across the top to reveal more tip definition.
	FRIEZE Yarn is tightly twisted to display a curled, or kinked, effect.
	TIP-SHEARED Tops of tufted loops are shaved to create a cut or uncut texture or pattern.
	RANDOM-SHEARED Similar to multilevel loop, but the higher loops are sheared off to reveal a combination cut and uncut texture.
	MULTI-LEVEL LOOP Yarns that are looped at varying heights to create patterns or effects.
	LEVEL LOOP Tops of the yarn loops are level with each other.

NOTE

Critical radiant flux refers to the rate of flow of radiant energy through a material. In the case of resilient flooring, it is measured in terms of watts per square centimeter (W/cm^2) because it is a direct reflection of resistance to the spread of fire in this type of material. Class I materials (the most fire resistant) have a 0.45 W/cm^2 CRF, and Class II materials are 0.22 W/cm^2.

counted across a specific dimension of the width), pitch (number of surface yarn ends in 27 in. of width), total weight, face weight, tuft density (total tufts for each 1 in.), and yarn count. Each of these properties affects the look, feel, and performance of a given carpet.

Carpets are also evaluated for flammability, as regulated by building codes. Flame spread is tested by the flooring radiant panel test, which determines a carpet's ability to spread fire; this is identified as critical radiant flux (CRF). Another measure of the ability of a carpet to promulgate a flame from a small source, such as a match or cigarette, is determined from the pass-or-fail methenamine pill test, where an ignited tablet is dropped on the carpet and flammability is then measured.

In addition, the performance of carpets can be enhanced by treating them to resist staining, static electricity, and bacteria growth.

Carpet Cushion

Some, but certainly not all, broadloom or rolled goods carpets require a cushion between the subfloor and the carpet to achieve its desired performance. Cushions tend to be used in hospitality areas, hotel rooms, and most residential buildings. Carpet cushions are not required for carpet tiles. Several types of carpet cushions are commonly available, including fiber, sponge rubber, and polyurethane foam.

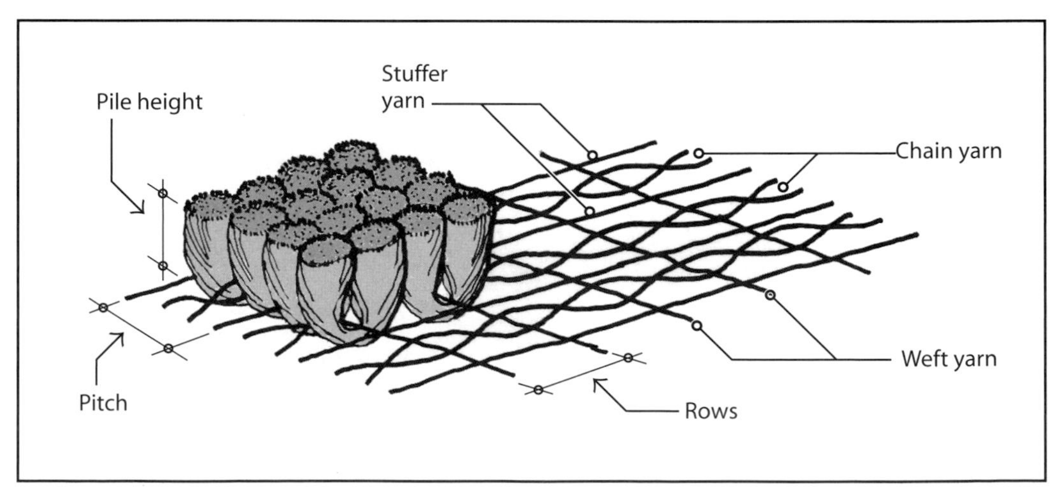

FIGURE 34.18 General characteristics of knitted and woven carpets.

Installation

There are various methods for installing carpet and carpet cushion, and the method depends on the carpet type, the end use of the areas, and the type of subfloor.

- *Stretch-In:* Used for broadloom carpets, the carpet is stretched into tension and then securely hooked to tackless nail strips at the perimeter of the room. This method is generally limited to those areas that experience low traffic, and it requires an independent carpet cushion. Replacement of worn or outdated carpet can be easily accomplished. If used in heavy or extra-heavy traffic areas, the carpet and the cushion will tend to "travel" in the direction of the dominant foot traffic and will become wrinkled or detached from the nail strips, and the cushion may bunch up, or buckle, underneath the carpet.
- *Direct Glue-Down:* Used for broadloom carpet and carpet tile, a compatible adhesive is used to attach the carpet to the subfloor; it will prevent buckling under traffic load and will also resist carpet movement due to rolling loads as well as changes due to temperature and humidity.
- *Double Glue-Down:* This is essentially the same as the direct-glue method; however, the carpet cushion is also glued to the subfloor.
- *Other Methods:* While not frequently used, there are several other methods available, including the hook-and-loop method and a preapplied-adhesive system.

34.7 WOOD FLOORING

Wood and manufactured wood products were discussed in Chapters 11 and 12, and much of the same information also applies to wood flooring. Wood flooring is valued in residential applications for the warmth and beauty it brings to a room. It is also used extensively for athletic floors, such as basketball, volleyball, and racquetball courts, as well as for theater and dance floors, where durability, hardness, and resiliency are highly desired. Although most wood species can be fabricated for wood flooring, all are not equal in performance.

Characteristics of Wood Flooring

Virtually any hardwood or softwood species that can be domestically harvested or imported and provided to a fabricator can be used for wood flooring. Secondary to availability, any wood that can be finished suitably for the application can be used for wood flooring. The characteristics of most importance for evaluating wood flooring are species, appearance (including grain, texture, and color), cut, durability, dimensional stability, and ability to accept finishes.

According to manufacturers, oak and maple are the most commonly used species; however, there are currently as many as 50 domestic and exotic species available, including the following:

Ash: White
Beech: American and red
Birch: Yellow
Cherry: African, black, and Brazilian
Hickory
Maple: Black and hard
Oak: Northern red, red Southern, and white
Pecan
Pine: Eastern white, Southern yellow, and heart
Walnut: African, black, Brazilian, and Peruvian

Generally, wood floorings are classified as either solid wood flooring or engineered wood flooring. As the name implies, in solid wood flooring the same wood species is used throughout the entire piece. It is susceptible to changes in size due to moisture content. It can be plain sawn, quartersawn, or rift sawn and is available in different appearance grades. Like plywood, engineered wood flooring is a combination of a surface veneer, usually a hardwood, which is laminated to one or more plies of a wood veneer from a less expensive wood species that provides dimensional stability and added strength.

Either material can be fabricated into a variety of shapes and sizes. They are categorized into three groups: strips (usually $1\frac{1}{2}$ to $2\frac{1}{4}$ in. in width and random in length), planks (usually 3 to 8 in. in width and random in length), and parquet (a patterned floor, parquet can be strips or planks or can be blocks or square tiles).

The typical profile for wood flooring is known as a tongue-and-groove profile, Figure 34.19. This allows for accurate alignment during installation and—because of its uniformity—tight

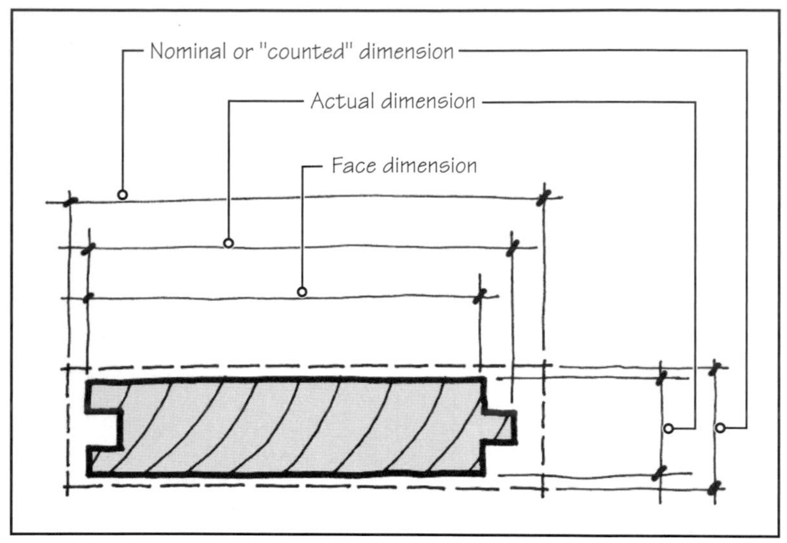

FIGURE 34.19 Solid wood flooring cross-sectional dimension terminology.

and level joints. The thickness of solid wood flooring varies according to usage: $\frac{5}{16}$, $\frac{3}{8}$, $\frac{1}{2}$, and $\frac{5}{8}$ in. thick for light use; $\frac{3}{4}$ and $\frac{25}{32}$ in. thick for normal use; and $\frac{33}{32}$, $1\frac{1}{4}$ and $1\frac{1}{2}$ in. thick for heavy use.

Wood floorings are finished to protect the wood from dirt, abrasion, wear, oxidation, and moisture (to some extent). They can be finished by the manufacturer, resulting in higher-performing finishes than can be achieved when finished in place. When finished in place, the floor is sanded, stained, and waxed. Wood flooring can be factory impregnated with an acrylic, which increases its density, hardness, and resistance to wear. Because the acrylic seals the wood from moisture, the wood's dimensional stability is greatly improved.

Physical Properties of Wood Flooring

The two most important physical properties to consider for wood flooring are density and hardness. A wood species with a high density is required for wood flooring. Hardness is necessary to resist indentation and marring. The side hardness of the wood species range from 380 ft-lb for eastern white pine (softest) to 3,680 ft-lb for Brazilian walnut (hardest).

The species of the wood used and its grade and cut are important considerations that affect the dimensional stability, durability, and appearance of wood flooring.

Wood flooring used for athletic purposes should possess the physical properties of absorption of impact shock, ball rebound, deflection resistance, and control of surface friction. Hard maple is favored because it meets these requirements well.

Installation

Unlike the other floor coverings discussed in this chapter, controlling the moisture content of wood flooring during fabrication and installation is imperative. As wood absorbs and releases moisture (each wood species and cut has unique rates), a hygroscopic material, it changes dimensionally by swelling and shrinking. Controlling moisture content is more important for solid wood flooring than for engineered wood flooring. To control moisture, wood is kiln dried to establish a baseline of moisture content; then during fabrication and installation it should be protected from changes in moisture. It is common to place the wood flooring in the place it will be installed to allow it to acclimate to the temperature and relative humidity.

Another unique factor for wood flooring is that an expansion space is necessary around the perimeter to accommodate the small amount of movement the flooring may encounter due to changes in temperature and relative humidity during the life of the installation.

There are several methods of installation for those wood floorings not installed for athletic purposes. Historically, strips and planks were installed over wood sleepers. Today, strips, planks, and parquet can be installed over a plywood subbase, Figure 34.20, or directly to the subfloor. Nails, staples, and water-based adhesives are used to attach to wood subfloors, whereas adhesives are required for concrete.

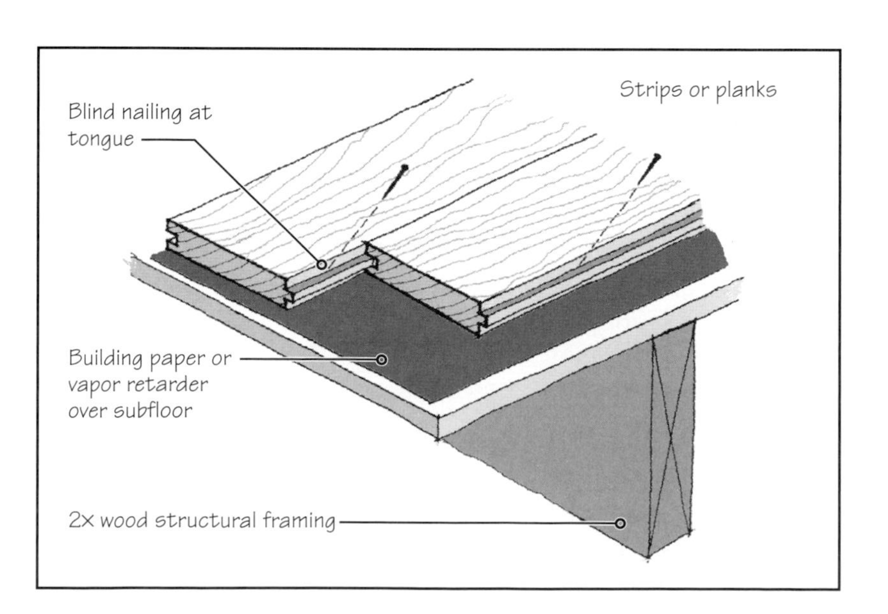

FIGURE 34.20 Installation of wood strip or plank flooring over a plywood subfloor.

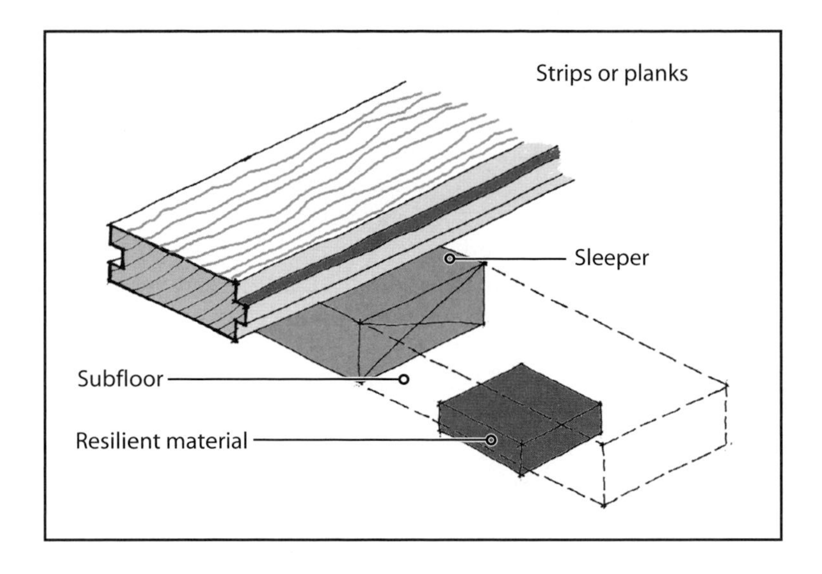

FIGURE 34.21 Basic athletic floating-wood-flooring components.

A more sophisticated method of installation, in which an air space is established between the wood flooring and the substrate, is used for athletic wood flooring. Usually a vapor retarding membrane is placed over the concrete substrate; then one of a variety of support systems may be used:

- *Floating Systems:* Resilient materials such as neoprene, rubber, or polyvinyl chloride sheets may be adhered to the substrate or a foam underlayment to isolate the wood flooring from the subfloor, Figure 34.21.
- *Fixed-Wood-Sleeper Systems:* One or two layers of wood strips, at opposing directions, are attached to the subfloor.
- *Fixed-Metal Sleeper Systems:* Like the wood sleeper system, metal tubes, channels, or other shapes are used to create a grid on the subfloor.

PRACTICE QUIZ

Each question has only one correct answer. Select the choice that best answers the question.

11. Terrazzo flooring is like concrete in that it
a. is composed of a binding agent and aggregate.
b. is liquid when placed and dries hard.
c. is sawcut to control cracking.
d. all the above.
e. (a) and (b).

12. Cementitions terrazzo floor coverings are approximately $\frac{1}{4}$ in. thick.
a. True
b. False

13. Carpet is a relatively new flooring material.
a. True
b. False

14. The major distinguishing physical characteristic that differentiates carpets is
a. color.
b. fabric.
c. construction.
d. size.
e. cost.

15. The desired carpet color is achieved by
a. dying the fiber.
b. dying the yarn.
c. dying the carpet after it is constructed.
d. all the above.
e. none of the above.

16. In the face construction of cut pile carpet,
a. the yarn is twisted, curled, and kinked.
b. a multilevel effect is created by shearing the yarn.
c. the yarn is looped at varying heights.
d. the tops of yarn loops are cut to reveal the tips of the fabers.

17. The majority of broadloom carpets are
a. fusion bonded.
b. woven.
c. tufted.
d. knitted.
e. needle punched.

18. All carpet installations include a cushion between the subfloor and the carpet.
a. True
b. False

19. The standard size for wood floor pieces is
a. strips $1\frac{1}{2}$ to $2\frac{1}{4}$ in. wide.
b. planks 3 to 8 in. wide.
c. parquet blocks or tiles.
d. no standard size for wood floor.

20. The highest-performing floor finishes are
a. waxed in place.
b. acrylic impregnated.
c. finished in place with water-based sealant.
d. applied by the manufacturer prior to placing the wood.

21. It is essential to control the moisture content of wood flooring, especially during installation and fabrication.
a. True
b. False

22. Wood flooring is installed using
a. adhesives.
b. nails.
c. screws.
d. all the above.
e. none of the above.

Answers: 11-e, 12-b, 13-b, 14-c, 15-d, 16-d, 17-c, 18-b, 19-d, 20-d, 21-a, 22-d.

34.8 RESILIENT FLOORING

One of the two most commonly used floor coverings is resilient flooring (the other being carpet). Resilient flooring is characterized as those products that spring back into shape after being compressed as a result of impacts due to walking or other activities normally found in buildings. It can be easily cleaned without special knowledge or specialized products, and it can withstand general use without permanent deformations or damage. Resilient flooring is typically used for more utilitarian purposes in rooms that require regular wet cleaning and where there is a probability water will be on the floor.

Forms of Resilient Flooring

There are two forms of resilient flooring: tiles and sheets. Resilient tiles are uniformly sized after manufacturing, most commonly 12 in. square; however, other sizes and shapes are available. Tiles are usually the most economical and are used when joints are not objectionable or there is a desire to create a particular pattern with different color tiles. Resilient sheets are available in rolls of continuous length, which are typically 6 or 12 ft in width. Sheets are used when there is either a desire to cover as much of the floor as possible without joints or there is a desire to create a larger pattern that cannot be accomplished by tiles.

Physical Properties of Resilient Flooring

As with the floor covering materials discussed previously, the physical properties that must be considered when selecting a covering for a particular use include the following:

- *Appearance:* Among all building materials and products available, few exceed the vast variety of colors, patterns, textures, styles, and designs that are commonly available. Common appearances include the traditional speckled pattern, solid colors in earth tones and bright colors, natural stone imitations, and transparent finished woods.
- *Chemical Resistance:* Many resilient flooring products are resistant to such chemicals as alkalis, alcohols, oils, grease, and aliphatic hydrocarbons; however, others may cause permanent damage, such as ketones, esters, and chlorinated and aromatic hydrocarbons.
- *Static and Dynamic Load Resistance:* Permanent indention of the surface can result from furniture and equipment loading if the appropriate floor covering is not selected.
- *Hardness:* Some resilient floorings are measured for their hardness or softness. Soft materials may be required in some uses to resist cuts, punctures, tears, or heavy traffic. In other instances, the materials may need to be hard to endure its use.
- *Cigarette Burn Resistance:* Resilient floorings have varying degrees of resistance to damage due to a dropped, lighted cigarette.
- *Light Stability:* Exterior application of resilient flooring is usually not recommended by manufacturers because ultraviolet light can cause fading, shrinking, and blistering. When used in rooms flooded with exterior light, degradation potential should be evaluated.
- *Flammability:* Resilient floorings are tested for critical radiant flux (CRF), a measure of the ability to resist the spread of fire.
- *Electrostatic Discharge Resistance:* Reserved for when static control is necessary, product types include static dissipative, static generation resistant, and static decay resistant.
- *Nonasbestos Content:* Unlike the past, resilient floor covering materials are no longer formulated with asbestos.

NOTE

Fire-rated rooms and corridors may restrict the use of resilient floor coverings based on CRF (class).

Solid Vinyl Tiles

Solid vinyl tiles are composed of three primary ingredients—pigments, fillers, a vinyl chloride polymer or copolymer binder—and other modifying resins, plasticizers, and stabilizers. The thickness is approximately $\frac{3}{32}$ in., and sizes range between 12 to 36 in. square. Solid vinyl is also available in planks that replicate the appearance of transparent finished wood and are 36 in. in length and 3 to 6 in. in width. Tile classification includes monolithic, surface-decorated, and printed film (all smooth or embossed surface).

Vinyl Composition Tiles (VCT)

Widely popular and similar to solid vinyl tiles, vinyl composition tiles are a less expensive version, Figure 34.22. Thickness is usually $\frac{1}{8}$ in. However, $\frac{3}{32}$ in. is also available, and the size is usually 12 in. square. Classifications include solid color, through-pattern, and surface-pattern tiles.

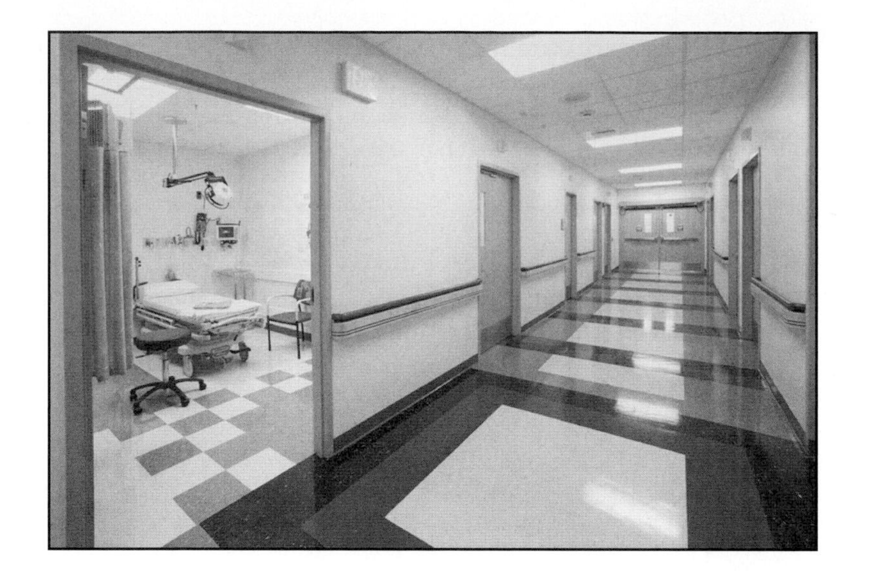

FIGURE 34.22 A multicolored pattern of vinyl composition tiles installed in a hospital. (Photo courtesy of HKS Inc.)

Rubber Tiles

Rubber tiles are manufactured of a vulcanized compound of natural or synthetic rubber (no minimum content required) along with pigments, fillers, and plasticizers. Thicknesses range from $\frac{3}{16}$ to $\frac{5}{16}$ in., and sizes range between 12 to 24 in. square. Rubber tiles are commonly used for stair treads and riser covers. Tile classifications include homogeneous (solid color and mottled) and laminated (solid color and mottled wear-layer). Recycled-rubber material content is very popular and is formed into mats with thicknesses ranging from 1 to $2\frac{1}{2}$ in.

Sheet Vinyl

Similar to solid vinyl and vinyl composition tiles, sheet vinyl is ideally suited for those installations that are regularly or frequently cleaned with water and cleaners, Figure 34.23. It provides minimal joints and the sheets can be bent up the wall to form an integral base, known as a flash cove base. Additionally, the joints may be sealed using a vinyl rod of the same color and pattern as the sheet vinyl. The joint is either chemically bonded or welded with hot air, making the joint watertight and giving a seamless appearance. The wear layer can be clear, translucent, or opaque, it can have a background that is either printed or otherwise prepared, and the surface can be smooth, embossed, or textured. Thickness is approximately $\frac{3}{32}$ in., and widths are 6 ft 0 in., 6 ft 7 in., and 12 ft 0 in. Sheet classifications include backed (nonasbestos fibrous formulations, nonfoamed plastic, and foamed plastic, each graded according to application) and unbacked.

Linoleum Tiles and Sheets

Linoleum is the original resilient sheet after which all other products have patterned themselves. Invented in England in 1860, linoleum consists of a binding agent (a mixture of linseed oil combined with either fossil, pine, or other resins or rosins or an equivalent oxidized

NOTE

Resin is a naturally occurring, sticky, organic substance that is discharged from pine, fir, or other plants. Rosin is a hard, amber-colored by-product of the distillation of turpentine or naptha extract from pine trees. The "resin" in resinous flooring is a liquid synthetic polymer or epoxy that forms into a sheet as it cures.

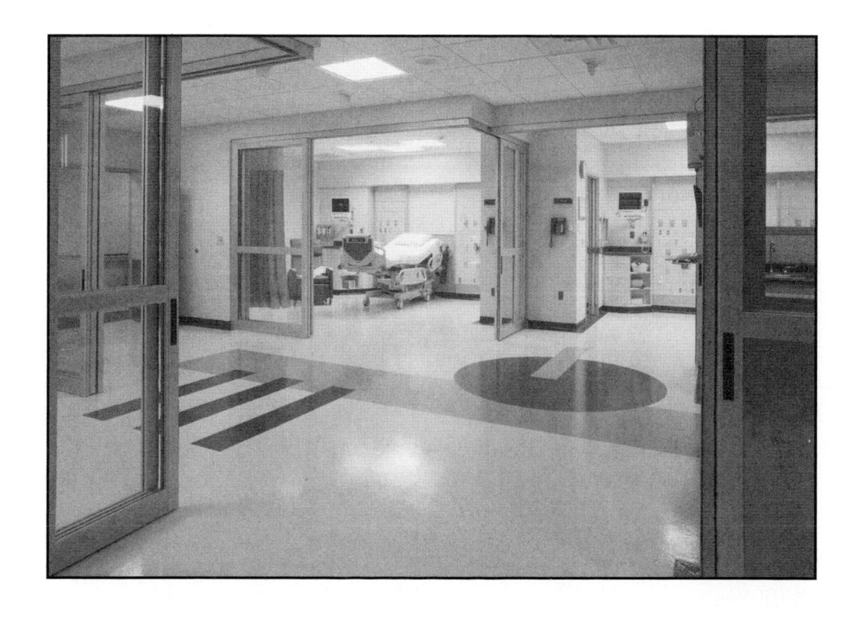

FIGURE 34.23 A creative design in sheet vinyl installed in a hospital. (Photo courtesy of HKS Inc.)

oleoresinous material) and filler (ground cork, wood flour, mineral fillers, and pigments). This mixture is then solidified, bonded, and keyed to a fibrous backing such as burlap (jute). Joints of sheets and tiles can be made watertight by sealing them with heat-welded rods of linoleum. Thicknesses range from $\frac{3}{32}$ to $\frac{1}{8}$ in., and sheets are as wide as 6 ft 7 in.

CORK TILES

The bark of cork trees and recycled cork are ground into small granules to form a synthetic resin matrix, which, when formed into a tile, becomes a flexible product that is homogenous and uniform in composition. Product options include solid and homogeneous tiles and engineered tiles—where a patterned cork veneer is laminated to a cork base and combined with rubber (30% by volume) to form cork rubber or laminated around a medium-density fiberboard core to form planking with tongue-and-groove edges (that can be constructed into a floating floor). The surface can be sanded, left unsanded, waxed, stained, coated with polyurethane or acrylic, or finished in the factory or in the building with other finishes. Thicknesses range from $\frac{3}{16}$ to $\frac{5}{16}$ in., and sizes range between 12 to 24 in. square.

STATIC-CONTROL RESILIENT FLOORING

There are instances where floor coverings are required to control electrostatic discharge (ESD), such as in sensitive manufacturing environments, computer and electronics rooms, and explosive environments. A static-control floor covering system is a vital component to resisting electrostatic charges generated by people, furniture casters, and equipment movements. Essentially, static-control resilient flooring directs an electrical charge to a reliable grounding source. These systems are composed of floor-covering products (like those described earlier) with conductivity elements within the body of the material and require the use of a static-control adhesive and grounding strips.

INSTALLATION

Preparation of the subfloor is critically important for a successful installation of resilient flooring, and smoothness is the most essential factor. Ridges, high spots, and other imperfections extending above the surrounding surface should be removed. Depressions, scratches, gouges, and other imperfections extending below the surrounding surface should be filled. Latex-modified portland cement based products can be applied with a trowel to level and patch imperfections. In some instances, a self-leveling product may be necessary for an extensive area.

Once the imperfections are corrected and the subfloor is free of dust and debris, the resilient flooring is applied by a full-spreading of adhesive using a notched trowel. There are a variety of adhesives available, and the adhesive used should be appropriate for the floor covering product and the subfloor material.

The layout of the floor covering should be considered in order to achieve the desired design expression and joint pattern and to minimize waste. Consideration should be given to alignment with the principal wall while discounting minor offsets. Floor coverings should be scribed and cut to fit around, and butt tightly to, vertical surfaces and permanent fixtures, furniture, and cabinetry. Cuts should be neat and straight, and joints should be tight.

34.9 RESINOUS FLOORING

In recent years, the dramatic advancement in chemical technology has led to improvements in liquid-applied resinous flooring. Resinous flooring, also identified as polymer flooring or epoxy flooring, is applied in its liquid form; when cured, it provides a flexible, seamless, and uniform surface (applied as a liquid that cures into a sheet). The flooring is thin and has excellent bonding capability, mechanical strength, and abrasion and impact resistance. Resinous flooring systems fall into the following categories:

- *Decorative Systems:* Decorative aggregates (ceramic-coated silica, known as colored quartz, marble, granite, dyed stone, vinyl flakes) are combined in an epoxy resin matrix. Resinous (epoxy) terrazzo is actually resinous flooring.
- *General Commercial and Industrial Systems:* Similar to decorative systems, these are usually thicker and have higher performance capabilities.
- *High-Performance or Special-Use Systems:* These systems are formulated for specific environmental exposures or for specific purposes such as resistance to severe and corrosive

FIGURE 34.24 A clear coating applied to exposed concrete flooring. (Used with permission of RetroPlate System; photo by Rhonda Clinton.)

chemicals, acids, and solvents; resistance to extreme cyclic changes in temperature, or for static-dissipative or conductive purposes. Epoxy, epoxy-novolac, urethane, and vinyl-ester systems are the primary systems available.

34.10 OTHER FLOOR-COVERING MATERIALS

In addition to the floor-covering materials discussed before, there are other materials that are commonly available and can be used:

- *Brick:* Full-size brick or split-face brick can be used for interior and exterior floor coverings. They can be thin-set or thick-set mortared for interior floors, with or without grouted joints, generally following the same methods as used for ceramic and stone tile flooring or stone panel flooring.
- *Concrete Staining:* A popular finish is to apply an acid stain to concrete, which penetrates down into the concrete surface and reduces visibility to wear. The finish has a mottled appearance and can be left as is or can be covered with a clear sealing product.
- *Clear Finished Concrete:* Another popular finish is the application of a clear or tinted sealing product that can be polished to a high gloss finish, Figure 34.24.
- *Specialty Floor Covering:* There are many floor-covering products that can be used for specialty purposes. These include rubber mats around swimming pools and in athletic locker rooms and elevated access flooring used in computer equipment and television broadcast rooms where there is a considerable amount of wiring on the floor. Also, increasingly popular in large office buildings is elevated access flooring, which is combined with the heating-cooling systems to distribute air below the flooring instead of through overhead ducts.

34.11 UNDERLAYMENTS

An underlayment is a thin material that is adhered or applied to the subfloor prior to installing the final floor covering. It either provides protection or prepares the subfloor to receive the flooring. There are three types of underlayments: membrane, fill, and solid.

- *Membrane Underlayments:* There are three performance functions that can be accomplished by membrane underlayments. Some products may simultaneously provide two or all three of these functions.
  - *Waterproofing:* These products are single or multiple components (some contain internal reinforcing fabrics) that are dimensionally stable, load bearing, and resistant to mold growth and possess strength in breaking (at seams and in shear). Product types include liquid-applied components (which cure into a continuous membrane), plastic sheets (polyvinyl chloride, chlorinated polyethylene, and polyethylene (applied with mortar), Figure 34.25, and self-adhering bituminous sheets.
  - *Crack Isolation (also known as crack suppression):* Because subfloor structures expand, contract, and deflect ever so slightly over the course of the life of a building, it is sometimes necessary to provide crack-isolation membranes. They are

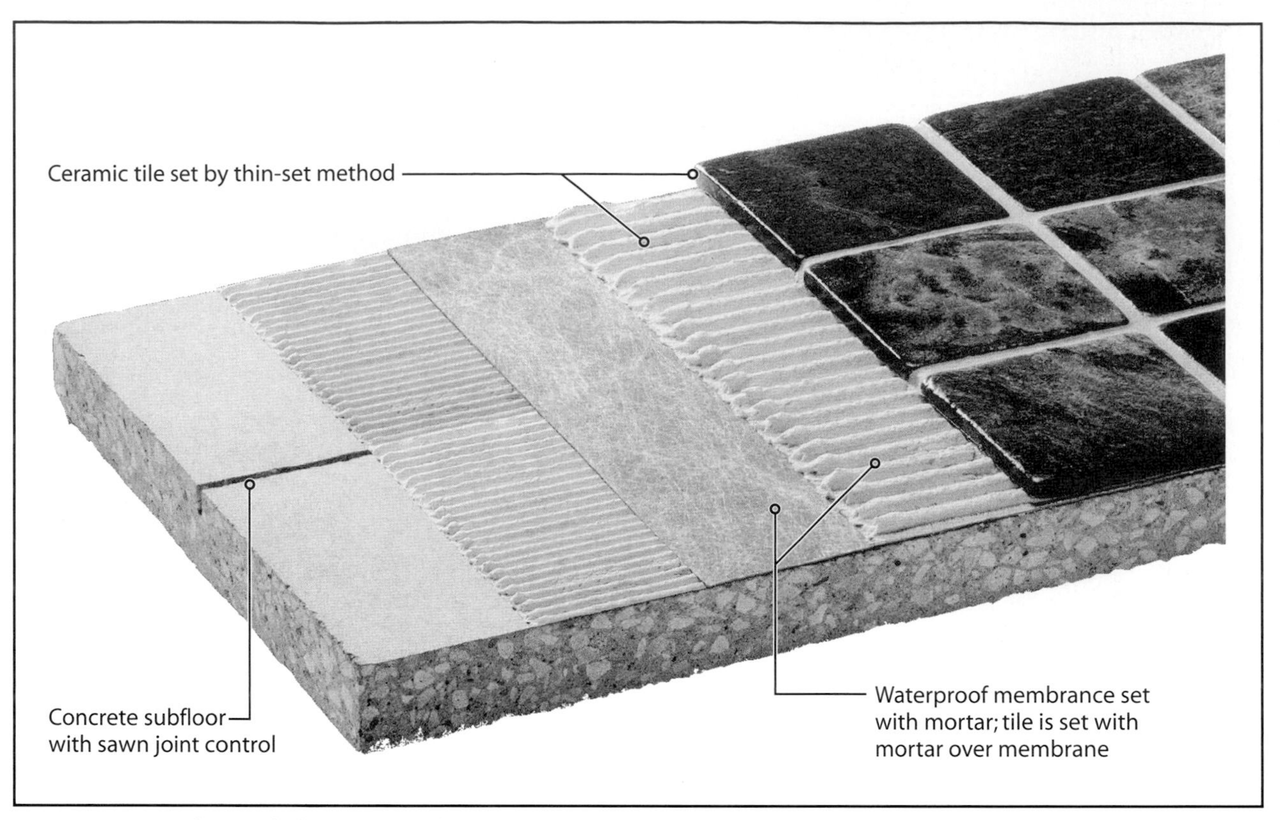

FIGURE 34.25 A waterproofing underlayment in a thin-set ceramic tile assembly. (Used with permission of Noble Company.)

designed to either eliminate the transfer of cracks or greatly reduce the likelihood of cracks. They are fungus and microorganism resistant and have good shear strength, good point-load resistance, and crack resistance. It should be noted that crack-isolation membranes are not a substitute for properly located and designed expansion and control joints in the building structure.

- *Sound Reduction:* These membranes mitigate, or reduce, the noise between floors, such as between two stacked residential units. It is common for building codes to require certain performance levels.

- *Fill Underlayments:* As discussed earlier, it is important that the subfloor be smooth before a floor covering is installed. Imperfections extending below the surrounding surface are filled with underlayments that can be troweled smooth and feather-edged for a smooth transition. They are also used for leveling or resurfacing a subfloor. If the surface of the subfloor has to be raised, the self-leveling versions can be used. Fill underlayments are not intended as wear surfaces. They are typically hydraulic cement based (high compressive strengths and used for commercial applications) and gypsum based (low compressive strengths and used for residential applications) materials.
- *Solid Underlayments:* When the elevation of the subfloor has to be raised slightly or the subfloor is significantly damaged, solid underlayments may be used. Common materials used include medium-density fiberboard, cement fiberboards, and cement backer units.

34.12 RESILIENT ACCESSORIES—WALL BASE AND MOLDINGS

Just as woodwork is trimmed with narrow strips of wood to close gaps between surfaces, narrow strips of resilient materials are used to trim around flooring installations and to cover small gaps, Table 34.5. Resilient accessories are typically manufactured of polyvinyl chloride (PVC). This thermoplastic material can be softened by heating, shaped to fit, and hardened. It retains its shape when cooled. It can also be fused together. Rubber is a thermoset material that has a memory and cannot be shaped.

Wall base, as the name indicates, is the narrow strip of material that is installed at the base of the wall adjoining the floor covering. If manufactured of PVC, it is typically available in long rolls, which allow long installations without joints. If manufactured of rubber, it is typically available in 48-in.-long strips. Premolded interior and exterior corners are also available. There are a variety of heights available in three styles.

TABLE 34.5 TYPICAL RESILIENT WALL BASE AND MOLDING PROFILES.

WALL BASE Left: Toeless, or straight, base Middle: Cove base Right: Trimmed base (where floor covering extends up to, but not under base) Typical heights are $2\frac{1}{2}$, 4, $4\frac{1}{2}$, and 6 in.
STAIR TREADS AND RISER COVERS Left: Riser cover Middle: Tread cover Right: Alternate tread nosing Options include an abrasive strip just behind the nosing (shown); surface patterns on treads (smooth, dots, grooved diamond, longitudinal grooves), glow-in-the-dark colored nosings
STAIR NOSINGS Options include surface patterns on tread surface abrasive strip, longitudinal grooves; glow-in-the-dark colors
CARPET-TO-CARPET MOLDING Resilient exposed trim that snaps into an aluminum retainer; options include transitions into other floor-covering materials
TRANSITION MOLDINGS

Moldings cover small gaps between floor coverings as well as transitions from one covering to another. They are available in a multitude of shapes and profiles to accommodate the many situations that occur.

PRACTICE QUIZ

Each question has only one correct answer. Select the choice that best answers the question.

23. Resilient flooring is available in
- **a.** blankets.
- **b.** sheets.
- **c.** tiles.
- **d.** panels.
- **e.** (b) and (c).

24. Floor coverings that return to their original shape after normal impacts are called
- **a.** ceramic tile.
- **b.** stone panels.
- **c.** terrazzo.
- **d.** carpet.
- **e.** resilient flooring.

25. All vinyl tile is available only in. 12-in. × 12-in. squares.
- **a.** True
- **b.** False

26. The binding agent in linoleum is.
- **a.** portland cement.
- **b.** PVC (polyvinyl chloride).
- **c.** linseed oil and organic resins.
- **d.** epoxy.

27. Electrostatic discharges generated by people, furniture casters, and the movement of equipment can be hazardous in
- **a.** rooms or buildings that house computers.
- **b.** some manufacturing facilities.
- **c.** storage rooms for explosives.
- **d.** all the above.
- **e.** none of the above.

28. Resinous flooring is a site-applied, synthetic seamless sheet material.
- **a.** True
- **b.** False

29. The purpose of a membrane underlayment is to provide
- **a.** resistance to cracking.
- **b.** mitigation of the transmission of sound.
- **c.** waterproofing.
- **d.** any of the above.
- **e.** none of the above.

30. A fill underlayment is a solid, hard material that is nailed directly to the subfloor.
- **a.** True
- **b.** False

31. Most floor coverings require moldings or trim where the flooring meets the wall or one type of flooring meets a different type of flooring.
- **a.** True
- **b.** False

Answers: 23-e, 24-e, 25-b, 26-c, 27-d, 28-a, 29-d, 30-b, 31-a.

PRINCIPLES IN PRACTICE

Showers and Tile

Setting tile in a shower is one of the most challenging installations in the tile-setting industry. The challenge is to form the necessary surfaces on which the tile is set, to set multiple tile shapes into a consistently smooth surface with uniform joint spacing, to ensure that water will drain off the walls and floor as well as not leak through the installation, and, finally, to have it look good, Figure 1.

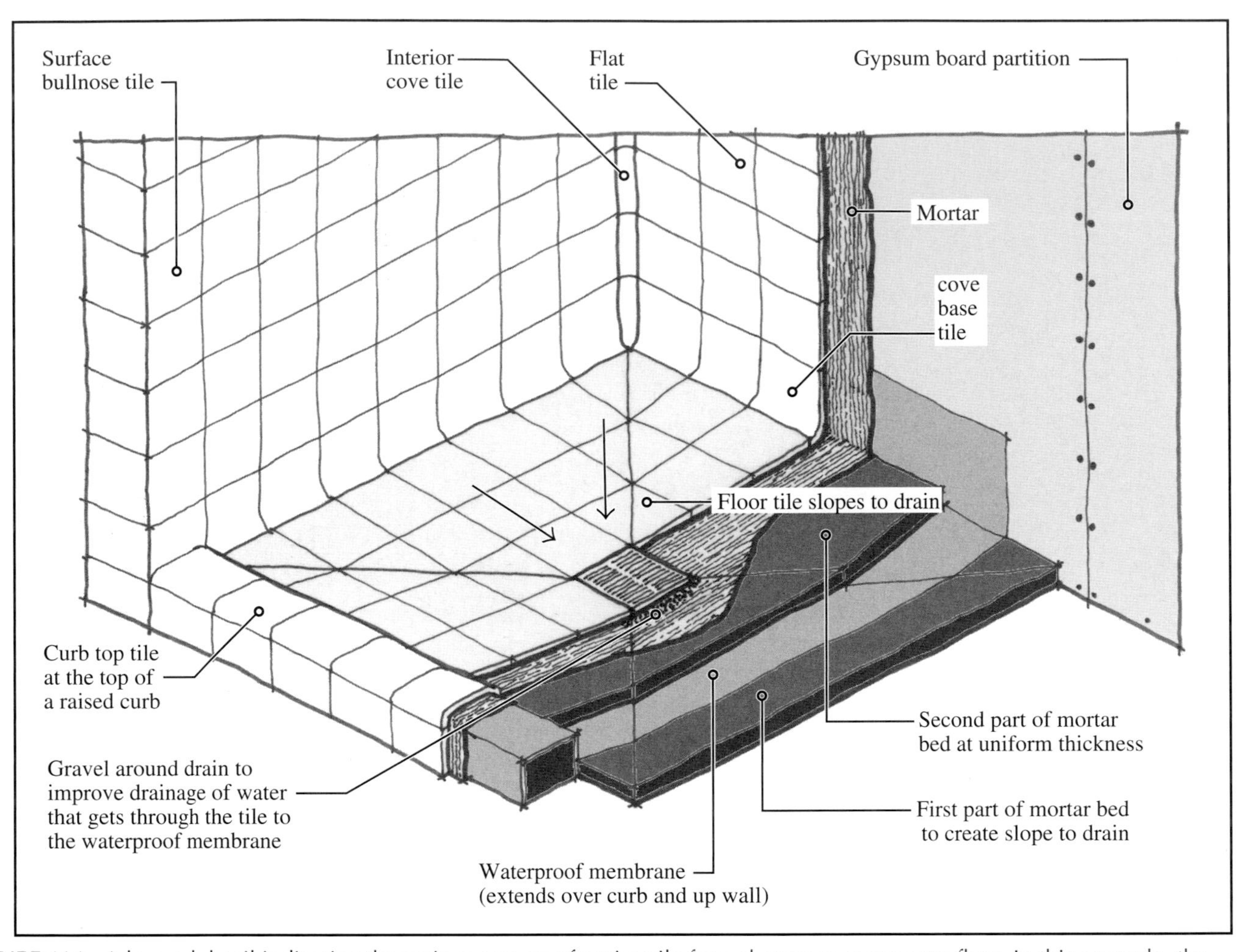

FIGURE 1(a) A layered detail indicating the various aspects of setting tile for a shower on a concrete floor. In this example, the waterproof membrane is placed within the mortar bed.

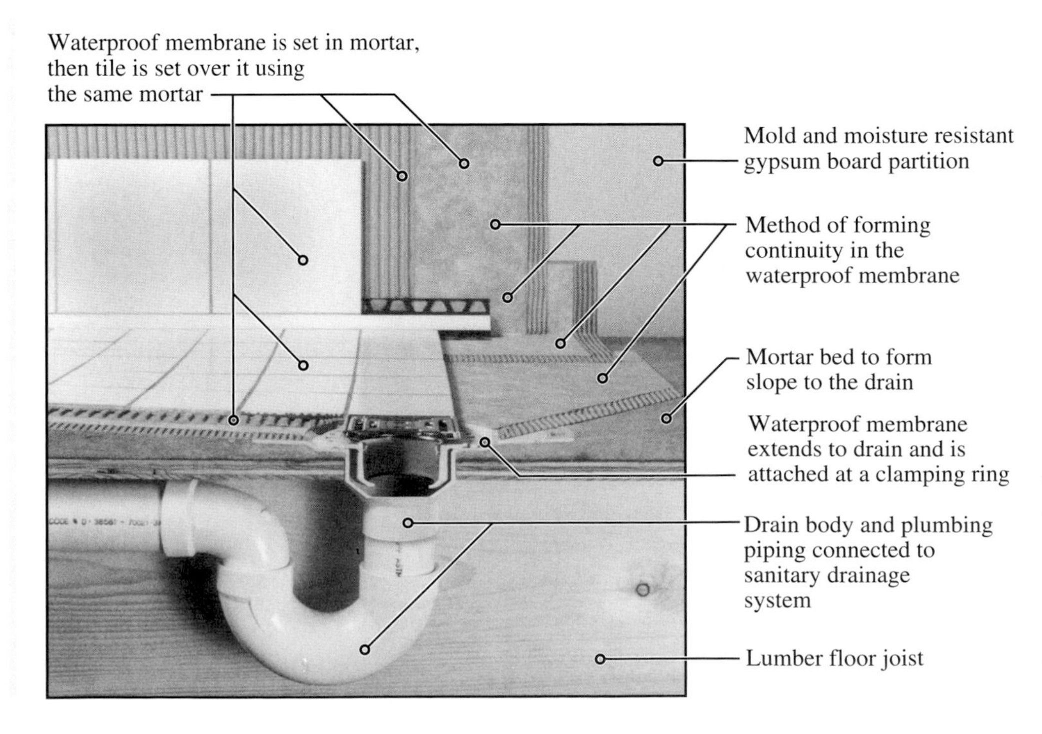

FIGURE 1(b) A layered section through a shower floor that is thick-set ceramic tile, indicating the various materials. In this example, the waterproof membrane is set on top of the mortar bed and the tile is set over the membrane (there is only one setting of a mortar bed, not two as in Figure 1(a). (Photo courtesy of Schluter.)

PRINCIPLES IN PRACTICE

Showers and Tile (*Continued*)

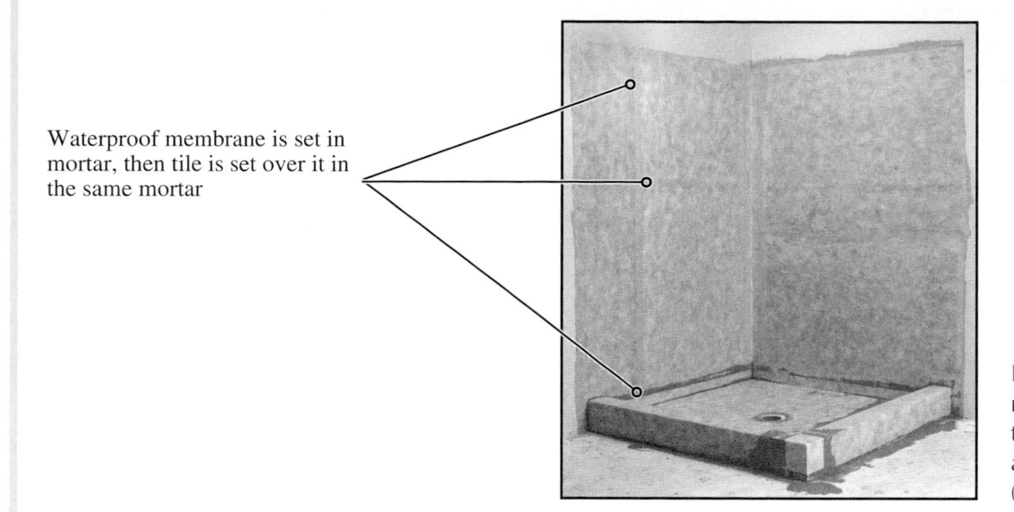

FIGURE 1(c) A waterproofing underlayment installed over the shower curb, at the shower floor for thick-set ceramic tile, and at the walls for thin-set ceramic tile. (Photo courtesy of Schluter.)

KEY TERMS AND CONCEPTS

Carpet
Carpet tile
Floating floor systems
Grout
Linoleum
Mortar
Nylon
Patching compound
Polypropylene
Resilient flooring
Resinous flooring
Rubber flooring
Setting materials
Stone panel flooring
Subfloor
Terrazzo
Tile
Tile flooring
Underlayment
Vinyl flooring
Wood flooring

REVIEW QUESTIONS

1. Use a sketch and notes to illustrate the difference between the thin-set and thick-set methods of laying tiles. Describe when each of these systems should be used.
2. Describe the differences and similarities between terrazzo and resinous flooring.
3. Draw plans and cross sections of a ceramic tile, stone tile, and stone panel that illustrate the differences between the materials.
4. Use a sketch and notes to describe the difference between a conventional wood plank floor and a floating wood plank floor assembly.
5. Using the information provided in this chapter, select an appropriate type of flooring and state why it should be used for the following applications: busy corridor in a school, hospital kitchen, residential living room, and office with a raised-floor system.

SELECTED WEB SITES

Canadian Stone Association, CSA (www.stone.ca)

Carpet Cushion Council, CCC (www.carpetcushion.org)

Carpet and Rug Institute, CRI (www.carpet-rug.com)

Ceramic Tile Institute of America, CTIA (www.ctioa.org)

Floor Covering Installation Contractors Association, FCICA (www.fcica.com)

Indiana Limestone Institute of America, ILI (www.iliai.com)

Marble Institute of America, MIA (www.marble_institute.com)

The National Terrazzo & Mosaic Association, Inc. (www.ntma.com)

Tile Council of America, TCA (www.tileusa.com)

CHAPTER 35 Ceilings

CHAPTER OUTLINE

35.1 Selection Criteria for Ceiling Finish Materials

35.2 No Ceiling Finish—Exposed to Above

35.3 Ceilings Attached to Building Structure

35.4 Ceilings Suspended from Building Structure

A standard suspension acoustical ceiling used in the large entry Foyer of Garland (Texas) Independent School District Special Event Center. (Photo courtesy of HKS Inc.)

(a) (b)

FIGURE 35.1 The smooth suspended gypsum ceiling (curved in plan, narrows in width from one end to the other, and barrel vault in profile) in a contemporary university foundation building on a college campus stands in contrast to as well as is reminiscent of St. Peter's Cathedral in Rome. (Photo of foundation building courtesy of HKS Inc.)

Ceilings can be brought low and articulated to create intimate spaces, or they can be pushed high to create awe-inspiring and dramatic spaces. They can be subtle in design or they can be highly expressive. They can impede sound or amplify it, and they can direct light by reflection or they can black it out. Whatever the requirements, ceilings can be used to accomplish any design intent.

Historically, ceilings were the underside of the floor or roof structure above; the ceilings were either finished or the structure was left exposed. The several architects of St. Peter's in Rome, Figure 35.1, placed an exquisitely detailed dome at the center of the building and consequently created one of the most awe-inspiring buildings in the history of architecture. The ceiling of another awe-inspiring ancient building, the Pantheon in Rome, is coffered concrete that was originally covered in stucco (Figure 18.5). Buildings today are far more complex and require space for the building utilities (ductwork for the distribution of heat and cooling, electrical power, lighting, communications wiring, fire protection piping, etc). In most buildings, the space overhead is the preferred location for many of these components, and for that reason, ceiling systems were developed to conceal these from view. Today there are a vast number of products and materials available for this purpose.

In this chapter, we discuss the three primary strategies for constructing ceilings. First, and the simplest to accomplish, is when the underside of the structure above forms the ceiling and any overhead mechanical and electrical components are left exposed. Second, the ceiling finish materials are attached directly to the overhead building structure to form a cover. Third, and most popular, ceiling finish materials are suspended from the overhead building structure. In the second and third strategies, the space between the ceiling and the floor or roof above is known as a *plenum.* When the ceiling is on the exterior of a building, it is called a *soffit.*

35.1 SELECTION CRITERIA FOR CEILING FINISH MATERIALS

Unlike many other building-product assemblies, the architect or owner decides which strategy will be used for the overhead space. Unless the ceiling is part of a fire-rated assembly (which is not very often), the reasons for a ceiling are discretionary. In residential buildings, the ceiling is attached directly to the structure, and in most commercial buildings the suspended ceiling strategy is used. When ceilings are not included, it may be to save building cost or to express the building utilities as part of the design. Other times, suspended ceilings can be used creatively to enhance the building design.

There are several factors that must be considered when determining the appropriate ceiling type and finishes, and many of the principles discussed in Part I of the text apply. Factors unique to ceilings include the following:

- *Aesthetic Expectations:* Ceiling finish materials, heights, and profiles can create inviting environments and can influence the way light interacts with the building interior. Unlike floors, ceilings are not limited to a single horizontal plane; they can be articulated at several elevations or can include vertical or sloped surfaces. Ceilings are used to introduce a sense of scale and proportion to an interior space. For example, if the area of the room is large, a high ceiling can create a grand space; however, if the area is small and the ceiling is high, the room may simply feel cavernous.
- *Conceal the Building Utilities Overhead (mechanical and electrical equipment and components):* As previously discussed, the most common reason for a ceiling is to conceal the building structure and overhead utilities, Figure 35.2.
- *Wind Loading:* Wind can impose an uplift load on the exterior soffit surface; therefore, wind-uplift resistance becomes an important consideration. In circumstances when large areas of the building enclosure are regularly opened to the exterior, wind can enter the building and exert a force on the ceilings.
- *Volume of Occupied Space:* Ceiling height is related directly to the amount of volume required to be heated and cooled.
- *Humidity:* Ceilings adjacent to openings in exterior walls, especially major openings such as at a bank of frequently used doors, may be subjected to higher levels of humidity than the ceilings in the remainder of the building. Because many materials, when oriented horizontally, are vulnerable to sagging due to moisture absorption, the choice of materials must be carefully considered.
- *Flammability:* Building codes sometimes require ceilings to be resistant to fire propagation and spread.
- *Seismic:* In seismically active areas, building codes require products, materials, and equipment installed overhead to resist the movements caused by an earthquake. During seismic activity, support structures and ceiling finish materials must not fall down, which would cause injuries to people trying to vacate the building.
- *Sound Absorption:* Ceiling finish materials and the method of application can help absorb sound generated within a room or space.
- *Sound Isolation:* Building codes sometimes require that a floor/ceiling assembly help isolate sound from above. See impact isolation class (IIC) in Chapter 8.
- *Sustainability:* Many ceiling products contain recycled content and can make contributions to sustainability goals.
- *Antimicrobial Resistance:* In addition to normal maintenance, some applications may require ceilings to be resistant to bacteria growth and mold and mildew development.
- *Light Reflectance:* Ceilings can be used to reflect and/or diffuse light from other sources in order to uniformly distribute it throughout a space. Sometimes certain ceiling finish materials are selected specifically to give the space a brighter appearance because indirect lighting is the preferred lighting source. This is also an important consideration when daylighting an interior space is part of the design.

FIGURE 35.2 In most buildings, there are usually many building utilities located overhead as shown in this image. Because many of these utilities require periodic access, an acoustical suspended ceiling is frequently selected. (Photo courtesy of HKS Inc.)

FIGURE 35.3 The building structure, utilities, and lapidaries of a professional basketball and hockey stadium are exposed as part of the architectural design. Note contrast with the interior atmosphere in Figure 35.6. (Photo courtesy of HKS Inc.)

- *Maintenance:* Some occupancies, such as health care facilities, require ceiling finishes that can be regularly cleaned and scrubbed to remove possible contaminates. Other occupancies may require ceilings to resist soiling, scratching, and impact.

35.2 NO CEILING FINISH—EXPOSED TO ABOVE

One method of treating an interior space is not to have a ceiling, but rather to expose the building structure and utilities. Although this may seem simple, it requires careful design, detailing, and coordination of all exposed elements to accomplish a satisfactory result, Figures 35.3, 35.4 and 35.5. This approach can make it easier to access some elements for routine maintenance, but it also creates a difficult environment for routine cleaning. In applications such as a museum, all the systems may be painted black so they tend to disappear, or they may be painted and expressed as a design feature. This ceiling is especially common in commercial buildings.

35.3 CEILINGS ATTACHED TO BUILDING STRUCTURE

In most residential buildings and some light-commercial buildings, lightweight ceiling finish materials are attached directly to the building structure. This is the most economical approach in residential applications, especially when there are few mechanical, plumbing, or electrical systems to be accommodated within the plenum. The most important consideration for this type of ceiling is that the underside of the building structure be specifically designed and located to provide surfaces for attaching the finished ceiling. Ceilings attached directly to the building structure do not usually require another structure, as is required for suspended ceilings.

The most common material attached to the building structure is gypsum board. It is either (a) nailed to the underside of wood framing, Figure 35.6, or (b) screwed to light-gauge steel framing (as at the underside of the framing in Figure 17.15). The gypsum board is subsequently taped, bedded, textured, and painted. In addition, acoustical tiles can be attached with adhesive to the gypsum board finish.

FIGURE 35.4 Interior of an architect's office with exposed wood structural frame and building utilities. (Photo courtesy of HKS Inc.)

FIGURE 35.5 The exposed building structure of the entry to HKS Architects, Dallas, is painted black to contrast with the white suspended gypsum ceilings.

FIGURE 35.6 A typical gypsum board ceiling attached to the wood framing in a residential application.

PRACTICE QUIZ

Each question has only one correct answer. Select the choice that best answers the question.

1. All ceilings are formed by suspending a lightweight material from the underside of the structure to provide the interior finish.
   a. True
   b. False
2. The most common reason for a ceiling is to
   a. control heat and humidity.
   b. provide a sculpted or multilevel surface overhead.
   c. reflect light.
   d. conceal the building structure and overhead utilities.
   e. isolate sound.
3. Antimicrobial-resistant ceiling materials are most likely to be used in a
   a. university classroom.
   b. hospital or medical laboratory.
   c. department store.
   d. pharmacy.
   e. warehouse.
4. In a building where no ceiling is provided and the structure and utilities are exposed,
   a. mechanical and electrical subcontractors are allowed easy access and freedom of choice in locating their work.
   b. all systems must be carefully designed.
   c. on-site coordination of construction is critical.
   d. all the above.
   e. (b) and (c).
5. The ceiling provides an excellent opportunity to modulate natural and artificial light.
   a. True
   b. False

Answers: 1-b, 2-b, 3-b, 4-e, 5-a.

35.4 CEILINGS SUSPENDED FROM BUILDING STRUCTURE

By far the most common ceiling in nonresidential buildings is one that is suspended from the building structure overhead.

PRINCIPLES OF CEILING SUSPENSION SYSTEMS

A suspended ceiling system is a relatively simple concept. A lightweight metal grid structure is suspended from the building structure to provide support for the ceiling finish products as well as to establish the ceiling height. This grid structure consists of main runners, suspended from above by steel wire, and interconnecting cross runners spaced uniformly to support the ceiling finish products. The main runners are usually larger in cross section and thicker than the cross runners.

There are three types of metal suspension systems, Figure 35.7. The type of suspension system used for a particular application depends on the structural requirements to support the ceiling finish products and other components that will be mounted or attached to the grid structure, Table 35.1.

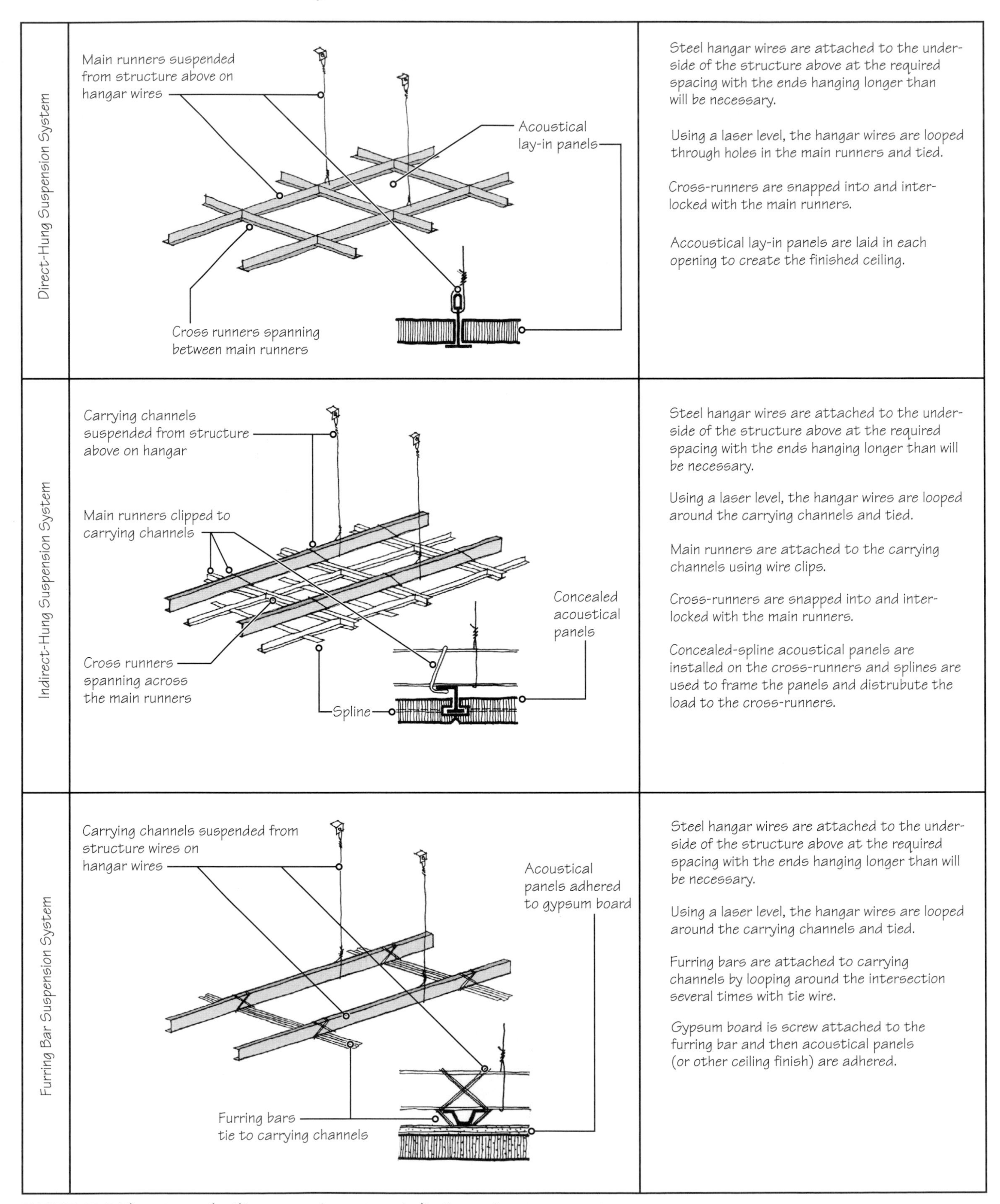

FIGURE 35.7 Three types of ceiling suspension systems indicating various components.

TABLE 35.1 METAL SUSPENSION SYSTEMS—DUTY CLASSIFICATIONS AND STRUCTURAL CAPACITIES

Duty	Carrying capacity	Minimum load-carrying capacity (pounds per square foot)		
		Direct-hung	Indirect-hung	Furring bar
Light	Supports only the finish material itself	5.0	2.0	4.5
Intermediate	For ordinary ceilings; supports finish material and lightweight components such as light fixtures and ceiling diffusers/grilles	12.0	3.5	6.5
Heavy	Greatest capacity to support finish material and other ceiling-mounted components	16.0	8.0	N/A

Suspended Acoustical Ceilings

The most common commercial ceiling consists of modular acoustical panels laid in a direct-hung suspension system known as an inverted-tee grid. The acoustical panels are typically 24 in. square, or 24 by 48 in., and are laid on the horizontal legs of the inverted tees, Figure 35.8.

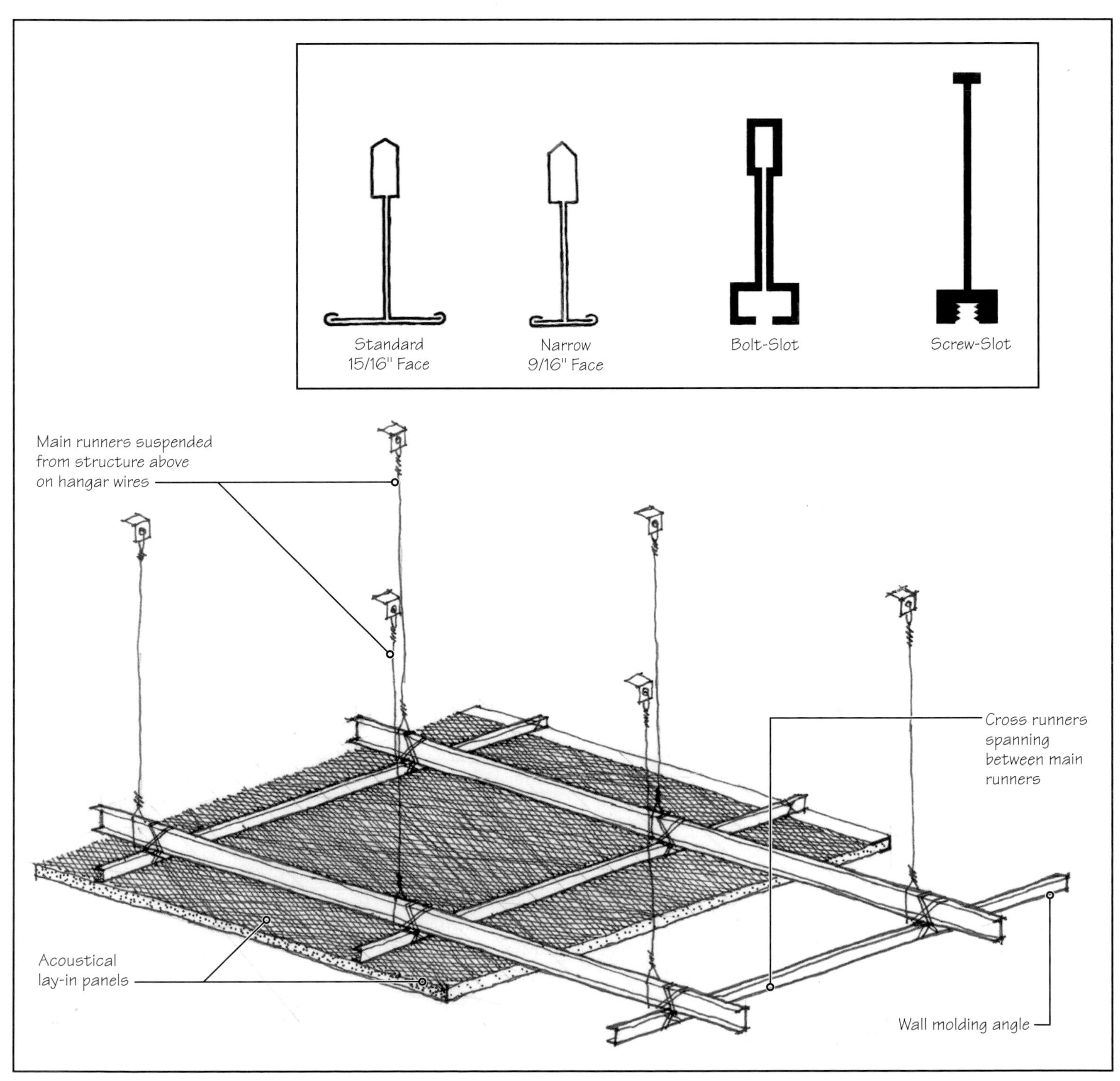

FIGURE 35.8 Suspended acoustical ceiling components and typical profiles for inverted tees.

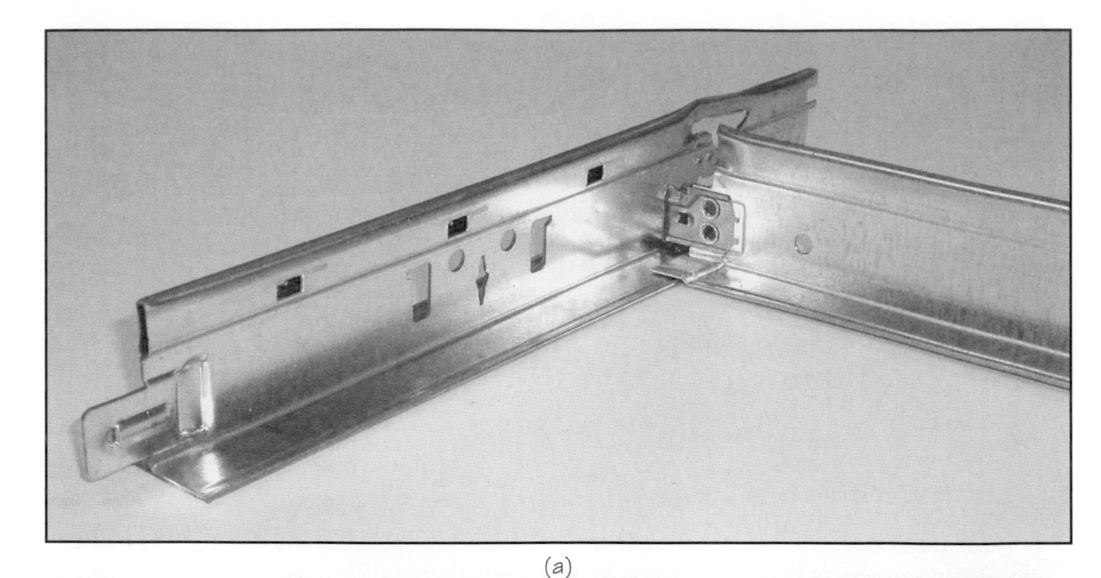

(a)

(b) (c)

FIGURE 35.9 Suspended acoustical grid components. **(a)** Typical intersection of cross runner into main runner. **(b)** Typical profiles of inverted tees. **(c)** Close-up of intersection. Notice crimped hole at top of main runner for attachment of hanger wire.

This grid structure is composed of corrosion-resistant steel or extruded aluminum inverted tees (upside-down T-shape). The main runners and cross runners are in the same plane and are assembled by interlocking the pieces to form a unified, modular grid structure, usually on some increment of 12 in. The main runners have slots and holes that have been prepunched in the web at regular spacings to receive the end-clips that are on each end of the cross tees, Figures 35.9 and 35.10. Figure 35.11 shows three examples of acoustical ceilings.

Acoustical panels and tiles are composed of a variety of materials, including mineral wool, mineral fibers, fiberglass, and/or perlite, which are combined with fillers, binders, and water and then formed into sheets (and cut into panels or tiles) or molded in pans. There are several surface finishes available, including paint, membrane overlay, fabric, and thin metal sheets. Surface textures and patterns include perforations, fissures, embossings, printings, and scorings, Figure 35.12. Finally, panels can have one of several edge treatments, Figure 35.13. Once the grid structure is assembled and the panels and tiles are installed, no other finishing work is necessary.

When an acoustical ceiling without an exposed grid is desired, a variation of the inverted-tee grid structure is used. In addition to a suspended grid, metal splines are inserted in the edges of acoustical tiles, usually 12 in. square, which are, in turn, supported on a suspended grid.

SUSPENDED GYPSUM BOARD CEILINGS

The second most commonly used commercial suspended ceiling consists of screw-attaching gypsum board sheets to an inverted-tee grid structure. The gypsum board is then finished

FIGURE 35.10 Installation of a suspended grid. (Photos of Vaden's Acoustical & Drywall, Inc. by Gary Yancy of USG Corp). **(a)** Attaching steel hanger wire to the steel structural joist above. In this case, the hanger wire is tied to the steel joist. When a concrete slab or metal deck is present, a hanger wire, with integral clip angel, is usually powder-fastener attached. Hanger wires are always attached to the building structure and not to building utilities. **(b)** Layout begins by establishing the location of the main runners. For most ceilings, installers will use stilts to assist in installation or will use a rolling scaffold. **(c)** Using string and laser level see (e) installers establish main runner locations. **(d)** Until permanent attachments can be made, installers clamp intersecting tees together. **(e)** Continuing with installation; notice laser level at wall. **(f)** Final suspended grid installation, including openings for light fixtures.

Beveled tubular edge acoustical panel with narrow 9/16 inch wide screw-slot face suspended grid.

Beveled tubular edge acoustical panel with narrow 9/16 inch wide face suspended grid.

Square edge acoustical panel with standard 15/16 inch wide suspended grid.

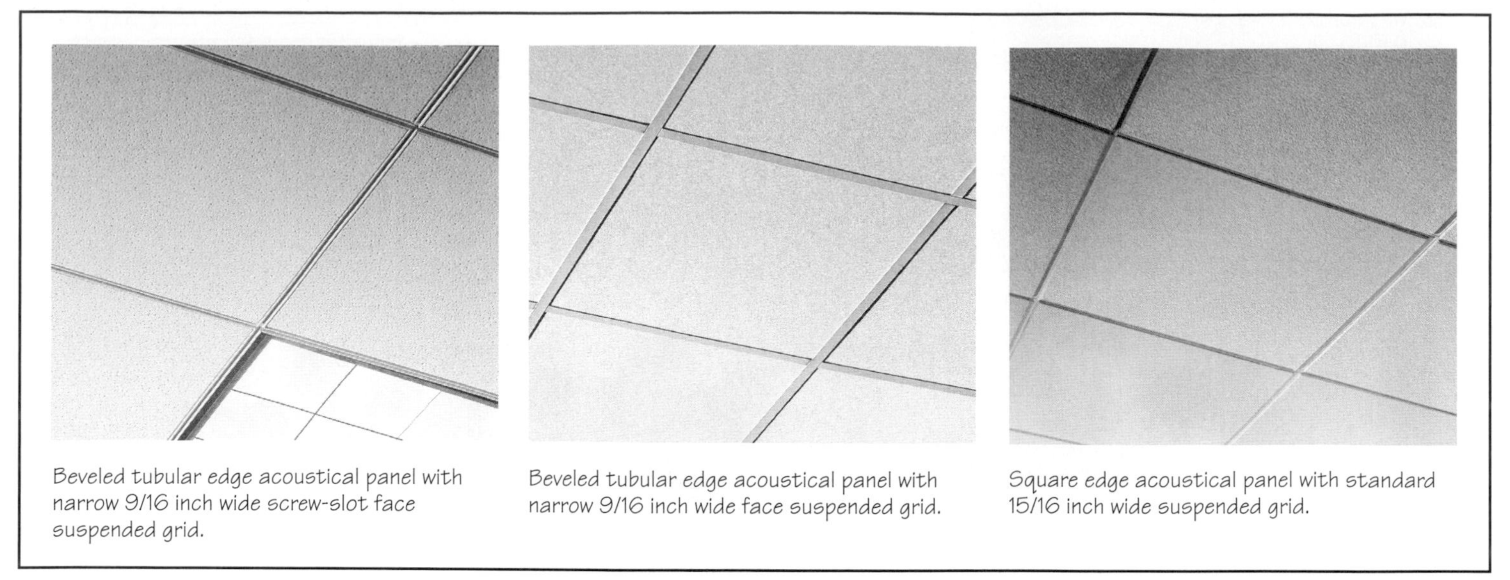

FIGURE 35.11 Examples of suspended acoustical panels show a range of panel types and grid profiles. (Photo courtesy of Armstrong Ceiling Systems.)

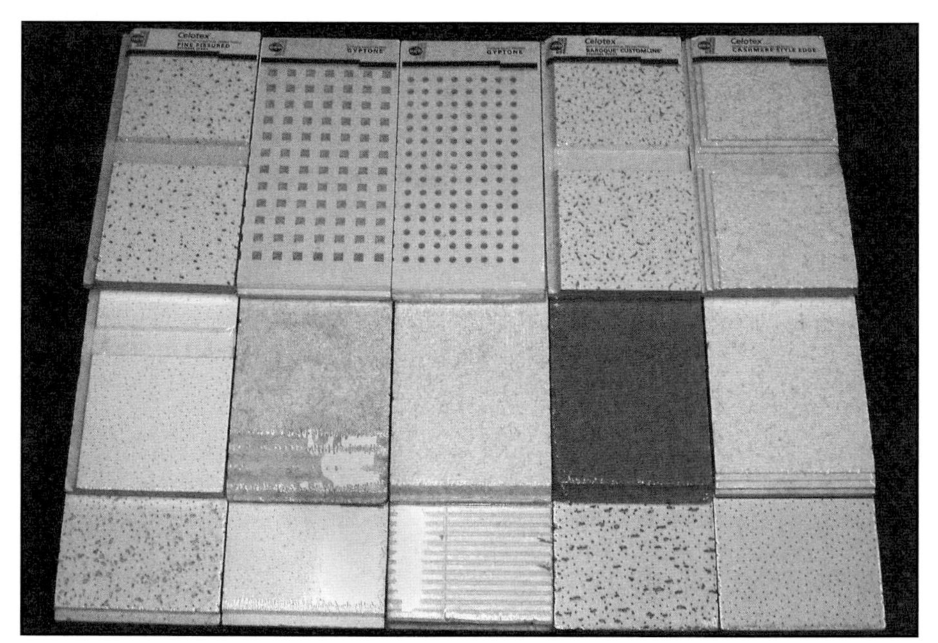

FIGURE 35.12 Just a few of the many acoustical panels available (see also Figure 8.11).

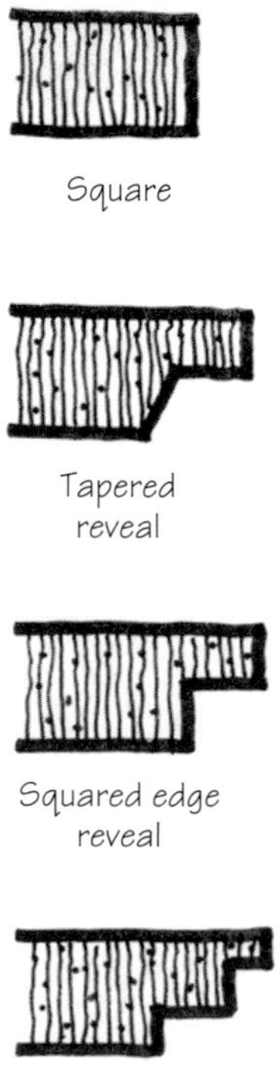

Square
Tapered reveal
Squared edge reveal
Stepped edge reveal

FIGURE 35.13 Various acoustical lay-in panel edge types.

like other gypsum board surfaces. Suspended gypsum board ceilings are versatile because they can be integrated with stud framing hung from the building structure above to form ceilings that have various profiles vertically and horizontally.

The suspended inverted-tee grid structure for gypsum board ceilings is very similar to the grid structure used for suspended acoustical ceilings, except the bottom surface of the horizontal tee flange is embossed to improve the grip on the gypsum board, Figures 35.14, 35.15, and 35.16.

Suspended Gypsum Plaster Ceilings and Portland Cement Plaster Soffits

Portland cement and gypsum plasters are heavier than acoustical products and materials and thus require the use of a channel grid structure. Wire lath is attached horizontally to the main and cross runners using steel wire; then the plaster is applied as described in Chapter 27, Figure 35.17.

The channel grid structure is used for ceilings that are heavier than acoustical or gypsum board materials. Using galvanized-steel channel shapes, main runners and cross runners are

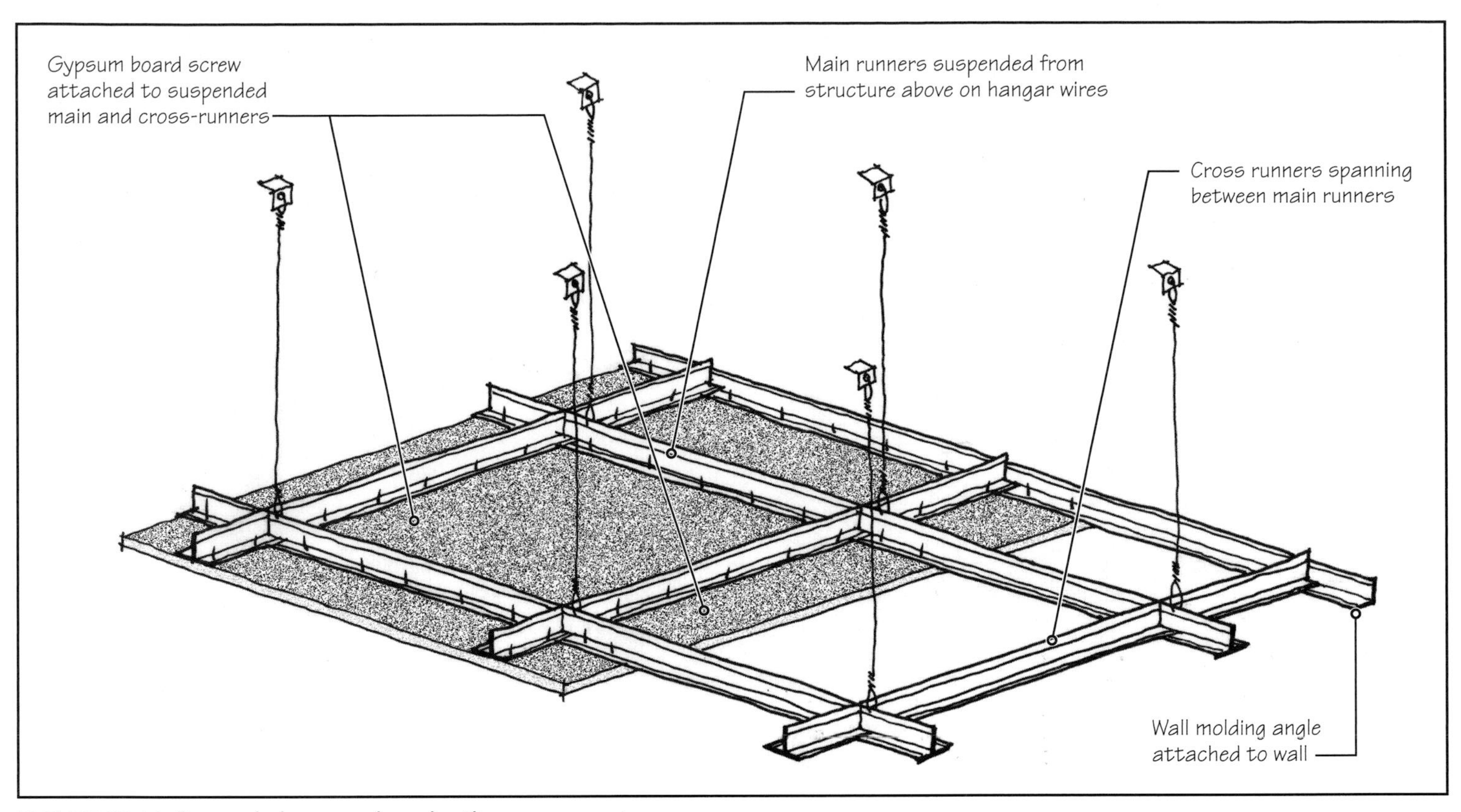

FIGURE 35.14 Suspended gypsum board ceiling components.

overlapped, and each intersection is tied with steel wire. Unlike the inverted-tee grid, the spacing of the channels can be adjusted to fit the application conditions, the applied finish, and the component loading.

Suspended Specialty Ceilings and Soffits

Like so many other building products, a wide variety of specialty ceiling and soffit finishes and materials is available, limited only by the designer's imagination. These systems can be flat planes, squared with corners, curved vertically or horizontally, or formed into the appearance of waves. They can be continuous or broken into sections. Lighting fixtures can be above, within, or below the systems. The following discussion describes just a few suspended specialty ceilings and soffits.

- Canopy ceilings appear as floating ceilings because they emphasize the negative space above and at the sides of suspended ceiling shapes. The materials involved are usually metals, and an almost infinite number of shapes and combinations can be achieved, Figure 35.18.
- Decorative grids use rectangular main runners and cross runners that are combined and formed into an interlocking grid that is uniform in appearance, Figure 35.19.

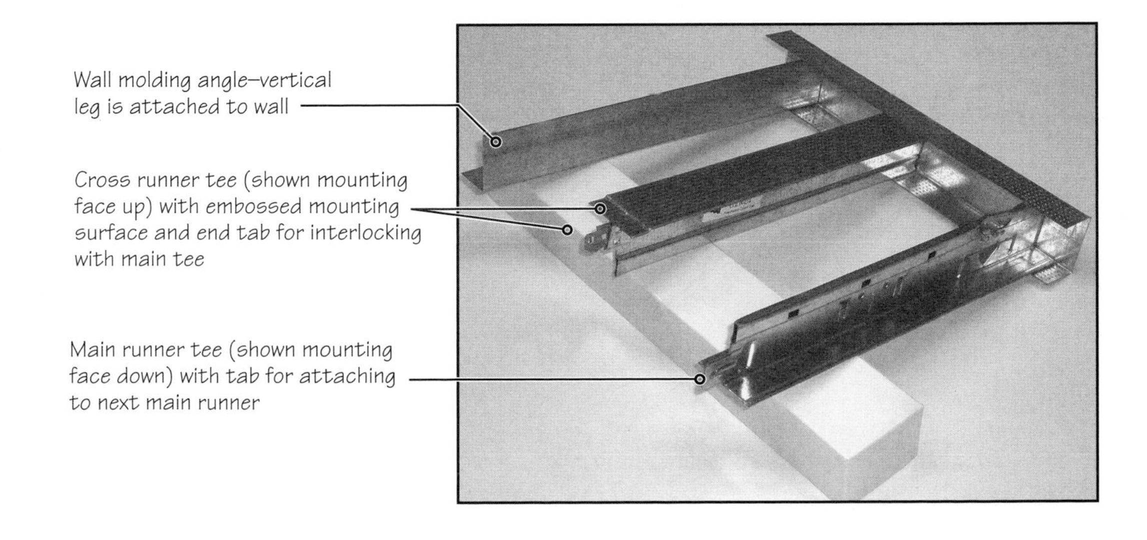

FIGURE 35.15 Samples of the various framing members of a suspended gypsum board ceiling.

FIGURE 35.16 A sculpted suspended gypsum board ceiling in a dining room of an office building. (Photo courtesy of HKS Inc.)

- Transparent-finished wood panels (very much like wall paneling) can be formed into panels laid into an exposed grid structure or into planks attached to a specialized suspension system, Figure 35.20.
- Acoustical metal facings, with or without acoustical backings, can be formed into panels or tiles and placed in suspension systems, Figure 35.21. The surface can be perforated in a variety of patterns, or it can be unperforated.
- Sheet metals can be formed into linear strips that resemble planks, which are attached to a specialized suspension system to achieve a directional look within a space, Figure 35.22. The spaces between the strips can be left open or covered.
- Luminous ceilings combine overhead lighting with either open or translucent (opaque) ceilings to create dramatic effects, Figure 35.23.

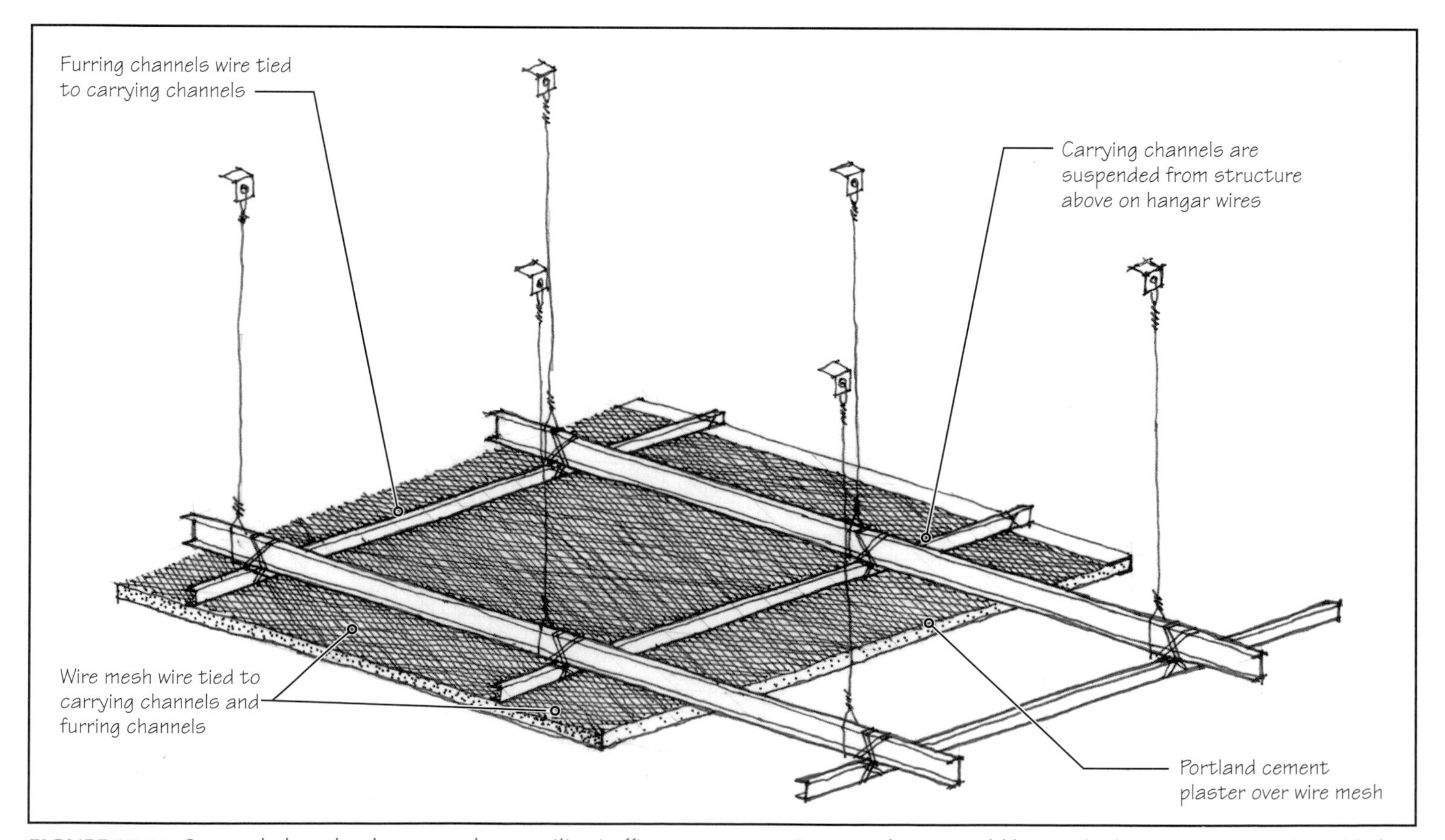

FIGURE 35.17 Suspended portland cement plaster ceiling/soffit components. Gypsum plaster would be similar however, gypsum board lath would be used in lieu of wire mesh lath.

FIGURE 35.18 Curved, perforated metal panel canopy ceiling in a conference room at HKS Architects. Portions around the canopy are open to the building structure, above which is painted black.

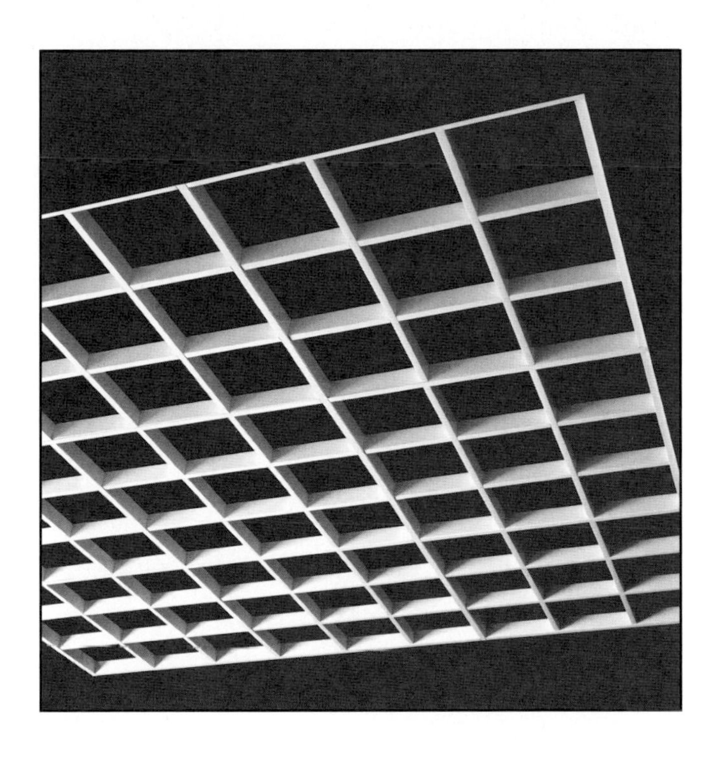

FIGURE 35.19 Simple open-cell, suspended metal grid ceiling. (Photo courtesy of Armstrong Ceiling Systems.)

FIGURE 35.20 Suspended wood panel ceiling. (Photo courtesy of Armstrong Ceiling Systems.)

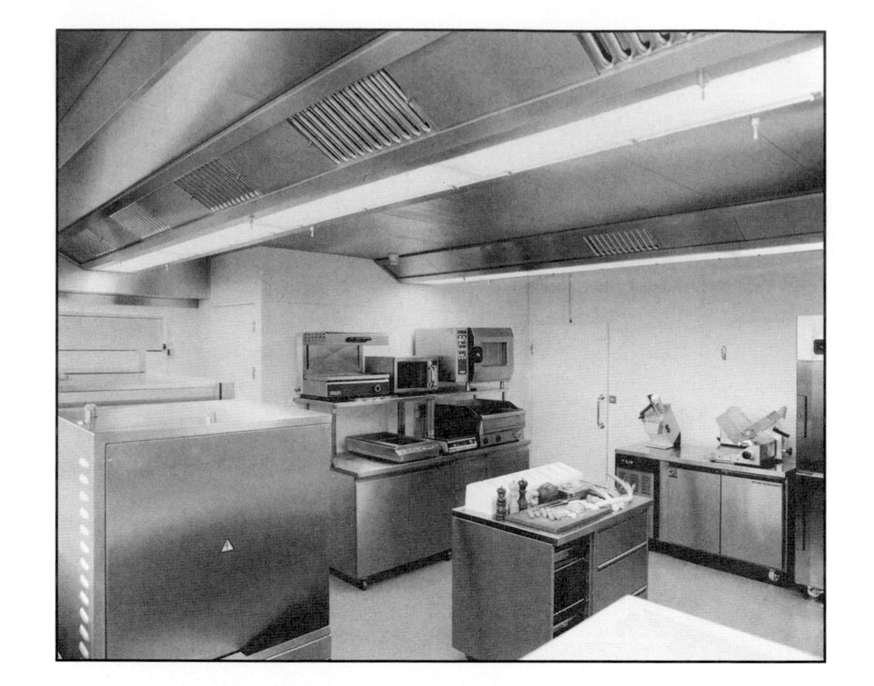

FIGURE 35.21 Suspended metal panel ceiling in a commercial kitchen. (Photo courtesy of Armstrong Ceiling Systems.)

FIGURE 35.22 A suspended linear metal ceiling in an area in which streets and drives pass under a building.

FIGURE 35.23 A luminous ceiling, imitating a basketball, in a professional basketball team's locker room. (Photo courtesy of HKS Inc.)

Each question has only one correct answer. Select the choice that best answers the question.

6. In a suspended ceiling system, steel hanger wires are supplied by the manufacturer already cut to length.
  a. True
  b. False

7. The strongest metal suspension system is always a properly installed direct-hung system.
  a. True
  b. False

8. In suspended acoustical ceilings, the grid structure is made of
  a. a hardwood such as oak or maple.
  b. extruded aluminum tees.
  c. corrosion resistant steel tees.
  d. all the above.
  e. (b) and (c).

9. Acoustical panels and tiles are fabricated using
  a. mineral wood, mineral fibers, fiberglass, and/or perlite.
  b. fillers, binders, and water.
  c. (a) and (b).
  d. none of the above.

10. The edges of acoustical panels
  a. are designed to rest on the metal suspension system.
  b. are available in three different profiles.
  c. are snapped into extruded plastic shapes.
  d. do not affect the overall pattern of the ceiling.
  e. all the above.

11. Gypsum board is a commonly used finish material. It is typically
  a. attached directly to a structural metal or wood light frame.
  b. attached directly to a suspended metal frame.
  c. designed to rest on the metal suspension system.
  d. all the above.
  e. (a) and (b).

12. The design of a ceiling in a large office building calls for a multilevel series of ceiling planes finished with gypsum board. The most likely method for attaching the ceiling to the structure is to
  a. nail gypsum board directly to the wood light frame structure.
  b. screw the gypsum board directly to a suspended metal framework.
  c. use lay-in acoustical tile.

13. It is not possible to use lath and plaster with a suspended ceiling structure.
  a. True
  b. False

14. Ceiling materials are essentially limited to drywall, lath and plaster, and acoustical tiles.
  a. True
  b. False

Answers: 6-b, 7-b, 8-e, 9-e, 10-a, 11-e, 12-b, 13-b, 14-b.

KEY TERMS AND CONCEPTS

Acoustical panel or tile
Canopy ceilings
Exposed ceiling
Luminous ceiling
Selecting ceiling finish materials
Suspended ceiling types
Suspended wood panel or grid ceiling

REVIEW QUESTIONS

1. Diagram and describe the difference between a ceiling that is attached to the building structure and one that is suspended from the building structure.
2. Examine the ceiling in one of your classrooms. What type of ceiling is it? Do a section sketch illustrating the parts of the system.
3. Use a section sketch and notes to illustrate a direct-hung suspended ceiling.
4. Use a sketch and notes to illustrate two methods for attachment when gypsum board is used as a finish ceiling material.

SELECTED WEB SITES

Ceilings & Interior Systems Construction Association, CISCA (www.cisca.org)
Gypsum Association, GA (www.gypsum.org)

APPENDIX A

SI System and U.S. System of Units

APPENDIX OUTLINE

Rules of Grammar in the SI System

Length, Thickness, Area, and Volume

Fluid Capacity

Mass, Force, and Weight

Pressure and Stress

Unit Weight of Materials

Temperature and Energy

Conversion from the U.S. System to the SI System

NOTE

The SI system, although commonly referred to as the International System, is, in fact, an acronym for Le Système International d'Unités, a name given by the thirty-six nations meeting at the eleventh General Conference on Weights and Measures (CGPM) held in Paris in 1960. (CGPM is an acronym for Conférence Générale des Poids et Mesures.)

TABLE 1 PREFIXES IN THE SI SYSTEM

Factor	Prefix	Symbol
10^{12}	tera	T
10^{9}	giga	G
10^{6}	mega	M
10^{3}	kilo	k
10^{2}	hecto	h
10	deka	da
10^{-1}	deci	d
10^{-2}	centi	c
10^{-3}	milli	m
10^{-6}	micro	μ
10^{-9}	nano	n
10^{-12}	pico	p

The system of measurement (or units) commonly used at the present time for the design and construction of buildings in the United States is the foot-pound-second (FPS) system. In this system, the length is measured using the foot or its multiples—the yard and the mile—or its submultiple—the inch. Weight is measured in pounds, kilo-pounds, or ounces. Time is measured in hours, minutes or seconds. Although the United States was the first country to establish the decimal currency in 1785, it is one of the two (or three) countries where (the nondecimal) FPS system is still used.* Therefore, the FPS system, earlier known as the Imperial System of units, is now commonly known as the *U.S. System of Units* or *U.S. Customary Units.*

The system of units used by the rest of the world is the meter-kilogram-second (MKS) system. A rationalized and more commonly used version of the MKS system is called the SI system, popularly known as the International System of Units.

The advantage of the SI system (or the MKS system) lies primarily in that the multiples and submultiples of each unit (the secondary units) have a decimal relationship with each other, which makes computations easier and less susceptible to errors. For instance, the secondary units of length, the centimeter and the kilometer, are related to the base unit, the meter, by 10^{-2} and 10^{3}, respectively. By contrast, the length unit in the U.S. system, the foot, does not bear a decimal relationship with its secondary units, the inch, yard, and the mile.

Twelve prefixes have been standardized in the SI system for use with the base unit to make the secondary units. These prefixes, along with their symbols, are given in Table 1. Note that the prefixes are uppercase for magnitudes 10^{6} and greater and lowercase for magnitudes 10^{3} and lower.

The SI system uses seven base quantities—length, mass, time, temperature, electric current, luminous intensity, and amount of substance. The units corresponding to the base quantities and their symbols are listed in Table 2. Of these, only the first four quantities are of interest to design and construction professionals—length, mass, time, and temperature.

Quantities other than the base quantities, such as force, stress, pressure, velocity, acceleration, and energy, and their units are derived from a combination of two or more base quantities. For example, acceleration is the rate of change of velocity, which, in turn, is the rate of change of distance. Consequently, the units for acceleration are meters per second squared (m/s^2).

Another advantage of the SI system is that there is one and only one unit for each quantity—the meter for length, kilogram for mass, second for time, and so on. This is not so in the U.S. system, where multiple units are often used. For instance, power is measured in Btu per hour and also in horsepower.

RULES OF GRAMMAR IN THE SI SYSTEM

The symbols for units in the SI system are always lowercase unless the unit is named after a person, such as Newton, Pascal, Hertz, or Kelvin. In this case, the first letter of the symbol is in uppercase and the second letter in the lowercase, as in Pa. The exception is, however, made in the case of the liter, which is given the symbol L.

Multiples of base units are given with a single prefix. Double prefixes are not allowed. Thus, megakilometer (Mkm) is incorrect; instead, we use gigameter (Gm). No space is left between a prefix and symbol. Thus, we use km, and not k m.

TABLE 2 BASE QUANTITIES AND THEIR SYMBOLS IN THE SI SYSTEM

Quantity	Unit name	Symbol
Length	meter	m
Mass	kilogram	kg
Time	second	s
Temperature	kelvin	K
Electric current	ampere	A
Luminous intensity	candela	cd
Amount of substance	mole	mol

*Although not yet officially adopted, the SI system is being increasingly used (together with the U.S. System—in a dual-unit format) in several important U.S. publications that regulate building design and construction.

The product of two or more units in symbolic form is given by using a multiplication dot between individual symbols (example N · m). Mixing symbols and names of units is incorrect (example, N · meter).

A space must be left between the numerical value and the unit symbol. Thus, we use 300 m, not 300m. No space, however, is left between the degree symbol and C (Celsius). Plurals are not used in symbols. For instance, we use 1 m and 50 m. Periods are not used after symbols except at the end of a sentence.

In architectural and engineering drawings, dimensions are generally given in millimeters, and when that is done, the use of mm is avoided. For instance, the measurements of a floor tile are given as 300 × 300, and not as 300 mm × 300 mm. Larger dimensions may be given in meters, if necessary. For instance, a building may be dimensioned as 30.800 m × 50.600 m in plan (elevation or section), but it is preferable to dimension it as 30,800 × 50,600.

LENGTH, THICKNESS, AREA, AND VOLUME

In the U.S. system, the standard unit of length is the foot: 1 ft = 12 in. Long distances are measured in miles. In the SI system, the standard unit of length is the meter. Most dimensions in the SI system are given in millimeters (mm), where 1 mm = 10^{-3} m. Long distances are given in kilometers (km).

1 ft = 0.3048 m; 1 m = 3.281 ft = 39.37 in; 1 in. = 25.4 mm
1 mi = 1.609 km

FLUID CAPACITY

Fluid capacity is usually given in gallons in the U.S. system and in liters (L) in the SI system. The imperial gallon is slightly larger than the U.S. gallon, but the imperial gallon is no longer used as a unit.

1 gal (U.S.) = 4 qt = 3.7854 L
1 L = 0.001 m^3 (by definition) = 0.264 gal

MASS, FORCE, AND WEIGHT

In the U.S. system, the unit of weight (and mass) is the pound (lb); the corresponding unit of mass in the SI system is the kilogram (kg).

1 lb = 16 oz = 7,000 gr (grains) = 0.4536 kg
1 kg = 2.205 lb

Force is defined as mass times acceleration. Because the unit of acceleration in the U.S. system is feet/second^2 (ft/s^2), the unit of force is (lb-ft/s^2). This unit is called pound-force, usually referred to as the pound. In the SI system, the unit of force is (kg · m)/s^2. This complex unit is called the newton (N), after the famous physicist Isaac Newton (1642–1727).

Because the weight of an object is the force exerted on it by the gravitational pull of the earth, the weight of an object is equal to mass × acceleration due to gravity. The acceleration due to gravity on Earth's surface is 9.8 m/s^2. Therefore, the weight of an object whose mass is 1 kg is 9.8 (kg · m)/s^2, or 9.8 N.

We see that in the SI system, there is a clear distinction between the units of mass and the weight of an object. The distinction between the mass and the weight in the U.S. system is rather obscure because the pound is used as the unit for both the mass and the weight of an object.

1 kilogram-force = 9.8 N; 1 lb = 4.448 N; 1 N = 0.2248 lb
1 kilopound (kip) = 1,000 lb = 4.448 kN; 1 kN = 0.2248 kip

NOTE

Some countries that have not fully adopted the SI system use kilogram as the unit for mass as well as force (in the same way as pound is used as a unit for mass and weight). In these countries, the unit for stress is kilogram per square centimeter (kg/cm^2) instead of Pa. Weight density is expressed as kilograms per cubic centimeter (kg/cm^3) instead of N/m^3.

PRESSURE AND STRESS

Because pressure, or stress, is defined as force per unit area, the unit for pressure, or stress, in the U.S. system is lb/ft^2 (psf). Other units commonly used are pounds per square inch (psi) and kilopounds per square inch (ksi). In the SI system, the unit of pressure, or stress, is N/m^2. This unit is called the pascal (Pa) after the physicist Blaise Pascal (1623–1662). Thus, 1 N/m^2 = 1 Pa.

1 psf = 47.880 Pa; 1 Pa = 0.208 85 psf
1 psi = 6.895 kPa; 1 kPa = 0.1450 psi

In weather-related topics, the unit of pressure is the atmosphere (atm); 1 atm is the standard atmospheric pressure at sea level.

1 atm = 760 mm of mercury (Hg) = 29.92 in. of Hg = 14.69 psi = 2,115.4 psf = 101.3 kPa. For all practical purposes, the atmospheric pressure may be taken as 2,100 psf or 101 kPa.

UNIT WEIGHT OF MATERIALS

Density is defined as the mass per unit volume. Its units are lb/ft^3 in the U.S. system and kg/m^3 in the SI system.

1 lb/ft^3 = 16.018 kg/m^3; 1 kg/m^3 = 0.064 243 lb/ft^3

In building construction, however, we are more interested in weight density of materials instead of the mass density. Weight density (or simply the unit weight) is defined as the weight per unit volume, and its units are lb/ft^3 and N/m^3 in the U.S. system and SI system, respectively.

1 lb/ft^3 = 157.1 N/m^3

TEMPERATURE AND ENERGY

In the U.S. system, temperature is measured in degrees Fahrenheit (°F). This scale was introduced during the early eighteenth century. Zero on the Fahrenheit scale was established based on the lowest obtainable temperature at the time, and 100°F, on the basis of human body temperature as it was then considered to be.

On the Celsius scale (°C), earlier known as the centigrade scale, the zero refers to the freezing point of water, and 100°C to the boiling point of water.

The unit of temperature in the SI system is the kelvin (K). The preference for Kelvin scale over the Celsius scale is due to the fact that on the Kelvin scale, the temperature is always positive. This is not so on the Celsius or the Fahrenheit scale, on which the temperature may be positive or negative. In other words, 0 K (as we know it now) is the lowest obtainable temperature and is, therefore, called *absolute zero.* The relationship between Kelvin and Celsius temperatures is T °C = (T + 273.15) K. In other words, 20°C = 293.15 K. Other relationships are

T °F = [(1.8)T + 32]°C; T °C = [(0.555 . . .)(T − 32)]°F.

The intervals in both the Kelvin and the Celsius scales are equal. Therefore, the Celsius and Kelvin scales start at different points, but their subdivisions are equal. Note that the word *degree* is not used on the Kelvin scale. (That is, we say 8 Kelvins, not 8 degrees Kelvin.) Although Kelvin is the appropriate scale to use in the SI system, degrees Celsius are also used because of the smaller numbers associated with the Celsius scale.

CONVERSION FROM THE U.S. SYSTEM TO THE SI SYSTEM

The conversion of a physical quantity such as length, weight, or stress from the U.S. system to the SI system simply involves using the appropriate conversion factor. A comprehensive list of conversion factors to convert from the U.S. system to the SI system is given in Table 3. However, when it comes to converting building products' sizes from the U.S. system to the SI system, three types of conversion are possible:

- *Exact Conversion:* This conversion is made simply by multiplying a value given in the U.S. system by the appropriate conversion factor to obtain the corresponding value in the SI system. For example, 12 in. is exactly equal to 304.8 mm.
- *Soft Conversion:* In this conversion, a product's size is not converted, only its description. For example, a manufacturer may decide to continue making 12 in. × 12 in. floor tiles but market them as 305 mm × 305 mm tiles. During the initial period of changes over to the SI system in the building industry, most product sizes will be soft-converted.

- *Hard Conversion:* In hard conversion, the physical size of the product is changed to a new metric equivalent. For example, 12 in. × 12 in. floor tiles will most probably be changed to 300 mm × 300 mm tiles. Hard conversion of a product requires a change in the manufacturing equipment and a great deal of coordination among various related products but has the advantage that the product sizes are rationalized.

TABLE 3 UNIT CONVERSION FACTORS

Quantity	To convert from	To	Multiply by
Length	mi	km	1.609 344*
	yd	m	0.9144*
	ft	m	0.304 8*
	ft	mm	304.8*
	in.	mm	25.4*
Area	mi^2	km^2	2.590 00*
	acre	m^2	4 046.87
	acre	ha**	0.404 687
	ft^2	m^2	0.092 903 04*
	in.^{2}	mm^2	645.16*
Volume	yd^3	m^3	0.764 555
	ft^3	m^3	0.028 3168
	100 board feet	m^3	0.235 974
	gal	L	3.785 41
	in.^{3}	cm^3	16.387 064
	in.^{3}	mm^3	16 387.064
Velocity	ft/s	m/s	0.3048
Rate of fluid flow, infiltration	ft^3/s	m^3/s	0.028 3168
	gal/h	mL/s	1.051 50
Acceleration	ft/s^2	m/s^2	0.3048
Mass	lb	kg	0.453 59
Mass per unit area	psf	kg/m^2	4.882 43
Mass density	pcf	kg/m^3	16.018 5
Force	lb	N	4.448 22
Force per unit length	plf	N/m	14.593 9
Pressure, stress	psf	Pa	47.880 26
	psi	kPa	6.894 76
	in. of mercury (in. Hg)	kPa	3.386 38
	in. of Hg (in. Hg)	psf	70.72
	atm***	kPa	101.325
Temperature	°F	°C	5/[9(°F − 32)]
	°F	K	(°F + 459.7)/1.8
Quantity of heat	Btu	J	1055.056
Power	ton (refrigeration)	kW	3.517
	Btu/h	W	0.293 07
	hp	W	745.7
	Btu/(h-ft^2)	W/m^2	3.154 59
Thermal conductivity	Btu-in/(ft^2-h-°F)	W/(m·°C)	0.144 2
Thermal conductance, or Thermal transmittance, U	Btu/(ft^2-h-°F)	W/(m^2·°C)	5.678 263
Thermal resistance	(ft^2-h-°F)/Btu	(m^2·°C)/W	0.176 110
Thermal capacity	Btu/(ft^2-°F)	kJ/(m^2·°C)	20.44
Specific heat	Btu/(lb-°F)	J/(kg·°C)	4.186 8
Vapor permeability	perm-in	ng/(Pa·m·s)	1.459 29
Vapor permeance	perm	ng/(Pa·m^2·s)	57.213 5
Angle	degree	radian	0.017 453

*Denotes exact conversion
**1 hectare (ha) = 100 m × 100 m
***1 atmosphere (atm) = 29.92 in. of mercury

APPENDIX B

Preliminary Sizing of Structural Members

APPENDIX OUTLINE

CONVENTIONAL WOOD LIGHT FRAME (WLF) BUILDINGS

CONVENTIONAL LIGHT-GAUGE STEEL FRAME (LGSF) BUILDINGS

STRUCTURAL STEEL FRAME BUILDINGS

SITE-CAST CONCRETE FRAME BUILDINGS

PRECAST-PRESTRESSED CONCRETE MEMBERS

LOAD-BEARING MASONRY AND CONCRETE BUILDINGS

The information provided in this appendix is valid only for approximate sizing of structural members during the sketch design (SD) or initial design development (DD) stages of conventional buildings. A structural engineer should be consulted for final member sizes.

CONVENTIONAL WOOD LIGHT FRAME (WLF) BUILDINGS

WALL FRAMING

Approximate Stud Size and Spacing

Number of stories	Stud size	Stud spacing
3 stories	2 × 6	16 in. o.c.
2 stories	2 × 6 2 × 4	16 in. o.c. 12 in. o.c.
1 story	2 × 4	24 in. o.c.

1. Stud size and spacing are for approximately 10-ft-high studs. For tall walls, such as those used in double-height spaces, doubled studs, and (or) closer stud spacing may be required.
2. In cold climates, the stud size may be governed by insulation requirements. For example, 2 × 6 studs may be necessary where 2 × 4 studs are structurally adequate.
3. In high-wind regions, the exterior walls may require larger studs and (or) closer spacing or doubled studs.

FLOOR FRAMING

Lumber Joists—Approximate Span Capabilities

	Joist spacing		
Joist size	**12 in. o.c.**	**16 in. o.c.**	**24 in. o.c.**
2 × 6	10 ft	9 ft	8 ft
2 × 8	13 ft	12 ft	10 ft
2 × 10	17 ft	15 ft	12 ft
2 × 12	20 ft	17 ft	14 ft

I-Joists—Approximate Span Capabilities

	I-joist spacing		
I-joist depth	**12 in. o.c.**	**16 in. o.c.**	**24 in. o.c.**
$9\frac{1}{2}$ in.	18 ft	16 ft	14 ft
$11\frac{7}{8}$ in.	21 ft	19 ft	15 ft
14 in.	24 ft	20 ft	17 ft
16 in.	28 ft	24 ft	19 ft

Trussed Joists—Span Capabilities Unlike lumber or I-joists, trussed joists are custom manufactured for a project and are not trimmable. Typical spacing of trussed joists = 24 in. o.c.

$$\textbf{Approx. joist depth} = \frac{\textbf{joist span}}{\textbf{18}}$$

Example

If joist span = 30 ft, approximate joist depth = (30 × 12)/18 = 20 in.
Because trussed joists are made of 2 × 4 lumber, width of joists is $3\frac{1}{2}$ in.

ROOF FRAMING

Sawn Lumber Rafters—Approximate Span Capabilities

	Rafter spacing		
Rafter size	**12 in. o.c.**	**16 in. o.c.**	**24 in. o.c.**
2 × 6	14 ft	13 ft	11 ft
2 × 8	19 ft	17 ft	14 ft
2 × 10	24 ft	22 ft	18 ft

The above table applies to a roof live load (or snow load) ≤20 psf and a light roof cover (such as asphalt shingles). If snow load exceeds 20 psf and (or) roof cover is heavier (such as clay or concrete tiles), the span capability of a given rafter size will be smaller.

Sawn Lumber Ceiling Joists—Approximate Span Capabilities

	Joist spacing		
Joist size	**12 in. o.c.**	**16 in. o.c.**	**24 in. o.c.**
2 × 6	18 ft	16 ft	14 ft
2 × 8	24 ft	22 ft	18 ft
2 × 10	—	—	22 ft

The table above applies to an uninhabitable attic without storage.

A dash (—) indicates that the span capability exceeds 26 ft—the maximum sawn lumber length available.

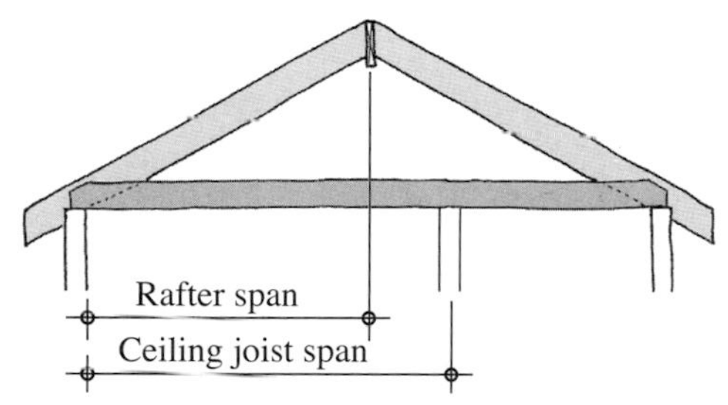

Because the span of a hip or valley rafter is larger than the common rafters, hip or valley rafters are generally one size larger than the common rafters. Thus, for 2 × 8 common rafters, a hip or valley rafter is generally 2 × 10.

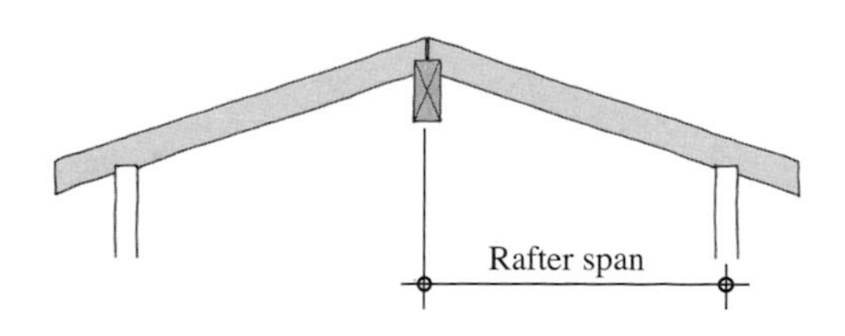

CONVENTIONAL LIGHT-GAUGE STEEL FRAME (LGSF) BUILDINGS

WALL FRAMING

Approximate Stud Size and Spacing

Number of stories	Stud size	Stud spacing
2 stories	550S162-33 350S162-43	24 in. o.c. 16 in. o.c.
1 story	350S162-33	24 in. o.c.

Stud size and spacing is based on sheet steel with a yield strength of 33 ksi. The other commonly used yield strength is 50 ksi (Chapter 17).

550S162-33 implies a stud (joist or rafter) with a web depth of 5.5 in. and flange width of 1.625 in., made of 33-mil- (0.033-in.) thick sheet steel (Chapter 17).

1. Stud size and spacing are for approximately 10-ft-high studs. For tall walls, such as those used in double-height spaces, larger stud size and (or) thicker sheet steel or closer spacing of studs may be required.
2. In cold climates, the stud size may be governed by insulation requirements. For example, 550S162 studs may be necessary where 350S162 studs are structurally adequate.
3. In high-wind regions, the exterior walls may require larger studs, studs made of thicker sheets, or higher yield strength.

FLOOR FRAMING

Joists—Approximate Span Capabilities

	Joist spacing	
Joist size	**16 in. o.c.**	**24 in. o.c.**
550S162-54	11 ft	10 ft
800S162-54	15 ft	13 ft
1000S162-54	18 ft	17 ft
1200S162-54	21 ft	18 ft

1. For a given joist size, the span capability can be increased by increasing sheet thickness. Commonly used sheet thickness for joists are 33, 43, 54, 68, and 97 mil.
2. The span capabilities given here are for the intermediate sheet thickness of 54 mil.

ROOF FRAMING

Rafters—Approximate Span Capabilities

	Rafter spacing	
Rafter size	**16 in. o.c.**	**24 in. o.c.**
550S162-54	18 ft	15 ft
800S162-54	24 ft	20 ft
1000S162-54	28 ft	23 ft
1200S162-54	30 ft	25 ft

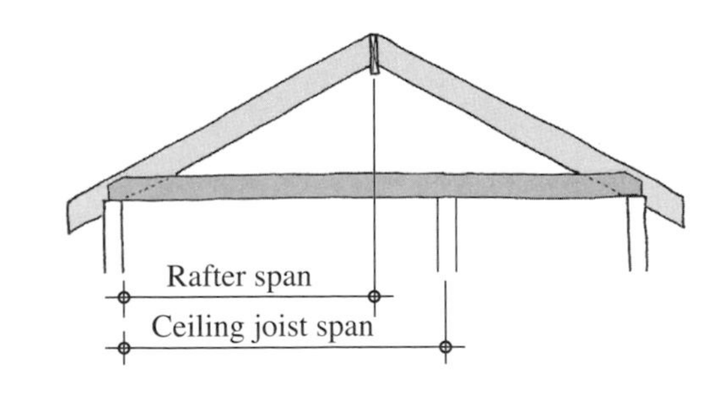

1. The table above applies to roof live load (or snow load) $\leq$20 psf and a light roof cover (such as asphalt shingles). If snow load exceeds 20 psf and (or) roof cover is heavier (such as clay or concrete tiles), the span capability of a given rafter size will be smaller.
2. For a given rafter size, the span capability can be increased by increasing sheet thickness. Commonly used sheet thickness for rafters are 33, 43, 54, 68 and 97 mil.
3. The span capabilities given here are for the intermediate sheet thickness of 54 mil.

Ceiling Joists—Approximate Span Capabilities

	Joist spacing	
Joist size	**16 in. o.c.**	**24 in. o.c.**
550S162-54	18 ft	15 ft
800S162-54	20 ft	18 ft
1000S162-54	21 ft	19 ft
1200S162-54	23 ft	20 ft

1. The above table applies to an uninhabitable attic without storage.
2. For a given joist size, the span capability can be increased by increasing sheet thickness. Commonly used sheet thickness for rafters are 33, 43, 54, 68 and 97 mil.
3. The span capabilities given here are for the intermediate sheet thickness of 54 mil.

STRUCTURAL STEEL FRAME BUILDINGS

Floors and Roofs

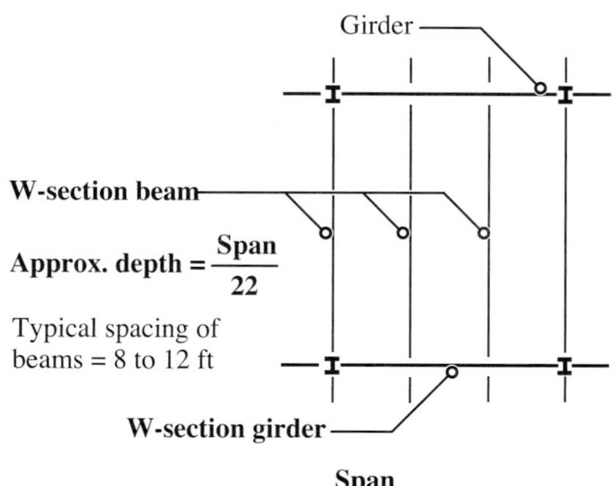

Nominal Depths of W-Sections in inches
44, 40, 36
33, 30, 27, 24, 21
18, 16, 14, 12, 10, 8, 6, 4

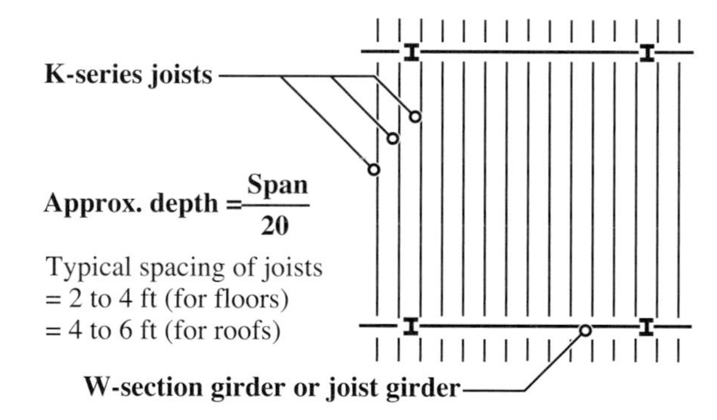

Approx. depth = Span/16 (for W-section girder)

Approx. depth = Span/14 (for joist girder)

Standard Depths of K-Series Joists in inches
8, 10, 12, 14, 16, ...30

Example 1

Determine the approximate depths of beams and girders for the floor of Figure 16.23.

Solution

The maximum span of girder is 26.5 ft. Therefore, approximate girder depth = $(26.5 \times 12)/16 = 19.9$ in., or 20 in. The nearest W-section is W21. Hence, use a W21 girder.

The maximum beam span is 30 ft. Therefore, approximate beam depth = $(30 \times 12)/22 = 16.4$ in. Hence, use a W16 beam.

COLUMNS

Interior Columns To approximate the size of an interior column, compute the total floor area supported by the column on all floors and then select the column size from the following table.

Column size	Maximum floor area on all floors
8 in. × 8 in.	2,000 sq. ft
10 in. × 10 in.	3,000 sq. ft
12 in. × 12 in.	4,500 sq. ft
14 in. × 14 in.	6,000 sq. ft

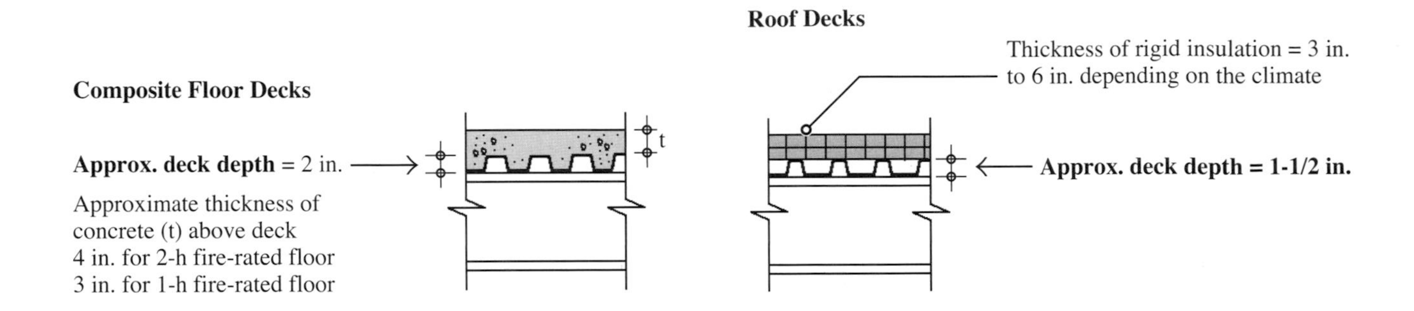

Exterior Columns Although the floor area supported by an exterior column is less than that of an interior column, exterior columns will generally support (exterior) walls, generally not supported by interior columns. Therefore, exterior columns may be assumed to be the same size as interior columns.

Example 2

If the building of Figure 16.23 is five-stories tall, determine the approximate size of the columns.

Solution

We will base the column size on the maximum floor area supported by an interior column. The largest bay size is (30.0 × 26.5) = 795 ft^2. The total area on all 5 floors is 5(795) = 3,975 ft^2. From the table above, an approximate column size is a W12 column (web depth approximately 12 in.).

SITE-CAST CONCRETE FRAME BUILDINGS

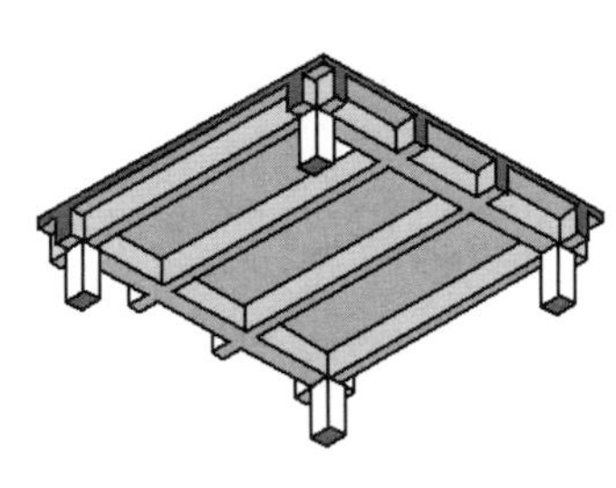

Beam and Girder-Supported One-Way Solid Slab

Reinforced Concrete

Girder

Approx. depth $= \dfrac{\textbf{span}}{\textbf{12}}$

Width = 0.6 (depth)

Beam

Approx. depth $= \dfrac{\textbf{span}}{\textbf{15}}$

Width = 0.6 (depth)

Slab

Approx. thickness $= \dfrac{\textbf{span}}{\textbf{24}}$

Typical distance between beams = 8 ft to 15 ft

Post-Tensioned Concrete

Girder

Approx. depth $= \dfrac{\textbf{span}}{\textbf{18}}$

Width = 0.6 (depth)

Beam

Approx. depth $= \dfrac{\textbf{span}}{\textbf{20}}$

Width = 0.6 (depth)

Slab

Approx. thickness $= \dfrac{\textbf{span}}{\textbf{40}}$

Typical distance between beams = 15 ft to 25 ft

Span for girder is clear distance between columns, and span for beams is clear distance between girders.

Depth of girder or beam includes thickness of slab.

Span for slab is clear distance between beams.

As far as possible, girder and beam depths should be the same. Therefore, the girder should span along the shorter direction.

Round beam and girder widths and depths to whole inches. Round slab thickness to $\frac{1}{2}$ inch, not less than 4 in. Slab thickness may be governed by fire-resistance requirements.

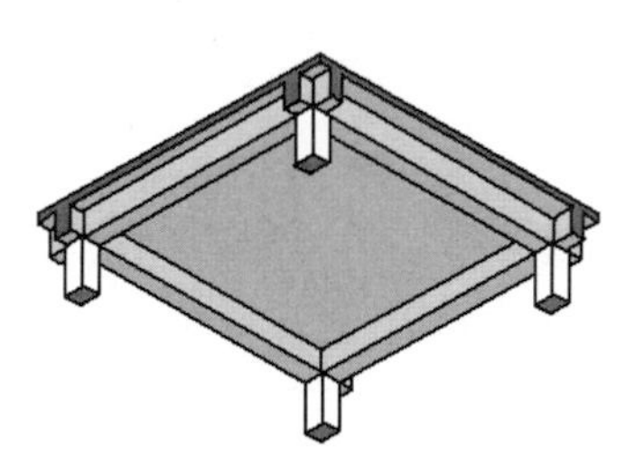

Two-way Solid Slab

Reinforced Concrete

Beam

Approx. depth $= \dfrac{\textbf{span}}{\textbf{15}}$

Width = 0.6 (depth)

Slab

Approx. thickness $= \dfrac{\textbf{span}}{\textbf{36}}$

Typical column spacing = 8 ft to 25 ft

Post-Tensioned Concrete

Beam

Approx. depth $= \dfrac{\textbf{span}}{\textbf{20}}$

Width = 0.6 (depth)

Slab

Approx. thickness $= \dfrac{\textbf{span}}{\textbf{48}}$

Typical column spacing = 25 ft to 30 ft

Span for beams is clear distance between columns.

Span for slab is the longer of the two clear distances between beams.

Round beam width and depth to whole inches. Round slab thickness to $\frac{1}{2}$ inch, not less than 4 in. Slab thickness may be governed by fire-resistance requirements.

As far as possible, distance between columns in both directions should be the same.

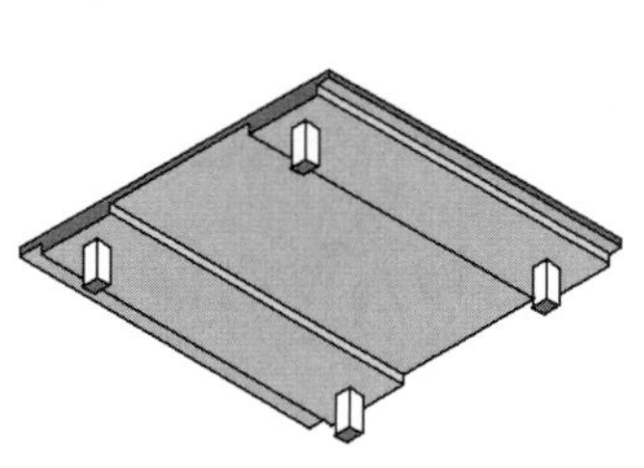

One-Way Band Beam Slab

Reinforced Concrete

Beam

Approx. depth = 2.0 to 2.5 (slab thickness)

Width = 0.25 to 0.3 (center-to-center beam spacing)

Slab

Approx. thickness $= \dfrac{\textbf{span}}{\textbf{24}}$

Typical column spacing = 25 ft to 30 ft

Post-Tensioned Concrete

Beam

Approx. depth = 2.0 to 2.5 (slab thickness)

Width = 0.25 to 0.3 (center-to-center beam spacing)

Slab

Approx. thickness $= \dfrac{\textbf{span}}{\textbf{40}}$

Typical column spacing = 30 ft to 40 ft

Span for slab is clear distance between beams.

Round beam width and depth to whole inches. Round slab thickness to $\frac{1}{2}$ inch. Slab thickness should not be less than 4 in. Slab thickness may be governed by fire-resistance requirements.

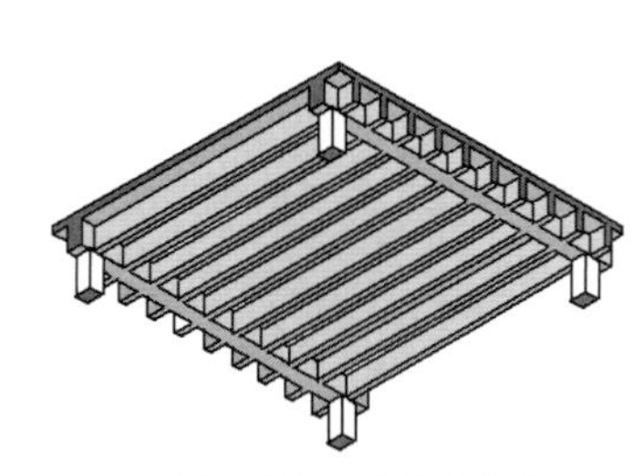

One-Way Joist Slab

Reinforced Concrete

Beam

Depth = same as joists

Width = 1.25 (joist depth)

Joist

Approx. depth $= \dfrac{\textbf{span}}{\textbf{18}}$

Width = 5 in. (standard-module pans), 7 in. for wide-module pans

Slab

Thickness = 3 in. (standard-module pans), **4 in.** (wide-module pans)

Typical column spacing = 25 ft to 40 ft

Post-Tensioned Concrete

Beam

Depth = same as joists

Width = 1.25 (joist depth)

Joist

Approx. depth $= \dfrac{\textbf{span}}{\textbf{25}}$

Width = 5 in. (standard-module pans), 7 in. for wide-module pans

Slab

Thickness = 3 in. (standard-module pans), **4 in.** (wide-module pans)

Typical column spacing = 35 ft to 50 ft

Span for joists is clear distance between beams.

Depth of joists or beams includes slab thickness. Round joist depth to pan depth + slab thickness.

Slab thickness may be governed by fire-resistance requirements.

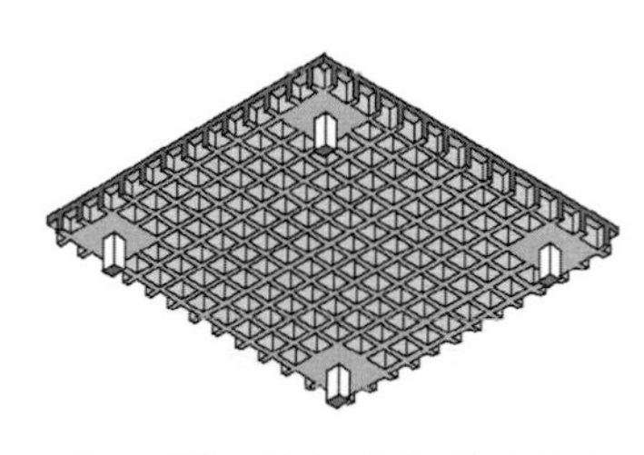

Two-Way Joist (Waffle) Slab

Reinforced Concrete

Joist

Approx. depth $= \dfrac{\textbf{span}}{\textbf{22}}$

Width = 5 in. to 6 in. depending on dome size

Slab

Thickness = 3 in. to 4 in.

Typical column spacing = 30 ft to 45 ft

Post-Tensioned Concrete

Joist

Approx. depth $= \dfrac{\textbf{span}}{\textbf{30}}$

Width = 5 in. to 6 in. depending on dome size

Slab

Thickness = 3 in. to 4 in.

Typical column spacing = 40 ft to 60 ft

Depth of joists includes slab thickness. Round joist depth to dome depth + slab thickness.

Slab thickness may be governed by fire-resistance requirements.

The number of domes filled around columns is a function of column spacing, floor load and dome size. Beams may be used in place of filled domes.

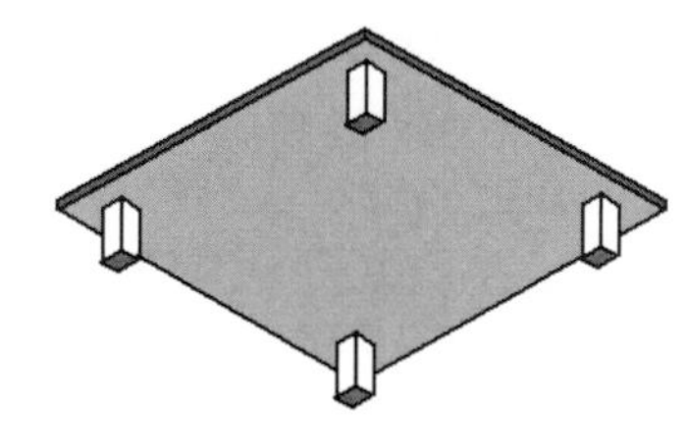

Flat Plate

Reinforced Concrete

Slab

Approx. thickness $= \dfrac{\textbf{span}}{\textbf{30}}$

Typical column spacing = 15 ft to 20 ft

Post-Tensioned Concrete

Slab

Approx. thickness $= \dfrac{\textbf{span}}{\textbf{40}}$

Typical column spacing = 20 ft to 25 ft

Slab span is (longer) clear distance between columns.

As far as possible, distance between columns in both directions should be the same.

Slab thickness may be governed by fire-resistance requirements.

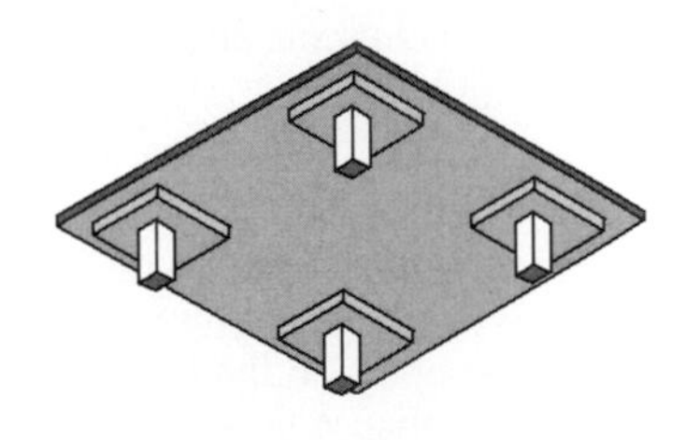

Flat Slab

Reinforced Concrete Slab

Approx. thickness $= \frac{\textbf{span}}{\textbf{35}}$

Typical column spacing = 20 ft to 30 ft

Post-Tensioned Concrete Slab

Approx. thickness $= \frac{\textbf{span}}{\textbf{45}}$

Typical column spacing = 30 ft to 45 ft

Slab span is (longer) clear distance between columns.

As far as possible, distance between columns in both directions should be the same.

Slab thickness may be governed by fire-resistance requirements.

Columns

Column size depends on various factors such as the total floor area supported by the column, concrete strength, amount of reinforcement, column height and whether the column is part of the lateral load resistance system of the building. For conventional buildings in which the lateral load is resisted by shear walls, the following rule of thumb may be used for the approximate size of an interior column.

$$\text{Area of column} = \frac{\text{total floor area supported by column}}{\text{10 to 20 (depending on concrete strength)}} \text{ not less than 10 in. in any direction}$$

As far as possible, column size should be the same for interior and exterior columns and from floor to floor. Note that the amount of reinforcement and concrete strength in a column can be increased toward lower floors.

Example: Determine the column size required for a square reinforced concrete column at the ground floor of a building supporting 1,000 ft^2 at each floor. Number of floors = 4.

Solution:

$$\text{Area of column} = \frac{4{,}000}{10} = 400 \text{ in.}^2$$

Hence, approx. column size = 20 in. × 20 in.

PRECAST-PRESTRESSED CONCRETE MEMBERS

Double-T Units

Approx. depth, h $= \frac{\textbf{Span}}{\textbf{28}}$

Hollow-core slabs

Approx. depth, h $= \frac{\textbf{Span}}{\textbf{40}}$

LOAD-BEARING MASONRY AND CONCRETE BUILDINGS

Reinforced-Concrete Masonry Bearing Walls in Residential (e.g., Apartment and Hotel) Buildings

8-in.-thick, CMU walls for up to 8-floor-high buildings.

10-in.-thick CMU walls for 11- to 15-floor-high buildings.

12-in.-thick CMU walls for 16- to 20-floor-high buildings.

Upper floors in 10-in.- or 12-in.-thick wall structures may be constructed of 8-in.-thick walls.

Reinforced-Concrete Masonry Bearing Walls in Single-Story, Long-Span Structures (e.g., Gymnasiums)

10-in.- or 12-in.-thick CMU walls, depending on the span.

Site-Cast Reinforced-Concrete Bearing Walls in Residential Buildings

6-in.-thick site-cast reinforced concrete walls for up to 20-floor-high buildings.

Precast-Concrete Tilt-Up Walls

The thickness of tilt-up walls in a single-story building can be approximated by dividing wall height by 48 (not less than 6 in.). For two- or three-story buildings, slightly thicker walls may be needed.

GLOSSARY

A

Abrasive blasting A method of roughening a concrete or steel surface by an abrasive medium, such as sand, under high air pressure.
Acid etching A method used to expose the aggregate on a concrete surface through the use of a dilute acid wash.
Acoustical materials See *sound-absorbing materials.*
ACQ See *alkaline copper quat.*
ADA Americans with Disabilities Act.
Adhered veneer A thin finish material, such as thin brick or tile, that is adhered directly to a backup wall with mortar or adhesive.
Adobe Sun-dried clay masonry units, generally molded on site.
Aggregate Granular materials, including gravel, sand, crushed stone, expanded shale, and slag, that serve as filler in a concrete mix.
Air retarder A microperforated plastic sheet that prevents the passage of outside air through a wall or roof assembly but allows water vapor to pass through.
Air seasoning A method of drying lumber in which lumber members are stacked to allow free circulation of air around them.
Alkaline copper quat (ACQ) A chemical used for wood preservation.
Alloy A chemical combination of two or more metals.
Anchored veneer A cladding system (usually exterior) in which masonry units are mechanically anchored to the backup wall.
Anisotropic material A material that has different properties in different directions (e.g., wood). See also *isotropic material.*
Annealed glass Flat glass obtained by heating and then gradually cooling it to relieve internal stresses that develop during the early stage of its manufacturing process. Annealed glass is the basic form of flat glass.
Arch A curved structural member that spans over a wall opening and carries loads in compression.
Arching action A structural action in which a vertical load on a wall or beam translates into two inclined loads; generally occurs in a deep beam or wall.
Architectural concrete Concrete that is obtained from a specially designed formwork to give a smooth, patterned, textured or otherwise detailed surface finish on concrete.
Architectural precast concrete Nonstructural precast concrete, as in a concrete curtain wall.
Architectural sheet-metal roof A roof covering made of sheet metal (e.g., standing seam roof, batten seam roof, etc.), applied over a structural roof deck and underlayment.
Ashlar Stone dressed with square-cut edges and laid in a coursed or random pattern in a wall.
Asphalt A tacky black liquid derivative of the petroleum-distillation process, used in roofing and waterproofing applications.
Asphalt-saturated felt An organic paper felt that is saturated with asphalt, typically used as an underlayment in steep roofs (sometimes in walls).
Asphalt shingles A roofing unit composed of heavy organic paper or fiberglass felt saturated (or treated and coated) with asphalt and faced with mineral granules.
Asphalt-treated felt A fiberglass felt that is treated with asphalt, typically used in built-up roofs.
ASTM American Society for Testing and Materials.
Attactic polypropylene (APP) A polymer used to obtain an APP modified-bitumen roof membrane.
Auger A tool for drilling holes in wood or soil. Large mechanized augers drill holes in the ground for piers, piles, or caissons.
Axial A direction that is parallel to the length of a structural member. For a column, the axial direction is vertical; for a horizontal beam, the axial direction is horizontal.

B

Backer rod A compressible spherical foam rod used to control the depth of an elastomeric sealant, which allows the tooling of the sealant.
Backfill Soil used to fill the space between the foundation and the boundaries of an excavation. Generally, the soil excavated from foundation trenches and pits is used as backfill, unless found unsuitable.
Backup wall Load-bearing or non-load-bearing wall to which exterior cladding is adhered or anchored.
Balance point temperature Exterior temperature at which the heat loss through the building envelope equals heat gain from interior activities such as cooking, lighting, and human occupancy.
Ballast (in roofing) Aggregate or concrete pavers used over a loose-laid, single-ply roof membrane to resist wind uplift and protect the membrane from degradation by solar radiation.
Balloon frame One of the two wood light frame construction systems in which the studs extend from the foundation to the roof, bypassing the intermediate floor(s). See also *platform frame.*
Ballooning Interior pressurization of building due to high wind. It typically occurs in buildings that have large openings on only one exterior face, such as an aircraft hangar.
Baluster Vertical support of a staircase handrail.
Barrier wall An exterior wall that resists water leakage by providing an impervious barrier or by serving as a water reservoir.
Basement Below-ground portion of a building.
Basic wind speed Peak 3-second gust wind speed with a 50-year recurrence interval, used to determine wind loads on a building.
Batt Precut units of fiberglass insulation sized to fit between wall studs, ceiling joists, or rafters.
Batten A long, narrow strip of wood.
Batten bar A long metal or plastic bar used in mechanically fastened single-ply roof membrane.
Bay In a building structure, regular, repeated space defined by four adjacent columns and beams (and/or girders) that span between them.

Beamless floor A reinforced-concrete slab supported directly on columns (without supporting beams).
Bearing pad A block of high-strength plastic, metal, or rubber placed under a beam or curtain wall to distribute the load on a larger area of the supporting member.
Bearing plate A steel plate welded to the base of a steel column to distribute the column load on a larger area of the footing. Also a steel plate welded to the ends of a steel joist or beam.
Bearing wall A wall that carries gravity loads from the floor(s) and/or roof.
Bed joint Horizontal mortar joint in a masonry wall. See also *head joint.*
Bending Deformation of a structural member caused by a force acting perpendicular to the member axis.
Bent glass Flat glass bent to shape by heating.
Bentonite clay A highly expansive clay typically used as slurry to fill the excavation for a concrete wall.
Billet A large rectangular bar of cast steel used to roll finished shapes, such as smaller bars and rods.
Bioclimatic glass curtain wall See *double-skin glass curtain wall.*
Bitumen A waterproofing compound (asphalt or coal tar) used for roofing and waterproofing.
Blast furnace A furnace used to melt iron ore.
Blocking Short pieces of lumber that fit between adjacent wood joists, rafters, or studs to provide lateral stability and additional nailing members.
Bloom Steel cast in a large rectangular section, which is then used to make wide-flange sections, angles, and channels by the hot-rolling process.
Board foot The amount of lumber contained in a 1-in.-thick board that measures 1 ft by 1 ft.
Board lumber Lumber that is less than 2 in. (nominal) thick.
Bolt A threaded steel rod with a fixed head at one end and closed with a nut at the other end.
Bond beam A continuous reinforced beam embedded in a masonry wall (generally at the floor and roof levels) formed either by bond beam masonry units or site-cast concrete.
Bond breaker A material used to prevent the adhesion of an elastomeric sealant to a backup surface.
Bond pattern The pattern used to arrange masonry units in a wall.
Brick A small masonry unit, generally manufactured from clay. Concrete bricks are also available.
Brick ledge Depressed portion of a concrete foundation to support the first story of brick veneer.
Bridging Structural members, laid perpendicular to wood or steel framing members, to stabilize them against overturning and brace them against buckling.
Brittle material A material that deforms little before failure. It is generally stronger in compression than tension.
Brown coat The second layer of base coat in portland cement plaster (stucco) applied on metal lath.
Buckling A type of failure that results in the sudden bending of a slender structural member subjected to excessive axial loading.
Building brick A type of clay brick recommended for interior applications, with less-stringent durability specifications than facing brick.
Building code A legal document that regulates the design and construction of buildings to ensure that the buildings meet minimum standards of health, safety, and welfare.
Building component A part of a building that performs a specific function, that is, a window, door, or wall assembly.
Building envelope All building components that separate the building's interior from the exterior environment.
Building expansion joint See *building separation joint.*
Building movement joint See *building separation joint.*
Building separation joint A continuous joint that extends through all floors and roof of a building, dividing the building into smaller buildings that can move independently of each other, generally $1\frac{1}{2}$ to 2 in. wide.
Built-up beam A beam made by combining two or more standard structural members.
Built-up roof (BUR) membrane A multi-ply roof membrane composed of alternating layers of felt and moppings of bitumen with a top cover of aggregate.

C

Caisson A large-diameter, deep reinforced-concrete foundation element made by drilling a hole into the ground and filling it with concrete; an enclosure that permits excavation work to be carried out under water.
Calcium silicate masonry unit Sand-lime bricks or blocks cured in an autoclave to enhance the pozzolanic reaction between sand and lime.
Camber An upward curvature introduced in a beam to ensure that it will be flat under dead loads.
Cant strip A triangular strip of perlite board or pressure-treated wood to provide a smooth transition between a horizontal and vertical surface on a roof, required with a built-up or modified-bitumen roof membrane.
Capital Upper part of a column, generally of a larger cross section than the column.
Carriage An inclined structural member (beam) that supports a stair.
Cast-in-place (CIP) concrete See *site-cast concrete.*
Cast iron An iron-carbon alloy formed by casting molten iron in a sand mold and milling to the final shape—relatively brittle and nonmalleable.
Caulk Soft, pliable material (such as polyisobutyline and acrylic) used to fill narrow, nonmovement joints between building components.
Ceiling The visible, overhead interior surface of a room.
Cells Voids in concrete masonry units; voids in foamed insulation; microscopic voids in wood.
Centering Temporary formwork for constructing an arch, vault, or dome.
Ceramic material A material produced by firing clay in a high-temperature kiln, such as brick, tile, or porcelain.
Certified wood Wood obtained from sustainable forestry practices.
Chair A small support to raise steel reinforcing bars above the surface of the formwork or ground.
Chimney A hollow vertical structure lined with an internal flue to carry smoke and other effluents from wood or charcoal fires used for heating a building.
Chord Member of a truss; can be the top or bottom chord of a truss.
CIP concrete See *cast-in-place concrete.*
Cladding Exterior weather-resistant layer of a wall assembly.
Clad window Framing members of a wood window clad in aluminum, PVC, or fiberglass on the outside to increase wood's durability.
Clapboard A type of horizontal lap siding

CMU See *concrete masonry unit.*
Coal tar A tacky black liquid derivative of the coal-distillation process, used in roofing and water-proofing applications.
Coefficient of friction A measure of the resistance to sliding between contact surfaces of two components.
Cohesive soil A soil, such as clay, whose particles tend to adhere (cohere) to each other in the presence of water.
Cold joint See *construction joint.*
Column An upright vertical structural member that supports a slab, beam, or truss.
Column cover Preformed exterior cladding element that covers a column for aesthetic purposes.
Column splice A method of connecting the lower part of a tall steel column with that of the upper part. Column splices are typically provided every two stories.
Combustible A material that will ignite in the presence of an open flame or high temperature.
Commissioning A process of adjusting the performance of individual components of the mechanical and electrical systems of a building to achieve energy efficiency.
Compaction A decrease in the amount of void space in a soil mass brought about by mechanical action (e.g., tamping) or natural process, resulting in a densified soil mass.
Composite deck Corrugated-steel floor deck, which acts as formwork and primary reinforcement for a concrete slab.
Compression Stress created in a structural member as a result of bending or axial compressive force.
Compressive strength Measure of the ability of a material to resist compressive force.
Concave joint A type of mortar joint profile that is most weather resistant.
Concrete A composite material consisting of portland cement, coarse aggregate (crushed stone), fine aggregate (sand), and water.
Concrete admixture Material added to concrete mix to influence its performance.
Concrete block See *concrete masonry unit.*
Concrete masonry unit Standardized masonry unit made of concrete and consisting of face shells and webs surrounding two or three voids, called cells.
Condensation resistance factor A measure of the potential for condensation to occur on a glazed assembly (window or glass curtain wall).
Conduction A mode of heat transfer in a solid.
Conduit A hollow metal or plastic tube used as protective cover for electric wires.
Construction documents Documents used by the contractor to construct a building, including drawings and specifications.
Construction drawings Drawings used by a contractor to construct a building.
Construction joint A nonmovement joint resulting when fresh concrete is placed against previously placed concrete.
Construction manager A manager hired by the owner to coordinate and supervise the work of multiple contractors. A project with a construction manager does not generally have a general contractor.
Control joint A sawed or tooled joint on the top surface of a concrete slab-on-ground; a continuous vertical joint in a concrete masonry wall.
Convection A mode of heat transfer in liquids and gases.
Conveying equipment Mechanical equipment that is used to move people or products vertically and/or horizontally through a building.
Coping A protective cap or cover used on the exposed top of a wall, typically sloped to shed water.
Core holes Holes in an extruded brick, to provide uniform drying and firing.
Corrosion An electrochemical process that deteriorates metal surfaces exposed to air, water, or excessive humidity.
Counterflashing A removable sheet-metal flashing that laps over membrane flashing to prevent water from penetrating between the membrane and the wall.
Course A single horizontal layer of units in a masonry wall.
Coursed rubble Irregularly shaped and sized stone laid with periodically aligned bed joints.
Crane A type of construction equipment used to hoist and place heavy building components or materials during construction.
Crawl space Air space between an elevated ground floor and the ground.
Creep Permanent deformation of a component under sustained loads, generally of concern in concrete and wood.
CRF See *condensation resistance factor.*
Cricket A pyramidal formation on a roof to divert water to drains or on an easily drained area of the roof.
CSI Construction Specification Institute.
CSMU See *calcium silicate masonry unit.*
Curtain wall Exterior wall cladding system suspended from or supported by the structural frame of the building.

Dampproofing A material applied to the exterior of a wall to resist the intrusion of capillary water.
Dead load Loads created by the weight of building components.
Decibel (dB) A measure of the loudness of sound.
Deep foundation A foundation element that extends deep below the ground to reach bedrock or higher-bearing-capacity soil.
Deformation Change in size or shape of a structural member as a result of an applied load.
Dew point The temperature at which the relative humidity of an air mass reaches 100% and the water vapor in air converts to (liquid) water, that is, condenses.
Diagonal brace A linear, diagonal stiffening element against lateral loads.
Die A tool with an orifice (hole) through which a soft, heated metal or plastic or a column of wet clay is pushed to give a long, continuous element.
Dimension lumber Lumber that ranges from 2 in. to 4 in. (nominal dimensions) thick.
Dimension stone Stone that has been fabricated to specific dimensions, texture, and finish for use in buildings.
Dimmable glass A glass that changes from transparent to translucent or opaque condition when exposed to sun (photochromic glass) or electrical current (electrochromic glass).
Dome A curved roof structure over a round or rectangular space.
Double-skin glass curtain wall A glass curtain wall assembly composed of two layers of glazing separated by 1 to 5 ft of air space, also called double-skin facade or bioclimatic glass curtain wall.
Double-strength (DS) glass $\frac{1}{8}$-in.-thick flat glass.
Double-Tee A precast concrete floor or roof element to span long distances.

Dowel A steel or wood pin used with an adhesive to connect adjacent pieces of wood; a steel reinforcing bar that projects from a foundation to form a splice with reinforcing bars in a concrete column or wall above the foundation.
Drainage wall A cavity wall with an air space between the exterior cladding and the inner backup wall, allowing water to drain out through weep holes.
Drip edge Perimeter or edge flashing with an angled lip to keep water away from a vertical surface below.
Drop panels A widening of the underside of a concrete slab at a column location.
Dry glazing Use of preformed compression gaskets to seal the glass against the metal frame of a window or metal-glass curtain wall.
Drywall See *gypsum board.*
Ductile material A material that produces large deformations under load before failure.
Dynamic joint A joint that allows a predetermined amount of unrestrained movement between components.

E

Eave The low end of a sloped roof.
Edge block A resilient material used to prevent the vertical edge of glass from touching the frame of a window or curtain wall.
Efflorescence A white deposit of water-soluble salts on an exterior masonry or concrete wall caused by the wetting of the wall.
Egress window A window in a habitable room of a dwelling that opens to a minimum size for use as a fire escape.
EIFS See *exterior insulation and finish system.*
Elasticity The ability of a deformed material to return to its original shape and size after the removal of the load, such as rubber.
Elastomeric A material that exhibits elastic (rubberlike) properties.
Elastomeric sealant A synthetic semiliquid material to seal the joints between adjacent building components, which becomes an elastic material after curing.
Elevated slab An above-ground floor or roof slab, supported on columns and/or beams and forming an integral part of a structural frame.
Embodied energy Total amount of energy required to extract raw materials and manufacture a building material.
Emissivity A property of the surface of a material that governs its potential to emit radiation; value lies between 0 and 1.0.
End-bearing pile A deep foundation element within the ground that transfers load to bedrock in contrast to a friction pile that transfers load to the surrounding soil. See also *friction pile.*
Engineered fill Soil and/or aggregate with specified proportions and properties compacted into place so that the resulting product has predictable physical characteristics.
Engineered wood products A manufactured wood product rated for structural applications, such as plywood, oriented strandboard, or laminated veneer lumber; see also *manufactured wood products* and *industrial wood products.*
Epoxy A polymer-based adhesive.
Ethylene propylene diene monomer (EPDM) A stretchable polymeric membrane used as single-ply roof membrane.
Expanded polystyrene Open-cell foamed plastic insulating material formed into rigid boards.
Expansion anchor An anchor that allows expansion and contraction while connecting the vertical framing members of a glass curtain wall to the building structure.
Expansion joint A narrow space between two building components that allows their unrestrained expansion and contraction.
Exposed aggregate A type of concrete finish with coarse aggregate exposed at the surface.
Exposed ceiling A ceiling where all structural components and mechanical and electrical services are left exposed to below.
Exposure The exposed portion of lapped roofing felt or shingles.
Exterior insulation and finish system A stuccolike exterior finish that includes a layer of foam insulation, fiberglass reinforcing mesh, and one or two coats of a polymer-based finish, also called synthetic stucco.
Extrusion See *die.*
Extruded brick Wet clay extruded through a die and cut to size before drying and firing. See also *molded brick.*
Extruded polystyrene Closed-cell foamed plastic insulating material formed as rigid boards by extrusion process.

F

Face shell Longitudinal walls of a concrete masonry unit.
Facing brick Brick used in an exposed exterior masonry wall, controlled for dimensional variations, warpage, and durability.
Factor of safety A safety margin added to structural calculations to account for our imprecise knowledge of material strength, loads and theories pertaining to structural analysis and design, defined as the actual strength of a component divided by the strength required to carry the load.
Ferrous metal A metal that contains mainly iron.
Ferrule A threaded cylindrical sleeve that accepts a bolt, typically embedded in a precast-concrete member.
Fiberglass Mineral fiber spun from glass.
Fibrous insulation An insulating material that has a high R-value due to the air contained between the fibers.
Fill Soil or aggregate that fills the area under the foundations or between the exterior of a foundation and the boundaries of an excavation. See also *engineered fill and backfill.*
Filter fabric A porous filter fabric for filtering out silt, sand, or debris in a protected membrane roof, French drain or foundation drain.
Finger joint A method of joining short lengths of lumber using interlocking "fingers" at the end of each piece and gluing the members together.
Finish coat The last layer of material that provides final color and texture on a surface, such as the Portland cement based layer on stucco or polymer-based layer on EIFS.
Fire-rated glazing A glazing system that will resist standard fire for the measured duration.
Fire rating The ability of a building assembly to endure fire, measured in hours or minutes of time and determined from standardized full-scale tests.
Fire resistance rating See *fire rating.*
Fire retardant lumber Wood that is pressure treated with chemicals to increase its fire resistance.
Firestopping A noncombustible material used to seal the space around a penetration or joint in a fire-rated wall, floor, or roof.
Fixed window A window unit where the glass is permanently fixed into the frame.
Flame finish A type of finish on a stone surface (generally granite) using a torch on wetted stone to break the surface, creating a rough, slip-resistant finish; also called thermal finish.
Flame spread index (FSI) A measure of the rate at which flames spread on the surface of an interior finish.

Flange Top and bottom components of an I-section or C-section steel beam.
Flashing A flexible metal or plastic used at roof terminations, edges, or penetrations to increase the strength and water resistance of the roof; a flexible metal or plastic sheet used as drainage channel at the base of a cavity wall.
Flat glass Planar glass sheet (window glass), as opposed to glass block or other glass products.
Flat plate A reinforced concrete slab of uniform thickness supported directly on columns.
Flat-sawn lumber See *plain-sawn lumber.*
Flat slab A reinforced concrete slab, like flat plate, but the slab is thickened at the columns.
Flex anchor An L-shaped steel pin that connects to the GFRC skin with a bonding pad and is welded to the supporting light-gauge steel frame.
Float glass Flat glass manufactured using the float glass process.
Floating The process of smoothing a freshly placed concrete surface after it has been struck (leveled). See also *striking.*
Floating floor A floor assembly consisting of a resilient underlayment to isolate the finished floor from the subfloor; can also be a double floor with an intervening air space.
Fluid-applied roof membrane A combination of sprayed polyurethane foam insulation topped by a fluid that cures to a membrane.
Flush joint Troweled masonry mortar joint, where the mortar is finished flush with the masonry face.
Flux A material added to the primary material to lower the primary material's melting-point temperature, saving energy.
Fly ash A waste product obtained from the combustion of coal in a thermal power plant (a pozzolanic material), used as concrete admixture.
Flying form A large, prefabricated formwork for concrete floors, which can be lifted (flown) in position by a crane and reused multiple times.
Foamed concrete See *insulating concrete.*
Foamed plastic Plastic insulating material that contains small volumes of air or other gas.
Footing The bottom portion of a foundation resting directly on the supporting soil with an area larger than that of the supported wall or column.
Form ties Specially shaped steel wires that are used to connect and separate opposing panels of formwork for a concrete wall.
Formwork Temporary support system for fresh concrete to give it the desired shape (form) until it cures to hardness.
Frame structure Building structure that is composed of columns and beams, called a skeleton frame in the case of a steel frame structure.
Framing plan Architectural plan that delineates the organization and dimensions of columns, beams, and load-bearing walls (if any).
Freeze-thaw A cycle of freezing and thawing of water within water-absorptive materials, such as soil, masonry walls, or concrete.
Friction pile A load-bearing pile that carries load by friction developed between the surface area of the pile and the soil. See also *end-bearing pile.*
Fritted glass Glass made semiopaque or opaque by the application of patterns of tiny dots or lines of ceramic material on one surface of flat glass.
Frog A small depression in a molded brick.
Frost line The depth of soil below which groundwater will not freeze.
Furring Narrow wood or formed sheet-metal sections applied to a wall or ceiling, used for supporting and attaching a finish material on the wall.

Gable Triangular portion of an exterior wall between the eave and the ridge in a gable roof.
Gable roof A roof with two sloping planes that meet at a point at the top, called a ridge.
Gable vent Passive exhaust ventilation unit located high on the gable end, used to exhaust attic air.
Galvanic corrosion Corrosion that occurs when two dissimilar metals are in direct contact with each other.
Galvanizing Method of coating steel with zinc to provide a corrosion-resistant (sacrificial) coating.
Gasket A shaped piece of resilient material that provides a weatherproof seal between the glass and the frame in a window or metal-glass curtain wall. See also *dry glazing.*
General contractor A contractor who has the final responsibility for the construction of the project including site safety, supervision, and coordination of the work of all subcontractors.
Geodesic dome A hemispherical dome formed by short, crisscross linear members that run along great circles of the sphere, creating triangles or polygons on the surface of the dome.
Geotechnical report A document prepared by a geotechnical (soils) engineer that identifies and describes the properties of soil and underlying geology of the building site.
GFRC Glass fiber–reinforced concrete.
GFRP Glass fiber–reinforced plastic.
Girder A large beam (also called a primary beam) that carries the loads from secondary beams or joists to columns in a frame structure.
Glass block See *glass masonry unit.*
Glass fiber–reinforced concrete Glass fibers reinforce a mix of portland cement, sand, and water that is sprayed on formwork to form a panel that is integrated with a light-gauge steel frame and used in curtain wall assemblies.
Glass fiber–reinforced plastic A plastic material reinforced with fiberglass, molded to the desired shape.
Glass masonry unit A hollow glass unit used in non-load-bearing wall applications to provide a translucent wall.
Glazing An assembly of flat glass and its supporting frame.
Glazing compound A semisolid compound mastic, which cures to hardness, used to bed small panes of glass in a window frame, not commonly used in contemporary buildings.
Glue laminated Lengths of dimension lumber, glued and laminated together to create a structural member of a large cross section.
Glulam See *glue laminated.*
GMU Glass masonry unit.
Gradation Distribution by size of soil particles within a soil mass; also applies to aggregates in a concrete mix.
Grade beam A reinforced-concrete beam constructed at ground level.
Graded lumber Lumber with a grade stamp.
Grade stamp Stamp affixed by a grading (inspection) agency on lumber or a wood panel that describes its in-service performance.
Grading Mechanically moving soil on a site to predetermined ground elevations.

Grain structure Pattern and density of fibers that compose wood.
Granite Strongest and densest of stones; it takes good polish and weathers more slowly than other stones, obtained from igneous rock.
Gravity load Building loads caused by gravity and acting in the vertical direction; also called vertical loads.
Green roof A landscaped roof.
Grille A grating or screen that protects an opening while allowing the passage of air.
Grout A high-slump (semiliquid) mortar that flows well enough to be placed or pumped between the voids in a masonry wall; a mortarlike, high-strength cementitious material for filling the space between the steel base plate of a steel column and concrete footing; a cementitious material to fill joints between ceramic or quarry floor tiles or panels after they have been laid.
Growth ring A nearly circular ring of wood fibers; a tree adds approximately one per year.
Guard rail A horizontal rail and associated supports to prevent people falling from a stair or balcony.
Gunite See *shotcrete.*
Gusset plate A plate that connects two structural steel members, commonly used in a large steel truss.
Gypsum board A building panel faced on two sides by paper (or vinyl) with a gypsum core. Typically used on fire-rated interior walls. It is also called drywall, wallboard, or gypsum wallboard.

H

Handrail Diagonal rail at hand level adjacent to a stairway to provide support for people using the stairs, as distinguished from a guardrail.
Hardwood Wood obtained from trees that are deciduous and have broad leaves. See also *softwood.*
Head Top of a window or door.
Header The short face of a brick when laid in a horizontal position in a course of bricks, historically used to tie together two wythes of a thick brick wall; a beam (lintel) above a door or window opening in a wood frame or light-gauge steel construction.
Head joint Vertical mortar joint in a masonry wall. See also *bed joint.*
Headroom Minimum code-required clearance between a floor and an overhead projection or ceiling.
Heartwood The central portion of a tree trunk that no longer conducts nutrients, generally darker than sapwood. See also *sapwood.*
Heat-absorbing glass See *tinted glass.*
Heat-soaked glass A type of tempered glass obtained from a process that reduces (or eliminates) the possibility of spontaneous breakage of tempered glass during its service life. See also *tempered glass.*
Heat-strengthened glass A glass obtained by heating annealed (basic) glass to a high temperature and then suddenly cooling it; approximately twice as strong as annealed glass. See also *annealed glass* and *tempered glass.*
Heavy timber Sawn lumber with both cross-sectional dimensions greater than 5 in. (nominal).
Heavy-timber construction A construction type where the floors, roofs, and structural frame are composed of heavy sections of wood without added fire protection.
Hip The edge of a hip roof where adjacent slopes meet; shows as a diagonal line in the plan of the roof.
Hip roof A roof formed by four sloping planes that intersect to form a pyramidical or elongated pyramidal shape.
Hollow-core door Wood or metal veneer placed on both sides of a frame to build a door that is partially or almost entirely hollow within.
Hollow structural section (HSS) Square or rectangular tubular steel section used as columns or beams in a steel frame structure or as components of a steel truss.
Honed finish A type of smooth finish on stone or metal created by using abrasives.
Horizontal diaphragm A structural engineering term that refers to the floors and roof in a building when they function as part of the lateral load resisting system.
Hydrated lime Calcium hydroxide, produced by combining quick lime and water to produce a relatively benign lime, used in powder form in masonry mortar. See also *slaked lime.*
Hydration Chemical reaction between portland cement and water causing the mix to become hard.
Hydraulic cement A cement (glue) that, after setting, does not chemically react with or dissolve in water, such as portland cement.

I

IBC International Building Code.
Ice dam protection membrane A self-adhering waterproof membrane that prevents water from backing up under roof shingles when an ice dam forms at the eave.
IGU Insulating glass unit.
IMG Intumescent multilaminate glass
Impact insulation class (IIC) Measure of the ability of an assembly to resist the transmission of structure-borne sounds.
Impact sound Structure-borne sound produced by an impact on a building component.
Impact wrench A tool for tightening bolts and nuts through rapidly repeated torque impulses produced by electric or pneumatic power.
Industrial wood products Manufactured wood products used for nonstructural applications, including particle board, medium-density fiberboard (MDF), and high-density fiberboard (HDF). See also *manufactured wood products* and *engineered wood products.*
Infrared radiation Long-wave radiation emitted by objects at low temperature
Infrastructure Basic organizational systems that provide services to a building, including roads, power, water, and electricity.
Initial rate of absorption (IRA) A measure of the ability of clay bricks to absorb water, used to determine if bricks need to be wetted before laying.
Inside glazed Glazing system where the glass is set into the frame from the inside of a building. See also *outside glazed.*
Insulated metal panels Metal panels consisting of polyurethane foam sandwiched between and bonded to two metal sheets, used in curtain wall applications.
Insulating concrete Lightweight concrete consisting of portland cement, water, and expanded aggregate (perlite or vermiculite), primarily used as low-slope roof insulation. A type of insulating concrete (called foamed concrete) does not contain aggregate but contains air particles. A fire-resistant insulation material for roofs and floors.
Insulating glass unit A sealed assembly of two layers of glass separated by an air or gas-filled chamber. The assembly has a higher R-value than a single sheet of glass.

Insulation See *thermal insulation* and *sound insulation.*
Interior finish class Rating of interior finish materials as Class A, B, or C, based on flame-spread index (FSI) and smoke-developed index (SDI).
Intumescent paint A paint that swells when exposed to heat, creating an insulating layer that protects steel from short-term exposure to fire.
Inverted roof See *protected membrane roof.*
Inverted T-beam A precast concrete beam with flanges at the bottom, generally used to support double-tee concrete floors.
Isolation joint A joint in a concrete slab-on-ground that penetrates the entire thickness of the slab, used to separate the slab from the structure (walls and columns).
Isotropic material A material that has the same properties in all directions, such as steel. See also *anisotropic material.*

J

Jack stud A shorter stud length attached to a longer king stud to support a header in the opening of a wood light frame wall.
Jack rafter A shorter rafter (than a common rafter) that joins to a hip or valley rafter.
Jamb Side of a window, door opening, or frame.
Joint compound Soft material used together with tape to finish a joint or cover nail dents in a gypsum wallboard assembly.
Joint cover Cover for an expansion or seismic joint in buildings to provide a continuous protective surface.
Joint reinforcement Continuous horizontal steel wire reinforcement laid in mortar joints of concrete masonry to control shrinkage cracks.
Joint sealants See *elastomeric sealants.*
Joist floor An elevated reinforced concrete slab with integral narrow ribs (like wood floor joists), also called a one-way joist floor or ribbed floor.
Joists Slender, closely spaced, parallel beams in a wood light (or light-gauge steel) frame floor.
Junction box Rectangular or circular metal or plastic box used to protect connections between electrical wires.

K

Kerf Slit cut into a surface, used to form a capillary break (drip-edge) at the bottom edge of an exterior surface; a slit cut into building stone (used as wall cladding) to serve as a receptacle for a mechanical attachment system.
Key See *shear key.*
Key way A gap between two adjacent precast-concrete floor units, typically reinforced and filled with portland cement grout.
Kiln A furnace used to dry materials such as green lumber; a furnace for drying and firing bricks; a furnace for making quicklime.
Knot An approximately circular formation of wood fibers formed where a limb branches from a tree trunk. A knot generally lowers the strength of wood.
Kraft paper Asphalt-treated paper that serves as a vapor retarder, commonly used as lamination on fiberglass insulation.
Kilopascal (kPa) A measure of load (force) in the SI system of units.

L

Laminated glass Two pieces of glass laminated under heat and pressure to a plastic interlayer to form a fused unit.
Laminated veneer lumber (LVL) Dried wood veneers laminated in layers, all oriented in the same direction, to form a large structural member.
Landing A horizontal platform between two flights of stairs.
Landing frame A structural frame that forms a landing in a prefabricated steel stair.
Lap splice A connection between two steel or wood members, where the ends of each member are overlapped and connected so that the two members function as one continuous member.
Lateral load Load that acts predominantly in the horizontal direction, such as wind or earthquake load or soil pressure on a basement wall.
Lath Expanded sheet metal attached to a backup wall and used as a mechanical key to bond and reinforce portland cement plaster (stucco).
LEED Leadership in Energy and Environmental Design.
Light-gauge steel framing A framing system that mimics wood light frame construction, but the elements are made of cold-formed, galvanized sheet steel.
Light-gauge stud Generally a C-shaped member made of light-gauge steel.
Light-transmitting plastic Transparent or opaque polycarbonate or acrylic sheet used as glazing in some limited situations.
Lime See *hydrated lime* and *quicklime.*
Limestone A sedimentary rock composed primarily of calcium carbonate and used as exterior and interior wall cladding.
Linoleum A resilient sheet or tile flooring made from linseed oil, organic resins, and fillers bonded to a fibrous backing.
Lintel A beam that spans over a door or window opening.
Lite A pane or sheet of glass.
Live load A type of gravity load that changes in magnitude and placement over time, due to factors such as human occupancy or storage for roof repair or maintenance materials, subdivided into floor live load and roof live load. See also *dead load.*
Load-bearing-wall A wall that supports superimposed gravity loads, such as from floors and roof. See also *non-load-bearing wall.*
Longitudinal bars Steel reinforcing bars placed in the long direction of a reinforced concrete beam or column.
Louvers Fixed or operable, closely spaced, angled slats attached to a frame to direct the passage of air or light.
Low-emissivity (low-E) coating A surface coating that reflects most of the long-wave radiation.
Low-emissivity glass A glass treated with a low-E coating, called low-E glass.
Low-emissivity (low-E) material A material that is an efficient reflector (poor absorber) of long-wave radiation (heat), generally obtained through a low-E lamination (such as aluminum foil) or a low-E coating.
Low-slope roof A roof with a slope less than 3:12.
Lumber Solid wood products derived directly from logs by sawing and planing only, with no additional machining.
LVL See *laminated veneer lumber.*

M

Machine grading A method of grading lumber using a machine, also called machine stress-rated lumber (MSR), in contrast with the more commonly used method of visual grading.
Machine stress-rated (MSR) lumber See *machine grading.*
Malleability A property of metals that indicates their ability to be formed to shape by hammering, bending, forging, pressing, rolling, and so on.
Manufactured wood products Any of a variety of wood products made by bonding smaller pieces of wood veneers, fibers, strands, wafers, or particles to produce a composite wood material. See also *engineered wood products* and *industrial wood products.*
Marble A metamorphic rock geologically formed from limestone under high pressure and heat.
Masonry Materials such as brick, stone, concrete masonry units (concrete blocks), and glass masonry units that are stacked and adhered using mortar.
Masonry cement A preblended mix of portland cement, lime, and/or pulverized limestone (typically manufacturer specific) used in masonry mortar in place of portland cement and lime.
Masonry grout A high-slump concrete used to fill the voids in a masonry wall.
Masonry unit Single brick, concrete (or glass) block, or stone used in masonry construction.
MasterFormat The standard organizational format for construction specifications (note the absence of space between r and F in "MasterFormat").
Mat foundation A type of concrete foundation where one large, combined footing is used for several columns and load-bearing walls, often for the entire building. A concrete slab-on-ground used as foundation for light frame buildings is the simplest type of mat foundation.
Means of egress The route by which one exits a building in the case of a fire.
Metal A highly refined (generally solid) material that is typically hard, malleable, ductile, and capable of carrying an electrical current. Steel, copper, and aluminum are commonly used metals in building construction.
Metal panel roof A roof covering composed of thin sheets of metal, laid over rigid insulation and mechanically attached to the roof structure.
Mission tile A clay or concrete tile with a half-barrel profile.
Model code Building code developed and updated regularly by an independent agency and adopted and adapted for use by federal, state, or local jurisdictions. The International Building Code, International Residential Code, and International Plumbing Code are examples of a model code.
Modified bitumen A type of polymer-modified bitumen (generally asphalt) used in making modified-bitumen roof membranes.
Modular brick The most commonly used size of extruded brick (4 in. × $2\frac{2}{3}$ in. × 8 in. nominal dimensions).
Modulus of elasticity A property of a material that measures its stiffness against deformation under loads.
Moisture movement Expansion or contraction of a material that occurs as a result of changing moisture content.
Moisture retarder A membrane that retards (prevents) liquid water from passing through.
Molded brick A brick formed by pressing small clumps of wet clay into molds. See also *extruded brick.*
Moment connection A connection between a beam and its support (e.g., a wall or column) that transfers the bending of the beam to the support, or vice versa.
Mortar A mixture of portland cement, lime, sand, and water used to bond masonry units; cementations mixture used to adhere tiles to a subfloor.
Mortar-capturing device A synthetic mesh located just above the flashing in the air space in a brick veneer wall used to prevent mortar from clogging the weep holes.
Mortar cement A preblended (typically manufacturer-specific) mix of portland cement, lime, and admixtures.
Mortar joint The layer of mortar placed between masonry units, generally about $\frac{3}{8}$ in. thick.
Movement joint Same as *dynamic joint.*
Mud sill See *sill plate.*
Mullion A vertical member in a metal-glass curtain wall; a vertical member between two adjacent windows or doors.

N

Nail Metal fastener made from a wire with a pointed tip (bottom) and a (generally flat) head.
Nailing flange Continuous flat metal or plastic extrusion from a window frame used to nail the unit to the exterior of a wall opening in a wood frame building.
Nail plate A metal plate with several sharp, flat, nail-like projections.
Neoprene A synthetic rubber primarily used in glazing gaskets.
NFRC rating Window-performance rating system developed by the National Fenestration Rating Council (NFRC).
Noise-reduction coefficient (NRC) A measure of the ability of a material to absorb sound; lies between 0 and 1.0.
Noncombustible A material that will not ignite when subjected to an open flame or high temperature. See also *combustible.*
Non-load-bearing wall A wall that does not support floor or roof loads.
Nonmovement joint Same as *static joint.* See also *dynamic joint* or *movement joint.*
Nosing The leading edge of a tread of a stair.
Nylon A versatile polymer whose fibers are used in carpet and fabrics; also produced in sheets or molded into fittings used in mechanical devices.

O

Occupancy load Floor live load based on its occupancy.
One-way joist floor See *joist floor.*
One-way slab An elevated reinforced concrete slab where most of the load on the slab is carried to the supporting beams in one direction; a four-sided, supported rectangular slab whose length is greater than or equal to twice its width.
Open-web steel joist A standardized, prefabricated steel parallel chord truss used to span between beams, larger joists, or trusses.
Operable window A window unit that opens to allow the passage of air, sound, and so on. See also *fixed window.*
Organic soil A soil containing decayed plant material.
Oriented strandboard A wood-based panel made by gluing several layers of wood strands under heat and pressure so that the adjacent layers are oriented in opposite directions.
OSB See *oriented strandboard.*

Outside glazed Glazing system where the glass is set into the frame from the outside of a building.
Oxidation Chemical weathering process by which the source material combines with atmospheric oxygen, generally leading to the material's degradation, such as oxidation of iron or steel, called corrosion, or rusting.

P

Pan Glass-reinforced plastic pans used as an economical formwork for one-way concrete joist floors.
Pane A sheet of glass.
Panel A sheet of plywood, OSB, particleboard, and so on; a prefabricated building component whose surface dimensions are much greater than its thickness, such as a curtain wall panel.
Panel door A wood door constructed by inserting solid wood panels or glass panes in horizontal and vertical frame members (rails and stiles, respectively).
Panelized stone cladding Multiple stone slabs attached edge to edge to a prefabricated steel frame work which is then attached to the exterior of a steel or concrete frame building.
Panel points Points on a truss where members of a truss connect.
Parallel-strand lumber Manufactured wood product composed of narrow strands of veneered lumber glued together, all oriented in the same direction to form a member of large cross section.
Parapet Portion of an exterior wall that extends above the roof line.
Partition wall A non-load-bearing interior wall that separates spaces, but does not carry floor or roof loads.
Patching compound A liquid or semisolid compound that can be troweled on to correct surface irregularities.
Paving bricks Bricks used for paving, graded for abrasion resistance and freeze-thaw damage.
Permanent formwork Material used as formwork (typically metal deck) and retained as a permanent part of a reinforced concrete floor slab.
Permeability Property of a material (e.g., soil) or assembly that allows water or water vapor to pass through.
Permeance A measure of the rate at which water vapor flows through a material or assembly.
Pier A relatively short column or a deep, vertical foundation element (typically of cast-in-place concrete).
Pilaster A column formed by thickening an area of a masonry or concrete wall.
Pile Driven or drilled long, slender foundation element.
Pile cap A concrete cap that transfers foundation loads to multiple piles beneath.
Placing concrete The operation of pouring fresh concrete from a concrete mixer into the formwork.
Plain concrete Concrete without steel reinforcement.
Plain masonry A masonry wall without (vertical) steel reinforcement.
Plain-sawn lumber Method of sawing a log by sawing in one or two directions only. The grain pattern varies from nearly parallel to the wide face to perpendicular.
Planing Operation that uses high-speed knives to smooth the surface of rough-sawn lumber.
Plaster Pastelike cementitious material that, when applied to the surface of a wall or ceiling, cures to a hard surface.
Plasterboard Specific type of gypsum board used as substrate for plaster.
Plastic A synthetic material composed of polymers that can be molded or shaped when soft; it cures into a rigid or semielastic form and is a material that does not return to its original shape after being deformed.
Plate girder A heavy steel beam fabricated from steel plates.
Platform frame One of the two types of wood light frame constructions, the other being balloon frame. In a platform frame, the studs are one floor high and extend between adjacent floors or from the floor or roof. See also *balloon frame.*
Plywood A panel made from multiple, thin layers of wood veneer adhered with glue under heat and pressure so that the grain direction of a veneer is perpendicular to that of the adjacent veneer.
Pointing Raking an existing (generally defective) masonry mortar joint to sufficient depth and then finishing it with new mortar.
Point-supported curtain wall Mullionless glass curtain wall with glass panes supported at its corners by a metal connector and glass panes sealed at vertical and horizontal joints with a sealant. See also *spider connector.*
Polished finish A type of stone finish, attained by buffing to give a reflective sheen.
Polyethylene A polymer, typically used in sheets as vapor retarder.
Polyisocyanurate Closed-cell foamed plastic insulating material, sandwiched between two facing layers, available as rigid (flat or tapered) boards, typically used as roof insulation.
Polypropylene Synthetic fiber, commonly used in carpets and fabrics.
Polyvinyl chloride (PVC) A versatile, thermoplastic polymer, used in single-ply roof membranes, rigid plastic pipes, and so on.
Porosity The property of a material to allow liquid or air to pass through.
Portland cement A noncombustible, hydraulic cement produced by burning limestone and clay. See also *hydraulic cement.*
Portland cement plaster Multilayered application of a cementitious mix of portland cement, lime, sand, and water to a thickness ranging from $\frac{5}{8}$ in. to 1 in., commonly used on exterior walls.
Posttensioning Subjecting a concrete or masonry member to compressive stresses by tensioning high-strength steel strands (wires) after the concrete has developed sufficient strength.
Pour To cast or place concrete; an increment of concrete placement carried out without a long interruption.
Pour stop An L-shaped steel member welded to a steel spandrel or edge beam to stop fresh concrete from flowing beyond the edge of a floor or roof.
Pozzolanic reaction Reaction between lime and amorphous silica (such as in fly ash, silica fume, etc.) that converts the mixture into a hydraulic cement. See also *hydraulic cement.*
Precast concrete Concrete members typically cast in a precasting plant and transported to the construction site for erection. Precasting can also be done at construction site, such as with concrete tilt-up walls.
Precast, prestressed concrete A precast-concrete member that has been subjected to compressive stresses by tensioning high-strength steel wires before casting concrete for the member.
Preformed tape An elastomeric material (available in rolls) that provides a seal between the glass and the supporting frame in a window or metal-glass curtain wall.
Pressure-equalized wall A drainage wall where the air space between the exterior cladding and the backup wall is at the same pressure as the outside air so that water penetration through the wall is minimized.

Pressure-treated wood Wood into which preservatives have been pressure injected to retard termite infestation and fungal decay. Using a different preservative, the pressure treatment can also be used for increasing the fire resistance of wood.
Prestressed concrete See *precast, prestressed concrete.*
Pretensioning Same as prestressing. See also *posttensioning.*
Primary reinforcement Steel reinforcement in a one-way concrete slab oriented in the direction that carries most of the loads; see also *one-way slab.*
Primer A liquid that improves the adhesion of a sealant or paint to the substrate.
Probability of breakage The probability that a pane of glass will break unpredictably, measured as the number of broken panes (out of 1,000) when the panes are subjected to a given load.
Protected membrane roof (PMR) A roof membrane directly attached to the roof deck and covered with rigid insulation held in place by ballast.
PSL See *parallel-strand lumber.*
Punched window A window surrounded by opaque portions of the wall.
Punch list A list of outstanding problems that must be corrected before the architect certifies that the building is complete.
Putty See *glazing compound.*
Pylon A heavy vertical support member, commonly used in suspension bridges.

Q

Quarry An excavation from which stone is obtained.
Quarry tile A clay floor tile, generally unglazed.
Quarter-sawn lumber A method of sawing lumber where the log is cut radially into four quarters and then sawed along radial lines.
Quicklime Lime obtained by heating limestone. Chemically, it is calcium oxide, which is a fairly caustic substance and is, therefore, not used in construction. When quicklime is mixed with water, it becomes calcium hydroxide, also called hydrated lime. Hydrated lime is used in building construction.

R

Racking Angular deformation of a structure under lateral loads, such as the deformation caused by a horizontal force on a book rack.
Radiant barrier Layer of material that slows the transmission of heat due to its low emissivity. Aluminum foil is a commonly used radiant barrier.
Radiation Emission of electromagnetic waves by an object. All objects emit radiation at all times.
Rain screen wall Same as pressure-equalized wall.
Rake Sloping edge of a gable roof.
Raked joint Tooled, inset masonry mortar joint that creates a strong shadow line.
Random rubble Irregularly shaped and sized stone laid in a random pattern in a wall.
Rebar An abbreviation for reinforcing bar (deformed steel bar used as concrete reinforcement).
Reciprocating saw Fixed or handheld saw that uses a linear blade, which moves horizontally and/or vertically at the same time.
Reflective glass A glass that reflects incoming visible radiation due to a very thin metal oxide coating on one surface.
Reglet A slot in a concrete or masonry wall in which a flashing or roof membrane is inserted and held in place by inserting a compressible material in the slot.
Reinforced concrete Concrete with integral steel reinforcing bars.
Reinforced masonry A masonry wall with vertical reinforcing bars; see also *plain masonry.*
Reinforcement cage A preassembled unit of reinforcing bars used in concrete members.
Relative humidity Amount of water vapor present in the air, measured as a percentage of the maximum water vapor the air can hold under the same temperature.
Reshoring A method of supporting a concrete floor after the formwork has been stripped and before it has gained strength required to support the superimposed loads.
Resilient channel Light-gauge steel channel used for attaching gypsum boards so that the gypsum boards are acoustically decoupled from the backup wall on which the channel is attached.
Resilient flooring Synthetic flooring material, such as linoleum, vinyl, or rubber tiles or sheets, that returns to its original shape after being compressed by impacts caused by walking, dropped objects, and so on.
Resinous flooring A liquid-applied flooring material that dries to form a thin, strong floor surface, also called epoxy flooring. Includes epoxy-based terrazzo.
Ribbed floor See *joist floor.*
Ribbed slab A concrete slab that is stiffened by edge and interior ribs (beams) in both directions, used as an elevated or ground-supported slab. Where the ribs are closely spaced (in both directions), it is also called a waffle slab or two-way joist floor.
Ridge beam A structural beam that supports the top ends of roof rafters at the ridge line and forms a triangular shape without creating lateral thrust.
Ridge board A nonstructural board used to align and join rafters at ridge line of a sloping roof.
Ridge line The line of intersection of two oppositely sloping roof planes.
Ridge vent Passive air-exhaust unit located along the ridge of a sloping roof to exhaust attic air.
Rigid connection See *moment connection.*
Rigid insulation Rigid boards of foamed, granular, or cellular materials.
Rim joist Outer joist that surrounds the wood frame floor structure.
Riser Vertical component of a stair.
Roof deck A load-bearing surface on which a roof membrane is attached.
Roofing felt Organic (typically paper-based) or inorganic (typically fiberglass) fibers pressed into a thin flat sheet and saturated with bitumen (asphalt or coal tar).
Roof membrane The entire waterproofing layer on the roof.
Rubber flooring Rubber sheets or tiles used as floor finish.
Running bond Masonry bond pattern that staggers the units in each course, creating staggered head joints.
R-value See *thermal resistance.*

S

Safety glass Glass that can resist impact (per the safety glass test standard) and, upon breaking, falls into pieces that are small and blunt enough not to cause injury.
Safety margin See *factor of safety.*

Sapwood The outer portion of a tree trunk that conducts nutrients, typically lighter in color than heartwood. See also *heartwood.*
Sash The (operable) part of a window, which holds the glass.
SBS See *styrene butadiene styrene.*
Scab A small piece of lumber nailed to two butt-jointed lumber members to give them continuity.
Scaffold An elevated temporary platform (movable or unmovable) on which workers (such as brick layers, plasterers and painters) stand to perform their work. The term scaffolding is used for a system of scaffolds.
Scratch coat First of two base coats in stucco.
Seasoning Process of drying of lumber until the moisture content reaches the desired percentage.
Secondary reinforcement Reinforcement in a one-way concrete slab placed perpendicular to primary reinforcement. See *one-way slab.*
Security glazing Glazing system that can withstand various levels of assault from handheld weapons, ballistic weapons, and so on.
Seismic joint Continuous joint that divides large or complex buildings into smaller buildings that can move independently of each other to reduce damage from earthquake; generally wider than a building separation joint. See also *building separation joint.*
Self-drilling fastener A fastener that drills its own hole.
Self-furring metal lath A metal lath with dimples that holds the lath away from the substrate to which the lath is applied, allowing the plaster to go through the lath and embed the lath in the plaster.
Setting block A strong, resilient block that supports the glass pane in its frame and cushions it against damage, located at the lower edge of a glazing pocket.
Setting materials A material used to adhere ceramic and stone floor tiles to the subfloor.
Settlement Natural compaction of a soil mass, or shifts in a structure resulting from compaction or instability of soil underlying the foundation.
S4S A lumber member surfaced on all four sides.
Shading coefficient Solar heat gain through a glass divided by the solar heat gain through clear $\frac{1}{8}$-in.-thick glass.
Shaft A continuous vertical enclosure around stairs, elevators, ducts, pipes, and so on.
Shear Stress that acts tangential to the cross section of a member, causing various planes of the member to slide relative to each other.
Shear block A small metal block used to attach horizontal rails to mullions in a metal-glass curtain wall.
Shear key A continuous groove in the first concrete pour that is filled with concrete during the second pour to prevent differential movement between the two elements, such as between two adjacent concrete slabs or a foundation wall and its footing.
Shear stud A short, headed steel rod welded to the top of a steel beam (girder or joist) that is embedded in the concrete slab over the beam so that the beam and the slab act as one structural unit.
Shear wall See *vertical diaphragm.*
Sheathing A panelized material applied to the exterior surfaces of wood or light-gauge steel frame members to add rigidity to the frame and to serve as a base for (wall) cladding or roofing.
Shed roof A roof that slopes to one side.
Sheet bracing A panelized material applied to a structural frame to prevent the frame from racking. See *racking.*
Sheeting Sheet piles; a thin sheet material such as polyethylene sheet.
Sheet pile Piles made of interlocking sheet steel driven into the ground to support an excavation.
Shelf angle Steel angle attached to a spandrel beam or a load-bearing wall to support masonry veneer.
SHGC See *solar heat-gain coefficient.*
Shim A thin, flat piece of metal, plastic, or wood placed between two components to adjust their relative positions during construction.
Shingles Small roofing units of asphalt-treated felts, wood, metal, concrete, or clay tiles, slate, and so on, that lap over each other to waterproof steep roofs.
Shop drawings Detailed drawings of a building component developed by its fabricator based on construction drawings and specifications.
Shores Temporary vertical or inclined supports used in concrete formwork or excavation.
Shotcrete Concrete mix that is deposited through a nozzle at high pressure in combination with a stream of compressed air; also called gunite.
Siding Exterior wall-finish material applied to a wood light frame or light-gauge steel frame building.
Sill Bottom horizontal portion of a window, the exterior portion of which is typically sloped away from the window.
Sill plate The bottom, horizontal member in a wood light frame (or light-gauge steel) wall that is directly in contact with the foundation.
Simple (Shear) connection A connection between a beam and its support (e,g., a wall or column) that allows the beam to bend without transferring its bending to the support.
Simply supported beam A single-span beam with simple connections at both supports.
Single-ply roof (SPR) membrane A single-layer synthetic roof membrane.
Single-strength (SS) glass A $\frac{3}{32}$-inch-thick flat glass
SIPS See *structural insulated panels.*
Sitecast concrete Concrete cast in the position where it will remain in the final structure.
Slab-on-grade A concrete slab that is supported directly on the ground.
Slab-on-ground Same as *slab-on-grade.*
Slag A noncombustible waste product obtained from the manufacture of iron in a blast furnace, commonly used as aggregate on built-up roofs.
Slag wool Slag converted to fibrous insulation.
Slip connection A connection that allows movement between connected element in at least one direction.
Slip-critical connection A connection between two structural steel members bolted together in which the load on one member is transferred to the other member mainly by friction.
Slump test A test that measures the workability of fresh concrete by filling a cone-shaped mold with concrete, removing the mold, and measuring the height to which the concrete settles below its original height.
Slurry A liquid mixture of an insoluble material and water, such as portland cement and water (called portland cement slurry) or bentonite clay and water (called bentonite slurry).
Slurry wall A method of constructing a reinforced concrete basement wall by temporarily stabilizing the wall excavation with bentonite slurry and then pumping concrete in it. As the concrete is pumped, the slurry is displaced.
Smoke-developed index (SDI) A measure of the visibility through smoke in a building fire.
Smooth, off-the-form concrete The smooth finish on concrete revealed when formwork is removed, without any further surface treatment.

Snap ties A steel wire tie used in formwork for concrete walls.
Soffit The underside of a horizontal building element, such as of a roof overhang or of a stair.
Soffit vent Air intake ventilation unit located in the soffit of a roof overhang.
Softwood Wood obtained from trees that have thin conical leaves and are typically evergreen. See *hardwood.*
Solar heat-gain coefficient Solar heat transmitted through a glass divided by the solar heat that falls on the glass.
Solid-core door Flush wood or metal veneers laminated to both sides of a solid interior, in contrast to a hollow-core door.
S1S2E A lumber member surfaced on one side and two edges.
Sound-absorbing material A material that absorbs more sound than most other materials.
Sound frequency The number of to-and-fro movements of air particles in 1 s in a sound wave.
Sound insulation A material that retards the transmission of sound.
Space truss A three-dimensional truss.
Spall Loss of the surface of a material, such as concrete or masonry, resulting from stresses caused by expansion or contraction of the material, such as freeze-thaw cycles.
Spandrel area The area of the exterior facade of a building at the level of the spandrel beam.
Spandrel beam A beam that spans between columns on the exterior face of a frame structure.
Spandrel glass Glass (usually opaque) used in a metal-glass curtain wall in spandrel area of the facade.
Spandrel panel A panel used in spandrel area of a glass curtain wall.
Specific gravity Density of a material divided by the density of water.
Specifications Written technical description of materials and assemblies shown in construction drawings, organized in three parts: general, products, and execution.
Spider connector Four-pronged stainless steel connector used at the corner of four lites of glass in a mullionless glass curtain wall.
SPR See *single-ply roof membrane.*
Spray-applied fire protection A mix composed of portland cement, water, and mineral fibers (or lightweight aggregate) sprayed on a steel member to increase its fire resistance.
Stack bond Masonry bond pattern, where units are stacked vertically with continuous head joints.
Standing-seam metal roof Sheet metal roof panels joined by folding and/or interlocking seams.
Static joint Same as *nonmovement joint.*
Steel A strong and malleable metal (alloy of iron and carbon) used for structural and nonstructural shapes.
Steel decking A ribbed-sheet steel formed into panels, commonly used in floors and roofs.
Steep roof A roof with a slope greater than or equal to 3:12. See *low-slope roof.*
Stem wall A short foundation wall, generally used to form a crawl space below the ground floor of a light-frame building.
Step flashing Small lengths of metal sheets placed between roof shingles, used at locations where a sloping roof meets a wall.
Stick-built curtain wall A metal-glass curtain wall whose framing members are installed at the site member by member.
Stiffback A member attached to a frame to increase its stiffness.
Stirrup A loop or U-shaped steel reinforcement used to increase the shear resistance of a concrete beam.
Stone cladding Stone panels attached to a backup wall or curtain wall frame.
Stone honeycomb panel A lightweight panel made of thin stone veneer laminated to an aluminum honeycomb backing.
Stone masonry Stone laid with mortar unit by unit.
Storefront system Glazed facade, generally one or two stories high from the ground, with a framing system similar to a metal-glass curtain wall but with less stringent performance requirements.
Story drift Horizontal deflection of an upper-story floor with respect to a lower floor caused by lateral loads, as a result of racking.
Strain Change in the dimension of a member divided by its original dimension.
Strands High-strength steel wires used in prestressing cables; thin wafers of wood.
Stress Internal resistance created by a member in response to an applied external force, expressed as force divided by the cross-sectional area of the member.
Stretcher Long face of a brick as laid in its usual way in a wall.
Striking Leveling a freshly placed concrete surface, generally followed by floating. See *floating.*
Stringer See *carriage.*
Strip flooring Wood finish flooring made of long, narrow tongue-and-groove boards.
Stripping time Length of time between placing concrete and removing the formwork.
Strip windows An array of continuous window units placed side by side horizontally or vertically.
Structural composite lumber Manufactured wood products produced by gluing together longitudinally oriented wood veneers or stands.
Structural insulated panels Structural panels composed of rigid insulation glued on each side with OSB or plywood panels.
Stucco See *portland cement plaster.*
Studs Closely spaced vertical members that constitute a wood light frame or light-gauge steel frame wall.
Styrene butadiene styrene A polymer with rubberlike properties generally used as a modifier of asphalt, giving an SBS-modified roof membrane.
Subfloor Structural (rough) floor beneath a floor finish, such as carpet or floor tiles.
Substrate Underlying member on which other material is placed.
Substructure Below-ground portion of a structure, including basement and foundation.
Sump A pit used to collect water, from where it is pumped out.
Superstructure Above-ground portion of a structure.
Surety bond A type of insurance required of a contractor by the owner to ensure that the obligations of the construction contract will be met.
Suspended ceiling A ceiling hung from the overlying floor or roof structure.
Sustainable Comprehensively addressing aspects of building design and construction in addition to health, safety, welfare, and aesthetic aspects that account for resource consciousness and stewardship of the environment and ensure that the needs of future generations are not compromised by today's generation.

T

T&G See *tongue-and-groove.*
Tabular area Allowable area per floor as obtained from the building code table.
Tempered glass A glass obtained by heating annealed glass to a high temperature and then suddenly cooling it, which makes

it four times stronger than annealed (basic) glass; used as safety glass because it breaks into pieces that are small and blunt enough not to cause injury.
Tendon The combination of high-strength steel strands, sleeves, and end anchorages used for post-tensioning concrete.
Terrazzo A concretelike material (binder, filler, and aggregate mix) that is poured over a subfloor in a thin layer, ground and polished after it cures to form a floor finish.
Thermal break A section of a material with high thermal resistance epoxied between two parts of a material with low thermal resistance.
Thermal bridge A portion of building envelope or a building assembly with a much lower thermal resistance than the rest.
Thermal capacity Ability of a building material or component to store heat.
Thermal conductor Material that conducts heat rapidly
Thermal finish See *flame finish.*
Thermal insulation A material or assembly that retards the flow of heat.
Thermal movement Changes in the dimensions of a material that occur as a result of a change in its temperature.
Thermal resistance (R-value) Measure of the ability of a building material or assembly to resist the flow of heat.
Thermal transmittance Inverse of thermal resistance.
Thermoplastic polymer A polymer that softens when heated and hardens when cooled.
Thermoplastic polyolefin A thermoplastic polymer used as a single-ply roof membrane.
Thermosetting polymer A polymer that will not soften when heated.
Thick-set mortar Thick bed of cementitious mortar on which floor tiles are set, used where the subfloor surface or the bottom of the floor tiles (or slabs) are irregular or too large for the use of thin-set mortar.
Thin-set mortar Method for attaching tile to the subfloor or underlayment using a thin layer of adhesive.
Threshold Floor surface immediately below an exterior door, typically sloped to the exterior.
Tie-back A fastener that connects an exterior cladding to the supporting frame to resist lateral loads.
Ties Closed-loop reinforcement used in columns to prevent the buckling of longitudinal bars.
Tile Small, thin flooring or ceiling unit.
Tilt-up wall Reinforced concrete wall, precast on the ground-floor slab of the building and lifted by a crane to its required position.
Timber See *heavy timber.*
Tinted glass A type of glass made by adding a metallic pigment during manufacture.
Tongue-and-groove A joint between two members in which the edge of one member is milled with a groove and that of the other member is milled with a projecting tongue.
Tooled joint Joint in masonry mortar or sealant that is compressed by a specific tool.
Torched membrane SBS modified-bitumen roofing sheet adhered using an open-flame torch.
TPO See *thermoplastic polyolefin.*
Travertine A type of natural stone pitted with voids.
Tread Horizontal (walking) surface of a stair.
Tread pan A steel pan in a prefabricated steel stair that is later filled with concrete at the site to give a finished tread.
Tributary area Area of a building that contributes to the load on a building component.
Truss A structural member with triangulated, linear elements, typically used for large spans.
Trussed joist Parallel chord truss used as a floor joist.
Two-way joist floor See *ribbed slab* and *waffle slab.*
Two-way slab An elevated, reinforced-concrete slab, where the load on the slab is carried to the supporting beams in both directions; a four-side-supported rectangular slab whose length is less than twice its width.

U

Ultraviolet radiation (UV) Shortwave solar radiation that can damage human skin and eyes as well as fade colors of materials and art.
Ultraviolet transmittance A measure of the transmission of ultraviolet radiation through glass.
Underlayment (floor) A panel attached to the subfloor to create a smooth, rigid surface for the application of finish flooring.
Underlayment (roof) Water-resistant sheet applied as a second layer of defense under shingles in a steep roof.
Unit-and-mullion curtain wall Preassembled metal-glass curtain units attached to mullions that have already been attached to the building frame.
Unitized curtain wall Preassembled metal-glass curtain wall units attached to building frame.
UVT See *ultraviolet transmittance.*

V

Valley A trough formed at the intersection of two adjacent sloping roofs.
Valley flashing Flashing material used in a steep roof valley to increase the water resistance of valley trough.
Vapor drive analysis A numerical analysis that determines the amount of water vapor flow through the building envelope to ascertain condensation potential.
Vapor pressure Pressure exerted by water vapor on the building envelope. Vapor pressure is independent of the air pressure.
Vapor retarder A material that restricts the flow (transmittance) of water vapor.
Vault A continuous, curved roof member.
Vaulted ceiling High ceiling that follows the shape of the roof structure. If the gypsum board is attached directly to the rafters in a wood frame building (with no attic space), such a ceiling is called a vaulted ceiling.
Veneer A thin layer of material over a back-up component. See also *adhered veneer* and *anchored veneer.*
Vertical diaphragm Walls that serve to transfer lateral loads in a building to the foundation by interacting with horizontal diaphragms (floor and roof structures).
Vinyl Abbreviation for polyvinyl chloride, used in window frames, plumbing pipes, and so on; sheet vinyl used in floor tiles, siding, and trim, and so on.
Visible transmittance A measure of the transmission of visible light through glass.
Vision glass Transparent glass used between spandrel areas of a curtain wall.
Visual grading A method of grading lumber based on visual inspection of each piece by trained (graders) inspectors.

Vitreous Glasslike substance with zero vapor permeability.
VT See *visible transmittance.*

Waffle slab Also called two-way joist floor or ribbed slab that has closely spaced ribs (beams) in both directions. See also *ribbed slab.*
Waler Horizontal member used to stiffen and support concrete formwork.
Wall anchor A steel fastener that connects a veneer wall to the backup wall.
Wallboard See *gypsum board.*
Wall cavity Air space between the veneer and the backup wall.
Waterproofing Liquid or sheet material applied to the exterior of a basement wall or floor to resist the intrusion of water under hydrostatic pressure, as distinct from dampproofing.
Waterstop A rubber or polymeric strip of material, used to seal joints between two concrete pours.
Water table The level below ground where the water pressure due to the water in the soil equals the atmospheric pressure; the level below ground to which water will fill an excavation.
Wavelength Distance between adjacent air pressure maxima (compression peaks) or adjacent minima (rarefaction crests) at a given point in time caused by sound.
Weatherstripping Strip of material (wood, metal, foam, etc.) applied to an exterior door or window, which allows the unit to open and close but prevents the passage of air and rainwater when it is in the closed position.
Web Transverse wall in a CMU; the vertical part of a steel wide-flange section; members in a steel truss between the top and bottom chords.
Weep hole A small opening in a veneer wall directly above the flashing to drain water.
Weld Method of attaching two steel members by heating them until they fuse together.
Welded-wire reinforcement A prefabricated rectangular grid of steel wires spot-welded together at intersections, used as reinforcement in concrete slabs.
Wet glazing Use of a semiliquid elastomeric sealant to set glass in a frame. See also *dry glazing.*
White cement White portland cement, commonly used in architectural concrete, stucco, and terrazzo flooring.
Wind load Difference between inside and outside air pressures on a building component caused by the wind.
Wind uplift Upward pressure caused by wind on a building component, generally on the roof.
Wired glass Rolled glass sheet with embedded thin steel wire mesh, used in fire-rated glazed openings.
Wood flooring Solid or manufactured wood available in strips, planks, and tilelike units and used as a floor finish.
Wood light frame A structural frame assembly composed primarily of dimension lumber studs, floor joists, and roof rafters and panels of wood-based sheathing materials.
Workability Ease with which fresh concrete can be placed and compacted, roughly measured through a slump test; ease with which fresh masonry mortar or portland cement plaster can be troweled on.
Worked lumber Lumber that has been machined beyond surfacing to obtain a specific profile or cross section.
Wrought iron Iron-carbon alloy that is hammered into shape.
Wythe Vertical section of masonry that is 1 unit thick.

Yield strength Maximum stress in steel before it experiences excessive deformation.

Z

Zero-slump concrete A stiff concrete that will not settle at all in a slump test, generally used in making concrete masonry units.
Zoning ordinance A document that describes regulations for the use of land in a particular jurisdiction.

INDEX

Accessibility, building, 44–49
Accessory-use space, 39
Acheson, Edward, 87
Acid etching, 456
Acoustical properties. *See* Sound
Acoustical suspended ceilings, 939–940, 941, 942
Acoustical units, 574
Acrolein, 155
Acrylic glazing, 750
Adhesives, for ceramic and stone tiles, 909
Aggregate interlock, 470
Air barrier, 136
Air conditioning, invention of, 198
Air infiltration, glass-aluminum curtain walls and, 811
Air leakage
  indoor air quality and, 137
  rate, 134–135
Air quality, air leakage and indoor, 137
Air retarders, 135–137
  perm ratings, 143
Air space, veneer walls and, 645–646
Airborne sound-borne, 169
Air-to-air heat exchanger, 137
Air-to-air heat transfer, 110
Air-weather retarders, 136
Allowable stress, 97
Aluminum
  anodized, 787
  doors, 774
  extrusions, 786
  for formwork and shores, 451
  painted, 787
  siding, 313
  windows, 764–766
Aluminum foil
  R-values, 110–111
  as vapor retarder, 142
American Architectural Manufacturers Association (AAMA), 792, 811
American bond, 560–562
American Concrete Institute (ACI), 183, 423
American Institute of Steel Construction (AISC), 348, 365
American National Standards Institute (ANSI), 41
American Plywood Association, 248
American Society for Testing and Materials (ASTM), 42
  building brick, 558
  calcium silicate masonry units, 584
  combustible and noncombustible materials, 154
  facing brick, 557
  paving brick, 559
  roofing asphalt types, 823–824
  wind-load resistance of glass, 747, 748
  World Trade Center fires, 161
American Society of Civil Engineers (ASCE), 41
American Society of Heating, Refrigeration and Air Conditioning Engineers (ASHRAE), 41, 201
American standard shapes, 347
Americans with Disabilities Act (ADA), 44–45, 906
American Wood Preservers' Association (AWPA), 41
Anchors, 645, 654
  bolt, 712–713
  dead-load and expansion, for glass-aluminum curtain walls, 794–799
  flex, 681–682
  installation of stone and tieback, 710–714
Annealed glass, 732
Annealing lehr, 731
Anodized aluminum, 787
Antiwalk blocks, 753
APA—The Engineered Wood Association, 248–249, 260, 261
APP. *See* Attactic polypropylene (APP)-modified bitumen roof membrane
Appeals
  board of, 31
  zoning, 44
Approved assembly, 845
Arch, 80
Arching action, in masonry walls, 573
Architect, 5
  field observation, 17
  inspections, 17–18
  relationships between general contractor and, 12
  summary of functions, 18
Architectural Barriers Act (ABA), 1969, 45
Area, allowable, 46–49
As-built drawings, 19
Ashlar masonry, 591, 592
Asphalt
  amounts needed, for roofs, 824
  flashings, 854, 856, 863–866
  modifiers, 825–829
  ridge and hip treatment, 862–863
  shingles and underlayment for steep roofs, 852, 854–858
  treated fiberglass felt, 823
  types of roofing, 823–824
  valley treatment, 859–862
Aspidin, Joseph, 418
Assemblies, fire-rated requirements, 156–157, 161
Atmospheric corrosion, 376
Atmospheric pressure equalization, 638
Attactic polypropylene (APP)-modified bitumen roof membrane, 825, 827–829
Attic space, R-values, 111
Attic ventilation
  importance of, 143–144
  vapor-tight construction and, 144
Audible frequency range, 168
A-units, 573, 574
Autoclaving, 584
Autogenous healing, 544, 545
Automatic sprinklers, 36, 47, 48–49, 164
Axial stress, 88

Backer rods, 191, 192
Backfill, 509
Backup rods, 191, 192
Backwrapping, 706
Balloon frame, 270–272
Band joists, 274, 283–285
Basalt, 586
Base coats, 694, 696
Base flashing, 837, 838–840
Base sheet, 826
Basic Building Code, 34
Basic wind speed, 59
Batten bar, 833
Bayonet clips, 712–713
Bead boards, 126
Beam(s)
  balanced and unbalanced glulam, 248
  column-to-beam connection, 365–368
  concrete floors, 480–488
  continuous, 457
  doubly reinforced, 457–459
  simply-supported, 457
  spandrel, 188
  -to-beam connection, 368
Bearing capacity, 94
Bearing connection, 363
Bearing plates, 94
  steel, 380
Bearing strength, 93–94
Bearing stress, 93
Bearing supports, 673, 674
Bed joints, 550
Below-grade waterproofing, 531–534
Benched excavations, 510
Bending failure, 94
Bending strength, 88–91
Bending stress, 89
Bent glass, 735
Bentonite slurry, 516
Bent-plate angles, 348
Bessemer, Henry, 344
Bidding, selecting a general contractor and
  bond, 13, 14
  competitive, 12–13
  documents, 12–13
  invitational, 13
Bid negotiation (preconstruction phase), 11–15
Bimetallic corrosion, 376
Bioclimatic glass curtain walls, 812
Biodegradability, 205
Biodeterioration, 234
Bird's mouth, 288
Bitumen
  asphalt versus coal tar, 823
  built-up roof membrane and, 819, 820–824
  modified roof membrane and, 819, 820, 825–829
Blast furnace, 343–344
Blast-resistant glass, 746
Blind-side waterproofing, 533
Blocking, 283–285
Board foot measure, 227–228
Board of adjustment, 44
Board of appeals, 31
BOCA National Building Code, 34
Boiling water absorption, 557
Bolts, 265, 362, 363–364, 712–713
Bond beams, 581
  in masonry walls, 605–606
Bond breaker, 192
Bond strength, 544, 545
Bond patterns
  in masonry walls, 559–562
  in stone walls, 591–592

Bonding pad, 681
Bonds
  bid, 13, 14
  payment, 14
  performance, 14
  pros and cons of, 14
  surety, 14
Boring field log, 507, 508
Bottom plate, 276, 277
Boulders, 504
Braced frame, pin connections, 385, 386–387
Braces (bracing)
  bundled tubes, 57–58
  cross (X), 385
  diagonal, 57, 58
  K, 385
  multibay single-story frames, 387–388
Brick(s)
  building, 558–559
  clay, 556–559
  construction of, 563–564
  core holes, 553–554
  efflorescence, 565
  expansion control, 565–566
  extruded (modular), 552–554, 556
  facing, 557–558
  floor covering, 927
  freeze-thaw resistance, measuring, 557
  initial rate of absorption (IRA), 562–563
  ledge, 647
  manufacture of, 552
  masonry, 543
  molded, 554–555
  orientations on wall elevation, 560
  paving, 559
  physical requirements for grades of, 558
  precast curtain wall, 675–677
Brick Industry Association (BIA), 41, 183, 669
Brick veneer. *See* Veneer walls
Bridging channels, light-steel, 400
British Standards Institution (BSI), 41
Brittleness, 83–85
Brown coat, 696, 697
Brundtland Report, 199
Buckling, 94–96
Building brick, 558–559
Building codes
  administration, 35–36
  application of, 36–39
  building accessibility, 44–49
  construction standards, 40–42
  contents of, 35–36
  enforcement of, 29–31
  historical, 27
  model, 33–35
  objectives of, 26–29
  other major governmental constraints, 42–43
  prescriptive and performance, 32–33
  special, 35
  zoning ordinance, 43–44
Building envelope, defined, 102
Building expansion joints, 179
Building joints, 178, 179
Building official, 29
Building Officials and Code Administrators (BOCA), 34
Building paper, 142
Building permits, 30
Building planning, 36
Building separation joints, 179–180, 380–381
Building stone, 585
Built-up roof (BUR) membrane, 819, 820–824, 846–847
Built-up sections, 349
Bun, 126
Bundled tubes, 57–58
Burnished units, 575, 576
Caissons, 526
Calcining, 414
Calcite limestone, 587
Calcium silicate masonry units (CSMUs), 584
Calibrated wrench method, 363
Cambering, 370
Canadian inspection agencies, for lumber, 242
Cantilevered footings, 521
Cantilevered frame, pin connections, 383–384
Cant strips, 839
Cap sheet, 826
Capillary break, 633
Carbon dioxide, 155
Carbon monoxide, 155
Carpet and carpet tile flooring
  characteristics of, 918
  construction methods, 919–920
  cushions, 920
  fibers and yarns, 918–919
  installation, 921
Carpet and Rug Institute (CRI), 207
Carrier, 825
Carrier, Willis, 198
Cast iron, 343–344
Cathedral ceilings, 291
Caulks, 193
Ceilings
  attached to building structure, 936–937
  cathedral or vaulted, 291
  finish materials, selection criteria, 934–936
  joists, 289
  no ceiling finish, 936
Ceilings, suspended, 379
  acoustical, 939–940, 941, 942
  carrying capacity, 939
  gypsum board, 940, 942, 943, 944
  gypsum plaster and portland cement plaster, 942–943, 944
  principles of, 937–939
  specialty, 943–946
  types of, 938
Cells, 570
Cellular concrete, 129, 838
Cellular-type plan, 606
Cement. *See* Portland cement
Ceramic frit, 735
Ceramic tile
  characteristics of, 906
  grouts, 910–911
  installation, 911–913
  mortars, adhesives and epoxies, 909–910
  permeability, 907
  physical properties of, 907
  setting methods, 907–908
  types, shapes and sizes of, 907, 908, 909, 910
Certificate of final completion, 19
Certificate of occupancy, 19, 31
Certified Wood Label, 206–207
Chairs (bolsters), 462
Change orders, 18
Chemical weathering, 504
Chimney, flashing around a masonry, 864–866
Cladding, exterior rain-screen, 636–640
Cladding, exterior wall
  brick veneer aesthetics, 669–670
  brick veneer with concrete masonry units or concrete backup wall, 654–660
  brick veneer with steel stud backup wall, 661–668
  concrete masonry units backup versus steel stud backup, 668–669
  curtain walls, glass fiber-reinforced concrete, 680–687
  curtain walls, precast concrete, 670–679
  curtain walls, prefabricated brick, 687–689
  exterior insulation and finish system, 702–709
  masonry veneer assembly, 644–652
  metal panels, insulated, 722–725
  stone, 709–722
  stucco, 694–702
Clapboard, 313–315
Clay, 504
  bricks, types of, 556–559
  difference between shale and, 552
  masonry, 543
  surface, 552
Clay and concrete roof tiles, 853, 866–872
Clinkers, 419
Closure strip, 623
CM at risk (CMAR) method, 21
CM (construction management) method, 19–21
Coal tar-treated fiberglass felt, 823
Cobbles, 504
Coefficient of friction, 500, 906
Coefficient of thermal expansion, 184
Coke, 343
Cold forming, 355, 399
Cold joints, 178, 470–472
Collar ties, 289–291
Column-beam system, 616
Columns
  concrete reinforced, 463–464
  defined, 574
  footings, 520–521
  forms, 463
  spiral, 463
  tied, 463
Column-to-beam connection, 365–368
Column-to-column (column splices) connection, 369
Combustible materials, 154
Combustion
  complete, 155
  products of, 154–155
Commissioning authority (agent), 201–202
Competitive bidding, 12–13
Complete combustion, 155
Component joints, 178
Composite steel floor deck, 357–359
Composition shingles, 852
Compression, 168
Compression gaskets, 754–755
Compression reinforcement, 458
Compressive failures, 94
Compressive strain, 84
Compressive stress, 80–83
Concrete
  See also Portland cement
  admixtures, 436
  aggregate, sizes of, 422–424
  building codes and, 448
  cast-in-place, 448
  cellular, 129, 836
  consolidation, 430–431
  curing, 433–435
  defined, 421
  fire resistance, 498
  foamed, 129, 836
  form liners, 456
  high-strength, 435–437
  ingredients of, 421–424
  insulating, 126–127
  insulating, and roof membrane attachment, 835–836
  making, 428–429
  moisture movement in, 186
  placing and finishing, 429–433
  properties of, 425–428
  Roman, 415–416
  roof tiles, 853, 866–872
  smooth, off-the-form, 455–456
  stairs, 898–899
  stucco on substrates, 700–701
  sustainability, 441
  textured surfaces, 456

water quality, 424
water-reducing admixtures, 435
Concrete, precast, 447
architectural and structural, 493
curtain walls, 670–679
double-tee units, 495–498
hollow-core slabs and solid planks, 494–495
inverted-tee beams, 498
mixed, 494–498
single-tee units, 498
Concrete, reinforced, 446
columns, 463–464
corrosion protection, 461–462
couplers, 459–460
defined, 448
form release agents, 455
formwork and shores, 448–455
hooks, 460–461
prestressed, 446–447
principles of, 456–459
splices, 459, 460
strands, 446, 447
versatility of, 448
walls, 464–467
Concrete, site-cast, 447
building separation joint, 499–500
defined, 448
elevated (framed), 468, 480
reinforced bearing walls, 617–618
Concrete blocks. *See* Concrete masonry units (CMUs)
Concrete floors, beamless, 488–490
Concrete floors, beam-supported
band beam, 482
beam and girder, 481–482
one-way joist, 482
one-way joist, standard-module, 484
one-way joist, wide-module, 484–486
one-way solid, 480
two-way joist (waffle), 486–488
two-way solid, 480
Concrete floors, posttensioned elevated, 491–492
Concrete floors, subfloors, 904–905
Concrete masonry units (CMUs), 543
A, H, and bond beam, 573–574
acoustical, 574
burnished and glazed, 575
construction of, 578–579
control joints, 582–583
fire resistance of, 595–596
grout for, 583–584
horizontal reinforcement, 580–582
lintel, 572
shapes, 571–575
shrinkage control in walls, 579–583
sizes of, 570–571
specifications, 576–577
split-face and ribbed, 574–575
stack-bond, 581–582
veneer wall with, 654–660, 668–669
Concrete slab-on-ground, 468
fibrous, 469
insulation under, 119–120
joints, 470–472
steel reinforcement amounts and location, 469–470
subbase, 469
vapor retarder, 144, 468–469, 905
Concrete slabs
constant-thickness, 475
elevated (framed), 468, 480
flat, 490
flat plate, 489–490
ground-supported isolated, 468–472
ground-supported stiffened, 472–476
posttensioned, 475
prestressing tendons, 473–475
ribbed, 472, 475
Concrete walls
air leakage in, 136
reinforced, 464–467, 516, 617–626
reinforced-concrete tilt-up, 619–626
Condensation of water vapor, 140–141
Condensation resistance factor (CRF), 745, 757, 811
Conduction, 104
Connections
beam-to-beam, 368
column-to-beam, 365–368
column-to-column (column splices), 369
interchangeability of terms, 365
moment-resisting, 385
pin, 382, 383–385
rigid, 382, 385–387
shear, 383
simple, 383
tilt-up walls, 623–626
Connectors, sheet metal, 265–266
Consolidation settlement, 524
Construction documents (CD)
set, 7–8
stage, 6–8
Construction drawings
defined, 7
relationship between specifications and, 8
Construction joints, 178, 470–472
Construction management (CM) method, 19–21
at risk (CMAR) method, 21
Construction phase, 15–18
Construction Specification Institute (CSI) MasterFormat, 9–11
Construction type, 36, 39
fire and, 157–160
Consultants, types of, 5
Contiguous bored concrete piles (CBPs), 514–515
Continuous beams, 457
Continuous flight auger, 514–515
Continuous vertical expansion joints, 651
Contract documents, 15
Control joints, 179, 186, 190, 470, 582–583
stucco and, 696, 697–699, 701
Convection, 104
Core holes, 553–554
Cork tiles, 926
Corrosion, types of, 376
Corrosion protection, of steel, 375–376
of light-gauge steel, 401
of reinforced, 461–462
Counterflashing, 839
Couplers, 459–460
Cover board, 832
Crack isolation (suppression), 927–928
Creep deformation, 182, 188
Creosote, 237
Critical load, 95
Cross-grain lumber dimensions, equalizing, 294–295
Crown glass process, 728
C-shapes, structural steel, 348
CSI (Construction Specification Institute) MasterFormat, 9–11
Curb flashing, 837, 840
Curtain walls
defined, 636, 637
prefabricated brick, 687–689
prefabricated stone, 717–719
prefabricated stone-honeycomb, 719–722
Curtain walls, glass-aluminum
anchors, dead-load and expansion, 794–799
connecting rails to mullions, 800–801
details of, 806–808
inside-glazed, 801, 806
nontraditional, 812
outside-glazed, 801–806
performance criteria 809–812
pressure plate-captured glass, 801–803, 804–805
standard and custom, 793–794
structural silicone sealant-adhered glass, 803, 806
types of, 792–793
Curtain walls, glass fiber-reinforced concrete (GFRC)
anchors, 681–682
components of, 680
details of, 685–687
fabrication of panels, 683–684
panel shapes, 682
surface finishes, 684
Curtain walls, precast concrete (PC)
advantages of, 670–671
brick and stone-face, 675–677
concrete strength, 671–672
connecting, to a structure, 673–675
insulated, sandwiched panels, 679
joints between panels, 678–679
mock-ups, 673
panel shapes and size, 671
panel thickness, 672–673
steel frame structure, 675
Cut stone. *See* Stone tile
Cylindrical glass process, 728

Darby, Abraham, 343
Dead loads, 54, 55
anchors, for glass-aluminum curtain walls, 794–799
on a beam, 71–72
supports for slabs, 710–711
veneer walls and, 646–647
on wood frame residential floor, 72–73
on wood frame residential floor with concrete topping, 74
Decibel scale, 169
Decks
masonry walls and floor and roof, 608–609, 610–612
roof, 820, 854
Deep foundations. *See* Foundations, deep
Design-bid-build method, 15
Design-build (DB) method, 22
Design-Build Institute of America (DBIA), 22
Design development (DD) stage, 6
Design phase, 5–8
Design team, members of, 5
Design wind speed, 59
Deutsches Institute fur Normung (DIN), 41
Dew point temperature, 140
location of, in an assembly, 146–148
temperature gradient across an assembly, 146
Diagonal bridging, 283–285, 354–355
Diamonds, artificial, 87
Diaphragm action, 306
Diaphragm compression, 306
Diaphragm tension, 306
Dicing, 733
Die skin texture, 553
Dimension stone, 585
Dimensions, of masonry units, 555–556
Dimmable (electrochromic) glass, 751
Direct-glaze windows, 764
Directory of Fire Resistance, 161
Direct-tension indicator (DTI) method, 363
Disabled, 44–45
Documents
bidding, 12–13
construction, 6–8
contract, 15
record, 19
Dolomitic limestone, 587
Dome, 80

Doors
classification of, 773–779
fire-rated, 783
frames, 780–782
sealing, 136
terminology, 785
Double-skin glass-aluminum curtain walls, 812
Double-tee units, 495–498
Dragon wagon, 828
Drainable EIFS, 708–709
Drainage, roof, 845
Drainage plane, 696, 701
Drainage system, 532–533
Drawings, shop, 16
Drip edges for eaves and rakes, 856, 870
Drop panels, 490
Dry glazing, 753–755
Drywall, 320
installing and finishing interior, 323–326
Dual-carrier sheet, 825
Ductility, 83–85
Durability, 206
Dust marking, 408
Dynamic joints, 178
Dynamic state, 120–121

Earthquake load, 64
factors that affect, 68–69
soil liquefaction, 65
tsunamis, 65
Earthquakes
frequency of occurrence and location of, 66–67
geological explanations of, 66
intraplate, 66
list of major, 67
Richter scale, 67–68
Eaves, 318, 319
clay and concrete tile, 870–872
drip edges for, 856
Ecolabeling
of building products, 206–207
of buildings, 200–202
Edge blocks, 753
Efflorescence, 565
EIFS. *See* Exterior insulation and finish system
EIFS Industry Members Association (EIMA), 703
Elastic deformation, 188
Elasticity, 86–88
Elastic-plastic materials, 86
Elastomeric sealants, 136, 189, 193
Electrochromic glass, 751
Electrostatic discharge, controlling, 926
Elevators, invention of, 198
Embodied energy, 198, 205
Emergency shelters, 63
Emissivity, surface, 110–112
Enclosure classification, wind pressure and, 63
End-bearing piers, 527
End-bearing piles, 526
End dams, 647
Energy, units of, 103–104
Energy performance ratings of windows, 767
Energy Star Label, 206, 767
Engineered fill, 509
Engineered wood (EW) products, 246, 266
English bond, 560–562
Envelope. *See* Building envelope, defined
Environmental Protection Agency (EPA), 207
EPDM (ethylene propylene diene monomer), 829, 830, 831
Epoxies, for ceramic and stone tiles, 909
EPS boards, 108
Erection, steel, 372–375
Evaporation, 137
Excavation(s)
benched, 510
defined, 507–508
keeping them dry, 518–519
open, 508, 509–510
pits, 508
supports for deep, 511–517
trenches, 508
Exfiltration, 134
Exhaust ventilation, 144
Exit, 28
Exit access, 28
Exit discharge, 28
Expanded metal lath, 696
Expanded polystyrene (EPS) boards, 126, 330
Expansion anchors, for glass-aluminum curtain walls, 794–799
Expansion control in brick walls, 565–566
Expansion joints, 179, 186, 190
stucco and, 696, 697–699
Expansive soils, 537
Exposure
category, 61–62
felt, 846
siding, 314
Exterior insulation and finish system (EIFS)
application of polymer-based, 703–707
backwrapping, 706
basics, 702–703
categories of, 703
drainable, 708–709
impact-resistant, 707–708
installation of insulation, 704–705
movement control, 706–707
Exterior rain-screen cladding. *See* Cladding, exterior rain-screen
Exterior wall cladding. *See* Cladding, exterior wall
Extruded polystyrene board (XEPS), 127–129, 832–835, 842

Fabrication, steel, 369–371
Face mix, 684
Face-shell mortaring, 578–579
Face shells, 570
Facing brick, 557–558
Factor of safety, 97
Factory Mutual Research Corporation (FMRC), 845
Factory square, 824
Failure to detect, 17
Failure warning, ductility and, 84
Failures, structural, 94–96
Fair Housing Act, 43
Fasteners for connecting wood members, 261–265
Fast-track project delivery, 15
Feldspar, 586
Felt
built-up roof membrane and, 819, 820–824
exposure, 846
fiberglass roof, 822, 823
organic versus inorganic, 822, 855
shingles, 846–847
use of term, 826
Fiber-cement siding, 316
Fiberglass, 123, 124
doors, 773
felt, 822, 823
windows, 766
Fibrous insulation, 123–124
Field observation, 17
Fill, 509
Film resistance, 110
Filter fabric, 842
Finger-jointed lumber, 227
Fire
codes, 153–154
combustible and noncombustible materials, 154
construction types and, 157–160
factors that affect, 153
gases produced in a fire, 155
interior finishes, 162–164
products of combustion, 154–155
protection, 36
protection of steel, 376–379
safety, 27–27
smoke, effects of, 155
sprinklers, 36, 47, 48–49, 164
Fire Prevention Code, 35, 153–154
Fire-rated
assemblies, requirements of, 156–157, 161
doors, 783
roofs, 844–845
windows, 784
Fire resistance
concrete members, 498
glass, 747–749
gypsum board, 161, 320, 377–378
of masonry walls, 595–596
ratings, 157, 161, 326
of roofs, 844–845
Fire Resistance Design Manual, 161
Fire-retardant-treated wood (FRTW), 154, 239–240
Fire-stopping, for penetrations and joints, 161–162
Fixed windows, 763, 764
Flame-spread index (FSI), 162–163, 164
Flange flashing, 837, 840–841
Flashing(s), 636–637
apron, 864
around a masonry chimney, 864–866
base, 837, 838–840
counter, 839, 865–866
cricket, 864–865
curb, 837, 840
flange, 837, 840–841
low-slope roof, 820, 837–841
purpose of roof, 820, 837
for steep roofs, 854, 856, 863–866
step, 863, 864
veneer walls and, 647–650
Flat glass, manufacture of, 730–732
Flat slabs, 490
Flat-plate slabs, 489–490
Flemish bond, 560–562
Flex anchors, 681–682
Flexible insulation, 123
Flexural stress, 89
Flexural tensile strength, 544, 545
Flexural tension/compression, 544
Flexure, 88
Float bath, 731
Float glass process, 730–732
Floor area ratio (FAR), 43–44
Floor live load, 55–56
Floors
beamless concrete, 488–490
beam-supported concrete, 480–488
decks, masonry walls and, 608–609, 610–612
frame configuration and spacing of members, 274
framing, essentials of, 281–285
framing in light-gauge steel, 403–405
joists, 274, 283–285
over unheated basements or crawl spaces, insulation for, 117, 119
posttensioned elevated concrete, 491–492
sheathing, 272, 293
steel decks, 357–359
structural insulated panels (SIP), 334
Sturd-I-Floor sheathing, 259
sub-, 272, 293, 904–905
trusses, 252–254

Floors, coverings
brick, 927
carpet and carpet tile, 918–921
ceramic and stone tile, 906–914
concrete, 927
cork tiles, 926
linoleum, 925–926
resilient, 924–926
resinous, 926–927
rubber tiles, 925
selection criteria for, 905–906
sheet vinyl, 925
stone panel, 914–915
terrazzo, 916–918
underlayments, 927–928
vinyl tiles, 924
wall bases and moldings, 928–929
wood, 921–923
Fluid-applied membrane roofs, 820
Flux, 730
Fly ash, 437
Flying forms, 453
Foamed concrete, 129, 836
Foamed glass, 129
Foamed insulation, 123, 128–129
Foamed plastics, 108
Foamed-in-place insulation, 129
Footings
design parameters of, 522–523
settlement of, 524
tilt-up walls, 621–622
types of, 520–522
Forced convection, 104
Forest Stewardship Council (FSC), 206–207, 266
Form steel floor deck, 357–359
Formaldehyde, 137, 266
Formwork, 448–455
Foundations
below-grade waterproofing, 531–534
defined, 519
for a two-story wood light frame building, 296
Foundations, deep
construction details of, 528
defined, 520
piers, 526–527
piles, 526
Foundations, shallow
allowable bearing capacity of soil, 524–525
construction details of, 525
defined, 519
design parameters of footings, 522–524
frost protection for, 120, 528, 531
settlement of footings, 524
types of footings, 520–522
Frame structure
defined, 344
fundamentals of, 382
Freeze-thaw resistance of brick, measuring, 557
Friction
coefficient of, 500, 906
piers, 527
piles, 526
Frog, 555
Frontage, 36, 47
Frost-protection for shallow foundations, 120, 528, 531
Fully tempered (FT) glass, 732–733
Fungal decay (rotting) of wood, 234–235
Furring channels, light-steel, 400

Gable fans, 144, 145
Gable ventilators, 144, 145
Galvanic corrosion, 376
Galvanizing, hot-dip, 375–376
Gang forms, 453
Gases produced in a fire, 155
Gasket reglet, 755
Gaskets, 754–755
General contractor(s)
relationship between architect and, 12
role of, 11, 12
selecting a, 12–15
use of multiple, 14–15
Geological cycle, 504
Ghosting, 408
Girders, joist, 352–355
Glass
annealed, 732
bent, 735
ceramic frit on, 735
crown process, 728
cylindrical process, 728
dimmable (electrochromic), 751
double-strength, 732
fire-resistant, 747–749
glazing and, 743–745
glazing (plastic), 750
glazing pocket, anatomy of, 752–755
heat-modified, types of, 732–735
heat-strengthened, 734–735
historical use and development of, 728
insulating, 740–741
intumescent gel-filled units, 749
intumescent multilaminate, 749
laminated, 746
long-wave radiation and, 739–740, 756–757
low-emissivity (low-E) glass, 739–740, 742
manufacture of flat, 730–732
radiation-shielding, 751
R-values, 742–743
safety, 745–746
selection of, 752
self-cleaning, 751
single-strength, 732
solar radiation and, 736–737, 738, 756
for special purposes, 750–751
structural performance of, 747
surface designations for, 741
tempered, 732–733
tempered, heat-soaked, 733–734
tinted and reflective, types of, 737–738
transparency of, 730
U-values, 114, 742–743
wired, 748–749
Glass-aluminum walls
See also Curtain walls, glass-aluminum
punched and strip glazing, 812–813
storefront system, 813
Glass blocks. *See* Glass masonry units (GMUs)
Glass fiber-reinforced concrete (GFRC)
curtain walls. *See* Curtain walls, glass fiber-reinforced concrete (GFRC)
Glass fiber-reinforced plastic (GFRP), 453, 482
Glass masonry units (GMUs)
applications, 592–593
construction of, 594–595
sizes, 594
Glazed openings, U-values, 114
Glazed units, 575, 576
Glazing
acrylic versus polycarbonate, 750
compound, 753
glass, R-values and, 743–745
plastic, 750
pocket, anatomy of, 752–755
punched, 812–813
putty, 753
rain screen principle, 755
seals, 753–755
security, 750–751
strip (ribbon), 813
Global warming, 199
Glulam (glue-laminated wood) members
applications and sizes, 246–247
balanced and unbalanced beams, 248
grades, 248–249
Gothic architecture, 80, 627–628
Grades
glulam member, 248–249
lumber, 231–234
veneer, of softwood plywood, 257
wood panels, 260, 261
Granite, 585–587
versus marble, 590
Granular insulation, 123, 124–127
Gravel, 423, 504
Gravity
loads, 54, 74, 303–306
water infiltration control and, 632–633
Greek architecture, 80
Green building products, characteristics of, 202–206
Green development, 198
Green Label and Green Label Plus, 207
Green roofs, 845
Green Seal Label, 207
Greenhouse effect, 757
Ground coverage, 43
Ground-supported isolated concrete slabs, 468–472
Ground-supported stiffened concrete slabs, 472–476
Grout
for ceramic and stone tiles, 910–911
types of, 583–584
Gypsum Association, 161, 326
Gypsum board
encasement, 377–378
fire resistance, 161, 320, 377–378
installing and finishing interior, 323–326
other names for, 320
panels, types of, 321–323
sizes for, 321
suspended ceilings, 940, 942, 943, 944
Gypsum plaster and portland cement plaster, suspended ceilings, 942–943, 944

Hail, roofs and, 845
Hardboard, 246
siding, 316
Hardwoods, 217–220
veneers, 257
Header, 278, 279–280
Head joints, 550
Head lap, 846
Heal bead, 755
Health and welfare, 29
green building and, 206
Heat-absorbing glass, 737
Heating degree days (HDD), 116, 117
Heat-modified glass, types of, 732–735
Heat-soaked tempered glass, 733–734
Heat-strengthened glass, 734–735
Heavy timber (HT) construction, 158–159
Height, allowable, 36, 46
Height above ground, wind pressure and, 61
Hertz, 168
Hollow structural section (HSS), 349
Hollow-core slabs, 494–495
Hooks, in bars, 460–461
Horizontal bridging members, 353
Horizontal reinforcement, 580–582
Hot rolling, 355
HP-shapes, structural steel, 347–348
H-units, 573, 574
Hurricanes, 58–59
glass-aluminum curtain walls and, 811
resistant glass, 746
Hurricane straps for roof, 307
Hydro-chloro-fluoro-carbon (HCFC) gas, entrapment of, 108

Hydrogen chloride, 155
Hydrogen cyanide, 155

Ice dams, 143–144
protection membrane, 853, 869
Igenous rock, 585, 586
I-joists, 250–251
Impact insulation class (IIC), 173–174
Impact sound, 169
Impact-resistant EIFS, 707–708
Incidental-use space, 39
Industrial wood products, 246
Infill panels, 675
Infill walls, 636, 637
Infiltration, 134
Initial rate of absorption (IRA), 562–563
In-line framing, 403, 404
Inside-glazed curtain walls, 801, 806
Inspection(s)
failed, 30
process, 16–18
stages of, 30
substantial completion, 18–19
Insulating concrete, 126–127
and roof membrane attachment, 835–836
Insulating concrete forms (ICF), 126
Insulating glass unit (IGU), 740–741
Insulation, 106
reflective, 111
requirements for residential structures, 117–120
rigid-board, and roof membrane attachment, 832–835
roof, 820
thermal capacity versus, 121–122
types of, 122–129
warm-in-winter side of, 143
where and how much, 116–120
Intake ventilation, 144
Interior finishes, fire testing of, 162–164
Interior partitions, framing in light-gauge steel, 409
International Building Code (IBC), 34, 35, 595
International Code Council (ICC), 34, 35, 154
International Conference of Building Officials (ICBO), 34
International Electrical Code, 35
International Energy Conservation Code (IECC), 35, 116
International Fire Code, 35, 154
International Mechanical Code, 35
International Plumbing Code, 35
International Residential Code (IRC), 35
International Standards Organization (ISO), 207
Interstitial condensation, 140, 141
Intumescent gel-filled glass, 749
Intumescent multilaminate, glass 749
Intumescent paints, 378–379
Inverted roof, 842
Inverted-tee beams, 498
Invitational bidding, 13
Iron
cast, 343–344
pig, 344
wrought, 343
I-sections, structural steel, 347–348
Iso boards, 128–129, 832–835
Isolation joints, 179, 470

Jack studs, 278–279
John Hancock Center, 57, 58
Joinery methods for connecting wood members, 261–265
Joint compound, 324–325
Joints
bed, 550
between precast curtain wall panels, 678–679
building, 178, 179
building separation, 179–180
ceiling, 289
cold, 178, 470–472
component, 178
components of sealed, 191–192
concrete slab-on-ground, 470–472
construction, 178, 470–472
continuous vertical expansion, 651
control, 179, 186, 190, 470, 582–583
covers for, 181–182
creep deformation, 182, 188
dovetail, 261
dynamic, 178
elastic deformation, 188
expansion, 179, 186, 190
head, 550
isolation, 179, 470
moisture movement, 186–187
mortar joint thickness and profiles, 550–551
mortise-and-tenon, 261
movement, 178–179, 182–183
principles of detailing, 190
reinforcement, 580–581
repointing, 564–565
sealants, 192–194, 634
seismic, 179, 181–182
sizing and spacing requirements, 182–183
slip, 380
static, 178
stucco and control and expansion, 696, 697–699
in stucco-finished masonry and concrete walls, 701
thermal movement, 184–186
total joint dimension, 189–190
Joist floor
one-way, 482
one-way, standard-module, 484
one-way, wide-module, 484–486
two-way (waffle), 486–488
Joist girders, 352–355
Joists
band, 274, 283–285
bracing, 353–355
floor, 283–285
header, 281
I-, 250–251
rim, 274, 283–285
steel, 350–355, 380–381
Joules, 103

Kerfs, 710
Kraft paper, 124, 142

Ladder-type joint reinforcement, 580
Lagging, 512
Lag screws, 265
Laminated glass, 746
Laminated veneer lumber (LVL), 249–250
Lateral buckling, 95
Lateral loads, 54, 304–306, 809–810
Lath, expanded metal, 696
LEED (Leadership in Energy and Environmental Design) rating system, 201
Let-in lumber brace, 292
Life cycle assessment (LCA), 206
Life safety, 26–27
Light-gauge steel. *See* Steel, light-gauge
Light-to-solar gain (LSG) index, 737
Light-transmitting plastics, 750
Lime
applications, 417
carbonation of, 414–415
chemistry of, 415
hydrated, 414
mortar and use of, 543–544, 546–547
pozzolanic reaction, 415–416
quick, 414
slaked, 414
types of, 416–417
Limestone, 587
Liner blocks, 710
Lining the air space, 110
Linoleum, 925–926
Lintel angles, 647
Lintel unit, 572
Liquid limit, 537, 538
Lites, 792
Live loads, 54, 55–56
Load resistance
bearing strength of materials, 92–94
bending strength of materials, 88–91
compressive and tensile strengths of materials, 80–83
ductility and brittleness, 83–85
elasticity and plasticity, 86–87
modulus of elasticity, 87–86
shear strength of materials, 91–92
structural failures, 94–96
structural safety, 96–98
yield strength of metals, 85–86
Load-bearing walls, 281–282, 308
Loads
dead, 54, 55, 71–74
earthquake, 64–69
factors, 97–98
gravity, 54, 74, 303–304
K-series joists, 353
lateral, 54, 304–306
live, 54, 55–56
rain, 56–57
snow, 63–64
two-story wood light frame building, 303–308
units of measurement for, 54
vertical, 54
wind, 57–63
Local crushing, 93
Long-span steel buildings, 392–394
Long-wave radiation, 739–740, 756–757
Low-carbon steel, 85
Low-emissivity (low-E) glass, 739–740, 742
L-shapes, structural steel, 349
Lumber
See also Wood
air versus kiln seasoning, 224
board foot measure, 227–228
boards, 229
checks, shakes, and splits, 230–231
classification, 229
conversion from logs to, 221–223
dimension, 229
drying of, 223–226
end tags, 239
finger-jointed, 227
flat-sawn and radial-sawn, 222–223
grades, 231–234, 241
grading agencies, 241, 242
graphic symbols for framing and worked, 226
green versus dry, 224
knots, 230
laminated veneer, 249–250
nominal and actual dimensions of, 227
parallel strand, 250
plain and quarter sawing, 222
seasoning, 223
shrinkage and swelling of, 224–226

slope of grain, 230
softwood and hardwood species, 220
strength and appearance, 229–231
structural composite, 250
surfacing, 226–227
sustainability features of, 239
timbers, 229
use of term, 212
worked, 226

Machine stress-rated (MSR) lumber, 233–234
Magnetic sputtering, 738
Malleability, ductility and, 84
Manufactured wood products. *See* Wood products, manufactured
Marble, 587
granite versus, 590
Masonry
See also Brick(s); Concrete masonry units (CMUs); Glass masonry units (GMUs); Mortar; Stone
ashlar, 591, 592
cement, 547–548
clay, 543, 556–559
course of, 560
dimensions of units, 555–556
historical use of, 542–543
moisture movement in, 186
rubble, 591, 592
stucco on substrates, 700–701
sustainability features, 596
units, 543
wythe, 560
Masonry walls. *See* Walls, masonry bearing
Mason's sand, 546
Master-builder method, 22
MasterFormat, 9–11
Mat footings, 521
Means of egress, 28
Measurement
for loads, 54
permeance, 141
Mechanical weathering, 504
Medium-set (bed) method, 907–908
Membranes
See also Roof(s); Roofs, low-slope
base flashing, 838–839
use of term, 826
Metal back pan, 807
Metal coping, 839–840
Metal doors, 779
Metal panel roofs, 820
Metal panels, insulated, 722–724
Metal roofs, 820, 872–876
Metals
corrosion of, 376
ferrous versus nonferrous, 342
precious, 344
Metamorphic rock, 586, 587, 588
Mica, 586
Middle-third rule, 601–602, 627–628
Mild steel, 85
Mill finish, 787
Minimum Design Loads for Buildings and Other Structures, 41
Mission tile, 867, 872
Mixed-occupancy building, 38–39
Mock-up samples, 16
Model building codes, 33–35
Modified bitumen roof (MBR) membrane, 819, 820, 825–829
Modulus of elasticity, 87–88
Moisture coefficient, 186
Moisture movement, 186–187
Moldings, 928–929
Mortar
capturing device, 651
cement, 548
for ceramic and stone tiles, 909–910
contents of, 543
joint thickness and profiles, 550–551
materials and specifications, 546–549
repointing, 564–565
retempering, 563
strength, 544–546, 548
white and colored, 549
workability and water retentivity of, 543–544
Movement joints
in building components, 182–183
types of, 178–179
M-shapes, structural steel, 347–348
Mud sill, 277
Mullions, 794
connecting rails to, 800–801
Multistory glass curtain wall, 806–808
Multistory steel frames, 389–392

Nailing
blind, 263
power-driven, and stapling, 264
Nailing flanges, installing windows with, 769–771
Nails
brite, 262
connections, 262–263
popping, 264
sinker, 262
sizes, 262
types of, 261–262
National Building Code of Canada, 33
National building codes, 33–34
National Concrete Masonry Association (NCMA), 41
National Fenestration Rating Council (NFRC), 767
National Fire Protection Association (NFPA), 35, 153
National Institute of Standards and Technology (NIST), World Trade Center fires, 161
National Roofing Contractors' Association (NRCA), 41
National Terrazzo and Mosaic Association (NTMA), 917
Natural convection, 104
Negative-side waterproofing, 533
Negotiated contract, 13–14
Neutral axis, 89
NFPA 5000: Building Construction and Safety Code, 35
Nocturnal cooling, 184
Noise-reduction coefficient (NRC), 174
Noncombustible materials, 154

Occupancy
certificate of, 19, 31
classification, 36, 37–38
load, 55
mixed-occupancy building, 38–39
One-way joist floor, 482
standard-module, 484
wide-module, 484–486
One-way solid slab, 480
Open excavations, 508, 509–510
Operable windows, 764
Organic soil, 505
Oriented strandboard (OSB), 255, 257–258, 292, 293, 330
Otis, Elisha Graves, 198
Outgasing, 192
Outside-glazed curtain walls, 805–808
Owner's program, 5
Owner's role, in design phase, 8
Ozone depletion, 199

Painted aluminum, 787
Panels, wood
oriented strandboard (OSB), 255, 257–258
plywood, 255–257, 258
rating, 259–261
types of, 255
Parallel chord truss, 252
Parallel strand lumber (PSL), 250
Parallel-path heat flow, 114
Particle board, 246, 255
Paving brick, 559
Payment bonds, 14
Payment certifications, 18
Peat, 505
Performance bonds, 14
Performance codes, 32–33
Performance-rated engineered wood panels, 259–260
Perlite boards, 125, 832
Perm ratings, 141–142
of air retarders, 143
Permeability, ceramic tile and, 907
Permeance, 141–142
Perrot, Bernard, 728
PFS, 248
Phenolformaldehyde, 266
Picture window, 764
Piers, 526–527, 574
Pig iron, 344
Pilasters, 574
Piles, 526
Pilkington Co., 812
Pin connection, 382, 383–385
Pitch, 823
Pits, 508
Plain-masonry walls, 604
Planning phase, 4–5
Plaster. *See* Stucco
Plaster of paris, 321
Plasterboard, 320
Plastic glazing, 750
radiation-shielding, 751
Plastic limit, 537, 538
Plasticity, 86–88, 417
Plate girders, 349
Platform frame, 272–274
Plywood, 246
for formwork and shores, 451
panels, 255–257, 258
siding, 315–316
Point-supported glass-aluminum curtain walls, 812
Polycarbonate glazing, 750
Polyicynene, 129
Polyisocyanurate boards (iso boards), 128–129, 832–835
Polymer-based, EIFS, 703–707
Polymer-dispersed liquid crystals (PDLC), 751
Polymers
single-ply roof (SPR) membrane and, 819, 820, 829–831
thermosetting, 829
Polysulfide, 193
Polyurethane, 193
Polyvinyl chloride (PVC)
room membranes, 829, 830–831
wall bases and moldings, 928–929
windows, 766
Porous sound absorbers, 174
Portland cement, 323
See also Concrete
composition of, 420
entrapped air versus entrained air in, 420–421
hydration of, 433–435
manufacturing process of, 418–419
for masonry mortar, 546–547
sustainability, 441
water reaction and, 433–435
white, 421
types of, 419–420

Portland Cement Association (PCA), 183
Portland cement plaster. *See* Stucco
Positive-side waterproofing, 533
Post-and-beam (lintel) frame structure, 382
Postconstruction phase, 18–19
Posttensioned concrete slabs, 475
Posttensioned elevated concrete floors, 491–492
Pour stops, 360
Power, energy and, 103–104
Pozzolanic reaction, 415–416
Precast concrete. *See* Concrete, precast
Precast/Prestressed Concrete Institute (PCI), 675
Preconstruction phase, 11–15
Preconstruction services, 14
Predesign phase, 4–5
Prefabricated brick curtain walls, 687–689
Prefabricated stone curtain walls, 717–719
Prefabricated stone-honeycomb curtain walls, 719–722
Preformed tapes, 192
Prescriptive codes, 32–33
Preservative-treated wood, 237–239
Pressure plate-captured glass, 801–803, 804–805
Pressure-treated wood, 237–239
Prestressing tendons, 473–475
Primary panels, 516
Prime contractors, 14–15, 20
Primer, 191
Procurement and Contracting Group division, 9, 11
Products of combustion, 154–155
Professional societies, 41
Project closeout, 18–19
Project deliver
  construction phase, 15–18
  design phase, 5–8
  postconstruction phase, 18–19
  preconstruction phase, 11–15
  predesign phase, 4–5
Project manual, 8
Project superintendent, 16
Property protection, 29
Protected membrane roof (PMR), 841–842
Punch list, 18
Punched glazing, 812–813
Purlins, 289, 290
p-values
  of commonly used building materials, 107
  difference between R-values and, 106–107
  effect of density and moisture content on, 107–108
PVC. *See* Polyvinyl chloride (PVC)
Pyrolytic disposition, 738

Quarter-point supports, 710
Quartz, 586
Quicklime, 414

Rabbet, 753
Rack (racking), 306, 384
Radiant barrier, 111
Radiant flux, 922
Radiation, 104–105
  important facts about, 756
  long-wave, 739–740, 756–757
  shielding glass and plastic, 751
  solar, 736–737, 738, 756
Radon, 137
Raft foundation, 521
Rafters, 287–288
Raggle, 866
Rails, 794
  connecting mullions to, 800–801
Rain loads, 56–57
Rain-screen exterior cladding, 636–640
  glazing pocket and, 755
Rainwater infiltration control. *See* Water infiltration control
Rake, 318, 319
  clay and concrete tile, 871–872
  drip edges for, 856
Rarefaction, 168
Rasping, 705
Rebars, 414, 446
Record documents, 19
Record drawings, 19
Record specifications, 19
Recovery and reusability, 203–204
Recycling, 204–205
  difference between renewal and, 204
  internal and external, 204
  postconsumer waste, 204
Reflective glass, 737–738
Reflective insulation, 111
Reinforced concrete. *See* Concrete, reinforced
Reinforced steel. *See* Steel, reinforced
Reinforcement
  compression, 458
  horizontal, 580–582
  ladder-type joint, 580
  mesh, 706
  shear, 457
  tension, 457
  truss-type joint, 580
  vertical, in masonry walls, 604
Relative humidity of air, 138–139
Relieving angles, 647
Renewability, 202–203
Renewal, difference between recycling and, 204
Repointing, mortar joints, 564–565
Reshoring, 454
Resilient channels or clips, 172–173
Resilient flooring
  forms of, 924
  installation, 926
  physical properties of, 924
  static-control, 926
Resin, 922
Resinous flooring, 926–927
Resistance factor, 98
Retempering mortar, 563
Ribband, 270
Ribbed units, 575
Richter scale, 67–68
Ridge beam, 291
Ridge board, 287, 288
Ridge straps, 289, 290
Ridge ventilation, 318–319
Ridge vents, 144, 145
Rigid body, 83
Rigid connection, 382, 385–387
Rigid insulation, 123
Rigid-board insulation and roof membrane attachment, 832–835
Rim joists, 274, 283–285
Rivets, 362
Rocks
  chemical classification of, 586
  classification of, 504
  geological classification of, 586
  -soil cycle, 504
Rock wool, 123
Rodding, 426
Roman concrete and mortar, 415–416
Roof(s)
  built-up surfaces on, 824
  decks, masonry walls and, 608–609, 610–612
  frame configuration and spacing of members, 274
  framing, essentials of, 287–291
  framing in light-gauge steel, 405–408
  green, 845
  inverted, 842
  live load, 56
  rise-to-run ratio, 285
  seismic or hurricane straps for, 307
  shapes and terminology, 286
  sheathing, 293–294
  snow load, 63–64
  square, 824
  steel decks, 355–357
  structural insulated panels (SIP), 334–336
  trusses, 252
  types and slope, 285–287
Roofs, low-slope
  built-up membrane, 819, 820–824, 846–847
  decks, 820
  design considerations, 842–847
  difference between steep roofs and, 818–819
  drainage design, 845
  fire safety, 844–845
  flashings, 820, 837–841
  fundamentals, 819–820
  insulating concrete and membrane attachment, 835–836
  insulation, 820
  membranes, 819–820
  modified bitumen membrane, 819, 820, 825–829
  protected membrane, 841–842
  recyclable products, 845
  rigid-board insulation and membrane attachment, 832–835
  single-ply membrane, 819, 820, 829–831
  surface reflectivity, 845
  torched membrane, use of, 828
  wind uplift and hail impact, 845
Roofs, steep (water-shedding)
  asphalt shingles and underlayment, 854–858
  clay and concrete tiles, 866–872
  deck, 854
  difference between low-slope roofs and, 818–819
  drip edges for eaves and rakes, 856, 870–872
  flashings, 854, 856, 863–866
  fundamentals, 852–854
  ice dam protection, 853
  metal, 820, 872–876
  ridge and hip treatment, 862–863
  shingles, types of, 852–853
  standard slope, 855
  underlayment, 853, 854–856
  valley treatment, 859–862
Rooting of wood, 234–235
Rubber tiles, 925
Rubble masonry, 591, 592
R-values
  of a building component, 105–108
  difference between *p*-values and, 106–107
  of glass, 742–743
  glazing and, 743–745
  of a multilayer component, 109–110

Safety
  factor of, 97
  glass, 745–746
  margin, 97
  structural, 28–29, 96–98
Sand, 504
Sanded plywood panels, 260–261
Sand-lime bricks and blocks, 584
Sandstone, 587
Saturated air, 138
Saturation coefficient, 557
Saturation vapor pressure, 138
SBS. *See* Styrene butadiene styrene (SBS)-modified bitumen roof membrane
Schematic design (SD) stage, 6
Scratch coat, 696, 697
Screw splines, 764
Screws, 264–265
Sealants
  backup, 191

caulks, 193
comparison of commonly used, 193
elastomeric, 136, 189, 193
joint, 192–194
life expectancy of, 194
movement ability of, 193
one-part versus two-part, 194
preformed tapes, 192
tooling time, cure time, and temperature range, 194
Sealed joints, components of, 191–192
Sears Center, 57–58
Secant piles, 515–516
Second-tier subcontractor, 12
Security glazing, 750–751
Sedimentary rock, 586, 587
Seismic clips, 654, 656
Seismic joints, 179, 181–182
Seismic resistance, glass-aluminum curtain walls and, 811
Seismic straps for roof, 307
Seismic use group, 69
Select fill, 509
Self-adhering tape, 136
Self-cleaning glass, 751
Self-furring lath, 696
Setbacks, 44
Setting blocks, 752–753
Setting materials, for ceramic and stone tile, 909–911
Setting methods, for ceramic and stone tile, 907–908
Settlement of footings, 524
Shading coefficient (SC), 736–737
Shale, difference between clay and, 552
Shallow foundations. *See* Foundations, shallow
Shear blocks, 800
Shear connection, 383
Shear reinforcement, 457
Shear strength, 92
Shear walls, 306
Sheathing
floor, 272, 293
roof, 293–294
wall, 292
Sheet
base, 826
cap, 826
piles, 511
Sheet metal
connectors, 265–266
roofs, 820, 872–876
Sheet vinyl, 925
Sheetrock, 320
Shelby tube, 506
Shelf angles, 646–647
Shingles
asphalt, 854–858
clay and concrete tile, 866–872
composition, 852
fasteners for, 858
felt, 846–847
ridge and hip treatment, 862–863
types of, 852–853
valley treatment, 859–862
Shop drawings, 16
Shores, 448–455
Showers, tile in, 930–931
Sick building syndrome, 137
Siding
aluminum and steel, 313
fiber-cement, 316
hardboard, 316
horizontal, 313–317
plywood, 315–316
vertical, 317–318
vinyl, 317
wood (clapboard), 313–315
Silica fume, 437
Silica glass, 730
Silicone, 193
Sill plate, 276, 277
Silt, 504
Simple connection, 383
Simply-supported beams, 457
Single-ply roof (SPR) membrane, 819, 820, 829–831
fully adhered system, 833
loose-laid, ballasted system, 834–835
mechanically fastened system, 833–834
SI system of units
loads in, 54
permeance, 141
Site-cast concrete. *See* Concrete, site-cast
Skeleton frame. *See* Frame structure
Skyscrapers, early, 198
Slab-on-grade, 119
Slab-on-ground, concrete
insulation under, 119–120
tilt-up wall, 621
vapor retarder under, 144
Slag, 345
Slag wool, 123, 345
Slate, 588
shingles, 852
Slip-critical bolted connection, 363
Slump, 425–426
Smoke-developed index (SDI), 163–164
Smooth, off-the-form concrete, 455–456
Snow, George Washington, 270
Snow load, 63–64
Snug-tight connection, 363
Soda-lime glass, 730
Soffit vents, 144, 145
Soffits, 318, 319
Softening-point temperature (SPT), of asphalt, 823–824
Soft story, 607
Softwoods, 217–220
Soil
bearing capacity, 94, 523, 524–525
below-grade waterproofing, 531–534
classification of, 504–505, 535–538
coarse-grained versus fine-grained, 504, 505
cohesive and noncohesive, 504–505
earthwork and excavations, 507–519
expansive, 537
foundation systems, 519–531
laboratory testing, 507
nailing, 516–517
organic, 505
plastic and liquid limit, 537, 538
sampling and testing, 505–507, 538
Solar heat-gain coefficient (SHGC), 736, 811
Solar radiation, 736–737, 738, 756
Soldier piles, 511–514
Sole plate, 276, 277
Sound
absorption, 174
airborne and structure-borne, 169
decibel scale, 169
frequency, speed, and wavelength of, 168
glass-aluminum curtain walls and transmission of, 811
impact, 169
impact insulation class (IIC), 173–174
pressure level, 169
-transmission class (STC), 171–172
-transmission loss (TL), 170–171
wave, 168
Southern Building Code Congress International (SBCCI), 34
Spandrel beam, 188
Spandrel panels, 675
Specialities division, difference between special construction division and, 11
Specifications
CSI MasterFormat and, 9–11
defined, 7
Group division, 9
record, 19
relationship between construction drawings and, 8
Specific gravity, 217
Spider-shaped connectors, 812
Splices, 459, 460
Split-face units, 574, 575, 590
Split-spoon, 506
SPR. *See* Single-ply roof (SPR) membrane
Sprinklers, 36, 47, 48–49, 164
S-shaped tiles, 867–868
S-shapes, structural steel, 347–348
Stabilizer angle, 366
Stack casting, 619
Stack-bond walls, 581–582
Stairs
balusters, 884
carriages (rough stringers), 886
concrete, 898–899
drawing conventions, 885
freestanding, self-supporting, 896–898
guard units, 883
handrails, 884
head room, 883
helical, 897–898
landing frame, 893–895
layout and plan, 884–885
newel post, 884
open or closed, 882
open-riser, 882
shafts or shaft enclosures, 886
shapes, 882
steel, 888–898
stringers, 886, 891–893
tread, drop-in, 889
tread, riser, and nosing, 881–882, 929
width of, 883
wood, 886–888
Standard Building Code, 34
Standard penetration resistance, 506
Standard penetration test, 506
Standard slope, 855
Standards, 40–42
Standards Association of Australia (SAA), 41
Stapling, power, 264
Static joints, 178
Static-control resilient flooring, 926
Steady state, 120
Steel
for formwork and shores, 451
stairs, 888–898
stucco on steel-stud walls, 695–700
windows, 766
Steel, light-gauge
advantages and disadvantages of, 408
applications, 398–399
corrosion protection, 401
difference between reinforced, structural and, 346
floor framing, 403–405
framing members, 399–400
gauge versus mil, 399
load-bearing applications, 400–408
non-load-bearing applications, 409–410
roof framing, 405–408
stud size and spacing, 401
thickness and depth of studs and joists, 398
wall framing, 401–403
Steel, reinforced
bar identification markings, 439–440
bending and tying bars, 440
diameter and length of bars, 438–439
difference between structural, light-gauge and, 346
epoxy coated bars, 440
grades, 438

Steel, structural
  bolts and welds, 362–364
  building separation joints, 380–381
  commonly-used structural, 345
  composite action between floor slab and supporting members, 359
  corrosion protection, 375–376
  detailing and fabrication, 369–371
  development of, 343–344
  difference between reinforced, light-gauge and, 346
  erection, 372–375
  fire protection, 376–379
  floor decks, 357–359
  framing members, connections between, 364–369
  framing members, preliminary layout of, 360–362
  importance of, 342
  joists and joist girders, 350–355
  long-span buildings, 392–394
  low-carbon (mild), 85
  making of modern, 344–345
  multistory frames, 389–392
  pour stops, 360
  roof decks, 355–357
  shapes and designations, 346–349
  siding, 313
  slag, 345
  sustainability of, 379
Steel Joist Institute (SJI), 350, 351
Steel stud backup walls, 644, 661–669
Steel Stud Manufacturers Association (SSMA), 398, 399
Steel wire designation, 581
Steiner Tunnel Test, 162
Step flashing, 863, 864
Stiffbacks, 465
Stirrups, 457
Stone
  applications, 591
  bond patterns, 591–592
  building, 585
  dimension, 585
  -faced precast curtain walls, 675–677
  finishes on slabs and panels, 589–590
  from blocks to finished, 588–590
  honeycomb panels, 719–722
  masonry, 591
  selection, 590–591
  veneer-faced precast panels, 677
  types of, 585–588
Stone, exterior cladding with
  bayonet clips and bolt anchors, 712–713
  prefabricated stone curtain walls, 717
  standard-set installation, 710–714
  thin, 719–722
  types of stone used, 709
  vertical channel support installation, 714–717
Stone panel flooring
  installation, 914–915
  patterns and finishes for, 914
  physical properties of, 914
Stone tile
  characteristics of, 907
  grouts, 910–911
  installation, 911–913
  mortars, adhesives and epoxies, 909–910
  permeability, 907
  physical properties of, 907
  setting methods, 907–908
  types, shapes and sizes of, 907, 908, 909, 910
Stop work order, 30
Storefront systems, 813
Strain, 83–84
Strain hardening, 86
Strands, 446, 447
Strap footings, 521
Stress, 80–83
  allowable, 97
  axial, 88
  bearing, 93
  bending or flexural, 89
  normal, 91
  shear, 91–92
  -strain diagram, 85
Stretcher face, 560
Strip footings, 520
Strip (ribbon) glazing, 813
Strongbacks, 620
Structural composite lumber (SCL), 250
Structural efficiency, 89–91
Structural failures, 94–96
Structural insulated panels (SIP), 212
  advantages of, 336–337
  basics of, 330–331
  disadvantages of, 338
  floors, 334
  panel sizes, 331
  roofs, 334–336
  walls, 331–334
Structural safety, 28–29, 96–98
Structural silicone sealant-adhered glass, 803, 806
Structural steel. *See* Steel, structural
Structure-borne sound, 169
Stucco
  accessories and trims for, 700
  advantages and disadvantages of, 701
  applications for, 694
  control and expansion joints, 696, 697–699
  drainage plane, 696, 701
  how to apply, 696–697
  on masonry and concrete substrates, 700–701
  mix composition for base and finish coats, 694
  one-coat, 700
  rigidity of stud wall assembly, 699
  on steel- or wood-stud walls, 695–700
  stop, 698
  stud wall deflection, 700
  suspended ceilings, 942, 944
  synthetic versus conventional, 703
Studs
  jack, 278–279
  light-steel, 399, 401
  precut length, 280
  at wall corners or at T-junctions, 277–278
Sturd-I-Floor sheathing, 259
Styrene butadiene styrene (SBS)-modified bitumen roof membrane, 825, 826–827
Subcontractors
  role of, 11
  second-tier, 12
Subfloors, 272, 293, 904–905
Substantial completion inspection, 18–19
Substrates, 191
  stucco on masonry and concrete, 700–701
Substructure, 519
Sump pumps, 518
Superstructure, 519
Supports for deep excavations, 511–517
Surety bonds, 14
Surface color, 112
Surface condensation, 140
Surface emissivity, 110–112
Surface resistance, 110
Surface splines, 331–332
Surface tension
  defined, 633
  water infiltration control and, 633
Suspended ceilings. *See* Ceilings, suspended
Sustainable development
  characteristics of green building products, 202–206
  ecolabeling of building products, 206–207
  ecolabeling of buildings, 200–202
  fundamentals of, 200
  use of term, 198–199

Tar, 823
Taylor, Augustine, 270
TECO, 248
Temperature
  building envelope versus ambient (air), 184
  gradient, 146, 185
Tempered glass, 732–733
  heat-soaked, 733–734
Tendons, prestressing, 473–475
Tensile failures, 94
Tensile strain, 84
Tensile strength, 80–83
Tension reinforcement, 457
Termite control, 235–237
Terrazzo flooring
  characteristics of, 916
  installation, 918
  types of, 917–918
Test boring method, 506–507
Test pit method, 506
Thermal break, 744
Thermal bridging, 115–116
Thermal capacity, 120–122
Thermal conductor, 106
Thermal inertia, 121
Thermal insulator, 106
Thermal mass, 121
Thermal movement, 184–186
Thermal properties
  conduction, 104
  convection, 104
  importance of, 102–103
  insulation, types of, 122–129
  insulation, where and how much, 116–120
  radiation, 104–105
  R-value of a building component, 105–108
  R-value of a multilayer component, 109–110
  surface emissivity, 110–112
  thermal capacity, 120–122
  units of energy, 103–104
  U-value of an assembly, 113–116
Thermal resistance. *See* R-values
Thermal resistivity, 106
Thermal short circuiting, 115
Thermal transmittance, 113
Thermoplastic, 829
Thermoplastic polyolefin (TPO) membrane, 829, 831
Thermosetting polymer, 829
Thick-set (bed) method, 907
Thin-set (bed) method, 908
Threshold of audibility, 169
Thumbprint hard, 563
Tiebacks, 512–514, 673, 674–675, 710–714
Ties, 463, 465
Tile
  See also Ceramic tile
  in showers, 930–931
Tile Council of America (TCA), 41, 909, 910, 911
Tile Council of North America (TCNA), 909
Tilt-up Concrete Association (TCA), 621
Tilt-up walls, reinforced-concrete, 619
  aesthetics, 626
  connections, 623–626
  foundations, 621–623
  shape and sizes of panels, 620–621
  thickness and concrete strength, 620, 621
Tinted glass, 737
Top plates
  double, 275–276
  triple, 276–277
Topography of site, wind pressure and, 62–63
Tornadoes, 58–59

Touchsanded plywood panels, 260–261
TPO. *See* Thermoplastic polyolefin (TPO) membrane
Tracks, light-steel, 399
Trade associations, 41
Travertine, 587
Tremie pipes, 516
Trenches, 508
Triaxial compression, 93
Tributary area, 71
Trussed joists, 252–253
Trusses, wood, 291
  floor, 252–254
  roof, 252
  specifying, 254
  terminology use with, 251
Truss-type joint reinforcement, 580
T-shapes, structural steel, 348
Tsunamis, 65
Tunnel form, 618
Turbine ventilators, 144, 145
Turnkey method, 22
Turn-of-nut method, 363
Twist-off bolt method, 363
Two-story wood light frame building, step-by-step process
  ceiling joist and roof framing, 299–300
  exterior and interior finishes, 301–302
  first-floor walls framing, 296–297
  floor sheathing, 299
  foundation preparation, 296
  loads, 303–308
  second floor framing, 298
  second-floor wall framing, 299
  wall and roof sheathing, 300–301
Two-way joist floor (waffle), 486–488
Two-way solid slab, 480
Type of construction, 36, 39

Ultimate strain, 84
Ultimate strength or stress, 82, 86
Ultraviolet transmittance (UVT), 737
Underlayments, 853, 927–928
Underwriters Laboratories (UL), 42, 161, 595, 844
Underwriters Laboratory of Canada, 161
Unified Soil Classification System (USCS), 504–505, 535–538
Uniform Building Code (UBC), 34
Uniform Federal Accessibility Standard (UFAS), 45
U.S. Department of Housing and Urban Development (HUD), 43
U.S. Green Building Council (USGBC), 200–201
U.S. model codes, 34–35
U.S. System of Measurement
  for loads, 54
  permeance, 141
Unit masonry, 543
Urea formaldehyde, 137, 266
U-values
  of an assembly, 113–116
  of glass, 742–743, 811

Valley treatment, asphalt shingle roofs and, 859–862
Valley underlayment, 853
Vaporretarders
  location of, 143–144
  materials used as, 141–142
  under concrete slab-on-ground, 144, 468–469
  warm climates and, 143
Vault, 80
Vaulted ceilings, 291
Veneer walls
  aesthetics of, 669–670
  air space, 645–646
  anchors, 645, 654
  concrete masonry units and, 654–660
  continuous vertical expansion joints, 651, 668–669
  dead load support, 646–647
  defined, 636, 637
  flashings and end dams, 647–650
  lintel angles, 647
  mortar droppings, 651
  mortar type and joint profile, 651–652
  shelf angles, 646–647
  steel stud backup walls, 644, 661–669
  weep holes, 650–651
Veneers, hardwood, 257
Ventilation ridge, 318–319
Vertical loads, 54
Vertical reinforcement, masonry walls and, 604
Vinyl composition tiles (VCT), 924
Vinyl siding, 317
Vinyl tiles, 924
Vinyl windows, 766
Visible transmittance (VT), 737
Visual grading, 231, 233
Volatile organic compounds (VOC), 137

Waffle slab, 486–488
Walers, 465
Wallboard, 320
Walls
  See also Cladding, exterior wall; Curtain walls; Veneer walls
  barrier, 635
  bond patterns in stone, 591–592
  concrete reinforced, 464–467, 516, 617–626
  drainage, 635–636
  footings, 520
  framing in light-gauge steel, 401–403, 409–410
  glass masonry, 594–595
  infill, 636, 637
  insulation for, 118
  insulation for basement, 119
  overhanging, 634
  reinforced-concrete tilt-up, 619–626
  stack-bond, 581–582
  steel stud backup walls, 644, 661 668
  structural insulated panels (SIP), 331–334
  stucco on steel- or wood-stud walls, 695–700
  tilt-up, 619–626
  U-values, 114, 115
Walls, masonry bearing
  advantages of, 600–601
  air leakage in, 136
  arching action, 573
  bond beams, 605–606
  bond patterns, 559–562
  column-beam system, 616
  construction of, 601–604
  coordination between trades during construction, 614–616
  fire resistance of, 595–596
  floor and roof decks, role of, 608–609, 610–612
  height limitations, 602–603
  layouts, 606–609
  limitations of, 613–615
  middle-third rule, 601–602, 627–628
  mortarless, 543
  in reinforced concrete, 601
  vertical reinforcement, 604
  water-resistive exteriors, 614
  watertightness in, 544
Walls, wood light frame (WLF) construction
  exterior finishes, 312–313
  frame configuration and spacing of members, 274
  framing, essentials of, 275–278
  load-bearing, 281–282, 308
  openings, framing around, 278–280
  shear, 306
  sheathing, 292
  sidings, horizontal, 313–317
  sidings, vertical, 317–318
  studs at corners or at T-junctions, 277–278
Warm climates, vapor retarders and, 143
Warm edge technology (WET), 745
Warranties, 18–19
Water retentivity of mortar, 543–544
Water vapor
  in air, 137–140
  condensation of, 140–141
  movement of, through assemblies, 139–140
  retarders, 141–145
Waterborne preservatives, 237–239
Water-holding curvature, 88
Water-infiltration control
  air space, compartmentalization of, 639–640
  airtight backup wall, 639
  barrier walls, 635
  cladding, 636–640
  drainage walls, 635–636
  gravitational force-induced, 632–633
  joint sealants, importance of, 634
  overhanging walls, 634
  surface tension-induced, 633
  wind-induced, 633–634
Water-penetration control, glass-aluminum curtain walls and, 811
Waterproofing, below-grade, 531–534
Water-reducing admixtures (WRA), 435
Water-resistive finishes, exterior walls, 614
Water-shedding curvature, 88
Watertightness in masonry walls, 544
Wavelength, of sound, 168
Weathering, mechanical and chemical, 504
Weathering index (WI), 559
Webs, 570
Weep holes, 649–651
Weight, of aggregate surfacing on roofs, 825
Welded wire reinforcement (WWR), 414
  designation of, 441
Welds, 362, 364
Well points, 518
Wet glazing, 753–755
Wide-flange section, 347
Wind
  -bracing elements, 57–58
  direction, 59
  hurricanes, 58–59
  important facts about loads, 62
  loads, 57–63, 638, 639, 748
  pressure and suction, 60–61
  speed, basic versus design, 59
  storms, 59
  straight-line, 59
  tornadoes, 58–59
  uplift and roofs, 845
  water infiltration control and, 633–634
Windows
  building codes and, 768–769
  fire-rated, 785
  installation and surrounding details, 769–773
  materials for, 764–766
  performance ratings for, 767–769
  sealing exterior, 136
  styles, 762–764, 768
  terminology, 763, 784–785
Wind pressures, 60
  factors that affect, 61–63
Wired glass, 748–749

Wood
See also Lumber
anisotropicity, 219–220
Certified Wood Label, 206–207
compared with other structural materials, 213
connectors, sheet metal, 265–266
doors, 773
durability of, 212, 234
earlywood and latewood, 216
fasteners for connecting wood members, 261–265
fire-retardant, 154, 239–240
for formwork and shores, 451
fungal decay (rotting), 234–235
growth rings, 215–216
heartwood and sapwood, 216–217
moisture content and relative humidity of air in, 226
moisture movement in, 186–187
nailers, 839
photosynthesis and respiration, 215
preservative-treated, 237–239
renewability, 212–213
shingles and shakes, 853
siding, 313–315
softwoods and hardwoods, 217–220
stairs, 886–888
stucco on wood-stud walls, 695–700
subfloors, 904–905
termite control, 235–237
windows, 764
Wood flooring
characteristics of, 921–922
installation, 922–923
physical properties of, 922
Wood I-joists, 250–251
Wood light frame (WLF) construction
building codes, 294, 295
configuration and spacing of members, 274–275
contemporary, 272–274
drywall, installing and finishing interior, 323–326
eaves, 318
equalizing cross-grain lumber dimensions, 294–295
evolution of, 270–272
example of two-story, 296–308
exterior wall finishes, 312–313
fire-resistance ratings, 326
floor framing, essentials of, 281–285
foundation, 296
gypsum board, 320–323
ridge ventilation, 318–319
roof framing, essentials of, 287–291
roof types and slope, 285–287
sheathing, applying to frame, 292–294
sidings, horizontal, 313–317
sidings, vertical, 317–318
vaulted ceilings, 291
wall framing, essentials of, 275–278
wall openings, framing around, 278–280
Wood nailers, 839
Wood panels
oriented strandboard (OSB), 255, 257–258
plywood, 255–257, 258
rating, 259–261
types of, 255
Wood products, manufactured
examples of, 212
glulam (glue-laminated wood), 246–249
I-joists, 250–251
laminated veneer lumber (LVL), 249–250
parallel strand lumber (PSL), 250
plywood, 246
structural composite lumber (SCL), 250
trusses, 251–254
types of, 246
Workability, 417, 425–426, 543–544
Working drawings, 7
World Trade Center fires, 161
Wrought iron, 343
W-shapes, structural steel, 347–348

XEPS (extruded polystyrene boards), 127–129, 832–835, 842

Yield point, 85
Yield strength, 85–86
Yield stress, 85, 86

Zoning
appeals, 44
map, 43
ordinance (code), 43–44
text, 43–44
Zoning and planning commission, 44